THIRD EDITION

CHEMISTRY

John C. Bailar, Jr.
University of Illinois

Therald Moeller
Arizona State University

Jacob Kleinberg
University of Kansas

Cyrus O. Guss
University of Nevada at Reno

Mary E. Castellion
Norwalk, Connecticut

Clyde Metz
College of Charleston

Harcourt Brace Jovanovich, Publishers
and its subsidiary, Academic Press
San Diego New York Chicago Austin Washington, D.C.
London Sydney Tokyo Toronto

Front Cover: Paul Silverman, *Test Tube Abstract*. Light striking test tubes filled with potassium permanganate, sodium dichromate, and cobalt(III) fluoride. © Fundamental Photographs.

ISBN: 0-15-506456-8
Library of Congress Catalog Card Number: 88-80622
Printed in the United States of America

Illustration Credits: Appear on page G16, which constitutes a continuation of the copyright page.

PREFACE TO THE STUDENT

You are holding in your hands a textbook written with your needs in mind. The authors of this textbook have been listening to the questions of chemistry students for many years. We have learned what causes excitement and what causes boredom and distress. This book includes many features designed to give you help where we think you are most likely to need it.

We invite you to use this book actively rather than passively. As you read each chapter, look at the figures and examine the tables. The information in the figures and tables supports and expands on the subject matter. Pause for a moment at each chemical or mathematical equation and think over the meaning of the symbols. Take special note of the features listed below, which are there to help you. Several of these features focus on two major areas in chemistry where we want to assist you: arithmetical problem solving and the nonquantitative part of chemistry known as descriptive chemistry.

- **General problem-solving method**—a method for deciding how to solve *any* type of problem (introduced in Chapter 2)
- **Specific problem-solving methods**—stepwise guides to solving major types of problems in chemistry (e.g., stoichiometry, Section 5.5; equilibrium problems, Section 19.12)
- **Solutions to sample problems**—demonstrations of how to solve problems in every topic presented, quantitative and nonquantitative
- **Problem exercises**—matched to solved problems and including answers
- **Notes on common errors**—warnings of potential pitfalls
- **Chemical reaction classification scheme**—a guide to similarities among a seemingly wide variety of chemical reactions (Table 17.8)
- **Summaries of outstanding properties of elements and compounds**—tables designed to answer the question students always ask about descriptive chemistry: "But what am I supposed to know?" (e.g., Tables 16.6, 28.8)
- **Study aids**—
  - *Significant terms*—boldface when first given in defining sentence; listed at end of each chapter
  - *Glossary*—alphabetical list (at back of book) of all significant terms
  - *Summaries*—at end of each major subject area

We wish you well in your study of chemistry. You are about to discover why chemistry has sometimes been called "the science of everything."

The Authors

PREFACE TO THE INSTRUCTOR

What is the role of a chemistry textbook? We often ask ourselves that question. Is it a tool for teaching? A stimulant for curiosity? A source of homework assignments? A reference book? It may play any or all of these roles, for a textbook should be adaptable to many different teaching strategies.

It is much easier to say what a chemistry textbook is not—it is not a mandatory syllabus. Textbook authors are not arbiters of course content, nor do we wish to be. We expect you to mold the course to the needs of your students, colleagues, and institution. Moreover, the course should reflect your own interests. If you are more excited about crystal structure than kinetics, teach more of one and less of the other. Conveying your personal enthusiasm and a view of how science works is considerably more important than conveying additional facts.

In this book, topics most commonly taught in the first semester are covered mainly in Chapters 1–15. For many instructors, Chapters 18–24 form the core of the second semester. Chapters 26–33 cover chemistry of the elements, organic chemistry, polymers, and biochemistry. We do not expect you to teach all of Chapters 26–33, although we know a few hardy souls who do so. Choose the topics you most enjoy teaching and/or those your students are most likely to need in the future. Those of you who teach inorganic qualitative analysis will want to include Chapters 34 and 35, found only in our alternate edition (Moeller et al., *Chemistry with Inorganic Qualitative Analysis*).

As textbook authors, we are obligated to respond to teaching trends. Today everyone is striving to find better, different, or more interesting ways to "integrate" descriptive chemistry and the principles of chemistry—more difficult to do in writing than in lectures. Our approach is to combine them in a few chapters where we believe the fit is natural.

Chapter 14 combines the chemistry of water with the introduction to ions, acids, bases, and equilibrium in aqueous solution. Chapter 16 combines the chemistry of hydrogen and oxygen with the introduction to redox reactions. If you wish to delay all descriptive chemistry to later in the course, it would be best to include the non-descriptive portions of these chapters earlier. Chapter 17 is a qualitative presentation of chemical reactions that can be a foundation for the quantitative topics that follow.

We believe a textbook should be "user friendly." Our Preface to the Student lists the ways in which we try to achieve this goal. Furthermore, every subject should be covered clearly, with sufficient depth to allow for understanding, and in a scientifically honest manner. Users of previous editions lead us to believe we have had some success in achieving these goals.

As another response to the needs of instructors today, this edition is shorter than the previous one. Eight color-photo segments, uniquely designed to be independent of the text, illustrate various aspects of chemical principles, descriptive chemistry, and applied chemistry. A few changes in organization have been made as well (e.g., uniting organic chemistry in one chapter, reorganizing Chapters 9–11 on bonding and periodic properties), and we have provided questions and problems paired as to topic, with answers to one of each selected pair in Appendix VII. Other adjustments too numerous to mention have been made. Suffice it to say that we have continued our efforts to produce a textbook that supports your efforts to teach and those of your students to learn.

The Authors

ACKNOWLEDGMENTS

We are grateful to the reviewers for their helpful comments: David Brown, College of Du Page; Edward Carberry, Southwest State University; Lawrence Conroy, University of Minnesota; Raymond Cozort, Northeast Mississippi Junior College; Robert V. Dilts, Vanderbilt University; Emily Dudek, Brandeis University; David Dunham, Mount Hood Community College; Robert Freeman, Oklahoma State University; Michael Golde, University of Pittsburgh; Charles H. Henrickson, Western Kentucky University; Thomas Lubin, Cypress College; Magla Niazy, Drexel University; Miriam Rossi, Vassar College. In addition, we would like to thank the following individuals: Arnold Drucker, University of Connecticut—Stamford; Kip Peticolas, Fundamental Photographs; Allan L. Smith, Drexel University; Walter Thomas, American Cyanamid Company. And we sincerely thank the chemistry faculty at College of Charleston, and our production and editorial staff at Harcourt Brace Jovanovich.

The Authors

CONTENTS

ASIDES

To everyone who helped.

CHAPTER

1

The Nature of Chemistry

(We open this book with a letter to a friend from John C. Bailar, Jr., who has been a member of the chemistry department faculty at the University of Illinois for 61 years.)

Dear Chris:

This letter is an answer to your questions about just what chemistry is and what chemists do. I'm glad that you asked, for many people have a distorted, or at least superficial, view of what the subject is all about. Whether I can give you a clear picture of it in a letter like this, I am not sure, but I shall try.

You know, of course, that chemistry is one of the **physical sciences**, along with physics, geology, and astronomy. Closely related, but in a somewhat different category, are the **biological sciences**, such as botany, physiology, ecology, and genetics. There is no sharp distinction between the two groups of sciences, or between those in either group, for they overlap each other. Often it is difficult to decide whether a specific topic belongs in one area or another. Many important subjects fall within the boundaries of several different disciplines. [Definitions of terms given in boldface type in this letter are presented at the end of the letter.]

All of the sciences overlap extensively with chemistry; they depend upon it and, in large measure, are based upon it. By that I mean that chemistry is really a part of all of the natural sciences, and a person cannot go very far in any science without some knowledge of chemistry. It would be possible to be a chemist without much knowledge of astronomy or physiology, but certainly, one could not make great progress in astronomy or physiology without some understanding of chemistry. A knowledge of chemistry is essential in other scientific fields as well. Agriculturists, engineers, and medical doctors use chemical concepts constantly.

Chemistry is concerned with the composition of **matter** and the changes in composition which matter undergoes—in brief, chemistry is the science of matter. Physics is concerned chiefly with energy and with the interactions of matter and energy, including energy in such forms as heat, light, sound, electricity, mechanical energy, and nuclear energy. All changes in the composition of matter either release or absorb energy and for this reason the relationship between chemistry and physics is a most intimate one.

We think of any change in which the composition of matter changes as a **chemical change.** For example, if you pour vinegar on baking soda in a glass vessel, you will see bubbles of gas escaping and the liquid will become warm as energy is released. When the bubbling stops, you can evaporate the liquid by boiling it, until finally only a white powder remains. But this white powder is not the original baking soda. It is a new substance with new characteristics. For example, it won't give off bubbles if you pour vinegar on it. This new material is different in composition from either of the materials which you originally mixed together. A chemical change has taken place.

By contrast, a **physical change** does not involve a change in the composition of matter. The melting of ice or the stretching of a rubber band are physical changes. It is often impossible to say whether a particular change is chemical or physical. Happily, it is not usually necessary to make a clear distinction between the two.

You must not assume that in your first course in chemistry you'll learn about the chemistry of the digestion of food or how a mixture of cement and water sets and hardens. These are complex processes, and before one can understand them one must first learn the chemistry of simpler substances. In learning to play the piano, a student does not start with Rachmaninoff's *Prelude in C# Minor*. A music student must first learn to play scales, and then simple pieces. It is only after months or years of practice that an individual can play the music of the masters. So it is with chemistry. You must first learn the fundamental principles and something about simple substances such as water and oxygen. A good understanding of the behavior of such substances will then allow you to understand the chemical behavior of more complex materials.

The science of chemistry is so broad that no one can be expert in all of its aspects. It is necessary to study the different branches of chemistry separately, and, if you become a chemist, to specialize in one or two branches of the subject.

Until about 150 years ago, it was believed that inanimate matter and living matter were of entirely different natures and had different origins. The inanimate matter was referred to as "inorganic" (meaning "without life") and the living matter and material derived from living matter were called "organic." However, in 1828, a German chemist named Friedrich Wöhler heated a material that was known to be inorganic and obtained a substance that all chemists recognized to be a product formed in life processes. So the distinction between "inorganic" and "organic" broke down. We still use these terms, but they now have different meanings from those they had in the early days. All living matter contains carbon chemically combined with hydrogen, so the chemistry of chemical compounds of carbon and hydrogen, whatever their origin, is called **organic chemistry**. Substances that do not contain carbon combined with hydrogen are "inorganic," and their chemistry is called **inorganic chemistry.** Carbon is very versatile in its behavior and is a key substance in a great many compounds, including most of the compounds essential to life.

There are other branches of chemistry, too. *Analytical chemistry* is concerned with the detection or identification of the substances that are present in a material (**qualitative analysis**) and how much of each is present (**quantitative analysis**). *Physical chemistry* is the application of the methods and theories of physics to the study of chemical changes and the properties of matter. Physical chemistry really forms the foundation for all of the other branches of the subject. *Biochemistry*, as the name implies, is concerned with the chemistry of the processes that take place in living things.

Inorganic, organic, analytical, physical chemistry, and biochemistry are the main branches of chemistry, but it is possible to combine portions of them, or to elaborate on them in many ways. For example, *bioinorganic chemistry* deals with the function of the metals that are present in living matter and that are essential to life. *Pharmaceutical chemistry* is concerned with drugs: their manufacture, their composition, and their effects upon the body. *Clinical chemistry* is concerned chiefly with the analysis of blood, urine, and other biological materials. *Polymer chemistry* deals with the formation and behavior of such substances as rayon, nylon, and rubber. (Some people would include inorganic polymers such as glass and quartz.) *Environmental chemistry*, of course, deals with the composition of the atmosphere and the purity of water supplies—essentially, with the chemistry of our surroundings. *Agricultural chemistry* is concerned with fertilizers, pesticides, plant growth, the nutrition of farm animals, and every other chemical topic that is involved in farming.

One more topic should be mentioned—*chemical engineering*—which is concerned with the applications of chemistry on a large scale. Chemical engineers design and operate chemical factories; they deal with the economics of making chemicals on a commercial scale. They are also concerned with such processes as distilling, grinding, and drying materials in large amounts—even the study of the friction of liquids and gases flowing through pipes.

Before you can undertake the study of any of these broad fields of chemistry, you will need to take a course, usually called "General Chemistry," which is the basis for more specialized study. You will quickly learn that general chemistry consists of two interrelated parts: **descriptive chemistry** and **principles of chemistry.**

Descriptive chemistry generally deals with the "What . . . ?" questions: What does that substance look like? What happens when it is heated? What happens when an electric current flows through it? What occurs when it is mixed with another specific substance? Chemistry is an experimental science and chemists work with a great many substances. It is important that they know the nature of these substances: their solubility in water or other liquids, their flammability, their toxicity, whether they undergo chemical changes in damp air, and many other characteristics. Sometimes the availability and cost of a substance are also important. The descriptive part of the general chemistry course is concerned chiefly with the behavior of some of the simpler inorganic substances, but may include brief discussions of organic and biochemical materials as well.

The *principles* part of the course is concerned with theories of chemical behavior. That is, it attempts to answer the "Why . . . ?" questions: Why won't a substance dissolve in water? Why did an explosion take place when a mixture was heated? Why was a particular substance and not a different one formed in a chemical change? Why do some chemical changes speed up dramatically if a tiny amount of something else is added?

The study of chemical principles is of great practical as well as intellectual interest. We can, for example, calculate how much heat is given off when a particular fuel burns, and determine how to speed up or slow down its combustion. When we know why certain substances behave as they do we can often modify their behavior to achieve desirable or useful results.

Chemistry is an experimental science. By this statement, I do not mean that chemists do not have theories about changes in chemical composition—under what conditions they will or will not take place, how they take place, and what the products will be. There are always theories. But **theory** must always be subject to experiment. If one's theory is not in accordance with carefully executed experiments, then the theory, not the experiment, must be wrong. The theory must then be abandoned or modified. In this regard, chemistry is quite different from the social sciences, such as sociology and economics. People who work in those fields may have theories about the causes of inflation or unemployment or marital unhappiness, and they may carry out experiments to test their theories. But these experiments can never be repeated and checked under the same conditions, for in the act of doing the experiment the conditions have been irretrievably changed. This is true to some extent also in the biological sciences. A pharmacologist may test the effect of a given drug on a mouse and draw some conclusions from what happens to the mouse. But the experiment cannot be repeated with that same mouse, for there is no certainty that the health of the mouse has not been changed by the first administration of the drug. And although the experiment can be done with another mouse, there is no certainty that the second mouse will respond exactly as the first one did. Chemists are more fortunate: under the same conditions, pure chemicals will always react with each other in exactly the same way. The trick is to make sure that the chemicals are pure and that the conditions of the experiment are exactly the same.

But, you will ask, "Just what do chemists do?" That is a difficult question to answer, for chemists do many different things. About half the chemists in the United States work in laboratories. Some of them are "quality control" chemists. By a variety of laboratory techniques (some simple and some complex), they analyze or otherwise test materials that are to be used or are products of a chemical factory (be it a drug manufacturer, a food factory, or a steel mill) to ensure that these products are uniform and pure. Some chemists do laboratory research hoping to discover new chemicals or new uses for known chemicals, or to improve methods of making useful chemicals. Some seek to unearth new principles of chemical behavior, and their activities may range from laboratory work to using only pure mathematics. None of this, of course, is hit-and-miss experimentation. A chemist is always guided by a background of both chemical theory and practical experience, and the broader these are the more successful the chemist will be.

But what about those who do not work in laboratories? Are they still chemists? Indeed they are, though they may combine their chemical activity with some other professional work. Some spend their time looking for new uses and markets for substances that the research chemists have discovered; some are teachers (or divide their time between teaching and research); some become writers of scientific articles for newspapers and magazines.

You may be wondering whether you should study chemistry at all. I hope that you will do so, for as I indicated earlier, a knowledge of chemistry is useful, no matter what profession you follow. If you decide to become a mechanical engineer, you'll need to know something about fuels and alloys and corrosion; if a civil engineer, you must have a knowledge of cement, plaster, steel, and other building materials; if an electrical engineer, you'll need a knowledge of how a battery produces electrical energy, and the changes that take place in it when it is recharged, as well as a knowledge of transistors and lasers. Should you become a medical doctor, you'll be dealing with the most complex chemical plant of all—the human body—and the multitude of chemicals in it. Although my own son, John, studied chemistry for three years as an undergraduate, after one year in medical school he returned to his undergraduate college to take a summer course in physical chemistry, for he had discovered that he needed that extra chemical knowledge in his medical studies.

If you decide to go into agriculture, you'll need to know about fertilizers and pesticides, as well as animal nutrition. Even if you enter some profession that seems to have no connection with chemistry, such as law, you'll find a knowledge of chemistry very useful. Lawyers frequently must deal with patents that concern chemical inventions. Some members of the U.S. Congress have had extensive chemical training, which gives them a great advantage in discussions of environmental pollution, nuclear energy, the regulations of the Food and Drug Administration, and in other legislation that concerns scientific matters.

The chemical profession is so broad that persons of many different interests and temperaments find satisfaction in it. A person who studies chemistry for very long develops habits of thinking logically and clearly. Once that is accomplished, he or she can do almost any sort of work.

I hope that you will enjoy your study of chemistry. I have found it to be a fascinating subject, because of its history, the beauty of its logic, and its multitude of applications.

Sincerely,

John C. Bailar Jr.

SIGNIFICANT TERMS

biological sciences the study of living matter

chemical change a change in the composition of matter

chemistry the study of matter, the changes that matter can undergo, and the laws that describe these changes

descriptive chemistry the description of the elements and their compounds, their physical states, and how they behave

inorganic chemistry the chemistry of all the elements and their compounds, except compounds of carbon with hydrogen and their derivatives

matter everything that has mass

organic chemistry the chemistry of compounds of carbon with hydrogen, and of their derivatives

physical change a change in which the composition of matter is unaltered

physical sciences the study of natural laws and processes other than those peculiar to living matter

principles of chemistry the explanations of chemical facts, for example, by theories and mathematics

qualitative analysis the identification of substances, often in a mixture

quantitative analysis the determination of the amounts of substances in a compound or mixture

theory a unifying principle or group of principles that seeks to explain a body of facts or phenomena

CHAPTER

2

Units, Measurements, and Numbers

CHEMISTRY AS A QUANTITATIVE SCIENCE

Every object in the world around you can be described in terms of chemistry, and many events that you can see occurring involve chemical changes. The changing color of the leaves in the fall, the transformation of a pond into a swamp, the rusting of iron, the explosion of fireworks—these are all chemical changes.

Let's consider what is seen in one specific chemical change. Two substances are involved. One is a black powdery solid. The other is a colorless liquid that causes irritation if spilled on the skin. If some of the black solid is placed in a container and the liquid slowly added, things happen. The black solid begins to dissolve, and the solution that is formed is not black but very pale green. At the same time, a gas begins to bubble out of the solution and the air is filled with a terrible smell, like that of rotten eggs.

What a multitude of questions can be asked here. What are these substances? Why did the black solid dissolve? What was formed in its place? *How much* of the liquid does it take to dissolve all of the black solid? *How much* of the gas can be produced? *How long* did the change take? Will events speed up if the mixture is heated? If so, by *how much per degree of temperature*?

Notice how many of these are quantitative questions. Eventually, the study of almost every science arrives at numbers and measurements. In chemistry, to learn about the composition of matter requires experiments on carefully weighed amounts of material under controlled conditions.

In studying chemistry you will be presented with facts accumulated during hundreds of years of observation and measurement. You will also learn how the principles of chemistry are used to explain what has been observed. To test your understanding of chemical principles, you will solve problems, frequently problems that utilize the results of measurements of physical properties. Therefore, right from the beginning in studying chemistry, it is best to be prepared to solve problems—the goal of this chapter is to provide you with the problem-solving skills that you will need.

This chapter may be frustrating to some of you. Here you are ready to study chemistry and your first encounter is with topics that appear very nonchemical, more like the beginning of a mathematics course. We assure you that mastering the subject matter of this chapter is going to make everything that follows easier.

2.1 MEASUREMENT AND SIGNIFICANT FIGURES

The result of measuring a physical property is expressed by a numerical value together with a unit of measurement. For example

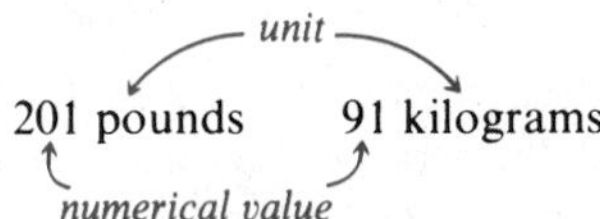

Here and in the next four sections we concentrate on the numerical part of such physical quantities—on how many digits should be included and on the use of scientific notation. In later sections, we discuss the units.

If you count the pencils on your desk, there is no uncertainty in the number that results. You may have, say, five pencils, *exactly*. Also, there is no uncertainty in the number of inches in one foot. By definition 12 inches = 1 foot. These are **exact numbers**—numbers with no uncertainty. Exact numbers arise by directly counting whole items or by definition.

Numbers that result from measurements are never exact. There is always some degree of uncertainty due to experimental errors: limitations of the measuring instrument, variations in how each individual makes measurements, or other conditions of the experiment. For example, with a finely constructed, sensitive balance, it is possible to read the mass of a sample as 3.1267 grams. A simple balance might only allow the mass of the same sample to be read as 3.13 grams—such a balance cannot detect differences in mass beyond hundredths of a gram (Figure 2.1). In both of these measurements there is a degree of uncertainty in the final digit—the 7 in 3.1267 and the final 3 in 3.13. The preceding digits are definitely known, but the final digits are near the limits of what the balances can detect.

Physical quantities are reported to include only the first uncertain digit. The **significant figures,** or **significant digits,** in a number include all the digits that are known with certainty, plus the first digit to the right that has an uncertain value.

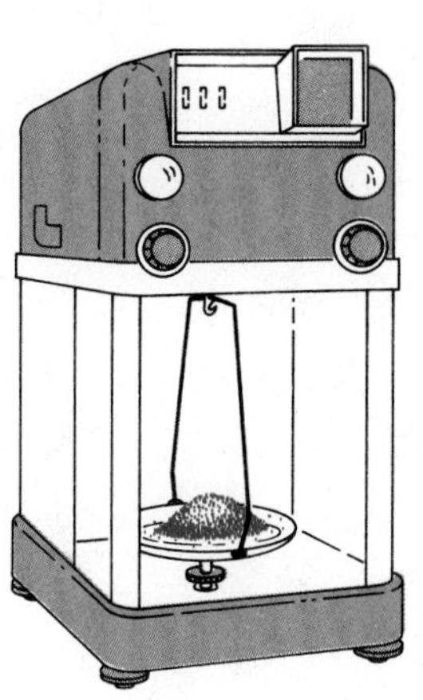

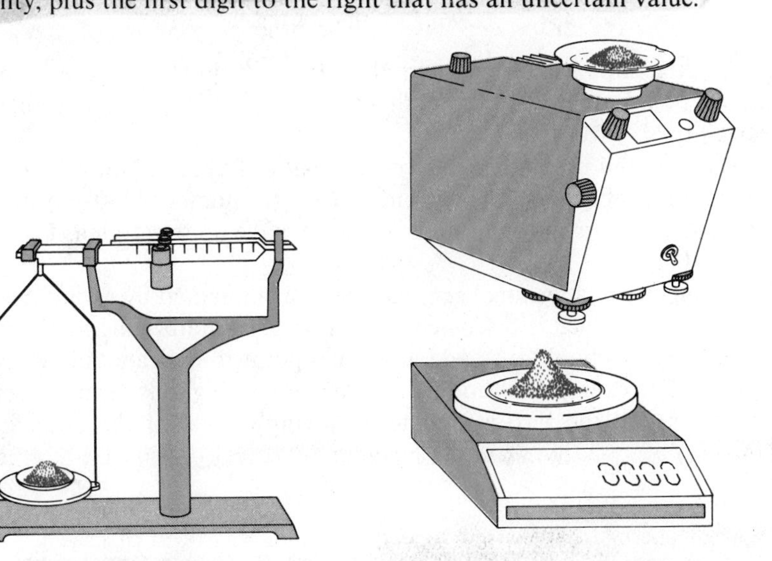

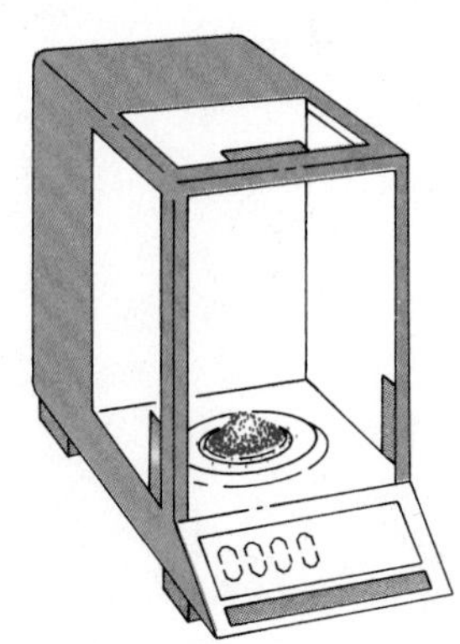

Single-pan balance **Top-loading balances** **Automatic single-pan balances**

Figure 2.1

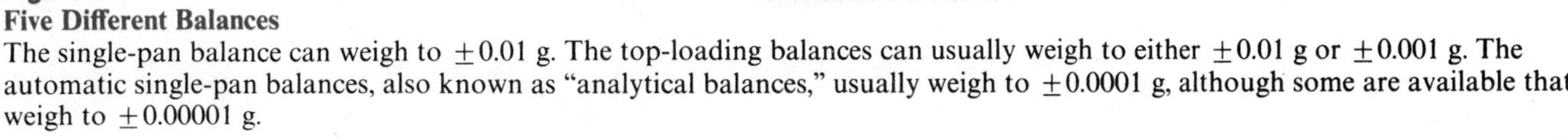
Five Different Balances
The single-pan balance can weigh to ±0.01 g. The top-loading balances can usually weigh to either ±0.01 g or ±0.001 g. The automatic single-pan balances, also known as "analytical balances," usually weigh to ±0.0001 g, although some are available that weigh to ±0.00001 g.

Consider the following results reported after the same piece of candy was weighed by the members of a class (g is the symbol for grams).

23.1 g	23.3 g	23.0 g	23.3 g	23.2 g	23.2 g
23.3 g	23.4 g	23.2 g	23.1 g	23.3 g	23.2 g

As you can see, there is some uncertainty in the measurements. Everyone agreed on the first two digits, but the third digit from the left—the tenths of a gram place—differs. There is no doubt that the piece of candy weighs 23 g plus something more, but the uncertainty lies in how much more.

The average of the measured masses is 23.2 g. To reflect the uncertainty in the measurement, the mass could be reported as 23.2 ± 0.2 g, which includes all the available data. (The highest value is 23.4 g and the lowest is 23.0 g.) However, data are usually given without any indication of the amount of the specific uncertainty. Consequently, the mass of the piece of candy would be reported as 23.2 g.

It is the responsibility of one reporting the results of experiments to include only the number of digits that are significant. In using measured data reported by others, we assume, therefore, that the uncertainty lies only in the last digit to the right. Without specific information, we assume that the uncertainty is at least ± 1. The decimal place of the digit does not matter. For example, the uncertainty in 3285 lies in the 5, and in 0.042 in the 2.

uncertainty of ±1 → 3285   *uncertainty of ±0.001* → 0.042

2.2 FINDING THE NUMBER OF SIGNIFICANT FIGURES

In order to use physical quantities correctly in arithmetic, one must recognize how many significant figures are included in the numerical values. The last digit to the right is assumed to be uncertain. The number of significant figures is found by counting from left to right, beginning with the first nonzero digit and ending with the digit that has the uncertain value. Each of the following numbers has three significant figures.

123	*123*	*1 23*	*123*	*12 3*
454	0.296	7.31	0.00846	10.7

Zeros at the end of a number and also after a decimal point are significant. For example, the final zero in 1.520 shows that this number has four significant figures and an uncertainty of ±0.001.

Zeros at the end of a number given *without* a decimal point present a problem because they are ambiguous. The terminal zero in the number 1520 may or may not be significant; unless more information is given, we have no way to tell. The uncertainty might be ±10 or ±1.

The ambiguity in a terminal zero can best be eliminated by giving the number in scientific notation, which is discussed in Section 2.4. Another method, which we use in some cases in this book, is to add a decimal point to indicate that a terminal zero is significant. For example, although 1520 has an ambiguous terminal zero, writing 1520. indicates that the zero is significant, giving a total of four significant digits.

How to count the significant figures in a number is summarized in the following rules:

1. All nonzero digits are significant.
2. To count significant digits, begin with the first nonzero digit in the number as read from left to right. This means that zeros that *precede* the first nonzero digit are not significant.

12	*1*	*1*
0.096	0.000003	0.8

3. Zeros are significant when they appear (a) in the *middle* of a number or (b) at the *end* of any number that includes a decimal point.

1234	*12 345*	*1234*	*12*	*12 3456*
1003	70,204	0.6900	50.	50.0000

4. Zeros at the end of a number given without a decimal point are ambiguous. Adding a decimal point indicates their significance, according to rule 3(b).

	12??	*1234*
two or more significant digits	3600	3600.

[In this book, worked examples are followed by exercises that pose similar problems. The exercises include only the answers. By doing the exercises immediately after reading the examples, you can be sure that you understand how to solve each type of problem.]

EXAMPLE 2.1 Significant Figures

How many significant figures are there in the numbers (a) 57, (b) 82.9, (c) 10.000, (d) 0.000002, (e) 0.0402, (f) 0.04020?

(a) Beginning with the 5 and counting gives **two** significant figures — *12* 57

(b) Beginning to count with the 8 gives **three** significant figures — *12 3* 82.9

(c) All zeros here are significant [rule 3(b)], giving **five** significant figures — *12 345* 10.000

(d) None of the zeros here are significant (rule 2), giving **one** significant figure — *1* 0.000002

(e) The zeros at the left are not significant, but the zero in the middle is [rules 2 and 3(a)], giving **three** significant figures — *123* 0.0402

(f) The zero at the end is significant as well as the zero in the middle (rule 3), giving **four** significant figures — *1234* 0.04020

EXERCISE How many significant figures are in the numbers (a) 234.7, (b) 0.0300, (c) 630, (d) 700., (e) 1036? *Answers* (a) 4, (b) 3, (c) 2 (or more), (d) 3, (e) 4.

2.3 ARITHMETIC USING SIGNIFICANT FIGURES

The result of a calculation involving physical quantities should not be given with a smaller uncertainty than the original quantities. It is poor practice, for example, to divide 7.8 g by exactly 7 and give more than two significant figures in the answer. The uncertainty represented in the original measurement has not changed.

Paying attention to the correct number of digits in the result of a calculation is especially important when working with electronic calculators. With the touch of a button, 7.8 divided by 7 is seen to be 1.114285714. *Do not assume that all the numbers in the digital display are significant.* Limit the number of digits in your answer as explained in the following sections.

a. Addition and Subtraction Adding 23.2 + 6.052 + 139.4 gives 168.652. However, this is not the correct answer because the implied uncertainty, ±0.001, is smaller than it should be. The greatest uncertainty lies in 23.2 and 139.4, both of which are uncertain to ±0.1. Therefore, the answer should be rounded to 168.7, which also has an uncertainty of ±0.1.

To maintain the correct number of significant figures in addition or subtraction, round the answer to the place (before or after the decimal point) with the greatest uncertainty. An easy way to apply this rule is to draw a vertical line after the digit of the greatest uncertainty in the numbers being added or subtracted, and then round the answer to the same place, as shown.

```
  23.2¦      ← rule follows the place of greatest uncertainty
   6.0¦52
 139.4¦
 ---------
 168.6¦52
 168.7¦ (±0.1)  ← rounded to same uncertainty as 23.2 and 139.4
```

In this book we use the following rule for rounding off an answer: If the numbers following the desired place are from 000 through 499, drop these numbers and do nothing to the number in the desired place; if the numbers following are from 500 through 999, drop these numbers and increase the number in the desired place by one.

For example, to round 6.713, 6.785, and 6.75 to the one-tenths place:

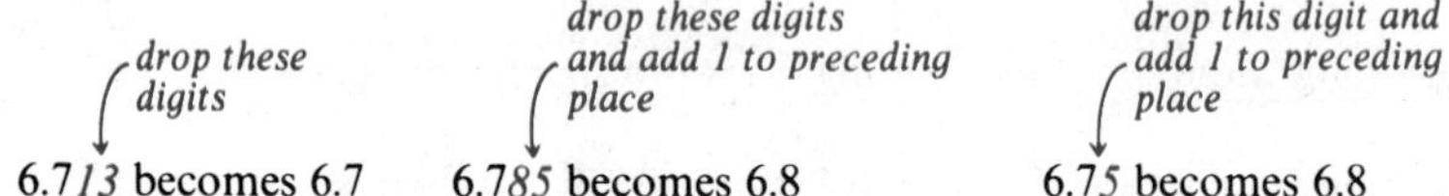

b. Multiplication and Division Multiplying the two numbers (23.2) (0.1257) gives 2.91624. Here again, this is not the correct answer, because of the difference in the uncertainty of the numbers. To maintain the correct number of significant figures in multiplication and division, perform the multiplication or division and then round the answer to the same number of significant figures as the number with the fewest significant figures. For the preceding problem, the answer should have three significant figures:

$$\overset{123}{(23.2)}\overset{1234}{(0.1257)} = [2.91624] = \overset{123}{2.92}$$

An exact number does not limit the number of significant figures in the result of a calculation. For example, if an experiment is to be performed exactly six times and 6.35 g of a substance is needed each time, the total amount needed is 38.1 g.

$$\text{exact number } (6)\overset{123}{(6.35\text{ g})} = \overset{123}{38.1\text{ g}}$$

need not be rounded to one significant figure

Note that when an exact number has more digits than an inexact number by which it is multiplied or divided, the number of significant digits is determined by the inexact number. For example

$$\overset{12}{(2.54)}(33) = [83.82] = \overset{12}{84}$$

exact number

EXAMPLE 2.2 Significant Figures—Arithmetic

Perform the following calculations and express the answers to the proper number of significant figures:

(a) $14 - 0.072$ (b) 0.0097×0.0563 (c) $\dfrac{26.9}{2.69}$

(d) (0.0729)(10) (e) $(13.65)^2$ (f) $\dfrac{14.89}{0.0003}$

In (d), assume that the factor 10 is exact.

(a)

$$\begin{array}{r|l} 14 & \\ -\ 0 & 072 \\ \hline 13 & 928 \end{array}$$

14

(b)

$$\begin{array}{r|} 0.0097 \\ 0.0563 \\ \hline 0.0660 \end{array}$$

0.0660

(c) $\dfrac{\overset{12\ 3}{26.9}}{\underset{1\ 23}{2.69}} = [\overset{12\ 3}{10}] = \mathbf{10.0}$

(d) $(\overset{123}{0.0729})(\overset{\text{exact number}}{10}) = \overset{123}{\mathbf{0.729}}$

(e) $(\overset{12\ 34}{13.65})^2 = [\overset{123\ 4}{186.3225}] = \mathbf{186.3}$

(f) $\dfrac{\overset{12\ 34}{14.89}}{\underset{1}{0.0003}} = [\overset{1}{49{,}633}] = \mathbf{50{,}000}$

(a) The numbers to the right of the vertical line are greater than 500, so they are dropped and 1 is added. (b) The uncertainty is in the same decimal place in both numbers added and the answer need not be rounded off. (c) A decimal point and a zero must be added to the calculated answer to obtain the correct number of significant figures. (d) The exact factor of 10 does not limit the number of significant figures in the answer. (e) Because $(13.65)^2$ is equivalent to (13.65)(13.65), the answer has four significant figures. (f) Despite all the decimal places in the factors, the answer should have only one significant figure. [Although many students at first are troubled by this big difference between the calculated answer and the properly significant answer, it is perfectly acceptable.]

EXERCISE Perform the following calculations and express the answers to the proper number of significant figures: (a) (16.3) – (10.02), (b) (6000.) – (32), (c) (16.3)/(10.02), (d) (6000.)(32), (e) (0.0327)/(0.0147) *Answers* (a) 6.3, (b) 5968, (c) 1.63, (d) 190,000, (e) 2.22.

The question arises of how to determine the number of significant figures in the answer to a problem that includes multiplication or division, as well as addition or subtraction. One approach, illustrated in Example 2.3, is to carry out the complete calculation without any rounding off and then to go back and determine which step or steps may limit the number of significant figures. Because addition, subtraction, multiplication, and division can be combined in many ways, each problem must be examined individually to determine which steps limit the significant figures. For example, because the number of significant figures can change in subtraction, the answer to a subtraction step (as illustrated in Example 2.3) must be examined. In a problem like that in the exercise following Example 2.3, the answer to each division step must be calculated separately and the subtraction then performed.

EXAMPLE 2.3 Significant Figures—Arithmetic

Perform the following calculation and express the answer to the proper number of significant figures.

$$x = \frac{(95.316)}{(2.303)(1.987)}\left(\frac{1}{298} - \frac{1}{308}\right)$$

Solving the expression in one step with a calculator gives

$$x = \left(\frac{(95.316)}{(2.303)(1.987)}\right)\left(\frac{1}{298} - \frac{1}{308}\right) = [2.269378954]$$

The answer to the first factor in the calculation (shown in the first set of braces above) is limited to four significant figures. To find the number of significant figures in the second factor requires carrying out the following subtraction:

$$\frac{1}{298} = 0.00336 \qquad \frac{1}{308} = 0.00325$$

$$0.00336 - 0.00325 = 0.00011$$

Because the second factor in the problem is limited to two significant figures, the calculated value of x must also have two significant figures, giving $\boldsymbol{x = 2.3}$.

EXERCISE Perform the following calculation and express the answer to the proper number of significant figures.

$$x = \frac{(44.008)}{(0.0820568)(273.15)(1.976757)} - \frac{1}{1.3}$$

Answer $x = 0.22$

2.4 SCIENTIFIC NOTATION

In chemistry, because very large and very small numbers are used often, we must be able to represent such numbers in some convenient fashion. Scientific notation (also called exponential notation) is the answer. In **scientific notation** the significant figures of a number are retained, usually as a factor from 1 to 9.99... and the location of the decimal point is indicated by a power of 10.

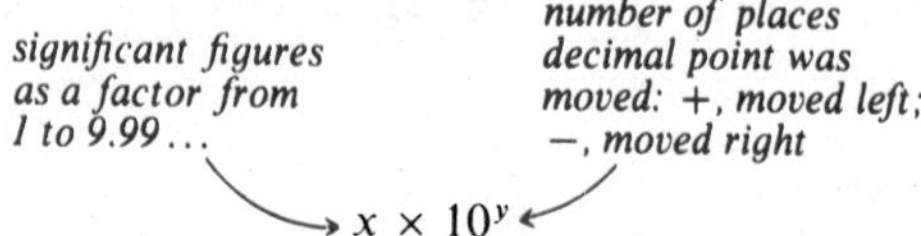

For example, the numbers 0.0063 and 900,000,000 are expressed in scientific notation as follows:

factor, two significant figures → 6.3×10^{-3} ← *exponent*

factor, one significant figure → 9×10^{8} ← *exponent*

As a rule, a number is not written in scientific notation unless it is easier to use (or less ambiguous) than the original number. Numbers such as 1.32, 0.9624, and 42 usually would not be written in scientific notation.

To change a number greater than 1 to scientific notation, the decimal point is moved to the left until only one digit remains to the left of it. The power of 10 is equal to the number of places that the decimal point was moved. For example, to rewrite 2,000,000 in scientific notation, the decimal point is moved left six places, giving 10 to the sixth power.

2,000,000 becomes 2×10^6

654 321

[This operation is equivalent to dividing the number by 1 million and then multiplying it by 1 million expressed as 10^6.]

To change a number less than 1 to scientific notation, the decimal point is moved to the right until only one digit appears to the left of it. The power of 10 is equal to the number of digits that the decimal point was moved and a minus sign appears in the exponent. For example

$$0.0006 \text{ becomes } 6 \times 10^{-4} \qquad 0.0263 \text{ becomes } 2.63 \times 10^{-2}$$

1234   *12*

The problem of the ambiguous significance of the zeros at the end of a number such as 200 can be solved with scientific notation. The number can be written to clearly show one, or two, or three significant digits.

$$2 \times 10^2 \qquad 2.0 \times 10^2 \qquad 2.00 \times 10^2$$

Table 2.1 gives some examples of physical quantities of various magnitudes.

Table 2.1
Physical Quantities

Masses	
Sun	2.0×10^{30} kg
Earth	6.0×10^{24} kg
Person	59 kg
Basketball	0.6 kg
Virus	2.3×10^{-13} kg
Electron	9.1×10^{-31} kg
Lengths	
Radius of our galaxy	5×10^{20} m
Radius of sun	7.0×10^{8} m
Radius of Earth	6.4×10^{6} m
Height of Mt. Everest	8.9×10^{3} m
Person	1.8 m
Virus	1.7×10^{-8} m

EXAMPLE 2.4 Scientific Notation

Express the following numbers in standard scientific notation: (a) 6100 and (b) 0.000870. Write out in full the following numbers: (c) 4.33×10^4, (d) 6.1234×10^2, (e) 2.0×10^{-2}.

The decimal point must be moved until only one digit is to the left of it; the number of places the decimal point was moved gives the power of 10 (positive for number >1; negative for number <1).

(a) $6100 = \mathbf{6.1 \times 10^3}$ (*321*)   (b) $0.000870 = \mathbf{8.70 \times 10^{-4}}$ (*1234*)

To write out a number given in scientific notation, the reverse procedure is carried out. For positive powers of 10, the decimal point in the factor is moved to the right a number of places equal to the power of 10; for negative powers of 10, the decimal point is similarly moved to the left.

(c) $4.33 \times 10^4 = 43{,}300 = \mathbf{43{,}300}$ (*1 234*)   (d) $6.1234 \times 10^2 = 612.34 = \mathbf{612.34}$ (*12*)

(e) $2.0 \times 10^{-2} = 0.020 = \mathbf{0.020}$ (*21*)

EXERCISE Express the following numbers in standard scientific notation: (a) 0.0203, (b) 5,260,000, (c) 0.0010. Write out in full the following numbers: (d) 6.3×10^{-6}, (e) 1.01×10^4, (f) 3.0×10^{-2}.

Answers (a) 2.03×10^{-2}, (b) 5.26×10^6, (c) 1.0×10^{-3}, (d) 0.0000063, (e) 10,100, (f) 0.030.

2.5 ARITHMETIC USING SCIENTIFIC NOTATION

When doing arithmetic with numbers expressed in scientific notation, the rules for significant figures are applied to the factors exactly as for any numbers. For addition and subtraction using pencil and paper, all the numbers must have the same exponent, even if this produces some numbers with factors not from 1 to 9.99.... For example, to add 9.9×10^6 to 1.23×10^8, the 9.9×10^6 must be expressed with the eighth power of 10:

$$\begin{array}{l} 0.099 \times 10^8 \\ 1.23 \times 10^8 \\ \hline 1.329 \times 10^8 \\ 1.33 \times 10^8 \end{array}$$

The power of 10 in the answer is then the same as that in the numbers added or subtracted. Today, to do addition or subtraction with numbers in scientific notation, most of us need only enter the numbers into our calculators.

To multiply and divide numbers in scientific notation using pencil and paper, the factors are multiplied or divided according to the rules of significant figures. Then, the power of 10 for the answer is found by adding powers of 10 for multiplication and subtracting them for division. If necessary, the answer is rewritten with a factor from 1 to 9.99...

equal to 23 + 3

$$(6.022 \times 10^{23})(4.2 \times 10^{3}) = 25 \times 10^{26} = 2.5 \times 10^{27}$$

restated with factor from 1 to 9.99...

equal to 11 − 6 *equal to (−2) − (−1)*

$$\frac{9.9 \times 10^{11}}{5.1 \times 10^{6}} = 1.9 \times 10^{5} \qquad \frac{4.3 \times 10^{-2}}{2.01 \times 10^{-1}} = 2.1 \times 10^{-1} = 0.21$$

Of course, calculators directly give the answer for multiplication and division complete with the correct exponent. Where necessary, the number should be restated with the correct number of significant figures and a factor from 1 to 9.99....

EXAMPLE 2.5 Scientific Notation—Arithmetic

Perform the following calculations and express the answers in scientific notation.

(a) $(6.38 \times 10^{4}) + (5.2 \times 10^{5})$ (b) $(4.77 \times 10^{-4}) - (2.66 \times 10^{-3})$

(c) $(5.3 \times 10^{8})(9.62 \times 10^{-4})$ (d) $\dfrac{7.11 \times 10^{-3}}{4.26 \times 10^{4}}$ (e) $\dfrac{(6.7 \times 10^{-5})(1.2 \times 10^{10})^2}{(8.1 \times 10^{-3})^{11}(2.5 \times 10^{2})}$

(a)
$$\begin{array}{r} 0.638 \times 10^{5} \\ 5.2 \times 10^{5} \\ \hline 5.838 \times 10^{5} \\ \mathbf{5.8} \mathbf{\times 10^{5}} \end{array}$$

(b)
$$\begin{array}{r} 0.477 \times 10^{-3} \\ -2.66 \times 10^{-3} \\ \hline -2.183 \times 10^{-3} \\ \mathbf{-2.18} \mathbf{\times 10^{-3}} \end{array}$$

(c) $(5.3 \times 10^{8})(9.62 \times 10^{-4}) = 51 \times 10^{4} = \mathbf{5.1 \times 10^{5}}$

(d) $\dfrac{7.11 \times 10^{-3}}{4.26 \times 10^{4}} = \mathbf{1.67 \times 10^{-7}}$ (e) $\dfrac{(6.7 \times 10^{-5})(1.2 \times 10^{10})^2}{(8.1 \times 10^{-3})^{11}(2.5 \times 10^{2})} = \mathbf{3.9 \times 10^{36}}$

EXERCISE Perform the following calculations and express the answers in scientific notation: (a) $(3.62 \times 10^{-3}) + (2.68 \times 10^{-4})$, (b) $(3.14 \times 10^{-4})(5.1 \times 10^{-5})$, (c) $(4.58 \times 10^{4})/(5.44 \times 10^{-14})$, (d) $(7.39 \times 10^{5}) + (7.39 \times 10^{-5})$.

Answers (a) 3.89×10^{-3}, (b) 1.6×10^{-8}, (c) 8.42×10^{17}, (d) 7.39×10^{5}.

SUMMARY OF SECTIONS 2.1–2.5

Physical quantities must be used correctly in dealing with the results of measurements. Based upon the uncertainty in the measurement, the numerical value of a physical quantity must be expressed to the correct number of significant figures. In doing arithmetic with physical quantities, the uncertainty in the answer must be no smaller than that in the numbers involved in the calculation.

Very large and very small numbers are best expressed in scientific notation. A factor from 1 to 9.99... indicates the significant figures in the number (allowing the ambiguity of terminal zeros to be avoided) and the location of the decimal point is indicated by a power of 10. A positive power of 10 indicates that the number is greater than 1 and gives the number of places the decimal point was moved to the left. A negative power of 10 indicates that the number is less than 1 and gives the number of places the decimal point was moved to the right.

The rules for doing arithmetic with significant figures and scientific notation are summarized in Table 2.2.

Table 2.2
Rules for Arithmetic

Significant figures
Addition and subtraction: add or subtract; round the answer to the place (before or after the decimal point) with the greatest uncertainty in the numbers being added or subtracted.
Multiplication and division: multiply or divide; round the answer to the same number of significant figures as in the number with the fewest significant figures.

Scientific notation
Addition and subtraction: write all numbers in the same power of 10; add or subtract factors; answer has same power of 10.
Multiplication and division: multiply or divide factors; get power of 10 by adding (multiplication) or subtracting (division) powers of 10.

UNITS OF MEASUREMENT

2.6 SYSTEMS OF MEASUREMENT

If you ask an Englishman his weight, he might reply, "14 stone." An American might say, "180 pounds," and a Canadian might say "91 kilograms," in answer to the same question. These answers leave you in the dark about who is the heaviest person unless you are familiar with three different systems of measurement and the relationships among them.

Easy comparison of measurements made by different people, in different laboratories, or in different countries, is only possible when measurements of the same quantities are expressed in the same units. If the Englishman had used the unit kilograms instead of stones, he would have replied "89 kilograms." The American, using the same system of measurement, would have replied that he weighs 82 kilograms. Without any effort, we can now see that the Canadian is the heavyweight of the group at 91 kilograms and the American is the lightweight at 82 kilograms.

The metric system was devised by the French National Academy of Sciences in 1793 to replace the profusion of units handed down from medieval times. Since then, various international bodies have been defining and redefining units of measurement and attempting to gain widespread uniformity in their use. For many years people in most European countries and scientists everywhere have used the metric system. The United States, until 1975, officially stayed with a weights and measures system based upon the English system of inches and feet, ounces and pounds, pints and quarts, and so on. The Metric Conversion Act of 1975 committed this country to a policy of voluntary conversion to the metric system. Although the metric system is gaining acceptance, the changeover from the English system is proceeding very slowly because of conversion costs and general public resistance to change.

In 1960, the International Bureau of Weights and Measures adopted the International System of Units, known as the "SI system" (for Système International). The SI system is a revision and extension of the metric system. Scientists and engineers throughout the world in all disciplines are now being urged to use *only* the SI system of units. However, the scientific community is as resistant to change as the general public, and some older metric units remain in use. At the present time, everyone working in science must be familiar with both the metric and the SI systems of measurement.

2.7 THE SI SYSTEM OF MEASUREMENT

In the SI system there are seven base physical quantities. The units for these quantities are defined in terms of constant physical phenomena, rather than other units (see Appendix II). The seven base physical quantities are given in Table 2.3, together with their units. To express, for example, the length of a piece of pipe 2 meters long, the unit and quantity would be used as follows: $l = 2$ m.

Table 2.3
SI Base Physical Quantities and Units
Luminous intensity and the candela are not commonly encountered in chemistry.

Quantity (symbol)	Name of unit	Unit symbol
Length (l)	meter (or metre)	m
Mass (m)	kilogram	kg
Time (t)	second	s
Electric current (I)	ampere	A
Temperature (T)	kelvin	K
Luminous intensity (I_v)	candela	cd
Amount of substance (n)	mole	mol

In the SI system, all physical quantities other than the base quantities and the units for all physical quantities are derived from the base units. For example, volume is a derived physical quantity. To find the volume of a box, the lengths of its three sides are multiplied together. Length is a base physical quantity in the SI system, and the base unit of length is the *meter*. With the length of the sides in meters, the volume of a box has units of meters cubed, or cubic meters.

$$(\text{meters})(\text{meters})(\text{meters}) = (\text{meters})^3 \text{ or cubic meters}$$

Cubic meters is the derived unit for volume in the SI system.

Table 2.4 lists some derived units that are used in chemistry. Many combine several base units. Such derived units are simplified by assigning a single unit to represent the physical quantity. We do not, for example, have to recall each time we use it that pressure is given in "kilograms per meter per second squared," kg/m s^2. The SI approved unit for pressure is the *pascal*, abbreviated Pa. By definition, 1 Pa = 1 kg/m s^2.

Often the SI unit for a physical quantity is inconvenient because it leads to a numerical value that is very large or very small. To solve this problem, multiples and fractions of the base or derived units are indicated by using the prefixes given in Table 2.5. For example, one type of virus is about 0.00000001 m or 1×10^{-8} m, long. The meter is too big a unit for the length of viruses, and it is more convenient to discuss viruses in units of nanometers. One nanometer is equal to 0.000000001 m, and the virus is about 10 nm long. (Note that the prefix "nano" in nanometer is written in front of the unit name without a hyphen and that the prefix abbreviation "n" in nm is similarly written in front of the unit abbreviation.)

2.8 UNITS OF MEASUREMENT IN CHEMISTRY

In the sections that follow, many of the units for physical quantities important to chemistry are discussed individually. Table 2.6 compares some of the SI units to units with which you may be more familiar. We suggest that you refer back to this table as you read the following sections.

Table 2.4
SI Derived Physical Quantities and Units

Quantity (symbol)	Name of unit (symbol)	Derived unit
Area (A)	square meter	m^2
Volume (V)	cubic meter	m^3
Density (ρ)	kilogram per cubic meter	kg/m^3
Velocity (u)	meter per second	m/s
Pressure (P)	pascal (Pa)	kg/m s^2
Energy (E)	joule (J)	kg m^2/s^2
Frequency (ν)	hertz (Hz)	1/s
Quantity of electricity (Q)	coulomb (C)	A s
Electromotive force (E)	volt (V)	kg m^2/A s^3
Force	newton (N)	kg m/s^2

Table 2.5
Prefixes for Multiples and Fractions of SI Units
The symbol for the prefix "micro" is the Greek letter mu, μ.

Decimal location	Prefix	Prefix symbol
$1{,}000{,}000{,}000{,}000 = 10^{12}$	tera	T
$1{,}000{,}000{,}000 = 10^{9}$	giga	G
$1{,}000{,}000 = 10^{6}$	mega	M
$1{,}000 = 10^{3}$	kilo	k
$100 = 10^{2}$	hecto	h
$10 = 10^{1}$	deka	da
$0.1 = 10^{-1}$	deci	d
$0.01 = 10^{-2}$	centi	c
$0.001 = 10^{-3}$	milli	m
$0.000\,001 = 10^{-6}$	micro	μ
$0.000\,000\,001 = 10^{-9}$	nano	n
$0.000\,000\,000\,001 = 10^{-12}$	pico	p
$0.000\,000\,000\,000\,001 = 10^{-15}$	femto	f
$0.000\,000\,000\,000\,000\,001 = 10^{-18}$	atto	a

a. Length In both the older metric and the newer SI systems, the meter (m) is the base unit of length. Many very small lengths in chemistry are conveniently expressed in nanometers (1 nm $= 10^{-9}$ m) or picometers (1 pm $= 10^{-12}$ nm). An older unit for very small distances is the angstrom, abbreviated Å (1 Å $= 10^{-10}$ m $= 0.1$ nm).

b. Volume Because of the relationship of length to volume (as illustrated above), the SI derived unit for volume is the cubic meter (m^3). (The SI unit for area is the square meter, m^2.) On this basis, the recommended replacement for the liter (L), the volume unit in the older metric system, is the cubic decimeter (1 L = 1 dm^3). However, the liter and the milliliter (mL) remain the most common volume units in chemical laboratory work, and we use these units in this book. One milliliter exactly equals one cubic centimeter; one liter exactly equals 1000 milliliters or 1000 cubic centimeters.

c. Mass and Weight The distinction between mass and weight should be clear to anyone who has seen pictures of the astronauts bounding over the surface of the moon. The gravity of the moon is smaller than that of the Earth, and so the weight of the astronauts there was less. Their bodies, however, were unchanged and had the same mass as on Earth. **Mass** is a physical property that represents the quantity of matter in a body. **Weight** is the force exerted on a body by the pull of gravity on the mass of that body.

Both the SI and metric systems rely on the gram, and the multiples and fractions of the gram, as the units for mass. The kilogram is the base unit for mass in the SI system. Strictly speaking, weight should be expressed in units of force (Section 2.8g). In practice, however, the distinction between weight and mass is often ignored.

d. Density Mass and volume are physical quantities that by themselves disclose nothing about the identity of the substance measured (both depend upon the amount of material). However, when mass and volume are combined in a ratio, they yield one of the distinctive properties of different substances (Table 2.7). **Density** is the mass per unit volume of a substance. The SI unit for density is derived from the base units of kilograms and meters, and is kilograms per cubic meter (kg/m^3). The most common unit for density is grams per cubic centimeter (g/cm^3), which is equivalent to grams per milliliter (g/mL). For example, the density of aluminum metal might be given as 2.7×10^3 kg/m^3 or 2.7 g/cm^3, the density of water as about 1 g/mL, and the density of oxygen gas (at room temperature and pressure) as 1.3 g/L. Note that the volume of a substance (especially a gas) varies with temperature and pressure. In careful work, therefore, the temperature and pressure at which a density was measured must be stated.

Table 2.6
Equivalence between Units
The equivalences marked by * are exact (see Section 2.1). Appendix II gives additional conversion factors and larger numbers of significant digits for certain ones.[a]

Length	
1 km	= 0.621 mile
1 m	= 3.281 ft
1 cm	= 0.3937 inch
1 nm	= 10 Å* = 1 mμ*
Volume	
1 L	= 1 dm^3* = 10^{-3} m^3*
1 cm^3	= 1 mL*
1 L	= 1.0567 qt
Mass	
1 kg	= 2.205 lb
1 g	= 0.0353 oz
1 metric ton	= 10^3 kg
Energy	
1 J	= 1 kg m^2/s^2*
1 J	= 0.239 cal
1 erg	= 10^{-7} J*
1 L atm	= 101.325 J*
1 cal	= 4.184 J*
1 eV	= 1.6022×10^{-19} J
Force	
1 dyne	= 1 g cm/s^2*
1 N	= 1 kg m/s^2*
1 N	= 10^5 dyne
1 N	= 0.225 pound (force)
Pressure	
1 Pa	= 1 N/m^2* = 1 kg/m s^2*
1 atm	= 101,325 Pa*
1 atm	= 760 Torr* = 760 mmHg
1 bar	= 1×10^5 Pa*

[a] Å, angstrom; L, liter; eV, electron volt; atm, atmosphere; mmHg, millimeters of mercury.

Table 2.7
Densities of Some Common Substances
Approximate values for ordinary room temperature and pressure.

Substance	Density (g/cm^3)
Ammonia, dilute solution (4%)	0.96
Balsa wood	0.1
Bone	1.9
Chalk	2.4
Chloroform	1.5
Cork	0.24
Corn oil	0.91
Creosote	1.1
Diamond	3.3
Limestone	2.7
Maple wood	0.69
Mercury	13.5
Water	0.99
Whiskey	0.92

e. Temperature You should be familiar with three temperature scales: the SI scale, measured in kelvin units (K); the Celsius scale, measured in degrees Celsius (°C); and the Fahrenheit scale, measured in degrees Fahrenheit (°F). The Fahrenheit scale has been in common use in the United States. However, most scientific measurements are reported in degrees Celsius and some are given in kelvins, the SI unit (Figure 2.2).

On the Celsius scale, the freezing point of water is set at 0 °C and the boiling point at 100 °C (at 1 atm pressure). Anders Celsius described a thermometer using such a scale in 1742; the same scale has been called the centigrade scale. Fortunately, both are represented by °C.

On the Fahrenheit scale, water freezes at 32 °F and boils at 212 °F (at 1 atm). Thus, there are 100 degrees between these two points on the Celsius scale and 180 degrees between them on the Fahrenheit scale, giving rise to the exact relationships 180 Fahrenheit degrees = 100 Celsius degrees, or 1.8 Fahrenheit degrees = 1 Celsius degree. To convert between temperatures on the Fahrenheit and Celsius scales, account must be taken of the different locations of zero (0 °C = 32 °F).

$$\frac{1.8\ \text{F}^\circ}{1\ \text{C}^\circ}\,T(^\circ\text{C}) = T(^\circ\text{F}) - 32 \tag{2.1}$$

To use Equation (2.1), one inserts the known temperature and solves for the unknown temperature. For example, to convert the temperature in a cold room of 16 °C:

$$\left(\frac{1.8\ \text{F}^\circ}{1\ \text{C}^\circ}\right)(16\ ^\circ\text{C}) = T(^\circ\text{F}) - 32$$

$$T(^\circ\text{F}) = \left(\frac{1.8\ \text{F}^\circ}{1\ \text{C}^\circ}\right)(16\ ^\circ\text{C}) + 32 = 61\ ^\circ\text{F}$$

In the SI system, the kelvin is used without a degree sign. One kelvin is the same size as one degree Celsius. On the Kelvin temperature scale, absolute zero, the lowest possible temperature, is equal to 0 kelvin, or 0 K. The freezing point of water is 273.15 K and the boiling point of water is 373.15 K. (The physical explanation for the relationship of these two temperature scales is discussed in Section 6.5.) A Celsius temperature can be converted to a Kelvin temperature by adding 273.15.

$$T(\text{K}) = T(^\circ\text{C})\left(\frac{1\ \text{K}}{1\ \text{C}^\circ}\right) + 273.15\ \text{K}$$

or, as usually written

$$T(\text{K}) = T(^\circ\text{C}) + 273.15 \tag{2.2}$$

Example 2.6 illustrates how to convert temperatures.

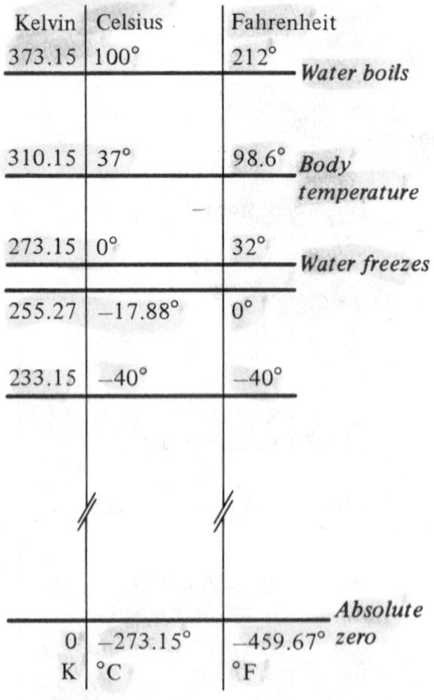

Kelvin	Celsius	Fahrenheit	
373.15	100°	212°	*Water boils*
310.15	37°	98.6°	*Body temperature*
273.15	0°	32°	*Water freezes*
255.27	−17.88°	0°	
233.15	−40°	−40°	
0	−273.15°	−459.67°	*Absolute zero*
K	°C	°F	

Figure 2.2
Comparison of Temperature Measured in Kelvin Units and on the Celsius and Fahrenheit Scales

EXAMPLE 2.6 Temperature Conversion

Convert (a) 72 °F to both Celsius and Kelvin temperatures and (b) 64 K to Celsius temperature.

(a) First use Equation (2.1) to find the Celsius temperature from the Fahrenheit temperature that is given, $T(^\circ\text{F}) = 72\ ^\circ\text{F}$.

$$\frac{1.8\ \text{F}^\circ}{1\ \text{C}^\circ}\,T(^\circ\text{C}) = 72\ ^\circ\text{F} - 32$$

$$T(^\circ\text{C}) = \frac{1\ \text{C}^\circ}{1.8\ \text{F}^\circ}(72\ ^\circ\text{F} - 32) = 22\ ^\circ\text{C}$$

Then use Equation (2.2) to find the Kelvin temperature from the Celsius temperature. (Note that 273 may be used, because it has enough significant figures.)

$$T(\text{K}) = T(^\circ\text{C}) + 273$$
$$= 22 + 273 = \mathbf{295\ K}$$

(b)

$$64 \text{ K} = T(°\text{C}) + 273$$
$$T(°\text{C}) = \mathbf{-209\ °C}$$

EXERCISE Convert −15 °F to both Celsius and Kelvin temperatures.

Answer −26 °C, 247 K.

f. Heat and Energy In the SI system the unit for energy of all types is the joule (pronounced "jool"), a derived unit defined as $1 \text{ J} = 1 \text{ kg m}^2/\text{s}^2$, where m is the symbol for meters and s is that for seconds. At the present time both joules and kilojoules, as well as the older metric units of calories and kilocalories, are seen often. The calorie is now defined in terms of the joule (1 cal = 4.184 J, exactly). The calorie and the joule are rather small units for the heat exchanged in many chemical processes. Most frequently, the kilocalorie or the kilojoule is used for such purposes. The "calories" counted by dieters are really kilocalories.

1 kilocalorie (kcal) = 4.184 kilojoules (kJ)

g. Force and Pressure If you push or pull an object, you are exerting a force on the object. A **force** is any interaction that can cause a change in the motion or state of rest of a body. Chemistry often deals with forces that are interactions between two bodies. When such an interaction occurs, the state of motion of the bodies may be changed, or the shape or size of the bodies altered. The SI unit of force is the newton, N ($1 \text{ N} = 1 \text{ kg m/s}^2$).

Pressure is a force exerted per unit area. The SI unit of pressure is the pascal ($1 \text{ Pa} = 1 \text{ N/m}^2 = 1 \text{ kg/m s}^2$). Other, more commonly used units for pressure such as atmospheres, bars, Torr, and millimeters of mercury (mmHg) are discussed in Aside 6.1.

SUMMARY OF SECTIONS 2.6–2.8

Comparisons of physical quantities are only possible when the quantities are expressed in the same units. Scientific measurements are often expressed in either metric units or in the units of the newer SI (Système International) system. The SI system defines seven base physical quantities and their units; all other physical quantities and units are derived from these base quantities and units. Prefixes are used to designate the multiples or fractions of SI units. The following quantities and units are common in chemistry: length in nanometers and picometers; volume in milliliters and liters; mass in grams, milligrams, and kilograms; density in grams per cubic centimeter; temperature in degrees Celsius or in kelvins; heat and energy in kilojoules or kilocalories.

THE DIMENSIONAL METHOD AND PROBLEM SOLVING

2.9 THE DIMENSIONAL METHOD

The unit should *always* accompany the numerical value of a measurement in writing about the measurement, talking about it, or using it in any kind of calculation. In a numerical problem, units are included in setting up the calculation and are treated exactly as numbers would be. Such treatment of units as numbers is the basis for what is called the dimensional method of calculation.

Because only numbers with the same units can be added or subtracted, units do not change in these operations. For example

$$9.0 \text{ V} + 3.29 \text{ V} = 12.3 \text{ V} \qquad 635 \text{ nm} - 91 \text{ nm} = 544 \text{ nm}$$

In multiplication, the answer has as its units the product of the units multiplied.

$$(6\ \text{L})(0.3\ \text{atm}) = 2\ \text{L atm} \qquad (29.0\ \text{cm})^2 = 841\ \text{cm}^2$$

read "liter atmospheres"

In division the units may appear in the answer or they may cancel.

$$\frac{3.0\ \text{cm}}{2.0\ \text{s}} = 1.5\ \text{cm/s} \qquad \frac{203\ \text{kcal}}{69\ \text{kcal}} = 2.9$$

read "centimeters per second" — *units cancelled; a dimensionless quantity*

(Often a unit that would appear in the denominator of an expression is instead written with a negative exponent. For example, for "centimeters per second," cm s^{-1} would be written instead of cm/s.)

Units are canceled exactly as numbers (or algebraic variables) would be in calculating the answer to a problem. For example, suppose you wanted to know how many liters of gasoline you would need to travel 650 km each day for 10 days in a car that requires 21 L of gasoline for each 160 km. The correct solution to the problem looks like this:

$$\left(\frac{650\ \cancel{\text{km}}}{1\ \cancel{\text{day}}}\right)(10\ \cancel{\text{days}})\left(\frac{21\ \text{L}}{160\ \cancel{\text{km}}}\right) = 850\ \text{L}$$

cancel *cancel*

The dimensional method of calculation is powerful and useful for several reasons. It is an excellent guide in deciding how to solve some problems. And it makes many errors instantly recognizable. A correct setup usually leads to an answer in the desired units, as is shown by the answer in liters just above. A wrong setup usually leads to an answer in the wrong units. A good question to ask yourself before solving a dimensional problem is, What must the units be in the final answer? The dimensional method is of great help in solving all kinds of problems, not just those in chemistry. Try the method on a physics problem or on an everyday domestic problem. It works!

2.10 CONVERSION FACTORS

The conversion of a physical quantity in one unit to the same quantity expressed in another unit is a common necessity in solving problems in chemistry. To carry out such a transformation, conversion factors based on the relationship between the two units are used. The conversion factors are derived from equalities (see Table 2.6) such as 1 cal = 4.184 J. Each such equality yields two conversion factors:

$$1\ \text{cal} = 4.184\ \text{J} \qquad\qquad 1\ \text{cal} = 4.184\ \text{J}$$

$$\frac{1\ \cancel{\text{cal}}}{1\ \cancel{\text{cal}}} = \frac{4.184\ \text{J}}{1\ \text{cal}} \qquad\qquad \frac{1\ \text{cal}}{4.184\ \text{J}} = \frac{4.184\ \cancel{\text{J}}}{4.184\ \cancel{\text{J}}}$$

$$1 = \frac{4.184\ \text{J}}{1\ \text{cal}} \longleftarrow \textit{two conversion factors; reciprocals} \longrightarrow \frac{1\ \text{cal}}{4.184\ \text{J}} = 1$$

As you can see, these conversion factors are reciprocals of each other and are both equal to 1. They are read as "4.184 joules per calorie" and "1 calorie per 4.184 joules." These two conversion factors, like all such pairs of conversion factors, allow for the interconversion of the two units.

$$\text{Calories} \underset{\times\left(\frac{1\ \text{cal}}{4.184\ \text{J}}\right)}{\overset{\times\left(\frac{4.184\ \text{J}}{1\ \text{cal}}\right)}{\rightleftarrows}} \text{Joules} \qquad (2.3)$$

For example

$$30.0\ \cancel{\text{cal}} \times \frac{4.184\ \text{J}}{1\ \cancel{\text{cal}}} = 126\ \text{J} \qquad 30.0\ \cancel{\text{J}} \times \frac{1\ \text{cal}}{4.184\ \cancel{\text{J}}} = 7.17\ \text{cal}$$

How do you choose which conversion factor to use? It's easy—*choose the conversion factor that eliminates the unit that you do not want*. In each case shown above, the unit to be eliminated is in the denominator (on the *bottom*) in the conversion factor. Of course, it is possible that the reverse might occur and the unit to be eliminated would have to be in the numerator in the conversion factor.

The number of significant figures in a conversion factor depends on whether or not the number is an exact number. Note that in part (a) of Example 2.7, the factor 10 Å does not limit the answer to two digits. This is because there are *exactly* 10 Å in exactly 1 nm. The multiples and fractions of units indicated by the prefixes of Table 2.5 are exact numbers; for example, 1 dL is exactly equal to 0.1 L.

EXAMPLE 2.7 Unit Conversion

Convert 8160 Å to (a) nanometers and (b) meters.

(a) Table 2.6 shows that 1 nm = 10 Å, giving the conversion factors

$$\frac{1\ \text{nm}}{10\ \text{Å}} \qquad \frac{10\ \text{Å}}{1\ \text{nm}}$$

Choosing the conversion factor that eliminates Å

$$(8160\ \cancel{\text{Å}})\left(\frac{1\ \text{nm}}{10\ \cancel{\text{Å}}}\right) = \mathbf{816\ nm}$$

(b) The unit equivalence (Table 2.6) and the possible conversion factors are

$$1\ \text{nm} = 1 \times 10^{-9}\ \text{m} \qquad \frac{1\ \text{nm}}{1 \times 10^{-9}\ \text{m}} \qquad \frac{1 \times 10^{-9}\ \text{m}}{1\ \text{nm}}$$

The conversion is accomplished as follows:

$$(816\ \cancel{\text{nm}})\left(\frac{1 \times 10^{-9}\ \text{m}}{1\ \cancel{\text{nm}}}\right) = 816 \times 10^{-9}\ \text{m} = \mathbf{8.16 \times 10^{-7}\ m}$$

EXERCISE Convert 265 μL to (a) liters and (b) milliliters.

Answers (a) 2.65×10^{-4} L, (b) 0.265 mL.

EXAMPLE 2.8 Unit Conversion

While on vacation in France, the Jones family made a shopping list that included 2 qt of milk, 2 gal of wine, 6.5 oz of cheese, and 2.5 lb of beef. Convert this into a shopping list for liters of milk and wine, grams of cheese, and kilograms of beef. (See Table 2.6.) How far in miles is it to the store, which is 19 km away?

Choosing (Table 2.6) a conversion factor that eliminates the unwanted unit in each case, and limiting the conversion factors to two significant figures gives

$$(2\ \cancel{\text{qt}}\ \text{milk})\left(\frac{1\ \text{L}}{1.1\ \cancel{\text{qt}}}\right) = \mathbf{2\ L\ milk} \qquad (2\ \cancel{\text{gal}}\ \text{wine})\left(\frac{4\ \cancel{\text{qt}}}{1\ \cancel{\text{gal}}}\right)\left(\frac{1\ \text{L}}{1.1\ \cancel{\text{qt}}}\right) = \mathbf{7\ L\ wine}$$

$$(6.5\ \cancel{\text{oz}}\ \text{cheese})\left(\frac{1\ \text{g}}{0.035\ \cancel{\text{oz}}}\right) = \mathbf{190\ g\ cheese} \qquad (2.5\ \cancel{\text{lb}}\ \text{beef})\left(\frac{1\ \text{kg}}{2.2\ \cancel{\text{lb}}}\right) = \mathbf{1.1\ kg\ beef}$$

$$(19\ \cancel{\text{km}})\left(\frac{0.62\ \text{mi}}{1\ \cancel{\text{km}}}\right) = \mathbf{12\ mi}$$

EXERCISE A package weighs 362 g. What is the mass expressed in ounces?

Answer 12.8 oz.

When supplying conversion factors or values of physical constants in a calculation, it is important not to decrease the number of significant figures in the answer by giving too few figures in the number supplied. In addition and subtraction, this means supplying at least as many decimal places as are in the number in the problem with the fewest decimal places. In multiplication and division, this means supplying at least the same number of significant figures as are in the other numbers being multiplied or divided:

express conversion factor to at least two significant figures

$$(1.2\,\text{eV})\left(\frac{1.6 \times 10^{-19}\,\text{J}}{1\,\text{eV}}\right) = 1.9 \times 10^{-19}\,\text{J}$$

two significant figures (1.2 eV) → *two significant figures* (answer)

$$(99.94\,\text{eV})\left(\frac{1.602 \times 10^{-19}\,\text{J}}{1\,\text{eV}}\right) = 1.601 \times 10^{-17}\,\text{J}$$

four significant figures (99.94 eV); *express conversion factor to at least four significant figures*; *four significant figures* (answer)

Once you understand the use of the dimensional method, many steps in a unit conversion or any other problem can be combined in a single expression, as shown in Example 2.9.

EXAMPLE 2.9 Unit Conversion

How many seconds are there in exactly one day? Use the following equalities to derive the necessary conversion factors:

$$1\text{ day} = 24\text{ h} \qquad 1\text{ h} = 60\text{ min} \qquad 1\text{ min} = 60\text{ s}$$

Days must be converted to hours (1), then hours to minutes (2), then minutes to seconds (3). The individual steps in such a unit conversion can be strung together and the final answer calculated all at once as follows:

$$(1\,\text{day}) \times \underbrace{\left(\frac{24\,\text{h}}{1\,\text{day}}\right)}_{\text{step 1}} \times \underbrace{\left(\frac{60\,\text{min}}{1\,\text{h}}\right)}_{\text{step 2}} \times \underbrace{\left(\frac{60\,\text{s}}{1\,\text{min}}\right)}_{\text{step 3}} = \mathbf{86{,}400\,s}$$

EXERCISE How many seconds are there in exactly one hour? *Answer* 3600 s.

The conversion factors that we have used so far all convert between different units for the same physical property: distance in miles or kilometers, heat in kilocalories or kilojoules, and so on. The dimensional method is also effective in the use of factors that allow conversion between related but different physical properties. For example, density—the mass per unit volume—is a factor that allows conversion from volume to mass or mass to volume:

$$\text{Mass} = (\text{density})\,(\text{volume}) \qquad \textbf{(2.4a)}$$

$$\textit{grams} \quad \frac{\textit{grams}}{\textit{cubic centimeter}} \quad \textit{cubic centimeters}$$

$$\text{Volume} = \left(\frac{1}{\text{density}}\right)(\text{mass}) \qquad \textbf{(2.4b)}$$

$$\textit{cubic centimeters} \quad \frac{\textit{cubic centimeters}}{\textit{grams}} \quad \textit{grams}$$

Such conversion factors are ratios between two different physical quantities.

EXAMPLE 2.10 Unit Conversion

What is the mass in grams of a 9.00 cm^3 piece of lead? The density of lead is 11.3 g/cm^3.

The solution to the problem is set up using the dimensional method

$$\text{Mass} = (9.00\ \cancel{cm^3})\left(\frac{11.3\ g}{1\ \cancel{cm^3}}\right) = \mathbf{102\ g}$$

EXERCISE What is the volume in cubic centimeters of a 6.35 g piece of lead? *Answer* 0.562 cm^3.

As illustrated in Examples 2.11 and 2.12, the ratios needed to solve a problem can be derived from the problem itself, as well as from a defined quantity such as density. Many chemical problems are of this type.

EXAMPLE 2.11 Unit Conversion

A 250. cm^3 volume of a liquid weighs 312 g. What volume of the liquid will weigh 4.5 g?

The information given provides a ratio of mass to volume for this liquid, which can be used in a dimensional calculation as follows:

$$\left(\frac{250.\ cm^3}{312\ \cancel{g}}\right)(4.5\ \cancel{g}) = \mathbf{3.6\ cm^3}$$

EXERCISE A 100. mL sample of an oil weighs 86.2 g. What is the mass of 16.3 mL of the oil? *Answer* 14.1 g.

EXAMPLE 2.12 Unit Conversion

A chemical plant can produce 5.4 kg of a substance each day at a cost of \$1300 per kilogram. What will it cost to run the plant for 5 days?

The conversion factors for this problem are derived from the data given: the mass produced per day and the cost per unit mass.

$$\text{Total cost} = \left(\frac{5.4\ \cancel{kg}}{1\ \cancel{day}}\right)\left(\frac{\$1300}{1\ \cancel{kg}}\right)(5\ \cancel{days}) = \mathbf{\$35{,}000}$$

EXERCISE Starting with \$8300, how many days can the chemical plant described above operate until the funds are depleted? *Answer* 1.2 days.

EXAMPLE 2.13 Unit Conversion

What is the mass of the snow (in tons) on a 150 ft × 45 ft flat roof after a 6.0 inch snowfall? Assume that 11 inches of snow is equivalent to 1.0 inch of water. Density of water = 1.0 g/cm^3; 1 lb = 454 g; 1 ton = 2000 lb.

The volume of water equivalent to the volume of snow on the roof is

$$(6.0\ \cancel{\text{inches snow}})\left(\frac{1.0\ \cancel{\text{inch}}\ \text{water}}{11\ \cancel{\text{inches snow}}}\right)(150\ \cancel{ft})(45\ \cancel{ft})\left(\frac{12\ \cancel{\text{inches}}}{1\ \cancel{ft}}\right)^2\left(\frac{2.54\ cm}{1\ \cancel{\text{inch}}}\right)^3$$

$$= 8.7 \times 10^6\ cm^3\ \text{water}$$

The mass of this volume of water, which corresponds to the mass of the snow, is found by using the density

$$(8.7 \times 10^6 \not{cm^3})\left(\frac{1.0 \not{g}}{1 \not{cm^3}}\right)\left(\frac{1 \not{lb}}{454 \not{g}}\right)\left(\frac{1 \text{ ton}}{2000 \not{lb}}\right) = \mathbf{9.6\ tons}$$

EXERCISE Astronomical distances are often measured in units of light years (the distance that light travels in a year). What is the number of meters in a light year? The speed of light is 3.00×10^8 m/s. *Answer* 9.46×10^{15} m.

2.11 A PROBLEM-SOLVING METHOD

Success in problem solving is greatly aided by applying the following general method, which is intended as a guide to problem analysis. The goal is to make sure that you understand the problem sufficiently to choose an approach to solve it. The method outlined can be applied to both quantitative and qualitative problems, and to simple and highly complex problems. When applied to complex problems, the method is especially helpful in sorting out the available information and deciding how to use it.

We are not suggesting that you always write out problem solutions as illustrated (although it might be helpful if you are stuck). Instead, let the steps lead you through a logical approach to analyzing problems. If you practice the method for a while, its use will become a habit—one of value in many types of problems, not just those of chemistry.

In later sections of this book as new types of problems are introduced, from time to time we shall present problem solutions according to the plan outlined here. The crucial part of the solution is always in recognizing the connection between the known and the unknown. Many types of chemistry problems can be characterized by how the connection is made.

The following four general steps form the basis for the general problem-solving method we recommend. As you apply this method, ask yourself questions about each step. Although the questions will vary with the nature of the problem, they should always include an analysis of what is known and unknown in the problem and what conversion factor, mathematical relationship, or other information will provide a connection between the known and the unknown parts of the problem.

1. Study the problem and be sure you understand it. To understand a problem, you must know what facts are given—the known information about the system. Read the problem carefully to determine what "knowns" are included. This information might be a single physical quantity or it might be several quantities. In some cases these quantities will be related to each other and can be combined into conversion factors or ratios. Read the problem carefully again to be sure you know what the question is. What is the "unknown" that must be determined from the known information? The unknown might be a single physical quantity or several; it also might be nonquantitative information.
2. Decide how to solve the problem. With the known and the unknown clearly in mind, the next step is to determine the connection between them. What have you learned that applies to the physical quantities and the kind of system under consideration? A simple conversion factor or ratio might be all that is needed. A standard mathematical expression might have to be solved for the unknown. Information about the chemical or physical properties of the system might be needed.

   In deciding how to solve a quantitative problem, analyze the units of the known and the units desired in the unknown. It might be necessary to find the unknown physical quantity in one unit and then convert it to another unit before the problem is solved. Or it might be necessary to convert the known into other units before it can be used in the appropriate equation. If solving the problem requires conversion factors or equations not in the statement of the problem, be sure you use them correctly. Check back in the book or some other source if your memory is hazy.

3. Set up the problem and solve it. At this step the dimensional method is of primary importance in quantitative problems. With practice, many problems can be set up as single expressions to be solved. Check the setup by checking the unit cancellation. Will the answer be in the desired units? In doing the arithmetic, do not forget the rules for significant figures and scientific notation.
4. Check the result. Solving a problem *is not finished* until you check the result. Do this in two ways. First, check both the numerical part of the answer and the units. Is the answer OK with respect to significant figures, decimal places, scientific notation, and units? Second, *think* about the answer. Is it reasonable? If the answer is a physical quantity, is it much larger or much smaller than is realistically possible?

To illustrate the problem-solving steps, the analysis of a simple problem is presented in Example 2.14.

EXAMPLE 2.14 Unit Conversion; Problem Solving

A small airplane traveled 128 km in 48 min. What is the speed of the airplane in kilometers per hour?

1. Study the problem and be sure you understand it.
   (a) What is unknown?
   The speed of the plane in kilometers per hour.
   (b) What is known?
   The distance traveled = 128 km; the time to travel that distance = 48 min.
2. Decide how to solve the problem.
   (a) What is the connection between the known and the unknown?
   Speed is a ratio, distance/time.
   (b) What is necessary to make the connection?
   First, the calculation of speed in km/min. Second, a conversion factor based on the prior knowledge that 60 min = 1 h.
3. Set up the problem and solve it.
   The solution can be set up in two steps:

$$\frac{128\text{ km}}{48\text{ min}} = 2.7\,\frac{\text{km}}{\text{min}} \qquad \left(2.7\,\frac{\text{km}}{\cancel{\text{min}}}\right)\left(\frac{60\,\cancel{\text{min}}}{1\text{ h}}\right) = 160\text{ km/h}$$

   or in one step:

$$\left(\frac{128\text{ km}}{48\,\cancel{\text{min}}}\right)\left(\frac{60\,\cancel{\text{min}}}{1\text{ h}}\right) = \mathbf{160\ km/h}$$

4. Check the result.
   (a) Are significant figures used correctly?
   Yes.
   (b) Did the answer come out in the correct units?
   Yes.
   (c) Is the answer reasonable?
   Yes, this is a reasonable speed for a small airplane. [When SI units are more widely used we shall know that a highway speed limit for a car might be about 80 km/h. It is reasonable that a small plane travels somewhat faster than a car.]

Aside 2.1 THOUGHTS ON CHEMISTRY: PROBLEM SOLVING

The following extract is from the widely known book on mathematical problem solving by G. Polya, which provided the inspiration for the problem-solving method introduced in Section 2.11.

It would be a mistake to think that solving problems is a purely "intellectual affair"; determination and emotions play an important role. Lukewarm determination and sleepy consent to do a little something may be enough for a routine problem in the classroom. But, to solve a serious scientific problem, willpower is needed that can outlast years of toil and bitter disappointments.

Determination fluctuates with hope and hopelessness, with satisfaction and disappointment. It is easy to keep on going when we think that the solution is just around the corner; but it is hard to persevere when we do not see any way out of the difficulty. We are elated when our forecast comes true. We are depressed when the way we have followed with some confidence is suddenly blocked, and our determination wavers....

Incomplete understanding of the problem, owing to lack of concentration, is perhaps the most widespread deficiency in solving problems. With respect to devising a plan and obtaining a general idea of the solution two opposite faults are frequent. Some students rush into calculations and constructions without any plan or general idea; others wait clumsily for some idea to come and cannot do anything that would accelerate its coming. In carrying out the plan, the most frequent fault is carelessness, lack of patience in checking each step. Failure to check the result at all is very frequent; the student is glad to get an answer, throws down his pencil, and is not shocked by the most unlikely results.

Source: *How To Solve It: A New Aspect of Mathematical Method*, pp. 93–95. Copyright 1945 by Princeton University Press; © 1957 by G. Polya.

SUMMARY OF SECTIONS 2.9–2.11

The numerical value of a measurement must always be expressed together with the correct unit. In solving problems by the dimensional method, units are multiplied, divided, and canceled exactly as are numbers. If the problem is correctly set up and solved, the answer obtained will be in the correct units. The relationship between any two units expressed as an equality (e.g., 1 cal = 4.184 J) provides two conversion factors that permit the conversion of values given in one unit to values given in the other. Choosing the correct conversion factor permits cancellation of the unwanted unit. Conversion factors are also derived from the relationships between two different physical quantities (e.g., density = mass/volume, which is often expressed as grams per cubic centimeter).

To solve a problem, first be certain that you understand what is known and what is unknown. Then determine how to solve the problem based upon whatever permits the connection to be made between the known and unknown. Next, set up and solve the problem, using the dimensional method and making sure to follow the rules for significant figures. Finally, check the answer to make sure that the units are correct and that the answer is reasonable.

SIGNIFICANT TERMS (with section references)

(Significant terms—terms that should be understood—are listed at the end of each chapter, with the number of the section in which the term is introduced given in parentheses. An alphabetized Glossary containing all significant terms is presented at the back of the book.)

density (2.8)
exact numbers (2.1)
force (2.8)
mass (2.8)
pressure (2.8)
scientific notation (2.4)
significant figures, significant digits (2.1)
weight (2.8)

QUESTIONS AND PROBLEMS

(The more challenging questions and problems are indicated by asterisks.)

Numbers in Physical Quantities—Significant Figures

2.1 Assume that the uncertainty in a number is ± 1 in the last significant figure; for example, the uncertainty in 1.36 is ± 0.01. What is the uncertainty of each of the following numbers: (a) 273, (b) 0.5649, (c) 470, (d) 12.529, (e) 6000, (f) 0.0006, (g) 12.00, (h) 1,300,020?

2.2 Repeat Question 2.1 for (a) 1432, (b) 632.2, (c) 710, (d) 0.92, (e) 500, (f) 0.09, (g) 8.0, (h) 3.14159.

2.3 How many significant figures are there in each of the following numbers: (a) 454, (b) 2.2, (c) 2.205, (d) 0.0353, (e) 1.0080, (f) 14.00, (g) 1030?

2.4 How many significant figures are there in each of the numbers in Question 2.2?

2.5 Perform the following calculations and express each answer to the proper number of significant figures:

(a) $\begin{array}{r} 423.1 \\ 0.256 \\ 100 \\ \hline \end{array}$ (b) $\begin{array}{r} 52.987 \\ 9.3545 \\ 6.12 \\ \hline \end{array}$ (c) $\begin{array}{r} 14.3920 \\ -4.4 \\ \hline \end{array}$

(d) (5183)(2.2) (e) $\dfrac{14.000}{6.1}$ (f) $(6.11)(\pi)$

(g) (14.3)(60) (h) $\dfrac{1020}{1.2}$ (i) $\dfrac{(3.2)(454)}{(8.6214)}$

(j) $(4/3)\pi(2.16)^3$
(k) $(6.0 + 9.57 + 0.61)(1.113)$
(l) $(2.93)(14.7) + (1203)(0.0296) + (9.38)(5.2)$

2.6 Repeat Problem 2.5 for the following calculations:

(a) $\begin{array}{r} 1900 \\ -6.25 \\ \hline \end{array}$ (b) $\begin{array}{r} 963.2 \\ 1.46 \\ 10.5 \\ \hline \end{array}$ (c) $\begin{array}{r} 16.3256 \\ -49.3 \\ \hline \end{array}$

(d) (1492)(6.3) (e) $\dfrac{0.25}{137}$ (f) $\dfrac{(9.4)(16)}{(9.354)}$

(g) $\pi(8.2)^2$
(h) $(6.35 + 2.9 + 163)(7.5 + 6.3)$

2.7* A group of students reported the following measurements for the diameter of a quarter: 2.50 cm, 2.42 cm, 2.43 cm, 2.40 cm, and 2.41 cm. (a) Calculate the class average for the diameter. (b) What is the uncertainty in the measurement?

To check the accuracy of the result, the "% error" was calculated.

$$\%\ \text{error} = \frac{(\text{experimental value}) - (\text{accepted value})}{(\text{accepted value})} \times 100$$

Using the accepted diameter as 2.44 cm, (c) calculate the % error of the class average.

2.8* The calibration of a thermometer was checked by placing it into an ice-water bath at exactly 0 °C. A student made the following temperature readings: 0.02 °C, 0.01 °C, 0.02 °C, 0.03 °C, and 0.02 °C. (a) Calculate the average reading for the thermometer in the bath. (b) What is the uncertainty in the measurement? (c) How could the % error (see Problem 2.7) in the thermometer calibration be calculated?

Numbers in Physical Quantities—Scientific Notation

2.9 Express the following numbers in scientific notation: (a) 6500, (b) 0.0041, (c) 0.003050, (d) 810., (e) 0.0000003, (f) 9,352,000, (g) 42×10^3.

2.10 Express the following numbers in scientific notation: (a) 0.0516, (b) 1420, (c) 1260., (d) 0.0002, (e) 0.010, (f) 6,925,300, (g) 0.28×10^{-5}.

2.11 Express the following numbers using ordinary notation (for example, $5.2 \times 10^{-2} = 0.052$): (a) 5.26×10^3, (b) 4.10×10^{-6}, (c) 5×10^5, (d) 0.3×10^4, (e) 16.2×10^{-3}, (f) 9.346×10^3.

2.12 Express the following numbers using ordinary notation: (a) 6.90×10^{-4}, (b) 1.426×10^5, (c) 4×10^{-3}, (d) 52.3×10^3, (e) 3.200×10^3, (f) 0.2×10^{-2}.

2.13 Perform the following calculations and express each answer in scientific notation to the proper number of significant figures:
(a) $(5.29 \times 10^3) - (1.609 \times 10^2)$
(b) $(2.547 \times 10^2)(3.2 \times 10^{-1})$
(c) $(6.1 \times 10^{-2})(5.800 \times 10^{-6})$
(d) $\dfrac{(3.261 \times 10^{-3}) + (2.58 \times 10^4)}{1.2 \times 10^{-7}}$

2.14 Repeat Problem 2.13 for the following calculations:
(a) $(6.057 \times 10^3) + (9.35)$
(b) $(2.35 \times 10^{-14}) - (7.1 \times 10^{-15})$
(c) $\dfrac{4.51 \times 10^{-3}}{8.78 \times 10^4}$ (d) $\dfrac{(1812)(1492)}{1979}$
(e) $\dfrac{(7.33 \times 10^{-3}) + (4.29 \times 10^3)}{(5.88 \times 10^{-3}) + (4.29 \times 10^3)}$

2.15* At a temperature T between 0 °C and 100 °C, the density in g/cm^3 of liquid water at one atmosphere pressure is given by the equation

$$d = \frac{[(0.99983952) + (1.6945176 \times 10^{-2})T - (7.9870401 \times 10^{-6})T^2 - (4.6170461 \times 10^{-8})T^3 + (1.0556032 \times 10^{-10})T^4 - (2.8054253 \times 10^{-13})T^5]}{1.0000000000 + (1.6879850 \times 10^{-2})T}$$

Calculate the density at 25.00 °C to the proper number of significant figures.

2.16* The average radius of the orbit of the planet Pluto is 5.91×10^{12} m and the radius of the sun is 6.95×10^8 m. (a) Calculate the ratio of the radius of Pluto's orbit to the radius of the sun. The radius of a hydrogen atom is 5.29×10^{-11} m and the radius of the proton at the center is 1.5×10^{-15} m. (b) Calculate the ratio of the radius of the atom to the radius of the proton. (c) Which ratio is greater?

Units of Measurement—SI System

2.17 List the seven base physical quantities as specified by the SI system. How are all other physical quantities treated in this system?

2.18* Many physical properties are related to other physical properties by simple equations; for example, (force) = (mass) × (acceleration) where (acceleration) = (velocity)/(time). The derived SI unit for the property can be determined by using the same equation and substituting the units for the other properties; for example, the derived unit for force is (kg)(m/s^2) = kg m/s^2. Given that (work) = (force) × (distance) and (pressure) = (force)/(area), derive the SI units for work and pressure.

2.19 Which of the following units can be used to express (a) mass, (b) energy, (c) length, (d) volume, and (e) temperature: (i) erg, (ii) cm, (iii) cm^3, (iv) K, (v) Å, (vi) J, (vii) km, (viii) mL, (ix) g, (x) °C, (xi) nm, (xii) L, (xiii) cal, (xiv) m, (xv) dm^3, (xvi) kg, (xvii) mg?

2.20 Which of the following combinations of property measured and unit of measurement are acceptable: (a) the area of a football field in m^2, (b) the volume of an apple juice bottle in L^3, (c) the density of wood in kg/m^3, (d) the length of an eraser in mL, (e) the radius of a basketball in kg, (f) the length of time of a TV commercial in Ms, (g) the height of an evergreen tree in cm^3?

2.21 Choose the largest unit from each group: (a) cm, pm, or km; (b) MV, mV, or nV; (c) fm, mm, pm, or nm; (d) TJ, kJ, or μJ; (e) cm^3, dm^3, or km^3.

2.22 Write each of the following values, first using scientific notation and then using a prefix: (a) 0.0001 V, (b) 13,500 Pa, (c) 0.0000000005 m.

Units of Measurement—Temperature Conversion

2.23 Convert the following temperatures, which are commonplace in our daily lives, to the Celsius scale: (a) normal body temperature, 98.6 °F; (b) the temperature on a cold, wintry day, −10. °F; (c) the temperature on a warm fall day, 78 °F; (d) the running temperature of a modern auto engine, 250 °F, (e) the melting point of ice, 32.00 °F.

2.24 The electrical wires in a certain furnace cannot be used at temperatures above 950 °C. The only available temperature-measuring device was calibrated in °F. What is the maximum "safe" operating temperature on this scale?

2.25 Convert the temperatures given in Problem 2.23 to kelvins.

2.26 Convert each of the following boiling point temperatures to values on the Celsius scale: (a) water, 373.15 K; (b) nitric oxide, 121.4 K; (c) sulfur, 717.8 K; (d) iron, 3020 K; (e) sulfuric acid, 611 K.

2.27 Convert the temperatures given in Problem 2.26 to values on the Fahrenheit scale.

2.28* At what temperature will a Fahrenheit thermometer give (a) the same reading as a Celsius thermometer, (b) a reading that is twice that on the Celsius thermometer, (c) a reading that is numerically the same but opposite in sign from the Celsius scale?

The Dimensional Method and Problem Solving—Unit Conversion

2.29 Derive two conversion factors for pressure based on the following equality: 1 atm = 760 Torr.

2.30 Repeat Problem 2.29 for the equality: 1 bar = 101,325 Pa.

2.31 Water depth can be measured in units of fathoms (1 fathom = 6 ft) and leagues (1 league = 3040 fathoms). What is the depth in meters corresponding to (a) 2.00 fathoms (the source of the pen name of Samuel Clemens, which was Mark Twain), (b) 20,000 leagues (the distance traveled by a submarine in a story by Jules Verne)?

2.32 The speed of sound in air at 25 °C is 346 m/s. This speed is called one "mach number." (a) Write two conversion factors relating m/s and mach number. (b) Derive a conversion factor relating mach number and miles per hour. (c) Express the speed of an "Indy-500" race car going 215 miles per hour in mach numbers.

2.33 Make each of the following conversions: (a) 10.3 Å to nm, (b) 635 cal to J, (c) 14.6 L to dm^3, (d) 14.6 kg to g, (e) 14.6 atm to Pa, (f) 1.2 eV to J, (g) 735.2 Torr to atm, (h) 21.65 mL to cm^3.

2.34 Make each of the following conversions: (a) 75 yards to cm, (b) 16 pounds to g, (c) 45.6 L to gallons, (d) 450 miles to km, (e) 1.429 g/L to pounds per cubic foot, (f) 45 miles per hour to m/s, (g) 5.0 grains of aspirin to mg (1 grain = 0.06479891 g).

2.35 Make each of the following conversions: (a) 25 m to yards, (b) 13.7 g/mL to kg/m^3, (c) 3.2 Torr to Pa, (d) 4.6 atm to bars, (e) 14.6 lb to kg.

2.36 Determine which quantity is larger: (a) 1.0 mg or 1.0 cg, (b) 325 kcal or 95 J, (c) 50 nm or 0.5 m, (d) 0.8 nm or 8 Å, (e) 75 Pa or 747 Torr, (f) 5 L or 3.2 m^3.

2.37 A constant used frequently in chemistry is known as the ideal gas constant. It is numerically equal to 8.314 J/K mol. Express the value of this constant in (a) erg/K mol, (b) cal/K mol, (c) L atm/K mol.

2.38 The density of water is 1.00 g/cm^3 at room temperature. Express this value in (a) g/mL, (b) kg/m^3, (c) mg/mL, (d) lb/ft^3.

2.39 Each molecule of sucrose (ordinary sugar) contains 12 carbon atoms. How many carbon atoms are present in 5×10^{21} molecules?

2.40 An atom of heavy hydrogen (deuterium) contains 1 electron, 1 proton, and 1 neutron. How many protons are present in 3.2×10^{25} atoms? How many protons and neutrons are present in 3.2×10^{25} atoms?

2.41 A chemical plant releases 5.0 tons of gas into the atmosphere each day. The gas contains 5% by mass of sulfur dioxide (i.e., for every 100 parts by mass of gas, 5 parts are sulfur dioxide). What mass of sulfur dioxide is released in a period of one week (7 days)?

2.42 A certain chemical process required 75 gallons of pure water each day. The available water contained 11 parts per million by mass of salt (i.e., for every 1,000,000 parts by mass of impure water, 11 parts are salt). What mass of salt must be removed each day? A gallon of water weighs 3.78 kg.

2.43 The radius of a hydrogen atom is about 0.58 Å and the distance between the sun and the Earth is about 1.5×10^8 km. Find the ratio of the radius of the hydrogen atom to the sun-Earth distance so that the units cancel.

2.44 A molecule of palmitic acid has a volume of 110 Å^3. When a drop of the acid is placed on water, the molecules spread out on the surface of the water producing a layer that is one molecule thick. The height of the molecule is 4.6 Å in this layer. (a) Calculate the cross sectional area of the molecule. (b) What area in m^2 will 6.022×10^{23} molecules occupy? The area of 1 Å^2 is equivalent to 10^{-20} m^2.

2.45* The value of 273.15 used to convert Celsius temperatures to absolute temperatures in kelvins is exact. What is the minimum number of significant figures to which this conversion factor (represented by T_0) in each of the calculations below should be expressed so that the uncertainty in the answer is the result of the uncertainty in the other temperature and not from the conversion factor?

(a) $16 + T_0$ (b) $0.094 + T_0$
(c) $103.7 + T_0$ (d) $729.65 - T_0$
(e) $(2.303)(5.26)T_0$ (f) $\dfrac{1}{305} - \dfrac{1}{T_0}$

2.46* The mass of an electron, m_e, is 9.109534×10^{-28} g (known only to seven significant figures). What is the minimum number of significant figures to which the mass should be expressed in each of the calculations below so that the uncertainty in the answer is the result of the uncertainty in the other numbers and not from the value of m_e (if possible)?

(a) $\dfrac{m_e}{1.6726485 \times 10^{-24}}$
(b) $(1.6726485 \times 10^{-24}) + m_e$
(c) $(6 \times 10^3)^2 m_e$
(d) $(6.635 \times 10^{12}) m_e + (1.602 \times 10^{-19})$

The Dimensional Method and Problem Solving—Density

2.47 Vinegar has a density of 1.0056 g/cm^3. What is the mass of 1.0000 L of vinegar?

2.48 What is the mass of a cylinder of aluminum that has a 0.50 m radius and is 0.35 m in height? The density of aluminum is 2.702 g/cm^3. The volume of a cylinder is $V = \pi r^2 h$.

2.49 A small crystal of sucrose (table sugar) had a mass of 2.236 mg. The dimensions of the box-like crystal were $1.11 \times 1.09 \times 1.12$ mm. What is density of sucrose expressed in g/cm^3?

2.50 The radius of a neutron (a particle smaller than the smallest atom and present in most atoms) is approximately 1.5×10^{-15} m. Calculate the volume of a neutron using the formula $V = 4\pi r^3/3$. Calculate the density of a neutron given that its mass is 1.675×10^{-24} g.

2.51* A container has a mass of 68.31 g empty and 93.34 g filled with water. Calculate the volume of the container using a density of 1.0000 g/cm^3 for water. The container when filled with an unknown liquid had a mass of 88.42 g. Calculate the density of the unknown liquid.

2.52* The mass of an empty container is 66.734 g. The mass of the container filled with water is 91.786 g. (a) Calculate the volume of the container, using a density of 1.0000 g/cm^3 for water. A piece of metal was placed in the empty container and the combined mass was 87.807 g. (b) Calculate the mass of the metal. The container with the metal was filled with water and the mass of the entire system was 105.408 g. (c) What mass of water was added? (d) What volume of water was added? (e) What is the volume of the metal? (f) Calculate the density of the metal.

The Dimensional Method and Problem Solving—Problem Solving

2.53* Apply the four steps of the general problem-solving method described in the text to the following problem: A "double-sided, double-density $5\frac{1}{4}$-inch floppy" diskette for a computer has 80 tracks, each with 9 sectors holding 512 bytes. A "$5\frac{1}{4}$-inch fixed" disk has 1227 tracks, each with 17 sectors holding 512 bytes. From approximately how many diskettes can the information be transferred to the fixed disk?

2.54* Apply the four steps of the general problem-solving method described in the text to the following problem: What mass of sodium fluoride is needed for a year for one person living in a large city so that the water contains 1.0 ppm fluoride ion by mass? Sodium fluoride is 45% fluoride ion by mass. The daily per capita consumption of water in large cities is 145 gal.

CHAPTER

3

Chemistry: The Science of Matter

ATOMS AND ELEMENTS

3.1 ELEMENTS

Long ago, before the existence of chemistry as a science, it was observed that some matter is composed of other kinds of matter. For example, when a rock was split, the colors and textures of many substances within it could frequently be observed. Experiments with matter later showed that some substances, such as iron, could *not* be broken down into other substances.

From ancient to medieval times, philosophers used the word *element* to refer to the simple substances of which all matter was thought to be composed. Eventually, substances that cannot be broken down into other kinds of matter came to be called "elements." Today chemists have methods by which the elements combined in any type of matter can be identified.

The elements that have been known since ancient times are listed in Table 3.1. One by one other elements were discovered, sometimes only after long and tedious experiments.

More than one hundred substances are now recognized as elements. Some, such as gold and sulfur, can be found in the crust of the Earth in their uncombined, elemental forms. Others, such as chlorine and uranium, are only found combined with other elements. And with the advent of our understanding of nuclear chemistry has come a string of man-made elements—elements not found naturally on Earth.

Table 3.1
Elements Known in Antiquity

Antimony
Carbon
Copper
Gold
Iron
Lead
Mercury
Silver
Sulfur
Tin

3.2 SYMBOLS FOR THE ELEMENTS

In talking and writing about chemistry, we come upon a recurring problem: How do we clearly communicate with each other about elements and the more complex substances derived from them? So many kinds of chemicals are now known that the problem can only be solved by a systematic approach. The language of chemistry is really *two* languages. One is based on the names of the elements and more complex chemical substances; the other is based upon symbols for the elements.

Table 3.2 gives the names and symbols for some of the more familiar elements. Thirteen elements have as their symbols the first letter of their modern names or older names. All other commonly encountered elements have two-letter symbols. Single-letter symbols are always capital letters. Other symbols are written with the first letter capitalized and the second lower case.

Some of the elements have symbols that do not appear to be related to their names. Most of these elements have been known since ancient times and their symbols are derived from their Latin names, for example

Cu	Pb	Fe
copper (*Latin* cuprum)	*lead* (*Latin* plumbum)	*iron* (*Latin* ferrum)

All elements with symbols not based on their modern names are listed in Table 3.3.

The names of the known elements are given in alphabetical order inside the front cover of this book. Opposite the alphabetical list of the elements is a *periodic table,* in which the elements are organized into groups based upon similarities in their chemical behavior and properties. (The format of the periodic table is explained in Chapter 8.) For more information about the names of the elements, see Aside 3.1.

Table 3.2
Symbols for Some Common Elements

Aluminum	Al
Arsenic	As
Barium	Ba
Boron	B
Bromine	Br
Calcium	Ca
Carbon	C
Chlorine	Cl
Chromium	Cr
Cobalt	Co
Fluorine	F
Helium	He
Hydrogen	H
Iodine	I
Lithium	Li
Magnesium	Mg
Nickel	Ni
Nitrogen	N
Oxygen	O
Phosphorus	P
Platinum	Pt
Silicon	Si
Sulfur	S
Uranium	U
Zinc	Zn

3.3 ATOMS

Early philosophers wondered whether matter that could be seen was composed of smaller bits of matter. Perhaps the rock, or the iron derived from it, were both "made of" something else—something so tiny that it couldn't be seen. The name *atom* was suggested by the Greeks for such small particles.

All matter is now understood to be composed of atoms. The description of matter in a scientific manner—what chemistry is all about—began with an understanding of atoms. The period from 1879 to 1932 was an exciting time, as chemists and physicists found out more and more about atoms. Many surprises were part of the story, the first being that atoms themselves have "structure"—atoms contain smaller particles within themselves.

Our modern picture of atomic structure, known as the nuclear model, is based upon a series of classic experiments designed to "see the unseeable and know the unknowable." (These experiments are described in Sections 3.6–3.10.) Atoms are *very* small. Their diameters are on the order of 1×10^{-10} m or, in more common units, 0.1 nm (nanometer). It has been estimated that 3.8×10^{13} (thirty-eight trillion) atoms of iron can "dance" on the head of a pin.

Table 3.3
Elements with Symbols Not Based on Their Modern Names
All of these elements except tungsten (discovered in 1783) and potassium and sodium (both discovered in 1807) have been known since antiquity. All of the symbols are based on Latin names except W, for tungsten, which derives from the name of an ore, wolfram.

Modern name	Symbol	Derivation of symbol
Antimony	Sb	*stibium*
Copper	Cu	*cuprum*
Gold	Au	*aurum*
Iron	Fe	*ferrum*
Lead	Pb	*plumbum*
Mercury	Hg	*hydrargyrum*
Potassium	K	*kalium*
Silver	Ag	*argentum*
Sodium	Na	*natrium*
Tin	Sn	*stannum*
Tungsten	W	*wolfram*

Aside 3.1 THE NAMES OF THE ELEMENTS

The names of the elements provide a fascinating glimpse into the history of chemistry. Gold, silver, and the other elements listed in Table 3.1 have been known since ancient times, and their names reflect what the Romans observed about them. For example, gold was called *aurum*, meaning "shining dawn"; mercury was *hydrargyrum*, meaning "liquid silver"; lead was *plumbum*, which means "heavy."

In the eighteenth century, chemists were intrigued with studies of the atmosphere, and the gases they discovered were given names based on what was then known of their chemistry. "Hydrogen" is from the Greek words meaning water-former, and "nitrogen" and "oxygen" are from the Greek words meaning soda-former and acid-former, respectively.

The names of metals, except for those known since antiquity, all end in *ium*. "Aluminium" was the name first given to the metal called "aluminum" in the United States. It is still called "aluminium" in England and many other parts of the world. We are blessed with the four tongue twisters terbium, erbium, ytterbium, and yttrium because these metals were all isolated from ores found in Ytterby, Sweden.

Names of some elements honor the places where the elements were discovered: californium, berkelium, europium, americium, francium, germanium. In recent years a series of man-made elements have been named in honor of famous scientists: einsteinium (Albert Einstein), fermium (Enrico Fermi), mendelevium (Dimitri Mendeleev), nobelium (Alfred Nobel), and lawrencium (Ernest Lawrence).

When faced with the challenge of naming new elements, chemists have turned to the heavens for plutonium, uranium, and cerium (for Ceres, an asteroid discovered at about the same time as the element). They have looked to mythology for thorium (Thor, the Scandinavian god of war), and promethium (Prometheus, the bringer of fire).

The honor of naming an element has traditionally gone to its discoverer. The prestige associated with discovering an element is reflected in an argument that has lasted for years over who should name elements 104 and 105. In the 1960s both American and Russian teams of scientists claimed discovery of elements 104 and 105. The Russians suggested the name khurchatovium (after a Russian scientist) for element 104 and the Americans suggested rutherfordium (Ernest Rutherford).

The International Union of Pure and Applied Chemistry (IUPAC) has been given the responsibility for making rules for naming elements and other chemical substances. In 1978 the conflict over naming element 104, and also the need for establishing the "discoverers" of this and other elements, was partially resolved. It was recommended that, at least for now, elements above 103 be named by a system based on numerical root words. Element 104 becomes unnilquadium ("un" for 1; "nil" for 0, "quad" for 4, plus the "ium" ending). Element 105 becomes unnilpentium, 106 becomes unnilhexium, and the system can cope with elements all the way up to atomic number 999, which will be ennennennium if and when it is discovered. These elements would become the first to have three-letter symbols (104, Unq; 105, Unp; 106, Unh, and so on).

A suggested alternate method is to represent these elements solely by their numbers until permanent names of the usual type are assigned to them. For example, in words "element 105" would be used; where a symbol is required for this element, simply the atomic number "105" would be used.

As knowledge from experiments accumulated, theoretical explanations of the behavior and structure of atoms were developed. Modern atomic theory views an atom as having a dense, central core (a *nucleus*) containing positively charged particles (*protons*) and neutral particles (*neutrons*). Negatively charged particles (*electrons*) are scattered in a relatively large space around the nucleus. Nuclei have diameters on the order of 1×10^{-6} nm, one hundred thousand times smaller than atoms. If an atom were expanded to the size of one of our largest football stadiums, the nucleus would be about the size of a marble at the center.

In the periodic table (inside the front cover), the elements appear in the order of their *atomic numbers* (shown in the upper left-hand corners of the boxes). The atomic number (Section 3.10) is equal to the number of positively charged particles in the atoms of a given element. As a result of the different numbers of particles in their nuclei, atoms of different elements, as we shall see, have different masses. An **atom** is the smallest particle of an element. Aside 3.2 explains further the concept of the atom.

Aside 3.2 THE CONCEPT OF THE ATOM

If you divide a drop of water into smaller drops, and one of those into still smaller drops, and so on, will you ultimately reach a point where the drop cannot be further subdivided, even in imagination, without destroying the substance of the water itself? The ancient Greek philosophers debated this question for many years.

Democritus of Abdera, in about 400 B. C., came close to the modern theory of matter—the atomic theory. He argued that all matter is composed of tiny, homogeneous particles that are hard and impenetrable, differ in size and shape, can come together in different combinations, and are constantly in motion. Democritus named the particles "atoms," from the Greek word meaning "indivisible." Several hundred years later, Lucretius in Rome said it this way: "... all nature as it is in itself consists of two things—for there are atoms and there is the void."

The remarkable way in which atomic theory can explain the behavior of matter gradually came to be fully appreciated as chemistry became a quantitative science. Let's look at one example of the kind of evidence that led John Dalton, an English chemist (1766–1844), to formulate his atomic theory (Table 3.4).

Many compounds containing only the elements carbon and hydrogen are known. One of the simplest of these contains 85.7% carbon and 14.3% hydrogen (by mass). Another common one contains 75.0% carbon and 25.0% hydrogen.

As they stand, such figures are of scant interest. But if we calculate from them the mass of hydrogen combined with one gram of carbon in each case, we find that in the first compound it is 0.167 g and in the second, 0.333 g. The second of these numbers is almost exactly twice the first. Your reaction to this may be that it is an interesting coincidence. However, there are many other cases like it. The amount of hydrogen that combines with a given amount of carbon is always 1 or 2 or 3 or 4 ... times a specific amount of hydrogen. And the same type of relationship is found for the combination of other elements. So we must conclude that this is not a coincidence. How, then, shall we explain it? The most logical explanation is that the elements exist in the form of extremely small, discrete units—atoms—and that these combine in simple ratios. In other words, one atom of A will combine with one atom of B, or with two atoms of B, and so on. The smallest possible particle of water is, we now know, that in which two hydrogen atoms are combined with one oxygen atom.

The importance and usefulness of Dalton's atomic theory are not diminished by the modern knowledge that some of the statements are not correct under all circumstances. [The second and third statements do not apply to reactions in which the nucleus undergoes a change (Section 12.2), and the fourth statement was found to be incorrect when isotopes were discovered (Section 3.11).]

Table 3.4
Dalton's Atomic Theory (1808)

1. All matter consists of tiny particles. Dalton, like the Greeks, called these particles atoms.
2. Atoms of one element can neither be subdivided nor changed into atoms of any other element.
3. Atoms can neither be created nor destroyed.
4. All atoms of the same element are identical in mass, size, and other properties.
5. Atoms of one element differ in mass and other properties from atoms of other elements.
6. Chemical combination is the union of atoms of different elements; the atoms combine in simple, whole-number ratios to each other.

SUMMARY OF SECTIONS 3.1–3.3

Substances that cannot be broken down chemically into simpler substances are referred to as elements. More than 100 elements are now known, most of which occur naturally on Earth but some of which are man-made. Chemical elements are symbolized by one- or two-letter abbreviations derived from their modern names or, in some cases, from their Latin names. All matter is composed of atoms, and an atom is the smallest particle of an element. Atoms themselves are composed of smaller particles. A dense central core, known as the nucleus, contains positively charged protons and uncharged neutrons. Much lighter, negatively charged electrons occupy a relatively large space surrounding the nucleus.

KINDS OF MATTER

3.4 PURE SUBSTANCES AND MIXTURES

Every kind of matter can be classified as a pure substance or a mixture. The nature of a sample of matter is determined by examining its properties and its composition. Color, melting point, density, and hardness are among the properties of matter. These are **physical properties**: they can be measured or observed without changing the composition and identity of a substance. By contrast, any process in which the *identity* and composition of at least one substance is changed is a **chemical reaction**. **Chemical properties** can be observed only in chemical reactions, which result in changes in the identities of substances.

A **pure substance** is a form of matter that has the same composition and the same properties, no matter what its source. For example, pure water has the composition 88.8% oxygen and 11.2% hydrogen (by mass). Pure water is colorless and odorless, and at atmospheric pressure boils at 100 °C and freezes at 0 °C. Water weighs 1 gram per milliliter at 4 °C and it does not burn. Water has this composition and these properties whether it is distilled from seawater or from melted snow, or prepared in a chemical reaction by the union of hydrogen and oxygen.

All pure substances are either elements or chemical compounds. **Elements** are composed solely of atoms of the same atomic number (Sections 3.10 and 3.11). A pure sample of the element iron is 100% iron atoms.

A **chemical compound** is a substance of definite, fixed composition in which atoms of two or more elements are chemically combined. "Chemically combined" means that the atoms are held together by forces strong enough to keep them together under normal conditions, giving a substance that has properties different from those of the original elements. For example, in water there are *always* two hydrogen atoms for every oxygen atom, and the mass ratio of hydrogen to oxygen is 1 to 8, *always*. The properties of water are very different from those of hydrogen and oxygen.

A **mixture** is composed of two or more substances that retain their separate identities. The substances in a mixture (unlike the elements in a chemical compound) can be present in *any* proportions, and the components can be retrieved intact from the mixture without a chemical change (although sometimes only with difficulty). A **heterogeneous mixture is** a mixture in which the individual components of the mixture remain physically separate and can be seen as separate components, although in some cases a microscope is needed. Concrete and granite are heterogeneous mixtures. Powdered iron and powered sulfur, no matter how well stirred, form a heterogeneous

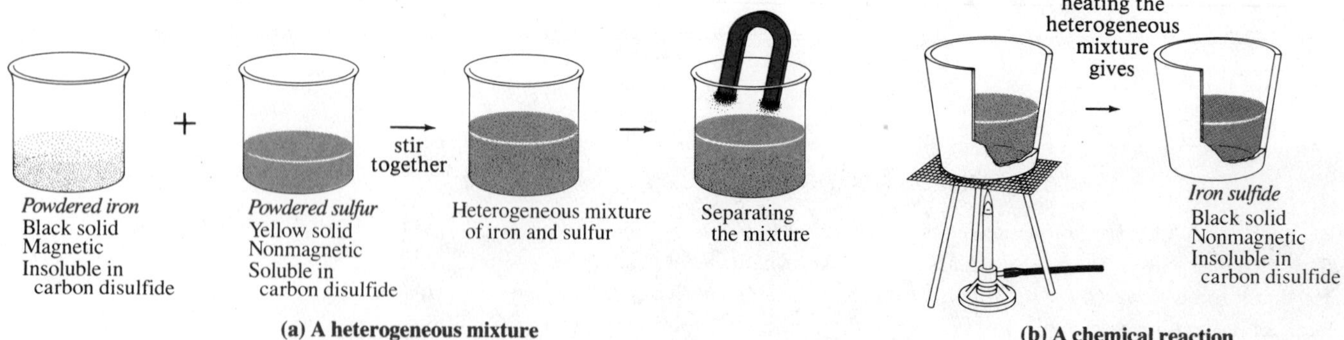

Figure 3.1
Iron Plus Sulfur: (a) A Mixture or (b) a Chemical Compound
(a) If iron and sulfur powders are thoroughly mixed but not heated, a mixture results. The iron can be completely removed from the sulfur by a magnet. Conversely, the sulfur can be completely removed from the iron by dissolving it in carbon disulfide, a liquid in which iron is not soluble. (b) When heated together, powdered iron and powdered sulfur give a chemical compound, iron sulfide.

mixture. They can be separated by using a magnet to attract the powdered iron (Figure 3.1a).

By contrast, when iron and sulfur are heated together, a chemical reaction occurs and iron sulfide, a chemical compound, is formed. The properties of iron sulfide (Figure 3.1b) are clearly different from those of iron and sulfur.

The substances in a **homogeneous mixture** are thoroughly intermingled, and the composition and appearance of the mixture are uniform throughout. Air is a homogeneous mixture of gases, and motor oil is a homogeneous mixture of liquid petroleum derivatives.

We commonly think of solutions as homogeneous mixtures of something with water. Strictly speaking, any homogeneous mixture of two or more substances is a **solution**. A solution of any substance in water is an **aqueous solution**.

Solutions are spoken of as having two components: the solvent and the solute (or solutes). If a handful of salt is dissolved in a bucket of water, the salt is the solute and the water is the solvent. The **solvent** is the component of a solution usually present in the larger amount. The solvent is the medium in which the **solute**—the component of a solution usually present in the smaller amount—has dissolved. The terms "solvent" and "solute," while convenient, are often imprecise and do not have fixed scientific meaning. The process of one substance dissolving in another is called **dissolution**.

The general classification of matter is summarized in Figure 3.2.

3.5 STATES OF MATTER

There are three different **states of matter**: the gaseous state, the liquid state, and the solid state. A substance is generally described as a gas if it is found in the gaseous state at ordinary temperatures and pressures (at roughly room temperature and pressure). The elements oxygen, nitrogen, and hydrogen, and the compounds carbon dioxide, ammonia, and methane (the major component of natural gas) are all gases. The compounds water and ethyl alcohol, and the element mercury are liquids at ordinary temperatures and pressures. And solids, of course, are everywhere we look. The majority of pure substances are solids. Common pure substances that are solids at

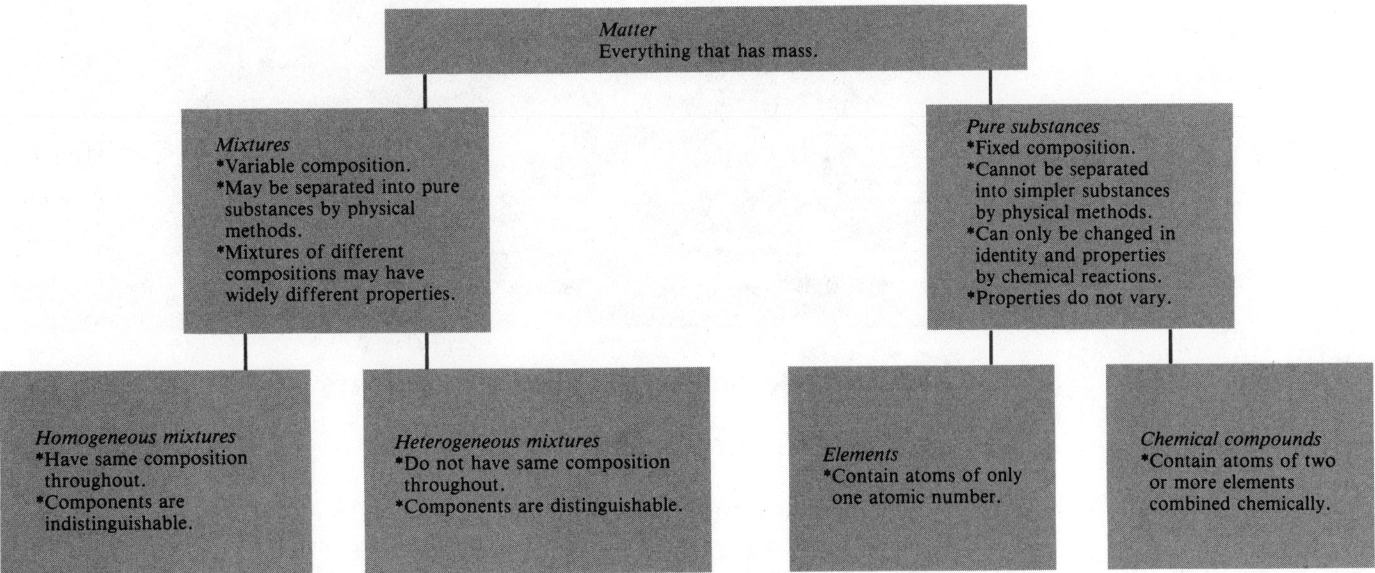

Figure 3.2
The Classification of Matter

ordinary temperatures and pressures include most metallic elements; carbon, as either diamond or graphite; and the compounds sodium chloride (table salt) and sucrose (table sugar). Like sugar and salt, many pure chemical compounds are crystalline solids.

Some substances, at different temperatures and/or pressures, can exist in all three states. Water is known in the solid state as ice, in the liquid state (at room temperature), and in the gaseous state as steam or water vapor. Interconversions between the solid, liquid, and gaseous states are referred to as **changes of state**. Many metals, which are usually solid, can be melted, and if heated to even higher temperatures, can become gaseous. Some substances, however, cannot exist in the gaseous state; others cannot exist in the liquid state; and some cannot exist in either the gaseous or the liquid state. For example, calcium carbonate, a solid, cannot be melted or vaporized: upon heating it decomposes into calcium oxide, a different solid chemical compound, and carbon dioxide, a gas. Upon gentle heating, sugar melts to the liquid state, but upon heating to higher temperatures sugar does not become gaseous. Instead, it decomposes into a variety of products that contain carbon. However, all gases and liquids can be condensed to the solid state.

The term **phase** refers to a homogeneous part of a system in contact with but separate from other parts of the system. A glass of iced tea—we might think of the iced tea as the system—has a solid phase (ice, a pure substance) and a liquid phase (tea, a solution), for example. The iced tea includes substances in the solid and the liquid states. A bottle of oil and vinegar contains only substances in the liquid state, but it also has two phases, because the oil phase and the vinegar phase remain in contact with each other but do not mix. It is also possible to have a completely solid substance in which several phases that are different crystalline forms are in contact with each other.

SUMMARY OF SECTIONS 3.4–3.5

Every kind of matter can be classified as a pure substance or a mixture. A pure substance always has the same chemical properties (observed in chemical reactions, in which the identities of substances are changed) and physical properties (observed or measured with no change in the composition and identity of substances). All pure substances are either elements, which contain only atoms of one kind, or chemical compounds, which contain chemically combined atoms of different elements.

In a mixture, two or more substances are intermingled but retain their separate identities. Mixtures can be homogeneous (of uniform composition and appearance) or heterogeneous (of nonuniform composition and appearance). The components of either type of mixture can be retrieved from the mixture without any change in identity.

The three states of matter are the gaseous state, the liquid state, and the solid state. Interconversions between these states are known as changes of state. A phase is a homogeneous part of a system that is in contact with but separate from other parts of the system—such as the solid phase of the ice that floats in the liquid phase in a glass of iced tea.

ATOMIC STRUCTURE: FIVE CLASSIC EXPERIMENTS

3.6 CATHODE RAYS: THE ELECTRON

At atmospheric pressure, neither air nor other gases conduct electricity very well. However, a discovery important to the progress of chemistry was made in 1821 when it was found that at very low pressure, gases conduct electricity readily. Numerous studies were made on the passage of electrical current through different gases contained in **gas-discharge tubes**—glass tubes that can be evacuated and into which electrodes are sealed (e.g., see Figure 3.3). (**Electrodes** are conductors through which electrical current enters or leaves a conducting medium. For a brief introduction to electricity, see Aside 3.3 on page 40.)

As the pressure is gradually lowered by pumping gas out of a gas-discharge tube, a glow appears, the color of the glow being different for different gases. Eventually, current flows between the electrodes, the glow fades out, and a greenish fluorescence is produced on the glass wall at the far end of the tube. (A **fluorescent** substance emits radiation during exposure to light or some other form of energy.)

William Crookes, an English editor, inventor, and scientist confirmed the important discovery (made earlier by Julius Plücker in Bonn) that the fluorescent spot can be moved by a magnet. Crookes concluded, in 1879, that rays of particles were flowing from the negative electrode, or *cathode*, in a gas-discharge tube. He called them "cathode rays." Cathode rays are the same, Crookes found, no matter what material is used for the negative electrode or what gas is in the tube.

The *cathode ray tube* is the ancestor of neon signs and television tubes. What we call "neon" signs are cathode-ray tubes that may contain neon (which gives a red glow) or other gases that produce other colors. In television tubes, a rapidly moving cathode ray continuously sweeps across the surface of the tube, which is coated with a material that fluoresces when struck by the cathode ray. The picture is created by variations in the intensity of the ray.

George Johnstone Stoney, a physicist, proposed in 1881 that electricity is carried by individual, negatively charged particles. Stoney named the proposed particle the "electron" (from the Greek for amber, a material known to acquire an electric charge when rubbed with silk).

J. J. Thomson, a remarkable man who at age 28 (in 1884) had been named the director of the Cavendish Laboratory at Cambridge University in England and seven of whose students went on to win Nobel prizes, studied cathode rays intensively. Thomson found that cathode rays are deflected by both electric and magnetic fields. This and the other properties of cathode rays (Table 3.5) convinced Thomson that cathode rays are streams of negatively charged particles of extremely small mass, much smaller than the masses of any atoms—particles identical with Stoney's proposed "electrons." Furthermore, Thomson concluded that because cathode rays are always the same, electrons must be present in *all* matter.

To prove that cathode rays consist of such particles, Thomson measured the ratio of the charge (e) of the particles to their mass (m). He did this by an ingenious method of balancing the forces of electric and magnetic fields, as described in Figure 3.3. The value

Table 3.5
Properties of Cathode Rays

1. Travel in straight lines from the cathode to the anode
2. Cast shadows when metal objects are placed in their path
3. Produce fluorescence where they strike the glass walls of the tube
4. Heat thin metal foils to incandescence
5. Cause ionization of gas molecules
6. "Expose" photographic films or plates
7. Produce highly penetrating radiation (x-rays) when directed against a target
8. Impart a negative charge to such a target
9. Undergo deflection parallel to an applied electrostatic field (away from the negative electrode) and perpendicular to an applied magnetic field

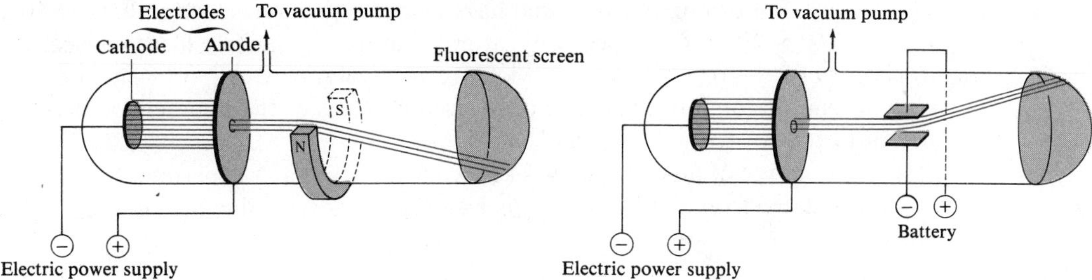

(a) Deflection of cathode ray by magnetic field

(b) Deflection of cathode ray by electric field

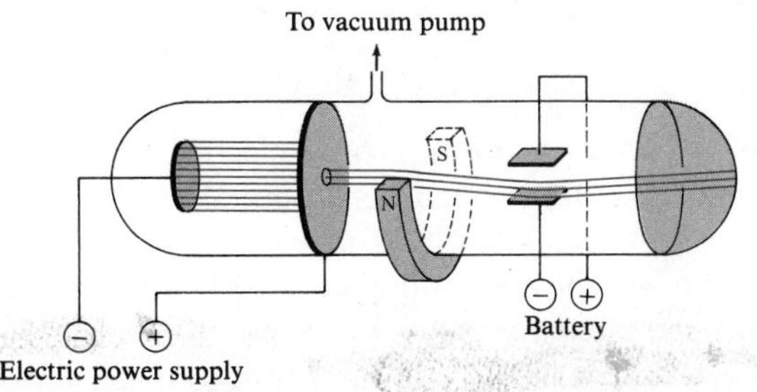

(c) Balanced deflection of cathode ray by electric and magnetic fields

Figure 3.3
Determination of e/m of the Electron
The magnetic and electric fields deflect the cathode ray in opposite directions. By varying these two opposing forces until they balance (an often-used principle in the design of experiments), Thomson was able to calculate from the field strengths, the e/m value of the electron.

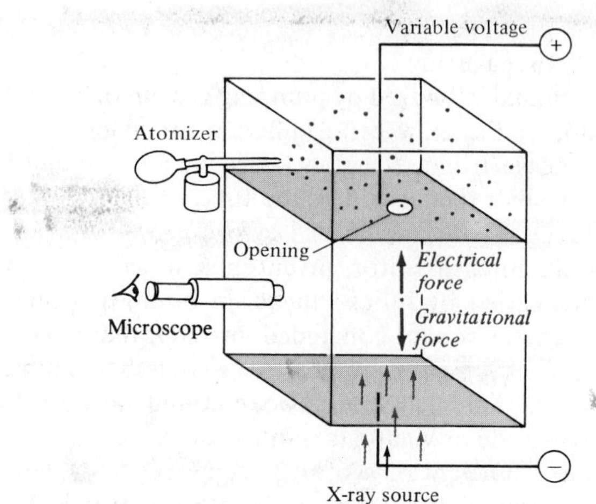

Figure 3.4
Millikan Oil-Drop Experiment
The velocity of the rise and fall of charged oil drops between the + and − plates is observed through the microscope. Both the charge on the drops and the electric field can be varied. Here again (see Figure 3.3) an experiment makes use of the balancing of opposing forces.

of e/m was the same for different gases and different electrode materials, evidence that electrons are indeed present in all matter. The modern value for e/m is 1.76×10^8 C/g. (C is the abbreviation for the SI unit of electrical charge, the coulomb.)

In 1909 Robert A. Millikan at the University of Chicago accomplished what Thomson had been unable to do: he experimentally determined the charge on the electron. In Millikan's experiment (Figure 3.4), tiny oil droplets, some of them charged by friction, were released between two horizontal plates between which an electrical field (see Aside 3.3) could be created. The rate of fall of the droplets under the influence of gravity alone, and then their rate of rise and fall when the electric field was on, were closely observed. Sometimes the rate changed abruptly, showing that the droplet had captured a charged particle from the air.

In some experiments x-rays were passed through the air to assure the formation of charged particles. After many measurements on droplets with different charges, Millikan found that the charge, whether positive or negative, was always a whole-number multiple of the same value: 1.60×10^{-19} C. This, therefore, is the charge of the smallest charged particle, the electron. During the experiment, **ions**—atoms (or other species) that have become positively or negatively charged by the loss or gain of electrons—had been formed from atoms in the air. The oil droplets became charged when they captured ions, and therefore had charges equal to those of one or more electrons. For example, atoms that have gained or lost one electron have charges of − or $+1.60 \times 10^{-19}$ C, respectively; atoms that have gained or lost two electrons have charges of − or $+3.20 \times 10^{-19}$ C, respectively. (Instead of using units of charge, the charges on ions are commonly given in terms of the numbers of electrons gained or lost: −1, +1, −2, +2, and so on.)

Once the charge of an electron was known from Millikan's experiments, the mass could be calculated from the value of e/m. Using the modern value of e/m and solving for the mass gives

$$\frac{e}{m} = \frac{1.60 \times 10^{-19}\ \text{C}}{m} = 1.76 \times 10^{8}\ \text{C/g}$$

$$m = \frac{1.60 \times 10^{-19}\ \cancel{\text{C}}}{1.76 \times 10^{8}\ \cancel{\text{C}}/\text{g}} = 9.09 \times 10^{-28}\ \text{g}$$

As predicted by Thomson, the mass of an electron is much smaller than the mass of any atom. It would take 1837 electrons to equal the mass of one hydrogen atom, the smallest atom.

The electron was the first fundamental, subatomic particle discovered. A **subatomic particle** is a particle smaller than the smallest atom. A **fundamental particle** is one that is present in all matter. The **electron** is a fundamental, subatomic, negatively charged

Particle	Symbol	Mass[a]		Charge	
		In grams	In atomic mass units[b]	In Coulombs	Relative
Electron	e^-	9.109534×10^{-28} g	0.0005485802 u	-1.602×10^{-19}	-1
Proton	p^+	$1.6726485 \times 10^{-24}$ g	1.0072764 u	$+1.602 \times 10^{-19}$	$+1$
Neutron	n	$1.6749543 \times 10^{-24}$ g	1.0086650 u	0	0

[a] Strictly speaking, we should use the term *rest mass.* Physicists have shown that moving particles have greater mass than particles that are "resting." The distinction has little bearing here, and throughout this book we use "mass" to refer to the rest mass of particles.

[b] The "u" is the abbreviation for "atomic mass unit," which is defined in Section 3.12.

Table 3.6
Fundamental Particles of Importance to Chemistry
Electrons and protons are both stable outside of an atom. Neutrons eventually decompose spontaneously.

particle; **cathode rays,** as Thomson thought, are streams of electrons flowing from the cathode toward the anode in a gas-discharge tube. Modern values for the properties of the electron and other fundamental particles are given in Table 3.6.

3.7 CANAL RAYS: THE PROTON

Atoms are electrically neutral. Therefore, the identification of positive charges to balance the negative charge of the electrons was essential. Also, since electrons have such small mass, something else had to be found to account for the much greater mass of atoms.

The first observations of charged particles other than cathode rays were also made in gas-discharge tubes. In 1886, Eugen Goldstein discovered that rays with a positive charge were flowing in the opposite direction from the cathode rays in such tubes. These were named *canal rays,* because they pass through "canals"—openings cut into the cathode.

In 1898, Wilhelm Wien, a German physicist, succeeded in measuring e/m for canal rays by a method similar to that used by Thomson for the electron. The work of Wien and others showed that canal ray particles are much heavier than electrons and that, unlike cathode ray particles, they vary in mass according to the gas present in the tube. Canal rays have positive charges that are small-whole-number multiples of $+1.60 \times 10^{-19}$ C (a charge equal but opposite to that on the electron). The properties of canal rays are summarized in Table 3.7.

Table 3.7
Properties of Canal Rays
These rays are so named because they pass through holes, or "canals," in the cathode.

1. **Travel in straight lines toward the cathode**
2. **Produce fluorescence when they strike the walls of the tube**
3. **Are deflected in the opposite direction from cathode rays by both electric and magnetic fields**
4. **Are deflected less than cathode rays by fields of equal strength**
5. **Expose photographic plates**
6. **Differ for different gases in the tube**

After the discovery of the canal rays, it was possible to explain completely what happens in a gas-discharge tube (Figure 3.5). Electrons from the cathode collide with atoms of whatever gas is present in the tube, knocking out of each of these atoms one or more additional electrons. These collisions leave behind positive ions formed by the loss of electrons from the atoms of the gas. Most of the positive ions strike the cathode, but a few pass through the holes ("canals") in the cathode. **Canal rays** are positively charged ions flowing from the anode to the cathode in a gas-discharge tube. They can be observed as luminous rays and also can be detected as they impinge on a fluorescent material on the inside of the tube (Figure 3.6; page 41).

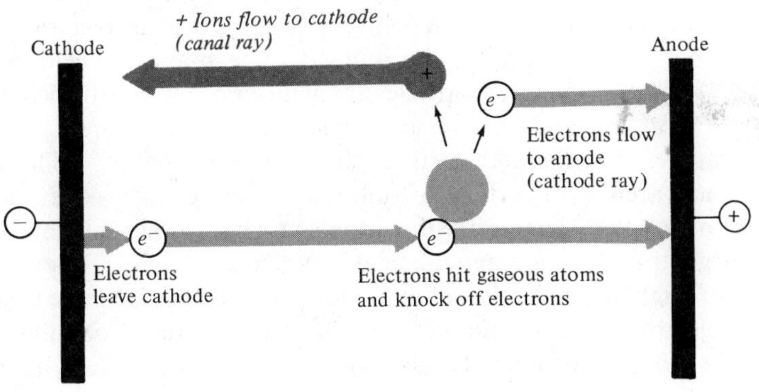

Figure 3.5
Cathode Rays and Canal Rays
A cathode ray consists of electrons from the atoms of the cathode material flowing toward the anode. These electrons collide with gaseous atoms, knocking off other electrons and leaving behind positively charged ions. Some of the excess energy that the gaseous atoms acquire is given off as radiation, causing a glow in the tube. A canal ray consists of the positive ions flowing toward the anode. As the pressure becomes lower, the electrons encounter fewer and fewer atoms, and eventually the glow disappears. However, the cathode ray continues to flow from the cathode to the anode.

The lightest and simplest canal-ray particles are formed when a gas-discharge tube contains hydrogen. Ernest Rutherford (a student of Thomson's, who succeeded him as director of the Cavendish Laboratory) focused on these particles in his search for a fundamental, positively charged particle. Eventually he showed that particles identical to hydrogen atoms with one electron missing (hydrogen ions, H^+), are present in all matter. He named these particles protons. A **proton** is a fundamental subatomic particle with a positive charge equal in magnitude to the negative charge of the electron (see Table 3.5).

Aside 3.3 TOOLS OF CHEMISTRY: ELECTRICITY AND MAGNETISM

The study of electricity and magnetism is one of the major areas of physics. By observing the behavior of matter under the influence of electricity or magnetism, much has been learned about the properties of matter. For the purpose of studying chemistry, therefore, it helps to understand in a general way some of the concepts and terms related to electricity and magnetism.

What is commonly thought of as "electricity" is the flow of electrons through a wire. The essential property of electrons is their negative charge. Positively charged objects are attracted to negatively charged objects, and objects of the same charge repel each other. Protons are positively charged, and the attraction between electrons and protons holds the particles together in atoms. The charges on electrons and protons, and the attraction and repulsion of charged objects are not *proven* or *explained* by theory. These are fundamental properties that are accepted because they are observed to exist.

The force of electrical attraction or repulsion between charged particles is called the **Coulomb force**. The magnitude of the Coulomb force between two particles depends upon the charges on the particles and the distance between them.

$$\text{Coulomb force} = k\frac{q_1 q_2}{r^2} \tag{3.1}$$

charge on particles 1 and 2 ($q_1 q_2$); *proportionality constant* (k); *distance between particles 1 and 2* (r)

The region around one charged particle or bit of matter in which another charged particle will be attracted or repelled, depending on their charges, is an *electric field.* A charged particle in an electric field has *electrical potential energy,* just as the water in a reservoir has gravitational potential energy. **Potential energy** is energy that an object has by virtue of its position, that is, because of a force acting on it that could cause it to move. In flowing out of the reservoir, the water flows from an area of higher gravitational potential energy to one of lower gravitational potential energy. Similarly, a charged particle in an electric field moves from an area of higher electrical potential energy to one of lower electrical potential energy.

Charged particles in an electric field are described as moving in response to a *potential difference.* There is a potential difference between the cathode and anode in a gas-discharge tube (see Figure 3.3) and between the parallel plates in the Millikan experiment (see Figure 3.4). In a diagram of an electrical apparatus or circuit, you can recognize by the location of the + and − signs the presence of a potential difference that will cause charged particles to flow. The *volt* is the unit of electrical potential difference. To "put a voltage across" something means to apply a potential difference.

The movement of charged particles constitutes an electric current. The metal from which wires are made, and all other substances through which current can flow, are called *conductors.* Substances that do not conduct electricity are called *insulators.* To be conductive, a substance must contain charge carriers—charged particles that are free to move through the material. For example, in a gas-discharge tube (see Figure 3.5), electrons and positive ions flow through a gas.

The ability of different substances to allow charge carriers to move varies widely. The *electrical conductivity* of a substance is a measurable quantity used to compare the ability of different substances to carry current, or the ability of the same substance to carry current under different conditions (for example, at different temperatures). In later chapters, we discuss how the chemical constitution of a substance influences its electrical conductivity.

For a long time electricity and magnetism were thought to be different phenomena, and we still frequently refer to them as though they are different. Actually, electricity and magnetism are two aspects of the same phenomenon. A wire through which an electric current is moving is surrounded by a magnetic field. Whenever charged particles are in motion, a magnetic field is created and, conversely, where a magnetic field is detected, charged particles are in motion nearby. (The magnetic properties of materials derive from the motion of electrons within atoms.) A charged particle that is moving in a magnetic field experiences a force. This explains the deflection of cathode rays in a gas-discharge tube when a magnetic field is applied (see Figure 3.3) and the deflection of ions in a mass spectrometer (see Aside 3.4).

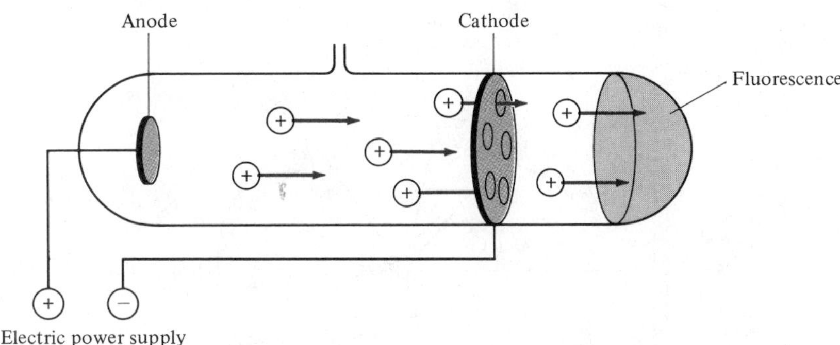

Figure 3.6
Canal Rays
Positive ions are produced in the gas-discharge tube when electrons from the cathode collide with gaseous atoms. The canal ray consists of the resulting positive ions moving toward the cathode. Some ions pass through the "canals" in the cathode and cause fluorescence when they strike the end of the tube.

3.8 α-PARTICLE SCATTERING: THE NUCLEUS

a. Radioactivity and α-Particles During the years that Rutherford was studying atomic structure, he was also studying radioactive elements—elements whose atoms spontaneously break down because they are unstable. (Chapter 12 is devoted to radioactivity and nuclear chemistry.) One result of this work was Rutherford's identification of α-particles ("alpha-particles"), which are given off by some radioactive elements. An **α-particle** is a helium ion with a charge of +2, He^{2+}. (The symbol for an ion is written by adding the charge as a superscript to the element symbol.) Each α-particle has a positive charge twice that of a proton and a mass four times that of a proton.

The α-particle, which is not a fundamental particle, was used in some pioneering experiments. α-Particles are emitted from radioactive elements with very high speed, and therefore carry very high energy. Rutherford and others recognized the value of such particles. Because of their high energy, they could be fired like bullets at atoms and, from what happened when the particles collided, information about the structure of atoms could be obtained.

b. α-Particle Scattering and the Nuclear Atom The third of the classic experiments on atomic structure was carried out in 1911 in Rutherford's laboratory by his colleagues Hans Geiger and Ernest Marsden. At the time, the prevailing picture of an atom was that proposed by J. J. Thomson after his determination of e/m for the electron. Thomson suggested that an atom is a sphere of positive charge in which electrons are embedded like raisins in a plum pudding. The electrons would be distributed at maximum distances from each other because of the repulsion of their negative charges. This model was tested by the experiment of Geiger and Marsden.

A narrow beam of α-particles from a radioactive source was aimed at a thin foil of a metal such as gold, platinum, or copper (Figure 3.7). Rutherford wrote of what happened as follows: "I remember ... Geiger coming to me in great excitement and saying, 'We have been able to get some of the alpha particles coming backwards....' It was quite the most incredible event that has ever happened to me in my life. It was almost as incredible as if you fired a 15-inch shell at a piece of tissue paper and it came back and hit you."

If the mass and positive charge were uniformly distributed throughout each atom in the metal foil, as suggested by the Thomson model, there would be no concentration of charge or mass large enough to deflect the α-particles. Faced with this disagreement between fact and theory, Rutherford did what a good scientist must do. He abandoned the old theory and devised a better one. He proposed that each atom has a dense central core, which he called the nucleus. The **nucleus** is a central region, very small by comparison with the total size of an atom, in which virtually all of the mass and positive charge of the atom are concentrated.

In α-particle bombardment of a foil of such nuclear atoms, the majority of the α-particles encounter only empty space and pass through undeflected (see Figure 3.7b).

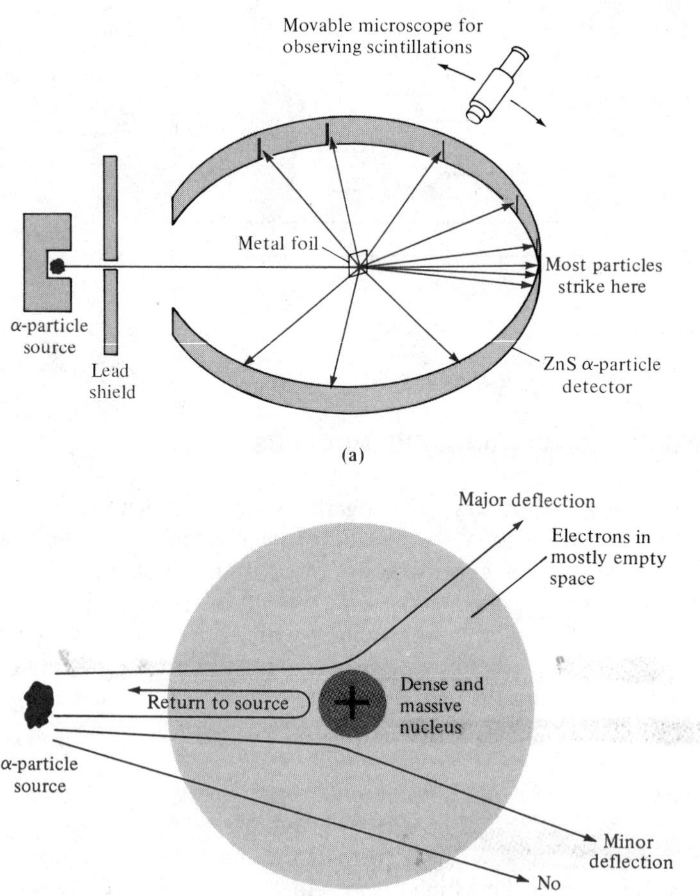

Figure 3.7
α-Particle Scattering
(a) Where an α-particle strikes the zinc sulfide screen, a flash of light (a "scintillation") can be observed through the microscope. Most α-particles pass through the metal foil undeflected. Some are deflected through varying angles and a few are deflected back toward the source.
(b) This diagram shows how α-particles are scattered by a single atom with a dense positive nucleus.

But those few that come very close to the nucleus experience maximum repulsion and are returned in the direction from which they came.

The nucleus, according to Rutherford, contained closely packed protons. The positive charge of the protons had to be balanced by the negative charge on the electrons outside the nucleus. However, this model was not completely satisfactory because the mass of protons and electrons equal in number to the nuclear charge did not account for the observed atomic masses. For example, a nucleus of two protons—to balance the charge of its two electrons—could not explain the helium atom, which has a mass *four* times that of a proton. For a while it was thought that the nucleus contained enough additional protons to account for the observed masses of atoms, plus enough electrons inside the nucleus to maintain electrical neutrality. After Chadwick's discovery of the neutron in 1932, a more satisfactory model of the nucleus emerged.

3.9 20 YEARS LATER: THE NEUTRON

In 1920 Rutherford had proposed that the nucleus might contain an uncharged particle with a mass close to that of the hydrogen atom. The search for this "neutron" lasted another 12 years. James Chadwick of Rutherford's laboratory in Cambridge saw the answer in some α-particle experiments reported by other workers. They had produced a "highly penetrating" radiation from beryllium (Figure 3.8). This radiation knocked protons out of paraffin with great force.

Chadwick felt that the "highly penetrating radiation" must be a beam of uncharged particles, each with the mass expected for a neutron. Chadwick performed his own experiments and in 1932 was able to prove conclusively the existence of the

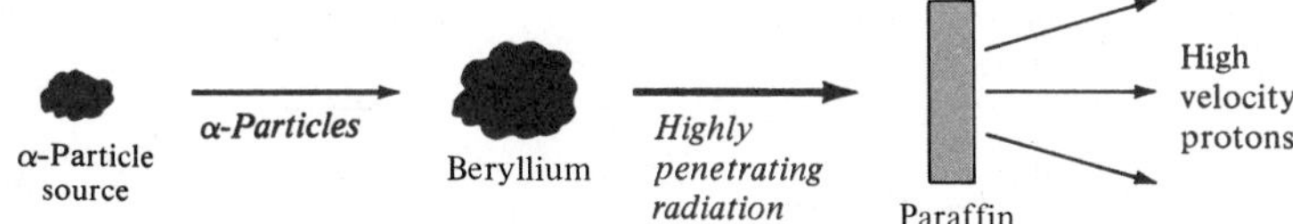

Figure 3.8
Experimental Discovery of the Neutron
Such experiments were first performed by Irène and Frédéric Joliot-Curie in Paris, who interpreted the "highly penetrating" radiation as an x-ray. Chadwick recognized that, based on what was known of energy and momentum, only a neutral particle with mass close to that of the proton could knock protons out of paraffin.

neutron. The **neutron** is a fundamental, subatomic particle that has a mass almost the same as the mass of the proton and has no charge (see Table 3.5). [Note that a neutron is considered a fundamental particle even though there is one type of matter—most hydrogen atoms—in which there are no neutrons.]

With the discovery of the neutron, it could be assumed that the nucleus contains protons equal in number to the electrons, plus enough neutrons to account for the total mass of the atom. For example, the helium atom contains in its nucleus two protons and two neutrons that account for its mass (Figure 3.9). All the electrons were assumed to be outside of the nucleus, although their exact arrangement was yet to be understood.

3.10 X-RAY SPECTRA: ATOMIC NUMBER

In 1913 Henry G. J. Moseley, a young Englishman, performed a series of experiments crucial to the understanding of atomic numbers and the atomic nucleus. It had been found that when electrons from the cathode in a gas-discharge tube hit a metal target, the metal emits x-rays (Figure 3.10). When the spectra from these x-rays were photographed, series of lines that varied with the metal could be seen. (A **spectrum**—the plural is spectra—is an array of radiation or particles spread out according to the increasing or decreasing magnitude of some physical property. In the case of x-ray spectra, this property is wavelength. See Aside 8.1). Moseley examined the x-ray spectra of 38 elements from aluminum to gold. Several series of lines appear in the spectrum of each element. Moseley found that, with one or two exceptions, as the mass of the target atoms increased, the position of any specific line in a specific series moved by regular intervals toward shorter wavelengths (Figure 3.11). There was a simple mathematical relationship between the position of these lines and a property of the target elements. Here, in his own words, are his conclusions:

1. Every element from aluminum to gold is characterized by an integer *N* which determines its x-ray spectrum. Every detail in the spectrum of an element can therefore be predicted from the spectra of its neighbors [in the periodic table].
2. This integer *N*, the atomic number of the element, is identified with the number of positive units of electricity contained in the atomic nucleus. [H. G. J. Moseley, in *The World of the Atom*, H. A. Boorse and L. Motz, eds. (New York: Basic Books, 1966), p. 883.]

The **atomic number** of an element, now symbolized by *Z*, is equal to the number of protons in the nucleus of each atom of that element. Inside the front cover of this book, in the table that lists the elements alphabetically, is a column labeled "atomic number." Look at the periodic table opposite the alphabetical table and note that the atomic numbers correspond to the order in which the elements appear in the periodic table. The periodic table was devised before the significance of this arrangement was fully understood. (The history of the periodic table is discussed in Aside 9.2.)

Rutherford's concept of the nucleus is now universally accepted. Rutherford did not know the whole story of what was inside the nucleus. We still do not know it. Nor did Rutherford have a clear understanding of the location and behavior of electrons in atoms. But we now believe he was correct in picturing an atom as mostly empty space occupied by electrons moving around a very small, dense central core.

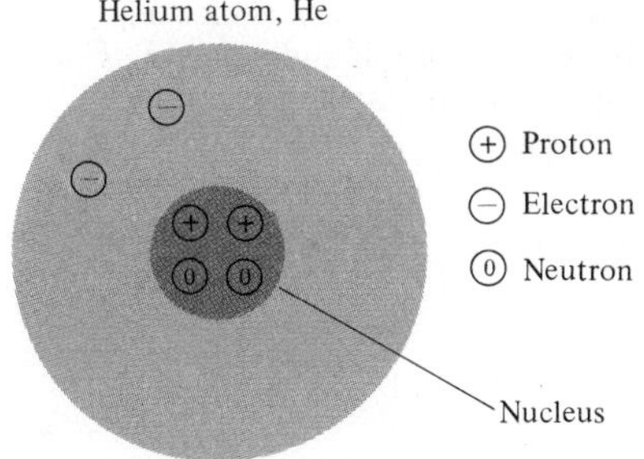

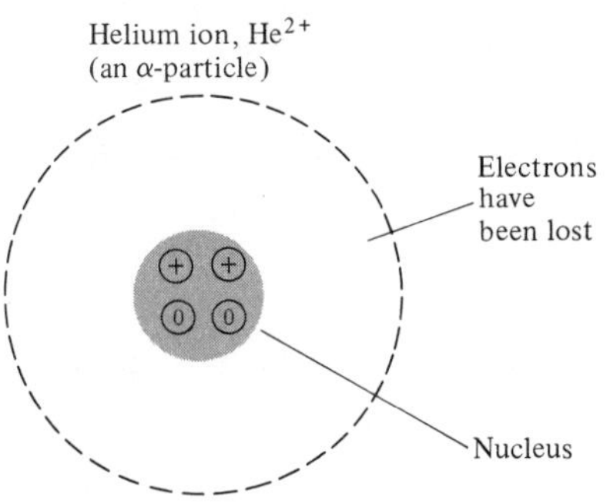

Figure 3.9
The Helium Atom and the Helium +2 Ion
This is a schematic picture of the location of the electrons, protons, and neutrons in helium (not to scale).

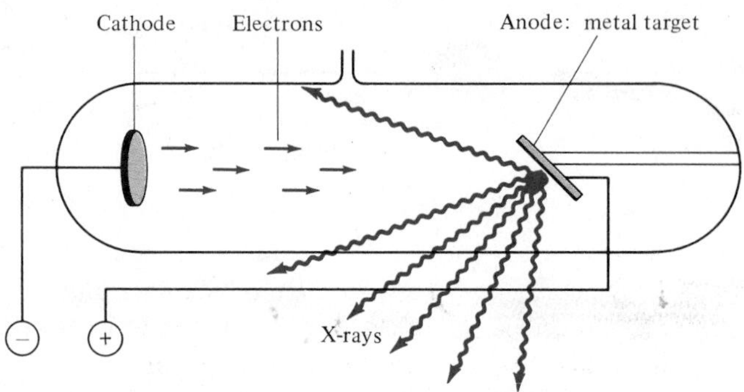

Figure 3.10
An X-Ray Tube.

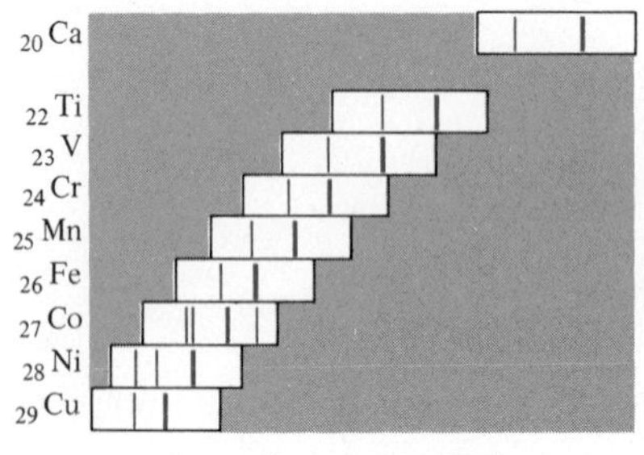

Figure 3.11
Atomic X-Ray Spectra
A sketch of the relative positions of emission lines in some of Moseley's spectra. The element scandium, not available to Moseley, is missing between titanium and calcium. (For a discussion of wavelength and radiation, see Aside 8.1.)

SUMMARY OF SECTIONS 3.6–3.10

Dalton's original picture of atoms as hard and indivisible evolved into the nuclear model of the atom used today as a result of a number of classic experiments. First, the electron—a very light, negatively charged particle—was identified in cathode rays by J. J. Thomson, who measured its charge-to-mass ratio. The Millikan oil-drop experiment gave a value for the charge on the electron, allowing determination of its mass also. The study of the canal rays in gas-discharge tubes proved the existence of a second fundamental particle, the proton—a positively charged particle (an H^+ ion) with much greater mass than the electron.

An experiment by Geiger and Marsden on α-particle scattering led Rutherford to propose that the mass of an atom is concentrated in a small, dense central nucleus, with electrons moving outside the nucleus in mostly empty space. The basic model of the nuclear atom was completed in 1932 with the discovery of the neutron by Chadwick. Neutrons and protons together in the nucleus account for the mass of each atom. The significance of the number of protons in the nucleus, called the atomic number, became clear when Moseley found in x-ray studies that atoms of each element have different and characteristic numbers of protons.

NUCLEAR ARITHMETIC

3.11 ATOMIC NUMBER, ISOTOPES, AND MASS NUMBERS

The atomic number—the number of protons in the nucleus—determines the identity of an atom. Every atom with an atomic number of 8 is an oxygen atom and every oxygen atom contains 8 protons in its nucleus. Atoms with atomic numbers of 9 or 7 are atoms of fluorine or nitrogen, and contain 9 or 7 protons, respectively. Because of the necessity for charge balance in all matter, the atomic number also equals the number of electrons normally present in every atom of the same element. So far, then, we see that atoms of the same element each contain the same numbers of protons and electrons.

The number of neutrons, however, can vary for atoms of the same element. For example, oxygen as it occurs naturally in the atmosphere and elsewhere includes atoms that contain 8, 9, or 10 neutrons.

The **mass number** is the sum of the number of neutrons and the number of protons in an atom. Where *A* is the mass number, *Z* is the atomic number, and *N* is the **neutron number**—the number of neutrons in an atom—

$$\underset{\text{mass number}}{A} = \overset{\text{atomic number } (= \text{no. of protons})}{Z} + \underset{\text{neutron number } (= \text{no. of neutrons})}{N} \tag{3.2}$$

In the symbolism used for atoms and ions, the left superscript position is reserved for the mass number and the left subscript position for the atomic number (Figure 3.12). For the naturally occurring oxygen atoms the symbols and their meaning are as follows:

Symbol	Name	Mass number A	=	Atomic number Z	+	Neutron number N
$^{16}_{8}O$	oxygen-16	16	=	8	+	8
$^{17}_{8}O$	oxygen-17	17	=	8	+	9
$^{18}_{8}O$	oxygen-18	18	=	8	+	10

These are atoms of different **isotopes**—forms of the same element with different mass numbers. In other words, isotopes have the same number of protons but different numbers of neutrons. (Note that the formation of ions by loss or gain of electrons does not change the identity of an element or its isotopes.)

Natural hydrogen is almost entirely a mixture of isotopes of mass numbers 1 and 2, $^{1}_{1}H$ and $^{2}_{1}H$. Each hydrogen atom contains a single proton and a single electron. The nucleus of the heavier hydrogen isotope contains in addition a single neutron. A third and much scarcer isotope of hydrogen has a mass number of 3 (Figure 3.13) and contains two neutrons. Hydrogen-2 is commonly known as **deuterium** or "heavy hydrogen," and hydrogen-3 is known as **tritium.** Most elements, although not all, occur in nature as mixtures of isotopes (see Aside 3.4).

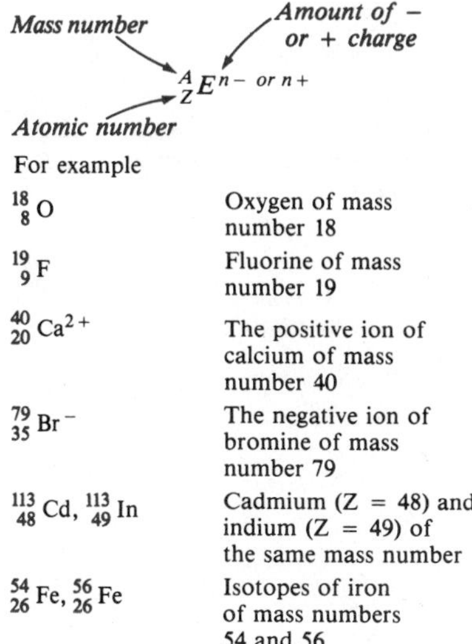

Figure 3.12
Notation for an Atom or Ion of Element E
Only the subscripts or superscripts needed for the purpose at hand are used. Often the atomic number is not included. (This is the notation as agreed upon today. Some older publications place the mass number at the right, e.g., O^{18}.)

EXAMPLE 3.1 Nuclear Arithmetic

What are the compositions of the nuclei of $^{35}_{17}Cl$ and $^{35}_{17}Cl^{-}$?

The atomic number, Z, of chlorine (found from the table inside the front cover) is 17; thus the nucleus contains 17 protons.

The mass number, A, of this isotope is 35; thus the total number of protons and neutrons in the nucleus is 35.

The neutron number, N, is the difference between the mass number and the atomic number.

$$N = A - Z = 35 - 17 = 18$$

The nucleus of $^{35}_{17}Cl$ consists of 17 protons and 18 neutrons. The electrical charge on an ion is the result of the gain or loss of electrons, which takes place outside the nucleus. Therefore, **the nuclear composition of $^{35}_{17}Cl^{-}$ is the same as that of $^{35}_{17}Cl$.**

EXERCISE What is the composition of the nucleus of $^{23}_{11}Na$? Will this composition be any different from that of the nucleus of $^{23}_{11}Na^{+}$? *Answers* 11 protons, 12 neutrons; no.

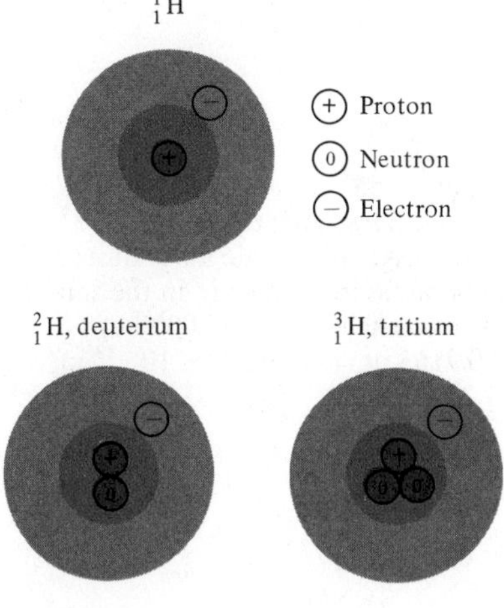

Figure 3.13
The Isotopes of Hydrogen
This is a schematic picture (not to scale) of the number of electrons, protons, and neutrons in the hydrogen isotopes. Deuterium is sometimes symbolized D and tritium is sometimes symbolized T.

3.12 ATOMIC MASS

Today we know that one atom of bismuth, one of the heaviest of the elements, has a mass of 3.470256×10^{-22} g, and that one atom of hydrogen ($^{1}_{1}H$), the lightest element, has a mass of 1.673559×10^{-24} g. There was a time, however, when knowing the mass of an atom, or even knowing whether an atom of one element was heavier than an atom of another element, seemed an insurmountable problem. Early chemists had to establish a relative atomic mass scale by assigning a mass to an atom of one kind and then determining all other atomic masses relative to that one. The process is like deciding that the king's foot has a length of one "foot." Anything equal in length to five king's feet is then 5 "feet" long, and so on. If the queen's foot is chosen as the relative length scale standard, then all objects in the kingdom have a different measured length (Figure 3.16).

Figure 3.14
Length Measured Relative to Royal Feet

1 King's foot

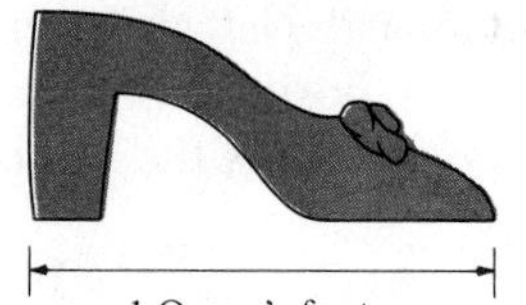

1 Queen's foot

A 1.25 foot snake (King's foot standard)

A 1.66 foot snake (Queen's foot standard)

We still use a mass scale for atoms based on the relative masses of atoms of different elements, rather than a scale of conventional mass units such as grams. It is much easier to think of an atom of bismuth as about 209 times heavier than a hydrogen atom, than to deal with numbers with more than 20 zeros after the decimal point. By international agreement, the standard for atomic masses is now the isotope of carbon of mass number 12, ${}^{12}_{6}C$, which is *assigned* an atomic mass of *exactly* 12 units. An **atomic mass unit**, u, is defined as $\frac{1}{12}$ of the mass of one carbon-12 atom. One atomic mass unit is equal to $1.6605655 \times 10^{-24}$ g, often rounded off to 1.6606×10^{-24} g (Figure 3.17).

On this atomic mass scale, the mass of any atom is *close* to the whole number that is its mass number. This is so because on this scale the masses of protons and neutrons are each close to 1 (see Table 3.5). For example, for the three naturally occurring isotopes of neon of mass numbers 20, 21, and 22 the masses of the atoms are

$${}^{20}_{10}\text{Ne } 19.992 \text{ u} \qquad {}^{21}_{10}\text{Ne } 20.994 \text{ u} \qquad {}^{22}_{10}\text{Ne } 21.991 \text{ u}$$

The masses of atoms deviate slightly from whole numbers for several reasons. First, the neutron and proton masses are not *exactly* equal to 1 (see Table 3.6). Second, the mass includes the masses of electrons, which make a tiny contribution to the mass. And third, a relatively small amount of mass is converted into the energy that holds the particles together (known as the binding energy; Section 12.2).

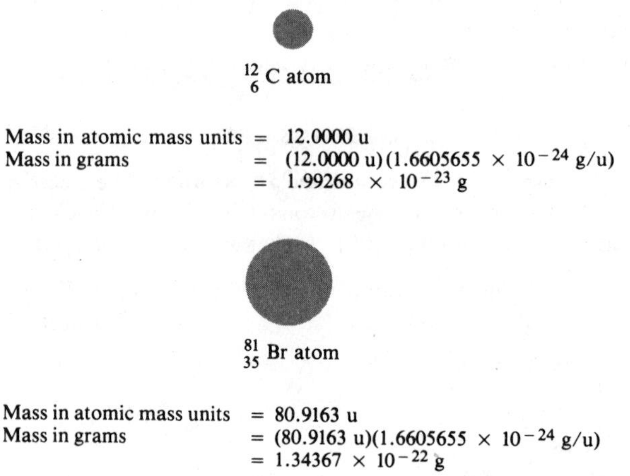

${}^{12}_{6}$C atom

Mass in atomic mass units = 12.0000 u
Mass in grams = $(12.0000 \text{ u})(1.6605655 \times 10^{-24} \text{ g/u})$
= 1.99268×10^{-23} g

${}^{81}_{35}$Br atom

Mass in atomic mass units = 80.9163 u
Mass in grams = $(80.9163 \text{ u})(1.6605655 \times 10^{-24} \text{ g/u})$
= 1.34367×10^{-22} g

Figure 3.15
Atomic Mass Relationships
The masses in atomic mass units and the masses in grams are in the same ratio to each other: $(12.0000 \text{ u})/(80.9163 \text{ u}) = (1.99268 \times 10^{-23} \text{ g})/(1.34367 \times 10^{-22} \text{ g}) = 0.148301$.

EXAMPLE 3.2 Masses of Atoms

An atom of the most abundant isotope of uranium, ${}^{238}_{92}U$, has a mass of 238.051 u. What is the mass in grams of an atom of ${}^{238}_{92}U$?

To find the mass of an atom in grams, knowing its mass in atomic mass units, requires a unit conversion based on the knowledge that 1 u = $1.6605655 \times 10^{-24}$ g. Using the conversion factor rounded to six significant digits gives

$$(238.051 \text{ u})\left(\frac{1.66057 \times 10^{-24} \text{ g}}{1 \text{ u}}\right) = \mathbf{3.95300 \times 10^{-22} \text{ g}}$$

EXERCISE What is the mass in atomic mass units of an atom of $^{20}_{10}Ne$, which has a mass in grams of 3.3198×10^{-23} g? *Answer* 19.992 u.

As they are found naturally, either as elements or in chemical compounds, most elements are composed of mixtures of isotopes. For every element, the composition of the naturally occurring mixture of isotopes is very nearly the same everywhere (on this planet). Except in specialized applications, chemists are not concerned with individual isotopes or individual isotopic masses. Therefore, what is most useful is the *average* atomic mass of the naturally occurring mixture of isotopes. Tables like those inside the front cover of this book give as the **atomic mass** of each element the average mass in atomic mass units of the atoms of the naturally occurring mixture of isotopes. (Note that the term "atomic weight" is often used instead of "atomic mass.") When we speak of the atomic mass of, say, bromine, we mean 79.904 u—in a bottle of bromine, this is the average mass of the atoms of the different isotopes of bromine that are present.

The atomic mass of an element is a weighted average based on the percentage of atoms of each isotope that are present. The general expression for calculation of atomic mass from the isotopic masses is as follows:

$$\begin{aligned}\text{Atomic mass} &= (\text{fraction of isotope } i)(\text{mass of isotope } i) \\ &+ (\text{fraction of isotope } ii)(\text{mass of isotope } ii) + \ldots\end{aligned} \tag{3.3}$$

EXAMPLE 3.3 Atomic Mass

Naturally occurring iron contains 5.82% $^{54}_{26}Fe$, 91.66% $^{56}_{26}Fe$, 2.19% $^{57}_{26}Fe$, and 0.33% $^{58}_{26}Fe$. The respective atomic masses are 53.940 u, 55.935 u, 56.935 u, and 57.933 u. Calculate the average atomic mass of iron.

To find the weighted average, multiply the mass of each isotope by the fraction of that isotope present and add the results.

$$\begin{aligned}\text{Atomic mass} &= (0.0582)(53.940\ \text{u}) + (0.9166)(55.935\ \text{u}) \\ &+ (0.0219)(56.935\ \text{u}) + (0.0033)(57.933\ \text{u}) \\ &= 3.14\ \text{u} + 51.27\ \text{u} + 1.25\ \text{u} + 0.19\ \text{u} \\ &= \mathbf{55.85\ u}\end{aligned}$$

(You can check this value in the tables inside the front cover.)

EXERCISE Naturally occurring neon gas contains 90.92% $^{20}_{10}Ne$, 0.257% $^{21}_{10}Ne$, and 8.82% $^{22}_{10}Ne$. The respective atomic masses are 19.992 u, 20.994 u, and 21.991 u. Calculate the average atomic mass of neon. *Answer* 20.17 u.

SUMMARY OF SECTIONS 3.11–3.12

The identity of an atom is determined by the number of protons in the nucleus, which is given by the atomic number. Isotopes are forms of the same element with different numbers of neutrons. The mass number of an isotope equals the number of protons plus the number of neutrons. The relative atomic mass scale is based upon carbon-12, which is assigned an atomic mass of exactly 12 u. "Atomic mass" refers to the average mass of the naturally occurring mixture of isotopes of an element.

Aside 3.4 TOOLS OF CHEMISTRY: MASS SPECTROMETER

Measurements with mass spectrometers are the principal source of our modern, highly accurate information about the mass and natural distribution of isotopes. The modern mass spectrometer is a direct descendant of the gas-discharge tube in which Goldstein studied canal rays.

In a mass spectrometer, positive ions (most having +1 charge) are produced from the substance being studied under vacuum in the gas phase. The positive ion beam is lined up ("collimated") by a slit system and accelerated by an electrical field of variable strength. The ions then pass through a magnetic field of variable strength that is perpendicular to their path. This force deflects the ions into curved paths. How far the path of an ion is deflected depends upon its mass-to-charge ratio: ions with the smallest mass are deflected the most. As a result, the beam of positive ions is spread out according to the mass-to-charge ratio of the ions.

The spectrum can be recorded in a photograph; the instrument that produces such a record is called a mass spectrograph (Figure 3.16). The locations of the lines show the relative masses of the ions and the intensities of the lines show the relative numbers, or abundances, of ions of each mass.

In a mass spectrometer, the flow of ions is controlled by controlling the electric and magnetic fields so that ions of the same mass and charge arrive at a detector at the same time and are recorded as peaks on a graph. The positions of the peaks give the mass numbers of the ions present. The relative heights of the peaks give the relative abundances of ions of each mass number. Mass spectrometers are very sensitive and can separate ions that differ in mass by as little as one unit. With the aid of a computer coupled to a mass spectrometer, mass spectra are provided directly as printouts of the masses and abundances of the ions present.

Figure 3.17 is a simple mass spectrum showing the relative abundances of the two naturally occurring isotopes of antimony. Mass spectra have many uses in addition to the determination of isotopic masses and abundances. The masses of chemical compounds and information about their structures can be found. Also, chemical compounds can be identified by their "fingerprints," which are the distinctive patterns of ions formed when compounds break down in the spectrometer.

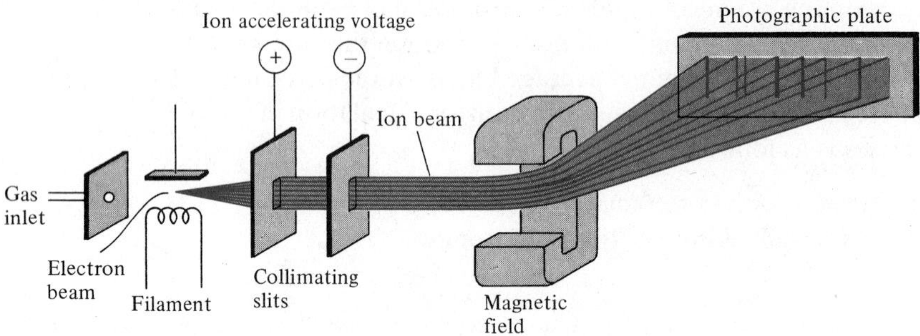

Figure 3.16
Mass Spectrograph
The electron beam ionizes molecules of the gas, which are given a high velocity by the ion accelerators. As the ions pass through the magnetic field, they are deflected from their straight path, the lighter ones being deflected the most. In the instrument shown, the arrival of ions of different e/m is recorded on a photographic plate. In a mass spectrometer, they are recorded on a graph or chart.

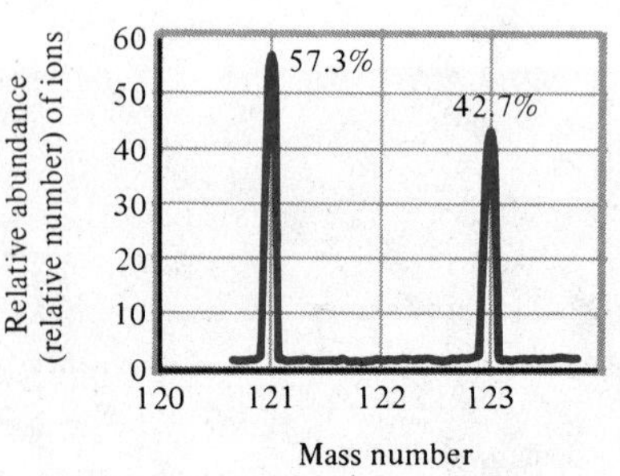

Figure 3.17
Mass Spectrum of Antimony
The peak heights show the abundances of antimony 121 and antimony 123 in the naturally occurring mixture.

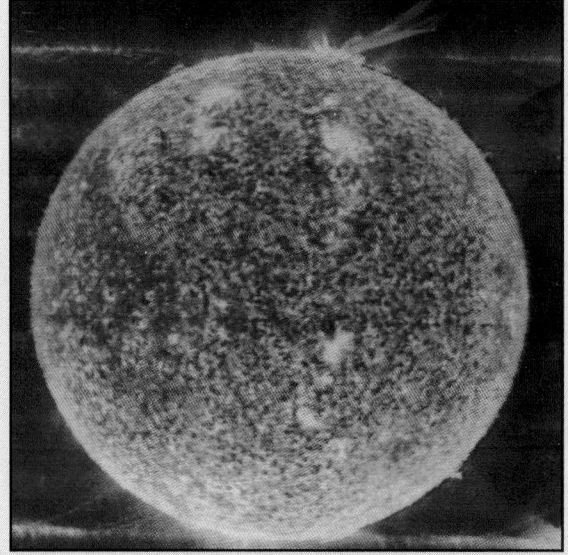

■ **THE SUN** *above* Nuclear fusion in the Sun creates an interior temperature of 15 million °C and is the source of energy for Earth. This photo of the Sun was taken by Skylab 2 in 1973. The image was created by recording the 304 angstrom ultraviolet radiation emitted by He^+ ions.

■ **EXPLOSION OF A THERMONUCLEAR BOMB** *below*

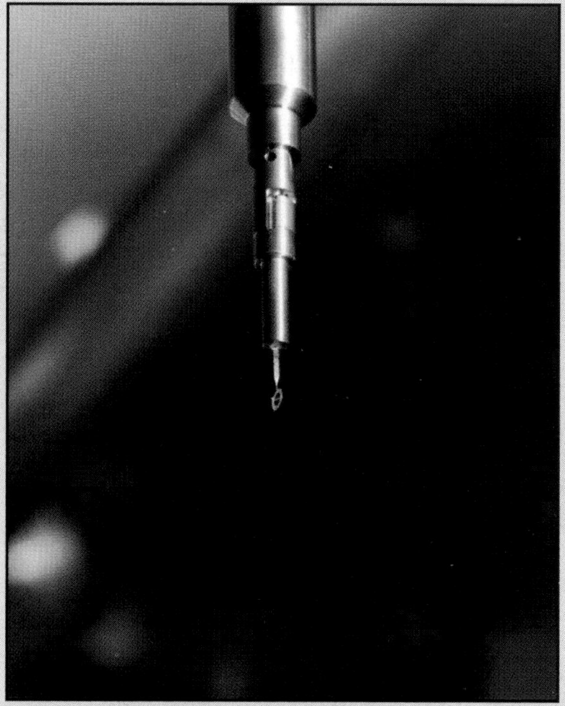

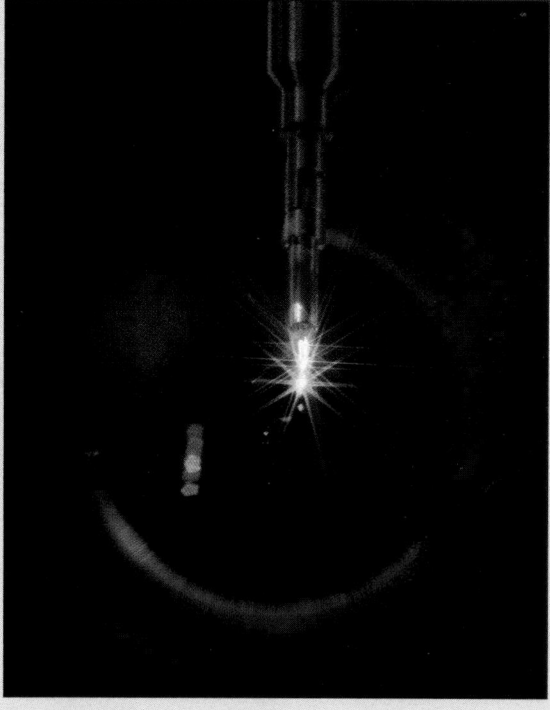

■ **NUCLEAR FUSION BY INERTIAL CONFINEMENT IN A LASER BEAM** *above* *left:* A 1 mm diameter glass capsule containing deuterium and tritium set up as a target for Nova, the world's most powerful laser. *right:* Star-like fusion created by laser-induced implosion of the target capsule. In the 0.5 picosecond burst, 10^{13} neutrons were produced by the reaction ${}^{2}_{1}H + {}^{3}_{1}H \rightarrow {}^{4}_{2}He + {}^{1}_{0}n$.

CONTINUOUS VISIBLE SPECTRUM

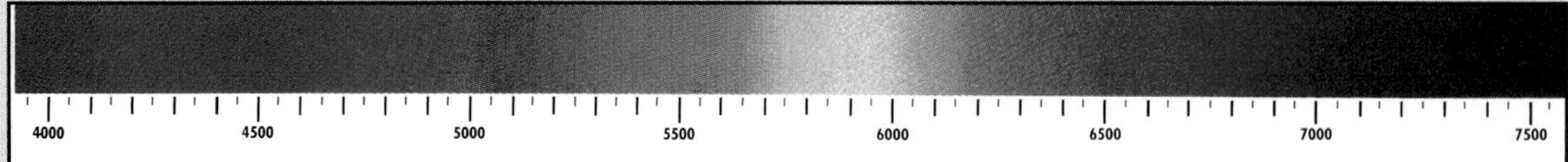

EMISSION SPECTRA

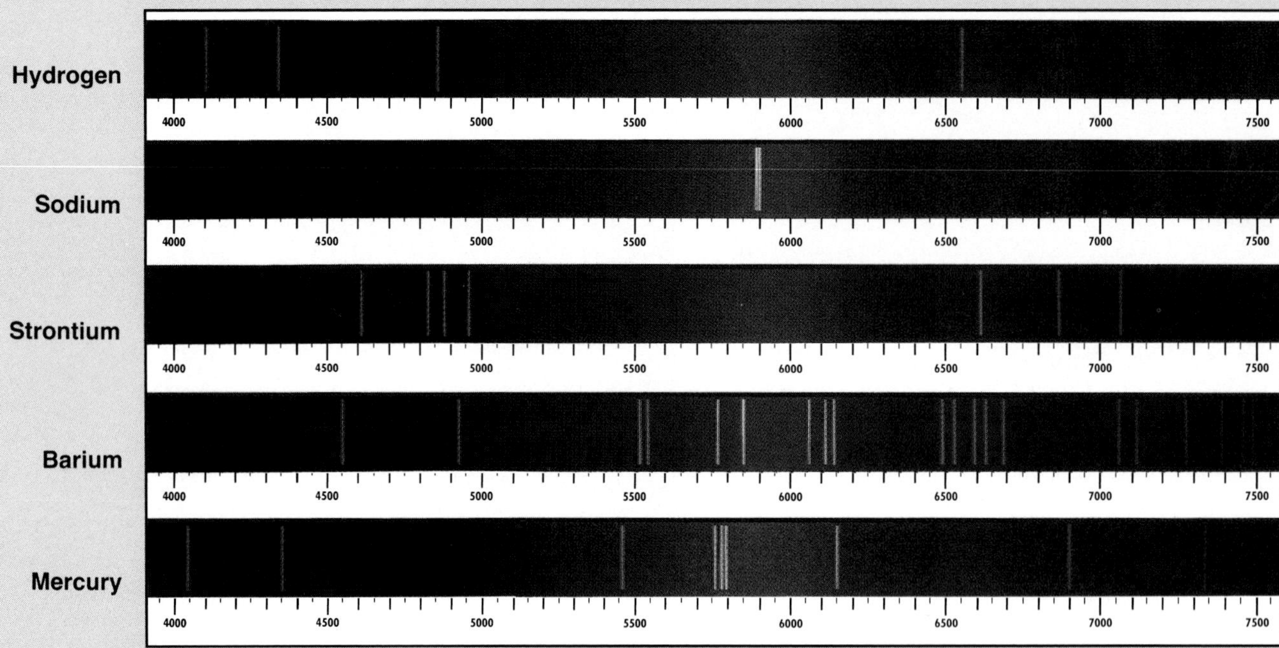

■ SPECTRA *above* White light (e.g., from an incandescent light bulb), when it passes through a prism, is dispersed into a continuous spectrum. Line emission spectra — shown for hydrogen (H), sodium (Na), strontium (Sr), barium (Ba), and mercury (Hg) — are emitted by excited atoms in the gaseous state, vary with electron structure, and are different for each element. Sodium-vapor and mercury-vapor streetlights emit at the wavelengths shown, an intense blue radiation dominating in the light from mercury lamps. (Wavelengths in angstroms.)

■ FLAME TESTS AND FIREWORKS *below and right* Many metal cations impart characteristic colors to flames, and flame tests can be used to detect their presence. *left to right:* the flame colors produced by ignition of methanol-salt mixtures — lithium ion (Li^+), crimson; sodium ion (Na^+), yellow; potassium ion (K^+), purple; and boron ion (B^{3+}), green. These and other cation flame colors are used in fireworks.

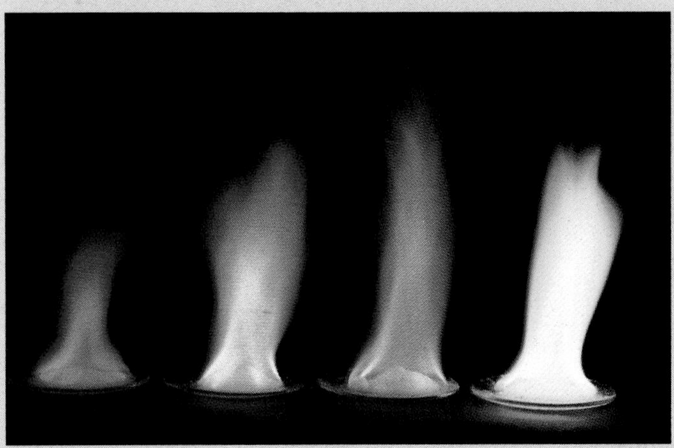

COMBUSTION — IGNITION OF A STRIKE-ANYWHERE MATCH *left* Friction causes ignition of tetraphosphorus trisulfide, and combination of phosphorus and sulfur with oxygen in the following exothermic reaction: $P_4S_3\,(s) + 8O_2\,(g) \longrightarrow P_4O_{10}\,(s) + 3SO_2\,(g)$

COMBUSTION — MAGNESIUM BURNS IN AIR *below* The intense white light produced when magnesium combines with oxygen is utilized in fireworks and flashbulbs. In the photo, white finely divided magnesium oxide, formed by the reaction $2Mg\,(s) + O_2\,(g) \longrightarrow 2MgO(s)$, is seen above and below the flame.

COMBUSTION — OIL ACCIDENT *above* Flames emerging from oil burning in a ruptured storage tank as the oil is converted to carbon dioxide, water, and carbon-containing soot. When gasoline or oil burns, as in any combustion reaction, chemical energy is converted to thermal energy and light energy.

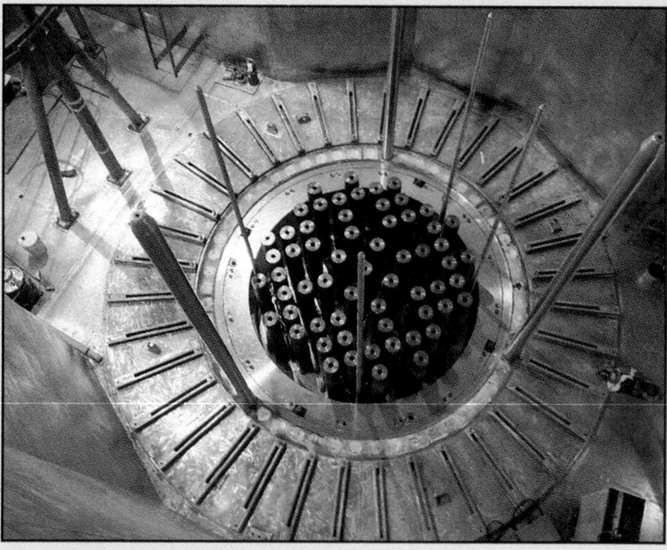

■ **NUCLEAR FISSION REACTOR IN ELECTRIC POWER GENERATING STATION** *above left:* The top of the empty nuclear reactor, which is about 12 m deep and has 22-cm thick stainless steel walls, is seen at the bottom of the pit. During operation water flows to and from the steam generator through the three round openings on the far wall. The same water serves as moderator, slowing neutrons emitted during uranium fission. At the back of the pit, the head that will close the reactor during operation is visible. This photo was taken before the initial insertion of fuel rods. During refueling of an operating reactor, the pit is flooded with water to protect workers from radiation and rods are replaced by a crane from above the pit. *right:* A close-up of the open top of the reactor vessel with the control rod guide tubes in place. During shutdown, control rods slide down these tubes into spaces between the fuel rods (about 5 m beneath the opening shown).

■ **AN OIL-FUELED ELECTRIC POWER GENERATING STATION** *below*
Many generating stations, like the one pictured in Norwalk, Connecticut, are near large bodies of water. This plant draws 200,000 gal/min of cooling water from the harbor and burns 325 gal of oil/min. Ash is removed from stack gases by a high-voltage precipitator, mixed with water, and piped from the small gray building at the base of the stack to the tanks and ponds of the water treatment area seen in the foreground, where the water is cleaned before recycling. Note the conveyor belt just beyond the three oil storage tanks. This station was designed to burn coal, which was delivered on barges. If economic conditions require it, the plant could be converted back to coal burning.

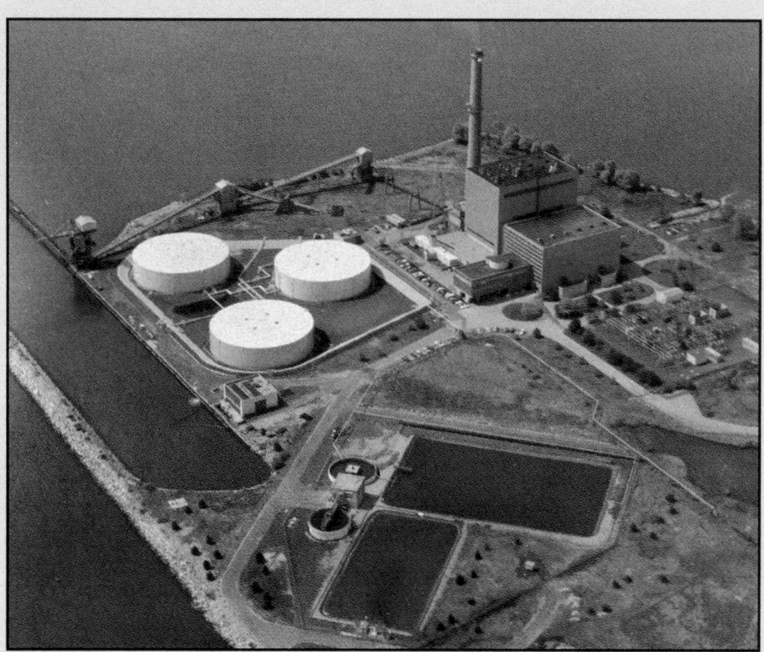

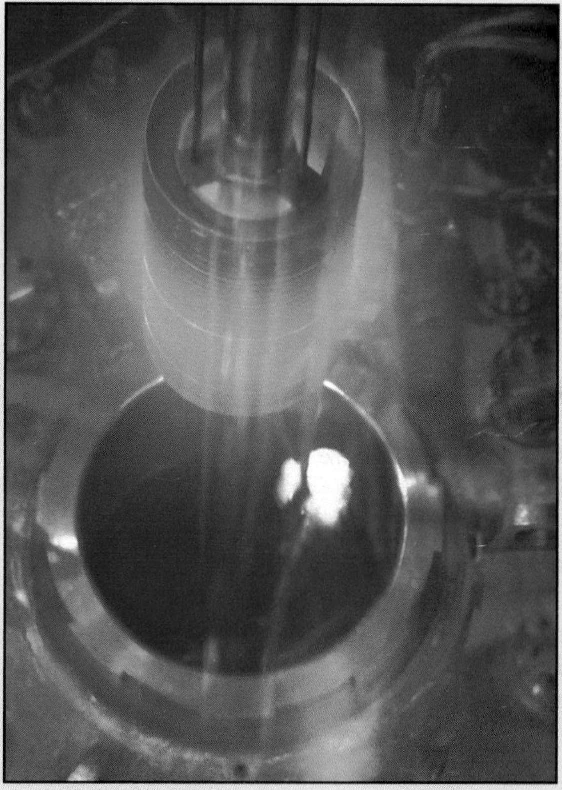

■ **NUCLEAR FISSION REACTOR IN RESEARCH** *above*
Removal of fuel element from a small reactor used in research on man-made elements. The blue glow, due to Cerenkov radiation, is produced when charged particles, in this case β^- particles, lose energy as they move through a transparent medium (like water) faster than the speed of light in that medium.

SIGNIFICANT TERMS (with section references)

α-particle (3.8)	heterogeneous mixture (3.4)
aqueous solution (3.4)	homogeneous mixture (3.4)
atom (3.3)	ions (3.6)
atomic mass (3.12)	isotopes (3.11)
atomic mass unit (3.12)	mass number (3.11)
atomic number (3.10)	mixture (3.4)
canal rays (3.7)	neutron (3.9)
cathode rays (3.6)	neutron number (3.11)
changes of state (3.5)	nucleus (3.8)
chemical compound (3.4)	phase (3.5)
chemical properties (3.4)	physical properties (3.4)
chemical reaction (3.4)	potential energy (Aside 3.3)
Coulomb force (Aside 3.3)	proton (3.7)
deuterium (3.11)	pure substance (3.4)
dissolution (3.4)	solute (3.4)
electrodes (3.6)	solution (3.4)
electron (3.6)	solvent (3.4)
elements (3.4)	spectrum (3.10)
fluorescent (3.6)	states of matter (3.5)
fundamental particle (3.6)	subatomic particle (3.6)
gas-discharge tubes (3.6)	tritium (3.11)

QUESTIONS AND PROBLEMS

Atoms and Elements

3.1 Name the element represented by each of the following symbols: (a) Be, (b) B, (c) V, (d) As, (e) Ba.

3.2 Select the symbols that are not derived from the modern names of the respective elements: (a) Ac, (b) Cs, (c) Cu, (d) Au, (e) Fe, (f) Mn, (g) Th. Name these elements.

3.3 Write the symbol for each of the following elements: (a) xenon, (b) nickel, (c) magnesium, (d) cobalt, (e) silicon, (f) lead, (g) potassium, (h) silver, (i) fluorine, (j) radon.

3.4 Name the element that corresponds to each of the following symbols: (a) Al, (b) At, (c) Ca, (d) Cl, (e) Ir, (f) Re. What are the origins of the names of these elements?
[Hint: Use a suitable reference, such as the *Handbook of Chemistry and Physics* (The Chemical Rubber Co.).]

Kinds of Matter

3.5 Using the terminology and classifications given in Figure 3.2, describe each of the following: (a) a drop of mercury, (b) a mound of sugar crystals, (c) an ice cube, (d) a melting ice cube, (e) a puddle of water, (f) a scoop of "smooth" peanut butter, (g) a scoop of "crunchy" peanut butter. Identify the number of phases present in each sample.

3.6 A chemical change has taken place if the identity and composition of at least one substance has been changed by the process. Choose from the following list of processes those that are chemical changes: (a) a nail is magnetized, (b) grape juice is fermented, (c) gasoline is burned, (d) meat is cut, (e) meat is cooked, (f) a cake is baked, (g) a leaf changes color in the autumn, (h) concrete becomes hard, (i) a balloon is inflated.

3.7 A student found the following list of properties of molecular iodine (I_2) in an encyclopedia: (a) grayish-black granules, (b) metallic luster, (c) characteristic odor, (d) forms a purple vapor, (e) density = 4.93 g/cm^3, (f) melting point = 113.5 °C, (g) soluble in alcohol, (h) insoluble in water, (i) noncombustible, (j) forms $[I_3]^-$ ions in aqueous solutions, (k) poisonous. Which of these properties are chemical properties?

3.8 Name two changes of state. Give a specific example of each change.

Atomic Structure—Five Classic Experiments

3.9 Prepare a sketch of a gas-discharge tube showing the production of cathode rays. Are the cathode rays produced in a gas-discharge tube different for different gases in the tube? What conclusion can be drawn from this fact?

3.10 Prepare a sketch of a gas-discharge tube showing the production of canal rays. Are the canal rays produced in a gas-discharge tube different for different gases in the tube? What conclusion can be drawn from your answer?

3.11 Describe the experiment in which the charge-to-mass ratio of an electron was determined. Once the charge on an electron was determined, the mass could be calculated. What did the value of the mass prove?

3.12 Describe the experiment in which the neutron was discovered. How does this subatomic particle fit into Rutherford's concept of a nucleus?

3.13 If the mass and electrical charge were uniformly distributed throughout an atom, what would be the expected results of an α-particle scattering experiment? What was the major conclusion drawn from the results of the α-particle scattering experiments?

3.14 What property of each element did Moseley use to assign atomic numbers to the elements?

3.15 The approximate radius of a hydrogen atom is 0.0529 nm and of a proton is 1.5×10^{-15} m. Assuming both the hydrogen atom and the proton to be spherical, calculate the fraction of the space in an atom of hydrogen that is occupied by the nucleus. $V = (4/3)\pi r^3$ for a sphere.

3.16 The approximate radius of a neutron is 1.5×10^{-15} m and the mass is 1.675×10^{-27} kg. Calculate the density of a neutron. $V = (4/3)\pi r^3$ for a sphere.

3.17 Two charged particles are attracted to each other by a Coulombic force. What will be the effect on the strength of the force produced by the following changes: (a) the distance between the particles is increased, (b) the charge on one particle is increased, (c) the charge on both particles is increased, (d) the distance between the particles is halved and the charge on each particle is doubled?

3.18* The following data are measurements of the charges on oil droplets using an apparatus similar to that used by Millikan:

11.215×10^{-19} C	14.423×10^{-19} C
12.811×10^{-19} C	24.037×10^{-19} C
14.419×10^{-19} C	9.621×10^{-19} C
12.815×10^{-19} C	16.012×10^{-19} C

Each of these charges should be some integral multiple of a fundamental charge. (a) Calculate the fundamental charge by trial-and-error. The value of e/m for an electron is 1.76×10^8 C/g. Using your value of the charge calculated above, (b) calculate the mass of an electron.

Nuclear Arithmetic

3.19 What is an isotope? What are the special names given to the isotopes of hydrogen?

3.20 Which symbol gives us more information about an unidentified isotope: ^{40}X or $_{40}X$? Why is the atomic number of primary importance to a chemist?

3.21 How many protons are present in each atom of (a) ^{100}Rh, (b) ^{146}Nd, (c) ^{79}Br, (d) ^{7}Li, (e) ^{159}Tb? How many neutrons are present in each atom?

3.22 Calculate the mass number for each of the following isotopes: (a) 74 protons and 106 neutrons, (b) 92 protons and 144 neutrons, (c) 38 protons and 46 neutrons. Write the complete symbol for each isotope.

3.23 Complete the following table for the atoms or ions as illustrated for $^{13}_{6}C$:

Symbol	Z	N	A	Number of electrons	Electrical charge
$^{13}_{6}C$	6	7	13	6	0
	14	15			0
S		18		18	
			56	24	+2
Au			188	76	

3.24 Repeat Question 3.23 for the following table:

Symbol	Z	N	A	Number of electrons	Electrical charge
$^{229}_{90}Th$	90	139	229	90	0
$^{235}_{91}Pa^{4+}$					
	92	143			+6
		146	238	89	
		146		90	+3

Identify the pairs of atoms or ions that have the same (a) atomic number, (b) neutron number, (c) mass number, (d) number of electrons, (e) electrical charge.

3.25 Choose from the following list the symbols that represent (a) groups of isotopes of the same element, (b) atoms with the same number of neutrons, (c) atoms with the same mass number (4 different sets): (i) ^{12}N, (ii) ^{13}B, (iii) ^{13}N, (iv) ^{14}C, (v) ^{14}N, (vi) ^{15}N, (vii) ^{16}N, (viii) ^{16}O, (ix) ^{17}N, (x) ^{17}F, (xi) ^{18}Ne.

3.26 Place the following atomic symbols in order of increasing (a) mass number, (b) neutron number, (c) atomic number: (i) $^{49}_{24}Cr$, (ii) $^{59}_{27}Co$, (iii) $^{50}_{23}V$, (iv) $^{56}_{26}Fe$, (v) $^{64}_{28}Ni$, (vi) $^{58}_{25}Mn$.

Nuclear Arithmetic—Atomic Masses

3.27 Calculate the mass in grams of an atom of each of the following isotopes: (a) ^{78}Kr, (b) ^{80}Kr, and (c) ^{82}Kr. The respective atomic masses are 77.9204 u, 79.9164 u, and 81.9135 u.

3.28 Repeat Problem 3.27 for (a) ^{6}Li and (b) ^{7}Li. The respective atomic masses are 6.01512 u and 7.01600 u.

3.29 Calculate the atomic mass in atomic mass units of an atom of each of the following isotopes: (a) ^{151}Eu and (b) ^{153}Eu. The respective masses are 2.506119×10^{-22} g and 2.539352×10^{-22} g.

3.30 Repeat Problem 3.29 for (a) ^{231}Np and (b) ^{233}Np. The respective masses are 3.836542×10^{-22} g and 3.869792×10^{-22} g.

3.31 Calculate the mass in grams of an atom of each of the following: (a) ^{56}Mn, 55.93904 u; (b) ^{56}Fe, 55.9349 u; and (c) ^{56}Co,

55.94002 u. Why is there a difference in the masses even though the mass number is 56 in all three cases?

3.32 On the average, a silver atom weighs 107.02 times as much as an atom of hydrogen. Using 1.0079 u as the atomic mass of hydrogen, calculate the atomic mass of silver in atomic mass units.

3.33 Calculate the atomic mass of magnesium using the following data for the percent of natural abundance and mass of each isotope: 78.99% of ^{24}Mg (23.98504 u), 10.00% of ^{25}Mg (24.98584 u), and 11.01% of ^{26}Mg (25.98259 u).

3.34 Calculate the atomic mass of strontium using the following data for the percent of natural abundance and mass of each isotope: 0.5% of ^{84}Sr (83.9134 u), 9.9% of ^{86}Sr (85.9094 u), 7.0% of ^{87}Sr (86.9089 u), and 82.6% of ^{88}Sr (87.9056 u).

3.35 Only two isotopes of copper are present in naturally occurring copper: ^{63}Cu (62.9298 u) and ^{65}Cu (64.9278 u). Calculate the percent composition of naturally occurring copper using the atomic mass of 63.546 u.

3.36 The atomic mass of chlorine is 35.453 u. There are only two isotopes in naturally occurring chlorine: ^{35}Cl (34.96885 u) and ^{37}Cl (36.96712 u). Calculate the percent composition of naturally occurring chlorine.

3.37* There are two naturally occurring isotopes of hydrogen (^{1}H, >99% and ^{2}H, <1%) and two of chlorine (^{35}Cl, 76% and ^{37}Cl, 24%). (a) How many different HCl molecules can be formed from these isotopes? (b) What are the approximate masses of each of the molecules expressed in atomic mass units? (c) List the molecules in order of decreasing relative abundance—as would be observed on a mass spectrometer—assuming only the formation of $(HCl)^+$ ions.

3.38* In a suitable reference such as the Table of Isotopes in the *Handbook of Chemistry and Physics* (The Chemical Rubber Co.), look up the following information for selenium: (a) the total number of known isotopes, (b) the atomic mass, (c) the percentage of natural abundance and mass of each of the stable isotopes. (d) Calculate the atomic mass of selenium.

CHAPTER

4

Atoms, Molecules, and Ions

ATOMS AND IONS IN COMBINATION

For someone living today, it is almost impossible to imagine how difficult it was for the early chemists to understand the nature of matter. Many of us—even if we understand no other chemistry—know that water is "H-2-oh," and unquestioningly accept the fact that all little bits of water contain two H's and one O combined in some way. Even this simple basic assumption—that in a pure substance the elements are combined in definite proportions—was first arrived at with great difficulty. From our modern perspective, we can easily see that the existence of atoms explains this fact.

One important step in understanding the nature of a chemical compound is to know its composition. At the atomic level, this means knowing the numbers of atoms of each kind that have combined. This knowledge is used to write chemical formulas. For example, water has the chemical formula H_2O, and the bit of matter in which two hydrogen atoms and one oxygen atom are joined firmly together is called a water molecule.

To begin a study of chemical compounds, it is first necessary to learn a few of the rules about how chemical formulas are written and how chemical compounds are named. Then, it is possible to proceed to an essential aspect of the subject: atoms, molecules, and ions are tiny bits of matter that cannot be directly seen or weighed. Ordinarily, however, one must deal with quantities of matter large enough to be handled easily. The most important thing to learn from this chapter is how to relate seeable, weighable quantities of chemical compounds to the masses of individual atoms, ions, and molecules.

4.1 MOLECULAR AND IONIC COMPOUNDS

a. Molecular Compounds A **molecule** is the smallest particle of a pure substance that has the composition of that substance and is capable of independent existence. Many chemical compounds and some elements are composed of molecules. Crystal-

Table 4.1
Chemical Formulas for Molecules of Elements
The subscript gives the number of atoms in a molecule of the element. Other elements exist in large aggregates of atoms not thought of as molecules.

Monatomic molecules		Diatomic molecules		Polyatomic molecules	
The following elements ordinarily exist as gases:		**The following elements (except for bromine, a liquid and iodine, a solid) ordinarily exist as gases:**		**The following elements ordinarily exist as solids:**	
He	Helium	H_2	Hydrogen	P_4	Phosphorus
Ne	Neon	O_2	Oxygen	As_4	Arsenic
Ar	Argon	N_2	Nitrogen	Sb_4	Antimony
Kr	Krypton	F_2	Fluorine	S_8	Sulfur
Xe	Xenon	Cl_2	Chlorine	Se_8	Selenium
Rn	Radon	Br_2	Bromine		
		I_2	Iodine		

line sulfur, for example, is made up of individual sulfur molecules, each containing eight sulfur atoms. The chemical formula of this molecule is written as S_8; the subscript 8 shows that eight sulfur atoms are chemically combined. The S_8 molecule is a **polyatomic molecule**—a molecule containing more than two atoms.

Oxygen, another element, usually exists as **diatomic molecules**—molecules made of two atoms. Like oxygen, the elements hydrogen, nitrogen, fluorine, chlorine, bromine, and iodine exist under ordinary conditions as diatomic molecules (Table 4.1). A few elements—helium, neon, argon, krypton, xenon, and radon—ordinarily exist as single atoms. (We might think of these as *monatomic* molecules.) Elements other than those listed in Table 4.1 usually are composed of large aggregates of atoms, and such aggregates are *not* thought of as "molecules."

When one atom of element A combines with one atom of element B, they form a molecule of a compound, symbolized AB, or A—B. A molecule may contain only two atoms or it may contain many thousands of atoms. There are simple molecules, such as the carbon dioxide molecule, in which three atoms are combined (Figure 4.1a). There are very large molecules, such as the chlorophyll *a* molecule, a natural compound that traps solar energy in green plants. One chlorophyll *a* molecule is composed of 137 atoms of the elements carbon, hydrogen, nitrogen, oxygen, and magnesium. And there are **polymers**—very large molecules formed by linking many smaller molecules together. Polymer molecules may contain from thousands to hundreds of thousands of atoms. For example, a single molecule of high-density polyethylene—familiar for its use in such everyday items as toys and chairs—may contain hundreds of thousands of ethylene molecules joined together, each ethylene molecule being composed of two carbon atoms and four hydrogen atoms.

The line sometimes drawn between the symbols for the atoms in a molecule, A—B, represents a chemical bond. Until we provide a more formal definition (Chapter 9), simply think of the chemical bond as the force that holds the atoms together.

Chemical compounds composed of molecules are referred to as "molecular compounds." Water and carbon dioxide are molecular compounds. A block of *dry ice*, which is solid carbon dioxide, contains *only* carbon dioxide molecules, which have the formula CO_2 (Figure 4.1b).

b. Ionic Compounds Ions are formed from atoms by the loss or gain of electrons (Sections 3.6, 3.7). Atoms (or other species) that have lost electrons to form positively charged ions are called **cations;** atoms that have gained electrons to form negatively charged ions are called **anions.** For example, a sodium cation, represented by Na^+, is formed from a sodium atom by the loss of one electron and therefore has a single positive charge. A barium cation, Ba^{2+}, is formed from a barium atom by the loss of two electrons and therefore has twice the positive charge of a sodium cation. A chlorine atom gains one electron to form Cl^-, the chloride anion; an oxygen atom gains two electrons to form O^{2-}, the negatively charged oxide anion.

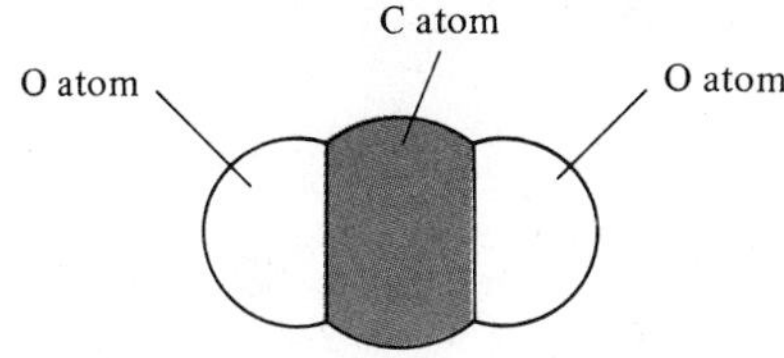

(a) A carbon dioxide molecule

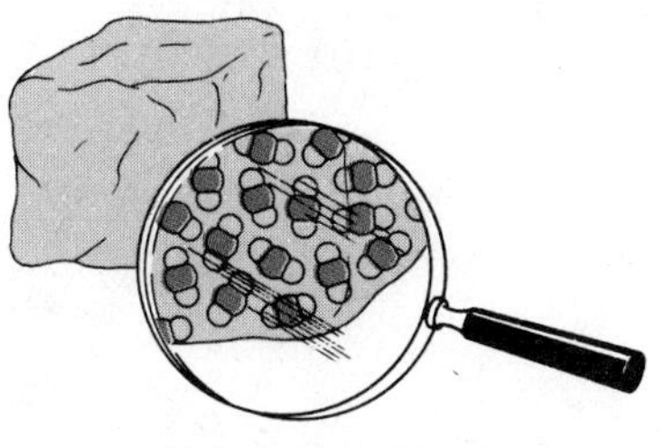
(b) A block of dry ice

Figure 4.1
Carbon Dioxide, a Molecular Compound
Dry ice is frozen carbon dioxide.

Table 4.2
Some Common Monatomic Ions
Anions, shown in color, are generally formed by nonmetals and cations by metals.

Li^+	Mg^{2+}	Ag^+	Zn^{2+}	Al^{3+}	Sn^{2+}	N^{3-}	O^{2-}	F^-
Na^+	Ca^{2+}		Cd^{2+}		Pb^{2+}	Bi^{3+}	S^{2-}	Cl^-
K^+	Ba^{2+}		$Hg^{2+,a}$					Br^-
								I^-
Ions of different charges from same elements								
Cr^{2+}, Cr^{3+}		Fe^{2+}, Fe^{3+}		Cu^+, Cu^{2+}				
Mn^{2+}, Mn^{3+}		Co^{2+}, Co^{3+}						

[a] Mercury (Hg) is unusual in that it also commonly forms a diatomic ion, Hg_2^{2+}.

Although ions have only a brief independent existence in the gas phase, positive and negative ions together form stable chemical compounds known as "ionic compounds." Table 4.2 lists some of the common **monatomic ions**—cations or anions formed by single atoms of elements. These monatomic ions are found combined in numerous simple ionic compounds. The types of ions formed by an element are characteristic of that element. Some elements form no stable ions, some form only one type of ion, and some form ions of two different charges.

At the top of Table 4.2, the ions formed from elements in the same vertical column in the periodic table are grouped together. For example, the lithium, sodium, and potassium cations (Li^+, Na^+, K^+) are all formed from elements in the first vertical column of the table on the left (see inside front cover). And the fluoride, chloride, bromide, and iodide anions (F^-, Cl^-, Br^-, I^-) are all formed from elements in the next-to-last vertical column on the right. You can see that ions from the same column tend to form ions with the same charges. Note that nitrogen (a nonmetal) forms an anion (N^{3-}) and bismuth (a metal) forms a cation (Bi^{3+}). In general, nonmetals form anions and metals form cations. (After we have studied the arrangements of electrons in atoms, the basis for the similarities in the ions formed by different types of elements will become apparent.) At the bottom of Table 4.2 are listed elements that commonly form ions with two different charges.

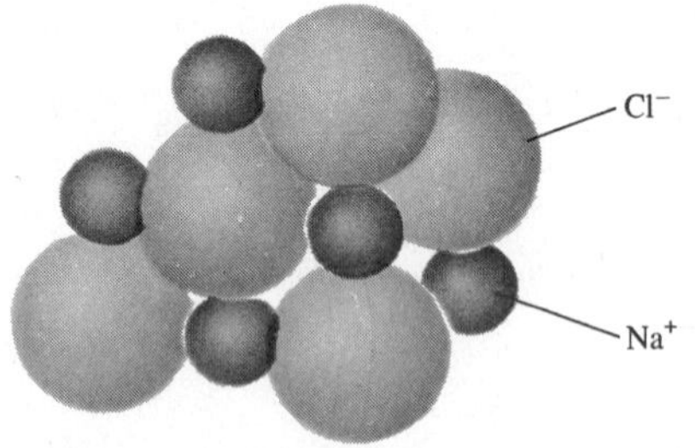

A portion of a crystal of NaCl

In ionic compounds positive and negative ions combine in sufficient numbers to maintain electrical neutrality. This means, for example, that in an ionic compound there is one Na^+ ion for every one Cl^- ion, or one Ba^{2+} ion for every two Cl^- ions. The resulting ionic compounds are sodium chloride (NaCl) and barium chloride ($BaCl_2$). Most ionic compounds are crystalline solids like sodium chloride, which is table salt. A solid ionic compound is a collection of ions held together by the mutual attraction of positive and negative charges. Independent molecules of NaCl and other ionic compounds do not ordinarily exist. A salt shaker holding pure table salt contains *only* crystals composed of sodium and chloride ions combined in a 1-to-1 ratio (Figure 4.2).

A salt shaker filled with NaCl crystals

Figure 4.2
Sodium Chloride, an Ionic Compound

4.2 FORMULAS FOR CHEMICAL COMPOUNDS

Chemical compounds are symbolized by chemical formulas. A **chemical formula** gives the symbols for the elements combined, with subscripts indicating how many atoms of each element are included. The chemical formulas of a few simple molecular compounds are given in Table 4.3. To write the correct chemical formula for a molecular compound, it is necessary to have information about how many atoms are combined in each molecule.

By contrast, the formula of an ionic compound shows only the *ratios* in which the ions are combined, as there are no molecules present. For example, the formula $BaCl_2$ indicates that in this substance, barium and chloride ions are present in the ratio 1:2. The term **formula unit** refers to the simplest unit indicated by the formula of a nonmolecular compound. The formula unit of barium chloride is one barium ion plus two chloride ions. As illustrated below, once the identities of the ions present are

Table 4.3
Some Simple Molecular Compounds

Water	H_2O	Odorless liquid
Carbon monoxide	CO	Odorless, flammable, toxic gas
Carbon dioxide	CO_2	Odorless, nonflammable, suffocating gas
Sulfur dioxide	SO_2	Nonflammable gas; suffocating odor
Ammonia	NH_3	Colorless, nonflammable gas; pungent odor
Methane	CH_4	Odorless, flammable gas
Carbon tetrachloride	CCl_4	Nonflammable, dense liquid

known, a correct formula can be written for a simple ionic compound based on the requirement that the total positive and negative charges be equal to each other.

Chemical formulas are used to represent ions as well as neutral compounds. For example, one sulfur atom and four oxygen atoms form an ion with a charge of −2, the sulfate ion, SO_4^{2-}. Ions that incorporate more than one atom are called **polyatomic ions**. Table 4.4 lists the formulas and names of some of the common polyatomic ions, most of which are negatively charged. (Memorizing these names and formulas will be *very* helpful.)

Like the monatomic anions, polyatomic anions also combine with cations to give ionic compounds in which electrical neutrality is maintained. Here are the formulas of two of the many compounds that are formed between the cations of Table 4.2 and the polyatomic anions of Table 4.4.

read "N-A-oh-H"
Na^+OH^-
1 Na⁺ plus 1 OH⁻

read "M-G-N-oh-three taken twice"
$Mg^{2+}(NO_3^-)_2$
parentheses to avoid confusion about what is taken twice

Note that when a chemical formula includes more than one polyatomic ion of the same type, parentheses are placed around the formula of that ion.

Remember that in an ionic compound, the total positive charge must equal the total negative charge. Determining how many ions of each charge must combine is obvious when the charges are equal or when one is two or three times the other. For example, Ca^{2+} plus F^-, Ca^{2+} plus SO_4^{2-}, and Na^+ plus PO_4^{3-} obviously give $Ca^{2+}(F^-)_2$, $Ca^{2+}SO_4^{2-}$, and $(Na^+)_3PO_4^{3-}$. In other cases, a number of positive ions equal to the charge on the negative ion and a number of negative ions equal to the charge on the positive ion often gives a correct formula. For example, Al^{3+} plus SO_4^{2-} gives $(Al^{3+})_2(SO_4^{2-})_3$ and Sn^{2+} plus PO_4^{3-} gives $(Sn^{2+})_3(PO_4^{3-})_2$. To check for charge balance, multiply the subscript times the charge on each ion:

total + charge of 2 × (+3) = +6
$(Al^{3+})_2(SO_4^{2-})_3$
total − charge of 3 × (−2) = −6

total + charge of 3 × (+2) = +6
$(Sn^{2+})_3(PO_4^{3-})_2$
total − charge of 2 × (−3) = −6

The charges of ions are not ordinarily included in formulas. The formulas for the compounds described above would normally be written as follows:

$$NaOH \qquad Mg(NO_3)_2 \qquad Al_2(SO_4)_3 \qquad Sn_3(PO_4)_2$$

Sometimes, molecules or polyatomic ions are represented by *structural formulas* in which the symbols for the elements are arranged on the page so that they give information about which atoms are connected to each other. For example, for SO_4^{2-} and NH_3, the structural formulas are

```
 [    O    ]2-
 [    |    ]          H       H
 [ O—S—O  ]    and     \    /
 [    |    ]             N
 [    O    ]             |
                         H
```

Table 4.4
Polyatomic Ions
These are some of the more common polyatomic ions.

NH_4^+	Ammonium ion
CN^-	Cyanide ion
CO_3^{2-}	Carbonate ion
ClO_3^-	Chlorate ion
ClO_4^-	Perchlorate ion
CrO_4^{2-}	Chromate ion
$Cr_2O_7^{2-}$	Dichromate ion
MnO_4^-	Permanganate ion
NO_2^-	Nitrite ion
NO_3^-	Nitrate ion
O_2^{2-}	Peroxide ion
OH^-	Hydroxide ion
PO_4^{3-}	Phosphate ion
SO_3^{2-}	Sulfite ion
SO_4^{2-}	Sulfate ion
CH_3COO^-	Acetate ion

Ethyl alcohol contains two C atoms, six H atoms, and one O atom. Instead of C_2H_6O, the formula is frequently written as either C_2H_5OH (which shows that the C_2H_5 forms a "group" of atoms and so does the OH) or as CH_3CH_2OH (which shows that an OH group is attached to a CH_2 group that is attached to a CH_3 group). The structural formula for ethyl alcohol is

```
    H   H
    |   |
H — C — C — O — H
    |   |
    H   H
```

EXAMPLE 4.1 Formulas for Ionic Compounds

Write the formulas for the nine compounds that might form between the cations Na^+, Zn^{2+}, Al^{3+} and the anions Cl^-, CO_3^{2-}, and PO_4^{3-}.

The total positive charge in each case must be equal to the total negative charge. For example, for Zn^{2+} with Cl^-

$$\left.\begin{array}{ll} \text{total positive charge} = (1)(+2) = +2 & Zn^{2+} \\ \text{total negative charge} = (2)(-1) = -2 & Cl^-, Cl^- \end{array}\right\} ZnCl_2$$

To check the formulas for simple ionic compounds, be sure that the subscript on the positive ion times the charge on the positive ion equals the subscript on the negative ion times the charge on the negative ion.

	Cl^-	CO_3^{2-}	PO_4^{3-}
Na^+	$NaCl$	Na_2CO_3	Na_3PO_4
Zn^{2+}	$ZnCl_2$	$ZnCO_3$	$Zn_3(PO_4)_2$
Al^{3+}	$AlCl_3$	$Al_2(CO_3)_3$	$AlPO_4$

EXERCISE Write the formulas for the six compounds that can form between the cations NH_4^+, Ca^{2+}, and Fe^{3+}, and the anions NO_3^- and SO_4^{2-}.

Answers NH_4NO_3, $(NH_4)_2SO_4$, $Ca(NO_3)_2$, $CaSO_4$, $Fe(NO_3)_3$, $Fe_2(SO_4)_3$.

4.3 MOLECULAR MASS

The **molecular mass** (or molecular weight) of a chemical compound is the sum of the atomic masses, in atomic mass units, of all the atoms in the formula of the compound. To calculate a molecular mass, the correct formula of the compound and the atomic mass of each element in the compound must be known. The term "molecular mass" is commonly used to refer to the mass of any species for which a chemical formula can be written, whether or not it is a molecular compound. In the case of an ionic compound, the term refers to the sum of the atomic masses in one *formula unit*. [The mass of an electron is so small that electrons gained or lost need not be considered in finding the molecular masses of polyatomic ions.]

EXAMPLE 4.2 Molecular Mass

Calculate the molecular masses of (a) N_2O_5 and (b) SO_4^{2-}.

The values of the atomic masses are taken from the tables like those inside the front or back covers of this book. For most calculations, we need only consider the atomic and molecular masses to the nearest 0.01 u.

(a) N_2O_5

	number of atoms		atomic mass (u/atom)		mass (u)
N	2	×	14.01	=	28.02
O	5	×	16.00	=	80.00
molecular mass of N_2O_5				=	**108.02 u**

(b) SO_4^{2-}

	number of atoms		atomic mass (u/atom)		mass (u)
S	1	×	32.06	=	32.06
O	4	×	16.00	=	64.00
molecular mass of SO_4^{2-}				=	**96.06 u**

EXERCISE Calculate the molecular mass of KBH_4. *Answer* 53.95 u.

4.4 NAMING CHEMICAL COMPOUNDS

Ideally, every chemical compound should have a unique name. **Chemical nomenclature** is the collective term for the rules and regulations that govern naming chemical compounds. The nomenclature of some simple types of ions and compounds is discussed in the following sections.

a. Monatomic Cations Monatomic cations are formed by the removal of one or more electrons from an atom. The Stock system of nomenclature is recommended for naming monatomic cations. (The International Union of Pure and Applied Chemistry, IUPAC, sets standards for nomenclature.) In the Stock system, the name of a cation consists of (1) the name of the element, (2) the charge on the ion given inside parentheses as a Roman numeral, and (3) the word "ion." For example

Cr^{2+}	chromium(II) ion	Fe^{2+}	iron(II) ion
Cr^{3+}	chromium(III) ion	Fe^{3+}	iron(III) ion

For elements that form only one cation (see Table 4.2), the Roman numeral is omitted. Also, the 1 is not written in the superscript for ions of +1 (or −1) charge, for example

H^+ hydrogen ion Na^+ sodium ion

EXAMPLE 4.3 Chemical Nomenclature—Monatomic Cations

(a) Name the monatomic cations K^+ and Cr^{2+}. (b) Write the symbols for gallium(III) ion and nickel(II) ion.

(a) K is the symbol for potassium and, as shown in Table 4.2, potassium commonly forms only one ion, so K^+ is named the **potassium ion.** Cr is the symbol for chromium, the Roman numeral for a +2 charge is II, so Cr^{2+} is called the **chromium(II) ion.**

(b) The symbols for gallium and nickel are Ga and Ni. These ions are written $\mathbf{Ga^{3+}}$ and $\mathbf{Ni^{2+}}$.

EXERCISE (a) Name the monatomic cations Cr^{3+} and Sc^{3+}. (b) Write the symbols for the cesium ion and the lead ion.

Answers (a) chromium(III) ion, scandium(III) ion; (b) Cs^+, Pb^{2+}.

An older system of naming cations uses word endings instead of numerals. The cation with the lower charge has the suffix "ous" and the cation with the higher charge has the suffix "ic" following the stem of the name of the element.

	Older system	Stock system
Fe^{2+}	Ferrous ion	Iron(II) ion
Fe^{3+}	Ferric ion	Iron(III) ion
Mn^{2+}	Manganous ion	Manganese(II) ion
Mn^{3+}	Manganic ion	Manganese(III) ion
Cu^{+}	Cuprous ion	Copper(I) ion
Cu^{2+}	Cupric ion	Copper(II) ion

Frequently, the ancient name of the metal provides the root in this system, as, for example, in cuprous and cupric, which are based on "cuprum." Although you may encounter this older system for naming cations, the Stock system is now preferred.

b. Monatomic Anions Monatomic anions are formed by the addition of one or more electrons to an atom. Such anions are named by writing (1) the root of the name of the element modified by (2) the ending "ide" followed by (3) the word "ion." Most elements form only one anion and it is not necessary to include the charge in the name. Some examples of monatomic anions are

Cl^-	chloride ion *(from chlorine)*	O^{2-}	oxide ion *(from oxygen)*	N^{3-}	nitride ion *(from nitrogen)*

EXAMPLE 4.4 Chemical Nomenclature—Monatomic Anions

(a) Name the anion Te^{2-}. (b) Write the symbol for the astatide anion, which has a charge of -1.

(a) Te is the symbol for tellurium. Dropping the "ium" and modifying the name with the ending "ide" gives the name **telluride ion** for Te^{2-}.
(b) The name "astatide" is derived from the element "astatine," which has the symbol At; therefore, the astatide ion is written $\mathbf{At^-}$.

EXERCISE (a) Name the anion F^-. (b) Write the symbols for the iodide ion, which has a charge of -1, and the sulfide ion, which has a charge of -2.

Answers (a) fluoride ion; (b) I^-, S^{2-}.

c. Ionic Compounds **A binary compound** is a compound that contains atoms or ions of only two elements. A binary compound formed by two ions is named by giving the cation name first, followed by the anion name. The word "ion" does not appear in the name of the compound. For example

NaF	sodium fluoride	FeO	iron(II) oxide
Al_2O_3	aluminum oxide	Fe_2O_3	iron(III) oxide

With the older system of naming cations by word endings, names such as the following were used:

CuCl	cuprous chloride	MnF_2	manganous fluoride
$CuCl_2$	cupric chloride	MnF_3	manganic fluoride

Compounds of polyatomic ions are also named by giving the cation name followed by the anion name. For example

Cu_2SO_4	copper(I) sulfate	$Al_2(SO_4)_3$	aluminum sulfate
$Mg(NO_3)_2$	magnesium nitrate	$ZnCO_3$	zinc carbonate

EXAMPLE 4.5 Chemical Nomenclature—Ionic Compounds

(a) Write the formulas for the following compounds: (i) iron(III) bromide, (ii) calcium sulfide, (iii) calcium permanganate. (b) Name the following compounds: (iv) CaI_2, (v) CuO, (vi) $(NH_4)_2SO_4$, (vii) $K_2Cr_2O_7$.

(a) There are two steps to writing the formulas: (1) Identify the symbols for the ions and the charges of the ions (check Tables 4.2 and 4.4) and (2) combine the ions into a formula based on electrical neutrality.

(i) Fe^{3+} and Br^- give $\mathbf{FeBr_3}$

(ii) Ca^{2+} and S^{2-} give **CaS**

(iii) Ca^{2+} and MnO_4^- give $\mathbf{Ca(MnO_4)_2}$

(b) (iv) Calcium forms *only* an ion with a + 2 charge (see Table 4.2) and CaI_2 is therefore named **calcium iodide.**

(v) The copper in CuO obviously has a +2 charge, since oxygen has a −2 charge. The compound is named **copper(II) oxide.** (Copper also forms Cu^+.)

(vi) $(NH_4)_2SO_4$ is simply named **ammonium sulfate.**

(vii) Potassium forms only one ion and $K_2Cr_2O_7$ is therefore named **potassium dichromate.**

EXERCISE (a) Write the formulas for potassium nitride and strontium carbonate. (b) Name the compounds that have the formulas $Na(CH_3COO)$ and $NiCl_2$.

Answers (a) K_3N, $SrCO_3$; (b) sodium acetate, nickel(II) chloride.

d. Acids and Bases At this point you should become familiar with the names of the common acids and bases. Acids and bases are chemical compounds that are often used in the laboratory, and they are mentioned in many of the following chapters. The common acids, some of which are listed in Table 4.5, all contain hydrogen and when dissolved in water give solutions that contain hydrogen ions (H^+) and anions. Most common bases contain OH^- ions and dissolve to give these ions plus cations. The three most common bases of this type are sodium hydroxide, NaOH; potassium hydroxide, KOH; and calcium hydroxide, $Ca(OH)_2$, which is much less soluble in water than the other two common bases.

The anions formed when the common acids are dissolved in water include many of the simple anions and the polyatomic anions listed in Tables 4.2 and 4.4. In addition, anions that contain hydrogen are formed by acids that contain more than one hydrogen atom. For example, H_2SO_4 can yield both SO_4^{2-} (sulfate ion) and HSO_4^-. Anions that include hydrogen are named by using either the word "hydrogen" or, in some cases, the prefix "bi" as follows:

HSO_4^- hydrogen sulfate ion *or* bisulfate ion
HCO_3^- hydrogen carbonate ion *or* bicarbonate ion

Table 4.5
Some Common Acids
Carbonic acid, nitrous acid, and sulfurous acid have never been isolated and are known only in aqueous (*aq*) solutions; HCl *not* in aqueous solution is the gas hydrogen chloride.

Acid		Anion	
Hydrochloric acid	HCl(*aq*)	Chloride ion	Cl^-
Carbonic acid	$H_2CO_3(aq)$	Carbonate ion	CO_3^{2-}
		Hydrogen carbonate ion	HCO_3^-
Nitric acid	HNO_3	Nitrate ion	NO_3^-
Nitrous acid	$HNO_2(aq)$	Nitrite ion	NO_2^-
Perchloric acid	$HClO_4$	Perchlorate ion	ClO_4^-
Phosphoric acid	H_3PO_4	Phosphate ion	PO_4^{3-}
		Hydrogen phosphate ion	HPO_4^{2-}
		Dihydrogen phosphate ion	$H_2PO_4^-$
Phosphorous acid	H_3PO_3	Hydrogen phosphite ion	HPO_3^{2-}
Sulfuric acid	H_2SO_4	Sulfate ion	SO_4^{2-}
		Hydrogen sulfate ion	HSO_4^-
Sulfurous acid	$H_2SO_3(aq)$	Sulfite ion	SO_3^{2-}
		Hydrogen sulfite ion	HSO_3^-

Table 4.6
Multiplying Prefixes

Number indicated	Prefix
1	mono
2	di
3	tri
4	tetra
5	penta
6	hexa
7	hepta
8	octa
9	nona
10	deca

The formulas and names of anions are included in Table 4.5. Take a moment right now to look at this table. (Memorizing these names and formulas is *very* helpful.)

Ionic compounds formed between cations (except H^+) and the anions of acids are called **salts**. For example, KCl, Na_2SO_4, $Mg_3(PO_4)_2$, and $NaHSO_4$ are all referred to as salts.

e. Binary Molecular Compounds The classical system for naming binary molecular compounds is based on the prefixes listed in Table 4.6. The appropriate prefix is placed before the name of the first element in the formula of the compound. For example, for N_2O_3 the first part of the name is *di*nitrogen, showing that there are two nitrogen atoms. The second part of the name consists of the appropriate prefix before the name of the second element, which has been modified by adding "ide." For N_2O_3 this would be *tri*oxide, showing that there are three oxygen atoms. Thus, the complete name of N_2O_3 is dinitrogen trioxide.

The prefix "mono" for one atom of an element in a compound is omitted for the first element named and usually for the second, except where there is more than one compound formed between the two elements.

Following are some additional examples of the names of binary molecular compounds:

N_2O	dinitrogen monoxide	ICl	iodine monochloride	SO_2	sulfur dioxide
N_2O_5	dinitrogen pentoxide	ICl_3	iodine trichloride	SO_3	sulfur trioxide

(Note that in many binary molecular compounds, atoms of nonmetals have combined.)

EXAMPLE 4.6 Chemical Nomenclature—Binary Molecular Compounds

(a) Write the formulas for phosphorus triiodide and oxygen difluoride. (b) Write names for BrF_3, S_2O_7 and S_2Br_2.

(a) The "tri" indicates three iodine atoms, and therefore the formula of phosphorus triiodide is $\mathbf{PI_3}$. The "di" indicates two fluorine atoms, and therefore the formula of oxygen difluoride is $\mathbf{OF_2}$.

(b) The name for BrF_3 must include the prefix "tri" to show the presence of three fluorine atoms, but need not include a prefix to indicate that there is one bromine atom; the name is **bromine trifluoride.** S_2O_7 is named **disulfur heptoxide** and S_2Br_2 is named **disulfur dibromide.**

EXERCISE (a) Write the formulas for dialuminum hexachloride and iodine trifluoride. (b) Write names for IF_7 and P_4O_{10}.

Answers (a) Al_2Cl_6, IF_3; (b) iodine heptafluoride, tetraphosphorus decoxide.

SUMMARY OF SECTIONS 4.1–4.4

A molecule is the smallest particle of a pure substance that is capable of independent existence. The chemical formula for a molecular compound includes the symbols for the elements combined and shows by subscripts the number of atoms present in each molecule. Binary molecular compounds are named by combining the names of the elements with prefixes that indicate how many atoms of each element are present in each molecule (e.g., CO_2, carbon dioxide).

In an ionic compound, sufficient numbers of positive and negative ions combine so that electrical neutrality is maintained. The formula of an ionic compound shows the ratio of positive ions (cations) to negative ions (anions). In the Stock system, a cation is given the name of the element followed by the charge given as a Roman numeral in parentheses (but usually omitted for elements that form only one cation); an anion is given the name of the element, modified with the ending "ide." The name of an ionic compound consists of the name of the cation followed by the name of the anion (e.g., $BaSO_4$, barium sulfate; Cu_2O, copper(I) oxide).

The common acids all contain hydrogen and give aqueous solutions that contain H^+ and anions. Most common bases contain OH^- and give aqueous solutions that contain cations and OH^-. It is helpful to memorize the names and formulas of the common acids given in Table 4.5 and the common ions in Tables 4.2, 4.4, and 4.5.

The "molecular mass" of a chemical species of any type is the sum of the atomic masses, in atomic mass units, of the total number of atoms in the formula of the species.

CHEMICAL EQUATIONS

4.5 READING AND WRITING CHEMICAL EQUATIONS

a. The Conservation of Mass Substances that undergo chemical reactions may be composed of atoms, molecules, or ions. They may be in the gaseous, liquid, or solid states, or in solution. The substances that are changed in a chemical reaction are called the **reactants**. The substances that are produced in a chemical reaction are called the **products**.

Chemical equations are the sentences of the symbolic language of chemistry. A **chemical equation** represents with symbols and formulas the total chemical change that occurs in a chemical reaction. The general scheme for writing a chemical equation is

$$\text{reactant}_1 + \text{reactant}_2 + \text{reactant}_3 + \cdots \longrightarrow \text{product}_1 + \text{product}_2 + \text{product}_3 + \cdots$$

For example, the equation for the chemical reaction between the elements phosphorus and chlorine to form the binary compound phosphorus trichloride is as follows (the arrow is read "yields"):

$$P_4 + 6Cl_2 \longrightarrow 4PCl_3 \qquad \textbf{(4.1)}$$

Phosphorus plus chlorine yields phosphorus trichloride

The interpretation of Equation (4.1) at the micro, or atomic and molecular level, is as follows:

$$P_4 + 6Cl_2 \longrightarrow 4PCl_3$$

One P_4 molecule plus six Cl_2 molecules yields four molecules of PCl_3

Like all correctly written chemical equations, this equation is *balanced*. All chemical equations must be balanced—the number of atoms of each kind must be the same in the products and in the reactants. There is a reason for this—it is the law of conservation of mass: In ordinary chemical reactions matter is neither created nor destroyed. This law means that all of the atoms present in the reactants, however their arrangements or combinations are changed, must still be present when the reaction is completed. Without the coefficients of 6 and 4, Equation (4.1) would be incorrect, because it would show the destruction of three phosphorus atoms and the creation of one chlorine atom. (Equation balancing is discussed in Section 4.6.)

b. States of Matter and Reaction Conditions Information about the physical states of the reactants and products, or about some of the conditions under which a reaction occurs, is often added to chemical equations. The states of the substances involved are shown by placing in parentheses after the formulas, the symbols *g* for gas, *l* for liquid, and *s* for solid (Table 4.7). For example

$$P_4(s) + 6Cl_2(g) \longrightarrow 4PCl_3(l) \qquad \textbf{(4.2)}$$

Solid phosphorus plus gaseous chlorine yields liquid phosphorus trichloride

Table 4.7
Symbols for the States of Reactants and Products

Symbol	State
(*s*)	Solid
(*l*)	Liquid
(*g*)	Gas
(*aq*)	Aqueous solution

The capital Greek letter delta, Δ, is often placed over the arrow to indicate that heat has been found necessary to make a reaction take place, for example

$$\underset{\text{Solid calcium carbonate}}{CaCO_3(s)} \xrightarrow[\text{when heated yields}]{\Delta} \underset{\text{solid calcium oxide}}{CaO(s)} \underset{\text{plus}}{+} \underset{\text{gaseous carbon dioxide}}{CO_2(g)} \quad (4.3)$$

Other information about the conditions under which a reaction takes place is sometimes written over the arrow, such as the temperature or the pressure, or the formula of a **catalyst**—a substance that increases the rate of a reaction but can be recovered chemically unchanged after the reaction is complete. For example, the industrial preparation of methyl alcohol (CH_3OH) is carried out at high temperature and pressure in the presence of a catalyst that is a mixture of zinc and chromium oxides.

$$CO(g) + 2H_2(g) \xrightarrow[ZnO-Cr_2O_3]{350°C,\ 200-300\ atm} CH_3OH(g)$$

Equations are sometimes used to represent the dissolution of a substance in water or another solvent. The presence of solvent can be indicated by writing its formula or name over the arrow. The ions in ionic compounds separate from each other in aqueous solutions, and this is indicated as follows:

$$\underset{\text{solid sodium hydroxide}}{NaOH(s)} \xrightarrow{H_2O} \underset{\text{sodium ion in aqueous solution}}{Na^+(aq)} + \underset{\text{hydroxide ion in aqueous solution}}{OH^-(aq)} \quad (4.4)$$

The **dissociation of an ionic compound** is the separation of the ions of a neutral ionic compound, usually by dissolution in water. The term **ionization** is reserved for the formation of ions from a molecular compound or from atoms. For example, ionization occurs when hydrogen chloride, a gas, is dissolved in water to give hydrogen ions and chloride ions. The solution of hydrogen chloride in water is hydrochloric acid.

$$\underset{\text{hydrogen chloride}}{HCl(g)} \xrightarrow{H_2O} \underset{\text{hydrochloric acid}}{H^+(aq) + Cl^-(aq)} \quad (4.5)$$

4.6 BALANCING CHEMICAL EQUATIONS

Once the reactants and products in a chemical reaction are known, the first step in writing the equation is to write the correct formulas on the appropriate sides of the arrow. Suppose we have found that when solid aluminum sulfide is mixed with water, the products are solid aluminum hydroxide and hydrogen sulfide gas. Starting with the formulas, we write

$$\underset{\text{aluminum sulfide}}{Al_2S_3(s)} + \underset{\text{water}}{H_2O(l)} \xrightarrow{\text{not balanced}} \underset{\text{aluminum hydroxide}}{Al(OH)_3(s)} + \underset{\text{hydrogen sulfide}}{H_2S(g)} \quad (4.6)$$

The next step is to balance, one by one, the atoms of each element and also any polyatomic ions that appear on *both sides of the equation.* It is often easiest to begin with atoms that appear in only one formula on each side of the equation. In Equation (4.6), there are three such atoms: Al, S, and O. Also, it often helps to begin with the most complicated formula, which in this case is Al_2S_3, and, if water is present, to balance O and H atoms last.

Starting with Al, inspect the formula of the reactant in which it appears: aluminum sulfide, Al_2S_3. The subscript shows that two Al atoms are present, and so there must also be two Al atoms on the right in the equation. To increase the number of Al atoms on the right from one in aluminum hydroxide, $Al(OH)_3$, to two, the *coefficient* 2 must be added.

$$Al_2S_3 + H_2O \xrightarrow{\text{not balanced}} 2Al(OH)_3 + H_2S \quad (4.7)$$

It would be *incorrect* to write $Al_2(OH)_3$ because this would be a different chemical compound (if it existed, which it doesn't).

In the same manner, inspection shows that there must be three S atoms on the right.

$$Al_2S_3 + H_2O \xrightarrow{\text{not balanced}} 2Al(OH)_3 + 3H_2S \quad \textbf{(4.8)}$$

Going on to oxygen, there are now six oxygen atoms on the right (2 × 3, the coefficient times the subscript), so six are needed on the left.

$$Al_2S_3(s) + 6H_2O(l) \longrightarrow 2Al(OH)_3(s) + 3H_2S(g) \quad \textbf{(4.9)}$$

Now it is necessary to check the number of hydrogen atoms to see if they balance. On the left there are (6)(2) = 12 hydrogen atoms and on the right there are (2)(3) = 6 plus (3)(2) = 6 hydrogen atoms, also giving a total of 12 hydrogen atoms. Equation (4.9) is balanced.

It is important to emphasize that the formulas for the reactants and products in a chemical reaction must represent the species that have been shown, by experience, to take part in the reaction. Any change in the correct chemical formulas of these species makes the equation *wrong*, even if it is balanced. Balancing an equation must never be achieved by changing the chemical formula of a reactant or product. Do not make the common error of changing subscripts to balance an equation.

Some chemical equations can be balanced by using fractional coefficients; for example

$$Zn(s) + \tfrac{1}{2}O_2(g) \longrightarrow ZnO(s) \quad \textbf{(4.10)}$$

Fractional coefficients are used most often when it is desired to write a chemical equation that shows one molecule or one formula unit of a product, without any coefficients. For example, Equation (4.10) would be used to represent the formation of one formula unit of zinc oxide from the elements. Otherwise, fractional coefficients are avoided, and Equation (4.10) normally would be written as follows:

$$2Zn(s) + O_2(g) \longrightarrow 2ZnO(s) \quad \textbf{(4.11)}$$

EXAMPLE 4.7 Balancing Chemical Equations

An important step in the treatment of municipal water supplies and sewage is the reaction between $Al_2(SO_4)_3$ and $Ca(OH)_2$:

$$Al_2(SO_4)_3(aq) + Ca(OH)_2(aq) \xrightarrow{\text{not balanced}} CaSO_4(s) + Al(OH)_3(s)$$

Balance this equation.

First balance the two Al atoms on the left by placing the coefficient 2 in front of $Al(OH)_3$ on the right:

$$Al_2(SO_4)_3 + Ca(OH)_2 \xrightarrow{\text{not balanced}} CaSO_4 + 2Al(OH)_3$$

Recognizing that there are (2)(3) = 6 hydroxide ions on the products side and only 2 on the reactants side, write a 3 in front of $Ca(OH)_2$ to give 6 hydroxide ions on the reactants side.

$$Al_2(SO_4)_3 + 3Ca(OH)_2 \xrightarrow{\text{not balanced}} CaSO_4 + 2Al(OH)_3$$

The balancing is completed by placing a 3 before the formula for $CaSO_4$, which equalizes both the number of Ca atoms and the number of sulfate ions on both sides of the equation:

$$\mathbf{Al_2(SO_4)_3(s) + 3Ca(OH)_2(s) \longrightarrow 3CaSO_4(s) + 2Al(OH)_3(s)}$$

Next it is necessary to be sure that the equation is balanced.

On the left:	2 Al	3 SO_4	3Ca	(3)(2) = 6 OH
On the right:	2 Al	3 SO_4	3Ca	(3)(2) = 6 OH

The equation is balanced.

EXERCISE Solid phosphorus pentabromide, PBr_5, reacts with water to form phosphoric acid, H_3PO_4, and hydrobromic acid, HBr. Write the balanced chemical equation for this reaction. *Answer* $PBr_5(s) + 4H_2O(l) \longrightarrow H_3PO_4(aq) + 5HBr(aq)$

SUMMARY OF SECTIONS 4.5–4.6

Chemical reactions are represented in writing by chemical equations. In accord with the law of conservation of mass, atoms are neither created nor destroyed in ordinary chemical reactions. To reflect this, a chemical equation must be balanced—it must show equal numbers of atoms of each type in the reactants and products.

To write a chemical equation, the formulas of the reactants and products are first written on the left and right sides of the arrow, respectively. The equation is then balanced by adding coefficients to the formulas of the reactants and products. For simple equations, it is best to balance the atoms of each element or any unchanged polyatomic ions one by one. Equations must not be balanced by changing the subscripts in chemical formulas.

Information about the states of reactants and products (gas, liquid, or solid), the presence of substances in aqueous solution, and the conditions of a reaction may be included in chemical equations. Chemical equations can be written to show the dissolution of ionic compounds to give ions in aqueous solution, or the ionization of molecular compounds.

ATOMIC, MOLECULAR, AND MOLAR MASS RELATIONSHIPS

4.7 AVOGADRO'S NUMBER, THE MOLE, AND MOLAR MASS

How can masses of matter that can be seen and weighed be related to the numbers and masses of individual atoms and molecules, which cannot be seen, weighed, or counted? Three quantities are at the heart of the matter in dealing with these relationships:

Avogadro's number (6.022×10^{23}) is the number of atoms in exactly 12 g of carbon-12.

The **mole** represents a number of anything equal to Avogadro's number.

The **molar mass** of a substance is the mass in grams of one mole of that substance.

a. Avogadro's Number The masses of substances in chemistry are most commonly measured in grams, milligrams, or kilograms. It would be useful to know how many atoms, ions, molecules, or formula units are present in a sample of matter—say, 0.2 g of sodium chloride or 500 g of sulfuric acid. This information is desirable because chemical formulas and chemical equations show how atoms combine, not in terms of their masses, but in terms of their relative numbers. The formula SO_2, for example, does *not* tell us what *mass* of sulfur will combine completely with a given mass of oxygen; it tells us only that there is one sulfur atom for every two oxygen atoms in this compound. But in weighing out a few grams of sulfur, we are weighing trillions upon trillions of atoms which are both too small to see and too numerous to count.

What is needed is a simple way to connect macroscopic masses with numbers of atoms, molecules, or ions. There is such a simple connection and it is made possible by the relative atomic mass scale.

To understand the connection, consider an analogy: Nails used in carpentry are ordinarily sold by the pound. The 2-inch nails weigh twice as much as the $1\frac{1}{2}$-inch nails. If the $1\frac{1}{2}$-inch nails are assigned a relative mass of 1, the 2-inch nails have a relative mass of 2. To get exactly the same number of nails of each size, *it is not necessary to count the nails.* All that a carpenter need do is buy 1 pound of $1\frac{1}{2}$-inch nails and 2 pounds of 2-inch nails. As long as the *relative* masses are the same, the number of nails will be the same. Furthermore, the unit of mass does not make any difference. One kilogram of $1\frac{1}{2}$-inch nails will also contain the same number of nails as 2 kg of 2-inch nails.

Now consider the masses of two elements on the relative atomic mass scale:

$$\text{atomic mass of O} = 15.9994 \text{ u} \qquad \text{atomic mass of S} = 32.06 \text{ u}$$

Because a sulfur atom weighs about twice as much as an oxygen atom, 1 g of oxygen and 2 g of sulfur will contain roughly the same number of atoms of each element, as will 16 g of oxygen and 32 g of sulfur.

To weigh out exactly the same number of atoms of each element, it is only necessary to take 15.9994 g of oxygen and 32.06 g of sulfur. Just as we saw for the nails, once a relative mass scale is established, the relative number of individual items can be determined from the masses, regardless of what units of mass are used.

By the same reasoning, 1.0079 g of hydrogen or 238.03 g of uranium must contain *the same number of atoms* as 15.9994 g of oxygen or 32.06 g of sulfur. Why? Because these are also masses in grams numerically equal to the relative atomic masses of these elements. For every element, a mass in grams numerically equal to its atomic mass in atomic mass units contains the same number of atoms; this is the number known as Avogadro's number.

The relative atomic mass scale is based on the assigned mass of 12 for carbon-12; therefore, Avogadro's number is the number of atoms in exactly 12 g of carbon-12. Avogadro's number is the connecting link between the macro and micro scales of mass. It is also the connecting link between macroscopic masses and specific numbers of atoms.

The same relationship can be extended from atoms to molecules. For every molecular compound, a mass in grams equal to its molecular mass contains an Avogadro's number of molecules. For example, the molecular mass of sulfur dioxide is 64.06 u, the sum of the atomic masses of one sulfur atom and two oxygen atoms. A mass of 64.06 g of sulfur dioxide contains an Avogadro's number of SO_2 molecules.

Avogadro's number has been determined by many methods and with great accuracy. The most up-to-date value is 6.022045×10^{23}. In most calculations in this book, it will be sufficient to use 6.022×10^{23}. (To grasp the magnitude of this number, consider that 6×10^{23} baseballs would cover the entire surface of the earth to a depth of 97 km, or that a digital electronic timer that can register 10 million counts per second would take almost 2 billion years to count Avogadro's number of seconds.)

b. The Mole Avogadro's number enables us to define an extremely convenient unit, the *mole*, which represents an Avogadro's number of anything. The mole is a unit like the dozen. The dozen can be used to represent 12 of anything. Similarly, the mole represents an Avogadro's number of "entities" of any kind whatsoever—ions, atoms, or molecules. We could also refer to a mole of eggs or a mole of butterflies. The mole allows for counting atoms, molecules, or ions by weighing them.

The mole is one of the seven SI base units. As defined formally, one mole of a substance contains the same number of "elementary entities" as there are atoms in 12 g of carbon-12 (Figure 4.3). The SI definition notes that the "elementary entities," which may be "atoms, ions, electrons, other particles, or groups of such particles," must be specified. The SI symbol for the mole is "mol."

The mole gives us an equality

$$6.022 \times 10^{23} \text{ entities} = 1 \text{ mol}$$

from which, as usual, two conversion factors can be derived.

$$\frac{6.022 \times 10^{23} \text{ entities}}{1 \text{ mol}} \quad \text{and} \quad \frac{1 \text{ mol}}{6.022 \times 10^{23} \text{ entities}}$$

Avogadro's number provides the conversion factor between the number of entities on the micro scale and the number of moles. For example, when the entities are molecules

$$(\text{No. of moles})\left(\frac{6.022 \times 10^{23} \text{ molecules}}{1 \text{ mol}}\right) = \text{no. of molecules} \tag{4.12}$$

and

$$(\text{No. of molecules})\left(\frac{1 \text{ mol}}{6.022 \times 10^{23} \text{ molecules}}\right) = \text{no. of moles} \tag{4.13}$$

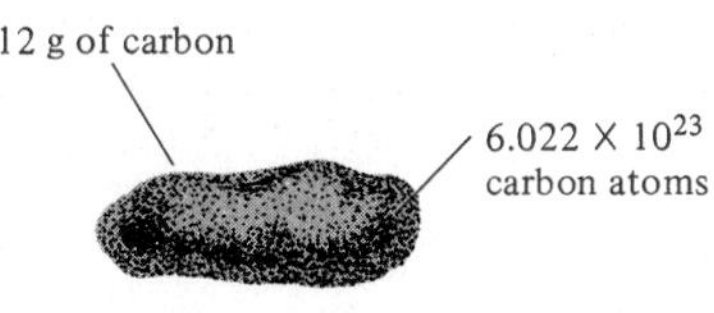

Figure 4.3
The Mole
The word "mole" comes from the Latin *moles*, meaning a mass or a pile of something.

EXAMPLE 4.8 Avogadro's Number and Moles

A 1 L flask of air contains 0.040 mol of N_2. How many molecules of nitrogen are present?

This problem is a mole → molecules conversion. The number of molecules is given by

$$(0.040\ \cancel{\text{mol}\ N_2})\left(\frac{6.022 \times 10^{23}\ \text{molecules}\ N_2}{1\ \cancel{\text{mol}\ N_2}}\right) = \mathbf{2.4 \times 10^{22}\ molecules\ N_2}$$

EXERCISE How many moles of CsBr are present in a sample containing 7.8×10^{24} formula units of CsBr? *Answer* 13 mol.

c. Molar Mass The molar mass of a substance is the mass in grams of one mole of that substance. Here is where the quantities that we have been discussing—relative atomic mass, Avogadro's number, and the mole—come together beautifully.

For an element, the molar mass is the mass in grams numerically equal to the atomic mass in atomic mass units.

For a chemical compound, the molar mass is the sum of the molar masses of all of the atoms that have combined in one molecule or formula unit.

One mole of a substance contains the same number of molecules (or whatever other "entity" is under discussion) as one mole of any other substance. From the following chemical equation

$$N_2(g) + 3H_2(g) \longrightarrow 2NH_3(g)$$

it can be seen that one molecule of nitrogen combines with three molecules of hydrogen to give two molecules of ammonia. This also means that 1 mol of nitrogen will combine with 3 mol of hydrogen to give 2 mol of ammonia. Here is the connection that we have been looking for. The masses of reactants and products can be related to each other by the numbers of moles given by the balanced chemical equation. The very important use of such relationships in determining the masses of substance that take part in chemical reactions is discussed in Chapter 5.

[Note that molar amounts of substances are sometimes expressed in terms of "millimoles," with masses given in milligrams, or "kilomoles," with masses given in kilograms. Also note that in an older terminology that you may encounter in other books, the molar masses of atomic, molecular, and ionic substances are called the gram atomic weight, gram molecular weight, and gram formula weight, respectively.]

Table 4.8 gives the molar masses of a variety of chemical "entities." Think for a moment about how the molar masses are related to the atomic masses of the elements involved.

From Table 4.8 it should be clear why it is necessary to specify what "entity" is referred to when the number of moles is used. For example, one mole of oxygen *molecules* contains two moles of oxygen *atoms*. One mole of $NaNO_3$ contains one mole of Na^+ ions and one mole of NO_3^- ions, or a total of two moles of ions.

Molar mass, as commonly used, has the units of grams per mole; for example, the molar mass of aluminum is 26.98 g/mol. Molar mass therefore provides the conversion factors that relate the mass in grams and the number of moles of a substance.

Table 4.8
Molar Masses of Some Different Chemical "Entities"

	Molar mass (g/mol)
Atoms	
O atoms	16.00
Fe atoms	55.85
S atoms	32.06
N atoms	14.01
Molecules	
O_2	32.00
H_2O	18.02
SO_2	64.06
Ions	
Na^+	22.99
Fe^{3+}	55.85
NO_3^-	62.01
Ionic compounds	
$NaNO_3$	85.00
$Fe(NO_3)_3$	241.88

$$(\text{Mass})\left(\frac{1}{\text{molar mass}}\right) = \text{no. of moles} \qquad \textbf{(4.14)}$$

$$\cancel{grams} \quad \frac{moles}{\cancel{grams}} \quad moles$$

$$(\text{No. of moles})(\text{molar mass}) = \text{mass} \qquad \textbf{(4.15)}$$

$$\cancel{moles} \quad \frac{grams}{\cancel{mole}} \quad grams$$

EXAMPLE 4.9 Molar Mass

What is the molar mass of chlorophyll *a*, $C_{55}H_{72}MgN_4O_5$?

The molar mass is found by taking the sum of the molar masses of all the atoms present in the compound.

	moles of atoms		molar mass of atoms (g/mol)		mass (g)
C	55	×	12.01	=	660.6
H	72	×	1.01	=	72.7
Mg	1	×	24.31	=	24.31
N	4	×	14.01	=	56.04
O	5	×	16.00	=	80.00
molar mass of $C_{55}H_{72}MgN_4O_5$				=	**893.7 g**

EXERCISE What is the molar mass of benzoic acid, C_6H_5COOH?

Answer 122.13 g.

EXAMPLE 4.10 Molar Mass and Moles

What is the mass of the nitrogen sample described in Example 4.8, which contained 0.040 mol of N_2?

The molar mass of N_2 is 28.02 g. This problem is a moles → mass conversion.

$$(0.040\ \cancel{\text{mol N}_2})\left(\frac{28.02\text{ g N}_2}{1\ \cancel{\text{mol N}_2}}\right) = \mathbf{1.1\ g\ N_2}$$

EXERCISE How many moles are equivalent to 34 kg of sucrose, $C_{12}H_{22}O_{11}$?

Answer 99 mol $C_{12}H_{22}O_{11}$.

The relationships among Avogadro's number, the mole, and molar mass can be used separately or in combination to solve a wide variety of problems. As an illustration of how to analyze solving such problems, Example 4.11 is worked by the problem-solving method introduced in Section 2.11.

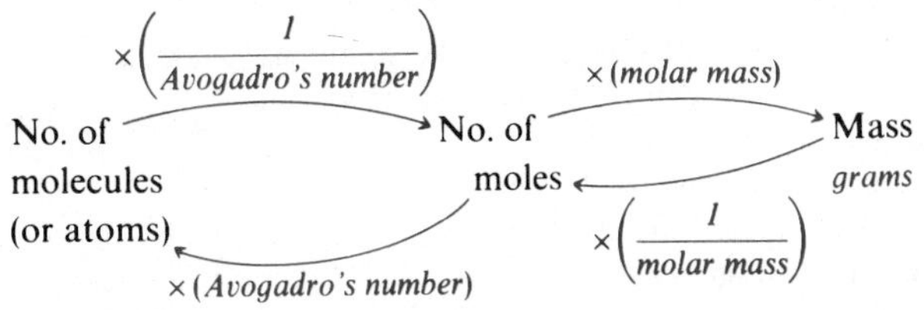

$$\text{No. of molecules (or atoms)} \xrightarrow{\times\left(\frac{1}{\text{Avogadro's number}}\right)} \text{No. of moles} \xrightarrow{\times(\text{molar mass})} \text{Mass (grams)}$$

$$\text{Mass (grams)} \xrightarrow{\times\left(\frac{1}{\text{molar mass}}\right)} \text{No. of moles} \xrightarrow{\times(\text{Avogadro's number})} \text{No. of molecules (or atoms)} \quad \textbf{(4.16)}$$

EXAMPLE 4.11 Molar Mass, Avogadro's Number, and Moles: Problem Solving

What mass of zirconium contains the same number of atoms as 16 g of calcium?

1. Study the problem and be sure you understand it.
   (a) What is unknown?
   The mass of zirconium containing the same number of atoms as 16 g of calcium.

(b) What is known?
The sample size—16 g of calcium. Also the molar masses of calcium and zirconium (from the tables).

2. Decide how to solve the problem.
(a) What is the connection between the known and the unknown?
The mole provides the connection. *Equal molar amounts* of calcium and zirconium will contain equal numbers of atoms.
(b) What is necessary to make the connection?
It is necessary to do a mass ⟶ moles conversion and then a moles ⟶ mass conversion. First, from the known mass of calcium, find the number of moles of calcium

$$16 \text{ g Ca} \xrightarrow{\times\left(\frac{1}{\text{molar mass}}\right)} \text{no. of moles of Ca}$$

Then, taking an equal number of moles of zirconium (because this amount contains an equal number of atoms), find the equivalent mass of zirconium, which is the unknown.

$$(\text{No. of } \cancel{\text{moles Ca}})\left(\frac{1 \text{ mol Zr}}{1 \cancel{\text{ mol Ca}}}\right) = \text{no. of moles Zr}$$

$$\text{No. of moles Zr} \xrightarrow{\times \text{(molar mass Zr)}} \text{mass Zr (grams)}$$

3. Set up the problem and solve it.
First, find the number of moles of calcium in 16 g of calcium.

$$(16 \cancel{\text{ g Ca}})\left(\frac{1 \text{ mol Ca}}{40.08 \cancel{\text{ g Ca}}}\right) = 0.40 \text{ mol Ca}$$

Then, find the number of grams of zirconium that represents the same number of moles:

$$(0.40 \cancel{\text{ mol Ca}})\left(\frac{1 \cancel{\text{ mol Zr}}}{1 \cancel{\text{ mol Ca}}}\right)\left(\frac{91.22 \text{ g Zr}}{1 \cancel{\text{ mol Zr}}}\right) = 36 \text{ g Zr}$$

Or, in a single setup:

$$(16 \cancel{\text{ g Ca}})\left(\frac{1 \cancel{\text{ mol Ca}}}{40.08 \cancel{\text{ g Ca}}}\right)\left(\frac{1 \cancel{\text{ mol Zr}}}{1 \cancel{\text{ mol Ca}}}\right)\left(\frac{91.22 \text{ g Zr}}{1 \cancel{\text{ mol Zr}}}\right) = \mathbf{36 \text{ g Zr}}$$

The answer is that 36 g of zirconium contains the same number of atoms as 16 g of calcium.

4. Check the result.
(a) Are significant figures used correctly?
Yes, the answer can have only two significant figures.
(b) Did the answer come out in the correct units?
Yes.
(c) Is the answer reasonable?
Yes. The molar mass of zirconium (91.22 g/mol) is slightly more than two times the molar mass of calcium (40.08 g/mol). Therefore, it is reasonable that 36 g of zirconium contains the same number of atoms as 16 g of calcium, because 36 g is roughly two times 16 g.

EXAMPLE 4.12 Molar Mass and Moles

Which is equivalent to the greater number of moles: 3.5 g of carbon dioxide, CO_2, or 3.5 g of sodium chloride, NaCl?

To answer this question, the conversion required is mass → moles. The molar masses are 44.01 g for CO_2 and 58.44 g for NaCl. The numbers of moles of the substances are

$$(3.5\ \cancel{g\ CO_2})\left(\frac{1\ mol\ CO_2}{44.01\ \cancel{g\ CO_2}}\right) = \mathbf{0.080\ mol\ CO_2}$$

$$(3.5\ \cancel{g\ NaCl})\left(\frac{1\ mol\ NaCl}{58.44\ \cancel{g\ NaCl}}\right) = \mathbf{0.060\ mol\ NaCl}$$

The number of moles of CO_2 is larger than the number of moles of NaCl.

EXERCISE Which is equivalent to the greater number of moles: 2.1 g of iron(II) oxide, FeO, or 2.9 g of iron(III) oxide, Fe_2O_3? *Answer* 2.1 g FeO.

EXAMPLE 4.13 Molar Mass, Avogadro's Number, and Moles

The nutritional recommended daily allowance (rda) of iron is 18 mg for an adult. How many atoms of iron is this?

First mass must be converted to moles and then moles to number of atoms, or mass → moles → atoms. Note that in molar calculations all masses must be converted to *grams*. This problem can also be solved in one step:

$$(18\ \cancel{mg\ Fe})\left(\frac{1\ \cancel{g}}{1000\ \cancel{mg}}\right)\left(\frac{1\ \cancel{mol\ Fe}}{55.85\ \cancel{g\ Fe}}\right)\left(\frac{6.022 \times 10^{23}\ Fe\ atoms}{1\ \cancel{mol\ Fe}}\right) = \mathbf{1.9 \times 10^{20}\ Fe\ atoms}$$

EXERCISE How many atoms of carbon are present in 14.6 g of $CaCO_3$? *Answer* 8.78×10^{22} C atoms.

4.8 MOLARITY: MOLAR MASS IN SOLUTIONS

The mole is used to express the amounts of *pure* substances. Frequently, chemical reactions are performed in aqueous solutions. Such solutions are mixtures and can contain widely varying proportions of dissolved substances and water. To know the amount of a substance present in a given amount of solution, we must know the **concentration** of the solution, which is a quantitative statement of the amount of solute in a given amount of solvent or solution.

The concentration of aqueous solutions is often most conveniently given as the number of moles of solute per liter of solution, which is called the **molarity** of the solution:

$$\text{Molarity} = \frac{\text{no. of moles of solute}}{\text{volume of solution in liters}} \quad \textbf{(4.17)}$$

A one-molar solution, written 1 M, contains 1 mol of a substance dissolved in enough water to give exactly 1 L of solution. For example, 1 L of a 1 M NaCl solution contains 58.5 g of dissolved sodium chloride. And 1 L of a 0.1 M NaCl solution contains 5.85 g of dissolved sodium chloride. In finding the molarity of a solution, do not make the common error of using liters of solvent rather than liters of solution.

Molarity provides the connection between the molar amount of a substance in solution and the volume of the solution.

$$\underset{\cancel{liters}}{\text{(Volume of solution)}}\underset{\frac{moles}{\cancel{liter}}}{\text{(molarity)}} = \underset{moles}{\text{no. of moles of solute}} \quad \textbf{(4.18)}$$

$$\underset{\cancel{moles}}{\text{(No. of moles of solute)}}\underset{\frac{liters}{\cancel{mole}}}{\left(\frac{1}{\text{molarity}}\right)} = \underset{liters}{\text{volume of solution}} \quad \textbf{(4.19)}$$

EXAMPLE 4.14 Molarity

What is the concentration of a solution which contains 4.03 g of NaOH dissolved in sufficient water to give 500.0 mL of solution?

The number of moles of NaOH is

$$(4.03\ \cancel{\text{g NaOH}})\left(\frac{1\ \text{mol NaOH}}{40.00\ \cancel{\text{g NaOH}}}\right) = 0.101\ \text{mol NaOH}$$

The molarity of the solution is found by dividing the number of moles by the volume of the solution in liters.

$$\frac{(0.101\ \text{mol NaOH})}{(500.0\ \cancel{\text{mL}})(1\ \text{L}/1000\ \cancel{\text{mL}})} = \mathbf{0.202\ mol\ NaOH/L}$$

EXERCISE A solution contains 9.68 g of $Ca(NO_3)_2$ dissolved in 250.0 mL of solution. What is the molarity of this solution? *Answer* 0.236 M.

EXAMPLE 4.15 Molarity and Moles

A student transferred 25.00 mL of a 0.0839 M solution of hydrochloric acid solution from a stockroom container into a flask. How many moles of HCl were transferred?

This is a volume of solution → moles of solute problem. The conversion requires multiplying the volume of the solution by the concentration of the solution:

$$(25.00\ \cancel{\text{mL}})\left(\frac{1\ \cancel{\text{L}}}{1000\ \cancel{\text{mL}}}\right)\left(\frac{0.0839\ \text{mol HCl}}{1\ \cancel{\text{L}}}\right) = \mathbf{0.00210\ mol\ HCl}$$

EXERCISE How many moles of NaOH are present in 10.5 mL of a 6.0 M solution of NaOH? *Answer* 0.063 mol NaOH.

EXAMPLE 4.16 Molarity and Moles

An experiment called for the addition of 1.50 mol of NaOH in the form of a dilute solution. The only sodium hydroxide solution that could be found in the laboratory was labeled "0.1035 M NaOH." What volume of this solution would be required for the 1.50 mol of NaOH?

A moles → volume of solution conversion is required and the key to this conversion is multiplying the number of moles by liters per mole:

$$(1.50\ \text{mol}\ \cancel{\text{NaOH}})\left(\frac{1\ \text{L}}{0.1035\ \text{mol}\ \cancel{\text{NaOH}}}\right) = \mathbf{14.5\ L}$$

EXERCISE What volume of 6.0 M HCl is needed to obtain 3.0 mmol of HCl? *Answer* 0.50 mL.

4.9 COMPOSITION OF A CHEMICAL COMPOUND

The composition of a chemical compound can be stated in terms of atoms, moles, atomic mass units, or grams (Table 4.9). All of this information is known once the correct chemical formula of a compound is known.

A chemical formula also provides the information necessary to calculate the **percentage composition** of a chemical compound—the percentage by mass of each element present in the compound. It is most convenient to do this by calculating the

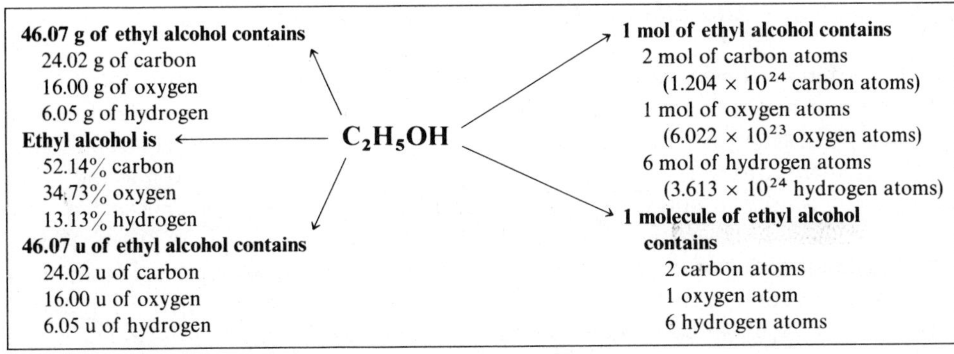

Table 4.9
Information from a Chemical Formula
The information given here is derived from the formula of ethyl alcohol, C_2H_5OH. (The percentage composition is by mass.)

percentage of each element present in one mole of the compound, as illustrated in Example 4.17.

The percentage composition of a compound for which the chemical formula is not known must be determined experimentally. The mass of each element present in a sample of the compound of known mass is determined by carrying out chemical reactions in which the masses of the reactants and products are measured, as illustrated by Example 4.18.

EXAMPLE 4.17 Percentage Composition from Formula

What is the percentage composition of sulfur dichloride, SCl_2?

The molar mass of sulfur dichloride is 102.96 g and one mole of SCl_2 contains one mole, or 32.06 g, of sulfur atoms and two moles, or 70.90 g, of chlorine atoms. The percentage composition is found as follows:

$$\%\,S = \frac{32.06\text{ g S}}{102.96\text{ g SCl}_2}(100) = \mathbf{31.14\%} \qquad \%\,Cl = \frac{70.90\text{ g Cl}}{102.96\text{ g SCl}_2}(100) = \mathbf{68.86\%}$$

The percentage composition of sulfur dichloride is 31.14% sulfur and 68.86% chlorine.

EXERCISE What is the percentage composition of DDT, $(ClC_6H_4)_2CHCCl_3$?

Answer 47.42% C, 2.56% H, 50.01% Cl.

EXAMPLE 4.18 Percentage Composition—Experimental Data

A 3.91 g sample of potassium metal when burned in oxygen formed a compound weighing 7.11 g and containing only potassium and oxygen. What is the percentage composition of this compound?

The percentage composition is the percent by mass of each element in the compound. Thus

$$\%\,K = \frac{(3.91\text{ g K})}{(7.11\text{ g compound})}(100) = \mathbf{55.0\%}$$

$$7.11\text{ g compound} - 3.91\text{ g K} = 3.20\text{ g O}$$

$$\%\,O = \frac{(3.20\text{ g O})}{(7.11\text{ g compound})}(100) = \mathbf{45.0\%}$$

The percentage composition of the compound is 55.0% potassium and 45.0% oxygen.

EXERCISE A 13.73 g sample of a compound of phosphorus and chlorine contains 3.10 g P. What is the percentage composition of this compound?

Answer 22.6% by mass of P, 77.42% by mass of Cl.

4.10 SIMPLEST AND EMPIRICAL FORMULAS

Determining the correct formula of a chemical compound newly prepared in the laboratory or newly discovered in nature is an essential part of its identification. The first step in the study of such a compound is frequently the experimental determination of the elements present and the percentage composition of the compound. This information is used to find the simplest, or empirical, formula of the compound.

The **simplest formula** of a compound gives the simplest whole-number ratio of atoms in the compound. For example, the simplest formula of diborane, a compound of boron and hydrogen, is BH_3. The simplest formula of ammonia is NH_3. The actual formula of a chemical compound is either the same as the simplest formula or is a multiple of the simplest formula (Section 4.11). The actual formula of diborane is B_2H_6, which has twice the molar mass of BH_3, but, of course, the same percentage composition.

For an ionic compound, the simplest formula is usually the only one that is needed. Recall our examples of ionic compounds, NaCl and $BaCl_2$. These are simplest formulas that show the smallest whole-number ratios of the ions present.

An experimentally determined simplest formula is called an **empirical formula**, implying that the correct formula is not yet known. (An **empirical relationship** is a relationship based solely on experimental facts or derived without the use of any theory or explanation of the facts.)

To find an empirical formula from experimentally determined masses of the elements present requires (1) converting the mass of each element to the number of moles, (2) finding the mole ratios of the atoms present, and (3) finding the simplest whole number ratios of the atoms present.

EXAMPLE 4.19 Empirical Formula

A 1.19 g sample of an oxide contains 0.52 g of phosphorus and 0.67 g of oxygen. What is the empirical formula of this oxide?

First, it is necessary to determine from the experimental data the number of moles of atoms of each element present.

$$(0.52 \text{ g P})\left(\frac{1 \text{ mol P}}{30.97 \text{ g P}}\right) = 0.017 \text{ mol P} \qquad (0.67 \text{ g O})\left(\frac{1 \text{ mol O}}{16.00 \text{ g O}}\right) = 0.042 \text{ mol O}$$

Because for a given compound the ratios of numbers of atoms and numbers of moles are the same, we could write an empirical formula of $P_{0.017}O_{0.042}$. However, this is unrealistic since whole atoms combine, not fractions of atoms. To proceed to a more reasonable empirical formula, the simplest ratio of moles is found by dividing the number of moles of each element by the number of moles of the element present in the smallest amount:

$$\frac{0.042 \text{ mol O}}{0.017 \text{ mol P}} = 2.5 \qquad \frac{0.017 \text{ mol P}}{0.017 \text{ mol P}} = 1.0$$

Thus for every mole of P atoms, there are 2.5 moles of O atoms in the compound. This means that for each P atom in the compound there are 2.5 O atoms. The empirical formula could be written as $PO_{2.5}$. However, fractional numbers of atoms are usually avoided in formulas. The relative numbers are doubled, and the empirical formula becomes $\mathbf{P_2O_5}$.

EXERCISE A 13.74 g sample of a compound contains 3.10 g of P and 10.64 g of Cl. What is the empirical formula of the compound? *Answer* PCl_3.

If the experimentally determined composition of a compound is obtained as the mass percentage of each element present, it is convenient to take 100 g of the compound as the basis for calculation. Each mass percentage is then taken as the mass in grams of an element. (This is a convenient way to deal with percentages in any dimensional calculation.)

EXAMPLE 4.20 Empirical Formula

The mineral cryolite contains 33% by mass of Na, 13% by mass of Al, and 54% by mass of F. Determine the empirical formula of the compound.

Choose exactly 100 g of cryolite as a basis to solve the problem. The 100 g of cryolite contains 33 g of Na, 13 g of Al, and 54 g of F. The entries in a table summarizing the calculations are as follows (*n* represents numbers of moles):

	Na	**Al**	**F**
No. of moles	$\frac{33\text{ g}}{22.99\text{ g/mol}} = 1.4\text{ mol}$	$\frac{13\text{ g}}{26.98\text{ g/mol}} = 0.48\text{ mol}$	$\frac{54\text{ g}}{19.00\text{ g/mol}} = 2.8\text{ mol}$
Mole ratio, n/n_{Al}	$\frac{1.4}{0.48} = 2.9$	$\frac{0.48}{0.48} = 1.0$	$\frac{2.8}{0.48} = 5.8$
Relative no. of atoms	3	1	6

The empirical formula of cryolite is $\mathbf{Na_3AlF_6}$.

EXERCISE Sodium thiosulfate contains 29.1% by mass of Na, 40.6% by mass of S, and 30.4% by mass of O. Determine the empirical formula for this compound.

Answer $Na_2S_2O_3$.

4.11 MOLECULAR FORMULAS

For molecular compounds, it is important to know how many atoms are present in each molecule. The **molecular formula** of a compound represents the actual number of atoms of each element that are combined in each molecule. To find the molecular formula from the empirical formula, the molecular or molar mass of the compound in question must be determined. This mass will be equal or nearly equal to some multiple of the mass calculated from the empirical formula. For example, for diborane, which has an empirical formula of BH_3, the molar mass is two times the empirical formula mass and therefore the molecular formula of the compound is B_2H_6.

EXAMPLE 4.21 Molecular Formula

The empirical formula for a substance was determined to be CH. The approximate molar mass of the substance was experimentally found to be 79 g. What is the molecular formula of this molecular compound? What is the exact molar mass?

The molar mass of the compound based on its empirical formula is 12.01 g + 1.01 g = 13.02 g. Dividing the experimental molar mass by the empirical formula molar mass

$$\frac{79\ \cancel{\text{g/mol}}}{13.02\ \cancel{\text{g/mol}}} = 6.1$$

shows that the empirical formula should be multiplied by six to obtain the molecular formula, $\mathbf{C_6H_6}$. The exact molar mass is

$$(13.02\text{ g/mol})(6) = \mathbf{78.12\text{ g/mol}}$$

EXERCISE The empirical formula of enneaborane was determined to be B_3H_5. The approximate molar mass of the compound is 115 g. What is the molecular formula of this compound? *Answer* B_9H_{15}.

Aside 4.1 THOUGHTS ON CHEMISTRY: A PRESIDENT'S STATEMENT

The following comments to students, made in 1982 when Robert W. Parry was President of the American Chemical Society, are no less relevant today.

Is chemistry really important to our future as a nation or is it a source of our problems? Why should anyone study chemistry? ... These are all fair and common questions demanding a reasoned answer. Here is the way I see it.

As we all know, the population of the world continues to increase while the resources of planet Earth remain constant. Concentrated mineral ores are being scattered. The entropy of the system climbs. Clearly our hopes for a better life must involve better and more enlightened use of our resources. We must get needed materials from lower and lower grade raw materials. Our processes must get more and more efficient. Undesirable waste materials must be returned in suitable form to the use-cycle. Knowledge is the key to our hopes and dreams—knowledge about science and about the immutable natual laws that govern the world in which we live.

Chemistry is central to our quest for this scientific knowledge. It looks toward physics and mathematics on one hand and toward biology and medicine on the other. Almost any serious effort in science ... must involve chemistry at some level. For example, the new field of molecular biology, with its promise in medicine, involves the chemistry of giant molecules that are taken apart and reconstructed to achieve a given goal. New techniques are being used, but standard chemical principles of structure and dynamics underlie the entire operation. Looking toward physics, the new field of optical communication is in commercial development today because of spectacular advances in the preparation of ultra-clear glass. Chemistry is truly the central science that offers fantastic opportunities for the future.

Those of us who have elected to study and practice chemistry have both a responsibility and an opportunity to serve mankind in a crucial period of the world's technological history. The benefits to humanity are large; failure is unthinkable.

Source: Robert W. Parry, President, American Chemical Society 1982, "Some Thoughts on Chemistry and the American Chemical Society," *The Philter*, Student Affiliate Newsletter, American Chemical Society, 14, No. 3 (Spring, 1982).

SUMMARY OF SECTIONS 4.7–4.11

Avogadro's number (6.022×10^{23}) is equal to the number of atoms in exactly 12 g of carbon-12; this number provides the connection between the numbers of atoms, molecules, or ions and the weighable masses of substances. A mole represents a number of anything equal to Avogadro's number, and the molar mass is the mass of one mole. The molar mass of an element or a chemical compound is a mass in grams numerically equal to the atomic mass or molecular mass in atomic mass units, respectively. The concentrations of solutions are often given as the number of moles of solute per liter of solution, known as the molarity of the solution.

The percentage by mass of each element in a chemical compound is its percentage composition. The simplest formula of a compound gives the simplest whole-number ratio of the atoms it contains. An experimentally determined simplest formula, called an empirical formula, can be determined from the percentage composition and the molar masses of the elements present. The molecular formula of a compound represents the actual number of atoms of each element present in a molecule of that compound. The molecular formula can be determined from the empirical formula by finding the molar mass, which is usually some whole-number multiple of the mass calculated from the empirical formula.

SIGNIFICANT TERMS (with section references)

anions (4.1)
Avogadro's number (4.7)
binary compound (4.4)
catalyst (4.5)
cations (4.1)
chemical equation (4.5)
chemical formula (4.2)
chemical nomenclature (4.4)
concentration (4.8)
diatomic molecules (4.1)
dissociation of an ionic compound (4.5)
empirical formula (4.10)

empirical relationship (4.10)
formula unit (4.2)
ionization (4.5)
molar mass (4.7)
molarity (4.8)
mole (4.7)
molecular formula (4.11)
molecular mass (4.3)
molecule (4.1)
monatomic ions (4.1)
percentage composition (4.9)
polyatomic ions (4.2)
polyatomic molecule (4.1)
polymers (4.1)
products (4.5)
reactants (4.5)
salts (4.4)
simplest formula (4.10)

QUESTIONS AND PROBLEMS

Molecular and Ionic Compounds

4.1 Classify each of the following species as (i) atomic, (ii) molecular, or (iii) ionic: (a) SO_4^{2-}, (b) S_8, (c) Na^+Cl^-, (d) CO_2, (e) NH_4^+, (f) Fe.

4.2 Repeat Question 4.1 for (a) Cl, (b) Cl_2, (c) Fe^{3+}, (d) Pt, (e) Ca^{2+} (CO_3^{2-}).

4.3 Classify each of the following substances as (i) monatomic, (ii) diatomic, or (iii) polyatomic: (a) CO, (b) CO_2, (c) Ne, (d) Cl_2, (e) PCl_5, (f) N_2O_4.

4.4 Repeat Question 4.3 for (a) H_2O_2, (b) SF_6, (c) He, (d) P_4, (e) ICl, (f) C_6H_5OH.

4.5 Classify each of the following ions as (i) a cation or (ii) an anion: (a) Sn^{2+}, (b) N^{3-}, (c) Cu^{2+}, (d) I^-, (e) K^+, (f) Fe^{2+}.

4.6 Repeat Question 4.5 for (a) SO_4^{2-}, (b) H_3O^+, (c) Ca^{2+}, (d) F^-, (e) NH_4^+, (f) NO_3^-.

4.7 Write the chemical formula for the ionic compound formed between (a) Ca^{2+} and ClO_3^-, (b) Al^{3+} and SO_4^{2-}, (c) Au^{3+} and Br^-, (d) NH_4^+ and CN^-, (e) K^+ and PO_4^{3-}, (f) Na^+ and $Cr_2O_7^{2-}$, (g) Ca^{2+} and MnO_4^-.

4.8 Repeat Question 4.7 for (a) Cu^{2+} and SO_4^{2-}, (b) Ba^{2+} and O_2^{2-}, (c) K^+ and CrO_4^{2-}, (d) Cs^+ and OH^-, (e) La^{3+} and CO_3^{2-}, (f) NH_4^+ and SO_4^{2-}, (g) Zn^{2+} and N^{3-}.

4.9 Calculate the molecular mass for each of the following: (a) Cl_2, (b) Fe, (c) Fe^{3+}, (d) $C_{12}H_{22}O_{11}$, (e) $KClO_3$, (f) $CoWO_4$, (g) $Pt_2(CO)_3Cl_4$, (h) $Cu_2(CO_3)(OH)_2$, (i) H_3PO_4, (j) $(NH_4)_3AsO_4$.

4.10 Repeat Problem 4.9 for (a) H_2SO_4, (b) $Ca(OH)_2$, (c) $Ca_3(PO_4)_2$, (d) $(NH_4)_2SO_4$, (e) $Ba(OH)_2$, (f) $Cr_2(SO_4)_3$, (g) Fe_3O_4, (h) $UO_2(SO_4)_2$, (i) Hg_2Br_2.

Naming Chemical Compounds

4.11 Name the following monatomic cations: (a) Li^+, (b) Cd^{2+}, (c) Fe^{2+}, (d) Mn^{2+}, (e) Al^{3+}. Use the Stock system of nomenclature.

4.12 Repeat Question 4.11 for (a) Au^+, (b) Au^{3+}, (c) Ba^{2+}, (d) Sn^{2+}, (e) Ag^+.

4.13 Write the chemical symbol for each of the following: (a) sodium ion, (b) zinc ion, (c) silver ion, (d) mercury(II) ion, (e) iron(III) ion.

4.14 Repeat Question 4.13 for (a) lithium ion, (b) bismuth(III) ion, (c) iron(II) ion, (d) chromium(III) ion, (e) potassium ion.

4.15 Name the following ions: (a) N^{3-}, (b) O^{2-}, (c) Se^{2-}, (d) F^-, (e) Br^-.

4.16 Write the chemical symbol for each of the following: (a) nitride ion, (b) sulfide ion, (c) telluride ion, (d) chloride ion, (e) iodide ion.

4.17 Name the following as ionic compounds: (a) Li_2S, (b) SnO_2, (c) RbI, (d) Li_2O, (e) UO_2, (f) Ba_3N_2, (g) NaF.

4.18 Repeat Question 4.17 for (a) NaI, (b) Hg_2S, (c) Li_3N, (d) $MnCl_2$, (e) $ZrBr_4$, (f) $AlCl_3$, (g) CuF_2, (h) FeO.

4.19 Write the chemical formula for each of the following compounds: (a) sodium fluoride, (b) zinc oxide, (c) barium peroxide, (d) magnesium bromide, (e) hydrogen iodide, (f) copper(I) chloride, (g) potassium iodide.

4.20 Repeat Question 4.19 for (a) sodium peroxide, (b) calcium phosphide, (c) iron(II) oxide, (d) silver fluoride, (e) manganese(IV) oxide, (f) iron(III) oxide.

4.21 Name the following salts containing polyatomic anions: (a) $(NH_4)_2SO_4$, (b) $K_2Cr_2O_7$, (c) $Fe(ClO_4)_2$, (d) $CaCO_3$, (e) $NaNO_2$, (f) K_2CrO_4, (g) Na_2SO_3.

4.22 Repeat Question 4.21 for (a) NH_4CN, (b) $Al(NO_3)_3$, (c) $Ca_3(PO_4)_2$, (d) Li_2CO_3, (e) BaO_2, (f) $FeCO_3$, (g) $Fe_2(SO_4)_3$.

4.23 Write the chemical formula for each of the following salts containing polyatomic anions: (a) potassium sulfite, (b) calcium permanganate, (c) barium phosphate, (d) copper(I) sulfate, (e) ammonium acetate, (f) silver nitrate, (g) uranium(IV) sulfate.

4.24 Repeat Question 4.23 for (a) iron(II) perchlorate, (b) potassium nitrite, (c) sodium peroxide, (d) ammonium dichromate, (e) sodium carbonate, (f) aluminum acetate, (g) manganese(II) phosphate.

4.25 Name the following common acids: (a) HCl, (b) H_3PO_4, (c) $HClO_4$, (d) HNO_3, (e) H_2SO_3, (f) H_3PO_3.

4.26 Write the chemical formula for each acid or base: (a) nitrous acid, (b) sulfuric acid, (c) carbonic acid, (d) sodium hydroxide, (e) potassium hydroxide.

4.27 What is the name of the acid with the formula H_2CO_3? Write the formulas of the two anions derived from it and name these ions.

4.28 What is the name of the acid with the formula H_3PO_3? What is the name of the HPO_3^{2-} ion?

4.29 Name the following binary molecular compounds: (a) CO, (b) CO_2, (c) SF_6, (d) $SiCl_4$, (e) IF.

4.30 Repeat Question 4.29 for (a) AsF_3, (b) Br_2O, (c) BrO_2, (d) CSe_2, (e) Cl_2O_7.

4.31 Write the chemical formula for each of the following compounds: (a) diboron trioxide, (b) silicon dioxide, (c) phosphorus trichloride, (d) sulfur tetrachloride, (e) bromine trifluoride, (f) hydrogen telluride, (g) diphosphorus trioxide.

4.32 Repeat Question 4.31 for (a) iodine monobromide, (b) dinitrogen pentasulfide, (c) phosphorus triiodide, (d) silicon monosulfide, (e) tetrasulfur dinitride, (f) diiodine pentoxide, (g) xenon tetrafluoride, (h) hydrogen sulfide.

Chemical Equations

4.33 The chemical equation describing the decomposition of dinitrogen monoxide is

$$2N_2O(g) \xrightarrow{\Delta} 2N_2(g) + O_2(g)$$

(a) What is the reactant? (b) What are the products? (c) What are the physical states of the substances involved in the reaction? (d) What does the "Δ" represent?

4.34 The chemical equation describing the reaction of aluminum with iodine is

$$2Al(s) + 3I_2(s) \xrightarrow{1\text{ mL } H_2O} 2AlI_3(s)$$

(a) What are the reactants? (b) What is the product? (c) What are the special conditions of the experiment?

4.35 For each of the following chemical equations, write a sentence that describes the chemical reaction:

(a) $N_2(g) + 3H_2(g) \xrightarrow{400\,°C,\,250\,atm,\,FeO} 2NH_3(g)$

(b) $2CO(g) + O_2(g) \xrightarrow{\Delta} 2CO_2(g)$

(c) $SiO_2(s) + 2C(s) \xrightarrow{3000\,°C} Si(l) + 2CO(g)$

4.36 Repeat Question 4.35 for

(a) $SiI_4(s) + 2H_2O(l) \xrightarrow{H_2O} SiO_2(s) + 4HI(aq)$
(b) $2H_3AsO_3(aq) + 3H_2S(g) \longrightarrow As_2S_3(s) + 6H_2O(l)$
(c) $(NH_4)_2Cr_2O_7(s) \xrightarrow{\Delta} N_2(g) + Cr_2O_3(s) + 4H_2O(g)$

4.37 Balance each of the following chemical equations:
(a) $Cl_2O_7(g) + H_2O(l) \longrightarrow HClO_4(aq)$
(b) $Br_2(l) + H_2O(l) \longrightarrow HBr(aq) + HBrO(aq)$
(c) $Ca_3(PO_4)_2(s) + H_2SO_4(aq) \longrightarrow CaSO_4(s) + H_3PO_4(aq)$
(d) $V_2O_4(s) + HClO_4(aq) \longrightarrow VO(ClO_4)_2(aq) + H_2O(l)$
(e) $Fe_2O_3(s) + H_2(g) \longrightarrow Fe(s) + H_2O(l)$
(f) solid potassium reacts with water to give aqueous potassium hydroxide and gaseous hydrogen
(g) upon heating, solid magnesium carbonate decomposes to form solid magnesium oxide and gaseous carbon dioxide
(h) solid aluminum sulfide reacts with liquid water to give solid aluminum hydroxide and gaseous hydrogen sulfide
(i) solid barium peroxide reacts with aqueous sulfuric acid to give aqueous hydrogen peroxide and solid barium sulfate

4.38 Repeat Question 4.37 for
(a) $Fe_3O_4(s) + H_2(g) \longrightarrow Fe(s) + H_2O(l)$
(b) $KClO_3(s) \longrightarrow KCl(s) + O_2(g)$
(c) $H_2O_2(aq) \longrightarrow H_2O(l) + O_2(g)$
(d) $Cl_2(aq) + KI(aq) \longrightarrow KCl(aq) + I_2(aq)$
(e) steam and hot carbon react to form gaseous hydrogen and gaseous carbon monoxide
(f) liquid phosphorus tribromide reacts with liquid water in the presence of excess water to produce aqueous phosphorous acid and aqueous hydrogen bromide
(g) gaseous ozone reacts with gaseous nitrogen monoxide to produce gaseous nitrogen dioxide and gaseous oxygen

Atomic, Molecular, and Molar Mass Relationships

4.39 Calculate the number of moles equivalent to each of the following: (a) 9.5×10^{21} atoms of Cs, (b) 4.7×10^{27} molecules of CO_2, (c) 1.63×10^{23} formula units of $BaCl_2$, (d) 1.2×10^{22} atoms of Cu.

4.40 Repeat Problem 4.39 for (a) 5.5×10^{16} atoms of Fe, (b) 1.5×10^{24} atoms of Tc, (c) 3.92×10^{18} molecules of CH_4, (d) 4.61×10^{25} molecules of O_3, (e) 4.6×10^{25} formula units of $Fe(NO_3)_3$.

4.41 A sample contains 13.4 mol of a substance. (a) How many atoms are present in the sample if it is gallium, Ga? (b) How many molecules are present in the sample if it is toluene, $C_6H_5CH_3$? (c) How many formula units are present in the sample if it is silver nitrate, $AgNO_3$?

4.42 You are given a sample containing 0.37 mol of a substance. (a) How many atoms are present in the sample if it is uranium metal, U? (b) How many molecules are present in the sample if it is acetylene, C_2H_2? (c) How many formula units are present in the sample if it is silver chloride, AgCl?

4.43 Calculate the molar mass for each of the following: (a) NO_2, (b) $Ba(OH)_2$, (c) XeF_6, (d) Mn^{2+}, (e) $H_2PO_4^-$, (f) N_2, (g) $KAuI_4$, (h) $C_6H_5N_3O_4$, (i) $Cu(IO_3)_2$.

4.44 Calculate the molar mass for each of the following: (a) H_3PO_4, (b) $(NH_4)_3AsO_4$, (c) $UO_2(SO_4)$, (d) $HgBr_2$.

4.45 What mass corresponds to (a) 5.3 mol of C, (b) 0.1273 mol of N_2O_5, (c) 1.3 μmol of $AmBr_3$, (d) 1×10^{-10} mol of HCl?

4.46 What mass corresponds to (a) 0.50 mol of phenol, C_6H_5OH; (b) 1.01 mol of quartz, SiO_2; (c) 3 mol of quicksilver, Hg; (d) 0.42 mol of saccharin, $C_6H_4(CO)(SO_2)NH$; (e) 0.25 mol of saltpeter, KNO_3?

4.47 Calculate the number of moles equivalent to each of the following samples: (a) 7.9 mg of Tc, (b) 16.8 g of NH_3, (c) 3.25 kg of NH_4Br, (d) 5.6 g of PCl_5.

4.48 Repeat Problem 4.47 for (a) 10.03 g of calcium carbonate, $CaCO_3$; (b) 14 g of iron, Fe; (c) 24.5 g of formaldehyde, H_2CO; (d) 33.5 g of acetic acid, CH_3COOH.

4.49 How many ozone molecules and how many oxygen atoms are contained in 48.00 g of ozone, O_3?

4.50 How many nitrogen dioxide molecules are contained in 46.01 g of NO_2? How many nitrogen atoms? How many oxygen atoms? What is the total number of atoms contained in the sample?

4.51 How many formula units are contained in 222.99 g of AuCN? How many gold(I) ions? How many cyanide ions? How many atoms are contained in the sample?

4.52 How many formula units are contained in 238.1 g of K_2MoO_4? How many potassium ions? How many MoO_4^{2-} ions? How many atoms are contained in the sample?

Atomic, Molecular, and Molar Mass Relationships—Molarity

4.53 Solutions containing (a) 10.0 g of Na_2SO_4, (b) 56 g of $CaCl_2$, and (c) 42.6 g of $Al(NO_3)_3$, each dissolved in sufficient water to make a total volume of 1.00 liter of solution were prepared. What are the concentrations expressed in molarity that should be written on the respective bottles?

4.54 These solutions containing (a) 5.2 g of H_2O_2 in 100.0 mL of solution, (b) 5.26 g of KOH in 250.0 mL of solution, and (c) 15.26 g of NaOH in 500.0 mL of solution were prepared. What concentrations expressed in molarity should be written on their respective bottles?

4.55 What mass of solid sodium hydroxide is needed to prepare 250.0 mL of a 0.100 M NaOH solution?

4.56 What masses of solutes are needed to prepare 250 mL of 1.0 M solutions of each of the following: (a) KCl, (b) $NiCl_2$, (c) $FeSO_4$?

4.57* How many moles of acid are contained in 108 mL of 0.62 M solution? If we add enough water to make 0.300 L of acid solution, how many moles of acid will the resulting solution contain? What is the molarity of the final solution?

4.58* What is the molarity of chloride ion in a solution prepared by dissolving 16.7 g of $CaCl_2$ in sufficient water to obtain 400. mL of solution? Note that 2 mol of Cl^- are released when 1 mol of $CaCl_2$ dissolves in water.

Atomic, Molecular, and Molar Mass Relationships—Percentage Composition

4.59 A 3.56 g sample of iron powder was heated in gaseous chlorine and 10.39 g of a dark substance presumed to be an iron chloride was formed. What is the percentage composition of this compound?

4.60 A 0.2360 g sample of a white compound was analyzed and found to contain 0.0944 g of Ca, 0.0283 g of C, and 0.1133 g of O. What is the percentage composition of this compound?

4.61 Calculate the percentage composition of (a) KClO, (b) $KClO_2$, (c) $KClO_3$, (d) $KClO_4$.

4.62 Calculate the percentage composition of (a) acetone, CH_3COCH_3; (b) corundum, Al_2O_3; (c) aspirin, $CH_3COOC_6H_4$-COOH; (d) beryl, $Be_3Al_2(SiO_3)_6$; (e) carborundum, SiC; (f) LSD, $C_{20}H_{25}N_3O$.

4.63 What mass of oxygen is contained in 5.5 g of $KClO_3$?

4.64 What mass of silver is contained in (a) 0.263 g of AgF, (b) 5.92 g of AgCl, (c) 136.9 g of AgBr, (d) 1.6 μg of AgI?

4.65* Copper is obtained from ores containing the following minerals: azurite, $Cu_3(CO_3)_2(OH)_2$; chalcocite, Cu_2S; chalcopyrite, $CuFeS_2$; covellite, CuS; cuprite, Cu_2O; and malachite, $Cu_2(CO_3)(OH)_2$. Which mineral has the highest copper content on a percentage basis?

4.66* A sample contained 50.0% NaCl and 50.0% KCl by mass. What is the mass percent of Cl^- in this sample? A second sample of NaCl and KCl contained 50.0% Cl^- by mass. What is the mass percent of NaCl in this sample?

Atomic, Molecular, and Molar Mass Relationships—Chemical Formulas

4.67 The hormone epinephrine is released in the human body during stress and increases the body's metabolic rate. Like many biochemical compounds, epinephrine is composed of carbon, hydrogen, oxygen, and nitrogen. The percentage composition of the hormone is 56.8% C, 6.56% H, 28.4% O, and 8.28% N. Determine the empirical formula.

4.68 Determine the empirical formula for (a) copper(II) tartrate, which contains 30.03% Cu, 22.70% C, 1.91% H, and 45.37% O, and (b) nitrosyl fluoroborate, which contains 11.99% N, 13.70% O, 9.25% B, and 65.06% F.

4.69 A compound containing 52.2% C, 13.0% H, and 34.8% O has an observed molar mass of 91.6 g. Determine (a) the empirical formula, (b) the molecular formula, (c) the exact molar mass of the compound.

4.70 A compound containing 12.3% N, 3.5% H, 28.0% S, and 56.1% O has an observed molar mass of 228 g. Determine (a) the empirical formula, (b) the molecular formula, (c) the exact molar mass.

4.71* A 2.00 g sample of compound gave 4.86 g of CO_2 and 2.03 g of H_2O upon combustion in oxygen. The compound is known to contain only C, H, and O. What is its empirical formula?

4.72* A 1.000 g sample of an alcohol was burned in oxygen and produced 1.913 g of CO_2 and 1.174 g of H_2O. The alcohol contained only C, H, and O. What is the empirical formula of the alcohol?

4.73* Determine the empirical formula from the given molecular formula for each of the following compounds: (a) octane, C_8H_{18}; (b) caffeine, $C_8H_{10}N_4O_2$; (c) dextrose, $C_6H_{12}O_6$; (d) ascorbic acid (vitamin C), $H_2C_6H_6O_6$.

4.74* Find the empirical formula of each of the following minerals: (a) talc (used for talcum powder, ceramics, and laundry tubs), which contains 19.23% Mg, 29.62% Si, 42.18% O, and 8.97% OH; and (b) borax (used for softening water and washing clothes), which contains 12.1% Na, 11.3% B, 29.4% O, and 47.3% H_2O. Note that the percentages of OH in (a) and H_2O in (b) are given separately from the percentages of O. This is the manner in which geological analyses are stated.

ADDITIONAL PROBLEMS

4.75 The molar mass of hemoglobin is about 65,000 g. A molecule of hemoglobin contains 0.35% Fe by mass. How many iron atoms are in a hemoglobin molecule?

4.76 Carbon tetrachloride, CCl_4, was formerly used as a dry cleaning fluid. What are the (a) molecular mass and (b) molar mass of this substance? (c) Calculate the mass of one molecule of CCl_4. How many (d) moles and (e) molecules of CCl_4 are contained in 17.93 g of the compound? How many (f) carbon atoms and (g) chlorine atoms are contained in the 17.93 g sample?

4.77 Test tubes containing 1.00 g samples of (a) lime, CaO; (b) slaked lime, $Ca(OH)_2$; (c) magnesium, Mg; (d) ethylene glycol, $HOCH_2CH_2OH$; and (e) water, H_2O, are placed in front of you. Which contains the largest number of moles? Which contains the largest number of atoms?

4.78* A highly purified sample of carbon tetrabromide, CBr_4, contains 96.379% bromine and 3.621% carbon by mass. Using the atomic mass of carbon as 12.011 u, find the exact atomic mass of bromine.

4.79* Calculate the atomic mass of a metal that forms an oxide having the empirical formula M_2O_3 and contains 68.4% of the metal by mass. Identify the metal.

4.80* A 10.0 g sample of an orange powder was piled on an asbestos sheet and ignited. A chemical reaction took place that produced a miniature volcano. As sparks flew, a fluffy green powder was formed, gaseous nitrogen was released, and steam was generated. (a) The orange powder contained 11.1% N, 3.2% H, 41.3% Cr, and 44.4% O; (b) the green powder contained 68.4% Cr and 31.6% O. Find the empirical formula for each of these compounds. (c) How many moles of orange powder were used? (d) How many atoms were contained in the sample of orange powder? (e) Write a chemical equation describing the reaction.

CHAPTER

5

Chemical Reactions and Stoichiometry

CHEMICAL REACTIONS

With the first few sections of this chapter, we begin the study of descriptive chemistry, wherein the properties of the elements and their compounds are examined. The study of chemical reactions is an essential aspect of descriptive chemistry. Which substances can react with each other? What kinds of chemical reactions are possible between given reactants? What products are likely to form? To answer such questions, one may turn to qualitative or to quantitative information. Commonly, the first approach is qualitative and is based on general knowledge of the properties of substances. In the early days of chemistry, qualitative information based on observation was all that was available. The second approach is based on whatever quantitative data and theoretical explanations apply to the expected type of reaction.

Here we begin a qualitative examination of how chemical reactions can be classified and the possible products predicted. Our next encounter with descriptive chemistry occurs in Chapters 9 and 10, where the properties of different types of elements are examined with respect to their positions in the periodic table and the types of chemical bonds they form.

5.1 SOME SIMPLE TYPES OF CHEMICAL REACTIONS

Four simple, general types of chemical reactions are introduced in the following sections. A surprising number of reactions fit into these four categories, and it is not too early in your study of chemistry to begin to examine chemical reactions for similarities and differences.

We suggest that you practice *reading* the chemical equations in the following sections. Unfortunately, it is all too easy to let your eyes quickly pass over a chemical equation without learning anything from it. Think about what the equations in the following sections mean in terms of the conservation of mass and note the physical states of the reactants and products. The equations given here can be written with confidence because they represent reactions that have been carried out many times with the same results. Never forget that *writing* a balanced equation does not necessarily mean that the reaction described will occur as indicated. Only experimentation can prove that an equation is correct.

a. Combination Reactions Sodium is a metallic element that is silvery and, unlike most metals, is very soft. It has such a high **chemical reactivity**—a tendency to undergo chemical reactions—that it cannot be stored in contact with the atmosphere, for it immediately reacts with oxygen and moisture in the air. Therefore, pure sodium is rarely encountered outside a laboratory. Chlorine is a yellow-green, poisonous gas with a suffocating odor. Like sodium, chlorine is a quite reactive element. It is often stored in tanks adjacent to water treatment plants, for it is used in purifying the water. When sodium and chlorine are brought together, they undergo a chemical reaction, sometimes with an explosive release of heat and light:

$$2Na(s) + Cl_2(g) \longrightarrow 2NaCl(s) \quad \textbf{(5.1)}$$

sodium / *silvery, soft metal* — *chlorine* / *yellow-green gas, nonmetal* — *sodium chloride* / *white, crystalline solid (table salt)*

This is an example of a **combination reaction**—the combination of two reactants to form a single product (Figure 5.1).

$$A + B \longrightarrow C$$

Many elements undergo combination reactions, and such reactions frequently occur between metals and nonmetals (Section 9.3). In the case of less chemically reactive elements, the reactions are not as spectacular and may not occur as readily. For example, aluminum, a silvery white metal, and sulfur, a brittle, yellow solid nonmetal, are less reactive than sodium and chlorine, respectively. However, when heated together they undergo a combination reaction:

$$2Al(s) + 3S(s) \xrightarrow{\Delta} Al_2S_3(s) \quad \textbf{(5.2)}$$

aluminum / *silvery white metal* — *sulfur* / *yellow solid, nonmetal* — *aluminum sulfide* / *white, crystalline solid*

Chemical compounds also take part in combination reactions. Whenever bottles of hydrochloric acid and ammonia in aqueous solution are placed near each other, the fumes escaping from the bottles combine in the air, forming a finely divided white powder that looks like smoke.

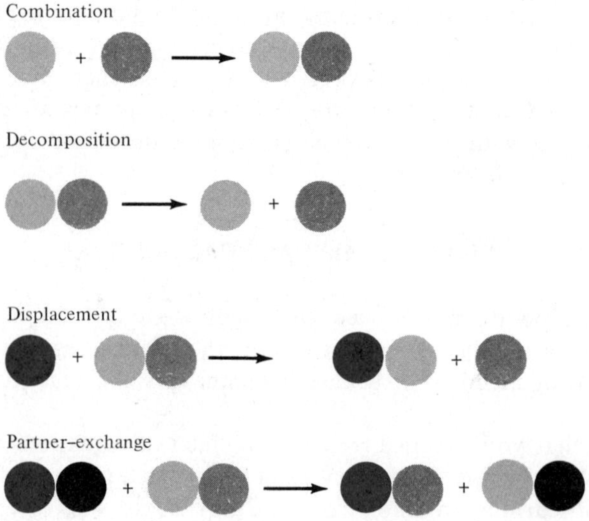

Figure 5.1
Four General Types of Chemical Reactions
The circles may represent atoms or groups of atoms.

■ **THE WHOLE EARTH** *above* The Earth as photographed by a NASA Applications Technology Satellite. The raw materials for all human activities must be drawn from the atmosphere, waters, and crust of the Earth.

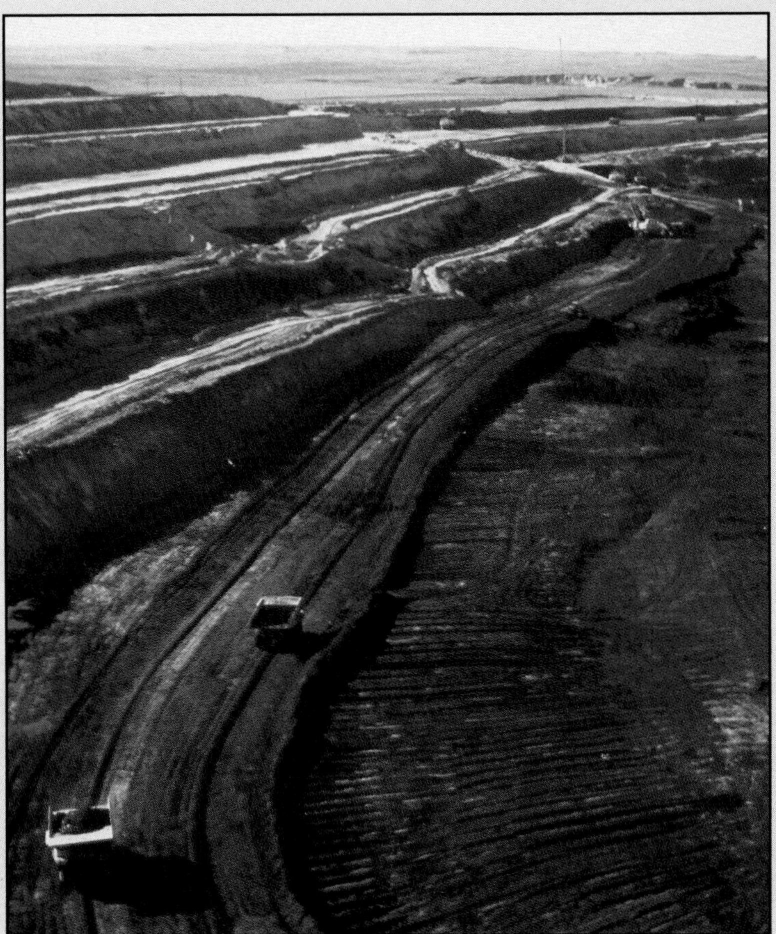

■ **SURFACE MINING** *above* The Belle Ayr surface coal mine in Wyoming's Powder River Basin. This mine, one of the largest in the United States, can process more than 6,000 tons of coal per hour.

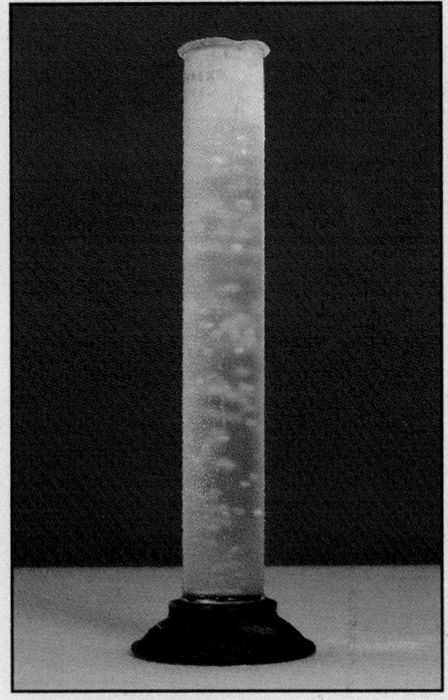

■ **GASES FROM THE ATMOSPHERE** *above* Liquid air, which is fractionally distilled to produce nitrogen, oxygen, and the noble gases neon, argon, krypton, and xenon.

■ **BRINE EVAPORATION** *right* Brine from the Great Salt Lake is pumped into giant ponds where solar evaporation increases the concentration of magnesium chloride naturally present in the brine. Magnesium is produced from the concentrated brine by electrolysis. Lithium, sodium, potassium, and calcium salts can also be obtained from brines.

■ **HEAP LEACHING** *below* Leaching — separating a metal or a metal compound from an ore by washing with water or water solution — is useful for ores that contain a low percentage of a metal. At the Sleeper Mine in Nevada, crushed ore containing native gold is piled on impermeable pads covering 2 million sq. ft. The ore is soaked with dilute cyanide (CN^-) solution, and the gold is washed out as the $[Au(CN)_2]^-$ ion in solution.

■ **UNDERGROUND MINING** *above* Molybdenum ore (primarily molybdenite, MoS_2) is loaded at Henderson mine in Empire, Colorado for transportation by a 9.6-mile underground railroad to a processing plant on other side of the Continental Divide.

■ **POURING REFINED GOLD** *right* Pure gold, produced by reduction of $[Au(CN)_2]^-$, is seen being poured into ingots at the Sleeper Mine.

■ **NATIVE GOLD** *below* Gold (Au) does not combine easily with other elements, and therefore can be found as the free element in the crust of the Earth. Gold is shown embedded in quartz.

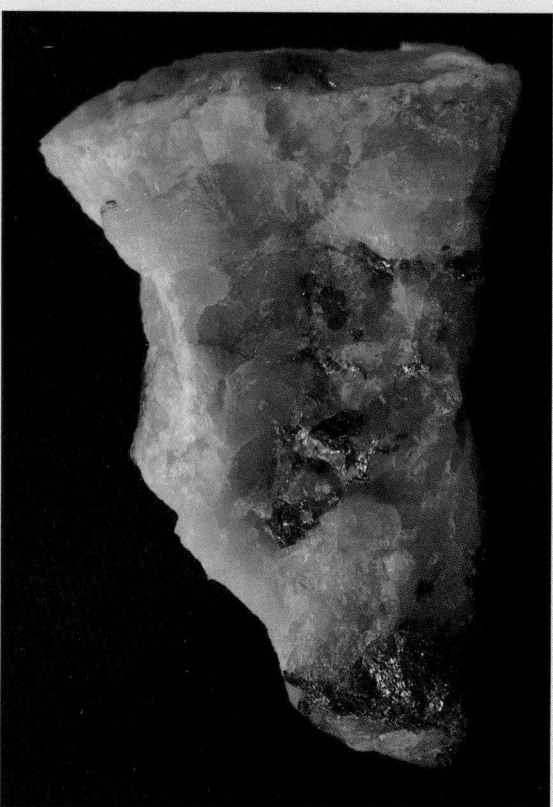

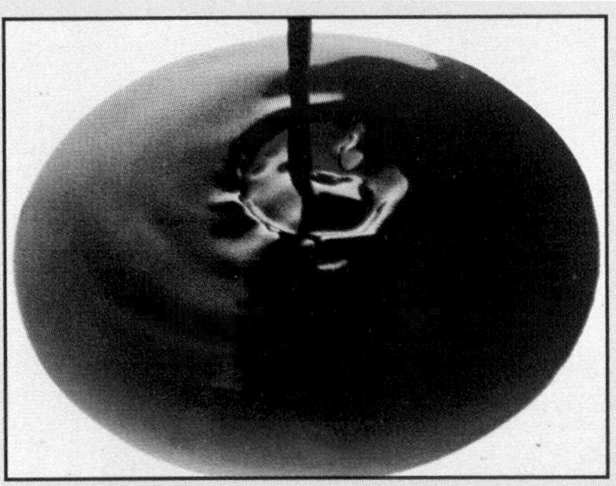

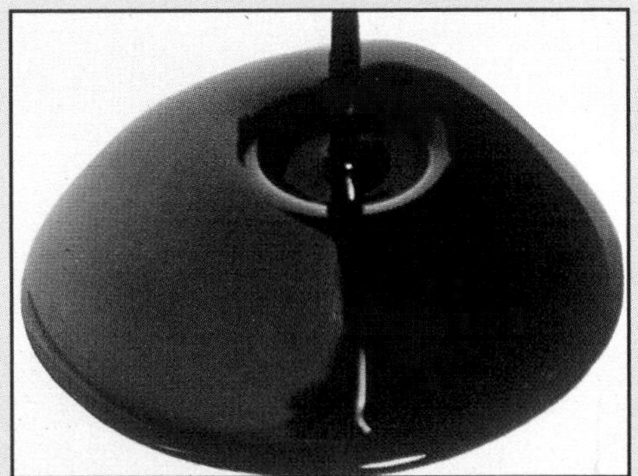

■ **CRUDE OIL** *left* As it is pumped from underground, crude oil (or petroleum) is a viscous, dark brown to black liquid of unpleasant odor. *left to right:* The samples shown are Arabian Light and Gulf Coast Empire Mix crude oil. Petroleum is a complex mixture of saturated and aromatic hydrocarbons with a small percentage of sulfur compounds and smaller amounts of nitrogen and oxygen compounds.

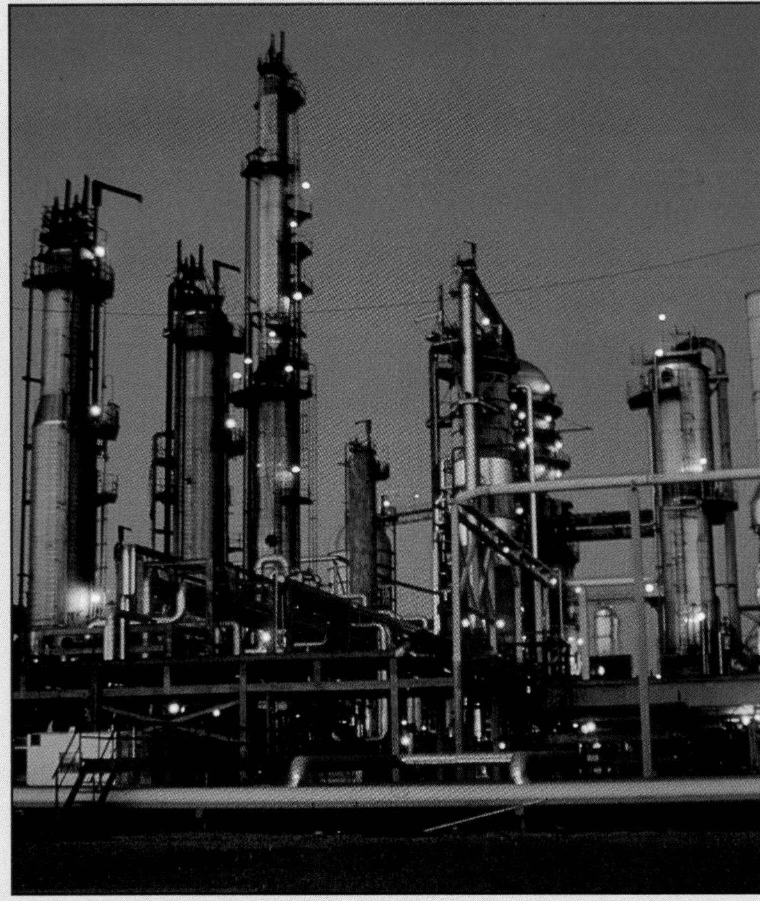

■ **CATALYTIC CRACKING** *left* Heavy oil distillate is converted by passage over catalysts to lighter oil containing the smaller molecules used in gasoline. The catalytic cracking unit shown is at a refinery in Texas City, Texas.

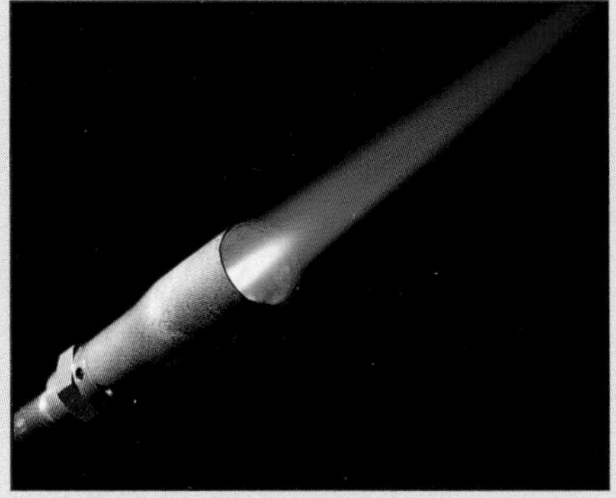

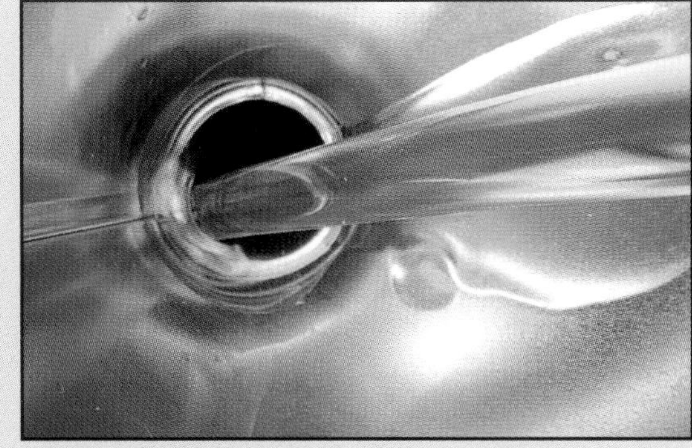

■ **SOME PETROLEUM PRODUCTS** *above* *from the left:* Propane (propane torch flame) is a three-carbon hydrocarbon. Mineral oil, petroleum jelly, and motor oil (shown being poured into a funnel) contain 16–20 carbon atoms per molecule and boil above 350 °C. Higher boiling, heavier petroleum products include waxes and asphalt.

A CLUSTER OF METAL ORES *below* The photo shows 21 minerals that are metal ores and 2 samples of native metals (silver and bismuth). The minerals and metals are identified in the sketch and list below.

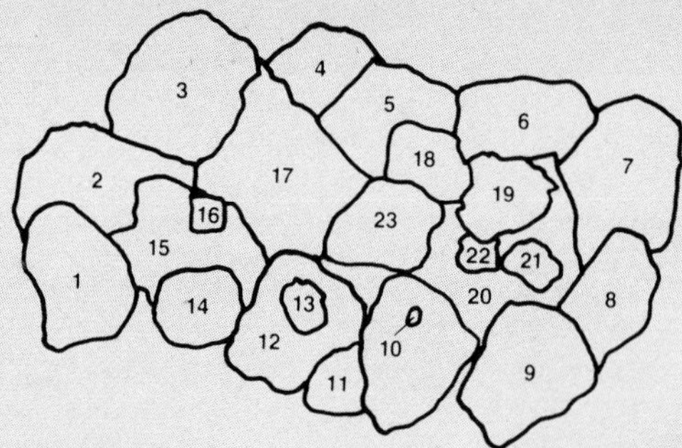

1. Bornite (iridescent) — COPPER
2. Dolomite (pink) — MAGNESIUM
3. Molybdenite (grey) — MOLYBDENUM
4. Skutterudite (grey) — COBALT, NICKEL
5. Zincite (mottled red) — ZINC
6. Chromite (grey) — CHROMIUM
7. Stibnite (*top right*, grey) — ANTIMONY
8. Gummite (yellow) — URANIUM
9. Cassiterite (rust, *bottom right*) — TIN
10. Vanadinite crystal on Goethite (red crystal) — VANADIUM
11. Cinnabar (red) — MERCURY
12. Galena (grey) — LEAD
13. Monazite (white) — RARE EARTHS: Cerium, Lanthium, Neodymium, Thorium
14. Bauzite (gold) — ALUMINUM
15. Strontianite (white, spiny) — STRONTIUM
16. Cobaltite (grey cube) — COBALT
17. Pyrite (gold) — IRON
18. Columbinite (tan, grey stripe) — NIOBIUM, TANTALUM
19. Native Bismuth (shiny)
20. Rhodochrosite (pink) — MAGNESIUM
21. Rutile (shiny twin crystal) — TITANIUM
22. Native Silver (filigree on quartz)
23. Pyrolusite (black, powdery) — MANGANESE

$$HCl(g) + NH_3(g) \longrightarrow NH_4Cl(s) \quad (5.3)$$

hydrogen chloride — *ammonia* — *ammonium chloride*

The result is a frosting of ammonium chloride that settles on everything nearby. Additional examples of combination and other types of reactions discussed below are given in Table 5.1.

Table 5.1
Some Examples of Chemical Reactions

Combination	
$C(s) + O_2(g) \longrightarrow CO_2(g)$ (*carbon dioxide*)	Occurs when coal is burned in an excess of air; large amounts of heat released
$2C(s) + O_2(g) \longrightarrow 2CO(g)$ (*carbon monoxide*)	CO, a poisonous gas, is formed when coal is burned in a limited amount of air
$N_2(g) + 3H_2(g) \longrightarrow 2NH_3(g)$ (*ammonia*)	Major industrial preparation of NH_3, which is used as a fertilizer and source of nitric acid, HNO_3
$S(s) + O_2(g) \longrightarrow SO_2(g)$ (*sulfur dioxide*)	Important reaction in the industrial production of H_2SO_4; reaction also occurs when coal and other sulfur-containing fossil fuels are burned; SO_2 is an atmospheric pollutant
$CaO(s) + H_2O(l) \longrightarrow Ca(OH)_2(s)$ (*calcium oxide (lime)* → *calcium hydroxide (slaked lime)*)	Preparation of "slaked lime," used in mortar and plaster
Decomposition	
$NH_4NO_3(s) \xrightarrow{\Delta} N_2O(g) + 2H_2O(g)$ (*ammonium nitrate* → *dinitrogen monoxide*)	N_2O is "laughing gas," the first synthetic anesthetic discovered
$CaCO_3(s) \xrightarrow{\Delta} CaO(s) + CO_2(g)$ (*calcium carbonate (limestone)* → *calcium oxide (lime)*)	Source of lime, which is used in the manufacture of mortar and plaster (see above)
$2Al_2O_3(s) \xrightarrow{\text{electric current}} 4Al(l) + 3O_2(g)$ (*aluminum oxide*)	A major step in the production of aluminum
$SiI_4(g) \xrightarrow{\Delta} Si(s) + 2I_2(g)$ (*silicon(IV) iodide* → *silicon*)	Reaction by which ultrapure silicon is obtained for the electronics industry
Displacement	
$Fe(s) + H_2SO_4(aq) \longrightarrow H_2(g) + FeSO_4(aq)$ (*iron(II) sulfate*)	First preparation of hydrogen, in 1671
$Na(g) + KCl(l) \xrightarrow{\Delta} K(g) + NaCl(l)$	Industrial preparation of potassium
Partner Exchange	
$BaCl_2(aq) + H_2SO_4(aq) \longrightarrow BaSO_4(s) + 2HCl(aq)$ (*barium chloride* → *barium sulfate*)	Used in analysis for sulfur
$AgNO_3(aq) + NaCl(aq) \longrightarrow AgCl(s) + NaNO_3(aq)$ (*silver nitrate* → *silver chloride*)	Used in analysis for chloride ion and to make AgCl for photography

b. Decomposition Reactions Mercury(II) oxide is a red crystalline solid—an oxide of mercury, which is a much less reactive metal than sodium or aluminum. Oxygen was discovered (in 1774) by heating mercury(II) oxide. Heat causes the oxide to decompose, as shown by the evolution of gaseous oxygen and the condensation of elemental mercury on the walls of the container.

$$\underset{\substack{\textit{mercury(II) oxide}\\ \textit{red powder}}}{2HgO(s)} \xrightarrow{\Delta} \underset{\substack{\textit{mercury}\\ \textit{silvery liquid}}}{2Hg(l)} + \underset{\substack{\textit{oxygen}\\ \textit{colorless gas}}}{O_2(g)} \qquad \textbf{(5.4)}$$

The production of oxygen and mercury from mercury(II) oxide is an example of a **decomposition reaction**—a reaction in which a single compound breaks down to give two or more other substances.

$$C \longrightarrow A + B$$

In most cases, energy, usually in the form of heat, must be provided to cause a compound to decompose. When the decomposition occurs as the result of heating, it is referred to as *thermal decomposition.*

Compounds that contain three or more elements can decompose to give several different products. For example, the white crystalline solid, sodium hydrogen carbonate—familiarly known as bicarbonate of soda or baking soda—decomposes by losing water and carbon dioxide, leaving behind another white crystalline substance, sodium carbonate.

$$\underset{\substack{\textit{sodium hydrogen carbonate}\\ \textit{white, crystalline solid}}}{2NaHCO_3(s)} \xrightarrow{\Delta} \underset{\substack{\textit{sodium carbonate}\\ \textit{white, crystalline solid}}}{Na_2CO_3(s)} + \underset{\textit{water}}{H_2O(g)} + \underset{\substack{\textit{carbon dioxide}\\ \textit{colorless gas}}}{CO_2(g)} \qquad \textbf{(5.5)}$$

c. Displacement Reactions: Activity Series of Metals Iron is a metal familiar to everyone as a major component of cast iron and steel. It is much less reactive than sodium, but more reactive than mercury. Dropping a bit of iron powder into a solution of dilute sulfuric acid results in the evolution of hydrogen gas; this is the reaction by which hydrogen was discovered (in 1671). The metallic iron is converted to Fe^{2+} ions and a pale-green aqueous solution of iron(II) sulfate is formed.

$$\underset{\textit{iron}}{Fe(s)} + \underset{\textit{sulfuric acid}}{H_2SO_4(aq)} \longrightarrow \underset{\substack{\textit{iron(II) sulfate}\\ \textit{pale green solution}}}{FeSO_4(aq)} + \underset{\textit{hydrogen}}{H_2(g)} \qquad \textbf{(5.6)}$$

Trying the same reaction with mercury would produce no evolution of gas and no visible evidence of change, for no chemical reaction would take place.

Reaction (5.6) is a **displacement reaction**—a reaction in which the atoms or ions of one substance take the place of other atoms or ions in a compound.

$$A + BC \longrightarrow AC + B$$

The iron in reaction (5.6) has displaced hydrogen from the acid. In all such displacement reactions between a metal and an acid, the products are a salt formed from the cation of the metal and the anion of the acid, plus hydrogen. Sodium is so active a metal that it not only displaces hydrogen from acids—sometimes with the generation of enough heat to cause the highly flammable hydrogen to burst into flame—

$$2Na(s) + \underset{\textit{hydrochloric acid}}{2HCl(aq)} \longrightarrow \underset{\textit{sodium chloride}}{2NaCl(aq)} + H_2(g) \qquad \textbf{(5.7)}$$

but it also displaces hydrogen from water (HOH).

$$2Na(s) + 2H_2O(l) \longrightarrow \underset{\substack{\textit{sodium hydroxide}\\ \textit{(a base)}}}{2NaOH(aq)} + H_2(g) \qquad \textbf{(5.8)}$$

The additional product of this reaction is an aqueous solution of sodium hydroxide, which is a strong base.

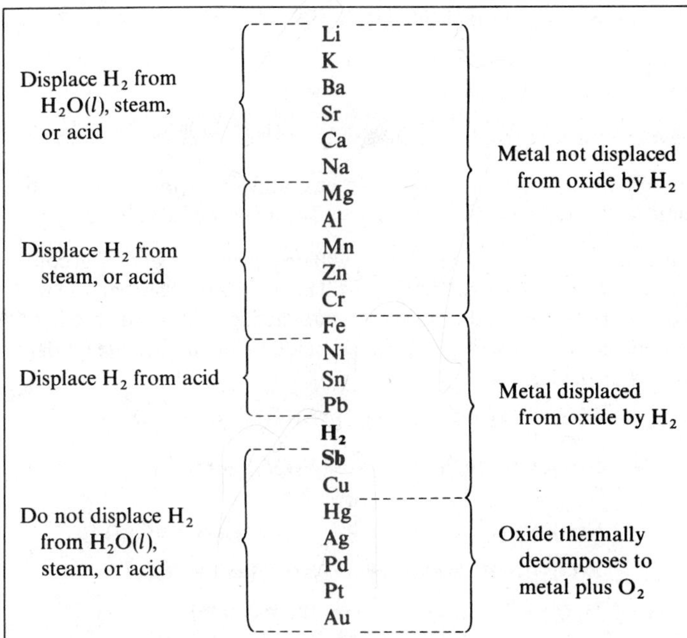

Hydrogen displacement	Metal	Oxide behavior
Displace H_2 from $H_2O(l)$, steam, or acid	Li	Metal not displaced from oxide by H_2
	K	
	Ba	
	Sr	
	Ca	
	Na	
Displace H_2 from steam, or acid	Mg	
	Al	
	Mn	
	Zn	
	Cr	
	Fe	Metal displaced from oxide by H_2
Displace H_2 from acid	Ni	
	Sn	
	Pb	
	H_2	
Do not displace H_2 from $H_2O(l)$, steam, or acid	**Sb**	
	Cu	
	Hg	Oxide thermally decomposes to metal plus O_2
	Ag	
	Pd	
	Pt	
	Au	

Table 5.2
Activity Series of Metals
The most reactive metals are at the top and activity decreases down the series. The metals at the top form cations most readily. The displacement of hydrogen occurs with hydrochloric acid and dilute sulfuric acid, but not with nitric acid, which leads to the formation of different types of products.

In displacement reactions in which the metals are converted to ions in aqueous solution, the ions formed are usually the following: Li^+, K^+, Ba^{2+}, Sr^{2+}, Ca^{2+}, Na^+, Mg^{2+}, Al^{3+}, Mn^{2+}, Zn^{2+}, Cr^{2+}, Fe^{2+}, Ni^{2+}, Sn^{2+}, Pb^{2+}.

By experimenting with the chemical reactions of many pure metals it is possible to order the metals in an *activity series* like that in Table 5.2. The chemical reactivity of the metals in displacement reactions decreases down the list. All metals above hydrogen, which is included for reference, displace hydrogen from various acids, for example, hydrochloric acid. The most active metals also displace hydrogen from water. Metals of intermediate activity only displace hydrogen from water at the temperature of steam, giving hydrogen and either the metal oxide or the metal hydroxide as products. If the hydroxide is not stable to thermal decomposition at the reaction temperature, the product will be the oxide. For example

$$\underset{}{Mg(s)} + \underset{\textit{steam}}{H_2O(g)} \xrightarrow{\Delta} \underset{\substack{\textit{magnesium}\\ \textit{oxide}}}{MgO(s)} + H_2(g) \qquad \textbf{(5.9)}$$

The metals below hydrogen in the activity series do not displace hydrogen in this manner.

The activity series can also be used to predict the outcome of other types of reactions, including the displacement of one metal by another. An elemental metal will displace a metal below it in the series from a solution of a salt of that metal. For example, placing a piece of zinc metal in a solution of a copper salt such as copper(II) chloride produces copper metal and a solution of zinc chloride. The progress of this reaction can be observed as the blue color of the copper(II) chloride solution gradually fades to the colorless zinc chloride solution, the piece of zinc grows smaller, and the red-brown to black copper metal can be seen to form.

$$Zn(s) + \underset{\substack{\textit{copper(II) chloride}\\ \textit{blue solution}}}{CuCl_2(aq)} \longrightarrow \underset{\substack{\textit{zinc(II) chloride}\\ \textit{colorless solution}}}{ZnCl_2(aq)} + Cu(s) \qquad \textbf{(5.10)}$$

The ease with which metals can be regained from their oxides is also indicated by the activity series. As shown on the right in Table 5.2, it is the reverse of the ease with which hydrogen displacement takes place. The oxides of the least active metals, like mercury (Equation 5.4), need only be heated to give the metal and oxygen. Metals of intermediate reactivity can be displaced from their oxides by hydrogen. For example

$$H_2(g) + \underset{\substack{\textit{nickel(II)}\\ \textit{oxide}}}{NiO(s)} \xrightarrow{\Delta} H_2O(g) + Ni(s) \qquad \textbf{(5.11)}$$

Oxides of the most active metals undergo neither reaction.

EXAMPLE 5.1 Activity Series of Metals

Compare the activity of strontium, an active metal, and lead, a moderately active metal, by writing chemical equations for the information summarized in Table 5.2.

Strontium, near the top of the activity series, will displace hydrogen from cold water, from steam, or from acids such as hydrochloric acid. The product of the reaction with water will be the hydroxide. However, without specific knowledge, we cannot predict whether the product in the reaction with steam will be the oxide or the hydroxide [it depends on the thermal stability of the hydroxide].

$$\mathbf{Sr}(s) + 2\mathbf{H_2O}(l) \longrightarrow \mathbf{Sr(OH)_2}(aq) + \mathbf{H_2}(g)$$

$$\mathbf{Sr}(s) + \mathbf{H_2O}(g) \xrightarrow{\Delta} \mathbf{SrO}(s) + \mathbf{H_2}(g)$$

or

$$\mathbf{Sr}(s) + 2\mathbf{H_2O}(g) \xrightarrow{\Delta} \mathbf{Sr(OH)_2}(s) + \mathbf{H_2}(g)$$

$$\mathbf{Sr}(s) + 2\mathbf{HCl}(aq) \longrightarrow \mathbf{SrCl_2}(aq) + \mathbf{H_2}(g)$$

$$\mathbf{H_2}(g) + \mathbf{SrO}(s) \not\longrightarrow \text{(no reaction)}$$

Lead displaces hydrogen from acids, but not from water or steam, and hydrogen displaces lead from its oxide.

$$\mathbf{Pb}(s) + \mathbf{H_2O}(l \text{ or } g) \not\longrightarrow \text{(no reaction)}$$

$$\mathbf{Pb}(s) + 2\mathbf{HCl}(aq) \longrightarrow \mathbf{PbCl_2}(aq) + \mathbf{H_2}(g)$$

$$\mathbf{H_2}(g) + \mathbf{PbO}(s) \xrightarrow{\Delta} \mathbf{Pb}(s) + \mathbf{H_2O}(g)$$

Neither SrO nor PbO will undergo thermal decomposition.

EXERCISE Nickel is a moderately active metal. Write chemical equations describing the chemical behavior of this metal as summarized in Table 5.2.

Answers $Ni(s) + H_2O(g \text{ or } l) \not\longrightarrow$ (no reaction), $Ni(s) + 2HCl(aq) \longrightarrow NiCl_2(aq) + H_2(g)$, $NiO(s) + H_2(g) \longrightarrow Ni(s) + H_2O(g)$.

EXAMPLE 5.2 Activity Series of Metals

Use the activity series given in Table 5.2 to predict whether or not reactions will occur between the following substances:

(a) $Fe(s) + CuCl_2(aq) \longrightarrow$

(b) $Zn(s) + MgCl_2(aq) \longrightarrow$

(c) Manganese + hydrochloric acid $\longrightarrow$

Complete the equations if reactions occur.

(a) Iron is above copper in the activity series. Therefore, it will displace copper ions from a solution of copper(II) chloride to give free copper.

$$\mathbf{Fe}(s) + \mathbf{CuCl_2}(aq) \longrightarrow \mathbf{FeCl_2}(aq) + \mathbf{Cu}(s)$$

(b) Zinc is a less active metal than magnesium. Therefore, it cannot displace magnesium from the solution.

$$\mathbf{Zn}(s) + \mathbf{MgCl_2}(aq) \not\longrightarrow \text{(no reaction)}$$

(c) Manganese is above hydrogen in the activity series and will displace hydrogen from an acid. The products of the reaction with hydrochloric acid will be hydrogen and the salt formed between the manganese and the chloride ions. The charges of the manganese and chloride ions are, respectively, $+2$ and -1, and the product is $MnCl_2$.

$$\mathbf{Mn}(s) + 2\mathbf{HCl}(aq) \longrightarrow \mathbf{MnCl_2}(aq) + \mathbf{H_2}(g)$$

EXERCISE Use the activity series in Table 5.2 to predict whether or not the following reactions will occur:

(a) $2Al(s) + 3CuSO_4(aq) \longrightarrow Al_2(SO_4)_3(aq) + 3Cu(s)$
(b) $Sn(s) + FeCl_2(aq) \longrightarrow SnCl_2(aq) + Fe(s)$
(c) $Pt(s) + 2HCl(aq) \longrightarrow PtCl_2(aq) + H_2(g)$

Answers (a) yes, (b) no, (c) no.

d. Partner-Exchange Reactions In **partner-exchange reactions** two compounds interact as follows:

$$AC + BD \longrightarrow AD + BC \qquad (5.12)$$

where A, B, C, and D may be atoms, monatomic ions, or polyatomic ions. Many different names have been given to this type of reaction—double decomposition, double displacement, and, often, metathesis—you may see these terms in other books. We feel that "partner exchange" is more descriptive of what takes place in reactions that correspond to the pattern of Equation (5.12).

In many cases partner exchange takes place between two ionic compounds that are both dissolved in water. Evidence that a reaction is taking place will often be the evolution of heat, the formation of a gas that bubbles out of the solution, or the formation of a solid product. A solid that forms during a reaction in solution is called a **precipitate.** The formation of a solid when a reaction takes place in solution is called **precipitation.** For example, the partner-exchange reaction between barium chloride and potassium chromate, both in solution, results in the precipitation of barium chromate:

$$\underset{\textit{barium chloride}}{BaCl_2(aq)} + \underset{\textit{potassium chromate}}{K_2CrO_4(aq)} \longrightarrow \underset{\textit{barium chromate}}{BaCrO_4(s)} + \underset{\textit{potassium chloride}}{2KCl(aq)} \qquad (5.13)$$

In the next section you will learn how solubility rules can be used to predict the products of partner-exchange reactions in which precipitates are formed.

The reaction between an acid and base (see Section 4.4d) is also a partner-exchange reaction, often detectable by the evolution of heat. The products are water and a salt:

$$\underset{\textit{hydrochloric acid}}{HCl(aq)} + \underset{\textit{potassium hydroxide (a base)}}{KOH(aq)} \longrightarrow \underset{\textit{water}}{H_2O(l)} + \underset{\textit{potassium chloride (a salt)}}{KCl(aq)} \qquad (5.14)$$

The first step in obtaining magnesium from sea water is the following partner-exchange reaction in which magnesium ions are removed from solution as solid magnesium hydroxide:

$$\underset{\textit{magnesium chloride}}{MgCl_2(aq)} + \underset{\textit{calcium hydroxide}}{Ca(OH)_2(s)} \longrightarrow \underset{\textit{magnesium hydroxide}}{Mg(OH)_2(s)} + \underset{\textit{calcium chloride}}{CaCl_2(aq)} \qquad (5.15)$$

EXAMPLE 5.3 Classification of Reactions

Classify each of the following chemical reactions as combination, decomposition, displacement, or partner exchange.

(a) $HNO_3(aq) + NH_3(aq) \longrightarrow NH_4NO_3(aq)$
(b) $NH_4Cl(s) \xrightarrow{\Delta} NH_3(g) + HCl(g)$
(c) $ZrCl_4(s) + 2Mg(s) \xrightarrow{\Delta} Zr(s) + 2MgCl_2(s)$
(d) $CaCl_2(aq) + 2AgNO_3(aq) \longrightarrow 2AgCl(s) + Ca(NO_3)_2(aq)$
(e) $Fe_3O_4(s) + 4H_2(g) \xrightarrow{\Delta} 3Fe(s) + 4H_2O(g)$

In reaction (a) two compounds combine to form a single compound. This is a **combination reaction.**

In decomposition reactions, a single compound decomposes to give other substances. Reaction (b) is a **decomposition.**

In reaction (c), the Mg atoms replace the Zr atoms in $ZrCl_4$ and in reaction (e), the H atoms replace the Fe atoms in Fe_3O_4. These are **displacement reactions.**

Reaction (d) fits the pattern of Equation (5.12) and is therefore a **partner-exchange reaction.**

EXERCISE Classify each of the following chemical reactions as combination, displacement, decomposition, or partner exchange:

(a) $ZnCO_3(s) \longrightarrow ZnO(s) + CO_2(g)$

(b) $P_4O_6(s) + 6H_2O(l) \longrightarrow 4H_3PO_3(aq)$

(c) $Al_2(SO_4)_3(aq) + 3Ca(OH)_2(aq) \longrightarrow 2Al(OH)_3(s) + 3CaSO_4(s)$

Answers (a) decomposition, (b) combination, (c) partner exchange.

5.2 NET IONIC EQUATIONS; PRECIPITATION REACTIONS

In the reaction between barium chloride and potassium chromate (Equation 5.13), three of the four chemical compounds involved are soluble and are present as ions in aqueous solution. Writing out separately all the ions in solution in Equation (5.13) gives

$$Ba^{2+}(aq) + 2Cl^-(aq) + 2K^+(aq) + CrO_4^{2-}(aq) \longrightarrow BaCrO_4(s) + 2K^+(aq) + 2Cl^-(aq) \quad (5.16)$$

Inspection of this equation shows that two of the species present—the K^+ and Cl^- ions—actually undergo no chemical change. These are **spectator ions**—ions that are present during a reaction in aqueous solution, but are unchanged in the reaction.

Equal numbers of spectator ions that appear on both sides of an equation can be cancelled. Doing so for Equation (5.16) leaves

$$Ba^{2+}(aq) + CrO_4^{2-}(aq) \longrightarrow BaCrO_4(s) \quad (5.17)$$

Equation (5.17) is a **net ionic equation**—an equation that shows only the species involved in a chemical change and excludes spectator ions. Equation (5.17) tells us that barium chromate can precipitate whenever barium and chromate ions are brought together in solution. The barium and chromate ions can be from any source. They need not be only from the compounds we started with in Equation (5.16).

A net ionic equation can be written for any reaction that has ions in solution among the reactants and products. For the displacement of copper ion from aqueous $CuCl_2$ by zinc

$$Zn(s) + Cu^{2+}(aq) + \cancel{2Cl^-(aq)} \longrightarrow Zn^{2+}(aq) + \cancel{2Cl^-(aq)} + Cu(s) \quad (5.18)$$
$$Zn(s) + Cu^{2+}(aq) \longrightarrow Zn^{2+}(aq) + Cu(s)$$

In writing net ionic equations, only ions in solution are written as ions. Complete formulas are given for reactants or products that are insoluble solids, even if they are ionic compounds. Also, complete formulas are written for gases, for water, or for other molecular compounds. The equations for the reaction between solid manganese sulfide and hydrochloric acid are written as follows:

$$MnS(s) + 2HCl(aq) \longrightarrow MnCl_2(aq) + H_2S(g) \quad (5.19)$$
$$MnS(s) + 2H^+(aq) \longrightarrow Mn^{2+}(aq) + H_2S(g) \quad (5.20)$$

Note that although one product is a gas rather than a precipitate, this reaction fits the pattern for partner exchange.

In net ionic equations, the sum of the charges of the reactants and those of the products must balance. Electrical charge, like mass, must be conserved. In summing the charges, keep in mind that you are taking an algebraic sum of positive and negative numbers. For Equation (5.17), for example, the charge on the left is zero (+2 plus −2 = 0), as is the charge for the compound on the right. As long as charge balance is maintained, the total charge on each side of an ionic equation can have any value. In the following equation the charge on each side is +17.

$$\underset{(-1)}{MnO_4^-(aq)} + \underset{5(+2)}{5Fe^{2+}(aq)} + \underset{8(+1)}{8H^+(aq)} \longrightarrow \underset{5(+3)}{5Fe^{3+}(aq)} + \underset{(+2)}{Mn^{2+}(aq)} + 4H_2O(l) \qquad (5.21)$$
$$(-1) + 5(+2) + 8(+1) = +17 \quad 5(+3) + (+2) = +17$$

Precipitation can often be predicted on the basis of generalizations, or rules, about the solubilities of ionic compounds. Table 5.3 summarizes some general solubility rules. For example, consider what might happen when solutions of the soluble salts potassium carbonate (K_2CO_3) and calcium chloride ($CaCl_2$) are mixed. The following ions are present in the solution:

$$2K^+(aq) + CO_3^{2-}(aq) + Ca^{2+}(aq) + 2Cl^-(aq)$$

Possible precipitates—if the compounds are insoluble—would be potassium chloride (KCl) or calcium carbonate ($CaCO_3$). Table 5.3 shows that chlorides are generally soluble, and KCl is not one of the exceptions. Carbonates are generally insoluble, and $CaCO_3$ is not an exception to this rule. Therefore, we can predict that $CaCO_3$ would precipitate when solutions of potassium carbonate and calcium chloride are mixed. The complete equation, the equation showing all of the ions, and the net ionic equation are given below. In these equations and throughout the remainder of this book, we usually omit the designation (*aq*) after ions. You may always assume that ions are in aqueous solution unless it is explicitly stated that they are not.

$$K_2CO_3(aq) + CaCl_2(aq) \longrightarrow 2KCl(aq) + CaCO_3(s) \qquad (5.22)$$
$$\cancel{2K^+} + CO_3^{2-} + Ca^{2+} + \cancel{2Cl^-} \longrightarrow \cancel{2K^+} + \cancel{2Cl^-} + CaCO_3(s) \qquad (5.23)$$
$$Ca^{2+} + CO_3^{2-} \longrightarrow CaCO_3(s) \qquad (5.24)$$

Table 5.3
Solubility of Common Ionic Compounds
These generalities apply to aqueous solutions at room temperature. *Soluble* indicates > 0.1 mol/L; *moderately soluble* (mod. sol.) indicates 0.1 − 0.01 mol/L; *insoluble* (insol.) indicates < 0.01 mol/L.

Generally soluble	Exceptions
Na^+, K^+, NH_4^+ compounds	—
Chlorides	Insol.: AgCl, Hg_2Cl_2 Sol. in hot water: $PbCl_2$
Bromides	Insol.: AgBr, $PbBr_2$, Hg_2Br_2 Mod. sol.: $HgBr_2$
Iodides	Insol.: heavier metal iodides
Sulfates	Insol.: $SrSO_4$, $BaSO_4$, $PbSO_4$, Hg_2SO_4 Mod. sol.: $CaSO_4$, Ag_2SO_4
Nitrates, nitrites	Mod. sol.: $AgNO_2$
Chlorites, perchlorates, permanganates	Mod. sol.: $KClO_4$
Acetates	Mod. sol.: $AgCH_3COO$
Generally insoluble	**Exceptions**
Sulfides	Sol.: NH_4^+, Na^+, K^+
Oxides, hydroxides	Sol.: Li_2O^a, LiOH, Na_2O^a, NaOH, K_2O^a, KOH, BaO^a, $Ba(OH)_2$ Mod. sol.: CaO^a, SrO^a, $Ca(OH)_2$, $Sr(OH)_2$
Carbonates, phosphates, cyanides, sulfites	Sol.: those of NH_4^+, Li^+, Na^+, K^+

[a] Dissolve with evolution of heat and formation of hydroxides.

EXAMPLE 5.4 Net Ionic Equations

Based on the solubility rules given in Table 5.3, how would you write the formulas for (a) Na_2S, (b) AgBr, (c) $NH_4(CH_3COO)$, and (d) $Fe(OH)_3$ in the net ionic equations for reactions in aqueous solution?

(a) Na_2S is soluble in water and should be written as $2Na^+(aq) + S^{2-}(aq)$, or $\mathbf{2Na^+ + S^{2-}}$.
(b) AgBr is insoluble in water. Thus the formula should be written as **AgBr(*s*)**.
(c) $NH_4(CH_3COO)$ is soluble in water and should be written as $NH_4^+(aq) + CH_3COO^-(aq)$, or $\mathbf{NH_4^+ + CH_3COO^-}$.
(d) $Fe(OH)_3$ is an insoluble hydroxide and the formula should be **$Fe(OH)_3(s)$**.

EXERCISE Based on the solubility rules given in Table 5.3, write the formulas for (a) KNO_2, (b) $BaCO_3$, and (c) Hg_2I_2 that would be used in the equations for reactions in aqueous solution. *Answers* (a) $K^+ + NO_2^-$, (b) $BaCO_3(s)$, (c) $Hg_2I_2(s)$.

EXAMPLE 5.5 Net Ionic Equations

Chlorine gas can be prepared in the laboratory by the reaction of manganese dioxide with concentrated hydrochloric acid (which is completely ionized in aqueous solution).

$$MnO_2(s) + 4HCl(aq) \longrightarrow MnCl_2(aq) + 2H_2O(l) + Cl_2(g)$$

Write the net ionic equation for this reaction.

It is first necessary to identify which species exist as ions in aqueous solution and which do not. We know that HCl is completely ionized. From Table 5.3 we can determine that MnO_2 is not soluble, and that metal chlorides are generally soluble ionic compounds, and that $MnCl_2$ is not an exception. Therefore, HCl and $MnCl_2$ should be represented in the equation as ions, and MnO_2 as a solid. We know water to be a molecular compound and chlorine to be a gas. Therefore, the equation showing all of the ions is as follows:

$$MnO_2(s) + 4H^+ + 4Cl^- \longrightarrow Mn^{2+} + 2Cl^- + 2H_2O(l) + Cl_2(g)$$

Two chloride ions are spectator ions and can be canceled from each side, giving the net ionic equation

$$\mathbf{MnO_2(s) + 4H^+ + 2Cl^- \longrightarrow Mn^{2+} + 2H_2O(l) + Cl_2(g)}$$

A check reveals that the atoms are balanced (1Mn, 2O, 4H, and 2Cl atoms on each side) and that the charge is balanced (+2 on each side).

EXERCISE One industrial method for preparing sodium hydroxide is the reaction of sodium carbonate with calcium hydroxide:

$$Na_2CO_3(aq) + Ca(OH)_2(aq) \longrightarrow 2NaOH(aq) + CaCO_3(s)$$

Write the net ionic equation for this reaction.
Answer $Ca^{2+} + CO_3^{2-} \longrightarrow CaCO_3(s)$.

EXAMPLE 5.6 Net Ionic Equations

Using the solubility rules in Table 5.3, determine whether or not precipitation reactions will occur when aqueous solutions of the following compounds are combined. If the reactions occur, write complete and net ionic equations.

(a) $ZnI_2(aq) + Pb(NO_3)_2(aq) \longrightarrow$ (b) $MgBr_2(aq) + Na_2SO_4(aq) \longrightarrow$

(a) Of the two possible products of a partner-exchange reaction, zinc nitrate is soluble and lead iodide is insoluble and would precipitate.

$$ZnI_2(aq) + Pb(NO_3)_2(aq) \longrightarrow Zn(NO_3)_2(aq) + PbI_2(s)$$
$$Pb^{2+} + 2I^- \longrightarrow PbI_2(s)$$

(b) The two possible products, magnesium sulfate and sodium bromide, are both soluble salts. Therefore, **no reaction** will occur and the ions will all remain in solution.

EXERCISE Predict whether or not precipitation reactions will occur between the following pairs of soluble compounds and, if they do occur, write the complete and net ionic equations: (a) magnesium nitrate plus sodium hydroxide, (b) potassium chloride plus sodium nitrate.

Answers (a) $Mg(NO_3)_2(aq) + 2NaOH(aq) \longrightarrow Mg(OH)_2(s) + 2NaNO_3(aq)$, $Mg^{2+} + 2OH^- \longrightarrow Mg(OH)_2(s)$; (b) no reaction.

SUMMARY OF SECTIONS 5.1–5.2

Four simple types of chemical reactions form the basis for the classification of many reactions and the prediction of their products [further explored in Chapter 17]. In combination reactions, two substances (often a metal and a nonmetal) combine to form a third. Many elements undergo combination, the vigor of the reaction being dependent upon the reactivity of the elements. In decomposition reactions, a single reactant breaks down to two or more products, frequently as the result of heating the reactant.

Displacement reactions occur when the atoms or ions of one substance take the place of other atoms or ions in a compound. Metals can be arranged in an activity series based upon the ease with which they displace or are displaced by other metals. Such an activity series allows the prediction (on a qualitative basis) of (1) displacement of hydrogen from water and acids by the metals, (2) production of free metals from oxides by displacement by hydrogen or by thermal decomposition, and (3) displacement of metals from solutions of their salts by more active metals. [Placing such predictions on a quantitative basis is discussed in Chapter 24.]

Partner-exchange reactions result in two chemical compounds exchanging "partners" to form two other chemical compounds: $AC + BD \longrightarrow AD + BC$. For example, partner-exchange reactions occur when two salts in solution are combined and a precipitate is formed; general solubility rules allow the prediction of when such reactions will occur. [A quantitative basis for such reactions is introduced in Chapter 22.] Precipitation reactions are one type of reaction conveniently represented by net ionic equations, in which ions that participate in the reaction are included but spectator ions are excluded.

STOICHIOMETRY

5.3 INFORMATION FROM CHEMICAL EQUATIONS

The calculation of the quantitative relationships in chemical changes is called **stoichiometry** (from the Greek words meaning "something that cannot be divided" and "to determine relative magnitudes"). The information that can be derived from a chemical formula—information about atomic, molar, and mass relationships in chemical compounds—is summarized in Table 4.9. Stoichiometry uses all this information and adds to it the information provided by chemical equations about the ratios of atoms, molecules, and ions and their masses in chemical reactions.

Consider the information given by the equation for the combination of oxygen with sulfur dioxide to give sulfur trioxide.

$$2SO_2(g) + O_2(g) \xrightarrow{\text{catalyst}} 2SO_3(g) \qquad \textbf{(5.25)}$$

Table 5.4
Information from a Chemical Equation
The reaction between sulfur dioxide and oxygen accounts for the presence of sulfur trioxide in the air as a pollutant. The molecular masses of the reactants and products are SO_2, 64 u; O_2, 32 u; SO_3, 80 u.

$2SO_2(g)$ *sulfur dioxide*	+	$O_2(g)$ *oxygen*	$\longrightarrow$	$2SO_3(g)$ *sulfur trioxide*
Each		**can react with**		**to yield**
2 molecules of SO_2		1 molecule of O_2		2 molecules of SO_3
2 moles of SO_2		1 mole of O_2		2 moles of SO_3
128 u of SO_2		32 u of O_2		160. u of SO_3
128 g of SO_2		32 g of O_2		160. u of SO_3
2 volumes of SO_2[a]		1 volume of O_2[a]		2 volumes of SO_3[a]

[a] For gaseous reactants and products, coefficients also give relative volumes (measured at same temperature and pressure), as discussed in Section 6.7.

The coefficients indicate that for every two molecules of SO_2 that react, one molecule of O_2 is required and two molecules of SO_3 form. Therefore, based upon the relationships between numbers of molecules and Avogadro's number (Section 4.7b), we know that every 2 mol of SO_2 requires 1 mol of O_2 and yields 2 mol of SO_3. By using the molar masses determined from the formulas in the equation, the masses of the reactants and products involved in the complete reaction between SO_2 and O_2 to form SO_3 can also be found. Table 5.4 summarizes the information derived from Equation (5.25).

The relative masses of reactants and products based upon the coefficients in a chemical equation hold for *any unit of mass*. (Recall our discussion of counting out nails by weighing them; Section 4.7.) Sometimes it is convenient to deal with the relative masses of reactants and products in units other than grams. For example, for large amounts ton-moles might be used. A ton-mole of oxygen is the mass in tons equal to the mass in atomic mass units: 1 ton-mol O_2 = 32 tons O_2. From the chemical equation and the information given in Table 5.4, it can be concluded that 128 tons of SO_2 and 32 tons of O_2 would similarly give 160 tons of SO_3.

5.4 USING MOLE RATIOS

In order to carry out a chemical reaction, many questions must be answered about the amounts of reactants and products. How much of each reactant has to be weighed out and allowed to react in order to produce the desired amount of product? If only a few grams of one reactant are available, how much of the other reactant will be needed? How much product can be prepared from a given amount of starting materials? Will any amount of the reactants be left unchanged? These are all questions about stoichiometry.

To answer such questions, it is necessary to know the balanced chemical equation for the reaction. If the equation is not known, finding it becomes the first step in solving the problem.

The known and unknown facts in stoichiometry problems are different kinds of information about the quantities of reactants and products. In *every* stoichiometry problem, the mole ratios in the balanced chemical equation provide the connection between the known and the unknown information. The mole ratios show the relative molar amounts of any pair of reactants and/or products. For example, in the reaction of sulfur dioxide with oxygen given in Table 5.4, the coefficients of the reactants and products show that every 2 mol of SO_2 requires 1 mol of O_2 and yields 2 mol of SO_3, giving the mole ratios

$$\frac{2\text{ mol }SO_2}{1\text{ mol }O_2} \qquad \frac{2\text{ mol }SO_2}{2\text{ mol }SO_3} \qquad \frac{1\text{ mol }O_2}{2\text{ mol }SO_3}$$

(Note that mole ratios are exact conversion factors and do not limit the number of significant figures in a calculation.)

EXAMPLE 5.7 Mole Ratios

Calcium carbide, a commercial source of acetylene, is produced by the reaction of calcium oxide with carbon at high temperatures:

$$CaO(s) + 3C(s) \xrightarrow{\Delta} CaC_2(s) + CO(g)$$

What are the mole ratios that give (a) the amount of calcium carbide produced by each mole of calcium oxide that reacts, (b) the amount of carbon required by each mole of calcium oxide that reacts, and (c) the amount of calcium carbide produced by each mole of carbon that reacts?

(a) The balanced chemical equation shows that 1 mol of calcium carbide is produced by each mole of calcium oxide that reacts, giving the mole ratio shown below. (b) Similarly, 3 mol of carbon is needed for each mole of calcium oxide that reacts. (c) And 1 mol of calcium carbide is produced by every 3 mol of carbon that reacts.

$$\text{(a)} \; \frac{\textbf{1 mol CaC}_2}{\textbf{1 mol CaO}} \qquad \text{(b)} \; \frac{\textbf{3 mol C}}{\textbf{1 mol CaO}} \qquad \text{(c)} \; \frac{\textbf{1 mol CaC}_2}{\textbf{3 mol C}}$$

EXERCISE A beautiful yellow suspension of solid As_2S_3 in water may be prepared as shown by the following equation:

$$2H_3AsO_3(aq) + 3H_2S(g) \longrightarrow As_2S_3(s) + 6H_2O(l)$$

What are the mole ratios of (a) H_3AsO_3 to H_2S, (b) As_2S_3 to H_3AsO_3, and (c) As_2S_3 to H_2S?

Answers (a) 2 mol H_3AsO_3/3 mol H_2S, (b) 1 mol As_2S_3/2 mol H_3AsO_3, (c) 1 mol As_2S_3/3 mol H_2S.

EXAMPLE 5.8 Mole Ratios

The overall process for producing elemental phosphorus from phosphate rock can be represented by the following chemical equation:

$$2Ca_3(PO_4)_2(s) + 6SiO_2(s) + 10C(s) \xrightarrow{\Delta} 6CaSiO_3(l) + 10CO(g) + P_4(g)$$

How many moles of silicon dioxide (SiO_2) would be required to produce 2.6 mol of P_4?

To solve this problem, the mole ratio of SiO_2 to P_4 given by the chemical equation must be used to solve for moles of SiO_2:

$$(2.6 \text{ mol } P_4)\left(\frac{6 \text{ mol } SiO_2}{1 \text{ mol } P_4}\right) = \textbf{16 mol SiO}_2$$

EXERCISE How many moles of $CO_2(g)$ and $H_2O(g)$ will be produced by the combustion of 15.0 mol of methane, $CH_4(g) + 2O_2(g) \longrightarrow CO_2(g) + 2H_2O(l)$?

Answer 15.0 mol CO_2, 30.0 mol H_2O.

5.5 SOLVING STOICHIOMETRY PROBLEMS

Many stoichiometry problems, no matter what kind of information is known and what kind of information is unknown, can be solved by using the following four steps:

1. Write the balanced chemical equation.
2. Convert the known quantity to number of moles.
3. Use a mole ratio from the balanced chemical equation to find the unknown in terms of number of moles.
4. Convert from number of moles to the unknown quantity that is desired.

You have already studied everything that is necessary to solve a great many types of stoichiometry problems. Often, balanced equations are known. When only the identities of the reactants and products are known for a simple reaction, the equation can be balanced as discussed in Section 4.6.

Among the possible types of known or unknown information about reactants and products are masses, molarity and volumes for reactions in aqueous solution, or possibly numbers of molecules. The methods for converting each of these types of information to numbers of moles (Step 2) have already been presented (Sections 4.7c, 4.8, and 4.7b, respectively).

We cannot emphasize too strongly that the essential step in every stoichiometry problem is the use of a mole ratio to connect the known and unknown quantities (Step 3).

The following three examples represent a simple type of stoichiometry problem in which the known and unknown are both masses of reactants and/or products. How one might analyze a problem to be solved by our four-step method for solving stoichiometry problems is illustrated in Example 5.9.

EXAMPLE 5.9 Stoichiometry; Problem Solving

Hydrogen fluoride gas for industrial use is produced by the partner-exchange reaction between calcium fluoride in the mineral fluorspar and excess sulfuric acid.

$$CaF_2(s) + H_2SO_4(aq) \longrightarrow CaSO_4(s) + 2HF(g)$$

What mass of hydrogen fluoride can be produced by the reaction of 10.0 g of calcium fluoride with excess sulfuric acid (in other words, with more than enough sulfuric acid for all the calcium fluoride to react)?

1. Study the problem and be sure you understand it.
   (a) What is unknown?
   The mass of the hydrogen fluoride that can be produced.
   (b) What is known?
   The chemical equation and the mass of calcium fluoride that reacts.
2. Decide how to solve the problem.

Step 1. Write balanced chemical equation. The equation is given in the problem.
Step 2. Convert known quantity to number of moles. The known information is the mass of a reactant, so the conversion to number of moles will be a mass-to-moles conversion:

$$(\text{mass of } CaF_2) \times \left(\frac{1}{\text{molar mass } CaF_2}\right) = \text{no. of moles of } CaF_2.$$

Step 3. Use mole ratio to find unknown in number of moles.

$$(\text{no. of moles of } CaF_2) \times (\text{mole ratio}) = \text{no. of moles of HF}$$

Step 4. Convert from number of moles to desired quantity. The unknown is the mass in grams; this step is essentially the reverse of Step 2.

$$(\text{no. of moles of HF}) \times (\text{molar mass of HF}) = \text{mass of HF}$$

3. Set up the problem and solve it.
   Steps 2, 3, and 4 can be done separately, as follows:

Step 2. $$(10.0 \text{ g } CaF_2)\left(\frac{1 \text{ mol } CaF_2}{78.1 \text{ g } CaF_2}\right) = 0.128 \text{ mol } CaF_2$$

Step 3. $$(0.128 \text{ mol } CaF_2)\left(\frac{2 \text{ mol HF}}{1 \text{ mol } CaF_2}\right) = 0.256 \text{ mol HF}$$

Step 4. $$(0.256 \text{ mol HF})\left(\frac{20.0 \text{ g HF}}{1 \text{ mol HF}}\right) = \mathbf{5.12 \text{ g HF}}$$

Or the problem can be solved in a single expression:

$$\underbrace{}_{\text{Step 2}}\quad \underbrace{}_{\text{Step 3}}\quad \underbrace{}_{\text{Step 4}}$$

$$(10.0\ \text{g CaF}_2)\left(\frac{1\ \text{mol CaF}_2}{78.1\ \text{g CaF}_2}\right)\left(\frac{2\ \text{mol HF}}{1\ \text{mol CaF}_2}\right)\left(\frac{20.0\ \text{g HF}}{1\ \text{mol HF}}\right) = \mathbf{5.12\ g\ HF}$$

4. Check the result.
   (a) Are significant figures used correctly? Yes. In each calculation the three significant figures in the numbers of moles allow three significant figures in the answer. (The numbers of moles in the mole ratios are exact.)
   (b) Is the answer reasonable?
   Yes. Although 2 mol of HF are produced for each mole of CaF_2, the molar mass of CaF_2 is almost three times that of HF; therefore, it is reasonable that the mass of HF produced is less than the mass of the initial CaF_2.

EXAMPLE 5.10 Stoichiometry

Isooctane (C_8H_{18})—which undergoes combustion very efficiently in automobile engines and does not cause knocking—is the standard for octane numbers for fuels. A fuel that undergoes combustion with as little knocking as isooctane is assigned an octane number of 100. What mass of oxygen is consumed in the complete combustion of 1.00 g of isooctane to form carbon dioxide and water?

$$2C_8H_{18}(l) + 25O_2(g) \longrightarrow 16CO_2(g) + 18H_2O(g)$$

The equation is given in the statement of the problem (Step 1). The known mass of isooctane must be converted to number of moles of isooctane by using the molar mass (Step 2). Next, the number of moles of isooctane required by the calculated moles of oxygen must be found by using the mole ratio 25 mol O_2/2 mol C_8H_{18} (Step 3). Finally, moles of oxygen must be converted to mass of oxygen by using the molar mass of oxygen (Step 4).

Steps 2–4 may be combined into a single calculation.

$$\underbrace{}_{\text{Step 2}}\quad \underbrace{}_{\text{Step 3}}\quad \underbrace{}_{\text{Step 4}}$$

$$(1.00\ \text{g C}_8\text{H}_{18})\left(\frac{1\ \text{mol C}_8\text{H}_{18}}{114.26\ \text{g C}_8\text{H}_{18}}\right)\left(\frac{25\ \text{mol O}_2}{2\ \text{mol C}_8\text{H}_{18}}\right)\left(\frac{32.00\ \text{g O}_2}{1\ \text{mol O}_2}\right) = \mathbf{3.50\ g\ O_2}$$

EXERCISE Hydrogen chloride can be prepared by the reaction of sodium chloride with warm concentrated sulfuric acid

$$NaCl(s) + H_2SO_4(conc) \xrightarrow{\Delta} HCl(g) + NaHSO_4(aq)$$

What mass of HCl can be obtained if 120 g of NaCl reacts with excess H_2SO_4?

Answer 77 g HCl.

EXAMPLE 5.11 Stoichiometry

Each day a power plant burns 4.0×10^3 tons of coal that contains 1.2% sulfur by mass. During the combustion process, the sulfur is completely converted to sulfur dioxide. Calculate the mass of sulfur dioxide produced each day by the plant.

Step 1. Write balanced chemical equation

$$S(s) + O_2(g) \longrightarrow SO_2(g)$$

Step 2. Convert known quantity to number of moles. The amount of sulfur in 4.0×10^3 tons of coal is

$$(4.0 \times 10^3\ \text{tons coal})\left(\frac{1.2\ \text{tons S}}{100.0\ \text{tons coal}}\right) = \mathbf{48\ tons\ S}$$

The problem can be solved for the mass in tons by using ton-moles of sulfur:

$$(48 \text{ tons S})\left(\frac{1 \text{ ton-mol S}}{32.06 \text{ tons S}}\right) = \mathbf{1.5\ ton\text{-}mol\ S}$$

Steps 3 and 4. Use mole ratio to find unknown in number of moles; convert from number of moles to desired quantity

$$(1.5 \text{ ton-mol S})\left(\frac{1 \text{ ton-mol S}}{1 \text{ ton-mol } SO_2}\right)\left(\frac{64.06 \text{ tons } SO_2}{1 \text{ ton-mol S}}\right) = \mathbf{96\ tons\ SO_2}$$

EXERCISE The sulfur dioxide released by the power plant described above is removed from the stack gases (the gases that go up the chimney) by passing these gases through a series of scrubbers containing powdered limestone, $CaCO_3(s)$. In the following reaction SO_2 is completely converted to $CaSO_3$

$$SO_2(g) + CaCO_3(s) \longrightarrow CaSO_3(s) + CO_2(g)$$

What mass (in tons) of limestone is needed to remove the daily effluent of SO_2 from this plant? *Answer* 150 tons $CaCO_3$ daily.

The following examples illustrate two common but slightly more complex situations for which stoichiometry calculations are required. In each case our method for solving stoichiometry problems is easily applied.

Example 5.12 illustrates the situation in which two or more consecutive reactions make up a single process. The equations for the individual steps can be added together, if necessary multiplying one or more equations by factors that will allow all intermediate products containing the elements of interest to be canceled out. The mole ratios from the resulting overall equation can then be used to make the connection between the amounts of the original reactants and the final products. The amounts of intermediate products need not be taken into account *so long as they have been canceled out of the overall equation.* (It should be noted that the overall equation may not represent a reaction which would take place as written.)

EXAMPLE 5.12 Stoichiometry—Consecutive Reactions

The production of potassium permanganate requires two separate reactions. In the first, manganese dioxide (the naturally occurring mineral known as pyrolusite) is converted to potassium manganate by reaction with molten KOH and oxygen:

(a) $$\underset{\textit{manganese dioxide}}{2MnO_2(s)} + \underset{\textit{potassium hydroxide}}{4KOH(l)} + O_2(g) \xrightarrow{\Delta} \underset{\textit{potassium manganate}}{2K_2MnO_4(s)} + 2H_2O(l)$$

In the second reaction, potassium manganate is converted to potassium permanganate.

(b) $$2K_2MnO_4(aq) + Cl_2(g) \longrightarrow \underset{\textit{potassium permanganate}}{2KMnO_4(aq)} + 2KCl(aq)$$

What mass (in grams) of $KMnO_4$ will be produced from 100.0 g of MnO_2?

Step 1. Write balanced chemical equation.

Adding Equations (a) and (b) as given allows cancellation of the intermediate potassium manganate, giving the overall chemical equation

$$\begin{array}{rcl} 2MnO_2 + 4KOH + O_2 & \longrightarrow & \cancel{2K_2MnO_4} + 2H_2O \\ \cancel{2K_2MnO_4} + Cl_2 & \longrightarrow & 2KMnO_4 + 2KCl \\ \hline 2MnO_2 + 4KOH + O_2 + Cl_2 & \longrightarrow & 2KMnO_4 + 2KCl + 2H_2O \end{array}$$

Steps 2–4. Convert known quantity to number of moles; use mole ratio; convert to desired quantity.

$$(100.0 \text{ g } MnO_2)\underbrace{\left(\frac{1 \text{ mol } MnO_2}{86.94 \text{ g } MnO_2}\right)}_{\textit{Step 2}}\underbrace{\left(\frac{2 \text{ mol } KMnO_4}{2 \text{ mol } MnO_2}\right)}_{\textit{Step 3}}\underbrace{\left(\frac{158.04 \text{ g } KMnO_4}{1 \text{ mol } KMnO_4}\right)}_{\textit{Step 4}}$$

$$= \mathbf{181.8\ g\ KMnO_4}$$

Reactions (a) and (b) will yield a maximum of 181.8 g of potassium permanganate for each 100.0 g of manganese dioxide.

EXERCISE Antimony can be prepared from its sulfide ore by the following two-step process:

$$2Sb_2S_3(s) + 9O_2(g) \xrightarrow{\Delta} Sb_4O_6(s) + 6SO_2(g)$$

$$Sb_4O_6(s) + 6C(s) \xrightarrow{\Delta} 4Sb(l) + 6CO(g)$$

What mass of elemental antimony will be produced from 1.00 kg of an ore that contains 23.2 mass % Sb_2S_3? *Answer* 166 g Sb.

Example 5.13 illustrates finding the composition of a mixture by carrying out a chemical reaction that changes one component of the mixture but not another.

EXAMPLE 5.13 Stoichiometry—Composition of a Mixture

A by-product of an industrial process was a mixture of sodium sulfate (Na_2SO_4) and sodium hydrogen carbonate ($NaHCO_3$). To determine the composition of the mixture, a sample weighing 8.00 g was heated until constant mass was achieved, indicating that the heat-induced reaction was complete. Under these conditions the sodium hydrogen carbonate undergoes the following decomposition reaction

$$\underset{\text{sodium hydrogen carbonate}}{2NaHCO_3(s)} \longrightarrow \underset{\text{sodium carbonate}}{Na_2CO_3(s)} + \underset{\text{carbon dioxide}}{CO_2(g)} + \underset{\text{water}}{H_2O(g)}$$

and the sodium sulfate is unchanged. The mass of the sample after heating was 6.02 g. What was the mass percent of $NaHCO_3$ in the original by-product?

Before we can find the mass of $NaHCO_3$ in the sample before heating, we must determine either the mass of CO_2 or the mass of H_2O that was present in the mixture of gases released during the decomposition. The molar ratio of CO_2 to H_2O in the mixture of gaseous products is (1 mol CO_2/1 mol H_2O). This means that for each 44.02 g of CO_2, 18.02 g H_2O was produced. The mass percent of CO_2 in the gaseous products is

$$\left(\frac{44.02 \text{ g } CO_2}{44.02 \text{ g } CO_2 + 18.02 \text{ g } H_2O}\right) \times (100) = 70.95\%$$

The mass of the gaseous mixture released by the decomposition was

$$8.00 \text{ g} - 6.02 \text{ g} = 1.98 \text{ g}$$

The mass of CO_2 in 1.98 g of gaseous products was

$$(1.98 \text{ g mixture})\left(\frac{70.95 \text{ g } CO_2}{100.0 \text{ g mixture}}\right) = 1.40 \text{ g } CO_2$$

Using the mole ratio of $NaHCO_3$ to CO_2, the mass of $NaHCO_3$ originally present can now be found:

$$(1.40 \text{ g } CO_2)\underbrace{\left(\frac{1 \text{ mol } CO_2}{44.02 \text{ g } CO_2}\right)}_{\text{Step 2}}\underbrace{\left(\frac{2 \text{ mol } NaHCO_3}{1 \text{ mol } CO_2}\right)}_{\text{Step 3}}\underbrace{\left(\frac{84.01 \text{ g } NaHCO_3}{1 \text{ mol } NaHCO_3}\right)}_{\text{Step 4}} = 5.34 \text{ g } NaHCO_3$$

The mass of $NaHCO_3$ in 8.00 g of the original mixture was 5.34 g, which corresponds to

$$\left(\frac{5.34 \text{ g}}{8.00 \text{ g}}\right) \times (100) = \mathbf{66.8\%} \text{ (by mass)}$$

EXERCISE A Cu_2O-CuO mixture was analyzed by heating a sample of the mixture with gaseous hydrogen to produce copper metal and water.

$$Cu_2O(s) + H_2(g) \xrightarrow{\Delta} 2Cu(s) + H_2O(g)$$

$$CuO(s) + H_2(g) \xrightarrow{\Delta} Cu(s) + H_2O(g)$$

A 1.351 g sample of the mixture gave 1.152 g of copper metal. What is the composition of this mixture? *Answer* 61% Cu_2O, 39% CuO (by mass).

5.6 REACTIONS IN AQUEOUS SOLUTION

When chemical reactions are carried out in aqueous solution, the amounts of reactants and products are often expressed as concentrations in moles per liter. The amounts are found from the volumes of the solutions and the known molarities, as illustrated in Examples 5.14 and 5.15.

EXAMPLE 5.14 Stoichiometry—Solutions

What volume of 0.10 M barium chloride must be added to 25 mL of 0.23 M solution of sodium sulfate to completely precipitate barium sulfate?

Step 1. Write balanced chemical equation. This is a partner-exchange reaction.

$$BaCl_2(aq) + Na_2SO_4(aq) \longrightarrow BaSO_4(s) + 2NaCl(aq)$$

Step 2. Convert known quantity to number of moles.

$$(25 \text{ mL})\left(\frac{1 \text{ L}}{1000 \text{ mL}}\right)\left(\frac{0.23 \text{ mol } Na_2SO_4}{1 \text{ L}}\right) = 5.8 \times 10^{-3} \text{ mol } Na_2SO_4$$

Step 3. Use mole ratio to find unknown in number of moles.

$$(5.8 \times 10^{-3} \text{ mol } Na_2SO_4)\left(\frac{1 \text{ mol } BaCl_2}{1 \text{ mol } Na_2SO_4}\right) = 5.8 \times 10^{-3} \text{ mol } BaCl_2$$

Step 4. Convert from number of moles to desired quantity.

$$(5.8 \times 10^{-3} \text{ mol } BaCl_2)\left(\frac{1 \text{ L}}{0.10 \text{ mol } BaCl_2}\right)\left(\frac{1000 \text{ mL}}{1 \text{ L}}\right) = \mathbf{58 \text{ mL}}$$

EXERCISE What volume of 0.103 M HCl will react with 25.00 mL of 0.112 M NaOH?

$$HCl(aq) + NaOH(aq) \longrightarrow H_2O(l) + NaCl(aq)$$

Answer 27.2 mL.

EXAMPLE 5.15 Stoichiometry—Solutions

The concentration of iodine dissolved in aqueous potassium iodide was determined by allowing the solution to react with 25.00 mL of 0.00397 M arsenous acid (H_3AsO_3).

$$H_3AsO_3(aq) + I_2(aq, KI) + H_2O(l) \longrightarrow H_3AsO_4(aq) + 2H^+ + 2I^-$$

Find the concentration of the iodine solution if 199 mL of the iodine solution was used.

Step 1. Write balanced chemical equation. The equation is given above.
Step 2 and 3. Convert known quantities to number of moles; use mole ratios to find unknown in number of moles.

$$(25.00 \text{ mL})\underbrace{\left(\frac{1 \text{ L}}{1000 \text{ mL}}\right)\left(\frac{0.00397 \text{ mol } H_3AsO_3}{1 \text{ L}}\right)}_{\textit{Step 2}}\underbrace{\left(\frac{1 \text{ mol } I_2}{1 \text{ mol } H_3AsO_3}\right)}_{\textit{Step 3}} = 9.93 \times 10^{-5} \text{ mol } I_2$$

Step 4. Convert from number of moles to desired quantity.

$$\left(\frac{9.93 \times 10^{-5} \text{ mol } I_2}{199 \text{ mL}}\right)\left(\frac{1000 \text{ mL}}{1 \text{ L}}\right) = \mathbf{4.99 \times 10^{-4} \text{ mol/L}}, \text{ or } 4.99 \times 10^{-4} \text{ M}$$

EXERCISE A 0.863 g sample of impure potassium acid phthalate ($KHC_8H_4O_4$) was dissolved in water and allowed to react with a dilute solution of sodium hydroxide.

$$KHC_8H_4O_4(aq) + NaOH(aq) \longrightarrow NaKC_8H_4O_4(aq) + H_2O(l)$$

What is the purity of the sample if 22.7 mL of 0.106 M NaOH was used?

Answer 57.0 mass % $KHC_8H_4O_4$.

5.7 LIMITING REACTANTS

Sometimes only a limited amount of one of the reactants needed for a chemical reaction is available. Or perhaps it is easier to carry out a reaction by adding an excess of one of the reactants. The maximum amount of product that can be formed is determined by the amount of the reactant that is used up first. The situation is not unlike trying to put together new bicycles in a toy store. Suppose that each bicycle requires six nuts, six bolts, and six washers. If one shipment of parts includes 60 washers, 60 nuts, but only 59 bolts (as is so often the case), then bolts become the limiting factor. Only nine bicycles can be assembled.

The exact amount of a substance required or produced as determined by a balanced chemical equation is the **stoichiometric amount**. For example, in the combination of sodium with chlorine

$$2Na(s) + Cl_2(g) \longrightarrow 2NaCl(s)$$

1 mol of Na will react with 0.5 mol of Cl_2 to produce 1 mol of NaCl. These are the stoichiometric amounts required for the reaction of 1 mol of sodium according to the balanced chemical equation.

When more than the stoichiometric amount of one reactant is present, some of it will remain unreacted. If 1 mol of Na is mixed with 2 mol of Cl_2, then 1.5 mol of the chlorine will be left over and the same amount of sodium chloride as before—1 mol—will be produced. In this case, sodium is the **limiting reactant**—the reactant that determines the amount of product that can be formed. Even if 100 mol of chlorine is available, 1 mol of sodium will limit the reaction to the formation of 1 mol of sodium chloride.

In a limiting reactant problem, information is known about two of the reactants and it is necessary to determine which reactant is limiting. First, the number of moles of each reactant is found. Next, the stoichiometric amount of one reactant *required* by the known amount of the other reactant is calculated. If more of a reactant is required than is available, then that is the limiting reactant. If less of a reactant is required than is available, then that it is not the limiting reactant.

EXAMPLE 5.16 Stoichiometry—Limiting Reactant

What mass of hydrogen will be produced from the displacement reaction of 6.0 g of zinc with 25 mL of 6.0 M by hydrochloric acid?

Step 1. Write balanced chemical equation.

$$Zn(s) + 2HCl(aq) \longrightarrow ZnCl_2(aq) + H_2(g)$$

Step 2. Convert known quantities to number of moles.

$$(6.0\ \text{g Zn})\left(\frac{1\ \text{mol Zn}}{65.38\ \text{g Zn}}\right) = 0.092\ \text{mol Zn}$$

$$(25\ \text{mL})\left(\frac{1\ \text{L}}{1000\ \text{mL}}\right)\left(\frac{6.0\ \text{mol HCl}}{1\ \text{L}}\right) = 0.15\ \text{mol HCl}$$

Step 3. Use mole ratios to find unknown in number of moles. For all of the Zn to react, the number of moles of HCl must be

$$(0.092 \text{ mol Zn})\left(\frac{2 \text{ mol HCl}}{1 \text{ mol Zn}}\right) = 0.18 \text{ mol HCl}$$

The amount of HCl available, 0.15 mol, is less than the required amount. Therefore, HCl is the limiting reactant. (If we had checked the amount of Zn, we would have found that more Zn is available than required to react with all of the HCl.) The amount of H_2 formed is therefore determined by the amount of HCl:

$$(0.15 \text{ mol HCl})\left(\frac{1 \text{ mol } H_2}{2 \text{ mol HCl}}\right) = 0.075 \text{ mol } H_2$$

Step 4. Convert from number of moles to desired quantity.

$$(0.075 \text{ mol } H_2)\left(\frac{2.02 \text{ g } H_2}{1 \text{ mol } H_2}\right) = \mathbf{0.15 \text{ g } H_2}$$

EXERCISE A 10.0 g sample of H_2 was mixed with a 10.0 g sample of O_2. An electrical spark ignited this mixture to produce water. Calculate the mass of water produced.

Answer 11.3 g H_2O.

5.8 YIELDS

The maximum amount of a product that can, *according to the balanced chemical equation*, be obtained from known amounts of reactants is called the **theoretical yield.** It is the stoichiometric amount of the product. For many reasons, the amount of a product actually obtained is often less than the theoretical amount. Perhaps some product was lost or spilled in handling, or perhaps not all of the reactant was converted to product. Also, sometimes undesired further reactions occur and use up some of the reactant or product.

The **actual yield** in a reaction is the weighed mass or measured volume of product formed. The extent to which product has been formed is given as the **percent yield:**

$$\% \text{ yield} = \left(\frac{\text{actual yield}}{\text{theoretical yield}}\right)(100) \tag{5.26}$$

EXAMPLE 5.17 Stoichiometry—Percent Yield

The final step in the industrial production of aspirin (acetylsalicylic acid) is the reaction of salicylic acid with acetic anhydride:

$$\underset{\text{salicylic acid}}{HOC_6H_4COOH(s)} + \underset{\text{acetic anhydride}}{(CH_3CO)_2O(l)} \longrightarrow \underset{\text{acetylsalicylic acid (aspirin)}}{CH_3COOC_6H_4COOH(s)} + \underset{\text{acetic acid}}{CH_3COOH(l)}$$

To test a new method of handling the materials, a chemist ran the reaction on a laboratory scale with 25.0 g of salicylic acid and excess acetic anhydride (over 20 g). The actual yield was 24.3 g of aspirin. What was the percent yield?

First calculate the theoretical yield, which is the stoichiometric amount of the product. Setting up the dimensional calculation in one expression gives

$$(25.0 \text{ g sal. acid})\underbrace{\left(\frac{1 \text{ mol sal. acid}}{138.13 \text{ g sal. acid}}\right)}_{\textit{Step 2}}\underbrace{\left(\frac{1 \text{ mol aspirin}}{1 \text{ mol sal. acid}}\right)}_{\textit{Step 3}}\underbrace{\left(\frac{180.17 \text{ g aspirin}}{1 \text{ mol aspirin}}\right)}_{\textit{Step 4}} = 32.6 \text{ g aspirin}$$

The percent yield is given by

$$\% \text{ yield} = \frac{\text{actual yield}}{\text{theoretical yield}} \times (100) = \left(\frac{24.3 \text{ g}}{32.6 \text{ g}}\right)(100) = \mathbf{74.5\%}$$

EXERCISE The equation for the decomposition of $NH_4Cl(s)$ to $NH_3(g)$ and $HCl(g)$ is

$$NH_4Cl(s) \xrightarrow{\Delta} NH_3(g) + HCl(g)$$

What is the theoretical yield of NH_3 for 1.00 g NH_4Cl? If the percent yield is 78%, what is the actual yield? *Answers* 0.319 g NH_3, 0.25 g NH_3.

SUMMARY OF SECTIONS 5.3–5.8

Stoichiometric calculations are based on balanced chemical equations. The known and unknown information is always the quantities of reactants and/or products, which may be expressed in many different ways. The connection between the known and unknown information is always provided by mole ratios from the chemical equations. Most stoichiometry problems, simple or complex, can be solved by applying four general steps: (1) write the balanced chemical equation, (2) convert the known quantity to number of moles, (3) use mole ratios to find the unknown quantity in terms of number of moles, (4) convert the answer from number of moles to the desired quantity.

The stoichiometric amount of a substance required or produced in a chemical reaction is the exact amount calculated according to the chemical equation. When one of two reactants is present in less than the stoichiometric amount required by the other reactant, that reactant is the limiting reactant—it determines how much product can be formed. The maximum amount of a product that can be formed is the theoretical yield of a reaction. Frequently the percent yield, determined by the amount of product actually obtained, is less than the theoretical yield.

SIGNIFICANT TERMS (with section references)

chemical reactivity (5.1)
combination reaction (5.1)
decomposition reaction (5.1)
displacement reaction (5.1)
limiting reactant (5.7)
net ionic equation (5.2)
partner-exchange reaction (5.1)
precipitate (5.1)
precipitation (5.1)
spectator ions (5.2)
stoichiometric amount (5.7)
stoichiometry (5.3)
yield, actual (5.8)
yield, percent (5.8)
yield, theoretical (5.8)

QUESTIONS AND PROBLEMS

Chemical Reactions—Types

5.1 Classify each of the following equations as describing a (i) combination reaction, (ii) decomposition reaction, (iii) displacement reaction, or (iv) partner-exchange reaction:

(a) $Cl_2O_7(g) + H_2O(l) \longrightarrow 2HClO_4(aq)$
(b) $Br_2(l) + H_2O(l) \longrightarrow HBr(aq) + HBrO(aq)$
(c) $Ca_3(PO_4)_2(s) + 3H_2SO_4(aq) \longrightarrow 3CaSO_4(s) + 2H_3PO_4(aq)$
(d) $2K(s) + 2H_2O(l) \longrightarrow 2KOH(aq) + H_2(g)$
(e) $MgCO_3(s) \longrightarrow MgO(s) + CO_2(g)$
(f) $Fe_3O_4(s) + 4H_2(g) \longrightarrow 3Fe(s) + 4H_2O(l)$
(g) $2KClO_3(s) \longrightarrow 2KCl(s) + 3O_2(g)$
(h) $H_2O(g) + C(s) \xrightarrow{\Delta} H_2(g) + CO(g)$

5.2 Repeat Question 5.1 for

(a) $V_2O_4(s) + 4HClO_4(aq) \longrightarrow 2VO(ClO_4)_2(aq) + 2H_2O(l)$
(b) $Fe_2O_3(s) + 3H_2(g) \longrightarrow 2Fe(s) + 3H_2O(l)$
(c) $2H_2O_2(aq) \longrightarrow 2H_2O(l) + O_2(g)$
(d) $Cl_2(aq) + 2KI(aq) \longrightarrow 2KCl(aq) + I_2(aq)$
(e) $NH_4HSO_4 \longrightarrow NH_3 + H_2SO_4$
(f) $2NaI + Br_2 \longrightarrow 2NaBr + I_2$
(g) $Zn(NO_3)_2 + Na_2S \longrightarrow ZnS + 2NaNO_3$
(h) $4Fe + 3O_2 \longrightarrow 2Fe_2O_3$
(i) $2HAuCl_4 \longrightarrow 2Au + 3Cl_2 + 2HCl$
(j) $Xe + 2F_2 \longrightarrow XeF_4$

5.3 Use the activity series to predict whether or not the following reactions will occur:
(a) $Fe(s) + Mg^{2+} \longrightarrow Mg(s) + Fe^{2+}$
(b) $Ni(s) + Cu^{2+} \longrightarrow Ni^{2+} + Cu(s)$
(c) $Cu(s) + 2H^+ \longrightarrow Cu^{2+} + H_2(g)$
(d) $Mg(s) + H_2O(g) \longrightarrow MgO(s) + H_2(g)$

5.4 Repeat Question 5.3 for
(a) $Sn(s) + Ba^{2+} \longrightarrow Sn^{2+} + Ba(s)$
(b) $Al_2O_3(s) + 3H_2(g) \xrightarrow{\Delta} 2Al(s) + 3H_2O(g)$
(c) $Ca(s) + 2H^+ \longrightarrow Ca^{2+} + H_2(g)$
(d) $Cu(s) + Pb^{2+} \longrightarrow Cu^{2+} + Pb(s)$

Chemical Reactions—Net Ionic Equations

5.5 Based on the solubility rules given in Table 5.3, how would you write the formulas for the following substances in a net ionic equation: (a) $PbSO_4$, (b) $Na(CH_3COO)$, (c) $(NH_4)_2CO_3$, (d) MnS, (e) $BaCl_2$?

5.6 Repeat Question 5.5 for (a) $(NH_4)_2SO_4$, (b) NaBr, (c) $Ba(CN)_2$, (d)$Mg(OH)_2$, (e) K_2CO_3.

5.7 Write the net ionic equations for each of the following reactions:
(a) $2FeCl_3(aq) + 2KI(aq) \longrightarrow 2FeCl_2(aq) + I_2(s) + 2KCl(aq)$
(b) $2NaIO_3(aq) + 6NaHSO_3(aq) \longrightarrow 2NaI(aq) + 3Na_2SO_4(aq) + 3H_2SO_4(aq)$
(c) $BrCl(g) + H_2O(l) \longrightarrow HCl(aq) + HBrO(aq)$
(d) $PCl_5(g) + 4H_2O(l) \longrightarrow H_3PO_4(aq) + 5HCl(aq)$
(e) $2Na_2S_2O_3(aq) + I_2(s) \longrightarrow Na_2S_4O_6(aq) + 2NaI(aq)$

5.8 Repeat Question 5.7 for
(a) $AgNO_3(aq) + KCN(aq) \longrightarrow AgCN(s) + KNO_3(aq)$
(b) $2Al(s) + 2NaOH(aq) + 6H_2O(l) \longrightarrow 2Na[Al(OH)_4](aq) + 3H_2(g)$
(c) $K_2Cr_2O_7(aq) + 6KBr(aq) + 14HBr(aq) \longrightarrow 2CrBr_3(aq) + 3Br_2(aq) + 7H_2O(l) + 8KBr(aq)$
(d) $Na_2S(aq) + H_2O(l) \longrightarrow NaHS(aq) + NaOH(aq)$
(e) $NH_4NO_3(aq) + Na_2CO_3(aq) \longrightarrow NH_3(aq) + NaHCO_3(aq) + NaNO_3(aq)$

5.9 Using the solubility rules given in Table 5.3, determine whether or not reactions will occur when aqueous solutions of the compounds are mixed:
(a) $Hg(NO_3)_2(aq) + Na_2S(aq) \longrightarrow$
(b) $Al(NO_3)_3(aq) + LiOH(aq) \longrightarrow$
(c) $Li_2SO_3(aq) + NaCl(aq) \longrightarrow$
(d) $Fe(OH)_3(s) + KNO_3(aq) \longrightarrow$
Write net ionic equations for those reactions that occur.

5.10 Repeat Question 5.9 for
(a) $Al(OH)_3(s) + NaNO_3(aq) \longrightarrow$
(b) $NaBr(aq) + NH_4I(aq) \longrightarrow$
(c) $AgNO_3(aq) + HCl(aq) \longrightarrow$
(d) $CaCl_2(aq) + Na_2CO_3(aq) \longrightarrow$

Stoichiometry—Information from Chemical Equations

5.11 What is the relationship shown by the equation

$$C_7H_{16}(l) + 11O_2(g) \longrightarrow 7CO_2(g) + 8H_2O(g)$$

between C_7H_{16} and each of the other substances in terms of (a) moles, (b) molecules, (c) mass?

5.12 Repeat Question 5.11 for SO_2 and each of the other substances in the following equation

$$SO_2(g) + Br_2(g) + 2H_2O(g) \xrightarrow{\Delta} 2HBr(g) + H_2SO_4(aq)$$

5.13 The equation which describes the commercial "roasting" of zinc sulfide is

$$2ZnS(s) + 3O_2(g) \xrightarrow{\Delta} 2ZnO(s) + 2SO_2(g)$$

What is the mole ratio of (a) O_2 to ZnS, (b) ZnO to ZnS, (c) SO_2 to ZnS?

5.14 The reaction between dilute nitric acid and copper is given by the equation:

$$3Cu(s) + 8HNO_3(aq) \longrightarrow 3Cu(NO_3)_2(aq) + 2NO(g) + 4H_2O(l)$$

What is the mole ratio of (a) HNO_3 to Cu, (b) NO to Cu, (c) $Cu(NO_3)_2$ to Cu?

5.15 How many moles of oxygen can be obtained by the decomposition of 1.00 mol of reactant in each of the following reactions?
(a) $2KClO_3(s) \longrightarrow 2KCl(s) + 3O_2(g)$
(b) $2H_2O_2(aq) \longrightarrow 2H_2O(l) + O_2(g)$
(c) $2HgO(s) \longrightarrow 2Hg(l) + O_2(g)$
(d) $2NaNO_3(s) \longrightarrow 2NaNO_2(s) + O_2(g)$
(e) $KClO_4(s) \longrightarrow KCl(s) + 2O_2(g)$

5.16 For the formation of 1.00 mol of water, which reaction uses the most nitric acid?
(a) $3Cu(s) + 8HNO_3(aq) \longrightarrow 3Cu(NO_3)_2(aq) + 2NO(g) + 4H_2O(l)$
(b) $Al_2O_3(s) + 6HNO_3(aq) \longrightarrow 2Al(NO_3)_3(aq) + 3H_2O(l)$
(c) $4Zn(s) + 10HNO_3(aq) \longrightarrow 4Zn(NO_3)_2(aq) + NH_4NO_3(aq) + 3H_2O(l)$

5.17 Consider the reaction

$$NH_3(g) + O_2(g) \xrightarrow{\text{not balanced}} NO(g) + H_2O(l)$$

For every 1.50 mol of NH_3, (a) how many moles of O_2 are required, (b) how many moles of NO are produced, (c) how many moles of H_2O are produced?

5.18 Consider the reaction

$$2NO(g) + Br_2(g) \longrightarrow 2NOBr(g)$$

For every 3.00 mol of bromine that reacts, how many moles of (a) NO react and (b) NOBr are produced?

5.19 It has been proposed that electricity could be generated in a fuel cell by the following reaction:

$$4NH_3(g) + 3O_2(g) \longrightarrow 2N_2(g) + 6H_2O(l)$$

For every 1.00 mol of ammonia that reacts, how many (a) moles of oxygen are needed and how many (b) moles of nitrogen and (c) moles of water are produced?

5.20 Consider the following decomposition reaction

$$2Pb(NO_3)_2(s) \xrightarrow{\Delta} 2PbO(s) + 4NO_2(g) + O_2(g)$$

For every 0.050 mol of $Pb(NO_3)_2$ that reacts, how many moles of (a) PbO, (b) NO_2, (c) O_2, (d) gaseous products are produced?

Stoichiometry—General

5.21 Calculate the mass of sodium required to produce 80.0 g of sodium hydroxide by direct reaction with water

$$2Na(s) + 2H_2O(l) \longrightarrow 2NaOH(aq) + H_2(g)$$

(This reaction is rather dangerous because the hydrogen can form an explosive mixture with the oxygen in the air and bits of molten sodium metal may be released and start a fire.)

5.22 Find the mass of chlorine that will combine with 1.38 g of hydrogen to form hydrogen chloride

$$H_2(g) + Cl_2(g) \longrightarrow 2HCl(g)$$

5.23 What mass of solid AgCl will precipitate from a solution containing 1.50 g of $CaCl_2$ if an excess amount of $AgNO_3$ is added?

$$CaCl_2(aq) + 2AgNO_3(aq) \longrightarrow 2AgCl(s) + Ca(NO_3)_2(aq)$$

5.24 A sample of magnetic iron oxide, Fe_3O_4, reacted completely with hydrogen at red heat. The water vapor formed by the reaction

$$Fe_3O_4(s) + 4H_2(g) \xrightarrow{\Delta} 3Fe(s) + 4H_2O(g)$$

was condensed and found to weigh 7.5 g. Calculate the mass of Fe_3O_4 that reacted.

5.25 What mass of cobalt(II) chloride and of hydrogen fluoride are needed to prepare 10.0 g of cobalt(II) fluoride by the following reaction?

$$CoCl_2(s) + 2HF(g) \longrightarrow CoF_2(s) + 2HCl(g)$$

5.26 Zirconium is obtained industrially using the Kroll process

$$ZrCl_4(s) + 2Mg(s) \longrightarrow Zr(s) + 2MgCl_2(s)$$

Calculate the mass of Zr obtainable for each ton of Mg consumed.

5.27 Gaseous chlorine and gaseous fluorine undergo a combination reaction to form the interhalogen compound $ClF(g)$. Write the chemical equation for this reaction and calculate the mass of fluorine needed to react with 3.27 g of Cl_2.

5.28 Dinitrogen pentoxide, N_2O_5, undergoes a decomposition reaction to form nitrogen dioxide, NO_2, and oxygen. Write the chemical equation for this reaction. A 0.165 g sample of O_2 was produced by the reaction. What mass of NO_2 was produced?

5.29 Gaseous chlorine will displace bromide ion from an aqueous solution of potassium bromide to form aqueous potassium chloride and aqueous bromine. Write the chemical equation for this reaction. What mass of bromine will be produced if 0.289 g of chlorine undergoes reaction?

5.30 Solid zinc sulfide will undergo a partner-exchange reaction with hydrochloric acid to form a mixture of aqueous zinc chloride and hydrogen sulfide, $H_2S(g)$. Write the chemical equation for this reaction. What mass of zinc sulfide is needed to react with 10.65 g of HCl?

5.31 An impure sample of $CuSO_4$ weighing 5.52 g was dissolved in water and allowed to react with excess zinc.

$$CuSO_4(aq) + Zn(s) \longrightarrow ZnSO_4(aq) + Cu(s)$$

What is the percent $CuSO_4$ in the sample if 1.49 g of Cu was produced?

5.32 You are designing an experiment for the preparation of hydrogen. For the production of equal amounts of hydrogen, which metal, Zn or Al, is less expensive if Zn costs about half as much as Al on a mass basis?

$$Zn(s) + 2HCl(aq) \longrightarrow ZnCl_2(aq) + H_2(g)$$
$$2Al(s) + 6HCl(aq) \longrightarrow 2AlCl_3(aq) + 3H_2(g)$$

Stoichiometry—Consecutive Reactions

5.33 Consider the two-step process for the formation of tellurous acid described by the following equations:

$$TeO_2(s) + 2OH^- \longrightarrow TeO_3{}^{2-} + H_2O(l)$$
$$TeO_3{}^{2-} + 2H^+ \longrightarrow H_2TeO_3(s)$$

What mass of H_2TeO_3 would be formed from 62.1 g of TeO_2, assuming 100% yield?

5.34 Consider the formation of cyanogen, C_2N_2, and its subsequent decomposition in water given by the equations:

$$2Cu^{2+} + 6CN^- \longrightarrow 2[Cu(CN)_2]^- + C_2N_2(g)$$
$$C_2N_2(g) + H_2O(l) \longrightarrow HCN(aq) + HOCN(aq)$$

What is the theoretical yield of hydrocyanic acid, HCN, assuming 100% yield, that can be produced from 10.00 g of KCN?

5.35 What mass of potassium chlorate would be required to supply the proper amount of oxygen needed to burn 35.0 g of methane, CH_4?

$$2KClO_3(s) \longrightarrow 2KCl(s) + 3O_2(g)$$
$$CH_4(g) + 2O_2(g) \longrightarrow CO_2(g) + 2H_2O(g)$$

5.36 Hydrogen, obtained by the electrical decomposition of water, was combined with chlorine to produce 51.0 g of hydrogen chloride. Calculate the mass of water decomposed.

$$2H_2O(l) \longrightarrow 2H_2(g) + O_2(g)$$
$$H_2(g) + Cl_2(g) \longrightarrow 2HCl(g)$$

5.37* (a) Limestone, $CaCO_3$, is heated to produce unslaked lime (or quicklime), CaO, and carbon dioxide

$$CaCO_3(s) \longrightarrow CaO(s) + CO_2(g)$$

What mass of CaO will be produced from the decomposition of 2.25 tons of impure limestone? Assume the limestone is 45% pure. (b) Slaked lime, $Ca(OH)_2$, is produced from CaO by the reaction

$$CaO(s) + H_2O(l) \longrightarrow Ca(OH)_2(aq)$$

What mass of slaked lime will be produced from the CaO produced from the impure limestone?

5.38* What mass of sulfuric acid can be obtained from 1.00 kg of sulfur by the following series of reactions?

$$S(s) + O_2(g) \xrightarrow{98\%\text{ yield}} SO_2(g)$$
$$2SO_2(g) + O_2(g) \xrightarrow{96\%\text{ yield}} 2SO_3(g)$$
$$SO_3(g) + H_2SO_4(l) \xrightarrow{100\%\text{ yield}} H_2S_2O_7(l)$$
$$H_2S_2O_7(l) + H_2O(l) \xrightarrow{97\%\text{ yield}} 2H_2SO_4(aq)$$

Stoichiometry—Solutions

5.39 What volume of 0.50 M HBr is required to react completely with 0.75 mol of $Ca(OH)_2$?

$$2HBr(aq) + Ca(OH)_2(aq) \longrightarrow CaBr_2(aq) + 2H_2O(l)$$

5.40 What volume of 0.324 M HNO_3 solution is required to react completely with 22.0 mL of 0.0612 M $Ba(OH)_2$?

$$Ba(OH)_2(aq) + 2HNO_3(aq) \longrightarrow Ba(NO_3)_2(aq) + 2H_2O(l)$$

5.41 What is the concentration in mol/L of an HCl solution if 23.65 mL reacts completely with (neutralizes) 25.00 mL of a 0.1037 M solution of NaOH?

$$HCl(aq) + NaOH(aq) \longrightarrow NaCl(aq) + H_2O(l)$$

5.42 An excess of $AgNO_3$ reacts with 100.0 mL of an $AlCl_3$ solution to give 0.275 g of AgCl. What is the concentration in mol/L of the $AlCl_3$ solution?

$$AlCl_3(aq) + 3AgNO_3(aq) \longrightarrow 3AgCl(s) + Al(NO_3)_3(aq)$$

5.43 An impure sample of solid Na_2CO_3 was allowed to react with 0.1026 M HCl.

$$Na_2CO_3(s) + 2HCl(aq) \longrightarrow 2NaCl(aq) + CO_2(g) + H_2O(l)$$

A 0.1247 g sample of sodium carbonate required 14.78 mL of HCl. What is the purity of the sodium carbonate?

5.44 Calculate the theoretical yield of AgCl formed from the reaction of an aqueous solution containing excess $ZnCl_2$ with 35.0 mL of 0.325 M $AgNO_3$.

$$ZnCl_2(aq) + 2AgNO_3(aq) \longrightarrow Zn(NO_3)_2(aq) + 2AgCl(s)$$

Stoichiometry—Limiting Reactants

5.45 What is the maximum mass of sodium chloride that can be formed by the reaction of 5.00 g of sodium with 7.10 g of chlorine? Which substance is the limiting reactant? Which substance is in excess?

$$2Na(s) + Cl_2(g) \longrightarrow 2NaCl(s)$$

5.46 What mass of potassium can be produced by the reaction of 100.0 g of Na with 100.0 g of KCl?

$$Na(g) + KCl(l) \xrightarrow{\Delta} NaCl(l) + K(g)$$

5.47 A reaction mixture contained 25.0 g of PCl_3 and 45.0 g of PbF_2. What mass of $PbCl_2$ can be obtained from the following partner-exchange reaction?

$$3PbF_2(s) + 2PCl_3(l) \longrightarrow 2PF_3(g) + 3PbCl_2(s)$$

How much of which reactant will be left unchanged?

5.48 What mass of $BaSO_4$ will be produced by the reaction of 33.2 g of Na_2SO_4 with 43.5 g of $Ba(NO_3)_2$?

$$Ba(NO_3)_2(aq) + Na_2SO_4(aq) \longrightarrow BaSO_4(s) + 2NaNO_3(aq)$$

5.49* Consider the reaction

$$3HCl(g) + 3HNF_2(g) \longrightarrow 2ClNF_2(g) + NH_4Cl(s) + 2HF(g)$$

A mixture of 8.00 g of HCl and 10.00 g of HNF_2 is allowed to react. If only 15% of the limiting reactant does react, what is the composition of the final mixture?

5.50* Consider the following reaction:

$$HNO_3(aq) + Cu(s) \longrightarrow Cu(NO_3)_2(aq) + NO_2(g) + H_2O(l)$$

(a) Balance the equation. A piece of Cu metal 3.31 cm × 1.84 cm × 1.00 cm reacts with 157 mL of 1.35 M nitric acid solution. The density of copper is 8.92 g/cm^3. (b) Find the number of moles of each reactant. (c) Describe what would happen if the amount of Cu were doubled.

Stoichiometry—Percent Yield

5.51 The percent yield for the reaction

$$PCl_3(g) + Cl_2(g) \longrightarrow PCl_5(g)$$

is 85%. What mass of PCl_5 would be expected from the reaction of 38.5 g of PCl_3 with excess chlorine?

5.52 The percent yield for the following reaction carried out in carbon tetrachloride solution

$$Br_2(CCl_4) + Cl_2(CCl_4) \longrightarrow 2BrCl(CCl_4)$$

is 57%. (a) What amount of BrCl would be formed from the reaction of 0.0100 mol Br_2 with 0.0100 mol Cl_2? (b) What amount of Br_2 is left unchanged?

5.53 Solid silver nitrate undergoes thermal decomposition to form silver metal, nitrogen dioxide, and oxygen. Write the chemical equation for this reaction. A 0.362 g sample of silver metal was obtained from the decomposition of a 0.575 g sample of $AgNO_3$. What is the percent yield of the reaction?

5.54 Gaseous nitrogen and hydrogen undergo a combination reaction to form gaseous ammonia (the Haber process). Write the chemical equation for this reaction. At a temperature of 400 °C and a total pressure of 250 atm, 0.72 g of NH_3 was produced by the reaction of 2.80 g of N_2 with excess H_2. What is the percent yield of the reaction?

5.55* When sulfuric acid dissolves in water, the following reactions take place:

$$H_2SO_4(aq) \longrightarrow H^+ + HSO_4^-$$
$$HSO_4^- \longrightarrow H^+ + SO_4^{2-}$$

The first reaction is 100.0% complete and the second reaction is 10.0% complete. Calculate the concentrations of the various ions in a 0.100 M aqueous solution of H_2SO_4.

5.56* The chief ore of zinc is the sulfide, ZnS. The ore is concentrated by flotation and then heated in air, which converts the ZnS to ZnO.

$$2ZnS(s) + 3O_2(g) \longrightarrow 2ZnO(s) + 2SO_2(g)$$

The ZnO is then treated with dilute H_2SO_4

$$ZnO(s) + H_2SO_4(aq) \longrightarrow ZnSO_4(aq) + H_2O(l)$$

to produce an aqueous solution containing the zinc as $ZnSO_4$. An electrical current is passed through the solution to produce the metal.

$$2ZnSO_4(aq) + 2H_2O(l) \longrightarrow 2Zn(s) + 2H_2SO_4(aq) + O_2(g)$$

What mass of Zn will be obtained from an ore containing 100. kg of ZnS? Assume the flotation process to be 91% efficient, the electrolysis step to be 98% efficient, and the other steps to be 100.% efficient.

CHAPTER

6

The Gaseous State

THE NATURE OF GASES

6.1 GENERAL PROPERTIES OF GASES

Strictly speaking, any substance in which the molecules are far apart and move independently of each other, and which expands infinitely to fill any container completely, is a gas. Although gases can exist at any temperature, the word "gas" is commonly used to refer to a substance that at room temperature and pressure exists in the gaseous state. The names and formulas of some chemical compounds that are gases in this sense are given in Table 6.1. Eleven of the elements are also gases at room temperature and pressure: H_2, O_2, N_2, F_2, Cl_2, He, Ne, Ar, Kr, Xe, Rn.

The word "vapor" is usually applied to a gas that is evaporating from a solid or liquid. For example, on a hot day the smell of gasoline *vapor* is obvious near a gasoline pump. A substance that very readily forms a vapor is referred to as **volatile.** Volatile liquids have relatively low boiling points. For example, ether (b.p. 34.5 °C) is a *volatile* liquid and its *vapor* is highly flammable. Water (b.p. 100 °C) is not a very volatile liquid.

CO	Carbon monoxide	Odorless, poisonous
CO_2	Carbon dioxide	Odorless, nonpoisonous
NH_3	Ammonia	Pungent odor, poisonous
PH_3	Phosphine	Garlic-like odor, very poisonous
CH_4	Methane	Odorless, flammable
C_2H_2	Acetylene	Mild odor, flammable
HCl	Hydrogen chloride	Choking odor, harmful and poisonous
SO_2	Sulfur dioxide	Suffocating odor, irritating to eyes, poisonous
NO_2	Nitrogen dioxide	Red-brown, irritating odor, very poisonous
H_2S	Hydrogen sulfide	Rotten egg odor, very poisonous

Table 6.1
Gases
This is a list of some compounds that are gases at ordinary temperatures and pressures. Pranksters often ask for a harmless gas with a very bad odor, but there is no such thing—all evil-smelling gases are poisonous. Of course, some odorless gases, such as CO, are also poisonous.

Table 6.2
Observable Physical Properties of Gases

Fill whatever space is available and take shape of container
Are infinitely expandable
Are very compressible
Flow rapidly
Expand and contract with changes in temperature
Exert greater pressure in closed container when temperature rises
Have low density

Progress in the understanding of gases provides a superb example of how science works. Observations led to quantitative experiments. Generalizations known as the "gas laws" were formulated to summarize the results of the experiments. Ultimately, a theory was developed that explained the related observations, facts, and laws. (Sometimes in science, of course, it goes the other way. A theory can come first and then gradually be accepted as more and more experiments support its predictions.)

During the latter part of the eighteenth and the early part of the nineteenth century, chemistry advanced rapidly, largely through the study of gases. At first glance this may appear strange, for to most of us gases seem rather nebulous. Most common gases are invisible and we cannot hold a gas in our hands.

Yet the properties of gases are exactly what made the study of gases so fruitful. Gases, although they differ as widely as solids and liquids in their chemical properties, have many physical properties in common (Table 6.2).

Eventually, it became clear that these properties all depend upon one thing—the behavior of molecules flying through space. This realization led to the kinetic-molecular theory (the theory of moving molecules). The kinetic-molecular theory explained what had already been observed about gas behavior and predicted much of what has since been observed. From a historical viewpoint, the gas laws preceded the kinetic-molecular theory. But in terms of the present understanding of atoms and molecules, it is more direct to begin our study of gases with the kinetic-molecular theory. Then, as the gas laws are introduced, their relationship to the theory can be explained.

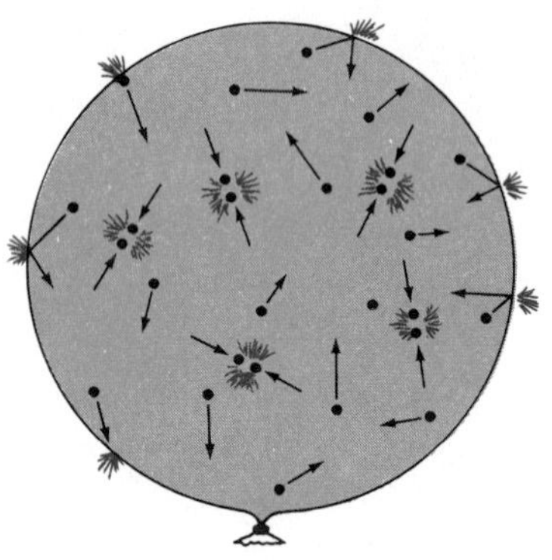

Figure 6.1
Molecular Motion in a Gas
Gas molecules are far apart, constantly moving in random straight lines, and colliding with each other and the walls of the container. The pressure on the container walls is the sum of the force of the individual impacts.

6.2 KINETIC-MOLECULAR THEORY OF GASES

The essential concept of the kinetic-molecular theory is that molecules are always in motion. The molecules in a gas are continuously and freely moving about, colliding with each other and whatever surfaces are nearby (Figure 6.1).

The following five statements summarize the kinetic-molecular theory of gases. Keep in mind that these statements form the basis for a mathematical model. As is often necessary, to develop such a model certain simplifying assumptions must be made. In the case of the gas laws and kinetic-molecular theory, the assumption is made that the gas under consideration is an **ideal gas**—a gas composed of molecules that occupy negligible volume and exert no forces of attraction on each other. An ideal gas exactly obeys the gas laws and exhibits the behavior predicted by the kinetic-molecular theory. Under the conditions at which real gases normally are encountered, many are very much like "ideal" gases.

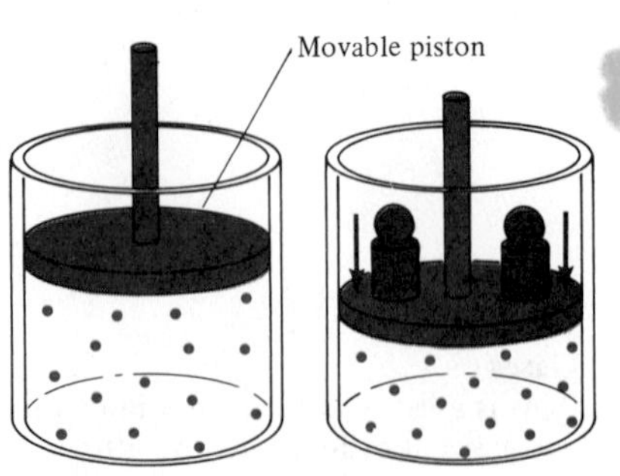

Figure 6.2
Compression of a Gas
Gas molecules are far apart relative to their size and gases are easily compressed. As gas pressure increases, volume decreases.

1. Gases are made up of molecules that are relatively far apart; in comparison with their size, the spaces between them are large. A substance in the gaseous state occupies a much larger volume than the same amount of that substance in the liquid or solid state. For example, when liquid water at 100 °C and 1 atm pressure changes to steam at the same temperature and pressure, the volume of the steam is about 1700 times greater than the volume of the same amount of water. The same number of water molecules have simply spread out over a much larger space. Gases can be compressed easily because the distances between neighboring molecules decrease when pressure is applied (Figure 6.2).
2. The molecules of a gas are in constant motion in random straight lines and their collisions with surfaces cause gas pressure. The molecules change direction only when they collide with each other or with the walls of their container. The gas pressure is the result of impacts with the walls. In an inflated balloon, it is the impact of multitudes of gas molecules with the inside of the balloon that keeps it stretched out and prevents it from collapsing (see Figure 6.1). The pressure does not vary with each blow because the number of gas molecules is very large, the molecules move rapidly, and the impacts are so

frequent that they cannot be detected individually. Gases fill any container by constantly moving about through its entire volume. (Think of how quickly the smell of frying onions fills a house.)

3. The average speed and average kinetic energy of gas molecules are proportional to the temperature. The energy of a moving body—**kinetic energy**, E_k—is a function of its body mass, m, and its velocity, u. As the velocity of a body increases, its kinetic energy increases.

$$E_k = \tfrac{1}{2} m u^2 \quad (6.1)$$

(kinetic energy; mass; velocity)

Temperature in kelvins and kinetic energy are directly proportional to each other. (Proportionality is discussed in the next section.) Therefore, the average kinetic energy of the molecules in a sample of gas is greater at higher temperatures and lower at lower temperatures. This is one very important outcome of the kinetic-molecular theory—the understanding that temperature is dependent solely upon the kinetic energy of atoms, molecules, and ions.

Note that not all the molecules in a gas sample are flying about at the same speed (Figure 6.3). Some go much faster than others. Collisions between molecules continually cause exchanges in energy, slowing down some molecules and speeding up others. It is the *average* speed of all the molecules in a gas that increases as the temperature is raised. The effect of this increase is seen in changes in the pressure and/or the volume of a given gas sample (Figure 6.4).

4. At the same temperature, the molecules of every gas have the same average kinetic energy. If $E_k = \frac{1}{2} mu^2$ has the same average value for every gas at a given temperature, then molecules with larger mass must, on the average, move more slowly at that temperature than molecules with smaller mass. In short, the larger the m, the smaller the u. As a consequence, a gas with a lower molar mass released into a container will spread throughout the container more quickly than will a gas with a higher molar mass.

5. Collisions of gas molecules with each other or with the walls of a container are perfectly elastic. According to the kinetic-molecular theory, when gas molecules collide with each other or with the walls of a container, they bounce off with no loss in energy—they undergo frictionless, or perfectly elastic, collisions. Energy may be exchanged between colliding molecules, but the *total* energy of the molecules that have collided is the same after the collision as it was before. Ideal gas molecules must therefore be assumed to exert no forces of attraction or repulsion on each other because, if the gas molecules attracted each other, their collisions would not be perfectly elastic.

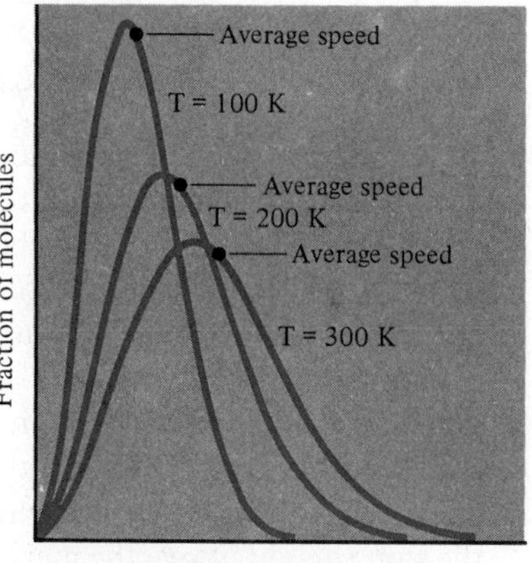

Figure 6.3
Distribution of Molecular Speeds
Both the average speed and the range of speeds increases with increasing temperature. At any given temperature, very few molecules have the highest or the lowest speeds.

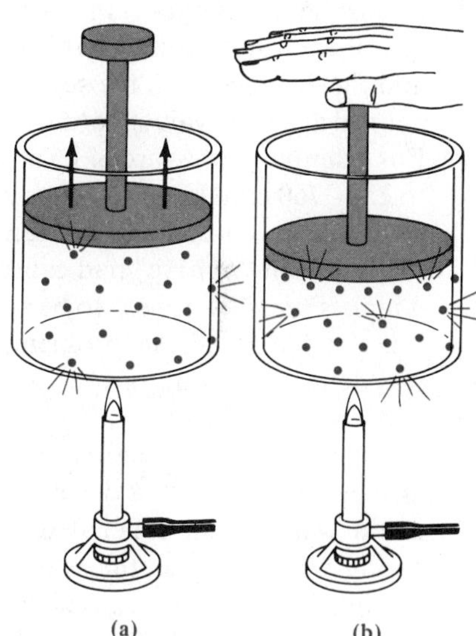

Figure 6.4
Effect on a Gas of Increasing Temperature
Gas in a cylinder with a movable piston. (a) With increasing temperature, the molecules move faster, hit the piston with greater force, and push it up. The pressure remains the same; the volume increases. (b) When the piston cannot move, as the molecules move faster they hit the piston with greater force and more often, and the pressure increases.

SUMMARY OF SECTIONS 6.1–6.2

Gases have in common many easily observable physical properties. The kinetic-molecular theory explains these properties by assuming that molecules in a gas are constantly in motion in random straight-line paths. Gas pressure is the result of the collisions of gas molecules with the walls of a container. In an ideal gas, the gas molecules occupy no volume, collisions are perfectly elastic, and the molecules exert no forces of attraction or repulsion on each other. The average kinetic energy of gas molecules increases with the temperature, causing increases in pressure and/or volume. (Average kinetic energy and average speed are both proportional to temperature.) A mathematical model based on these assumptions is successful at explaining gas behavior.

Aside 6.1 TOOLS OF CHEMISTRY: PRESSURE UNITS AND PRESSURE MEASUREMENT

Pressure is defined as force per unit area. The atmosphere exerts pressure on the surface of the Earth and on all sides of all objects on the Earth's surface. The first **barometer**—an instrument that measures atmospheric pressure—was made in about 1643 by Evangelista Torricelli, who worked with Galileo.

Torricelli filled a tube with mercury and inverted it in a dish of mercury. He found that whatever the length or diameter of the tube, the mercury always fell in the tube to the same height above the mercury level in the dish, about 760 mm (Figure 6.5) At this height, the weight of the mercury is balanced by the constant pressure exerted by the atmosphere on the surface of the mercury in the dish. Pressure is still sometimes expressed in millimeters of mercury (e.g., a pressure of 20 mmHg).

The height of the liquid column in the barometer depends on the density of the liquid, as well as on the atmospheric pressure. Since it is very dense, mercury is best for barometers, because the column is of a convenient height. The atmosphere supports a column of water about 34 ft high, not too useful for barometers.

High pressures are frequently expressed in units of atmospheres. Because the actual pressure of the atmosphere varies somewhat with latitude, temperature, and the distance from the surface of the Earth, the unit is defined at a fixed value: 1 atmosphere (atm) = 760 mmHg. For example, a pressure of 25 atm means a pressure equal to 25 × 760 mmHg, which is roughly 25 times the pressure of the Earth's atmosphere at sea level.

The unit "mmHg" had caused some distress because it makes pressure appear to be a function of length, which it is not. To avoid this problem, the Torr (for Torricelli) was defined as a unit of pressure: 1 Torr = 1 mmHg and 760 Torr = 1 atm.

Atmospheric pressure is now often measured with an aneroid barometer—a device in which two thin, flexible metal walls of an evacuated box are held apart by a spring, which opposes the external pressure of the air. The motion of the walls caused by changes in pressure is transmitted by a series of mechanical linkages to a dial calibrated in pressure units. The gauges on tanks containing gases and those used for measuring tire pressure operate on the same principle.

The pressure in closed systems is measured with a device known as a **manometer**. In an open-end manometer (Figure 6.6), the pressure in the system is compared with

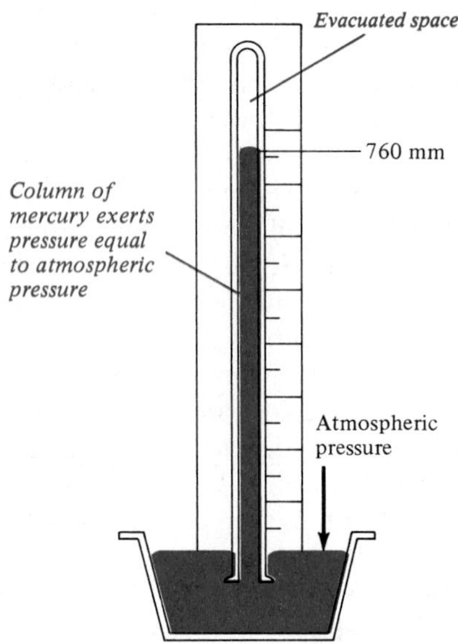

Figure 6.5
Mercury Barometer
The height of the column of mercury in millimeters is read as the atmospheric pressure. The space above the mercury is not a complete vacuum, for a little mercury evaporates into it. But mercury is not very volatile and at ordinary temperatures the pressure exerted by this vapor is so small that it can be disregarded.

VOLUME, PRESSURE, AND TEMPERATURE RELATIONSHIPS

6.3 VARIABLES AND PROPORTIONALITY

Variables are measurable properties that can change. The pressure, volume, temperature, and the amount of a gas are all variables. To investigate the effects of

atmospheric pressure. A closed-end manometer (Figure 6.7) is commonly used to measure a pressure less than atmospheric in a closed system.

Uniformity in the choice of units for the measurement of gas pressures seems particularly far off. Each branch of science and engineering has its own preferences. Engineers, in particular, frequently express pressure in units with the dimensions of force per unit area, such as pounds per square inch (psi), dynes per square centimeter, or newtons per square meter. A distinction is sometimes made between "absolute pressure" and "gauge pressure." When the pressure in a tire is given as, for example, 30 psi or 30 psig, it is gauge pressure—pressure *greater* than atmospheric pressure. Absolute pressure includes the atmospheric pressure, and for this tire would be 44.7 psi (30 psi + 14.7 psi). In chemistry, pressures are usually expressed as absolute pressures, but without any special notation to designate absolute pressure. The SI unit for pressure is the pascal (Pa), a derived unit related to the base units and the newton (N), the derived unit for force:

SI unit of force $1\ \text{N} = 1\ \text{kg m/s}^2$

SI unit of pressure $1\ \text{Pa} = 1\ \text{N/m}^2 = 1\ \text{kg/s}^2\ \text{m}$

The SI system includes the definition of a unit called the *bar*, which is close in magnitude to the standard atmosphere, as shown below. Following are the equivalences between some of the various pressure units:

1 atm = 760 mmHg (exactly)
= 760 Torr (exactly)
= 29.92 inches Hg
= 14.7 psi
= 101,325 Pa (exactly)
= 1.01325 bar (exactly)

1 mmHg = 1 Torr (exactly)
1 Torr = 133.32 Pa
1 bar = 10^5 Pa (exactly)
= 750.06 mmHg
1 psi = 6894.76 Pa

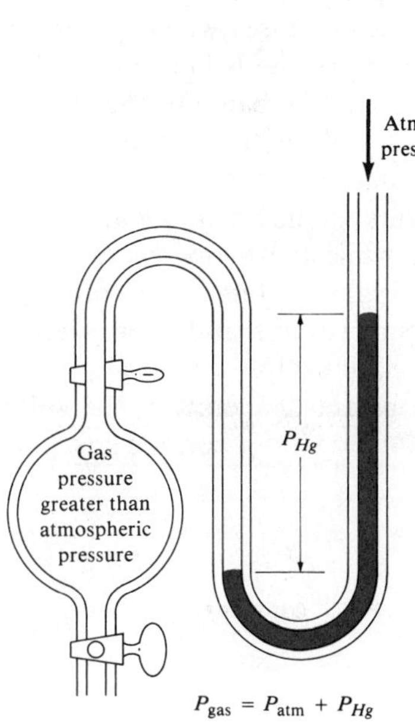

Figure 6.6
Open-End Manometer
Atmospheric pressure (P_{atm}) is read from a barometer and P_{Hg} is the height of the mercury column between the two horizontal lines. Gas pressure greater than atmospheric pressure is calculated as shown. If pressure in the system is less than atmospheric pressure, the mercury is higher in the tube on the left and pressure in the system is equal to atmospheric pressure minus the height of the mercury column ($P_{gas} = P_{atm} - P_{Hg}$).

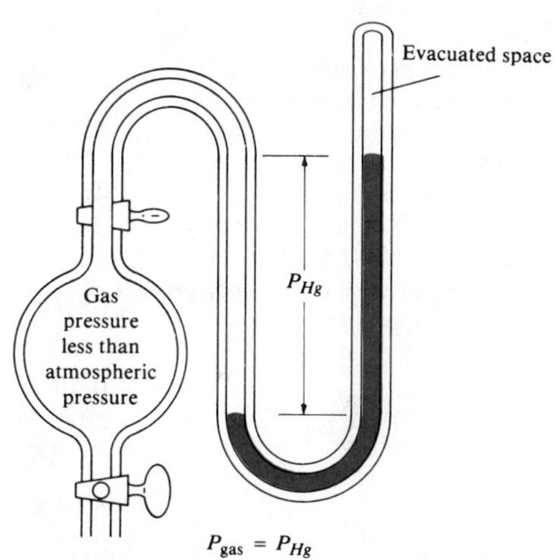

Figure 6.7
Closed-End Manometer
The pressure exerted by the gas is equal to P_{Hg}, the height of the mercury column between the two horizontal lines.

changes in one variable on changes in a related variable, other quantities must be held constant. The study of the gas laws illustrates this common experimental method.

Each of the gas laws is based upon one or more proportional relationships originally found by holding all but two variables constant. Proportional quantities are related to each other mathematically by multiplication by a constant.

In a *direct proportion* one variable is equal to a constant times a second variable. A **constant** has a numerical value that does not change.

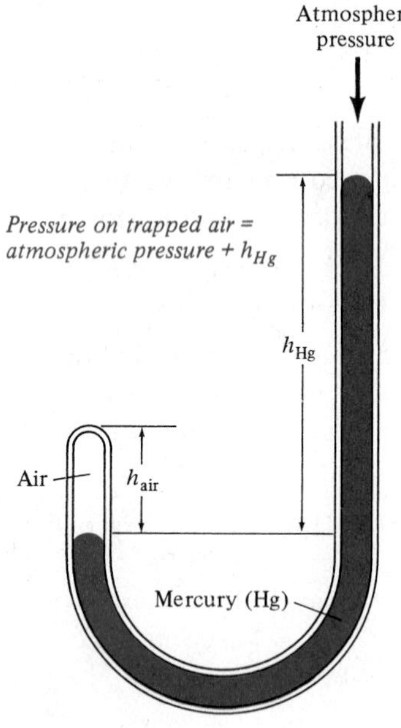

Figure 6.8
Boyle's Apparatus for Measuring Volume Changes with Changing Pressure
The height, h_{air}, is proportional to the volume of the trapped gas. The total pressure on the trapped gas is the atmospheric pressure plus the pressure of the mercury column, which is varied by adding mercury to the open side of the J tube. Boyle said that the particles of air behaved like coiled springs.

a direct proportion

$$\underbrace{a}_{\text{variable}} = \underbrace{k}_{\text{constant}}\underbrace{b}_{\text{variable}} \quad \text{or} \quad \frac{a}{b} = k \tag{6.2}$$

Directly proportional quantities increase and decrease together. For example, if the value of a doubles, the value of b doubles, so that their ratio a/b remains equal to k.

In an *inverse proportion*, one quantity is equal to a constant times the *reciprocal* of the other quantity (the reciprocal of b is $1/b$).

an inverse proportion

$$\underbrace{a}_{\text{variable}} = \underbrace{k}_{\text{constant}} \times \frac{1}{\underbrace{b}_{\text{variable}}} \quad \text{or} \quad ab = k \tag{6.3}$$

If one of two inversely proportional quantities increases, the other decreases. For example, if the value of a doubles, the value of b is halved, so that their product remains constant and equal to k.

6.4 VOLUME VERSUS PRESSURE: BOYLE'S LAW

Boyle's law applies to volume and pressure changes when the temperature and amount of a gas are constant. The volume of the gas decreases when the pressure exerted on the gas increases, and vice versa. For example, the helium in a balloon vendor's tank is under high pressure. The pressure within the balloons that the vendor fills is much lower. Therefore, the total volume of all of the balloons that the vendor can fill from the tank will be much greater than that of the tank itself.

The apparatus used by Boyle in his experiments is shown in Figure 6.8. He observed that if the pressure on a gas is doubled while the temperature remains unchanged, the gas volume decreases to one-half the original volume; if the pressure is tripled, the gas volume decreases to one-third the original volume; and so on. These are changes characteristic of an inversely proportional relationship.

Boyle's law is stated in words, as follows: At constant temperature, the volume of a given mass of gas is inversely proportional to the pressure upon the gas. Stated mathematically

At constant temperature and mass of gas:

$$\underbrace{V}_{\text{volume of the gas}} = \text{constant} \times \frac{1}{\underbrace{P}_{\text{pressure of the gas}}} \quad \text{or} \quad PV = \text{constant} \tag{6.4}$$

These relationships hold for any given quantity of gas at a fixed temperature, whether the gas is a pure substance or a mixture. Volume is plotted versus pressure in Figure 6.9, showing the types of curves always obtained for inversely proportional quantities.

The explanation of Boyle's law in terms of kinetic-molecular theory is quite simple: When gas volume is decreased, the space between gas molecules becomes smaller. However, because the temperature is unchanged, there is no change in the speeds with which the molecules are moving. Therefore, it is inevitable that the molecules collide with the walls more often, leading to higher pressure.

For a given amount of a gas at constant temperature, the Boyle's law relationship can be written as $PV = k$, where k represents a constant. Mathematically, where P_1 and V_1 represent the initial conditions, and P_2 and V_2 represent the new conditions

$$P_1V_1 = k \qquad P_2V_2 = k$$

and therefore, P_1V_1 and P_2V_2 must equal each other

$$P_1V_1 = P_2V_2 \tag{6.5}$$

If any three of the quantities in Equation (6.5) are known, the equation can be solved to find the value of the fourth quantity. For example, if a gas of known P_1 and V_1 undergoes a change to a known new volume of V_2, then the new pressure can be found from the following equation (provided that the temperature is unchanged):

$$P_2 = \frac{P_1V_1}{V_2} \tag{6.6}$$

The factor V_1/V_2 in Equation (6.6) can be thought of as a correction factor: Multiplying the original pressure, P_1, by this correction factor gives the new pressure. Use of this reasoning either to solve a problem or to provide a commonsense check of the answer to a problem is illustrated in Example 6.1.

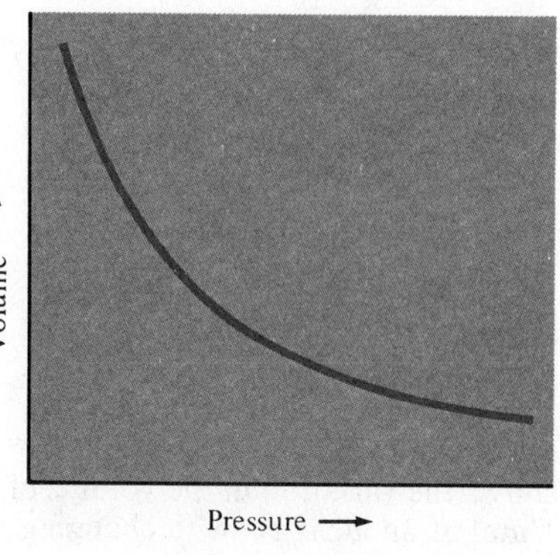

(a)

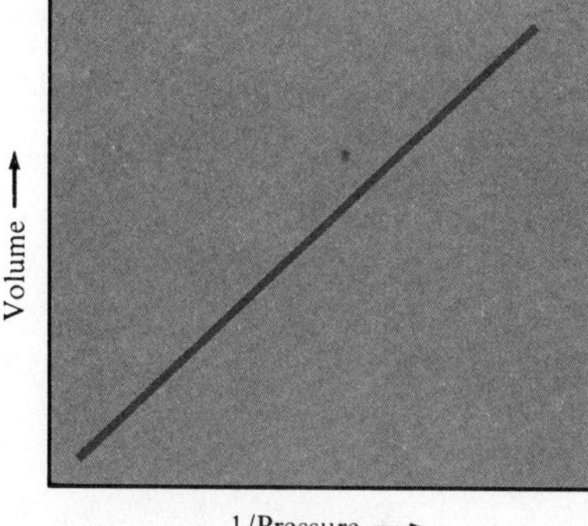

(b)

Figure 6.9
Boyle's Law
In (a) volume is plotted versus pressure for a gas at constant temperature. If volume is plotted versus the reciprocal of the pressure ($1/P$) as in (b), a straight line is obtained. Plots like these are always obtained when the two plotted quantities are inversely proportional to each other.

EXAMPLE 6.1 *P–V–T* **Relationships:** *V* **and** *P*

A sample of an ideal gas at 0.93 atm and 25 °C occupied a volume of 17.3 L. This gas was transferred to a 3.7 L container without a temperature change. What was the pressure of the gas under the new conditions?

Because the process involved a fixed mass of gas and was done at a constant temperature, the Boyle's law relationship, $P_1V_1 = P_2V_2$, can be used. The original conditions of the gas were

$$P_1 = 0.93 \text{ atm} \qquad V_1 = 17.3 \text{ L}$$

and the final conditions are

$$P_2 = ? \qquad V_2 = 3.7 \text{ L}$$

Using Equation (6.5), solving for P_2, and substituting the above values gives

$$P_2 = (P_1)\left(\frac{V_1}{V_2}\right) = (0.93 \text{ atm})\left(\frac{17.3 \text{ L}}{3.7 \text{ L}}\right) = \mathbf{4.3 \text{ atm}}$$

This problem could also be solved by analyzing what must happen to the original pressure. Because the volume has decreased, the pressure must increase. Therefore, the correction factor for P_1 based on the known volumes must be greater than one, that is, it must be (17.3 L/3.7 L). As a commonsense check on the answer obtained above, we can see that the new pressure found is greater than the original pressure, and this is reasonable, for the volume decreased.

EXERCISE What will be the volume occupied by an ideal gas at 75 kPa after it has expanded from 10.0 L and 145 kPa at constant temperature? *Answer* 19 L.

6.5 VOLUME VERSUS TEMPERATURE: CHARLES' LAW

Charles' law applies to volume and temperature changes when the pressure and amount of a gas are constant. As a gas is heated, it expands, and as it is cooled, it contracts. Have you ever left a balloon in the bright sunlight and seen it burst? As the temperature rises the balloon expands. Eventually its maximum volume is reached, and as the temperature increases still further the increasing pressure within causes it to burst.

Plotting the volume versus the temperature for any quantity of an ideal gas at constant pressure gives a straight line, showing that volume and temperature have a directly proportional relationship. Such a plot for an ideal gas is drawn in Figure 6.10. The dashed portion of the plot illustrates a "logical" conclusion from the observations:

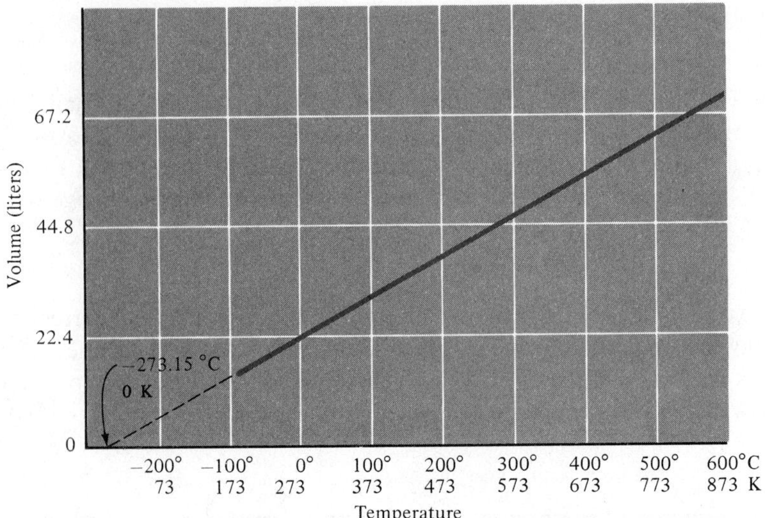

Figure 6.10
Charles' Law
Volume versus temperature is plotted at constant pressure. This graph shows the variation in the volume of 1 mol of an ideal gas with changing temperature at 1 atm. The equivalent temperatures on the Celsius and Kelvin scales are shown. Note that when the Kelvin temperature doubles (2×273 K), the volume occupied by 1 mol doubles (from 22.4 L to 44.8 L).

When an ideal gas is cooled to −273.15 °C, it should disappear entirely. We know from experiment that this does not happen to real gases. Instead, Charles' law becomes a less and less accurate description of gas behavior as the temperature decreases. In other words, at lower temperatures real gases become less like ideal gases. The velocities of the molecules decrease, spaces between them become smaller, and the real forces of attraction that do exist between molecules take effect. Before −273.15 °C is reached, the gas molecules get close enough together for the forces of attraction between them to overcome the energy of their random motion; as a result, the gas liquefies or solidifies.

When extended, a plot of volume versus temperature for any ideal gas will reach zero volume at the same temperature. This temperature is **absolute zero,** −273.15 °C, the lowest possible temperature. According to kinetic-molecular theory, temperature drops as energy is removed. At absolute zero no further energy can be removed and this means that no lower temperature is possible. Absolute zero has been approached very closely in the laboratory, but never attained.

An **absolute temperature scale** takes absolute zero as its zero point. The **Kelvin temperature scale** (see Section 2.8e and Figure 2.2) is an absolute temperature scale with units equal to Celsius degrees; 0 K is equal to −273.15 °C. Kelvin's scale allows **Charles' law** to be stated as follows: At constant pressure, the volume of a given mass of gas is directly proportional to the absolute temperature. Mathematically, Charles' law takes the following forms.

At constant pressure and for a given mass of gas:

temperature of the gas in kelvins

$$V = \text{constant} \times T$$

or **(6.7)**

$$\frac{V}{T} = \text{constant}$$

By the same reasoning as used for Boyle's law, since $V_1/T_1 = k$ and $V_2/T_2 = k$, then

$$\frac{V_1}{T_1} = \frac{V_2}{T_2} \qquad \textbf{(6.8)}$$

EXAMPLE 6.2 P–V–T Relationships: V and T

A cylinder with a movable piston is filled at 24 °C with a gas that occupies 36.2 cm³. If the maximum capacity of the cylinder is 65.2 cm³, what is the highest temperature to which the cylinder can be heated at constant pressure without having the piston come out?

The initial and final conditions of the gas are

$$V_1 = 36.2\ \text{cm}^3 \qquad T_1 = 24° + 273 = 297\ \text{K}$$
$$V_2 = 65.2\ \text{cm}^3 \qquad T_2 = ?$$

(Note that temperature must always be given on the Kelvin scale in gas law problems.) Solving Charles' law, Equation (6.8), for T_2 and substituting the known values gives

$$T_2 = (T_1)\left(\frac{V_2}{V_1}\right) = (297\ \text{K})\left(\frac{65.2\ \text{cm}^3}{36.2\ \text{cm}^3}\right) = \mathbf{535\ K} \text{ or } \mathbf{262\ °C}$$

(Let's apply the commonsense check of the solution to make sure that we have not made an error. The volume of the gas increases in this problem. According to Charles' law, for this to occur the temperature must have increased. Mathematically, V_2/V_1, the correction factor that must be applied to the temperature, is greater than 1. Thus the answer of 535 K seems reasonable.)

EXERCISE The temperature of a 35.2 L sample of a gas was changed from 25 °C to 75 °C while keeping the pressure constant. What is the new volume? *Answer* 41.1 L.

6.6 *P, V,* AND *T* CHANGES IN A FIXED AMOUNT OF GAS

a. Combined Gas Law The relationships among pressure, volume, and temperature expressed in Boyle's law and Charles' law (Equations 6.5 and 6.8) can be combined mathematically:

For a fixed mass of gas:

$$\frac{P_1V_1}{T_1} = \frac{P_2V_2}{T_2} \tag{6.9}$$

This equation, sometimes known as the **combined gas law**, represents the interrelated changes in temperature, pressure, and volume for a gas sample of fixed mass. From the viewpoint of the kinetic-molecular theory, as the molecules of a gas confined in a vessel move faster and faster with increasing temperature, they collide with the walls more often and with greater force. If the walls are flexible, they are pushed back and the volume increases. If the walls are fixed, the pressure on the walls increases (see Figure 6.4). With lower temperatures, the molecular velocities are lower and the opposite effects occur: pressure decreases and, if the walls are flexible, volume decreases.

The combined gas law can be used to solve any problems involving changes in the variables *P*, *V*, and *T* for a fixed amount of a gas. Knowing any five of the quantities in Equation (6.9), the sixth can be calculated by solving the equation for that quantity (as illustrated in Example 6.4). When one variable remains unchanged ($P_1 = P_2$, or $V_1 = V_2$, or $T_1 = T_2$), that variable can be cancelled from both sides of the equation (Table 6.3).

Example 6.3 is solved by using our method for analyzing a problem (Section 2.11). Note that because variables which do not change may be canceled, one need only remember the combined gas law, rather than also remembering Boyle's and Charles' laws separately.

Table 6.3
***P, V, T,* Changes in a Fixed Mass of Gas**
The subscript 1 indicates the initial conditions and the subscript 2, the new conditions.

At constant *P*,	$\frac{V_1}{T_1} = \frac{V_2}{T_2}$
At constant *V*,	$\frac{P_1}{T_1} = \frac{P_2}{T_2}$
At constant *T*,	$P_1V_1 = P_2V_2$

EXAMPLE 6.3 *P-V-T* **Relationships: Problem Solving**

A sample of gas confined in a 2.0 L container at 930 Torr and 25 °C was heated to 45 °C. What was the final pressure?

1. Study the problem and be sure you understand it.
   (a) What is known?
   The initial and final conditions for a gas that has been heated at constant volume:

$$P_1 = 930 \text{ Torr} \qquad T_1 = 25\ ^\circ\text{C} \qquad V_1 = 2.0 \text{ L}$$
$$T_2 = 45\ ^\circ\text{C} \qquad V_2 = 2.0 \text{ L}$$

   (b) What is unknown?
   The final pressure, P_2.
2. Decide how to solve the problem.
   (a) What is the connection between the known and the unknown?
   This is a gas law problem for a fixed mass of gas. As for all problems of this type, the connection can be made by using

$$\frac{P_1V_1}{T_1} = \frac{P_2V_2}{T_2}$$

   (Alternatively, the correction factors that must be applied to the initial pressure can be employed.)
   (b) What is necessary to make the connection?
   First, recognizing that $V_1 = V_2$, cancel these terms from the equation. Then solve for the unknown, P_2.

$$\frac{P_1}{T_1} = \frac{P_2}{T_2} \qquad P_2 = \frac{P_1T_2}{T_1}$$

   This is the same expression that would have been obtained by utilizing the other approach—adjusting the initial pressure (P_1) by a correction factor (T_2/T_1) based on the temperature change. Before the problem can be solved, the temperature, which is given in Celsius degrees, must be converted to temperature on the Kelvin scale for use in a gas law calculation.

$$T_1 = 25^\circ + 273 = 298 \text{ K}$$
$$T_2 = 45^\circ + 273 = 318 \text{ K}$$

3. Set up the problem and solve it.

$$P_2 = (930 \text{ Torr})\left(\frac{318 \text{ K}}{298 \text{ K}}\right) = \mathbf{990\ Torr}$$

4. Check the result.
   (a) Are significant figures used correctly?
   Yes; the limit is two significant figures.
   (b) Is the answer in the correct units?
   Yes; the temperature units in the correction factor cancel.
   (c) Is the answer reasonable?
   An increase in temperature should cause an increase in pressure at constant volume, and the correction factor should be larger than 1, which it is (318 K/298 K). The final pressure is greater than the initial pressure, and the answer is reasonable.

EXAMPLE 6.4 *P-V-T* **Relationships:** *P*, *V*, **and** *T*

A sample of hydrogen occupied a volume of 4.00 L at 760. Torr and 31 °C. What volume would the gas occupy at 205 °C and 382 Torr?

All three variables change for this gas sample. The initial and final conditions are

$$P_1 = 760.\text{ Torr} \qquad T_1 = 31^\circ + 273 = 304 \text{ K} \qquad V_1 = 4.00 \text{ L}$$
$$P_2 = 382 \text{ Torr} \qquad T_2 = 205^\circ + 273 = 478 \text{ K} \qquad V_2 = ?$$

Solving Equation (6.9) for V_2 and substituting into it the above values gives

$$V_2 = (V_1)\left(\frac{P_1}{P_2}\right)\left(\frac{T_2}{T_1}\right) = (4.00\text{ L})\left(\frac{760.\text{ Torr}}{382\text{ Torr}}\right)\left(\frac{478\text{ K}}{304\text{ K}}\right) = \mathbf{12.5\text{ L}}$$

(To check the answer, consider that the factor P_1/P_2 should be larger than 1 because the pressure on the hydrogen gas has decreased, which means volume increases. Also, the factor T_2/T_1 should be larger than 1 because an increase in the temperature on the gas means the volume increases. The factors of 760. Torr/382 Torr and 478 K/304 K, both larger than 1, do seem reasonable.)

EXERCISE An ideal gas that initially had a volume of 175 L at 15 atm and 298 K was compressed to 75 L at 5.0 atm. What was the new temperature of this gas?

Answer 43 K.

b. Standard Temperature and Pressure (STP) Because the volume of a gas changes with both pressure and temperature, stating only the volume of a gas does not indicate the quantity of gas present. The temperature and pressure at which the volume was measured must also be given. For example, 4 L of hydrogen at 760 Torr and 30 °C is about three times as much hydrogen as 4 L of hydrogen at 380 Torr and 200 °C.

To simplify comparisons among quantities of gases, scientists have agreed to state gas volume at a specified temperature and pressure. The **standard temperature and pressure** (abbreviated STP) universally used for this purpose are 0 °C (273 K) and 760 Torr.

The volume of a gas is often reported as, for example, 25 L (STP). When the pressure and temperature are not given with a gas volume, you can assume that the conditions are STP. Measurements of gas volumes need not, of course, always be made at STP conditions. But the volume measured under any conditions of temperature and pressure may be converted to the value at STP by using the combined gas law.

EXAMPLE 6.5 *P-V-T* Relationships: *P*, *V*, and *T*

What is the volume at STP of a sample of gas that occupies 16.5 L at 352 °C and 0.275 atm?

This problem is solved in the same way as Example 6.4.

$$P_1 = 0.275\text{ atm} \qquad V_1 = 16.5\text{ L} \qquad T_1 = 352° + 273 = 625\text{ K}$$

$$P_{STP} = 1.000\text{ atm} \qquad V_{STP} = ? \qquad T_{STP} = 0° + 273 = 273\text{ K}$$

$$V_{STP} = V_1\left(\frac{P_1}{P_{STP}}\right)\left(\frac{T_{STP}}{T_1}\right) = (16.5\text{ L})\left(\frac{0.275\text{ atm}}{1.000\text{ atm}}\right)\left(\frac{273\text{ K}}{625\text{ K}}\right) = \mathbf{1.98\text{ L}}$$

EXERCISE What will be the volume at STP of a sample of gas that occupies 35.2 mL at 27 °C and 742 Torr? *Answer* 31.3 mL.

SUMMARY OF SECTIONS 6.3–6.6

For a fixed amount of gas, Boyle's law expresses the relationship of volume to pressure at constant temperature, $V = k \times (1/P)$, and Charles' law expresses the relationship of volume to temperature at constant pressure, $V = kT$. In all calculations dealing with gas laws, temperature is expressed on an absolute temperature scale, usually the Kelvin scale. The combined gas law, Equation (6.9), combines the relationships of pressure, temperature, and volume for a fixed amount of gas. At constant temperature or constant pressure, the combined gas law reduces to Boyle's law or Charles' law, respectively. Gas volumes are reported at STP, standard temperature and pressure, which are 0 °C and 760 Torr.

Aside 6.2 GASES AND ATOMIC MASSES

A chapter on gases in a chemistry textbook is overrun with laws named for scientists of the past. A chronology of the study and discovery of gases is given in Table 6.4. What was learned from gases contributed to an understanding of the elements and atomic theory, and to the beginnings of physical chemistry.

The study of gases led directly to an understanding of the distinctions among "atoms," "molecules," and "elements." Only after these distinctions were clarified was it possible to develop a valid relative atomic mass scale (see Section 3.12).

Joseph-Louis Gay-Lussac had observed a remarkably simple relationship: In chemical reactions involving gases as either reactants or products, the ratios of the gas volumes (at constant temperature and pressure) are small whole numbers. For example

$$\underset{\textit{2 volumes}}{\text{Hydrogen}} + \underset{\textit{1 volume}}{\text{oxygen}} \longrightarrow \underset{\textit{2 volumes}}{\text{steam}}$$

$$\underset{\textit{1 volume}}{\text{Hydrogen}} + \underset{\textit{1 volume}}{\text{chlorine}} \longrightarrow \underset{\textit{2 volumes}}{\text{hydrogen chloride}}$$

These simple ratios reflect the existence of atoms and molecules. However, explanation of these ratios was a problem that neither Gay-Lussac nor John Dalton could solve. Dalton considered the obvious possibility that equal volumes of gases contain equal numbers of reacting particles (at the same temperature and pressure), but had to reject the idea because he thought that water was HO. In this case, the formation of two volumes of steam would require two volumes of oxgyen as well as two volumes of hydrogen.

Avogadro resolved the dilemma by concluding that, at the same temperature and pressure, equal volumes of gases contain equal numbers of *molecules*. Furthermore, he recognized that not only do atoms of *different* elements combine, but that atoms of the *same* element can combine to form molecules.

Consider the reaction of hydrogen with chlorine as shown above. If Avogadro's conclusion that equal volumes of gases contain equal numbers of molecules is correct, then the number of molecules of hydrogen chloride gas formed is twice the number of hydrogen molecules consumed. But every molecule of hydrogen chloride must contain at least one atom of hydrogen, so each molecule of the original hydrogen molecule must have split in two. The formula of hydrogen should therefore be H_2, H_4, H_6, or some other multiple. Many experiments showed that when gaseous hydrogen is a reactant, the volume of a gaseous product is never more than twice the volume of the hydrogen consumed. Therefore, hydrogen must be H_2. By exactly the same reasoning, it could be found that chlorine is Cl_2, showing that the reaction of hydrogen with chlorine is correctly represented as

$$\underset{\textit{1 volume}}{H_2} + \underset{\textit{1 volume}}{Cl_2} \longrightarrow \underset{\textit{2 volumes}}{2HCl}$$

Once the relationship between the numbers of molecules and the volumes of gases was understood, a means was available for measuring atomic and molecular masses. By comparing the masses of equal volumes of gases (at the same temperature and pressure), the masses of individual molecules could be compared. For example, suppose the hydrogen atom has been assigned an atomic mass of one. Measurements would show that (at the same temperature and pressure) a given volume of oxygen is 16 times heavier than an equal volume of hydrogen and that a given volume of steam is nine times heavier than an equal volume of hydrogen. By using the correct molecular formulas, the molecular masses of 2 for hydrogen, 32 for oxygen, and 18 for water, are found.

$$\underset{\textit{mol. mass 2}}{2H_2} + \underset{\textit{mol. mass 32}}{O_2} \longrightarrow \underset{\textit{mol. mass 18}}{2H_2O}$$

MASS, MOLECULAR, AND MOLAR RELATIONSHIPS

6.7 GAY-LUSSAC'S LAW OF COMBINING VOLUMES AND AVOGADRO'S LAW

As discussed in Aside 6.2, the relationships among volumes of reacting gases were first observed by Gay-Lussac and then interpreted by Avogadro (see Table 6.4). The results of their efforts are now known as the following laws:

Gay-Lussac's law of combining volumes: When gases react and/or gaseous products are formed, the ratios of the volumes of the gases involved, measured at the same temperature and pressure, are small whole numbers. In other words, the relative volumes of gaseous reactants and/or products are the same as the coefficients of their formulas in the balanced equation (Figure 6.11, page 115). The relative volumes may

Obviously, the atomic mass of oxygen is 16 on this scale. Eventually, methods like this were used to develop the first accurate tables of relative atomic masses.

Sadly for the progress of chemistry, however, it took 50 years before the significance of Avogadro's law came to the attention of the scientific community. A Congress was called in 1860 in Karlsruhe, Germany, for the purpose of resolving the confusion among the many different "atomic" and "molecular" mass scales and systems of naming chemical compounds then in use. At this congress, Cannizzaro made a vigorous speech stressing the importance of Avogadro's work. Cannizzaro's reception was not enthusiastic, but the seeds had been sown. As a direct result of Cannizzaro's speech and his distribution of a paper at the congress, the validity of Avogadro's conclusions eventually came to be accepted. One chemist, after reading Cannizzaro's paper, wrote, "The scales seemed to fall from my eyes. Doubts disappeared and a feeling of quiet certainty took their place."

Table 6.4
A Chronology of Gases

Event	Year	Year	Event
		1620	Pilgrims land at Plymouth
Boyle's law (V–P; Robert Boyle)	1662		
		1672	Marquette and Joliet explore the Mississippi
Guillaume Amontons (V–T)	1701		
		~1715	"Age of Reason" begins
Daniel Bernoulli relates pressure to molecular motion	1738		
		1760	George III begins 60-year reign in England
Henry Cavendish discovers hydrogen	1766		
		1769	James Watt invents steam engine; industrial revolution begins in England
Daniel Rutherford discovers nitrogen	1772		
Joseph Priestley discovers oxygen	1774		
		1776	U.S. Declaration of Independence
Charles' law (P–T; Jacques-Alexandre-César Charles)	1787		
		1789	French revolution
Dalton's law (partial pressures; John Dalton)	1801		
		1804	Napoleon I becomes Emperor in France
Gay-Lussac's law (combining volumes; Joseph-Louis Gay-Lussac)	1808		
Avogadro's law (Amadeo Avogadro)	1811		
Graham's law of diffusion (Thomas Graham)	1830		
		1837	Queen Victoria begins 54-year reign in England
Graham's law of effusion	1846		
Stanislao Cannizarro presents Avogadro's work at Karlsruhe	1860	1860	Abraham Lincoln elected U.S. President
J. Clerk Maxwell and Ludwig Boltzmann, mathematics of kinetic theory	1860, 1868		
		1885	Louis Pasteur, first vaccine
Noble gases discovered	1894–1898		
		1904	Theodore Roosevelt elected U.S. President

be expressed in any units. For example, 1 L of H_2 and 1 L of Cl_2 will react to give 2 L of HCl; 0.25 mL of H_2 and 0.25 mL of Cl will give 0.50 mL of HCl.

Avogadro's law: Equal volumes of gases, measured at the same temperature and pressure, contain equal numbers of molecules. Therefore, equal volumes of gases at the same temperature and pressure must also contain equal numbers of moles of gases. Under identical conditions, a large gas sample contains more molecules and therefore a larger molar amount than a small gas sample. Thus, at a given temperature and pressure, gas volume is directly proportional to number of moles.

Avogadro's law applies to all gases, whether they are elements, simple chemical compounds, or the vapors from solids or liquids. In terms of kinetic-molecular theory, this is easy to understand. A gas is mostly empty space occupied by molecules moving around with an average energy that varies only with the temperature. At any given temperature a mole of gas A and a mole of gas B both contain the same number of

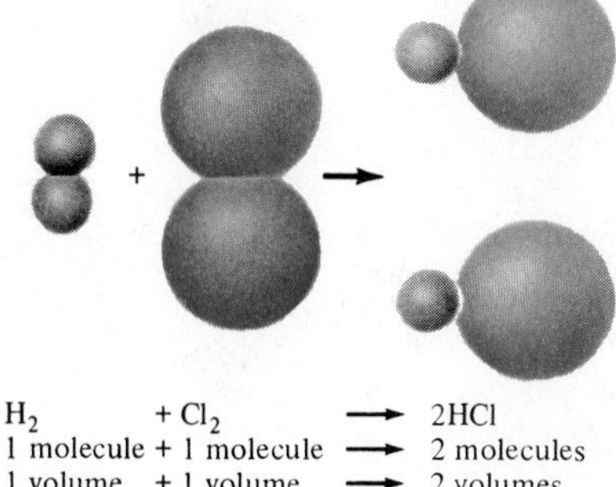

Figure 6.11
Reaction of Hydrogen with Chlorine
The relative gas volumes are given by the coefficients in a chemical equation.

molecules and the average kinetic energy of the molecules in both samples is the same. Therefore if the pressure is the same, the two gases should occupy the same volume.

Since the number of moles is directly proportional to the volume, Avogadro's law may be stated mathematically as follows:

At constant temperature and pressure:

$$V = \text{constant} \times n \qquad \text{or} \qquad \frac{V_1}{n_1} = \frac{V_2}{n_2} \tag{6.10}$$

number of moles of gas

when n_1 and n_2 represent the number of moles of the same gas in different volumes, or the number of moles of different gases in different volumes, both at the same temperature and pressure.

EXAMPLE 6.6 *P-V-T-n* **Relationships: Avogadro's Law**

Incandescent light bulbs contain inert gases, such as argon, so that the filament will last longer. The approximate volume of a 100-watt bulb is 130 cm^3, and the bulb contains 3×10^{-3} mol of argon. How many moles of argon would be contained in a 150-watt bulb under the same pressure and temperature conditions if the volume of the larger wattage bulb is 185 cm^3?

For the 100-watt bulb

$$n_1 = 3 \times 10^{-3}\ \text{mol} \qquad V_1 = 130\ \text{cm}^3$$

and for the 150-watt bulb

$$n_2 = ? \qquad V_2 = 185\ \text{cm}^3$$

Solving the mathematical statement of Avogadro's law (Equation 6.10) for n_2 and substituting the above values gives

$$n_2 = (n_1)\left(\frac{V_2}{V_1}\right) = (3 \times 10^{-3}\ \text{mol})\left(\frac{185\ \text{cm}^3}{130\ \text{cm}^3}\right) = \mathbf{4 \times 10^{-3}\ mol}$$

(Again we can check our answer by the commonsense method. Because the 150-watt bulb has a larger volume, according to Avogadro's law, there should be a larger quantity of gas in it—our answer of 4×10^{-3} mol seems reasonable.)

EXERCISE A flask contains 0.116 mol of nitrogen at a given temperature and pressure. Under the same temperature and pressure conditions, a 6.25 ft^3 flask contains 1.522 mol of nitrogen. What is the volume of the first flask? *Answer* 0.476 ft^3.

6.8 MOLAR VOLUME

According to Avogadro's law, at the same temperature and pressure, one mole of any gas should occupy the same volume as one mole of any other gas. **Standard molar volume** is defined as the volume occupied by one mole of a substance at standard temperature and pressure (STP). For gases that are reasonably close to "ideal," the standard molar volume averages 22.4 L/mol (Figure 6.12). The molar volume gives the factor 22.4 L/mol, which allows conversion between the number of moles of a gas and its volume at STP.

$$\text{Volume (liters; STP)} \underset{\times\left(\frac{22.4\ L}{1\ mol}\right)}{\overset{\times\left(\frac{1\ mol}{22.4\ L}\right)}{\rightleftarrows}} \text{No. of moles} \tag{6.11}$$

If one mole of gas at STP always occupies 22.4 L, then the molar mass of the gas is simply the mass of 22.4 L of that gas at STP. This is easy to find from the density of the gas at STP.

$$\text{(Gas density)(molar volume)} = \text{molar mass} \qquad \textbf{(6.12)}$$

$$\frac{\textit{grams}}{\cancel{\textit{liter}}} \qquad \frac{\cancel{\textit{liters}}}{\textit{mole}} \quad \frac{\textit{grams}}{\textit{mole}}$$

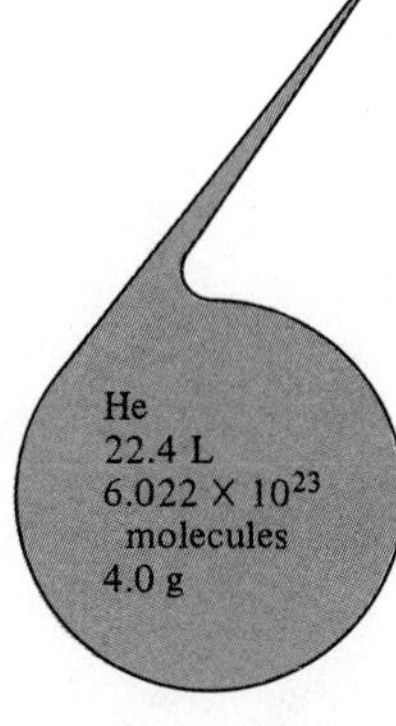

In solving some problems, 22.4 L/mol is a useful conversion factor; however, remember that it applies *only* to gases *at standard temperature and pressure.* Do not make the common error of applying the factor 22.4 L/mol to gases at other than standard temperature and pressure, or to solids or liquids.

Example 6.7 illustrates the use of molar volume and Example 6.8 shows how to analyze solving a problem that requires several steps and brings together concepts presented earlier. Note the technique of going backward from the unknown to determine what information is needed.

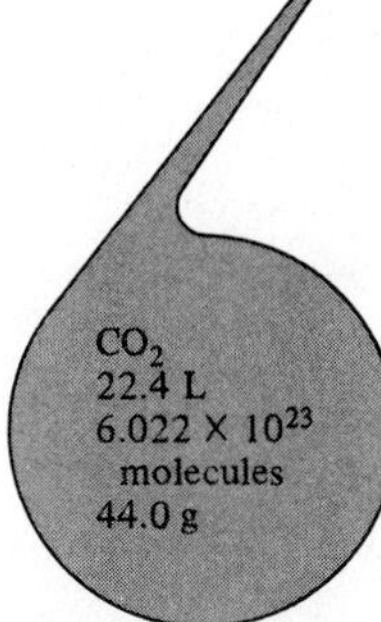

EXAMPLE 6.7 *P-V-T-n* Relationships: Molar Volume

What volume would be occupied at STP by 3.25 mol of an ideal gas?

The volume can be found by simply multiplying the number of moles of gas by the molar volume of an ideal gas at STP (22.4 L/mol), giving

$$(3.25 \text{ mol})\left(\frac{22.4 \text{ L}}{1 \text{ mol}}\right) = \mathbf{72.8\ L}$$

EXERCISE What volume at STP would be occupied by 0.0036 mol of an ideal gas?

Answer 81 mL.

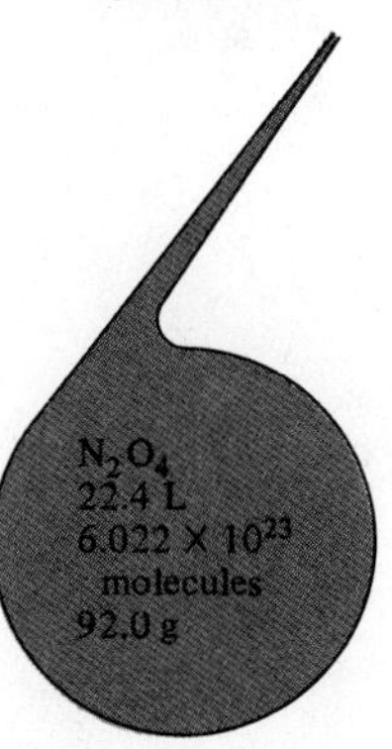

Figure 6.12
Avogadro's Law and Molar Volume

EXAMPLE 6.8 *P-V-T-n* Relationships: Molar Volume

A pure gas containing 92.3% carbon and 7.7% hydrogen by mass has a density of 1.16 g/L at STP. What is the molecular formula of the gas?

1. Study the problem and be sure you understand it.
   (a) What is known?
   The percentage composition of a gas and its density at STP.
   (b) What is unknown?
   The molecular formula of the gas—the number of carbon and hydrogen atoms combined in each molecule.
2. Decide how to solve the problem.
   (a) What is the connection between the known and the unknown?
   Making the connection in this problem is going to require several steps. First, we must ask what is directly needed to find the unknown. To find the molecular formula requires knowing the simplest formula and the molar mass (Section 4.11)—both also unknown. Then examine the known data: How do they apply? The percentage composition data can be used to find the simplest formula—the molar masses of carbon and hydrogen provide the connection. This leaves the molar mass of the gas in question unknown. It can be found from the gas density—here the standard molar volume

provides the connection. Then, knowing the simplest formula and the molar mass of the compound, it will be possible to determine the molecular formula.

(b) What is necessary to make the connection?
There are three steps to solving this problem. (i) Find the simplest formula. To do this, take 100 g of the compound as a basis for calculation and find the number of moles of carbon and hydrogen represented by 92.3 g of carbon and 7.7 g of hydrogen. The simplest formula is then found by determining the ratio of carbon to hydrogen atoms (Section 4.10). (ii) Find the molar mass by multiplying the density by the molar volume. (iii) Decide upon the molecular formula based on the simplest formula and molar mass.

3. Set up the problem and solve it.

(i) Find the simplest formula.

$$(92.3\text{ g C})\left(\frac{1\text{ mol C}}{12.0\text{ g}}\right) = 7.69\text{ mol C}$$

$$(7.7\text{ g H})\left(\frac{1\text{ mol H}}{1.0\text{ g H}}\right) = 7.7\text{ mol H}$$

It is apparent that the C to H ratio is 1 to 1 and the simplest formula is CH.

(ii) Find the molar mass.

$$\left(1.16\,\frac{\text{g}}{\text{L}}\right)\left(\frac{22.4\text{ L}}{\text{mol}}\right) = 26.0\text{ g/mol}$$

(iii) Find the molecular formula.
The molar mass based on the simplest formula of CH is 13.0 g/mol (12.0 g for C plus 1.0 g for H). The molar mass of the compound in question is twice this value and so the molecular formula is clearly $\mathbf{C_2H_2}$. [The compound is acetylene.]

4. Check the result.

(a) Are the calculations correct?
Yes. Significant figures have been used correctly, and the molar mass units are correct.

(b) Is the answer reasonable?
This is difficult to judge here. Your later studies will show that it is reasonable for a gaseous compound of carbon and hydrogen to have a low molar mass.

6.9 IDEAL GAS LAW

Boyle's law, Charles' law, and Avogadro's law establish relationships for gases among four variables—the volume, the pressure, the temperature, and the amount of gas expressed in moles, *n*.

$$V = \text{constant} \times \frac{1}{P}, \text{ at fixed } T \text{ and } n$$

$$V = \text{constant} \times T, \text{ at fixed } P \text{ and } n$$

$$V = \text{constant} \times n, \text{ at fixed } P \text{ and } T$$

These proportionalities combine to give a single relationship

$$V = \text{constant} \times \frac{1}{P} \times T \times n \qquad \text{or} \qquad PV = \text{constant} \times T \times n$$

The constant is represented by the symbol *R* and, written in its usual form, this gives the **ideal gas law**

$$PV = nRT \tag{6.13}$$

Based as it is on the individual gas laws, which assume ideal behavior of gases, this equation summarizes the behavior of ideal gas.

The gas constant, R, is the same for all gases under conditions where their behavior is close to ideal. The value and units for R depend upon the units of P, V, and T. The value of R can be calculated from molar volume and the standard temperature and pressure. In conventional units, using the most accurate value for the molar volume of an ideal gas, 22.41383 L

$$R = \frac{PV}{nT} = \frac{(1\ \text{atm})(22.41383\ \text{L})}{(1\ \text{mol})(273.15\ \text{K})} = 0.082057\ \text{L atm/K mol}$$

Usually, this value of R is rounded off to 0.0821 L atm/K mol.

As Table 6.5 shows, pressure and volume can be expressed in many different units, and for each combination of units R has a different numerical value. In working problems with the ideal gas law, it is *essential* that the units of P, V, and T and the units of the gas constant be in agreement. (It is convenient to memorize one value for the gas constant and, in solving a problem, to convert the pressure and volume to the appropriate units.)

Table 6.5
Ideal Gas Constant Values

0.0821 L atm/K mol
62.36 L Torr/K mol
8.314×10^3 L Pa/K mol
8.314 m^3 Pa/K mol
1.987 cal/K mol
8.314 J/K mol

The original gas laws were based on observation. From kinetic-molecular theory it can be shown that PV is a function solely of the number of molecules, their mass, and their velocity. By combining this relationship with the proportionalities between kinetic energy and temperature, and between the number of molecules and the number of moles, it is possible to arrive at $PV = nRT$ from theory rather than experiment.

Example and Exercise 6.9 illustrate some of the many types of problems that can be solved by using the ideal gas law. Knowing any three of the four quantities P, V, n, and T, the ideal gas law can be used to find the fourth.

EXAMPLE 6.9 *P-V-T-n* Relationships: Ideal Gas Law

The statement was made in Section 6.6b that 4.0 L of hydrogen at 760 Torr and 30. °C (sample 1) contains about three times as much hydrogen as does 4.0 L of hydrogen at 380 Torr and 200. °C (sample 2). Prove this statement by using the ideal gas law and comparing the number of moles of each gas.

For sample 1:

$$P = (760\ \text{Torr})(1\ \text{atm}/760\ \text{Torr}) = 1.0\ \text{atm} \qquad V = 4.0\ \text{L}$$
$$T = 30.^\circ + 273 = 303\ \text{K} \qquad n = ?$$

which, upon substituting into the ideal gas law, gives

$$n = \frac{PV}{RT} = \frac{(1.0\ \text{atm})(4.0\ \text{L})}{(0.0821\ \text{L atm/K mol})(303\ \text{K})} = 0.16\ \text{mol}$$

For sample 2:

$$P = (380\ \text{Torr})(1\ \text{atm}/760\ \text{Torr}) = 0.50\ \text{atm} \qquad V = 4.0\ \text{L}$$
$$T = 200.^\circ + 273 = 473\ \text{K} \qquad n = ?$$

which gives

$$n = \frac{(0.50\ \text{atm})(4.0\ \text{L})}{(0.0821\ \text{L atm/K mol})(473\ \text{K})} = 0.052\ \text{mol}$$

The ratio of the number of moles in sample 1 to the number of moles in sample 2 is

$$\frac{(0.16\ \text{mol})}{(0.052\ \text{mol})} = 3.1$$

Sample 1 contains a little over three times as much hydrogen as sample 2.

EXERCISE What pressure would be necessary to confine 1.38 mol of an ideal gas at 375 °C in a volume of 5.2 L? *Answer* 14 atm.

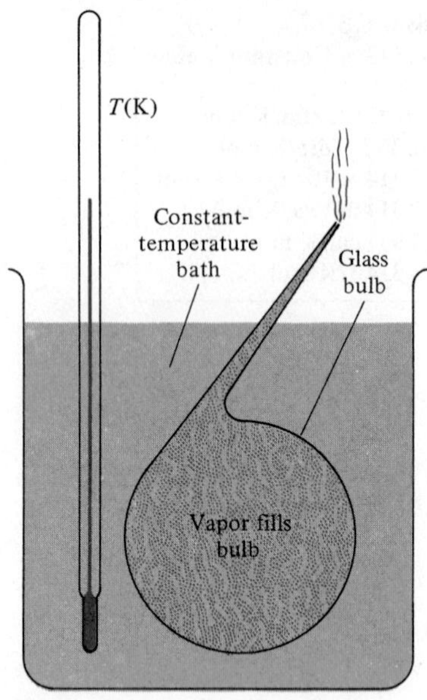

Figure 6.13
Molar Mass Determination (the Dumas Method)

6.10 MASS AND DENSITY

Molar mass, gas density, and sample mass can all be determined for gases from P, V, T data and the ideal gas law. The number of moles (n) equals the mass (m) divided by the molar mass (M).

$$\text{No. of moles} = \frac{\text{mass}}{\text{molar mass}} \qquad \text{or} \qquad n = \frac{m}{M}$$

$$\textit{moles} \qquad \frac{\textit{grams}}{\textit{grams/mole}}$$

Substitution of this relationship into the ideal gas equation gives a form convenient for use in problems where masses are involved.

$$PV = \frac{m}{M}RT \tag{6.14}$$

Equation (6.14) is the basis for a method of molar mass determination (the Dumas method, Figure 6.13). In this method, an empty glass bulb with a known volume is weighed. A few grams of a volatile liquid or solid are placed in the bulb. The sample is heated until it is completely vaporized and no excess vapor escapes from the thin neck of the bulb. The bulb is then cooled and weighed. The molar mass is determined from $M = mRT/PV$; $m =$ the mass of the bulb plus the cooled vapor minus the mass of the empty bulb; $T =$ the temperature of the bath; $P =$ the atmospheric pressure, read from a barometer; $V =$ the volume of the flask.

It is also sometimes useful to introduce density into the ideal gas law. Density (d) is mass per unit volume (m/V) and can be used to convert volume into mass:

$$d = \frac{m}{V} \qquad \text{or} \qquad m = dV$$

Substituting $m = dV$ into Equation (6.14), cancelling V from both sides, and rearranging gives a relationship which allows calculation of the density of a gas at a given pressure and temperature if its molar mass is known.

$$PV = \frac{dV}{M}RT \qquad d = \frac{PM}{RT} \tag{6.15}$$

EXAMPLE 6.10 Molar Mass of Ideal Gas

A liquid was known to be either methyl alcohol (CH_3OH) or ethyl alcohol (CH_3CH_2OH). The Dumas method for determining molar masses (see Figure 6.13). was used to obtain an approximate molar mass, and this value was used to identify the alcohol. The gaseous alcohol at 98 °C and 740 Torr had a mass of 0.276 g in a Dumas bulb of volume equal to 270 mL. Which alcohol was present?

For the alcohol

$$P = (740\ \text{Torr})\left(\frac{1\ \text{atm}}{760\ \text{Torr}}\right) = 0.97\ \text{atm} \qquad T = 98° + 273 = 371\ \text{K}$$

$$V = (270\ \text{mL})\left(\frac{1\ \text{L}}{1000\ \text{mL}}\right) = 0.27\ \text{L} \qquad m = 0.276\ \text{g}$$

Solving Equation (6.14) for the molar mass and substituting the above values, we get

$$M = \frac{mRT}{PV} = \frac{(0.276\ \text{g})(0.0821\ \text{L atm/K mol})(371\ \text{K})}{(0.97\ \text{atm})(0.27\ \text{L})} = \mathbf{32\ g/mol}$$

The molar mass of the unknown alcohol is 32 g/mol, which agrees with the value calculated from atomic masses for **methyl alcohol.**

EXERCISE An ideal gas had a mass of 0.0218 g and occupied a volume of 1.111 L at 0 °C and 0.0100 atm. What is the molar mass of this gas? *Answer* 44.0 g/mol.

EXAMPLE 6.11 Density of Ideal Gases

What is the density of acetone vapor, CH_3COCH_3 at 95 °C and 650 Torr?

For the gas:

$$P = (650\text{ Torr})(1\text{ atm}/760\text{ Torr}) = 0.86\text{ atm} \qquad M = 58\text{ g/mol}$$
$$T = 95° + 273 = 368\text{ K}$$

The density of the gas can be calculated using Equation (6.15).

$$d = \frac{PM}{RT} = \frac{(0.86\text{ atm})(58\text{ g/mol})}{(0.0821\text{ L atm/K mol})(368\text{ K})} = \mathbf{1.7\text{ g/L}}$$

EXERCISE The density of ozone at 1.013 atm and 26.5 °C is 1.979 g/L. What is the molar mass of this gas? *Answer* 48.04 g/mol.

6.11 PRESSURE IN GAS MIXTURES: DALTON'S LAW

In a mixture of ideal gases that do not interact chemically with each other, each molecule moves about independently just as it would in the absence of molecules of other kinds. Each gas distributes itself uniformly throughout the entire space available as though no other gas were present. The molecules strike the walls as frequently, with the same energy, and therefore with the same pressure, as they do when no other gas is present.

Among the studies which led John Dalton to the atomic theory were experiments on gas pressures in mixtures. He summarized his conclusions in 1803 in what is known as **Dalton's law of partial pressures:** In a mixture of gases, the total pressure exerted is the sum of the pressures that each gas would exert if it were present alone under the same conditions.

If several different gases (1, 2, 3, ...) are placed in the same container they form a homogeneous mixture. The pressure of a single gas in a mixture is called its **partial pressure.** The total pressure of a gas mixture is given by

partial pressures of gas₁, gas₂, gas₃,...

$$P_{\text{total}} = P_1 + P_2 + P_3 + \cdots \qquad \textbf{(6.16)}$$

Consider a volume of air under a pressure of 1000 Torr. Approximately one-fifth of the molecules are oxygen molecules and approximately four-fifths are nitrogen molecules. The approximate partial pressures are, therefore, 200 Torr for oxygen and 800 Torr for nitrogen.

The fraction of molecules of each gas present in a mixture is given by the **mole fraction,** X—the number of moles of a component in a mixture divided by the total number of moles of all components in the mixture.

mole fraction of component 1

$$X_1 = \frac{n_1}{n_1 + n_2 + \cdots} \qquad \textbf{(6.17)}$$

Because each gas in a mixture exerts the same pressure that it would exert by itself, the mole fraction of a gas is also the fraction of the total pressure exerted by that gas.

partial pressure of component 1

$$P_1 = X_1 P_{\text{total}} \quad \text{or} \quad X_1 = \frac{P_1}{P_{\text{total}}} \qquad \textbf{(6.18)}$$

EXAMPLE 6.12 Dalton's Law

A 6.2 L sample of N_2 at 738 Torr is mixed (at constant temperature) with 15.2 L of O_2 at 325 Torr. The gaseous mixture is placed in a 12.0 L container. (a) What is the pressure of the mixture? (b) What is the composition of the mixture in mole fractions?

(a) For each gas, Boyle's law applies as follows
For N_2:

$$P_1 = 738 \text{ Torr} \qquad V_1 = 6.2 \text{ L}$$
$$P_2 = ? \qquad V_2 = 12.0 \text{ L}$$
$$P_2 = (P_1)\left(\frac{V_1}{V_2}\right) = (738 \text{ Torr})\left(\frac{6.2 \text{ L}}{12.0 \text{ L}}\right) = 380 \text{ Torr}$$

For O_2:

$$P_1 = 325 \text{ Torr} \qquad V_1 = 15.2 \text{ L}$$
$$P_2 = ? \qquad V_2 = 12.0 \text{ L}$$
$$P_2 = (325 \text{ Torr})\left(\frac{15.2 \text{ L}}{12.0 \text{ L}}\right) = 412 \text{ Torr}$$

According to Dalton's law of partial pressure, the pressure of the mixture is

$$P_{\text{total}} = P_{O_2} + P_{N_2} = 412 \text{ Torr} + 380 \text{ Torr} = \mathbf{790\ Torr}$$

(b) The mole fraction of each component in the mixture can be found by using Equation (6.18).

$$X_{N_2} = \frac{P_{N_2}}{P_{\text{total}}} = \frac{380 \text{ Torr}}{790 \text{ Torr}} = \mathbf{0.48} \qquad X_{O_2} = \frac{P_{O_2}}{P_{\text{total}}} = \frac{412 \text{ Torr}}{790 \text{ Torr}} = \mathbf{0.52}$$

EXERCISE The total pressure of a mixture of water vapor and helium is 0.893 atm. The partial pressure of water is 27.3 Torr at this temperature. For helium, what are (a) the partial pressure and (b) the mole fraction in the mixture?

Answers (a) 0.857 atm and (b) 0.960.

In a closed container partly filled with water, the air over the liquid soon becomes mixed with water vapor. Eventually, the air holds all of the water vapor that it can, and some of the water vapor condenses. For example, in a terrarium (Figure 6.14), water is constantly evaporating from the surface of the plants and soil, and in an equal but opposite change, water is condensing on the walls and lid and running back into the soil. In the terrarium, as in any other closed vessel containing a liquid, the processes of evaporation and condensation eventually reach a state of **dynamic equilibrium**—a state of balance between exactly opposite changes occurring at the same rate.

$$H_2O(l) \rightleftharpoons H_2O(g)$$

Figure 6.14
Dynamic Equilibrium in a Terrarium
Liquid water and water vapor are in equilibrium.

The **equilibrium vapor pressure** (often referred to as simply the "vapor pressure") is the pressure exerted by the vapor over a liquid (or solid) once evaporation and condensation have reached equilibrium. The total pressure over a liquid is the sum of the pressure of any gas present plus the vapor pressure of the liquid at the existing temperature.

Frequently a gas is collected in the laboratory over some liquid in which it is not very soluble (Figure 6.15). The procedure is always basically the same: The vessel in which the gas is to be confined is filled with the liquid and inverted in a container of the same liquid. The gas is led through a tube from the apparatus in which it is generated, under the liquid, and to the mouth of the vessel, where it bubbles up into the vessel through the liquid. Once the gas is collected, the bottle is moved up or down until the liquid levels inside and outside the bottle are the same. At this point the gas inside the bottle is at atmospheric pressure.

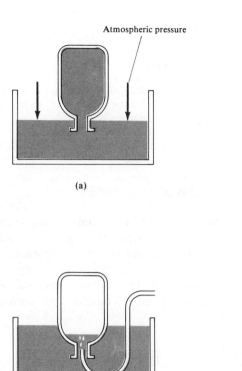

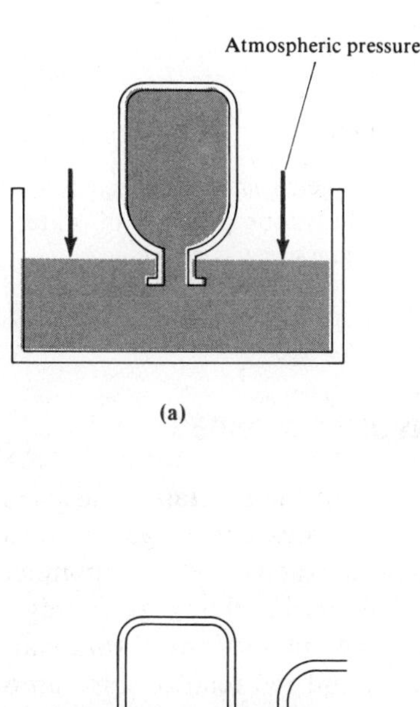

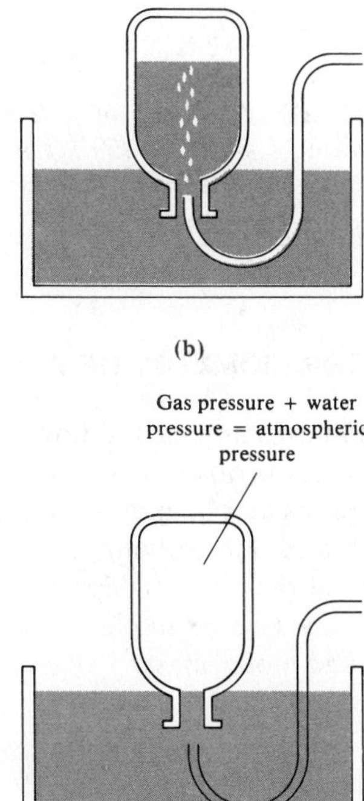

Figure 6.15
Collecting a Gas over a Liquid
Atmospheric pressure on the surface in the dish keeps the bottle full (a) until the liquid in it is displaced by the gas (b). When all of the gas is collected its pressure may be smaller (b) or greater (c) than that of the atmosphere. By raising or lowering the bottle, the liquid levels are made equal (d). In this way, the gas is brought to atmospheric pressure (which can be read from a barometer). The gas volume is determined by marking the level of the liquid in the bottle, setting the bottle upright, and measuring the volume of water required to fill the bottle to the mark.

A gas collected over water is "wet"—it contains water vapor. For a gas collected at atmospheric pressure, the total gas pressure is the sum of the pressure of the water vapor and the pressure of the gas collected, and equals the atmospheric pressure.

$$P_{atm} = P_{H_2O} + P_{gas} \tag{6.19}$$

The volume of the gas collected is measured at atmospheric pressure (see Figure 6.15). The value of P_{atm} is found by reading a barometer. The value of P_{H_2O} is read from a table of water vapor pressures at various temperatures (Appendix IV). With this information, the partial pressure of the gas collected and the amount of gas collected can be found as shown in Example 6.13.

EXAMPLE 6.13 Dalton's Law

Potassium chlorate ($KClO_3$) was heated to give oxygen.

$$2KClO_3(s) \xrightarrow[MnO_2]{\Delta} 2KCl(s) + 3O_2(g)$$

A volume of 550 mL of gas was collected over water at 21 °C and an atmospheric pressure of 743 Torr. The vapor pressure of water at 21 °C is 19 Torr. How many moles of oxygen were collected?

The pressure of oxygen is the total pressure less the partial pressure of the water vapor. Thus

$$P = (743 \text{ Torr} - 19 \text{ Torr})(1 \text{ atm}/760 \text{ Torr}) = 0.953 \text{ atm}$$

$$V = (550 \text{ mL})\left(\frac{1 \text{ L}}{1000 \text{ mL}}\right) = 0.55 \text{ L}$$

$$T = 21° + 273 = 294 \text{ K}$$

Substituting these values into the ideal gas law gives us

$$n = \frac{PV}{RT} = \frac{(0.953\ \text{atm})(0.55\ \text{L})}{(0.0821\ \text{L atm/K mol})(294\ \text{K})} = \mathbf{0.022\ mol}$$

EXERCISE A volume of 45.2 mL of "wet" hydrogen gas was collected by the displacement of water at 759.3 Torr and 23.8 °C. The vapor pressure of water at this temperature is 22.1 Torr. What mass of "dry" hydrogen gas was collected?

Answer 0.00364 g H_2.

6.12 STOICHIOMETRY OF REACTIONS INVOLVING GASES

The known and/or unknown information about the reactants and products in a stoichiometry problem may include the volumes or pressures of gases. In such cases, volume ratios can be used to solve the problem in exactly the same manner as mole ratios. In addition, solving a stoichiometry problem involving gases might require using any of the relationships discussed in the preceding sections of this chapter—the ideal gas law (see Example 6.15), molar volume, and the relationships among mass, density, and molar mass.

EXAMPLE 6.14 Stoichiometry—Gases

The chemical reaction in old-fashioned laboratory alcohol burners is

$$\underset{\textit{ethyl alcohol}}{CH_3CH_2OH(l)} + 3O_2(g) \longrightarrow 2CO_2(g) + 3H_2O(g)$$

What volume of carbon dioxide will be produced by this combustion reaction if 6.2 L of oxygen has been consumed (gas volumes measured at the same temperature and pressure)?

The volume ratio needed is that which relates CO_2 volume to O_2 volume (2 volumes CO_2/3 volumes O_2). In terms of liters, the problem is solved as follows.

$$(6.2\ \text{L}\ O_2)\left(\frac{2\ \text{L}\ CO_2}{3\ \text{L}\ O_2}\right) = \mathbf{4.1\ L\ O_2}$$

EXERCISE Bromine trifluoride can be prepared by the direct combination of the elements:

$$Br_2(g) + 3F_2(g) \xrightarrow{\Delta} 2BrF_3(g)$$

What volumes of Br_2 and of F_2 measured at the same temperature and pressure are needed to prepare 50. mL of BrF_3? *Answers* 25 mL Br_2, 75 mL F_2.

EXAMPLE 6.15 Stoichiometry—Gases

Metallic tungsten is obtained by displacement of tungsten from tungsten(VI) oxide by hydrogen at high temperatures. What volume of H_2 (measured at 0.975 atm and 25 °C) is required to react quantitatively with 46.4 g of WO_3?

The stepwise method for solving stoichiometry problems is applied to this problem as follows, utilizing the ideal gas law to convert from moles to volume of hydrogen.

Step 1. Write balanced equation. The displacement reaction is

$$3H_2(g) + WO_3(s) \xrightarrow{\Delta} 3H_2O(g) + W(s)$$

Steps 2 and 3. Convert known quantity to number of moles, and use mole ratio to find unknown in number of moles. The mole ratio must connect H_2, about which information is needed, to WO_3, about which information is known.

$$(46.4 \text{ g WO}_3)\overbrace{\left(\frac{1 \text{ mol WO}_3}{231.85 \text{ g WO}_3}\right)}^{\textit{Step 2}}\overbrace{\left(\frac{3 \text{ mol H}_2}{1 \text{ mol WO}_3}\right)}^{\textit{Step 3}} = 0.600 \text{ mol H}_2$$

Step 4. Convert from number of moles to desired quantity.

$$V = \frac{nRT}{P} = \frac{(0.600 \text{ mol})(0.0821 \text{ L atm/K mol})(298 \text{ K})}{(0.975 \text{ atm})}$$

$$= \mathbf{15.1 \text{ L}}$$

EXERCISE Hydrazine, H_2NNH_2, is used as a rocket fuel.

$$H_2NNH_2(l) + O_2(g) \longrightarrow N_2(g) + 2H_2O(g)$$

What volume of N_2 will be produced at 850 °C and 0.23 atm for each 15.0 g of hydrazine that reacts? What volume of steam will be produced?

Answers 190 L N_2, 380 L $H_2O(g)$.

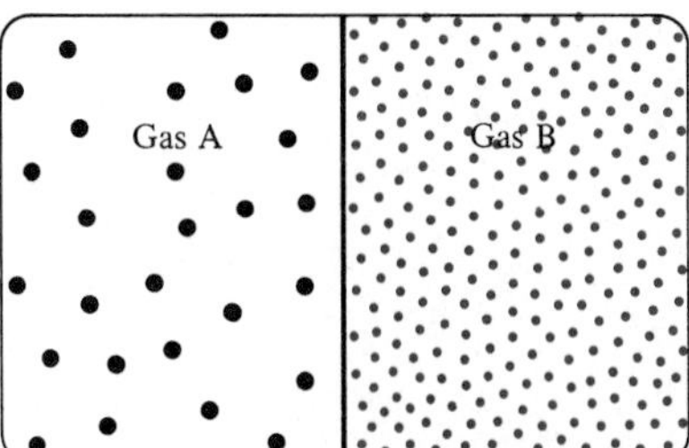

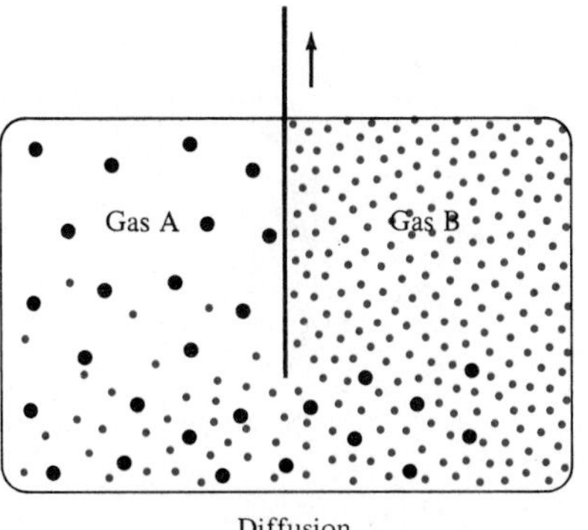

Figure 6.16
Diffusion

SUMMARY OF SECTIONS 6.7–6.12

Under the same conditions of temperature and pressure, equal volumes of gases contain equal numbers of molecules (Avogadro's law). As a result, in chemical reactions the relative volumes of gaseous reactants and products are given by the coefficients in the balanced chemical equation. In solving stoichiometry problems, therefore, the volume ratios given by the coefficients of gaseous reactants and/or products may be used in the same manner as mole ratios.

For any ideal gas, the ideal gas law ($PV = nRT$) summarizes the relationships among pressure, volume, temperature, and amount of gas. Knowing any three of these variables permits calculation of the fourth. Mass, density, and molar mass relationships can also be found by introducing these quantities into the ideal gas law (Equation 6.15). The sum of the pressures that individual gases would exert if they were present alone under the same conditions is the total pressure of a gaseous mixture (Dalton's law). The ideal gas law, molar volume, or any of the relationships involving the mass, density, or molar mass of gases may be needed to solve stoichiometry problems for reactions that include gases among the reactants and/or products.

BEHAVIOR OF GAS MOLECULES

6.13 DIFFUSION AND EFFUSION: GRAHAM'S LAW

Both diffusion and effusion are processes based on the random motion of gas molecules. When a bottle of perfume is opened on one side of a room and the aroma is soon noticed on the other side of the room, it is because molecules from the bottle have diffused into the air within the room. **Diffusion** is the mixing of molecules by random motion and collisions until a mixture becomes homogeneous (Figure 6.16). **Effusion** is the escape of molecules in the gaseous state one by one, without collisions, through a hole of molecular dimensions (Figure 6.17). The rate of effusion is controlled by the rate with which the random motion of the molecules leads them through the hole. Effusion is a simpler process than diffusion, because it does not involve collisions.

In studies conducted between 1831 and 1846, Scottish chemist Thomas Graham found that the relative rates of both the diffusion and effusion of two gases are inversely proportional to the square roots of the densities of the gases—a relationship known as **Graham's law.**

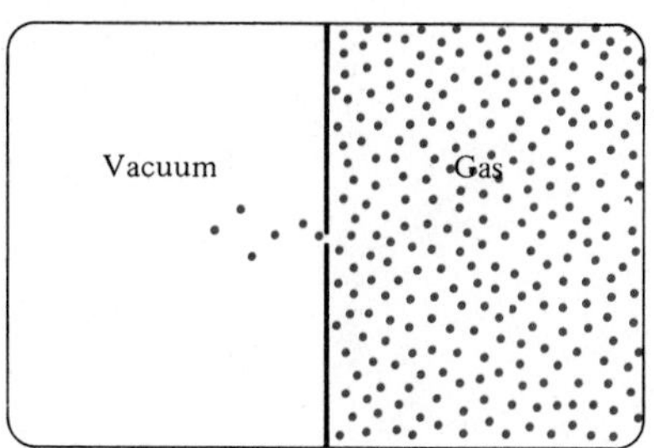

Figure 6.17
Effusion

$$\frac{\text{rate}_1}{\text{rate}_2} = \frac{\sqrt{d_2}}{\sqrt{d_1}} \tag{6.20}$$

By using the ideal gas equation stated in terms of density (Equation 6.15), the rates of diffusion and effusion are found also to be inversely proportional to the square roots of the molar masses, M. For gases 1 and 2

$$\frac{d_1}{d_2} = \frac{PM_1/RT}{PM_2/RT} = \frac{M_1}{M_2} \tag{6.21}$$

Combining this relationship with Graham's law gives

$$\frac{\text{rate}_1}{\text{rate}_2} = \frac{\sqrt{d_2}}{\sqrt{d_1}} = \frac{\sqrt{M_2}}{\sqrt{M_1}} \tag{6.22}$$

It can be seen qualitatively that Graham's law provides support for the assumptions of kinetic-molecular theory. According to the theory, at the same temperature gases of different masses have the same average kinetic energy, and therefore gases of higher molar mass must have slower average speeds than gases of lower molar mass. A heavier gas should collide with the hole in an effusion experiment less often and have a slower rate of effusion. Suppose gas 1 is a heavier gas and gas 2 is a lighter gas. From Equation (6.22) you can see that the larger is M_1, the molar mass of the heavier gas, the smaller must be its rate of effusion, rate 1.

Diffusion and effusion have several practical applications. The diffusion of a gaseous mixture through a membrane can be used to separate atoms or molecules of different masses. This technique is especially valuable in separating isotopes, which have very small mass differences. Such isotope separation is of crucial importance to the nuclear energy industry. Naturally occurring uranium contains only 0.72% of uranium-235 in a mixture with uranium-238—not a high enough percentage for practical use of the uranium in nuclear reactors. The uranium is converted to a mixture of the gaseous hexafluorides, $^{235}UF_6$ and $^{238}UF_6$. The $^{235}UF_6$, being lighter, diffuses faster than the compound containing the other isotope. By successive passage of the mixture through a series of porous membranes from higher to lower pressure in what is called the uranium enrichment process, the percentage of uranium-235 can be raised to a useful level. Example 6.16 illustrates the use of an effusion experiment to determine molar mass.

EXAMPLE 6.16 Graham's Law

It is more convenient to measure the time for a gas sample to effuse than to measure its actual rate of effusion. The rate is defined as the number of molecules moving through a molecular-sized hole per unit time. Therefore, rate is inversely proportional to time, giving for the effusion of equal numbers of molecules

$$\frac{\text{rate}_1}{\text{rate}_2} = \frac{\text{time}_2}{\text{time}_1} = \frac{\sqrt{M_2}}{\sqrt{M_1}}$$

Find the molar mass of a gas that takes 33.5 s to effuse from a porous container. An identical number of moles of CO_2 takes 25.0 s.

Solving the above equation for the molar mass of the unknown gas and substituting the respective times and M = 44.0 g/mol for CO_2 gives

$$M_2 = M_{CO_2}\left(\frac{\text{time}_2}{\text{time}_{CO_2}}\right)^2 = \left(44.0\frac{\text{g}}{\text{mol}}\right)\left(\frac{33.5\text{ s}}{25.0\text{ s}}\right)^2 = \mathbf{79.0\ g/mol}$$

EXERCISE A sample of methane, molar mass = 16.04 g, required 23.2 s to effuse from a container. What time was required for a sample of ethane, molar mass = 30.07 g?

Answer 31.8 s.

6.14 DEVIATIONS FROM THE GAS LAWS

Real gases differ from ideal gases because some of the assumptions that are made in describing ideal gas behavior are not entirely correct. For example, although it is assumed that the molecules in a gas exert no forces on each other, this is not quite true. There are forces of attraction that act between molecules of all gases. As pressure is increased and the molecules come closer together, real gas molecules experience these forces and PV is less than predicted for an ideal gas until the pressure becomes very high. In Figure 6.18 it is seen that N_2, CH_4, and O_2 are gases of this type. The greater the forces of attraction between molecules, the greater the deviation from ideal behavior.

Moreover, in deriving the gas laws it is assumed that changes in temperature and pressure affect the entire volume occupied by a gas. This is also not quite correct. The *spaces* between the molecules change with temperature, but the molecules themselves do not expand or contract. The volume occupied by molecules is ordinarily so small a fraction of the total that the error introduced by this assumption can be disregarded. However, at very high pressures or low temperatures, this error becomes significant. As the pressure is increased or the temperature decreased, the fraction of the total volume occupied by molecules becomes larger. As a result, the volume decreases less than the gas laws predict. Hydrogen and helium, which are virtually unaffected by attraction between molecules, deviate from ideal behavior in this manner at all pressures. As shown in Figure 6.18, when pressure becomes sufficiently high, all gases show this effect.

These two sources of error tend to offset each other, but in exact work, the gas law equation must be corrected to eliminate the inaccuracies. One of several more nearly exact ways of expressing the gas law is the van der Waals equation

$$\left(P + \frac{an^2}{V^2}\right)(V - nb) = nRT \tag{6.23}$$

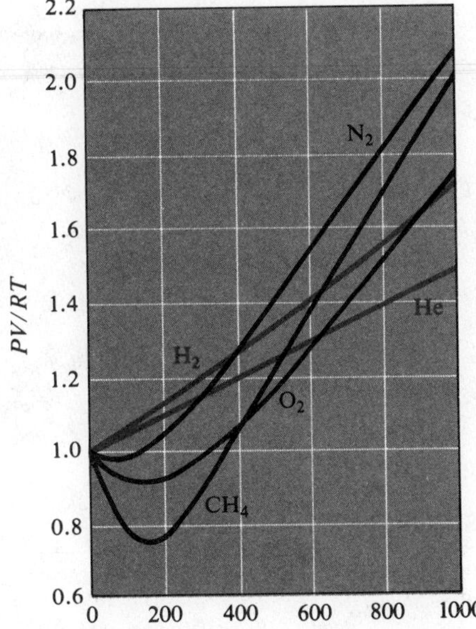

Figure 6.18
Deviations from Ideality with Increasing Pressure for One Mole of Gas (at 0 °C)
For one mole of an ideal gas, $PV/RT = 1$. For N_2, CH_4, and O_2, attractions between molecules cause PV/RT to drop below 1 as pressure increases. For all gases, the actual volume of the molecules makes an increasingly greater contribution to V as pressure increases and PV/RT eventually becomes increasingly greater than 1. For H_2 and He, this latter effect is the major cause of deviation from ideality at this temperature.

Table 6.6
Van der Waals Constants

Gas	*a* (L^2 atm/mol^2)	*b* (L/mol)
H_2	0.2444	0.02661
He	0.03412	0.02370
O_2	1.360	0.03183
CO_2	3.592	0.04267
H_2O	5.464	0.03049

in which a is a constant that takes into account the molecular attractions and b is a constant related to the volume actually occupied by the molecules of the substance. The corrections represented by a and b are different for different gases. Therefore, standard tables must be consulted to find the values of a and b needed for various gases. Values of a and b for some common gases are given in Table 6.6.

EXAMPLE 6.17 Real Gases

Using the van der Waals equation, calculate the pressure for a 1.25 mol sample of xenon contained in a volume of 1.000 L at 75 °C; $a = 4.194$ L^2 atm/mol^2 and $b = 0.05105$ L/mol for Xe. Compare this result to that predicted by the ideal gas law.

The pressure for the real gas is

$$P = \frac{nRT}{(V - nb)} - \frac{an^2}{V^2}$$

$$= \frac{(1.25\text{ mol})(0.0821\text{ L atm/K mol})(384\text{ K})}{(1.000\text{ L}) - [(1.25\text{ mol})(0.05105\text{ L/mol})]} - \frac{(4.194\text{ L}^2\text{ atm/mol}^2)(1.25\text{ mol})^2}{(1.00\text{ L})^2}$$

$$= \mathbf{31.6\ atm}$$

The ideal gas law would predict

$$P = \frac{nRT}{V} = \frac{(1.25\text{ mol})(0.0821\text{ L atm/K mol})(348\text{ K})}{(1.000\text{ L})} = \mathbf{35.7\ atm}$$

The pressure of the gas is 31.6 atm. The ideal gas law predicts a value of 35.7 atm, which is about 13% too high.

EXERCISE A 1.75 L container holds 32.5 mol of Xe at 75 °C. Calculate the pressure of the gas (a) assuming ideality and (b) using the van der Waals equation. For Xe, $a = 4.194$ L^2 atm/mol^2 and $b = 0.05105$ L/mol. *Answers* (a) 531 atm, (b) 9000 atm.

SUMMARY OF SECTIONS 6.13–6.14

Effusion is the escape of gas molecules, one by one, through a hole of molecular dimensions. Diffusion is the mixing of different gases by random molecular motions and collisions. According to Graham's law, at the same temperature and pressure, the rates of both effusion and diffusion are inversely proportional to the square roots of the densities (and therefore to the square roots of the molecular masses) of the gases.

Molecules of all gases occupy a finite volume and interact with each other to some extent. These factors cause some deviations from the ideal gas law. The van der Waals equation, which introduces corrections that take into account the volumes and interactions of gas molecules, describes more accurately the behavior of real gases.

SIGNIFICANT TERMS (with section references)

absolute temperature scale (6.5)
absolute zero (6.5)
Avogadro's law (6.7)
barometer (Aside 6.1)
Boyle's law (6.4)
Charles' law (6.5)
combined gas law (6.6)
constant (6.3)
Dalton's law of partial pressures (6.11)
diffusion (6.13)
dyanamic equilibrium (6.11)
effusion (6.13)
equilibrium vapor pressure (6.11)
Gay-Lussac's law of combining volumes (6.7)
Graham's law (6.13)

ideal gas (6.2)
ideal gas law (6.9)
Kelvin temperature scale (6.5)
kinetic energy (6.2)
manometer (Aside 6.1)
mole fraction (6.11)
partial pressure (6.11)
standard molar volume (6.8)
standard temperature and pressure (STP) (6.6)
variables (6.3)
volatile (6.1)

Aside 6.3 THOUGHTS ON CHEMISTRY: THE SCIENTIFIC METHOD

The following often-quoted excerpt comes from a non-fiction story about more than motorcycles.

When I think of formal scientific method an image sometimes comes to mind of an enormous juggernaut, a huge bulldozer—slow, tedious, lumbering, laborious, but invincible. It takes twice as long, five times as long, maybe a dozen times as long as informal mechanic's techniques, but you know in the end you're going to get it. There's no fault isolation problem in motorcycle maintenance that can stand up to it. When you've hit a really tough one, tried everything, racked your brain and nothing works, and you know that this time Nature has really decided to be difficult, you say, "Okay, Nature, that's the end of the nice guy," and you crank up the formal scientific method.

For this you keep a lab notebook. Everything gets written down, formally, so that you know at all times where you are, where you've been, where you're going and where you want to get. In scientific work and electronics technology this is necessary because otherwise the problems get so complex you get lost in them and confused and forget what you know and what you don't know and have to give up. In cycle maintenance things are not that involved, but when confusion starts it's a good idea to hold it down by making everything formal and exact. Sometimes just the act of writing down the problems straightens out your head as to what they really are....

The real purpose of scientific method is to make sure Nature hasn't misled you into thinking you know something you don't actually know. There's not a mechanic or scientist or technician alive who hasn't suffered from that one so much that he's not instinctively on guard. That's the main reason why so much scientific and mechanical information sounds so dull and so cautious. If you get careless or go romanticizing scientific information, giving it a flourish here and there, Nature will soon make a complete fool out of you. It does it often enough anyway even when you don't give it opportunities. One must be extremely careful and rigidly logical when dealing with Nature: one logical slip and an entire scientific edifice comes tumbling down. One false deduction about the machine and you can get hung up indefinitely.

Source: Robert Pirsig, Zen and the Art of Motorcycle Maintenance (New York: William Morrow & Co., 1975), p. 100.

QUESTIONS AND PROBLEMS

The Nature of Gases

6.1 List the five statements that summarize the kinetic-molecular theory for an ideal gas. Explain how these might not be valid for a liquid or a solid.

6.2 The radius of a typical molecule of a gas is 0.2 nm. (a) Find the volume of a molecule assuming it to be spherical. ($V = \frac{4}{3}\pi r^3$ for a sphere.) (b) Calculate the volume actually occupied by 1.00 mol of these molecules. (c) If 1.00 mol of this gas actually occupies 22.4 L, find the fraction of the volume actually occupied by the molecules. (d) Comment on your answer to (c) in view of the first statement summarizing the kinetic-molecular theory of an ideal gas.

6.3 The temperature of an ideal gas was changed from 75 °C to 25 °C. What is the ratio of kinetic energy of the gas at the lower temperature to that at the higher temperature?

6.4* The average speed of oxygen molecules at 25 °C is 4.44×10^2 m/s. What is the average speed of nitrogen molecules at this temperature?

Volume, Pressure and Temperature Relationships

6.5 Classify the relationship between the variables (a) P and V, (b) V and T, (c) P and T, (d) V and n as either (i) directly proportional or (ii) inversely proportional.

6.6* Prepare sketches of plots of (a) P vs. V, (b) P vs. $1/V$, (c) V vs. T, (d) P vs. T for an ideal gas.

Volume, Pressure, and Temperature Relationships—*V* and *P*

6.7 What pressure is needed to confine an ideal gas at 75 L after it has expanded from 25 L and 1.00 atm at constant temperature?

6.8 An ideal gas underwent an expansion at constant temperature from 2.55 bar to 1.00 bar. The final volume of the gas is 32.7 L. What was the initial volume of the gas?

6.9* A flask of unknown volume was filled with air to a pressure of 3.6 atm. This flask was then attached to an evacuated flask having a known volume 4.9 L and the air was allowed to expand into the flask. The final pressure of the air (in both flasks) was 2.5 atm. Calculate the volume of the first flask.

6.10* Demonstrate that the following data for the experimental apparatus depicted in Figure 6.8 are consistent with Boyle's law.

P_{atm} = 745 Torr

h_{Hg} in cm	1.0	2.0	3.0	4.0	5.0
h_{air} in cm	25.00	24.68	24.34	24.05	23.75

Volume, Pressure, and Temperature Relationships—*V* and *T*

6.11 What will be the new volume after 1.000 L of an ideal gas is heated from 0.0 °C to 1.0 °C under constant pressure?

6.12 What will be the final temperature after 10.0 L of an ideal gas at 950 K is reduced to a volume of 1.0 L under constant pressure?

6.13 The volume of an ideal gas sample was 102.7 cm^3 at 25 °C and changed to 126.4 cm^3 when the temperature was changed under constant pressure conditions. What is the new temperature of the gas sample?

6.14* Calculate the volume of an ideal gas at dry ice (−78.5 °C), liquid N_2 (−195.8 °C), and liquid He (−268.9 °C) temperatures if it occupies 10.00 L at 25.0 °C. Assume constant pressure. Plot your results and extrapolate to zero volume. At what temperature is zero volume reached?

Volume, Pressure, and Temperature Relationships—*P* and *T*

6.15 To what temperature must an ideal gas at 830 °C and 14.0 psi be cooled under constant volume conditions so that the final pressure is 1.0 psi?

6.16 A steel cylinder of fixed volume has an attached pressure gauge which reads 298 Torr at 25 °C. At what temperature should it register 1000. Torr? Could you use this sort of device as a thermometer? Why?

6.17 What will be the final pressure of an ideal gas that is heated from 0 °C and 1.00 atm to 135 °C in a fixed volume?

6.18 A basketball was inflated to 8 psig in a garage at 20 °C. While playing basketball on a driveway at a temperature of −5 °C, the ball seemed "flat." Calculate the pressure of the air in the cold ball assuming the atmospheric pressure to be 14.4 psi.

Volume, Pressure, and Temperature Relationships—*P*, *V*, and *T*

6.19 What is the volume of an ideal gas at 1246 K and 5.30 atm if it occupied 16.3 L at 273 K and 0.937 atm?

6.20 What is the volume of an ideal gas at −14 °C and 367 Torr if it occupied 3.65 L at 25 °C and 745 Torr?

6.21 What pressure is necessary to contain an ideal gas at a volume of 37.5 ft^3 and 78 °F if it originally occupied 375 ft^3 at 135 psi and 85 °F?

6.22 What is the pressure of an ideal gas confined to 15.9 L at 63 °C if it occupied 22.4 L at 25 °C and 757 Torr?

6.23 What is the temperature of 83 L of an ideal gas at 425 Torr if it occupied a volume of 75 L at 763 Torr and 15.0 K?

6.24 What temperature would be necessary to double the volume of an ideal gas initially at STP if the pressure decreased by 25%?

Mass, Molecular, and Molar Relationships—Gay-Lussac's Law

6.25 What volume of chlorine under the same temperature and pressure conditions will react with 2 L of each of the following gases: H_2, C_2H_4, CO, C_2H_2? The equations are

(a) $H_2(g) + Cl_2(g) \longrightarrow 2HCl(g)$
(b) $C_2H_4(g) + Cl_2(g) \longrightarrow C_2H_4Cl_2(g)$
(c) $CO(g) + Cl_2(g) \longrightarrow COCl_2(g)$
(d) $C_2H_2(g) + 2Cl_2(g) \longrightarrow C_2H_2Cl_4(g)$

6.26 Which reaction requires the smallest volume of gaseous oxygen for the reaction of 1 volume of the other gas?

(a) $CH_4(g) + 2O_2(g) \longrightarrow CO_2(g) + 2H_2O(l)$
(b) $2CH_3OH(g) + 3O_2(g) \longrightarrow 2CO_2(g) + 4H_2O(l)$
(c) $2C_2H_2(g) + 5O_2(g) \longrightarrow 4CO_2(g) + 2H_2O(l)$
(d) $CH_3CH_2OH(g) + 3O_2(g) \longrightarrow 2CO_2(g) + 3H_2O(l)$

6.27 Assuming the volumes of all gases in the reaction are measured at the same temperature and pressure conditions, calculate the volume of water vapor obtainable by the explosion of a mixture of 250 mL of hydrogen gas and 125 mL of oxygen gas.

6.28 One liter of sulfur vapor at 500 °C and 1 atm is burned in pure oxygen to give 8 L of sulfur dioxide gas, SO_2, measured at the same temperature and pressure. How many atoms are there in a molecule of sulfur in the gaseous state?

Mass, Molecular, and Molar Relationships—Avogadro's Law

6.29 A 503 mL flask contains 0.0179 mol of an ideal gas under a given set of temperature and pressure conditions. Another flask contains 0.0256 mol of the gas under the same temperature and pressure conditions. What is the volume of the second flask?

6.30 A "one-liter" beaker (actual volume = 1.08 L) contains 0.0411 mol of air under a given set of temperature and pressure conditions. How many moles of air are in a "two-liter" beaker (actual volume = 2.23 L) under these same temperature and pressure conditions?

Mass, Molecular, and Molar Relationships—Molar Volume

6.31 How many moles of an ideal gas are contained in 1.00 m^3 at STP?

6.32 What is the volume occupied by 2.25 mol of an ideal gas at STP?

6.33 How many molecules of an ideal gas are contained in a 1.00 L flask at STP?

6.34 The limit of sensitivity for the analysis of carbon monoxide, CO, in air is 1 ppb (ppb = parts per billion) by volume. What is the number of CO molecules that can be detected in 10 L of air at STP?

6.35 The density of an ideal gas at STP is 6.13 g/L. What is the molar mass of the gas?

6.36 What is the density of gaseous fluorine, F_2, at STP?

Mass, Molecular, and Molar Relationships—Ideal Gas Law

6.37 For an ideal gas, calculate the number of moles in 1.00 L at 25 °C and 1.00 atm.

6.38 A sample of an ideal gas is confined in a 3.0 L container at a pressure of 2280 Torr and a temperature of 27 °C. How many moles of gas are there?

6.39 How many moles of Cl_2 are there in a 26.5 mL sample of the gas measured at STP?

6.40 Choose the sample that contains the larger number of moles of gas: (a) 52.9 L measured at −12 °C and 0.255 atm or (b) 32.6 mL measured at 37 °C and 37.6 atm.

6.41 Calculate the pressure needed to contain 5.29 mol of an ideal gas at 45 °C in a volume of 3.45 L.

6.42 A 10.0 mol sample of oxygen is confined in a vessel with a capacity of 8.0 L. If the temperature is 0 °C, what is the pressure?

6.43 A 1.3 mol sample of an ideal gas occupies 74 L at 45 atm. What is the temperature of this gas?

6.44 A 6.00 mol sample of helium is confined in a 4.5 L vessel. What is the temperature if the pressure is 3.0 atm?

6.45 What volume is occupied by 1.00 mol of an ideal gas at −75 °C and 12.5 atm?

6.46* A barge containing 640 tons of liquid chlorine was involved in an accident on the Ohio River. What volume would this amount of chlorine occupy if it were all converted to a gas at 740 Torr and 15 °C? Assume that the chlorine is confined to a width of 0.5 mile and an average depth of 50 ft. How long would this chlorine "cloud" be?

6.47 How many gaseous molecules are in a one-liter container if the pressure is 1.6×10^{-9} Torr and the temperature is 1475 K?

6.48 The molar volume of an ideal gas at STP is 0.790 ft^3. Calculate the gas constant, R, in units of ft^3 atm/K mol and in terms of ft^3 atm/K pound-mol.

Mass, Molecular, and Molar Relationships—Molar Mass

6.49 An ideal gas has a molar mass of 95 g. What volume will 5.0 g of the gas occupy at 25 °C and 776 Torr?

6.50 What is the molar mass of an ideal gas if 0.52 g of the gas occupies 610 mL at 385 Torr and 45 °C?

6.51 The Dumas method was used to determine the molar mass of a liquid. The vapor occupied a 103 mL volume at 99 °C and 721 Torr. The condensed vapor has a mass of 0.800 g. Calculate the molar mass of the liquid.

6.52* A highly volatile liquid was allowed to vaporize completely into a 250 mL flask immersed in boiling water. From the following data, calculate the molar mass of the liquid: mass of empty flask = 65.347 g, mass of flask filled with water at room temperature = 327.4 g, mass of flask and condensed liquid = 65.739 g, atmospheric pressure = 743.3 Torr, temperature of boiling water = 99.8 °C, density of water at room temperature = 0.997 g/mL.

Mass, Molecular, and Molar Relationships—Density

6.53 What is the density of an ideal gas at 25 °C and 10.0 atm if the molar mass is 18 g?

6.54 A pure gas contains 81.71% carbon and 18.29% hydrogen by mass. The density of this gas is 1.97 g/L at STP. What is the molecular formula of this gas?

Mass, Molecular, and Molar Relationships—Dalton's Law

6.55 A sample of molecular oxygen of mass 24.0 g is confined in a vessel at 0 °C and 1000. Torr. If 6.00 g of molecular hydrogen is now pumped into the vessel at constant temperature, what will be the final pressure in the vessel (assuming only simple mixing)?

6.56 A gaseous mixture contains 4.18 g of chloroform, $CHCl_3$, and 1.95 g of ethane, C_2H_6. What pressure is exerted by the mixture inside a 50.0 mL metal bomb at 375 °C? What pressure is contributed by the $CHCl_3$?

6.57 A cyclopropane-oxygen mixture can be used as an anesthetic. If the partial pressures of cyclopropane and oxygen are 150 Torr and 550 Torr, respectively, what is the ratio of the number of moles of cyclopropane to the number of moles of oxygen in this mixture?

6.58 What are the mole fractions of each gas in a mixture that contains 0.267 atm of He, 0.369 atm of Ar, and 0.394 atm of Xe?

6.59 A sample of hydrogen was collected over water at 25 °C. The vapor pressure of water at this temperature is 23.8 Torr. A dehydrating agent (something that absorbs water) was added to remove the water. If the original volume of wet hydrogen was 33.3 L and the original pressure of the wet hydrogen was 738 Torr, what volume would the dry hydrogen occupy at 743 Torr?

6.60 A 5.00 L flask containing He at 5.00 atm was connected to a 4.00 L flask containing N_2 at 4.00 atm. Using Boyle's law for each gas, (a) find the partial pressures of the gases after they are allowed to mix; using Dalton's law, (b) find the total pressure of the mixture. (c) What is the mole fraction of the helium?

Mass, Molecular, and Molar Relationships—Stoichiometry

6.61 What volume of N_2 is required to convert 15.0 L of H_2 to NH_3? Assume all gases to be at the same temperature and pressure, and that the reaction is complete.

$$N_2(g) + 3H_2(g) \longrightarrow 2NH_3(g)$$

6.62 Many woodsmen use small propane stoves to cook meals. What volume of air (assumed to be 20.% O_2 by volume) will be required to burn 10.0 L of propane, C_3H_8? Assume all gas volumes are measured at the same temperature and pressure. The equation is

$$C_3H_8(g) + 5O_2(g) \longrightarrow 3CO_2(g) + 4H_2O(g)$$

6.63 What volume of carbon dioxide, measured at STP, can be obtained by the reaction of 50.0 g of $CaCO_3$ with excess hydrochloric acid?

$$CaCO_3(s) + 2HCl(aq) \longrightarrow CaCl_2(aq) + CO_2(g) + H_2O(l)$$

What would be the volume of CO_2 had it been measured at 25 °C and 0.975 atm?

6.64 Aqueous ammonium nitrite decomposes upon heating to form water and nitrogen. What volume of N_2, at STP, will be released by the decomposition of 80.0 g of NH_4NO_2?

$$NH_4NO_2(aq) \xrightarrow{\Delta} N_2(g) + 2H_2O(g)$$

6.65 What volume of hydrogen fluoride at 743 Torr and 24 °C will be released by the reaction of 47.2 g of xenon difluoride with a stoichiometric amount of water? The unbalanced equation is

$$XeF_2(s) + H_2O(l) \longrightarrow Xe(g) + O_2(g) + HF(g)$$

What volumes of oxygen and xenon will be released under these conditions?

6.66 Hydrogen reacts with some of the more active metals to form crystalline ionic hydrides. For example, Li forms LiH

$$2Li(s) + H_2(g) \longrightarrow 2LiH(s)$$

(a) What mass of LiH would be produced by allowing 10.0 g of Li to react with 10.0 L of H_2 (measured at STP)? (b) If the actual yield was 6.7 g of LiH, what was the percent yield?

6.67* A common laboratory preparation of oxygen is

$$2KClO_3(s) \xrightarrow[\Delta]{MnO_2} 2KCl(s) + 3O_2(g)$$

If you were designing an experiment to generate four bottles (each containing 250 mL) of O_2 at 25 °C and 723 Torr and allowing for 50% waste, what mass of potassium chlorate would be required? As a laboratory instructor, how would you explain to your students the symbol Δ and MnO_2 written near the arrow?

6.68* Sheet iron is galvanized (plated with zinc) on both sides to protect it from rust. The thickness of the zinc coating was determined by allowing hydrochloric acid to react with the zinc and collecting the resulting hydrogen. (Note: The acid solution contained an "inhibitor" ($SbCl_3$) which prevented the iron from reacting.)

$$Zn(s) + 2HCl(aq) \longrightarrow ZnCl_2(aq) + H_2(g)$$

Determine the thickness of the zinc plate from the following data: sample size = 1.50 cm × 2.00 cm, volume of dry hydrogen = 30.0 mL, temperature = 25 °C, pressure = 747 Torr, density of zinc = 7.11 g/cm^3.

Behavior of Gas Molecules—Graham's Law

6.69 How much faster would neon effuse than argon?

6.70 A sample of unknown gas flows through the wall of a porous cup in 39.9 min. An equal volume of molecular hydrogen, measured at the same temperature and pressure, flows through in 9.75 min. What is the molar mass of the unknown gas?

6.71 What would be the rate of diffusion for $^{235}UF_6$ relative to that for $^{238}UF_6$ under similar conditions? Assume atomic masses of 235 u for ^{235}U, 238 u for ^{238}U, and 19 u for F.

6.72 What would be the relative rates of effusion of gaseous H_2, HD, and D_2? (D is a chemical symbol used to represent deuterium, an isotope of hydrogen that has an atomic mass of 2.0140 u as compared to that of 1.0078 u for H.)

Behavior of Gas Molecules—Real Gases

6.73 A sample of gas has a molar volume of 10.3 L at a pressure of 745 Torr and a temperature of −138 °C. Is the gas acting ideally?

6.74 The van der Waals constants for carbon tetrachloride, CCl_4, are $a = 20.39$ L^2 atm/mol^2 and $b = 0.1383$ L/mol. Find the pressure of a sample of CCl_4 if one mole occupies 30.0 L at 77 °C (just slightly above the boiling point). Assume CCl_4 to obey the (a) ideal gas law and (b) van der Waals gas law.

6.75 Repeat the calculations of Problem 6.74 using a 3.25 mol gas sample confined to 6.25 L at 115 °C.

6.76 Values of molar mass calculated using the ideal gas law are good only to the extent that the gas behaves as an ideal gas. However, all real gases approach ideal gas behavior at very low pressures, so a common technique for obtaining very accurate molar masses is to measure the density of a gas at various low pressures, calculate d/P from the data, plot d/P against P, extrapolate the curve to $P = 0$ to find the intercept, and calculate the molar mass, M, using $M = (\text{intercept})RT$ where $R =$ 0.0820568 L atm/K mol. Find the molar mass for SO_2 from the following data at 0 °C:

P in atm	0.1	0.01	0.001	0.0001
(d/P) in g/L atm	2.864974	2.858800	2.858183	2.858121

Assume that the temperature and pressure values are exact.

ADDITIONAL QUESTIONS AND PROBLEMS

6.77 A typical laboratory atmosphere pressure reading is 745 Torr. Convert this value to (a) psi, (b) mmHg (c) inches Hg, (d) Pa, (e) atm, (f) ft H_2O.

6.78 A laboratory technician forgot what the color coding on some commercial cylinders of gas meant, but remembered that each of two specific tanks contained one of the following gases: He, Ne, Ar, or Kr. Density measurements at STP were made on samples of the gases from these cylinders and were found to be 0.178 g/L and 0.900 g/L. Which of these gases was present in each tank?

6.79 Cyanogen is 46.2% carbon and 53.8% nitrogen by mass. At a temperature of 25 °C and a pressure of 750 Torr, 1.00 g of cyanogen occupies 0.476 L. Determine the empirical formula and the molecular formula of cyanogen.

6.80* The density of dry air at STP is 1.2929 g/L and that of molecular nitrogen is 1.25055 g/L. (a) Find the average molar mass of air from these data. The mole fractions of the major components of air are 0.7808 for N_2, 0.2095 for O_2, 0.0093 for Ar, and 0.0003 for CO_2. (b) Calculate the average molar mass of air and (c) compare your answers from (a) and (b). (d) Why can't Graham's law be used to find the average molar mass of air?

What is the density of (e) dry air at 745 Torr and 25 °C and of (f) wet air at 745 Torr and 25 °C if it contains water vapor at a partial pressure of 13 Torr? (g) Explain why the answer to (f) should be less than the answer to (e).

(h) What is the partial pressure of CO_2 in a room containing dry air at 1.00 atm? (i) What mass of CO_2 will be present at 25 °C in a closet 1.5 m by 3.4 m by 3.0 m if the air pressure is 1.00 atm?

6.81* About 25 years ago it was discovered that the "inert" gases are not really inert after all. In particular, xenon reacts with fluorine under various conditions to form a series of compounds.

Colorless crystals of XeF_4 can be prepared by heating a 1-to-5 mixture by volume of Xe to F_2 in a nickel can at 400 °C and 6 atm pressure for a few hours and cooling. The equation for the reaction is

$$Xe(g) + 2F_2(g) \longrightarrow XeF_4(s)$$

After the reaction the nickel container contains gaseous F_2 and a little XeF_4 vapor above the crystals of XeF_4.

According to Gay-Lussac's law for combining volumes, (a) what volume of F_2 would react for every milliliter of Xe that reacts? (b) Which law allows us to deduce that the partial pressure of Xe is 1 atm and the partial pressure of F_2 is 5 atm in the original reaction mixture? Predict what will happen to the gases in the nickel container after the reaction if (c) the pressure is increased at constant temperature (use Boyle's law) and (d) the temperature is increased at constant pressure (use Charles' law). (e) Using Graham's law, predict the increasing order for the rate of effusion of the gases Xe, F_2, and XeF_4. (f) Basing your argument on the actual volumes of molecules, which of the three gases—Xe, F_2 or XeF_4—do you predict would deviate most from ideal gas behavior at high pressures?

CHAPTER

7

Thermochemistry

ENERGY

7.1 ENERGY IN CHEMICAL REACTIONS

Exactly what is it that happens in chemical reactions? So far, we have learned that reactants are transformed into products. The products have different compositions and properties than the reactants, and during the reactions mass is neither lost nor gained.

Let's consider what can be observed in two simple chemical reactions that produce similar products from familiar materials. Limestone, chalk, and marble are all composed mainly of calcium carbonate. When sulfuric acid is poured over any of these substances, bubbles appear in the liquid as gaseous carbon dioxide is formed and a finely divided white solid appears. The balanced chemical equation for this reaction is

$$CaCO_3(s) + H_2SO_4(aq) \longrightarrow CO_2(g) + H_2O(l) + CaSO_4(s) \qquad \textbf{(7.1)}$$

Careful observation reveals that something more than the transformation of reactants into products is taking place. During the course of the reaction, heat is evolved, as shown by an increase in the temperature of the mixture.

Sodium hydrogen carbonate is also a familiar substance; you know it as "baking soda," or "bicarbonate of soda." It can be used in the home for cleaning teeth, deodorizing refrigerators, and even putting out small fires, as well as for baking biscuits. Gentle heating transforms sodium hydrogen carbonate into products similar to those in the reaction described above—carbon dioxide, water, and a finely divided white salt, in this case sodium carbonate.

$$2NaHCO_3(s) \xrightarrow{\Delta} CO_2(g) + H_2O(g) + Na_2CO_3(s) \qquad \textbf{(7.2)}$$

However, baking soda remains unchanged in a box on the kitchen or bathroom shelf without decomposing. Only when the compound is heated does the reaction take place.

Have you noticed an essential difference between these two reactions, both of which produce carbon dioxide, water, and a solid salt? In the first reaction, heat is *produced*. In the second, the reaction does not take place unless heat is *added*. Heat, we know, is associated with energy. Changes in energy apparently accompany both reactions. Is it possible that this is the case for all chemical reactions? "Yes, always," is the answer. Experience has shown that energy changes accompany all chemical reactions.

The atoms or ions in chemical compounds are held together by the forces referred to as chemical bonds. When new substances are produced by a chemical reaction, existing chemical bonds are broken and new ones are formed. In general, breaking chemical bonds requires energy and the formation of new bonds releases energy. Whether a specific reaction requires energy or releases energy depends upon the final balance between the energy used to break old bonds and the energy released as new bonds are formed. Despite the similarity of the products in reactions (7.1) and (7.2), the overall result in one case is the release of energy as heat whereas, in the other case the addition of heat is required. The amount of heat absorbed or released in a given chemical reaction is always the same when the amounts of the reactants and the reaction conditions are identical.

7.2 THERMODYNAMICS

Thermodynamics is the study of energy transformations. The question that immediately arises is, What is energy? This is a hard question to answer because we cannot see energy or directly measure it. But we *accept* the existence of energy because we can see and measure the effects of energy changes. When energy is available, things happen.

Kinetic energy (Section 6.2) is energy of motion, and depends upon both the mass and the velocity of the moving object. A train roaring down the tracks has kinetic energy, as do the molecules of a gas. An object has *potential energy* (Aside 3.3) by virtue of its position with respect to a force that could cause it to move. A boulder poised at the top of a cliff, for example, has potential energy with respect to the Earth's gravity. An electron in an atom has potential energy with respect to the force of electrical attraction of the nucleus.

In the study of thermodynamics, it is convenient to discuss a process in terms of a system and its interaction with the surroundings. The system and the surroundings are defined according to what is being studied. A **system** is the portion of the universe under study. A system can be simple or complex—the boulder on the cliff, the contents of a test tube, or an entire train with its engine and all of its cars. A system in chemistry is usually the substances undergoing a physical or chemical change. In the decomposition of sodium hydrogen carbonate, the sample of the compound would be the system.

The **surroundings** is everything in the universe that is not part of the system. The flask in which the sodium hydrogen carbonate was heated, the hot plate on which it was heated, and everything else in the universe would be the surroundings. In this case we can say that heat was added to the system from the rest of the universe and caused a change in the system.

Most commonly, chemical systems exchange energy with their surroundings in the form of heat—thermal energy. **Thermochemistry**, which is the study of the thermal energy changes that accompany chemical and physical changes, is the subject of this chapter.

Heat is the energy transferred between objects or systems at different temperatures. Note that energy is a property of a system or a substance, but heat is not. Heat, which is detected by temperature changes (Section 7.5), is something happening—it is energy in transit between objects or systems that are at different temperatures and are in contact with each other.

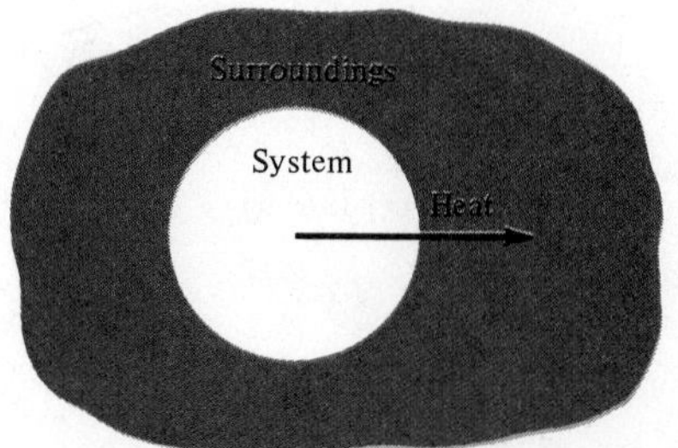

Figure 7.1
Exothermic Processes

Any process that releases heat from the system to the surroundings is referred to as **exothermic** (Figure 7.1). When an exothermic chemical reaction occurs, the temper-

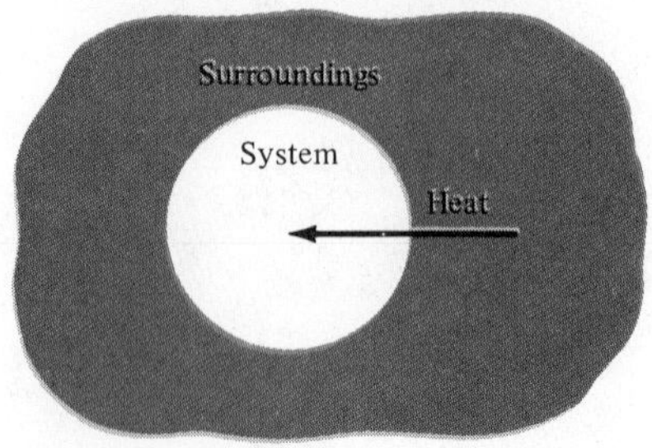

Endothermic process

Figure 7.2
Endothermic Processes

ature of the reacting substances increases and heat is transferred to the surroundings. The reaction of calcium carbonate with sulfuric acid (Equation 7.1) is exothermic. Any process in which the system absorbs heat from the surroundings is **endothermic** (Figure 7.2). For an endothermic chemical reaction to occur, the necessary amount of heat must be available from surroundings that are at a higher temperature than the reacting substances. The decomposition of sodium hydrogen carbonate (Equation 7.2) is endothermic.

7.3 INTERNAL ENERGY

All of the energy contained within a chemical system is **internal energy.** It is the internal energy of a substance that participates in thermal energy exchange with the surroundings. A chemical system might also have gravitational potential energy (a beaker might be poised on the lab bench) or it might have kinetic energy (a tankcar full of chlorine might be speeding along as part of a freight train), but these types of energy on a macro scale are usually not of interest in the study of chemistry.

Internal energy is the sum of all the energy of all of the atoms, molecules, or ions within a system. For example, if the system is a gas, there is a large contribution to the internal energy from the kinetic energy of the molecules moving about randomly. Additional energy is contributed by the molecules rotating in space and the atoms in the molecules moving with respect to each other (Figure 7.3). Potential energy due to attraction and repulsion among the atoms, molecules, or ions in a substance is also included in the internal energy.

An *increase* in the internal energy of a chemical system has three possible results:

1. The temperature can increase. The increased internal energy of a system when the temperature rises is demonstrated visibly by the expansion of an ideal gas at constant pressure.
2. Melting or vaporization or a change in crystalline form can occur. All molecules are influenced by forces known collectively as **intermolecular forces**—the forces of attraction and repulsion between individual molecules. (By contrast, forces that act between atoms in the same molecule are known as **intramolecular forces.**) The molecules in a liquid are attracted to each other by intermolecular forces. A liquid boils and changes into a gas when the heat added provides sufficient energy for the molecules to overcome the intermolecular forces. Similarly, added energy can cause a solid to melt. Some substances are able to crystallize in two or more forms that differ in the arrangements of the atoms or ions in the crystal, a phenomenon known as *polymorphism* (Section 13.10). For such substances, a change in internal energy may result in a change from one crystalline form to another.
3. A chemical reaction can occur. When the increase in internal energy of a system is sufficient to cause chemical bonds to break, allowing new ones to form, a chemical reaction takes place. A chemical reaction that results from the absorption by the system of energy from the surroundings is an endothermic reaction.

A *decrease* in the internal energy of a system can result in a decrease in the temperature of the system or a change of state—freezing or condensation in this case (or a change in crystalline form). Although a decrease in internal energy rarely initiates a chemical reaction, the outcome of a chemical reaction can be that the system has a lower internal energy than it had before the reaction.

The total internal energy, E, of a system cannot be determined. However, changes in internal energy

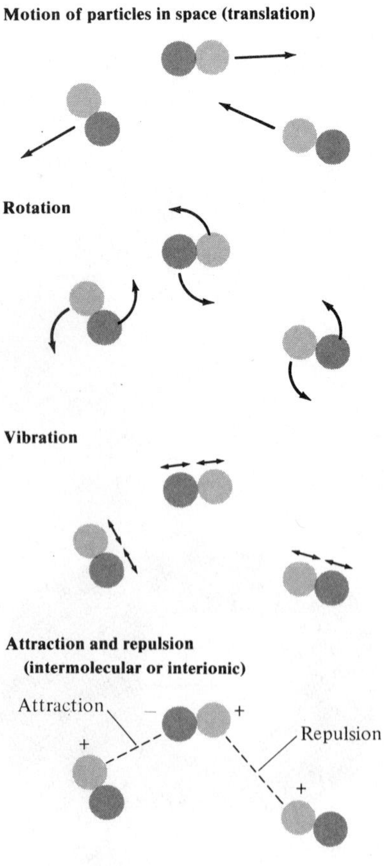

change in energy → $$\Delta E = E_{\text{final}} - E_{\text{initial}}$$

Motion of particles in space (translation)

Rotation

Vibration

Attraction and repulsion (intermolecular or interionic)

Attraction − + + Repulsion +

Figure 7.3
Contributions to Internal Energy

can be both measured and calculated. The **internal energy change** of a system, ΔE, is the amount of energy exchanged with the surroundings during a chemical or physical change of the system.

The symbol Δ is used to represent a change in a variable that is a property of a system. For most variables, it is customary to find the change by subtracting the *initial* value from the *final* value.

7.4 ENERGY, HEAT, AND WORK

The **first law of thermodynamics** is called the law of conservation of energy. It is often stated as follows: The energy of the universe is constant. (Other ways of stating the first law are given in Table 7.1.) The first law is accepted because no observation contrary to it has ever been made. The energy changes in any system and in its surroundings are always observed to offset each other so that the total energy of the universe remains constant.

In the interaction between a system and its surroundings, the entire energy change must be accounted for by heat, q, and/or work, w.

$$\Delta E = \underset{\text{heat}}{q} + \underset{\text{work}}{w} \qquad (7.3)$$

This equation is a mathematical statement of the first law.

We choose to define the terms in Equation (7.3) so that ΔE is negative when a system loses energy and positive when a system gains energy. To do this, heat that has been added to the system and work that has been done *on* a system—both of which can increase the internal energy of the system—are given positive values. Heat lost by a system or work done *by* a system on the surroundings—both of which can decrease the internal energy of the system—are given negative values (Figure 7.4). [Until the mid-1970s, a sign convention for work opposite to that just described was used. It results in Equation 7.3 taking the form $\Delta E = q - w$. Always check the author's definitions when reading about heat and work in other books and publications.]

The SI unit for heat, for work, or for energy in any form is the joule (Section 2.8f).

$$1 \text{ calorie (cal)} = 4.184 \text{ joules (J), exactly}$$

The magnitude of energy changes in chemical reactions frequently makes it more convenient to use the kilojoule (1 kJ = 1000 J).

Work is performed when a force moves an object over a distance. (In the SI system, $1 \text{ J} = 1 \text{ N m} = 1 \text{ kg m}^2/\text{s}^2$.) Because all energy transfer other than heat is defined as work, there are many different kinds of work. And because of the way it is calculated, work of different kinds is traditionally expressed in a variety of units, but these can all be converted to joules (see Appendix II). The only kinds of work relevant to our study of chemistry are work done when a gas expands or contracts (pressure–volume work) and work associated with electrochemical changes (Sections 24.6 and 24.10).

We have already encountered pressure–volume work in the expansion of a gas that pushes back a piston. Similarly, in a system open to the atmosphere, an expanding gas does work by pushing back the atmosphere. For a gas *expanding* against a constant external pressure

$$\underset{\text{work}}{w} = -\underbrace{\overset{\text{constant pressure}}{P}\ \overset{\text{final volume minus initial volume}}{\Delta V}}_{\text{work done by system}} \qquad (7.4)$$

If the volume of a system does not change ($\Delta V = 0$), no work is done. From the first law (Equation 7.3), it can be seen that at constant volume, with $w = 0$, the change in internal energy of a system is equal to the heat exchanged with the surroundings.

Table 7.1
Statements of the First Law of Thermodynamics

The energy of the universe is constant.
Energy can be converted from one form to another, but cannot be destroyed.
The change in energy of a system equals the heat exchanged by the system plus the work done on or by the system.
$\Delta E = q + w$
You can't get something for nothing.
There is no such thing as a free lunch.

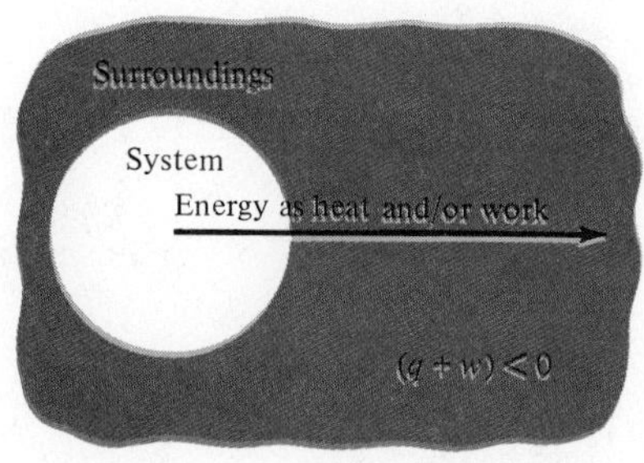

(a) Energy of system decreases, $\Delta E < 0$

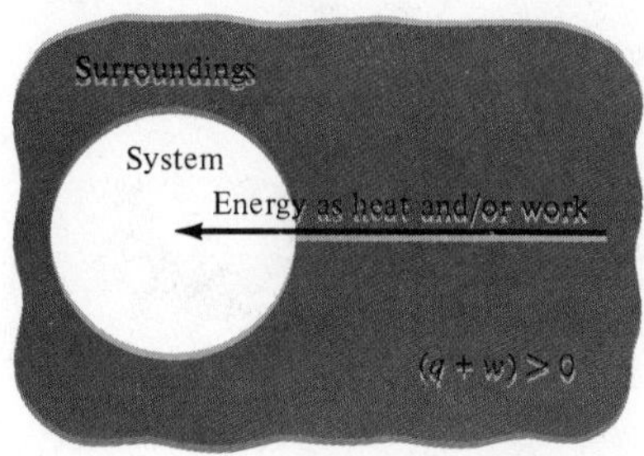

(b) Energy of system increases, $\Delta E > 0$

Figure 7.4
Energy Flow, Heat, and Work

$$\Delta E = q_V \quad \text{(heat at constant volume)} \tag{7.5}$$

This is a useful relationship—heat can be measured and this equation shows that the value of ΔE can be found by measuring q_V.

Chemicals are commonly handled in the laboratory and industrial plants at constant external pressure, most often atmospheric pressure. Therefore, we are often interested in q_P—the heat at constant external pressure—rather than q_V. For a system at constant external pressure, which can exchange energy with the surroundings only as heat or work

$$\Delta E = q + w$$
$$\Delta E = q_P - P\,\Delta V \quad (q_P\text{: heat at constant external pressure}) \tag{7.6}$$

or

$$E_2 - E_1 = q_P - (PV_2 - PV_1) \tag{7.6a}$$

Rearranging gives

$$q_P = (E_2 + PV_2) - (E_1 + PV_1) \tag{7.6b}$$

Because of the interest in q_P and the ability to measure it , $E + PV$ is defined as a quantity called *enthalpy*, H (from the Greek *enthalpo*, warming up). **Enthalpy** is a thermodynamic property of a system defined so that a change in enthalpy is equal to the amount of heat gained or lost in processes that occur at constant external pressure.

$$\text{enthalpy} \rightarrow H = E + PV \tag{7.7}$$
$$\text{enthalpy change} \rightarrow \Delta H = \Delta E + \Delta(PV) \tag{7.8}$$

At constant external pressure

$$\Delta H = \Delta E + P\,\Delta V \tag{7.9}$$

and, from Equation (7.6b)

$$\Delta H = H_2 - H_1 = (E_2 + PV_2) - (E_1 + PV_1) \tag{7.10}$$
$$\Delta H = q_P$$

Both ΔE and $\Delta(PV)$ or $P\,\Delta V$ have units of energy, and therefore ΔH is also expressed in energy units, usually joules or kilojoules. Frequently, even in reactions involving gases, the contribution of $\Delta(PV)$ or $P\,\Delta V$ to ΔH is so small that it can be neglected.

EXAMPLE 7.1 Energy, Heat, and Work

In a single process, a system does 125 J of work on its surroundings while 75 J of heat is added to the system. What is the internal energy change for the system?

A system loses energy when it performs work on its surroundings, therefore, $w = -125$ J. The energy of the system is increased by the flow of heat into the system, so q is given a positive value, $q = 75$ J. To find ΔE

$$\Delta E = q + w = 75\text{ J} + (-125\text{ J}) = \mathbf{-50.\ J}$$

The system has lost energy.

EXERCISE The internal energy of a system increased by 323 kJ while the system performed 111 kJ of work on its surroundings. How much heat was transferred between the system and its surroundings during this process? In which direction did the heat flow? *Answers* 434 kJ of heat; from surroundings to system.

EXAMPLE 7.2 Energy, Heat, and Work

For each of the following chemical and physical changes at constant pressure, is work done by the system (the substances undergoing the change) on the surroundings or by the surroundings on the system, or is the amount of work negligible?

(a) $Sn(s) + 2F_2(g) \longrightarrow SnF_4(s)$
(b) $AgNO_3(aq) + NaCl(aq) \longrightarrow AgCl(s) + NaNO_3(aq)$
(c) $C(s) + O_2(g) \longrightarrow CO_2(g)$
(d) $SiI_4(g) \xrightarrow{\Delta} Si(s) + 2I_2(g)$

(a) There are 2 mol of gaseous reactant and no gaseous product. The volume of the system decreases and, therefore, **work is done *on* the system by the surroundings** (w is positive).
(b) No gaseous reactants or products are involved and therefore the **work is negligible.**
(c) With one mole of gaseous reactant and one mole of gaseous product, $\Delta V = 0$ and the amount of **work is negligible.**
(d) One mole of gaseous reactant yields two moles of gaseous product. Therefore, the volume increases as **work is done by the system** (w is negative).

EXERCISE In which of the following chemical and physical changes is work being done by the system (the substances undergoing the change) on the surroundings: (a) $I_2(g) \longrightarrow I_2(s)$, (b) $CaCO_3(s) \xrightarrow{\Delta} CaO(s) + CO_2(g)$, (c) $H_2(g) + Cl_2(g) \longrightarrow 2HCl(g)$, and (d) $H_2O(l) \longrightarrow H_2O(s)$? *Answer (b).*

SUMMARY OF SECTIONS 7.1–7.4

To examine energy changes in thermodynamics, it is customary to define a system and its surroundings. Heat is energy in transit between objects or systems at different temperatures, or between a system and its surroundings. Exothermic processes release heat to the surroundings and endothermic processes absorb heat from the surroundings. In either case, the total amount of heat gained and lost by the system plus the surroundings is equal to zero, a consequence of the first law of thermodynamics (the law of conservation of energy).

In chemistry, we deal mainly with substances and their internal energy—the total energy of all the atoms, molecules, or ions within the substances. Changes in internal energy (ΔE) are most often exhibited as temperature changes, changes of state, or chemical reactions. At constant volume, ΔE equals the heat exchanged with the surroundings ($\Delta E = q_V$). The thermodynamic quantity enthalpy, H, is defined as $H = E + PV$. At constant external pressure, $\Delta H = \Delta E + P\Delta V$, and ΔH equals the heat exchanged with the surroundings ($\Delta H = q_P$). For many processes, especially those involving solids and liquids, $P\Delta V$ is small and ΔH is approximately equal to ΔE.

MEASURING HEAT

7.5 HEAT CAPACITY

When a hotter object is placed in contact with a cooler object, energy in the form of heat flows between them until they reach the same temperature. Heat flow is detected by changes in temperature. The symbol ΔT represents a change in temperature. The value of ΔT is taken as the difference between the final and the initial temperatures of a substance or a system.

temperature

change in

$$\Delta T = T_{\text{final}} - T_{\text{initial}} \qquad (7.11)$$

Aside 7.1 HEAT AND CALORIC THEORY

In the eighteenth century, heat was thought to be due to the presence in all matter of a fluid substance called "caloric." When heat flowed from a hotter body to a cooler one, caloric was thought to be flowing from one to the other.

However, attempts to show the loss or gain of mass when heat flowed were never successful. The definitive word on the subject was written by one of the most fascinating characters in the history of science, Count Rumford (1753–1814). In his "Inquiry Concerning the Weight Ascribed to Heat" he reported meticulous experiments on the weight of water at various temperatures and in various states. He ended his paper with the statement, "I think we may safely conclude that all attempts to discover any effect of heat upon the apparent weights of bodies will be fruitless."

Rumford, who was born Benjamin Thompson in Woburn, Massachusetts, was a man of many parts—a politician, a military man, and a scientist. He was also an unscrupulous scoundrel. He managed to be a spy for both the British and the French, in each instance to his financial advantage. More than once, Rumford married for money.

While Rumford was supervising the boring of cannon barrels at an arsenal at Munich, he observed that a great amount of heat was generated in the process. He pursued this observation with experiments showing that work and heat were related. Rumford immersed the cannon-boring equipment in water and found, much to his delight, that after $2\frac{1}{2}$ hours of work by the borer, the water had absorbed enough heat to come to a boil.

Rumford himself recorded the outcome of this experiment as follows:

> *It would be difficult to describe the surprise and astonishment expressed in the countenances of the bystanders on seeing so large a quantity of cold water heated, and actually made to boil, without any fire. Though there was, in fact, nothing that could justly be considered as surprising in this event, yet I acknowledged fairly that it afforded me a degree of childish pleasure, which, were I ambitious of the reputation of a grave philosopher, I ought most certainly rather to hide than to discover.*
>
> *It is hardly necessary to add that anything which any insulated body, or system of bodies, can continue to furnish without limitation, cannot possibly be a material substance; and it appears to me to be extremely difficult, if not quite impossible, to form any distinct idea of anything capable of being excited and communicated in the manner in which heat was excited and communicated in these experiments, except it be MOTION.*

Source: Count Rumford, "An Inquiry Concerning the Source of the Heat Which is Excited by Friction," in *Philosophical Transactions*, Vol. 88, p. 80, 1798. Quoted from *Harvard Case Histories in Experimental Science*, Vol. 1 (Cambridge: Harvard University Press, 1966).

In thermodynamics, as in the gas law calculations (Chapter 6), temperature should be expressed in an absolute temperature unit, commonly the kelvin. However, *because the Celsius degree and the kelvin are the same size, ΔT has the same value in Celsius degrees as it does in kelvins.* Simply write the value of ΔT in kelvins, where these are needed. For example, for a temperature increase from 23 °C to 47 °C

$$\Delta T = T_{\text{final}} - T_{\text{initial}} = 47\ °\text{C} - 23\ °\text{C} = 24\ °\text{C, or } 24\ \text{K} \tag{7.12}$$

For a temperature decrease from 100. °C to 35 °C

$$T_{\text{final}} - T_{\text{initial}} = 35\ °\text{C} - 100.\ °\text{C} = -65\ °\text{C, or } -65\ \text{K} \tag{7.13}$$

As Equations (7.12) and (7.13) show, when temperature increases, ΔT is positive; when temperature decreases, ΔT is negative.

The magnitude of the heat flow that accompanies an increase in temperature depends upon the mass and the identity of the substance involved. To determine the amount of heat associated with a known ΔT for a given substance, it is necessary to know the **heat capacity** of that substance—the amount of heat required to raise the temperature of a given amount of a substance by one kelvin.

Heat capacity values are usually tabulated in one of two ways: (1) as **molar heat capacity,** the amount of heat required to raise the temperature of one *mole* of substance by one kelvin (J/K mol) or (2) as *specific heat capacity,* commonly referred to as **specific heat,** the amount of heat required to raise the temperature of one *gram* of a substance by one kelvin (J/K g).

The molar heat capacity varies with the complexity of the structure of a substance. When a monatomic gas is heated, all the added energy goes into increasing the random, straight-line motion (the translational kinetic energy) of the gaseous atoms and thereby directly raises the temperature of the substance. As illustrated in Table 7.2 for argon and helium, monatomic gases have identical molar heat capacities. For the same reason, elemental metals, which are composed of identical atoms, have very similar molar heat capacities. When more complex substances are heated, some of the added energy also goes into increasing the rotation of the molecules and/or the motion of the atoms or ions with respect to each other (see Figure 7.3). Therefore, more energy must be added to raise the temperature by one kelvin. For example, in Table 7.2 note the increasing heat capacities in the series of carbon-containing molecular compounds and in the series of calcium salts. (Additional molar heat capacity values are given in Appendix IV.)

Molar heat capacity and specific heat provide the connections between heat, q, and the amount of substance and the change in temperature.

$$q = (\text{no. of moles})(\text{molar heat capacity})(\Delta T) \quad \textbf{(7.14)}$$

$$\textit{joules} \quad \cancel{\textit{moles}} \quad \frac{\textit{joules}}{\cancel{\textit{kelvin}}\ \cancel{\textit{mole}}} \quad \cancel{\textit{kelvins}}$$

$$q = (\text{mass})(\text{specific heat})(\Delta T) \quad \textbf{(7.15)}$$

$$\textit{joules} \quad \cancel{\textit{grams}} \quad \frac{\textit{joules}}{\cancel{\textit{kelvin}}\ \cancel{\textit{gram}}} \quad \cancel{\textit{kelvins}}$$

Table 7.2
Some Molar Heat Capacity Values at Constant Pressure and 298 K[a,b]

Substance	Molar heat capacity (J/K mol)
Ar(g)	20.786
He(g)	20.786
Ne(g)	20.786
Ag(s)	25.351
Al(s)	24.35
Fe(s)	25.10
Pb(s)	26.44
CO(g)	29.142
$CO_2(g)$	37.11
$CH_4(g)$	35.309
$CH_3CH_2OH(l)$	111.46
$CH_3COOH(l)$	124.3
CaO(s)	42.80
$CaCl_2(s)$	72.59
$Ca_3(PO_4)_2(s)$	227.82

[a] The values given here are heat capacities under constant pressure, as for a substance at the standard state pressure of 1 bar. Values of heat capacity when the volume is constant and the pressure can vary slightly are different for solids and liquids and appreciably different for gases.
[b] Additional values given in Appendix IV.

EXAMPLE 7.3 Heat Flow–Specific Heat

How much heat is needed to raise the temperature of 21 g of aluminum from 25 °C to 161 °C (no phase changes occur)? The specific heat of aluminum is 0.902 J/K g.

The amount of heat needed is found from the known mass of aluminum, the change in temperature and the specific heat.

$$\Delta T = 161\ ^\circ\text{C} - 25\ ^\circ\text{C} = 136\ ^\circ\text{C, or } 136\ \text{K}$$
$$q = (\text{mass})(\text{sp. heat})(\Delta T)$$
$$= (21\ \text{g})\left(\frac{0.902\ \text{J}}{\text{K g}}\right)(136\ \text{K})$$
$$= \mathbf{2600\ J}$$

Note that the temperature increased and q is positive.

EXERCISE Exactly 100 J of heat was absorbed by 1.00 mol of iron and the resulting temperature change was 3.98 °C. Calculate the molar heat capacity of iron.

Answer 25.1 J/K mol.

7.6 CALORIMETRY

In all calorimetry, the objective is to measure ΔT for the calorimeter and its contents, and to calculate from ΔT the amount of heat that has been exchanged between the calorimeter and its contents as the contents undergo a chemical reaction or a change of state. The heat capacity of the calorimeter itself and also in some cases that of the contents of the calorimeter must be known. These heat capacities are used to connect the measured temperature change with the amount of heat gained or lost during the process under study. This value is equal to the internal energy change if the measurement is made in a constant-volume calorimeter and it is equal to the enthalpy change if the measurement is made in a constant-pressure calorimeter ($\Delta E = q_V$, $\Delta H = q_P$; Section 7.4).

The heat capacity of a calorimeter is determined experimentally by one of several standard methods. For example, a known amount of heat is introduced into the

Aside 7.2 TOOLS OF CHEMISTRY: CALORIMETERS

A **calorimeter** is a device for measuring the heat absorbed or released during a chemical or physical change. The measurement is made by determining the temperature change (ΔT) that occurs when the process under study takes place within the calorimeter. As illustrated in the next section (Section 7.6), calculation of the heat absorbed or released requires knowing the heat capacity of the calorimeter vessel.

Many types of calorimeters have been designed for different purposes. One of the most common is a bomb calorimeter (Figure 7.5). The physical or chemical change takes place in a "bomb," a chamber strong enough to withstand high temperatures and pressures. The heat flow measured in a bomb calorimeter is a direct measure of the change in internal energy, ΔE, because the measurement is made at constant volume ($\Delta E = q_V$).

Bomb calorimeters are frequently used for measuring heats of combustion or the caloric value of foods, each of which is the heat released in the combination of a substance with oxygen. Although **combustion** is a general term for any chemical change in which heat and light are produced, the term usually refers to burning in the presence of oxygen.

The substance to be burned is weighed and placed in the reaction chamber, which is then sealed and filled with oxygen under pressure in order to assure complete combustion. The reaction chamber is placed in the water bath. The reaction is then initiated by a small, electrically heated wire in the calorimeter. The stirrer keeps the water in motion so that heat evolved by the reaction is evenly distributed and the temperature is uniform. The temperature of the water, once it has become constant, is accurately measured.

The total heat produced by an exothermic reaction is equal to the heat absorbed by the water plus the heat absorbed by all of the parts of the calorimeter within the insulated chamber. The heat evolved is calculated from the recorded temperature increase by using the known specific heat of water and the heat capacity of the calorimeter, which is determined experimentally (see Example 7.12).

A calorimeter for use with processes that take place in solution is shown in Figure 7.6. The pressure in this apparatus is constant and is equal to atmospheric pressure. Therefore, the heat flow in such a solution calorimeter is equal to ΔH ($\Delta H = q_P$). The sealed glass ampule holds a substance that will react with the solution or dissolve in it. After the calorimeter is tightly sealed and the temperature recorded, the ampule is broken to initiate the process to be studied. As in a bomb calorimeter, the temperature change is recorded. The heat released or absorbed is calculated from the heat capacity of the calorimeter and, in this case, also the heat capacity of the solution, for both the calorimeter and the solution undergo the same change in temperature.

A simple, constant-pressure calorimeter can be made from a Styrofoam coffee cup (Figure 7.7). Styrofoam is a good insulator and prevents heat exchange with the surroundings. The calorimeter has four parts—the coffee cup holding a solution, a lid, preferably of an insulating substance such as Styrofoam or cardboard, a thermometer, and a rod for stirring.

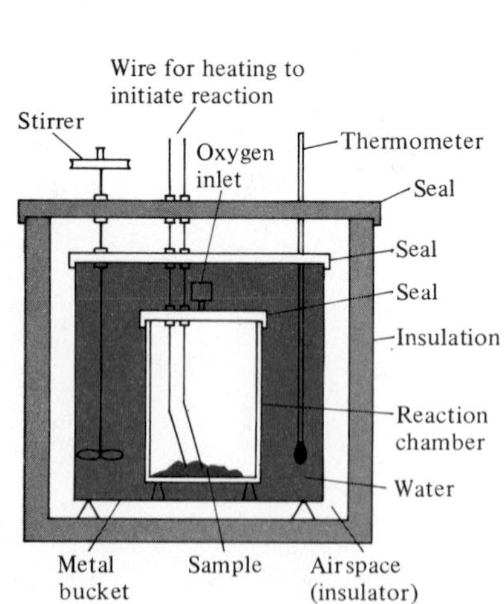

Figure 7.5
Bomb Calorimeter
Volume is constant and ΔE is measured.

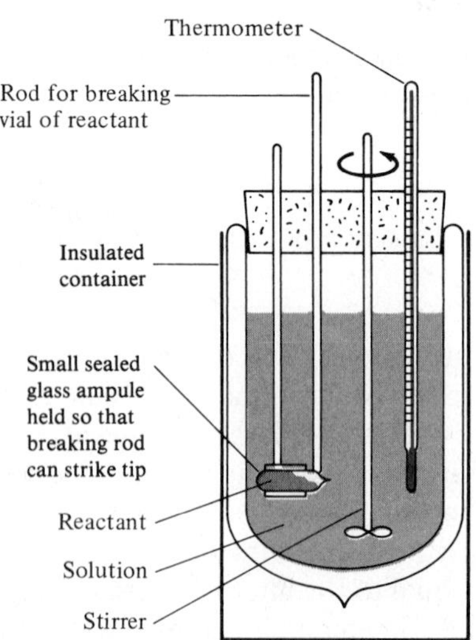

Figure 7.6
Solution Calorimeter
Pressure is constant and ΔH is measured.

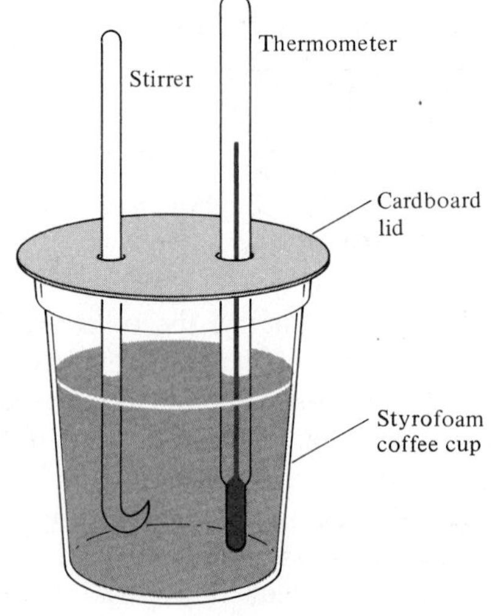

Figure 7.7
Coffee Cup Calorimeter

calorimeter with an electric heater and the temperature rise measured. Or a known amount of heat is introduced by means of a heated metal of known specific heat (see Example 7.4). A reaction for which the heat of reaction is known can also be used. The heat capacity of the calorimeter is found from the relationship

$$q_{\text{known}} = (\text{calorimeter heat capacity})(\Delta T) \qquad \textbf{(7.16)}$$
$$\textit{joules} \qquad \frac{\textit{joules}}{\cancel{\textit{kelvin}}} \qquad \cancel{\textit{kelvins}}$$

Examples 7.4–7.6 illustrate calculations necessary in calorimetry.

EXAMPLE 7.4 Calorimetry

To determine the heat capacity of a solution calorimeter (like that in Figure 7.6), 75.3 g of copper at 370.5 K was placed into 100.3 g of water in the calorimeter. The initial temperature of the water was 294.4 K; the final temperature of the system was 298.8 K. What is the heat capacity of the calorimeter? For Cu, specific heat = 0.385 J/K g; for H_2O, specific heat = 4.184 J/K g.

The copper loses energy and decreases in temperature; the water and the calorimeter gain energy and increase in temperature.

$$\Delta T_{\text{Cu}} = 298.8\text{ K} - 370.5\text{ K} = -71.7\text{ K}$$
$$\Delta T_{\text{calorimeter}} = \Delta T_{\text{H}_2\text{O}} = 298.8\text{ K} - 294.4\text{ K} = 4.4\text{ K}$$

The amount of heat lost by the copper was

$$\begin{aligned} q_{\text{lost, Cu}} &= (\text{mass, Cu})(\text{sp. heat, Cu})(\Delta T_{\text{Cu}}) \\ &= (75.3\text{ g})(0.385\text{ J/K g})(-71.7\text{ K}) = -2080\text{ J} \end{aligned}$$

The heat lost is transferred to the water and the calorimeter, for both of which $\Delta T = 4.4$ K

$$\begin{aligned} q_{\text{gained, H}_2\text{O}} &= (\text{mass, H}_2\text{O})(\text{sp. heat, H}_2\text{O})(\Delta T_{\text{H}_2\text{O}}) \\ &= (100.3\text{ g})(4.184\text{ J/K g})(4.4\text{ K}) = 1800\text{ J} \\ q_{\text{gained, calorimeter}} &= (\text{calorimeter heat capacity})(\Delta T_{\text{calorimeter}}) \\ &= (\text{calorimeter heat capacity})(4.4\text{ K}) \end{aligned}$$

The sum of the heat lost and the heat gained must be zero. The heat capacity of the calorimeter can therefore be calculated as follows:

$$q_{\text{lost, Cu}} + q_{\text{gained, H}_2\text{O}} + q_{\text{gained, calorimeter}} = 0$$
$$-2080\text{ J} + 1800\text{ J} + (\text{calorimeter heat capacity})(4.4\text{ K}) = 0$$
$$\text{Calorimeter heat capacity} = \frac{(2080\text{ J} - 1800\text{ J})}{4.4\text{ K}} = \mathbf{60\ J/K}$$

EXERCISE A calorimeter was calibrated by mixing together dilute solutions of HCl and NaOH, which undergo an exothermic reaction, $HCl(aq) + NaOH(aq) \longrightarrow NaCl(aq) + H_2O(l)$. The amount of heat released by the reaction was 5583.5 J. The 200.7 g of solution that resulted from the reaction had a specific heat of 3.97 J/K g. A temperature increase of 6.5 °C was observed for the calorimeter and the contents. Calculate the heat capacity of the calorimeter. *Answer* 60 J/K.

EXAMPLE 7.5 Calorimetry

The calorimeter described in Example 7.4 was used to find the heat flow associated with the dissolution of 24 g of NaCl in 176 g of water. The observed ΔT (for both the calorimeter and the solution) was −1.6 K. The calorimeter heat capacity is 60 J/K. The specific heat of the resulting NaCl solution is 3.64 J/K g. (This value is found from a standard table of specific heats.) What is the heat flow for this process?

The total heat flow in this experiment is accounted for as follows:

$$q_{\text{dissoln}} + q_{\text{calorimeter}} + q_{\text{NaCl soln}} = 0$$

The negative ΔT shows that the dissolution process is endothermic; the solution and the calorimeter lose energy during the dissolution.

$$\begin{aligned} q_{\text{lost, NaCl soln}} &= (\text{mass, NaCl soln})(\text{sp. heat, NaCl soln})(\Delta T_{\text{NaCl soln}}) \\ &= (176 \text{ g} + 24 \text{ g})(3.64 \text{ J/K g})(-1.6 \text{ K}) = -1200 \text{ J} \\ q_{\text{lost, calorimeter}} &= (\text{calorimeter heat capacity})(\Delta T_{\text{calorimeter}}) \\ &= (60 \text{ J/K})(-1.6 \text{ K}) = -100 \text{ J} \end{aligned}$$

Solving the first law expression above gives the heat absorbed in dissolving 24 g of NaCl

$$\begin{aligned} q_{\text{dissoln}} - 100 \text{ J} - 1200 \text{ J} &= 0 \\ q_{\text{dissoln}} &= \mathbf{1300 \text{ J}} \end{aligned}$$

EXERCISE The heat capacity of a calorimeter is 7.3 J/K. This calorimeter was used to determine the amount of heat gained or lost in the reaction between 100. mL of 0.0100 M Ag^+ and 100. mL of 0.0100 M Cl^-. The observed temperature change was +0.077 K for the 200. g of solution (specific heat = 4.35 J/K g) and the calorimeter. What is the value of q for the reaction, which is $Ag^+ + Cl^- \longrightarrow AgCl(s)$? Was heat released or absorbed in this reaction?

Answer $q = -68$ kJ; as shown by the minus sign, heat was released.

EXAMPLE 7.6 Calorimetry

A bomb calorimeter was used to measure the heat released in the combustion of naphthalene ($C_{10}H_8$) in oxygen (at 298 K). The reaction of a 0.640 g sample of naphthalene raised the temperature of the calorimeter and its contents by 2.54 K. The calorimeter was of a type that has a relatively large heat capacity (10.13 kJ/K). The heat gain of such a calorimeter is so much larger than the heat gain of its contents that the heat gain of the contents can be neglected. Calculate the heat released in the combustion of one mole of naphthalene ($\Delta E = q_V$).

In this experiment the total heat flow, neglecting the heat absorbed by the contents of the calorimeter, is

$$q_{\text{rxn}} + q_{\text{calorimeter}} = 0$$

The heat gained by the calorimeter is

$$q_{\text{calorimeter}} = (\text{calorimeter heat capacity})(\Delta T) = (10.13 \text{ kJ/K})(2.54 \text{ K}) = 25.7 \text{ kJ}$$

giving for the combustion of the sample

$$\begin{aligned} q_{\text{rxn}} + 25.7 \text{ kJ} &= 0 \\ q_{\text{rxn}} &= -25.7 \text{ kJ} \end{aligned}$$

The amount of naphthalene that reacted was

$$(0.640 \text{ g } C_{10}H_8)\left(\frac{1 \text{ mol } C_{10}H_8}{128.17 \text{ g } C_{10}H_8}\right) = 0.00499 \text{ mol}$$

giving the heat of reaction on a molar basis as

$$\frac{-25.7 \text{ kJ}}{0.00499 \text{ mol}} = \mathbf{-5150 \text{ kJ/mol}}$$

The value of -5150 kJ/mol, because it was measured in a constant volume calorimeter, is the internal energy change (ΔE) for the combustion of naphthalene.

EXERCISE The same bomb calorimeter was used to measure the heat of combustion of urea, $(NH_2)_2CO(s)$. The temperature increase of the calorimeter and contents was 1.05 K for the combustion of 1.013 g of urea. Calculate the molar heat of combustion of urea. *Answer* $\Delta E = -629$ kJ/mol.

The data of the experiment described in Example 7.6 can be used to demonstrate the relationship between the internal energy change and the heat of a reaction involving gases. The reaction was

$$C_{10}H_8(s) + 12O_2(g) \longrightarrow 10CO_2(g) + 4H_2O(l) \qquad \Delta E = -5150 \text{ kJ}$$

Recall that the heat of reaction, ΔH, is defined as $\Delta H = \Delta E + \Delta(PV)$. Using the ideal gas equation for a change from state 1 to state 2 shows that we can replace $\Delta(PV)$ by $\Delta(RTn)$:

$$P_1V_1 = n_1RT_1 \qquad P_2V_2 = n_2RT_2$$

$$P_2V_2 - P_1V_1 = n_2RT_2 - n_1RT_1$$

$$\Delta(PV) = \Delta(nRT)$$

At constant temperature

$$\Delta(PV) = RT(\Delta n_g) \quad \text{change in no. of moles of gas}$$

giving

$$\Delta H = \Delta E + RT(\Delta n_g)$$

This expression permits calculation of the difference between ΔH and ΔE. For example, for the reaction of Example 7.6, which was carried out at 298 K and for which the change in the number of moles of gas is 10 mol − 12 mol = −2 mol, the difference is

$$\Delta H - \Delta E = RT(\Delta n_g) = (8.314 \times 10^{-3} \text{ kJ/K mol})(298 \text{ K})(-2 \text{ mol}) = -4.96 \text{ kJ}$$

This reaction involving pressure–volume work has a difference between ΔH and ΔE of about 0.1% of the value of ΔE. You can see that in reactions involving only solids and liquids, the difference would be negligible.

SUMMARY OF SECTIONS 7.5–7.6

The heat capacity of a substance is the amount of heat needed to raise the temperature of the substance by 1 K. Molar heat capacity is the heat capacity per mole; specific heat is the heat capacity per gram. Knowing the change in temperature and the heat capacity of a substance permits calculation of the amount of heat that the substance has released or absorbed. Measurement of the change in temperature of a calorimeter of known heat capacity permits calculation of the heat absorbed or released during a chemical reaction or a change of state. Bomb calorimeters measure heat exchanged at constant volume, which is equal to ΔE; solution calorimeters measure heat exchanged at constant pressure, which is equal to ΔH.

HEATS OF REACTION AND OTHER ENTHALPY CHANGES

7.7 HEATS OF REACTION

In studying thermodynamics, a chemical reaction is considered complete when no further changes in composition take place and the substances have returned to their original temperature, usually room temperature. The total amount of heat released or absorbed between the beginning of a reaction and the return of the substances present to the original temperature is referred to as the **heat of reaction.** When the reaction occurs at constant pressure, the heat of reaction is equal to the enthalpy change. Thus, ΔH is correctly referred to as either the enthalpy change or the heat of reaction. The value of ΔH depends on the specific reaction that has taken place, the amounts of the substances involved, and the temperature. Enthalpy changes, therefore, must be expressed as the quantity of heat per quantity of the substance or substances in question, and the temperature must be specified.

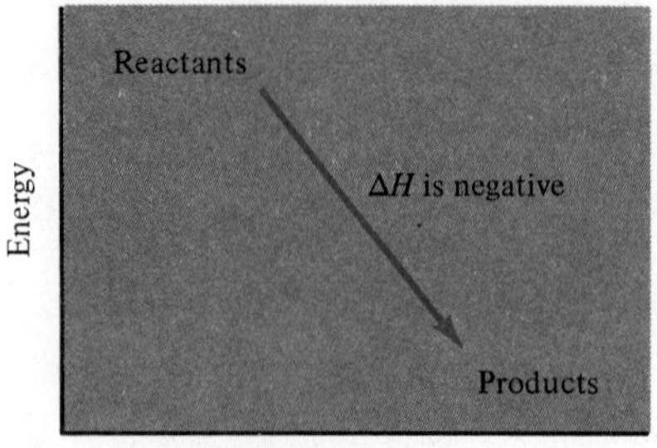

(a) Exothermic reaction: heat is evolved; products have lower energy than reactants.

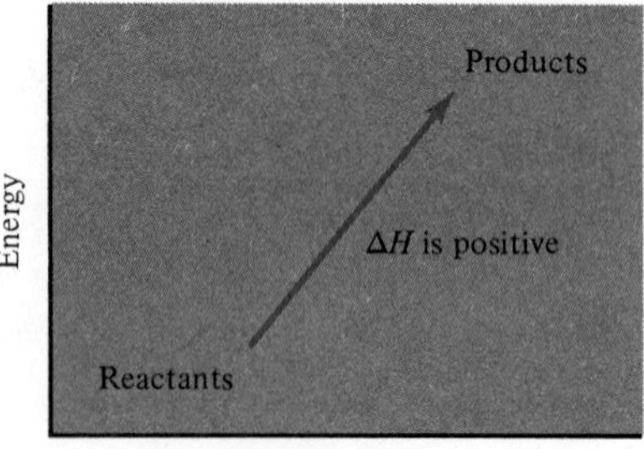

(b) Endothermic reaction: heat is absorbed; products have higher energy than reactants.

Figure 7.8
Energy of Reactants and Products in Exothermic and Endothermic Reactions

The value of ΔH is usually given as the heat released or absorbed in the reaction of the molar amounts of the reactants represented by the balanced equation. A **thermochemical equation** includes ΔH for the balanced equation as written. For example, the following thermochemical equation

$$2HgO(s) \xrightarrow{\Delta} 2Hg(l) + O_2(g) \qquad \Delta H_{298} = 181.67 \text{ kJ} \tag{7.17}$$

tells us that 181.67 kJ of thermal energy must be *added* (because ΔH is positive) to convert 2 mol of solid mercury(II) oxide to 2 mol of liquid mercury and 1 mol of gaseous oxygen (at 298 K). The equation

$$K_2O(s) + CO_2(g) \longrightarrow K_2CO_3(s) \qquad \Delta H_{298} = -391.1 \text{ kJ} \tag{7.18}$$

tells us that when one mole of solid potassium oxide reacts with one mole of gaseous carbon dioxide to produce one mole of solid potassium carbonate, 391.1 kJ of energy is *released* as heat (at 298 K). Note that the value of ΔH is positive for an endothermic reaction and negative for an exothermic reaction.

It might be helpful to think of the heat exchanged in a reaction as one of the reactants or products. In an exothermic reaction such as the combustion of carbon disulfide

$$CS_2(l) + 3O_2(g) \longrightarrow CO_2(g) + 2SO_2(g) + 1075 \text{ kJ} \tag{7.19}$$

energy has been released and therefore, the total internal energy of the products is less than that of the reactants (Figure 7.8a). For an endothermic reaction such as the thermal decomposition of mercury(II) oxide

$$2HgO(s) + 181.67 \text{ kJ} \longrightarrow 2Hg(l) + O_2(g) \tag{7.20}$$

energy has been added to the system and therefore the total internal energy of the products is greater than that of the reactants (Figure 7.8b).

7.8 STANDARD STATE AND STANDARD ENTHALPY CHANGES

To compare heats of reaction, it is necessary to know the temperature at which ΔH was measured and the physical state of each of the reactants and products. Chemists have agreed to report heats of reaction for reactions carried out with reactants in what is called the standard state. The **standard state** of any substance is the physical state at which it is most stable at 1 bar pressure (slightly less than 1 atm) and a specified temperature. [The standard state pressure was formerly 1 atm, but was changed because the atmosphere is not a unit accepted in the SI system. One bar is equal to 10^5 Pa or 100 kPa.] The usual specified temperature for the standard state is 298 K (25 °C, roughly room temperature). For substances that are solid at this temperature and pressure, and can also have several different crystalline forms, it is necessary to specify for which form the heat of reaction is given.

Standard enthalpy changes are enthalpy changes expressed for chemical reactions or other transformations of substances in their standard states. (For example, the **standard enthalpy of reaction** is the enthalpy change for the transformation of reactants in their standard states to products in their standard states.) The symbol for a standard enthalpy change is ΔH°, and the specified temperature is often given as a subscript, ΔH°_{298}. We have chosen to omit specifying the temperature if it is 298 K. Unless otherwise stated, all ΔH° values in this book are given for 298 K. When changes in internal energy are reported for substances in their standard states, the symbol ΔE° is used. (Note that the standard state of a substance is a specific physical state and is in no way related to the "standard conditions" of temperature and pressure, STP, used most often in discussing gases.) Many chemical reactions, of course, do not actually take place at 298 K, but require higher or lower temperatures. The standard enthalpy changes at 298 K reported for such reactions are calculated from ΔH values measured at other temperatures.

EXAMPLE 7.7 Standard States

What would be the standard states at 25 °C of the following substances?

Ca	$GeCl_4$	GeH_4	$CH_3(CH_2)_6CH_3$
Calcium	Germanium tetrachloride	Germanium hydride	Octane
m.p. 839 °C	m.p. −49.5 °C	m.p. −165 °C	m.p. −56.8 °C
b.p. 1484 °C	b.p. 84 °C	b.p. −88.5 °C	b.p. 125.7 °C

By examining the melting and boiling points, it can be seen that in their standard states at 25 °C **calcium is a solid (m.p. > 25 °C), germanium tetrachloride and octane are liquids (m.p. < 25 °C, b.p. > 25 °C), and germanium hydride is a gas (b.p. < 25 °C).**

EXERCISE What would be the standard states of the substances listed above at 100 °C?

Answers Ca(*s*), $GeCl_4(g)$, $GeH_4(g)$, $CH_3(CH_2)_6CH_3(l)$.

7.9 HEATS OF FORMATION

One of the simplest types of chemical reactions is the formation of a compound from its elements, a combination reaction. The **standard enthalpy of formation**, or heat of formation of a compound, ΔH_f°, is the heat of formation of one mole of the compound in its standard state by combination of the elements in their standard states at a specified temperature. The thermochemical equation for the formation of water at 298 K by combination of hydrogen and oxygen is

$$H_2(g) + \tfrac{1}{2}O_2(g) \longrightarrow H_2O(l) \qquad \Delta H_f^\circ = -285.83 \text{ kJ}$$

Hydrogen and oxygen are gases, and water is a liquid at this temperature and therefore the standard enthalpy of formation of water is given for the reaction with the substances in these states.

Some standard heats of formation are given in Table 7.3. (A more comprehensive table is given in Appendix IV.) To write thermochemical equations for the combination

Table 7.3
Standard Heats of Formation at 298 K
The values of ΔH_f° are for the formation of one mole of the given compound in its standard state by combination of the elements in their standard states at 298 K. A compound that is a gas or a liquid at 298 K and 1 bar (slightly less than 1 atm) has that form as its standard state. For solids that can exist in several different crystalline forms under these conditions, such as carbon as diamond or graphite, it is necessary to specify which is the most stable form (for which $\Delta H_f^\circ = 0$). For additional thermodynamic data, see Appendix IV.

Substance	ΔH_f° (kJ/mol)	Substance	ΔH_f° (kJ/mol)
Standard state of all elements	0.00		
C(graphite)	0.00	$PCl_3(l)$	−319.7
P(white)	0.00	$PCl_5(s)$	−443.5
		AgCl(*s*)	−127.068
		NaCl(*s*)	−411.153
$H_2O(l)$	−285.830	KCl(*s*)	−436.747
$SO_2(g)$	−296.830		
$SO_3(g)$	−395.72		
NO(*g*)	90.25	HF(*g*)	−271.1
$NO_2(g)$	33.18	HCl(*g*)	−92.307
CO(*g*)	−110.525	HBr(*g*)	−36.40
$CO_2(g)$	−393.509	HI(*g*)	26.48
		$H_2S(g)$	−20.63
PbO(*s*, red)	−218.99	$NH_3(g)$	−46.11
$PbO_2(s)$	−277.4		
Al_2O_3(α-solid)	−1675.7		
$Fe_2O_3(s)$	−824.2	$CS_2(l)$	89.70
$Fe_3O_4(s)$	−1118.4	$CH_4(g)$	−74.81
CaO(*s*)	−635.09	$C_2H_2(g)$	226.73
Ca(OH)(*s*)	−986.09	$CF_4(g)$	−925
		$PbSO_4(s)$	−919.94
		$CaCO_3$(calcite)	−1206.92

reactions to which these heats of formation apply, the reactants must be elements in their standard states at 298 K and the product must be *one mole* of the substance in its standard state. For example,

$$\tfrac{1}{2}H_2(g) + \tfrac{1}{2}I_2(s) \longrightarrow HI(g) \qquad \Delta H_f^\circ = +26.48\text{ kJ}$$
$$2Al(s) + \tfrac{3}{2}O_2(g) \longrightarrow Al_2O_3(\alpha\text{-solid}) \qquad \Delta H_f^\circ = -1675.7\text{ kJ}$$

It is important to note that the standard enthalpy of formation of an element in its standard state is taken to be zero. This convention allows all of the heat absorbed or released in such combination reactions to be assigned to formation of the products.

EXAMPLE 7.8 Heats of Reaction

Using the data given in Table 7.3, write the thermochemical equations for the formation of (a) PbO(*s*, red) and (b) $PCl_3(l)$.

(a) The equation must give the elements in their standard states at 298 K as reactants and one mole of PbO(*s*, red) in its standard state as the only product. Oxygen in its standard state is $O_2(g)$ and lead is a solid.

$$Pb(s) + O_2(g) \xrightarrow{\text{(unbalanced)}} PbO(s, \text{red})$$

The equation is balanced by putting a coefficient of $\tfrac{1}{2}$ before the O_2 term

$$Pb(s) + \tfrac{1}{2}O_2(g) \longrightarrow PbO(s, \text{red})$$

The heat of formation of PbO(*s*, red) is given as −218.99 kJ/mol in Table 7.3, so the complete equation is

$$\mathbf{Pb(s) + \tfrac{1}{2}O_2(g) \longrightarrow PbO(s, red) \qquad \Delta H_f^\circ = -218.99\text{ kJ}}$$

(b) Chlorine, Cl_2, is gaseous in its standard state and phosphorus is a solid. The table shows that ΔH° is zero for white phosphorus, which is considered the standard state. The complete equation is

$$\mathbf{P(s, white) + \tfrac{3}{2}Cl_2(g) \longrightarrow PCl_3(l) \qquad \Delta H_f^\circ = -319.7\text{ kJ}}$$

EXERCISE Use the data given in Table 7.3 to write the thermochemical equations for the formation of (a) iron(III) oxide, Fe_2O_3, and (b) calcite, $CaCO_3$.

Answers (a) $2Fe(s) + \tfrac{3}{2}O_2(g) \longrightarrow Fe_2O_3(s) \quad \Delta H^\circ = -824.2$ kJ; (b) $Ca(s) + C(\text{graphite}) + \tfrac{3}{2}O_2(g) \longrightarrow CaCO_3(\text{calcite}) \quad \Delta H^\circ = -1206.92$ kJ.

7.10 FINDING ENTHALPY CHANGES

The following sections illustrate how known enthalpy changes are used to find other enthalpy changes.

a. Reversing Reactions Suppose we want to know the heat of reaction for the decomposition of one mole of silver oxide, Ag_2O.

$$Ag_2O(s) \xrightarrow{\Delta} 2Ag(s) + \tfrac{1}{2}O_2(g) \qquad \Delta H^\circ = ? \qquad \textbf{(7.21)}$$

What is available is the heat of reaction for the formation of one mole of silver oxide from the elements, which is the reverse reaction.

$$2Ag(s) + \tfrac{1}{2}O_2(g) \longrightarrow Ag_2O(s) \qquad \Delta H_f^\circ = -31.05\text{ kJ} \qquad \textbf{(7.22)}$$

The reverse of an exothermic reaction is always an endothermic reaction (and vice versa). As expected, based on the first law of thermodynamics, ΔH for a reaction in one direction is equal in magnitude to ΔH for the reaction in the reverse direction, but is opposite in sign. The heat of reaction for reaction (7.21) is, therefore, $\Delta H^\circ = 31.05$ kJ. The energy required to decompose a compound to its constituent elements is always equal to the energy released in formation of the compound from the elements (with the reactions compared under identical conditions).

b. Changing Quantities If the quantities of reactants are changed, the heat of reaction also changes. For example, to find the heat of reaction for the decomposition of 2 mol of Ag_2O, the heat of reaction for 1 mol must be multiplied by 2, as are the coefficients for the reactants and products.

$$2[Ag_2O(s) \longrightarrow 2Ag(s) + \tfrac{1}{2}O_2(g)] \qquad \Delta H° = (2)(31.05\ \text{kJ})$$
$$2Ag_2O(s) \longrightarrow 4Ag(s) + O_2(g) \qquad \Delta H° = 62.10\ \text{kJ}$$

This illustrates the second important principle of thermochemistry: For a given chemical reaction or change of state, ΔH is directly proportional to the quantities of reactants or products.

To find the heat of reaction for any amount of a substance, known in mass units, the mass is first converted to moles as in any stoichiometry problem. Then the enthalpy change is multiplied by the number of moles of the substance. Example 7.9 illustrates how to analyze a stoichiometry problem involving thermochemistry.

EXAMPLE 7.9 Heats of Reaction: Problem Solving

Aluminum displaces chromium from chromium(III) oxide.

$$2Al(s) + Cr_2O_3(s) \longrightarrow Al_2O_3(s) + 2Cr(s) \qquad \Delta H° = -536\ \text{kJ}$$

How much heat will be released in the reaction (under standard state conditions at 298 K) of 10.0 g of aluminium with 25.0 g of Cr_2O_3?

1. Study the problem and be sure you understand it.
   (a) What is unknown?
   The heat released in the reaction of known amounts of Al and Cr_2O_3 according to the equation given.
   (b) What is known?
   The thermochemical equation for the reaction under consideration and the masses of the two reactants.
2. Decide how to solve the problem.
   (a) What is the connection between the known and the unknown?
   Several steps will be necessary to make the connection. In such a case, it is helpful to analyze the problem beginning with the unknown and working backwards. To find the unknown heat, the number of moles of reactant that are completely consumed must be known. Because masses of both reactants are given, it will first be necessary to determine which is the limiting reactant. To find the limiting reactant, the known masses must be converted to moles and mole ratios then used to determine which reactant is limiting.
   (b) What is necessary to make the connection?
   (i) Convert masses of Al and Cr_2O_3 to amounts in moles.
   (ii) Check amount of one reactant, say Cr_2O_3, required by the other, Al, to see which is limiting.
   (iii) Use molar amount of limiting reactant to find how much heat will be released.
3. Set up the problem and solve it.

(i)
$$(10.0\ \cancel{\text{g Al}})\left(\frac{1\ \text{mol Al}}{26.98\ \cancel{\text{g Al}}}\right) = 0.371\ \text{mol Al}$$

$$(25.0\ \cancel{\text{g } Cr_2O_3})\left(\frac{1\ \text{mol } Cr_2O_3}{151.99\ \cancel{\text{g } Cr_2O_3}}\right) = 0.164\ \text{mol } Cr_2O_3$$

(ii)
$$(0.371\ \cancel{\text{mol Al}})\left(\frac{1\ \text{mol } Cr_2O_3}{2\ \cancel{\text{mol Al}}}\right) = 0.186\ \text{mol } Cr_2O_3$$

More Cr_2O_3 is required than is available. Therefore, Cr_2O_3 is the limiting reactant.

(iii) The value of ΔH° is the amount of heat for the reaction as written and can be expressed, for this reaction, as either -536 kJ/1 mol Cr_2O_3 or -536 kJ/2 mol Al. Since Cr_2O_3 is the limiting reactant, it is used to find the amount of heat released.

$$(0.164 \text{ mol } Cr_2O_3)\left(\frac{-536 \text{ kJ}}{1 \text{ mol } Cr_2O_3}\right) = \mathbf{-87.9 \text{ kJ}}$$

4. Check the result.
The answer is reasonable. The units came out correctly, significant digits are used correctly (limited to three throughout by 10.0 g and 25.0 g), and the magnitude of the answer is reasonable (25.0 g of Cr_2O_3 is roughly 0.2 mol and 87.9 kJ is roughly two-tenths of the value of ΔH°).

EXAMPLE 7.10 Heats of Reaction

Write the thermochemical equation for the combustion of acetylene, $C_2H_2(g)$, in oxygen, for which the standard enthalpy of reaction is $-1300.$ kJ per mole of acetylene. The reaction products are carbon dioxide and water. What is the heat of reaction for burning 65 g of acetylene in oxygen (at 298 K)?

The thermochemical equation is

$$C_2H_2(g) + \tfrac{5}{2}O_2(g) \longrightarrow 2CO_2(g) + H_2O(l) \qquad \Delta H^\circ = -1300. \text{ kJ}$$

To find the heat of reaction for 65 g of acetylene, the value of ΔH° is multiplied by the number of moles of acetylene. Doing the calculation in one step gives

$$(65 \text{ g } C_2H_2)\left(\frac{1 \text{ mol } C_2H_2}{26.04 \text{ g } C_2H_2}\right)\left(\frac{-1300. \text{ kJ}}{1 \text{ mol } C_2H_2}\right) = \mathbf{-3200 \text{ kJ}}$$

[This highly exothermic reaction is used in welding.]

EXERCISE Write the thermochemical equation for the reaction of benzoic acid, $C_6H_5COOH(s)$, with oxygen to give CO_2 and H_2O. The heat of reaction is -3227 kJ per mole of C_6H_5COOH. What is the heat of reaction for the combustion of 1.00 g of benzoic acid?

Answer $C_6H_5COOH(s) + \tfrac{15}{2}O_2(g) \longrightarrow 7CO_2(g) + 3H_2O(l) \qquad \Delta H^\circ = -3227$ kJ; heat of reaction is -26.4 kJ.

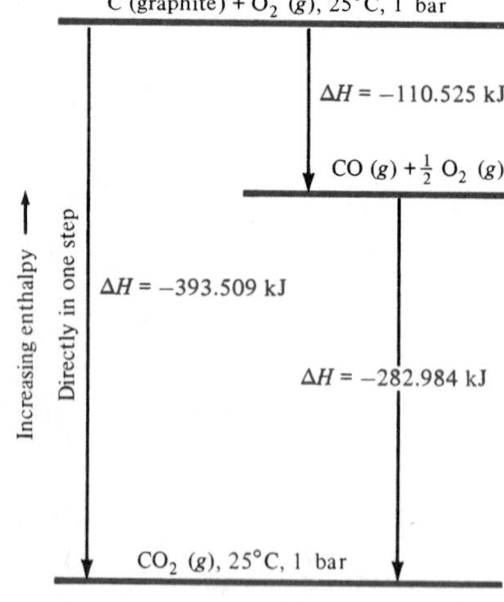

Figure 7.9
Hess's Law
The total amount of heat released in the combustion of graphite is the same whether the reaction takes place in one step ($C + O_2 \longrightarrow CO_2$) or in two steps ($C + \tfrac{1}{2}O_2 \longrightarrow CO$, $CO + \tfrac{1}{2}O_2 \longrightarrow CO_2$).

c. Hess's Law: Combining ΔH Values **Hess's law,** discovered experimentally in 1840, is the third important principle of thermochemistry: The enthalpy change of a chemical reaction is the same whether the reaction takes place in one step or several steps. In other words, the heat absorbed or evolved in going from the initial state to the final state is the same no matter by what route the reaction takes place (Figure 7.9).

Hess's law and the other thermochemical principles are all consequences of the law of conservation of energy. It is impossible to reverse a reaction and get more heat out than was put in, or to go from the same reactants to the same products by different intermediate steps and get more heat out by one route than by another.

The very useful result of Hess's law is that thermochemical equations can be dealt with algebraically to find enthalpy changes that are not known or are difficult to measure. The desired ΔH value is calculated from a sequence of thermochemical equations that can be combined algebraically to give the desired equation. *It doesn't matter whether the sequence of reactions can actually take place or exists only on paper.*

A method for using known enthalpies to calculate unknown enthalpies can be summed up in five general steps. In all Hess's law problems, the unknown is the enthalpy change of a particular chemical reaction; the knowns are thermochemical equations for changes involving the substances in the process in question; and the connection between the unknown and the knowns is an algebraic manipulation of the known thermochemical equations. The objective is to write the known thermochemical equations in such a way that they will add up to the desired equation. This can be accomplished by the steps outlined below and used to solve the following problem: Given the enthalpy changes for the following two reactions:

$$\tfrac{1}{2}N_2(g) + \tfrac{1}{2}O_2(g) \longrightarrow NO(g) \qquad \Delta H^\circ = 90.25 \text{ kJ} \qquad \textbf{(7.23)}$$

$$2NO_2(g) \longrightarrow 2NO(g) + O_2(g) \qquad \Delta H^\circ = 114.14 \text{ kJ} \qquad \textbf{(7.24)}$$

find the standard heat of formation of NO_2.

1. Write the desired thermochemical equation. The equation for the formation of NO_2 from the elements in their standard states is

$$\tfrac{1}{2}N_2(g) + O_2(g) \longrightarrow NO_2(g) \qquad \textbf{(7.25)}$$

2. If necessary, reverse some of the known thermochemical equations so that the major reactants and products of the desired reaction are on the reactants and products sides. The reactants are on the desired side in Equation (7.23), but Equation (7.24) must be reversed so that the product NO_2 will be on the appropriate side.

$$2NO(g) + O_2(g) \longrightarrow 2NO_2(g) \qquad \Delta H^\circ = -114.14 \text{ kJ} \qquad \textbf{(7.24a)}$$

3. If necessary, multiply the known equations by appropriate coefficients so that the major reactants and products have the same coefficients as they do in the desired equation. Because NO_2 has the coefficient of 1 in the desired reaction, Equation (7.24) must be multiplied by $\tfrac{1}{2}$.

$$\begin{aligned} \tfrac{1}{2}[2NO(g) + O_2(g) &\longrightarrow 2NO_2(g)] \qquad \Delta H^\circ = \tfrac{1}{2}(-114.14 \text{ kJ}) \\ NO(g) + \tfrac{1}{2}O_2(g) &\longrightarrow NO_2(g) \qquad \Delta H^\circ = -57.07 \text{ kJ} \end{aligned} \qquad \textbf{(7.24b)}$$

Nitrogen, N_2, has the same coefficient in both Equations (7.23) and (7.25), so no further change is necessary.

4. If unwanted reactants or products appear in the equations after Step 3, further equations must be included so that the unwanted species will cancel out when all of the equations are added. This is accomplished by having the unwanted species, with the same coefficients, on opposite sides of the equations. In this case, the only unwanted species is NO and it already appears on opposite sides of Equations (7.23) and (7.24a) with the same coefficient (see Example 7.11, where this step must be followed).

5. Add the known equations and the values of ΔH° as they appear after Steps 2, 3, and 4.

$$\begin{aligned} \tfrac{1}{2}N_2(g) + \tfrac{1}{2}O_2(g) &\longrightarrow \cancel{NO(g)} \qquad \Delta H^\circ = 90.25 \text{ kJ} \\ \cancel{NO(g)} + \tfrac{1}{2}O_2(g) &\longrightarrow NO_2(g) \qquad \Delta H^\circ = -57.07 \text{ kJ} \\ \hline \tfrac{1}{2}N_2(g) + O_2(g) &\longrightarrow NO_2(g) \qquad \Delta H^\circ = 33.18 \text{ kJ} \end{aligned}$$

The standard heat of formation of NO_2 is 33.18 kJ/mol.

EXAMPLE 7.11 Heats of Reaction

Use the following thermochemical equations

$$\begin{aligned} Ca(s) + 2C(\text{graphite}) &\longrightarrow CaC_2(s) & \Delta H^\circ &= -59.8 \text{ kJ} \\ Ca(s) + \tfrac{1}{2}O_2(g) &\longrightarrow CaO(s) & \Delta H^\circ &= -635.09 \text{ kJ} \\ CaO(s) + H_2O(l) &\xrightarrow{H_2O} Ca(OH)_2(aq) & \Delta H^\circ &= -653.1 \text{ kJ} \\ C_2H_2(g) + \tfrac{5}{2}O_2(g) &\longrightarrow 2CO_2(g) + H_2O(l) & \Delta H^\circ &= -1300. \text{ kJ} \\ C(\text{graphite}) + O_2(g) &\longrightarrow CO_2(g) & \Delta H^\circ &= -393.509 \text{ kJ} \end{aligned}$$

to find ΔH° for

$$CaC_2(s) + 2H_2O(l) \xrightarrow{H_2O} Ca(OH)_2(aq) + C_2H_2(g)$$

The complete equation for the desired reaction is given in the statement of the problem (Step 1). The first, third, and fourth known equations include the major reactants and products, and the first and fourth equations must be reversed, giving the following three equations that include the main reactants and products (Step 2).

$$
\begin{array}{lcll}
CaC_2(s) & \longrightarrow & Ca(s) + 2C(\text{graphite}) & \Delta H^\circ = 59.8\text{ kJ} \\
CaO(s) + H_2O(l) & \xrightarrow{H_2O} & Ca(OH)_2(aq) & \Delta H^\circ = -653.1\text{ kJ} \\
2CO_2(g) + H_2O(l) & \longrightarrow & C_2H_2(g) + \frac{5}{2}O_2(g) & \Delta H^\circ = 1300.\text{ kJ}
\end{array}
$$

No changes in coefficients are needed (Step 3).

The $CaO(s)$ and $2CO_2(g)$ are unwanted reactants and $Ca(s)$, $2C(\text{graphite})$, and $\frac{5}{2}O_2(g)$ are unwanted products. These can be eliminated by using the second and fifth equations (Step 4). The objective now is to have formulas on the *opposite* sides from their positions in the first three equations. By including the second equation as it is and the fifth equation multiplied by 2 to allow the elimination of $2CO_2$, the desired equation is obtained.

$$
\begin{array}{lcll}
CaC_2(s) & \longrightarrow & \cancel{Ca(s)} + \cancel{2C(\text{graphite})} & \Delta H^\circ = 59.8\text{ kJ} \\
\cancel{CaO(s)} + H_2O(l) & \xrightarrow{H_2O} & Ca(OH)_2(aq) & \Delta H^\circ = -653.1\text{ kJ} \\
\cancel{2CO_2(g)} + H_2O(l) & \longrightarrow & C_2H_2(g) + \cancel{\frac{5}{2}O_2(g)} & \Delta H^\circ = 1300.\text{ kJ} \\
\cancel{Ca(s)} + \cancel{\frac{1}{2}O_2(g)} & \longrightarrow & \cancel{CaO(s)} & \Delta H^\circ = -635.09\text{ kJ} \\
\cancel{2C(\text{graphite})} + \cancel{2O_2(g)} & \longrightarrow & \cancel{2CO_2(g)} & \Delta H^\circ = -787.018\text{ kJ} \\
\hline
CaC_2(s) + 2H_2O(l) & \xrightarrow{H_2O} & Ca(OH)_2(aq) + C_2H_2(g) & \Delta H^\circ = \mathbf{-715\text{ kJ}}
\end{array}
$$

EXERCISE The following thermochemical equations can be written for the oxides of iron:

$$
\begin{array}{lcll}
Fe(s) + \frac{1}{2}O_2(g) & \longrightarrow & FeO(s) & \Delta H^\circ = -272.0\text{ kJ} \\
3Fe(s) + 2O_2(g) & \longrightarrow & Fe_3O_4(s) & \Delta H^\circ = -1118.4\text{ kJ}
\end{array}
$$

Find ΔH° for $4FeO(s) \longrightarrow Fe(s) + Fe_3O_4(s)$. *Answer* $\Delta H^\circ = -30.4$ kJ.

When standard heats of formation are known for all of the reactants and products in a given reaction, the heat of reaction can be found by a more simple method than that illustrated above (but also based on Hess's law). Subtraction of the sum of the ΔH_f values of all of the *reactants* from the sum of the ΔH_f values of all of the *products* gives the desired ΔH° value.

$$\Delta H^\circ = (\text{sum of } \Delta H_f^\circ \text{ products}) - (\text{sum of } \Delta H_f^\circ \text{ reactants}) \qquad \textbf{(7.26)}$$

In using Equation (7.26) keep in mind that for elements in their standard states, $\Delta H_f^\circ = 0$.

To find ΔH° for the reaction

$$2A + B \longrightarrow 3C + 2D \qquad \Delta H^\circ = ?$$

ΔH_f° values would be combined according to Equation (7.26) as follows:

$$\Delta H^\circ = [(3\text{ mol})\Delta H_f^\circ(C) + (2\text{ mol})\Delta H_f^\circ(D)] - [(2\text{ mol})\Delta H_f^\circ(A) + (1\text{ mol})\Delta H_f^\circ(B)]$$

Note that each value must be multiplied by the number of moles that appear in the desired reaction.

EXAMPLE 7.12 Heats of Reaction

Find the heat of reaction for

$$CH_4(g) + 4F_2(g) \longrightarrow CF_4(g) + 4HF(g)$$

using the heat of formation data given in Table 7.3.

The heats of formation of all of the reactants and products are known. Therefore

$$\begin{aligned}\Delta H^\circ &= [(1\ \text{mol})\Delta H_f^\circ(CF_4) + (4\ \text{mol})\Delta H_f^\circ(HF)] \\ &\quad - [(1\ \text{mol})\Delta H_f^\circ(CH_4) + (4\ \text{mol})\Delta H_f^\circ(F_2)] \\ &= [(1\ \text{mol})(-925\ \text{kJ/mol}) + (4\ \text{mol})(-271.1\ \text{kJ/mol})] \\ &\quad - [(1\ \text{mol})(-74.81\ \text{kJ/mol}) + (4\ \text{mol})(0)] \\ &= \mathbf{-1935\ kJ}\end{aligned}$$

EXERCISE Use the heat of formation data given in Table 7.3 to find the heat of reaction for $2Fe_3O_4(s) + \frac{1}{2}O_2(g) \longrightarrow 3Fe_2O_3(s)$. *Answer* $\Delta H^\circ = -235.8$ kJ.

7.11 HEATS OF OTHER CHEMICAL AND PHYSICAL CHANGES

The heat gained or lost in any chemical or physical change, when measured at constant pressure, is an enthalpy change. Although often referred to by various names other than as enthalpies, all such values can be included in thermochemical equations, which can be used to find other enthalpy changes as illustrated in Section 7.9. For example, many chemical compounds, particularly organic compounds containing carbon and hydrogen, react exothermically with oxygen. The **standard enthalpy of combustion,** or heat of combustion, is the heat released in the reaction of one mole of a substance in its standard state with oxygen. When compounds containing only carbon, hydrogen, and possibly oxygen burn completely, the products are only carbon dioxide and water. The thermochemical equation for the combustion of such a compound must show the reaction of one mole of the compound with sufficient oxygen to convert all of the carbon and hydrogen present to gaseous carbon dioxide and liquid water. The combustion of table sugar is represented as follows:

$$\underset{\substack{\textit{sucrose}\\(\textit{table sugar})}}{C_{12}H_{22}O_{11}(s)} + 12O_2(g) \longrightarrow 12CO_2(g) + 11H_2O(l) \qquad \Delta H_c^\circ = -5641\ \text{kJ}$$

The caloric values of foods are heats of combustion, usually reported as kilocalories (but often referred to as "calories" or "Calories") in a given mass or in a normal serving. For example, three teaspoons of table sugar has a caloric value of 45 kcal (about 190 kJ) and one raw apple has a caloric value of 80 kcal (about 300 kJ).

Table 7.4 gives the names and general equations for a number of other commonly encountered types of enthalpy changes, some of which are introduced in later chapters.

Table 7.4
Types of Enthalpy Changes

Name of enthalpy change (section where discussed)	Physical or chemical change
Heat of reaction, ΔH° (Section 7.7)	Reactants $\longrightarrow$ products
Heat of formation, ΔH_f° (Section 7.9)	Elements $\longrightarrow$ 1 mol substance
Heat of combustion, ΔH_c° (Section 7.11)	1 mol substance + $nO_2(g)$ $\longrightarrow$ combustion products
Heat of change of state (Section 7.11 and Figure 7.11)	1 mol substance in state A $\longrightarrow$ 1 mol substance in state B
Heat of neutralization, $\Delta H_{\text{neutralization}}^\circ$ (per 1 mol of acid or base)	Acid + base $\longrightarrow$ salt(aq) + $H_2O(l)$
Heat of solution, $\Delta H_{\text{soln}}^\circ$	1 mol solute $\xrightarrow{n\ \text{solvent}}$ 1 mol solute (in n mol solvent)
Heat of dilution, $\Delta H_{\text{dil}}^\circ$	1 mol solute (in n mol solvent) $\xrightarrow{m\ \text{solvent}}$ 1 mol solute (in $n + m$ mol solvent)
Bond dissociation energy (Section 10.6)	$X_2(g) \longrightarrow 2X(g)$
Ionization energy (Section 9.7)	$X(g) \longrightarrow X^+(g) + e^-$
Electron affinity (Section 9.9)	$X(g) + e^- \longrightarrow X^-(g)$

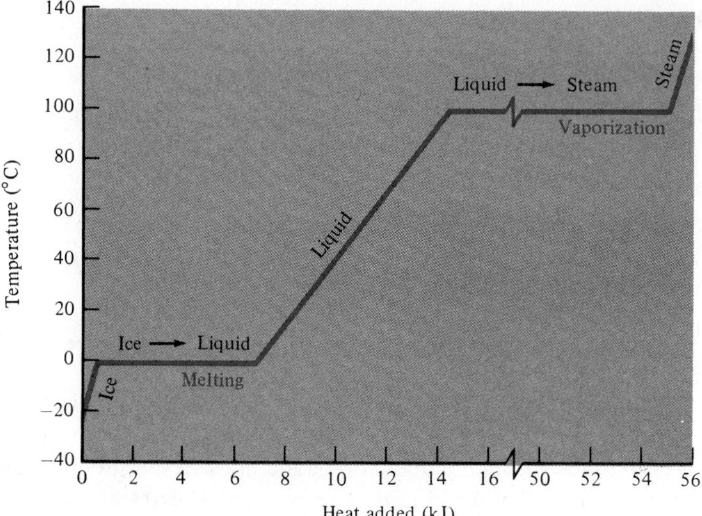

Figure 7.10
Heating Curve from −25 °C to 125 °C for One Mole of Water
Note that more heat is required for vaporization than for melting.

To understand the heats for changes of states, consider the changes as one mole of ice at −25 °C is slowly heated at a constant rate while the temperature of the system is carefully measured. The ice gradually increases in temperature until the temperature reaches 0 °C (Figure 7.10). The temperature then remains constant, even as more heat is added, until all of the ice has melted into liquid water. All of the added thermal energy has gone into melting the ice.

Once the ice is melted and only liquid water is present, the system again increases in temperature, this time until the boiling point of water is reached (100 °C at 1 atm pressure). Again, the temperature remains constant as more heat is added, this time until the liquid water has completely vaporized into steam. All of the thermal energy has gone into breaking up clusters of water molecules held together by the intermolecular forces. After the vaporization is complete, the steam can again increase in temperature.

The enthalpy for a change of state is the amount of heat absorbed or released in the process (at constant pressure) of completing the transformation from one state to the other, without any change in the temperature. The enthalpies for changes of state (Figure 7.11) are usually referred to as heats of vaporization, heats of fusion, and so on. (Definitions of these terms are given in Section 13.4. Note that it is common to refer to something that has melted and then solidified as "fused," but in scientific terminology "fusion" means melting.) Some further examples of enthalpy changes for changes of state and also a few for changes in crystalline form are given in Table 7.5.

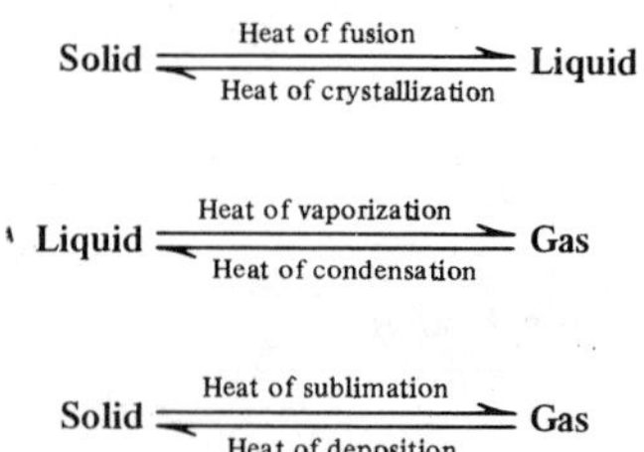

Figure 7.11
Changes of State
Enthalpy values for changes of state are usually given in kilojoules per mole of substance changed at constant temperature. The heat added to raise the substance to the temperature at which vaporization, for example, begins is not included. Sometimes values for changes of state are given in joules per gram. Usually the heats of fusion, vaporization, and sublimation are tabulated in handbooks of chemistry. For a further discussion of changes of state, see Section 13.4.

SUMMARY OF SECTIONS 7.7–7.11

The heats of chemical and physical changes carried out at constant external pressure are expressed as changes in enthalpy, ΔH. The enthalpy change for a chemical reaction (the heat of the reaction) is included in a thermochemical equation, and the value given is for the equation as written. Exothermic reactions have negative ΔH values; endothermic reactions have positive ΔH values. Standard enthalpy changes, ΔH°, apply to reactants and products in their standard states—the states most stable at 1 bar (slightly less than 1 atm) and a specified temperature (most often 298 K).

For any chemical reaction, ΔH has the same value whether the reaction takes place in one step or in several steps (Hess's law). Therefore, thermochemical equations can be added or

Table 7.5
Enthalpy Changes for Changes of State or Crystalline Form

Substance	Transformation	T(K)	ΔH(kJ/mol)
CO	solid II $\longrightarrow$ solid I	61.52	0.632
	fusion	68.09	0.837
	vaporization	81.65	6.042
CO_2	sublimation	194.67	25.23
CH_3OH(methyl alcohol)	solid II $\longrightarrow$ solid I	157.4	0.644
	fusion	175.25	3.17
	vaporization	337.9	35.27
$(CH_3)_2O$(dimethyl ether)	fusion	131.65	4.94
	vaporization	248.33	21.51
Cl_2	fusion	172.15	6.41
	vaporization	239.09	20.41
HF	fusion	190.08	4.58
	vaporization	293.1	7.5
H_2	fusion	13.95	0.117
	vaporization	20.38	0.904
H_2O	fusion	273.15	6.01
	vaporization	373.15	40.66
Na	fusion	371	2.64
NaCl	fusion	1081	28.5
	vaporization	1738	170.7

subtracted algebraically to find unknown heats of reaction from known heats of reaction. When the direction of a reaction is reversed, the sign of ΔH must be reversed. When a chemical equation is multiplied or divided by a factor, the value of ΔH changes by the same factor, permitting the value of ΔH for the reaction of any quantity of a reactant to be calculated.

Standard heats of formation are ΔH° values for the formation of one mole of a compound by combination of the elements (with the heat of formation of elements in their standard states taken to be zero). Heats of reaction can be found from tabulated standard heats of formation by subtracting the sum of the heats of formation of all reactants from the sum of those for all products.

Table 7.4 lists some of the many types of enthalpy changes utilized in various aspects of chemistry, all of which can be combined according to Hess's law.

SIGNIFICANT TERMS (with section references)

calorimeter (Aside 7.2)
combustion (Aside 7.2)
endothermic (7.2)
enthalpy (7.4)
exothermic (7.2)
first law of thermodynamics (7.4)
heat (7.2)
heat capacity (7.5)
heat of reaction (7.7)
Hess's law (7.10)
intermolecular forces (7.3)
internal energy (7.3)
internal energy change (7.3)
intramolecular forces (7.3)
molar heat capacity (7.5)
specific heat (7.5)
standard enthalpy changes (7.8)
standard enthalpy of combustion (7.11)
standard enthalpy of formation (7.9)
standard enthalpy of reaction (7.8)
standard state (7.8)
surroundings (7.2)
system (7.2)
thermochemical equation (7.7)
thermochemistry (7.2)
thermodynamics (7.2)
work (7.4)

QUESTIONS AND PROBLEMS

Energy

7.1 What happens to ΔE for a system during a process in which (a) $q < 0$ and $w < 0$, (b) $q = 0$ and $w > 0$, (c) $q > 0$ and $w < 0$?

7.2 What happens to ΔE for a system during a process in which (a) $q > 0$ and $w > 0$, (b) $q = w = 0$, (c) $q < 0$ and $w > 0$?

7.3 A system performs 460. L atm of pressure–volume work (1 L atm = 101.325 J) on its surroundings and absorbs 6300 J of heat from its surroundings. What is the change in internal energy of the system?

7.4 A system receives 45 J of electrical work, delivers 165 J of pressure–volume work, and releases 212 J of heat. What is the change in internal energy of the system?

7.5 What will be the volume change if 125 J of work is done on a system containing an ideal gas? The surroundings exert a constant pressure of 5.2 atm.

7.6 A system performs 1.5×10^5 J of work on the surroundings while undergoing an increase of volume equal to 47.9 L. What constant external pressure is being exerted on the system?

7.7 For each of the following chemical and physical changes carried out at constant pressure, decide whether work is done by the system (the substances undergoing the change) on the surroundings or by the surroundings on the system or whether the amount of work is negligible.

(a) $C_6H_6(l) \longrightarrow C_6H_6(s)$
(b) $\frac{1}{2}N_2(g) + \frac{3}{2}H_2(g) \longrightarrow NH_3(g)$
(c) $3H_2S(g) + 2HNO_3(g) \longrightarrow 2NO(g) + 4H_2O(l) + 3S(s)$

7.8 Repeat Question 7.7 for

(a) $4HNO_3(l) + P_4O_{10}(s) \longrightarrow 2N_2O_5(s) + 4HPO_3(s)$
(b) $2MnO_4^- + 5H_2O_2(aq) + 6H^+ \longrightarrow 2Mn^{2+} + 5O_2(g) + 8H_2O(l)$
(c) $CO_2(g) + H_2O(l) + CaCO_3(s) \longrightarrow Ca^{2+} + 2HCO_3^-$

7.9 Discuss the relationships among ΔE, ΔH, q_V, and w for the formation of ozone carried out under constant pressure conditions

$$3O_2(g) \xrightarrow{\Delta} 2O_3(g)$$

7.10 Repeat Question 7.9 for the equation for the combustion of hydrazine under constant volume conditions

$$H_2NNH_2(l) + O_2(g) \longrightarrow N_2(g) + 2H_2O(l)$$

Measuring Heat—Heat Capacity

7.11 How much energy must be removed from 15.0 g of red phosphorus to cool it from 25 °C to 23 °C? The molar heat capacity of red phosphorus is 21.21 J/K mol.

7.12 Which system—one mole of diamond or one mole of graphite—must absorb the greater amount of heat for an increase in temperature from 298 K to 371 K? How much more heat is required? The molar heat capacities are 6.11 J/K mol for diamond and 8.53 J/K mol for graphite.

7.13 One mole samples of chlorine (molar heat capacity = 33.91 J/K mol), sodium (molar heat capacity = 28.24 J/K mol), and sodium chloride (molar heat capacity = 50.50 J/K mol) are each heated from 25 °C to 35 °C. Which sample must absorb the most heat?

7.14* A 12-fluid-ounce can of diet cola contains 360.1 g of cola (specific heat = 4.2 J/K g) and the aluminum can weighs 19.3 g (specific heat = 0.90 J/K g). (a) What amount of thermal energy must be removed to cool a six-pack from 25 °C to 0 °C? Ice requires 334 J/g to melt. (b) What mass of ice is needed to cool the six-pack of soft drinks?

7.15 The absorption of 413 J of heat by 200.59 g of liquid mercury caused a temperature increase of 14.76 °C. Calculate the molar heat capacity and the specific heat of mercury.

7.16* A 3.72 g sample of magnesium absorbed enough heat to produce a temperature increase of 13.6 °C. The same amount of heat was absorbed by a 5.63 g sample of silver producing a temperature increase of 37.6 °C. The molar heat capacity of silver is 25.351 J/K mol. Calculate the molar heat capacity of magnesium.

7.17 The molar heat capacity of copper is 24.435 J/K mol. What temperature change would take place if 260 J of thermal energy was removed from a 0.30 mol block of copper?

7.18 The specific heat for aluminum is 0.900 J/K g and for lead 0.130 J/K g. Which substance would have the greater decrease in temperature if 65 J of heat was removed from (a) one gram of each and (b) one mole of each?

7.19* In 1819, Pierre Dulong and Alexis Petit recognized that for all solid metallic elements at room temperature (except for a few with very small atomic masses) the product of the atomic mass of the element and its specific heat is approximately a constant

$$(\text{atomic mass})(\text{sp. heat}) = 26 \text{ J/K mol}$$

Using the following specific heat data, show that this empirical relationship is true.

Al	0.900 J/K g	Fe	0.452 J/Kg
Be	1.824	Pb	0.130
Cr	0.460	Sn	0.226

Suggest the basis for this relationship.

7.20* A temperature increase of 54.0 °C was observed upon the addition of 76.5 J of heat to a 2.71 g sample of an unknown metal. Calculate the specific heat of the metal. Using the rule of Dulong-Petit (see Problem 7.19), identify the unknown metal.

Measuring Heat—Calorimetry

7.21 A calorimeter contained 75.0 g of water at 16.95 °C. A 75.2 g sample of iron at 63.14 °C was placed in it giving a final temperature of 19.68 °C for the system. Calculate the heat capacity of the calorimeter. The specific heats are 4.184 J/K g for H_2O and 0.450 J/K g for Fe.

7.22 A calorimeter was calibrated by pouring 50.0 g of hot water into 100.0 g of cold water in the calorimeter. The original temperature of the hot water was 83.7 °C and of the cold water and calorimeter was 26.3 °C. The final temperature of the combined system was 41.2 °C. What is the value of the heat capacity of the calorimeter? The specific heat of water is 4.184 J/K g.

7.23 A bomb calorimeter was constructed so that the contribution of the calorimeter contents to the heat flow could be neglected. A 1.298 g sample of benzoic acid was burned, and the observed temperature change was 4.32 °C. The heat of combustion of benzoic acid is −26,440 J/g. What is the heat capacity of the calorimeter?

7.24 A calorimeter was calibrated by adding 100.0 g of dilute HCl to 100.0 g of dilute NaOH. The resulting temperature change of the calorimeter and of the NaCl solution formed as a result of the reaction (specific heat = 4.06 J/K g) was 0.627 °C. The amount of energy released by the reaction was 558.4 J. Find the heat capacity of the calorimeter.

7.25 A 0.241 g piece of magnesium metal was dropped into 100.0 g of dilute hydrochloric acid in a calorimeter. A temperature increase of 10.89 °C is observed as a result of the reaction

$$Mg(s) + 2HCl(aq) \longrightarrow MgCl_2(aq) + H_2(g)$$

The specific heat of the $MgCl_2$-HCl solution is 4.21 J/K g, and the heat capacity of the calorimeter is 2.74 J/K. Calculate the heat of reaction per mole of Mg.

7.26 A 1.02 g sample of urea, $(NH_2)_2CO$, was burned in a bomb calorimeter having a heat capacity of 28.1 kJ/K, and the observed temperature increase was 0.381 °C. The calorimeter was designed so that the contribution of the contents to the heat flow is negligible. Find the molar heat of combustion of urea.

7.27* (a) A student heated a sample of a metal weighing 32.6 g to 99.83 °C and put it into 100.0 g of water at 23.62 °C in a calorimeter. The final temperature was 24.41 °C. The student calculated the specific heat of the metal, but neglected to use the heat capacity of the calorimeter. The specific heat of water is 4.184 J/K g. What was his answer? The metal was known to be either chromium, molybdenum, or tungsten. By comparing the value of the specific heat to those of the metals (Cr, 0.460; Mo, 0.250; W, 0.135 J/K g), the student identified the metal. What was the metal?

(b) A student at the next laboratory bench did the same experiment, obtained the same data, and used the heat capacity of the calorimeter in his calculations. The heat capacity of the calorimeter was 410 J/K. Was his identification of the metal different?

7.28* A student measured the heat of solution of sodium iodide by dissolving 1.00 g of solid NaI in 20.00 mL of water in a styrofoam cup calorimeter. In the dissolution process, the temperature rose from 25.00 °C to 25.61 °C. In another experiment, the student placed 30.00 g of water at 20.00 °C in the calorimeter and then added 20.00 g of water at 96.50 °C. Upon mixing the two portions of water, the final temperature was found to be 50.00 °C. The specific heat of water is 4.184 J/K g and that of the NaI solution is 4.046 J/K g. Calculate (a) the heat capacity of the calorimeter and (b) the molar enthalpy of solution of NaI(*s*) under these conditions.

Heats of Reaction and Other Enthalpy Changes

7.29 What is the standard state of water at (a) −5 °C, (b) 5 °C, (c) 93 °C, (d) 103 °C?

7.30 What is the standard state of chlorine at (a) −175 °C, (b) −100 °C, (c) −65 °C, (d) 25 °C, (e) 1275 °C? The melting and boiling points of Cl_2 are 172.15 K and 239.09 K, respectively.

7.31 Choose the formula or symbol from each set that represents the substance that will have an enthalpy of formation at 25 °C equal to zero: (a) $O_3(g)$, $O_2(g)$, $O_2(l)$, $O(g)$; (b) $C(g)$, C(*s*, graphite), C(*s*, diamond); (c) $Hg(g)$, $Hg_2(g)$, $Hg(l)$, $Hg(s)$; (d) $I_2(g)$, $I_2(l)$, $I_2(s)$, $I(s)$.

7.32 Following is a list of the standard states at 25 °C of several elements: C(*s*, graphite), $Cl_2(g)$, $Na(s)$, $Br_2(l)$, $Ca(s)$, $S_8(s)$, $O_2(g)$. Write chemical equations for the formation from the elements of (a) $CCl_4(l)$, (b) $NaBr(s)$, (c) $CaSO_4(s)$.

7.33 At 800 K

$$Ag(s) + \tfrac{1}{2}Cl_2(g) \longrightarrow AgCl(l) \qquad \Delta H^\circ = -106.12 \text{ kJ}$$

Calculate ΔH° for each of the following thermochemical equations

(a) $AgCl(l) \longrightarrow Ag(s) + \tfrac{1}{2}Cl_2(g)$
(b) $2Ag(s) + Cl_2(g) \longrightarrow 2AgCl(l)$
(c) $2AgCl(l) \longrightarrow 2Ag(s) + Cl_2(g)$

7.34 Compare the enthalpy change per mole of iron formed when the oxides Fe_3O_4 and Fe_2O_3 react with aluminum.

$$3Fe_3O_4(s) + 8Al(s) \longrightarrow 4Al_2O_3(s) + 9Fe(s) \qquad \Delta H^\circ = -3347.6 \text{ kJ}$$

$$Fe_2O_3(s) + 2Al(s) \longrightarrow Al_2O_3(s) + 2Fe(s) \qquad \Delta H^\circ = -851.4 \text{ kJ}$$

7.35 When a welder uses an acetylene torch, it is the combustion of acetylene that liberates the intense heat for welding metals together. The equation for this process is

$$2C_2H_2(g) + 5O_2(g) \longrightarrow 4CO_2(g) + 2H_2O(g)$$

The heat of combustion of acetylene is −1300. kJ/mol. What is the enthalpy change when 0.260 kg of C_2H_2 is burned?

7.36 What amount of energy is released as 1.00 g of Ag reacts with excess chlorine at (a) 298 K and (b) 800 K?

$$Ag(s) + \tfrac{1}{2}Cl_2(g) \longrightarrow AgCl(s) \qquad \Delta H^\circ_{298} = -127.068 \text{ kJ}$$
$$Ag(s) + \tfrac{1}{2}Cl_2(g) \longrightarrow AgCl(l) \qquad \Delta H^\circ_{800} = -106.12 \text{ kJ}$$

7.37 What is the enthalpy change during the formation of ozone as (a) 1.00 mol O_2 reacts, (b) 1.00 mol O_3 is formed, (c) 1.00 g O_2 reacts, (d) 1.00 g O_3 is formed?

$$3O_2(g) \longrightarrow 2O_3(g) \qquad \Delta H^\circ = 285.4 \text{ kJ}$$

7.38* The Thermit reaction used for welding iron is the reaction of Fe_3O_4 with Al:

$$8Al(s) + 3Fe_3O_4(s) \longrightarrow 4Al_2O_3(s) + 9Fe(s) \qquad \Delta H^\circ = -3347.6 \text{ kJ}$$

[Because this large amount of heat cannot be rapidly dissipated to the surroundings, the reacting mass may reach temperatures near 3000 °C.] What is the change in enthalpy generated by the reaction of 10.0 g of Al with 25.0 g of Fe_3O_4?

7.39 Combine the following thermochemical equations

$$N_2O_4(g) \longrightarrow 2NO_2(g) \qquad \Delta H^\circ = 57.20 \text{ kJ}$$
$$2NO(g) + O_2(g) \longrightarrow 2NO_2(g) \qquad \Delta H^\circ = -114.14 \text{ kJ}$$

to find the heat of reaction for

$$2NO(g) + O_2(g) \longrightarrow N_2O_4(g)$$

7.40 Use the following thermochemical equations

$$H^+ + OH^- \longrightarrow H_2O(l) \qquad \Delta H^\circ = -55.836 \text{ kJ}$$
$$HCN(aq) + OH^- \longrightarrow H_2O(l) + CN^- \qquad \Delta H^\circ = -10.5 \text{ kJ}$$

to find ΔH° for

$$HCN(aq) \longrightarrow CN^- + H^+$$

7.41 Use the following thermochemical equations to find ΔH_f° for $CuCl_2(s)$.

$$2Cu(s) + Cl_2(g) \longrightarrow 2CuCl(s) \qquad \Delta H^\circ = -274.4 \text{ kJ}$$
$$2CuCl(s) + Cl_2(g) \longrightarrow 2CuCl_2(s) \qquad \Delta H^\circ = -165.8 \text{ kJ}$$

7.42 Find the heat of formation of liquid hydrogen peroxide at 25 °C from the following thermochemical equations.

$$H_2(g) + \tfrac{1}{2}O_2(g) \longrightarrow H_2O(g) \qquad \Delta H^\circ = -241.818 \text{ kJ}$$
$$2H(g) + O(g) \longrightarrow H_2O(g) \qquad \Delta H^\circ = -926.919 \text{ kJ}$$
$$2H(g) + 2O(g) \longrightarrow H_2O_2(g) \qquad \Delta H^\circ = -1070.60 \text{ kJ}$$
$$2O(g) \longrightarrow O_2(g) \qquad \Delta H^\circ = -498.340 \text{ kJ}$$
$$H_2O_2(l) \longrightarrow H_2O_2(g) \qquad \Delta H^\circ = 51.46 \text{ kJ}$$

7.43 Use the data given in Table 7.3 to find the enthalpy of reaction for

(a) $P(s, \text{white}) + \tfrac{5}{2}Cl_2(g) \longrightarrow PCl_5(s)$
(b) $2Fe_3O_4(s) + \tfrac{1}{2}O_2(g) \longrightarrow 3Fe_2O_3(s)$
(c) $KCl(s) + Na(s) \longrightarrow K(s) + NaCl(s)$

7.44 Repeat Problem 7.43 for

(a) $C(s, \text{graphite}) + CO_2(g) \longrightarrow 2CO(g)$
(b) $2HI(g) + F_2(g) \longrightarrow 2HF(g) + I_2(s)$
(c) $2SO_2(g) + O_2(g) \longrightarrow 2SO_3(g)$

7.45 The enthalpy of formation for H^+ is taken as zero. This allows enthalpies of formation to be assigned to individual ions in aqueous solutions. For example, the enthalpy of formation for OH^- is -229.994 kJ/mol. For $H_2O(l)$, the enthalpy of formation is -285.830 kJ/mol. What is the enthalpy of reaction for the addition of a strong acid such as $HCl(aq)$ to a strong base such as $NaOH(aq)$?

$$H^+ + OH^- \longrightarrow H_2O(l)$$

7.46 Aragonite, a mineral, undergoes the following reaction in a very dilute solution of carbonic acid (CO_2 in H_2O).

$$CaCO_3(s, \text{aragonite}) + H_2CO_3(aq) \rightleftharpoons Ca(HCO_3)_2(aq)$$

This means that the $CaCO_3$ can be dissolved in one place and transported to another in the form of $Ca(HCO_3)_2$—a very important step in the formation of stalagmites and stalactites. Calculate the enthalpy of reaction at 25 °C given that the enthalpies of formation are -1207 kJ/mol for $CaCO_3(s$, aragonite), $-700.$ kJ/mol for $H_2CO_3(aq)$, and -1925 kJ/mol for $Ca(HCO_3)_2(aq)$.

7.47 Write chemical equation for the formation of one mole of methane, $CH_4(g)$, from the elements. Find the enthalpy of formation of CH_4 at 20 °C from the following thermochemical equations and values of enthalpies of combustion at 20 °C:

$$H_2(g) + \tfrac{1}{2}O_2(g) \longrightarrow H_2O(l) \qquad \Delta H^\circ = -286.10 \text{ kJ}$$
$$C(s) + O_2(g) \longrightarrow CO_2(g) \qquad \Delta H^\circ = -394.89 \text{ kJ}$$
$$CH_4(g) + 2O_2(g) \longrightarrow CO_2(g) + 2H_2O(l) \qquad \Delta H^\circ = -882.0 \text{ kJ}$$

7.48 Write the chemical equation for the combination reaction between ethene, C_2H_4, and hydrogen (a hydrogenation reaction) to form ethane, C_2H_6. Find the enthalpy of hydrogenation using the following thermochemical equations and values of enthalpies of combustion of 25 °C:

$$C_2H_4(g) + 3O_2(g) \longrightarrow 2CO_2(g) + 2H_2O(l) \qquad \Delta H^\circ = -1410.94 \text{ kJ}$$
$$C_2H_6(g) + \tfrac{7}{2}O_2(g) \longrightarrow 2CO_2(g) + 3H_2O(l) \qquad \Delta H^\circ = -1559.83 \text{ kJ}$$
$$H_2(g) + \tfrac{1}{2}O_2(g) \longrightarrow H_2O(l) \qquad \Delta H^\circ = -285.830 \text{ kJ}$$

7.49 The heat of formation at 1000 K for Al(*s*) is -10.519 kJ/mol, for Al(*l*) it is 0.000 kJ/mol, and for Al(*g*) it is 310.114 kJ/mol. Find the heat of (a) fusion, (b) vaporization, and (c) sublimation at this temperature.

7.50 A system consists of several cubes of ice in liquid water at 0 °C. What will happen to the system if the surroundings performs more work on the system than the system can dissipate to the surroundings as heat?

7.51 Which is more exothermic, the combustion of one mole of methane to form $CO_2(g)$ and liquid water or to form $CO_2(g)$ and steam? Why?

7.52 Which is more exothermic, the combustion of one mole of gaseous benzene or one mole of liquid benzene? Why?

ADDITIONAL QUESTIONS AND PROBLEMS

7.53 Arrange the following amounts of energy in increasing order: (a) 3 kcal, (b) 2 MJ, (c) 14 L atm, (d) 7.2 erg, (e) 15 J, (f) 4.0 cal.

7.54 Convert the following values of molar heat capacity to specific heat: (a) $O_2(g)$, 29.355 J/K mol; (b) $Br_2(l)$, 75.689 J/K mol; (c) $Ca_3(PO_4)_2(s)$, 227.82 J/K mol.

7.55* A system consisting of 1.0 kg of water falls a distance of 1.5 m toward the ground. The energy change of the system is given by *mgd*, where *m* is the mass in kg, *g* is the gravitational constant ($g = 9.81$ m/s^2), and *d* is the distance that the system falls in meters. Assuming that this energy is dissipated by heating the water upon contact with the ground, what will be the increase in temperature? The specific heat of water is 4.184 J/K g.

7.56* As a person walks along a level sidewalk, approximately 2.0 kJ of energy is dissipated for each mile traveled for each pound the person weighs. How far should a 120 lb person walk to overcome the weight-gaining effects of a 1 lb box of candy? Assume that the candy is 100% sucrose and that the reaction

$$\underset{\textit{sucrose}}{C_{12}H_{22}O_{11}(s)} + 12O_2(g) \longrightarrow 12CO_2(g) + 11H_2O(l)$$

is 12% efficient in the body and that $\Delta H_f^\circ = -2221.7$ kJ/mol for sucrose.

7.57 Calculate the enthalpy change for heating 10.0 g of water from -5.0 °C to 25.0 °C. The heat of fusion is 6009.5 J/mol and the specific heat is 2.1 J/K g for ice and 4.184 J/K g for water.

7.58* The heat of formation of $HCl(g)$ is -92.307 kJ/mol at 25 °C. The value of ΔH_f° at 500. K can be found by (a) calculating the enthalpy change, ΔH_i°, for cooling $\frac{1}{2}$ mol of $H_2(g)$ and $\frac{1}{2}$ mol of $Cl_2(g)$ from 500. K to 298 K; (b) adding the heat of reaction at 298 K, ΔH_{ii}°, to the answer to part (a); and (c) adding the enthalpy change, ΔH_{iii}°, for heating one mole of $HCl(g)$ from 298 K to 500. K to the result of part (b).

$$\begin{array}{ccc} \frac{1}{2}H_2(g) + \frac{1}{2}Cl_2(g) & \xrightarrow{\Delta H_{500}^\circ = ?} & HCl(g) \\ \downarrow \Delta H_i^\circ & & \uparrow \Delta H_{iii}^\circ \\ \frac{1}{2}H_2(g) + \frac{1}{2}Cl_2(g) & \xrightarrow{\Delta H_{ii}^\circ} & HCl(g) \end{array}$$

Find ΔH_f° at 500. K, given the heat capacities of 33.907 J/K mol for $Cl_2(g)$, 29.12 J/K mol for $HCl(g)$, and 28.824 J/K mol for $H_2(g)$. Is the reaction more or less exothermic at 500. K than at 298 K?

7.59* The heat of reaction at 25 °C for

$$CO(g) + \tfrac{1}{2}O_2(g) \longrightarrow CO_2(g)$$

is −282.984 kJ. Using the heat capacity data given in Table 7.2 and 29.355 J/K mol for $O_2(g)$, calculate the heat of reaction at 500. K.

7.60* A series of experiments was designed to determine the enthalpy of formation of $KBr(s)$. These experiments were all carried out in the same Styrofoam coffee cup calorimeter.

The first experiment consisted of pouring 10.0 g of water at 37.42 °C into the cup containing 10.0 g of water at 25.16 °C, giving a final temperature of 29.64 °C. (a) Find the heat capacity of the calorimeter. The specific heat of water is 4.184 J/K g.

The second experiment consisted of allowing 1.0 millimole of HBr in dilute solution to react with 1.0 millimole of KOH in dilute solution in the cup. A temperature increase of 0.49 °C was observed for the 20.00 g of water and products. (b) Find the enthalpy change for

$$KOH(aq) + HBr(aq) \longrightarrow KBr(aq) + H_2O(l)$$

Assume that the specific heat is 4.2 J/K g for the solution. (c) Find the enthalpy of formation of $KBr(aq)$ from your answer to part (b) and the following enthalpies of formation: −482.37 kJ/mol for $KOH(aq)$, −121.55 kJ/mol for $HBr(aq)$, and −285.830 kJ/mol for $H_2O(l)$.

The third experiment consisted of dissolving 1.0 millimole of $KBr(s)$ in 20.00 g of water at 24.33 °C to give a final temperature of 24.16 °C. (d) Find ΔH° for the process which can be represented as

$$KBr(s) \xrightarrow{\text{solvent}} KBr(aq)$$

Assume that the specific heat is 4.2 J/K g for the dilute solution. Using the enthalpy of formation for $KBr(aq)$ determined in part (c) and your answer to part (d), (e) find the enthalpy of formation for $KBr(s)$.

CHAPTER

8

Electronic Structure and the Periodic Table

QUANTUM THEORY

8.1 LIGHT, ELECTRONS, AND THE PERIODIC TABLE

Imagine the time when new elements were being discovered in nature and their properties investigated. Some elements were found to be very similar to each other, and others were very different. Lithium and sodium, for example, both have very similar properties. Fluorine and chlorine are also very similar to each other, but they are very different from lithium and sodium. Following the natural human desire to arrange things into categories, many individuals sought an organizing principle for the properties of the elements.

When the first 17 known elements were arranged in the order of their atomic masses, six elements were found to fall between lithium and sodium. What an exciting moment it must have been when it was recognized for the first time that six elements also separated fluorine and chlorine, and several other pairs of similar elements.

The periodic table, which is the most meaningful guide that we have to the properties of the elements and their compounds, originated in observations like these. In the **periodic table**, the elements are arranged in the order of increasing atomic number so that elements with similar properties fall near each other. In this chapter we first examine how the reasons for this arrangement came to be understood.

The early experiments that led to the nuclear model of the atom were described in Chapter 3. The studies described in Sections 8.1–8.10 extended the atomic model to include the location of the electrons and their role in the properties of the elements and their compounds. It is now believed that most properties of matter that are of interest in chemistry are intimately related to **electronic structure**—the arrangement of electrons in atoms, ions, or molecules. Understanding the electronic structure of atoms, which is described in Sections 8.11–8.13, is essential for understanding the properties of the elements and their compounds.

The basis for today's model of electronic structure lies in the modern theory of the nature of light (electromagnetic radiation; Aside 8.1) and in some aspects of the interaction of light with matter. You may wonder why it is important to study light in order to understand the periodic table. It turns out that light and electrons in atoms have many properties in common. Only when what physicists had been learning about light and what chemists had been learning about atoms were brought together in the 1920s were both the arrangement of electrons in atoms and the arrangement of the elements in the periodic table fully explained.

Physics, like all of the sciences, began with observations of the world around us. The major areas of interest that developed in what is now called classical physics were mechanics (the motion of objects), electricity and magnetism (the interactions of charged particles and charged particles in motion, which result in magnetic forces), thermodynamics (the study of heat and energy), and optics (the study of light). Experiments with large, easily observed objects led to laws explaining phenomena in each of these fields. Classical mechanics, for example, is based on Isaac Newton's laws of motion from the seventeenth century.

When something is described as "classical" it is almost certain that in modern times it has either been replaced or had added to it something new and quite different. The essential difference between classical physics and modern physics lies in the scale of what they deal with. In the late nineteenth century, physics and chemistry began to look at the behavior of tiny particles and at phenomena that are *not* part of daily life. The old laws were not in harmony with the new observations and a change became inevitable—a change that has been called "the greatest and most radical revolution in natural philosophy since the time of Newton." In the early 1900s a new physics was born.

Classical physics was not replaced. An additional set of theories and laws was developed to deal with small particles like electrons that move at great speeds. The classical laws are not "wrong"—they are fine for apples falling out of trees. For electrons in atoms, modern physics is better.

8.2 LIGHT AS WAVES

Early physicists found that the description of light as a series of waves traveling through space successfully accounts for many of the properties of light. Generally, waves are set in motion by a vibrating body, just as you set water waves in motion by moving your hand in a pond. In the case of light, the vibrations were presumed to be those of electrons, atoms, or groups of atoms. The waves that result are waves of electromagnetic energy, in the form of fluctuating electric and magnetic fields.

The terminology and units of measurement that are used to describe light are based on the picture of waves traveling through space. **Wavelength** (λ, Greek lambda) is the distance between any two similar points on adjacent waves (Figure 8.1). The **frequency** (ν, Greek nu) of light is the number of complete waves, also known as the number of *cycles*, passing a given point in a unit of time (see Figure 8.1).

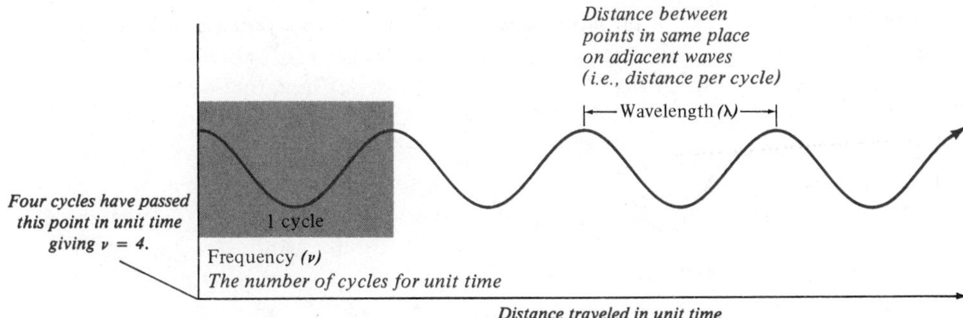

Figure 8.1
Wavelength and Frequency

Table 8.1
Units for Wavelength
The preferred units for wavelength are those based on the meter. Although angstroms, millimicrons, and microns are no longer recommended, they are still seen in older publications.

Unit abbreviation	Unit	Unit equal to: Meters	Unit equal to: Nanometers
m	meter	—	1×10^9 nm
cm	centimeter	1×10^{-2} m	1×10^7 nm
μm	micrometer	1×10^{-6} m	1×10^3 nm
μ	micron	1×10^{-6} m	1×10^3 nm
nm	nanometer	1×10^{-9} m	—
mμ	millimicron	1×10^{-9} m	—
Å	angstrom	1×10^{-10} m	0.1 nm

Wavelength has units of length. For light, wavelengths are now often given in nanometers, nm (see Table 8.1 for other units that are sometimes encountered). Frequency has units of cycles per unit time. Frequency is usually given in cycles per second. Since cycles is a dimensionless quantity, the unit is written as s^{-1}, reciprocal seconds, with the word "cycles" left out. In the SI system one cycle per second is a hertz, abbreviated Hz; 1 Hz = 1 s^{-1}. The speed of light in a vacuum (2.9979×10^8 m/s) is one of the fundamental constants of nature and does not vary with the wavelength or any other properties of the light. The speed of light in air is slightly less than the speed in a vacuum, but the difference is so small that it can be ignored in most cases.

Wavelength and frequency are related to the speed of light by the equation

$$\underset{\text{speed of light}}{c} = \overset{\text{wavelength}}{\lambda}\underset{\text{frequency}}{\nu} \qquad \textbf{(8.1)}$$

or, with the wavelength expressed in meters, the frequency in reciprocal seconds, and the speed of light given to three significant figures

$$\underset{\text{meters/second}}{c = 3.00 \times 10^8 \text{ m/s}} = \underset{\text{meters}}{\lambda}\ \underset{\text{seconds}^{-1}}{\nu} \qquad \textbf{(8.1a)}$$

Equation (8.1) allows the calculation of frequency from wavelength or of wavelength from frequency. Another quantity used to characterize waves is the wave number.

$$\underset{\text{wave number}}{\bar{\nu}} = \frac{1}{\lambda}$$

The **wave number** is the number of wavelengths per unit of length. Its unit is the reciprocal of the wavelength unit, and it is often given in cm^{-1}.

(a) Constructive interference

(b) Intermediate between constructive and destructive interference

(c) Destructive interference

Figure 8.2
Wave Interference
(a) The waves are in phase and reinforce one another. (b) The waves are partly out of phase. (c) The waves are completely out of phase and cancel out one another.

Two properties characteristic of light waves are interference and diffraction, both of which can be seen in water waves on the surface of a pond. If you throw two stones into the pond, the two sets of wave patterns interfere with each other in the area where they meet. When identical waves meet in phase, that is, so that their high and low points match up, they reinforce each other. This is called *constructive interference* (Figure 8.2a). In the opposite situation, identical waves meet out of phase, so that the high points of one exactly match the low points of the other, and they cancel each other. This is *destructive interference* (Figure 8.2c).

Diffraction is the spreading of waves as they pass obstacles or openings comparable in size to their wavelength (Figure 8.3). You can, for example, see waves spreading out as they pass between rocks at the edge of the pond. In circumstances where light passes through two slits, the diffracted light waves interfere with each other and produce a pattern of light and dark—a diffraction pattern—on a screen placed in their path (Figure 8.4).

When interference and diffraction are observed, they are interpreted as proof that the passage of waves has taken place.

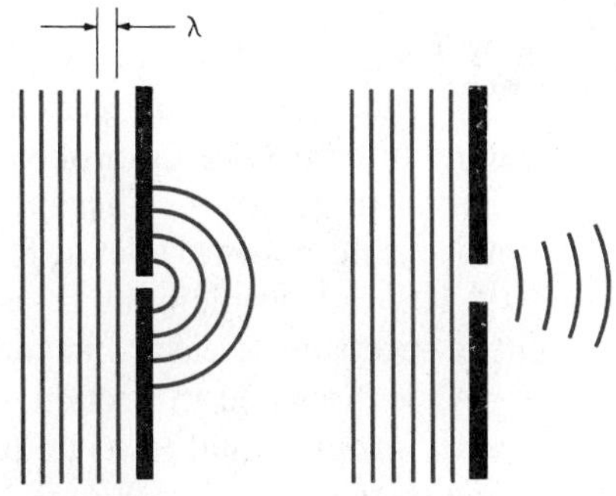

Figure 8.3
Diffraction
The amount of spreading depends upon the size of the opening relative to the wavelength. If the opening is very much larger than the wavelength (as with a beam of visible light passing through a keyhole), the effect is undetectable.

EXAMPLE 8.1 Wave Nature of Light

Blue-green light has a wavelength of about 520 nm. What are the frequency (in Hz) and wave number (in cm^{-1}) of light of this color?

Given the wavelength and wanting to know the frequency, the relationship between these two quantities and the speed of light given by Equation (8.1a) is used

$$\nu = \frac{c}{\lambda} = \frac{3.00 \times 10^8 \text{ m/s}}{(520 \text{ nm})(1 \text{ m}/10^9 \text{ nm})} = 5.8 \times 10^{14} \text{ s}^{-1} = \mathbf{5.8 \times 10^{14}\ Hz.}$$

Note that the conversion factor based on 1 m = 10^9 nm must be used to convert the wavelength to meters so that the answer is in the correct units of seconds^{-1}.

The wave number is simply the reciprocal of the wavelength

$$\bar{\nu} = \frac{1}{\lambda} = \frac{1}{(520 \text{ nm})(1 \text{ m}/10^9 \text{ nm})(10^2 \text{ cm/m})} = \mathbf{1.9 \times 10^4\ cm^{-1}}$$

EXERCISE What are the wavelength (in nm) and the wave number (in cm^{-1}) of yellow light with a frequency of 5.2×10^{14} s^{-1}? *Answer* 580 nm, 1.7×10^4 cm^{-1}.

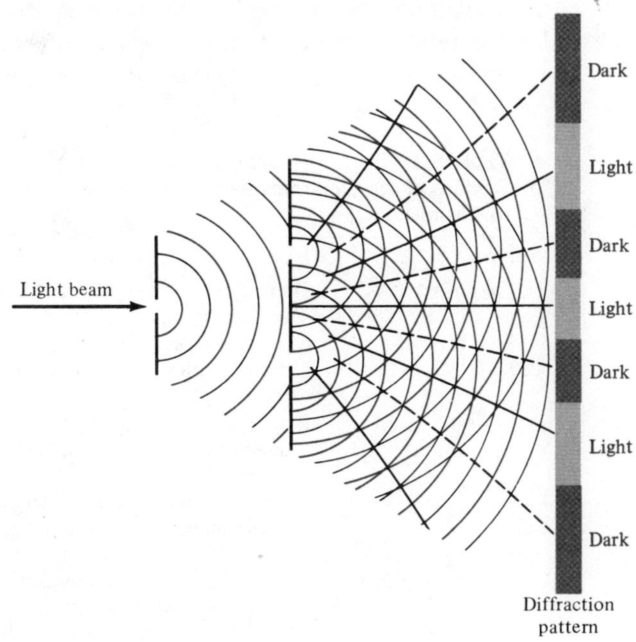

Figure 8.4
A Diffraction Pattern
Such a pattern is produced by the interference of light waves that have passed though two or more openings close in size to the wavelength of the light. Dark bands appear where waves arrive out of phase and cancel out one another; light bands appear where waves arrive in phase and reinforce one another.

Aside 8.1 TOOLS OF CHEMISTRY: ELECTROMAGNETIC RADIATION AND SPECTRA

When a rainbow forms in the sky or gasoline produces colored patterns as it floats on a puddle of water, sunlight is being broken up into colored light of different frequencies. The spectrum of light that we can see ranges from violet at 8×10^{14} Hz to red at 4×10^{14} Hz. But visible light is just a small portion of the entire spectrum of electromagnetic radiation.

As is discussed in Section 8.3, the energy of light is directly proportional to its frequency ($E = h\nu$). The electromagnetic spectrum (Figure 8.5) extends from short-wavelength, high-frequency gamma rays of 10^{20} Hz and beyond at the high-energy end, to long-wavelength, low-frequency radiowaves of 10^8 Hz and less at the low-energy end.

When electromagnetic radiation interacts with matter, energy is exchanged, and atoms and molecules may absorb this energy. In returning to their normal, more stable lower energy states, the atoms and molecules may then emit radiation in various parts of the electromagnetic spectrum.

The spectrum of radiation emitted by a substance that has absorbed energy is called an **emission spectrum.** Atoms, molecules, or ions that have absorbed radiation are spoken of as "excited." To produce an emission spectrum, energy is supplied to a sample by heating it or passing radiation through it, and the wavelength (or frequency) of the radiation emitted as the sample gives up the absorbed energy is recorded.

An absorption spectrum is like the photographic negative of an emission spectrum. A continuum of radiation is passed through a sample, which absorbs radiation of certain wavelengths. The missing wavelengths leave dark spaces in the bright continuous spectrum. The radiation *not* absorbed by the sample is recorded to produce an **absorption spectrum.**

The study of spectra is referred to as **spectroscopy.** An instrument that records the intensity and frequency of absorbed or emitted radiation is called a spectrophotometer, or simply a **spectrometer.** A spectrometer usually has five basic sections (Figure 8.6): (1) a source of the beam of radiation; (2) an analyzer, which divides up the beam according to the property or properties being analyzed; (3) a sample holder; (4) a detector, which measures the quantity or quantities being determined (e.g., the intensity of radiation absorbed or emitted); (5) a display device, which makes the results visible as a graph, chart, or photograph.

Some of the applications of spectroscopy in the various regions of the electromagnetic spectrum are listed in Table 8.2.

The sun and heated solids (such as the filament in a light bulb) produce **continuous spectra**—spectra in which radiation is emitted at all of the wavelengths in a region. The rainbow is a continuous spectrum. **Line spectra** are produced when radiation is emitted (or absorbed) only at

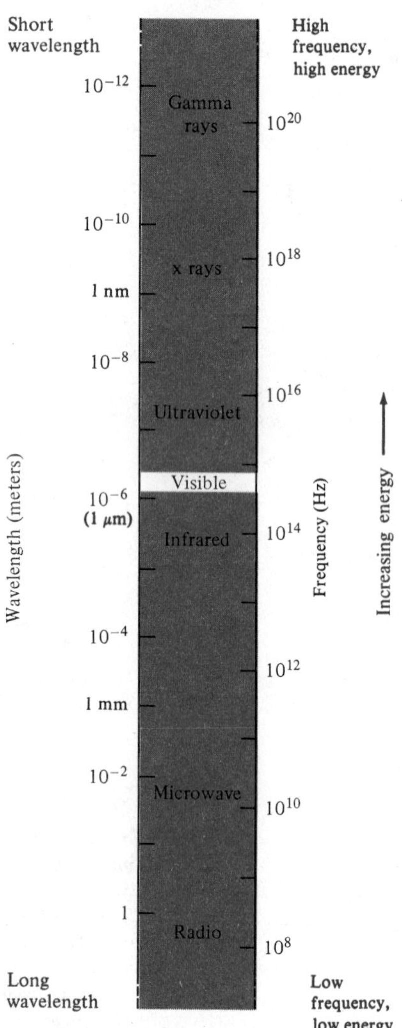

Figure 8.5
The Electromagnetic Spectrum

Table 8.2
Regions of the Electromagnetic Spectrum of Chief Interest to Chemistry

Region	Use
Gamma rays	Emitted in radioactive decay
x-Rays	Diffraction of x-rays used in determining crystal structure
Ultraviolet	Absorption spectra used for structure determination in organic and inorganic molecules
Visible	Emission and absorption spectra used for qualitative and quantitative identification of elements
Infrared	Absorption spectra used for structure determination in organic and inorganic molecules
Radiofrequency	Absorption at accurately measured frequencies used in structure determination

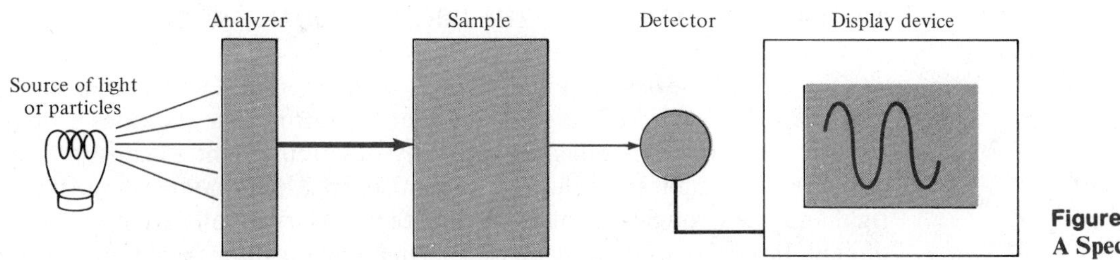

Figure 8.6
A Spectrometer

specific wavelengths. Line emission spectra are of greatest interest to studies of electronic structure, for they are produced by electrons that have absorbed energy. Moseley was studying line emission spectra of atoms in the x-ray region when he discovered the significance of atomic numbers (see Section 3.10). Sodium vapor gives a line emission spectrum in the visible region, as does hydrogen in a gas-discharge tube (Figure 8.7). The origin and production of line spectra is discussed in Section 8.6. In **band spectra,** which are produced by molecules that have become excited, groups of closely spaced lines form bands.

Much of our knowledge of the structure of matter is derived from spectroscopy. In addition, many theories of the structure of matter have had experimental support from the accurate prediction of where spectral absorption or emission will occur.

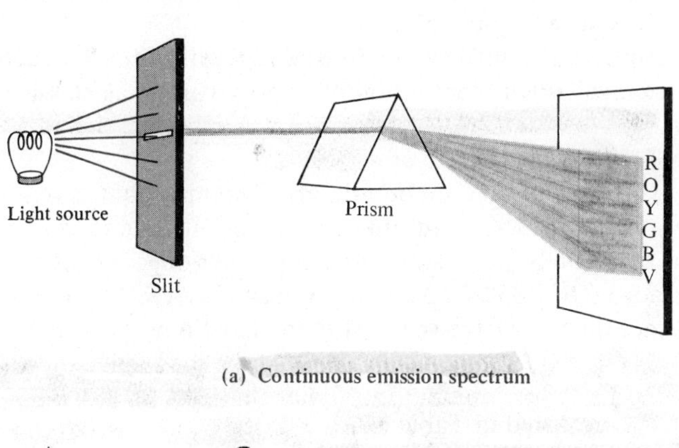

(a) Continuous emission spectrum

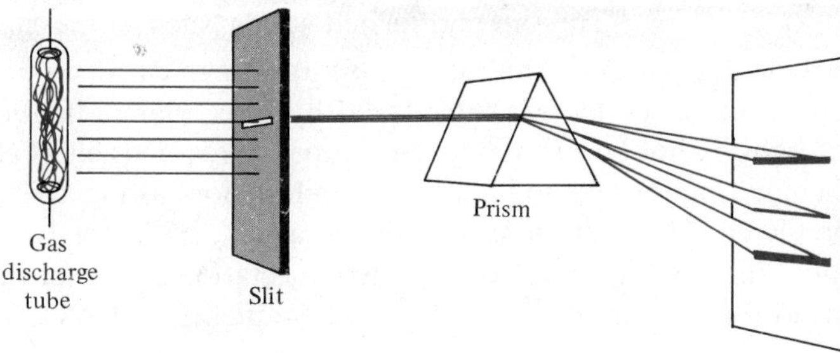

(b) Line emission spectrum

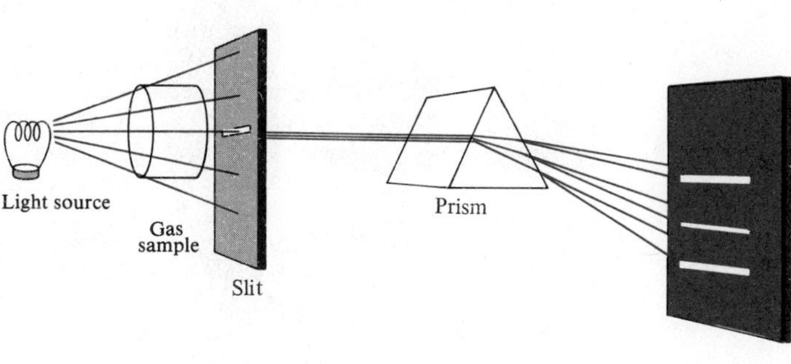

(c) Line absorption spectrum

Figure 8.7
Continuous and Line Spectra
(a) Continuous spectra are emitted when solids, liquids, or very highly compressed gases have been heated to incandescence. (b) Line spectra are emitted by atoms in the gaseous state that have absorbed extra energy. (Molecules under similar conditions give band spectra—groups of lines spaced very closely together.) (c) When a continuum of wavelengths is passed through a sample of unexcited atoms or molecules, certain wavelengths are absorbed, producing dark lines (or bands) in the continuous spectrum. This is known as an absorption spectrum. The wavelengths absorbed are the same ones that the sample would radiate if it were excited.

8.3 LIGHT AS PARTICLES: THE BIRTH OF QUANTUM THEORY

For a time, the description of light as waves was satisfactory in answering most of the questions that were being asked about light. But then questions arose that could only by answered in a different way—by describing light as consisting of separate particles. You might think, as many scientists once did, that one description had to be right and one wrong. Not true, for both are necessary to fully account for the behavior of light. It was later found that electrons in atoms, as discussed in the next section, can also be described as both waves and particles. This existence of two ways of describing the behavior of light and small particles is referred to as *wave–particle duality:* In some ways light behaves like continuous waves and in some ways it behaves like a stream of individual particles. Similarly, in some ways matter behaves like individual particles and in some ways it behaves like waves. Wave–particle duality should not be thought of as something strange in the behavior of the natural world, but as a result of the way that we observe and describe the world.

The shortcomings in the classical wave model of light first became apparent when physicists began to study the energy of light. In classical theory, the energy of a wave can have any value along a continuum. To make an analogy, your car can be driven at any speed along a continuum. From standing still you can gradually accelerate to the speed at which you choose to travel. This speed may be 40 km/h, 80 km/h, 80.5 km/h, or 80.501 km/h—any speed is possible.

The new description of light began with the realization that at each wavelength light does not have available a continuum of energy values, as classical theory had assumed. Rather, light can be regarded as made up of particles each of which carries a definite amount of energy, referred to as a **quantum.**

A quantum is like a package or bundle of something that is available only in specific and separate amounts. A quantum is somewhat like the scoops of ice cream in an ice cream cone that you buy at a store. You can order one scoop, or two scoops, or three scoops, but not 1.5 or 2.35 scoops. The amount of ice cream you get must be one, two, three, or n times the size of the scoop (Figure 8.8). Something that is **quantized** is restricted to amounts that are *whole-number multiples* of the basic unit, or quantum, for the particular system. A whole-number multiplier that specifies an amount of energy (or anything else that is quantized) is called a **quantum number. Quantum theory** is a general term for the idea that energy is quantized and the consequences of that idea. Like wave–particle duality, quantum theory also applies to electrons in atoms.

The quantization of energy was introduced by Max Planck in 1900 to explain the emission of light by hot bodies at various temperatures. This effect can be seen in the change of color from orange to "red hot" to white hot" as metal is heated. Classical theory predicted that with increasing temperature, the intensity of light of all frequencies would increase equally. Contrary to expectations, most of the increase in intensity occurred for light in a narrow range of frequencies. Planck found that he

Figure 8.8
If Ice Cream Were Quantized . . .

could account mathematically for this distribution of frequencies by assuming that *energy* was quantized. It was thought at the time that the radiation was given off by vibrating atoms. Planck proposed that the energies and frequencies of such atoms are related by the following equation:

$$E = h\nu \qquad (8.2)$$

(energy: E; Planck's constant: h; frequency: ν)

Furthermore, he proposed that the energies were quantized—the energy could have only values that are small whole-number multiples of $h\nu$, that is $1h\nu$, $2h\nu$, $3h\nu$, and so on.

Planck's constant, h, is a fundamental, universal constant. It has the units of energy × time and appears in the mathematical statements of many laws. The modern value for h is $h = 6.626 \times 10^{-34}$ J s. The units in Equation (8.2) would be

$$\underset{\text{joules}}{E} = \underset{\text{joule }\cancel{\text{second}}}{h}\ \underset{\cancel{\text{seconds}}^{-1}}{\nu}$$

In 1905 Albert Einstein extended the quantum concept devised by Planck for the special case of radiation by hot bodies. Einstein proposed that all electromagnetic radiation, whatever its source, is quantized in accordance with the relationship $E = h\nu$. Here was the first suggestion that radiation could behave like a wave, yet also be divided into packets of energy like a stream of particles.

By assuming that light is quantized, Einstein was able to explain the **photoelectric effect,** in which electrons are released by certain metals (particularly Cs and the other alkali metals, Li, Na, K, and Rb) when light shines on them. (The photoelectric effect is used in practical devices such as automatic door openers; Figure 8.9.) Previous attempts to explain the photoelectric effect had encountered a dilemma. According to classical theory, light of any frequency should deliver more energy to a surface as the intensity of the light increases (that is, as more light per unit area reaches the surface). Electrons should be able to accumulate energy from the light until they have acquired enough to break away from the metal. But, in reality, electrons do not leave the surface when exposed to light of just any frequency. For each metal there is a characteristic minimum frequency—the *threshold frequency*—below which the photoelectric effect does not occur.

A red light ($\nu = 4.3$–4.6×10^{14} Hz) of any brightness can shine on a piece of potassium for hours and no photoelectrons will be released. But as soon as even a very weak yellow light ($\nu = 5.1$–5.2×10^{14} Hz) shines on potassium, the photoelectric effects begins. The threshold frequency of potassium is 5×10^{14} Hz.

A quantum of radiant energy is called a **photon,** and each photon has energy equal to $h\nu$. Individual photons of high-intensity light of a given frequency have no greater energy than individual photons of a low-intensity light of the same frequency. High-intensity light just has *more* photons, each with the *same* energy (Figure 8.10).

Each photon can act like a particle in a collision with a single electron and deliver to that electron a maximum of 1 $h\nu$ of energy. For example, one photon of the blue-green light described in Example 8.1 ($\nu = 5.8 \times 10^{14}\ \text{s}^{-1}$) has energy $E = h\nu = (6.626 \times 10^{-34}\ \text{J s})(5.8 \times 10^{14}\ \text{s}^{-1}) = 3.8 \times 10^{-19}$ J. The amount of energy delivered by one photon can increase only if the frequency increases. This explains the threshold frequency. When the frequency, and therefore the energy of the incoming light, is too low, not even one electron can acquire enough energy to escape. Red light can never initiate the photoelectric effect in potassium, because each photon has too little energy. (The energy is dissipated by an increase in the temperature of the metal.)

8.4 ELECTRONS AS WAVES

After Planck and Einstein established that waves of light also behave like particles, Louis de Broglie, in 1923, turned up the other side of the coin. If radiation has particle-like properties, then particles in motion should have wave-like properties.

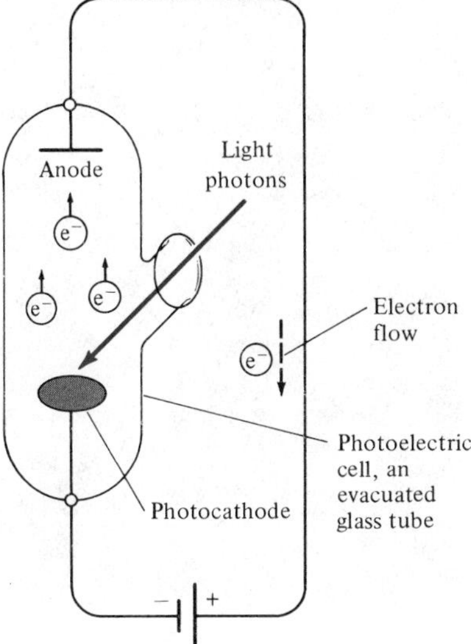

Figure 8.9
A Photoelectric Cell
The photocathode is made of a metal that exhibits the photoelectric effect. Such a photoelectric cell is an essential part of an "electric eye" device used, for example, in automatic door openers. When the light beam is interrupted, the photoelectric current no longer flows and this triggers opening of the door.

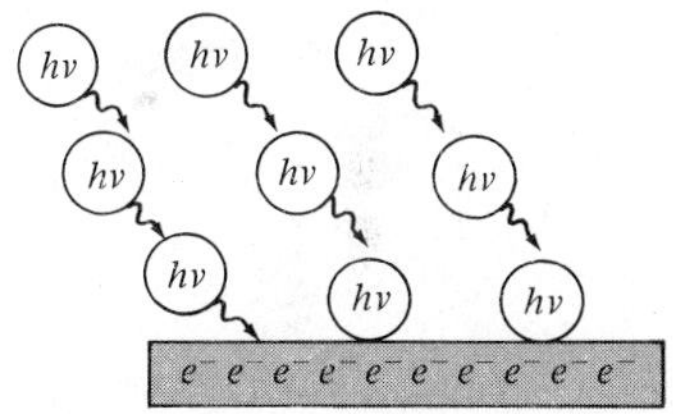

(a) Low-energy photons: below threshold frequency, so no electrons dislodged

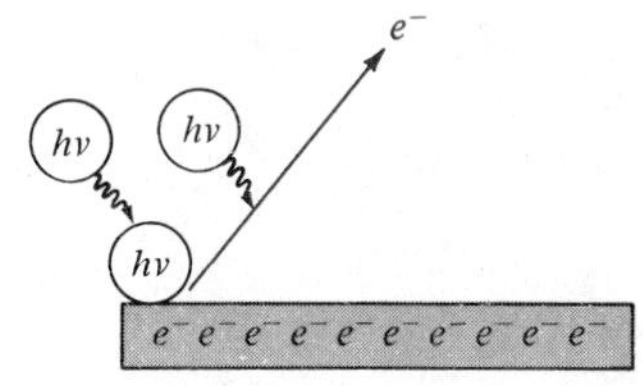

(b) Higher-energy photons: above threshold frequency, so electrons dislodged

Figure 8.10
The Photoelectric Effect
It takes light of energy equal to or greater than the $h\nu$ threshold to knock electrons out of the metal. (a) A high-intensity light of low energy cannot initiate the photoelectric effect. (b) A low-intensity light of sufficiently high energy immediately initiates the photoelectric effect.

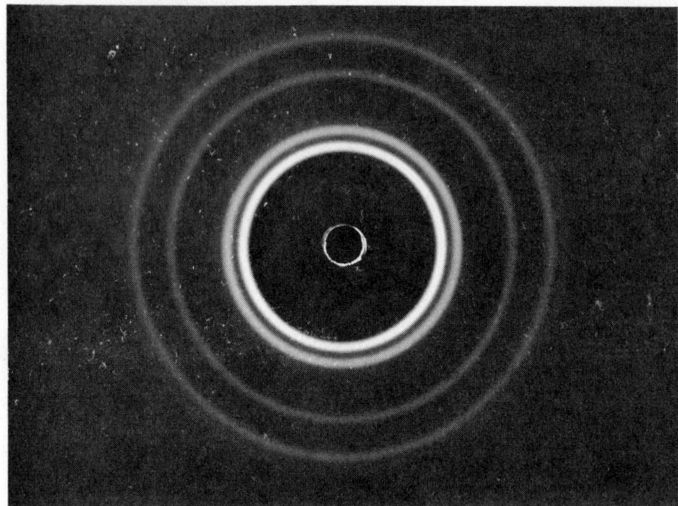

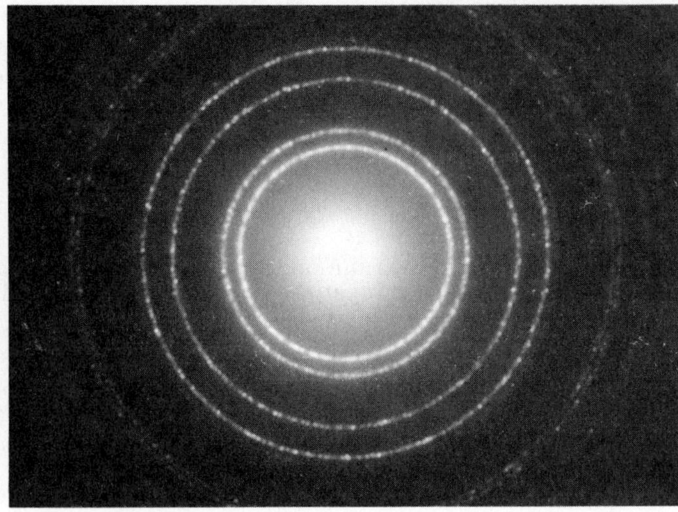

Figure 8.11
Diffraction of (a) X-rays and (b) An Electron Beam by the Same Thin Aluminum Foil
The identical patterns of the rings demonstrate the wavelike nature of both x-rays and the electron beam. Source: PSSC Physics film *Matter Waves* (Cambridge, MA: Education Development Center).

De Broglie predicted that the wavelength of a moving particle could be calculated from the equation

$$\lambda = \frac{h}{mu} \tag{8.3}$$

(wavelength of particle: λ; Planck's constant: h; mass of particle: m; velocity of particle: u)

Two American physicists at the Bell Telephone Laboratory, C. J. Davisson and L. H. Germer, soon did an experiment that proved that de Broglie was correct. In 1927 Davisson and Germer found an unexpected result in an electron-scattering experiment—the diffraction of the electron beam. Diffraction, as we have explained, is clearly a wave property. When their electron beam was aimed at a nickel crystal, they obtained a diffraction pattern similar to the pattern obtained from the diffraction of x-rays by a crystal (Figure 8.11). The wavelength of the electron beam calculated from the diffraction pattern agreed to within 1% with the wavelength calculated from de Broglie's equation. Electrons can indeed behave like waves. (See Aside 13.2.)

The wavelength of a moving body of any mass can be calculated with the de Broglie equation, and presumably moving bodies of any mass have wavelengths. However, the wave properties of matter are observable only for particles of very small mass. An α particle of mass 6.6×10^{-27} kg which is emitted from a radioactive source at a speed of 1.9×10^{7} m/s has a wavelength of 5.3×10^{-6} nm, which is observable. A 2 g hummingbird flying at 160 km/h has a wavelength of 7×10^{-24} nm, much too small to be observed.

8.5 QUANTUM MECHANICS: THE HEISENBERG UNCERTAINTY PRINCIPLE

"Mechanics" is the study of motion, and **quantum mechanics** refers to the study of the motion of entities that are small enough and move fast enough to have both observable wavelike and particlelike properties. When quantum mechanics is applied to large-scale, familiar phenomena, the effects are too small to be significant and we are left with the laws of classical mechanics intact. Classical mechanics and quantum mechanics are in a sense at the opposite ends of the same continuum.

A principle of central significance in the difference between classical mechanics and quantum mechanics is named for Werner Heisenberg, who first stated it. Classical physics incorporates the assumption that if we are clever enough and careful enough we can continue indefinitely to improve the precision with which anything can be measured. Werner Heisenberg discovered that in the realm of photons and electrons this assumption does not hold. What is called the **Heisenberg uncertainty principle** may be stated as follows: It is impossible to know simultaneously both the exact momentum and the exact position of a particle such as an electron (or any particle that exhibits quantum-mechanical behavior). **Momentum** is mass times velocity. It expresses not only the tendency of a moving body to keep moving but also, since velocity is a directional quantity, to maintain the direction of its motion. The Heisenberg uncertainty principle, as we shall see, continues to have a profound influence on the modern picture of the distribution of electrons in atoms. This principle is viewed by some as an intrinsic property of nature, by some as a statement of the limits of our knowledge, and by others as a matter for philosophical thought.

The Heisenberg principle, theoretically, holds for all objects, including those that we see around ourselves all the time. However, like the wavelength of particles, it becomes significant only at the subatomic level. Suppose you shine a flashlight into a dark closet and observe a mouse running across the floor. The light from the flashlight does not noticeably slow the mouse down or change the direction of his flight. Trying to look at an electron is a different story. The electron is so small that no matter what kind of radiation we shine on it, the speed and direction of the electron will be changed the instant the radiation hits it.

SUMMARY OF SECTIONS 8.1–8.5

Electromagnetic radiation is characterized by its wavelength (λ) and frequency (ν), which are related by $c = \lambda\nu$ (c is the speed of light), and displays the wave properties of interference and diffraction. On the other hand, the energy of radiation is quantized according to $E = h\nu$ (h is Planck's constant). Light and small particles such as electrons exhibit observable wave–particle duality. To describe their behavior requires both wave properties and particle properties, and their energy is quantized according to the mathematics of quantum mechanics. Fundamental to quantum mechanics, the Heisenberg uncertainty principle states that is not possible to know simultaneously both the exact position and the exact momentum of a particle.

QUANTUM THEORY AND THE ATOM

8.6 ATOMIC SPECTRA

In the early 1900s atomic spectra seemed as hard to explain as the photoelectric effect. Atoms that have absorbed energy over and above their normal energy content radiate that energy at specific wavelengths. All atoms in the gaseous state give line emission spectra, and there is a characteristic line spectrum for each element. This regularity in the line spectra, so puzzling at first, was the key to understanding electronic structure.

The simplest line spectrum is that of hydrogen, the lightest element. Atomic spectra become more and more complex for heavier atoms. The hydrogen spectrum consists of five series of lines, named for the men who discovered them (Figure 8.12). The first interpretation of the hydrogen spectrum was an empirical one based on the observed spectrum. It was found that the wave numbers of all of the lines in the hydrogen spectrum can be calculated accurately from the equation

$$\bar{\nu} = R\left(\frac{1}{n_1^2} - \frac{1}{n_2^2}\right) \tag{8.4}$$

where R is a constant called the Rydberg constant (named for spectroscopist Johannes Rydberg), and n_1 and n_2 are whole numbers. There is a characteristic single value of n_1

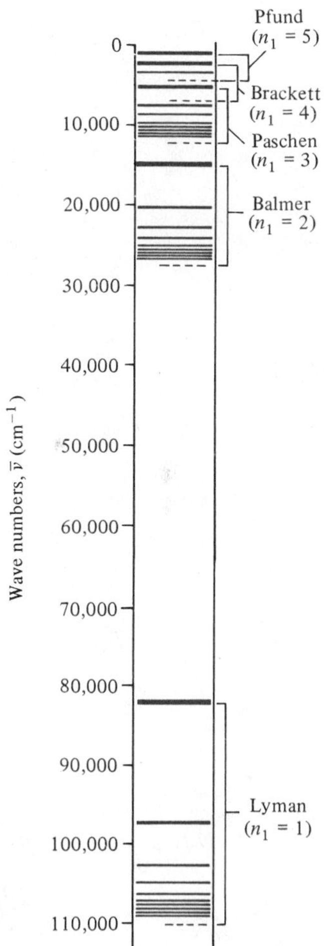

Figure 8.12
The Hydrogen Emission Spectrum
A schematic drawing showing the five series of lines. The dashed lines indicate the limits of each series. The Balmer series, which is in the visible region, was the first to be discovered.

for each series of lines in the hydrogen emission spectrum, and for each series $n_2 = n_1 + 1, n_1 + 2, \ldots$. In 1885 J. J. Balmer first developed this empirical relationship for the series bearing his name. In the Balmer series, which lies in the visible region of the spectrum, $n_1 = 2$ and $n_2 = 3, 4, 5 \ldots$. For the four later-discovered series n_1 has the values 5, 4, 3, and 1. The experimentally determined value of the Rydberg constant for the hydrogen spectrum is 109,678 cm^{-1}.

Such regularity in the spectrum had to be related to some kind of regularity in the structure of the atom. Here was another problem that begged to be solved.

8.7 THE BOHR MODEL OF THE HYDROGEN ATOM

The Rutherford nuclear atom (Section 3.8) left an inexact picture of how electrons are distributed about the nucleus. Rutherford knew that the electrons must be in motion or they would be pulled into the nucleus by the attraction between opposite charges. He suggested that the electrons might orbit the nucleus as planets orbit the sun. But this was not satisfactory. Electrons are charged, while planets are not. Classical physics predicted that charged particles moving in circles should emit radiation. As a result, the electrons would very quickly radiate away all of their energy and spiral into the nucleus.

The difficulties of the Rutherford atomic model were overcome in 1913 by Niels Bohr. The Bohr model of the hydrogen atom was the first atomic model based on the quantization of energy. Bohr made the following three assumptions:

1. The electron in the hydrogen atom can move about the nucleus in any one of several fixed circular orbits, but *only* in these orbits.
2. The angular momentum of the electron in a hydrogen atom is quantized; it is a whole-number multiple of $h/2\pi$. This is why only certain fixed orbits are allowed. **Angular momentum,** which is given by mass times velocity times the radius of a body's motion, is a measure of the tendency of a body to keep moving on a curved path. For an electron in a Bohr atom

$$mur = n\left(\frac{h}{2\pi}\right) \tag{8.5}$$

By using relationships from classical physics together with the assumption that angular momentum is quantized, Bohr showed that the energy of an electron in a fixed orbit is also quantized. This electron energy is dependent only on the mass and charge of the electron and the value of n, according to following equation:

$$E \propto \frac{me^4}{n^2} \tag{8.6}$$

where the symbol $\propto$ means "is proportional to." No energies other than those governed by Equation (8.6) are possible. The lowest energy orbit, the one in which the single electron in a hydrogen atom normally resides, is the **ground state** for that electron. The higher energy orbits (reached by the hydrogen electron only when the atom has absorbed extra energy) are referred to as **excited states.** It can also be shown that the distance of each Bohr orbit from the nucleus in an atom is quantized ($r \propto n^2/me^2$, where r is the radius). As

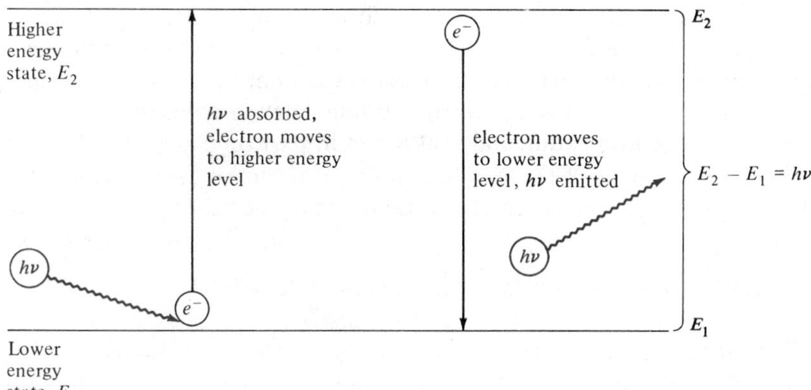

Figure 8.13
Absorption and Emission of Photons by an Electron in an Atom

n increases, the radius increases and electrons are increasingly farther from the nucleus.

3. The electron does not radiate energy as long as it remains in one of the orbits. When the electron drops from a higher energy state (E_2) to a lower one (E_1), a definite quantity of energy is emitted as one photon of radiation. The energy change and the frequency of this radiation are proportional to each other according to Planck's relationship:

$$\Delta E = E_2 - E_1 = h\nu \qquad \textbf{(8.7)}$$

If the electron is to be raised from state 1 to state 2, the same quantity of energy must be absorbed (Figure 8.13).

In the normal (or ground) state, the electron in a hydrogen atom is in the lowest energy level. The quantum number n for the ground state is 1. When the atom absorbs energy (e.g., from the electric current in a gas-discharge tube), the atom becomes "excited," and the electron jumps to higher energy levels ($n > 1$). As it falls back to the ground state, the extra energy is emitted as electromagnetic radiation. This produces a line in the hydrogen spectrum. The wavelength of the line depends on the size of the energy jump, in accordance with the basic quantum relationship, $E = h\nu$ (or, using Equation 8.1, $E = hc/\lambda$).

From the mathematics of his theory, Bohr was able to calculate the wavelengths of the hydrogen spectral lines. How each of the five series of lines was explained by the Bohr theory of the hydrogen atom is shown in Figure 8.14. Bohr's calculated values agreed with the observed values. Furthermore, his theory led directly to Equation (8.4) and a value of the Rydberg constant close to that found experimentally.

When de Broglie proposed that particles can have wavelike properties (Section 8.3), he had in mind the Bohr model of the atom. He saw that the quantized energy of an electron moving in a circle about the nucleus could be explained as a wave property. To see how, picture a vibrating guitar string of length l, which is fixed at both ends. Only a certain number of wavelengths can fit into a fixed length. Thus the wavelength of the vibration of the string (and consequently of the sound that it produces) is quantized. The possible wavelengths are given by the expression $\lambda = 2l/n$, where n is a quantum number. When $n = 1$ (Figure 8.15a), the vibration of the string has the longest allowed wavelength. The fixed points, where there is no motion, are known as *nodes*. With $n = 1$, nodes exist only at the ends of the string. At higher values of n, additional nodes appear along the length of the string (see Figure 8.15a). Only waves with such fixed points—standing waves—are allowed for guitar strings.

The same conditions apply to electrons in atoms. For an electron in a Bohr orbit to have wavelike motion, the wave must be a standing wave (Figure 8.15b) quantized according to $\lambda = 2\pi r/n$, where n is a whole number. Otherwise the wave would interfere with itself and cancel itself (Figure 8.15c).

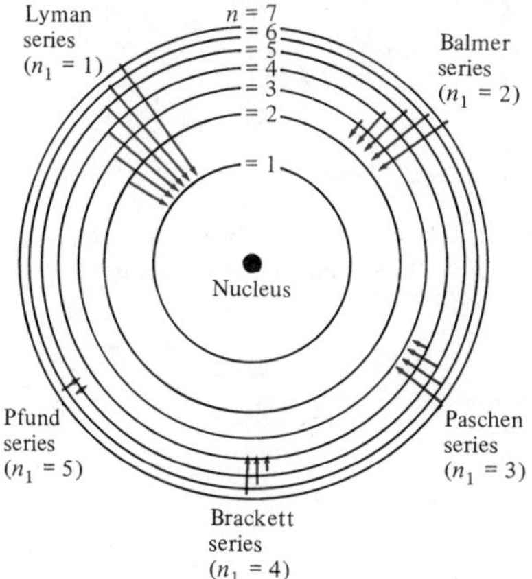

Figure 8.14
Energy Levels in the Bohr Hydrogen Atom (not to scale)
The single electron can reside in any circular orbit without emitting energy. The electron reaches orbits with $n > 1$ only when the atom is excited. The series of lines in the hydrogen spectrum arise from transitions of the electron from orbit to orbit. The n value for each series refers to the n_1 in Equation (8.4). The n_2 values are all of the n values greater than n_1 in each series. The Lyman series lines represent all the transitions in which an electron drops from a higher energy level to level 1 (the ground state). Similarly, the Balmer series lines represent transitions down to the second energy level, and so on.

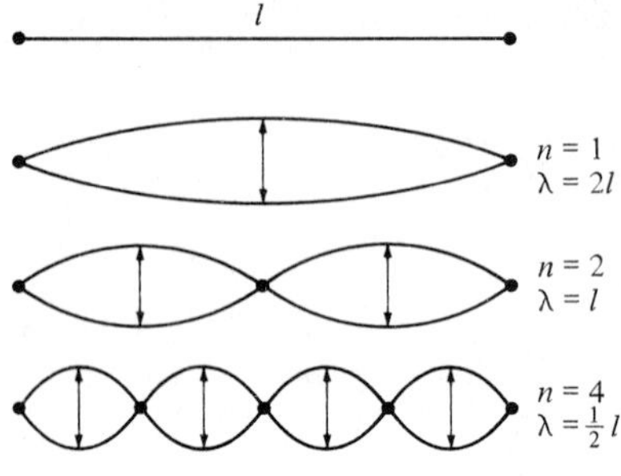

(a) Standing waves in a guitar string, $\lambda = 2l/n$

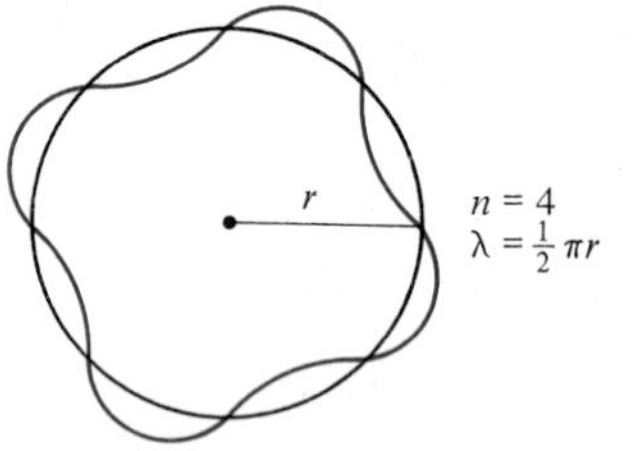

(b) Standing wave for an electron in a Bohr orbit, $\lambda = 2\pi r/n$

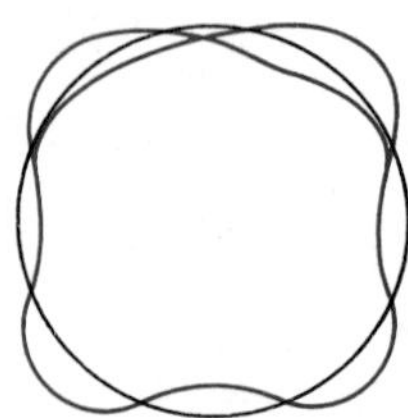

(c) Unstable wave for an electron in an atom; $n \neq$ whole number and wave interferes with itself

Figure 8.15
Standing Waves

Two of Bohr's ideas—quantization of energy levels for electrons in atoms, and energy radiation by the electron only when changing energy levels—remain as part of modern atomic theory. However, the Bohr model did not fully explain the spectra of atoms containing more than one electron. While starting from the "new" ideas of quantized energy, the Bohr atom was still based in part on classical physics. Also, it presumed exact knowledge of the location and momentum of the electron at the same moment, a violation of Heisenberg's uncertainty principle.

8.8 THE QUANTUM-MECHANICAL MODEL OF THE ATOM

Erwin Schrödinger, in developing a quantum-mechanical model for the atom, began with a classical equation for the properties of waves. He modified this equation to take account of the mass of a particle and the de Broglie relationship between mass and wavelength. The outcome—the Schrödinger equation—is the basis for a purely mathematical description of electrons in atoms. The electrons are treated as having wavelike properties as they move in the three-dimensional space surrounding the nucleus, not just in planar orbits as in the Bohr atom. The wavelike motion, just as for the guitar string, naturally imposes the requirement for standing waves, nodes, and quantized energy. But when the waves associated with electrons in atoms are extended into three dimensions, *three* quantum numbers are needed in the equation.

The Schrödinger equation for the hydrogen atom successfully predicts all aspects of the hydrogen atom spectrum, including some phenomena that could not be explained by the Bohr model. In addition, the quantization of energy of the electrons arises as a natural consequence of the mathematics—it is not necessary to begin from an assumption of quantization. Application of the hydrogen atom wave equation to many-electron atoms requires modifications for the effect of electrons on each other and the effect of the nuclear charge on electrons. The mathematics is difficult, but accurate approximations carried out with the aid of computers permit the description of atoms other than hydrogen with reasonable success (e.g., the frequencies of spectral lines can be predicted to within 1%).

The important consequences of the quantum-mechanical view of atoms are the following:

1. The energy of electrons in atoms is quantized.
2. The existence of quantized energy levels is a direct result of the wavelike properties of electrons.
3. The exact position and exact momentum of an electron cannot both be determined simultaneously (the Heisenberg uncertainty principle).
4. The region in space around the nucleus in which an electron is most probably located can be predicted for each electron in an atom. Electrons of different energies are found in different regions. The region in which an electron with a specific energy will *most probably* be located is called an **atomic orbital.**

Note the distinction between an orbit, as in the Bohr atom, and an orbital. An orbit is a clearly defined path. An orbital is a region within which an electron of a given energy is most likely to be found. Quantum mechanics can predict only the *probable* location of an electron. Suppose at different instants many, many snapshots of the location of an electron of a given energy could be taken. Superimposing these snapshots would show the greatest "electron density" in the region designated as the atomic orbital for that electron (see Section 8.10).

8.9 ORBITALS AND QUANTUM NUMBERS

The designation of the orbital "location" of an electron requires four quantum numbers. Introducing the quantum numbers is difficult because there is no satisfactory physical description of them. So let's start by explaining why quantum numbers are

Aside 8.2 THOUGHTS ON CHEMISTRY: QUANTUM BEHAVIOR

Some comments on quantum behavior by Nobel-prize-winning physicist, Richard P. Feynman.

Because atomic behavior is so unlike ordinary experience, it is very difficult to get used to, and it appears peculiar and mysterious to everyone—both to the novice and to the experienced physicist. Even the experts do not understand it the way they would like to, and it is perfectly reasonable that they should not, because all of direct, human experience and of human intuition applies to large objects. We know how large objects will act, but things on a small scale just do not act that way. So we have to learn about them in a sort of abstract or imaginative fashion and not by connection with our direct experience. . . .

We would like to emphasize a very important difference between classical and quantum mechanics. We have been talking about the probability that an electron will arrive in a given circumstance. We have implied that in our experimental arrangement (or even in the best possible one) it would be impossible to predict exactly what would happen. We can only predict the odds! This would mean, if it were true, that physics has given up on the problem of trying to predict exactly what will happen in a definite circumstance. Yes! physics has given up. We do not know how to predict what would happen in a given circumstance, and we believe now that it is impossible—that the only thing that can be predicted is the probability of different events. It must be recognized that this is a retrenchment in our earlier ideal of understanding nature. It may be a backward step, but no one has seen a way to avoid it. . . . So at the present time we must limit ourselves to computing probabilities. We say "at the present time," but we suspect very strongly that it is something that will be with us forever—that it is impossible to beat that puzzle—that this is the way nature really is.

Source: Richard P. Feynman, Robert B. Leighton, and Matthew Sands, *The Feynman Lectures on Physics*, Vol. 3 (Reading, MA: Addison-Wesley, 1965).

needed. Each electron in a many-electron atom has its own unique set of quantum numbers. By using quantum numbers we can identify which electron we are talking about and know to which atomic orbital the electron belongs. To make a rough analogy (rough because for electrons we deal only with their *probable* locations), suppose it is necessary to locate a person who is attending a performance in a large theater. We know that his ticket reads, "Second balcony, A104." Three "quantum numbers" identify his location. To find him we must look first for balcony number 2, then in row A, and then in seat 104.

Being able to specify the "location" of electrons is important. Many of the properties of atoms are determined by how many electrons are present and to which orbitals they belong. The orbitals may have different energies and occupy different regions in space—both properties that profoundly affect the way atoms combine and what the geometry of molecules will be.

The first quantum number, the *principal* quantum number, identifies the main energy levels (like the balconies). The second, the *subshell* quantum number (traditionally called either the angular momentum or azimuthal quantum number) identifies sublevels of energy within the main energy level (like the rows in each balcony). The third quantum number is the *orbital* quantum number (traditionally called the magnetic quantum number)—it pins down the location of individual electrons in orbitals (like the seats in each row). A fourth quantum number is needed because the electron can occupy the orbital in two different orientations.

1. Principal quantum number, n. The principal quantum number is roughly equivalent to the n of the energy levels in the Bohr atom. It has whole number values, $n = 1, 2, 3, 4, \ldots$, and designates what are called the *main energy levels* in an atom. As n increases, electrons are generally farther from the nucleus and have higher total potential energy. Values of n range from 1 to 7 in the unexcited states of the known elements. Values of n from 1 to ∞ are possible in the excited states of atoms (∞, infinity, corresponds to the complete removal of an electron, to form an ion).
2. Subshell quantum number, l. The second quantum number, l, designates the different energy subshells, or sublevels, within the main, or n, level and indicates the general shape of different types of orbitals. Only certain values of l are allowed. These depend on n: $l = 0, 1, 2, \ldots, (n - 1)$. *The total number of*

Table 8.3
Subshell Quantum Number (l) Values
The letters *s*, *p*, *d*, and *f* come from the designation of lines in emission spectra as belonging to the sharp, principal, diffuse, or fundamental series. The number of sublevels in each quantum level is equal to *n*.

For $n =$	$l = 0, 1, 2, \ldots, (n-1)$ (subshell letters)
1	0 (*s*)
2	0, 1 (*s*, *p*)
3	0, 1, 2 (*s*, *p*, *d*)
4	0, 1, 2, 3 (*s*, *p*, *d*, *f*)

possible subshells in each level is equal to n. For example, the $n = 3$ shell has three l levels, $l = 0, 1$, and 2. The first four subshells are identified by the letters *s*, *p*, *d*, and *f* (Table 8.3), which correspond to l values of 0, 1, 2, and 3. Higher values of l represent subshells occupied only in excited states of atoms. Transitions between the subshells can be identified in spectra and, in general, the energies of the subshells within the same main energy level increase from *s*, to *p*, to *d*, to *f*.

3. Orbital quantum number, m_l. Within the subshell, it is possible for electrons to occupy different regions of space governed by the values of m_l. The number of allowed values of m_l is limited and depends upon l (and therefore also on n): $m_l = -l$ to $+l$. The number of orbitals in each subshell equals the number of values of m_l, which is given by $2l + 1$, resulting in one *s* orbital, three *p* orbitals, five *d* orbitals, and seven *f* orbitals, as shown in Table 8.4. Each of the different *p*, *d*, or *s* orbitals has a different orientation in space (see Section 8.10).
4. Spin quantum number, m_s. An electron within an orbital has angular momentum. Although the picture should not be taken too literally, we can imagine the electron as spinning on its own axis like the Earth as it orbits the sun. This spinning causes each electron to behave like a tiny magnet. Electron spin has two possible orientations (Figure 8.16) and the two values of m_s are $m_s = +\frac{1}{2}, -\frac{1}{2}$.

Table 8.4
Orbital Quantum Number (m_l) Values

For $l =$	$m_l = -l, \ldots, -1, 0, 1, \ldots, +l$ (total)
0 (*s*)	0 (one *s* orbital)
1 (*p*)	+1, 0, −1 (three *p* orbitals)
2 (*d*)	+2, +1, 0, −1, −2 (five *d* orbitals)
3 (*f*)	+3, +2, +1, 0, −1, −2, −3 (seven *f* orbitals)

Because only certain values of l and m_l are allowed, only a specific number of atomic orbitals are allowed within each main energy level. For the lowest energy level, where $n = 1$, l has only one possible value, $l = 0$, and m_l has only one possible value, $m_l = 0$. Therefore, at the $n = 1$ level, only a single atomic orbital is present, an *s* orbital.

Atomic orbitals are designated by the following notation:

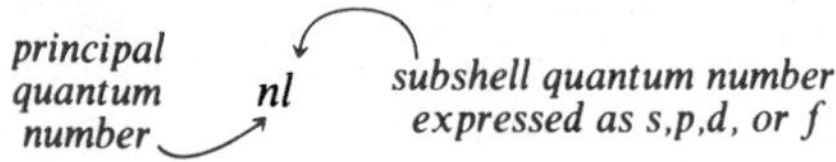

The single orbital in the *s* subshell of the lowest energy level is a 1*s* orbital. The one electron in a hydrogen atom normally occupies a 1*s* orbital.

Study of Tables 8.3 and 8.4 and a little thought show how the quantum numbers determine the number of available orbitals. When $n = 2$, four orbitals are possible—one *s* orbital (for $l = 0$, the 2*s* orbital) and three *p* orbitals (for $l = 1$, the 2*p* orbitals). When $n = 3$, nine orbitals are possible, and when $n = 4$, sixteen orbitals are possible.

$n = 1, l = 0$	one *s* orbital
$n = 2, l = 0, 1$	one *s* orbital + three *p* orbitals
$n = 3, l = 0, 1, 2$	one *s* orbital + three *p* orbitals + five *d* orbitals
$n = 4, l = 0, 1, 2, 3$	one *s* orbital + three *p* orbitals + five *d* orbitals + seven *f* orbitals

8.10 PICTURING ORBITALS

Unfortunately, drawing pictures of atomic orbitals is difficult. Artists' best efforts at representing orbitals still convey the idea of distinct shapes. These "shapes" are only the regions of probable electron location, and the probability gradually trails off to zero in every direction. The shapes have no *independent* physical existence, and *the overall, total distribution of electron density for all the orbitals in each principal quantum level is close to a sphere for any isolated atom.* However, the differences among the regions of greatest electron density for each different type of orbital—*s*, *p*, *d*, and *f*—help to determine the shapes of molecules (Chapter 11).

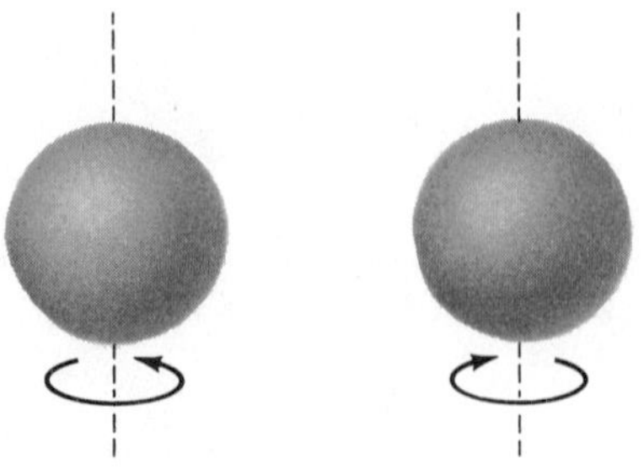

Figure 8.16
Electron Spin
Two possible orientations of the spin of an electron about its own axis.

One way to picture the locations of electrons is to look at the relative distances from the nucleus of electrons in different orbitals. A plot of the probability of finding

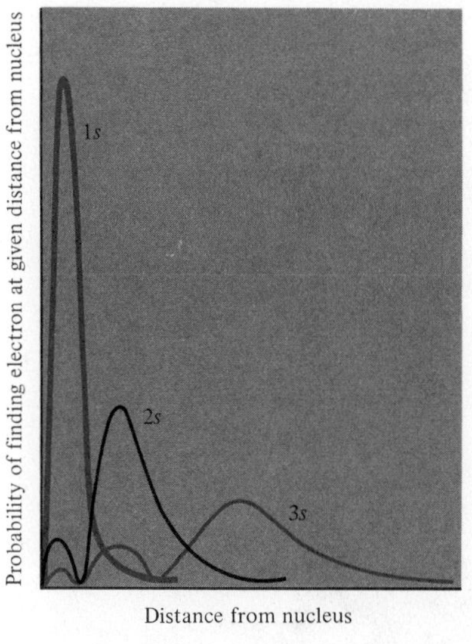

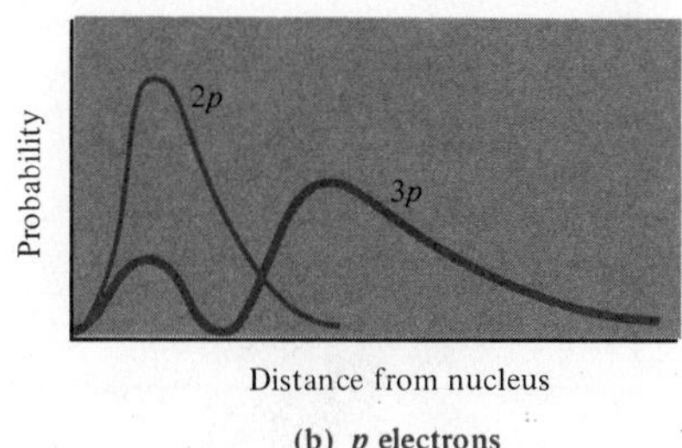

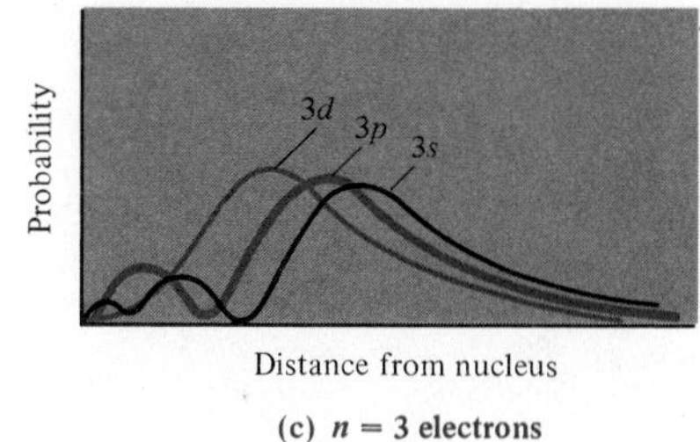

Figure 8.17
Probable Distances of $n = 1, 2,$ and 3 Electrons from the Nucleus
(a) and (b) Plots for the $n = 1$–3 *s* and *p* electrons, which show that as *n* increases the electrons are generally farther from the nucleus. (c) Plots for the *s*, *p*, and *d* electrons in the $n = 3$ energy level, which show that they occupy roughly the same region.

1*s*, 2*s*, and 3*s* electrons at varying distances from the nucleus is shown in Figure 8.17a. A 1*s* electron is most likely to be found quite close to the nucleus; the 2*s* and 3*s* electrons are successively farther away. Similar plots for the 2*p* and 3*p* electrons (Figure 8.17b) reveal the same increasing distance from the nucleus with the value of *n*. The probability plots for the 3*s*, 3*p*, and 3*d* orbitals are superimposed in Figure 8.17c to show that electrons with the same *n* value are roughly the same distance from the nucleus, and therefore have similar potential energies.

Another approach to picturing orbitals is to draw boundary contours—surfaces that are like solid shapes viewed from a distance. The surface indicates the region within which the electron has a certain probability (usually 90%) of being found. Such a contour surface for an *s* orbital has the shape of a spherical shell centered on the nucleus (Figure 8.18). For each value of *n*, there is one *s* orbital. As *n* increases, the spherical shells increase in size, like the successive layers in an onion.

The *p* subshell at each energy level has three orbitals. Each *p* orbital has two lobes, one on each side of the nucleus. The lobes of the three different *p* orbitals can be thought of as directed along the *x*, or *y*, or *z* axes of a set of coordinates. Essentially, the electrons in each *p* orbital stay as far from the electrons in the other *p* orbitals as possible (Figure 8.19).

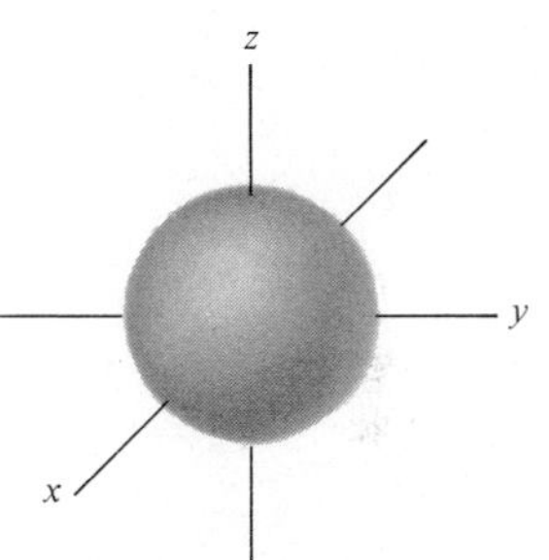

Figure 8.18
An *s* Orbital

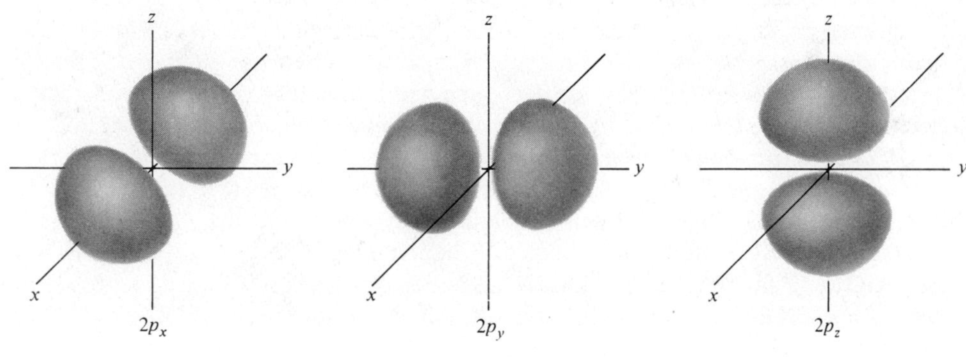

Figure 8.19
The *p* Orbitals
Contour surfaces for the three 2*p* orbitals.

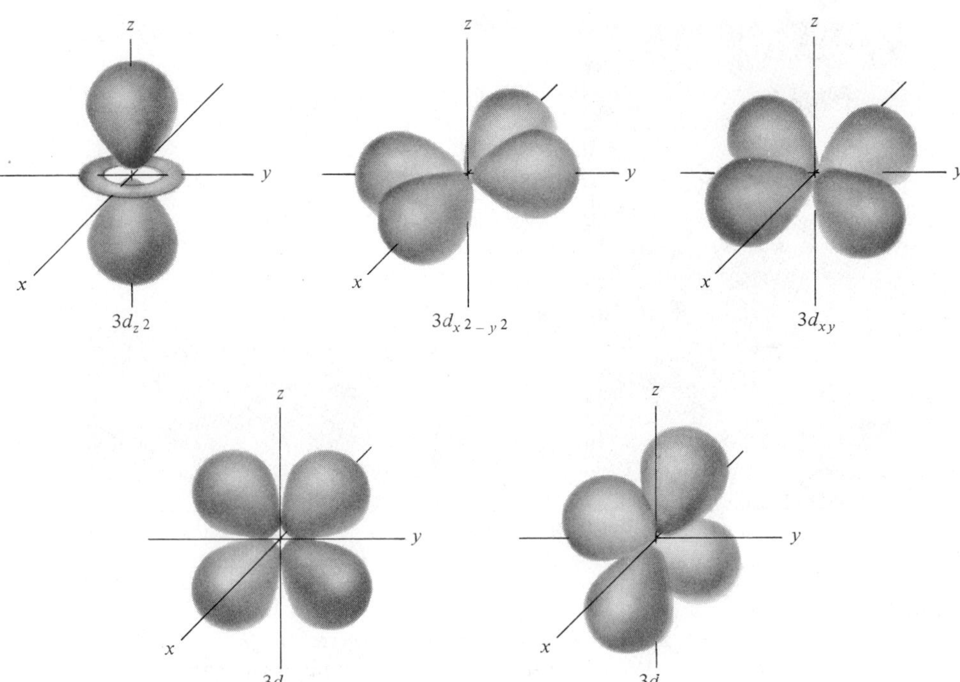

Figure 8.20
The 3*d* Orbitals

The *d* orbitals—five of them for the five possible values of m_l—are more complex (Figure 8.20). [The *f* orbitals are still more complex.] One *d* orbital resembles the *p* orbital along the *z* axis, but with an additional doughnut-shaped volume in the center. The other four have four lobes each, extending in one case along the *x* and *y* axes and in the other cases, between the axes.

SUMMARY OF SECTIONS 8.6–8.10

Line-emission spectra result when electrons that have gained extra energy fall back to their normal energy levels in atoms; each element has a characteristic line-emission spectrum (atomic spectra). The partially successful Bohr model of the atom, devised to explain the hydrogen spectrum, pictures electrons as moving in fixed orbits of quantized energy and radiating energy only when moving from a higher energy orbit to a lower one. In the modern, quantum-mechanical view of the atom, the energy of electrons is also quantized. The electrons have wave-like motion and occupy three-dimensional regions known as orbitals.

Four quantum numbers are needed to describe the location of each electron in an atom—*n*, which designates the main energy level; *l*, which designates the subshell within the main energy level; m_l, which designates the orbital; and m_s, which designates one of two possible spin orientations. An atomic orbital is the region in which an electron with a specific set of quantum numbers is most likely to be found. Only the probability of the location of an electron at a given position within the orbital can be known.

The four types of orbitals (*s*, *p*, *d*, and *f*) occupy different and distinctive regions surrounding an atom (see Figures 8.18–8.20). Within a main energy level, the number of types of orbitals equals *n* ($n = 1$, *s* orbitals; $n = 2$, *s* and *p* orbitals; $n = 3$, *s*, *p*, and *d* orbitals; $n = 4$, *s*, *p*, *d*, and *f* orbitals). A maximum of one *s*, three *p*, five *d*, and seven *f* orbitals are possible in a given main energy level (see Tables 8.3 and 8.4).

Aside 8.3 THE QUANTUM REVOLUTION

Toward the end of the nineteenth century, some scientists were ready to conclude that the era of discovery in physics was over. They felt that everything there was to know had been found out—an attitude that proved to be far from correct. The first 25 years of the twentieth century saw a revolution in physics that also had a profound effect upon chemistry. It gradually became clear that the laws that explain readily observable phenomena do not apply on the scale of electrons and radiation. The revolution was founded on the work of a number of European physicists who dared to depart from classical physics.

Table 8.5 relates the classical experiments on atomic structure discussed in Chapter 3 and the major events of the quantum revolution to historical events. The political and social turmoil caused in Europe by the rise to power of the Nazi party in Germany touched the lives of Planck, Einstein, and many of the physicists who were carrying forward the quantum revolution. Einstein was horrified by the Nazi preparations for war and, although a pacifist, urged that Europe take up arms. In 1933 when Hitler became Chancellor, Einstein renounced his German citizenship and moved to the United States. Bohr, a brilliant physicist, played a role in the development of the atomic bomb during World War II. He had fled from Denmark to Sweden in a fishing boat when it was learned that Hitler had ordered that he be kidnapped. Still in danger, he was then flown to England in a British bomber. In later years, Bohr was deeply troubled by the awful destructive potential of the atomic bomb and was outspoken on the need for peaceful coexistence among all nations.

Table 8.5
Historical Perspective—Atomic Structure

		1876	Custer's Last Stand
Cathode rays, William Crookes	1879		
"Electron," G. Johnstone Stoney	1881		
Hydrogen spectrum, Johann Balmer	1885		
Canal rays, Eugen Goldstein	1886		
Electron e/m, J. J. Thomson	1897	1897	Klondike gold rush
Canal ray e/m, W. Wien	1898	1898	Spanish-American War
$E = h\nu$, Max Planck	1900		
		1901	William McKinley assassinated: Theodore Roosevelt becomes U.S. President
		1903	Wright Brothers flight at Kitty Hawk
Quantized radiation, Albert Einstein	1905		
Charge on electron, Robert A. Millikan	1909		
Nuclear atom model, Ernest Rutherford	1911		
		1912	Woodrow Wilson elected U.S. President
Hydrogen atom model, Niels Bohr Meaning of atomic number, Henry Moseley	1913	1914	World War I begins
		1917	Russian Bolshevik Revolution
		1919	Treaty of Versailles ends World War I
$\lambda = \frac{h}{m\mu}$, Louis de Broglie	1923		
Wave-mechanical atomic model, Erwin Schrödinger Uncertainty principle, Werner Heisenberg	1926		
Electron diffraction, C. J. Davisson and L. H. Germer	1927	1927	Transatlantic flight by Charles Lindbergh
		1929	U.S. stock market crash; beginning of Depression
Neutron, James Chadwick	1932		
		1933	Adolf Hitler becomes Chancellor of Germany

ELECTRON CONFIGURATIONS

8.11 SYMBOLS FOR ELECTRON CONFIGURATIONS

The **electron configuration** of an atom is the distribution among the subshells of all of the electrons in the atom. Electron configurations are designated by using the following subshell notation:

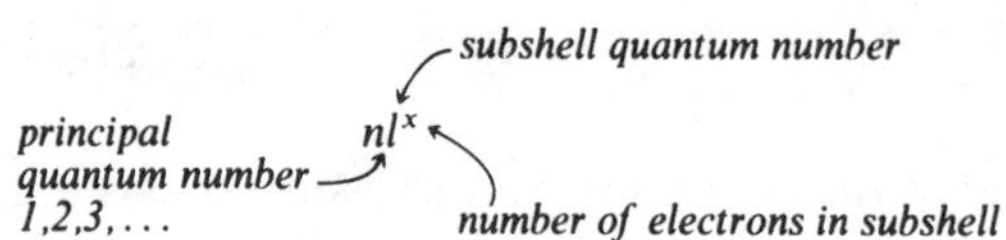

Here n represents the principal quantum level as a number $(1, 2, 3, \ldots)$; l represents the subshell as a letter (s, p, d, f); and x indicates the number of electrons in the subshell. For example

$3d^8$

In the n = *3 level* — *the* d *orbitals* — *are occupied by 8 electrons*

The complete electron configuration for an atom is given by a series of symbols representing each occupied subshell. For example, the complete ground-state configuration of an argon atom is written

$$\text{Ar} \qquad 1s^2 2s^2 2p^6 3s^2 3p^6$$

showing that the $1s$, $2s$, and $3s$ subshells each contain two electrons, and that the $2p$ and $3p$ subshells each contain six electrons. For a neutral atom, the sum of all the x values is the atomic number (Z) of the element. Thus for argon (atomic number 18)

$$Z = 2 + 2 + 6 + 2 + 6 = 18$$

8.12 WRITING ELECTRON CONFIGURATIONS

A good approach to understanding electron configurations is to start with hydrogen and consider where each successive electron is added as the atomic numbers increase from element to element. This is known as the *aufbau* process (from the German for "building up").

The single electron in hydrogen is a $1s$ electron. In which orbital will the second electron that is present in a helium atom be located? There are three rules for building up electron configurations, and two of them apply here. The first is the *Pauli exclusion principle* (named for Wolfgang Pauli, who suggested it in 1925 on the basis of spectra): No two electrons can have the same four quantum numbers (n, l, m_l, and m_s). In other words, each electron must have a unique set of quantum numbers.

The maximum number of electrons that can go into each atomic orbital is set by this principle. Once all of its possible quantum number combinations have been used, an energy level, and each of its orbitals, is full. With this rule the spin quantum number takes on great significance, for it allows two electrons of opposite spin to occupy each atomic orbital. The maximum numbers of electrons for each principal quantum level and each subshell are summarized in Table 8.6. (Note that the maximum number of electrons in each n level is $2n^2$.)

Based on the Pauli exclusion principle, the second electron needed to form a helium atom ($Z = 2$) should be a $1s$ electron, but of opposite spin from the first. (The two sets of quantum numbers would be $n = 1, l = 0, m_l = 0, m_s = +\frac{1}{2}$ and $n = 1, l = 0, m_l = 0, m_s = -\frac{1}{2}$.)

But couldn't the second electron instead be an $n = 2$ electron, say a $2s$ electron, and still obey the Pauli exclusion principle? This is where the next rule applies, the *lowest energy principle:* Electrons occupy the lowest-energy orbitals available to them; they enter higher energy orbitals only when the lower energy orbitals are filled. This rule establishes that the second electron in helium must be a $1s$ electron, not an electron in an $n = 2$ orbital, which is of higher energy. The electron configuration of the helium atom is written

$$\text{He} \qquad 1s^2 \qquad (Z = 2)$$

The configurations for atoms of the next three elements follow easily from the two rules given:

Table 8.6
Electron Distribution and Maximum Electron Population
The number of *s*, *p*, *d*, and *f* orbitals is determined by the number of m_l values for each subshell (see Table 8.4). Two electrons of opposite spin can reside in each orbital.

n	l	Number of l orbitals	Maximum number of electrons per l subshell	Maximum number of electrons per n level ($=2n^2$)
1	0(*s*)	One *s* orbital	2	2
2	0(*s*)	One *s* orbital	2	
	1(*p*)	Three *p* orbitals	6	8
3	0(*s*)	One *s* orbital	2	
	1(*p*)	Three *p* orbitals	6	
	2(*d*)	Five *d* orbitals	10	18
4	0(*s*)	One *s* orbital	2	
	1(*p*)	Three *p* orbitals	6	
	2(*d*)	Five *d* orbitals	10	
	3(*f*)	Seven *f* orbitals	14	32
5	0(*s*)	One *s* orbital	2	
	1(*p*)	Three *p* orbitals	6	
	2(*d*)	Five *d* orbitals	10	
	3(*f*)	Seven *f* orbitals	14	
	4("*g*")[a]	Nine "*g*" orbitals[a]	18[a]	50[a]

[a] No element has yet been found that uses an $l = 4$ ("*g*") energy sublevel in the ground state.

Li	$1s^2 2s^1$	($Z = 3$)
Be	$1s^2 2s^2$	($Z = 4$)
B	$1s^2 2s^2 2p^1$	($Z = 5$)

What about the sixth electron to form a carbon atom? It should be another *p* electron. Does it make any difference to which of the three 2*p* atomic orbitals it belongs? They are of equal energy, so the lowest energy principle is of little help. The third rule for electron configurations, *Hund's principle*, applies here: Orbitals of equal energy are each occupied by a single electron before a second electron, which will have the opposite spin quantum number, enters any of them. For a carbon atom ($Z = 6$) this means that the sixth electron enters a *different p* orbital from the first electron. Single electrons in orbitals in the same sublevel spin in the same direction. By using a box for each orbital and an arrow for each electron, with opposite directions indicating opposite spins, the boron and carbon configurations are written

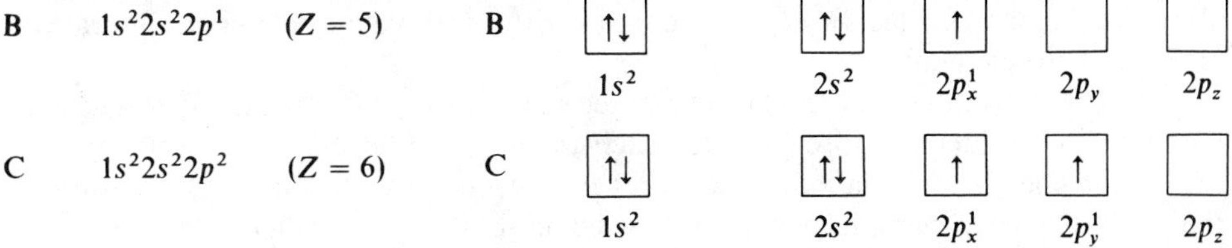

In nitrogen there is one electron in each of the three *p* orbitals.

N $1s^2 2s^2 2p^3$ ($Z = 7$) N [↑↓] [↑↓] [↑] [↑] [↑]
$1s^2$ $2s^2$ $2p_x^1$ $2p_y^1$ $2p_z^1$

The next three electrons pair up one by one with the unpaired 2*p* electrons, and at neon ($Z = 10$) all of the orbitals in the $n = 2$ level are full.

Ne $1s^2 2s^2 2p^6$ ($Z = 10$) Ne [↑↓] [↑↓] [↑↓] [↑↓] [↑↓]
$1s^2$ $2s^2$ $2p_x^2$ $2p_y^2$ $2p_z^2$

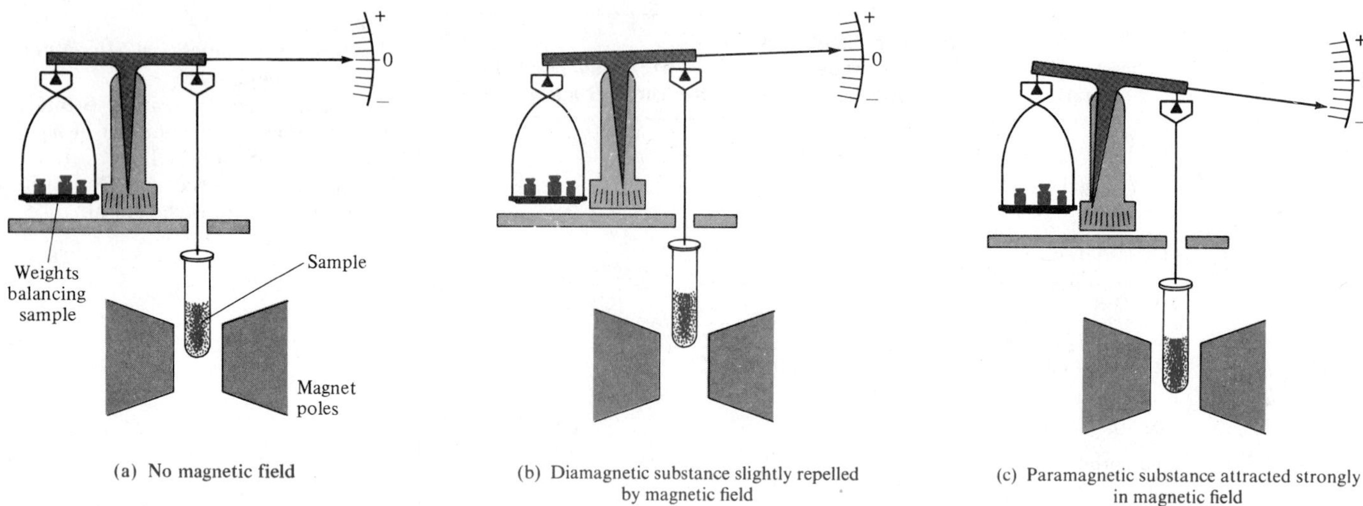

(a) No magnetic field

(b) Diamagnetic substance slightly repelled by magnetic field

(c) Paramagnetic substance attracted strongly in magnetic field

Figure 8.21
Experimental Determination of Magnetic Properties
(a) The sample is first balanced exactly by weights in the absence of a magnetic field. (b) A diamagnetic substance—one with only paired electrons—is weakly repelled by the field, as shown by the pointer moving in the + direction. (c) A paramagnetic substance—one with one or more unpaired electrons—is more strongly attracted by the magnetic field, as shown by the pointer moving in the − direction.

The eleventh electron, in sodium, will be a 3*s* electron and the process will begin again.

According to Hund's principle, atoms or ions (or molecules) can have varying numbers of unpaired electrons. The presence or absence of unpaired electrons is found experimentally by the behavior of a substance in the presence of a magnetic field (Figure 8.21), and such measurements verify the presence of unpaired electrons as predicted by Hund's rule. **Paramagnetism** is the property of attraction to a magnetic field shown by substances containing unpaired electrons. **Diamagnetism** is the property of repulsion by a magnetic field and shows the absence of unpaired electrons. There can be one unpaired *s* electron and as many as three unpaired *p* electrons, or five unpaired *d* electrons, or seven unpaired *f* electrons. (Ferromagnetism is the more familiar property of permanent magnets; Section 30.3)

With increasing atomic number, electrons occupy the subshells available in each main energy level in the following order:

$$1s\,2s\,2p\,3s\,3p\,4s\,3d\,4p\,5s\,4d\,5p\,6s\,4f\,5d\,6p\,7s\,5f\,6d\,7p$$

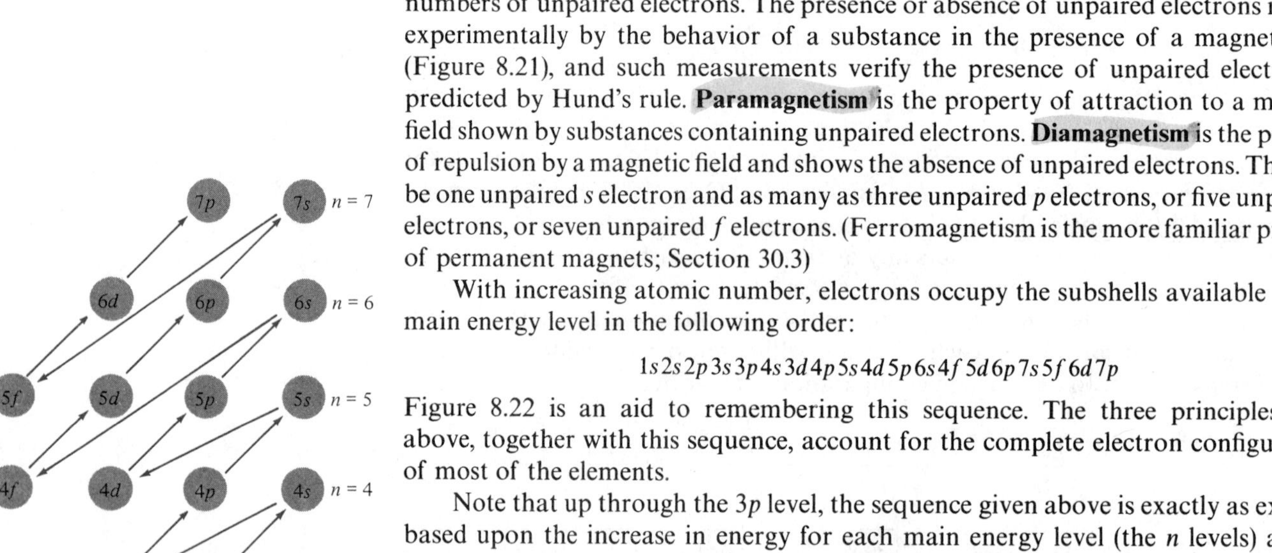

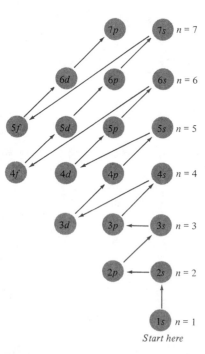

Figure 8.22
Orbital Occupancy Sequence—A Diagram

Figure 8.22 is an aid to remembering this sequence. The three principles given above, together with this sequence, account for the complete electron configurations of most of the elements.

Note that up through the 3*p* level, the sequence given above is exactly as expected based upon the increase in energy for each main energy level (the *n* levels) and the increase in energy within a main energy level for each subshell. After 3*p*, variations in the filling sequence begin to occur, the first being that the 4*s* subshell is fully occupied before electrons enter the 3*d* subshell. These variations result from the greater complexity of atoms with higher atomic number. As the atomic number increases, nuclear charge increases. The electrons are drawn closer together and also closer to the nucleus, allowing electrons in different energy levels to influence each other to an increasing extent. The influences on each electron are unique. How strongly an electron is attracted to the nucleus and how strongly it is repelled by other electrons affect the energy of that electron. The configuration of each atom is the one that gives the lowest energy to the atom as a whole. The result is that the 4*s* subshell is filled before the 3*d* subshell begins to fill, the 5*s* subshell before the 4*d* subshell, the 5*p and* 6*s* subshells before the 4*f* subshell, and so on.

In a few cases, there are exceptions to the *ideal* configurations predicted by the sequence in Figure 8.22, notably in a trade-off between the expected occupancy of an outer *s* subshell and a next-to-outer *d* subshell. In Table 8.7 the elements with configurations that vary from the expected configurations are marked with asterisks. Of these elements the most common are chromium ($Z = 24$), copper ($Z = 29$), and silver ($Z = 47$). In Table 8.7 (page 183) each of these elements has one less electron in the *s* subshell and one more in the *d* subshell than expected. These variations have been attributed to the possible greater stability of fully occupied or one-half occupied subshells (e.g., $3d^5$ occupancy rather than $3d^4$ for Cr; $3d^{10}$ rather than $3d^9$ for Cu).

EXAMPLE 8.2 Writing Electron Configurations

What is the complete electron configuration of the zirconium atom ($Z = 40$)?

Forty electrons must be accommodated. The $n = 1$ and $n = 2$ energy levels are filled to maximum capacity first.

$$1s^2 2s^2 2p^6 \qquad \text{(10 electrons)}$$

The next subshells in the filling sequence (see Figure 8.20) are 3*s*, 3*p*, 4*s*, and 3*d*. There are enough electrons to fill up all of these subshells, also. Now the $n = 1, 2,$ and 3 levels have their maximum populations and 4*s* is also full.

$$1s^2 2s^2 2p^6 3s^2 3p^6 3d^{10} 4s^2 \qquad \text{(30 electrons)}$$

[Note that complete configurations are written in this book in the order of the principal quantum number, not the filling order.]

Ten electrons remain to be placed. The next sublevel in the sequence is 4*p*, which takes a maximum of six electrons; followed by 5*s*, which takes two electrons. The remaining two electrons go into 4*d* orbitals. The total electron configuration of Zr is

$$\textbf{Zr} \qquad \mathbf{1s^2 2s^2 2p^6 3s^2 3p^6 3d^{10} 4s^2 4p^6 4d^2 5s^2} \qquad \text{(40 electrons)}$$

The two 4*d* electrons are in separate orbitals and spin in the same direction.

EXERCISE Write the complete electron configuration for the arsenic atom.

Answer $1s^2 2s^2 2p^6 3s^2 3p^6 3d^{10} 4s^2 4p^3$.

8.13 ELECTRON CONFIGURATIONS AND THE PERIODIC TABLE

A modern periodic table is given inside the front cover of this book, and the relationship between the subshell filling sequence and the periodic table is shown in Figure 8.23. The arrangement of the elements in the table, first devised on the basis of

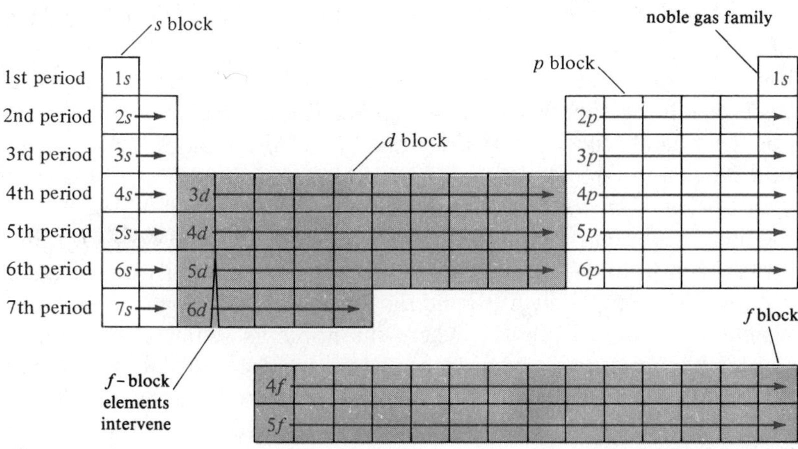

Figure 8.23
Orbital Filling Sequence in the Periodic Table

the observed properties of the elements, is fully explained by the electron configuration of atoms of the elements. The length of each horizontal row—called a **period**—is determined by the number of electrons accommodated by the subshells that become filled as atomic number increases. The elements in a vertical column in the table, referred to as members of a **group** or **family**, have similar properties because they have identical (or, in a few cases, almost identical) outer electron configurations.

Earlier we noted that lithium ($Z = 3$) and sodium ($Z = 11$) are elements with very similar properties, as are fluorine ($Z = 9$) and chlorine ($Z = 17$). Examination of the electron configurations of these elements in Table 8.7 shows that the **outer electron configurations**—the configurations in the occupied energy level of highest principal quantum number (and, for the transition elements, the next-to-highest principal quantum number)—are identical. For lithium and sodium, these outer configurations are a single electron in an s orbital ($2s^1$ and $3s^1$, respectively) and, for fluorine and chlorine, seven electrons in the s and p orbitals ($2s^2 2p^5$ and $3s^2 p^5$, respectively). Furthermore, all the elements in the respective groups with these elements also have the same outer configurations (see periodic table inside the front cover, where outer configurations are given for all elements).

Hydrogen and helium, in which electrons occupy the $n = 1$ main energy level, stand alone in the first period of the table. The next main energy level to be filled has $n = 2$ and accommodates eight electrons, accounting for the eight elements from lithium through neon (see Table 8.7 and Figure 8.23). There are also eight third-period elements (Na–Ar), accounted for by the filling of the s and p subshells in the $n = 3$ level. Note that n is commonly used as shorthand in representing the occupied orbitals of highest n value. For example, one could say lithium and sodium both have ns^1 outer configurations.

To maintain elements with similar outer configurations in the same columns, beginning with the fourth period the lengths of the periods must be extended to accommodate the elements in which the d subshells are being filled. For example, in the fourth period, after the $4s$ subshell is filled by two electrons in potassium ($Z = 19$) and calcium ($Z = 20$), the $3d$ subshell is gradually filled in the next 10 elements (Sc through Zn, for which $Z = 21$ to $Z = 30$), as shown in Table 8.7. The fourth and fifth periods each contain 18 elements; in each an ns subshell (2 electrons), the d subshell of the next lowest energy level, referred to as the $(n - 1)d$ subshell (10 electrons), and the np subshell (6 electrons) are being occupied.

The first three elements in the sixth period—cesium, barium, and lanthanum—like the comparable elements in the fifth period, represent filling of an ns subshell followed by the first electron to enter an $(n - 1)d$ subshell. However, with cerium ($Z = 58$; see Table 8.7) the first electron enters an $(n - 2)f$ subshell, and filling of this $4f$ subshell accounts for the 14 elements from cerium through lutetium ($Z = 71$). The seventh period similarly includes 14 elements in which the $(n - 2)f$ subshell is being occupied. To save space and to confine similar elements to the same vertical columns, the elements in which f subshells are being filled are usually placed at the bottom of the periodic table.

As shown in Figure 8.23, elements with electrons that have been added to subshells of the same type fall together in the periodic table. It is often convenient to refer to the elements of the s block, the d block, the p block, and the f block, as indicated in the figure. The d-block and f-block elements are also referred to as the *d-transition elements* and the *f-transition elements*, respectively (Section 9.2).

The elements at the end of each period are known as the **noble gases** (helium, neon, argon, krypton, xenon, and radon). In atoms of these elements, all outermost subshells that are occupied are completely filled and the outer configuration is $ns^2 np^6$, except for helium, which is $1s^2$. (See Table 8.7, where the noble gas configurations are shown against a color background.) All the noble gases are normally present in the atmosphere, although in such small concentration that they are sometimes called the "rare gases." Helium and argon were the first noble gases discovered, and at the time of

Table 8.7
Electron Configurations of Atoms of the Elements
Noble gases are shown against a color background. Incompletely filled orbitals are shown in color. Transition elements are shown against a gray background. Asterisks mark those elements with configurations that differ from the expected configurations.

Group	Atomic number	Symbol	Core	Electron configuration
	1	H		$1s^1$
	2	He		$1s^2$
	3	Li		$1s^2 2s^1$
	4	Be		$1s^2 2s^2$
	5	B		$1s^2 2s^2 2p^1$
	6	C		$1s^2 2s^2 2p^2$
	7	N		$1s^2 2s^2 2p^3$
	8	O		$1s^2 2s^2 2p^4$
	9	F		$1s^2 2s^2 2p^5$
	10	Ne		$1s^2 2s^2 2p^6$
	11	Na		$1s^2 2s^2 2p^6\ 3s^1$
	12	Mg		$1s^2 2s^2 2p^6\ 3s^2$
	13	Al		$1s^2 2s^2 2p^6\ 3s^2 3p^1$
	14	Si		$1s^2 2s^2 2p^6\ 3s^2 3p^2$
	15	P		$1s^2 2s^2 2p^6\ 3s^2 3p^3$
	16	S		$1s^2 2s^2 2p^6\ 3s^2 3p^4$
	17	Cl		$1s^2 2s^2 2p^6\ 3s^2 3p^5$
	18	Ar		$1s^2 2s^2 2p^6\ 3s^2 3p^6$
	19	K		$1s^2 2s^2 2p^6\ 3s^2 3p^6\ 4s^1$
	20	Ca		$1s^2 2s^2 2p^6\ 3s^2 3p^6\ 4s^2$
Transition Elements Period 4	21	Sc		$1s^2 2s^2 2p^6\ 3s^2 3p^6 3d^1\ 4s^2$
Transition Elements Period 4	22	Ti		$1s^2 2s^2 2p^6\ 3s^2 3p^6 3d^2\ 4s^2$
Transition Elements Period 4	23	V		$1s^2 2s^2 2p^6\ 3s^2 3p^6 3d^3\ 4s^2$
Transition Elements Period 4	24	Cr*		$1s^2 2s^2 2p^6\ 3s^2 3p^6 3d^5\ 4s^1$
Transition Elements Period 4	25	Mn		$1s^2 2s^2 2p^6\ 3s^2 3p^6 3d^5\ 4s^2$
Transition Elements Period 4	26	Fe		$1s^2 2s^2 2p^6\ 3s^2 3p^6 3d^6\ 4s^2$
Transition Elements Period 4	27	Co		$1s^2 2s^2 2p^6\ 3s^2 3p^6 3d^7\ 4s^2$
Transition Elements Period 4	28	Ni		$1s^2 2s^2 2p^6\ 3s^2 3p^6 3d^8\ 4s^2$
Transition Elements Period 4	29	Cu*		$1s^2 2s^2 2p^6\ 3s^2 3p^6 3d^{10} 4s^1$
Transition Elements Period 4	30	Zn		$1s^2 2s^2 2p^6\ 3s^2 3p^6 3d^{10} 4s^2$
	31	Ga		$1s^2 2s^2 2p^6\ 3s^2 3p^6 3d^{10} 4s^2 4p^1$
	32	Ge		$1s^2 2s^2 2p^6\ 3s^2 3p^6 3d^{10} 4s^2 4p^2$
	33	As		$1s^2 2s^2 2p^6\ 3s^2 3p^6 3d^{10} 4s^2 4p^3$
	34	Se		$1s^2 2s^2 2p^6\ 3s^2 3p^6 3d^{10} 4s^2 4p^4$
	35	Br		$1s^2 2s^2 2p^6\ 3s^2 3p^6 3d^{10} 4s^2 4p^5$
	36	Kr		$1s^2 2s^2 2p^6\ 3s^2 3p^6 3d^{10} 4s^2 4p^6$
	37	Rb	Krypton Core	$5s^1$
	38	Sr	Krypton Core	$5s^2$
Transition Elements Period 5	39	Y	Krypton Core	$4d^1\ 5s^2$
Transition Elements Period 5	40	Zr	Krypton Core	$4d^2\ 5s^2$
Transition Elements Period 5	41	Nb*	Krypton Core	$4d^4\ 5s^1$
Transition Elements Period 5	42	Mo*	Krypton Core	$4d^5\ 5s^1$
Transition Elements Period 5	43	Tc	Krypton Core	$4d^5\ 5s^2$
Transition Elements Period 5	44	Ru*	Krypton Core	$4d^7\ 5s^1$
Transition Elements Period 5	45	Rh*	Krypton Core	$4d^8\ 5s^1$
Transition Elements Period 5	46	Pd*	Krypton Core	$4d^{10}$
Transition Elements Period 5	47	Ag*	Krypton Core	$4d^{10}\ 5s^1$
Transition Elements Period 5	48	Cd	Krypton Core	$4d^{10}\ 5s^2$
	49	In	Krypton Core	$4d^{10}\ 5s^2 5p^1$
	50	Sn	Krypton Core	$4d^{10}\ 5s^2 5p^2$
	51	Sb	Krypton Core	$4d^{10}\ 5s^2 5p^3$
	52	Te	Krypton Core	$4d^{10}\ 5s^2 5p^4$
	53	I	Krypton Core	$4d^{10}\ 5s^2 5p^5$
	54	Xe	Krypton Core	$4d^{10}\ 5s^2 5p^6$
	55	Cs	Krypton Core	$4d^{10}\ 5s^2 5p^6\ 6s^1$
	56	Ba	Krypton Core	$4d^{10}\ 5s^2 5p^6\ 6s^2$
Lanthanides	57	La*	Krypton Core	$4d^{10}\ 5s^2 5p^6 5d^1\ 6s^2$
Lanthanides	58	Ce*	Krypton Core	$4d^{10} 4f^1\ 5s^2 5p^6 5d^1\ 6s^2$
Lanthanides	59	Pr	Krypton Core	$4d^{10} 4f^3\ 5s^2 5p^6\ 6s^2$
Lanthanides	60	Nd	Krypton Core	$4d^{10} 4f^4\ 5s^2 5p^6\ 6s^2$
Lanthanides	61	Pm	Krypton Core	$4d^{10} 4f^5\ 5s^2 5p^6\ 6s^2$
Lanthanides	62	Sm	Krypton Core	$4d^{10} 4f^6\ 5s^2 5p^6\ 6s^2$
Lanthanides	63	Eu	Krypton Core	$4d^{10} 4f^7\ 5s^2 5p^6\ 6s^2$
Lanthanides	64	Gd*	Krypton Core	$4d^{10} 4f^7\ 5s^2 5p^6 5d^1\ 6s^2$
Lanthanides	65	Tb	Krypton Core	$4d^{10} 4f^9\ 5s^2 5p^6\ 6s^2$
Lanthanides	66	Dy	Krypton Core	$4d^{10} 4f^{10} 5s^2 5p^6\ 6s^2$
Lanthanides	67	Ho	Krypton Core	$4d^{10} 4f^{11} 5s^2 5p^6\ 6s^2$
Lanthanides	68	Er	Krypton Core	$4d^{10} 4f^{12} 5s^2 5p^6\ 6s^2$
Lanthanides	69	Tm	Krypton Core	$4d^{10} 4f^{13} 5s^2 5p^6\ 6s^2$
Lanthanides	70	Yb	Krypton Core	$4d^{10} 4f^{14} 5s^2 5p^6\ 6s^2$
Lanthanides	71	Lu	Krypton Core	$4d^{10} 4f^{14} 5s^2 5p^6 5d^1\ 6s^2$
Transition Elements Period 6	72	Hf	Krypton Core	$4d^{10} 4f^{14} 5s^2 5p^6 5d^2\ 6s^2$
Transition Elements Period 6	73	Ta	Krypton Core	$4d^{10} 4f^{14} 5s^2 5p^6 5d^3\ 6s^2$
Transition Elements Period 6	74	W	Krypton Core	$4d^{10} 4f^{14} 5s^2 5p^6 5d^4\ 6s^2$
Transition Elements Period 6	75	Re	Krypton Core	$4d^{10} 4f^{14} 5s^2 5p^6 5d^5\ 6s^2$
Transition Elements Period 6	76	Os	Krypton Core	$4d^{10} 4f^{14} 5s^2 5p^6 5d^6\ 6s^2$
Transition Elements Period 6	77	Ir	Krypton Core	$4d^{10} 4f^{14} 5s^2 5p^6 5d^7\ 6s^2$
Transition Elements Period 6	78	Pt*	Krypton Core	$4d^{10} 4f^{14} 5s^2 5p^6 5d^9\ 6s^1$
Transition Elements Period 6	79	Au*	Krypton Core	$4d^{10} 4f^{14} 5s^2 5p^6 5d^{10}\ 6s^1$
Transition Elements Period 6	80	Hg	Krypton Core	$4d^{10} 4f^{14} 5s^2 5p^6 5d^{10}\ 6s^2$
	81	Tl	Krypton Core	$4d^{10} 4f^{14} 5s^2 5p^6 5d^{10}\ 6s^2 6p^1$
	82	Pb	Krypton Core	$4d^{10} 4f^{14} 5s^2 5p^6 5d^{10}\ 6s^2 6p^2$
	83	Bi	Krypton Core	$4d^{10} 4f^{14} 5s^2 5p^6 5d^{10}\ 6s^2 6p^3$
	84	Po	Krypton Core	$4d^{10} 4f^{14} 5s^2 5p^6 5d^{10}\ 6s^2 6p^4$
	85	At	Krypton Core	$4d^{10} 4f^{14} 5s^2 5p^6 5d^{10}\ 6s^2 6p^5$
	86	Rn	Krypton Core	$4d^{10} 4f^{14} 5s^2 5p^6 5d^{10}\ 6s^2 6p^6$
	87	Fr	Krypton Core	$4d^{10} 4f^{14} 5s^2 5p^6 5d^{10}\ 6s^2 6p^6\ 7s^1$
	88	Ra	Krypton Core	$4d^{10} 4f^{14} 5s^2 5p^6 5d^{10}\ 6s^2 6p^6\ 7s^2$
Actinides	89	Ac*	Krypton Core	$4d^{10} 4f^{14} 5s^2 5p^6 5d^{10}\ 6s^2 6p^6 6d^1 7s^2$
Actinides	90	Th*	Krypton Core	$4d^{10} 4f^{14} 5s^2 5p^6 5d^{10}\ 6s^2 6p^6 6d^2 7s^2$
Actinides	91	Pa*	Krypton Core	$4d^{10} 4f^{14} 5s^2 5p^6 5d^{10} 5f^2\ 6s^2 6p^6 6d^1 7s^2$
Actinides	92	U*	Krypton Core	$4d^{10} 4f^{14} 5s^2 5p^6 5d^{10} 5f^3\ 6s^2 6p^6 6d^1 7s^2$
Actinides	93	Np*	Krypton Core	$4d^{10} 4f^{14} 5s^2 5p^6 5d^{10} 5f^4\ 6s^2 6p^6 6d^1 7s^2$
Actinides	94	Pu	Krypton Core	$4d^{10} 4f^{14} 5s^2 5p^6 5d^{10} 5f^6\ 6s^2 6p^6\ 7s^2$
Actinides	95	Am	Krypton Core	$4d^{10} 4f^{14} 5s^2 5p^6 5d^{10} 5f^7\ 6s^2 6p^6\ 7s^2$
Actinides	96	Cm*	Krypton Core	$4d^{10} 4f^{14} 5s^2 5p^6 5d^{10} 5f^7\ 6s^2 6p^6 6d^1 7s^2$
Actinides	97	Bk	Krypton Core	$4d^{10} 4f^{14} 5s^2 5p^6 5d^{10} 5f^9\ 6s^2 6p^6\ 7s^2$
Actinides	98	Cf	Krypton Core	$4d^{10} 4f^{14} 5s^2 5p^6 5d^{10} 5f^{10} 6s^2 6p^6\ 7s^2$
Actinides	99	Es	Krypton Core	$4d^{10} 4f^{14} 5s^2 5p^6 5d^{10} 5f^{11} 6s^2 6p^6\ 7s^2$
Actinides	100	Fm	Krypton Core	$4d^{10} 4f^{14} 5s^2 5p^6 5d^{10} 5f^{12} 6s^2 6p^6\ 7s^2$
Actinides	101	Md	Krypton Core	$4d^{10} 4f^{14} 5s^2 5p^6 5d^{10} 5f^{13} 6s^2 6p^6\ 7s^2$
Actinides	102	No	Krypton Core	$4d^{10} 4f^{14} 5s^2 5p^6 5d^{10} 5f^{14} 6s^2 6p^6\ 7s^2$
Actinides	103	Lr	Krypton Core	$4d^{10} 4f^{14} 5s^2 5p^6 5d^{10} 5f^{14} 6s^2 6p^6 6d^1 7s^2$
	104	Unq	Krypton Core	$4d^{10} 4f^{14} 5s^2 5p^6 5d^{10} 5f^{14} 6s^2 6p^6 6d^2 7s^2$
	105	Unp	Krypton Core	$4d^{10} 4f^{14} 5s^2 5p^6 5d^{10} 5f^{14} 6s^2 6p^6 6d^3 7s^2$
	106	Unh	Krypton Core	$4d^{10} 4f^{14} 5s^2 5p^6 5d^{10} 5f^{14} 6s^2 6p^6 6d^4 7s^2$
	107	Uns	Krypton Core	$4d^{10} 4f^{14} 5s^2 5p^6 5d^{10} 5f^{14} 6s^2 6p^6 6d^5 7s^2$
	108	Uno	Krypton Core	$4d^{10} 4f^{14} 5s^2 5p^6 5d^{10} 5f^{14} 6s^2 6p^6 6d^6 7s^2$
	109	Une	Krypton Core	$4d^{10} 4f^{14} 5s^2 5p^6 5d^{10} 5f^{14} 6s^2 6p^6 6d^7 7s^2$

their discovery there was no place for them in the periodic table. William Ramsay, a Scottish chemist, was convinced that they represented a new periodic table group. By carefully separating the components of liquid air, in 1898 Ramsay and his assistant, William Travers, isolated from the atmosphere and identified with the then-new spectroscope the additional noble gases—krypton, neon, and xenon.

As a group the noble gases are the least reactive of all the elements. This **chemical stability**—resistance to chemical change—is credited to the completely filled outer *s* and *p* subshells of these elements.

With the element following each noble gas, electrons begin to fill the next-highest main energy level (e.g., Na, $1s^2 2s^2 2p^6 3s^1$ follows Ne, $1s^2 2s^2 2p^6$). As atomic number and nuclear charge increase, the electrons in the lower filled energy levels are drawn closer to the nucleus by its charge. These inner electrons become unavailable for interaction with the surroundings and have a decreasing influence on properties. It is, therefore, the outer electron configurations that are of greatest interest.

To recognize the outer configuration of an element from its position in the periodic table is most useful. For example, consider phosphorus. Because phosphorus is in the third period, its inner, or *core,* electrons have the same configuration as the last element in the second period—neon. Writing electron configurations can be simplified by using the symbol of the appropriate noble gas to represent the configuration of the inner electrons. Phosphorus is the fifth element in the third period. Moving across the period from left to right (see Figure 8.22) shows that phosphorus has two 3*s* electrons and, because it is the third element from the left in the *p* block, it has three *p* electrons. Using the symbol for neon in square brackets to represent the inner core electrons, the configuration of phosphorus is written as follows

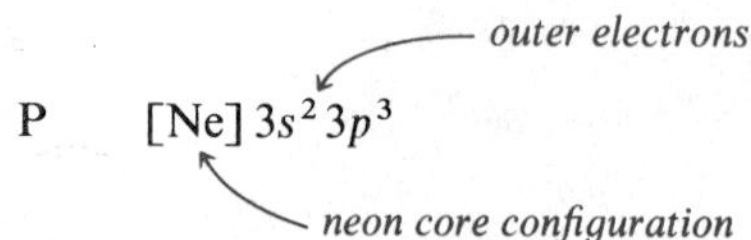

As we have noted, certain *d*-block elements (those marked with asterisks in Table 8.7) have electron configurations that differ slightly from the expected configurations. However, for all *d*-block elements, *the total number of s plus d electrons is equal to two (for the s subshell) plus one electron for each element counted from the left in the d block.*

EXAMPLE 8.3 Writing Electron Configurations

What are the electron configurations of rubidium, molybdenum, and iodine? Write configurations using the noble gas symbols for the core electrons.

These three elements are all in the fifth period, and therefore all have an inner core of electrons in the krypton configuration. Rubidium is the first element in the *s* block and therefore has one *s* electron

$$\textbf{Rb} \quad [Kr]5s^1$$

Molybdenum is in the *d* block; because it is the fourth element from the left the sum of its 5*s* and 4*d* electrons must be six. The "ideal" configuration is

$$\textbf{Mo} \quad [Kr]4d^4 5s^2$$

(The actual configuration is $[Kr]4d^5 5s^1$.) Iodine is the fifth *p*-block element in the fifth period. As such, it has completely filled *d* sublevel and the following configuration

$$\textbf{I} \quad [Kr]4d^{10} 5s^2 5p^5$$

EXERCISE Write electron configurations using the noble gas symbols for the core electrons for (a) Se, (b) Fe, and (c) Hg.

Answers (a) $[Ar]3d^{10} 4s^2 4p^4$, (b) $[Ar]3d^6 4s^2$, (c) $[Xe]4f^{14} 5d^{10} 6s^2$.

EXAMPLE 8.4 Writing Electron Configurations

Write the electron configurations for elements A, B, and C.

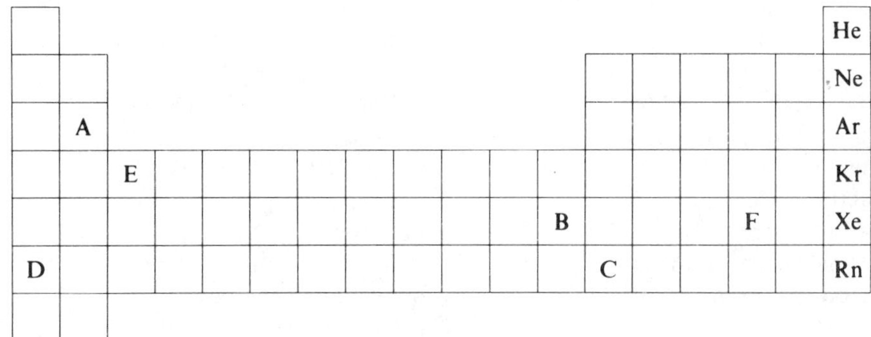

Element A is a third-period element and it is second in the *s* block. It has two *s* level electrons, giving

$$\text{A} \quad [\text{Ne}]3s^2$$

As a fifth-period, *d*-block element, B has electrons in the 5*s* and 4*d* subshells (note that the *d* electrons are in the next lowest main energy level). Because it is the last *d*-block element, it has 10 *d* level electrons, giving a configuration of

$$\text{B} \quad [\text{Kr}]4d^{10}5s^2$$

Element C is a sixth-period element and the first in the *p* block. It has two 6*s* electrons, ten 5*d* electrons, and one 6*p* electron, giving

$$\text{C} \quad [\text{Xe}]5d^{10}6s^2 6p^1$$

EXERCISE Write the electron configurations for the elements (a) D, (b) E, and (c) F, given in the above periodic table.

Answers (a) $[\text{Xe}]6s^1$, (b) $[\text{Ar}]3d^1 4s^2$, (c) $[\text{Kr}]4d^{10}5s^2 5p^4$.

SUMMARY OF SECTIONS 8.11–8.13

The electron configuration of an atom is the distribution among the subshells of all the electrons in the atom. The notation for each occupied subshell is of the general form nl^x, where n is the principal quantum number, l is the subshell quantum number (shown by *s, p, d, f*), and x is the number of electrons in the subshell.

Three principles govern the sequence in which orbitals are filled as atomic number increases: (1) the Pauli exclusion principle: no two electrons can have the same four quantum numbers; (2) the principle of lowest energy: electrons occupy the lowest-energy orbitals available; (3) Hund's principle: orbitals of equal energy are each occupied by a single electron before the second electron (with opposite spin quantum number) enters the orbital.

In the periodic table, atoms of elements in the same vertical column have identical or almost identical outer electron configurations and similar properties. The length of each horizontal row is determined by the number of electrons that can be accommodated in the subshells being occupied. Each period ends with a noble gas. The chemical stability of the noble gases is attributed to their filled outer subshells ($1s^2$ for He; ns^2np^6 for others).

In the *s*-block elements (see Figure 8.23), the outermost configurations are ns^1 and ns^2. In the fourth through the sixth periods, there are 10 *d*-block elements in which $(n-1)d$ subshells are being occupied. For atoms of these elements, the number of outer electrons equals two plus one electron for each family counted from the left in the *d* block. In the six groups of the *p* block at the right, np subshells are being occupied. Fourteen *f*-block elements, in which $(n-2)f$ subshells are being filled, belong in the sixth and seventh periods, but are usually placed at the bottom of the periodic table.

SIGNIFICANT TERMS (with section references)

absorption spectrum (Aside 8.1)
angular momentum (8.7)
atomic orbital (8.8)
band spectra (Aside 8.1)
chemical stability (8.13)
continuous spectra (Aside 8.1)
diamagnetism (8.12)
diffraction (8.2)
electron configuration (8.11)
electronic structure (8.1)
emission spectrum (Aside 8.1)
excited states (8.7)
frequency (8.2)
ground state (8.7)
group, family (periodic table) (8.13)
Heisenberg uncertainty principle (8.5)
line spectra (Aside 8.1)
momentum (8.5)
noble gases (8.13)
outer electron configurations (8.13)
paramagnetism (8.12)
period (periodic table) (8.13)
periodic table (8.1)
photoelectric effect (8.3)
photon (8.3)
quantized (8.3)
quantum (8.3)
quantum mechanics (8.5)
quantum number (8.3)
quantum theory (8.3)
spectrometer (Aside 8.1)
spectroscopy (Aside 8.1)
wave number (8.2)
wavelength (8.2)

QUESTIONS AND PROBLEMS

Quantum Theory—Light

8.1 What two phenomena that could not be explained by classical physics were successfully explained by quantum theory?

8.2 What made it difficult for early scientists to accept the particle nature of light?

8.3 Excited lithium ions emit radiation at a wavelength of 670.8 nm in the visible range of the spectrum. [This characteristic red color is often used as a qualitative analysis test for the presence of Li^+.] Calculate (a) the frequency, (b) the wave number, (c) the energy of a photon of this radiation.

8.4 Carbon dioxide absorbs energy at $\nu = 2.001 \times 10^{13}\ s^{-1}$, $4.017 \times 10^{13}\ s^{-1}$, and $7.043 \times 10^{13}\ s^{-1}$. Calculate the wavelengths for these absorptions. In what spectral range do these absorptions occur?

8.5 Ozone in the upper atmosphere absorbs ultraviolet radiation which induces the following chemical reaction:

$$O_3(g) \longrightarrow O_2(g) + O(g)$$

What is the energy of a 340 nm photon that is absorbed? What is the energy of a mole of these photons?

8.6 During photosynthesis, chlorophyll-*a* absorbs light of wavelength 440 nm and emits light of wavelength 670 nm. What is the energy available for photosynthesis from the absorption–emission of a mole of photons?

8.7 Assume that 10^{-17} J of light energy is needed by the interior of the human eye to "see" an object. How many photons of green light (wavelength = 550 nm) are needed to generate this minimum energy?

8.8 Water absorbs microwave radiation of wavelength 3 mm. How many photons are needed to raise the temperature of a cup of water (250 g) from 25 °C to 85 °C in a microwave oven using this radiation? This specific heat of water is 4.184 J/K g.

8.9 Describe what happens to a photoelectric surface when a low-intensity light of a frequency below the threshold energy shines on the surface. What happens if the intensity of the low-frequency light is increased by a factor of 1000? What happens if a low-intensity light of a frequency *above* the threshold frequency is used? What happens if the intensity of this light is increased?

8.10 The energy of a photon of light must be 2.26 eV in order to produce the photoelectric effect in potassium. What is the wavelength of a photon having this energy? (1 eV = 1.602 $\times 10^{-19}$ J.) Would a photon having $\lambda = 700$ nm produce the photoelectric effect in potassium?

Quantum Theory—Particles

8.11 Why are wave properties important for particles that have very small masses? Why can we neglect the wave properties of objects that have large masses?

8.12 State the Heisenberg uncertainty principle. Is this principle significant at the atomic level? Why don't we worry about this principle when discussing everyday phenomena?

8.13 What is the wavelength corresponding to a neutron of mass 1.67×10^{-27} kg moving at 2200 m/s?

8.14* The energy of a photon in the x-ray region of the spectrum is 7×10^{-16} J. According to de Broglie's equation, what is the mass of this photon?

Quantum Theory—Spectroscopy

8.15 Name the five basic parts of a spectrometer. Briefly describe the function of each.

8.16* On a relative scale, very little energy is needed to excite the rotational motion of a molecule, an intermediate amount of energy is needed to excite vibrations in a molecule, and the largest amount of energy is needed to excite electrons to higher energy levels in a molecule. Which type(s) of transitions will be observed in (a) microwave spectroscopy, (b) infrared spectroscopy, (c) visible-ultraviolet spectroscopy?

8.17 Briefly discuss the interaction that occurs between radiation and matter during the recording of an emission spectrum.

8.18 Prepare a sketch similar to Figure 8.13 that shows a ground energy state and two excited energy states. Using vertical arrows, indicate the transitions which would correspond to the absorption spectrum for this system.

8.19 Briefly discuss how helium was first discovered. The *Handbook of Chemistry and Physics* is an example of a suitable reference for the necessary information.

8.20 Each of the ions formed by the elements in the lithium family of the periodic table has a spectral line in the visible region of the spectrum which can be used to identify the element. Using the following data for the most intense lines, predict what color would be observed (violet, 400–450 nm; blue, 450–510 nm; green, 510–550 nm; yellow, 550–590 nm; orange, 590–620 nm; red, 620–700 nm); (a) Li, $\lambda = 6708$ Å for $2p \longrightarrow 2s$; (b) Na, $\bar{\nu} = 16980\ \text{cm}^{-1}$ for $3p \longrightarrow 3s$; (c) K, $\nu = 3.90 \times 10^{14}$ Hz for $4p \longrightarrow 4s$ and 7.41×10^{14} Hz for $5p \longrightarrow 4s$; (d) Rb, $\lambda = 7.9 \times 10^{-7}$ m for $5p \longrightarrow 5s$ and 4.2×10^{-7} m for $6p \longrightarrow 5s$; (e) Cs, $\nu = 3.45 \times 10^{14}$ Hz for $6p \longrightarrow 6s$ and 6.53×10^{14} Hz for $7p \longrightarrow 6s$.

Quantum Theory and the Atom—Atomic Spectroscopy

8.21 Use Equation 8.4 to calculate $\bar{\nu}$ for atomic hydrogen as an electron changes from the $n = 3$ level to the $n = 1$ level. To what wavelength does this value of $\bar{\nu}$ correspond? Is this wavelength in the visible region of the spectrum?

8.22 Calculate the wave numbers of the first six lines in the Lyman series for atomic hydrogen.

8.23 Choose the electronic transition that emits more energy than the $n = 3$ to $n = 1$ transition: (a) $n = 1$ to $n = 25$, (b) $n = 3$ to $n = 2$, (c) $n = 5$ to $n = 3$, (d) $n = 4$ to $n = 1$.

8.24* An atom absorbs light of wavelength λ_1 and emits light of wavelength λ_2 followed by a second emission of light of wavelength λ_3 as it returns to its ground state. What can be concluded about the relative values of the three wavelengths?

8.25* As an electron is added by a hydrogen ion, $H^+(g) + e^- \longrightarrow H(g)$, light energy is emitted. Assuming that the resulting hydrogen atom is in the ground electronic state after the capture, what is the wavelength of the emitted radiation?

8.26* The following lines were observed in the spectrum of atomic hydrogen: $5331.5\ \text{cm}^{-1}$, $7799.3\ \text{cm}^{-1}$, $9139.8\ \text{cm}^{-1}$, and $9948.1\ \text{cm}^{-1}$. These lines all belong to the same spectroscopic series. Identify this series.

Quantum Theory and the Atom—Bohr Theory

8.27 State the three basic principles that make up the Bohr theory. Why is his theory no longer used as the working model of the atom?

8.28 How does our solar system resemble and differ from a Bohr atom having nine electrons (the fluorine atom)?

8.29 The radii of the orbits of a Bohr hydrogen atom are given by

$$r = (0.0529\ \text{nm})n^2$$

where n is the principal quantum number. Calculate the radii of the first five orbits and prepare a scale drawing. Comment on the relative distances.

8.30 The energy of the Bohr orbit having quantum number n is

$$E_n = -\frac{2.178720 \times 10^{-18}\ \text{J}}{n^2}$$

(The negative sign simply means that the atom is more stable than a proton and an electron separated by an infinite distance.) (a) Calculate the energies of the first five orbits. (b) Calculate the energy change as an electron changes from $n = 5$ to $n = 2$, $n = 4$ to $n = 2$, and $n = 3$ to $n = 2$. (The negative sign simply means that energy is released.) (c) Convert these energies to wave numbers and compare your answers to Figure 8.12. To which series do these spectral lines belong?

Quantum Theory and the Atom—Quantum Numbers and Orbitals

8.31 What are the permitted values of the principal quantum number? What is the range of values of this quantum number for the ground states of the known elements? What does $n = \infty$ mean? What physical interpretation can be given for n?

8.32 Your laboratory partner asks you to explain the subshell quantum number to him. You decide that the important things to cover include symbol, permitted values, and a physical interpretation. What do you tell him?

8.33 Briefly discuss the orbital quantum number. What are the permitted values of this quantum number?

8.34 What values of the spin quantum number are permitted?

8.35 What values of the subshell quantum number correspond to the (a) d, (b) f, (c) s, (d) p, (e) g subshells?

8.36 Write the subshell notations that correspond to (a) $n = 2$, $l = 0$; (b) $n = 4$, $l = 2$; (c) $n = 7$, $l = 0$; (d) $n = 5$, $l = 3$.

8.37 What values can m_l take for (a) a $4d$ orbital, (b) a $1s$ orbital, (c) a $3p$ orbital?

8.38 How many orbitals are in the (a) s, (b) p, (c) d, (d) f subshells? How many electrons can each orbital hold? How many electrons can each subshell hold?

8.39 Choose the set of quantum numbers that correctly describes an electron in an atom: (a) $n = 4$, $l = 4$, $m_l = 3$, $m_s = +\frac{1}{2}$; (b) $n = 3$, $l = 2$, $m_l = -3$, $m_s = -\frac{1}{2}$; (c) $n = 0$, $l = 0$, $m_l = 0$, $m_s = +\frac{1}{2}$; (d) $n = 3$, $l = 1$, $m_l = 0$, $m_s = -\frac{1}{2}$.

8.40 Choose the set of quantum numbers that represents the electron of the lowest energy: (a) $n = 2$, $l = 0$, $m_l = 0$, $s = -\frac{1}{2}$;

(b) $n = 2$, $l = 1$, $m_l = 0$, $s = +\frac{1}{2}$; (c) $n = 4$, $l = 0$, $m_l = 0$, $s = +\frac{1}{2}$: (d) $n = 4$, $l = 0$, $m_l = 0$, $s = -\frac{1}{2}$.

8.41 Using Figure 8.17, place the following subshells in order of increasing distance from the nucleus (use the maximum in the curve for the basis of your ordering): 1*s*, 2*s*, 2*p*, 3*s*, 3*p*, and 3*d*.

8.42 Prepare sketches of the (a) 1*s*, (b) $2p_x$, (c) $2p_y$, (d) $2p_z$, (e) $3d_{z^2}$, (f) $3d_{x^2-y^2}$, (g) $3d_{xy}$, (h) $3d_{yz}$ (i) $3d_{xz}$ atomic orbitals.

8.43 The Bohr theory places the 1*s* electron in a hydrogen atom in a fixed orbit having a radius of 0.0529 nm. Discuss this result based on the plots given in Figure 8.17.

8.44* Plot the atomic radii (see Table III.2 of Appendix III) for the elements from hydrogen through calcium against atomic number. Why is there such a large jump in the values between helium and lithium, neon and sodium, and argon and potassium?

Electron Configurations

8.45 State the Pauli exclusion principle. Do any of the following electron configurations violate this rule: (a) $1s^2$, (b) $1s^2 2p^1$, (c) $1s^3$? Explain.

8.46 State the lowest energy principle. Do any of the following electron configurations violate this rule: (a) $1s^1 2s^1$, (b) $1s^2 2p^1$, (c) $1s^2 2s^2 2p_x^1 2p_y^1$? Explain.

8.47 State the Hund principle. Do any of the following electron configurations violate this rule: (a) $1s^2$, (b) $1s^2 2s^2 2p_x^2$, (c) $1s^2 2s^2 2p_x^1 2p_y^1$, (d) $1s^2 2s^2 2p_x^1 2p_z^1$, (e) $1s^2 2s^2 2p_x^2 2p_y^1 2p_z^1$? Explain.

8.48 Classify each of the following atomic electron configurations as (i) a ground state, (ii) an excited state, or (iii) a forbidden state:

(a) $1s^2 2s^2 2p^5 3s^1$,
(b) [Kr] $4d^{10} 5s^3$,
(c) $1s^2 2s^2 2p^6 3s^2 3p^6 3d^8 4s^2$,
(d) $1s^2 2s^2 2p^6 3s^2 3p^6 3d^1$,
(e) $1s^2 2s^2 2p^{10} 3s^2 3p^5$.

8.49 Using the rules given in Section 8.12, write reasonable ground state electron configurations for atoms of (a) K, (b) Sc, (c) Si, (d) F, (e) U, (f) Ag, (g) Mg, (h) Fe, (i) Pr. Compare your answers to the known configurations given in Table 8.7. Which of these elements are paramagnetic?

8.50 Repeat Question 8.49 for (a) C, (b) Cr, (c) Zn, (d) P, (e) Ne, (f) Au, (g) Sn, (h) Ga, (i) W.

8.51 Which elements are represented by the following electron configurations

(a) $1s^2 2s^2 2p^6 3s^2 3p^6 3d^{10} 4s^2 4p^3$
(b) [Kr] $4d^{10} 4f^{14} 5s^2 5p^6 5d^{10} 5f^{14} 6s^2 6p^6 6d^2 7s^2$
(c) [Kr] $4d^{10} 4f^{14} 5s^2 5p^6 5d^{10} 6s^2 6p^4$
(d) [Kr] $4d^5 5s^2$
(e) $1s^2 2s^2 2p^6 3s^2 3p^6 3d^3 4s^2$

8.52 Repeat Question 8.51 for (a) $1s^2 2s^2 2p^6 3s^2 3p^6 3d^5 4s^1$, (b) [Kr] $4d^{10} 4f^{14} 5s^2 5p^6 5d^{10} 6s^2 6p^1$, (c) $1s^2 2s^2 2p^6 3s^2 3p^6$, (d) [Kr] $4d^{10} 4f^{14} 5s^2 5p^6 5d^{10} 6s^2 6p^6 7s^2$.

8.53* Write the ground state electron configuration for the Cu atom using the order of filling predicted by the lowest energy principle. Because configurations of d^5 and d^{10} are very favorable, one of the 4*s* electrons is often promoted to the 3*d* subshell. Write the electron configuration for this alternative arrangement of electrons. Do the different configurations predict a difference in the magnetic properties of Cu?

8.54* Repeat Question 8.53 for a Cr atom.

8.55 How many unpaired electrons do atoms of each of the following elements contain: (a) Ca, (b) W, (c) Ge, (d) Ce, (e) Tc?

8.56 Repeat Question 8.55 for (a) Nd, (b) Pb, (c) V, (d) Cl, (e) Kr.

8.57 Write the ground state electron configurations for elements A–E.

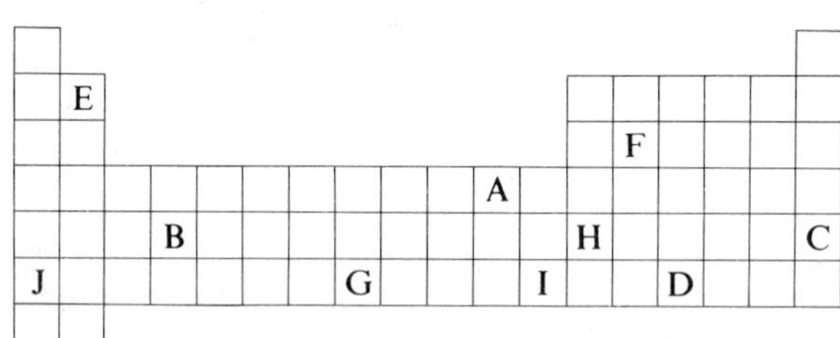

8.58 Repeat Question 8.57 for elements F–J.

8.59 Predict and write the electron configuration for the unknown element 119. In what family of the periodic table would it fall?

CHAPTER

9

Periodic Perspective I: Atoms and Ions

CLASSIFICATION OF THE ELEMENTS

Information about the elements and their compounds can be organized in a variety of ways. In Chapter 6 the study of descriptive chemistry was begun with the introduction of four simple types of chemical reactions. Recognizing reaction types is one of the ways in which information about the properties of the elements and their compounds can be organized. Another is by relating the general properties of elements and their compounds to positions in the periodic table. In this chapter, we begin to examine this latter aspect of descriptive chemistry.

The elements are classified in two major ways relative to the periodic table. In the first classification, every element is either a representative element or a transition element. In the second, every element is a metal, a semiconducting element, or a nonmetal.

9.1 REPRESENTATIVE ELEMENTS

a. Representative Element Groups The elements for which all inner subshells are fully occupied and the outer *s* and *p* subshells are filling are called **representative elements** (Figure 9.1)—the *s*-block elements on the left and the *p*-block elements on the right. The representative elements are also often referred to as **main group elements.**

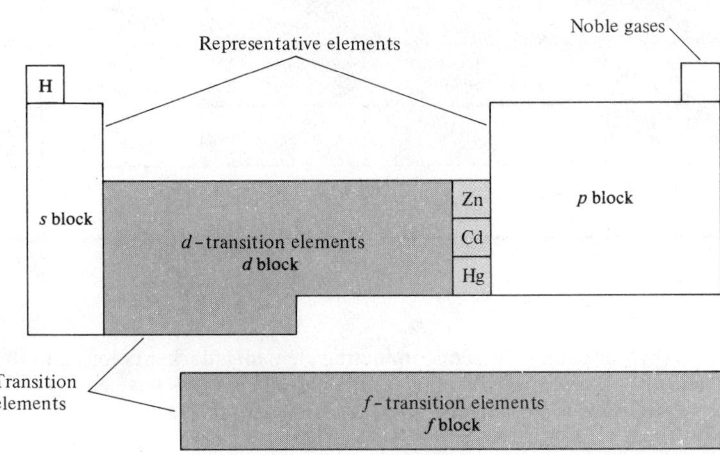

Figure 9.1
Transition and Representative Elements
The locations of transition elements are shown in gray, the representative elements in white.

The groups of representative elements, with the exception of the noble gases, are sometimes assigned Roman numerals, starting at the left in the periodic table with I for lithium through francium. (Note that hydrogen is a unique element and is not included in Representative Group I.) The result of this numbering, illustrated in Figure 9.2, is that the group number is equal to the number of electrons in the highest occupied energy levels of atoms of the elements in each group. Representative Groups I and II include those with one and two outer *s* electrons, respectively. The elements in vertical columns in the periodic table are also referred to collectively as families, named by the first element in the family—Representative Groups I and II are the lithium and beryllium families. Representative Groups III–VII (the boron family through the fluorine family), include elements with two outer *s* electrons plus one to five outer *p* electrons in their atoms.

The group configurations are given in Figure 9.2, where *n* represents the quantum number of the highest occupied main energy level. For example, atoms of the Representative Group III elements each have three outer electrons in the following configurations: B, $2s^2 2p^1$; Al, $3s^2 3p^1$; Ga, $4s^2 4p^1$; In, $5s^2 5p^1$; Tl, $6s^2 6p^1$. Take a moment to study the configurations of the elements in the other groups of representative elements.

Several groups of representative elements (in addition to the noble gases) have distinctive names. The elements of the first two groups are known collectively as the **alkali metals** (the lithium family) and the **alkaline earth metals** (the beryllium family). The elements of Representative Group VII, the fluorine family, are virtually always referred to as the **halogens.**

b. Valence Electrons and Lewis Symbols The electrons available to take part in chemical bonding are known as **valence electrons.** In chemical compounds (as is discussed in Chapter 10) valence electrons are either lost, gained, or shared between atoms, with the result that atoms are held together by the attraction between their nuclei and the electrons.

For atoms of representative elements, the outermost electrons are the valence electrons. Therefore, with groups of elements numbered as in Figure 9.2, the total

	I (except H) ns^1	II ns^2	III ns^2np^1	IV ns^2np^2	V ns^2np^3	VI ns^2np^4	VII ns^2np^5	Noble gases ns^2np^6 (except He)
Period 1	$_{1}$H $1s^1$							$_{2}$He $1s^2$
Period 2	$_{3}$Li $2s^1$	$_{4}$Be $2s^2$	$_{5}$B $2s^2 2p^1$	$_{6}$C $2s^2 2p^2$	$_{7}$N $2s^2 2p^3$	$_{8}$O $2s^2 2p^4$	$_{9}$F $2s^2 2p^5$	$_{10}$Ne $2s^2 2p^6$
Period 3	$_{11}$Na $3s^1$	$_{12}$Mg $3s^2$	$_{13}$Al $3s^2 3p^1$	$_{14}$Si $3s^2 3p^2$	$_{15}$P $3s^2 3p^3$	$_{16}$S $3s^2 3p^4$	$_{17}$Cl $3s^2 3p^5$	$_{18}$Ar $3s^2 3p^6$
Period 4	$_{19}$K $4s^1$	$_{20}$Ca $4s^2$	$_{31}$Ga $3d^{10} 4s^2 4p^1$	$_{32}$Ge $3d^{10} 4s^2 4p^2$	$_{33}$As $3d^{10} 4s^2 4p^3$	$_{34}$Se $3d^{10} 4s^2 4p^4$	$_{35}$Br $3d^{10} 4s^2 4p^5$	$_{36}$Kr $3d^{10} 4s^2 4p^6$
Period 5	$_{37}$Rb $5s^1$	$_{38}$Sr $5s^2$	$_{49}$In $4d^{10} 5s^2 5p^1$	$_{50}$Sn $4d^{10} 5s^2 5p^2$	$_{51}$Sb $4d^{10} 5s^2 5p^3$	$_{52}$Te $4d^{10} 5s^2 5p^4$	$_{53}$I $4d^{10} 5s^2 5p^5$	$_{54}$Xe $4d^{10} 5s^2 5p^6$
Period 6	$_{55}$Cs $6s^1$	$_{56}$Ba $6s^2$	$_{81}$Tl $4f^{14} 5d^{10} 6s^2 6p^1$	$_{82}$Pb $4f^{14} 5d^{10} 6s^2 6p^2$	$_{83}$Bi $4f^{14} 5d^{10} 6s^2 6p^3$	$_{84}$Po $4f^{14} 5d^{10} 6s^2 6p^4$	$_{85}$At $4f^{14} 5d^{10} 6s^2 6p^5$	$_{86}$Rn $4f^{14} 5d^{10} 6s^2 6p^6$
Period 7	$_{87}$Fr $7s^1$	$_{88}$Ra $7s^2$	Transition elements, all metals, intervene					
	s-block elements		*p*-block elements					

Figure 9.2
The Representative Elements and the Noble Gases
The heavy lines separate the elements classified as metals (lighter color), the semiconducting elements (darker color), and the nonmetals (white). First and second period elements have only *s*, or *s* and *p* orbitals. Third-period elements have empty $3d$ orbitals available. Fourth and higher period elements have $(n-1)d$ electrons close in energy to the ns electrons. (The valence electrons are shown in color.)

Table 9.1
Lewis Symbols for the Second-Period Elements

Element	Representative group number	Outer electron configuration	Number of outer electrons
Li·	I	$2s^1$	1
Be:	II	$2s^2$	2
·B:	III	$2s^2 2p^1$	3
·C:	IV	$2s^2 2p^2$	4
·N:	V	$2s^2 2p^3$	5
·O:	VI	$2s^2 2p^4$	6
·F:	VII	$2s^2 2p^5$	7
:Ne:	VIII	$2s^2 2p^6$	8

number of valence electrons in an atom of a representative element is equal to the group number of that element.

Lewis symbols provide a convenient notation for outer electron configurations. In a **Lewis symbol** the outer electrons are indicated by dots arranged around the atomic symbol. For example, the Lewis symbol for an atom of nitrogen, which is from Representative Group V and has five valence electrons, is

$_7N$	$1s^2 2s^2 2p^3$	·N:
	complete configuration	*Lewis symbol—dots represent the five valence electrons*

Table 9.1 gives the Lewis symbols for atoms of the second-period elements.

EXAMPLE 9.1 Valence Electrons of Representative Elements

How many valence electrons are present in (a) antimony, (b) bromine, and (c) barium atoms? Write the outer electron configurations and the Lewis symbols for atoms of these elements.

(a) Antimony is in the nitrogen family, which is Representative Group V, and therefore each atom has **five valence electrons:** the two *s* and the three *p* electrons in the $n = 5$ energy level, giving $\mathbf{5s^2 5p^3}$ and **·Sb:**

(b) Bromine is in the fluorine family, which is Representative Group VII. Each atom has **seven valence electrons:** the two *s* and the five *p* electrons in the $n = 4$ level, giving $\mathbf{4s^2 4p^5}$ and **·Br:**

(c) Barium is in the beryllium family, which is Representative Group II. Each barium atom has **two valence electrons** in the $n = 6$ energy level, giving $\mathbf{6s^2}$ and **Ba:**

EXERCISE How many valence electrons are present in (a) cesium, (b) silicon, and (c) sulfur atoms? Write the outer electron configurations and the Lewis symbols for these elements. *Answers* (a) one, $6s^1$, Cs·; (b) four, $3s^2 3p^2$, ·Si:; (c) six, $3s^2 3p^4$, ·S:

9.2 TRANSITION ELEMENTS

The elements in which electrons enter *d* or *f* subshells as atomic numbers increase are referred to as **transition elements.** Atoms of all but one of the transition elements possess one or two electrons in the outermost *s* subshell. (See Table 9.2, where the representative and transition element configurations are compared.) The elements of the *d* block are known as the *d-transition elements* (see Figure 9.1). There are three complete series of *d*-transition elements, in which the 3*d*, 4*d*, and 5*d* subshells are being occupied.

Table 9.2
The Three Basic Types of Electron Configurations

	Inner levels	Outermost level (n)	Outer configuration
Representative elements e.g., Na, Sr, Br	Filled	s and p being filled	$ns^{1,2}$, ns^2np^{1-6}
***d*-Transition elements** e.g., Sc, Pd, W	$(n-1)d$ being filled, others filled	s^1 or s^2	$(n-1)d^{1-10}ns^{1,2}$
***f*-Transition elements** e.g., Nd, Tm, Pu	$(n-2)f$ being filled, $(n-1)d^{0-2}$, others filled	s^2	$(n-2)f^{1-14}(n-1)d^{0-2}ns^2$

The elements of the f block are known as the *f-transition elements,* or sometimes as the *inner transition elements.* In the two series of f-transition elements, the $4f$ and $5f$ subshells, respectively, are being filled. Take a moment to examine the configurations of the d and f transition elements in the periodic table inside the front cover.

In the transition elements, the outer level s electrons and the d or f electrons in the subshell that is being filled have almost the same energy. As a result, the d and f electrons as well as the outer s electrons influence the properties of these elements. By contrast, in the representative elements the d and f electrons in filled subshells have little effect on properties. The outstanding characteristic of the transition elements is that the similarities in their electron configurations cause them to have many similarities in properties.

In the periodic table, zinc, cadmium, and mercury fall within the d block and are in a sense the last elements of the d-transition series. However, atoms of each of these three elements have two outer electrons in the s subshell as well as a completely filled $(n-1)d$ subshell, which is a configuration more like those of the representative elements. Because these elements also have many properties in common with the representative elements, they are often included in that category.

The elements of the first f-transition series (cerium through lutetium) are named for the first element *preceding* this group, lanthanum, and are known as the **lanthanides.** They are members of the sixth period. Similarly, the elements of the second f-transition series (thorium through lawrencium) are known as the **actinides** and belong to the seventh period. The elements in the seventh period of atomic number greater than uranium, which includes the actinides beyond uranium and the d-transition elements of atomic number 104 and higher, are known collectively as the **transuranium elements.** These elements are all man-made, unstable, and radioactive.

One grouping of transition elements bears a distinctive name based on their origin in the same types of minerals. The members of the scandium family, plus all the f-transition elements to lutetium, are known as the **rare earth elements.**

The grouping of elements bearing distinctive names are summarized in Table 9.3.

Table 9.3
Named Element Groupings

Representative elements	
Alkali metals (Group I)	Li, Na, K, Rb, Cs, Fr
Alkaline earth metals (Group II)	Be, Mg, Ca, Sr, Ba, Ra
Halogens (Groups VII)	F, Cl, Br, I, At
Noble gases	He, Ne, Ar, Kr, Xe, Rn
Transition elements	
Rare earth elements	Sc, Y, La–Lu
Lanthanides	Ce–Lu
Actinides	Th–Lr
Transuranium elements	Np, Pu, Am...

Aside 9.1 GROUP NUMBERS IN THE PERIODIC TABLE

In recent years efforts have been under way to eliminate differences in how groups of elements in the periodic table are designated in western Europe and in the United States. Both the European and U.S. periodic tables use Roman numerals for the representative element groups, as described in Section 9.1a (except in those cases where the noble gases are named as Group VIII). However, the letters A and B are often appended to the group numbers, and the usage of these letters is opposite in the two systems. The U.S. system has labeled the representative groups as A groups (e.g., the lithium family is Group IA). Certain transition-element groupings that traditionally have been numbered are labeled B groups in the U.S. system (e.g., the copper family is known as Group IB, based on the early discovery that these elements have some properties similar to those of the lithium family elements, Group IA). The European usage of A and B has been exactly the opposite. You will note that in this textbook, letters are not used with representative group numbers and transition elements are not given group numbers or letters. Where necessary, groupings of transition elements are referred to as "families"—a simpler and less ambiguous approach.

Recently it has been recommended that a periodic table be introduced in which the vertical columns are numbered all the way across the table from left to right with Arabic numbers (the lithium family would be Group 1, the zinc family would be Group 12, and the noble gases would be Group 18). In this system one must subtract 10 (for the transition elements) as a reminder that, for example, the Group 13, or boron family, elements (Representative Group III) have three valence electrons. At the present time, the controversy generated by this suggestion need be of little concern to chemistry students. However, it is likely that you may encounter periodic tables numbered in this new system (or in alternative new systems).

9.3 METALS, NONMETALS, AND SEMICONDUCTING ELEMENTS

Based on their physical and chemical properties, the elements are classified as metals, nonmetals, and semiconducting elements. As might be expected, the elements of each of these types fall near one another in the periodic table and have similarities in electron configurations.

How the elements are divided into metals, nonmetals, and semiconducting elements is shown in Figure 9.3, where heavy lines separate the types of elements. All the *s*-block elements, all the transition elements, and some of the *p*-block elements are metals. The most characteristic property of **metals**, the property that distinguishes elemental metals most clearly from the other elements, is the ease with which they conduct electricity. The basis for this property, and many other properties of metals, is the presence of a small number of *s* or *s* and *p* electrons that are easily removed. The

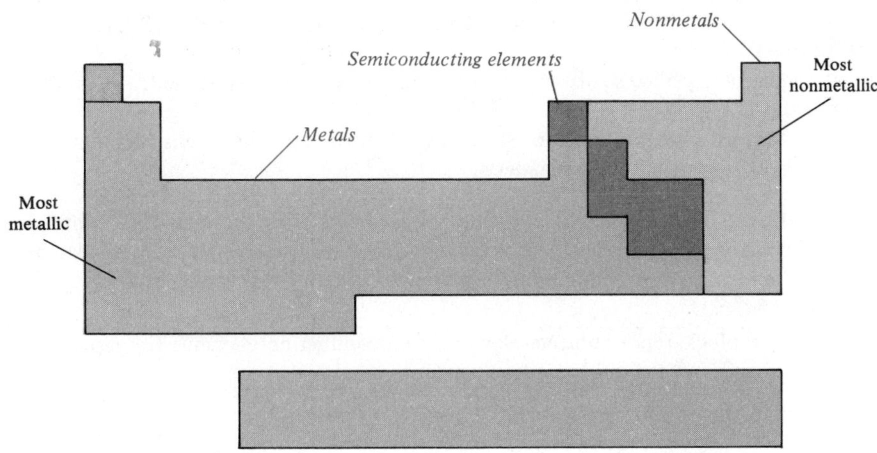

Figure 9.3
Metals, Semiconducting Elements, and Nonmetals
Elements with the most strongly metallic properties fall at the bottom left-hand corner of the periodic table; elements with the most strongly nonmetallic properties are at the top right-hand corner.

two complete groups of representative metals (Representative Groups I and II) and all but one transition metal have outer configurations of ns^1 or ns^2. The heavier *p*-block elements in Representative Groups III or VI are also metals (see Figure 9.2).

As to both position in the periodic table and properties, the **semiconducting elements** fall between the metals and the nonmetals. In appearance, the seven semiconducting elements—boron, silicon, germanium, arsenic, selenium, antimony, and tellurium—resemble metals. However, in chemical behavior and in the properties of their compounds, the semiconducting elements are more like the nonmetals.

At the right in the periodic table are two complete families of nonmetals—the noble gases and the halogens. In the carbon family, just the first element, carbon, is a nonmetal. In both the nitrogen and oxygen families, the first two elements are nonmetals (nitrogen and phosphorus, and oxygen and sulfur, respectively). Only 17 of the elements are nonmetals. The most distinguishing property of **nonmetals** compared to other elements is that they do not conduct electricity (except under special conditions). All but five of the nonmetals (carbon, phosphorus, sulfur, bromine, and iodine) are gases at room temperature and pressure.

At the time the elements were first successfully organized into a periodic table, it was recognized that certain properties of the elements vary periodically across the rows in the table; that is, the properties are similar at regular intervals in the sequence (see Aside 9.2). Such variation is summarized by the **periodic law:** The properties of the elements vary periodically with their atomic numbers. In the following sections, we examine several periodic properties that are explained by the variation in atomic number and electron configuration across the table.

EXAMPLE 9.2 Electron Configurations and the Periodic Table

In the periodic table, locate the elements with the following configurations and classify each as either a representative or a transition element, and also as a metal, a nonmetal, or a semiconducting element.

(a) $[Xe]5d^{10}6s^2 6p^2$ (c) $[Kr]4d^7 5s^1$
(b) $1s^2 2s^2 2p^6 3s^2$ (d) $[Kr]4d^{10} 5s^2 5p^6$

(a) The highest level configuration of s^2p^2, and the Xe core with filled *d* sublevel, both indicate that this is a **representative element.** The s^2p^2 configuration identifies this as an element from the carbon family, in the *p* block (Representative Group IV). The outer electrons are in the $n = 6$ energy level, showing that this element is in the sixth period and should be a **metal,** as metallic properties extend across to Representative Group IV in this period. [See Figure 9.2. The element is lead.]

(b) All occupied subshells below the highest main energy level are filled, showing this also to be a **representative element.** The two *s* electrons in the highest energy level indicate that this element is from the *s* block and is an alkaline earth **metal** (Representative Group II).

(c) The incompletely filled *d* subshell identifies this as a ***d*-transition element,** and therefore a **metal.**

(d) The outermost s^2p^6 configuration shows this element to be a noble gas, and therefore a **representative element** and a **nonmetal.**

EXERCISE In the periodic table, locate the elements with the following configurations and classify each as either a representative or a transition element, and also as a metal, a nonmetal, or a semiconducting element. (a) $[Ne]3s^2 3p^2$, (b) $[Rn]5f^6 7s^2$, (c) $[Xe]4f^{14} 5d^9 6s^1$.

Answers (a) *p*-block representative element, semiconducting element; (b) *f*-transition element, actinide, metal; (c) *d*-transition element, metal.

SUMMARY OF SECTIONS 9.1–9.3

All elements can be classified on the basis of their electron configurations and properties as either representative elements or transition elements (see Table 9.2) and also as metals, semiconducting elements, or nonmetals. The representative elements are those of the *s* block (Representative Groups I and II; see Figure 9.2) and the *p* block (Representative Groups III–VII and the noble gases). In *d*-transition elements, the $(n - 1)d$ subshell is filling and in *f*-transition elements the $(n - 2)f$ subshell is filling. Certain groupings of elements have distinctive names (see Table 9.3)

All *s*-block elements and transition elements, and some *p*-block elements are metals, with the distinctive property of readily conducting electricity. The seventeen nonmetals, which do not conduct electricity and many of which are gases, are all *p*-block elements except for hydrogen and helium, which have no *p* electrons. The seven semiconducting elements (B, Si, Ge, As, Sb, Se, Te) fall between the metals and nonmetals and are intermediate in their properties. Many properties of the elements, like metallic behavior, vary periodically and recur for elements that fall together in the periodic table.

Aside 9.2 THE EVOLUTION OF THE PERIODIC TABLE

The pathway to the periodic table began with a search for numerical relationships among the atomic masses of similar elements. In 1829 the German chemist Johann Döbereiner (1780–1849) called attention to several sets of three elements—triads—which are similar in properties, which form similar compounds, and for which the atomic mass of one is approximately the average of the atomic masses of the other two (e.g., the average mass of Cl and I is 81 and the mass of Br is 80). Döbereiner drew the logical conclusion that the properties of an element depend upon its atomic mass.

The English chemist John A. R. Newlands (1837–1898) proposed in 1864 that if the elements are arranged in the order of increasing atomic mass, the eighth is like the first, the ninth is like the second, and so on. He called this rule the "law of octaves" and set up a table to illustrate it. Unfortunately, the value of Newlands's work was not appreciated at first and he was even ridiculed when he presented his ideas at a meeting of the Chemical Society. Later, however, he was highly honored for his work.

The form of the periodic table most common today originated in 1869. A German chemist, Lothar Meyer, and a Russian chemist, Dmitri Mendeleev, published similar tables, although they had worked independently of each other. Meyer also illustrated the periodicity of properties by plotting properties that could be expressed as numbers versus atomic masses (similar to those in Figures 9.8–9.10).

Mendeleev, like Newlands, proposed a table with columns of related elements. But Mendeleev's table was more nearly complete and left spaces for undiscovered elements. For example, in 1869, the next heaviest known element after zinc was arsenic. Arsenic is not much like aluminum or silicon—it is much more like phosphorus. So Mendeleev placed arsenic next to phosphorus and left spaces next to aluminum and silicon in his table for elements that he felt sure would be discovered later (Figure 9.4). He was even able to predict the properties of these elements with considerable accuracy. Both were discovered a few years later and fulfilled his predictions.

Eventually, the growth of our knowledge of atomic structure indicated that after the first two horizontal rows, the next rows in the table should be made longer. This could have been predicted from Lothar Meyer's curves.

Even with the addition of the later-discovered noble gases and the expansion of the table into eighteen columns, some difficulties remained. In three cases, the order of increasing atomic masses did not put the elements in the proper chemical sequence (cobalt and nickel, argon and potassium, tellurium and iodine). Also, the rare earth elements (elements 57–71) all seemed to belong to a single space between barium and hafnium. The first of these difficulties was resolved when Henry Moseley found that the atomic number, rather than the atomic mass, is the fundamental property for the arrangement of the elements. The second was resolved in a practical way by relegating the rare earth elements to a space outside the table proper. Knowledge of electronic structure explained the similarities of these elements and showed that they do indeed all belong between barium and hafnium as the 4*f* subshells are filled.

The four heaviest elements known for many years (elements 89–92, actinium–uranium) were assigned positions in the four columns of the table following radium. In terms of the properties of the elements, this assignment was only moderately satisfactory, but it was accepted. On the basis of this plan, elements 93 and 94 should resemble rhenium and osmium. However, when these elements were synthesized, it was found that they are not like rhenium and osmium, but instead *both* resemble uranium.

This similarity in properties brought to mind the rare earth series. In 1944 Glenn Seaborg at the University of California, where the new elements had been made, proposed that the elements after radium were misplaced in the periodic table. Instead they are part of a second "rare earth" series, which like the first, would contain fourteen elements, the last being element number 103. All of these elements have now been synthesized and found to fit into the scheme very well. Moreover, elements 104–109 have been synthesized. So far, chemical studies have shown that 104 has properties that relate it to hafnium, as the periodic table would lead us to expect. All of the synthetic elements are radioactive and, except for plutonium, have been prepared in only very small amounts.

Figure 9.4
Mendeleev's Periodic Table (as published in *Zeitschrift für Chemie* in 1869)
The tables of Mendeleev and Meyer were very similar, although Mendeleev based his primarily on chemical properties and Meyer based his primarily on physical properties. Note the spaces for unknown elements, as indicated by question marks in the symbol column. Tellurium and iodine (symbolized here by J for German Jod), are out of atomic mass order because Mendeleev recognized that their chemical properties related them to selenium and bromine, respectively. In a later version, Mendeleev rearranged his table so that the vertical columns listed the groups as in the modern periodic table.

			Ti = 50	Zr = 90	? = 180
			V = 51	Nb = 94	Ta = 182
			Cr = 52	Mo = 96	W = 186
			Mn = 55	Rh = 104,4	Pt = 197,4
			Fe = 56	Ru = 104,4	Ir = 198
			Ni = Co = 59	Pd = 106,6	Os = 199
H = 1			Cu = 63,4	Ag = 108	Hg = 200
	Be = 9,4	Mg = 24	Zn = 65,2	Cd = 112	
	B = 11	Al = 27,4	? = 68	Ur = 116	Au = 197?
	C = 12	Si = 28	? = 70	Sn = 118	
	N = 14	P = 31	As = 75	Sb = 122	Bi = 210?
	O = 16	S = 32	Se = 79,4	Te = 128?	
	F = 19	Cl = 35,5	Br = 80	J = 127	
Li = 7	Na = 23	K = 39	Rb = 85,4	Cs = 133	Tl = 204
		Ca = 40	Sr = 87,6	Ba = 137	Pb = 207
		? = 45	Ce = 92		
		?Er = 56	La = 94		
		?Yt = 60	Di = 95		
		?In = 75,6	Th = 118?		

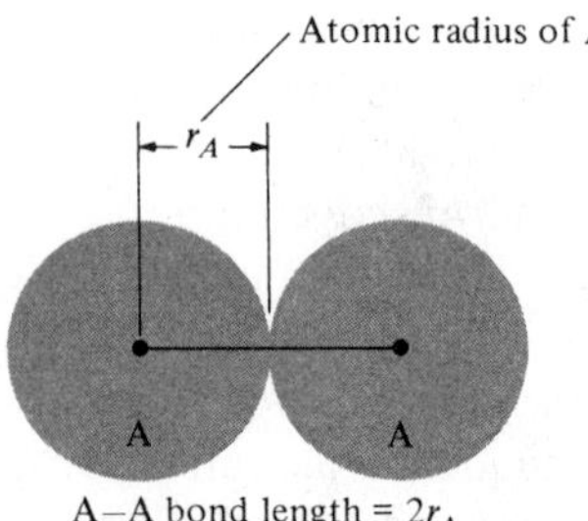

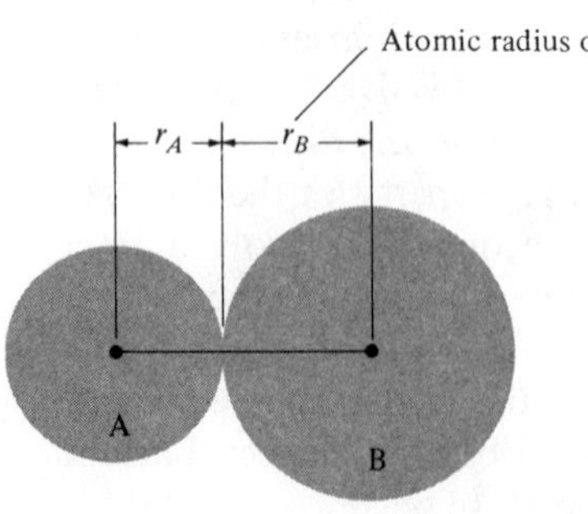

Figure 9.5
Atomic Radii

THE SIZES OF ATOMS AND IONS

The sizes of atoms and ions are compared by comparing their radii. This is not so simple as it seems, because measuring the radius of an atom or ion is not like measuring the radius of, say, a basketball. Atoms and ions are not rigid and they do not have clearly defined boundaries. Instead, the electron density drops off gradually in all directions. Also, atoms or ions cannot be isolated for measurement of their radii. Only the distance between nuclei in elements or compounds can be measured. This distance depends upon the environment—the type and strength of the bonding in the substance under investigation.

The approach used, therefore, is to define sets of radii for different bonding situations. Our concern here is primarily with what are called atomic radii and ionic radii.

9.4 ATOMIC AND IONIC RADII

The term **atomic radii** refers to an internally consistent set of radii based on the distance between atoms in comparable chemical bonds. Atomic radii are found by measuring the distances between bonded atoms and assigning part of the distance to each atom.

Bond length is defined as the distance between the nuclei of two bonded atoms. One-half the length of the bond between two atoms of the same element is taken as the radius of atoms of that element (Figure 9.5). For example, the length of the chlorine–chlorine bond in Cl_2 is 0.198 nm. The atomic radius of chlorine is therefore taken as

$\frac{1}{2}$ (0.198 nm) = 0.099 nm. Radii assigned in this way are used to determine the radii of atoms of other elements. Suppose the Sn—Cl bond length was measured to be 0.240 nm. Subtracting the known chlorine atomic radius of 0.099 nm from this value gives a radius of 0.141 nm for the tin atom. The atomic radii of the representative and *d*-transition elements, together with the radii of their most common ions, are given in Figure 9.6, where the radii of ions are shown by dashed lines.

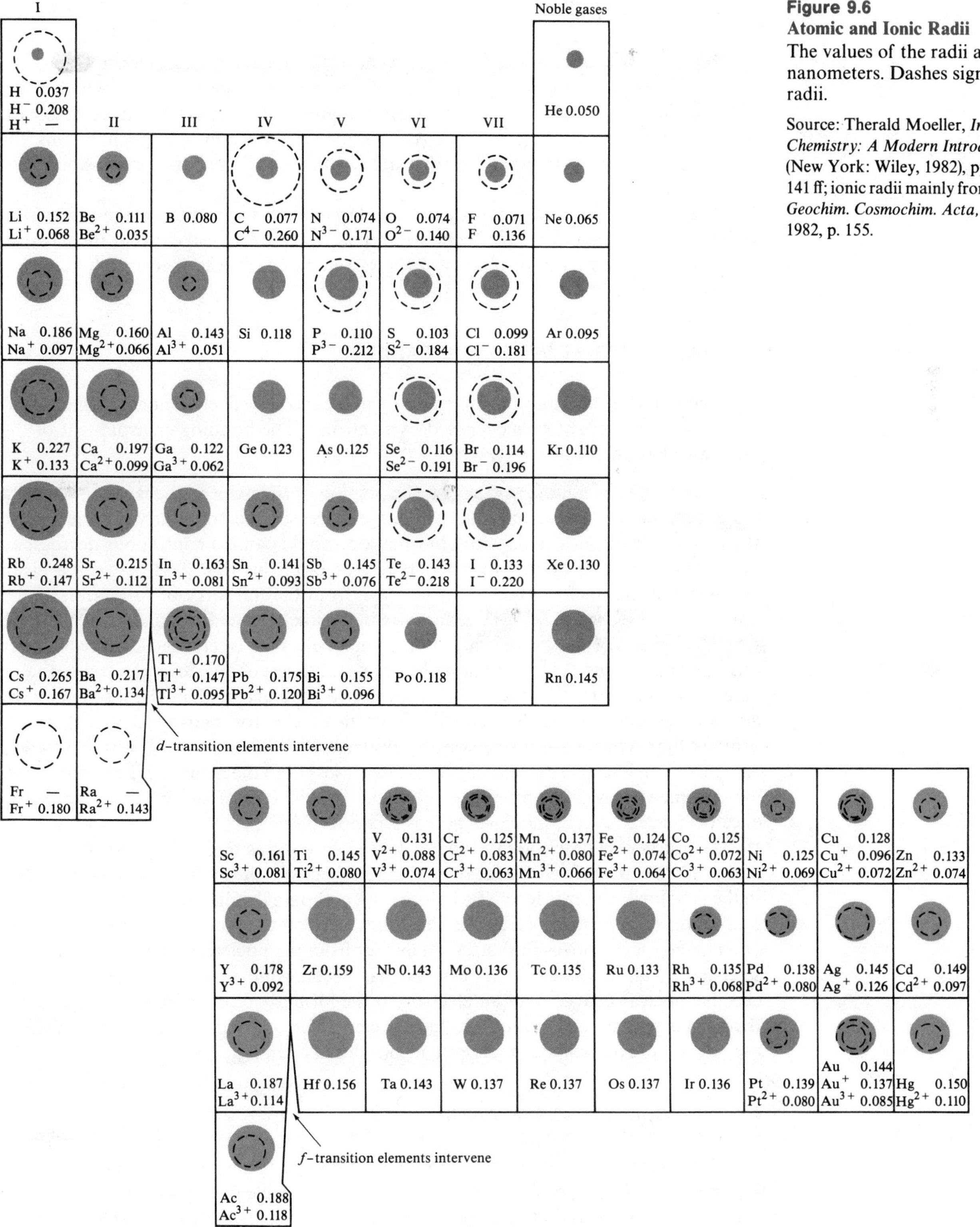

Figure 9.6
Atomic and Ionic Radii
The values of the radii are given in nanometers. Dashes signify ionic radii.

Source: Therald Moeller, *Inorganic Chemistry: A Modern Introduction* (New York: Wiley, 1982), pp. 70 ff. and 141 ff; ionic radii mainly from L. H. Ahrens, *Geochim. Cosmochim. Acta*, Vol. 2, 1982, p. 155.

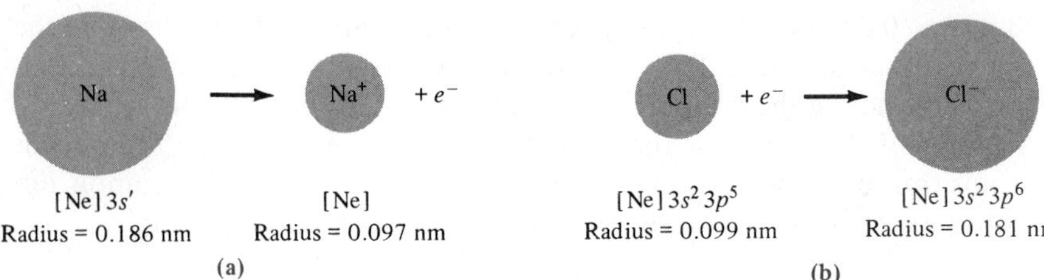

Figure 9.7
Variations in Radii with Cation and Anion Formation

Ionic radii are the radii of anions and cations in crystalline ionic compounds. The radius of a positive ion is always smaller than the radius of the atom from which it was formed. The loss of one electron permits the remaining electrons to be pulled in toward the nucleus a bit more (Figure 9.7a). Also, in many of the common positive ions, the atom has lost all of its valence electrons and is smaller because it has one less occupied energy level.

The radius of a negative ion is always larger than the radius of the atom from which it was formed. An added electron repels other electrons, causing the electron cloud to spread out, and also adds to the screening effect described in the following section (Figure 9.7b).

9.5 FACTORS THAT INFLUENCE RADII

The radius of an atom or ion is mainly influenced by effective nuclear charge and the number of energy levels occupied by electrons. (The bonding environment of the atom also has an effect.)

1. The **effective nuclear charge** is the portion of the nuclear charge that acts on a given electron. As the actual nuclear charge increases, electrons are attracted more strongly to the nucleus. This effect, in isolation, would cause a continuous decrease in radii with increasing atomic number. However, electrons that pass between a given electron and the nucleus shield that electron from the full force of attraction of the nucleus. The **screening effect** is the decrease in the nuclear charge acting on an electron due to the effects of other electrons. A given outer electron is screened slightly by other outer level electrons, somewhat more by the next highest energy level electrons, and to an even greater extent by the electrons closer to the nucleus than the $n - 1$ level. In addition, because of the different distributions of electron density in *s*, *p*, *d*, and *f* orbitals, the effectiveness of screening by electrons in different types of subshells varies. As a result, the effective nuclear charge is always less than the actual nuclear charge and varies somewhat with the number of electrons in each type of subshell. In general, the effective nuclear charge increases across a period.
2. Earlier we described the energy levels of atoms as resembling the layers in an onion. The electrons in each energy level are further from the nucleus than the electrons in the next lowest energy level. (In Figure 8.17, compare the distances from the nucleus of 2*s* and 3*s* or 2*p* and 3*p* electrons.) The first electron to enter the *s* subshell in a new outer energy level moves into a region farther from the nucleus and also, because it is screened by all of the electrons in what has become the $n - 1$ level, feels a smaller effective nuclear charge than an electron in the formerly outermost level. As a result, there is a big increase in radius from the last element in one period to the first element in the next period as well as an increase in size down a family.

9.6 PERIODIC TRENDS IN RADII

The representative elements decrease in atomic radii across a period, as shown in Figures 9.6 and 9.8. Across a period of representative elements, the effective nuclear charge is increasing, pulling electrons closer to the nucleus, but there is no change in the

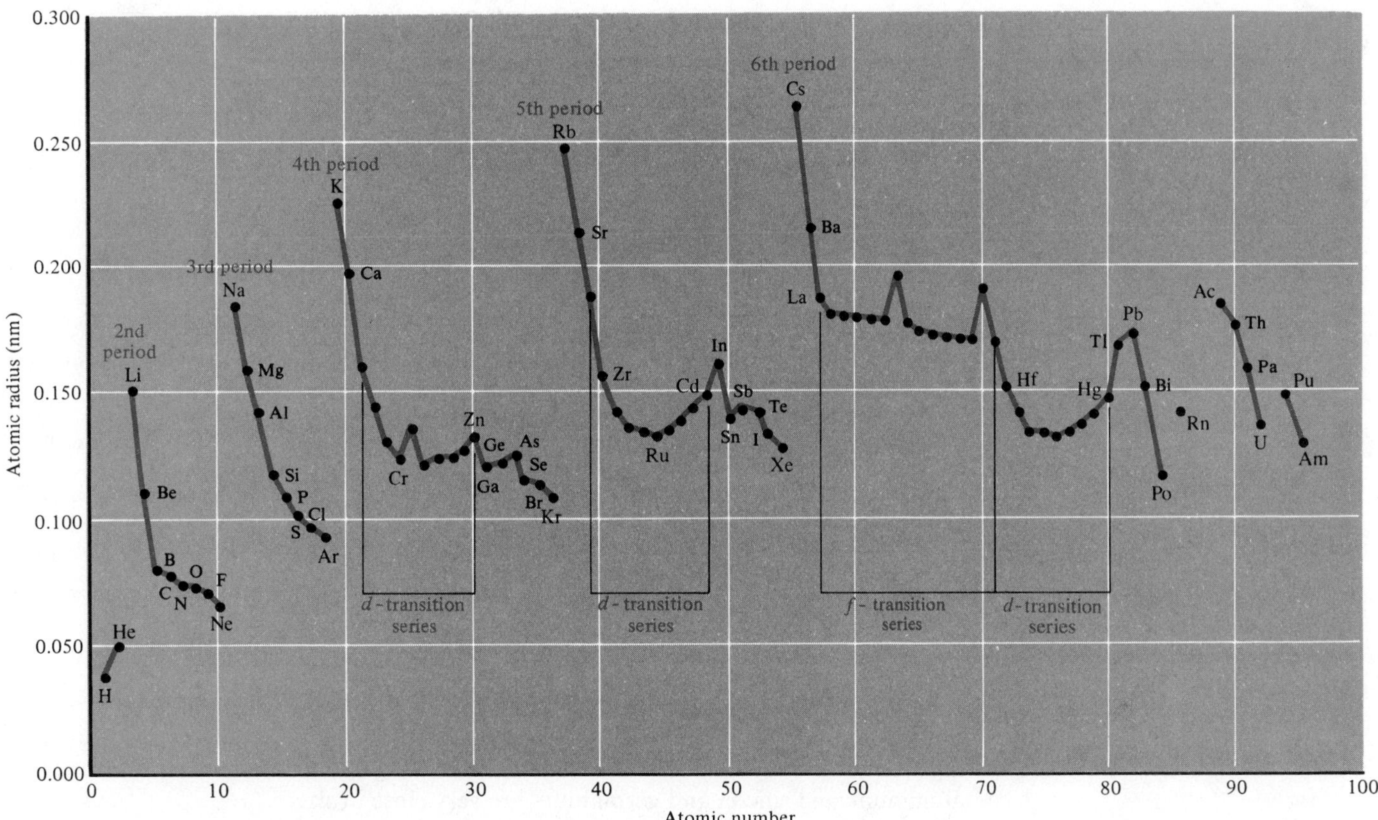

Figure 9.8
Variation of Atomic Radii with Atomic Number

highest main energy level that is occupied. Therefore, the decrease in radii due to increased nuclear charge predominates.

In going *down* a family of representative elements, the atomic radii increase. Here the occupancy of higher energy levels has a greater influence on size than the increasing nuclear charge.

In each family of representative elements the first member differs in many respects from the other elements in the family, partly due to its much smaller size. Lithium, beryllium, and boron bear many resemblances to the elements diagonally below each of them in the next family. (In Figure 9.8, compare the radii of these elements and the radii of those in their families.)

$$
\begin{array}{cccc}
{}_{3}\text{Li} & {}_{4}\text{Be} & {}_{5}\text{B} & {}_{6}\text{C} \\
{}_{11}\text{Na} & {}_{12}\text{Mg} & {}_{13}\text{Al} & {}_{14}\text{Si}
\end{array}
$$

The transition elements decrease less in size across the periods than do representative elements. Here *d* or *f* electrons are being added to inner shells, where they are very effective at screening the outer *s* electrons from the nuclear charge. Thus the outer *s* electrons feel only a slightly increased nuclear charge, and the decrease in size across a transition element period is much more gradual than for the representative elements. The gradual decrease in size across the first *f*-transition element series from lanthanum to lutetium is referred to as the **lanthanide contraction.**

The intervention of the transition elements and the accompanying increase in nuclear charge have a significant effect upon the radii of the *p*-block elements. In Figures 9.6 or 9.8 compare the radii of, for example, magnesium and calcium, with the radii of aluminum and gallium or silicon and germanium. Atoms of the fourth-period *s*-block elements are larger than those of the third-period elements. However, in the

Table 9.4
Atomic and Ionic Radii

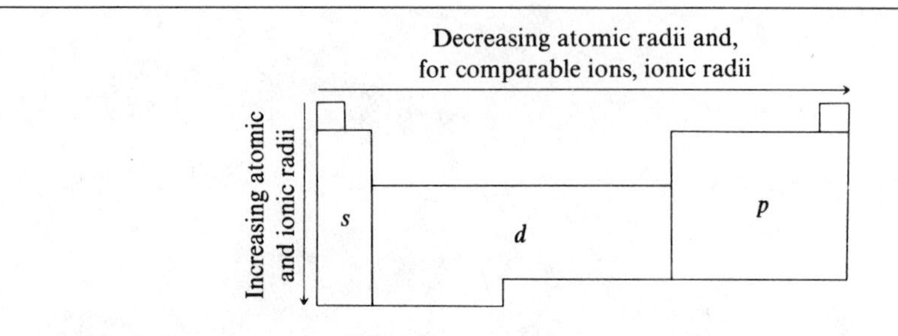

Radii generally decrease across the periods.
Radii generally increase down the families.
Cations are smaller and anions are larger than neutral atoms of the elements from which they are formed.
Cations are generally smaller than anions.
Effective nuclear charge increases across each period.
In *p*-block families, the second and third members are similar in size and properties, because of the intervention of the *d*-transition elements.
In *d*-transition element families, the two heaviest members are similar in size and properties because of the intervention of the lanthanides.
The first element in each representative element family differs from other elements in the family because of the much smaller size of its atoms.

p block, the relatively greater nuclear charge of the elements in the fourth period pulls in the electrons more strongly and counteracts the presence of electrons in a higher main energy level. As a result, in the *p* block gallium is somewhat smaller than aluminum, and silicon and germanium are very close in size.

A similar effect on the sizes of atoms of the *d*-transition elements results from the intervention of the lanthanides (La–Lu) between barium and hafnium. The expected increase in radii due to occupation by electrons of a higher energy level is just about balanced by the additional nuclear charge due to the lanthanides. Consequently, the two heaviest elements in each family of *d*-transition elements have almost identical radii and very similar chemical properties.

Ionic radii follow general trends similar to those of atomic radii, increasing down families for ions of the same charge. The effect of increasing nuclear charge on radius can be seen by comparing **isoelectronic** ions, that is, ions that have identical electron configurations. For example, N^{3-}, O^{2-}, Na^+, and Mg^{2+} are an isoelectronic series of ions. The N^{3-} and O^{2-} ions have gained electrons to give the neon configuration and the Na^+ and Mg^{2+} ions have lost their valence electrons to leave the neon configuration. In such a series, the decreased radii for the heavier ions are solely the result of greater attraction by the more highly charged nuclei.

Isoelectronic ions	N^{3-}	O^{2-}	Na^+	Mg^{2+}
Z	7	8	11	12
Radii (nm)	0.171	0.140	0.097	0.066

The trends in atomic and ionic radii are summarized in Table 9.4.

EXAMPLE 9.3 Atomic Radii

Consulting only the periodic table (inside front cover), decide whether the first atom in each of the following pairs is larger, smaller, or similar in atomic radius to the second atom: (a) Si, Pb; (b) Cs, Pb; (c) Rh, Ir; (d) Ti, V.

(a) Silicon is the second element in the carbon family, while lead is the fifth. Silicon atoms should be **smaller** as size increases down a representative element family.
(b) Cesium and lead are in the same period, with lead further to the right in the period. Size decreases across the periods, and therefore cesium atoms should be **larger** than lead atoms.
(c) Rhodium and iridium are the second and third members of a *d*-transition metal family. Because of the lanthanide contraction, the atomic radii of these elements should be **similar.**
(d) Titanium and vanadium are adjacent elements in the same *d*-transition period. Atoms of these two elements should be **similar** in atomic radius, for the decrease in size across the periods for transition elements is gradual.

[Check the conclusions reached here by consulting Figure 9.6.]

EXERCISE Choose the element from each set that has the larger atomic radius: (a) K, Na; (b) Na, Mg; and (c) As, S. *Answers* (a) K, (b) Na, (c) As.

SUMMARY OF SECTIONS 9.4–9.6

Atomic radii decrease with increasing effective nuclear charge (the portion of the nuclear charge that acts on the outer electrons after screening by other electrons is taken into account) and increase when a higher main energy level is occupied. As a result, atomic radii generally increase down representative element groups and decrease across periods. The combination of these effects causes the first elements in several representative groups to resemble in properties the second member of the next family to the right.

The radii of cations are smaller than those of neutral atoms of the same element and the radii of anions are larger. Ionic radii are also influenced by effective nuclear charge, as shown in series of isoelectronic ions.

The atomic radii of the transition elements decrease less across the periods than those of the representative elements, because inner electrons screen outer electrons from the nuclear charge. Due to the intervention of transition elements and the resulting greater nuclear charges, the third- and fourth-period *p*-block elements are similar in size. Also, the second and third members of each *d*-transition element family are very similar in size and properties.

ELECTRON GAIN AND LOSS

9.7 IONIZATION ENERGY

The **ionization energy** (sometimes called the ionization potential) is the enthalpy change for the removal of the least tightly bound electron from an atom or an ion in the gaseous state. Ionization energies are given per mole of atoms or ions of a given type. The *first* ionization energy is that required for the removal of one electron from a neutral atom. For example, for sodium

$$Na(g) \longrightarrow Na^+(g) + e^- \qquad \Delta H^\circ_0 = 495.8 \text{ kJ}$$

1st ionization energy

Such reactions are always endothermic and ionization energies are always positive—energy must be put into the system to pull an electron away from the nuclear charge. The energies for removal of successive electrons are the second ionization energy, the third ionization energy, and so on. For example

$$Al(g) \longrightarrow Al^{+}(g) + e^{-} \qquad \Delta H^\circ_0 = 578 \text{ kJ} \quad \textit{1st ionization energy} \tag{9.1}$$

$$Al^{+}(g) \longrightarrow Al^{2+}(g) + e^{-} \qquad \Delta H^\circ_0 = 1817 \text{ kJ} \quad \textit{2nd ionization energy} \tag{9.2}$$

$$Al^{2+}(g) \longrightarrow Al^{3+}(g) + e^{-} \qquad \Delta H^\circ_0 = 2745 \text{ kJ} \quad \textit{3rd ionization energy} \tag{9.3}$$

Removing electrons from ions of increasing positive charge is increasingly difficult.

Note that the ionization energy is defined as the energy required for the removal of *the least tightly bound* electron. In the total environment of the atom or ion under consideration, the electron removed is at the highest energy level. You might expect this to be the last electron that was added according to the building up order across the periodic table. This is not necessarily what happens. Electrons *always* come off in the order of the principal energy levels; for example, $n = 3$ electrons before $n = 2$ electrons, and $n = 2$ electrons before $n = 1$ electrons. Within the same main energy level, the order of electron loss in ionization is f, then d, then p, then s electrons. As a consequence, in the d-transition elements the $(n - 1)d$ electrons enter last, but the ns electrons leave first. For example, the ions of iron have the following configurations:

Fe [Ar]$3d^6 4s^2$ (*last electron added here* → $3d$; *first electrons to be lost* → $4s$)

Fe^{2+} [Ar]$3d^6$

Fe^{3+} [Ar]$3d^5$

9.8 PERIODIC TRENDS IN IONIZATION ENERGIES

Ionization energies, like radii, are influenced by the effective nuclear charge and the electron configuration. In general, a greater effective nuclear charge makes it more difficult to remove an electron, resulting in a higher value for the ionization energy.

Table 9.5
First Ionization Energies
The values are given in kilojoules per mole at 0 K. To use these values in calculations with other enthalpies that are given for 298 K, the values in this table must be converted to approximate values at 298 K by adding 6.19 kJ.

Ionization energies increase across periods → ; Ionization energies decrease down groups ↓

I	II							
H 1312								
Li 520	Be 899							
Na 496	Mg 738							
K 419	Ca 590	Sc 631	Ti 658	V 650	Cr 653	Mn 717	Fe 759	
Rb 403	Sr 550	Y 616	Zr 660	Nb 664	Mo 685	Tc 702	Ru 711	
Cs 376	Ba 503	La 538	Hf 680	Ta 761	W 770	Re 760	Os 840	
Fr	Ra 509	Ac 670						

lowest first ionization energy (Cs 376); *lanthanides intervene* (between La and Hf)

Source: Adapted from Therald Moeller, *Inorganic Chemistry: A Modern Introduction* (New York: Wiley, 1982), pp. 76–79.

Table 9.6
Ionization Energies of Na–Ar Period Elements

Energies for the removal of successive outer electrons are given in black, and for removal of inner shell electrons, in color. Values are in kilojoules per mole at 0 K. In the $n = 1$ and $n = 2$ energy levels each of these elements has a neon-like ($1s^2 2s^2 2p^6$) configuration. The screening effect across the period is approximately constant. To convert to approximate values at 298 K, add 6.19 kJ per electron removed.

Ionization energy	Na ($3s^1$)	Mg ($3s^2$)	Al ($3s^2\ 3p^1$)	Si ($3s^2 3p^2$)	P ($3s^2 3p^3$)	S ($3s^2 3p^4$)	Cl ($3s^2 3p^5$)	Ar ($3s^2 3p^6$)
1st	496	738	578	786	1,012	1,000.	1,251	1,521
2nd	4,562	1,451	1,817	1,577	1,903	2,251	2,297	2,666
3rd	6,912	7,733	2,745	3,232	2,912	3,361	3,822	3,931
4th	9,543	10,540.	11,577	4,355	4,956	4,564	5,158	5,771
5th	13,353	13,629	14,831	16,091	6,274	7,013	6,540	7,238
6th	16,610.	17,994	18,377	19,784	21,268	8,495	9,362	8,781
7th	20,114	21,703	23,294	23,776	25,397	27,106	11,018	11,995
8th	25,489	25,655	27,459	29,251	29,853	31,669	33,604	13,841

Source: Adapted from Therald Moeller, *Inorganic Chemistry: A Modern Introduction* (New York: Wiley, 1982), pp. 76–79.

However, all other factors being equal, it is easier to remove an electron from a larger atom or ion than from a smaller one. In addition, since certain electron configurations are more stable than others, more energy is required for electron removal.

Because the **noble gas configuration**—an outer configuration of $ns^2 np^6$ (or $1s^2$)—is very stable, removal of an electron from an atom or ion with this type of configuration is especially difficult. The result, as shown in Table 9.5 and Figure 9.9, is that the noble gases have high ionization energies. Note that when the term "ionization energy" is used in this general way, it usually means the first ionization energy.

The relatively greater stability of the noble gas configuration is dramatically illustrated by the successive ionization energies of the elements. In Table 9.6, the black

				III	IV	V	VI	VII	Noble gases
									He 2372 ← *highest first ionization energy*
				B 801	C 1086	N 1402	O 1314	F 1681	Ne 2081
				Al 578	Si 786	P 1012	S 1000	Cl 1251	Ar 1521
Co 758	Ni 737	Cu 745	Zn 906	Ga 579	Ge 762	As 947	Se 941	Br 1140	Kr 1351
Rh 720	Pd 805	Ag 731	Cd 868	In 558	Sn 709	Sb 834	Te 869	I 1008	Xe 1170
Ir 880	Pt 870	Au 890	Hg 1007	Tl 589	Pb 716	Bi 703	Po 818	At	Rn 1037

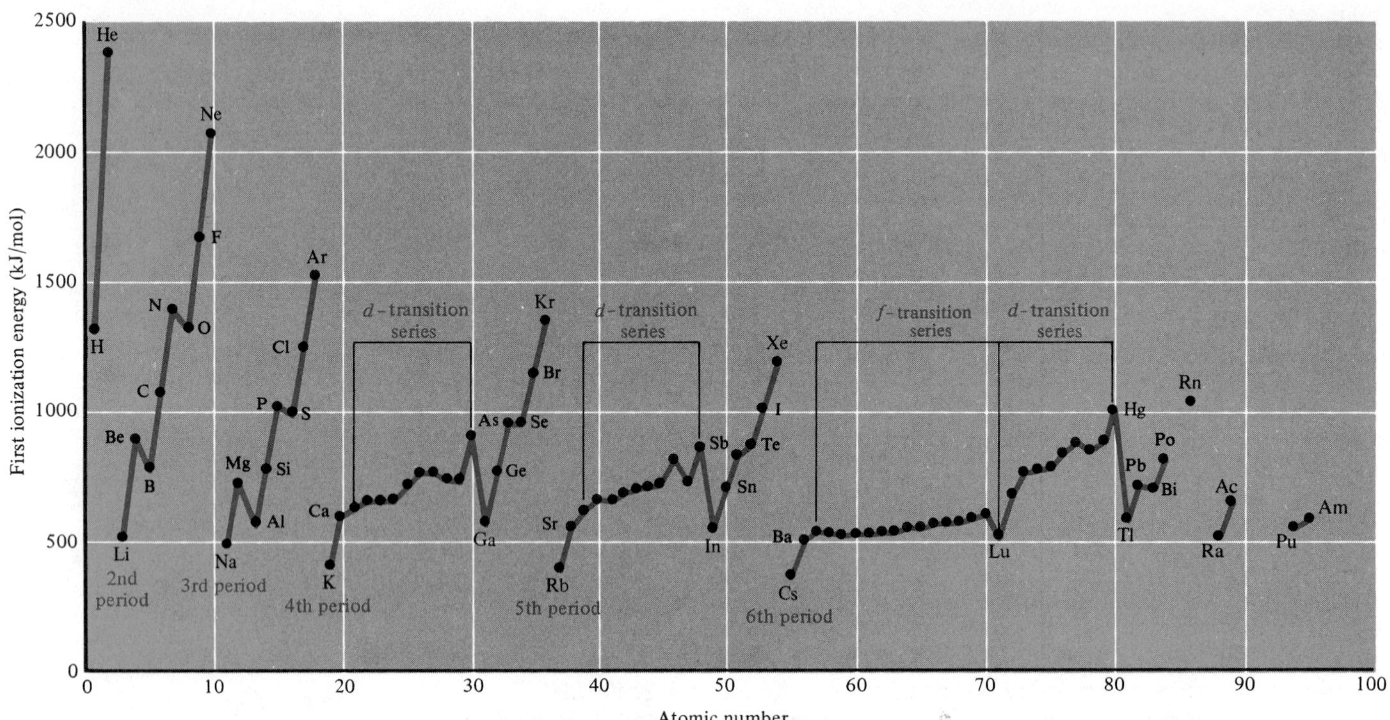

Figure 9.9
Variation of First Ionization Energies with Atomic Number
In each period the lithium family element has the lowest ionization energy and the noble gas the highest ionization energy. Note the lack of variation for the transition elements.

Source: Therald Moeller, *Inorganic Chemistry: A Modern Introduction* (New York: Wiley, 1982), pp. 76–79.

numbers represent removal of outer electrons and the color numbers represent removal of inner electrons from completely filled subshells. Compare the first and second ionization energies of sodium, the second and third ionization energies of magnesium, and so on. The successive ionization energies also provide a satisfying experimental verification of the existence of energy levels and the numbers of electrons expected to be in each level.

First ionization energies generally increase across the periods and decrease down the families of the periodic table. Comparison of Figure 9.9 with Figure 9.8 confirms that less energy is required to remove electrons from larger atoms than from smaller atoms. The locations of the metals on both curves show that, in general, metal atoms are larger than nonmetal atoms and lose electrons more easily. These differences account for many of the differences in properties between the metals and nonmetals. Note that, as expected from the small variation in their radii, the ionization energies of the transition metals do not vary greatly.

The ionization energies for the third-period elements (see Figure 9.9) illustrate variations in the general trend caused by the extra stability of electron configurations in which subshells are completely filled or one-half filled. It is easier to remove a $3p$ electron from an aluminum atom ([Ne]$3s^2 3p^1$) than to remove a $3s$ electron from the pair in an s subshell in a magnesium atom ([Ne]$3s^2$). Also, it is easier to remove the fourth p electron from a sulfur atom ([Ne]$3s^2 3p^4$) than to remove a p electron from the half-filled p subshell in a phosphorus atom ([Ne]$3s^2 3p^3$). Similar variations can be seen for each period. The trends in ionization energies are summarized in Table 9.7.

Table 9.7
Ionization Energy

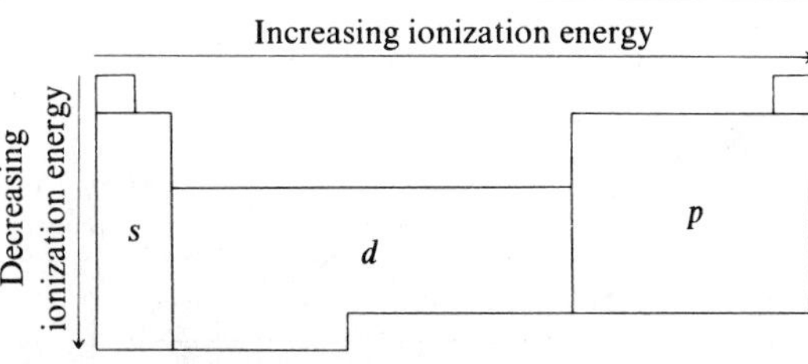

Ionization energies generally increase across the periods.
Ionization energies generally decrease down the families.
Removal of electrons from filled and half-filled subshells requires higher energy.
In successive ionization energies, a large increase occurs for removal of electrons that lie beneath the outermost main energy level.
Metals have low ionization energies.
Nonmetals have high ionization energies.

EXAMPLE 9.4 Ionization Energy

The pairs of elements, Mg and Ca, and Al and Ga, occupy corresponding positions in the beryllium and boron families. The ionization energy for the removal of all of the valence electrons is much larger for Mg than for Ca [$\Delta H^\circ_0 = 2188$ kJ/mol for $Mg(g) \longrightarrow Mg^{2+}(g) + 2e^-$; 1735 kJ/mol for $Ca(g) \longrightarrow Ca^{2+}(g) + 2e^-$], while the similar energy for Al is smaller than that for Ga [$\Delta H^\circ_0 = 5139$ kJ/mol for $Al(g) \longrightarrow Al^{3+}(g) + 3e^-$; 5521 kJ/mol for $Ga(g) \longrightarrow Ga^{3+}(g) + 3e^-$]. Account for this difference.

Magnesium and calcium atoms have the same outer electron configurations. However, calcium atoms are also larger. The ionization energies show that the **size factor predominates,** making it easier to remove the valence electrons from calcium than from magnesium.

The intervention of the 3*d*-transition elements between aluminum and gallium causes atoms of these two elements to be much closer in size than magnesium and calcium atoms, but gives gallium atoms a relatively much larger effective nuclear charge. It is this **greater charge** that causes the ionization energy for the formation of Ga^{3+} to be larger than that for Al^{3+}.

EXERCISE The first ionization energy of Y is larger than that of La, but for the rest of the *d*-transition elements in these two periods the order is reversed; for example, the first ionization energy of Hf is larger than that of Zr and the first ionization energy of Ta is larger than that of Nb. Explain this trend.

Answer Intervention of lanthanides makes atoms in these two periods virtually identical in size and the much larger nuclear charge of Hf through Hg produces a larger ionization energy.

9.9 ELECTRON AFFINITY

The **electron affinity** is the enthalpy change for the addition of one electron to an atom or ion in the gaseous state. Electron affinity values are given per mole of atoms or ions. For example, for chlorine

$$Cl(g) + e^- \longrightarrow Cl^-(g) \qquad \Delta H^\circ_0 = -349 \text{ kJ}$$

The chlorine atom, which is one electron short of a noble gas configuration, adds an electron readily in an exothermic reaction.

Unlike ionization energies, electron affinities can be either positive or negative. Energy is *required* to add an electron to an already stable configuration. For example,

Table 9.8
Electron Affinities of the Representative Elements
Values are given in kilojoules/mole at 0 K. Values in parentheses have been calculated from theory; others are from experimental measurements. (Note that in some older publications an opposite sign convention is used.)

I	II	III	IV	V	VI	VII	Noble gases
H −73							He (21)
Li −60	Be (240)	B −83	C −123	N 0.0	O −141	F −322	Ne (29)
Na −53	Mg (230)	Al (−50)	Si −120	P −74	S −200	Cl −349	Ar (35)
K −48	Ca (156)	Ga (−36)	Ge −116	As −77	Se −195	Br −325	Kr (39)
Rb −47	Sr (168)	In −34	Sn −121	Sb −101	Te −183	I −295	Xe (41)
Cs −46	Ba (52)	Tl −50	Pb −101	Bi −101	Po (−170)	At (−270)	Rn (41)
Fr (−44)	Ra						

Source: Therald Moeller, *Inorganic Chemistry: A Modern Introduction* (New York: Wiley, 1982), p. 81.

the electron affinity values for the beryllium family elements, which have ns^2 configurations, and for the noble gases, with their ns^2np^6 configurations, are positive (Table 9.8). The ability of such nuclei to bind electrons in unoccupied energy levels is extremely small, and the positive electron affinities suggest that negative ions of these elements would be unstable. Second electron affinities are *all* positive, for it takes energy to overcome the repulsion between an electron and the already-negative ion.

As expected, the halogens have the most highly negative electron affinities and metals generally have large positive electron affinities. The plot of first electron affinities in Figure 9.10 reveals some periodic trends. For example, examine the values

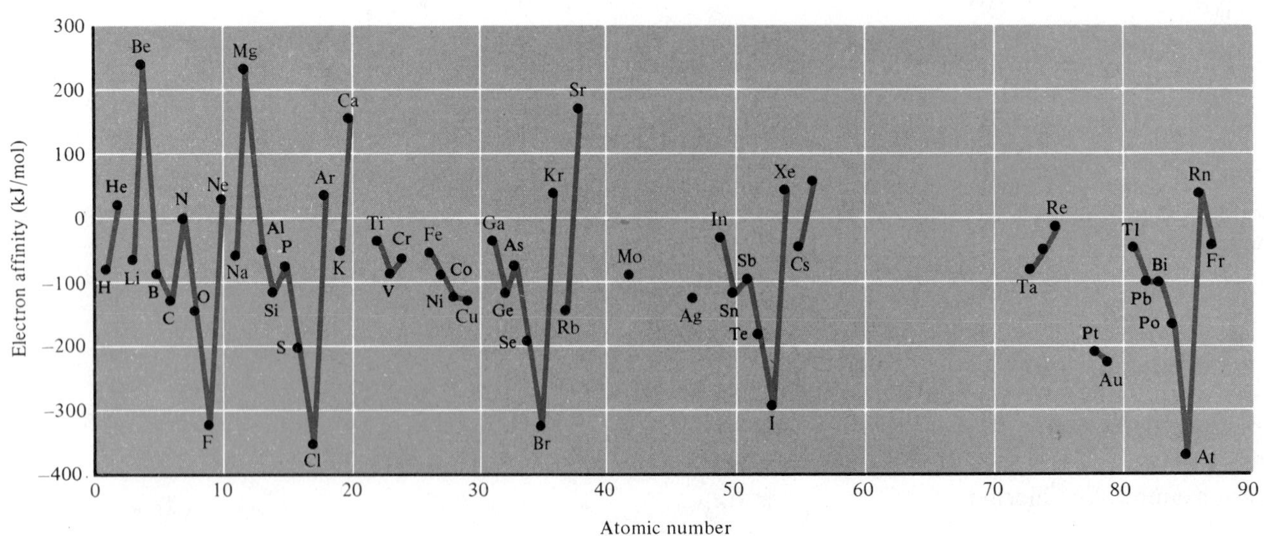

Figure 9.10
Variation of Electron Affinity with Atomic Number
The halogens, which add electrons most readily of any elements, have the most strongly negative electron affinities.

Source: Therald Moeller, *Inorganic Chemistry: A Modern Introduction* (New York: Wiley, 1982), p. 81.

for the third-period elements. The large increase in electron affinity between sodium and magnesium occurs because the ns^2 subshell of magnesium is filled, and the next electron must enter the p subshell. The small increase between silicon and phosphorus occurs because an electron added to a phosphorus atom enters a p orbital already occupied by one electron. This occurs less readily than addition of an electron to an unoccupied p orbital. Similar variations in the general trend for electron affinities for elements of other periods can be seen in Figure 9.10. Variations in electron affinity down a family are not so easily explained.

SUMMARY OF SECTIONS 9.7–9.9

The ionization energy is the enthalpy change for removal of the least tightly bound electron from an atom or ion in the gaseous state; similarly, electron affinity is the enthalpy change for the addition of one electron. First ionization energies, commonly referred to simply as ionization energies, increase across periods (because of increasing effective nuclear charge) and decrease down families (because of increasing radii). Metals generally have low ionization energies and nonmetals high ionization energies. Periodic trends in electron affinities are less clear-cut, but nonmetals have large negative electron affinities, and the beryllium family elements and the noble gases have positive electron affinities.

SIGNIFICANT TERMS (with section references)

actinides (9.2)
alkali metals (9.1)
alkaline earth metals (9.1)
atomic radii (9.4)
bond length (9.4)
effective nuclear charge (9.5)
electron affinity (9.9)
halogens (9.1)
ionic radii (9.4)
ionization energy (9.7)
isoelectronic (9.6)
lanthanide contraction (9.6)
lanthanides (9.2)
Lewis symbol (9.1)
metals (9.3)
noble gas configuration (9.8)
nonmetals (9.3)
periodic law (9.3)
rare earth elements (9.2)
representative elements, main group elements (9.1)
screening effect (9.5)
semiconducting elements (9.3)
transition elements (9.2)
transuranium elements (9.2)
valence electrons (9.1)

QUESTIONS AND PROBLEMS

Classification of the Elements

9.1 Identify the group, family, and/or other periodic table location of each element with the following outer electron configuration: (a) ns^2np^3, (b) ns^1, (c) $ns^2(n-1)d^{0-2}(n-2)f^{1-14}$.

9.2 Repeat Question 9.1 for (a) ns^2np^5, (b) ns^2, (c) $ns^2(n-1)d^{1-10}$, (d) ns^2np^1.

9.3 Write the outer electron configurations for the (a) alkaline earth metals, (b) d-transition metals, (c) halogens.

9.4 Repeat Question 9.3 for the (a) noble gases, (b) alkali metals, (c) f-transition metals, (d) vanadium family.

9.5 Identify the elements and the part of the periodic table in which the elements with the following configurations are found: (a) $1s^22s^22p^63s^23p^64s^2$, (b) $[Kr]4d^85s^1$, (c) $[Xe]4f^{14}5d^66s^2$, (d) $[Xe]4f^{12}6s^2$, (e) $[Kr]4d^{10}5s^25p^6$, (f) $[Kr]4d^{10}4f^{14}5s^25p^65d^{10}6s^26p^2$.

9.6 Repeat Question 9.5 for (a) $1s^22s^22p^63s^23p^63d^84s^2$, (b) $[Xe]4f^{11}6s^2$, (c) $[Kr]4d^{10}5s^25p^3$, (d) $[Rn]7s^1$, (e) $[Rn]5f^77s^2$, (f) $[Rn]7s^2$.

9.7 Which of the elements in the following periodic table is/are (a) alkali metals, (b) an element with the outer configuration of d^8s^2, (c) lanthanides, (d) p-block representative elements, (e) elements with incompletely filled f-subshells, (f) halogens, (g) s-block representative elements, (h) actinides, (i) d-transition elements, (j) noble gases?

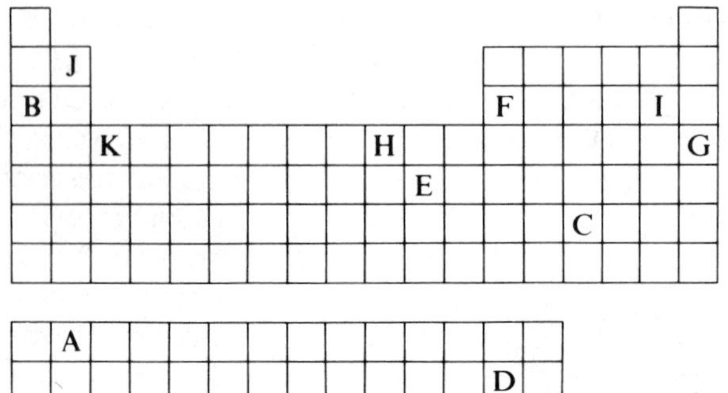

9.8 The following "message" was found in an alien spacecraft:

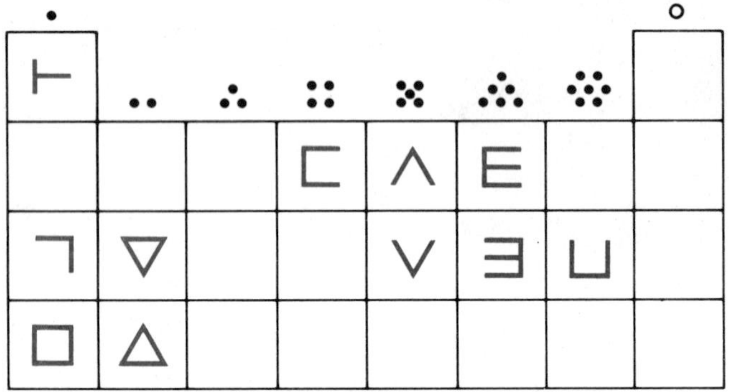

A chemist on Earth quickly recognized the "message" as a portion of the periodic table showing the elements important to the alien life form. List the (a) metals and (b) nonmetals that appear in the table. (c) What is the alien symbol for the halogen that appears in the table? (d) How many alkali metals are listed? (e) Is their life form based on carbon or silicon?

9.9 How many valence electrons are present in an atom of each of the following elements: (a) Rb, (b) As, (c) S, (d) Ne, (e) I?

9.10 Repeat Question 9.9 for (a) Na, (b) Ga, (c) Xe, (d) O, (e) Sr.

9.11 Write the Lewis symbols for (a) Tl, (b) As, (c) Mg, (d) F, (e) O.

9.12 Repeat Question 9.11 for (a) Al, (b) Li, (c) Ar, (d) Te, (e) S.

The Size of Atoms and Ions

9.13 What are two factors that influence atomic radii? What is the trend in atomic radii across a period for the representative elements? What is the trend down a family?

9.14 How does the radius of a cation compare to that of the atom from which it was formed? How does the radius of an anion compare to that of the atom from which it was formed?

9.15 Why are the atomic radii of Al and Ga nearly the same? How does this similarity affect the chemical behaviors of these elements?

9.16 What is the lanthanide contraction? How does this contraction affect the radii of the two heaviest elements in each d-transition element family?

9.17 Using only the arrangement of elements in the periodic table, match the following atomic radii—(i) 0.095 nm, (ii) 0.099 nm, (iii) 0.141 nm, and (iv) 0.175 nm—to the following elements: (a) Sn, (b) Ar, (c) Cl, (d) Pb.

9.18 Repeat Question 9.17 for (a) F, (b) Sr, (c) Br, (d) Li using (i) 0.114 nm, (ii) 0.071 nm, (iii) 0.152 nm, (iv) 0.215 nm.

9.19 Place the following species in order of increasing radius: Cl, Cl^-, Cl^+.

9.20 Repeat Question 9.19 for Fe, Fe^{2+}, and Fe^{3+}.

9.21 Arrange the following ions in order of increasing radius: F^-, Mg^{2+}, Cl^-, Be^{2+}, S^{2-}, Na^+.

9.22 Arrange the following species in order of decreasing radius: Li, O, I, Li^+, O^{2-}, I^-.

9.23 The P—Cl bond length in PCl_3 is 0.204 nm. The bond length in Cl_2 is 0.198 nm. Calculate the atomic radii for these elements. Using the atomic radius for F given in Figure 9.6, predict the P—F bond length in PF_3.

9.24 The bond lengths in F_2 and Cl_2 molecules are 0.142 nm and 0.198 nm, respectively. Calculate the atomic radii for these elements. Predict the Cl—F bond length. [The actual Cl—F bond length is 0.164 nm.]

9.25 The atoms in crystalline nickel are arranged so that they are touching each other in a plane as shown in the sketch:

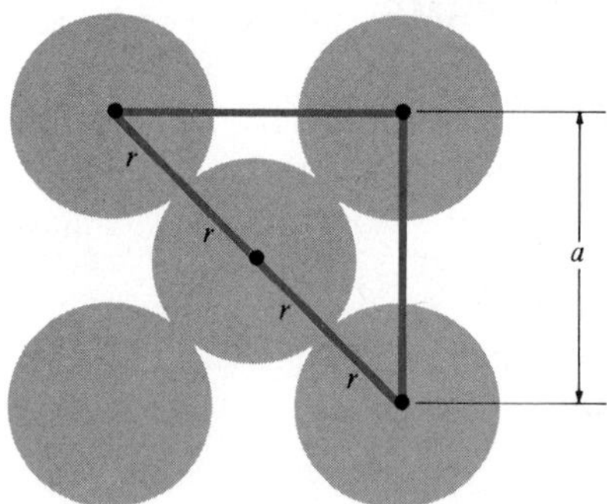

From plane geometry, we can see that $4r = a\sqrt{2}$. Calculate the radius of a nickel atom given that $a = 0.35238$ nm.

9.26* The ions in a plane in crystalline KBr, KCl, and LiCl are arranged as shown in the sketch:

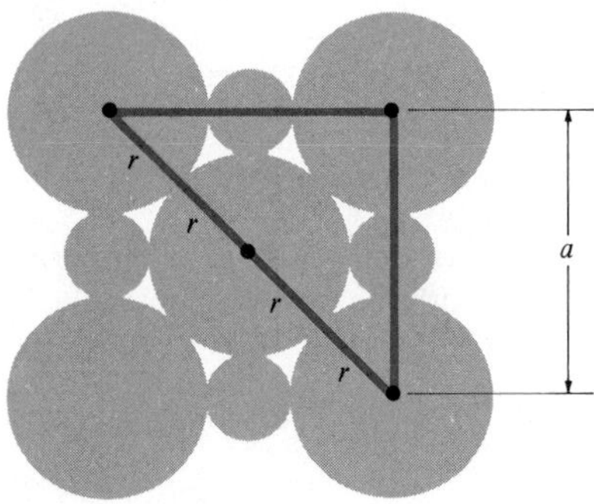

What is the relationship between a and the ionic radii? Using $a = 0.6578$ nm for KBr, 0.62931 nm for KCl, and 0.514 nm for LiCl, and 0.196 nm for the ionic radius for Br^-, calculate the ionic radii for K^+, Cl^-, and Li^+.

Electron Gain and Loss

9.27 Write the electron configurations for (a) Mg, (b) Al, (c) Sc. Predict the cations that these elements might form by the loss of the valence electrons and write the electron configurations for these ions.

9.28 Repeat Question 9.27 for (a) K, (b) Zn, (c) Sn.

9.29 Write the electron configurations for (a) N, (b) S, (c) P. Predict the anions that these elements might form by the addition of electrons and write the electron configurations for these ions.

9.30 Repeat Question 9.29 for (a) F, (b) C, (c) I.

9.31 Write the Lewis symbols for Mg^{2+}, Al^{3+}, Sc^{3+}, N^{3-}, S^{2-}.

9.32 Repeat Question 9.31 for K^+, Zn^{2+}, F^-, C^{4-}, I^-.

9.33 What is the general trend of first ionization energies for atoms in the same period? What is the general trend within a family of representative elements?

9.34 What would you predict as the general trend for values of electron affinity across a period? What is the general trend within a family of representative elements?

9.35 How do the values of the ionization energies of the metals compare to those of the nonmetals? How do these values determine the chemical behaviors of the metals and nonmetals?

9.36 Compare the respective values of the first ionization energy (Table 9.5) and electron affinity (Table 9.8) for several elements. Which energy is greater? Why?

9.37 Compare the respective values of the first ionization energy (Table 9.5) and electron affinity (Table 9.8) for nitrogen to that for carbon and oxygen. Explain why the nitrogen values are considerably different.

9.38 Based on general trends, the electron affinity of fluorine would be expected to be greater than that of chlorine; however, the value is less and is similar to the value for bromine. Explain.

9.39 The ΔH_f° at 25 °C is 78.99 kJ/mol for F(*g*) and −255.39 kJ/mol for $F^-(g)$. Calculate the electron affinity for fluorine at this temperature.

9.40 The ΔH_f° at 25 °C is 159.37 kJ/mol for Li(*g*) and 685.783 kJ/mol for $Li^+(g)$. Calculate the ionization energy of lithium at this temperature.

9.41 The first ionization energy of oxygen is 1313.9 kJ/mol at 0 K. The second ionization energy is 3388.1 kJ/mol. Why is this second value much larger than the first?

9.42 The enthalpy change for the addition of one electron to an oxygen atom is −148 kJ/mol, and for the addition of a second electron, 880 kJ/mol. Explain the difference between these numbers.

ADDITIONAL QUESTIONS AND PROBLEMS

9.43 An atom of element "E" has 15 electrons. Answer as many of the following questions about element E as you can without looking at a periodic table: (a) What is the approximate atomic mass? (b) What is the atomic number? (c) What is the total number of *s* electrons? (d) Is the element a metal, nonmetal, or semiconducting element?

9.44 Pretend we know nothing about element 52, tellurium. From our working knowledge of the periodic law and atomic structure, we can attempt to predict many of its properties. (a) Write a reasonable electron configuration for the 52 electrons and (b) identify the period and family in which the element would appear in the periodic table. (c) Would we expect this element to have properties similar to those of metals, nonmetals, or semiconducting elements? Will the value of the atomic radius be larger or smaller than that of (d) I and (e) Se? (f) Are the atoms paramagnetic?

9.45* Prepare plots of the densities of (a) Cr, Mn, Fe and (b) W, Re, Os against atomic number. Use a reference such as the *Handbook of Chemistry and Physics*. Observe the trend in density. Assume the same trend for Mo, Tc, Ru and (c) predict the density of Tc.

9.46* Using the data given in Tables 9.5 and 9.8 for As, Se, Br, Sb, I, Bi, Po, predict the (a) electron affinity and (b) first ionization energy for tellurium. (c) Compare your answer for the ionization energy to that calculated using the following heat of formation data: 196.73 kJ/mol for Te(g) and 1022.19 kJ/mol for $Te^+(g)$.

CHAPTER

10

Periodic Perspective II: Chemical Bonds

TYPES AND PROPERTIES OF CHEMICAL BONDS

10.1 DEFINITION OF THE CHEMICAL BOND

Few terms in chemistry are as difficult to define as "the chemical bond." However, in 55 B.C. even to Roman poet and philospher Lucretius it was apparent that a force of some kind holds atoms together. He wrote that certain atoms, when they collide, "do not recoil far, being driven into a closer union and held there by the entanglement of their own interlocking shapes." This was an early attempt to define the chemical bond.

Historically, the most difficult problem was in understanding the nature of the force of attraction between atoms in chemical compounds. To solve this problem, it was necessary to discover how electrons are arranged in atoms of different elements (the topic of Chapter 9). It is now believed that all types of chemical bonding are accounted for by the distribution of electrons around the nuclei of bonded atoms.

In modern terms, a **chemical bond** is a force that acts strongly enough between two atoms or groups of atoms to hold them together in a different species that has measurable properties. This definition excludes the weaker forces that attract molecules to each other (Sections 11.10–11.13), but normally do not result in the formation of independent species that can be isolated and studied.

The types of chemical bonds formed by atoms of different elements can be related to the periodic table positions of the elements. Given the role of electrons in chemical bonding, this is not surprising. In this chapter we define the major types of chemical bonds and explore their properties.

10.2 BOND TYPE–PROPERTY RELATIONSHIPS: AN EXAMPLE

Gaseous chlorine and solid sodium react to form sodium chloride:

$$2\text{Na}(s) + \text{Cl}_2(g) \longrightarrow 2\text{NaCl}(s)$$

This is a common type of reaction—the formation of an ionic compound by the

Metals	Nonmetals
Solids of high melting point	Gases or low-melting solids
Lustrous, reflecting light of many wavelengths	Dull, reflecting light poorly or absorbing strongly
High density	Low density
Often hard	Often soft
Malleable, ductile, strong	Brittle and weak if solid
Conductors of heat and electricity	Insulators

Table 10.1
General Physical Properties of Metallic and Nonmetallic Elements
The properties of many elements differ from these general properties. Sodium, e.g., melts below 100 °C, Hg melts at −38.87 °C; C as diamond melts at about 3700 °C and is the hardest of all substances; Sn is brittle above 161 °C; Li has a low density; C as graphite is a conductor; and $I_2(s)$ conducts electricity (weakly).

combination of a metal (Na) and a nonmetal (Cl_2). As we noted in discussing combination reactions (Section 5.1a), both sodium and chlorine are highly reactive elements, and the product of their combination is the high-melting, white crystalline solid that we know as table salt.

Some of the general properties of metals and nonmetals are listed in Table 10.1. Although sodium is softer and less dense than most metals, it does have a high boiling point and the characteristic metallic luster. Like many of the nonmetals, chlorine is a gas composed of molecules in which two atoms are joined together.

The properties of sodium, sodium chloride, and chlorine illustrate the properties associated with the three basic types of chemical bonds—metallic bonds, ionic bonds, and bonds in molecules, known as covalent bonds. Experiments on the electrical conductivities of these substances provide information about their bonding. Recall that for an electrical current to pass through a substance, charged particles, either electrons or ions, must be available to move through the substance (Aside 3.3). The following observations can be made:

1. Chlorine—whether in the gaseous, liquid, or solid state—is an extremely poor conductor of electricity, behavior that is characteristic of nonmetallic substances.
2. Solid sodium apparently offers almost no resistance to the flow of electrical current, for it has very high electrical conductivity. With increasing temperature, the electrical conductivity gradually decreases and at the melting point there is a sharp drop. However, sodium conducts electrical current very well at all temperatures.
3. Solid sodium chloride is a very poor conductor of electricity, but at its melting point the conductivity increases dramatically, as shown in Figure 10.1b. In comparing the conductivities of sodium and sodium chloride in Figure 10.1, note the difference in the electrical conductivity scales. The conductivity of solid sodium is 10^5 times larger than that of molten sodium chloride. Another noteworthy difference is that the conductivity of sodium chloride increases with increasing temperature, whereas that of sodium decreases.
4. If direct current is passed through molten sodium chloride, the elements that have combined—sodium and chlorine—are regenerated at the electrodes.

What conclusions about the distribution of electrons in metals, nonmetals, and ionic compounds can be drawn from these observations? Clearly, chlorine—whether in the gaseous, liquid, or solid states—contains neither freely mobile electrons nor freely mobile ions. Somehow, the electrons succeed in holding two chlorine atoms together in a molecule in such a way that neither ions nor free electrons are present at any temperature.

By contrast, the high electrical conductivity of sodium suggests that it contains highly mobile, charged particles in both the solid and liquid states. Because electrons are considerably lighter than ions, and therefore able to move much more freely, it is reasonable to assume that electrons rather than ions carry the current in sodium.

Furthermore, the electrons in either solid or liquid sodium apparently are free from the atoms and are ready to flow when electrons enter from an external source—

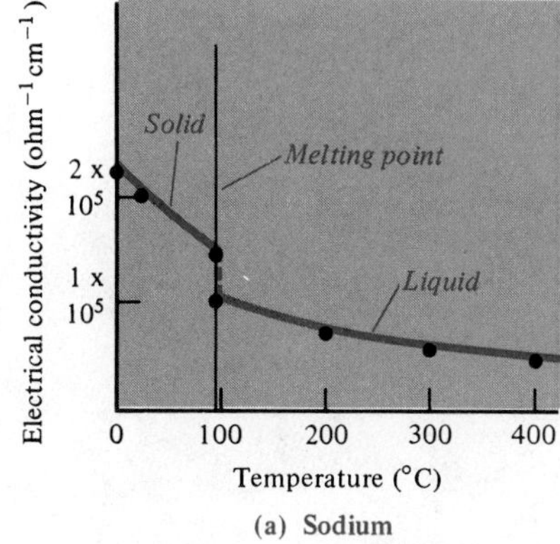

(a) Sodium

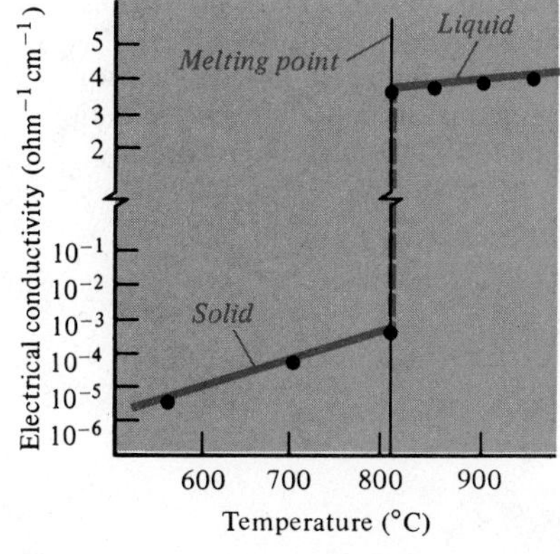

(b) Sodium chloride

Figure 10.1
Electrical Conductivity of Sodium and Sodium Chloride
Note that the scales of these two graphs are quite different. At the melting point, the conductivity of sodium chloride increases 10,000-fold. However, at its lowest point in (a), the conductivity of sodium is still 100,000 times greater than the conductivity of molten sodium chloride.

somewhat like the water in a full water pipe. The presence of such mobile electrons implies that positive metal ions are also present. The decrease in conductivity of a metal with increasing temperature is attributed to increasing vibration of the metal cations, permitting them to interfere with the flow of the electrons.

Solid sodium chloride apparently contains neither mobile electrons nor mobile ions. However, at the melting point current carriers are set free. Because the electrical conductivity is low relative to that of metallic sodium it seems likely that ions rather than electrons have become mobile. As the temperature increases, the ions flow more easily and conductivity increases.

The chemical change that occurs when current flows through the molten sodium chloride confirms the presence of both sodium and chloride ions. The sodium and chloride ions are converted to sodium atoms ($Na^+ + e^- \longrightarrow Na$) and to chlorine molecules ($2Cl^- \longrightarrow Cl_2 + 2e^-$).

In the following sections, metallic, ionic, and covalent bonding—the types of bonding in sodium, sodium chloride, and chlorine, respectively—are described.

10.3 THE METALLIC BOND

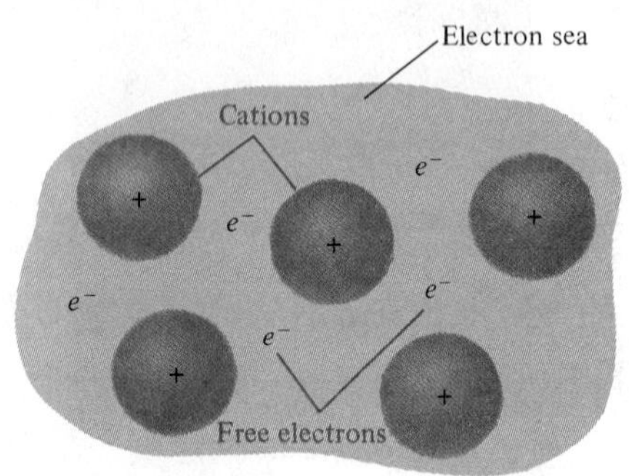

Figure 10.2
Metallic Bonding
Metallic bonding is the attraction between positive ions and surrounding freely mobile electrons. Most metals contribute more than one mobile electron per atom.

a. Bonding in Metals Metals other than those from the *p* block have only one or two electrons in their highest energy levels. In general, metals give up their outer electrons more easily than nonmetals. Each atom in an elemental metal contributes one or more valence electrons to what has been called an "electron sea." The freed valence electrons no longer "belong" to specific atoms. **Metallic bonding** is the attraction between positive metal ions and surrounding, freely mobile electrons (Figure 10.2). Representative metal atoms contribute their outermost *s* electrons or *s* and *p* electrons to the sea of free electrons. In transition metals some of the $(n - 1)d$ electrons can also join the electron sea.

b. Properties Imparted by the Metallic Bond Most substances that have metallic bonding are either elemental metals or alloys. Their general properties are those given in Table 10.1 for the metallic elements. Some compounds, such as titanium nitride (TiN), also conduct electricity and apparently contain free electrons.

Many metals are dense, but when a mechanical force is applied to a metal, the cations can move, sliding along on a "cushion" of free electrons (see Figure 10.3). No specific bonds need be broken, the forces between the cations and the free electrons need not be disrupted, and no additional forces of repulsion are encountered. This explains the ease with which metals are hammered into shape (*malleability*) or drawn out into wires (*ductility*).

The free electrons occupy energy levels so close together that they are essentially merged. Because the electrons have a wide distribution of energies, they can absorb and reemit visible light of many wavelengths, leading to the characteristic luster of metals and many alloys. As is demonstrated by sodium, pure metals (and also many alloys) retain the properties of high electrical conductivity and metallic luster in the liquid state. When the surface of a metal is dull rather than lustrous, it is usually because a layer of metal oxide, sulfide, or some other compound has formed on the surface by reaction of the metal with substances in the atmosphere.

The high electrical conductivity of metals is provided by the mobile electrons, which begin to flow when an electric potential is applied across a piece of metal. Metals

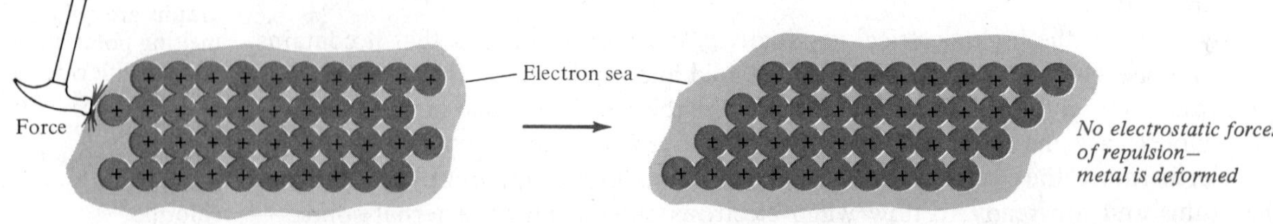

Figure 10.3
Bending a Metal

also conduct heat very well, 10 to 10^5 times better than do most other substances. Unlike the ions in an ionic compound, which are held back by the attraction of oppositely charged ions, the ions in a metal are quite free to vibrate in place. Therefore, when an increase in temperature causes an increase in the random motion of ions and also of free electrons, this kinetic energy is easily passed along to other ions and electrons, explaining the high thermal conductivity of metals.

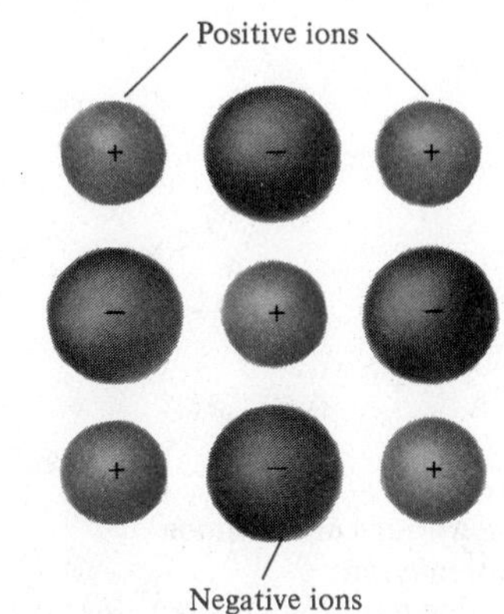

Figure 10.4
Ionic Bonding
Ionic bonding is the attraction between positive and negative ions.

10.4 THE IONIC BOND

a. Bonding in Ionic Compounds **Ionic bonding** is the attraction between positive and negative ions (Figure 10.4). Ionic compounds are simply collections of ions held together by the force of electrostatic attraction between cations and anions. Independent molecules do not exist in ionic compounds, and in the solid state the ions occupy fixed positions in a crystal.

Many common ionic compounds, like sodium chloride, are composed of metal cations and nonmetal anions. In combination reactions like that between sodium and chlorine (Equation 10.1), the overall result is that the electrons given up by sodium atoms to form sodium cations have been gained by chlorine atoms to form chloride ions:

$$\underset{(1s^2 2s^2 2p^6 3s^1)}{\text{Na}\cdot} + \underset{(1s^2 2s^2 2p^6 3s^2 3p^5)}{\cdot\ddot{\underset{..}{\text{Cl}}}:} \longrightarrow \underset{(1s^2 2s^2 2p^6)}{[\text{Na}]^+} \quad \underset{(1s^2 2s^2 2p^6 3s^2 3p^6)}{\left[:\ddot{\underset{..}{\text{Cl}}}:\right]^-} \qquad \textbf{(10.1)}$$

Metals that have large atomic radii and that easily give up their valence electrons undergo such combination reactions most readily. The halogens, with their small radii and high electron affinities, combine with many metals in such reactions.

Chemical bonds tend to form and chemical reactions tend to take place so that substances more stable than the original substances are formed. As we have pointed out (Section 8.13), the noble gases are the least reactive (most stable) of the elements. This lack of reactivity is attributed to the presence of the ns^2np^6 electron configuration (or $1s^2$ in helium). Note in the preceding equation that by the respective loss and gain of an electron, ions that have noble gas configurations have been formed by both sodium and chlorine. In many ionic compounds this is the case—atoms have gained and lost electrons to form ions with eight outer electrons in a noble gas configuration.

A rule that accounts for the formation of many chemical compounds is based upon the observed stability of compounds in which atoms are associated with eight valence electrons. According to the **octet rule,** atoms tend to combine by gain, loss, or sharing of electrons so that the outer energy level of each atom holds or shares four pairs of electrons. As we shall see, the octet rule applies to many ionic and molecular compounds, notably those of elements that closely precede or follow the noble gases in atomic numbers.

b. Electron Configurations of Ions The electron configurations and charges of the ions most common in ionic compounds are summarized in periodic table form in Figure 10.5. Note that the charges on the Representative Group I and II metal ions are the same as the group number. Atoms of these *s*-block elements lose their *s* subshell electrons to give singly or doubly charged cations with noble gas configurations. By losing two *s* electrons and one *p* electron, aluminum can also form an ion with a noble gas configuration and a charge equal to its group number (Al^{3+}).

The Group III metals of periods 4–6 form both +3 and +1 cations by the loss of their single *p* electron, or all three outer electrons. For example, for gallium—which has the configuration $[Ar]3d^{10}4s^24p^1$—the ions are

$$Ga^+ \; [Ar]3d^{10}4s^2 \qquad Ga^{3+} \; [Ar]3d^{10}$$

A noble gas core with an outer $\boldsymbol{d^{10}}$ **configuration**, as in Ga^{3+}, is sometimes known as a **pseudo-noble gas configuration**. A "d^{10} cation" is stable because the highest principal

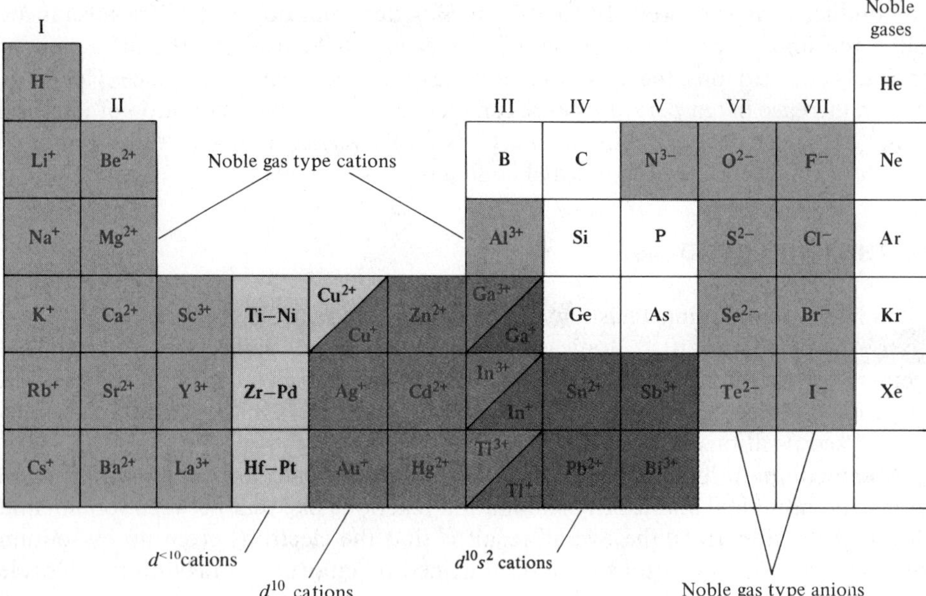

Figure 10.5
Configurations of Common Monatomic Ions
The noble gas type ions have ns^2 or ns^2np^6 configurations. The d^{10} cations form by loss of s electrons to leave filled d subshells. The $d^{10}s^2$ cations form by loss of p electrons to leave $(n-1)d^{10}ns^2$ configurations. Atoms of transition elements that form $d < 10$ cations do so by loss of s or s plus d electrons.

energy level is completely filled by 18 electrons, with 10 electrons in the d subshell.

The metals of Groups IV and V do not form d^{10} cations, but form only the +2 and +3 ions, respectively, that result from loss of the valence electrons from the outer p subshells. For example, Bi^{3+} $[Xe]4f^{14}5d^{10}6s^2$, is what might be called a "$d^{10}s^2$ cation."

Among the p-block elements, boron, carbon, silicon, germanium, phosphorus, arsenic and the noble gases are rarely found as monatomic ions in ordinary compounds. For the nonmetals that do form ions, it is easier to gain electrons to achieve noble gas configurations than it is to give up electrons. A sulfur atom, for example, would have to lose six electrons to form an ion with the neon configuration. Removing successive electrons from positive ions becomes increasingly more difficult and the energy needed to remove this many electrons is not available in ordinary chemical reactions. (Adding successive electrons to anions is also increasingly difficult. Compounds containing ions with charges greater than +3 or −3 are rare.)

The nonmetals each form only one monatomic anion. These anions have negative charges equal to the number of electrons needed to give eight outer electrons. For example, for sulfur from Group VI and iodine from Group VII

$$\text{S} \quad [\text{Ne}]3s^2 3p^4 \qquad \text{S}^{2-} \quad [\text{Ne}]3s^2 3p^{4+2}$$

added electrons

$$\text{I} \quad [\text{Kr}]4d^{10}5s^2 5p^5 \qquad \text{I}^- \quad [\text{Kr}]4d^{10}5s^2 5p^{5+1}$$

Hydrogen gives an anion with the helium configuration (the hydride ion, H^-). The semiconducting elements selenium and tellurium also form anions with noble gas configurations (see Figure 10.5).

To predict the charges of ions formed by the d-transition elements on the basis of electron configurations alone is difficult. Because the ns and $(n-1)d$ subshells are similar in energy, the d electrons are often involved in bonding. Most d-transition elements give cations with +2 charges, in some cases by losing one s and one d electron. As illustrated in Table 10.2 and Figure 10.5, some d-transition elements also form +3 or +1 ions. Note, however, that the s electrons are always the first to be lost in the ionization of d-transition metal atoms.

Table 10.2
Some Fourth-Period Transition Metal Ions
The Cu^{2+} ion is more common than the Cu^+ ion, and the Ti^{2+} ion is rare.

Atom	Ion
Sc$[Ar]3d^1 4s^2$	Sc^{3+} [Ar]
Ti$[Ar]3d^2 4s^2$	Ti^{2+} $[Ar]3d^2$
	Ti^{3+} $[Ar]3d^1$
Cr$[Ar]3d^5 4s^1$	Cr^{2+} $[Ar]3d^4$
	Cr^{3+} $[Ar]3d^3$
Fe$[Ar]3d^6 4s^2$	Fe^{2+} $[Ar]3d^6$
	Fe^{3+} $[Ar]3d^5$
Cu$[Ar]3d^{10} 4s^1$	Cu^+ $[Ar]3d^{10}$
	Cu^{2+} $[Ar]3d^9$

EXAMPLE 10.1 Electron Configurations of Ions

Write the electron configurations of the following ions, using the noble gas notation for their inner electrons: (a) Se^{2-}, (b) Ni^{2+}, (c) Cs^+, (d) Mo^{2+}.

(a) Selenium is the fourth-period Group VI element and therefore has an argon core, ten 4*d* electrons, and six valence electrons. Adding two electrons to account for the −2 charge gives **[Ar]$4d^{10}5s^25p^6$**.

(b) Nickel is the eighth *d*-transition metal in the fourth period and therefore has a total of ten 4*s* and 3*d* electrons, and an expected outer configuration for the neutral atoms of $3d^84s^2$ (Sections 8.12 and 8.13). Checking the periodic table inside the front cover shows that this is its configuration. To form a +2 ion, the two *s* electrons would be lost, **[Ar]$3d^8$**.

(c) Cesium is the sixth-period Group I element and therefore has a xenon core and one 6*s* electron, [Xe]$6s^1$. The electron configuration of the +1 ion is simply **[Xe]**.

(d) Molybdenum is the fourth transition metal in the fifth period and therefore has a total of six 5*s* and 4*d* electrons, giving an expected outer configuration of $4d^45s^2$. Checking shows that its outer configuration is actually $4d^55s^1$. To form a + 2 ion, molybdenum must lose one *s* and one *d* electron, to give **[Kr]$4d^4$**.

EXERCISE Write the electron configurations for the following ions: (a) Ba^{2+}, (b) Ag^+, (c) Cd^{2+}, (d) Pb^{2+}. *Answers* (a) [Xe], (b) [Kr]$4d^{10}$, (c) [Kr]$4d^{10}$, (d) [Xe]$4f^{14}5d^{10}6s^2$.

EXAMPLE 10.2 Electron Configurations of Ions

Show with Lewis symbols and electron configurations the changes that occur in the individual atoms (a) when strontium and oxygen combine to give strontium oxide and (b) when magnesium combines with nitrogen to give magnesium nitride. Both strontium oxide and magnesium nitride are ionic compounds. What are the formulas of the products?

(a) Strontium is in Representative Group II and forms Sr^{2+}. Oxygen is in Representative Group VI and forms O^{2-}.

$$\underset{[Kr]5s^2}{\text{Sr:}} + \underset{[He]2s^22p^4}{\cdot\ddot{\text{O}}\text{:}} \longrightarrow \underset{[Kr]}{Sr^{2+}} \quad \underset{[He]2s^22p^6 \text{ or } [Ne]}{:\ddot{\text{O}}:^{2-}}$$

Strontium oxide has the formula **SrO**.

(b) Magnesium is in Representative Group II and forms Mg^{2+}. Nitrogen is in Representative Group V and forms N^{3-}. The charges on these ions are not equal, and it takes three magnesium atoms to provide the six electrons needed to form two N^{3-} ions

$$\underset{[Ne]3s^2}{3\text{Mg:}} + \underset{[He]2s^22p^3}{2\cdot\dot{\ddot{\text{N}}}\cdot} \longrightarrow \underset{[Ne]}{3Mg^{2+}} \quad \underset{[He]2s^22p^6 \text{ or } [Ne]}{2:\ddot{\text{N}}:^{3-}}$$

The product, magnesium nitride, has the formula $\mathbf{Mg_3N_2}$. [Note that Mg^{2+} and N^{3-} both have the same electron configuration, but, because of their different molar masses and charges, they are very different species.]

EXERCISE Barium reacts with chlorine to produce an ionic compound. What is the formula of the compound? Show with Lewis symbols and electron configurations the changes that occur in the individual atoms during this reaction.

Answers $BaCl_2$;

$$\underset{[Xe]6s^2}{\text{Ba:}} + \underset{[Ne]3s^2\,3p^5}{2\cdot\ddot{\text{Cl}}\text{:}} \longrightarrow \underset{[Xe]}{Ba^{2+}} \quad \underset{[Ar]}{2[:\ddot{\text{Cl}}:]^-}$$

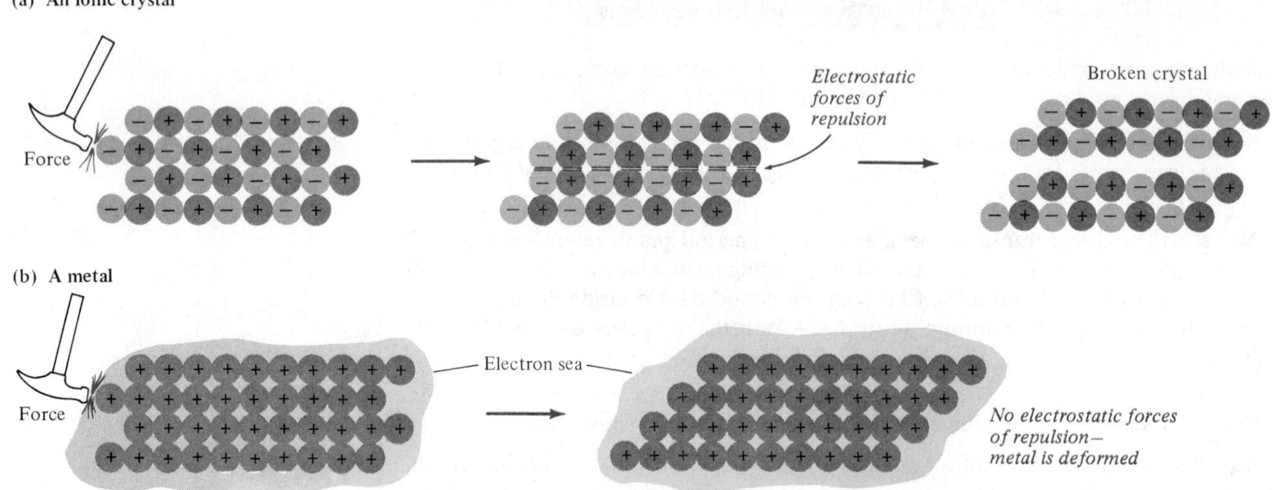

Figure 10.6
Shattering an Ionic Crystal

c. Properties Imparted by the Ionic Bond Each ion in a solid ionic crystalline substance is surrounded by other ions of opposite charge. For example, in a sodium chloride crystal (see Figure 13.25) each Na^+ ion is surrounded by six Cl^- ions and each Cl^- ion is surrounded by six Na^+ ions. The electrostatic force of attraction holding the ions together is strong. Any changes that require disrupting the arrangement of ions in a crystalline ionic compound, therefore, require a large amount of energy. As a result, ionic compounds have high melting points and boiling points and high heats of vaporization and fusion. For the same reasons, ionic crystalline substances are hard—a strong force is needed to break up the crystal lattice. However, such substances are brittle and when struck with sufficient force shatter along the planes between rows of ions (Figure 10.6).

The ratio of the numbers of cations and anions in a crystal is the ratio shown in the formula of the compound—the ratio necessary for electroneutrality. The geometrical arrangement of the ions is determined by the number of ions of each kind (Section 13.10) and by their sizes. The density of ionic compounds varies with the spacing of ions in the crystal and also, of course, with the masses of the ions. Ionic solids are less dense than most metals.

Most solid ionic compounds are poor conductors of electricity because the ions are rigidly fixed in their positions. The ions become free to move and conduct electricity when ionic compounds melt or when they dissolve in water. Ionic solids are also not very good conductors of heat, for the ions do not easily pass kinetic energy along to their neighbors.

10.5 THE COVALENT BOND

a. Bonding in Molecular Compounds When a chemical bond forms between two nonmetal atoms, a molecule is produced, as in H_2 or Cl_2 or HCl. The bond that holds atoms together in such molecules is the result of the sharing of valence electrons, leading to the name "covalent" for such bonding. Like ionic bonds, covalent bonds often form so that the bonded atoms achieve octet configurations. For example, a chlorine atom is one electron short of an octet configuration. In the Cl_2 molecule, two $3p$ electrons are shared by the two chlorine atoms, permitting each atom to achieve a stable octet:

the shared pair of electrons

:C̤̈l:Cl:

$$1s^22s^22p^63s^23p^23p^2 \; \boxed{3p^1 \quad 3p^1} \; 3p^23p^23s^22p^62s^21s^2$$

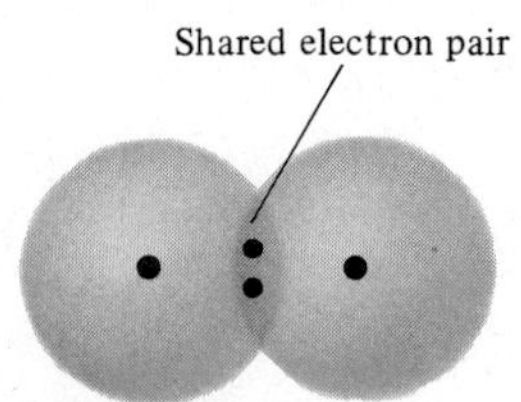

Figure 10.7
Covalent Bonding
The bonded atoms come together sufficiently for their electron clouds to overlap.

Structures in which Lewis symbols are combined so that the bonding and nonbonding outer electrons are indicated are called **Lewis structures** (or electron dot structures). The symbols are placed so that bonding electrons are represented by dots between the symbols. Sometimes, as above, both dots and Xs are used in Lewis structures to show the atomic origin of the electrons. However, the electrons in the bond are equivalent and indistinguishable from each other.

In an H_2 molecule, the two 1*s* electrons are shared to give a stable two-electron configuration to each hydrogen atom. The sharing of a 1*s* electron from one hydrogen atom with the unpaired 3*p* electron from one chlorine atom gives a hydrogen chloride molecule in which the hydrogen and chlorine atoms both have noble gas configurations

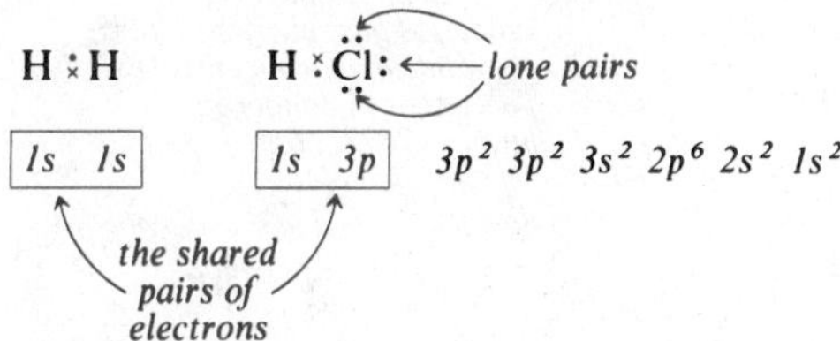

Pairs of valence electrons not involved in bonding are called **lone electron pairs** (also nonbonding electron pairs or unshared electron pairs). The chlorine atoms in Cl_2 and HCl each have three lone electron pairs.

Covalent bonding is based upon electron-pair sharing and is the attraction between two atoms that share electrons (Figure 10.7). A **single covalent bond** is a bond in which two atoms are held together by sharing two electrons. The electrons in the bond spend enough time in the space between the two atoms to provide the "glue" that holds the atoms together. Each positive nucleus is attracted toward the region of high electron density between them.

Covalent bonds form between two atoms when the sharing of electrons requires less energy than electron loss (for metallic bonding) or electron loss and gain (for ionic bonding). The most obvious example is the bonding of nonmetal atoms with themselves (as in, e.g., Cl_2, H_2, O_2) and with each other (as in, e.g., HCl, H_2O, SO_3). In bond formation, the semiconducting elements tend to behave like the nonmetals, forming covalent bonds in many compounds with nonmetals (e.g., BCl_3, SiF_4). Covalent bonds also form between metal atoms and nonmetal atoms. For example, tin tetrabromide ($SnBr_4$) is a molecular compound—the loss of four electrons from the tin atom requires more energy than normally is available.

The number of valence electrons and the octet rule govern the formation of many covalent compounds. For second-period elements (Li to F), which have only *s* and *p* orbitals available, eight is the *maximum number* of valence electrons that can be accommodated.

Carbon, nitrogen, and oxygen form a great many molecular compounds that follow the octet rule. The carbon atom, which has four valence electrons, forms four covalent bonds to achieve an octet. A nitrogen atom (five valence electrons) forms three covalent bonds and retains a lone pair of electrons to achieve an octet configuration, whereas oxygen (six valence electrons) forms two covalent bonds and retains two lone pairs of electrons. The halogen atoms, with seven valence electrons, join in single covalent bonds in many compounds to complete their octets.

The compounds listed in Table 10.3 illustrate simple covalent bonding to carbon, nitrogen, and oxygen atoms by hydrogen and chlorine atoms. As is customary, a line is drawn to indicate each shared electron pair. The nonbonding electron pairs, shown in the table to illustrate how each atom is surrounded by an octet, are often not included. In many compounds, as illustrated in the table, nitrogen and carbon atoms are the central atoms and are surrounded by hydrogen or halogen atoms.

More than one pair of electrons can be shared between the same two atoms, resulting in what is called a **multiple covalent bond.** In a **double covalent bond** two electron pairs are shared between the same two atoms.

Table 10.3
Some Simple Molecular Compounds
All atoms in these compounds, except for hydrogen atoms, follow the octet rule.

C (Group IV) 4 valence electrons 4 shared pairs	N (Group V) 5 valence electrons 3 shared pairs	O (Group VI) 6 valence electrons 2 shared pairs
methane, CH_4 *colorless gas, nonpoisonous and flammable* *m.p. −183 °C, b.p. −161 °C*	*ammonia, NH_3* *colorless gas, pungent odor (inhalation of concentrated vapor is dangerous)* *m.p. −78 °C, b.p. −33 °C*	*water, H_2O* *colorless liquid* *m.p. 0 °C, b.p. 100 °C*
carbon tetrachloride, CCl_4 *colorless liquid, dense and nonflammable, poisonous* *m.p. −23 °C, b.p. 77 °C*	*nitrogen trichloride, NCl_3* *yellow, oily liquid, very unstable and explosive* *m.p. −91.7 °C, b.p. 71 °C*	*chlorine monoxide, Cl_2O* *yellowish brown gas, explosive* *m.p. −116 °C, b.p. 2 °C*

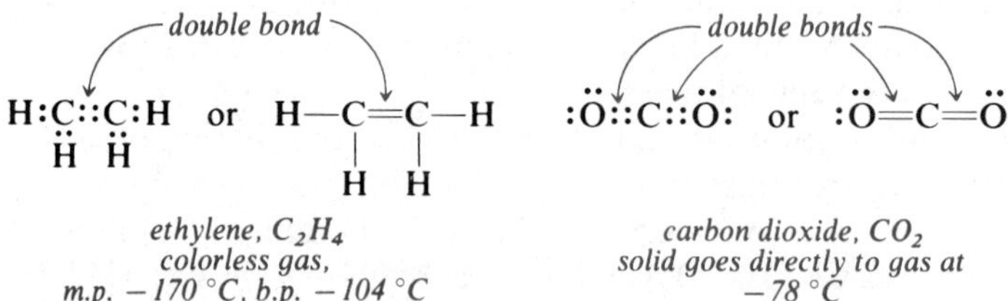

ethylene, C_2H_4
colorless gas,
m.p. −170 °C, b.p. −104 °C

carbon dioxide, CO_2
solid goes directly to gas at −78 °C

In a **triple covalent bond** three electron pairs are shared between the same two atoms.

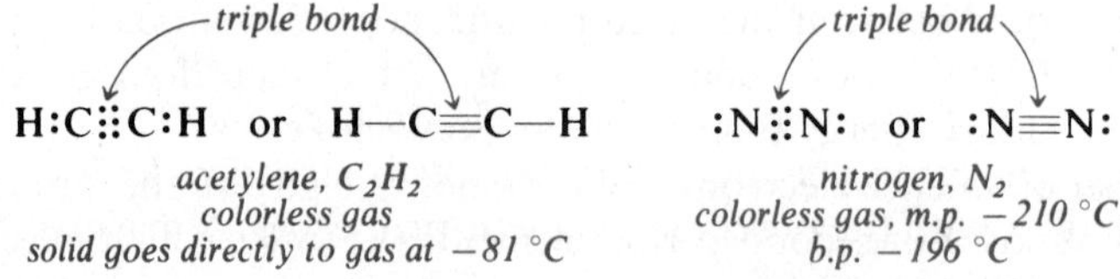

acetylene, C_2H_2
colorless gas
solid goes directly to gas at −81 °C

nitrogen, N_2
colorless gas, m.p. −210 °C, b.p. −196 °C

In the compounds above with double and triple covalent bonds, note that multiple-bonded atoms share electrons so that each is surrounded by eight electrons.

Carbon, nitrogen, and oxygen are the most common participants in multiple covalent bonds. Carbon and nitrogen can form either double or triple bonds. Oxygen is doubly bonded in many compounds and triply bonded in a few cases, most notably, carbon monoxide, :C≡O:. Phosphorus, sulfur, and selenium can also form multiple bonds in some compounds.

b. Exceptions to the Octet Rule Many compounds are known in which the central atom is *not* surrounded by eight electrons. Most of these compounds fall into one of the three categories described below.

1. Certain molecular compounds of beryllium, boron, and aluminum: Beryllium, from Representative Group II, has two valence electrons and boron and aluminum, from Group III, each has three valence electrons. Under certain circumstances (mainly in the gaseous state) these elements are found in compounds in which they have only two or three covalent bonds, respectively, and therefore less than an electron octet.

:F̤—Be—F̤:
beryllium difluoride

:C̤l: / B / :C̤l: :C̤l:
boron trichloride

2. Certain molecular compounds of phosphorus, sulfur, chlorine, or other elements from the third or higher periods: Atoms of elements from the third and higher periods have valence electrons in the outermost *s* and *p* subshells, but no electrons in the outermost *d* subshell. (Recall that it is the $(n - 1)d$ subshell that is occupied in the *d*-transition metals.) Therefore, if electrons occupy these empty *d* subshells, atoms of such elements can be surrounded by more than eight electrons. For example, in sulfur hexafluoride (SF_6) the sulfur atom shares six electron pairs. An isolated sulfur atom has six valence electrons in *s* and *p* subshells ($3s^2\,3p^4$), and therefore room for only two more electrons in these occupied subshells. Of necessity, some of the shared electron pairs in the SF_6 molecule must utilize empty 3*d* subshells.

   In some cases, compounds with fewer than four covalent bonds to the central atom can also have more than an octet of electrons. In chlorine trifluoride, for example, the chlorine atom is surrounded by two unshared pairs of electrons and three shared pairs of electrons. To accommodate these 10 electrons, the empty *d* subshells must obviously also be utilized. (How the *d* subshells are involved in covalent bonding in such compounds is discussed in Section 11.8.)

12 electrons in outermost level of S

10 electrons in outermost level of Cl

sulfur hexafluoride
extremely stable colorless gas, m.p. −50 °C b.p. 64 °C

chlorine trifluoride
very reactive colorless gas, m.p. 83 °C b.p. 11.3 °C

3. Compounds with unpaired electrons: Some compounds exist in which one or more electrons remain unpaired. In most the total of the valence electrons of the central atom and the atoms bonded to it is an odd number. Chlorine dioxide (ClO_2), for example, has a total of 19 valence electrons (6 from each of the two oxygen atoms and 7 from the chlorine atom). Measurements of the magnetism of this compound (see Figure 8.21) show the presence of one unpaired electron per molecule. The Lewis structure for such a compound must agree with the observed magnetic properties. A reasonable Lewis structure for chlorine dioxide is

:Ö:C̤l:Ö:

chlorine dioxide
unstable and explosive yellow-red gas m.p. −59.5 °C b.p. 9.9 °C

c. Coordinate Covalent Bonds So far, we have considered single covalent bonds to which each atom contributes one electron. Sometimes, both electrons are provided to a bond by the same atom. A single covalent bond in which both electrons in the shared pair come from the same atom is called a **coordinate covalent bond.**

The **donor atom** provides both electrons to a coordinate covalent bond, and the **acceptor atom** accepts an electron pair for sharing in a coordinate covalent bond. For a coordinate covalent bond, as for any other kind of bond, it is impossible to distinguish among the electrons once the bond has formed. For example, a hydrogen ion unites with an ammonia molecule by a coordinate covalent bond to form the ammonium ion (NH_4^+)

donor, acceptor

$$H:\ddot{N}(H)(H): + H^+ \longrightarrow [H:\ddot{N}(H)(H):H]^+$$

but all four hydrogen atoms and all four nitrogen–hydrogen bonds in the ammonium ion are found by experiment to be equivalent.

d. Polyatomic Ions In polyatomic ions such as the ammonium ion or the sulfate ion ($SO_4{}^{2-}$)

```
        ..      2-
       :O:
    .. .. ..
  [ :O:S:O: ]
    .. .. ..
       :O:
        ..
```

sulfate ion, SO_4^{2-}

a central atom is covalently bonded to other atoms. The ions are charged because they have fewer or more electrons than are needed to balance the positive charges of the nuclei present in the ion. In the ammonium ion, one valence electron for each of four hydrogen atoms and the normal number of five valence electrons for the nitrogen atom would give a total of nine electrons. With eight electrons present, the ammonium ion is short one electron and thus has a + 1 charge. The sulfate ion, in addition to the 30 valence electrons from four oxygen atoms and one sulfur atom, has 2 electrons gained from some other species that easily gives up electrons. As a result, the ion as a whole has a −2 charge.

e. Properties Imparted by the Covalent Bond Covalent bonds hold atoms together in discrete molecules (or in polyatomic ions). The covalent bonds to a given atom have a specific spatial arrangement determined by the types of orbitals occupied by the bonding electrons. Therefore, individual molecules have distinctive geometries.

Molecules usually remain intact when a molecular compound melts, evaporates, or dissolves. As a result, it is not the bonding forces but the weaker forces that act between molecules—the intermolecular forces (Sections 11.11–11.13)—that determine many of the properties of molecular compounds. For example, liquid and solid molecular compounds generally have relatively low melting and boiling points, because in these changes of state it is only the intermolecular forces of attraction that must be overcome. For similar reasons, solids composed of molecular compounds are usually either soft and waxy, or brittle and easily broken up. The geometry of molecules and the various types of intermolecular forces are explored in the next chapter, which is fully devoted to the properties of molecules and molecular compounds. The general properties of ionic and molecular compounds are compared in Table 10.4.

Certain substances such as diamond (crystalline carbon), borazon (one form of boron nitride, BN), carborundum (silicon carbide, SiC), and quartz (silicon dioxide, SiO_2) resemble molecular compounds, except that they are very hard and have very

Table 10.4
General Properties of Ionic and Molecular Compounds
Melting and boiling points, and heats of vaporization and fusion all tend to increase with increasing molecular mass for similar molecular compounds.

Ionic compounds	Molecular compounds[a]
Crystalline solids	Gases, liquids, or solids
Hard and brittle	Solids brittle and weak, or soft and waxy
High melting points	Low melting points
High boiling points (approx. range, 700 °C to 3500 °C)	Low boiling points (approx. range, −250 °C to 600 °C)
High heats of vaporization	Low heats of vaporization
High heats of fusion	Low heats of fusion
Good conductors of electricity when molten; poor conductors of heat and electricity when solid	Poor conductors of heat and electricity
Many soluble in water	Many insoluble in water but soluble in organic solvents
Many formed by combination of reactive metals with reactive nonmetals	Many formed by combination of nonmetals with other nonmetals or with less reactive metals

[a] Many of these properties do not apply to network covalent compounds.

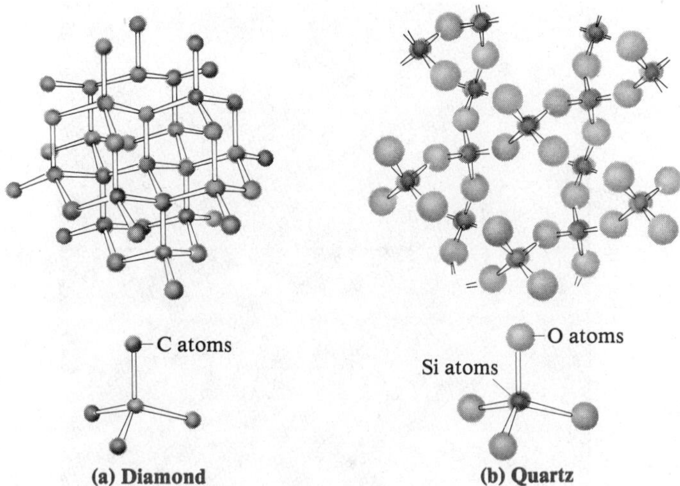

Figure 10.8
Network Covalent Substances
These are three-dimensional arrays of covalently bonded atoms. Diamond and quartz structures are shown as examples. On a hardness scale from 1 to 10, diamond has a hardness of 10 and quartz of 7. For comparison, rock salt has a hardness of 2; marble, of 3; and iron, of 4–5 on the same scale.

high melting points (approximate range 2000 °C–6000 °C). In these materials no discrete molecules are present. The diamond and the other materials like it are **network covalent substances**—three-dimensional arrays of covalently bonded atoms.

In diamond (Figure 10.8a) each carbon atom is covalently bonded to four other carbon atoms, each of which is bonded to three more carbon atoms, and so on throughout the entire diamond crystal. In quartz (Figure 10.8b), a form of silica (SiO_2), each Si atom is bonded to four O atoms and each O atom is bonded to two Si atoms in a three-dimensional array. Because many covalent bonds must be broken to disrupt the crystals, hardness, high melting and boiling points, and high heats of fusion and vaporization are characteristic of these substances.

For network covalent substances as for ionic substances, the simplest formula represents only the stoichiometric composition. A piece of the substance is a giant molecule of molecular mass determined only by the size of the piece.

10.6 LENGTH AND STRENGTH OF COVALENT BONDS

Suppose that two atoms sufficiently far apart to have no detectable influence on one another approach along a straight line. The nucleus of each atom is gradually attracted more and more by the electron cloud of the other. But there is also repulsion between the two nuclei and between the two electron clouds. As the distance between the two atoms decreases (from right to left on the graph in Figure 10.9), the attraction is initially stronger than the repulsion and the potential energy of the atoms decreases. At a certain distance, however, the repulsion becomes stronger and the potential energy of the atoms increases again. Where the attraction and repulsion balance, the atoms are at the bottom of an "energy well" and their potential energy is at a minimum. The average distance between the nuclei of the two atoms at this point is the bond length. The energy that would be needed to pull them apart from this distance is the bond strength, or bond energy.

The sum of the atomic radii of two atoms gives a reasonable estimate of the length of a single covalent bond between the two atoms. However, exact bond lengths must be determined experimentally, for they vary somewhat with the type of compound.

The heats of the reactions in which bonds are broken are measures of the strengths of chemical bonds. Bond energies are determined by measuring heats of reactions in the gaseous state, where the atoms formed are distinctly separate and free from attraction or repulsion by other atoms, molecules, or ions.

In some reactions it is possible to rupture a single chemical bond to give the free atoms.

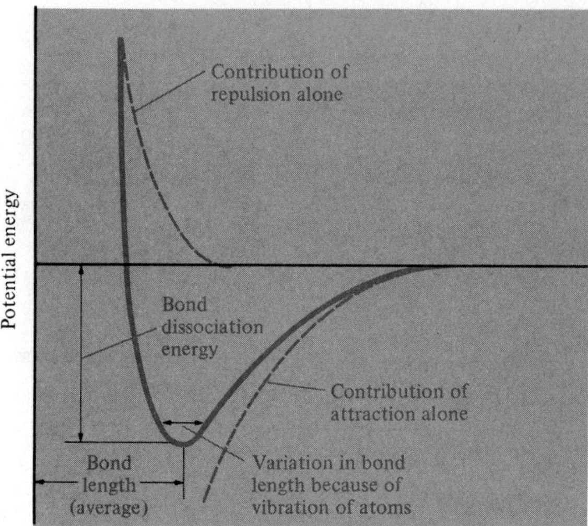

Figure 10.9
Energy–Distance Relationship in Bond Formation
The values of the bond dissociation energy and the bond length vary for different molecules. However, if bond formation is to occur, the general shape of the curve is as shown.

$$H_2(g) \longrightarrow H(g) + H(g) \qquad \Delta H^\circ = 436 \text{ kJ}$$
$$HI(g) \longrightarrow H(g) + I(g) \qquad \Delta H^\circ = 298 \text{ kJ}$$

Bond dissociation energy is the enthalpy change per mole required to break exactly one bond of the same type per molecule. When it is only possible to rupture several bonds in a single reaction, bond energies that are averages must be used. For example, four carbon–hydrogen bonds are broken in the gas-phase dissociation of methane.

$$CH_4(g) \longrightarrow C(g) + 4H(g) \qquad \Delta H^\circ = 1663 \text{ kJ}$$

One-fourth of the enthalpy change in the preceding reaction is the average bond energy of the C—H bond, 415.8 kJ/mol. **Bond energy** is the *average* enthalpy change per mole for breaking one bond of the same type per molecule. Bond breaking is an endothermic process, and bond energies are positive enthalpies. In general, the larger the bond energy, the stronger the bond.

The heat of a reaction in the gas phase is equal to the sum of the energy required to break all of the bonds in the reactants and the energy released by the formation of all of the bonds in the products.

$$\Delta H^\circ = \begin{pmatrix} \text{sum of heats of} \\ \text{breaking all bonds} \\ \text{in reactants} \end{pmatrix} + \begin{pmatrix} \text{sum of heats of} \\ \text{forming all bonds} \\ \text{in products} \end{pmatrix} \qquad \textbf{(10.2)}$$

The first term in Equation (10.2) is the sum of the bond energies for all of the bonds in the reactants. Bond formation is an exothermic process and is the reverse of bond breaking, meaning that the heat of forming a bond has the same value as the bond energy, but is negative rather than positive. The second term in Equation (10.2) is, therefore, equivalent to −(sum of bond energies for all of the products). These relationships allow Equation (10.2) to be rewritten in the following useful form

$$\Delta H^\circ = \begin{pmatrix} \text{sum of bond energies} \\ \text{of all bonds in reactants} \end{pmatrix} - \begin{pmatrix} \text{sum of bond energies} \\ \text{of all bonds in products} \end{pmatrix} \qquad \textbf{(10.3)}$$

The accuracy of the calculation of ΔH° by using Equation (10.3) is limited, because the strength of a chemical bond varies somewhat depending upon its environment in a given molecule. Also, note that Equations (10.2) and (10.3) apply only when all reactants and products are gases. When other states of matter are involved, the contribution to ΔH° of the heats of vaporization or sublimation must be included.

For single covalent bonds, bond lengths range from roughly 0.05 to 0.2 nm and bond energies, given in Table 10.5, from about 160 to 600 kJ/mol. The relationship between

Table 10.5
Bond Energies (in kJ/mol)

	C	N	S	O	I	Br	Cl	F	H
H—	414	389	368	464	297	368	431	569	435
F—	490	280	343	213	280	285	255	159	
Cl—	326	201	272	205	209	218	243		
Br—	272	163	209	—	176	192			
I—	218	—	—	—	151				
O—	326	230	423	142					
O=	803[a]	590[b]	523	498					
O≡	1075	—	—	—					
S—	289	—	247						
S=	582	—	—						
N—	285	159							
N=	515	473							
N≡	858	946							
C—	331								
C=	590[c]								
C≡	812								

[a] Value for CO_2; 728 if —C(—)=O

[b] 406 if $—NO_2$
368 if $—NO_3$

[c] 506 if alternating — and =

bond length and bond strength is illustrated in Table 10.6. In general, shorter bonds are stronger. Multiple bonds, which have much greater electron density between the atoms, are always stronger and shorter than single bonds. In Table 10.6, compare the values for C—C, C=C, and C≡C.

Table 10.6
Some Bond Lengths and Bond Strengths

	Bond length (nm)	Bond energy (kJ/mol)
H—H	0.074	435
H—Cl	0.127	431
Cl—Cl	0.198	243
H—C	0.109	414
C—C	0.154	331
C=C	0.134	590
C≡C	0.120	812
C—O	0.143	326
C=O	0.120	803
C≡O	0.113	1075
N—N	0.145	159
N=N	0.125	473
N≡N	0.110	946

EXAMPLE 10.3 Bond Energies

Estimate the heat released as one mole of *n*-butane burns. The necessary bond energies are given in Table 10.5.

$$\mathrm{H{-}\underset{H}{\overset{H}{C}}{-}\underset{H}{\overset{H}{C}}{-}\underset{H}{\overset{H}{C}}{-}\underset{H}{\overset{H}{C}}{-}H}(g) + \tfrac{13}{2}\mathrm{O{=}O}(g) \longrightarrow 4\mathrm{O{=}C{=}O}(g) + 5\mathrm{H{-}O{-}H}(g)$$

In this reaction we find that 3 C—C, 10 C—H and $\frac{13}{2}$ O=O bonds are being broken and 8 C=O and 10 O—H bonds are being formed. Thus

$$\Delta H^\circ = \left[(3\text{ mol})\left(\frac{331\text{ kJ}}{1\text{ mol}}\right) + (10\text{ mol})\left(\frac{414\text{ kJ}}{\text{mol}}\right) + \left(\frac{13}{2}\text{ mol}\right)\left(\frac{498\text{ kJ}}{1\text{ mol}}\right)\right]$$
$$- \left[(8\text{ mol})\left(\frac{803\text{ kJ}}{1\text{ mol}}\right) + (10\text{ mol})\left(\frac{464\text{ kJ}}{1\text{ mol}}\right)\right]$$
$$= \mathbf{-2694\ kJ}$$

[The experimental value of ΔH° is -2660 kJ for the reaction as written.]

EXERCISE Use the bond energies given in Table 10.5 to calculate the heat of reaction for

$$NF_3(g) + H_2(g) \longrightarrow NF_2H(g) + HF(g)$$

Answer −243 kJ.

SUMMARY OF SECTIONS 10.1–10.6

The three principal types of bonding are metallic bonding (in which valence electrons form a sea of electrons surrounding metal cations), ionic bonding (in which cations and anions are held together by the attraction of their opposite charges), and covalent bonding (in which pairs of electrons are shared between bonded atoms, leading to the formation of molecules). In many ionic and molecular compounds each atom or ion is surrounded by eight electrons (the octet rule). Exceptions to the octet rule include compounds of B, Be, and Al; some compounds of elements from the third period and higher; and compounds with unpaired electrons.

Metallic bonding imparts the properties of electrical conductivity, luster, malleability, and ductility. Ionic compounds are crystalline solids that conduct electricity poorly when solid but quite well when molten or dissolved in water. The common cations of the representative elements have charges equal to the group number or the group number minus two (see Figure 10.5); the common monatomic anions contain enough added electrons to give an octet configuration. Most *d*-transition elements form cations with $+2$ charges and in many cases also cations of $+1$ or $+3$ charge. In forming cations, transition metals lose their outer *s* electrons first.

In single covalent bonds, a single electron pair is shared by two atoms; in double and triple covalent bonds, two or three electron pairs are shared, respectively. In a coordinate covalent bond, both shared electrons come from the same atom. Molecular compounds, which may be gases, liquids, or solids, are poor conductors of heat and electricity. Certain substances—such as diamond—resemble molecular compounds, except that they are very strong and very hard, being held together by a continuous network of covalent bonds.

The general properties of metals, nonmetals, ionic compounds, and molecular compounds are summarized in Tables 10.1 and 10.4.

Bond energy is the average enthalpy change per mole for breaking one bond of the same type per molecule. The breaking of bonds is always endothermic. In general, the stronger a bond is, the shorter it is. Multiple bonds are always stronger and shorter than single bonds between the same types of atoms.

THE CONTINUA OF BOND TYPES

10.7 VARIATIONS IN BOND TYPES

(a) Nonpolar covalent compound

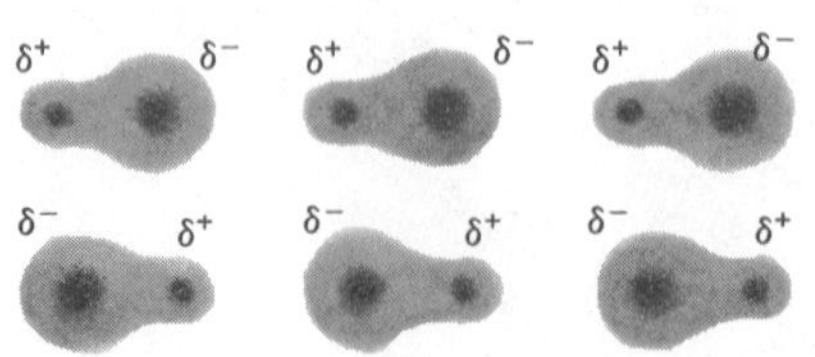

(b) Polar covalent compound

Figure 10.10
Polar and Nonpolar Covalent Bonding

a. Nonpolar and Polar Covalent Bonds In molecules such as H_2, Cl_2, and N_2, electron density (the probability of finding the valence electrons in a given region) is equally divided between the two bonded atoms. In a covalent bond of this type—a **nonpolar covalent bond**—the electrons are shared equally (Figure 10.10a). Atoms of different elements, however, do not have exactly the same electron-attracting ability. Consequently, whenever atoms of two different elements are covalently bonded, the sharing of the electrons becomes unequal and the electron density around one atom becomes greater than that around the other. How unequally the electrons are shared depends on the relative abilities of the two different atoms to attract electrons.

A covalent bond in which electrons are shared unequally is called a **polar covalent bond** (Figure 10.10b); one atom acquires a *partial* negative charge ($\delta-$; Greek delta and a minus sign) and the other acquires a *partial* positive charge ($\delta+$). These are not unit charges, but only represent a reorientation, a sort of pushing around, of the total electron density of the two bonded atoms.

A polar molecule is a **dipole**—a pair of opposite charges of equal magnitude at a specific distance from each other. The entire molecule remains electrically neutral.

H—H *nonpolar covalent bonds* F—F $\delta+$ H—F $\delta-$ *polar covalent bonds* $O\delta-$ (bonded to two H $\delta+$)

In the hydrogen fluoride and water molecules, electrons are attracted away from hydrogen atoms and toward fluorine and oxygen atoms, respectively.

Polar covalent bonds are also found between atoms of the same element, if other parts of the molecule differ in electron-attracting ability. For example, the carbon–carbon bond in trifluoroethane is polar (although less so than a carbon–fluorine bond) because of the strong electron-attracting ability of fluorine atoms:

$$\overset{\delta+}{H_3C}—\overset{\delta-}{CF_3}$$

trifluoroethane

As is discussed further in Chapter 11, whether or not a molecule is a dipole depends upon molecular geometry in addition to the presence of polar bonds.

b. Bonding Continua The nonpolar covalent bond is at one end of a continuum of variation in bond polarity. At the other end of this continuum, one atom attracts electrons so strongly that the electrons depart completely from the other atom and the bonding is ionic.

Although we often speak and write about "ionic compounds," or "covalent bonds," or "metallic bonding," the bonding in most chemical species is *not* 100% ionic, or 100% covalent, or 100% metallic. Instead, the bonding is usually somewhere in between—anywhere along a continuum of bonding from covalent to ionic, or from ionic to metallic, or from metallic to covalent.

The bonds in H_2 or Cl_2 are certainly 100% "covalent." And the bonds in many elemental metals are certainly 100% "metallic." However, the semiconducting elements have properties that are partly metallic and partly covalent. Also, some molecular substances have properties that are partly those expected for metals. For example, iodine (I_2) crystals have a metallic luster and conduct electricity in the same manner as semiconductors. In the following series, metallic character increases as covalent character decreases:

$$F_2 \quad I_2 \quad Te \quad Sn \quad Ag \quad Li$$

In the next series, metallic character increases as ionic character decreases:

$$CsF \quad Na_3N \quad Na_3As \quad Na_3Bi \quad Li$$

The existence of a chemical compound in which the bonding is 100% ionic is not likely. A cation has an attraction for the electron density of an anion, causing it to be somewhat unsymmetrical, or *polarized* (Figure 10.11). The more unsymmetrical the electron density distribution, the greater is the covalent character of the bond. Viewed another way, a covalent bond is polar if the atoms attract electrons unequally. Along the ionic-to-covalent bonding continuum, there eventually comes an area in which there is only a semantic difference between a polar covalent bond and a polarized ionic bond (see Figure 10.11).

Various approaches are used to describe or predict the degree of covalent bonding in an "ionic" compound or the degree of polarity in a molecular compound. In Sections 10.8 and 10.9 two concepts that are useful for this purpose are introduced, one based on the influence of ions on each other (polarizability) and the other based on the ability of atoms in covalent bonds to attract electrons (electronegativity).

(a) Idealized 100% ionic bond

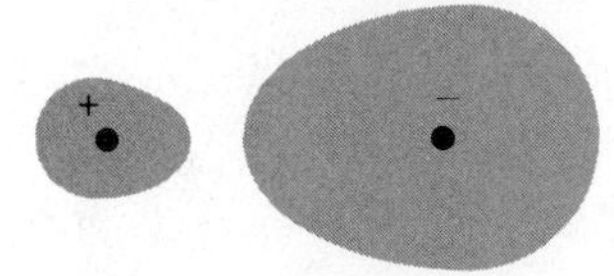

(b) Polarized, partly covalent, ionic bond

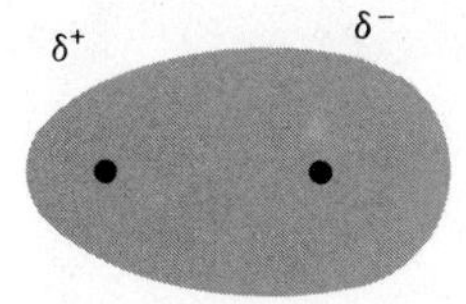

(c) Polar, partly ionic, covalent bond

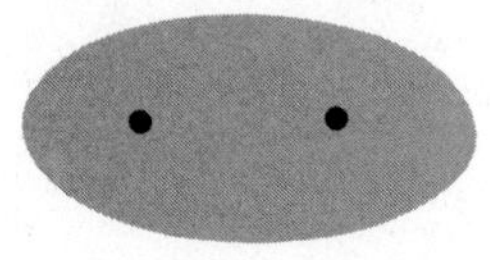

(d) 100% covalent bond

Figure 10.11
The Ionic–Covalent Bonding Continuum

10.8 POLARIZATION

The **polarization of an ion** is the distortion of its electron cloud by an ion of opposite charge. Because cations are small and have a high charge density, they tend to attract the electron clouds of anions. The smaller a cation and the larger its charge, the greater is its polarizing ability and the more it draws electron density into the region between itself and an anion. The result is an increase in the covalent character of the bond (see Figure 10.11).

The charge-to-size ratio of a cation—the absolute value of its charge divided by its ionic radius—is a *relative* measure of the polarizing ability of a cation, if it were to form. The larger the charge-to-size ratio of an ion, the greater the degree of covalent character in its bonds. The charge-to-size ratios for the beryllium and calcium +2 ions

$$Be^{2+} \quad 2/0.035 = 57 \qquad Ca^{2+} \quad 2/0.099 = 20$$

reflect what is known from experiment to be true. The bonds to beryllium in most of its compounds are highly covalent, while calcium forms many ionic compounds. Note that the significance of the charge-to-size ratios is not in their numerical values, but in the relative magnitudes of the absolute values.

The relative effects of size and charge contribute to the diagonal similarity of the representative elements and their compounds mentioned in Section 9.6. The Be^{2+} ion is smaller than the Al^{3+} ion, but the size difference is compensated for by the larger charge of the aluminum ion (Be^{2+}, $2/0.035 = 57$; Al^{3+}, $3/0.051 = 59$). As a result, the ions have nearly identical polarizing abilities, and many analogous aluminum and beryllium compounds have very similar properties.

The charge-to-size ratio concept is frequently used qualitatively in rationalizing the ionic or covalent bonding of the elements. For example, boron is the first element in Representative Group III and boron atoms are quite small. A noble gas type cation of boron would have a +3 charge. Based on what would be the high charge and small size of a boron +3 ion, we would expect to find, for example, that BCl_3 is not a highly ionic compound. (It isn't. It is a molecular compound.)

There is one significant effect on the polarizing ability of cations that is not reflected in the charge-to-size ratio. Cations with d^{10} electron configurations (see Section 10.5) cause greater polarization than cations with noble gas configurations. For example, the Na^+ and Cu^+ cations are of the same charge and almost identical size. However, Cu^+ has only outer *d* level electrons, which shield the nuclear charge much less than the outer *s* and *p* level electrons of Na^+. Therefore, the Cu^+ ion exerts a much stronger polarizing effect and forms compounds with greater covalent character.

Table 10.7
The Complete Electronegativity Scale

	I	II						
	H 2.1							
	Li 1.0	Be 1.5						
	Na 0.9	Mg 1.2						
	K 0.8	Ca 1.0	Sc 1.3	Ti 1.5	V 1.6	Cr 1.6	Mn 1.5	Fe 1.8
	Rb 0.8	Sr 1.0	Y 1.2	Zr 1.4	Nb 1.6	Mo 1.8	Tc 1.9	Ru 2.2
least electronegative →	Cs 0.7	Ba 0.9	La-Lu 1.1–1.2	Hf 1.3	Ta 1.5	W 1.7	Re 1.9	Os 2.2
least electronegative →	Fr 0.7	Ra 0.9						

Source: Adapted from Linus Pauling, *The Nature of the Chemical Bond*, 3rd ed. (Ithaca, New York: Cornell University Press, 1960), p. 43.

EXAMPLE 10.4 Polarization

Based on the polarization of ions, predict the order of decreasing ionic character of the bonding in the following three compounds: $CaCl_2$, $ZnCl_2$, and KCl. The cation radii are Ca^{2+}, 0.099 nm; Zn^{2+}, 0.074 nm; K^+, 0.133 nm.

The K^+ cation is the least polarizing cation in the group—it is the largest and has only a +1 charge (1/0.133 nm = 7.5); therefore, KCl is more highly ionic than the other two compounds. Because it is both smaller and of a higher charge than K^+, the Ca^{2+} ion (2/0.099 nm = 20.) is more polarizing than K^+ and the bonding in $CaCl_2$ is more covalent than in KCl. The Zn^{2+} cation, in addition to its still smaller radius and its +2 charge (2/0.074 nm = 27), also has a d^{10} configuration, making it the most polarizing cation in the group. The order of decreasing ionic bond character is therefore, **KCl > $CaCl_2$ > $ZnCl_2$.**

EXERCISE Which one of the following bonds will be the most ionic and which the least ionic: (a) Na—Cl, (b) Mg—Cl, (c) Be—Cl? *Answers* (a) most, (c) least.

10.9 ELECTRONEGATIVITY

Electronegativity is the ability of an atom in a covalent bond to attract electrons to itself. Many different approaches have been made to assigning numerical values to electronegativity. Linus Pauling originated the term and derived an electronegativity scale based upon calculations using bond energies. The Pauling scale, which we use in this book, is adjusted so that fluorine, the most electronegative element, has an electronegativity of 4, the highest value (Table 10.7).

Like the charge-to-size ratio, the significance of electronegativity lies mainly in its usefulness for predicting the types and properties of bonds, not in the specific numerical values. Pairs of atoms with moderate differences in electronegativity form polar covalent bonds. The magnitude of the electronegativity difference reflects the degree of polarity. In a polar bond, the more electronegative atom has the partial negative charge. With a large enough difference in electronegativity, an ionic bond can be

				III	IV	V	VI	VII
				B 2.0	C 2.5	N 3.0	O 3.5	F 4.0 (most electronegative)
				Al 1.5	Si 1.8	P 2.1	S 2.5	Cl 3.0
Co 1.8	Ni 1.8	Cu 1.9	Zn 1.6	Ga 1.6	Ge 1.8	As 2.0	Se 2.4	Br 2.8
Rh 2.2	Pd 2.2	Ag 1.9	Cd 1.7	In 1.7	Sn 1.8	Sb 1.9	Te 2.1	I 2.5
Ir 2.2	Pt 2.2	Au 2.4	Hg 1.9	Tl 1.8	Pb 1.8	Bi 1.9	Po 2.0	At 2.2

Table 10.8
Electronegativity

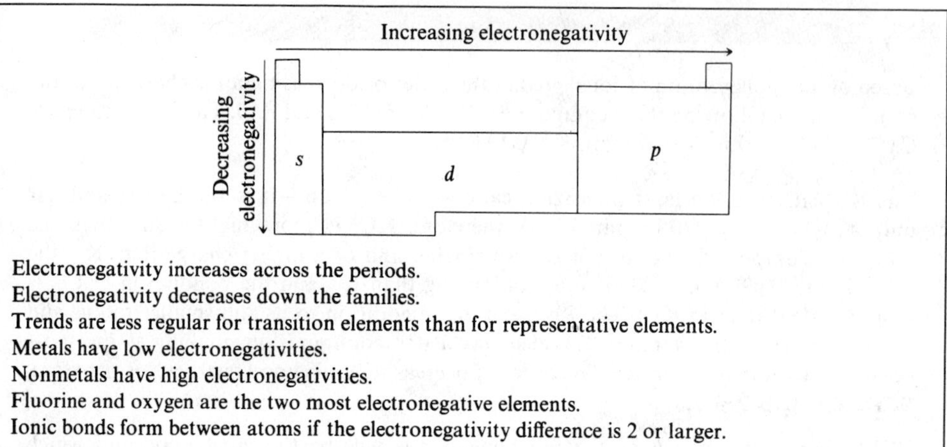

Electronegativity increases across the periods.
Electronegativity decreases down the families.
Trends are less regular for transition elements than for representative elements.
Metals have low electronegativities.
Nonmetals have high electronegativities.
Fluorine and oxygen are the two most electronegative elements.
Ionic bonds form between atoms if the electronegativity difference is 2 or larger.

expected to form. A commonly used rule of thumb is that if the electronegativity difference is two or greater, the bond between two atoms will be more ionic than covalent in character.

As shown in Tables 10.7 and 10.8, electronegativity is a periodic property. Electronegativities are lowest for metals at the bottom left corner in the periodic table and highest for the nonmetals in the top right corner. Among the representative elements, electronegativities increase regularly across each period and decrease, with a few exceptions, down each family. Comparison of Table 10.7 with Tables 9.5 and 9.8 shows that, as might be expected, increasing electronegativity is associated with increasingly difficult electron removal (increasing ionization energy) and increasing ease of electron addition (increasing electron affinity). Ionic compounds form most readily between large metal atoms of low electronegativity and small nonmetal atoms of high electronegativity.

Frequently "electronegative" and "electropositive" are used in a general way to describe the properties of atoms. An **electronegative atom** tends to acquire a partial negative charge in a covalent bond or to form a negative ion. Nonmetals are generally electronegative. An **electropositive atom** tends to acquire a partial positive charge in a covalent bond or to form a positive ion. Metals are generally electropositive. Of course, the terms are relative—in ICl the chlorine atom is the more electronegative atom and the iodine atom is the more electropositive atom.

Trends in electronegativity are not as regular among the transition metals, all of which have electronegativity values between 1.1 and 2.4. This smaller variation in electronegativity is in line with the smaller variations of atomic radii and ionization energies of these elements. Table 10.8 summarizes trends in electronegativity.

EXAMPLE 10.5 Electronegativity

With the use of the electronegativities in Table 10.7, arrange the following covalent bonds in order of increasing polarity: O—H, I—Br, C—F, P—H, S—Cl.

The polarity of a bond increases with increasing difference in electronegativity of the bonded atoms. The covalent bonds in question have the following order of increasing polarity, with the electronegativity differences shown:

P—H	I—Br	S—Cl	O—H	C—F
0.0	*0.3*	*0.5*	*1.4*	*1.5*

EXERCISE Arrange the following bonds in order of increasing polarity: (a) C—N, (b) S—O, (c) Si—N, (d) B—O. *Answer* C—N, S—O, Si—N, B—O.

EXAMPLE 10.6 Electronegativity

By consulting only the periodic table, arrange the following elements in order of increasing electronegativity: P, Al, Mg, Rb, Cl, Br. Give reasons for your arrangement.

For the representative elements, electronegativity decreases down a family and increases across a period. Rubidium, magnesium, and aluminum are metals and are less electronegative than the remaining elements, which are nonmetals. Rubidium, a member of Representative Group I and the first element in the fifth period, is clearly the least electronegative element on the list. Magnesium, aluminum, phosphorus, and chlorine fall in the third period—magnesium in Representative Group II, aluminum in Group III, phosphorus in Group V, and chlorine in Group VII. Effective nuclear charge increases in the same order and so does electronegativity. Bromine is also in Group VII, but in the fourth period, and therefore is less electronegative than chlorine. Even though bromine is one period farther down than phosphorus in the periodic table and its atomic size is slightly greater (0.114 nm vs 0.110 nm), it is the more electronegative element. The much greater nuclear charge on the bromine atom ($Z = 35$ for Br vs. $Z = 15$ for P) gives it a substantially larger effective nuclear charge. The order of increasing electronegativity is **Rb < Mg < Al < P < Br < Cl.**

EXERCISE Arrange the following elements in order of increasing electronegativity: K, N, As, F, and Sn. *Answer* K < Sn < As < N < F.

SUMMARY OF SECTIONS 10.7–10.9

Except for 100% covalent bonds (e.g., in H_2) and 100% metallic bonds in elemental metals, bonds fall along continua with varying contributions from ionic or covalent or metallic bonding. Bonds with partial ionic and partial covalent character can be thought of as polar covalent bonds or as polarized ionic bonds. The charge-to-size ratio of a cation (or a hypothetical cation) is a relative measure of the polarizing ability of the cation—the ability to attract electrons and induce covalent character in a bond with an anion. The smaller a cation and the greater its charge, the greater is its polarizing ability. Also, cations with d^{10} configurations cause greater polarization than cations with noble gas configurations.

Electronegativity is the ability of an atom in a covalent bond to attract electrons to itself. Electronegativity increases across the periods and decreases down the families in the periodic table (see Table 10.8). A difference of two or more in the Pauling electronegativity values is taken as an indication that a given bond will be more ionic than covalent.

OXIDATION STATE

10.10 ASSIGNING OXIDATION NUMBERS

The extent to which bonded atoms have changed their share of "ownership" of their valence electrons is useful information. A simple bookkeeping system for electrons permits assignment of numbers that, in many cases, indicate the relative electron ownership of the bonded atoms.

Oxidation numbers are equal either to the charges of ions or to the charges atoms *would* have if the compound were ionic are assigned to atoms in compounds. The term **oxidation state** has the same meaning as "oxidation number."

Table 10.9
Rules for Assigning Oxidation Numbers
In (6) and (7) the oxidation numbers calculated using the rules are shown in black.

(1) Free elements, oxidation no. = 0

$\overset{0}{Ca}$ $\overset{0}{O_2}$ $\overset{0}{S_8}$ $\overset{0}{O_3}$

(2) Monatomic ions, oxidation no. = ionic charge

$\overset{+1}{Na}\overset{-1}{Cl}$ $\overset{+2}{Ca}\overset{-1}{F_2}$ $\overset{+1}{Cu}\overset{-1}{Br}$ $\overset{+2}{Cu}\overset{-1}{Br_2}$

(3) Fluorine, oxidation no. = −1, *always* in compounds

$Ba\overset{-1}{F_2}$ $H\overset{-1}{F}$ $BaSi\overset{-1}{F_6}$ $Br\overset{-1}{F}$

(4) Oxygen, oxidation no. = −2

$H_2\overset{-2}{O}$ $Ba\overset{-2}{O}$ $H_2S\overset{-2}{O_4}$ $C\overset{-2}{O_2}$

except, e.g. $H_2\overset{+1-1}{O_2}$ $\overset{+2-1}{OF_2}$

(5) Hydrogen, oxidation no. = +1

$\overset{+1}{H_2}O$ $\overset{+1}{H_2}SO_4$ $Ba(O\overset{+1}{H})_2$ $C\overset{+1}{H_4}$

except in metal hydrides, e.g. $\overset{+1}{Na}\overset{-1}{H}$

(6) Neutral compounds, sum of oxidation nos. = 0

$\overset{2(+5)}{As_2}\ \overset{5(-2)}{O_5}$ $\overset{+1}{K}\ \overset{+7}{Cl}\ \overset{4(-2)}{O_4}$

$\overset{-3}{N}\ \overset{3(+1)}{H_3}$ $\overset{+4}{C}\ \overset{4(-1)}{Cl_4}$

(7) Polyatomic ions, sum of oxidation nos. = charge of ion

$\left(\overset{-2}{O}\ \overset{+1}{H}\right)^-$ $\left(\overset{+6}{S}\ \overset{4(-2)}{O_4}\right)^{2-}$

$\left(\overset{+3}{Cl}\ \overset{2(-2)}{O_2}\right)^-$ $\left(\overset{2(+5)}{P_2}\ \overset{7(-2)}{O_7}\right)^{4-}$

In many cases oxidation numbers do show the relative shift of electrons toward or away from each atom, but in some cases they have no such significance. Nevertheless, oxidation numbers are very useful in balancing equations, in predicting possible reaction products, and in predicting the properties of chemical compounds.

The rules for assigning oxidation numbers are given in the following paragraphs. In each case, the objective of the rule is to assign a negative oxidation number to the most electronegative element in a compound. Table 10.9 summarizes the rules and gives examples. Study the table while you study the rules.

1. The oxidation number of any element in the free state is zero. For example, Cu in elemental copper and H in elemental hydrogen (H_2) are both assigned oxidation numbers of zero.
2. The oxidation number of a monatomic ion is equal to the charge on the ion. For example, in $CaCl_2$ the oxidation number of the chlorine (Cl^-) is -1 and the oxidation number of the calcium (Ca^{2+}) is $+2$.
3. Fluorine (the most electronegative element) has the oxidation number -1 in all its compounds.
4. Oxygen, with only a few exceptions, has the oxidation number of -2 in all its compounds. Oxygen is assigned an oxidation number of -2 in all ionic and nonionic compounds except peroxides, superoxides, and ozonides, which contain O—O bonds, and the few compounds that contain O—F bonds (in which F has the negative oxidation number).
5. Hydrogen has an oxidation number of $+1$ in all compounds except metal hydrides. In metal hydrides, such as NaH, hydrogen is assigned an oxidation number of -1.
6. For a neutral compound, the algebraic sum of the oxidation numbers of all the atoms must equal zero. For example, in MnO_2, the Mn oxidation number is $+4$ and the O oxidation number is -2, giving $+4 + (2 \times -2) = 0$.
7. For a polyatomic ion, the algebraic sum of the oxidation numbers of the atoms must equal the charge on the ion. For example, in the carbonate ion, CO_3^{2-}, oxygen contributes a total oxidation number of $3 \times (-2)$, or -6; therefore, to have a total charge of -2, the oxidation number of carbon must be $+4$.

Usually, it is wise to assign all oxidation numbers that you are sure of and then solve mathematically for the one you are least sure of by using rules 6 and 7. For example, to find the oxidation numbers of all atoms in $HClO_4$, use the assigned values of $+1$ for H and -2 for O and solve for the oxidation number of Cl:

$$\begin{aligned} &(\text{No. of H atoms})(\text{oxidation no. of H}) + (\text{no. of O atoms})(\text{oxidation no. of O}) \\ &\qquad + (\text{no. of Cl atoms})(\text{oxidation no. of Cl}) = 0 \\ &(1)(+1) + (4)(-2) + (1)(\text{oxidation no. of Cl}) = 0 \\ &\text{Oxidation no. of Cl} = (-1) + (+8) = +7 \end{aligned}$$

The oxidation number of Cl in $HClO_4$ is $+7$. Although we have written out an equation to illustrate this process, an equation is not necessary. Just by examining the formula and the known oxidation states, what is needed to maintain neutrality can be seen.

$$\overset{+1}{H}\ \overset{+7}{Cl}\ \overset{4(-2)}{O_4}$$

In binary compounds for which the rules do not specifically indicate the oxidation states of both of the atoms present, the element that usually forms negative ions or is more electronegative is assigned an oxidation number equal to the charge it would have *if* it were present as a negative ion (whether it is ionic or not). The other element is given the positive oxidation number that makes the molecule neutral. For example, in PCl_3 and PCl_5 chlorine is assigned an oxidation number of -1, resulting in oxidation numbers of $+3$ and $+5$ for phosphorus, respectively.

For some compounds, following the rules leads to fractional oxidation numbers. For example, Fe in Fe_3O_4 must have an oxidation number of $+2\frac{2}{3}$ ($3 \times 2\frac{2}{3} = +8$, to balance the -8 of the 4 O atoms). This seemingly strange result is explained by the fact that Fe_3O_4 contains Fe atoms of both $+2$ and $+3$ oxidation numbers, combined in what may be thought of as $FeO \cdot Fe_2O_3$. The oxidation number found in such a case is an average.

EXAMPLE 10.7 Oxidation Numbers

Using the rules discussed above, find the oxidation states of sulfur in the following species: (a) S_8, (b) S, (c) S^{2-}, (d) H_2S, (e) SO_2, (f) SO_3, (g) HSO_4^-, (h) H_2SO_4, (i) SO_3^{2-}

(a) and (b) The oxidation number of S is **zero in both S_8 and S,** for these symbols both represent the free element (rule 1).

(c) The oxidation number of S in S^{2-} **is -2,** the charge on the ion (rule 2).

(d) Based on the oxidation number of $+1$ for hydrogen (rule 5) and the maintenance of an oxidation number sum of zero for a neutral compound (rule 6), sulfur has an oxidation number of -2.

$$\overset{2(+1)-2}{H_2S}$$

(e) and (f) Based on an oxidation number of -2 for O (rule 4) and maintaining oxidation number sums of zero for a neutral compound (rule 6), S has oxidation numbers of **$+4$ and $+6$** in these compounds.

$$\overset{+4\ 2(-2)}{SO_2} \qquad \overset{+6\ 3(-2)}{SO_3}$$

(g), (h), and (i) Applying rules 4, 5, 6 and 7 (O, -2; H, $+1$; oxidation number total $= 0$ for neutral compound; oxidation number total $=$ total charge of polyatomic ion) gives oxidation numbers of **$+6$, $+6$, and $+4$** to sulfur in these species.

check

$$\overset{+1\ +6\ 4(-2)}{HSO_4^-} \qquad (+1) + (+6) + 4(-2) = -1$$

$$\overset{2(+1)\ +6\ 4(-2)}{H_2SO_4} \qquad 2(+1) + (+6) + 4(-2) = 0$$

$$\overset{+4\ 3(-2)}{SO_3^{2-}} \qquad (+4) + 3(-2) = -2$$

EXERCISE Find the oxidation states of nitrogen in the following species: (a) HNO_3, (b) HNO_2, (c) NO^+, (d) N^{3-}, (e) N_2, (f) NH_3.

Answers (a) $+5$, (b) $+3$, (c) $+3$, (d) -3, (e) 0, (f) -3.

10.11 PERIODIC TRENDS IN OXIDATION STATE

Knowing which elements can display only one, or only two, or more than two oxidation states, and knowing the more common oxidation states is useful in many ways. For example, possible products of chemical reactions can be predicted from possible oxidation states of the elements involved. Also, in their higher oxidation states many elements are more electronegative, with a resulting influence on properties.

Table 10.10 summarizes the variability of the oxidation states of the representative elements. Table 10.11 lists common oxidation states and gives examples of compounds of the representative elements. The following guidelines are helpful in remembering the common oxidation states of the common elements:

1. For the representative elements, the maximum possible positive oxidation state equals the representative group number. What we might call the *group oxidation state* in a sense represents the involvement of all of the *s* and *p* electrons in bonding.
2. The members of the *s* block and the scandium family show only one oxidation state in their compounds.
3. Most *p*-block elements show more than one oxidation state. Halogens other than fluorine have either oxidation states of −1 or they have positive oxidation states when combined with more electronegative elements. Many *p*-block elements have positive oxidation states that decrease from the group oxidation state by increments of 2 (see Table 10.11).
4. Variability of oxidation states is the rule for most transition elements. In most cases it is difficult to predict from the electron configurations which are the most common oxidation states of transition elements.

I	II	III	IV
H +1 HCl −1 Na^+H^-			
Li +1 Li^+Cl^-	**Be** +2 $BeCl_2$	**B** +3 BCl_3	**C** −4 CH_4 +2 CO +4 CO_2 CCl_4
Na +1 Na^+Cl^-	**Mg** +2 $Mg^{2+}(Cl^-)_2$	**Al** +3 Al_2Cl_6	**Si** −4 SiH_4 +4 $SiCl_4$
K +1 K^+Cl^-	**Ca** +2 $Ca^{2+}(Cl^-)_2$	**Ga** +3 Ga_2Cl_6	**Ge** +2 $GeCl_2$ +4 $GeCl_4$
Rb +1 Rb^+Cl^-	**Sr** +2 $Sr^{2+}(Cl^-)_2$	**In** +3 $InCl_3$	**Sn** +2 $Sn^{2+}(F^-)_2$ $SnCl_2$ +4 $SnCl_4$ SnO_2
Cs +1 Cs^+Cl^-	**Ba** +2 $Ba^{2+}(Cl^-)_2$	**Tl** +1 Tl^+Cl^- +3 TlF_3	**Pb** +2 $Pb^{2+}(F^-)_2$ +4 PbO_2

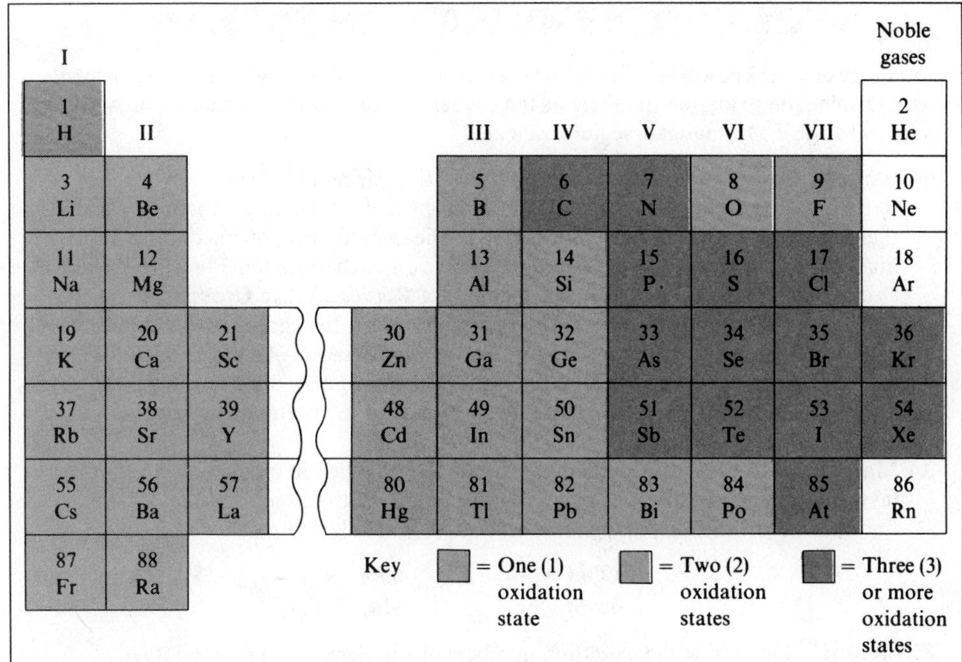

Table 10.10
Variability of Common Oxidation States in Compounds
This table is based upon the commonly encountered oxidation states of these elements. In some compounds, boron has oxidation state +3; in others, the bonding is unique and assigning oxidation states has little meaning. As far as is now known, He, Ne, Ar, and Rn form no chemical compound.

Table 10.11
Oxidation States and Typical Compounds for the Representative Elements
Except for oxygen and fluorine, the most electronegative elements, the maximum oxidation state for each element is equal to its group number.

V	VI	VII	Noble gases
			He
N −3 $(Li^+)_3N^{3-}$ NH_3 +2 NO +3 N_2O_3 +4 NO_2 +5 N_2O_5	**O** −2 $(Na^+)_2O^{2-}$ H_2O −1 $(Na^+)_2O_2^{2-}$ H_2O_2	**F** −1 Na^+F^- CF_4	**Ne**
P −3 PH_3 Na_3P +3 PCl_3 +5 PCl_5 P_4O_{10}	**S** −2 $(Na^+)_2S^{2-}$ H_2S +4 SO_2 +6 SO_3	**Cl** −1 Na^+Cl^- CCl_4 +1 HOCl +3 $HClO_2$ +5 $HClO_3$ +7 $HClO_4$	**Ar**
As −3 AsH_3 Na_3As +3 $AsCl_3$ +5 As_4O_{10}	**Se** −2 H_2Se +4 SeO_2 +6 SeO_3	**Br** −1 Na^+Br^- HBr +1 HOBr +5 $HBrO_3$ +7 $HBrO_4$	**Kr** +2 KrF_2
Sb −3 SbH_3 +3 $Sb^{3+}(F^-)_3$ $SbCl_3$ +5 $SbCl_5$	**Te** −2 H_2Te +4 TeO_2 +6 TeO_3	**I** −1 Na^+I^- HI +1 HOI +5 HIO_3 +7 HIO_4	**Xe** +2 XeF_2 +4 XeF_4 +6 XeF_6 +8 Na_4XeO_6
Bi −3 BiH_3 +3 $Bi^{3+}(F^-)_3$ +5 Bi_2O_5	**Po** +4 PoO_2	**At**	**Rn**

EXAMPLE 10.8 Oxidation Numbers

Make use of your knowledge of periodic positions and trends in properties of the elements to determine the oxidation numbers of the elements in (a) $BaSiF_6$, barium fluorosilicate, and (b) $Mn_2(SiF_6)_3$, manganese fluorosilicate.

(a) Barium, silicon, and fluorine are representative elements. Barium, a member of Representative Group II, is a metal and forms the Ba^{2+} ion in its compounds. Barium, therefore, has an oxidation number of +2. The combination of silicon and fluorine must have a charge of −2 to balance the charge on the barium ion. Fluorine, the most electronegative element known, is a member of Representative Group VII and each fluorine atom is assigned an oxidation number of −1, the charge it would have if it were present in the compound as an ion. For the combination of silicon and fluorine to have a charge of −2, the silicon atom must be assigned an oxidation number of +4. This value is consistent with the position of silicon in the table, in Representative Group IV.

(b) Manganese is a transition metal that exhibits several different oxidation states. In part (a) we found that the SiF_6 group has a charge of −2. With this information at hand, we find that in $Mn_2(SiF_6)_3$, manganese must have an oxidation number of +3.

(a) $\overset{+2}{\mathbf{Ba}}\ \overset{+4}{\mathbf{Si}}\ \overset{6(-1)}{\mathbf{F_6}}$ (b) $\overset{2(+3)}{\mathbf{Mn_2}}\ \overset{3(-2)}{\mathbf{(SiF_6)_3}}$

EXERCISE Determine the oxidation numbers of the elements in $CaMg(SiO_3)_2$.

Answer Ca is +2, Mg is +2, Si is +4, O is −2.

10.12 OXIDATION STATES IN FORMULAS AND NAMES

In Section 4.4 we introduced the Stock system for naming chemical compounds and illustrated its use for cations and binary ionic compounds such as

Mn^{2+}	manganese(II) ion	CuCl	copper(I) chloride
Mn^{3+}	manganese(III) ion	$CuCl_2$	copper(II) chloride

The Stock system is used to name molecular as well as ionic compounds. For molecular compounds the Roman numeral is simply the oxidation number. Binary molecular compounds may be named by the classical system *or* by the Stock system.

Classical system		Stock system	
N_2O	dinitrogen monoxide	nitrogen(I) oxide	$\overset{+1}{N_2}\overset{-2}{O}$
N_2O_3	dinitrogen trioxide	nitrogen(III) oxide	$\overset{+3}{N_2}\overset{-2}{O_3}$
N_2O_5	dinitrogen pentoxide	nitrogen(V) oxide	$\overset{+5}{N_2}\overset{-2}{O_5}$
ICl	iodine monochloride	iodine(I) chloride	$\overset{+1}{I}\overset{-1}{Cl}$
ICl_3	iodine trichloride	iodine(III) chloride	$\overset{+3}{I}\overset{-1}{Cl_3}$

One shortcoming of the Stock system is its inability to distinguish between different binary compounds of the same two elements in which the oxidation number is the same. For example, the name *nitrogen(IV) oxide* does not distinguish between NO_2 and N_2O_4, both of which exist. In such cases the classical names, which for these compounds are *nitrogen dioxide* and *dinitrogen tetroxide,* are more descriptive.

EXAMPLE 10.9 Chemical Nomenclature

Manganese forms oxides in which the metal exhibits oxidation states of +2, +3, +4, and +7. Write the formulas and give the names (Stock system) for these compounds.

Oxygen exhibits an oxidation number of −2 in these oxides. The formulas of the manganese compounds and their names are

MnO	**manganese(II) oxide**	**MnO_2**	**manganese(IV) oxide**
Mn_2O_3	**manganese(III) oxide**	**Mn_2O_7**	**manganese(VII) oxide**

EXERCISE Determine the oxidation state of gold in each of the following compounds and name the compounds using the Stock system of nomenclature: (a) $AuCl_3$, (b) Au_2S, (c) Au_2O_3.

Answers (a) +3, gold(III) chloride; (b) +1, gold(I) sulfide; (c) +3, gold(III) oxide.

EXAMPLE 10.10 Chemical Nomenclature

Write the formulas of the following binary compounds: (a) iodine(V) fluoride, (b) tin(IV) sulfide, (c) selenium(VI) oxide.

The oxidation numbers of the more electropositive elements are indicated in the names of the compounds. Based on their positions in the periodic table, the oxidation numbers of the other elements are −1 for fluorine (Group VII) and −2 for sulfur and oxygen (both Group VI). Knowing the oxidation numbers allows the correct formulas to be written:

$$\text{(a)}\ \overset{+5\,5(-1)}{\mathbf{IF_5}} \qquad \text{(b)}\ \overset{+4\,2(-2)}{\mathbf{SnS_2}} \qquad \text{(c)}\ \overset{+6\,3(-2)}{\mathbf{SeO_3}}$$

EXERCISE Write the formulas for the following compounds: (a) arsenic(III) fluoride, (b) bismuth(V) oxide, (c) cerium(IV) sulfate, (d) chromium(II) chloride.

Answers (a) AsF_3, (b) Bi_2O_5, (c) $Ce(SO_4)_2$, (d) $CrCl_2$.

SUMMARY OF SECTIONS 10.10–10.12

The oxidation state (oxidation number) of an atom in a compound is equal to its charge (if it is a monatomic ion) or to the charge it would have if the compound were ionic. The rules for assigning oxidation numbers are summarized in Table 10.9. For the representative elements, the maximum possible oxidation state is given by the representative group number. Most *p*-block and transition elements exhibit more than one oxidation state. The common oxidation states of the representative elements are listed in Table 10.11. When the Stock system (Section 4.4) is used to name molecular compounds, the oxidation number is given as a Roman numeral in parentheses for elements of variable oxidation number.

* * *

10.13 IN SUMMARY: TRENDS IN THE PROPERTIES OF THE ELEMENTS

The study of the descriptive chemistry of the elements and their compounds is aided by an understanding of the periodic variation of elemental properties. The chemical properties of the elements are determined to a large extent by their electron configurations and radii. Across the periods of representative elements, except for hydrogen and helium, the outermost configurations build up from s^1 to s^2p^5, and each period (except the first) ends with a noble gas of the s^2p^6 configuration. The effective nuclear charge increases across each period, causing a continuously greater attraction by the nuclei for the outer electrons and a decrease in atomic radii.

The trend across the periods to decreasing atomic radii is accompanied by a general increase in the energy needed to remove electrons (increasing ionization energy). There are also increases in the ease with which isolated atoms add electrons (increasing electron affinity) and in the ability of covalently bonded atoms to attract electrons and increase the polarity of bonds (increasing electronegativity). The most electronegative elements are fluorine and oxygen; the least electronegative elements as a group are the highly reactive metals of the lithium family (Group I, outer configuration ns^1.)

Atoms of the transition elements have one or two outer *s* electrons (except Pd). The incomplete inner-level *d* or *f* subshells are close in energy to the outer energy level, and electrons in the *d* subshells can participate in bonding of the *d*-transition elements. Because of strong screening by electrons in the incompletely filled *d*- and *f*-subshell, effective nuclear charge increases only slightly across the periods. For the transition elements, this results in a greater uniformity of radii (which decrease slightly) and ionization energy and electronegativity (both of which increase slightly) across the periods. Because of the intervention of the 4*f* series of transition elements, atoms of the two heaviest members of each transition metal family are very similar in size and chemical properties.

Metallic properties in general are associated with large radii, a small number of valence electrons, easy electron removal, difficulty of electron addition, and low electronegativity (Table 10.12). Nonmetallic properties are associated with small radii, large numbers of valence electrons, difficult electron removal, easy electron addition, and high electronegativity. As a result of the periodic trends in radii and other properties, there is a progression from metallic to nonmetallic character across the periods for the representative elements. The noble gases, with their very stable outer electron configurations of ns^2p^6 ($1s^2$ for helium), are the least reactive elements—they neither gain nor lose electrons readily.

In each family of representative elements, there is a general increase in atomic radii and decrease in ionization energy with increasing atomic number. As a result, the transition from metallic to nonmetallic character across a period moves farther to the right with increasing period number. In Group IV, tin and lead (fifth and sixth periods, respectively) are metals; in Groups V and VI, bismuth and polonium (sixth period) are the only metals; in Group VII and the noble gas group, none of the elements are metals. The seven *p*-block elements that fall between the metals and the nonmetals are known as the semiconducting elements (B, Si, Ge, As, Sb, Se, Te) and have intermediate properties between those of the metals and the nonmetals.

Table 10.12
Comparisons of Ranges of Ionization Energies, Electron Affinities, and Electronegativities
For electron affinities we have considered only the experimentally determined values (see Table 9.8). The Representative Group II metals have calculated positive values ranging from 52 to 240 kJ/mol.

First ionization energies (kJ/mol)	
Nonmetals	1000–2372
Semiconducting elements	762–947
Metals	376–1007
Electron affinities (kJ/mol)	
Nonmetals	−349 to 0
Semiconducting elements	−195 to −77
Metals	−120 to −14
Electronegativities	
Nonmetals	2.1–4.0
Semiconducting elements	1.8–2.4
Metals	0.7–2.4

When representative metals combine with the nonmetals of Groups V, VI, and VII, ionic compounds are usually formed. As a generalization, combination of elements with large differences in electronegativity yields ionic compounds; combination of elements with relatively small differences in electronegativity yields molecular compounds.

In forming ionic compounds, atoms of representative elements often achieve octet configurations by the gain or loss of electrons. The ions of the transition metals are formed by the loss of the outer *s* electrons and also, in some cases, by the loss of one or two $(n-1)$ *d* electrons.

Covalent bonds and molecular compounds generally result from bonding between nonmetals and nonmetals or between nonmetals and the less electropositive metals of smaller radii. Covalent bonds between atoms of differing electronegativity are polar bonds, and the greater the difference in electronegativity the greater the polarity. Carbon, nitrogen, and oxygen form many compounds in which they follow the octet rule.

The properties of the first member of each representative family differ significantly from those of the remaining family members due to the smaller atomic size of the first family member. As a result, first members of families tend to resemble second members of the families to their right. This diagonal relationship is strongest for beryllium and aluminum, which have approximately the same charge-to-size ratio and the same electronegativity.

For the representative elements, the maximum positive oxidation states (except for oxygen and fluorine) equal the group numbers. Metals have *only* positive oxidation states, and those in Groups I and II show only the group oxidation state. The maximum negative oxidation states of the nonmetals correspond to the number of electrons required to attain noble gas configurations. The *p*-block metals other than aluminum (Ga, In, Tl, Sn, Pb, Bi) can exhibit two oxidation states, one that corresponds to the loss of the *p* valence electrons (e.g., Tl^{+}, Sn^{2+}, Bi^{3+}) and the other that corresponds to the use of all valence electrons (e.g., $TlCl_3$, $SnCl_4$, BiF_5). Note that in higher oxidation states, these elements are more likely to form molecular compounds.

All of the nonmetals (except fluorine) and many of the semiconducting elements show a variety of oxidation states, but correlations with configurations for these states are not always possible.

In atoms of the *d*-transition elements, the $(n-1)d$ and *ns* levels differ relatively little in energy, electrons in both levels are available for chemical bonding, and multiple oxidation states are common for the *d*-transition metals. In fact, only the members of the scandium family commonly show a single oxidation state.

SIGNIFICANT TERMS (with section references)

acceptor atom (10.5)
bond dissociation energy (10.6)
bond energy (10.6)
chemical bond (10.1)
coordinate covalent bond (10.5)
covalent bonding (10.5)
d^{10} configuration, pseudo-noble gas configuration (10.4)
dipole (10.7)
donor atom (10.5)
double covalent bond (10.5)
electronegative atom (10.9)
electronegativity (10.9)
electropositive atom (10.9)
ionic bonding (10.4)
Lewis structure (10.5)
lone electron pairs (10.5)
metallic bonding (10.3)
multiple covalent bond (10.5)
network covalent substance (10.5)
nonpolar covalent bond (10.7)
octet rule (10.4)
oxidation numbers, oxidation states (10.10)
polar covalent bond (10.7)
polarization of an ion (10.8)
single covalent bond (10.5)
triple covalent bond (10.5)

QUESTIONS AND PROBLEMS

Types and Properties of Chemical Bonds

10.1 Give a suitable definition of a chemical bond. What type of force holds the atoms together in a chemical bond?

10.2 List the three basic types of chemical bonding. Give an example of a substance with each type of bonding. What are the differences among these types? What are some of the general properties associated with the three types of bonding?

10.3 The outermost electron configuration atoms of the alkali metals (lithium family elements) is ns^1. How can an atom of each of these metals attain a noble gas electron configuration?

10.4 Repeat Question 10.3 for the halogens (fluorine family elements); these have an outer electron configuration of ns^2np^5.

10.5 Describe the types of bonding in barium hydroxide, $Ba(OH)_2$

$$[Ba]^{2+} \quad 2[:\ddot{\underset{..}{O}}-H]^-$$

10.6 Repeat Question 10.5 for potassium chlorate, $KClO_3$

$$[K]^+ \left[\begin{array}{c} :\ddot{O}: \\ | \\ :\ddot{\underset{..}{O}}-\underset{..}{Cl}-\ddot{\underset{..}{O}}: \end{array}\right]^-$$

Types and Properties of Chemical Bonds—Metallic Bonds

10.7 Describe how a metallic bond is formed. Why don't nonmetals form metallic bonds?

10.8 Which metal would you predict to have the stronger metallic bonding: K or Ca? Why?

10.9 Why are metals malleable and ductile?

10.10 Why are metals excellent conductors of heat and electricity? What happens to the electrical conductivity of a metals as the temperature is increased? Why?

Types and Properties of Chemical Bonds—Ionic Bonds

10.11 Describe what happens to the valence electron(s) as an ionic bond is formed between a metal atom and a nonmetal atom.

10.12 State the octet rule. How is this rule modified for elements having atomic numbers less than 5?

10.13 Show with Lewis symbols and electron configurations the changes that occur when (a) zinc and fluorine combine to form ionic zinc fluoride and (b) calcium and oxygen combine to form ionic calcium oxide.

10.14 Repeat Question 10.13 when (a) lithium and chlorine combine to form ionic lithium chloride, (b) magnesium and sulfur combine to form ionic magnesium sulfide, (c) magnesium and chlorine combine to form ionic magnesium chloride.

10.15 Write the formulas for the ionic compounds that form between (a) La and Cl_2, (b) Cu and F_2, (c) Cs and Br_2.

10.16 Repeat Question 10.15 for (a) Ca and O_2, (b) Sc and O_2, (c) Na and S.

10.17 Write formulas for any ionic compounds that might form between (a) calcium and nitrogen, (b) aluminum and silicon, (c) potassium and selenium, (d) iron and sulfur. Consult Figure 10.5 to determine which of the elements will form ions.

10.18 Repeat Question 10.17 for (a) F and Cl, (b) Na and F, (c) Mg and As.

10.19 Describe an ionic crystal. What determines the geometrical arrangement of the ions?

10.20 Why are most solid ionic compounds rather poor conductors of electricity? Why does conductivity increase when an ionic compound is melted or dissolved in water?

Types and Properties of Chemical Bonds—Covalent Bonds

10.21 How many electrons are shared between two atoms in (a) a single covalent bond, (b) a double covalent bond, (c) a triple covalent bond?

10.22 What is the maximum number of covalent bonds that second-period elements can form? Why can the representative elements beyond the second period form more than this number of covalent bonds?

10.23 What is the term for a single covalent bond in which both electrons in the shared pair come from the same atom? Identify the donor and acceptor atoms in the following

$$\begin{array}{c} H \\ | \\ H-N: \\ | \\ H \end{array} + \begin{array}{c} :\ddot{F}: \\ | \\ B-\ddot{\underset{..}{F}}: \\ | \\ :\underset{..}{F}: \end{array} \longrightarrow \begin{array}{c} H \quad :\ddot{F}: \\ | \qquad | \\ H-N-B-\ddot{\underset{..}{F}}: \\ | \qquad | \\ H \quad :\underset{..}{F}: \end{array}$$

10.24 Identify the donor and acceptor atoms in each of the following

(a) $$\begin{array}{c} H \\ | \\ H-\underset{..}{O}: \end{array} + H^+ \longrightarrow \left[\begin{array}{c} H \\ | \\ H-\underset{..}{O}-H \end{array}\right]^+$$

(b) $$6\left[:\ddot{\underset{..}{Cl}}:\right]^- + Pt^{4+} \longrightarrow \left[\begin{array}{c} :\ddot{Cl}\quad \ddot{Cl}: \\ \diagdown \quad \diagup \\ :\ddot{\underset{..}{Cl}}-Pt-\ddot{\underset{..}{Cl}}: \\ \diagup \quad \diagdown \\ :\underset{..}{Cl}\quad \underset{..}{Cl}: \end{array}\right]^{2-}$$

10.25 Identify the species that obey the octet rule:

(a) $:\ddot{\underset{..}{F}}-\ddot{\underset{..}{F}}:$ (b) $:\ddot{\underset{..}{O}}-\dot{\underset{..}{Cl}}-\ddot{\underset{..}{O}}:$ (c) $:\ddot{\underset{..}{F}}-\ddot{\underset{..}{Xe}}-\ddot{\underset{..}{F}}:$

(d) $$\left[\begin{array}{c} :\ddot{O}: \\ | \\ :\ddot{\underset{..}{O}}-S-\ddot{\underset{..}{O}}: \\ | \\ :\underset{..}{O}: \end{array}\right]^{2-}$$ (e) $$\begin{array}{c} :\ddot{\underset{..}{Cl}} \\ \diagdown \\ C=\ddot{\underset{..}{O}} \\ \diagup \\ :\underset{..}{\ddot{Cl}} \end{array}$$ (f) $$\begin{array}{c} :\ddot{F}: \\ | \\ :\ddot{\underset{..}{F}}-\underset{..}{P}-\ddot{\underset{..}{F}}: \\ \diagup \quad \diagdown \\ :\underset{..}{F}: \quad :\underset{..}{F}: \end{array}$$

10.26 Repeat Question 10.25 for

(a) P_4 (tetrahedral: :P̈ at the apex bonded to P, :P and P:, each P bonded to the other three) (b) $:\ddot{O}{=}C{=}\ddot{O}:$ (c) $[:\ddot{O}{-}\ddot{Cl}(-\ddot{O}:){-}\ddot{O}:]^-$

(d) $:\ddot{Cl}{-}B(-\ddot{Cl}:){-}\ddot{Cl}:$ (e) $[H{-}\ddot{O}(-H){-}H]^+$

10.27 Explain the statement: "The properties of molecular substances are not so much determined by the covalent bonding within the molecule, but by the bonding between the molecules (intermolecular bonding)."

10.28 Name two compounds that contain network covalent bonds. What kinds of properties are associated with these substances?

Types and Properties of Chemical Bonds—Bond Energies and Bond Lengths

10.29 What does the term "bond length" refer to in a covalently bonded substance? Why is it necessary to discuss an average value for the bond length? How does the bond length change as the bonding increases from a single to a double to a triple covalent bond?

10.30 What does the term "bond energy" mean? How does the value change as the bonding increases from a single to a double to a triple covalent bond?

10.31 Calculate the O—H bond energy given $\Delta H_f^\circ =$ 249.170 kJ/mol for $O(g)$, 217.965 kJ/mol for $H(g)$, and -241.818 kJ/mol for $H_2O(g)$.

10.32 Calculate the Br—F bond energy given $\Delta H_f^\circ =$ -428.9 kJ/mol for $BrF_5(g)$, 111.884 kJ/mol for $Br(g)$, 78.99 kJ/mol for $F(g)$.

10.33 Use the following heat-of-reaction data to calculate the S—H bond energy:

$$H_2S(g) \longrightarrow H(g) + HS(g) \qquad \Delta H^\circ = 381.27 \text{ kJ}$$
$$HS(g) \longrightarrow H(g) + S(g) \qquad \Delta H^\circ = 354.10 \text{ kJ}$$

10.34 Use the following heat-of-reaction data to calculate the N—H bond energy:

$$NH_3(g) \longrightarrow NH_2(g) + H(g) \qquad \Delta H^\circ = 449 \text{ kJ}$$
$$NH_2(g) \longrightarrow NH(g) + H(g) \qquad \Delta H^\circ = 385 \text{ kJ}$$
$$NH(g) \longrightarrow N(g) + H(g) \qquad \Delta H^\circ = 339 \text{ kJ}$$

10.35 Calculate the C=O bond energy using

$$CH_4(g) + 2O_2(g) \longrightarrow CO_2(g) + 2H_2O(g) \qquad \Delta H^\circ = -802.335 \text{ kJ}$$

and the values of the O=O, O—H, and C—H bond energies given in Table 10.5.

10.36 Calculate the O—H bond energy using

$$H_2(g) + \tfrac{1}{2}O_2(g) \longrightarrow H_2O(g) \qquad \Delta H^\circ = -241.818 \text{ kJ}$$

and the values of the H—H and O=O bond energies given in Table 10.5.

10.37 Using the bond energies given in Table 10.5, predict the heats of reaction for

(a) $2{:}C{\equiv}O{:}(g) + O{=}O(g) \longrightarrow 2{:}\ddot{O}{=}C{=}\ddot{O}{:}(g)$

(b) $:C{\equiv}O{:}(g) + :\ddot{Cl}{-}\ddot{Cl}{:}(g) \longrightarrow :\ddot{Cl}{-}C(-\ddot{Cl}:){=}\ddot{O}(g)$

(c) $2H{-}C(H)(H){-}H(g) + 2H{-}\ddot{N}(H){-}H(g) + 3O{=}O(g) \longrightarrow 2H{-}C{\equiv}N{:}(g) + 6H{-}\ddot{O}{-}H(g)$

10.38 Using the bond energies given in Table 10.5, predict the heats of reaction for

(a) $H{-}CH_2{-}CH_2{-}H(g) \longrightarrow H{-}C(H){=}C(H){-}H(g) + H{-}H(g)$

(b) $H{-}C(H){=}C(H){-}H(g) + :\ddot{Cl}{-}\ddot{Cl}{:}(g) \longrightarrow H{-}C(H)(:\ddot{Cl}:){-}C(H)(:\ddot{Cl}:){-}H(g)$

(c) $H{-}C(H)(H){-}H(g) + :\ddot{Cl}{-}\ddot{Cl}{:}(g) \longrightarrow H{-}C(H)(H){-}\ddot{Cl}{:}(g) + H{-}\ddot{Cl}{:}(g)$

10.39 Would the value of ΔH° be the same for both of the following reactions if you used average bond energies to calculate the enthalpy change? Give a reason for your answer.

$$CH_3{-}CH(CH_3){-}CH_3 + \tfrac{13}{2}O_2(g) \longrightarrow 4CO_2(g) + 5H_2O(l)$$
$$CH_3CH_2CH_2CH_3 + \tfrac{13}{2}O_2(g) \longrightarrow 4CO_2(g) + 5H_2O(l)$$

10.40 Even though values of bond energies are given for molecules and atoms in the gaseous phase, they are often used for calculations involving reactions in solid, liquid, and solution phases. What additional information is needed to perform such calculations?

The Continua of Bond Types

10.41 What reservation should we keep in mind when we classify the bonding in a substance as metallic, covalent, or ionic?

10.42 Briefly describe the concept of polarization of an ion by another ion.

10.43 What causes a covalent bond to be polar? Choose the polar covalent bonds from the following list: (a) F—Xe, (b) O—O, (c) O=O, (d) C=O, (e) C—O, (f) I—I.

10.44 Arrange the following bonds in order of increasing polarity: (a) C—F, (b) H—F, (c) F—F, (d) O—F.

10.45 Choose the cation that would be most effective in polarizing a given anion: Na^+, ionic radius = 0.097 nm; Mg^{2+}, ionic radius = 0.066 nm; or Al^{3+}, ionic radius = 0.051 nm.

10.46 Using the data for radii of the ions given in Figure 9.6, calculate the charge to radius ratio for the following cations: (a) K^+, (b) Ca^{2+}, (c) Sc^{3+}, (d) Fe^{2+}, (e) Zn^{2+}, (f) Ga^{3+}. Discuss the trend across this period. Which ion would polarize a given anion most?

10.47 Repeat the calculation in Problem 10.46 for (a) Cu^+ and (b) Cu^{2+}. Discuss the trend for cations of the same element. Which ion would polarize a given anion more?

10.48 Repeat the calculation in Problem 10.46 for (a) Al^{3+}, (b) Ga^{3+}, (c) In^{3+}, and (d) Tl^{3+}. Discuss the trend for equally charged ions of elements in the same family of the periodic table. Which ion would polarize a given anion most?

10.49 What is meant by the term "electronegativity"? How does the difference in electronegativity between atoms of different elements affect the bonding between these atoms? What value of the electronegativity difference is usually considered necessary for the formation of an ionic bond?

10.50 In which part of the periodic table would we find the elements with the smallest values of electronegativity? Where would the elements with the largest values be?

10.51 Arrange the following elements in order of increasing electronegativity: (a) I, (b) Te, (c) Bi, (d) Ra.

10.52 Repeat Question 10.51 for (a) Rb, (b) Sn, (c) Si, (d) O.

10.53 Classify the bonds that would form between the following pairs of atoms as (a) ionic, (b) polar covalent, (c) nonpolar covalent: (i) Si, O; (ii) N, O; (iii) Sr, F, (iv) As, As.

10.54 Repeat Question 10.53 for (i) Li, O; (ii) Br, I; (iii) Ca, H; (iv) O, O; (v) H, O.

10.55 A common laboratory technique to remove oxygen from a gas is to pass the gas over hot, finely divided copper metal:

$$2Cu(s) + O_2(g) \xrightarrow{\Delta} 2CuO(s)$$

The copper can be regenerated for further use by allowing the CuO to react with hydrogen:

$$CuO(s) + H_2(g) \xrightarrow{\Delta} Cu(s) + H_2O(g)$$

What type of bonding would be found in (a) Cu, (b) O_2, (c) CuO, (d) H_2, (e) H_2O? Write the Lewis structures for (f) O_2, (g) CuO, (h) H_2, (i) H_2O. (j) Which of these substances would have relatively low melting and boiling points? (k) Which of these substances should be good conductors of heat and electricity?

10.56 Beryllium and magnesium are both reactive metals and form compounds with similar empirical formulas. Yet in many cases there are marked differences in the properties of such compounds. For example, beryllium chloride ($BeCl_2$) melts at 405 °C and boils at 488 °C, whereas magnesium chloride ($MgCl_2$) melts at 712 °C and boils at about 1400 °C. Molten beryllium chloride has an electrical conductance about one-thousandth that of a fully ionized salt; molten magnesium chloride exhibits the conductance of a typical salt. (a) Account for the difference in properties of the two compounds. (b) Write Lewis structures for $BeCl_2$ and $MgCl_2$.

Oxidation State

10.57 What are oxidation numbers? Do oxidation numbers represent the actual charges on atoms in molecular compounds?

10.58 What is the maximum possible positive oxidation state for the representative elements? Which two representative elements do not form species with the maximum possible values of their oxidation numbers?

10.59 Which elements in the periodic table have only one oxidation number (other than 0)? Which elements have oxidation numbers equal to the group number and the group number minus 2 as the two possible oxidation states?

10.60 What oxidation numbers might be expected for a nonmetal?

10.61 Find the oxidation numbers of the atoms in the following species: (a) KH, (b) $MnCl_2$, (c) NH_4^+, (d) P_4, (e) Cl^-, (f) SO_3^{2-}, (g) Na_2O_2, (h) MnF_3, (i) ICl_3, (j) H_2Se.

10.62 Repeat Question 10.61 for (a) Al^{3+}, (b) $Cr_2O_7^{2-}$, (c) $Mg(NO_3)_2$, (d) Al_2O_3, (e) P_4O_{10}, (f) $KHCO_3$, (g) $(NH_4)_2SO_4$, (h) $Fe(ClO_4)_3$, (i) H_2O_2, (j) N_3^-.

10.63 Determine the oxidation numbers of each of the atoms in the reactants and products in each of the following reactions and decide whether or not changes in oxidation numbers have occurred as a result of the reaction:

(a) $2NH_3(g) + 3CuO(s) \longrightarrow N_2(g) + 3Cu(s) + 3H_2O(g)$
(b) $H_2SO_4(aq) + 2NaOH(aq) \longrightarrow Na_2SO_4(aq) + 2H_2O(l)$

10.64 Repeat Question 10.63 for:

(a) $2O_3(g) \longrightarrow 3O_2(g)$
(b) $SnO_2(s) + 2C(s) \longrightarrow Sn(l) + 2CO(g)$
(c) $AgCl(s) \rightleftharpoons Ag^+ + Cl^-$

10.65 Name the following substances using the Stock system: (a) N_2O_3, (b) ICl_3, (c) CO_2, (d) SO_2, (e) BF_3, (f) N_2O_5, (g) $SiCl_4$, (h) CCl_4.

10.66 Repeat Question 10.65 for (a) CO, (b) IF_3, (c) I_2O_5, (d) NO, (e) SF_6, (f) SO_3, (g) ICl.

10.67 Write the formulas for the following compounds: (a) boron(III) nitride, (b) carbon(IV) selenide, (c) bromine(I) chloride, (d) nitrogen(III) oxide, (e) oxygen(II) fluoride, (f) sulfur(IV) fluoride, (g) nitrogen(II) oxide, (h) phosphorus(V) chloride.

10.68 Repeat Question 10.67 for (a) nitrogen(I) oxide, (b) silicon(IV) oxide, (c) phosphorus(III) chloride, (d) iodine(V) oxide, (e) sulfur(II) chloride, (f) bromine(III) fluoride, (g) silicon(IV) sulfide.

ELEMENTS AND COMPOUNDS

■ **THE ELEMENTS OF GROUP V** *left, top to bottom* A main group in the *p* block, Group V includes the nonmetals nitrogen and phosphorus, the semiconducting elements arsenic and antimony, and the metal bismuth.

■ **NITROGEN (N_2)** *above* As liquid nitrogen (b.p. $-195\,°C$) boils, water vapor in the air is frozen into clouds.

■ **PHOSPHORUS: RED AND WHITE ALLOTROPES** *above* White phosphorus (P_4, *above right*), which burns on exposure to air, is stored under water, with which it does not react. Red phosphorus (a polymer of P_4 units, *above left*) is much less reactive.

■ **ARSENIC AND ANTIMONY: SEMICONDUCTING ELEMENTS** *above* These allotropes of antimony (Sb, *left*) and arsenic (As, *right*), like other semiconducting elements, resemble metals in appearance but nonmetals in many properties.

■ **BISMUTH** *above* Bismuth is the heaviest main group metal, other than radioactive polonium.

■ **THE MOLE — 6.022×10^{23} FORMULA UNITS** *above* Shown are one mole each of aluminum and copper atoms (*left and right dishes in foreground*); oxygen molecules (in balloons, at room temperature and pressure); ionic formula units of sodium chloride, potassium chromate, and potassium dichromate (NaCl, K_2CrO_4, $K_2Cr_2O_7$, *left to right in beakers*); and water and ethyl alcohol molecules (H_2O and CH_3CH_2OH, *left and right graduated cylinders*).

■ **CRYSTALS** *three across top; two at bottom left* The five photomicrographs (*a–e*) show crystals formed by an element (sulfur), two inorganic compounds (copper(II) chloride and ammonium dichromate), and two organic compounds (aspirin, which is acetylsalicylic acid, and urea, which is present in urine and was the first organic compound to be synthesized). (*a*)–(*d*) Crystals grown by cooling melted substances between microscope slides and photographed in polarized light, which is diffracted by differently ordered planes of atoms to give different pure spectral colors. (*e*) Three-dimensional crystals photographed in nonpolarized light against a dark field. (Magnifications from 50 × to 100 ×.)

(*a*) **SULFUR** *above*

(*b*) **ASPIRIN ($CH_3COOC_6H_4COOH$)** *above*

(*c*) **UREA (H_2NCONH_2)** *above*

(*d*) **COPPER(II) CHLORIDE ($CuCl_2$)** *above*

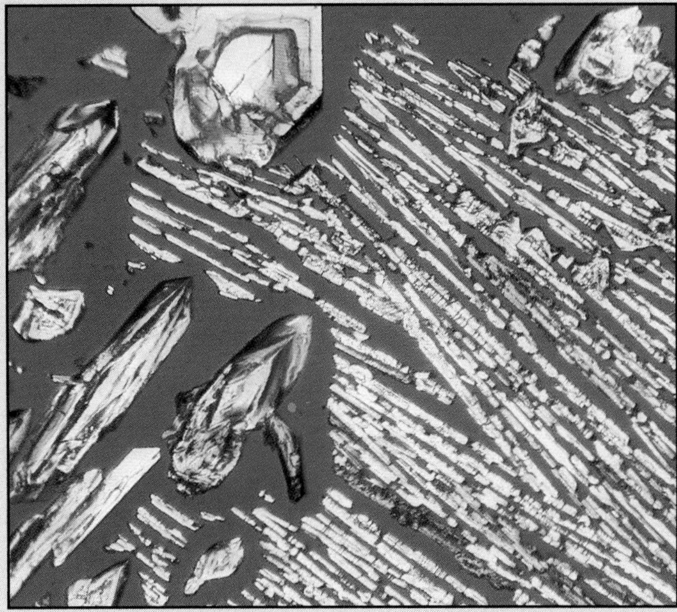

(*e*) **AMMONIUM DICHROMATE ($(NH_4)_2Cr_2O_7$)** *above*

■ **AMORPHOUS SUBSTANCES** *right and below* Amorphous substances form when liquids solidify with their atoms or molecules in random arrangements, rather than in the ordered arrangements of crystals. Opal and amber are two naturally occurring amorphous materials, both used in jewelry. Opal is hydrated silica ($SiO_2 \cdot nH_2O$); it is found in cavities in igneous and sedimentary rocks, and in fossilized wood. Amber, a resin from prehistoric pine trees, is of variable carbon–hydrogen–oxygen ratio, but usually contains 3–8% of succinic acid.

ROUGH PRECIOUS OPAL *above*

AMBER WITH PRESERVED INSECTS *above*

■ **SOME REPRESENTATIVE ELEMENTS AND THEIR CHLORIDES** *below from the left:* In the beakers, stored under oil because of their reactivity with the atmosphere, are lithium and sodium metals and, in the watch glasses, their chlorides (LiCl, NaCl). Next are pieces of magnesium with magnesium chloride ($MgCl_2$) and a coil of aluminum with aluminum chloride ($AlCl_3$). These four metals all form white, crystalline, ionic chlorides. Silicon, the shiny gray substance in the next watch glass, forms a molecular, liquid chloride ($SiCl_4$, not shown) that fumes as it reacts with water in the air. The closed flask contains carbon tetrachloride (CCl_4), a molecular compound, and the dish at the right contains carbon as graphite.

■ **SOME 4TH PERIOD *d*-BLOCK METALS** *below from the upper left, clockwise:* Copper (Cu), zinc (Zn), nickel (Ni), manganese (Mn), titanium (Ti), chromium (Cr), and iron (Fe) as wire and powder. Scandium, vanadium, and cobalt are also members of this *d*-block period.

■ **CHLORIDES OF SOME 4TH PERIOD *d*-BLOCK METALS** *right from upper left, clockwise:* Chlorides of iron ($FeCl_3 \cdot 6H_2O$, bright yellow), cobalt ($CoCl_2 \cdot 6H_2O$, ruby red), nickel ($NiCl_2 \cdot 6H_2O$, green), copper ($CuCl_2 \cdot 2H_2O$, green), and zinc ($ZnCl_2$, white). All but the zinc compound have characteristic colors due to the transition metal cations.

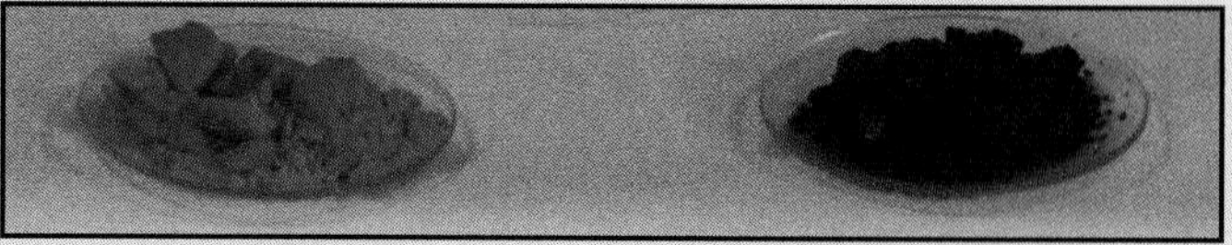

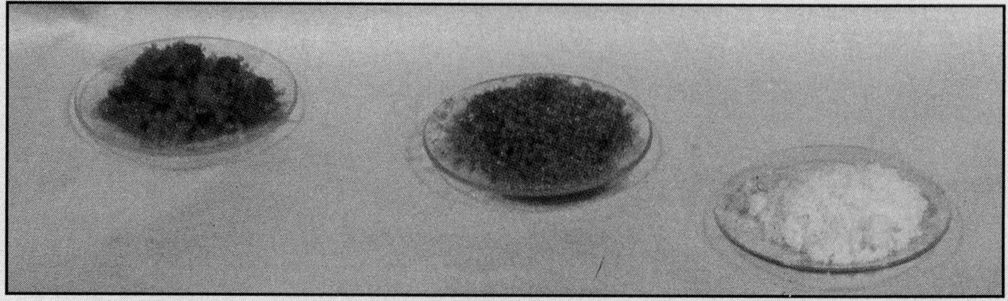

CHAPTER

11

Covalent Bonding and Properties of Molecules

LEWIS STRUCTURES AND VSEPR

11.1 THE SHAPES OF MOLECULES

Almost everyone has played with building toys. There are wooden balls to be connected by sticks and there are plastic parts of many kinds to be snapped together. The structural shapes one can build are governed by the number of connecting points on each block or part, and by the angles at which such connections can be made.

Atoms are built up into molecules in somewhat the same way. The angles between atoms are determined by the electron configurations of the atoms. Some atoms have more "connecting points" than others and therefore can form molecules with a greater variety of shapes. **Molecular geometry** is the spatial arrangement of the atoms in a molecule. Molecules, however, differ from building blocks because they are flexible rather than rigid, and because their geometry is influenced by attraction and repulsion between areas of greater or lesser electron density.

The first step in investigating the shape of a molecule is often determining the Lewis structure of the molecule. Lewis structures (introduced in Section 11.2) represent on paper which atoms in a molecule are bonded to each other and the locations of any unshared electrons. Lewis structures alone give no information about molecular geometry. However, unshared electrons, as we shall see, have an important influence on molecular geometry.

The geometry of a molecule is described by the lengths and angles of the bonds. A **bond angle** is the angle between the bonds that join one atom to two other atoms. The relationships among atomic radii (Section 9.4), bond length (Section 9.4), and bond angle are shown in Figure 11.1.

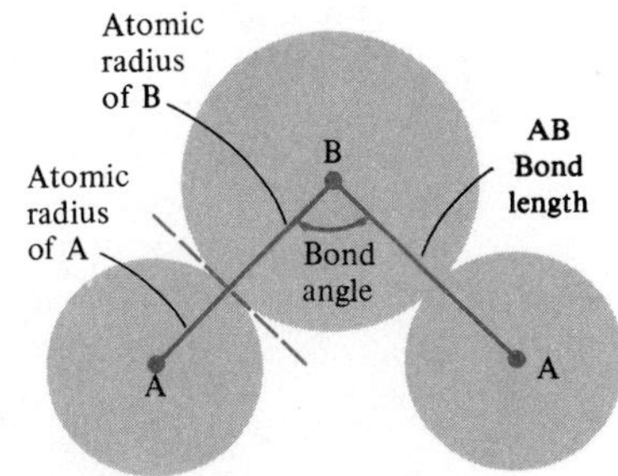

Figure 11.1
Bond Length, Bond Angle, and Atomic Radii

Experimental methods permit the determination of bond lengths and bond angles. Given the knowledge that, for example, the mercury(II) chloride molecule is linear but the water molecule is bent,

Cl—Hg—Cl H/O\H

we then want to know why. Why should three atoms form a linear molecule in one case and a bent molecule in another case? The valence-shell electron-pair repulsion theory (VSEPR), which is presented in Sections 11.4 and 11.5, provides a simple approach to predicting the geometry of molecules such as these.

Molecular geometry greatly influences the properties of molecular substances. For example, molecular geometry directly affects the strength of intermolecular forces (Sections 11.10–11.13), which in turn influence melting and boiling points, chemical reactivity, and other molecular properties.

11.2 LEWIS STRUCTURES FOR MOLECULAR COMPOUNDS AND POLYATOMIC IONS

To write a Lewis structure, one must know the arrangement of the atoms—which atoms are directly connected to each other. Are the three atoms in SO_2 arranged like this, S—O—O, or like this, O—S—O? For many simple molecules, the arrangement can be predicted using the following guidelines. In many other cases, however, only by consulting reference books or performing experiments can one determine how the atoms in a molecule are connected to each other.

The steps for writing Lewis structures are as follows:

1. Write the correct arrangement of the atoms using single bonds. Where necessary, apply the following guidelines:
   (a) Smaller, more electronegative nonmetal atoms surround larger, less electronegative nonmetal atoms.
   (b) Oxygen, hydrogen, and/or halogen atoms often surround a central metal or nonmetal atom in a symmetrical arrangement.
   (c) Carbon atoms are usually bonded to each other.
   (d) Oxygen atoms are bonded to each other only in peroxides (or superoxides; Section 16.10a).
   (e) In most acids, such as H_2SO_4, and in many other compounds that contain both oxygen and hydrogen atoms, the hydrogen atoms are all bonded to oxygen atoms.
2. Find the total number of valence electrons. Add together the number of valence electrons contributed by each atom. If the species is an ion, subtract one electron for each unit of positive charge or add one electron for each unit of negative charge.
3. Assign two electrons to each covalent bond.
4. Distribute the remaining electrons so that each atom has the appropriate number of nonbonded electrons. For elements from the second period, other than beryllium and boron, this is the number of electrons needed so that each atom is surrounded by an octet. For elements of the third period and beyond, except for aluminum, this is often the number of electrons needed to complete an octet, although extra electrons can also be placed around atoms of these elements *when they are the central atoms in compounds.* Remember that atoms bonded to a central atom usually obey the octet rule.
5. If there are not enough electrons to go around, change some single bonds to multiple bonds. Multiple bonds can be written between carbon, nitrogen, oxygen, sulfur, selenium, and phosphorus atoms. (Note that beryllium, boron, and aluminum do *not* form multiple bonds.)

EXAMPLE 11.1 Lewis Structures; Problem Solving

Predict the arrangement of the atoms in a molecule of chlorous acid ($HClO_2$) and write the Lewis structure for this molecule. Note that, as is apparent from its name, this compound is not a peroxide.

1. Study the problem and be sure you understand it.
   (a) What is unknown?
   How the atoms are connected to each other, whether the bonds are single or multiple, and where lone pairs of electrons are located are unknown.
   (b) What is known?
   The molecular formula of the compound and its chemical name are known.
2. Decide how to solve the problem.
   (a) How can the connection be made between the known and the unknown?
   First, the guidelines given in Step 1a–e above must be applied to determine how the atoms are connected. Then, the total number of valence electrons must be determined and distributed among the bonded atoms according to Steps 3–5 above.
   (b) What is necessary to make the connection?
   It is necessary to recognize that the chlorine atom, which is larger and less electronegative than the oxygen atoms, will be the central atom.
3. Solve the problem.
   The smaller, more electronegative oxygen atoms, which we know from the name of the compound are not bonded to each other as in a peroxide, will both be bonded to the central chlorine atom. The hydrogen atom will be bonded to one of the oxygen atoms, as is the case for most oxygen-containing acids. The atoms are connected as follows

$$\mathrm{H{-}O{-}Cl{-}O}$$

   The total number of valence electrons is 1 for the hydrogen atom, 6 for each oxygen atom (from Group VI), and 7 for the chlorine atom (Group VII); $1 + (2 \times 6) + 7 = 20$. Six electrons are used in the three covalent bonds shown above, leaving 14 to be distributed. If the electron octets are completed for the oxygen and chlorine atoms by adding lone pairs, all 20 electrons have been used.

$$\mathrm{H{-}\ddot{\underset{..}{O}}{-}\ddot{\underset{..}{Cl}}{-}\ddot{\underset{..}{O}}{:}}$$

4. Check the result.
   Is the answer reasonable?
   Yes. Each atom has a noble gas configuration and also, chlorine and oxygen are both atoms that can accommodate eight valence electrons.

EXAMPLE 11.2 Lewis Structures

Write Lewis structures for (a) chlorate ion (ClO_3^-), (b) nitrosyl ion (NO^+), (c) bromine trifluoride (BrF_3).

As a first step, determine the arrangement of the atoms in each molecule:

$$\text{(a)}\ \mathrm{O{-}\underset{\underset{\displaystyle O}{|}}{Cl}{-}O} \qquad \text{(b)}\ \mathrm{N{-}O} \qquad \text{(c)}\ \mathrm{F{-}\underset{\underset{\displaystyle F}{|}}{Br}{-}F}$$

The arrangements in (a) and (c) follow the rule that the larger, less electronegative atom is the central atom and is surrounded by the smaller, more electronegative atoms. In (b) there is no other choice.

(a) ClO_3^-, $7e^-$ [from Cl] + ($3 \times 6e^-$) [from O_3] + $1e^-$ [-1 charge] = 26 valence electrons. The three bonds in structure (a) account for 6 valence electrons. If the octets on all atoms are completed, all 20 remaining electrons are used.

(b) NO^+, $5e^-$ [from N] + $6e^-$ [from O] − $1e^-$ [+1 charge] = 10 valence electrons. The single bond in (b) above uses 2 electrons and leaves both the N and the O atoms 6 electrons short of octets. A double bond would use 4 electrons instead, leaving both atoms 4 electrons short of octets. Only a triple bond will permit an octet of electrons on each atom and also accommodate all 10 valence electrons.

(c) BrF_3, $7e^-$ [from Br] + ($3 \times 7e^-$) [from F_3] = 28 valence electrons. If all the octets on the Br and F atoms are completed, 26 electrons are used, 6 in the bonds and 20 in the lone pairs, leaving 2 more electrons to be accounted for. Because bromine is in the fourth period, it can accommodate more than 8 valence electrons, and placing a second lone pair on bromine gives the correct Lewis structure.

(a) [:Ö—C̈l—Ö: with :Ö: below Cl]⁻ (b) [:N≡O:]⁺ (c) :F̈—B̈r—F̈: with :F̈: below Br

EXERCISE Write Lewis structures for (a) iodate ion (IO_3^-), (b) hydrogen peroxide (H_2O_2), (c) tetrahydroxoborate(III) ion ($[B(OH)_4]^-$), (d) iodine pentafluoride (IF_5), (e) carbon disulfide (CS_2).

Answers

(a) [:Ö—Ï—Ö: with :Ö: below I]⁻ (b) H—Ö—Ö—H (c) [H—Ö—B—Ö—H with :Ö:—H above and :Ö:—H below B]⁻

(d) :F̈—Ï—F̈: with :F̈: above I and :F̈: :F̈: below (e) S̈=C=S̈

11.3 RESONANCE

The representation of bonding by Lewis structures runs into difficulty when more than one structure that agrees with the electronic requirements and the properties of a compound can be written. A typical example is nitrogen(I) oxide, N_2O, a diamagnetic molecule in which the two nitrogen atoms are bonded to each other. The 16 valence electrons of N_2O can be arranged in two reasonable ways:

:N̈=N=Ö: :N≡N—Ö:

Each of these formulas gives each atom eight valence electrons, each is electrically neutral, and in each, the atoms form numbers of bonds that might reasonably be expected (see Section 11.9). Two structures can also be written for ozone, O_3

Ö=Ö—Ö: :Ö—Ö=Ö

In writing molecular structures, we must not lose sight of an essential point—the structures are our way of trying to put down on paper what we have learned by experiment. Experimental investigation of properties of N_2O, O_3, and many other such molecules invariably reveals not one structure *or* another, but a structure that is a composite of the possible structures. For example, the two O—O bonds in ozone are identical—there is no evidence for both single and double bond properties in the molecule.

Many organic compounds also have structures intermediate between those that can be written with the lines and dots of Lewis formulas. The classic example is that of the benzene molecule, a six-membered ring of carbon atoms with the molecular formula C_6H_6. Each carbon atom in benzene has an octet of electrons in each of the first pair of formulas below (called Kekulé structures, after the man who first devised them). In one structure there are double covalent bonds between three pairs of adjacent carbon atoms, and in the other structure there are single covalent bonds between the same pairs of carbon atoms. All experimental evidence, however, indicates that all of the carbon–carbon bonds in benzene are the same.

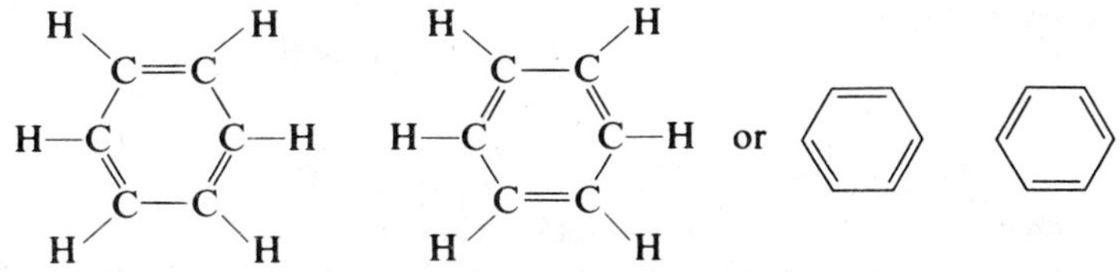

```
 H       H        H       H
  \     /          \     /
   C═C              C─C
  /     \          //     \\
H─C      C─H    H─C      C─H   or
  \\     //        \     /
   C─C              C═C
  /     \          /     \
 H       H        H       H
```

The second pair of structures above illustrates the customary method of writing structural formulas for carbon atoms in a ring. The symbols C and H are omitted, and it is understood that at each corner is a carbon atom bonded to the correct number of hydrogen atoms to give the carbon atom four covalent bonds. Often benzene is represented by just one of these structures, with the understanding that both are needed for a complete representation.

Obviously, no single structure accounts for the characteristic properties of substances such as benzene, N_2O, and O_3. To deal with this problem, if two or more structures can be written for a molecule or ion that differ only in the position of the valence electrons, the molecule is said to exhibit resonance. **Resonance** refers to the arrangement of valence electrons in molecules or ions for which more than one Lewis structure can be written. That two Lewis structures are resonance structures of the same compound is indicated by a double-headed arrow, for example

$$:\ddot{N}{=}N{=}\ddot{O}: \longleftrightarrow :N{\equiv}N{-}\ddot{\underset{\cdot\cdot}{O}}:$$

[Note that this differs from the arrows used to indicate equilibrium, $\rightleftharpoons$.]

"Resonance" does *not* mean that the molecule constantly flips from one structure to another. The bonds actually have lengths and strengths intermediate between those of single and double bonds, or double and triple bonds. For example, the carbon–carbon bonds in benzene have a bond length of 0.140 nm, compared to 0.154 nm for C—C bonds and 0.134 nm for C═C bonds. The concept of resonance is necessary because of limitations in the way we *write* structures. The structure of a molecule or ion for which resonance structures can be written is called a **resonance hybrid.** The *actual* molecule is always the same. A resonance hybrid is similar to the tangelo, a hybrid citrus fruit. You can pick up a tangelo and examine its characteristic properties—it is not a tangerine at one moment and a grapefruit at another moment.

Sometimes a dashed or dotted line in a single structure is used to indicate resonance, for example

O┄O┄O

This structure shows that the oxygen–oxygen bonds are equivalent and are intermediate in their properties between single bonds and double bonds. The structure of benzene is often written in one of the following ways:

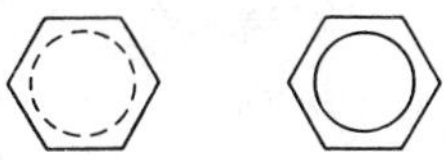

to emphasize that benzene has only one structure, the hybrid in which all carbon–carbon bonds are identical.

In writing resonance structures, the following rules apply:

1. The sequence of atoms in each resonance structure must be the same. That is, the same atoms must be connected to the same other atoms. For example

   $$:N{\equiv}C{-}\ddot{\underset{\cdot\cdot}{O}}{-}H \quad \text{and} \quad H{-}\ddot{N}{=}C{=}\ddot{O}:$$

   are not resonance structures for the same compound; they are the structures of different compounds [cyanic acid and isocyanic acid, respectively].
2. All resonance structures for the same molecule must have the same total number of valence electrons.

EXAMPLE 11.3 Resonance

Write the possible resonance forms of nitryl chloride, NO_2Cl, in which nitrogen is the central atom.

Putting in single bonds gives

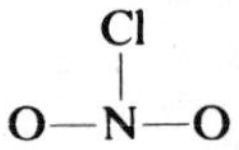

$$\begin{array}{c} Cl \\ | \\ O-N-O \end{array}$$

The total number of valence electrons available is 24 [7 + 5 + (2 × 6)]. With 6 electrons in the single bonds, 18 electrons remain to be distributed. Putting enough lone pairs for complete octets into each atom would require 20 electrons (6 for Cl, 2 for N, and 6 each for the O atoms), two more than we have. Therefore, one double bond must be used. Nitrogen–oxygen double bonds are to be expected, while Cl—N double bonds are not. There are two equivalent possibilities for nitrogen–oxygen double bonds, giving the two resonance forms, which can also be written as a hybrid:

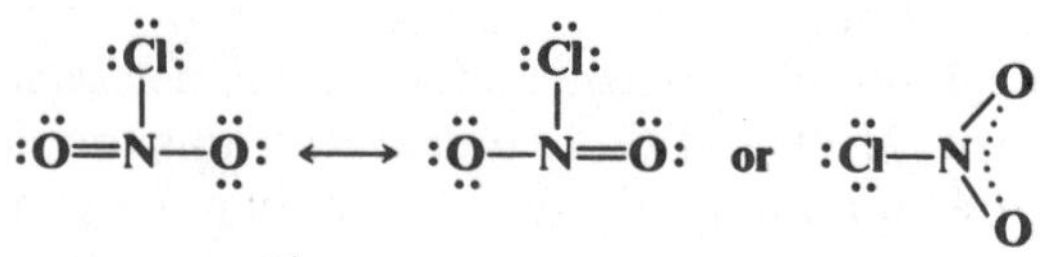

$$:\ddot{O}=N(-:\ddot{Cl}:)-\ddot{O}: \longleftrightarrow :\ddot{O}-N(-:\ddot{Cl}:)=\ddot{O}: \quad \text{or} \quad :\ddot{Cl}-N\overset{O}{\underset{O}{\vdots}}$$

EXERCISE Write the Lewis structures for the possible resonance forms of the nitrite ion, NO_2^-.

Answer $[:\ddot{O}-\ddot{N}=\ddot{O}]^- \longleftrightarrow [\ddot{O}=\ddot{N}-\ddot{O}:]^-$

11.4 VSEPR: MOLECULES WITHOUT LONE-PAIR ELECTRONS ON THE CENTRAL ATOM

Valence-shell electron-pair repulsion theory (VSEPR) uses repulsion between electron pairs as the basis for prediction of molecular geometry. Many simple molecules or polyatomic ions consist of a central atom to which a number of other atoms or groups are bonded, for example

Formulas as usually written

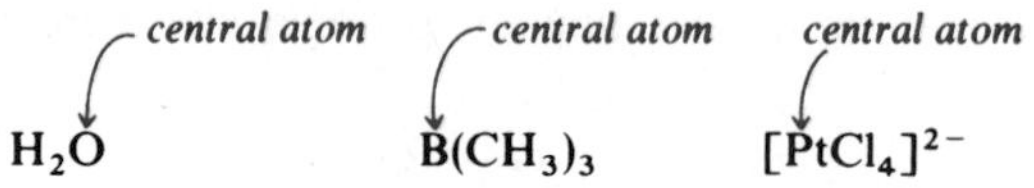

H_2O (central atom: O) $B(CH_3)_3$ (central atom: B) $[PtCl_4]^{2-}$ (central atom: Pt)

Structural formulas

H—O—H (central atom: O)

$$\begin{array}{c} CH_3 \\ | \\ CH_3-B-CH_3 \end{array}$$ (central atom: B)

$$\left[\begin{array}{c} Cl \\ | \\ Cl-Pt-Cl \\ | \\ Cl \end{array}\right]^{2-}$$ (central atom: Pt)

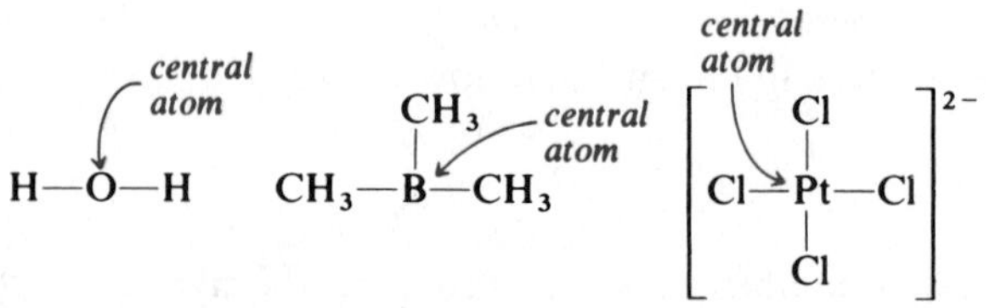

To apply VSEPR theory, molecules and ions are classified according to the number of bonding electron pairs and lone electron pairs surrounding central atoms. The predicted molecular geometry is the one that places the atoms or groups bonded to the central atom, as well as the lone pairs, as far apart as possible.

Consider a molecule AB_n in which A is the central atom. For $n = 1$ there is only one possible structure—a linear AB molecule (whether or not lone electron pairs are present).

$n = 1$ A—B e.g., $H-\ddot{\underset{..}{Cl}}:$

hydrogen chloride

An AB_2 molecule contains two shared electron pairs. Obviously, they achieve the greatest distance from each other if the A—B bonds are on opposite sides of A in a linear molecule. The BAB bond angle is 180°. Beryllium hydride (BeH_2) has such a linear molecule.

When three covalent bonds surround the central molecule (AB_3), BAB angles of 120° place the bonding electron pairs at the greatest distance from each other. This gives a planar molecule with A at the center of an equilateral triangle and the B atoms at the corners, as shown in Table 11.1. Boron trifluoride forms such a molecule.

When $n = 4$ we reach molecules of the common type AB_4, with four substituents around a central atom that has an octet of bonding electrons. (The word "substituent" refers to atoms or groups bonded to another atom or group, in this case B substituents on A.) The four substituents remain as far apart as possible, at the corners of a regular tetrahedron, as shown in Table 11.1. Tetrahedral geometry occurs often, particularly in carbon compounds. Methane (CH_4) is a simple carbon compound that has a central tetrahedral carbon atom.

Table 11.1 summarizes the ideal geometries and ideal bond angles for AB_n molecules with 2 to 6 covalent bonds and no lone-pair electrons on the central atom. Each of these arrangements places the bonding electron pairs at maximum distances from each other. In real molecules, the bond angles often vary from the ideal values given in the table, but the overall geometry of the molecule is based on that shown.

In each structure in Table 11.1 except that for AB_5 molecules, the ideal BAB angles are equal to each other. A triangular bipyramidal molecule has two different types of BAB bond angles. The three B substituents in the same plane with the central atom occupy what are called the *equatorial* positions and lie at 120° angles from each other. The two B substituents at opposite ends of the molecule, in what are referred to as the *axial* positions, are at angles of 90° from the B substituents in the equatorial positions. (In the next section we shall see that this difference influences the geometry of molecules with lone-pair electrons.)

In predicting molecular geometry by VSEPR theory, double and triple bonds are treated like single bonds. For example, the formaldehyde molecule, $H_2C{=}O$, is classified as an AB_3 molecule. And each carbon atom in ethylene, $H_2C{=}CH_2$, is the A

Table 11.1
Geometry of Covalent Molecules with No Unshared Electron Pairs on the Central Atom
In the molecules shown here, all valence electrons on the central atom (A) participate in covalent bonds. Geometry could possibly be limited by the size of A and B. Only in AB_5 molecules are there different BAB angles within the molecule. Each B^e is separated by 120° from other B^e's. Each B^a is separated by 90° from the B^e's. Therefore, the B^e's have more room. Note: The geometry summarized here and in Table 11.2 applies to most cases where A is a representative element and often when A is a transition element.

Formula type	AB_2	AB_3	AB_4	AB_5	AB_6
Shared electron pairs	2	3	4	5	6
Arrangement of B atoms relative to A atoms (ideal BAB bond angle)	Linear (180°)	Triangular planar (120°)	Tetrahedral (109.47°)	Triangular bipyramidal (B^eAB^e, 120°) (B^eAB^a, 90°)	Octahedral (90°)
Molecular structures	B——A——B	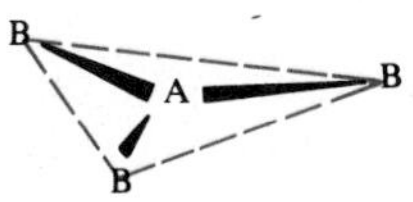	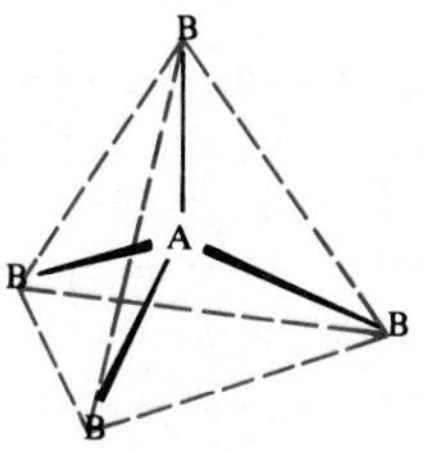	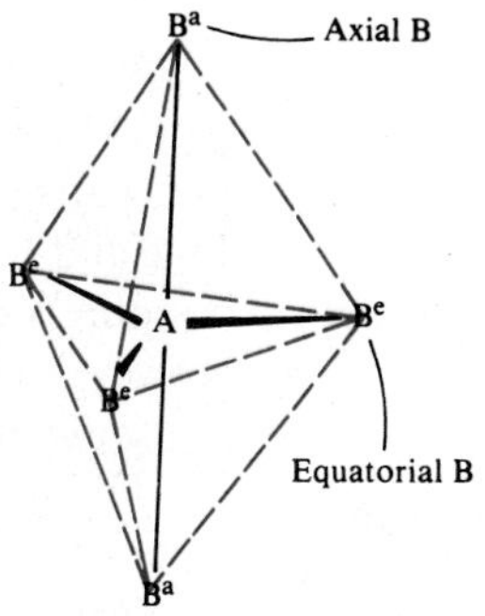 Axial B; Equatorial B	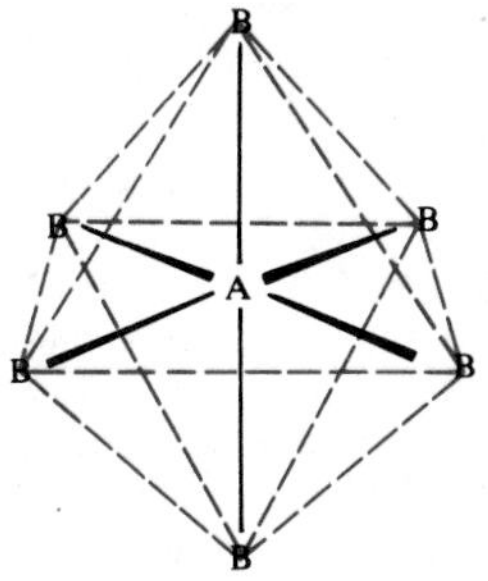

atom in an AB_3 situation (one of the B's is $=CH_2$). Double bonds take up more space around A than single bonds because their greater electron density repels the other bonding electrons. As a result, the bond angles are distorted from the ideal angles of Table 11.1, for example

H, H, C=O, 118°, 121°
formaldehyde
a triangular planar molecule

H, H, C=C, H, H, 117.4°, 121.3°
ethylene
a planar molecule

To use VSEPR theory, it is necessary to write the correct Lewis structure for the molecule and determine whether or not there are any lone electron pairs present on the central atom. If there are no lone pairs, the molecule is of the AB_n type and the geometry is expected to be based on that shown in Table 11.1.

EXAMPLE 11.4 VSEPR

Use VSEPR to predict the geometry of (a) PF_5, (b) SO_4^{2-}, and (c) $BeCl_2(g)$.

(a) Phosphorus is the third-period element of Group V and has five valence electrons. Because some of its bonding electrons can occupy vacant *d* orbitals, the phosphorus atom can accommodate more than an octet of valence electrons, as shown by the Lewis structure below. There are no lone-pair electrons on the phosphorus atom. Therefore, PF_5 is an AB_5 molecule and should have **triangular bipyramidal** geometry.

(b) Sulfur is a third-period element of Group VI. The Lewis structure shows no lone pairs on the sulfur atom, and therefore this AB_4 ion has **tetrahedral** geometry. Note that the VSEPR theory applies equally well to both ions and molecules.

(c) The beryllium atom has two valence electrons ($2s^2$) and the Lewis structure shows no lone-pair electrons on the central beryllium atom. Therefore, $BeCl_2$ is a **linear,** AB_2 type molecule (in the gas phase).

An AB_5 molecule
(a)

An AB_4 ion
(b)

:Cl—Be—Cl:
An AB_2 molecule
(c)

EXERCISE Using VSEPR, predict the geometry of (a) CO_2, (b) SF_6, (c) NH_4^+.

Answers (a) Linear, AB_2 type; (b) octahedral, AB_6 type; (c) tetrahedral, AB_4 type.

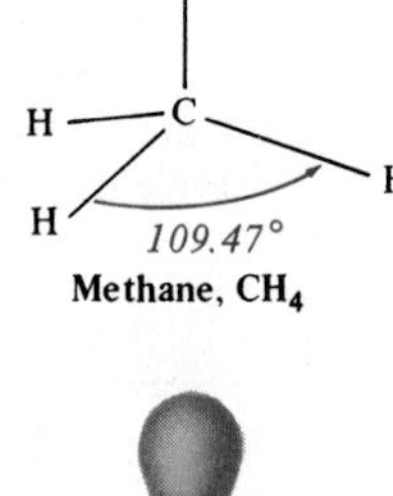

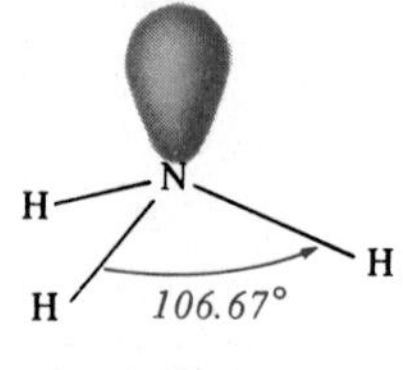

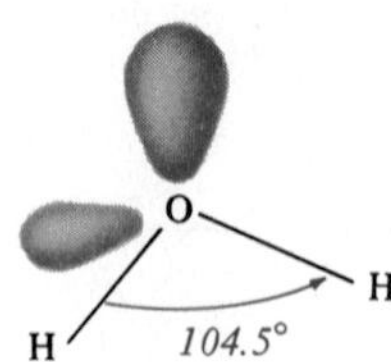

Figure 11.2
Methane, Ammonia, and Water Molecules
The methane molecule has tetrahedral geometry (see AB_4 in Table 11.1); ammonia is a triangular pyramidal molecule; and water is a bent, or angular, molecule.

11.5 VSEPR: MOLECULES WITH LONE-PAIR ELECTRONS ON THE CENTRAL ATOM

What happens when the central atom in a molecule has one or more lone electron pairs? Like bonding pairs, lone pairs repel each other. There is also repulsion between lone pairs and bonding pairs. This leads to arrangements in which all electron pairs—lone and bonding—are as far as possible from each other. Taken together, therefore, bonding and lone pairs assume the same general arrangements shown in Table 11.1. However, because in molecular geometry we are looking only at the positions of *atoms*, replacing an atom with a lone pair changes the shape of the molecule.

For example, methane, ammonia, and water molecules each have four electron pairs around a central atom. Each has a different molecular geometry, but all derive their geometry from the AB_4 tetrahedron (Figure 11.2). The ammonia molecule has the geometry of a triangular pyramid, rather than a tetrahedron, because one place in AB_4 is filled by a lone pair. The general formula for such a molecule is AB_3E, where E represents a lone pair. Water, with two lone pairs, is a bent molecule represented by the general formula AB_2E_2. Note that one unpaired electron has the same effect on geometry as a lone pair.

Lone pairs occupy more space than electron pairs in bonds, because lone pairs are attracted by only one nucleus, not two. A lone pair on atom A repels the shared electron pairs of A—B bonds. Therefore, in most cases the BAB bond angles in compounds where atom A has one or more lone pairs are compressed and are smaller than the ideal angles given in Table 11.1. The effect of one lone pair is shown by comparing the bond angles in methane and ammonia molecules (see Figure 11.2). Methane, with no lone pair, has the ideal tetrahedral angle, 109.47°. In ammonia, with one lone pair, the extra repulsion of the lone pair decreases the H—N—H angle to 106.67°.

Two lone pairs repel each other more than a lone pair and a bonding pair. As a result, the bond angles in a molecule that has two lone pairs are more compressed than the bond angles in a similar molecule that has one lone pair. For example, compare the bond angle in ammonia (106.67°) with that in water (104.5°), which has two lone pairs rather than one. A *single* unpaired electron on a central atom requires *less* space than either a bonding pair or a lone pair of electrons.

Because the lone, or nonbonding, electron pairs need more space, they enter AB_5 and AB_6 molecules at specific positions. In Figure 11.3, which summarizes the effects of lone pairs on molecular shape, look at the rows beginning AB_5 and AB_6. As pointed out above, AB_5 molecules have two different bond angles between nearest neighbors.

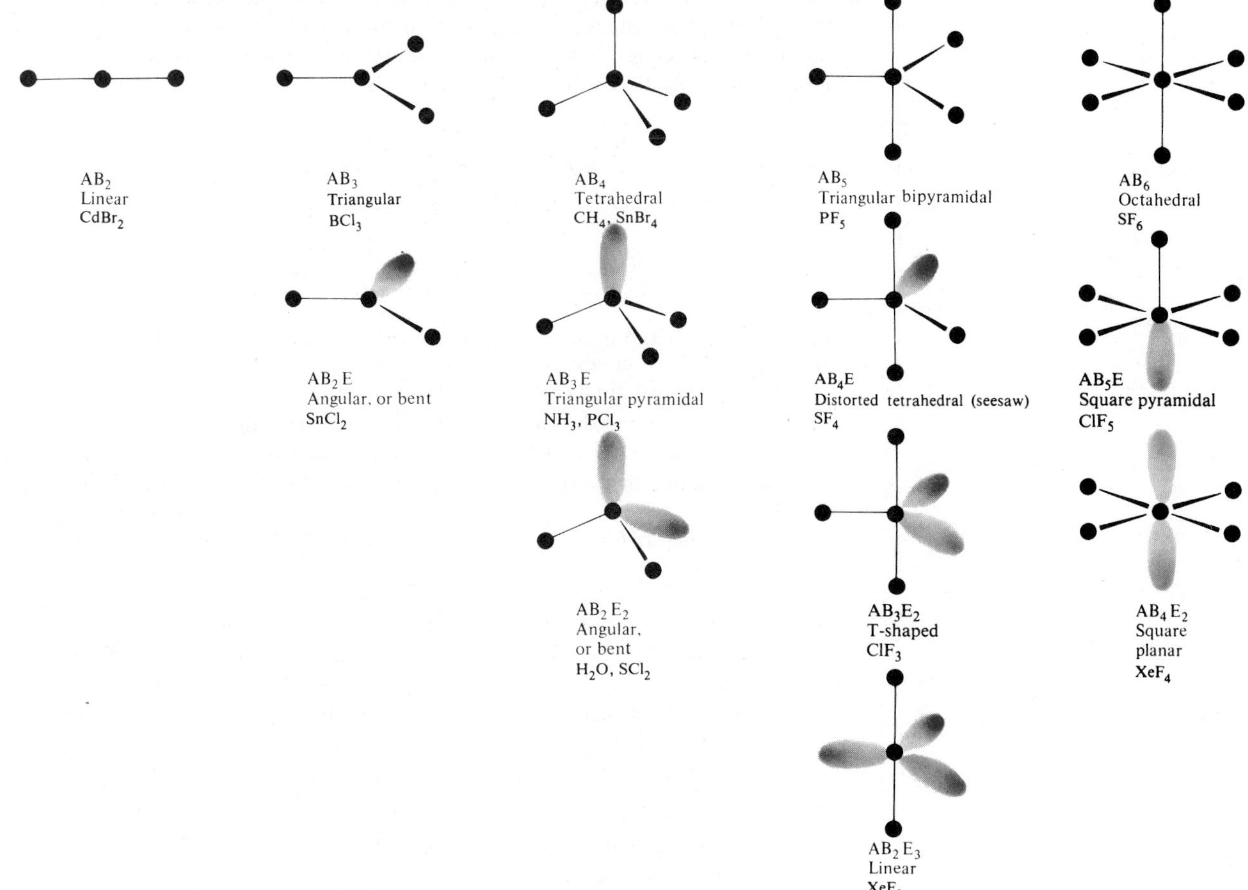

Figure 11.3
General Shapes of Molecules with Bonding and Nonbonding Electron Pairs
A nonbonding electron pair is symbolized by E. With a nonbonded electron pair on A, the BAB bond angles tend to be smaller than ideal. Source: R. J. Gillespie, *J. Chem. Educ.*, **40**, 1963, p. 295

The equatorial positions have more space than the axial positions, and lone pairs occupy these equatorial positions (see Table 11.1). In AB_6 molecules, all the B positions are equivalent; the lone pairs can best stay separated from each other by entering positions opposite each other. Note that the smallest bond angle in AB_5 and AB_6 molecules is 90°. Thus in each case, the lone pairs occupy the positions that minimize the number of lone pairs separated by 90°. Examples of molecules with each type of geometry are included in Figure 11.3.

The following steps are necessary in using VSEPR to predict molecular geometry:

1. Write the Lewis structure of the molecule.
2. Determine the number of bonding pairs and lone pairs of electrons around the central atom.
3. From Table 11.1 determine the ideal geometry. Then, if necessary taking into account the presence of lone pairs, use Table 11.1 and Figure 11.3 to predict the actual shape of the molecule. (Remember that multiple bonds are treated like single bonds, and unpaired electrons are treated like lone electron pairs.)
4. Keep in mind that lone pairs occupy larger sites (equatorial in molecules derived from AB_5 geometry) or, when sites are equal, occupy sites opposite rather than next to each other.

EXAMPLE 11.5 VSEPR

Use VSEPR to predict the molecular geometry of (a) BrF_3 and BrF_5, (b) NO_2. Discuss in general the bond angles in these molecules.

(a) Bromine is the fourth-period element in Group VII. The bromine atom has seven valence electrons and can accommodate more than an octet of electrons in its outermost energy level. The Lewis structures show the number of lone pairs and bonding pairs in BrF_3 and BrF_5 as follows

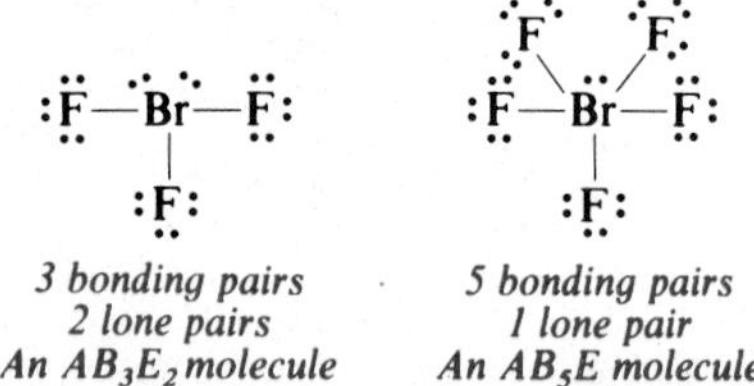

From Table 11.1 and Figure 11.3 it can be seen that BrF_3 will be a **T-shaped** molecule and BrF_5 will have **square pyramidal** geometry.

The F—Br—F bond angles in BrF_3 should be less than the ideal 90° (see AB_5 in Table 11.1) because of the two lone pairs of electrons. Similarly, **the angles in BrF_5 should also be compressed from the ideal 90°** (AB_6 in Table 11.1) because there is one lone pair. [Experiment confirms this prediction—the F—Br—F bond angles are about 83.5°.]

(b) The Lewis structure for NO_2, which has two resonance structures

$$:\ddot{O}{=}\dot{N}{-}\ddot{\underset{..}{O}}: \longleftrightarrow :\ddot{\underset{..}{O}}{-}\dot{N}{=}\ddot{O}$$

shows that there are two substituents and one lone electron on the central nitrogen atom. Thus NO_2 fits the AB_2E pattern and should be an **angular, or bent,** molecule.

The repulsion between *one* lone electron and bonding electron pairs is not nearly as great as that between a lone pair or a bonding pair and other bonding pairs. Therefore, **the O—N—O bond angle should be somewhat *greater* than 120°** because of the repulsion between the bonding pairs. [Experiment confirms this prediction—the O—N—O bond angle is 134.25°.]

EXERCISE Use VSEPR to predict the shape of $[I_3]^-$ and ClO_2. Explain the observed O—Cl—O bond angle of 112°.

Answer $[I_3]^-$ is a linear ion, AB_2E_3 type. ClO_2 is a bent molecule, AB_2E_2. Angle greater than ideal tetrahedral angle because the central Cl atom has one unshared pair and one unpaired electron.

SUMMARY OF SECTIONS 11.1–11.5

Molecular geometry refers to the spatial arrangement of atoms in a molecule. It is determined by the number of bonds to each atom and the angles between them. The first step in examining the geometry of a molecule or a polyatomic ion is to write its Lewis structure. To do this, one must know how the atoms are connected to each other and then distribute the valence electrons among them. Oxygen, hydrogen, or halogen atoms often surround a central atom in symmetrical arrangements. If several different correct Lewis structures can be written for a single compound, the compound is said to exhibit resonance. In such cases the actual structure is always an average of the structures that can be written.

VSEPR explains and predicts the geometries of molecules on the basis of repulsion between electron pairs. The geometry of a molecule in which atoms or groups are bonded to a central atom places the atoms or groups and the lone pairs as far apart as possible. This principle gives rise to the molecular geometries shown in Table 11.1 for molecules in which there are no lone pairs on the central atom. When lone electron pairs take the place of one or more bonded pairs, the geometry of the resulting molcule is altered, as shown in Figure 11.3. Because lone pairs strongly repel each other, they occupy locations as far apart as possible (equatorial positions in AB_5 structures, positions opposite one another in AB_6 structures). Lone pairs compress adjacent bond angles, making them less than their ideal values.

VALENCE BOND THEORY

11.6 BOND FORMATION

The VSEPR theory presented in the preceding sections is a simple, practical tool that works well in predicting molecular geometry. However, there are several things that it does *not* do. It does not relate the shapes of molecules to the orbitals and energy levels of atoms. And it gives no picture of when, how, or why bonds form.

Two approaches are used to understand bonding in these terms—the valence bond theory and the molecular orbital theory. In both theories, nuclei are pictured as attracted to an area of high electron density located along the line between the two nuclei—the **bond axis.** At the same time, the bonding electrons are attracted by both nuclei.

So far we have pictured covalent bonding as the result of sharing of electron pairs. **Valence bond theory** describes bond formation as the interaction, or overlap, of atomic orbitals. The concept of shared electron pairs remains important in valence bond theory. The sharing occurs when the atomic orbitals from two atoms overlap so that a region of high electron density is possible between the bonded atoms. Two electrons jointly occupy this region and form the bond. In molecular orbital theory, which is discussed in Chapter 25, atomic orbitals are pictured as combining to form molecular orbitals—new orbitals that "belong" to the entire molecule or to groups of atoms, rather than to individual atoms. Both valence bond and molecular orbital theories are based on the mathematics of quantum mechanics. Both seek to explain experimental facts, such as the *observed* geometry of molecules, their molecular spectra, or their bond energies, and to correctly predict these properties for other molecules.

Figure 11.4 presents a comparison of the valence bond and molecular orbital pictures of bonding. In the valence bond picture, the atomic orbitals maintain their identity and overlap to give an area of greater electron density between A and B. In the molecular orbital picture, the overlapping orbitals have combined and rearranged to give a single bonding molecular orbital with greater electron density between A and B. (An antibonding orbital, not pictured in the figure, is also formed; see Section 25.3.)

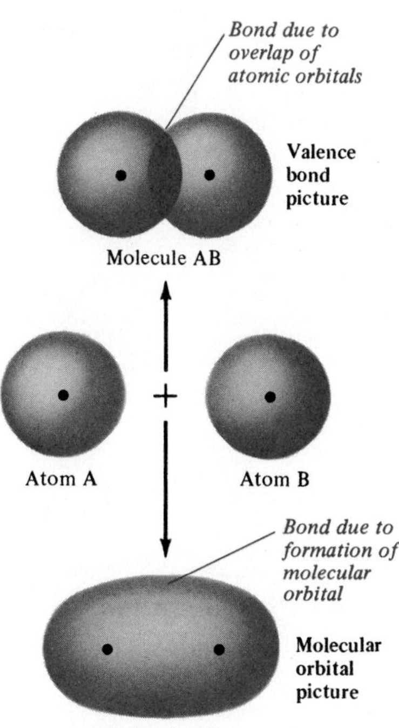

Figure 11.4
Comparison of Valence Bond and Molecular Orbital Pictures of Bond Formation

11.7 SINGLE BONDS IN DIATOMIC MOLECULES

To describe when and how bonding will occur according to valence bond theory, the electron configurations and orbitals of the atoms that are about to combine must be considered. The possibility for bond formation exists when two atoms can approach each other in such a way that occupied orbitals with similar energies are able to come into contact, or overlap. The greater the amount of overlap, the stronger the bond.

The simplest example of bond formation by atomic orbital overlap is given by the H_2 molecule. The single electron associated with each hydrogen atom occupies a spherical 1*s* orbital. Overlap of the two 1*s* orbitals allows pairing of the electron spins and formation of a single covalent bond.

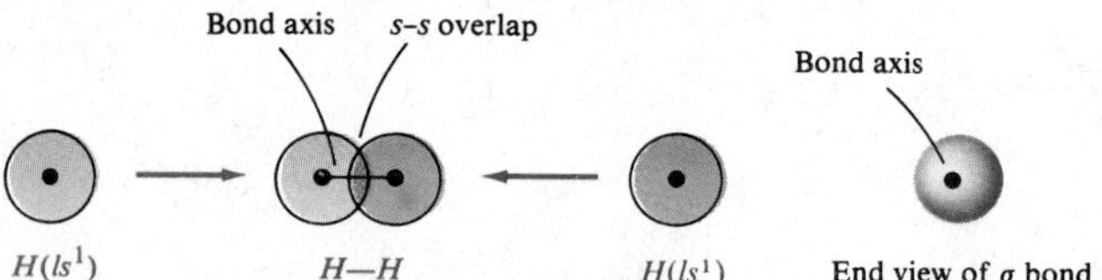

All bonds such as this, in which the region of highest electron density surrounds the bond axis, are called **σ bonds** (sigma bonds). Only one σ bond can form between any two atoms.

An *s* orbital and a *p* orbital (each of which contains a single electron) can also overlap to yield a σ bond. For example, a hydrogen atom ($1s^1$) can combine with a chlorine atom ([Ne]$3s^2 3p^2 3p^2 3p^1$) to form an HCl molecule by overlap of the 1*s* orbital of the hydrogen atom and the 3*p* orbital of the chlorine atom that holds a single electron.

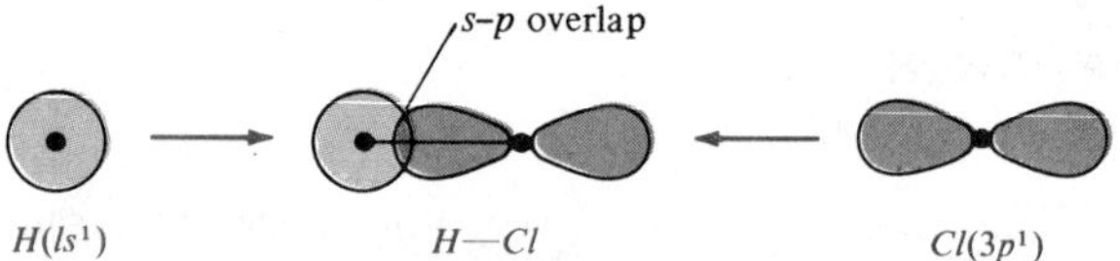

Similarly, two *p* orbitals can overlap to produce a σ bond, *if* the overlapping orbitals lie along the same axis. For example, the chlorine–chlorine bond in the Cl_2 molecule can be described by the overlap of two 3*p* orbitals, each containing one electron. To form a bond, the orbitals must approach end-on.

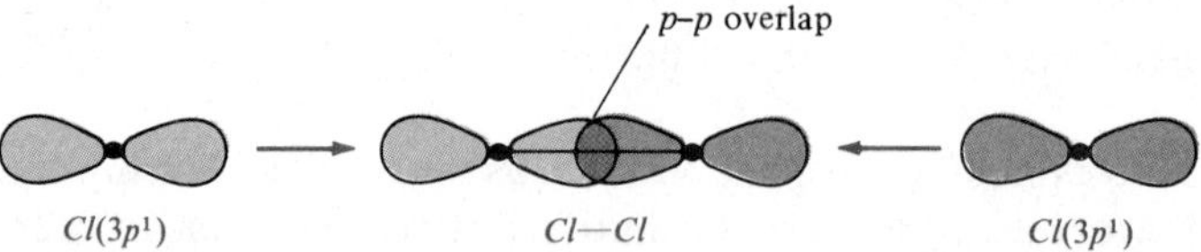

In general, overlapping orbitals, in order to form a bond, must have the same symmetry with respect to the bond axis. The three *p* orbitals on a single atom are oriented in space at 90° to each other. Therefore, geometry does not allow bond formation between *s* and *p* orbitals when two atoms approach each other in certain ways. A *p* orbital cannot overlap an *s* or another *p* orbital sufficiently to form a σ bond when they approach as follows:

No overlap

No overlap

11.8 SINGLE BONDS IN POLYATOMIC MOLECULES; HYBRIDIZATION

Overlap of *s* and *p* orbitals as pictured thus far cannot explain the bond lengths and bond angles in most molecules with more than two atoms. For example, consider the O—H and N—H bonds in H_2O and NH_3. If each bond resulted from overlap of the hydrogen $1s^1$ orbital with either a *p* orbital of an oxygen atom ($1s^2 2s^2 2p^4$) or a *p* orbital of a nitrogen atom ($1s^2 2s^2 2p^3$), the HOH or HNH bond angles should be about the same as the angles between the *p* orbitals, which are 90°. But in both compounds the angles are closer to the tetrahedral angle, as we have seen in our discussion of VSEPR theory (see Figure 11.2).

Furthermore, there are difficulties in explaining the observed geometry of a molecule that contains bonds derived from the overlap of different types of orbitals. In methane (CH_4), two *s* electrons and two *p* electrons from a C atom ($1s^2 2s^2 2p^2$) must each form a bond with a 1*s* electron from an H atom. Differences in bond length and bond angle might be expected between C—H bonds from *s–p* overlap and those from *s–s* overlap. However, experimental measurements show that the four C—H bonds in CH_4 are identical in length and form equal tetrahedral angles with each other.

The concept of hybridization was introduced to allow an explanation of molecular geometry in terms of atomic orbitals and valence bond theory. **Hybridization** is the mixing of the atomic orbitals on a single atom to give a new set of orbitals, called *hybrid orbitals,* on that atom.

Hybridization applies only to the orbitals of a single, covalently bonded atom in a molecule. In terms of quantum theory, hybridization represents a combination of the mathematical functions that describe the atomic orbitals involved in bonding. The result is a new description of the probable electron density about the atom. In terms of energy, hybridization represents the blending of higher energy and lower energy orbitals to form orbitals of intermediate energy. The hybrid orbitals are still atomic orbitals and remain oriented around a single nucleus.

Hybridization provides a connection between observed molecular geometry and the electron configurations of the combining atoms. Consider, for example, the $BeCl_2$ molecule, which is known to be linear in the gas phase. A beryllium atom has two 2*s* electrons which are, of course, paired. How can *two* equivalent covalent bonds be formed by an atom that has only paired electrons? The $2s^2$ valence electrons must in some way become unpaired. Hybridization pictures this as happening by combination of the *s* and *p* orbitals to give two equivalent hybrid orbitals, each containing one electron. These are known as *sp* hybrid orbitals, or simply *sp* orbitals. (The inner electrons, which are not involved in the hybridization, are not included in the following diagram.)

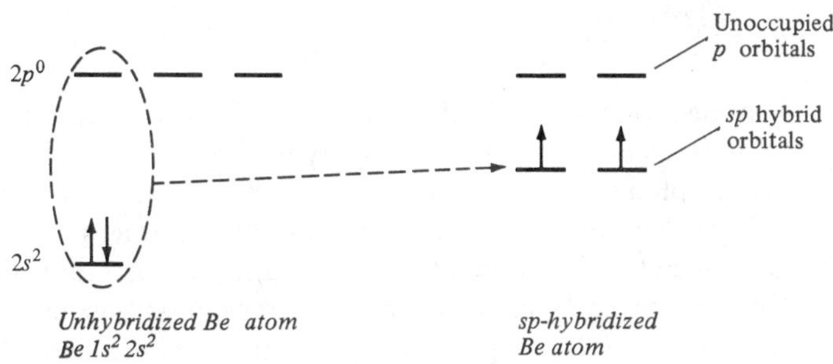

Each of the two *sp* hybrid orbitals can be pictured as overlapping with a *p* orbital of a chlorine atom to give the two covalent bonds of $BeCl_2$ (Figure 11.5).

The number of hybrid orbitals formed always equals the number of atomic orbitals that have combined. Each *sp* hybrid orbital has a large lobe on one side of the nucleus and a smaller lobe on the other (Figure 11.6). For simplification, the smaller lobe is often omitted in drawings. *sp* Hybrid orbitals form somewhat stronger bonds than *p* orbitals because they allow greater overlap.

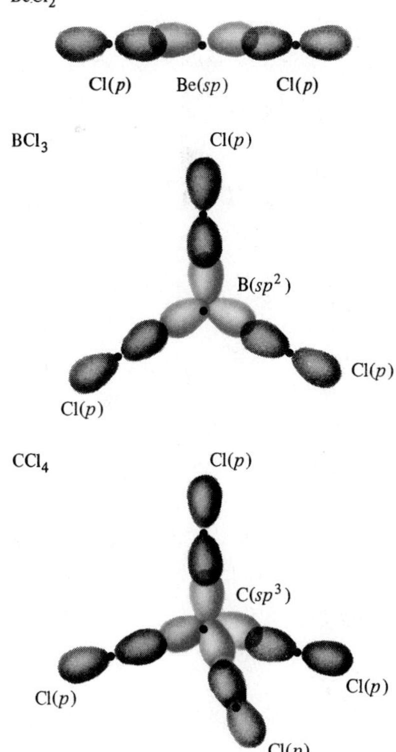

Figure 11.5
$BeCl_2$, BCl_3, and CCl_4
Valence bond pictures of bonding. The hybrid orbitals are shown in color.

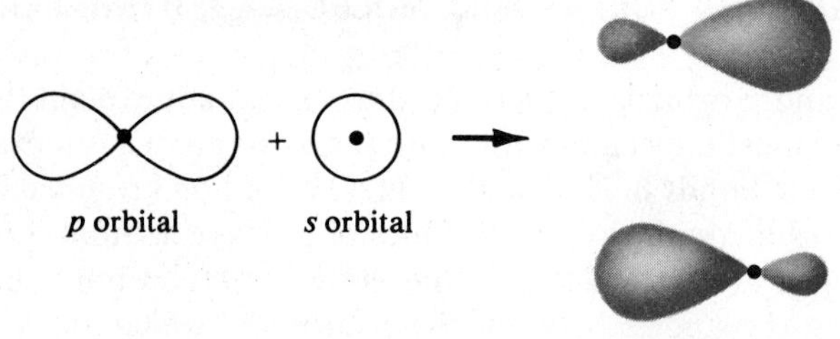

Figure 11.6
Hybridization to Give sp Orbitals
The sp orbitals, shown separately for clarity, are from the same atom.

Boron atoms, with the $2s^2 2p^1$ configuration, form three equivalent bonds to chlorine in BCl_3. As predicted by VSEPR and as shown by experiment, this is a triangular planar molecule. In order to give three equivalent orbitals, hybridization in this case involves one s orbital and two p orbitals and is termed sp^2 hybridization.

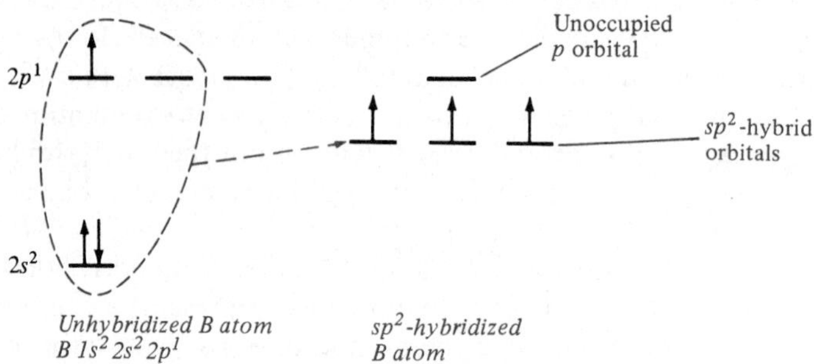

The three sp^2 orbitals of the boron atom can be pictured as overlapping with orbitals from chlorine atoms to form BCl_3 (see Figure 11.5).

A carbon atom, which has the $2s^2 2p^2$ configuration, forms four equivalent covalent bonds in CCl_4 (see Figure 11.5). This is accomplished by sp^3 hybridization, the combination of one s and three p orbitals.

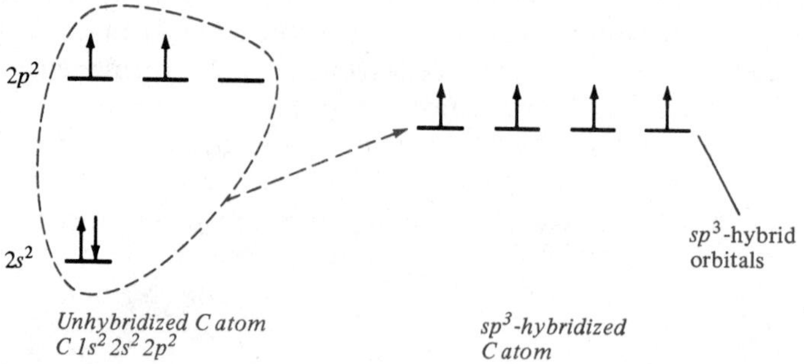

As is discussed in connection with multiple bonds in the next section, carbon atoms in various bonding situations can be sp, sp^2, or sp^3 hybridized.

Linear, triangular planar, and tetrahedral geometries (see Figure 11.5) are all accounted for by hybridization of s and p orbitals. Where lone-pair electrons are present on atoms with hybridized orbitals, they occupy one or more of the hybrid orbitals. For example, the carbon, nitrogen, and oxygen atoms in the methane, ammonia, and water molecules (see Figure 11.2) are all pictured in valence bond theory as having sp^3-hybridized orbitals. In the ammonia molecules, one sp^3 hybrid orbital is occupied by a lone pair, and in the water molecules two hybrid orbitals are occupied by lone pairs.

In this discussion of hybridization we have focused on the central atom as in VSEPR theory. The question arises as to whether or not the orbitals on other atoms in a molecule, for example, the chlorine atoms in $BeCl_2$, BCl_3, or CCl_4, should be viewed as hybridized. In Figure 11.5 we have shown the bonds as formed by overlap of hybrid orbitals with the unhybridized p orbitals on chlorine. These bonds could equally well be pictured as formed between hybrid orbitals on the central atom and hybrid orbitals on the chlorine atoms. There are some valid arguments for doing so, because the overlap, and hence the bond strengths, would be greater. The mathematical techniques upon which hybridization is based allow either complete or partial hybridization, so that the percentage of hybridization can be adjusted to match the observed bond strengths. Because our concern here is mainly with providing some insight into how observed molecular geometries are explained by bonding theories, we have, in most cases, considered only hybridization of the central atoms.

The *s* and *p* orbitals of any atom give a maximum of four hybrid orbitals. Atoms of second-period elements, which have *only s* and *p* orbitals, can form no more than four hybrid orbitals and no more than four covalent bonds. Beyond the second period, atoms can form a larger number of covalent bonds by involving *d* orbitals in hybridization. In Example 11.3, we found that PF_5 is a triangular bipyramidal molecule. This geometry for PF_5 is explained by sp^3d hybridization. A phosphorus atom has five valence electrons in four orbitals ($3s^2 3p^3$). Hybridization that includes one vacant 3*d* orbital allows formation of five equivalent sp^3d hybrid orbitals.

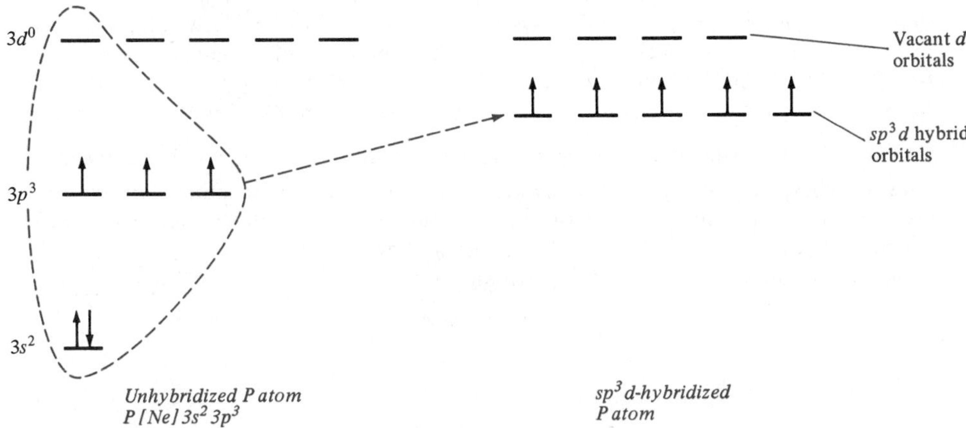

The octahedral geometry of molecules with six covalent bonds is, in most cases, accounted for by sp^3d^2 hybridization.

To explain bonding by using hybrid orbitals, first decide how many σ bonds and lone pairs surround each atom. This indicates how many hybrid orbitals are needed and the type of hybridization. For example, an atom with two σ bonds and two lone pairs needs four hybrid orbitals and must be sp^3-hybridized. Table 11.2 summarizes the

Table 11.2
Hybridized Orbitals
These orbitals form σ bonds by overlap with other hybrid orbitals, and with *s* and *p* orbitals.

Type	sp	sp^2	sp^3	sp^3d, dsp^3	sp^3d^2, d^2sp^3
Constituent orbitals	One *s* + one *p* orbital	One *s* + two *p* orbitals	One *s* + three *p* orbitals	One *s* + three *p* + one *d* orbital	One *s* + three *p* + two *d* orbitals
Ideal bond angle	180°	120°	109.47°	180°, 120°, 90°[a]	90°
Hybrid orbitals					
Geometry	Linear	Triangular planar	Tetrahedral	Triangular bipyramidal	Octahedral

[a] See AB_5 in Table 11.1, where axial and equatorial positions are shown.

formation and geometry of hybrid orbitals. Comparison with Table 11.1 shows the equivalence of VSEPR and hybridization in describing molecular geometry.

EXAMPLE 11.6 Valence Bond Theory

Describe the bonding in BrF_3 in terms of the valence bond theory.

```
:F̤—B̈r—F̤:
      |
     :F̤:
```

The Lewis structure shows that BrF_3 is an AB_3E_2 type molecule and the bromine atom is surrounded by three σ bonds and two lone pairs. To explain bonding in this molecule, five equivalent hybridized orbitals on the bromine atom are necessary. The outer electron configuration for the bromine atom is $4s^2 4p^5$. By using the $4s$ orbital, the three $4p$ orbitals, and one of the empty $4d$ orbitals in **sp^3d hybridization, five hybrid orbitals** can be formed. **Two of these hybrid orbitals contain lone pairs of electrons and three of the hybrid orbitals contain single electrons which will form the σ bonds with the fluorine atoms.**

The five hybridized orbitals will be arranged in the shape of a **triangular bipyramid** (see Table 11.2). With lone pairs in two of the hybrid orbitals in the equatorial positions, the T shape predicted by VSEPR for BrF_3 is explained in terms of orbitals.

EXERCISE Describe the bonding in H_2O in terms of the valence bond theory.

Answer There are four equivalent hybrid orbitals on the oxygen atom, formed by sp^3 hybridization and directed toward the four corners of the tetrahedron (see Table 11.2); two contain lone pairs of electrons and two contain a single electron each, which will form the σ bonds with the hydrogen atoms.

EXAMPLE 11.7 Valence Bond Theory

Describe and sketch the bonding in ethane, C_2H_6, in terms of the valence bond theory.

```
    H  H
    |  |
  H—C——C—H
    |  |
    H  H
```

Each carbon atom must have four equivalent hybridized orbitals. The orbitals are formed by **sp^3 hybridization.** Three of the hybridized orbitals on each carbon atom contain a single electron which will form a σ bond with a hydrogen atom and the fourth contains a single electron which will form a σ bond with the other carbon atom.

H H
H C C H
H H

EXERCISE Describe and sketch the bonding in hydrogen peroxide, H_2O_2, in terms of the valence bond theory.

Answer Each oxygen atom has four equivalent hybrid orbitals formed by sp^3 hybridization; two contain lone pairs of electrons, one contains a single electron which will form a σ bond with a hydrogen atom, and one contains a single electron which will form a σ bond with the other oxygen atom.

11.9 MULTIPLE COVALENT BONDS

a. π Bonds So far, we have dealt only with single covalent bonds, which are σ bonds and are symmetrical about the bond axis. Only one σ bond can form between any two atoms. Multiple covalent bonds arise when p or d orbitals on atoms that are σ-bonded to each other also overlap.

When p orbitals approach parallel to each other, both lobes can overlap.

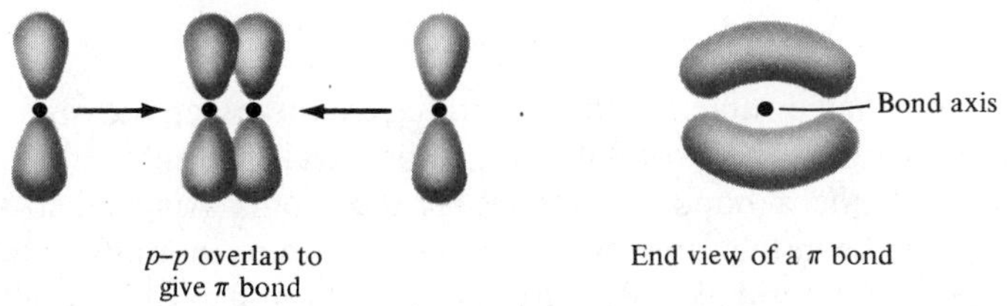

A bond formed in this way is called a π bond (pi bond). **π Bonds** concentrate electron density above and below the bond axis and always have a plane of zero electron density passing through the bond axis. π-Bond formation can also result from the interaction of p and d orbitals, or d and d orbitals.

Multiple covalent bonds are the result of σ- and π-bond formation between the same two atoms. One σ bond plus one π bond form a double covalent bond, and one σ bond plus two π bonds form a triple covalent bond. The σ bonds are stronger and provide most of the force holding the atoms together. The overall geometry of the molecule is also determined by the σ bonds, for the π bonds simply lie above and below the geometrical framework of the σ-bonded atoms. For this reason double bonds can be treated like single bonds in using VSEPR theory.

Carbon–carbon multiple covalent bonds are explained by the hybridization of carbon atoms to form either sp or sp^2 hybrid orbitals. In ethylene ($CH_2{=}CH_2$) and in other molecules that contain carbon–carbon double bonds, each of the two carbon atoms in the double bonds is sp^2-hybridized (Figure 11.7a) and is able to form three σ bonds. This provides a planar skeleton of σ bonds for the molecule, with each carbon atom at the center of three σ bonds arranged in a planar triangle (Figure 11.7a). One electron occupies each of the three sp^2 orbitals on each carbon atom, leaving one electron on each carbon atom in an unhybridized p orbital, which we picture as the p_z orbital. With the carbon–hydrogen skeleton in the x-y plane, the two valence electrons in the p_z orbitals can be shared by parallel overlap, creating a π bond.

In the acetylene molecule ($HC{\equiv}CH$) each of the carbon atoms is sp-hybridized and can form two σ bonds. Each forms a σ bond with the other carbon atom and with one hydrogen atom. Sidewise overlap of the p_z and p_y orbitals leads to sharing of the remaining two electrons from each carbon atom to form a pair of π bonds (Figure 11.8). All carbon–carbon triple bonds, like the triple bond in acetylene, are combinations of one σ bond and two π bonds.

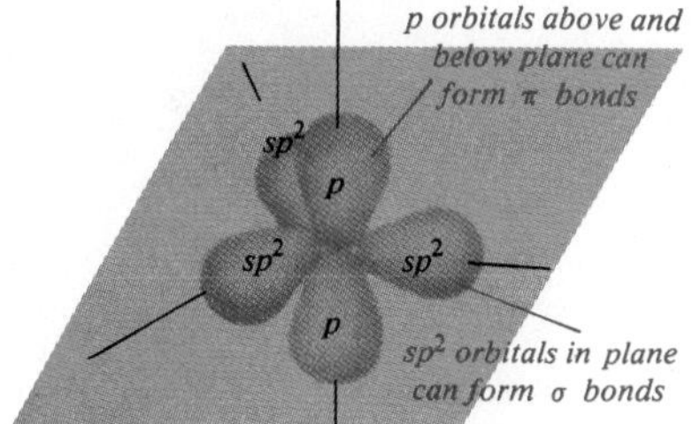

(a) An sp^2-hybridized carbon atom

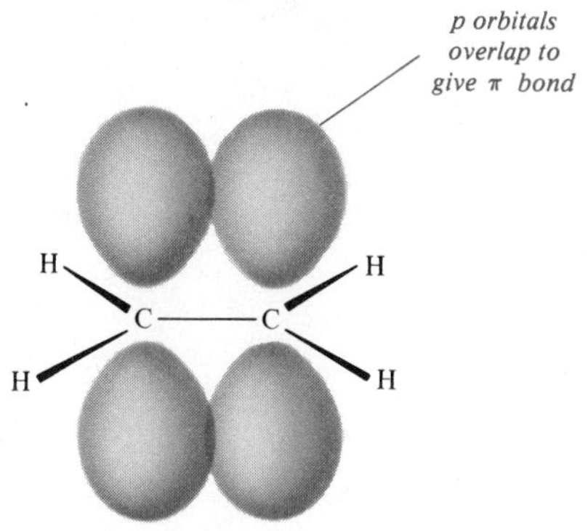

(b) Bonds in ethylene formed by two sp^2-hybridized carbon atoms

Figure 11.7
Hybridization and Bonding in Ethylene
The C—H bonds in ethylene are formed by overlap of sp^2 orbitals from C atoms and s orbitals from H atoms. The C—C bond is formed by sp^2–sp^2 overlap and the π bond by p–p overlap.

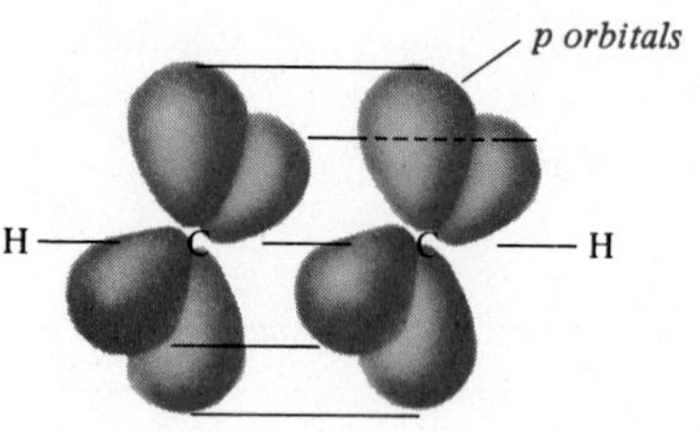

(a) p orbitals approaching to form π orbitals in acetylene

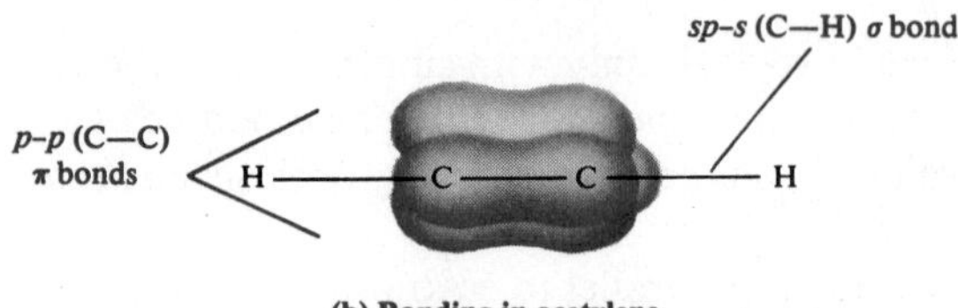

(b) Bonding in acetylene

Figure 11.8
Acetylene, H—C≡C—H

For the π bonds in a double covalent bond to remain intact, the orientation of the atoms must not change—the p orbitals have to be parallel to each other. Turning the atoms would break the π bond. Therefore, (1) the atoms attached to the multiply bonded atoms must lie in the same plane, and (2) the atoms in multiple bonds are not free to rotate about the bond axis. Because rotation about a double bond is restricted, the way in which other atoms or groups are attached to the double-bonded atoms makes a difference in the structure of the molecule (Section 31.4).

EXAMPLE 11.8 Valence Bond Theory

Describe and sketch the bonding in formaldehyde, H_2CO, using the valence bond theory.

$$\begin{array}{c} :\ddot{O}: \\ \| \\ H—C—H \end{array}$$

The Lewis structure for the formaldehyde molecule shows that around the carbon atom there are two carbon–hydrogen single bonds and a carbon–oxygen double bond. The carbon atom is ***sp^2*-hybridized** to produce the three hybrid orbitals needed for the formation of the three σ bonds. The $2p_z$ orbital on the carbon atom is unhybridized and contains a single electron which forms the π bond by parallel overlap with the unhybridized $2p_z$ orbital on the oxygen atom. The oxygen atom may also be pictured as ***sp^2*-hybridized,** with lone pairs in two of the hybrid orbitals. The third hybrid orbital, which contains a single electron, forms the σ bond by overlap with one of the sp^2 hybrid orbitals on carbon, and overlap of the unhybridized p orbital forms the π bond.

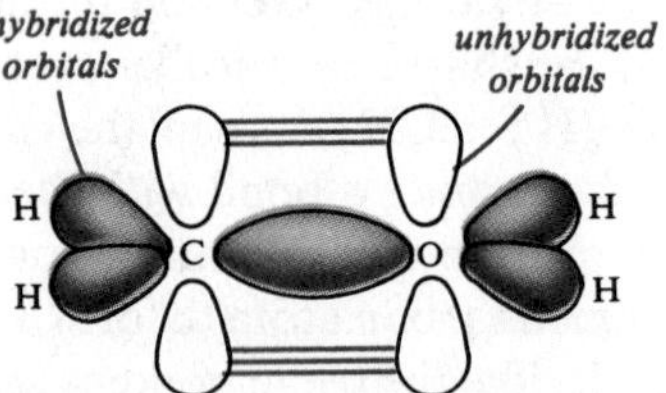

EXERCISE Describe and sketch the bonding in nitroxyl hydride, HNO, using the valence bond theory. The Lewis structure is H—N̈=Ö:

Answer The nitrogen atom is sp^2-hybridized; the three hybridized orbitals are used for the N—H σ bond, the lone pair of electrons, and the N—O σ bond; the unhybridized $2p_z$ orbital is used to form the π bond by the parallel overlap of the unhybridized $2p_z$ orbital of the oxygen atom. The oxygen may be pictured as also sp^2-hybridized.

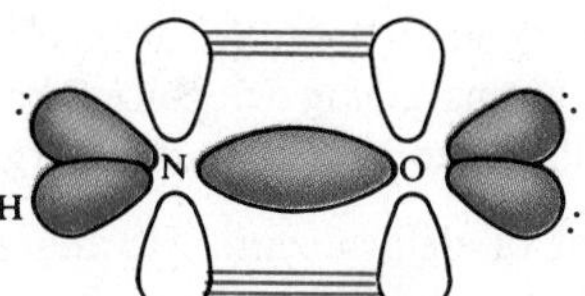

b. Delocalization of Electrons The valence bond picture of bonds formed by the overlap of orbitals on adjacent atoms must be modified to account for the bonding in molecules that exhibit resonance (Section 11.3). Consider the resonance forms of the nitrate ion and the single resonance hybrid that can be drawn

It is known from experiment that the three nitrogen–oxygen bonds are equivalent (e.g., the bond lengths are the same). To achieve a single structure for the nitrate ion, or any other such ion or molecule, the electrons in the unhybridized p orbitals are described as **delocalized electrons**—electrons that occupy a space spread over three or more atoms. Merging the electrons in the π bonds shown above gives a picture of the delocalization of electrons in the nitrate ion.

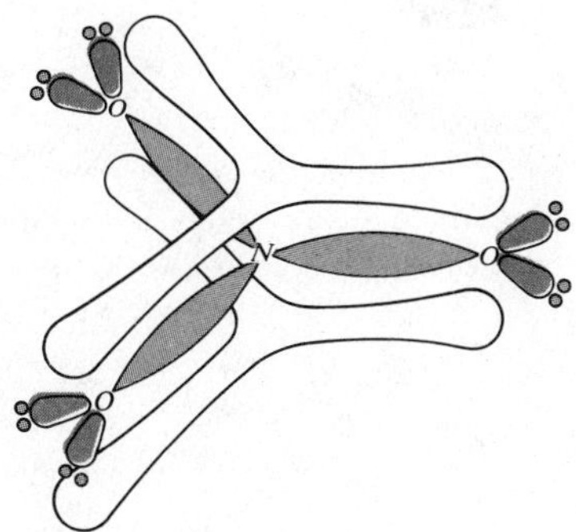

The delocalized electrons occupy a larger region of space than they do in localized bonds and their repulsion for each other is decreased. For this reason, molecules with delocalized electrons gain in stability.

Mathematically, in aspects of bonding theory that we need not pursue, delocalized electrons are a more natural consequence of molecular orbital theory (Chapter 25) than of valence bond theory. The important point is that the properties of equivalent bonds in molecules for which several resonance structures can be written are accounted for by the spreading of electron density over three or more atoms.

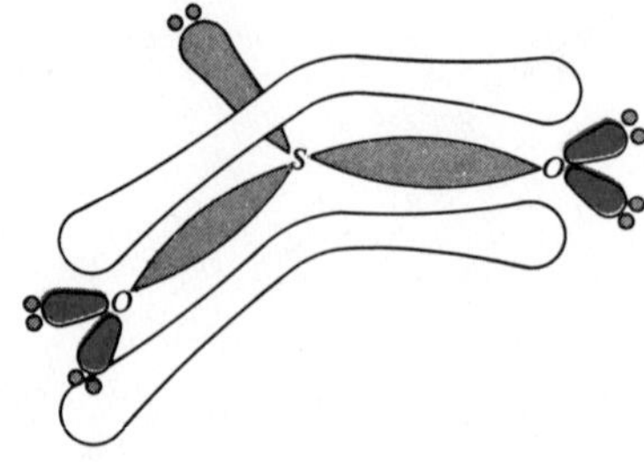

EXAMPLE 11.9 Delocalized Electrons

Use the concept of delocalization of electrons to describe the bonding in sulfur dioxide, SO_2.

$$:\ddot{\underset{..}{O}}-\ddot{S}=\ddot{\underset{..}{O}} \longleftrightarrow \ddot{\underset{..}{O}}=\ddot{S}-\ddot{\underset{..}{O}}:$$

The sulfur atom is sp^2-hybridized. Two of the three sp^2 orbitals are used to form σ bonds between the sulfur atom and the oxygen atoms and the third, for the lone pair of electrons on sulfur. The unhybridized p orbital on the sulfur atom overlaps with the unhybridized p orbitals on both of the oxygen atoms, giving delocalized bonding of the π electrons. Thus both S—O bonds are identical, as can be seen in the margin figure.

EXERCISE Use the concept of delocalization to describe the bonding in benzene, C_6H_6.

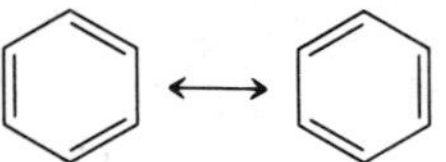

Answer Each carbon atom is sp^2-hybridized; one hybrid orbital is used for the C—H σ bond; two hybrid orbitals are used for the C—C σ bonds; the parallel unhybridized p orbitals allow delocalization of the π bonding over the entire carbon ring (see Figure 25.11c).

SUMMARY OF SECTIONS 11.6–11.9

In valence bond theory, bonds are pictured as formed by the overlap of atomic orbitals that retain their identity. For bond formation to occur, two orbitals must have similar energies and the same symmetry with respect to the bond axis, so that they can overlap enough to allow interaction. The combination of orbitals by hybridization is used to account for observed molecular geometry and properties. In sp hybridization, one s and one p orbital combine to give two sp hybrid orbitals that lie at 180° to each other. In sp^2 hybridization, one s and two p orbitals combine to give three sp^2 hybrid orbitals that lie in a plane with 120° angles between them. In sp^3 hybridization, one s and three p orbitals combine to give four sp^3 hybrid orbitals arranged tetrahedrally.

Second-period elements can form up to four covalent bonds by hybridization of s and p orbitals. Elements of the third period and beyond have d orbitals available for hybridization and can form five and six equivalent orbitals by sp^3d and sp^3d^2 hybridization; these equivalent orbitals account for triangular bipyramidal and octahedral geometry, respectively.

σ Bonds give regions of high electron density that surround the bond axis. π Bonds, which are weaker than σ bonds, give regions of high electron density on opposite sides of the bond axis, but have no electron density along the bond axis. Double and triple covalent bonds include one σ bond and one and two π bonds, respectively. The concept of bonding by delocalized electrons is applied to molecules for which resonance forms can be written.

INTERMOLECULAR FORCES

Intermolecular forces act between molecules, causing them to be attracted to each other in varying degrees. The strength of the intermolecular forces at a particular temperature determines whether a molecular substance is a gas, a liquid, or a solid at that temperature.

Three principal types of intermolecular forces are discussed in the following sections: dipole–dipole forces, London forces, and hydrogen bonding. Dipole–dipole forces act between molecules with an uneven distribution of charge, that is, molecules with dipole moments. London forces are often weak, but act between all types of molecules. Collectively, dipole–dipole forces and London forces (and certain other types of intermolecular forces) are sometimes referred to as **van der Waals forces.** Hydrogen bonding, the strongest type of intermolecular attraction, acts between highly polar molecules that contain hydrogen atoms bonded to very electronegative atoms.

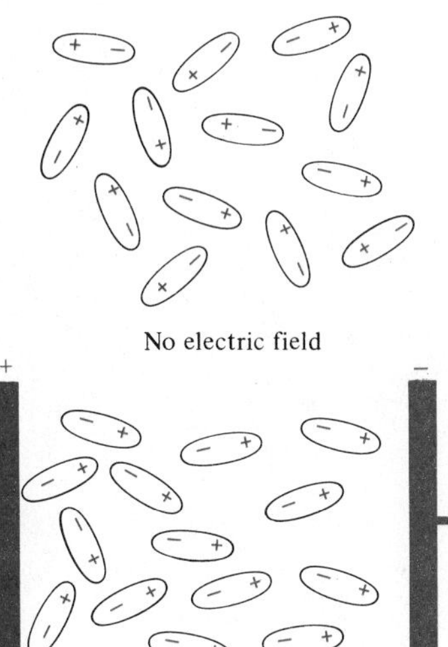

Figure 11.9
Polar Molecules in an Electric Field
The degree of alignment of polar molecules in an electric field in the gas phase allows the dipole moment of the molecules to be measured.

11.10 DIPOLE MOMENT

Covalent bonds between atoms of differing electronegativity are dipoles—they have a partial negative charge at one end and a partial positive charge at the other (Section 10.7). Whether or not a *molecule* is a dipole depends not only upon bond polarity, but also upon molecular geometry and the presence of lone-pair electrons.

The degree of polarity of a molecule is measured by its **dipole moment,** μ (Greek mu, pronounced "mew")—defined as the magnitude of the plus and minus charges times the distance between them. The common unit for dipole moments is the debye, D (pronounced "de-buy"). In SI units the dipole moment is given in coulomb meters (1 D = 3.336×10^{-30} C m). For a diatomic molecule in the gaseous state, the dipole moment (Figure 11.9) is a direct indication of the polarity of the bond. The decrease in bond polarity with decreasing electronegativity of the halogen atom is shown by the dipole moments of the hydrogen halides.

	$\overset{\delta+}{H}—\overset{\delta-}{F}$	$\overset{\delta+}{H}—\overset{\delta-}{Cl}$	$\overset{\delta+}{H}—\overset{\delta-}{Br}$	$\overset{\delta+}{H}—\overset{\delta-}{I}$
Dipole moment	*1.9 D*	*1.04 D*	*0.79 D*	*0.38 D*
Electronegativity of halogen atom	*4.0*	*3.0*	*2.8*	*2.5*

The effect of molecular geometry on dipole moment is illustrated in Figure 11.10. The colored arrows show individual bonds that are polar. The water molecule is highly polar. The two lone pairs of electrons concentrate negative charge on the oxygen atom. In addition, the hydrogen–oxygen bonds are highly polar, with the negative charge also concentrated on the oxygen atom. The positive end of the dipole that represents the entire bent molecule is centered between the two hydrogen atoms. The ammonia molecule is also polar because both the lone pair on the nitrogen atom and the polar nitrogen–hydrogen bonds concentrate negative charge on the nitrogen atom.

In some molecules internal compensation of partial charges can lead to an overall absence of polarity even though the individual bonds are themselves polar. For example, in the linear arrangement of atoms in the carbon dioxide molecule, one strongly polar carbon–oxygen double bond cancels the other (see Figure 11.10).

The symmetrical regular tetrahedral geometry of carbon tetrachloride (CCl_4) provides the same internal compensation. However, in chloroform ($CHCl_3$), the hydrogen atom is both less electronegative than the carbon atom and markedly

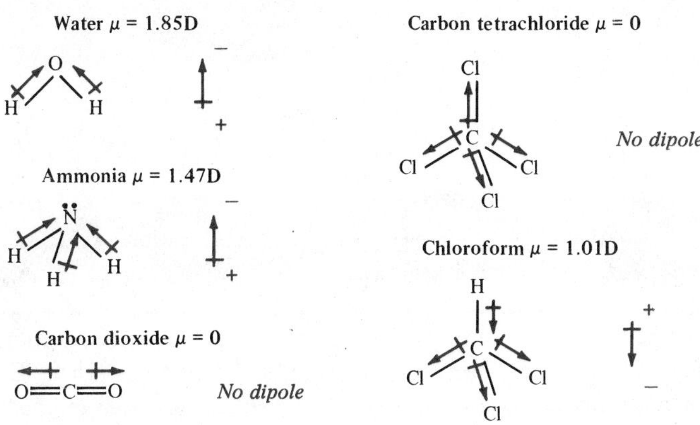

Figure 11.10
Dipole Moments of Some Gaseous Polyatomic Molecules
The direction of the dipole for each bond and each molecule is shown by the arrow. The arrowhead points to the negative end; the crossed tail indicates the positive end. In CO_2 and CCl_4, the individual bonds are polar. In both cases, however, the atoms with the partial negative charges (O and Cl) are symmetrically placed about the central carbon atom. Consequently the *center* of the negative charge coincides with the carbon atom, where the positive charge is also localized. With no net separation of charge, these molecules have no dipole moments.

different in electronegativity, size, and electronic atmosphere from the chlorine atoms. Thus, a partial negative charge is concentrated among the chlorine atoms, and the molecule has a dipole moment.

EXAMPLE 11.10 Intermolecular Forces

Identify which of the following molecules would be polar: (a) SO_2, (b) BCl_3, (c) ClF_3. The structures of the molecules are shown below:

(a) SO_2 is an angular AB_2E molecule. The lone pair concentrates electron density on the sulfur atom, but the polarity of the S—O bonds partially offsets it. The electron density is not symmetrically distributed and the molecule is **polar.**

(b) BCl_3 is a triangular AB_3 molecule. The planar symmetrical arrangement of the B—Cl bonds means that the overall effect of the polarity of the individual bonds is canceled and the molecule is **not polar.**

(c) ClF_3 is a T-shaped AB_3E_2 molecule. The polarities of the two axial Cl—F bonds cancel each other. However, the electron density distribution in the plane around Cl that includes the two lone pairs is uneven because the effect of the lone pairs is not canceled by the one polar Cl—F bond. Therefore, the molecule as a whole is **polar.**

EXERCISE Identify which of the following molecules would be polar: (a) SO_3, (b) BCl_2F, (c) XeF_2. *Answer* (b).

11.11 DIPOLE–DIPOLE FORCES

In **dipole–dipole interaction,** molecules with dipole moments attract each other electrostatically; the positive end of one molecule attracts the negative end of another molecule, and so on, leading to an alignment of the molecules (Figure 11.11a). The dipole–dipole attraction, along with other forces, must be overcome in melting a solid or vaporizing a liquid; it thereby influences the melting point, heat of fusion, boiling point, and heat of vaporization.

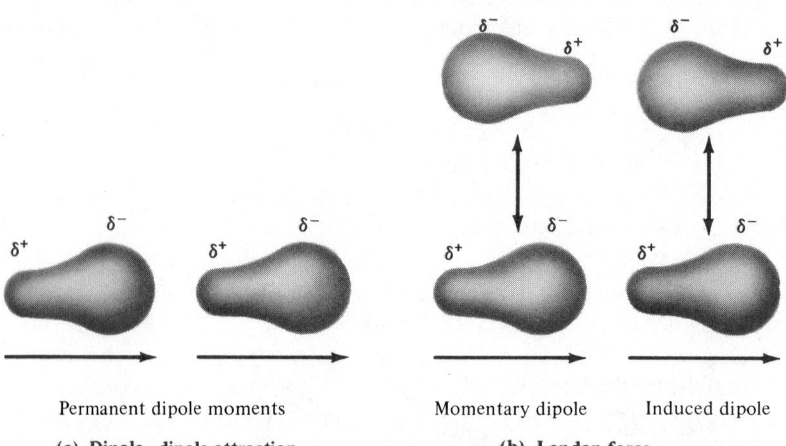

Figure 11.11
Intermolecular Forces
(a) Dipole–dipole attraction (b) London force

Table 11.3
Properties of Some Nonmetal Hydrides
Data for the second-period elements are given in black, those for the third-period elements in color.

Property	Methane CH_4 (nonpolar)	Silane SiH_4 (nonpolar)	Ammonia NH_3 (polar, H bonded)	Phosphine PH_3 (polar)	Water H_2O (polar, H bonded)	Hydrogen sulfide H_2S (polar)	Hydrogen fluoride HF (polar, H bonded)	Hydrogen chloride HCl (polar)
Appearance	Colorless gas	Colorless gas	Colorless gas	Colorless gas	Colorless liquid	Colorless gas	Colorless fuming gas, or liquid	Colorless gas
Molecular mass (u)	16.04	32.09	17.03	34.00	18.02	34.08	20.01	36.46
Melting point (°C)	−182	−185	−78	−134	0.00	−85.6	−83.0	−114
Boiling point (°C)	−161	−111	−33	−87.8	100.	−60.8	19.5	−84.9
Density of liquid (g/mL)	0.42 (−164 °C)	0.68 (−185 °C)	0.68 (−33.4 °C)	0.75 (−87.78 °C)	0.96 (100 °C)	0.95 (−61.3 °C)	0.99 (19.5 °C)	1.19 (−84.9 °C)
Heat of fusion at m.p. (kJ/mol)	0.941	0.665	5.52	1.13	6.02	2.38	4.56	1.99
Heat of vaporization at b.p. (kJ/mol)	8.20	13	23.3	14.6	40.7	18.7	25.6	16.1
Dipole moment (D)	0	0	1.47	0.55	1.85	1.10	1.9	1.04

Comparison of the properties of compounds of similar molecular mass illustrates the effect of dipole–dipole forces. Look at the properties of SiH_4, PH_3, and H_2S in Table 11.3. Among these compounds of similar molecular mass, SiH_4 is nonpolar and has the lowest melting and boiling points, and the lowest heats of fusion and vaporization. Hydrogen sulfide, H_2S, has a dipole moment twice that of PH_3 and, as expected, has the highest melting and boiling points, and the highest heat of fusion and vaporization of the three compounds.

11.12 LONDON FORCES

London forces act on *all* atoms and *all* molecules, whether polar or nonpolar. London forces are responsible for the condensation, at low enough temperatures, of even the monatomic noble gases, which come closer to ideal behavior than any other gases. (London forces are among the forces represented by the constant *a* in the van der Waals equation for nonideal gases, Section 6.14. Sometimes the phrase "van der Waals attraction" is used for London forces alone.)

London forces (also known as *dispersion forces*) are the result of momentary shifts in the symmetry of the electron cloud of a molecule. Recall that we spoke of the polarizability of the electron cloud—the ease with which it is distorted. In a large collection of molecules, at any given moment collisions are taking place, with resulting polarization of the molecules. As soon as a slight positive charge is produced at one end of one molecule, it induces a slight negative charge in one end of the molecule next to it—an *induced dipole*. For an instant, a force of attraction exists between these molecules. **London forces** are the forces of attraction between fluctuating dipoles in atoms and molecules that are very close together (see Figure 11.11b).

The strength of London forces is influenced by the size and geometry of the molecules involved and by the ease of polarization of the electron clouds. Close contact between the molecules over a larger region gives greater opportunity for dipole interaction than when only a small amount of contact is possible. In the series of pentanes shown in Table 11.4a, the London forces get stronger (as shown by the higher boiling points) as the molecules get less spherical in shape, because spheres can make contact at only one point. Among molecules of similar geometry (Table 11.4b), London forces increase with increasing number of electrons, that is, with increasing molecular

Table 11.4
London Forces and Boiling Points
The increase in boiling point in each series of compounds is due to increasing London forces.

(a) Same molecular mass, decreasingly compact shape	(b) Similar molecular geometry, increasing molecular mass
$CH_3C(CH_3)_2CH_3$ Neopentane, C_5H_{12} b.p. 9.5 °C	CH_4 Methane, CH_4 b.p. −161 °C
	CH_3CH_3 Ethane, C_2H_6 b.p. −88.6 °C
$CH_3CH_2CH(CH_3)CH_3$ Isopentane, C_5H_{12} b.p. 28 °C	$CH_3CH_2CH_3$ *n*-Propane, C_3H_8 b.p. −44.5 °C
$CH_3CH_2CH_2CH_2CH_3$ *n*-Pentane, C_5H_{12} b.p. 36 °C	$CH_3CH_2CH_2CH_3$ *n*-Butane, C_4H_{10} b.p. −0.5 °C

mass. The difference in boiling point between CH_4 and SiH_4 (Table 11.3), which are both nonpolar, is due to the different strengths of the London forces. Melting points are affected by crystal geometry and other factors in addition to London forces, and therefore may not show as good a correlation with molecular mass and shape as do boiling points.

11.13 HYDROGEN BONDS

When a hydrogen atom is covalently bonded to an electronegative atom that strongly attracts the shared electron pair, the small hydrogen atom has little electron density around it. Under these circumstances, the hydrogen atom carries a partial positive charge and can act as a bridge to another electronegative atom. A **hydrogen bond** is the attraction of a hydrogen atom covalently bonded to an electronegative atom for a second electronegative atom. Strong hydrogen bonds form between hydrogen atoms and fluorine, nitrogen, or oxygen atoms, which are small and have their negative charges highly concentrated in a small volume. Hydrogen bonds are the strongest intermolecular forces. It is the strength of the hydrogen bonds in water that is largely responsible for the unique properties of water and the resulting influence of water on the environment and the living things on Earth (Sections 14.1 and 14.2).

The hydrogen bond between hydrogen and the most electronegative element, fluorine, is the strongest type of hydrogen bond (100 kJ/mol). Hydrogen fluoride crystals contain long chains in which each hydrogen atom is believed to be covalently bonded to one fluorine atom and hydrogen-bonded to another.

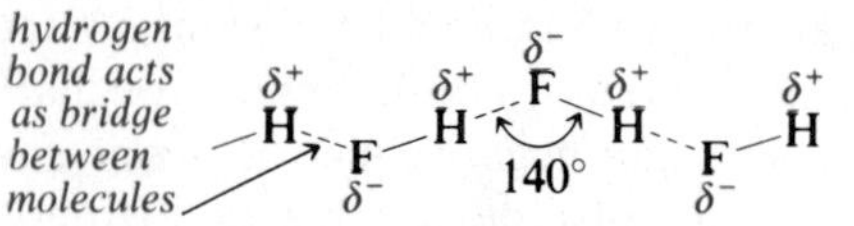

A hydrogen bond is essentially a partial coordinate covalent bond. The bond forms between a hydrogen atom with very little electron density (because it is bonded to an electronegative atom) and the lone electron pair on a highly electronegative oxygen, nitrogen, or halogen atom.

The effect of hydrogen bonding on the properties of compounds is clearly illustrated by the nonmetal hydrides. In Table 11.3, compare the properties of the hydrogen-bonded, second-period hydrides ammonia, water, and hydrogen fluoride, with the non-hydrogen-bonded third-period hydrides phosphine, hydrogen sulfide, and hydrogen chloride. In each of the comparable pairs from the same periodic table families (NH_3–PH_3, H_2O–H_2S, HF–HCl), the compound with the smaller molecular

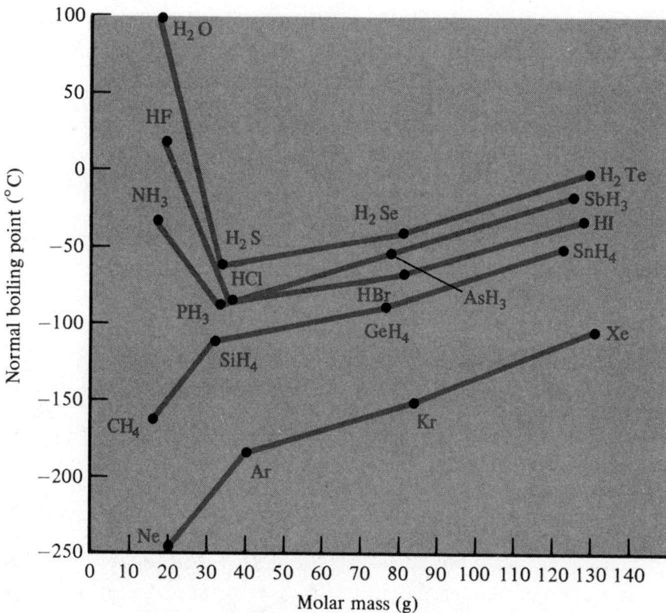

Figure 11.12
Boiling Points of Simple Hydrides and Noble Gas Elements

mass is less volatile and has the higher melting point and the larger heats of fusion and vaporization. These differences, which are illustrated in Figure 11.12, are due to the extra energy needed to break the intermolecular hydrogen bonds in NH_3, H_2O, and HF.

Hydrogen bonds can form between atoms in different molecules or in the same molecule. The large molecules in living systems are often held in exactly the shape they need for their specific biochemical function by hydrogen bonds. Except for the very strong H---F hydrogen bonds, the energies of hydrogen bonds are about 20–40 kJ/mol. By comparison, covalent bond energies range from 150 to 900 kJ/mol. More energy would be needed than the body could easily supply to quickly break covalent bonds in molecules undergoing biochemical reactions. For example, DNA (deoxyribonucleic acid), the molecule that carries the genetic code, must untwine each time a cell is duplicated (Figure 11.13). Hydrogen bonds help to hold the strands in place until the molecule reacts.

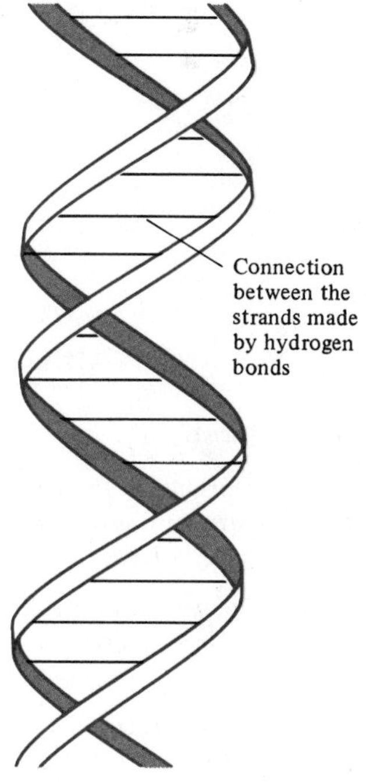

Figure 11.13
The DNA Double Helix

EXAMPLE 11.11 Intermolecular Forces

Which intermolecular forces influence the properties of (a) bromine(V) fluoride (BrF_5) and (b) acetone (CH_3COCH_3)? The Lewis structures of these molecules are

```
 :F̤    F̤:          H  :O:  H
   \ ‥ /            |   ‖   |
:F̤—Br—F̤:        H—C—C—C—H
     |              |       |
    :F̤:             H       H
```

(a) Bromine(V) fluoride is an AB_5E molecule and has square pyramidal geometry. The molecule is a dipole because of the unsymmetrical molecular geometry, and therefore its properties are influenced by **dipole–dipole attraction.** In addition, BrF_5 would have **London forces** (as do all molecules). There is no hydrogen bonding between molecules of BrF_5.

(b) The central carbon atom in the acetone molecule is surrounded by three σ bonds, which should be separated by two approximately 120° angles. The molecule will be a dipole because of the electronegativity of the oxygen atom and the geometry of the molecule. There is no hydrogen atom bonded to oxygen, and therefore there is no hydrogen bonding in this molecule. As in all molecules, London forces will be present. Therefore, the intermolecular forces that are important in determining the properties of acetone are **dipole–dipole interactions** and **London forces.**

EXERCISE List the intermolecular forces that influence the properties of (a) ethyl alcohol (CH_3CH_2OH) and (b) phosphine (PH_3).

Answers (a) Dipole–dipole interactions, London forces, and hydrogen bonding. (b) Dipole–dipole interactions and London forces.

SUMMARY OF SECTIONS 11.10–11.13

Bond polarity and molecular geometry both influence whether or not a given molecule has a permanent dipole moment and is therefore a polar molecule. The forces that act between polar molecules are referred to as dipole–dipole forces. All molecules, polar or not, are subject to London forces, which result from the attraction between transient dipoles in molecules that are very close together. London forces vary in strength with the size of molecules and the amount of contact possible between them. A hydrogen bond is the attraction of a hydrogen atom covalently bonded to an electronegative atom for another electronegative atom. The strongest hydrogen bonds join hydrogen atoms to the unpaired electrons on fluorine, nitrogen, or oxygen atoms. Hydrogen bonds can form between atoms in different molecules or in the same molecule; they have a crucial influence on the properties of water and many biologically important molecules.

SIGNIFICANT TERMS (with section references)

bond angle (11.1)
bond axis (11.6)
delocalized electrons (11.9)
dipole moment (11.10)
dipole–dipole interactions (11.11)
hybridization (11.8)
hydrogen bond (11.13)
London forces (11.12)
molecular geometry (11.1)
π bonds (11.9)
resonance (11.3)
resonance hybrid (11.3)
σ bonds (11.7)
valence bond theory (11.6)
valence-shell electron-pair repulsion theory (VSEPR) (11.4)
van der Waals forces (11.10, intro.)

QUESTIONS AND PROBLEMS

Lewis Structures

11.1 What information about chemical bonding can a correctly drawn Lewis structure give? What information about bonding cannot be given by a Lewis structure?

11.2 What do we mean by the term "resonance"? Do the resonance structures that we draw actually represent the bonding in the substance? Explain your answer.

11.3 Write Lewis structures for (a) ClO_4^-, (b) NOF, (c) XeF_4, (d) $CrCl_6^{3-}$, (e) $COCl_2$, (f) ClF_3.

11.4 Repeat Question 11.3 for (a) H_2S_2, (b) IO_4^-, (c) BeH_2, (d) NCl_3, (e) CH_3COOH, (f) HClO.

11.5 Repeat Question 11.3 for (a) H_2NOH; (b) S_8, ring of eight atoms; (c) SiH_4; (d) F_2O_2, oxygen atoms in center and fluorine atoms on outside; (e) CO; (f) $SeCl_6$.

11.6 Repeat Question 11.3 for (a) BCl_3, (b) SF_6, (c) CN^-, (d) AlH_4^-, (e) N_2H_4, (f) H_2SO_3.

11.7* Write the Lewis structure for molecular $AlCl_3$. Note that the aluminum atom is an exception to the octet rule in this molecule. In the gaseous phase, two molecules of $AlCl_3$ are joined together (dimerized) to form Al_2Cl_6. (The two molecules are joined by two Al—Cl—Al bonds.) Write the Lewis structure for this molecule.

11.8* Write the Lewis structure for molecular ClO_2. Note that there is a single unshared electron on the chlorine atom in this molecule. Two of these molecules will dimerize to form Cl_2O_4. (The two molecules are joined by a Cl—Cl bond.) Write the Lewis structure for this molecule.

11.9 Write the Lewis structures for the formate ion, $HCOO^-$.

11.10 Repeat Question 11.9 for the sulfur dioxide molecule, SO_2.

11.11 There are two resonance structures for toluene, $C_6H_5CH_3$, which can be written

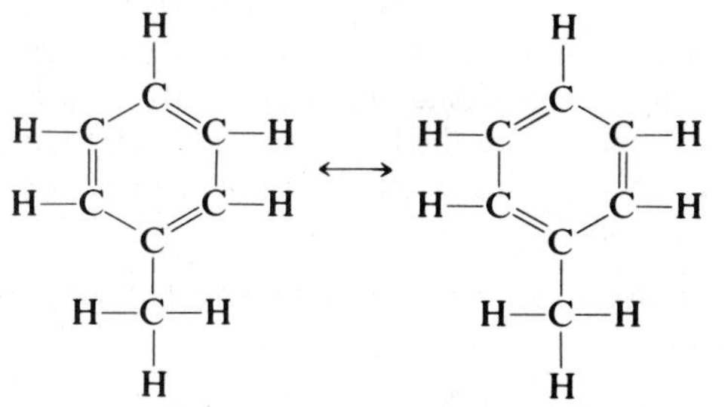

```
         H                      H
         |                      |
         C                      C
      //   \                 /    \\
H—C         C—H      H—C          C—H
   ||       |     ⟷      |         ||
H—C         C—H      H—C          C—H
      \   //                \\    /
         C                      C
         |                      |
      H—C—H                  H—C—H
         |                      |
         H                      H
```

How would you expect the carbon–carbon bond length in the six-member ring to compare with the carbon–carbon bond length between the CH_3 group and the carbon atom on the ring?

11.12 Draw the Lewis structures for the nitric acid molecule (HNO_3) that are consistent with the following bond length data: 0.1405 nm for the bond between the nitrogen atom and the oxygen atom that is attached to the hydrogen atom; 0.1206 nm for the bonds between the nitrogen atom and the two remaining oxygen atoms.

VSEPR

11.13 What geometrical structures would be predicted for molecules having the formulas (a) AB, (b) AB_2, (c) AB_3, (d) AB_5?

11.14 When using VSEPR theory to predict molecular geometry, how are double and triple bonds treated? How is a single nonbonding electron treated?

11.15 List, in decreasing order, the strength of electron-pair repulsions involving lone pairs (LP) and bonded pairs (BP) of electrons: LP—BP, LP—LP, BP—BP.

11.16 How does the presence of lone pairs of electrons on an atom influence the bond angles around that atom?

11.17 Sketch the three different possible arrangements of the two B atoms around the central atom A for the molecule AB_2E_3. Which of these structures correctly describes the molecular geometry? Why?

11.18 Sketch the three different possible arrangements of the three B atoms around the central atom A for the molecule AB_3E_2. Which of these structures correctly describes the molecular geometry? Why? What are the predicted ideal bond angles? How would the actual bond angles deviate from these values?

11.19 Use VSEPR to predict the geometries of each of the following species: (a) F_2, (b) H_2S, (c) F_2O, (d) H_3O^+, (e) XeF_4, (f) CO, (g) $[I_3]^-$.

11.20 Repeat Question 11.19 for (a) H_2Be, (b) molecular $AlCl_3$, (c) SiH_4, (d) SF_6, (e) IO_4^-, (f) NCl_3, (g) $[ICl_4]^-$.

11.21 Repeat Question 11.19 for (a) CO_2, (b) $[AlH_4]^-$, (c) NH_4^+, (d) $[Cr(H_2O)_6]^{3+}$, (e) ClO_3^-, (f) SeF_6, (g) ONCl.

11.22 Repeat Question 11.19 for (a) Cl_2CO, (b) PCl_5, (c) BCl_3, (d) ClO_4^-, (e) molecular $MgCl_2$, (f) ClO_2, (g) XeF_2.

11.23 Briefly discuss the bond angles in the hydroxylamine molecule

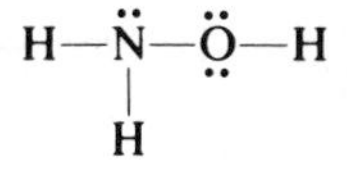

```
    ••  ••
H—N—O—H
   |  ••
   H
```

in terms of the ideal geometry and the small changes caused by electron pair repulsions.

11.24 Repeat Question 11.23 for nitric acid.

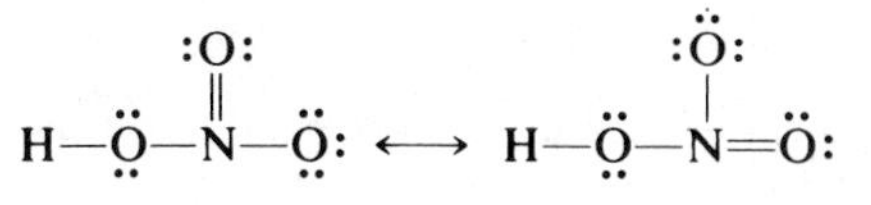

```
        :O:                  :Ö:
         ||                   |
  H—Ö—N—Ö:   ⟷   H—Ö—N=Ö:
```

Valence Bond Theory

11.25 Prepare sketches of the overlap of the following atomic orbitals: (a) *s* with *s*, (b) *s* with *p* along the bond axis, (c) *p* with *p* along the bond axis, (d) *p* with *p* perpendicular to the bond axis.

11.26 Prepare a sketch of the cross section taken between two atoms that have formed (a) a single σ bond, (b) a double bond consisting of a σ and a π bond, and (c) a triple bond consisting of a σ bond and two π bonds.

11.27 Describe the bonding in the F_2 molecule in terms of simple valence bond theory.

11.28 Repeat Question 11.27 for the HCl molecule.

11.29 What are hybridized atomic orbitals? Why was the theory of hybridized orbitals introduced?

11.30 Prepare sketches of the orbitals around atoms which are (a) *sp*, (b) sp^2, (c) sp^3, (d) sp^3d, (e) sp^3d^2 hybridized. Show in the sketches any unhybridized *p* orbitals that might participate in multiple bonding.

11.31 What types of hybridization would you predict for molecules having the following general formulas: (a) AB_3, (b) AB_2E_2, (c) AB_3E, (d) ABE_4, (e) ABE_3?

11.32 Repeat Question 11.31 for (a) ABE_5, (b) AB_2E_4, (c) AB_4, (d) AB_3E_2, (e) AB_5.

11.33 What is the hybridization of the central atom in each of the following: (a) H_2Be, (b) $AlCl_3$, (c) SiH_4, (d) IO_4^-, (e) NCl_3, (f) ClO_3^-, (g) PCl_5, (h) BCl_3, (i) ClO_4^-?

11.34 Repeat Question 11.33 for (a) $[AlH_4]^-$, (b) $AsCl_5$, (c) NH_4^+, (d) $[Cr(H_2O)_6]^{3+}$, (e) SeF_4, (f) ClO_2, (g) molecular $MgCl_2$, (h) $[I_3]^-$, (i) XeF_2.

11.35 What type of hybridization will the carbon atom have in (a) CO, (b) CO_2, (c) CO_3^{2-}? In which species are there π bonds that are not delocalized? In which species will there be delocalized π bonding?

11.36 Prepare a sketch of the molecule $CH_3CH{=}CH_2$ showing orbital overlaps. Identify the type of hybridization of atomic orbitals for each carbon atom.

11.37 Draw the Lewis structures for molecular oxygen and ozone. Assuming that all of the oxygen atoms are hybridized, what will be the hybridization of the oxygen atoms in each substance? Prepare sketches of the molecules.

11.38 What type of hybridization do the nitrogen atoms have in the following molecules: (a) NO; (b) N_2O_2, the dimer of NO; (c) N_2O_5; (d) NO_2; (e) N_2O_4, the dimer of NO_2; (f) N_2O_3?

Intermolecular Forces

11.39 What are van der Waals forces? What are the two types of van der Waals forces that are considered in this chapter?

11.40 Briefly describe the interaction between two dipoles.

11.41 What causes London forces? What factors determine the strength of London forces between molecules?

11.42 What is a hydrogen bond? Which atoms can form strong hydrogen bonds?

11.43 Choose the polar molecules: (a) SiH_4, (b) molecular $MgCl_2$, (c) NOCl, (d) NCl_3, (e) F_2O.

11.44 Repeat Question 11.43 for (a) molecular $AlCl_3$, (b) molecular Al_2Cl_6, (c) SF_6, (d) Cl_2CO, (e) NO, (f) SeF_4.

11.45 Indicate which of the following molecular species would exhibit hydrogen bonding: (a) CH_4, (b) N_2H_4, (c) CH_3CH_2OH, (d) H_2Se.

11.46 Repeat Question 11.45 for (a) H_2Be, (b) SiH_4, (c) NH_3, (d) HI, (e) molecular $AlCl_3$.

11.47 Why does HF have a lower boiling point and lower heat of vaporization than H_2O even though the molecular masses are nearly the same and the hydrogen bonds between molecules of HF are stronger?

11.48 Draw a likely structure for the dimer of acetic acid, which is formed as a result of the existence of two hydrogen bonds between the two individual molecules. The structural formula of the acetic acid molecule is

```
        :O:
         ‖   ..
  CH3—C—O—H
             ..
```

11.49 List the intermolecular forces that would be important for (a) CO_2, (b) $AsCl_5$, (c) Cl_2CO, (d) molecular $MgCl_2$, (e) SeF_4, (f) BCl_3, (g) NOCl.

11.50 Repeat Question 11.49 for (a) H_2Be, (b) molecular $AlCl_3$, (c) SiH_4, (d) SF_6, (e) F_2, (f) XeF_4.

11.51 Choose the substance within each group that has the largest intermolecular force: (a) P_4, S_8, or Cl_2 (considering mass); (b) CO_2 or SO_2 (considering dipole moment); (c) F_2 or Ar (considering shape); (d) *n*-octane or isooctane—

```
                                            CH3   CH3
                                             |     |
CH3CH2CH2CH2CH2CH2CH2CH3           CH3CCH2CHCH3
                                             |
                                            CH3
        n-octane                         isooctane
```

(considering shape); (e) CH_4 or CCl_4 (considering mass).

11.52* Assign each of the boiling points to the respective substances within each group on the basis of intermolecular forces: (a) Ne, Ar, Kr, −246.048 °C, −185.7 °C, −152.30 °C; (b) N_2, HCN, C_2H_6, −195.8 °C, −88.63 °C, 26 °C; (c) NH_3, H_2O, HF, −33.35 °C, 19.54 °C, 100.00 °C.

11.53 List the intermolecular forces that are present in liquid ammonia, NH_3, and liquid methane, CH_4. Which of these compounds should have the lower freezing and boiling points? Which substance would you expect to be a liquid over a larger temperature range?

11.54 The structures for three molecules having the formula $C_2H_2Cl_2$ are

```
:Cl:       H        :Cl:       H        :Cl:      :Cl:
    \     /             \     /             \     /
     C=C                 C=C                 C=C
    /     \             /     \             /     \
:Cl:       H        H       :Cl:        H          H
```

Describe the intermolecular forces present in each of these compounds and predict which has the lowest boiling point.

11.55 There are two different molecules having the formula C_4H_{10}.

```
                                           H
                                           |
   H  H  H  H                    H  H—C—H  H
   |  |  |  |                    |     |     |
H—C—C—C—C—H               H—C————C————C—H
   |  |  |  |                    |     |     |
   H  H  H  H                    H     H     H
```

Which would you expect to have the higher boiling point? Why?

11.56* The van der Waals constants (see Section 6.14) are $a = 19.01\ L^2\ atm/mol^2$, $b = 0.1460$ L/mol for *n*-pentane, and $a = 18.05\ L^2\ atm/mol^2$, $b = 0.1417$ L/mol for isopentane.

```
                                                   CH3
                                                    |
CH3—CH2—CH2—CH2—CH3        CH3—C—CH2—CH3
                                                    |
                                                    H
        n-pentane                            isopentane
```

(a) Basing your reasoning on intermolecular forces, why would you expect *a* for *n*-pentane to be larger? (b) Basing your reasoning on molecular size, why would you expect *b* for *n*-pentane to be larger?

ADDITIONAL QUESTIONS AND PROBLEMS

11.57 Sulfur reacts with fluorine to form two gaseous compounds: SF_4 and SF_6. The respective heats of formation at 25 °C are −774.9 kJ/mol and −1209 kJ/mol. The heat of formation is 278.805 kJ/mol for S(*g*) and 78.99 kJ/mol for F(*g*). Calculate the S—F bond energy in each compound.

11.58 The enthalpy of vaporization is 18.673 kJ/mol for H_2S. 19.33 kJ/mol for H_2Se, and 23.22 kJ/mol for H_2Te. (a) Explain this general trend. (b) Plot the heat of vaporization against the atomic number of the central atom. (c) Extrapolate your plot and predict the heat of vaporization of water. (d) Why is there a large difference between your answer for part (c) and the measured value of40.6563 kJ/mol?

11.59* Consider the following proposed Lewis structures for ozone (O_3):

(i) :Ö—Ö=Ö: ⟷ :Ö=Ö—Ö: (ii) triangular :O: bonded to :O—O: (iii) :Ö—Ö—Ö:

(a) Which of these structures correspond to a polar molecule? (b) Which of these structures predict covalent bonds of equal lengths and strengths? (c) Which of these structures predict a diamagnetic molecule? The properties listed in (a), (b), and (c) are those observed for ozone. (d) Which structure correctly predicts all three? (e) Which of these structures contains a considerable amount of "strain" in it?

11.60* What hybridization is predicted for the central atom in molecules having the formulas AB_2E_2 and AB_3E? What are the predicted bond angles for these molecules? The actual bond angles for representative substances are

H_2O	104.45°	NH_3	106.67°
H_2S	92.2°	PH_3	93.67°
H_2Se	91.0°	AsH_3	91.8°
He_2Te	89.5°	SbH_3	91.3°

What would be the predicted bond angle if no hybridization occurred? What conclusion can you draw concerning the importance of hybridization for molecules of compounds involving elements with large atomic numbers?

11.61* The following fluorides of xenon have been well characterized: XeF_2, XeF_4, and XeF_6. (a) Write Lewis structures for these substances and decide what type of hybridization of the Xe atomic orbitals has taken place. (b) Draw all of the isomers of XeF_2 and discuss your choice of molecular geometry. (c) What shape do you predict for XeF_4? (d) What are the important intermolecular forces in XeF_2 and XeF_4?

11.62* Iodine and fluorine form a series of interhalogen molecules and ions. Among these are IF (minute quantities observed spectroscopically), IF_3, IF_4^-, IF_5, IF_6^-, and IF_7. (a) Draw Lewis structures for each of these species. (b) Identify the type of hybridization that the orbitals of the iodine atom have undergone in each substance. (c) Identify the shape of the molecule or ion. (d) What intermolecular forces are important for IF, IF_3, IF_5, and IF_7?

CHAPTER

12

Nuclear Chemistry

NUCLEAR STABILITY AND RADIOACTIVITY

12.1 THE NUCLEUS

To review briefly (Section 3.11), the mass number (A) is the sum of the atomic number (Z, equal to the number of protons in the nucleus) and the number of neutrons in the nucleus (N), that is, $A = Z + N$. Atoms with the same number of protons but different numbers of neutrons are isotopes—atoms of the same element that have different masses. **Nuclide** is a general term used to refer to any isotope of any element.

Three nuclides

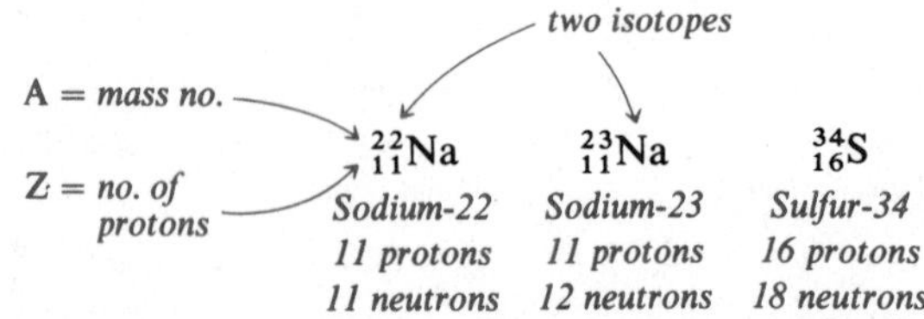

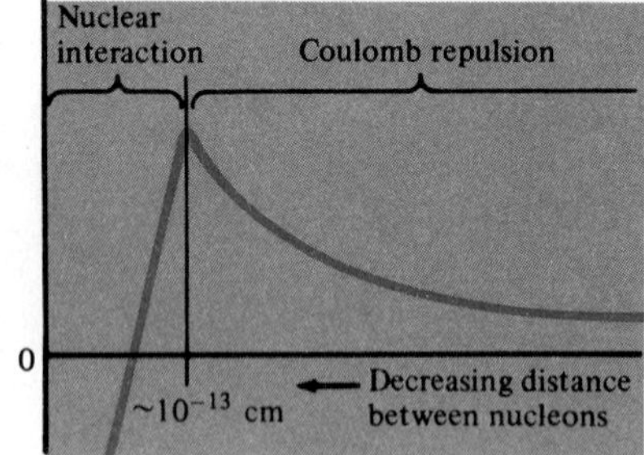

Figure 12.1
Potential Energy Barrier between Protons
Coulombic repulsion acts between two protons until they reach a distance of about 10^{-13} cm. At this point the nuclear force of attraction takes over. No forces act between a proton and a neutron (or between two neutrons) until a distance of 10^{-13} cm is reached.

Protons and neutrons, collectively called **nucleons**, are packed tightly together in the nucleus. All nuclei have roughly the same extremely high density of about 10^{14} g/cm^3. By comparison, platinum, one of the densest metals, has a density of about 22 g/cm^3. A piece of platinum the size of an average egg weighs about 1.5 kg. A comparable volume of matter with the density of nuclei would weigh about 6.5×10^{12} kg.

The **nuclear force** (also called the "strong force") is the force of attraction between nucleons; it acts between protons, between neutrons, and between protons and neutrons. The nuclear force, the strongest force yet discovered, is 30–40 times stronger than the Coulomb force, but has a very short range of only about 10^{-13} cm. Figure 12.1 shows the potential energy changes as a proton approaches another proton.

12.2 MASS, ENERGY, AND NUCLEAR BINDING ENERGY

In "ordinary" chemical reactions, atoms are rearranged but always maintain their identity. Two aspects of nuclear chemistry make it notably different from other chemistry. First, the composition of nuclei, and therefore the *identities* of atoms, change in nuclear reactions. Second, in nuclear reactions, the relationship between mass and energy takes on a significance that it does not have in ordinary chemical reactions. **Nuclear chemistry** deals with nuclei and reactions that cause changes in nuclei. The uses of nuclear chemistry in weapons and in power plants is constantly debated in the media. For this reason, if for no other, everyone should have some understanding of nuclear chemistry.

The law of conservation of mass and the law of conservation of energy are dealt with separately in ordinary chemistry, and this causes no problems. In reality, however, mass and energy are equivalent to each other and are connected by a more fundamental law, sometimes called the law of conservation of mass–energy: The total of the mass and energy in the universe is a constant. This relationship was first expressed by Albert Einstein (in 1905) in his now-famous equation (part of the special theory of relativity)

energy → E; *mass* → m; *speed of light* → c

$$E = mc^2 \qquad \text{(12.1)}$$

joules   *kilograms* $\dfrac{\textit{meters}^2}{\textit{seconds}^2}$

[Recall that in the SI system, $1\ \text{J} = 1\ \text{kg m}^2/\text{s}^2$.]

With the discovery of the mass–energy relationship it became clear that mass and energy are interconvertible. For the energy of the universe to remain constant, every energy change must be accompanied by a change in mass ($\Delta E = \Delta m \times c^2$). This means that when energy is released during a chemical reaction, mass must be lost by the substances involved.

The difference between the mass–energy changes in ordinary chemical reactions and in nuclear reactions is one of magnitude. In ordinary chemical reactions, the changes in mass that are equivalent to the amount of energy gained or lost are too small to be detected, permitting observations to be made that are in agreement with the law of conservation of mass. In nuclear reactions, the changes in energy are determined by the nuclear force and are a million or more times greater than those in ordinary chemical reactions. The mass changes that are equivalent to this energy *can* be detected.

Evidence for the mass changes associated with rearranging protons and neutrons can be found in the masses of atoms. The sum of the masses of the individual neutrons, protons, and electrons combined in any given atom is always greater than the mass of the atom. The difference between the total mass of its nucleons and electrons and the mass of an atom is called the **mass defect**. The missing mass represents the **nuclear binding energy**—the energy that *would be* released in the combination of nucleons to form the nucleus (Figure 12.2). (The mass change due to the binding of the electrons is too small to be detected.)

Although no way has been found to directly combine protons and neutrons, nuclear binding energy can be calculated from the mass defect. Nuclear binding energy is similar in nature to the standard enthalpy of formation of a chemical compound. Just as the enthalpy of formation applies to the heat of formation of one mole of the compound from the elements, the binding energy is the energy of formation of a single nucleus from the nucleons. Note that binding energy is expressed *per nucleus*, and not per mole. In this book we give nuclear binding energy as a negative number, showing that energy is *released*. (In some publications, nuclear binding energy is reported as a positive number.) The following example illustrates calculation of the mass defect and the nuclear binding energy.

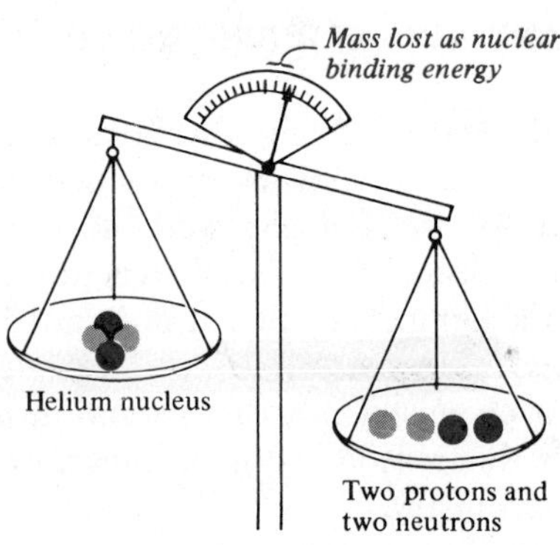

Figure 12.2
Nuclear Mass versus Nucleon Mass
The mass of a nucleus is always less than the mass of the uncombined neutrons and protons. The difference in mass has been converted to nuclear binding energy. (This difference could not, of course, be weighed on a balance.)

EXAMPLE 12.1 Nuclear Binding Energy

The atomic mass of $^{39}_{19}K$ is 38.96371 u. Calculate the binding energy for this nuclide using 1.008665 u for the mass of a neutron, 1.007276 u for the mass of a proton, 0.00054858 u for the mass of an electron, 2.9979×10^8 m/s for the speed of light, and 1 u = $1.6605655 \times 10^{-27}$ kg. Calculate the total binding energy of one mole of $^{39}_{19}K$ atoms.

There are 19 protons and 39 − 19 = 20 neutrons in a $^{39}_{19}K$ nucleus, giving as the calculated mass of one atom of $^{39}_{19}K$

$$(19 \text{ protons})(1.007276 \text{ u/proton}) + (19 \text{ electrons})(0.00054858 \text{ u/electron}) + (20 \text{ neutrons})(1.008665 \text{ u/neutron}) = 39.32197 \text{ u}$$

The mass defect is the difference between the actual mass and the calculated mass:

$$\text{Mass defect} = 38.96371 \text{ u} - 39.32197 \text{ u} = -0.35826 \text{ u}$$

This loss of mass is equivalent to

$$(-0.35826 \text{ u})\left(\frac{1.6606 \times 10^{-27} \text{ kg}}{1 \text{ u}}\right) = -5.9493 \times 10^{-28} \text{ kg}$$

According to the Einstein mass–energy relationship, the energy equivalent to this mass is

$$\begin{aligned} E &= mc^2 \\ &= (-5.9493 \times 10^{-28} \text{ kg})(2.9979 \times 10^8 \text{ m/s})^2\left(\frac{1 \text{ J}}{1 \text{ kg m}^2/\text{s}^2}\right) \\ &= \mathbf{-5.3469 \times 10^{-11} \ J} \end{aligned}$$

For one mole of $^{39}_{19}K$ atoms, the total nuclear binding energy is this energy multiplied by Avogadro's number:

$$(-5.3469 \times 10^{-11} \text{ J/atom})(6.0220 \times 10^{23} \text{ atom/mol}) = \mathbf{-3.2199 \times 10^{13} \ J/mol}$$

EXERCISE Calculate the total binding energy of one mole of $^{81}_{35}Br$ atoms. The atomic mass of $^{81}_{35}Br$ is 80.9163 u. *Answer* -6.799×10^{13} J/mol.

Energies are often expressed in millions of electron volts (MeV) instead of joules when dealing with nuclear processes. Using the conversion factor 1 MeV = $1.6021892 \times 10^{-13}$ J gives the following value for the binding energy of $^{39}_{19}K$:

$$(-5.3469 \times 10^{-11} \text{ J})\left(\frac{1 \text{ MeV}}{1.6022 \times 10^{-13} \text{ J}}\right) = -333.72 \text{ MeV}$$

For the direct conversion of mass changes in atomic mass units to energy in millions of electron volts the conversion factor is 1 u = 931.5017 MeV. For the $^{39}_{19}K$ nuclide

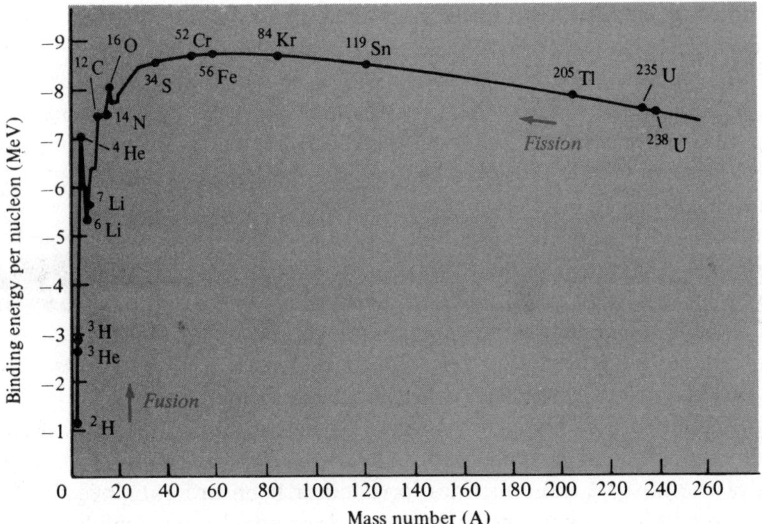

Figure 12.3
The Binding Energy per Nucleon versus the Mass Number
Nuclear fission and nuclear fusion release energy and form nuclei with greater binding energies per nucleon.

$$(-0.35826\text{ u})\left(\frac{931.50\text{ MeV}}{1\text{ u}}\right) = -333.72\text{ MeV}$$

The **binding energy per nucleon** is the nuclear binding energy of a nucleus divided by the number of nucleons in that nucleus. For example, for $^{39}_{19}K$ from Example 12.1 the binding energy per nucleon is

$$\frac{-333.72\text{ MeV}}{39\text{ nucleons}} = \frac{-8.5569\text{ MeV}}{\text{nucleon}}$$

Binding energy per nucleon is more useful than the total nuclear binding energy for comparing one nucleus with another. For the lighter elements, the binding energy per nucleon increases rapidly (Figure 12.3), showing that as the number of nucleons increases for these elements, the nucleons are held together more strongly. Once carbon-12 is reached, there is much less variation in the binding energy per nucleon. Apparently, the strength of interaction of each additional nucleon with its neighbors is very similar beyond this point. A maximum in the curve occurs in the region of iron, showing that in iron nuclei, the nucleons are most strongly bound together.

Combination of two of the lightest nuclei to give a heavier nucleus further along on the curve releases energy. **Nuclear fusion** is the combination of two light nuclei to give a heavier nucleus.

Nuclei at the heavier end of the curve have slightly *lower* binding energies per nucleon than those in the middle. Thus the splitting of a heavy nucleus to give two lighter nuclei also releases energy. **Nuclear fission** is the splitting of a heavy nucleus into two lighter nuclei of intermediate mass numbers (and sometimes other particles as well). In both fission and fusion, nuclei are transformed into other nuclei in which the nucleons are more strongly bound together—those in the middle of the curve in Figure 12.3. The total mass of the nuclei involved decreases, and energy is released in both processes. In Sections 12.15 and 12.16, the uses of fission and fusion as sources of energy are discussed.

12.3 RADIOACTIVITY: NATURAL AND ARTIFICIAL

The early discoveries that pitchblende ore and the elements uranium and thorium are radioactive (see Aside 12.1) led Ernest Rutherford and others to study the properties of the radiation. Soon three different types of radiation from such materials were identified and named after the first three letters of the Greek alphabet: α rays (alpha rays), β rays (beta rays), and γ rays (gamma rays). Each type of ray responds differently to an electric field. The γ rays are undeflected by the field, showing that they have no

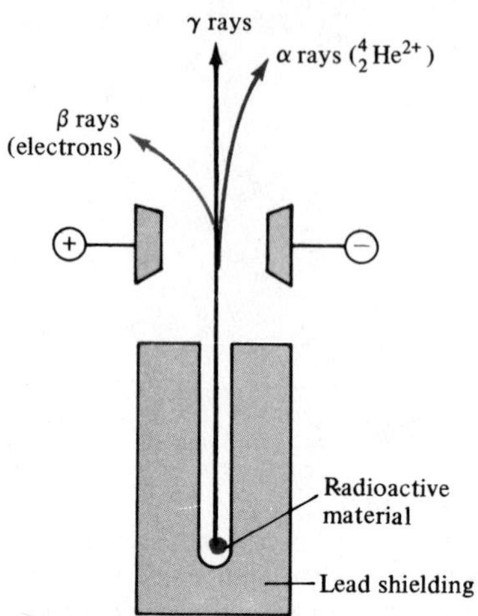

Figure 12.4
α, β, and γ Rays in an Electric Field
The α rays have a positive charge and are heavier (they are less deflected) than β rays, which have a negative charge. The γ rays are neutral and are unaffected by the field. While this sketch illustrates the behavior of all three types of rays, they are not necessarily produced simultaneously by the same radioactive source.

charge. The α rays are attracted to the negative pole, indicating a positive charge. The β rays are attracted to the positive pole, indicating a negative charge (Figure 12.4).

Although we now know the true nature of α, β, and γ rays, their names and letter symbols continue to be used in nuclear chemistry. The α rays are helium nuclei ($^4_2He^{2+}$) and the β rays are electrons. The γ rays are not beams of particles, but beams of photons—electromagnetic radiation at the high-energy, short-wavelength end of the spectrum (see Aside 8.1).

When radiation is emitted, unstable nuclei are converted to more stable nuclei. **Radioactivity** is the spontaneous emission by unstable nuclei of particles, or electromagnetic radiation, or both. Nuclides that spontaneously break down, or *decay*, are called radioisotopes, radioactive nuclides, or **radionuclides.** Radioactive substances and their radioactivity are often referred to as "natural" if the substance occurs naturally on Earth or as "artificial" if the substance is man-made.

Roughly one-third of the elements have natural radioisotopes. All of the known isotopes of elements heavier than bismuth ($Z > 83$) are radioactive.

Most of the natural radionuclides have either been present since the Earth was formed or are products of the decay of other, long-lived radionuclides. Some radionuclides are continuously produced on Earth by the cosmic ray bombardment of stable natural nuclides. (Primary cosmic rays are high-energy particles from outer space—mainly protons—that continually bombard the Earth. In the atmosphere they collide with atoms to produce secondary cosmic rays composed of photons, electrons, neutrons, and other particles that in turn collide with materials at the surface of the Earth.)

More than 350 artificial radionuclides have been identified in the environment. Most of them were created during the period (1955–1962) when nuclear bombs were being tested in the atmosphere by many nations. Radionuclides are also sometimes introduced into the environment during the operation of nuclear power plants, research laboratories, and facilities such as hospitals that use radionuclides in their many applications.

Radioactive nuclides of the same element—radioisotopes—have essentially the same physical and chemical properties. This is to be expected, for the number of electrons and the electron structure are the same. Because of their similar properties, isotopes are difficult to separate. The radioactivity of an isotope is largely unaffected by its chemical environment. For example, carbon-14 decays by the same nuclear reactions and at the same rate whether it is in a sample of coal or is incorporated in a complex biological molecule.

12.4 NEUTRON–PROTON RATIO

Why are some nuclides radioactive and others not radioactive? An answer to this question is suggested by a plot of the number of neutrons versus the number of protons in stable nuclei. In Figure 12.5 the solid line shows where nuclei with an equal number of neutrons and protons would fall. Along this line $Z = N$ and the neutron-to-proton ratio is 1:1.

Isotopes of the lighter elements, up to $^{40}_{20}Ca$, fall on or quite close to the line. For heavier elements, the number of neutrons increases faster than the number of protons and the neutron–proton ratio eventually reaches about 5:3. The additional neutrons apparently provide the additional nuclear force necessary to hold larger numbers of protons close together within the nucleus. Once the atomic number reaches 84, even extra neutrons are not sufficient to maintain stability, with the result that all known nuclides of $Z > 83$ are unstable and radioactive.

For each nuclear charge, only isotopes with a neutron–proton ratio within a specific range—the **band of stability**—are stable and not radioactive. Essentially, radioactivity is the spontaneous transformation of unstable nuclei to nuclei with more favorable neutron–proton ratios. Nuclides with too many protons fall below the stable nuclei band in Figure 12.5. Such nuclides decay so that the net result is a decrease

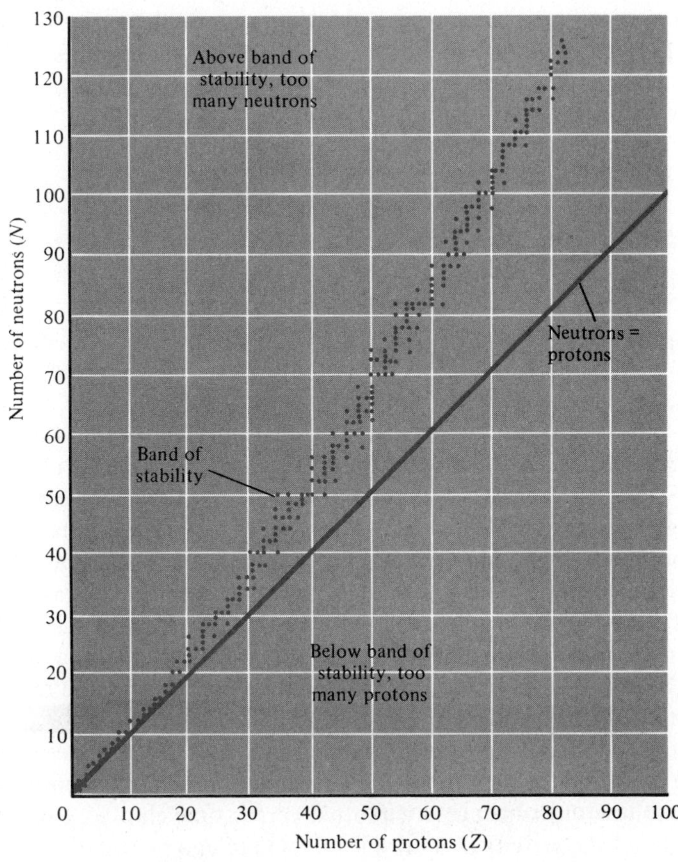

Figure 12.5
Number of Neutrons versus Number of Protons for Stable Nuclei

in the number of protons relative to the number of neutrons (greater n/p ratio). Nuclides with too many neutrons fall above the stable nuclei band in Figure 12.5. Such nuclides decay so that the net result is a decrease in the number of neutrons relative to the number of protons (smaller n/p ratio).

12.5 HALF-LIFE

The decay of an unstable nuclide is the spontaneous disintegration of individual nuclei. Each disintegration both decreases the mass of the original radionuclide sample and contributes to its radioactivity. The type of radiation emitted and the rate of decay are characteristic of a given nuclide. Although it is not possible to predict when an individual nucleus will disintegrate, the overall rate of decay can be expressed as the fraction of nuclei that will disintegrate in a given time interval. Alternatively, the rate of decay is given as the time necessary for a given fraction of the nuclei to disintegrate. Unlike a chemical reaction rate, the rate of radioactive decay is not influenced by temperature or any other factor that we can control.

To compare rates of decay, values of half-lives are usually compared. The **half-life** ($t_{\frac{1}{2}}$) of a radionuclide is the time it takes for one-half of the nuclei in a sample of that nuclide to decay. For example, consider a radionuclide with a half-life of ten years. Assume 100 g of this nuclide is put away today. Ten years from today, 50 g of the original nuclide will remain, along with some of the more stable decay products. Ten years after that, 25 g of the original nuclide will remain, and so on. This characteristic rate of decay is the same for a sample of this nuclide of *any size*.

Half-lives range from a few microseconds to 10^{15} yr (Table 12.1). The less stable the nuclide, the shorter is the half-life. The more stable nuclides persist longer, but they are less radioactive—fewer nuclei are disintegrating per unit time.

One way to observe the rate of decay of a radionuclide is to measure the amounts of the original nuclide remaining after given time intervals. The decays of all radioactive

Table 12.1
Some Radioisotope Half-Lives
Sodium-24, iron-52, and cobalt-60 are produced in accelerators for use in medical treatment or diagnosis. Strontium-90, iodine-131, and cesium-137 have been introduced to the atmosphere as fallout from nuclear weapons testing.

Isotope	Half-life
Radioisotopes produced on Earth by cosmic rays	
$^{3}_{1}H$ (tritium)	12.26 yr
$^{14}_{6}C$	5730 yr
Natural radioisotopes	
$^{40}_{19}K$	1.28×10^{9} yr
$^{144}_{60}Nd$	5×10^{15} yr
$^{232}_{90}Th$	1.41×10^{10} yr
$^{235}_{92}U$	7.1×10^{8} yr
$^{238}_{92}U$	4.51×10^{9} yr
Artificial radioisotopes	
$^{24}_{11}Na$	15.0 h
$^{52}_{26}Fe$	8.2 h
$^{60}_{27}Co$	5.26 yr
$^{90}_{38}Sr$	28.1 yr
$^{87}_{35}Br$	55 s
$^{131}_{53}I$	8.070 day
$^{137}_{55}Cs$	30.23 yr
$^{239}_{94}Pu$	24,400 yr

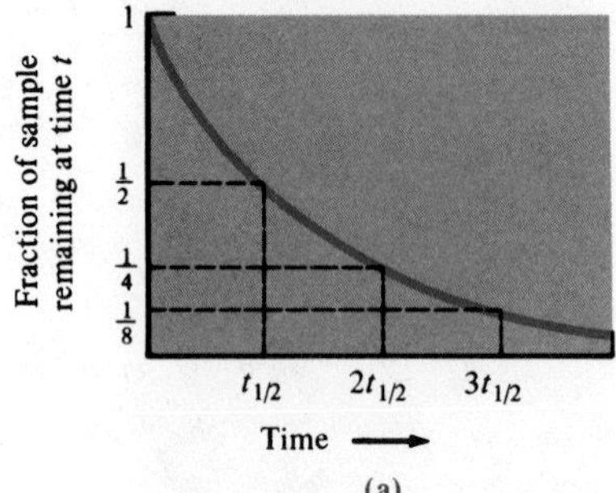

(a)

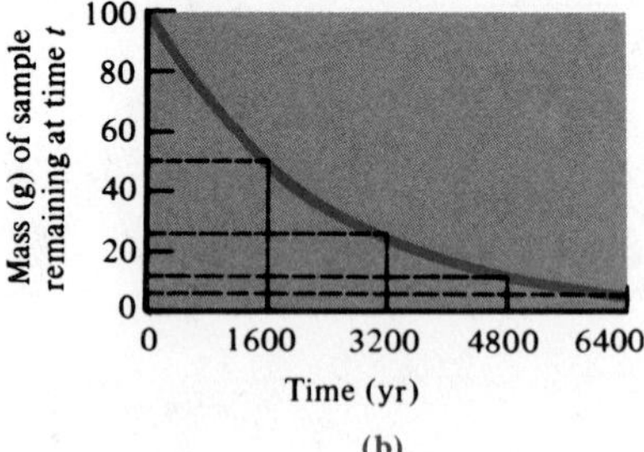

(b)

Figure 12.6
Radioactive Decay Curves
(a) A general curve for radioactive decay. (You may recognize this as an exponential decay curve.) (b) Radioactive decay of a 100-g sample of $^{226}_{88}Ra$, which has a half-life of 1,600 years.

nuclides, whatever the rate, are governed by the same mathematical relationship. In terms of half-lives and taking 1 as the amount of the original sample of a nuclide

Number of half-lives that have elapsed	Amount of sample remaining
$1\,t_{\frac{1}{2}}$	$\frac{1}{2}$
$2\,t_{\frac{1}{2}}$	$\frac{1}{2} \times \frac{1}{2} = \frac{1}{4}$
$3\,t_{\frac{1}{2}}$	$\frac{1}{2} \times \frac{1}{2} \times \frac{1}{2} = \frac{1}{8}$
$4\,t_{\frac{1}{2}}$	$\frac{1}{2} \times \frac{1}{2} \times \frac{1}{2} \times \frac{1}{2} = \frac{1}{16}$
⋮	⋮

or, with n = the number of half-lives elapsed

$$\text{Fraction of original mass remaining} = (\tfrac{1}{2})^n$$

A general plot of the amount of sample remaining versus the time elapsed—a radioactive decay curve—is shown in Figure 12.6a and the curve for a specific nuclide is shown in Figure 12.6b. From the known half-life of a nuclide and the fractional relationships given above or curves such as those in Figure 12.6, estimates can be made of (1) the amount of a given sample that will remain after a given time interval, (2) the time necessary for a nuclide to decay, or (3) the age of a sample. For example, given that the half-life of $^{60}_{27}Co$ is 5.26 years, how long would it take for $\frac{7}{8}$ of a sample to decay, leaving $\frac{1}{8}$ of the sample unchanged? As can be seen from Figure 12.6a or the equation given above, three half-lives $[\frac{1}{8} = (\frac{1}{2})^3]$, which is 15.8 years, would be required.

The **activity** of a radionuclide is the number of disintegrations in a given unit of time. This activity is directly proportional to the mass of the substance that is present. Therefore, the mass and the activity of a given radioisotope both decrease at the same rate, and both follow the same curve if plotted versus time elapsed. For example, if it has been found that the activity of a sample of $^{68}_{31}Ga$ after a period of 136.6 min is one-fourth that of the original sample, then from Figure 12.6a or the relationship above it can be determined that two half-life periods must have passed $[\frac{1}{4} = (\frac{1}{2})^2]$ and that the half-life of this nuclide is 68.3 min.

The rate of radioactive decay is found by experiment to be directly proportional to the number of radioactive nuclei in the sample, N

$$\text{rate} = kN \tag{12.2}$$

where k is a proportionality constant known as a rate constant. For each nuclide, k has a characteristic value. The initial quantity of a radionuclide, q_0 (in any mass units or as the number of nuclei) is related to the quantity, q, after *any* period of time, t, has passed by the following equation:

rate constant
initial quantity *time elapsed*

$$\log\left(\frac{q_0}{q}\right) = \frac{kt}{2.303} \tag{12.3}$$

quantity at time t

Equation (12.3) is also valid for the variation in the activity of a sample of a radionuclide with time.

$$\log\left(\frac{a_0}{a}\right) = \frac{kt}{2.303} \tag{12.4}$$

(At this point you may want to review the use of logarithms discussed in Appendix I.)

By substituting into Equation (12.3) the half-life, $t_{\frac{1}{2}}$, and the amount of an original sample remaining after one half-life, which is one-half of q_0, or $0.5q_0$, it can be shown that the half-life is independent of the size of the sample and is dependent only on the rate constant.

$$\log\left(\frac{q_0}{0.5q_0}\right) = \frac{kt_{\frac{1}{2}}}{2.303}$$

$$\log 2 = \frac{kt_{\frac{1}{2}}}{2.303}$$

$$0.301 = \frac{kt_{\frac{1}{2}}}{2.303}$$

$$t_{\frac{1}{2}} = \frac{0.693}{k} \qquad \textbf{(12.5)}$$

(The relationships given by Equations 12.3 and 12.5 also apply to certain types of chemical reactions and are further discussed in Sections 18.8 and 18.9.)

EXAMPLE 12.2 Half-Life

What percentage of a $^{68}_{31}$Ga sample, which has a half-life of 68.3 min, would remain at the end of 3.00 h?

The rate constant for the radioactive decay of this nuclide is

$$k = \frac{0.693}{t_{\frac{1}{2}}} = \frac{0.693}{68.3 \text{ min}} = 0.0101/\text{min}$$

With q_0 the initial amount of the isotope and q the amount remaining after 3.00 h, Equation (12.3) for $^{68}_{31}$Ga is, using $t = (3.00 \text{ h})(60 \text{ min/h}) = 180.$ min and $k = 0.0101/\text{min}$,

$$\log\left(\frac{q_0}{q}\right) = \frac{kt}{2.303}$$

$$= \frac{(0.0101/\text{min})(180. \text{ min})}{(2.303)} = 0.789$$

The fraction of the isotope remaining after 3.00 h is q/q_0, which can be found from this equation as follows:

$$\log\left(\frac{q_0}{q}\right) = 0.789$$

$$\frac{q_0}{q} = 6.15 \text{ and therefore } \frac{q}{q_0} = \frac{1}{6.15} = 0.163$$

The percentage remaining is therefore **16.3%**.

EXERCISE What percentage of a $^{60}_{27}$Co sample remains at the end of 3.0 yr? The half-life of $^{60}_{27}$Co is 5.26 yr. *Answer* 67%.

12.6 RADIOISOTOPE DATING

Ingenious use has been made of the natural radioactive isotope carbon-14 to determine the age of plant and animal relics. Carbon-14 is formed in the upper atmosphere by cosmic ray bombardment of nitrogen and is incorporated into CO_2. The amount of carbon-14 in the Earth's environment is assumed to have been relatively constant over the centuries.

As long as a plant or animals lives, it contains the same proportion of $^{14}_{6}$C as its surroundings. But as soon as a plant stops utilizing CO_2 or an animal stops eating carbon-containing plants, its supply of $^{14}_{6}$C is no longer replenished and the ratio of radioactive $^{14}_{6}$C to nonradioactive $^{12}_{6}$C begins to decrease. From the half-life of carbon-14, which is 5730 yr, and the measured ratio of $^{14}_{6}$C to $^{12}_{6}$C, the age of the plant or animal remains can be calculated. Such a procedure is known as radiocarbon dating.

Radiocarbon dating is limited to objects less than about 50,000 to 75,000 years old. In older objects, the activity is too low for reliable measurements. Rocks that are older can be dated by measuring their ratio of uranium-238 to lead-206. It is assumed that the radioactive uranium and its decay products have been trapped in the rock since the time that the rock solidified. Ultimately, all uranium-238 decay products are converted to the stable isotope, lead-206 (Section 12.10).

Highly refined mass spectrometers and very careful experimental techniques have also been used in radioisotope dating. The actual relative numbers of atoms of the parent nuclide and its decay products are counted. Use of this technique for moon rocks established the age of the moon and the Earth to be the same, 4.5×10^9 years.

EXAMPLE 12.3 Half-Life

Ptolemy V reigned in Egypt during the period 203 to 181 BC (about 2180 years ago). Could a piece of wood from an artifact that had a $^{14}_{6}C$ activity of 11.8 disintegrations per minute per gram of carbon have come from this same time period? The half-life of $^{14}_{6}C$ is 5730 years and the current $^{14}_{6}C$ activity is 15.3 disintegrations per minute per gram of carbon.

The rate constant for the decay of $^{14}_{6}C$ is

$$k = \frac{0.693}{t_{\frac{1}{2}}} = \frac{0.693}{5730 \text{ yr}} = 1.21 \times 10^{-4}/\text{yr}$$

Solving Equation (12.4) for the time and substituting $a = 11.8/\text{min g C}$, $a_0 = 15.3/\text{min g C}$, and $k = 1.21 \times 10^{-4}/\text{yr}$ gives the age of the wood as

$$t = \frac{(2.303)\log\left(\frac{a_0}{a}\right)}{k} = \frac{(2.303)\log\left(\frac{15.3/\text{min g C}}{11.8/\text{min g C}}\right)}{1.21 \times 10^{-4}/\text{yr}} = 2150 \text{ yr}$$

The wooden artifact is 2150 years old and could have come from the reign of Ptolemy, about 2180 years ago.

EXERCISE A bone found in a cave was "dated" using carbon-14 analysis. A sample of the bone had an activity of 2.93 disintegrations per minute per gram of carbon. What is the age of the bone specimen? *Answer* 1.37×10^4 yr.

12.7 COSMIC ABUNDANCE AND NUCLEAR STABILITY

A plot of the cosmic abundance of the elements (Figure 12.7)—their abundance not on Earth but in the universe—reveals some interesting relationships. Hydrogen and helium atoms are far more abundant in the universe than any other elements. Except for hydrogen, elements with even atomic numbers are more abundant than their neighbors with odd atomic numbers. And in general, light elements are more abundant than heavy elements.

If abundance is a reflection of nuclear stability—a reasonable assumption—then even numbers of protons and nuclear stability must be related in some way. Further examination reveals that of the ten most abundant elements in the cosmos (H, He, O, C, N, Ne, Mg, Si, Fe, S), all except hydrogen have even numbers of both neutrons and protons. In addition, of all the nuclides in the Earth's crust, 86% have even mass numbers.

Certain numbers of neutrons or protons—called **magic numbers**—impart particularly great nuclear stability. The magic numbers are 2, 8, 20, 28, 50, 82, and 126. Nuclides with magic numbers of *both* neutrons and protons are very stable and are

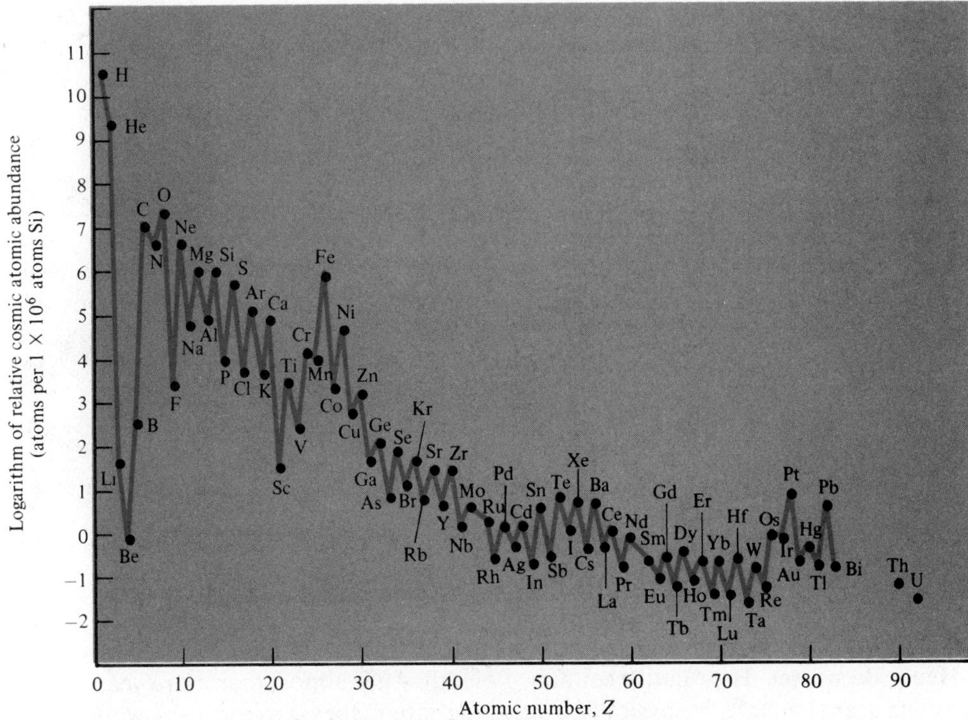

Figure 12.7
Relative Cosmic Abundances of the Elements
Silicon is widespread in the crust of the Earth and its abundance is used as a standard of comparison. Source: Therald Moeller, *Inorganic Chemistry: A Modern Introduction* (New York: Wiley, 1982), p. 28.)

abundant, for example, $^{4}_{2}He$, $^{16}_{8}O$, $^{40}_{20}Ca$, and $^{208}_{82}Pb$. The magic numbers were so named before enough was known about the nucleus to interpret their significance. Now, these relationships are explained by the existence of energy levels within the nucleus similar to the electron energy levels outside the nucleus. The magic numbers represent energy levels filled by nucleons. Nuclides with filled levels have increased nuclear stability, just as atoms with energy levels filled by electrons have increased chemical stability.

The elements originated, it is believed, in nuclear reactions within the stars. In stars like our sun, hydrogen is converted to helium by nuclear fusion. By successive fusion reactions, helium then yields the elements through iron and nickel. Because nuclear binding energy is at a maximum with the iron and nickel nuclides, continued fusion to produce heavier elements does not occur. The heavier elements are formed in smaller quantities by subsequent nuclear reactions of other types.

After the Earth was formed and before it cooled, natural geochemical processes concentrated iron and nickel in its core and the light elements in its crust and atmosphere. Therefore, the portion of the Earth available to chemists and geologists offers an abundance and distribution pattern of the elements quite different from that in the universe as a whole. (The ten most abundant elements in the Earth's crust are O, Si, Na, Al, Mg, Ca, K, Fe, C, Ti, F.)

SUMMARY OF SECTIONS 12.1–12.7

Nuclear chemistry deals with nuclei and reactions that cause changes in nuclei. In nuclear chemistry, as opposed to "ordinary" chemistry, the interconversion of mass and energy is of significance.

The mass difference between the sum of the individual nucleons present and the mass of an atom represents the nuclear binding energy. The most stable nuclei are those with the highest binding energy per nucleon.

Light nuclides are generally stable when their nuclei have about equal numbers of protons and neutrons. The neutron–proton ratio needed for stability is larger for heavier nuclides. Radioactivity is the decay of unstable nuclei to form more stable nuclei. All known nuclei with $Z > 83$ are unstable. α Rays ($^{4}_{2}He^{2+}$), β rays (electrons), and γ rays (electromagnetic radiation) are emitted by naturally radioactive isotopes. The chemical properties of radioisotopes are essentially the same as those of nonradioactive isotopes, and the radioactivity of an isotope is unaffected by its chemical environment.

The half-life of a radionuclide is the time it takes for one-half of the nuclei in a sample of that nuclide to decay. The less stable the nuclide, the shorter the half-life. The fraction of the mass of an original radionuclide sample and the activity (the number of disintegrations per unit time) both change with time in the same way (see Figure 12.6) for all radionuclides. The decay of carbon-14 incorporated in living things is the basis for finding the age of plant and animal relics.

Nuclides with even numbers of nucleons are more stable and abundant than those with odd numbers of nucleons; certain numbers of nucleons, called magic numbers, confer an especially high degree of stability. The existence of nucleon energy levels within the nucleus is assumed to explain these facts.

Aside 12.1 RADIOACTIVITY

Radioactivity was discovered quite by accident in 1896 by a French physicist, Henri Becquerel. He was studying a uranium salt (potassium uranyl sulfate, $K_2SO_4 \cdot UO_2SO_4 \cdot 2H_2O$) that was known to be phosphorescent—it reemitted radiation when it was exposed to light. Becquerel was looking for a relationship between x-rays, which had been discovered the year before by Röntgen, and the phenomenon of phosphorescence.

X-rays had been found to penetrate the black paper wrappers that kept ordinary light from exposing photographic plates. Becquerel had already discovered that radiation from the phosphorescing uranium salt exposed photographic plates in the same way. During several cloudy days some samples were left waiting in a drawer. Being curious, Becquerel developed the plates. He had quite a surprise. The image of the uranium salt appeared sharp and clear on the plates. The emission of radiation was obviously unrelated to exposure of the salt to any other radiation, but was a property of the uranium salt itself.

In 1897 Marie Curie began a systematic search for naturally radioactive elements. Among the elements known at that time, and with the methods of measurement available, she found only uranium and thorium to be radioactive. But a uranium-containing ore, pitchblende, was more radioactive than either the uranium or the thorium contained in it. Marie Curie correctly suspected that an undiscovered radioactive element might also be present in pitchblende. Using chemical methods, Marie and her physicist husband, Pierre, demonstrated the presence of two previously unknown radioactive elements in the ore. The first, which resembled barium in its chemical behavior, they named radium (from the Latin for "ray"). The other, which resembled bismuth in its chemical behavior, they named polonium (for Marie's native country, Poland). After almost four more years of physically exhausting labor, they succeeded in winning one-tenth of a gram of radium chloride from several tons of pitchblende ore.

Pierre Curie was killed tragically in 1906 when he was struck by a heavy horse-drawn wagon while crossing the street. Marie carried on their work with courage. In 1921 she traveled to the United States, where President Harding presented her with 1 gram of pure radium to aid in her ongoing research. The women of the United States, to honor this great woman scientist, had contributed \$100,000 to finance the mining and purification of this radium.

The mystery of radioactivity deepened when it was found that not only radiation, but also atoms of other elements, were continuously produced in radioactive decay. It was Rutherford and Frederick Soddy who concluded in 1902 that in the process of radioactive decay, one element is transformed, or *transmuted,* into another element. To announce this conclusion to the scientific community must have taken a special form of courage. It was contrary to what had been firmly believed since the time of Dalton—that atoms of one element cannot be transformed into atoms of another element. (Rutherford is reported to have remarked "For Mike's sake, Soddy, don't call it transmutation. They'll have our heads off as alchemists.") A further milestone in the understanding of radioactivity and its relationship to nuclear instability came in 1934 when Marie Curie's daughter, Irène, and son-in-law, Frédéric Joliot-Curie, discovered that radioactive substances not only existed in nature, but could be produced artificially (Section 12.11).

NUCLEAR REACTIONS

12.8 WRITING EQUATIONS FOR NUCLEAR REACTIONS

The various types of **nuclear reactions**—reactions that lead to changes in the atomic number, mass number, or energy states of nuclei—are described below. Nuclear reactions result from (1) the spontaneous decay of radioactive nuclides, either artificial or natural, (2) **bombardment reactions**, in which electromagnetic radiation or fast-moving particles are captured by a nucleus to form an unstable nucleus that subsequently decays, (3) the fission of unstable heavy nuclei, or (4) the fusion of light nuclei, which occurs naturally only in the sun and other stars.

The symbols used for the particles produced in nuclear reactions are summarized in Table 12.2. In writing equations for nuclear reactions, the mass numbers of all reactant and product nuclides are included in the nuclide symbols (upper left). Inclusion of mass numbers and atomic numbers (lower left) for all reactants and products is optional, but is helpful in writing and balancing nuclear equations.

Following are the equations for some nuclear reactions of historical interest:

Table 12.2
Particles in Nuclear Reactions
These are the particles and the symbols most commonly encountered in writing nuclear reactions. The superscripts are the mass numbers (neutrons + protons). The subscripts are the number of protons or, for the β particles, values that compensate for the change in the number of protons (see Sections 12.9c, d).

Name	Symbols
Alpha particle (helium nucleus)	α, ${}^{4}_{2}\alpha$, ${}^{4}_{2}\mathrm{He}$
Electron (beta-minus particle)	β^-, ${}^{0}_{-1}\beta$, ${}^{0}_{-1}e$
Positron (beta-plus particle)	β^+, ${}^{0}_{+1}\beta$, ${}^{0}_{+1}e$
Proton (hydrogen nucleus)	p, ${}^{1}_{1}p$, ${}^{1}_{1}\mathrm{H}$
Neutron	n, ${}^{1}_{0}n$

First recognized natural transmutation of an element (Rutherford and Soddy, 1902)

$$\underset{\text{an atom of radium-226}}{{}^{226}_{88}\mathrm{Ra}} \underset{\text{decays to}}{\longrightarrow} \underset{\text{an }\alpha\text{ particle}}{{}^{4}_{2}\mathrm{He}} + \underset{\text{an atom of radon-222}}{{}^{222}_{86}\mathrm{Rn}} \quad (12.6)$$

First artificial transmutation of an element (Rutherford, 1919)

$$\underset{\text{an atom of nitrogen-14}}{{}^{14}_{7}\mathrm{N}} + \underset{\text{bombarded by an }\alpha\text{ particle}}{{}^{4}_{2}\mathrm{He}} \underset{\text{yields}}{\longrightarrow} \underset{\text{an atom of oxygen-17}}{{}^{17}_{8}\mathrm{O}} + \underset{\text{a proton}}{{}^{1}_{1}p} \quad (12.7)$$

Discovery of the neutron (Chadwick, 1932)

$$\underset{\text{an atom of beryllium-9}}{{}^{9}_{4}\mathrm{Be}} + \underset{\text{bombarded by an }\alpha\text{ particle}}{{}^{4}_{2}\mathrm{He}} \underset{\text{yields}}{\longrightarrow} \underset{\text{an atom of carbon-12}}{{}^{12}_{6}\mathrm{C}} + \underset{\text{a neutron}}{{}^{1}_{0}n} \quad (12.8)$$

Some of the species in nuclear reactions may be ions, for example, the α particle, which is ${}^{4}_{2}\mathrm{He}^{2+}$. However, the charges are not included in the symbols. As a nuclear reaction proceeds, excited electrons within an atom lose energy so as to return the atom to a stable state, and positive ions eventually pick up electrons.

A nuclear equation is correctly written when the following two rules are obeyed:

1. Conservation of mass number. The sum of the number of protons and neutrons in the reactants must equal the sum of the number of protons and neutrons in the products. This can be checked by comparing the sum of the mass numbers of the products with that of the reactants. For example, in Equation (12.8), $9 + 4 = 12 + 1$.
2. Conservation of nuclear charge. The total nuclear charge of the products must equal the total nuclear charge of the reactants. This can be checked by comparing the sum of the atomic numbers (equal to the nuclear charge) of the products with the sum of the atomic numbers of the reactants. For example, in Equation (12.8), $4 + 2 = 6 + 0$.

If the atomic numbers and mass numbers of all but one of the atoms or particles in a nuclear reaction are known, the unknown particle can easily be identified by using the rules given.

EXAMPLE 12.4 Nuclear Equations

A nuclide of element 104 (for which the name unnilquadium has been suggested) with a mass number of 257, $^{257}_{104}Unq$, is formed by the nuclear reaction of $^{249}_{98}Cf$ and $^{12}_{6}C$, with the emission of four neutrons. This new nuclide has a half-life of about 5 s and decays by emitting an α particle. Write the equations for these nuclear reactions and identify the nuclide formed as $^{257}_{104}Unq$ undergoes decay.

The problem gives all the information needed to write the equation for the formation of $^{257}_{104}Unq$.

$$^{249}_{98}\mathbf{Cf} + {}^{12}_{6}\mathbf{C} \longrightarrow {}^{257}_{104}\mathbf{Unq} + 4\,{}^{1}_{0}\boldsymbol{n}$$

Note that the totals of the mass numbers of the reactants and products are equal, $249 + 12 = 261 = 257 + 4(1)$, as are the totals of the atomic numbers, $98 + 6 = 104 = 104 + 4(0)$.

The equation for the radioactive decay of $^{257}_{104}Unq$ can be written as

$$^{257}_{104}\mathrm{Unq} \longrightarrow {}^{4}_{2}\mathrm{He} + {}^{A}_{Z}E$$

where E represents the unidentified element. To conserve mass number (A) and nuclear charge

$$257 = 4 + A \quad \text{and} \quad 104 = 2 + Z$$

or

$$A = 257 - 4 = 253 \quad \text{and} \quad Z = 104 - 2 = 102$$

The unidentified element has atomic number 102 and the periodic table (inside front cover) shows it to be an isotope of nobelium. The complete equation is

$$^{257}_{104}\mathbf{Unq} \longrightarrow {}^{4}_{2}\mathbf{He} + {}^{253}_{102}\mathbf{No}$$

EXERCISE A $^{10}_{5}B$ nucleus reacts with a neutron to form a $^{10}_{4}Be$ nucleus and another particle. Write the equation for this nuclear process and identify the unknown particle.

Answer $^{10}_{5}B + {}^{1}_{0}n \longrightarrow {}^{10}_{4}Be + {}^{1}_{1}H$, proton.

12.9 SPONTANEOUS RADIOACTIVE DECAY

a. γ Decay Frequently, radioactive decay by particle emission leaves the nucleus above its ground-state nuclear energy level. To return to this most stable energy level the nucleus undergoes **γ decay**—the emission of a γ ray—thereby giving up the excess energy in the form of electromagnetic radiation. The wavelength range for both x-rays and γ rays is from roughly 1 to 0.001 nm. X-rays are at the longer wavelength end of this range and γ rays at the shorter wavelength end of the range. However, the terms refer not to wavelength, but to the origin of the radiation—"x-rays" from energy changes of electrons and "γ rays" from nuclear energy changes.

γ Rays travel with the speed of light and only certain specific frequencies of γ rays are emitted by specific radionuclides. There is no change of mass number or nuclear charge in γ decay.

indicates excited state

$$^{A}_{Z}X^{*} \longrightarrow {}^{A}_{Z}X + \gamma \tag{12.9}$$

γ-Ray emission usually occurs within a nanosecond following particle emission—virtually at the same time. An excited nuclide that survives for a longer time is called an "isomer" and its decay by delayed γ-ray emission is called an **isomeric transition** (see Table 12.8). These longer-lived excited nuclides are usually designated by adding the letter "m" to the name or symbol for the nuclide, for example, technetium-99m, ^{99m}Tc.

b. α Decay **α Decay** is the emission of an α particle by a radionuclide. The mass number is decreased by 4 units and the atomic number by 2.

$$^{A}_{Z}X \longrightarrow {}^{A-4}_{Z-2}Y + {}^{4}_{2}\mathrm{He} \tag{12.10}$$

Heavy nuclides for which the neutron–proton ratio is too low and those with $Z > 83$ undergo α decay. For example, you can see that the gold isotope with 79 protons and a mass number of 185, or 106 neutrons, lies below the stability band in Figure 12.5, and that the product nuclide with 77 protons and 104 neutrons is closer to the band and therefore more stable.

$$^{185}_{79}\text{Au} \longrightarrow {}^{181}_{77}\text{Ir} + {}^{4}_{2}\text{He}$$

Emission of an α particle can leave a nuclide in an excited state; it then emits γ radiation to reach the stable state. The total energy of the emitted α particle and the γ radiation is equal to the reaction energy. For example, in Figure 12.8 the energy of α_1 plus γ_2 plus γ_3, or α_1 plus γ_1, equals $E_1 - E_2$, the total reaction energy.

c. β Decay: Electron Emission **β Decay** includes electron emission, positron emission, and electron capture. It is thought that the electrons emitted in β decay are created at the moment of emission by decay of neutrons within the nucleus to give one proton plus one electron each (see equation in Table 12.3).

Emission of an electron causes no change in the mass number of the nuclide, but *increases* the positive charge on the nucleus by 1 because a proton is left behind. To compensate for this increase in nuclear charge, a -1 is written for the "nuclear charge" of an electron, allowing conservation of nuclear charge to be indicated in writing the equations for nuclear reactions.

$$^{A}_{Z}X \longrightarrow {}_{Z+1}^{A}Y + {}_{-1}^{0}e \quad \textbf{(12.11)}$$

The net effect of electron emission is the conversion of a neutron to a proton (Table 12.3). Thus, in electron emission, the neutron–proton ratio decreases. Electron emission is the most common mode of decay for nuclides that are radioactive because they lie above the band of stability (too many neutrons; Figure 12.5). For example, carbon-14 undergoes electron emission to form nitrogen-14, which is not radioactive.

$$^{14}_{6}\text{C} \longrightarrow {}^{14}_{7}\text{N} + {}_{-1}^{0}e$$

You can see that nuclear charge is conserved: $6 = 7 - 1$.

d. β Decay: Positron Emission A **positron** is a particle identical to an electron in all of its properties except for the charge, which is $+1$ rather than -1. Positron emission causes a decrease of 1 in atomic number, but no change in mass number. In the equations for nuclear reactions, the positron is written with a $+1$ "nuclear charge" to compensate for the decrease in atomic number.

$$^{A}_{Z}X \longrightarrow {}_{Z-1}^{A}Y + {}_{+1}^{0}e \quad \textbf{(12.12)}$$

The net effect of positron emission is the conversion of a proton to a neutron (see Table 12.3). This increases the neutron–proton ratio.

Table 12.3
Net Effect of β-Decay Reactions
An isolated neutron decays by the reaction shown here with a half-life of about 12 min. The other reactions do not occur in isolation from nuclei. The net effect of β decay is either the conversion of a neutron to a proton or the conversion of a proton to a neutron.

Electron emission ($n \longrightarrow p$)
$^{1}_{0}n \longrightarrow {}^{1}_{1}p + {}_{-1}^{0}e$
Positron emission ($p \longrightarrow n$)
$^{1}_{1}p \longrightarrow {}^{1}_{0}n + {}_{+1}^{0}e$
Electron capture ($p \longrightarrow n$)
$^{1}_{1}p + {}_{-1}^{0}e \longrightarrow {}^{1}_{0}n$

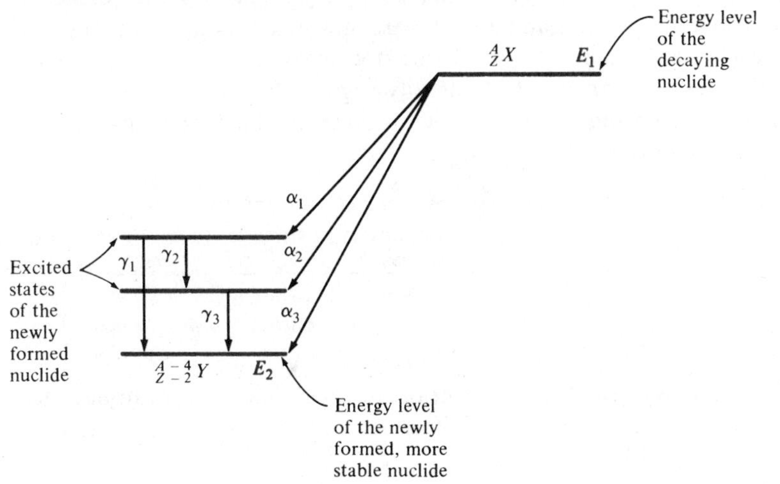

Figure 12.8
An α-Decay Scheme
Isotope X yields α particles with three different energies plus three different γ rays.

Isotopes with too many protons (below the band in Figure 12.5) decay by positron emission. Only artificial radionuclides have been observed to undergo positron emission. When a positron and an electron interact, they annihilate each other—all of their mass is converted to energy in the form of two 0.51 MeV γ rays traveling in opposite directions.

e. β Decay: Electron Capture The positive charge of an unstable nucleus can also be decreased by **electron capture**—the capture by the nucleus of one of its own inner orbital electrons. The mass number is again unchanged while the atomic number decreases by 1.

$$^{A}_{Z}X + {}^{0}_{-1}e \longrightarrow {}_{Z-1}^{\ \ \ A}Y \qquad \textbf{(12.13)}$$

Here a proton captures an electron to produce a neutron (Table 12.3). The result is an increase in the neutron–proton ratio. As the electrons rearrange themselves to compensate for the electron pulled into the nucleus, x-rays are emitted.

12.10 NATURAL RADIOACTIVE SERIES

The heavy radioactive elements found in nature are members of three different series of elements (Table 12.4). Each series starts with a long-lived radionuclide and ends with a stable isotope of lead. Decay occurs by a sequence of reactions in which intermediate radionuclides of varying half-lives are produced by α and/or β^- decay. The series that begins with uranium-238 is shown in Figure 12.9, where diagonal lines indicate α decay and horizontal lines indicate β^- decay. For example, radium-226 (half-life 1600 yr) undergoes α decay to give radon-222 (half-life 3.823 days), which by α decay gives polonium-218 (half-life 3.05 min), which can decay by alternate routes, and so on.

Since the 1970s, concern has grown over exposure of individuals to radon in homes and other buildings. Because radon is a gas, it can percolate from uranium-bearing rocks, subsoil, or building materials into the air within a building. Radon itself is not a health threat because it is an unreactive noble gas. It enters and exits the lungs without remaining in the body. However, its decay product polonium, from the oxygen family, is more reactive.

Polonium can adhere to dust particles in the air that then become trapped in the lungs. Being an α emitter, polonium and its decay products are potential cancer-causing agents. As is always the case, evaluating the seriousness of the health threat from such a substance in the environment is difficult. Data must be gathered from many sources, and the effects of other substances and the habits of individuals, notably smoking, must be separated from the effects of radon.

At first it was thought that newer, more energy-efficient buildings might pose a greater threat of radon exposure because they are more airtight. Further studies have begun to show that the age and airtightness of a structure do not necessarily correlate with radon levels in the inside air. More extensive data are being gathered throughout the United States in an effort to identify regions that might be most at risk. Also, methods for preventing circulation of radon from the ground into foundations are under development.

Table 12.4
The Natural Radioactive Decay Series

Designation	Parent	Half-life of parent (yr)	Final product
Thorium series	$^{232}_{90}Th$	1.39×10^{10}	$^{204}_{82}Pb$
Uranium series	$^{238}_{92}U$	4.49×10^{9}	$^{206}_{82}Pb$
Actinium series	$^{235}_{92}U$	7.13×10^{8}	$^{207}_{82}Pb$

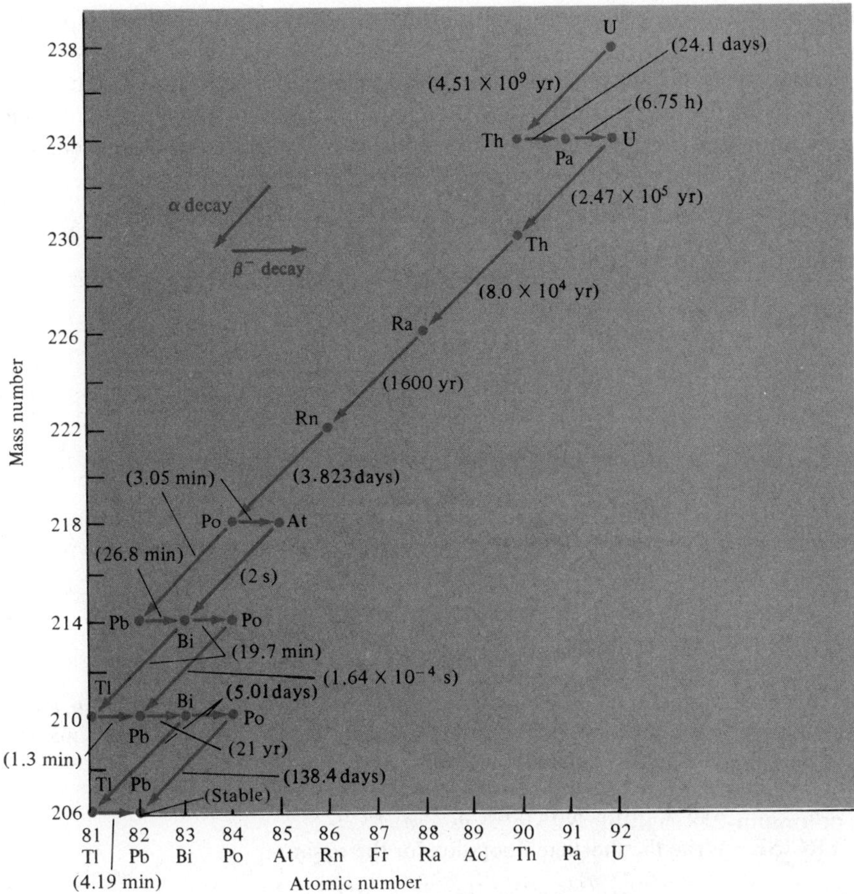

Figure 12.9
Uranium-238 Decay Series
Some of the isotopes decay by both α- and β-particle emission. The half-lives of the nuclides are given in parentheses.

12.11 BOMBARDMENT

In *bombardment* a nucleus is struck by a moving particle. The particles used to bombard nuclei usually have energies of 10 MeV or more. The particle combines with the nucleus to form an unstable compound nucleus that decays either instantaneously or with a measurable half-life (Figure 12.10). If the bombarding particle has enough energy to overcome the nuclear force (in the billion electron volt range), it may smash the atom to bits. Physicists search among the debris from such reactions for information about nuclear particles and forces.

Early investigators were limited to bombardment with α particles from naturally radioactive sources. To enter a nucleus, a particle must have enough energy to overcome any repulsive forces between itself and the nucleus. The α particles from natural sources have energies of less than 10 MeV and can be captured only by relatively light nuclei.

Two English physicists, Sir John Cockcroft and E. T. S. Walton, in 1932 succeeded in accelerating protons in a vacuum tube. The first nuclear reaction produced by artificially accelerated particles was the splitting of lithium-7 into two α particles.

$$^{7}_{3}Li + ^{1}_{1}p \longrightarrow ^{4}_{2}He + ^{4}_{2}He$$

Since the experiments of Cockroft and Walton, accelerators have continued to grow in size and complexity. The cost of the big accelerators is now so great that only national governments can afford to pay for them.

The synthesis of the transuranium elements ($Z > 92$) was made possible by the advent of the big accelerators. Most of these elements were discovered at the famous Lawrence Radiation Laboratory of the University of California at Berkeley. Weighable amounts of elements 93–99 have been prepared. The remaining transur-

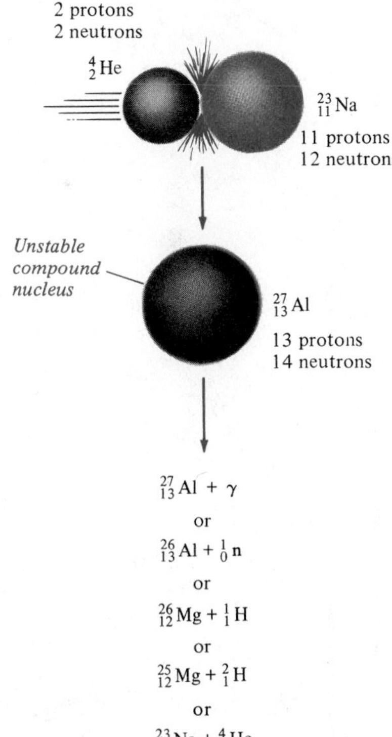

Figure 12.10
Bombardment of Sodium-23 by α Particles
The compound nucleus may decay by one or several of the routes shown.

anium elements have been identified literally atom by atom, an impressive achievement.

A shorthand notation is used for bombardment reactions. The bombarding particle and the product particle are written in parentheses between the symbols for the reactant and product nuclides. For example, the first transuranium element, neptunium, was synthesized by the neutron bombardment of uranium-238

$$^{238}_{92}U + ^{1}_{0}n \longrightarrow ^{239}_{93}Np + ^{0}_{-1}e$$

a reaction represented as follows:

bombarding particle ↘ ↙ *product particle*

$$^{238}_{92}U(n, e^-)^{239}_{93}Np \quad \text{or} \quad ^{238}_{92}U(n, \beta^-)^{239}_{93}Np$$

EXAMPLE 12.5 Nuclear Equations

Many neutrons, α particles, and other particles are produced in reactions between cosmic rays and nuclei near the top of the Earth's atmosphere. Most of the neutrons produced then react with ^{14}N to produce ^{14}C, as shown by the shorthand notation

$$^{14}N(n, p)^{14}C$$

Write the nuclear equation for this reaction.

The shorthand notation indicates that the bombarding particle is a neutron, $^{1}_{0}n$, and the product particle is a proton, $^{1}_{1}H$. The reactant nuclide is $^{14}_{7}N$ and the product nuclide is $^{14}_{6}C$. The complete nuclear equation is

$$^{14}_{7}N + ^{1}_{0}n \longrightarrow ^{14}_{6}C + ^{1}_{1}H$$

EXERCISE Write the shorthand notation for the equation

$$^{242}_{96}Cm + ^{4}_{2}He \longrightarrow ^{245}_{98}Cf + ^{1}_{0}n$$

Answer $^{242}Cm(\alpha, n)^{245}Cf$.

12.12 RADIATION AND MATTER

During the decay of a radioactive nuclide, mass is lost and energy is released. The energy is given up mainly as the kinetic energy of the particles and nuclei produced in the reaction. These particles and nuclei collide with the atoms and molecules of their surroundings, gradually losing energy. What happens to the atoms and molecules of the irradiated matter depends on several things, including the state and type of matter, the energy of the radiation, and the type of radiation. Valence electrons may be knocked out of atoms and molecules to produce ions, or they may be pushed up to excited states and then emit x-rays as they return to the ground state. Highly reactive atoms or molecules with unpaired electrons may be produced. In some cases, anions are formed or chemical bonds are broken. Electromagnetic radiation (x-rays and γ rays) and particle radiation (α rays, β rays, neutrons, etc.) from radioactive sources are collectively referred to as *ionizing radiation*. The study of the interaction of ionizing radiation with matter is the concern of *radiation chemistry*.

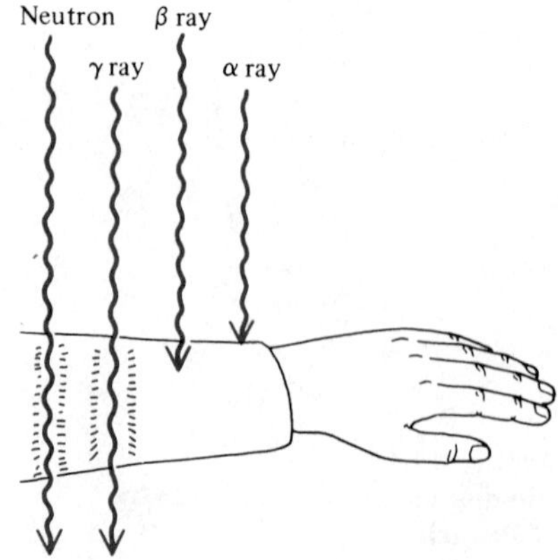

Figure 12.11
Penetration of Radiation

The distance that radiation travels depends upon the medium that it encounters and the type of radiation. γ Rays are highly penetrating, and neutrons even more so. Aluminum 5 to 11 cm thick is required to stop most γ rays, which can penetrate deeply into the human body or even pass through it. β Rays are considerably less penetrating than γ rays. In air, β rays can travel for several meters, but they penetrate only a few millimeters into human tissue. α Rays, with their heavier, doubly charged particles, are the least penetrating natural radiation. Typical α particles travel only a few centimeters in air and are stopped by a sheet of paper or a layer of clothing (Figure 12.11).

Table 12.5
Radiation Units

Unit	Quantity measured	Definition of unit
Roentgen (R)	Exposure to x-rays or γ rays	Radiation that produces 2×10^9 ion pairs in 1 cm^3 of dry air at normal temperature and pressure
Rad (radiation absorbed dose)	Energy absorbed by tissue	1×10^{-5} J absorbed per gram of tissue
Gray (Gy) (SI unit)	Energy absorbed by tissue	1 J/kg tissue (1 gray = 100 rad)
RBE (relative biological effectiveness)	Factor relating effects of different types of radiation	Comparison to γ rays from cobalt-60
Rem (roentgen equivalent man)	Dose to human beings	Rems = rads × RBE (1 rem = same effect in man as 1 roentgen)
Sievert (SI unit)	Dose to human beings	1 sievert = 100 rem
Curie (Ci)	Activity	3.7×10^{10} disintegrations/s
Becquerel (Bq) (SI unit)	Activity	1 disintegration/s
Linear energy transfer (LET)	Energy transferred	Energy lost per unit path length (varies with material irradiated)

If radiation originates outside the body, γ radiation and neutron radiation are most hazardous; the body is easily shielded from α particles and β particles. Neutron radiation is very damaging because the neutrons, being uncharged, collide with nuclei. High-energy neutrons cause recoil of nuclei that can break bonds; lower energy neutrons are captured to give radionuclides that decay further. If radioactive material is created in the body or enters the body through a wound or by being inhaled or swallowed, α particles become more dangerous. Internally, they give up all of their energy to the tissue in a very small distance, causing great damage.

A variety of units are used to measure the interaction of radiation with matter. For reference they are summarized in Table 12.5. The roentgen measures x- or γ radiation by the number of ions it produces while passing through the air. Geiger counters (Figure 12.12), commonly used to detect the presence of radioactive materials, are often calibrated in roentgens.

Comparison of the amounts of radiation to which human beings are exposed is best done in terms of the rem (roentgen equivalent man). This unit takes into account not only the amount of energy delivered by the radiation, but also the variation in the biological effects of different types of radiation.

Everyone is exposed to radiation from the environment. Those who live at high elevations or travel often in jet planes receive more exposure to cosmic rays than those who pass their lives at sea level. Uranium and its decay products are present in natural building materials, and living in a stone or brick house increases radiation exposure compared to living in a wooden house. Smoking one pack of cigarettes a day yields 40 mrem/yr of exposure to radiation from radioactive isotopes that concentrate

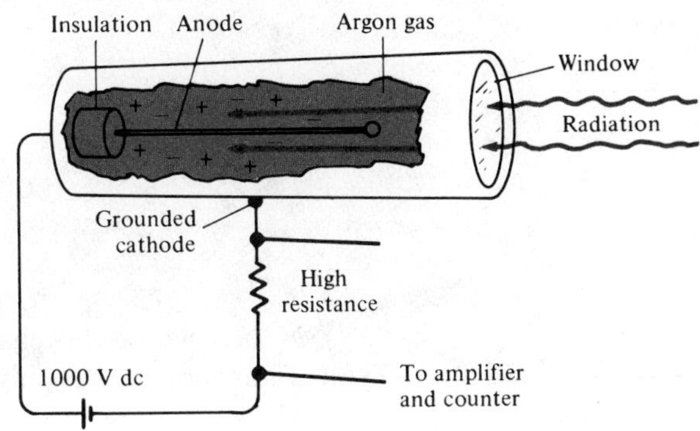

Figure 12.12
Geiger Counter
The radiation ionizes the argon gas in the tube and the resulting electric current is measured.

in tobacco leaves. Potassium is an essential element for plants and we take in small amounts of potassium-40 with all foods that contain potassium. (We note in passing that Brazil nuts and caviar have been reported to contain abnormally high levels of radioactive isotopes.) Table 12.6 summarizes radiation exposure as averaged over the whole population of the United States.

How seriously an organism is damaged by ionizing radiation depends on many factors and varies from individual to individual. An irradiated organism or plant may undergo *somatic effects*—changes in its own cell structure, immediate or delayed, that may be damaging but will not be passed on to future generations. It may also suffer *genetic effects*—changes in the genes that will produce physical changes in future generations.

Intense short-term exposure to radiation, as in an accident with radioactive materials, leads to immediate somatic damage. A 25 rem exposure, about the lowest level detectable by blood tests, raises the white blood cell count. In a few months the blood will return to normal. With exposure to 200 rem, nausea and fatigue are induced and infection-fighting capability is reduced. This level of radiation can cause death in six weeks or so. A dose of 400 rem will be fatal to 50% of those exposed by causing damage to the bone marrow and spleen.

Radionuclides that are incorporated into the body cause damage by exposure over a long time span. For example, cancer may be initiated by radioactive iodine that concentrates in the thyroid gland, or by radioactive calcium, strontium, and radium that concentrate in bones. At one time, luminescent radium-containing paint was widely used in watch and instrument dials. However, the radiation from radium is hazardous and other materials have been substituted for it. In the 1920s, many women painting watch dials developed bone cancer because they were in the habit of pointing the tips of their paint brushes by licking them.

Longer term exposure to low levels of background radiation can lead to mutations—changes in the structure of the DNA molecule that carries the code of heredity. What percentage of mutations are due to the background radiation is difficult to estimate, as there are other natural causes of such effects.

Figure 12.13 is a generalized summary of the possible effects of radiation exposure on plants and/or animals.

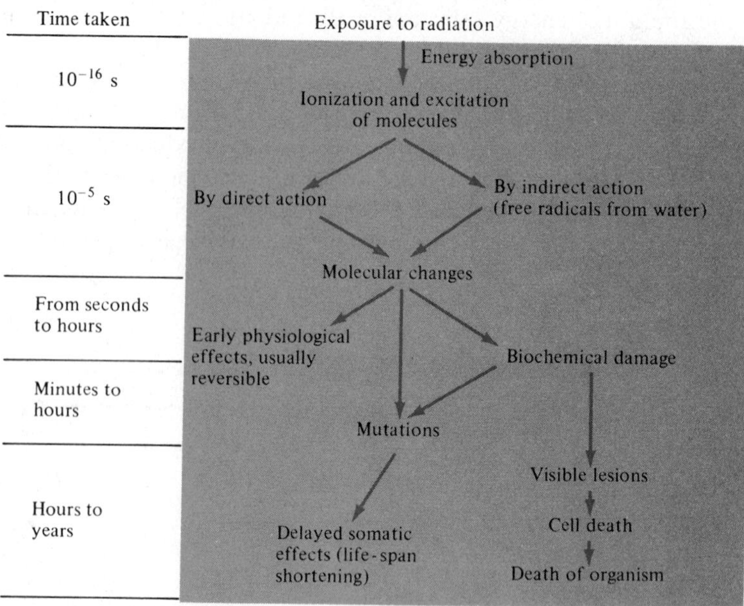

Figure 12.13
Effects of Radiation Exposure
The period of time following radiation exposure during which various effects, depending on the degree of exposure, can occur. Molecular changes in cells are caused both directly by radiation and indirectly by free radicals formed by irradiation of water molecules. Source: P. N. Tiwari, *Fundamentals of Nuclear Science* (New Delhi, India: Wiley Eastern Limited, 1974), p. 131.

Aside 12.2 USES OF RADIOISOTOPES

Each distinctive property of radioisotopes has been put to use in dozens of ways in fields ranging from chemical synthesis to medicine, to agriculture, to manufacturing and engineering.

Many applications arise simply because a radioisotope is continuously emitting radiation that can be detected and measured, making it possible to monitor the location of the isotope. Radioactive tracers have helped to follow the course of underground water supplies, and a large number of isotopes are used as diagnostic tools in medicine. Specific isotopes find their way in the body to specific organs, and by emitting radiation from these locations, make the organ visible to a radiation-detecting camera. For example, thallium-201 concentrates in normal heart muscle, allowing determination by the dark areas on the film of the amount of damage caused by a heart attack. Complementary information is obtained with technetium-99m, which is taken up by damaged heart cells, but not by healthy ones.

One of the first applications of radioisotopes was based on the ability of radiation to destroy cells—radiation therapy for cancer, in which the malignant cells are wiped out. In another application of this property, male insects have been irradiated to destroy their reproductive cells. For instance, millions of sterile flies were released near cattle farms, where they mated with female flies, who then produced no offspring. In this way, the screwworm fly, which does great damage to livestock, has been eliminated in several areas.

In chemistry and biochemistry, radioactive labeling of compounds has revealed much information about how reactions take place. By determining in which of the reaction products the radioactive atom winds up, the route of the atom through intermediate chemical and metabolic processes can be determined. For example, use of a radioactive isotope of oxygen as a "tracer" revealed that, contrary to what had been believed, protein molecules in the liver, although their number remains unchanged, are constantly being synthesized and destroyed at equal rates.

The heat generated by radioactive decay can be converted to electricity in nuclear-powered batteries for use in remote locations, such as the moon. Plutonium-238 powers a tiny battery used in heart pacemakers implanted in the human body. The long life of the isotope allows the battery to function for almost 10 years before an operation must be performed to replace it.

The regular variation in the penetration by radiation of materials of various thicknesses has found many industrial and engineering applications. The thickness of a film of plastic passing over a set of rollers as it is formed can be monitored by radiation passing through the film. This information is then fed back electronically to control the machinery and regulate the thickness of the product.

Neutron activation analysis is a valuable technique for measuring the concentration of small amounts of elements. The great advantage of neutron activation analysis is that the sample is not destroyed in the process. For example, irradiating an old Dutch painting with neutrons forms silver-110m from the traces of silver in white pigment. The amount of silver present can be found from the γ-ray spectrum of the isotope. A change in silver extraction methods in the mid-nineteenth century decreased the amount of silver impurity in the white pigment made after that time. Therefore, the amount of silver-110m present in the pigment reflects the date when a painting was made.

Table 12.6
Background Radiation
Estimated annual average whole-body dose rates from all sources of ionizing radiations in the United States.

Source	Dose (mrem/yr)
Natural background	
Cosmic rays	30
Terrestrial radiation	
External	50
Internal	25
Artificial sources	
Fallout from weapons tests	10–15
Medical and dental	150–300
Occupational	5
Nuclear power generation	5
	275–430

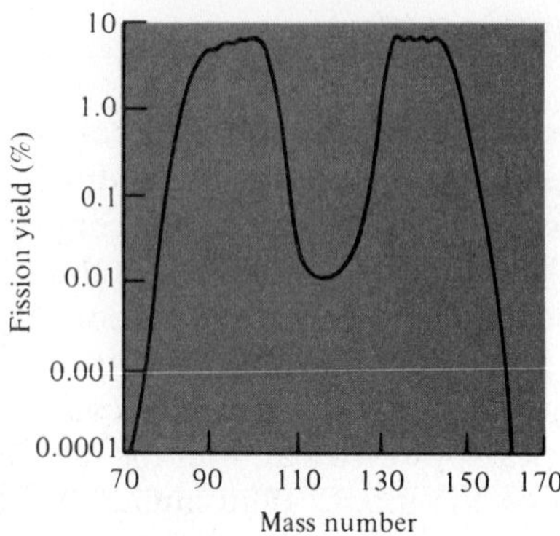

Figure 12.14
Distribution of Isotopes in the Slow-Neutron-Induced Fission of Uranium-235
On the average, 2.5 neutrons and 200 MeV per fission are also produced.

12.13 NUCLEAR FISSION

In a single nuclear fission event, one atom of a heavy fissionable isotope splits into two atoms of intermediate mass and several neutrons. Fission can be initiated by bombardment with many types of particles or by γ radiation, and many heavy nuclides can undergo fission. Some do so spontaneously. Most commonly, nuclear fission reactions are initiated by bombardment with what are called "slow neutrons"—neutrons that move slowly enough to be captured by the target nuclei rather than causing them to break up immediately.

The history of the discovery of nuclear fission is intertwined with the history of World War II (see Aside 12.3). Uranium-235 and plutonium-239 were the fissionable isotopes used in the first atomic bombs. Uranium-235 fission produces a wide array of nuclides, with many of mass numbers near 95 and 139 (Figure 12.14), together with varying amounts of energy. Most of the nuclides produced are radioactive and continue to decay through as many steps as are necessary to reach a stable neutron–proton ratio. These nuclides provide an ongoing source of radiation in the environment.

A series of reactions in which the products of the reactions initiate other similar reactions, so that the series can be self-sustaining, is called a **chain reaction.** The single fact of greatest significance for the use of nuclear energy is that fission produces more than one neutron per fission event. Enough neutrons are thus available to keep the fission going. Nuclear fission becomes a self-sustaining chain reaction when the number of neutrons emitted equals or is greater than the number of neutrons absorbed by fissioning nuclei plus those lost to the surroundings. The **critical mass** of a fissionable material is the smallest mass that will support a self-sustaining chain reaction under a given set of conditions.

Critical mass is affected by many factors, including the geometry of the sample of fissionable material. In the first atomic bomb to be exploded (see Aside 12.4), a sphere of plutonium was compressed to a critical mass by the force of exploding TNT. The rapid chain reaction that is set off by assembly of a critical mass continues until the force of the explosion disperses the fuel (Figure 12.15a).

By controlling the rate of a chain reaction, energy can be produced slowly enough to allow its use for constructive rather than destructive purposes. The term **nuclear reactor** is usually applied to the equipment in which fission is carried out at a controlled rate (Figure 12.15b).

12.14 NUCLEAR FUSION

So far, continuous nuclear fusion has been observed only in the sun and other stars. The conversion of hydrogen to helium in the sequence of fusion reactions shown in Table 12.7 is the main source of solar energy.

Nuclei must approach each other with a large amount of energy if they are to overcome the electrostatic repulsion between their positive charges and be able to fuse. Particles can be given enough energy in an accelerator to fuse with target atoms, and fusion was known and understood several years before the discovery of fission. However, in an accelerator bombardment reaction, only a random few atoms collide and fuse, while a large amount of energy is expended by the accelerator. Useful extraction of energy from nuclear fusion is not possible under these conditions.

The temperature in the center of the sun is about 1.5×10^7 K. At such a high temperature, atoms are stripped of electrons, forming a *plasma*—a neutral mixture of ions and electrons at a high temperature. Within a dense, very hot plasma, light nuclei are moving fast enough for large numbers of them to collide and fuse, releasing large amounts of energy in the process. Nuclear fusion reactions, which occur at very high temperatures (roughly $> 10^6$ K), are called **thermonuclear reactions.**

Table 12.7
Nuclear Reactions in the Stars

Overall reaction:
$4\,^1_1\text{H} \longrightarrow \,^4_2\text{He} + 2\,^0_{+1}e + 26.7\text{ MeV}$
Individual reactions:
$^1_1\text{H} + \,^1_1\text{H} \longrightarrow \,^2_1\text{H} + \,^0_{+1}e$
$^2_1\text{H} + \,^1_1\text{H} \longrightarrow \,^3_2\text{He} + \gamma$
$^3_2\text{He} + \,^3_2\text{He} \longrightarrow \,^4_2\text{He} + 2\,^1_1\text{H}$

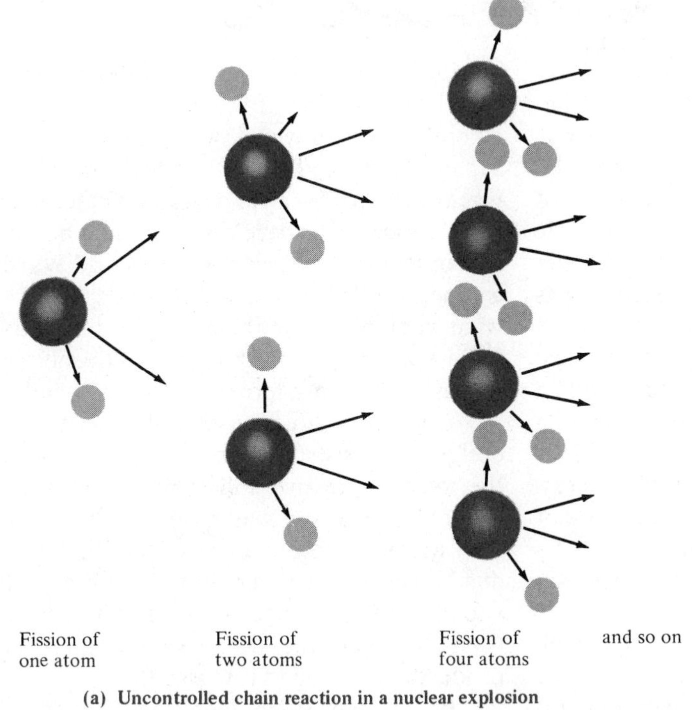

(a) Uncontrolled chain reaction in a nuclear explosion

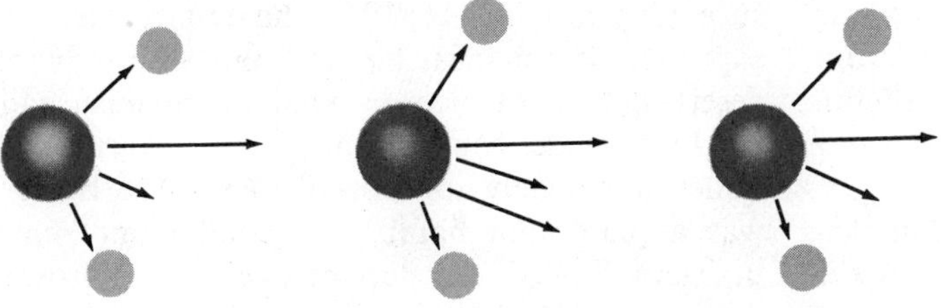

(b) Controlled nuclear reaction in a nuclear reactor

Figure 12.15
Chain Reactions
(a) In an uncontrolled chain reaction, the fission of each atom releases enough high-energy neutrons to initiate the fission of more than one new atom. Fission of all the fuel available takes place rapidly and with explosive force. (b) In a nuclear reactor, the chain reaction is controlled so that each fission initiates the fission of only one more atom and energy is released at a controlled rate.

To produce the temperature needed for thermonuclear fusion on Earth once seemed impossible. But man-made nuclear fission, in the form of the atomic bomb, provided the match needed to light a large-scale fusion reaction. The result was the "H bomb" first tested at Bikini atoll in the Pacific Ocean in 1952.

In very general terms, a hydrogen, or thermonuclear, bomb can be pictured as a fission bomb surrounded by the compound formed by deuterium (hydrogen-2) and lithium-6, ${}^{6}_{3}Li{}^{2}_{1}H$. The sequence of reactions is

$$\text{fission bomb} \longrightarrow \text{heat} + \text{neutrons} \quad (12.14)$$

$${}^{6}_{3}\text{Li} + {}^{1}_{0}n \longrightarrow {}^{4}_{2}\text{He} + {}^{3}_{1}\text{H} + 4.78\text{ MeV} \quad (12.15)$$

$${}^{2}_{1}\text{H} + {}^{3}_{1}\text{H} \longrightarrow {}^{4}_{2}\text{He} + {}^{1}_{0}n + 17.6\text{ MeV} \quad (12.16)$$

where the energy released is equivalent to the loss in mass in the reaction.

The reaction of lithium and neutrons is used to produce tritium for the fusion reaction between tritium and deuterium. The energy release per nuclear fusion is roughly one-tenth that released per nuclear fission event. (Compare 200 MeV per uranium-235 fission with 17.6 MeV per deuterium–tritium reaction.) However, the yield from a thermonuclear explosion for a given quantity of fuel is much greater than that from a fission explosion for several reasons: because of the smaller mass of lithium and deuterium compared to that of uranium, because new fusionable nuclei are produced in the explosion, and possibly also because a larger proportion of the fuel is consumed. Fortunately, thermonuclear weapons have so far been used only in tests.

Aside 12.3 THE ATOMIC BOMB

At the end of the 1930s, turmoil had come to Europe. Refugees were fleeing troubled countries and national energies were being turned to war and defense.

In 1938 Otto Hahn, a chemist at the Kaiser Wilhelm Institute in Berlin, identified barium in a sample of uranium that had been bombarded by slow neutrons. He dared to write of the discovery to Lise Meitner, a former colleague who had fled from Germany to Sweden. It was Christmas time and Lise Meitner's nephew Otto Frisch, a physicist working at Niels Bohr's Institute of Theoretical Physics in Copenhagen, was visting her. They walked together in the snow near Gothenburg; while sitting on a log they figured out how fission of a heavy nucleus into smaller fragments could release a large amount of energy. Frisch went back to his laboratory, measured the speed of a barium nucleus released in uranium bombardment by neutrons, and found that it was what they had thought it would be—very fast.

Nuclear fission was announced on February 11, 1939. The news caused great excitement among physicists, and about a month later Leo Szilard made clear the implications of the discovery for a world heading for war. Szilard showed that in each fission of a uranium atom, two or three neutrons were released. This meant one thing—a terrible explosive force could be released in a nuclear fission chain reaction.

An extraordinary group of refugees in America, physicists all, drafted a letter to President Franklin Roosevelt. Szilard, Edward Teller, and Eugene Wigner, all Hungarians; Victor Weisskopf, an Austrian; and the brilliant Italian, Enrico Fermi, explained the potential of nuclear fission, and warned that Germany might already be working to build a nuclear bomb. So that their warning would carry more weight, they persuaded Albert Einstein, the world's most famous and respected scientist, to sign the letter.

President Roosevelt was impressed enough by the letter to authorize financial assistance to American universities studying nuclear reactions. The major problem was the production of the fissionable isotopes needed for a bomb. Progress was slow. Then, in December, 1941, the Japanese attacked Pearl Harbor and the United States was at war with Japan as well as with Germany. An all-out crash program to develop an atomic bomb—the Manhattan Project—was launched. World War II had thrust a generation of physicists, chemists, and engineers into what was for many of them a moral dilemma—rushing to build a weapon that, while certain to end the war, would unleash a force too terrible to comprehend.

Under the direction of J. Robert Oppenheimer the design and construction of the bomb was undertaken by the most impressive collection of scientific talent ever assembled. The site was a ranch school in Los Alamos, New Mexico. A group of physicists that included numerous Nobel prize winners found themselves facing a new and different challenge. As one put it, they had to change from "probing the deep-seated secrets of nature to producing a device that worked."

Proceeding at an increasingly feverish pace, they succeeded. On July 16, 1945, the Trinity Test—the first explosion of a nuclear fission bomb—took place in the desert in New Mexico, in what the Spanish had called the Jornada del Muerto Valley—the Journey of Death Valley, a strangely appropriate name. The Trinity device was a plutonium bomb. A second bomb made from uranium leveled the Japanese city of Hiroshima on August 6, 1945, six years almost to the day after delivery of the letter to President Roosevelt. A third bomb was dropped on Nagasaki on August 9, 1945. Five days later Japan surrendered.

Many benefits to mankind have developed from the application of radioactivity in medicine and industry, and from the utilization of nuclear energy for power. But the fear persists that the horror of nuclear weapons may fade from the minds of those who control their possible use.

SUMMARY OF SECTIONS 12.8–12.14

In nuclear reactions—which result in changes in atomic number, mass number, or energy states of nuclei—mass number and nuclear charge are conserved. Spontaneous decay of radionuclides generally occurs by α decay or β decay, which may be electron emission, positron emission, or electron capture. γ Decay permits nuclei in excited states to give up energy. Nuclides that fall above the band of stability (Figure 12.5) undergo electron emission (net effect, $n \longrightarrow p$); those that fall below it undergo positron emission ($p \longrightarrow n$), electron capture ($p \longrightarrow n$), or α decay (the heavier nuclides). Natural radioactive nuclides are members of three series with long-lived parents and stable lead isotopes as the end products. Table 12.8 summarizes the reactions of spontaneous radioactive decay.

In bombardment reactions, moving particles of high energy combine with target nuclei, temporarily forming unstable nuclei that subsequently decay. Most transuranium elements ($Z > 92$) have been synthesized in bombardment reactions.

Radiation varies in the degree of penetration into matter and the degree of damage that can be done to living things. Neutrons and γ rays are the most harmful radiation from external sources because they are most penetrating. α Particles are very harmful when they come from an internal source, for they give up all their energy to surrounding tissue. Collectively, radiation that causes damage to tissue is referred to as ionizing radiation. Both somatic and genetic damage can result from exposure to ionizing radiation, depending upon the intensity and duration of the exposure.

Nuclear fission is the splitting of a fissionable nucleus into two lighter nuclei. The fission of ^{238}U, initiated by thermal neutrons, becomes a self-sustaining chain reaction when a critical mass of material is present. Uncontrolled nuclear fission results in ex[illegible] nuclear reactors, fission is carried out under controlled conditions. Nuclear fusion (which occurs in the sun and stars, and in thermonuclear weapons) is the combination of lighter nuclei to form heavier nuclei; the initiation of fusion requires very high temperatures.

Table 12.8
Spontaneous Radioactive Decay
An m following the mass number superscript, as in the isomeric transition reaction, indicates an excited, or isomeric, state of an isotope. (Sometimes an asterisk is used to designate an excited state.)

Reaction	Reacting nuclide	Reaction symbol	Change in A	Change in Z	Change in n/p	Example
γ Decay (isomeric transition)	Excited nuclear state	IT	0	0	—	$^{77m}_{34}Se \longrightarrow ^{77}_{34}Se + \gamma$ ray
Spontaneous fission	Too heavy ($Z > 92$)	SF	—	—	—	$^{254}_{98}Cf \longrightarrow$ intermediate mass nuclides and neutrons
α Decay	$Z > 83$; n/p too low	α	−4	−2	Increase	$^{210}_{84}Po \longrightarrow ^{206}_{82}Pb + ^{4}_{2}He$
β Decay						
Electron emission	n/p too high	β^-	0	+1	Decrease	$^{227}_{89}Ac \longrightarrow ^{227}_{90}Th + ^{0}_{-1}e$
Positron emission	n/p too low	β^+	0	−1	Increase	$^{13}_{7}N \longrightarrow ^{13}_{6}C + ^{0}_{+1}e$
Electron capture	n/p too low	EC	0	−1	Increase	$^{73}_{33}As + ^{0}_{-1}e \longrightarrow ^{73}_{32}Ge$

NUCLEAR ENERGY

12.15 FISSION REACTORS

Virtually all commercial nuclear power in the United States is generated in light water reactors of two types, either pressurized water or boiling water reactors. A nuclear fission reactor cooled by light water has five essential parts (Figure 12.16):

1. Fuel The fuel is a core of fissionable material. Light water reactors use enriched uranium oxide, U_3O_8, in which the natural 0.7% of uranium-235 has been raised to the 2%–3% uranium-235 needed for efficient operation. The fuel is formed into solid elements with a protective coating.
2. Moderator The fast neutrons produced in fission must be slowed down to speeds at which they produce the fission reaction most efficiently. Neutrons must be able to lose energy by colliding with the nuclei of the moderator, without being absorbed by or reacting with the moderator. Materials with a

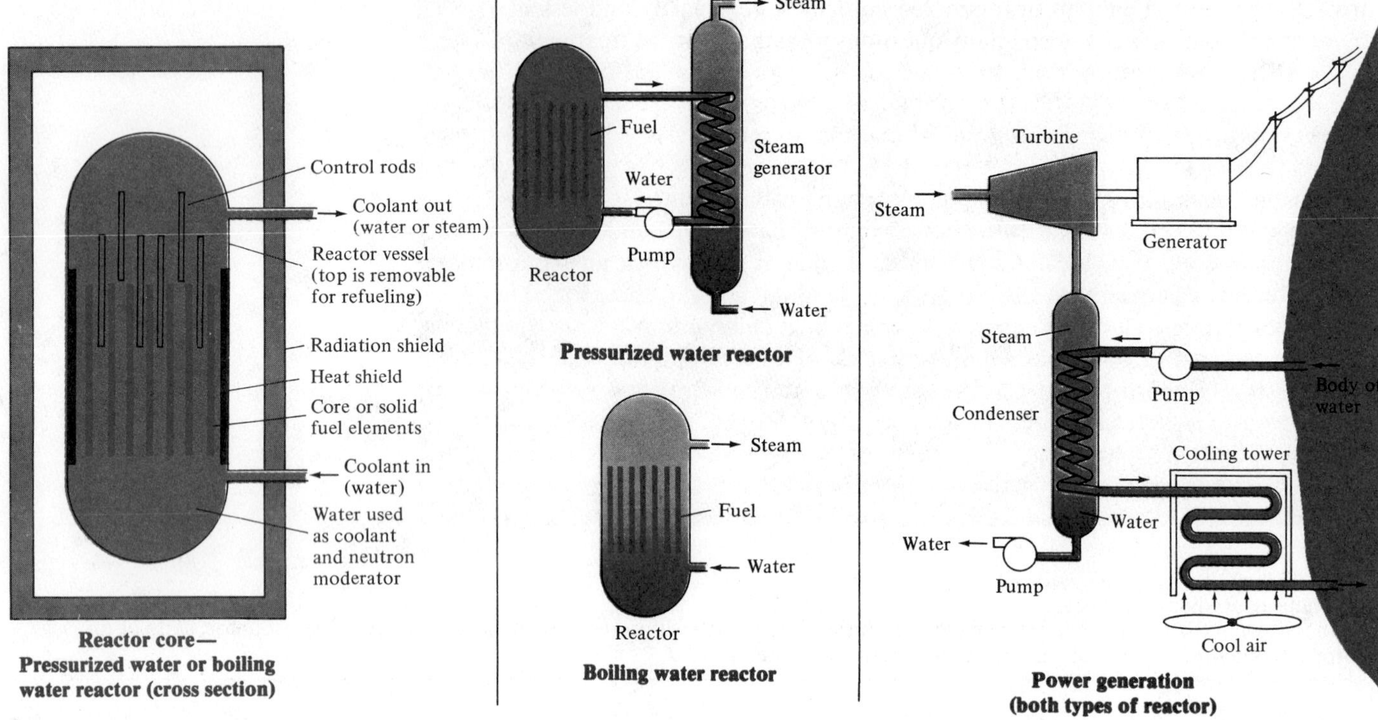

Figure 12.16
Light Water Nuclear Reactors
Both boiling water and pressurized water reactors generate electricity from steam in the same way electricity is generated in a coal-fired power station. Note that the coolant is in a closed system and does not come in contact with the outside body of water. Both nuclear and coal-fired power stations need cooling water; the thermal pollution (temperature rise in a natural body of water used for cooling) is somewhat greater for nuclear reactors.

large proportion of hydrogen atoms are good moderators. Light water reactors are so called because they use ordinary water as the moderator (as opposed to "heavy" water, D_2O). Graphite is also a good moderator.

3. Control system Just enough free slow neutrons are needed to carry on the chain reaction at a safe rate (see Figure 12.15). If too many neutrons initiate fission, more heat would build up than could be carried away. This condition could lead to the most serious type of nuclear reactor accident, a meltdown, in which the fuel core, and eventually its container, would melt.

   In a light water reactor control rods containing boron are raised and lowered in between the fuel elements so that unneeded neutrons are removed by the reaction

   $$^{10}_{5}B + ^{1}_{0}n \longrightarrow ^{7}_{3}Li + ^{4}_{2}He$$

   Enough control rods are available to completely stop the reaction by lowering all of them into the spaces between the radioactive rods.

4. Cooling system The energy of the fission reaction must be carried away for transformation into electrical power and to keep the reactor from overheating as well. In *pressurized water reactors,* liquid water under pressure is the coolant; in *boiling water reactors* steam is the coolant (see Figure 12.16).

5. Shielding Both the walls of the reactor and the personnel operating the reactor must be protected from heat and radiation. In addition, the entire nuclear reactor is enclosed in a heavy steel or concrete dome that is intended to contain the radioactive materials that might be set free in a serious reactor accident.

Some major technological problems facing the present-day nuclear power industry lie in the processing of fuel. The enrichment of natural uranium to give a product with a high enough uranium-235 content is very costly and less expensive methods are being sought. Also, many questions remain about the treatment of spent fuel once it is removed from reactors—reprocessing may be possible but has not yet been achieved economically. Ultimately, waste radioactive material must be stored; however, for both technical and political reasons no method for permanent storage has yet been agreed upon.

It is interesting to note that the first nuclear fission reactor was most likely not man-made. In 1972 it was discovered that apparently between 1.7 and 1.9 million years ago in what is now an open-pit uranium mine at Oklo in West Africa, an extraordinary series of events occurred that created a natural fission reactor. Concentrated uranium deposits containing a greater than usual abundance of uranium-235 are thought to have accumulated as water washed into the area. A seam of uranium thick enough to sustain a flow of neutrons and a natural mechanism for moderating, or slowing down, the neutrons (possibly by reaction with the hydrogen atoms of water), apparently were available. The existence of the natural reactor was proposed to explain the significantly lower uranium-235 content of the ore in the region today.

A **breeder reactor** can produce at least as many fissionable atoms as it consumes. The primary advantages of a breeder reactor are that it is able to convert uranium-238 or thorium-232, both naturally more plentiful than uranium-235, into fissionable fuel, and that it is thermally more efficient than the light water reactors because it can operate at higher temperatures. By using breeder reactors and also recycling fuel, up to 70% of the available energy might be set free. The excess fissionable artificial isotopes (uranium-233 or plutonium-239) produced in a breeder can be used as fuel for other reactors. Prototype breeder reactors are in operation in many countries. However, their possible widespread use raises serious questions. First, do we need the power and fuel economy of the breeders enough to risk having to handle large amounts of plutonium-239, a material so poisonous that inhaling a tiny speck of it can be fatal? Second, how great a risk is incurred by making available plutonium-239, which is much more readily used in nuclear weapons than enriched uranium?

12.16 FUSION REACTORS

The problem of igniting, controlling, and maintaining a fusion reaction in a power plant has been called the greatest technological challenge taken up by man. Light atoms must be held together in a vacuum at high density, at a high enough temperature, and for a long enough time for the thermonuclear fusion to take place.

Confinement of nuclear fusion in a vessel is not possible for two reasons: (1) the reacting nuclei and electrons, which have such high energy that they are traveling at several hundred miles per second, would lose their energy as soon as they struck the walls and before fusion could occur; (2) in striking the walls the particles would knock atoms out of the walls and these impurities would contribute to a loss of energy in the reaction system. Two major approaches to containing nuclear fusion are being investigated—magnetic confinement and inertial confinement.

If a large enough current is passed through a plasma parallel to the axis of a magnetic field, the plasma is pinched in and pulled away from the walls of the container, forming a so-called magnetic bottle. Plasma behavior in such a field has turned out to be very complex, but gradually it is being understood, and experiments are coming closer to plasma confinement under the conditions needed for fusion.

The basic concept of inertial confinement is that a small sphere of fuel is made to react before it has time to fly apart. In laser fusion, a powerful laser beam strikes a sphere of fuel with such energy that some of the surface material is vaporized. (A laser is a device that produces an intense beam of coherent radiation—radiation of a single wavelength with all of the waves in step with one another.) The resulting shock wave compresses the fuel at the center of the sphere, both bringing the nuclei close enough together and raising the temperature to the point where fusion can occur (Figure 12.17).

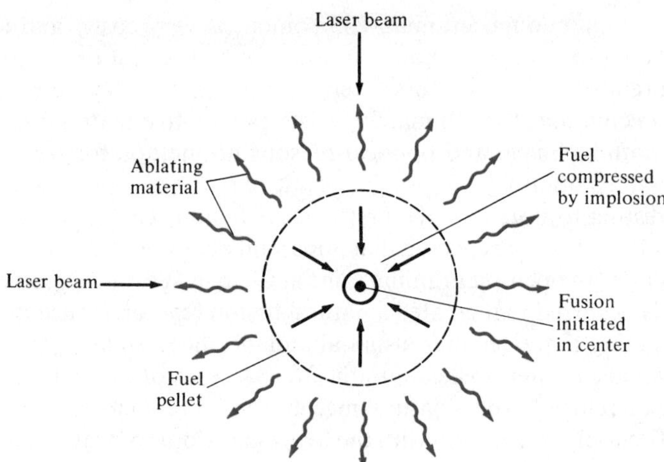

Figure 12.17
Nuclear Fusion by Inertial Confinement in a Laser Beam
The synchronized laser beam rapidly heats the pellet. In ablating, or burning away of the surface, the center of the pellet is compressed to a pressure great enough to cause fusion (10^{12} atm).

The fusion of deuterium and tritium to give helium (the reaction used in the hydrogen bomb; Equation 12.16) is the prime candidate for the major reaction in thermonuclear power plants (Table 12.9). Deuterium is available in huge quantities in the oceans and obtaining it would be easy and relatively inexpensive. Two reaction schemes have been proposed for fusion reactors, one involving lithium as a breeder for tritium, as in reactions (12.15) and (12.16), and the other involving only deuterium:

$$\begin{array}{lcl} {}^2_1\text{H} + {}^2_1\text{H} & \longrightarrow & {}^3_1\text{H} + {}^1_1\text{H} + 4.0\text{ MeV} \\ {}^2_1\text{H} + {}^2_1\text{H} & \longrightarrow & {}^3_2\text{He} + {}^1_0n + 3.3\text{ MeV} \\ {}^2_1\text{H} + {}^3_1\text{H} & \longrightarrow & {}^4_2\text{He} + {}^1_0n + 17.6\text{ MeV} \\ \hline 5\,{}^2_1\text{H} & \longrightarrow & {}^4_2\text{He} + {}^3_2\text{He} + {}^1_1\text{H} + 2\,{}^1_0n + 24.9\text{ MeV} \end{array}$$

Fusion reactors would have a number of advantages over fission reactors, including (1) the lower cost and unlimited supply of the fuel; (2) the worldwide availability of the fuel, thereby avoiding international tensions over obtaining it; (3) the potential for production of energy at a lower cost; and (4) the production of by-products that are not radioactive.

Table 12.9
Some Fusion Reactions
The energy ratio compares the potential of the reactions for use in commercial energy production. The deuterium–tritium reaction is most favorable. The other combinations are spoken of as "advanced fuels."

Reaction	Energy produced (keV)	Energy required for ignition (keV)	Energy ratio (energy produced/ energy required)
${}^2_1\text{H} + {}^3_1\text{H} \longrightarrow {}^4_2\text{He} + {}^1_0n$	17,600	10	1,760
${}^2_1\text{H} + {}^2_1\text{H} \longrightarrow {}^3_2\text{He} + {}^1_0n$	3,300	50	66
${}^2_1\text{H} + {}^2_1\text{H} \longrightarrow {}^3_1\text{H} + {}^1_1p$	4,000	50	80
${}^2_1\text{H} + {}^3_2\text{He} \longrightarrow {}^4_2\text{He} + {}^1_1p$	18,300	100	183
${}^1_1p + {}^{11}_5\text{B} \longrightarrow 3\,{}^4_2\text{He}$	8,700	300	29

SUMMARY OF SECTIONS 12.15–12.16

In a nuclear fission reactor, neutrons produced by reaction of the uranium-235 fuel are slowed by a moderator (often D_2O or graphite) to speeds at which they produce fission most effectively. A breeder reactor produces as much new fissionable material as it consumes, or more. It is difficult to confine the reacting nuclei and electrons, and to sustain the high temperatures needed for controlled nuclear fusion. The two most promising approaches to utilizing nuclear fusion as a source of energy involve magnetic confinement of the reacting materials, or the use of lasers.

Aside 12.4 THOUGHTS ON CHEMISTRY: THE ATOMIC AGE BEGINS

William L. Laurence, a journalist, was invited to view the first atomic bomb explosion—the test explosion in the desert in New Mexico on July 16, 1945. His description of what he saw, from his book *Men and Atoms*, is reprinted here.

Suddenly, at 5:29:50, as we stood huddled around our radio, we heard a voice ringing through the darkness, sounding as though it had come from above the clouds: "Zero minus ten seconds!" A green flare flashed out through the clouds, descended slowly, opened, grew dim, and vanished into the darkness.

The voice from the clouds boomed out again: "Zero minus three seconds!" Another green flare came down. Silence reigned over the desert. We kept moving in small groups in the direction of Zero. From the east came the first faint signs of dawn.

And just at that instant there rose as if from the bowels of the earth a light not of this world, the light of many suns in one. It was a sunrise such as the world had never seen, a great green supersun climbing in a fraction of a second to a height of more than eight thousand feet, rising even higher until it touched the clouds, lighting up earth and sky all around with a dazzling luminosity.

Up it went, a great ball of fire about a mile in diameter, changing colors as it kept shooting upward, from deep purple to orange, expanding, growing bigger, rising as it expanded, an elemental force freed from its bonds after being chained for billions of years. For a fleeting instant the color was unearthly green, such as one sees only in the corona of the sun during a total eclipse. It was as though the earth had opened and the skies had split.

A huge cloud rose from the ground and followed the trail of the great sun. At first it was a giant column, which soon took the shape of a supramundane mushroom. Up it went, higher and higher, quivering convulsively, a giant mountain born in a few seconds instead of millions of years. It touched the multicolored clouds, pushed its summit through them, and kept rising until it reached a height of 41,000 feet, 12,000 feet higher than the earth's highest mountain.

All through the very short but long-seeming time interval not a sound was heard. I could see the silhouettes of human forms motionless in little groups, like desert plants in the dark. The newborn mountain in the distance, a giant among the pygmies of the Sierra Oscuro range, stood leaning at an angle against the clouds, like a vibrant volcano spouting fire to the sky.

Then out of the great silence came a mighty thunder. For a brief interval the phenomena we had seen as light repeated themselves in terms of sound. It was the blast from thousands of blockbusters going off simultaneously at one spot. The thunder reverberated all through the desert, bounced back and forth from the Sierra Oscuro, echo upon echo. The ground trembled under our feet as in an earthquake. A wave of hot wind was felt by many of us just before the blast and warned us of its coming.

The big boom came about a hundred seconds after the great flash—the first cry of a newborn world.

Source: William L. Laurence, *Men and Atoms.* pp. 116, 117.

SIGNIFICANT TERMS (with section references)

α decay (12.9)
activity (radioactive) (12.5)
β decay (12.9)
band of stability (12.4)
binding energy per nucleon (12.2)
bombardment reactions (12.8)
breeder reactor (12.15)
chain reaction (12.13)
critical mass (12.13)
electron capture (12.9)
γ decay (12.9)
half-life (12.5)
isomeric transition (12.9)
magic numbers (12.7)
mass defect (12.2)
nuclear binding energy (12.2)
nuclear chemistry (12.2)
nuclear fission (12.2)
nuclear force (12.1)
nuclear fusion (12.2)
nuclear reactions (12.8)
nuclear reactor (12.13)
nucleons (12.1)
nuclide (12.1)
positron (12.9)
radioactivity (12.3)
radionuclides (12.3)
thermonuclear reactions (12.14)

QUESTIONS AND PROBLEMS

Nuclear Stability and Radioactivity

12.1 What are nucleons? What is the relationship between the number of protons and the atomic number? What is the relationship among the number of protons, the number of neutrons, and the mass number?

12.2 Describe what happens to the energy of the system as a proton approaches another proton until they are less than 10^{-13} cm apart. Repeat your description for a neutron approaching a proton and for a neutron approaching a neutron.

12.3 What is the nuclear binding energy? How is this energy calculated?

12.4 Define the term "binding energy per nucleon." How can this quantity be used to compare the stabilities of nuclei?

12.5 Calculate the nuclear binding energy for each of the following: (a) $^{14}_{7}N$, (b) $^{56}_{26}Fe$, (c) $^{130}_{52}Te$. This respective atomic masses are 14.00307 u, 55.9349 u, and 129.9067 u. Which of these nuclides has the largest binding energy per nucleon?

12.6 Repeat Problem 12.5 for (a) $^{20}_{10}Ne$, (b) $^{59}_{27}Co$, (c) $^{106}_{46}Pd$. The atomic masses are 19.99244 u, 58.9332 u, and 105.9032 u, respectively.

12.7 Describe the general shape of the plot of binding energy per nucleon against mass number.

12.8 Briefly describe a plot of the number of neutrons against the number of protons for the stable nuclides.

12.9 Compare the behaviors of α, β, and γ radiation (a) in an electrical field, (b) in a magnetic field, (c) with various shielding materials such as a piece of paper and concrete. What is the composition of each type of radiation?

12.10 Describe how (a) nuclear fission and (b) nuclear fusion generate more stable nuclei.

Nuclear Stability and Radioactivity—Half-Life and Dating

12.11 Why do the various isotopes of an element (such as $^{1}_{1}H$, $^{2}_{1}H$, and $^{3}_{1}H$) undergo the same chemical reactions?

12.12 What does the half-life of a radionuclide represent? How do we compare the relative stabilities of radionuclides in terms of half-lives?

12.13 The half-lives of nuclides that undergo, α, β^-, β^+, and γ decay do not depend on whether the atom is in the free state, is a monatomic ion, or is in a molecule. However, recent experimental results indicate that the half-life of a nuclide that decays by electron capture may be influenced by the chemical environment of the atom. Briefly explain why this might be true.

12.14 Briefly describe radiocarbon dating. What is the source of the radioactive carbon? What is the limit of its use?

12.15 The half-life of $^{19}_{8}O$ is 29 s. What fraction of the isotope originally present would be left after 5.0 s?

12.16 The half-life of $^{11}_{6}C$ is 20.3 min. How long will it take for exactly 90% of a sample to decay? How long will it take for exactly 99% of the sample to decay?

12.17 The activity of a sample of tritium decreased by 5.5% over the period of a year. What is the half-life of $^{3}_{1}H$?

12.18 The rate constant for the radioactive decay of $^{193}_{81}Tl$ is 1.81 h^{-1}. How many half-lives will be required for a 1.00 g sample of $^{193}_{81}Tl$ to decay to 0.01 g?

12.19 Potassium chloride serves as a common salt substitute in low sodium diets. Potassium-40 ($t_{\frac{1}{2}} = 1.28 \times 10^9$ yr) is a naturally occurring radioisotope that is present to the extent of 1.17% in potassium. (a) Calculate the rate constant for the decay of ^{40}K. (b) Calculate the number of potassium ions in 1.00 g of KCl. (c) Calculate the activity of the KCl sample in terms of the number of K^+ ions disintegrating per second by multiplying the number of K^+ ions by the rate constant.

12.20 A very unstable isotope of beryllium, ^8Be, undergoes α emission with a half-life of 0.07 fs. How long does it take for 99.99% of a 1.0 μg sample of ^8Be to undergo decay? What would be the volume of helium measured at STP generated from the decay of the 1.0 μg sample?

12.21 The $^{14}_{6}C$ activity of an artifact from the tomb of Hemaka (2930 ± 200 BC) was 8.3/(min)(g C). Using the data given in Example 12.3 for $^{14}_{6}C$ dating, determine the age of this artifact.

12.22 A piece of wood from a burial site was analyzed using $^{14}_{6}C$ dating and was found to have an activity of 13.4/(min)(g C). Using the data given in Example 12.3 for $^{14}_{6}C$ dating, determine the age of this piece of wood.

12.23* The activity of a sample of ^{131}I was monitored over a period of time and the following data obtained:

a, cpm	*t*, day	*a*, cpm	*t*, day
10,800	0	2,600	16
9,100	2	1,700	21
7,200	4	280	42
6,000	7	230	44
3,400	13	160	49
3,200	14		

(a) Calculate the respective values of $\log a$ and plot $\log a$ against t. Equation (12.4) implies that the slope of the best straight line through the data points is equal to $-k/(2.303)$. (b) Determine the value of k from your graph. (c) From the value of k, determine the half-life of ^{131}I.

12.24* A current theory suggests that the ratio of $^{235}_{92}U$/$^{238}_{92}U$ was nearly unity at the time of the formation of the elements. The current ratio is 7.25×10^{-3}. The half-life of $^{235}_{92}U$ is 7.1×10^8 yr and of $^{238}_{92}U$ is 4.51×10^9 yr. Calculate when the elements were formed using these data. Hint: (i) solve Equation (12.3) for $\log q_{0,235}$ in terms of $\log q_{235}$, k_{235}, and t; (ii) solve Equation (12.3) for $\log q_{238}$ in terms of $\log q_{0,238}$, k_{238}, and t; (iii) substitute equation (i) into equation (ii) for $\log q_{0,238}$ because $q_{0,238} = q_{0,235}$; (iv) solve equation (iii) for t in terms of k_{235}, k_{238}, and $\log(q_{235}/q_{238})$; (v) substitute the data into equation (iv) to find t.

12.25 Briefly describe a plot of relative cosmic atomic abundance against atomic number.

12.26 What are nuclear "magic numbers"? What are some of the values? What do these numbers actually represent?

Nuclear Reactions—Nuclear Equations

12.27 Write the nuclear equations for the following processes: (a) ${}^{63}_{28}Ni$ undergoing β^- emission, (b) two deuterium ions undergoing fusion to give ${}^{3}_{2}He$ and a neutron, (c) a nuclide being bombarded by a neutron to form ${}^{7}_{3}Li$ and an α particle (identify the unknown nuclide), and (d) ${}^{14}_{7}N$ being bombarded by a neutron to form 3 α particles and an atom of tritium.

12.28 Write the nuclear equations for the following processes: (a) ${}^{228}_{90}Th$ undergoing α decay, (b) ${}^{110}_{49}In$ undergoing positron emission, (c) ${}^{127}_{53}I$ being bombarded by a proton to form ${}^{121}_{54}Xe$ and 7 neutrons, (d) tritium and deuterium undergoing fusion to form an α particle and a neutron, (e) ${}^{95}_{42}Mo$ being bombarded by a proton to form ${}^{95}_{43}Tc$ and radiation (identify this radiation).

12.29 "Radioactinium" is produced in the actinium series from ${}^{235}_{92}U$ by the successive emission of an α particle, a β^- particle, an α particle, and a β^- particle. What are the symbol, atomic number, and mass number for "radioactinium"?

12.30 An alkaline earth element (Representative Group II) is radioactive. It undergoes decay by emitting three α particles in succession. In what periodic table group is the resulting element found?

12.31 A proposed series of reactions (known as the carbon–nitrogen cycle) that could be important in the very hottest region of the interior of the Sun is

$$\begin{aligned} {}^{12}C + {}^{1}H &\longrightarrow A + \gamma \\ A &\longrightarrow B + {}^{0}_{+1}e \\ B + {}^{1}H &\longrightarrow C + \gamma \\ C + {}^{1}H &\longrightarrow D + \gamma \\ D &\longrightarrow E + {}^{0}_{+1}e \\ E + {}^{1}H &\longrightarrow {}^{12}C + F \end{aligned}$$

Identify the species labeled A–F.

12.32 A proposed series of reactions (known as the proton–proton chain) that could be important in the cooler region of the sun is

$$\begin{aligned} {}^{1}H + {}^{1}H &\longrightarrow A + {}^{0}_{+1}e \\ A + {}^{1}H &\longrightarrow B + \gamma \\ 2B &\longrightarrow C + 2{}^{1}H \end{aligned}$$

Identify A, B, C.

Nuclear Reactions—Natural Radioactivity

12.33 Describe what happens to the (i) atomic number, (ii) mass number, (iii) n/p ratio during (a) isomeric transition, (b) α decay, (c) β^- decay. Write a nuclear equation for each of the following processes: (d) ${}^{91m}_{39}Y$ undergoing isomeric transition, (e) ${}^{205}_{84}Po$ undergoing α decay, (f) ${}^{215}_{83}Bi$ undergoing β^- decay.

12.34 Repeat Question 12.33 for (a) β^+ emission and (b) electron capture. (c) Compare the product nuclides formed by ${}^{148}_{63}Eu$ as it undergoes β^+ emission and electron capture.

12.35 Consider a radioactive nuclide with a neutron–proton ratio that is larger than those for the stable isotopes of that element. What mode(s) of decay might be expected for this nuclide and why?

12.36 Repeat Question 12.35 for a nuclide with a neutron–proton ratio that is smaller than those for the stable isotopes.

12.37 Calculate the neutron–proton ratio for each of the following radioactive nuclides and predict how each of the nuclides might decay: (a) ${}^{13}_{5}B$ (stable mass numbers for B are 10 and 11), (b) ${}^{81}_{38}Sr$ (stable mass numbers for Sr are between 84 and 88), (c) ${}^{212}_{82}Pb$ (stable mass numbers for Pb are between 204 and 208).

12.38 Repeat Problem 12.37 for (a) ${}^{193}_{79}Au$ (stable mass number for Au is 197), (b) ${}^{184m}_{75}Re$ (stable mass numbers for Re are 185 and 187), (c) ${}^{142}_{59}Pr$ (stable mass number for Pr is 141).

12.39* Predict the mode of decay for ${}^{159}_{64}Gd$. The energies of the radiation are 0.59 MeV, 0.95 MeV, and 0.89 MeV. The energies of the accompanying γ radiation are 0.362 MeV and 0.058 MeV. Prepare a decay scheme consistent with these data.

12.40* The nuclide ${}^{19}_{8}O$ is radioactive. (a) Predict the mode of decay for this isotope. (b) Particles with a maximum energy of 4.60 MeV are emitted followed by 0.20 MeV γ radiation, along with particles with a maximum energy of 3.25 MeV followed by 1.37 MeV and 0.20 MeV γ radiation. Prepare a decay scheme for this nuclide.

12.41* Calculate the binding energy per nucleon for the following isotopes: (a) ${}^{15}_{8}O$ with a mass of 15.00300 u; (b) ${}^{16}_{8}O$ with a mass of 15.99491 u; (c) ${}^{17}_{8}O$ with a mass of 16.99913 u; (d) ${}^{18}_{8}O$ with a mass of 17.99915 u; (e) ${}^{19}_{8}O$ with a mass of 19.0035 u. Which of these would you expect to be most stable?

12.42* The masses of several nuclides having mass number 56 are 55.940671 u for ${}^{56}Cr$, 55.938906 u for ${}^{56}Mn$, 55.934939 u for ${}^{56}Fe$, 55.939842 u for ${}^{56}Co$, and 55.942134 u for ${}^{56}Ni$. Prepare a plot of binding energy per nucleon for these nuclides and identify the stable nuclide. Assuming that the unstable nuclides will change by either positron emission or electron capture or by electron emission to more stable nuclides, predict the mode of decay for each of the unstable nuclides.

Nuclear Reactions—Bombardment Reactions

12.43 What are bombardment reactions? Explain the shorthand notation used to describe bombardment reactions.

12.44 What particles were used by early investigators for bombardment reactions? Why were accelerated protons more useful than α particles for bombardment reactions? What other types of particles can be used for bombarding particles?

12.45 Write the nuclear equation for each of the following bombardment processes: (a) ${}^{14}_{7}N(\alpha, p){}^{17}_{8}O$, (b) ${}^{106}_{46}Pd(n, p){}^{106}_{45}Rh$, (c) ${}^{23}_{11}Na\ (n, \beta^-)X$, and identify X.

12.46 Repeat Question 12.45 for (a) $^{113}_{48}\text{Cd}(n,\gamma)^{114}_{48}\text{Cd}$, (b) $^{6}_{3}\text{Li}(n,\alpha)^{3}_{1}\text{H}$, (c) $^{2}_{1}\text{H}(\gamma,p)\text{X}$, and identify X.

12.47 Write the shorthand notation for each of the following nuclear equations:
(a) $^{6}_{3}\text{Li} + ^{1}_{0}n \longrightarrow ^{4}_{2}\text{He} + ^{3}_{1}\text{H}$,
(b) $^{31}_{15}\text{P} + ^{2}_{1}\text{H} \longrightarrow ^{32}_{15}\text{P} + ^{1}_{1}\text{H}$,
(c) $^{238}_{92}\text{U} + ^{1}_{0}n \longrightarrow ^{239}_{93}\text{Np} + ^{0}_{-1}e$.

12.48 Repeat Question 12.47 for
(a) $^{253}_{99}\text{Es} + ^{4}_{2}\text{He} \longrightarrow ^{256}_{101}\text{Md} + ^{1}_{0}n$,
(b) $^{27}_{13}\text{Al} + ^{1}_{0}n \longrightarrow ^{26}_{13}\text{Al} + 2^{1}_{0}n$,
(c) $^{37}_{17}\text{Cl} + ^{1}_{1}\text{H} \longrightarrow ^{1}_{0}n + ^{37}_{18}\text{Ar}$.

12.49 The first nuclear transformation (discovered by Rutherford) can be represented by the shorthand notation $^{14}_{7}\text{N}(\alpha,p)^{17}_{8}\text{O}$. (a) Write the corresponding nuclear equation for this process. The respective atomic masses are 14.00307 u for $^{14}_{7}\text{N}$, 4.00260 u for $^{4}_{2}\text{He}$, 1.007825 u for $^{1}_{1}\text{H}$, 16.99913 u for $^{17}_{8}\text{O}$. (b) Calculate the energy change of this reaction.

12.50 Consider the following reactions which are possible when $^{27}_{13}\text{Al}$ is bombarded with neutrons:

$$^{27}_{13}\text{Al} + ^{1}_{0}n \longrightarrow ^{28}_{13}\text{Al} \longrightarrow \begin{cases} ^{26}_{13}\text{Al} + 2\,^{1}_{0}n \\ ^{28}_{13}\text{Al} + \gamma \\ ^{27}_{12}\text{Mg} + ^{1}_{1}\text{H} \\ ^{24}_{11}\text{Na} + ^{4}_{2}\text{He} \end{cases}$$

The nuclide masses are 26.98153 u for $^{27}_{13}\text{Al}$, 1.008665 u for $^{1}_{0}n$, 25.9858 u for $^{26}_{13}\text{Al}$, 27.98193 u for $^{28}_{13}\text{Al}$, 26.98437 u for $^{27}_{12}\text{Mg}$, 1.007825 u for $^{1}_{1}\text{H}$, 23.99102 u for $^{24}_{11}\text{Na}$, 4.00260 u for $^{4}_{2}\text{He}$. Which reaction releases the most energy and is therefore most favorable?

Radiation and Matter

12.51 Compare the general penetrating abilities of α particles, β particles, γ rays, and neutrons. Why are α particles that are absorbed internally by the body particularly dangerous?

12.52 What types of processes could take place in a substance that absorbs radiation? In what two ways can an organism be damaged by radiation? In which of these ways might the giant insects that so often threaten to destroy the world in science fiction movies be produced?

Nuclear Fission and Fusion

12.53 Briefly describe a nuclear fission process. What are the two most important fissionable materials?

12.54 What is a "chain reaction"? Why are nuclear fission processes considered chain reactions? What is the "critical mass" of a fissionable material?

12.55 Where have continuous nuclear fusion processes been observed? What is the main reaction that occurs in these sources?

12.56 The reaction that occurred in the first fusion bomb was $^{7}_{3}\text{Li}(p,\alpha)\text{X}$. (a) Write the complete equation for the process and identify the product, X. (b) The atomic masses are 1.007825 u for $^{1}_{1}\text{H}$, 4.00260 u for α, 7.01600 u for $^{7}_{3}\text{Li}$. Find the energy for the reaction.

Nuclear Energy

12.57 List the five primary components of a nuclear reactor and briefly describe their functions. Do any of these components present ecological or environmental problems?

12.58 Name and briefly discuss the major technological problems involved in the "fuel cycle" for a nuclear reactor.

12.59 How does a breeder reactor operate? Write the nuclear equations for the production of $^{239}_{94}\text{Pu}$ from $^{238}_{92}\text{U}$.

12.60 Name some of the advantages that fusion reactors could have over fission reactors. Name two methods for containment of fusion materials. Briefly describe these methods.

12.61 Which reaction produces the larger amount of energy per atomic mass unit of material reacting?

$$\text{fission: } ^{235}_{92}\text{U} + ^{1}_{0}n \longrightarrow ^{94}_{40}\text{Zn} + ^{140}_{58}\text{Ce} + 6\,^{0}_{-1}e + 2\,^{1}_{0}n$$
$$\text{fusion: } 2\,^{2}_{1}\text{H} \longrightarrow ^{3}_{1}\text{H} + ^{1}_{1}\text{H}$$

The atomic masses are 235.0439 u for $^{235}_{92}\text{U}$, 1.00867 u for $^{1}_{0}n$, 93.9061 u for $^{94}_{40}\text{Zn}$, 139.9053 u for $^{140}_{58}\text{Ce}$, 0.00055 u for $^{0}_{-1}e$, 3.01605 u for $^{3}_{1}\text{H}$, 1.007825 u for $^{1}_{1}\text{H}$, 2.0140 u for $^{2}_{1}\text{H}$.

12.62 The separation of isotopes is commonly done by gaseous diffusion processes. (a) Using Graham's law, calculate the ratio of the rates of effusion for $^{1}_{1}\text{H}_2(g)$ and $^{2}_{1}\text{H}_2(g)$. The atomic masses are 1.007825 u and 2.0140 u for $^{1}_{1}\text{H}$ and $^{2}_{1}\text{H}$, respectively. (b) Repeat the calculation for $^{235}_{92}\text{UF}_6(g)$ and $^{238}_{92}\text{UF}_6(g)$. The atomic masses of the uranium isotopes are 235.0439 u and 238.0508 u. respectively. (c) In which pair of isotopes are the nuclides easier to separate?

CHAPTER

13

Liquid and Solid States; Changes of State

LIQUIDS AND SOLIDS

13.1 KINETIC-MOLECULAR THEORY FOR LIQUIDS AND SOLIDS

The atoms, molecules, or ions in liquids and solids, like those in gases, are in constant motion. In a gas the molecules are far apart, are relatively independent of each other, and move in straight lines until they collide with something. In solids and liquids the particles are close together and there is comparatively little empty space (Figure 13.1). As a result, the motion of the particles is restricted. However, except for the smaller distance between molecules (see Section 6.2), the kinetic-molecular theory applies to liquids and solids as well as to gases. The distribution of the kinetic energy (or speed) of the particles in a liquid is represented by a curve similar to that for the molecules in a gas (Figure 13.2; see also Figure 6.3). The higher the temperature, the greater the average kinetic energy of the particles. Within the limits of their rigidly fixed positions, the particles in a solid have a similar distribution of energies.

The physical state in which a substance exists is dependent on the balance between the kinetic energies of the particles, which tend to keep them apart, and the intermolecular and/or bonding forces, which tend to pull them together.

With decreasing temperature, particles move more slowly. Eventually the interparticle forces of attraction exceed the kinetic energy and pull the particles together—a gas becomes a liquid or a liquid becomes a solid. Raising the temperature adds kinetic energy and eventually gives the particles sufficient energy to overcome the forces of attraction, leading to a phase change in the opposite direction.

In a liquid, the motion of a particle is restricted and the particle often collides with other particles. Although each of the particles rebounds, the distance that any particle can move away from another is so small that a liquid retains its volume. However, individually or in random groups the particles in a liquid can easily slide past each other, permitting a liquid to take the shape of its container.

Figure 13.1
The Three States of Matter

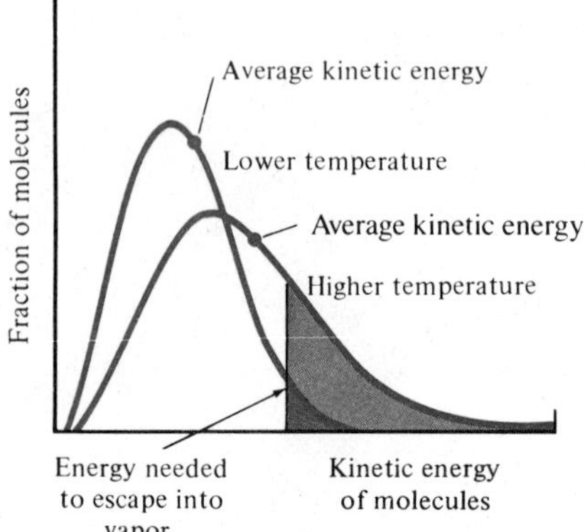

Figure 13.2
Change in Kinetic Energy with Temperature
The color areas represent the relative numbers of molecules of a substance that can enter the vapor at higher and lower temperatures.

The motion of the particles in a liquid, although less free than the motion of the molecules in a gas, is sufficient to allow liquids to mix together by diffusion. This effect can be seen in the gradual spreading of color when a drop of ink or food coloring is placed in a glass of water. Closeness of the particles accounts for the low compressibility of liquids, a property that permits liquids to transmit pressure in hydraulic systems—for example, in automobile brake systems.

Particles in a solid, unlike those in liquids and gases, are held rigidly in place by surrounding particles and can only vibrate about their fixed positions. This rigid structure prevents solids from spreading out; consequently, solids are only very slightly compressible. The particles of two solids placed in contact with each other can diffuse together at the surfaces, but the process is exceedingly slow.

13.2 GENERAL PROPERTIES AND TYPES OF LIQUIDS

Most substances that come to mind as examples of liquids—such as water, cleaning fluid, or antifreeze—are composed of rather simple molecules with up to about 20 atoms per molecule. Larger molecules generally have stronger intermolecular forces, if only because they come into contact with each other over larger areas. Therefore, substances composed of larger molecules are likely to be solids at ordinary temperatures. Such molecular solids become liquids when the temperature is raised sufficiently (if they do not decompose before they melt). Metals and some salts also melt at sufficiently high temperatures. The types of liquids are summarized in Table 13.1.

The difference in density between the solid and liquid states of pure substances is not great, and pressure changes have only a small effect on the volume and density of liquids and solids. For most liquids, and for water above 4 °C (Section 14.1), raising the temperature causes a slight increase in volume and a slight decrease in density.

The resistance of a substance to flow is **viscosity**—the opposite of fluidity. Most often viscosity is a property of concern for liquids. "Molasses in January" has a high viscosity; it will not run downhill easily or quickly. Internal forces that keep liquid particles from flowing past each other are responsible for viscosity.

Like most other properties of liquids, viscosity is dependent upon the size, shape, and chemical nature of the molecules. In general, greater intermolecular forces mean higher viscosity, and for similar types of compounds, greater molar mass also means higher viscosity. As illustrated by molasses, viscosity increases as the temperature decreases.

Lubricating oils are classified by their viscosity. The Society of Automotive Engineers assigns numbers to motor oils based on their relative viscosities. The SAE numbers range from 5 to 50 at a given temperature—0 °F for winter (W) oils and 210 °F for oils that are usable at summer temperatures. The larger the number, the greater the viscosity. Oils with a double number contain additives that moderate the change of viscosity with temperature so that they are neither too thin in summer nor

Table 13.1
Types of Liquids

Substances	Examples	Particles in liquid	Major interparticle forces
Monatomic species at low temperatures	Noble gases at low temperatures (e.g., He, Ne)	Atoms	London forces
Hydrogen-bonded liquids	H_2O, C_2H_5OH, NH_3	Associated molecules	Hydrogen bonding (plus other intermolecular forces)
Non-hydrogen-bonded liquids	Benzene, Br_2, chloroform ($CHCl_3$)	Molecules	London forces; dipole–dipole forces for polar molecules
Molten salts	Molten NaCl, $Na_3[AlF_6]$ (cryolite)	Ions	Attraction between + and − ions
Molten metals and Hg	Molten Fe, Cu	Atoms[a]	Metallic bond

[a] Neither "atoms" nor "ions" is an exact description, for each electron in the "electron sea" is associated with several electron-deficient atoms.

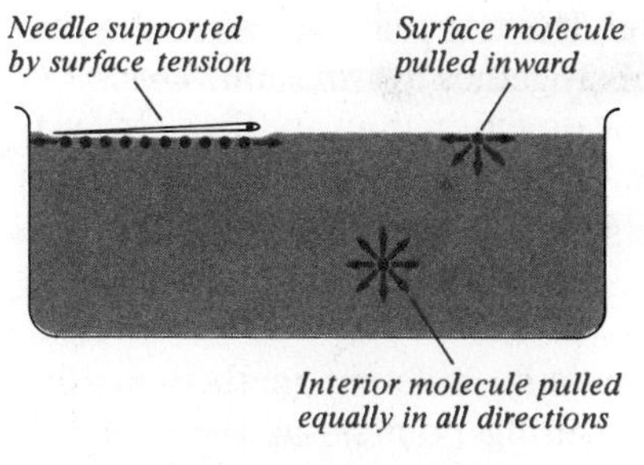

(a) Surface tension of a liquid

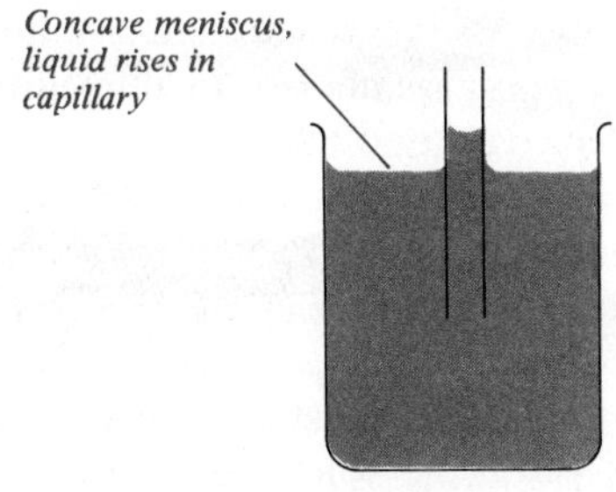

(b) Liquid wets solid (e.g., H_2O on unwaxed, dirty car)

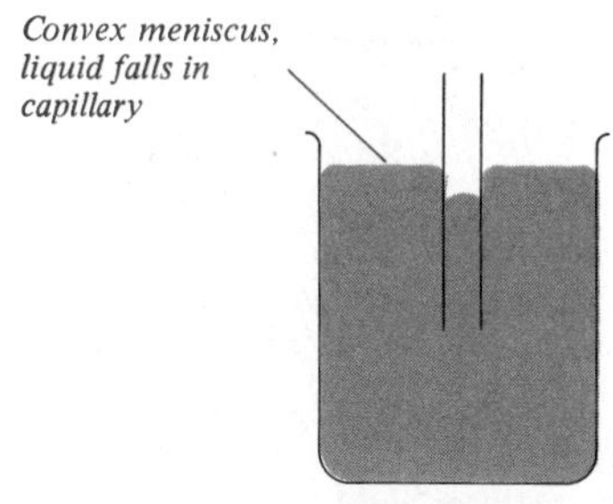

(c) Liquid does not wet solid (e.g., H_2O on freshly waxed car; Hg in glass tube)

Figure 13.3
Surface Tension and Wetting
The curved shape of the surface of a liquid in a narrow tube is called a *meniscus.*

too thick in winter. For example, SAE 10W/40 oil has an SAE number of 10 at 0 °F and of 40 at 210 °F.

In the interior of a liquid each molecule is subjected to equal forces of attraction from other molecules on all sides. Molecules at the surface feel these forces only on the liquid side, however, and as a result are pulled inward and closer together (Figure 13.3). The surface of a liquid can be thought of as behaving like a stretched membrane trying to contract to the smallest possible surface area. **Surface tension** is the property of a surface that imparts membrane-like behavior to the surface; it is formally defined as the amount of energy required to expand the surface of a liquid by a unit area.

Surface tension causes a tiny suspended liquid droplet to take the shape of a sphere—the shape with the smallest surface-to-volume ratio. It is surface tension that supports a steel needle on the surface of a glass of water (Figure 13.3a). (The needle is denser than water, and if pushed under, it sinks.) The surface tension at a liquid–solid interface is responsible for whether or not the liquid "wets" the solid. "Wetting" occurs when the attraction between the liquid and solid molecules is greater than the internal cohesive forces in the liquid (Figure 13.3b,c). The rise of a liquid in a narrow tube due to wetting of the walls—*capillary action*—contributes to the rise of water in the stems of plants.

13.3 GENERAL PROPERTIES AND TYPES OF SOLIDS

Solids were once described as "those parts of the material world which support when sat on, which hurt when kicked, which kill when shot." In terms of their chemical and physical properties, there are two types of solids—crystalline solids and amorphous solids. A **crystalline solid** is a substance in which the atoms, molecules, or ions assume a characteristic, regular, and repetitive three-dimensional arrangement. Sugar and salt are crystalline solids. An **amorphous solid** is a substance in which the atoms, molecules, or ions assume a random and nonrepetitive three-dimensional arrangement. Tar is an amorphous solid. The gemstone opal and volcanic glass (obsidian) are two of the very few parts of the Earth's crust that are amorphous.

Everyone is familiar with the beautiful and varied crystalline forms of the gemstones formed in nature and collected by mineralogists (Figure 13.4). A **crystal** is a solid that has a shape bounded by plane surfaces intersecting at fixed angles. To a chemist, a crystal is an array of atoms, molecules, or ions in which a structural pattern is repeated periodically in three dimensions. The regular geometrical shape of a crystal is the large-scale expression of the internal order of its atoms, molecules, or ions. Perfectly symmetrical single crystals of pure substances form only when the crystals have the opportunity to grow slowly in all directions. Most crystalline materials are *polycrystalline*—composed of small regions with plane faces and the characteristic internal order, but randomly oriented toward each other (Figure 13.5).

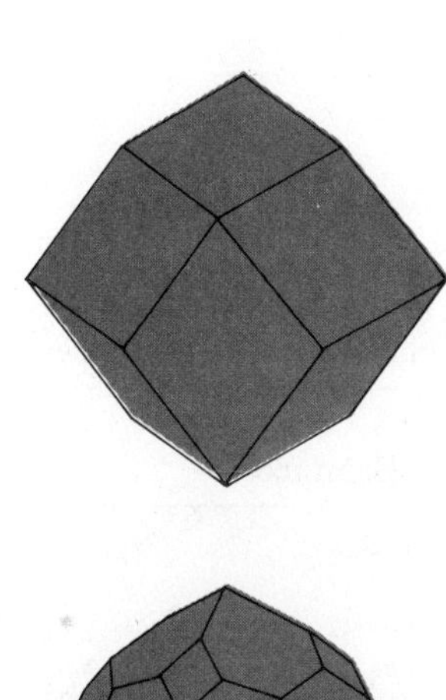

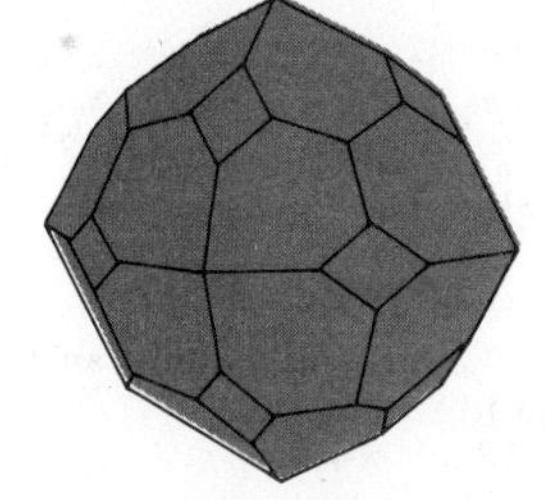

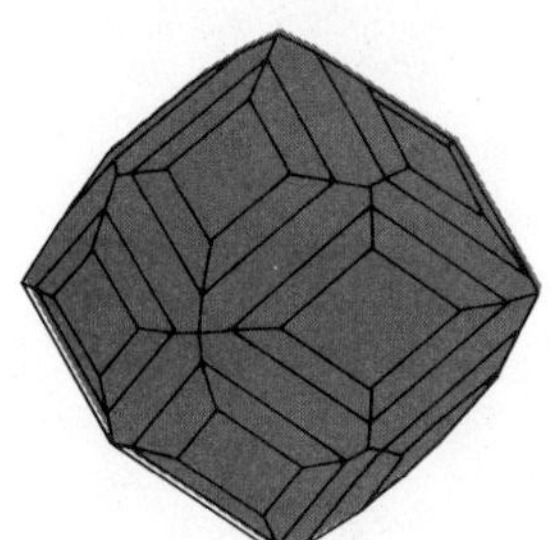

Figure 13.4
Some of the Natural Crystal Forms of Garnet

Source: C. S. Hurlburt, Jr., *Dana's Manual of Mineralogy*, 18th ed. (New York: Wiley & Sons, 1971).

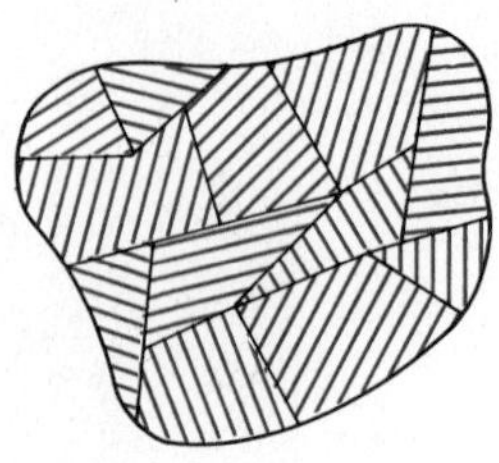

Figure 13.5
Polycrystalline Solid
In a polycrystalline material, many single crystal regions, called *crystallites*, lie adjacent to each other in different orientations.

Crystalline substances, because of their internal order, may be *anisotropic*—they may have certain physical and chemical properties that vary with direction. For example, electrical conductivity may be greater in one direction than another. One way to identify a crystalline substance is by its cleavage, when struck sharply, into smaller pieces with planar faces. Amorphous substances are generally *isotropic*—their physical properties are the same in all directions. Cleavage of an amorphous material gives smaller pieces with nonplanar faces.

Amorphous and crystalline materials are also quite different in their melting behavior. Crystalline solids generally have sharp melting points—as the crystal is heated, all of the particles are released from their fixed positions at the same temperature. Amorphous solids soften gradually as the temperature rises. Glass, tar, and many polymers are, in fact, often classified not as solids, but as liquids of such high viscosity at ordinary temperatures that their flow is not normally observable. In a sheet of glass taken from an old window, there is a significant (though small) difference in thickness between the bottom and the top. The bottom is thicker because the glass has flowed slightly over the years. Most glasses are mixtures of oxygen-containing compounds that have cooled from a molten state without crystallizing.

Table 13.2 summarizes the types of crystalline solids. Molecular crystals include almost all organic compounds in the solid state, held together by intermolecular forces that include hydrogen bonding, dipole–dipole attraction, and London forces (Sections 11.11–11.13).

Table 13.2
Types of Crystalline Solids
Keep in mind that both covalent and ionic contributions to bonding often exist in the same substance.

Substances	Examples	Particles in crystal lattice	Major interparticle forces
Pure metals	Fe, Cu, Al	Atoms[a]	Metallic bond
Alloys	Bronze (Cu–Sn)	Atoms[a]	Metallic bond
Ionic crystals	All salts (e.g., NaCl, $BaSO_4$)	Monatomic or polyatomic ions	Ionic bond
Molecular crystals	All molecular compounds (e.g., CO_2, ice, organic compounds)	Neutral molecules	Intermolecular forces
Network covalent materials[b]	Silicon, diamond, quartz, germanium, boron nitride	Atoms	Covalent bonds

[a] Neither "atoms" nor "ions" is an exact description, for each electron in the "electron sea" is associated with many electron-deficient atoms.
[b] Section 10.5.

SUMMARY OF SECTIONS 13.1–13.3

In solids and liquids, the particles are close together and motion is restricted. In other respects, however, the kinetic-molecular theory applies to solids and liquids as it does to gases, as demonstrated by the diffusion of liquids into each other and also the diffusion of solids, although the latter is very slow. The physical state in which a substance exists depends upon the balance between the kinetic energies of the particles and the intermolecular and/or bonding forces acting between them.

In a liquid the arrangement of the particles is neither completely random nor completely ordered, whereas in a true crystalline solid the particles have a characteristic, regular, and repetitive spatial arrangement. Intermolecular forces in liquids are responsible for the properties of viscosity and surface tension. An amorphous solid is one in which the particles have a random and nonrepetitive arrangement. Crystalline solids (as opposed to amorphous solids) can be cleaved along planes, have sharp melting points, and sometimes possess anisotropic properties, that is, properties that vary with direction.

THE CRYSTALLINE STATE

If you have ever looked at the patterns in a brick wall, a tile floor, or a produce display at a grocery store, you have an insight to our model of the crystalline state. The major differences between the patterns of brick walls or tile floors and the patterns of crystals are that a very large number of atoms, ions, or molecules make up the latter and that their patterns are three dimensional.

To help in visualizing this three dimensional aspect of the crystalline state, the unit cells of several simple solids and two types of packing that can occur in metals are illustrated on the following pages. The illustrated systems include a primitive cubic unit cell (e.g., elemental polonium); a body-centered cubic unit cell (e.g., elemental sodium); a face-centered cubic unit cell (e.g., elemental platinum); a cesium chloride unit cell, in which the ions are similar in size; and a sodium chloride unit cell, in which the ions are considerably different in size. The illustrations also include diagrams representing hexagonal closest packing (e.g., elemental zinc) and cubic closest packing. A view from the top and a perspective front view are shown for most of the illustrated structures.

As you turn the page and rotate the book, you will see a page of captions and a page consisting of a text page with three transparent overlays on it. Refer to the captions and to the drawings on the text page and overlays at the same time. Each caption first gives a description of the atoms or ions that make up each layer and then describes how the layers are stacked to give the crystal structure. In all cases the layer drawn on the text page is referred to as the first layer, the layer drawn on the overlay next to the text page is the second layer, the center overlay is the third layer, and the top overlay is the fourth layer. (Most of the crystal structures require the use of the first three layers.) By analyzing each of the layers separately and then repeatedly stacking and separating the layers, you should acquire a good understanding of the model of the crystalline state.

Primitive Cubic Unit Cell

Both layers consist of four atoms arranged in a square. Each atom touches two other atoms in the square. The cube is formed by placing the second layer directly over the first layer so that the atoms touch.

Body-Centered Cubic Unit Cell

The first and third layers consist of four atoms arranged in a square. The atoms are separated slightly from each other. The single atom that makes up the second layer is placed in the center of the depression formed by the four atoms in the first layer so that it touches all four atoms. The cube is formed by placing the third layer directly over the first layer so that the four atoms touch the body-centered atom.

Face-Centered Cubic Unit Cell

The first and third layers consist of four atoms arranged in a square plus a fifth atom located in the center. The second layer consists of four atoms in a square with the atoms touching. The second layer is placed on the first layer so that an atom in the second layer touches three atoms in the first layer. The cube is formed by placing the third layer directly over the first layer.

Hexagonal Closest Packing

Each layer consists of an atom surrounded by six atoms in a hexagon. The second layer (layer B) is placed on the first layer (layer A) so that each atom of the second layer fits into the center of a depression made by three atoms in the first layer. The third layer is placed directly over the first layer, with the atoms fitting into the depressions of the second layer. Because the third layer is directly above the first layer, it is also known as layer A. Hexagonal closest packing is called ''ABABABA . . .'' packing, after the arrangement of the layers. There are three complete unit cells shown in this model.

Cubic Closest Packing

Each layer consists of an atom surrounded by six atoms in a hexagon. The placement of layer B on layer A is identical to that described for the hexagonal closest packing model. Note that only one-half of the depressions in the first layer are filled by the atoms in the second layer. The third layer is placed so that the atoms of the third layer fit into the centers of the depressions made by the atoms in the second layer, but so that the atoms of the third layer are directly above the depressions in the first layer that were not filled by the atoms in the second layer. Because this third layer is placed differently than the others, it is known as layer C. The fourth layer is placed directly over the first layer, with the atoms fitting into the depressions of the third layer. Because this fourth layer is directly above the first layer, it is also known as layer A. Cubic closest packing is called ''ABCABCA . . .'' packing, after the arrangement of the layers, and is the same as the packing in the face-centered cubic unit cell.

Cesium Chloride Unit Cell

The first and third layers consist of four Cl^- ions (0.181 nm) arranged in a square. The Cl^- ions are separated slightly from each other. The single Cs^+ ion (0.167 nm) that makes up the second layer is placed in the center of the depression formed by the four Cl^- ions in the first layer so that it touches all four Cl^- ions. The cube is formed by placing the third layer directly over the first layer so that the four Cl^- ions touch the body-centered Cs^+ ion.

Sodium Chloride Unit Cell

The first and third layers consist of five Cl^- ions (0.181 nm) and four Na^+ ions (0.097 nm) arranged in a square. The fifth Cl^- ion is in the center of the square and touches the four Na^+ ions that are along the edges and nearly touches the other four Cl^- ions that are at the corners. The second layer is similar to the other layers, except the positions of the ions are reversed—Na^+ ions in the center and at the corners, Cl^- ions along the edges of the square. The second layer is placed directly over the first layer so that the Na^+ ions in the second layer are directly over the Cl^- ions in the first layer and the Cl^- ions in the second layer are directly over the Na^+ ions in the first layer. The cube is formed by placing the third layer directly over the first layer so that the Cl^- ions in the third layer are directly over the Na^+ ions in the second layer and the Na^+ ions in the third layer are directly over the Cl^- ions in the second layer.

Primitive Cubic Unit Cell

Body-Centered Cubic Unit Cell

Face-Centered Cubic Unit Cell

Hexagonal Closest Packing

A A

Cubic Closest Packing

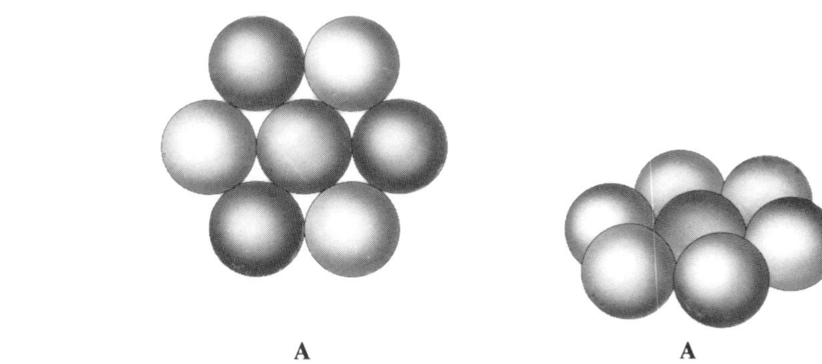

A A

Cesium Chloride Unit Cell

Sodium Chloride Unit Cell

Aside 13.1 LIQUID CRYSTALS

The liquid crystalline state is intermediate between the solid state and the liquid state. Liquid crystals have become familiar through the advertising for LCD—liquid crystal display—wrist watches, calculators and other items with digital displays (i.e., displays of numbers). A liquid crystal display usually shows black numbers against a silvery background.

Liquid crystals flow like liquids, but exhibit some of the optical properties of crystals. In the liquid crystalline state, substances have lost the three-dimensional order typical of crystals, but have retained two-dimensional order and are therefore more ordered than liquids. The optical, mechanical, and electrical properties of liquid crystals are anisotropic, that is, they vary with the direction in which they are observed. This variation is essential to the use of liquid crystals in display devices.

Most substances that have liquid crystalline phases are composed of long, narrow organic molecules. The molecules must have polar groups arranged so that as the solid is melted or in some cases dissolved in a solvent, the molecules are held together with some degree of order. Three basic types of molecular arrangement have been identified in liquid crystals. *Smectic liquid crystals* are turbid, viscous substances with parallel molecules in planes that can slip past each other (Figure 13.6a). (The curious name comes from the Greek word for soap; soap films are smectic substances.) *Nematic liquid crystals* are less turbid and more mobile than smectic liquid crystals. In the nematic state the molecules are parallel to each other but are not ordered into planes (Figure 13.6b). ("Nematic" comes from the Greek word meaning thread and refers to a threadlike optical pattern seen in films of such substances.) Frequently a smectic substance is transformed into a nematic substance—a further decrease in order—during heating to a higher temperature.

Cholesteric liquid crystals, sometimes called twisted nematic crystals, have order similar to that of the nematic substances. However, the molecules in successive layers are not randomly oriented with respect to each other. Instead, the molecules in successive layers are rotated about axes perpendicular to their centers, with the result that they form a helical coil (Figure 13.6c). A complete 360° turn in such a substance frequently occurs within the distance of the wavelength of visible light. As a result, cholesteric substances have the remarkable property of reflecting light of different colors at different temperatures because of the sensitivity of this helical structure to temperature. The color observed is the result of reflection of visible light by the layers of ions or molecules in a crystal (Bragg reflection; see Aside 13.2). The variation in color of cholesteric liquid crystals with changing temperature can be exploited in a variety of ways. For example, thin films incorporating these substances are used to map

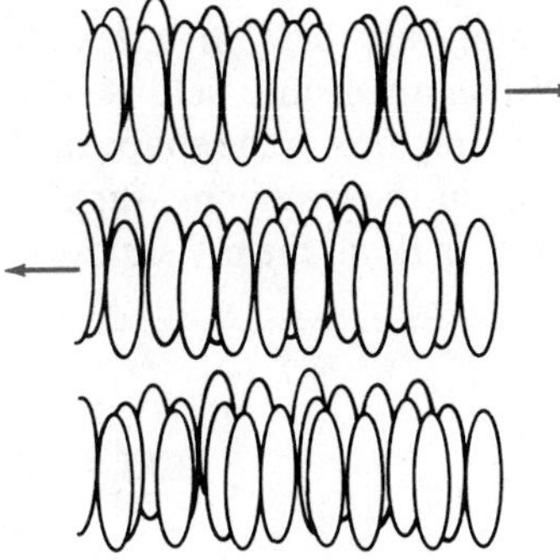
Molecules remain in
their own planes,
planes slide past each other
each other

(a) Smectic

Parallel molecules slide
past each other

(b) Nematic

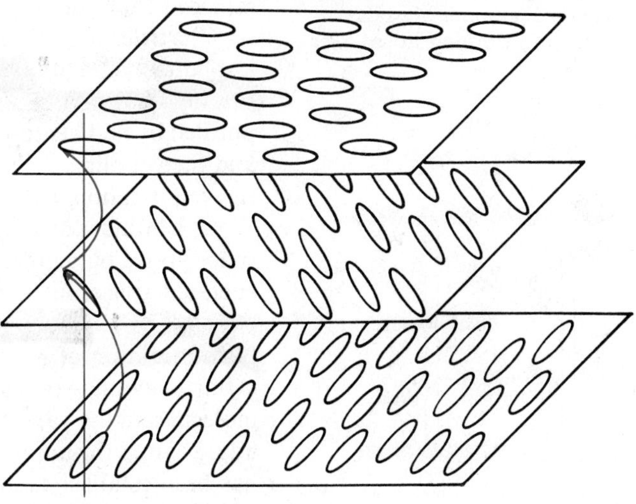
Molecules in successive planes are
twisted about vertical axes

(c) Cholesteric

Figure 13.6
Liquid Crystal Structures

skin temperature (as an aid to the diagnosis of diseased tissue) and electrical circuit temperatures. Liquid crystalline substances have been found in numerous living systems, and a better understanding of their behavior is therefore also of importance in biochemistry.

A typical LCD consists of a thin flat cell in which the long liquid crystals are held in a specific orientation (Figure 13.7). When an electric field is applied to certain regions of the cell (regions in the shape of digits), various optical properties in those regions, such as reflection of light, are changed, making them visible.

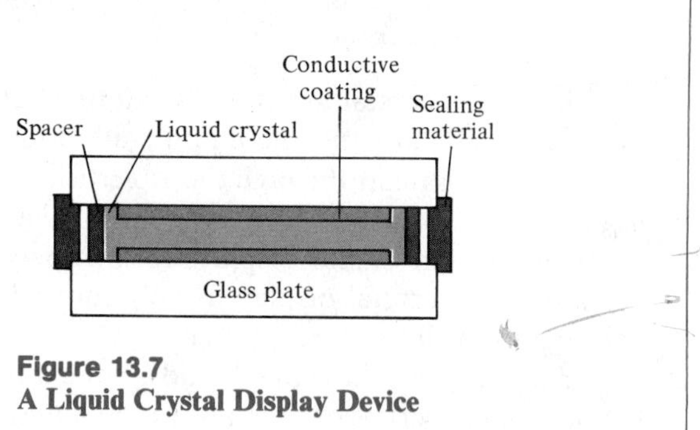

Figure 13.7
A Liquid Crystal Display Device

RELATIONSHIPS BETWEEN PHASES

13.4 CHANGES OF STATE

Molecules near the surface of a liquid can escape into the gaseous phase if they have sufficient kinetic energy to overcome the attraction of neighboring molecules. Many such escaping molecules strike molecules of the gas above the liquid, lose energy, and bounce back into the liquid phase. However, if the vessel is open to the air, all the molecules eventually escape from the liquid. This process is called **evaporation**—the escape of molecules from the surface of a liquid in an open container to the gas phase (see Figure 13.10a). (**Vaporization** is the more general term for escape of molecules from the liquid or solid phase to the gas phase.)

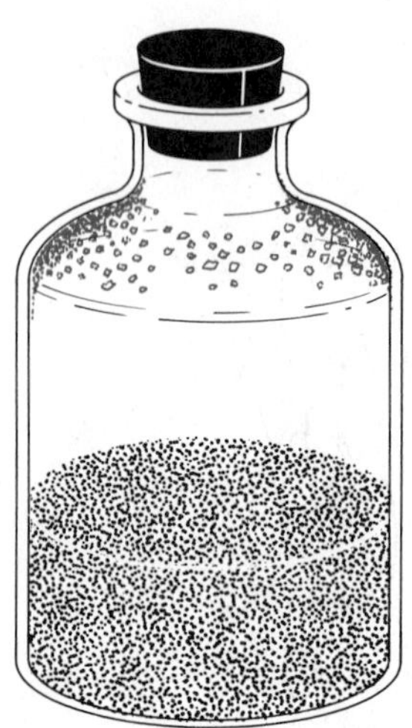

Figure 13.8
Camphor—A Solid That Sublimes at Room Temperature
Camphor is an organic compound used in some mothballs. Frequently you will find camphor crystals sublimed onto clothing stored over mothballs.

The escaping molecules are the "hottest" molecules—the ones at the high-kinetic-energy end of the curve (see Figure 13.2). Therefore, evaporation is faster when the temperature is higher because a larger fraction of molecules have sufficient energy to escape. The rate of evaporation is also dependent on surface area—the larger the surface area, the larger the number of "hot" molecules that reach the surface. Escape of the highest energy molecules leaves behind a collection of molecules with a lower average kinetic energy and, consequently, a lower temperature. When evaporation is rapid—for example, when you get out of a swimming pool on a windy day—the cooling effect of evaporation is obvious. It is less noticeable during the slow evaporation of a liquid from a noninsulated container because the lost heat is gradually replaced from the surroundings.

Particles with a sufficiently high kinetic energy can also escape from the surface of a solid. **Sublimation** is the vaporization of a solid (Figure 13.8). The reverse transition, from the gas phase directly to the solid phase, is called "deposition." A solid can be purified by first heating it, so that the pure material sublimes away from the impurities, and then cooling the vapor. (The term "sublimation" is also sometimes used to mean the vaporization *and* redeposition of the same solid.)

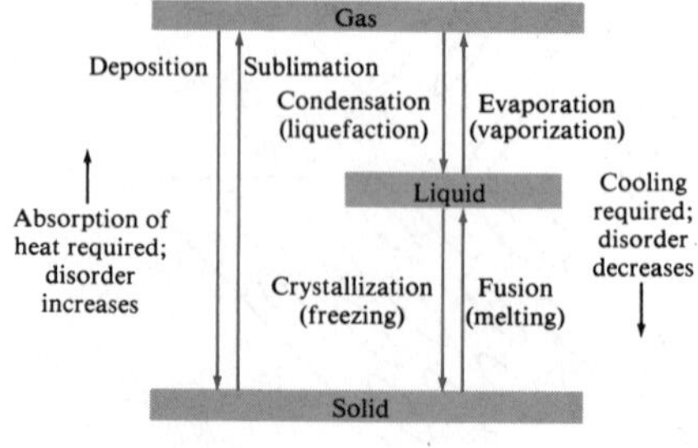

Figure 13.9
Changes of State
The alternate names for changes of state are given in parentheses.

If a liquid is in a tightly closed container, molecules escape from the liquid into the space above but can go no farther. Some of the molecules, while bouncing within the enclosed space, hit the liquid surface and reenter the liquid phase. This process is **condensation**—the movement of molecules from the gas phase to the liquid phase. The transformation of a gas into a liquid is also referred to as **liquefaction.** (Note the *e* in liquefaction—to use *i* in its place is wrong.) Eventually, the concentration of molecules in the vapor is so great that the number of molecules moving back into the liquid equals the number escaping from the liquid. At this point, a dynamic equilibrium has been established and the partial pressure of the vapor over the liquid equals its vapor pressure (Section 6.11).

When a solid is heated, the particles vibrate faster and faster in their fixed positions until they are no longer held firmly in place and are thus sufficiently free to form a liquid. The **melting point** of a solid is the temperature at which the solid and liquid phases of a substance are at equilibrium. **Fusion** is a term also used to mean melting. (Remember that "fusion" means melting, *not* solidification.) The freezing point is identical to the melting point but is thought of as a property of the liquid, whereas melting point is a property of the solid.

The terminology for changes of state is summarized in Figure 13.9. Fusion and vaporization always occur with the absorption of heat and the ΔH values for these processes are positive. The ΔH values for evaporation, ΔH_{vap}, reflect the strength of intermolecular forces in a liquid—the larger ΔH_{vap} the stronger the intermolecular forces. For example, in the following series of nonpolar hydrocarbons, the increasing heats of evaporation and boiling points show the increasing London forces in the larger molecules:

	CH_4	C_2H_6	C_3H_8
b.p. (°C)	−164	−88.6	−42.1
ΔH_{vap} (kJ/mol)	8.22	14.68	18.83

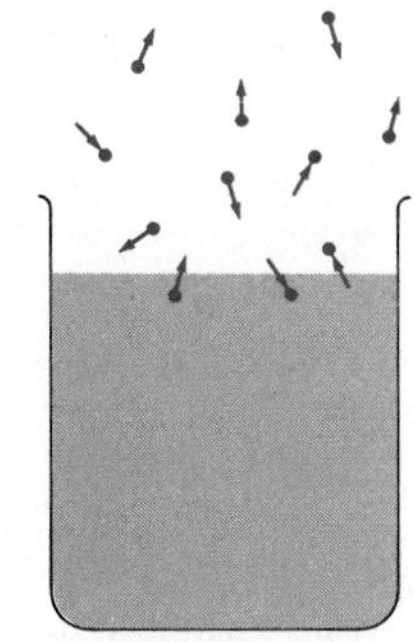

(a) Evaporation

The reverse processes of condensation and crystallization have ΔH values of equal magnitude but opposite sign from heats of vaporization and fusion. (Some values for enthalpies for changes of state were given in Table 7.5.) Changes of state all occur at constant temperatures, as reflected in cooling curves like that for water (see Figure 7.10).

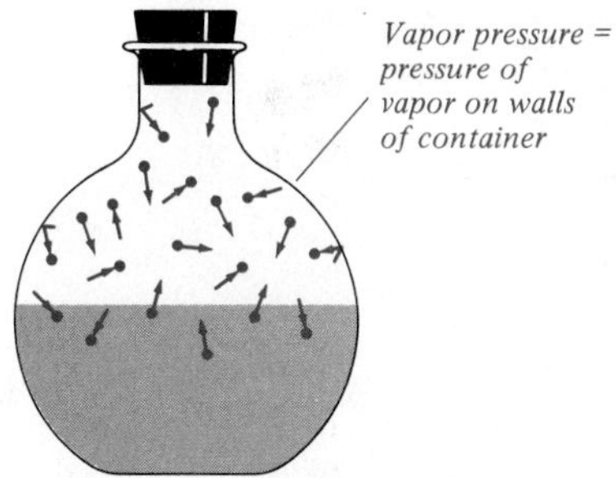

(b) Vaporization ⇌ Condensation Equilibrium

13.5 VAPOR PRESSURE OF LIQUIDS

The *equilibrium vapor pressure* (Section 6.11), usually referred to simply as the "vapor pressure," is the pressure exerted by a vapor in contact with a liquid once vaporization and condensation have come to equilibrium (Figure 13.10b). Vapor pressure increases with temperature, as more "hot" molecules are able to leave the surface. At a given temperature, a substance with strong intermolecular forces will have a lower vapor pressure than a substance with weak intermolecular forces. Hydrogen-bonded substances and those with large dipole moments usually have lower vapor pressures than substances of comparable molecular mass in which such strong intermolecular forces are absent. Even in the absence of hydrogen bonding and dipole–dipole forces, vapor pressure decreases with increasing molecular mass, due to increasing London forces. Volatile substances are those with high vapor pressures.

Figure 13.11 (p. 308) shows the vapor pressure–temperature curves of several simple molecular compounds. For each, the vapor pressure increases with temperature in the same general way. Compare the vapor pressures of water, chloroform, and diethyl ether at 20 °C. Water, with its hydrogen bonds, has the strongest intermolecular forces and the lowest vapor pressure. Although chloroform and diethyl ether have similar dipole moments, the possibility for London forces is greater in chloroform because it has a greater molecular mass. Thus, chloroform has a lower vapor pressure than diethyl ether, which is an extremely volatile liquid. Such substances as nitrous oxide, carbon dioxide, and cyclopropane, which have low molecular masses and either no dipole moments or very small dipole moments, have such high vapor pressures that they remain entirely in the gaseous state until cooled to quite low temperatures. The relative vapor pressures of liquids can be demonstrated by introducing the liquids into barometer-like tubes of mercury, as shown in Figure 13.12 (p. 308).

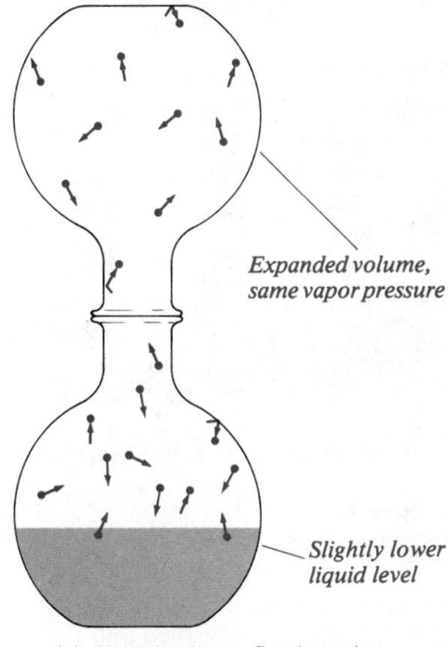

(c) Vaporization ⇌ Condensation Equilibrium reestablished

Figure 13.10
Evaporation, Vaporization, Condensation, and Vapor Pressure

Each vapor pressure–temperature curve ends at the **critical point**—the point at which the densities of a liquid and its vapor become equal, and the boundary between the phases disappears. The nature of the critical point can be demonstrated by heating (carefully) a liquid sealed in a closed glass vessel. (*Do not attempt this experiment.*) With increasing temperature the pressure of the vapor increases as more molecules leave

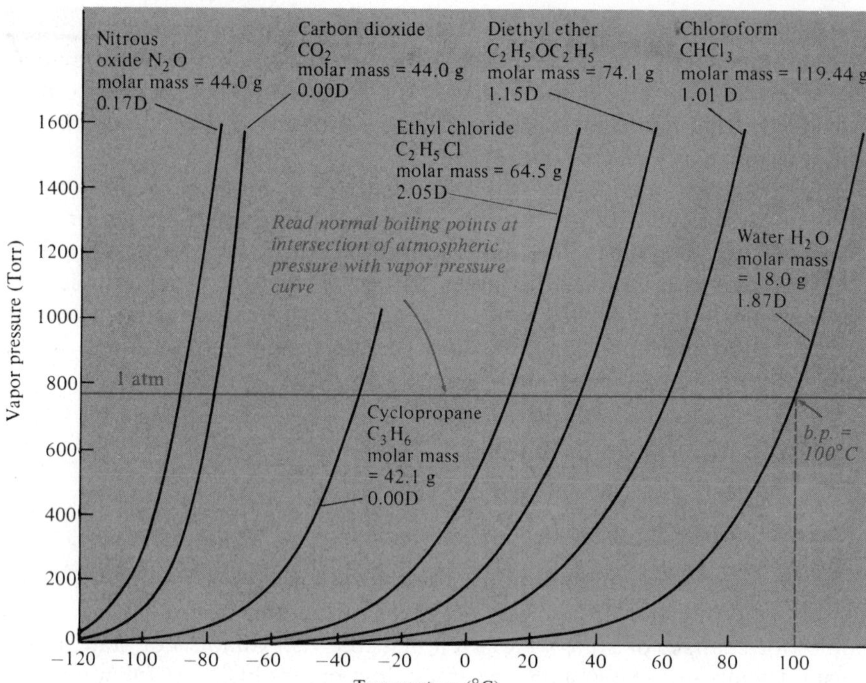

Figure 13.11
Vapor Pressure–Temperature Curves
With the exception of water and carbon dioxide, which are included for comparison, each of these liquids is an anesthetic. The molar mass and dipole moment of each substance are given on the curves.

the liquid phase and enter the gas phase. Eventually, the critical point is reached and the boundary between the vapor and liquid phases disappears—all of the molecules present have acquired sufficient kinetic energy to prevent them from remaining close together in the liquid phase. The temperature at which this occurs is the **critical temperature**—the temperature *above* which a substance cannot exist as a liquid no matter how great the pressure. The pressure that will liquefy a substance at its critical temperature is known as the **critical pressure**.

The equilibrium vapor pressure is unaffected by the volume of the liquid. As long as even the smallest amount of liquid is present, the vapor pressure at a given temperature is the same. Similarly, equilibrium is established at the same vapor pressure for a given substance no matter what the size of the container (see Figure 13.10c).

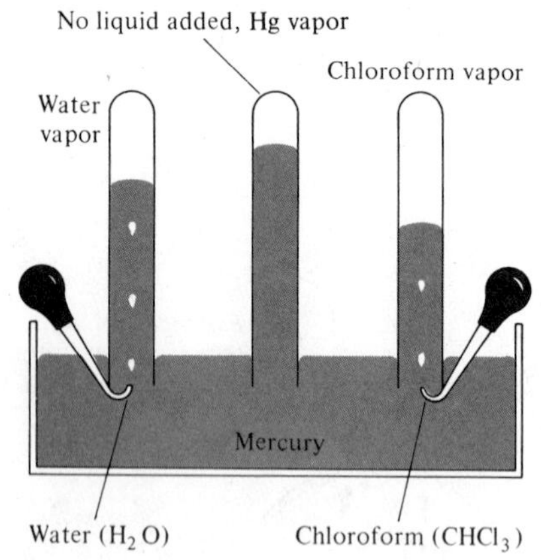

Figure 13.12
Relative Vapor Pressures
The liquids added are less dense than mercury and rise to the surface. Equilibrium is established and the vapor pressure pushes the mercury down. The distance the mercury falls is a measure of the vapor pressure of the liquids added (at the temperature of the experiment).

We have seen that both vapor pressure and heat of evaporation (or sublimation) are strongly influenced by intermolecular forces. As you might expect, there is a relationship between these two properties. If the logarithm of the vapor pressure is plotted versus $1/T$, data like those plotted in Figure 13.11 yield straight lines. The slope of such lines is equal to $-\Delta H^\circ_{\text{vap or sub}}/2.303\,R$, where R is the gas constant and ΔH° is the heat of evaporation or sublimation of the substance for which the data are plotted. The following equation, in which P_1 and P_2 are the vapor pressures of the same substance at two different temperatures T_1 and T_2, gives the relationship shown by the straight-line plots:

$$\log\left(\frac{P_2}{P_1}\right) = \frac{-\Delta H^\circ_{\text{vaporization or sublimation}}}{2.303\,R}\left(\frac{1}{T_2} - \frac{1}{T_1}\right) \quad \textbf{(13.1)}$$

Equation (13.1) (known as the Clausius-Clapeyron equation) can be used (1) to predict the value of the vapor pressure of a substance at a given temperature if the heat of vaporization and the vapor pressure at another temperature are known, (2) to calculate the heat of vaporization if two vapor pressures at two temperatures are known, or (3) to predict the temperature at which the vapor pressure has a certain value if the heat of vaporization and the vapor pressure at another temperature are known.

EXAMPLE 13.1 Vapor Pressure

Each of the following substances is a liquid at −100 °C:

```
   H  H              H       H              H  H
   |  |              |  ..   |              |  |  ..
H—C—C—H          H—C—O—C—H          H—C—C—O—H
   |  |              |  ..   |              |  |  ..
   H  H              H       H              H  H
```

ethane *dimethyl ether* *ethyl alcohol*

Place these liquids in order of increasing vapor pressure, basing your arrangement on the strengths of intermolecular forces (Sections 11.11–11.13).

London forces are expected in all molecules and are present in each substance. Both dimethyl ether and ethyl alcohol molecules should be bent at the electronegative oxygen atoms (AB_2E_2; Figure 11.3), which means that dipole–dipole attraction will be significant. In addition, ethyl alcohol has a polar OH group and will be relatively strongly hydrogen-bonded in the liquid state. Hydrogen bonding is of no significance in either dimethyl ether or ethane. The intermolecular attraction should be strongest in ethyl alcohol and weakest in ethane, giving the predicted order of increasing vapor pressures at −100 °C as

ethyl alcohol < dimethyl ether < ethane

which is in excellent agreement with the respective measured vapor pressure values of 4×10^{-4} Torr, 6 Torr, and 360 Torr at −100 °C.

EXERCISE Which of the following compounds

```
(a)      Cl  H          (b)      H  H
         |   |                   |  |
     Cl—C—C—H              Cl—C—C—Cl
         |   |                   |  |
         H   H                   H  H
```

will have the higher vapor pressure at a given temperature? *Answer* compound (b).

EXAMPLE 13.2 Vapor Pressure

The vapor pressure of liquid bromine at room temperature is 168 Torr. Suppose that bromine is introduced drop by drop into a closed system containing air at 745 Torr and room temperature. [The liquid bromine volume is negligible compared to the gas volume.] (a) If the bromine is added until no more vaporizes and a few drops of liquid are present in the flask, what will be the total pressure in the flask? (b) What will be the total pressure if the volume of this closed system is decreased to one-half its original value?

(a) The total pressure of the mixture of gases in the flask will be that given by Dalton's law as

$$P_{total} = P_{air} + P_{Br_2}$$
$$= 745 \text{ Torr} + 168 \text{ Torr} = \mathbf{913\ Torr}$$

(b) To solve this problem, the air and the bromine must be treated differently. The new pressure for the air can be calculated by using Boyle's law:

$$P_{air} = (P_1)\frac{V_1}{V_2} = (745 \text{ Torr})\left(\frac{V_1}{0.5\ V_1}\right) = 1490 \text{ Torr}$$

The vapor pressure of the bromine will remain constant at 168 Torr because the temperature has not changed and because the vapor remains in equilibrium with the liquid. However, some of the gaseous bromine will condense to form additional liquid bromine because the volume of the system has decreased.

The total pressure of the mixture in the flask after the volume change will be

$$P_{total} = P_{air} + P_{Br_2}$$
$$= 1490 \text{ Torr} + 168 \text{ Torr} = \mathbf{1660 \text{ Torr}}$$

EXERCISE Suppose the volume of the closed system containing air at 745 Torr and bromine can be decreased to one-fifth its original value. What will be the pressure in the system after the change? *Answer* 3.90×10^3 Torr.

EXAMPLE 13.3 P–T–$\Delta H_{vap\,or\,sub}$ Relationship

The heat of sublimation of carbon dioxide is 25,800 J/mol and the vapor pressure is 35 Torr at −110. °C. Assuming that the heat of sublimation is constant, calculate the vapor pressure of dry ice at −90. °C.

The vapor pressure at −90.0 °C (= 183.2 K) is found from Equation (13.1).

$$\log\left(\frac{P}{35 \text{ Torr}}\right) = \frac{-(25{,}800 \text{ J/mol})}{(2.303)(8.314 \text{ J/K mol})}\left(\frac{1}{183 \text{ K}} - \frac{1}{163 \text{ K}}\right) = 0.90$$

Taking antilogarithms of both sides gives

$$\frac{P}{35 \text{ Torr}} = 7.9$$

which gives

$$P = (7.9)(35 \text{ Torr}) = \mathbf{280 \text{ Torr}}$$

EXERCISE The vapor pressure of liquid tungsten is 1.00 Torr at 3990. °C and 100.0 Torr at 5168 °C. Calculate the heat of vaporization. *Answer* 754 kJ/mol.

13.6 BOILING POINT

We have all seen bubbles rising to the surface in a pot of boiling water. The first small bubbles, when heating begins, are bubbles of air which is driven out of solution by the rising temperature. When the water has come to a boil, larger bubbles continuously form and rise to the surface. These bubbles consist of water vapor.

As water (or any other liquid) is heated in an open container, the average kinetic energy of the molecules increases, the rate of evaporation increases, and the vapor pressure increases. Eventually the point is reached where the vapor pressure of the liquid equals the pressure of the atmosphere pushing down on the surface of the liquid and a great many molecules have acquired enough energy to escape from the attraction by their neighbors. At this point bubbles of vapor can form in the liquid, whereas previously they would have been collapsed by the pressure from above. As the bubbles form they rise to the surface and boiling commences. The temperature at which this happens is the **boiling point**—the temperature at which the vapor pressure of a liquid in an open container equals the pressure of the gases above the liquid and bubbles of vapor form throughout the liquid. A boiling point given without any mention of pressure can be assumed to be a **normal boiling point**—the boiling point at 760 Torr, the average atmospheric pressure at sea level.

Plots of the variation of the vapor pressure with the temperature like those in Figure 13.11 also show the variation of the boiling point with pressure. Below atmospheric pressure, boiling points are lower than the normal boiling point. For example, you can see from Figure 13.11 that at 600 Torr ethyl chloride boils at about 0 °C. At 9000 ft above sea level, where the pressure is about 550 Torr, water boils at 91 °C. At this altitude an egg must be boiled for about 5 min to reach the same consistency as a 3 min egg at sea level, and a hard-boiled egg requires about 18 minutes of

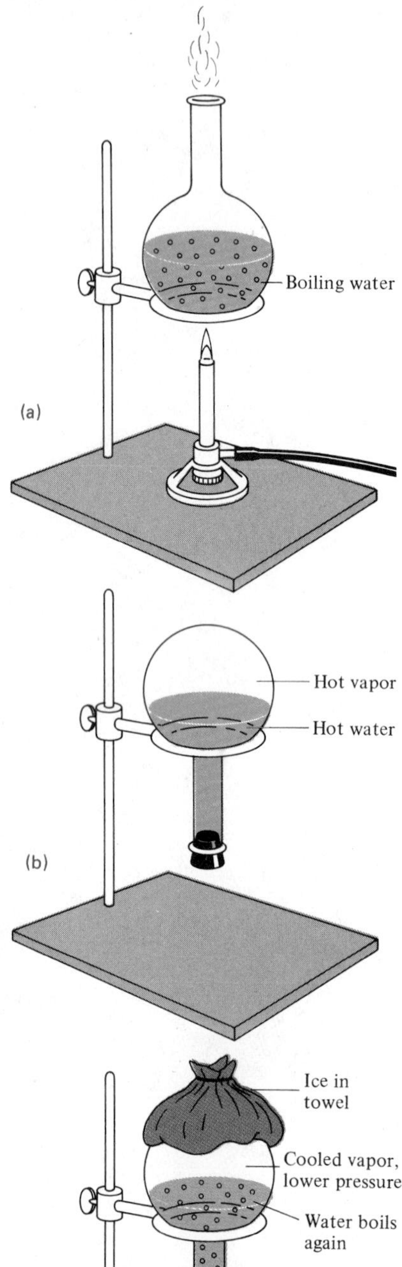

Figure 13.13
Lowering of Boiling Point by Lowering Vapor Pressure
(a) Water is brought to a boil. (b) Heat is removed, stopper added, and flask inverted. (c) Vapor is cooled, lowering pressure and bringing water back to boil. Water will boil until pressure builds up and temperature is below boiling point at new pressure.

cooking time. An experiment that demonstrates the boiling of water at lower temperatures under reduced pressure is shown in Figure 13.13.

It is possible for a liquid to be **superheated**—heated to a temperature above the boiling point without the occurrence of boiling. Superheating occurs when it is difficult for molecules with enough kinetic energy to get together to form a bubble. Bubble formation is encouraged by the presence of "boiling chips"—pieces of broken porcelain, or something similar, that release tiny air bubbles during heating. The boiling liquid evaporates into the bubble, which grows in size, and smooth boiling is encouraged. *A superheated liquid is hazardous, for it can start to boil suddenly and with great violence* (called "bumping").

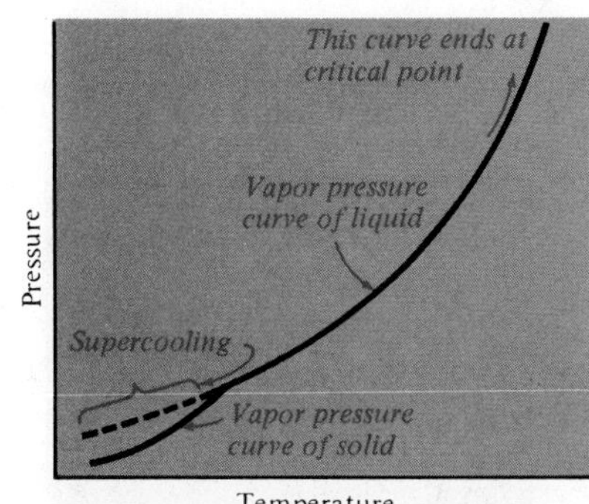

(a)

13.7 PHASE DIAGRAMS

A phase diagram is used to show the relationships between the physical state of a substance and the pressure and temperature. To construct a phase diagram for a three-phase solid–liquid–gas system, the vapor pressure–temperature curve for the liquid phase (like those in Figure 13.11) is combined with the vapor pressure–temperature curve for the solid phase (Figure 13.14a). This second curve represents the temperatures and pressures at which solid and vapor are in equilibrium.

The curve for the vapor pressure of the solid ends at its intersection with the liquid curve—at this point the solid melts. A liquid can be **supercooled**—cooled below its freezing point without the occurrence of freezing—as shown by the dashed line in Figure 13.14a. In this condition, crystallization will sometimes start if a "seed crystal" of the substance is added to provide a surface on which crystals begin to form. Once crystallization is triggered in this way, supercooled liquids often crystalize very rapidly, with the noticeable evolution of heat, and the solid-liquid equilibrium is restored.

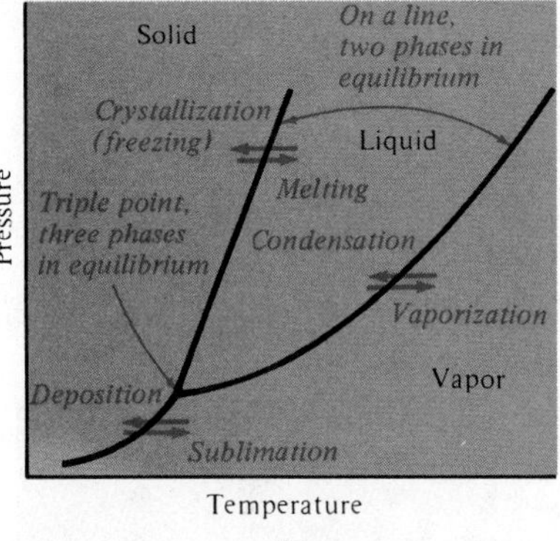

(b)

Figure 13.14
Phase Diagram for a Solid–Liquid–Gas System
To the vapor pressure curves of the solid and liquid (a) are added the melting point–vapor pressure curve of the solid (b). The color arrows show the phase changes at constant pressure that occur in crossing the lines at various points.

To complete the simple phase diagram, the line representing the relationship between pressure and melting point is added (Figure 13.14b). In most cases, the melting point increases with pressure; either way, increase or decrease, the effect is usually quite small, so this line is usually almost vertical.

At any point not on a line in the phase diagram, only a single phase of the substance exists. At any point along a line, two phases are in equilibrium. At the point where three lines intersect—called a **triple point**—three phases are in equilibrium. Phase diagrams for carbon dioxide and water are given in Figure 13.15 (see page 312). (The scale and shape of the lines have been adjusted to make the relationships more easily visible.) Note the different slopes of the solid–liquid equilibrium lines. The freezing point of water *decreases* with increasing pressure.

The **normal freezing point** of a liquid is the temperature at which the liquid freezes at 760 Torr pressure, that is, the temperature at which solid and liquid are in equilibrium at 760 Torr pressure. The temperatures of the normal freezing point and the triple point are usually not the same. At the triple point the solid and liquid phases are both also in equilibrium with the *pure* vapor. Therefore, the pressure at the triple point can be quite different from 760 Torr.

The triple point for water lies *below* atmospheric pressure, while the triple point for carbon dioxide lies above atmospheric pressure. The significance of this difference is apparent when a block of ice and a block of dry ice, which is solid carbon dioxide, stand side by side on a hot day. The water ice melts to a liquid; the dry ice disappears—it sublimes directly to a gas. For the few substances like CO_2, with triple points above atmospheric pressure, the vapor pressure reaches atmospheric pressure *before* a liquid forms and therefore sublimation occurs. Liquid carbon dioxide exists only at pressures greater than 5.11 atm, the pressure at the triple point.

The boiling point of water (see Figure 13.15) decreases with decreasing pressure along the line connecting the normal boiling point and the triple point. Below the triple point, the temperature of sublimation decreases with decreasing pressure. Practical use is made of this effect. At lower pressure, it is possible to remove water safely from a substance that would be damaged by heating for a long time at 100 °C. Coffee and many

The critical point is at 304.2 K and 72.8 atm; the triple point is at 216.6 K and 5.11 atm; the sublimation temperature at 1 atm is 194.6 K.

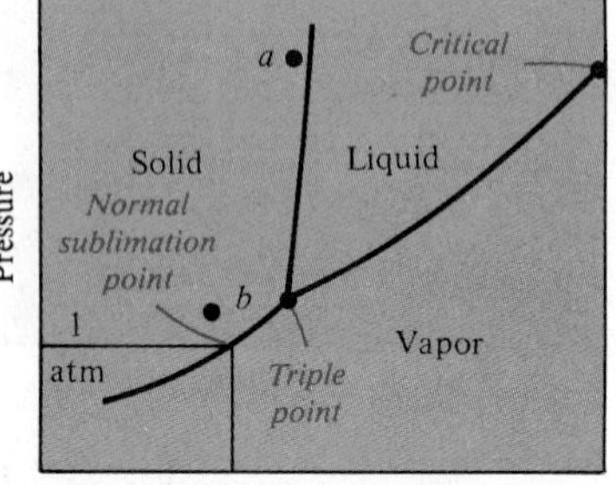

CO_2 (not to scale)

The critical point is at 647.3 K and 217.7 atm; the triple point is at 273.16 K and 0.00603 atm; the normal freezing point is 273.15 K; the normal boiling point is 373.15 K.

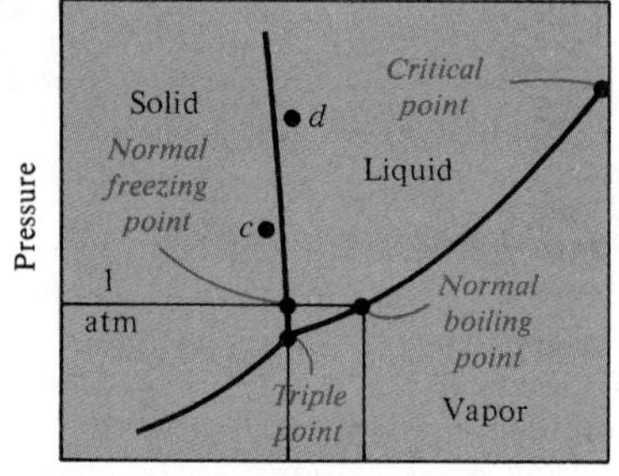

H_2O (not to scale)

Figure 13.15
Phase Diagrams for CO_2 and H_2O
The temperature and pressure scales are not linear and the slope of the solid–liquid line is greatly exaggerated. (The lettered points are used in various examples and exercises in this chapter.)

kinds of food are preserved by "freeze-drying." The coffee or food is first frozen and the ice then vaporized by lowering the pressure. The organic compounds that give food its flavor and texture are unharmed at the low temperature and pressure.

Note that it is the *partial pressure* of the vapor that is plotted in a phase diagram, not the total pressure above a substance. On a dry winter day the partial pressure of water vapor in the air can drop low enough to allow ice to sublime. The gradual disappearance of snow during a bitter cold period is due to sublimation.

EXAMPLE 13.4 Interpretation of a Phase Diagram

Describe what happens to a sample of CO_2 originally at point *a* in the phase diagram (Figure 13.15) as the pressure is decreased at constant temperature.

For a change at constant temperature, the CO_2 will follow a vertical path on the diagram. (See the figure for this example.) Along this path, five different conditions occur: **(1) The CO_2 remains a solid as the pressure decreases. At the most, a very slight expansion in volume might occur. (2) The solid becomes a liquid. (3) The CO_2 remains liquid as the pressure decreases. As for the solid, a very slight expansion might occur. (4) The liquid CO_2 vaporizes. (5) With further decrease in pressure, the gas volume increases in accordance with the gas laws.**

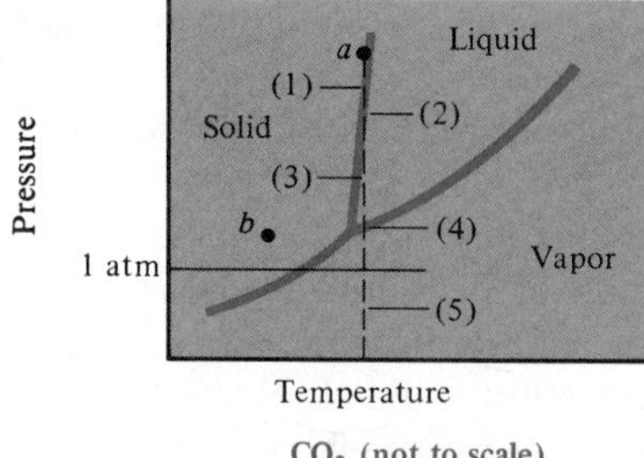

CO_2 (not to scale)

EXERCISE Would liquid CO_2 form upon heating at constant pressure a sample of dry ice originally at point *a* in the accompanying figure? What about point *b*?

Answers yes, no.

SUMMARY OF SECTIONS 13.4–13.7

Changes of state (Figure 13.9) occur at constant temperature as energy is absorbed or released by a substance. In evaporation and condensation, particles escape from or reenter the liquid phase, respectively. Melting (technically known as fusion) and crystallization are the transformation of a solid to a liquid or vice versa.

Particles with sufficiently high kinetic energy can leave the surface of a liquid or a solid and enter the gas phase. At a given temperature, the equilibrium vapor pressure of a liquid is constant. Vapor pressure increases with increasing temperature. The boiling point of a liquid is the temperature at which the vapor pressure equals the pressure of the gases above the liquid; boiling point thus varies with the pressure. The normal boiling point is the temperature at which a substance boils at 760 Torr. Stronger intermolecular forces cause lower vapor pressures, higher boiling points, and higher heats of vaporization. Above the critical temperature, no amount of pressure is sufficient to cause liquefaction of a vapor.

A phase diagram shows the relationships between the physical state of a substance and the pressure and temperature. A simple phase diagram combines the vapor pressure curves for the liquid and solid with the pressure–melting point curve. A liquid can, under certain conditions, be supercooled, that is, cooled below its melting point. At a triple point in a phase diagram, three phases are in equilibrium.

THE SOLID STATE

Skip

13.8 CLOSEST PACKING

A pure metal in its solid state is crystalline and is composed of atoms that are identical in size. The high densities of many metals show that the atoms are close together and the amount of unoccupied space is small. To visualize the crystal structures of metals, think of each metal atom as a solid sphere. Spheres of the same radius pack together most efficiently in a single layer in a hexagonal arrangement. Each sphere touches six other spheres (Figure 13.16a). This arrangement, called closest packing, can continue indefinitely in a single layer, and a crystal can be built up by superimposing one such layer over another.

Two types of superimposition maintain closest packing in three dimensions. In each, the second layer (B) is placed over the first layer (A) so that three spheres of layer B touch the same sphere of layer A (Figure 13.16b).

The third layer can then be superimposed in one of two ways. In the first, each atom in the third layer lies directly above an atom in the first layer. In this arrangement, called **hexagonal closest packing,** closest-packed layers of atoms are arranged in an ABABAB... sequence shown at the left in Figure 13.17.

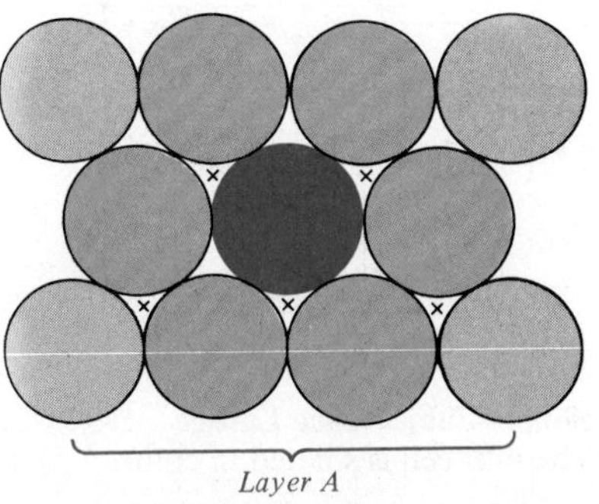

(a) Closest packing of spheres of equal radius in a single layer. Here and in (b), all of the spheres that touch the dark central spheres are shown in light color.

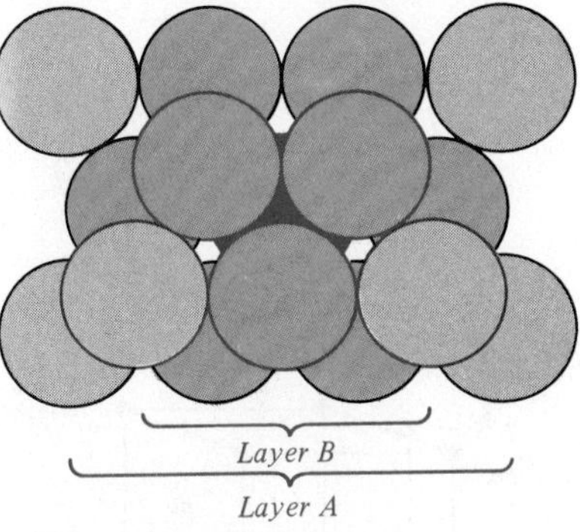

(b) Another closest-packed layer, layer B, has been added above layer A. Layer B spheres cover the layer A spaces marked x in (a). Three spheres in layer B touch the dark central sphere of layer A.

Figure 13.16
Closest Packing in Layers
The light-colored spheres all touch the dark-colored spheres.

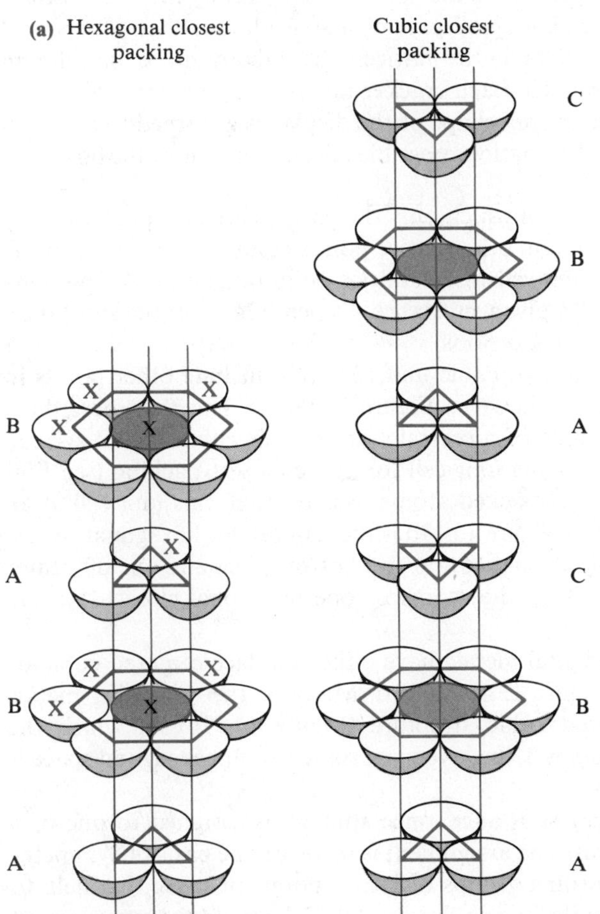

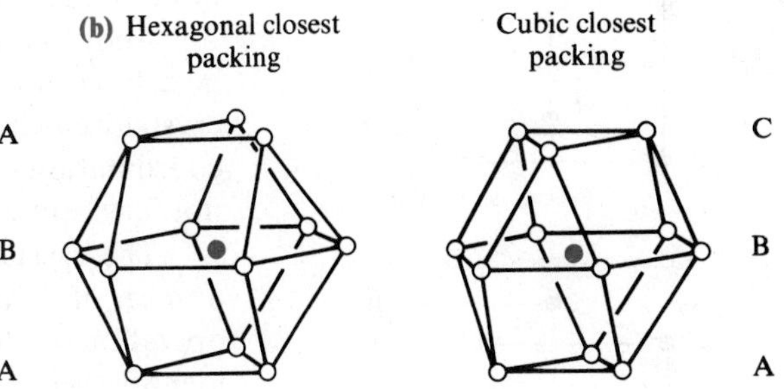

Figure 13.17
Hexagonal and Cubic Closest Packing
(a) Exploded views show stacking of planes of spheres (one-half of each sphere shown). In both types of packing, spheres in layers with the same letters are directly above one another. In hexagonal packing, two layers repeat; in cubic packing, three layers repeat—note the dark-colored spheres one above the other in both cases. (Xs mark the atoms of one hexagonal closest-packed unit cell.)
(b) Arrangement of the twelve nearest neighbors about one sphere in each type of packing. The dark spheres in (a) are equivalent to the dark spheres in (b).

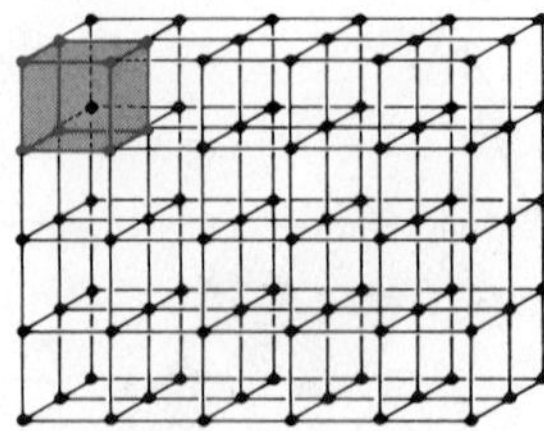

Figure 13.18
Simple Cubic Space Lattice
The unit cell is shaded in color.

In the second way of adding the third layer (layer C), the layer is displaced so that its atoms are not directly above those of either layer A or layer B. In this arrangement, called **cubic closest packing,** closest-packed layers of atoms are arranged in an ABCABCABC... sequence shown at the right in Figure 13.17.

In both hexagonal and cubic closest packing, each sphere touches six other spheres in its own layer, plus three in the layer above and three in the layer below. This gives each sphere twelve nearest neighbors (see Figure 13.17b). The **coordination number** of an atom, ion, or molecule in a particular crystal structure is the number of nearest neighbors of that atom, ion, or molecule. In metals with either of the two closest-packed structures, 74% of the available space is occupied and the coordination number of each atom is 12.

13.9 CRYSTAL SYSTEMS; UNIT CELLS

The infinite repetition of an orderly arrangement of particles in a crystal means that beginning at any given atom and moving in a specific direction will lead to another atom with an identical environment. A **space lattice** is an orderly array of points representing sites with identical environments in the same orientation in a crystal. A simple cubic space lattice is shown in Figure 13.18. The **crystal structure** of a substance is the complete geometrical arrangement of the particles that occupy the space lattice.

A **unit cell** is the most convenient small part of a space lattice that, if repeated in three dimensions, will generate the entire lattice. The cube outlined in color in Figure 13.18 is a simple cubic unit cell. Each cracker box in a supermarket display is like a unit cell in the display. The overall shape of the display is governed to a certain extent by the shape of the box, with variations possible depending on how the boxes are stacked.

A **primitive unit cell** is a unit cell in which only the corners are occupied. In some cases, unit cells are chosen that contain other lattice points in addition to those at the corners; these are called **multiple unit cells.** Each lattice point in a unit cell represents the location of a single particle of a given substance. Depending upon the substance, the particles may be atoms, molecules, or ions.

A primitive cubic unit cell and two cubic unit cells that include other points in addition to those at the corners are shown in Figure 13.19a–c. Together these three cubic unit cells comprise an example of a crystal system—the cubic crystal system. The face-centered cubic unit cell is the unit cell for cubic closest packing (see Figure 13.17); parallel planes of closest-packed atoms pass through this unit cell at an angle. The fourth unit cell shown in Figure 13.19d is the unit cell for hexagonal closest packing; it differs in having diamond-shaped top and bottom faces and one off-center interior lattice point. (In Figure 13.17 the atoms of one hexagonal closest packing unit cell are marked with Xs.)

About two-thirds of all elemental metals have either the face-centered cubic or hexagonal unit cells shown in Figure 13.19c and d. Many of the remaining metals crystallize with the body-centered cubic structure (Figure 13.19b), in which the coordination number of each atom is 8, the layers are not closest packed, and space is not filled quite as efficiently.

The crystal structure of every substance, once studied, is assigned to one of a number of different crystal systems. The assignment is made on the basis of symmetry operations—different ways of turning groups of lattice points in space that help to identify their symmetry. For example, in a perfect crystal of diamond the arrangement of the carbon atoms has the symmetry characteristics of a cube, and diamond belongs to the cubic crystal system. Information from x-ray analysis (see Aside 13.2) shows that, in addition to the atoms that occupy the face-centered lattice points, the diamond unit cell contains four interior carbon atoms arranged in a tetrahedron (Figure 13.20). (We need not consider assignment of crystal systems and unit cells. It is a matter best left to those with experience in the field or to more advanced study.)

The unit cell diagrams in Figure 13.19 are misleading because they suggest that

Cubic crystal system

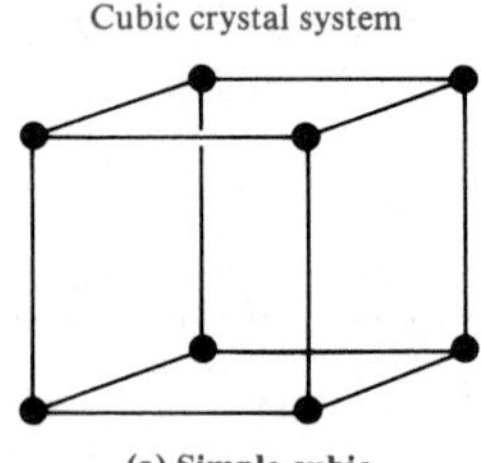

(a) Simple cubic

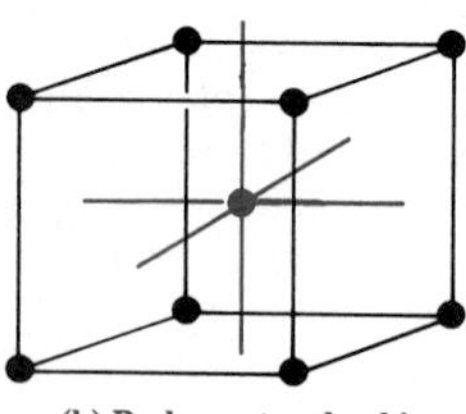

(b) Body-centered cubic

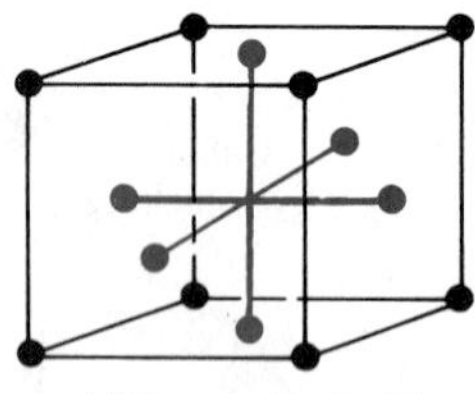

(c) Face-centered cubic

Hexagonal closest-packed unit cell

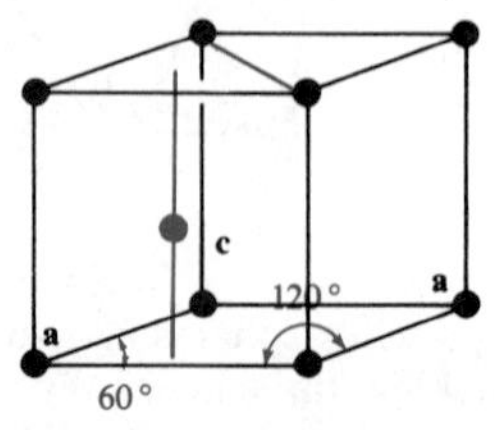

(d) Hexagonal closest packed

Figure 13.19
Four Unit Cells

each cell contains a lot of empty space. The packing of atoms in a crystal is more realistically shown by the sphere-based packing models given in Figure 13.21.

The concepts of space lattices and unit cells are applied to all types of crystals. When the points of a space lattice are occupied by ions of different sizes (as discussed in the next section) or by molecules, the total structure of the unit cell becomes more complex. The unit cell of carbon dioxide (dry ice) is a simple example of the structure of a molecular crystal (Figure 13.22).

Crystallographers specify the number of atoms per unit cell for elements. At first glance, you might think that there are eight atoms occupying the simple cubic unit cell shown in Figures 13.19 and 13.21 but this is not the case. Even though eight lattice points outline the cubic unit cell, an atom at each point is part of seven other units cells that are adjacent to the cell under consideration (Figure 13.23). To find the number of atoms that belong to *each* unit cell, the atoms must be sliced along the faces of the cube. In the simple cubic unit cell, only one-eighth of each corner atom is contained within the individual cell. Therefore, the simple, or primitive, cubic unit cell "contains" one atom:

$$(8 \text{ corners})\left(\frac{1/8 \text{ atom}}{\text{corner}}\right) = 1 \text{ atom}$$

The body-centered cubic unit cell has one complete atom in the interior in addition to the corner atoms, to give a unit cell with two atoms:

$$(8 \text{ corners})\left(\frac{1/8 \text{ atom}}{\text{corner}}\right) + (1 \text{ center atom}) = 2 \text{ atoms}$$

An atom at a face-centered point is divided between two unit cells, giving the face-centered cubic unit cell a total of four atoms.

$$(8 \text{ corners})\left(\frac{1/8 \text{ atom}}{\text{corner}}\right) + (6 \text{ faces})\left(\frac{1/2 \text{ atom}}{\text{face}}\right) = 4 \text{ atoms}$$

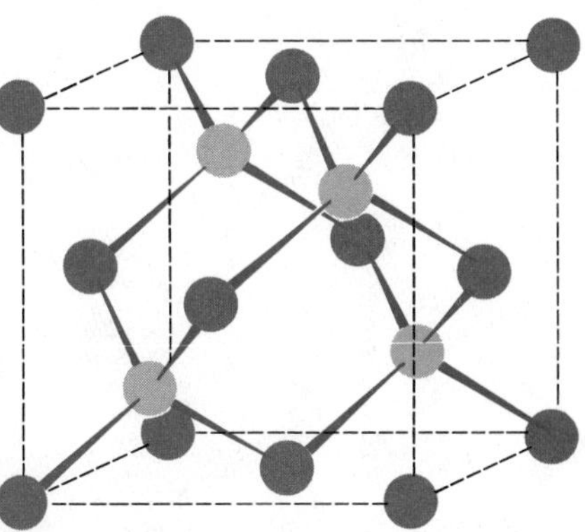

Figure 13.20
A Diamond Unit Cell
Each sphere represents a carbon atom. The unit cell belongs to the cubic crystal system; the atoms forming the face-centered cube that defines this unit cell are shown in the dark color. The solid lines are the C—C bonds. This unit cell also contains four interior carbon atoms, shown in the lighter color.

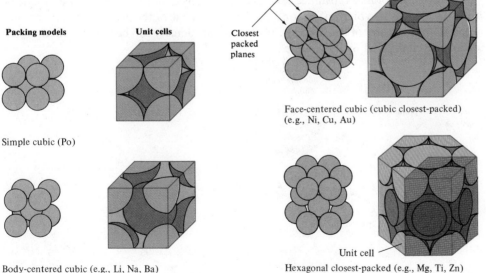

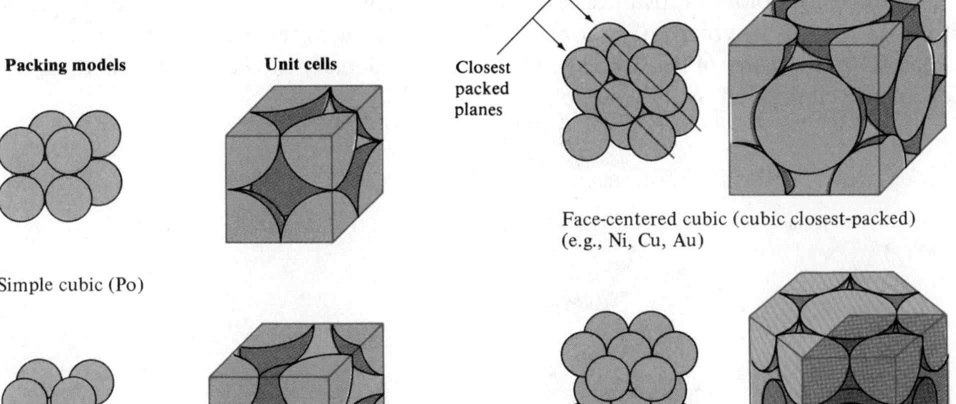

Figure 13.21
Crystal Structure of Metals: Packing Models and Unit Cells for Cubic and Hexagonal Crystal Systems
One metal, polonium, has a simple cubic structure. Most other metals have one of the remaining three structures shown. The number of atoms per unit cell for hexagonal closest packing (two) is not so obvious as it is in the other cases. The unit cell for hexagonal closest packing is the hexagonal cell with one additional interior point.

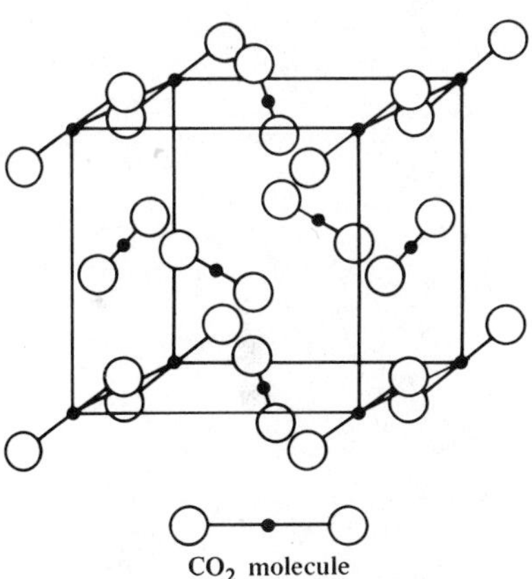

Figure 13.22
The CO_2 Unit Cell

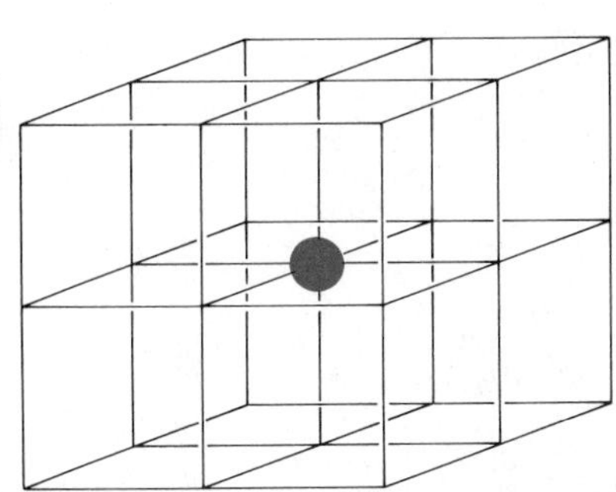

Figure 13.23
Sharing of a Corner Atom by Eight Unit Cells

Once the number of atoms in the unit cell is known, the mass of the unit cell can be calculated. To do this, the mass of one atom is multiplied by the number of atoms in the unit cell. (The mass of one atom can be found by dividing the molar mass of the substance by Avogadro's number.)

EXAMPLE 13.5 Unit Cell—Number of Atoms

(a) How many atoms occupy the diamond unit cell shown in Figure 13.20? (b) What is the mass of a diamond unit cell?

(a) The sketch of the unit cell shows that there are eight corner atoms, six atoms at face-centered points, and four atoms at points in the interior of the cell. The number of atoms in the unit cell is

$$(8 \text{ corners})\left(\frac{1/8 \text{ atom}}{\text{corner}}\right) + (6 \text{ faces})\left(\frac{1/2 \text{ atom}}{\text{face}}\right) + (4 \text{ atoms in interior}) = \mathbf{8\ atoms}$$

(b) The mass of a carbon atom is found by dividing the molar mass by Avogadro's number.

$$\frac{(12.01 \text{ g/mol})}{(6.022 \times 10^{23} \text{ C atoms/mol})} = 1.994 \times 10^{-23} \text{ g/C atom}$$

Thus, for the eight carbon atoms in the unit cell,

$$(8 \text{ C atoms})(1.994 \times 10^{-23} \text{ g/C atom}) = \mathbf{1.595 \times 10^{-22}\ g}$$

EXERCISE The unit cell of metallic gold is face-centered cubic. (a) How many atoms occupy the gold unit cell? (b) What is the mass of a gold unit cell?

Answers (a) 4; (b) 1.308×10^{-21} g.

The lengths of the sides of a unit cell can be found by x-ray diffraction (see Aside 13.2). For a cubic unit cell, the lengths of the sides are identical. The volume of the cubic unit cell can be calculated by cubing the length of the edge of the unit cell.

The density of any substance can be calculated by dividing the mass of a sample of the substance by its volume (Section 2.8d). Density is one of the properties that does not depend upon the size of the sample. For this reason, density can be found from the mass of a unit cell divided by its volume.

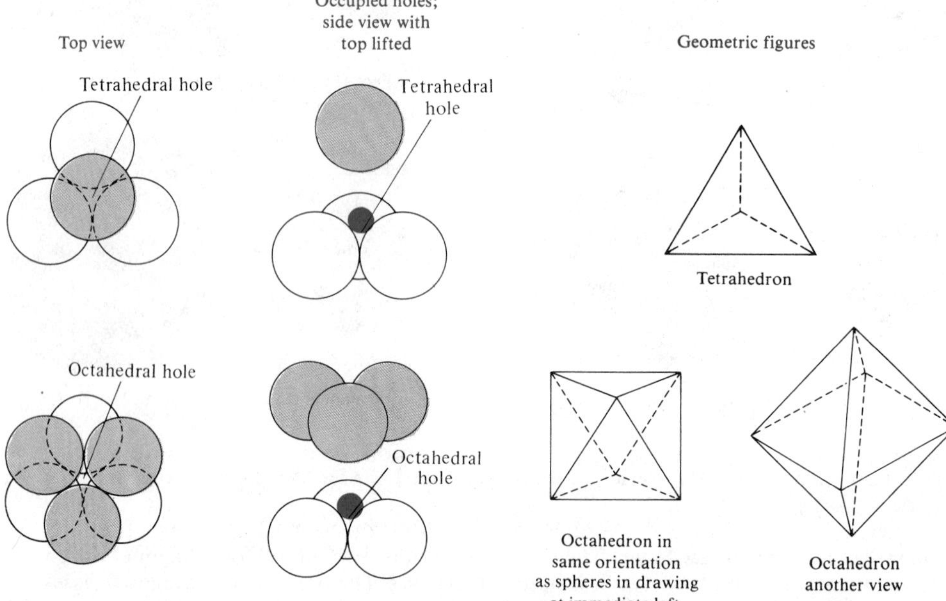

Figure 13.24
Tetrahedral and Octahedral Holes

Density calculated in this way for a unit cell is sometimes called the **theoretical density.** The theoretical density differs from the actual density because most crystals have defects (Section 13.12). Vacancies, for example, would lead to an actual density less than the theoretical density. The presence of impurities could make the actual density either greater or less than the theoretical density, depending on the relative masses of the metal atoms and the impurity atoms.

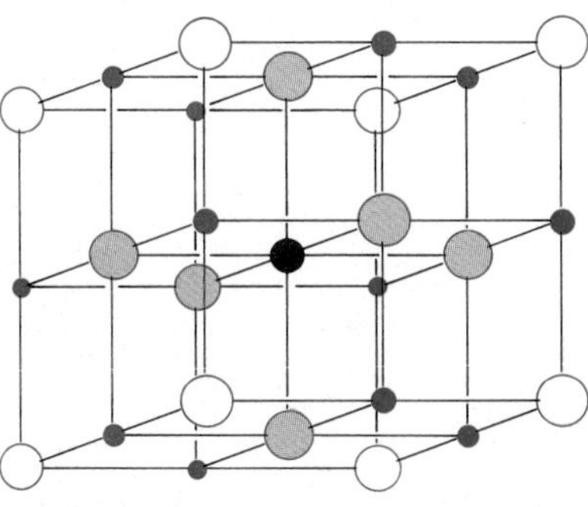

Figure 13.25
Ions in a Sodium Chloride Crystal
The larger Cl^- ions outline the cube. The colored Cl^- ions form one octahedron. The Na^+ ion in the hole at the center of this octahedron is shown in black.

EXAMPLE 13.6 Unit Cell—Volume and Density

The unit cell length of diamond was measured as 0.3567 nm. (a) Calculate the volume of this cubic unit cell in cubic centimeters. (b) Using the mass found in Example 13.5 for the diamond unit cell, find the theoretical density of diamond.

(a) To find the unit cell length expressed in centimeters, the conversion factors based on the meaning of the prefixes "nano" and "centi" must be used:

$$(0.3567\ \text{nm})\left(\frac{1\ \text{m}}{10^9\ \text{nm}}\right)\left(\frac{10^2\ \text{cm}}{\text{m}}\right) = 3.567 \times 10^{-8}\ \text{cm}$$

The volume of the cubic unit is found by cubing the length of the side

$$(3.567 \times 10^{-8}\ \text{cm})^3 = \mathbf{4.538 \times 10^{-23}\ cm^3}$$

(b) Dividing the mass found in Example 13.5b by the volume gives the density as

$$\frac{1.595 \times 10^{-22}\ \text{g}}{4.538 \times 10^{-23}\ \text{cm}^3} = \mathbf{3.515\ g/cm^3}$$

[The measured density of diamond is 3.513 g/cm^3 at 25 °C.]

EXERCISE The unit cell length of gold is 0.4079 nm. (a) Calculate the volume of this cubic unit cell in cubic centimeters. (b) The mass of the gold unit cell is 1.308×10^{-21} g. Calculate the theoretical density of gold.

Answers (a) 6.787×10^{-23} cm^3, (b) 19.27 g/cm^3.

13.10 CRYSTAL STRUCTURE OF IONIC COMPOUNDS

Closest packing of spheres of the same size leaves two types of open spaces, or "holes," between the layers. One approach to visualizing the crystal structure of ionic compounds is to picture a closest-packed array of larger ions, with smaller ions occupying the holes. Usually anions, which are commonly larger than cations, form the closest-packed array, with cations in the holes, but in some cases the situation is reversed.

A *tetrahedral hole* lies at the center of a cluster of four spheres that form a tetrahedron (Figure 13.24). In a closest-packed array there are two tetrahedral holes per closest-packed sphere. An *octahedral hole,* larger than the tetrahedral hole, lies at the center of a cluster of six spheres that form an octahedron (Figure 13.24). One octahedral hole is present for every sphere in a closest-packed array.

An array of anions may open up and depart somewhat from closest packing in order to accommodate cations in the holes. For example, in a sodium chloride crystal, the Na^+ cations occupy octahedral holes in a slightly expanded Cl^- cubic closest-packed array. There is one octahedral hole per Cl^- ion; all of these holes are occupied by Na^+ ions, and thus the 1-to-1 NaCl stoichiometry is achieved. The six Cl^- ions that comprise one octahedron are shown in color in Figure 13.25. Each Na^+ ion in an octahedral hole is surrounded by six Cl^- ions and has a coordination number of six. Similarly, each Cl^- ion has a coordination number of six. Figure 13.26 shows how the sodium chloride crystal structure fits into the cubic crystal system. All ionic crystal structures can be assigned to crystal systems as discussed in the preceding section.

Ionic crystal structures are influenced by the relative charges and sizes of ions. A crystal is most stable when each cation just touches the surrounding anions (and vice

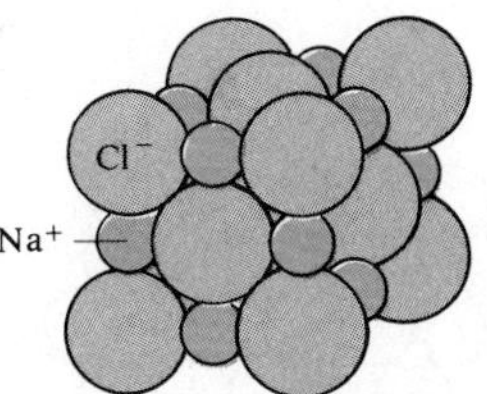

(a) NaCl packing model

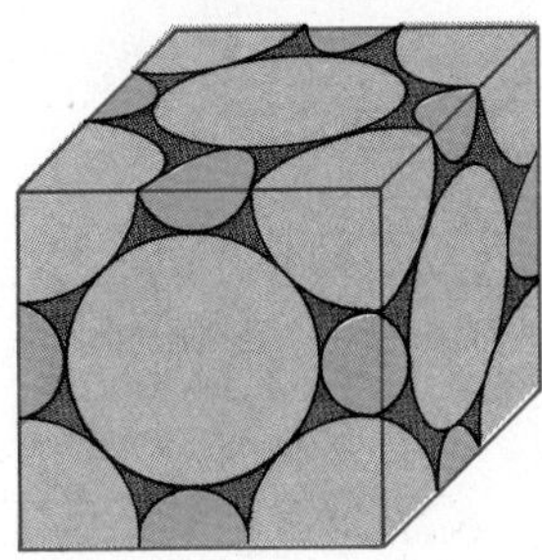

(b) NaCl unit cell (one Na^+ ion is concealed in the center of the cell)

Figure 13.26
Sodium Chloride Crystal Structure
You can see that the Na^+ ions are each surrounded by six Cl^- ions and the Cl^- ions are each surrounded by six Na^+ ions.

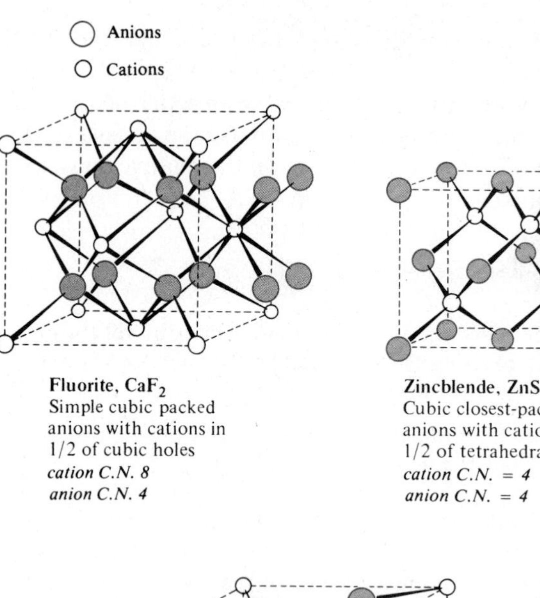

Fluorite, CaF_2
Simple cubic packed anions with cations in 1/2 of cubic holes
cation C.N. 8
anion C.N. 4

Zincblende, ZnS
Cubic closest-packed anions with cations in 1/2 of tetrahedral holes
cation C.N. = 4
anion C.N. = 4

Wurtzite, ZnS
Hexagonal closest packed anions with cations in 1/2 of tetrahedral holes
cation C.N. = 4
anion C.N. = 4

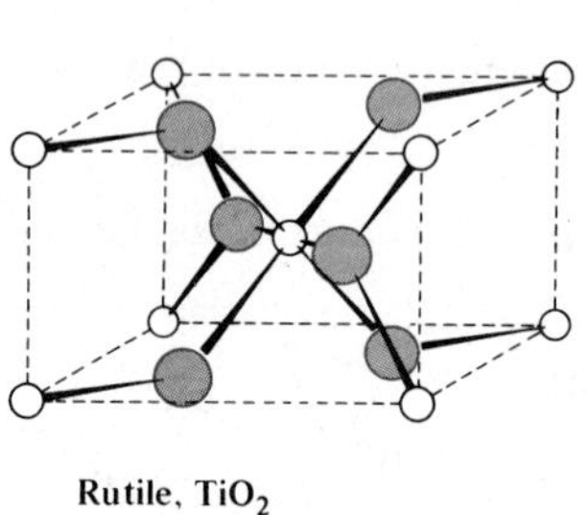

Rutile, TiO_2
Hexagonal closest packed anions with cations in 1/2 of octahedral holes
cation C.N. = 6
anion C.N. = 3

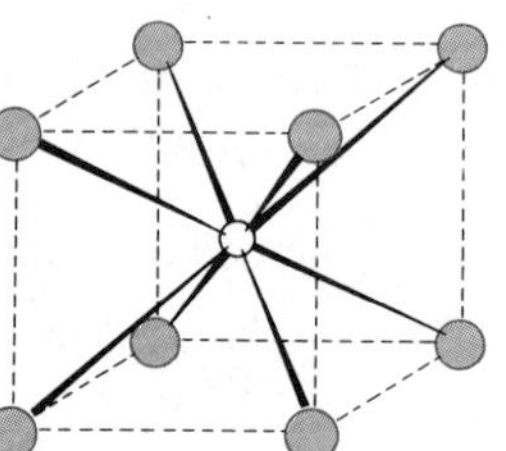

Cesium chloride, CsCl
Simple cubic packed anions with cations in all cubic holes
cation C.N. = 8
anion C.N. = 8

Figure 13.27
Some Simple Crystal Structures
These are common types of crystal structures adopted by many different compounds and referred to by the name of the compound given here. The large color circles represent anions; the smaller open circles represent cations. An antifluorite structure in which the cations and anions are reversed is adopted in some compounds where cations are larger than anions.

versa). Smaller cations best make such contact when they have four nearest neighbors (coordination number 4) and occupy tetrahedral holes, which are smaller than octahedral holes. There are two tetrahedral holes per anion in a closest-packed array of anions. In compounds with a 2-to-1 stoichiometry, such as Li_2O and Na_2S, every tetrahedral hole is occupied by a cation. In compounds with a 1-to-1 stoichiometry, such as ZnS, one-half the tetrahedral holes are occupied.

Zinc sulfide is an example of a compound that is **polymorphous**—able to crystallize in more than one crystal structure. Both zinc sulfide structures are shown in Figure 13.27, which illustrates some of the more common crystal structures and their descriptions in terms of anion packing, hole geometry, and hole occupation. Titanium dioxide (TiO_2) is also polymorphous. In anatase TiO_2 (Figure 13.28a), the oxygen anions form a cubic close-packed array and the titanium cations occupy one-half the octahedral holes. In rutile TiO_2 (Figure 13.28b), the anions are hexagonally closest packed and one-half the octahedral holes are also occupied, but in a different pattern. In both forms of TiO_2 the cation coordination number is 6 and the anion coordination number is 3.

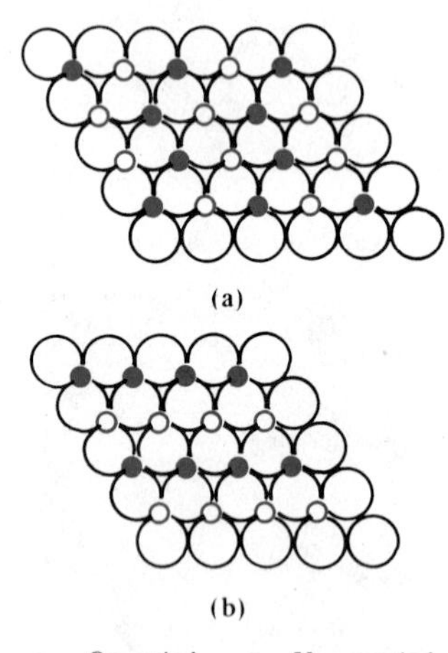

Figure 13.28
Occupation of $\frac{1}{2}$ the Octahedral Holes
(a) Pattern followed in the anatase form of TiO_2; (b) pattern followed in the rutile form of TiO_2. Other patterns are possible.

When the cation is large relative to the anion, it may not fit into either the tetrahedral or octahedral holes in a closest-packed array of anions. In such a case, the anions assume a simple cubic array (see Figure 13.21, top), which leaves cubic holes that provide sufficient space for the larger cations. A cation in a cubic hole has a coordination number of 8. Cesium chloride (see Figure 13.27) has this type of structure.

Substances that have the same crystal structure are said to be **isomorphous.** Some such substances can crystallize together to give a mixed product. For example, sodium nitrate ($NaNO_3$) and calcium carbonate ($CaCO_3$) form mixed crystals, although in terms of other physical properties and all chemical properties, they are quite different.

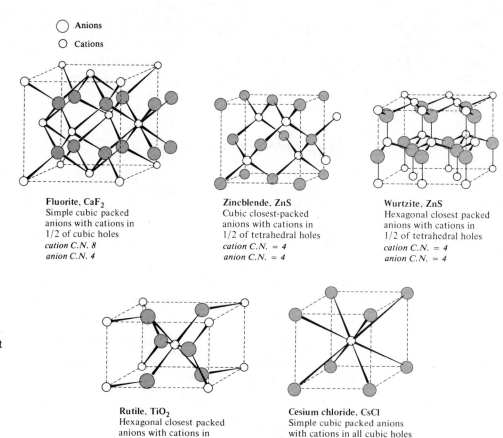

13.11 LATTICE ENERGIES; THE BORN–HABER CYCLE

Lattice energy is a thermodynamic quantity defined as the energy liberated when gaseous ions combine to give one mole of a crystalline ionic compound. Lattice energy is essentially the enthalpy of formation of an ionic compound from ions in the gas phase. As illustrated by the values given in Table 13.3, lattice energies vary with interionic distances in the crystal and the charges of the ions. The closer together the ions and the larger their charges, the greater the lattice energy.

Lattice energies may be calculated by the application of Hess's law (Section 7.10c) to a standard cycle of reactions known as the Born–Haber cycle. The reactions in the cycle for the formation of a metal halide of formula MX represent conversion of the solid metal to cations in the gas phase (steps 1 and 2 in Figure 13.29), conversion of halogen molecules to anions in the gas phase (steps 3 and 4), and combination of the gaseous ions to give the solid compound (step 5). The enthalpy for the last step (step 5) is the lattice energy.

Sublimation energies (step 1), ionization energies (step 2), and bond energies (step 3) are generally known from experimental measurements. Electron affinities (step 4) and lattice energies (step 5), both of which are difficult to measure experimentally, are often determined by Born–Haber cycle calculations.

Steps (1), (2), and (3) all require the input of energy and have positive ΔH values. Electron affinity (step 4) is negative for the halogens and ranges from -349 kJ to -295 kJ. Lattice energy values (step 5) are always negative. When the lattice energy and electron affinity combined provide the energy required by steps (1) to (3), the formation of the ionic compound in question is exothermic and therefore is more favorable than if it were endothermic.

In Example 13.7, the Born–Haber cycle is used to answer a question frequently posed by students of chemistry.

Table 13.3
Lattice Energies
In the series of halides, the lattice energy decreases as the size of the anion or the cation increases. The pairs of Fe and Na salts show the effect of higher charges in increasing the lattice energy.

Halide	Lattice energy (kJ/mol)	Interionic distance (nm)
NaF	−910	0.2317
NaCl	−769	0.282
NaBr	−732	0.297
NaI	−682	0.323
LiCl	−834	0.257
NaCl	−769	0.282
KCl	−701	0.315
RbCl	−680	0.329
CsCl	−657	—
$FeCl_2$	−2525	—
$FeCl_3$	−5364	—
Na_2SO_4	−1827	—
$FeSO_4$	−2983	—
Na_2CO_3	−2301	—
$FeCO_3$	−3121	—

EXAMPLE 13.7 Born–Haber Cycle

Even though much less energy is required to remove one valence electron from a magnesium atom than both electrons ($_{12}$Mg, $1s^2 2s^2 2p^6 3s^2$), the element forms compounds containing Mg^{2+} ion rather than Mg^+. With the use of the thermochemical information given below (at 25 °C) show that the reaction

$$Mg(s) + Cl_2(g) \longrightarrow MgCl_2(s)$$

is more favorable energetically than

$$Mg(s) + \tfrac{1}{2}Cl_2(g) \longrightarrow MgCl(s)$$

Assign the known data to Born–Haber cycle steps according to Figure 13.29 for the formation of MgCl. Note that the stoichiometry of the steps must be modified for the formation of $MgCl_2$.

Reaction	$\Delta H°$(kJ)
$Mg(s) \longrightarrow Mg(g)$	150.2
$Mg(g) \longrightarrow Mg^+(g) + e^-$	743.8
$Mg(g) \longrightarrow Mg^{2+}(g) + 2e^-$	2200.3
$Cl_2(g) \longrightarrow 2Cl(g)$	243.4
$Cl(g) + e^- \longrightarrow Cl^-(g)$	−367.8
$Mg^+(g) + Cl^-(g) \longrightarrow MgCl(s)$	−676.1
$Mg^{2+}(g) + 2Cl^-(g) \longrightarrow MgCl_2(s)$	−2500.1

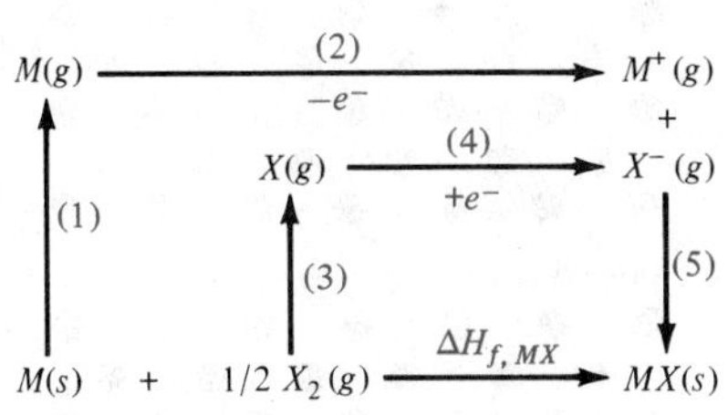

ΔH_1 = sublimation energy
ΔH_2 = ionization energy
ΔH_3 = 1/2 (bond dissociation energy)
ΔH_4 = electron affinity
ΔH_5 = lattice energy
$\Delta H_{f,MX} = \Delta H_1 + \Delta H_2 + \Delta H_3 + \Delta H_4 + \Delta H_5$

Figure 13.29
Born–Haber Cycle for the Formation of MX(s) from M(s) and $\frac{1}{2}X_2(g)$
This type of thermochemical cycle was introduced by M. Born, K. Fajans, and F. Haber in 1919. For reasons unknown, the cycle is commonly called the "Born–Haber" cycle rather than, more correctly, the Born–Fajans–Haber cycle.

For the formation of MgCl(*s*) from the elements, the Born–Haber cycle steps in Figure 13.29 apply as follows:

Reaction	$\Delta H°$(kJ)	
$Mg(s) \longrightarrow Mg(g)$	150.2	(Step 1, sublimation energy)
$Mg(g) \longrightarrow Mg^+(g) + e^-$	743.8	(Step 2, ionization energy)
$\frac{1}{2}Cl_2(g) \longrightarrow Cl(g)$	121.7	(Step 3, $\frac{1}{2}$ × bond dissociation energy)
$Cl(g) + e^- \longrightarrow Cl^-(g)$	−367.8	(Step 4, electron affinity)
$Mg^+(g) + Cl^-(g) \longrightarrow MgCl(s)$	−676.1	(Step 5, lattice energy)
$Mg(s) + \frac{1}{2}Cl_2(g) \longrightarrow MgCl(s)$	**−28.2 kJ = $\Delta H_f°$, MgCl(*s*)**	

For the formation of $MgCl_2(s)$, the thermochemical steps are as follows:

Reaction	$\Delta H°$(kJ)	
$Mg(s) \longrightarrow Mg(g)$	150.2	(Step 1, sublimation energy)
$Mg(g) \longrightarrow Mg^{2+}(g) + 2e^-$	2200.3	(Step 2, ionization energy)
$Cl_2(g) \longrightarrow 2Cl(g)$	243.4	(Step 3, bond dissociation energy)
$2Cl(g) + 2e^- \longrightarrow 2Cl^-(g)$	−735.6	(Step 4, 2 × electron affinity)
$Mg^{2+}(g) + 2Cl^-(g) \longrightarrow MgCl_2(s)$	−2500.1	(Step 5, lattice energy)
$Mg(s) + Cl_2(g) \longrightarrow MgCl_2(s)$	**−641.8 kJ = $\Delta H_f°$, $MgCl_2(s)$**	

The energy released in the formation of $MgCl_2(s)$ from its elements is much greater than that which would be released if MgCl(*s*) were formed. Thermochemically, in terms of $\Delta H_f°$ values, $MgCl_2(s)$ is the much more stable species. It is clear from the data that the much greater lattice energy of $MgCl_2(s)$ is the major factor contributing to this greater stability. The smaller size and greater charge of Mg^{2+} compared to Mg^+ contribute to the greater lattice energy of $MgCl_2$.

EXERCISE The heat of formation of KF(*s*) at 25 °C is −562.58 kJ/mol. Use the following thermodynamic data at 25 °C to calculate the lattice energy of KF(*s*): $\Delta H° = 424.93$ kJ/mol for the ionization of K(*g*), $\Delta H° = -349.7$ kJ/mol for the electron affinity of F(*g*), $\Delta H° = 90.00$ kJ/mol for the sublimation of K(*s*), and $\Delta H° = 157.99$ kJ/mol for the bond dissociation energy of $F_2(g)$.

Answer −806.8 kJ/mol.

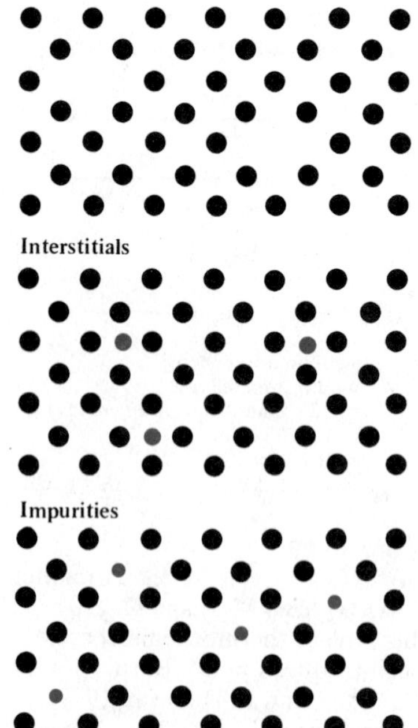

Figure 13.30
Point Defects

13.12 DEFECTS

Except for single crystals grown under special conditions, crystalline substances are seldom perfect. A perfect crystal is chemically pure and structurally perfect, with every lattice point occupied as specified by the unit cell.

Many physical and chemical properties of solids *depend* upon the presence of defects in the solid state. Perfect crystals are very strong, whereas most solids contain enough defects to allow them to yield more easily to mechanical forces. Also, chemical reactions in the solid state require the motion of atoms or ions through a solid. In a perfect crystal there is no available pathway for such motion, whereas in real crystals, atoms or ions can move from defect to defect. The defect structure plays an essential role in determining the properties of semiconductors (Section 30.8).

Point defects, one of several types of defects defined by solid-state chemists, are variations in the occupation of the lattice or interstitial sites in the crystal. Three basic types of point defects occur (Figure 13.30): (1) vacancies—unoccupied lattice sites; (2) interstitial atoms or ions—atoms or ions in the spaces among lattice sites; and (3) impurity defects—foreign ions in regular lattice sites or interstitial sites. In ionic crystals neutrality must be maintained, in some cases by the establishment of an equilibrium between positively and negatively charged defects.

The tendency of some substances to incorporate point defects in their crystal structures accounts for the occurrence of **nonstoichiometric compounds**—compounds in which atoms of different elements combine in other than whole-number ratios. Such compounds exist only in the solid state and in some cases have variable compositions. For example, in wüstite, $Fe_{<1}O$, up to 14% of the normal cation sites can be empty. To maintain neutrality, two Fe^{2+} ions are converted to Fe^{3+} for every missing Fe^{2+}. As prepared in the laboratory at atmospheric pressure, neither iron(II) oxide nor copper(II) sulfide has a stoichiometric composition. And in titanium oxide of the stoichiometric composition TiO, 15% of the sites of each type are vacant. Therefore, there exists a wide range of nonstoichiometric titanium oxide compositions, TiO_x, where x can be less than or greater than one depending upon the oxygen pressure during preparation of the sample.

SUMMARY OF SECTIONS 13.8–13.12

The atoms, molecules, or ions in crystalline substances have a regular, repetitive three-dimensional arrangement. In a single layer, atoms or ions of the same size are arranged most efficiently in a closest-packed layer with each atom surrounded by six others. Most metals have crystal structures based on the stacking of closest-packed layers in either hexagonal closest packing (AB AB... layers) or cubic closest packing (ABC ABC... layers).

Crystal structures are classified by assigning the spatial arrangement of atoms, ions, or molecules to crystal systems identified by the type of space lattice and unit cell that are occupied. Examples are given in Section 13.10 of the simple cubic lattice and some unit cells based upon it. Other similar crystal systems, not discussed here, are used in classifying all types of crystals, whether made up of atoms, molecules, or ions.

Ionic crystals can be pictured as arrays of the larger ions, usually anions, with the smaller ions occupying holes. In cubic and hexagonal closest-packed arrays there are two types of holes—tetrahedral holes and octahedral holes. The octahedral holes are the larger of the two. Still larger holes—cubic holes—are formed in simple cubic arrays of anions, which are not closest packed.

The lattice energy—the enthalpy of formation of an ionic compound from ions in the gas phase—varies with the charges of ions and the distance between them. Lattice energy can be calculated by applying Hess's law to a sequence of reactions known as the Born–Haber cycle, in which the heat of formation of an ionic compound is set equal to the sum of the appropriate sublimation and ionization energies, bond energy, electron affinity, and lattice energy.

Crystals are seldom perfect. They incorporate defects of various types that contribute to the physical and chemical properties of substances.

Aside 13.2 TOOLS OF CHEMISTRY: DIFFRACTION

Two beams of light of the same wavelength that come from two different sources and meet at the same point can reinforce or cancel each other. These effects represent constructive and destructive interference, respectively (see Figure 8.2). If the waves are in phase and reinforce each other, a bright light appears. If the waves are partially or completely out of phase, a dimmer light or no light at all is visible. A diffraction pattern of light and dark lines is produced when light is split by passing through two or more slits (see Figure 8.4). To be diffracted, waves must have a wavelength comparable in size to the width of the slits.

The regularly spaced rows of atoms, ions, or molecules in a crystal can diffract radiation with wavelengths on the order of 0.1 nm. X-rays, neutron beams, and electron beams that fulfill this requirement are used in studying both crystal structure and molecular structure.

X-ray diffraction is a valuable tool in the determination of crystal structure. Only after 1912, when the German physicist Max von Laue found that x-rays could be diffracted by crystals, was proof of the orderly structures of crystals possible. X-rays interact with the orbital electrons of the atoms in a crystal and x-ray diffraction shows where electron density is concentrated. X-ray diffraction patterns are interpreted by a method developed by the English scientists William and Lawrence Bragg. If x-rays of the same wavelength arrive at a given point in phase, the distance they have traveled must be a whole-number multiple of their wavelength. Picture two x-ray beams as reflected by successive planes of atoms in a crystal (Figure 13.31). The rays from the lower plane travel a distance $2l$ farther than the rays reflected by the upper plane. If this distance is a whole-number multiple of the wavelength, beams 1′ and 2′ are in phase and reinforce each other. Mathematically this condition is stated

$$n\lambda = 2l$$

whole number → n; *wavelength* → λ

The angle of incidence of the beam θ is equal to the angle BAC and the triangle BAC is a right triangle, with AC equal to d, the distance between the planes. Therefore

$$l = d \sin \theta$$

Combining the two preceding equations gives the *Bragg equation*, or Bragg's law

$$n\lambda = 2d \sin \theta$$

In the original Bragg method a single crystal is rotated through many values of θ. Using x-rays of a known single wavelength, the value of d and the dimensions of the unit cell in a crystal can be found. The intensity of the diffracted radiation varies with the type and location of atoms or ions in the unit cell. Information about the structure of molecules in a molecular crystal or the arrangement of ions in an ionic crystal is derived by studying these variations. Polycrystalline samples can also be studied, using the technique called x-ray powder diffraction.

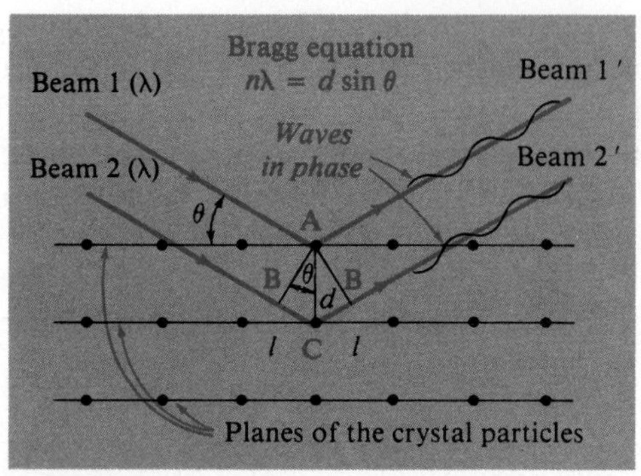

Figure 13.31
X-Ray Diffraction as Interpreted by the Bragg Equation
Beams 1 and 2 of monochromatic x-radiation are reflected by the successive planes of atoms, molecules, or ions in the crystal to give beams 1′ and 2′ with their waves in phase. Beam 2 travels $2l$ farther than beam 1, where l = BC. The distance between the planes is designated by d, where d = AC.

An electron does not penetrate as far past a crystalline surface as an x-ray beam does. Therefore, low-energy electron diffraction (LEED) is useful for studying atomic arrangements and electron densities at surfaces. In electron diffraction, scattering of the radiation is the result of interaction with both orbital electrons and nuclei. Electron diffraction studies of thin foils yield information both about the regularity of the crystal structure and about its defects.

In neutron diffraction, an incident neutron beam is scattered solely by the nuclei, not by the orbital electrons. This makes neutron diffraction useful in cases where x-ray diffraction is not effective.

SIGNIFICANT TERMS (with section references)

amorphous solid (13.3)
boiling point (13.6)
condensation, liquefaction (13.4)
coordination number (crystal) (13.8)
critical point (13.5)
critical pressure (13.5)
critical temperature (13.5)
crystal (13.3)
crystal structure (13.9)
crystalline solid (13.3)
cubic closest packing (13.8)
evaporation (13.4)
fusion (13.4)
hexagonal closest packing (13.8)
isomorphous (13.10)
lattice energy (13.11)
melting point (13.4)
multiple unit cells (13.9)
nonstoichiometric compounds (13.12)
normal boiling point (13.6)
normal freezing point (13.7)
polymorphous (13.10)
primitive unit cell (13.9)
space lattice (13.9)
sublimation (13.4)
supercooled (13.7)
superheated (13.6)
surface tension (13.2)
theoretical density (13.9)
triple point (13.7)
unit cell (13.9)
vaporization (13.4)
viscosity (13.2)

QUESTIONS AND PROBLEMS

Liquids and Solids

13.1 Why does a gas completely fill its container, a liquid spread to take the shape of its container, and a solid retain its shape?

13.2 The density of solid lead is 11.288 g/cm^{3}at 20 °C, that of liquid lead is 10.43 g/cm^3 at 500 °C, and that of gaseous lead is 1.110 g/L at 2000 °C and 1 atm pressure. (a) Calculate the volume occupied by one mole of lead in each state. The radius of a lead atom is 0.175 nm. Calculate (b) the volume actually occupied by one mole of lead atoms and (c) the fraction of volume of each state actually occupied by the atoms.

13.3 Name a property of a liquid that is related to the viscosity. What factors determine how viscous a liquid is? How does viscosity change with an increase in temperature?

13.4 What is the surface tension of a liquid? What causes this phenomenon? How does surface tension change with an increase in temperature?

13.5 What part of the meniscus is "read" for a liquid that "wets" glass, such as an aqueous solution of NaOH in a graduated cylinder? What part is "read" for a liquid that does not "wet" glass, such as mercury in a barometer tube?

13.6 What happens inside a capillary tube if a liquid "wets" the tube? What happens if a liquid does not "wet" the tube?

13.7 Compare several properties of crystalline and amorphous substances.

13.8 In the solid state, argon, calcium, sodium fluoride, and carbon dioxide have quite different physical properties such as melting points and heats of fusion, yet all four substances have similar molar masses (about 40 g/mol) and have similar crystal structures. Account for the differences in their properties.

Relationships between Phases—Changes of State and Boiling Point

13.9 Is the equilibrium that is established between two physical states of matter an example of static or dynamic equilibrium? Explain your answer.

13.10 Choose the changes of state that are exothermic: (a) fusion, (b) liquefaction, (c) sublimation, (d) deposition.

13.11 The heat of formation of BeF_2 at 700 K is -1023 kJ/mol for the solid and -797 kJ/mol for the gas. Calculate the heat of sublimation of BeF_2 at 700 K.

13.12 The heat of formation of ICl at 25 °C is -35.1 kJ/mol for the solid, -23.89 kJ/mol for the liquid, and 17.78 kJ/mol for the gas. Calculate the heats of fusion, vaporization, and sublimation for ICl at this temperature.

13.13 Water can be cooled in hot climates by the evaporation of water from the surfaces of canvas bags. What mass of water (heat capacity = 75 J/K mol) can be cooled from 35 °C to 20 °C by the evaporation of one gram of water (heat of vaporization = 44 kJ/mol)?

13.14 What is the enthalpy change for preparing 1.00 g of ice at 0 °C from water at 25 °C? If this energy is to be absorbed by Freon-12 (a common household refrigerant, CCl_2F_2), what mass of refrigerant must be vaporized? The heat of vaporization of Freon-12 is 165.1 J/g. The heat of fusion of water is 6.01 kJ/mol and the specific heat of liquid water is 4.18 J/K g.

Relationships between Phases—Vapor Pressure

13.15 Choose from each pair the substance that in the liquid state would have the larger vapor pressure at a given temperature: (a) C_6H_6 or C_6Cl_6, (b) $H_2C{=}O$ or CH_3OH, (c) Ga or Cu, (d) He or H_2. Base your choice on intermolecular-force strengths.

13.16 Repeat Question 13.15 for (a) $BiBr_3$ or $BiCl_3$, (b) CO or CO_2, (c) Li or Au, (d) N_2 or NO, (e) CH_3COOH or $HCOOCH_3$.

13.17 A closed flask contained liquid water at 73.2 °C. The total pressure of the air and water vapor mixture was 627.4 Torr. The vapor pressure of water at this temperature is 268.0 Torr. What was the partial pressure of the air?

13.18 The vapor pressure of solid iodine at 25 °C is 0.466 Torr. What mass of iodine will sublime into a 1.0 L container? If the atmospheric pressure in the container is 763 Torr, what will be the total pressure in the container?

13.19 The total pressure in a flask containing dry air and silicon tetrachloride, $SiCl_4$, was 988 Torr at 25 °C. The volume of the flask was halved and the pressure changed to 1742 Torr while the temperature was held constant. What is the vapor pressure of $SiCl_4$ at this temperature? Assume that a small amount of liquid $SiCl_4$ is present in the container at all times.

13.20 The vapor pressure of $CH_3CH_2CH_2Cl$ at room temperature is 385 Torr. The total pressure of $CH_3CH_2CH_2Cl$ and air in a container is 745 Torr. What will happen to the pressure if the volume of the container is doubled at constant temperature? Assume that a small amount of liquid $CH_3CH_2CH_2Cl$ is present in the container at all times.

13.21 The apparatus shown below can be used to determine the vapor pressure of a liquid. A very small amount of liquid is injected and allowed to evaporate. More liquid is added until a small amount of liquid remains and establishes equilibrium with the vapor. The difference between the liquid heights in the U-shaped tube, *h*, measured in mm is the vapor pressure in Torr if the tube contains mercury. At 15 °C the values at *a* and *b* were 26.9 and 22.5 mm, and at 32 °C the values were 47.6 and 1.7 mm. Find the vapor pressure at each temperature.

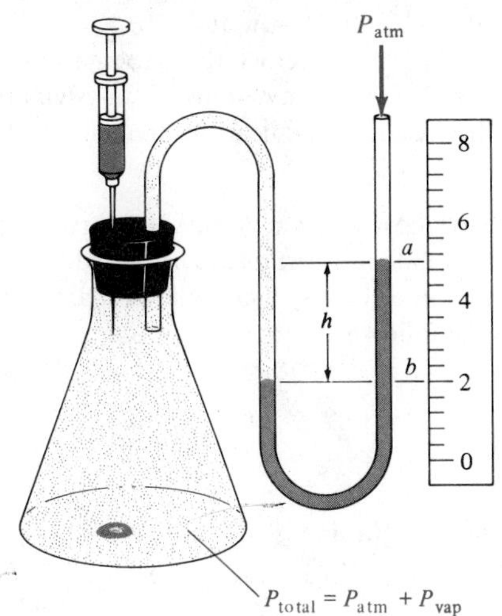

13.22 The vapor pressure of a liquid can be determined by slowly bubbling an inert gas through the liquid to be analyzed (making sure equilibrium has been established) and measuring the total pressure of the gaseous mixture (P_t), the number of moles of inert gas (n_1), and the number of moles of liquid that vaporized (n_2). Dalton's law of partial pressures gives the vapor pressure of the liquid (P_2) as

$$P_2 = P_t\left[\frac{n_2}{n_1 + n_2}\right]$$

After bubbling nitrogen through formic acid, HCOOH, a gaseous mixture collected at 745 Torr contained 30.28 g of N_2 and 2.35 g of HCOOH. Find the vapor pressure of formic acid at this temperature.

13.23 The relative humidity is defined as the actual partial pressure of water present in air divided by the equilibrium vapor pressure at that temperature. What is the relative humidity of air containing 14.3 Torr water vapor at 22 °C, given that the equilibrium vapor pressure of water is 19.8 Torr at this temperature? Express your answer as a percentage.

13.24 What is the partial pressure of water in air that has a relative humidity (see Problem 13.23) equal to 82% at 28 °C? The vapor pressure of water at 28 °C is 28.3 Torr. What would happen as this air sample is cooled to 22 °C? The vapor pressure of water at 22 °C is 19.8 Torr.

13.25 The vapor pressure of solid carbon dioxide is 35 Torr at −110 °C and 1485 Torr at −70 °C (see Figure 13.11). A student predicted that at −90 °C, which is halfway between −110 °C and −70 °C, the vapor pressure would be (35 Torr + 1485 Torr)/2 = 760. Torr. The student was amazed to find that the value was only 280. Torr at this temperature. What did the student overlook in making this prediction?

13.26 At the normal boiling point the heat of vaporization of water (100.00 °C) is 40,656 J/mol and of heavy water (101.41 °C) is 41,606 J/mol. Use these data to calculate the vapor pressure of each liquid at 75.00 °C.

13.27 What is the boiling point of water at a typical laboratory pressure of 745 Torr? The heat of vaporization of water at 100.00 °C and 760.0 Torr is 40,656 J/mol.

13.28 The sublimation pressure of tungsten is 9.08×10^{-8} atm at 3000. K and 7.42×10^{-6} atm at 3500. K. Find the heat of sublimation of tungsten. Predict the sublimation pressure of tungsten at 3200. K. How many atoms of tungsten are in a cubic centimeter in the gaseous state near the filament in a light bulb at 3200. K?

13.29* A plot of the logarithm of the vapor pressure against the inverse of the absolute temperature is predicted by the Clausius-Clapeyron equation to be a straight line. Prepare such a plot for mercury from the following data:

P, Torr	1.00	10.0	40.0	100.0	400.0	760.0
T, °C	126.2	184.0	228.8	261.7	323.0	357.0

The slope of the straight line is equal to $-\Delta H_{\text{vaporization}}/2.303\ R$. Calculate the value of $\Delta H_{\text{vaporization}}$ from your graph.

13.30* Prepare plots of the logarithm of the vapor pressure against the inverse of the absolute temperature for the solid and liquid forms of ammonia using the following data:

	solid			liquid		
P, Torr	1.00	10.0	40.0	100.0	400.0	760.0
T, °C	−109.1	−91.9	−79.2	−68.4	−45.4	−33.6

The slopes of the lines through these data are equal to $-\Delta H_{\text{sublimation}}/2.303\ R$ and $-\Delta H_{\text{vaporization}}/2.303\ R$, respectively. Calculate these heats of transformation. Predict the vapor pressure of ammonia at the melting point, −77.8 °C.

13.31 The three major components of air are N_2 (b.p. − 196 °C, O_2 (b.p. − 183 °C), and Ar (b.p. − 186 °C). Suppose we had a sample of liquid air at −200 °C. In what order would these gases evaporate as the temperature is raised?

13.32 Why is it necessary to specify the atmospheric pressure over a liquid when measuring a boiling point? What is the definition of the normal boiling point?

Relationships between Phases—Phase Diagrams

13.33 What is the critical point? Will a substance always be a liquid below the critical temperature? Why or why not?

13.34 Are samples of supercooled and superheated liquids examples of stable states? Explain.

13.35 How many phases exist at a triple point? Describe what would happen if a small amount of heat were added under constant-volume conditions to a sample of water at the triple point. Assume a negligible volume change during fusion.

13.36 Describe what would happen to a sample of CO_2 originally at point *b* in Figure 13.15 if it undergoes a sudden isothermal (constant temperature) decrease in pressure. Compare your answer to that for a sample originally at point *b* being heated at constant pressure.

13.37 The pressure on a sample of CO_2 at point *a* and on a sample of H_2O at point *c* in Figure 13.15 is greatly increased under constant-temperature conditions. Compare the effects on the samples.

13.38 Describe the physical state of a sample of water originally at point *d* in Figure 13.15. (a) Describe the changes on the sample as the pressure is decreased at constant temperature. (b) Prepare a heating curve (see Figure 7.10) for a second sample of water originally at point *d* as it is heated at constant pressure. (c) Prepare a cooling curve for a third sample of water originally at point *d* as it is cooled at constant pressure.

The Solid State—Metals and Unit Cells

13.39 Briefly describe how layers of closest-packed atoms can be arranged to give hexagonal and cubic closest packing.

13.40 What does the term "crystal coordination number" mean? What is the value of the coordination number for an atom in either hexagonal or cubic closest packing?

13.41 Define the term "unit cell." Prepare sketches of the primitive, body-centered, and face-centered cubic unit cells.

13.42 Choose two different unit cells in the two-dimensional lattice shown. Make one as simple as possible and the other orthogonal (containing right angles). Which of these is simpler to use to calculate area, etc.?

13.43 Polonium crystallizes in a primitive cubic unit cell with an edge length of 0.336 nm. (a) What is the mass of the unit cell? (b) What is the volume of the unit cell? (c) What is the theoretical density of Po?

13.44 Calculate the theoretical density of Na. The length of the body-centered cubic unit cell is 0.424 nm.

13.45 Silver crystallizes in a face-centered cubic unit cell. The density of Ag is 10.5 g/cm^3. Calculate the length of the edge of the unit cell.

13.46 Platinum crystallizes in a cubic unit cell with an edge length of 0.39231 nm. The density of Pt is 21.45 g/cm^3. Determine the type of cubic unit cell that Pt forms and calculate the value of Avogadro's number from these data.

13.47 Crystalline silicon has the same structure as diamond, with a unit-cell edge length of 0.54305 nm. What is the theoretical density of Si?

13.48* Magnesium crystallizes in the hexagonal closest-packed unit cell with $a = 0.3203$ nm and $c = 0.5196$ nm, as shown in Figure 13.19. (a) What is the mass contained in the unit cell? The volume of the unit cell is given by $a^2c \sin 60°$. (b) Calculate the volume of the unit cell. (c) Calculate the theoretical density of Mg.

The Solid State—Ionic Compounds

13.49 A unit cell consists of a cube in which there are cations at each corner and anions at the center of each face. (a) Sketch the unit cell. How many (b) cations and (c) anions are present? (d) What is the simplest formula of the compound?

13.50 A unit cell consists of a cube in which there are anions at each corner and one at the center of the unit cell and cations at the center of each face. How many cations and anions make up the unit cell? What is the simplest formula for this compound?

13.51* The unit cell length of the sodium chloride crystal (see Figure 13.25) is 0.56402 nm. Calculate the theoretical density of NaCl.

13.52* The ions in crystalline LiI are arranged as shown in the sketch

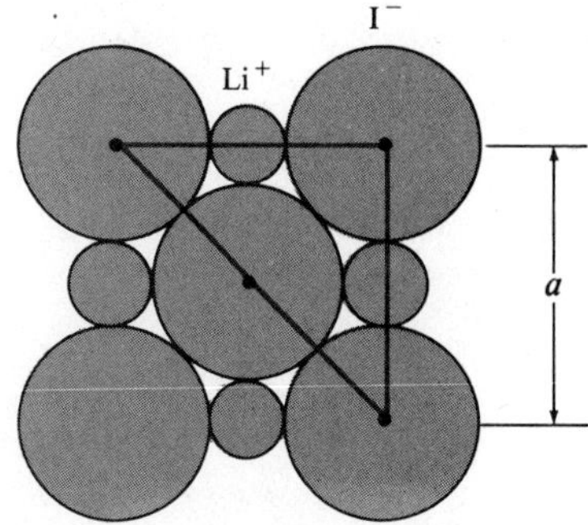

What is the relationship between a and the radius of the iodide ion? Calculate the ionic radius of the iodide ion given that $a = 0.600$ nm for LiI.

13.53 Define the term "lattice energy." Write a chemical equation illustrating the reaction between M^{2+} and X^{2-} that fits your definition. How does the value of the lattice energy change with increasing ionic charge and with increasing ionic radius?

13.54 Sketch the Born–Haber cycle for the formation of MO(*s*) from M(*s*) and $O_2(g)$. Write the expression for ΔH_f° in terms of the various enthalpy changes involved.

13.55 Use the following heat of formation data at 25 °C to calculate the lattice energies for NaCl and KCl: −233.13 kJ/mol for $Cl^-(g)$, 514.26 kJ/mol for $K^+(g)$, 609.358 kJ/mol for $Na^+(g)$, −411.153 kJ/mol for NaCl(*s*), and −436.747 kJ/mol for KCl(*s*). Both of these ionic compounds have the same charges on the ions and have the same crystal geometry. What conclusion can you make, based on your results, about the value of lattice energy and interionic distance?

13.56 The ΔH_f° at 25 °C is 2348.504 kJ/mol for $Mg^{2+}(g)$, 984 kJ/mol for $O^{2-}(g)$, and −601.70 kJ/mol for MgO(*s*). (a) Calculate the lattice energy for MgO. The ΔH_f° at 25 °C is 609.358 kJ/mol for $Na^+(g)$, −255.39 kJ/mol for $F^-(g)$, and −573.647 kJ/mol for NaF(*s*). (b) Calculate the lattice energy for NaF. (c) Why is there such a large difference between these lattice energies even though both substances crystallize in the same pattern and there is only about a 10% difference in the distance between the ions?

The Solid State—Defects and Diffraction

13.57 Compare the theoretical density that can be calculated for a crystalline substance from the unit cell data to that actually observed for a macroscopic-sized sample of the substance.

13.58 Name the three basic types of point defects. How do these defects account for nonstoichiometric compounds?

13.59 What is diffraction? What must be the relationship between the wavelength of incident radiation and the spacing of the particles in a crystal in order for diffraction to occur? What types of waves are suitable for diffraction studies of crystals?

13.60* Write the Bragg equation. Identify each symbol. X-rays from a palladium source ($\lambda = 0.0576$ nm) were reflected by a sample of copper at an angle of 9.40°. This reflection corresponds to the unit cell length ($d = a$) with $n = 2$ in the Bragg equation. Calculate the length of the copper unit cell.

CHAPTER

14

Water and Solutions in Water

THE CHEMISTRY OF WATER

To study the properties of a chemical compound, we review its molecular geometry and/or its crystal structure, its physical properties, and its chemical reactions. The properties of the compound are compared with those of similar compounds, especially those of elements from similar general areas in the periodic table. It also may be helpful to review the role of the compound in the environment or in practical applications.

In this chapter, the properties of that most important of chemical compounds, water, are explored. In addition, you will be introduced to some substances that are soluble in water—electrolytes and nonelectrolytes, acids and bases, and complex ions—all to be studied further in later chapters.

14.1 THE PROPERTIES OF WATER

Water (H_2O) consists of simple triatomic molecules, but the behavior of water is complex and in some ways quite different from that of any other substance. As you might imagine, because of the abundance of water in our surroundings and the essential nature of water for all living things, considerable effort has gone into studying the properties of water. Yet, it is surprising that the behavior of water at the molecular level in either living or nonliving systems is far from completely understood.

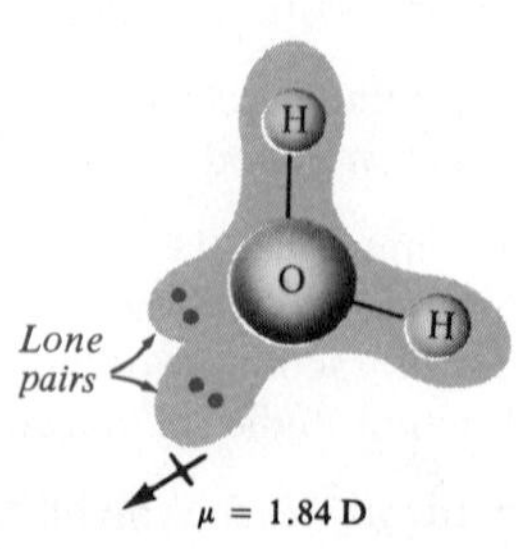

Figure 14.1
The Water Molecule
Compare the water dipole moment of 1.84 D with that of H_2S, which is 0.89 D.

The unique properties of water arise mainly from its molecular structure and the resultant intermolecular forces. The oxygen atom in the water molecule has what are described as sp^3-hybridized outer orbitals. Two of the four hybrid orbitals are occupied by lone-pair electrons and two by electron pairs that form covalent bonds to the two hydrogen atoms (Figure 14.1). The water molecule is bent and in the VSEPR classification is an AB_2E_2 molecule (Section 11.5). Because the covalent bonds to the hydrogen atoms are compressed by the lone pairs, the H—O—H angle is 104.5°, as compared to the tetrahedral angle of 109.47° (see Figure 11.2).

The presence of two lone electron pairs on the oxygen atom and the polar nature of the oxygen–hydrogen bonds combine to make the water molecule highly polar. The negative charge is centered on the oxygen atom and the positive charge midway between the two hydrogen atoms (see Figure 11.10).

In addition to the resultant strong dipole–dipole interaction between water molecules, hydrogen bonding plays a very important role in determining the properties of water. The hydrogen atoms, with their partial positive charges, are strongly attracted to the oxygen atoms of adjacent molecules, forming strong hydrogen bonds in liquid and solid water. Each oxygen atom can form two hydrogen bonds, one through each lone pair.

Although much remains to be learned about the structure of liquid water, it is certain that aggregates of varying numbers of water molecules held together by hydrogen bonds are always present, together with some uncombined water molecules. The "flickering cluster" model of liquid water structure pictures a dynamic equilibrium between aggregates of various sizes and single molecules, with the aggregates continually forming and breaking up.

In ice, water molecules are held in fixed positions by hydrogen bonding. Each oxygen atom is surrounded by four hydrogen atoms, two of which are connected to the oxygen atom by covalent bonds and two by hydrogen bonds (Figure 14.2a). These bonds through the hydrogen atoms join each oxygen atom to four other oxygen atoms. Repetition of this grouping in three dimensions builds a honeycomb-like structure containing large open spaces. Water molecules joined in this way occupy more space than they do in the less-ordered form of liquid water. The result is that water expands when it freezes, a phenomenon observed when the freezing of water in a bottle breaks the bottle. Another familiar effect of this increased volume is that ice floats, because it is less dense than liquid water (compare the density values in Table 14.1).

When air-free water is warmed from 0 °C to 3.98 °C, the water gradually becomes *more* dense (Figure 14.3). Few liquids become denser upon heating in any temperature range. Apparently, when ice melts not all hydrogen bonds are broken and the honeycomb structure only partially collapses. As temperature is increased from 0 °C, additional hydrogen bonds break, volume decreases, and there is a corresponding increase in density until maximum density is reached at 3.98 °C. Above this temperature the expected expansion due to the increased energy of individual molecules takes over. The transformation of solid ice to liquid water and then to water vapor passes through the following stages:

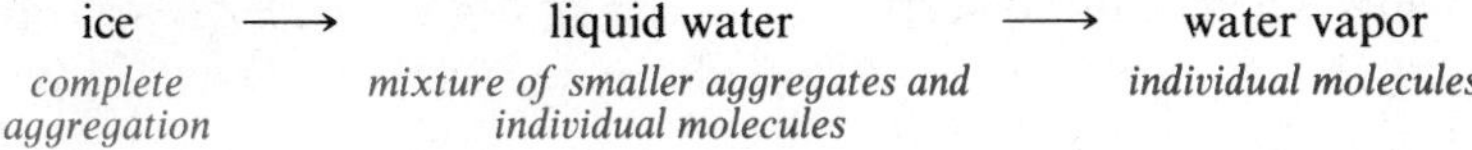

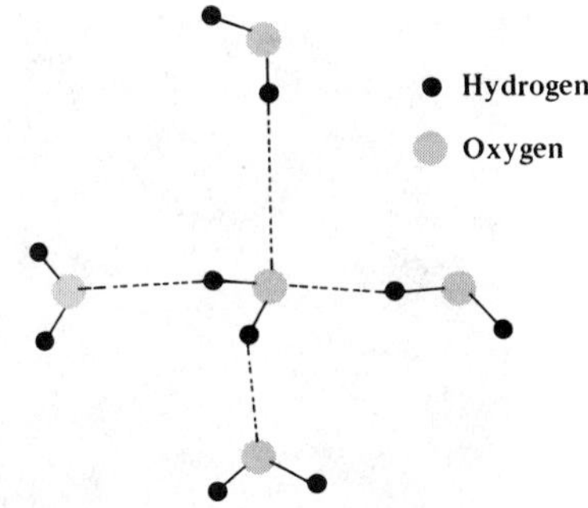

(a) Hydrogen bonds to one water molecule in ice

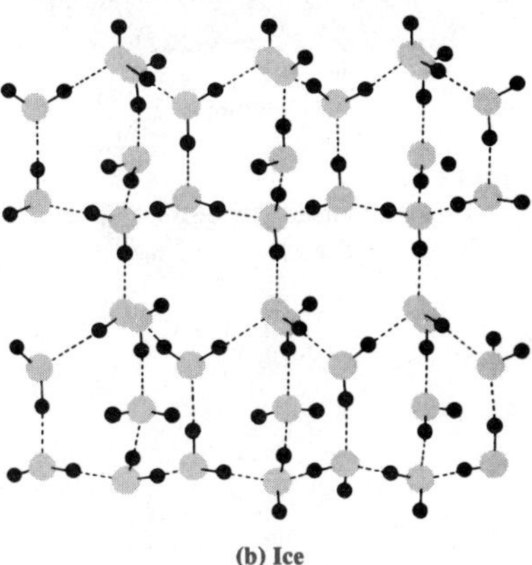

(b) Ice

Figure 14.2
Aggregate of Water Molecules
The O---H hydrogen bonds in ice are about 0.177 nm in length.

Table 14.1
The Properties of Water

Molar mass	18.02 g/mol
Melting point	0.0 °C
Boiling point (1 atm)	100.0 °C
Density (g/cm^3)	
0 °C	0.9168 g/cm^3 (ice)
	0.99984 g/cm^3 (liquid)
3.98 °C	0.99997 g/cm^3
25 °C	0.99704 g/cm^3
Vapor pressure (25 °C)	23.756 Torr
Triple point	0.0100 °C and 4.58 Torr
Dipole moment	1.84 D
Surface tension (25 °C)	0.07197 N/m
Heat capacity (25 °C)	75.2 J/K mol
Heat of formation (of liquid) (25 °C)	−286 kJ/mol
Heat of fusion (0 °C)	6.02 kJ/mol
Heat of vaporization (100 °C)	40.7 kJ/mol

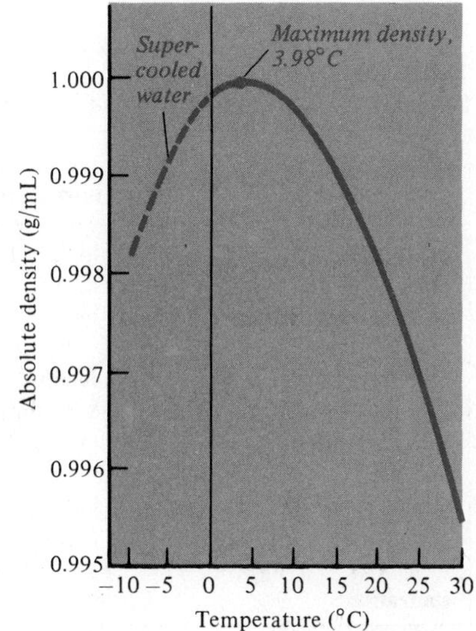

Figure 14.3
The Effect of Temperature on the Density of Water

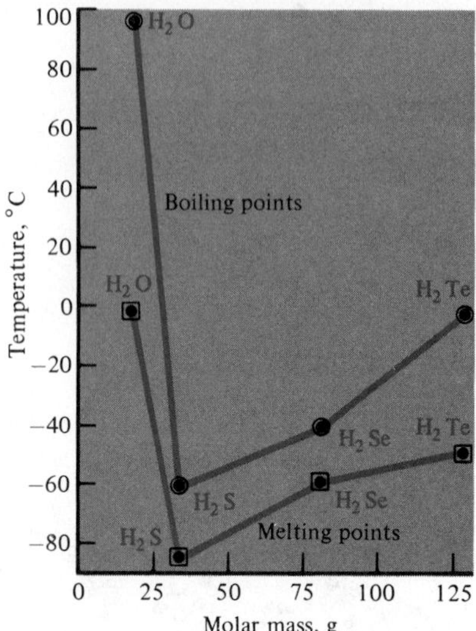

Figure 14.4
Melting and Boiling Points of the Hydrides of the Oxygen Family

The weathering of rocks and the formation of holes in streets are hastened by the expansion of freezing water inside cracks. More importantly, because ice stays at the top of a lake, pond, or other body of water, and also because it acts as an insulator, fish can live through the winter at the bottom, where the water does not freeze. If water behaved like most liquids, lakes and streams would freeze solid as ice that formed at the top continually sank to the bottom. As a result aquatic life would be nonexistent—or very different. The seasonal changes in temperature in temperate climates and the resulting changes in the density of water allow the water in many lakes to turn over completely each spring and fall. During the spring thaw, surface water sinks as it is warmed to 3.98 °C and becomes denser. During the fall freeze, warm surface water sinks as it is cooled to 3.98 °C and becomes denser. This leads to the uniform distribution of oxygen and nutrients throughout the water.

A relatively large increase in kinetic energy is necessary before water molecules can break free from the strong forces that hold them together. Therefore, as shown in Table 14.1, the heats of fusion and vaporization of water, as well as the melting and boiling points, are high relative to those of other simple molecular substances that have no hydrogen bonding. Look again at Table 11.3 and compare the values of these quantities for water with those for the other simple hydrides. In particular, note the effect of hydrogen bonding on the differences between water and the most closely comparable molecule, hydrogen sulfide. As a further comparison, the melting and boiling points of the hydrides of the oxygen family elements are plotted in Figure 14.4.

The uniqueness of water extends to other properties as well; for example, water has a higher heat capacity and surface tension than almost any other liquid. Table 14.2 summarizes some of the ways that the unusual properties of water contribute to its role in the environment. Because of the high heat capacity of water, the lakes, oceans, and body fluids can absorb or release large quantities of heat, while undergoing only small changes in temperatures. This property permits large bodies of water to moderate the temperature of the atmosphere and thereby affect the climate. The hottest and coldest regions on Earth are all inland regions, those farthest from the moderating effects of the oceans. Warm-blooded animals maintain their necessary narrow ranges of body temperatures with the aid of water. Excess heat is expended in evaporating water from the skin. Also, because of the high heat of vaporization of water, less liquid evaporates with the addition of a given amount of heat than for most other liquids. This keeps water loss to a minimum, making it easier for plants and animals to stay alive in environments where water is scarce.

Table 14.2
Some of the Unusual Physical–Chemical Properties of Water and Their Environmental and Biological Significance

Property	Comparison with normal liquids	Significance
Heat capacity	Very high	Moderates environmental temperatures, good heat transport medium
Heat of fusion	Very high	Has moderating effect, tends to stabilize liquid state
Heat of vaporization	Very high	Has moderating effect important in atmospheric physics and in precipitation–evaporation balance
Density	Anomalous maximum at 3.98 °C (for pure water)	Allows freezing from the surface and controls temperature distribution and circulation in bodies of water
Surface tension	Very high	Is important in surface phenomena, droplet formation in the atmosphere, and many physiological processes including transport through biomembranes
Dipole moment	Very high	Provides good solvent properties
Hydration[a]	Very extensive	Provides good solvent properties and allows mobilization of environmental pollutants and alteration of the biochemistry of solutes
Ionization	Very small	Provides a neutral medium but with some availability of both H^+ and OH^- ions

[a] See Section 14.3.

Source: Adapted from R. A. Horne, *The Chemistry of Our Environment* (New York: Wiley-Interscience, 1978), p. 237. Copyright © 1978 by Wiley and Sons.

EXAMPLE 14.1 Properties of Water

Why is hydrogen bonding stronger in H_2O than in H_2S?

The electronegativity (3.5) of an oxygen atom is considerably greater than the electronegativity (2.5) of a sulfur atom. As a result, **the covalent H—O bonds in water are much more polar than are the H—S bonds in hydrogen sulfide,** and the partial positive charges on the hydrogen atoms in H_2O are larger than those in H_2S. Therefore, stronger hydrogen bonds form between H_2O molecules than between H_2S molecules. [Hydrogen bonding in liquid H_2S is very weak, one of the reasons why H_2S is a gas at room temperature.]

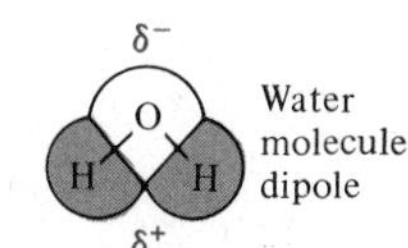

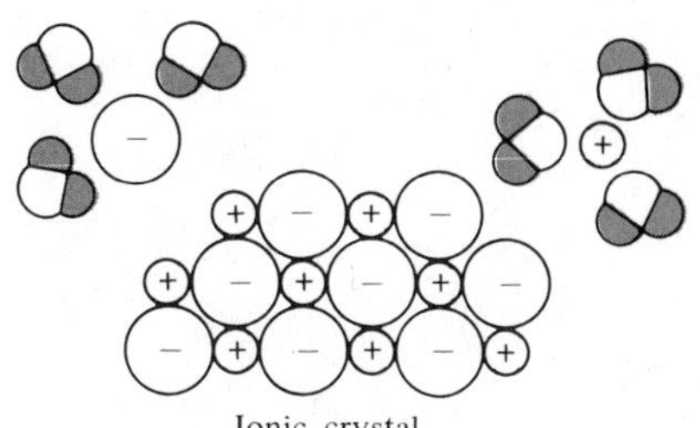

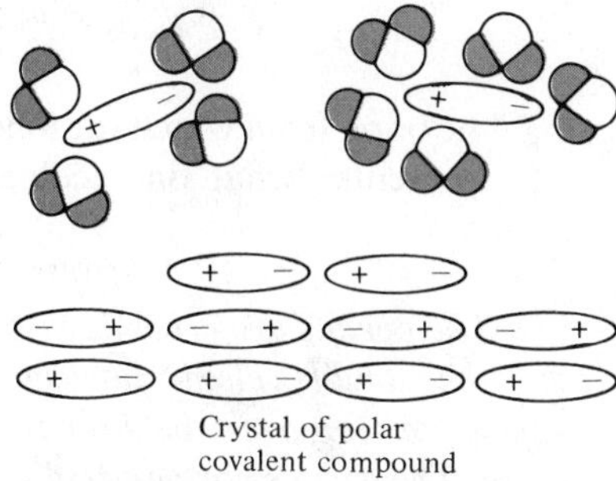

Figure 14.5
Dissolution of Ionic and Polar Covalent Substances in Water

14.2 WATER AS A SOLVENT

Water is a remarkably versatile solvent, making it an excellent medium for transporting ions and molecules in the environment and in living things. The substances most likely to be soluble in water are those with ionic or polar covalent bonds. Ionic substances dissolve when the ion–dipole attraction between the ions and the water molecules is stronger than the attraction of the ions for each other (Figure 14.5). Polar molecular substances such as ethyl alcohol similarly dissolve if the combined dipole–dipole attraction and hydrogen bonding between the solute molecules and the water molecules is greater than the attraction of the solute molecules for each other. The majority of nonpolar substances do not dissolve in water to any great extent because water molecules are more strongly attracted to each other than to nonpolar molecules.

The association of water molecules with the ions or molecules of another substance is called **hydration.** A more general term for the association of solvent molecules with the ions or molecules of a solute is **solvation.**

All ions in aqueous solution are assumed to be **hydrated**—surrounded by water molecules. A hydrated ion and its associated water molecules may be written together within square brackets, for example, $[Cu(H_2O)_4]^{2+}$. However, water molecules are generally omitted from the formulas of ions in solution unless their role is under discussion.

Ions of high charge and small radius (i.e., high charge-to-size ratio) attract water molecules more strongly than those of lower charge and larger radius. The *heat of hydration*—the heat released when an ion becomes hydrated—reflects these trends. For example, the fluoride ion has a higher heat of hydration than the much larger iodide ion.

	Ionic radius	Charge/size ratio	ΔH°_{hydr}
F^-	0.136 nm	1/0.136 = 7.35	−483 kJ/mol
I^-	0.220 nm	1/0.220 = 4.55	−273 kJ/mol

The effect of charge is shown by comparing the heat of hydration of the magnesium(II) ion with that of the lithium ion, which is comparable in size.

	Ionic radius	Charge/size ratio	ΔH°_{hydr}
Li^+	0.068 nm	1/0.068 = 15	−543 kJ/mol
Mg^{2+}	0.066 nm	2/0.066 = 30.	−1921 kJ/mol

You should not assume that all ionic or highly polar substances are highly soluble in water. (Recall the variation in the solubilities of salts shown by the solubility rules given in Table 5.3.) Solubility varies with both the solute–solvent attraction and the attraction of solute particles for each other, as we shall see in the next section.

EXAMPLE 14.2 Properties of Water

If water molecules were linear instead of bent, would H_2O be as good a solvent as it is? Why or why not?

Water definitely would not be as good a solvent if the molecules were linear. A linear H—O—H molecule, even though it had polar hydrogen–oxygen bonds, would not be a dipole and could not dissolve substances by the mechanisms shown in Figure 14.5. The nonlinear shape of the water molecule produces a polar molecule which will interact with either positively or negatively charged parts of the solute. In fact, life on Earth would be quite different if the water molecule were linear, for many of its unusual properties depend on the polar nature of the molecule.

Aside 14.1 THOUGHTS ON CHEMISTRY: THE OCEAN SYSTEM

The book from which the following quotation is taken presents "what one ecologist thinks his subject is about."

The oceans are cruel deserts, and the things they lack are the soluble plant nutrients. But this is a strange conclusion for one who broods about the history of the earth. The oceans have existed since almost the beginnings of earthly time, changing in shape, shoved from one part of the planet to another before the drifting continents, but always present in roughly the volume we know. And all the time soluble nutrients have been washed into them by the rivers coming from the land. The sea has been made salty by this process. And yet it lacks the nutrients needed for life. Strange.

Some details of the answer to this riddle still escape us, but the outlines are clear. The chemistry of the sea is controlled by its mud. Even as the endless-flowing rivers discharge their chemicals into the ocean, so the mud at the bottom soaks them up. The mud is selective. It has complicated minerals similar to the clays of temperate soils that hold metallic cations like calcium, potassium, and sodium. In its medium of salt, the surface of the mud allows slow crystals to grow, like the nodules of manganese that some mining corporations plan to dredge for. There are sites where calcium carbonate precipitates and collects into reefs of limestone, dragging with it other elements such as magnesium. The mud contains organic debris on which bacteria do their strange feeding, fixing some elements to their corpses and discarding others. . . .

Chemicals are removed from the ocean basins as fast as they arrive from the rivers. They flow back to the land with the writhing of the earth's crust. Every thrust of a mountain range and every emergence of a coastline from the sea brings the chemical-rich mud back to the land. All the sedimentary rocks, from limestone and sandstone to shale and schist, were once part of the ocean's mud. When they were lifted out of the sea, they took with them the chemical nutrients stored in them. At once these nutrients began to be washed out of the rocks again by rain, to be caught in the roots of land plants and held for a while in land ecosystems, then to escape in a slow leak to the rivers, and to be sent on another journey to the sea, another spreading through the fluid mass, and another sorting on the sea's bottom.

It is an enormous chemical machine that keeps the sea a desert. All the chemicals in the sea are cycled slowly through it. They come from the rivers, they spend a time in the oceans, diluted and in suspense, then they are taken in by the mud, held for a brief few million years, and then thrust back onto the land in a prison of rock. This is a system that holds the chemistry of the sea constant from eon to eon. It is not an ecosystem, for all that bacteria and other forms of life do some of the chemical things on the way, particularly influencing the deposition of carbonates. It is a passive physical and chemical system, driven by the sun because the sun is there, but not organized by life.

Source: Paul Colinvaux, *Why Big Fierce Animals Are Rare; An Ecologist's Perspective.*

14.3 HEATS OF SOLUTION IN WATER FOR IONIC COMPOUNDS

The process of dissolution can be exothermic or endothermic. Before solute and solvent particles can mingle, the solute–solute and solvent–solvent forces of attraction must be overcome. Like bond breaking, these processes are endothermic. Once dissolution begins, solute and solvent particles are pulled together in a process that, like

Compound	Lattice energy (kJ/mol)	Heat of hydration (kJ/mol)	Calculated heat of solution (kJ/mol)	Experimental heat of solution (kJ/mol)
LiCl	−834	−884	−50	−37.15
NaCl	−769	−769	0	3.891
KCl	−701	−685	16	17.24
RbCl	−680	−664	16	16.7
CsCl	−657	−640	17	18.0
KF	−808	−827	−19	−17.74
KBr	−671	−658	13	20.04
KI	−632	−617	15	20.5

Table 14.3
Heats of Solution, Calculated and Experimental
Lattice energy, hydration energy, and heats of solution are given for selected alkali metal halides. The calculated heats of solution were found by adding the hydration energy and the heat for the formation of gaseous ions from the solid, which has the same value as the lattice energy, but the opposite sign (see Equations 14.1 and 14.2).

bond formation, is exothermic. The overall heat of solution depends upon the relative magnitudes of these endothermic and exothermic processes. The **heat of solution** is the enthalpy change for the dissolution of one mole of a substance in a given amount of solvent.

To calculate the heats of solution of ionic crystalline substances, dissolution is divided into two theoretical steps. First, the ions of the solute are set free from their positions in a crystal to give ions in the gas phase. This dissociation is the reverse of crystal formation and has an enthalpy change of the same value as the lattice energy (Section 13.11), but of the opposite sign. Dissociation is always endothermic; it takes energy to separate ions (e.g., Equation 14.1). Second, the gaseous ions become hydrated. Equation (14.2) gives the heat of hydration of K^+ and Cl^- ions in a large amount of water (sufficiently large that the ions do not influence each other). Combination of the endothermic dissociation of the solid compound into gaseous ions and the exothermic hydration of ions gives the heat of solution, for example, for potassium chloride

$$KCl(s) \xrightarrow{\Delta} K^+(g) + Cl^-(g) \qquad \Delta H^\circ = 701 \text{ kJ} \qquad (14.1)$$

$$K^+(g) + Cl^-(g) \xrightarrow{H_2O} K^+(aq) + Cl^-(aq) \qquad \Delta H^\circ_{hydr} = -685 \text{ kJ} \qquad (14.2)$$

$$KCl(s) \xrightarrow{H_2O} K^+(aq) + Cl^-(aq) \qquad \Delta H^\circ_{soln} = 16 \text{ kJ}$$

Heats of solution vary with the concentration of the solution and are recorded either for a specific number of moles of solvent per mole of solute, as shown in the following equation, or, as calculated above, for a very large quantity of solvent (the heat of solution at infinite dilution)

$$AgNO_3(s) \xrightarrow[\ ]{\text{50 moles } H_2O} Ag^+(aq) + NO_3^-(aq) \qquad \Delta H^\circ = 20.17 \text{ kJ}$$

Lattice energies and hydration energies tend to be of similar magnitude. Therefore, their combined effect results in heats of solution that are often of small magnitude, as shown in Table 14.3. Note the effect on the lattice energies and heats of hydration of increasing ionic charge/size ratios for the series of cations (Li^+ through Cs^+) and anions (F^- through I^-) of increasing size and the same charge. Also, salts of dipositive ions tend to have higher heats of solution than those of monopositive ions. For example, compare the heats of solution of $MgCl_2$, −155.06 kJ/mol, and $CaCl_2$, −82.9 kJ/mol, with the values in Table 14.3.

14.4 HYDRATES

Water molecules are often incorporated into chemical compounds, which are then referred to as **hydrates.** Hydrates can be found in all three physical states; the most familiar are crystalline solids.

Table 14.4
Some Hydrates
The common names are given in parentheses.

$CuSO_4 \cdot 5H_2O(s)$	Copper sulfate pentahydrate (blue vitriol)
$KAl(SO_4)_2 \cdot 12H_2O(s)$	Potassium aluminum sulfate dodecahydrate (alum)
$Na_2B_4O_7 \cdot 10H_2O(s)$	Sodium tetraborate decahydrate (borax)
$FeSO_4 \cdot 7H_2O(s)$	Iron(II) sulfate heptahydrate (green vitriol)
$H_2SO_4 \cdot H_2O(l)$	Sulfuric acid monohydrate, m.p. 8.6 °C

The formulas of hydrates are often written with a centered dot between the water molecules and the compound that has been hydrated, for example,

$$Ba(OH)_2 \cdot 8H_2O \qquad CuSO_4 \cdot 5H_2O$$

barium hydroxide octahydrate *copper sulfate pentahydrate*

Each hydrate listed in Table 14.4 contains a fixed quantity of water and has a definite composition. Each may lose its water of hydration upon heating and may be re-formed by reaction of the *anhydrous* (water-free) substance with water. The forces holding the water molecules in hydrates are often not very strong, as shown by the ease with which water is lost and regained.

Not all of the water molecules in a given hydrate need be held in the same way. The pentahydrate of copper(II) sulfate is a good example of a crystalline hydrate in which water molecules are bound in different ways. If solid anhydrous copper(II) sulfate is dissolved in water and the solution allowed to evaporate by standing in air, blue crystals are deposited. After careful drying at room temperature, these crystals have the chemical composition $CuSO_4 \cdot 5H_2O$, and remain unchanged in normally moist air. Careful heating causes loss of four water molecules to give a monohydrate, while stronger heating drives off the fifth water molecule.

$$CuSO_4 \cdot 5H_2O(s) \xrightarrow{140\ ^\circ C} CuSO_4 \cdot H_2O(s) + 4H_2O(g)$$

$$CuSO_4 \cdot H_2O(s) \xrightarrow{400\ ^\circ C} CuSO_4(s) + H_2O(g)$$

Studies of the crystal structure of $CuSO_4 \cdot 5H_2O$ have shown that four water molecules are held to Cu^{2+} by coordinate covalent bonds between the copper ion and the oxygen atoms of water and one is held by hydrogen bonds to the oxygen atoms of SO_4^{2-}. A better representation of this compound might be $[Cu(H_2O)_4][SO_4(H_2O)]$. In some hydrates, water molecules simply occupy fixed random positions in the crystal structure.

When a hydrated compound dissolves in water, the hydrate water molecules merge with the solvent water molecules and are, of course, indistinguishable. The concentration of the solution is, therefore, based on the mass of the nonhydrated compound that has dissolved. In weighing the hydrate, however, the mass of the water molecules must be included. There are two common errors here: (1) Do not include the mass of the water molecules with the mass of the solute in calculating the concentration of the solution of a hydrate and (2) be sure to include the mass of the water molecules in calculating the desired mass of a hydrate solute.

The loss of water by a hydrate on exposure to air is called **efflorescence.** Many hydrates have appreciable water vapor pressure because the water molecules are held loosely. If the vapor pressure of a hydrate is greater than that of the water vapor in the air, the hydrate will effloresce until a state of equilibrium has been reached. For example, $Na_2SO_4 \cdot 10H_2O$ has a vapor pressure of 30.8 Torr at room temperature. The partial pressure of water vapor at room temperature and average humidity is about 20 Torr. Therefore, $Na_2SO_4 \cdot 10\,H_2O$ is normally an efflorescent hydrate. With the loss of water, the crystal structure of a hydrated substance collapses and a powder appears on the surface.

Some compounds are **hygroscopic**—they take up water from the air. In some cases the result may be the formation of a hydrate. Compounds that take up enough water from the air to dissolve in the water they have taken up are called **deliquescent.** For example, calcium chloride ($CaCl_2$) and sodium hydroxide (NaOH) are deliquescent.

Water is often removed from gases or liquids by "drying agents" that are anhydrous salts, such as Na_2SO_4, $CaCl_2$, or $MgSO_4$. Of course, the drying agent should not react with the substance being dried.

EXAMPLE 14.3 Hydrates

A 1.83 g sample of a hydrate of $Al_2(SO_4)_3$ was heated until no more water vapor was driven off. The anhydrous material weighed 0.94 g. Determine the formula of the hydrate.

The mass of water driven off by the heating process was

$$1.83\text{ g} - 0.94\text{ g} = 0.89\text{ g}$$

The number of moles of water and of anhydrous $Al_2(SO_4)_3$ are

$$(0.89\text{ g } H_2O)\left(\frac{1\text{ mol } H_2O}{18.02\text{ g } H_2O}\right) = 0.049\text{ mol } H_2O$$

$$(0.94\text{ g } Al_2(SO_4)_3)\left(\frac{1\text{ mol } Al_2(SO_4)_3}{342.14\text{ g } 1\text{ mol } Al_2(SO_4)_3}\right) = 0.0027\text{ mol } Al_2(SO_4)_3$$

The ratio of moles of water to moles of $Al_2(SO_4)_3$ is

$$\frac{0.049\text{ mol } H_2O}{0.0027\text{ mol } Al_2(SO_4)_3} = \frac{18\text{ mol } H_2O}{1\text{ mol } Al_2(SO_4)_3}$$

The formula of the hydrate is $\mathbf{Al_2(SO_4)_3 \cdot 18H_2O}$.

EXAMPLE 14.4 Hydrates

How would 500. mL of 0.10 M $CuSO_4$ be prepared from the hydrate, $CuSO_4 \cdot 5H_2O$?

For 500. mL of 0.10 M solution, the required number of moles of $CuSO_4$ is

$$(0.500\text{ L})(0.10\text{ mol/L}) = 0.050\text{ mol}$$

From the formula of the hydrate, we can see that each mole of hydrate contains 1 mol of $CuSO_4$. Therefore, the solution can be prepared from 0.050 mol of $CuSO_4 \cdot 5H_2O$. The mass of the hydrate needed is

$$(0.050\text{ mol } CuSO_4 \cdot 5H_2O)\left(\frac{249.68\text{ g } CuSO_4 \cdot 5H_2O}{1\text{ mol } CuSO_4 \cdot 5H_2O}\right) = 12\text{ g } CuSO_4 \cdot 5H_2O$$

The solution is prepared by adding sufficient water to 12 g of $CuSO_4 \cdot 5H_2O$ to make 500. mL of solution.

EXERCISE A dilute solution of sodium tetraborate (borax) was prepared by adding sufficient water to 10. g of the decahydrate ($Na_2B_4O_7 \cdot 10H_2O$) to give 1.0 L of solution. What is the molarity of the solution? *Answer* 0.026 M.

14.5 THE IONIZATION OF WATER: A CHEMICAL EQUILIBRIUM

Experiments show that a very small percentage of hydrogen ions and hydroxide ions are always present in liquid water. The ions form by ionization of the water

$$H_2O \longrightarrow H^+ + OH^-$$

and can also combine to give water molecules

$$H^+ + OH^- \longrightarrow H_2O$$

Left undisturbed, these two reactions establish a dynamic chemical equilibrium. To indicate that the changes in both directions are taking place simultaneously, a double arrow is used in writing the equation.

$$H_2O(l) \underset{\text{reverse reaction}}{\overset{\text{forward reaction}}{\rightleftharpoons}} H^+ + OH^-$$

When **chemical equilibrium** has been reached, the rates of the forward and the reverse chemical reactions are the same and the amounts of the species present do not change with time. Chemical equilibria are possible in solution, in all phases, and between species in different phases.

A chemical equation like that above gives *no* indication of the amounts of the substances present at equilibrium. In pure water, only $1.8 \times 10^{-7}\%$ of the water molecules (at 25 °C) are ionized at any given moment. The amounts of the species present at equilibrium vary greatly for different types of reactions, different substances, and different reaction conditions. (The quantitative aspects of chemical equilibrium reactions are examined in Chapters 19–22.)

The question arises of how to represent the hydrogen ion when it is involved in reactions in aqueous solution. The hydrogen ion is the smallest positive ion. It is very strongly attracted to the negative end of the water dipole and therefore the presence of hydrogen ions free from association with water molecules is unlikely. In aqueous solution, each hydrogen ion—whether from the ionization of water or from some other source—is always strongly bonded to one or more water molecules. To emphasize the association of the hydrogen ion with water molecules, the hydronium ion, H_3O^+, is frequently written instead of H^+. This allows the ionization of hydrogen-containing molecules or ions to be pictured as the transfer of a hydrogen ion to a water molecule, for example

$$H_2O(l) + H_2O(l) \rightleftharpoons H_3O^+ + OH^-$$
$$H_2O(l) + HSO_4^- \rightleftharpoons H_3O^+ + SO_4^{2-}$$

The structure of the hydronium ion has been investigated by various spectroscopic methods. It has the shape (a) of a triangular pyramid (an AB_3E ion; see Figure 11.3) and is similar in overall size to the K^+ ion, which is one of the larger monatomic ions. In aqueous solution the hydronium ion is surrounded by four water molecules (b), three held by hydrogen bonding and a fourth attracted to the positive charge on the hydronium ion through the negative end of the water dipole (ion–dipole interaction).

(a) H_3O^+ structure 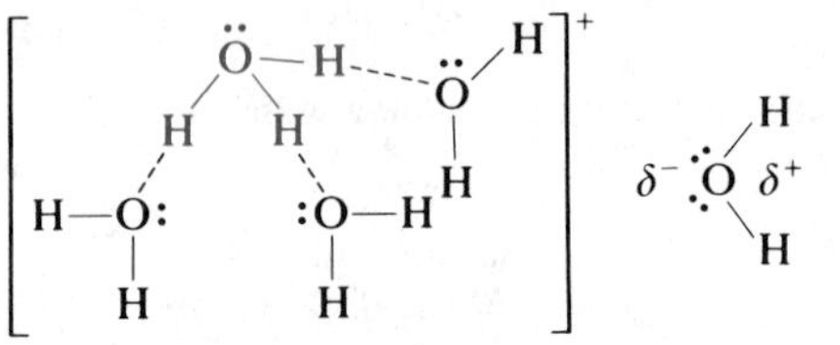 (b) H_3O^+ in aqueous solution

"Hydrogen ions" are known to move very rapidly through aqueous solutions, accounting in part for the rapid rate of reactions involving them. If H_3O^+ ions were the mobile species, the ions should move at about the same speed as the K^+ ion of similar dimensions. This is not the case—the hydrogen ions move about twice as fast. Therefore, it has been proposed that H^+ moves rapidly through aqueous solutions by a "jump," or "pass the proton," mechanism.

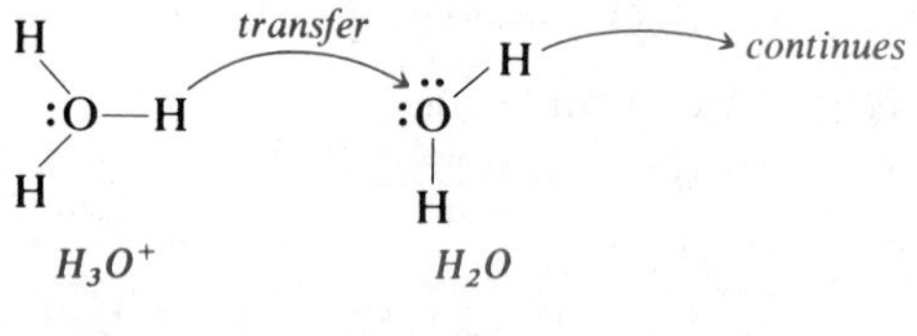

14.6 THE CHEMICAL REACTIONS OF WATER

The chemical reactions in which water participates are many and varied. Often the reactants and/or products are ions in aqueous solution, and in many cases H^+ and OH^- ions are involved. In some reactions, the oxygen atom of the polar water molecule is so strongly attracted to another species that the hydrogen–oxygen bond breaks, yielding ions or molecules in which oxygen is bonded to atoms of other elements (e.g., Equations 14.5 and 14.6, below). **Hydrolysis** is a general term for reactions in which the water molecule is split. The hydrolysis of ions (e.g., Equations 14.3 and 14.4) or molecular compounds (Equations 14.5 and 14.6) often results in bonding of the hydrogen atom to other atoms. In other reactions, reactants known as reducing agents (Section 16.3) yield hydrogen by reaction with water (Equation 14.7). Water also undergoes combination reactions (e.g., Equations 14.8 and 14.9).

The reactions of water are discussed in conjunction with the chemistry of the other reactants wherever appropriate in later chapters. Following are examples of some common types of reactions of water.

Equilibria with ions in aqueous solution Such reactions, which are hydrolysis reactions of ions, are discussed in Section 21.1.

$$CN^- + H_2O(l) \rightleftharpoons HCN(aq) + OH^- \quad (14.3)$$
$$CO_3^{2-} + H_2O(l) \rightleftharpoons HCO_3^- + OH^- \quad (14.4)$$

Reactions of molecular halides and other binary compounds in which one or both hydrogen–oxygen bonds are broken Such reactions are hydrolysis reactions of molecular compounds

$$Al_2S_2(s) + 6H_2O(l) \longrightarrow 2Al(OH)_3(s) + 3H_2S(g) \quad (14.5)$$
$$SiI_4(s) + 2H_2O(l) \longrightarrow SiO_2(s) + 4HI(aq) \quad (14.6)$$

Reactions in which elemental hydrogen is produced This type of reaction occurs between active metals and water (Section 5.1c), for example

$$2Na(s) + 2H_2O(l) \longrightarrow 2NaOH(aq) + H_2(g) \quad (14.7)$$

Reactions in which water combines with another substance The reactions of oxides with water, a common type of combination reaction of water, are discussed in Section 16.1.

$$SO_3(g) + H_2O(l) \longrightarrow H_2SO_4(aq) \quad (14.8)$$
$$CaO(s) + H_2O(l) \longrightarrow Ca(OH)_2(aq) \quad (14.9)$$

EXAMPLE 14.5 Chemical Reactions of Water

Several simple types of chemical reactions were introduced in Section 5.1: combination, decomposition, displacement, and partner exchange. Into which categories do the reactions represented by (a) Equation (14.3), (b) Equation (14.5), and (c) Equation (14.7) fall?

(a) To classify this reaction from the net ionic equation, first add back the spectator ion, a metal cation (M^+)

$$M^+CN^-(aq) + H_2O(l) \rightleftharpoons HCN(aq) + M^+OH^-(aq)$$

This reaction is a **partner-exchange reaction.**

(b) In this reaction, sulfur becomes bonded to hydrogen and aluminum to oxygen—a **partner-exchange reaction.**

(c) By recognizing that sodium has formed sodium hydroxide, which is a soluble ionic compound, it can be seen that sodium has displaced hydrogen from the water—a **displacement reaction.**

EXERCISE Place the reactions of (a) Equation (14.6) and (b) Equation (14.9) into the categories described in this Example. *Answers* (a) partner exchange, (b) combination.

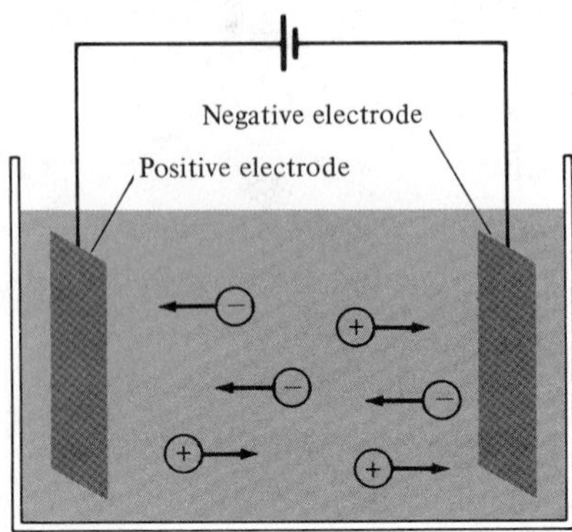

Figure 14.6
Conduction of Electricity by a Solution Containing Ions

SUMMARY OF SECTIONS 14.1–14.6

The water molecule is angular with a large dipole moment. Water molecules are strongly hydrogen-bonded to each other, and this is the basis for many of the unusual properties of water. Ice, which has an open crystal structure, floats on liquid water, which attains its maximum density at 3.98 °C. Water has a high heat capacity and high heat of vaporization, properties important to the role of water in the environment and in living things. A great many ionic and polar covalent substances are soluble in water, being attracted in varying degrees to the polar water molecules. Water is very slightly ionized to H^+ and OH^- in a dynamic chemical equilibrium, $H_2O \rightleftharpoons H^+ + OH^-$. Often the hydronium ion, H_3O^+, is written instead of the hydrogen ion to emphasize that hydrogen ions are always associated with water molecules in aqueous solution.

Ions and molecules dissolved in water are hydrated—surrounded by water molecules. Compounds in which intact water molecules are incorporated are called hydrates. The water of hydration may be bound by coordinate covalent or by hydrogen bonds, or may simply occupy spaces in the crystal structure. Efflorescence is the loss of water of hydration. Hygroscopic substances take up water from the air, and deliquescent substances dissolve in the water that they take up from the air.

Common types of chemical reactions of water are reactions in which elemental hydrogen is produced; reactions in which water combines with another substance, and hydrolysis reactions of ions, molecular halides, or other compounds in which the water molecule is split.

IONS IN AQUEOUS SOLUTION

14.7 ELECTROLYTES AND NONELECTROLYTES

An ionic compound such as sodium chloride when melted conducts electricity because mobile ions are available to carry the current (Section 10.2). Dissolving an ionic compound in water, like melting, frees the ions from their fixed positions in the solid. As a result, aqueous solutions of ionic compounds also conduct electricity. When a direct current is passed through a solution that contains ions, the positive ions are attracted to the negative electrode and the negative ions to the positive electrode (Figure 14.6). Pure substances and substances in solution that conduct electricity by the movement of ions are called **electrolytes.** All compounds that yield ions in aqueous solution are electrolytes. Compounds that are 100% dissociated or ionized in aqueous solution are called **strong electrolytes.** Whatever the *amount* of a strong electrolyte dissolved in a given volume of water, virtually all of the compound yields ions in solution.

Table 14.5
Ions Present in Common Strong Electrolytes
Salts containing the ions listed are 100% dissociated into ions in aqueous solution, to the extent that they are soluble.

Ions
Li^+
Na^+
K^+
Mg^{2+}
Ca^{2+}
Sr^{2+}
Ba^{2+}
La^{3+}
NH_4^+
Cl^-
Br^-
I^-
ClO_4^-
BrO_3^-
IO_3^-
SO_4^{2-}
HSO_4^-
NO_3^-
CO_3^{2-}
HCO_3^-

Most salts are strong electrolytes. If a salt that is a strong electrolyte is only slightly soluble in water, a chemical equilibrium is established between any remaining undissolved salt and the ions in aqueous solution. For example, for mercury(II) iodide, one of the *least soluble* of salts, the equilibrium is written

$$HgI_2(s) \overset{H_2O}{\rightleftharpoons} Hg^{2+} + 2I^- \qquad \textbf{(14.10)}$$

The only species dissolved in the water are ions. Table 14.5 is a list of the ions most commonly found in strong electrolytes. For example, salts containing Na^+, whatever the anion, or salts containing IO_3^-, whatever the cation, are strong electrolytes.

Molecular compounds can also become strong electrolytes in aqueous solution. The molecular electrolytes are generally highly polar compounds such as hydrogen chloride. The oppositely charged ends of the HCl and H_2O molecules attract each other strongly enough to break the H—Cl bond

$$HCl(g) + H_2O(l) \longrightarrow H_3O^+ + Cl^- \qquad \textbf{(14.11)}$$

Hydrogen chloride and other molecular compounds that are strong electrolytes are 100% ionized when they are present in any concentration in aqueous solutions.

Many polar molecular compounds dissolve in water with only partial, rather than complete, ionization. An equilibrium is established between the molecules and ions, all of which are dissolved in the water. For example, acetic acid is soluble in water, but only four out of every 1000 molecules of the acetic acid are ionized in a 1 M solution at 25 °C.

$$\underset{\text{acetic acid}}{CH_3COOH(aq)} + H_2O(l) \rightleftharpoons \underset{\text{acetate ion}}{CH_3COO^-} + H_3O^+ \quad \textbf{(14.12)}$$

Here the solution contains both ions and molecules. Acetic acid and other substances like it that are only partially ionized in aqueous solution are called **weak electrolytes.** The extent to which different weak electrolytes ionize under similar conditions varies widely.

Substances that dissolve in water to give solutions that do not conduct electricity are called **nonelectrolytes.** Nonelectrolytes remain as molecules in solution. Some examples are ethyl alcohol (C_2H_5OH), acetone (CH_3COCH_3, a common laboratory solvent), and sucrose (table sugar).

Figure 14.7 illustrates a demonstration of the conduction of electricity by electrolytes. Pure water is so slightly ionized that its conductance is detected only by very sensitive instruments. Tapwater, however, is seldom pure; it usually contains various ions (Section 14.13) and so conducts electricity. (That is why you should never touch electrical equipment when your hands are wet or while you are standing on a wet floor.)

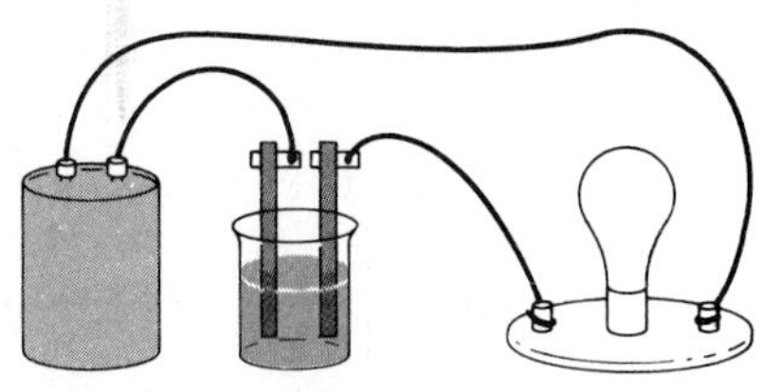

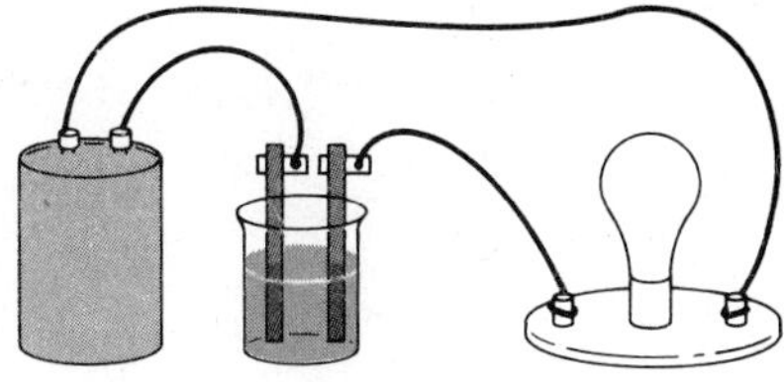

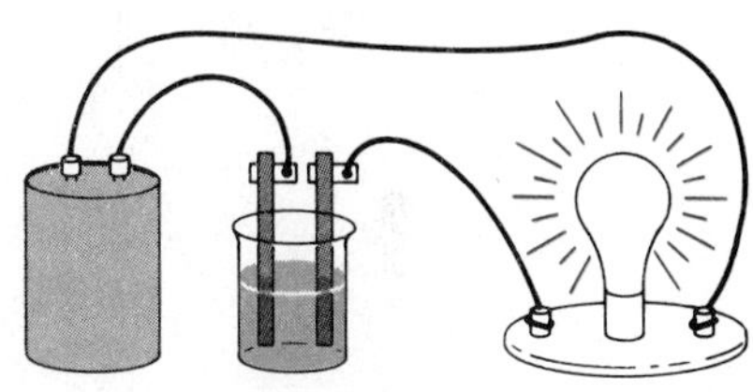

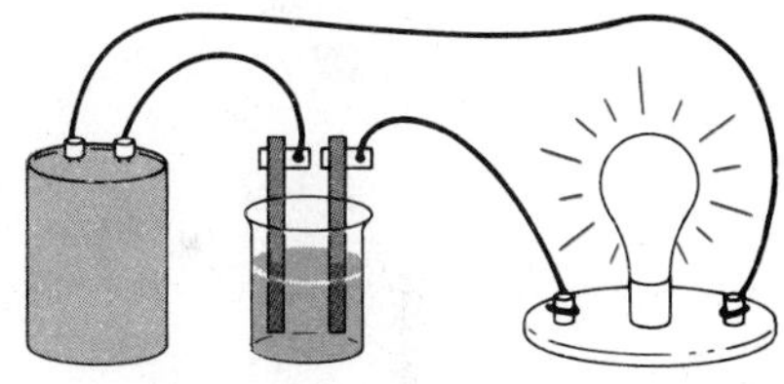

Figure 14.7
Electrolytes and Nonelectrolytes
Pure water contains so few ions that it is a nonconductor in this type of experiment. Tap water usually contains many ions in solution and is conductive.

EXAMPLE 14.6 Electrolytes and Nonelectrolytes

Write chemical equations for what happens when each of the following substances is mixed with water: (a) gaseous hydrogen iodide; (b) solid calcium nitrate; (c) solid magnesium hydroxide; and (d) *n*-propyl alcohol (a liquid with the structure $CH_3CH_2CH_2OH$).

(a) Based on what we know about HCl—that it is a strong electrolyte (Equation 14.11)—we would expect HI(g) to behave in a similar fashion and be 100% ionized in aqueous solution.

$$HI(g) + H_2O(l) \longrightarrow H_3O^+ + I^-$$

(b) Nitrates (see Table 5.3) are generally soluble salts. Also, Ca^{2+} and NO_3^- are ions found in strong electrolytes (see Table 14.4). Therefore, calcium nitrate should be a strong electrolyte and should be soluble and 100% dissociated in aqueous solution.

$$Ca(NO_3)_2(s) \xrightarrow{H_2O} Ca^{2+} + 2NO_3$$

(c) Only the alkali metal hydroxides are soluble (see Table 5.3). Therefore, solid $Mg(OH)_2$ most likely establishes an equilibrium with ions in solution.

$$Mg(OH)_2(s) \overset{H_2O}{\rightleftharpoons} Mg^{2+} + 2OH^-$$

(d) By analogy with what we know of ethyl alcohol, we would expect *n*-propyl alcohol to be soluble in water and to be a **nonelectrolyte** (which indeed it is).

$$CH_3CH_2CH_2OH(l) \xrightarrow{H_2O} CH_3CH_2CH_2OH(aq)$$

EXERCISE Write chemical equations describing the dissolution of the following substances in water: (a) gaseous oxygen, O_2; (b) solid potassium bromide, KBr; (c) solid sodium hydroxide, NaOH; and (d) solid silver chloride, AgCl.

Answers (a) $O_2(g) \xrightarrow{H_2O} O_2(aq)$, (b) $KBr(s) \xrightarrow{H_2O} K^+ + Br^-$, (c) $NaOH(s) \xrightarrow{H_2O} Na^+ + OH^-$, (d) $AgCl(s) \overset{H_2O}{\rightleftharpoons} Ag^+ + Cl^-$.

14.8 ACIDS AND BASES: H^+ AND OH^-

Acids and bases were first recognized as specific classes of compounds because of the distinctive properties exhibited by their aqueous solutions. The classic acid and base properties and a few common acidic and basic substances are listed in Table 14.6.

In the 1880s Svante Arrhenius, a Swedish chemist, developed a theory for ionization in aqueous solutions. Part of his theory explains that the classic acidic properties are imparted to an aqueous solution by the hydrogen ion, H^+, and the basic properties by the hydroxide ion, OH^-. An **acidic aqueous solution** contains a greater concentration of H^+ ions than of OH^- ions. A basic aqueous solution, better called an **alkaline aqueous solution**, contains a greater concentration of OH^- ions than H^+ ions. (The term "alkaline" is preferable because "basic" also has other meanings. "Alkaline" derives from "alkali," an old name for substances with the classic properties of bases.)

The Arrhenius definitions of acids and bases can be called the *water-ion* definitions. An **acid (water-ion)** is a substance that contains hydrogen and yields hydrogen ions in aqueous solution. The common *strong acids* are all water-ion acids. They are molecular compounds that are strong electrolytes and are 100% ionized in aqueous solution. Some are binary acids like hydrochloric acid (Equation 14.11) and others are oxoacids, such as nitric acid:

$$HNO_3(l) + H_2O(l) \longrightarrow H_3O^+ + NO_3^- \qquad \textbf{(14.13)}$$

A **base (water-ion)** is a compound that contains hydroxide ions and when it dissolves in water dissociates to give hydroxide ions. The bases defined in this way are all ionic compounds and strong electrolytes. Those hydroxides that are soluble or moderately soluble (the alkali metal hydroxides, and calcium, strontium, and barium hydroxide) give high concentrations of hydroxide ions in solution and are referred to as *strong bases,* for example

$$NaOH(s) \xrightarrow{H_2O} Na^+ + OH^- \qquad \textbf{(14.14)}$$

$$KOH(s) \xrightarrow{H_2O} K^+ + OH^- \qquad \textbf{(14.15)}$$

Many hydrogen-containing molecular compounds are weak electrolytes and therefore *weak acids,* for they are less than 100% ionized. For example, acetic acid (Equation 14.12) and hydrogen cyanide, known in its aqueous solution as hydrocyanic acid, are both weak acids and weak electrolytes.

$$HCN(aq) + H_2O(l) \rightleftharpoons H_3O^+ + CN^- \qquad \textbf{(14.16)}$$

Table 14.6
Acids and Bases
Early chemists blithely tasted their chemicals. Experience has shown that this can be a fatal mistake. *Never* taste any laboratory chemicals. The acidity of vinegar is due to acetic acid (CH_3COOH). Fruits and vegetables derive their acidity from various organic acids. Gastric juice contains hydrochloric acid (HCl). Carbonated beverages are acidic because of the reaction of CO_2 with water. Aspirin and vitamin C are organic acids, acetylsalicylic acid and ascorbic acid, respectively. Household ammonia is alkaline because of the reaction of ammonia gas (NH_3) with water, and milk of magnesia contains magnesium hydroxide [$Mg(OH)_2$].

Acids	Bases
Sour taste	Bitter taste
Change colors of indicators, e.g., litmus turns from blue to red; phenolphthalein turns from red to colorless	Slippery feeling
React with active metals to give hydrogen, e.g., $Fe(s) + 2H^+ \longrightarrow Fe^{2+} + H_2(g)$	Change colors of indicators, e.g., litmus turns from red to blue; phenolphthalein turns from colorless to red
Acidic properties disappear in reaction with a base	Basic properties disappear in reaction with an acid
Some Acidic Substances	**Some Alkaline Substances**
Vinegar	Household ammonia
Tomatoes	Baking soda
Citrus fruits	Soap
Carbonated beverages	Detergents
Black coffee	Milk of magnesia
Gastric fluid	Oven cleaners
Vitamin C	Lye
Aspirin	Drano
Saniflush	

You should become familiar with the names and formulas of the common acids and bases listed in Table 14.7.

Sulfuric acid and phosphoric acid, and other acids like them that contain more than one ionizable hydrogen atom, are known as **polyprotic acids**. Such acids ionize in steps, for example, for phosphoric acid

$$H_3PO_4(aq) \rightleftharpoons H_2PO_4^- + H^+ \quad (14.17)$$
$$H_2PO_4^- \rightleftharpoons HPO_4^{2-} + H^+ \quad (14.18)$$
$$HPO_4^{2-} \rightleftharpoons PO_4^{3-} + H^+ \quad (14.19)$$

Broader definitions of acids and bases than those of Arrhenius have been introduced over the years as it became desirable to categorize more types of substances as acids and bases. Consider ammonia, which is clearly a base although not an ionic hydroxide:

$$NH_3(g) + H_2O(l) \rightleftharpoons NH_4^+ + OH^- \quad (14.20)$$

Ammonia is the most common weak base. (A useful way of defining acids and bases that includes ammonia as a base is discussed at length in Chapter 20.)

Table 14.7
Common Acids and Bases
Nitrous acid is known only in solution.

Strong acids	
HCl	Hydrochloric acid
HBr	Hydrobromic acid
HI	Hydroiodic acid
HNO_3	Nitric acid
H_2SO_4	Sulfuric acid
$HClO_4$	Perchloric acid
Weak acids	
CH_3COOH	Acetic acid
HF	Hydrofluoric acid
H_3PO_4	Phosphoric acid
HNO_2	Nitrous acid
Strong bases	
$NaOH$	Sodium hydroxide
KOH	Potassium hydroxide
$Ca(OH)_2$	Calcium hydroxide
Weak base	
NH_3	Ammonia

EXAMPLE 14.7 Acids and Bases

Write equations for the interaction of water with (a) perchloric acid; (b) nitrous acid (*aq*); and (c) the weak base aniline, $C_6H_5NH_2$.

(a) Perchloric acid, a strong acid, is completely ionized.

$$HClO_4(l) + H_2O(l) \longrightarrow H_3O^+ + ClO_4^-$$

(b) Nitrous acid is a weak acid and establishes an equilibrium with water molecules.

$$HNO_2(aq) + H_2O(l) \rightleftharpoons H_3O^+ + NO_2^-$$

(c) Given that aniline is a weak base, we can assume that it enters into an equilibrium that yields OH^- ions.

$$C_6H_5NH_2(aq) + H_2O(l) \rightleftharpoons C_6H_5NH_3^+ + OH^-$$

EXERCISE Write equations showing the interaction of water with (a) formic acid, HCOOH, a weak acid like acetic acid; (b) hydrogen bromide; and (c) cesium hydroxide.

Answers (a) $HCOOH(aq) + H_2O(l) \rightleftharpoons HCOO^- + H_3O^+$; (b) $HBr(g) + H_2O(l) \longrightarrow H_3O^+ + Br^-$; (c) $CsOH(s) \xrightarrow{H_2O} Cs^+ + OH^-$.

14.9 NEUTRALIZATION

Neutralization is the reaction of an acid with a base. Neutralization reactions between water-ion acids and bases are partner-exchange reactions in which the products are water and an ionic compound—a salt. ("An acid plus a base yields a salt plus water" is an old mnemonic phrase describing neutralization reactions.)

$$HCl(aq) + NaOH(aq) \longrightarrow NaCl(aq) + H_2O(l) \quad (14.21)$$
$$HNO_3(aq) + KOH(aq) \longrightarrow KNO_3(aq) + H_2O(l) \quad (14.22)$$

The term "neutralization" is commonly applied to any acid–base reaction. Strictly speaking, only the reactions of stoichiometric amounts of strong acids with strong bases should be referred to as neutralization. Only in such reactions do acidic and basic properties completely disappear. How this occurs is made apparent by writing the net ionic equations for the two reactions above, in which the acids, bases, and salts are all strong electrolytes. Writing these compounds as ions and canceling

$$H_3O^+ + \cancel{Cl^-} + \cancel{Na^+} + OH^- \longrightarrow \cancel{Na^+} + \cancel{Cl^-} + 2H_2O(l)$$
$$H_3O^+ + OH^- \longrightarrow 2H_2O(l)$$
$$H_3O^+ + \cancel{NO_3^-} + \cancel{K^+} + OH^- \longrightarrow \cancel{K^+} + \cancel{NO_3^-} + 2H_2O(l)$$
$$H_3O^+ + OH^- \longrightarrow 2H_2O(l)$$

reveals that the result is the same in both cases. In the reaction between a strong acid and a strong base, the net ionic reaction is the reverse of the ionization of water. When the H^+ and OH^- ions combine to give water, the original acidic and basic properties of the solutions can no longer be detected. (Other types of acid–base reactions are discussed in Chapters 20 and 21.)

14.10 FORMATION OF COMPLEX IONS

Frequently nonmetal atoms that are already present in molecules or ions form coordinate covalent bonds with metal atoms or ions, usually those of transition metals. For example, the nitrogen atom in ammonia can "coordinate" with a silver cation to form what is called a *complex ion*

$$2H{:}\ddot{\underset{\cdot\cdot}{N}}{:} + Ag^+ \longrightarrow [H{:}\ddot{\underset{\cdot\cdot}{N}}{:}Ag{:}\ddot{\underset{\cdot\cdot}{N}}{:}H]^+ \quad \text{or} \quad [Ag(NH_3)_2]^+$$

a complex ion (pointing to $[Ag(NH_3)_2]^+$)

A **complex ion**, or "complex," consists of a central metal atom or cation to which are bonded one or more molecules or anions. The molecules or ions bonded to the central atom or cation are called **ligands.** The ligands are referred to as "coordinated to" the central atom or ion.

The charge of a complex ion equals the sum (algebraic) of the charge of the cation and the charges of any ligands that are anions. For example, with ammonia, a neutral ligand, the charge of the complex ion equals that of the cation.

$[Ni(NH_3)_4]^{2+}$ — *cation charge = +2*; *charge of complex ion*; *ligand charge = 0*

With anionic ligands, the charge of a complex ion is found as follows:

$[Cd(CN)_4]^{2-}$ — *cation charge = +2*; *charge of complex ion = (+2) + (−4) = −2*; *total ligand charge = (4 × −1) = −4*

$[CoF_6]^{3-}$ — *cation charge = +3*; *charge of complex ion = (+3) + (−6) = −3*; *total ligand charge = (6 × −1) = −6*

Anionic and molecular ligands can be present in the same complex, and in some cases the total charge of a complex is zero.

$[Co(NH_3)_4(CO_3)]^+$ — *cation charge = +3*; *charge of complex ion = (+3) + (−2) = +1*; *total ligand charge = 0 + (−2) = −2*

$[Pt(NH_3)_2(Cl)_2]^0$ — *cation charge = +2*; *charge of complex ion = (+2) + (−2) = 0*; *total ligand charge = 0 + (2 × −1) = −2*

A hydrated metal cation is a complex ion with water molecules as ligands, for example, $[Cu(H_2O)_4]^{2+}$. The stepwise displacement of water molecules leads to the formation of other complex ions, for example

$$[Cu(H_2O)_4]^{2+} + NH_3(aq) \rightleftharpoons [Cu(H_2O)_3(NH_3)]^{2+} + H_2O(l)$$
$$[Cu(H_2O)_3(NH_3)]^{2+} + NH_3(aq) \rightleftharpoons [Cu(H_2O)_2(NH_3)_2]^{2+} + H_2O(l)$$

Like slightly soluble salts, complex ions in solution are in equilibrium with their cations and ligands. In writing equations, the water molecules are generally omitted, and the overall equilibrium in the solution of a complex ion is represented, for example, as follows:

$$Ag^+ + 2NH_3(aq) \rightleftharpoons [Ag(NH_3)_2]^+ \qquad \textbf{(14.23)}$$

Coordination compounds contain metal ions or atoms and their surrounding ligands. The metal plus its ligands may be a cation, an anion, or have no charge. For example, a compound between a complex ion that is a cation and a simple anion, such as $[Ni(NH_3)_4]Cl_2$, would be described as a coordination compound. Coordination compounds can be acids, bases, salts, or nonelectrolytes, as shown by the following examples:

$H_2[PtCl_4]$	$[Co(NH_3)_6](OH)_3$	$[Co(NH_3)_6]Cl_3$	$[Co(NH_3)_3(NO_2)_3]$
acid	*base*	*salt*	*nonelectrolyte*

Many metal coordination compounds are soluble in water. The exceptions are certain nonelectrolytes and compounds that contain numerous repeating units (i.e., polymers). Most coordination compounds that are acids are extremely strong acids because there is little attraction between the anions in the completely self-contained complex ion and the acidic hydrogen ions. Similarly, coordination compounds with hydroxide ions are extremely strong bases (the cobalt compound shown above is as strong a base as sodium hydroxide). Hydrated metal cations are mildly acidic because, as the electrons on the oxygen atom are drawn toward the metal ion, the oxygen–hydrogen bonds are weakened.

$$[Cu(H_2O)_4]^{2+} + H_2O(l) \longrightarrow [Cu(H_2O)_3OH]^+ + H_3O^+$$

(Complex ion equilibria are covered quantitatively in Chapter 22; bonding and other aspects of the chemistry of coordination compounds are discussed in Chapter 31.)

Table 14.8
Electrolytes and Nonelectrolytes in Aqueous Solutions

Type of compound	Species present in pure compound	Species present in aqueous solution	Examples
Strong electrolytes			
Strong acids (water-ion)	Polar molecules	Ions only	$HCl(g) + H_2O(l) \longrightarrow H_3O^+ + Cl^-$
Strong bases (water-ion)	Ions	Ions only	$NaOH(s) \xrightarrow{H_2O} Na^+ + OH^-$
Salts	Ions	Ions only	$NaCl(s) \xrightarrow{H_2O} Na^+ + Cl^-$
Weak electrolytes			
Weak acids	Polar molecules	Ions and molecules in equilibrium	$HF(aq) + H_2O(l) \rightleftharpoons H_3O^+ + F^-$
Weak bases	Polar molecules	Ions and molecules in equilibrium	$NH_3(aq) + H_2O(l) \rightleftharpoons NH_4^+ + OH^-$
Nonelectrolytes			
Water-soluble molecular compounds	Molecules	Molecules only	$CH_3CH_2OH(l) \xrightarrow{H_2O} CH_3CH_2OH(aq)$

SUMMARY OF SECTIONS 14.7–14.10

Substances that conduct electricity by the movement of ions are called electrolytes, and all compounds that dissolve in water with the formation of ions are electrolytes. Strong electrolytes are 100% ionized or dissociated into ions in aqueous solution. Weak electrolytes establish equilibria between ions and molecules in aqueous solution. Compounds that dissolve in water but form no ions and give solutions that do not conduct electricity are nonelectrolytes. The simple types of compounds that are strong or weak electrolytes, or nonelectrolytes, are summarized in Table 14.8.

If total world water supply = 4 liters

Total fresh water supply = 100 mL

Total fresh water that's not ice = 25 mL

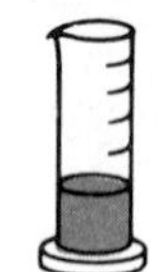

And available fresh water = 1 drop

Figure 14.8
Availability of Fresh Water

Strong acids (water-ion) and strong bases (water-ion) are strong electrolytes and in aqueous solutions yield hydrogen ions and hydroxide ions, respectively, that impart acidic and basic properties to their solutions. In the reaction between stoichiometric amounts of such strong acids and bases, the acidic and basic properties of the solutions are "neutralized." The products of such neutralizations are a salt and water. Aqueous solutions of weak electrolytes, such as weak acids or weak bases, contain varying proportions of molecules and ions that depend upon individual compounds and conditions.

In complex ions, nonmetal atoms in molecules or anions called ligands form coordinate covalent bonds to metal cations. The charge of a complex ion equals the algebraic sum of the charge of the metal cation and the charges of any ligands that are anions. Coordination compounds contain metal cations surrounded by ligands.

WATER: PURE AND IMPURE

14.11 NATURAL WATERS AND WATER POLLUTION

The Earth's water supply is estimated to be approximately 1.4×10^9 km^3. For comparison, Lake Ontario contains 1.6×10^3 km^3 of water, or roughly one-millionth of the total water on our planet. The oceans plus a few inland saltwater bodies hold 97.3% of the total water, leaving only 2.7% as fresh water. Of this, most is locked up in polar ice and glaciers, lies deep in the ground, or resides in the atmosphere or the soil. Furthermore, many freshwater lakes and rivers are not close to populated areas. This inaccessibility reduces the portion of the world's water available for human use to approximately 0.003% of the total (Figure 14.8).

All natural waters contain ions in solution. The oceans contain soluble salts that have been there since the era of volcanic activity and the original formation of these bodies of water. The highest concentrations of ions in the oceans are chloride (19,000 mg/L), sodium (10,600 mg/L), and magnesium ions (1300 mg/L). Fresh water contains lower concentrations of ions accumulated as the water courses over the land, dissolving soluble salts from rocks and minerals. In addition, natural water always contains carbon dioxide, which in solution is in equilibrium with hydrogen carbonate ions, causing a slight acidity in such waters:

$$CO_2(g) + 2H_2O(l) \rightleftharpoons HCO_3^- + H_3O^+$$

The most abundant ions in natural fresh water are usually calcium and hydrogen carbonate ions (Table 14.9).

Table 14.9
Ions in Natural Fresh Waters
These are the ions found most frequently.

Cations	Anions
Ca^{2+}	HCO_3^-, CO_3^{2-}
	OH^-
Na^+	SO_4^{2-}
Mg^{2+}	Cl^-
K^+	NO_3^-
Fe^{2+}, Fe^{3+}	F^-
NH_4^+	PO_4^{3-}

As long as the quantities of substances dissolved in water do not become excessive, water purification occurs naturally. Dissolved gases and volatile impurities are flushed out by air mixed with the water as it trickles over shallow stream beds. Solids settle out in quiet pools and lakes, and are filtered out as water seeps through the soil. And a host of bacteria and other microorganisms help to decompose the by-products of plants and animals, living and dead.

A **pollutant** is an undesirable substance added to the environment, usually by the activities of Earth's human inhabitants. Substances considered to be water pollutants are generally (1) poisonous or cause disease (viruses and bacteria); (2) decrease the oxygen content of water during their decay, thereby leading to the death of aquatic life; or (3) are indirectly harmful by making water unpleasant to use or by destroying the natural beauty and health of lakes, rivers, and oceans. More than fifteen thousand scientists and engineers in the United States are continually working to purify water either for human consumption or so that it can be returned safely to natural bodies of water after industrial or domestic use.

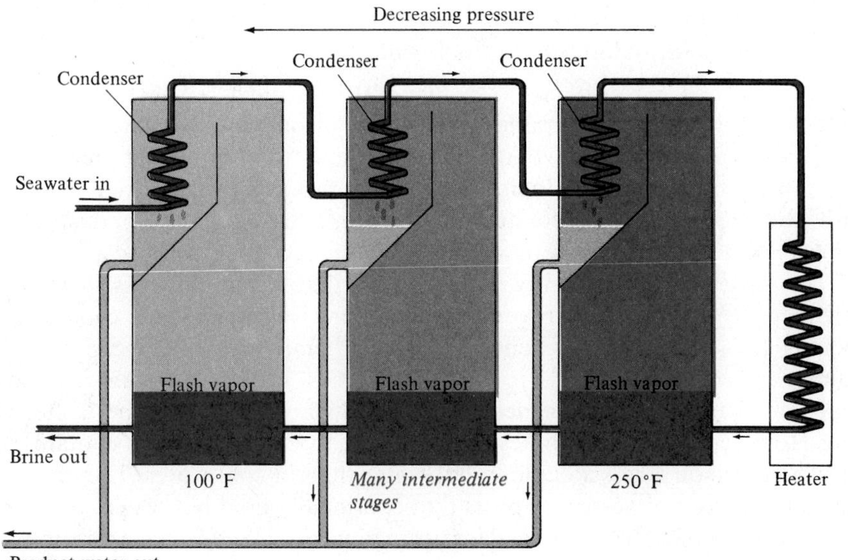

Figure 14.9
Seawater Desalination by Multistage Flash Distillation
This is the process most widely used for seawater desalination. A plant at Key West, Florida, produces 2.6 million gallons/day by this method. Hot seawater enters a chamber where the pressure is low enough to cause some water to vaporize instantly—to "flash" into steam. The temperature of the remaining seawater drops as it moves over to the next chamber, where low pressure causes more water to flash distill. The Key West plant takes water from 38 °C to 121 °C in 50 stages.

14.12 DESALINATION OF WATER

The removal of ions, especially Na^+ and Cl^-, from water is called *desalination.* Desalination is necessary to obtain very pure water from ordinary tap water, to obtain drinking water from seawater or brackish water (less salty than seawater), and to remove excessively high concentrations of ions in purifying industrial or municipal wastewater.

Desalination methods utilize one of several different principles. In *distillation processes,* the salty water is boiled and the pure water vapor condensed. Distillation has long been used to make small quantities of water of high purity in laboratories (Aside 15.2). It is also the basis for the majority of the large-scale seawater desalination plants. To get the maximum efficiency out of the energy input needed to vaporize the water, the vapor is passed back over the incoming seawater and the heat of condensation given up by the vapor aids in heating the seawater (Figure 14.9).

Desalination by *crystallization* also utilizes a phase change. Salty water is partially frozen to a slurry, then the ice crystals are separated and melted. Seawater desalination by crystallization is still in the experimental stage. Several processes utilize membranes that in essence filter out the ions (see description of reverse osmosis, Section 15.13). In *electrodialysis,* ion-selective membranes—membranes that permit only positive or only negative ions pass through—are combined with an electrochemical cell.

14.13 HARD WATER

The presence of certain metal ions in water destined for domestic or industrial use causes practical problems. Water containing these ions is called "hard" water. It can be recognized by the grayish white scum it forms with soap, the dull appearance of clothes washed in it using soap, the hard scale it forms in boilers and tea kettles, and sometimes, by an unpleasant taste. **Hard water** contains metal ions (principally Ca^{2+}, Mg^{2+}, and Fe^{2+}) that form precipitates with soap or upon boiling.

The properties of hard water are partly determined by what anions accompany the metal cations. To the extent that water contains HCO_3^- anions together with the Ca^{2+}, Mg^{2+}, and Fe^{2+} ions, it displays **carbonate hardness** (or **temporary hardness**). Boiling hard water that contains hydrogen carbonate ions drives off carbon dioxide, leaving behind carbonate ions that form insoluble salts.

$$M^{2+} + 2HCO_3^- \xrightarrow{\Delta} MCO_3(s) + CO_2(g) + H_2O(g) \quad \textbf{(14.24)}$$

Carbonate hardness is undesirable because the insoluble carbonates form a hard coating known as "boiler scale" that collects on the walls of hot water pipes and other vessels. (Have you ever seen a domestic hot water pipe that was choked off by such a deposit?) Boiler scale is a poor conductor of heat, and the efficiency of industrial boilers (and tea kettles) is decreased by its presence.

Water softening is the removal of the ions that cause hardness in water. If a sufficient concentration of hydrogen carbonate ions is present, water can be softened by boiling (as shown above); hence the name "temporary" hardness. Water that contains Ca^{2+}, Mg^{2+}, and Fe^{2+} ions but little HCO_3^- ion has what is referred to as **noncarbonate hardness** (or **permanent hardness**). Such water cannot be "softened" by boiling. However, it also contributes to boiler scale to the extent that it contains sulfate ion. Calcium sulfate is less soluble in hot water than in cold water. Therefore, it too collects on the walls of vessels or pipes that continually hold hot water.

Hard water is effectively softened for use in homes or in industrial operations either by removing the metal ions or by tying them up chemically in complexes. Chemicals may be added to precipitate the metal ions. In home laundries water was formerly softened by adding washing soda ($Na_2CO_3 \cdot 10H_2O$) before putting in the soap. This causes Ca^{2+}, Mg^{2+}, Fe^{2+} ions to precipitate out as carbonates. Modern synthetic detergents (Section 15.17) contain *builders*—compounds that, among other functions, prevent the formation of precipitates in hard water by tying up the metal ions as soluble complex ions.

Another method for softening water involves the replacement of the objectionable cations by sodium ions, which do not give a precipitate with soap or form scales. The process of replacement of one ion by another is known as **ion exchange.** Water is passed over naturally occurring zeolites or similar synthetic materials. The zeolites, which are complex sodium aluminum silicates (Section 29.13), possess a three-dimensional open structure, with a negatively charged aluminosilicate framework and sodium ions in the openings of the framework. When hard water is passed through a column of a zeolite, the Ca^{2+}, Mg^{2+}, and Fe^{2+} ions present in the water change places with the sodium ions and are thus removed from the liquid phase.

$$M^{2+} + 2NaZ(s) \longrightarrow 2Na^+ + M(Z)_2(s) \qquad (Z = \text{anion of zeolite})$$

The zeolite can be regenerated by passing a concentrated NaCl solution through it. Thus, the zeolite can be used repeatedly. People on restricted diets should be aware that water softened in this way contains a high concentration of sodium ions.

Synthetic organic materials known as *ion-exchange resins* are even better water softeners than the zeolites. The resins have a hydrocarbon framework to which are chemically bonded either negatively charged groups, such as sulfonate, $—SO_3^-$, or positively charged groups, such as $—NR_3^+$ (where R = a group containing only hydrogen and carbon). The former are counterbalanced by positive ions, usually Na^+ or H^+, and the latter by anions, for example, OH^-. It is possible to remove virtually all the unwanted ions from water by passing it through a mixture of a cation-exchange resin containing H^+ as the positive ion and an anion-exchange resin having OH^- as the negative ion. The H^+ replaced from one resin by cation impurities in the water combines with the OH^- replaced from the other by anionic impurities. Water with almost all ions removed (except, of course, the H^+ and OH^- ions normally present from the H_2O ionization equilibrium) is called *deionized water*. (Deionized water is not "pure"—it still contains dissolved gases and molecular compounds, and bacteria.)

14.14 WATER POLLUTION AND SEWAGE TREATMENT

Municipal sewage is what drains from our sinks, washing machines, bathtubs, and toilets, combined with industrial wastewater and, in some places, with the water that runs into storm sewers. Surprisingly, average domestic sewage is 99.94% pure water and only 0.06% dissolved and suspended solids. Taking out as much of that 0.06% as possible before returning the water to the hydrosphere is increasingly important as the population increases.

The pollutants in sewage, as well as in many polluted natural bodies of water, include (1) pathogens—the viruses and bacteria that can cause illness, (2) oxygen-demanding organic wastes, and (3) the by-products of our advanced industrial society—heavy metals, pesticides, fertilizers, and detergents.

The oxygen-demanding wastes are organic materials that come from plants and animals, living and dead. The large organic molecules, which contain primarily carbon, hydrogen, nitrogen, oxygen, sulfur, and phosphorus, are broken down in both natural waters and sewage treatment plants by microorganisms to simple, harmless molecules and ions. Decomposition of organic matter by bacteria in the presence of oxygen is called *aerobic decomposition.*

aerobic decomposition

$$\text{Organic molecules containing C, H, N, O, S, P} \xrightarrow[O_2]{\text{aerobic bacteria}} CO_2, H_2O, NO_3^-, SO_4^{2-}, HPO_4^{2-}, H_2PO_4^-$$

As long as enough oxygen is available to decompose all of the organic matter present, natural microorganisms can keep a body of water sparkling and clean.

If the oxygen supply is cut down, or if the supply of organic material increases to the point where aerobic decomposition cannot keep up with it, drastic changes take place. Bacteria that depend upon oxygen die, other bacteria switch to oxygen-containing ions such as NO_3^- for their oxygen supply, and anaerobic bacteria, which require oxygen-free conditions, thrive. Decomposition by bacteria in the absence of oxygen is called *anaerobic decomposition.* The products are as simple as those in aerobic decomposition, but much less pleasant.

anaerobic decomposition

$$\text{Organic molecules containing C, H, N, O, S, P} \xrightarrow[\text{no } O_2]{\text{anaerobic bacteria}} CH_4, NH_3, NH_4^+, H_2S, HPO_4^{2-}, H_2PO_4^-, \text{sometimes } PH_3$$

In a body of water that has "gone anaerobic," gas bubbles are visible and the smell of rotten eggs characteristic of hydrogen sulfide (H_2S) is in the air, sometimes combined with the smell of phosphine (PH_3), which is equally noxious. The water appears black and is filled with slime. Fish and other oxygen-consuming residents of ponds or lakes that have gone anaerobic eventually die.

Nitrogen- and phosphorus-containing wastes, which include fertilizers and detergents, play a role in upsetting the balance of living organisms and destroying oxygen-demanding pollutants in natural waters. The outcome is what is referred to as eutrophication, or the "death" of, say, a lake or pond. *Eutrophication* is a natural process in which a lake grows rich in nutrients and subsequently gradually fills with organic sediment and aquatic plants. The lake becomes shallower as undecomposed sediment builds up, and plant life thrives within the lake and at its boundaries. The end result of eutrophication, a very slow process when unaffected by man's activities, is the transformation of the lake first to a marsh and then to dry land.

Before water that has been used in households is returned to natural bodies of water, it must be purified. Sewage treatment, with the overall goal of reducing all types of pollutants, takes place in three stages—primary, secondary, and advanced, or tertiary, wastewater treatment.

The first step in *primary sewage treatment* (Figure 14.10) is to filter out obvious debris such as pieces of paper and wood, and then to get rid of sand, cinders, and gravel (collectively called "grit") that might clog or damage pipes and machinery farther along in the treatment plant. Next, organic and inorganic solids are allowed to settle out in large sedimentation tanks. Sometimes chemicals are added to speed up settling and sometimes provision is made for further purification of the *sludge* (the solids that settle out of sewage) by sludge digestion, which utilizes anaerobic decomposition. At the end

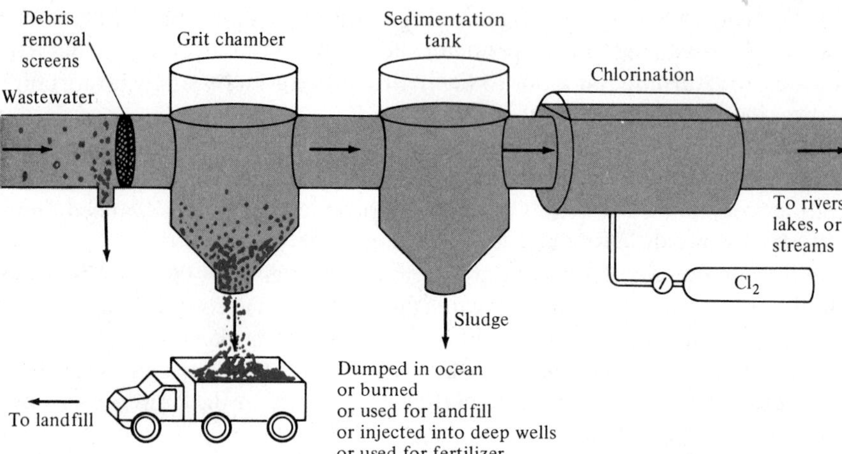

Figure 14.10
Primary Sewage Treatment
The sludge is partially digested and the water partly removed before the sludge is disposed of.

of primary treatment 40%–60% of the suspended solids and 25%–35% of the oxygen-demanding wastes have been removed. Before its return to natural bodies of water, the treated water is disinfected, usually by chlorine, to destroy any remaining pathogens.

Secondary sewage treatment (Figure 14.11) goes on to remove up to 90% of the oxygen-demanding wastes. Two methods are used, both of which expose the sewage to a vigorous population of aerobic bacteria and a plentiful supply of oxygen. In the trickling filter method, which doesn't really filter anything, the water runs over a bed of stones 5–10 cm in diameter. Aerobic bacteria and other microorganisms attach themselves to the stones and take organic matter out of the water as it trickles past, while oxygen from the air mixes into the running water. The other method relies on vigorously mixing air (or oxygen generated on site) with the sewage and adding *activated sludge*—sludge from previous batches of sewage that has developed a high population of microorganisms. This is necessary because one batch of sewage cannot produce enough microorganisms during a short stay in the plant to purify itself sufficiently.

Primary plus secondary sewage treatment does a good job of removing solids, oxygen-demanding wastes, and disease-causing organisms. Dissolved ions, including heavy metals and plant nutrients, still remain, as do organic compounds, such as pesticides that are nonbiodegradable and also radioactive materials. Sewage treatment designed to remove specific pollutants that remain after secondary treatment is called *tertiary*, or *advanced, wastewater treatment*. The major problem currently being

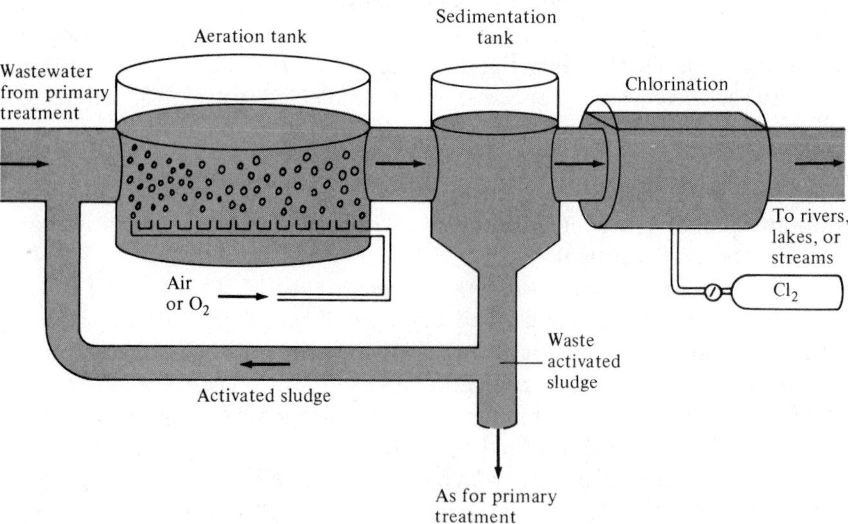

Figure 14.11
Secondary Sewage Treatment

attacked by tertiary treatment is the removal of nitrogen- and phosphorus-containing chemicals; both are growth accelerators for microorganisms, algae, and aquatic plants, and nitrogen compounds are also oxygen-demanding pollutants.

There eventually will be about 22,000 sewage treatment plants in the United States. Communities that once discharged untreated sewage are now being forced by Federal Government regulations to build treatment plants. Also, all existing treatment plants must, if necessary, upgrade their treatment to meet Federal water pollution standards. When all plants are in conformance with the government requirements, 57% of the plants will have secondary treatment and 43% will have tertiary treatment. Primary treatment alone is no longer sufficient.

SUMMARY OF SECTIONS 14.11–14.14

In the oceans Cl^-, Na^+, and Mg^{2+} ions are present in highest concentrations; in fresh water Ca^{2+} and HCO_3^- are usually most abundant. Bodies of water that are not overly polluted are purified by natural processes. Water pollutants introduced by human activities are (1) poisonous or disease causing, (2) oxygen demanding, (3) or indirectly harmful. Desalination, the removal of ions from water, is carried out by various techniques on sea water or on industrial or municipal wastewater.

Hard water contains metal ions (chiefly Ca^{2+}, Mg^{2+}, and Fe^{2+}) that form precipitates with soap or upon boiling. Carbonate hardness can be decreased by boiling, which drives off CO_2. Noncarbonate hardness is overcome by water softeners that precipitate the metal ions or tie them up in soluble complexes. Hard water is also softened by ion exchange, in which the metal cations are replaced by Na^+.

Bacteria in natural waters and sewage treatment plants remove organic wastes from water by breaking down large molecules to smaller ones either with (aerobic decomposition) or without (anaerobic decomposition) the consumption of oxygen. Lakes or ponds undergo eutrophication (gradual transformation to marshes and then dry land), a process that is speeded up by large concentrations of substances that contain nitrogen and/or phosphorus (e.g., fertilizers and detergents).

Sewage treatment requires removal of pathogens, oxygen-demanding wastes, and man-made pollutants. Primary sewage treatment includes removal of solids by filtration and sedimentation, and removal of a portion of the oxygen-demanding wastes. Secondary sewage treatment removes up to 90% of the oxygen-demanding wastes by exposing the sewage to large populations of aerobic microorganisms.

SIGNIFICANT TERMS (with section references)

acid (water-ion) (14.8)
acidic aqueous solution (14.8)
alkaline aqueous solution (14.8)
base (water-ion) (14.8)
carbonate hardness, temporary hardness (14.13)
chemical equilibrium (14.5)
complex ion (14.10)
coordination compounds (14.10)
deliquescent (14.4)
efflorescence (14.4)
electrolytes (14.7)
hard water (14.13)
heat of solution (14.3)
hydrated (14.2)
hydrates (14.4)
hydration (14.2)
hydrolysis (14.6)
hygroscopic (14.4)
ion exchange (14.13)
ligands (14.10)
neutralization (14.9)
noncarbonate hardness, permanent hardness (14.13)
nonelectrolytes (14.7)
pollutant (14.11)
polyprotic acids (14.8)
solvation (14.2)
strong electrolytes (14.7)
water softening (14.13)
weak electrolytes (14.7)

QUESTIONS AND PROBLEMS

The Chemistry of Water—Properties of Water

14.1 Draw the Lewis structure of a water molecule. What type of bonding exists between the oxygen and hydrogen atoms within the molecule?

14.2 Draw a three-dimensional sketch of a water molecule. Describe the geometry of the molecule. Describe the hybridization of the atomic orbitals on the oxygen atom.

14.3 What is the value of the predicted bond angle in water? Is the actual value larger or smaller than this value? Why is there a difference?

14.4 What types of intermolecular forces are present in water?

14.5 Why is liquid water less dense at temperatures both higher and lower than 3.98 °C? Why is liquid water more dense than ice?

14.6 What properties of water make it useful as a cooling agent, for example, in automobile radiators and in industrial processes?

14.7 The heat of combustion of a typical coal is 31 kJ/g. Utilizing this heat, calculate the mass of steam at 125 °C that can be prepared from water at 25 °C for each gram of coal burned. Assume the heat of vaporization of water to be 2260 J/g, the specific heat of water to be 4.184 J/K g, and the specific heat of steam to be 2.1 J/K g.

14.8 What will be the final temperature of the mixture prepared from 100.0 g of ice at 0 °C and 100.0 g of steam at 100 °C? For water the heat of vaporization is 2256.2 J/g, the heat of fusion is 333.5 J/g, and the specific heat is 4.184 J/K g.

14.9* The crystal structure of ordinary ice is a hexagonal unit cell with volume given by $(0.4535\ \text{nm})^2(0.741\ \text{nm})(\sin 60°)$. The number of molecules in the unit cell is four. Calculate the theoretical density of ice.

14.10* What is the pressure predicted by the ideal gas law for one mole of steam at 31.0 L and 100.00 °C? What is the pressure predicted by the van der Waals equation given that $a = 5.464\ \text{L}^2\ \text{atm/mol}^2$ and $b = 0.03049$ L/mol? What is the percent difference between these values? Does steam deviate from ideality significantly at 100 °C?

The Chemistry of Water—Solutions

14.11 Why is water a much better solvent for ionic and polar substances than for nonpolar substances? Are all ionic and highly polar substances highly soluble in water?

14.12 What does the term "solvation" mean? What do we call the solvation process that occurs in an aqueous solution?

14.13 What two factors determine the strength of the attraction of water molecules to ions during hydration? Will Na^+ or Zn^{2+} have the larger energy of hydration?

14.14 Describe the processes by which NaCl, HCl, and CH_3OH dissolve in water. Prepare sketches showing how these compounds exist in aqueous solution.

14.15 Compare the relative magnitudes of the solvent–solvent, solute–solute, and solvent–solute interactions (a) for a solution with a positive heat of solution (endothermic) and (b) for a solution with a negative heat of solution (exothermic).

14.16 We can see from Table 14.3 that for the dissolution of ionic solids, the lattice energy is a large negative number, the hydration energy is a large negative number, and the calculated heat of solution is a relatively small number. Describe the effect on the calculated heat of solution that would be caused by a small error (say, 5%) in either the lattice energy or the hydration energy.

14.17 The heat of formation at 25 °C is −366 kJ/mol for $NH_4NO_3(s)$ and −341 kJ/mol for NH_4NO_3 (1 M). Calculate the heat of solution for preparing a 1 M solution

$$NH_4NO_3(s) \xrightarrow{H_2O} NH_4NO_3(1\ M)$$

If the solution cannot gain or lose this heat fast enough to remain at 25 °C, what sensation would you feel if you touched the beaker in which this reaction was performed?

14.18 The enthalpy change for dissolving $MgBr_2(s)$ in 25 mol of H_2O is −179.9 kJ/mol and the enthalpy change for dissolving $MgBr_2{\cdot}6H_2O(s)$ in 25 mol of H_2O is −5.0 kJ/mol. Compare the relative magnitudes of the hydration and lattice energies for these salts.

The Chemistry of Water—Hydrates

14.19 Name the following compounds: (a) $Cr(CH_3COO)_3{\cdot}H_2O$, (b) $Cd(NO_3)_2{\cdot}4H_2O$, (c) $(NH_4)_2C_2O_4{\cdot}H_2O$ ($C_2O_4^{2-}$ is oxalate ion), (d) $LiBr{\cdot}2H_2O$.

14.20 Write the formulas of the following compounds: (a) sodium acetate trihydrate, (b) aluminum iodide hexahydrate, (c) magnesium bromate hexahydrate (bromate ion is BrO_3^-), (d) thallium(III) chloride tetrahydrate.

14.21 Calculate the mass percent of water of hydration in (a) epsomite, $MgSO_4{\cdot}7H_2O$ and (b) turquoise, $CuAl_6(PO_4)_4(OH)_8{\cdot}4H_2O$.

14.22 Repeat Problem 14.21 for (a) gypsum, $CaSO_4{\cdot}2H_2O$ and (b) natrolite, $Na_2Al_2Si_3O_{10}{\cdot}2H_2O$.

14.23 A student needed anhydrous calcium chloride, $CaCl_2$, to use as a drying agent in an experiment. The stockroom had only one-pound bottles of the hexahydrate. Upon consulting a reference book, the student found that the hexahydrate could be converted to the anhydrous compound by heating at 200 °C. The student heated the contents of one of the bottles at that temperature until the mass no longer decreased to ensure that all of the water had been driven off. What was the final mass of anhydrous salt obtained?

14.24 One liter of a 0.100 M aqueous solution of lead acetate, $Pb(CH_3COO)_2$, was being prepared by a stockroom attendant. The only available source of lead acetate was the decahydrate. What mass of the hydrated salt did the attendant use for the solution?

14.25 A 23.4 g sample of a hydrate of $CuCl_2$ was heated at 100 °C until no additional water was driven off. The anhydrous sample weighed 18.5 g. What is the formula of the hydrate?

14.26 A sample of a pale-green hydrated salt was dissolved in water. Standard qualitative analysis tests showed that it contained Fe^{2+} and Cl^-. A 4.04 g sample of the solid hydrate was heated in a stream of HCl, producing 2.18 g of anhydrous salt. What is the formula of the hydrate?

The Chemistry of Water—Chemical Reactions

14.27 How do we usually write the chemical formula for the hydrated hydrogen ion? Draw a Lewis structure for this cation.

14.28 Would you expect OH^- to be highly hydrated? Draw a Lewis structure for $H_3O_2^-$.

14.29 What is wrong with the following statement: Aqueous solutions of acids are characterized by the presence of H^+ or H_3O^+ and the absence of OH^-?

14.30 One student classified the reaction

$$H^+ + OH^- \longrightarrow H_2O(l)$$

as a combination reaction, whereas a second student classified the reaction as partner-exchange. Comment on both statements.

14.31 Write chemical equations for (a) the thermal dissociation of steam at high temperature into atomic hydrogen and hydroxyl radicals (OH), (b) the self-ionization of liquid water, (c) the hydration of $CaCl_2 \cdot H_2O(s)$ to form calcium chloride hexahydrate.

14.32 Write chemical equations for (a) the decomposition of water into hydrogen and oxygen, (b) the reaction of water with magnesium oxide to give magnesium hydroxide, (c) the reaction of carbon monoxide with steam at high temperatures to produce hydrogen and carbon dioxide.

14.33 Classify each of the following reactions as (i) combination, (ii) decomposition, (iii) displacement, or (iv) partner exchange:
(a) $H^+ + OH^- \longrightarrow H_2O(l)$
(b) $S^{2-} + H_2O(l) \rightleftharpoons HS^- + OH^-$
(c) $SO_3(g) + H_2O(l) \longrightarrow H_2SO_4(aq)$
(d) $CuSO_4 \cdot 5H_2O(s) \xrightarrow{\Delta} CuSO_4(s) + 5H_2O(g)$
(e) $[Cu(H_2O)_4]^{2+} + 4NH_3(aq) \longrightarrow [Cu(NH_3)_4]^{2+} + 4H_2O(l)$

14.34 Repeat Question 14.33 for
(a) $Na_2CO_3(s) + 10H_2O(l) \longrightarrow Na_2CO_3 \cdot 10H_2O(s)$
(b) $C(s) + H_2O(g) \longrightarrow H_2(g) + CO(g)$
(c) $CaO(s) + H_2O(l) \longrightarrow Ca(OH)_2(s)$
(d) $CaC_2(s) + 2H_2O(l) \longrightarrow Ca(OH)_2(s) + C_2H_2(g)$
(e) $SO_4^{2-} + H_2O(l) \rightleftharpoons HSO_4^- + OH^-$

Ions in Aqueous Solution—Electrolytes and Nonelectrolytes

14.35 Is water a very weak electrolyte? Why?

14.36 Are slightly soluble ionic compounds considered to be strong or weak electrolytes? Why?

14.37 Identify the species present in an aqueous solution of each of the following compounds: (a) potassium nitrate, a salt; (b) perchloric acid, a strong acid; (c) potassium hydroxide, a strong base; (d) silver chloride, a slightly soluble salt; (e) formic acid (HCOOH), a weak acid; (f) methanol (CH_3OH), a nonelectrolyte.

14.38 Repeat Question 14.37 for (a) isopropyl alcohol ($CH_3CHOHCH_3$), a nonelectrolyte; (b) barium sulfate, a slightly soluble salt; (c) lithium hydroxide, a strong base; (d) hydrosulfuric acid, a weak diprotic acid; (e) hydrobromic acid, a strong acid; (f) sodium iodide, a salt.

14.39 Write chemical equations describing what happens as each of the following substances is mixed with water: (a) $H_2SO_4(l)$, (b) $NH_4I(s)$, (c) $Sr(OH)_2(s)$, (d) $HCN(g)$, (e) $AgCl(s)$.

14.40 Repeat Question 14.39 for (a) $HNO_3(l)$; (b) $KOH(s)$; (c) $KF(s)$; (d) $MgCO_3(s)$, a slightly soluble salt; (e) $Ca(ClO_4)_2(s)$.

Ions in Aqueous Solution—Acids, Bases, and Neutralization

14.41 Define "acid" and "base" in terms of the water-ion theory.

14.42 Name four classic properties of aqueous solutions of acids. Contrast these with the properties of bases.

14.43 What does the term "polyprotic" mean? Write the formula for a common polyprotic acid. Write the chemical equations showing the stepwise ionization of the acid.

14.44 What is a neutralization reaction? Write a chemical equation for the neutralization of any strong acid by any strong base. Write the net ionic equation for the reaction.

14.45 Write the chemical equations for the interaction with water of (a) oxalic acid, HOOCCOOH(s); (b) hydrogen cyanide, HCN(g); (c) nitric acid, $HNO_3(l)$.

14.46 Repeat Question 14.45 for (a) hydrogen bromide, HBr; (b) hypobromous acid, HBrO; (c) phosphorous acid, H_3PO_3, in which only two of the protons are ionizable.

14.47 Which requires the greater mass of NaOH for neutralization—1.00 g of HBr or 1.00 g of $HClO_4$?

14.48 Will a smaller mass of HCl or HNO_3 be needed to neutralize 1.00 g of KOH?

14.49 What volume of 0.1123 M HCl is needed to neutralize 16.9 g of $Ca(OH)_2$?

14.50 What mass of NaOH is needed to neutralize 50.0 mL of 0.1036 M HCl? If the NaOH is available as a 0.0937 M aqueous solution, what volume will be required?

Ions in Aqueous Solution—Complex Ions

14.51 What is a complex ion? What is the term for the molecules or ions that bond to the central metal atom or cation?

14.52 Why do coordination compounds that are acids act as very strong acids and those that contain hydroxide ions act as very strong bases? Why are solutions of many hydrated metal cations slightly acidic?

14.53 Determine the charges on the following complex ions:
(a) $[AuCl_4]^?$, where the ionic charge on Au is +3;
(b) $[Cr(H_2O)_4Cl_2]^?$, where the ionic charge on Cr is +3;
(c) $[Al(OH)_4]^?$, where the ionic charge on Al is +3.

14.54 Repeat Question 14.53 for (a) $[Cr(H_2O)_6]^?$, where the ionic charge on Cr is +3; (b) $[Fe(H_2O)_5Cl]^?$, where the ionic charge on Fe is +3; (c) $[Co(NH_3)_5Cl]^?$, where the ionic charge on Co is +3.

Water: Pure and Impure

14.55 Describe how "nature's water treatment system" functions to purify water containing (a) dissolved gases and other volatile impurities, (b) solids, (c) dissolved ions, (d) plant and animal by-products in the presence and in the absence of oxygen.

14.56 Name the process in which (a) a lake grows rich in nutrients and subsequently fills in gradually with organic sediment and aquatic plants, (b) bacteria and other microorganisms break down organic molecules in the presence ofair, (c) bacteria and other microorganisms break down organic molecules in the absence of air.

14.57 Seawater contains 19,000 mg of chlorine per liter. Assuming the chlorine is in the form of Cl^-, express this concentration in molarity.

14.58 One of the sources of elemental magnesium is seawater, in which the magnesium occurs to the extent of 1300 mg/L as the chloride and sulfate. The magnesium is recovered from the seawater as $MgCl_2 \cdot H_2O$ and the molten hydrate is converted to the metal. What volume of seawater would have to be processed in order to produce 1.00 g of the metal? Assume that the entire process is 80% efficient.

14.59 What is meant by the term "hard water"? What are the two types of hard water and how do they differ from each other?

14.60 What does the term "ion exchange" mean? Briefly describe how ion-exchange resins are used in a home water softener.

14.61 An analysis of water from the Mississippi River showed that it contains 34 mg Ca^{2+}/L and 7.6 mg Mg^{2+}/L. What masses of $Ca(OH)_2$ and Na_2CO_3 are needed to remove these ions from one liter of river water using the "soda-lime process" for softening, which can be represented by the equation

$$Mg^{2+} + Ca(OH)_2 + Na_2CO_3 \longrightarrow Mg(OH)_2 + CaCO_3 + 2Na^+$$

$$Ca^{2+} + 2HCO_3^- + Ca(OH)_2 \longrightarrow 2CaCO_3 + 2H_2O$$

14.62* The scum-producing reaction between soap and Ca^{2+} represented by

$$Ca^{2+} + 2Na(C_{17}H_{35}COO) \longrightarrow 2Na^+ + Ca(C_{17}H_{35}COO)_2$$

can be used to determine water hardness quantitatively. A standardized soap solution is slowly added to a water sample until a lasting sudsing effect is observed upon shaking. A soap solution was treated with 20.0 mL of a solution that contained 495 ppm of Ca^{2+} (1 ppm = 10^{-6} g Ca^{2+}/mL of hard water). Express the concentration of the soap solution in terms of grams of Ca^{2+}/mL soap solution if 35.2 mL of the soap solution were required for the reaction.

A 25.0 mL sample of well water required 3.0 mL of this soap solution before a lasting suds was formed. What is the concentration of Ca^{2+} in the well water (expressed in ppm)?

14.63 Name three common classes of pollutants that are present in sewage and in many polluted natural waters. Cite an example or two for each class.

14.64 Describe the processes occurring in a municipal sewage treatment plant that performs primary, secondary, and tertiary treatment. Identify the types of materials that are removed during each step.

CHAPTER

15

Solutions and Colloids

GENERAL PROPERTIES OF SOLUTIONS

15.1 SOME DEFINITIONS

Any mixture with only a single phase is a solution. We are used to dealing with liquid solutions, but solutions are also found in the gas phase (e.g., air) or in the solid phase (some types of metal alloys; Section 26.8). Most of this chapter is devoted to solutions in the liquid phase, especially to those in which various solutes are dissolved in water.

Suppose a soluble salt is gradually added to water at a given temperature. Beyond a certain point, no more seems to dissolve. In reality, it is still dissolving, but at the same time, an equal amount of salt is coming out of solution and crystallizing on the surface of the solid salt or on the sides of the container. A condition of dynamic equilibrium has been reached—the processes of dissolution and crystallization are proceeding at the same rate, so that each exactly offsets the other (Figure 15.1).

$$\text{Solid salt} \rightleftharpoons \text{salt in solution}$$

This equilibrium can be demonstrated by placing a broken, irregular crystal of salt in the solution. As ions from the crystal go into solution, ions from the solution precipitate on the crystal surface. These ions precipitate onto areas which will give a more symmetrical crystal. Over a period of several days (at constant temperature), the broken corners fill in and a more nearly perfect crystal of the same mass as the broken crystal is produced.

Figure 15.1
Crystals in a Saturated Solution
The total mass of the crystals remains the same as long as the temperature is unchanged. Molecules or ions join and leave the crystal surface in a dynamic equilibrium; irregular crystals become more nearly perfect.

A solution in which no more solute will dissolve is described as saturated. A **saturated solution** contains the amount of dissolved solute that would be in equilibrium with undissolved solute at a given temperature. If a solution can still dissolve more solute, it is described as an **unsaturated solution**.

For every solute and solvent combination, there is a characteristic amount of solute that will dissolve under the given conditions. The **solubility** of a substance is the

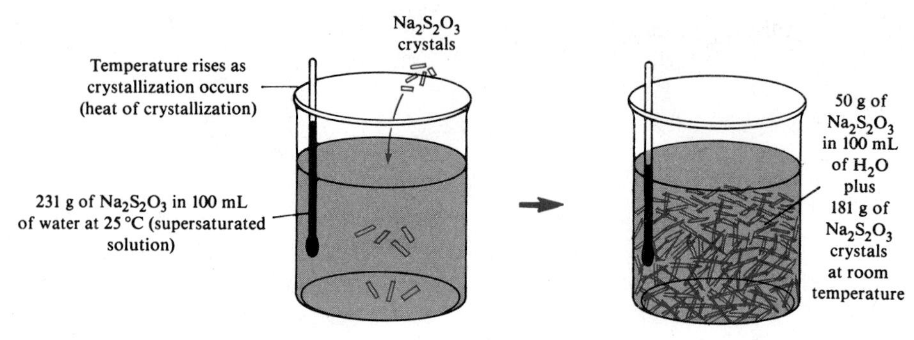

Figure 15.2
Crystallization from a Supersaturated Solution
The solubility of sodium thiosulfate is 50 g/100 mL of water at 25 °C and 231 g/100 mL of water at 100 °C. A supersaturated solution is prepared by dissolving the salt at 100 °C and then carefully cooling it. Seeding the supersaturated solution with a few $Na_2S_2O_3$ crystals causes the excess solute to precipitate. Due to heat of crystallization, the temperature first rises (the reverse of heat of solution) and then returns to room temperature, leaving a saturated solution.

amount that will dissolve in a given solvent to produce a saturated solution (at a given temperature). When the temperature is not specified, it may be assumed that solubility data are given for 25 °C.

The terms "dilute" and "concentrated" are used loosely to indicate solutions with relatively little solute or relatively abundant solute. It is a common error to confuse "dilute" and "concentrated" with "unsaturated" and "saturated". These terms refer to quite different properties of solutions. For example, barium sulfate ($BaSO_4$) is not very soluble—a saturated aqueous solution of barium sulfate is very dilute (0.002 g/L at 18 °C). On the other hand, sodium thiosulfate ($Na_2S_2O_3$) is very soluble—a saturated solution is very concentrated (500 g/L at 25 °C).

Students often make the common error of defining a saturated solution as one that "holds all the solute that it can." This is not correct. It is quite possible to have a **supersaturated solution**—a solution that holds more dissolved solute than would be in equilibrium with undissolved solute. Many solids are more soluble at higher temperatures than at lower temperatures. Sometimes when such a material is dissolved at a higher temperature and the solution then allowed to cool slowly and without disturbance, all of the solute remains dissolved. Honey is essentially a supersaturated solution of various sugars in water. It is produced in the beehive when some of the water of an unsaturated solution evaporates. Honey does not quickly revert to saturation, but on long standing it may do so and become at least partly crystalline.

The more complex the crystal structure of a particular compound and the more soluble the compound, the more apt that compound is to form a supersaturated solution. If a crystal with a similar or identical crystal structure is introduced into a supersaturated solution, the solute molecules or ions can orient themselves on the surface of the crystal and precipitation begins (Figure 15.2). The solid precipitates until equilibrium is established and a saturated solution is formed. Inducing crystallization in a supersaturated solution by adding a crystal is called "seeding."

15.2 THE NATURE OF SOLUTIONS IN THE LIQUID PHASE

a. Liquid–Liquid Solutions The mutual solubility of two substances in the same phase (both liquids, both solids, or both gases) is called **miscibility**. Substances that are completely soluble in each other are **infinitely miscible**. All gases are infinitely miscible. Some liquids and solids are also infinitely miscible.

There is an old adage that "like dissolves like," meaning that substances with similar bonding, structures, and intermolecular forces will be mutually soluble. Like most adages, "like dissolves like" must be applied with caution, for there are many exceptions. However, the adage is true for many liquids. The more similar the intermolecular forces within each of the two liquids, the more likely it is that their molecules will be able to mingle together. Ethyl alcohol (CH_3CH_2OH) and water

(HOH) are infinitely miscible because the primary intermolecular force in each pure liquid is hydrogen bonding. On mixing, the CH_3CH_2OH and HOH molecules readily form strong hydrogen bonds with each other.

Another infinitely miscible pair is benzene and toluene.

benzene *toluene* ($-CH_3$)

Here London forces are the most important forces of attraction between the molecules in the pure liquids, meaning that the attraction of benzene and toluene molecules for each other (the solute–solvent interaction) will be similar in strength to that of benzene molecules for each other and that of toluene molecules for each other. However, benzene and water are only partially miscible because the intermolecular forces in each are different (London forces in benzene, and the stronger forces of hydrogen bonding in water; Figure 15.3).

Liquids that are mutually insoluble, or very nearly so, are referred to as **immiscible.** Toluene and water are considered to be immiscible, as are vinegar and oil. No matter how hard you shake a bottle of vinegar and oil salad dressing, the vinegar droplets are soon drawn back together by the strong intermolecular force in water, squeezing out the nonpolar oil droplets.

Infinitely miscible

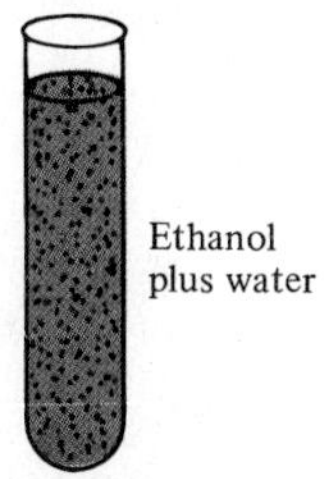

Partially miscible

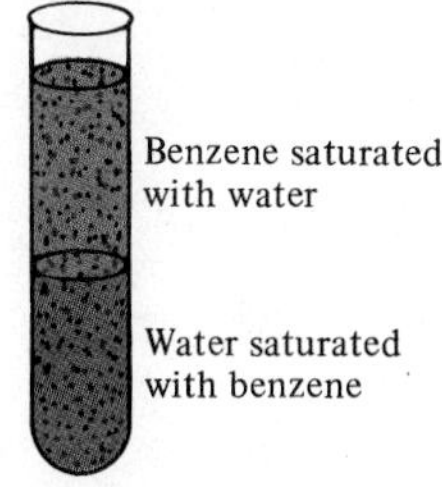

Immiscible

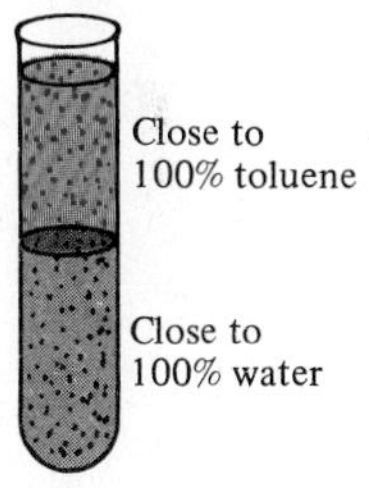

Figure 15.3
Miscibility of Liquids
Very few liquids are totally immiscible.

b. Solid–Liquid Solutions The dissolution of a solid occurs at the interface between the solid and the liquid. Sodium chloride will not dissolve in liquids like benzene, because the benzene is nonpolar and does not attract the sodium and chloride ions. Even though alcohol is somewhat polar, sodium chloride will not dissolve in it to any great extent because the polar attractions are not great enough to overcome the attraction of the sodium ions and chloride ions in the crystal for each other.

A finely divided solid presents a much larger surface area to a solvent than a large piece of the same solid. Therefore, crushing a solid will hasten dissolution. For example, confectioner's sugar dissolves much more rapidly than rock candy. Shaking a mixture of a solid solute and a solvent will also hasten dissolution. But note that while the *rate* of dissolution varies with the shaking or size of the pieces of solid, the *solubility* of the solid is unaffected. Whether it takes a few seconds with shaking or weeks without shaking, dissolution will appear to stop when saturation for that solute–solvent combination at that temperature is reached.

Do not make the common error of confusing the solubility of a substance with its rate of dissolution. For example, in the laboratory do not immediately assume that a substance is insoluble if a large chunk of it does not immediately dissolve.

EXAMPLE 15.1 Mutual Solubility

Decide whether each of the following substances is more likely to be soluble in water or in benzene: (a) KCl; (b) C_8H_{18}(*n*-octane; the *n* indicates that the carbon atoms are attached to each other in a straight chain; Section 32.1); (c) $[Ag(NH_3)_2]Cl$; and (d) $HOCH_2CH_2OH$ (ethylene glycol).

(a) Potassium chloride is an ionic compound that is soluble in **water** (most chlorides are) and is unlikely to be soluble in a nonpolar solvent like benzene.
(b) Hydrocarbons like *n*-octane, which contain no strongly electronegative atoms, are essentially nonpolar. Octane is more likely to be soluble in **benzene** than in water.
(c) This is an ionic compound that contains the complex ion $[Ag(NH_3)_2]^+$ combined with Cl^-. It is more likely to be soluble in **water** than in benzene.
(d) With two —OH groups in the molecule, ethylene glycol [the main ingredient in antifreeze] should be able to form hydrogen bonds with water. It is more likely to be soluble in **water** than in benzene.

EXERCISE Decide whether each of the following substances is more likely to be soluble in water or in *n*-hexane (n-C_6H_{14}): (a) $K_3[Fe(CN)_6]$; (b) CCl_4 (carbon tetrachloride); (c) $C_{10}H_{22}$ (*n*-decane); and (d) $Ca(NO_3)_2$.

Answers (a) water; (b) *n*-hexane; (c) *n*-hexane; (d) water.

15.3 IDEAL VERSUS NONIDEAL SOLUTIONS

Real atoms, ions, and molecules influence each other, and their behavior rarely is exactly what theories predict. It is useful to have a convenient way to compare real to theoretical behavior.

Ideal gas molecules are thought of as far enough apart to be independent of each other's influence. A different approach to ideality is needed for liquid solutions, where the solvent and solute particles are intimately mixed and in contact with each other. In an **ideal solution of a molecular solute**, the forces between all particles of both solvent and solute are identical. In other words, in an ideal solution of A and B the forces between particles of A and A, or A and B, or B and B are the same. If A and B form an ideal solution and both substances are liquids, interparticle forces in pure A are so similar to those in pure B that particles of one added to the other find themselves in virtually identical environment. Benzene and toluene (Section 15.2a) are so similar that they form solutions that are close to ideal.

An **ideal solution of an ionic solute** is defined as one in which the ions in solution are independent of each other and attracted only to solvent molecules. Very dilute solutions of ionic substances approach ideal behavior because the ions are so far apart that electrostatic attraction is weak.

Components in an ideal solution make a fully effective contribution to the concentration. Think of a lone swimmer in a swimming pool. He can go anywhere in the pool he wants to at any speed he chooses. He makes a fully effective contribution to the concentration of swimmers in the pool (1 swimmer/pool). Now put 24 other swimmers in the pool with him. The effectiveness of each is decreased by bumping into or going around the others. The effective concentration is less than 25 swimmers/pool. Furthermore, if three sea lions were thrown into the pool, the effective concentrations of the swimmers would change drastically.

In a nonideal solution, forces between atoms, ions, and molecules must be taken into account. Consider, for example, the electrical conductivity of an aqueous solution of a strong electrolyte, say, a salt. If the electrolyte solution is very dilute—about 0.01 M or less—the conductivity is just about that expected on the basis of complete dissociation of a salt into ions. But the difference between the expected and observed values becomes larger and larger as the concentration of the solution is increased. For many years, such behavior was interpreted to mean that dissociation or ionization of strong electrolytes was not complete except in dilute solutions. We know now, however, that in solution a strong electrolyte is fully dissociated into ions. Another explanation has been found, and it is quite simple—ions of opposite charge attract each other, and this mutual attraction increases as the ions come closer together, as they must in more concentrated solutions. Each ion becomes surrounded by ions of opposite charge; these tend to hold the original ion back and so retard its journey toward the electrode, causing a lower conductivity than expected. The effect is larger if the solution is more concentrated or if the ions have charges greater than one, for example, for $MgSO_4$, as compared with NaCl.

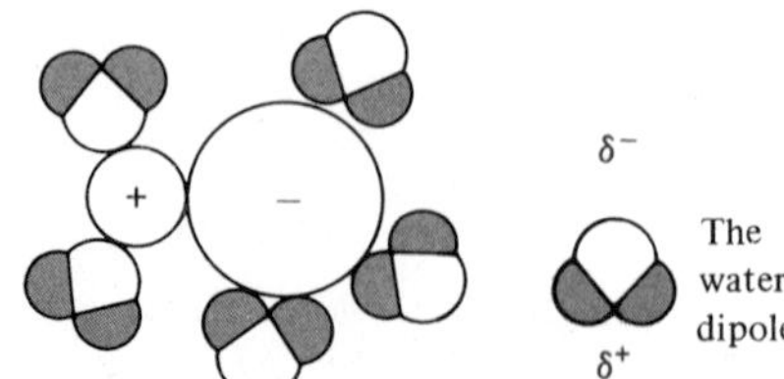

Figure 15.4
Pairing of Ions in Aqueous Solution
The formation of such ion pairs contributes to the nonideal behavior of solutions.

In quite concentrated solutions, among highly charged ions, or among ions in solvents other than water, ion pairing may also occur. Individual pairs of ions held together by electrostatic attraction contribute to the nonideal behavior of such solutions (Figure 15.4).

For extremely accurate experimentation, for the mathematics of physical chemistry, or for work with quite concentrated solutions, it is necessary to deal with effective concentrations rather than actual concentrations. Physical chemists define a quantity called "activity" for this purpose. However, in our study of chemistry and in many aspects of everyday laboratory work in chemistry, this is not necessary and actual concentrations may be used.

15.4 THE EFFECT OF TEMPERATURE AND PRESSURE ON SOLUBILITY

a. Temperature and Solubility The solubility in water of most inorganic salts increases with temperature. However, there are many exceptions, and one study has noted the presence of the following anions in most salts that become less soluble with increasing temperature: SO_4^{2-}, SeO_4^{2-}, SO_3^{2-}, AsO_4^{3-}, and PO_4^{3-}. Sodium chloride (see Figure 15.5) is unusual because the various forces balance out so that there is almost no change in solubility with changing temperature.

The effect of temperature on the solubility of gases is, by contrast, much more predictable. The solubility of gases in liquids decreases with rising temperature in most cases (Figure 15.6), and most gases that dissolve without ionizing can be driven out of solution through boiling.

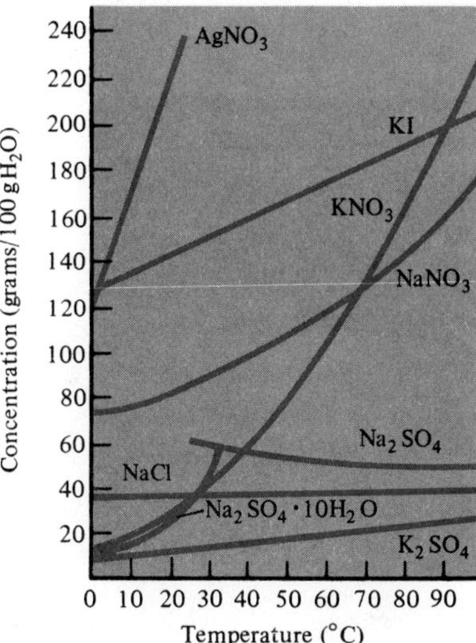

Figure 15.5
The Effect of Temperature on the Solubility of Several Inorganic Substances in Water

b. Pressure and the Solubility of Gases The solubility of gases, unlike that of solids or liquids, varies not only with the temperature but also with the pressure. The relationship between the partial pressure of a solute gas (Section 6.11) and its solubility is expressed by **Henry's law:** At constant temperature, the partial pressure of a gas over a solution is directly proportional to the concentration of the gas in that solution. Doubling the oxygen pressure, for example, doubles the amount of oxygen that will dissolve in a given amount of solvent. Henry's law is followed most closely by dilute solutions of gases that do not react with the solvent.

Mathematically, Henry's law is expressed as

partial pressure of solute gas → $P_A = kC$ ← *constant*; *gas concentration in solution* **(15.1)**

$$P_A = kC \qquad (15.1)$$

where k is a constant characteristic of the specific combination of solvent and gas, P_A is the partial pressure of the solute gas in the gas phase over the solution, and C is the concentration of the gas in the solution.

A variety of units can be used in Henry's law. The value and units of the constant differ for different concentration units. The constant is often given in either liter atmospheres per mole or in atmospheres, to be used with concentrations in moles per liter and mole fractions, respectively.

$$P_A = kC \qquad (15.2a)$$

atmospheres $\quad \dfrac{\text{liter atmospheres}}{\text{mole}} \quad \dfrac{\text{moles}}{\text{liter}}$

$$P_A = kX \qquad (15.2b)$$

atmospheres *atmospheres* *mole fraction*

The mole fraction is a dimensionless number (Section 6.11).

Probably the most familiar gas–liquid solution is carbon dioxide in water. In manufacturing soft drinks, the sweetened and flavored water is saturated at a carbon dioxide pressure greater than that in the atmosphere. The carbon dioxide does not escape immediately when the pressure is relieved, for carbon dioxide readily forms supersaturated solutions in water. If the solution is stirred or shaken, however, most of the gas escapes rapidly, leaving a solution that is saturated at the newly established partial pressure of the carbon dioxide above the solution. This is in accordance with Henry's law and Dalton's law of partial pressures ($P_{total} = P_1 + P_2 + P_3 + \ldots$; Section 6.11). Because carbon dioxide in the atmosphere has a partial pressure of only 4×10^{-4} atm, a soft drink tastes flat when it has come to equilibrium with the carbon dioxide in the atmosphere.

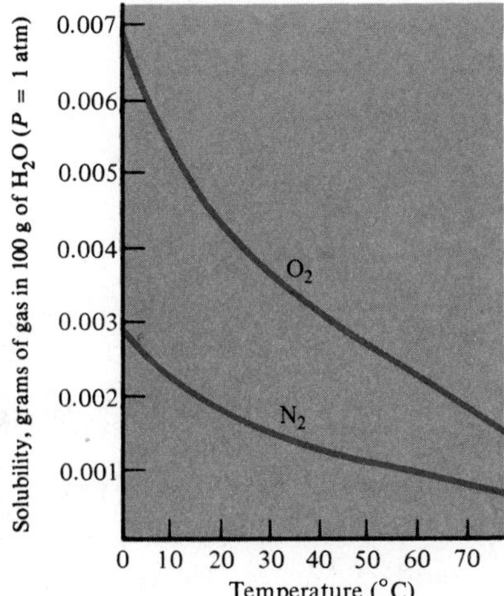

Figure 15.6
Solubility of Nitrogen and Oxygen in Water at 1 atm
Data for solubility at pure oxygen or pure nitrogen pressure.

EXAMPLE 15.2 Solubility: Henry's Law

During bottling, a carbonated beverage was made by saturating water with CO_2 at 0 °C and a pressure of 4.0 atm. Later, the bottle was opened and the soft drink allowed to come to equilibrium at 25 °C with air containing CO_2 at a partial pressure of 4.0×10^{-4} atm. Find the concentration of CO_2 in the freshly bottled soda and in the soda after it had stood open and come to equilibrium with the air. The Henry's law constants for aqueous solutions of CO_2 are $k = 13.0$ L atm/mol at 0 °C and 32 L atm/mol at 25 °C.

The partial pressures of CO_2 over the aqueous solutions during the bottling process (4.0 atm) and after the bottle was opened (4.0×10^{-4} atm) are known. These pressures can be used in Henry's law to find the concentration of the CO_2 before and after the bottle was opened.

$$C = \frac{P_{CO_2}}{k} = \frac{4.0 \text{ atm}}{13.0 \text{ L atm/mol}} = \mathbf{0.31\ mol/L}$$

$$C = \frac{4.0 \times 10^{-4} \text{ atm}}{32 \text{ L atm/mol}} = \mathbf{1.3 \times 10^{-5}\ mol/L}$$

The concentration of CO_2 decreased from 0.31 mol/L to 1.3×10^{-5} mol/L, a decrease of more than 99.99%. [The drink had definitely become flat.]

EXERCISE Calculate the solubility of argon in water at 0 °C at a pressure of 2.16 atm. The value of the Henry's law constant for Ar in H_2O at 0 °C is 396 L atm/mol.

Answer 5.45×10^{-3} mol/L.

SUMMARY OF SECTIONS 15.1–15.4

Solutions can form in all three phases. Solutions in the liquid phase are governed by the strength of solute–solute, solvent–solvent, and solute–solvent forces. The first two must be disrupted by the latter in the process of dissolution. Liquids with similar intermolecular forces are most likely to be miscible. Solutions are considered ideal if forces between all particles of both molecular solute and solvent are identical, or if ions in solution are independent of each other.

A saturated solution has in solution the amount of solute that would be in equilibrium with undissolved solute. This amount, at a given temperature, is the solubility of a given solute in a specific solvent. More solute will dissolve in an unsaturated solution and a supersaturated solution holds more solute than a saturated solution.

Pressure has only a small effect on the solubility of liquids and solids. The solubility of a gas varies with pressure according to Henry's law, $P_A = kC$, where P_A is the partial pressure of solute gas A and C is the concentration of the gas in solution. The majority of inorganic salts increase in solubility with increasing temperature (but there are many exceptions).

Aside 15.1 THOUGHTS ON CHEMISTRY: SIGNS OF SOLUBILITY

Michael Faraday (1791–1867) was an entirely self-taught yet skilled experimentalist in both physics and chemistry. He is most famous for his discoveries about electricity and matter (Section 24.5). The following excerpt is from his book, *Chemical Manipulation*, which he described as "Instructions to students in chemistry, on the methods of performing experiments of demonstration or of research with accuracy and success." The quoted instructions are as useful today as they must have been for nineteenth-century students.

333. *There are two great and general objects to be gained by solution, which render it a process of constant occurrence in the laboratory. The first is that of preparing substances for the exertion of chemical action; and from the perfect manner in which it separates the particles one from another,*

every obstacle dependant upon the attraction of aggregation is removed, at the same time that other advantages are obtained. The second object is that of separating one substance from another; this being continually effected by the use of such fluids as have a solvent power over one or more of the substances present.

338. *If the substance appear to be insoluble, or if it be necessary to know whether it be soluble in alcohol, ether, oils, or any other body, for the purpose of selecting a solvent from among them, a portion should be pulverized finely, and introduced into a small tube with a little of the fluid to be tried, and heated; if the substance disappear, it is of course soluble. But if it be supposed to be a mixed body, and partly soluble, though not altogether so, then the presumed solution should be poured from the tube into an earthenware or platina capsule, and evaporated carefully and slowly; if any substance remain, it of course indicates a degree of solubility....*

339. *Indications of solution dependant upon chemical action are obtained from the changed appearance of the substance. A body not soluble in water except by the use of acids or alkalies, is generally, though not always, rendered so by chemical action, and has its properties changed. It is rarely that these chemical agents are applied with any other liquid than water. Indications of their power may be obtained in the manner before directed.*

341. *It may here be proper to warn the student of an appearance which sometimes takes place during mere solution, which has been frequently misconstrued into an indication of chemical action. Common salt, and several other salts, as well as their strong solutions, and also alcohol or spirit added to common or even distilled water, which has been exposed to air, frequently causes the evolution of a number of small air bubbles; these have the appearance of being the direct result of some chemical action, but are occasioned merely by the expulsion of the air dissolved in the water, which being incompatible with the substances added, is separated. Some of those solutions which dissolve air, produce the same appearances when added to the bodies mentioned.*

Source: Michael Faraday, *Chemical Manipulation* (London, 1827). Quoted from Halsted Press edition (New York: Wiley, 1974).

CONCENTRATION OF SOLUTIONS

15.5 UNITS OF CONCENTRATION

The concentration of a solution is the amount of solute (usually expressed as mass, moles, or volume) in a given amount (mass or volume) of solvent or solution. The concentration units commonly encountered in chemistry are summarized in Table 15.1. Molarity, the most useful and convenient unit for the stoichiometry of chemical reactions in solution, was introduced in Section 4.8. A 1 M solution contains 1 mol of solute in 1 L of solution. For example, 1 L of a 1 M solution is prepared by dissolving 1 mol of a substance in enough water to make 1 L of solution (Figure 15.7). Knowing the molarity of a solution permits calculation of the number of moles of solute or the number of grams of solute in a given volume of solution.

$$\text{(Molarity)(volume)} = \text{moles of solute}$$
$$\frac{\textit{moles}}{\cancel{\textit{liter}}} \quad \cancel{\textit{liters}} \qquad \textit{moles}$$

$$\text{(Molarity)(volume)(molar mass of solute)} = \text{mass of solute}$$
$$\frac{\cancel{\textit{moles}}}{\cancel{\textit{liter}}} \quad \cancel{\textit{liters}} \quad \frac{\textit{grams}}{\cancel{\textit{mole}}} \qquad \textit{grams}$$

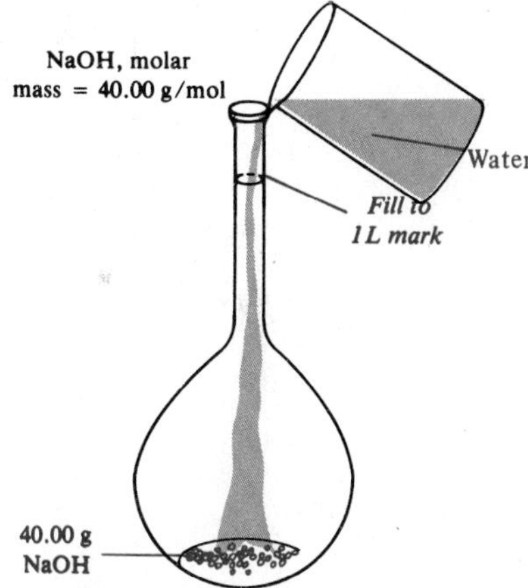

Figure 15.7
Preparation of a 1.000 M Solution of NaOH

Table 15.1
Summary of Concentration Units
A is the solvent and B is the solute. The second column shows how to calculate the concentration from known quantities of A and B.

Unit	Calculation
Mass %	$\frac{\text{mass of B}}{\text{mass of A + mass of B}} \times 100$
Molarity (moles/liter)	$\frac{\text{moles of B}}{\text{volume of solution (L)}}$
Mole fraction, X	$\frac{\text{moles of B}}{\text{moles of A + moles of B}}$
Molality (moles/kg of solvent)	$\frac{\text{moles of B}}{\text{mass of A (kg)}}$

The mole fraction was introduced in Section 6.11 in connection with Dalton's law of partial pressures. The mole fraction is useful because it gives the fractions or percentages of molecules of different types in a mixture. For gases, this means that the mole fraction of a gas in a mixture is also the fraction of the total pressure exerted by that gas. The sum of the mole fractions of all of the components of a mixture is one.

Mass percent and molality, the other two concentration units with which one should be familiar in the study of chemistry, are introduced in the following sections.

15.6 MASS PERCENT

The concentration of a solution can be given in **mass percent (weight percent).**

$$\text{Mass percent} = \frac{\text{mass of solute}}{\text{mass of solution}} \times 100 \tag{15.3}$$

The mass of the solution is equal to the sum of the masses of the solute (or solutes) and the solvent. Concentrations in units based only upon masses, such as mass percent (and mole fraction), do not vary with temperature—an advantage in some applications.

EXAMPLE 15.3 Solution Concentration: Mass Percent

What is the mass percent of methanol in a solution of 2.0 L of methanol in 3.5 L of diethyl ether? The density of methanol is 0.79 g/mL and the density of diethyl ether is 0.71 g/mL.

To find the mass percent of methanol, we must know the masses of all of the components in the solution. The known densities allow conversion of the known volumes to the masses of methanol and diethyl ether.

$$(2.0\text{ L methanol})\left(\frac{1000\text{ mL}}{\text{L}}\right)\left(\frac{0.79\text{ g}}{\text{mL}}\right) = 1600\text{ g methanol}$$

$$(3.5\text{ L diethyl ether})\left(\frac{1000\text{ mL}}{\text{L}}\right)\left(\frac{0.71\text{ g}}{\text{mL}}\right) = 2500\text{ g diethyl ether}$$

The mass of the solution is

$$\text{Mass of solution} = 1600\text{ g} + 2500\text{ g} = 4100\text{ g}$$

The mass percent of the methanol in the solution is

$$\text{Mass }\% = \frac{\text{mass methanol}}{\text{mass solution}} \times 100 = \frac{1600\text{ g}}{4100\text{ g}} \times 100 = \mathbf{39\%}$$

EXERCISE The solubility of $CoSO_4$ in water is 36.2 g $CoSO_4$/100 g H_2O at 20 °C. What is the mass percent of $CoSO_4$ in this solution? *Answer* 26.6 mass %.

15.7 MOLALITY

Molality is defined as the number of moles of solute per *kilogram* of *solvent*.

$$\text{Molality} = \frac{\text{moles of solute}}{\text{kilogram of solvent}} \tag{15.4}$$

A 1 molal solution of NaOH would be prepared by dissolving 40 g (1 mol) of NaOH in 1000 g of water, or 20 g (0.5 mol) of NaOH in 500 g of water, or any other combination that maintains this proportion. The bottle containing this solution would be labeled 1 *m* NaOH. (Note the different abbreviations—lowercase *m* for "molal" and uppercase M for "molar".) Molality is another concentration unit independent of temperature because it is defined only in terms of masses. Usually molality is chosen as the concentration unit for solutions in which one component is clearly the solvent and another the solute.

The molarity and molality of *dilute* aqueous solutions have very similar values because a dilute solution has a density close to that of pure water and contains close to 1 kg of solvent per liter of solution (density of $H_2O = 1$ kg/L). But as the concentration increases, the mass of the solute begins to contribute significantly to the density of the solution. (Remember that molarity is moles per liter of *solution*, not per liter of solvent.) Therefore, in concentrated solutions the density of the solution differs significantly from 1 kg/L, and the molarity and molality no longer have similar values.

EXAMPLE 15.4 Solution Concentration: Molality

A 6.0 M nitric acid solution contains 95 g, or 1.5 mol, of HNO_3 in 298 g of solution. What is the molality of this solution?

To find the molality we must know the mass in kilograms of the solvent, water, which is the difference between the mass of solution and that of HNO_3:

$$298\text{ g} - 95\text{ g} = 203\text{ g H}_2\text{O} \quad \text{or} \quad 0.203\text{ kg H}_2\text{O}$$

The molality is

$$\text{Molality} = \frac{\text{moles HNO}_3}{\text{kg H}_2\text{O}} = \frac{1.5\text{ mol}}{0.203\text{ kg}} = 7.4\text{ mol/kg}$$

and the concentration of this solution is **7.4 *m*.**

EXERCISE An aqueous solution contains 30.0 g of table sugar (sucrose, $C_{12}H_{22}O_{11}$) and 70.0 g of water. Calculate the concentration of the solution in terms of molality.

Answer 1.25 *m*.

Example 15.5 permits comparison of different ways of expressing the same solution concentration. Note that of the common units for solution concentration, molarity is the only one based on solution volume (see Table 15.1). To convert from other concentration units to molarity requires knowing solution density.

When there are no volumes as either knowns or unknowns, as in Example 15.5, one mass or volume is chosen as the basis for converting from one concentration unit to another.

EXAMPLE 15.5 Solution Concentration

Commercial vinegar is an aqueous solution that must contain at least 4 mass percent of acetic acid (CH_3COOH, molar mass, 60.05 g). The density of such a solution is 1.0058 g/mL. Express the concentration of acetic acid in this vinegar in terms of (a) mol fraction, (b) molality, and (c) molarity.

This problem does not specify a given mass or volume as a known quantity. We are free to choose any convenient sample of solution for our calculations. One obvious choice is 100.00 g of solution, which contains 4.00 g of CH_3COOH and 96.00 g of H_2O.

(a) The numbers of moles of solute and solvent in 100.00 g of solution are

$$(4.00\text{ g CH}_3\text{COOH})\left(\frac{1\text{ mol CH}_3\text{COOH}}{60.05\text{ g CH}_3\text{COOH}}\right) = 0.0666\text{ mol CH}_3\text{COOH}$$

$$(96.00\text{ g H}_2\text{O})\left(\frac{1\text{ mol H}_2\text{O}}{18.015\text{ g H}_2\text{O}}\right) = 5.329\text{ mol H}_2\text{O}$$

The mole fraction of acetic acid in vinegar is

$$X_{\text{CH}_3\text{COOH}} = \frac{0.0666\text{ mol}}{0.0666\text{ mol} + 5.329\text{ mol}} = \mathbf{0.0123}$$

(b) The molality of the solution is

$$\frac{0.0666 \text{ mol } CH_3COOH}{0.09600 \text{ kg } H_2O} = \mathbf{0.694 \text{ mol/kg } H_2O}$$

(c) To calculate the solution molarity, the solution volume must be found. The known density of the solution provides the connection. The volume of the 100.00 g of solution is

$$(100.00 \text{ g})\left(\frac{1 \text{ mL}}{1.0058 \text{ g}}\right) = \mathbf{99.42 \text{ mL}}$$

giving the molarity as

$$\frac{0.0666 \text{ mol } CH_3COOH}{0.09942 \text{ L}} = \mathbf{0.670 \text{ mol/L}}$$

To summarize, the concentration of acetic acid in commercial vinegar which is 4 mass percent acetic acid can be expressed as $X_{CH_3COOH} = 0.0123$, 0.694 *m*, or 0.670 M.

EXERCISE A solution of acetone, CH_3COCH_3, in water contains 0.500 g of CH_3COCH_3 dissolved in 99.500 g of H_2O. The density of this solution is 0.9993 g/mL. Express the concentration of this solution in terms of (a) molarity, (b) molality, and (c) mole fraction of acetone. *Answers* (a) 0.0860 M, (b) 0.0865 *m*, (c) 1.56×10^{-3}.

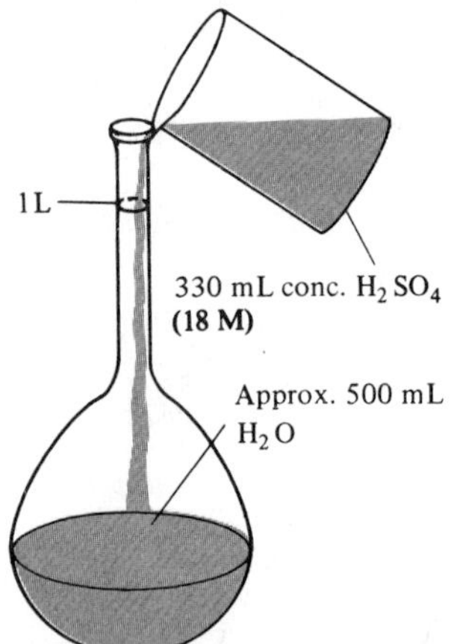

(a) Add acid to water

15.8 DILUTION OF SOLUTIONS

In the laboratory it is often necessary to prepare a solution of a desired concentration from a more concentrated solution, or to know the concentration of a solution that has been diluted. When solvent is added to a solution, the *concentration* of the solution changes, but the *amount* of solute remains the same. For example, suppose it is necessary to prepare 1.00 L of 6.0 M H_2SO_4 by diluting concentrated H_2SO_4, which has a concentration of 18 M. In deciding how to dilute a concentrated solution, it is necessary to know the amount of solute needed in the new solution. One liter of a 6 M H_2SO_4 solution will contain 6.0 mol of the acid. The volume of the concentrated solution that contains 6.0 mol of the acid is

$$(6.0 \text{ mol } H_2SO_4)\left(\frac{1 \text{ L}}{18.0 \text{ mol } H_2SO_4}\right) = 0.33 \text{ L, or } 330 \text{ mL}$$

The procedure for this dilution is illustrated in Figure 15.8. Sulfuric acid is always added to water because a considerable amount of heat is released when H_2SO_4 is dissolved and the bulk of the water helps to dissipate the heat.

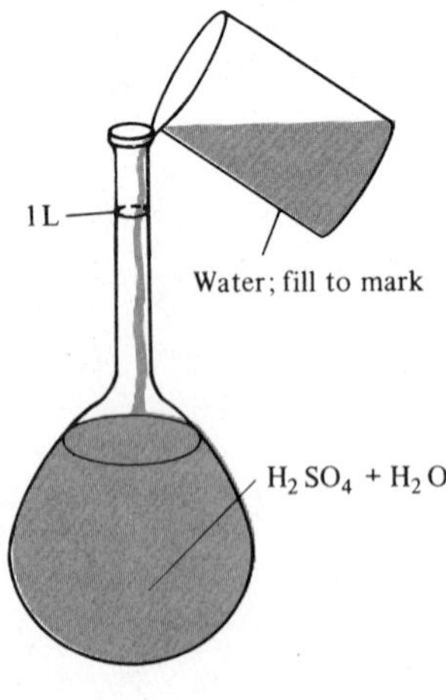

(b) Dilute to exact volume desired

Figure 15.8
Dilution of Concentrated Sulfuric Acid

EXAMPLE 15.6 Solution Concentration: Dilution

What volume of 12 M HCl must be must be diluted to give 1.5 L of 2.8 M HCl?

The amount of HCl needed in the new solution is

$$(1.5 \text{ L})\left(\frac{2.8 \text{ mol HCl}}{1 \text{ L}}\right) = 4.2 \text{ mol HCl}$$

The volume of 12 M HCl that will contain 4.2 mol of HCl is

$$(4.2 \text{ mol})\left(\frac{1 \text{ L}}{12 \text{ M}}\right) = \mathbf{0.35 \text{ L}}$$

To prepare the desired solution, 0.35 L of 12 M HCl must be diluted to give 1.5 L of solution.

EXERCISE What volume of commercially available concentrated nitric acid, which has a concentration of 15.6 M, must be diluted to give 1.75 L of 6.0 M HNO_3? *Answer* 0.71 L.

EXAMPLE 15.7 Solution Concentration: Dilution

An analytical procedure required that 13.2 mL of a 0.1016 M solution of sodium hydroxide be diluted to exactly 25.00 mL. What was the molarity of the diluted solution?

The number of moles of NaOH in the original solution is

$$(13.2\ \text{mL})\left(\frac{0.1016\ \text{mol NaOH}}{1000\ \text{mL}}\right) = 0.00134\ \text{mol NaOH}$$

giving a new molarity in the dilute solution of

$$\left(\frac{0.00134\ \text{mol NaOH}}{25.00\ \text{mL}}\right)\left(\frac{1000\ \text{mL}}{1\ \text{L}}\right) = \mathbf{0.0536\ mol\ NaOH/L}$$

Alternate Solution Many dilution problems involving concentrations in molarity (or normality, Section 20.6), including this problem, can be solved by using the relationship $C_1V_1 = C_2V_2$, where C_1 and V_1 are the concentration and volume before dilution, and C_2 and V_2 are the concentration and volume after dilution. In this case, C_2 is the unknown and the solution would take the following form:

$$C_2 = \frac{C_1V_1}{V_2} = \frac{(0.1016\ \text{mol NaOH/L})(13.2\ \text{mL})}{(25.00\ \text{mL})} = 0.0536\ \text{mol NaOH/L}$$

EXERCISE A 0.250 *m* NaCl solution contained 14.6 g NaCl and 996 g H_2O. An additional 225 g of H_2O was added. What is the concentration of the diluted solution?

Answer 0.205 *m*.

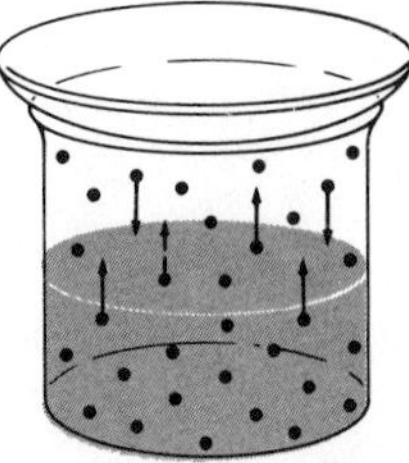

(a) Pure solvent (A)

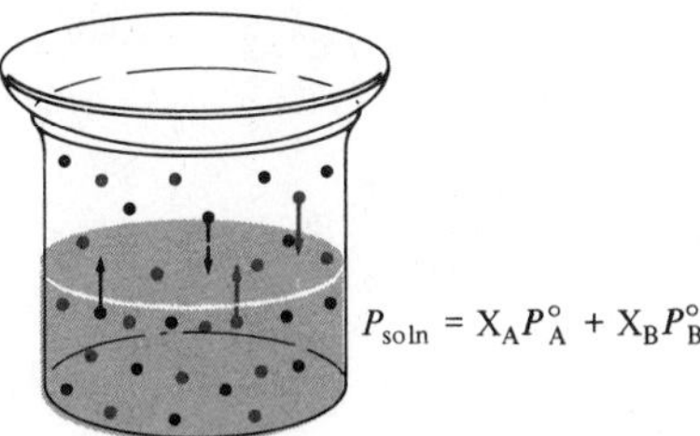

(b) Liquid (B) in solution in solvent (A)

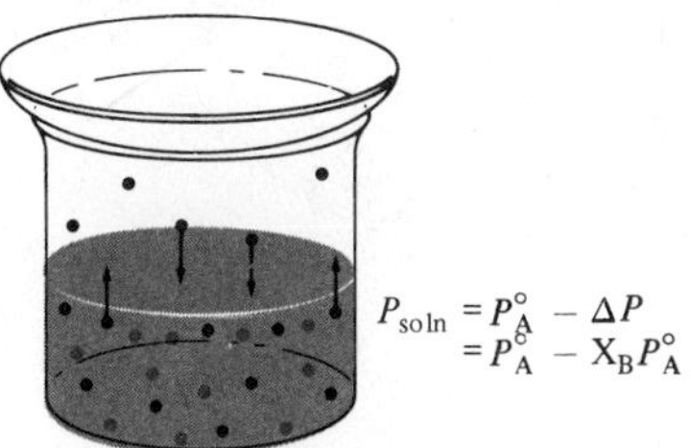

(c) Nonvolatile solute (B) in solution in solvent (A)

Figure 15.9
Vapor Pressure over Solutions
In (b) and (c), fewer solvent molecules are available at the surface to evaporate.

SUMMARY OF SECTIONS 15.5–15.8

Solution concentration is commonly expressed as mass, moles, or volume of solute in a given mass or volume of solvent or solution (see Table 15.1). Molality is concentration in moles of solute per kilogram of solvent. Concentrations in units that do not include volume do not vary with temperature, but molarity does so. Conversion from other concentration units to molarity can be effected by using solution density to find solution volume. The key to dilution problems is recognizing that the amount of solute remains unchanged during dilution.

VAPOR PRESSURES OF LIQUID SOLUTIONS AND RELATED PROPERTIES

15.9 LIQUID–LIQUID SOLUTION VAPOR PRESSURES: RAOULT'S LAW

The molecules in a liquid are in rapid constant motion. This motion causes molecules to escape from the surface into the vapor phase, creating a vapor pressure above the liquid. If more than one liquid is present in a solution, the total vapor pressure of the solution is the sum of the vapor pressures of each liquid. Molecules of each liquid are present at the solution surface. For an ideal liquid–liquid solution, where the forces between all molecules are identical, the tendency of either type of molecule to escape depends on the relative numbers of each (Figure 15.9b), which are given by their mole fractions.

The effect on vapor pressure of mixing two volatile liquids to form an ideal solution is shown in Figure 15.10. Adding either liquid to the other (i.e., starting at either side of Figure 15.10) causes a decrease from the vapor pressure of that pure liquid because fewer molecules of the original liquid are at the surface. The partial vapor pressure of either liquid in an ideal solution of liquids A and B is dependent upon its concentration

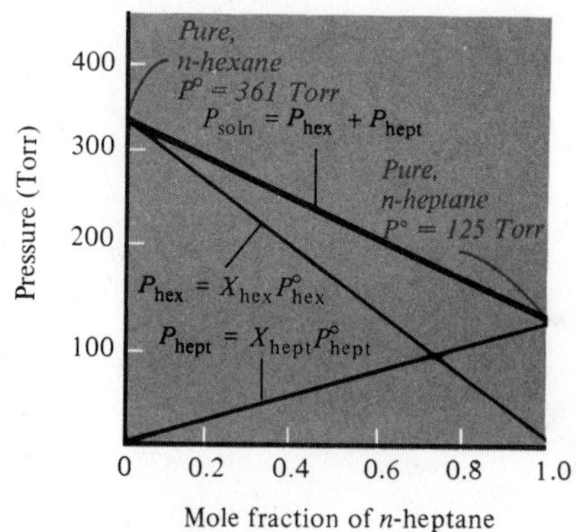

Figure 15.10
Vapor Pressure of a Solution That Follows Raoult's Law
The top line is the total vapor pressure of the hexane–heptane solution. The other two lines are the partial vapor pressures of each component of the solution.

in the solution, expressed as the mole fraction, and the vapor pressure of the pure liquid as follows:

mole fraction of A in solution

partial vapor pressure of A over solution

$$P_A = X_A P_A^\circ \qquad P_B = X_B P_B^\circ \qquad \textbf{(15.5)}$$

vapor pressure of pure A

Equation (15.5) is a mathematical statement of **Raoult's law:** The vapor pressure of a liquid in a solution is equal to the mole fraction of that liquid in the solution times the vapor pressure of the pure liquid.

The total vapor pressure of the solution is the sum of the partial pressures of all of the components of the solution (see Dalton's law, Section 6.11). For a solution of A and B

total vapor pressure of solution

$$P_{soln} = P_A + P_B \qquad \textbf{(15.6)}$$

Substituting the relationship of partial pressure to vapor pressure given by Raoult's law (Equation 15.5) into Dalton's law gives

$$P_{soln} = X_A P_A^\circ + X_B P_B^\circ \qquad \textbf{(15.7)}$$

An ideal solution of two liquids would obey Raoult's law (Equations 15.3 and 15.7) perfectly. There are a few miscible liquids that form solutions that come quite close to ideality. To do so, the liquids must be very similar in molecular mass, in structure, and in types of possible intermolecular forces, as illustrated by the following pairs of substances that form close-to-ideal solutions:

CH_3CH_2Br *ethyl bromide*	*and*	CH_3CH_2I *ethyl iodide*
$CH_3(CH_2)_4CH_3$ *n-hexane*	*and*	$CH_3(CH_2)_5CH_3$ *n-heptane*

EXAMPLE 15.8 Solution Properties: Vapor Pressure

What is the vapor pressure (at 25 °C) of a benzene–toluene solution of composition $X_{benz} = 0.30$ and $X_{tol} = 0.70$? The vapor pressures of the pure substances are 73 Torr for benzene (C_6H_6) and 27 Torr for toluene (C_7H_8) (at 25 °C). Assume that the benzene and toluene form an ideal solution.

The partial pressure of each component in the vapor phase is directly proportional to its mole fraction in the solution, according to Raoult's law.

$$P_{benz} = X_{benz} P_{benz}^\circ = (0.30)(73\ \text{Torr}) = 22\ \text{Torr}$$
$$P_{tol} = X_{tol} P_{tol}^\circ = (0.70)(27\ \text{Torr}) = 19\ \text{Torr}$$

and the total vapor pressure of the solution is the sum of the partial pressures.

$$P_{soln} = P_{benz} + P_{tol} = 22\ \text{Torr} + 19\ \text{Torr} = \mathbf{41\ Torr}$$

EXERCISE Calculate the vapor pressure at 25 °C of the benzene–toluene solution of composition $X_{benz} = 0.70$ and $X_{tol} = 0.30$. *Answer* 59 Torr.

The composition of the vapor over a solution of two liquids is not the same as the composition of the solution. The vapor composition depends upon the relative vapor pressures of the two liquids. The vapor over a solution of any composition and at any temperature is always richer in the more volatile (i.e., lower boiling) component than is the solution.

Table 15.2
Properties of Real Solutions of Two Liquids, A and B
Positive deviation is the more common situation.

Positive deviation from Raoult's law	Negative deviation from Raoult's law
A----B forces less than A----A or B----B forces	A----B forces greater than A----A or B----B forces
Dissolution process endothermic	Dissolution process exothermic
Heating increases solubility	Heating decreases solubility

For example, benzene (b.p. 80.1 °C) is more volatile than toluene (b.p. 110.6 °C). Using the relationship between the partial and total pressures of a mixture of gases, $P_A = X_A P^\circ_{total}$ (Section 6.11), the composition of the vapor over the solution described in Example 15.8 can be found by taking the total pressure as the solution vapor pressure:

$$P_{benz} = X_{benz} P_{soln}$$
$$22 \text{ Torr} = X_{benz}(41 \text{ Torr})$$
$$X_{benz} = \frac{22 \text{ Torr}}{41 \text{ Torr}} = 0.54$$

The vapor contains almost twice as great a concentration of benzene as the liquid phase.

Nonideal liquid–liquid solutions show either positive or negative deviations from Raoult's law, as summarized in Table 15.2. When the different molecules of two liquids (A----B) attract each other less than they attract other molecules of the same kind (A----A, B----B), molecules of either A or B escape from solution more easily than from the pure liquids. This causes *positive* deviation from Raoult's law Figure 15.11a. The partial vapor pressures of both A and B are greater than those predicted by Raoult's law. Ethanol and *n*-hexane, and methyl alcohol and water form such solutions. Most liquid–liquid solutions behave in this way. The dissolution process is endothermic for solutions that deviate positively from Raoult's law because energy is required to disrupt the A----A and B----B attractive forces. Heating such a solution will increase solubility.

Negative deviations from Raoult's law occur when the different molecules attract each other more strongly than they attract other molecules of the same kind (Figure 15.11b). The molecules have a smaller tendency to escape from the solution surface than from their pure liquid surfaces and the partial vapor pressures of both A and B are lower than predicted by Raoult's law. Formic acid (HCOOH) and water form this type of solution. The dissolution proccess is exothermic for solutions that deviate negatively from Raoult's law because energy is released when the stronger A----B attraction replaces the weaker A----A and B----B attractions. Heating such a solution will decrease solubility.

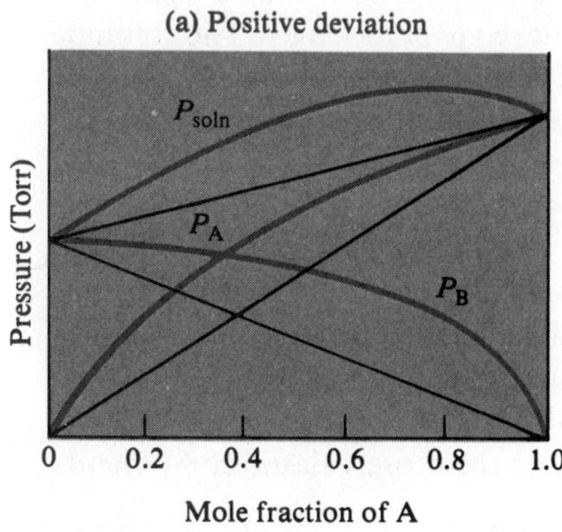

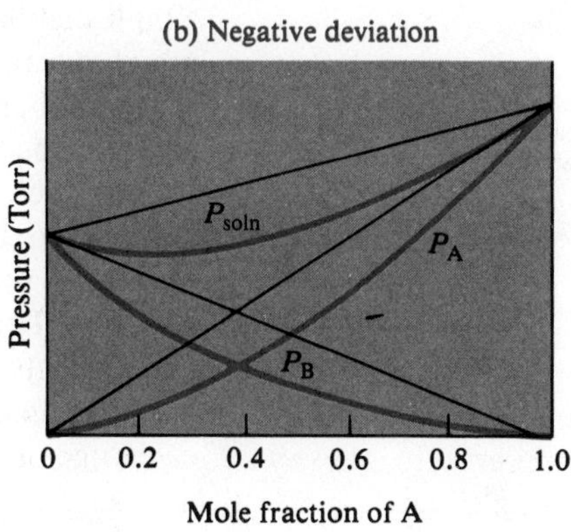

Figure 15.11
Deviations from Raoult's Law
The black lines represent ideal Raoult's law behavior. The colored lines show the deviations from ideal behavior. The degree of deviation, and therefore the shapes of the real solution vapor pressure curves, can vary widely. Methyl alcohol (CH_3OH) and water form a solution that deviates positively from Raoult's law. Nitric acid (HNO_3) and water solutions deviate negatively.

Aside 15.2 TOOLS OF CHEMISTRY: DISTILLATION

The liquids in a solution can be separated from each other by making use of the fact that the vapor above the solution is always richer in the more volatile component of the mixture than is the solution. **Distillation** involves heating a liquid or a solution to boiling, and collecting and condensing the vapors. A laboratory scale set up for simple distillation is shown in Figure 15.12a. This type of apparatus might be used, for example, to prepare distilled water. The mixture to be distilled is heated, the vapor condenses in the water-cooled condenser, and the **distillate**—the product of distillation—is collected in a receiving flask.

As a mixture of liquids is distilled, the composition of the liquid phase, the composition of the vapor phase, and the boiling point of the solution change continuously. Figure 15.13 illustrates these changes for an ideal mixture of liquids A and B. The three dots mark the temperature, vapor composition, and liquid composition at the beginning of the distillation of a 50–50 mixture of A and B in which B is the more volatile component. The color arrows mark the directions of change during distillation. The composition of the vapor at any temperature is read from the top curve and the composition of the liquid solution is read from the bottom curve.

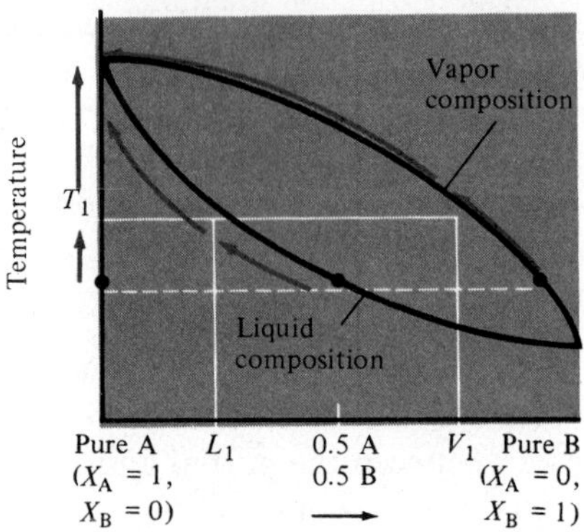

Figure 15.13
Boiling Point Diagram for a Solution of Two Liquids A and B (at atmospheric pressure)
The dots mark the beginning of distillation of a 50–50 mixture of A and B. For a mixture boiling at T_1, the composition of the liquid is given by the mole fractions of A and B at L_1 and the composition of the vapor is given by the mole fractions of A and B at V_1.

From the boiling point diagram in Figure 15.13 it is apparent that a single simple distillation cannot completely separate two or more volatile liquids from each other. The vapor collected as the temperature rises, while richer in the more volatile component (B) of the mixture, is still a mixture of the two vapors at all temperatures up to the boiling point of A. A *series* of simple distillations would provide distillates richer and richer in the more volatile component, but the repeated distillations would be tedious.

Fractional distillation is a process that separates liquid mixtures into fractions that differ in boiling points. A fractionating column is designed so that a series of simple distillations is achieved over its length. The column may have an internal structure of projections or trays or perforated plates, or it may be packed with small pieces of ceramic, metal, or glass. The purpose of all these items is to provide surfaces at which the rising vapor meets descending liquid and exchanges heat with it. Vapor moves up the column and some of it condenses on the packing. This condensate is richer in the less volatile components of the mixture. The condensate trickles down the column, where it meets rising hotter vapor. A heat exchange occurs—more of the more volatile component

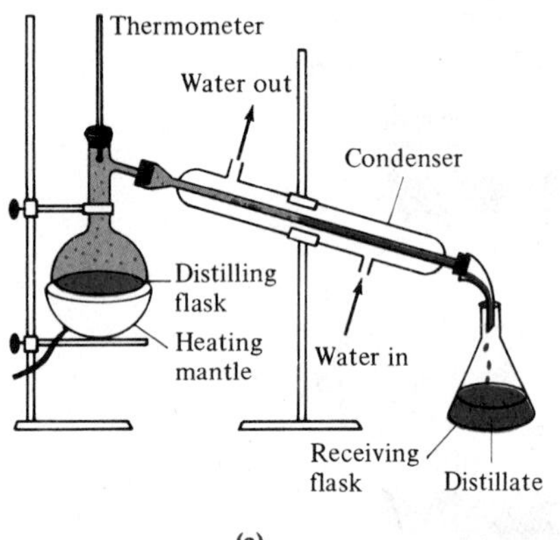

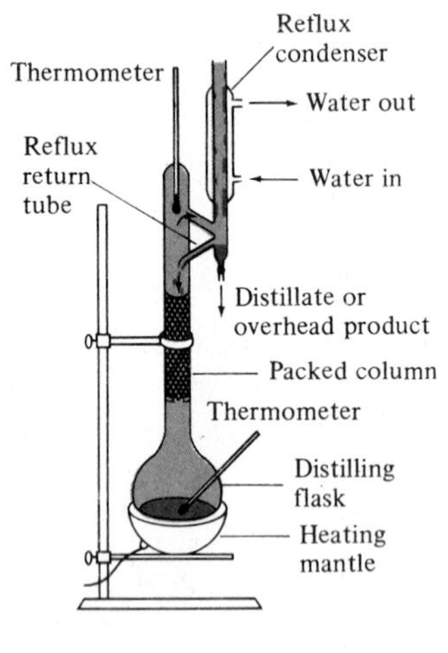

Figure 15.12
Laboratory Distillation
(a) Apparatus for simple distillation. (b) Apparatus for batch fractional distillation.

leaves the condensate and joins the rising vapor, while more of the less volatile component leaves the vapor and joins the descending liquid. The net result is that pure vapor can be collected at the top of the column and pure liquid at the bottom of the column. The process of vapor moving up the column, condensing, and trickling down the column is called **refluxing.**

A laboratory-scale fractional distillation apparatus is shown in Figure 15.12b. During distillation some of the condensing vapor is returned to the column and some is collected as product. Batch distillation—the distillation of one batch of the mixture at a time—is carried out in such an apparatus.

In most industrial-scale distillations, continuous operation is more economical than batch distillation. A simple continuous distillation column for separating two liquids is shown in Figure 15.14. The material to be distilled (the feed) is added continuously at the middle of the column, at a point where the composition of the liquid phase in the column is similar to that of the feed. The more volatile product is collected continuously at the top (the overhead product), and the less volatile product at the bottom (the bottoms product). Distillation may be carried out at atmospheric pressure, or at either reduced or increased pressure, depending on the properties of the materials involved.

When a several-component mixture is fractionally distilled, fractions that differ in boiling point are collected in the order of increasing boiling point, either one by one at the top of the column or continuously at successively higher points along the column.

Solutions that deviate sufficiently from Raoult's law cannot be totally separated by fractional distillation. At some definite composition they form **azeotropes**—constant-boiling mixtures that distill without change in composition. A minimum boiling point azeotrope has a boiling point below that of either component, and a maximum boiling point azeotrope has a boiling point above that of either component (Figure 15.15). If A and B can form an azeotrope, the best separation possible from distillation of a mixture of A and B is to obtain pure A *or* pure B plus the azeotrope. Distillation of a mixture with *exactly* an azeotropic composition would yield no separation at all.

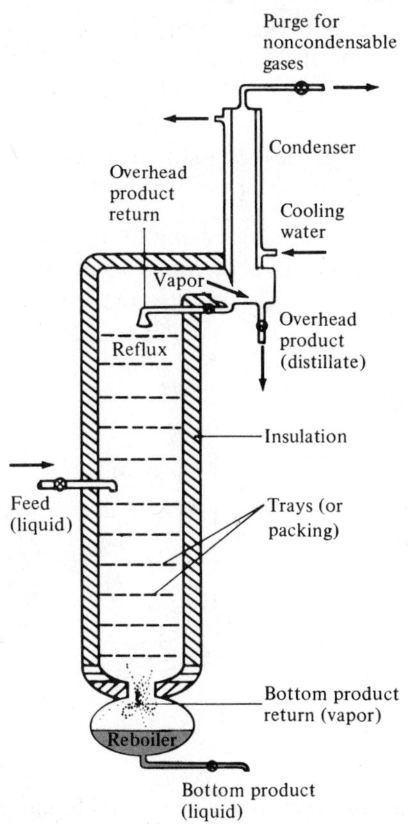

Figure 15.14
Column for Continuous Fractional Distillation
Source: Gerhard A. Cook, *Survey of Modern Industrial Chemistry* (Ann Arbor, MI: Ann Arbor Science, 1977), p. 29.

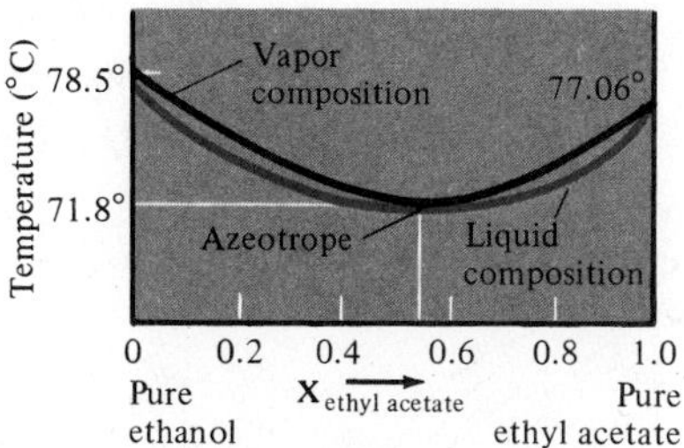

(a) Minimum boiling point azeotrope – ethyl acetate ($CH_3CO_2C_2H_5$) + ethanol (C_2H_5OH)

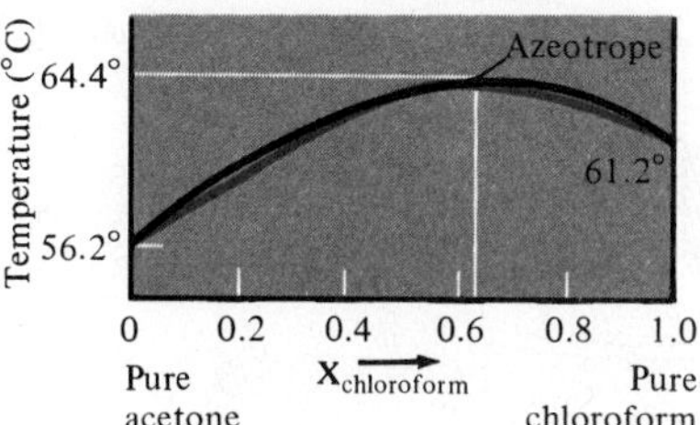

(b) Maximum boiling point azeotrope – acetone (CH_3COCH_3) + chloroform ($CHCl_3$)

Figure 15.15
Boiling Point Diagrams for Azeotropic Solutions (at 760 Torr)

15.10 VAPOR PRESSURE LOWERING

Solutes with essentially no vapor pressure of their own (nonvolatile solutes) give solutions over which the vapor pressure is less than that of pure solvent. At the surface of such a solution, the nonvolatile solute molecules mingle with the solvent molecules, allowing fewer solvent molecules to escape. That the vapor pressure will be lowered by a nonvolatile nonelectrolyte solute can be seen from Raoult's law and Equation (15.7), where A is the solvent and $P^\circ_B = 0$, giving

$$P_{soln} = X_A P^\circ_A \qquad \textbf{(15.8)}$$

In the presence of a solute, X_A is less than one, and therefore the vapor pressure of the solution is less than P°_A, the vapor pressure of the pure solvent.

The **vapor pressure lowering,** ΔP, is the difference between the vapor pressure of a pure solvent and the vapor pressure over a solution of a nonvolatile solute. Introducing the mole fraction of the solute B into Equation (15.8) and rearranging gives an expression for the vapor pressure lowering in terms of the mole fraction of the *solute* (B).

$$X_A + X_B = 1, \quad \text{so} \quad X_A = 1 - X_B$$
$$P_{soln} = X_A P^\circ_A = (1 - X_B)P^\circ_A$$

$$\Delta P = P^\circ_A - P_{soln} = P^\circ_A - (P^\circ_A - X_B P^\circ_A) \qquad \textbf{(15.9)}$$

$$\underset{\text{vapor pressure lowering}}{\Delta P} = \underset{\text{mole fraction of solute}}{X_B}\ \underset{\text{vapor pressure of pure solvent}}{P^\circ_A} \qquad \textbf{(15.10)}$$

If the solute molecules have no tendency to escape from the solution (a nonvolatile solute) and do not interact with each other (true only for nonelectrolyte solutes), the vapor pressure lowering is dependent only on the relative numbers of solute and solvent molecules.

The effect of vapor pressure lowering can be seen in the transfer of a solvent in the closed system illustrated in Figure 15.16. The vapor pressure of the solution is always

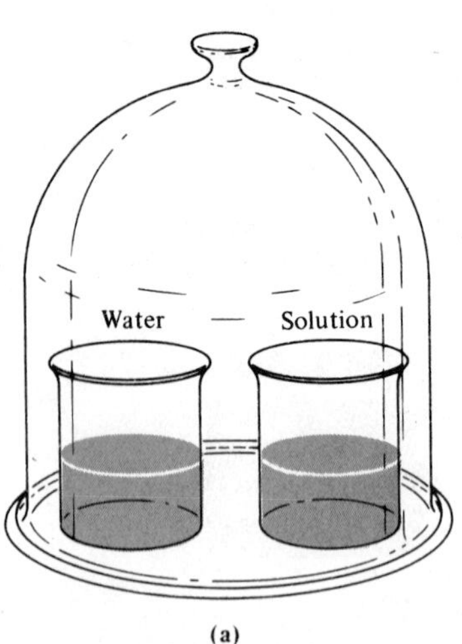

(a)

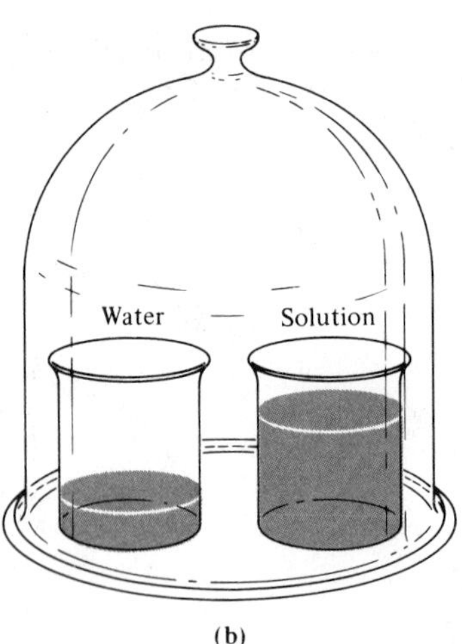

(b)

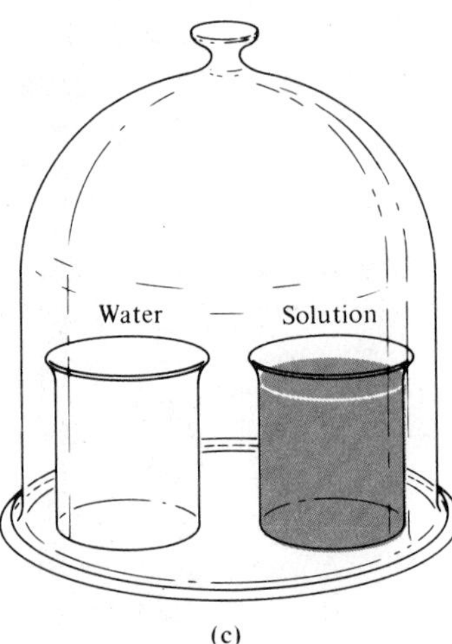

(c)

Figure 5.16
Demonstration of Vapor Pressure Lowering
(a) Molecules evaporate from the surface of the pure water. If the solution were not also present, a state of equilibrium would eventually be attained. (b) The vapor pressure of the solution, however, is always *lower* than that of pure water (Equation 15.8). As soon as the partial pressure of water vapor in the bell jar rises higher than the vapor pressure of the *solution,* water molecules start to enter the solution beaker more rapidly than they evaporate from it. There is a continuous net movement of water molecules into the solution. (c) Since equilibrium can never be attained, *all* of the water eventually ends up in the solution beaker except for those molecules that remain in the gaseous state.

lower than that of the pure water, and equilibrium between water molecules in the gas phase and those in the two beakers is not possible. More molecules leave the surface of the water than reenter it, because some enter the solution beaker. *All* of the water molecules eventually migrate to the solution beaker.

Any property of a solution that depends on the relative numbers of solute and solvent particles is called a **colligative property** (from the Latin *colligare,* to bind together). Vapor pressure lowering is a colligative property. The three phenomena discussed in the following sections—boiling point elevation, freezing point depression, and osmosis—are also colligative properties.

The vapor pressure over a solution of a nonvolatile solute can be found from Equation (15.8) using the mole fraction of the solvent, or from the following equation if the value of the vapor pressure lowering is known.

$$P_{\text{soln}} = P_A^\circ - \Delta P \qquad \textbf{(15.11)}$$

EXAMPLE 15.9 Solution Properties: Vapor Pressure

An aqueous solution contains 0.0220 mole fraction of dissolved sucrose. (a) What is the vapor pressure lowering of this solution? (b) By what percent has the vapor pressure of pure water been lowered by the addition of sucrose to this solution? (c) What is the vapor pressure over this solution? Assume that the vapor pressure of the sucrose is negligible and that the vapor pressure of pure water at 25 °C is 23.756 Torr.

(a) The vapor pressure lowering, ΔP, is found from the mole fraction of the solute, sugar, which is 0.0220.

$$\Delta P = X_{\text{sucrose}} P_{H_2O}^\circ = (0.0220)(23.756\ \text{Torr}) = \mathbf{0.523\ Torr}$$

(b) The percent by which the vapor pressure has been lowered from that of pure water is

$$\frac{0.523\ \text{Torr}}{23.756\ \text{Torr}} \times 100 = \mathbf{2.20\%}$$

(c) The solution vapor pressure is the difference between the vapor pressure of the pure water and the vapor pressure lowering.

$$P_{\text{soln}} = P_{H_2O}^\circ - \Delta P = 23.756\ \text{Torr} - 0.523\ \text{Torr} = \mathbf{23.233\ Torr}$$

EXERCISE Find the vapor pressure at 45 °C of a solution which contains 0.083 mol of urea (NH_2CONH_2) in 1000.0 g H_2O. The vapor pressure of water at 45 °C is 71.88 Torr. *Answer* 71.77 Torr.

15.11 BOILING POINT ELEVATION AND FREEZING POINT DEPRESSION

Figure 15.17 shows the effect of vapor pressure lowering on the boiling point and also the freezing point of pure water. The solution–vapor equilibrium curve lies *below* the curve for pure water at all points. As a result of this vapor pressure lowering, the solution must be heated to a higher temperature to reach a vapor pressure equal to atmospheric pressure, and therefore to boil. The boiling point has been elevated—the solution boils at a higher temperature than the solvent. The lowering of the vapor pressure also causes the freezing point to be lowered—the solution freezes at a lower temperature than the solvent.

Ideally one mole of any nonvolatile nonelectrolyte solute in a given amount of solvent will have the *same* effect on colligative properties as one mole of any other nonvolatile nonelectrolyte solute. Each contains the *same* number of particles—equal to Avogadro's number. In working with colligative properties *molality* (or mole fraction) is used as the concentration unit because it gives the ratio of molecules of solute to a fixed number of molecules of solvent and also because it is independent of temperature changes.

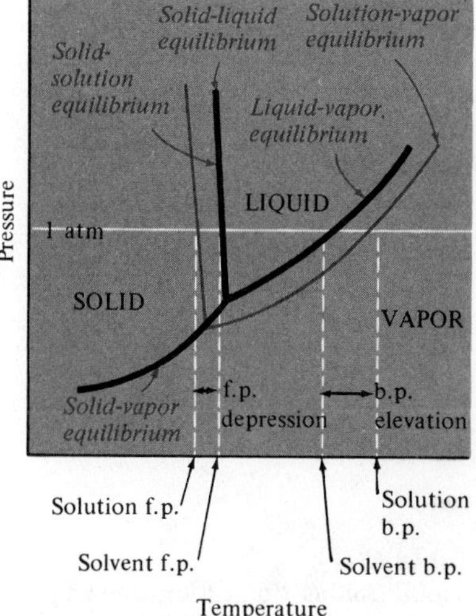

Figure 15.17
Comparison of Phase Diagrams of Pure Water and an Aqueous Solution of a Nonvolatile Solute
The solution freezing point is the point at which crystals of pure *solvent* appear. If a compound or solid solution formed between the solvent and solute as cooling occurred, the phase diagram would be more complex.

The boiling point elevation, ΔT_b (= b.p. solution − b.p. solvent), and the freezing point depression, ΔT_f (= f.p. solvent − f.p. solution), are directly proportional to the molality of a solution.

$$\underset{\text{b.p. elevation}}{\Delta T_b} = \underset{\text{molal b.p. elevation constant}}{K_b} m \qquad \underset{\text{f.p. depression}}{\Delta T_f} = \underset{\text{molal f.p. depression constant}}{K_f} m \tag{15.12a,b}$$

A 1 molal aqueous solution of any nonvolatile nonelectrolyte boils at 100.512 °C, a boiling point elevation of 0.512 °C. This same solution freezes at −1.86 °C, a freezing point lowering of 1.86 °C. These values provide the constants used in Equations (15.12a) and (15.12b) for aqueous solutions. The units for the constants are derived as follows:

molal b.p. elevation constant

$$K_b = \frac{0.512\ ^\circ\text{C}}{1\ m\ \text{solution}} = 0.512\ ^\circ\text{C}\left(\frac{1\ \text{kg}}{1\ \text{mol}}\right) = 0.512\,\frac{^\circ\text{C kg}}{\text{mol}} \tag{15.13a}$$

molal f.p. depression constant

$$K_f = \frac{1.86\ ^\circ\text{C}}{1\ m\ \text{solution}} = 1.86\ ^\circ\text{C}\left(\frac{1\ \text{kg}}{1\ \text{mol}}\right) = 1.86\,\frac{^\circ\text{C kg}}{\text{mol}} \tag{15.13b}$$

and the complete equations for aqueous solutions are

$$\Delta T_b = \left(0.512\,\frac{^\circ\text{C kg}}{\text{mol}}\right) m \qquad \Delta T_f = \left(1.86\,\frac{^\circ\text{C kg}}{\text{mol}}\right) m \tag{15.14a,b}$$

$$^\circ\text{C} \quad \frac{^\circ\text{C}\ \cancel{\text{kg}}}{\cancel{\text{mol}}} \quad \frac{\cancel{\text{mol}}}{\cancel{\text{kg}}} \qquad ^\circ\text{C} \quad \frac{^\circ\text{C}\ \cancel{\text{kg}}}{\cancel{\text{mol}}} \quad \frac{\cancel{\text{mol}}}{\cancel{\text{kg}}}$$

The values of K_b and K_f are different for each solvent (Table 15.3).

To find the boiling or freezing point of a solution, the tabulated value of K_b or K_f for the solvent and the known molality of the solution are used in Equation (15.12a) or (15.12b) to find the boiling point elevation or the freezing point lowering. The boiling point (T_b) or freezing point (T_f) of the given solution is then found from the following relationships:

$$T_{b,\,\text{solution}} = T_{b,\,\text{solvent}} + \Delta T_b \qquad T_{f,\,\text{solution}} = T_{f,\,\text{solvent}} - \Delta T_f \tag{15.15a,b}$$

Because ΔT_b represents a boiling point *elevation* it is *added* to the boiling point of the solvent. Because ΔT_f represents a *depression,* or lowering of the freezing point, it is *subtracted* from the freezing point of the solvent. Think of it this way—the freezing point moves toward a colder temperature, the boiling point moves toward a warmer temperature. Do not make the common error of disregarding or confusing the + and − signs in Equations (15.15a) and (15.15b).

Table 15.3
Molal Boiling Point Elevation (K_b) and Molal Freezing Point Lowering (K_f) Constants for Several Solvents

Solvent	K_b (°C kg/mol)	K_f (°C kg/mol)
Water	0.512	1.86
Benzene	2.53	4.90
Nitrobenzene ($C_6H_5NO_2$)	5.24	7.00
Biphenyl	—	8.00
Ethylene dibromide	—	11.80
Naphthalene	—	6.8
Carbon disulfide	2.34	—
Carbon tetrachloride	5.03	32
Ethyl alcohol	1.22	—
Methyl alcohol	0.83	—

CHEMICAL PRINCIPLES

■ **SOLUBILITY** *above* Each beaker holds the amount of a crystalline, ionic compound that will dissolve in the amount of hot water (100 mL) in the cylinder at the top right. The compounds are (*top row, left to right*) 39.12 g sodium chloride (NaCl), 102 g potassium dichromate ($K_2Cr_2O_7$), 340.7 g nickel sulfate hexahydrate ($NiSO_4 \cdot 6H_2O$); (*bottom row, left to right*) 79.2 g potassium chromate (K_2CrO_4), 190.7 g cobalt(II) chloride hexahydrate ($CoCl_2 \cdot 6H_2O$), and 203.3 g copper sulfate pentahydrate ($CuSO_4 \cdot 5H_2O$).

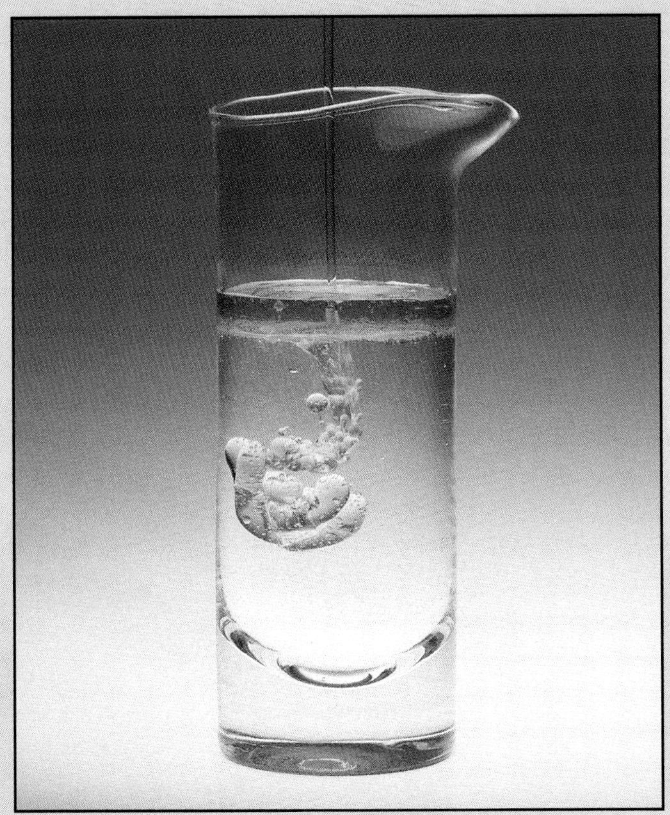

■ **IMMISCIBILITY** *above* The differences in the strengths of their intermolecular forces prevent oil and water molecules from intermingling. Because oil is less dense than water, it rises to the surface.

■ **DISSOLUTION** *above* Sugar molecules are sufficiently polar that they are attracted to polar water molecules and are pulled from the crystal surface. As the hydrated sugar molecules diffuse away, regions in the solution of different concentration bend light rays differently (i.e., the index of refraction varies with concentration).

■ **SUPERSATURATION** *above* The test tube originally held a supersaturated solution of sodium acetate ($NaCH_3COO$). Seeding with a few crystals of the acetate caused rapid crystallization.

■ **CRYSTAL FORMS** *above* Very pure sodium chloride (NaCl) forms cubic crystals that reflect in shape the cubic unit cell of the sodium chloride crystal lattice. This scanning electron micrograph shows the crystals magnified 305 times and reveals dislocations in the uniformity of the planar surfaces.

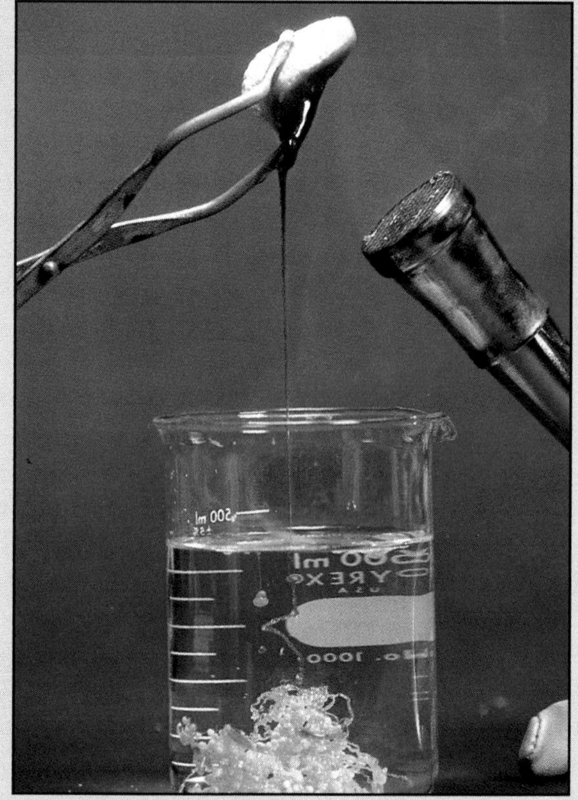

■ **ALLOTROPY: SULFUR** *right* Yellow rhombic sulfur, the stable form of sulfur at room temperature, is converted above 160 °C to a brown viscous liquid. Rapid cooling in water converts this liquid into plastic sulfur, which is rubbery because it contains coiled chains of sulfur atoms.

SUBLIMATION *above* Iodine sublimes — goes directly from the solid state to the vapor state. Sublimation occurs if the triple point of a substance (equilibrium of solid, liquid, and gas phases) is above atmospheric pressure, meaning that its vapor pressure can reach atmospheric pressure before a liquid forms.

DIFFUSION *above* Bromine diffuses into air (from the watch glass) and into an aqueous solution (in the beaker). Bromine vapor has come to equilibrium with liquid bromine in the closed flask.

DELIQUESCENCE *left* Sodium hydroxide (NaOH) exposed to the atmosphere takes up water to such an extent that it dissolves in the water it has accumulated.

EFFLORESCENCE AND DEHYDRATION *left and above* On exposure to dry air, blue copper sulfate pentahydrate ($CuSO_4 \cdot 5H_2O$) (*left*) slowly effloresces: it spontaneously loses water molecules to form white anhydrous copper sulfate. Heating speeds up conversion of copper sulfate pentahydrate to the anhydrous salt (*above*).

■ **SURFACE TENSION AND WETTING** *above* The attraction between water molecules and glass (*left*) overcomes the surface tension of water; water ''wets'' glass and forms a concave surface (meniscus). The attraction between mercury atoms and glass (*right*) is weaker than the surface tension of mercury; mercury does not ''wet'' glass and forms a convex surface.

■ **HYDROLYSIS** *above* The beakers contain dilute aqueous solutions of the salts shown on the watch glasses. The indicator bromothymol blue — which is yellow in acid solution, greenish in neutral solution, and blue in alkaline solution — has been added to the solutions. *left to right:* Ammonium chloride (NH_4Cl) gives an acidic solution due to NH_4^+ hydrolysis; ammonium acetate (NH_4CH_3COO) gives a neutral solution because both ions hydrolyze equally; potassium chloride (KCl) gives a neutral solution because neither ion hydrolyzes; potassium acetate (KCH_3COO) gives an alkaline solution because CH_3COO^- hydrolyzes.

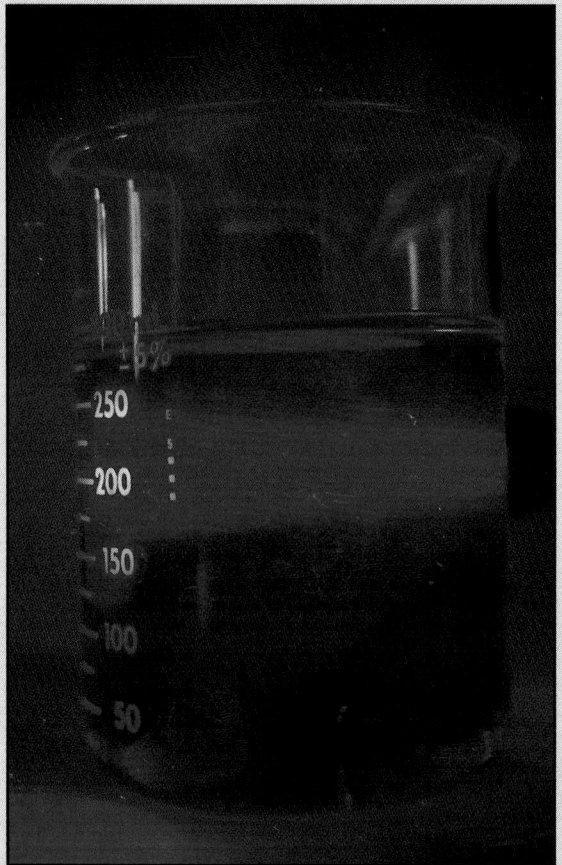

■ **EQUILIBRIUM DISPLACEMENT — BY ACID** *left* Addition of acid from the dropper displaces the equilibrium from yellow chromate ion (CrO_4^{2-}) in alkaline solution to orange dichromate ion ($Cr_2O_7^{2-}$) in acidic solution.

$$CrO_4^{2-} \underset{OH^-}{\overset{H^+}{\rightleftharpoons}} HCrO_4^- \underset{OH^-}{\overset{H^+}{\rightleftharpoons}} Cr_2O_7^{2-}$$

■ **A COLLOID AND THE TYNDALL EFFECT** *left* A sol is a type of colloid composed of solid particles suspended in a liquid. The beaker contains a clear, bright red sol composed of hydrated iron(III) oxide particles ($Fe_2O_3 \cdot xH_2O$). Light is scattered by the suspended particles, making the light beam visible as it passes through the beaker (the Tyndall effect).

EXAMPLE 15.10 Solution Properties: Colligative Properties

A solution contained a mixture of sugars: 0.50 mol of fructose, 0.50 mol of glucose, 0.50 mol of lactose, 0.50 mol of maltose, and 0.50 mol of sucrose dissolved in 1.00 kg of water. What are the freezing point and boiling point of this solution?

The total number of moles of solute is 2.50 mol (for the five sugars) and the total concentration of solute is

$$\text{Molality} = \frac{\text{moles of solute}}{\text{kg of solvent}} = \frac{2.50\ \text{mol}}{1.00\ \text{kg H}_2\text{O}} = 2.50\ \text{mol/kg}$$

From the relationships among freezing point lowering, concentration, and solution freezing point (Equations 15.13b and 15.14b)

$$\Delta T_f = K_f m = (1.86\ ^\circ\text{C kg/mol})(2.50\ \text{mol/kg}) = 4.65\ ^\circ\text{C}$$
$$T_{f,\,\text{solution}} = T_{f,\,\text{solvent}} - \Delta T_f = 0.00\ ^\circ\text{C} - 4.65\ ^\circ\text{C} = \mathbf{-4.65\ ^\circ C}$$

Similarly, for boiling point elevation

$$\Delta T_b = K_b m = (0.512\ ^\circ\text{C kg/mol})(2.50\ \text{mol/kg}) = 1.28\ ^\circ\text{C}$$
$$T_{b,\,\text{solution}} = T_{b,\,\text{solvent}} + \Delta T_b = 100.00\ ^\circ\text{C} + 1.28\ ^\circ\text{C} = \mathbf{101.28\ ^\circ C}$$

EXERCISE What are the boiling and freezing points of a 0.125 m aqueous solution of sucrose? *Answer* 100.0640 °C, −0.233 °C.

15.12 MOLAR MASS DETERMINATION

Any of the colligative properties can be used to determine the molar masses of nonvolatile nonelectrolytes. A weighed amount of a substance of unknown molar mass is dissolved in a weighed amount of solvent. The colligative property, say boiling point elevation, is measured. The molality of the solution can then be calculated from the relationship $\Delta T_b = K_b m$ rearranged to give $m = \Delta T_b/K_b$. Once the molality is known, the molar mass can be found.

Experimentally, freezing point lowering is the colligative property easiest to measure and it is used most often in molar mass determinations of relatively low molecular mass substances. However, molar masses over 1000 are difficult to determine by using vapor pressure lowering, ΔT_b, or ΔT_f because of the large mass of material that must be dissolved in order to produce significant changes in the vapor pressure, freezing point, or melting point. Osmotic pressure measurements (next section) offer a better method for determining high molar masses. It should be pointed out that the boiling point elevation constant and the freezing point depression constant, like many so-called scientific constants, are not really constant. The values of K_b and K_f vary slightly with the nature of the solute and the concentration of the solution. A great many scientific "constants" are in actuality subject to minor variations with changing conditions, and the results of calculations involving such constants are therefore only approximate, as shown by the result in Example 15.11.

EXAMPLE 15.11 Solution Properties: Colligative Properties

Either camphor ($C_{10}H_{16}O$, molar mass = 152.24 g) or naphthalene ($C_{10}H_8$, molar mass = 128.19 g) can be used in mothballs. A 5.0 g sample of mothballs was dissolved in 100.0 g of ethyl alcohol and the resulting solution had a boiling point of 78.91 °C. Determine whether the mothballs were made of camphor or naphthalene. Pure ethyl alcohol has a boiling point of 78.41 °C and $K_b = 1.22$ °C kg/mol for this solvent.

The boiling point elevation is

$$\Delta T_b = T_{b,\text{solution}} - T_{b,\text{solvent}} = 78.91\ ^\circ\text{C} - 78.41\ ^\circ\text{C} = 0.50\ ^\circ\text{C}$$

Using $\Delta T_b = 0.50\ ^\circ\text{C}$ and $K_b = 1.22\ ^\circ\text{C kg/mol}$, gives the molality of the solution as

$$\text{Molality} = \frac{\Delta T_b}{K_b} = \frac{0.50\ ^\circ\text{C}}{1.22\ ^\circ\text{C kg/mol}} = 0.41\ \text{mol/kg}$$

The number of moles of solute in the 100.0 g, or 0.1000 kg, of solvent is

$$(0.41\ \text{mol/kg})(0.1000\ \text{kg}) = 0.041\ \text{mol}$$

The molar mass of the solute is found by dividing the known mass by the number of moles of solute

$$\text{Molar mass} = \frac{5.0\ \text{g}}{0.041\ \text{mol}} = 120\ \text{g/mol}$$

The value of 120 g/mol for the molar mass indicates that **naphthalene** was used to make mothballs.

EXERCISE A solution contained 15.00 g of urea dissolved in 85.00 g of water. The freezing point of this solution was $-5.02\ ^\circ\text{C}$. What is the molar mass of urea? $K_b = 1.86\ ^\circ\text{C kg/mol}$ for H_2O. *Answer* 65.2 g/mol (actual molar mass = 60.06 g).

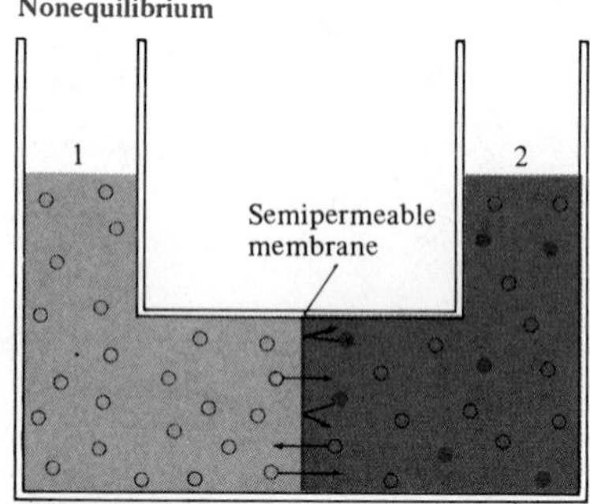

Equilibrium

1

2

Osmotic pressure, Π

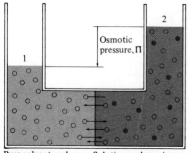

Pure solvent, volume decreased because of diffusion through membrane

Solution, volume increased and solution concentration decreased because of solvent diffusion

Figure 15.18
Osmosis
(a) Diffusible molecules of solvent pass through a semipermeable membrane separating two chambers. Because chamber 2 contains molecules of a nondiffusible solute (solid circles), the effective solvent concentration there is lower than in chamber 1. Consequently more molecules diffuse from left to right than from right to left.
(b) Equilibrium is reached when the rate of diffusion from right to left equals that in the opposite direction. The additional pressure needed to equalize the rates of diffusion is the *osmotic pressure* of the solution in chamber 2.

15.13 OSMOTIC PRESSURE

Liquids can diffuse through skin and other biological membranes, as well as through, for example, parchment, cellophane, and poly (vinyl chloride) membranes. **Semipermeable membranes** are membranes that allow the passage of some molecules but not others. The passage of *solvent* molecules through a semipermeable membrane from a more dilute solution into a more concentrated solution is called **osmosis.** Osmosis is one of the ways in which water molecules move in and out of living cells. If red blood cells are placed in a solution containing a lower concentration of inorganic ions than the fluid inside the cells, the cells swell up and burst as more and more water molecules enter the cells from the solution.

[Note that *osmosis* specifically refers to conditions under which only *solvent* molecules pass through a membrane. There are other processes in which membranes selectively allow passage of particles in a particular size range while retaining larger particles. For example, in the purification of commercially produced enzymes (biological catalysts; Section 33.9c), the enzymes are retained by membranes that allow the passage of water, small organic molecules, and inorganic impurities.]

Osmosis, like vapor pressure, is dependent upon the relative numbers of solute and solvent molecules in a solution. To visualize the process of osmosis, consider a vessel divided by a semipermeable membrane (Figure 15.18). With the same pure liquid in both halves of the vessel, molecules of the liquid diffuse through the membrane with equal freedom in both directions, and there is no apparent change. When a solute that will not pass through the membrane is dissolved in the solvent on one side of the membrane, the effective concentration of the solvent on that side is lowered. Fewer solvent molecules make contact with the membrane and diffuse through it. (The situation is similar to the evaporation of solvent molecules shown in Figure 15.7.) The molecules on the other side of the membrane can still diffuse freely, however, with the result that the rates of diffusion of liquid in the two directions are unequal.

The liquid level on the solution side rises as more water molecules enter than leave until rates of diffusion of solvent molecules in the two directions are once again equal. The pressure exerted by the solution at this point is equal to the osmotic pressure—a colligative property exhibited when a pure solvent and a solution, or solutions of two different concentrations, are separated by a semipermeable membrane. **Osmotic pressure** is defined as the external pressure exactly sufficient to oppose osmosis and stop it.

Osmosis can be reversed by applying a pressure *greater* than the osmotic pressure, a principle utilized in desalination. In reverse osmosis, a pressure greater than the osmotic pressure is applied on the saltwater side of a semipermeable membrane. The water molecules diffuse through the membrane to the freshwater side. (The pressures used are from 200 to 1500 psig.) This is the opposite of the direction in which the water molecules move in osmosis. Reverse osmosis is being used in small-scale units for the desalination of brackish water. The process will retain solutes of molecular mass below 500. (It does not work well with seawater; the ion concentration is too high.)

The mathematical expression for osmotic pressure, Π, in dilute solution is as follows:

osmotic pressure → Π (atm); *molarity of solution* → M

$$\underset{\text{atm}}{\Pi} = \left(\frac{n_B}{V}\right) RT = MRT \quad \left(\frac{\text{mol}}{\text{L}}\;\frac{\text{L atm}}{\text{K mol}}\;\text{K}\right) \qquad \textbf{(15.16)}$$

where n_B is moles of solute, V is solution volume in liters, R is ideal gas constant (0.0821 L atm/K mol for Π in atmospheres), T is the absolute temperature, and M is the molarity of the solution. [You may have noticed that this equation has the same form as the ideal gas law. The behavior of solute molecules in dilute solutions is not unlike the behavior of gas molecules.]

EXAMPLE 15.12 Solution Properties: Colligative Properties

An aqueous solution contained 77.1 g of inulin, $(C_6H_{10}O_5)_x$, per liter of solution. (Inulin is a high molecular mass sugar.) The osmotic pressure at 20 °C of this solution was 0.58 atm. Calculate the molar mass of inulin. What is the value of x in the formula for inulin?

The molarity of the solution can be found from the osmotic pressure.

$$\Pi = MRT$$

$$M = \frac{\Pi}{RT} = \frac{(0.58\text{ atm})}{(0.0821\text{ L atm/K mol})(293\text{ K})} = 0.024\text{ mol/L}$$

The molar mass of the solute, as found from the mass and molarity, is

$$\left(\frac{77.1\text{ g}}{1\text{ L}}\right)\left(\frac{1\text{ L}}{0.024\text{ mol}}\right) = 3200\text{ g/mol}$$

The mass of the formula unit, $C_6H_{10}O_5$, is 162 g/mol, so the value of x is found by dividing 3200 g/mol by 162 g/mol and rounding off to the nearest integer.

$$x = \frac{3200\text{ g/mol}}{162\text{ g/mol}} = 20$$

The molar mass of inulin is **3200 g** and with $\mathbf{x = 20}$ the formula for inulin is $(C_6H_{10}O_6)_{20}$.

EXERCISE What pressure would have to be applied to the solution side of a semipermeable membrane separating pure water from a 0.25 M aqueous solution of sucrose to prevent solvent flow from taking place? Assume the temperature to be 25 °C.

Answer 6.1 atm.

15.14 COLLIGATIVE PROPERTIES OF ELECTROLYTE SOLUTIONS

In a 0.1 *m* solution of sodium chloride (NaCl)—a strong electrolyte—there are twice as many solute particles (Na^+, Cl^-) as in a 0.1 *m* solution of a nonelectrolyte such as sucrose. The effect of sodium chloride on the colligative properties of a 1 *m* solution—if such a solution were ideal—would be twice that of sucrose. Experimentally, it is found that the value of the freezing point depression of a 0.1 *m* NaCl solution is 1.87 times that predicted from Equation (15.12b) for a nonelectrolyte, and for a 0.001 *m* solution it is 1.97 times the expected value.

These data illustrate the general trend in the colligative properties of solutions of strong electrolytes. As the solution becomes more dilute, and therefore approaches ideality, the values of the colligative properties approach those expected based on the number of ions per formula unit of solute. For NaCl, the values approach two times the expected values; for K_2SO_4, they approach three times the expected values; and so on. The following example illustrates how the effect on colligative properties of a weak electrolyte can be used to find the extent of ionization of the electrolyte.

EXAMPLE 15.13 Solution Properties: Colligative Properties

Hydrofluoric acid (HF) is a weak electrolyte—only a small number of the HF molecules ionize to form H^+ ions and F^- ions. The observed freezing point depression of a 0.100 *m* HF solution is 0.197 °C. What percentage of the HF molecules are ionized in the solution? $K_f = 1.86$ °C kg/mol for H_2O.

In this solution there are three types of solute particles: H^+ ions, F^- ions, and non-ionized HF molecules. If we let X represent the fraction of molecules that are ionized, the concentration of each species is

$$\left(\frac{0.100 \text{ mol HF}}{\text{kg}}\right)\left(\frac{X \text{ mol H}^+}{1 \text{ mol HF}}\right) = (0.100\,X) \text{ mol H}^+/\text{kg}$$

$$\left(\frac{0.100 \text{ mol HF}}{\text{kg}}\right)\left(\frac{X \text{ mol F}^-}{1 \text{ mol HF}}\right) = (0.100\,X) \text{ mol F}^-/\text{kg}$$

$$\left(\frac{0.100 \text{ mol HF}}{\text{kg}}\right)\left(\frac{(1 - X) \text{ mol non-ionized HF}}{1 \text{ mol HF}}\right) = (0.100 - 0.100\,X) \text{ mol non-ionized HF/kg}$$

The total concentration of solute particles is

$$(0.100\,X) + (0.100\,X) + (0.100 - 0.100\,X) = (0.100 + 0.100\,X) \text{ mol/kg}$$

The freezing point depression for this solution in terms of the total concentration of solute particles is

$$\Delta T_f = K_f m = (1.86 \text{ °C kg/mol})[(0.100 + 0.100\,X) \text{ mol/kg}] = (1.86)(0.100 + 0.100\,X) \text{ °C}$$

Equating this predicted freezing point depression to the observed value gives

$$(1.86)(0.100 + 0.100\,X) \text{ °C} = 0.197 \text{ °C}$$

$$\mathbf{X = 0.06, \text{ or } 6\%}$$

EXERCISE Benzoic acid, C_6H_5COOH, is a weak electrolyte—only 2.6% of it ionizes in a 0.100 *m* aqueous solution to form H^+ ions and $C_6H_5COO^-$ ions. Calculate the total concentration of solute particles in this solution. What is the predicted freezing point depression? $K_f = 1.86$ °C kg/mol for H_2O. *Answer* 0.1026 mol/kg, 0.191 °C.

SUMMARY OF SECTIONS 15.9–15.14

The total vapor pressure for a mixture of volatile liquids is governed by Raoult's law, $P_{soln} = X_A P_A^\circ + X_B P_B^\circ$, where P_A° and P_B° are the vapor pressures of the pure liquids. Nonvolatile solutes lower the vapor pressure from that of the pure solvent. Vapor pressure lowering and the properties dependent upon it—freezing point depression, boiling point elevation, and osmotic pressure—are called colligative properties. These properties are dependent on the number of nonvolatile particles.

Osmosis is the passage of solvent molecules through a semipermeable membrane, and the external pressure sufficient to halt osmosis is known as osmotic pressure. Any of the colligative properties can be used to find the molality of a solution and from the molality, the molar mass of the solute; for high molar mass solutes, osmotic pressure measurements are best.

As electrolyte solutions become more dilute, the closer they approach ideality, in which condition each ion counts fully in determining colligative properties. The more concentrated an electrolyte solution, the more the ions interact with each other, and the further the colligative properties of the solution deviate from those expected on the basis of the number of ions per formula unit of solute.

COLLOIDS

15.15 PROPERTIES OF COLLOIDS

A sample of mud can be kept in suspension in water by stirring. As soon as the stirring is stopped, the larger particles will fall to the bottom of the vessel. Smaller particles may remain in suspension for a while, but most of them will settle out before long. There are usually some particles, however, that remain suspended in the water for several days, or perhaps indefinitely. These make the mixture cloudy, but the individual particles cannot be seen even under powerful magnification, and the mixture cannot be clarified by passing it through a conventional filter.

A mixture in which small particles remain dispersed almost indefinitely is called a **colloidal dispersion,** or simply a colloid. Rather than use the terms "solute" and "solvent" for a colloid, we speak of the dispersing medium and the dispersed substance or dispersed phase. As illustrated in the next section, all three states of matter may serve in either capacity.

Strictly speaking, a **colloid** is a substance made up of suspended particles larger than most molecules, but too small to be seen in an optical microscope. The size range of colloidal particles is not clearly defined, but is usually set as somewhere between 1 nm and 1000 nm. Colloidal dispersions can be produced by suspending particles, or by breaking up larger aggregates, or by condensation of smaller particles.

When a colloidal dispersion is brightly illuminated by a beam of light at right angles to the line of sight and examined under a microscope, the individual particles cannot be seen, but they are detected as tiny flashes of light dancing in the liquid. This motion, called *Brownian motion* after its discoverer, Robert Brown (1773–1858), is constant but quite irregular. For many years after its discovery, the cause of the Brownian motion was not known. Several theories were advanced to explain it, Brown himself suggesting that the dancing particles were alive. Recognition that the movement of the suspended particles is due to bombardment by rapidly moving molecules and that the speed increases with temperature led to the development of the kinetic-molecular theory.

When a strong beam of light passes through a colloid, the scattering of light by the suspended particles makes the beam clearly visible. This is called the *Tyndall effect* and

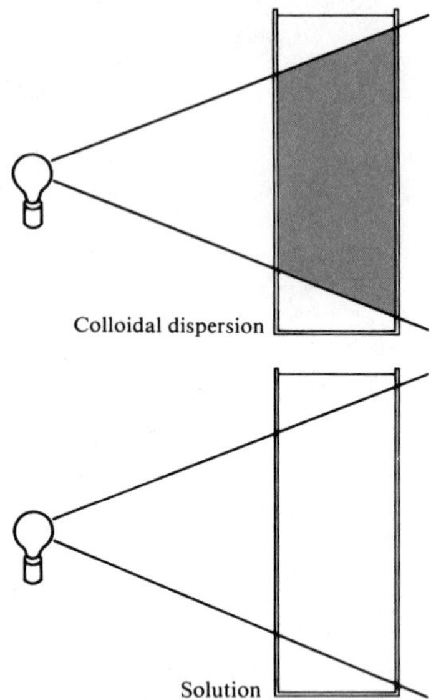

Figure 15.19
Light and Colloids: The Tyndall Effect

is frequently used to distinguish between true solutions and colloids (Figure 15.19). The Tyndall effect can be seen when a beam of light passes through a cloud of smoke or a glass of soapy water.

The properties of colloids in which water is the dispersing medium, such as clays, paints, gelatins, and the colloids present in many biological systems, are of great interest. To get a general picture of such colloids, consider a solid suspended in water which, as is usually the case, also has dissolved in it varying concentrations of strong electrolytes.

Like all macroscopic substances, the colloidal dispersion as a whole is neutral. However, in the presence of an electrical field it is found that all of the colloidal particles migrate to one electrode or the other. The particles themselves are obviously all of the same charge. How does this come about?

At the centers of the particles are chemically similar molecules which attract from the surrounding medium layers of ions all of the same charge. (The attraction of a substance to the surface of a solid is called **adsorption.** An adsorbed layer can be held to a surface by intermolecular forces, the forces of chemical bonding, or a combination of the two.) As a colloidal particle with its charged surface layer drifts through the water, it attracts a more loosely held "atmosphere" made up of ions bearing the opposite charge. The charge on the particle is exactly neutralized by these ions, which cluster around the particle, forming a thin, diffuse layer of ions intermingled with water molecules. The ions are kept from adhering to the surface of the particle by the motion of the intervening water molecules. As the colloidal particles approach each other, their surrounding diffuse layers merge and strong repulsion arises between the charged surface layers. It is this electrical repulsion that prevents colloidal particles from coalescing and thus keeps them in suspension. If the central molecules in the colloidal particles could collide, they would adhere to each other as a result of van der Waals attractions.

The thickness of the diffuse layer containing the ions required for electrical neutrality is determined by the concentration of ions in the suspending medium. If there are few ions, the diffuse layer is large and the colloidal particles experience maximum repulsion. When the concentration of ions is large, the diffuse layer becomes very thin and two colliding particles can collide with sufficient energy to overcome all forces of repulsion, allowing them to come close enough to each other to stick together because of van der Waals attractions. For this reason, colloidal particles may be coagulated by adding electrolytes. Deltas form where rivers meet the ocean partly because the electrolyte concentration in seawater is higher than that in the river (Figure 15.20), causing the colloidal clay particles to coagulate and settle.

Colloidal particles have very large surface areas in comparison with their diameters, a geometrical property characteristic of small objects. As a result, colloidal particles are very efficient at adsorbing other substances on their surfaces, a property useful in the role of many colloidal substances as catalysts.

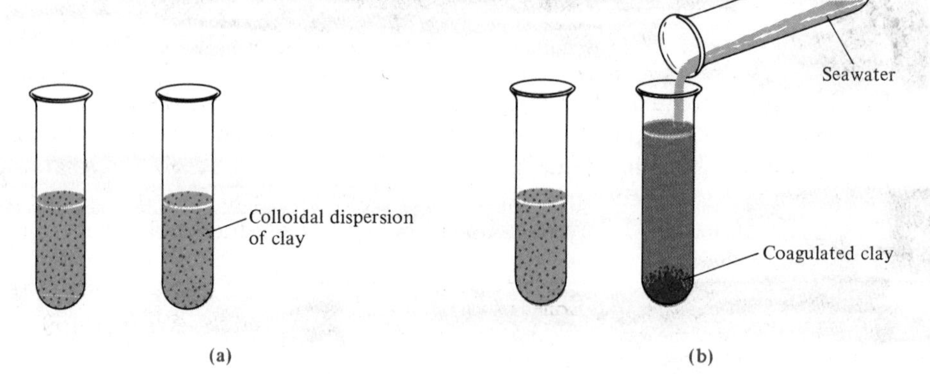

Figure 15.20
Coagulation of a Colloid
Clay is dispersed in water by vigorously shaking the test tubes. Seawater (or 0.5 M NaCl + 0.05 M $MgCl_2$—"artifical" seawater) added to one test tube causes the colloidal dispersion to coagulate as the negative charge on the colloid particles is neutralized by Na^+ and Mg^{2+} ions.

15.16 TYPES OF COLLOIDS

Colloids are categorized by the states of the dispersed phase and the dispersing medium. Examples of the various types of colloids are given in Table 15.4.

Mud or clay consists of solid particles suspended in a liquid—a type of colloid called a **sol**. Sols are very common and examples of them are easily prepared in the laboratory. If, for example, a small amount of a solution of iron(III) chloride is poured slowly into a large volume of boiling water, and the boiling is continued for some time, a clear red sol of hydrated iron(III) oxide particles is formed:

$$2FeCl_3(aq) + (x + 3)H_2O(l) \longrightarrow Fe_2O_3 \cdot xH_2O(\text{red sol}) + 6HCl(aq)$$

A **gel** is a special type of colloid in which solid particles, usually very large molecules, unite in a random and intertwined structure that gives rigidity to the mixture. Pectin, the carbohydrate from fruit, is the dispersed substance that stiffens, for example, grape jelly. Many gels can be converted into sols by changes in temperature, pH, or other conditions. Jello dessert is a colloid that undergoes sol–gel interconversion—the gel form from the refrigerator "melts" to a sol in a warm room. Many of the unusual properties of proteins in aqueous media are due to sol–gel interconversion.

A colloid in which particles of a liquid are dispersed in another liquid is called an **emulsion.** Colloidal suspensions of immiscible liquids, such as oil in water, and water in oil, are emulsions. Such colloids form only in the presence of an emulsifying agent which forms a protective layer on the surface of each colloidal particle. Without the emulsifying agent the droplets would combine and the two liquids would separate. Soap molecules (next section) emulsify droplets of oily dirt. Mayonnaise is a familiar example of an emulsion—the dispersed phase is oil, the dispersing medium is water, and egg yolk is the emulsifying agent.

Foam is a colloid consisting of tiny bubbles of gas dispersed in a liquid. Most foams in which the liquid phase is pure water are short-lived, but if the surface tension of the water is reduced by the addition of a surface-active agent, very stable foams may be generated. Common substances that can be added to water to produce this effect are soap and licorice, both of which lower the surface tension of water. Whipped cream is both an emulsion and a foam, for in it both butterfat and air bubbles are colloidally dispersed.

Smoke is a colloidal dispersion of solid particles in air, and fog is a colloid consisting of droplets of water in the air. A colloidal dispersion in which air (or any gas) is the dispersing medium is called an **aerosol.**

Table 15.4
Types of Colloids

Dispersing medium	Dispersed substance	Type of colloid	Examples
Liquid	Gas	Foam[a]	Soap suds, whipped cream, beer foam
	Liquid	Emulsion[a]	Mayonnaise, milk, face cream
	Solid	Sols, gels[b]	Protoplasm, starch, gelatin and jelly, clay
Gas	Liquid	Liquid aerosol	Fog, mist, aerosol spray
	Solid	Solid aerosol	Smoke, airborne bacteria and viruses
Solid	Gas	Solid foam	Aerogels, polyurethane foam
	Liquid	Solid emulsions, some gels	Cheese
	Solid	Solid sol	Ruby glass, some alloys

[a] Stable foams and emulsions are usually formed only when an emulsifying agent, such as a soap, is present in addition to the pure liquid and gas.

[b] Sols contain individual dispersed particles; in gels the particles link together in a structure of some strength.

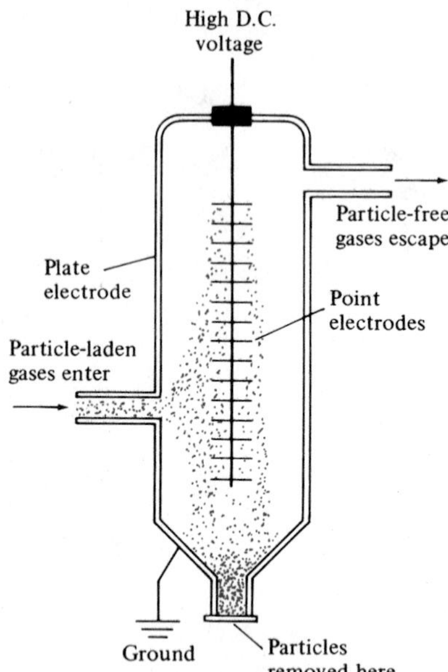

Figure 15.21
The Cottrell Precipitator
Installed near the top of the smokestack, the precipitator removes charged colloidal particles from the smoke.

The precipitation of colloidal particles is illustrated by the operation of electrostatic smoke and dust removers. In the Cottrell precipitator, for example, metal plates carrying a strong electrostatic charge are mounted in a smokestack. Because of their electrical charge, the colloidal smoke particles that pass through the stack are attracted to the plate of opposite charge and, upon contact with it, lose their charge and precipitate (Figure 15.21).

Many colloids are known in which the dispersing medium is a liquid, and the dispersed particles are either gaseous or solid. A very efficient device for cleaning grease spots from clothes consists of a can containing colloidal silica suspended in a mixture of a volatile oil and a pressurized gas. When the pressure is released by opening the valve, a jet of fluid is ejected from the can and is directed on the grease spot. The liquefied gas immediately evaporates; the oil dissolves the grease in the spot and this solution is adsorbed on the silica. The oil evaporates in a few minutes, and the silica with the adsorbed grease is brushed from the garment.

15.17 THE SOAP AND DETERGENT OPERA

Hydrolysis of natural fats and oils in the presence of a base, usually sodium hydroxide, produces a soap. Long before the chemical structure of the reactants was known, soaps were made by the treatment of animal fats with lye. What we call soap, once used in bar or flake form for all household cleaning and laundry, is a mixture of the salts of straight-chain organic acids. A soap has two parts: (1) a long, straight hydrocarbon chain that is oil soluble and (2) the water-soluble sodium or potassium salt of an acid. For example, a common soap is sodium stearate

$$CH_3CH_2CH_2CH_2CH_2CH_2CH_2CH_2CH_2CH_2CH_2CH_2CH_2CH_2CH_2CH_2CH_2\overset{\overset{\displaystyle O}{\|}}{C}—O^-Na^+$$

Oily or greasy dirt adheres firmly to clothing and is undisturbed by washing in tap water, in which oil and grease are insoluble. Soap acts as an emulsifying agent and brings the dirt into colloidal dispersion. The hydrocarbon end of the soap molecule is soluble in the oil or grease. The anionic end is not soluble in the oil or grease, but protrudes from the colloidal droplets, making them water soluble and repelling other similar droplets, thus preventing coalescence (Figure 15.22).

The great disadvantage of soap is that in hard water (Section 14.13) the sodium ion is replaced by magnesium or calcium ions from the water to give insoluble compounds. These compounds precipitate to form bathtub scum and a dull film on laundry washed with soap in hard water.

Synthetic detergents, often simply called detergents, are synthetic cleaning agents (not made directly from animal fats) used as substitutes for soap. Most laundry detergents contain a surface-active agent that, like a soap, consists of a long hydrocarbon chain with a water-soluble group at the end of the chain; for example

$$CH_3CH_2CH_2CH_2CH_2CH_2CH_2CH_2CH_2CH_2CH_2CH_2—C_6H_4—SO_3^-Na^+$$

a straight-chain sodium alkylbenzenesulfonate

The calcium and magnesium salts of such compounds are soluble and do not precipitate.

Synthetic detergents were introduced in 1945 and quickly became very popular. These compounds are more effective cleaning agents than soap, and they can be used in hard water as well. In the first generation of detergents, however, the hydrocarbon side chains were not straight (as they are in soap and in the sodium alkylbenzenesulfonate written above), but branched; for example

$$CH_3\underset{\displaystyle CH_3}{\underset{|}{C}H}CH_2\underset{\displaystyle CH_3}{\underset{|}{C}H}CH_2\underset{\displaystyle CH_3}{\underset{|}{C}H}CH_2\underset{\displaystyle CH_3}{\underset{|}{C}H}—C_6H_4—SO_3^-Na^+$$

a branched-chain sodium alkylbenzenesulfonate

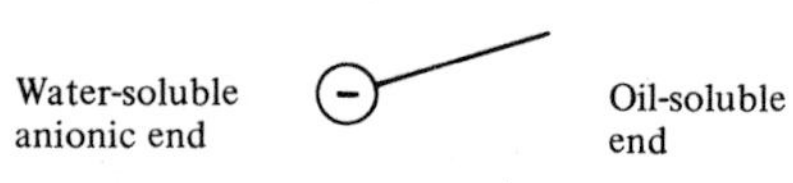

(a) A molecule of soap or synthetic detergent

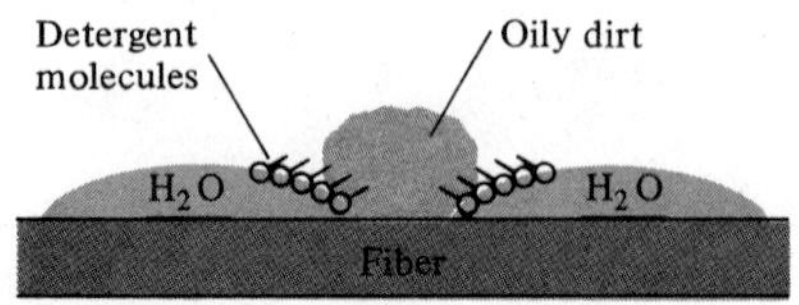

(b) Action of soap or synthetic detergent on oily dirt on clothing

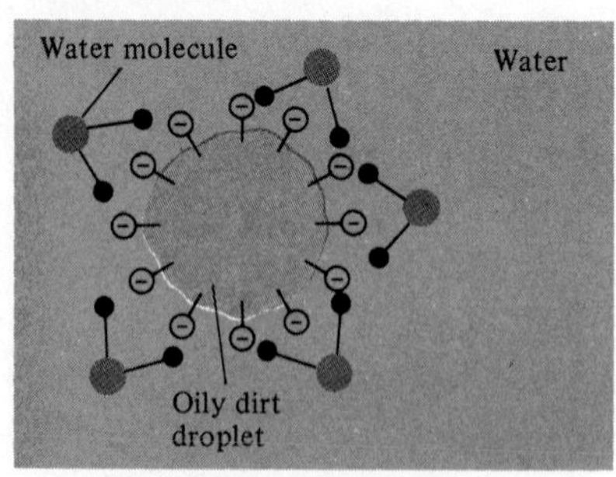

(c) Oily dirt in wash water

Figure 15.22
The Action of Soap and Synthetic Detergents on Oily and Greasy Dirt

It was soon discovered that these compounds are not biodegradable, that is, they cannot be broken down by organisms in the environment. A major reason is that the bacteria in natural bodies of water and in municipal wastewater treatment plants reject a branched-chain diet. The detergents remained effective, and sudsy water was found in streams and even ran out of faucets. This problem was solved by a second generation of detergents with straight hydrocarbon chain surface-active agents, a more acceptable diet to bacteria.

SUMMARY OF SECTIONS 15.15–15.17

A mixture in which particles generally between 1 nm and 1000 nm in diameter remain dispersed almost indefinitely is called a colloidal dispersion or colloid. Colloidal particles in aqueous suspension are usually surrounded by layers of ions which prevent them from coalescing.

A sol consists of solid particles suspended in a liquid. In a gel the solid particles, usually very large molecules, unite in a random, intertwined structure, producing greater rigidity. Particles of a liquid colloidally dispersed in another liquid constitute an emulsion. A foam consists of tiny gas bubbles suspended in a liquid. A colloid in which a gas is the suspending medium is an aerosol.

Soaps are mixtures of the salts of straight-chain organic acids. The long hydrocarbon chains dissolve in droplets of oil or grease; the anions at the ends of the chains protrude from the droplets and interact with water molecules, preventing coalescence of the droplets and creating a colloidal dispersion. Synthetic detergents also consist of long hydrocarbon chains with water-soluble groups at the end. The Ca and Mg salts of synthetic detergents, unlike those of soaps, are water soluble.

SIGNIFICANT TERMS (with section references)

adsorption (15.15)
aerosol (15.16)
azeotropes (Aside 15.2)
colligative property (15.10)
colloid (15.15)
colloidal dispersion (15.15)
distillate (Aside 15.2)
distillation (Aside 15.2)
emulsion (15.16)
foam (15.16)
fractional distillation (Aside 15.2)
gel (15.16)
Henry's law (15.4)
ideal solution of a molecular solute (15.3)
ideal solution of an ionic solute (15.3)
immiscible (15.2)
infinitely miscible (15.2)
mass percent, weight percent (15.6)
miscibility (15.2)
molality (15.7)
osmosis (15.13)
osmotic pressure (15.13)
Raoult's law (15.9)
refluxing (Aside 15.2)
saturated solution (15.1)
semipermeable membranes (15.13)
sol (15.16)
solubility (15.1)
supersaturated solution (15.1)
unsaturated solution (15.1)
vapor pressure lowering (15.10)

QUESTIONS AND PROBLEMS

General Properties of Solutions

15.1 Give an example of solutions that contain (a) a solid dissolved in a liquid, (b) a gas dissolved in a gas, (c) a gas dissolved in a liquid, (d) a liquid dissolved in a liquid, (e) a solid dissolved in a solid. Identify the substances that are the solvent and solute in each case.

15.2 Why is dry HCl in the liquid state a nonelectrolyte, whereas in aqueous solution it is a good electrical conductor?

15.3 Briefly describe the equilibrium between a saturated solution and undissolved solute. How might you prove that your description is correct?

15.4 Two liquids, A and B, do not react chemically and are completely miscible. What would be observed as one is poured into the other? What would be observed in the case of two completely immiscible liquids and in the case of two partially miscible liquids?

15.5 An old saying is that "oil and water don't mix." Why is this true?

15.6 Both methanol (CH_3OH) and ethanol (CH_3CH_2OH) are completely miscible with water at room temperature because of strong solvent–solute intermolecular forces. Predict the trend in solubility in water for 1-propanol ($CH_3CH_2CH_2OH$), 1-butanol ($CH_3CH_2CH_2CH_2OH$), and 1-pentanol ($CH_3CH_2CH_2CH_2CH_2OH$).

15.7 On what basis would you choose the components to prepare an ideal solution of a molecular solute? Which of the following combinations would be the closest to acting ideally: (a) $CH_4(l)$ and $CH_3OH(l)$, (b) $CH_3OH(l)$ and $NaCl(s)$, (c) $CH_4(l)$ and $CH_3CH_3(l)$?

15.8 Do aqueous solutions of ionic solutes act more nearly ideally at high or low concentrations? Which of the following would be closest to acting ideally: (a) 0.001 M $Al_2(SO_4)_3$, (b) 0.001 M KNO_3, (c) 0.010 M KNO_3?

15.9 A handbook lists the value of the Henry's law constant as 3.02×10^4 atm for ethane, C_2H_6, dissolved in water at 25 °C. The absence of concentration units on k means that the constant is meant to be used with concentration expressed as mole fraction. Calculate the mole fraction of ethane in water at an ethane pressure of 0.15 atm.

15.10 The mole fraction of methane, CH_4, in water can be calculated from the Henry's law constants of 4.13×10^4 atm at 25 °C and 5.77×10^4 atm at 50 °C. Calculate the solubility of methane at these temperatures for a methane pressure of 15 atm above the solution. Does the solubility increase or decrease with increasing temperature?

15.11 Henry's law constants at 25 °C for N_2 and O_2 in water are 4.34×10^5 Torr/(g N_2/100 g H_2O) and 1.93×10^5 Torr/(g O_2/100 g H_2O), respectively. Assume the partial pressure of N_2 to be 608 Torr and of oxygen to be 152 Torr above the solution. Find the solubility of each of these gases in water. Is the ratio of the mass of O_2 to N_2 larger or smaller in water than in air?

15.12 Use the following solubility data for SO_2 in water at 0 °C to calculate the Henry's law constant for this system:

P_{SO_2} (Torr)	0.25	0.60	1.2	1.9
solubility (g SO_2/100 g H_2O)	0.02	0.05	0.10	0.15

15.13 What is the general trend in the solubility of inorganic salts in water with increasing temperature? Which anions do not follow this general trend?

15.14 (a) Does the solubility of a solid in a liquid exhibit an appreciable pressure dependence? (b) Is the same true for the solubility of a liquid in a liquid?

Concentration of Solutions

15.15 Concentration units that are based on the volume of solution, solvent, solute, etc., are dependent on temperature. Which of the following units depends on temperature: molality, molarity, mole fraction, mass %?

15.16 What physical quantity provides the connection between concentration units based on mass (mole fraction, molality, mass %, etc.) and those based on volume (molarity, etc.)?

15.17 Many handbooks list solubilities in units of (g solute/100 g H_2O). How would you convert from this unit to mass percent?

15.18 Under what conditions are the molarity and molality of a solution nearly the same? Which concentration unit is more useful when measuring volume with burets, pipets, and volumetric flasks in the laboratory? Why?

15.19 A 40.0 mL sample of ethyl ether, $(C_2H_5)_2O$, is dissolved in enough methyl alcohol. CH_3OH, to make 250. mL of solution. The density of the ether is 0.714 g/mL. What is the molarity of this solution?

15.20 Hydrogen chloride and water form an azeotrope at 1 atm containing 20.2 mass % HCl. This azeotrope has a density of 1.102 g/cm^3. What is the molarity of this solution?

15.21 Find the mole fraction of table sugar (sucrose, $C_{12}H_{22}O_{11}$) in an aqueous solution containing 30.0 g of the sugar (molar mass = 342.3 g) and 70.0 g of water.

15.22 A solution contained 5.0 g NaCl, 15.0 g KCl, and 80.0 g H_2O. Calculate the mole fraction of water.

15.23 What masses of NaCl and H_2O are present in 160. g of a 12 mass % aqueous solution of NaCl?

15.24 Sodium fluoride has a solubility of 4.22 g in 100.0 g of water at 18 °C. Express the concentration in mass percent.

15.25 The solubility of K_2ZrF_6 at 100 °C in 100 g of H_2O is 25 g. Express this concentration in terms of molality.

15.26 A solution contains 5.0 g KI and 10.0 g KCl dissolved in 85.0 g H_2O. What is the concentration of KCl expressed in molality?

15.27 Describe how to prepare 1.000 L of 0.250 m NaCl. The density of this solution is 1.011 g/mL.

15.28 A piece of jewelry is marked "14 carat gold," meaning that on a mass basis the jewelry is 14/24 pure gold. What is the molality of this alloy—assuming the other metal is the solvent?

15.29 Given a sufficient quantity of 4.38 M NaOH, describe how you would prepare exactly 250 mL of 0.876 M NaOH.

15.30 How must a 2.56 M potassium permanganate solution be diluted to prepare 500. mL of a 0.15 M solution?

15.31 The heat of formation at 25 °C is −887 kJ/mol for H_2SO_4 (1 M) and −866 kJ/mol for H_2SO_4(16 M). As for chemical reactions and changes of state, the process of dilution of a solution can be exothermic or endothermic. How much heat is absorbed or released in the dilution of 16 M H_2SO_4 to 1 M H_2SO_4?

$$H_2SO_4(16\text{ M}) \xrightarrow{H_2O} H_2SO_4(1\text{ M})$$

If the solution cannot gain or lose this heat fast enough to remain at 25 °C, what sensation would you feel if you touched the beaker in which this reaction was performed?

15.32 The heat of formation at 25 °C is −341 kJ/mol for NH_4NO_3(1 M) and −340. kJ/mol for NH_4NO_3(0.1 M). Calculate the heat absorbed or released in preparing a 0.1 M solution from a 1 M solution

$$NH_4NO_3(1\ M) \xrightarrow{H_2O} NH_4NO_3(0.1\ M)$$

15.33 An aqueous ammonium chloride solution contains 6.50 mass % NH_4Cl. The density of the solution is 1.0201 g/mL. Express the concentration of this solution in molarity, molality, and mole fraction of solute.

15.34 The density of a solution containing 10.00 g K_2SO_4 in 100.00 g solution is 1.0825 g/mL. Calculate the concentration of this solution in molarity, molality, mass percent of K_2SO_4, and mole fraction of solvent.

15.35 The density of a sulfuric acid solution taken from a car battery is 1.225 g/cm^3. This corresponds to a 3.75 M solution. Express the concentration of this solution in molality, mole fraction of H_2SO_4, and mass % of water.

15.36 Which aqueous solution has the largest Cl^- concentration: 0.05 *m* HCl, 15 mass % NaCl, or a $CaCl_2$ solution that has a mole fraction of 0.10 for $CaCl_2$?

Colligative Properties—Vapor Pressure

15.37 A solution contains two volatile components, A and B. What will be the relative compositions of the vapor and liquid if (a) $P^\circ_A = P^\circ_B$, (b) $P^\circ_A > P^\circ_B$, (c) $P^\circ_A < P^\circ_B$?

15.38 A solution shows positive deviation from Raoult's law for both components A and B. What can be inferred about the intermolecular forces between A and A, B and B, and A and B?

15.39 Toluene and water are essentially immiscible. If a flask containing the two liquids is heated, the system begins to boil at a temperature that is below the boiling point of either liquid. Why?

15.40 Describe what happens to the composition of the liquid phase of a solution of two volatile components as boiling occurs. Assume that the vapor pressures of the solvent and solute are different.

15.41 Using Raoult's law, predict the partial pressures over a solution containing 0.300 mol acetone (P° = 345 Torr) and 0.200 mol chloroform (P° = 295 Torr). What is the total pressure over this solution?

15.42 At −100 °C the vapor pressure of ethane (CH_3CH_3) is 394 Torr and of propane ($CH_3CH_2CH_3$) is 22 Torr. What is the vapor pressure over a solution containing equal molar amounts of these substances? What is the composition of the vapor?

15.43 What is the vapor pressure (at 25 °C) above a solution containing 100.00 g of water and 10.0 g of urea, $CO(NH_2)_2$, a nonvolatile solute? The vapor pressure of pure water (at 25 °C) is 23.76 Torr.

15.44 What mass of a nonvolatile solute having a molar mass of 250. g would be required to decrease the vapor pressure of 1.00 kg of water by 1.00 Torr at 100 °C?

Colligative Properties—Freezing Point Lowering and Boiling Point Elevation

15.45 What are the boiling and freezing points of a 0.125 *m* aqueous solution of urea, a nonelectrolyte? K_b = 0.512 °C kg/mol and K_f = 1.86 °C kg/mol for water.

15.46 The molal freezing point constant for copper is 23 °C kg/mol. If pure copper melts at 1083 °C, what will be the melting point of a brass made of 10. mass % Zn and 90. mass % Cu?

15.47 Methyl alcohol, CH_3OH, and ethylene glycol, HOH_2C-CH_2OH, are both used to prevent the freezing of water in automobile radiators in cold weather. (a) Which will be more effective in a given radiator, 25 g of methyl alcohol or 25 g of ethylene glycol? (b) Which will be more effective, a 5 *m* solution of methyl alcohol or a 5 *m* solution of ethylene glycol?

15.48 An aqueous solution containing 0.502 g of a nonvolatile nonelectrolyte organic compound (molar mass = 60.0 g) dissolved in 7.50 g of water had a boiling point of 100.57 °C at 1 atm pressure. What is the molality of the solution? Calculate the boiling point constant for water.

15.49 A solution was made by dissolving 3.75 g of a nonvolatile solute in 95.0 g of acetone. The solution boiled at 56.58 °C. The boiling point of pure acetone is 55.95 °C and K_b = 1.71 °C kg/mol. Calculate the molar mass of the solute.

15.50 The molar mass of an organic compound was determined by measuring the freezing point depression of a benzene solution. A 0.500 g sample was dissolved in 50.0 g of benzene (K_f = 2.53 °C kg/mol), and the resulting depression was 0.42 °C. What is the approximate molar mass? The compound gave the following elemental analysis: 40.0 mass % C, 6.67 mass % H, 53.3 mass % O. Determine the formula and exact molar mass of the substance.

Colligative Properties—Osmotic Pressure

15.51 What is the osmotic pressure associated with a 0.001 M aqueous solution of a nonvolatile nonelectrolyte solute at 75 °C?

15.52 The osmotic pressure of an aqueous solution of a nonvolatile nonelectrolyte solute is 14.7 atm at 0 °C. What is the molarity of the solution?

15.53 At what temperature would a 1.00 M aqueous solution of sugar have an osmotic pressure of 1.00 atm? Is this answer reasonable?

15.54 A solution containing 3.6 g of a nonelectrolyte polymer per liter of benzene has an osmotic pressure of 0.66 Torr at 20 °C. Calculate the molar mass of the polymer. Assuming the molarity and molality of this dilute solution to be equal, what would be the freezing point depression for this solution (K_{fp} = 4.90 °C kg/mol for benzene)? Why are boiling point elevations and freezing point depressions difficult to use to measure molar masses of polymers?

15.55* A sample of a drug ($C_{21}H_{23}O_5N$, molar mass = 369 g) mixed with lactose (a sugar, $C_{12}H_{22}O_{11}$, molar mass = 342 g) was analyzed by osmotic pressure to determine the amount of sugar present. If 100.00 mL of solution containing 1.00 g of the drug-sugar mixture has an osmotic pressure of 539 Torr at 25 °C, what is the percent sugar present?

15.56* If a sugar maple tree grows to a height of 45 feet, what must be the concentration of sugar in its sap so that osmotic pressure forces the sap to the top of the tree at 0 °C? The density of mercury is 13.6 g/cm^3 and the density of the sap can be considered to be 1.00 g/cm^3.

Colligative Properties—Solutions of Electrolytes

15.57 Compare the number of solute particles present in solutions of equal concentrations of strong electrolytes, weak electrolytes, and nonelectrolytes.

15.58 Four beakers contain, respectively, 0.010 *m* aqueous solutions of CH_3OH, NaOH, $CaCl_2$, and CH_3COOH. Which of these solutions has the lowest freezing point?

15.59 The saline solution used for intraveneous injections contains 8.5 g of NaCl dissolved in 1.00 kg of water. What is the osmotic pressure of this solution at 37 °C? Assume that the molarity is equal to the molality for this solution.

15.60 Upon dissolution, Na_2HPO_4 forms two Na^+ ions for every HPO_4^{2-} ion. Assuming an ideal solution, what is the predicted freezing point depression of a 0.010 *m* aqueous solution of Na_2HPO_4? $K_f = 1.86$ °C kg/mol for H_2O.

15.61 A 0.050 *m* aqueous solution of $K_3Fe(CN)_6$ has a freezing point of −0.2800 °C. Calculate the total concentration of solute particles in this solution and interpret your results.

15.62 A compound was known to be either K_2SO_4, which forms two K^+ ions for each SO_4^{2-} ion upon dissolution, or $KHSO_4$, which forms one K^+ ion for each HSO_4^- ion upon dissolution. (Assume that the HSO_4^- does not ionize further.) The osmotic pressure of a 0.010 M solution of this compound at 25 °C was 0.50 atm. Which substance was present?

15.63 One gram each of NaCl, NaBr, and NaI was dissolved in 100.0 g water. What is the vapor pressure above the solution at 100 °C?

15.64 You are designing an experiment for a general chemistry laboratory in which the molar mass will be determined for some simple inorganic salts having the empirical formula MX. The procedure calls for measuring the boiling point elevation of an aqueous solution containing 10.0 g of water. If the average molar mass of the salts is 75 g, approximately what mass of salt should be used to give a boiling point elevation of 2.0 °C? $K_b = 0.512$ °C/kg mol for water?

15.65 The freezing point of a 1.00 mass % aqueous solution of acetic acid, CH_3COOH, is −0.31 °C. What is the approximate molar mass of acetic acid in water? A 1.00 mass % solution of acetic acid in benzene ($K_f = 4.90$ °C kg/mol) has a freezing point depression of 0.441 °C. What is the molar mass of acetic acid in this solvent? Explain the difference.

15.66* The freezing point depression for a 1.00 *m* aqueous solution of NH_3 is 1.94 °C. What fraction of ammonia molecules react with water in this solution?

$$NH_3(aq) + H_2O(l) \rightleftharpoons NH_4^+ + OH^-$$

15.67* The major components of seawater are 18,980. ppm Cl^-, 10,561 ppm Na^+, 2652 ppm SO_4^{2-}, 1272 ppm Mg^{2+}, 400. ppm Ca^{2+}, 380. ppm K^+, 142 ppm HCO_3^-, 65 ppm Br^-. A ppm (part per million) means 1 g solute in 10^6 g of solution. Calculate the molality of seawater with respect to each ion. What would be the predicted freezing and boiling points? Assuming molarity and molality to be equal, what pressure would be required at 25 °C in order to obtain pure water by reverse osmosis?

15.68* An aqueous solution contained 1.00 g each of acetic acid (CH_3COOH, a weak electrolyte that is 4% ionized into H^+ ions and CH_3COO^- ions), ammonium sulfate [$(NH_4)_2SO_4$, a strong electrolyte], and formic acid (HCOOH, a weak electrolyte that is 6% ionized into H^+ ions and $HCOO^-$ ions) in 100.0 g H_2O. What is the freezing point of this solution?

Colloids

15.69 How does a colloidal dispersion differ from a true solution?

15.70 The terms "condensation" and "dispersion" are often used to describe two common ways in which colloids can be prepared. How do these methods differ?

15.71 Suspensions of gases in gases do not exist. Why? How can the dispersed phase be removed from colloids in which the dispersing phase is a gas?

15.72 What is the usual reason that colloidal particles do not coalesce into larger particles? How can colloids be coagulated?

15.73 What special name is given to colloids (a) in which the *dispersing* phase is a gas and (b) in which the *dispersed* phase is a gas?

15.74 The cleansing action of soaps and detergents (see Section 15.17) illustrates how an insoluble material can be suspended in a solvent. Prepare a sketch showing this mechanism.

ADDITIONAL PROBLEMS

15.75 At 1.000 atm the solubility of CO in water at 25 °C is 0.002603 g/100 g H_2O and of SO_2 in water is 9.41 g/100 g H_2O. What is the Henry's law constant for each gas? What is the molality of each solution? What would be the freezing point of each solution? Why is there such a large difference in the solubilities of these gases?

15.76* The density of one brand of commerical vinegar (5.00 mass % CH_3COOH) is 1.0055 g/cm^3. (a) Express the concentration of the solute in molarity, molality, and mole fraction.

The freezing point of vinegar −1.65 °C. (b) Calculate the fraction of molecules that ionize. (c) Calculate the osmotic pressure of the solution at 25 °C.

Assuming an ideal solution, use Raoult's law to calculate (d) the total vapor pressure of the vinegar. The vapor pressure of water is 23.8 Torr and of acetic acid is 15.5 Torr at 25 °C.

The heat of formation is −484.5 kJ/mol for $CH_3COOH(l)$ and −485.54 kJ/mol for CH_3COOH (5.00 mass %) at 25 °C. (e) Calculate the heat of solution.

CHAPTER

16

Hydrogen and Oxygen; Oxidation–Reduction Reactions

16.1 HYDROGEN AND OXYGEN

Oxygen is the most abundant element on Earth and hydrogen forms more chemical compounds than any other element. Water, the most abundant compound of each of these elements, covers 70% of the Earth's surface. Consequently, it is not surprising that hydrogen and oxygen played significant roles in the experiments of early chemists. For many years, an atomic mass scale based on an assigned mass of 1 for hydrogen was used. The chemistry of hydrogen and oxygen continues to be studied and to be of great practical importance.

Although hydrogen was the most abundant element in the atmosphere of the primitive Earth, very little remains today (Table 16.1). Because many hydrogen molecules have sufficiently high velocities to escape from the Earth's atmosphere, the concentration of hydrogen has steadily decreased. However, large quantities of hydrogen are produced each year from inexpensive and widely available natural sources—water and hydrocarbons (Section 16.8).

A hydrogen atom can, like alkali metal atoms, lose an electron to form a hydrogen ion (H^+). However, the hydrogen ion, having no surrounding electrons, is an unshielded proton; no other positive ion is so small or has such a high charge density. Therefore, in chemical reactions hydrogen atoms give up electrons only when the resulting H^+ ions can share two electrons from other atoms by forming coordinate covalent bonds (as in H_3O^+ or NH_4^+).

A hydrogen atom can, like halogen atoms, also gain an electron to form the hydride ion (H^-). However, unlike halide ions, the hydride ion is not stable in aqueous solution—it reacts immediately with water to form hydrogen and hydroxide ion.

$$H^- + H_2O(l) \longrightarrow H_2(g) + OH^- \qquad \textbf{(16.1)}$$

Because of its unique properties, hydrogen is not classified as a member of any family in the periodic table.

Table 16.1
Average Composition of Clean, Dry Air

Symbol or formula	Name	Percent by volume
N_2	Nitrogen	78.084
O_2	Oxygen	20.946
Ar	Argon	0.934
CO_2	Carbon dioxide	0.0325
Ne	Neon	1.818×10^{-3}
He	Helium	5.24×10^{-4}
Kr	Krypton	1.14×10^{-4}
Xe	Xenon	8.7×10^{-6}
CH_4	Methane	1.6×10^{-4}
H_2	Hydrogen	5×10^{-5}
CO	Carbon monoxide	8×10^{-6} to 5×10^{-5}
N_2O	Nitrous oxide	2 to 4×10^{-7}
SO_2	Sulfur dioxide	7×10^{-7} to $>1 \times 10^{-4}$
NO	Nitric oxide	10^{-6} to 10^{-4}
NO_2	Nitrogen dioxide	10^{-6} to 10^{-4}
HCHO	Formaldehyde	$\leq 10^{-5}$
NH_3	Ammonia	$\leq 10^{-4}$
O_3	Ozone	0 to 5×10^{-5}

The atmosphere of the Earth today contains about 20% oxygen (see Table 16.1), and air is the major industrial source of elemental oxygen (Section 16.12). The primitive atmosphere probably contained very little oxygen. Our oxygen has been generated from water by photosynthesis in green plants. The rocks and soil on the Earth's surface are composed largely of oxygen-containing oxides, silicates, and carbonates.

Oxygen is the first element in Representative Group VI. Oxygen atoms have six valence electrons and can gain two electrons to form the oxide ion (O^{2-}), or can form two covalent bonds. In aqueous solution, the oxide ion reacts immediately and completely to form hydroxide ion.

$$O^{2-} + H_2O(l) \longrightarrow 2OH^- \qquad \textbf{(16.2)}$$

An excess of hydroxide ions in aqueous solution imparts the classical base properties to a solution; an excess of hydrogen ions imparts the classical acid properties.

Almost all elements form binary oxides. As might be expected, the bonding and properties of the oxides vary greatly. Oxygen also is present in many of the common acids and polyatomic ions.

The properties of water, hydrogen ions, hydroxide ions, hydrogen, oxygen, and chemical compounds containing hydrogen and/or oxygen are fundamental to many areas in chemistry. Two such areas of importance are acid–base chemistry, which was introduced earlier (Section 14.8), and oxidation–reduction chemistry, which is introduced in the following sections. The concepts of oxidation and reduction originated from studies of the chemical properties of hydrogen and oxygen.

OXIDATION AND REDUCTION

16.2 OXIDATION–REDUCTION REACTIONS

All chemical reactions can be classified as oxidation–reduction reactions or as reactions in which oxidation and reduction do *not* occur. Just as for "acids" and "bases," the meanings of the terms "oxidation" and "reduction" have been expanded over the years. "Oxidation" was first known as the addition of oxygen to another substance, which was thus "oxidized," for example

$$2\overset{0}{Cu}(s) + O_2(g) \longrightarrow 2\overset{+2}{Cu}O(s) \qquad \textbf{(16.3)}$$

In this reaction copper is oxidized, as shown by the increase in the oxidation number of copper (0 to +2). Any process in which an oxidation number increases is **oxidation.**

"Reduction" first meant the production of a free metal from its ore. Many ores are oxides, and "reduction" came to mean the *removal* of oxygen from a substance—the opposite of oxidation. Tin, for example, is prepared by reduction of the mineral cassiterite, SnO_2, by reaction with coke (coal that has been heated in the absence of air to leave behind mostly carbon).

$$\overset{+4}{Sn}O_2(s) + 2C(s) \xrightarrow{\Delta} \overset{0}{Sn}(l) + 2CO(g) \qquad \textbf{(16.4)}$$

The tin(IV) oxide is reduced by carbon. In modern terms, the reduction of the tin is shown by its decrease in oxidation number (+4 to 0). Any process in which an oxidation number decreases is **reduction.**

What happens to the carbon in reaction (16.2)?

$$SnO_2(s) + 2\overset{0}{C}(s) \xrightarrow{\Delta} Sn(l) + 2\overset{+2}{C}O(g) \qquad \textbf{(16.5)}$$

In both the original and the modern meanings of the term, carbon has been oxidized—oxygen has been added and the oxidation number increased (0 to +2).

Equations (16.4) and (16.5) illustrate an important point. Oxidation and reduction *always* occur together. If the oxidation number of one atom or ion increases in a reaction, the oxidation number of another atom or ion *must* decrease. Reactions in which oxidation and reduction take place are known as **oxidation–reduction reactions,** or frequently as **redox reactions.**

Often, oxidation is accompanied by electron loss and reduction by electron gain. For example, in the combination of copper and oxygen (Equation 16.3), copper loses electrons to form Cu^{2+} and oxygen gains electrons to form O^{2-}. However, the definitions in terms of oxidation number changes are the most generally useful. Note that the changes in oxidation numbers are algebraic increases and decreases. [An algebraic increase is a change in the direction of a more positive number. Oxidation number changes of -4 to -2, or -1 to $+1$, or $+2$ to $+3$ represent oxidation. Similarly, changes of $+5$ to $+2$, or $+2$ to -1, or -1 to -2 represent reduction. Here is a simple mnemonic trick: **O**xidation and **i**ncrease in oxidation number both begin with vowels; **r**eduction and **d**ecrease in oxidation number both begin with consonants.]

As illustrated in Example 16.1, the distinction between redox and nonredox reactions can be made by determining whether or not any oxidation number changes occur. (The rules for assigning oxidation numbers were given in Section 10.10.)

EXAMPLE 16.1 Redox Reactions

Are either or both of the following, redox reactions?

$$Zn(s) + HgO(s) \longrightarrow ZnO(s) + Hg(l)$$

$$NaOH(aq) + HCl(aq) \longrightarrow NaCl(aq) + H_2O(l)$$

To determine whether or not these are redox reactions we can put in all of the oxidation numbers.

$$\overset{0}{Zn}(s) + \overset{+2-2}{HgO}(s) \longrightarrow \overset{+2-2}{ZnO}(s) + \overset{0}{Hg}(l)$$

This is **a redox reaction**—zinc has been oxidized (0 to +2) and mercury has been reduced (+2 to 0).

$$\overset{+1-2+1}{NaOH}(aq) + \overset{+1-1}{HCl}(aq) \longrightarrow \overset{+1-1}{NaCl}(aq) + \overset{+1-2}{H_2O}(l)$$

This is **not a redox reaction**—no changes in oxidation number have occurred. [It is a neutralization reaction and also a partner-exchange reaction.]

EXERCISE Determine whether or not the following reactions are redox reactions: (a) $I^- + ClO^- \longrightarrow IO^- + Cl^-$, (b) $H_2O(l) \rightleftharpoons H^+ + OH^-$, (c) $4HNO_3(l) \longrightarrow 2N_2O_4(g) + O_2(g) + 2H_2O(l)$, and (d) $C_6H_6(l) \longrightarrow C_6H_6(s)$.

Answer Reactions (a) and (c) are redox reactions.

16.3 OXIDIZING AGENTS AND REDUCING AGENTS

Oxygen, in its reaction with copper (Equation 16.3), is referred to as the oxidizing agent—it has caused the copper to undergo an increase in oxidation number. The oxygen itself has gone from oxidation state 0 in molecular oxygen to oxidation state -2 in the copper(II) oxide—oxygen has been reduced.

$$2\overset{0}{Cu}(s) + \overset{0}{O_2}(g) \longrightarrow 2\overset{+2-2}{CuO}(s) \qquad \textbf{(16.6)}$$

An atom, molecule, or ion that causes an *increase* in the oxidation state of another substance and is itself reduced is called an **oxidizing agent.** Oxygen rarely has anything but negative oxidation states, and therefore molecular oxygen is an oxidizing agent in almost all of its reactions.

The original meaning of "oxidation" as a reaction with oxygen was expanded as it was recognized that other substances could cause similar reactions. For example, compare the reaction of copper and fluorine with the reaction of copper and oxygen:

$$\overset{0}{Cu}(s) + \overset{0}{F_2}(g) \longrightarrow \overset{+2-1}{CuF_2}(s) \qquad \textbf{(16.7)}$$

Although oxygen has not been added, this reaction is clearly of the same type as that in Equation (16.6). Fluorine is the oxidizing agent in this reaction; like oxygen, it has caused the copper to be oxidized and has itself been reduced (from oxidation state 0 to oxidation state -1).

In the reaction of copper(II) oxide with hydrogen, hydrogen is the reducing agent—it causes the reduction of copper from oxidation state $+2$ to oxidation state 0.

$$\overset{+2}{\mathrm{Cu}}\mathrm{O}(s) + \overset{0}{\mathrm{H}}_2(g) \xrightarrow{\Delta} \overset{0}{\mathrm{Cu}}(s) + \overset{+1}{\mathrm{H}}_2\mathrm{O}(g) \tag{16.8}$$

The hydrogen itself has gone from oxidation state 0 to oxidation state $+1$—the hydrogen has been oxidized. An atom, molecule, or ion that causes a *decrease* in the oxidation state of another substance and is itself oxidized is called a **reducing agent.** In all of its compounds except the ionic hydrides, hydrogen is assigned oxidation state $+1$, and therefore in all of its reactions except the formation of ionic hydrides, molecular hydrogen is a reducing agent. Note that in redox reactions, the oxidizing agent is always reduced and the reducing agent is always oxidized.

The activity series introduced earlier (Section 5.1c) is a list of metals in the order of decreasing strength as reducing agents. Metals rarely have negative oxidation states, and therefore elemental metals are reducing agents in most of their reactions. The active metals of the sodium family and also calcium, barium, and strontium are very strong reducing agents.

Obviously, because oxidation and reduction always occur together, each redox reaction includes both an oxidizing agent and a reducing agent. In the reaction between ammonia and oxygen to give nitrogen(II) oxide and water

$$4\,\overset{-3}{\mathrm{N}}\mathrm{H}_3(g) + 5\,\overset{0}{\mathrm{O}}_2(g) \longrightarrow 4\,\overset{+2-2}{\mathrm{NO}}(g) + 6\,\mathrm{H}_2\overset{-2}{\mathrm{O}}(l) \tag{16.9}$$

(oxidizing agent: O_2; reducing agent: NH_3)

ammonia is the reducing agent and oxygen is the oxidizing agent. (Note that the entire molecule or ion—not just the atom that changes oxidation number—is called the oxidizing or reducing agent.) Whether a reaction is commonly thought of as "oxidation" or "reduction," or as caused by the oxidizing agent or by the reducing agent, usually depends only on the aspect of the reaction in which we are most interested.

How can an atom, molecule, or ion be recognized as a *potential* oxidizing or reducing agent? A clue lies in the statement made above—oxidizing agents are reduced and reducing agents are oxidized. An oxidizing agent must contain an atom that can have a lower oxidation number (i.e., is in an oxidized state). Otherwise, no reduction is possible. Similarly, a reducing agent must contain an atom that can have a higher oxidation number (i.e., is in a reduced state). Ammonia is the reducing agent in the reaction with oxygen (Equation 16.9). Ammonia contains nitrogen in oxidation state -3 and nitrogen can have many different oxidation states more positive than -3. Elements like nitrogen that can have many different oxidation states are found in both oxidizing and reducing agents.

16.4 BALANCING REDOX EQUATIONS: OXIDATION NUMBER METHOD

The equations for many redox reactions are difficult to balance by inspection. Step-by-step methods must be used to balance such equations. The method discussed here is based on the total oxidation number changes in the reaction. (A second method, based on electron "gain" and "loss," is discussed in Section 24.2.)

In a redox reaction, the total increase in oxidation number for all oxidized atoms or ions must equal the total decrease in oxidation number for all reduced atoms or ions. This is reasonable, for oxidation numbers are assigned so that they generally indicate a greater or lesser share in "ownership" of valence electrons. If one atom gains in the ownership of electrons, then another atom must lose ownership to an equal extent.

The *total* increase in oxidation number in a redox reaction is the increase in oxidation number times the number of atoms that undergo the increase. In the reaction

$$2\overset{0}{H_2}(g) + \overset{0}{O_2}(g) \longrightarrow 2\overset{+1}{H_2}\overset{-2}{O}(l)$$

each H atom is oxidized from 0 to +1 and the total oxidation number increase is (4)(+1) = +4. Similarly, for the two oxygen atoms that are reduced, the total oxidation number decrease is (2)(−2) = −4.

To balance the equation for a redox reaction by the oxidation number method requires five steps. These steps are given below and are illustrated for the reaction in an acidic solution between dichromate ion and iron(II) ion to give chromium(III) ion and iron(III) ion.

Step A. Write the unbalanced equation for major reactants and products and then determine the oxidation number changes.

$$\overset{+6}{Cr_2}O_7^{2-} + \overset{+2}{Fe^{2+}} \longrightarrow \overset{+3}{Cr^{3+}} + \overset{+3}{Fe^{3+}}$$

To find the oxidation number changes write two partial "equations" and balance the *numbers of the atoms* oxidized and those reduced. Here, since there are two Cr atoms in $Cr_2O_7^{2-}$, there must also be two Cr^{3+} ions produced.

$$\overset{+6}{Cr_2}O_7^{2-} \longrightarrow 2\overset{+3}{Cr^{3+}} \qquad \textit{decrease in oxidation number is } (2)(-3) = -6$$

$$\overset{+2}{Fe^{2+}} \longrightarrow \overset{+3}{Fe^{3+}} \qquad \textit{increase in oxidation number is } (1)(+1) = +1$$

Step B. Add coefficients to balance change in oxidation number. The total increase in oxidation number must equal the total decrease in oxidation number. If necessary, multiply the species oxidized and reduced by factors that equalize the change in oxidation number. In this case, with a decrease of −6 (for Cr) and an increase of +1 (for Fe), both Fe^{2+} and Fe^{3+} must be multiplied by 6 to equalize the change.

$$\overbrace{Cr_2O_7^{2-} + 6Fe^{2+} \longrightarrow 2Cr^{3+}}^{(2)(-3) = -6} + 6Fe^{3+}$$

$$(6)(+1) = +6 \quad (Fe^{2+} \rightarrow Fe^{3+})$$

Step C. If necessary, balance the charges by adding H^+ for reactions in acidic solution and OH^- for reactions in alkaline solution. After Step B, the equation has a total charge of (−2) + (6)(+2) = +10 on the left and (2)(+3) + (6)(+3) = +24 on the right. To balance the charge for this reaction in acidic solution, 14 H^+ ions must be added on the left.

$$Cr_2O_7^{2-} + 6Fe^{2+} + 14H^+ \longrightarrow 2Cr^{3+} + 6Fe^{3+}$$

Note: In balancing the charge in a redox equation, and also in checking the charge balance, be sure that you use the *charges on ions.* Do not make the common error of confusing oxidation numbers and ionic charges.

Step D. Complete balancing by inspection, using water to balance oxygen for reactions in aqueous solution. Seven oxygen atoms are needed on the right in this case, so seven water molecules are added.

$$Cr_2O_7^{2-} + 6Fe^{2+} + 14H^+ \longrightarrow 2Cr^{3+} + 6Fe^{3+} + 7H_2O(l)$$

Inspection shows that the hydrogen atoms are now balanced and no other atoms need to be balanced.

Step E. Check to be sure that both atoms and charge are balanced. On each side of the equation there are 2 Cr, 7 O, 6 Fe, and 14 H atoms. On each side the total charge is +24. The equation is balanced.

As Examples 16.2 and 16.3 demonstrate, this method applies equally well to reactions in alkaline solution and to molecular rather than net ionic equations.

EXAMPLE 16.2 Balancing Redox Equations

Write the balanced chemical equation for the oxidation of iodate ion (IO_3^-) to periodiate ion (IO_4^-) by permanganate ion (MnO_4^-) in alkaline solution. The permanganate ion is reduced to manganese(IV) oxide (MnO_2).

Step A. *Write the unbalanced equation for major reactants and products, and then determine the oxidation number changes.*

$$\overset{+7}{Mn}O_4^- + \overset{+5}{I}O_3^- \longrightarrow \overset{+4}{Mn}O_2 + \overset{+7}{I}O_4^-$$

$$\overset{+7}{Mn}O_4^- \longrightarrow \overset{+4}{Mn}O_2 \qquad (1)(-3) = -3$$

$$\overset{+5}{I}O_3^- \longrightarrow \overset{+7}{I}O_4^- \qquad (1)(+2) = +2$$

Step B. *Add coefficients to balance change in oxidation number.*

$$(2)(-3) = -6$$

$$2MnO_4^- + 3IO_3^- \longrightarrow 2MnO_2 + 3IO_4^-$$

$$(3)(+2) = +6$$

Step C. *If necessary, balance the charges.* The total charge on the left is $(2)(-1) + (3)(-1) = -5$ and the total charge on the right is $(3)(-1) = -3$. Hydroxide ion must be added on the right to balance the charge, since this is a reaction in alkaline solution.

$$2MnO_4^- + 3IO_3^- \longrightarrow 2MnO_2 + 3IO_4^- + 2OH^-$$

Step D. *Complete balancing by inspection, using water to balance oxygen.* There are $(2)(4) + (3)(3) = 17$ oxygen atoms on the left and $(2)(2) + (3)(4) + (2)(1) = 18$ oxygen atoms on the right. One water molecule must be added on the left to balance oxygen.

$$\mathbf{2MnO_4^- + 3IO_3^- + H_2O(\mathit{l}) \longrightarrow 2MnO_2(\mathit{s}) + 3IO_4^- + 2OH^-}$$

Step E. *Check atom and charge balance.* On each side there are 2 Mn, 18 O, 3 I, and 2 H atoms. On each side the charge is -5. The equation is balanced.

EXERCISE Balance the following redox equation:

$$MnO_2(s) + PbO_2(s) \xrightarrow{H^+} MnO_4^- + Pb^{2+}$$

Answer $2MnO_2(s) + 3PbO_2(s) + 4H^+ \longrightarrow 2MnO_4^- + 3Pb^{2+} + 2H_2O(l)$.

EXAMPLE 16.3 Balancing Redox Equations

The products of the thermal decomposition of nickel(II) nitrate are nickel(II) oxide, nitrogen dioxide, and oxygen. Write a balanced equation for this reaction.

Step A. *Write the unbalanced equation for major reactants and products, and then determine the oxidation number changes.*

$$Ni(\overset{+5}{N}\overset{-2}{O}_3)_2(s) \xrightarrow{\Delta} NiO(s) + \overset{+4}{N}O_2(g) + \overset{0}{O}_2(g)$$

$$Ni(\overset{+5}{N}O_3)_2 \longrightarrow 2\overset{+4}{N}O_2 \qquad (2)(-1) = -2$$

$$Ni(N\overset{-2}{O}_3)_2 \longrightarrow 3\overset{0}{O}_2 \qquad (6)(+2) = +12$$

Step B. *Add coefficients to balance change in oxidation number.*

$$(6)(-2) = -12$$

$$6Ni(NO_3)_2(s) \xrightarrow{\Delta} NiO(s) + 12NO_2(g) + 3O_2(g)$$

$$(6)(+2) = +12$$

(Not all of the oxygen atoms are oxidized.)

Step C. *If necessary, balance the charges.* Not needed.

Step D. *Complete balancing by inspection.* In this case, water is not involved in balancing the equation, for the reaction does not take place in aqueous solution. It is apparent that 6 Ni atoms and 6 O atoms are needed on the right.

$$6Ni(NO_3)_2(s) \xrightarrow{\Delta} 6NiO(s) + 12NO_2(g) + 3O_2(g)$$

Step E. *Check atom and charge balance.* There are 6 Ni, 12 N, and 36 O atoms on each side and there are no ions present. The equation is balanced. Occasionally a stepwise method leads, as in this case, to an equation balanced with coefficients that have a common denominator. This equation would ordinarily be written as

$$\mathbf{2Ni(NO_3)_2(s) \xrightarrow{\Delta} 2NiO(s) + 4NO_2(g) + O_2(g)}$$

EXERCISE Sulfite ion (SO_3^{2-}) reacts with chromate ion (CrO_4^{2-}) in alkaline solution to give $[Cr(OH)_4]^-$ and SO_4^{2-}. Write the balanced chemical equation for this redox reaction.

Answer $3SO_3^{2-} + 2CrO_4^{2-} + 5H_2O(l) \longrightarrow 3SO_4^{2-} + 2[Cr(OH)_4]^- + 2OH^-$.

SUMMARY OF SECTIONS 16.2–16.4

In oxidation, oxidation numbers increase; in reduction, oxidation numbers decrease. Oxidation and reduction always occur together, and in all redox reactions the total oxidation number increase equals the total oxidation number decrease. Oxidizing and reducing agents cause other species to be oxidized or reduced, respectively; in the process, the oxidizing agent is reduced and the reducing agent is oxidized. The vocabulary of oxidation and reduction is summarized in Table 16.2.

The equations for many redox reactions are too complex to be balanced by inspection. A stepwise approach to balancing such equations, based upon oxidation number changes, is given in Section 16.4. In a balanced redox equation, both the number of atoms and the change in oxidation number must be balanced. If necessary, water is added to balance oxygen atoms for reactions in aqueous solution. If the equation is a net ionic equation, charge must also be balanced, if necessary by adding H^+ ions for reactions in acidic solution or OH^- ions for reactions in alkaline solution.

Table 16.2
Vocabulary of Oxidation and Reduction
Oxidizing agents are also sometimes referred to as *oxidants* and reducing agents as *reductants.*

In:	Oxidation number:
oxidation	increases
reduction	decreases
the oxidized substance	increases
the reduced substance	decreases
the oxidizing agent	decreases
the reducing agent	increases

HYDROGEN

16.5 PROPERTIES OF HYDROGEN

Molecular hydrogen (H_2) is a colorless, odorless gas. It can be liquefied and solidified only at temperatures close to absolute zero, showing that there are only weak intermolecular forces acting between the hydrogen molecules.

The molecular and atomic properties of hydrogen are summarized in Table 16.3. The atomic radius of hydrogen is about one-half that of oxygen. Because of the considerable energy needed to rupture the H—H bond, high temperatures and/or catalysts are required for many hydrogen reactions. Often, the hydrogen molecule is split into atoms as the reaction proceeds. Hydrogen *atoms* are highly reactive. (Note that oxygen has a bond energy similar to that of hydrogen and is of similar reactivity.)

In bonding, hydrogen atoms achieve the stable helium configuration either by sharing electrons in covalent bonds or by gaining electrons to form hydride ions (H^-). The hydrogen atom shares electrons easily and forms many covalent compounds. It gains electrons less readily, but does so in combination with metals that easily lose electrons. Hydrogen has a +1 oxidation state in all of its compounds, except for the ionic hydrides (e.g., Na^+H^-), where it is assigned an oxidation state of −1.

Table 16.3
Properties of Hydrogen

		For comparison	
Molecular properties			
Melting point (°C)	−260	He	−272
Boiling point (°C)	−253	He	−269
Density (g/L, at STP)	0.090	He	0.18
Bond length (nm)	0.074	Cl_2	0.198
		N_2	0.110
Bond dissociation energy (kJ/mol)	436	O_2	498
		N_2	946
Atomic properties			
Ionization energy (kJ/mol)	1312	Na	496
Electron affinity (kJ/mol)	−73	F	−322
Electronegativity	2.1	O	3.5
Atomic radius (nm)	0.037	O	0.074

Hydrogen forms compounds with almost all of the other elements and these compounds range over the entire continuum from ionic to covalent. Molecules in which hydrogen is bonded to small, highly electronegative atoms (e.g., F, O, N) exhibit hydrogen bonding (Section 11.13).

The high flammability of hydrogen requires that it be handled with care. Hydrogen burns with a very hot, nearly invisible flame in mixtures with air of from 4% to 74% H_2 by volume. Air containing from 18% to 60% of hydrogen by volume is potentially explosive and can be ignited by a spark or by static electricity. However, hydrogen, like many other hazardous materials (for example, natural gas), is routinely handled and transported with safety. It is shipped by rail as a liquid held at low temperatures in tank cars equipped with specially designed insulation.

The density of gaseous hydrogen is low, about $\frac{1}{14}$ that of air, and it provides excellent buoyancy to balloons. In 1937, a hydrogen-filled airship, the Hindenburg, on its twenty-second Atlantic crossing, caught fire after passing through an electrical storm and was destroyed with the loss of many lives. This disaster permanently impressed upon the public that hydrogen is a dangerous material and led to the use of helium instead of hydrogen in airships.

16.6 REACTIONS OF HYDROGEN

Hydrogen is a reducing agent in all the reactions described below except those with active metals.

a. Reduction of Oxides High temperatures are generally required for the reduction of metals oxides to free metals by reaction with hydrogen (see Equation 16.8). Hydrogen is not, however, a sufficiently strong reducing agent to reduce the oxides of the most electropositive metals (those at the top of the activity series, Table 5.2).

Hydrogen can also reduce certain nonmetal oxides. Note that the sulfur in SO_2 is reduced to free sulfur in the presence of a limited amount of hydrogen, but to a lower oxidation state if hydrogen is in excess.

$$\overset{+4}{S}O_2(g) + 2H_2(g) \longrightarrow \overset{0}{S}(s) + 2H_2O(g)$$

$$\overset{+4}{S}O_2(g) + 3H_2(g) \longrightarrow H_2\overset{-2}{S}(g) + 2H_2O(g)$$

$$2\overset{+2}{N}O(g) + 2H_2(g) \longrightarrow \overset{0}{N_2}(g) + 2H_2O(g)$$

b. Combination of Hydrogen with Oxygen Like many reactions, the combination of hydrogen and oxygen, although it has a favorable, negative $\Delta H°$

$$2H_2(g) + O_2(g) \longrightarrow 2H_2O(g) \qquad \Delta H° = -484 \text{ kJ}$$

proceeds at an insignificant rate at ordinary temperatures and pressures. Left undisturbed, a mixture of hydrogen and oxygen will stand without reacting. However, once the reaction is initiated, the combination occurs with explosive violence. The primary engines in the U.S. space shuttle are fueled by this reaction. The flame temperature of hydrogen burning in air is about 2000 °C, and temperatures up to 2800 °C are achieved in the oxyhydrogen torch, which is used in cutting metals. Even higher temperatures (up to 5000 °C) are achieved if hydrogen molecules are split into hydrogen atoms in an electrical arc before they are burned. The higher torch temperature is possible because none of the energy from the combination of hydrogen with oxygen in the flame is taken up by breaking the hydrogen–hydrogen bonds.

c. Combination of Hydrogen with Nonmetals Other than Oxygen Hydrogen combines directly with all of the halogens:

$$\begin{array}{llll} H_2(g) + F_2(g) & \longrightarrow 2HF(g) & & \Delta H^\circ = -542\ \text{kJ} \\ H_2(g) + Cl_2(g) & \longrightarrow 2HCl(g) & & \Delta H^\circ = -185\ \text{kJ} \\ H_2(g) + Br_2(l) & \longrightarrow 2HBr(g) & & \Delta H^\circ = -73\ \text{kJ} \\ H_2(g) + I_2(s) & \longrightarrow 2HI(g) & & \Delta H^\circ = +53\ \text{kJ} \end{array}$$

(Note how ΔH° varies with the size and mass of the halogen atoms.) The reaction with fluorine is rapid, and it was once considered useful in rocket propulsion. Hydrogen will burn in chlorine, and an explosive reaction of hydrogen and chlorine is initiated by light. (Hydrogen–chlorine mixtures should be kept in the dark or under red light, since they are very sensitive to short-wavelength radiation and are thus extremely dangerous to handle.)

Ammonia is produced by the combination of nitrogen with hydrogen at elevated temperature and pressure, in the presence of a catalyst:

$$N_2(g) + 3H_2(g) \xrightleftharpoons[\text{pressure}]{\Delta,\ \text{catalyst}} \underset{\textit{ammonia}}{2NH_3(g)}$$

Also at elevated temperature, hydrogen unites with sulfur vapor to form hydrogen sulfide. The reaction is much less exothermic than that with oxygen, which precedes sulfur in the same periodic table family

$$H_2(g) + S(g) \xrightarrow{\Delta} H_2S(g) \qquad \Delta H^\circ = -20.6\ \text{kJ}$$

d. Combination of Hydrogen with Active Metals The active metals of the sodium family and also calcium, barium, and strontium are such strong reducing agents that in reacting with molecular hydrogen, they cause the hydrogen itself to be reduced to hydride ion (H^-) in reactions such as

$$\overset{0}{2Li}(s) + \overset{0}{H_2}(g) \longrightarrow \underset{\textit{lithium hydride}}{2\overset{+1}{Li}\overset{-1}{H}(s)}$$

$$\overset{0}{Ba}(s) + \overset{0}{H_2}(g) \xrightarrow{\Delta} \underset{\textit{barium hydride}}{\overset{+2}{Ba}\overset{-1}{H_2}(s)}$$

(Contrary to the situation in most reactions, hydrogen is the "oxidizing agent" in combination reactions with active metals.)

e. Interaction of Hydrogen with Transition Metals The interaction of hydrogen with transition metals has several different outcomes. A number of the transition metal hydrides, particularly those of *d* transition metals, are nonstoichiometric compounds. For example, zirconium reacts with hydrogen to form compounds with two different crystal structures, one with the composition range $ZrH_{1.73}$–$ZrH_{2.00}$ and the other with the composition range $ZrH_{1.50}$–$ZrH_{2.00}$.

The hydrides of the 4*f*-transition metals and some 5*f*-transition metals, referred to as "metallic hydrides," have considerable ionic character, but also resemble metals in their appearance and electrical conductivity. Such hydrides are less dense than the parent metals and tend to be brittle. The formation of brittle metal hydrides contributes to the deterioration of metal pipes and vessels that are exposed to hydrogen.

Hydrogen is absorbed by many metals. **Absorption** is the incorporation of one substance into another at the molecular level. Palladium, amazingly, can absorb 800 to 900 times its own volume of hydrogen. The hydrogen enters the crystal structure of the metal to form what is essentially a solution. Apparently, H_2 dissociates into atoms at the metal surface. The atoms then diffuse into the metal and occupy interstitial or vacant lattice positions. One method of purifying hydrogen consists of putting it into a thin-walled palladium vessel under slight pressure. The hydrogen dissolves in the metal and then escapes on the other side of the wall, where the pressure is lower. Other gases are not dissolved by palladium and remain behind.

f. Hydrogenation **Hydrogenation** is the addition of hydrogen atoms to molecular compounds. Hydrogenation reactions generally require catalysts and are usually carried out under pressure. Often the purpose of hydrogenation is to add hydrogen to the carbon–carbon multiple bond in an organic compound or a mixture of organic compounds. A simple example is the hydrogenation of ethylene.

$$\underset{\textit{ethylene}}{H_2C{=}CH_2(g)} + H_2(g) \xrightarrow[\text{pressure}]{\text{Pt or Ni}} \underset{\textit{ethane}}{H_3C{-}CH_3(g)}$$

Vegetable oils, such as soybean, cottonseed, and coconut oils, are hydrogenated to produce solid fats for shortening and margarine. (When organic chemists speak of a "reduction reaction," they are frequently referring to the addition of hydrogen to an organic molecule.)

16.7 HYDRIDES OF THE REPRESENTATIVE ELEMENTS

The formulas of the representative element hydrides show the numbers of hydrogen atoms to be expected based on electron configurations (Table 16.4).

The hydrides of the Group I and II metals are highly reactive ionic compounds that are strong reducing agents and that form strong bases in aqueous solution, for example

$$NaH(s) + H_2O(l) \longrightarrow NaOH(aq) + H_2(g) \qquad \textbf{(16.10)}$$

Table 16.4
Examples of Hydrides of Representative Elements[a]
Electrons of the H^- ion or contributed to covalent bonding by H atoms are shown in color to illustrate how the occurrence of bonding is pictured.

Group I	Group II	Group III[b]	Group IV	Group V	Group VI	Group VII
$Li^+H:^-$	$Ca^{2+}(H:^-)_2$	H H:Al:H	H H:Sn:H H	H :P:H H	H:Se:H	H:F:
lithium hydride (LiH)	*calcium hydride (CaH_2)*	*aluminum hydride (AlH_3)*	*stannane (SnH_4)*	*phosphine (PH_3)*	*hydrogen selenide (H_2Se)*	*hydrogen fluoride (HF)*

[a] The Group I–III compounds, and also SbH_3, BiH_3, SeH_2, and TeH_2 are named as hydrides, although SeH_2 and TeH_2 are also called hydrogen selenide and hydrogen telluride. The hydrides of elements in other groups are named as follows: Group IV, CH_4, methane; SiH_4, silane; GeH_4, germane; SnH_4, stannane; PbH_4, plumbane. Group V, NH_3, ammonia; PH_3, phosphine; AsH_3, arsine. Group VI, H_2O, water; H_2S, hydrogen sulfide. Group VII, the hydrogen halides, e.g., HF, hydrogen fluoride.

[b] Aluminum hydride is polymeric.

The reaction of calcium hydride (CaH_2) with water is utilized both for removing the last traces of water from organic liquids and to generate hydrogen for filling weather balloons in the field. The ionic hydrides decompose upon heating, in some cases even at room temperature, to give hydrogen and the metal, for example

$$2NaH(s) \xrightarrow{\Delta} 2Na(s) + H_2(g) \qquad \textbf{(16.11)}$$

The hydrides of the Group III–VII elements are all molecular compounds, most of which are gases at room temperature. Some are quite stable thermally and are chemically unreactive (e.g., methane, CH_4). Others are very reactive (e.g., silane, SiH_4, and germane, GeH_4, are spontaneously flammable in air). Some hydride molecules (e.g., those of Be and Al) are so strongly associated with each other by bonding through the hydrogen atoms that the simplest molecular formulas do not correctly represent the molecular structures. The many hydrides formed by boron (Section 29.8) have structures with unique types of bonding.

Methane (CH_4) is the simplest compound of hydrogen and carbon. It is the first in the vast series of **hydrocarbons**—compounds of carbon and hydrogen—in which increasing numbers of carbon atoms are joined to each other in various ways and are bonded only to other carbon atoms or to hydrogen atoms. Many of the compounds present in or derived from the **fossil fuels**, which are natural gas, coal, and petroleum, are hydrocarbons.

In addition to the binary hydrides, several of the representative elements form more complex hydrides. Sodium borohydride ($NaBH_4$) and lithium aluminum hydride ($LiAlH_4$), for example, are well-known compounds widely used as reducing agents in organic chemistry. These are ionic compounds containing BH_4^- and AlH_4^- ions, respectively.

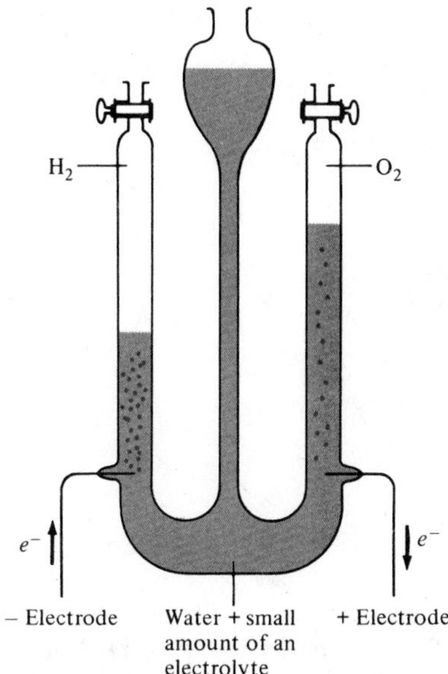

Figure 16.1
Electrolysis of Water
In electrolysis (Section 24.9), electrical energy is used to cause a redox reaction to occur. Hydrogen is produced at one electrode and oxygen at the other. The volume of H_2 produced is twice that of O_2. The volumes of each gas collected depend upon the relative solubilities of H_2 and O_2 at the temperature of the experiment.

16.8 PREPARATION AND USES OF HYDROGEN

Hydrogen is produced industrially from hydrocarbons. In the presence of various combinations of catalysts and elevated temperatures, hydrocarbons react with steam and/or oxygen to form hydrogen plus carbon dioxide and/or carbon monoxide (Section 29.4). For example, in the industrial process called *steam reforming,* gaseous or vaporized hydrocarbons react with steam over a bed of nickel catalyst at temperatures of 600° to 1000 °C and pressures of 8–50 atm. The product is a mixture of hydrogen, carbon monoxide, and carbon dioxide. At the present time, the steam reforming of natural gas, which is mostly methane, is the major industrial source of hydrogen.

$$CH_4(g) + H_2O(g) \xrightarrow[\text{pressure}]{\Delta,\ \text{catalyst}} CO(g) + 3H_2(g) \qquad \textbf{(16.12)}$$

$$CH_4(g) + 2H_2O(g) \xrightarrow[\text{pressure}]{\Delta,\ \text{catalyst}} CO_2(g) + 4H_2(g) \qquad \textbf{(16.13)}$$

The partial oxidation of hydrocarbons from petroleum also yields mixtures of hydrogen and the oxides of carbon. An older method for producing hydrogen was by the action of steam on coke.

$$C(s) + H_2O(g) \xrightarrow{1000\ ^\circ C} H_2(g) + CO(g) \qquad \textbf{(16.14)}$$

Hydrogen is also produced in large quantities as a by-product of the manufacture of gasoline.

Pure hydrogen can be obtained by the electrolysis of water, as shown in Figure 16.1. Because of the large amounts of electricity required, electrolysis is an expensive preparative method.

Table 16.5 summarizes the uses of hydrogen. Most commercially produced hydrogen is used within the chemical industry. Well over 100 billion cubic feet of hydrogen is made each year (not including hydrogen used directly where it is produced). The single largest use of hydrogen is in the manufacture of ammonia, which

Table 16.5
Uses of Hydrogen

Major
Ammonia manufacture
Hydrogenation
of petroleum products
of vegetable oils
of CO to give CH_3OH (methanol)
of CO and CO_2 to give hydrocarbons
Manufacture of plastics, pesticides, and industrial chemicals
Others
Fuel (including rocket fuel)
Reduction of oxide-containing ores to give free metals
Welding

requires a 1:3 mixture (by volume) of nitrogen and hydrogen. Large amounts are also used in methanol manufacture, which requires a 1:2 mixture (by volume) of carbon monoxide and hydrogen (Section 32.5b). These mixtures are obtained directly from the steam reforming of hydrocarbons (air is added as the source of nitrogen).

SUMMARY OF SECTIONS 16.5–16.8

Hydrogen, the lightest element, is unique in its properties because of its small atomic size and the high charge density of the hydrogen ion. Hydrogen is a reducing agent and an important industrial chemical; it is produced mainly by the steam reforming of methane (Equations 16.12 and 16.13). The properties of hydrogen and some of its compounds are summarized in Table 16.6.

Table 16.6
Outstanding Properties of Hydrogen and Some of Its Compounds

H 1*s*	Lightest atom; one proton, one electron
H—H (*strong bond*)	Most reactions of H_2 require high temperature and/or catalysts
$\overset{+1}{H}$—	Oxidation state +1 in all compounds except ionic hydrides
$\overset{0}{H_2} \longrightarrow \overset{+1}{H}$	Reducing agent
H_3O^+ (*hydronium ion*)	H^+ hydrated in aqueous solution
$M^+H^-, M^{2+}(H^-)_2$	Ionic hydrides
$H^- + H_2O(l) \longrightarrow OH^- + H_2(g)$ (*hydride ion*)	H^- is a strong base and a very strong reducing agent
EH, EH_2, EH_3, EH_4	Molecular hydrides formed with representative elements of Groups III–VII
M_aH_b (*transition metal*)	*a, b* not always whole numbers and may vary for same hydride
Combination with other elements	
$2H_2(g) + O_2(g) \longrightarrow 2H_2O(l)$	Can occur with explosive force; used in oxyhydrogen torch
$H_2(g) + X_2(g) \longrightarrow 2HX(g)$	$X_2 = F_2, Cl_2, Br_2, I_2$; reactivity decreases $F_2 \rightarrow I_2$
$3H_2(g) + N_2(g) \rightleftharpoons 2NH_3(g)$	An important industrial process; catalyst, high pressure, and moderately high temperature needed
$H_2(g) + M(s) \longrightarrow M^+H^-, M^{2+}(H^-)$	M = Li, Na, K, Rb, Cs; Ca, Sr, Ba
Hydrogenation	
$H_2C{=}CH_2(g) + H_2(g) \xrightarrow[\text{pressure}]{\Delta,\ \text{catalysts}} H_3C{-}CH_3$ (*ethylene* → *ethane*)	With many compounds containing C=C bonds. Preparation of saturated fats

Aside 16.1 THE HYDROGEN ECONOMY

An economy based upon the chemical energy stored in hydrogen is one possible alternative to our present fossil fuel-based economy. Only about 10% of the energy needs in the United States are met directly by electricity. For the remainder, we require a fuel that can be transported, stored, and consumed at the site where energy is needed. Hydrogen might fill this role and might also replace electricity in some uses.

It is proposed that energy from nuclear power plants or conventional power plants be used to convert water to hydrogen during periods when demand for electricity is low. Solar energy might also be used. The hydrogen could be transported directly to homes and factories in pipelines as is done with natural gas. In processes that regenerate water, the energy stored in the hydrogen would be utilized to produce electricity in fuel cells (Section 24.17).

$$2H_2(g) + O_2(g, \text{from air}) \xrightarrow{\text{fuel cell}} 2H_2O(l) + \text{electrical energy}$$

or it could be burned.

$$2H_2(g) + O_2(g, \text{from air}) \longrightarrow 2H_2O(l) + \text{thermal energy}$$

Hydrogen-burning automobile engines are also under development. The hydrogen supply might be carried as a liquid or stored in a metal alloy. As noted earlier, many transition metals absorb hydrogen. This hydrogen can be released as needed by heating the alloy. One hydrogen storage system utilizes iron–titanium hydride, $FeTiH_{1-2}$.

The hydrogen for a hydrogen-based economy might be produced by improved methods of electrolysis. Another possibility is the use of thermal energy in a thermochemical cycle—a series of chemical reactions in which only heat and water are consumed and only hydrogen and oxygen are produced. All reactants other than hydrogen and oxygen are recycled. Two possible thermochemical cycles are given in Table 16.7. (Check for yourself by canceling like reactants and products that the overall reaction in each case is $H_2O \quad H_2 + \frac{1}{2}O_2$.)

Table 16.7
Two Proposed Thermochemical Cycles for Hydrogen Production

(1) $CaBr_2 + 2H_2O \xrightarrow{730\,°C} Ca(OH)_2 + 2HBr$

(2) $2HBr + Hg \xrightarrow{250\,°C} HgBr_2 + H_2$

(3) $HgBr_2 + Ca(OH)_2 \xrightarrow{100\,°C} CaBr_2 + H_2O + HgO$

(4) $HgO \xrightarrow{600\,°C} Hg + \frac{1}{2}O_2$

(1) $2HI \xrightarrow{425\,°C} I_2 + H_2$

(2) $2H_2O + SO_2 + I_2 \xrightarrow{90\,°C} H_2SO_4 + 2HI$

(3) $H_2SO_4 \xrightarrow{850\,°C} SO_2 + H_2O + \frac{1}{2}O_2$

A hydrogen economy would have numerous advantages. Hydrogen can be continuously generated, stored, and called upon when needed. Water, the source of the fuel, is widely distributed—nations would not have to fight over it. Moreover, water is regenerated—a homeowner could collect more than 10 L of pure water as the by-product of an average day's production of electricity by fuel cells (Section 24.17). Hydrogen combustion and the use of fuel cells produce far less pollution than burning fossil fuels. And the hydrogen would also be available for use in the chemical industry. How, when and whether or not a hydrogen economy becomes operative depends primarily upon the costs of the processes involved relative to the cost of using other possible fuels.

OXYGEN

16.9 PROPERTIES OF OXYGEN

Molecular oxygen (O_2) is a colorless, odorless, tasteless gas that is slightly more dense than air. Its properties are given in Table 16.8. Liquid oxygen, which boils at −183 °C, is pale blue. Oxygen is essential to most life on Earth. Although it is not very soluble in water—at 20 °C and 1 atm a saturated solution is about 0.01 M—this slight solubility makes possible both aquatic life and the destruction of organic waste in bodies of water (Section 14.14).

Table 16.8
Properties of Oxygen

		For comparison	
Molecular properties			
Melting point (°C)	−219	H_2	−260
Boiling point (°C)	−183	H_2	−253
Density (g/L)	1.43	Air	1.29
Bond length (nm)	0.12	H_2	0.0742
Bond dissociation energy (kJ/mol)	498	H_2	436
		N_2	946
Atomic properties			
Ionization energy (kJ/mol)	1314	Na	496
Electron affinity (kJ/mol)	−141	F	−322
Electronegativity	3.5	F	4.0
Atomic radius (nm)	0.074	F	0.071
Ionic radius (O^{2-}) (nm)	0.140	F^-	0.136

The oxygen–oxygen bond in O_2 is comparable in strength to the hydrogen–hydrogen bond in H_2 (see Table 16.8). Some reactions of oxygen, notably its reaction with iron in the presence of water (rusting), are spontaneous, but slow at ordinary temperatures. As for hydrogen, elevated temperatures and pressures are necessary to carry out effectively most reactions of oxygen. The oxygen molecule, despite its even number of electrons (16), has two unpaired electrons and is thus paramagnetic (this exception to the usual pairing of electrons is discussed in Section 25.7).

Oxygen forms compounds with all of the elements except (so far as is known) helium, neon, argon, and krypton. In bonding, the oxygen atom readily shares electrons with nonmetal atoms and with metal atoms that do not lose electrons easily. Oxygen forms single, double, and triple covalent bonds. With metals that lose electrons easily, oxygen gains electrons to form ionic compounds. The following compounds illustrate different types of bonding by oxygen.

$Ca^{2+}\,:\ddot{\underset{..}{O}}:^{2-}$	$H—\ddot{\underset{..}{O}}—H$	$H—\ddot{\underset{..}{O}}—Cl(—\ddot{O}:)_3$	$CH_3C(H)=\ddot{O}$	$:C\equiv O:$
calcium oxide	*water*	*perchloric acid*	*acetaldehyde*	*carbon monoxide*

The oxidation state of oxygen in such compounds is -2.

Oxygen is second only to fluorine in electronegativity and therefore is assigned a positive oxidation number only in its compounds with fluorine (e.g., OF_2, in which fluorine has oxidation number -1 and oxygen therefore has $+2$). Oxygen has an oxidation number other than -2 in the small number of compounds in which two or three oxygen atoms are bonded to each other (peroxides, e.g., H_2O_2; superoxides, e.g., KO_2; and ozonides, e.g., KO_3; see Sections 16.10a, 16.13, and 16.14).

We have noted that the first element in a representative element family is frequently quite different in its properties from the other family members. Atoms of these first elements are lighter and smaller, and they do not have *d* orbitals available for octet expansion. Oxygen is followed in Group VI by the nonmetal sulfur, the two semiconducting elements selenium and tellurium, and the radioactive metal, polonium. The significant differences between oxygen and these elements are summarized in Table 16.9, which we urge you to study thoroughly, for it summarizes some very interesting chemistry.

Table 16.9
Differences between the Chemistry of Oxygen and That of Other Members of the Oxygen Family (S, Se, Te)

Property	Comments
The only element of the group that is a gas at room temperature.	The free element is a diatomic molecule, O_2, at room temperature. The common forms of sulfur and selenium are octa-atomic molecules, S_8 and Se_8, whereas tellurium has metallic bonding.
Exhibits positive oxidation states only when bound to fluorine.	Positive oxidation states of $+4$, and $+6$, e.g., SO_2, H_2SO_4, SF_6, SeO_2, H_2SeO_4, H_6TeO_6, are common for other elements in the group.
Next to fluorine, oxygen has the highest electronegativity of any element.	The high electronegativity accounts for the hydrogen bonding of H_2O in the liquid and solid states. The analogous hydrides of S, Se, and Te are not hydrogen-bonded.
Formation of compounds with more than two covalent bonds to O is rare.	Because O has six valence electrons and only *s* and *p* orbitals available for bonding, it generally forms two covalent bonds. (One exception is H_3O^+.) Compounds with four and six covalent bonds to S, Se, and Te are common.
The oxide ion, O^{2-}, and covalent species where oxygen exhibits a -2 oxidation state, e.g., H_2O, OH^-, are oxidized only with difficulty.	A reflection of the high electronegativity of oxygen. S, Se, and Te in the -2 oxidation state are easily oxidized (good reducing agents).
Forms no chains containing more than three oxygen atoms, e.g., O_3 (ozone), O_3^- (ozonide ion).	Salts containing polysulfide (e.g., S_6^{2-}) and polyselenide (Se_5^{2-}) chains are readily prepared. Polytellurides are also known, but are not very stable. Relative ability to form chains is consistent with element–element bond energies: O—O, 138; S—S, 264; Se—Se, 193; Te—Te, 138 kJ/mol.

16.10 REACTIONS OF OXYGEN

All of the reactions discussed in the following sections are redox reactions in which molecular oxygen is the oxidizing agent and is reduced.

a. Combination of Oxygen with Lithium Family Elements The reaction of oxygen with lithium gives lithium oxide, which contains the oxide ion, O^{2-}. Reaction with sodium yields a salt of the less common peroxide ion, O_2^{2-}, and the larger and heavier family elements, potassium, rubidium, and cesium give salts that contain the still less common superoxide ion, O_2^-.

$$4Li(s) + O_2(g) \longrightarrow \underset{\text{lithium oxide}}{2Li_2O(s)}$$

$$2Na(s) + O_2(g) \longrightarrow \underset{\text{sodium peroxide}}{Na_2O_2(s)}$$

$$\underset{\text{(also Rb, Cs)}}{K(s)} + O_2(g) \longrightarrow \underset{\text{potassium superoxide}}{KO_2(s)}$$

b. Combination of Oxygen with Other than Lithium Family Elements Oxygen combines directly with all of the elements except the noble gases, the halogens, and some of the less active metals such as silver, gold, and platinum. Except in the reactions with the lithium family elements described above, the products always contain oxygen in the -2 oxidation state.

The compounds formed with nonmetals contain covalent oxygen–nonmetal bonds. The compounds of oxygen with metals may be anywhere on the continuum from ionic to covalent, depending upon the metal and its oxidation state.

Combination with oxygen of elements that form two oxides can yield either oxide, depending upon the amount of oxygen available (and other reaction conditions). For example, carbon monoxide forms when the oxygen supply is limited but carbon dioxide forms when more oxygen is available.

$$C(s) + \tfrac{1}{2}O_2(g) \longrightarrow CO(g)$$

$$C(s) + O_2(g) \longrightarrow CO_2(g)$$

c. Oxidation of Compounds Exposure to heat and an ample supply of oxygen can be used to oxidize the lower oxides of elements that can exhibit more than one oxidation state. For example

$$\underset{\text{copper(I) oxide}}{2Cu_2O(s)} + O_2(g) \xrightarrow{\Delta} \underset{\text{copper(II) oxide}}{4CuO(s)}$$

$$\underset{\text{sulfur dioxide}}{2SO_2(g)} + O_2(g) \underset{\text{catalyst}}{\overset{\Delta}{\rightleftharpoons}} \underset{\text{sulfur trioxide}}{2SO_3(g)}$$

The combustion of fossil fuels is the reaction of hydrocarbons with oxygen to produce water and carbon dioxide (Section 7.11). For example, methane, the major component of natural gas, yields almost 900 kJ of energy for each mole that is burned.

$$CH_4(g) + 2O_2(g) \longrightarrow CO_2(g) + 2H_2O(l) \qquad \Delta H° = -890.\ \text{kJ}$$

The reactions of oxygen with ammonia and carbon disulfide are further examples of reactions in which, as in combustion of hydrocarbons, each of the elements of the compounds combines with oxygen.

$$4NH_3(g) + 5O_2(g) \xrightarrow{\text{catalyst}} 4NO(g) + 6H_2O(g)$$

$$CS_2(g) + 3O_2(g) \longrightarrow CO_2(g) + 2SO_2(g)$$

The human body uses about 500 L of oxygen each day to "burn" fats and carbohydrates with the production of energy, CO_2, and H_2O. Many materials that will not burn, such as glass, ceramics, clays, and water, contain elements that are already fully combined with oxygen and are at their highest oxidation states.

16.11 OXIDES AND HYDROXIDES

With elements to the left in the periodic table, oxygen forms ionic oxides. With elements to the right, the oxides are molecular compounds. With elements between these extremes, the bonding in the oxides is intermediate between ionic and covalent. Many oxides are classified as either "acidic" or "basic" by the products formed in their reactions with water.

a. Nonmetal Oxides: Acidic Oxides Oxygen forms covalent bonds with the nonmetals, as shown by the low melting and boiling points of nonmetal oxides, many of which are gases. Most nonmetal oxides combine with excess water in nonredox, hydrolysis reactions to give oxygen-containing acids, for example

$$SO_2(g) + H_2O(l) \rightleftharpoons \underset{\textit{sulfurous acid}}{H_2SO_3(aq)}$$

$$SO_3(g) + H_2O(l) \longrightarrow \underset{\textit{sulfuric acid}}{H_2SO_4(aq)}$$

$$P_4O_6(s) + 6H_2O(l) \longrightarrow \underset{\textit{phosphorous acid}}{4H_3PO_3(aq)}$$

A nonmetal oxide that combines with water to give an acid is known as an **acidic anhydride** (anhydride meaning "without water"), or an **acidic oxide.** Acidic oxides react with aqueous bases to give salts plus water, for example

$$CO_2(g) + 2NaOH(aq) \longrightarrow Na_2CO_3(aq) + H_2O(l)$$
$$SiO_2(s) + 4NaOH(aq) \longrightarrow Na_4SiO_4(aq) + 2H_2O(l)$$

b. Metal and Semiconducting Element Oxides: Basic and Amphoteric Oxides Oxygen forms ionic compounds with the Group I and II metals that have large atomic radii and that readily lose electrons. These ionic crystalline compounds react with water to yield metal hydroxides either in solution or as solids, depending upon the solubilities of the hydroxides, for example

$$K_2O(s) + H_2O(l) \longrightarrow 2KOH(aq)$$
$$MgO(s) + H_2O(l) \longrightarrow Mg(OH)_2(s)$$

A metal oxide that yields a hydroxide base with water is known as a **basic anhydride,** or a **basic oxide.** Basic oxides, including many that are only slightly soluble in water, react with aqueous acids to form cations and water, for example

$$Na_2O(s) + 2H_3O^+ \longrightarrow 2Na^+ + 3H_2O(l)$$
$$MnO(s) + 2H_3O^+ \longrightarrow Mn^{2+} + 3H_2O(l)$$
$$Fe_2O_3(s) + 6H_3O^+ \longrightarrow 2Fe^{3+} + 9H_2O(l)$$

Not all oxides of metals or semiconducting elements are ionic, basic oxides. Whether a particular oxide might be ionic, covalent, or intermediate in its bonding is generally indicated by the charge-to-size ratio of the cation that would be present if the compound were ionic. Consider, for example, the elements sodium, magnesium, aluminum, and silicon.

	Valence electrons	Ion formed by loss of valence electrons	Atomic radius (nm)	Ionic radius (nm)	Charge-to-size ratio of ion
Na	$3s^1$	Na^+	0.186	0.097	10.
Mg	$3s^2$	Mg^{2+}	0.160	0.066	30.
Al	$3s^23p^1$	Al^{3+}	0.143	0.051	59
Si	$3s^23p^2$	$[Si^{4+}]$	0.118	0.026	154

+4 ion not likely to form

Recall that the larger the charge-to-size ratio (Section 10.8) the greater the polarizing ability of the cation and the greater the covalent character of the bond. On the basis of the information given above, silicon(IV) oxide (SiO_2) must surely have covalent bonding and thus must be an acidic oxide (which it is). The large charge-to-size ratio indicates that aluminum oxide (Al_2O_3) should have a greater degree of covalent character in its bonding than either sodium oxide (Na_2O) or magnesium oxide (MgO). This prediction is borne out by the chemical properties of the oxides. Both Na_2O and MgO are basic oxides, showing that they are ionic compounds. On the other hand, Al_2O_3 exhibits basic properties in the presence of strong acids and acidic properties in the presence of strong hydroxide bases

$$Al_2O_3(s) + 6H_3O^+ \longrightarrow 2Al^{3+} + 9H_2O(l) \quad \textbf{(16.15)}$$

$$Al_2O_3(s) + 2OH^- + 3H_2O(l) \longrightarrow 2[Al(OH)_4]^- \quad \textbf{(16.16)}$$

Substances like aluminum oxide that can act as either acids or bases are described as **amphoteric.** The common amphoteric oxides are listed in Table 16.10. [Formulation of the aluminate ion as a complex ion, $[Al(OH)_4]^-$, best represents its properties. However, this anion is also sometimes written AlO_2^-, which is related to $[Al(OH)_4]^-$ by the loss of two molecules of water. As further explained in the next section, this type of dehydration formula can be written for the anion formed by any of the amphoteric oxides.]

The covalent character of oxides of the same metal increases with the oxidation state of the metal, as shown by an increase in the acidic nature of the oxide. For example, in each of the following pairs of oxides, the second oxide is more acidic than the first: FeO, Fe_2O_3; MnO, MnO_2; UO_2, UO_3. The metal atoms in their higher oxidation states are assumed to be more highly polarizing, explaining the more covalent nature of the bonding.

Most metal oxides are quite stable and are unaffected by high temperatures. The few that do decompose upon heating are those of the least active metals—mercury, silver, gold, and platinum (see Table 5.2), for example

$$2HgO(s) \xrightarrow{\Delta} 2Hg(l) + O_2(g)$$

Some oxides of metals in their less stable, higher oxidation states decompose with heat to give oxygen and an oxide of the metal in a lower oxidation state

$$2PbO_2(s) \xrightarrow{>500\,°C} 2PbO(s) + O_2(g)$$

c. Metal Hydroxides In writing chemical equations, the precipitates produced when solutions containing metal ions are treated with hydroxide ions are usually formulated as hydroxides; for example

$$Al^{3+} + 3OH^- \longrightarrow Al(OH)_3(s)$$

Many metal ions yield precipitates with hydroxide ions, and the compounds formed are referred to as insoluble hydroxides. The precipitates—for example, the substances

Table 16.10
Amphoteric Oxides and Hydroxides

Compound	Reaction with H^+ produces	Reaction with OH^- produces[a]
BeO or $Be(OH)_2$	Be^{2+}	$[Be(OH)_4]^{2-}$ or BeO_2^{2-}
Sb_2O_3 or $Sb(OH)_3$	Sb^{3+} or SbO^+	$[Sb(OH)_4]^-$ or SbO_2^-
SnO or $Sn(OH)_2$	Sn^{2+}	$[Sn(OH)_4]^{2-}$ or SnO_2^{2-}
PbO or $Pb(OH)_2$	Pb^{2+}	$[Pb(OH)_4]^{2-}$ or PbO_2^{2-}
SnO_2 or $Sn(OH)_4$	Sn^{4+}	$[Sn(OH)_6]^{2-}$ or SnO_3^{2-}
Al_2O_3 or $Al(OH)_3$	Al^{3+}	$[Al(OH)_4]^-$ or AlO_2^-
Cr_2O_3 or $Cr(OH)_3$	Cr^{3+}	$[Cr(OH)_4]^-$ or CrO_2^-
ZnO or $Zn(OH)_2$	Zn^{2+}	$[Zn(OH)_4]^{2-}$ or ZnO_2^{2-}
Ga_2O_3 or $Ga(OH)_3$	Ga^{3+}	$[Ga(OH)_4]^-$ or GaO_2^-
TiO_2 or $Ti(OH)_4$	TiO^{2+}	$[Ti(OH)_6]^{2-}$ or TiO_3^{2-}

[a] Written as hydroxo complex ions or as their dehydration products.

written as $Cu(OH)_2$, $Fe(OH)_3$, $Mg(OH)_2$, and $Cr(OH)_3$—are almost always gelatinous in nature. All such precipitates contain varying numbers of water molecules. Some, such as $Mg(OH)_2$, are true insoluble hydroxides accompanied by water molecules. Others are simply insoluble *oxides* with water molecules trapped in the solid in stoichiometry equivalent to that of hydroxides. For example, "$Cr(OH)_3$" and "$Ti(OH)_4$" are equivalent to Cr_2O_3 plus three water molecules and to TiO_2 plus two water molecules, respectively. Still other gelatinous "hydroxide" precipitates are oxides combined with water molecules, but *not* in the correct stoichiometry for simple hydroxides; $Fe_2O_3 \cdot H_2O$, sometimes written FeOOH, is such a compound ($Fe_2O_3 \cdot H_2O = 2FeOOH$). Although representing a precipitate as a hydroxide may not always be correct, it is convenient, especially because the actual structures of many "hydroxide" precipitates are not known.

Whatever their structures, however, the hydroxides are chemically equivalent to the oxides. The "hydroxides" are acidic, basic, or amphoteric, like the oxides from which they are derived. Therefore, the following equations

$$Al(OH)_3(s) + 3H_3O^+ \longrightarrow Al^{3+} + 6H_2O(l) \qquad \textbf{(16.17)}$$

$$Al(OH)_3(s) + OH^- \longrightarrow [Al(OH)_4]^- \qquad \textbf{(16.18)}$$

represent the same chemical reactions as Equations (16.15) and (16.16).

EXAMPLE 16.4 Properties of Oxides

Of the three oxides BaO, SeO_3, and Ga_2O_3, one is acidic, one amphoteric, and one basic. Predict which oxide fits into each category and describe simple tests that will either confirm or contradict your predictions.

The highest oxidation state of the elements bonded to oxygen in these three oxides is +6 for selenium. The oxide **SeO_3 is most likely to be the acidic one**—the bonding of Se to O must be covalent, for a +6 ion is not likely to form. The lowest oxidation state is +2 for barium and the charge-to-size ratio for Ba^{2+} is 2/0.134 = 14.9. **Barium oxide is probably ionic and basic.** This leaves **Ga_2O_3 as the amphoteric oxide,** as might be predicted from the large charge-to-size ratio for Ga^{3+}, 3/0.062 = 48. Barium oxide should react with acids such as aqueous HCl, but not with aqueous bases. Gallium(III) oxide should react with solutions of both strong acids and strong bases. Selenium(VI) oxide should react with solutions of strong bases but not with solutions of acids.

EXERCISE Of the three oxides P_4O_{10}, Cs_2O, and ZnO, one is acidic, one amphoteric, and one basic. Identify which oxide fits each category.

Answers P_4O_{10}, acidic; Cs_2O, basic; ZnO, amphoteric.

16.12 PREPARATION AND USES OF OXYGEN

A large percentage of the annual output of oxygen (more than 400 billion cubic feet per year) is produced by liquefaction of air on the site of its use. Modern liquid-air plants utilize a combination of cooling (by permitting compressed gas to do expansion work), fractional distillation, and other techniques to fully separate air into its components. Such plants are located beside steel mills, the single largest users of oxygen, and beside metal fabricating plants. The oxygen goes directly to the mill or plant, where it is used to speed combustion and raise the combustion temperature (Section 26.7). A relatively new use for on-site-generated oxygen is in wastewater treatment, where it replaces air in the activated sludge process (see Section 14.14).

Oxygen is also transported to industrial users in cylinders, tanks, or tank cars. The greatest outlet for such oxygen is the chemical industry, where oxygen is consumed in oxidation reactions and the production of oxygen-containing compounds. Other uses are summarized in Table 16.11.

Table 16.11
Uses of Oxygen

Uses of Oxygen
Major
Combustion aid
In steelmaking
In metallurgy and metal fabricating
Production of oxygen-containing chemicals
Others
Life support
Medical
Aerospace
Wastewater treatment
Oxidizer
Liquid oxygen (LOX) in rockets
Oxyacetylene torch (3000 °C)

At one time, oxygen, hydrogen, and all gases needed in the laboratory had to be generated at the time of their use. The common laboratory method for oxygen production was the thermal decomposition of potassium chlorate ($KClO_3$), a redox reaction aided by a catalyst.

$$\underset{\textit{potassium chlorate}}{2KClO_3(l)} \xrightarrow{\Delta,\ MnO_2 \text{ or } Fe_2O_3} 3O_2(g) + 2KCl(s) \qquad \textbf{(16.19)}$$

This is a hazardous reaction and can lead to explosions, particularly if the hot reaction mixture comes in contact with rubber stoppers, hoses, wood, or other combustible materials. Gases are now available in small tanks that are safer and more convenient to use than the classical gas-producing reactions.

Chlorates are fabricated into "chlorate candles" that generate oxygen by reactions like the one above. The reaction mixture is enclosed in a sturdy container equipped with an ignition device to set off the reaction. These candles have a long storage life and serve as emergency sources of oxygen for breathing support on aircraft and submarines.

16.13 OZONE AND OZONIDES

Ozone (O_3) is oxygen in the form of gaseous, triatomic molecules. Oxygen and ozone are allotropes—different forms of the same element in the same state (in this case, both gases). Most of the nonmetals tend to "self-link" in this way—that is, to form chains. (The bonding of many atoms of the same element to each other in chains or rings is called *catenation.*)

Earlier we discussed the molecular and electronic structure of ozone. It is a bent, or angular, molecule that can be pictured as having resonance forms (Section 11.3).

Pure ozone is a pale blue gas which condenses to a dark blue liquid that boils at −111.5 °C. Ozone is formed when energy in the form of radiation, electricity, or heat is supplied to gaseous oxygen.

$$3O_2(g) \longrightarrow 2O_3(g) \qquad \Delta H^\circ = 285 \text{ kJ} \qquad \textbf{(16.20)}$$

The characteristic odor of ozone is often noticeable near a sparking electric motor.

Usually, ozone is prepared by passing gaseous oxygen through an electrical field under high voltage. The yield of ozone is low. Equation (16.20) shows that ozone has a much higher energy content than molecular oxygen. As might be expected, the stability and reactivity of ozone differ from those of oxygen. Ozone is more reactive than oxygen. In general, substances formed in endothermic reactions are reactive, readily releasing the energy stored in their formation. Ozone tends to decompose to oxygen, in a reversal of reaction (16.20). The decomposition, slow at low temperatures, increases rapidly with rising temperature, is catalyzed by a variety of substances such as platinum and manganese dioxide (MnO_2), and can be explosive.

Ozone is a much more powerful oxidizing agent than ordinary oxygen. Usually, when ozone functions as an oxidizing agent only one oxygen atom from each molecule is reduced; the other two form molecular oxygen, as, in the oxidation of lead sulfide.

$$PbS(s) + 4\overset{0}{O}_3(g) \longrightarrow PbS\overset{-2}{O}_4(s) + 4\overset{0}{O}_2(g)$$

In this reaction, sulfur is oxidized from its lowest oxidation state (−2) to its highest (+6).

The oxidation of potassium iodide by ozone is often used as a test for ozone. The iodine liberated from KI is detected by the deep blue color it gives to a suspension of starch in water.

$$2KI(aq) + O_3(g) + H_2O(l) \longrightarrow 2KOH(aq) + I_2(aq) + O_2(g)$$

$$I_2(aq) + \text{starch} \longrightarrow \text{blue complex}$$

(Other oxidizing agents also liberate iodine from KI.)

Ozone performs a vital role in protecting living things on Earth from ultraviolet radiation from the sun. In the stratosphere (11 – 50 km above the Earth's surface) ozone is formed from oxygen in a two-step process initiated by radiation.

$$O_2(g) \xrightarrow{h\nu} O(g) + O(g)$$
$$O(g) + O_2(g) + M(g) \longrightarrow O_3(g) + M(g)$$

(M represents a "third body," usually a nonreactive atom, needed to carry excess energy away from the reactions so that O and O_2 can "stick" together when they collide; see Figure 18.5.)

The ozone cycle in the stratosphere is completed when ozone absorbs ultraviolet radiation from the Sun and breaks down into an oxygen molecule and an oxygen atom.

$$O_3 \xrightarrow{h\nu} O_2 + O + \text{heat}$$

Concern has been growing in recent years that gases added to the atmosphere by the activities of mankind are finding their way to the stratosphere and destroying ozone there. In particular, nitrogen oxides (Section 28.9) and compounds containing chlorine, fluorine, and carbon (Section 27.12) are suspect. As ozone concentration decreases, the amount of ultraviolet radiation reaching the surface of the Earth increases. It is known that such an increase will cause a rise in human skin cancer and eye diseases. Changes are likely in plant, animal, and marine life, and possibly also in the properties of materials exposed to sunlight, although less is known about what these changes might be. (The ozone depletion problem is further discussed in Section 27.12.)

Most uses of ozone are dependent upon its strength as an oxidizing agent or its reactions with various organic compounds. Ozone bleaches paper or cloth and destroys harmful chemicals and bacteria in water. In large amounts, ozone is toxic. Restaurant kitchens once used machines to generate ozone; the ozone could react with and destroy airborne particles of fat and smoke from the cooking. This practice was stopped years ago when it was discovered that ozone is toxic.

The ozone molecule can gain an electron to form the ozonide ion (O_3^-). Potassium, rubidium, and cesium ozonides are formed by the reaction of ozone with the hydroxides, for example

$$6CsOH(s) + 4O_3(g) \longrightarrow \underset{\textit{cesium ozonide}}{4CsO_3(s)} + 2CsOH \cdot H_2O(s) + O_2(g)$$

16.14 HYDROGEN PEROXIDE AND PEROXIDES

Hydrogen peroxide (H_2O_2) is a colorless, somewhat unstable liquid in which the two oxygen atoms are joined by a single covalent bond (Figure 16.2). The oxygen–oxygen bond is weak (142 kJ/mol) compared to that in O_2, and hydrogen peroxide is quite reactive. The pure compound and its concentrated aqueous solutions can decompose, often explosively, with heat or in the presence of a catalyst (e.g., metal ions such as Fe^{2+}, finely divided metals, metal oxides, blood, or saliva).

$$2H_2O_2(l) \xrightarrow{\text{heat or catalyst}} 2H_2O(l) + O_2(g) \qquad \Delta H^\circ = -196 \text{ kJ} \qquad \textbf{(16.21)}$$

The hydrogen peroxide molecules are polar and associated through hydrogen bonding. The compound is completely miscible with water and is an extremely weak acid that yields two hydrogen ions per molecule.

$$H_2O_2(aq) \rightleftharpoons H^+ + \underset{\textit{hydroperoxide ion}}{HO_2^-}$$
$$HO_2^- \rightleftharpoons H^+ + \underset{\textit{peroxide ion}}{O_2^{2-}}$$

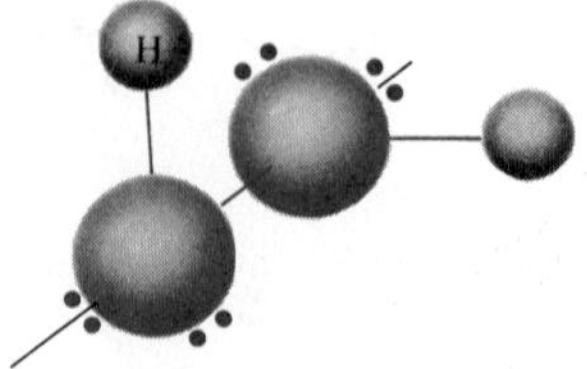

Figure 16.2
The H_2O_2 Molecule
Hydrogen peroxide has m.p. 0.4 °C and b.p. 151 °C.

The oxygen in a peroxide ion, which is assigned an oxidation state of −1, can be both oxidized to the zero state (in O_2) and reduced to the −2 state (in H_2O), as in reaction (16.21). Therefore, hydrogen peroxide is both an oxidizing and a reducing agent. Actually it is an extremely powerful oxidizing agent and a rather weak reducing agent, usually in acidic solution. The reaction with lead(II) sulfide illustrates the behavior of hydrogen peroxide as an oxidizing agent.

$$\overset{-2}{PbS}(s) + 4H_2\overset{-1}{O_2}(aq) \longrightarrow Pb\overset{+6}{S}O_4(s) + 4H_2\overset{-2}{O}(l)$$

The conversion of permanganate ion (MnO_4^-) to manganous ion (Mn^{2+}) is an example of reduction by hydrogen peroxide.

$$2\overset{+7}{Mn}O_4^- + 5H_2\overset{-2}{O_2}(aq) + 6H^+ \longrightarrow 2\overset{+2}{Mn^{2+}} + 5\overset{0}{O_2}(g) + 8H_2O(l)$$

In general, when hydrogen peroxide acts as a reducing agent, *both* water and oxygen are products, with one mole of oxygen formed for each mole of hydrogen peroxide consumed. When hydrogen peroxide acts as an oxidizing agent, water is a product, but oxygen is not.

Commercial hydrogen peroxide solutions have a concentration of about 30% H_2O_2 by mass. A 3% solution of H_2O_2 is sometimes used medically as an antiseptic. A stabilizer must be added to such solutions to prevent the decomposition of H_2O_2 during storage.

Reactions between strong bases and hydrogen peroxide yield metal peroxides, usually as the hydrates; for example

$$Ba(OH)_2(aq) + H_2O_2(aq) + 6H_2O(l) \longrightarrow \underset{\text{barium peroxide octahydrate}}{BaO_2 \cdot 8H_2O(s)}$$

Peroxides are known for all of the lithium family elements and for calcium, strontium, and barium—all very active metals. The peroxides react with acid to produce hydrogen peroxide. This reaction can be used to prepare hydrogen peroxide in the laboratory.

$$BaO_2(s) + H_2SO_4(aq) \longrightarrow H_2O_2(aq) + BaSO_4(s)$$

To demonstrate that analyzing a problem according to what is known and unknown is useful for purely chemical problems as well as those involving mathematics, the third of the following three examples is worked by the problem-solving method introduced in Chapter 2. At first glance, this may look like a very long solution to a simple problem. Remember that this method illustrates a way of *thinking* about the problem. (Also, without mathematics, more words are necessary.) We have set the problem up assuming a knowledge only of facts that we have discussed thus far. (With greater or different knowledge of descriptive chemistry, other approaches would, of course, be possible.)

EXAMPLE 16.5 Oxygen Anions

Oxygen is known to form four anions: O^{2-}, oxide ion; O_2^{2-}, peroxide ion; O_2^-, superoxide ion; and O_3^-, ozonide ion. (a) Show how the oxidation number of oxygen in each of the anions is determined. (b) Predict which of them can behave as oxidizing agents, knowing that these are the only anions oxygen forms. (c) Write the Lewis structures for the peroxide, superoxide, and ozonide ions.

(a) **In the oxide ion, O^{2-}, the oxidation number of oxygen is simply the charge, -2. For the peroxide ion, O_2^{2-}, it is the charge divided by 2, or -1; for the superoxide ion, O_2^-, it is the charge divided by 2, which gives an oxidation state of $-\frac{1}{2}$; and for the ozonide ion, O_3^-, it is the charge divided by 3, or $-\frac{1}{3}$.**

(b) **The peroxide, superoxide, and ozonide ions can all act as oxidizing agents.** In each of them the oxidation number of oxygen lies above -2, the lowest state known for the element. Therefore, in a redox reaction oxygen in these ions can decrease in oxidation state, allowing the increase in oxidation state—the oxidation—of another substance. [In fact, all three ions are strong oxidizing agents.] Oxide ion, which is oxygen in its lowest oxidation state, cannot be an oxidizing agent.

(c) Each oxygen atom contributes six valence electrons. Taking into account the charges on the ions, the peroxide ion, O_2^{2-}, has $(2 \times 6) + 2 = 14$ electrons; the superoxide ion, O_2^-, has 13 electrons, one less electron than the peroxide ion; and the ozonide ion, O_3^- has $(3 \times 6) + 1 = 19$ electrons. Placing single bonds and lone pairs of electrons gives the correct Lewis structure for the peroxide ion, and shows that, because of their odd numbers of electrons, resonance must be used to account for the Lewis structures of the superoxide and ozonide ions.

$(:\ddot{\underset{..}{O}}-\ddot{\underset{..}{O}}:)^{2-}$ $(:\ddot{\underset{..}{O}}-\ddot{\underset{.}{O}}:)^- \rightleftharpoons (:\ddot{\underset{.}{O}}-\ddot{\underset{..}{O}}:)^-$

peroxide ion   *superoxide ion*

$(:\ddot{\underset{.}{O}}-\ddot{\underset{..}{O}}-\ddot{\underset{..}{O}}:)^- \rightleftharpoons (:\ddot{\underset{..}{O}}-\ddot{\underset{.}{O}}-\ddot{\underset{..}{O}}:)^- \rightleftharpoons (:\ddot{\underset{..}{O}}-\ddot{\underset{..}{O}}-\ddot{\underset{.}{O}}:)^-$

ozonide ion

EXAMPLE 16.6 Properties of Oxides; Problem Solving

Based on what you have learned, how do $Na_2O_2(s)$ and $MnO_2(s)$ differ in structure and bonding?

1. Study the problem and be sure you understand it.
   (a) What is unknown?
   The structure and bonding of the two oxides. Are they ionic or covalent compounds? Which atoms shown in the formulas are connected to which other atoms? Are the compounds similar or different in structure and bonding?
   (b) What is known?
   What the formulas show—that in each compound a metal and oxygen have combined—two oxygen atoms with two sodium atoms, and two oxygen atoms with one manganese atom.
   Whatever you know about the chemistry of sodium, manganese, and oxygen.
2. Decide how to solve the problem.
   It is necessary to review the facts that you know about the chemistry of the elements involved and choose those facts that apply to the problem at hand.
3. Set up the problem and solve it.
   (a) Review the known chemistry that is relevant to structure and bonding.
   Sodium is a lithium family element of large atomic radius and is an active metal. Sodium atoms lose electrons readily to give Na^+ ions, and $+1$ is the only common oxidation state of sodium, which is one of the least electronegative elements (0.9).
   Manganese is a transition metal and as such has several oxidation states. It is more electronegative (1.5) than sodium.
   Oxygen is the second most electronegative element (3.5). It is assigned the oxidation state of -2 in most compounds, but also forms the peroxide ion, $(O—O)^{2-}$, where it has oxidation state -1. Peroxides form with active metals, which include sodium. Normal oxides of a given metal tend to be ionic for the lower oxidation state of the metal and covalent for the higher oxidation states.
   (b) Choose the appropriate structures and give the best prediction for the type of bonding.
   First, we must decide whether Na_2O_2 and MnO_2 are oxides or peroxides. If Na_2O_2 were a normal oxide, sodium would have an oxidation state of $+2$. Sodium is a Group I metal and has only one oxidation state, which is $+1$. For this reason, and also because sodium is one of the most active metals, which we know form peroxides, we can conclude that **Na_2O_2 is sodium peroxide, and is an ionic compound that contains Na^+ and O_2^{2-} ions.**
   On the other hand, because manganese is a transition metal and therefore not one of the most active metals, the formation of manganese peroxide seems unlikely. Moreover, if MnO_2 is a normal oxide, the manganese must have an oxidation state of $+4$, which for a transition metal

of variable oxidation state is reasonable. Therefore, we conclude that MnO_2 is probably a normal oxide.

It remains to decide whether the bonding in MnO_2 is ionic or covalent, or intermediate between the two. We have learned several different ways to examine this problem. If the compound is ionic, the formation of Mn^{4+} is necessary. Removal of four electrons requires a large amount of energy and the formation of +4 ions is not common. Also, the Mn^{4+} ion would have quite a large charge-to-size ratio, as transition metal atoms are not very large to begin with (see Figure 9.8). Therefore, if it existed, Mn^{4+} would be a highly polarizing ion. This suggests that the manganese–oxygen bond is probably covalent. The electronegativity difference between manganese and oxygen is 2, which is right on the borderline between covalent (<2) and ionic (>2) bonding, according to the usual rule of thumb. **The best prediction, then, is that in MnO_2 the two oxygen atoms are joined to the manganese atom by highly polar covalent bonds.**

SUMMARY OF SECTIONS 16.9–16.14

Oxygen is an abundant element essential to life. It is an important industrial chemical and an oxidizing agent. As the first member of Representative Group VI, oxygen differs in properties from the other group members (Table 16.8). The outstanding properties of oxygen and some of its compounds are summarized in Table 16.12.

Table 16.12
Outstanding Properties of Oxygen and Its Compounds

O $2s^22p^4$	A member of Representative Group VI; six valence electrons
O_2	Oxygen molecule has two unpaired electrons
O_3	Ozone, an allotrope of oxygen and a strong oxidizing agent
$\overset{-2}{O}$	Oxidation state −2 except in compounds with F (e.g., +2 in OF_2), peroxides (−1), superoxides ($-\frac{1}{2}$), and ozonides ($-\frac{1}{3}$)
$\overset{0}{O_2} \longrightarrow \overset{-2}{O}$	Oxidizing agent
polar covalent bond $\overset{\delta-}{O} - \overset{\delta+}{E}$	Second most electronegative element
M_2O, MO	Ionic oxides with Li and Be family elements
$E_2O, E_2O_3, \ldots, E_2O_5, E_2O_7$	Binary oxides with all elements except He, Ne, Ar, Kr
H_2O_2	Both oxidizing and reducing agent. As oxidizing agent, product is H_2O. As reducing agent, products are H_2O and O_2
Fossil fuels, carbohydrates, fats $\xrightarrow{O_2}$ $CO_2 + H_2O$	Essential to combustion of fuels and to living things
Combination with other elements	
Metal + $O_2 \longrightarrow$ metal oxide	Metal = all except least active, e.g., Ag, Au, Pt; peroxide with Na; superoxides with K, Rb, Cs
Nonmetal + $O_2 \longrightarrow$ nonmetal oxide	Nonmetal = all except noble gases, some halogens. Oxidation state of metal or nonmetal governed by O_2 supply—highest oxidation state when O_2 is in excess
Reactions of oxides	
$2HgO(s) \xrightarrow{\Delta} 2Hg(l) + O_2(g)$	Oxides of least active metals decompose with heat
$2PbO_2(s) \xrightarrow{>500\,^\circ C} 2PbO(s) + O_2(g)$	Reduction of oxides of metals in their less stable, higher oxidation states
$Na_2O(s) + H_2O(l) \longrightarrow 2NaOH(aq)$	Oxides of many metals are basic
$FeO(s) + 2HCl(aq) \longrightarrow FeCl_2(aq) + H_2O(l)$	
$SO_3(g) + H_2O(l) \longrightarrow H_2SO_4(aq)$	Oxides of many nonmetals are acidic
$Cr_2O_3(s) \xrightarrow{+H^+} Cr^{3+} + H_2O(l)$; $Cr_2O_3(s) \xrightarrow{+OH^-} [Cr(OH)_4]^-$	Oxides of many metals in intermediate oxidation states are amphoteric

SIGNIFICANT TERMS (with section references)

absorption (16.6)
acidic anhydride, acidic oxide (16.11)
allotropes (16.13)
amphoteric (16.11)
basic anhydride, basic oxide (16.11)
fossil fuels (16.7)
hydrocarbons (16.7)
hydrogenation (16.6)
oxidation (16.2)
oxidation–reduction reactions, redox reactions (16.2)
oxidizing agent (16.3)
reducing agent (16.3)
reduction (16.2)

QUESTIONS AND PROBLEMS

Oxidation and Reduction

16.1 Which of the following changes in oxidation number represent oxidation: (a) $0 \rightarrow +2$, (b) $-3 \rightarrow -2$, (c) $+1 \rightarrow -1$, (d) $0 \rightarrow -2$, (e) $+3 \rightarrow 0$, (f) $-2 \rightarrow +1$, (g) $+2 \rightarrow +1$?

16.2 How can the formula of an atom, molecule, or ion indicate whether or not the substance could serve as a potential oxidizing or reducing agent?

16.3 Which of the following are redox reactions? Identify the oxidizing and reducing agents in each of the redox reactions.

(a) $2Na_2O_2(s) + 2H_2O(l) \longrightarrow O_2(g) + 4NaOH(aq)$
(b) $Sn(s) + 2HCl(aq) \longrightarrow SnCl_2(aq) + H_2(g)$
(c) $CaO(s) + H_2O(l) \longrightarrow Ca(OH)_2(aq)$
(d) $N_2O_5(s) + H_2O(l) \longrightarrow 2HNO_3(aq)$

16.4 Repeat Question 16.3 for

(a) $LiAlH_4(s) + 4H^+ \longrightarrow Li^+ + Al^{3+} + 4H_2(g)$
(b) $Cr_2O_7^{2-} + 2OH^- \longrightarrow 2CrO_4^{2-} + H_2O(l)$
(c) $3KClO(s) \longrightarrow 2KCl(s) + KClO_3(s)$
(d) $24Cu_2S(s) + 128H^+ + 32NO_3^- \longrightarrow 48Cu^{2+} + 32NO(g) + 3S_8(s) + 64H_2O(l)$

16.5 For each of the following balanced equations choose the oxidizing agent and the reducing agent. Show the change in oxidation number which occurs for each substance:

(a) $Sn^{2+} + 2Fe^{3+} \longrightarrow Sn^{4+} + 2Fe^{2+}$
(b) $MnO_2(s) + 4HCl(aq) \longrightarrow MnCl_2(aq) + Cl_2(g) + 2H_2O(l)$
(c) $2XeF_2(s) + 2H_2O(l) \longrightarrow 2Xe(g) + O_2(g) + 4HF(g)$
(d) $I^- + ClO^- \longrightarrow IO^- + Cl^-$
(e) $4HNO_3(l) \longrightarrow 2N_2O_4(g) + O_2(g) + 2H_2O(l)$

16.6 Repeat Question 16.5 for

(a) $4Al(s) + 3O_2(g) \longrightarrow 2Al_2O_3(s)$
(b) $Cr_2O_7^{2-} + 3SO_3^{2-} + 8H^+ \longrightarrow 2Cr^{3+} + 3SO_4^{2-} + 4H_2O(l)$
(c) $Fe_3O_4(s) + 4H_2(g) \longrightarrow 3Fe(s) + 4H_2O(g)$
(d) $3PbO_2(s) \xrightarrow{\Delta} Pb_3O_4(s) + O_2(g)$
(e) $I_2(s) + H_2S(aq) \longrightarrow S(s) + 2HI(aq)$
(f) $2HNO_3(aq) + SO_2(g) \longrightarrow H_2SO_4(aq) + 2NO_2(g)$

16.7 Using the oxidation number method, balance the following equations:

(a) $SO_2(g) + H_2S(g) \longrightarrow S_8(s) + H_2O(g)$
(b) $Ca_3(PO_4)_2(s) + C(s) + SiO_2(s) \longrightarrow CaSiO_3(l) + P_4(g) + CO(g)$
(c) $SO_2(g) + Cr_2O_7^{2-} + H^+ \longrightarrow Cr^{3+} + H_2SO_4(aq) + H_2O(l)$
(d) $Cl_2(g) + OH^- \longrightarrow ClO_3^- + Cl^- + H_2O(l)$
(e) $S(s) + OH^- \longrightarrow S^{2-} + SO_3^{2-} + H_2O(l)$
(f) $MnO_4^- + IO_3^- + H_2O(l) \longrightarrow MnO_2(s) + IO_4^- + OH^-$
(g) $Sn(s) + OH^- + H_2O(l) \longrightarrow [Sn(OH)_4]^{2-} + H_2(g)$

16.8 Repeat Question 16.7 for

(a) $SiO_2(s) + Al(s) \longrightarrow Si(s) + Al_2O_3(s)$
(b) $I_2(s) + H_2S(aq) \longrightarrow I^- + S(s) + H^+$
(c) $PbO_2(s) + HCl(aq) \longrightarrow Cl_2(g) + PbCl_2(s) + H_2O(l)$
(d) $Cu(s) + Br_2(aq) + OH^- \longrightarrow Cu_2O(s) + Br^- + H_2O(l)$
(e) $S^{2-} + Cl_2(g) + OH^- \longrightarrow SO_4^{2-} + Cl^- + H_2O(l)$
(f) $H_2O_2(aq) + HI(aq) \longrightarrow I_2(s) + H_2O(l)$
(g) $H_2S(g) + O_2(g) \longrightarrow SO_2(g) + H_2O(g)$
(h) $NH_3(g) + O_2(g) \longrightarrow NO(g) + H_2O(g)$

Hydrogen

16.9 Write the electron configuration for atomic hydrogen. What oxidation states would you predict for this element?

16.10 In many periodic tables hydrogen is placed in Representative Group I or in both Representative Groups I and VII. On what bases are these placements justified? What is wrong with these placements?

16.11 The average atomic mass of naturally occurring hydrogen is 1.00797 u. Assuming only ${}^1_1H$ (1.007825 u) and ${}^2_1H$ (2.0140 u) to be present, find the percentage of natural abundance of ${}^2_1H$.

16.12 The half-life of tritium is 12.26 yr. What fraction of a sample of tritium will remain at the end of one year?

16.13 Molecular hydrogen can act as either an oxidizing agent or a reducing agent. Write a chemical equation illustrating each type of reaction.

16.14 Many of the reactions of molecular hydrogen described in this chapter do not occur at ordinary temperatures. Why?

16.15 When hydrogen reacts with lithium, an ionic substance is formed. When hydrogen reacts with chlorine, a molecular substance results. Explain this difference.

16.16 What type of element reacts with molecular hydrogen to form ionic hydrides? What types of hydrides are formed when hydrogen combines with transition metals? In addition to compound formation, in what other ways can transition elements interact with molecular hydrogen?

16.17 Complete and balance chemical equations for the reaction of $H_2(g)$ with (a) $N_2(g)$, (b) $I_2(g)$, (c) $Ca(s)$, (d) $SnO(s)$. In which reactions is hydrogen acting as an oxidizing agent and in which is it acting as a reducing agent?

16.18 Write the chemical equations for the reaction of $H_2(g)$ as (a) a reducing agent with $WO_3(s)$, $Br_2(g)$, and $C_2H_4(g)$ and (b) an oxidizing agent with $Na(s)$.

16.19 Currently the major industrial sources of hydrogen are water and natural gas. Describe how the element is obtained from each of these raw materials. Write chemical equations to illustrate these methods.

16.20 Write the chemical equation for the production of molecular hydrogen from water by electrolysis. What is the relationship between the volumes of gases collected? Why are methods other than electrolysis used to obtain hydrogen industrially?

16.21 Torches using hydrogen and oxygen can produce very high temperatures. Those that electrically decompose the hydrogen molecules into atomic hydrogen before mixing the fuel with the oxygen generate even higher temperatures. The heat of formation at 25 °C is −242 kJ/mol for $H_2O(g)$ and 218 kJ/mol for $H(g)$. Calculate $\Delta H°$ for the reaction

$$2H(g) + \tfrac{1}{2}O_2(g) \longrightarrow H_2O(g)$$

and compare the value to that for

$$H_2(g) + \tfrac{1}{2}O_2(g) \longrightarrow H_2O(g)$$

16.22 Water gas is produced by the reaction between steam and hot carbon

$$C(s) + H_2O(g) \xrightarrow{\Delta} CO(g) + H_2(g)$$

What volume of steam at 1.53 atm and 184 °C is needed to produce 100.0 L of water gas at 820 °C and 1.02 atm?

Oxygen

16.23 Write the electron configuration for atomic oxygen. What oxidation states would you predict for this element?

16.24 Draw a Lewis structure for the oxygen molecule using a double bond between the atoms. Does this structure correctly show that the O_2 molecule is paramagnetic?

16.25 Oxygen is assigned a positive oxidation state when it is covalently bonded to only one specific element. What element is this? Why is oxygen positive?

16.26 List the negative oxidation states of oxygen as they appear in oxides, peroxides, superoxides, and ozonides. For each, give a chemical formula of a compound having oxygen in this oxidation state and name that compound.

16.27 Describe what happens when a limited amount of oxygen combines with C, S, and Fe. What will happen to the compounds formed upon reaction with additional oxygen?

16.28 Write chemical equations for the reactions of the elements in the lithium family with molecular oxygen. Name the compounds that are formed.

16.29 Write the chemical equations for the reactions between molecular oxygen (in excess) and (a) phosphorus, (b) carbon, (c) sulfur dioxide, (d) tin(II) oxide.

16.30 Repeat Question 16.29 for (a) $S_8(s)$, (b) $Mg(s)$, (c) $Cu_2O(s)$, (d) $CH_4(g)$.

16.31 Each of the following compounds is a basic anhydride or an acidic anhydride: (a) $MgO(s)$, (b) $P_4O_6(s)$, (c) $CO_2(g)$, (d) $SO_2(g)$, (e) $Na_2O(s)$. Write the chemical equation for the reaction between each substance and water.

16.32 Classify each of the following oxides as (i) probably basic, (ii) probably acidic, (iii) probably amphoteric: (a) Cl_2O_7, (b) SrO, (c) MnO, (d) Mn_2O_7, (e) Sb_2O_3, (f) Cl_2O.

16.33 The metallic element M forms a series of oxides: MO, M_2O_3, and M_2O_5. One of these oxides will dissolve only in an acidic solution, whereas the others will dissolve in alkaline or neutral solutions. How can you choose the oxide that is different?

16.34 A few metal oxides decompose upon heating. Write chemical equations for the decompositions of (a) $HgO(s)$, (b) $Ag_2O(s)$, (c) $PbO_2(s)$.

16.35 What volume of dry O_2 at 37 °C and 1.007 atm can be produced from the thermal decomposition of 25 g of $KClO_3$ described by the following equation?

$$2KClO_3(s) \xrightarrow{MnO_2} 2KCl(s) + 3O_2(g)$$

16.36 When heated, lead(IV) oxide loses half its oxygen content, potassium chlorate loses all its oxygen content, and potassium nitrate loses one-third its oxygen content. Write a balanced equation for each reaction. Disregarding the cost of heat energy, determine by calculation which compound would be the least expensive laboratory source of oxygen given the following prices per pound: PbO_2, \$24.94; $KClO_3$, \$18.81; KNO_3, \$12.60.

16.37 Write the Lewis structure(s) for the ozone molecule. Describe the geometry of the molecule.

16.38 What types of intermolecular forces are present in liquid ozone? Would you expect the boiling point of ozone to be higher or lower than that of oxygen? Why?

16.39 Write chemical equations representing the photochemical formation and decomposition of ozone in the stratosphere. Why is a change in the ozone concentration in the stratosphere a cause for concern?

16.40 Can ozone be an oxidizing and/or a reducing agent? What usually happens to the oxygen atoms in ozone during an oxidation–reduction reaction?

16.41 A mixture of ozone and oxygen was bubbled through an excess of potassium iodide solution

$$2KI(aq) + O_3(g) + H_2O(l) \longrightarrow 2KOH(aq) + I_2(aq) + O_2(g)$$

The mass of the liberated elemental iodine was 101.6 mg. What mass of ozone reacted with the potassium iodide? Assume that the iodide ion does not react with molecular oxygen.

16.42 Silver will not react directly with oxygen at room temperature to form Ag_2O (even though the reaction is slightly exothermic), but will react with ozone according to the equation

$$2Ag(s) + O_3(g) \longrightarrow Ag_2O(s) + O_2(g)$$

Calculate the enthalpy of reaction given that the $\Delta H°$ of formation at 25 °C is −31.05 kJ/mol for $Ag_2O(s)$ and 142.7 kJ/mol for $O_3(g)$.

16.43 Write the Lewis structure for the hydrogen peroxide molecule. Describe the geometry of the molecule. What intermolecular forces are present in hydrogen peroxide?

16.44 Write chemical equations showing the stepwise ionization of hydrogen peroxide as a weak acid. Name all ions formed.

16.45 Is the decomposition of H_2O_2 an example of an oxidation–reduction reaction? What is undergoing oxidation and what is undergoing reduction? Write the chemical equation for the reaction.

16.46 Why can hydrogen peroxide act as both an oxidizing and a reducing agent? Which behavior is dominant? Write the equations for the reactions of hydrogen peroxide with (a) $SO_3^{2-}(aq)$ and (b) acidic $MnO_4^-(aq)$.

16.47 Following are some heats of formation at 25 °C: $H_2O(g)$, −241.818 kJ/mol; $H_2O(l)$, −285.83 kJ/mol; $H_2O_2(g)$, −136.31 kJ/mol; $H_2O_2(l)$, −187.78 kJ/mol. Calculate the heat of vaporization of both compounds at this temperature and compare the strengths of the intermolecular forces present.

16.48 The analysis of an ionic compound of sodium showed that it contained 59% by mass of Na and 41% by mass of O. Determine the simplest formula of the compound. Using your knowledge of the oxidation states of sodium and oxygen, determine the true formula and name the compound.

ADDITIONAL QUESTIONS AND PROBLEMS

16.49 A student was given two unmarked test tubes, one containing hydrogen and one containing oxygen. A glowing wooden splint was inserted into the mouth of each tube. It burst into flame in the first tube and was extinguished in the second tube. A burning splint was held near the mouth of the second tube, and a minor explosion occurred which sounded similar to a sharp bark. Which gas did the student decide was in each of the test tubes?

16.50 Classify each of the following reactions involving hydrogen or oxygen as (i) combination, (ii) decomposition, (iii) displacement, (iv) partner exchange:

(a) $SnO_2(s) + C(s) \longrightarrow Sn(s) + CO_2(g)$
(b) $H_2(g) + Cl_2(g) \longrightarrow 2HCl(g)$
(c) $2NaH(s) \xrightarrow{\Delta} 2Na(s) + H_2(g)$
(d) $NaH(s) + H_2O(l) \longrightarrow NaOH(aq) + H_2(g)$
(e) $P_4O_6(s) + 6H_2O(l) \longrightarrow 4H_3PO_3(aq)$
(f) $O_3(g) \xrightarrow{h\nu} O_2(g) + O(g)$
(g) $BaO_2(s) + H_2SO_4(aq) \longrightarrow H_2O_2(aq) + BaSO_4(s)$

16.51 A 20.0-L vessel contains 9.90 mol of H_2 and 2.00 mol of O_2 at 25 °C. Calculate the total pressure exerted on the inner walls of the container by the gaseous mixture. The mixture is ignited by a spark and the vessel is subsequently cooled at 25 °C. Calculate the pressure in the container after the reaction has occurred. (The vapor pressure of water at 25 °C is 23.756 Torr.)

16.52* Aqueous solutions of H_2O_2 undergo the following decomposition:

$$2H_2O_2(aq) \longrightarrow 2H_2O(l) + O_2(g)$$

This is the reaction that makes H_2O_2 an important disinfectant and bleach.

The oxygen formed from the decomposition of 1.00 g of H_2O_2 is confined to a volume of 25 mL at 25 °C. (a) What will be the pressure?

The enthalpy of formation at 25 °C is −191.17 kJ/mol for $H_2O_2(aq)$ and −285.830 kJ/mol for $H_2O(l)$. (b) What is the heat of reaction?

The heat of formation of $H_2O_2(g)$ is −136.31 kJ/mol and of $H_2O(g)$ is −241.818 kJ/mol. The bond energy of H—H is 435.93 kJ/mol and of O═O is 498.17 kJ/mol. Calculate the (c) O—H and (d) O—O bond energies.

CHAPTER

17

Chemical Reactions in Perspective

CHEMICAL REACTIONS

17.1 THE STUDY OF CHEMICAL REACTIONS

Many students look upon the prediction of what might happen when chemicals are brought together as a formidable task. With more than 100 elements, all of their possible compounds, and all of the possible reactions of those compounds, how can the study of chemical reactions be organized?

We cannot offer a systematic classification of all possible chemical reactions. There is nothing in chemistry comparable to the kingdoms, phyla, classes, and orders that biologists use to classify living things. What we do have is a body of knowledge based on years of observation of chemical reactions. We also have generalities about chemical properties based on the positions of the elements in the periodic table. We come to expect that compounds of certain elements or molecules with certain structures will react in certain ways. For example, we expect an oxide of a metal from the lithium family to give a strongly alkaline solution in water, and we would be quite surprised if it did not do so.

A person studying or using chemistry every day becomes familiar with the properties of the substances that he or she works with regularly. An experienced chemist acquires knowledge based upon direct experience and also develops a good "chemical intuition" that allows the prediction of what might happen in untried circumstances. One of the authors of this book has spent many years studying the chemistry of complex ions and their compounds. He can predict with reasonable certainty the outcome of hundreds of reactions involving complex ions. Another author is an organic chemist. He can make predictions about hundreds of reactions of organic compounds. But given the need to predict, say, the outcome of a reaction between free atoms in the upper atmosphere, each would have to consult books in the library or experts on atmospheric chemistry.

At the beginning of a study of chemical reactions, limits must be set. Our goal is to review *the most common types of reactions of the simplest types of common compounds.* The emphasis is on inorganic chemistry. Within inorganic chemistry, the emphasis is on reactions in aqueous solution, particularly the behavior of ions in aqueous solution. As you learn more of the facts of chemistry, you will find it easier and easier to make reasonable and correct predictions. What is important here is to learn to recognize the possible simple *types* of chemical reactions and then, by analogy with some examples, conclude what products are possible in a given reaction.

You have already learned more about chemical reactions and the properties of chemical compounds that govern chemical reactions than you may realize. At this point you may wish to review briefly this chemistry, for it is the foundation for this chapter. [For review: general types of reactions, Table 5.1; the activity series of metals and the solubility rules, Tables 5.2 and 5.3 (repeated in this chapter as Tables 17.1 and 17.3); periodic trends in the properties of the elements, Section 10.13; rules for assigning oxidation numbers, Table 10.9; ions present in common strong electrolytes, Table 14.5; common strong acids and bases, Table 14.7; the chemistry of water, Sections 14.2 and 14.5; the properties of hydrogen and oxygen, Tables 16.6 and 16.12.]

For success in studying this chapter, it is essential to understand that **this is not an exercise in the memorization of chemical reactions.** It is, rather, an exercise in recognizing what is possible—first, whether a reaction is or is not a redox reaction; second, which type of redox or nonredox reaction can occur between the given reactants (four types of nonredox reactions and six types of redox reactions are covered). The chemical equations in each section are meant to be read and analyzed as examples of the type of reaction under discussion. At the end of the chapter, after each of the types of reactions has been considered individually, we present a flowchart for the systematic analysis of what reaction is possible for a given set of reactants. Intelligent use of the flowchart can lead to predicting the outcomes of a substantial number of common, simple inorganic reactions.

17.2 EQUILIBRIUM AND STABILITY

a. Is Chemical Equilibrium Established in All Reactions? You have been introduced to the chemical equilibria of ions in aqueous solution (e.g., the self-ionization of water in Section 14.5 and the ionization of weak acids in Section 14.8). Equilibrium is possible, under the appropriate conditions, in all phases. Any chemical reaction that can proceed in either direction is thought of as reversible and is potentially capable of establishing equilibrium.

The classic example of a reversible reaction involves iron and iron oxide (Fe_3O_4), and hydrogen and water. In 1766, Sir Henry Cavendish discovered the element hydrogen by passing steam through a red-hot iron gun barrel.

$$3Fe(s) + 4H_2O(g) \xrightarrow{\Delta} Fe_3O_4(s) + 4H_2(g) \qquad \textbf{(17.1)}$$

When the hydrogen is removed from the reaction system, the reaction proceeds in the direction shown by the arrow in Equation (17.1).

If instead, hydrogen gas is passed over hot iron oxide, the reaction that occurs is exactly the reverse of the one that Cavendish carried out.

$$Fe_3O_4(s) + 4H_2(g) \xrightarrow{\Delta} 3Fe(s) + 4H_2O(g) \qquad \textbf{(17.2)}$$

Here also the reaction proceeds in the direction shown when one of the products—in this case the steam—is removed from the reaction system.

However, if iron and steam are heated together in a closed vessel, the situation is different (Figure 17.1). As soon as traces of iron oxide and hydrogen have formed, they begin to react with each other to form iron and steam; reactions (17.1) and (17.2) proceed simultaneously.

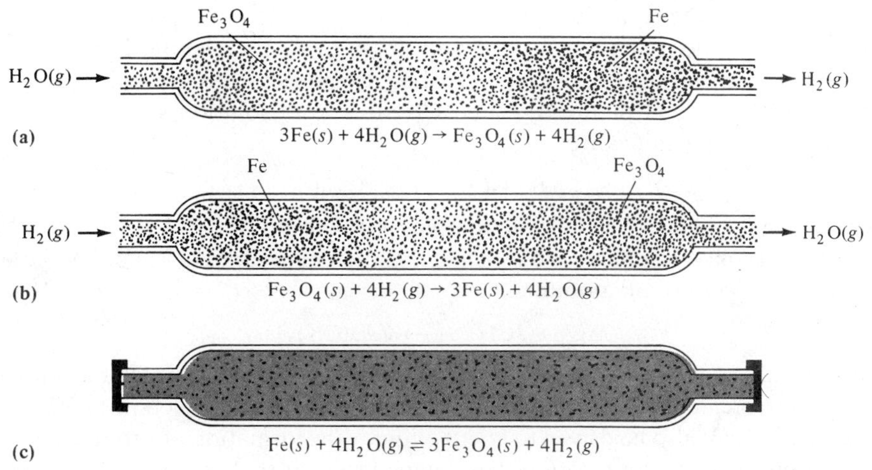

Fe_3O_4 Fe

$H_2O(g)$ → → $H_2(g)$

(a) $3Fe(s) + 4H_2O(g) \rightarrow Fe_3O_4(s) + 4H_2(g)$

Fe Fe_3O_4

$H_2(g)$ → → $H_2O(g)$

(b) $Fe_3O_4(s) + 4H_2(g) \rightarrow 3Fe(s) + 4H_2O(g)$

(c) $Fe(s) + 4H_2O(g) \rightleftharpoons 3Fe_3O_4(s) + 4H_2(g)$

Figure 17.1
A Reversible Reaction
The direction of the reaction in (a) and (b) is determined by the removal of a component of the reaction system. Equilibrium is reached in the closed system (c).

At first, the concentration of steam is relatively high and reaction (17.1) is faster than reaction (17.2). As steam is consumed, this reaction slows down. At the same time, since a supply of hydrogen is building up, the reverse reaction (17.2) between H_2 and Fe_3O_4 speeds up. Eventually, both reactions proceed with equal speed. Once this point is reached, the amounts of the substances present do not change. The reactions have not ceased, but each exactly opposes the other. A state of chemical equilibrium has been established.

The correct answer to the question posed in the heading to this section is, Yes, every chemical reaction is an equilibrium reaction. A finite chance always exists that under the appropriate conditions some product atoms, molecules, or ions will undergo the reverse of the reaction by which they were formed. A more practical question is: Are there any circumstances in which chemical equilibrium is relatively unimportant in considering the outcome of a reaction?

To some extent, of course, whether equilibrium is "important" or not depends upon the situation. For example, cadmium sulfide (CdS) is sparingly soluble in water. If it is to be prepared in the laboratory by precipitation from a solution that contains cadmium ion, an almost 100% yield can be expected and the reaction is thought of as "going to completion."

$$Cd^{2+} + S^{2-} \longrightarrow CdS(s) \qquad \textbf{(17.3)}$$

However, cadmium compounds are poisonous. If the effect of dumping cadmium sulfide near a public water supply is under consideration, the concentration of cadmium ions that will dissolve due to the reverse of reaction (17.3) and the establishment of the equilibrium

$$CdS(s) \overset{H_2O}{\rightleftharpoons} Cd^{2+} + S^{2-} \qquad \textbf{(17.3a)}$$

though, exceedingly small, might be *very* important.

Reactions that appear to "go to completion" continue until practically all of the reactants have been converted to products (as governed by the stoichiometry of the reaction). Such reactions *do* reach equilibrium, but one set of reactants is present in so much smaller concentration than the other that it appears that the reaction has gone only one way.

Many reactions considered in this chapter appear to go to completion. These reactions proceed because a product is formed that by its nature becomes unavailable for the reverse reaction or undergoes it only very slightly. For example, if a reaction takes place in the liquid or solid phase and a product escapes as a gas, the reverse reaction simply cannot take place. The following reaction is driven in the forward direction by the escape of hydrogen fluoride.

$$CaF_2(s) + H_2SO_4(conc) \xrightarrow{\Delta} CaSO_4(s) + 2HF(g) \qquad \textbf{(17.4)}$$

Where the reactants are ions in solution, the formation of a product that is only slightly ionized causes the reaction to appear to go to completion. For example, if solutions of silver nitrate and sodium chloride are mixed, silver chloride precipitates, thereby removing most of the Ag^+ and Cl^- ions from solution so that the reverse reaction is of little significance.

$$Ag^+ + Cl^- \longrightarrow AgCl(s)$$

The situation is similar when a stable complex ion such as $[Ag(CN)_2)]^-$ is formed. The complex ion is only very slightly dissociated, and therefore the Ag^+ and CN^- ions are present only in very small amounts.

$$Ag^+ + 2CN^- \longrightarrow [Ag(CN)_2]^-$$

Products such as water or a weak electrolyte that is soluble in water but only very slightly ionized can also essentially remove ions from solution (Section 17.3d).

A reaction can also be driven to completion by the formation of products so stable with respect to the reverse reaction that the tendency for that reaction to occur is very small. The reactions of strong oxidizing agents or strong reducing agents (Section 17.5) are often "irreversible." For example, lead sulfate and water, the products of the reaction of the oxidizing agent hydrogen peroxide with lead(II) sulfide

$$\underset{\text{reducing agent}}{PbS(s)} + 4\underset{\text{oxidizing agent}}{H_2O_2(aq)} \longrightarrow PbSO_4(s) + 4H_2O(l) \qquad \textbf{(17.5)}$$

are such poor oxidizing and reducing agents that the reverse reaction is of little significance. In other words, a mixture of lead(II) sulfate and water is very stable.

A general principle that applies to all chemical equilibria, as well as to all equilibria of other types, is known as **Le Chatelier's principle:** If a system at equilibrium is subjected to a stress, the system will react in a way that tends to relieve the stress. Suppose you had a saturated solution of sodium chloride (see Figure 15.1) with ions from the salt crystal entering and leaving the solution at equal rates. What do you think would happen if you removed some of the chloride ions from the solution (e.g., by reaction with Ag^+ as above)? The original equilibrium would respond to this stress—more chloride ions would dissolve from the salt crystal. As you might suspect, Le Chatelier's principle can be applied purposefully to upset an equilibrium in order to form more of a desired product. Later chapters give numerous examples of the application of Le Chatelier's principle.

One general objective in studying chemical reactions is to recognize reactions that proceed so far toward the formation of products that they can be viewed as going to "completion." To do so in a qualitative way, we look for the following:

1. Gaseous products from a reaction in the liquid or solid phase
2. Products that are insoluble from a reaction in solution
3. Products that remove ions from aqueous solution (e.g., precipitates, weak electrolytes, or complex ions)
4. Reactions of strong oxidizing and reducing agents

EXAMPLE 17.1 Equilibrium in Chemical Reactions

Which of the following reactions are likely to "go to completion"? For what reason would they do so?

(a) $Pb(NO_3)_2(aq) + 2NaOH(aq) \longrightarrow 2NaNO_3(aq) + Pb(OH)_2(s)$

(b) $2NaHCO_3(s) \xrightarrow[\text{in open vessel}]{\Delta} Na_2CO_3(s) + H_2O(g) + CO_2(g)$

(c) $2KCl(aq) + MgSO_4(aq) \longrightarrow K_2SO_4(aq) + MgCl_2(aq)$

(d) $H_2O_2(aq) + 2Fe^{2+} + 2H^+ \longrightarrow 2Fe^{3+} + 2H_2O(l)$

(a) This is a partner-exchange reaction driven by the formation of a slightly soluble product, lead(II) hydroxide. Formation of a solid phase removes lead and hydroxide ions from solution. **Completion is likely.**
(b) In this thermal decomposition reaction, two gaseous products can escape, preventing the reverse reaction. **Completion is likely.**
(c) The reactants and possible products are all strong electrolytes. **No reaction will occur and all ions will remain in solution.**
(d) This is a redox reaction—hydrogen peroxide is a strong oxidizing agent (Section 16.14). For the reverse reaction to occur, water would have to act as a reducing agent and be oxidized to hydrogen peroxide, which is unlikely. **The reaction should go to completion.**

EXERCISE Which of the following reactions are likely to go to completion? For what reasons?

(a) $Zn(CN)_2(s) + 2HCl(aq) \longrightarrow ZnCl_2(aq) + 2HCN(g)$
(b) $K_2CrO_4(aq) + 2AgNO_3(aq) \longrightarrow 2KNO_3(aq) + Ag_2CrO_4(s)$
(c) $2KCl(aq) + Mg(NO_3)_2(aq) \longrightarrow 2KNO_3(aq) + MgCl_2(aq)$

Answers (a) Completion. HCN is a gaseous product. (b) Completion. Ag_2CrO_4 is only slightly soluble. (c) No reaction. All reactants and products are strong electrolytes.

b. Stability in Chemistry A few comments are in order about the word "stable" as used in chemistry. Earlier, in connection with Equation (17.5), we stated that "a mixture of lead(II) sulfate and water is very stable." The reference was to stability toward oxidation and reduction. Water is too weak a reducing agent and lead(II) sulfate is too weak an oxidizing agent for these compounds to undergo a redox reaction with each other.

A description of something as "stable" means that it is likely to remain unchanged under a certain set of conditions. Whether or not the conditions are specified, we must always ask ourselves: Stable in what way? Stable with respect to what? The conditions might be temperature, pressure, acidity, alkalinity, oxidation–reduction, concentration, or the presence of certain other substances. Often a chemical compound is described as "stable" simply because it can be handled and stored without special precautions to protect it from reacting with water or oxygen in the air.

Clearly, understanding chemical reactions requires not only knowing the possible products of reactions, but also knowing whether or not a chemical reaction will actually occur when substances are brought together under a given set of conditions. Two areas of physical chemistry are devoted to providing a quantitative basis for predicting whether or not a reaction will occur—chemical thermodynamics and chemical kinetics.

Strictly speaking, there are two kinds of stability in chemistry—thermodynamic stability and kinetic stability. Thermodynamics allows the prediction of whether or not a reaction is favorable so far as energy changes are concerned. A *thermodynamically stable* substance is not likely to undergo chemical change readily because it is already in a state of low energy. In discussing chemical properties, general references to "stability" almost always refer to thermodynamic stability.

Thermodynamics, however, does not reveal how *fast* a reaction takes place; reaction rates are the concern of chemical kinetics. A *kinetically stable* substance might have the potential for undergoing a given reaction, but the reaction would take place too slowly to be useful or interesting.

The approach to studying chemical reactions in this chapter is qualitative. The following six chapters build upon the qualitative aspects of chemical reactions. Quantitative methods are presented for examining the rates and mechanisms of chemical reactions (chemical kinetics, Chapter 18), the nature and extent of equilibria in aqueous solutions (Chapters 19–22), the spontaneity of chemical reactions (thermodynamics, Chapter 23, which was introduced in our discussion of heats of reaction in Chapter 7), and electrochemistry and redox reactions (Chapter 24).

SUMMARY OF SECTIONS 17.1–17.2

In theory, all chemical reactions reach equilibrium. In practice, many reactions "go to completion"—they proceed so far that the reactants are almost entirely converted to products. "Completion" occurs when one or more products become unavailable for the reverse reaction (e.g., when a gas or a solid is formed; when ions are removed from aqueous solution; when a product stable to the reverse reaction is formed, often the case in reactions of strong oxidizing or reducing agents). A reaction is driven to "completion" in accordance with Le Chatelier's principle—if a system at equilibrium is subjected to a stress, the system will react in a way that tends to relieve the stress. The removal of a product or its unavailability for the reverse reaction stresses a system and results in the formation of additional product.

Not all reactions that appear possible actually take place. Both the energy change (thermodynamics) and the rate (kinetics) of the reaction must be favorable.

NONREDOX AND REDOX REACTIONS

17.3 A REVIEW OF NONREDOX REACTIONS

In the following sections several simple types of nonredox reactions are reviewed, all of which fit into the reaction categories introduced in Section 5.1—combination, decomposition, displacement, and partner exchange. Labeled examples of the types of reactions discussed are grouped at the end of each section and are followed by an unlabeled group for practice in recognizing the different reaction types.

Note that free elements are never reactants or products in nonredox reactions—to undergo a chemical change an element must be oxidized or reduced. [The sole exception is the formation of a complex between a metal and a ligand, such as $Ni(CO)_4$, in which the metal maintains its oxidation state of zero.] Many of the reactions of ions in aqueous solution are nonredox reactions.

a. Combination Reactions of Acidic and Basic Compounds Substances that have acidic properties and substances that have basic properties are likely to react with each other. We already have seen that nonmetal oxides such as SO_3 are acidic and metal oxides such as CaO are basic. The reactions of such oxides with water to give acids and bases, as illustrated below, are nonredox combination reactions. Acidic and basic oxides also can combine with each other to form salts. The combination of ammonia, a weak base, with acids to form ammonium salts is another example of this type of reaction.

An acidic oxide such as SO_2 combines with a limited amount of base to give an **acid salt**—a salt with the hydrogen-containing anion of a polyprotic acid, like $NaHCO_3$ or $NaHSO_3$. (With excess base the normal salts, Na_2CO_3 or Na_2SO_3, are formed.)

Nonredox combination—acidic and basic compounds

$SO_3(g) + H_2O(l) \longrightarrow H_2SO_4(l)$ — Acidic oxide + water $\longrightarrow$ acid

$CaO(s) + H_2O(l) \longrightarrow Ca(OH)_2(s)$ — Basic oxide + water $\longrightarrow$ base

$SO_3(g) + CaO(s) \longrightarrow CaSO_4(s)$ — Acidic oxide + basic oxide $\longrightarrow$ salt

$NH_3(g) + HCl(g) \longrightarrow NH_4Cl(s)$ — Ammonia + acid $\longrightarrow$ salt

$SO_2(g) + NaOH(aq) \longrightarrow NaHSO_3(aq)$ — Acidic oxide + limited base $\longrightarrow$ acid salt

Take a moment to think about the further examples given below. Read them thoroughly. Which are the acidic and the basic oxides?

Further examples of nonredox combination

$SO_2(g) + H_2O(l) \longrightarrow H_2SO_3(aq)$

$P_4O_{10}(s) + 6H_2O(l) \longrightarrow 4H_3PO_4(aq)$

$CO_2(g) + NaOH(aq) \longrightarrow NaHCO_3(aq)$

$Na_2O(s) + SO_2(g) \longrightarrow Na_2SO_3(s)$

$BaO(s) + SiO_2(s) \xrightarrow{\Delta} BaSiO_3(l)$

$2CO_2(g) + Ca(OH)_2(aq) \longrightarrow Ca(HCO_3)_2(aq)$

b. Thermal Decomposition to Give Compounds The driving force for thermal decomposition is likely to be the breakdown of a compound to form a gas. The clue that a compound will decompose thermally is often the presence of an anion that can be converted into a simple gaseous compound.

In some cases thermal decomposition is the reverse of combination. For example, many carbonates, which can be formed by the combination of basic metal oxides and carbon dioxide, undergo thermal decomposition to yield carbon dioxide and the metal oxide. The exceptions to this reaction are the carbonates of the active metals sodium, potassium, rubidium, and cesium.

Similarly, ammonium salts can decompose to give ammonia. However, this reaction occurs only if the ammonium salt contains an anion that cannot be reduced—a *nonoxidizing anion*. For example, the chloride ion is a nonoxidizing anion because the chlorine atom is already in its most negative oxidation state. (Decompositions of ammonium salts with oxidizing anions are redox reactions; Section 17.4b.)

Acid salts decompose with the production of water, the normal salt, and a gaseous product in which the oxidation number of the central atom is unchanged. For example, HSO_3^- yields SO_2 and HCO_3^- yields CO_2. Hydroxides, as might be expected, decompose thermally to yield water and oxides. Here again, compounds of the active metals are an exception—the hydroxides of sodium, potassium, rubidium, and cesium do not decompose upon heating.

Nonredox thermal decomposition

Reaction	
$CaCO_3(s) \xrightarrow{\Delta} CaO(s) + CO_2(g)$	Except carbonates of Na, K, Rb, Cs, which form the most basic oxides.
$NH_4Cl(s) \xrightarrow{\Delta} NH_3(g) + HCl(g)$	Ammonium salts containing *nonoxidizing* anions
$2NaHSO_3(s) \xrightarrow{\Delta} Na_2SO_3(s) + H_2O(g) + SO_2(g)$	All hydrogen sulfites
$2NaHCO_3(s) \xrightarrow{\Delta} Na_2CO_3(s) + H_2O(g) + CO_2(g)$	All hydrogen carbonates
$Mg(OH)_2(s) \xrightarrow{\Delta} MgO(s) + H_2O(g)$	Except hydroxides of Na, K, Rb, Cs

Take a moment to think about the further examples given below. Read them thoroughly. Which are the volatile products?

Further examples of nonredox thermal decomposition

$$(NH_4)_2S(s) \xrightarrow{\Delta} 2NH_3(g) + H_2S(g) \qquad 2Al(OH)_3(s) \xrightarrow{\Delta} Al_2O_3(s) + 3H_2O(g)$$

$$NH_4HCO_3(s) \xrightarrow{\Delta} NH_3(g) + CO_2(g) + H_2O(g) \qquad Ca(OH)_2(s) \xrightarrow{\Delta} CaO(s) + H_2O(g)$$

c. Nonredox Displacement Reactions Two types of nonredox displacement reactions are common. The first type is the displacement from a compound of one oxide by another oxide that is less volatile. In other words, the displaced oxide is more readily converted to a gas. For example, silicon dioxide (b.p. 2000 °C) displaces carbon dioxide (sublimes at −78 °C), as shown below.

The second type is the formation of one complex ion (Section 14.10) from another by displacement of ligands by other ligands. Such reactions occur when the product is more stable to dissociation than the reactant complex ion. Note the use of "stable" in referring to complex ions and their equilibria. Complex ions are present in solution in equilibrium with their dissociation products, for example

$$[HgCl_4]^{2-} \rightleftharpoons Hg^{2+} + 4Cl^-$$
$$[Hg(CN)_4]^{2-} \rightleftharpoons Hg^{2+} + 4CN^-$$

The $[Hg(CN)_4]^{2-}$ ion is the "more stable" of these two complex ions because it dissociates less. In solutions of these two complex ions of comparable concentration, the concentration of simple Hg^{2+} ions would be lower in the solution of the cyanide complex.

In general, CN^- forms quite stable complex ions; therefore, CN^- ions displace many other ligands, as illustrated below. [At this point in the study of chemical reactions you should recognize ligand displacement as a possible reaction; however, only with further study will prediction of which is a more stable complex ion be possible (Section 31.6).]

Nonredox displacement

$CaCO_3(s) + SiO_2(s) \xrightarrow{\Delta} CaSiO_3(l) + CO_2(g)$	Displacement of a more volatile oxide by a less volatile oxide
$[HgCl_4]^{2-} + 4CN^- \longrightarrow [Hg(CN)_4]^{2-} + 4Cl^-$	Displacement of a ligand in a complex ion to give a more stable complex ion.

Take a moment to think about the further examples given below. Read them thoroughly. Which are the displaced ligands or volatile oxides?

Further examples of nonredox displacement

$$Na_2CO_3(s) + SiO_2(s) \xrightarrow{\Delta} Na_2SiO_3(l) + CO_2(g)$$
$$2Ca_3(PO_4)_2(s) + 6SiO_2(s) \xrightarrow{\Delta} 6CaSiO_3(s \text{ or } l) + P_4O_{10}(g)$$
$$[PtCl_4]^{2-} + 2NH_3(aq) \longrightarrow [PtCl_2(NH_3)_2](s) + 2Cl^-$$
$$[Cu(H_2O)_4]^{2+} + 4NH_3(g) \longrightarrow [Cu(NH_3)_4]^{2+} + 4H_2O(l)$$

d. Partner-Exchange Reactions Partner-exchange reactions follow the pattern

$$AC + BD \longrightarrow AD + BC$$

and occur with no oxidation or reduction of A, B, C, or D. Partner-exchange reactions are most often thought of in connection with the reactions of ions in aqueous solution. In fact, when aqueous solutions of *soluble* ionic compounds are combined and no redox reaction can occur, there are only two possible results: no reaction or partner exchange. For partner exchange to occur between ionic compounds in aqueous solution, at least one product must remove ions from solution.

1. Formation of a gas Carbonates, sulfites, and many sulfides react with acids to produce gases—CO_2, SO_2, and H_2S, respectively. The first products of the reactions of carbonates and sulfites are carbonic acid (H_2CO_3) and sulfurous acid (H_2SO_3), which are *unstable* acids. In this case, "unstable" refers to the immediate decomposition of these acids to give water plus carbon dioxide or sulfur dioxide. The formation of gases drives these reactions toward completion (in an open container). For example

sulfite *acid* *unstable acid*

$$CaSO_3(s) + 2HCl(aq) \longrightarrow CaCl_2(aq) + H_2SO_3(aq) \quad \textbf{(17.6a)}$$

gas

$$H_2SO_3(aq) \longrightarrow H_2O(l) + SO_2(g) \quad \textbf{(17.6b)}$$

The overall equation is

$$CaSO_3(s) + 2HCl(aq) \longrightarrow CaCl_2(aq) + SO_2(g) + H_2O(l) \quad \textbf{(17.6c)}$$

2. Formation of an insoluble product The prediction of precipitate formation by use of the general solubility rules (reproduced in Table 17.1) was discussed in Section 5.2.

3. Formation of water or a weak electrolyte All reactions between water-ion acids and bases are irreversible due to removal of H^+ and OH^- through the formation of water. Similarly, reactions in which slightly soluble salts dissolve in strong aqueous acids are driven by formation of molecules of weak acids (illustrated below by the reaction of magnesium borate ($MgBO_3$) with

Table 17.1
Solubility of Common Ionic Compounds
These generalities apply to aqueous solutions at room temperature. *Soluble* indicates > 0.1 mol/L; *moderately soluble* (mod. sol.) indicates 0.1 – 0.01 mol/L; *insoluble* (insol.) indicates < 0.01 mol/L.

Generally soluble	Exceptions
Na^+, K^+, NH_4^+ compounds	—
Chlorides	Insol.: AgCl, Hg_2Cl_2 Sol. in hot water: $PbCl_2$
Bromides	Insol.: AgBr, $PbBr_2$, Hg_2Br_2 Mod. sol.: $HgBr_2$
Iodides	Insol.: heavier metal iodides
Sulfates	Insol.: $SrSO_4$, $BaSO_4$, $PbSO_4$, Hg_2SO_4 Mod. sol.: $CaSO_4$, Ag_2SO_4
Nitrates, nitrites	Mod. sol.: $AgNO_2$
Chlorites, perchlorates, permanganates	Mod. sol.: $KClO_4$
Acetates	Mod. sol.: $AgCH_3COO$
Generally insoluble	**Exceptions**
Sulfides	Sol.: NH_4^+, Na^+, K^+
Oxides, hydroxides	Sol.: Li_2O^a, LiOH, Na_2O^a, NaOH, K_2O^a, KOH, BaO^a, $Ba(OH)_2$ Mod. sol.: CaO^a, SrO^a, $Ca(OH)_2$, $Sr(OH)_2$
Carbonates, phosphates, cyanides, sulfites	Sol.: those of NH_4^+, Li^+ (except Li_3PO_4), Na^+, K^+

[a] Dissolve with evolution of heat and formation of hydroxides.

hydrochloric acid (HCl)). Neither the water nor the weak acids are physically removed from the solutions. However, water and weak acids are so slightly ionized that the ions from which they were formed are no longer freely available.

Partner-exchange reactions are not limited to ionic compounds in aqueous solution. Many other reactions follow the same pattern. For example, many hydrolysis reactions are partner-exchange reactions in which the OH group from the water molecule combines with the less electronegative atom of the reactant and the remaining hydrogen atom from water combines with the more electronegative atom. The products are usually two acids. In the reaction of boron trichloride, shown below, the products are boric acid, which could be written $B(OH)_3$ to show partner exchange more clearly, and HCl.

Partner-exchange reactions

$Na_2CO_3(aq) + 2HCl(aq) \longrightarrow 2NaCl(aq) + CO_2(g) + H_2O(l)$ $Na_2SO_3(aq) + 2HCl(aq) \longrightarrow 2NaCl(aq) + SO_2(g) + H_2O(l)$ $Na_2S(aq) + 2HCl(aq) \longrightarrow 2NaCl(aq) + H_2S(g)$	**Formation of a gas**—reaction of metal carbonates, sulfites, and sulfides with acids
$Al_2S_3(s) + 6HCl(aq) \longrightarrow 2AlCl_3(aq) + 3H_2S(g)$	**Formation of a gas**—slightly soluble sulfide + acid → metal salt + H_2S (except with *very* slightly soluble sulfides, e.g., HgS)
$NaCl(s) + H_2SO_4(conc) \longrightarrow NaHSO_4(s) + HCl(g)$	**Formation of a gas**—solid chloride + conc. sulfuric acid → metal hydrogen sulfate + HCl(*g*)
$AgNO_3(aq) + NaCl(aq) \longrightarrow AgCl(s) + NaNO_3(aq)$	**Formation of a precipitate**
$HCl(aq) + NaOH(aq) \longrightarrow NaCl(aq) + H_2O(l)$	**Formation of water**—all neutralizations of H^+ acid and OH^- base
$Mg_3(BO_3)_2(s) + 6HCl(aq) \longrightarrow 3MgCl_2(aq) + 2H_3BO_3(aq)$	**Formation of a weak electrolyte**—salt of weak acid + stronger acid → soluble salt + weaker acid
$BCl_3(l) + 3H_2O(l) \longrightarrow H_3BO_3(aq) + 3HCl(aq)$	**Hydrolysis**—compound of two nonmetals → two acids.

Take a moment to think about the further examples given below. Read them thoroughly. Which is the product that by its formation drives the reaction to completion?

Further examples of partner-exchange reactions

$$BaCl_2(aq) + (NH_4)_2CO_3(aq) \longrightarrow BaCO_3(s) + 2NH_4Cl(aq)$$
$$\text{or} \quad Ba^{2+} + CO_3^{2-} \longrightarrow BaCO_3(s)$$
$$PbCl_2(aq) + Na_2CrO_4(aq) \longrightarrow PbCrO_4(s) + 2NaCl(aq)$$
$$BaS(aq) + ZnSO_4(aq) \longrightarrow BaSO_4(s) + ZnS(s)$$
$$Na_2SO_3(aq) + 2HCl(aq) \longrightarrow 2NaCl(aq) + SO_2(g) + H_2O(l)$$
$$Ba(OH)_2(aq) + 2HNO_3(aq) \longrightarrow Ba(NO_3)_2(aq) + 2H_2O(l)$$
$$CuS(s) + 2HCl(aq) \longrightarrow CuCl_2(aq) + H_2S(g)$$
$$SiBr_4(l) + 3H_2O(l) \longrightarrow H_2SiO_3(s) + 4HBr(aq)$$

EXAMPLE 17.2 Nonredox Reactions

Will the following partner-exchange reactions take place? Give reasons for your answers.

(a) $H_2SO_4(aq) + 2KOH(aq) \longrightarrow K_2SO_4(aq) + 2H_2O(l)$

(b) $H_2SO_4(aq) + 2NaNO_3(aq) \longrightarrow 2HNO_3(aq) + Na_2SO_4(aq)$

(c) $HNO_3(aq) + NaF(aq) \longrightarrow HF(aq) + NaNO_3(aq)$

(d) $Cs_2CO_3(aq) + H_2SO_4(aq) \longrightarrow Cs_2SO_4(aq) + CO_2(g) + H_2O(l)$

(a) Water, a very slightly ionized substance, is formed, thus effectively removing H^+ (from H_2SO_4) and OH^- (from KOH). **The reaction will take place.**

(b) The reactants and products here are all strong electrolytes and are soluble in water. **No reaction will occur.**

(c) The formation of the weak acid HF ties up the H^+ and F^- ions from the reactants. **The reaction will take place.**

(d) The carbonic acid decomposes to yield a gas that escapes, driving the reaction to continue to form carbonic acid. **The reaction will take place.**

EXERCISE Will the following partner-exchange reactions occur?

(a) $K_2SO_3(aq) + 2HCl(aq) \longrightarrow SO_2(g) + H_2O(l) + 2KCl(aq)$

(b) $SnSO_4(aq) + H_2S(aq) \xrightarrow{H^+} SnS(s) + H_2SO_4(aq)$

(c) $BaSO_4(s) + 2CH_3COOH(aq) \longrightarrow Ba(CH_3COO)_2(aq) + H_2SO_4(aq)$

Explain your reasoning.

Answers (a) Yes, formation of a gas; (b) yes, formation of a precipitate; (c) no, neither product removes ions from availability for reverse reaction, $Ba^{2+} + SO_4^{2-} \longrightarrow BaSO_4(s)$.

EXAMPLE 17.3 Nonredox Reactions

Predict the products of each of the following nonredox reactions, write the balanced equations, and identify the type of reaction that occurs. If no reaction can occur, give a reason.

(a) $NH_3(g) + HNO_3(aq) \longrightarrow$

(b) $Na_2SO_3(s) + SiO_2(s) \xrightarrow{\Delta}$

(c) $CO_2(g) + MgO(s) \longrightarrow$

(d) $Na_2O(s) + H_2O(l) \longrightarrow$

(e) $Pb(NO_3)_2(aq) + Na_2S(aq) \longrightarrow$

(a) Ammonia, a basic substance, combines with acids to give ammonium salts:

$$\mathbf{NH_3(g) + HNO_3(aq) \longrightarrow NH_4NO_3(aq)}$$

This is a **combination reaction.**

(b) Silicon dioxide, which is not volatile, displaces the volatile oxide SO_2 from sodium sulfite:

$$\mathbf{Na_2SO_3(s) + SiO_2(s) \longrightarrow Na_2SiO_3(s) + SO_2(g)}$$

This is a **displacement reaction.**

(c) Sulfur dioxide, the acidic oxide of a nonmetal, and magnesium oxide, the basic oxide of a metal, will undergo a **combination reaction** to form a salt.

$$\mathbf{CO_2(g) + MgO(s) \longrightarrow MgCO_3(s)}$$

(d) The oxides of the active metals are basic and react with water to give hydroxides in solution:

$$\mathbf{Na_2O(s) + H_2O(l) \longrightarrow 2NaOH(aq)}$$

This is a **combination reaction.**

(e) A reaction between two soluble salts can occur only if ions are removed from solution. In this case, a precipitate of an insoluble sulfide will form.

$$\mathbf{Pb(NO_3)_2(aq) + Na_2S(aq) \longrightarrow PbS(s) + 2NaNO_3(aq)}$$

This is a **partner-exchange reaction.**

EXERCISE Predict the products of each of the following nonredox reactions:
(a) $H_2SO_4(conc) + NaCl(s) \longrightarrow$ (c) $[Zn(NH_3)_4]^{2+} + CN^- \longrightarrow$
(b) $PbSO_4(s) + HCl(aq) \longrightarrow$
Write the balanced equations and identify the type of reaction that occurs. If no reaction occurs, give a reason.

Answers (a) $H_2SO_4(conc) + NaCl(s) \longrightarrow NaHSO_4(s) + HCl(g)$, partner exchange, formation of a gas; (b) no reaction (the reverse reaction is favorable because $PbSO_4$ is slightly soluble); (c) $[Zn(NH_3)_4]^{2+} + 4CN^- \longrightarrow [Zn(CN)_4]^{2-} + 4NH_3(aq)$, partner exchange, formation of more stable complex ion.

17.4 A REVIEW OF REDOX REACTIONS

One way to recognize a redox reaction is by the presence of a free element as a reactant or product. Another way is by recognizing a reactant that is an oxidizing or reducing agent. Recall that an oxidizing agent contains an element that can have a lower oxidation state and a reducing agent contains an element that can have a higher oxidation state. Of course, the most certain way to recognize a redox reaction, *if* all of the reactants and products are known, is by determining whether changes in oxidation number have occurred.

A knowledge of the common oxidation states of the common elements is of great value in understanding redox reactions and in predicting their products. Table 17.2 summarizes these common oxidation states, and we recommend that you learn them. For the representative elements (except for oxygen and fluorine, which are the most electronegative elements), the highest positive oxidation state equals the group number. In general, the *p*-block elements have more than one positive oxidation state, and these states decrease by increments of 2 from the group oxidation state (except for nitrogen). The *d*-transition elements also tend to have more than one common oxidation state.

Metals, because they have only positive oxidation states, can function only as reducing agents. The activity series used earlier in our discussion of displacement reactions (Section 5.1c) is a list of metals in decreasing order of reactivity as reducing agents. (For convenience Table 17.3 duplicates Table 5.2.)

The nonmetals and semiconducting elements, which can have either positive or negative oxidation states, can function as either oxidizing or reducing agents. All are oxidizing agents in at least some of their reactions.

Simple redox reactions are often combination, decomposition, or displacement reactions, and many of these involve free elements as reactants or products. Such reactions are reviewed below. Many redox reactions are quite complex, however, and

Table 17.2
Common Oxidation States of Common Elements

	One common oxidation state (other than zero): Metals or semiconducting elements		Nonmetals	
s block	Na	+1	F	−1
	K	+1		
	Mg	+2		
	Ca	+2		
d transition	Ag	+1		
	Zn	+2		
	Ni	+2		
p block	Al	+3		
	Si	+4		
	More than one common oxidation state (other than zero): Metals or semiconducting elements		**Nonmetals**	
d transition	Cr	+3, +6	N	−3, +1, +2, +3, +4, +5
	Mn	+2, +4, +7	P	−3, +1, +3, +5
	Fe	+2, +3	O	−2, −1 (in peroxides)
	Co	+2, +3	S	−2, +4, +6
	Cu	+1, +2	Cl	−1, +1, +3, +5, +7
	Au	+1, +3	Br	−1, +5, +7
	Hg	+1, +2	I	−1, +5, +7
p block	Sn	+2, +4		
	Pb	+2, +4		
	As	−3, +3, +5		

Table 17.3
Activity Series of Metals
The most reactive metals are at the top, and activity in reactions in water decreases down the series. The metals at the top form cations most readily. Of the common acids, displacement of hydrogen occurs with hydrochloric acid and dilute sulfuric acid, but not with nitric acid, which forms nitrogen-containing reduction products instead of H_2. In displacement reactions where the metals are converted to ions in aqueous solution, the ions formed are usually the following: Li^+, K^+, Ba^{2+}, Sr^{2+}, Ca^{2+}, Na^+, Mg^{2+}, Al^{3+}, Mn^{2+}, Zn^{2+}, Cr^{2+}, Fe^{2+}, Ni^{2+}, Sn^{2+}, Pb^{2+}. (Note that all these ions have the lowest oxidation states of these elements.)

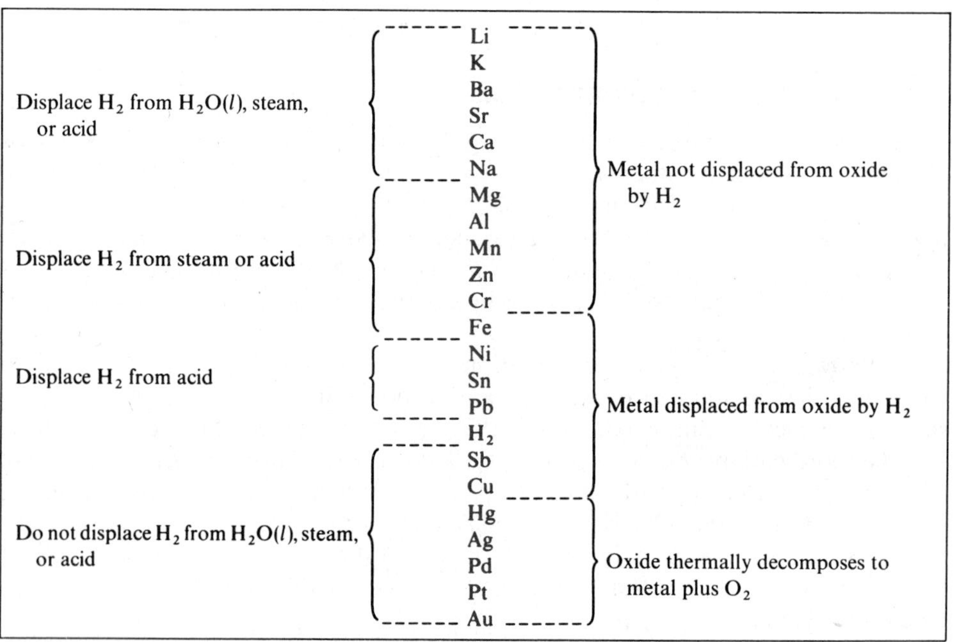

Displacement of H_2	Metal	Oxide
Displace H_2 from $H_2O(l)$, steam, or acid	Li, K, Ba, Sr, Ca, Na	Metal not displaced from oxide by H_2 (Li through Cr)
Displace H_2 from steam or acid	Mg, Al, Mn, Zn, Cr, Fe	
Displace H_2 from acid	Ni, Sn, Pb	Metal displaced from oxide by H_2 (Fe through Cu)
	H_2	
Do not displace H_2 from $H_2O(l)$, steam, or acid	Sb, Cu, Hg, Ag, Pd, Pt, Au	Oxide thermally decomposes to metal plus O_2 (Hg through Au)

there is no value in attempting to categorize them here. The products of such more complex redox reactions can, nevertheless, often be predicted based on a knowledge of the products usually formed by the more common oxidizing and reducing agents, which are discussed in Section 17.5.

a. Combination Reactions of Elements The simplest combination reactions are those between two different elements. The more electronegative element of the two, usually a nonmetal, is reduced. The more electropositive element of the two, which may be a metal or a nonmetal, is oxidized. You should be able to write equations for all possible combinations with each other of the elements in Table 17.2.

A number of metals have two oxidation states. A stronger oxidizing agent will yield the product with the metal in the higher oxidation state. For example, in the combination of iron with chlorine, either $FeCl_2$ or $FeCl_3$ can be formed. Because chlorine is quite a strong oxidizing agent, the formation of $FeCl_3$ is favored. However, the reactions of iron with iodine or sulfur, which are weaker oxidizing agents than chlorine (see Table 17.4), lead to FeI_2 and FeS, compounds of iron in its lower oxidation state. Which of two possible products will form may also depend upon the relative amounts of reactants, temperature, and other conditions.

Many compounds of an element that can have a higher oxidation state will combine with fluorine, chlorine, or oxygen to give a product in which that element has been oxidized (illustrated below for compounds of iron and carbon).

Table 17.4
Activity Series of Nonmetals
The strength of the nonmetals as oxidizing agents decreases down this series.

F_2
Cl_2
O_2
Br_2
I_2
S
Se
P(red)

Combination of two elements to give a compound

$2Mg(s) + O_2(g) \longrightarrow 2MgO(s)$	Most metals with O_2
$Ti(s) + 2Cl_2(g) \longrightarrow TiCl_4(l)$	Most metals with halogens
$2H_2(g) + O_2(g) \longrightarrow 2H_2O(l)$	Oxygen with nonmetals except halogens and noble gases
$S(s) + 3F_2(g) \longrightarrow SF_6(g)$	Two nonmetals; more electronegative nonmetal reduced

Combination of an element with a compound to give another compound

$2FeCl_2(s) + Cl_2(g) \longrightarrow 2FeCl_3(s)$	Element oxidized is in lower of two positive oxidation states
$4FeO(s) + O_2(g) \longrightarrow 2Fe_2O_3(s)$	
$2CO(g) + O_2(g) \longrightarrow 2CO_2(g)$	
$CO(g) + Cl_2(g) \longrightarrow COCl_2(g)$	

Take a moment to think about the further examples given below. Read them thoroughly. Which elements are oxidized and which reduced?

Further examples of combination reactions of elements

$ZnS(s) + 2O_2(g) \longrightarrow ZnSO_4(s)$	$Zn(s) + S(s) \longrightarrow ZnS(s)$
$SnS(s) + S(s) \longrightarrow SnS_2(s)$	$2SO_2(g) + O_2(g) \longrightarrow 2SO_3(g)$
$H_2(g) + Cl_2(g) \longrightarrow 2HCl(g)$	$SO_2(g) + Cl_2(g) \longrightarrow SO_2Cl_2(g)$
$2Al(s) + N_2(s) \longrightarrow 2AlN(s)$	$SO_2(g) + I_2(s) \longrightarrow$ no reaction

EXAMPLE 17.4 Redox Reactions

Describe what happens in each of the following combination reactions in terms of oxidation and reduction. Predict the products of the reactions by referring to the common oxidation states given in Table 17.2. Write the balanced equations.

(a) $Al(s) + S(s) \xrightarrow{\Delta}$ (b) $Na(s) + P(s) \xrightarrow{\Delta}$ (c) $Sn(s) + O_2(g) \longrightarrow$

In each case, the metal is the reducing agent and the nonmetal is the oxidizing agent. The metal increases in oxidation state and the nonmetal decreases in oxidation state. Note that in a binary compound between a metal and a nonmetal, the nonmetal has the negative oxidation state.

(a) Aluminum will have a +3 oxidation state and sulfur a −2 oxidation state.

$$\mathbf{2Al(s) + 3S(s) \xrightarrow{\Delta} Al_2S_3(s)}$$

(b) Sodium can have only a + 1 oxidation state and phosphorus will have a − 3 oxidation state.

$$\mathbf{3Na(s) + P(s) \xrightarrow{\Delta} Na_3P(s)}$$

(c) Tin is a metal that has more than one common oxidation state, +2 and +4. There are two possible products, SnO and SnO_2.

$$2Sn(s) + O_2(g) \longrightarrow 2SnO(s)$$
$$Sn(s) + O_2(g) \longrightarrow SnO_2(s)$$

Without specific knowledge of the chemistry of tin, it would be difficult to predict which oxide forms. If tin were in excess, SnO might form; if oxygen were in excess SnO_2 might form.

EXERCISE Describe what happens in each of the following combination reactions in terms of oxidation and reduction. Predict the products and write the balanced equations.

(a) $Mg(s) + N_2(g) \xrightarrow{\Delta}$ (b) $K(s) + Br_2(l) \longrightarrow$ (c) $P(s) + Cl_2(g) \longrightarrow$

Answers (a) and (b) The metals are the reducing agents (undergoing oxidation) and the nonmetals are the oxidizing agents (undergoing reduction):

$$3Mg(s) + N_2(g) \xrightarrow{\Delta} Mg_3N_2(s) \qquad 2K(s) + Br_2(l) \longrightarrow 2KBr(s)$$

(c) The more electronegative of the two nonmetals, Cl_2, is the oxidizing agent; P is the reducing agent and would be oxidized to either the +3 or +5 oxidation state:

$$2P(s) + 3Cl_2(g) \xrightarrow{\Delta} 2PCl_3(l) \qquad 2P(s) + 5Cl_2(g) \longrightarrow 2PCl_5(s)$$

b. Decomposition with Oxidation–Reduction Many of the oxides and halides (except fluorides) of the least active metals decompose when heated to give the metal plus oxygen or the free halogen, as shown below for mercury(II) oxide and platinum(IV) chloride. Such decomposition reactions are **internal redox reactions**—the oxidized and reduced elements originate in the same compound.

Metal nitrates undergo thermal decomposition to give oxygen and other products which depend upon the nature of the metal. Lithium and beryllium family nitrates are reduced to nitrites, whereas nitrates of less active metals yield oxides plus nitrogen dioxide and oxygen.

Heating an ammonium salt of an **oxidizing anion**—an anion capable of being reduced—causes an internal redox reaction in which the ammonium ion is oxidized and the anion is reduced. For example, the nitrate ion is an oxidizing anion because it contains nitrogen in the +5 oxidation state, from which it can be reduced.

$$\overset{-3}{N}H_4\overset{+5}{N}O_3(s) \xrightarrow{\Delta} \overset{+1}{N_2}O(g) + 2H_2O(g) \qquad \textbf{(17.7)}$$

Ammonium nitrite undergoes a similar reaction:

$$\overset{-3}{N}H_4\overset{+3}{N}O_2(s) \xrightarrow{\Delta} \overset{0}{N_2}(g) + 2H_2O(g) \qquad \textbf{(17.8)}$$

Redox decomposition

$2HgO(s) \xrightarrow{\Delta} 2Hg(l) + O_2(g)$ $PtCl_4(s) \xrightarrow{\Delta} Pt(s) + 2Cl_2(g)$	Of oxides and halides (except fluorides) of least active metals to give the metal and elemental O_2 or halogen
$2KNO_3(s) \xrightarrow{\Delta} 2KNO_2(s) + O_2(g)$	Of Li and Be family nitrates (except $Be(NO_3)_2$) to give nitrites
$2Pb(NO_3)_2(s) \xrightarrow{\Delta} 2PbO(s) + 4NO_2(g) + O_2(g)$	Of nitrates of less active metals to give oxide, NO_2, and O_2
$(NH_4)_2Cr_2O_7(s) \xrightarrow{\Delta} N_2(g) + Cr_2O_3(s) + 4H_2O(g)$	Of ammonium salts with oxidizing anions

Take a moment to think about the further examples given on the next page. Read them thoroughly. Which elements are oxidized and which are reduced?

Further examples of redox decomposition

$$2HgO(s) \xrightarrow{\Delta} 2Hg(l) + O_2(g)$$
$$2AuCl_3(s) \xrightarrow{\Delta} 2Au(s) + 3Cl_2(g)$$
$$Mg(NO_3)_2(s) \xrightarrow{\Delta} Mg(NO_2)_2(s) + O_2(g)$$

c. Displacement of One Element from a Compound by Another Element Reactions in which one element is displaced from a compound by another element make up a large group of simple redox reactions. The displacement reactions discussed in connection with the activity series of metals in aqueous solution are all redox reactions of this type (Section 5.1c). A stronger reducing agent can displace a weaker one from a compound. For example, as shown below, zinc, the stronger reducing agent, displaces copper from a solution of copper(II) sulfate. Other examples of the displacement reactions of metals and hydrogen discussed earlier are also given below.

Within limits, the short nonmetal activity series can be used to predict the displacement in aqueous solution of nonmetals by other nonmetals that are stronger oxidizing agents. The halogens, in general, displace each other in the order shown in the table. For example, chlorine displaces bromine from bromides.

$$\overset{0}{Cl_2}(g) + 2K\overset{-1}{Br}(aq) \longrightarrow 2K\overset{-1}{Cl}(aq) + \overset{0}{Br_2}(aq) \tag{17.9}$$

stronger oxidizing agent than Br_2 (Cl₂) — *weaker oxidizing agent than Cl_2* (Br₂)

Also, as is predicted by the activity series, oxygen and iodine both displace sulfur.

Redox displacement

Reaction	
$2Na(s) + 2H_2O(l) \longrightarrow H_2(g) + 2NaOH(aq)$	Of H_2 from H_2O by an active metal
$Zn(s) + 2HCl(aq) \longrightarrow H_2(g) + ZnCl_2(aq)$	Of H_2 from a nonoxidizing acid by metals above H_2 in the activity series
$Zn(s) + CuSO_4(aq) \longrightarrow ZnSO_4(aq) + Cu(s)$	Of a metal from a compound by a more active metal
$H_2(g) + CuO(s) \xrightarrow{\Delta} Cu(s) + H_2O(g)$	Of a metal from an oxide by hydrogen
$Br_2(aq) + 2NaI(aq) \longrightarrow 2NaBr(aq) + I_2(aq)$	Of a halogen by a more active halogen
$I_2(aq) + H_2S(aq) \longrightarrow S(s) + 2HI(aq)$	Of a nonmetal by a more active nonmetal
$O_2(g) + 2H_2S(aq) \longrightarrow 2H_2O(l) + 2S(s)$	

Reactions that follow the displacement pattern occur not only in solution, but in other phases as well, for example

$$2\overset{0}{Mg}(l) + \overset{+4}{Ti}Cl_4(g) \xrightarrow{\Delta} \overset{0}{Ti}(s) + 2\overset{+2}{Mg}Cl_2(l) \tag{17.10}$$

However, note that the reactivity of the nonmetals and metals under conditions other than reactions in aqueous solution can sometimes be quite different from that reflected in the activity series. For example, from Table 17.4 we would predict that the displacement of chlorine by oxygen is not possible because chlorine stands above oxygen in the activity series. However, the reaction does occur at a high temperature in the presence of a catalyst and indeed at one time was the basis for an industrial method for the production of chlorine.

$$4HCl(g) + O_2(g) \xrightarrow[\text{catalyst}]{350\text{–}400\ ^\circ C} 2H_2O(g) + 2Cl_2(g)$$

Take a moment to think about the further examples given on the next page. Read them thoroughly. Which elements are oxidized and which reduced? Which elements are more active oxidizing and reducing agents?

Further examples of redox displacement

$$H_2(g) + NiO(s) \longrightarrow H_2O(l) + Ni(s)$$
$$Ni(s) + H_2O(l) \longrightarrow \text{no reaction}$$
$$Ni(s) + 2HCl(aq) \longrightarrow NiCl_2(aq) + H_2(g)$$
$$Zn(s) + Na_2SO_4(aq) \longrightarrow \text{no reaction}$$
$$Ca(s) + 2H_2O(l) \longrightarrow Ca(OH)_2(s) + H_2(g)$$
$$Cd(s) + NiCl_2(aq) \longrightarrow Ni(s) + CdCl_2(aq)$$

EXAMPLE 17.5 Redox Reactions

Will the following reactions occur? Justify your answers.

(a) $Cu(s) + Sn^{2+} \longrightarrow Cu^{2+} + Sn(s)$

(b) $Fe(s) + 2HCl(aq) \longrightarrow FeCl_2(aq) + H_2(g)$

(c) $2Al(s) + 3CaO(s) \xrightarrow[\text{low pressure}]{\Delta} Al_2O_3(s) + 3Ca(g)$

(d) $Zn(s) + CaCl_2(aq) \longrightarrow ZnCl_2(aq) + Ca(s)$

(e) $2NaF(aq) + Br_2(l) \longrightarrow 2NaBr(aq) + F_2(g)$

(f) $H_2S(aq) + Cl_2(g) \longrightarrow S(s) + 2HCl(aq)$

These are all reactions involving displacement of one element from a compound by another element. Examination of the activity series for metals in Table 17.3 indicates that **reaction (a) will not occur** (copper is not more active than tin) and **reaction (b) will occur** (iron is above hydrogen in the series). The important thing to note first about reaction (c) is that it does *not* take place in aqueous solution, and therefore the activity series may not apply. We can predict that **reaction (c) will occur,** however, because the temperature is apparently meant to be high enough for the calcium to escape as a vapor at a lowered pressure. By contrast, **reaction (d),** a displacement of calcium by zinc, **will not take place** in aqueous solution because zinc is a less active metal than calcium [also because Ca is too reactive to form in H_2O]. **Reaction (e) will not take place** because it would be a displacement of fluorine, the most active nonmetal, by bromine, which is less active than fluorine. **Reaction (f), however, will take place,** for chlorine is a more active nonmetal than sulfur.

EXERCISE Will the following reactions occur? Justify your answers.

(a) $2Cl^- + I_2(aq) \longrightarrow Cl_2(aq) + 2I^-$

(b) $ZnCl_2(aq) + Pb(s) \longrightarrow Zn(s) + PbCl_2(aq)$

(c) $Cu(s) + 2Ag^+ \longrightarrow Cu^{2+} + 2Ag(s)$

Answers (a) No, Cl_2 is a stronger oxidizing agent than I_2; (b) no, Zn is a stronger reducing agent than Pb; (c) yes, Cu is a stronger reducing agent than Ag.

EXAMPLE 17.6 Redox Reactions

Some information about the reactions of four metals is given below. On the basis of this information, arrange the metals in the order of decreasing strength as reducing agents.

(a) $Sn(s) + H_2O(g) \longrightarrow \text{no reaction}$

(b) $Zn(s) + H_2O(g) \xrightarrow{\Delta} ZnO(s) + H_2(g)$

(c) $Ca(s) + 2H_2O(l) \longrightarrow Ca(OH)_2(s) + H_2(g)$

(d) $Zn(s) + H_2O(l) \longrightarrow \text{no reaction}$

(e) $3Fe(s) + 4H_2O(g) \xrightarrow{\Delta} 4H_2(g) + Fe_3O_4(s)$

(f) $Fe(s) + H_2O(l) \longrightarrow \text{no reaction}$

(g) $4H_2(g) + Fe_3O_4(s) \xrightarrow{\Delta} 3Fe(s) + 4H_2O(g)$

(h) $H_2(g) + ZnO(s) \longrightarrow \text{no reaction}$

(i) $H_2(g) + SnO(s) \longrightarrow Sn(s) + H_2O(l)$

The most reactive metals displace hydrogen from cold water. Calcium (reaction c) is thus the most reactive metal in the group. The least active metals will not displace hydrogen from water even at the temperature of steam. Tin (reaction a) therefore appears to be the least reactive metal in the group. Metals of activities intermediate between calcium and tin would displace hydrogen from steam but not from cold water. Zinc (reactions b and d) and iron (reactions e and f) fall into this category. To distinguish between them we can examine the reactions of their oxides with hydrogen. Hydrogen is able to reduce the oxides of less active metals, but not those of more active metals. Reactions (g), (h), and (i) show that zinc is a more active metal than either tin or iron. The order of decreasing strength as reducing agents for these metals is therefore

$$\mathbf{Ca > Zn > Fe > Sn}$$

[For a long period in the history of chemistry, reasoning like that demonstrated above was the *only* way to derive information on the relative reactivity of chemical substances.]

EXERCISE On the basis of the information given in the following reactions, arrange the metals in order of decreasing strength as reducing agents.

(a) $Cu(s) + Pd^{2+} \longrightarrow Pd(s) + Cu^{2+}$ (c) $Pd(s) + Rh^{3+} \longrightarrow$ no reaction
(b) $Pt(s) + Pd^{2+} \longrightarrow$ no reaction (d) $Rh(s) + Cu^{2+} \longrightarrow$ no reaction

Answer Cu > Rh > Pd > Pt.

d. Disproportionation Reactions In a **disproportionation reaction** an element in one oxidation state is both oxidized and reduced. One reactant must contain an element that is capable of having at least three oxidation states—that in the reactant plus one higher and one lower oxidation state. The halogens, with their many common oxidation states, can undergo disproportionation in many ways. For example, the reaction of chlorine with water to give hydrochloric acid and hypochlorous acid is a disproportionation reaction.

can have both higher and lower ox. states (0) — *lower ox. state* (−1) — *higher ox. state* (+1)

$$\overset{0}{Cl_2}(g) + H_2O(l) \longrightarrow H\overset{-1}{Cl}(aq) + HO\overset{+1}{Cl}(aq) \qquad (17.11)$$

hypochlorous acid

The elements most commonly encountered in disproportionation reactions are listed in Table 17.5. Generally it is necessary to know the chemistry of a specific element to predict the products of a disproportionation reaction involving that element. Sulfites, copper(I) compounds, and manganates are common examples of compounds that disproportionate. Soluble copper(I) compounds, when they dissolve, immediately disproportionate to give the free metal and a copper(II) compound, and manganate ion disproportionates in acidic solution.

$$4Na_2\overset{+4}{S}O_3(s) \xrightarrow{\Delta} Na_2\overset{-2}{S}(s) + 3Na_2\overset{+6}{S}O_4(s)$$ Sulfite ion disproportionation

$$\overset{+1}{Cu_2}SO_4(s) \xrightarrow{H_2O} \overset{0}{Cu}(s) + \overset{+2}{Cu}SO_4(aq)$$ Copper(I) disproportionation

$$3\overset{+6}{Mn}O_4^{2-} + 4H^+ \longrightarrow \overset{+4}{Mn}O_2(s) + 2\overset{+7}{Mn}O_4^- + 2H_2O(l)$$ Manganate ion disproportionation

Substances that disproportionate immediately upon their formation are often referred to as "unstable." For example, chlorous acid, which is formed when an alkaline solution of chlorite ion (ClO_2^-) is treated with acid, is an *extremely unstable substance*. When "unstable" is used with such strong emphasis, it usually means "watch out." In this case, chlorous acid ($HClO_2$) disproportionates as soon as it is formed, a reaction that often is accompanied by the explosion of the chlorine dioxide (ClO_2) produced.

$$ClO_2^- + H^+ \rightleftharpoons HClO_2(aq)$$

$$5H\overset{+3}{Cl}O_2(aq) \longrightarrow 4\overset{+4}{Cl}O_2(g) + H\overset{-1}{Cl}(aq) + 2H_2O(l)$$

Table 17.5
Elements Most Commonly Involved in Disproportionation Reactions
Note that the maximum oxidation states of the nonmetals equal their periodic table group numbers.

	Possible oxidation states
N	−3, 0, +1, +2, +3, +4, +5
P	−3, 0, +3, +5
S	−2, 0, +4, +6
Cl	−1, 0, +1, +3, +5, +7
Br	−1, 0, +1, +3, +5, +7
I	−1, 0, +1, +3, +5, +7
Mn	0, +2, +3, +4, +5, +6, +7
Cu	0, +1, +2
Au	0, +1, +3
Hg	0, +1, +2

e. Electron Transfer between Monatomic Ions in Aqueous Solution Direct transfer of electrons from one ion to another can occur in aqueous solution. For example

$$2FeCl_3(aq) + SnCl_2(aq) \longrightarrow 2FeCl_2(aq) + SnCl_4(aq) \qquad \textbf{(17.12a)}$$

The net ionic equation for this or any reaction between a soluble iron(III) compound and a soluble tin(II) compound is

$$\underset{\textit{oxidizing agent}}{\overset{\textit{gains electrons}}{2Fe^{3+}}} + \underset{\textit{reducing agent}}{\overset{\textit{loses electrons}}{Sn^{2+}}} \longrightarrow 2Fe^{2+} + Sn^{4+} \qquad \textbf{(17.12b)}$$

Nonmetal ions can also participate in electron-transfer reactions, for example

$$\underset{\textit{oxidizing agent}}{2Cu^{2+}} + \underset{\textit{reducing agent}}{2I^-} \longrightarrow \overset{\textit{electrons gained}}{2Cu^+} + \overset{\textit{electrons lost}}{I_2(aq)}$$

To predict with certainty when such reactions will occur requires the use of quantitative methods for evaluating the oxidizing and reducing strengths of ions (Section 24.8).

f. More Complex Redox Reactions Many redox reactions do not fit any of the "classes" of reactions described in the preceding sections. In general, redox reactions involving oxygen-containing redox agents in acidic or alkaline solution are not easily categorized. In most such reactions, the participation of water and hydrogen ion or hydroxide ion as reactants or products is a necessity. One example is the oxidation of copper by nitric acid

$$3\overset{0}{Cu}(s) + 8H\overset{+5}{N}O_3(dil) \longrightarrow 3\overset{+2}{Cu}(NO_3)_2(aq) + 2\overset{+2}{N}O(g) + 4H_2O(l)$$

or, written as a net ionic equation

$$3Cu(s) + 8H^+ + 2NO_3^- \longrightarrow 3Cu^{2+} + 2NO(g) + 4H_2O(l)$$

The participation of water and H^+ or OH^- in redox reactions contributes to the complexity of the equations and the need for a stepwise method for balancing them (Sections 16.4 and 24.2). It also means that the concentration of an acid or a base can influence the course of a reaction. For example, if concentrated instead of dilute nitric acid is used to dissolve copper, the reduction product formed from the nitric acid is different from that shown above.

$$\overset{0}{Cu}(s) + 4H\overset{+5}{N}O_3(conc) \longrightarrow \overset{+2}{Cu}(NO_3)_2(aq) + 2\overset{+4}{N}O_2(g) + 2H_2O(l)$$

Some further examples of redox reactions that do not fit into simple categories are the combination of oxygen with each of the elements in a binary compound:

$$2Zn\overset{-2}{S}(s) + 3\overset{0}{O}_2(g) \overset{\Delta}{\longrightarrow} 2ZnO(s) + 2\overset{+4-2}{SO_2}(g)$$

$$\overset{-4}{C}H_4(g) + 2\overset{0}{O}_2(g) \longrightarrow \overset{+4}{C}O_2(g) + 2H_2\overset{-2}{O}(l)$$

and specific reactions between gaseous, liquid, or solid compounds such as

$$2\overset{-3}{N}H_3(g) + 3\overset{+2}{Cu}O(s) \longrightarrow \overset{0}{N}_2(g) + 3\overset{0}{Cu}(s) + 3H_2O(g)$$

$$3\overset{+4}{C}O_2(g) + \overset{-4}{C}H_4(g) \overset{\Delta}{\longrightarrow} 4\overset{+2}{C}O(g) + 2H_2O(g)$$

EXAMPLE 17.7 Redox Reactions

Predict the possible products of the following reactions. Explain your reasoning.

(a) $Mn(s) + HCl(aq) \longrightarrow$
(b) $Al(s) + O_2(g) \longrightarrow$
(c) $Cl_2(g) + NaBr(aq) \longrightarrow$
(d) $Ni(s) + CaCl_2(aq) \longrightarrow$

(a) Manganese lies above hydrogen in the activity series and should displace it. Manganese would probably be oxidized to its lowest oxidation state of +2. Products: $\mathbf{MnCl_2}$**(*aq*)** + $\mathbf{H_2}$**(*g*)**
(b) Aluminum is an active metal and will combine with oxygen. The +3 oxidation state is most common for aluminum. Product: $\mathbf{Al_2O_3}$**(*s*)**
(c) Chlorine is a more active nonmetal than bromine and should displace it. Products: $\mathbf{Br_2}$ **+ NaCl(*aq*)**
(d) Nickel is a less active metal than calcium. **No reaction will occur.**

EXERCISE Predict the possible products of the following reactions. Explain your reasoning.

(a) $PCl_3(l) + Cl_2(g) \longrightarrow$
(b) $Fe(s) + Br_2(l) \longrightarrow$
(c) $Br_2(l) + NaCl(aq) \longrightarrow$

Answers (a) PCl_5—phosphorus can be oxidized to the +5 oxidation state; (b) $FeBr_2$ or $FeBr_3$—iron has two common oxidation states; (c) no reaction—bromine is less active than chlorine.

EXAMPLE 17.8 Nonredox and Redox Reactions

Classify each of the following reactions as a nonredox or a redox reaction. Further classify each reaction according to type of nonredox or redox reaction where possible.

(a) $3NaOH(aq) + H_3PO_4(aq) \longrightarrow Na_3PO_4(aq) + 3H_2O(l)$
(b) $3NaOCl(s) \xrightarrow{\Delta} NaClO_3(s) + 2NaCl(s)$
(c) $PCl_3(l) + 3H_2O(l) \longrightarrow H_3PO_3(aq) + 3HCl(aq)$

(a) This is the reaction of an acid and a base to give a salt and water. **Nonredox. Partner-exchange reaction**—neutralization, driven by the formation of water.
(b) Heating a single substance causes a decomposition reaction. Chlorine, an element of variable oxidation state, is +1 in NaOCl, +5 in $NaClO_3$, and −1 in NaCl. **Redox. A decomposition reaction that is also a disproportionation reaction.**
(c) Phosphorus trichloride is a compound between two nonmetals. The water molecule is split in this reaction. **Nonredox. Partner-exchange**—a hydrolysis reaction driven by formation of a weak electrolyte.

EXERCISE Classify each of the following reactions as a nonredox or a redox reaction. Further classify each reaction according to type of nonredox or redox reaction where possible.

(a) $KCl(l) + Na(g) \xrightarrow{\Delta} K(g) + NaCl(l)$
(b) $2Na(s) + O_2(g) \xrightarrow{\Delta} Na_2O_2(s)$
(c) $2B_2O_3(s) + Na_2CO_3(s) \xrightarrow{\Delta} Na_2B_4O_7(s) + CO_2(g)$
(d) $Cr_2O_7^{2-} + 6I^- + 14H^+ \longrightarrow 2Cr^{3+} + 3I_2(s) + 7H_2O(l)$

Answers (a) Redox, displacement of one element from a compound by another element; (b) redox, combination of elements; (c) nonredox, displacement of a more volatile oxide by a less volatile oxide; (d) redox, a reaction of an oxygen-containing anion (a "more complex" redox reaction).

17.5 OXIDIZING AND REDUCING AGENTS

a. Common Oxidizing Agents Some common oxidizing agents are listed in Table 17.6, with their usual reduction products. Fluorine is the strongest of the common oxidizing agents. Where more than one product is listed, the one formed depends upon the relative oxidizing and reducing strengths of the reactants and the other conditions of the reaction.

Table 17.6
Common Oxidizing Agents
In the presence of a sufficiently strong oxidizing agent, the species on the right can be oxidized to give the species on the left. It is not possible to predict the products of specific reactions in which nitric acid, nitrous acid, nitrate ion, or concentrated sulfuric acid are the oxidizing agents.

Oxidizing agent	Usual reduction product
F_2	F^-
Cl_2	Cl^-
Br_2	Br^-
O_2	O^{2-}
S	S^{2-}
Cu^+, Cu^{2+}	Cu
Ag^+	Ag
Fe^{3+}	Fe^{2+}, Fe
conc. H_2SO_4	SO_2, H_2S, S
conc. HNO_3	NO_2, various others[a]
dil. HNO_3	NO, NO_2, N_2O, various others[a]
HNO_2 (nitrous acid)	NO, N_2O
HClO (hypochlorous acid)	Cl^-, Cl_2
NO_3^- (nitrate ion, in acidic solution)	HNO_2, NO, N_2O_4
ClO_3^- (chlorate ion)	Cl^-, Cl_2
ClO^- (hypochlorite ion)	Cl^-, Cl_2
BrO_3^- (bromate ion)	Br^-, Br_2
IO_3^- (iodate ion)	I^-, I_2
H_2O_2 (hydrogen peroxide)	H_2O
MnO_4^- (permanganate ion)	
in acidic solution	Mn^{2+}
in neutral or alkaline solution	$MnO_2(s)$
$Cr_2O_7^{2-}$ (dichromate ion, in acidic solution)	Cr^{3+}
BiO_3^- (bismuthate ion, in acidic solution)	Bi(III)
$PbO_2(s)$ (lead(IV) oxide, in acidic solution)	Pb^{2+}
$MnO_2(s)$ (manganese(IV) oxide, in acidic solution)	Mn^{2+}

[a] Other possible products, such as NH_3, NH_4^+, and HNO_2, depending upon acid concentration and nature of the reducing agent.

Many oxidizing agents contain oxygen combined with other elements in their higher oxidation states. In any reaction of a compound or anion containing oxygen plus an element of variable oxidation state, it is reasonable to consider the possibility of an oxidation–reduction reaction.

Among the nonmetals, nitrogen (Group V), sulfur (Group VI), and chlorine, bromine, and iodine (Group VII) are found in common oxygen-containing oxidizing agents. In addition, hydrogen peroxide is an oxidizing agent (Section 16.14).

Acids that are oxidizing agents are referred to as **oxidizing acids.** Two common strong acids—sulfuric acid and nitric acid—are oxidizing acids, sulfuric acid only when it is concentrated. In predicting the possible products of reactions of these acids, oxidation–reduction must be considered as well as acid–base interaction. Concentrated perchloric acid ($HClO_4$) is, as you might surmise, also an oxidizing acid. At room temperature concentrated perchloric acid is a weak oxidizng agent, but when concentrated and hot it is a very strong oxidant. **The reactions of perchloric acid can be violently explosive,** particularly if the acid comes into contact with any organic material such as rubber. Therefore, it is not commonly used as an oxidizing agent.

Among the metals, those more stable in lower oxidation states are oxidizing agents when in their higher oxidation states, usually in combination with oxygen. For example, the representative metals lead and bismuth are most stable as lead(II) and bismuth(III). Lead(IV) appears on the list of common oxidizing agents in PbO_2 and bismuth(V) in the bismuthate ion, BiO_3^-.

The 3*d*-transition metals with the highest oxidation states are chromium (+6), manganese (+7), and iron (+6). The oxoanions of these elements in these states (chromate ion, CrO_4^{2-}; dichromate ion, $Cr_2O_7^{2-}$; permanganate ion, MnO_4^-; and ferrate ion, FeO_4^{2-}) are all powerful oxidizing agents, whether in solid compounds or in solutions. Permanganate ion, usually from potassium permanganate solutions (which are a beautiful deep purple), and dichromate ion, usually from potassium dichromate solutions (which are deep orange), are frequently used as oxidizing agents in the laboratory.

b. Common Reducing Agents Free metals dominate the list of common reducing agents (Table 17.7). Lithium is the strongest common reducing agent, and all the other metals of the lithium and beryllium families are also strong reducing agents. Of the metals above hydrogen in the activity series (Table 17.3), all of which are reducing agents, aluminum and tin (representative metals) and iron and zinc (transition metals) are most often utilized for their reducing properties. The Fe^{2+} ion is also a useful reducing agent.

Several nonmetals in their lowest oxidation states are also good reducing agents. The hydride ion (H^-) is the strongest reducing agent of this type. The reducing strength of the halide ions increases down the fluorine family and only I^- is a reducing agent of significance. Sulfur in its -2 oxidation state, while not a strong reducing agent, is useful in this capacity, and sulfide ion (S^{2-}) and hydrogen sulfide (H_2S) are encountered as reducing agents. The sulfite ion (SO_3^{2-}), also a common reducing agent, contains sulfur in the $+4$ state and is oxidized to sulfate ion, with sulfur in the $+6$ state.

Table 17.7
Common Reducing Agents

Reducing agent	Usual oxidation product
Li, Na, K	Li^+, Na^+, K^+
Mg, Ca	Mg^{2+}, Ca^{2+}
Al, Sn	Al^{3+}, Sn^{2+}, Sn^{4+} (most often)
Fe,Zn	Fe^{2+} (sometimes Fe^{3+}), Zn^{2+}
Fe^{2+}	Fe^{3+}
Sn^{2+}	Sn^{4+}
H_2	H^+
H^-	H_2 or H^+
I^-	I_2
S^{2-}, H_2S	S
SO_3^{2-}	SO_4^{2-}
HPO_3^{2-}	H_3PO_4
NH_3	N_2
N_2H_4	N_2

Some examples of redox reactions of common oxidizing and reducing agents

Take a moment to think about the examples of redox reactions below. Read them thoroughly and compare them with Tables 17.6 and 17.7. Which are the common oxidizing and reducing agents? What are their products? Note that many of these reactions occur in acidic solutions.

$$SO_3^{2-} + Br_2(l) + H_2O(l) \longrightarrow SO_4^{2-} + 2Br^- + 2H^+$$
$$IO_3^- + 3SO_2(g) + 3H_2O(l) \xrightarrow{H^+} I^- + 3SO_4^{2-} + 6H^+$$
$$5PbO_2(s) + 2Mn^{2+} + 4H^+ \longrightarrow 2MnO_4^- + 5Pb^{2+} + 2H_2O(l)$$
$$4I^- + 2Cu^{2+} \longrightarrow 2CuI(s) + I_2(s)$$
$$6Fe^{2+} + Cr_2O_7^{2-} + 14H^+ \longrightarrow 6Fe^{3+} + 2Cr^{3+} + 7H_2O(l)$$
$$2MnO_4^- + 5SO_2(g) + 2H_2O(l) \xrightarrow{H^+} 2Mn^{2+} + 5SO_4^{2-} + 4H^+$$

EXAMPLE 17.9 Redox Reactions

Referring to Tables 17.6 and 17.7, predict the oxidation and reduction products of each of the following reactions. Do not be concerned with other possible products or with writing balanced equations. (These reactions all take place.)

(a) $Al(s) + H^+ \longrightarrow$

(b) $Fe^{2+} + MnO_4^- \xrightarrow{H^+}$

(c) $SO_3^{2-} + Cl_2 \longrightarrow$

(d) $Cr_2O_7^{2-} + Sn^{2+} \xrightarrow{H^+}$

(e) $BiO_3^- + Mn^{2+} \xrightarrow{H^+}$

(f) $H_2S(g) + BrO_3^- \xrightarrow{H^+}$

(a) Aluminum can act as a reducing agent and will displace hydrogen ion from an acid in solution.

$$Al(s) + H^+ \xrightarrow{\text{not balanced}} Al^{3+} + H_2$$

(b) Permanganate ion is a common oxidizing agent that is reduced to Mn^{2+} in acidic solution. The Fe^{2+} ion is commonly oxidized to Fe^{3+}.

$$Fe^{2+} + MnO_4^- \xrightarrow{\text{not balanced}} Fe^{3+} + Mn^{2+}$$

(c) Sulfite ion is a common reducing agent and chlorine is a common oxidizing agent.

$$SO_3^{2-} + Cl_2 \xrightarrow{\text{not balanced}} SO_4^{2-} + Cl^-$$

(d) Dichromate ion in acidic solution is an oxidizing agent and yields Cr^{3+} as its reduction product. The Sn^{2+} ion should go to the next highest oxidation state, Sn^{4+}.

$$Cr_2O_7^{2-} + Sn^{2+} \xrightarrow{\text{not balanced}} Cr^{3+} + Sn^{4+}$$

(e) The bismuthate ion is a very strong oxidizing agent in acidic solution. The common species in which manganese occurs in higher oxidation states are MnO_2(Mn + 4) or MnO_4^-(Mn + 7). Because bismuthate is a strong oxidizing agent we predict that it would give MnO_4^-.

$$BiO_3^- + Mn^{2+} \xrightarrow{\text{not balanced}} Bi^{3+} + MnO_4^-$$

(f) Hydrogen sulfide is a reducing agent and gives free sulfur as its usual oxidation product. Bromate ion is an oxidizing agent which in acidic solution is reduced to bromide ion or bromine.

$$H_2S(g) + BrO_3^- \xrightarrow{\text{not balanced}} S(s) + Br^- \text{ or } Br_2$$

EXERCISE Predict the oxidation and reduction products of each of the following reactions:

(a) $HNO_2(aq) + HI(aq) \longrightarrow$

(b) $Cr_2O_7^{2-} + Fe^{2+} \xrightarrow{H^+}$

(c) $PbO_2(s) + SO_3^{2-} \xrightarrow{H^+}$

Answers (a) $NO(g) + I_2(aq)$, (b) $Cr^{3+} + Fe^{3+}$, (c) $PbSO_4(s)$.

17.6 A METHOD FOR CLASSIFYING CHEMICAL REACTIONS AND PREDICTING REACTION PRODUCTS

The types of chemical reactions that we have described in this chapter are organized into a flowchart in Table 17.8. By examining the reactants and identifying them in the order of the numbered questions in Table 17.8, it is possible to narrow down the choice of the types of reactions that might occur. Once this is done, general and specific information can be utilized to predict reaction products. For example, if a partner-exchange reaction is found to be a possible type of reaction, general information about solubilities (see Table 17.1) can be used to predict whether or not a precipitate might form.

To use Table 17.8, answer the questions in numerical order until you come to a box listing the types of reactions that are possible for the reactants under consideration. At that point, call upon your knowledge and the tables in this book to predict what the reaction products might be. The result may be a reasonably firm prediction of the reaction products (or that no reaction can occur); prediction of the type of reaction that is possible, if not the exact products; or the conclusion that no prediction can be made without additional knowledge.

In all cases, note must be taken of the reaction conditions, the states of the reactants, whether or not water is present as a solvent, and how the reactants are formulated. For reactions in aqueous solution, it is important to recognize soluble strong electrolytes, as well as slightly soluble compounds and weak electrolytes. To illustrate these points, and the use of Table 17.8 in general, consider the reaction of barium chloride and sodium sulfate in aqueous solution.

$$BaCl_2(aq) + Na_2SO_4(aq) \longrightarrow$$

The answers to the first two questions are "no"—there is not just a single reactant and there is no element among the reactants. To answer question 3, you must know whether or not the reactants are soluble strong electrolytes, which they indeed are (Table 17.1). Therefore, the species that are possible reactants are the Ba^{2+}, Na^+, Cl^-, and SO_4^{2-} ions in aqueous solution.

Next, in order to answer question 4, you must determine whether the formation of a gas, a precipitate, a weak electrolyte, or water is possible. Of the ions present, Ba^{2+} and SO_4^{2-} can combine to give a precipitate of barium sulfate (Table 17.1). Therefore, it is possible to predict that the reaction will be a nonredox, partner-exchange reaction driven by the formation of a precipitate.

Table 17.8
A Guide to Some Simple Types of Chemical Reactions

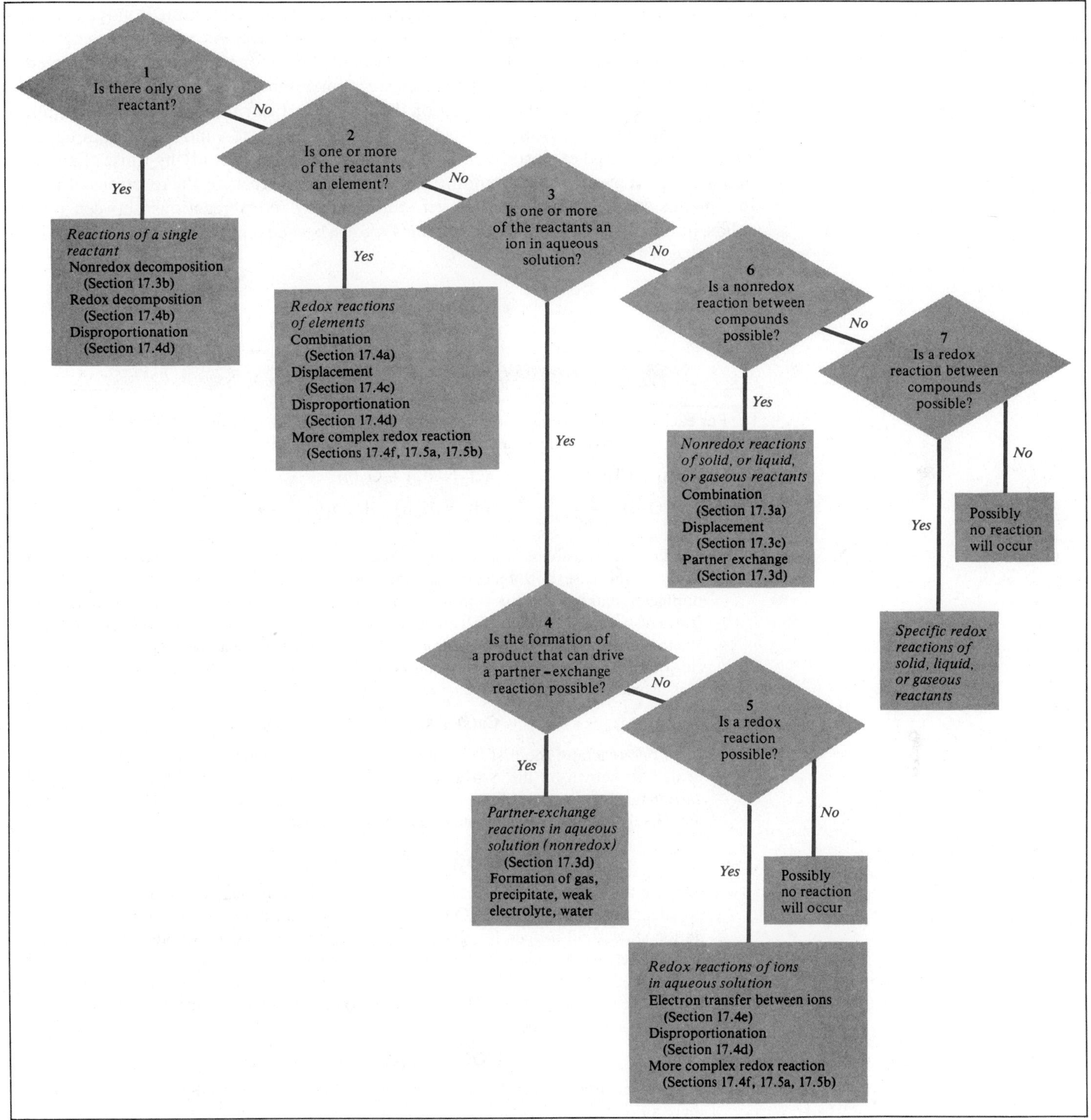

$$BaCl_2(aq) + Na_2SO_4(aq) \longrightarrow BaSO_4(s) + 2NaCl(aq)$$

Keep in mind that the patterns of partner-exchange and displacement reactions in aqueous solution are based on the complete formulas of the compounds. It takes a moment's thought to recognize the partner-exchange pattern from the net ionic equation

$$Ba^{2+} + SO_4^{2-} \longrightarrow BaSO_4(s)$$

The use of the scheme outlined in Table 17.8 is further illustrated in the next two examples. Like the problem-solving method that we have used from time to time, the table is a guide meant to help you in deciding how to answer a question. It may not lead you to the right answer every time, nor is it all-inclusive. As your knowledge of chemistry increases, more and more questions will arise for which you will not need such a guide. (Perhaps you did not need it to recognize immediately that a precipitate of barium sulfate would be the product of the reaction of barium chloride and sodium sulfate in aqueous solution.) The behavior of ions and acids and bases in aqueous solution is discussed extensively in Chapters 19–22, and after studying those chapters you will have a greater understanding of the possible products of the reactions of ions in aqueous solution. We have not yet studied many redox reactions between pure solids, liquids, or gases—reactions of this type are best covered in connection with the chemistry of the specific elements.

EXAMPLE 17.10 Predicting Reaction Products

For each of the following reactions, identify the most likely type of reaction and predict which are the most likely reaction products. Follow the guide in Table 17.8.

(a) $Cu(s) + Cl_2(g) \longrightarrow$ (c) $NO(g) + O_2(g) \longrightarrow$

(b) $ZnCO_3(s) \xrightarrow{\Delta}$ (d) $N_2H_4(l) + H_2O_2(l) \longrightarrow$

(a) The first "yes" answer is for question 2, indicating that a redox reaction of an element is likely. In this case, both reactants are elements, one a metal and the other a reactive nonmetal, making a redox combination reaction likely. Copper can have oxidation states of +1 or +2 (Table 17.2) and chlorine is an active oxidizing agent (Table 17.6). We therefore predict that the reaction will be a **combination reaction** to give copper(II) chloride because chlorine is a strong enough oxidizing agent to take copper to its higher oxidation state.

$$Cu(s) + Cl_2(g) \longrightarrow CuCl_2(s)$$

(b) With one reactant, the first "yes" answer is for question 1. The compound is to be heated, as shown by the Δ over the arrow. In a thermal decomposition reaction, carbonates form carbon dioxide and the metal oxide (Section 17.3b). Therefore, the most likely reaction is a **nonredox thermal decomposition.**

$$ZnCO_3(s) \xrightarrow{\Delta} ZnO(s) + CO_2(g)$$

(c) Because the first "yes" answer is for question 2, a redox reaction of an element is possible. Nitrogen, which in NO is in the +2 oxidation state, has several higher oxidation states, and oxygen is a good oxidizing agent, making a **combination reaction** appear likely. Without specific knowledge of the chemistry of nitrogen compounds, we cannot predict with certainty to which oxidation state nitrogen will be oxidized. One reasonable guess would be that it is oxidized to the next highest multiple of 2 state, to give nitrogen(IV) oxide.

$$2NO(g) + O_2(g) \longrightarrow 2NO_2(g)$$

(d) The answers to questions 1, 2, and 3 are all "no." There is not a single reactant, nor is there an element or an ion in aqueous solution present. This leads us to question 6. None of the simple nonredox reactions between compounds that we have studied—combination, displacement, or partner exchange—appears likely. To answer question 7, we consider whether the reactants are known as oxidizing and reducing agents. This is the case: N_2H_4 (hydrazine) is a reducing agent (Table 17.7) and H_2O_2 is an oxidizing agent (Table 17.6). We predict that a **redox reaction** will occur between these two liquids to give nitrogen and water.

$$N_2H_4(l) + 2H_2O_2(l) \longrightarrow N_2(g) + 4H_2O(g)$$

EXERCISE For each of the following, identify the most likely type of reaction and predict the most likely products.

(a) $BrO_3^- + Br^- \xrightarrow{H^+}$ (b) $P_4(s) + I_2(g) \xrightarrow{\Delta}$ (c) $KMnO_4(aq) + H_2S(g) \xrightarrow{OH^-}$

Answers (a) complex redox reaction of ions in aqueous solution, $Br_2(aq) + H_2O(l)$; (b) redox combination of elements, $PI_3(s)$; (c) complex redox reaction (common oxidizing agent plus common reducing agent), $MnO_2(s) + S(s)$.

EXAMPLE 17.11 Predicting Reaction Products

Will reactions occur when the following combinations of reactants are brought together in aqueous solution? If so, what type of reaction can take place and what might the products be? Follow the guide in Table 17.8.

(a) $ZnS + NaNO_3 \longrightarrow$ (c) $KMnO_4(aq) + Na_2SO_3(aq) \xrightarrow{OH^-}$
(b) $I^- + Br_2 \longrightarrow$ (d) $Na_2SO_3 + HCl \longrightarrow$

(a) Both questions 1 and 2 are answered by "no" and it is next necessary to decide whether ions in aqueous solution are present. Sodium nitrate is soluble (Table 17.1), so the answer is "yes." However, zinc sulfide is not soluble (Table 17.1). Therefore no partner exchange will occur, because the least soluble product is already present as a solid. Of the species present, $ZnS(s)$, Na^+, NO_3^-, none is likely to act as an oxidizing or reducing agent (Table 17.6; NO_3^- is commonly an oxidizing agent only in acid solution), and therefore we predict that **no chemical reaction will occur.**

(b) One reactant is an element (question 2), making a redox reaction of an element possible. Bromine is a more active nonmetal than iodine (Table 17.3), and thus it can oxidize iodide ion. Although the cation is not written, it is apparent that this will be what we have categorized as a **redox displacement reaction.**

$$2I^- + Br_2(aq) \longrightarrow I_2(aq) + 2Br^-$$

(c) Here again, "no" answers to questions 1 and 2 lead us to question 3. We are shown that these are soluble compounds by the (*aq*) in the formulas, and therefore the answer to question 3 is "yes." Partner exchange is unlikely, since Na^+ and K^+ salts are generally water soluble, and the formation of a gas or a weak electrolyte does not appear likely. However, the answer to question 5 is "yes," because permanganate ion is an oxidizing agent (Table 17.6) and sulfite ion is a reducing agent (Table 17.7). We predict that **the reaction will occur and that the major products will be $MnO_2(s)$ and SO_4^{2-}.** [The net ionic equation for this reaction is $2MnO_4^- + 3SO_3^{2-} + H_2O(l) \longrightarrow 2MnO_2(s) + 3SO_4^{2-} + 2OH^-$.]

(d) Questions 1 and 2 are answered by "no." Next, we must decide whether ions are present. They are, for sodium sulfite is a soluble salt (Table 17.1) and an aqueous solution of hydrogen chloride is the strong acid, hydrochloric acid. To answer question 4, we must determine whether any possible products could drive a partner-exchange reaction. The answer is "yes." **Partner exchange will result in the formation of sulfurous acid, H_2SO_3, which immediately decomposes to give gaseous sulfur dioxide, thereby driving the reaction to completion.**

$$Na_2SO_3(aq) + 2HCl(aq) \longrightarrow 2NaCl(aq) + SO_2(g) + H_2O(l)$$

EXERCISE Will reactions occur when the following combinations of reactants are brought together in aqueous solution? If so, what types of reactions can take place and what might the products be?

(a) $AgCl + Cu$ (b) $Fe^{3+} + I^-$ (c) $SiCl_4(l) + H_2O(l)$

Answers (a) No reaction, AgCl is not soluble; (b) can't be sure, possible electron transfer to give Fe^{2+} and I_2; (c) nonredox partner exchange (hydrolysis), HCl plus oxide or hydroxide of Si(IV).

SUMMARY OF SECTIONS 17.3–17.6

It is possible to predict the products of many chemical reactions by classifying the reactions as one of several simple types of nonredox or redox reactions. Memorization of vast numbers of reactions is not necessary. The flowchart in Table 17.8 aids in such classification and prediction of products.

Many nonredox reactions of simple substances (but never elements, which undergo only redox reactions) can be classified as combination, decomposition, displacement, or partner-exchange reactions.

Compounds may combine to give other compounds. Compounds may decompose with heat to give other compounds, particularly if the products are gases. Less volatile oxides displace more volatile oxides, and ligands displace other ligands to form more stable complex ions. Partner-exchange reactions between ions in solution take place when ions are removed from solution as gases, precipitates, weak electrolytes, or water.

Some redox reactions may also be classified as combination (of elements with other elements or with compounds), decomposition, or displacement (of one element from a compound by another element, according to the activity series for metals or nonmetals). In combination of elements, the more electronegative element is reduced and the more electropositive element is oxidized.

Other common types of simple redox reactions are disproportionation (in which an element in one oxidation state is both oxidized and reduced) and electron transfer between monatomic ions in aqueous solution. More complex redox reactions that are not easily classified further include the reactions of oxygen-containing oxidizing or reducing agents in acidic or alkaline solutions and specific reactions between gaseous, solid, or liquid compounds. Recognizing the common oxidizing and reducing agents and knowing the products that they usually form are extremely useful in identifying redox reactions and in predicting the major products of the more complex redox reactions.

SIGNIFICANT TERMS (with section references)

acid salt (17.3)
disproportionation reaction (17.4)
internal redox reactions (17.4)
Le Chatelier's principle (17.2)
oxidizing acids (17.5)
oxidizing anion (17.4)

QUESTIONS AND PROBLEMS

Chemical Equilibrium and Stability

17.1 When a chemical reaction reaches equilibrium, are the amounts of reactants and products equal? Have all chemical reactions stopped at equilibrium? Explain your answers.

17.2 Name the general principle that governs the displacement of equilibria. State this principle in words.

17.3 Which of the following reactions are likely to "go to completion"? Explain your reasoning.

(a) $2NaCl(s) + H_2SO_4(conc) \longrightarrow Na_2SO_4(aq) + 2HCl(g)$
(b) $2Fe^{3+} + 3Cu^{2+} \longrightarrow 2Fe(s) + 3Cu(s)$
(c) $UO_2(NO_3)_2(aq) + (NH_4)_2SO_3(aq) \longrightarrow UO_2SO_3(s) + 2NH_4NO_3(aq)$
(d) $H_3AsO_3(aq) + H_2O_2(aq) \longrightarrow H_3AsO_4(aq) + H_2O(l)$

17.4 Repeat Question 17.3 for

(a) $CaC_2(s) + 2H_2O(l) \longrightarrow C_2H_2(g) + Ca(OH)_2(aq)$
(b) $H_2AsO_3^- + 3Ag^+ \longrightarrow Ag_3AsO_3(s) + 2H^+$
(c) $CaCO_3(s) + 2HCl(aq) \longrightarrow CaCl_2(aq) + CO_2(g) + H_2O(l)$
(d) $AgCl(s) \xrightarrow{H_2O} Ag^+ + Cl^-$

17.5 Compare the relative stabilities of AgCl and $[Ag(NH_3)_2]^+$ given that AgCl dissolves in aqueous ammonia:

$$AgCl(s) + 2NH_3(aq) \xrightarrow{\text{excess } NH_3} [Ag(NH_3)_2]^+ + Cl^-$$

17.6 Compare the relative stabilities of ozone and oxygen, given the following data:

$$2O_3(g) \longrightarrow 3O_2(g) \qquad \Delta H^\circ = -285.3 \text{ kJ}$$

CHEMICAL REACTIONS

$NH_3(g) + HCl(g) \rightarrow NH_4Cl(s)$

■ **NONREDOX COMBINATION** *above* Ammonia and hydrogen chloride gases escape from the surfaces of their aqueous solutions and diffuse into the atmosphere, where they react to form finely divided white ammonium chloride.

$Ag^+ + Cl^- \rightarrow AgCl$ (s, white) $Ag^+ + Br^- \rightarrow AgBr$ (s, cream) $Ag^+ + I^- \rightarrow AgI$ (s, yellow)

■ **NONREDOX, PARTNER-EXCHANGE, PRECIPITATE FORMATION** *above* Each beaker originally was filled with an aqueous solution of silver ion. As solutions containing (*left to right*) chloride, bromide, and iodide ion are added with stirring, the finely divided silver halides are seen to precipitate.

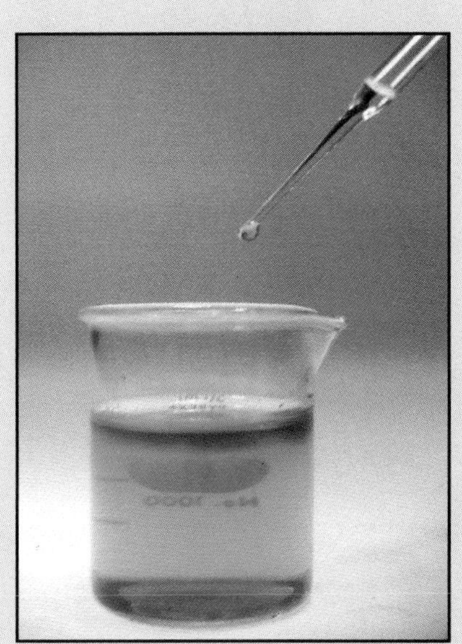

$[Ni(H_2O)_6]^{2+} + 6NH_3(aq) \rightarrow [Ni(NH_3)_6]^{2+} + 6\,H_2O(l)$

■ **NONREDOX DISPLACEMENT** *above* In aqueous solution, the hydrated nickel ion is green (at bottom of beaker). As aqueous ammonia is added, the blue nickel-ammonia complex ion forms at the surface of the solution.

$[Fe(H_2O)_6]^{3+} + NH_4SCN(aq) \rightarrow [Fe(H_2O)_5SCN]^{2+} + NH_4^+ + H_2O(l)$

■ **NONREDOX DISPLACEMENT** *right* Hydrated iron(III) ion is converted to bright-red pentaaquothiocyanato complex ion when an aqueous solution of ammonium thiocyanate (NH_4SCN) is added. The intensity of the red color can be used as a measure of Fe^{3+} concentration.

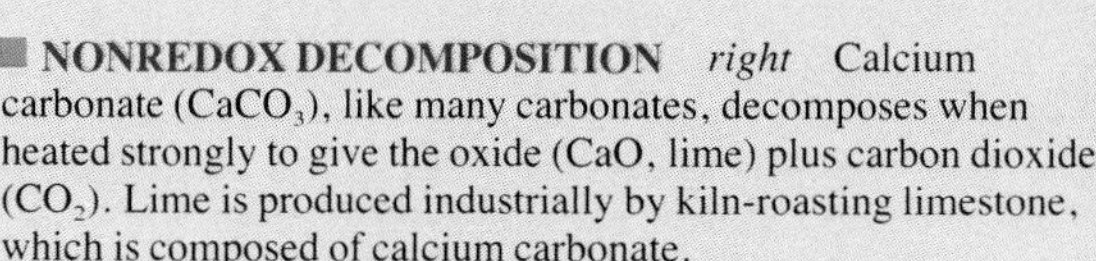

■ **NONREDOX DECOMPOSITION** *right* Calcium carbonate ($CaCO_3$), like many carbonates, decomposes when heated strongly to give the oxide (CaO, lime) plus carbon dioxide (CO_2). Lime is produced industrially by kiln-roasting limestone, which is composed of calcium carbonate.

$$CaCO_3(s) \rightarrow CaO(s) + CO_2(g)$$

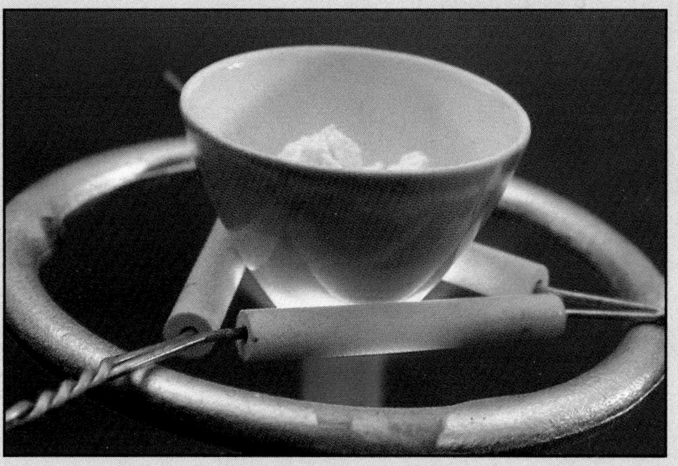

$CH_3COOH(aq) + NaHCO_3(s) \rightarrow NaCH_3COO(aq) + CO_2(g) + H_2O(l)$

■ **NONREDOX, PARTNER-EXCHANGE, GAS FORMATION** *left* Vinegar, which contains acetic acid (CH_3COOH), is being added from the dropper to solid sodium hydrogen carbonate ($NaHCO_3$) in the test tube. Bubbles of carbon dioxide form where the acid is in contact with the solid. Carbonic acid, the initial product of partner exchange, decomposes to carbon dioxide and water.

$NaOH(aq) + HCl(aq) \rightarrow NaCl(aq) + H_2O(l)$

■ **NONREDOX, PARTNER-EXCHANGE, WATER FORMATION** *above* In titration of an acid with a base, the first drop of base in excess is shown (*left*) by pink color of phenolphthalein indicator (pink in alkaline solution). Additional base gives more alkaline, deeper pink solution (*right*).

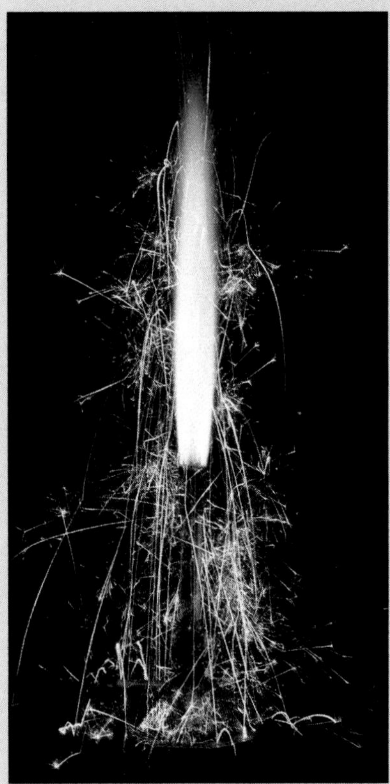

$3Fe(s) + 2O_2(g) \rightarrow Fe_3O_4(s)$

■ **REDOX COMBINATION OF ELEMENTS** *above* Iron powder sprinkled into the flame creates a shower of sparks as iron combines with oxygen from the air to form iron(II, III) oxide.

$S(s) + O_2(g) \rightarrow SO_2(g)$

■ **REDOX COMBINATION OF ELEMENTS** *above* Sulfur burns in oxygen with a bright blue flame. Traces of sulfur trioxide (SO_3) may form in addition to the major product, which is sulfur dioxide (SO_2)

$2Al(s) + 3I_2(s) \rightarrow 2AlI_3(s)$

■ **REDOX COMBINATION OF ELEMENTS** *above* Aluminum and iodine combine in an exothermic reaction. The heat of reaction produces a swirl of iodine vapor as some of the solid reactant iodine sublimes.

$$2HgO(s) \longrightarrow 2Hg(l) + O_2(g)$$

■ **REDOX DECOMPOSITION** *above* Heating red mercury(II) oxide (HgO) causes decomposition to mercury, which can be seen as silver droplets on the side of the test tube, and oxygen. This is the reaction by which Joseph Priestley prepared oxygen in 1774. Some of the red mercury oxide has undergone the solid-state phase transition to yellow mercury oxide.

$$(NH_4)_2Cr_2O_7(s) \longrightarrow N_2(g) + Cr_2O_3(s) + 4H_2O(g)$$

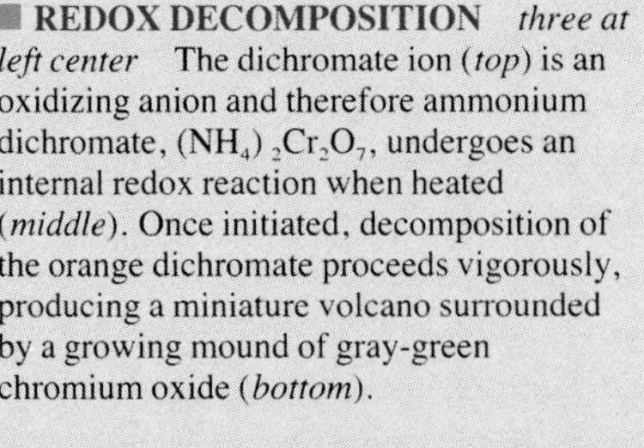

■ **REDOX DECOMPOSITION** *three at left center* The dichromate ion (*top*) is an oxidizing anion and therefore ammonium dichromate, $(NH_4)_2Cr_2O_7$, undergoes an internal redox reaction when heated (*middle*). Once initiated, decomposition of the orange dichromate proceeds vigorously, producing a miniature volcano surrounded by a growing mound of gray-green chromium oxide (*bottom*).

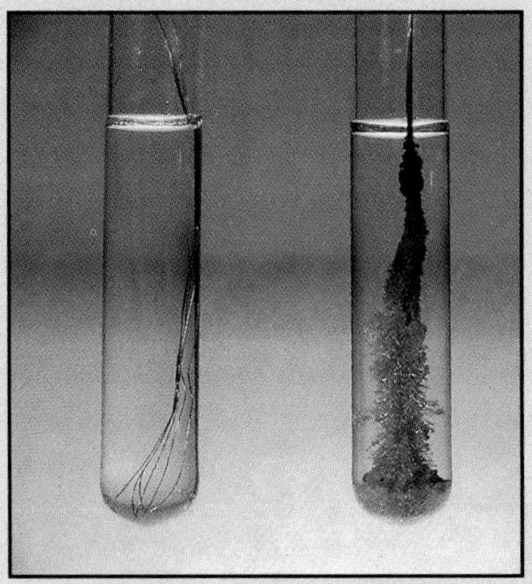

$$Ni(s) + 2Ag^+ \longrightarrow Ni^{2+} + 2Ag(s)$$
$$Cu(s) + 2Ag^+ \longrightarrow Cu^{2+} + 2Ag(s)$$

■ **REDOX DISPLACEMENT** *above* In the left-hand test tube, a nickel wire was immersed in a solution containing silver ion. The displacement of silver ion by nickel is shown by the shiny silver coating on the wire and the faint green color of hydrated nickel ion in the solution. In the right-hand test tube, a copper wire was immersed in a solution containing silver ion. The displacement of silver ion by copper is shown by the silver crystals attached to the wire and the blue-green layer containing hydrated copper ion.

$$Cu(s) + 4HNO_3(conc) \longrightarrow Cu(NO_3)_2(aq) + 2NO_2(g) + 2H_2O(l)$$

■ **NITRIC ACID AS AN OXIDIZING AGENT** *left* A copper penny reacts with concentrated nitric acid to produce an aqueous solution of copper(II) nitrate, as shown by the blue color of the hydrated copper ion (*left*). The other reaction products are nitrogen dioxide, a brown gas, and water, which is seen condensing on the side of the test tube (*right*). This is the reaction made famous in a historical account by Ira Remsen of his quest to discover what was meant by the statement, "Nitric acid acts upon copper." Note that the penny in this experiment is a pre-1982 penny. After that year, pennies have been made of zinc coated with copper, rather than of a 95% copper–5% zinc alloy.

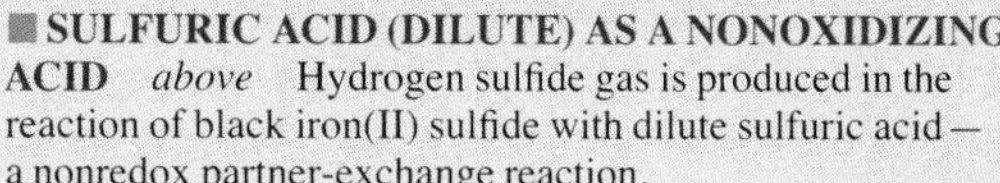

$$FeS(s) + H_2SO_4(aq) \longrightarrow FeSO_4(aq) + H_2S(g)$$

■ **SULFURIC ACID (DILUTE) AS A NONOXIDIZING ACID** *above* Hydrogen sulfide gas is produced in the reaction of black iron(II) sulfide with dilute sulfuric acid—a nonredox partner-exchange reaction.

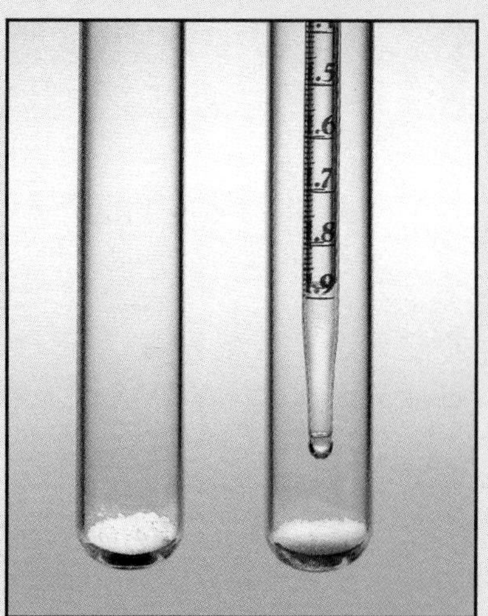

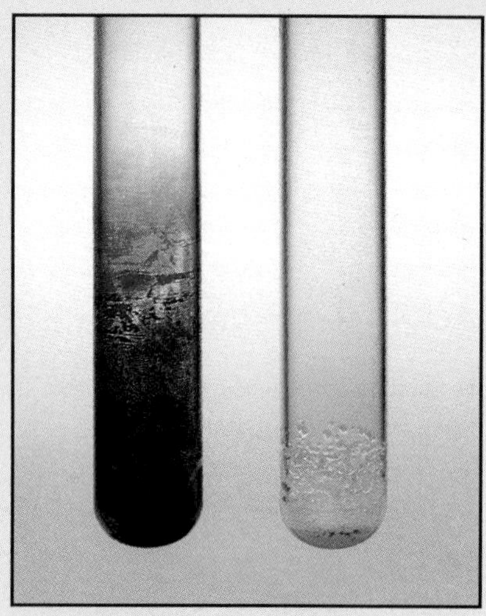

$$8NaI(s) + 9H_2SO_4(conc) \longrightarrow 8NaHSO_4(s) + H_2S(g) + 4I_2(g) + 4H_2O(l)$$
$$2NaBr(s) + 2H_2SO_4(conc) \longrightarrow Na_2SO_4(s) + SO_2(g) + Br_2(l) + 2H_2O(l)$$

■ **SULFURIC ACID (CONCENTRATED) AS AN OXIDIZING AGENT** *above* Addition of concentrated sulfuric acid to sodium iodide (*left test tube*) and sodium bromide (*right test tube*) causes oxidation to iodine and bromine.

■ **SULFURIC ACID (CONCENTRATED) AS A DEHYDRATING AGENT** *above* Treatment with concentrated sulfuric acid (*left*) converts sugar (sucrose, $C_{11}H_{22}O_{11}$) to black carbon by removal of water (*right*).

Nonredox Reactions

17.7 Write chemical equations for the nonredox combination of (a) $NH_3(g)$ with $H_2SO_4(aq)$ and (b) $CaO(s)$ with $SiO_2(l)$. (The first reaction produces a useful nitrogen fertilizer and the second occurs during the production of glass.)

17.8 Write chemical equations for the nonredox thermal decomposition of (a) $CaCO_3(a)$ and (b) $Ni(CO)_4(g)$. (The first reaction is used to produce quicklime and the second to produce very pure nickel.)

17.9 Complete and balance the equations for the following nonredox displacement reactions:

(a) $Na_2CO_3(l) + SiO_2(s) \longrightarrow$
(b) $[Cu(H_2O)_4]^{2+} + NH_3(aq) \longrightarrow$
(c) $[AgCl_2]^- + NH_3(aq) \longrightarrow$

17.10 Choose the nonredox partner-exchange reactions that will take place:

(a) $FeS(s) + 2HCl(aq) \longrightarrow FeCl_2(aq) + H_2S(g)$
(b) $BaSO_4(s) + 2KCl(aq) \longrightarrow BaCl_2(aq) + K_2SO_4(aq)$
(c) $2NaOH(aq) + H_2S(aq) \longrightarrow Na_2S(aq) + 2H_2O(l)$

17.11 Predict the products of each of the following nonredox partner-exchange reactions:

(a) $Na_2S(aq) + HCl(aq) \longrightarrow$
(b) $CdSO_4(aq) + Na_2S(aq) \longrightarrow$
(c) $(NH_4)_2CrO_4(aq) + Pb(CH_2COO)_2(aq) \longrightarrow$
(d) $AsCl_3(aq) + H_2S(aq) \longrightarrow$

17.12 Repeat Question 17.11 for

(a) $Ba(NO_3)_2(aq) + H_2SO_4(aq) \longrightarrow$
(b) $CaI_2(aq) + AgNO_3(aq) \longrightarrow$
(c) $Ca_3(PO_4)_2(s) + H_2SO_4(l) \longrightarrow$
(d) $NH_4Cl(aq) + KOH(aq) \longrightarrow$
(e) $CaCl_2(aq) + Na_2C_2O_4(aq) \longrightarrow$

17.13 Predict the products of each of the following nonredox reactions, classify each reaction as to type, and write the balanced chemical equation for the reaction.

(a) $Ca(OH)_2(s) \xrightarrow{\Delta}$
(b) $[Cu(H_2O)_4]Cl_2(aq) + NH_3(aq) \longrightarrow$
(c) $CuCl_2(aq) + NaOH(aq) \longrightarrow$
(d) $PCl_3(l) + H_2O(l) \longrightarrow$
(e) $SO_2(g) + CaO(s) \longrightarrow$
(f) $Be(OH)_2(s) \xrightarrow{\Delta}$
(g) $CO_2(g) + LiOH(aq) \longrightarrow$

17.14 Repeat Question 17.13 for

(a) $BaCO_3(s) + HCl(aq) \longrightarrow$
(b) $MgSO_4(aq) + BaCl_2(aq) \longrightarrow$
(c) $CaCO_3(s) \xrightarrow{\Delta}$
(d) $Pb(NO_3)_2(aq) + Na_2CrO_4(aq) \longrightarrow$
(e) $CaO(s) + H_2O(l) \longrightarrow$
(f) $Na_2CO_3(aq) + CaCl_2(aq) \longrightarrow$
(g) $Ba(OH)_2(aq) + H_2SO_4(aq) \longrightarrow$
(h) $K_2CO_3(s) + SiO_2(s) \xrightarrow{\Delta}$

Redox Reactions

17.15 Determine which of the following are redox reactions. Identify the oxidizing and reducing agent in each of the redox reactions.

(a) $NaOH(aq) + H_3PO_4(aq) \longrightarrow NaH_2PO_4(aq) + H_2O(l)$
(b) $NH_3(g) + CO_2(g) + H_2O(l) \longrightarrow NH_4HCO_3(aq)$
(c) $TiCl_4(g) + 2Mg(l) \xrightarrow{\Delta} Ti(s) + 2MgCl_2(l)$
(d) $NaCl(s) + NaHSO_4(s) \xrightarrow{\Delta} HCl(g) + Na_2SO_4(s)$

17.16 Repeat Question 17.15 for

(a) $Ti(s) + 2Cl_2(g) \xrightarrow{\Delta} TiCl_4(l)$
(b) $NH_4Br(s) \longrightarrow NH_3(g) + HBr(g)$
(c) $3CuO(s) + 2NH_3(g) \xrightarrow{\Delta} 3Cu(s) + N_2(g) + 3H_2O(g)$
(d) $K_2Cr_2O_7(aq) + 14HCl(aq) \longrightarrow 2KCl(aq) + 2CrCl_3(aq) + 7H_2O(l) + 3Cl_2(g)$

17.17 Identify the products of the following redox combination reactions:

(a) $C(s) + O_2(g, \text{excess}) \longrightarrow$
(b) $P_4(s) + Cl_2(g, \text{excess}) \longrightarrow$
(c) $Ti(s) + Cl_2(g) \longrightarrow$
(d) $Mg(s) + N_2(g) \longrightarrow$
(e) $FeO(s) + O_2(g) \longrightarrow$
(f) $NO(g) + O_2(g) \longrightarrow$
(g) $P_4O_6(s) + O_2(g) \longrightarrow$

17.18 Complete and balance the following equations for redox displacement reactions:

(a) $K(s) + H_2O(l) \longrightarrow$
(b) $Mg(s) + HBr(aq) \longrightarrow$
(c) $NaBr(aq) + Cl_2(aq) \longrightarrow$
(d) $WO_3(s) + H_2(g) \longrightarrow$
(e) $H_2S(aq) + Cl_2(aq) \longrightarrow$

17.19 Using the activity series, predict whether or not the following reactions will occur in aqueous solution:

(a) $Mg(s) + Ca(s) \longrightarrow Mg^{2+} + Ca^{2+}$
(b) $2Al^{3+} + 3Pb^{2+} \longrightarrow 2Al(s) + 3Pb(s)$
(c) $H_2(g) + Zn^{2+} \longrightarrow 2H^+ + Zn(s)$
(d) $2KI(aq) + Br_2(aq) \longrightarrow 2KBr(aq) + I_2(aq)$

Explain your reasoning.

17.20 Repeat Question 17.19 for

(a) $Mg(s) + Cu^{2+} \longrightarrow Mg^{2+} + Cu(s)$
(b) $Pb(s) + 2H^+ \longrightarrow H_2(g) + Pb^{2+}$
(c) $2Ag^+ + Cu(s) \longrightarrow 2Ag(s) + Cu^{2+}$
(d) $2Al^{3+} + 3Zn(s) \longrightarrow 3Zn^{2+} + 2Al(s)$

17.21 The sulfite ion is a good reducing agent. Predict the products of the redox reactions between SO_3^{2-} and (a) $Br_2(aq)$, (b) $O_2(g)$, (c) $H_2O_2(aq)$, (d) MnO_4^- in acidic solution. Write balanced equations.

17.22 Lithium metal is a good reducing agent. Predict the products of the reactions between $Li(s)$ and (a) $H_2O(l)$, (b) $O_2(g)$, (c) $F_2(g)$, (d) $H_2(g)$, (e) $N_2(g)$.

17.23 Gaseous chlorine is a good oxidizing agent. Predict the products of the redox reactions between $Cl_2(g)$ and (a) $P_4(s)$, (b) $PCl_3(l)$, (c) $CuCl(s)$, (d) I^-.

17.24 The permanganate ion is a good oxidizing agent. Predict the products of the redox reactions between MnO_4^- and (a) $SO_2(g)$, (b) Cl^-, (c) Sn^{2+}, assuming all reactions take place in an acidic solution.

17.25 Predict the oxidation and reduction products of each of the following reactions, classify each reaction as to type, and write the balanced chemical equation describing the reaction:

(a) $MnO_2(s) + Cl^- + H^+ \longrightarrow$
(b) $Sn^{2+} + Fe^{3+} \longrightarrow$
(c) $K(s) + S_8(s) \longrightarrow$
(d) $N_2H_4(l) + O_2(g) \longrightarrow$
(e) $CuSO_4(s) \xrightarrow{H_2O}$
(f) $Hg_2Cl_2(s) \xrightarrow{\Delta}$

17.26 Repeat Question 17.25 for

(a) $Cl_2(aq) + Br^- \longrightarrow$
(b) $PbO_2(s) + Mn^{2+} + H^+ \longrightarrow$
(c) $Cr_2O_7^{2-} + I^- + H^+ \longrightarrow$
(d) $ClO_3^- + Cl^- + H^+ \longrightarrow$
(e) $HNO_2(aq) + Sn^{2+} + H^+ \longrightarrow$
(f) $KClO_3(l) \xrightarrow{\Delta}$
(g) $Cl_2(g) + H_2O(l) \longrightarrow$

17.27 What mass of $K_2Cr_2O_7$ is required to oxidize 50.0 g of HCHO to HCOOH? Assume Cr^{3+} is the reduction product of $Cr_2O_7^{2-}$ in acidic medium.

17.28 A 0.500 g sample of a powder containing $SnCl_2$ was oxidized using 0.103 M Ce^{4+} (in HCl solution). Find the mass % $SnCl_2$ in the powder, given that 35.72 mL of the Ce^{4+} solution was used. The reduction product of Ce^{4+} is Ce^{3+}. (HCl is not reduced.)

Classifying Chemical Reactions and Predicting Reaction Products

17.29 Classify each of the following reactions:

(a) $MgO(s) + 2HCl(aq) \longrightarrow MgCl_2(aq) + H_2O(l)$
(b) $WO_3(s) + 3H_2(g) \longrightarrow W(s) + 3H_2O(l)$
(c) $2NaHCO_3(s) \xrightarrow{\Delta} Na_2CO_3(s) + CO_2(g) + H_2O(g)$
(d) $CaO(s) + SO_2(g) \longrightarrow CaSO_3(s)$
(e) $2KMnO_4(aq) + 16HCl(aq) \longrightarrow 2KCl(aq) + 2MnCl_2(aq) + 5Cl_2(g) + 8H_2O(l)$
(f) $Na_2CO_3(l) + SiO_2(s) \xrightarrow{\Delta} Na_2SiO_3(l) + CO_2(g)$
(g) $3Cu(s) + 8HNO_3(aq) \longrightarrow 3Cu(NO_3)_2(aq) + 2NO(g) + 4H_2O(l)$
(h) $2NO(g) + O_2(g) \longrightarrow 2NO_2(g)$
(i) $(NH_4)_2Cr_2O_7(s) \xrightarrow{\Delta} N_2(g) + Cr_2O_3(s) + 4H_2O(g)$
(j) $SnCl_4(aq) + Fe(s) \longrightarrow FeCl_2(aq) + SnCl_2(aq)$

17.30 Repeat Question 17.29 for

(a) $BaO(s) + H_2O(l) \longrightarrow Ba(OH)_2(aq)$
(b) $CaH_2(s) + 2H_2O(l) \longrightarrow Ca(OH)_2(aq) + 2H_2(g)$
(c) $2SO_2(g) + O_2(g) \longrightarrow 2SO_3(g)$
(d) $2KClO_3(s) \xrightarrow{\Delta} 2KCl(s) + 3O_2(g)$
(e) $H_2SO_3(aq) + Br_2(aq) + H_2O(l) \longrightarrow H_2SO_4(aq) + 2HBr(aq)$
(f) $Zn(s) + 2AgNO_3(aq) \longrightarrow 2Ag(s) + Zn(NO_3)_2(aq)$
(g) $BaO(s) + SO_3(g) \longrightarrow BaSO_4(s)$
(h) $[Ni(H_2O)_4]^{2+} + 4CN^- \longrightarrow [Ni(CN)_4]^{2-} + 4H_2O(l)$
(i) $CaCO_3(s) + 2HCl(aq) \longrightarrow CaCl_2(aq) + H_2O(l) + CO_2(g)$
(j) $2NH_3(g) + 3CuO(s) \xrightarrow{\Delta} N_2(g) + 3Cu(s) + 3H_2O(g)$

17.31 Classify the following reactions and predict the major products formed.

(a) $Li(s) + H_2O(l) \longrightarrow$
(b) $Ag_2O(s) \xrightarrow{\Delta}$
(c) $Li_2O(s) + H_2O(l) \longrightarrow$
(d) $H_2O(l) \xrightarrow{\text{electrolysis}}$
(e) $I_2(s) + Cl^-(aq) \longrightarrow$
(f) $Cu(s) + HCl(aq) \longrightarrow$
(g) $NaNO_3(s) \xrightarrow{\Delta}$

17.32 Repeat Question 17.31 for

(a) $SO_3(g) + H_2O(l) \longrightarrow$
(b) $Sr(s) + H_2(g) \longrightarrow$
(c) $Mg(s) + H_2SO_4(aq, \text{dilute}) \longrightarrow$
(d) $Na_3PO_4(aq) + AgNO_3(aq) \longrightarrow$
(e) $Li_2O(s) + SO_2(g) \longrightarrow$
(f) $Ca(HCO_3)_2(s) \xrightarrow{\Delta}$
(g) $P_4O_6(s) + O_2(g) \longrightarrow$

17.33 Identify the products and write balanced chemical equations for the following reactions:

(a) $NH_4NO_3(s) \xrightarrow{\Delta}$
(b) $H_2S(g) + Pb(CH_3COO)_2(aq) \longrightarrow$
(c) $Ba(OH)_2(aq) + H_2SO_4(aq) \longrightarrow$
(d) $Fe^{3+} + Sn^{2+} \longrightarrow$
(e) $FeCl_2(aq) + Cl_2(g) \longrightarrow$
(f) $Hg(l) + O_2(g) \xrightarrow{\Delta}$
(g) $CaCl_2(aq) + H_2SO_4(aq) \longrightarrow$
(h) $SO_2(g) + NaOH(aq) \longrightarrow$

17.34 Repeat Question 17.33 for

(a) $Al(s) + H_2SO_4(aq) \longrightarrow$
(b) $NO_2(g) + H_2O(l) \longrightarrow$
(c) $Pb_3O_4(s) \xrightarrow{\Delta}$
(d) $N_2(g) + H_2(g) \longrightarrow$
(e) $BaCl_2(aq) + H_3PO_4(aq) \longrightarrow$
(f) $MnO_2(s) + HCl(aq) \longrightarrow$
(g) $Mg(s) + N_2(g) \longrightarrow$
(h) $Mg(s) + HCl(aq) \longrightarrow$

17.35 Will the following reactions take place?

(a) $BaSO_4(s) + NaCl(aq) \longrightarrow$
(b) $FeSO_4(aq) + Cu(s) \longrightarrow$
(c) $Ba(NO_3)_2(aq) + HCl(aq) \longrightarrow$
(d) $Ba(OH)_2(aq) + HCl(aq) \longrightarrow$
(e) $Al_2(SO_4)_3(aq) + Zn(s) \longrightarrow$

Identify the products of those reactions that occur and give a reason for each answer.

17.36 Repeat Question 17.35 for

(a) $ZnSO_4(aq) + BaS(aq) \longrightarrow$
(b) $MgO(s) + H_2SO_4(aq) \longrightarrow$
(c) $Al_2(SO_4)_3(aq) + Hg(l) \longrightarrow$
(d) $Ag(s) + H_2SO_4(aq) \longrightarrow$

17.37 A piece of magnesium ribbon was placed in a crucible and was strongly heated in air to give a mixture of magnesium oxide and magnesium nitride. A small amount of water was added to convert the magnesium nitride to gaseous ammonia and magnesium hydroxide. The sample was heated again to convert the magnesium hydroxide to magnesium oxide. Write chemical equations for each of these reactions. Classify each reaction as nonredox or redox. Within each classification, identify the type of reaction.

17.38* A piece of copper was allowed to react in hot concentrated sulfuric acid to give a solution. This solution was added slowly to a large volume of water, giving a blue solution. Addition of aqueous sodium sulfide to the blue solution produced a black precipitate. Addition of nitric acid dissolved the precipitate, forming a blue solution. Addition of a small amount of ammonia gave a light blue precipitate which dissolved upon addition of more ammonia, giving a blue-purple solution. A blue solution was formed upon the addition of sulfuric acid. A blue precipitate was produced by the addition of sodium hydroxide solution. This precipitate was removed by filtering and heated to give a black solid. The black solid dissolved in sulfuric acid. Upon addition of zinc, copper was regenerated. Write a net ionic equation for each of these reactions. Classify each as nonredox or redox. Within each classification, identify the type of reaction.

CHAPTER

18

Chemical Kinetics

CHEMICAL KINETICS AT THE MOLECULAR LEVEL

18.1 CHEMICAL KINETICS: RATES AND MECHANISMS

Information about the molar amounts and masses of reactants and products can be derived from a balanced chemical equation. Also, bond energies and heats of formation can be used to determine the amount of heat absorbed or released by the reaction. All this information can be derived from the chemical equation without knowing anything about what happens to individual atoms, molecules, or ions during the course of the reaction. The pathway from reactants to products is of no concern.

Chemical kinetics, by contrast, deals with what happens while a reaction is under way, not just with the final result. Kinetics provides information about how fast reactions take place, information of great practical value. For example, chemical engineers must consult data on reaction rates in deciding how best to design chemical plants. Chemical kinetics also provides information about how individual molecules interact with each other, a topic of intense interest to those in pursuit of a deeper understanding of chemical reactions.

Ideally, to study kinetics we would like to observe the progress of each atom from its position in the reactants to its position in the products. Because chemical reactions take place by the transfer of electrons or by changes in the way electrons are shared, we would also like to know the fate of every valence electron. At one time, such close observation seemed an impossibility. Today, ingenious experimental techniques permit studies of individual reacting molecules, transient intermediate states of reactants, and very fast reactions (with reaction times of nanosecond and even picosecond duration). In addition, computer modeling aids the study of complex systems, and computers can solve the complex mathematical equations describing reaction kinetics.

However, such studies are difficult to perform, time-consuming, and expensive. Therefore, chemical reactions are still often studied by observing the overall behavior of quantities of reactants and products large enough to handle easily. Then mathematical models and prior knowledge about other reactions are used to deduce reasonable descriptions of reaction pathways at the molecular and electron levels.

The best way to obtain information about the pathway of a chemical reaction is to study how fast the reaction proceeds—the *rate* of the reaction—generally expressed as, for example, the amount of a given reactant that is consumed per unit of time. Once the reaction rate is known, pathways can be proposed that are consistent with the rate information.

Perhaps an analogy will demonstrate the relationship between how fast a reaction takes place and what the pathway might be. Suppose you had several hundred textbooks in the basement of your house and wanted to move them to your study on the second floor. One possible pathway, or "reaction mechanism," might require the following steps: (1) put ten books (the maximum number you can lift) into a carton, (2) walk up one flight of stairs to the first floor, (3) rest a bit, (4) walk up another flight of stairs to the second floor, (5) empty the carton, and (6) return for another load. Another "reaction mechanism" might involve a different series of steps: (1) fill three cartons with books, (2) push the three full cartons into the elevator, (3) ride up to the second floor, (4) push the cartons off the elevator, (5) empty the cartons and (6) ride down for another load. The rate of the reaction might be expressed as the number of books moved per hour. The reaction rate and how fast the complete reaction—moving all the books—is accomplished are clearly dependent upon the pathway.

A typical pathway for a chemical reaction includes several simple steps called elementary reactions. An **elementary reaction** is a reaction that occurs in a single step exactly as written. The equation for an elementary reaction shows the actual species that must interact at the molecular level for the reaction to take place. It often does not represent what can be observed as the overall final result of a chemical reaction.

A **reaction mechanism** consists of all of the elementary steps in a single reaction pathway. These steps show all of the changes that take place at the molecular level. The sum of the reaction steps is the stoichiometric equation for the reaction. For example, the reaction of iodine monochloride and hydrogen,

$$2ICl(g) + H_2(g) \longrightarrow I_2(g) + 2HCl(g)$$

is proposed to have a two-step mechanism. One ICl molecule and one H_2 molecule first react to give HI and HCl. The HI molecule then reacts with another molecule of ICl. The HI molecule in this mechanism is an **intermediate**—a reactive species that is produced during the course of a reaction but always reacts further and is not among the final products.

$$\begin{array}{rcl} ICl(g) + H_2(g) & \longrightarrow & HI(g) + HCl(g) \\ HI(g) + ICl(g) & \longrightarrow & I_2(g) + HCl(g) \\ \hline 2ICl(g) + H_2(g) & \longrightarrow & I_2(g) + 2HCl(g) \end{array}$$

Chemical kinetics is the study of the rates and mechanisms of chemical reactions. We can use the book-moving analogy to illustrate another important point—an essential difference between chemical kinetics and chemical thermodynamics. The reaction rate, a kinetic property, is dependent upon the mechanism, but the thermodynamic properties of a reaction are independent of the mechanism. Earlier we studied the heats of chemical reactions without needing to know anything about reaction rates or mechanisms. Similarly, we could calculate the change in potential energy of the books as a result of their move from the basement to the second floor without knowing by which "mechanism" they got there. Thermodynamics (Chapters 7 and 23) can tell us whether or not a reaction is possible, but kinetics is needed to tell us if the reaction happens *fast enough* to be of any interest or value.

18.2 HOW REACTANTS GET TOGETHER

What is necessary for two molecules to react with each other? In considering this question, let's look at the usual simplest case—some isolated molecules in the gas phase. Ozone and nitric oxide react with each other in the upper atmosphere as follows:

$$O_3(g) + NO(g) \longrightarrow NO_2(g) + O_2(g) \quad \textbf{(18.1)}$$

This is one of the reactions that may contribute to depletion of the ozone layer.

Suppose we have mixed small amounts of these two gases and can study the reaction by watching the individual molecules. We see the molecules flying about in random paths, as expected from kinetic-molecular theory. Some molecules move faster and have more kinetic energy than others—the distribution of speeds and energies is determined by the temperature. The molecules also have internal energy, represented by the vibrations of the individual atoms and the rotations of the molecules.

Ozone and nitric oxide molecules often collide with each other as they fly about. The frequency of collisions between O_3 and NO clearly depends on their relative *concentrations*. At 1 atm and 0 °C, for equal concentrations of O_3 and NO, there are about 10^{29} O_3–NO collisions per cubic centimeter per second. Each collision presents an opportunity for a reaction to occur between the two molecules that collide. We can think of a "collision" as having three parts: (1) the approach of the molecules, (2) interaction between the molecules, and (3) separation.

When two approaching molecules get close enough together, their electrons "see" each other and repulsion builds up between them. If their kinetic energies are not great enough to overcome this repulsion, the molecules just veer away from each other. The only results of this interaction are changes in the kinetic energy and direction of the molecules after they separate. No chemical reaction occurs.

On the other hand, the energy of the approaching molecules may be sufficient to drive them together in spite of the forces of repulsion, allowing the molecules to stay together for a brief time. In our imagined observation of such a collision, we see the electron clouds of the two molecules interpenetrate each other, allowing a redistribution of electron density to take place. The molecules merge into a temporary intermediate "molecule." During this phase of the collision, the kinetic energy of the colliding molecules is also available for redistribution within the temporary intermediate. The internal energy of this "molecule" can increase to the point where the vibrations of some atoms become so strong that bonds break.

Suppose the nitrogen atom from the NO molecule collides with an oxygen atom at one end of the O_3 molecule and their electron clouds merge. Enough energy from the collision can be converted into vibrational energy in the oxygen–oxygen bond in O_3 to weaken it. For a brief moment there is a partial bond between the oxygen atoms as electron density shifts away from the bond that is about to break. At the same time, electron density shifts toward the nitrogen atom, and a partial bond forms between the oxygen and nitrogen atoms. Electron density in the other bonds in these two molecules is also shifted.

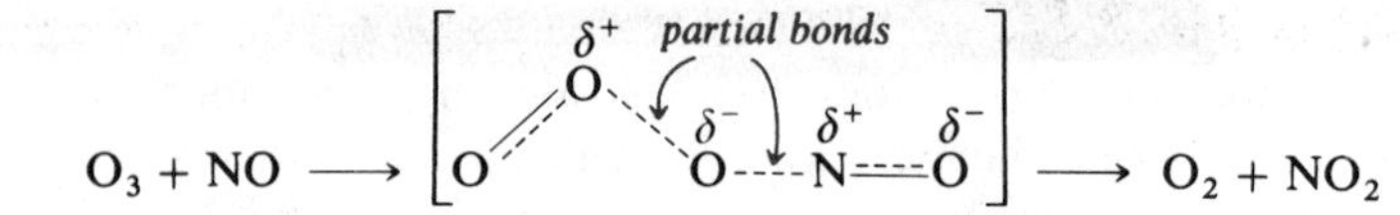

$$O_3 + NO \longrightarrow \left[O \overset{\delta^+}{\underset{}{O}} \cdots \overset{\delta^-}{O} \cdots \overset{\delta^+}{N} = \overset{\delta^-}{O} \right] \longrightarrow O_2 + NO_2$$

partial bonds

transition state

The short-lived combination of reacting atoms, molecules, or ions that is intermediate between reactants and products is called the **transition state**, or **activated complex.** The transition state breaks up either to give the products of the reaction, which then separate, or to give back the reactants. The overall result of an *effective* collision of an NO molecule and an O_3 molecule—a collision that leads to a reaction—is the breaking of an O—O bond, the formation of an N—O bond, and distribution of the internal and kinetic energies of the reactant molecules among the internal and kinetic energies of the product molecules. (The *total* energy of the system is unchanged.)

Of all the collisions that occur, only a small fraction are effective and result in a chemical change. There are three requirements for effective collisions:

1. The orientation of ions or molecules when they collide must be such that the atoms directly involved in the transfer or sharing of electrons come into contact.
2. The collision must take place with enough energy that the outer electron shells of the atoms concerned penetrate each other to some extent and bonding electrons are rearranged.
3. The transfer or sharing of electrons must give a structure that is capable of existence under the conditions of the collision; that is, stable bonds or new stable species must be formed.

We have already discussed the second condition: that the collisions must take place with sufficient energy so that the molecules do not fly apart unchanged. The third condition is straightforward—no reaction can occur if there is no possible product.

Condition 1 becomes more important as the reacting species become larger. In the reaction of hydrogen chloride (HCl) with tri-*n*-propylamine [$(CH_3CH_2CH_2)_3N$] to give tri-*n*-propylammonium chloride [$(CH_3CH_2CH_2)_3NH^+Cl^-$], the bulky hydrocarbon groups hinder the approach of the HCl molecule. The number of directions from which the HCl molecule can approach for an effective collision to take place is limited (Figure 18.1).

The process of collision that we have described is fundamental to kinetics. Two major mathematical models for the advanced study of kinetics are the "collision theory" and the "transition state theory." You may see these terms in other books. The differences between these theories lie in the emphasis placed on different aspects of the collision and in the mathematics of the models. There is no essential difference between these two theories in what is proposed as happening to individual atoms, molecules, or ions during a reaction.

Because of the three requirements that we have discussed, only a small fraction of all collisions are effective for most reactions. In a few types of chemical reactions, however, almost every collision produces products. Mostly, these are reactions of ions in aqueous solution, such as

$$H^+ + OH^- \longrightarrow H_2O(l) \quad (18.2)$$

$$H^+ + HS^- \longrightarrow H_2S(aq) \quad (18.3)$$

or they are reactions between radicals. A radical, or **free radical**, is a highly reactive species that contains an unpaired electron. For example, hydrogen and chlorine atoms are radicals, as is any fragment of a compound containing a carbon atom with three bonds and one unpaired electron. A reaction between a radical and a molecule, such as reaction (18.4), almost always yields another radical and a molecule.

$$H\cdot(g) + Br_2(g) \longrightarrow HBr(g) + Br\cdot(g) \quad (18.4)$$

A reaction between two radicals generally produces a nonradical.

$$CH_3(g) + \cdot CH_3(g) \longrightarrow CH_3CH_3(g) \quad (18.5)$$

Reactions such as these are often very fast.

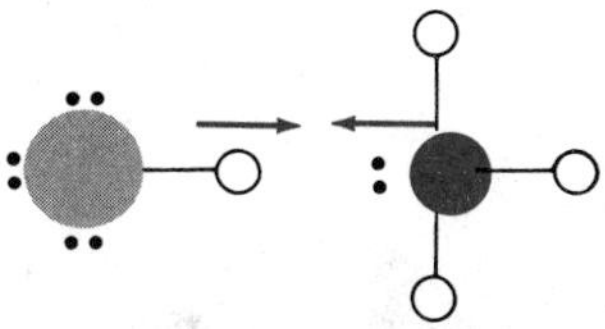

(a) Effective collision, HCl + NH_3

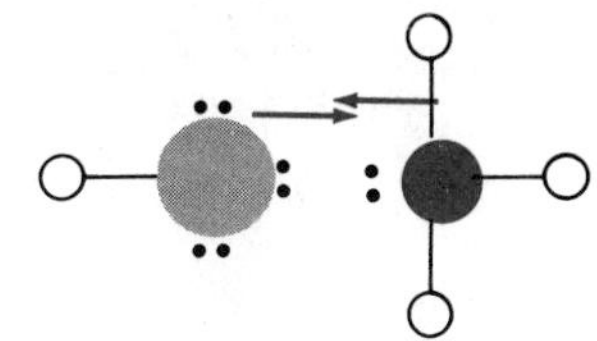

(b) Ineffective collision, HCl + NH_3

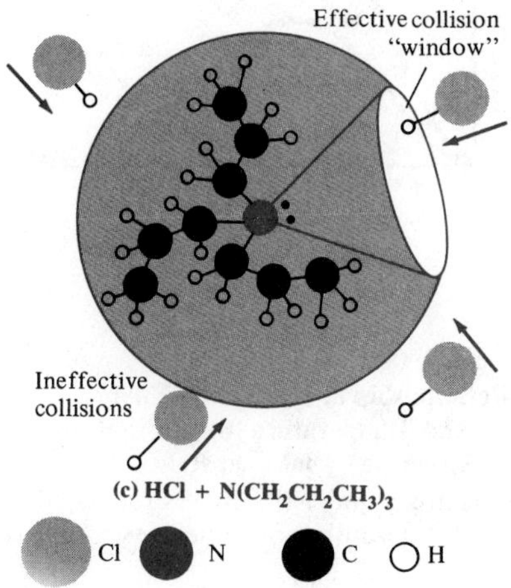

(c) HCl + $N(CH_2CH_2CH_3)_3$

Figure 18.1
Collisions of Hydrogen Chloride with Ammonia and Tripropylamine

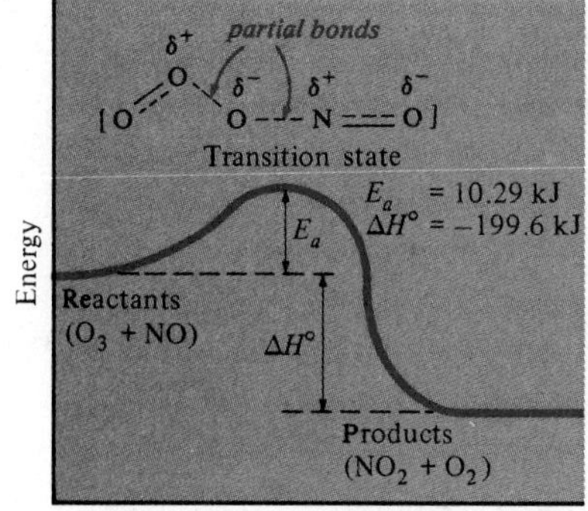

Figure 18.2
Energy Pathway in the Reaction of O_3 with NO (not drawn to scale)
The heat of reaction, $\Delta H°$ (with no $P\Delta V$ work) is the difference between the energy of the reactants and the energy of the products.

18.3 THE ENERGY PATHWAY OF A CHEMICAL REACTION

The energy relationships *during* the course of a chemical reaction can be represented by a diagram like that in Figure 18.2. At the left, the curve begins with the internal energy possessed by the reactants, in this case ozone and nitric oxide. At the right, the curve ends with the internal energy possessed by the products. The reaction

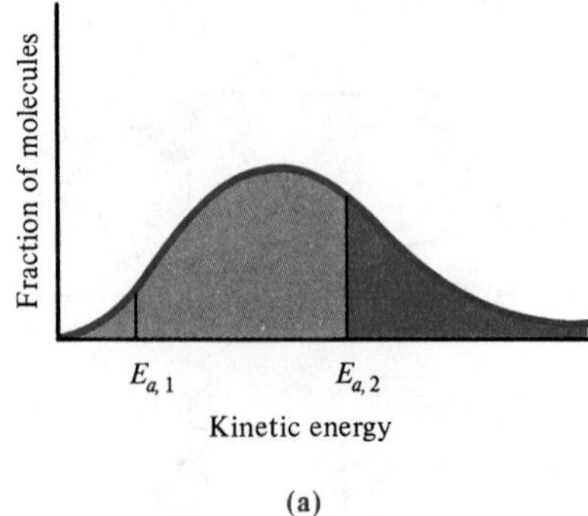

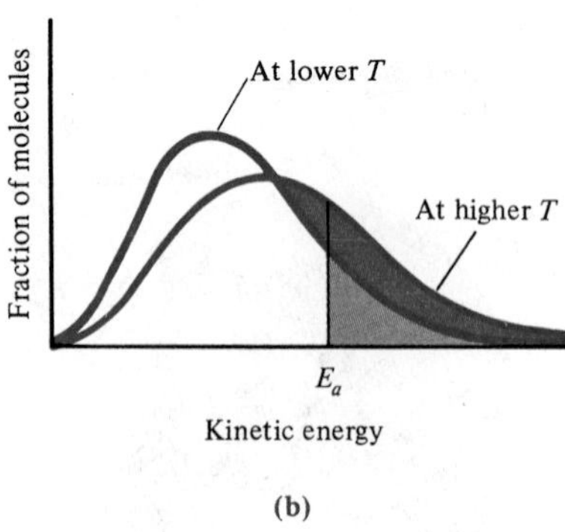

Figure 18.3
Relationship of Activation Energy (a) and Temperature (b) to Fraction of Molecules That Can React
The areas under the curves represent the total number of molecules.

between ozone and nitric oxide is exothermic. The products possess less energy than the reactants and heat (kinetic energy) is released during the reaction.

$$O_3(g) + NO(g) \longrightarrow O_2(g) + NO_2(g) \qquad \Delta H^\circ = -199.6 \text{ kJ}$$

In Figure 18.2 the total energy change in the reaction is represented by the difference between the reactant and product energies. [For this reaction $\Delta E^\circ = \Delta H^\circ$ because there is no change in the number of moles of gases present, that is, no PV work is done. Recall that $\Delta H = \Delta E + \Delta(PV)$, or at constant pressure, $\Delta H = \Delta E + P\,\Delta V$; Section 7.4. For the purpose of the general discussion in this section, we assume that $P\,\Delta V$ is negligible and $\Delta E = \Delta H$ in all cases.]

The horizontal axis in Figure 18.2 is called the reaction coordinate. The progress of the reaction is represented by moving from left to right along this coordinate. As the reactants approach each other, the potential energy rises—repulsion is taking effect. The highest point in the curve represents formation of the transition state. Only collisions that have enough energy to bring the reactants to this point can lead to reaction. It takes about 1.7×10^{-23} kJ to form the transition state between one O_3 molecule and one NO molecule. The **activation energy**, E_a, is defined as the minimum energy that reactants must have for reaction to occur. It is the difference in energy between the transition state and the reactants, and it is always positive (or zero). Customarily, activation energy is reported as the energy needed for the formation of one mole of the transition state. The activation energy for the ozone–nitric oxide reaction is 10.29 kJ. (This is a relatively low activation energy and the reaction is a relatively fast one.)

Activation energy is a characteristic property of a reaction and is found by experiment (Section 18.13). Whenever bonds are broken in a chemical reaction, there must be an activation energy. In only a few types of reactions, such as the rapid reactions of ions or radicals mentioned above, is there no activation energy. Most values of E_a fall in the range of 40 to 200 kJ. The magnitude of E_a depends on the nature of the reacting atoms, molecules, or ions—their bond energies, structures, and so on.

The larger the activation energy, the slower a reaction will tend to be. The smaller the activation energy, the faster a reaction will tend to be. These relationships can be explained by referring to the curve in Figure 18.3a. The total area under the curve represents the total number of molecules of all possible kinetic energies. Suppose $E_{a,1}$ is the activation energy for a particular reaction. The entire shaded area represents the number of molecules that have enough energy to react. However, if the activation energy is higher, say $E_{a,2}$, then only a much smaller number of the molecules, shown by the darkly shaded area, have sufficient energy to react. In other words, there will be fewer effective collisions.

In general, reaction rates increase with temperature. An old rule of thumb states that the rate of a chemical reaction is doubled for each 10 Celsius degree rise in temperature. This is a rough approximation. Reactions with very small activation energies have only a small temperature dependence, while the rates of some reactions are much more than doubled by a 10 Celsius degree temperature increase.

The effect of changing the temperature of a reacting system is twofold. As the temperature is increased, the molecules move more rapidly, so collisions are more frequent. However, calculations show that only a small percentage of the rate increase with temperature can be accounted for by the larger number of collisions. The energy distribution curve provides the explanation for most of the increase in rate. Figure 18.3b shows how the curve changes as the temperature increases. The area under both curves is the same, but as the curve flattens out with increasing temperature, a larger number of molecules occupy the high-energy end of the distribution, as shown by the darkly shaded portion of the area under the curve. More molecules have higher energy and can undergo effective collisions.

The energy changes for exothermic and endothermic reactions are compared in Figure 18.4. The heat of reaction in both cases is the difference between the energy of the products and the energy of the reactants. Take a moment to think about the relationships represented by these curves. Note that the reverse of an exothermic reaction is an endothermic reaction, and that the endothermic reaction always has the larger activation energy.

As we have pointed out, thermodynamics is concerned only with the initial and final states of a chemical reaction, which are represented in Figure 18.4 by the energy levels of the reactants and products. The existence of the activation energy explains why some reactions that are spontaneous according to the calculations of thermodynamics do not take place rapidly as soon as the reactants are mixed. Look at Figure 18.4a and imagine a situation in which, although the products have lower energy than the reactants (which is thermodynamically favorable), very little reaction occurs because the activation energy "hill" is too high and very few effective collisions can occur.

The first step in the mechanism for the oxidation of carbon monoxide is such a reaction.

$$CO(g) + O_2(g) \longrightarrow CO_2(g) + O(g) \qquad \Delta H° = -33 \text{ kJ}, \quad E_a = 213 \text{ kJ} \qquad (18.6)$$

The reaction is exothermic—the products have less energy than the reactants. But carbon monoxide is certainly not oxidized by oxygen in the air as soon as it is formed. You can die from breathing the carbon monoxide that collects when an automobile engine is left running in a closed garage. The activation energy of reaction (18.6) is high and the temperature must be raised to over 2000 °C before CO and O_2 molecules collide with enough energy to get over the energy barrier. (One purpose of the catalytic converter in an automobile exhaust system is to allow this reaction to take place at a lower temperature.)

Some reactions keep going once a few molecules gather the energy needed to react. This occurs when the overall reaction is exothermic. The products carry away much of the energy released by the reaction in the form of kinetic energy, raising the temperature of the entire system. As a result, more and more reactants have sufficient energy to react as time proceeds. The classic example of this type of reaction is the combination of hydrogen and oxygen. In a clean, undisturbed flask, a mixture of hydrogen and oxygen will stand indefinitely virtually without reaction. A single spark which heats only a few molecules can set off an explosive reaction of the entire contents of the flask.

Exothermic reaction

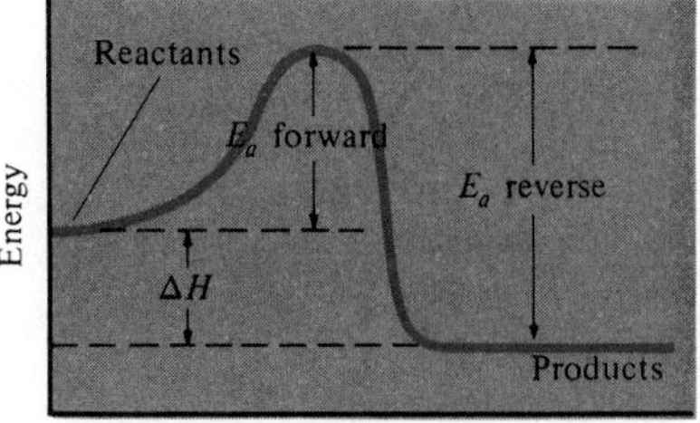

Endothermic reaction

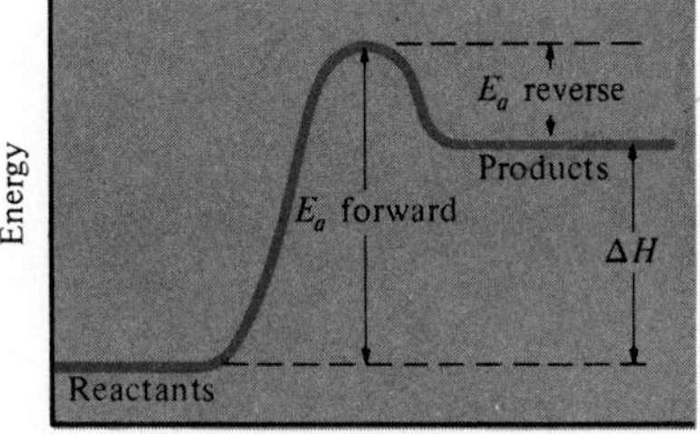

Figure 18.4
Energy Pathways for Exothermic (a) and Endothermic (b) Reactions
Note that the activation energy for an endothermic reaction is always larger than that for the reverse, exothermic reaction.

18.4 ELEMENTARY REACTIONS

Elementary reactions, which represent what happens at the molecular level, are categorized in terms of their **molecularity**—the number of reactant particles involved in an elementary reaction. The reactants in elementary reactions are often highly reactive intermediates—atoms and free radicals, or ionic fragments of larger molecules.

The reaction of nitric oxide and ozone (Equation 18.1) is a **bimolecular reaction**—an elementary reaction that has two reactant particles. The reactants in a bimolecular reaction can be the same or different—the point is that two particles must collide.

elementary bimolecular reactions

$$A + A \longrightarrow \text{products} \qquad (18.7)$$

$$A + B \longrightarrow \text{products} \qquad (18.8)$$

By far the majority of elementary reactions are bimolecular. Such reactions are common in the gas phase—for example, the reaction between carbon monoxide and molecular oxygen (Equation 18.6) or the reaction between ozone and an oxygen atom to give two oxygen molecules

$$O_3(g) + O(g) \longrightarrow O_2(g) + O_2(g)$$

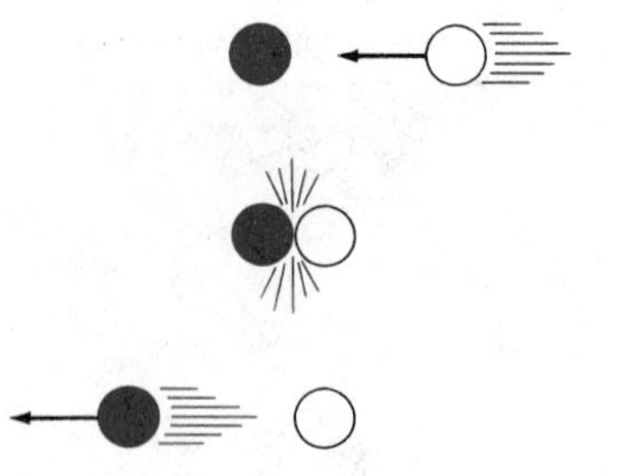

Two bodies cannot stay together after impact

(a) Two-body collision

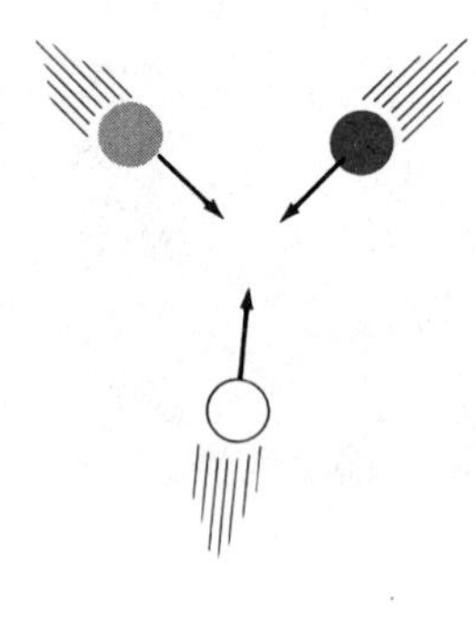

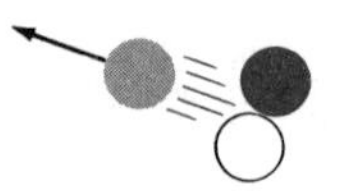

Two bodies can stay together after impact because third body carries away excess energy

(b) Three-body collision

Figure 18.5
Two-Body and Three-Body Collisions (Such collisions between molecules have similarities to collisions between billiard balls.) A successful three-body collision between molecules occurs when the third molecule arrives before the other two have flown apart.

Bimolecular elementary reactions are also common in the solution phase—for example, the combination of the hydrogen ion with hydroxide ion or hydrogen sulfide (HS^-) ion (Equations 18.2, 18.3) or the displacement of the bromide ion in methyl bromide by the chloride ion, which occurs in acetone solution

$$CH_3Br + Cl^- \xrightarrow{\text{acetone}} CH_3Cl + Br^-$$

A **termolecular reaction** is an elementary reaction that requires the interaction of three reactant particles. Termolecular reactions are rare, for the probability of three reacting particles coming together at the same time to form a transition state is low. The following reactions are thought to be termolecular elementary reactions:

$$2NO(g) + O_2(g) \longrightarrow 2NO_2(g)$$
$$2NO(g) + Br_2(g) \longrightarrow 2NOBr(g)$$

In the recombination of atoms in the gas phase to give diatomic molecules, however, a third participant is *necessary*. Suppose two atoms come together with sufficient energy to combine in an activated complex, [A—A]*. To form the molecule A—A, the activated complex must lose some energy. Otherwise, all of the kinetic energy of the two original atoms will be converted to vibrational energy in the A—A bond, and this bond will break. Instead of settling down into a stable molecule, the atoms will fly apart again. Only if a third body carries away some of the energy can the two atoms remain combined (Figure 18.5). Following are some examples of recombination reactions in the gas phase that require a third body (M):

$$Br(g) + Br(g) + M(g) \longrightarrow Br_2(g) + M(g)$$
$$O(g) + O_2(g) + M(g) \longrightarrow O_3(g) + M(g)$$

The question arises as to whether or not there are unimolecular reactions. These would be elementary reactions in which one reactant gives one or more products. But there is a problem. We know that most reactions have an activation energy. How can a single reactant in a unimolecular reaction become activated? The answer is that in most cases, reactions involving only a single reactant are *not* really elementary unimolecular reactions. The reactions proceed by a simple mechanism in which the first step is activation of the reactant: by radiation or by collision with something that does not become part of the product. The collision might be with an inert atom or molecule or even with the wall of the container. The activated molecule does not react immediately. The energy it has gained must be distributed among the various rotational and vibrational motions of the molecule. The reaction occurs when sufficient energy reaches the bond that will break, usually the weakest bond in the molecule, and reaction occurs. The mechanism is

activated species

$$A + M \longrightarrow A^* + M$$
$$A^* \longrightarrow B$$

At the molecular level, reactions in solution differ somewhat from gas-phase reactions because of the solvation of the reactants. Reactants cannot simply come together, interact briefly, and then separate. Their coming together is slowed down by surrounding solvent molecules, called the "solvent cage," and the frequency of collisions between isolated molecules is lower than in the gas phase. But once the reactants come close together, they are trapped in the same solvent cage and collide with each other frequently within a short period of time (Figure 18.6). The reactions of ions in aqueous solution are, as noted, generally very fast, because the activation energy is close to zero. The rate of such reactions is limited only by the time required for the ions to diffuse through the water until they encounter each other.

SUMMARY OF SECTIONS 18.1–18.4

Chemical kinetics deals with the rates and mechanisms of chemical reactions and, as opposed to thermodynamics, deals with what occurs during the course of a reaction at the molecular level, rather than only the overall outcome. A chemical reaction occurs when atoms, molecules, or ions undergo effective collisions; this requires proper orientation of the colliding particles, energy greater than a certain minimum energy (known as the activation energy), and the possibility of formation of a stable product. The larger the activation energy, the slower a chemical reaction, and vice versa. In general, reaction rates increase with temperature, because higher temperature increases both the frequency of collisions and the number of molecules possessing enough energy to undergo effective collisions. The reverse of an exothermic reaction is always an endothermic reaction, which always has a larger activation energy.

A chemical equation represents an elementary reaction only if the reactants in the equation represent the actual atoms, molecules, or ions that must interact for the reaction to occur. Bimolecular elementary reactions (the most common) have two reactant species and termolecular reactions have three. In unimolecular reactions, a single reactant must be activated by collision with a nonreactant or by some other means.

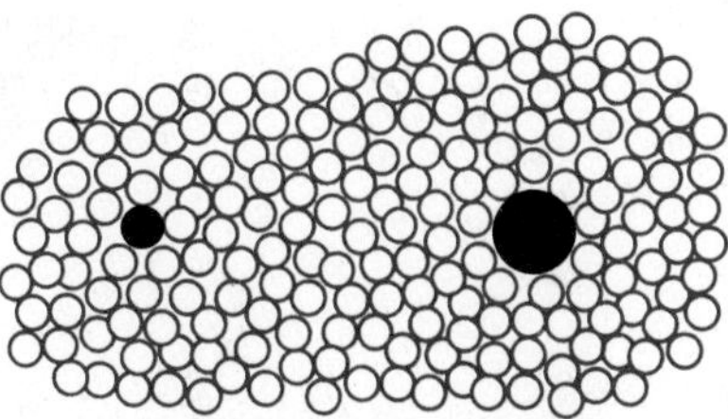

Separated reactant molecules

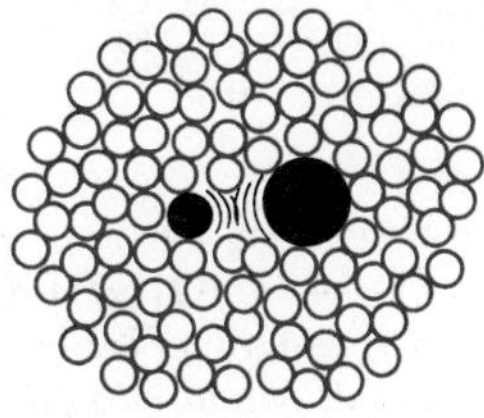

Reactant molecules in same solvent cage

Figure 18.6
Reactants in Solution
The reactant molecules collide frequently while they are in the same solvent cage.

REACTION RATES AND RATE EQUATIONS

18.5 DEFINITION OF REACTION RATE

The rate of a process is expressed as the amount of change per unit of time. A car may travel at a rate of 88 km/h. Water may flow through a pipe at a rate of 32 L/min. In a chemical reaction the quantities that are changing are the amounts of reactants and products. The concentration of reactants decreases with time and the concentration of products increases (Figure 18.7). **Reaction rate** is defined as the change in concentration of a reactant or a product in a unit of time.

$$\text{Rate} = \frac{\text{change of concentration}}{\text{time period of change}}$$

The change of concentration can be that of any one of the reactants or products. For the reaction A + B $\longrightarrow$ C, the rate could be expressed as follows

$$\text{Rate} = \frac{\Delta[\text{A}]}{\Delta t} \quad \text{or Rate} = \frac{\Delta[\text{B}]}{\Delta t} \quad \text{or Rate} = \frac{\Delta[\text{C}]}{\Delta t}$$

where $\Delta[\text{A}]$, for example, indicates the change in the concentration of A and the square brackets indicate concentrations in moles per liter; Δt represents the change in time. It is convenient to express the reaction rate as a positive number. This is achieved by defining the rate of reaction as

$$\text{Rate} = \pm \frac{\Delta[\text{X}]}{\Delta t} \tag{18.9}$$

change in reactant concentration (in moles/liter) (points to $\Delta[\text{X}]$); *change in time* (points to Δt)

where [X] is the concentration of any reactant or product. The negative sign applies if X is a reactant ($\Delta[\text{X}]$ is negative) and the positive sign if X is a product ($\Delta[\text{X}]$ is positive).

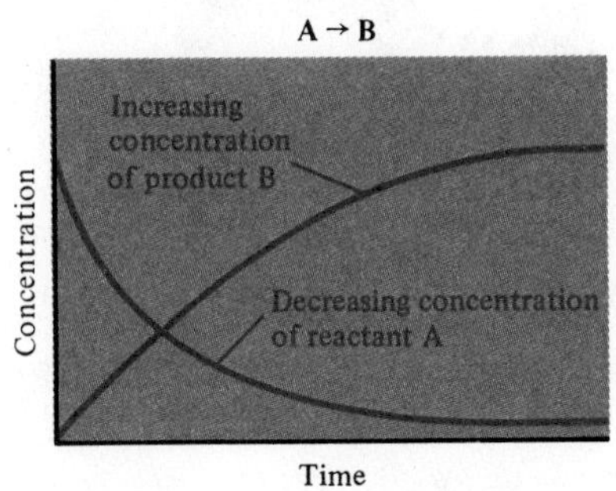

Figure 18.7
Change in Reactant and Product Concentrations with Time

Reaction rates are expressed in units of concentration per time unit, most commonly moles per liter per time unit—seconds, minutes, days, and so on, depending

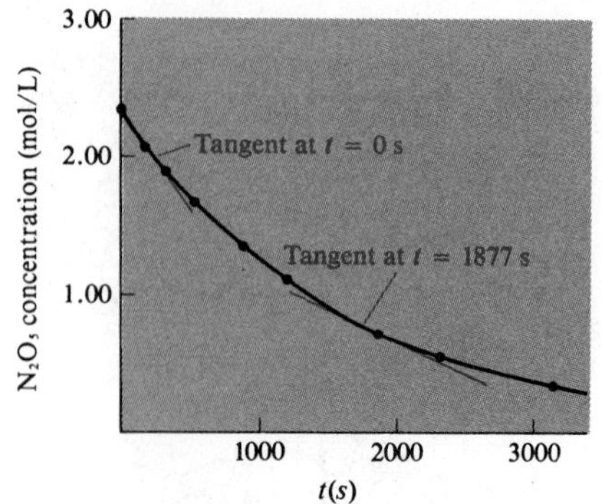

Figure 18.8
Plot of Kinetic Data for $2N_2O_5(g) \longrightarrow 4NO_2(g) + O_2(g)$ at 45 °C in CCl_4 Solution (Table 18.1)
Tangents are drawn at two points. The slope of the tangent to such a concentration versus time plot at any value of t equals the reaction rate at that instant.

upon how slow the reaction is. The units are written, for example, mol/L s (s is the abbreviation for seconds), or mol L^{-1} s^{-1} (where the -1 superscript indicates the reciprocal of the unit: $s^{-1} = 1/s$), or sometimes M/s (where M is the abbreviation for mol/L). In this book we write mol/L s, mol/L h, and so on. For reactions of gases, reaction rates are sometimes expressed in units of partial pressure change per unit time. (Reaction rates are seldom defined in terms of the amounts of solids or pure liquids.)

A significant difference exists between, say, the rate of flow of water through a pipe and the rate of a chemical reaction. The flow rate of water can be held constant, but in most cases a reaction rate is not constant. The rate of a chemical reaction is a smoothly changing quantity because it is dependent upon concentration, which changes as reactants are consumed and products are formed.

18.6 DETERMINATION OF REACTION RATE

The study of the kinetics of a chemical reaction begins with the experimental determination of how the concentration of reactants or products varies with time. A typical set of data for the following reaction is shown in Table 18.1 and plotted in Figure 18.8.

$$2N_2O_5 \xrightarrow[45\ °C]{\text{in } CCl_4} 4NO_2 + O_2$$

The experimental data give the concentration of N_2O_5 measured at the beginning of the reaction ($t = 0$) and then at a number of time intervals during the course of the reaction. For example, 184 s after the reaction began, the concentration of N_2O_5 had dropped from 2.33 mol/L to 2.07 mol/L. The average reaction rate is the calculated average rate during the time period between a pair of concentration measurements. You can see from the data in the table that it decreases as the concentration of the reactant decreases.

The average reaction rate, in terms of the reactant N_2O_5, is found by using Equation (18.9) as follows:

reactant concentration at t_2 ↘ ↙ *reactant concentration at t_1*

$$\text{Rate} = -\frac{\Delta[X]}{\Delta t} = -\frac{[N_2O_5]_2 - [N_2O_5]_1}{t_2 - t_1}$$

The reactant concentration decreases with time, making $[N_2O_5]_2 - [N_2O_5]_1$ a negative number. Thus, the minus sign gives the rate as a positive number. For example, the average reaction rate between the 319th second and the 526th second—a period

Table 18.1
Kinetic Data for the Decomposition of N_2O_5 at 45 °C in CCl_4 Solution

t (s)	Concentration of N_2O_5 (mol/L)	Average reaction rate in each time interval (mol/L s)	Instantaneous reaction rate at t (mol/L s)
0	2.33		1.42×10^{-3}
		1.14×10^{-3}	
184	2.07		1.27×10^{-3}
		1.19×10^{-3}	
319	1.91		1.17×10^{-3}
		1.11×10^{-3}	
526	1.68		1.02×10^{-3}
		0.97×10^{-3}	
867	1.35		0.83×10^{-3}
		0.76×10^{-3}	
1198	1.11		0.68×10^{-3}
		0.57×10^{-3}	
1877	0.72		0.44×10^{-3}
		0.39×10^{-3}	
2315	0.55		0.34×10^{-3}
		0.25×10^{-3}	
3144	0.34		0.21×10^{-3}

during which the concentration of the reactant N_2O_5 decreased from 1.91 mol/L to 1.68 mol/L—was found as follows

$$\text{Rate} = -\frac{\Delta[N_2O_5]}{\Delta t} = -\frac{(1.68 - 1.91)\ \text{mol/L}}{(526 - 319)\ \text{s}} = 1.11 \times 10^{-3}\ \frac{\text{mol}}{\text{L s}}$$

Keep in mind that the reaction rate is a constantly changing quantity (except for zero-order reactions, Section 18.10). The **instantaneous reaction rate**—the rate at any specific instant of time during a chemical reaction—is a more useful quantity in chemical kinetics than the average rate. It is not calculated, but must be found from a plot of the experimental data. The variation in concentration of a reactant or product with time is first plotted as illustrated in Figure 18.8.

The slope of the tangent to the curve at any given time gives the value of the instantaneous reaction rate at that time. Tangents are drawn in Figure 18.8 at two points during the reaction, $t = 0$ s and $t = 1877$ s. The instantaneous reaction rate at $t = 0$, s, the instant when the reaction begins, is called the **initial reaction rate**. This and the instantaneous reaction rate for each data point in Figure 18.8 are given in the last column in Table 18.1. Like the average rate, the instantaneous rate decreases with time (except for zero-order reactions). (The method for calculating the slope of the curve is described in Appendix I.B.)

EXAMPLE 18.1 Reaction Rate

How might the rate of reaction be expressed for the following reaction?

$$H_2(g) + I_2(g) \longrightarrow 2HI(g)$$

The rate of reaction can be defined in terms of the changes in concentration with time of either reactants or products. In terms of the reactants, the rate of this reaction could be expressed as

$$\textbf{Rate} = -\frac{\Delta[H_2]}{\Delta t} = -\frac{\Delta[I_2]}{\Delta t}$$

Because one molecule of I_2 and one molecule of H_2 react with each other, the concentrations of these two reactants change at the same rate. In terms of the product, the rate would be

$$\textbf{Rate} = \frac{\Delta[HI]}{\Delta t}$$

Note that two molecules of HI are produced for each molecule of H_2 and I_2 that react. Therefore, if the rate of reaction is defined as the change in the concentration of HI, it will be twice as large as the rate of reaction defined in terms of either H_2 or I_2.

EXERCISE How might the rate of reaction be defined for the following reaction: $2O_3(g) \longrightarrow 3O_2(g)$? *Answer* $-\Delta[O_3]/\Delta t$ or $\Delta[O_2]/\Delta t$.

18.7 SIMPLE RATE EQUATIONS AND REACTION ORDER

Having gathered concentration versus time data, the next step in studying the kinetics of a reaction is to fit the data to a **rate equation**, which gives the mathematical relationship between the reaction rate and the concentration of one or more of the reactants. Remember that in most kinetics experiments and in the determination of rate equations, it is the overall macroscopic result of the reaction that is under study.

The rate of a reaction is dependent upon the number of collisions between reactant particles. The number of collisions, in turn, is dependent upon the concentration of the reactants. A rate equation expresses the relationship between rate and concentration; it is always some function of one or more constants and the concentrations of reactants and/or products during the reaction raised to various powers. Frequently, although by no means always, experimentally determined rate equations have the following form:

$$\text{Rate} = k[\text{A}]^m[\text{B}]^n[\text{C}]^p\cdots \quad \textbf{(18.10)}$$

rate constant — *reactant concentrations in moles/liter*

where the square brackets in [A], [B], [C], . . . denote reactant concentrations in moles per liter and k is the standard notation for a proportionality constant. The exponents $m, n, p, \ldots$ are most often 1 or 2. Sometimes they have values of 3 or of 0, or they may have values that are fractions. On occasion they may be negative (showing that increasing the concentration of that reactant would actually slow down the reaction). For reactions in the gas phase, partial pressures can be used instead of molar concentrations.

The proportionality constant k that relates reaction rate to some function of reactant concentrations raised to various powers is called the **rate constant**. The value of a rate constant is characteristic of a specific reaction. It reflects the fraction of collisions that are effective in producing a reaction at a given temperature. Because k varies with the temperature (Section 18.13), the temperature at which a rate constant applies must always be stated. For fast reactions k is large; for slow reactions k is small. Both the value of the rate constant and the form of the "reactant concentrations" part of the rate equation must be determined by experiment.

The kinetics of a chemical reaction is categorized by the "order" of the reaction, a term that refers to the exponents on the concentration terms in the rate equation. The **overall reaction order** is the sum of the exponents of the concentration terms in the rate equation. For example, a reaction with the rate equation

$$\text{Rate} = k[\text{A}]^2[\text{B}]$$

is described as a third-order reaction overall, because the sum of the exponents on [A] and [B] is 3. The **reactant reaction order** is the exponent on the term for a given reactant in the rate equation. The reaction with the rate equation given above is described as "second order in A" and "first order in B."

It is often convenient to describe a reaction as "first order" or "second order" overall. However, for reactions that have complex rate equations, overall reaction order has little meaning. Some further examples of the order of generalized rate equations for simple reactions are given in Table 18.2.

Table 18.2
Rate Equations and Reaction Order for Simple Reactions
Because any number raised to the zero power equals 1, $k[\text{A}]^0 = k$. For complex rate equations, the concept of overall reaction order has little meaning.

Rate equation rate =	Overall reaction order
$k[\text{A}]^0 = k$	zero
$k[\text{A}]$	first
$k[\text{A}]^2$	second
$k[\text{A}][\text{B}]$	second
$k[\text{A}]^3$	third
$k[\text{A}]^2[\text{B}]$	third
$k[\text{A}][\text{B}][\text{C}]$	third
$k[\text{A}]^{3/2}[\text{B}]$	5/2

EXAMPLE 18.2 Reaction Order

The rate equation for the oxidation of SO_2 to SO_3 in excess O_2 is as follows:

$$2SO_2(g) + O_2(g) \xrightarrow{\text{Pt}} 2SO_3(g) \qquad \text{Rate} = k[SO_2][SO_3]^{-1/2}$$

What is the order of the reaction with respect to each substance in the rate equation? What happens to the rate of reaction as $[SO_3]$ increases?

The order of the reaction for each substance is given by the exponent on the concentration term in the rate equation. The reaction is **first order with respect to SO_2 and negative one-half order with respect to SO_3.** As the concentration of SO_3 increases, the reaction rate decreases.

EXERCISE The decomposition of hydrogen peroxide

$$2H_2O_2(aq) \xrightarrow{\text{I}^-} 2H_2O(l) + O_2(g)$$

is known to be first order with respect to H_2O_2 and first order with respect to I^-. Write the rate equation for this reaction. What is the overall order of the reaction?

Answer Rate = $k[H_2O_2][I^-]$, second order overall.

18.8 RATE EQUATIONS FOR FIRST-ORDER REACTIONS

The rate of a first-order reaction is dependent on the concentration of only one reactant.

$$\underset{\frac{mole}{(liter)(time\ unit)}}{\text{Rate}} = \underset{\frac{1}{time\ unit}\ \frac{mole}{liter}}{k[\text{A}]} \qquad (18.11)$$

For rate expressed in the usual units of moles per liter per unit of time, the rate constant of a first-order reaction has reciprocal time units, for example

$$\frac{1}{\text{s}} \text{ or } \text{s}^{-1} \qquad \frac{1}{\text{min}} \text{ or } \text{min}^{-1} \qquad \frac{1}{\text{h}} \text{ or } \text{h}^{-1}$$

The high-temperature (1000 °C) decomposition of carbon dioxide to give carbon monoxide and atomic oxygen (Equation 18.12) and the decomposition of cyclobutane to give two molecules of ethylene (Equation 18.13) are both first-order reactions.

$$CO_2(g) \xrightarrow{1000\ °C} CO(g) + O(g) \qquad k = 1.8 \times 10^{-13}\ \text{s}^{-1} \qquad (18.12)$$

$$\begin{matrix} H_2C{-}CH_2 \\ | \quad\ \ | \\ H_2C{-}CH_2 \end{matrix}(g) \xrightarrow{1000\ °C} 2CH_2{=}CH_2(g) \qquad k = 87\ \text{s}^{-1} \qquad (18.13)$$

[Because gas-phase reactions are easier to study than condensed-phase reactions, the most thoroughly understood examples of reaction kinetics are for gas-phase reactions.]

The rate of a first-order reaction can be expressed as the change in concentration of the reactant with time (Section 18.5), giving

$$\text{Rate} = -\frac{\Delta[\text{A}]}{\Delta t} \qquad (18.14)$$

By equating Equations (18.11) and (18.14), we find that for a first-order reaction

$$k[\text{A}] = -\frac{\Delta[\text{A}]}{\Delta t} \qquad (18.15)$$

Using the mathematical methods of calculus, this equation can be transformed into the following equation that relates concentration and time for a first-order reaction, where $[\text{A}]_0$ is the concentration at $t = 0$ and $[\text{A}]$ is the concentration at time t.

$$\log[\text{A}] = \log[\text{A}]_0 - \frac{kt}{2.303} \qquad (18.16)$$

Combining the log terms gives

$$\log\frac{[\text{A}]}{[\text{A}]_0} = -\frac{kt}{2.303} \quad \text{or} \quad \log\frac{[\text{A}]_0}{[\text{A}]} = \frac{kt}{2.303} \qquad (18.17)$$

Equations (18.16) and (18.17) are known as *integrated rate equations.*

[Calculus gives these and other similar equations in terms of the natural logarithm, usually abbreviated, for example, $\ln[\text{A}]$. The factor 2.303 enters into the equation in order to convert natural logarithms to base 10 logarithms, $\log[\text{A}] = \ln[\text{A}]/2.303$. It is customary to work with logarithms to base 10, but with an appropriate calculator, there is no difference in the ease of using either type of logarithm. It is necessary to use logarithms frequently in this and later chapters. At this point, you may want to review logarithms, which are discussed in Appendix I.A.]

If k is known, Equations (18.16) and (18.17) allow the calculation of the time needed to reach a given concentration of reactant or the concentration of reactant remaining after a given amount of time. Alternatively, k can be found by measuring these times and concentrations. Because the units of $[\text{A}]$ and $[\text{A}]_0$ cancel out in Equation (18.17), for a first-order reaction the concentrations of $[\text{A}]$ and $[\text{A}]_0$ can be replaced by any quantity that is proportional to the concentration in moles per liter.

For example, if the volume remains constant, the concentrations might be replaced by mass in grams, or amount of absorbed radiation, or for radioactive substances, by number of nuclei or even counts per minute per gram measured by a Geiger counter (see Figure 12.12).

EXAMPLE 18.3 Rate Equations

The decomposition of N_2O_5 in CCl_4 at 45 °C,

$$2N_2O_5 \xrightarrow{CCl_4} 4NO_2 + O_2$$

is first order with $k = 6.32 \times 10^{-4}\ s^{-1}$. Find the concentration of N_2O_5 remaining after 1.0 h (3600 s) when the initial concentration of N_2O_5 was 0.40 mol/L. What percentage of the N_2O_5 has reacted?

The equation relating time and concentration for a first-order reaction, Equation (18.16), is used with $[N_2O_5]_0 = 0.40$ mol/L and $t = 3600$ s to find $[N_2O_5]$.

$$\log [N_2O_5] = \frac{-kt}{2.303} + \log [N_2O_5]_0$$

$$= \frac{(-6.32 \times 10^{-4}\ s^{-1})(3600\ s)}{(2.303)} + \log (0.40)$$

$$= (-0.988) + (-0.40) = -1.39$$

Taking antilogarithms of both sides gives the concentration of N_2O_5 remaining after 1.0 h

$$[N_2O_5] = \mathbf{0.041\ mol/L}$$

The difference between this and the initial concentration (0.40 mol/L − 0.041 mol/L) = 0.36 mol/L, is the amount that has reacted, which gives

$$\left(\frac{0.36\ \text{mol/L}}{0.40\ \text{mol/L}}\right)(100) = \mathbf{90.\%}$$

EXERCISE In the reaction described above, given the same initial concentration of N_2O_5, find the time it would take for 95% of the N_2O_5 to decompose.

Answers 4.74×10^3 s, or 79 min.

18.9 HALF-LIFE OF FIRST-ORDER REACTIONS

Half-life, which is often used to characterize the relative stabilities of radioactive materials, is the time it takes for one-half of a reactant to undergo a reaction. In terms of kinetics, radioactive decay is a first-order process. The equation that relates the initial mass (q_0) of a radioactive nuclide and the mass (q) at time t (see Equation 12.3)

$$\log \frac{q_0}{q} = \frac{kt}{2.303} \qquad \textbf{(18.18)}$$

is the integrated rate equation for a first-order reaction (Equation 18.17) with concentration replaced by mass. The expression for half-life derived from this relationship in Section 12.5

$$t_{1/2} = \frac{0.693}{k} \qquad \textbf{(18.19)}$$

applies to all first-order reactions, from which it is apparent that the half-life of a first-order reaction is *independent* of the concentration of the reactant.

EXAMPLE 18.4 Rate Equations: Half-Life

After 2.00 h, a solution originally containing 1.30×10^{-6} mol/L of $^{240}AmCl_3$ contained only 1.27×10^{-6} mol/L of this radioactive substance. What is the half-life of ^{240}Am?

Before the half-life can be found by using Equation (18.19), the first-order rate constant must be found by solving the integrated first-order rate equation, Equation (18.16), for k.

$$k = \frac{(2.303)(\log[^{240}\text{Am}]_0 - \log[^{240}\text{Am}])}{t}$$

$$= \frac{(2.303)[\log(1.30 \times 10^{-6}) - \log(1.27 \times 10^{-6})]}{2.00\ \text{h}}$$

$$= 1.2 \times 10^{-2}\ \text{h}^{-1}$$

$$t_{1/2} = \frac{0.693}{k} = \frac{(0.693)}{(1.2 \times 10^{-2}\ \text{h}^{-1})} = \mathbf{58\ h}$$

EXERCISE The decomposition of gaseous PH_3 at 600 °C is known to be first order. After exactly 2 min, the partial pressure of PH_3 decreased from 262 Torr to 16 Torr. Using these data, calculate the half-life of this reaction at this temperature.

Answer 29.7 s.

18.10 RATE EQUATIONS FOR ZERO-ORDER AND SECOND-ORDER REACTIONS

The rate of a second-order reaction can be dependent upon the concentration of either one or two reactants.

$$\text{Rate} = k[\text{A}]^2 \qquad \textbf{(18.20)}$$

$$\text{Rate} = k[\text{A}][\text{B}] \qquad \textbf{(18.21)}$$

$$\frac{\textit{mole}}{(\textit{liter})(\textit{time unit})} \qquad \frac{\cancel{\textit{liter}}}{(\cancel{\textit{mole}})(\textit{time unit})} \frac{\cancel{\textit{mole}}}{\cancel{\textit{liter}}} \frac{\textit{mole}}{\textit{liter}}$$

Second-order rate constants have units of reciprocal concentration (liters/mole) times reciprocal time. For example, the decomposition of hydrogen iodide in the gas phase at 25 °C is quite a slow second-order reaction

$$2\text{HI}(g) \longrightarrow \text{H}_2(g) + \text{I}_2(g) \qquad k = 2.4 \times 10^{-21}\ \text{L/mol s}$$

and the reaction of ozone with nitric oxide that was discussed earlier is a quite rapid second-order reaction

$$\text{O}_3(g) + \text{NO}(g) \longrightarrow \text{O}_2(g) + \text{NO}_2(g) \qquad k = 2.2 \times 10^7\ \text{L/mol s}$$

The rate of a zero-order reaction is not dependent on the reactant concentration, and therefore the rate is a constant.

$$\text{Rate} = k[\text{A}]^0 = k(1) = k$$

This means that the rate is controlled by something other than collisions that involve reactants. The controlling factor might be the amount of light available in a photochemical reaction or the amount of catalyst present. Many reactions that occur on the surface of a metal (heterogeneous catalysis; Section 18.15) are zero order—the rate is controlled by the interaction of the reactant with the surface and the availability of active sites on the surface, for example

$$2\text{NH}_3(g) \xrightarrow{\text{tungsten}} \text{N}_2(g) + 3\text{H}_2(g) \qquad \text{Rate} = k$$

By using the methods of calculus, equations relating the variation of concentration with time (integrated rate equations) can be derived for zero- and second-order

Table 18.3
Equations for the Variation of Rate with Concentration, the Variation of Concentration with Time, and Half-Life
Concentration at $t = 0$ is shown by $[A]_0$, $[B]_0$; concentration at t is shown by $[A]$, $[B]$.

Type of reaction	Rate equation	Integrated rate equation	Half-life
Zero order	Rate $= k$	$[A] = [A]_0 - kt$	$\frac{[A]_0}{2k}$
First order	Rate $= k[A]$	$\log\frac{[A]}{[A]_0} = -\frac{kt}{2.303}$ or $\log[A] = \log[A]_0 - \frac{kt}{2.303}$	$\frac{0.693}{k}$
Second order	Rate $= k[A]^2$ or Rate $= k[A][B]$ with $[A] = [B]$	$\frac{1}{[A]} = \frac{1}{[A]_0} + kt$	$\frac{1}{k[A]_0}$
	Rate $= k[A][B]$ with $[A] \neq [B]$	$\log\frac{[A][B]_0}{[A]_0[B]} = ([A]_0 - [B]_0)\frac{kt}{2.303}$	—

reactions. These equations and also the half-life equations are given in Table 18.3. Note that second-order reactions include those in which the rate is dependent upon the concentration of one reactant, upon the equal concentrations of two different reactants, or upon the different concentrations of two reactants. The latter type of second-order reaction gives a different form for the integrated rate equation.

The half-life equations indicate the interesting fact that only the half-lives of first-order reactions are independent of the concentration.

The equations in Table 18.3 for zero- and second-order reactions can be used to solve the types of problems illustrated for first-order reactions in Examples 18.3 and 18.4.

18.11 FINDING RATE EQUATIONS

With a set of data at hand giving the concentrations of reactants or products at different times, what can be done to find the rate equation? The first step is to determine if the data fit any of the rate equations given in Table 18.3. In this section we describe two simple methods for doing this. First, however, the relationships among the elementary reactions, the overall reaction, and the rate equations must be discussed.

a. Elementary Reactions, Overall Reactions, and Rate Equations *If* a reaction is known to be an elementary reaction, its rate equation can be written based solely on the chemical equation for the reaction. In the general rate equation, Rate $= k[A]^m[B]^n\cdots$, A and B are the reactants and m and n are the coefficients of those reactants in the equation for an *elementary reaction*. For example

$$A + 2B \longrightarrow C \qquad \text{Rate} = k[A][B]^2$$

or for the bimolecular reaction of ozone with nitric oxide

$$O_3(g) + NO(g) \longrightarrow O_2(g) + NO_2(g) \qquad \text{Rate} = k[O_3][NO_2]$$

Writing a rate equation by inspection of the chemical equation is possible *only* for an elementary reaction, because the reactants in this case represent the actual species whose concentrations determine the reaction rate. The most common misunderstanding of kinetics by students is to assume that the exponents in a rate equation are necessarily equal to the coefficients in the chemical equation for the reaction. The exponents show how rate varies with collisions of individual reactant particles, and this is determined by the reaction mechanism and what is happening at the molecular level.

Experimental rate equations usually apply to overall reactions, *not* elementary reactions.

Consider the reaction

$$2NO_2(g) + F_2(g) \longrightarrow 2NO_2F(g) \quad \textbf{(18.22)}$$

for which the experimentally determined rate equation is

$$\text{Rate} = k[NO_2][F_2] \quad \textbf{(18.23)}$$

Note that the exponent of $[NO_2]$ is *not* 2. A proposed mechanism for reaction (18.22) includes two steps:

$$NO_2(g) + F_2(g) \xrightarrow{\text{slow}} NO_2F + F \quad \textbf{(18.24)}$$

$$NO_2(g) + F(g) \xrightarrow{\text{fast}} NO_2F \quad \textbf{(18.25)}$$

$$2NO_2(g) + F_2(g) \longrightarrow 2NO_2F(g)$$

Reaction (18.24) is the **rate-determining step**—the slowest step in the reaction mechanism. Like all steps in a mechanism, reaction (18.24) is an elementary reaction. In this case of a reaction with a two-step mechanism, the rate is determined solely by collisions between NO_2 and F_2 and the formation of an activated complex between them. Further reaction depends on the presence of a product of this reaction, atomic fluorine, and the overall reaction cannot go any faster than the rate at which atomic fluorine is produced. Knowing only the experimentally determined rate equation (18.23), we could suspect that the rate of the reaction might depend on the formation of an activated complex between one NO_2 molecule and one F_2 molecule. However, although experimental rate equations do provide clues to the species involved in the activated complex, the relationship between the rate equation and the mechanism is not always as simple as in this example.

For *elementary* reactions, the exponents in the rate equation *do* equal the coefficients in the chemical reaction. Equation (18.23) is the rate equation for the overall reaction represented by Equation (18.22), which is the sum of the separate steps of the reaction mechanism. Equation (18.23) is also the rate equation for reaction (18.24), which happens to be the elementary reaction step that governs the rate of the overall reaction. As we have said, the relationship is not always so simple.

The essential point is that neither the rate equation nor the mechanism of a reaction can be deduced from the overall stoichiometric chemical equation. Do not make the common error of assuming that the exponents in the rate equation are the same as the coefficients in the overall equation.

EXAMPLE 18.5 Rate Equations

Each of the reactions listed below is thought to occur in the upper atmosphere. *Assuming that these are all elementary reactions,* write rate equations for these reactions.

(a) $CH_3 + O_2 + M \longrightarrow CH_3O_2 + M$ (c) $2O + M \longrightarrow O_2 + M$
(b) $CH_3 + O_3 \longrightarrow CH_2O + O_2 + H$ (d) $2NO_3 \longrightarrow 2NO_2 + O_2$

For *elementary* reactions, the rate equations are of the form Rate $= k[A]^m[B]^n[C]^p \cdots$ with exponents equal to the coefficients.

(a) **Rate = $k[CH_3][O_2][M]$** (c) **Rate = $k[O]^2[M]$**
(b) **Rate = $k[CH_3][O_3]$** (d) **Rate = $k[NO_3]^2$**

These rate equations can be assumed to apply to reaction (a)–(d) *only* if these are elementary reactions. Rate equations cannot be written based on overall chemical equations.

EXERCISE Each of the following elementary reactions is part of a mechanism by which Fe^{3+} catalyzes the decomposition of aqueous hydrogen peroxide: (a) $Fe^{3+} + HO_2^- \rightarrow Fe^{2+} + HO_2(aq)$, (b) $Fe^{3+} + HO_2(aq) \rightarrow Fe^{2+} + O_2(g) + H^+$, (c) $Fe^{3+} + O_2^- \rightarrow Fe^{2+} +$

$O_2(g)$, (d) $Fe^{2+} + H_2O_2(aq) \rightarrow Fe^{3+} + OH^- + OH(aq)$, (e) $Fe^{2+} + OH(aq) \rightarrow Fe^{3+} + OH^-$. Write the rate equations for these reactions.

Answers (a) Rate $= k[Fe^{3+}][HO_2^-]$, (b) Rate $= k[Fe^{3+}][HO_2]$, (c) Rate $= k[Fe^{3+}] \times [O_2^-]$, (d) Rate $= k[Fe^{2+}][H_2O_2]$, (e) Rate $= k[Fe^{2+}][OH]$.

b. Method of Initial Rates for Finding Rate Equations To find the experimental rate equation by the method of initial rates, it is necessary to gather concentration–time data in a series of separate experiments. Each experiment must have a different initial concentration of one or more reactants. (The experiments must be run at the same temperature.) The initial rate for each experiment can be found from the concentration versus time curve, as described in Section 18.6.

Suppose a reaction of the type A → products is under study. The data are inspected to find how the initial rate varies with $[A]_0$, the initial concentration of A. The objective is to find the value of the exponent in the rate equation, rate $= k[A]^m$. For a first-order reaction, $m = 1$ and the rate varies directly with $[A]_0$—if $[A]_0$ is doubled, the rate is doubled; if $[A]_0$ is tripled, the rate is tripled; and so on (Table 18.4). For a second-order reaction, $m = 2$ and the rate is increased by a factor of 2^2, or 4, if $[A]_0$ is doubled, and so on. If the reaction is zero order in A, changes in the concentration of A have no effect on the reaction rate.

The order for each of several reactants can be found by varying one initial concentration at a time, while keeping the others constant. For example, the kinetics of the reaction

$$2NO(g) + O_2(g) \longrightarrow 2NO_2(g) \qquad \textbf{(18.26)}$$

was studied (at 80 K) by varying the initial pressures of the reactants individually and finding the initial reaction rates. The following data are typical of such a study.

Experiment	$P_{init}(NO)$ (Torr)	$P_{init}(O_2)$ (Torr)	Rate of NO_2 Formation (Torr/s)
(a)	1630	1630	6.13×10^{-8}
(b)	3260	1630	24.5×10^{-8}
(c)	1630	3260	12.2×10^{-8}

Holding the O_2 pressure constant while doubling the NO pressure quadrupled the rate, indicating that the reaction is second order in NO. With constant NO pressure, doubling the O_2 pressure caused the rate to double, showing that the reaction is first order in O_2. The rate equation is, therefore

$$\text{Rate of } NO_2 \text{ formation} = kP_{NO}{}^2P_{O_2}$$

The experimental data can be used in the rate equation to find the value of k. For the data from experiment (a), for example

$$k = \frac{\text{Rate of } NO_2 \text{ formation}}{P_{NO}{}^2P_{O_2}} = \frac{6.13 \times 10^{-8}\ \text{Torr/s}}{(1630\ \text{Torr})^2(1630\ \text{Torr})} = 1.42 \times 10^{-17}\ \text{Torr}^{-2}\ \text{s}^{-2}$$

The best value for k is taken as an average of the k values calculated for each run; in this case, it is 1.41×10^{-17} Torr^{-2} s^{-1}.

In studying reaction (18.26) and the experiments described above, the value of the exponent in the rate equation was found by inspecting the data. Where this is not

Table 18.4
Change in Initial Rate with Change in $[A]_0$ for Rate $= k[A]^m$

For a reaction of	If $[A]_0$ is doubled	If $[A]_0$ is tripled
Zero order ($m = 0$)	rate is unchanged	rate is unchanged
First order ($m = 1$)	rate is doubled	rate is tripled
Second order ($m = 2$)	rate is quadrupled ($\times 2^2$)	rate is 9 times greater ($\times 3^2$)

possible, other methods, such as the graphic method described in the next section, must be tried. If it is proven that the data do not fit any of the simple rate equations, then a more complex rate equation must be sought.

EXAMPLE 18.6 Reaction Order and Rate Constant

The rate equation for the reaction of iodide ion with hypochlorite ion in an alkaline solution

$$I^- + OCl^- \xrightarrow{OH^-} Cl^- + IO^-$$

was thought to be of the form

$$\text{Rate} = k[I^-]^m[OCl^-]^n[OH^-]^p$$

The reaction was studied using the method of initial rates and the following data collected at 25 °C:

Experiment	$[ClO^-]_0$	$[I^-]_0$	$[OH^-]_0$	Initial rate (mol/L s)
(a)	2.00×10^{-3}	2.00×10^{-3}	1.00	2.42×10^{-4}
(b)	4.00×10^{-3}	2.00×10^{-3}	1.00	4.82×10^{-4}
(c)	2.00×10^{-3}	4.00×10^{-3}	1.00	5.02×10^{-4}
(d)	2.00×10^{-3}	2.00×10^{-3}	0.500	4.64×10^{-4}

What is the order of reaction with respect to each reactant and what is the value of the rate constant?

The data for experiments (a) and (b) show that the reaction rate essentially doubles as the concentration of ClO^- doubles. Thus the reaction is **first order with respect to $[OCl^-]$ and $n = 1$.** Likewise, the data for experiments (a) and (c) show that the reaction rate essentially doubles as the concentration of I^- doubles. Thus the reaction is **first order with respect to $[I^-]$ and $m = 1$.**

Upon comparison of the rate for experiment (d) with that of experiment (a), we can see that the rate essentially doubles as the concentration of OH^- is halved. This means that there is an inverse relationship between the rate and $[OH^-]$. The reaction is **negative first order with respect to $[OH^-]$ and $p = -1$.**

The complete rate equation is

$$\text{Rate} = k[I^-][OCl^-][OH^-]^{-1} = k\frac{[I^-][OCl^-]}{[OH^-]}$$

The value of k is found by substituting the values of the rate and the respective concentrations into the rate law. For example, for experiment (a)

$$k_1 = \frac{\text{rate}_1[OH^-]_1}{[I^-]_1[OCl^-]_1} = \frac{(2.42 \times 10^{-4}\ \text{mol/L s})(1.00\ \text{mol/L})}{(2.00 \times 10^{-3}\ \text{mol/L})(2.00 \times 10^{-3}\ \text{mol/L})} = 60.5\ \text{s}^{-1}.$$

Likewise, for the other experiments, $k_2 = 60.3\ \text{s}^{-1}$, $k_3 = 62.8\ \text{s}^{-1}$, and $k_4 = 58.0\ \text{s}^{-1}$, giving an average of $k = 60.4\ \text{s}^{-1}$.

EXERCISE The data below were obtained for the following reaction: $A + B \rightarrow$ products.

Experiment	(a)	(b)	(c)
Initial Rate (mol/L s)	0.030	0.059	0.060
$[A]_0$	0.10	0.20	0.20
$[B]_0$	0.20	0.20	0.30

Write the rate equation for this reaction. Evaluate k.

Answer Rate $= k[A]$, $k = 0.30\ \text{s}^{-1}$.

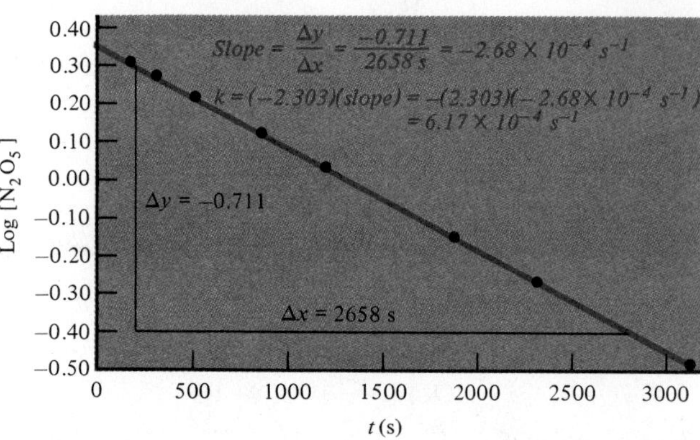

Figure 18.9
Plot of Kinetic Data for Decomposition of N_2O_5 at 45 °C in CCl_4 Solution (Table 18.1), a First-Order Reaction

c. Graphic Method for Finding Rate Equations Whether a reaction is zero, first, or second order in a specific reactant can be found by plotting data appropriately (Appendix I.B). The general equation for a straight-line plot of y versus x is

$$y = mx + b \qquad (18.27)$$

(m: constant = slope of y vs. x plot; b: constant, intercept of y axis (y value at x = 0))

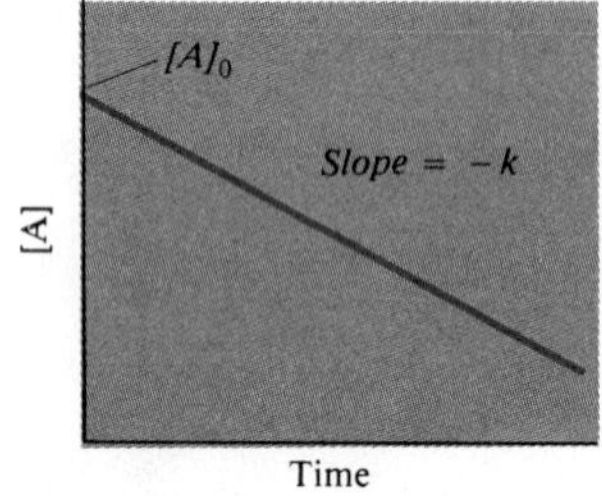

Figure 18.10
Plot of Concentration versus Time for a Zero-Order Reaction

The integrated rate equations of zero-, first-, and second-order reactions can be arranged into this form. For example, for a first-order reaction

$$\log[A] = -\frac{kt}{2.303} + \log[A]_0 \qquad (18.28)$$

(log [A]: plot on y axis; t: plot on x axis; −k/2.303: slope; log [A]₀: y intercept)

Using the data for the decomposition of N_2O_5 given in Table 18.1 and plotting log $[N_2O_5]$ versus t gives a straight line, as shown in Figure 18.9. Therefore, it can be concluded that this reaction is first order in N_2O_5. From the slope of the line, the value of k is found to be $6.17 \times 10^{-4}\ s^{-1}$ (at 45 °C).

As shown in Table 18.5, straight lines are obtained for a plot of [A] versus t for a zero-order reaction (Figure 18.10), of log [A] versus t for a first-order reaction (Figure 18.11), and of 1/[A] versus t for a second-order reaction (Figure 18.12). In each case the value of k can be found from the slope of the straight line.

Table 18.5
Equations for Plotting Kinetic Data

Type of reaction	Equation for straight line $y = ax + b$	Linear plot	Slope	y Intercept
Zero order (rate = k)	$[A] = -kt + [A]_0$	[A] vs. t	$-k$	$[A]_0$
First order (rate = k[A])	$\log[A] = -\frac{kt}{2.303} + \log[A]_0$	log [A] vs. t	$-\frac{k}{2.303}$	$\log[A]_0$
Second order (rate = $k[A]^2$)	$\frac{1}{[A]} = kt + \frac{1}{[A]_0}$	$\frac{1}{[A]}$ vs. t	k	$\frac{1}{[A]_0}$

18.12 REACTION MECHANISMS AND RATE EQUATIONS

Once an experimental rate equation has been established, it is then possible to propose a mechanism for the reaction. In order to check the validity of a proposed mechanism, a rate equation based on the mechanism is derived. Remember that rate equations can be written by inspection of a chemical equation for *elementary* reactions (Section 18.11). Algebraic combination of the rate equations for the steps in a reasonably simple reaction mechanism gives an overall rate equation.

If the rate equation based on a proposed mechanism *is not* the same as the experimental rate equation, then the mechanism is wrong. If the rate equation based on a mechanism *is* the same as the experimental rate equation, the proposed mechanism may be the correct one. The remaining uncertainty arises because two quite different mechanisms can lead to the same rate equation. A chemist studying kinetics tries to eliminate as many possible explanations for a reaction as he can. A reaction mechanism is often arrived at by a negative approach that first proves what the mechanism cannot be. One hopes that after the elimination process, there will be only one remaining mechanism. If there is more than one, the most probable mechanism must be selected. Chemists may have different opinions as to which mechanism is most probable, a fact that sometimes leads to lively controversy in chemical publications. Writing possible reaction mechanisms based on experimental rate equations is an exercise best left to the experts. What we can do here is explore a few generalities about the relationship between mechanisms and overall rate equations.

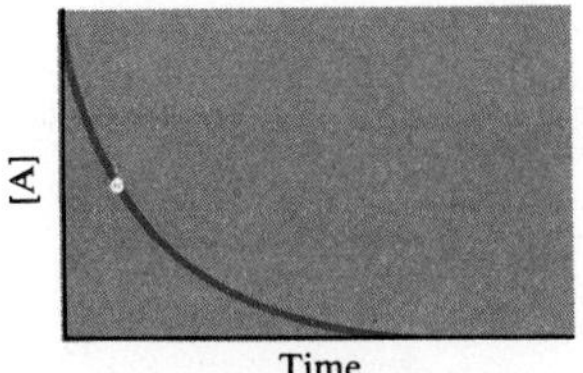

(a) Concentration versus time for a first-order reaction

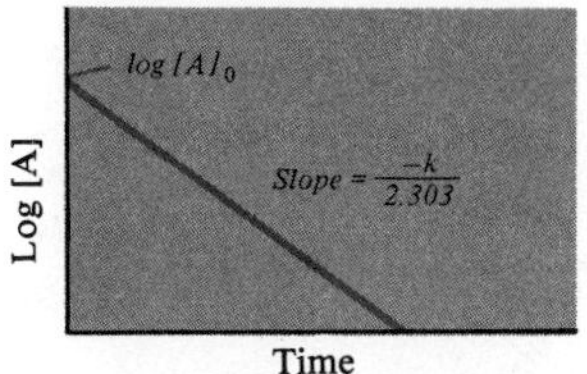

(b) Plot of the equation log [A] = (−kt/2.303)+ log [A]$_0$ for a first-order reaction

Figure 18.11
Plots for a First-Order Reaction

One of the simplest types of reaction mechanism involves a single, rate-determining step upon which the rate equation is based. This occurs when the first step in the mechanism is the rate-determining step, as illustrated earlier for the reaction between nitrogen dioxide and fluorine (Equations 18.22–18.25). In general, the rate of a reaction may be affected by reaction steps that precede the rate-determining step, but it is not influenced by steps that follow it. To give another example, the overall third-order reaction of nitrogen(II) oxide (NO) with hydrogen

$$2NO(g) + 2H_2(g) \longrightarrow N_2(g) + 2H_2O(g) \qquad \text{Rate} = k[NO]^2[H_2] \qquad \textbf{(18.29)}$$

is believed to occur by a two-step mechanism in which the first step is rate determining.

$$2NO(g) + H_2(g) \xrightarrow{\text{slow}} N_2(g) + H_2O_2(g) \qquad \textbf{(18.29a)}$$

$$H_2O_2(g) + H_2(g) \xrightarrow{\text{fast}} 2H_2O(g) \qquad \textbf{(18.29b)}$$

Note that the single elementary reaction step that determines the overall reaction order for Equation (18.29) is a third-order reaction (Equation 18.29a). As we have seen, all collisions are either bimolecular or (in a few cases) termolecular. This means that a reaction for which the experimental overall rate equation is other than second order or third order must proceed by a more complex mechanism, one in which the rate is not dependent on a single elementary reaction.

As an example of a reaction with a fairly common type of more complex rate equation and mechanism, consider the decomposition of acetic anhydride in the gas phase into acetic acid and ketene.

$$\underset{\textit{acetic anhydride}}{(CH_3CO)_2O(g)} \longrightarrow \underset{\textit{acetic acid}}{CH_3COOH(g)} + \underset{\textit{ketene}}{CH_2{=}C{=}O(g)} \qquad \textbf{(18.30)}$$

The proposed mechanism is as follows (where the asterisk indicates the activated acetic anhydride molecule, and k_{-1} is used to show that the second step is the reverse of the first step):

step 1

$$(CH_3CO)_2O(g) + (CH_3CO)_2O(g) \xrightarrow{k_1} (CH_3CO)_2O^*(g) + (CH_3CO)_2O(g) \qquad \textbf{(18.30a)}$$

step −1

$$(CH_3CO)_2O^*(g) + (CH_3CO)_2O(g) \xrightarrow{k_{-1}} (CH_3CO)_2O(g) + (CH_3CO)_2O(g) \qquad \textbf{(18.30b)}$$

step 2

$$(CH_3CO)_2O^*(g) \xrightarrow{k_2} CH_3COOH(g) + CH_2{=}C{=}O(g) \qquad \textbf{(18.30c)}$$

The reaction can only begin when two molecules of acetic anhydride collide and one of the molecules is activated by the collision (step 1). Once this occurs, there are two possible outcomes. The activated molecule can be deactivated by another collision before it has a chance to react (step − 1); or it can react to give the products before being

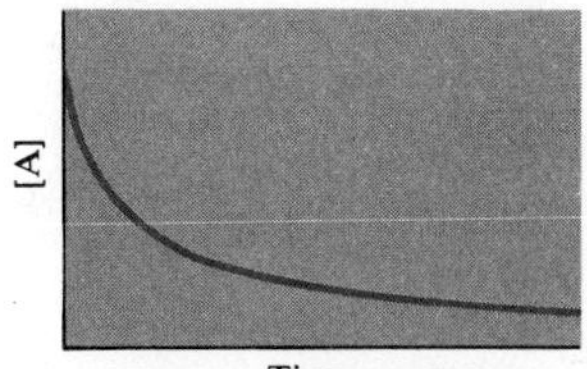

(a) Concentration versus time for a second-order reaction

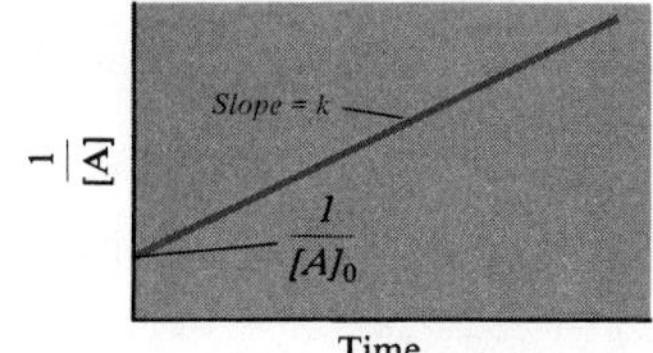

(b) Plot of the equation $\frac{1}{[A]} = kt + \frac{1}{[A]_0}$ for a second-order reaction

Figure 18.12
Plots for a Second-Order Reaction

deactivated (step 2). The rate equation derived for this reaction from the mechanism agrees with experimental observations (in terms of the k's in Equation 18.30a, b, c):

acetic anhydride

$$\text{Rate} = \frac{k_1 k_2 [\text{A}]^2}{k_{-1}[\text{A}] + k_2} \qquad \textbf{(18.31)}$$

At high pressure, step 2 is the rate-determining step. Because k_2 is small relative to the values of k_1, k_{-1}, and [A], it can be neglected in the denominator of the rate equation. This expression therefore reduces to a first-order rate equation:

at high pressure $$\text{Rate} = \frac{k_1 k_2}{k_{-1}}[\text{A}] = k'[\text{A}]$$

Since k_1, k_2, and k_{-1} are all constants, $k_1 k_2/k_{-1}$ is also a constant. Note that the *overall* first-order rate constant, k', includes the rate constants of steps that *precede* the rate-determining step.

At low pressure, [A] is small; the $k_{-1}[\text{A}]$ term in the denominator in Equation (18.31) is therefore small compared to k_2 and can be dropped. The rate equation thus becomes

at low pressure $$\text{Rate} = k_1[\text{A}]^2$$

The reaction is second order, with a rate determined solely by step 1.

A limited number of generalities can be drawn about the relationship between reaction mechanisms and rate equations:

1. The sum of the steps in a reaction mechanism must be the equation for the overall chemical reaction.
2. The concentrations of short-lived, activated intermediates do not appear in overall rate equations.
3. The experimental overall rate equation is derivable by some combination of the rate equations for the steps in a possible reaction mechanism.
4. It is not always feasible to eliminate all but one proposed mechanism for a reaction.
5. Steps that follow the rate-determining step in a mechanism do not generally influence the rate equation.
6. Reactions that are other than second order or third order overall *must* have rates dependent on more than one elementary step.

SUMMARY OF SECTIONS 18.5–18.11

Reaction rate is expressed as the change in concentration of a reactant or product per unit time. The rate of a chemical reaction is not constant, but changes continually as the amounts of reactants change (except for zero-order reactions). Both reaction rates and rate equations *must* be determined experimentally, often by gathering data on the variation of concentration with time. Rate equations often, though not always, take the form of Rate = $k[\text{A}]^m[\text{B}]^n[\text{C}]^p$.... The sum of m, n, p... is the overall reaction order; the order for a specific reactant is given by m, or n, or p.... The exponents in a rate equation are often 1, 2, or 3, but may also be fractions or negative numbers. Rates of first-order and second-order reactions are dependent upon the concentrations of one or two reactants, respectively, whereas rates of zero-order reactions are independent of reactant concentrations. The half-life of a first-order reaction is independent of reactant concentration. Rate equations, integrated rate equations, and half-life equations for zero-, first-, and second-order reactions are summarized in Table 18.3.

If a reaction is elementary, the exponents in the rate equation are given by the coefficients of the reactants in the chemical equation, but this *does not apply* to the overall equations for reactions that proceed by more than one step. Rate equations for such reactions are found experimentally, for example, by the method of initial rates in which initial amounts of reactants are varied one at a time or by the graphic method by making plots of time versus concentration. In both methods the order of reaction for individual reactants is determined by comparing the experimental outcome with what would be expected for different reaction orders.

The sequence in a study of the kinetics of a chemical reaction is (1) experiments that yield concentration versus time data, (2) determination of the experimental rate equation and evaluation of the rate constant, and (3) the proposal of possible reaction mechanisms. Some generalities about the relationship between rate equations and reaction mechanisms are listed at the end of Section 18.12.

FACTORS THAT INFLUENCE REACTION RATES

18.13 EFFECT OF TEMPERATURE ON REACTION RATE

In 1889 Svante Arrhenius, the same Swedish chemist who gave us the classical, water-ion, definitions of acids and bases, observed that there is a mathematical relationship that connects activation energy, temperature, and the rate constant. This relationship, now known as the Arrhenius equation, is

activation energy

$$k = Ae^{-E_a/RT} \quad \text{ideal gas constant} \tag{18.32}$$

rate constant at T — natural log base — a constant

or, taking logarithms of both sides

$$\log k = -\frac{E_a}{2.303RT} + \log A \tag{18.33}$$

These equations show that reactions with larger activation energies have smaller values of k (the exponent is negative) and are, therefore, slower. This is what we would expect from our earlier discussion of effective collisions and energy relationships. Also, for a given value of the activation energy, the equations show that as the temperature increases, the value of the rate constant increases, indicating that the reaction is faster, as expected, because at higher temperatures more molecules can collide effectively.

In terms of collision theory, $e^{-Ea/RT}$ in Equation (18.32) is the fraction of collisions in an elementary reaction in which the particles have enough energy to react. The constant A corrects for the frequency of collisions, the necessity for the proper orientation in effective collisions, and all factors other than activation energy that are significant in effective collisions.

The Arrhenius equation and the equations derived from it can be applied both to data for overall reactions and to data for elementary reactions. (Some complex reactions, such as explosions and certain catalyzed reactions, have non-Arrhenius temperature dependence.) When found by using the rate constant for an overall reaction, the activation energy value represents the net effect of the activation energies of *all of the steps* in the reaction mechanism. In this way, *effective* activation energies can be found even though reaction mechanisms are unknown.

Equation (18.33) is the equation for a straight line (see Table 18.5).

plot on y axis — slope

$$\log k = \left(-\frac{E_a}{2.303R}\right)\left(\frac{1}{T}\right) + \log A \tag{18.34}$$

plot $1/T$ on x axis — y intercept

The value of E_a can be found from the slope of a plot of log k versus $1/T$, $E_a = -2.303R \times$ (slope).

Also, by combining the equations for values of k at two different temperatures for the same reaction, a relationship is obtained that allows for the calculation of E_a

$$\log\frac{k_2}{k_1} = \frac{-E_a}{2.303R}\left(\frac{1}{T_2} - \frac{1}{T_1}\right) \quad \text{or} \quad \log\frac{k_2}{k_1} = \frac{E_a}{2.303R}\left(\frac{T_2 - T_1}{T_1 T_2}\right) \tag{18.35}$$

Alternatively, if E_a and three of the four values k_1, k_2, T_1, and T_2 are known, the fourth can be found. In using the Arrhenius equation, the energy is customarily expressed in joules, temperature must be in kelvins, and R is 8.314 J/K mol or 1.987 cal/K mol.

EXAMPLE 18.7 Activation Energy and Temperature

The thermal decomposition of $CH_3N{=}NCH_3$ in the gas phase to give nitrogen and methyl radicals

$$CH_3N{=}NCH_3(g) \xrightarrow{\Delta} N_2(g) + 2{\cdot}CH_3(g)$$

has an activation energy of 2.14×10^5 J/mol and, at 600. K, $k = 1.99 \times 10^8\ s^{-1}$. Does the rule of thumb that a reaction rate doubles with a ten degree rise in temperature hold for this reaction?

To answer the question, we must calculate k at 610. K by using the appropriate form of the Arrhenius equation, Equation (18.35). The knowns are E_a, k_1 at $T_1 = 600.$ K, and $T_2 = 610.$ K. The unknown is k_2 at 610. K.

$$\log\frac{k_2}{k_1} = \frac{E_a}{2.303R}\left(\frac{T_2 - T_1}{T_1 T_2}\right)$$

$$\log\frac{k_2}{1.99 \times 10^8\ s^{-1}} = \left(\frac{2.14 \times 10^5\ \text{J/mol}}{(2.303)(8.314\ \text{J/K mol})}\right)\left(\frac{610.\ \text{K} - 600.\ \text{K}}{(610.\ \text{K})(600.\ \text{K})}\right) = 0.31$$

Taking antilogarithms of both sides gives

$$\frac{k_2}{1.99 \times 10^8\ s^{-1}} = 2.0$$

$$k_2 = (2.0)(1.99 \times 10^8\ s^{-1}) = 4.0 \times 10^8\ s^{-1}$$

Comparison of k_1 with k_2 shows that **for this reaction, the rate constant is doubled by a ten degree rise in temperature.** (This is not always the case.)

EXERCISE The thermal decomposition of CH_3CHF_2,

$$CH_3CHF_2(g) \xrightarrow{\Delta} CH_2CHF(g) + HF(g)$$

has an activation energy of 260 kJ/mol. The reported value of the rate constant at 462 °C is $5.8 \times 10^{-6}\ s^{-1}$. What is the value of the rate constant at 487 °C?

Answer $2.4 \times 10^{-5}\ s^{-1}$.

EXAMPLE 18.8 Activation Energy and Temperature

The rate constant for the decomposition of N_2O_5 in chloroform

$$2N_2O_5 \xrightarrow{CHCl_3} 4NO_2 + O_2$$

was measured at two different temperatures: $T_1 = 25$ °C, $k_1 = 5.54 \times 10^{-5}\ s^{-1}$ and $T_2 = 67$ °C, $k_2 = 9.30 \times 10^{-3}\ s^{-1}$. Find the activation energy for this reaction.

The solution to this problem requires solving Equation (18.35) for E_a

$$E_a = \left(\log \frac{k_2}{k_1}\right)(2.303R)\left(\frac{T_1 T_2}{T_2 - T_1}\right)$$

and then just substituting the known values of k_1 and k_2, and T_1 and T_2.

$$E_a = \left[\log\left(\frac{9.30 \times 10^{-3}\ \text{s}^{-1}}{5.54 \times 10^{-5}\ \text{s}^{-1}}\right)\right](2.303)\left(8.314 \frac{\text{J}}{\text{K mol}}\right)\frac{(298\ \text{K})(340.\ \text{K})}{(340.\ \text{K} - 298\ \text{K})}$$

$$= (2.245)(2.303)\left(8.314 \frac{\text{J}}{\text{K mol}}\right)(2400\ \text{K})$$

$$= \mathbf{1.0 \times 10^5\ J/mol}$$

EXERCISE The rate constant for the decomposition of gaseous CH_3CHF_2,

$$CH_3CHF_2(g) \longrightarrow CH_2CHF(g) + HF(g)$$

is $7.9 \times 10^{-7}\ \text{s}^{-1}$ at 429 °C and $1.7 \times 10^{-4}\ \text{s}^{-1}$ at 522 °C. Calculate the activation energy for this reaction. *Answer* 270 kJ/mol.

18.14 HOMOGENEOUS AND HETEROGENEOUS REACTIONS

A reaction between substances in the same gaseous or liquid phase is referred to as a **homogeneous reaction.** The question of contact between the reactive molecules in a homogeneous reaction is not an important one, for the molecules and ions are free to move and collisions are frequent. However, for a reaction between substances in different phases—a **heterogeneous reaction**—bringing the reacting molecules or ions together may be difficult. For example, the reaction between steam and red hot iron,

$$3Fe(s) + 4H_2O(g) \longrightarrow Fe_3O_4(s) + 4H_2(g)$$

proceeds very slowly if the iron is in one large block, but goes rapidly if the metal is powdered and spread out so as to expose a large surface to the steam. [Heterogeneous reactions involving the catalysis of reactions of gases by solids (Section 18.15) have been studied extensively.]

The rates at which solids react with each other are often limited by the amount of contact between them. The reaction of mercury(II) chloride and potassium iodide is a good illustration.

$$\underset{\textit{white}}{HgCl_2(s)} + \underset{\textit{white}}{2KI(s)} \longrightarrow \underset{\textit{red}}{HgI_2(s)} + \underset{\textit{white}}{2KCl(s)}$$

If large crystals of the two solids are shaken together, there is no sign of reaction; if the crystals are ground together, reaction takes place fairly rapidly, as shown by the appearance of the bright red mercury(II) iodide. If the reactants are dissolved separately in water and the solutions are mixed, the insoluble red mercury iodide precipitates almost instantly. In general, for reactions between solids the reaction rate is proportional to the area of surface in contact.

There are many practical examples of how the degree of contact influences the rate of reaction, such as dust explosions in flour mills, grain elevators, and coal mines, or the effect of tetraethyllead in preventing "knock" in automobile engines. To eliminate knock, tetraethyllead was once added to almost all gasoline in the United States. When the volatile liquid tetraethyllead is drawn into the hot cylinder, it decomposes, giving an extremely fine suspension of metallic lead. This exposes a very large surface area to the burning gases. Reaction on the surface of the metallic lead tends to slow down the combustion process and distribute the burning evenly throughout the cylinder, preventing knock. Placing a large sheet of lead in the cylinder would not have the same effect, because the sheet would have much less surface area in contact with the gases in the cylinder. Pollution control requirements have forced the phasing out of the use of

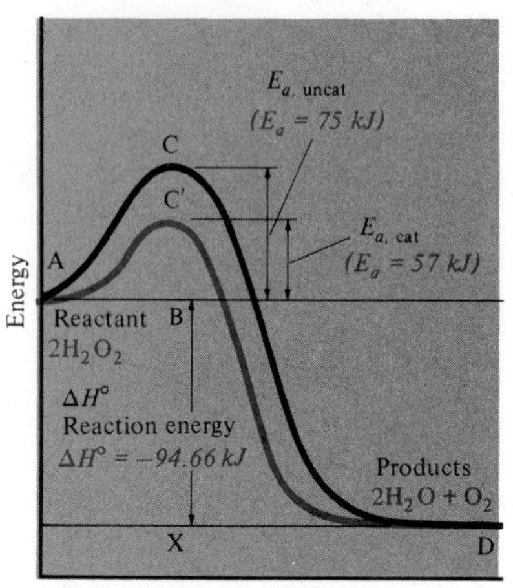

Figure 18.13
Energy Path of the Decomposition of H_2O_2, Catalyzed by I^- and Uncatalyzed
The catalyst decreases the activation energy from $E_{a,\text{uncat}}$ to $E_{a,\text{cat}}$.

lead compounds in gasoline; instead, engines are being redesigned to burn gasoline more smoothly and the use of other fuel additives, such as ethanol, is spreading.

18.15 CATALYSIS

A catalyst increases the rate of a chemical reaction, but can be recovered in its original form when the reaction is finished. A catalyst cannot assist a reaction that could not proceed in its absence. Nor can a catalyst change the concentrations in an equilibrium mixture; it can only lessen the *time* needed to reach equilibrium.

A catalyst works by providing an alternate and easier pathway from reactants to products. Catalysts accomplish this by a variety of mechanisms, but in each case the sole function of a catalyst is to lower the activation energy of a reaction. The role of a catalyst is highly specific to a particular reaction. A substance that catalyzes one reaction may well have no effect on another reaction, even if that reaction is very similar. Many of the most highly specific catalysts are those designed by nature. The chemical reactions in living things are controlled by biochemical catalysts—the enzymes.

Catalysts are classified as homogeneous or heterogeneous. A **homogeneous catalyst** is present in the same phase as the reactants and functions by combining with reacting molecules or ions to form unstable intermediates. These intermediates combine with other reactants to give the desired product and to regenerate the catalyst.

The energy pathway from reactants to products for the decomposition of hydrogen peroxide

$$2H_2O_2(aq) \longrightarrow 2H_2O(l) + O_2(g) \qquad \Delta H° = -94.66 \text{ kJ}$$

is shown in Figure 18.13. The upper curve is for the uncatalyzed reaction. The lower curve shows the change in the energy pathway after the addition of iodide ion, I^-, which serves as a homogeneous catalyst by the mechanism

$$\begin{array}{rcl} H_2O_2 + I^- & \longrightarrow & H_2O + IO^- \\ IO^- + H_2O_2 & \longrightarrow & H_2O + O_2 + I^- \\ \hline 2H_2O_2 & \longrightarrow & 2H_2O + O_2 \end{array}$$

The distance BC represents the activation energy of the uncatalyzed reaction and the distance BC′, the activation energy of the catalyzed reaction. Note that a catalyst decreases the energies of activation of both the forward and reverse reactions by an equal amount. Note also that the overall energy change in the reaction (BX) remains the same. The catalyst I^- takes part in the mechanism of the reaction, but is regenerated in its original state and can be recovered. Do not make the common error of saying that a catalyst "changes the rate of a reaction without taking part in it."

A **heterogeneous catalyst** is present in a phase different from that of the reactants and products. Such catalysts are usually solids in the presence of gaseous or liquid reactants. Reactions occur at the surfaces of heterogeneous catalysts. For this reason the catalysts are usually finely divided solids or have particle shapes that provide a high surface-to-volume ratio—a property of importance in heterogeneous catalysis. Table 18.6 lists catalysts and conditions that are effective in controlling the products of the reaction between carbon monoxide and hydrogen.

One or both reactants are chemically adsorbed at the surface of a heterogeneous catalyst (Figure 18.14). Adsorption is the adherence of atoms, molecules, or ions (the adsorbate) to a surface. It is called **physical adsorption**, or physisorption, when the forces between surface and adsorbate are van der Waals forces, and it is called **chemical adsorption**, or chemisorption, when the forces between adsorbate and surface are of the magnitude of chemical bond forces. The process of chemisorption on a catalyst can alter the structure and activity of a reactant in the same way as the formation of an activated complex—by weakening some bonds and allowing the formation of others.

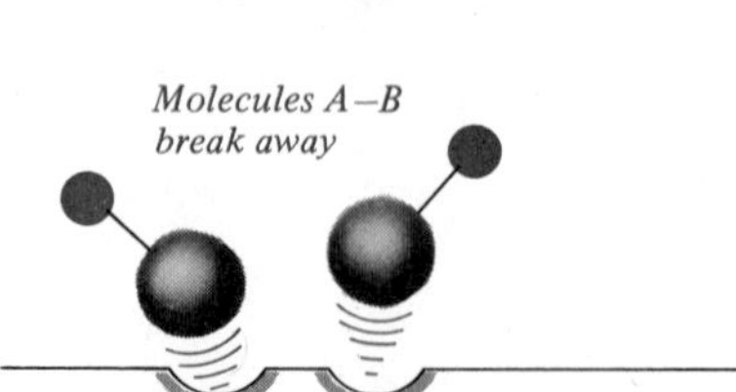

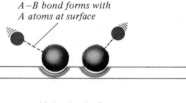

Figure 18.14
How the Reaction A—A + 2B ⟶ 2AB Might Take Place at a Catalyst Surface

Table 18.6
Reactions between Carbon Monoxide and Hydrogen

Catalyst	Conditions	Products
Ni[a]	100–200 °C, 1–10 atm	$CH_4 + H_2O$
$ZnO \cdot Cr_2O_3$	400 °C, 500 atm	$CH_3OH + H_2O$
Co/ThO_2	190 °C, 1–20 atm	CH_4, C_2H_6, and higher hydrocarbons + H_2O
Ru	200 °C, 200 atm	High-molecular-mass hydrocarbons + H_2O
ThO_2	400 °C, 200 atm	Branched-chain hydrocarbons + H_2O

[a] This is the *methanation* reaction of current interest in the manufacture of *synthetic natural gas*. It converts carbon monoxide, which has a relatively low heat of combustion, to methane, which has a high one. (Source: R. L. Burwell, Jr., "Heterogeneous Catalysts," in *Survey of Progress in Chemistry*, A. F. Scott (ed.), Vol. 8, Academic Press, New York, 1977, p. 2.)

Figure 18.15 shows some of the ways in which small molecules are chemisorbed at catalyst surfaces. A heterogeneously catalyzed reaction occurs in the following steps: (1) Reactants diffuse from the fluid phase to the catalyst surface; (2) one or more reactants are chemisorbed; (3) reaction occurs between adsorbed reactants or between a reactant in the fluid phase and an adsorbed reactant; (4) products are desorbed; and (5) products diffuse back into the fluid phase.

One difficulty in the use of heterogeneous catalysts is that most of them are readily "poisoned"—that is, impurities in the reactants coat the catalyst with unreactive material or modify its surface, so that the catalytic activity is lost. Frequently, but not always, the poisoned catalyst can be purified and used again.

Inhibitors are substances that slow down a catalyzed reaction—usually by tying up the catalyst, possibly also by tying up a reactant. For example, the development of rancidity in butter is greatly retarded by the addition of very small amounts of certain organic substances. The added organic materials apparently combine with the traces of copper ion that get into nearly all butter from the processing equipment. The copper ion catalyzes the development of rancidity, but in combination with the organic material, the copper ion does not exert a catalytic effect.

Promoters are substances that make a catalyst more effective. In the case of solid catalysts, a small amount of a promoter may encourage the formation of lattice defects, which are often the active sites on a catalyst surface.

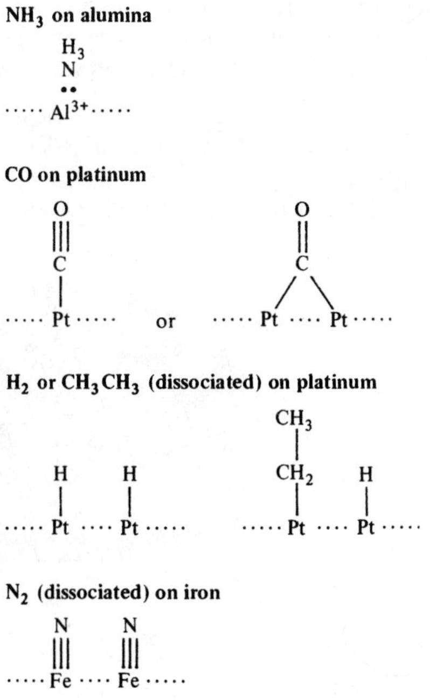

NH_3 on alumina

H_3
N
••
····· Al^{3+}·····

CO on platinum

O ≡ C — ····· Pt ····· or O = C bridging ····· Pt ···· Pt ·····

H_2 or CH_3CH_3 (dissociated) on platinum

H H on ····· Pt ···· Pt ····· CH_3–CH_2 H on ····· Pt ···· Pt ·····

N_2 (dissociated) on iron

N ≡ N ≡ on ····· Fe ···· Fe ·····

Figure 18.15
Some Examples of Chemisorption on Catalyst Surfaces
Source: Burwell, *op. cit.* Table 18.6.

SUMMARY OF SECTIONS 18.13–18.15

Reaction rates are influenced by the concentrations of reactants, as represented in rate equations (Sections 18.8 and 18.10). They are also influenced by temperature, catalysts, and the degree of contact between reactants.

Reaction rates and rate constants vary with temperature. The Arrhenius equation (see Equations 18.32–18.35) gives the relationships among the temperature, the rate constant, and the activation energy for a given reaction, either an elementary reaction or an overall reaction. This equation can be used, for example, to evaluate E_a or calculate a rate constant at one temperature from the known rate constant at another temperature.

In a homogeneous reaction, all reactants are in the same phase, either gaseous or liquid. In a heterogeneous reaction, the reactants are in different phases, and the rates of such reactions are limited by the amount of contact between reactants. Catalysts increase the rates of chemical reactions but can be recovered unchanged at the end of a reaction. All catalysts function by lowering the activation energy; they do not change the amounts of products present once a reaction is complete or equilibrium is reached. Heterogeneous catalysts, often used in industrial applications, are present in a different phase from the reactants and reaction often occurs at the surface of a solid heterogeneous catalyst.

SIGNIFICANT TERMS (with section references)

activation energy (18.3)
bimolecular reaction (18.4)
chemical adsorption (18.15)
chemical kinetics (18.1)
elementary reaction (18.1)
free radical (18.2)
heterogeneous catalyst (18.15)
heterogeneous reaction (18.14)
homogeneous catalyst (18.15)
homogeneous reaction (18.14)
inhibitors (18.15)
initial reaction rate (18.6)
instantaneous reaction rate (18.6)
intermediate (18.1)
molecularity (18.4)
overall reaction order (18.7)
physical adsorption (18.15)
promoters (18.15)
rate constant (18.7)
rate equation (18.7)
rate-determining step (18.11)
reactant reaction order (18.7)
reaction mechanism (18.1)
reaction rate (18.5)
termolecular reaction (18.4)
transition state, activated complex (18.2)

QUESTION AND PROBLEMS

Chemical Kinetics at the Molecular Level

18.1 What is an "elementary reaction"? How are elementary reactions used to construct a reaction mechanism?

18.2 What is meant by the term "intermediate" as applied to one or more of the species in a reaction mechanism?

18.3 What are the three conditions for effective collisions between reacting species? Which of these conditions is influenced most by (a) temperature, (b) molecular geometry, (c) relative bond energies?

18.4 Define "activation energy" for a chemical reaction. How does the value of the activation energy affect the rate of reaction? How does the rate of reaction vary with temperature? Why?

18.5 Using the numbers from the diagram shown below, identify (a) the reactant(s), (b) the product(s), (c) the energy change of reaction, (d) the activation energy for the forward reaction, (e) the activation energy for the reverse reaction.

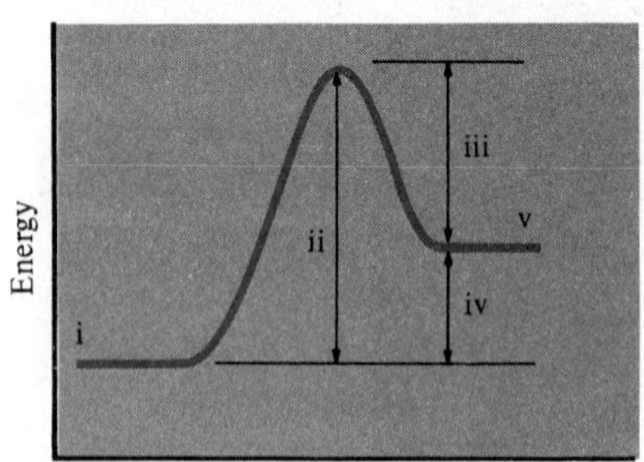

18.6 From the plots of energy versus extent of reaction shown below, identify the diagram for (a) a highly endothermic reaction with a low activation energy, (b) a reaction in which a stable intermediate is formed, (c) an exothermic reaction with a low activation energy, (d) an endothermic reaction with a high activation energy, (e) a highly exothermic reaction with a high activation energy.

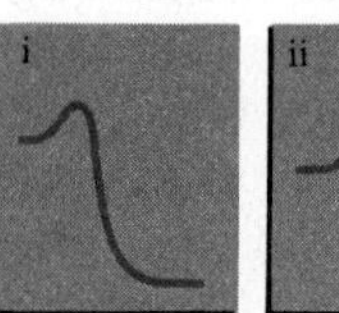

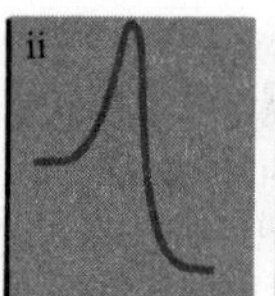

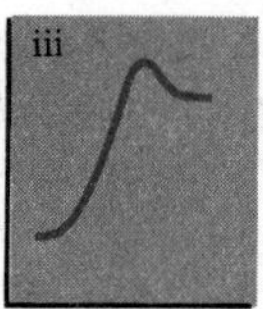

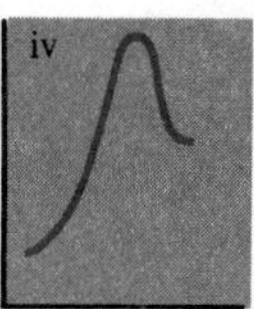

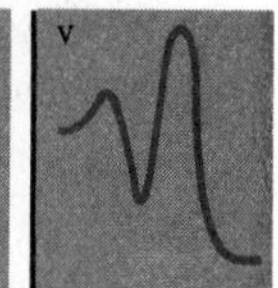

18.7 The activation energy for the reaction

$$2HI(g) \longrightarrow H_2(g) + I_2(g) \qquad \Delta E^\circ = 9.478 \text{ kJ}$$

is 179 kJ. Construct a diagram similar to Figure 18.4 for this reaction.

18.8 The activation energy for the reaction between O_3 and NO is 9.6 kJ/mol.

$$O_3(g) + NO(g) \longrightarrow NO_2(g) + O_2(g)$$

At 25 °C, the heat of formation is 33.2 kJ/mol for $NO_2(g)$, 90.25 kJ/mol for $NO(g)$, and 142.7 kJ/mol for $O_3(g)$. Calculate the heat of reaction. Prepare an activation energy plot similar to Figure 18.4 for this reaction. (For this reaction, $\Delta E^\circ = \Delta H^\circ$.)

18.9 A proposed mechanism for the neutralization of acetic acid by the hydroxide ion:

$$CH_3COOH(aq) + OH^- \longrightarrow CH_3COO^- + H_2O(l)$$

consists of the following elementary reactions:

$$CH_3COOH(aq) + H_2O(l) \longrightarrow CH_3COO^- + H_3O^+$$
$$H_3O^+ + OH^- \longrightarrow 2H_2O(l)$$

What is the molecularity of each of these elementary reactions?

18.10 A proposed mechanism for the decomposition of hydrogen peroxide in the presence of H_3O^+ and Br^-

$$2H_2O_2(aq) \xrightarrow{H_3O^+,\ Br^-} 2H_2O(l) + O_2(g)$$

consists of the following elementary reactions:

$$H_2O_2(aq) + H_3O^+ + Br^- \longrightarrow HOBr(aq) + 2H_2O(l)$$
$$H_2O_2(aq) + HOBr(aq) \longrightarrow H_3O^+ + O_2 + Br^-$$

Discuss these elementary reactions in terms of their molecularity.

Reaction Rates and Rate Equations

18.11 Express the rate of reaction in terms of the reactants and/or products for each of the following:

(a) $2HgO(s) \longrightarrow 2Hg(l) + O_2(g)$
(b) $CaCO_3(s) + 2HCl(aq) \longrightarrow CaCl_2(aq) + CO_2(g) + H_2O(l)$
(c) $S^{2-} + H_2O(l) \longrightarrow HS^- + OH^-$
(d) $2KClO_3(l) \longrightarrow 2KCl(s) + 3O_2(g)$
(e) $2H_2O(g) \xrightarrow{\Delta} 2H_2(g) + O_2(g)$

18.12 Repeat Question 18.11 for

(a) $2H_2O_2(aq) \longrightarrow 2H_2O(l) + O_2(g)$
(b) $CH_3COOH(aq) + OH^- \longrightarrow CH_3COO^- + H_2O(l)$
(c) $I^- + ClO^- \longrightarrow IO^- + Cl^-$
(d) $2K^+ + 2Fe^{3+} + 2[Fe(CN)_6]^{3-} + 2I^- \longrightarrow 2KFe[Fe(CN)_6](s) + I_2(aq)$
(e) $C_2H_4(g) + Br_2(g) \longrightarrow C_2H_4Br_2(g)$

18.13 Write the rate equation for the overall first-order reaction A $\longrightarrow$ products. What are the units of k? What is the integrated rate equation? What would you plot so that the concentration–time data would be linear?

18.14 Repeat Question 18.13 for the overall second-order reaction 2A $\longrightarrow$ products.

18.15 The reaction between $NO(g)$ and $H_2(g)$

$$2NO(g) + 2H_2(g) \longrightarrow N_2(g) + 2H_2O(g)$$

is second order with respect to NO and first order with respect to H_2. Write the rate equation for this reaction. What is the overall order of this reaction?

18.16 The oxidation of nitrogen(II) oxide

$$2NO(g) + O_2(g) \longrightarrow 2NO_2(g)$$

is known to be second order with respect to NO and first order with respect to O_2. Write the rate equation for this reaction. What is the overall order of the reaction?

18.17 The hydrogenation of ethene

$$C_2H_4(g) + H_2(g) \xrightarrow{Cu} C_2H_6(g)$$

is known to be first order with respect to H_2 and negative first order with respect to C_2H_4. Write the rate equation for this reaction.

18.18 Under certain conditions, the rate equation for

$$H_2(g) + Br_2(g) \longrightarrow 2HBr(g)$$

is Rate $= k[H_2][Br_2]^{3/2}[HBr]^{-1}$. What is the order of the reaction with respect to each reactant? What happens to the rate of reaction as the concentrations of hydrogen and bromine decrease and the concentration of hydrogen bromide increases?

18.19 The rate equation for the decomposition of N_2O_5 in CCl_4 is Rate $= k[N_2O_5]$, where $k = 6.32 \times 10^{-4}\ \text{min}^{-1}$ at 45 °C. What is the initial rate of decomposition of N_2O_5 in (a) a 0.10 M and (b) a 0.010 M solution of N_2O_5?

18.20 The rate equation for the decomposition of hydrogen peroxide in 0.02 M KI is Rate $= k[H_2O_2]$, where $k = 5.21 \times 10^{-3}\ \text{min}^{-1}$ at 25 °C. What is the initial rate of decomposition of hydrogen peroxide in (a) a 0.10 M and (b) a 0.010 M solution of hydrogen peroxide?

Reaction Rates and Rate Equations—Integrated Rate Equations

18.21 The first-order rate constant for the decomposition

$$CS_2(g) \longrightarrow CS(g) + S(g)$$

is $2.8 \times 10^{-7}\ \text{s}^{-1}$ at 1000 °C. What will be the concentration of CS_2 at $t = 2.0$ days given that the initial concentration is 1.30×10^{-2} mol/L? What is the half-life of this reaction?

18.22 The decomposition of gaseous PH_3 at 600 °C

$$4PH_3(g) \longrightarrow P_4(g) + 6H_2(g)$$

is first order with $k = 0.023\ \text{s}^{-1}$. (a) Find the partial pressure of PH_3 after 1.00 min if the initial partial pressure of PH_3 was 275 Torr. What percentage of the PH_3 has reacted? (b) Find the time required for the partial pressure to drop to 1 Torr.

18.23 For the reaction

$$CH_3COOCH_2CH_3(aq) + OH^- \longrightarrow CH_3COO^- + CH_3CH_2OH(aq)$$

the rate equation is Rate $= k[CH_3COOCH_2CH_3][OH^-]$, where $k = 6.49$ L/mol min at 25 °C. Calculate the half-life for the reaction given $[OH^-]_0 = [CH_3COOCH_2CH_3]_0 = 0.0100$ mol/L. How long will it take for the reaction to go to 95% completion?

18.24* The second-order rate constant for the reaction

$$2HI(g) \longrightarrow H_2(g) + I_2(g)$$

is 2.4×10^{-21} L/mol s at 25 °C. What will be the concentration of HI at $t = 1.0$ yr, given that the initial concentration is 5.2×10^{-3} mol/L? What is the half-life of the reaction for this initial value of concentration?

Reaction Rates and Rate Equations—Determining the Order of Reaction

18.25 Determine the rate equation for the gaseous reaction

$$2NO(g) + 2H_2(g) \longrightarrow N_2(g) + 2H_2O(g)$$

from the following data:

Experiment	(a)	(b)	(c)	(d)
Initial rate (Torr/s)	1.50	0.39	1.60	0.80
Initial P_{NO} (Torr)	359	182	378	376
Initial P_{H_2} (Torr)	295	299	289	146

18.26 Determine the rate equation for the reaction

$$(C_6H_5)_2(NH)_2 + I_2 \longrightarrow (C_6H_5)_2N_2 + 2HI$$

from the following data:

Experiment	(a)	(b)	(c)
Initial rate (mol/L s)	2.0	4.0	12.0
$[(C_6H_5)_2(NH)_2]_0$ (mol/L)	0.01	0.01	0.03
$[I_2]_0$ (mol/L)	0.01	0.02	0.02

18.27 Consider the reaction $A + B \longrightarrow$ products. The following table gives the initial rates of reaction at various initial concentrations of A and B:

Experiment	(a)	(b)	(c)	(d)
Initial rate (mol/L s)	0.0090	0.036	0.018	0.027
$[A]_0$(mol/L)	0.10	0.20	0.10	0.10
$[B]_0$(mol/L)	0.10	0.10	0.20	0.30

Find the order of the reaction with respect to A and B and the overall order.

18.28 The following initial rate data were obtained at 0 °C for the reaction described by the equation

$$2ClO_2(aq) + 2OH^- \longrightarrow ClO_3^- + ClO_2^- + H_2O(l)$$

Experiment	(a)	(b)	(c)	(d)
Initial rate (mol/L s)	3.88×10^{-4}	1.55×10^{-3}	7.76×10^{-4}	3.11×10^{-3}
$[ClO_2]_0$ (mol/L)	1.5×10^{-2}	3.0×10^{-2}	1.5×10^{-2}	3.0×10^{-2}
$[OH^-]_0$ (mol/L)	1.5×10^{-2}	1.5×10^{-2}	3.0×10^{-2}	3.0×10^{-2}

Determine the order of reaction with respect to $[ClO_2]$ and $[OH^-]$. Calculate k at this temperature.

18.29 The following data were obtained for the decomposition of diazomethane, CH_2N_2, at 600 °C:

t(min)	0	5	10	15	20	25
$[CH_2N_2]$(mol/L)	0.100	0.076	0.058	0.044	0.033	0.025

Show graphically that the reaction is first order and determine the rate constant.

18.30 By graphical means, determine the order of reaction and the rate constant for the reaction described by the equation

$$2Br(g) \xrightarrow{SF_6} Br_2(g)$$

from the following data:

t(μs)	120	220	320	420	520	620
[Br](μmol/L)	25.8	15.1	10.4	8.0	6.7	5.6

18.31* The following concentration–time data were obtained for the decomposition of N_2O_5 at 45 °C:

t(s)	0	423	753	1116	1582	1986	2343
$[N_2O_5]$(mol/L)	1.40	1.09	0.89	0.72	0.54	0.43	0.35

(a) Prepare a plot of concentration against time. (b) Estimate the concentrations of N_2O_5 at 600 s, 1200 s, 1800 s from your plot. (c) Determine the rate of reaction at 600 s by calculating the slope of a line drawn tangent to the curve at 600 s. (d) Repeat the calculation of (c) at 1200 and 1800 s. Does the rate seem to depend on concentration? (e) Divide your answers for (c) and (d) by the respective values of $[N_2O_5]$ and $[N_2O_5]^2$. Is the reaction first or second order?

18.32* A convenient experimental method for determining the order of reaction involves measuring the time necessary for a reaction to proceed to a predetermined extent. This procedure is known as the "clock method." For example, the decomposition of gaseous PH_3 was studied at 600 °C:

$$4PH_3(g) \longrightarrow P_4(g) + 6H_2(g)$$

using the clock method. The time required for the partial pressure of PH_3 to decrease by 5.00 Torr was recorded at three different initial partial pressures of PH_3. The data were

$P_{PH_3,0}$(Torr)	262.40	65.60	16.31
t(s)	0.83	3.42	15.81

The equation relating the time to the initial pressure (or concentration) is $\log(1/t) = K + n \log P_0$, where K is a constant and n is the order of reaction. Prepare a plot of $\log(1/t)$ versus $\log P_0$ and determine the reaction order.

Reaction Rates and Rate Equations—Reaction Mechanisms

18.33 The rate equation for the decomposition of $NO_2Cl(g)$

$$2NO_2Cl \longrightarrow 2NO_2 + Cl_2$$

is Rate $= k[NO_2Cl]$. A proposed mechanism is

$$NO_2Cl \xrightarrow{k_1} NO_2 + Cl$$

$$NO_2Cl + Cl \xrightarrow{k_2} NO_2 + Cl_2$$

Based on the experimental rate equation, which is the rate-determining step?

18.34 The rate equation for the reaction of NO and Cl_2

$$2NO(g) + Cl_2(g) \longrightarrow 2NOCl(g)$$

is Rate $= k[Cl_2][NO]^2$. Rather than proposing a termolecular mechanism to describe this reaction, the following two-step mechanism is proposed:

$$NO + Cl_2 \xrightarrow{\text{fast}} NOCl_2$$

$$NOCl_2 + NO \xrightarrow{\text{slow}} 2NOCl$$

Show that the proposed mechanism gives the experimentally determined rate equation.

18.35* Given the following steps in the mechanism for a chemical reaction:

$$A + B \xrightarrow{\text{fast}} C$$
$$B + C \xrightarrow{\text{slow}} D + E$$
$$D + F \xrightarrow{\text{fast}} A + E$$

Identify which species, if any, are (a) catalysts and (b) intermediates in this reaction. (c) Write the rate equation for the rate-determining step. Given that the concentration of C is directly proportional to the concentration of A, (d) rewrite the rate equation so that it includes concentration terms for only reactants and products, not intermediates. (e) What is the overall order of the reaction?

18.36* A proposed mechanism for the aqueous reaction

$$I^- + ClO^- \longrightarrow IO^- + Cl^-$$

consists of the following steps:

$$ClO^- + H_2O \xrightarrow{\text{rapid}} HClO + OH^-$$
$$HClO + OH^- \xrightarrow{\text{rapid}} ClO^- + H_2O$$
$$I^- + HClO \xrightarrow{\text{slow}} HIO + Cl^-$$
$$OH^- + HIO \xrightarrow{\text{rapid}} H_2O + IO^-$$
$$H_2O + IO^- \xrightarrow{\text{rapid}} OH^- + HIO$$

(a) Write the rate equation for the rate-determining step. The concentration of HClO in a dilute aqueous solution is known to be proportional to $[ClO^-]/[OH^-]$. (b) Rewrite the rate equation in terms of $[ClO^-]$, $[OH^-]$, and $[I^-]$. (c) What is the overall order of the reaction? (d) What would happen to the reaction rate if the concentration of OH^- were doubled?

Factors That Influence Reaction Rates

18.37 The rate constant for the reaction

$$\begin{matrix} H_2C-C{=}O \\ | \quad\ \ | \\ H_2C-CH_2 \end{matrix}(g) \longrightarrow C_2H_4(g) + H_2C{=}C{=}O(g)$$

at 361 °C is $4.6 \times 10^{-4}\ s^{-1}$ and at 371 °C is $7.2 \times 10^{-4}\ s^{-1}$. What is the energy of activation for this reaction?

18.38 The activation energy for the reaction

$$2C_6H_5CHO(aq) \xrightarrow{CN^-} C_6H_5CH(OH)COC_6H_5$$

is 54.8 kJ. At 25 °C, the rate constant is $1.40 \times 10^{-4}\ L^2/mol^2\ s$. Calculate the rate constant at 60. °C.

18.39 The rate constant for the decomposition of N_2O

$$2N_2O(g) \longrightarrow 2N_2(g) + O_2(g)$$

is $2.6 \times 10^{-11}\ s^{-1}$ at 300. °C and $2.1 \times 10^{-10}\ s^{-1}$ at 330. °C. What is the activation energy for this reaction? Prepare a reaction coordinate diagram like Figure 18.4 using −164.1 kJ as the energy of reaction.

18.40 For the reaction

$$N_2O_5(g) \longrightarrow 2NO_2(g) + \tfrac{1}{2}O_2(g) \qquad \Delta E° = 51.51\ kJ$$

$k = 8.0 \times 10^{-7}\ s^{-1}$ at 0.0 °C and $8.9 \times 10^{-4}\ s^{-1}$ at 50.0 °C. Prepare a reaction coordinate diagram like Figure 18.4 for this reaction.

18.41 How much faster would a reaction proceed at 25 °C than at 0 °C if the activation energy is 65 kJ?

18.42 "A temperature change affects the reaction rates for reactions with high energies of activation more than the rates for reactions with small energies of activation." Show that this statement is true by calculating k_2/k_1 for (a) a reaction with $E_a = 45$ kJ and for (b) a reaction with $E_a = 5$ kJ as the temperature changes from 275 K to 775 K.

18.43 The energy of activation for the decomposition of ammonia

$$2NH_3(g) \longrightarrow N_2(g) + 3H_2(g) \qquad \Delta H° = 92.22\ kJ$$

is 350 kJ for the uncatalyzed reaction and 162 kJ for the reaction catalyzed by tungsten. Draw a reaction coordinate diagram similar to Figure 18.13 for this reaction. Assume $\Delta E° = \Delta H°$.

18.44 Revise your drawing in Question 18.7 to include curves for the reaction in the presence of different catalysts: $E_a = 105$ kJ for Au and $E_a = 59$ kJ for Pt. Compare the rates of reaction for the uncatalyzed reaction and the Pt-catalyzed reaction by calculating k_{Pt}/k_{uncat} at 355 K using Equation (18.32).

ADDITIONAL PROBLEMS

18.45 For a hypothetical reaction

$$A + 2B \xrightarrow{C} D + E$$

the rate equation is

$$\text{Rate} = \frac{k[A][B][C]}{[D]}$$

What will happen to the reaction rate if (a) [A] is doubled, (b) [B] is doubled, (c) more catalyst is added, (d) D is removed as it is formed to keep [D] at a small value, (e) [A] and [B] are both doubled, (f) the temperature is increased?

18.46 The rate constant for the decomposition of A in solvent X is $4.7 \times 10^{-5}\ s^{-1}$ at 60.0 °C and $1.60 \times 10^{-4}\ s^{-1}$ at 70.0 °C; in solvent Y it is $7.0 \times 10^{-5}\ s^{-1}$ at 65.0 °C and $2.30 \times 10^{-4}\ s^{-1}$ at 75.0 °C. Calculate the energy of activation for the reaction in each solvent. Is one solvent acting as a stronger catalyst than the other (as determined by a lower value of E_a)?

18.47* For the reaction $2A + B \longrightarrow$ products, the following data were obtained for a solution containing a large excess of B:

t(s)	0	10	20	30	40	50	60	70
[A] (mol/L)	10.00	7.95	6.31	5.00	3.92	3.16	2.50	1.99

(a) Calculate log [A] and 1/[A] for the data. (b) Plot your results from part (a) against t and determine the order of reaction with respect to [A]. (c) What is the value of the rate constant? (d) Which of the following mechanisms are acceptable possible mechanisms?

(1) $2A + B \longrightarrow$ products

(2) $A + B \xrightarrow{\text{slow}} C$
$C + A \longrightarrow$ products

(3) $A \rightleftharpoons C$
$C + A + B \xrightarrow{\text{slow}}$ products

(4) $A \xrightarrow{\text{slow}} C$
$C + A + B \longrightarrow$ products

18.48* Unless the reaction is first order, the half-life of a reaction depends on concentration. The general relationship between half-life and initial concentration is

$$\log t_{1/2} = \log\left[\frac{2^{n-1} - 1}{(n-1)k}\right] - (n-1)\log[A]_0$$

where n is the order of reaction. Thus, a plot of $\log t_{1/2}$ against $\log[A]_0$ will be a straight line with slope equal to $-(n-1)$. For the reaction

$$2Br(g) \xrightarrow{SF_6} Br_2(g)$$

the following data were obtained:

$[Br]_0$ (mol/L)	2.58×10^{-5}	1.51×10^{-5}	1.04×10^{-5}
$t_{1/2}(s)$	1.4×10^{-4}	2.2×10^{-4}	3.4×10^{-4}

Prepare a plot of $\log t_{1/2}$ against $\log[Br]_0$ and determine the reaction order.

18.49* An important gas-phase reaction, which occurs in the atmosphere, is described by the equation

$$2O_3(g) \longrightarrow 3O_2(g) \qquad \Delta H^\circ_{298} = -285.3 \text{ kJ}$$

A set of experiments gave the following data.

Initial P_{O_3} (Torr)	0.20	0.20	0.40
Initial P_{O_2} (Torr)	0.50	1.0	1.0
Initial rate (Torr/s)	6.0	3.0	12.0

(a) What are the values of x and y in the rate equation

$$\text{Rate} = kP_{O_3}^x P_{O_2}^y$$

(b) The reaction was studied in the presence of CO_2 to investigate the catalytic effect of CO_2 on the reaction rate. The results of two such experiments at 50 °C are as follows:

P_{CO_2} **= 100 Torr**

$t(s)$	0	1800	3600	7200
P_{O_3} **(Torr)**	200	140	100	50

P_{CO_2} **= 180 Torr**

$t(s)$	0	1800	3600
P_{O_3} **(Torr)**	220	120	70

The complete rate equation for the catalyzed reaction is

$$\text{Rate} = k'P_{O_3}^m P_{CO_2}$$

Find the order of reaction with respect to P_{O_3} for each experiment and the rate constant in each case. (c) Does the catalyzed reaction have a mechanism different from that of the uncatalyzed reaction? The rate constants found in part (b) are equal to $k'P_{CO_2}$. (d) Find k'.

18.50* Consider the decomposition of acetic anhydride in the gas phase

$$(CH_3CO)_2O(g) \longrightarrow CH_3COOH(g) + CH_2{=}C{=}O(g)$$

The proposed mechanism is

$$2A \xrightarrow{k_1} A^* + A$$
$$A^* + A \xrightarrow{k_{-1}} 2A$$
$$A^* \xrightarrow{k_2} \text{products}$$

where the * indicates the activated acetic anhydride molecule, k_{-1} signifies the second step is the reverse of the first step, and A represents the acetic anhydride molecule. The rate equation derived for this reaction from the mechanism (that agrees with experimental observations) is

$$\text{Rate} = \frac{k_1 k_2 P_A^2}{k_{-1}P_A + k_2}$$

Show that at high pressures, when the second step is the rate-determining step, the expression reduces to a first-order rate equation and at low pressure, where $k_2 > k_{-1}P_A$, the expression reduces to a second-order rate equation (that is, the rate is determined by the first step).

18.51* The rate constant for the decomposition of N_2O_5 was studied as a function of temperature:

T(°C)	0	25	35	45	55	65
$k(s^{-1})$	7.36×10^{-7}	3.33×10^{-5}	1.29×10^{-4}	4.58×10^{-4}	1.51×10^{-3}	4.64×10^{-3}

Prepare a plot of $\log k$ against $1/T$. The slope of the straight line through the data is related to E_a by $E_a = -2.303R \times$ (slope). Calculate E_a for this reaction.

CHAPTER

19

Chemical Equilibrium

THE LAW OF CHEMICAL EQUILIBRIUM

Picture a large chicken house divided into two parts by a chicken-wire wall. On one side live 200 white chickens. On the other side live 200 brown chickens. What would happen if someone left the connecting door open? Gradually the brown and white chickens would mix together as their random paths took them through the door. How fast this happens— the "reaction rate"—would depend upon how fast the chickens are moving. The process might be quicker in the morning than just before sundown when the chickens are drowsy.

Eventually, we can assume, a state of dynamic equilibrium would be reached. The brown and white chickens would be pretty well mixed together on each side, and at any given moment some brown and some white chickens would be going through the door in each direction. There would be a roughly unchanging number of chickens on each side.

If we knew how long it would take to reach this equilibrium, we would still know nothing about how many chickens were on each side at equilibrium. It might be an equal number, or it might be any combination adding up to 400, depending upon the "conditions" of the "reaction." Suppose the water buckets on one side were empty. Then at equilibrium there might be only 50 chickens on that side and 350 on the other.

The purpose of this analogy, about which you are probably wondering by now, is to point out that the "rate" of a reaction and the equilibrium state of a reaction are different concepts. A thorough understanding of any chemical reaction must include quantitative knowledge about both the rate of the reaction and the conditions at which it reaches equilibrium.

$A + B \rightarrow C$

[C]

[A], [B]

Concentration

Time

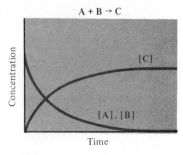

(a) Reaction goes to completion

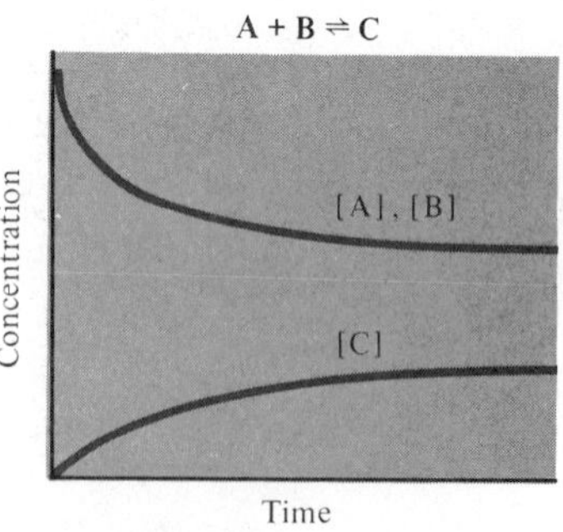

(b) Equilibrium with many reactant molecules unchanged

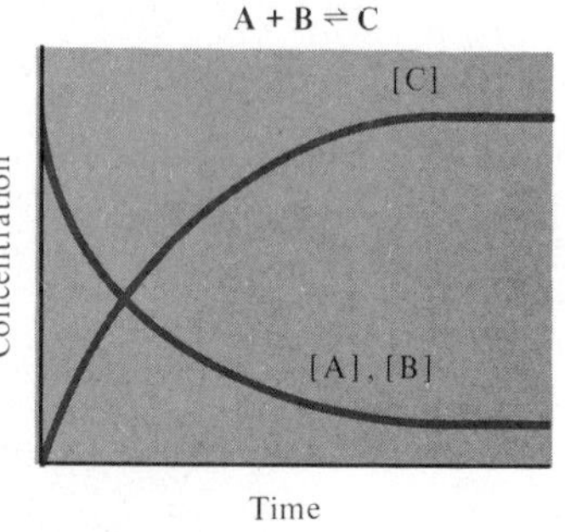

(c) Equilibrium with few reactant molecules remaining

Figure 19.1
Concentration Change with Time for Reactions That Go To Completion or Reach Equilibrium
The initial concentrations of A and B are equal.

19.1 A CLOSE LOOK AT AN EQUILIBRIUM REACTION

As a chemical reaction proceeds, the concentration of reactants decreases and the concentration of products increases (assuming that nothing escapes from the system). If the reaction goes to "completion," the concentration of reactants, say A and B in the combination reaction

$$A + B \longrightarrow C$$

approaches zero, while the concentration of product C approaches the amount determined by the stoichiometry of the reaction and the initial amounts of A and B (Figure 19.1a).

If, instead, a dynamic equilibrium is reached, some molecules of A, B, and C will all remain at equilibrium. Equilibrium might be reached at a point where the number of reactant molecules remaining is much larger than the number of product molecules (Figure 19.1b). In such a case the reaction can be described as "not going very far toward completion." On the other hand, the reaction might come to equilibrium at a point where few reactant molecules remain and the system contains mainly products (Figure 19.1c). Such a reaction is described as "favorable in the forward direction" or as going "to the right."

Suppose some PCl_3 and Cl_2 are confined in a closed vessel at 500 K. As soon as the gases are mixed, the formation of PCl_5—the forward reaction—begins.

$$PCl_3(g) + Cl_2(g) \longrightarrow PCl_5(g)$$

At this temperature PCl_5 can decompose. As soon as some PCl_5 has formed, the reverse reaction begins—PCl_5 molecules decompose to give PCl_3 and Cl_2 molecules.

$$PCl_5(g) \longrightarrow PCl_3(g) + Cl_2(g)$$

The concentrations of the three gases will continue to change in one of the two general ways shown in Figure 19.1b and c. As the concentrations of PCl_3 and Cl_2 decrease, whatever collisions precede the formation of PCl_5 occur less often, and the rate of the forward reaction decreases. The rate of the reverse reaction depends upon the concentration of PCl_5 molecules; the more of these that are present in a given volume, the more there are that decompose. Thus, as more PCl_5 is formed, the rate of the reverse reaction increases (Figure 19.2). The rates of the forward and the reverse reactions, and the concentrations of the three gases continue to change until equilibrium is reached. At this point the rates of the forward and reverse reactions are equal

$$PCl_3(g) + Cl_2(g) \rightleftharpoons PCl_5(g)$$

with the net result that the concentrations of the three gases remain constant.

The system is in a state of dynamic, chemical equilibrium. Each time one PCl_5 molecule is formed, one PCl_3 and one Cl_2 molecule disappear. Each time a PCl_5 molecule decomposes, one PCl_3 and one Cl_2 molecule are formed. As long as the temperature and pressure remain constant and nothing is added to or taken from the mixture, the equilibrium state remains unchanged. This is true for all chemical equilibria.

19.2 EQUILIBRIUM CONSTANTS

Our understanding of dynamic chemical equilibrium grew out of observing the concentrations of reactants and products. Changing the concentration of one reactant or product was found to cause changes in the concentrations of the other reactants and products. Both experimental and theoretical approaches eventually led to the same general mathematical relationship for the equilibrium concentrations of reactants and products. In this section we present some experimental results and show how they reveal this relationship.

Nitrogen dioxide (NO_2) and dinitrogen tetroxide (N_2O_4) form an equilibrium mixture by the following reaction:

$$\underset{\textit{brown}}{2NO_2(g)} \rightleftharpoons \underset{\textit{colorless}}{N_2O_4(g)} \qquad \textbf{(19.1)}$$

The concentrations of these two gases have been measured after equilibrium was reached in a closed vessel at constant temperature (about 21 °C). In each of four experiments the initial concentrations of the two gases were different. Once equilibrium was reached, the following concentrations were found:

Experiment no.	At equilibrium $[NO_2]$ (mol/L)	$[N_2O_4]$ (mol/L)
1	0.052	0.595
2	0.024	0.127
3	0.068	1.02
4	0.101	2.24

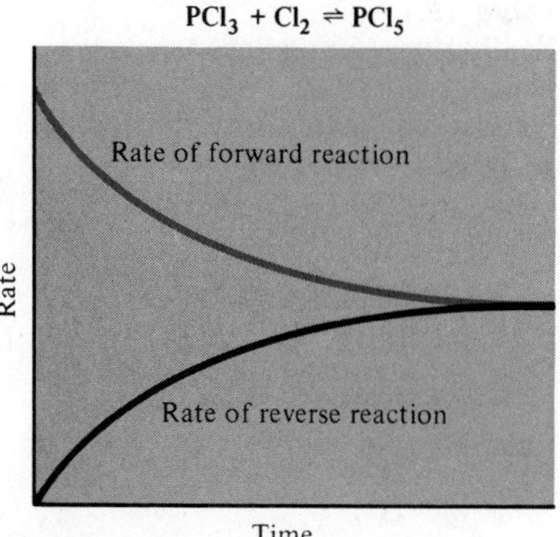

Figure 19.2
Variation of Forward and Reverse Reaction Rates in an Equilibrium Reaction

It would be satisfying to find a mathematical relationship that brings some meaning to these numbers. The table shows that the concentrations of the reactant and product are *not* equal to each other at equilibrium. A little arithmetic shows that these concentrations do not maintain a constant ratio, either.

Experiment no.	$[N_2O_4]/[NO_2]^2$
1	11
2	5.3
3	15
4	22

We might try several dozen manipulations with such data before coming up with a meaningful relationship. Let's move ahead by skipping directly to the right one. Dividing the concentration of N_2O_4 by the *square* of the concentration of NO_2 gives a constant.

Experiment no.	$[N_2O_4]/[NO_2]^2$
1	220
2	220
3	220
4	220

Remember that the initial concentrations of the gases were different in each experiment. The existence of this mathematically constant relationship shows that whatever the initial conditions, the forward and reverse reactions adjust in reaching equilibrium so that $[N_2O_4]/[NO_2]^2$ remains the same.

Experiments with many different equilibrium reactions over the years and a theoretical derivation based on the principles of thermodynamics both give the same result. It is possible to formulate an expression for the relationships among equilibrium concentrations for any reaction. For the general reaction

$$aA + bB + \cdots \qquad rR + sS + \cdots \qquad \textbf{(19.2)}$$

Table 19.1
Equilibrium Expressions for Homogeneous Gas-Phase Reactions
(Brackets indicate concentration in moles per liter.)

$N_2(g) + 3H_2(g) \rightleftharpoons 2NH_3(g)$ (a)

$$K = \frac{[NH_3]^2}{[N_2][H_2]^3}$$

$2NO_2(g) \rightleftharpoons N_2O_4(g)$ (b)

$$K = \frac{[N_2O_4]}{[NO_2]^2}$$

$COCl_2(g) \rightleftharpoons CO(g) + Cl_2(g)$ (c)

$$K = \frac{[CO][Cl_2]}{[COCl_2]}$$

the "equilibrium constant," K, is given by the following expression:

$$K = \frac{[R]^r[S]^s\cdots}{[A]^a[B]^b\cdots} \quad \begin{matrix}\leftarrow \textit{product concentrations} \\ \leftarrow \textit{reactant concentrations}\end{matrix} \tag{19.3}$$

where [R], [S], [A], [B], and so on represent the concentrations of the products and reactants, at equilibrium, in moles per liter. It is more difficult to define the equilibrium constant in words than with an equation: The **equilibrium constant** is equal to the product of the concentrations of the reaction products, each raised to the power equal to its stoichiometric coefficient, divided by the product of the concentrations of the reactants, each raised to the power equal to its stoichiometric coefficient. (Note that K is used for the *equilibrium constant expression,* or the equilibrium constant, and k for the *rate constant.*)

The NO_2–N_2O_4 reaction fits into the general scheme of Equations (19.2) and (19.3) as

$$aA \rightleftharpoons rR \qquad 2NO_2 \rightleftharpoons N_2O_4$$

and

$$K = \frac{[R]^r}{[A]^a} = \frac{[N_2O_4]}{[NO_2]^2} \tag{19.4}$$

Table 19.1 gives a few other examples.

Some carefully gathered experimental data on the reaction between hydrogen and iodine

$$H_2(g) + I_2(g) \rightleftharpoons 2HI(g) \tag{19.5}$$

illustrate several additional important points about chemical equilibrium. In one set of experiments, the amount of hydrogen in the initial mixture was held relatively constant (at about 0.01 mol/L), while the amount of iodine was varied. The concentrations of H_2, I_2, and HI at equilibrium were measured (at 425.4 °C). These concentrations and the values of K calculated from

$$K = \frac{[HI]^2}{[H_2][I_2]} \tag{19.6}$$

were as follows:

At equilibrium

$[H_2]$	$[I_2]$	$[HI]$	$K = \frac{[HI]^2}{[H_2][I_2]}$
0.001831	0.003129	0.01767	54.5
0.003560	0.001250	0.01559	54.6
0.004565	0.0007378	0.01354	54.4

The small variations in the value of K are due to small variations in the precision of the experiment and are not significant. Clearly, changes in the initial concentration of one of two reactants do not alter the constancy of K. Additional experiments show that K remains constant with simultaneous variations in the concentrations of both reactants. An infinite number of combinations of the concentrations of reactants and products is possible, but at equilibrium at a specific temperature and pressure, there is only one value for K.

However, similar experiments show that the value of K does vary with *temperature.* For example, at a temperature of 490.65 °C (instead of 425.4 °C) the value of K is constant at 45.6 (instead of 54.5).

In further experiments with this same system, hydrogen iodide was placed in a sealed vessel and allowed to reach equilibrium with its decomposition products hydrogen and iodine—the same reaction but approached from the reverse direction.

Similar data and calculations (for 425.4 °C also) show by the constancy of K that the equilibrium state is the same, no matter from which direction it is approached.

EXAMPLE 19.1 Equilibrium Constant Expressions

Write the equilibrium constant expression for each of the following gas-phase reactions:

(a) $N_2(g) + O_2(g) \rightleftharpoons 2NO(g)$

(b) $2NO(g) + Br_2(g) \rightleftharpoons 2NOBr(g)$

For chemical equation (a), there is only one product and it has a stoichiometric coefficient of 2; therefore the numerator of the equilibrium constant expression (Equation 19.3) consists of one term, $[NO]^2$. There are two reactants, each with a stoichiometric coefficient of 1; therefore the denominator of the equilibrium constant expression consists of the product of these concentrations, each raised to the first power, $[N_2][O_2]$. The complete expression for K is

$$K = \frac{[NO]^2}{[N_2][O_2]}$$

For chemical equation (b), the numerator is $[NOBr]^2$ and the denominator is $[NO]^2[Br_2]$, giving

$$K = \frac{[NOBr]^2}{[NO]^2[Br_2]}$$

EXERCISE Write the equilibrium constant expression for each of the following gas-phase reactions:

(a) $2NO(g) + O_2(g) \rightleftharpoons 2NO_2(g)$

(b) $2HF(g) \rightleftharpoons H_2(g) + F_2(g)$

Answers (a) $K = [NO_2]^2/[NO]^2[O_2]$, (b) $K = [H_2][F_2]/[HF]^2$.

EXAMPLE 19.2 Chemical Equilibrium

Describe what happens at the molecular level when more PCl_3 is added to a system containing PCl_3, Cl_2, and PCl_5 that have come to equilibrium as described in Section 19.1.

$$PCl_3(g) + Cl_2(g) \rightleftharpoons PCl_5(g)$$

Assume that the temperature and volume of the system remain unchanged. What can be concluded about the new concentrations of the reactants and product once equilibrium is reached?

With more molecules of reactant PCl_3 available, more collisions with Cl_2 molecules will occur in a given time, resulting in an increase in the rate of formation of PCl_5. However, as more PCl_5 molecules are formed, the number decomposing to give the reactants in a given time—and therefore the rate of the reverse reaction—will also increase. The amounts of the reactants and products as well as the rates of the forward and reverse reactions will continue to change until equilibrium is reached once again. **At equilibrium, the concentrations of the reactants and product will be different from those before the additional PCl_3 was added. There will be more PCl_5 than originally, but the value of $[PCl_5]/[PCl_3][Cl_2]$ will be unchanged.**

EXERCISE What would happen if some of the chlorine escaped from the system described above?

Answer Additional PCl_5 will decompose, rates will become equal again, and $[PCl_5]/[PCl_3][Cl_2]$ will have the same value as before.

Table 19.2
Some Equilibrium Constants

	K_c at 298 K
Gas-phase reactions	
$2H_2(g) + O_2(g) \rightleftharpoons 2H_2O(g)$	3.1×10^{81}
$H_2(g) + I_2(g) \rightleftharpoons 2HI(g)$	8.7×10^{2}
$2CO_2(g) \rightleftharpoons 2CO(g) + O_2(g)$	3.0×10^{-92}
Ionization of weak acids in water	
$CH_3COOH(aq) \rightleftharpoons CH_3COO^- + H^+$	1.8×10^{-5}
$HCN(aq) \rightleftharpoons H^+ + CN^-$	6.2×10^{-10}
Solubility of precipitates	
$Ag_2SO_4(s) \rightleftharpoons 2Ag^+ + SO_4^{2-}$	1.5×10^{-5}
$AgCl(s) \rightleftharpoons Ag^+ + Cl^-$	1.8×10^{-10}
$Ag_2S(s) \rightleftharpoons 2Ag^+ + S^{2-}$	7.1×10^{-50}

19.3 UNITS AND EQUILIBRIUM CONSTANT VALUES

In the discussion of K for the hydrogen–iodine reaction, we bypassed the subject of units. Like other relationships in chemistry (e.g., the ideal gas law), the equilibrium constant expression was first found empirically and later derived from theory. In the derivation of the equilibrium constant expression from theoretical thermodynamics relationships, effective concentrations of the solutes (Section 15.3) replace actual concentrations in moles per liter or as partial pressures. The result is that K is a dimensionless number. It has become customary to omit units for K, even when concentrations rather than effective concentrations are used.

The values of equilibrium constants are usually given in scientific notation because they range greatly in magnitude. When K is greater than 1, products are favored; when K is less than 1, reactants are favored. In Table 19.2, compare the K values for the gas-phase reactions of hydrogen with oxygen and the decomposition of carbon dioxide. The combination of hydrogen and oxygen, once initiated, can proceed with explosive violence. When the reaction is finished, very little hydrogen and oxygen remain unreacted. The high concentration of product when this reaction has come to equilibrium leads to a very large value for K. On the other hand, carbon dioxide decomposes to only a very small extent. The very low concentrations of the products at equilibrium lead to a very small value for K (at 298 K).

EXAMPLE 19.3 Interpreting Values of *K*

It was necessary to remove silver ion from an aqueous solution as completely as possible by causing it to form a precipitate. Which would be the better method: to add a solution containing sulfate ion or one containing an equivalent amount of sulfide ion? (See Table 19.2.)

Silver sulfate and silver sulfide would be the precipitates formed upon addition of sulfate ion or sulfide ion, respectively. Comparison of the equilibria that would result after precipitate formation and the K values

$$Ag_2SO_4(s) \rightleftharpoons 2Ag^+ + SO_4^{2-} \qquad K = 1.5 \times 10^{-5}$$
$$Ag_2S(s) \rightleftharpoons 2Ag^+ + S^{2-} \qquad K = 7.1 \times 10^{-50}$$

shows that the **sulfide would leave a far lower concentration of Ag^+ in solution after the precipitation reaction was complete.**

EXERCISE A solution contained equal concentrations of acetate ion and cyanide ion. Which ion would decrease most in concentration as the H^+ ion concentration of the solution is increased? *Answer* Cyanide ion.

SUMMARY OF SECTIONS 19.1–19.3

Experimental data show the following:

1. The equilibrium concentrations of reactants and products in a reversible reaction are related by the law of chemical equilibrium

$$K = \frac{[R]^r[S]^s\cdots}{[A]^a[B]^b\cdots} \qquad \text{where } a\text{A} + b\text{B} + \cdots \rightleftharpoons r\text{R} + s\text{S} + \cdots$$

and [ ] indicates concentration in moles per liter at equilibrium.
2. An initial mixture of reactants of any concentrations will reach equilibrium (at a specific temperature and pressure) so that the concentrations of reactants and products are related by the law of chemical equilibrium.
3. The value of K is dependent upon the temperature; at a given temperature, the value of K for a particular reaction is always the same (provided the reaction is represented by identical equations).
4. Approaching equilibrium from either direction, at the same temperature, leads to equilibrium concentrations determined by the same value of K.

When K is greater than 1, products are favored over reactants; the larger the value of K, the higher the concentrations of products. When K is less than 1, reactants are favored over products; the smaller the value of K, the lower the concentrations of products. It is customary to omit units with equilibrium constant values.

EQUILIBRIUM CONSTANT EXPRESSIONS

19.4 EQUILIBRIUM CONSTANTS FOR REACTIONS OF GASES

If the reactants and products to be represented in an equilibrium constant expression are all gases, the expression can be written in two different ways—with concentrations in moles per liter (K_c) as shown in Section 19.2, or with partial pressures (K_p). For the general reaction of Equation (19.2)

$$K_p = \frac{P_R{}^r P_S{}^s\cdots}{P_A{}^a P_B{}^b\cdots} \tag{19.7}$$

partial pressures

indicates K for partial pressures of reactants and products

The partial pressures are given in bars [with K_p values based on current standard state definition; Section 7.8] or atmospheres [with K_p values based on older standard state definition]. The K_p values in this book are for partial pressures in bars.

Partial pressures are proportional to concentrations in moles per liter, and K_c and K_p for different reactions have different values. For example, for the following reaction carried out at 1000 °C

$$PCl_3(g) + Cl_2(g) \rightleftharpoons PCl_5(g)$$

$$K_c = \frac{[PCl_5]}{[PCl_3][Cl_2]} = 9.7 \times 10^{-4} \qquad K_p = \frac{P_{PCl_5}}{P_{PCl_3}P_{Cl_2}} = 1.2 \times 10^{-5}$$

indicates K for gas concentrations in moles/liter

indicates K for partial pressures of gas

The value of K_p for a given reaction is related to the value of K_c by the ideal gas law. For substance [A]

$$P_AV_A = n_ART \qquad \text{and therefore} \qquad P_A = \frac{n_ART}{V_A}$$

Because n/V is moles per liter, this expression can be written as

$$P_A = [A]RT \tag{19.8}$$

Substituting relationships such as that in Equation (19.8) into the expression for K_p gives

$$K_p = \frac{([R]RT)^r([S]RT)^s\cdots}{([A]RT)^a([B]RT)^b\cdots} = \frac{[R]^r[S]^s\cdots}{[A]^a[B]^b\cdots} \times (RT)^{(r+s+\cdots)-(a+b+\cdots)}$$

which yields the general expression

$$K_p = K_c(RT)^{\Delta n} \tag{19.9}$$

where Δn for the balanced chemical equation is given by

$$\begin{aligned}\Delta n &= (\text{moles of gaseous products}) - (\text{moles of gaseous reactants})\\ &= (r + s + \cdots) - (a + b + \cdots)\end{aligned}$$

For example, for the reaction of PCl_3 with Cl_2, $\Delta n = 1 - 2 = -1$ and $K_p = K_c(RT)^{-1}$. If $\Delta n = 0$, that is, when the number of moles of reactants equals the number of moles of products in the balanced equation for the reaction being considered, then K_c has the same value as K_p [because $(RT)^0 = 1$]. For use in Equation (19.9) with pressure in bars, $R = 0.08314$ L bar/K mol.

EXAMPLE 19.4 Evaluation of K

Methyl alcohol can be prepared commercially by the reaction of hydrogen with carbon monoxide:

$$CO(g) + 2H_2(g) \rightleftharpoons CH_3OH(g)$$

Under equilibrium conditions at 700 K, $[H_2] = 0.072$ mol/L, $[CO] = 0.020$ mol/L, and $[CH_3OH] = 0.030$ mol/L. What are the values of K_c and K_p at this temperature?

Substituting the equilibrium concentrations in moles per liter into the equilibrium constant expression gives the value of K_c:

$$K_c = \frac{[CH_3OH]}{[CO][H_2]^2} = \frac{(0.030)}{(0.020)(0.072)^2} = \mathbf{290}$$

In order to convert K_c to K_p, the value of Δn must be found:

$$\begin{aligned}\Delta n &= (\text{moles of gaseous products}) - (\text{moles of gaseous reactants})\\ &= (1) - (1 + 2) = -2\end{aligned}$$

Thus

$$K_p = K_c(RT)^{\Delta n} = (290)[(0.08314)(700)]^{-2} = \mathbf{0.086}$$

EXERCISE $K_c = 2.6 \times 10^{-9}$ at 1500 K for the reaction

$$2BrF_5(g) \rightleftharpoons Br_2(g) + 5F_2(g)$$

Evaluate K_p at this temperature. *Answer* 0.63.

19.5 HETEROGENEOUS AND SOLUTION EQUILIBRIA

Reversible reactions can take place and equilibrium is established no matter what the physical state of the materials involved. For example, in the iron–steam reaction (Section 17.2)

$$3Fe(s) + 4H_2O(g) \rightleftharpoons Fe_3O_4(s) + 4H_2(g) \tag{19.10}$$

a solid and a gas are in equilibrium with another solid and gas. The solid substances, lead(II) oxide and tin, when both are finely powdered and heated to just below the melting point, slowly reach equilibrium with lead and tin(II) oxide

$$PbO(s) + Sn(s) \rightleftharpoons Pb(s) + SnO(s) \tag{19.11}$$

In *homogeneous equilibria* all of the reactants and products are present in the same phase, as in the gas-phase reactions given in Table 19.1. In *heterogeneous equilibria* the reactants and products are *not* all in the same phase.

The question arises of how to deal with solids and pure liquids in writing equilibrium constant expressions. Consider the iron–steam reaction (Equation 19.10). As the reaction proceeds, the concentrations of water and hydrogen in the gas phase vary as the reaction approaches equilibrium. What about the "concentrations" of iron and iron oxide, the solids? As long as *some* of a solid reactant is always present, equilibrium is reached in the same way. For a solid (or a pure liquid), whether the amount is large or small, the *concentration*—the *mass per unit volume*—is always the same. How much iron or iron oxide is present during the time the reaction is coming to equilibrium has no influence on the equilibrium state that is eventually reached. Furthermore, adding or taking away some of either solid will cause no change in the equilibrium state once it is established.

For these reasons (and also for theoretical reasons), the concentration terms for solids or pure liquids do not appear in equilibrium constant expressions. For the iron–steam reaction, K_c and K_p are given by

$$K_c = \frac{[H_2]^4}{[H_2O]^4} \quad \text{or} \quad K_p = \frac{P_{H_2}{}^4}{P_{H_2O}{}^4}$$

For the reaction

$$CaCO_3(s) \rightleftharpoons CaO(s) + CO_2(g)$$

the equilibrium constant expressions are

$$K_c = [CO_2] \quad \text{or} \quad K_p = P_{CO_2}$$

These expressions indicate that as long as both $CaCO_3$ and CaO are present, the pressure of carbon dioxide cannot vary (at a given temperature). If the CO_2 pressure is increased, then CO_2 combines with CaO until the equilibrium pressure of CO_2 is restored.

Many reactions with which we are concerned occur in dilute aqueous solutions. Frequently, water is a "participant" in the reactions in the sense that ions in solution are hydrated or formed by reaction with water. Water is present in such solutions in great excess compared to the other "reactants." The solvent water is considered to be virtually a pure liquid, and changes in the amount of water have little influence on the position of the equilibrium. Therefore, for reactions in dilute aqueous solution, the concentration of water need not appear in the equilibrium constant expression. For example, for the equilibria of weak electrolytes in water, such as ammonia and acetic acid

$$NH_3(aq) + H_2O(l) \rightleftharpoons NH_4{}^+ + OH^-$$
$$CH_3COOH(aq) + H_2O(l) \rightleftharpoons CH_3COO^- + H_3O^+$$

we write

$$K_c = \frac{[NH_4{}^+][OH^-]}{[NH_3]} = 1.6 \times 10^{-5} \quad \text{and} \quad K_c = \frac{[CH_3COO^-][H_3O^+]}{[CH_3COOH]} = 1.8 \times 10^{-5}$$

Note, however, that when water is a reactant or product but is *not* present in excess as the solvent, its concentration must be included in the equilibrium constant expression, as was done above for the iron–steam reaction. Do not make the common error of leaving $[H_2O]$ out of an equilibrium expression for a reaction in which the variation in the concentration of water is of significance.

It is possible to have both pure substances and solutions as reactants and products in the same reaction. Consider, for example, the reaction of hydrochloric acid with calcium carbonate:

$$CaCO_3(s) + 2HCl(aq) \rightleftharpoons CaCl_2(aq) + CO_2(g) + H_2O(l)$$

The equilibrium constant expression for this reaction can be written as follows, where $[CO_2]$ is the concentration of CO_2 in the gas phase:

$$K_c = \frac{[CaCl_2][CO_2]}{[HCl]^2}$$

Although at first it seems strange, an equilibrium constant expression for this heterogeneous reaction can also be written as follows [because concentration in moles per liter and partial pressures are both approximations to effective concentration]

$$K_p = \frac{[CaCl_2]P_{CO_2}}{[HCl]^2}$$

In dealing with heterogeneous reactions that include a gas, it must be made clear how the expression for K has been formulated.

The partial dissolution of any slightly soluble salt in water is another example of a heterogeneous equilibrium. The participants in the equilibrium are the solid salt, the solvent water, and the ions in aqueous solution. Only the concentrations of the ions in solution need appear in the equilibrium constant expression, for example

$$Ag_2S(s) \xrightleftharpoons{H_2O} 2Ag^+ + S^{2-} \qquad K_c = [Ag^+]^2[S^{2-}]$$

EXAMPLE 19.5 Equilibrium Constant Expressions

Sulfur reacts with fluorine to form sulfur hexafluoride:

$$S(s) + 3F_2(g) \rightleftharpoons SF_6(g)$$

Write the equilibrium constant expressions for this reaction.

The concentration of solid sulfur need not appear in K_c and K_p for this reaction. The expressions are

$$K_c = \frac{[SF_6]}{[F_2]^3} \qquad K_p = \frac{P_{SF_6}}{P_{F_2}{}^3}$$

EXERCISE Hydrogen gas can be produced from the action of an acid on magnesium metal.

$$2H^+ + Mg(s) \rightleftharpoons Mg^{2+} + H_2(g)$$

Write the equilibrium constant expressions for this reaction.

Answers $K_c = [Mg^{2+}][H_2]/[H^+]^2$, $K_p = [Mg^{2+}]P_{H_2}/[H^+]^2$.

EXAMPLE 19.6 Equilibrium Concentrations

Anhydrous calcium chloride is often used as a desiccant—a substance that removes water vapor from an enclosed volume of air. In the presence of excess calcium chloride, the amount of water taken up is governed by $K_p = 1.28 \times 10^{85}$ for the following reaction, at room temperature.

$$CaCl_2(s) + 6H_2O(g) \rightleftharpoons CaCl_2 \cdot 6H_2O(s)$$

What is the equilibrium vapor pressure of water in a closed vessel that contains $CaCl_2(s)$?

The equilibrium expression is

$$K_p = \frac{1}{P_{H_2O}{}^6}$$

Rearranging gives

$$P_{H_2O} = \sqrt[6]{1/K_p} = \sqrt[6]{1/1.28 \times 10^{85}} = \mathbf{6.54 \times 10^{-15}\ bar}$$

The partial pressure of water in the enclosed volume will be reduced until it reaches 6.54×10^{-15} bar.

EXERCISE For the reaction

$$2Cs(s) + F_2(g) \longrightarrow 2CsF(s)$$

$K_p = 1.24 \times 10^{184}$ at 298 K. What is the partial pressure of fluorine in equilibrium with solid Cs and CsF?

Answer 8.06×10^{-185} bar. [This is a *very* favorable reaction.]

19.6 EQUILIBRIUM CONSTANTS AND REACTION EQUATIONS

In order to write the equilibrium constant expression for a given reaction, the balanced chemical equation for the reaction must be known. How the reaction equation is written determines the form of the expression for K. For example, two different equations for the combination of carbon monoxide and oxygen give two different K_c expressions:

$$2CO(g) + O_2(g) \rightleftharpoons 2CO_2(g) \qquad K_{12} = \frac{[CO_2]^2}{[CO]^2[O_2]} \tag{19.12}$$

$$CO(g) + \tfrac{1}{2}O_2(g) \rightleftharpoons CO_2(g) \qquad K_{13} = \frac{[CO_2]}{[CO][O_2]^{1/2}} \tag{19.13}$$

Equation (19.12) has been divided by 2 to give Equation (19.13). The value of the equilibrium constant for reaction (19.13), K_{13}, is the square root of the value of the equilibrium constant for reaction (19.12), K_{12}

$$K_{13} = \frac{[CO_2]}{[CO][O_2]^{1/2}} = \left(\frac{[CO_2]^2}{[CO]^2[O_2]}\right)^{1/2} \qquad \text{or} \qquad K_{13} = K_{12}^{\ 1/2}$$

In general, if a chemical equation has been divided by n, the new K value is the $\dfrac{1}{n\text{th}}$ root of the original K value.

If an equation is multiplied by n, its equilibrium constant becomes the original equilibrium constant raised to the nth power. In the following example, for instance, $n = 2$:

$$\tfrac{1}{2}N_2(g) + \tfrac{3}{2}H_2(g) \rightleftharpoons NH_3(g) \tag{19.14}$$

$$N_2(g) + 3H_2(g) \rightleftharpoons 2NH_3(g) \tag{19.15}$$

$$K_{15} = \frac{[NH_3]^2}{[N_2][H_2]^3} = \left(\frac{[NH_3]}{[N_2]^{1/2}[H_2]^{3/2}}\right)^2 \qquad \text{or} \qquad K_{15} = K_{14}^{\ 2}$$

If an equation is reversed, the new equilibrium constant is the reciprocal of the equilibrium constant for the original equation:

$$2NO_2(g) \rightleftharpoons N_2O_4(g) \tag{19.16}$$

$$N_2O_4(g) \rightleftharpoons 2NO_2(g) \tag{19.17}$$

$$K_{17} = \frac{[NO_2]^2}{[N_2O_4]} = \left(\frac{[N_2O_4]}{[NO_2]^2}\right)^{-1} \qquad \text{or} \qquad K_{17} = \frac{1}{K_{16}}$$

When the equations for two reactions are added together, the equilibrium constant for the total reaction is the product of the equilibrium constants of the original reactions. The addition of two equations, and the resulting overall equilibrium constant, are illustrated below for a two-step mechanism for the formation of sulfuric acid in the gas phase:

$$2SO_2 + O_2 \rightleftharpoons 2SO_3 \tag{19.18}$$

$$2SO_3 + 2H_2O \rightleftharpoons 2H_2SO_4 \tag{19.19}$$

$$2SO_2 + O_2 + 2H_2O \rightleftharpoons 2H_2SO_4 \tag{19.20}$$

$$K_{20} = K_{18}K_{19} = \left(\frac{\cancel{[SO_3]^2}}{[O_2][SO_2]^2}\right) \times \left(\frac{[H_2SO_4]^2}{[H_2O]^2\cancel{[SO_3]^2}}\right) = \frac{[H_2SO_4]^2}{[SO_2]^2[O_2][H_2O]^2}$$

This brings us to a very important point: The overall equilibrium constant is independent of the number of steps in the reaction mechanism. Whether reaction (19.20) proceeds by two steps or by 50 steps, its equilibrium expression remains the same. For this reason, equilibrium constant expressions can be written from the balanced overall equations. Rate equations, on the other hand, *do* depend upon the reaction pathway and, it bears saying once more, must be found by experiment.

EXAMPLE 19.7 Evaluation of *K*

A handbook gives the equilibrium constant for the formation of lithium iodide by combination of the elements in the gas phase at 3000 K as follows:

$$Li(g) + \tfrac{1}{2}I_2(g) \rightleftharpoons LiI(g) \qquad K_p = 644$$

(a) What is the equilibrium constant value for this reaction if it is represented by the equation given above multiplied by 2 in order to clear the fractional coefficient? (b) What is the equilibrium constant value for the reverse of this reaction as written above?

(a) To find the value of K_p for the equation multiplied by 2, the given value must be squared:

$$2Li(g) + I_2(g) \rightleftharpoons 2LiI(g) \qquad K_p = (644)^2 = \mathbf{4.15 \times 10^5}$$

(b) When an equation is reversed, the equilibrium constant value becomes the reciprocal of the original value:

$$LiI(g) \rightleftharpoons Li(g) + \tfrac{1}{2}I_2(g) \qquad K_p = \frac{1}{644} = \mathbf{1.55 \times 10^{-3}}$$

EXERCISE For the reactions

$$Cl_2(g) + F_2(g) \rightleftharpoons 2ClF(g)$$
$$Cl_2(g) + 3F_2(g) \rightleftharpoons 2ClF_3(g)$$

$K_p = 2.09 \times 10^{18}$ and 4.57×10^{41}, respectively, at 25 °C. Find the value of K_p for the reaction

$$ClF_3(g) \rightleftharpoons ClF(g) + F_2(g)$$

Answer 2.14×10^{-12}.

19.7 THE REACTION QUOTIENT

The equilibrium constant expression and the numerical value of K provide a means of predicting what will happen when substances with the potential for reacting with each other are mixed. To do this, nonequilibrium concentrations are used to calculate the **reaction quotient,** Q, a value found from an expression which takes the same form as the equilibrium constant but is used for a reaction not at equilibrium.

$$aA + bB + \cdots \longrightarrow rR + sS + \cdots \qquad \textbf{(19.21)}$$

$$Q = \frac{[R]^r[S]^s\cdots}{[A]^a[B]^b\cdots}$$

The value of Q indicates what changes will occur in reaching equilibrium:

1. Q less than K. To establish equilibrium, the concentrations of the reactants will decrease and those of the products will increase. The reaction will proceed toward equilibrium in the forward direction.
2. Q greater than K. To establish equilibrium the concentrations of the products will decrease and those of the reactants will increase. The reaction will approach equilibrium from the direction of the reverse reaction.

For example, at 425.4 °C, $K = 54.5$ for the combination of hydrogen and iodine to give hydrogen iodide (Equation 19.5). Suppose that 0.02 mol of H_2, 0.02 mol of I_2, and

0.04 mol of HI are introduced into a 1 L flask at this temperature. Would more HI form or would the reverse reaction occur to give more $H_2 + I_2$? The reaction quotient is

$$Q = \frac{[HI]^2}{[H_2][I_2]} = \frac{[0.04]^2}{[0.02][0.02]} = 4$$

This value of Q is smaller than K. Equilibrium will be established by the forward reaction proceeding, allowing [HI] to increase and $[H_2]$ and $[I_2]$ to decrease until $Q = K$.

EXAMPLE 19.8 Reaction Quotient

The concentration equilibrium constant for the reaction

$$ClF_3(g) \rightleftharpoons ClF(g) + F_2(g)$$

is 8.77×10^{-14} at 25 °C. Describe what happens when a solution is prepared so that it is 17.25 M in ClF_3, 1.3×10^{-6} M in ClF, and 4.62×10^{-7} in F_2.

The reaction quotient for the solution is

$$Q = \frac{[ClF][F_2]}{[ClF_3]} = \frac{(1.3 \times 10^{-6})(4.62 \times 10^{-7})}{(17.25)} = 3.5 \times 10^{-14}$$

Because $Q < K_c$, ClF_3 will decompose until equilibrium is reached.

EXERCISE For the reaction

$$BrF_3(g) \rightleftharpoons BrF(g) + F_2(g)$$

$K_p = 64$ at 2000 K. Calculate the value of the reaction quotient and describe what will happen if a mixture containing $P_{F_2} = 1.36$ bar, $P_{BrF} = 0.01$ bar, and $P_{BrF_3} = 0.52$ bar is prepared. *Answer* $Q = 0.03$, some of the BrF_3 will decompose.

SUMMARY OF SECTIONS 19.4–19.7

The concentrations of solids, pure liquids, and the solvents in dilute solutions need not appear in equilibrium constant expressions.

The form of the equilibrium constant expression is determined by how the chemical equation is written. The value of K and the equilibrium constant expression for chemical equations for the same reaction written in different ways are related mathematically as shown in Table 19.3.

The equilibrium constant in terms of partial pressures, K_p, may be used when all reactants and products represented in the K expression are present in the gas phase. The values of K_c and K_p are related by the ideal gas law (Equation 19.9).

The reaction quotient, Q, takes the same form as K but is written for reactions not at equilibrium. If Q is greater than K, the reverse reaction is favored; if Q is less than K, the forward reaction is favored. When $Q = K$ the reaction is at equilibrium.

Table 19.3
Variation of Equilibrium Constant with How the Reaction Equation Is Written
K_0 is the original equilibrium constant.

When the equation is	The new equilibrium constant is
Multiplied by n	K_0^n
Divided by n	$K_0^{1/n}$
Reversed	K_0^{-1}
Divided into 2 steps (a and b)	$K_0 = K_aK_b$

FACTORS THAT INFLUENCE EQUILIBRIA

19.8 LE CHATELIER'S PRINCIPLE

The general principle that underlies all changes in equilibria is Le Chatelier's principle (Section 17.2): If a system at equilibrium is subjected to a stress, the system will react in a way that tends to relieve the stress. In the chemical sense, the stresses that can be applied to a system at equilibrium are changes in temperature, concentration, and pressure.

Table 19.4
Comparison of Variations in Equilibrium Constants, Rate Constants, and Rates
The effects summarized here occur in most cases, although exceptions can be found.

When	Equilibrium constant (K)	Rate constant (k)	Rate
Temperature			
Increases	Changes	Increases	Increases
Decreases	Changes	Decreases	Decreases
Catalyst is added	Does not change	Increases	Increases
Concentration of reactants			
Increases	Does not change	Does not change	Increases
Decreases	Does not change	Does not change	Decreases

Catalysts can increase the speed with which equilibrium is reached, but they cause no change in the equilibrium constant or in the concentrations at equilibrium. The rates of both the forward and reverse reactions are increased equally by a catalyst (by lowering E_a; see Figure 18.13).

The effects of changes in reaction conditions on equilibrium constants, rate constants, and rates are summarized in Table 19.4. We suggest you consult this table as you read Sections 19.9–19.11.

White BiOCl precipitate

$BiCl_3(aq) + H_2O(l) \rightleftharpoons BiOCl(s) + 2HCl(aq)$

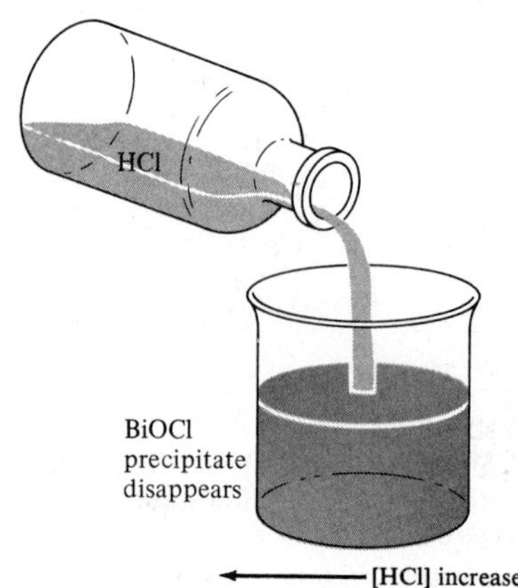

[HCl] increase
$BiCl_3(aq) + H_2O(l) \rightleftharpoons BiOCl(s) + 2HCl(aq)$

[HCl] decrease
$BiCl_3(aq) + H_2O(l) \rightleftharpoons BiOCl(s) + 2HCl(aq)$

Figure 19.3
Demonstration of Change in Equilibrium with Change in Concentration of Reactant or Product

19.9 CONCENTRATION

When the concentration of a reactant or product in a system in chemical equilibrium is changed, the system is put out of balance and the rates of the forward and reverse reactions become temporarily unequal. The system is stressed and must change to relieve the stress.

The reaction of bismuth(III) chloride with water to give bismuth(III) oxochloride visibly demonstrates the effect of changing concentration on chemical equilibrium. The experiment begins (Figure 19.3) with a mixture in which the white precipitate of BiOCl is in equilibrium with $BiCl_3$ in solution.

$$\underset{\textit{bismuth(III) chloride}}{BiCl_3(aq)} + H_2O(l) \rightleftharpoons \underset{\textit{bismuth oxochloride}}{BiOCl(s)} + 2HCl(aq)$$

As hydrochloric acid is added, the BiOCl disappears, showing that the reverse reaction is occurring at a greater rate than the forward reaction. The increase in concentration of HCl drives the reaction toward the reactants. More $BiCl_3$ and H_2O are formed until equilibrium is reestablished.

If the concentration of HCl is decreased (e.g., by adding base), the opposite change occurs. The BiOCl precipitate reappears as the forward reaction increases in rate until equilibrium is reached again.

$$\xleftarrow{\textit{[HCl] increase}}$$
$$BiCl_3(aq) + H_2O(l) \rightleftharpoons BiOCl(s) + 2HCl(aq)$$
$$\xrightarrow{\textit{[HCl] decrease}}$$

A *decrease* in a concentration causes a shift in the direction that increases that concentration. Thus the continuous removal of a product—a *continuous decrease* in concentration—allows a reaction to be driven to completion. The forward reaction continues to occur in order to relieve the stress of the decreasing concentration. (The continuous displacement of solubility equilibria is utilized in separating substances by chromatography; see Aside 19.1.)

EXAMPLE 19.9 Factors That Influence Equilibria: Concentration

A mixture of $K_2CrO_4(aq)$ and $HCl(aq)$ was allowed to come to equilibrium:

$$\underset{\textit{chromate ion yellow}}{2CrO_4^{2-}} + 2H^+ \rightleftharpoons \underset{\textit{dichromate ion orange}}{Cr_2O_7^{2-}} + H_2O(l)$$

0.04 mol of HI are introduced into a 1 L flask at this temperature. Would more HI form or would the reverse reaction occur to give more $H_2 + I_2$? The reaction quotient is

$$Q = \frac{[HI]^2}{[H_2][I_2]} = \frac{[0.04]^2}{[0.02][0.02]} = 4$$

This value of Q is smaller than K. Equilibrium will be established by the forward reaction proceeding, allowing [HI] to increase and $[H_2]$ and $[I_2]$ to decrease until $Q = K$.

EXAMPLE 19.8 Reaction Quotient

The concentration equilibrium constant for the reaction

$$ClF_3(g) \rightleftharpoons ClF(g) + F_2(g)$$

is 8.77×10^{-14} at 25 °C. Describe what happens when a solution is prepared so that it is 17.25 M in ClF_3, 1.3×10^{-6} M in ClF, and 4.62×10^{-7} in F_2.

The reaction quotient for the solution is

$$Q = \frac{[ClF][F_2]}{[ClF_3]} = \frac{(1.3 \times 10^{-6})(4.62 \times 10^{-7})}{(17.25)} = 3.5 \times 10^{-14}$$

Because $Q < K_c$, ClF_3 will decompose until equilibrium is reached.

EXERCISE For the reaction

$$BrF_3(g) \rightleftharpoons BrF(g) + F_2(g)$$

$K_p = 64$ at 2000 K. Calculate the value of the reaction quotient and describe what will happen if a mixture containing $P_{F_2} = 1.36$ bar, $P_{BrF} = 0.01$ bar, and $P_{BrF_3} = 0.52$ bar is prepared. *Answer* $Q = 0.03$, some of the BrF_3 will decompose.

SUMMARY OF SECTIONS 19.4–19.7

The concentrations of solids, pure liquids, and the solvents in dilute solutions need not appear in equilibrium constant expressions.

The form of the equilibrium constant expression is determined by how the chemical equation is written. The value of K and the equilibrium constant expression for chemical equations for the same reaction written in different ways are related mathematically as shown in Table 19.3.

The equilibrium constant in terms of partial pressures, K_p, may be used when all reactants and products represented in the K expression are present in the gas phase. The values of K_c and K_p are related by the ideal gas law (Equation 19.9).

The reaction quotient, Q, takes the same form as K but is written for reactions not at equilibrium. If Q is greater than K, the reverse reaction is favored; if Q is less than K, the forward reaction is favored. When $Q = K$ the reaction is at equilibrium.

Table 19.3
Variation of Equilibrium Constant with How the Reaction Equation Is Written
K_0 is the original equilibrium constant.

When the equation is	The new equilibrium constant is
Multiplied by n	K_0^n
Divided by n	$K_0^{1/n}$
Reversed	K_0^{-1}
Divided into 2 steps (a and b)	$K_0 = K_a K_b$

FACTORS THAT INFLUENCE EQUILIBRIA

19.8 LE CHATELIER'S PRINCIPLE

The general principle that underlies all changes in equilibria is Le Chatelier's principle (Section 17.2): If a system at equilibrium is subjected to a stress, the system will react in a way that tends to relieve the stress. In the chemical sense, the stresses that can be applied to a system at equilibrium are changes in temperature, concentration, and pressure.

Table 19.4
Comparison of Variations in Equilibrium Constants, Rate Constants, and Rates
The effects summarized here occur in most cases, although exceptions can be found.

When	Equilibrium constant (K)	Rate constant (k)	Rate
Temperature			
Increases	Changes	Increases	Increases
Decreases	Changes	Decreases	Decreases
Catalyst is added	Does not change	Increases	Increases
Concentration of reactants			
Increases	Does not change	Does not change	Increases
Decreases	Does not change	Does not change	Decreases

Catalysts can increase the speed with which equilibrium is reached, but they cause no change in the equilibrium constant or in the concentrations at equilibrium. The rates of both the forward and reverse reactions are increased equally by a catalyst (by lowering E_a; see Figure 18.13).

The effects of changes in reaction conditions on equilibrium constants, rate constants, and rates are summarized in Table 19.4. We suggest you consult this table as you read Sections 19.9–19.11.

19.9 CONCENTRATION

When the concentration of a reactant or product in a system in chemical equilibrium is changed, the system is put out of balance and the rates of the forward and reverse reactions become temporarily unequal. The system is stressed and must change to relieve the stress.

The reaction of bismuth(III) chloride with water to give bismuth(III) oxochloride visibly demonstrates the effect of changing concentration on chemical equilibrium. The experiment begins (Figure 19.3) with a mixture in which the white precipitate of BiOCl is in equilibrium with $BiCl_3$ in solution.

$$\underset{\text{bismuth(III) chloride}}{BiCl_3(aq)} + H_2O(l) \rightleftharpoons \underset{\text{bismuth oxochloride}}{BiOCl(s)} + 2HCl(aq)$$

As hydrochloric acid is added, the BiOCl disappears, showing that the reverse reaction is occurring at a greater rate than the forward reaction. The increase in concentration of HCl drives the reaction toward the reactants. More $BiCl_3$ and H_2O are formed until equilibrium is reestablished.

If the concentration of HCl is decreased (e.g., by adding base), the opposite change occurs. The BiOCl precipitate reappears as the forward reaction increases in rate until equilibrium is reached again.

$$\xleftarrow{\text{[HCl] increase}}$$
$$BiCl_3(aq) + H_2O(l) \rightleftharpoons BiOCl(s) + 2HCl(aq)$$
$$\xrightarrow{\text{[HCl] decrease}}$$

A *decrease* in a concentration causes a shift in the direction that increases that concentration. Thus the continuous removal of a product—a *continuous decrease* in concentration—allows a reaction to be driven to completion. The forward reaction continues to occur in order to relieve the stress of the decreasing concentration. (The continuous displacement of solubility equilibria is utilized in separating substances by chromatography; see Aside 19.1.)

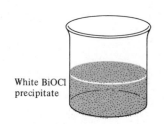

White BiOCl precipitate

$BiCl_3(aq) + H_2O(l) \rightleftharpoons BiOCl(s) + 2HCl(aq)$

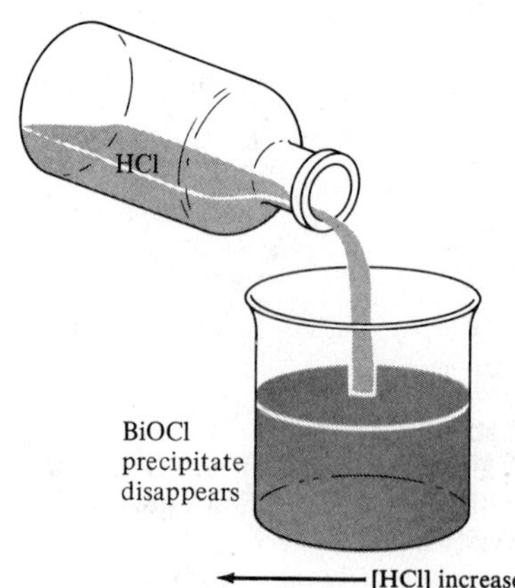

← [HCl] increase
$BiCl_3(aq) + H_2O(l) \rightleftharpoons BiOCl(s) + 2HCl(aq)$

[HCl] decrease →
$BiCl_3(aq) + H_2O(l) \rightleftharpoons BiOCl(s) + 2HCl(aq)$

Figure 19.3
Demonstration of Change in Equilibrium with Change in Concentration of Reactant or Product

EXAMPLE 19.9 Factors That Influence Equilibria: Concentration

A mixture of $K_2CrO_4(aq)$ and $HCl(aq)$ was allowed to come to equilibrium:

$$\underset{\text{chromate ion, yellow}}{2CrO_4^{2-}} + 2H^+ \rightleftharpoons \underset{\text{dichromate ion, orange}}{Cr_2O_7^{2-}} + H_2O(l)$$

Describe the changes that will occur in the equilibrium system as (a) additional acid is added; (b) additional $K_2CrO_4(s)$ is added; (c) $K_2Cr_2O_7(s)$ is added; (d) Zn^{2+} is added—note that $ZnCrO_4$ is highly insoluble and $ZnCr_2O_7$ is very soluble; (e) NaOH is added.

(a) Some of the additional H^+ will react with some of the remaining chromate ion to produce more dichromate ion and water; **$[CrO_4^{2-}]$ will decrease and $[Cr_2O_7^{2-}]$ will increase.**
(b) Some of the additional chromate ion will react with some of the H^+ to **produce more dichromate ion, and water and the concentration of H^+ will decrease.**
(c) Some of the **additional dichromate ion will react to form chromate ion, and the concentration of CrO_4^{2-} will increase.**
(d) Addition of Zn^{2+} will remove some of the chromate ion as $ZnCrO_4$, **lowering the concentration of chromate ion. Thus some of the dichromate ion will react to form additional chromate ion and H^+.**
(e) Addition of OH^- will remove some of the H^+ as a result of an acid–base reaction. The **decrease in H^+ concentration will cause the reaction of additional $Cr_2O_7^{2-}$ with H_2O; $[Cr_2O_7^{2-}]$ will decrease and $[CrO_4^{2-}]$ will increase.**

EXERCISE Describe the changes that will occur in the equilibrium system

$$2BrF_5(g) \rightleftharpoons Br_2(g) + 5F_2(g)$$

as (a) P_{BrF_5}(or$[BrF_5]$) is increased, (b) P_{F_2} (or$[F_2]$) is decreased, (c) P_{Br_2} and P_{F_2} (or$[Br_2]$ and $[F_2]$) are both increased, (d) a catalyst is added.

Answers (a) Additional Br_2 and F_2 are formed; (b) additional Br_2 is formed by the further decomposition of BrF_5; (c) additional BrF_5 is formed; (d) no changes are observed.

Aside 19.1 TOOLS OF CHEMISTRY: CHROMATOGRAPHY

Biochemical fluids, inorganic salts, organic compounds, polymers—mixtures of all kinds of substances can be separated by chromatography. It is a technique of special value because it can be used to analyze very complex mixtures and it works effectively with very small amounts of material. In addition, the separated substances can be isolated, allowing chromatography to be used as a preparative method.

The basis of chromatography is the distribution of a solute between two immiscible solvents, as illustrated by the simple separations performed using separatory funnels. Suppose we have an aqueous solution of A, a fairly nonpolar substance. Most likely A is also soluble in a nonpolar solvent such as hexane. When the aqueous solution of A and hexane are shaken together in a separatory funnel (Figure 19.4), an equilibrium is established, as some A is transferred from the aqueous solution to the hexane.

$$A(aq) \rightleftharpoons A(hexane)$$

The equilibrium constant for the distribution of a solute between two immiscible solvents is called the **distribution coefficient.** For the equilibrium above

$$K = \frac{[A]_{hexane}}{[A]_{aq}}$$

The value of the distribution coefficient expresses the relative affinity of the solute for hexane and water. If K is large, most of the A dissolved in the water can be transferred to hexane by extraction in a separatory funnel as shown in Figure 19.4. If some other solute, B, that is less soluble in hexane than A were also present in the aqueous solution, A and B could be separated from each other by extraction.

In chromatography a *series* of extractions like the one just described is set up and equilibrium is constantly displaced by fresh solvent. **Chromatography** is the distribution of a solute between a stationary phase and a mobile phase. The stationary phase may be a solid or a liquid that is supported as a thin film on the surface of an inert solid—the *support*. The mobile phase flowing over the surface of the stationary phase may be a gas or a liquid.

Substances are separated in chromatography according to their relative affinities for the stationary and mobile phases. Suppose a mixture of A and B is adsorbed on a solid support and a solvent flows over the solid. The solutes will continuously move into the solvent because fresh solvent is continuously available and equilibrium can never be reached. The solute which has the greater affinity for the solvent will dissolve more quickly in the fresh solvent as it flows by than will the other solute. This more soluble solute, say A, will move down the column faster than B and thus be separated from it.

Chromatographic methods in which the stationary phase is a solid are classified as adsorption chromatography. A substance leaves the mobile phase to become adsorbed on the surface of the solid phase, called the adsorbent (Figure 19.5).

The equilibria in adsorption chromatography are governed by the weak intermolecular forces of physical adsorption such as London forces and hydrogen bonding. The adsorbent must have a large surface area. Alumina (Al_2O_3) and silica (SiO_2) in various forms are common adsorbents.

The many different types of separation methods that are classified as chromatography are summarized in Table 19.5. For example, in column chromatography, which includes most gas chromatography, the stationary phase is held in a tube. In paper chromatography, a stationary liquid phase is supported in the pores of a piece of paper. The end product of paper chromatography is a dried piece of paper with spots of the components of the mixture spread out across the paper. In thin-layer chromatography, the adsorbent is spread in a thin layer on an inert support such as a piece of glass.

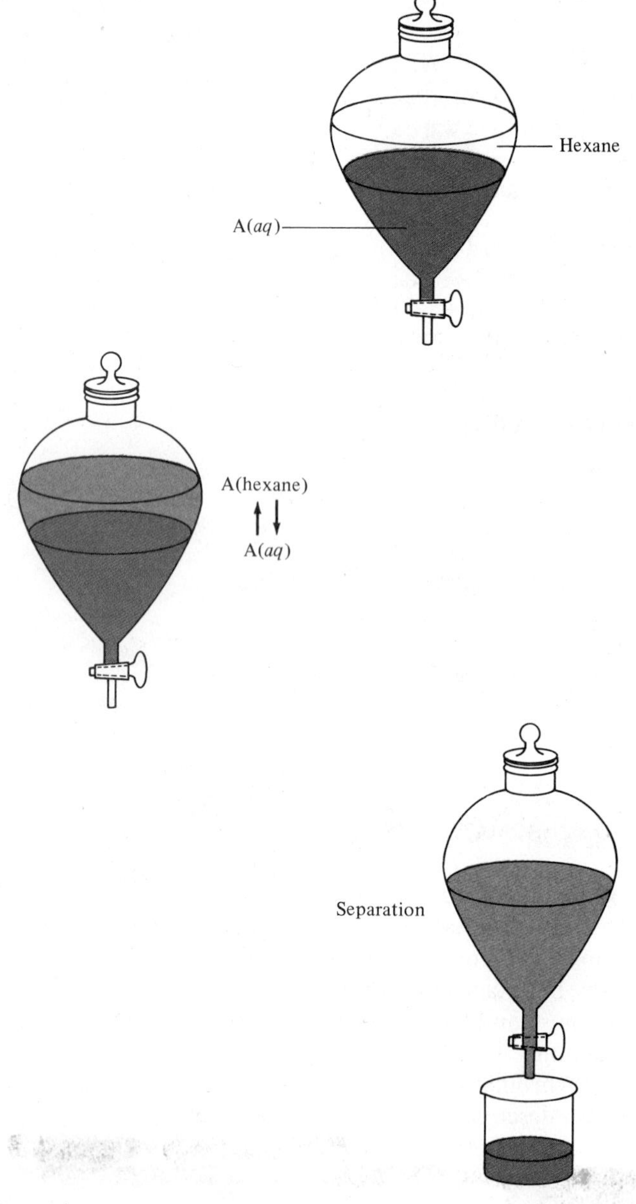

Figure 19.4
Extraction in a Separatory Funnel
In this method of separation, a solute is distributed between two immiscible solvents.

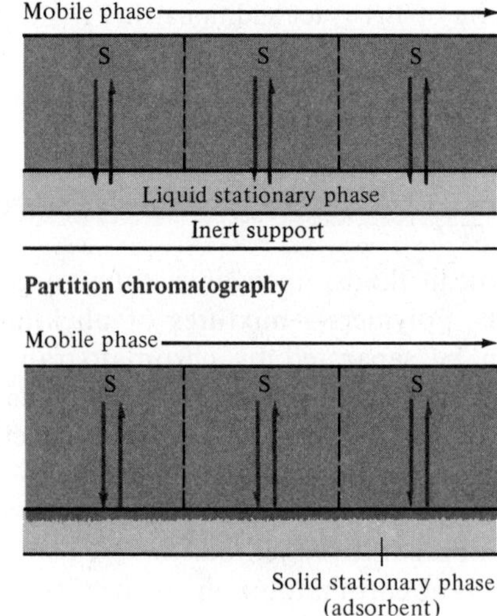

Figure 19.5
Equilibria in Partition and Adsorption Chromatography
S represents the solute.

Table 19.5
Chromatographic Methods

Method	Description
Adsorption column chromatography	Distribution of solute between a solid and liquid phase on a column
Partition column chromatography	Distribution of a solute between two liquids on a column
Thin-layer chromatography	Adsorption or partition on an open thin sheet
Paper chromatography	Partition on a paper sheet
High-pressure liquid chromatography	Column liquid chromatography under high inlet pressure
Ion-exchange chromatography	Exchange of ions
Gas chromatography	Distribution of a gaseous solute between a gas and liquid or solid phase
Zone electrophoresis	Separation on a sheet in the presence of an electrical field
Molecular sieves	Separation is based on size of solute and absorption
Gel permeation (filtration)	Separation is based on size of solute and absorption

19.10 PRESSURE

Changes in pressure displace equilibria in gas-phase reactions in which the number of moles of reactant is different from the number of moles of product. An increase in pressure shifts the reaction equilibrium in the direction that produces the smaller number of molecules in the gas phase. For example, in the phosphorus(III) chloride–chlorine reaction

$$\underset{\textit{2 moles of gas}}{PCl_3(g) + Cl_2(g)} \overset{\textit{pressure}\ \longrightarrow}{\rightleftharpoons} \underset{\textit{1 mole of gas}}{PCl_5(g)}$$

the stress of greater pressure can be relieved by an increase in the rate of the forward reaction, which decreases the number of molecules in the gas phase by forming one mole of gas from two. On the other hand, for the decomposition of iodopropane to give propene plus hydrogen iodide we have the opposite situation.

$$\underset{\textit{1 mole of gas}}{CH_3CH_2CH_2I(g)} \overset{\longleftarrow\ \textit{pressure}}{\rightleftharpoons} \underset{\textit{2 moles of gas}}{CH_3CH{=}CH_2(g) + HI(g)}$$

The reverse reaction is favored by increasing pressure. A change in pressure is essentially a change in concentration and it does not change the value of K. The effect of pressure changes on reactions involving only liquids and solids is very small.

The displacement of an equilibrium in the gas phase can be demonstrated visibly with the reaction of the oxides of nitrogen discussed in Section 19.2. Nitrogen dioxide (NO_2) is a brown gas that is always in equilibrium with dinitrogen tetroxide (N_2O_4) a colorless gas.

$$\underset{\textit{brown}}{2NO_2(g)} \rightleftharpoons \underset{\textit{colorless}}{N_2O_4(g)}$$

Increasing pressure favors the formation of N_2O_4, and the change in the concentration at equilibrium is observable by the decrease in intensity of the brown color of NO_2.

Note that only an increase or decrease in total pressure caused by a change in volume can displace equilibria as described above. If the pressure is increased by the addition of an ideal gas that does not participate in the chemical reaction, the partial pressures of the reactant gases remain unchanged. Therefore, the dynamic chemical equilibrium is undisturbed.

19.11 TEMPERATURE

Every equilibrium reaction involves one exothermic reaction and one endothermic reaction (see Figure 18.4). When the temperature is increased, the equilibrium shifts so that the endothermic reaction—the one that requires heat—is favored. It can be the forward or the reverse reaction. Therefore, unlike changes in concentration and pressure, changes in temperature cause changes in the *value* of K.

If we think of heat as a reactant or a product, it is easy to visualize, based on Le Chatelier's principle, which direction is favored by a temperature change. For example, the reaction between hydrogen and iodine to give hydrogen iodide is an overall exothermic reaction; that is, heat is a product ($\Delta H° = -52.7$ kJ). This means that the reverse reaction is endothermic. In Section 19.2 we saw that K varied with temperature:

$$H_2(g) + I_2(g) \overset{\longleftarrow\ \textit{heat}}{\rightleftharpoons} 2HI + 52.7\text{ kJ} \qquad K = \frac{[HI]^2}{[H_2][I_2]}$$

$$K = 54 \text{ at } 425.4\ °C$$
$$K = 45.6 \text{ at } 490.65\ °C$$

An increase in temperature causes the decomposition of additional hydrogen iodide. The concentration of hydrogen iodide becomes smaller and the concentrations of hydrogen and iodine become larger. The result is a decrease in the value of K.

In an overall endothermic reaction, the reverse reaction is exothermic. Here we may think of heat as a reactant. The endothermic reaction of oxygen to form ozone ($\Delta H^\circ = 285.3$ kJ) is at equilibrium at room temperature with very little ozone present. Increasing the temperature increases ozone formation, and the equilibrium constant becomes larger.

$$\text{285.3 kJ} + 3O_2(g) \xrightleftharpoons[]{\textit{heat} \longrightarrow} 2O_3(g) \qquad K = 6.2 \times 10^{-58} \text{ at } 25\ ^\circ\text{C}$$
$$K = 2.6 \times 10^{-56} \text{ at } 35\ ^\circ\text{C}$$

In general, K for exothermic reactions increases with lower temperatures and decreases with higher temperatures. The opposite is true for endothermic reactions, for which K decreases with lower temperatures and increases with higher temperatures.

EXAMPLE 19.10 Factors That Influence Equilibria: Temperature and Pressure

For the reaction

$$2Hg(l) + O_2(g) \rightleftharpoons 2HgO(s)$$

$\Delta H^\circ = -180.7$ kJ over the temperature range 298 K to 500 K and $K_p = 3.2 \times 10^{20}$ at 298 K. (a) Is the value of K_p at 500 K greater or less than this value? (b) What effect will a decrease in pressure have on this reaction?

(a) For this exothermic reaction, we can write

$$2Hg(l) + O_2(g) \rightleftharpoons 2HgO(s) + 180.7 \text{ kJ}$$

An increase in temperature favors the endothermic reaction—the reverse reaction in this case—and the equilibrium will be shifted toward the reactants side as the temperature is increased. **The value of K_p at 500 K should be less than 3.2×10^{20}.** [The value at 500 K is 5.1×10^7.]

(b) A decrease in pressure favors an increase in the amount of gas present. In this case **the reverse reaction would be favored.**

EXERCISE For the reaction

$$2BrF_5(g) \rightleftharpoons Br_2(g) + 5F_2(g)$$

$\Delta H^\circ = 858$ kJ over the temperature range 1000 K to 1500 K and $K_p = 7.4 \times 10^{-16}$ at 1000 K. (a) Is the value of K_p at 1500 K greater or less than this value? (b) What would be the effect of an increase in the total pressure of the system?

Answers (a) The value would be greater [$K_p = 0.60$]. (b) Additional BrF_5 would form.

SUMMARY OF SECTIONS 19.8–19.11

Le Chatelier's principle predicts the response of a system at equilibrium to a stress that upsets the equilibrium. An increase in the concentration of a specific substance causes a change that decreases the concentration of that substance (Table 19.6). A decrease in the

Table 19.6
Effect of Changing Conditions on a Chemical Reaction at Equilibrium
Note that the *value* of K changes only with changes in temperature.

Change	Effect
Concentration	
Increase reactant concentration or decrease product concentration	Shift toward product formation
Decrease reactant concentration or increase product concentration	Shift toward reactant formation
Temperature	
Increase	**Exothermic reaction**—shift toward reactant formation. K decreases. **Endothermic reaction**—shift toward product formation. K increases.
Decrease	**Exothermic reaction**—shift toward product formation. K increases. **Endothermic reaction**—shift toward reactant formation. K decreases.
Pressure (for gas-phase reactions with unequal numbers of product and reactant molecules)	
Increase	Shift toward decrease in number of molecules
Decrease	Shift toward increase in number of molecules

concentration favors a change that increases that concentration. Therefore, continuous removal of a product drives a reaction toward completion.

An increase in temperature favors whichever reaction requires heat (the endothermic reaction, which may be either the forward or the reverse reaction) and a decrease in temperature favors the reaction that liberates heat (the exothermic reaction). The value of the equilibrium constant varies with changes in temperature, but not with pressure. Changes in pressure cause significant changes only in equilibria in which the number of moles of gaseous products and reactants differ. An increase in pressure shifts an equilibrium in the direction that produces the smaller number of molecules in the gas phase.

Aside 19.2 THE HABER PROCESS FOR THE MANUFACTURE OF AMMONIA

The Haber process for the manufacture of ammonia is the classic example of the role of kinetics and equilibrium in an industrial process. The reaction of nitrogen and hydrogen to give ammonia is exothermic, occurs in the gas phase, and involves a decrease in pressure at a given temperature.

$$\underset{4\ moles}{N_2(g) + 3H_2(g)} \rightleftharpoons \underset{2\ moles}{2NH_3(g)}$$

$$\Delta H_{25^\circ C} = -92.22\ \text{kJ} \qquad K_{p,25^\circ C} = 5.5 \times 10^5$$
$$\Delta H_{400^\circ C} = -108\ \text{kJ} \qquad K_{p,400^\circ C} = 1.8 \times 10^{-4}$$

Le Chatelier's principle indicates that higher pressure will favor the formation of ammonia. Since the overall equilibrium is exothermic, higher temperatures will favor the decomposition of ammonia, an undesirable result, and lower temperatures will favor ammonia production.

However, a low temperature and a high pressure are not the easy answer to industrial production of ammonia by this reaction. The rate of the reaction at room temperature is very, very small. In commercial practice, both yield and rate must be considered, for a process is practical only if the plant produces a reasonable amount of the product per day.

Successful ammonia manufacture relies on a suitable catalyst, which increases the rate of attainment of equilibrium and permits the use of moderate temperatures. In modern industry the compromise among temperature, pressure, and yield of ammonia is struck by carrying out the reaction at about 250 atm (a pressure obtained economically with efficient centrifugal compressors) and 400 °C, the lowest temperature at which the catalyst is sufficiently effective. Under these conditions about 20%

ammonia is present in the gases that come off the catalytic reactor on each pass. Unconverted reactants are recycled, as shown in Figure 19.6, a simplified schematic diagram of NH_3 production from N_2 and H_2.

The catalyst is iron oxide, "doubly promoted" (that is, with the activity increased by two additives) by the addition of about 0.4% potassium oxide and 0.8% aluminum oxide. As is so often the case, the mechanism of the catalytic reaction is not known exactly. However, ammonia is formed at the catalyst surface, for the rate of ammonia formation is approximately proportional to the rate of nitrogen adsorption by the catalyst.

Fritz Haber, a German chemist, developed the industrial ammonia process named for him just before World War I. Had he not done so the war might have ended much sooner. Germany's trade was cut off by a blockade, and sodium nitrate from Chile was not available. Traditionally, nitric acid for explosives had been made from sodium nitrate, but ammonia from the Haber process provided another route to nitric acid. The success of the Haber process allowed the German chemical industry to produce the explosives and fertilizers needed for the war effort despite the blockade. (The importance of the Haber process in fertilizer manufacture is discussed in Aside 28.3.)

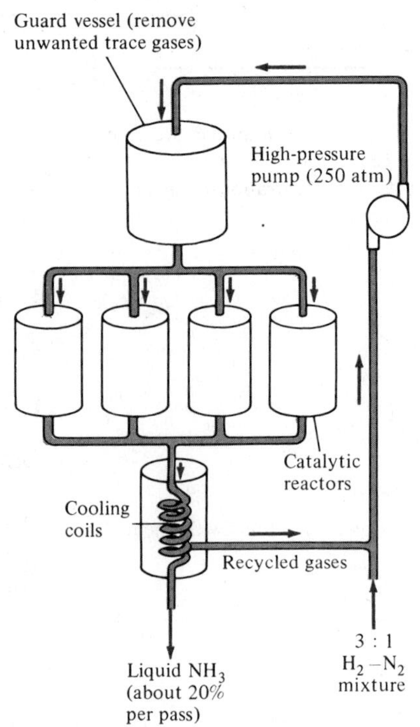

Figure 19.6
The Haber Process for NH_3 Manufacture

EQUILIBRIUM PROBLEMS

19.12 HOW TO SOLVE EQUILIBRIUM PROBLEMS

a. General Method The relationship between K and the equilibrium concentrations allows answers to be found for many questions about equilibrium reactions, such as, What are the concentrations of all species present at equilibrium? Given one concentration at equilibrium, what are the others? How must one concentration be changed to obtain a desired change in another concentration? The unknowns in equilibrium problems are most often the concentrations of one or more reactants or products. Whatever the unknown quantity in an equilibrium problem, it is most successfully obtained by careful and accurate bookkeeping. An excellent way to organize the solution is outlined below and illustrated for the following typical problem:

For the reaction $A \rightleftharpoons B + C$, what is the concentration of C at equilibrium when at the beginning of the reaction only 5.0 mol/L of A was present? Assume that $K = 3.0 \times 10^{-6}$.

Write the balanced chemical equation.

$$A \rightleftharpoons B + C$$

Write the corresponding equilibrium expression.

$$K = \frac{[B][C]}{[A]}$$

Identify what is unknown. Often this will be a single concentration, or the equal concentrations of several substances, or several concentrations that are related to each other algebraically. Select a single quantity for which to solve the problem. This might be an equilibrium concentration or a change in concentration. (Make a note of the unknown—you may forget what you solved for.) It is customary to let x equal this unknown. For this problem, let x equal the equilibrium concentrations of B and C; $x = [B] = [C]$.

Make a table. The table should have four lines: (1) the chemical equation; (2) initial concentrations, written under the reactants and products; (3) changes in concentration; and (4) equilibrium concentrations—the sum of lines (2) and (3). In line 3, use your knowledge of stoichiometry. In this example, to reach equilibrium [A] must decrease while [B] and [C] increase. Therefore, the change in [A] is $-x$ and the changes in [B] and [C] are $+x$

(1)	A	$\rightleftharpoons$	B	+	C
(2) Initial	5.0		0		0
(3) Change	$-x$		$+x$		$+x$
(4) Equilibrium	$5.0 - x$		x		x

Substitute the equilibrium concentrations from the table (line 4) into the equilibrium expression.

$$K = \frac{(x)(x)}{5.0 - x} = 3.0 \times 10^{-6} \tag{19.22}$$

Solve for the quantity represented by x. In some cases, solving for x is simple; in others, simplifying assumptions are made; in still others, simplifying assumptions are not possible and exact solutions are necessary. The problem here is solved by assuming that $5.0 - x$ is approximately equal to 5.0

$$K = \frac{x^2}{5.0 - x} \approx \frac{x^2}{5.0} = 3.0 \times 10^{-6}$$

$$x^2 = (5.0)(3.0 \times 10^{-6}) = 15 \times 10^{-6}$$
$$x = 3.9 \times 10^{-3} \text{ mol/L}$$

How to choose the method of solving for x is discussed further just below.

If an approximation was made, check it for validity. That $5.0 - x \approx 5.0$ is shown by substituting back the value of x:

$$5.0 - (3.9 \times 10^{-3}) = [4.9961] = 5.0$$

To the correct number of significant digits, the answer is 5.0. The assumption that the value of x could be neglected was valid.

Answer the questions in the problem. At equilibrium, $[C] = 3.9 \times 10^{-3}$ mol/L.

b. Solving for *x* Solving equilibrium expressions for unknowns requires the use of algebra. In actual laboratory and industrial situations, the algebra can become extremely complex. In this book we restrict ourselves to problems that can be solved by simple algebra and approximations, or to problems involving quadratic equations that can be solved exactly.

Let's look further at the expression that was solved for x in the problem just above

$$\frac{x^2}{5.0 - x} = 3.0 \times 10^{-6}$$

Because it includes both x and x^2, the problem can be solved only by simplifying the expression or by using the quadratic formula, which would require rearranging the equation into the quadratic form

$$ax^2 + bx + c = 0 \tag{19.23}$$

and solving for x with the quadratic formula

$$x = \frac{-b \pm \sqrt{b^2 - 4ac}}{2a} \tag{19.24}$$

The usual simplification is to neglect x in addition or subtraction (e.g., in $5.0 + x$ or $5.0 - x$) when the value of x will be small relative to the known number. If the value of K is significantly smaller than the known concentration values, it is usually safe to conclude that x will be small enough to neglect. This choice was made above based on the K value of 3.0×10^{-6} and the known concentration of 5.0 mol/L, and it was easily shown to be the correct choice.

If the validity of a simplification is in doubt, it can be checked by calculating the percent error. If x is neglected in $(A + x)$ or $(A - x)$

$$\%\text{ error} = \left(\frac{\text{value of } x}{\text{value of A}}\right) \times 100 \tag{19.25}$$

In this case

$$\%\text{ error} = \left(\frac{3.9 \times 10^{-3}}{5.0}\right) \times (100) = 0.078\%$$

Up to a 5% error is usually considered acceptable.

Suppose that instead in the same problem $K = 0.30$, a value closer to the initial concentration of 5.0 mol/L. If we assume that $(5.0 - x) \approx 5.0$

$$K = \frac{x^2}{5.0 - x} \approx \frac{x^2}{5.0} = 0.30 \tag{19.26}$$

$$x^2 = (5.0)(0.30) = 1.5$$

$$x = \sqrt{1.5} = 1.2$$

and then check the percent error by using Equation (19.25)

$$\%\text{ error} = \left(\frac{1.2}{5.0}\right) \times (100) = 24\%$$

we find that the assumption was not valid.

Using the quadratic equation, instead, requires first rearranging Equation (19.26) into the form of Equation (19.23)

$$\frac{x^2}{5.0 - x} = 0.30$$

$$x^2 = (0.30)(5.0 - x) = 1.5 - 0.30x$$

$$\overset{a}{x^2} + \overset{b}{0.30}x - \overset{c}{1.5} = 0$$

where $a = 1$, $b = 0.30$, and $c = -1.5$ (see Equation 19.23). Then, using the quadratic formula (Equation 19.24)

$$x = \frac{-0.30 \pm \sqrt{(0.30)^2 - (4)(1)(-1.5)}}{2(1)}$$

$$= \frac{-0.30 \pm 2.5}{2}$$

$$x = 1.1 \text{ or } -1.4$$

The negative value of x is discarded because a negative concentration has no physical meaning. The equilibrium concentration of C, found exactly with no approximation, is 1.1 mol/L.

(If K is very large, it is best to work backwards, assuming that a reaction goes to completion and using x for the small amount that does not react. For example, if $K = 3 \times 10^5$ for the $A \rightleftharpoons B + C$ reaction described above, assume that reaction of 5.0 mol of A gives equilibrium concentrations of x for A, $5.0 - x$ for B, and $5.0 - x$ for C. In this case, $x = 8 \times 10^{-5}$ mol.)

19.13 SOME SIMPLE EQUILIBRIUM PROBLEMS

Examples 19.11 through 19.14 represent some simple types of equilibrium problems. Example 19.11 illustrates the application of the problem-solving method of Section 2.11 to an equilibrium problem. The equilibrium constant expression provides the connection between the known and the unknown information for almost all equilibrium problems.

EXAMPLE 19.11 Equilibrium Problems; Problem Solving

For the reaction

$$H_2(g) + I_2(g) \rightleftharpoons 2HI(g)$$

$K_c = 29.1$ at 1000 K. What is the concentration of $I_2(g)$ under equilibrium conditions if the system originally contained only $[HI] = 10.0$ mol/L?

1. Study the problem and be sure you understand it.
   (a) What is unknown?
   The concentration of $I_2(g)$ at equilibrium.
   (b) What is known?
   Four things are known:
   (1) The chemical equation

$$H_2(g) + I_2(g) \rightleftharpoons 2HI(g)$$

   and from the equation the mole ratios of reactants and products.
   (2) The value of the equilibrium constant, $K = 29.1$.
   (3) The initial concentration of HI, 10.0 mol/L.
   (4) The initial concentrations of H_2 and I_2, which are zero.
2. Decide how to solve the problem.
   (a) What is the connection between the knowns and the unknown?
   The equilibrium expression, which is based on the known chemical equation.

$$K_c = \frac{[HI]^2}{[H_2][I_2]}$$

   (b) What is necessary to make the connection?
   It is necessary to substitute the known concentration and the unknown concentrations into the equilibrium constant expression. To do this, the information given by the chemical equation and the known concentration must be used to assign values to all of the terms in the equilibrium constant expression. In this problem, the equilibrium concentrations of H_2 and I_2 are both unknown. From the stoichiometry of the reaction we can see that H_2 and I_2 will be present in equal concentrations at equilibrium, and so we set $x = [H_2] = [I_2]$. The equilibrium concentration of HI will be the initial concentration *minus* the change due to conversion to H_2 and I_2. Because two moles of HI are required for every mole of H_2 and I_2 that form, the change in the concentration of HI will be $-2x$. The table format is best for using all of this information to determine what substitutions to make in the equilibrium constant expression

	H_2	$+ I_2$	$\rightleftharpoons$ 2HI
Initial	0	0	10.0
Change	$+x$	$+x$	$-2x$
Equilibrium	x	x	$10.0 - 2x$

the unknowns (x, x); *the known* (10.0); *found from stoichiometry* ($-2x$); *found from the known* ($10.0 - 2x$)

3. Set up the problem and solve it.

$$K_c = \frac{(10.0 - 2x)^2}{(x)(x)} = 29.1$$

At this point, inspect the expression to determine how it can be solved. Because all of the terms containing x are squared, the problem can be solved by taking the square root of each side.

$$\frac{(10.0 - 2x)^2}{x^2} = 29.1$$

$$\frac{10.0 - 2x}{x} = \sqrt{29.1} = 5.39$$

$$10.0 - 2x = 5.39x$$
$$-2x - 5.39x = -10.0$$
$$7.39x = 10.0$$
$$x = \mathbf{1.35\ mol/L}$$

At equilibrium 1.35 mol/L of I_2 (and also of H_2) is present.

4. Check the result. The result can be checked by substituting the value found for x back into the equilibrium constant expression that was solved in Step 3 to show that the result is equal to 29.1

$$\frac{[10.0 - 2(1.35)]^2}{(1.35)^2} = 29.1$$

No approximations were made, so no check of an introduced error is needed. Since K_c is greater than 1, we can expect a relatively large amount of HI to remain at equilibrium, and this is the case; $[HI] = 10.0 - 2x = 10.0 - 2.70 = 7.3$ mol/L. The answer is reasonable.

EXAMPLE 19.12 Equilibrium Concentrations

What is the concentration of CO in equilibrium at 25 °C in a sample of gas originally containing 1.00 mol/L of CO_2? For the dissociation of CO_2 at 25 °C (see Table 19.2), $K_c = 2.96 \times 10^{-92}$.

The reaction and K expression are

$$2CO_2(g) \rightleftharpoons 2CO(g) + O_2(g) \qquad K_c = \frac{[CO]^2[O_2]}{[CO_2]^2}$$

The equilibrium concentrations of both CO and O_2 are unknown; $[O_2]$ at equilibrium is a good choice for x.

	$2CO_2(g)$ $\rightleftharpoons$	$2CO(g)$ +	$O_2(g)$
Initial	1.00	0	0
Change	$-2x$	$+2x$	$+x$
Equilibrium	$1.00 - 2x$	$2x$	x

$$K_c = \frac{(2x^2)(x)}{(1.00 - 2x)^2} = 2.96 \times 10^{-92}$$

Because the value of K_c is much smaller than the known concentration, we can assume $1.00 - 2x \approx 1.00$.

$$\frac{(2x)^2(x)}{(1.00 - 2x)^2} \approx \frac{4x^3}{1.00} = 2.96 \times 10^{-92}$$

$$4x^3 = 2.96 \times 10^{-92}$$

$$x = \left(\frac{2.96 \times 10^{-92}}{4}\right)^{1/3} = 1.95 \times 10^{-31} \text{ mol/L}$$

The approximation is valid because $1.00 - 2(19.5 \times 10^{-31}) \approx 1.00$. At equilibrium

$$[CO] = 2(1.95 \times 10^{-31}) = \mathbf{3.90 \times 10^{-31}\ mol/L}$$

EXERCISE What is the concentration of Br_2 in equilibrium at 1500 K in a sample of gas originally containing 1.00 mol/L of BrF_5?

$$2BrF_5(g) \rightleftharpoons Br_2(g) + 5F_2(g) \qquad K_c = 2.6 \times 10^{-9}$$

Answer $[Br_2] = 9.7 \times 10^{-3}$ mol/L.

EXAMPLE 19.13 Equilibrium Concentrations

The concentration equilibrium constant for the reaction

$$ClF_3(g) \rightleftharpoons ClF(g) + F_2(g)$$

is 8.77×10^{-14} at 25 °C. A mixture of ClF_3 and F_2 was prepared such that it was 2.50 M in ClF_3 and 1.00 M in F_2. Find the concentration of ClF once equilibrium is reached.

Let x be the amount of ClF formed.

	$ClF_3(g)$ $\rightleftharpoons$	$ClF(g)$ +	$F_2(g)$
Initial	2.50	0	1.00
Change	$-x$	$+x$	$+x$
Equilibrium	$2.50 - x$	x	$1.00 + x$

$$K_c = \frac{[ClF][F_2]}{[ClF_3]} = \frac{(x)(1.00 + x)}{(2.50 - x)} = 8.77 \times 10^{-14}$$

If we assume that $(1.00 + x) \approx 1.00$ and $(2.50 - x) \approx 2.50$

$$\frac{x(1.00)}{(2.50)} = 8.77 \times 10^{-14}$$

$$x = 2.2 \times 10^{-13} \text{ M}$$

The approximations are valid; thus **[ClF] = x = 2.2×10^{-13} mol/L.**

EXERCISE For the reaction

$$Cl_2(g) + F_2(g) \rightleftharpoons 2ClF(g)$$

$K_c = 4.78 \times 10^{-19}$ at 25 °C. If a mixture that is 1.00 M in Cl_2 and 2.00 M in F_2 is prepared, what will be the equilibrium concentration of ClF?

Answer 9.78×10^{-10}mol/L.

EXAMPLE 19.14 Equilibrium Concentrations

At 3000 K chlorine gas dissociates into chlorine atoms in an equilibrium reaction for which $K_c = 0.37$. What is the concentration of chlorine atoms in a vessel that originally contained 1.0 mol/L of molecular chlorine?

The reaction is

$$Cl_2(g) \rightleftharpoons 2Cl(g)$$

The unknown is [Cl] at equilibrium and we choose the concentration of Cl_2 that dissociates into Cl as x. The algebraic expression that must be solved is found by arranging the known and unknown information in the usual way.

	$Cl_2(g)$ $\rightleftharpoons$	$2Cl(g)$
Initial	1.0	0
Change	$-x$	$+2x$
Equilibrium	$1.0 - x$	$2x$

$$K_c = \frac{[Cl]^2}{[Cl_2]} = \frac{(2x)^2}{(1.0 - x)} = 0.37$$

Inspection shows that the values of K_c and the known concentration (1.0 mol/L) are similar in magnitude. Therefore, this problem should be solved by using the quadratic formula (see Equations 19.23 and 19.24).

$$4x^2 = (0.37)(1.0 - x)$$
$$4x^2 + 0.37x - 0.37 = 0$$

with $a = 4$, $b = 0.37$, and $c = -0.37$

$$x = \frac{-0.37 \pm \sqrt{(0.37)^2 - (4)(4)(-0.37)}}{(2)(4)}$$
$$= \frac{-0.37 \pm (2.5)}{8} = 0.27 \text{ or } -0.36$$

Therefore

$$\mathbf{[Cl] = 2x = (2)(0.27\ mol/L) = 0.54\ mol/L}$$

EXERCISE For the reaction

$$ClF_3(g) \rightleftharpoons ClF(g) + F_2(g)$$

$K_c = 8.77 \times 10^{-14}$ at 25 °C. In Example 19.8, we saw that ClF_3 will decompose if a mixture containing 17.25 M ClF_3, 1.3×10^{-6} M ClF, and 4.62×10^{-7} F_2 is prepared. What are the concentrations of the three gases once equilibrium is reached?

Answers $[ClF_3] = 17.25$ mol/L, $[ClF] = 1.7 \times 10^{-6}$ mol/L, $[F_2] = 9 \times 10^{-7}$ mol/L.

SUMMARY OF SECTIONS 19.12–19.13

The stepwise method presented in Section 19.12 is recommended for solving all equilibrium problems. In most cases, the equilibrium constant expression provides the connection between the known and the unknown information. To solve an equilibrium problem, you must begin with the balanced chemical equation and the equilibrium constant expression. Then, the known and the unknown information should be identified, and a table utilized to summarize initial concentrations, changes in concentrations, and equilibrium concentrations. Next, the equilibrium concentrations (expressed in terms of x, where necessary) are inserted into the K expression, and the expression solved for x. In solving the problem, it is possible to neglect x in addition or subtraction where x is small relative to the known concentrations. However, if such an approximation introduces too great an error, the problem must be solved exactly by using the quadratic formula (Equation 19.24).

SIGNIFICANT TERMS (with section references)

chromatography (Aside 19.1)
distribution coefficient (Aside 19.1)
equilibrium constant (19.2)
reaction quotient (19.7)

QUESTIONS AND PROBLEMS

The Law of Chemical Equilibrium

19.1 What type of equilibrium is established in chemical reactions? When does a chemical reaction reach a state of equilibrium? Illustrate with a specific example.

19.2 Consider the reaction described by the equation

$$H_2O(l) \rightleftharpoons H^+ + OH^-$$

What is the relationship between the rates of the forward and reverse reactions at equilibrium? Does this mean that there will be equal masses of reactants and products at equilibrium? Explain.

19.3 Write the concentration equilibrium constant expression for each of the following reactions:

(a) $SO_3(g) + H_2(g) \rightleftharpoons SO_2(g) + H_2O(g)$
(b) $4NH_3(g) + 5O_2(g) \rightleftharpoons 4NO(g) + 6H_2O(g)$
(c) $C_3H_8(g) + 5O_2(g) \rightleftharpoons 3CO_2(g) + 4H_2O(g)$

19.4 Repeat Question 19.3 for

(a) $H_2O(g) \rightleftharpoons H(g) + OH(g)$
(b) $H_2C{=}CH_2(g) + Cl_2(g) \rightleftharpoons H_2ClCCClH_2(g)$
(c) $2O_3(g) \rightleftharpoons 3O_2(g)$
(d) $IF(g) \rightleftharpoons \frac{1}{2}I_2(g) + \frac{1}{2}F_2(g)$

19.5 Write the equilibrium constant expression for the following chemical reaction:

$$H_3PO_4(aq) \rightleftharpoons H^+ + H_2PO_4^-$$

Describe what happens at the molecular level as the H^+ ion from a concentrated strong acid such as hydrochloric acid is added to a system in which the H_3PO_4, H^+ ion, and $H_2PO_4^-$ ion have reached equilibrium.

19.6 Write the equilibrium constant expression for the following chemical reaction:

$$Cl_2(aq) + 2Br^- \rightleftharpoons 2Cl^- + Br_2(aq)$$

Describe what happens at the molecular level as the Br^- ion from solid NaBr, which is soluble, is added to a system in which the four species have reached equilibrium.

19.7 On the basis of the equilibrium constant values, choose the reactions in which the *products* are favored:

(a) $NH_3(aq) + H_2O(l) \rightleftharpoons NH_4^+ + OH^-$ $K = 1.6 \times 10^{-5}$
(b) $Au^+ + 2CN^- \rightleftharpoons [Au(CN)_2]^-$ $K = 2 \times 10^{38}$
(c) $PbC_2O_4(s) \rightleftharpoons Pb^{2+} + C_2O_4^{2-}$ $K = 4.8$
(d) $HS^- + H^+ \rightleftharpoons H_2S(aq)$ $K = 4.8 \times 10^{-12}$

19.8 On the basis of the equilibrium constant values, chose the reactions in which the *reactants* are favored:

(a) $H_2O(l) \rightleftharpoons H^+ + OH^-$ $K = 1.0 \times 10^{-14}$
(b) $[AlF_6]^{3-} \rightleftharpoons Al^{3+} + 6F^-$ $K = 3 \times 10^{-20}$
(c) $Ca_3(PO_4)_2(s) \rightleftharpoons 3Ca^{2+} + 2PO_4^{3-}$ $K = 10^{-26}$
(d) $2Fe^{3+} + 3S^{2-} \rightleftharpoons Fe_2S_3(s)$ $K = 1 \times 10^{88}$

19.9 A beaker contains a 0.1 M solution of Ba^{2+} ion. Which anion, CO_3^{2-} or CrO_4^{2-}, would be more effective in removing the Ba^{2+} ion as a precipitate?

$$Ba^{2+} + CO_3^{2-} \rightleftharpoons BaCO_3(s) \quad K = 5.0 \times 10^8$$
$$Ba^{2+} + CrO_4^{2-} \rightleftharpoons BaCrO_4(s) \quad K = 8.3 \times 10^9$$

Assume that equimolar amounts of anion would be added to cause precipitation.

19.10 Solid $Cd(NO_3)_2$ is added to a solution containing 0.1 M Cl^- and solid $AgNO_3$ is added to a second solution containing 0.1 M Cl^-. In which solution will the resulting Cl^- ion concentration be less?

$$Cd^{2+} + 4Cl^- \rightleftharpoons [CdCl_4]^{2-} \quad K = 1.1 \times 10^2$$
$$Ag^+ + 4Cl^- \rightleftharpoons [AgCl_4]^{3-} \quad K = 2 \times 10^5$$

Assume that equimolar amounts of each cation are added to cause complexation.

The Law of Chemical Equilibrium—Evaluation of *K*

19.11 The reaction between nitrogen and oxygen to form NO(*g*) is represented by the chemical equation

$$N_2(g) + O_2(g) \rightleftharpoons 2NO(g)$$

The equilibrium concentrations of the gases at 1500 K are 1.7×10^{-3} mol/L for O_2, 6.4×10^{-3} mol/L for N_2, and 1.1×10^{-5} mol/L for NO. Calculate the value of K_c at 1500 K from these data.

19.12 At elevated temperatures, BrF_5 establishes the following equilibrium:

$$2BrF_5(g) \rightleftharpoons Br_2(g) + 5F_2(g)$$

The equilibrium concentrations of the gases at 1500 K are 0.0064 mol/L for BrF_5, 0.0018 mol/L for Br_2, and 0.0090 mol/L for F_2. Calculate the value of K_c.

19.13 The reaction described by the equation

$$PCl_3(g) + Cl_2(g) \rightleftharpoons PCl_5(g)$$

has come to equilibrium at a temperature at which the concentrations of PCl_3, Cl_2, and PCl_5 are 10, 9, and 12 mol/L, respectively. Calculate the value of K_c for this reaction at that temperature.

19.14 Nitrogen reacts with hydrogen as follows:

$$N_2(g) + 3H_2(g) \rightleftharpoons 2NH_3(g)$$

An equilibrium mixture at a given temperature was found to contain 0.31 mol/L N_2, 0.50 mol/L H_2, and 0.14 mol/L NH_3. Calculate the value of K_c at the given temperature.

19.15* The reversible reaction

$$2SO_2(g) + O_2(g) \rightleftharpoons 2SO_3(g)$$

has come to equilibrium in a vessel of specific volume and at a given temperature. Before the reaction, the concentrations of the reactants were 0.060 mol/L of SO_2 and 0.050 mol/L of O_2. After equilibrium was reached, the concentration of SO_3 was 0.040 mol/L. What is the concentration of O_2 at equilibrium? What is the value of K_c?

19.16* Following are the concentrations of ions at equilibrium in three different experiments:

	$[Fe^{2+}]$	$[SCN^-]$	$[Fe(SCN)_x^{2-x}]$
Expt. 1	0.012	0.012	0.0012
Expt. 2	0.027	0.056	0.012
Expt. 3	0.039	0.016	0.0050

Use these data to determine the value of *x* in the equation

$$Fe^{2+} + xSCN^- \rightleftharpoons [Fe(SCN)_x]^{2-x}$$

Equilibrium Constant Expressions—Homogeneous Gas Reactions

19.17 Write the pressure equilibrium constant expressions for each of the following reactions:

(a) $SO_3(g) + H_2(g) \rightleftharpoons SO_2(g) + H_2O(g)$
(b) $4NH_3(g) + 5O_2(g) \rightleftharpoons 4NO(g) + 6H_2O(g)$
(c) $C_3H_8(g) + 5O_2(g) \rightleftharpoons 3CO_2(g) + 4H_2O(g)$

19.18 Repeat Question 19.17 for

(a) $H_2O(g) \rightleftharpoons H(g) + OH(g)$
(b) $H_2C{=}CH_2(g) + Cl_2(g) \rightleftharpoons H_2ClCCClH_2(g)$
(c) $2O_3(g) \rightleftharpoons 3O_2(g)$
(d) $IF(g) \rightleftharpoons \frac{1}{2}I_2(g) + \frac{1}{2}F_2(g)$

19.19 At 425 °C, the equilibrium partial pressures of H_2, I_2, and HI are 0.06443 bar, 0.06540 bar, and 0.4821 bar, respectively. Calculate the value of K_p for the reaction described by

$$2HI(g) \rightleftharpoons H_2(g) + I_2(g)$$

at this temperature.

19.20* At 27 °C and 1.00 bar, N_2O_4 is 20.0% dissociated into NO_2. Calculate the value of K_p for

$$N_2O_4(g) \rightleftharpoons 2NO_2(g)$$

at this temperature.

19.21 For the reaction

$$H_2(g) + Cl_2(g) \rightleftharpoons 2HCl(g)$$

$K_c = 193$ at 2500. K. What is the value of K_p for this reaction?

19.22 For the reaction

$$Br_2(g) \rightleftharpoons 2Br(g)$$

$K_p = 2550$ at 4000. K. What is the value of K_c for this reaction?

Equilibrium Constant Expressions—Heterogeneous Reactions

19.23 Write the equilibrium constant expressions (both K_p and K_c, when appropriate) for the following reactions:

(a) $Ag_2S(s) \rightleftharpoons 2Ag^+ + S^{2-}$
(b) $CaCO_3(s) \rightleftharpoons CaO(s) + CO_2(g)$
(c) $BaCO_3(s) + C(s) \rightleftharpoons BaO(s) + 2CO(g)$
(d) $Br_2(aq) + 2I^- \rightleftharpoons 2Br^- + I_2(s)$

19.24 Repeat Question 19.23 for

(a) $CO(g) + 2H_2(g) \rightleftharpoons CH_3OH(l)$
(b) $Cl_2(g) + H_2O(l) \rightleftharpoons H^+ + Cl^- + HOCl(aq)$
(c) $MnO_4^- + 5Fe^{2+} + 8H^+ \rightleftharpoons 5Fe^{3+} + Mn^{2+} + 4H_2O(l)$
(d) $Ag^+ + Cl^- \rightleftharpoons AgCl(s)$

19.25 For the reaction

$$B(s) + \tfrac{3}{2}F_2(g) \rightleftharpoons BF_3(g)$$

$K_p = 6.41 \times 10^{50}$ at 1100. K. What is the value of K_c for this reaction?

19.26 For the reaction

$$W(s) + 3Cl_2(g) \rightleftharpoons WCl_6(g)$$

$K_c = 2.27 \times 10^{15}$ at 1000. K. What is the value of K_p for this reaction?

Equilibrium Constant Expressions—Constants and Equations

19.27 Use the following chemical reactions and values of K_p at 1000 K

$$C(s) + \tfrac{1}{2}O_2(g) \rightleftharpoons CO(g) \qquad K_p = 2.9 \times 10^{10}$$
$$C(s) + O_2(g) \rightleftharpoons CO_2(g) \qquad K_p = 4.8 \times 10^{20}$$

to predict K_p for the chemical reaction

$$CO(g) + \tfrac{1}{2}O_2(g) \rightleftharpoons CO_2(g)$$

19.28 Use the following chemical reactions and values of K_p at 2000 K

$$\tfrac{1}{2}Br_2(g) + \tfrac{1}{2}F_2(g) \rightleftharpoons BrF(g) \qquad K_p = 148$$
$$\tfrac{1}{2}Br_2(g) + \tfrac{3}{2}F_2(g) \rightleftharpoons BrF_3(g) \qquad K_p = 2.3$$

to calculate K_p for

$$BrF_3(g) \rightleftharpoons BrF(g) + F_2(g)$$

19.29* Express the equilibrium constant K for the reaction

$$2NH_3(aq) + CO_2(g) + H_2O(l) \rightleftharpoons 2NH_4^+ + CO_3^{2-}$$

in terms of the equilibrium constants for the following reactions:

(a) $NH_3(aq) + H_2O(l) \rightleftharpoons NH_4^+ + OH^-$
(b) $CO_2(g) + H_2O(l) \rightleftharpoons H_2CO_3(aq)$
(c) $H_2CO_3(aq) \rightleftharpoons 2H^+ + CO_3^{2-}$
(d) $H_2O(l) \rightleftharpoons H^+ + OH^-$

19.30* Citric acid reacts with water to lose H^+ in three steps:

$$(C_3H_4OH)(COOH)_3(aq) \rightleftharpoons (C_3H_4OH)(COOH)_2(COO)^- + H^+ \qquad K_1 = 7.10 \times 10^{-4}$$
$$(C_3H_4OH)(COOH)_2(COO)^- \rightleftharpoons (C_3H_4OH)(COOH)(COO)_2^{2-} + H^+ \qquad K_2 = 1.68 \times 10^{-5}$$
$$(C_3H_4OH)(COOH)(COO)_2^{2-} \rightleftharpoons (C_3H_4OH)(COO)_3^{3-} + H^+ \qquad K_3 = 4.11 \times 10^{-7}$$

What is K for the overall reaction

$$(C_3H_4OH)(COOH)_3 \rightleftharpoons (C_3H_4OH)(COO)_3^{3-} + 3H^+$$

Equilibrium Constant Expressions—Reaction Quotient

19.31 How does the form of the reaction quotient compare with that for the equilibrium constant? What is the difference between these two expressions?

19.32 If the reaction quotient is larger than the equilibrium constant, what will happen to the reaction? What will happen if $Q < K$?

19.33 For the reaction

$$Cl_2(g) + F_2(g) \rightleftharpoons 2ClF(g)$$

$K_c = 19.9$. What will happen in a reaction mixture originally containing $[Cl_2] = 0.2$ mol/L, $[F_2] = 0.1$ mol/L, and $[ClF] = 3.65$ mol/L?

19.34 The value of K_p at 25 °C for

$$2CO(g) \rightleftharpoons C(\text{graphite}) + CO_2(g)$$

is 1.11×10^{21}. What is the value of K_c? Describe what will happen if 2 mol of CO and 1 mol of CO_2 are mixed in a 1 liter container with a suitable catalyst to make the reaction "go" at this temperature.

19.35 A solution that contains $[Ba^{2+}] = 1.0 \times 10^{-3}$ mol/L is mixed with an equal amount of a solution that contains $[CO_3^{2-}] = 2.0 \times 10^{-6}$ mol/L. (a) Calculate the reaction quotient for the reaction

$$Ba^{2+} + CO_3^{2-} \rightleftharpoons BaCO_3(s)$$

(b) If $K = 5 \times 10^3$ at 25 °C for this reaction, will a precipitate form?

19.36 What must be the pressure of hydrogen so that the reaction

$$WCl_6(g) + 3H_2(g) \rightleftharpoons W(s) + 6HCl(g)$$

will occur if $P_{WCl_6} = 0.012$ bar and $P_{HCl} = 0.10$ bar? $K_p = 1.37 \times 10^{21}$ at 900 K.

Factors That Influence Equilibrium

19.37 State Le Chatelier's principle. What factors do we usually consider to have an effect on a system at equilibrium? How does the presence of a catalyst or an inhibitor affect a system at chemical equilibrium? Explain your answer.

19.38 What will be the effect of increasing the total pressure on the equilibrium conditions for (a) a reaction that has more moles of gaseous products than gaseous reactants, (b) a reaction that has more moles of gaseous reactants than gaseous products, (c) a reaction that has the same number of moles of gaseous reactants and gaseous products, (d) a reaction in which all reactants and products are pure solids, pure liquids, or are in solution?

19.39 What would be the effect on the equilibrium position of an equilibrium mixture of Br_2, F_2, and BrF_5 if the total pressure of the system were decreased?

$$2BrF_5(g) \rightleftharpoons Br_2(g) + 5F_2(g)$$

19.40 What would be the effect on the equilibrium position of an equilibrium mixture of carbon, oxygen, and carbon dioxide if the total pressure of the system was decreased?

$$2C(s) + O_2(g) \rightleftharpoons 2CO(g)$$

19.41 A weather indicator can be made by coating an object with a hydrate of cobalt(II) chloride, which changes color as a result of the following reaction

$$\underset{pink}{[Co(H_2O)_6]Cl_2(s)} \rightleftharpoons \underset{blue}{[Co(H_2O)_4]Cl_2(s)} + 2H_2O(g)$$

Does a pink color indicate "moist" or "dry" air? Explain.

19.42 Consider the reaction

$$CaCO_3(s) \rightleftharpoons CaO(s) + CO_2(g)$$

Will the mass of $CaCO_3$ at equilibrium (i) increase, (ii) decrease, or (iii) remain the same if (a) CO_2 is removed from the equilibrium system? (b) the pressure is increased? (c) solid CaO is added?

19.43 The reaction between NO and O_2 is exothermic.

$$2NO(g) + O_2(g) \rightleftharpoons 2NO_2(g) \qquad \Delta H^\circ_{1000} = -116.86 \text{ kJ}$$

Will the concentration of NO_2 at equilibrium (i) increase, (ii) decrease, or (iii) remain the same if (a) additional O_2 is introduced, (b) additional NO is introduced, (c) the total pressure is increased, (d) the temperature is decreased?

19.44 Predict whether the equilibrium for the photosynthesis reaction described by the equation

$$6CO_2(g) + 6H_2O(l) \rightleftharpoons C_6H_{12}O_6(s) + 6O_2(g) \qquad \Delta H^\circ = 2801.69 \text{ kJ}$$

would (i) shift to the right, (ii) shift to the left, or (iii) remain unchanged if (a) $[CO_2]$ was increased, (b) P_{O_2} was increased, (c) one-half of the $C_6H_{12}O_6$ was removed, (d) the total pressure was increased, (e) the temperature was increased, (f) a catalyst was added.

Equilibrium Problems

19.45 When solving an equilibrium constant expression for an unknown concentration x, under what circumstances may approximations be made to simplify the solution of the expression?

19.46 Shown below are complete equilibrium constant expressions and approximations made to simplify these respective expressions. Solve for x in each case using the approximation and decide whether or not the approximation is valid.

(a) $\dfrac{(x)(0.1 + x)}{(0.1 - x)} \approx x = 1.3 \times 10^{-5}$

(b) $\dfrac{(x)(x)}{(0.1 - x)} \approx \dfrac{x^2}{0.1} = 4.8 \times 10^{-11}$

(c) $\dfrac{(x)^4(x)}{(0.001 - x)} \approx \dfrac{x^5}{0.001} = 9.3 \times 10^{-3}$

(d) $\dfrac{(x)(x)}{(0.1 - x)} \approx \dfrac{x^2}{0.1} = 1.3 \times 10^{-5}$

19.47 The equilibrium constant for the reaction

$$I_2(g) + F_2(g) \rightleftharpoons 2IF(g)$$

is $K_p = 1.15 \times 10^7$ at 2000 K. If the system is at equilibrium, and the partial pressure of IF is 0.32 bar and the partial pressure of F_2 is 1.9×10^{-2} bar, find the partial pressure of I_2.

19.48 For the reaction

$$2BrF_5(g) \rightleftharpoons Br_2(g) + 5F_2(g)$$

the value of K_p at 1000 K is 7.4×10^{-16}. Calculate the partial pressure of F_2 at equilibrium with BrF_5 at a partial pressure of 1.2×10^{-3} bar and Br_2 at a partial pressure of 3.8×10^{-5} bar.

19.49 Exactly one mole of carbon dioxide was placed in a 1.00 L container, and the following equilibrium was established

$$2CO_2(g) \rightleftharpoons 2CO(g) + O_2(g) \qquad K_c = 2.96 \times 10^{-92}$$

What is the concentration of carbon monoxide at equilibrium?

19.50 What are the concentrations of H_2, I_2, and HI once equilibrium has been established, after placing 0.100 mol of HI into a 0.100 L container?

$$2HI(g) \rightleftharpoons H_2(g) + I_2(g) \qquad K_c = 0.01813$$

19.51 At 43.9 °C, $K = 3.7 \times 10^{-3}$ for the reaction

$$(C_6H_5COOH)_2(\text{in } C_6H_6) \rightleftharpoons 2C_6H_5COOH\ (\text{in } C_6H_6)$$

What is the concentration of C_6H_5COOH at equilibrium if the initial concentration of $(C_6H_5COOH)_2$ was 0.45 mol/L?

19.52 The equilibrium constant for the reaction

$$\alpha\text{-D-glucose}(aq) \rightleftharpoons \beta\text{-D-glucose}(aq)$$

is 1.75 at 25 °C. What are the concentrations of these sugars at equilibrium if the initial concentration of α-D-glucose was 0.0100 mol/L?

19.53* The equilibrium constant for the reaction

$$H^+ + BrO^- \rightleftharpoons HBrO(aq)$$

is 4.5×10^8 at 25 °C. What is the concentration of BrO^- at equilibrium if the original solution contained 0.32 M H^+ and 0.32 M BrO^-? (The calculations may be simplified by assuming that practically all of the H^+ ion and BrO^- ion react and only small portions of the H^+ ion and BrO^- remain, so that $x = [H^+] = [BrO^-]$ and $[HBrO] = 0.32 - x$).

19.54* Adenosine triphosphate (ATP) reacts with water to form adenosine diphosphate (ADP) and produces an inorganic phosphate (P_i) such as HPO_4^{2-}.

$$ATP(aq) + H_2O(l) \rightleftharpoons ADP(aq) + P_i \qquad K = 4 \times 10^6$$

Both ATP and ADP are found in living cells and are important in the storage of energy for many chemical reactions in the body. What is the concentration of ATP at equilibirum if the original solution contained 1×10^{-7} mol/L ATP? (The calculations may be simplified by assuming that practically all of the ATP reacts and only a small portion of ATP is left, so that $x = [ATP]$ and $[ADP] = [P_i] = [(1 \times 10^{-7}) - x]$.)

19.55 For the reaction described by the equation

$$N_2(g) + C_2H_2(g) \rightleftharpoons 2HCN(g)$$

$K_c = 2.3 \times 10^{-4}$ at 300 °C. What is the equilibrium concentration of hydrogen cyanide if the initial concentrations of N_2 and acetylene (C_2H_2) were 5.0 mol/L and 2.0 mol/L, respectively?

19.56 For the reaction described by the equation

$$CN^- + H_2O(l) \rightleftharpoons HCN(aq) + OH^-$$

$K = 1.6 \times 10^{-5}$ at 25 °C. What is the equilibrium concentration of the hydroxide ion if the initial concentrations of CN^- ion and HCN were 0.15 mol/L and 0.10 mol/L, respectively?

19.57 The equilibrium constant for the reaction

$$Br_2(g) + F_2(g) \rightleftharpoons 2BrF(g)$$

is 54.7. What are the equilibrium concentrations of all these gases if the initial concentrations of bromine and fluorine were both 0.125 mol/L?

19.58 For the reaction

$$PCl_3(g) + Cl_2(g) \rightleftharpoons PCl_5(g)$$

$K_c = 96.2$ at 400 K. What is the concentration of Cl_2 at equilibrium if the initial concentrations were 0.10 mol/L for PCl_3 and 3.5 mol/L for Cl_2?

19.59* The equilibrium constant for the reaction

$$Fe^{3+} + 3C_2O_4^{2-} \rightleftharpoons [Fe(C_2O_4)_3]^{3-}$$

is 1.67×10^{20}. What is the concentration of Fe^{3+} left in solution if equal volumes of 0.0010 M $Fe(NO_3)_3$ and 1.000 M $K_2C_2O_4$ are mixed? (The calculations are simplified if you assume that all of the Fe^{3+} has reacted and write the "changes" on the basis of $[Fe(C_2O_4)_3]^{3-}$ undergoing decomposition.) Assume that the volume of the final solution is equal to the sum of the volumes of the solutions mixed together.

19.60* Chloromethane, CH_3Cl, is produced by the reaction of Cl_2 with methane, CH_4, as follows:

$$CH_4(g) + Cl_2(g) \rightleftharpoons CH_3Cl(g) + HCl(g)$$

Assuming that equal amounts of CH_4 and Cl_2 were placed in a reaction vessel so that their initial partial pressures were identical and equilibrium was allowed to be established, what is P_{CH_3Cl}/P_{CH_4} if $K_p = 7.4 \times 10^{17}$? (It is not necessary to find the individual pressures of the gases at equilibrium to calculate this ratio.)

COLOR IN TRANSITION METAL COMPOUNDS

The colors of compounds of a given transition element can vary with anions, with oxidation state, with the ligands in complex ions, or in different isomers.

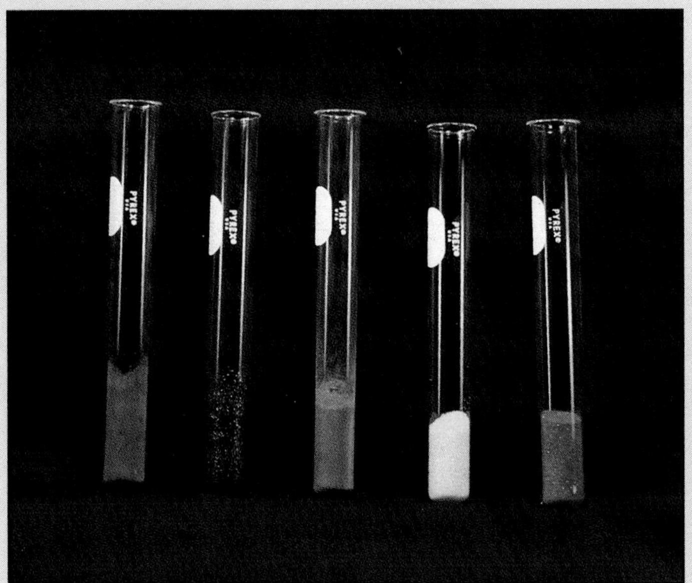

■ **CHROMIUM COMPOUNDS** *left* Chromium(III) compounds containing different anions (*left to right, first three*): gray-green chromium(III) acetate monohydrate ($Cr(CH_3COO)_3 \cdot H_2O$), deep violet chromium(III) chloride ($CrCl_3$), and green chromium(III) fluoride (CrF_3). Chromium(VI) compounds (*last two*): yellow potassium chromate (K_2CrO_4) and orange potassium dichromate ($K_2Cr_2O_7$), both used in making dyes and pigments, and potassium dichromate, a common laboratory oxidizing agent.

■ **IRON AND CHROMIUM COMPOUNDS IN DIFFERENT OXIDATION STATES** *right* *top:* Brownish yellow iron(III) chloride hexahydrate ($FeCl_3 \cdot 6H_2O$) (*left*) and greenish black iron(II) chloride tetrahydrate ($FeCl_2 \cdot 4H_2O$) (*right*). *bottom:* Green chromium(III) oxide (Cr_2O_3) (*left*), a pigment and the green granules in asphalt roofing, and dark red chromium(VI) oxide (CrO_3) (*right*).

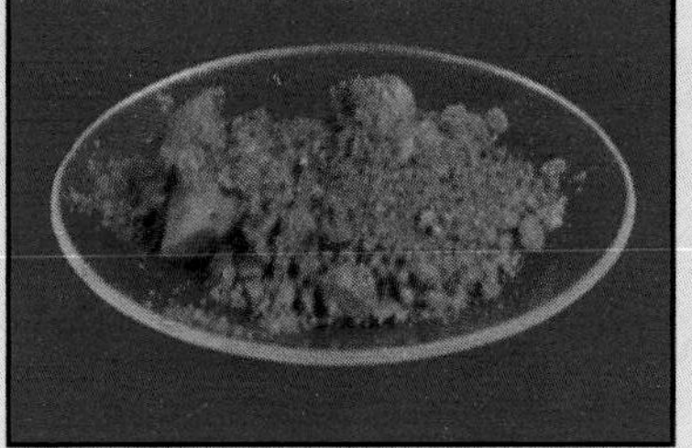

■ **COORDINATION COMPOUNDS OF COBALT** *above* The *cis* and *trans* isomers of the dichlorotetramminecobalt(III) complex ion: (*left*) *cis*-$[Co(NH_3)_4Cl_2]Cl$ and (*right*) *trans*-$[Co(NH_3)_4Cl_2]Cl$.

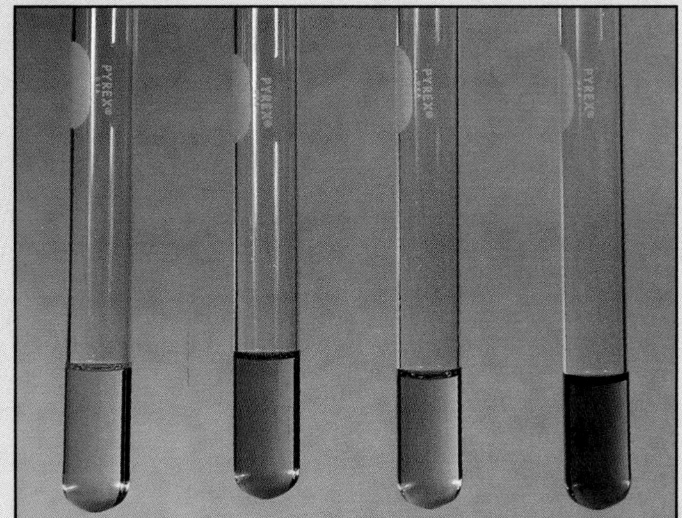

■ **HYDRATED IONS CONTAINING VANADIUM** *right* The test tubes contain solutions of vanadium ion in four different oxidation states (*left to right*): vanadium(V), VO_2^+; vanadium(IV), VO^{2+}; vanadium(III), V^{3+}; vanadium(II), V^{2+}.

■ **BALL-AND-STICK** (*left*) **AND SPACE-FILLING** (*right*) **MOLECULAR MODELS** *below* Small molecules are easily represented by ball-and-stick models or space-filling models (which more realistically show relative positions of atoms in molecules).

Methane (CH_4) *above*

Water (H_2O) *above*

■ **MOLECULAR MODELING BY COMPUTER — TRANSFER RIBONUCLEIC ACID (t-RNA)** *below* Programs that allow computers to draw and then rotate large biochemical molecules in three dimensions are increasingly valuable research tools. Pictured are a wireframe (*top*) and a space-filling (*bottom*) model of t-RNA. During protein synthesis, one end of the L-shaped t-RNA reads a nucleic acid codon for a specific amino acid, while the other end binds that amino acid and carries it to the growing protein. Each color represents a different molecular building block of t-RNA.

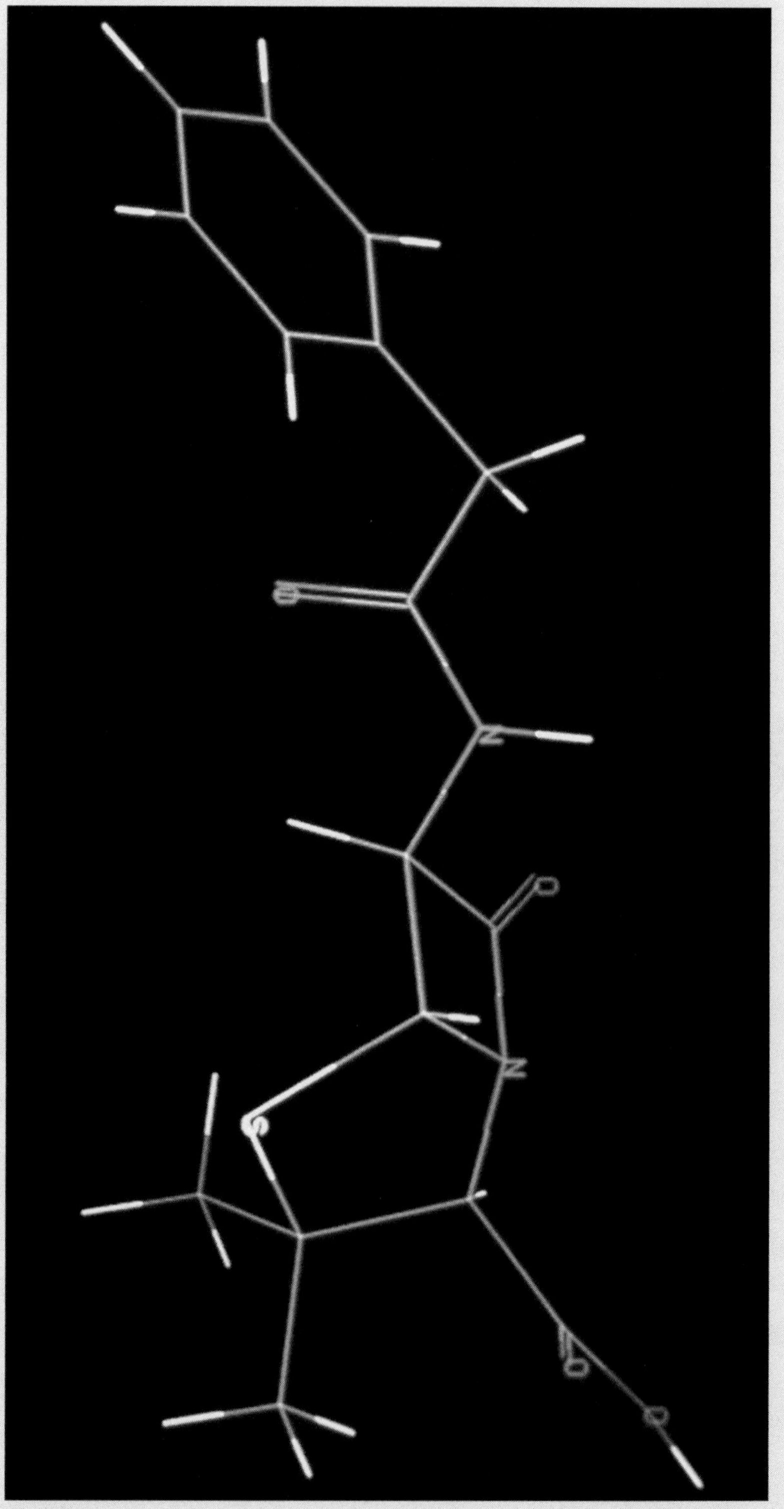

■ **MOLECULAR MODELING BY COMPUTER—PENICILLIN** *below* In "wireframe" models like that shown for the antibiotic, penicillin G, each straight line represents a bond, with the location of atoms other than carbon (pink) and hydrogen (yellow) indicated by symbols.

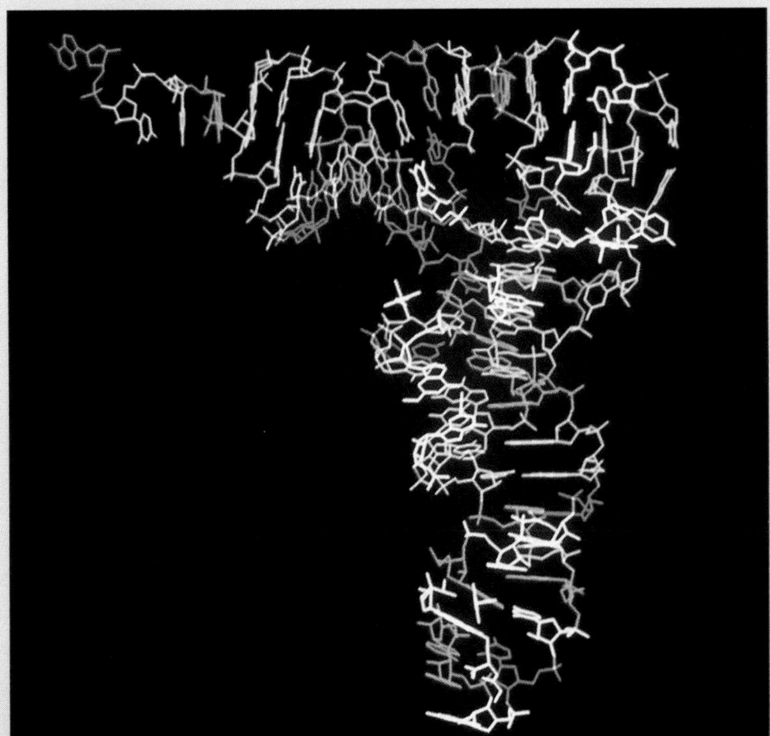

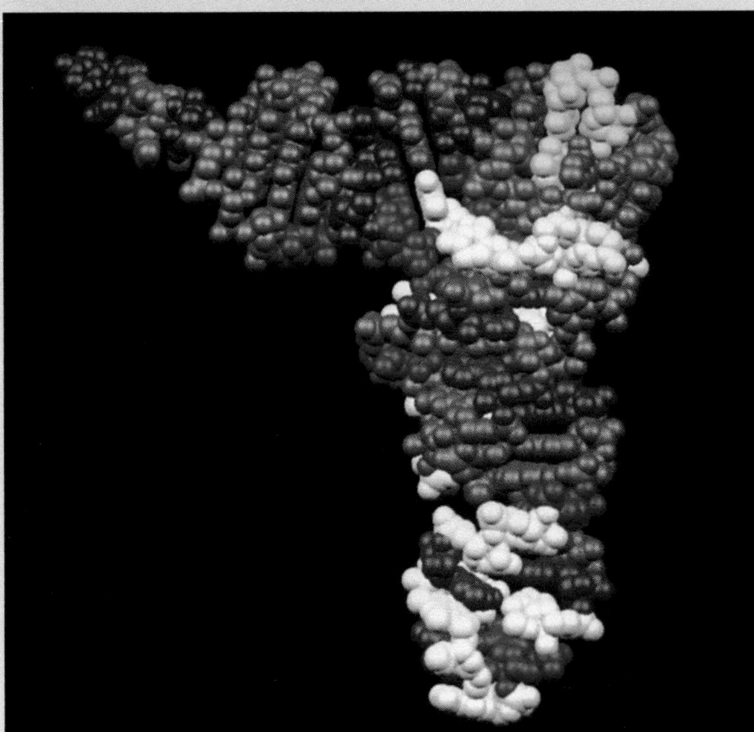

■ **MOLECULAR MODEL OF DEOXYRIBONUCLEIC ACID (DNA)** *below* Models like that shown are assembled by combining individual "atoms" to duplicate the molecular structure. The red oxygen atoms trace the turns of the sugar-phosphate chains of the double helix and the blue nitrogen atoms show the location of base pairs stacked in the center of the helix.

INORGANIC QUALITATIVE ANALYSIS

In inorganic "qual," cations in solution are separated into groups that form precipitates under similar conditions. Then individual cations are separated and each is identified by a characteristic reaction that gives a distinctive color, precipitate, or other visible result. The photos show these confirmatory tests for some ions of Cation Groups II and III.

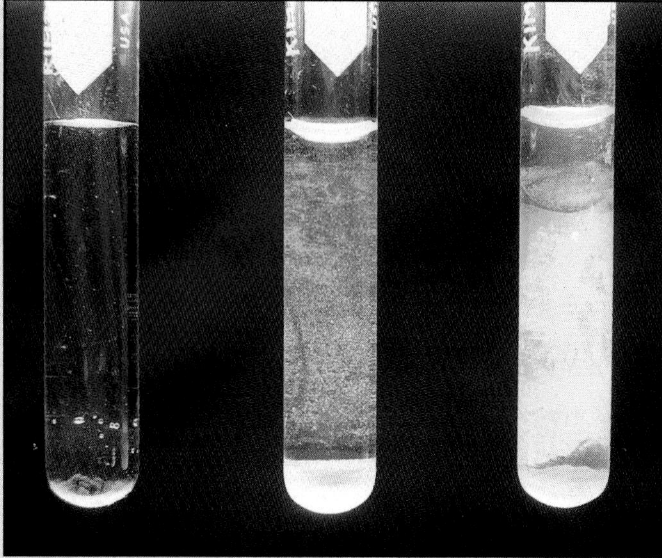

(*left to right*) *above*

■ Hg^{2+} — Reduction of $[HgCl_4]^{2-}$ by copper to give silvery Hg(*l*), which forms on the surface of unreacted copper.

■ Hg^{2+} — Reduction of mercury(II) in $[HgCl_4]^{2-}$ by Sn^{2+} in the presence of Cl^- yields a white precipitate of mercury(I) chloride (Hg_2Cl_2).

■ Pb^{2+} — Bright yellow lead chromate ($PbCrO_4$) precipitate forms by partner exchange when potassium chromate solution is added.

(*left to right*) *above*

■ Cu^{2+} — Addition of the hexacyanoferrate ion produces a dark reddish-brown fluffy precipitate of copper(II) hexacyanoferrate ($Cu_2[Fe(CN)_6]$).

■ Cu^{2+} — Addition of pyridine (C_5H_5N) and isothiocyanate ion (NCS^-) gives a green precipitate of $[Cu(NC_5H_5)_2(NCS)_2]$ (layer at the bottom of the test tube).

■ Sb^{3+} — An orange precipitate of antimonyl sulfide (Sb_2OS_2) forms in the presence of sulfide ion.

■ Sn^{2+} — Addition of excess mercury(II) results in a redox reaction and formation of a white precipitate of mercury(I) chloride (Hg_2Cl_2).

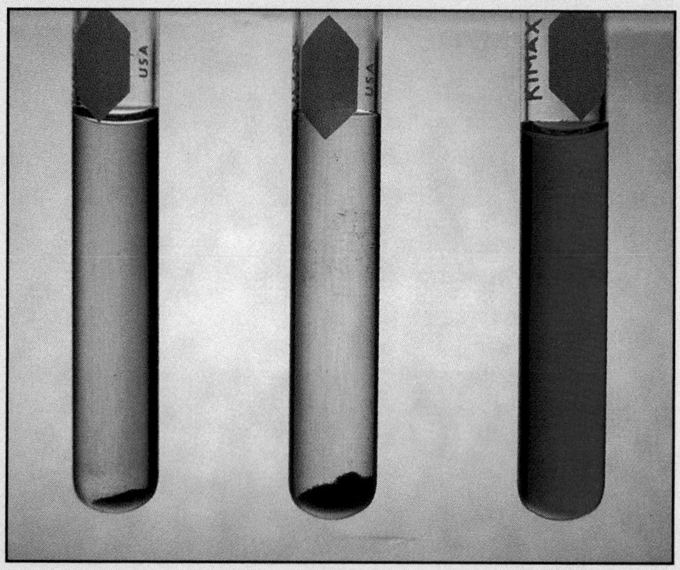

(*left to right*) *above*

■ Mn^{2+} — The strong oxidizing agent sodium bismuthate ($NaBiO_3$) converts Mn^{2+} to purple MnO_4^- in solution.

■ Fe^{3+} — Formation of a dark blue precipitate of the coordination compound $KFe[Fe(CN)_6]$ is a test for Fe^{3+}.

■ Fe^{3+} — An alternate test for Fe^{3+} is formation of the blood-red complex ion $[Fe(NCS)]^{2+}$ in solution.

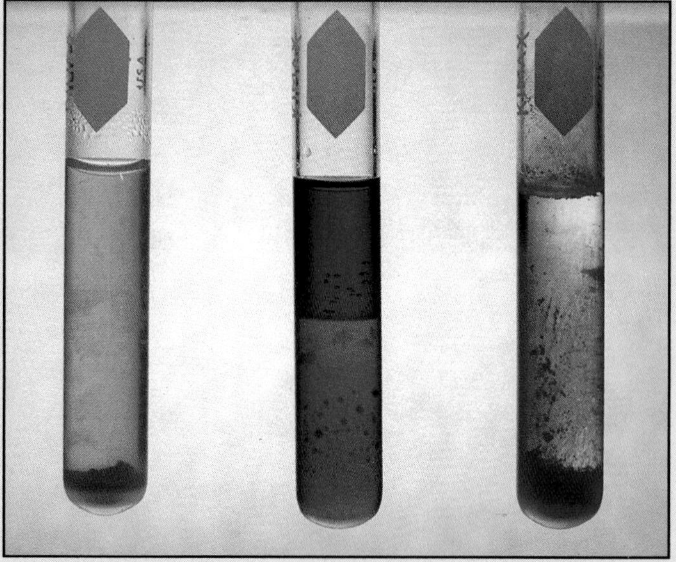

(*left to right*) *above*

■ Co^{2+} — Addition of potassium nitrite (KNO_2) to an acid solution produces the yellow precipitate $K_3[Co(NO_2)_6]$.

■ Co^{2+} — An alternate test for Co^{2+} is formation of blue $[Co(NCS)_4]^{2-}$, which dissolves preferentially in a layer of amyl alcohol.

■ Ni^{2+} — With dimethylglyoxime (DMG), Ni^{2+} produces a red precipitate of $[Ni(DMG)_2]$.

CHAPTER

20

Acids and Bases

BRØNSTED-LOWRY DEFINITIONS OF ACIDS AND BASES

20.1 PROTON DONORS AND ACCEPTORS

In Section 14.8 the classical definitions of acids and bases in terms of the ions of water were introduced. A water-ion acid yields hydrogen ions in aqueous solution and a water-ion base yields hydroxide ions in aqueous solution. An acid–base reaction results in the formation of water and a salt. Over the years many other definitions of acids and bases were introduced as it became desirable to include more types of molecules and ions in the categories of acids and bases. For reactions in aqueous solution, the aspect of chemistry upon which we concentrate in this book, the Brønsted-Lowry definitions are the most useful.

The Brønsted-Lowry acid–base concept emphasizes the role of the proton. Any molecule or ion that can act as a proton donor is a **Brønsted-Lowry acid.** Any molecule or ion that can act as a proton acceptor is a **Brønsted-Lowry base.** In the reaction of hydrogen chloride with water to form a solution of hydrochloric acid

proton acceptor (base)

proton donor (acid) → $HCl(g) + H_2O(l) \longrightarrow H_3O^+ + Cl^-$

HCl is a strong Brønsted-Lowry acid: Virtually all the HCl molecules give up their protons to water molecules. In this case the water molecule is the base and attracts H^+ to form an ion that we write as H_3O^+, the hydronium ion (see Section 14.5). A **Brønsted-Lowry acid–base reaction** is the transfer of a proton from a proton donor to a proton acceptor.

When a Brønsted-Lowry acid loses a proton, it forms the *conjugate base* of that acid, for example

acid (proton donor)		*proton*		*conjugate base*	
HCl	$\longrightarrow$	H^+	+	Cl^-	(20.1)
H_2SO_4	$\longrightarrow$	H^+	+	HSO_4^-	(20.2)
HSO_4^-	$\longrightarrow$	H^+	+	SO_4^{2-}	(20.3)
HNO_3	$\longrightarrow$	H^+	+	NO_3^-	(20.4)
H_2O	$\longrightarrow$	H^+	+	OH^-	(20.5)
NH_4^+	$\longrightarrow$	H^+	+	NH_3	(20.6)

water as an acid

When a Brønsted-Lowry base adds a proton, it forms the *conjugate acid* of that base, for example

base (proton acceptor)	*proton*		*conjugate acid*	
OH^-	$+ H^+$	$\longrightarrow$	H_2O	(20.7)
CN^-	$+ H^+$	$\longrightarrow$	HCN	(20.8)
CO_3^{2-}	$+ H^+$	$\longrightarrow$	HCO_3^-	(20.9)
HCO_3^-	$+ H^+$	$\longrightarrow$	H_2CO_3	(20.10)
NH_3	$+ H^+$	$\longrightarrow$	NH_4^+	(20.11)
H_2O	$+ H^+$	$\longrightarrow$	H_3O^+	(20.12)

water as a base

Note that water can be either an acid or a base.

An acid does not give up a proton unless another base is present to accept the proton. Similarly, a base cannot add a proton unless another acid is present to provide the proton. In the reaction of hydrogen chloride with water, as in all Brønsted-Lowry acid–base reactions, there are two acid–base pairs. In each of the following reactions, $acid_1$ gives up a proton to yield $base_1$, and $base_2$ adds a proton to form $acid_2$.

$acid_1$	*+ $base_2$*		*$base_1$*	*+ $acid_2$*	
HCl	$+ H_2O$	$\rightleftharpoons$	Cl^-	$+ H_3O^+$	(20.13)
HNO_3	$+ H_2O$	$\rightleftharpoons$	NO_3^-	$+ H_3O^+$	(20.14)
NH_4^+	$+ CO_3^{2-}$	$\rightleftharpoons$	NH_3	$+ HCO_3^-$	(20.15)
H_3O^+	$+ OH^-$	$\rightleftharpoons$	H_2O	$+ H_2O$	(20.16)
H_2O	$+ NH_3$	$\rightleftharpoons$	OH^-	$+ NH_4^+$	(20.17)

Most Brønsted-Lowry acid–base reactions that we discuss occur in aqueous solution. Note, however, that the definition includes any proton transfer reaction. For example, in the reaction of ammonia and hydrogen chloride in the gas phase, ammonia is a Brønsted-Lowry base and hydrogen chloride is a Brønsted-Lowry acid.

$$\underset{acid_1}{HCl(g)} + \underset{base_2}{NH_3(g)} \longrightarrow \underset{acid_2\ base_1}{NH_4^+Cl^-(s)}$$

Table 20.1 is a list of conjugate acid–base pairs. All substances that fit the classical water-ion definitions of acids are also Brønsted-Lowry acids, or what can be called *protonic acids*, because they give up protons. In addition, ions such as HSO_4^- or NH_4^+ that contain ionizable protons are included in the definition of Brønsted-Lowry acids.

In the presence of an acid stronger than itself, water is a base (Equation 20.14). In the presence of a base stronger than itself, water is an acid (Equation 20.17). Ammonia (see Table 20.1) can also act as either an acid or a base. Water, ammonia, and other such substances that can act as either acids or bases are called *amphoteric*. (We have discussed amphoteric oxides in Section 16.11.)

Protonic acids can be *monoprotic* (capable of losing one proton, e.g., HCl), or *diprotic* (capable of losing two protons, e.g., H_2SO_4), or *triprotic* (capable of losing three protons, e.g., H_3PO_4).

Table 20.1
Conjugate Acid–Base Pairs

	Acid		Base		
Increasing acid strength ↑					**Increasing base strength** ↓
Strong acids: 100% ionized	Perchloric acid	$HClO_4$	ClO_4^-	Perchlorate ion	**Virtually no reaction as bases**
	Sulfuric acid	H_2SO_4	HSO_4^-	Hydrogen sulfate ion	
	Hydriodic acid	HI	I^-	Iodide ion	
	Hydrobromic acid	HBr	Br^-	Bromide ion	
	Hydrochloric acid	HCl	Cl^-	Chloride ion	
	Nitric acid	HNO_3	NO_3^-	Nitrate ion	
	Hydronium ion	H_3O^+	H_2O	Water	
Weak acids	Trichloroacetic acid	Cl_3CCOOH	Cl_3CCOO^-	Trichloroacetate ion	**Weak bases**
	Hydrogen sulfate ion	HSO_4^-	SO_4^{2-}	Sulfate ion	
	Phosphoric acid	H_3PO_4	$H_2PO_4^-$	Dihydrogen phosphate ion	
	Nitrous acid	HNO_2	NO_2^-	Nitrite ion	
	Hydrofluoric acid	HF	F^-	Fluoride ion	
	Formic acid	$HCOOH$	$HCOO^-$	Formate ion	
	Acetic acid	CH_3COOH	CH_3COO^-	Acetate ion	
Very weak acids	Carbonic acid	H_2CO_3	HCO_3^-	Hydrogen carbonate ion	**Stronger weak bases**
	Hydrosulfuric acid	H_2S	HS^-	Hydrogen sulfide ion	
	Ammonium ion	NH_4^+	NH_3	Ammonia	
	Hydrogen cyanide	HCN	CN^-	Cyanide ion	
	Hydrogen sulfide ion	HS^-	S^{2-}	Sulfide ion	
	Water	H_2O	OH^-	Hydroxide ion	
	Ammonia	NH_3	NH_2^-	Amide ion	**Strong bases: 100% reaction with H_2O to yield OH^-**
	Hydrogen	H_2	H^-	Hydride ion	
	Methane	CH_4	CH_3^-	Methide ion	

The hydrogen-containing anions derived from diprotic and triprotic acids are amphoteric. They can be either proton donors or proton acceptors, for example, for HS^-

$$\underset{acid_1}{HS^-} + \underset{base_2}{OH^-} \rightleftharpoons \underset{base_1}{S^{2-}} + \underset{acid_2}{H_2O}$$

$$HCl + HS^- \rightleftharpoons Cl^- + H_2S$$

EXAMPLE 20.1 Identifying Acids and Bases

Identify the Brønsted-Lowry acid which is a reactant and its conjugate base which is a product in each of the following reactions:

(a) $HNO_3(l) + H_2O(l) \longrightarrow H_3O^+ + NO_3^-$

(b) $H_3O^+ + HS^- \rightleftharpoons H_2S(aq) + H_2O(l)$

(c) $HF(aq) + OH^- \rightleftharpoons H_2O(l) + F^-$

The acid is the species that donates the proton in each reaction: **(a) HNO_3, (b) H_3O^+, and (c) HF.** The conjugate base is the species formed by the removal of a proton from the acid: **(a) NO_3^-, (b) H_2O, and (c) F^-.**

EXERCISE Consider the following reactions of the hydrogen sulfide ion:

(i) $HS^- + H_2O(l) \rightleftharpoons H_3O^+ + S^{2-}$

(ii) $HS^- + H_3O^+ \rightleftharpoons H_2S(aq) + H_2O(l)$

In which reaction is HS^- acting as (a) an acid and as (b) a base? *Answers* (a) i, (b) ii.

20.2 RELATIVE STRENGTHS OF ACIDS AND BASES

A Brønsted-Lowry acid–base reaction results from the competition between two bases for a proton. In each of the reactions in Equations (20.13)–(20.17), $base_1$ and $base_2$ are competing for the proton. Whether a given acid–base reaction goes virtually to completion or reaches equilibrium with significant amounts of both reactants and products present depends on the relative strengths of the two bases. A chloride ion, for example, is a *very* weak base, with very little attraction for a proton. Because water is so much stronger a base than chloride ion, the reaction of hydrogen chloride with water (Equation 20.13) goes so far toward formation of H_3O^+ that for all practical purposes, ionization is complete. The acids and their conjugate bases listed above H_3O^+ and H_2O in Table 20.1 are all of this type—strong acids whose conjugate bases have virtually no attraction for protons. These acids are strong electrolytes—their aqueous solutions contain only hydrogen ions and anions. The anions of these acids are unable to attract hydrogen ions to re-form the acids. In the competition for the proton, water is the winner, driving reactions such as (20.13) and (20.14) to the right as written. (It is helpful to memorize the names and formulas of the strong acids above H_3O^+ in Table 20.1.)

In reactions of weak acids with water equilibrium is established with some undissociated acid still present. The anions of such weak acids are stronger bases than water and win back some of the protons, resulting in equilibria such as

$$HF(aq) + H_2O(l) \rightleftharpoons F^- + H_3O^+$$
$$HCN(aq) + H_2O(l) \rightleftharpoons CN^- + H_3O^+$$

All acids below H_3O^+ in Table 20.1 are of this type.

Looking down the list of bases in Table 20.1 you can see that all anions are Brønsted-Lowry bases, because they are potentially proton acceptors. The three anions below hydroxide ion in Table 20.1, because they are much stronger bases than the hydroxide ion, react immediately and virtually completely with water; they win the competition for the protons and give up practically none of them.

$$H_2O(l) + NH_2^- \longrightarrow OH^- + NH_3(g) \qquad \textbf{(20.18)}$$
$$H_2O(l) + H^- \longrightarrow OH^- + H_2(g) \qquad \textbf{(20.19)}$$
$$H_2O(l) + CH_3^- \longrightarrow OH^- + CH_4(g) \qquad \textbf{(20.20)}$$

Bases above hydroxide ion but below water in Table 20.1 are weaker bases than hydroxide ion, but strong enough to establish equilibria in the competition with hydroxide ion for the proton. These anions can, in the *absence* of H_3O^+ or other competing species, react with water establishing equilibria such as

$$H_2O(l) + F^- \rightleftharpoons OH^- + HF(aq) \qquad \textbf{(20.21)}$$
$$H_2O(l) + CN^- \rightleftharpoons OH^- + HCN(aq) \qquad \textbf{(20.22)}$$

As shown in Table 20.1, the weak bases fall into two groups, with those in one group significantly stronger than those in the other group. The anions above water in Table 20.1 are unable to take protons away from water molecules to any significant extent.

Examine the following reactions with reference to Table 20.1 and the competition between bases for the proton. Base$_2$ is stronger than base$_1$ in each case and the reactions are most favorable in the forward directions as written.

$$\underset{acid_1}{NH_4^+} + \underset{base_2}{OH^-} \rightleftharpoons \underset{acid_2}{H_2O(l)} + \underset{base_1}{NH_3(aq)}$$
$$HI(g) + CN^- \rightleftharpoons HCN(aq) + I^-$$

EXAMPLE 20.2 Identifying Acids and Bases; Relative Strengths of Acids and Bases

Write the equation for the equilibrium reaction between acetic acid and ammonia in aqueous solution. Identify the species acting as acids and bases. From the relative strengths of the bases (Table 20.1), in which direction is the reaction more favorable?

The chemical equation for the reaction is

$$CH_3COOH(aq) + NH_3(aq) \rightleftharpoons NH_4^+ + CH_3COO^-$$

The species that lose protons are **CH_3COOH and NH_4^+; therefore, these are the acids.** The species that gain protons are **NH_3 and CH_3COO^-; therefore, these are the bases.** As shown in Table 20.1, NH_3 is the stronger base and CH_3COO^- is the weaker base. Therefore, **the forward reaction should be favored.**

EXERCISE Write the equation for the equilibrium between hydrofluoric acid and sulfate ion in aqueous solution. Identify the species that act as acids. From experience we know that this reaction is not favorable in the forward direction as written. Which of the two bases present is stronger?

Answers $HF(aq) + SO_4^{2-} \rightleftharpoons HSO_4^- + F^-$, HF and HSO_4^- are the acids, F^- is the stronger base.

SUMMARY OF SECTIONS 20.1–20.2

The Brønsted-Lowry acid–base definitions emphasize the role of the proton. An acid is a proton donor; a base is a proton acceptor. According to these definitions, water can be either an acid or base (it is amphoteric). Anions are always Brønsted-Lowry bases, but those containing hydrogen (e.g., $H_2PO_4^-$) are amphoteric. Remembering the following simple concepts will help you understand and use the Brønsted-Lowry definitions of acids and bases:

1. A strong Brønsted-Lowry acid gives up its proton easily.
2. A strong Brønsted-Lowry base attracts a proton strongly.
3. An acid–base reaction results from a competition between two bases for the proton.
4. When one base is much stronger than the other, the attraction of the stronger base for a proton is so heavily favored that the reaction is virtually complete.
5. When both bases in a Brønsted-Lowry acid–base reaction are of moderate strength, equilibria are established in which significant amounts of both reactants and products are present.

BINARY ACIDS AND OXOACIDS

20.3 NAMING ACIDS AND OXOANIONS

Binary acids are compounds of hydrogen and another element, such as HCl or H_2S. These compounds are hydrides of nonmetals or certain semiconducting elements. Their names as acids are, for example

$HF(aq)$	$H_2S(aq)$	$HI(aq)$
hydrofluoric acid	*hydrosulfuric acid*	*hydroiodic acid*

Each name consists of three parts: (1) the prefix "hydro," which refers to the hydrogen atoms; (2) the name of the second element modified with the ending "ic"; (3) and the word "acid."

Oxoacids, for example, H_2SO_4 or HNO_3, contain hydrogen and oxygen atoms, and a central atom of a third element. Some oxoacids can be isolated as pure compounds. Others are known only in aqueous solution. Still others are known only by salts containing oxoanions.

The common names of the oxoacids were assigned when knowledge of chemistry was less developed than it is now. A name ending in "ic" was given to the acid that was best known or most common at the time. Other oxoacids of the same element were named according to the number of oxygen atoms surrounding the central element relative to this "most common" acid. The names assigned in this way remain in use, although in some cases one might argue that the "ic" named acid is not the most common one. To name the common series of oxoacids, it is necessary to know only the following formulas and names:

$HClO_3$	$HBrO_3$	HNO_3	H_3PO_4	H_2SO_4
chloric acid	*bromic acid*	*nitric acid*	*phosphoric acid*	*sulfuric acid*

As illustrated below for the oxoacids of chlorine and phosphorus, the acid with one less oxygen atom per molecule than the "ic" acid is identified by the ending "ous" and the acid with two less oxygen atoms is identified by the prefix "hypo" in addition to the "ous" ending.

no. of oxygen atoms	*name*	*examples*	
n	. . . ic acid	$HClO_3$ *chloric acid*	H_3PO_4 *phosphoric acid*
$(n - 1)$	. . . ous acid	$HClO_2$ *chlorous acid*	H_3PO_3 *phosphorous acid*
$(n - 2)$	hypo . . . ous acid	$HClO$ *hypochlorous acid*	H_3PO_2 *hypophosphorous acid*

In some cases, most notably for the halogens (except for fluorine), there also are acids that contain one oxygen atom per molecule more than the most common oxoacid. These acids are identified by adding the prefix "per" in addition to the suffix "ic," for example

no. of oxygen atoms	*name*	*examples*	
$(n + 1)$	per . . . ic acid	$HClO_4$ *perchloric acid*	$HBrO_4$ *perbromic acid*

The oxoanions formed by the removal of all acidic hydrogen atoms from the oxoacid are named as follows:

per . . . ic acid	gives	per . . . ate ion
. . . ic acid	gives	. . . ate ion
. . . ous acid	gives	. . . ite ion
hypo . . . ous acid	gives	hypo . . . ite ion

For example, the anions derived from the oxoacids of bromine are named as follows:

perbromic acid, $HBrO_4$	gives	perbromate ion, BrO_4^-
bromic acid, $HBrO_3$	gives	bromate ion, BrO_3^-
bromous acid, $HBrO_2$	gives	bromite ion, BrO_2^-
hypobromous acid, $HBrO$	gives	hypobromite ion, BrO^-

Occasionally, you may see acid names that begin with "ortho," "meta," or "pyro." These prefixes come from another naming system, which begins with the prefix "ortho" for the acid with the largest number of OH groups. For example, H_3PO_4, with its three OH groups, $OP(OH)_3$, contains more OH groups than any other phosphoric acid and is known as "orthophosphoric acid." An "ortho" acid that has "lost" one or more water molecules yields a "meta" acid, and an acid produced by the loss of one water molecule between two molecules of the acid is called a "pyro" acid (Table 20.2).

Table 20.2
"Ortho" Nomenclature for Acids
A better name for pyrophosphoric acid is diphosphoric acid.

Orthophosphoric acid	H_3PO_4
	H_3PO_4 minus H_2O
Metaphosphoric acid	HPO_3
	$2H_3PO_4$ minus H_2O
Pyrophosphoric acid	$H_4P_2O_7$

EXAMPLE 20.3 Nomenclature of Acids

Suppose we have discovered a remarkable new stable element and named it foxium, Fx. It readily forms a stable acid, $HFxO_3$, and three other less easily available acids, $HFxO_4$, $HFXO_2$, and HFxO. Salts of each of these acids can be isolated. Name the acids and their oxoanions.

Assuming that $HFxO_3$ is the most common acid and is named foxic acid, the names of the acids and the anions are as follows:

	oxoacid	*oxoanion*
$HFxO_4$	**perfoxic acid**	**perfoxate ion**
$HFxO_3$	**foxic acid**	**foxate ion**
$HFxO_2$	**foxous acid**	**foxite ion**
HFxO	**hypofoxous acid**	**hypofoxite ion**

EXERCISE Name the following oxoacids of sulfur and their respective oxoanions: (a) H_2SO_3, (b) H_2SO_4 (the most common sulfur oxoacid), (c) $H_2S_2O_7$.

Answers (a) sulfurous acid, sulfite ion; (b) sulfuric acid, sulfate ion, (c) pyrosulfuric acid, pyrosulfate ion.

20.4 STRENGTHS OF BINARY ACIDS

The strength of a binary acid depends upon the strength of the bond between the hydrogen atom and the second atom. The two most important factors influencing this bond strength are the radius of the second atom and its electronegativity.

In general, the larger an atom, the weaker are its covalent bonds. Therefore, the E—H bond, where E is the nonmetal atom in a binary acid, will become weaker as E becomes larger. Acid strength should increase with increasing radius of E in E—H bonds.

Electronegativity is the ability of an atom in a bond to attract electrons to itself. In the bond H—E, the greater the electronegativity of E, the more strongly E attracts electrons away from the hydrogen atom, thereby permitting the H^+ ion to separate more easily. Acid strength should also increase with the increasing electronegativity of E in H—E bonds.

In general, in going *down* a periodic table family, the acidity of the hydrides increases because the effect of increasing atomic radius predominates. For example, in the oxygen family, the strength of the binary acids increases with increasing radius down the family, despite the corresponding decrease in electronegativity.

Down a periodic table family

$$H_2O \quad H_2S \quad H_2Se \quad H_2Te$$
$$\longrightarrow$$

increasing acid strength with increasing radius

On the other hand, in going *across* a period in the periodic table, electronegativity becomes the predominating factor, despite the generally decreasing radii. For example, the hydrides of the second-row elements nitrogen, oxygen, and fluorine display increasing acid strength.

Across a periodic table period

$$HN_3 \quad H_2O \quad HF$$
$$\longrightarrow$$

increasing acid strength with increasing electronegativity

20.5 STRENGTHS OF OXOACIDS

As we saw in Chapter 16, compounds containing oxide or hydroxide groups can be acidic, basic, or amphoteric. Although we are not accustomed to thinking of the formulas in this way, sulfuric acid, H_2SO_4, for example, could be written $O_2S(OH)_2$ and H_3PO_4 could be written $OP(OH)_3$. Such oxoacids fit the general formula $O_nE(OH)_m$.

The strengths of oxoacids can vary with the size, electronegativity, and oxidation state of E, as well as the numbers of oxygen atoms or OH groups bonded to E. A most useful generality is that in a series of oxoacids of the same element E, acid strength increases with *n*, that is, with the number of oxygen atoms bonded to E but not to a hydrogen atom. Each such oxygen atom withdraws additional electron density from the bonds in the E—O—H part of the molecule, making it easier for the hydrogen atom to be attracted away as H^+. The chlorine oxoacids with an increasing number of oxygen atoms and the resultant increasing oxidation state of Cl are shown in Table 20.3.

Table 20.3
Oxochlorine Acids
Oxygen content, the oxidation state of chlorine, and the acidity all increase in this series.

Acid	Oxidation state of Cl	Acidity
Hypochlorous, HClO HO:C̈l: (with lone pairs)	+1	Very weak acid
Chlorous, $HClO_2$:Ö: HO:C̈l:	+3	Weak acid
Chloric, $HClO_3$:Ö: HO:C̈l:Ö:	+5	Strong acid
Perchloric, $HClO_4$:Ö: HO:C̈l:Ö: :Ö:	+7	Very strong acid

In comparing oxoacids of identical molecular structure, the acid strength is expected to be greater when E is more electronegative and/or smaller. This trend is illustrated by the hypohalous acids, which increase in acid strength with increasing electronegativity of E as follows: HOI < HOBr < HOCl.

In general, in an EOH group where E is of relatively large size and low electronegativity, the E—OH bond is ionic and the compound is basic. Where E is of intermediate electronegativity and size, amphoterism is encountered. For acids E is of small size and high electronegativity. These general trends are illustrated by the following three series of compounds. Note in the first and third series that the strongest acids are those with the largest number of oxygen atoms (not OH groups) bonded directly to the central atom.

(1) *Periodic table row: increasing acidity with increasing oxidation state and increasing electronegativity*

NaOH	$Mg(OH)_2$	$Al(OH)_3$	$Si(OH)_4$	$OP(OH)_3$	$O_2S(OH)_2$	$O_3Cl(OH)$
strongly basic	*basic*	*amphoteric*	*weakly acidic*	*acidic*	*strongly acidic*	*very strongly acidic*

(2) *Periodic table family: decreasing acidity with increasing radii (same oxidation state; similar electronegativity, except B)*

$B(OH)_3$	$Al(OH)_3$	$Ga(OH)_3$	$In(OH)_3$	$Tl(OH)_3$
acidic	*amphoteric*	*amphoteric*	*basic*	*basic*

(3) *Same element in different oxidation states: increasing acidity with increasing oxidation state*

+2	+3	+4	+5	+6	+7
$Mn(OH)_2$	$Mn(OH)_3$	MnO_2	$OMn(OH)_3$	$O_2Mn(OH)_2$	$O_3Mn(OH)$
basic	*weakly basic*	*amphoteric*	*acidic*	*strongly acidic*	*very strongly acidic*

EXAMPLE 20.4 Relative Strengths of Acids and Bases

Compare the acid strength of an aqueous solution of sulfurous acid (H_2SO_3) to the acid strengths of solutions of the following compounds: (a) sulfuric acid (H_2SO_4), (b) selenous acid (H_2SeO_3), (c) chloric acid ($HClO_3$).

To aid in the comparison, it is useful to write these acids according to the general formula for oxoacids. We want to compare $OS(OH)_2$ with $O_2S(OH)_2$, $OSe(OH)_2$, and $O_2Cl(OH)$.

(a) For oxoacids of a given element, the acid strength increases with the number of oxygen atoms bonded to the central atom. Therefore, **sulfurous acid will be a weaker acid than sulfuric acid.**

(b) For oxoacids of identical formula, the strength is greater for more electronegative and smaller central atoms. Therefore, **sulfurous acid should be a somewhat stronger acid than selenous acid.**

(c) Because the chlorine atom is smaller and more electronegative than the sulfur atom, and because chloric acid has two nonhydrogen-bearing oxygen atoms rather than on bonded to the central atom, **sulfurous acid should be weaker than chloric acid.** [It is.]

EXERCISE Compare the acid strengths of HClO(*aq*), HBrO(*aq*), and $HClO_2$(*aq*).

Answer HClO is stronger than HBrO and $HClO_2$ is stronger than HClO.

20.6 EQUIVALENT MASS AND NORMALITY FOR ACIDS AND BASES

The concept of equivalents played an important role in the early history of chemistry. The relative amounts of two elements or two compounds that reacted with each other in a particular reaction were considered "chemically equivalent." Information about chemical equivalents eventually led to our understanding of atomic mass,

mass in chemical combination, and stoichiometry. For example, Jeremias Richter and Ernst Fischer in the early 1800s published tables of equivalents of acids and bases. They knew that 1000 parts by mass of sulfuric acid or 1405 parts of nitric acid are each neutralized by 672 parts of ammonia. These quantities of sulfuric acid, nitric acid, and ammonia are chemically equivalent in neutralization reactions, although Richter and Fischer did not know why.

Our present knowledge of stoichiometry and the concept of the mole has made the use of equivalents unnecessary in most circumstances. The molar mass of a substance is more useful, for it is based solely on the identity of the substance and has only one value, no matter what the fate of the substance in a chemical reaction. Equivalent mass, however, is based on the behavior of a substance in a particular type of reaction. The definition of equivalent mass depends upon the type of reaction, and the equivalent mass of the same substance may be different in different reactions.

Equivalent masses are, however, sometimes still used in connection with acid–base reactions. The **equivalent mass of an acid** is the mass of that acid that donates one mole of H^+ ions. For a water–ion acid, the number of equivalents per mole is the number of ionizable protons in the acid: HCl has one equivalent per mole; H_2SO_4 has two equivalents per mole; and so on. The **equivalent mass of a base** is the mass of that base that reacts with one mole of H^+ ions. For a water–ion base containing OH^- ions, the number of equivalents per mole is equal to the number of OH^- ions. Thus, NaOH has one equivalent per mole; $Ca(OH)_2$ has two equivalents per mole; and so on.

The equivalent mass of an acid or base is found by dividing its molar mass by its number of equivalents per mole

$$\text{Equiv mass (g/equiv)} = \frac{\text{molar mass (g/mol)}}{\text{no. of equiv per mole}} \qquad \textbf{(20.23)}$$

For example, hydrochloric acid supplies one mole of H^+ for each mole of HCl.

$$HCl(aq) \longrightarrow H^+ + Cl^-$$

Therefore, the equivalent mass of HCl is the same as its molar mass and is equal to 36.46 g/equiv. Sodium hydroxide can react with one mole of H^+ and its equivalent mass is also the same as its molar mass, 40.00 g/equiv. Sulfuric acid and calcium hydroxide each have two equivalents per mole and their equivalent masses are

$$H_2SO_4 \longrightarrow 2H^+ + SO_4^{2-} \qquad \frac{98.07\ \text{g/mol}}{2\ \text{equiv/mol}} = 49.04\ \text{g/equiv} \quad \leftarrow \textit{equivalent mass}$$

$$Ca(OH)_2(s) \xrightarrow{H_2O} Ca^{2+} + 2OH^- \qquad \frac{74.09\ \text{g/mol}}{2\ \text{equiv/mol}} = 37.05\ \text{g/equiv} \quad \leftarrow \textit{equivalent mass}$$

In an acid–base reaction, equal numbers of equivalents of the acid and the base undergo complete reaction:

$$\begin{array}{lll} HCl(aq) & + \quad NaOH(aq) & \longrightarrow NaCl(aq) + H_2O(l) \\ (1\ mol)\left(\frac{1\ equiv}{mol}\right) = 1\ equiv & (1\ mol)\left(\frac{1\ equiv}{mol}\right) = 1\ equiv & \end{array}$$

$$\begin{array}{lll} H_2SO_4(aq) & + \quad 2NaOH(aq) & \longrightarrow Na_2SO_4(aq) + 2H_2O(l) \\ (1\ mol)\left(\frac{2\ equiv}{mol}\right) = 2\ equiv & (2\ mol)\left(\frac{1\ equiv}{mol}\right) = 2\ equiv & \end{array}$$

The number of equivalents per mole can vary with the reaction under consideration, as shown by these reactions of phosphoric acid.

$$\begin{array}{lll} H_3PO_4(aq) + 3NaOH(aq) & \longrightarrow & Na_3PO_4(aq) + 3H_2O(l) \\ \textit{3 equiv/mol} \quad \textit{1 equiv/mol} & & \\ H_3PO_4(aq) + 2NaOH(aq) & \longrightarrow & Na_2HPO_4(aq) + 2H_2O(l) \\ \textit{2 equiv/mol} \quad \textit{1 equiv/mol} & & \\ H_3PO_4(aq) + NaOH(aq) & \longrightarrow & NaH_2PO_4(aq) + H_2O(l) \\ \textit{1 equiv/mol} \quad \textit{1 equiv/mol} & & \end{array}$$

A solution containing one equivalent of a substance in one liter of solution is called a "one normal solution," written 1 N. **Normality** is the concentration of a solution expressed as the number of equivalents of solute per liter of solution. A solution containing 2 equivalents per liter is a 2 N solution, and 0.5 L of a 2 N solution contains one equivalent of solute.

The factor that connects the normality of a given solution to its molarity is the number of equivalents per mole in the reaction under consideration. The relationships are

$$(\text{Normality})\left(\frac{\text{moles}}{\text{equiv}}\right) = \text{molarity}$$
$$\frac{\textit{equiv}}{\textit{liter}} \quad \frac{\textit{moles}}{\textit{equiv}} \quad \frac{\textit{moles}}{\textit{liter}}$$

$$(\text{Molarity})\left(\frac{\text{equiv}}{\text{mole}}\right) = \text{normality}$$
$$\frac{\textit{moles}}{\textit{liter}} \quad \frac{\textit{equiv}}{\textit{mole}} \quad \frac{\textit{equiv}}{\textit{liter}}$$

EXAMPLE 20.5 Equivalents

What are the relationships between the equivalent masses and the molar masses of (a) HNO_3 and (b) $Ba(OH)_2$ in the following reaction?

$$2HNO_3(aq) + Ba(OH)_2(aq) \longrightarrow 2H_2O(l) + Ba(NO_3)_2(aq)$$

(a) Each mole of HNO_3 donates one mole of H^+ —1 equiv/mol—and so the **equivalent mass is the same as the molar mass.**

(b) Each mole of $Ba(OH)_2$ reacts with 2 mol of HNO_3—2 equiv/mol—and so the **equivalent mass is one-half of the molar mass** (Equation 20.23).

EXERCISE What are the relationships between the equivalent masses and the molar masses of H_2SO_3 and NaOH in the following reaction?

$$H_2SO_3(aq) + 2NaOH(aq) \longrightarrow 2H_2O(l) + Na_2SO_3(aq)$$

Answers H_2SO_3, equiv mass = $\frac{1}{2}$ molar mass; NaOH, equiv mass = molar mass.

EXAMPLE 20.6 Equivalents

A commercially available concentrated sulfuric acid is 35.9 N. Assuming H_2SO_4 to donate two moles of H^+ per mole of H_2SO_4, express this concentration in terms of molarity.

The connecting ratio between normality and molarity is 2 equiv/mol. The molarity of the solution is

$$\left(\frac{35.9 \text{ equiv}}{1 \text{ L}}\right)\left(\frac{1 \text{ mol}}{2 \text{ equiv}}\right) = \mathbf{18.0\ mol/L}$$

EXERCISE An old handbook listed various physical properties for a 0.5 N solution of lithium hydroxide. Express this concentration in terms of molarity. *Answer* 0.5 M.

SUMMARY OF SECTIONS 20.3–20.6

Binary acids are compounds of hydrogen and another element that yield hydrogen ions in aqueous solutions. They are identified as, for example, hydrofluoric acid (HF). The strength of a binary acid is dependent upon the strength of the bond between the hydrogen atom and the second atom. Down a periodic table family, the acidity of binary acids increases with the radius of the second atom; across a periodic table row, acidity increases with the electronegativity of the second atom.

Oxoacids contain oxygen, hydrogen, and a central atom of a third type; they have the general formula $O_nE(OH)_m$ (except H_3PO_3, which has one H—P bond). Oxoacids of an element are identified according to the number of oxygen atoms attached to the central element. One acid of the series is given the name of the central element modified with the ending "ic." Other acids are named according to the number of oxygen atoms they contain relative to the "ic" acid, as illustrated for the chlorine oxoacids in Table 20.3. In general, oxoacids are stronger with larger n, larger electronegativity of E, smaller size of E, and higher oxidation state of E.

The number of "equivalents" of an acid in a given reaction is the number of moles of H^+ ions it supplies; for a base, it is the number of moles of H^+ ions with which it reacts. The equivalent mass of an acid or base is the molar mass divided by the number of equivalents per mole. Normality is the concentration of a solution expressed as the number of equivalents of solute per liter of solution.

EXPRESSING THE STRENGTHS OF ACIDS AND BASES

20.7 IONIZATION EQUILIBRIUM OF WATER

Pure water is a slightly ionized substance (Section 14.5)

$$H_2O(l) \rightleftharpoons H^+ + OH^- \quad \text{or} \quad H_2O(l) + H_2O(l) \rightleftharpoons H_3O^+ + OH^- \qquad \mathbf{(20.24)}$$

One liter of water contains 55.5 mol of H_2O molecules. The ionization of water is so slight that in each liter of water there are roughly one H^+ and one OH^- for each 555 million H_2O molecules. Water is ionized to such a small extent that the nonionized water molecules can be considered as part of a pure liquid. Thus their concentration is constant and need not appear in the equilibrium constant expression (see Section 19.5). This expression for reaction (20.24), called the **ion product constant for water,** is therefore

$$K_w = [H^+][OH^-] \qquad \mathbf{(20.25)}$$

Measurements show that at 25 °C the value of K_w is 1.008×10^{-14}. The value increases with temperature; for example, at 60 °C it is 9.6×10^{-14}. With little introduction of error, the value of K_w at 25 °C (and frequently at other temperatures) is usually taken as

$$K_w = [H^+][OH^-] = 1.00 \times 10^{-14} \qquad \mathbf{(20.26)}$$

At equilibrium, the concentrations of H^+ and OH^- ions in pure water are the same. Letting $x = [H^+] = [OH^-]$ and solving the ion product constant expression for x shows that in pure water the concentration of each of these ions is equal to 1.00×10^{-7} mol/L (at 25 °C).

$$[H^+][OH^-] = 1.00 \times 10^{-14}$$
$$(x)(x) = 1.00 \times 10^{-14}$$
$$x^2 = 1.00 \times 10^{-14}$$
$$x = 1.00 \times 10^{-7}\ \text{mol/L}$$

A solution in which $[H^+] = [OH^-]$ is a **neutral solution.** In an **acidic aqueous solution** the concentration of H^+ ion is greater than 1×10^{-7} mol/L and in an **alkaline aqueous solution** the concentration of OH^- is greater than 1×10^{-7} mol/L (all at 25 °C). The relationship between $[H^+]$ and $[OH^-]$ for any aqueous solution, whatever other solutes may be present, is given by the expression for K_w. Based upon Le Chatelier's principle, it can be seen that adding H^+ to a neutral aqueous solution will drive the equilibrium of Equation (20.26) toward the formation of water and there will be a decrease in $[OH^-]$. Similarly, adding OH^- will decrease $[H^+]$.

By using the ion product constant expression, $[OH^-]$ can be found for any solution for which $[H^+]$ is known, and vice versa. The following example illustrates that, as predicted, increasing the hydrogen ion concentration decreases the hydroxide ion concentration from that in pure water.

EXAMPLE 20.7 Ion Product Constant of Water

What is the concentration of OH^- in a 0.01 M HCl solution? Because HCl is a strong acid and is 100% ionized, the H^+ concentration equals the molarity of the HCl solution.

The relationship between the concentrations of H^+ and OH^- in an aqueous solution is

$$K_w = [H^+][OH^-]$$

Solving this for $[OH^-]$ and substituting the known values of K_w and $[H^+]$ gives

$$[OH^-] = \frac{K_w}{[H^+]} = \frac{1.00 \times 10^{-14}}{0.01} = \mathbf{1 \times 10^{-12}\ mol/L}$$

EXERCISE What is the concentration of H^+ in a 0.5 M solution of KOH? Because KOH is a strong, soluble base and is 100% dissociated, the OH^- concentration equals the molarity of the solution. *Answer* 2×10^{-14} mol/L.

20.8 pH, pOH, AND pK_w

The concentrations of hydrogen ion and hydroxide ion in aqueous solutions are frequently of interest. These concentrations are often such small numbers that they would be best given in scientific notation. However, because they are in such common use, a more convenient way for expressing hydrogen and hydroxide ion concentrations has been devised. The expressions used, **pH** and **pOH**, are defined as the negative logarithms of the hydrogen ion and the hydroxide ion concentrations, respectively. (The use of logarithms is reviewed in Appendix I.)

$$pH = -\log[H^+] = \log\frac{1}{[H^+]} \tag{20.27}$$

$$pOH = -\log[OH^-] = \log\frac{1}{[OH^-]} \tag{20.28}$$

The notation "pH" derives from the French *pouvoir hydrogene,* meaning the "power of hydrogen" and referring to the exponent. For example, for pure water at 25 °C, for which $[H^+] = 1 \times 10^{-7}$, the pH is equal to 7.0. The values of pH or pOH for dilute solutions usually fall between 1 and 14, and it is for such solutions that pH or pOH values are most often used. [If the H^+ concentration is larger than 1 M, the pH is negative; if the OH^- concentration is larger than 1 M, the pH is larger than 14.]

The same notation (p...) can be used for the negative logarithm of any quantity, allowing pH and pOH to be combined with the ion product constant.

$$\begin{aligned} K_w &= [H^+][OH^-] \\ \log K_w &= \log([H^+][OH^-]) \\ &= \log[H^+] + \log[OH^-] \\ -\log K_w &= -\log[H^+] - \log[OH^-] \\ pK_w &= pH + pOH \end{aligned} \tag{20.29}$$

Numerically, from Equation (20.26)

$$pK_w = -\log(1.00 \times 10^{-14}) = 14.000$$

so that

$$pH + pOH = 14.000$$
$$pH = 14.000 - pOH \quad (20.30)$$
$$pOH = 14.000 - pH \quad (20.31)$$

(Note that in a logarithm the number of digits after the decimal point equals the number of significant digits in the original number. Thus, a pH of 2.00 represents $[H^+] = 1.0 \times 10^{-2}$.)

In a neutral solution

$$pH = pOH = 7.000$$

If the pH is smaller than 7, the solution is an *acidic aqueous solution;* if the pH is larger than 7, the solution is an *alkaline aqueous solution.* (Remember that this is strictly accurate only at 25 °C.) The larger the departure from 7, the more acidic or the more alkaline the solution is (Table 20.4). Although both pH and pOH are useful, it is common practice to use pH to indicate both the acidity and the alkalinity of an aqueous solution. The pH values of some well-known substances are given in Table 20.5.

Note that the pH and pOH scales are logarithmic, not linear, scales. For example, a solution of pH 1 has ten times the concentration of hydrogen ion of a solution of pH 2, not twice the concentration. A solution of pH 12 has 100 times the concentration of hydroxide ion of a solution of pH 10.

Examples 20.8–20.11 illustrate finding the pH and pOH values for various acidic or alkaline solutions. The solutions used as illustrations are all sufficiently more acidic or more alkaline than water itself that the effect of the ionization of water can be neglected. (*However*, in dealing with very dilute solutions, the contributions to $[H^+]$ or $[OH^-]$ of the ions from water itself cannot be ignored.)

Table 20.4
Acidity, Neutrality, and Alkalinity in Aqueous Solutions (at 25 °C)

$[H^+]$ (mol/L)	pH	$[OH^-]$ (mol/L)	
10^{1}	−1	10^{-15}	↑
10^{0}	0	10^{-14}	
10^{-1}	1	10^{-13}	
10^{-2}	2	10^{-12}	**Acidity increases**
10^{-3}	3	10^{-11}	
10^{-4}	4	10^{-10}	
10^{-5}	5	10^{-9}	
10^{-6}	6	10^{-8}	
10^{-7}	7	10^{-7}	**Neutral**
10^{-8}	8	10^{-6}	
10^{-9}	9	10^{-5}	
10^{-10}	10	10^{-4}	
10^{-11}	11	10^{-3}	**Alkalinity increases**
10^{-12}	12	10^{-2}	
10^{-13}	13	10^{-1}	
10^{-14}	14	10^{0}	
10^{-15}	15	10^{1}	↓

EXAMPLE 20.8 pH and pOH

What is the pH of a 0.01 M solution of HCl?

Hydrochloric acid, a strong acid, is 100% ionized and so $[H^+] = 0.01$ M. The pH for this concentration of H^+ is

$$pH = -\log[H^+]$$
$$= -\log(1 \times 10^{-2}) = -(-2.0)$$
$$= 2.0$$

The pH of 0.01 M HCl is 2.0, written **pH 2.0.**

EXERCISE The concentration of hydrogen ion in a cup of black coffee is 1.3×10^{-5} M. Find the pH of the coffee. Is this coffee acidic or alkaline? *Answer* 4.89, acidic.

Table 20.5
Approximate pH Values of Some Common Substances

Substance		pH
0.1 M HCl	↑	1.0
Gastric juice		1.4
Lemon juice		2.3
Vinegar	**Acidic substances**	2.9
Orange juice		3.5
Tomatoes		4.2
Coffee		5.0
Rainwater		6.2
Pure water		7.0
Blood		7.4
Seawater	**Alkaline substances**	8.5
Bar soap		11.0
Household ammonia		11.5
0.1 M NaOH	↓	13.0

EXAMPLE 20.9 pH and pOH

What is the hydrogen ion concentration in a solution of pH 1.5?

We need to rearrange the definition of pH

$$\log[H^+] = -pH = -1.5$$

To solve such an expression requires finding the antilogarithm of -1.5, which is equivalent to finding the antilogarithm of (0.5 − 2) or finding $10^{-1.5}$. The answer must be rounded to one significant figure. The answer is

$$[H^+] = 3 \times 10^{-2} \text{ mol/L, or } \mathbf{0.03 \text{ mol/L}}$$

EXERCISE The pH of blood serum is 7.4. What is the hydrogen ion concentration of blood serum? Is blood acid or alkaline? *Answer* 4×10^{-8} mol/L, alkaline.

EXAMPLE 20.10 pH and pOH

What are the pH and $[H^+]$ of a solution that has $[OH^-] = 3.5 \times 10^{-5}$ mol/L?

The pOH of the solution is

$$\text{pOH} = -\log[OH^-] = -\log(3.5 \times 10^{-5}) = 4.46$$

giving

$$\text{pH} = 14.00 - \text{pOH} = 14.00 - 4.46 = \mathbf{9.54}$$

which corresponds to

$$\log[H^+] = -\text{pH} = -9.54$$
$$[H^+] = \mathbf{2.9 \times 10^{-10} \text{ mol/L}}$$

EXERCISE The pH of an alkaline solution is 8.36. What are the values of (a) pOH, (b) $[OH^-]$, and (c) $[H^+]$ for this solution?
Answers (a) 5.64, (b) 2.3×10^{-6} mol/L, (c) 4.4×10^{-9} mol/L.

EXAMPLE 20.11 pH and pOH

Which is the most acidic, a solution with (a) $[H^+] = 0.3$ mol/L, (b) $[OH^-] = 0.5$ mol/L, (c) pH 1.2, or (d) pOH 5.9? Acidities can be compared by comparing any one of the four quantities, $[H^+]$, $[OH^-]$, pH, or pOH. (We shall convert all of the given information to $[H^+]$ to make the comparison.)

Solution (a) has $[H^+] = 0.3$ mol/L. The $[H^+]$ concentration for solution (b) can be found by using the expression for K_w:

$$K_w = [H^+][OH^-]$$
$$[H^+] = \frac{K_w}{[OH^-]} = \frac{1.00 \times 10^{-14}}{0.5} = 2 \times 10^{-14} \text{ mol/L}$$

Finding $[H^+]$ from the pH for solution (c) gives

$$\log[H^+] = -\text{pH} = -(1.2)$$
$$[H^+] = 0.06 \text{ mol/L}$$

For solution (d) we first find the pH

$$\text{pH} = 14.0 - \text{pOH} = 14.0 - 5.9 = 8.1$$

and then use this value to find $[H^+]$:

$$\log[H^+] = \text{pH} = -(8.1)$$
$$[H^+] = 8 \times 10^{-9} \text{ mol/L}$$

We see that **solution (a) is the most acidic.**

EXERCISE Which is the most alkaline, a solution with (a) $[H^+] = 0.0025$ mol/L, (b) $[OH^-] = 0.0025$ mol/L, (c) pH = 3.65, or (d) pOH = 9.26? *Answer* (b).

Aside 20.1 THOUGHTS ON CHEMISTRY: IRA REMSEN INVESTIGATES NITRIC ACID

You should learn a few practical lessons about acids and bases, if you have not already experienced them. Acids and bases can hurt you—from the stinging sensation when you accidentally squirt lemon juice (citric acid) in your eye to very severe and persistent burns if you spill battery acid (sulfuric acid) or lye (sodium hydroxide) on your skin and do not immediately flush it with water.

Acids and bases can also eat holes in clothing. You might first realize that an acidic or basic solution has splashed on you when your shirt or blouse comes out of the washing machine with little holes where the fabric was weakened. If you spill concentrated acid or base directly on your clothing, the effect will be immediately noticeable and the garment probably ruined. The conclusion is obvious—acids and bases must be handled with care.

How Ira Remsen, who later founded the chemistry department at Johns Hopkins University, discovered the properties of nitric acid in the 1800s is described in the following quotation:

"While reading a textbook of chemistry," said he, "I came upon the statement, 'nitric acid acts upon copper.' I was getting tired of reading such absurd stuff and I determined to see what this meant. Copper was more or less familiar to me, for copper cents were then in use. I had seen a bottle marked 'nitric acid' on a table in the doctor's office where I was then 'doing time!' I did not know its peculiarities, but I was getting on and likely to learn. The spirit of adventure was upon me. Having nitric acid and copper, I had only to learn what the words 'act upon' meant. Then the statement, 'nitric acid acts upon copper,' would be something more than mere words. All was still. In the interest of knowledge I was even willing to sacrifice one of the few copper cents then in my possession. I put one of them on the table; opened the bottle marked 'nitric acid'; poured some of the liquid on the copper; and prepared to make an observation. But what was this wonderful thing which I beheld? The cent was already changed, and it was no small change either. A greenish blue liquid foamed and fumed over the cent and over the table. The air in the neighborhood of the performance became colored dark red. A great colored cloud arose. This was disagreeable and suffocating—how should I stop this? I tried to get rid of the objectionable mess by picking it up and throwing it out of the window, which I had meanwhile opened. I learned another fact—nitric acid not only acts upon copper but it acts upon fingers. The pain led to another unpremeditated experiment. I drew my fingers across my trousers and another fact was discovered. Nitric acid acts upon trousers. Taking everything into consideration, that was the most impressive experiment, and, relatively, probably the most costly experiment I have ever performed. I tell of it even now with interest. It was a revelation to me. It resulted in a desire on my part to learn more about that remarkable kind of action. Plainly the only way to learn about it was to see its results, to experiment, to work in a laboratory."

Source: Frederick H. Getman, *The Life of Ira Remsen* (Easton, PA: Journal of Chemical Education, 1940), pp. 9, 10.

20.9 THE ACID IONIZATION CONSTANT, K_a

The strength of a protonic acid in aqueous solution is measured by the extent to which it ionizes

$$HA(aq) + H_2O(l) \rightleftharpoons H_3O^+ + A^- \quad \text{or} \quad HA(aq) \rightleftharpoons H^+ + A^- \tag{20.32}$$

The equilibrium constant for this reaction is called the **acid ionization constant**, K_a.

$$K_a = \frac{[H^+][A^-]}{[HA]} \tag{20.33}$$

The larger the value of K_a, the stronger is the acid. The strong acids that are virtually 100% ionized have K_a values much larger than 1 because [HA] is very small. Weak acids, on the other hand, are less than 100% ionized. For weak acids, the predominant direction of reaction (20.32) is toward the formation of HA, and the K_a values are less than 1. For example, consider the series of chlorine oxoacids listed in Table 20.3. Perchloric and chloric acids are both such strong acids that their K_a values are greater than 1; K_a values of such strong acids are usually reported in tables simply as "large." The predicted relative strengths of the other two chlorine oxoacids are shown by their K_a values: for the very weak hypochlorous acid, $K_a = 2.90 \times 10^{-9}$; for the somewhat stronger chlorous acid, $K_a = 1.1 \times 10^{-2}$.

Table 20.6
Ionization Constants of Weak Acids at 25 °C for the Reaction $HA(aq) \rightleftharpoons A^- + H^+$

	K_a	pK_a		K_a	pK_a
Hydrogen sulfate ion HSO_4^-	1.0×10^{-2}	2.00	Hydrosulfuric acid H_2S	1.0×10^{-7}	7.00
Phosphoric acid H_3PO_4	7.5×10^{-3}	2.12	Ammonium ion NH_4^+	6.3×10^{-10}	9.20
Nitrous acid HNO_2	7.2×10^{-4}	3.14	Hydrogen sulfide ion HS^-	3×10^{-13}	12.5
Hydrofluoric acid HF	6.5×10^{-4}	3.19	Chromium(III) hydroxide $Cr(OH)_3$	9×10^{-17}	16.0
Acetic acid CH_3COOH	1.754×10^{-5}	4.7560	Copper(II) hydroxide $Cu(OH)_2$	1×10^{-19}	19.0
Carbonic acid H_2CO_3	4.5×10^{-7}	6.35	Zinc hydroxide $Zn(OH)_2$	1.0×10^{-29}	29.00

Since K_a values for weak acids are frequently very small numbers, they can also be treated very conveniently as negative logarithms. The pK_a is defined as follows:

$$pK_a = -\log K_a \qquad \textbf{(20.34)}$$

Table 20.6 gives the K_a and pK_a values for some weak acids in the order of decreasing acid strength, as shown by the *decreasing* K_a values and *increasing* pK_a values. (A more extensive list of K_a values is given in Appendix VA.)

Equilibrium problems concerned with the concentrations of acids and their ions (or bases and their ions; Section 20.11) are solved by the same methods as any other equilibrium problems. If the concentration of any of the species present in the solution of an acid is changed, the equilibrium shifts according to Le Chatelier's principle, and the concentrations of the other species change so that the value of the equilibrium constant expression remains constant. Example 20.13 illustrates finding the concentration of an anion in the presence of additional hydrogen ion.

EXAMPLE 20.12 Acid Ionization Constants

A solution of iodic acid (HIO_3) was found to have, at equilibrium, a hydrogen ion concentration of 0.06 mol/L and an HIO_3 concentration of 0.02 mol/L. What is the value of K_a for the acid? Would HIO_3 be considered a "strong" or a "weak" acid? How does it compare in acid strength with the acids listed in Table 20.6?

The ionization equilibrium and the ionization constant expression for HIO_3 are

$$HIO_3(aq) \rightleftharpoons H^+ + IO_3^- \qquad K_a = \frac{[H^+][IO_3^-]}{[HIO_3]}$$

At equilibrium $[H^+] = [IO_3^-]$, allowing K_a to be found from the known $[H^+]$ and $[HIO_3]$.

$$K_a = \frac{[H^+][IO_3^-]}{[HIO_3]} = \frac{(0.06)(0.06)}{(0.02)} = \mathbf{0.18}$$

Because iodic acid has a K_a value less than 1, it is considered to be a **weak acid.** Comparison with the K_a values in Table 20.6 shows that it is **one of the stronger of the weak acids.**

EXERCISE The chemical equation for the ionization of benzoic acid (C_6H_5COOH) is

$$C_6H_5COOH(aq) + H_2O(l) \rightleftharpoons H_3O^+ + C_6H_5COO^-$$

A solution of benzoic acid was found to contain $[H^+] = 2.6 \times 10^{-3}$ mol/L and $[C_6H_5COOH] = 0.100$ mol/L at equilibrium. What is the value of K_a for this acid? *Answer* 6.8×10^{-5}.

EXAMPLE 20.13 Acid Ionization Constants

Hydrogen peroxide can act as a weak acid

$$H_2O_2(aq) + H_2O(l) \rightleftharpoons H_3O^+ + HO_2^- \qquad K_a = 2.2 \times 10^{-12}$$

Assuming that additional acid (e.g., hydrochloric acid) has been added to a hydrogen peroxide solution, giving, at equilibrium, $[H^+] = 0.036$ mol/L and $[H_2O_2] = 0.100$ mol/L, what is the concentration of HO_2^- present at equilibrium?

The equilibrium constant expression for this reaction is

$$K_a = \frac{[H^+][HO_2^-]}{[H_2O_2]}$$

The unknown is the $[HO_2^-]$ that is in equilibrium with known amounts of $[H^+]$ and $[H_2O_2]$. (Note that in this solution, because of the added acid, $[HO_2^-] \neq [H^+]$.) Solving the acid ionization constant expression for $[HO_2^-]$ and substituting the data gives

$$[HO_2^-] = \frac{K_a[H_2O_2]}{[H^+]} = \frac{(2.2 \times 10^{-12})(0.100)}{(0.036)} = \mathbf{6.1 \times 10^{-12}\ mol/L}$$

EXERCISE A solution contains formic acid (HCOOH) and hydrochloric acid. At equilibrium, $[H^+] = 0.15$ mol/L and $[HCOOH] = 0.037$ mol/L. What is the concentration of the formate ion under these conditions? $K_a = 1.77 \times 10^{-4}$ for formic acid. *Answer* 4.4×10^{-5} mol/L.

EXAMPLE 20.14 Acid Ionization Constants

Find the concentrations of the various species present in a 0.10 M solution of nitrous acid (HNO_2); $K_a = 7.2 \times 10^{-4}$ for HNO_2. What is the pH of this solution?

Nitrous acid establishes the equilibrium

$$HNO_2(aq) + H_2O(l) \rightleftharpoons H_3O^+ + NO_2^- \qquad K_a = \frac{[H^+][NO_2^-]}{[HNO_2]}$$

At equilibrium, $[H^+] = [NO_2^-]$. We choose $x = [H^+] = [NO_2^-]$.

	$HNO_2(aq)$ $\rightleftharpoons$	H^+	+ NO_2^-
Initial	0.10	0	0
Change	$-x$	$+x$	$+x$
Equilibrium	$0.10 - x$	x	x

Substituting into the expression for K_a gives

$$K_a = \frac{[H^+][NO_2^-]}{[HNO_2]} = \frac{(x)(x)}{(0.10 - x)} = 7.2 \times 10^{-4}$$

We first attempt to solve this expression for x by using the approximation $(0.10 - x) \approx 0.10$, which gives

$$x^2 = (7.2 \times 10^{-4})(0.10) = 7.2 \times 10^{-5}$$
$$x = 0.0085$$

The approximation is not valid because the error

$$\frac{0.0085}{0.10} \times 100 = 8.5\%$$

exceeds our permitted 5% error. Therefore we must use the quadratic formula to find the solution

$$x^2 = (7.2 \times 10^{-4})(0.10 - x) = 7.2 \times 10^{-5} - (7.2 \times 10^{-4})x$$
$$x^2 + (7.2 \times 10^{-4})x - 7.2 \times 10^{-5} = 0$$
$$x = \frac{-(7.2 \times 10^{-4}) \pm \sqrt{(7.2 \times 10^{-4})^2 - (4)(1)(-7.2 \times 10^{-5})}}{2}$$
$$= 0.0081$$

Thus at equilibrium $\mathbf{[H^+] = [NO_2^-] = x = 0.0081\ mol/L}$ and $\mathbf{[HNO_2] = (0.10 - x) = (0.10 - 0.0081) = 0.09\ mol/L}$. The pH of the solution is $\mathbf{pH = -log[H^+] = -log(0.0081) = 2.09}$.

EXERCISE Find the concentrations of the various species present in a 0.0108 M solution of hypochlorous acid. $K_a = 2.90 \times 10^{-8}$ for HClO.

Answer $[H^+] = [ClO^-] = 1.77 \times 10^{-5}$ mol/L, $[HClO] = 0.0108$ mol/L.

20.10 PERCENT IONIZATION OF WEAK MONOPROTIC ACIDS

The percent of ionization of a weak acid is found from the concentration of hydrogen ion at equilibrium and the *initial* concentration of the acid. In general, for a monoprotic acid

$$\text{Percent ionization} = \frac{[H^+]_{\text{equilibrium}}}{[HA]_{\text{initial}}} \times 100 \qquad \textbf{(20.35)}$$

For example, at equilibrium the hydrogen ion concentration in an initially 0.10 M acetic acid solution is 0.0013 M, giving a percent ionization of 1.3%:

$$\text{Percent ionization} = \left(\frac{0.0013\text{ mol/L}}{0.10\text{ mol/L}}\right) \times 100 = 1.3\%$$

The percentages of ionization for some weak acids are given in Table 20.7. Although it may not be immediately obvious, there is an important difference between K_a values and percent ionization values. For a given acid (at the same temperature), K_a is always the same. However, the percent of ionization of a weak acid or base varies with the initial concentration of the solution. The more dilute the solution, the larger is the percent ionization. For example, acetic acid in a 0.1 M solution is 1.3% ionized, but in a 0.01 M solution it is 4.3% ionized. This increase in ionization is a consequence of Le Chatelier's principle. The increasing dilution is equivalent to the addition of water to the following equilibrium system:

$$HA(aq) + H_2O(l) \rightleftharpoons H_3O^+ + A^-$$

The increase in the amount of water causes the equilibrium to shift in the direction of further ionization of HA. However, that is not the whole story. The greater dilution also decreases the *concentration* of the hydrogen, or hydronium, ions. Therefore, the

Table 20.7
Percent Ionization of Some Weak Acids in Aqueous Solution at 25 °C
The initial concentration of the acid is given in parentheses.

Reaction	Percent ionization
$H_3PO_4(aq) \rightleftharpoons H^+ + H_2PO_4^-$ ***phosphoric acid (0.5 M)***	17%
$HF(aq) \rightleftharpoons H^+ + F^-$ ***hydrofluoric acid (0.1 M)***	7.0%
$CH_3COOH(aq) \rightleftharpoons H^+ + CH_3COO^-$ ***acetic acid (0.1 M)***	0.4%
$H_2O + CO_2 \rightleftharpoons H^+ + HCO_3^-$ ***"carbonic acid" (H_2CO_3)(0.1 M)***	0.17%
$H_2S(aq) \rightleftharpoons H^+ + HS^-$ ***hydrosulfuric acid (0.1 M)***	0.07%

solution becomes *less* acidic as it is made more dilute, even though the *percentage of ionization* is increased. For acetic acid solutions:

$$CH_3COOH(aq) + H_2O(l) \rightleftharpoons CH_3COO^- + H_3O^+$$

Concentration	*% ionization*	*$[H^+]$*	*pH*
0.10 M	1.3%	0.0013 M	2.88
0.010 M	4.3%	0.00043 M	3.88
	more ionized		*less acidic*

The pH of a solution of a strong acid can be found directly from the concentration of the solution (see Example 20.8), but this is not the case for weak acids. Because weak acids are less than 100% ionized, it is necessary to know, in addition to the concentration of the solution, one of the three quantities K_a, pH, or percent ionization. Example 20.15 illustrates the relationships among these quantities.

EXAMPLE 20.15 Percent Ionization

Find the percent ionization of hypobromous acid (HBrO) in a 0.025 M solution. What is the pH of this solution?

$$\mathbf{HBrO(aq) \rightleftharpoons H^+ + BrO^- \qquad K_a = 2.2 \times 10^{-9}}$$

First it is necessary to find $[H^+]$ at equilibrium. Letting $x = [H^+]$, we have

	$HBrO(aq)$	$\rightleftharpoons H^+$	$+ \; BrO^-$
Initial	0.025	0	0
Change	$-x$	$+x$	$+x$
Equilibrium	$0.025 - x$	x	x

$$K_a \frac{[H^+][BrO^-]}{[HBrO]} = \frac{(x)(x)}{(0.025 - x)} = 2.2 \times 10^{-9}$$

Assuming that $(0.025 - x) \simeq 0.025$, we get

$$x^2 = (0.025)(2.2 \times 10^{-9}) = 5.5 \times 10^{-11}$$
$$x = 7.4 \times 10^{-6} = [H^+]$$

which corresponds to

$$pH = -\log[H^+] = -\log(7.4 \times 10^{-6}) = \mathbf{5.13}$$

The percent ionization is

$$\left(\frac{[H^+]_{equilibrium}}{[HBrO]_{initial}}\right)(100) = \left(\frac{7.4 \times 10^{-6}\ mol/L}{0.025\ mol/L}\right)(100) = \mathbf{0.030\%}$$

EXERCISE Find the percent ionization of hypobromous acid in a 0.020 M solution. What is the pH of this solution? *Answers* 0.033%, 5.18.

20.11 THE BASE IONIZATION CONSTANT, K_b

The equilibrium between a proton-accepting weak base, B, and water can be represented by

$$B(aq) + H_2O(l) \rightleftharpoons BH^+ + OH^-$$

and the **base ionization constant** for this equilibrium is

$$K_b = \frac{[BH^+][OH^-]}{[B]} \qquad \textbf{(20.36)}$$

Table 20.8
Ionization Constants for Some Weak Bases at 25 °C

	K_b	pK_b		K_b	pK_b
Hydroxide ion OH^-	1	0	Ammonia NH_3	1.6×10^{-5}	4.80
Sulfide ion S^{2-}	3×10^{-2}	1.5	Acetate ion CH_3COO^-	5.7×10^{-10}	9.24
Triethylamine $(C_2H_5)_3N$	5.2×10^{-4}	3.28	Aniline $C_6H_5NH_2$	4.2×10^{-10}	9.38
Carbonate ion CO_3^{2-}	2.1×10^{-4}	3.68	Fluoride ion F^-	1.5×10^{-11}	10.82
Trimethylamine $(CH_3)_3N$	6.3×10^{-5}	4.20	Nitrate ion NO_3^-	5×10^{-17}	16.3
Cyanide ion CN^-	1.6×10^{-5}	4.80	Chloride ion Cl^-	3×10^{-23}	22.5

The weak base B may be a neutral molecule, commonly ammonia

$$NH_3(aq) + H_2O(l) \rightleftharpoons NH_4^+ + OH^-$$

for which

$$K_b = \frac{[NH_4^+][OH^-]}{[NH_3]} = 1.6 \times 10^{-5} \qquad \textbf{(20.37)}$$

or an anion

$$F^- + H_2O(l) \rightleftharpoons HF(aq) + OH^-$$

for which

$$K_b = \frac{[HF][OH^-]}{[F^-]} = 1.5 \times 10^{-11} \qquad \textbf{(20.38)}$$

The larger the value of K_b, the stronger is the base; also, the larger the pK_b value, the weaker is the base. Table 20.8 gives the K_b and pK_b values of some common weak bases. You can see from the pK_b values that the anions of the weak acids H_2S, HCN, and H_2CO_3 are all relatively strong bases. As would be predicted from the Brønsted-Lowry acid–base definitions, anions of the strong acids such as HNO_3 and HCl are relatively weak bases.

EXAMPLE 20.16 Base Ionization Constants

What is the pH of a 0.15 M solution of NH_3? What is the percent ionization of the NH_3?

$$NH_3(aq) + H_2O(l) \rightleftharpoons NH_4^+ + OH^- \qquad K_b = 1.6 \times 10^{-5}$$

To find the pH from the known K_b, we first find $[OH^-]$ and then use this to find the pOH, from which we can find the pH. The concentration of OH^- can be found in the usual way.

	$NH_3(aq) + H_2O(l)$	$\rightleftharpoons NH_4^+$	+ OH^-
Initial	0.15	0	0
Change	$-x$	$+x$	$+x$
Equilibrium	$0.15 - x$	x	x

$$K_b = \frac{[NH_4^+][OH^-]}{[NH_3]} = \frac{(x)(x)}{(0.15 - x)} = 1.6 \times 10^{-5}$$

$$x = [OH^-] = 1.5 \times 10^{-3}\ \text{mol/L}$$

The percentage of ammonia that is ionized is

$$\frac{1.5 \times 10^{-3}\ \text{mol/L}}{0.15\ \text{mol/L}} \times 100 = \mathbf{1.0\%}$$

Now, solving for pOH and pH

$$pOH = -\log[OH^-] = -\log(1.5 \times 10^{-3}) = 2.82$$
$$pH = 14.00 - pOH = 14.00 - 2.82 = \mathbf{11.18}$$

EXERCISE What is the pH of a 0.015 M solution of NH_3? What is the percentage of ammonia that is ionized in this solution? Compare this to the 1.0% ionization of a 0.15 M NH_3 solution.

Answers pH 10.69, 3.3%, larger percentage of ionization in more dilute solution.

20.12 RELATIONSHIP OF K_a AND K_b TO K_w

The product of the ionization constant for an acid and the ionization constant of its conjugate base is the ion product constant of water.

$$K_a \times K_b = K_w \qquad (20.39)$$

For example, for hydrofluoric acid

$$HF(aq) + H_2O(l) \rightleftharpoons F^- + H_3O^+ \qquad K_a = \frac{[F^-][H^+]}{[HF]} = 6.5 \times 10^{-4}$$

and for fluoride ion, its conjugate base

$$F^- + H_2O(l) \rightleftharpoons HF(aq) + OH^- \qquad K_b = \frac{[HF][OH^-]}{[F^-]} = 1.5 \times 10^{-11}$$

The product of K_a and K_b is shown to be K_w as follows

$$K_a \times K_b = \frac{[\cancel{F^-}][H^+]}{[\cancel{HF}]} \times \frac{[\cancel{HF}][OH^-]}{[\cancel{F^-}]} = [H^+][OH^-] = K_w$$

EXAMPLE 20.17 K_a–K_b Relationship

The value of K_a for hydrocyanic acid (HCN) is 6.2×10^{-10}. What is the value of K_b for the conjugate base?

From the equation

$$HCN(aq) + H_2O(l) \rightleftharpoons H_3O^+ + CN^-$$

we can see that the conjugate base is the cyanide ion, CN^-. The value of K_b for the reaction of CN^- as a base with water

$$CN^- + H_2O(l) \longrightarrow HCN(aq) + OH^-$$
$$K_b = \frac{K_w}{K_a} = \frac{1.00 \times 10^{-14}}{6.2 \times 10^{-10}} = \mathbf{1.6 \times 10^{-5}}$$

EXERCISE The value of K_b for NH_3 is 1.6×10^{-5}. What is the value of K_a for the conjugate acid? *Answer* 6.3×10^{-10}.

SUMMARY OF SECTIONS 20.7–20.12

For water, K_w, the ion product constant, equals $[H^+][OH^-]$. The acidity or alkalinity of aqueous solutions is expressed as pH or pOH, which are the negative logarithms of $[H^+]$ and $[OH^-]$, respectively. Pure water is neutral and at 25 °C has pH 7. A pH > 7 indicates an alkaline solution; a pH < 7 indicates an acidic solution.

The relative strengths of acids and bases are shown by the K_a and K_b values—the equilibrium constants for the production of H^+ and OH^- ions in aqueous solution. For weak acids and bases, K_a and K_b are less then 1. The weaker the acid or base, the smaller is its value of K_a or K_b. The pK_a, the negative logarithm of the K_a, is also used to compare acid strengths; the pK_b, the negative logarithm of K_b, is also used to compare base strengths. The equilibrium constants K_a and K_b for conjugate acid–base pairs are related to K_w by $(K_a)(K_b) = K_w$.

LEWIS DEFINITIONS OF ACIDS AND BASES

20.13 ELECTRON-PAIR DONORS AND ACCEPTORS

The Brønsted-Lowry acid–base definitions are entirely satisfactory for interpreting acid–base reactions that involve proton transfer. And since we are mainly concerned with acid–base behavior in water as a solvent, those are the definitions that we have emphasized. However, there are many important reactions that do not involve proton transfer, which also can be looked upon as acid–base reactions. Of the many other acid–base definitions, the most generally useful has been that of G. N. Lewis.

Lewis observed that substances classified as acids and bases meet four experimental criteria:

1. Acids and bases combine rapidly with each other.
2. Acids and bases may be titrated against each other with the use of an indicator to identify an end point (Section 21.8).
3. An acid or base will usually displace a weaker acid or base from a compound.
4. Acids and bases often act as catalysts in chemical reactions.

Lewis applied these criteria (which are met by both water-ion and Brønsted-Lowry acids and bases) and introduced acid–base definitions that emphasize the role of electron pairs rather than protons.

A **Lewis acid** is a molecule or ion that can accept one or more electron pairs. A **Lewis base** is a molecule or ion that can donate an electron pair. The "donated" electron pair is shared between the acid and base (not physically transferred from one to the other). For example, in the following reaction, which occurs in the gas phase, boron trifluoride is an electron-pair acceptor and ammonia is an electron-pair donor

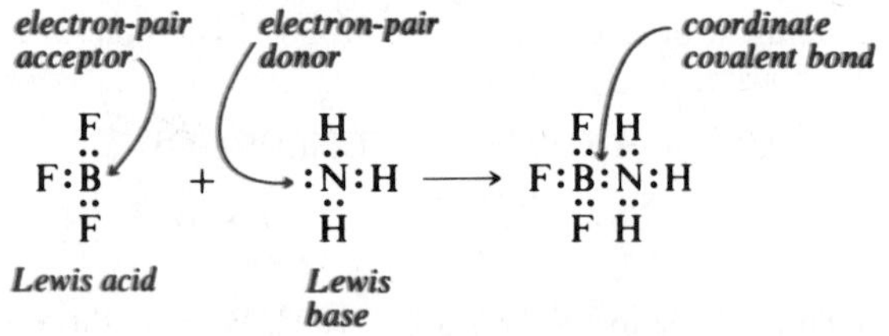

electron-pair acceptor — *electron-pair donor* — *coordinate covalent bond*

$$\underset{\text{Lewis acid}}{F{:}\overset{F}{\underset{F}{B}}} \; + \; \underset{\text{Lewis base}}{{:}\overset{H}{\underset{H}{N}}{:}H} \longrightarrow F{:}\overset{F}{\underset{F}{B}}{:}\overset{H}{\underset{H}{N}}{:}H$$

The result of this reaction is the formation of a coordinate covalent bond between the boron and nitrogen atoms. A **Lewis acid–base reaction** is the donation of an electron pair from one atom to a covalent bond formed with another atom.

Because boron and aluminum are Representative Group III elements, atoms of these elements are surrounded in their compounds by only six outer shell electrons. These elements can complete their octets by accepting an electron pair, allowing their compounds to function as Lewis acids. Boron and aluminum compounds are active as catalysts in many reactions, presumably by temporarily forming electron-pair bonds to produce reactive intermediates. Metal atoms in which vacant *d* orbitals are available for occupancy by electron pairs are also found in Lewis acids [see reaction (d) in Table 20.9].

Table 20.9
Lewis Acid–Base Reactions
The shared electron pair is shown in color.

Acid +	Base	⇌	Reaction product
Octet completion			
(a) BF_3 (:F:B with three F)	+ :N:H (NH_3)	⇌	$F_3B{:}NH_3$
(b) $AlCl_3$ (:Cl:Al)	+ $(:Cl:)^-$	⇌	$[:Cl:Al:Cl:]^-$ ($[AlCl_4]^-$)
(c) SO_3 (:O:S)	+ $(:O:)^{2-}$ Ba^{2+}	⇌	$(:O:S:O:)^{2-}$ Ba^{2+} (SO_4^{2-})
Occupation of empty d orbitals			
(d) $SnCl_4$	+ $2(:Cl:)^-$	⇌	$[Cl_4Sn(:Cl:)_2]^{2-}$
Reactions of cations			
(e) Fe^{3+}	+ 2:Ö—H (with H)	⇌	$[Fe{:}O{-}H]^{2+}$ + $[H{:}O{:}H$ (H)$]^+$
(f) Cu^{2+}	+ 4:N—H (with two H)	⇌	$[Cu(:N{-}H$ (with two H)$)_4]^{2+}$
Attraction of electron pair by atom in multiple bond			
(g) Ö::C::Ö	+ $(:Ö:H)^-$	⇌	$[$H, :Ö:, C, O, O$]^{2-}$

Several types of reactions that we have discussed earlier can also be classified as Lewis acid–base reactions. In the formation of complex ions, the ligands are Lewis bases and the metal cations are Lewis acids [reaction (f), Table 20.9]. The hydration of metal cations can also be viewed as Lewis acid–base reactions—an electron pair is donated from the oxygen atom of a water molecule to a cation. All M^{3+} cations and also Li^+, Be^{2+}, and Mg^{2+} are active as Lewis acids. The strength of a metal cation as a Lewis acid increases with its polarizing ability. The smaller cations with the higher charge-to-size ratios are the stronger Lewis acids.

The combination of an acidic oxide and a basic oxide is shown to be a Lewis acid–base reaction by writing out the Lewis structures of the two oxides [reaction (c), Table 20.9].

The Lewis definitions are especially useful in organic chemistry. In many reactions electron-deficient carbon atoms act as electron-pair acceptors. Carbon atoms are electron deficient when bonded to electronegative atoms and/or when electron density is withdrawn due to resonance in multiply bonded structures [reaction (g), Table 20.9].

Examination of the reactions given in Table 20.9 shows that all of the bases are also Brønsted-Lowry bases, but the Lewis definition includes many species that contain no protons and therefore are not Brønsted-Lowry acids. Protonic acids are traditionally also considered to be Lewis acids, and the hydrogen ion itself can be thought of as the electron-pair acceptor. (This causes no problems, although, strictly speaking, only free H^+ ions, which do not exist in aqueous solution, could *directly* accept electron pairs.)

EXAMPLE 20.18 Lewis Acids and Bases

Each of the following is a Lewis acid–base reaction:

(a) $Ni(s) + 4CO(g) \longrightarrow Ni(CO)_4(g)$ **[An Ni—C bond is formed.]**
(b) $NH_3(aq) + H^+ \longrightarrow NH_4^+$
(c) $BF_3(aq) + F^- \longrightarrow [BF_4]^-$
(d) $H^+ + OH^- \longrightarrow H_2O(l)$

Identify the base in each reaction.

(a) In this reaction the **carbon monoxide** is the base, for an electron pair is donated by the carbon atom to the nickel atom

$$Ni \longleftarrow :C{\equiv}O:$$

(b) In Lewis acid–base terms, the free hydrogen ion is thought of as the electron-pair acceptor and therefore NH_3 is the base and the electron-pair donor.

$$H^+ \longleftarrow :NH_3$$

(c) The **fluoride ion** donates a pair of electrons to the boron atom and is therefore the base.

$$F_3B \longleftarrow :\ddot{\underset{..}{F}}:^-$$

(d) The H^+ ion is an electron-pair acceptor, a Lewis acid; the OH^- ion is an electron-pair donor, a Lewis base, $H^+ \longleftarrow :OH^-$. Thus, you can see that the water-ion acids and bases are also Lewis acids and bases.

EXERCISE In each of the following Lewis acid–base reactions, identify the acid:
(a) $SO_3(g) + (C_2H_5)_2O(l) \longrightarrow (C_2H_5)_2OSO_3(l)$
(b) $SiF_4(g) + 2F^- \longrightarrow [SiF_6]^{2-}$
(c) $S(s) + S^{2-} \longrightarrow [S_2]^{2-}$

Answers (a) SO_3, (b) SiF_4, (c) S.

SUMMARY OF SECTION 20.13

A Lewis acid is an electron-pair acceptor; a Lewis base is an electron-pair donor; a Lewis acid–base reaction is the donation of an electron pair from one atom to a covalent bond formed with another atom. All Lewis bases are also Brønsted-Lowry bases. Lewis acids include small cations of high polarizing ability, compounds of boron and aluminum, and compounds of metal atoms with vacant *d* orbitals. The formation of complex ions and the hydration of cations are examples of Lewis acid–base reactions (others are given in Table 20.9).

SIGNIFICANT TERMS (with section references)

acid ionization constant (20.9)
acidic aqueous solution (20.7)
alkaline aqueous solution (20.7)
base ionization constant (20.11)
Brønsted-Lowry acid (20.1)
Brønsted-Lowry acid–base reaction (20.1)
Brønsted-Lowry base (20.1)
equivalent mass of a base (20.6)
equivalent mass of an acid (20.6)
ion product constant for water (20.7)
Lewis acid (20.13)
Lewis acid–base reaction (20.13)
Lewis base (20.13)
neutral solution (20.7)
normality (20.6)
oxoacids (20.3)
pH (20.8)
pOH (20.8)

QUESTIONS AND PROBLEMS

Brønsted-Lowry Acids and Bases

20.1 Define the term "Brønsted-Lowry acid." Why are all water-ion acids considered to be Brønsted-Lowry acids? Write the chemical equation for the ionization in water of each of the following Brønsted-Lowry acids: (a) HNO_3, (b) HSO_4^-, (c) H_2SO_4, (d) HCl, (e) H_2O. Identify the respective conjugate bases.

20.2 Define the term "Brønsted-Lowry base." Are all water-ion bases considered to be Brønsted-Lowry bases? Explain your answer. Write the chemical equation for the addition of a proton by each of the following Brønsted-Lowry bases: (a) NH_3, (b) H_2O, (c) HCO_3^-, (d) CO_3^{2-}, (e) CN^-, (f) OH^-. Identify the respective conjugate acids.

20.3 The bicarbonate ion, HCO_3^-, is amphoteric. (a) Illustrate this property by writing typical Brønsted-Lowry acid–base equations. (b) Identify the conjugate acid–base pairs.

20.4 Consider the following reactions of the hydrogen sulfate ion

$$HSO_4^- + H_3O^+ \rightleftharpoons H_2SO_4(aq) + H_2O^+$$
$$HSO_4^- + H_2O(l) \rightleftharpoons SO_4^{2-} + H_3O^+$$

In which reaction is HSO_4^- an acid and in which reaction is it a base?

20.5 List (a) all Brønsted-Lowry acids and (b) all Brønsted-Lowry bases from among the following: (i) SO_3^{2-}, (ii) $AlCl_3$, (iii) Cl^-, (iv) NH_4^+, (v) NH_3, (vi) H_2O, (vii) HBr.

20.6 Indicate which of the following substances—(i) HCl, (ii) $H_2PO_3^-$, (iii) H_2CaO_2, (iv) $ClO_3(OH)$, (v) $Sb(OH)_3$—are (a) Brønsted-Lowry acids or (b) Brønsted-Lowry bases. (Note: Do not be confused by the way the formulas are written.)

20.7 Define the term "Brønsted-Lowry acid–base reaction." Write a chemical equation illustrating the reaction in terms of the acid, base, conjugate acid, and conjugate base. Write the chemical equations for the following Brønsted-Lowry acid–base reactions: (a) H_2O and NH_3, (b) H_3O^+ and OH^-, (c) NH_4^+ and CO_3^{2-}, (d) HNO_3 and H_2O. Identify the respective conjugate acid–base pairs in each reaction.

20.8 Identify the Brønsted-Lowry base reactant and its conjugate acid product in each of the following reactions:
(a) $H_2PO_4^- + HSO_4^- \rightleftharpoons H_3PO_4(aq) + SO_4^{2-}$
(b) $H_3O^+ + HCOO^- \rightleftharpoons H_2O(l) + HCOOH(aq)$
(c) $HI(g) + H_2O(l) \rightleftharpoons H_3O^+ + I^-$

20.9 Based on the relative strengths of acid–base pairs given in Table 20.1, predict what will happen if (a) HF is added to Cl^-, (b) H_2S is added to CN^-, and (c) HCOOH is added to Cl^-. Write the appropriate equations.

20.10 Repeat Question 20.9 for (a) HCOOH and CH_3COO^- and (b) H_2S and OH^-.

Binary Acids and Oxoacids—Nomenclature

20.11 Name the following binary acids: (a) HF, (b) HBr, (c) H_2S, (d) H_2Se.

20.12 Write the formulas for the following binary acids: (a) hydrochloric acid, (b) hydroiodic acid, (c) hydrotelluric acid.

20.13 The "ic" oxoacids of phosphorus, arsenic, and antimony all have the general formula H_3EO_4. Name the following oxoacids: (a) H_3PO_4, H_3PO_3, and H_3PO_2; (b) H_3AsO_4 and H_3AsO_3; (c) H_3SbO_4 and H_3SbO_3.

20.14 The "ic" oxoacids of sulfur and selenium both have the general formula H_2EO_4. Write the formulas for the following oxoacids: (a) sulfuric acid and sulfurous acid and (b) selenic acid and selenous acid. The formula for tellurous acid is similar to those for sulfurous acid and selenous acid. (c) Write the formula for tellurous acid. The "ic" oxoacid of tellurium has six OH groups around a central tellurium atom. (d) Write the formula for telluric acid.

20.15 What is the relationship between "ortho" oxoacids and "meta" oxoacids? Write a chemical equation representing the formation of metaphosphoric acid from orthophosphoric acid.

20.16 What is the relationship between "pyro" oxoacids and "ortho" oxoacids? Write a chemical equation representing the formation of pyrosulfuric acid from sulfuric acid.

20.17 Name the following oxoanions: (a) BO_3^{3-}, (b) AsO_3^{3-}, (c) NO_2^-, (d) BiO_3^{3-}, (e) SeO_3^{2-}, (f) IO^-, (g) ClO_4^-.

20.18 Write the formulas for the following oxoanions: (a) aluminate ion, (b) carbonate ion, (c) antimonite ion, (d) hypophosphite ion, (e) sulfate ion, (f) chlorite ion, (g) hypobromite ion.

20.19 Name the following salts of oxoacids: (a) $Ga_2(SeO_4)_3$, (b) $Co_3(PO_4)_2$, (c) $Zn(BrO_3)_2$, (d) $TlNO_3$, (e) $Th(SO_4)_2$.

20.20 Name the following salts of oxoacids: (a) $NaIO_4$, (b) $AlPO_4$, (c) $Ca(NO_2)_2$, (d) $PbCrO_4$.

20.21 Write formulas for the following salts of oxoanions: (a) lead selenate, (b) cobalt(II) sulfite, (c) cadmium iodate, (d) cesium perchlorate, (e) copper(I) sulfate.

20.22 Write formulas for the following salts of oxoanions: (a) mercury(II) chromate, (b) manganese(III) phosphate, (c) nickel(II) carbonate, (d) silver sulfate, (e) iron(III) nitrate.

Binary Acids and Oxoacids—Strengths

20.23 Which effect—electronegativity or nonmetal radius—is more important for determining the acid strength of binary hydrides of elements belonging to the same family of the periodic table? Which is the stronger acid: H_2Se or H_2Te?

20.24 Which effect—electronegativity or nonmetal radius—is more important for determining the acid strength of binary hydrides of elements in the same row of the periodic table? Which is the stronger acid: H_2Se or H_3As?

20.25 Predict which acid is stronger in each pair (a) NH_3 or CH_4, (b) HI or HCl, (c) H_2S or HBr, (d) H_2S or HS^-.

20.26 List the following acids in order of increasing strength: (a) hydrosulfuric, hydroselenic, hydrotelluric; (b) HCl, H_2S, PH_3; (c) NH_3, H_2S, HBr.

20.27 How does the acid strength of a series of oxoacids of a given element depend on the oxidation state of the central atom? Which is the stronger acid, H_2SO_4 or H_2SO_3?

20.28 How does the acid strength of a series of oxoacids of elements in the same family of the periodic table in the same oxidation state depend on the electronegativities of the central atoms? Which is the stronger acid, $HBrO_2$ or $HClO_2$?

20.29 Predict which acid is stronger: (a) HIO_2 or HIO_3, (b) $As(OH)_3$ or $Sb(OH)_3$, (c) H_2S or HNO_3.

20.30 List the acids in each group in order of increasing strength: (a) sulfuric, phosphoric, perchloric; (b) HIO_3, HIO_2, HIO, HIO_4; (c) selenous, sulfurous, tellurous.

Binary Acids and Oxoacids—Equivalent Mass and Normality

20.31 How is the equivalent mass of an acid or base defined? Illustrate how the equivalent mass of a substance can vary from reaction to reaction.

20.32 Define the concentration term "normality." How can we convert from normality to molarity?

20.33 What is the relationship between the molar mass and the equivalent mass in acid–base reactions for (a) HBr, (b) H_3PO_4, (c) $Ca(OH)_2$? Assume complete reaction of all of the H^+ or OH^-.

20.34 What are the relationships between the molar masses and the equivalent masses of H_2CO_3 and $Ca(OH)_2$ in the following reaction?

$$2H_2CO_3(aq) + Ca(OH)_2(aq) \longrightarrow Ca(HCO_3)_2(aq) + 2H_2O(l)$$

20.35 What is the molarity of a 0.13 N solution of (a) H_2SO_4, (b) KOH, (c) HNO_3? Assume complete reaction of all of the H^+ or OH^-.

20.36 If 100.00 g of $Ba(OH)_2 \cdot 8H_2O$ is dissolved in enough water to make 1.000 L of solution, what is the normality of the solution? Assume that the $Ba(OH)_2$ will release both OH^- ions in an acid–base reaction.

Expressing the Strengths of Acids and Bases—K_w, pH, and pOH

20.37 Write a chemical equation showing the ionization of water. Write the equilibrium constant expression for this equation. What is the special symbol used for this equilibrium constant?

What is the relationship between $[H^+]$ and $[OH^-]$ in pure water? How can this relationship be used to define the terms "acidic" and "alkaline"?

20.38 Write mathematical definitions for pH and pOH. What is the relationship between pH and pOH? How can pH be used to define the terms "acidic" and "alkaline"?

20.39 Calculate $[OH^-]$ that is in equilibrium with (a) $[H^+] = 1.3 \times 10^{-4}$ mol/L and (b) $[H^+] = 7 \times 10^{-9}$ mol/L.

20.40 Calculate $[H^+]$ that is in equilibrium with $[OH^-] = 4.21 \times 10^{-6}$ mol/L.

20.41 Convert each of the following values of $[H^+]$ into pH: (a) 9.5×10^{-6} mol/L, (b) 0.0001472 mol/L, (c) 0.00203 mol/L, (d) 12×10^{-12} mol/L, (e) 1.0×10^{1} mol/L. Which solutions are acidic?

20.42 Convert each of the following pH values into $[H^+]$: (a) 4.67, (b) 9.05, (c) 0.0, (d) 15.97, (e) 0.153. Which solutions are acidic?

20.43 Convert each of the following values of $[OH^-]$ into pH: (a) 0.00035 mol/L, (b) 11×10^{-13} mol/L, (c) 4.5×10^{-7} mol/L, (d) 10 mol/L. Which solutions are alkaline?

20.44 Convert each of the following pOH values into $[H^+]$: (a) 13.97, (b) 0.12, (c) 5.04, (d) 11.021. Which solutions are acidic?

20.45 Calculate the pH of each of the following solutions of strong acids and bases: (a) 0.2374 M NaOH, (b) 0.0365 M HCl, (c) 4.2×10^{-5} M HNO_3, (d) 0.0826 M KOH.

20.46 Calculate the pH of each of the following solutions: (a) 0.0010 M HNO_3, (b) 0.103 M NaOH, (c) 0.5 M KOH, (d) 0.104 M HCl. Assume all solutes are completely ionized or dissociated.

Expressing the Strengths of Acids and Bases—K_a

20.47 Write a chemical equation representing the ionization of the acid HA. Write the equilibrium constant expression for this reaction. What is the special symbol used for this equilibrium constant?

20.48 What is the relationship between the strength of an acid and the numerical value of K_a? What is the relationship between the acid strength and the value of pK_a?

20.49 What is the concentration of IO^- in equilibrium with $[H^+] = 0.035$ mol/L and $[HIO] = 0.427$ mol/L? $K_a = 2.3 \times 10^{-11}$ for HIO.

20.50 What is the concentration of CN^- in a solution having an equilibrium concentration of 0.0123 M HCN and a pH of 3.56? $K_a = 6.2 \times 10^{-10}$ for HCN.

20.51 Find the concentrations of the various species present in a 0.025 M solution of hydrofluoric acid, HF. What is the pH of the solution? $K_a = 6.5 \times 10^{-4}$ for HF.

20.52 Calculate the concentrations of all the species present in a 0.35 M benzoic acid solution. $K_a = 6.6 \times 10^{-5}$ for C_6H_5COOH.

20.53 Electrical conductivity measurements show that 0.050 M acetic acid is 1.9% ionized at 25 °C. Calculate K_a for the acid.

20.54 A 0.0202 M solution of $ClCH_2COOH(aq)$, chloroacetic acid, is 24.1% ionized at 25 °C. Calculate K_a for this acid.

20.55 The ionization constant for acetic acid,

$$CH_3COOH(aq) + H_2O(l) \rightleftharpoons CH_3COO^- + H_3O^+$$

is 1.754×10^{-5} at 25 °C. What is the extent of ionization of the molecules in (a) a 0.100 M solution and (b) a 0.0100 M solution?

20.56 What is the percent ionization in a 0.0500 M solution of formic acid? $K_a = 1.772 \times 10^{-4}$ for HCOOH.

Expressing the Strengths of Acids and Bases—K_b

20.57 Write chemical equations representing (a) the dissociation of the base MOH and (b) the equilibrium between water and the proton-accepting base B. Write the equilibrium constant expressions for these reactions. What is the special symbol used for these equilibrium constants?

20.58 What is the relationship between base strength and the value of K_b? What is the relationship between base strength and the value of pK_b?

20.59 In a 0.0100 M aqueous solution of methylamine, CH_3NH_2, the equilibrium concentrations of the species are $[CH_3NH_2] = 0.0082$ mol/L and $[CH_3NH_3^+] = [OH^-] = 1.8 \times 10^{-3}$ mol/L. Calculate K_b for this weak base.

$$CH_3NH_2(aq) + H_2O(l) \rightleftharpoons CH_3NH_3^+ + OH^-$$

20.60 What is the concentration of NH_3 in equilibrium with $[NH_4^+] = 0.010$ mol/L and $[OH^-] = 1.2 \times 10^{-5}$ mol/L? $K_b = 1.6 \times 10^{-5}$ for NH_3.

20.61 Calculate the concentrations of NH_4^+ ion, OH^- ion, and NH_3 at equilibrium in a solution marked "0.10 M NH_4OH." $K_b = 1.6 \times 10^{-5}$ for NH_3.

20.62 Because K_b is larger for triethylamine

$$(C_2H_5)_3N(aq) + H_2O(l) \rightleftharpoons (C_2H_5)_3NH^+ + OH^- \qquad K_b = 5.2 \times 10^{-4}$$

than for trimethylamine

$$(CH_3)_3N(aq) + H_2O(l) \rightleftharpoons (CH_3)_3NH^+ + OH^- \qquad K_b = 6.3 \times 10^{-5}$$

an aqueous solution of triethylamine should have a larger concentration of OH^- ion than an aqueous solution of trimethylamine of the same concentration. Confirm this statement by calculating the $[OH^-]$ for 0.010 M solutions of both weak bases.

20.63 What is the value of K_b for CN^-, given the following equilibrium concentrations: $[CN^-] = 0.37$ mol/L, $[HCN] = 1.3 \times 10^{-3}$ mol/L, and $[OH^-] = 4.6 \times 10^{-3}$ mol/L?

20.64 What is the concentration of $[OH^-]$ in equilibrium with $[CH_3COO^-] = 7.5 \times 10^{-3}$ mol/L and $[CH_3COOH] = 0.092$ mol/L? $K_b = 5.701 \times 10^{-10}$ for CH_3COO^-.

20.65 What is the pH of a 0.13 M CN^- solution? $K_b = 1.6 \times 10^{-5}$ for CN^-.

20.66 Calculate the pH of a 0.50 M F^- solution. $K_b = 1.5 \times 10^{-11}$ for F^-. Neglect any $[OH^-]$ or $[H^+]$ from water.

20.67 A weak base, B, reacts only to the extent of 3.2% with water in a 0.100 M solution. Calculate K_b for this base.

20.68 Electrical conductance measurements at 18 °C show that the weak base $NH_3(aq)$ is 1.4% ionized in a 0.1000 M solution, 4.0% ionized in a 0.01000 M solution, and 12% ionized in a 0.0010 M solution. (a) Find the pH of each solution. (b) Calculate K_b for each solution.

Expressing the Strengths of Acids and Bases—K_a, K_b, and K_w

20.69 What is the relationship between the ionization constant for an acid and the ionization constant of its conjugate base?

20.70* Prove that $K_w = K_aK_b$ by making use of the expressions for K_a for the acid HA, K_b for the base A^-, and K_w, the ion product constant of water.

20.71 The value of K_a for hypobromous acid, HBrO, is 2.2×10^{-9}. What is the value of K_b for the conjugate base?

20.72 Calculate the values of K_b for the conjugate base of (a) HF, $K_a = 6.5 \times 10^{-4}$ for HF; (b) HNCS, $K_a = 69$ for the acid; (c) HIO_4, $K_a = 5.6 \times 10^{-9}$ for HIO_4.

20.73 Which ion, Cl^- or CN^-, has the larger value of K_b? $K_a = 3 \times 10^8$ for HCl and 6.2×10^{-10} for HCN.

20.74 The value of K_b for the sulfide ion is 3×10^{-2} and for the hydrogen sulfide ion it is 1.0×10^{-7}. What are the K_a values for the conjugate acids of these ions?

Lewis Acids and Bases

20.75 Define the term "Lewis acid." Prepare sketches showing how the following types of substances can act as Lewis acids: (a) cations, (b) molecules with unfilled octets, (c) nonmetal atoms which can accommodate more than an octet.

20.76 Define the term "Lewis base." Prepare sketches showing how the following types of substances can act as Lewis bases: (a) anions and (b) molecules with one or two pairs of unshared electrons.

20.77 Define the term "Lewis acid–base reaction." Which of the following are Lewis acid–base reactions? (a) $Ag^+ + 2NH_3 \longrightarrow [Ag(NH_3)_2]^+$, (b) $I^- + I_2 \longrightarrow I_3^-$, (c) $H_3O^+ + OH^- \longrightarrow 2H_2O$, (d) $H^- + H_2O \longrightarrow H_2 + OH^-$. Identify the Lewis acid in each of the acid–base reactions.

20.78 Which of the following are Lewis acid–base reactions? (a) $H_2O + SO_3 \longrightarrow H_2SO_4$, (b) $O^{2-} + SO_3 \longrightarrow SO_4^{2-}$, (c) $AlCl_3 + CH_4 \longrightarrow AlCl_3H^- + CH_3^+$. Identify the Lewis base in each of the acid–base reactions.

ADDITIONAL QUESTIONS AND PROBLEMS

20.79 The value of K_w is 9.614×10^{-14} at 60 °C. Find $[H^+]$ in water at this temperature and calculate the pH. Is pure water an acid at 60 °C? Explain.

20.80 The reagent bottles labeled "Dilute NaOH" and "Dilute HCl" that are commonly found in the laboratory usually contain 6 M solutions of the substances. Calculate the $[H^+]$, $[OH^-]$, pH, and pOH for each reagent, assuming that the solutes are completely ionized or dissociated.

20.81 Given the following heats of formation at 25 °C, 0 for H^+, −332.63 kJ/mol for F^-, −246.0 kJ/mol for Cl^-, −167.16 kJ/mol for HCl(*aq*), −320.08 kJ/mol for HF(*aq*), calculate ΔH° for the ionization of HCl(*aq*) and of HF(*aq*). Which acid would yield more heat if equimolar amounts were allowed to react with solutions of sodium hydroxide?

20.82 The enthalpy of neutralization of 1.00 mol of aqueous hydrocyanic acid by a strong base is −12.3 kJ. The enthalpy change for 1.00 mol of a strong acid undergoing neutralization by a strong base is −55.835 kJ. Calculate the enthalpy of reaction for

$$HCN(aq) + H_2O(l) \rightleftharpoons H_3O^+ + CN^-$$

20.83 At 25 °C, the value of p*K* for the following reaction of deuterium oxide (heavy water):

$$2D_2O(l) \rightleftharpoons D_3O^+ + OD^-$$

is 14.869. (a) What is the value of *K*? (b) Under equilibrium conditions, what are the values of $[D_3O^+]$ and $[OD^-]$? (c) What is the pD of pure D_2O? (d) What is the relation between pD and pOD? (e) Give a suitable definition in terms of pD for an acidic solution in this solvent system.

Consider the following acid–base reactions taking place in pure D_2O:

$$CH_3COOH + D_2O \rightleftharpoons HD_2O^+ + CH_3COO^-$$
$$NH_3 + D_2O \rightleftharpoons NH_3D^+ + OD^-$$

Which substances are acting as (f) water-ion acids, (g) classical bases, (h) Brønsted-Lowry acids, (i) Brønsted-Lowry bases?

The equilibrium constant for the reaction

$$D_2O(l) + H_2O(l) \rightleftharpoons 2HDO(l)$$

is 3.56 at 25 °C. (j) If we prepare an equimolar solution of D_2O and H_2O, which is essentially 27.7 M in each, what will be the concentration of HDO once equilibrium has been established?

CHAPTER

21

Ions and Ionic Equilibria: Acids and Bases

ACIDIC AND BASIC IONS

21.1 REACTIONS OF IONS WITH WATER

Consider for a moment what happens when ions are added to pure water. The ions enter an environment in which water molecules are in equilibrium with hydrogen ions and hydroxide ions. The added ions, depending upon their properties, may establish additional equilibria. All ions in aqueous solution are hydrated—surrounded by water molecules—as a result of the attraction between the ions and the water dipoles (Section 14.2). Hydration equilibria, such as the following, have no effect on the pH of a solution

$$K^+ + nH_2O(l) \rightleftharpoons [K(H_2O)_n]^+ \qquad \textbf{(21.1)}$$

Some ions establish only such hydration equilibria; others establish equilibria that produce acidic or alkaline solutions. The possible interactions between ions and water molecules are discussed in the following sections. Examples are given in Table 21.1. We suggest that you refer to this table as you read the following sections.

a. Cations or Anions That Simply Become Hydrated The hydration of a cation occurs by attraction between the positive charge of the cation and a lone electron pair on the oxygen atom of the water molecule. Cations that are simply hydrated, such as K^+, are the large cations of low charge—cations with low polarizing ability. Such cations hold their surrounding water molecules loosely, so that the water molecules can freely come and go in equilibria as shown in Equation (21.1). These are the cations found in strong bases.

Table 21.1
Types of Ions and Salts in Aqueous Solution

Common examples of each type of ion and salt are given.

	Cations that simply become hydrated Li^+, Na^+, K^+, Ca^{2+}, Ba^{2+}	**Cations that react with water to give H_3O^+** NH_4^+, Be^{2+}, Zn^{2+}, Al^{3+}, Fe^{3+}, Cr^{3+}, Cu^{2+}
Anions that simply become hydrated NO_3^-, Cl^-, Br^-, I^-, ClO_4^-, ClO_3^-	**Salts give neutral solutions** $NaCl$, $CaCl_2$, KNO_3	**Salts give acidic solutions** NH_4Cl, NH_4NO_3, $AlCl_3$, $Fe(NO_3)_3$
Anions that react with water to give OH^- S^{2-}, PO_4^{3-}, CO_3^{2-}, CN^-, SO_3^{2-}, HCO_3^-, CH_3COO^-, NO_2^-, F^-	**Salts give alkaline solutions** $Ba(CH_3COO)_2$, K_2CO_3, Na_2S, KCN	**pH of salt solution varies—weakly acidic or weakly alkaline** NH_4CH_3COO, NH_4NO_2, NH_4CN

Anions, on the other hand, are potential proton acceptors. Those that are simply hydrated are the anions found in strong acids (above H_2O as a base in Table 20.1). Such anions are so weak as bases that they cannot take away protons from water molecules. Instead, they become hydrated by attraction to the positive end of the water dipole.

$$NO_3^- + nH_2O(l) \rightleftharpoons [NO_3(H_2O)_n]^- \quad \textbf{(21.2)}$$

When salts containing cations or anions that do not react with water dissolve in water, such cations or anions are spectator ions.

b. Anions That React with Water to Give OH^- The anions that give alkaline aqueous solutions are stronger bases than water (below H_2O as a base in Table 20.1). They are the conjugate bases of weak acids and establish equilibria because they are able to attract protons away from water molecules. These anions are the winners in the competition for the protons, and in their presence water reacts as an acid. By taking protons from the water molecules they leave behind OH^- ions that make the solutions alkaline, for example

$$F^- + H_2O(l) \rightleftharpoons HF(aq) + OH^- \quad \textbf{(21.3)}$$

$$CN^- + H_2O(l) \rightleftharpoons HCN(aq) + OH^- \quad \textbf{(21.4)}$$

$$\underset{base_1}{S^{2-}} + \underset{acid_2}{H_2O(l)} \rightleftharpoons \underset{acid_1}{HS^-} + \underset{base_2}{OH^-} \quad \textbf{(21.5)}$$

c. Cations That React with Water to Give H_3O^+ Common cations that give acidic aqueous solutions are nitrogen-containing cations like NH_4^+ or metal cations of high polarizing ability. The ammonium ion gives up a proton to a water molecule to yield an acidic solution. The ammonium ion and all other cations that give acidic solutions are stronger acids than water, and in reacting with these cations water is the proton acceptor—the base.

$$\underset{acid_1}{NH_4^+} + \underset{base_2}{H_2O} \rightleftharpoons \underset{base_1}{NH_3} + \underset{acid_2}{H_3O^+} \quad \textbf{(21.6)}$$

A hydrated metal cation that gives an acidic aqueous solution does so by donating a proton from a coordinated water molecule, for example

$$\underset{acid_1}{[Al(H_2O)_6]^{3+}} + \underset{base_2}{H_2O(l)} \rightleftharpoons \underset{base_1}{[Al(H_2O)_5(OH)]^{2+}} + \underset{acid_2}{H_3O^+} \quad \textbf{(21.7)}$$

For simplicity, the bound water molecules are often omitted in writing the equations.

$$Al^{3+} + 2H_2O(l) \rightleftharpoons [AlOH]^{2+} + H_3O^+ \quad \textbf{(21.8)}$$

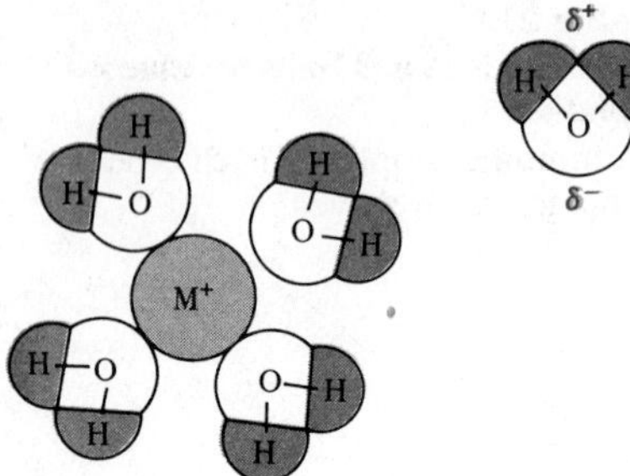

Figure 21.1
A Hydrated Metal Cation

The positive charge of the hydrated metal cation (Figure 21.1) draws electron density toward itself, thereby increasing the polarity of the oxygen–hydrogen bonds in the water molecule and making it easier for the water molecule to give up a hydrogen ion. Any property of a hydrated cation that increases its attraction for electrons favors its behavior as an acid. Most importantly, a small cation radius and a large positive charge, resulting in a high polarizing ability, favor acidic properties. Acidic properties are increasingly evident in metal ions to the right in the periodic table. Cations such as $[Cu(H_2O)_4]^{2+}$, $[Al(H_2O)_6]^{3+}$, and $[Fe(H_2O)_6]^{3+}$ behave as weak acids toward water.

EXAMPLE 21.1 Reactions of Ions with Water

How will the following ions react with water: (a) I^-, (b) NO_2^-, (c) Na^+, (d) Bi^{3+}, (e) Be^{2+}? Write chemical equations for the reactions that take place. Will the resulting solutions be neutral, acidic, or alkaline (in the absence of other ions that influence pH)?

(a) The I^- ion is the anion of HI, which is a strong acid. Therefore, I^- is a much weaker base than water, is **simply hydrated,** and gives a **neutral aqueous solution.**

(b) The NO_2^- ion is the conjugate base of a weak acid, HNO_2, and **reacts as a base to give an alkaline solution.**

$$NO_2^- + H_2O(l) \rightleftharpoons HNO_2(aq) + OH^-$$

(c) The large Na^+ ion has a small charge-to-size ratio [10] and does not react with water except to become **hydrated** and give a **neutral aqueous solution.**

(d) The smaller highly charged Bi^{3+} ion [charge-to-size ratio, 31] gives an **acidic solution.**

$$Bi^{3+} + 2H_2O(l) \rightleftharpoons [BiOH]^{2+} + H_3O^+$$

(e) Beryllium, the first element in its family, has the smallest cation in the family [charge-to-size ratio, 57] and will also give an **acidic solution.**

$$Be^{2+} + 2H_2O(l) \rightleftharpoons [BeOH]^+ + H_3O^+$$

EXERCISE Describe how the following ions will react with water: (a) CH_3COO^-, (b) NO_3^-, (c) Ba^{2+}, (d) Zn^{2+}. Write chemical equations showing the reactions that occur. Will these reactions produce neutral, acidic, or alkaline solutions?

Answers (a) $CH_3COO^- + H_2O(l) \rightleftharpoons CH_3COOH(aq) + OH^-$, alkaline; (b) hydration only; (c) hydration only; (d) $Zn^{2+} + 2H_2O(l) \longrightarrow [Zn(OH)]^+ + H_3O^+$, acidic.

21.2 THE BEHAVIOR OF SALTS TOWARD WATER

On the basis of the acidity or alkalinity of their solutions, there are four types of salts. (Examples of each type are given in Table 21.1.)

1. Salts that give neutral solutions because neither ion reacts with water to a significant extent Such salts contain both cations and anions that are simply hydrated in aqueous solution. Common salts of this type are the nitrates, chlorides, bromides, iodides, perchlorates, and chlorates of Li^+, Na^+, K^+, Ca^{2+}, and Ba^{2+}.
2. Salts that give alkaline solutions because only the anion reacts with water In such salts the anions are stronger bases than water. Common salts of this type are those carbonates, sulfides, phosphates, carbonates, cyanides, and acetates of Li^+, Na^+, K^+, Ca^{2+}, and Ba^{2+} that are soluble.
3. Salts that give acidic solutions because only the cation reacts with water Metal cations that are small and highly charged or ammonium ion (or similar nitrogen-containing cations) form such salts with nitrate, chloride, bromide, iodide, perchlorate, and chlorate anions.

4. Salts of cations and anions both of which react with water—pH varies with salt The pH of solutions of such salts, for example $(NH_4)_2CO_3$, depends upon the relative extents of the reactions of the two ions with water. The ion that reacts to the greater extent (Section 21.3) determines the pH.

EXAMPLE 21.2 Reactions of Ions with Water

Predict whether aqueous solutions of the following salts would be acidic, alkaline, or close to neutral: (a) KCl, (b) NH_4NO_2, (c) $AlBr_3$, (d) Na(HCOO), a salt of formic acid, HCOOH, a weak organic acid.

(a) Potassium chloride (KCl) is a salt of a cation, K^+, that is a weaker acid than water and an anion, Cl^-, that is a weaker base than water. Because neither ion reacts with water to any appreciable extent, a KCl solution will be **neutral.**

(b) Ammonium nitrite (NH_4NO_2) is the salt of a cation and an anion both of which react with water. A solution of this salt will be **close to neutral** and whether it is slightly alkaline or slightly acidic will depend upon the extent to which the individual ions react with water. [The solution will be slightly acidic.]

(c) Aluminum bromide ($AlBr_3$) is the salt of a cation that will react with water because it is a stronger acid than water and an anion that will not react because it is a weaker base than water. Therefore, the solution will be **acidic.**

(d) Sodium formate (Na(HCOO)) is the salt of a cation that does not react with water and an anion that is a stronger base than water (because it is the anion of a weak acid). The solution produced will therefore have an excess of OH^- and will be **alkaline.**

EXERCISE Predict whether aqueous solutions of the following salts would be acidic, alkaline, or close to neutral: (a) $Fe(NO_3)_3$, (b) NH_4CN, (c) KI, (d) $LiCH_3COO$.

Answers (a) acidic, (b) close to neutral, (c) close to neutral, (d) alkaline.

21.3 EQUILIBRIUM CONSTANTS FOR REACTIONS OF IONS WITH WATER: pH OF SALT SOLUTIONS

The reactions of ions or ionic compounds with water to give acidic or alkaline solutions traditionally have been called "hydrolysis" reactions. As we have seen (Section 14.6), "hydrolysis" is a general term for reactions in which the water molecule is split. The **hydrolysis of an ion** is the reaction of an ion with water to give either H_3O^+ or OH^-, plus whatever reaction product is formed by the ion.

Ion hydrolysis reactions are, however, merely the Brønsted-Lowry acid–base reactions between ions and water. The equilibrium constants for these reactions, although sometimes referred to as "hydrolysis constants," are simply the K_a and K_b values for acids and bases that happen to be ions. Some representative values are given in Table 21.2. For the reaction of an anion that gives an alkaline solution, the general expression is

$$A^- + H_2O(l) \rightleftharpoons HA(aq) + OH^- \qquad K_b = \frac{[HA][OH^-]}{[A^-]} \tag{21.9}$$

For an acetate ion, for example

$$CH_3COO^- + H_2O(l) \rightleftharpoons CH_3COOH(aq) + OH^- \qquad K_b = \frac{[CH_3COOH][OH^-]}{[CH_3COO^-]}$$

There is a wide variation in the strengths of anionic bases (see Table 21.2). The weaker the acid, the stronger its conjugate base. Of the anions in Equations (21.3)–(21.5), the fluoride ion is the conjugate base of the strongest acid and is the weakest base, as shown by its small K_b: F^-, $K_b = 1.5 \times 10^{-11}$; CN^-, $K_b = 1.6 \times 10^{-5}$; S^{2-}, $K_b = 3 \times 10^{-2}$. Note in Table 21.2 that the anions fall into two groups. Those below HCO_3^- are significantly weaker bases than the others.

Table 21.2
Some Equilibrium Constant Values for the Reactions of Ions with Water

Acidic cations	K_a
NH_4^+	6.3×10^{-10}
$C_6H_5NH_3^+$	2.4×10^{-5}
$CH_3NH_3^+$	2.6×10^{-11}
$CH_3CH_2NH_3^+$	2.1×10^{-11}
Bi^{3+}	1.0×10^{-2}
Fe^{3+}	4.0×10^{-3}
Al^{3+}	1.4×10^{-5}
Fe^{2+}	1.2×10^{-6}
Cu^{2+}	1.0×10^{-8}
Zn^{2+}	2.5×10^{-10}
Basic anions	K_b
S^{2-}	3.0×10^{-2}
CO_3^{2-}	2.1×10^{-4}
CN^-	1.6×10^{-5}
SO_3^{2-}	2.0×10^{-7}
HCO_3^-	2.2×10^{-8}
CH_3COO^-	5.7×10^{-10}
NO_2^-	1.4×10^{-11}
F^-	1.5×10^{-11}
SO_4^{2-}	1.0×10^{-12}

The equilibrium constants for the reactions of an acid and its conjugate base with water are related to the ion product constant of water by the expression $K_w = K_a K_b$ (Section 20.12). The K_b for acetate ion can be found from K_w and the value of K_a for acetic acid as follows:

for acetate ion

$$K_b = \frac{[CH_3COOH][OH^-]}{[CH_3COO^-]} = \frac{K_w}{K_a} = \frac{1.00 \times 10^{-14}}{1.75 \times 10^{-5}} = 5.70 \times 10^{-10} \quad \textbf{(21.10)}$$

for acetic acid

This type of relationship is particularly useful because K_a values are more likely to be found tabulated than K_b values for anions.

The K_a and K_b values for the reactions of an acidic cation and its conjugate base with water are related in the same way. For ammonium ion

for NH_4^+

$$NH_4^+ + H_2O(l) \rightleftharpoons NH_3(aq) + H_3O^+$$

$$K_a = \frac{[NH_3][H^+]}{[NH_4^+]} = \frac{K_w}{K_b} = \frac{1.00 \times 10^{-14}}{1.6 \times 10^{-5}} = 6.3 \times 10^{-10} \quad \textbf{(21.11)}$$

for NH_3

The pH of a solution of a salt in which only one ion reacts with water can be found from the known concentration of the solution and the K_a or K_b for the ion that reacts with water. Equations like (21.10) or (21.11) are used, if necessary, to find the equilibrium constant that is needed.

Whether the solution of a salt composed of ions both of which react with water will be slightly alkaline or slightly acidic can be predicted from the K_a and K_b values for the two ions. If $K_a > K_b$, the solution will be acidic; if $K_b > K_a$, the solution will be alkaline. For example, from the values of $K_a = 6.3 \times 10^{-10}$ for NH_4^+ and $K_b = 1.4 \times 10^{-11}$ for NO_2^-, it can be predicted that a solution of ammonium nitrite will be slightly acidic. Some ammonium salts give close to neutral solutions because the acid and base equilibrium constants for the ions are very close in value. Ammonium acetate (NH_4CH_3COO) is such a salt: for NH_4^+, $K_a = 6.3 \times 10^{-10}$ and for CH_3COO^-, $K_b = 5.7 \times 10^{-10}$.

EXAMPLE 21.3 Reactions of Ions with Water

(a) What is the pH of a 0.10 M KCN solution? (b) Would you expect a solution of NH_4CN to be acidic or alkaline? For HCN, $K_a = 6.2 \times 10^{-10}$.

(a) For KCN, the only reaction with water that will influence the pH is that of the cyanide ion. The solution should be alkaline due to the reaction

$$CN^- + H_2O(l) \rightleftharpoons HCN(aq) + OH^-$$

First, we find K_b for the CN^- ion from the K_a for HCN.

$$K_b = \frac{[HCN][OH^-]}{[CN^-]} = \frac{K_w}{K_a} = \frac{1.00 \times 10^{-14}}{6.2 \times 10^{-10}} = 1.6 \times 10^{-5}$$

Let $x = [OH^-]$.

	CN^-	+	H_2O	$\rightleftharpoons$	HCN	+	OH^-
Initial	0.10				0		0
Change	$-x$				$+x$		$+x$
Equilibrium	$0.10 - x$				x		x

Substitution into the equilibrium expression gives

$$K_b = \frac{(x)(x)}{0.10 - x} \simeq \frac{x^2}{0.10} = 1.6 \times 10^{-5}$$
$$x = 1.3 \times 10^{-3} \text{ mol/L} = [\text{OH}^-]$$

The approximation $0.10 - (1.3 \times 10^{-3}) \simeq 0.10$ is valid and the OH^- ion concentration is 1.3×10^{-3} mol/L. Next we find the pOH and the pH.

$$\text{pOH} = -\log[\text{OH}^-] = -\log(1.3 \times 10^{-3}) = 2.89$$
$$\text{pH} = 14.00 - \text{pOH} = 14.00 - 2.89 = \mathbf{11.11}$$

As we predicted, the solution is alkaline.

(b) Comparison of the K_a value for ammonium ion from Table 21.2, $K_a = 6.3 \times 10^{-10}$, with the K_b value for cyanide ion found in part (a) shows that the K_b value of the cyanide ion is larger. Therefore, an aqueous solution of NH_4CN should be **alkaline.**

EXERCISE What is the pH of a 0.10 M NH_4Cl solution? $K_b = 1.6 \times 10^{-5}$ for NH_3.

Answer 5.10.

SUMMARY OF SECTIONS 21.1–21.3

Anions that are the conjugate bases of strong acids and also cations of low polarizing ability simply become hydrated in aqueous solution and do not cause changes in the pH. Anions that are the conjugate bases of weak acids react to form OH^- ions in aqueous solution. The stronger the acid, the weaker its anion as a base. Ammonium ion and metal cations of high polarizing ability yield H_3O^+ upon reaction with water.

The equilibrium constants for the reactions of ions with water (sometimes called "hydrolysis constants") are the K_a and K_b values for acids and bases that are ions. They can be used, for example, to find the pH of a solution containing an ion that reacts with water. Such equilibrium constants are related to K_w by the equation $K_w = K_aK_b$, which shows that the larger the K_a of an acid, the smaller the K_b of its conjugate base.

Based upon the reactions of their ions with water there are four types of salts: (1) those that give neutral solutions because neither ion reacts with water to a significant extent, (2) those that give alkaline solutions because only the anion reacts with water, (3) those that give acidic solutions because only the cation reacts with water, and (4) those in which both ions react with water and the result depends upon the relative extent of the reactions of the two ions. (Examples of these four types of salts are given in Table 21.1.)

COMMON IONS AND BUFFERS

21.4 THE COMMON ION EFFECT

An ion added to a solution that already contains some of that ion is called a **common ion.** What happens if a small amount of sodium hydroxide is added to a saturated solution of magnesium hydroxide? The Na^+ ion is not a common ion, nor does it react with water; therefore, it is a spectator ion and has no effect on the equilibrium. The equilibrium in the solution is, however, displaced by the added hydroxide ions. As would be predicted on the basis of Le Chatelier's principle, the added hydroxide ions combine with magnesium ions until equilibrium is restored. The result is that the concentration of Mg^{2+} decreases and more solid magnesium hydroxide forms.

$$\xleftarrow[]{} \quad \textit{OH}^- \text{ \textit{a common ion}}$$

$$Mg(OH)_2(s) \rightleftharpoons Mg^{2+} + 2OH^-$$

This change in equilibrium is an example of the **common ion effect:** a displacement of an ionic equilibrium by an excess of one or more of the ions involved.

The common ion effect is frequently used to control the pH of a weak acid or base solution. If ammonium chloride is added to a solution of ammonia (with Cl^- as a spectator ion)

$$\xleftarrow[]{} \quad NH_4^+$$

$$\underset{\text{weak base}}{NH_3(aq)} + H_2O(l) \rightleftharpoons \underset{\text{conjugate acid}}{NH_4^+} + OH^- \qquad \textbf{(21.12)}$$

the NH_4^+ from the salt increases the total concentration of NH_4^+. The equilibrium is displaced toward the formation of NH_3, with an accompanying decrease in the concentration of OH^-, making the solution less alkaline.

The addition of sodium acetate, $Na(CH_3COO)$, to an acetic acid solution similarly displaces the equilibrium toward the formation of acetic acid. Here, adding a basic anion makes the solution less acidic.

$$\xleftarrow[]{} \quad CH_3COO^-$$

$$\underset{\text{weak acid}}{CH_3COOH(aq)} + H_2O(l) \rightleftharpoons \underset{\text{conjugate base}}{CH_3COO^-} + H_3O^+ \qquad \textbf{(21.13)}$$

The common ion effect can also be used to control the concentration of the anion of a weak acid. For example, addition of an acid to a solution of acetic acid will decrease the concentration of acetate ion.

$$\xleftarrow[]{} \quad H^+$$

$$CH_3COOH(aq) \rightleftharpoons H^+ + CH_3COO^- \qquad \textbf{(21.14)}$$

EXAMPLE 21.4 Common Ion Effect

What is the pH of a solution that is 0.10 M in NH_3 and 0.10 M in NH_4NO_3? $K_b = 1.6 \times 10^{-5}$ for NH_3.

The pH of the solution is controlled by the ammonia–water equilibrium:

$$NH_3(aq) + H_2O(l) \rightleftharpoons NH_4^+ + OH^- \qquad K_b = \frac{[NH_4^+][OH^-]}{[NH_3]}$$

Let $x = [OH^-]$. The ammonium ion concentration at equilibrium is the amount formed by the above reaction plus the amount added as ammonium nitrate (because ammonium nitrate is soluble and is a strong electrolyte).

	NH_3	+ $H_2O \rightleftharpoons$	NH_4^+	+ OH^-
Initial	0.10		0.10	0
Change	$-x$		$+x$	$+x$
Equilibrium	$0.10 - x$		$0.10 + x$	x

These values can be used in the equilibrium constant expression.

$$K_b = \frac{(0.10 + x)(x)}{(0.10 - x)} = 1.6 \times 10^{-5}$$

The equilibrium reaction is displaced toward the reactants by the added ammonium ion. Therefore, we can assume that x is small relative to the initial NH_3 and NH_4^+ concentrations, so that $(0.10 + x) \simeq 0.10$ and $(0.10 - x) \simeq 0.10$, giving

$$\frac{(0.10)(x)}{(0.10)} = 1.6 \times 10^{-5}$$

$$x = 1.6 \times 10^{-5}\ \text{mol/L} = [OH^-]$$

The approximation is valid. The pOH and pH of the solution are

$$pOH = -\log[OH^-] = -\log(1.6 \times 10^{-5}) = 4.80$$

$$pH = 14.00 - pOH = 14.00 - 4.80 = \mathbf{9.20}$$

(For comparison—the pH of a 0.1 M NH_3 solution is 11.1, showing that the alkalinity of the solution has been decreased by the added NH_4^+ ion.)

EXERCISE What is the pH of a solution which is 0.100 M in KClO and 0.050 M in HClO? $K_a = 2.90 \times 10^{-8}$ for HClO. *Answer* 7.82.

21.5 BUFFER SOLUTIONS

A **buffer solution** is a solution that resists changes in pH when small amounts of acid or base are added to it (or when it is diluted). The action of a buffer solution is dependent upon the common ion effect, as illustrated in Equations (21.12) and (21.13). The combination in solution of a weak acid and a weak base, often a conjugate acid–base pair, produces a buffer. For example, in a solution of weak acid HA and its conjugate base A^-, small amounts of added base react with the nonionized acid

$$OH^- + HA(aq) \rightleftharpoons A^- + H_2O(l)$$

and small amounts of added acid react with the basic anion

$$H_3O^+ + A^- \rightleftharpoons HA(aq) + H_2O(l)$$

A solution of a base, such as ammonia, plus the cation which is its conjugate acid (the ammonium ion in this case) can similarly act as a buffer:

$$OH^- + NH_4^+ \rightleftharpoons NH_3(aq) + H_2O(l)$$
$$H_3O^+ + NH_3(aq) \rightleftharpoons NH_4^+ + H_2O(l)$$

Such a combination of a weak acid and its conjugate base (HA/A^-) or of a weak base and its conjugate acid (e.g., NH_3/NH_4^+) is known as a *buffer pair*. The conjugate base or acid is added as a salt with an ion that does not react with water.

Rearranging the K_a expression for a weak acid and the K_b expression for ammonia shows that the ratio of the concentrations of the buffer pair determines the H^+ or OH^- concentration in a buffer solution

$$K_a = \frac{[H^+][A^-]}{[HA]} \qquad K_b = \frac{[NH_4^+][OH^-]}{[NH_3]}$$

$$[H^+] = \frac{[HA]}{[A^-]}K_a \qquad [OH^-] = \frac{[NH_3]}{[NH_4^+]}K_b \qquad \textbf{(21.15a,b)}$$

These equations also show that the pH of a buffer solution does not change if the solution is diluted. The concentrations of the buffer pair change by the same amount and their ratio remains the same.

A buffer solution is most effective when the concentrations of the buffer pair are roughly equal and are large relative to the amounts of added acid or base. If a small amount of acid is added to an HA/A^- buffer

$$\xleftarrow{\quad H^+ \quad}$$
$$HA(aq) + H_2O(l) \rightleftharpoons H_3O^+ + A^-$$

the concentration of A^- decreases ($-x$) and the concentration of HA increases by an equivalent amount ($+x$).

$$[H^+] = \frac{([HA] + x)}{([A^-] - x)} K_a$$

With x small relative to [HA] and $[A^-]$, you can see that the hydrogen ion concentration will not change greatly.

As illustrated in Example 21.5, Equations (21.15a) and (21.15b) can be used to find the ratio of concentrations of a buffer pair needed to maintain a given pH.

EXAMPLE 21.5 Common Ion Effect: Buffers

What ratio of $[HCOO^-]/[HCOOH]$ is needed to make a sodium formate–formic acid buffer solution of pH 3.80? $K_a = 1.77 \times 10^{-4}$ for formic acid, HCOOH.

The equilibrium is

$$HCOOH(aq) + H_2O(l) \rightleftharpoons H_3O^+ + HCOO^-$$

A pH of 3.80 corresponds to $[H^+] = 1.6 \times 10^{-4}$ mol/L. The ratio of acid to common ion in a buffer is given by Equation (21.15a). In this case

$$[H^+] = \frac{[HCOOH]}{[HCOO^-]} \times (1.77 \times 10^{-4}) = 1.6 \times 10^{-4}$$

$$\frac{[HCOO^-]}{[HCOOH]} = \frac{1.77 \times 10^{-4}}{1.6 \times 10^{-4}} = \mathbf{1.1}$$

The $[HCOO^-]/[HCOOH]$ ratio must be 1.1 to 1 to attain pH 3.80 with this buffer system. For example, if 1 M HCOOH is used, the solution must be 1.1 M in sodium formate.

EXERCISE What must be the concentration of fluoride ion in an NaF/HF buffer to give pH 4.00? Assume that solid NaF is added to a 0.10 M solution of HF without volume change. $K_a = 6.5 \times 10^{-4}$ for HF. *Answer* 0.65 mol/L

By taking the negative logarithm of both sides and by rearranging, Equation (21.15a) can be expressed in terms of pH and pK to give the following alternative to Equation (21.15a)

$$-\log[H^+] = -\log K_a - \log\frac{[HA]}{[A^-]}$$

$$pH = pK_a + \log\frac{[A^-]}{[HA]} \qquad \textbf{(21.16)}$$

This equation (known as the Henderson-Hasselbalch equation) is often encountered in the biological sciences, where the control of pH is essential. (A comparable equation can be derived for pOH.)

The narrow pH ranges necessary in the body fluids of animals are maintained by buffer systems. The H_2CO_3/HCO_3^- buffer pair plays a vital role in maintaining human blood at an almost constant pH of 7.4. Excess acid in the blood reacts with the HCO_3^- ion to form H_2CO_3. This acid decomposes into water and carbon dioxide, which is exhaled.

$$H^+ + HCO_3^- \rightleftharpoons H_2CO_3(aq) \rightleftharpoons H_2O(l) + CO_2(g)$$

Deep breathing, which is stimulated by excess acid in the blood, helps to drive the equilibria toward CO_2 formation and a decrease in $[H^+]$. Shallow breathing, which is stimulated by decreased acid in the blood, drives the equilibria in the opposite direction, causing an increase in $[H^+]$.

For a buffer pair of equal concentrations, $\log[A^-]/[HA] = 1$, showing from Equation (21.16) that pH = pK_a. Thus, a buffer pair for a desired pH range can be chosen according to the pK_a of the acid (or the pK_b of a base). Although a desired pH can be achieved by varying the concentration ratio of any buffer pair, as illustrated in Example 21.5, it is found that a given pair will best maintain pH in a range of $pK \pm 1$.

To calculate the change in pH when an acid or a base is added to a solution that contains a buffer pair requires (1) finding the original pH of the buffer solution, (2) finding the new concentrations of the buffer pair after acid or base is added, and (3) finding the new pH by using Equations (21.15a) or (21.15b).

Consider a buffer solution that is 0.100 M in acetic acid (CH_3COOH) and 0.100 M in potassium acetate (KCH_3COO). First we must find the pH of this solution. A relatively large amount of acetate ion has been added to slightly ionized acetic acid, shifting the equilibrium back toward the nonionized acid. The result is that the acid concentration is very close to its original concentration and the acetate ion concentration is due mainly to the added sodium acetate. Therefore, in this case (as for most buffer solutions) it is a valid assumption that the equilibrium concentrations of the members of the buffer pair are the same as their initial concentrations. (See Example 21.4 where concentrations of the NH_4^+/NH_3 buffer pair were similarly taken as equal to their initial concentrations.)

$$[H^+] = \frac{[CH_3COOH]}{[CH_3COO^-]} \times 1.75 \times 10^{-5}$$

$$= \frac{0.100}{0.100} \times 1.75 \times 10^{-5} = 1.75 \times 10^{-5}\ \text{mol/L}$$

$$\text{pH} = -\log(1.75 \times 10^{-5}) = 4.757$$

If 0.005 mol of potassium hydroxide is added to 1 L of this solution (assume the volume does not change), what happens? The added base reacts completely with 0.005 mol of acetic acid.

$$\underset{0.005\ mol}{CH_3COOH(aq)} + \underset{0.005\ mol}{OH^-} \longrightarrow \underset{0.005\ mol}{CH_3COO^-} + H_2O(l)$$

The result is to decrease the concentration of acetic acid to 0.095 mol/L,

$$[CH_3COOH] = (0.100 - 0.005)\ \text{mol/L} = 0.095\ \text{mol/L}$$

and increase the concentration of acetate ion to 0.105 mol/L,

$$[CH_3COO^-] = (0.100 + 0.005)\ \text{mol/L} = 0.105\ \text{mol/L}$$

Using these new concentrations of the buffer pair in Equation (21.15a) allows calculation of the pH:

$$[H^+] = \frac{0.095}{0.105} \times (1.75 \times 10^{-5}) = 1.6 \times 10^{-5}\ \text{mol/L}$$

$$\text{pH} = -\log(1.6 \times 10^{-5}) = 4.80$$

The addition of 0.005 mol of KOH to the buffer solution has changed the pH by only 0.04. The pH of pure water would have changed from 7.00 to 9.30 if the same amount of KOH were added. A similar calculation for the addition of 0.005 mol of hydrochloric acid to 1 L of this buffer solution shows a decrease in pH from 4.757 to 4.72. A series of similar calculations for the addition of acid and base to this system (Table 21.3) further illustrates its resistance to pH change.

Table 21.3
Effects of Added Potassium Hydroxide and Hydrochloric Acid on the pH of 1 L of 0.100 M CH_3COOH/0.100 M $K(CH_3COO)$ Solution

KOH added (mol)	$[H^+]$ (mol/L)	pH
0	1.75×10^{-5}	4.757
0.0001	1.75×10^{-5}	4.757
0.001	1.7×10^{-5}	4.77
0.005	1.6×10^{-5}	4.80
0.010	1.4×10^{-5}	4.85
0.050	0.58×10^{-5}	5.24
HCl added (mol)	**$[H^+]$ (mol/L)**	**pH**
0	1.75×10^{-5}	4.757
0.0001	1.75×10^{-5}	4.757
0.001	1.8×10^{-5}	4.74
0.005	1.9×10^{-5}	4.72
0.010	2.1×10^{-5}	4.68
0.050	5.3×10^{-5}	4.28

EXAMPLE 21.6 Common Ion Effect: Buffers

The 0.10 M NH_3/0.10 M NH_4NO_3 solution described in Example 21.4 is a buffer solution. Calculate the pH after 0.10 g (0.0025 mol) of solid NaOH has been added to 100.0 mL of this buffer solution. (Assume that the volume has not changed.) $K_b = 1.6 \times 10^{-5}$ for NH_3.

The first step in solving this problem—finding the original pH of the buffer solution—was completed in Example 21.4, where it was found to be 9.20. We must now determine the concentrations of the buffer pair, NH_3 and NH_4^+, after addition of the NaOH. Note that we are working with 100.0 mL, or 0.1000 L, of the solution, which therefore initially contains 0.010 mol of NH_4^+ and 0.010 mol of NH_3.

The 0.0025 mol of NaOH reacts with the NH_4^+, forming additional ammonia

$$\underset{0.010\ mol}{NH_4^+} + \underset{0.0025\ mol}{OH^-} \longrightarrow \underset{0.0025\ mol}{NH_3(aq)} + H_2O(l)$$

and leaving 0.008 mol of unreacted NH_4^+

$$(0.010 - 0.0025)\ mol\ NH_4^+ = 0.008\ mol\ NH_4^+$$

giving $[NH_4^+] = 0.08$ mol/L. The total amount of NH_3 in the solution is increased to 0.013 mol

$$(0.010 + 0.0025)\ mol\ NH_3 = 0.013\ mol\ NH_3$$

giving $[NH_3] = 0.13$ mol/L. Using these values of $[NH_4^+]$ and $[NH_3]$ in Equation (21.15b) gives an OH^- concentration of

$$[OH^-] = \frac{[NH_3]}{[NH_4^+]}(1.6 \times 10^{-5}) = \left(\frac{0.13}{0.08}\right)(1.6 \times 10^{-5})$$
$$= 3 \times 10^{-5}\ mol/L$$

or

$$pOH = -\log[OH^-] = -\log(3 \times 10^{-5}) = 4.5$$
$$pH = 14.00 - pOH = 14.00 - 4.5 = \mathbf{9.5}$$

The pH has changed from 9.20 to 9.5. [Pure water would have changed from pH 7.00 to pH 12.40.]

EXERCISE What is the pH of the 0.10 M NH_3/0.10 M NH_4NO_3 buffer described above after 10.0 mL of 0.10 M NaOH is dissolved in 100.0 mL of the buffer?

Answer 9.3.

SUMMARY OF SECTIONS 21.4–21.5

The common ion effect results from the addition of a common ion that shifts an ionic equilibrium in aqueous solution. Addition of a common ion can be used to control the pH of a solution of a weak acid or base, or the concentration of the anion of a weak acid. The common ion effect is the basis for the action of a buffer solution, which usually contains a weak acid or base and its conjugate base or acid. The pH of a buffer solution is controlled by the ratio of the concentrations of the buffer pair. Addition of a small amount of acid or base does not change the pH of a buffer solution, nor does it change with dilution. A buffer solution with roughly equal concentrations of the buffer pair that are large relative to the added acid or base is most effective. In addition pH is best controlled in range $pK \pm 1$.

ACIDS, BASES, AND pH

21.6 LIMITATIONS ON THE USE OF EQUILIBRIUM CONSTANTS

The equilibrium calculations in a book such as this one and in the practical applications of chemistry frequently have similar goals: to verify what chemical intuition tells us will happen in a given chemical reaction or to predict to what extent a chemical change will occur. For example, chemical intuition tells us that the anions of weak acids should give alkaline solutions. Equilibrium calculations using K_a and K_b values verify this prediction and allow comparison of the alkalinity of various solutions of various anions. Our focus is on the chemistry, and, like practicing chemists in many situations, we would prefer to avoid lengthy and complicated calculations.

The mathematics of equilibrium problems can be simplified in many cases by limiting the systems under consideration to those in which complications are minimized. One way to avoid the necessity for complex mathematics is to avoid very dilute or very concentrated solutions. Most examples and exercises in this book deal with ions at the 0.01 M to 1 M concentrations most often encountered in the laboratory.

As solutions become more concentrated, the ions are less and less isolated from each other, and the solutions become less than ideal. The use of concentrations rather than effective concentrations (activities; Section 15.3) in equilibrium expressions then gives results further and further from the true equilibrium concentrations of the ions. What is known as the *salt effect* is one result of this influence of ions on each other. The total concentration of ions in solution, common or not, has a small influence on the solubility of a salt. Silver chloride, for example, is slightly more soluble in a potassium nitrate solution than in pure water. In accurate analytical chemistry and in biochemistry the salt effect must be taken into account for solutions of *any* concentration.

In very dilute solutions of acidic or basic ions, the contribution of the H^+ and OH^- ions from the ionization of water cannot be ignored. For example, the rigorous calculation of the concentrations of *all* of the ions present in a very dilute solution of a weak acid requires solving four equations in four unknowns. Make no mistake—the exact equilibrium calculations based on purely mathematical considerations are always possible. One professor was fond of telling his general chemistry class that if one of them could earn a million dollars by such a calculation, he or she would be able to do it.

Without such motivation, we instead utilize our chemical knowledge to make assumptions that allow us to simplify the necessary mathematics. For example, when the ions in a given solution can participate simultaneously in several equilibria, each equilibrium constant must be satisfied. However, it is often possible to simplify the calculations by identifying one equilibrium as most important because it yields much larger concentrations of products. This allows us to treat contributions from other equilibria as negligible.

You will see in the following section, which deals with polyprotic acids, and later in dealing with complex ions (Section 22.1), that we simplify the problems both by specifying certain reaction conditions and by assuming that the contributions from certain reactions can be ignored.

21.7 POLYPROTIC ACIDS

A polyprotic acid contains more than one ionizable hydrogen atom per molecule. Such an acid ionizes or reacts with a base in a separate step for each proton, with a different equilibrium constant for each step. Each successive proton is released less readily than the preceding proton because each proton is held more strongly by the increasingly negative anion. (K_a values for some common polyprotic acids are given in Table 21.4.) For example, for carbonic acid

$$H_2CO_3(aq) + H_2O(l) \rightleftharpoons HCO_3^- + H_3O^+ \qquad K_{a1} = 4.5 \times 10^{-7}$$
$$HCO_3^- + H_2O(l) \rightleftharpoons CO_3^{2-} + H_3O^+ \qquad K_{a2} = 4.8 \times 10^{-11}$$

Equilibrium calculations for polyprotic acids are complex because the concentrations of the ions are determined by the overall result of the successive equilibria. This complexity is dealt with by making simplifying assumptions.

Table 21.4
K_a Values for Polyprotic Acids

Acid	K_a
Hydrosulfuric	
H_2S	1.0×10^{-7}
HS^-	3×10^{-13}
Sulfuric	
H_2SO_4	Large
HSO_4^-	1.0×10^{-2}
Carbonic	
H_2CO_3	4.5×10^{-7}
HCO_3^-	4.8×10^{-11}
Arsenic	
H_3AsO_4	6.5×10^{-3}
$H_2AsO_4^-$	1.1×10^{-7}
$HAsO_4^{2-}$	3×10^{-12}
Pyrophosphoric	
$H_4P_2O_7$	1.2×10^{-1}
$H_3P_2O_7^-$	7.9×10^{-2}
$H_2P_2O_7^{2-}$	2.0×10^{-7}
$HP_2O_7^{3-}$	4.8×10^{-10}
Phosphoric	
H_3PO_4	7.5×10^{-3}
$H_2PO_4^-$	6.6×10^{-8}
HPO_4^{2-}	1×10^{-12}

In general, a stepwise approach is used. The concentrations of the ions formed in the first ionization are found. Because the first step is always more extensive than the subsequent steps ($K_1 > K_2 > K_3$), whenever possible it is assumed that the subsequent steps make negligible changes in the concentrations of these ions. This assumption allows the concentrations found in the first step to be used in the equilibrium constant expressions for subsequent steps to find the concentrations of the other anions.

a. Sulfuric Acid Sulfuric acid is unique among the common polyprotic acids because it is a strong electrolyte in its first ionization and a moderately weak one in its second ionization.

$$H_2SO_4(aq) + H_2O(l) \longrightarrow H_3O^+ + HSO_4^- \qquad \text{complete in dilute solution}$$
$$HSO_4^- + H_2O(l) \rightleftharpoons H_3O^+ + SO_4^{2-} \qquad K_{a_2} = 1.0 \times 10^{-2}$$

The other common polyprotic acids are weak electrolytes in each ionization step, including the first.

As for all polyprotic acids, the H_3O^+ from the first reaction is a common ion for the second reaction:

$$HSO_4^- + H_2O(l) \xleftarrow{\quad H_3O^+ \quad} H_3O^+ + SO_4^{2-}$$

The result is to decrease the SO_4^{2-} concentration and increase the HSO_4^- concentration.

The pH in a concentrated sulfuric acid solution is less than 0. As shown in Figure 21.2, virtually the only anion in such a solution is HSO_4^-. If the concentration of H_3O^+ is decreased, more and more SO_4^{2-} is formed, until beyond pH 4, virtually all of the original HSO_4^- has been converted to SO_4^{2-}.

Because HSO_4^- is a weak acid, it is not fully dissociated in a sulfuric acid solution. As a result the concentration of H_3O^+ in a sulfuric acid solution is *not* equal to twice the molar concentration of the acid, as is shown in the following example. Do not make the common error of assuming that a sulfuric acid solution, because it is a strong acid, contains a concentration of hydrogen ion equal to twice the concentration of the sulfuric acid.

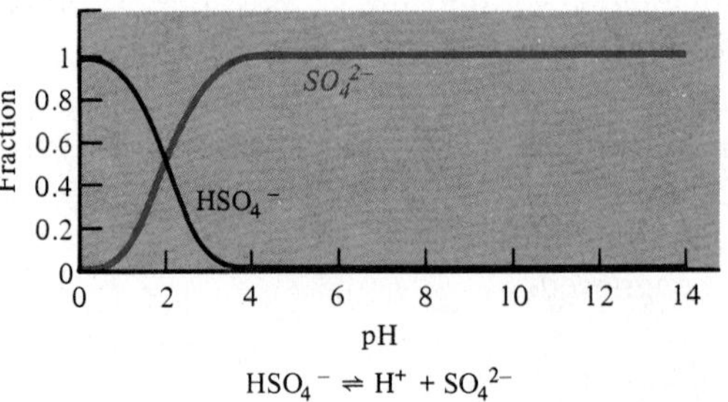

Figure 21.2
Effect of pH on Ionization of Sulfuric Acid in Aqueous Solution
Sulfuric acid is 100% ionized to HSO_4^-. The "fraction" plotted represents the fraction of the original HSO_4^- ion present as HSO_4^- and SO_4^{2-}. Beyond pH 4, virtually all of the original HSO_4^- has been converted to SO_4^{2-}. (Note that curves such as this do not show very low fractions of the species that are present.)

EXAMPLE 21.7 Polyprotic Acids

Find $[H^+]$, $[HSO_4^-]$, $[H_2SO_4]$, and $[SO_4^{2-}]$ in equilibrium in a 0.100 M solution of H_2SO_4. $K_{a_2} = 1.0 \times 10^{-2}$.

Because $H_2SO_4(aq)$ is a strong acid in the first ionization step

$$H_2SO_4(aq) + H_2O(l) \longrightarrow H_3O^+ + HSO_4^-$$

we can safely state that $[H_2SO_4] \simeq 0$ and that $[H_3O^+]$, or simply $[H^+] = [HSO_4^-] =$ 0.100 mol/L before the second ionization step occurs. For the second step, let $x = [SO_4^{2-}]$

	HSO_4^- $\rightleftharpoons$	H^+ +	SO_4^{2-}
Initial	0.100	0.100	0
Change	$-x$	$+x$	$+x$
Equilibrium	$0.100 - x$	$0.100 + x$	x

$$K_{a_2} = \frac{[H^+][SO_4^{2-}]}{[HSO_4^-]} = \frac{(0.100 + x)(x)}{(0.100 - x)} = 1.0 \times 10^{-2}$$

This equation must be solved by the exact method, using the quadratic equation, because the concentrations and K_{a_2} are of similar magnitude. Solving for x gives $[SO_4^{2-}] = 8.4 \times 10^{-3}$ mol/L. Thus

$$[HSO_4^-] = 0.100 - x = 0.100 - 0.0084 = 0.092 \text{ mol/L}$$

and

$$[H^+] = 0.100 + x = 0.100 + 0.0084 = 0.108 \text{ mol/L}$$

At equilibrium, $[H_2SO_4] \simeq 0$, $[HSO_4^-] = 0.092$ mol/L, $[SO_4^{2-}] = 0.0084$ mol/L, and $[H^+] = 0.108$ mol/L in a 0.100 M solution of H_2SO_4.

EXERCISE Find the molar concentrations of H^+, HSO_4^-, H_2SO_4, and SO_4^{2-} in equilibrium in a 0.0100 M solution of H_2SO_4. $K_{a_2} = 1.0 \times 10^{-2}$.

Answer $[H_2SO_4] \simeq 0$, $[HSO_4^-] = 0.006$ mol/L, $[SO_4^{2-}] = 0.004$ mol/L, $[H^+] = 0.014$ mol/L.

b. Hydrosulfuric Acid Hydrosulfuric acid (H_2S) is a weak diprotic acid:

$$H_2S(aq) + H_2O(l) \rightleftharpoons H_3O^+ + HS^- \qquad K_{a_1} = 1.0 \times 10^{-7}$$
$$HS^- + H_2O(l) \rightleftharpoons H_3O^+ + S^{2-} \qquad K_{a_2} = 3 \times 10^{-13}$$

The second ionization constant of the acid is so much smaller than the first that in calculating the hydrogen ion concentration in a solution of the acid alone, the second step ionization may be neglected, and $[H^+]$ and $[HS^-]$ equated to each other.

EXAMPLE 21.8 Polyprotic Acids

A saturated aqueous hydrosulfuric acid solution at 1 atm pressure and room temperature has a concentration of approximately 0.1 M H_2S. Find the concentrations of all of the ions present in this solution. For H_2S, $K_{a_1} = 1.0 \times 10^{-7}$; $K_{a_2} = 3 \times 10^{-13}$.

For the first ionization

$$H_2S(aq) + H_2O(l) \rightleftharpoons H_3O^+ + HS^-$$

Let $x = [H^+] = [HS^-]$.

	H_2S $\rightleftharpoons$	H^+ +	HS^-
Initial	0.1	0	0
Change	$-x$	$+x$	$+x$
Equilibrium	$0.1 - x$	x	x

$$K_{a_1} = \frac{[H^+][HS^-]}{[H_2S]} = \frac{(x)(x)}{(0.1 - x)} \simeq \frac{x^2}{0.1} = 1.0 \times 10^{-7}$$

Since K_{a_1} is very small, we can make the approximation that $(0.1 - x) \simeq 0.1$.

$$x^2 = 1 \times 10^{-8}$$
$$x = 1 \times 10^{-4} \text{ mol/L} = [H^+] = [HS^-]$$

The value found for the concentration of H^+ and HS^- ions must satisfy the second-step equilibrium, $HS^- \rightleftharpoons H^+ + S^{2-}$. Because K_{a_2} is much smaller than K_{a_1}, we assume that the contribution to $[H^+]$ from this equilibrium is negligible. Therefore, the S^{2-} ion concentration may be calculated using $[H^+]$ and $[HS^-]$ from the first ionization as follows:

	HS^-	$\rightleftharpoons$	H^+	+	S^{2-}
Initial	1×10^{-4}		1×10^{-4}		0
Change	$-x$		$+x$		$+x$
Equilibrium	$1 \times 10^{-4} - x$		$1 \times 10^{-4} + x$		x

$$K_{a_2} = \frac{[H^+][S^{2-}]}{[HS^-]} = \frac{(1 \times 10^{-4} + x)(x)}{(1 \times 10^{-4} - x)} \approx \frac{(1 \times 10^{-4})(x)}{(1 \times 10^{-4})} = 3 \times 10^{-13}$$

$$x = [S^{2-}] = 3 \times 10^{-13} \text{ mol/L}$$

In a 0.1 M aqueous H_2S solution, the concentrations of the ions present are **$[H^+] = [HS^-] = 1 \times 10^{-4}$ mol/L and $[S^{2-}] = 3 \times 10^{-13}$ mol/L.**

EXERCISE Find the molar concentrations of H^+, HS^-, S^{2-}, and H_2S in a 0.010 M solution of H_2S. $K_{a_1} = 1.0 \times 10^{-7}$ and $K_{a_2} = 3 \times 10^{-13}$ for H_2S.

Answers $[H^+] = [HS^-] = 3.2 \times 10^{-5}$ mol/L, $[S^{2-}] = 3 \times 10^{-13}$ mol/L, $[H_2S] \approx$ 0.010 mol/L.

Examination of the concentrations found in Example 21.8 and its Exercise shows that in a solution of H_2S alone, the molar concentration of the sulfide ion is equal to the value of K_{a_2}. Indeed, for most weak diprotic acids, the molar concentration of the ion formed by the loss of both protons is equal to the second ionization constant (in the absence of added common ions).

Overall K_a values for polyprotic acids can be found from the K_a values for individual steps. For example, multiplying the equilibrium constant expressions for the two ionization steps of aqueous H_2S gives

$$K_a(\text{overall}) = \frac{[H^+][HS^-]}{[H_2S]} \times \frac{[H^+][S^{2-}]}{[HS^-]} = (1.0 \times 10^{-7})(3 \times 10^{-13})$$

$$= \frac{[H^+]^2[S^{2-}]}{[H_2S]} = 3 \times 10^{-20} \qquad \textbf{(21.17)}$$

which represents the reaction

$$H_2S(aq) + 2H_2O(l) \rightleftharpoons 2H_3O^+ + S^{2-}$$

Ordinarily an overall equilibrium constant cannot be used to calculate the concentrations of the ions present in a solution of a polyprotic acid. In using such an equilibrium constant the assumption is implicit that the concentration of hydrogen ion produced in each step is the same. This is not correct, for at each step the acid ionizes to a lesser extent.

However, an expression that is very useful for finding concentrations in a saturated H_2S solution can be derived from Equation (21.17). For a 0.1 M saturated solution

$$\frac{[H^+]^2[S^{2-}]}{(0.1)} = 3 \times 10^{-20}$$

$$[H^+]^2[S^{2-}] = 3 \times 10^{-21} \qquad \textbf{(21.18)}$$

This equation allows for the calculation of the hydrogen ion concentration in a solution of known sulfide ion concentration, or of sulfide ion concentration in a solution of known hydrogen ion concentration.

The variation with pH of the fraction of the original H_2S present as HS^- and S^{2-} is shown in Figure 21.3. The sulfide ion concentration is high in alkaline solutions and

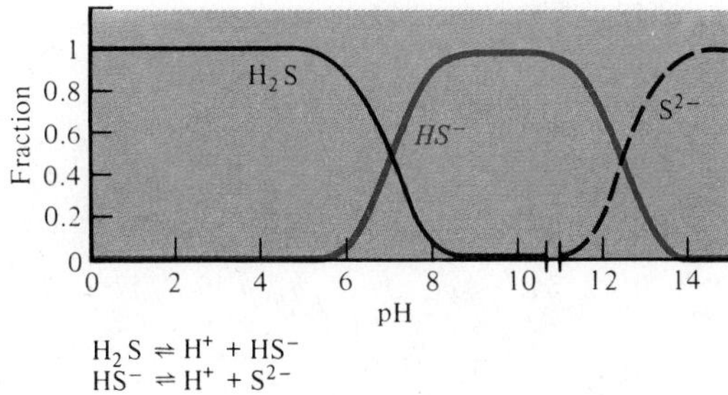

Figure 21.3
Effect of pH on Ionization of Hydrosulfuric Acid in Aqueous Solution
The fraction plotted is the fraction of the original H_2S present as HS^- and S^{2-}. Note that the sulfide ion concentration increases as the solution becomes more alkaline. In such plots for polyprotic acids, the curves cross at the point where the concentrations of the two species are equal and pH = pK_a (see Equation 21.16).

low in acidic solutions. By controlling the pH, the sulfide ion concentration can be controlled, an example of the common ion effect (Equation 21.14). This technique is used in qualitative analysis to separate cations by precipitation of their sulfides (Section 22.8).

EXAMPLE 21.9 Polyprotic Acids

To what pH must a saturated, 0.1 M H_2S solution be adjusted so that $[S^{2-}] = 1 \times 10^{-9}$ M?

For a saturated H_2S solution, the relationship

$$[H^+]^2[S^{2-}] = 3 \times 10^{-21}$$

is valid. This relationship can be solved for the hydrogen ion concentration necessary to achieve a desired sulfide concentration

$$[H^+]^2(1 \times 10^{-9}) = 3 \times 10^{-21}$$

$$[H^+] = \sqrt{\frac{3 \times 10^{-21}}{1 \times 10^{-9}}} = 2 \times 10^{-6} \text{ mol/L}$$

which corresponds to

$$\text{pH} = -\log[H^+] = -\log(2 \times 10^{-6}) = \mathbf{5.7}$$

EXERCISE To what pH must a 0.01 M solution of H_2S be adjusted so that $[S^{2-}] = 1 \times 10^{-9}$ mol/L? *Answer* 6.3.

c. Phosphoric Acid Phosphoric acid is a triprotic acid that releases protons in three steps.

$$H_3PO_4(aq) \rightleftharpoons H^+ + H_2PO_4^- \qquad K_{a_1} = 7.5 \times 10^{-3}$$
$$H_2PO_4^- \rightleftharpoons H^+ + HPO_4^{2-} \qquad K_{a_2} = 6.6 \times 10^{-8}$$
$$HPO_4^{2-} \rightleftharpoons H^+ + PO_4^{3-} \qquad K_{a_3} = 1 \times 10^{-12}$$

The ionization constant for the first step in the ionization of phosphoric acid is large relative to the ionization constants for the second and third steps. It can therefore be assumed that $[H^+]$ and $[H_2PO_4^-]$ at equilibrium in a phosphoric acid solution are determined by the first ionization. As in Example 21.8 for hydrosulfuric acid, the concentrations of the products of the first ionization step can be used in the equilibrium expressions for subsequent steps to find the concentrations of all ions in the solution. Such a calculation for 0.010 M H_3PO_4 gives the following results:

$$[H_3PO_4] = 4.3 \times 10^{-3} \text{ mol/L} \qquad [HPO_4^{2-}] = 6.6 \times 10^{-8} \text{ mol/L}$$
$$[H^+] = [H_2PO_4^-] = 5.7 \times 10^{-3} \text{ mol/L} \qquad [PO_4^{3-}] = 1 \times 10^{-17} \text{ mol/L}$$

The pH of this 0.010 M H_3PO_4 solution is 2.2. Figure 21.4 shows that most of the original H_3PO_4 in a solution at this pH is present as H_3PO_4 or $H_2PO_4^-$. (At this pH, the fractions of the other two anions are too small to appear in the figure.) The figure also illustrates that the concentration of PO_4^{3-} can be increased by making the solution alkaline.

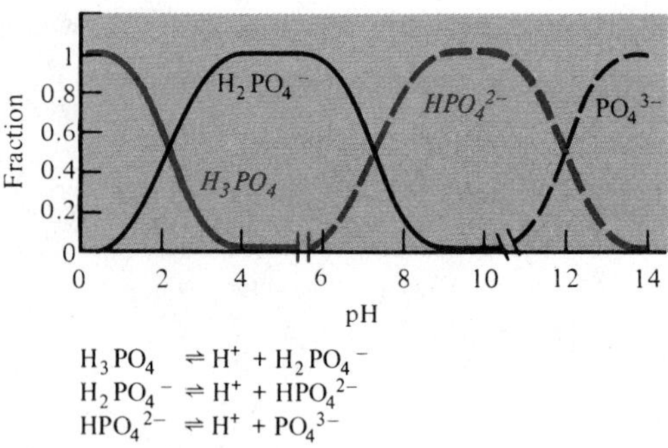

Figure 21.4
Effect of pH on Ionization of Phosphoric Acid
The "fraction" plotted is the fraction of the original H_3PO_4 converted to $H_2PO_4^-$, HPO_4^{2-}, and PO_4^{3-} (See Figure 21.3 caption.)

21.8 ACID–BASE REACTIONS AND TITRATION IN AQUEOUS SOLUTIONS

The need frequently arises to determine accurately the concentration of an acid or a base in aqueous solution. For example, a food chemist might need to know the concentration of citric acid in a fruit-flavored beverage. An environmental engineer might need to know the concentration of sulfuric acid in the wastewater from an industrial operation. One method for determining concentrations of acids or bases in solution is **titration**—the measurement of the volume of a solution of one reactant that is required to react completely with a measured amount of another reactant. (The technique can also be applied to redox reactions, so that the concentrations of oxidizing and reducing agents can be determined.)

A typical titration to determine the concentration of an acid or base in solution is conducted as follows: An exact volume of the solution of unknown concentration is measured out; let's say it's a solution of a base. The titration is performed by adding an acid solution of exactly known concentration (a standard solution) from a buret (Figure 21.5), which allows the volume of added solution to be measured accurately. The solution is added slowly until chemically equivalent, or stoichiometric, amounts of the added acid and the base in the unknown solution have reacted—at what is called the **equivalence point.**

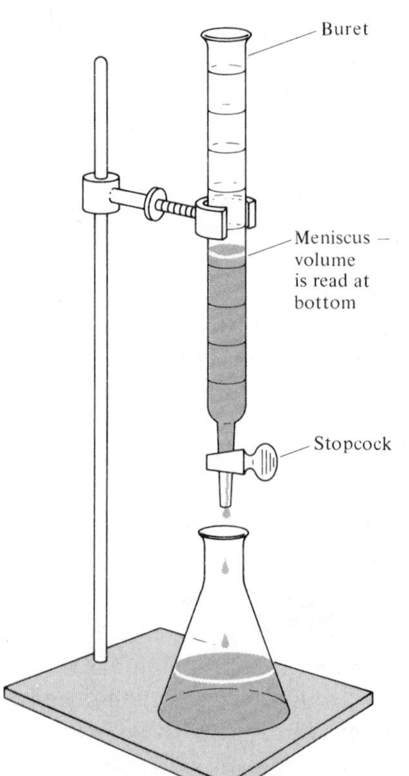

Figure 21.5
Titration
One solution is added from the buret to another in the flask. The volume of solution that must be added to reach the endpoint is carefully measured.

To detect the equivalence point, a small amount of an **indicator**—a compound that changes color in a specific pH range—may be added to the unknown solution at the beginning of the titration (Aside 21.1). In an acid–base reaction, the pH changes very little until the equivalence point is reached. At that point a small additional amount of acid or base causes a large and rapid change in pH, which in turn causes the indicator to change color, signaling the **endpoint** of the titration. An indicator is chosen so that the endpoint and the equivalence point are as close together as possible. There may be an endpoint error—a slight discrepancy between the endpoint and the exact equivalence point—but in most cases this error is negligible.

The goal of the titration is to measure the volume of acid of known concentration that must be added to reach the endpoint. The concentration of the base solution can then be calculated, as illustrated in Example 21.10, from the stoichiometry of the acid–base reaction, the measured volume, and the known concentration.

Essentially the same procedure described above is applied with variations to many different acid–base reactions. In each case, the reaction is carried out to the point where stoichiometric amounts of the acid and base have reacted. For a weak acid or base, the reaction is driven to completion according to Le Chatelier's principle—by continuous removal of H^+ or OH^- through the addition of base or acid. Example 21.11 illustrates titration of an unknown solution into a known solution and also the titration of the weak base, CO_3^{2-}, with hydrochloric acid, a strong acid.

All problems concerned with acid–base titrations can be solved with the concepts of stoichiometry and solution concentration based on moles, mole ratios, and molarity that we have studied earlier. (The problems could also be solved in terms of equivalents and normality, which were defined in Section 20.6. Although the use of these quantities is not necessary and has fallen from favor, some laboratories continue to express acid and base concentrations in normality. Be aware that you may encounter these units in some applications.)

EXAMPLE 21.10 Acid–Base Reactions

Exactly 23.6 mL of a 0.131 M HCl solution was required to completely react with 25.0 mL of an NaOH solution. What was the concentration of the NaOH solution?

The chemical equation for the acid–base reaction is

$$\mathrm{NaOH}(aq) + \mathrm{HCl}(aq) \longrightarrow \mathrm{H_2O}(l) + \mathrm{NaCl}(aq)$$

The number of moles of HCl that reacted is

$$(23.6\ \mathrm{mL})\left(\frac{1\ \mathrm{L}}{1000\ \mathrm{mL}}\right)\left(\frac{0.131\ \mathrm{mol\ HCl}}{\mathrm{L}}\right) = 0.00309\ \mathrm{mol\ HCl}$$

which corresponds to

$$(0.00309\ \mathrm{mol\ HCl})\left(\frac{1\ \mathrm{mol\ NaOH}}{1\ \mathrm{mol\ HCl}}\right) = 0.00309\ \mathrm{mol\ NaOH}$$

The concentration of NaOH is

$$\left(\frac{0.00309\ \mathrm{mol\ NaOH}}{25.0\ \mathrm{mL}}\right)\left(\frac{1\ \mathrm{L}}{1000\ \mathrm{mL}}\right) = 0.124\ \mathrm{mol/L}$$

The individual steps can be combined

$$(23.6\ \mathrm{mL})\left(\frac{1\ \mathrm{L}}{1000\ \mathrm{mL}}\right)\left(\frac{0.131\ \mathrm{mol\ HCl}}{\mathrm{L}}\right)\left(\frac{1\ \mathrm{mol\ NaOH}}{1\ \mathrm{mol\ HCl}}\right) \times \left(\frac{1}{25.0\ \mathrm{mL}}\right)\left(\frac{1000\ \mathrm{mL}}{1\ \mathrm{L}}\right) = \mathbf{0.124\ mol/L}$$

EXERCISE A volume of 42.6 mL of 0.0972 M NaOH was required to completely react with 50.0 mL of an $HClO_4$ solution. What was the concentration of the perchloric acid?

Answer 0.0828 mol/L.

EXAMPLE 21.11 Acid–Base Reactions

An approximately 0.1 M HCl solution was standardized (its exact concentration found) by titrating it into a solution containing 0.1223 g of 99.95% pure Na_2CO_3:

$$Na_2CO_3(aq) + 2HCl(aq) \longrightarrow 2NaCl(aq) + H_2O(l) + CO_2(g)$$

The equivalence point was reached when 22.65 mL of the HCl solution had been used. What was the exact concentration of the acid?

In this case, the known solution is that containing Na_2CO_3. The number of moles of Na_2CO_3 that reacted was

$$(0.1223 \text{ g impure } Na_2CO_3)\left(\frac{99.95 \text{ g } Na_2CO_3}{100.00 \text{ g impure } Na_2CO_3}\right)\left(\frac{1 \text{ mol } Na_2CO_3}{105.99 \text{ g } Na_2CO_3}\right)$$
$$= 0.001153 \text{ mol } Na_2CO_3$$

The chemical equation shows that 2 mol of HCl react for every mole of Na_2CO_3, so the stoichiometric amount of HCl required to react with the Na_2CO_3 was

$$(0.001153 \text{ mol } Na_2CO_3)\left(\frac{2 \text{ mol HCl}}{1 \text{ mol } Na_2CO_3}\right) = 0.002306 \text{ mol HCl}$$

The molarity of the solution is

$$\frac{0.002306 \text{ mol}}{0.02265 \text{ L}} = \mathbf{0.1018\ mol/L}$$

EXERCISE An approximately 0.1 M NaOH solution was standardized by titrating it into a solution of pure potassium acid phthalate (commonly abbreviated KHP), a monoprotic acid.

$$NaOH(aq) + C_6H_4(COOH)(COO^-K^+) \longrightarrow C_6H_4(COO^-Na^+)(COO^-K^+) + H_2O(l)$$

A 0.4963 g sample of the KHP was neutralized by 19.61 mL of the base. What is the exact concentration of the base? The molar mass of KHP is 204.23 g/mol.

Answer 0.1239 mol/L.

Aside 21.1 TOOLS OF CHEMISTRY: MEASURING pH

The pH of a solution or the endpoint in a titration is commonly found by the use of either an indicator or an instrument known as a pH meter. Because of their accuracy and speed, pH meters have superseded the older indicator method in many applications. However, the indicator method remains in use because it is simple and convenient.

An indicator is a weak organic acid or base that changes color with changes in the pH. Letting HInd represent an indicator acid, its ionization, like that of any acid, is

$$HInd(aq) \rightleftharpoons Ind^- + H^+$$

The HInd and Ind^- forms must have different colors. For the color change of an indicator to be visible, the concentration of one colored form must be about ten times greater than the concentration of the other. Most indicators have a useful color change over a 2 pH unit range. The commonly encountered indicator phenolphthalein, for example, changes from colorless to pink in the pH range from 8.3 to 10.0. Indicators are available for the entire pH range (Table 21.5).

One way to improve the accuracy of using indicators is by comparing the color of the indicator in a solution of unknown pH with the color of the indicator in a solution of known pH. Paper impregnated with an organic pH indicator is very convenient. Litmus paper and its pink and blue colors may be familiar to many of you. A "universal" pH paper combines several different indicators in the same paper. This paper will show a characteristic color for any point in the pH range.

Table 21.5
Common pH Indicators and Their Color Changes

Indicator	Color at lower pH	pH range	Color at higher pH
Methyl violet	Yellow	0–2	Violet
Malachite green (acidic)	Yellow	0–1.8	Blue-green
Thymol blue (acidic)	Red	1.2–2.8	Yellow
Bromphenol blue	Yellow	3.0–4.6	Purple
Methyl orange	Red	3.1–4.4	Yellow-orange
Bromcresol green	Yellow	3.8–5.4	Blue
Methyl red	Red	4.4–6.2	Yellow
Litmus	Red	4.5–8.3	Blue
Bromcresol purple	Yellow	5.2–6.8	Purple
Bromthymol blue	Yellow	6.0–7.6	Blue
Phenol red	Yellow	6.4–8.2	Red
m-Cresol purple	Yellow	7.6–9.2	Purple
Thymol blue (alkaline)	Yellow	8.0–9.6	Blue
Phenolphthalein	Colorless	8.3–10.0	Red
Thymolphthalein	Colorless	9.3–10.5	Blue
Alizarin yellow	Yellow	10.1–11.1	Lilac
Malachite green (alkaline)	Green	11.4–13.0	Colorless
Trinitrobenzene	Colorless	12.0–14.0	Orange

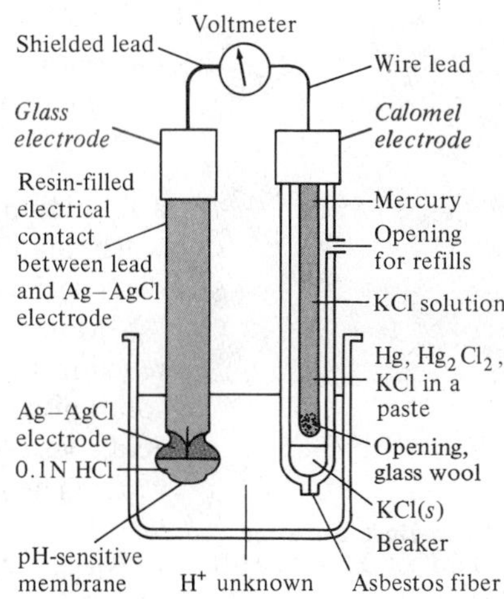

Figure 21.6
Electrodes Used for Measuring pH

A pH meter employs electrodes to measure the potential difference between solutions of known and unknown hydrogen ion concentration. The essential parts of a pH meter are (1) a glass electrode, (2) a reference electrode, and (3) a voltmeter calibrated to read in pH units rather than in volts. (In some instruments, for convenience the two electrodes are combined into one probe.) The glass electrode is based upon a unique property of a thin membrane (about 0.5 mm) made of special glass. An electric potential develops across the membrane when its two sides are in contact with solutions of different hydrogen ion concentrations. The potential arises because of ion exchange between Na^+ ions in the glass and H^+ ions in the solutions, aided by a thin layer of hydrated glass on the membrane surface.

The glass membrane is formed into a bulb that is filled with a solution of known and constant pH (Figure 21.6, left electrode). Sealed into the glass electrode and dipping into the solution with which the bulb is filled is an electrode, usually a silver wire coated with silver chloride. The entire glass electrode is placed in the solution of unknown pH during use, thereby putting the membrane in contact with the solutions of known and unknown pH. The reference electrode most often used is a calomel electrode, which contains a mixture of Hg, Hg_2Cl_2 (known as calomel), and KCl.

The arrangement is such that the potential difference measured by the voltmeter is due only to the difference in the hydrogen ion concentration of the two solutions. This potential is (see Section 24.12)

$$E = \text{a constant} + 0.0592\ \text{pH}$$

In practice, a pH meter must be calibrated before each use by placing the electrodes in a solution of known pH and setting the meter to read correctly. This is necessary because the value of the "constant" in the above equation varies slightly with conditions.

To find the titration equivalence point with a pH meter, after each addition of the known solution the pH is recorded and a titration curve is plotted, as illustrated in the following section (see, e.g., Figure 21.7). The point at which the largest pH change occurs is the equivalence point.

21.9 TITRATION CURVES

The change of the pH throughout the course of an acid–base titration can be represented by a **titration curve**—a plot of pH versus volume of acid or base added. The curves have different characteristic shapes that depend on the concentrations of reagents and on whether strong or weak acids and bases are involved. The calculation of the pH at various points in some representative titrations is discussed in the following sections and provides a review of pH calculations.

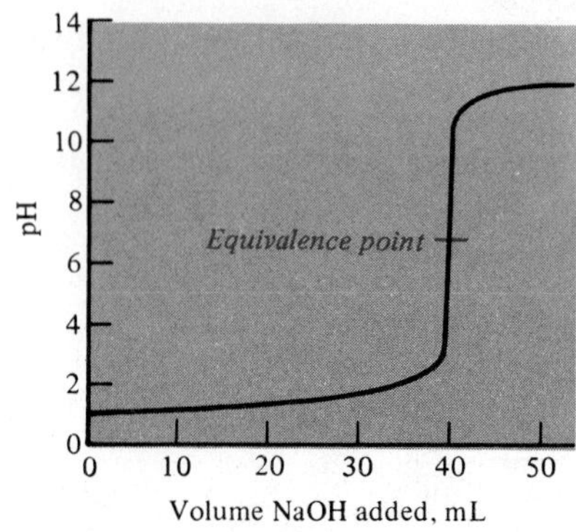

Figure 21.7
Strong Acid–Strong Base Titration
Titration of 40.0 mL of 0.100 M HCl by 0.100 M NaOH.

The pH at the equivalence point in any acid–base reaction depends upon which of the four types of salts discussed earlier (see Table 21.1) has been formed. For example, in the reaction of a hydrocyanic acid solution with sodium hydroxide

$$HCN(aq) + NaOH(aq) \rightleftharpoons NaCN(aq) + H_2O(l)$$

a solution of sodium cyanide in water is present at the equivalence point. The solution will be alkaline because sodium cyanide is a soluble compound of a cation that does not react with water and an anion that reacts to give hydroxide ion (Section 21.2).

Many students make the common error of assuming that in *any* acid–base reaction in aqueous solution, all acidic and basic character is destroyed, leaving a solution with pH 7. This is *not* the case. The solution at the equivalence point in an acid–base reaction is exactly the same as a solution produced by dissolving a salt in water. The solution will *not* necessarily be neutral.

a. Strong Acid–Strong Base Titration Figure 21.7 is the titration curve for the reaction of 40.0 mL of 0.100 M HCl with 0.100 M NaOH.

The pH of the solution before the addition of any base is determined by the initial concentration of the strong acid. In this titration, it is 1.000 ($pH = -\log[H^+] = -\log 0.100$).

The pH of the solution before the equivalence point is determined by how much of the strong acid has not yet reacted with base. The amount of HCl originally present in 40.0 mL of the 0.100 M solution is 0.00400 mol. After, say, 10.0 mL of base has been added, the amount of HCl that has reacted is

$$\left(\frac{0.100 \text{ mol NaOH}}{1 \text{ L}}\right)(10.0 \text{ mL})\left(\frac{1 \text{ L}}{1000 \text{ mL}}\right)\left(\frac{1 \text{ mol HCl}}{1 \text{ mol NaOH}}\right) = 0.00100 \text{ mol HCl}$$

and the amount of HCl that has not yet reacted, which determines the pH, is

$$0.00400 \text{ mol HCl} - 0.00100 \text{ mol HCl} = 0.00300 \text{ mol HCl}$$

In calculations involving titrations, it is important to remember that the volume of the solution is changing. The 0.00300 mol of HCl is present in (40.0 + 10.0) mL of solution, for a concentration and pH of

$$\left(\frac{0.00300 \text{ mol}}{40.0 + 10.0 \text{ mL}}\right)\left(\frac{1000 \text{ mL}}{\text{L}}\right) = 0.0600 \text{ mol/L, or pH } 1.222$$

Table 21.6 (center column) gives the pH of the solution as this titration proceeds.

At the equivalence point, there is no excess of HCl or NaOH and the solution is neutral because neither the Na^+ nor the Cl^- ions react with water. Thus the pH at the equivalence point is 7.000. [Suitable indicators for strong acid–strong base ti-

Table 21.6
pH during the Acid–Base Titrations

Volume of NaOH added, mL	Strong acid–strong base[a] pH		Weak acid–strong base[b] pH
0.0	1.000		2.879
10.0	1.222		4.279
20.0	1.477		4.756
30.0	1.845		5.233
39.0	2.898		6.347
39.3	3.076		6.505
39.7	3.378		6.878
40.0	7.000 ⟵	equivalence	⟶ 8.728
40.3	10.62	point	10.62
42.0	11.39		11.39
50.0	12.05		12.05

[a] 40.0 mL of 0.100 M HCl with 0.100 M NaOH.
[b] 40.0 mL of 0.100 M CH_3COOH with 0.100 M NaOH.

trations include litmus (pH 4.5–8.3), bromthymol blue (pH 6.0–7.6), and phenol red (pH 6.4–8.2).]

The pH of the solution after the equivalence point is determined by how much excess NaOH has been added. For example, after the addition of 45.0 mL, or 0.00450 mol, of base the amount of base in excess is

$$(0.00400 \text{ mol HCl})\left(\frac{1 \text{ mol NaOH}}{1 \text{ mol HCl}}\right) = 0.00400 \text{ mol NaOH reacted}$$

$$0.00450 \text{ mol NaOH} - 0.00400 \text{ mol NaOH} = 0.00050 \text{ mol NaOH}$$

which corresponds to a concentration of

$$\left(\frac{0.00050 \text{ mol NaOH}}{(40.0 + 45.0) \text{ mL}}\right)\left(\frac{1000 \text{ mL}}{1 \text{ L}}\right) = 0.0059 \text{ mol/L, pH } 11.77$$

As illustrated in Figure 21.8, the shape of a strong acid–strong base curve changes slightly with the concentration of the acid. However, the equivalence point remains at pH 7.

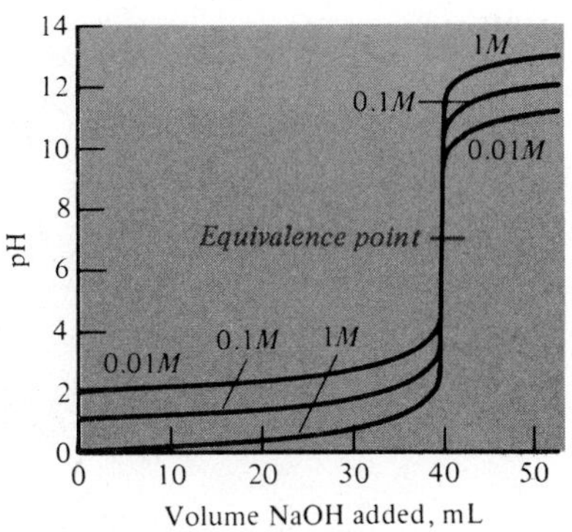

Figure 21.8
Strong Acid–Strong Base Titration
Titration of 40.0 mL of HCl by NaOH of equal concentrations.

b. Weak Acid–Strong Base or Strong Acid–Weak Base Titration Figure 21.9 illustrates the curve for the titration of 40.0 mL of 0.100 M CH_3COOH with 0.100 M NaOH.

$$CH_3COOH(aq) + NaOH(aq) \longrightarrow Na(CH_3COO)(aq) + H_2O(l)$$

The pH of the solution before the addition of any base is determined by the initial concentration of the weak acid. To find the pH requires finding $[H^+]$ by using the K_a of acetic acid. Let $x = [H^+]$.

	$CH_3COOH \rightleftharpoons$	CH_3COO^- +	H^+
Initial	0.100	0	0
Change	$-x$	$+x$	$+x$
Equilibrium	$0.100 - x$	x	x

$$K_a = \frac{[H^+][CH_3COO^-]}{[CH_3COOH]} = \frac{(x)(x)}{(0.100 - x)} \approx \frac{x^2}{0.100} = 1.75 \times 10^{-5}$$

$$x = 1.32 \times 10^{-3} \text{ mol/L} = [H^+]$$

$$\text{pH} = -\log(1.32 \times 10^{-3}) = 2.879$$

From the addition of the first drops of base and before the equivalence point, the solution in a weak acid–strong base titration contains a buffer pair—in this case, the solution contains acetic acid plus acetate ion. Equation (21.17) can be used to calculate the pH of the solution up to the equivalence point. As an example, we calculate the pH of the solution after the addition of 10.0 mL of base. The original amount of acetic acid present in 40.0 mL of 0.100 M CH_3COOH was 0.00400 mol. The amount of CH_3COOH that reacted was

$$\left(\frac{0.100 \text{ mol NaOH}}{1 \text{ L}}\right)(10.0 \text{ mL})\left(\frac{1 \text{ L}}{1000 \text{ mL}}\right)\left(\frac{1 \text{ mol } CH_3COOH}{1 \text{ mol NaOH}}\right) = 0.00100 \text{ mol } CH_3COOH$$

which means that the amount of CH_3COO^- formed is 0.00100 mol and the amount of unreacted CH_3COOH is

$$0.00400 \text{ mol } CH_3COOH - 0.00100 \text{ mol } CH_3COOH = 0.00300 \text{ mol } CH_3COOH$$

The concentrations of CH_3COO^- and unreacted CH_3COOH are

$$[CH_3COO^-] = \left(\frac{0.00100 \text{ mol } CH_3COO^-}{(40.0 + 10.0) \text{ mL}}\right)\left(\frac{100 \text{ mL}}{1 \text{ L}}\right) = 0.0200 \text{ mol/L}$$

$$[CH_3COOH] = \left(\frac{0.00300 \text{ mol } CH_3COOH}{(40.0 + 10.0) \text{ mL}}\right)\left(\frac{1000 \text{ mL}}{1 \text{ L}}\right) = 0.0600 \text{ mol/L}$$

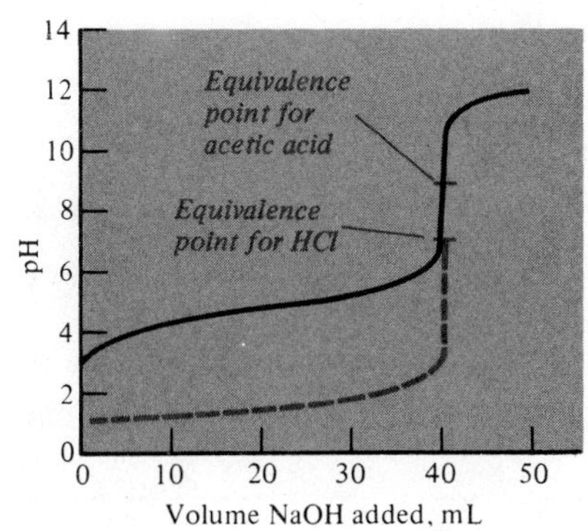

Figure 21.9
Weak Acid–Strong Base Titration
Titration of 40.0 mL of 0.100 M CH_3COOH by 0.100 M NaOH. Dashed line is for HCl–NaOH titration for comparison.

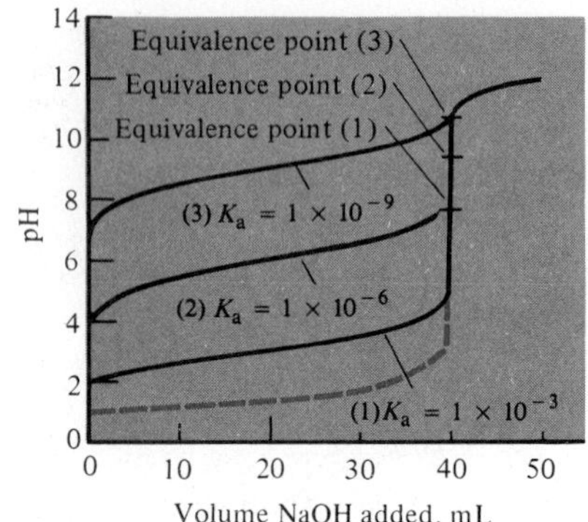

Figure 21.10
Weak Acid–Strong Base Titration
Titration of 40.0 mL of 0.100 M solutions of various weak acids with 0.100 M NaOH. Dashed line is for HCl for comparison.

The $[H^+]$ of the solution is (see Equation 21.15a)

$$[H^+] = \left(\frac{0.0600}{0.0200}\right)(1.75 \times 10^{-5}) = 5.25 \times 10^{-5} \text{ mol/L}$$

giving a pH of 4.279. Table 21.6 (right-hand column) gives the pH at various points throughout this titration.

At the equivalence point, the acetic acid has been neutralized by the sodium hydroxide and the 80.0 mL of solution contains 0.00400 mol of $Na(CH_3COO)$, for an acetate ion concentration of

$$[CH_3COO^-] = \left(\frac{0.00400 \text{ mol}}{80.0 \text{ mL}}\right)\left(\frac{1000 \text{ mL}}{1 \text{ L}}\right) = 0.0500 \text{ mol/L}$$

The solution is not neutral because the acetate ion reacts with water. Let $x = [OH^-]$.

	$CH_3COO^- + H_2O$	$\rightleftharpoons CH_3COOH$	$+ \; OH^-$
Initial	0.0500	0	0
Change	$-x$	$+x$	$+x$
Equilibrium	$0.0500 - x$	x	x

$$K_b = \frac{[CH_3COOH][OH^-]}{[CH_3COO^-]} = \frac{K_w}{K_a} = \frac{1.00 \times 10^{-14}}{1.75 \times 10^{-5}} = 5.71 \times 10^{-10}$$

$$\frac{(x)(x)}{(0.0500 - x)} \simeq \frac{x^2}{0.0500} = 5.71 \times 10^{-10}$$

$$x = 5.34 \times 10^{-6} \text{ mol/L} = [OH^-]$$

$$pOH = -\log(5.34 \times 10^{-6}) = 5.272 \qquad pH = 14.000 - 5.272 = 8.728$$

The pH at the equivalence point is in the alkaline range. [An indicator such as *m*-cresol purple (pH 7.6 – 9.2), thymol blue (pH 8.0 – 9.6), or phenolphthalein (pH 8.3 – 10.0) should be used for this titration.]

After the equivalence point the pH of the solution is controlled primarily by the excess NaOH. The calculation of the pH is the same as for points after the equivalence point in the titration of a strong acid with a strong base. From Table 21.6 you can see that beyond the equivalence point the titrations of a strong acid or a weak acid with a strong base are identical.

The pK_a of a weak acid can be determined experimentally from a titration curve. At the "half-equivalence" point, one-half of the acid has been neutralized and $[HA] = [A^-]$, which upon substitution into Equation (21.16) gives

$$pH = pK_a + \log\frac{[A^-]}{[HA]} = pK_a + \log(1) = pK_a$$

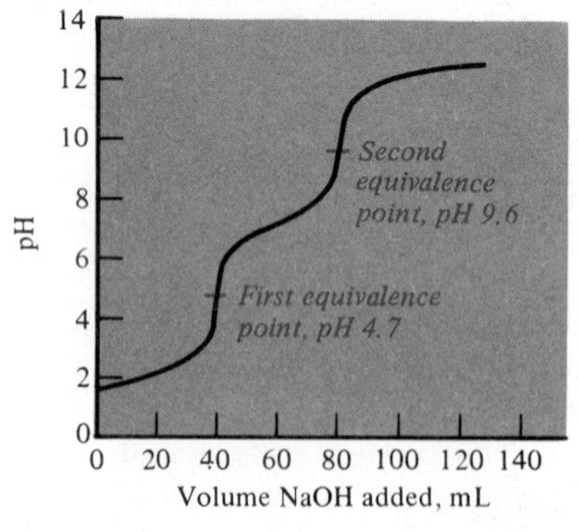

Figure 21.11
Triprotic Acid–Strong Base Titration
Titration of 40.0 mL of 0.100 M H_3PO_4 with 0.100 M NaOH.

The change in the pH at the equivalence point becomes less and less dramatic the weaker the acid is, as illustrated by Figure 21.10. The curves show that the equivalence point is very difficult to observe for acids with K_a values less than 10^{-8}. For this reason, it is not practical to titrate a weak acid with a weak base; the equivalence point is too difficult to detect.

c. Titration of a Polyprotic Acid with a Strong Base Figure 21.11 illustrates the titration curve for the reaction of 40.0 mL of 0.100 M H_3PO_4 with 0.100 M NaOH.

$$H_3PO_4(aq) + NaOH(aq) \longrightarrow NaH_2PO_4(aq) + H_2O(l)$$
$$NaH_2PO_4(aq) + NaOH(aq) \longrightarrow Na_2HPO_4(aq) + H_2O(l)$$
$$Na_2HPO_4(aq) + NaOH(aq) \longrightarrow Na_3PO_4(aq) + H_2O(l)$$

Only the equivalence points for the first two reactions appear in the titration curve. The third is not detectable (HPO_4^{2-} has $K_a = 1 \times 10^{-12}$). The pH before the first equivalence point is governed by the $H_3PO_4/H_2PO_4^-$ buffer pair and the pH between the first and second equivalence points is governed by the $H_2PO_4^-/HPO_4^{2-}$ buffer pair.

Aside 21.2 TOOLS OF CHEMISTRY: THE COMPUTER

It is widely believed that we are in the midst of an information revolution that will change society as profoundly as the industrial revolution did nineteenth century Europe and America. The digital computer is both the tool and the symbol of this revolution. Every scientific discipline has been affected by advances in computer technology, and chemistry is no exception.

Microcomputers can be used very effectively in the laboratory to control experimental equipment and to help record data. For example, the pH electrode you read about in Aside 21.1 can be interfaced with a microcomputer to produce an instrument that will record the pH every second during a titration. With another electronic interface and some flow control values, the microprocessor can be used to add specific volumes of titrant at regular intervals. With some additional programming, the two functions—data acquisition and experimental control—can be combined so that the volume of titrant added decreases as the endpoint is approached. It is then a simple matter to have the microcomputer calculate the molarity of the unknown acid. Now that a machine can do this much of your acid–base titration lab, what do you think the next step will be?

The real power of computer-based chemical instrumentation is seen in more complex experiments than titrations, however. For example, the tools of gas chromatography (Aside 19.1) and mass spectrometry (Aside 3.4) can be combined, with the aid of a minicomputer, to produce an analytical instrument of great sensitivity and specificity. A mixture of many different chemical compounds can be separated with chromatography, but simple detectors cannot identify the compounds as they are eluted from the column. A mass spectrometer can be used to read the "fingerprint" of each compound as it emerges. Instruments that combine mass spectrometers with chromatographs require delicate adjustments both before and during an experiment. Minicomputers are used to control the acquisition and analysis of the data, allowing specific compounds present in the original sample in concentrations of less than one part per million to be detected.

In the chemical industry, many chemical processes also require elaborate control of conditions such as temperature, pressure, flow rate, and catalyst condition, as well as immediate analyses of the process stream for the desired product. Computers are an essential tool used by chemical engineers in the design and operation of chemical plants.

The earliest applications of computational methods in chemistry were made by quantum theorists, who were trying to find accurate methods of calculating the electronic structure of molecules. Both the internal electronic energy of molecules and the distribution of electronic charge in a molecule are obtained by solving the Schrödinger equation (Aside 25.1). This equation can be solved exactly only for a few very simple problems of interest in chemistry, for example, the electronic structure of the hydrogen atom. It is from these exact solutions that our insight into chemical structure has developed; the concept of atomic orbitals comes directly from the solutions of the Schrödinger equation for the hydrogen atom. But even for the hydrogen molecule, an exact solution of the Schrödinger equation is impossible without lengthy calculations. Chemists during the 1930s were using analog computers (and even mechanical desk calculators) to solve the Schrödinger equation numerically, and they were able to determine the bond energy of the hydrogen molecule to within 10% of the presently accepted experimental value.

For any molecule heavier than the hydrogen molecule, numerical computations become very tedious if done by hand. As digital computers became available in universities during the 1950s, computational methods for solving Schrödinger's equation for molecules with many electrons were developed. There are now many research groups in both universities and industries that have computers dedicated entirely to electronic structural calculations, and accurate studies can be done on large, chemically interesting molecules. For example, it is possible to compute the interaction energies between two molecules such as the biochemically important cyclic bases adenine and thymine when they are hydrogen-bonded, as in the double-helix form of DNA (Section 33.10).

Of course, molecular structure can be determined by many experimental methods as well, for example, by X-ray diffraction (Aside 13.2) or molecular spectroscopy. In each of these experimental methods, the availability of computer-based methods of analyzing data has made possible the determination of structures far more complex than could have been done 20 years ago. For example, the complete three-dimensional structures of biologically significant molecules such as myoglobin, an oxygen-carrying protein that is similar to hemoglobin and contains more than one hundred atoms, has been completely worked out using X-ray diffraction and computer reduction of data.

Complex molecular structures can also be displayed on a variety of output devices, using the techniques of computer graphics. By representing the positions of all atoms within a molecule by vectors, each atom can be located in space with three numbers. Atoms themselves can be depicted in many ways: as points, as spheres with radii proportional to the van der Waals radii of the atoms or even as contour diagrams of electron density (Section 8.10). Devices such as high resolution graphics terminals or plotters can then be used to draw a picture of the molecule, viewed from any angle at any distance. The viewing perspective can be changed until a particular structural feature of interest comes into view. Examples of such computer-generated structures are given in Figure 21.12.

The ability of computers to search through huge data bases for desired information is also making significant changes in the field of chemistry. From the graphical input of the structure of a molecule, the data base of existing compounds can be checked to see if a molecule of interest has been made. It is also possible to search the data base for other compounds containing structural features similar to those of the compound originally entered. With the knowledge gained in learning to deal with complex chemical structures in this way, another interesting use of computers has emerged. There are now programs that will aid organic chemists in designing successful syntheses of compounds that have not yet been made. Several of these programs use concepts developed in the field of artificial intelligence to guide the synthesis design.

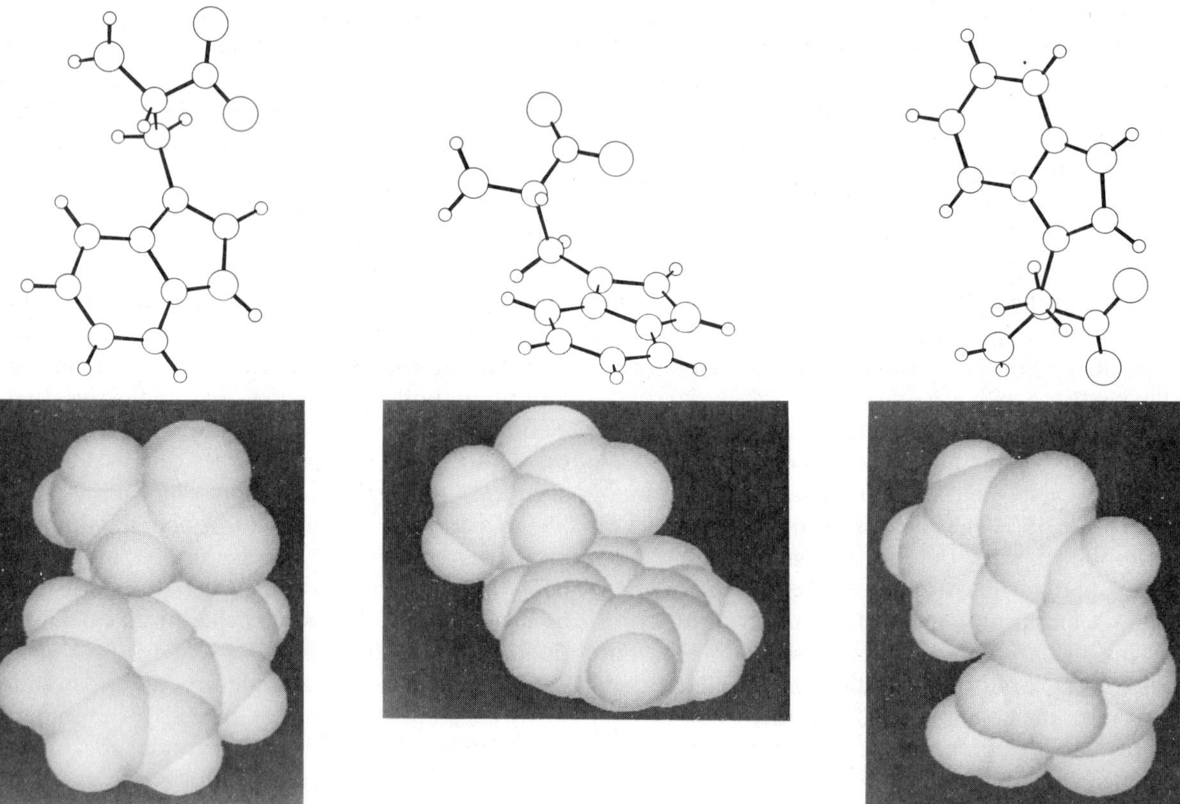

Figure 21.12
Computer-Generated Views of an Amino Acid
The tryptophan molecule

COO^-
$H_2N—CH_2$
C
CH
N
H

is shown in three different orientations in space. The structures on the bottom of each set are based on the van der Waals radii, which show the surfaces of the molecules. The structures on the top of each set are based on atomic radii. Can you identify the basic $—NH_2$ group and the acidic $—COO^-$ group in each view?

Source: Molecular graphics courtesy of Richard J. Feldmann, Division of Computer Research and Technology, National Institutes of Health.

SUMMARY OF SECTIONS 21.6–21.9

In considering chemical equilibria one should be aware that for highly concentrated solutions or very dilute solutions, the interactions of ions (the salt effect) or the concentrations of H^+ and OH^- from water influence concentrations. Also, where multiple equilibria are possible, the concentration of each ion must satisfy all possible equilibria. To simplify calculations, we deal mainly with concentrations in the 0.1–1 M range and make simplifying assumptions where possible.

For polyprotic acids, the concentrations of ions formed in the first ionization step are used to find the concentrations of ions formed in subsequent ionization steps. Important points to note about common polyprotic acids are (1) sulfuric acid is a strong acid in its first ionization but not its second, and therefore its H^+ concentration is not twice the solution concentration, and (2) for a saturated H_2S solution (at 1 atm and 25 °C), $[H_2S] = 0.1$ M and $[H^+]^2[S^{2-}] = 3 \times 10^{-21}$.

Titration permits determination of the concentration of an acid or base in solution. The change in pH as a titration proceeds is represented by a titration curve (Figures 21.7–21.11), in which pH is plotted versus volume of acid or base solution added. At the equivalence point there is a large change in pH, usually detected with a pH meter or the color change of an indicator.

SIGNIFICANT TERMS (with section references)

buffer solution (21.5)
common ion (21.4)
common ion effect (21.4)
endpoint (21.8)
equivalence point (21.8)
hydrolysis of an ion (21.3)
indicator (acid–base) (21.8)
titration (21.8)
titration curve (acid–base) (21.9)

QUESTIONS AND PROBLEMS

Acidic and Basic Ions—Cations

21.1 Some cations in aqueous solution undergo no significant reactions with water molecules. What is the relative acid strength of such a cation compared to water? What will be the effect on the pH of the solution upon dissolution of these cations?

21.2 Some cations in aqueous solution react with water molecules. What is the relative acid strength of such a cation compared to water? What will be the effect on the pH of the solution upon dissolution of these cations?

21.3 Choose the cations that will react with water to form H^+: (a) K^+, (b) NH_4^+, (c) $[Al(H_2O)_6]^{3+}$, (d) $[Fe(H_2O)_6]^{3+}$, (e) $[Sr(H_2O)_6]^{2+}$. Write chemical equations for the reactions.

21.4 Repeat Question 21.3 for (a) $[Zn(H_2O)_6]^{2+}$, (b) $[Ca(H_2O)_6]^{2+}$, (c) $[Be(H_2O)_4]^{2+}$, (d) $[Na(H_2O)_n]^+$. Write chemical equations for the reactions.

21.5 Calculate the equilibrium constant for the reaction of ammonium ion with water. $K_b = 1.6 \times 10^{-5}$ for NH_3.

21.6 Calculate the equilibrium constant for the reaction of $C_6H_5NH_3^+$ ion with water. $K_b = 4.2 \times 10^{-10}$ for aniline, $C_6H_5NH_2$.

Acidic and Basic Ions—Anions

21.7 Some anions, upon dissolution, undergo no significant reaction with water molecules. What is the relative base strength of such an anion compared to water? What will be the effect on the pH of the solution upon dissolution of these anions?

21.8 Some anions, upon dissolution, react with water molecules. What is the relative base strength of such an anion compared to water? What will be the effect on the pH of the solution upon dissolution of these anions?

21.9 Choose the anions that will react with water to form OH^-: (a) S^{2-}, (b) HS^- (c) F^-, (d) ClO_4^-, (e) NO_2^-. Write chemical equations for the hydrolysis reactions.

21.10 Repeat Question 21.9 for (a) I^-, (b) SO_4^{2-}, (c) HSO_4^-, (d) NO_3^-, (e) NH_2^-. Write chemical equations for the hydrolysis reactions.

21.11 Calculate the equilibrium constant for the reaction of fluoride ion with water. $K_a = 6.5 \times 10^{-4}$ for HF.

21.12 Calculate the equilibrium constant for the reaction of hypoiodite ion with water. $K_a = 2.3 \times 10^{-11}$ for HIO.

Acidic and Basic Ions—Salts

21.13 What determines whether the aqueous solution of a salt will be acidic, neutral, or alkaline?

21.14 If both the cation and anion of a salt react with water upon dissolution, what determines whether the solution will be slightly acidic or slightly alkaline? Classify aqueous solutions of the following salts as (a) slightly acidic or (b) slightly alkaline: (i) $NH_4F(aq)$ and (ii) $NH_4IO(aq)$. $K_b = 1.6 \times 10^{-5}$ for NH_3, $K_a = 6.5 \times 10^{-4}$ for HF, and $K_a = 2.3 \times 10^{-11}$ for HIO.

21.15 Classify aqueous solutions of the following salts as (a) acidic, (b) alkaline, or (c) essentially neutral: (i) $(NH_4)HSO_4$; (ii) $(NH_4)_2SO_4$, $K_a > K_b$; (iii) KCl; (iv) LiCN; (v) $Al(NO_3)_3$.

21.16 Repeat Question 21.15 for (i) $NaClO_4$; (ii) $K_2C_2O_4$; (iii) $(NH_4)_2S$, $K_a < K_b$; (iv) NaH_2PO_4; (v) NH_4CN, $K_a < K_b$.

21.17 What is the pH of a 0.10 M solution of KIO? $K_b = 4.3 \times 10^{-4}$ for IO^-.

21.18 What is the pH of a 0.10 M solution of KF? $K_b = 1.5 \times 10^{-11}$ for F^-.

Common Ions and Buffers—Common Ion Effect

21.19 Consider the ionization of formic acid, HCOOH

$$HCOOH(aq) + H_2O(l) \rightleftharpoons HCOO^- + H_3O^+ \qquad K_a = 1.772 \times 10^{-4}$$

What effect does the addition of sodium formate (NaHCOO) have on the fraction of formic acid molecules that undergoes ionization in aqueous solution?

21.20 Consider the reaction of water with sulfate ion

$$SO_4^{2-} + H_2O(l) \rightleftharpoons HSO_4^- + OH^- \qquad K_b = 1.0 \times 10^{-12}$$

Describe what "will happen" in a solution of Na_2SO_4 as solid sodium hydroxide is added.

21.21 Calculate the concentration of propionate ion, $CH_3CH_2COO^-$, in equilibrium with 0.010 M CH_3CH_2COOH (propionic acid) and 0.10 M H^+ from hydrochloric acid. $K_a = 1.3 \times 10^{-5}$ for CH_3CH_2COOH.

21.22 Calculate the concentration of $C_2H_5NH_3^+$ in equilibrium with 0.010 M $C_2H_5NH_2$ (ethylamine) and 0.0010 M OH^- ion from sodium hydroxide. $K_b = 4.7 \times 10^{-4}$ for ethylamine.

21.23* When chlorine gas is dissolved in water to make "chlorine water," HCl (a strong acid) and HClO (a weak acid) are produced in equal amounts

$$Cl_2(g) + H_2O(l) \longrightarrow HCl(aq) + HClO(aq)$$

What is the concentration of ClO^- ion in a solution containing 0.010 mol of each acid in 1.00 L of solution? $K_a = 2.90 \times 10^{-8}$ for HClO.

21.24* What is the pH of a solution that is a mixture of HBrO, HClO, and HIO, each at 0.10 M concentration? For HBrO, $K_a = 2.2 \times 10^{-9}$; for HClO, $K_a = 2.90 \times 10^{-8}$; and for HIO, $K_a = 2.3 \times 10^{-11}$.

Common Ions and Buffers—Buffer Solutions

21.25 What is the pH of a solution that is 0.10 M in $HClO_4$ and 0.10 M $KClO_4$? Is this a buffer solution? $HClO_4$ is a strong acid.

21.26 Briefly describe why the pH of a buffer solution remains nearly constant upon the addition of small amounts of acid or base. Over what pH range do we observe the best buffering action (nearly constant pH)?

21.27 What is the pH of a buffer solution that is 0.050 M in sodium acetate and 0.010 M in acetic acid? $K_a = 1.754 \times 10^{-5}$ for CH_3COOH.

21.28 What is the pH of a buffer solution that is 0.025 M in NH_4Cl and 0.010 M in NH_3? $K_b = 1.6 \times 10^{-5}$ for NH_3.

21.29 One liter of a buffer solution is prepared by dissolving 0.100 mol of $NaNO_2$ and 0.050 mol of HCl in water. What is the pH of this solution? If the solution is diluted twofold with water, what is the pH? $K_a = 7.2 \times 10^{-4}$ for HNO_2.

21.30 One liter of a buffer solution was made by mixing exactly 500 mL of 1.00 M acetic acid and 500 mL of 0.500 M calcium acetate solutions. What is the concentration of each of the following in the buffer solution: (a) CH_3COOH, (b) Ca^{2+}, (c) CH_3COO^-, (d) H^+? (e) What is the pH? $K_a = 1.754 \times 10^{-5}$ for CH_3COOH.

21.31 What concentration of benzoate ion, $C_6H_5COO^-$, should be added to a 0.010 M benzoic acid, C_6H_5COOH, solution to prepare a buffer that has pH 5.00? $K_a = 6.6 \times 10^{-5}$ for C_6H_5COOH.

21.32 What concentration of chloroacetic acid, $CH_2ClCOOH$, should be added to a 0.010 M $NaCH_2ClCOO$ solution to prepare a buffer that has pH 3.00? $K_a = 1.40 \times 10^{-3}$ for $CH_2ClCOOH$.

21.33 What concentration of NH_4^+ should be added to a 0.050 M NH_3 solution to prepare a buffer which has pH 8.80? $K_b = 1.6 \times 10^{-5}$ for NH_3.

21.34 Repeat the calculations of Problem 21.33 for a buffer solution having a pH of 7.00. Why might this buffer solution be impractical?

21.35 A buffer solution is 0.500 M in CH_3COOH and 0.500 M in $NaCH_3COO$. (a) What is the pH of this solution? What is the pH after the addition to 1.00 L of the buffer solution of (b) 0.001 mol HCl, (c) 0.001 mol NaOH, (d) 100 mL of water? $K_a = 1.754 \times 10^{-5}$ for CH_3COOH.

21.36 A buffer is 0.200 M in NH_4Cl and 0.100 M in NH_3. (a) What is the pH of this solution? $K_b = 1.6 \times 10^{-5}$ for NH_3. (b) What is the pH after the addition of 0.001 mol KOH to one liter of the buffer?

21.37* A beaker contains 1.00 L of 0.200 M lactic acid ($CH_3CHOHCOOH$); a second beaker contains 1.00 L of a buffer solution consisting of 0.200 M lactic acid and 0.200 M sodium lactate. What are the pH values of each solution before and after the addition of 0.005 mol of solid NaOH? $K_a = 1.36 \times 10^{-4}$ for lactic acid.

21.38* The Henderson-Hasselbalch equation can be used to calculate the pH of an acidic buffer solution consisting of HA and A^- only when K_a is small enough that the approximations

$[A^-] + [H^+] \approx [A^-]$ and $[HA] - [H^+] \approx [HA]$ are valid. Find the largest K_a that can be used with 0.100 M concentrations of acid and anion so that the error is less than 5%.

Acids, Bases, and pH—Polyprotic Acids

21.39 Write chemical equations showing the stepwise ionization of sulfuric acid. Which anion(s) of sulfuric acid will react with water to change the pH?

21.40 Repeat Question 21.39 for hydrosulfuric acid.

21.41 Find $[H^+]$, $[HSO_3^-]$, $[H_2SO_3]$, and $[SO_3^{2-}]$ at equilibrium in a 0.100 M solution of H_2SO_3. $K_{a_1} = 1.43 \times 10^{-2}$ and $K_{a_2} = 5.0 \times 10^{-8}$ for H_2SO_3.

21.42 Calculate the pH of a 0.10 M solution of $NaHSO_4$. $K_{a_2} = 1.0 \times 10^{-2}$ for H_2SO_4.

21.43 Find the pH of a solution of 0.25 M $H_2C_2O_4$ (oxalic acid). $K_{a_1} = 5.60 \times 10^{-2}$ and $K_{a_2} = 6.2 \times 10^{-5}$ for $H_2C_2O_4$.

21.44 What is the concentration of S^{2-} present in a saturated hydrosulfuric acid solution that has a pH of 5.3? $[H^+]^2[S^{2-}] = 3 \times 10^{-21}$ for a saturated solution of H_2S.

Acids, Bases, and pH—Acid–Base Reactions

21.45 Write the net ionic equation for the reaction between a strong monoprotic acid and a strong monohydroxo base that yields a completely soluble salt. Why is the heat of neutralization essentially constant for the reaction between 1.00 mol of a strong monoprotic acid and 1.00 mol of a strong monohydroxo base?

21.46 Write the net ionic equation for the neutralization of a weak monoprotic acid by a strong monohydroxo base. Will the heat of neutralization of a weak acid by a strong base be the same as that for the neutralization of a strong acid by a strong base? Explain your answer.

21.47 What volume of 0.106 M NaOH is needed to neutralize 25.00 mL of 0.263 M CH_3COOH?

21.48 The titration of a 10.00 mL sample of dilute sulfuric acid by 0.100 M NaOH was "followed" using a pH meter. A large change in pH was recorded after 13.74 mL of base was added and a second large pH jump was recorded after 27.42 mL of base was added. What is the concentration of the dilute sulfuric acid solution?

21.49 A 2.1734 g sample of Na_2CO_3 required 23.69 mL of a hydrochloric acid solution for complete reaction. What is the concentration of the acid solution?

21.50 Sulfamic acid, HSO_3NH_2, is a strong monoprotic acid that can be used to standardize a strong base:

$$HSO_3NH_2(aq) + KOH(aq) \longrightarrow KSO_3NH_2(aq) + H_2O(l)$$

A 0.179 g sample of HSO_3NH_2 required 19.35 mL of an aqueous solution of KOH for complete reaction. What was the molar concentration of the solution of the base?

21.51 The percent of Na_2CO_3 in a sample of impure soda ash was determined by titration with 0.107 M HCl. A 0.253 g sample of soda ash required 13.72 mL of HCl to reach the equivalence point. What was the percentage of Na_2CO_3 present? Assume that the impurities do not react with the acid.

21.52 A student prepared a solution of HCl having an approximate concentration of 0.15 M. To find the exact concentration, he observed that 22.37 mL of HCl was needed to react completely with 0.1870 g of pure Na_2CO_3. (a) Calculate the exact concentration of the acid. He then used his acid to determine the percent of Na_2CO_3 in an unknown, impure sample. From the data given: 0.3940 g of the unknown required 13.72 mL of HCl solution, (b) calculate the fraction of Na_2CO_3 present in the sample.

21.53 The percentage of acetic acid in a vinegar sample was determined by titration with 0.103 M NaOH:

$$CH_3COOH(aq) + NaOH(aq) \longrightarrow Na(CH_3COO)(aq) + H_2O(l)$$

A 10.13 g sample of vinegar required 67.43 mL of NaOH to reach the equivalence point. Did the vinegar meet the minimum specifications of 4 mass % acetic acid?

21.54* The molar mass of a metal carbonate, $MeCO_3$, was determined by adding excess acid to the carbonate and then titrating the excess acid with a strong base (called "back titrating"). (a) Write the chemical equations for the reaction of HCl with $MeCO_3$ and HCl with NaOH. Exactly 50.00 mL of 0.800 M HCl was poured into a 0.844 g sample of $MeCO_3$. The excess acid was determined by back titrating with 14.30 mL of 1.400 M NaOH. (b) Calculate the molar mass of the carbonate.

21.55 Calculate the pH of a solution of 0.125 M CH_3COOH after 0.100 mol of NaOH has been added to 1.00 L of the acid solution. Assume that no volume change occurs. $K_a = 1.754 \times 10^{-5}$ for CH_3COOH.

21.56 What is the pH of a solution of 0.0500 M NH_3 after 50.0 mL of 0.0500 M HNO_3 has been added to 100.0 mL of the solution of base? $K_b = 1.6 \times 10^{-5}$ for NH_3.

Acids, Bases, and pH—Titration Curves

21.57 Make a rough sketch of the titration curve expected for the titration of a strong acid with a strong base. What determines the pH of the solution at the following points: (a) no base added, (b) half-equivalence point, (c) equivalence point, (d) excess base?

21.58 Repeat Question 21.57 for the titration of a weak acid with a strong base.

21.59* Make a rough sketch of the titration curve expected for the titration of a diprotic acid with a strong base. What determines the pH of the solution at each of the following points: (a) no base added, (b) first half-equivalence point, (c) first equivalence point, (d) second half-equivalence point, (e) second equivalence point, (f) excess base?

21.60* A student obtained the following titration curve during the titration of 25.00 mL of a weak monoprotic acid with 0.100 M

NaOH. Calculate the original concentration and K_a of the weak acid.

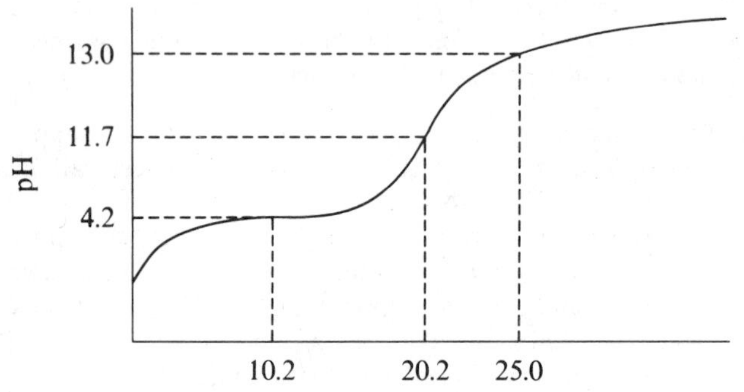

21.61 A 25.0 mL sample of 0.250 M HNO_3 is titrated with 0.100 M NaOH. Calculate the pH of the solution (a) before the addition of NaOH and after the addition of (b) 10.0 mL, (c) 25.0 mL, (d) 50.0 mL, (e) 62.5 mL, (f) 75.0 mL of NaOH.

21.62 A 15.00 mL sample of 0.1063 M KOH is titrated with 0.1077 M HCl. Calculate the pH of the solution (a) before the addition of HCl and after the addition of (b) 10.00 mL, (c) 15.00 mL, (d) 20.00 mL of HCl.

21.63* Repeat the calculations in Problem 21.61 assuming the acid was 0.250 M HCOOH. $K_a = 1.772 \times 10^{-4}$ for formic acid.

21.64* Repeat the calculations of Problem 21.62 assuming the base was aqueous NH_3 instead of KOH. $K_b = 1.6 \times 10^{-5}$ for NH_3.

Acids, Bases, and pH—Acid–Base Indicators

21.65 A "universal" indicator was prepared by mixing methyl red with thymolphthalein. What color would be observed at a pH value of (a) 3, (b) 7, (c) 11?

21.66 Using Table 21.5, choose one or more indicators that could be used to "signal" reaching a pH of (a) 2.4, (b) 7, (c) 10.3, (d) 5.1.

21.67 A solution of 0.01 M acetic acid is to be titrated with a 0.01 M NaOH solution. What is the approximate pH at the equivalence point? Choose an appropriate indicator for the titration. $K_b = 5.701 \times 10^{-10}$ for CH_3COO^-.

21.68 Repeat Problem 21.67 for a 0.01 M solution of formic acid with 0.01 M NaOH. $K_b = 5.643 \times 10^{-11}$ for $HCOO^-$.

21.69 A crime-lab technician wants to determine the barbital content in a sample by titration with NaOH. [Barbital is a sedative drug that can be addictive, and an overdose can lead to coma or death.]

$(C_2H_5)_2C(C(=O)NH)_2C{=}O + NaOH \longrightarrow Na^+ + H_2O + (C_2H_5)_2C[C(O^-){=}N][C(=O)NH]C{=}O$

From past experience the technician knows that the concentration of the $NaC_8H_{11}N_2O_3$ at the equivalence point will be approximately 0.001 mol/L. If $K_a = 3.7 \times 10^{-8}$ for barbital, at what approximate pH would the equivalence point be observed? What would be a good indicator for this titration?

CHAPTER

22

Ions and Ionic Equilibria: Complex Ions and Ionic Solids

COMPLEX ION EQUILIBRIA AND SOLUBILITY EQUILIBRIA

22.1 COMPLEX ION EQUILIBRIA

Complex ions result when coordinate covalent bonds form between metal ions and atoms in neutral or anionic ligands (Section 14.10). Like polyprotic acids, complex ions in aqueous solution form and dissociate by stepwise equilibria. Consider the dissociation of the silver–ammonia complex ion, $[Ag(NH_3)_2]^+$. The ligands separate from the metal ion one by one, and equilibrium constants can be determined for each dissociation step:

$$\textit{Step 1} \quad [Ag(NH_3)_2]^+ \rightleftharpoons [Ag(NH_3)]^+ + NH_3(aq) \qquad K_{d_1} = 1.45 \times 10^{-4} \qquad \textbf{(22.1)}$$

$$\textit{Step 2} \quad [Ag(NH_3)]^+ \rightleftharpoons Ag^+ + NH_3(aq) \qquad K_{d_2} = 4.3 \times 10^{-4} \qquad \textbf{(22.2)}$$

Equilibrium constants for the dissociation of complex ions are called **dissociation constants, K_d**. Some representative overall K_d values are given in Table 22.1 (similar to overall acid dissociation constants for polyprotic acids; Section 21.7). For the silver–ammonia complex ion

$$\textit{Overall} \quad [Ag(NH_3)_2]^+ \rightleftharpoons Ag^+ + 2NH_3(aq) \qquad \textbf{(22.3)}$$

$$K_d = \frac{[NH_3]^2[Ag^+]}{[Ag(NH_3)_2{}^+]} = K_{d_1} \times K_{d_2} = (1.45 \times 10^{-4})(4.3 \times 10^{-4}) = 6.2 \times 10^{-8}$$

Table 22.1
Dissociation Constants (K_d) for Complex Ions in Aqueous Solution (at 25 °C)[a]

Dissociation equilibria	K_d
$[Ag(NH_3)_2]^+ \rightleftharpoons Ag^+ + 2NH_3$	6.2×10^{-8}
$[AgCl_2]^- \rightleftharpoons Ag^+ + 2Cl^-$	9×10^{-6}
$[Au(CN)_2]^- \rightleftharpoons Au^+ + 2CN^-$	5×10^{-39}
$[Cd(CN)_4]^{2-} \rightleftharpoons Cd^{2+} + 4CN^-$	8.2×10^{-18}
$[CdI_4]^{2-} \rightleftharpoons Cd^{2+} + 4I^-$	8×10^{-7}
$[Cu(NH_3)_4]^{2+} \rightleftharpoons Cu^{2+} + 4NH_3$	1×10^{-13}
$[Cu(OH)_4]^{2-} \rightleftharpoons Cu^{2+} + 4OH^-$	7.6×10^{-17}
$[Fe(CN)_6]^{4-} \rightleftharpoons Fe^{2+} + 6CN^-$	1.3×10^{-37}
$[Fe(CN)_6]^{3-} \rightleftharpoons Fe^{3+} + 6CN^-$	1.3×10^{-44}
$[HgCl_4]^{2-} \rightleftharpoons Hg^{2+} + 4Cl^-$	2×10^{-16}
$[Zn(OH)_4]^{2-} \rightleftharpoons Zn^{2+} + 4OH^-$	5×10^{-21}

[a] Additional values given in Appendix V.4.

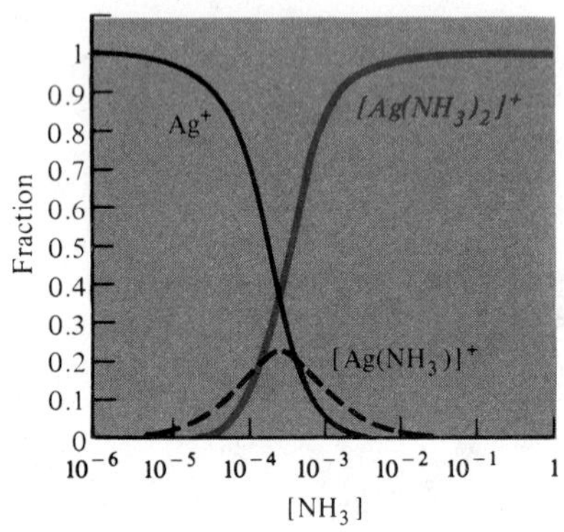

Figure 22.1
Effect of $[NH_3]$ on the Complexation of Ag^+ by NH_3
The "fraction" plotted is the fraction of the Ag^+ present as the uncomplexed ion and the ions with one and two ligands. Notice that the concentration of $[Ag(NH_3)]^+$ initially rises as $[NH_3]$ increases, and then falls as both Ag^+ and $[Ag(NH_3)]^+$ are converted to $[Ag(NH_3)_2]^+$.

The smaller the overall dissociation constant of a complex ion, the more stable it is and the lower the concentration of the metal ion in equilibrium with the complex ion.

Complex ion equilibria can also be described in terms of formation, the reverse of dissociation. The equilibrium constants for complex ion formation, known as **formation constants,** K_f, are the reciprocals of those for dissociation. For the silver–ammonia complex, the formation equilibria and the equilibrium constants are

Step 1 $$Ag^+ + NH_3(aq) \rightleftharpoons [Ag(NH_3)]^+ \qquad K_{f_1} = \frac{1}{K_{d_2}} = 2.3 \times 10^3 \tag{22.4}$$

Step 2 $$[Ag(NH_3)]^+ + NH_3(aq) \rightleftharpoons [Ag(NH_3)_2]^+ \qquad K_{f_2} = \frac{1}{K_{d_1}} = 6.9 \times 10^3 \tag{22.5}$$

Overall $$Ag^+ + 2NH_3(aq) \rightleftharpoons [Ag(NH_3)_2]^+ \qquad K_f = K_{f_1} \times K_{f_2} = 1.6 \times 10^7 \tag{22.6}$$

Figure 22.1 shows the variation in the amounts of Ag^+, $[Ag(NH_3)_2]^+$, and the intermediate ion $[Ag(NH_3)]^+$ with the concentration of the ligand, NH_3. As the concentration of NH_3 increases, the equilibria are displaced toward complex ion formation and the amount of uncomplexed Ag^+ decreases. Initially, the concentration of the intermediate ion with only one NH_3 ligand rises. Then, as the ion with two ligands is formed, the concentration of the intermediate ion decreases.

The stepwise equilibria of complex ions differ from those of polyprotic acids in a significant manner. For the acids, each successive ionization constant is much smaller than the preceding one. This difference allows for simplifying assumptions that make it possible to apply straightforward equilibrium calculations to one ionization step at a time. The stepwise equilibrium constants for complex ions, on the other hand, generally do not differ greatly from each other (for some examples, see Table 22.2). Consequently, similar simplifying assumptions are not valid because the concentrations of the ions produced by intermediate equilibria cannot be neglected. Consider the curves in Figure 22.2 for a complex ion that includes four ligands, $[Cu(NH_3)_4]^{2+}$. At any concentration of ammonia less than 0.1 mol/L there are from two to five intermediate ions present in significant concentrations.

In complex ion equilibria, the concentrations of intermediate ions can be neglected only in the presence of excess ligand. Therefore, overall dissociation constant values for complex ions can be used only for solutions that contain excess ligand. When excess ligand *is* present, the K_d values are used as in any other equilibrium problem. (If the ligand is *not* in excess, algebraic methods for solving *n* equations in *n* + 1 unknowns must be applied.)

Table 22.2
Some Stepwise Dissociation Constants for Complex Ions (25 °C)

	$[Cu(NH_3)_5]^{2+}$	$[AlF_6]^{3-}$
K_{d_1}	4	0.3
K_{d_2}	5×10^{-3}	3×10^{-2}
K_{d_3}	1×10^{-3}	2×10^{-3}
K_{d_4}	2×10^{-4}	1.7×10^{-4}
K_{d_5}	1×10^{-4}	1×10^{-5}
K_{d_6}	—	1×10^{-6}
K_d	3×10^{-13}	3×10^{-20}

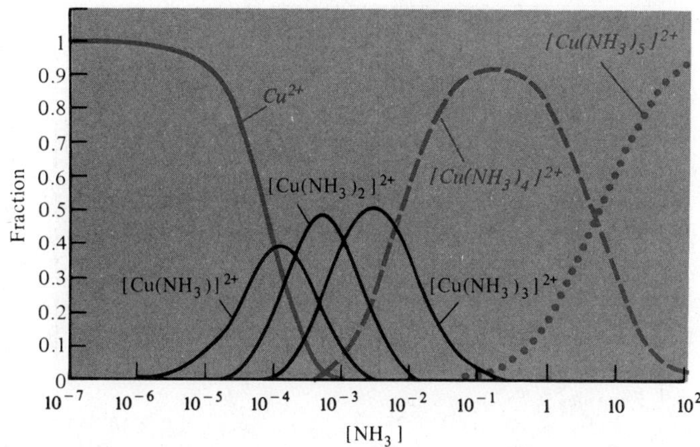

Figure 22.2
Effect of $[NH_3]$ on the Complexation of Cu^{2+} by NH_3
The "fraction" plotted is the fraction of the Cu^{2+} present as the uncomplexed ion and the complexes with one, two, three, four, and five ligands. $[Cu(NH_3)_5]^{2+}$ forms only when $[NH_3] > 1$ M and is not of concern at lower $[NH_3]$.

EXAMPLE 22.1 Complex Ions

A 0.0010 mol sample of solid silver nitrate was dissolved in 1.00 L of 6.0 M aqueous ammonia. What is the concentration of uncomplexed Ag^+ in this solution? $K_d = 6.2 \times 10^{-8}$ for $[Ag(NH_3)_2]^+$.

From the large excess of NH_3 and the small value of K_d (which shows that the complex ion forms readily and dissociates little), we can assume that the 0.0010 mol of Ag^+ will react completely with NH_3 to form 0.0010 mol of the $[Ag(NH_3)_2]^+$ complex ion. Therefore, at equilibrium the concentration of $[Ag(NH_3)_2]^+$ is 0.0010 mol/L and the concentration of NH_3 is essentially unchanged at 6.0 mol/L. The $[Ag^+]$ at equilibrium is found from the K_d expression

$$[Ag(NH_3)_2]^+ \rightleftharpoons Ag^+ + 2NH_3(aq) \qquad K_d = 6.2 \times 10^{-8}$$

$$\frac{[Ag^+][NH_3]^2}{[Ag(NH_3)_2^+]} = 6.2 \times 10^{-8}$$

$$[Ag^+] = \frac{(6.2 \times 10^{-8})[Ag(NH_3)_2^+]}{[NH_3]^2} = \frac{(6.2 \times 10^{-8})(0.0010)}{(6.0)^2}$$

$$= \mathbf{1.8 \times 10^{-12}\ mol/L}$$

EXERCISE A 0.001 mol sample of solid copper(II) nitrate was added to 1.00 L of 6.0 M aqueous ammonia. What is the concentration of Cu^{2+} in solution? $K_d = 1 \times 10^{-13}$ for $[Cu(NH_3)_4]^{2+}$. *Answer* 8×10^{-20} mol/L.

22.2 SOLUBILITY PRODUCT

The equilibrium constant expression for a solid electrolyte in equilibrium with its ions in solution is called the **solubility product,** or solubility product constant, K_{sp}. For example, for silver sulfate

$$Ag_2SO_4(s) \rightleftharpoons 2Ag^+ + SO_4^{2-}$$

the solubility product is

$$K_{sp} = [Ag^+]^2[SO_4^{2-}]$$

As for any equilibrium, the term for concentration of the solid need not appear in the equilibrium constant expression for solubility. Some representative K_{sp} values are given in Table 22.3. (Compare Table 22.3 with the general solubility rules given in Table 5.3.)

Table 22.3
Solubility Products, K_{sp}, for Ionic Solids in Saturated Aqueous Solution (at 25 °C)[a]

Equilibrium	K_{sp}
Bromides	
$PbBr_2 \rightleftharpoons Pb^{2+} + 2Br^-$	3.9×10^{-5}
$AgBr \rightleftharpoons Ag^+ + Br^-$	4.9×10^{-13}
Carbonates	
$MgCO_3 \rightleftharpoons Mg^{2+} + CO_3^{2-}$	1×10^{-5}
$FeCO_3 \rightleftharpoons Fe^{2+} + CO_3^{2-}$	2.1×10^{-11}
$Ag_2CO_3 \rightleftharpoons 2Ag^+ + CO_3^{2-}$	8.1×10^{-12}
Chlorides	
$PbCl_2 \rightleftharpoons Pb^{2+} + 2Cl^-$	2×10^{-5}
$AgCl \rightleftharpoons Ag^+ + Cl^-$	1.8×10^{-10}
Chromates	
$BaCrO_4 \rightleftharpoons Ba^{2+} + CrO_4^{2-}$	1.2×10^{-10}
$PbCrO_4 \rightleftharpoons Pb^{2+} + CrO_4^{2-}$	2.8×10^{-13}
Fluorides	
$BaF_2 \rightleftharpoons Ba^{2+} + 2F^-$	1.0×10^{-6}
$CaF_2 \rightleftharpoons Ca^{2+} + 2F^-$	2.7×10^{-11}
Hydroxides	
$Ba(OH)_2 \rightleftharpoons Ba^{2+} + 2OH^-$	1.3×10^{-2}
$Be(OH)_2 \rightleftharpoons Be^{2+} + 2OH^-$	4×10^{-13}
$Zn(OH)_2 \rightleftharpoons Zn^{2+} + 2OH^-$	1.2×10^{-17}
$Fe(OH)_3 \rightleftharpoons Fe^{3+} + 3OH^-$	3×10^{-39}
Iodides	
$PbI_2 \rightleftharpoons Pb^{2+} + 2I^-$	7.1×10^{-9}
$HgI_2 \rightleftharpoons Hg^{2+} + 2I^-$	3×10^{-26}
Phosphates	
$Li_3PO_4 \rightleftharpoons 3Li^+ + PO_4^{3-}$	3×10^{-13}
$AlPO_4 \rightleftharpoons Al^{3+} + PO_4^{3-}$	5.8×10^{-19}
$Ca_3(PO_4)_2 \rightleftharpoons 3Ca^{2+} + 2PO_4^{3-}$	1×10^{-26}
$Mg_3(PO_4)_2 \rightleftharpoons 3Mg^{2+} + 2PO_4^{3-}$	1×10^{-32}
Sulfates	
$Ag_2SO_4 \rightleftharpoons 2Ag^+ + SO_4^{2-}$	1.5×10^{-5}
$BaSO_4 \rightleftharpoons Ba^{2+} + SO_4^{2-}$	1.7×10^{-10}
Sulfides	
$CdS \rightleftharpoons Cd^{2+} + S^{2-}$	2×10^{-28}
$HgS \rightleftharpoons Hg^{2+} + S^{2-}$	4×10^{-53}

[a] A more extensive table of K_{sp} values is given in Appendix V.3.

22.3 K_{sp} AND SOLUBILITY

For salts of ions that do not react with water, it is possible to find K_{sp} values from solubility data or to find solubilities from known K_{sp} values. In doing so the relationships among solubility, concentrations of ions in solution, and the value of K_{sp} must be understood clearly.

Consider the solubility of barium sulfate in grams per liter of solution. Recall that solubility is the amount of a substance dissolved in a saturated solution, and therefore the amount in solution at equilibrium with the solid (Section 15.1). Solubility in grams per liter, or in any other combination of mass and volume units, can be converted easily to **molar solubility**—the solubility expressed in moles per liter of solution.

The molar solubility is related to the concentrations of ions at equilibrium in a saturated solution by the chemical equation. For barium sulfate, let x represent the molar solubility. The chemical equation shows that at equilibrium the concentrations of Ba^{2+} and SO_4^{2-} ions are both equal to the molar solubility:

$$\underset{}{BaSO_4(s)} \rightleftharpoons \underset{x}{Ba^{2+}} + \underset{x}{SO_4^{2-}} \qquad \textbf{(22.7)}$$

Next consider the molar solubility of cobalt(II) hydroxide. In this case, the concentration of the Co^{2+} ion is equal to the molar solubility. However, because each mole of the compound yields *two* moles of OH^-, the concentration of OH^- at equilibrium is twice the molar solubility.

$$Co(OH)_2(s) \rightleftharpoons \underset{x}{Co^{2+}} + \underset{2x}{2OH^-} \qquad \textbf{(22.8)}$$

For the two salts described above, observe closely what happens when molar solubilities are used in the equilibrium constant expressions:

$$K_{sp} = [Ba^{2+}][SO_4^{2-}] = (x)(x) = x^2$$

note that both *are required*

$$K_{sp} = [Co^{2+}][OH^-]^2 = (x)(2x)^2 = 4x^3$$

The K_{sp} values represent very different functions of the molar solubilities of these two salts. For $Co(OH)_2$, the stoichiometry of Equation (22.8) requires that $[OH^-]$ be squared in the K_{sp} expression. Also, because $[OH^-]$ is the *total* molar concentration of OH^-, the concentration of the OH^- ion is $2x$, *twice the solubility in moles per liter*. In calculations using K_{sp} values, do not make the common error of failing to take into account *both* the multiples of the molar solubility and the exponents on concentrations required by stoichiometry.

Obviously the form of K_{sp} expressed in terms of molar solubility varies with the general type of salt. Therefore, K_{sp} values permit the comparison of solubilities only for salts of the same type (an AB salt compared with an AB salt, an AB_2 salt compared with an AB_2 salt, and so on). For example, although iron(II) carbonate and calcium phosphate have very different K_{sp} values (see Table 22.3), their solubilities are of similar magnitude ($FeCO_3$, 0.0067 g/100 mL; $Ca_3(PO_4)_2$, 0.002 g/100 mL).

EXAMPLE 22.2 Solubility Product

The solubility of AgI at room temperature is 2.1×10^{-6} g/L. Calculate the solubility product of this salt from its solubility.

Letting x represent the molar solubility

$$AgI(s) \rightleftharpoons \underset{x}{Ag^+} + \underset{x}{I^-}$$

$$K_{sp} = [Ag^+][I^-] = (x)(x) = x^2$$

The value of the molar solubility is found from the solubility

$$x = \left(\frac{2.1 \times 10^{-6}\text{ g AgI}}{1\text{ L}}\right)\left(\frac{1\text{ mol AgI}}{234.77\text{ g AgI}}\right) = 8.9 \times 10^{-9}\text{ mol/L}$$

and the solubility product is found from the molar solubility

$$K_{sp} = x^2 = (8.9 \times 10^{-9})^2 = \mathbf{7.9 \times 10^{-17}}$$

EXERCISE The solubility of magnesium hydroxide is 0.0070 g/L. Calculate the K_{sp} of this base from its solubility. (Assume that the magnesium ion does not react appreciably with water to form H^+.) *Answer* 6.9×10^{-12}

EXAMPLE 22.3 Solubility Product

The solubility product of Ag_2SO_4 is 1.5×10^{-15}. Calculate the solubility of this salt from its solubility product.

Letting x represent the molar solubility gives, at equilibrium

$$Ag_2SO_4(s) \rightleftharpoons \underset{2x}{2Ag^+} + \underset{x}{SO_4^{2-}}$$

The K_{sp} expression can be solved to find the solubility as follows

$$K_{sp} = [Ag^+]^2[SO_4^{2-}] = (2x)^2(x) = 4x^3 = 1.5 \times 10^{-5}$$

$$x = \sqrt[3]{(1.5 \times 10^{-5})/4} = \mathbf{0.016\ mol/L\ or\ 5.0\ g/L}$$

EXERCISE The solubility product of strontium sulfate is 3.5×10^{-7}. Calculate the solubility of this salt in water from its K_{sp}. *Answer* 5.9×10^{-4} mol/L or 0.11 g/L.

22.4 SOLUBILITY: INFLUENCE OF COMMON IONS AND ION HYDROLYSIS

The presence in solution of common ions influences solubility equilibria just as it influences acid–base equilibria (Section 21.4). Any ionic compound will be *less* soluble in an aqueous solution in the presence of an additional amount of either of its ions (as long as no complex ion forms).

On the other hand, an ionic compound may be *more* soluble in aqueous solution than indicated by its K_{sp} value if either of its ions reacts appreciably with water. For example, due to hydrolysis of a basic anion such as CN^- or S^{2-}, a salt of a weak acid can be somewhat more soluble than expected. As anion concentration is decreased by reaction with water, more of the salt dissolves

$$\longrightarrow CN^- + H_2O \longrightarrow HCN + OH^-$$

$$Zn(CN)_2(s) \longrightarrow Zn^{2+} + 2CN^-$$

Consider lead(II) sulfide, which is the salt of an anion of a weak diprotic acid. When lead(II) sulfide dissolves,

$$PbS(s) \rightleftharpoons Pb^{2+} + S^{2-} \qquad K_{sp} = [Pb^{2+}][S^{2-}] = 1 \times 10^{-28}$$

the following additional equilibria are established in the same solution:

$$S^{2-} + H_2O(l) \rightleftharpoons HS^- + OH^- \qquad K_{b_2} = \frac{[HS^-][OH^-]}{[S^{2-}]} = 3 \times 10^{-2} \qquad \textbf{(22.9)}$$

$$HS^- + H_2O(l) \rightleftharpoons H_2S(aq) + OH^- \qquad K_{b_1} = \frac{[H_2S][OH^-]}{[HS^-]} = 1.0 \times 10^{-7} \qquad \textbf{(22.10)}$$

Using the K_{sp} alone, the solubility of lead(II) sulfide is found to be 1×10^{-14} mol/L. However, most of the sulfide ion goes on to react with water. If reactions (22.9) and (22.10) are taken into account, the solubility of PbS is found to be 7×10^{-12} mol/L, 700 times greater than the solubility found without considering the reaction of sulfide with water. A slightly greater solubility would be found if the reaction of Pb^{2+} with water to give H_3O^+ were also taken into account.

EXAMPLE 22.4 Solubility: Common Ion Effect

The solubility product of Ag_2SO_4 is 1.5×10^{-5}. Calculate the solubility of this salt in an aqueous solution that is 0.20 M Na_2SO_4. (Assume that none of the ions reacts appreciably with water to form H^+ or OH^-.)

Let x represent the molar solubility of Ag_2SO_4. From the chemical equation

$$\underset{}{Ag_2SO_4(s)} \rightleftharpoons \underset{2x}{2Ag^+} + \underset{x}{SO_4^{2-}}$$

we can see that the concentration of Ag^+ from the dissolution of Ag_2SO_4 will be equal to $2x$. However, there are two sources of SO_4^{2-} in this solution, the dissolved Ag_2SO_4 and the initial Na_2SO_4 in the solution. Therefore

$$K_{sp} = [Ag^+]^2[SO_4^{2-}] = (2x)^2(0.20 + x) = 1.5 \times 10^{-5}$$

Because the K_{sp} value is small relative to the 0.20 M added SO_4^{2-} ion, we can assume that $(0.20 + x) \approx 0.20$, giving

$$x = \sqrt{\frac{1.5 \times 10^{-5}}{(4)(0.20)}} = \mathbf{0.0043\ mol/L,\ or\ 1.3\ g/L}$$

The approximation is valid, so the molar solubility is 0.0043 mol/L or 1.3 g/L. The common ion effect has reduced the solubility from 5.0 g/L for Ag_2SO_4 in pure water (see Example 22.3) to 1.3 g/L in the Na_2SO_4 solution.

EXERCISE Calculate the solubility of $SrSO_4$ in a solution that is 0.10 M in Sr^{2+} $K_{sp} = 3.5 \times 10^{-7}$ for $SrSO_4$. Compare this solubility to 0.11 g/L, which represents the solubility of $SrSO_4$ in pure water. (Assume that neither of the ions reacts appreciably with water to form H^+ or OH^-.)

Answer 0.00064 g/L, solubility reduced to about 0.6% of the original value.

22.5 K_{sp} AND PRECIPITATION

The **ion product** (Q_i) is equivalent to the reaction quotient for the dissolution of an ionic solid—it is the product of the concentrations of ions that are not in equilibrium with the solid. Whether or not precipitation will occur is predicted by calculating the ion product and comparing it with the K_{sp} value. For example, $[Ca^{2+}][SO_4^{2-}]$ is the ion product for any solution containing calcium and sulfate ions. The ion product equals the K_{sp} when the solution is saturated. If the ion product is smaller than the K_{sp}, all the Ca^{2+} and SO_4^{2-} ions present are in solution. If the ion product is larger than the K_{sp}, solid calcium sulfate precipitates until equilibrium between the solid and a solution saturated with ions is established—in other words, precipitation continues until the ion product of the ions in solution equals the K_{sp} (Table 22.4).

In determining whether or not precipitation will occur when aqueous solutions are combined, the change in concentration that results from the change in volume must be taken into account. In any type of equilibrium calculation, do not make the common error of neglecting the changes in concentrations of ions that result when solutions are combined.

Table 22.4
Ion Product (Q_i) and Precipitation

$Q_i < K_{sp}$	All ions in solution
$Q_i = K_{sp}$	Equilibrium—a saturated solution
$Q_i > K_{sp}$	Precipitation occurs until $Q_i = K_{sp}$

EXAMPLE 22.5 Solubility: Ion Product

The concentration of chromate ion in an aqueous solution is 0.010 mol/L. (a) Will precipitation of silver chromate take place if a concentration of silver ion of 1×10^{-6} mol/L is introduced by adding solid silver nitrate to the solution? (b) If the answer to (a) is "no," find the concentration of silver ion at which precipitation will begin. $K_{sp} = 2.5 \times 10^{-12}$ for Ag_2CrO_4.

(a) The ion product for the reaction

$$Ag_2CrO_4(s) \rightleftharpoons 2Ag^+ + CrO_4^{2-}$$

is

$$Q_i = [Ag^+]^2[CrO_4^{2-}] = (1 \times 10^{-6})^2(0.010) = 1 \times 10^{-14}$$

Because the ion product is less than the solubility product, the forward reaction (the dissolution process) is favored and no precipitate forms.

(b) Precipitation would begin when $Q_i = K_{sp}$. The $[Ag^+]$ at which this occurs is found by substituting $[CrO_4^{2-}] = 0.010$ mol/L into the K_{sp} expression and solving for $[Ag^+]$.

$$K_{sp} = [Ag^+]^2[CrO_4^{2-}] = 2.5 \times 10^{-12}$$
$$[Ag^+]^2[0.010] = 2.5 \times 10^{-12}$$
$$[Ag^+] = \mathbf{1.6 \times 10^{-5}\ mol/L}$$

EXERCISE Suppose the solution described above contained 0.010 M Cl^- ion instead of CrO_4^{2-} ion. Would a precipitate form as the same amount of silver nitrate is added? $K_{sp} = 1.8 \times 10^{-10}$ for AgCl. *Answer* $Q_i = 1 \times 10^{-8}$; precipitation occurs.

EXAMPLE 22.6 Solubility: Ion Product

Will a precipitate of $Mg(OH)_2$ form as 10. mL of 0.010 M NaOH is added to 1.000 L of 0.010 M $MgCl_2$? (Assume that the volume of the resulting solution is 1.010 L.) $K_{sp} = 7.1 \times 10^{-12}$ for $Mg(OH)_2$.

The concentrations of each ion in the resulting solution are

$$[OH^-] = \frac{(0.010\ L)(0.010\ mol/L)}{(1.010\ L)} = 9.9 \times 10^{-5}\ mol/L$$

$$[Mg^{2+}] = \frac{(1.000\ L)(0.010\ mol/L)}{(1.010\ L)} = 9.9 \times 10^{-3}\ mol/L$$

and the ion product is

$$Mg(OH)_2(s) \rightleftharpoons Mg^{2+} + 2OH^-$$

$$Q_i = [Mg^{2+}][OH^-]^2 = (9.9 \times 10^{-3})(9.9 \times 10^{-5})^2 = 9.7 \times 10^{-11}$$

A precipitate of $Mg(OH)_2$ will form because Q_i is greater than K_{sp}.

EXERCISE Will $Mg(OH)_2(s)$ form if only 1 mL of 0.010 M NaOH is added to the original 0.010 M $MgCl_2$ solution described in the above example?

Answer No; $Q_i = 1 \times 10^{-12}$.

SUMMARY OF SECTIONS 22.1–22.5

Complex ions, like polyprotic acids, dissociate in aqueous solution by stepwise equilibria. Equilibrium constants for dissociation can be evaluated for individual steps and overall equilibria. The overall equilibrium constant expressions correctly apply only to solutions in which the ligand is in excess.

The solubility product is the equilibrium constant expression (K_{sp}) for a solid electrolyte in equilibrium with its ions in aqueous solution. Solubility and K_{sp} values can be found from each other; however, care must be taken in correctly using multiples and powers of the molar solubility in the K_{sp} expression. The presence of common ions decreases solubility and the occurrence of ion hydrolysis increases solubility. The ion product, found from the non-equilibrium ion concentrations, is used to predict whether or not precipitation will occur (see Table 22.4).

CONTROLLING SOLUBILITY AND PRECIPITATION

Often it is desirable to control whether or not a substance precipitates or dissolves.To do this, the concentrations of ions are adjusted so that the ion product either exceeds the solubility product or remains less than the solubility product. By controlling precipitation selectively, certain ions can be removed from a solution while other ions are left behind. Equilibrium constant values are used to calculate ion concentrations that will lead to the desired results. In the following sections, examples are given of the applications of solubility equilibria or solubility equilibria together with other types of equilibria.

22.6 DISSOLUTION OF IONIC PRECIPITATES

Ionic substances that are slightly soluble can be brought into solution by the application of Le Chatelier's principle. Any process that will remove, say, a calcium ion from a saturated calcium sulfate solution,

$$CaSO_4(s) \rightleftharpoons Ca^{2+} + SO_4^{2-} \quad \xrightarrow{\textit{remove } Ca^{2+}}$$

will cause more of the calcium sulfate to dissolve. Removal of an ion causes the ion product to drop to less than the K_{sp} value; more of the salt dissolves until the equilibrium is reestablished and $Q_i = K_{sp}$ once more.

Most often, when it is necessary to increase the solubility of a salt, ions in solution are "removed" by nonredox reactions that produce weak electrolytes such as water, weak acids, or complex ions. Redox reactions are also sometimes used to bring precipitates into solution. All of these ways of dissolving precipitates are important in the separation and identification of cations in qualitative analysis (Section 22.8; many examples given here are drawn from the qualitative analysis scheme).

a. Solubility and pH Many of the more common insoluble ionic compounds contain hydroxide, carbonate, or sulfide anions. These anions are all reasonably strong Brønsted-Lowry bases and combine with hydrogen ions to give weak electrolytes (H_2O, HCO_3^- and H_2CO_3, HS^- and H_2S, respectively). For example, most water-insoluble metal hydroxides that are either basic or amphoteric, such as iron(III) and aluminum hydroxides, dissolve readily in acidic solutions

$$Fe(OH)_3(s) + 3H_3O^+ \longrightarrow Fe^{3+} + 6H_2O(l)$$
$$Al(OH)_3(s) + 3H_3O^+ \longrightarrow Al^{3+} + 6H_2O(l)$$

In qualitative analysis and in other applications, the solubility and precipitation of compounds containing OH^-, CO_3^{2-}, and S^{2-} are controlled by varying the pH (Section 22.8).

No matter how many different ions are present in a single solution, all possible equilibria of those ions are established simultaneously. The concentration of any given ion must satisfy all equilibria involving that ion. Therefore, the hydroxide and hydrogen ion concentrations in every aqueous solution must satisfy the ion product constant for water. As a result, the OH^- ion concentration in any solution of known H^+ concentration can be found from the K_w expression. Suppose it is necessary to know whether any $Fe(OH)_3$ will precipitate from a solution that is 0.5 M each in Fe^{3+} and H^+. The OH^- concentration in the solution is found from the K_w expression and then is used to calculate the ion product.

$$K_w = [H^+][OH^-] = 1.00 \times 10^{-14}$$
$$[OH^-] = (1.00 \times 10^{-14})/0.5 = 2 \times 10^{-14}$$

$$Q_i = [Fe^{3+}][OH^-]^3 = (0.5)(2 \times 10^{-14})^3 = 4 \times 10^{-42}$$

Comparison with the K_{sp} of iron(III) hydroxide, which is 3×10^{-39}, shows that this compound will not precipitate in a solution that is 0.5 M in both Fe^{3+} and H^+.

Solubility equilibria and acid–base equilibria must be used together in solving many different types of problems. In deciding how to solve such problems, consider carefully what is known and unknown about concentrations and equilibria in the solution *at the point in question.* One general guideline is that *if* an intermediate ion is of negligible concentration, the equations and equilibrium constants for the two reactions involving that ion can be combined (see Example 22.8).

EXAMPLE 22.7 Solubility: pH

A 0.010 mol sample of $Fe(OH)_3(s)$ was added to 1.0 L of water, and a strong acid was added until the precipitate dissolved (assume negligible volume change due to added acid). At what pH was all of the solid $Fe(OH)_3$ dissolved? $K_{sp} = 3 \times 10^{-39}$ for $Fe(OH)_3$.

When all of the $Fe(OH)_3$ had dissolved, 0.010 mol of Fe^{3+} was present in 1.0 L of solution. The concentration of OH^- at the point where dissolution was complete can be found from the solubility product expression for $Fe(OH)_3$

$$Fe(OH)_3(s) \rightleftharpoons Fe^{3+} + 3OH^- \qquad K_{sp} = [Fe^{3+}][OH^-]^3 = 3 \times 10^{-39}$$

$$[OH^-] = \sqrt[3]{\frac{K_{sp}}{[Fe^{3+}]}} = \sqrt[3]{\frac{3 \times 10^{-39}}{0.010}} = 7 \times 10^{-13}\ \text{mol/L}$$

This $[OH^-]$ corresponds to

$$pOH = -\log[OH^-] = -\log(7 \times 10^{-13}) = 12.2$$
$$pH = 14.0 - pOH = 14.0 - 12.2 = \mathbf{1.8}$$

EXERCISE A 0.010 mol sample of $Fe(OH)_2(s)$ was added to 1.0 L of water and a strong acid was added until the solid dissolved (assume negligible volume change due to added acid). At what pH did all of the solid $Fe(OH)_2$ dissolve? $K_{sp} = 8 \times 10^{-16}$ for $Fe(OH)_2$.
Answer 7.5.

EXAMPLE 22.8 Solubility: pH

Calculate the molar solubility of BaF_2 in a solution buffered at pH 2.0. $K_a = 6.5 \times 10^{-4}$ for HF and $K_{sp} = 1.0 \times 10^{-6}$ for BaF_2.

Two equilibria are of concern here—the dissolution of BaF_2 and the reaction of the basic anion F^- with H^+.

$$BaF_2(s) \rightleftharpoons Ba^{2+} + 2F^- \qquad K = K_{sp} = 1.0 \times 10^{-6}$$
$$F^- + H^+ \rightleftharpoons HF(aq) \qquad K = 1/K_a = 1/(6.5 \times 10^{-4}) = 1.5 \times 10^3$$

Comparison of the K values for these two reactions shows that the concentration of the intermediate F^- ion can be considered negligible. Because the K value for the second reaction is so much larger than the first, as soon as any F^- is formed it will react immediately with H^+ ion. Therefore, the two equations and their equilibrium constants can be combined

$$BaF_2(s) \rightleftharpoons Ba^{2+} + 2F^- \qquad K = K_{sp}$$
$$2F^- + 2H^+ \rightleftharpoons 2HF(aq) \qquad K = (1/K_a)^2$$
$$BaF_2(s) + 2H^+ \rightleftharpoons Ba^{2+} + 2HF(aq) \qquad K = K_{sp}/K_a^2 = (1.0 \times 10^{-6})/(6.5 \times 10^{-4})^2 = 2.4$$

Letting x represent the molar solubility of BaF_2 gives

$$\underset{}{BaF_2(s)} + 2H^+ \rightleftharpoons \underset{x}{Ba^{2+}} + \underset{2x}{2HF(aq)}$$

The concentration of H^+ in this solution buffered at pH 2.0 is

$$-\log[H^+] = 2.0$$
$$[H^+] = 0.010\ \text{mol/L}$$

This value of $[H^+]$ can be used in the combined equilibrium expression to find the molar solubility of BaF_2 in the acid solution

$$K = \frac{[Ba^{2+}][HF]^2}{[H^+]^2} = \frac{(x)(2x)^2}{(0.010)^2} = 2.4$$

$$4x^3 = (2.4)(0.010)^2 = 2.4 \times 10^{-4}$$

$$x = \mathbf{0.039\ mol/L}$$

EXERCISE Calculate the solubility of $Ag(CH_3COO)$ in a solution buffered at pH 2.0. $K_a = 1.754 \times 10^{-5}$ for CH_3COOH and $K_{sp} = 4.4 \times 10^{-3}$ for $Ag(CH_3COO)$.

Answer 1.6 mol/L.

b. Solubility and Complex Ion Formation The formation and dissolution of precipitates can sometimes be controlled by complex ion formation. For example, the distinctive reactions of nickel and zinc ions with ammonia play a role in detecting the presence of these ions in solution. The hydroxides first precipitate upon addition of the aqueous ammonia solution, but then dissolve by the formation of complexes with the excess ammonia.

$$Ni^{2+} \xrightarrow{NH_3(aq)} \underset{\textit{pale green}}{Ni(OH)_2(s)} \xrightarrow{NH_3(aq,\ xs)} \underset{\textit{deep blue}}{[Ni(NH_3)_6]^{2+}}$$

$$Zn^{2+} \xrightarrow{NH_3(aq)} \underset{\textit{white}}{Zn(OH)_2(s)} \xrightarrow{NH_3(aq,\ xs)} \underset{\textit{colorless}}{[Zn(NH_3)_4]^{2+}}$$

Certain less soluble metal sulfides dissolve in hydrochloric acid, because both the cations and the sulfide ions are converted to weak electrolytes, the cations forming complex ions. For example

$$Sb_2S_3(s) + 6H^+ + 8Cl^- \longrightarrow 2[SbCl_4]^- + 3H_2S(aq)$$

$$SnS_2(s) + 4H^+ + 6Cl^- \longrightarrow [SnCl_6]^{2-} + 2H_2S(aq)$$

EXAMPLE 22.9 Solubility: Complex Formation

A qualitative analysis scheme calls for the precipitation of Zn^{2+} ion as $Zn(OH)_2(s)$ from an acidic aqueous solution by the addition of OH^- ion. However, there is a danger of redissolving the $Zn(OH)_2$ by adding too much OH^- ion because of the formation of the $[Zn(OH)_4]^{2-}$ complex ion. Find (a) the pH at which the concentration of Zn^{2+} falls below 1×10^{-4} mol/L due to precipitation of $Zn(OH)_2$ and (b) the pH at which the concentration of $[Zn(OH)_4]^{2-}$ increases above 1×10^{-4} mol/L. The pH range between these two points is the range in which essentially all the Zn^{2+} is present as the precipitate. $K_{sp} = 1.2 \times 10^{-17}$ for $Zn(OH)_2$ and $K_d = 5 \times 10^{-21}$ for $[Zn(OH)_4]^{2-}$.

(a) As the pH increases from acidic to neutral values, the formation of $Zn(OH)_2$ is important

$$Zn(OH)_2(s) \rightleftharpoons Zn^{2+} + 2OH^- \qquad K_{sp} = [Zn^{2+}][OH^-]^2$$

The OH^- ion concentration in equilibrium with $[Zn^{2+}] = 1 \times 10^{-4}$ mol/L is

$$[OH^-]^2 = \frac{K_{sp}}{[Zn^{2+}]} = \frac{1.2 \times 10^{-17}}{1 \times 10^{-4}} = 1 \times 10^{-13}$$

$$[OH^-] = 3 \times 10^{-7}\ \text{mol/L}$$

The pH of the solution is found from the pOH

$$pOH = -\log[OH^-] = -\log(3 \times 10^{-7}) = 6.5$$

$$pH = 14.0 - pOH = 14.0 - 6.5 = \mathbf{7.5}$$

(b) As the pH increases from neutral to alkaline values, the complexation of $Zn(OH)_2$ becomes more important. The solubility of zinc(II) is described by the equilibrium between the precipitate and the complex ion:

$$Zn(OH)_2(s) \rightleftharpoons Zn^{2+} + 2OH^- \qquad K_{sp}$$
$$Zn^{2+} + 4OH^- \rightleftharpoons [Zn(OH)_4]^{2-} \qquad 1/K_d$$

$$Zn(OH)_2(s) + 2OH^- \rightleftharpoons [Zn(OH)_4]^{2-} \qquad K = K_{sp}/K_d = \frac{[Zn(OH)_4{}^{2-}]}{[OH^-]^2}$$

The OH^- ion concentration in equilibrium with $[Zn(OH)_4{}^{2-}] = 1 \times 10^{-4}$ mol/L is

$$[OH^-]^2 = \frac{[Zn(OH)_4{}^{2-}]}{K_{sp}/K_d} = \frac{1 \times 10^{-4}}{(1.2 \times 10^{-17})/(5 \times 10^{-21})} = 4 \times 10^{-8}$$
$$[OH^-] = 2 \times 10^{-4}\ \text{mol/L}$$

The pH of the solution is found from the pOH

$$pOH = -\log(2 \times 10^{-4}) = 3.7$$
$$pH = 1.40 - 3.7 = \mathbf{10.3}$$

The pH of the solution should be controlled between 7.5 and 10.3 to effect the optimum precipitation of zinc(II).

EXERCISE During the precipitation of Ag^+ as AgCl in quantitative and qualitative analysis, it is possible to redissolve the AgCl as $[AgCl_2]^-$ if the concentration of Cl^- ion is too high. Find the concentration range of Cl^- in which the concentration of either Ag^+ or $[AgCl_2]^-$ is less than 1×10^{-4} mol/L. $K_{sp} = 1.8 \times 10^{-10}$ for AgCl(*s*) and $K_d = 9 \times 10^{-6}$ for $[AgCl_2]^-$. *Answer* 2×10^{-6} mol/L to 5 mol/L.

c. Solubility and Redox Reactions Some very slightly soluble metal sulfides (e.g., CuS, $K_{sp} = 6 \times 10^{-36}$) furnish so little S^{2-} in saturated solutions that they cannot be dissolved by aqueous HCl. In these cases S^{2-} can be effectively removed by oxidation to elemental sulfur. For PbS and CuS, 0.3 M nitric acid serves as the oxidizing agent. For example

$$3PbS(s) + 8H^+ + 2NO_3^- \longrightarrow 3Pb^{2+} + 3S(s) + 2NO(g) + 4H_2O(l)$$

The much less soluble HgS ($K_{sp} = 4 \times 10^{-53}$) does not furnish sufficient S^{2-} in saturated solution for oxidation with 0.3 M nitric acid to occur, and *aqua regia* (concentrated HCl and concentrated HNO_3 in a ratio of 3:1 by volume) is used. This reagent converts S^{2-} to the free element and at the same time ties up Hg^{2+} as the $[HgCl_4]^{2-}$ complex.

22.7 SELECTIVE PRECIPITATION

Consider the situation in which two different cations in a solution form precipitates with the same anion. If the precipitates have sufficiently different values of K_{sp}, it may be possible to separate the cations by **selective precipitation** (also called fractional precipitation), in which concentrations are controlled so that one compound precipitates while others do not. The separation of barium and strontium by selective precipitation of their sulfates is described in Example 22.10.

Often it is more successful to control the concentration of an ion with a chemical reaction, rather than by adding the necessary amount of an ion directly to the solution. The precipitation of sulfides in qualitative analysis is managed by this technique. The sulfide ion concentration is controlled by varying the hydrogen ion concentration. Recall that in a saturated hydrogen sulfide solution, the concentration of H_2S is approximately 0.1 M and that the overall equilibrium can be represented as

$$H_2S \rightleftharpoons 2H^+ + S^{2-} \qquad K_{H_2S} = [H^+]^2[S^{2-}] = 3 \times 10^{-21} \tag{22.11}$$

The larger the hydrogen ion concentration, the smaller is the sulfide ion concentration in a saturated H_2S solution. Based on their K_{sp} values, the slightly soluble sulfides fall roughly into two groups. In one group are sulfides so slightly soluble that no attainable hydrogen ion concentration is large enough to prevent their precipitation in the presence of sulfide ion. In the other group are sulfides sufficiently soluble that their precipitation can be prevented. The hydrogen ion concentration is increased to the point where the sulfide ion concentration is too small for precipitation of these more soluble sulfides to occur, as illustrated in Example 22.11.

EXAMPLE 22.10 Selective Precipitation

A solution is 0.1 M in Ba^{2+} and also 0.1 M in Sr^{2+}. Sulfate ion is introduced to the solution by the addition of sodium sulfate, which is soluble. (a) Will barium or strontium sulfate precipitate first, and at what concentration of sulfate ion will precipitation begin? (b) What percentage of the metal ion in the first precipitate is removed from solution before precipitation of the second compound begins? $K_{sp} = 1.7 \times 10^{-10}$ for $BaSO_4$ and $K_{sp} = 3.5 \times 10^{-7}$ for $SrSO_4$.

(a) The two equilibria under consideration are

$$BaSO_4(s) \rightleftharpoons Ba^{2+} + SO_4^{2-} \qquad K_{sp} = [Ba^{2+}][SO_4^{2-}] = 1.7 \times 10^{-10}$$
$$SrSO_4(s) \rightleftharpoons Sr^{2+} + SO_4^{2-} \qquad K_{sp} = [Sr^{2+}][SO_4^{2-}] = 3.5 \times 10^{-7}$$

Precipitation will begin when $Q_i = K_{sp}$ for one of these salts. Solving the K_{sp} expressions for $[SO_4^{2-}]$

$$[Ba^{2+}][SO_4^{2-}] = (0.1)[SO_4^{2-}] = 1.7 \times 10^{-10}$$
$$[SO_4^{2-}] = 1.7 \times 10^{-9} \text{ mol/L}$$
$$[Sr^{2+}][SO_4^{2-}] = (0.1)[SO_4^{2-}] = 3.5 \times 10^{-7}$$
$$[SO_4^{2-}] = 3.5 \times 10^{-6} \text{ mol/L}$$

shows that $Q_i = K_{sp}$ for barium sulfate at the lower sulfate ion concentration, 1.7×10^{-9} mol/L. Therefore, **barium sulfate will precipitate first at $[SO_4^{2-}] = 1.7 \times 10^{-9}$ mol/L.**

(b) Strontium sulfate precipitation will begin when $[SO_4^{2-}] = 3.5 \times 10^{-6}$ mol/L. The amount of Ba^{2+} remaining in solution at this point is found from the solubility product of barium sulfate:

$$1.7 \times 10^{-10} = [Ba^{2+}](3.5 \times 10^{-6})$$
$$[Ba^{2+}] = 4.9 \times 10^{-5} \text{ mol/L}$$

Thus, the percentage of the original Ba^{2+} remaining is

$$\frac{4.9 \times 10^{-5}\text{ M}}{0.10\text{ M}} \times 100 = \mathbf{0.049\%}$$

[You can see that the separation of these two ions from this solution by selective precipitation is almost complete.]

EXERCISE A solution is 0.00010 M in Ba^{2+} and 0.75 M in Sr^{2+}. $K_{sp} = 1.7 \times 10^{-10}$ for $BaSO_4$ and 3.5×10^{-7} for $SrSO_4$. (a) Which sulfate will precipitate first as solid Na_2SO_4 is added? (b) What percentage of the cation in the first precipitation remains in solution when precipitation of the second sulfate begins?

Answers (a) $SrSO_4$ begins to precipitate first, at $[SO_4^{2-}] = 4.7 \times 10^{-7}$ mol/L; $BaSO_4$ begins to precipitate second, at $[SO_4^{2-}] = 1.7 \times 10^{-6}$ mol/L. (b) 28% of the Sr^{2+} remains. [In this case the SO_4^{2-} anion concentrations at which precipitation of the two compounds begins are too close to allow effective separation by selective precipitation.]

EXAMPLE 22.11 Selective Precipitation

Zinc sulfide ($K_{sp} = 2 \times 10^{-24}$) is one of the more soluble of the slightly soluble sulfides and copper(II) sulfide ($K_{sp} = 6 \times 10^{-36}$) is one of the sulfides so slightly soluble that its precipitation in the presence of sulfide ion cannot be prevented by controlling $[H^+]$. A solution is 0.0010 M in Zn^{2+} and 0.0010 M in Cu^{2+}. It is desired to adjust the hydrogen ion concentration so that, upon saturation with H_2S, zinc(II) sulfide will not precipitate but copper(II) sulfide will precipitate. What is the necessary hydrogen ion concentration?

First, the concentration of sulfide ion at which zinc sulfide precipitation begins is found from the K_{sp}

$$[Zn^{2+}][S^{2-}] = 2 \times 10^{-24}$$
$$(0.0010[S^{2-}] = 2 \times 10^{-24}$$
$$[S^{2-}] = \frac{2 \times 10^{-24}}{(0.0010)} = 2 \times 10^{-21} \text{ mol/L}$$

Then, the pH at which the sulfide ion concentration is equal to this value is found from the equilibrium expression given in Equation (22.11).

$$[H^+]^2[S^{2-}] = 3 \times 10^{-21}$$
$$[H^+]^2(2 \times 10^{-21}) = 3 \times 10^{-21}$$
$$[H^+]^2 = 1.5$$
$$[H^+] = 1.2 \text{ mol/L}$$

The $[H^+]$ must be higher than 1.2 mol/L. Above this $[H^+]$, the concentration of sulfide ion will be too low for precipitation of ZnS to occur. [A similar calculation for CuS would show that $[H^+]$ greater than 7×10^5 mol/L is required to prevent CuS precipitation. This is not attainable—the concentration of concentrated hydrochloric acid, e.g., is only 12 mol/L.]

EXERCISE An acidic aqueous solution contains 1×10^{-4} M Zn^{2+} ion and 1×10^{-4} M Fe^{3+} ion. (a) As solid NaOH is added, will $Zn(OH)_2$ or $Fe(OH)_3$ precipitate first? (b) At what pH will the maximum amount of the first precipitate be formed before the second precipitate begins to form? *Answers* (a) $Fe(OH)_3$. (b) pH 7.5.

22.8 INORGANIC QUALITATIVE ANALYSIS

The properties of ions in aqueous solution are determined to a great extent by their strengths as Brønsted-Lowry acids and bases, the solubilities of their salts, the stabilities of their complex ions, and their strengths as oxidizing and reducing agents. By taking advantage of similarities and differences in such properties, the methods of **inorganic qualitative analysis** allow the identification of the cations and anions present in an aqueous solution of unknown composition.

You may have the opportunity to do some qualitative chemical analysis in the laboratory. Anions in solution are identified by their general and specific chemical properties. Cations are first separated into groups by precipitating those in one group while leaving the cations of other groups in solution. For example the addition of hydrochloric acid will precipitate Hg_2^{2+}, Pb^{2+}, and Ag^+ as chlorides, while other ions, for example Fe^{3+}, Cu^{2+}, Sb^{3+}, or Sn^{2+}, are left in solution. The cations precipitated as a group are further separated and then identified by specific reactions known as confirmatory tests.

To illustrate how the chemical properties of cations in aqueous solution are utilized in qualitative analysis, the reactions of Fe^{2+} during an analysis in the presence of other ions are shown below. At each step except the last (the confirmatory tests), the Fe^{2+} is being separated from other ions that react in different ways or do not react. At each step concentrations are adjusted so that the appropriate equilibrium is achieved.

Precipitation: governed by $K_{sp} = 8 \times 10^{-16}$ for $Fe(OH)_2$. The OH^- concentration is increased sufficiently to cause precipitation.

$$Fe^{2+} + 2OH^-(xs) \rightleftharpoons Fe(OH)_2(s)$$

Conversion to less soluble precipitate: governed by $K_{sp} = 4.2 \times 10^{-17}$ for FeS. The sulfide ion concentration necessary for precipitation is controlled by controlling the OH^- concentration in a saturated H_2S solution.

$$Fe(OH)_2(s) + S^{2-} \xrightleftharpoons{OH^- \text{ (high } [S^{2-}])} FeS(s) + 2OH^-$$

Dissolution in acid: driven by reaction of basic anion with acid to give H_2S.

$$FeS(s) + 2H_3O^+ \longrightarrow Fe^{2+} + H_2S(g) + 2H_2O(l)$$

Oxidation: driven by strength of NO_3^- in acid solution as oxidizing agent.

$$3Fe^{2+} + NO_3^- + 4H^+ \longrightarrow 3Fe^{3+} + NO(g) + 2H_2O(l)$$

Confirmatory tests: Only Fe^{3+} gives these characteristic colors.

$$Fe^{3+} + K^+ + [Fe(CN)_6]^{4-} \rightleftharpoons \underset{\textit{dark blue}}{KFe[Fe(CN)_6](s)}$$

$$Fe^{3+} + NCS^- \rightleftharpoons \underset{\textit{blood red}}{[Fe(NCS)]^{2+}}$$

SUMMARY OF SECTIONS 22.6–22.8

To dissolve an ionic solid, ions are removed from solution by the formation of weak electrolytes (water, weak acids, complex ions) or occasionally by a redox reaction, until the ion product is less than that required for precipitation. Solubility can be controlled by varying the pH, either directly or indirectly by means of a chemical equilibrium. In selective precipitation, ions in solution are separated by controlling concentrations so that one ion is removed as a precipitate while others remain in solution.

In solving problems involving the simultaneous occurrence of two or more equilibria, the equilibrium constant expressions for all possible equilibria must be satisfied simultaneously by the same ion concentrations. *If* the concentrations of intermediate ions are negligible, equations can be combined and overall equilibrium constant expressions used. (It is important to examine what is known and unknown about a solution at the point in question.)

All the methods of controlling precipitation and solubility described above are used in inorganic qualitative analysis, which has as its goal the identification of all cations and anions present in an aqueous solution. In particular, qualitative analysis uses the selective precipitation of sulfides, with the sulfide ion concentration being controlled by varying the hydrogen ion concentration.

SIGNIFICANT TERMS (with section references)

dissociation constants (22.1)
formation constants (22.1)
inorganic qualitative analysis (22.8)
ion product (22.5)
molar solubility (22.3)
selective precipitation (22.7)
solubility product (22.2)

QUESTIONS AND PROBLEMS

Complex Ion and Solubility Equilibria—Complex Ions

22.1 Write the chemical equations and the expressions for the dissociation constants for the stepwise dissociation of the $[Fe(C_2O_4)_3]^{3-}$ complex. What is the relationship between the overall dissociation constant and the stepwise constants?

22.2 Repeat Question 22.1 for $[Ag(CN)_2]^-$.

22.3 Use the overall dissociation constant for $[Ag(CN)_2]^-$ to find the concentration of Ag^+ in equilibrium with 0.25 M CN^- and 0.10 M $[Ag(CN)_2]^-$. $K_d = 1 \times 10^{-22}$ for $[Ag(CN)_2]^-$.

22.4 Use the overall dissociation constant for $[Ni(NH_3)_6]^{2+}$ to calculate the concentration of Ni^{2+} in equilibrium with 6 M NH_3 and 0.05 M $[Ni(NH_3)_6]^{2+}$. $K_d = 1 \times 10^{-9}$ for $[Ni(NH_3)_6]^{2+}$.

22.5 What is the concentration of Ag^+ ion remaining in a solution that originally contained 0.10 M Ag^+ and 1.3 M NH_3? $K_d = 6.2 \times 10^{-8}$ for $[Ag(NH_3)_2]^+$.

22.6 A solution is 0.010 M with respect to Cd^{2+} and also 0.010 M with respect to Pd^{2+}. Solid KBr is added to the solution (assume no change in volume) until $[Br^-] = 1.0$ M. Use the overall dissociation constants of the complexes shown below to calculate the concentrations of Cd^{2+} and Pd^{2+} once equilibrium is reached. $K_d = 2.0 \times 10^{-4}$ for $[CdBr_4]^{2-}$ and 7.7×10^{-14} for $[PdBr_4]^{2-}$.

Complex Ion and Solubility Equilibria—Solubility Product

22.7 For each of the following equilibria, write the expression for the solubility product: (a) $Ag_2SO_4(s) \rightleftharpoons 2Ag^+ + SO_4^{2-}$, (b) $Fe_2S_3(s) \rightleftharpoons 2Fe^{3+} + 3S^{2-}$, (c) $SrCrO_4(s) \rightleftharpoons Sr^{2+} + CrO_4^{2-}$.

22.8 Repeat Question 22.7 for (a) $Hg_2Br_2(s) \rightleftharpoons Hg_2^{2+} + 2Br^-$, (b) $CdCO_3(s) \rightleftharpoons Cd^{2+} + CO_3^{2-}$, (c) $Mn_3(PO_4)_2(s) \rightleftharpoons 3Mn^{2+} + 2PO_4^{3-}$.

22.9 Calculate the K_{sp} for each of the following substances from the given solubilities: (a) AgCl, 1.9×10^{-3} g/L; (b) $PbBr_2$, 7.8 g/L; (c) $Co(OH)_2$, 3.4×10^{-4} g/L.

22.10 Repeat Problem 22.9 for (a) AgBr, 1.3×10^{-4} g/L and (b) PbI_2, 0.56 g/L.

22.11 A saturated solution of $Mn(OH)_2$ has a pH of 9.57 at room temperature. Calculate the solubility product for $Mn(OH)_2$.

22.12 Magnesium hydroxide is a slightly soluble substance. The pH of a saturated solution of $Mg(OH)_2$ is 10.38. Find the K_{sp}. Calculate the solubility in units of g/100 g of water.

22.13 Calculate the solubility in grams per liter for each of the following substances from the given values of K_{sp}: (a) $Ba(OH)_2$, $K_{sp} = 1.3 \times 10^{-2}$, (b) PbI_2, $K_{sp} = 7.1 \times 10^{-9}$.

22.14 Repeat Problem 22.13 for (a) $Ni(OH)_2$, $K_{sp} = 3 \times 10^{-16}$ and (b) $Cu(OH)_2$, $K_{sp} = 1.3 \times 10^{-20}$.

22.15 The K_{sp} for AgBr is 4.9×10^{-13} at 25 °C. What is the concentration of Ag^+ in a solution in which the concentration of Br^- is 0.010 M?

22.16 The K_{sp} for $AgBrO_3$ is 3.9×10^{-5} at 25 °C. Calculate the solubility of silver ion in a 0.30 M bromic acid solution.

22.17 The solubility product constant for the reaction

$$Ag_2CrO_4(s) \rightleftharpoons 2Ag^+ + CrO_4^{2-}$$

is 2.5×10^{-12}. Find the solubility of Ag_2CrO_4 in (a) pure water, (b) 0.010 M $AgNO_3$, (c) 0.010 M K_2CrO_4.

22.18 Milk of magnesia is a suspension of the slightly soluble $Mg(OH)_2$ in water. $K_{sp} = 7.1 \times 10^{-12}$ for $Mg(OH)_2$. (a) What is the solublity of $Mg(OH)_2$ in 0.01 M NaOH solution? (b) What is the solubility of $Mg(OH)_2$ in 0.1 M Mg^{2+} solution?

22.19 A fluoridated water supply contains 1 mg/L of F^-. What is the maximum amount of Ca^{2+}, expressed in g/L, that can exist in this water supply? $K_{sp} = 2.7 \times 10^{-11}$ for CaF_2.

22.20 Which is more soluble in a 0.20 M CrO_4^{2-} solution: $BaCrO_4$, for which $K_{sp} = 1.2 \times 10^{-10}$ or Ag_2CrO_4, for which $K_{sp} = 2.5 \times 10^{-12}$?

22.21 If 1.0 g of $AgNO_3$ is added to 50 mL of 0.050 M NaCl, will a precipitate form? $K_{sp} = 1.8 \times 10^{-10}$ for AgCl.

22.22 Will a precipitate of $PbCl_2$ form as 5.0 g of solid $Pb(NO_3)_2$ is added to 1.00 L of 0.010 M NaCl? Assume that no volume change occurs. $K_{sp} = 2 \times 10^{-5}$ for $PbCl_2$.

22.23 Sodium bromide and lead nitrate are soluble in water. Will lead bromide precipitate as 1.03 g of NaBr and 0.332 g of $Pb(NO_3)_2$ are dissolved together in sufficient water to make 1.00 L of solution? $K_{sp} = 3.9 \times 10^{-5}$ for $PbBr_2$.

22.24 A solution contains 0.015 M CH_3COOH and 0.0030 M Ag^+. Will a precipitate form? $K_a = 1.754 \times 10^{-5}$ for acetic acid and $K_{sp} = 4.4 \times 10^{-3}$ for $Ag(CH_3COO)$.

Controlling Solubility and Precipitation

22.25 Why would MnS be expected to be more soluble in a 0.10 M HCl solution than in water? Would the same be true for $Mn(NO_3)_2$?

22.26 For each pair, choose the salt that would be expected to be more soluble in acidic solution than in pure water: (a) $Hg_2(CH_3COO)_2$ or Hg_2Br_2, (b) $Pb(OH)_2$ or PbI_2, (c) AgI or $AgNO_2$.

22.27 How can most water-insoluble metal hydroxides be dissolved? Write a chemical equation for the dissolution of $Fe(OH)_3$ in this way.

22.28 How does the presence of excess H_3O^+ aid in the dissolution of slightly soluble metal carbonates? Write the chemical equation for the dissolution of $MnCO_3$.

22.29 A concentrated, strong acid is added to a solid mixture of 0.010 mol samples of $Fe(OH)_2$ and $Cu(OH)_2$ placed in 1 L of water. At what values of pH will the dissolution of each hydroxide be complete? $K_{sp} = 8 \times 10^{-16}$ for $Fe(OH)_2$ and 1.3×10^{-20} for $Cu(OH)_2$. (Assume no change in volume.)

22.30 The solubility product of iron(III) hydroxide is 3×10^{-39}. (a) At what pH will a 0.001 M $Fe(NO_3)_3$ solution just begin to form the hydroxide precipitate? (b) If the source of the OH^- is a weak base having $K_b = 3.0 \times 10^{-6}$, what is the maximum concentration of that base which is permitted before $Fe(OH)_3$ forms?

22.31 A solution is 0.01 M in Pb^{2+} and 0.01 M in Ag^+. As Cl^- is introduced to the solution by the addition of solid NaCl: determine (a) which substance will precipitate first, AgCl or $PbCl_2$, and (b) the fraction of the metal ion in the first precipitate that remains in solution at the moment the precipitation of the second compound begins. $K_{sp} = 2 \times 10^{-5}$ for $PbCl_2$ and 1.8×10^{-10} for AgCl.

22.32 A solution is 0.01 M in I^- and 0.01 M in Br^-. As Ag^+ is introduced to the solution by the addition of solid $AgNO_3$: determine (a) which substance will precipitate first, AgI or AgBr, and (b) the percentage of the halide ion in the first precipitate that is removed from solution before the precipitation of the second compound begins. $K_{sp} = 8.3 \times 10^{-17}$ for AgI and 4.9×10^{-13} for AgBr.

22.33 Calculate the solubility of CaF_2 in a solution that is buffered at $[H^+] = 0.010$ M. $K_a = 6.5 \times 10^{-4}$ for HF and $K_{sp} = 2.7 \times 10^{-11}$ for CaF_2.

22.34 Calculate the solubility of AgCN in a solution that is buffered at $[H^+] = 0.010$ M. $K_a = 6.2 \times 10^{-10}$ for HCN and $K_{sp} = 2.3 \times 10^{-16}$ for AgCN.

22.35 A precipitate of $Cr(OH)_3$ forms as OH^- ion is added to an acidic solution of Cr^{3+} ion. However, the precipitate will redissolve as additional OH^- ion is added because of the formation of the $[Cr(OH)_4]^-$ complex ion. Find the pH range over which the concentration of Cr^{3+} or $[Cr(OH)_4]^-$ is less than 1×10^{-4} mol/L. $K_{sp} = 6 \times 10^{-31}$ for $Cr(OH)_3$ and $K_d = 1 \times 10^{-28}$ for $[Cr(OH)_4]^-$.

22.36 A solution containing Ag^+ was treated with CN^- ion to form the AgCN precipitate. However, in the presence of excess CN^- the $[Ag(CN)_2]^-$ complex ion will form. Over what range of $[CN^-]$ will the concentration of Ag^+ or $[Ag(CN)_2]^-$ remain below 1×10^{-4} mol/L? $K_{sp} = 2.3 \times 10^{-16}$ for AgCN and $K_d = 1.3 \times 10^{-22}$ for $[Ag(CN)_2]^-$.

22.37 What is the concentration of Ag^+ in a solution that is 0.10 M in KSCN and 0.10 M in $[Ag(SCN)_4]^{3-}$? Will Ag_2SO_4 precipitate if $[SO_4^{2-}] = 0.1$ M? $K_d = 2.1 \times 10^{-10}$ for $[Ag(SCN)_4]^{3-}$ and $K_{sp} = 1.5 \times 10^{-5}$ for Ag_2SO_4.

22.38 Zinc(II) ion forms a complex with $EDTA^{4-}$

$$Zn^{2+} + EDTA^{4-} \rightleftharpoons [Zn(EDTA)]^{2-} \qquad K_f = 3.9 \times 10^{16}$$

where $EDTA^{4-}$ represents the ethylenediaminetetraacetate ion. Will ZnS form in a solution originally 0.100 M in $EDTA^{4-}$, 0.01 M in S^{2-}, and 0.010 M in Zn^{2+}? $K_{sp} = 2 \times 10^{-24}$ for ZnS.

22.39 What pH could be used to separate Mn^{2+} from Zn^{2+} by selective precipitation of 0.010 mol of ZnS in 1 L of a saturated solution of H_2S? Assume that $[Mn^{2+}]$ is also 0.010 mol/L. $K_{sp} = 2.3 \times 10^{-13}$ for MnS and 2×10^{-24} for ZnS and $[H^+]^2[S^{2-}] = 3 \times 10^{-21}$ for a saturated solution of H_2S.

22.40 In a student-designed qualitative analysis scheme, Pb^{2+} is to be separated as PbC_2O_4 (lead oxalate) from Cu^{2+} by the addition of $Na_2C_2O_4$. To what value must the concentration of $C_2O_4^{2-}$ be adjusted so that only PbC_2O_4 will form in a solution that is roughly 0.01 M in each of these cations? $K_{sp} = 4.8 \times 10^{-12}$ for PbC_2O_4 and 3×10^{-8} for CuC_2O_4.

CHAPTER

23

Thermodynamics

DISORDER, SPONTANEITY, AND ENTROPY

The word "spontaneous" is applied to physical and chemical changes in a very specific way related to the thermodynamics of a system. The meaning is somewhat different from what most of us think of as "spontaneous." Perhaps we think of a crowd jumping up when the home team scores a touchdown, or the hug a mother gives a child who has fallen down. These are impulsive reactions—they are immediate and quick.

We know that the rate, or quickness, of a chemical reaction is dependent upon the reaction mechanism and must be studied by the methods of kinetics. "Spontaneity" as a thermodynamic concept is unrelated to speed. A "spontaneous" chemical reaction or physical change is one that can happen without any continuing outside influence. (We could say "thermodynamically favorable" instead, but this is cumbersome.) A spontaneous change, whether slow or fast, has a natural direction, like the melting of ice at room temperature, the reaction of an acid with a base, or the rusting away of a piece of iron.

The first law of thermodynamics—the law of conservation of energy—deals only with the total energy in chemical reactions. It tells only part of the story about spontaneity. In this chapter we introduce two additional thermodynamic quantities—entropy and free energy. Entropy and enthalpy together permit the calculation of free energy changes, which show whether chemical or physical changes will occur spontaneously.

23.1 ENTROPY: A QUANTITATIVE MEASURE OF RANDOMNESS

In Chapter 7 we found that when a chemical reaction or a change of state occurs, a change in the energy of the system occurs too. Although it is less obvious, something else also changes in most cases. When a liquid evaporates, what is happening to the molecules? They are gaining energy and moving into a larger space. When silver chloride precipitates from a solution, what is happening to the silver and chloride ions? Instead of moving randomly through the water, the ions assume fixed positions in a crystal.

In both of these changes, the amount of *order* among the particles in the system has changed. Molecules are more disordered in the gas phase—their motion and their arrangement in space are more random. Ions are more ordered in the solid phase—their motion is less random. In most physical and chemical changes there is a change in the order of the particles in the system as well as a change in energy.

A chemical or physical change that is exothermic is *often* spontaneous, but not always. A chemical or physical change that is endothermic is *most likely* not spontaneous, but not always.

To illustrate a case in which exothermic changes are not spontaneous, consider the three beakers of water shown in Figure 23.1. Each beaker holds a large reservoir of water maintained at a constant temperature. Ice cubes of equal size are dropped into each beaker. After a few minutes, the ice cube in the beaker containing water supercooled to −1 °C increases in size because some of the water freezes. No visible change occurs in the second beaker, and the ice cube in the third beaker decreases in size because some of the ice melts.

The enthalpy changes for the process

$$H_2O(l, 1\ \text{atm}) \longrightarrow H_2O(s, 1\ \text{atm})$$

at each of the three temperatures of the water in the beakers can be calculated. The values are

$$\Delta H^\circ_{272.15} = -5972.2\ \text{J/mol}$$
$$\Delta H^\circ_{273.15} = -6009.5\ \text{J/mol}$$
$$\Delta H^\circ_{274.15} = -6046.7\ \text{J/mol}$$

Based on the negative ΔH° values alone, the water should freeze in all three beakers, not just in one. Thus, we must conclude that the favorable enthalpy change is opposed by some other force. At −1 °C, this second force is not strong enough to counteract the enthalpy change. At 1 °C, it is strong enough to overcome the enthalpy change, and at 0 °C, it must be equal in magnitude to the enthalpy change, but opposite in effect, so that an equilibrium condition exists.

Could this second driving force toward spontaneity be related to order and disorder? You might think intuitively that it is, and your intuition would be correct. An

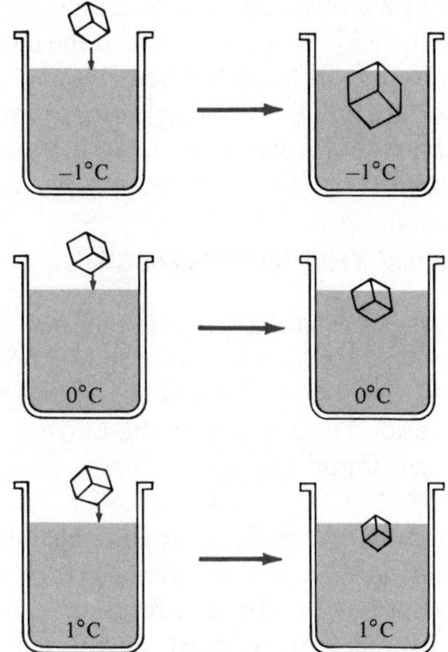

Figure 23.1
Changes in an Ice Cube at Different Temperatures
At −1 °C the ice cube grows larger, at 0 °C it is unchanged, and at 1 °C it melts a bit.

ordered state is not generally expected to occur spontaneously. If you dropped a piggy bank and the coins landed in separate stacks of pennies, nickels, and dimes, it would be quite a shock. Similarly, if two gases were mixed in a flask, you would not expect to find later that all the molecules of one had migrated to the left half of the container and all the molecules of the other occupied the right half.

In fact, just the opposite is true: It is *disorder* that tends to increase spontaneously. If two bulbs, each containing a different gas, were connected to each other and the gases did *not* eventually mix, you would conclude that the connection was stopped up.

Experience has also shown us that a process that is spontaneous in one direction is not spontaneous in the other. Left to themselves, the broken pieces of a piggy bank will not reunite, nor will the coins fly back into it. In Chapter 7 we discussed the spontaneous flow of heat from a hot block of metal to a reservoir of water. No one has yet seen a hot metal block get hotter when immersed in cold water.

Many practical circumstances arise in which it is desirable to know the natural direction of spontaneity of a physical or chemical change. To determine this natural direction of spontaneity it is necessary to take into account both the change in enthalpy and the change in the disorder of a system.

To deal quantitatively with the drive toward disorder, a thermodynamic property called **entropy,** which is a measure of disorder, is introduced. Entropy is symbolized by S; a change in entropy is symbolized in the usual way

$$\underbrace{\Delta S}_{\text{change in entropy}} = \underbrace{S_2}_{\text{final entropy}} - \underbrace{S_1}_{\text{initial entropy}}$$

As is done for enthalpy changes, entropy changes for substances in their standard states at 1 bar (slightly less than 1 atm) and a specified temperature are designated as standard state entropy changes, ΔS°. The units for entropy are most commonly joules per kelvin or calories per kelvin. When entropy is expressed on a molar basis, the units are, for example, joules per kelvin per mole (J/K mol). Note that because entropy is of a smaller magnitude than energy, joules or calories, rather than kilojoules or kilocalories, are used with entropy values. The value of ΔS is negative for a decrease in the entropy of a system and positive for an increase.

At this point students often ask, "So what is entropy, *really*?" On a microscopic, atom-for-atom, molecule-for-molecule level, entropy represents the number of possible ways the particles can be arranged—the statistical probability for disorder.

There are mathematical techniques for dealing with probability, and applied to the calculation of entropy they yield results that agree with what is observed. The entropy of a system is greater the more ways there are in which the individual atoms, molecules, or ions can be arranged with respect to each other. It is also greater the more possible states there are available to each individual particle. For example, the larger a molecule, the more possible ways there are for its atoms to rotate and vibrate relative to each other (see Figure 7.3); therefore, the greater is its entropy.

23.2 THE SECOND LAW OF THERMODYNAMICS

The role that the drive to disorder plays in every natural process is expressed in the second law of thermodynamics, which has profound implications that reach well beyond chemistry. Like the first law, **the second law of thermodynamics** is based on experience with the real world: The entropy of the universe is constantly increasing. Many other ways have been found to express the second law of thermodynamics (Table 23.1).

The second law means that *every* spontaneous chemical and physical change increases the entropy of the *universe as a whole.* A change can be spontaneous only when the entropy of the universe increases as a result of it. In a series of spontaneous changes over the course of time, entropy must always increase. The direction of all

Table 23.1
Statements of the Second Law of Thermodynamics

All systems tend to approach a state of equilibrium.
Entropy is time's arrow.[a]
The state of maximum entropy is the most stable state for an isolated system.[b]
Every system which is left to itself will, on the average, change toward a condition of maximum probability.[c]
It is impossible in any way to diminish the entropy of a system of bodies without thereby leaving behind changes in other bodies.[d]
Die Energie der Welt ist constant; die Entropie der Welt strebt einem Maximum zu.[e]
Things are getting more screwed up every day.[f]
You can't break even.

[a] Sir Arthur Eddington. [b] Enrico Fermi. [c] G. N. Lewis.
[d] Max Planck. This and three preceding statements are quoted from J. Arthur Campbell, *Chemical Systems: Energetics, Dynamics, Structure* (San Francisco: W. H. Freeman, 1970).
[e] R. J. E. Clausius, quoted by J. W. Gibbs to head his classic memoir, *The Equilibrium of Heterogeneous Substances.*
[f] Anonymous; with thanks to A. Truman Schwarz, *Chemistry: Imagination and Implication* (New York: Academic Press, 1973).

natural events is toward disorder—this is why entropy is sometimes referred to as "time's arrow."

Fortunately, it is not necessary to measure the entropy of the universe as a whole. Recall that it is customary in thermodynamics (Section 7.2) to divide the universe into the *system* (the part of the universe under study) and the *surroundings* (everything else). The condition expressed by the second law of thermodynamics for a spontaneous process becomes

$$\Delta S_{\text{universe}} = \Delta S_{\text{system}} + \Delta S_{\text{surroundings}} > 0 \qquad \textbf{(23.1)}$$

The entropy of a *system* can either increase or decrease. If ΔS_{system} decreases (has a negative sign) in a spontaneous change, this change *must* be accompanied by a simultaneous and larger increase in $\Delta S_{\text{surroundings}}$. (Can you see how the entropy of the universe increases in the natural processes in Figure 23.2?)

Expanding gases

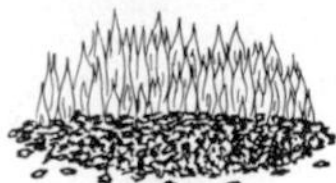
Chemical reactions

Formation of solutions

Boiling liquids

Falling bodies

Figure 23.2
Spontaneous Natural Processes in Which the Entropy of the Universe Increases

23.3 ENTROPY IN PHYSICAL CHANGES

With other conditions held constant, significant increases in the entropy of a chemical system accompany four types of physical change.

1. Increase in temperature (Figure 23.3a) With an increase in temperature, the kinetic energy of the atoms, molecules, or ions increases; they move around more rapidly, and there is greater disorder within the system. You might be wondering how it can be that an exothermic reaction, during which the internal energy of a system decreases, can be spontaneous. Remember that the second law deals with the entropy of the *universe.* The energy released in an exothermic reaction is transferred to the surroundings. The temperature of the surroundings increases, contributing to the overall increase of the entropy of the universe required for spontaneous change.
2. Changes of state: vaporization or melting (Figure 23.3b) An increase in the entropy of a system accompanies a change of state that allows greater freedom of motion to atoms, molecules, or ions. When a solid melts or a liquid evaporates, the *temperature* of the system does not increase, but the entropy increases because the particles have more freedom to move.

   Consider a sample of liquid argon in equilibrium at its normal boiling point with the gas. Each atom in the liquid can move around freely in a space about 0.32 nm in diameter. However, the atom has only a small chance of getting out of one space and into another one. An atom in the gas phase moves much further—about 20 nm—before hitting another atom. The difference

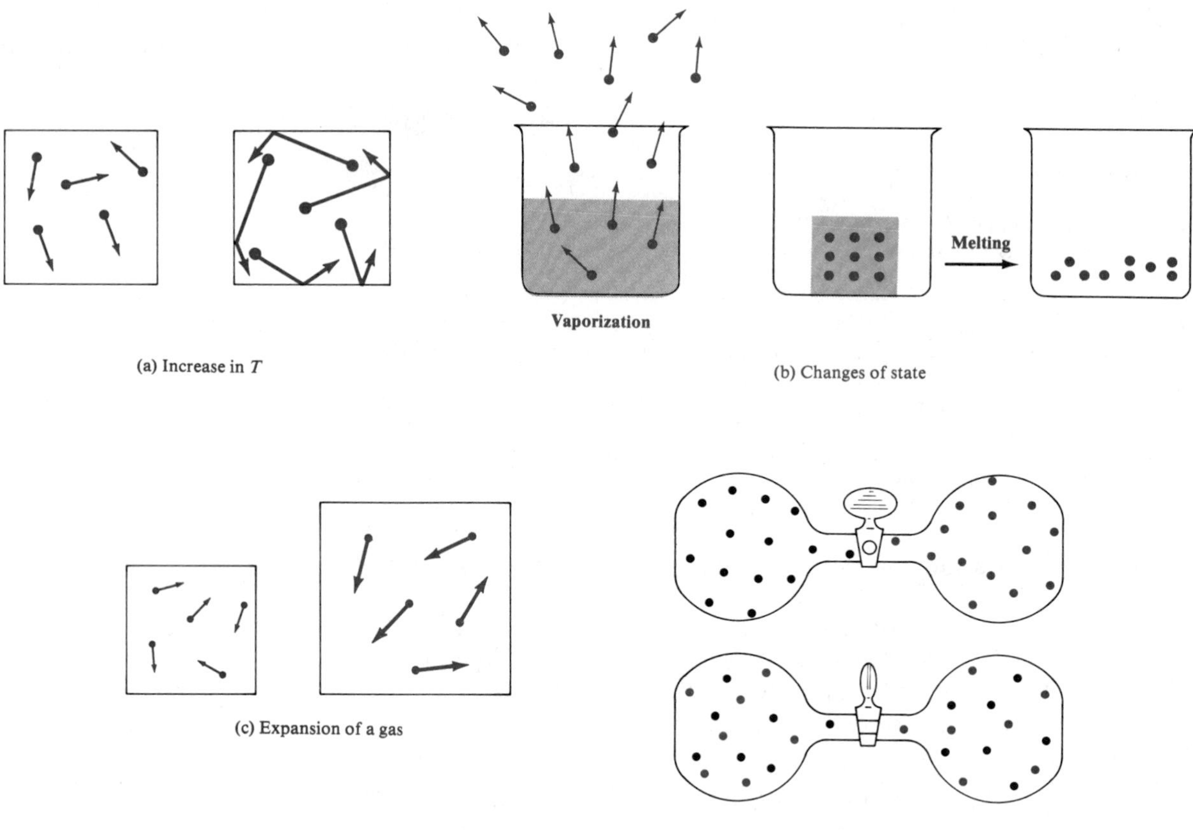

(a) Increase in T

(b) Changes of state

(c) Expansion of a gas

(d) Mixing

Figure 23.3
Changes in Which Entropy Increases

between these distances illustrates the significantly greater degree of randomness or disorder in the gaseous than in the liquid state. Keep in mind that the change in entropy alone is not a criterion for spontaneity. As we would intuitively predict, vaporization is accompanied by an entropy increase. But vaporization is not spontaneous—it takes place only as a result of the continuous absorption of heat from the surroundings. On a molar basis, the entropy change for vaporization or any change of state is given by

$$\Delta S = \frac{\Delta H}{T} \qquad \textbf{(23.2)}$$

enthalpy change for change of state (ΔH)

absolute temperature of the change of state (T)

$$\frac{\textit{joules}}{\textit{kelvin mole}} \qquad \frac{\textit{joules/mole}}{\textit{kelvins}}$$

3. Expansion of a gas (Figure 23.3c) When a given amount of a gas expands into a larger volume at a constant temperature, there is an increase in entropy. The molecules have a larger space in which to move about, and therefore the amount of disorder at the molecular level is greater.
4. Mixing of two or more substances (Figure 23.3d) Mixing is a physical process in which there is a change in entropy but no change in enthalpy. When two ideal gases that do not react with each other mix at constant temperature and pressure, the process is spontaneous and the *only* driving force is an entropy increase. When substances are mixed or dissolution occurs, entropy usually increases.

If a process results in an increase in the entropy of a system, the reverse of that process causes an equal *decrease* in entropy. Entropy decreases (1) when the temperature of a system becomes lower, (2) when a gas condenses or a liquid freezes, (3) when a gas is compressed into a smaller volume at a constant temperature, or (4) when a dissolved substance crystallizes out of solution.

The ice cubes of the experiment in Figure 23.1 demonstrate that entropy is not the sole criterion for spontaneity of a change in a chemical system. The entropy changes for freezing water at the three temperatures of the water in the beakers can be calculated and are found to be

$$\Delta S^\circ_{272.15} = -21.8640 \text{ J/K mol}$$
$$\Delta S^\circ_{273.15} = -22.0007 \text{ J/K mol}$$
$$\Delta S^\circ_{274.15} = -22.1369 \text{ J/K mol}$$

Based on entropy changes alone, water would not be expected to freeze in *any* of the three beakers, for the entropy value is negative in each case. In Section 23.6, we discuss how the effects of enthalpy and entropy changes can be combined to predict spontaneity.

EXAMPLE 23.1 Entropy in Physical Changes

Would you predict an increase or a decrease in entropy for the systems in which the following changes occur?

(a) The "disappearance" of the contents of a bottle of ethyl ether that was left open. (System = the ether)
(b) Making rock candy (crystalline sugar) from a saturated sugar solution. (System = candy + solution)
(c) Putting cream in your coffee. (System = coffee + cream)

(a) The evaporation of the ether would be an **increase** in the entropy, with entropy increasing due to both the phase change and expansion of the vapor.
(b) The crystallization of the sugar would be a **decrease** in entropy. The molecules are more ordered in a crystal than in solution.
(c) An **increase** in entropy would result from mixing the cream and the coffee solution. The randomness of the liquids combined is greater than that of the liquids in separate containers. (A slight decrease in entropy might result from cooling the coffee.)

EXERCISE Would you predict an increase or a decrease in entropy for the systems in which the following physical changes occur: (a) the formation of a rain drop in a cloud (system = water), (b) the crystallization of the metal alloy from the molten state in molding a bookend (system = alloy), and (c) the beating of an egg for an omelet (system = egg)? *Answers* (a) decrease, (b) decrease, (c) increase.

EXAMPLE 23.2 Entropy in Physical Changes

Calculate the standard state entropy change for the vaporization of one mole of argon at the normal boiling point. ΔH° (vaporization) = 6519 J/mol for Ar at 87.5 K. Is this an increase or decrease in entropy?

The entropy change for this change of state is

$$\Delta S^\circ_{87.5}(\text{vaporization}) = \frac{\Delta H^\circ_{87.5}(\text{vaporization})}{T} = \frac{6519 \text{ J/mol}}{87.5 \text{ K}} = \mathbf{74.5\ J/K\ mol}$$

The positive value of ΔS° tells us that the **entropy of argon has increased** by 74.5 J/K mol.

EXERCISE The normal boiling point of cesium is 690. °C and the heat of vaporization is 68.28 kJ/mol. Calculate ΔS° for the condensation of cesium vapor. Is this an increase or decrease in entropy for the cesium? *Answer* $\Delta S^\circ = -70.9$ J/K mol, decrease.

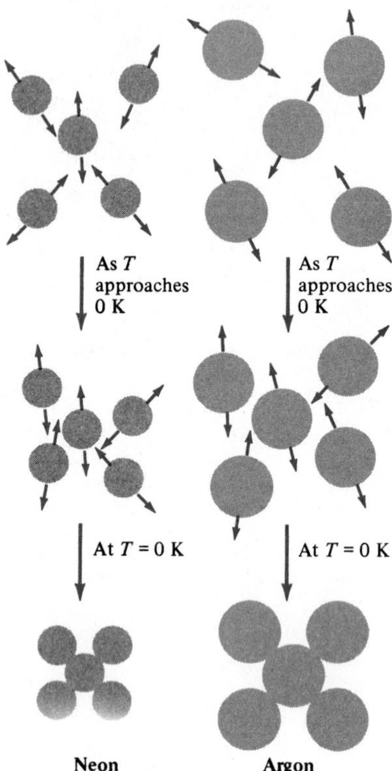

Figure 23.4
Decrease in Randomness as Absolute Zero Is Approached
The randomness of all perfect crystalline substances is the same at absolute zero and is $S_0^\circ = 0$. Note this does not mean that the crystal structures of all substances are the same at absolute zero.

23.4 ABSOLUTE ENTROPIES

Only *changes* in energy (ΔE) and enthalpy (ΔH) can be measured experimentally. Entropy is different. The disorder in a given amount of any pure substance can be determined experimentally. This is known as **absolute entropy,** S, the entropy of a pure substance. The determination of absolute entropies is possible because there is a natural reference state for the entropy of a pure substance—the entropy at absolute zero.

Consider two separate 1 mol samples of neon and argon that are cooled at 1 atm pressure to the point where they liquefy and then to the point where they solidify. Assume that they form perfect crystals. The amount of disorder in the crystals decreases in both samples as the temperature continues to decrease. What happens as the temperature approaches absolute zero? The vibrational kinetic energy of the atoms in both samples is decreasing (Figure 23.4), and the atoms are becoming more and more confined to their positions in the crystal. At absolute zero, the atoms no longer vibrate in their positions, and both substances have the same amount of disorder. The standard state entropy at absolute zero, S_0°, is zero for perfect crystalline substances. This is the **third law of thermodynamics:** The entropy of a perfect crystal at absolute zero is zero. In this context, a "perfect crystal" is one in which there are no crystal defects and, in addition, there is only one possible arrangement of particles and the particles are all of equivalent energy.

What about a substance more complicated than argon or neon—say, one composed of diatomic molecules AB? If the substance forms a perfect crystal with all molecules completely ordered (Figure 23.5a), again $S_0^\circ = 0$. On the other hand, if the substance forms a solid with the molecules oriented in a random way (Figure 23.5b), the entropy is not zero, but is larger than zero and is dependent upon the number of possible arrangements of the molecules. (The statistics of probability can be applied to calculate S_0° for such a crystal.)

The absolute entropy of a substance at a given temperature above absolute zero is the sum of the following contributions: (1) the entropy of the substance at absolute zero (calculated on the basis of the complexity of the molecule and the number of possible arrangements), (2) the entropy increases for any phase changes that the substance undergoes between absolute zero and the desired temperature, and (3) the entropy increases during the heating of each phase over the temperature range in which it is stable. Figure 23.6 illustrates the contributions to the entropy of a pure substance.

Table 23.2 includes the standard enthalpies of formation and the standard absolute entropies for a number of common substances at 25 °C (the ΔG° values are discussed in Section 23.6 and 23.7). The absolute entropies of substances in their standard states at 1 bar and a specified temperature are designated standard absolute entropies (S°). If no other temperature is indicated, the temperature can be assumed to be 25 °C.

Note in Table 23.2 that at 25 °C the standard state entropies of pure elements are *not* zero because the entropy is higher at the higher temperature. The standard state heats of formation [and also the values of ΔG_f°; Section 23.7] for the elements are *defined* as zero at 298 K because, since absolute values of these quantities cannot be measured, it is necessary to set a reference point. For ions in aqueous solution, the reference entropy is chosen to be $S^\circ = 0$ for $H^+(aq)$.

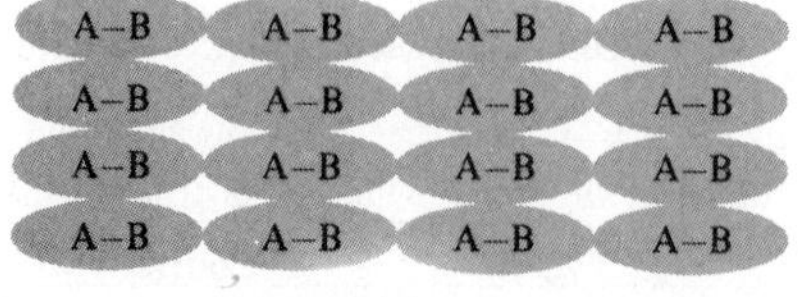

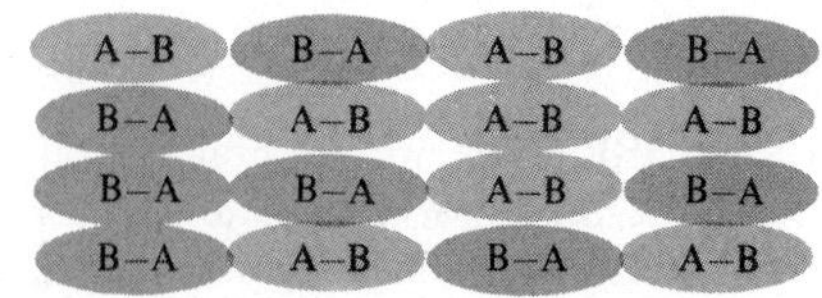

Figure 23.5
Order and Disorder in an A—B Crystal at Absolute Zero
(a) Perfect order in a perfect crystal.
(b) Random disorder.

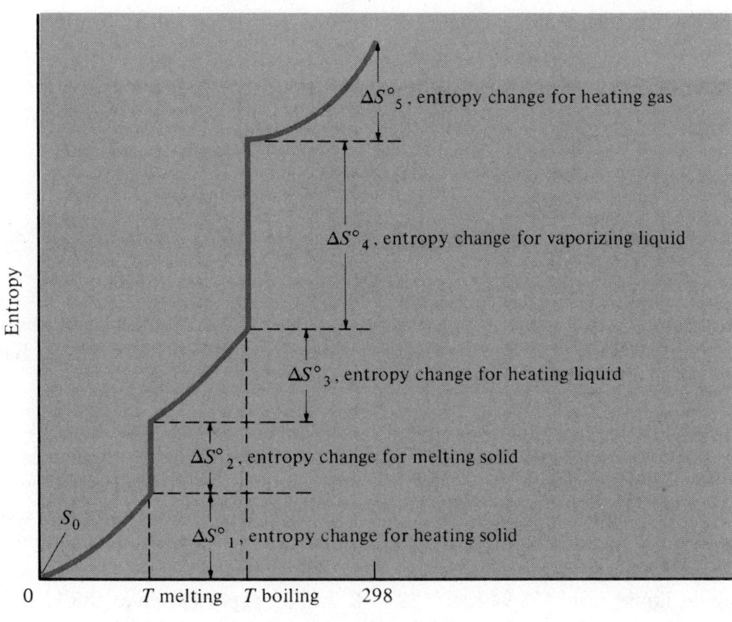

Figure 23.6
Contributions to the Standard Absolute Entropy of a Substance That Is a Gas at 298 K
For this substance, $S^\circ_{298} = S^\circ_0 + \Delta S^\circ_1 + \Delta S^\circ_2 + \Delta S^\circ_3 + \Delta S^\circ_4 + \Delta S^\circ_5$.

Table 23.2
Some Thermodynamic Data (at 25 °C)
See Appendix IV for additional values of thermodynamic quantities.

Substance	ΔH°_f (kJ/mol)	ΔG°_f (kJ/mol)	S° (J/K mol)
$O(g)$	249.17	231.73	161.06
$O_2(g)$	0	0	205.14
$O_3(g)$	142.7	163.2	238.93
$H(g)$	217.97	203.25	114.71
$H^+(g)$	1536.20	—	—
$H^+(aq)$	0[a]	0[a]	0[a]
$H_2(g)$	0	0	130.68
$OH^-(g)$	−143.5	—	—
$OH^-(aq)$	−229.99	−157.24	−10.75
$H_2O(l)$	−285.83	−237.13	69.91
$H_2O(g)$	−241.82	−228.57	188.83
$H_2O_2(l)$	−187.78	−120.35	109.6
$He(g)$	0	0	126.15
$Ne(g)$	0	0	146.33
$F_2(g)$	0	0	202.78
$HF(g)$	−271.1	−273.2	173.78
$Cl_2(g)$	0	0	223.07
$HCl(g)$	−92.31	−95.30	186.91
$HCl(aq)$	−167.16	−131.228	56.5
$Br_2(l)$	0	0	152.23
$HBr(g)$	−36.40	−53.45	198.70
Fe(α-solid)	0	0	27.28
$Fe^{2+}(aq)$	−89.1	−78.90	−137.7
$Fe^{3+}(aq)$	−48.5	−4.7	−315.9
$FeO(s)$	−272.0	—	—
$Fe_2O_3(s)$	−824.2	−742.2	87.40
$Fe_3O_4(s)$	−1118.4	−1015.4	146.4
$FeCl_2(s)$	−341.79	−302.30	117.95
$FeCl_2(aq)$	−423.4	−341.34	−24.7
$FeCl_3(s)$	−399.49	−334.00	142.3
$FeCl_3(aq)$	−550.2	−398.3	−146.4

[a] Chosen by convention to be zero. This choice sometimes gives negative values for S° for substances in aqueous solution, which does not mean that the entropy is negative, but that the entropy is negative with respect to the actual value for H^+.

In general, the data in Table 23.2 demonstrate the following: (a) Gases have greater entropies than liquids. [Compare $H_2O(l)$ and $H_2O(g)$.] (b) Substances with more complex molecular or crystalline structures have greater entropies than more simple substances. [Compare $H_2O(l)$ and $H_2O_2(l)$; or Fe, Fe_2O_3, and Fe_3O_4.]

EXAMPLE 23.3 Absolute Entropy

The absolute entropy of $HCl(g)$ at 25 °C is given in Table 23.2 as 186.91 J/K mol. Explain the contributions to this value, given that (a) at 0 K, HCl exists in a solid phase called solid-II (assume that this is not a perfect crystal); (b) at 98.38 K, solid-II changes to solid-I; (c) at 158.94 K, solid-I melts; and (d) at 188.11 K, the liquid boils.

There are eight contributions to the value of $S°$ for gaseous HCl. The first contribution is the result of the **disorder at absolute zero** because HCl does not form a perfect crystal. The second contribution is the entropy increase that results as **solid-II is heated** from 0 K to 98.38 K. The third contribution is the entropy increase that results as **solid-II changes to solid-I** at 98.38 K. The fourth and fifth contributions are for **heating solid-I** from 98.38 K to 158.94 K and for the **fusion process** at 158.94 K. The sixth and seventh contributions are for **heating the liquid** from 158.94 K to 188.11 K and for **vaporization** at 188.11 K. The last contribution is the entropy increase for **heating the gas** from 188.11 K to 298.15 K.

EXERCISE The absolute entropy of $F_2(g)$ at 25 °C is given in Table 23.2 as 202.78 J/K mol. Identify the contributions to this value given that (a) at 0 K, F_2 exists as a perfect crystal; (b) at 55.19 K, $F_2(s)$ melts; and (c) at 85.23 K and 1 atm, $F_2(l)$ boils.

Answer There are six contributions to the value of $S°$: (1) $S°_0 = 0$, (2) heating solid from 0 K to 55.19 K, (3) fusion at 55.19 K, (4) heating liquid from 55.19 K to 85.23 K, (5) vaporization at 84.23 K, and (6) heating gas from 85.23 K to 298 K.

23.5 ENTROPY IN CHEMICAL REACTIONS

Entropy can change in chemical reactions because of (1) a change in the number of uncombined atoms, ions, or molecules in the system; (2) a change in the phase in which the species in the system are present; or (3) a change in the complexity of the substances involved.

The thermal decomposition of potassium chlorate

$$2KClO_3(s) \xrightarrow{\Delta} 2KCl(s) + 3O_2(g)$$

can easily be predicted to result in a very large increase in entropy. Two moles of reactant have been converted into five moles of products, including three moles of gas where originally there were none. For this reaction, $\Delta S° = 808.14$ J/K.

In the combination of nitrogen and hydrogen to give ammonia

$$N_2(g) + 3H_2(g) \longrightarrow 2NH_3(g)$$

it is apparent that the entropy will decrease. Four moles of gas have been converted to 2 mol of gas. For this reaction, $\Delta S° = -198.55$ J/K.

In the following reaction, there is no change in the number of moles of substances, nor in the phases of the substances.

$$Cu^{2+} + Zn(s) \longrightarrow Zn^{2+} + Cu(s)$$

One mole of solid plus one mole of ions in solution have been converted to one mole of solid and one mole of ions in solution. For such reactions, it is difficult to predict whether entropy will increase or decrease. Contributions to the entropy change could be made by differences in the crystal structure of copper as compared to zinc, and differences in the hydration of the two ions. The best prediction here is that the entropy change will be small. In fact, it is −21.0 J/K.

EXAMPLE 23.4 Entropy in Chemical Reactions

Predict whether the entropy change for each of the following reactions will be large and negative, large and positive, or small:

(a) $H_2(g) + Cl_2(g) \longrightarrow 2HCl(g)$

(b) $2ZnS(s) + 3O_2(g) \xrightarrow{\Delta} 2ZnO(s) + 2SO_2(g)$

(c) $2N_2O(g) \longrightarrow 2N_2(g) + O_2(g)$

(a) Two moles of gas react to form two moles of gas. We would expect only a **small** change in randomness for this reaction. (The actual value is 20.07 J/K.)
(b) Two moles of solid react with three moles of gas to produce two moles of solid and two moles of gas. The net loss of one mole of gas represents a decrease in randomness, so we would predict a **large, negative** value for $\Delta S°$. (The actual value is −147.1 J/K.)
(c) Two moles of gas produce three moles of gas—a net gain of one mole of gas. Thus we would predict a **large, positive** value for $\Delta S°$. (The actual value is 148.54 J/K.)

EXERCISE Classify the entropy change for each of the following reactions as (i) large and negative, (ii) large and positive, or (iii) small:

(a) $C(\text{graphite}) + 2H_2(g) + \frac{1}{2}O_2(g) \longrightarrow CH_3OH(l)$

(b) $3NO_2(g) + H_2O(l) \longrightarrow 2HNO_3(aq) + NO(g)$

(c) $NH_3(g) + H_2O(l) \longrightarrow NH_4^+ + OH^-$

(d) $2C(\text{graphite}) + O_2(g) \longrightarrow 2CO(g)$

Answers (a) i, (b) i, (c) i, (d) ii.

The standard state entropy changes for chemical reactions can be calculated from known standard state entropies in two ways. In the first (Example 23.5) tabulated values for the *absolute* entropies of the reactants and products are used. The entropy change in the reaction is equal to the total entropy of the products (the final entropy) minus the total entropy of the reactants (the initial entropy).

$$\Delta S° = [\text{sum of } S°(\text{products})] - [\text{sum of } S°(\text{reactants})] \quad \textbf{(23.3)}$$

The second method is the same as that we have used for finding heats of reaction (Section 7.9). Like enthalpy changes, entropy changes depend only on the initial and final states, and not on the sequence of changes that have taken place. Known standard entropies of formation and/or known entropy changes for other chemical reactions, and their respective equations, can be combined algebraically as is done in Hess's law calculations of enthalpy changes.

EXAMPLE 23.5 Entropy in Chemical Reactions

Using values of absolute standard state entropies at 25 °C given in Table 23.2, find the $\Delta S°$ for each of the following reactions:

(a) $2O_3(g) \longrightarrow 3O_2(g)$ (b) $O_3(g) \longrightarrow O_2(g) + O(g)$

(c) Which reaction has the greater $\Delta S°$? Suggest why this is so.

To find the values of $\Delta S°$ from absolute standard state entropies, the sums of the entropies of reactants are subtracted from the sums of the entropies of the products.

(a) $\Delta S° = [(3\ \text{mol})S°(O_2)] - [(2\ \text{mol})S°(O_3)]$
$= [(3\ \text{mol})(205.14\ \text{J/K mol})] - [(2\ \text{mol})(238.93\ \text{J/K mol})]$
$= \mathbf{137.56\ J/K}$

(b) $\Delta S° = [(1\ \text{mol})S°(O_2) + (1\ \text{mol})S°(O)] - [(1\ \text{mol})S°(O_3)]$
$= [(1\ \text{mol})(205.14\ \text{J/K mol}) + (1\ \text{mol})(161.06\ \text{J/K mol})]$
$-[(1\ \text{mol})(238.93\ \text{J/K mol})]$
$= \mathbf{127.27\ J/K}$

(c) **Reaction (a) has the greater change in entropy.** Entropy increases in both reactions because in both there are increases in the number of particles present in the system. However, **the number and complexity of the particles produced in reaction (a) (and therefore the possibility for disorder) is greater.** In reaction (a) all products are molecules; in reaction (b) one-half the particles are atoms, which have no possibility for molecular vibrations or rotations.

EXERCISE Combine the equations and values of $\Delta S°$ from the preceding example to calculate $\Delta S°$ for the following reaction by means of a Hess's law type of calculation.

$$O_3(g) + O(g) \longrightarrow 2O_2(g)$$

Answer 10.29 J/K.

SUMMARY OF SECTIONS 23.1–23.5

Entropy changes and enthalpy changes are the two driving forces that influence the spontaneity of chemical reactions. Entropy is a quantitative measure of the degree of randomness, or disorder, in a system. A change in entropy of a system is a function both of the heat that flows into or out of a system and the temperature of the system. Unlike energy or enthalpy, for which only changes can be measured, the absolute standard state entropy ($S°$) of pure substances can be determined. Absolute entropies for substances at any given temperature above absolute zero include contributions from increases in entropy due to the increase in temperature and due to any changes of state that would occur between absolute zero and the given temperature.

The second law of thermodynamics states that the entropy of the universe must increase in any spontaneous process. The entropy of a specific *system* can either increase or decrease. Entropy increases as the number of possible ways for the atoms, molecules, or ions of a chemical system to be arranged increases. Both physical and chemical changes can cause increases or decreases in entropy. Entropy increases with increasing temperature, with the evaporation of a liquid, the melting of a solid, the expansion of a gas, or the increase of the number of particles during a chemical reaction. The reverse of these changes cause decreases in entropy. Some of the processes that cause entropy changes are listed in Table 23.3.

Table 23.3
Processes That Cause Entropy to Change

Entropy increases	Entropy decreases
Increasing temperature	Decreasing temperature
Melting a solid	Freezing a liquid
Evaporating a liquid	Condensing a gas
Expanding a gas	Confining a gas in a smaller volume
Increasing the number of moles of gas during a reaction	Decreasing the number of moles of gas during a reaction
Increasing the total moles of atoms, ions, or molecules during a reaction	Decreasing the total moles of atoms, ions, or molecules during a reaction
Dissolving a solid or a gas in a liquid	Precipitation of a product in solution
Mixing two substances in the same phase	

FREE ENERGY

23.6 FREE ENERGY CHANGE: THE CRITERION FOR SPONTANEITY

The natural direction of many physical and chemical changes, as we have seen, is toward minimum energy or toward maximum disorder. A **spontaneous chemical change** takes place without any continuing outside influence. When both the enthalpy change and the entropy change of a process are favorable, the process clearly will be spontaneous. And when changes in both enthalpy and entropy are unfavorable, the process clearly will be nonspontaneous. However, when these driving forces oppose each other, it is necessary to determine which predominates.

The combination of these two criteria for spontaneity into a single function was accomplished in 1876 by J. Willard Gibbs, a professor of mathematical physics at Yale University. Gibbs published a lengthy, highly abstract paper in the *Transactions of the Connecticut Academy of Sciences.* The paper went unnoticed for years, but today Gibbs is recognized as having been one of America's most brilliant scientists. In his honor, the important thermodynamic function that we are about to discuss is called the Gibbs free energy and given the symbol G.

For the type of system of greatest interest in chemistry, a system at constant temperature and pressure, the following relationship is valid

$$\Delta G = \Delta H - \underset{\text{absolute temperature}}{T}\,\Delta S \qquad (23.4)$$

The **free energy change,** ΔG, is the energy that is available, or *free,* to do useful work as the result of a chemical or physical change. Not all the energy transferred in a process can be harnessed for useful work because some is always lost to the entropy change. Here we are more interested in studying chemical reactions than in harnessing energy for useful work such as driving steam engines, although calculations of the changes in free energy can be used for that purpose.

Free energy has units of energy and is usually given in kilojoules or kilocalories, for example

$$\underset{\textit{kilojoules}}{\Delta G} = \underset{\textit{kilojoules}}{\Delta H} - \underset{\textit{kelvins}}{T}\,\underset{\frac{\textit{kilojoules}}{\textit{kelvin}}}{\Delta S} \qquad (23.5)$$

When dealing with changes on a molar basis (e.g., as for free energies of formation), the units are as follows:

$$\underset{\frac{\textit{kilojoules}}{\textit{mole}}}{\Delta G} = \underset{\frac{\textit{kilojoules}}{\textit{mole}}}{\Delta H} - \underset{\textit{kelvins}}{T}\,\underset{\frac{\textit{kilojoules}}{\textit{kelvin mole}}}{\Delta S} \qquad (23.6)$$

We noted earlier that reactions for which ΔH is negative are *often* spontaneous. From Equation (23.4) we can see why this is so. When $T\,\Delta S$ is quite small, the change in free energy is almost equal to the heat of a chemical reaction. In a spontaneous change, a system moves toward a state of minimum energy. Energy is released and the change in free energy, ΔG, has a negative value. Similarly, reactions for which ΔH is positive are *often* nonspontaneous. Here again, if $T\,\Delta S$ is small, the spontaneity of the reaction depends on ΔH and, with the reactants already at a state of low energy, no further change is likely. For nonspontaneous reactions, the value of ΔG is positive.

As in the case for enthalpy and energy, only *changes* in free energy can be measured. Furthermore, as for these other quantities, the change in free energy depends only on the initial and final states of the system, and is therefore independent of the number of steps in a chemical or physical change.

It is important to remember that free energy relationships are not in any way indicators of *kinetic* stability or instability—they give no indication of the *rate* of a chemical reaction. Do not make the common error of assuming that a thermodynamically spontaneous reaction occurs quickly.

Table 23.4
Free Energy and Spontaneity

$\Delta G < 0$	Spontaneous
$\Delta G = 0$	Equilibrium; no change
$\Delta G > 0$	Nonspontaneous; reverse change is spontaneous

To demonstrate how Equation (23.5) predicts spontaneity, let's return to the ice cubes of Figure 23.1. At all three temperatures the ΔH° values are negative (Section 23.1), favoring the process under consideration, the freezing of water

$$H_2O(l, 1\ \text{atm}) \longrightarrow H_2O(s, 1\ \text{atm})$$

but the ΔS° values are negative (Section 23.3), making the process unfavorable from the viewpoint of entropy changes. Combining the ΔH° and ΔS° values in the calculation of ΔG° shows which drive to spontaneity dominates at each temperature.

$$\Delta G^\circ \quad = \quad \Delta H^\circ \quad - \quad T \quad \Delta S^\circ$$

$$\Delta G^\circ_{272.15} = (-5972.2\ \text{J}) - (272.15\ \text{K})(-21.8640\ \text{J/K}) = -21.9\ \text{J} \quad \textbf{(23.7)}$$

$$\Delta G^\circ_{273.15} = (-6009.5\ \text{J}) - (273.15\ \text{K})(-22.0007\ \text{J/K}) = 0.0\ \text{J} \quad \textbf{(23.8)}$$

$$\Delta G^\circ_{274.15} = (-6046.7\ \text{J}) - (274.15\ \text{K})(-22.1369\ \text{J/K}) = 22.1\ \text{J} \quad \textbf{(23.9)}$$

At -1 °C, ΔH° is the most significant driving force, ΔG° has a negative value, and the process is spontaneous—water freezes. At 0 °C, the effects of enthalpy and entropy are in balance and $\Delta G^\circ = 0$. We know that at the freezing point ice and water are in equilibrium and no spontaneous change occurs. The value of the free energy change for a system at equilibrium is zero. The system is at the point of minimum free energy.

At $+1$ °C, ΔS° is the more important driving force, ΔG° has a positive value, and the process is nonspontaneous. Like ΔH°, the sign of ΔG° is reversed for a change in the reverse direction. A positive value of ΔG° indicates that under the given conditions a process is *spontaneous in the reverse direction.* Ice does indeed melt above 0 °C.

The three possible relationships between the free energy change and spontaneity illustrated by these calculations hold for *all* chemical and physical changes (Table 23.4). The free energy change is the unequivocal criterion for the spontaneity of a given chemical reaction or a physical change at constant temperature and pressure.

During all changes of state, two phases are in equilibrium, the temperature remains constant, and $\Delta G^\circ = 0$. Introducing $\Delta G^\circ = 0$ into Equation (23.4) and solving for ΔS° shows the origin of Equation (23.2), which we used earlier to calculate ΔS° for a change of state.

$$\Delta G^\circ = \Delta H^\circ - T\Delta S^\circ$$

$$0 = \Delta H^\circ - T\Delta S^\circ$$

$$\Delta S^\circ = \frac{\Delta H^\circ}{T}$$

23.7 ΔG° FOR CHEMICAL REACTIONS

Free energy changes for processes involving substances in their standard states at a specified temperature are designated as standard free energy changes, ΔG°, giving

$$\Delta G^\circ = \Delta H^\circ - T\Delta S^\circ \quad \textbf{(23.10)}$$

Like ΔH° and ΔS°, for a given chemical reaction or physical change ΔG° has a constant value at a given temperature and pressure.

Starting with reactants in their standard states at constant temperature and pressure, ΔG° represents the free energy change between the beginning of the reaction and the complete conversion of reactants to products in their standard states.

Free energy data are usually tabulated for the standard states of substances at 1 bar (slightly less than 1 atm) and a specified temperature. Where no temperature is given, the temperature is assumed to be 25 °C. Values for the standard free energies of formation, ΔG°_f, are included in Table 23.2. As for ΔH°_f, the reference point for the data is chosen by designating $\Delta G^\circ_f = 0$ for elements in their standard states.

The magnitude of the standard free energy of formation of a given species is an indicator of its *thermodynamic stability*. Species with large negative ΔG°_f values can be expected not to decompose into their elements unless they encounter vigorous reaction conditions. Look at the iron oxides in Table 23.2; it is not easy to win free iron from its oxide ores. On the other hand, a positive ΔG°_f is a good indicator that a compound will readily return to the free elements. Ozone, and hydrogen and oxygen atoms in the gas phase (the species in Table 23.2 that have positive free energies of formation) are highly reactive.

Depending upon the type of data available, there are several ways to calculate the value of ΔG° for a chemical reaction, for example: (1) If ΔH° and ΔS° are known, ΔG° can be found from Equation (23.10), which defines the free energy change (Example 23.6). The necessary values of ΔH° and ΔS° can, of course, be calculated from ΔH_f° and S° data, if they are available. (2) If ΔG° values are known for an appropriate series of reactions, the equations and ΔG° values can be combined by algebraic addition and subtraction according to Hess's law (Section 7.10c) to give ΔG° (See Example 23.8). (3) If standard free energies of formation of the reactants and products are available, ΔG° can be found (Example 23.7) by using the following relationship, similar to that for ΔH°.

$$\Delta G^\circ = (\text{sum of } \Delta G_f^\circ \text{ of products}) - (\text{sum of } \Delta G_f^\circ \text{ of reactants}) \qquad \textbf{(23.11)}$$

It is customary to tabulate ΔH_f° and ΔG_f° in units of *kilo*joules per mole (or kilocalories per mole). The S° values are customarily given in *joules* (or calories) per kelvin per mole. *This means* that when ΔH_f° and ΔG_f° data are used together with S°, it is usually necessary to convert kilojoules to joules or vice versa. Do not make the common error of assuming that ΔH° and S° data have the same energy units.

EXAMPLE 23.6 ΔG° of Reaction

Determine whether the reaction

$$CCl_4(l) + H_2(g) \longrightarrow HCl(g) + CHCl_3(l)$$

is spontaneous at 25 °C under standard state conditions. At 25 °C, $\Delta H^\circ = -91.34$ kJ and $\Delta S^\circ = 41.5$ J/K for this reaction.

The value of ΔG° is

$$\begin{aligned}\Delta G^\circ &= \Delta H^\circ - T\Delta S^\circ \\ &= (-91.34\ \text{kJ}) - (298.15\ \text{K})\left(41.5\frac{\text{J}}{\text{K}}\right)\left(\frac{1\ \text{kJ}}{1000\ \text{J}}\right) \\ &= \mathbf{-103.7\ kJ}\end{aligned}$$

The negative value of ΔG° indicates that the reaction will be **spontaneous** at this temperature and under standard state conditions. [Note the necessary conversion factor in the $T\Delta S^\circ$ term.]

EXERCISE For the reaction

$$O_3(g) + O(g) \longrightarrow 2O_2(g)$$

$\Delta H^\circ = -391.9$ kJ and $\Delta S^\circ = 10.29$ J/K at 25 °C. Calculate ΔG° at this temperature and state whether the reaction is spontaneous or not.

Answer $\Delta G^\circ = -395.0$ kJ; the reaction is spontaneous.

EXAMPLE 23.7 ΔG° of Reaction

The standard state free energies of formation at 25 °C are −78.90 kJ/mol for $Fe^{2+}(aq,\ 1\ M)$, −4.7 kJ/mol for $Fe^{3+}(aq,\ 1\ M)$, and 0 kJ/mol for Fe(*s*). Calculate ΔG° for the disproportionation reaction

$$3Fe^{2+}(aq, 1\ M) \longrightarrow 2Fe^{3+}(aq, 1\ M) + Fe(s)$$

Is the reaction spontaneous under these conditions?

The ΔG° for this reaction can be found by subtracting the free energy of formation of the reactant from the sum of the free energies of formation of the products:

$$\begin{aligned}\Delta G^\circ &= [(2 \text{ mol}) \times \Delta G^\circ_f(Fe^{3+}) + (1 \text{ mol}) \times \Delta G^\circ_f(Fe)] - [(3 \text{ mol}) \times \Delta G^\circ_f(Fe^{2+})] \\ &= [(2 \text{ mol})(-4.7 \text{ kJ/mol}) + (1 \text{ mol})(0 \text{ kJ/mol})] - [(3 \text{ mol})(-78.90 \text{ kJ/mol})] \\ &= \mathbf{227.3\ kJ}\end{aligned}$$

With $\Delta G^\circ = 227.3$ kJ, under standard state conditions at 25 °C, this disproportionation reaction is **not spontaneous.**

EXERCISE The standard state free energies of formation at 1000 K are -139.13 kJ/mol for $BrF_3(g)$, -78.58 kJ/mol for $BrF(g)$, and -144.85 kJ/mol for $BrF_5(g)$. Calculate ΔG° for the disproportionation reaction

$$2BrF_3(g) \longrightarrow BrF(g) + BrF_5(g)$$

Is the reaction spontaneous under these conditions?

Answers $\Delta G^\circ = 54.83$ kJ; the reaction is not spontaneous.

23.8 FREE ENERGY AND EQUILIBRIUM

In terms of thermodynamics, just what might take place when two substances are combined at constant temperature and pressure? No matter what else happens, there usually will be an increase in entropy due to an increase in disorder as the result of mixing. If the two substances can react with each other, the chemical reaction will begin, heat will be released or absorbed, and the concentrations of the substances in the mixture will change. A further change in entropy dependent upon changes in the nature of the reactants and products also will begin to take place. The evolution or absorption of heat, the changes in entropy, and the changes in concentrations will continue until equilibrium is reached. Equilibrium might occur with large amounts of reactants remaining, with virtually all reactants converted to products, or at any intermediate combination of concentrations.

From the equation defining the free energy change, $\Delta G = \Delta H - T\Delta S$, and the statements above, it is apparent that during a chemical or physical change at constant temperature the free energy of the system will continue to change until equilibrium is established. As you might guess, there is a clearly definable relationship between free energy and equilibrium. By a mathematical approach (that we need not explore), it can be shown that the free energy change for any reaction is given by

standard state free energy change

free energy change *reaction quotient*

$$\Delta G = \Delta G^\circ + RT \ln Q \qquad \textbf{(23.12)}$$

ideal gas constant

or expressed with the logarithm to the base 10

$$\Delta G = \Delta G^\circ + 2.303\ RT \log Q \qquad \textbf{(23.13)}$$

where the reaction quotient, as before (Section 19.7), represents the nonequilibrium concentrations (or partial pressures) of products and reactants. As a reaction proceeds, the free energy of the mixture and the concentrations change until at equilibrium $\Delta G = 0$ and the concentrations satisfy the equilibrium constant. At this point, Q has become equal to K. Inserting these values into Equation (23.13) gives the important and useful equation that relates thermodynamic quantities and the equilibrium constant:

$$0 = \Delta G^\circ + 2.303\ RT \log K$$

$$\Delta G^\circ = -2.303\ RT \log K \qquad \textbf{(23.14)}$$

Figure 23.7 illustrates the relationships between free energy and equilibrium. The left-hand ends of the curves represent the total free energy of the reactants and the

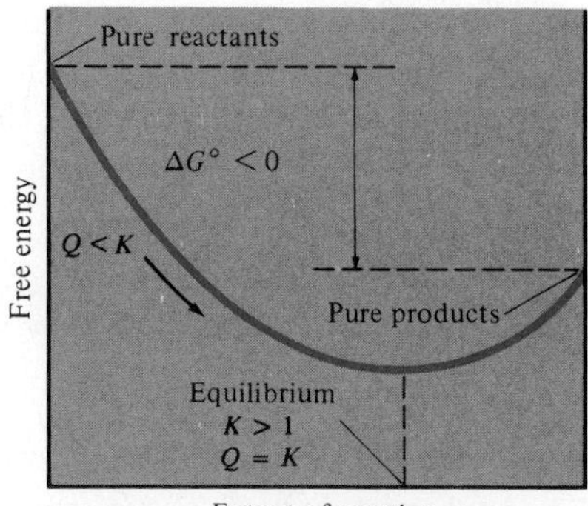

(a) Spontaneous in forward direction: $\Delta G^\circ < 0, K > 1$

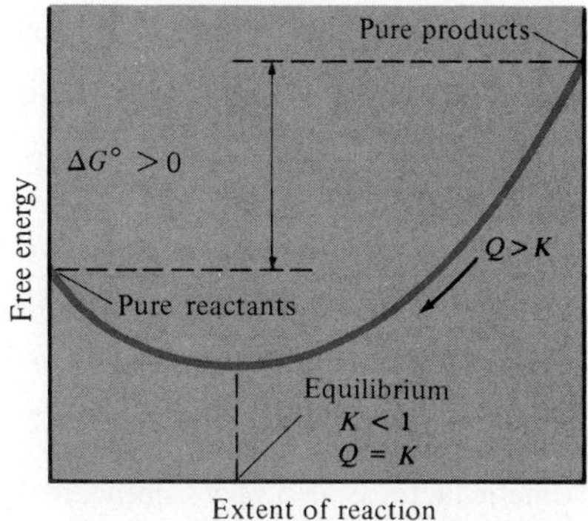

(b) Nonspontaneous in forward direction: $\Delta G^\circ > 0, K < 1$

Figure 23.7
Free Energy and Equilibrium in Chemical Reactions
The reactions progress from left to right. The equilibrium points correspond to the concentrations of reactants and products at which $\Delta G = 0$.

right-hand ends represent the total free energy of the products at standard state conditions. The difference between them is ΔG°, which, like K, depends only on temperature and is a constant for any given reaction.

From Equation (23.14) you can see that when ΔG° is negative, K is greater than 1 (because $\log K$ must be positive), showing that products are favored over reactants. This case is illustrated in Figure 23.7a. When ΔG° is positive, K is less than 1 (because $\log K$ must be negative), showing that reactants are favored over products at equilibrium, as illustrated in Figure 23.7b. In the relatively rare case of a chemical reaction for which $\Delta G^\circ = 0$, then $K = 1$ and in the equilibrium constant expression, the numerator and the denominator must be equal (i.e., $[R]^r[S]^s \cdots = [A]^a[B]^b \cdots$). These relationships are summarized in Table 23.5.

The direction of approach to equilibrium and the actual free energy change (ΔG) are not constants, but vary with the conditions and the initial concentrations. If at the beginning of the reaction, $Q < K$, equilibrium is approached from left to right on the curves in Figure 23.7. If $Q > K$, equilibrium is approached from right to left, and the reaction is spontaneous in the reverse direction.

The magnitude of ΔG°, as shown in Equation (23.14), is an indication of the *extent* to which a chemical reaction proceeds under standard state conditions—how far the reaction goes toward the formation of products before equilibrium is reached. The greater the magnitude of a negative ΔG° value, the larger the value of K and the

Table 23.5
Relationship between Standard Free Energy Change and Equilibrium Constant

ΔG°	K	Product formation
$\Delta G^\circ < 0$	$K > 1$	Products are favored over reactants at equilibrium
$\Delta G^\circ = 0$	$K = 1$	Equilibrium at $[R]^r[S]^s \cdots = [A]^a[B]^b \cdots$
$\Delta G^\circ > 0$	$K < 1$	Reactants are favored over products at equilibrium

more favorable the formation of products. The reactions that we think of as going to "completion" generally have large negative values of ΔG°. The greater the magnitude of a positive ΔG°, the smaller the value of K and the smaller the amounts of products at equilibrium.

EXAMPLE 23.8 Free Energy and Equilibrium

Use the following thermochemical equations

$$\text{S(rhombic)} + O_2(g) \longrightarrow SO_2(g) \qquad \Delta G^\circ = -300.19\text{ kJ}$$
$$\text{S(monoclinic)} + O_2(g) \longrightarrow SO_2(g) \qquad \Delta G^\circ = -300.29\text{ kJ}$$

to find ΔG° for the change of state

$$\text{S(monoclinic)} \longrightarrow \text{S(rhombic)}$$

Interpret ΔG° in terms of relative amounts of products and reactants.

The free energy change for the sulfur change of state can be found by adding the second equation to the reverse of the first equation

$$\text{S(monoclinic)} + O_2(g) \longrightarrow SO_2(g) \qquad \Delta G^\circ = -300.29\text{ kJ}$$
$$SO_2(g) \longrightarrow \text{S(rhombic)} + O_2(g) \qquad \Delta G^\circ = 300.19\text{ kJ}$$
$$\text{S(monoclinic)} \longrightarrow \text{S(rhombic)} \qquad \boldsymbol{\Delta G^\circ = -0.10\text{ kJ}}$$

This phase change is spontaneous at 25 °C under standard state conditions with $\Delta G^\circ = -0.10$ kJ/mol, but the small value of ΔG° indicates that the extent of the reaction is not great. **Monoclinic sulfur and rhombic sulfur will both be present in an equilibrium mixture in appreciable quantities.** [The kinetic factor is also very significant in this solid-state crystal structure change.]

EXERCISE Use the following thermochemical equations

$$2FeO(s) + \tfrac{1}{2}O_2(g) \longrightarrow Fe_2O_3(s) \qquad \Delta G^\circ = -252.7\text{ kJ}$$
$$3FeO(s) + \tfrac{1}{2}O_2(g) \longrightarrow Fe_3O_4(s) \qquad \Delta G^\circ = -281.1\text{ kJ}$$

to find ΔG° for the reaction

$$3Fe_2O_3(s) \longrightarrow 2Fe_3O_4(s) + \tfrac{1}{2}O_2(g)$$

Interpret ΔG° in terms of relative amounts of products and reactants, and the stability of Fe_2O_3 to decomposition at 25 °C.

Answers $\Delta G^\circ = 195.9$ kJ; reactant favored over products; quite stable.

EXAMPLE 23.9 Free Energy and Equilibrium

For the following reaction, $\Delta G^\circ = -103.7$ kJ at 25 °C

$$CCl_4(l) + H_2(g) \rightleftharpoons HCl(g) + CHCl_3(l)$$

(a) Calculate the equilibrium constant for this reaction. (b) Calculate ΔG for this reaction when $P_{H_2} = 10$ bar and $P_{HCl} = 0.1$ bar. Will the reaction be more or less favorable under these conditions than under standard state conditions?

(a) Equation (23.14) gives the relationship between the value of ΔG° and the equilibrium constant. Solving this equation for log K and substituting values gives

$$\log K = \frac{-\Delta G^\circ}{(2.303)\,RT} = \frac{-(-103.7\text{ kJ})(1000\text{ J/kJ})}{(2.303)(8.314\text{ J/K mol})(298\text{ K})}$$
$$= 18.18$$

Taking antilogarithms gives $\mathbf{K = 1.5 \times 10^{18}}$ Under standard state conditions at 25 °C this reaction is spontaneous and products are greatly favored over reactants at equilibrium. [For a reaction in which any reactants and products are gases, the equilibrium constant value calculated from thermodynamic data, the *thermodynamic equilibrium constant*, is the value of K found with partial pressures for the gases. If all reactants and products are gases, the value found is that of K_p.]

(b) The reaction quotient is

$$Q = \frac{P_{HCl}}{P_{H_2}} = \frac{0.1\ \text{bar}}{10\ \text{bar}} = 0.01$$

Using Equation (23.13), which relates ΔG, ΔG°, and Q, gives

$$\begin{aligned}\Delta G &= \Delta G^\circ + 2.303\ RT \log Q \\ &= (-103.7\ \text{kJ}) + (2.303)(8.314\ \text{J/K mol})(298\ \text{K})\left(\frac{1\ \text{kJ}}{1000\ \text{J}}\right)\log(0.01) \\ &= \mathbf{-115\ kJ}\end{aligned}$$

The value of ΔG has become slightly more negative, meaning that the reaction is **slightly more favorable** under these conditions. This is reasonable, for the hydrogen pressure is greater than it would be under standard state conditions and the hydrogen chloride pressure is less. Le Chatelier's principle predicts that the reaction will be more favorable in the forward direction.

EXERCISE Calculate ΔG at 1000. K for the reaction

$$2BrF_3(g) \rightleftharpoons BrF(g) + BrF_5(g)$$

given that $\Delta G^\circ_{1000} = 54.83$ kJ and $P_{BrF} = P_{BrF_5} = 0.0010$ bar and $P_{BrF_3} = 5.2$ bar. Will this reaction be more or less favorable under these nonstandard state conditions compared to standard state conditions?

Answer −87 kJ; the reaction is more favorable and under these conditions is spontaneous in the forward direction.

23.9 VARIATION OF REACTION SPONTANEITY WITH TEMPERATURE

a. Interpreting Thermodynamic Data The value of ΔG° varies with the temperature, as shown by the equation $\Delta G^\circ = \Delta H^\circ - T\Delta S^\circ$. How temperature changes affect the direction and extent of a reaction depends upon the signs of ΔH° and ΔS°. [Recall that the standard state is defined for any *specified* temperature. Because we commonly tabulate data for 25 °C, there is a tendency to forget that the standard state data can be calculated and reported for whatever temperature we choose to specify.]

When ΔH° and ΔS° have opposite signs, the outcome of a chemical reaction is predictable—it is either always spontaneous at all temperatures (negative ΔH°, positive ΔS°) or always nonspontaneous at all temperatures (positive ΔH°, negative ΔS°). Reactions (a) and (f) in Table 23.6 are reactions of these types. Consider the free energy variation with temperature for reaction (a):

$$\begin{aligned}\Delta G^\circ &= \Delta H^\circ - T\Delta S^\circ \\ \Delta G^\circ &= (-72.80\ \text{kJ}) - (T)(0.11448\ \text{kJ/K})\end{aligned} \qquad \textbf{(23.15)}$$

Table 23.6
Relations among ΔH°, ΔS°, and ΔG° of Reaction at 298 K, and Reaction Spontaneity

Reaction	ΔH°(kJ)	ΔS°(J/K)	ΔG°(kJ)
(a) $H_2(g) + Br_2(l) \longrightarrow 2HBr(g)$	−72.80 favorable	+114.48 favorable	−106.90 spontaneous
(b) $2H_2(g) + O_2(g) \longrightarrow 2H_2O(g)$	−483.36 favorable	−88.86 unfavorable	−457.14 spontaneous
(c) $Br_2(l) + Cl_2(g) \longrightarrow 2BrCl(g)$	+29.28 unfavorable	+104.90 favorable	−1.96 spontaneous
(d) $N_2(g) + 2F_2(g) \longrightarrow N_2F_4(g)$	−7.1 favorable	−295.98 unfavorable	+81.2 not spontaneous
(e) $H_2(g) \longrightarrow 2H(g)$	+435.93 unfavorable	+98.74 favorable	+406.49 not spontaneous
(f) $2HCl(g) \longrightarrow H_2(g) + Cl_2(g)$	+184.61 unfavorable	−20.07 unfavorable	+190.60 not spontaneous

In general, the value of ΔH° does not vary greatly with moderate temperature changes and ΔH° can be taken to be constant for most reactions. The entropy change, ΔS°, is also reasonably constant with changes in temperature. However, the $T\Delta S^\circ$ term is directly dependent upon temperature. From Equation (23.15) you can see that for a reaction of this type, no matter how the temperature changes, ΔG° will remain negative, and so the reaction will always be spontaneous. However, as illustrated below, the *extent* of such a reaction will increase with rising temperature.

For reaction (f), no matter what the temperature, ΔG° will always be positive

$$\Delta G^\circ = (+184.61 \text{ kJ}) - (T)(-0.02007 \text{ kJ/K}) \quad \textbf{(23.16)}$$

and the reaction will not be spontaneous.

When ΔH° and ΔS° are of the *same* sign, spontaneity depends upon the relative magnitudes of the ΔH° and $T\Delta S^\circ$ terms in the free energy equation. Under these circumstances, the value of ΔG° can change from positive to negative, or vice versa, with changes in temperature. As an example, look at reaction (e) in Table 22.6, for which both ΔH° and ΔS° are positive. A strong bond, the H—H bond, is broken, and as expected this is an endothermic process. However, the entropy change is favorable because two moles of gas are created from one mole of gas. At 25 °C the large value of ΔH° outweighs the $T\Delta S^\circ$ term, ΔG° is positive, and the reaction is not spontaneous. You can see, though, that increasing the value of T to the point where $T\Delta S^\circ > \Delta H^\circ$

$$\Delta G^\circ = (+435.93 \text{ kJ}) - (T)(0.09874 \text{ kJ/K})$$

will lead to a reversal in the sign of ΔG°. (This is reasonable: At a high enough temperature, we know that the bond will break.)

A similar argument would show that reaction (d), for which both ΔH° and ΔS° are negative, becomes spontaneous in the direction written at a *lower* temperature (unless phase changes counteract the free energy change). The effects of the relative signs of ΔH° and ΔS° with respect to changes in temperature are summarized in Table 23.7.

Table 23.7
Variation of ΔG° with Temperature

ΔH°	ΔS°	ΔG°	Examples in Table 23.6
−	+	− at all T	Reaction (a)
+	−	+ at all T	Reaction (f)
−	−	+ or −, depends on T Becomes more favorable at lower temperatures	Reactions (b) and (d)
+	+	+ or −, depends on T Becomes more favorable at higher temperatures	Reactions (c) and (e)

EXAMPLE 23.10 Reaction Spontaneity and Temperature

Discuss the meaning of the thermodynamic data for reaction (b) in Table 23.6.

Reaction (b) is the formation of water vapor by combination of the elements. We know that water is a thermodynamically stable compound. The large increase in the stability of the system when water is formed from its elements is shown by the large negative value of ΔH°. The drive to minimum energy is opposed in this reaction by a decrease in entropy due to the formation of two moles of gas from three moles of gas. However, the $T\Delta S$ term in $\Delta G^\circ = \Delta H^\circ - T\Delta S^\circ$ is small relative to the enthalpy change, and the reaction is spontaneous at 298 K. With negative ΔH° and negative ΔS°, this is the type of reaction that will remain spontaneous and become more favorable at lower temperatures (see Table 23.7).

EXERCISE For the reaction

$$2O_2(g) \longrightarrow O_3(g) + O(g)$$

$\Delta H^\circ = 391.9$ kJ and $\Delta S^\circ = -10.29$ J/K at 25 °C. Briefly discuss the temperature dependence of ΔG° for this reaction.

Answer ΔH° and ΔS° are both unfavorable; therefore ΔG° will remain positive and the reaction will remain nonspontaneous with changes in temperature.

b. Calculation of ΔG° and K at Given Temperatures; Other Thermodynamic Calculations The values of both ΔG° and K vary significantly with temperature. The values of ΔH° and ΔS°, however, can be assumed to be constant over reasonable temperature ranges. The use of this assumption makes possible the calculation of ΔG° and K at various temperatures from known values at other temperatures. There are a variety of approaches to solving such problems, all based on the equation that defines ΔG° ($\Delta G^\circ = \Delta H^\circ - T\Delta S^\circ$), the equation that relates ΔG° to K ($\Delta G^\circ = -2.303\ RT \log K$), or the combination of these equations:

$$\Delta H^\circ - T\Delta S^\circ = -2.303\ RT \log K \tag{23.17}$$

In fact, these equations provide many alternatives in the calculation of unknown thermodynamic data from known thermodynamic data. Consider the following equation, which is based upon Equation (23.17) solved for K, expressed for two different temperatures, and then combined. This equation can be used to find K at one temperature from K at another temperature, if ΔH° is known.

$$\log\left(\frac{K_2}{K_1}\right) = -\frac{\Delta H^\circ}{2.303\ R}\left(\frac{1}{T_2} - \frac{1}{T_1}\right) \tag{23.18}$$

Or if K is known for two different temperatures, this equation can be used to find ΔH°. A few calculations of thermodynamic quantities utilizing these equations are illustrated in Example 23.11.

EXAMPLE 23.11 Reaction Spontaneity and Temperature

Beryllium sulfate undergoes thermal decomposition at 400. K as follows

$$BeSO_4(s) \rightleftharpoons BeO(s) + SO_3(g) \qquad K_p = 3.87 \times 10^{-16}$$
$$\Delta H^\circ = 175\ \text{kJ}$$

(a) From the information given above, decide which of the reaction descriptions in Table 23.7 fits this reaction. (b) Calculate ΔG° at 400. K for this reaction. (c) At 600. K will this reaction be more or less favorable than at 400. K? Calculate K_p at this temperature. (d) Calculate the temperature above which this reaction will become spontaneous (under standard state conditions).

(a) The small value for K shows that at 400. K the value of ΔG° will be positive and rather large (see Figure 23.7). The value of ΔH° is positive. Because a gaseous product is formed from a solid, disorder will increase and the entropy change should be positive. From this information and Table 23.7, we can conclude that **this reaction will become more favorable as the temperature increases.**

(b) Knowing K, ΔG° can be found as follows

$$\Delta G^\circ_{400} = -2.303\ RT \log K$$
$$= (-2.303)\left(8.314 \frac{\text{J}}{\text{K mol}}\right)(400.\ \text{K}) \log(3.87 \times 10^{-16})\left(\frac{1\ \text{kJ}}{1000\ \text{J}}\right)$$
$$= \mathbf{118\ kJ/mol}$$

(c) This reaction should be more favorable at the higher temperature, meaning that the value of ΔG° should be smaller and the value of K larger. The value of K_p may be calculated for the higher temperature by using Equation (23.18) and assuming that the value of ΔH° remains unchanged.

$$\log\left(\frac{K_2}{K_1}\right) = \frac{-\Delta H^\circ}{(2.303)\ R}\left(\frac{1}{T_2} - \frac{1}{T_1}\right)$$

$$\log\left(\frac{K}{3.87 \times 10^{-16}}\right) = \frac{-(175\ \text{kJ/mol})(1000\ \text{J/kJ})}{(2.303)(8.314\ \text{J/K mol})}\left(\frac{1}{600.\ \text{K}} - \frac{1}{400.\ \text{K}}\right) = 7.62$$

Taking antilogarithms of both sides gives

$$\frac{K}{3.87 \times 10^{-16}} = 4.2 \times 10^{7}$$

$$\mathbf{K_p = 1.6 \times 10^{-8}}$$

(d) The reaction will become spontaneous in the forward direction, rather than in the reverse direction, above the temperature at which ΔG° becomes equal to zero. The temperature at which this occurs can be found by using Equation (23.10) and assuming that the values of ΔH° and ΔS° are constant. First, it is necessary to calculate ΔS°:

$$\Delta G^\circ = \Delta H^\circ - T\Delta S^\circ$$

$$\Delta S^\circ = \frac{\Delta H^\circ - \Delta G^\circ}{T} = \frac{175\ \text{kJ/mol} - 118\ \text{kJ/mol}}{400.\ \text{K}}$$

$$\Delta S^\circ_{400} = 0.14\ \text{kJ/K mol}$$

Then, the desired temperature can be found as follows:

$$\Delta G^\circ = \Delta H^\circ - T\Delta S^\circ$$

$$0 = \Delta H^\circ - T\Delta S^\circ$$

$$T = \frac{\Delta H^\circ}{\Delta S^\circ} = \frac{175\ \text{kJ/mol}}{0.14\ \text{kJ/mol K}} = \mathbf{1300\ K}$$

[Note that the transition from a "nonspontaneous" to a "spontaneous" reaction does *not* mean that suddenly there is product where there was none. What it does mean is that at this temperature (where $K = 1$), the transition occurs from less product than reactant ($K < 1$) to more product than reactant ($K > 1$) at equilibrium for reactants and products at standard state conditions.]

EXERCISE For the reaction

$$3Fe_2O_3(s) \longrightarrow 2Fe_3O_4(s) + \tfrac{1}{2}O_2(g)$$

$\Delta G^\circ = 195.8$ kJ, $\Delta H^\circ = 235.8$ kJ, and $K = 5.0 \times 10^{-35}$ at 25 °C. Assuming ΔH° to be independent of temperature, calculate (a) ΔG° at 1000. K, (b) K at 500. K, and (c) find the temperature above which the reaction will become spontaneous under standard state conditions. *Answers* (a) $\Delta G^\circ = 102$ kJ, (b) $K = 3 \times 10^{-18}$, (c) 1760 K.

SUMMARY OF SECTIONS 23.6–23.9

The free energy change ($\Delta G = \Delta H - T\Delta S$) of a chemical or physical process is the criterion for spontaneity of the process. For a spontaneous process (which occurs without any continuing outside influence), ΔG has a negative value; for a nonspontaneous process, ΔG has a positive value. For a process at equilibrium, $\Delta G = 0$. Values of ΔG°, the standard state free energy change, can be calculated from known values of ΔS° and ΔH°, from standard free energies of formation (Equation 23.11), or by combining equations algebraically.

The magnitude of ΔG° is an indication of how far a reaction proceeds toward the formation of products under standard state conditions. The possible change in spontaneity of a reaction with increasing temperature can be predicted based upon the signs of ΔH° and ΔS° (Tables 23.6 and 23.7).

Reaction quotients and equilibrium constants are related to free energy changes by two important and useful equations: $\Delta G = \Delta G^\circ + 2.303\,RT\log Q$ and $\Delta G^\circ = -2.303\,RT\log K$. Many types of thermodynamic calculations are made possible by these two equations, separately or combined, together with the equation $\Delta G^\circ = \Delta H^\circ - T\Delta S^\circ$ (e.g., finding any one of ΔH°, ΔG°, ΔS° when the other two are known; finding K at various temperatures; finding T at the point when a nonspontaneous reaction becomes spontaneous).

SIGNIFICANT TERMS (with section references)

absolute entropy (23.4)
entropy (23.1)
free energy change (23.6)
second law of thermodynamics (23.2)
spontaneous chemical change (23.6)
third law of thermodynamics (23.4)

QUESTIONS AND PROBLEMS

Disorder, Spontaneity, and Entropy

23.1 What do we call the quantitative measure of the randomness or disorder in a system? Place the following systems in order of increasing randomness: (a) 1 mol of gas A, (b) 1 mol of solid A, (c) 1 mol of liquid A.

23.2 Why would you expect a decrease in entropy as a gas condenses? Would this change be as large a decrease as when a liquid sample of the same substance crystallizes?

23.3 Briefly explain why heating a gas at constant pressure increases its entropy.

23.4 Would you expect a positive or negative ΔS for the system as the pressure on an ideal gas is increased under constant temperature conditions? Why?

23.5 When solid sodium chloride is cooled from 25 °C to 0 °C, the entropy change is -4.4 J/K mol. Is this an increase or decrease in randomness? Explain this entropy change in terms of what is happening in the solid at the molecular level.

23.6 When a one mole sample of argon gas at 0 °C is compressed to one-half its original volume, the entropy change is -5.76 J/K. Is this an increase or a decrease in randomness? Explain this entropy change in terms of what is happening in the gas at the molecular level.

23.7 Gallium undergoes a solid–solid phase transformation at 275.6 K for which $\Delta H^\circ_{275.6} = 2100$ J/mol. Calculate ΔS° for this change.

23.8 Gray tin changes at 291 K to white tin ($\Delta H^\circ_{291} = 2500$ J/mol), white tin changes to tin-I at 476.0 K ($\Delta H^\circ_{476} = 8$ J/mol), and tin-I melts at 505.1 K ($\Delta H^\circ_{505.1} = 7070$ J/mol). Calculate ΔS° for each phase change. Why are all three values relatively small?

23.9 Compare the entropy change for melting one mole of water at 273 K ($\Delta H^\circ_{273} = 6010$ J/mol for fusion) with that for vaporizing a mole of water at 373 K ($\Delta H^\circ_{373} = 40{,}660$ J/mol for vaporization). Why is the second value so much larger?

23.10* The value of ΔS° for vaporizing a sample of a substance at the normal boiling point is usually about 88 J/K mol, no matter what the substance (as long as there is no hydrogen bonding in the liquid and/or dimers are not formed in the gaseous phase). (a) Confirm this value by calculating ΔS° for the evaporation of the following substances:

(i) $Xe(l) \longrightarrow Xe(g)$ $\quad \Delta H^\circ_{165.1} = 12.64$ kJ/mol
(ii) $SO_2(l) \longrightarrow SO_2(g)$ $\quad \Delta H^\circ_{263.14} = 24.92$ kJ/mol
(iii) $CCl_4(l) \longrightarrow CCl_4(g)$ $\quad \Delta H^\circ_{349.9} = 30.0$ kJ/mol

(b) Would you predict ΔS° for vaporizing a mole of water to be greater or less than 88 J/K mol? Why? Confirm your answer using $\Delta H^\circ_{373.15} = 38{,}372$ J/mol. (c) Predict the boiling point of lead, given that ΔH° of vaporization is 179.9 kJ/mol.

23.11 Why is the value of S° for $Cl_2(g)$ given in Table 23.2 larger than that for $Br_2(l)$?

23.12 Why is the value of S° for $H(g)$ given in Table 23.2 smaller than that for $H_2(g)$?

23.13 Choose the substance having the larger entropy at 25 °C: (a) $O_3(g)$ or $H_2O(l)$, (b) $FeCl_2(s)$ or $FeCl_3(s)$, (c) $H_2O(l)$ or $H_2O(g)$.

23.14 Choose the substance having the larger entropy at 25 °C: (a) $H_2O_2(l)$ or $H_2O(l)$, (b) $HCl(aq)$ or $HCl(g)$, (c) $CH_3OH(g)$ or $CH_3OH(l)$.

23.15 Given the following information, specify the contributions to the absolute entropy of $O_2(g)$ at 298 K: (a) O_2 exists as the γ-solid at 0 K, (b) the γ-solid changes to the β-solid at 24 K, (c) the β-solid changes to the α-solid at 44 K, (d) the α-solid melts at 54 K, (e) the liquid boils at 90.2 K.

23.16 Repeat Question 23.15 for $H_2S(g)$ using the following data: (a) H_2S exists as solid-III, an imperfect crystal, at 0 K; (b) solid-III changes to solid-II at 104 K; (c) solid-II changes to solid-I at 126 K; (d) solid-I melts at 188 K; (e) the liquid boils at 213 K.

23.17* The entropy of one mole of solid N_2 at 0 K is 0.0 J/K. Entropy changes during heating the nitrogen to 298 K are the following: (a) heating the solid from 0 K to 35.61 K, 27.2 J/K mol; (b) changing crystal structures at 35.61 K, 6.4 J/K mol; (c) heating the solid from 35.61 K to 63.14 K, 23.4 J/K; (d) melting at 63.14 K, 11.4 J/K mol; (e) heating the liquid from 63.15 K to 77.32 K, 11.4 J/K mol; (f) evaporating the liquid at 77.32 K, 72.2 J/K mol; (g) heating the gas from 77.32 K to 298 K, 39.2 J/K mol. Calculate S°_{298} for nitrogen.

23.18* The stable form of $Na_2SO_4(s)$ at absolute zero is solid-V. At 298 K, solid-V changes to solid-III with $\Delta H^\circ = 3.00$ kJ/mol. The entropy change for heating solid-V from 0 K to 298 K is 149.58 J/K mol and for cooling solid-III from 298 K to 0 K it is -154.92 J/K mol. What is the value of S°_0 for solid-III?

Disorder, Spontaneity, and Entropy—Chemical Reactions

23.19 Classify the entropy change for each of the following reactions as (i) large and negative, (ii) large and positive, (iii) small:

(a) $2Au(s) + 3Cl_2(g) \longrightarrow 2AuCl_3(s)$
(b) $Cd(s) + Ni_2O_3(s) + 3H_2O(l) \longrightarrow Cd(OH)_2(s) + 2Ni(OH)_2(s)$
(c) $Cu^{2+} + H_2(g) \longrightarrow Cu(s) + 2H^+$
(d) $S(s) + 3F_2(g) \longrightarrow SF_6(g)$

(e) $2NaCl(aq) + 2H_2O(l) \xrightarrow[\text{energy}]{\text{electrical}} Cl_2(g) + H_2(g) + 2NaOH(aq)$

(f) $2HCl(aq) + Mg(s) \longrightarrow MgCl_2(aq) + H_2(g)$

(g) $SO_2(g) + Br_2(g) + 2H_2O(g) \xrightarrow{\Delta} 2HBr(g) + H_2SO_4(aq)$

Explain each answer.

23.20 Repeat Question 23.19 for

(a) $CaCO_3(\text{calcite}) \longrightarrow CaCO_3(\text{aragonite})$

(b) $Ag(s) + \frac{1}{2}Cl_2(g) \xrightarrow{\Delta} AgCl(l)$
(c) $ClO_4^- \longrightarrow Cl^- + 2O_2(g)$
(d) $2NH_4NO_3(s) \longrightarrow 2N_2(g) + 4H_2O(g) + O_2(g)$
(e) $Mg(s) + Br_2(l) \longrightarrow MgBr_2(s)$
(f) $2I^- + Cl_2(g) \longrightarrow I_2(s) + 2Cl^-$
(g) $NaHSO_4(s) + NaCl(s) \longrightarrow HCl(g) + Na_2SO_4(s)$
(h) $Cl_2(g) + H_2O(l) \longrightarrow HCl(aq) + HClO(aq)$

Explain each answer.

23.21 Calculate the entropy changes for the sublimation of iodine and for the formation of gaseous iodine atoms from gaseous iodine molecules

(a) $I_2(s) \longrightarrow I_2(g)$
(b) $I_2(g) \longrightarrow 2I(g)$

given $S° = 180.791$ J/K mol for $I(g)$, 116.135 J/K mol for $I_2(s)$, and 260.69 J/K mol for $I_2(g)$. Why are these entropy changes similar in sign and magnitude?

23.22 The standard state absolute entropy at 500 K in J/K mol is 219.681 for $F_2(g)$, 264.307 for $Br_2(g)$, 246.609 for $BrF(g)$, 329.394 for $BrF_3(g)$, 381.129 for $BrF_5(g)$. Calculate the entropy change for each of the following formation reactions:

(a) $Br_2(g) + F_2(g) \longrightarrow 2BrF(g)$
(b) $Br_2(g) + 3F_2(g) \longrightarrow 2BrF_3(g)$
(c) $Br_2(g) + 5F_2(g) \longrightarrow 2BrF_5(g)$

23.23 Using the data for absolute entropies given in Table 23.2 calculate $\Delta S°$ for the reactions

(a) $2Fe(\alpha\text{-solid}) + \frac{3}{2}O_2(g) \longrightarrow Fe_2O_3(s)$
(b) $3Fe(\alpha\text{-solid}) + 2O_2(g) \longrightarrow Fe_3O_4(s)$

Does either reaction have a favorable entropy change?

23.24 Repeat Problem 23.23 for

(a) $Fe_2O_3(s) + 6HCl(g) \longrightarrow 2FeCl_3(s) + 3H_2O(g)$
(b) $2HBr(g) + Cl_2(g) \longrightarrow 2HCl(g) + Br_2(l)$
(c) $H_2O_2(l) \longrightarrow H_2O(l) + \frac{1}{2}O_2(g)$

23.25 Combine your answers for Problem 23.21 to obtain $\Delta S°$ for the reaction

$$I_2(s) \longrightarrow 2I(g)$$

Is this a favorable entropy change?

23.26 Combine your answers for Problem 23.22 to obtain $\Delta S°$ for the reaction

$$2BrF_3(g) \longrightarrow BrF(g) + BrF_5(g)$$

Is this a favorable entropy change?

23.27 Combine your answers for Problem 23.23 to obtain $\Delta S°$ for the reaction

$$3Fe_2O_3(g) \longrightarrow 2Fe_3O_4(s) + \tfrac{1}{2}O_2(g)$$

Is this a favorable entropy change?

23.28 Using the $\Delta S°$ values at 25 °C given for the following reactions:

$$C(\text{graphite}) + 2Cl_2(g) \longrightarrow CCl_4(l) \qquad \Delta S° = -235.47 \text{ J/K}$$

$$C(\text{graphite}) + \tfrac{3}{2}Cl_2(g) + \tfrac{1}{2}H_2(g) \longrightarrow CHCl_3(l) \qquad \Delta S° = -204.0 \text{ J/K}$$

$$\tfrac{1}{2}H_2(g) + \tfrac{1}{2}Cl_2(g) \longrightarrow HCl(g) \qquad \Delta S° = 10.033 \text{ J/K}$$

find $\Delta S°$ for the reaction

$$CCl_4(l) + H_2(g) \longrightarrow HCl(g) + CHCl_3(l)$$

Free Energy

23.29 Define "free energy" in terms of enthalpy and entropy. Characterize ΔG for (a) a spontaneous process, (b) a nonspontaneous process, (c) a process at equilibrium.

23.30 What is chosen for the standard state free energy of formation of an element?

23.31 Which of the following conditions would predict a process that is (a) always spontaneous, (b) always nonspontaneous, or (c) spontaneous or nonspontaneous depending on the temperature and magnitudes of ΔH and ΔS: (i) $\Delta H > 0$, $\Delta S > 0$; (ii) $\Delta H > 0$, $\Delta S < 0$; (iii) $\Delta H < 0$, $\Delta S > 0$; (iv) $\Delta H < 0$, $\Delta S < 0$?

23.32 Why might we say that at absolute zero an exothermic reaction will always be spontaneous, but at temperatures above absolute zero we have to consider both enthalpy and entropy before we can predict spontaneity?

23.33 For the decomposition of $O_3(g)$ to $O_2(g)$

$$2O_3(g) \longrightarrow 3O_2(g)$$

$\Delta H° = -285.4$ kJ and $\Delta S° = 137.55$ J/K at 25 °C. Calculate $\Delta G°$ for the reaction. Is the reaction spontaneous? Are either or both the driving forces ($\Delta H°$ and $\Delta S°$) of the reaction favorable?

23.34 For the reaction

$$Ag^+ + Cl^- \longrightarrow AgCl(s)$$

$\Delta H° = -65.488$ kJ and $\Delta S° = -33.0$ J/K at 25 °C. Calculate $\Delta G°$ for this reaction. Is the reaction spontaneous? What is the controlling driving force ($\Delta H°$ or $\Delta S°$) for this reaction?

23.35 The enthalpy of formation of gaseous hydrogen bromide is -36.40 kJ/mol and the entropy of formation is 57.238 J/K mol under standard state conditions at 25 °C. Calculate the standard state free energy of formation.

23.36 The enthalpy of formation for $FeCl_2(aq)$ is -423.4 kJ/mol at 25 °C and the entropy of formation is -275.0 J/K mol. Calculate ΔG_f° for $FeCl_2(aq)$.

23.37 The free energy of formation of gaseous water at 25 °C under standard state conditions is −228.572 kJ/mol and the enthalpy of reaction is −241.818 kJ/mol. What is the entropy of formation?

23.38 The standard state free energy of formation of $O_2(g)$ is 0; that of $O(g)$ is 231.731 kJ/mol; that of $O_3(g)$ is 163.2 kJ/mol at 25 °C. Which allotrope of oxygen is the most stable at this temperature?

23.39 At 25 °C the standard state free energy of formation of $H_2O(g)$ is −228.572 kJ/mol and that of $H_2O(l)$ is −237.129 kJ/mol. Which form of water is more stable at this temperature?

23.40 The standard state free energy of formation at 25 °C of $Fe_2O_3(s)$ is −742.2 kJ/mol and that of $Fe_3O_4(s)$ is −1015.4 kJ/mol. Which oxide of iron would be formed under ideal stoichiometric conditions?

Free Energy—ΔG° for Chemical Reactions

23.41 Calculate $\Delta G°$ at 25 °C for the reaction

$$2NO_2(g) \longrightarrow N_2O_4(g)$$

given $\Delta H° = -57.20$ kJ and $\Delta S° = -175.83$ J/K. Is this reaction spontaneous? What is the driving force for spontaneity?

23.42 Calculate $\Delta S°$ at 25 °C for the reaction described by the equation

$$CuSO_4(aq) + Fe(s) \longrightarrow FeSO_4(aq) + Cu(s)$$

given $\Delta H° = -153.8$ kJ and $\Delta G° = -144.39$ kJ. Is this a favorable entropy change?

23.43 Use the following thermochemical equations:

$$HF(aq) \longrightarrow H^+ + F^- \qquad \Delta G° = 18.03 \text{ kJ}$$
$$H_2O(l) \longrightarrow H^+ + OH^- \qquad \Delta G° = 79.885 \text{ kJ}$$

to calculate $\Delta G°$ for the neutralization of hydrofluoric acid by a strong base:

$$HF(aq) + OH^- \longrightarrow H_2O(l) + F^-$$

23.44 Use the following thermochemical equations:

$$CH_4(g) + 2O_2(g) \longrightarrow CO_2(g) + 2H_2O(l) \qquad \Delta G° = -817.90 \text{ kJ}$$
$$CH_3OH(l) + \tfrac{3}{2}O_2(g) \longrightarrow CO_2(g) + 2H_2O(l) \qquad \Delta G° = -702.35 \text{ kJ}$$

to calculate $\Delta G°$ for the partial oxidation of methane to form methyl alcohol:

$$CH_4(g) + \tfrac{1}{2}O_2(g) \longrightarrow CH_3OH(l)$$

23.45 Use the following thermochemical equations:

$$H_2(g) + S(s) \longrightarrow H_2S(g) \qquad \Delta G° = -33.56 \text{ kJ}$$
$$H_2S(aq) \longrightarrow H_2S(g) \qquad \Delta G° = -5.73 \text{ kJ}$$
$$H_2S(aq) \longrightarrow HS^- + H^+ \qquad \Delta G° = 39.91 \text{ kJ}$$
$$HS^- \longrightarrow H^+ + S^{2-} \qquad \Delta G° = 73.7 \text{ kJ}$$

to find $\Delta G°$ for the reaction

$$H_2(g) + S(s) \longrightarrow 2H^+ + S^{2-}$$

23.46 Use the following thermochemical equations:

$$Br_2(l) \longrightarrow Br_2(g) \qquad \Delta G° = 3.110 \text{ kJ}$$
$$HBr(g) \longrightarrow H(g) + Br(g) \qquad \Delta G° = 339.09 \text{ kJ}$$
$$Br_2(g) \longrightarrow 2Br(g) \qquad \Delta G° = 164.792 \text{ kJ}$$
$$H_2(g) \longrightarrow 2H(g) \qquad \Delta G° = 406.494 \text{ kJ}$$

to find the $\Delta G°$ of formation of $HBr(g)$ at 25 °C.

23.47 The standard state free energy of formation at 25 °C is 56.9 kJ/mol for $SiH_4(g)$, −856.64 kJ/mol for $SiO_2(s)$, −237.129 kJ/mol for $H_2O(l)$. Calculate $\Delta G°$ for the reaction

$$SiH_4(g) + 2O_2(g) \longrightarrow SiO_2(s) + 2H_2O(l)$$

23.48 The standard state free energy of formation at 500 K is −212.5 kJ/mol for $KClO_4(s)$ and −389.0 kJ/mol for $KCl(s)$. Calculate $\Delta G°$ for the reaction represented by the equation

$$KClO_4(s) \longrightarrow KCl(s) + 2O_2(g)$$

23.49 The standard state free energy of formation is −286.06 kJ/mol for $NaI(s)$, −261.905 kJ/mol for Na^+, −51.57 kJ/mol for I^- at 25 °C. Calculate $\Delta G°$ for the reaction represented by the equation

$$NaI(s) \xrightarrow{H_2O} Na^+ + I^-$$

23.50 The standard state free energy of formation is 209.20 kJ/mol for acetylene, $C_2H_2(g)$; −394.359 kJ/mol for $CO_2(g)$; −237.129 kJ/mol for $H_2O(l)$. Calculate $\Delta G°$ for the combustion of acetylene.

23.51 Calculate $\Delta G°$ for the reduction of the oxides of iron and copper by carbon at 700 K represented by the equations

$$2Fe_2O_3(s) + 3C(s) \longrightarrow 4Fe(s) + 3CO_2(g)$$
$$2CuO(s) + C(s) \longrightarrow 2Cu(s) + CO_2(g)$$

given that the standard free energy of formation is −92 kJ/mol for $CuO(s)$, −637 kJ/mol for $Fe_2O_3(s)$, −395 kJ/mol for $CO_2(g)$. Which oxide can be reduced using carbon in a wood fire (which has a temperature of about 700 K) assuming standard state conditions?

23.52 A precipitate of AgCl will dissolve in an excess of aqueous NH_3, whereas AgI will not. Show that the stability of $[Ag(NH_3)_2]^+$ is large enough to permit AgCl to dissolve but not large enough to permit dissolution of the less soluble AgI by calculating $\Delta G°$ for each of the following reactions:

$$AgCl(s) \rightleftharpoons Ag^+ + Cl^-$$
$$AgI(s) \rightleftharpoons Ag^+ + I^-$$
$$Ag^+ + 2NH_3(aq) \rightleftharpoons [Ag(NH_3)_2]^+$$

The following standard state free energy of formation values are needed: 77.107 kJ/mol for $Ag^+(aq)$, −17.12 kJ/mol for $[Ag(NH_3)_2]^+$, −51.57 kJ/mol for I^-, −131.228 kJ/mol for Cl^-, −26.50 kJ/mol for $NH_3(aq)$, −109.789 kJ/mol for $AgCl(s)$, −66.19 kJ/mol for $AgI(s)$.

Free Energy—*K* and *Q*

23.53 What is the relationship between the enthalpy and the entropy changes for a process at equilibrium?

23.54 How is the free energy change related to the standard state free energy change and the reaction quotient? Will a large or a small value of the reaction quotient tend to make a chemical reaction more favorable?

23.55 For the reaction:

$$2NO_2(g) \rightleftharpoons N_2O_4(g)$$

$\Delta G^\circ = -4.73$ kJ at 25 °C. Calculate the value of K at 25 °C for this reaction. Would you have expected a value larger than unity? Why?

23.56 For this reaction

$$AgCl(s) \rightleftharpoons Ag^+ + Cl^-$$

$K = 1.8 \times 10^{-10}$ at 25 °C. Calculate the value of ΔG° for the reaction. Would you have expected a positive value? Why?

23.57 At 25 °C, $\Delta G^\circ = -95.299$ kJ/mol for the formation of $HCl(g)$:

$$\tfrac{1}{2}H_2(g) + \tfrac{1}{2}Cl_2(g) \longrightarrow HCl(g)$$

What is the value of ΔG for the process if the partial pressure of H_2 is 3.5 bar; of Cl_2, 1.5 bar; of HCl, 0.31 bar? Is the process more or less favorable under these conditions than under standard state conditions?

23.58 The standard state free energies of formation at 25 °C are 0 for H^+, -157.244 kJ/mol for OH^-, -237.129 kJ/mol for $H_2O(l)$. Find ΔG for the reaction

$$H_2O(l) \longrightarrow H^+ + OH^-$$

under conditions such that $[H^+] = [OH^-] = 1.0 \times 10^{-7}$ M.

23.59 The standard free energies of formation at 25 °C are -202.87 kJ/mol for $NH_4Cl(s)$, -16.45 kJ/mol for $NH_3(g)$, -95.299 kJ/mol for $HCl(g)$. (a) What ΔG° for the reaction

$$NH_4Cl(s) \rightleftharpoons NH_3(g) + HCl(g)$$

(b) Calculate the equilibrium constant for this decomposition. (c) Calculate the equilibrium partial pressure of HCl above a sample of NH_4Cl.

23.60 The standard state free energies of formation at 25 °C are -134.03 kJ/mol for $H_2O_2(aq)$, 0 for $O_2(g)$, -237.129 kJ/mol for $H_2O(l)$. Calculate ΔG° for the reaction

$$2H_2O_2(aq) \longrightarrow 2H_2O(l) + O_2(g)$$

This spontaneous reaction can be stopped by increasing the pressure of the oxygen. Find the value of the pressure necessary to stop the reaction.

Free Energy—Temperature Dependence of *K* and ΔG°

23.61 For the reaction

$$C(s) + O_2(g) \longrightarrow CO_2(g)$$

$\Delta H^\circ = -393.509$ kJ and $\Delta S^\circ = 2.86$ J/K at 25 °C. Does this reaction become more or less favorable as the temperature increases? For the reaction

$$C(s) + \tfrac{1}{2}O_2(g) \longrightarrow CO(g)$$

$\Delta H^\circ = -110.525$ kJ and $\Delta S^\circ = 89.365$ J/K at 25 °C. Does this reaction become more or less favorable as the temperature increases? Compare the temperature dependencies of these reactions.

23.62 How does the value of ΔG° change for the reaction

$$CCl_4(l) + H_2(g) \longrightarrow HCl(g) + CHCl_3(l)$$

if the reaction is carried out at 65 °C rather than at 25 °C? At 25 °C, $\Delta G^\circ = -103.75$ kJ and $\Delta H^\circ = -91.34$ kJ for the reaction.

23.63 The standard enthalpy of formation of $O_3(g)$ at 298 K is 142.7 kJ/mol and it varies negligibly with temperature up to 6000 K. Calculate the ΔG° of formation of ozone at 1000. K given that the ΔG° of formation is 163.2 kJ/mol at 298 K.

23.64 The enthalpy of reaction under standard state conditions at 25 °C for the combustion of CO,

$$CO(g) + \tfrac{1}{2}O_2(g) \longrightarrow CO_2(g)$$

is -282.984 kJ/mol and ΔG° for the reaction at 25 °C is -257.191 kJ/mol. At what temperature will this reaction no longer be spontaneous under standard state conditions?

23.65 At 25 °C, $K_a = 2.2 \times 10^{-9}$ and $\Delta H^\circ = 18.9$ kJ for the ionization of hypobromous acid, HOBr. What is the value of K_a at 75 °C?

23.66 At 25 °C, $K = 1.3 \times 10^{157}$ and $\Delta H^\circ = -1057$ kJ for the reaction

$$2Al(s) + 3Pb^{2+} \longrightarrow 2Al^{3+} + 3Pb(s)$$

What is the value of K at 75 °C?

23.67 The equilibrium constant for the reaction

$$Cl_2(g) + F_2(g) \rightleftharpoons 2\,ClF(g)$$

is 2.10×10^{18} at 298 K and 1.35×10^{11} at 500. K. Find ΔH° for the reaction.

23.68 The equilibrium constant is 3.66×10^{24} at 500. K and 2.58×10^{14} at 700. K for the reaction

$$P(s) + \tfrac{5}{2}Cl_2(g) \rightleftharpoons PCl_5(g)$$

Calculate ΔH_f° for $PCl_5(g)$ at 600. K from these data.

ADDITIONAL PROBLEMS

23.69 Calculate (a) ΔS° for the production of ozone from oxygen

$$3O_2(g) \longrightarrow 2O_3(g)$$

given that $S^\circ = 205.138$ J/K mol for $O_2(g)$ and 238.93 J/K mol for $O_3(g)$ at 25 °C. Is this an increase or decrease in randomness? The heat of formation at 25 °C is 142.7 kJ/mol for $O_3(g)$ and is 0 for $O_2(g)$. (b) Calculate ΔH° for the reaction. Is this a favorable or an unfavorable enthalpy change? (c) Calculate ΔG° for this reaction and the ΔG° of formation for $O_3(g)$. Is this reaction spontaneous under standard conditions?

23.70* Chlorine and fluorine react to form the interhalogen compound ClF according to the equation

$$Cl_2(g) + F_2(g) \longrightarrow 2ClF(g)$$

(a) What is ΔH° for this reaction, given that the heat of formation of $ClF(g)$ at 25 °C is −54.48 kJ/mol? (b) Calculate ΔG° for this reaction, given that the free energy of formation at 25 °C is −55.94 kJ/mol for $ClF(g)$. (c) Using your values of ΔH° and ΔG°, calculate ΔS° for the reaction at 25 °C. (d) Compare your answer for (c) to that calculated using absolute entropies of 217.89 J/K mol for $ClF(g)$, 223.066 J/K mol for $Cl_2(g)$, 202.78 J/K mol for $F_2(g)$. (e) What would be the value of ΔG if the partial pressure of Cl_2 is 1.52 bar, of F_2 is 1.78 bar, and of ClF is 0.10 bar? Is the reaction more or less favorable under these nonstandard state conditions?

23.71* The standard state heat of formation at 25 °C is −285.830 kJ/mol for $H_2O(l)$, 0 for $H^+(aq, 1\ M)$, and −229.994 kJ/mol for $OH^-(aq, 1\ M)$. (a) Calculate ΔH° for the reaction represented by the equation

$$H_2O(l) \rightleftharpoons H^+(aq, 1\ M) + OH^-(aq, 1\ M)$$

The standard state free energy of formation at 25 °C is −237.129 kJ/mol for $H_2O(l)$, 0 for $H^+(aq, 1\ M)$, and −157.244 kJ/mol for $OH^-(aq, 1\ M)$. (b) Calculate ΔG° for the above reaction. What are the values of K (c) at 25 °C and (d) at 35 °C? (e) What is the pH of water at 35 °C? Does this mean that water becomes acidic at 35 °C? Explain your answer.

23.72* Consider the conversion of L-aspartate ion into fumarate ion and ammonium ion:

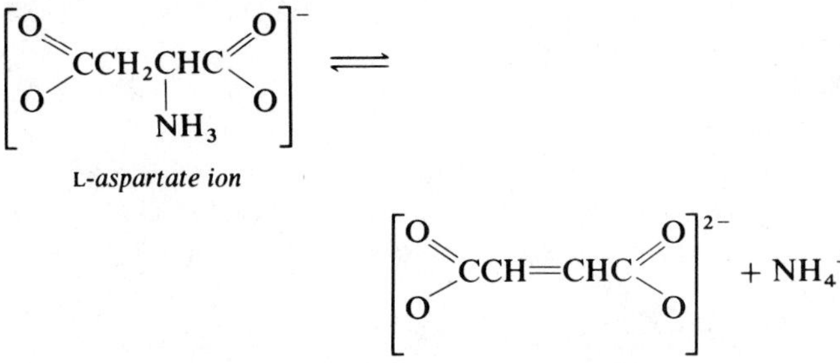

$$[O_2CCH_2CH(NH_3)CO_2]^- \rightleftharpoons [O_2CCH{=}CHCO_2]^{2-} + NH_4^+$$

L-aspartate ion *fumarate ion*

[This reaction, which is catalyzed by an enzyme, is part of a series of reactions that breaks down larger molecules in human and animal metabolism so that they can be eliminated as waste products. The overall result of the reaction cycle in which the aspartate–fumarate conversion takes part is the removal of NH_2 groups from amino acids and their conversion to urea, which is eliminated in urine.]

The equilibrium constant for the reaction is 7.4×10^{-3} at 29 °C and 1.60×10^{-2} at 39 °C. From this set of data, calculate ΔH°, K, ΔG°, ΔS° at 37 °C (approximate human body temperature, used as the standard state temperature for biochemical data).

CHAPTER

24

Oxidation–Reduction and Electrochemistry

REDOX REVISITED

24.1 ELECTRON LOSS AND GAIN IN REDOX REACTIONS

Oxidation and reduction were defined in Sections 16.2 and 17.4, in terms of changes in oxidation numbers. These definitions hold for substances with bonding of any type. It is also possible to define oxidation and reduction in terms of electron loss and gain, although *actual* electron loss and gain to form ions is not a requirement for oxidation and reduction to occur.

When a piece of zinc is dropped into a solution of copper(II) sulfate, a chemical reaction occurs (Figure 24.1). The deep blue color of the copper(II) sulfate solution grows lighter, the piece of zinc grows smaller, and solid elemental copper forms. Writing the chemical equation

$$Zn(s) + CuSO_4(aq) \longrightarrow Cu(s) + ZnSO_4(aq) \qquad \textbf{(24.1)}$$

shows this to be a redox displacement reaction. Because copper(II) sulfate and zinc sulfate are soluble strong electrolytes, the net ionic equation is

$$Zn(s) + Cu^{2+} \longrightarrow Cu(s) + Zn^{2+} \qquad \textbf{(24.2)}$$

To convert zinc metal to Zn^{2+} ions, each zinc atom must *lose* two electrons.

$$Zn(s) \longrightarrow Zn^{2+} + 2e^- \qquad \textbf{(24.3)}$$

The zinc has lost electrons and has thus been oxidized (its oxidation number has increased from 0 to +2). To convert copper ions into copper metal, each copper ion must *gain* two electrons.

$$Cu^{2+} + 2e^- \longrightarrow Cu(s) \qquad \textbf{(24.4)}$$

The copper ion has gained electrons and has thus been *reduced* (its oxidation number has decreased from +2 to 0).

Every oxidation–reduction reaction can be divided into two **half-reactions**—reactions representing either oxidation only or reduction only. The sum of the half-reactions of Equations (24.3) and (24.4)

$$\begin{array}{llll} \textit{oxidation} & Zn(s) & \longrightarrow & Zn^{2+} + 2e^- \\ \textit{reduction} & Cu^{2+} + 2e^- & \longrightarrow & Cu(s) \\ \hline & Zn(s) + Cu^{2+} & \longrightarrow & Cu(s) + Zn^{2+} \end{array}$$

is the net ionic equation for the redox reaction. In this reaction and in every redox reaction involving ions, oxidation is electron loss and reduction is electron gain.

Equations (24.3) and (24.4) are **ion-electron equations**—balanced equations that include only the electrons and other species directly involved in the oxidation or reduction of a given atom, molecule, or ion. Like net ionic equations, ion-electron equations must be balanced as to both number of atoms and charge.

The oxidized and reduced species that appear in an ion-electron equation are sometimes called a **redox couple,** or just a couple. In Equations (24.3) and (24.4) the redox couples are $Zn^{2+}/Zn(s)$ and $Cu^{2+}/Cu(s)$ (both written, as is done throughout the text, with the oxidized form first).

Like any other chemical equation, an ion-electron equation must always describe an actual chemical change. The element undergoing a change in oxidation number must appear as part of the chemical formula of the substance being oxidized or reduced. For reactions in aqueous solution, only species that exist as ions in solution during the reaction are written as ions. For example, when hydrogen sulfide in aqueous solution acts as a reducing agent, the sulfur in the −2 oxidation state is oxidized to free sulfur. The balanced ion-electron equation for this oxidation is

$$H_2S(aq) \longrightarrow S(s) + 2H^+ + 2e^- \tag{24.5}$$

not $S^{2-} \longrightarrow S(s) + 2e^-$, because sulfide ions are not involved.

Similarly, when nitrate ion acts as an oxidizing agent in acidic solution, one possible product is nitrogen(II) oxide. Nitrogen appears in the ion-electron equation as the nitrate ion and the molecular oxide.

$$NO_3^- + 4H^+ + 3e^- \longrightarrow NO(g) + 2H_2O(l) \tag{24.6}$$

not as $N^{5+} + 3e^- \longrightarrow N^{2+}$. Nitrogen never forms such ions. The error here is in confusing the *oxidation states* of nitrogen in NO_3^- and NO with the *charges* on monatomic ions. Do not make the common error of writing nonexistent species in ion-electron equations; in particular, do not confuse oxidation states and ionic charges.

Note that H^+ ions and water molecules appear in Equations (24.5) and (24.6). For redox reactions in aqueous systems, whatever hydrogen or oxygen atoms are required to balance the equations must come from water and either H_3O^+ or OH^-. Whether or not a reaction will occur and the type of product obtained often depend upon the pH of the system. Hydrogen ions and hydroxide ions never show up in the same ion-electron equation because, obviously, a solution cannot be both alkaline and acidic.

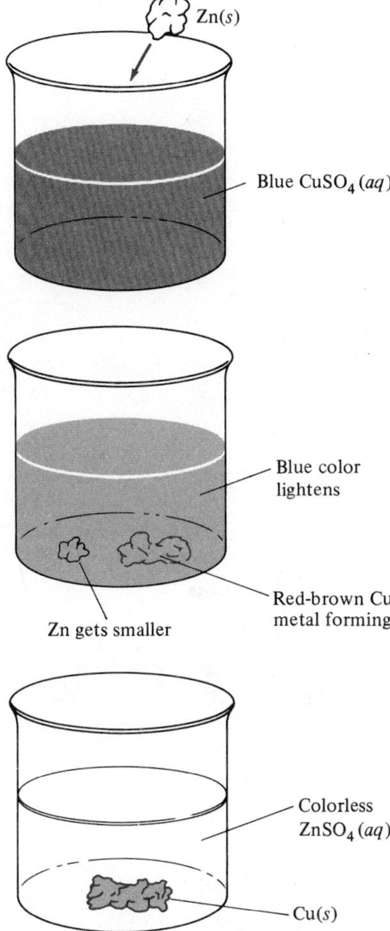

Figure 24.1
The Redox Displacement of Copper(II) by Zinc

EXAMPLE 24.1 Ion-Electron Equations

Write the ion-electron equations for the reduction of iron(III) ion to iron(II) ion by tin(II) ion:

$$Sn^{2+} + 2Fe^{3+} \longrightarrow 2Fe^{2+} + Sn^{4+}$$

What are the redox couples for these ion-electron equations?

Each iron(III) ion gains one electron during reduction and each tin(II) ion loses two electrons during oxidation.

$$Fe^{3+} + e^- \longrightarrow Fe^{2+} \qquad Sn^{2+} \longrightarrow Sn^{4+} + 2e^-$$

Remembering to write the oxidized form first, the redox couples for these equations are Fe^{3+}/Fe^{2+} and Sn^{4+}/Sn^{2+}.

EXERCISE Write the ion-electron equations for the oxidation of bromide ion by chlorine: $Cl_2(g) + 2Br^- \longrightarrow 2Cl^- + Br_2(l)$. Write the redox couples for these ion-electron equations.

Answers $Cl_2(g) + 2e^- \longrightarrow 2Cl^-$, $2Br^- \longrightarrow Br_2(l) + 2e^-$; $Cl_2(g)/Cl^-$, $Br_2(l)/Br^-$.

24.2 BALANCING REDOX EQUATIONS: HALF-REACTION METHOD

The oxidation number method of balancing redox equations (Section 16.4) has as its basis the equality of the increase and decrease in oxidation numbers. The half-reaction method is based on electron gain and loss. In a redox reaction, the number of electrons "gained" must equal the number of electrons "lost." This means that the number of electrons in two ion-electron equations must be made equal before the equations can be added to give an overall redox equation.

a. Balancing Ion-Electron Equations In order to balance a complete redox equation by the half-reaction method, it is necessary to first balance the ion-electron equations for the oxidation and reduction half-reactions that are involved. A procedure for doing so is outlined below and illustrated by balancing the ion-electron equations for the reduction of ClO_4^- to Cl_2. As illustrated, different approaches must be taken to balancing ion-electron equations for reactions in acidic and alkaline solutions.

1. Write the oxidized and reduced species on opposite sides of the equation and balance atoms other than oxygen and hydrogen by inspection.

$$2ClO_4^- \xrightarrow{\text{not balanced}} Cl_2 \qquad \textbf{(24.7)}$$

2. Balance oxygen atoms by adding H_2O to the side that is deficient in oxygen.

$$2ClO_4^- \xrightarrow{\text{not balanced}} Cl_2 + 8H_2O \qquad \textbf{(24.8)}$$

3. Balance hydrogen atoms. For reactions in acidic solution add H^+ ions to the side that is deficient in hydrogen.

$$2ClO_4^- + 16H^+ \xrightarrow{\text{not balanced}} Cl_2 + 8H_2O \qquad \textbf{(24.9)}$$

For reactions in alkaline solution, it is necessary to add OH^- to balance hydrogen atoms, but this also introduces more oxygen atoms. A method that often overcomes this problem is to add OH^- ions equal in number to the missing hydrogen atoms on the *same side* as the H_2O

$$2ClO_4^- \xrightarrow{\text{not balanced}} Cl_2 + 8H_2O + 16OH^- \qquad \textbf{(24.10)}$$

and then add the *same number* of H_2O molecules as OH^- ions on the opposite side

$$2ClO_4^- + 16H_2O \xrightarrow{\text{not balanced}} Cl_2 + 8H_2O + 16OH^-$$

and cancel the excess water molecules.

$$2ClO_4^- + 8H_2O \xrightarrow{\text{not balanced}} Cl_2 + 16OH^- \qquad \textbf{(24.11)}$$

Both Equations (24.9) and (24.11) are now balanced with regard to atoms.

4. Balance charge by adding electrons. The two balanced ion-electron equations are as follows:

$$2ClO_4^- + 16H^+ + 14e^- \longrightarrow Cl_2(g) + 8H_2O(l) \qquad \textbf{(24.12)}$$

$$2ClO_4^- + 8H_2O(l) + 14e^- \longrightarrow Cl_2(g) + 16OH^- \qquad \textbf{(24.13)}$$

EXAMPLE 24.2 Ion-Electron Equations

Write the balanced ion-electron equations for the reduction of NO_3^- to (a) $HNO_2(aq)$ in acidic solution and to (b) NO_2^- in alkaline solution.

(a) The oxidized and reduced species are

$$NO_3^- \longrightarrow HNO_2$$

This equation is already balanced with respect to the N atoms. The oxygen atoms are balanced by adding one H_2O molecule to the products side to obtain

$$NO_3^- \longrightarrow HNO_2 + H_2O$$

The H atoms are balanced in this acidic solution by adding three H^+ ions to the reactants side, which gives

$$NO_3^- + 3H^+ \longrightarrow HNO_2 + H_2O$$

The charge is balanced by adding two e^- to the reactants side.

$$\mathbf{NO_3^- + 3H^+ + 2e^- \longrightarrow HNO_2(aq) + H_2O(l)}$$

The total charge on the left in this equation is $(-1) + (3)(+1) + (2)(-1) = 0$, and the total charge on the right is also zero. The equation is balanced.

(b) For the reduction reaction in alkaline solution, the oxidized and reduced species are

$$NO_3^- \longrightarrow NO_2^-$$

This equation is already balanced with respect to the N atoms. The oxygen atoms are balanced by adding one H_2O molecule to the products side to obtain

$$NO_3^- \longrightarrow NO_2^- + H_2O$$

The H atoms are balanced in this alkaline solution by adding two OH^- ions to the product side (the same number as the H atom deficiency)

$$NO_3^- \longrightarrow NO_2^- + H_2O + 2OH^-$$

adding two H_2O molecules to the reactants side (the same number as the OH^- ions)

$$NO_3^- + 2H_2O \longrightarrow NO_2^- + H_2O + 2OH^-$$

and cancelling the excess H_2O molecules to obtain

$$NO_3^- + H_2O \longrightarrow NO_2^- + 2OH^-$$

The charge is balanced by adding two e^- to the reactants side.

$$\mathbf{NO_3^- + H_2O(l) + 2e^- \longrightarrow NO_2^- + 2OH^-}$$

The total charge on each side in this equation is $(-1) + (0) + (2)(-1) = (-1) + (2)(-1) = -3$. The equation is balanced.

EXERCISE Write the balanced ion-electron equations for the reduction of MnO_4^- to (a) Mn^{2+} in acidic solution and to (b) $MnO_2(s)$ in alkaline solution.

Answers (a) $MnO_4^- + 8H^+ + 5e^- \longrightarrow Mn^{2+} + 4H_2O(l)$, (b) $MnO_4^- + 2H_2O(l) + 3e^- \longrightarrow MnO_2(s) + 4OH^-$.

b. Using Ion-Electron Equations to Balance Redox Equations Earlier we gave the balanced ion-electron equations for the reactions of hydrogen sulfide in aqueous solution as a reducing agent and nitrate ion in acidic solution as an oxidizing agent (Equations 24.5 and 24.6). To find the balanced overall equation for the oxidation of

Table 24.1
Balancing Redox Equations by the Half-Reaction Method

1. Write the overall unbalanced equation, inlcuding all species that undergo change.
2. Identify oxidized and reduced substances, and write unbalanced ion-electron equations.[a]
3. Balance each ion-electron equation for atoms and charge:
   a. Balance for atoms other than O or H.
   b. Balance O atoms by adding H_2O on side deficient in O atoms.
   c. Balance H atoms by adding H^+ for acidic solutions or OH^- and H_2O for alkaline solutions.[b]
   d. Balance charge by adding electrons.
4. Multiply the ion-electron equations by appropriate factors so that electrons gained equals electrons lost.
5. Add the ion-electron equations, cancelling where appropriate.

[a] Oxidized and reduced atoms must be included in the actual ions, molecules, or solid ionic compounds that are present.

[b] For reactions in alkaline solution, add OH^- ions equal in number to the deficiency in hydrogen atoms to *same* side as H_2O was added in step b, then add equal number of H_2O molecules to opposite side and cancel excess H_2O.

hydrogen sulfide by nitrate ion in acidic solution, the two equations must be multiplied by 3 and 2, respectively, so that six electrons are gained and six are lost. Then the two ion-electron equations are added together and any species that appear on both sides are cancelled to obtain the overall equation.

$$(3)[H_2S(aq) \longrightarrow S(s) + 2H^+ + 2e^-] = 3H_2S(aq) \longrightarrow 3S(s) + \cancel{6H^+} + \cancel{6e^-}$$

$$(2)[NO_3^- + 4H^+ + 3e^- \longrightarrow NO(g) + 2H_2O(l)] =$$

$$2NO_3^- + \overset{2H^+}{\cancel{8H^+}} + \cancel{6e^-} \longrightarrow 2NO(g) + 4H_2O(l)$$

$$3H_2S(aq) + 2NO_3^- + 2H^+ \longrightarrow 3S(s) + 2NO(g) + 4H_2O(l)$$

The complete set of rules for balancing overall redox equations by the half-reaction method is given in Table 24.1 and illustrated in Example 24.3.

EXAMPLE 24.3 Half-Reaction Method

Bismuthate ion (BiO_3^-), a strong oxidizing agent, can oxidize Mn^{2+} to permanganate ion (MnO_4^-) in acidic solution. The BiO_3^- ion is reduced to Bi^{3+}. Balance the equation for this reaction by the half-reaction method.

Step 1. Write the overall unbalanced equation.

$$BiO_3^- + Mn^{2+} \longrightarrow Bi^{3+} + MnO_4^-$$

Step 2. Write the unbalanced ion-electron equations.

$$BiO_3^- \longrightarrow Bi^{3+} \qquad Mn^{2+} \longrightarrow MnO_4^-$$

Step 3. Balance each ion-electron equation for atoms and charge. The reaction occurs in acidic solution, so H^+ and H_2O may be used to balance atoms.

$$BiO_3^- + 6H^+ \longrightarrow Bi^{3+} + 3H_2O$$
$$Mn^{2+} + 4H_2O \longrightarrow MnO_4^- + 8H^+$$

Balancing for charge gives

$$BiO_3^- + 6H^+ + 2e^- \longrightarrow Bi^{3+} + 3H_2O$$
$$Mn^{2+} + 4H_2O \longrightarrow MnO_4^- + 8H^+ + 5e^-$$

Step 4. Multiply the ion-electron equations by appropriate factors so that electrons gained equal electrons lost.

$$(5)[BiO_3^- + 6H^+ + 2e^- \longrightarrow Bi^{3+} + 3H_2O]$$
$$= 5BiO_3^- + 30H^+ + 10e^- \longrightarrow 5Bi^{3+} + 15H_2O$$
$$(2)[Mn^{2+} + 4H_2O \longrightarrow MnO_4^- + 8H^+ + 5e^-]$$
$$= 2Mn^{2+} + 8H_2O \longrightarrow 2MnO_4^- + 16H^+ + 10e^-$$

Step 5. Add the ion-electron equations to obtain the overall equation, cancelling where appropriate.

$$\begin{array}{rcl} & & \\ 5BiO_3^- + \overset{14H^+}{\cancel{30H^+}} + \cancel{10e^-} & \longrightarrow & 5Bi^{3+} + \overset{7H_2O}{\cancel{15H_2O}} \\ 2Mn^{2+} + \cancel{8H_2O} & \longrightarrow & 2MnO_4^- + \cancel{16H^+} + \cancel{10e^-} \\ \hline \mathbf{5BiO_3^- + 2Mn^{2+} + 14H^+} & \longrightarrow & \mathbf{5Bi^{3+} + 2MnO_4^- + 7H_2O(\mathit{l})} \end{array}$$

Checking shows that the atoms are all balanced and the charge on the left, $(5)(-1) + (2)(+2) + (14)(+1) = +13$, equals the charge on right, $(5)(+3) + (2)(-1) = +13$. The equation is balanced.

EXERCISE In acidic solution, hydrogen peroxide is reduced to water by chlorate ion, which is oxidized to perchlorate ion. Write the ion-electron equations describing these processes and balance the overall equation for the redox process using the half-reaction method.

Answers $H_2O_2(aq) + 2H^+ + 2e^- \longrightarrow 2H_2O(l)$, $ClO_3^- + H_2O(l) \longrightarrow ClO_4^- + 2H^+ + 2e^-$, $H_2O_2(aq) + ClO_3^- \longrightarrow ClO_4^- + H_2O(l)$.

EXAMPLE 24.4 Half-Reaction Method

The nitrate ion is reduced to ammonia by elemental aluminum under alkaline conditions with the formation of $[Al(OH)_4]^-$. Using the half-reaction method, balance the overall equation.

Steps 1 and 2. Write the overall unbalanced equation and the unbalanced ion-electron equations.

$$Al(s) + NO_3^- + OH^- \longrightarrow NH_3(aq) + [Al(OH)_4]^-$$

$$\overset{0}{Al} \longrightarrow \overset{+3}{[Al(OH)_4]^-} \qquad \overset{+5}{NO_3^-} \longrightarrow \overset{-3}{NH_3}$$

Step 3. Balance each ion-electron equation for atoms and charge. Inspection shows that only OH^- ions need be added to balance atoms in the oxidation half-reaction.

$$Al + 4OH^- \longrightarrow [Al(OH)_4]^-$$

Balancing charge gives

$$Al(s) + 4OH^- \longrightarrow [Al(OH)_4]^- + 3e^-$$

Atoms in the reduction half-reaction in alkaline solution are balanced as follows:

$$\begin{array}{rcl} NO_3^- & \longrightarrow & NH_3 + 3H_2O \\ NO_3^- & \longrightarrow & NH_3 + 3H_2O + 9OH^- \\ NO_3^- + 9H_2O & \longrightarrow & NH_3 + 3H_2O + 9OH^- \\ NO_3^- + 6H_2O & \longrightarrow & NH_3 + 9OH^- \end{array}$$

The atoms are balanced. Balancing charge gives

$$NO_3^- + 6H_2O(l) + 8e^- \longrightarrow NH_3(aq) + 9OH^-$$

Step 4. Multiply the ion-electron equations by appropriate factors, so that electrons gained equal electrons lost.

$$\begin{array}{rcl} (8)[Al + 4OH^- & \longrightarrow & [Al(OH)_4]^- + 3e^-] \\ (3)[NO_3^- + 6H_2O + 8e^- & \longrightarrow & NH_3 + 9OH^-] \end{array}$$

Step 5. Add the ion-electron equations to get the overall equation, canceling where appropriate.

$$\begin{array}{rcl} 8Al(s) + \overset{5OH^-}{\cancel{32OH^-}} & \longrightarrow & 8[Al(OH)_4]^- + \cancel{24e^-} \\ 3NO_3^- + 18H_2O(l) + \cancel{24e^-} & \longrightarrow & 3NH_3(aq) + \cancel{27OH^-} \\ \hline \mathbf{8Al(\mathit{s}) + 3NO_3^- + 5OH^- + 18H_2O(\mathit{l})} & \longrightarrow & \mathbf{8[Al(OH)_4]^- + 3NH_3(\mathit{aq})} \end{array}$$

EXERCISE Magnesium metal is oxidized to magnesium hydroxide by chromate ion in an alkaline solution. The chromate ion is reduced to $Cr(OH)_3(s)$. Write the ion-electron equations describing these processes and balance the overall equation for the redox process using the half-reaction method.

Answer $Mg(s) + 2OH^- \longrightarrow Mg(OH)_2(s) + 2e^-$, $CrO_4^{2-} + 4H_2O(l) + 3e^- \longrightarrow Cr(OH)_3(s) + 5OH^-$, $3Mg(s) + 2CrO_4^{2-} + 8H_2O(l) \longrightarrow 3Mg(OH)_2(s) + 2Cr(OH)_3(s) + 4OH^-$.

EXAMPLE 24.5 Half-Reaction Method

In an alkaline solution, aqueous sodium hypochlorite disproportionates into sodium chloride and sodium chlorate. Write the ion-electron equations and use the half-reaction method to balance the overall redox equation for this process.

Once the stepwise method of balancing redox equations is learned, it is no longer necessary to write out each step. It is common practice to write the steps in the more condensed format shown below.

$$\begin{array}{rcl} ClO^- + 4OH^- & \longrightarrow & ClO_3^- + 2H_2O + 4e^- \\ (2)[ClO^- + H_2O + 2e^- & \longrightarrow & Cl^- + 2OH^-] \\ \hline 3ClO^- & \longrightarrow & ClO_3^- + 2Cl^- \end{array}$$

[By adding the spectator ions, the complete equation can be written from the net ionic equation: $3NaClO(aq) \longrightarrow NaClO_3(aq) + 2NaCl(aq)$.]

EXERCISE In an acidic solution, aqueous hydrogen peroxide disproportionates into water and oxygen. Write the ion-electron equations and overall redox equation for this process. Use the half-reaction method to balance the overall equation.

Answer $H_2O_2(aq) \longrightarrow O_2(g) + 2H^+ + 2e^-$, $H_2O_2(aq) + 2H^+ + 2e^- \longrightarrow 2H_2O(l)$, $2H_2O_2(aq) \longrightarrow 2H_2O(l) + O_2(g)$.

SUMMARY OF SECTIONS 24.1–24.2

An alternative to balancing redox equations by the oxidation number method (Section 16.4) is to balance them by combining ion-electron equations representing the oxidation and reduction half-reactions. The procedure for balancing by this method is given in Table 24.1.

FUNDAMENTALS OF ELECTROCHEMISTRY

24.3 ELECTROCHEMICAL CELLS

Electrons flowing through a wire constitute an electrical current, and ions flowing through an aqueous solution or a molten salt also constitute an electrical current. If the electrons lost and gained in a spontaneous reaction are able to flow through a wire on their pathway from the substance that is being oxidized to the substance that is being reduced, the energy of the reaction is released as electrical energy rather than thermal energy. Conversely, a *non*spontaneous redox reaction can be driven forward by the introduction of electrical energy into the system.

Electrochemistry deals with oxidation–reduction reactions that either produce or utilize electrical energy. Any device in which an electrochemical reaction occurs is called an **electrochemical cell.**

In Section 24.1 we used the spontaneous oxidation of metallic zinc by copper(II) ions as an example of a redox reaction in which actual electron gain and loss occur (see

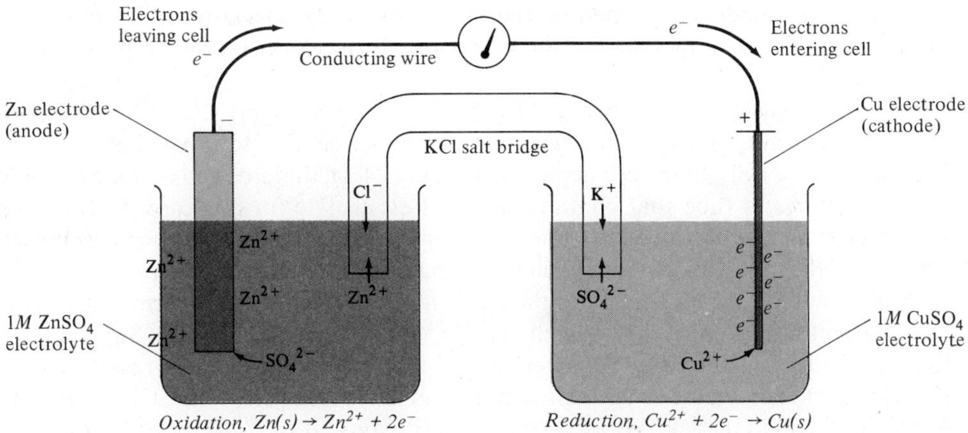

(a) Beginning of reaction, showing electron flow and ion flow

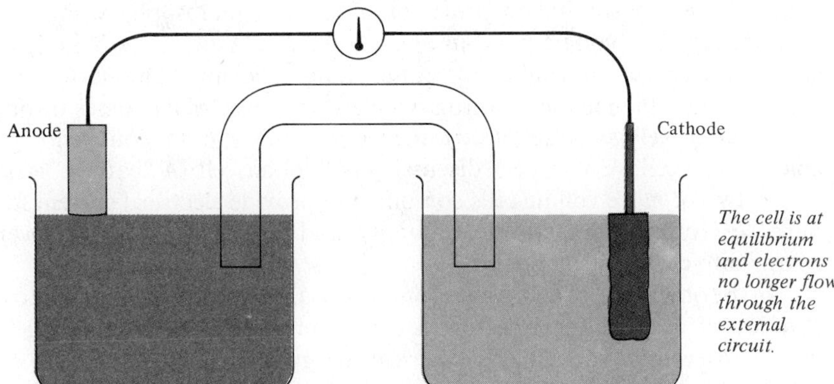

(b) Complete reaction; copper electrode covered with deposited copper and zinc electrode dissolved

Figure 24.2
The Zinc–Copper Voltaic Cell

Figure 24.1). An electrochemical cell in which this reaction takes place (Figure 24.2) illustrates the essential components of every electrochemical cell.

Two electrodes connected by an external conductor carry the electrons out of and into the cell. The electrodes are conductors and may or may not take part in the **cell reaction**—the overall chemical reaction that occurs in an electrochemical cell. The cell reaction is the sum of the half-reactions, as shown in Equation (24.3). In this cell the zinc of the electrode is a reactant.

Each electrode is immersed in an *electrolyte*—a medium through which ions can flow. The electrolytes in the zinc–copper cell are aqueous solutions of zinc sulfate and copper(II) sulfate. Electrolytes may also be molten salts, solids, or other media (for example, a paste).

An electrode and its surrounding electrolyte—for example, the copper electrode and the copper sulfate solution on the right in Figure 24.2—make up a **half-cell.** The half-reaction that occurs in a half-cell is an **electrode reaction.**

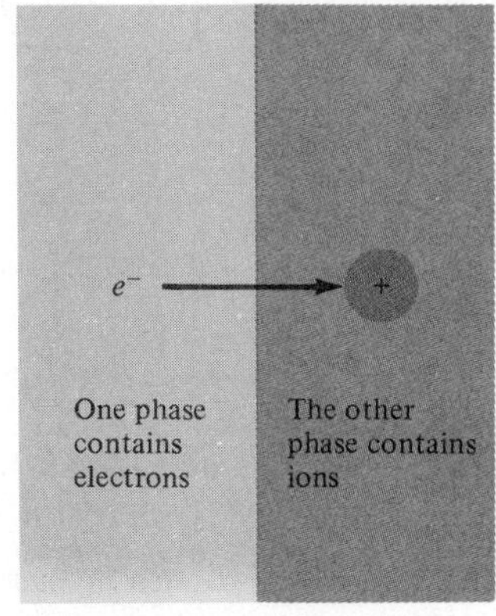

Figure 24.3
"The Fundamental Act in Electrochemistry"
Source: J. O. M. Bockris and A. K. N. Reddy, *Modern Electrochemistry* (New York: Plenum, 1970), p. 4.

When the electrical circuit is complete, current is carried in the electrodes and in the external circuit by electrons, and within the cell by ions. The chemical reaction occurs at the surfaces of the electrodes (Figure 24.3), where electron gain and loss must occur if current is to flow. The spontaneous reaction in an electrochemical cell, like every thermodynamically spontaneous reaction, continues until the reactants and products have come to equilibrium.

In the cell of Figure 24.2, oxidation occurs at the zinc electrode, where zinc atoms lose electrons to form Zn^{2+} ions. Reduction occurs at the copper electrode, where Cu^{2+} ions from the electrolyte gain electrons to form metallic copper. As the reaction proceeds, the zinc electrode erodes away, electrons flow from the zinc electrode through the external circuit to the copper electrode, and the copper(II) solution loses its blue

Table 24.2
Electrodes

Anode	Cathode
Oxidation	Reduction
Anions generally migrate to anode	Cations generally migrate to cathode
Electrons leave cell	Electrons enter cell
Negative in voltaic cell	Negative in electrolytic cell

color as fresh copper is deposited on the copper electrode. The current flowing in the external circuit could, in principle, be used to perform work—for example, to drive a motor.

Within the cell, current is carried by the flow of ions (Figure 24.2a). To maintain electrical neutrality, positive and negative ions must be able to cross the boundary between the two cells. In the cell depicted in Figure 24.2a, this crossing is made possible by a *salt bridge*, a tube that contains an inert electrolyte (in this case KCl) that is prevented from flowing out of the tube (e.g., by suspension in a gel or by porous plugs in the ends of the tube). In response to charge imbalance, ions are free to flow through the electrolyte in both the half-cells and the salt bridge. As Zn^{2+} ions are formed, Cl^- ions from the salt bridge enter the half-cell and Zn^{2+} ions leave the half-cell. Similarly, on the other side, K^+ and SO_4^{2-} ions enter and leave the half-cell. As a result, overall electrical neutrality in the cell is maintained by positive ions (Zn^{2+}, K^+, Cu^{2+}) flowing from left to right in the cell as drawn, while negative ions (SO_4^{2-}, Cl^-, SO_4^{2-}) flow from right to left.

There are two types of electrochemical cells. The first is called a "voltaic" cell after Alexander Volta, an Italian professor of natural philosophy who constructed the first battery, the "voltaic pile," in about 1800 (see Aside 24.1). Voltaic cells are also sometimes called "galvanic cells" after Luigi Galvani (famous for the accidental discovery that the muscle of a frog twitches when electricity passes through it).

A **voltaic cell** generates electrical energy from a spontaneous redox reaction. The zinc–copper cell we have been discussing is a voltaic cell. (A "battery" is defined properly as two or more voltaic cells combined to provide electrical current for a practical purpose. However, the term is commonly used for any device that converts chemical energy into electrical energy.)

In the other type of cell, a *non*spontaneous redox reaction is caused to occur by the addition of electrical energy from a direct current source such as a generator or a battery. An **electrolytic cell** uses electrical energy from outside the cell to cause a redox reaction to occur. The process of driving a nonspontaneous redox reaction to occur by means of electrical energy is called **electrolysis.**

By definition, in any cell the **anode** is the electrode at which oxidation occurs, and the **cathode** is the electrode at which reduction occurs (see Figure 24.2a). A good memory trick is this—**o**xidation and **a**node both begin with vowels, and **r**eduction and **c**athode with consonants. [Also by definition, in a voltaic cell the anode is labeled as the negative electrode and in an electrolytic cell the cathode is labeled as the negative electrode.]

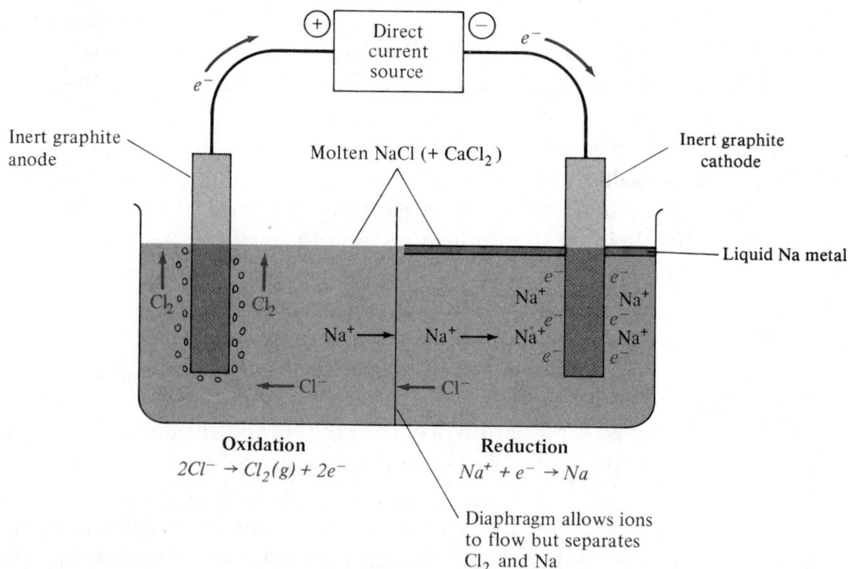

Figure 24.4
The Electrolysis of Molten Sodium Chloride
Addition of $CaCl_2$ to NaCl lowers the melting point enough to make the cell workable. (Pure NaCl melts at about the boiling point of metallic sodium.) The cell reaction is $2NaCl(l) \longrightarrow 2Na(l) + Cl_2(g)$.

Figure 24.4 is a simple diagram of a cell for the electrolysis of molten sodium chloride. In this cell, the electrodes are inert—they do not take part in the cell reaction. Electrons enter the cell from the outside source at the cathode, where they attract sodium ions and cause their reduction. As the sodium ions are consumed at the cathode, more cations are attracted into the vicinity of the cathode. At the anode, chloride ions are oxidized to give chlorine molecules, and the electrons that they lose leave the cell. More chloride ions move toward the anode and the reaction continues.

Note that in *both* the voltaic cell of Figure 24.2 and the electrolytic cell of Figure 24.4, cations move from left to right and anions move from right to left as the cells are drawn. In general, in both voltaic and electrolytic cells, *anions* move toward the *anode,* and *cations* move toward the *cathode*. Table 24.2 (top of page 604) summarizes what happens at the electrodes in both types of cells.

24.4 ELECTRODES AND CELL NOTATION

Figure 24.5 illustrates schematically some of the types of electrodes used in simple electrochemical cells. Electrodes of different types can be combined. The metal electrodes of the zinc–copper cell (see Figure 24.2) are of type a. A gas electrode (type b) requires that the gas be continuously bubbled into the half-cell, where it is oxidized or reduced at an inert electrode to give anions or cations in solution. For example, in a hydrogen electrode the redox couple is H^+/H_2, and in a chlorine electrode the redox couple is Cl_2/Cl^-.

A standard notation is used to describe voltaic cells. The pattern of the notation is as follows (with reduction on the right)

$$\text{anode}|\text{anode electrolyte}||\text{cathode electrolyte}|\text{cathode} \qquad \textbf{(24.14)}$$

A single vertical line indicates physical contact between species in different phases. A double line indicates a salt bridge, porous divider, or similar means of permitting ion flow while preventing the electrolytes from mixing. The zinc–copper cell would be written

phase boundary *salt bridge* *phase boundary*

$$\underbrace{Zn|Zn^{2+}}_{\textit{anode}}||\underbrace{Cu^{2+}|Cu}_{\textit{cathode}}$$

The symbol for an inert electrode, like the platinum electrode in the hydrogen half-cell, is often written in parentheses. For example

inert electrode

$$Mg|Mg^{2+}||H^+|H_2|(Pt)$$

The anode, cathode, and cell reactions for this cell are

anode reaction	$Mg(s)$	$\longrightarrow$	$Mg^{2+} + 2e^-$
cathode reaction	$2H^+ + 2e^-$	$\longrightarrow$	$H_2(g)$
cell reaction	$Mg(s) + 2H^+$	$\longrightarrow$	$Mg^{2+} + H_2(g)$ **(24.15)**

Species in the same phase in a cell are separated by commas, as in the notation for this zinc–iron(II)/iron(III) cell

in same phase *inert graphite electrode*

$$Zn|Zn^{2+}||\overbrace{Fe^{3+}, Fe^{2+}}|(C)$$

$$\textit{cell reaction} \quad Zn(s) + 2Fe^{3+} \longrightarrow Zn^{2+} + 2Fe^{2+} \qquad \textbf{(24.16)}$$

Frequently, the concentration of ions in solution and the pressure of gases are included, in parentheses, in the cell notation, for example

$$\underbrace{(Pt)|H_2(g, 1\ atm)|H^+(1\ M)}_{\textit{hydrogen gas anode}},\ \underbrace{Cl^-(1\ M)|AgCl(s)|Ag}_{\textit{metal-precipitate cathode}}$$

$$\textit{cell reaction} \quad 2\,AgCl(s) + H_2(g) \longrightarrow 2Ag(s) + 2H^+ + 2Cl^- \qquad \textbf{(24.17)}$$

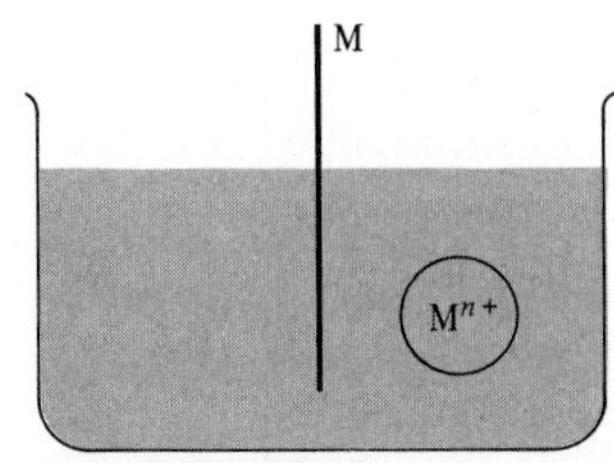

(a) Metal electrode, electrolyte containing the metal cation $M|M^{n+}$ (e.g., $Zn|Zn^{2+}$)

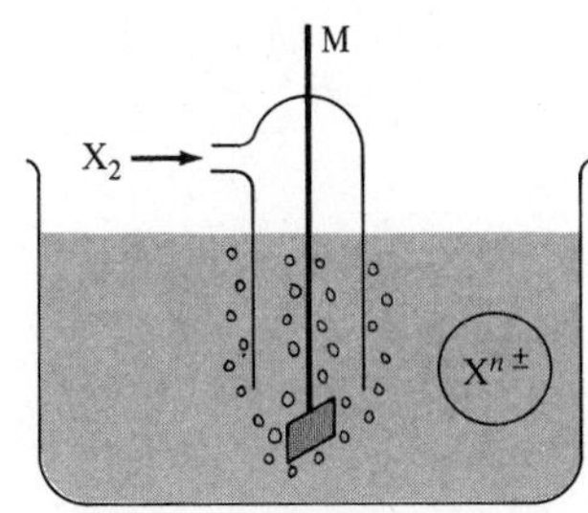

(b) Gas electrode with inert conductor, electrolyte containing anion or cation formed from the gas $(M)|X_2(g)|X^{n\pm}$ (e.g., $Cl_2(g)|Cl^-$)

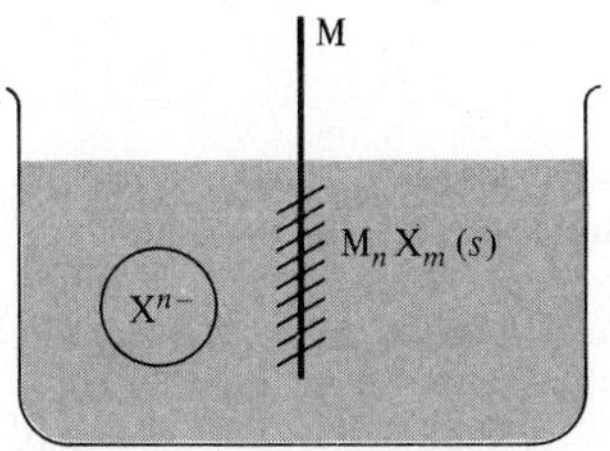

(c) Metal electrode coated with precipitate (e.g., Ag + AgCl), electrolyte contains anion of the salt $M|M_nX_m(s)|X^{n-}$ (e.g., $Ag|AgCl(s)|Cl^-$)

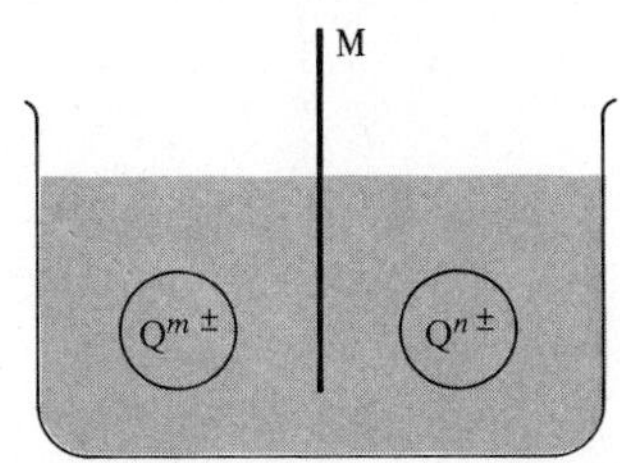

(d) "Redox" electrode with inert conductor, electrolyte containing two different species which include oxidized and reduced forms of same element (Q = monatomic or polyatomic species) $(M)|Q^{m\pm}, Q^{n\pm}$ (e.g., $(C)|Fe^{2+}, Fe^{3+}$)

Figure 24.5
Schematic Drawings of Some Simple Types of Electrodes
Species in circles are in solution.

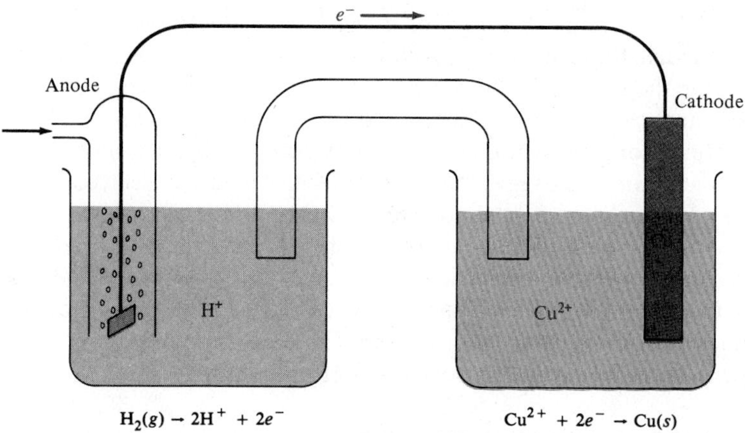

(a) Hydrogen electrode—Cu^{2+}/Cu cell

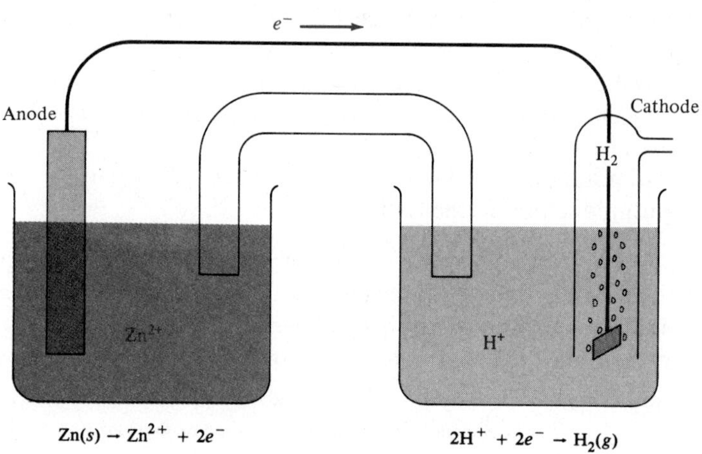

(b) Zn^{2+}/Zn—hydrogen electrode cell

Figure 24.6
Hydrogen Electrode as Anode or Cathode
In cell (a) oxidation occurs at the H_2 electrode and the cell reaction is $Cu^{2+} + H_2(g) \longrightarrow 2H^+ + Cu(s)$. In cell (b) reduction occurs at the H_2 electrode and the cell reaction is $Zn(s) + 2H^+ \longrightarrow Zn^{2+} + H_2(g)$.

No salt bridge is needed in this cell because the reactants are not soluble and do not mix.

Note that an electrode reaction may go in either direction and the electrode may be the anode or the cathode, depending upon the overall cell reaction. In Figure 24.6a, the hydrogen electrode is the anode, while in Figure 24.6b, the hydrogen electrode is the cathode. Table 24.3 lists some of the types of reactions most commonly encountered as anode reactions and cathode reactions.

EXAMPLE 24.6 Electrochemical Cell Notation

The notation for a voltaic cell is

$$Ag(s)|AgCl(s)|Cl^-||Br^-|Br_2(g)|(C)$$

Write the ion-electron equations for the electrode reactions and the overall chemical equation for the cell reaction.

The notation on the left represents the half-cell in which oxidation takes place—the anode (see Equation 24.14). The ion-electron equation for the anode reaction is

$$\mathbf{Ag(s) + Cl^- \longrightarrow AgCl(s) + e^-}$$

The notation on the right represents the half-cell in which reduction takes place—the cathode. The cathode reaction is

$$\mathbf{Br_2(g) + 2e^- \longrightarrow 2Br^-}$$

Table 24.3
Some Typical Electrode Reactions

Anode reactions

1. **Oxidation of an anion to a free element**

   (a) $2Cl^- \longrightarrow Cl_2(g) + 2e^-$
   (b) $4OH^- \longrightarrow O_2(g) + 2H_2O + 4e^-$

2. **Oxidation of an anion or a cation to another species in solution**

   (a) $ClO^- + 2H_2O \longrightarrow ClO_3^- + 4H^+ + 4e^-$
   (b) $Sn^{2+} \longrightarrow Sn^{4+} + 2e^-$
   (c) $2Cr^{3+} + 7H_2O \longrightarrow Cr_2O_7^{2-} + 14H^+ + 6e^-$

3. **Oxidation of a metal anode**

   (a) $Cu(s) \longrightarrow Cu^{2+} + 2e^-$
   (b) $Au(s) + 4Cl^- \longrightarrow AuCl_4^- + 3e^-$

4. **Oxidation of water**

   $2H_2O \longrightarrow O_2(g) + 4H^+ + 4e^-$ (Solution becomes acidic.)

Cathode reactions

5. **Reduction of a cation in solution, in solid salt, or in a complex ion to a free metal**

   (a) $Zn^{2+} + 2e^- \longrightarrow Zn(s)$
   (b) $AgCl(s) + e^- \longrightarrow Ag(s) + Cl^-$
   (c) $Ag(CN)_2^- + e^- \longrightarrow Ag(s) + 2CN^-$

6. **Reduction of an anion or cation**

   (a) $NO_3^- + 3H^+ + 2e^- \longrightarrow HNO_2 + H_2O$
   (b) $Ce^{4+} + e^- \longrightarrow Ce^{3+}$
   (c) $2H^+ + 2e^- \longrightarrow H_2$

7. **Reduction of an elemental nonmetal to an anion**

   $I_2 + 2e^- \longrightarrow 2I^-$

8. **Reduction of water**

   $2H_2O + 2e^- \longrightarrow H_2(g) + 2OH^-$ (Solution becomes alkaline.)

Combining these half-reactions gives the overall cell reaction as

$$\begin{array}{rcl} (2)[Ag + Cl^- & \longrightarrow & AgCl + e^-] \\ Br_2 + 2e^- & \longrightarrow & 2Br^- \\ \hline \mathbf{2Ag(s) + Br_2(g) + 2Cl^-} & \longrightarrow & \mathbf{2Br^- + 2AgCl(s)} \end{array}$$

[Note that the $Ag(s)|AgCl(s)|Cl^-$ electrode is the anode here, but was the cathode in the cell described above; Equation (24.17).]

EXERCISE The notation for a voltaic cell is

$$Fe(s)|Fe^{2+}||Sn^{2+}, Sn^{4+}|(Pt)$$

Write the ion-electron equations for the electrode reactions and the overall chemical equation for the cell reaction.

Answers anode: $Fe(s) \longrightarrow Fe^{2+} + 2e^-$, cathode: $Sn^{4+} + 2e^- \longrightarrow Sn^{2+}$, overall: $Fe(s) + Sn^{4+} \longrightarrow Fe^{2+} + Sn^{2+}$.

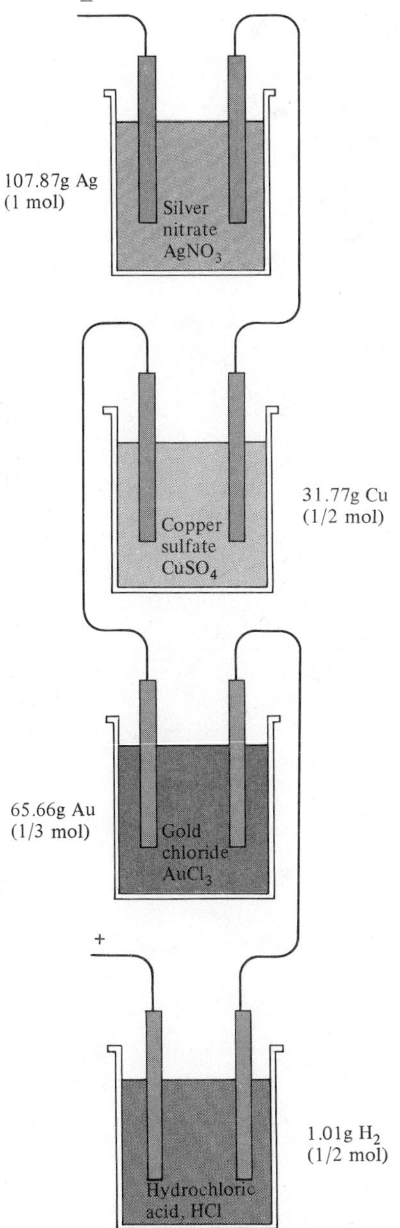

Figure 24.7
Simultaneous Electrolysis Experiment
This experiment illustrates chemical equivalence as the basis for Faraday's laws. The stoichiometric amount of each substance equivalent to 1 mol of electrons is produced.

24.5 ELECTROCHEMICAL STOICHIOMETRY

In an electrochemical reaction, there must obviously be a relationship between the amount of electricity that passes through a cell and the amount of chemical change that takes place. This relationship can be demonstrated by comparing the effect of a given amount of electrical current on several different electrolytic reactions.

Figure 24.7 illustrates an experiment for the simultaneous electrolyses of aqueous solutions of four different substances. Enough direct current is passed through the system to deposit 1 mol of silver (107.87 g). Using the mole ratios of the reduction half-reaction as for any chemical reaction shows that liberation of 1 mol of silver has required the passage of 1 mol of electrons.

$$Ag^+ + e^- \longrightarrow Ag(s) \qquad (1 \text{ mol Ag})\left(\frac{1 \text{ mol } e^-}{1 \text{ mol Ag}}\right) = 1 \text{ mol } e^-$$

Simultaneously with the deposition of the silver, it is found that $\frac{1}{2}$ mol of copper, $\frac{1}{3}$ mol of gold, and $\frac{1}{2}$ mol of hydrogen have been produced. The formation of these amounts of products is explained by the stoichiometry of the half-reactions and the passage of 1 mol of electrons through each solution, as follows:

$$Cu^{2+} + 2e^- \longrightarrow Cu(s) \qquad (1 \text{ mol } e^-)\left(\frac{1 \text{ mol Cu}}{2 \text{ mol } e^-}\right) = \frac{1}{2} \text{ mol Cu}$$

$$Au^{3+} + 3e^- \longrightarrow Au(s) \qquad (1 \text{ mol } e^-)\left(\frac{1 \text{ mol Au}}{3 \text{ mol } e^-}\right) = \frac{1}{3} \text{ mol Au}$$

$$2H^+ + 2e^- \longrightarrow H_2(g) \qquad (1 \text{ mol } e^-)\left(\frac{1 \text{ mol } H_2}{2 \text{ mol } e^-}\right) = \frac{1}{2} \text{ mol } H_2$$

The quantitative relationship between the amount of electricity and the amount of chemical change was first recognized by Michael Faraday, an English chemist, and is now known as Faraday's law. From his experiments in 1833 Faraday drew the following conclusion (in his own words): "The chemical power of a current of electricity is in direct proportion to the absolute quantity of electricity which passes." In modern terminology, "chemical power" is the extent to which the electrochemical reaction occurs, and the "absolute quantity of electricity" is the number of moles of electrons transferred.

The electrons involved in an electrochemical reaction can be dealt with in the same way as the stoichiometric amounts of reactants and products. However, electrical current is measured, not masses of electrons. Therefore, some unit conversions are necessary to perform electrochemical stoichiometry calculations.

The quantity of electrical charge is equal to the current that has flowed multiplied by the time. In the SI system of units, the coulomb (C) is the unit of electrical charge, the ampere (A) is the unit of current, and time is expressed in seconds (s).

$$\underset{\text{coulombs}}{\overset{\text{charge}}{q}} = \underset{\text{amperes}}{\overset{\text{current}}{I}}\,\underset{\text{seconds}}{\overset{\text{time}}{t}} \qquad \textbf{(24.18)}$$

The relationship among the units is

$$1 \text{ coulomb} = 1 \text{ ampere second} \qquad \textbf{(24.19)}$$

Or, as a conversion factor

$$\frac{1 \text{ coulomb}}{1 \text{ ampere second}}$$

Using the most accurate values, the charge on 1 mol of electrons is found to be

$$1.6021892 \times 10^{-19} \frac{\text{coulomb}}{\text{electron}} \times 6.022045 \times 10^{23} \frac{\text{electrons}}{\text{mol } e^-} = 96{,}484.56 \frac{\text{coulombs}}{\text{mol } e^-}$$

Because of its usefulness to chemistry, this quantity of electricity is defined as a unit, the **faraday** (*F*; also known as the faraday constant)—the amount of electrical charge represented by 1 mol of electrons, which for most calculations can be rounded to 96,500 C. The faraday provides the connection between the quantity of electrical charge and stoichiometry, as expressed in the numbers of moles of electrons:

$$(\text{moles of electrons})(\text{faraday}) = \text{amount of charge} \qquad \textbf{(24.20)}$$

$$(\text{moles } e^-)\left(\frac{96{,}500 \text{ coulombs}}{\text{mole } e^-}\right) \qquad \text{coulombs}$$

For example, the stoichiometric amount of charge required for the reduction of magnesium, $Mg^{2+} + 2e^- \longrightarrow Mg(s)$ is

$$(2 \text{ mol } e^-)\left(\frac{96{,}500 \text{ C}}{1 \text{ mol } e^-}\right) = 1.93 \times 10^5 \text{ C}$$

Because 1 coulomb = 1 ampere second, this amount of charge is equivalent to 1.93×10^5 ampere second (or 53.6 ampere hours).

To know the quantity of electrical charge that has flowed—for example, through an electrolytic cell—we must know the magnitude of the current in amperes and the length of time for which the current has flowed. For example, a 100 ampere current flowing for 60 s is equivalent to 6000 coulombs of charge:

$$q = It = (100 \text{ A})(60 \text{ s})\left(\frac{1 \text{ C}}{1 \text{ A s}}\right) = 6000 \text{ C}$$

The faraday allows conversion from coulombs to number of moles of electrons.

$$(6000 \text{ C})\left(\frac{1 \text{ mol } e^-}{96{,}500 \text{ C}}\right) = 0.06 \text{ mol } e^-$$

A 100 A direct current flowing for 60 s delivers 0.06 mol of electrons to an electrolytic cell. Conversely, the transfer of 0.06 mol of electrons in a voltaic cell would be required (ideally) to produce a 100 A current for 60 s.

EXAMPLE 24.7 Electrochemical Stoichiometry

What mass of elemental copper can be produced by the electrolysis of a copper(II) sulfate solution for 1.00 h at a steady direct current of 100. A?

First, we must use the current and the time to find the amount of electrical charge that has entered the cell.

$$(100. \text{ A})(1.00 \text{ h})\left(\frac{1 \text{ C}}{1 \text{ A s}}\right)\left(\frac{3600 \text{ s}}{1 \text{ h}}\right) = 3.60 \times 10^5 \text{ C}$$

The number of moles of electrons is found from the amount of charge.

$$(3.60 \times 10^5 \text{ C})\left(\frac{1 \text{ mol } e^-}{96{,}500 \text{ C}}\right) = 3.73 \text{ mol } e^-$$

Using the mole ratios from the reduction half-reaction

$$Cu^{2+} + 2e^- \longrightarrow Cu(s)$$

shows that the amount of copper produced will be

$$(3.73 \text{ mol } e^-)\left(\frac{1 \text{ mol Cu}}{2 \text{ mol } e^-}\right)\left(\frac{63.55 \text{ g Cu}}{1 \text{ mol Cu}}\right) = \mathbf{119 \text{ g Cu}}$$

The electrolysis will produce 119 g of copper metal. [This assumes, of course, that the reduction reaction is 100% efficient. In practice, as for example in a copper refinery, the reaction is not 100% efficient because of loss of current through competing electrochemical reactions.]

EXERCISE What mass of zinc must be oxidized (ideally) in a voltaic cell if the cell is to produce a direct current of 0.015 A for a period of 15 min? *Answer* 0.0049 g Zn.

24.6 CELL POTENTIAL

The spontaneous flow of electrons in the external circuit in a voltaic cell shows that electrons are attracted more strongly by one half-cell than by the other. There is an electrical potential difference between the two electrodes (see Aside 3.3). The anode has a higher negative potential than the cathode, and electrons flow from the anode to the cathode.

The **cell potential** or cell voltage (E or E_{cell}) is the potential difference between the two half-cells. This potential difference, also referred to as the **electromotive force** or **emf,** is sometimes described as the "driving force" of the cell reaction. The cell potential is a measure of the relative abilities of the species in each half-cell to gain and lose electrons. It is a thermodynamic quantity related to the amount of electrical work that a cell can perform and, as we shall see (Section 24.10), to the free energy change of the cell reaction.

Look again at Figure 24.6, which shows the hydrogen electrode acting as both an anode and a cathode. Relative to hydrogen, copper gives up electrons *less* easily (it is less easily oxidized). Therefore, the copper electrode is the cathode and the hydrogen electrode is the anode in the cell of Figure 24.6a. Relative to hydrogen, zinc gives up electrons *more* easily (it is more easily ozidized). Therefore, the zinc electrode is the anode and the hydrogen electrode is the cathode in the cell of Figure 24.6b.

Electrical potential is measured in volts (V). The cell potential is the voltage measured in the external circuit when no current is flowing (an "open" circuit). Under these conditions, the cell potential is a direct measure of the tendency of electrons to flow from one half-cell to the other. The potential varies with the nature of the chemical reaction, the temperature, and the concentration of the reactants (Section 24.11).

Recall that work is performed when an object moves over a distance. The motion of electrons through the external circuit in an electrochemical cell is electrical work, and one joule of work is performed when one coulomb of electrical charge moves through a potential difference of one volt.

$$1 \text{ joule} = (1 \text{ coulomb})(1 \text{ volt}) \quad \textbf{(24.21)}$$

or, as a single conversion factor

$$\frac{1 \text{ joule}}{1 \text{ volt coulomb}}$$

The amount of work done by a voltaic cell of a known electrical potential is given by the following relationship

electrical work → w, *cell potential* → E, *electrical charge delivered* → q

$$\underset{\text{joules}}{w} = -\underset{\text{volts}}{E}\underset{\text{coulombs}}{q} \quad \textbf{(24.22)}$$

(with a minus sign because this is work done *by* the system on the surroundings).

EXAMPLE 24.8 Thermodynamics of Electrochemical Cells

Calculate the electrical work done by a cell that has delivered 645 C of charge at a potential of 1.35 V.

Introducing $E = 1.35$ V and $q = 645$ C into the expression for work gives

$$w = -(1.35 \text{ V})(645 \text{ C})\left(\frac{1 \text{ J}}{1 \text{ V C}}\right) = \mathbf{-871 \text{ J}}$$

EXERCISE A voltaic cell is designed to deliver 1.0 kJ of electrical work during each minute of operation. If the operating potential of the cell is 1.52 V, what amount of electrical charge must be generated by the cell reaction each minute? *Answer* 660 C.

SUMMARY OF SECTIONS 24.3–24.6

The occurrence of a redox reaction in an electrochemical cell can generate electrical energy (in a voltaic cell) or be caused by electrical energy (in an electrolytic cell). Each of the two half-cells in such cells includes an electrode and its surrounding electrolyte; the electrodes are connected by an external circuit through which electrons enter and leave the cell. Four things happen simultaneously in the operation of an electrochemical cell:

1. An oxidation reaction occurs at the surface of one electrode (the anode).
2. A reduction reaction occurs at the surface of the other electrode (the cathode).
3. Ions flow in the electrolytes.
4. Electrons flow in an external circuit. A pathway for ions to cross the boundary between the half-cells must be provided (e.g., by a salt bridge).

In the notation for voltaic cells (Equation 24.14), single vertical lines indicate physical contact between different phases and double vertical lines indicate a barrier that permits ion flow but prevents mixing of electrolytes. Also, commas separate formulas of species in the same phase. Electrodes may be inert or may take part in the cell reaction (see Figure 24.5). The cell reaction is the redox reaction that is the sum of the oxidation and reduction half-reactions that occur at the anode and the cathode.

The charge of 1 mol of electrons, the faraday (96,500 C/mol e^-), provides the connection between the quantity of electrical charge that has entered or left a cell (equal to current, in amperes, multiplied by time) and the stoichiometry of the cell reaction. The cell potential (E, given in volts) is a measure of the relative abilities of the species in each half-cell to gain and lose electrons. The electrical work that a cell can generate is given by $w = -Eq$.

Aside 24.1 THOUGHTS ON CHEMISTRY: VOLTA DESCRIBES HIS DISCOVERIES

Written in 1800, the following excerpt is from a scientific publication in which Alexander Volta described in detail the construction of a "voltaic pile," the forerunner of voltaic cells as we know them.

I provide a few dozens of small round plates or disks of copper, brass, or rather silver, an inch in diameter more or less (pieces of coin, for example), and an equal number of plates of tin, or, what is better, of zinc, nearly of the same size and figure ... I prepare also a pretty large number of circular pieces of pasteboard, or any other spongy matter capable of imbibing and retaining a great deal of water or moisture, with which they must be well impregnated in order to ensure success to the experiments....

Having all these pieces ready in a good state ... I have nothing to do but to arrange them, a matter exceedingly simple and easy.... I place them horizontally, on a table or any other stand, one of the metallic pieces, for example one of silver, and over the first I adapt one of zinc; on the second I place one of the moistened disks, then another plate of silver followed immediately by another of zinc, over which I place another of the moistened disks. In this manner I continue ... to form, of several of these stories, a column as high as possible without any danger of its falling....

If, by means of an ample contact of the hand (well moistened) I establish on one side a good communication with one of the extremities of my electro-motive apparatus (we must give new names to instruments that are new not only in their form, but in their effects on the principle on which they depend); and on the other I apply the forehead, eye-lid, tip of the nose, also well moistened, or any other part of the body where the skin is very delicate: if I apply, I say, with a little pressure, any one of these delicate parts, well moistened, to the point of a metallic wire, communicating properly with the other extremity of the said apparatus, I experience, at the moment that the conducting circle is completed, at the place of the skin touched, and a little beyond it, a blow and a prick, which suddenly passes, and is repeated as many times as the circle is interrupted and restored....

What proof more evident of the continuation of the electric current as long as the communication of the conductors forming the circle is continued?—and that such a current is only suspended by interrupting that communication? This endless circulation of the electric fluid (this perpetual motion) may appear paradoxical and even inexplicable, but it is no less true and real, and you feel it, as I may say, with your hands.

Source: Alexander Volta, "On the Electricity Excited by the Mere Contact of Conducting Substances of Different Kinds," Alexander Volta's letter to the Royal Society of London, March 20, 1800. Quoted from an English translation, Philosophical Magazine, September 1800, as reproduced in Galvani-Volta, by Bern Dibner (Burndy Library, Norwalk, Connecticut, 1952), pp. 42 ff.

STANDARD REDUCTION POTENTIALS

24.7 $E°$ DEFINED

Just as a cell reaction can be divided into two half-reactions, a cell potential can be divided into two potentials. An **electrode potential** is the potential for an electrochemical half-reaction, and it is in this form that the potentials of electrochemical cells are tabulated. The cell potential is the sum of the electrode potentials, which can be thought of as anode and cathode potentials, or oxidation and reduction potentials:

$$E_{\text{cell}} = E_{\text{anode}} + E_{\text{cathode}} \quad \text{or} \quad E_{\text{cell}} = E_{\text{ox}} + E_{\text{red}} \qquad \textbf{(24.23a, b)}$$

As is customary for thermodynamic data, the electrode potentials are reported for substances in the standard states: all aqueous solutions at 1 M concentrations, all gases at 1 bar partial pressure, and all pure solids and liquids in their most stable forms at 1 bar. The symbol for potentials measured under these conditions is $E°$ and, as usual, we assume the temperature to be 25 °C unless stated otherwise.

Any redox couple can be described by an ion-electron equation written either as oxidation or reduction, for example

$$Fe^{2+} \longrightarrow Fe^{3+} + e^-$$
$$Fe^{3+} + e^- \longrightarrow Fe^{2+}$$

The potentials for half-reactions, which are the electrode potentials, by international agreement are tabulated for reduction—the addition of electrons.

In a by-now-familiar situation, a reference point must be chosen for electrode potential values because individual electrode potentials cannot be measured. (It is not possible to have oxidation without reduction or vice versa.) The proton–hydrogen gas redox couple is chosen as the reference point for the relative scale of electrode potentials. This electrode, referred to as the *standard hydrogen electrode (SHE)*, is assigned a standard potential of 0.000 V at 298 K.

$$2H^+(1\ M) + 2e^- \longrightarrow H_2(g, 1\ \text{bar}) \qquad E° = 0.000\ \text{V}$$

The entire potential of a cell that combines an H^+/H_2 electrode with a second electrode is assigned to the second electrode. The potential of 0.000 V for the H^+/H_2 couple and all other potentials established relative to it under the standard state conditions are called **standard electrode potentials,** or **standard reduction potentials.**

Consider the two cells of Figure 24.6. For the cell with the hydrogen and copper electrodes (Figure 24.6a)

anode reaction	$H_2(g) \longrightarrow 2H^+ + 2e^-$	
cathode reaction	$Cu^{2+} + 2e^- \longrightarrow Cu(s)$	*measured cell voltage*
cell reaction	$Cu^{2+} + H_2(g) \longrightarrow 2H^+ + Cu(s)$	0.337 V

The entire measured cell voltage is assigned to the Cu^{2+}/Cu half-cell. The cell reaction is spontaneous in the direction written, copper ion is *reduced,* and, assuming the measurements have been made at standard state conditions, the measured voltage is the standard reduction potential of copper:

$$Cu^{2+} + 2e^- \longrightarrow Cu(s) \qquad E° = 0.337\ \text{V}$$

In the cell in which the hydrogen and zinc electrodes are combined (Figure 24.6b), the hydrogen electrode acts as the *cathode:*

anode reaction	$Zn(s) \longrightarrow Zn^{2+} + 2e^-$	*measured*
cathode reaction	$2H^+ + 2e^- \longrightarrow H_2(g)$	*cell voltage*
cell reaction	$Zn(s) + 2H^+ \longrightarrow Zn^{2+} + H_2(g)$	0.763 V

Here again the reaction is spontaneous as written, and the entire cell voltage is assigned to the Zn^{2+}/Zn couple. Note that the cell and electrode potentials are positive for cell and electrode reactions that are spontaneous as written (equivalent to a negative $\Delta G°$).

However, because by convention it is chosen to report standard *reduction* potentials, the anode reaction and the sign of the measured voltage are reversed. Again assuming measurements at standard state conditions, the standard reduction potential of the Zn^{2+}/Zn couple is

$$Zn^{2+} + 2e^- \longrightarrow Zn(s) \qquad E^\circ = -0.763 \text{ V}$$

The result of reporting electrode potentials for reduction is that substances less easily reduced than hydrogen ion (such as Zn^{2+}) have negative reduction potentials and substances more easily reduced than hydrogen ion (such as Cu^{2+}) have positive reduction potentials.

24.8 USING E° VALUES

A table of standard reduction potentials can be used in many ways to provide both qualitative and quantitative information about redox reactions and electrochemical cells. Table 24.4 (page 614) illustrates the usual format for such tables (a more extensive table is given in Appendix VI). The table is a list of *reactants* in the order of increasing strength as oxidizing agents.

The reactants at the top of the table are the *weakest oxidizing agents;* the reactants at the bottom of the table are the *strongest oxidizing agents.* Compare the position of Na^+, which is very stable to reduction and is therefore a very weak oxidizing agent, with the positions of ozone and fluorine, which are strong oxidizing agents (and are easily reduced). Conversely, the products at the top of the tables are the *strongest reducing agents* (the common alkali and alkaline earth metals) and the products at the bottom are the *weakest reducing agents.*

Comparison of Table 24.4a with Table 17.3 shows the basis of the activity series of metals. The metals appear in the activity series in the order of increasingly positive standard reduction potentials, and therefore of decreasing activity as reducing agents.

Reactions in acidic and alkaline solutions are listed separately in tables of standard reduction potentials, because their standard state conditions differ. Where the hydrogen ion is a participant in the half-reaction, the potential must be measured at pH 0—that is, with a hydrogen ion concentration of 1 M, the standard state for all ions in solution. Where OH^- is a participant, its concentration must also be 1 M, giving a pH of 14 rather than 0. The potentials for reactions in which neither H^+ nor OH^- is a participant are usually listed together with reactions in acidic solution, although the standard potentials of such reactions are unaffected by pH.

A spontaneous redox reaction (at standard state conditions) occurs between the *product* in a given couple (which will be oxidized) and a *reactant* in any couple below it (which will be reduced) in a table such as Table 24.4. Zinc is a product and Cu^{2+} is a reactant in a couple farther along in the table. This information alone tells us that if zinc and copper electrodes are combined in an electrochemical cell, oxidation of zinc and reduction of the copper ion will occur. Furthermore, it indicates that this redox reaction will be thermodynamically spontaneous if these reactants are directly combined in aqueous solution, rather than in an electrochemical cell.

The standard potential for any redox reaction can be found by combining the potentials for the two half-cell reactions. A positive standard potential means that the reaction will be spontaneous in the direction written (under standard state conditions) *either* as a reaction in solution or as the cell reaction in a cell in which the two specific half-cells are combined. The calculated potential will be the cell potential for the reaction when it takes place in a voltaic cell (under standard state conditions).

Because electrode potentials are all given for half-reactions in the reduction direction, the oxidation half-reaction must be reversed in direction and the sign of its E° reversed before the half-reactions and E° values can be combined. The half-reaction that is higher in Table 24.4 (larger negative or smaller positive E° value) is reversed. This will be the oxidation (anode) half of the overall redox reaction. For example, to

Table 24.4
Standard State Reduction Potentials
A more extensive table is given in Appendix VI.

(a) Acidic solution	$E°$ (V)
$Li^+ + e^- \longrightarrow Li(s)$	−3.045
$K^+ + e^- \longrightarrow K(s)$	−2.925
$Ba^{2+} + 2e^- \longrightarrow Ba(s)$	−2.906
$Sr^{2+} + 2e^- \longrightarrow Sr(s)$	−2.888
$Ca^{2+} + 2e^- \longrightarrow Ca(s)$	−2.866
$Na^+ + e^- \longrightarrow Na(s)$	−2.714
$Mg^{2+} + 2e^- \longrightarrow Mg(s)$	−2.363
$H_2(g) + 2e^- \longrightarrow 2H^-$	−2.25
$Al^{3+} + 3e^- \longrightarrow Al(s)$	−1.662
$Mn^{2+} + 2e^- \longrightarrow Mn(s)$	−1.185
$Zn^{2+} + 2e^- \longrightarrow Zn(s)$	−0.763
$Cr^{3+} + 3e^- \longrightarrow Cr(s)$	−0.744
$Fe^{2+} + 2e^- \longrightarrow Fe(s)$	−0.440
$Cr^{3+} + e^- \longrightarrow Cr^{2+}$	−0.408
$Cd^{2+} + 2e^- \longrightarrow Cd(s)$	−0.403
$PbSO_4(s) + 2e^- \longrightarrow Pb(s) + SO_4^{2-}$	−0.359
$PbCl_2(s) + 2e^- \longrightarrow Pb(s) + 2Cl^-$	−0.268
$Ni^{2+} + 2e^- \longrightarrow Ni(s)$	−0.250
$Sn^{2+} + 2e^- \longrightarrow Sn(s)$	−0.136
$Pb^{2+} + 2e^- \longrightarrow Pb(s)$	−0.126
$2H^+ + 2e^- \longrightarrow H_2(g)$	0.000
$S(s) + 2H^+ + 2e^- \longrightarrow H_2S(aq)$	0.142
$Sn^{4+} + 2e^- \longrightarrow Sn^{2+}$	0.15
$Sb_2O_3(s) + 6H^+ + 6e^- \longrightarrow 2Sb(s) + 3H_2O(l)$	0.152
$Cu^{2+} + e^- \longrightarrow Cu^+$	0.153
$SO_4^{2-} + 4H^+ + 2e^- \longrightarrow H_2SO_3(aq) + H_2O(l)$	0.172
$AgCl(s) + e^- \longrightarrow Ag(s) + Cl^-$	0.222
$Cu^{2+} + 2e^- \longrightarrow Cu(s)$	0.337
$SO_4^{2-} + 8H^+ + 6e^- \longrightarrow S(s) + 4H_2O(l)$	0.357
$H_2SO_3(aq) + 4H^+ + 4e^- \longrightarrow S(s) + 3H_2O(l)$	0.450
$I_2(s) + 2e^- \longrightarrow 2I^-$	0.536
$MnO_4^- + e^- \longrightarrow MnO_4^{2-}$	0.564
$[PtCl_6]^{2-} + 2e^- \longrightarrow [PtCl_4]^{2-} + 2Cl^-$	0.68
$O_2(g) + 2H^+ + 2e^- \longrightarrow H_2O_2(aq)$	0.683
$Fe^{3+} + e^- \longrightarrow Fe^{2+}$	0.771
$Hg_2^{2+} + 2e^- \longrightarrow 2Hg(l)$	0.788
$Ag^+ + e^- \longrightarrow Ag(s)$	0.799
$2NO_3^- + 4H^+ + 2e^- \longrightarrow N_2O_4(g) + 2H_2O(l)$	0.803
$2Hg^{2+} + 2e^- \longrightarrow Hg_2^{2+}$	0.920
$NO_3^- + 3H^+ + 2e^- \longrightarrow HNO_2(aq) + H_2O(l)$	0.94
$NO_3^- + 4H^+ + 3e^- \longrightarrow NO(g) + 2H_2O(l)$	0.96
$Pd^{2+} + 2e^- \longrightarrow Pd(s)$	0.987
$Br_2(l) + 2e^- \longrightarrow 2Br^-$	1.065
$Br_2(aq) + 2e^- \longrightarrow 2Br^-$	1.087
$ClO_4^- + 2H^+ + 2e^- \longrightarrow ClO_3^- + H_2O(l)$	1.19
$2IO_3^- + 12H^+ + 10e^- \longrightarrow I_2(s) + 6H_2O(l)$	1.195

(a) Acidic solution (*continued*)	$E°$ (V)
$Pt^{2+} + 2e^- \longrightarrow Pt(s)$	~1.2
$ClO_3^- + 3H^+ + 2e^- \longrightarrow HClO_2(aq) + H_2O(l)$	1.21
$O_2(g) + 4H^+ + 4e^- \longrightarrow 2H_2O(l)$	1.229
$MnO_2(s) + 4H^+ + 2e^- \longrightarrow Mn^{2+} + 2H_2O(l)$	1.23
$2HNO_2(aq) + 4H^+ + 4e^- \longrightarrow N_2O(g) + 3H_2O(l)$	1.29
$Cr_2O_7^{2-} + 14H^+ + 6e^- \longrightarrow 2Cr^{3+} + 7H_2O(l)$	1.33
$Cl_2(g) + 2e^- \longrightarrow 2Cl^-$	1.360
$PbO_2(s) + 4H^+ + 2e^- \longrightarrow Pb^{2+} + 2H_2O(l)$	1.455
$Au^{3+} + 3e^- \longrightarrow Au(s)$	1.498
$MnO_4^- + 8H^+ + 5e^- \longrightarrow Mn^{2+} + 4H_2O(l)$	1.51
$2HClO(aq) + 2H^+ + 2e^- \longrightarrow Cl_2(g) + 2H_2O(l)$	1.63
$HClO_2(aq) + 2H^+ + 2e^- \longrightarrow HClO(aq) + H_2O(l)$	1.645
$H_2O_2(aq) + 2H^+ + 2e^- \longrightarrow 2H_2O(l)$	1.776
$O_3(g) + 2H^+ + 2e^- \longrightarrow O_2(g) + H_2O(l)$	2.07
$F_2(g) + 2e^- \longrightarrow 2F^-$	2.87
$F_2(g) + 2H^+ + 2e^- \longrightarrow 2HF(aq)$	3.06

(b) Alkaline solution	$E°$ (V)
$Mg(OH)_2(s) + 2e^- \longrightarrow Mg(s) + 2OH^-$	−2.690
$Al(OH)_3(s) + 3e^- \longrightarrow Al(s) + 3OH^-$	−2.30
$Zn(OH)_2(s) + 2e^- \longrightarrow Zn(s) + 2OH^-$	−1.245
$Fe(OH)_2(s) + 2e^- \longrightarrow Fe(s) + 2OH^-$	−0.877
$2H_2O(l) + 2e^- \longrightarrow H_2(g) + 2OH^-$	−0.828
$Cd(OH)_2(s) + 2e^- \longrightarrow Cd(s) + 2OH^-$	−0.809
$Ni(OH)_2(s) + 2e^- \longrightarrow Ni(s) + 2OH^-$	−0.72
$Fe(OH)_3(s) + e^- \longrightarrow Fe(OH)_2(s) + OH^-$	−0.56
$S(s) + 2e^- \longrightarrow S^{2-}$	−0.447
$Cu_2O(s) + H_2O(l) + 2e^- \longrightarrow 2Cu(s) + 2OH^-$	−0.358
$CrO_4^{2-} + 4H_2O(l) + 3e^- \longrightarrow Cr(OH)_3(s) + 5OH^-$	−0.13
$MnO_2(s) + 2H_2O(l) + 2e^- \longrightarrow Mn(OH)_2(s) + 2OH^-$	−0.05
$NO_3^- + H_2O(l) + 2e^- \longrightarrow NO_2^- + 2OH^-$	0.01
$HgO(s) + H_2O(l) + 2e^- \longrightarrow Hg(l) + 2OH^-$	0.098
$PbO_2(s) + H_2O(l) + 2e^- \longrightarrow PbO(s) + 2OH^-$	0.247
$IO_3^- + 3H_2O(l) + 6e^- \longrightarrow I^- + 6OH^-$	0.26
$ClO_3^- + H_2O(l) + 2e^- \longrightarrow ClO_2^- + 2OH^-$	0.33
$Ag_2O(s) + H_2O(l) + 2e^- \longrightarrow 2Ag(s) + 2OH^-$	0.345
$ClO_4^- + H_2O(l) + 2e^- \longrightarrow ClO_3^- + 2OH^-$	0.36
$O_2(g) + 2H_2O(l) + 4e^- \longrightarrow 4OH^-$	0.401
$NiO_2(s) + 2H_2O(l) + 2e^- \longrightarrow Ni(OH)_2(s) + 2OH^-$	0.490
$MnO_4^- + 2H_2O(l) + 3e^- \longrightarrow MnO_2(s) + 4OH^-$	0.588
$BrO_3^- + 3H_2O(l) + 6e^- \longrightarrow Br^- + 6OH^-$	0.61
$ClO^- + H_2O(l) + 2e^- \longrightarrow Cl^- + 2OH^-$	0.89
$O_3(g) + H_2O(l) + 2e^- \longrightarrow O_2(g) + 2OH^-$	1.24

Note: In some chemistry texts, particularly older American ones, you may find *standard oxidation potentials*. Such potentials are for the same standard state conditions relative to 0.000 V for the H_2/H^+ couple. The differences are that the ion-electron equations are written as oxidations (electrons on the other side) and the *signs* of the potentials are opposite. In 1953 the International Union of Pure and Applied Chemistry chose to recommend the uniform, worldwide use of standard reduction potentials, which had previously been more common in Europe. The numerical values for both sets of potentials are the same; only the signs are different.

Source: Therald Moeller, *Inorganic Chemistry: A Modern Introduction*, Appendix IV (New York: Wiley, 1982 © 1982 by John Wiley & Sons, Inc.).

calculate the standard state potential for a zinc-copper cell, the ion-electron equations and the potentials,

reduction	$Zn^{2+} + 2e^- \longrightarrow Zn(s)$	$E^\circ = -0.763$ V		**(24.24)**
reduction	$Cu^{2+} + 2e^- \longrightarrow Cu(s)$	$E^\circ = 0.337$ V		**(24.25)**

are combined as follows:

oxidation	$Zn(s) \longrightarrow Zn^{2+} + 2e^-$	$E^\circ = 0.763$ V
reduction	$Cu^{2+} + 2e^- \longrightarrow Cu(s)$	$E^\circ = 0.337$ V
	$Zn(s) + Cu^{2+} \longrightarrow Zn^{2+} + Cu(s)$	$E^\circ = 1.100$ V

Earlier, on the basis of qualitative knowledge of the chemistry of the substances involved, we predicted that the following reaction would occur (Example 17.5):

$$H_2S(aq) + Cl_2(g) \longrightarrow S(s) + 2HCl(aq)$$

By combining E° values we can verify this prediction by showing that the reaction potential is positive.

oxidation	$H_2S(aq) \longrightarrow S(s) + 2H^+ + 2e^-$	$E^\circ = -0.142$ V
reduction	$Cl_2(g) + 2e^- \longrightarrow 2Cl^-$	$E^\circ = 1.360$ V
	$H_2S(aq) + Cl_2(g) \longrightarrow S(s) + 2HCl(aq)$	$E^\circ = 1.218$ V

The displacement of sulfur from aqueous hydrogen sulfide by chlorine will be spontaneous under standard state conditions. With the appropriate electrodes and electrolytes, this reaction would take place in a voltaic cell, with a cell potential of 1.218 V.

In combining ion-electron equations it is often necessary to multiply the equations by whole numbers to equalize the number of electrons gained and lost. The E° values *are not* multiplied by these same factors. The standard reduction potentials are a measure of the *relative* tendencies of substances in their standard states to gain or lose electrons. The *amounts* of the substances have no effect on the size of the potential. Consider the common batteries: A, AA, C, and D. These batteries of the same type differ in size, but they each have a 1.5 V potential. Do not make the common errors of (1) multiplying the value of E° by the factors used to equalize the number of electrons in the ion-electron equations, or (2) forgetting to reverse the sign of the E^0 value (from Table 24.4) for the oxidation half-reaction.

Also, in using standard potential values it must always be remembered that, like ΔG° values, they only allow predictions of whether or not a reaction is thermodynamically spontaneous under standard state conditions. As for all thermodynamic quantities (e.g., ΔH or K values), the potentials have no bearing on *how fast* a possible reaction will proceed.

Examples 24.9, 24.10, and 24.11 illustrate how a table of standard reduction potentials can be used to provide qualitative information about the relative strengths of oxidizing and reducing agents, and also how values of E° for cell reactions can be calculated.

EXAMPLE 24.9 Interpreting E° Values

(a) Using Table 24.4, predict which species will be oxidized and which reduced in a voltaic cell in which the reactants in the half-cells are the Fe^{3+}/Fe^{2+} couple and the $I_2(s)/I^-$ couple. (b) Write the balanced equation and calculate E° for this reaction.

(a) Locating these couples in Table 24.4 shows that the $I_2(s)/I^-$ couple appears above the Fe^{3+}/Fe^{2+} couple and has the smaller positive E° value. Therefore, we would expect **I^- (the product in the higher couple in Table 24.4) to be oxidized at the anode and Fe^{3+} (the reactant in the lower couple) to be reduced at the cathode.**

(b) The oxidation and reduction half-reactions would be

$$\textit{oxidation} \quad 2I^- \longrightarrow I_2(s) + 2e^-$$
$$\textit{reduction} \quad Fe^{3+} + e^- \longrightarrow Fe^{2+}$$

To equalize the number of electrons, the reduction equation must be multiplied by two, but the E° value is *not* multiplied by this factor.

$$\begin{array}{rcll} 2I^- & \longrightarrow & I_2(s) + 2e^- & E^\circ = -0.536 \text{ V} \\ (2)[Fe^{3+} + e^- & \longrightarrow & Fe^{2+}] & E^\circ = 0.771 \text{ V} \\ \hline \mathbf{2Fe^{3+} + 2I^-} & \longrightarrow & \mathbf{2Fe^{2+} + I_2(s)} & \mathbf{E^\circ = 0.235 \text{ V}} \end{array}$$

The positive value of the emf confirms that this reaction will be spontaneous under standard state conditions.

EXERCISE Use the standard reduction potentials given in Table 24.4 to calculate E° for the reaction $3Zn(s) + IO_3^- + 3H_2O(l) \longrightarrow I^- + 3Zn(OH)_2(s)$.

Answer $E^\circ = 1.51$ V.

EXAMPLE 24.10 Interpreting E° Values

Earlier, we included nitric acid and the three chlorine-containing oxo anions—perchlorate ion, chlorate ion, and hypochlorite ion—in our list of common oxidizing agents (see Table 17.6). Using the E° values for the couples in the ion-electron equations in Table 24.4a, arrange these species in order of increasing strength as oxidizing agents at standard state conditions in acid solution (ClO^- is present in strongly acid solution as $HClO$).

The more positive the standard reduction potential values, the stronger the oxidizing agent. In the nitric acid solution we can assume that the nitrate ion is the reactant. Scanning Table 24.4a shows three possible reactions for NO_3^-. The order of the E° values for the possible reactions of the species under consideration is as follows:

Couples	E° (V)
$NO_3^-/NO(g)$, $NO_3^-/HNO_2(aq)$, $NO_3^-/N_2O_4(g)$	0.96, 0.94, 0.803
ClO_4^-/ClO_3^-	1.19
$ClO_3^-/HClO_2(aq)$	1.21
$HClO(aq)/Cl_2(g)$	1.63

At standard state conditions, nitrate ion is the weakest oxidizing agent and hypochlorous acid is the strongest. The order of strength as oxidizing agents is

$$\mathbf{NO_3^- < ClO_4^- < ClO_3^- < HClO(\mathit{aq})}$$

[The E° values for the three NO_3^- couples illustrate why prediction of the specific reduction products formed from nitric acid is difficult.]

EXERCISE Arrange ClO_4^-, ClO_3^-, ClO^-, and NO_3^- in order of increasing strength as oxidizing agents at standard state conditions in alkaline solution. Use the E° values given in Table 24.4b. *Answer* $NO_3^- < ClO_3^- < ClO_4^- < ClO^-$.

EXAMPLE 24.11 Calculating and Interpreting E° Values

There is a significant difference in the way some metals react with oxidizing and nonoxidizing acids. Using data from Table 24.4a, calculate E° and compare the reactions of zinc and copper with hydrochloric acid and nitric acid under standard state conditions.

To react with an acid, a metal must be oxidized. With hydrochloric acid, only H^+ is available to be reduced. Combining the electrode potentials for zinc and copper oxidation with that for the H^+/H_2 couple

$$\begin{array}{rcll} Zn(s) & \longrightarrow & Zn^{2+} + 2e^- & E^\circ = 0.763\text{ V} \\ 2H^+ + 2e^- & \longrightarrow & H_2(g) & E^\circ = 0.000\text{ V} \\ \hline Zn(s) + 2H^+ & \longrightarrow & Zn^{2+} + H_2(g) & E^\circ = 0.763\text{ V} \end{array}$$

$$\begin{array}{rcll} Cu(s) & \longrightarrow & Cu^{2+} + 2e^- & E^\circ = -0.337\text{ V} \\ 2H^+ + 2e^- & \longrightarrow & H_2(g) & E^\circ = 0.000\text{ V} \\ \hline Cu(s) + 2H^+ & \longrightarrow & Cu^{2+} + H_2(g) & E^\circ = -0.337\text{ V} \end{array}$$

shows that **under standard state conditions zinc will react with hydrochloric acid, but copper will not.**

In the reaction of nitric acid with metals, the nitrate ion is available to be reduced. [The H^+ ion is also available but is not of concern because it is not as easily reduced as NO_3^-.] Selecting from Table 24.4a the NO_3^- couple with the most positive E°, the reaction potentials are found to be

$$\begin{array}{rcll} (3)[Zn(s) & \longrightarrow & Zn^{2+} + 2e^-] & E^\circ = 0.763\text{ V} \\ (2)[NO_3^- + 4H^+ + 3e^- & \longrightarrow & NO(g) + 2H_2O(l)] & E^\circ = 0.96\text{ V} \\ \hline 3Zn(s) + 2NO_3^- + 8H^+ & \longrightarrow & 3Zn^{2+} + 2NO(g) + 4H_2O(l) & E^\circ = 1.72\text{ V} \end{array}$$

$$\begin{array}{rcll} (3)[Cu(s) & \longrightarrow & Cu^{2+} + 2e^-] & E^\circ = -0.337\text{ V} \\ (2)[NO_3^- + 4H^+ + 3e^- & \longrightarrow & NO(g) + 2H_2O(l)] & E^\circ = 0.96\text{ V} \\ \hline 3Cu(s) + 2NO_3^- + 8H^+ & \longrightarrow & 3Cu^{2+} + 2NO(g) + 4H_2O(l) & E^\circ = 0.62\text{ V} \end{array}$$

Both zinc and copper will dissolve in nitric acid at standard state conditions and 25 °C. The reaction with zinc is more favorable.

EXERCISE Can iodide ion be oxidized to iodate ion by reaction with chromate ion in alkaline solution under standard state conditions? Will the reaction occur using dichromate ion in acidic solution under standard state conditions? See Table 24.4 for values of E°.

Answers no, $E^\circ = -0.39$ V for $2CrO_4^{2-} + 5H_2O(l) + I^- \longrightarrow IO_3^- + 2Cr(OH)_3(s) + 4OH^-$; yes, $E^\circ = 0.79$ V for $6I^- + Cr_2O_7^{2-} + 14H^+ \longrightarrow 3I_2(s) + 2Cr^{3+} + 7H_2O(l)$ and $E^\circ = 0.14$ V for $5Cr_2O_7^{2-} + 34H^+ + 3I_2(s) \longrightarrow 6IO_3^- + 10Cr^{3+} + 17H_2O(l)$.

24.9 ELECTROLYSIS

In the electrolysis of molten sodium chloride (Figure 24.4), the only possible electrode reactions are the reduction of sodium ions and the oxidation of chloride ions. When electrical energy is introduced into an aqueous solution, however, the possibility of reactions of water, and (if their concentration is high) of H^+ and OH^-, must also be considered. In general, the anode reaction should be the one with the least positive reduction potential and the cathode reaction should be the one with the most positive reduction potential.

However, there are limitations to predicting the products of electrolysis. The voltage required to cause a particular reaction to occur is frequently greater than shown by the standard potential because of **overvoltage**—a collective term for several effects that add to the voltage required by an electrochemical reaction.

For example, the electrolysis of acidified zinc sulfate solution results in deposition of elemental zinc at the cathode ($Zn^{2+} + 2e^- \longrightarrow Zn$, $E^\circ = -0.763$ V) rather than the liberation of hydrogen gas ($2H^+ + 2e^- \longrightarrow H_2$, $E^\circ = 0.000$ V), even though the reduction potential data predict the opposite. Predictions are further complicated when several possible electrode reactions have the same or nearly the same potential values. Under these conditions, two or more reactions may occur simultaneously. This may be desirable (e.g., in the simultaneous electrodeposition of several metals) or undesirable (e.g., in the simultaneous liberation of hydrogen during deposition of a metal).

Electrolysis is used in many ways in industry. Metals are freed from their ores by electrolysis (Section 28.5), and objects are coated, or *plated*, with thin layers of metal. A major area of industrial chemistry—the chloralkali industry (Section 29.8a)—is based on the electrolysis of concentrated sodium chloride solutions. The two possible anode reactions in an aqueous sodium chloride solution are the oxidation of chloride ion or the oxidation of water:

$$2Cl^- \longrightarrow Cl_2(g) + 2e^- \qquad E^\circ = -1.36\ \mathrm{V}$$
$$2H_2O(l) \longrightarrow O_2(g) + 4H^+ + 4e^- \qquad E^\circ = -1.23\ \mathrm{V}$$

When the concentration of chloride ion is high the first reaction takes place to the greater extent, producing chlorine, which is a large-volume industrial chemical. In dilute solutions the second reaction is favored.

The two possible cathode reactions are the reduction of sodium ion or the reduction of water:

$$Na^+ + e^- \longrightarrow Na(s) \qquad E^\circ = -2.71\ \mathrm{V}$$
$$2H_2O(l) + 2e^- \longrightarrow H_2(g) + 2OH^- \qquad E^\circ = -0.83\ \mathrm{V}$$

The reduction of water is much more favorable and the cathode products in the electrolysis of concentrated sodium chloride are hydrogen and a solution of sodium hydroxide, both also valuable industrial chemicals.

SUMMARY OF SECTIONS 24.7–24.9

Electrochemical data are tabulated as standard reduction potentials—electrode potentials for reduction half-reactions at standard state conditions. The values are assigned relative to a value of zero for the standard hydrogen electrode, $H^+/H_2(g)$. The usual table of standard reduction potentials (see Table 24.4) is a list of reactants in the order of increasing strength as oxidizing agents. A spontaneous redox reaction (at standard state conditions) occurs between a species that is a product in a couple higher in the table (larger negative or smaller positive E°) and a species that is a reactant in a couple lower in the table. The E° for such a redox reaction has a positive value; a nonspontaneous reaction has a negative E° value.

The E° value of a redox reaction is found by adding the E° values of two half-cell reactions. The half-reaction of the couple that includes the weaker oxidizing agent (higher in a table like Table 24.4) and the sign of its E° value must be reversed. The E° values are *not* multiplied by the factors needed to equalize electrons gained and lost in the half-cell reactions.

In theory, in an electrolytic cell, the anode reaction should be that with the most negative reduction potential and the cathode reaction should be that with the most positive reduction potential. In practice, the outcome may be different.

THERMODYNAMICS OF REDOX REACTIONS

24.10 RELATIONSHIP BETWEEN ΔG° AND E°

The cell potential is a measure of whether a redox reaction is spontaneous. The free energy change of a reaction is also a measure of spontaneity. It is apparent that there must be some relationship between E and ΔG. Recall that ΔG is the energy free to do useful work (Section 23.6) and that electrical work is equal to $-Eq$, electrochemical potential times charge (Equation 24.22). The amount of electrical charge (in coulombs) available from a redox reaction is equal to the number of moles of electrons transferred times the faraday ($q = nF$). Equating the free energy change to electrical work and replacing q with nF gives the expression that relates the free energy change and electrochemical potential

$$\Delta G = -nFE$$

Because ΔG is usually given in kilojoules and E in volts, it is convenient to convert the units of the faraday as follows:

$$F = \left(\frac{96{,}500\ \text{C}}{\text{mol}\ e^-}\right)\left(\frac{1\ \text{J}}{1\ \text{V C}}\right)\left(\frac{1\ \text{kJ}}{1000\ \text{J}}\right) = \frac{96.5\ \text{kJ}}{\text{V mol}\ e^-} \qquad \textbf{(24.26)}$$

Therefore, when n moles of electrons are transferred in a redox reaction

$$\underset{\textit{kilojoules}}{\Delta G} = -\underset{\textit{moles } e^-}{n}\ \underset{\frac{\textit{kilojoules}}{\textit{volt moles } e^-}}{F}\ \underset{\textit{volts}}{E} \qquad \textbf{(24.27)}$$

When the cell potential is that for the usual standard state conditions, the standard state free energy change for the cell reaction can be found from the cell potential.

$$\Delta G^\circ = -nFE^\circ \qquad \textbf{(24.28)}$$

Values of E and E° can be interpreted in the same way as values of ΔG and ΔG° (see Table 23.4 and Figure 23.7), except that opposite signs are associated with spontaneity or nonspontaneity. The value of E° for a given redox reaction, like the value of ΔG°, is a constant and represents the extent of the reaction under standard state conditions at a specified temperature. The value of E, the electrical potential associated with the reaction as it takes place, changes during the course of a reaction until, at equilibrium, it is equal to zero.

When an electrochemical cell reaches equilibrium, electron flow ceases. If the cell is in a flashlight battery, we say that the battery is dead. At equilibrium, both the free energy change and the cell potential are equal to zero. These relationships are summarized in Table 24.5.

The value of n required in Equations (24.27) and (24.28) is most easily found from the oxidation and reduction half-reactions. This means essentially going backward in the half-reaction method for balancing redox equations.

EXAMPLE 24.12 Thermodynamics of Electrochemical Cells

Find the standard cell potential for an electrochemical cell in which the following reaction takes place spontaneously:

$$Cl_2(g) + 2Br^- \longrightarrow Br_2(aq) + 2Cl^- \qquad \Delta G^\circ = -50.61\ \text{kJ}$$

First, we must find the value of n, the moles of electrons transferred, by writing the half-reactions for reduction of Cl_2 and oxidation of Br^-.

$$\begin{array}{rcl} Cl_2(g) + 2e^- & \longrightarrow & 2Cl^- \\ 2Br^- & \longrightarrow & Br_2(aq) + 2e^- \\ \hline Cl_2(g) + 2Br^- & \longrightarrow & Br_2(aq) + 2Cl^- \end{array}$$

The value of n is 2.

The cell potential is found by solving Equation (24.28) for E° and substituting $n = 2$ and the known value of ΔG°.

$$E^\circ = \frac{-\Delta G^\circ}{nF} = \frac{-(-50.61\ \text{kJ})}{(2\ \text{mol}\ e^-)(96.5\ \text{kJ/V mol}\ e^-)} = \mathbf{0.262\ V}$$

EXERCISE The cell potential for an electrochemical cell in which the following reaction occurs, $2Fe^{3+} + 2I^- \longrightarrow 2Fe^{2+} + I_2(s)$, is $E^\circ = 0.236$ V. Calculate the value of ΔG° for this reaction. *Answer* −45.5 kJ.

Table 24.5
Electrical Potential and ΔG

	ΔG	E
Spontaneous reaction (voltaic cell)	Negative	Positive
Equilibrium	0	0
Non-spontaneous reaction (electrolytic cell)	Positive	Negative

24.11 CONDITIONS OTHER THAN STANDARD STATE

The free energy change at concentrations other than those chosen for the standard state (Section 23.8) is

$$\Delta G = \Delta G^\circ + 2.303\ RT \log Q$$

Substitution of $\Delta G = -nFE$ and $\Delta G^\circ = -nFE^\circ$ into this equation and rearranging gives the relationship between the potential at nonstandard and standard state conditions:

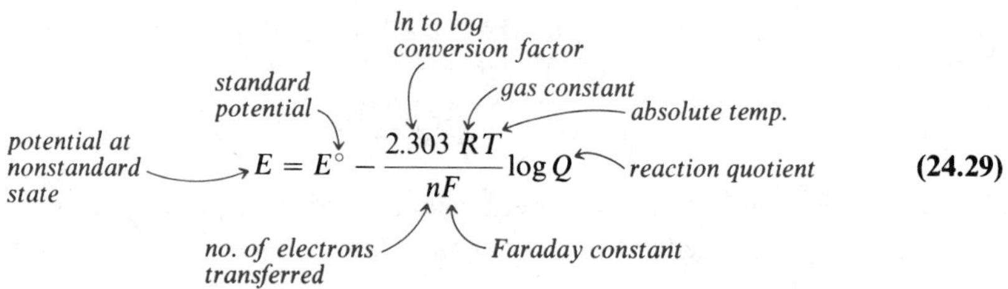

$$E = E^\circ - \frac{2.303\ RT}{nF} \log Q \qquad \textbf{(24.29)}$$

This is known as the Nernst equation, named for Walther Nernst, a German chemist who in the early 1900s studied electrochemistry and was also the first person to state the third law of thermodynamics. The Nernst equation is valid both for the potentials of half-reactions and for overall redox reactions. When $T = 298.15$ K, the value for the factor $2.303\ RT/F$ is 0.0592 and Equation (24.29) becomes

$$E = E^\circ - \frac{0.0592}{n} \log Q \qquad \textbf{(24.30)}$$

EXAMPLE 24.13 Thermodynamics of Electrochemical Cells

Under standard state conditions the reaction

$$3Zn(s) + 2Cr^{3+} \longrightarrow 3Zn^{2+} + 2Cr(s) \qquad E^\circ = 0.019\ \text{V}$$

is spontaneous. Will the above reaction occur spontaneously if $[Cr^{3+}] = 0.010$ mol/L and $[Zn^{2+}] = 5.3$ mol/L?

The value of the reaction quotient is

$$Q = \frac{[Zn^{2+}]^3}{[Cr^{3+}]^2} = \frac{(5.3)^3}{(0.010)^2} = 2.8 \times 10^5$$

and the cell potential is

$$E = E^\circ - \frac{0.0592}{n} \log Q$$

$$= 0.019 - \frac{0.0592}{6} \log(2.8 \times 10^5) = -0.035\ \text{V}$$

The negative cell potential implies that the reaction will not be spontaneous under these conditions. This is reasonable because the $[Zn^{2+}]$ concentration is larger than at standard state conditions and $[Cr^{3+}]$ has been reduced. From Le Chatelier's principle, the reverse reaction would be expected to be favored.

EXERCISE The standard reduction potential for the MnO_4^-/Mn^{2+} couple in acidic solution is 1.51 V at 25 °C. Will the permanganate ion become a stronger or weaker oxidizing agent in a solution which contains $[MnO_4^-] = 0.10$ mol/L, $[H^+] = 1.3 \times 10^{-2}$ mol/L, and $[Mn^{2+}] = 2.5 \times 10^{-5}$ mol/L?

Answer $E = 1.37$ V. Weaker oxidizing agent because E has smaller positive value.

24.12 REDOX EQUILIBRIA: FINDING *K* FROM *E°*

At equilibrium, $E = 0$ and $Q = K$. Introducing these values into the Nernst equation (Equation 24.30)

$$E = E^\circ - \frac{0.0592}{n} \log K = 0 \qquad \textbf{(24.31)}$$

and rearranging

$$\log K = \frac{n}{0.0592} E^\circ \qquad \textbf{(24.32)}$$

provides a simple relationship between the equilibrium constant and the standard potential, as well as an excellent experimental method for determining equilibrium constants. Equation (24.32) shows that the magnitude of K for a redox reaction is directly proportional to the number of electrons involved and to the standard potential for the reaction. Values of K can be calculated from the standard potential for an overall redox reaction.

EXAMPLE 24.14 Thermodynamics of Electrochemical Cells

The standard reduction potentials are 1.229 V for the $O_2(g)/H_2O(l)$ couple and 1.776 V for the $H_2O_2(aq)/H_2O(l)$ couple in acidic solution. Find E°, ΔG°, and K for the decomposition of hydrogen peroxide.

$$2H_2O_2(aq) \rightleftharpoons 2H_2O(l) + O_2(g)$$

The standard state potential of the cell is

$$\begin{array}{rcll} 2H_2O(l) & \longrightarrow & O_2(g) + 4H^+ + 4e^- & E^\circ = -1.229\text{ V} \\ (2)[H_2O_2(aq) + 2H^+ + 2e^- & \longrightarrow & 2H_2O(l)] & E^\circ = 1.776\text{ V} \\ \hline 2H_2O_2(aq) & \longrightarrow & O_2(g) + 2H_2O(l) & E^\circ = 0.547\text{ V} \end{array}$$

Using this value of E°, we obtain the following value of the standard state Gibbs free energy change:

$$\begin{aligned} \Delta G^\circ &= -nFE^\circ = -(4\text{ mol } e^-)(96.5\text{ kJ/V mol } e^-)(0.547\text{ V}) \\ &= -211\text{ kJ} \end{aligned}$$

and the equilibrium constant is

$$\begin{aligned} \log K &= \frac{n}{0.0592} E^\circ = \frac{4}{0.0592}(0.547) = 37.0 \\ K &= 1 \times 10^{37} \end{aligned}$$

For this spontaneous reaction (at standard state conditions), $E^\circ = 0.547$ V, $\Delta G^\circ = -211$ kJ, and $K = 1 \times 10^{37}$.

EXERCISE At 25 °C, $K = 9.9 \times 10^5$ for the reaction $Cl_2(aq) + I_2(aq) = 2ICl(aq)$. Determine E° for this reaction. *Answer* 0.177 V.

24.13 CALCULATION OF THE STANDARD REDUCTION POTENTIAL OF A HALF-REACTION

Often, not all of the oxidation states of a given element are represented in tables of standard reduction potentials. The electrode potentials for two couples can be combined to give the electrode potential for a different couple.

However, it is important to see that an unknown electrode potential *cannot* be found by simply adding known electrode potentials along with equations, as was done to find cell potentials. It is necessary to take into account the *number of electrons* transferred in each of the half-reactions. This can be demonstrated by combining two half-reactions and their ΔG° values in a Hess's law calculation.

$$\begin{array}{llll} A + 2e^- & \longrightarrow & B & \Delta G_1 = -2FE_1^\circ \\ B + 3e^- & \longrightarrow & D & \Delta G_2 = -3FE_2^\circ \\ \hline A + 5e^- & \longrightarrow & D & \Delta G_{1+2} = (-2FE_1^\circ) + (-3FE_2^\circ) \end{array}$$

The number of electrons in the new half-reaction is five, and the expression for the new electrode potential must be found as follows:

$$\Delta G_{1+2}^\circ = -5FE_{1+2}^\circ$$

$$-2FE_1^\circ - 3FE_2^\circ = -5FE_{1+2}^\circ$$

$$E_{1+2}^\circ = \frac{2E_1^\circ + 3E_2^\circ}{5}$$

To find an unknown electrode potential by combining two known electrode potentials, multiply each known E° by the number of electrons in the electrode reaction, add these two values together (algebraically) and divide by the total number of electrons in the new electrode reaction. [If you wish, you can prove for yourself by using—nFE° in place of ΔG° in a Hess's law calculation that the number of electrons cancels out when electrode potentials are combined to give a cell potential, although, as shown above, they do not when two half-reactions are combined to produce a third half-reaction.]

EXAMPLE 24.15 E° of a Half-Reaction

The standard reduction potentials are -1.798 V and -0.607 V for the couples $U^{3+}/U(s)$ and U^{4+}/U^{3+}, respectively. Calculate E° for the $U^{4+}/U(s)$ couple.

The number of electrons gained in the two known half-reactions and the new half-reactions are

$$\begin{array}{llll} U^{3+} + 3e^- & \longrightarrow & U(s) & E^\circ = -1.798 \text{ V} \\ U^{4+} + e^- & \longrightarrow & U^{3+} & E^\circ = -0.607 \text{ V} \\ \hline U^{4+} + 4e^- & \longrightarrow & U(s) & \end{array}$$

Multiplying each known E° by the number of electrons in the half-reaction, adding them, and dividing by the number of electrons in the new half-reaction gives

$$E^\circ = \frac{(3)(-1.798 \text{ V}) + (1)(-0.607 \text{ V})}{(4)} = \mathbf{-1.500 \text{ V}}$$

The standard reduction potential of the $U^{4+}/U(s)$ couple is -1.500 V.

EXERCISE The standard reduction potential is 0.536 V for the I_2/I^- couple and 1.195 V for the IO_3^-/I_2 couple. Calculate E° for the IO_3^-/I^- couple.

Answer $E^\circ = 1.085$ V.

24.14 EFFECT OF pH ON ELECTROCHEMICAL POTENTIAL

A change in the pH has a profound effect upon the value of the reduction potential for any couple that contains the H^+ or OH^- ion. For example, for the reference couple

$$2H^+ + 2e^- \longrightarrow H_2(g)$$

where $[H^+] = 1$ mol/L, E° is 0.000 V. A change to pure water, where $[H^+] = 1.00 \times 10^{-7}$ mol/L, while keeping the pressure of H_2 at 1.00 bar, decreases the potential for the reaction.

$$E = E^\circ - \frac{0.0592}{n}\log\frac{P_{H_2}}{[H^+]^2}$$

$$= 0.000 - \frac{0.0592}{n}\log\frac{1.00}{(1.00 \times 10^{-7})^2} = 0.000 - \frac{0.0592}{2}(14.00) = -0.414 \text{ V}$$

The negative E value shows that the reverse of the reaction as written is spontaneous. For the same couple in a 1.00 M NaOH solution, where $[OH^-] = 1.00$ mol/L and $[H^+] = 1.00 \times 10^{-14}$ mol/L,

$$E = 0.000 - \frac{0.0592}{2}\log\frac{1.00}{[1.00 \times 10^{-14}]^2} = -0.829 \text{ V}$$

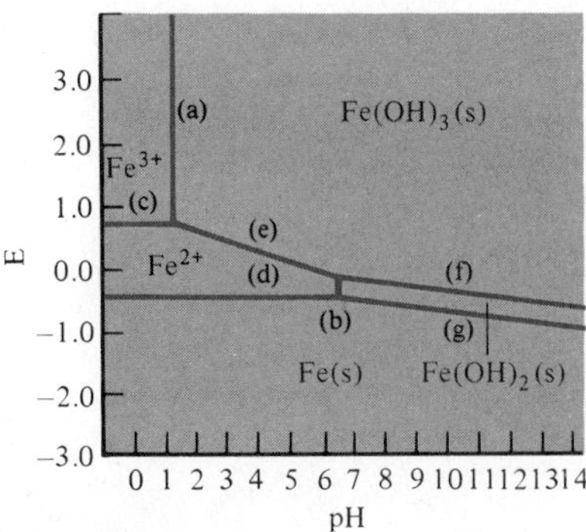

Figure 24.8
pH–Potential Diagram for the Iron–Water System
The various species are thermodynamically stable in the areas indicated.

Figure 24.8 illustrates the effects of variation of pH and potential on a system containing iron and water. Such pH–potential diagrams (also called Pourbaix diagrams) provide a very useful summary of the possible equilibria in a given system.

We can use the simple pH–potential diagram of Figure 24.8 to illustrate how pH–potential diagrams are interpreted. A vertical line represents an equilibrium that is independent of potential—a nonredox equilibrium (Figure 24.9). The iron–water system has two such equilibria (lines a and b), between the cations and the hydroxides that form in alkaline solution.

$$Fe^{3+} \underset{H^+}{\overset{OH^-}{\rightleftharpoons}} Fe(OH)_3(s) \qquad [a]$$

$$Fe^{2+} \underset{H^+}{\overset{OH^-}{\rightleftharpoons}} Fe(OH)_2(s) \qquad [b]$$

To the left of the vertical lines (lower pH) the cations are favored, to the right (higher pH) the hydroxides are favored.

Horizontal lines represent redox equilibria that are independent of pH. In a pH–potential diagram the oxidized species appear above the horizontal (or diagonal) lines and reduced species below the lines. For the iron–water system, two such equilibria (lines c and d) occur.

$$Fe^{3+} + e^- \rightleftharpoons Fe^{2+} \qquad [c]$$
$$Fe^{2+} + 2e^- \rightleftharpoons Fe(s) \qquad [d]$$

Both lie to the left of the points at which the solution becomes sufficiently alkaline for hydroxides to precipitate.

Diagonal lines show equilibria that depend upon both pH and potential—redox equilibria in which hydrogen or hydroxide ions are participants. Lines e, f, and g, respectively, represent the following equilibria of this type:

$$Fe(OH)_3(s) + e^- \rightleftharpoons Fe^{2+} + 3OH^- \qquad [e]$$
$$Fe(OH)_3(s) + e^- \rightleftharpoons Fe(OH)_2(s) + OH^- \qquad [f]$$
$$Fe(OH)_2(s) + 2e^- \rightleftharpoons Fe(s) + 2OH^- \qquad [g]$$

Within the spaces enclosed by the solid lines, the species indicated are thermodynamically stable. Species high in the diagram can react as oxidizing agents; those low in the diagram can react as reducing agents. It must be kept in mind that pH–potential diagrams are based on thermodynamic data and include only thermodynamically stable species. Also, a reminder about thermodynamic versus kinetic stability is in order with respect to pH–potential diagrams. In some cases, the reactions presented in the diagrams come to equilibrium too slowly to be of consequence.

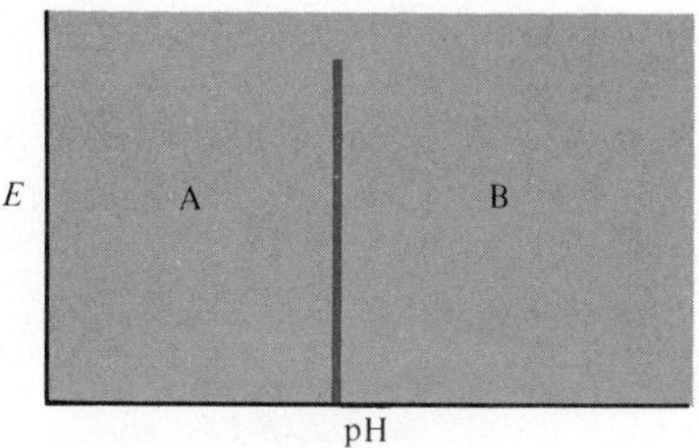

pH-dependent non-redox equilibrium

$$A \underset{OH^- \text{ or } H^+}{\overset{H^+ \text{ or } OH^-}{\rightleftharpoons}} B$$

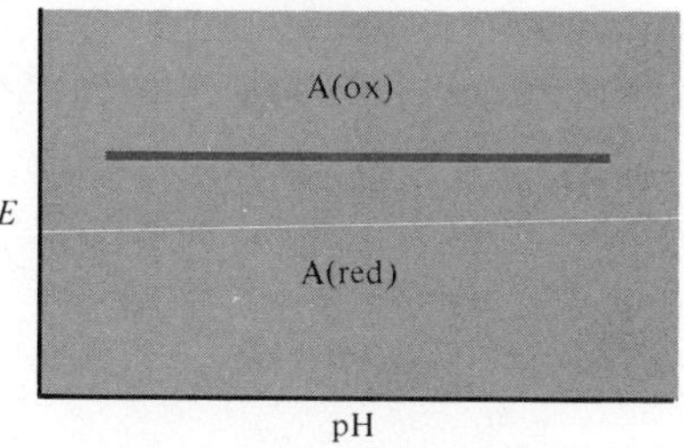

pH-independent redox equilibrium

$$A(ox) \rightleftharpoons A(red)$$

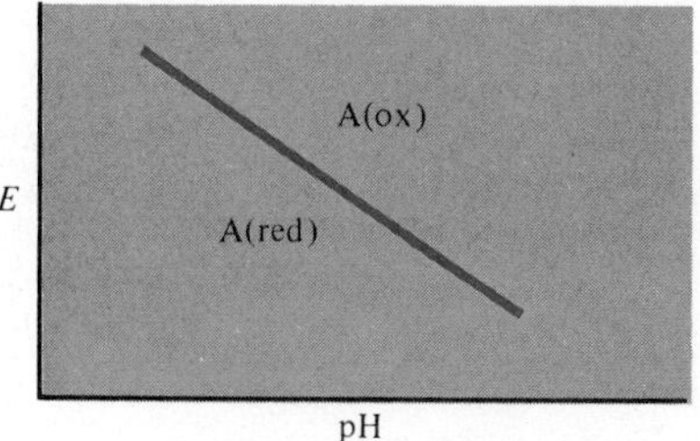

pH-dependent redox equilibrium

$$A(ox) \underset{OH^- \text{ or } H^+}{\overset{H^+ \text{ or } OH^-}{\rightleftharpoons}} A(red)$$

Figure 24.9
Interpretation of pH–Potential Diagrams

SUMMARY OF SECTIONS 24.10–24.14

Redox potentials and free energy changes are related according to $\Delta G = nFE$, where n is the number of moles of electrons transferred and F is the faraday constant. Values of E and E° can be interpreted in the same way as values of G and ΔG°, except that the signs have opposite meanings with respect to spontaneity. The value of E (nonstandard state conditions) and the value of Q, the reaction quotient, are related by the Nernst equation (Equation 24.30). At equilibrium, $E = 0$, $Q = K$, and the Nernst equation reduces to an equation that permits K to be found from $E°$ (Equation 24.32).

A standard reduction potential for a species containing an element in one oxidation state can be calculated from the potentials for species containing that element in two other oxidation states. However, this cannot be done by adding the potentials; it must be done by taking into account the numbers of electrons transferred in each half-reaction (as shown in Section 24.13). The reduction potential for any couple that includes H^+ or OH^- varies with the pH.

PRACTICAL ELECTROCHEMISTRY

24.15 DRY CELLS

Zinc–copper cells were used to provide direct current for many years. But sealed cells that contain no free liquid are more compact, portable, and practical. The best known and still the most common of these *dry cells* is the Leclanché cell, which has been in use for over 100 years.

The zinc anode in a Leclanché cell doubles as the cell container (Figure 24.10). There is a moist ammonium chloride–zinc chloride electrolyte which is thickened to a paste, and the reaction at the carbon cathode is the reduction of manganese dioxide. The anode reaction is the oxidation of zinc. The overall cell reaction is not fully understood, but an approximation to it that roughly agrees with the current-producing capacity of the cell is

$$Zn(s) + 2MnO_2(s) + 2H_2O(l) \longrightarrow Zn(OH)_2(s) + 2MnO(OH)(s)$$

The Leclanché cell has a potential of 1.5 V. Combining more than one of these cells in series (anode to cathode, cathode to anode) gives batteries with voltages that are multiples of 1.5 V. If energy is withdrawn too rapidly from a Leclanché cell, the voltage drops because by-products that hinder the cathode reaction accumulate. A cell in this condition regains its voltage on standing. Unused Leclanché cells deteriorate because of a slow reaction between zinc and the ammonium ion. Common flashlight batteries are Leclanché cells.

The alkaline manganese(IV) oxide–zinc cell, which is illustrated in Figure 24.11, has replaced the Leclanché cell in many applications because it has several advantages:

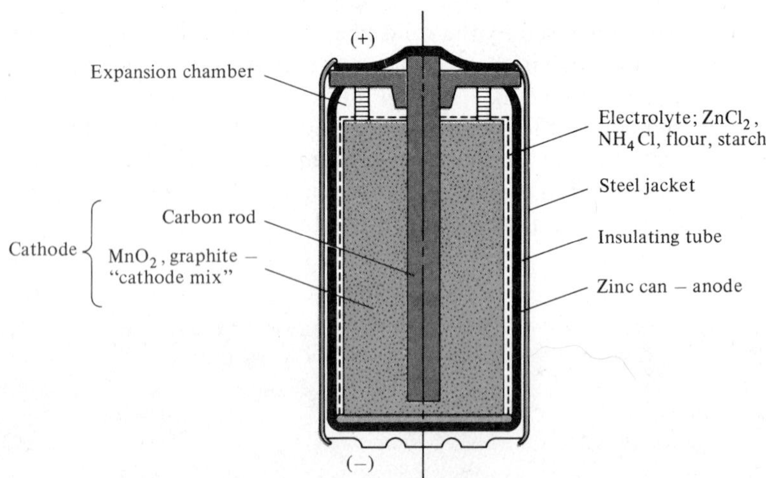

Figure 24.10
Leclanché Dry Cell

It yields a sustained operating voltage at large current drains, operates at low temperatures, has a longer shelf-life, and can, within limits, be recharged. Since the standard voltage of "alkaline batteries" is 1.54 V, they can be used in place of Leclanché cells without problems.

As in the Leclanché cell, the anode and cathode reactions in alkaline batteries are the oxidation of zinc and the reduction of MnO_2. The electrolyte is potassium hydroxide, and the anode is a gel-thickened mixture of zinc dust and potassium hydroxide. Either $[Zn(OH)_4]^{2-}$ or ZnO is formed at the anode and the final product of the complicated cathode reaction is $Mn(OH)_2$.

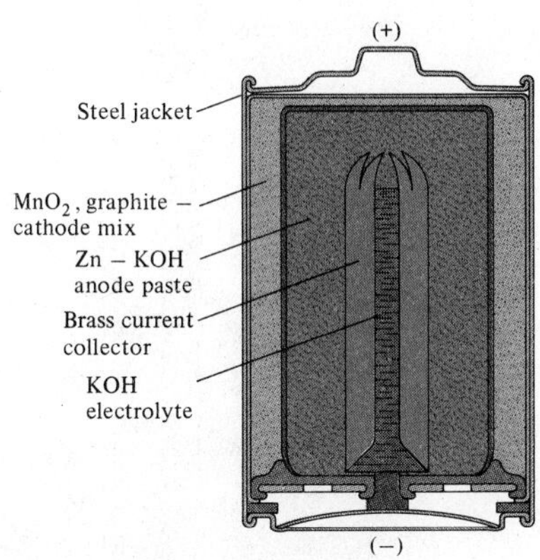

Figure 24.11
Alkaline Manganese(IV) Oxide–Zinc Cell

The solid-state revolution and the rapid growth in the availability of electronic watches and other small devices has stimulated the development of smaller cells that deliver larger currents at various potentials. A particularly successful cell of this type is the mercury cell, or Ruben-Mallory cell. This cell, also an alkaline cell, contains a zinc anode and a mercury(II) oxide cathode. The anode and cathode materials are both compacted powders. The space between these electrodes is filled with an absorbent material containing a sodium or potassium hydroxide electrolyte (Figure 24.12). The cell delivers electrical energy by the following reactions at a potential of 1.35 V

anode reaction $$Zn(s) + 2OH^- \longrightarrow ZnO(s) + H_2O(l) + 2e^-$$

cathode reaction $$HgO(s) + H_2O(l) + 2e^- \longrightarrow Hg(l) + 2OH^-$$

No gaseous products result. Spent cells should be reprocessed for mercury recovery or treated to prevent mercury or mercury compounds from entering the environment and causing contamination. A similar "button" type battery utilizes zinc and silver oxide.

24.16 STORAGE CELLS

Cells that can be recharged by using electrical energy from an external source to reverse the initial oxidation–reduction reactions are called **storage cells** or, if several are combined, **storage batteries** (or accumulators, or secondary cells). They are particularly useful in applications that require electrical energy at one time but generate energy at another time, notably in the automobile. A storage battery provides electrical energy for starting the automobile and is then recharged by energy from a generator or alternator while the engine is running.

The most widely used storage cell is the lead storage cell. It is constructed of electrodes that are lead alloy grids packed with either finely divided spongy elemental lead or finely divided lead(IV) oxide. These electrodes are arranged alternately, separated by thin wooden or fiberglass sheets, and suspended in dilute sulfuric acid solution (about 6 M).

When the external circuit is closed in the discharging cell, the spongy lead is oxidized and forms lead(II) sulfate that adheres to the electrode.

anode reaction $$Pb(s) + HSO_4^- \underset{\text{charge}}{\overset{\text{discharge}}{\rightleftharpoons}} PbSO_4(s) + H^+ + 2e^- \quad \textbf{(24.33)}$$

The released electrons reduce lead(IV) oxide preferentially, and in the presence of hydrogen ions and sulfate ion, adherent lead(II) sulfate forms here also, this time on the lead(IV) oxide cathode.

cathode reaction $$PbO_2(s) + 3H^+ + HSO_4^- + 2e^- \underset{\text{charge}}{\overset{\text{discharge}}{\rightleftharpoons}} PbSO_4(s) + 2H_2O(l) \quad \textbf{(24.34)}$$

The net results of discharge are the conversion of the active chemicals on both electrodes to lead(II) sulfate and a decrease in the concentration of the sulfuric acid solution, as indicated by a decrease in the density of the solution. (A service station attendant checks the charge of a battery by measuring the density, or specific gravity, of the sulfuric acid solution.) Reactions (24.33) and (24.34) are reversed when the

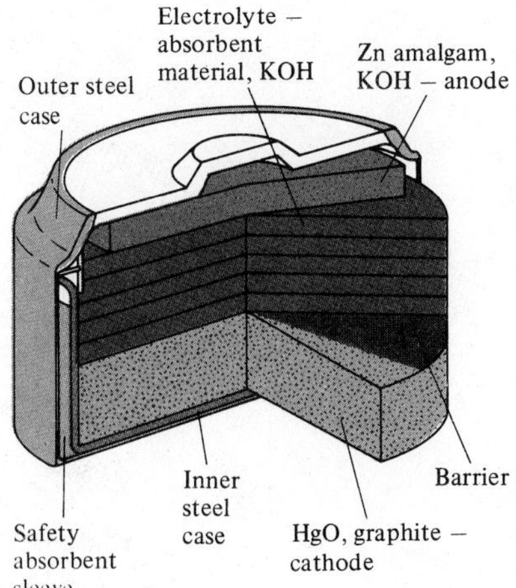

Figure 24.12
A "Button" Type Ruben-Mallory, Mercury Battery

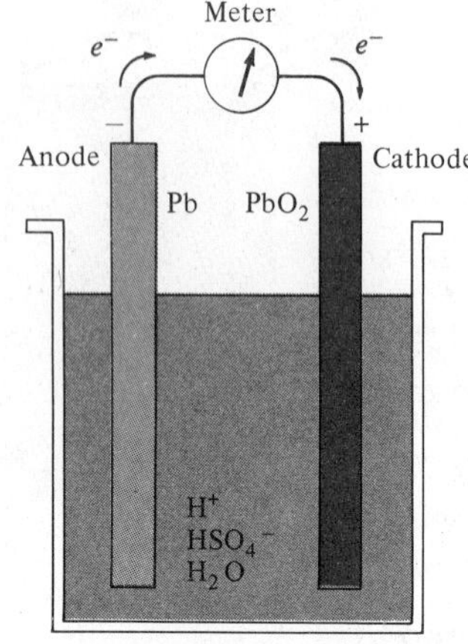

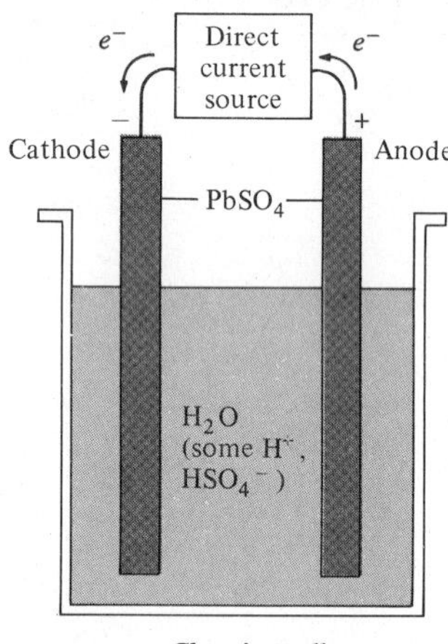

Figure 24.13
The Principles and Operation of the Lead Storage Cell during Discharge and Charge

battery is charged (Figure 24.13). Sulfuric acid is regenerated, and the specific gravity of the solution increases. During the charging operation, the cell becomes an electrolytic cell; as a side reaction to the regeneration of PbO_2 and Pb, some battery water is electrolyzed. The overall reaction in the lead storage cell is

$$Pb(s) + PbO_2(s) + 2H_2SO_4(aq) \underset{\text{charge}}{\overset{\text{discharge}}{\rightleftharpoons}} 2PbSO_4(s) + 2H_2O(l) \qquad \textbf{(24.35)}$$

Each lead storage cell develops a potential of 2 V. In practice, a number of these cells are connected in *parallel* (anode to anode, cathode to cathode) to increase the current that can be generated, and three or six such blocks of cells are connected in *series* to increase the voltage to 6 V or 12 V. The resulting lead storage batteries give excellent service for long periods of time, providing their liquid levels are maintained, they are never permitted to discharge completely, they are not allowed to freeze, and they are not subjected to frequent "quick charge" procedures. (A quick charge reverses the reactions so rapidly that the regenerated Pb and PbO_2 may not adhere completely to the electrodes and may fall off. The Pb and PbO_2 can build up to form a short-circuiting sludge if the quick charge procedure is repeated frequently. The same problem results if the car is driven over roads rough enough to dislodge the Pb and PbO_2.) In the new sealed automobile batteries, the lead plates have been modified so that during charging only a negligible amount of water is electrolyzed.

The nickel–cadmium alkaline storage cell (used in "ni-cad" batteries) has a longer life and delivers a more nearly constant potential than the lead storage cell. In this cell, a cadmium electrode and a metal grid containing a mixture of nickel oxides in various oxidation states are immersed in a potassium hydroxide solution. The chemical reactions during discharge and charge can be summarized by the equations

$$Cd(s) + 2OH^- \underset{\text{charge}}{\overset{\text{discharge}}{\rightleftharpoons}} Cd(OH)_2(s) + 2e^-$$

$$[NiOOH] + H_2O(l) + e^- \underset{\text{charge}}{\overset{\text{discharge}}{\rightleftharpoons}} Ni(OH)_2(s) + OH^-$$

where [NiOOH] represents a mixture of oxides and hydrates. No change in concentration of the electrolyte occurs in the overall cell reaction. This cell delivers current at a potential of about 1.4 V, is light, and is often used in cordless appliances. (Note that the words "cell" and "battery" are often used interchangably.)

24.17 FUEL CELLS

Burning fuel to obtain thermal energy that is then converted into electrical energy is a wasteful process. Heat is inevitably lost to the surroundings, and there are also thermodynamic limits to the amount of useful energy that can be produced. The best conventional power plant can operate at only about 40% efficiency. Theoretically, the electrochemical conversion of traditional fuels such as hydrocarbons to combustion products could be 100% efficient. Although perfect efficiency seems unattainable, 60% has been called possible.

Much research is being devoted to the development of **fuel cells,** cells that produce electrical energy directly from the air oxidation of a fuel continuously supplied to the cell. The technical, economic, and practical problems to be overcome are great. These problems include difficulties in providing contact among the three phases needed in a fuel cell (the gaseous fuel, the liquid electrolyte, and the solid conductor), the corrosiveness of acid electrolyte systems or the buildup of carbonates in alkaline systems (from CO_2 in the air), the high cost of catalysts needed for the electrode reactions (metals such as platinum, palladium, and silver), and the problems of handling gaseous fuels containing hydrogen at either low temperatures or high pressures.

A few test vehicles that run on energy from fuel cells have been operated. Fuel cells in which hydrogen is the fuel and oxygen is the oxidant have been built into a van by General Motors. Tanks of liquid hydrogen and liquid oxygen are carried in the back of the van and 32 fuel cells ride beneath the floorboards.

In alkaline hydrogen–oxygen fuel cells (Figure 24.14), the electrode reactions are

anode reaction $$2H_2(g) + 4OH^- \longrightarrow 4H_2O(l) + 4e^-$$

cathode reaction $$O_2(g) + 2H_2O(l) + 4e^- \longrightarrow 4OH^-$$

The cell reaction

$$2H_2(g) + O_2(g) \xrightarrow[70-140\ ^\circ C]{\text{catalyst}} 2H_2O(l)$$

is simply the same reaction as the combustion of hydrogen.

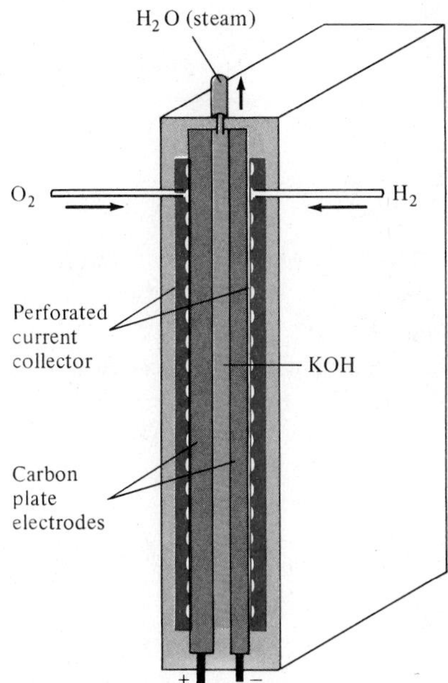

Figure 24.14
A Hydrogen–Oxygen Fuel Cell

24.18 CORROSION

Corrosion is the blanket name applied to a large variety of deterioration processes undergone by manufactured materials during exposure to their environments. We are interested here in the corrosion of metals, which is an electrochemical phenomenon.

The corrosion of metals may appear as a general material loss all over the surface (often with discoloration and scaling, as in rusted iron), deep gouge formation (usually found in bends in pipes), uncontrolled enlargement of cracks and crevices, or apparently inexplicable pinhole formation. Weakening may be due not only to material loss, but also to embrittlement and cracking under conditions of stress, or selective leaching of constituents from an alloy (e.g., the loss of zinc from brass).

Corrosion is the principal route by which manufactured materials are reclaimed by the ever-increasing entropy of the universe, and it has been likened to the death and decay of living things. Corrosion serves the useful function of making room for new generations of improved equipment, although by the same token it is desirable to postpone it as long as possible. Therefore, corrosion has been studied extensively.

All metallic corrosion involves an anodic process (oxidation of the metal to ions) that releases electrons into the metal and a cathodic process that consumes electrons at exactly the same rate. The cathodic process is usually the reduction of hydrogen ions or oxygen from the environment. Hydrogen ions are reduced to hydrogen atoms, which are then catalytically combined on the metal surface to give molecular hydrogen. Molecular oxygen is reduced to OH^- by reaction with water.

Corrosion often occurs by couple action, in which a voltaic cell is formed by direct contact between two different substances. While the presence of two different substances is not necessary for corrosion to occur, couple action does accelerate the process. Consider a piece of zinc—when placed in an acidic aqueous medium it tends to liberate electrons:

$$Zn(s) \longrightarrow Zn^{2+} + 2e^- \qquad \textbf{(24.36)}$$

This reaction proceeds only to a slight extent, for the electrons stay on the metal surface and the Zn^{2+} ions that have gone into solution remain close to the surface, preventing further reaction. If some of the electrons are removed from the zinc metal, then the Zn^{2+} ions can wander off into the solution and reaction (24.36), the corrosion of zinc, will continue to occur.

If a bit of graphite is embedded in the zinc surface (Figure 24.15), some of the electrons generated on the zinc move over to the graphite, which is a conductor. Hydrogen ions in the solution are attracted by the electrons, and the reaction

$$2H^+ + 2e^- \longrightarrow H_2(g)$$

occurs. The hydrogen seems to come from the carbon, but actually it is the result of the redox reaction, $Zn(s) + 2H^+ \longrightarrow Zn^{2+} + H_2(g)$, with zinc as the anode, and carbon as the inert cathode. If a piece of pure zinc immersed in an acid solution is touched with, say, a piece of platinum wire, the same thing happens. Some of the

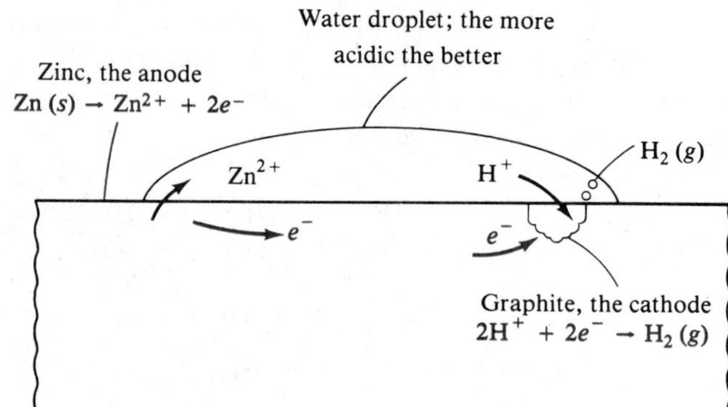

Figure 24.15
Couple Action and the Acceleration of Corrosion

electrons leave the zinc surface for the other metal, and hydrogen forms on the platinum wire. In this case, the platinum wire is the inert cathode.

The rusting of iron is a complex process that occurs in the presence of oxygen and water containing an electrolyte. Iron(III) cations (Fe^{3+}) are produced in the anode reaction and enter moisture on the metal surface. The electrons from the anode reaction can travel through the metal or through moisture on the metal surface to a cathodic region, which may be far from the site of the oxidation. An electrolyte dissolved in the moisture on the metal surface (e.g., salt from the sea, acids from the reaction of oxides with water in the air), like a salt bridge, aids in the flow of the electrons to the cathodic region. At the cathode, oxygen from the air is oxidized to OH^-. Ultimately iron ions, OH^- ions, and water unite to form rust—hydrated iron oxide, which may be written $Fe(OH)_3$ or $Fe_2O_3 \cdot 3H_2O$. Unlike many metals, iron forms no oxide that adheres tightly to the iron surface. Rust flakes off, exposing the free metal underneath to further corrosion.

Investigation of the relationship between pH and potential for metals in the presence of hydrogen ion and oxygen, and the construction of pH–potential diagrams (like Figure 24.8, but usually much more complex) reveal regions of potential and pH where the metal and one of (usually) several possible impervious oxides are the thermodynamically favored species. In these regions a metal will not corrode. A material with an impervious oxide coating is called *passivated.* Some materials, such as stainless steels, have exceptionally wide passive potential–pH regions. In pH–potential regions, where a metal is in equilibrium with its ions at finite concentrations, the anodic reaction will take place, and the metal will dissolve. Corrosion control consists of interfering with the anode or cathode reaction. Most commonly, this is done by applying coatings. Also, the potential can be forced into a more favorable region in the pH–potential relationship by an external power supply inserted between the equipment and an inert electrode. Such protection is usually cathodic—the potential is lowered, making the anodic reaction less favorable. For example, in pulp and paper mills, where strongly corrosive conditions prevail, expensive filter presses are continuously protected in this way. As a variation of this principle, protection is sometimes provided by what are called sacrificial anodes. For example, large magnesium blocks—the sacrificial anodes—are attached to the steel of an off-shore oil drilling platform or an underground oil tank (Figure 24.16). Because the magnesium is much more active than the steel, the anode reaction, oxidation of the metal, occurs at the magnesium instead of the steel, and the steel becomes the inert cathode. Massive sacrificial anodes are consumed fairly rapidly. Galvanizing—coating steel with zinc—works on the same principle. In the iron–zinc couple, zinc is the anode and iron the cathode.

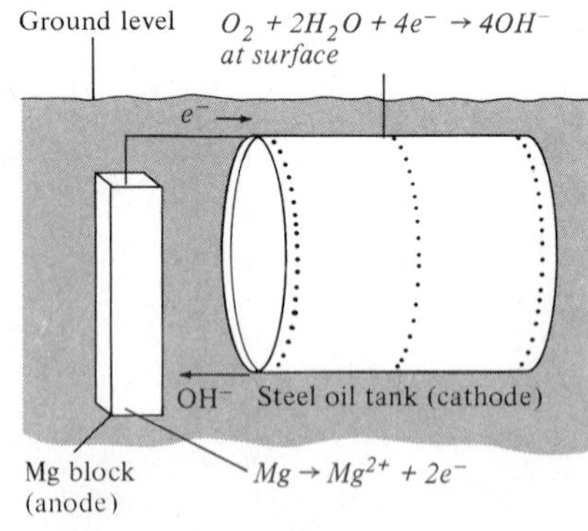

Figure 24.16
Cathodic Protection of an Oil Tank
The magnesium is embedded in sand to aid flow of ions away from the surface. The electrons used at the surface of the oil tank come through the conductor from the magnesium and the tank is undisturbed.

Under some conditions, a coating can accelerate rather than deter corrosion. Iron cans used to be coated with tin. However, tin is more noble than iron. Any imperfections in the coating provide a metal couple with a very large cathode (the tin) to anode (the iron) area ratio, excellent conditions for rapid localized corrosion. Organic lacquers and plastics are now used to protect the metal. Household water pipes provide another example. Given conventional hardness and pH in water, we can usually rely on a shell of calcium carbonate to form in iron or copper pipes and protect them from corrosion. But if the pH is not quite right or the water is a little too soft, the coating may not be complete. Because a calcium carbonate coating does not interfere with the cathode reaction, the result is a very small active anode (the metal exposed where the coating is incomplete) connected to a large cathode. Here again are the conditions for very rapid localized corrosion, the formation of a pinhole in the pipe, and a leak in the basement. Water that creates such conditions is referred to as *aggressive water*.

SUMMARY OF SECTIONS 24.15–24.18

Common flashlight batteries are Leclanché cells, the overall reaction of which can be approximated as $Zn(s) + 2MnO_2(s) + 2H_2O(l) \longrightarrow Zn(OH)_2(s) + 2MnOOH(s)$. Alkaline batteries also oxidize Zn and reduce MnO_2, but via somewhat different reactions. Mercury cells produce energy by oxidation of Zn and reduction of HgO to Hg and OH^-. Batteries that can be recharged by using electrical energy from an external source to reverse the redox reaction are called storage batteries. Among these is the lead storage battery (Equation 24.33). Nickel–cadmium alkaline storage batteries are light, long-lived, and deliver quite constant potential. Fuel cells produce electrical energy directly from the air oxidation of a fuel. Several such cells, utilizing fuels such as hydrogen, hydrocarbons, or ammonia, have been developed, but their technical and economic problems are still great.

The corrosion of metals is an electrochemical process. It involves oxidation of the metal in conjunction with reduction of hydrogen ions or oxygen from the environment, processes favored by low pH. Under certain conditions, some materials tend to form impervious oxide coatings, which prevent corrosion. Corrosion may be avoided by the use of an externally applied potential, or by a sacrificial anode—a more active metal which is consumed instead of the metal being protected.

SIGNIFICANT TERMS (with section references)

anode (24.3)
cathode (24.3)
cell potential, cell voltage, electromotive force, emf (24.6)
cell reaction (24.3)
electrochemical cell (24.3)
electrochemistry (24.3)
electrode potential (24.7)
electrode reaction (24.3)
electrolysis (24.3)
electrolytic cell (24.3)
faraday (24.5)
fuel cells (24.17)
half-cell (24.3)
half-reactions (24.1)
ion-electron equations (24.1)
overvoltage (24.9)
redox couple (24.1)
standard electrode potentials, standard reduction potentials (24.7)
storage cells, storage batteries (24.16)
voltaic cell, galvanic cell (24.3)

QUESTIONS AND PROBLEMS

Redox Revisited

24.1 Define oxidation and reduction in terms of electron gain or loss. What is the relationship between the number of electrons gained and lost in a redox reaction?

24.2 How is the symbol for a redox couple written?

24.3 Identify the two redox couples for each of the following unbalanced equations:

(a) $H_2O_2(aq) + I^- + H^+ \longrightarrow I_2(s) + H_2O(l)$
(b) $I^- + H^+ + NO_2^- \longrightarrow NO(g) + H_2O(l) + I_2(s)$
(c) $Al(s) + H_2SO_4(conc) + H^+ \longrightarrow Al^{3+} + H_2O(l) + SO_2(g)$
(d) $Zn(s) + OH^- + NO_3^- + H_2O(l) \longrightarrow NH_3(aq) + [Zn(OH)_4]^{2-}$

For each reaction, write the ion-electron equations for the half-reactions and balance the overall equations using the half-reaction method.

24.4 Repeat Question 24.3 for

(a) $I_2(s) + H_2S(aq) \longrightarrow I^- + S(s) + H^+$
(b) $HNO_3(aq) + Cu(s) \longrightarrow Cu(NO_3)_2(aq) + NO(g) + H_2O(l)$
(c) $I_2(s) + OH^- \longrightarrow I^- + IO_3^- + H_2O(l)$
(d) $H_2S(aq) + Cr_2O_7^{2-} + H^+ \longrightarrow Cr^{3+} + H_2O(l) + S(s)$

24.5 For each of the following unbalanced equations, write the ion-electron equations for the half-reactions and balance the overall equation using the half-reaction method:

(a) $I_2(s) + H_2S(aq) \longrightarrow I^- + S(s) + H^+$
(b) $PbO_2(s) + H^+ + Cl^- \longrightarrow Cl_2(g) + PbCl_2(s) + H_2O(l)$
(c) $Cu(s) + Br_2(aq) + OH^- \longrightarrow Cu_2O(s) + Br^- + H_2O(l)$
(d) $S^{2-} + Cl_2(g) + OH^- \longrightarrow SO_4^{2-} + Cl^- + H_2O(l)$
(e) $MnO_4^- + IO_3^- + H_2O(l) \longrightarrow MnO_2(s) + IO_4^- + OH^-$

24.6 Repeat Question 24.5 for

(a) $H_2O_2(aq) + H^+ + I^- \longrightarrow I_2(s) + H_2O(l)$
(b) $SO_2(g) + Cr_2O_7^{2-} + H^+ \longrightarrow Cr^{3+} + HSO_4^- + H_2O(l)$
(c) $Cl_2(g) + OH^- \longrightarrow ClO_3^- + Cl^- + H_2O(l)$
(d) $S(s) + OH^- \longrightarrow S^{2-} + SO_3^{2-} + H_2O(l)$
(e) $Sn(s) + OH^- + H_2O(l) \longrightarrow [Sn(OH)_4]^{2-} + H_2(g)$

24.7 Balance each of the following equations using the half-reaction method. [Note that in most cases H^+ or OH^- and H_2O are not shown. You should be able to decide whether the system is neutral, acidic, or alkaline with a little detective work, and add H^+ or OH^- and H_2O as needed to balance the equation.]

(a) $[Fe(CN)_6]^{4-} + CrO_4^{2-} \longrightarrow [Fe(CN)_6]^{3-} + Cr_2O_3(s)$
(b) $ClO^- + I_2(s) \longrightarrow Cl^- + IO_3^-$
(c) $CN^- + [Fe(CN)_6]^{3-} \longrightarrow CNO^- + [Fe(CN)_6]^{4-}$
(d) $N_2H_4(aq) + Cu(OH)_2(s) \longrightarrow N_2(g) + Cu(s)$

24.8 Repeat Question 24.7 for

(a) $Zn(s) + NO_3^- \longrightarrow Zn^{2+} + NH_4^+$
(b) $Cu(s) + NO_3^- \longrightarrow NO(g) + Cu^{2+}$
(c) $P_4(s) + NO_3^- \longrightarrow H_3PO_4(aq) + NO(g)$
(d) $H_2S(aq) + NO_3^- \longrightarrow S(s) + NO(g)$
(e) $Fe(s) + NO_3^- \longrightarrow Fe^{3+} + N_2(g)$

Fundamentals of Electrochemistry

24.9 What are the two classifications of electrochemical cells? What is the difference between them?

24.10 Distinguish clearly between the terms "anode" and "cathode." In writing the notation for an electrochemical cell, where do the anode and the cathode half-cells appear?

24.11 Prepare a simple sketch of a voltaic cell showing the anode, the cathode, the signs of the electrodes, and the direction of electron and ion flow for the cell represented by the notation

$$Ag(s)|AgCl(s)|HCl(aq)|Cl_2(g)|(C)$$

Is a salt bridge necessary for this reaction?

24.12 Repeat Question 24.11 for a voltaic cell in which the following reaction occurs:

$$Mg(s) + F_2(g) \longrightarrow MgF_2(l)$$

Write the cell notation for this electrochemical cell.

24.13 Write the ion-electron equations for the half-reactions and overall cell equation for the electrochemical cell represented by

$$(Pt)|H_2(g)|HCl(aq)||Fe^{3+}, Fe^{2+}|(Pt)$$

24.14 A storage battery based on the Edison cell and represented by

$$Fe(s)|Fe(OH)_2(s)|LiOH(aq), KOH(aq)|Ni(OH)_2(s)|[NiOOH](s)|(inert)$$

is sometimes used in place of a lead storage battery when weight considerations are important. Write the ion-electron equations for the half-reactions describing the oxidation and reduction processes and write the overall cell equation.

Fundamentals of Electrochemistry—Stoichiometry

24.15 What mass of zinc(II) ion will be reduced by one mole of electrons?

24.16 How many moles of electrons would be required to reduce 0.100 g of Eu^{3+} to the metal?

24.17 What mass of molten sodium would be produced by electrolyzing molten sodium bromide using a current of 15 A for 3.0 h?

24.18 The cells in an automobile battery were charged at a steady current of 5.0 A for exactly 5 h. What mass of Pb and PbO_2 were formed in each cell? The overall reaction is

$$2PbSO_4(s) + 2H_2O(l) \longrightarrow Pb(s) + PbO_2(s) + 2H_2SO_4(aq)$$

24.19 The chemical equation for the electrolysis of a fairly concentrated brine solution is

$$2NaCl(aq) + 2H_2O(l) \longrightarrow Cl_2(g) + H_2(g) + 2NaOH(aq)$$

What volume of gaseous chlorine would be generated at 745 Torr and 85 °C if the process is 75% efficient and if a current of 1.5 A flowed for 5.0 h?

24.20 The "life" of a certain voltaic cell is limited by the amount of Cu^{2+} in solution available to be reduced. If the cell contains 25 mL of 0.175 M Cu^{2+}, what is the maximum amount of electrical charge this cell can generate?

24.21 Calculate the current required to deposit 0.50 g of elemental platinum from a solution containing the $[PtCl_6]^{2-}$ ion within a period of 5.0 h.

24.22 What time would be required to plate an iron platter with 5.0 g of silver using a solution containing the $[Ag(CN)_2]^-$ ion and a current of 1.5 A?

24.23 The same quantity of electrical charge that deposited 0.583 g of silver was passed through a solution of a gold salt and 0.355 g of gold was formed. What is the oxidation state of gold in this salt?

24.24 A sample of Al_2O_3 dissolved in a molten fluoride bath is electrolyzed using a current of 1.00 A. (a) What is the rate of production of Al in grams per hour? (b) The oxygen liberated at the positive carbon electrode reacts with the carbon to form CO_2. What mass of CO_2 is produced per hour?

Standard Reduction Potentials

24.25 Standard reduction potentials are 2.9 V for $F_2(g)/F^-$, 0.8 V for $Ag^+/Ag(s)$, 0.5 V for $Cu^+/Cu(s)$, 0.3 V for $Cu^{2+}/Cu(s)$, −0.4 V for $Fe^{2+}/Fe(s)$, −2.7 V for Na^+/Na, and −2.9 V for $K^+/K(s)$. (a) Arrange the oxidizing agents in order of increasing strength. (b) Which of these oxidizing agents will oxidize Cu under standard state conditions?

24.26 Standard reduction potentials are 1.455 V for the $PbO_2(s)/Pb(s)$ couple, 1.360 V for $Cl_2(g)/Cl^-$, 3.06 V for $F_2(g)/HF(aq)$, and 1.776 V for $H_2O_2(aq)/H_2O(l)$. Under standard state conditions, (a) which is the strongest oxidizing agent, (b) which oxidizing agent(s) could oxidize lead to lead(IV) oxide, (c) which oxidizing agent(s) could oxidize fluoride ion in an acidic solution?

24.27 Arrange the following less commonly encountered metals in an activity series from the most active to the least active: radium [$Ra^{2+}/Ra(s)$, $E° = -2.9$ V]; rhodium [$Rh^{3+}/Rh(s)$, $E° =$ 0.80 V]; europium [$Eu^{2+}/Eu(s)$, $E° = -3.4$ V], plutonium [$Pu^{3+}/Pu(s)$, $E° = -2.0$ V]. How do these metals compare in reducing ability with the active metal lithium [$Li^+/Li(s)$, $E° =$ −3.0 V]; with hydrogen; and with gold [$Au^{3+}/Au(s)$, $E° =$ 1.5 V], which is a noble metal and one of the least active of the metals?

24.28 Arrange the following metals in an activity series from the most active to the least active: nobelium [$No^{3+}/No(s)$, $E° =$ −2.5 V]; cobalt [$Co^{2+}/Co(s)$, $E° = -0.28$ V]; gallium [$Ga^{3+}/Ga(s)$, $E° = -0.53$ V]; thallium [$Tl^+/Tl(s)$, $E° = -0.34$ V]; polonium [$Po^{2+}/Po(s)$, $E° = 0.65$ V].

24.29 What conclusion can be drawn from the $E°$ data for nitrate ion used in Example 24.10, plus the following additional data: $HNO_2(aq)/NO(g)$, $E° = 1.00$ V; $HNO_2(aq)/N_2O(g)$, $E° = 1.29$ V; $N_2O_4(g)/NO(g)$, $E° = 1.03$ V; $N_2O_4(g)/HNO_2(aq)$, $E° = 1.07$ V?

24.30 Use the data given in Table 24.4 to discuss the reduction of ClO_4^- in an acidic solution.

24.31 The standard reduction potential is −1.245 V for the $Zn(OH)_2(s)/Zn(s)$ couple, −2.690 V for $Mg(OH)_2(s)/Mg(s)$, −0.877 V for $Fe(OH)_2(s)/Fe(s)$, −2.30 V for $Al(OH)_3(s)/Al(s)$. Under standard state conditions, (a) which is the strongest reducing agent, (b) which reducing agent(s) could reduce $Zn(OH)_2(s)$ to $Zn(s)$, (c) which reducing agent(s) could reduce $Fe(OH)_2(s)$ to $Fe(s)$?

24.32 Using Table 24.4, identify an oxidizing agent for each case that under standard state conditions will oxidize (a) $I_2(s)$ to IO_3^-, but not Cl^- to $Cl_2(g)$; (b) $Fe(s)$ to Fe^{2+}, but not $Pb(s)$ to Pb^{2+}; (c) $Zn(s)$, but not $Cd(s)$, in alkaline solution.

Standard Reduction Potentials—$E°$ for Cells

24.33 The standard reduction potential is −0.76 V for the $Zn^{2+}/Zn(s)$ couple and 1.36 V for the $Cl_2(g)/Cl^-$ couple. Write the equation for the oxidation of $Zn(s)$ by $Cl_2(g)$. Calculate the potential of this reaction under standard state conditions. Is this a spontaneous reaction?

24.34 An electrochemical cell was needed in which hydrogen and oxygen would react to form water. Using the following standard reduction potentials for the couples given, determine which set of reactions gives the maximum output potential: $E° =$ −0.828 V for $H_2O(l)/H_2(g), OH^-$; $E° = 0.0000$ V for $H^+/H_2(g)$; $E° = 1.229$ V for $O_2(g), H^+/H_2O(l)$; and $E° = 0.401$ V for $O_2(g), H_2O(l)/OH^-$.

24.35 Using the standard reduction potentials given for the appropriate couples, calculate $E°$ for each reaction and indicate whether or not each of the reactions described below will be spontaneous under standard state conditions:

(a) $E° = -0.809$ V for $Cd(OH)_2(s)/Cd(s)$ and $E° = 0.098$ V for $HgO(s)/Hg(l)$

$$Cd(s) + HgO(s) + H_2O(l) \longrightarrow Hg(l) + Cd(OH)_2(s)$$

(b) $E° = -0.036$ V for $Fe^{3+}/Fe(s)$ and $E° = 0.0000$ V for $H^+/H_2(g)$

$$2Fe(s) + 6H^+ \longrightarrow 2Fe^{3+} + 3H_2(g)$$

(c) $E° = 1.36$ V for $Cl_2(g)/Cl^-$ and $E° = 1.33$ V for $Cr_2O_7^{2-}/Cr^{3+}$

$$6Cl^- + 14H^+ + Cr_2O_7^{2-} \longrightarrow 2Cr^{3+} + 3Cl_2(g) + 7H_2O(l)$$

(d) $E° = 0.17$ V for $SO_4^{2-}/H_2SO_3(aq)$ and $E° = 0.96$ V for $NO^{3-}/NO(g)$

$$3H_2SO_3(aq) + 2NO_3^- \longrightarrow 2NO(g) + 4H^+ + H_2O(l) + 3SO_4^{2-}$$

24.36 Using the standard reduction potentials given in Table 24.4, calculate $E°$ for each of the following reactions and indicate whether or not the reaction will be spontaneous under standard state conditions:

(a) $2MnO_4^- + 10HF(aq) + 6H^+ \longrightarrow 2Mn^{2+} + 5F_2(g) + 8H_2O(l)$

(b) $Mg(s) + Fe^{2+} \longrightarrow Fe(s) + Mg^{2+}$

(c) $H_2O_2(aq) + 2H^+ + 2Cl^- \longrightarrow Cl_2(g) + 2H_2O(l)$

(d) $2MnO_4^- + 3Ni(s) + 4H_2O(l) \longrightarrow 3Ni(OH)_2(s) + 2MnO_2(s) + 2OH^-$

(e) $Pb^{2+} + Mn(s) \longrightarrow Mn^{2+} + Pb(s)$

(f) $2Pb^{2+} + 2H_2O(l) \longrightarrow Pb(s) + PbO_2(s) + 4H^+$

(g) $Fe(s) + ClO_4^- + H_2O(l) \longrightarrow ClO_3^- + Fe(OH)_2(s)$

Standard Reduction Potentials—Electrolysis

24.37 Describe what would happen to an acidic aqueous sodium sulfate solution at standard state conditions during electrolysis between inert electrodes.

24.38 Write the balanced equation for the half-reaction that takes place at each electrode as an electrical current is passed through a 1 M aqueous solution of each of the following substances using inert electrodes: (a) $AgNO_3$, (b) $CuBr_2$, (c) H_2SO_4. (You may want to refer to Table 24.4 to decide which half-reactions are most favorable.)

24.39 On the basis of standard reduction potentials, the reaction

$$Cu(s) + 2K^+ \longrightarrow Cu^{2+} + 2K(s)$$

is predicted to be unfavorable (see Table 24.4). To make this reaction "go," a chemist supplied electrical energy from an external source, but to his amazement a flame appeared in the cell. What had the chemist forgotten to consider?

24.40 An examination question concerned the half-reactions in the electrolysis of a sodium chloride solution:

$$2NaCl(aq) + 2H_2O(l) \xrightarrow{\text{electrolysis}} 2NaOH(aq) + H_2(g) + Cl_2(g)$$

A student wrote that chlorine is liberated at the anode and sodium at the cathode, followed by a second reaction:

$$2Na(s) + 2H_2O(l) \longrightarrow 2NaOH(aq) + H_2(g)$$

What is wrong with this answer?

Thermodynamics of Redox Reactions

24.41 Calculate $\Delta G°$ for the half-reaction

$$\tfrac{1}{2}H_2O_2(aq) + H^+ + e^- \longrightarrow H_2O(l)$$

given that $E° = 1.776$ V for the $H_2O_2(aq)/H_2O(l)$ couple.

24.42 Given $E° = -1.529$ V for the reduction of Zr^{4+}

$$Zr^{4+} + 4e^- \longrightarrow Zr(s)$$

calculate $\Delta G°$ for this reaction.

24.43 The standard free energy of formation of water at 25 °C is −237.2 kJ/mol. What potential would be expected from a fuel cell that consumes hydrogen and oxygen and operates under standard state conditions?

24.44 At 1000 K, $\Delta G° = -81.06$ kJ for the reaction

$$Ag(s) + \tfrac{1}{2}Cl_2(g) \longrightarrow AgCl(l)$$

What is the potential of an electrochemical cell based on this reaction?

24.45 $E° = -0.036$ V for the $Fe^{3+}/Fe(s)$ reduction couple. Calculate $\Delta G°$ for the reactions

$$Fe^{3+} + 3e^- \longrightarrow Fe(s)$$
$$\tfrac{1}{3}Fe^{3+} + e^- \longrightarrow \tfrac{1}{3}Fe(s)$$

24.46 Show that $E° = -1.662$ V for the reduction of Al^{3+} to $Al(s)$, regardless of whether the equation for the reaction is written

(i) $\tfrac{1}{3}Al^{3+} + e^- \longrightarrow \tfrac{1}{3}Al(s) \qquad \Delta G° = 160.4\text{ kJ}$

or

(ii) $Al^{3+} + 3e^- \longrightarrow Al(s) \qquad \Delta G° = 481.2\text{ kJ}$

Thermodynamics of Redox Reactions—Nonstandard State Conditions

24.47 Calculate the potential associated with the half-reaction

$$Co(s) \longrightarrow Co^{2+} + 2e^-$$

given that the concentration of the cobalt(II) ion is 1×10^{-4} M. The standard reduction potential for the $Co^{2+}/Co(s)$ couple is −0.277 V.

24.48 Calculate the reduction potential for hydrogen ion in a system having a perchloric acid concentration of 1×10^{-4} M and a hydrogen pressure of 2 bar. (Recall that $HClO_4$ is a strong acid in aqueous solution.)

24.49 What is the concentration of Ag^+ in a half-cell if the reduction potential of the Ag^+/Ag couple is changed from $E° = 0.80$ V to 0.35 V?

24.50 The standard reduction potential for the $F_2(g)/F^-$ couple is 2.87 V. What must be the pressure of the fluorine gas in order to produce a reduction potential of 2.85 V in a solution that contains 0.10 M F^-?

24.51 The standard reduction potentials for the $H^+/H_2(g)$ and $O_2(g),H^+/H_2O(l)$ couples are 0.0000 V and 1.229 V, respectively. Write the half-reactions, the overall reaction, and calculate $E°$ for the reaction

$$2H_2(g) + O_2(g) \longrightarrow 2H_2O(l)$$

Calculate E for the cell when the pressure of H_2 is 5.0 bar and that of O_2 is 0.9 bar.

24.52 Consider the cell represented by the notation

$$Zn(s)|ZnCl_2(aq)|Cl_2(g, 1\text{ bar})|(C)$$

The standard reduction potentials are −0.763 V for $Zn^{2+}/Zn(s)$ and 1.360 V for $Cl_2(g)/Cl^-$ at 25 °C. Calculate $E°$ and then E for the cell when the concentration of the $ZnCl_2$ is 0.10 mol/L.

24.53 The potential of the cell for the reaction

$$M(s) + 2H^+(1.0\text{ M}) \longrightarrow H_2(g, 1.0\text{ bar}) + M^{2+}(0.10\text{ M})$$

is 0.500 V. What is the standard reduction potential for the $M^{2+}/M(s)$ couple?

24.54 Under standard state conditions the following reaction is not spontaneous:

$$Br^- + 2MnO_4^- + H_2O(l) \longrightarrow BrO_3^- + 2MnO_2(s) + 2OH^- \qquad E° = -0.022\text{ V}$$

The reaction conditions are adjusted so that $E = 0.100$ V by making $[Br^-] = [MnO_4^-] = 1.5$ mol/L and $[BrO_3^-] = 0.5$ mol/L. What is the concentration of hydroxide ion in this cell?

24.55 Consider the following unbalanced equation:

$$Hg(l) + Fe^{3+}(aq) \longrightarrow Hg_2^{2+}(aq) + Fe^{2+}(aq)$$

(a) Write the ion-electron equations for the half-reactions and the overall cell equation. (b) Write the shorthand notation for this cell. The standard reduction potential at 25 °C is 0.788 V for $Hg_2^{2+}/Hg(l)$ and 0.771 V for Fe^{3+}/Fe^{2+}. (c) Find $E°$ for the reaction. Is the reaction spontaneous under standard state conditions? (d) When $[Hg_2^{2+}] = 0.0010$ mol/L, $[Fe^{2+}] = 0.10$ mol/L, and $[Fe^{3+}] = 1.00$ mol/L, what is E for the cell? Is the reaction more, less, or identically favorable under these conditions compared to standard state conditions?

24.56* Find the potential of the cell in which identical iron electrodes are placed into solutions of Fe^{2+} of concentration 1.0 mol/L and 0.10 mol/L.

Thermodynamics of Redox Reactions—$E°$ and K

24.57 Calculate the value of the equilibrium constant for the reaction

$$2K(s) + 2H_2O(l) \rightleftharpoons 2K^+ + 2OH^- + H_2(g)$$

The standard reduction potential is -2.925 V for $K^+/K(s)$ and -0.8281 V for $H_2O(l)/H_2(g),OH^-$.

24.58 The standard reduction potential is 1.087 V for $Br_2(aq)/Br^-$ and 1.3597 V for $Cl_2(g)/Cl^-$. Calculate the equilibrium constant for the reaction

$$2Br^- + Cl_2(g) \rightleftharpoons Br_2(aq) + 2Cl^-$$

24.59 Using the following half-reaction and $E°$ data at 25 °C:

$$PbSO_4(s) + 2e^- \longrightarrow Pb(s) + SO_4^{2-} \qquad E° = -0.359 \text{ V}$$
$$PbI_2(s) + 2e^- \longrightarrow Pb(s) + 2I^- \qquad E° = -0.365 \text{ V}$$

calculate the equilibrium constant for the reaction

$$PbSO_4(s) + 2I^- \rightleftharpoons PbI_2(s) + SO_4^{2-}$$

24.60* Assuming $K = 1.00 \times 10^{-14}$ for the reaction

$$H_2O(l) \rightleftharpoons H^+ + OH^-$$

determine $E°$ for this reaction and write the notation for an electrochemical cell that could generate this value of emf.

24.61* At 25 °C, $E° = 0.071$ V for the reduction of AgBr and 0.799 V for the reduction of Ag^+. Calculate K_{sp} for AgBr(s).

24.62* At 25 °C, $E° = 0.142$ V for the reduction of elemental sulfur to the -2 state in acidic medium and -0.447 V in alkaline medium. Calculate K for the reaction

$$H_2S(aq) \rightleftharpoons 2H^+ + S^{2-}$$

Thermodynamics of Redox Reactions—Half-Reaction $E°$ Values

24.63 The element ytterbium forms both +2 and +3 cations in aqueous solution. $E° = -2.797$ V for $Yb^{2+}/Yb(s)$ and -2.267 V for $Yb^{3+}/Yb(s)$. What is the standard state reduction potential for the Yb^{3+}/Yb^{2+} couple?

24.64 The standard reduction potential for Cu^+ to Cu(s) is 0.521 V and for Cu^{2+} to Cu(s) is 0.337 V. Calculate $E°$ for the Cu^{2+}/Cu^+ couple.

24.65* Using $E° = 0.7991$ V for the reduction potential of Ag^+ at 25 °C and $K_{sp} = 1.8 \times 10^{-10}$ for AgCl at 25 °C, calculate $E°$ for the reduction of AgCl(s).

24.66* Using $E° = 2.87$ V for the $F_2(g)/F^-$ couple and $K_a = 6.5 \times 10^{-4}$ for HF(aq), calculate $E°$ for the $F_2/HF(aq)$ couple.

Thermodynamics of Redox Reactions—$E°$ and pH

24.67 A student wrote the equation

$$Fe(s) \longrightarrow Fe^{3+} + 3e^-$$

to describe the oxidation of iron at pH 4. Refer to Figure 24.8 and suggest what is wrong with this equation. Write the correct oxidation equation.

24.68 Prepare rough sketches for the pH–potential diagrams for chlorine and fluorine.

Practical Electrochemistry

24.69 Briefly describe how a storage cell operates.

24.70 How does a fuel cell differ from a dry cell or a storage cell?

24.71 Does the physical size of a commercial cell govern the potential that it will deliver? What will the size govern?

24.72 Many electronic calculators use rechargeable nickel–cadmium batteries. The overall equation for the spontaneous reaction in these cells is

$$Cd(s) + 2[NiOOH](s) + 2H_2O(l) \xrightarrow{KOH} Cd(OH)_2(s) + 2Ni(OH)_2(s)$$

What are the oxidation states of Cd in (a) Cd and (b) $Cd(OH)_2$ and of Ni in (c) [NiOOH] and (d) $Ni(OH)_2$? In the above reaction, what is the (e) oxidizing agent, (f) reducing agent, (g) substance oxidized, (h) substance reduced? Write the shorthand notation for (i) the reduction couple and (j) the overall cell. During the middle of an examination, the battery in a student's calculator failed. (k) What happened chemically? (l) Write the equation for the cell reaction for the recharge cycle of this cell.

ADDITIONAL PROBLEMS

24.73 A particular electrochemical cell consists of one half-cell in which a silver wire coated with AgCl(s) dips into a 1 M KCl solution and another half-cell in which a piece of platinum dips into a solution that is 0.1 M in $CrCl_3$, 0.001 M in $K_2Cr_2O_7$, and 1 M in HCl. In the cell described, the following reaction takes place:

$$Ag(s) + Cr_2O_7^{2-} + Cl^- + H^+ \longrightarrow AgCl(s) + Cr^{3+} + H_2O(l)$$

The standard reduction potentials are 1.33 V and 0.22 V for the $Cr_2O_7^{2-}/Cr^{3+}$ and AgCl(s)/Ag,Cl^- couples, respectively. Write (a) the ion-electron equations for the half-reactions for this cell and the overall cell equation. Determine (b) the standard state potential and (c) the potential of the cell under the above nonstandard state conditions. (d) Calculate the equilibrium constant for the reaction.

24.74 The standard state reduction potentials for $Au^{3+}/Au(s)$ and $Cl_2(g)/Cl^-$ are 1.498 V and 1.3597 V, respectively. (a) Will chlorine oxidize gold under standard state conditions? (b) Will the reaction occur if the concentration of $AuCl_3$ is 1.0×10^{-3} mol/L and $P_{Cl_2} = 1.0$ bar? (c) What will be the concentration of $AuCl_3$ once the cell reaches equilibrium if P_{Cl_2} is maintained at 1.0 bar?

24.75 The following reaction takes place in an electrochemical cell:

$$Sn(s) + 2AgCl(s) \longrightarrow SnCl_2(aq) + 2Ag(s)$$

For $AgCl(s)/Ag(s), Cl^-$, $E^\circ = 0.2222$ V and for $Sn^{2+}/Sn(s)$, $E^\circ = -0.136$ V. (a) Determine E° for this cell at 25 °C. (b) What is the cell potential, given that the concentration of the $SnCl_2$ is 0.10 mol/L? (c) What would be the resulting cell potential if the $SnCl_2$ solution were diluted tenfold? (d) At equilibrium, the cell potential assumes a unique value. What is this value? (e) From the result in (d), calculate the equilibrium constant. (f) What is the equilibrium concentration of Sn^{2+}?

24.76 Consider the electrochemical cell represented by $Zn(s)|Zn^{2+}||Fe^{3+}|Fe(s)$. (a) Write the ion-electron equations for the half-reactions and the overall cell equation. (b) The standard reduction potentials for $Zn^{2+}/Zn(s)$ and $Fe^{3+}/Fe(s)$ are -0.7628 V and -0.036 V, respectively, at 25 °C. Determine the standard potential for the reaction. (c) Determine E for the cell when the concentration of Fe^{3+} is 10 mol/L and that of Zn^{2+} is 1×10^{-3} mol/L. (d) If 150 mA is to be drawn from this cell for a period of 15 min, what is the minimum mass for the zinc electrode?

24.77* The production of uranium metal from purified uranium dioxide ore consists of the following steps:

$$UO_2(s) + 4HF(g) \longrightarrow UF_4(s) + 2H_2O(l)$$

$$UF_4(s) + 2Mg(s) \xrightarrow{\Delta} U(s) + 2MgF_2(s)$$

What is the oxidation number of U in (a) UO_2, (b) UF_4, and (c) U? Identify (d) the reducing agent and (e) the substance reduced. (f) What current could the second reaction produce if 1.00 g of UF_4 reacted each minute? (g) What volume of $HF(g)$ at 25 °C and 10.0 atm would be required to produce 1.00 g of U? (h) Would 1.00 g of Mg be enough to produce 1.00 g of U?

CHAPTER

25

Molecular Orbital Theory

25.1 VALENCE BOND THEORY VERSUS MOLECULAR ORBITAL THEORY

In Chapter 11 we introduced Lewis structures, valence-shell electron-pair repulsion theory, and valence bond theory to describe molecular structures and covalent bonds. We are about to introduce molecular orbital theory for the same purposes. Why are there so many ways of describing the same phenomena? As is often the case in science, each of several different approaches is useful, while each has its limitations. Lewis structures and VSEPR are simple and nonmathematical, while valence bond and molecular orbital theories relate bonding to electronic structure and quantum theory. Those who are interested in such things can push valence bond and molecular orbital theories to greater and greater levels of mathematical sophistication. For our purposes, it is sufficient to understand in a general way the differences between the valence bond and molecular orbital theories of bonding.

Molecular orbital theory explains bond formation as the occupation by electrons of orbitals characteristic of bonded atoms rather than individual atoms. As in valence bond theory, molecular orbital theory begins with atomic orbitals of similar energy that are able to overlap enough to allow interaction. One approach to simplifying the mathematics is to begin with the functions that describe the individual atomic orbitals and then combine them. The result is a mathematical description and a physical picture of a new orbital—a molecular orbital.

A **molecular orbital** is the space in which an electron with a specific energy is most likely to be found in the vicinity of two or more nuclei that are bonded together. The electron can be closer to one nucleus at one instant and closer to the other nucleus at another instant. The difference between bonds in the molecular orbital and valence bond theories is a little bit like the difference between scrambled eggs and two fried eggs that have run together at the edges in the frying pan. In forming molecular orbitals the atomic orbitals, like the scrambled eggs, lose their individuality.

Molecular structure and geometry are more easily pictured in terms of the VSEPR and valence bond theories (Chapter 11) than in terms of molecular orbitals. Atomic orbital overlap diagrams such as Figure 11.5 indicate how geometry and bonding are interrelated. However, the valence bond theory does not clarify the relative energy levels of the orbitals involved in the bonding—something that can be done by using molecular orbitals. An understanding of energy levels is important—for one reason, because we derive much of our information about molecules from spectroscopy. To interpret spectra, a knowledge of the number of possible excited states for electrons and of the energies of the transitions to and from these states is needed. One of the great successes of molecular orbital theory is the prediction and explanation of electronic spectra. Molecular orbital theory also permits a clearer understanding of multiple bonds, molecules for which resonance must be used to write Lewis structures, and molecules with unpaired electrons.

The molecular orbital approach, like the valence bond approach, is based on quantum theory and the use of the Schrödinger equation (Section 8.8 and Aside 25.1). The energies of electrons in bonds, like those in isolated atoms, are quantized. For molecules, not only the interactions of electrons with each other and with nuclei must be accounted for, but also the interactions of nuclei with each other. As for the atomic orbitals of many-electron atoms, approximations are used to reduce the complexity of the mathematical treatment of molecular orbitals.

Energy-level diagrams can be drawn for molecular orbitals. Then, as in the case of atoms (Section 8.13), electron configurations can be determined by distributing electrons among the energy levels. From the configuration of the electrons in the molecular orbitals, predictions can be made about whether or not a bond between two specific atoms will form and whether or not a molecule will be paramagnetic—information that is not easily available from valence bond theory. For example, molecular orbital theory explains why Be_2 molecules have only a brief existence and it can account for the magnetic properties of O_2 molecules, neither of which can be done easily by valence bond theory. In reading the following sections, keep in mind that we are describing a mathematical process for representing bonding as it is observed.

Aside 25.1 SCHRÖDINGER, HIS EQUATION, AND LIFE

Erwin Schrödinger was one of the architects of quantum mechanics (Chapter 8). The description of bonding by molecular orbital theory, discussed in this chapter, grows out of the mathematics of quantum mechanics. Schrödinger is most often remembered for the "Schrödinger equation," one of the first fundamental mathematical expressions of quantum theory (see Aside 21.2). For the hydrogen atom with its one electron the equation is

Planck's constant → h^2; *electron mass* → m; *variation in probability of finding electron in a three-dimensional space* → the bracketed term; *quantized energy levels of electron* → E; *potential energy of electron* → V

$$\frac{h^2}{8\pi^2 m}\left(\frac{\partial^2\psi}{\partial x^2}+\frac{\partial^2\psi}{\partial y^2}+\frac{\partial^2\psi}{\partial z^2}\right)+(V-E)\psi=0$$

Exploration of the meaning of this equation led to an exciting result. Aspects of the behavior of electrons and other small particles that had been introduced as assumptions resulted naturally from the mathematics.

Schrödinger was an individual with, in his own words, a "keen longing for unified, all embracing knowledge." In response to this longing, he set out, in a book called *What Is Life?* to explore the question, "How can the events in space and time which take place within the spatial boundary of a living organism be accounted for by physics and chemistry?"

Eventually, he felt, there was no doubt that physics and chemistry would achieve an understanding of life in terms of energy and molecules. But first these sciences would have to recognize an essential difference between living organisms and the systems they had been studying. Many of the laws of physics are based on the average behavior of large collections of identical particles. Living organisms, on the other hand, are unique. The code that leads, say, one egg to grow into a frog while another egg grows into a white rabbit must, Schrödinger concluded, lie in a "well-ordered association of atoms." Specific differences in the arrangement of the atoms in a small number of molecules must determine the distinctive features of different organisms. Schrödinger wrote of this conclusion as follows:

Well, this is a fantastic description, perhaps less becoming a scientist than a poet. However, it needs no poetical imagination but only clear and sober scientific reflection to recognize that we are here obviously faced with events whose regular and lawful unfolding is guided by a 'mechanism' entirely different from the 'probability mechanism' of physics. For it is simply a fact of observation that the guiding principle in every cell is embodied in a single atomic association existing only in one copy (or sometimes two)—and a fact of observation that it results in producing events which are a paragon of orderliness. Whether we find it astonishing or whether we find it quite plausible that a small but highly organized group of atoms be capable of acting in this manner, the situation is unprecedented, it is unknown anywhere else except in living matter.[a]

Before the mechanism of heredity was understood, Schrödinger had anticipated the field of molecular genetics. Francis H. C. Crick, who was a physicist, read *What Is Life?* a few years after it was written. The book helped to inspire a change in the direction of Crick's career: He moved into biology. In 1962, Crick shared a Nobel prize for the discovery, with James D. Watson and Maurice H. F. Wilkins, of the structure of DNA (Section 33.10), which carries the coded message of heredity from one generation to the next.

[a] Source: Erwin Schrödinger, *What Is Life?* 1944 (New York and London: Cambridge University Press combined reprint, 1980), p. 82.

25.2 MOLECULAR ORBITALS AND THEIR ENERGY LEVELS

The orbitals of two atoms that are about to form a bond are thought of as rearranging themselves as the electrons and nucleus of each atom come under the influence of the electrons and nucleus of the other. The number of molecular orbitals formed must equal the number of atomic orbitals that have combined. For example, when two atoms that each have two atomic orbitals combine, there will be four molecular orbitals in the area of the bonded atoms. This does not mean that all of these molecular orbitals will be filled with electrons. To use molecular orbital theory we first determine the number of possible molecular orbitals. Then we determine how many electrons are available in the atoms that have combined. Finally, we assign the electrons to molecular orbitals, writing an electron configuration for the molecule, somewhat as we wrote *spdf* configurations for atoms.

The process is a little like assigning dormitory rooms to freshmen in college. The person in charge of housing first has to figure out how many rooms are available and on what floors in a particular building. But the Office of Admissions has to decide how many freshmen there will be to occupy the rooms. Once this is known, the students are assigned to the empty rooms. Maybe there will be fewer students than rooms, but the empty rooms—like the vacant molecular orbitals—are still there to be occupied if someone else comes along.

The combination of two atomic orbitals must produce two molecular orbitals, and in the process energy must be conserved. The two molecular orbitals result from the combination of the atomic orbitals in such a way that the electron waves either reinforce each other (constructive interference; see Figure 8.2a) or cancel each other (destructive interference; see Figure 8.2c). In the first case, a **bonding molecular orbital**—an orbital in which most of the electron density is located between the nuclei of the bonded atoms—is formed. A bonding molecular orbital is always of lower energy than either of the atomic orbitals that have combined—formation of a bond creates a more stable situation. In the second case an **antibonding molecular orbital**—an orbital in which most of the electron density is located away from the space between the nuclei—is formed. The energy of the antibonding orbital is raised above the energy of the atomic orbitals that have combined by an amount equal to that by which the energy of the bonding orbital has been lowered. In this way, the two molecular orbitals have the

Atomic orbital | Molecular orbitals | Atomic orbital

Increasing energy

Antibonding orbital – higher energy than atomic orbitals

Bonding orbital – lower energy than atomic orbitals

Atom A | Molecule AA′ | Atom A′

Figure 25.1
Energy Levels in Formation of a Molecular Orbital

same total energy as the two original atomic orbitals. Mathematically, the formation of bonding and antibonding orbitals is most commonly described by the addition and subtraction of the wave functions of the individual atomic orbitals. A generalized energy-level diagram for the formation of a molecular orbital is given in Figure 25.1, which shows atomic orbitals of equal energy from two different atoms combining to give one antibonding and one bonding molecular orbital.

25.3 TYPES OF MOLECULAR ORBITALS

The redistribution of electron density in the formation of a bonding and an antibonding molecular orbital is illustrated in Figure 25.2, which depicts the combination of two 1*s* atomic orbitals. The boundary contours in the figure show the regions in which the electrons are most likely to be found. A *node,* which is a region of zero electron density, lies between the nuclei in the antibonding orbital. (A *nodal plane*—a plane that encounters no electron density—intersects the bond axis in an antibonding molecular orbital.) The nuclei are mutually repelled by each other's positive charges, rather than held together. Electrons in antibonding orbitals do not form bonds; rather, they can weaken bonds formed between the same two atoms by other electrons.

In Chapter 11, we learned that bonds in which the electron density surrounds the bond axis are called σ bonds and those that leave an area of zero electron density around the bond axis are called π bonds. The σ and π notation is used similarly for molecular orbitals.

The bonding molecular orbital formed from two 1*s* atomic orbitals (see Figure 25.2) is a bonding σ molecular orbital. It is symbolized as follows and the symbol is read as "sigma one ess":

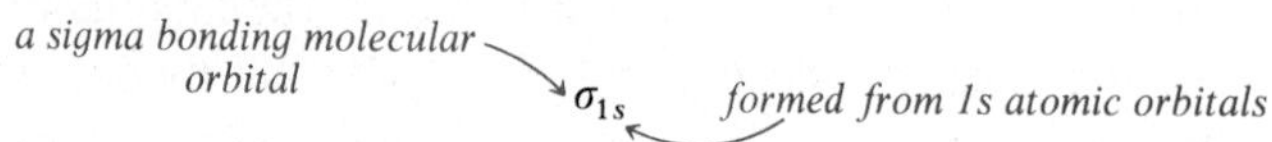

Antibonding molecular orbitals are symbolized by adding a "star" to the bonding molecular orbital symbol. The antibonding orbital from two 1*s* atomic orbitals is a "sigma star one ess" orbital:

In valence bond theory, π bonds are described as formed by parallel overlap of *p* orbitals. The molecular orbital picture is similar. The parallel combination of *p* atomic orbitals forms π bonding and π antibonding molecular orbitals.

The three *p* orbitals from each of two atoms must yield six molecular orbitals. Two pairs of *p* orbitals combine in parallel fashion to give four π molecular orbitals, two bonding (π_p) and two antibonding (π^*_p) orbitals. The two π_p orbitals lie at 90° to each other, as do the two π^*_p orbitals. The third pair of *p* orbitals combines in end-on fashion to give σ molecular orbitals (σ_p, σ^*_p).

When their energy levels are similar, an *s* and a *p* orbital can also combine to give

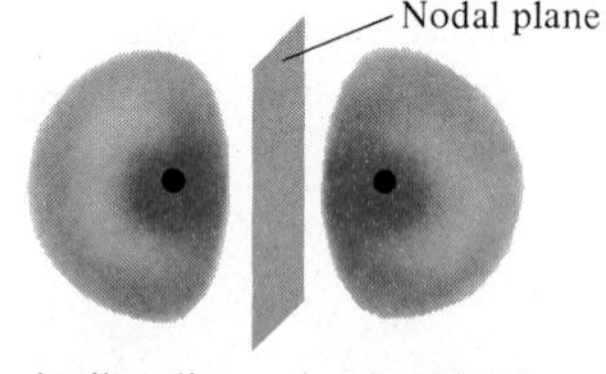

Antibonding molecular orbital

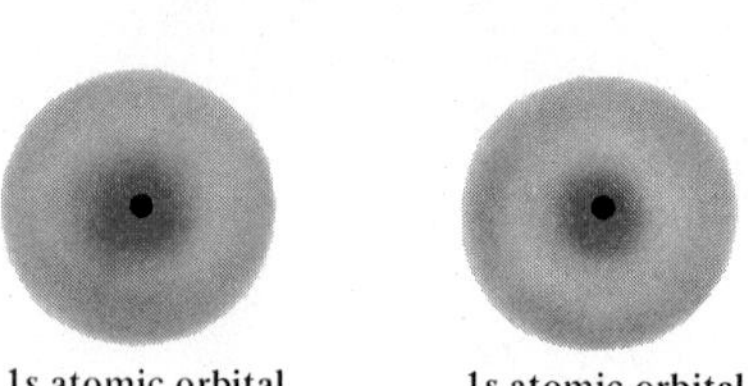

1*s* atomic orbital 1*s* atomic orbital

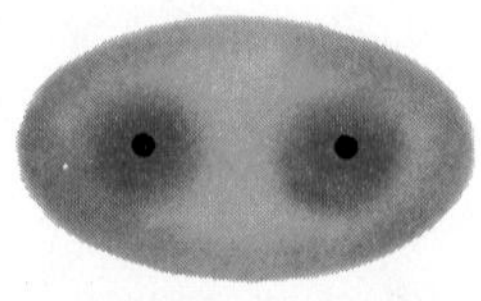

Bonding molecular orbital

Figure 25.2
Combination of *s* Atomic Orbitals to Give Antibonding and Bonding Molecular Orbitals

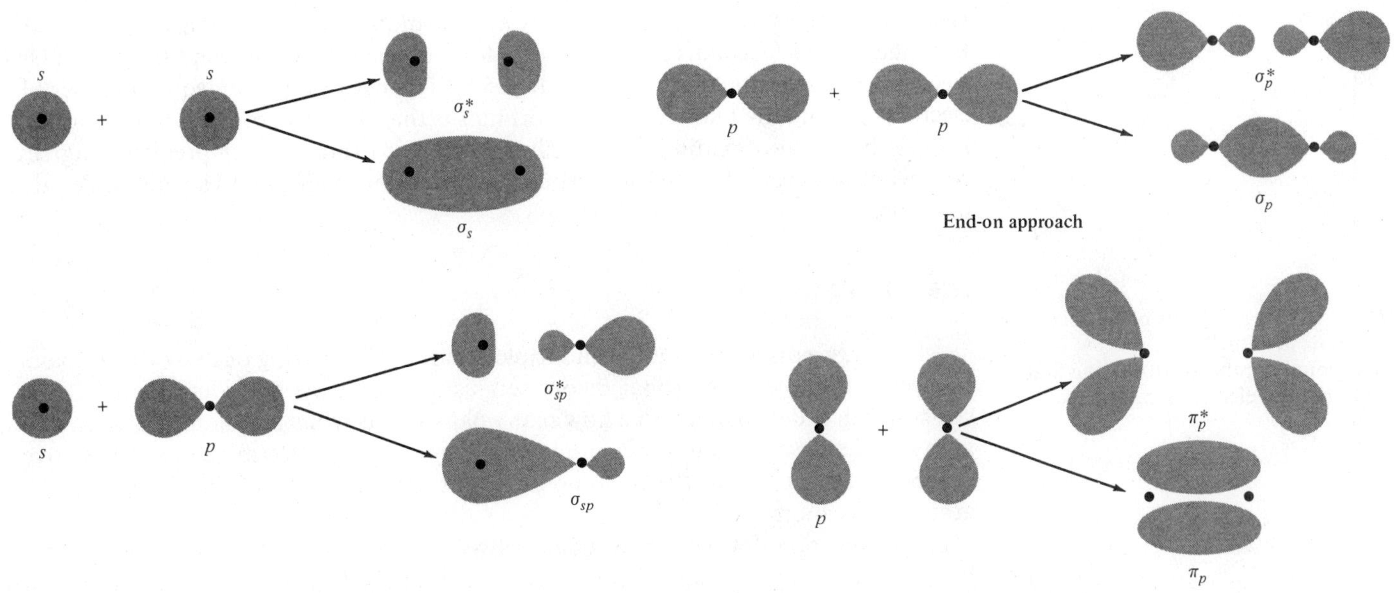

Figure 25.3
Molecular Orbitals from *s* and *p* Atomic Orbitals
The direction of approach is chosen as along the x axis. The antibonding molecular orbitals are higher in energy and the bonding molecular orbitals are lower in energy than the atomic orbitals from which they arise.

molecular orbitals ($\sigma_{sp}, \sigma_{sp}^*$). The formation of σ and π molecular orbitals from s and p atomic orbitals is summarized in Figure 25.3. Note that each antibonding molecular orbital has an area of zero electron density between the nuclei.

25.4 RULES FOR FILLING MOLECULAR ORBITALS

Just as for atomic orbitals, there are rules that determine the order in which molecular orbitals are filled. Electrons are pictured as entering the orbitals one by one according to the following rules:

1. Electrons first occupy the molecular orbitals of lowest energy; they enter higher energy molecular orbitals only when the lower energy orbitals are filled.
2. Each molecular orbital can accommodate a maximum of two electrons (Pauli exclusion principle).
3. Molecular orbitals of equal energy are occupied by single electrons before electron pairing begins (Hund principle).

25.5 H_2 AND "He_2"

As in our discussion of valence bond theory (Section 11.7), we look first at the simplest neutral, stable molecule, the hydrogen molecule. To derive the molecular orbital description of the hydrogen molecule we begin with two isolated hydrogen atoms, each with one electron in a $1s$ orbital. For this simple case, the two $1s$ atomic orbitals combine to give one lower energy, bonding σ molecular orbital and one higher energy, antibonding σ^* molecular orbital. The ground state H_2 molecule has a total of two molecular orbitals available to be filled. Following the rules given above, the two electrons pair up to occupy the lowest level molecular orbital available, the σ_{1s} orbital (Figure 25.4). Two electrons in a σ orbital formed from atomic $1s$ orbitals are symbolized as

$$H_2 \qquad \sigma_{1s}^2 \quad \textit{two electrons in a } \sigma_{1s} \textit{ orbital}$$

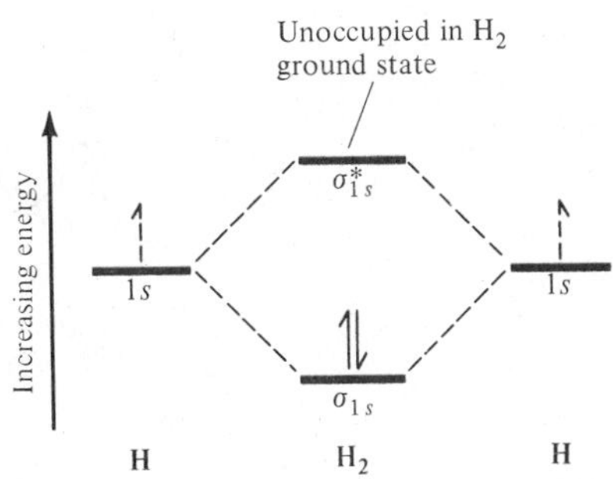

Figure 25.4
Relative Energy Levels in Formation of H_2

The antibonding σ_{1s}^* orbital is unoccupied in the ground state of the H_2 molecule.

The energy-level diagram for the combination of two He atoms into an He_2 molecule is the same as that for H_2. As for the hydrogen atoms, two atomic $1s$ orbitals

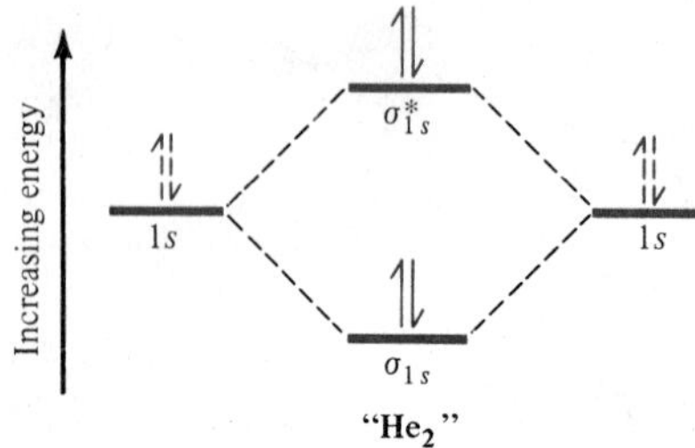

Figure 25.5
Relative Energy Levels of Molecular Orbitals in "He_2"
The equal numbers of bonding and antibonding electrons cancel each other.

yield a σ_{1s} and a σ^*_{1s} orbital. But here there are two more electrons to be accommodated. Both the σ and σ^* orbitals are filled by the four available electrons, and the molecular electron configuration is $\sigma_{1s}{}^2\sigma^{*2}_{1s}$ (Figure 25.5). The net result of an equal number of electrons in bonding and antibonding orbitals is that the effect of the bonding orbital is offset by the antibonding orbital. Thus molecular orbital theory prediction agrees with what has been observed—gaseous helium is monatomic and He_2 molecules are unstable.

25.6 BOND ORDER

Earlier, we discussed double and triple bonds as the sharing of two or three pairs of electrons. Molecular orbital theory sets us free from the consideration of Lewis structures in order to determine how many pairs of electrons are shared between two atoms. Whether a bond is, for example, single or double is determined by the relative numbers of electrons in the bonding and antibonding orbitals formed between the atoms in question.

The **bond order** of a covalent bond between two atoms is the number of effective bonding electron pairs shared between the two atoms. The repulsion due to electrons in antibonding orbitals cancels the attraction due to bonding electrons. Therefore, to find bond order, the number of electrons in antibonding orbitals must be subtracted from the number of electrons in bonding orbitals and the difference divided by 2.

$$\text{Bond order} = \tfrac{1}{2}[(\text{the number of bonding electrons}) - (\text{the number of antibonding electrons})] \quad \textbf{(25.1)}$$

For the H_2 molecule

$$\text{Bond order} = \tfrac{1}{2}(2 - 0) = 1$$

The hydrogen molecule has a single covalent bond.

For the "He_2" molecule

$$\text{Bond order} = \tfrac{1}{2}(2 - 2) = 0$$

A bond order of zero indicates that a stable bond does not form. In general, the strength of a bond increases with increasing bond order and the length of the bond decreases with increasing bond order. As shown in Example 25.1, it is possible to have fractional bond orders.

EXAMPLE 25.1 Bond Order

We have found that H_2 is a stable molecule and "He_2" is not. Would the formation of $H_2{}^+$ and $He_2{}^+$ be possible?

To form $H_2{}^+$ from H_2, which has the $\sigma_{1s}{}^2$ configuration, one electron would be removed to give one electron in a bonding orbital and a configuration of σ_{1s}. The bond order of $H_2{}^+$ is found as follows:

$$\text{Bond order} = \tfrac{1}{2}(1 - 0) = \tfrac{1}{2}$$

A bond order of $\frac{1}{2}$ indicates that **$H_2{}^+$ should exist, but would be less stable than H_2.**

The $He_2{}^+$ ion would have three electrons, one less than the "He_2" molecule, which has the $\sigma_{1s}{}^2\sigma^{*2}_{1s}$ configuration. This would give two electrons in the σ_{1s} orbital and one in the σ^*_{1s} orbital. Therefore, $He_2{}^+$ has a configuration of $\sigma_{1s}{}^2\sigma^*_{1s}$ and a bond order of

$$\text{Bond order} = \tfrac{1}{2}(2 - 1) = \tfrac{1}{2}$$

We conclude that the **$He_2{}^+$ ion can also exist.** [$H_2{}^+$ and $He_2{}^+$ have both been observed to exist in the gas phase; their bond dissociation energies are 255 kJ/mol and 300 kJ/mol, respectively, while that of H_2 is 435 kJ/mol.]

EXERCISE Write the molecular orbital configuration and calculate the bond order for $H_2{}^-$. Would the formation of $H_2{}^-$ be possible? *Answer* $\sigma_{1s}{}^2\sigma^*_{1s}$, $\frac{1}{2}$, yes.

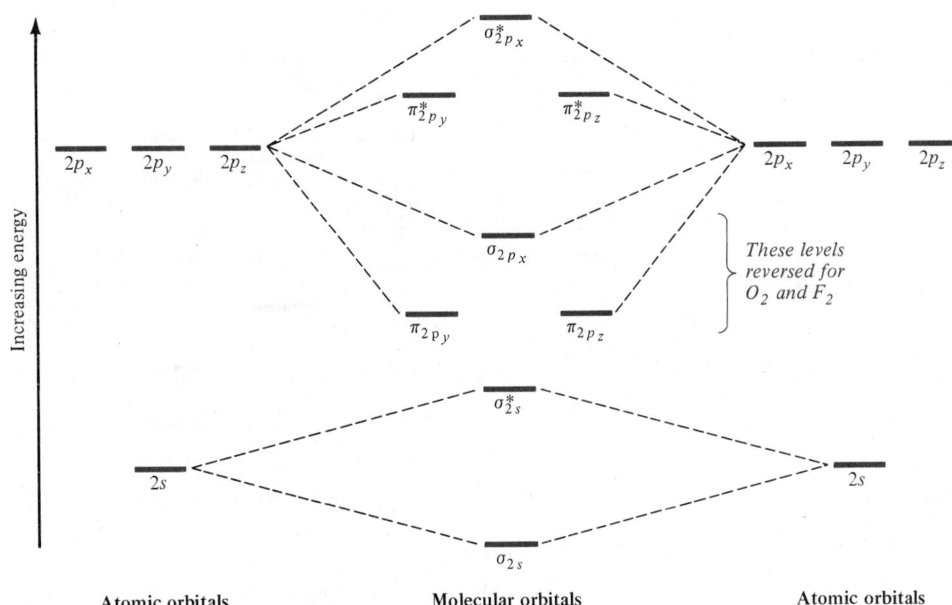

Figure 25.6
Relative Energy Levels for Atomic and Molecular Orbitals of Second-Period Elements
The $n = 1$ electrons are the nonbonding ("core") electrons and are not included in consideration of molecular orbitals. Each heavy line represents an orbital that can accommodate two electrons. (This and other energy-level diagrams are not drawn to scale.)

25.7 MOLECULAR ORBITALS FOR DIATOMIC MOLECULES OF SECOND-PERIOD ELEMENTS, Li–Ne

Atoms of second-period elements have two main energy levels ($n = 1, 2$) and five orbitals—one 1*s*, one 2*s*, and three 2*p*. The 1*s* electrons are core electrons. They are drawn close to the nucleus and remain essentially atomic in character. The 1*s* orbitals need not be considered in constructing molecular orbitals. In general, only valence electrons are placed into molecular orbitals.

In this section we consider how the molecular orbitals are filled for homonuclear diatomic molecules of the elements lithium through neon. **Homonuclear** means literally "the same nucleus"; the term refers to atoms of the same element. The value of molecular orbital theory in predicting whether or not a bond will form and in explaining the magnetic properties, bond energies, and bond lengths of molecules is well illustrated by these molecules.

The energy-level diagram for the molecular orbitals formed by the second-period element diatomic molecules is shown in Figure 25.6. Note that the *p* orbitals form two π molecular bonding orbitals (by parallel approach of $2p_y$ and $2p_z$ orbitals) and one σ molecular bonding orbital (by end-on approach of $2p_x$ orbitals; see Figure 25.3). For the molecules of the first five elements of the period (Li_2–N_2), experiment shows that the π_{2p} orbitals are of slightly lower energy than the σ_{2p} orbitals. For O_2 and F_2, the positions are reversed, and the σ_{2p} orbital is of lower energy than the π_{2p} orbitals.

What determines the relative energies of molecular orbitals? The situation is similar to that which we encountered with the *ns* and $(n - 1)d$ energy levels for electrons in many-electron atoms (Section 8.12). The relative energy levels in each atom or molecule depend upon the specific interactions possible in that atom or molecule.

Figure 25.7 (page 642) shows the relative positions of the molecular orbital energy levels for the diatomic molecules Li_2 to F_2. The electron configurations of the individual molecules (Table 25.1) are discussed in the following paragraphs. Note that in each case we are discussing orbitals and the possible formation of bonds in diatomic molecules in the *gaseous state*. *Elemental* lithium and beryllium are metals; *elemental* carbon is a network covalent substance.

Lithium, $1s^2 2s^1$. An Li_2 molecule will contain six electrons. The first four electrons, as for all of the second-period elements, are the nonbonding 1*s* electrons (the core electrons). The remaining two electrons come from the 2*s* orbitals of the two lithium atoms. Just as in H_2, the *s* orbitals combine to give one bonding and one antibonding σ

Table 25.1
Electron Configurations of the Diatomic Molecules of Second-Period Elements
The nonbonding 1*s* electrons are not included.

Li_2	${\sigma_{2s}}^2$
B_2	${\sigma_{2s}}^2 \sigma_{2s}^{*2} \pi_{2p_y}^1 \pi_{2p_z}^1$
C_2	${\sigma_{2s}}^2 \sigma_{2s}^{*2} \pi_{2p_y}^2 \pi_{2p_z}^2$
N_2	${\sigma_{2s}}^2 \sigma_{2s}^{*2} \pi_{2p_y}^2 \pi_{2p_z}^2 \sigma_{2p_x}^2$
O_2	${\sigma_{2s}}^2 \sigma_{2s}^{*2} \sigma_{2p_x}^2 \pi_{2p_y}^2 \pi_{2p_z}^2 \pi_{2p_y}^{*1} \pi_{2p_z}^{*1}$
F_2	${\sigma_{2s}}^2 \sigma_{2s}^{*2} \sigma_{2p_x}^2 \pi_{2p_y}^2 \pi_{2p_z}^2 \pi_{2p_y}^{*2} \pi_{2p_z}^{*2}$

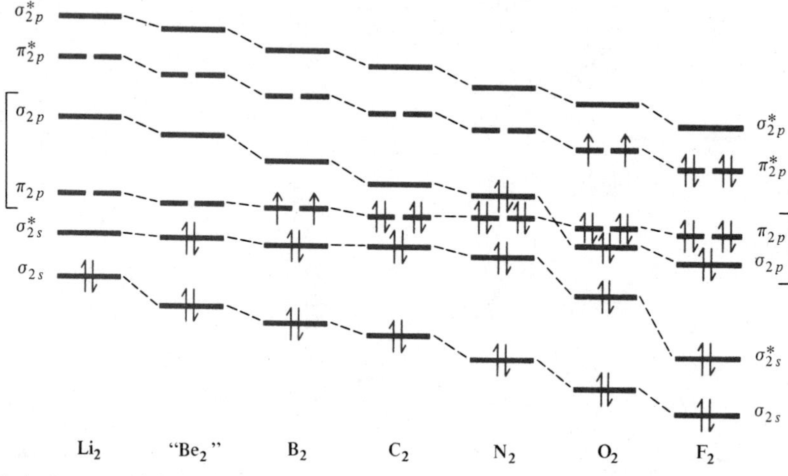

Figure 25.7
Molecular Orbital Diagram for the Second-Period Elements Li_2–F_2
Note the reversal of the σ_{2p} and π_{2p} orbitals for O_2 and F_2.

Source: F. A. Cotton and G. Wilkinson, *Advanced Inorganic Chemistry*, Fig. 3.12 (New York: Wiley, 1976) p. 66;

molecular orbital. The lower energy σ_{2s} orbital is occupied by the two electrons (see Figure 25.7), giving Li_2 a bond order of $\frac{1}{2}(2-0)=1$.

Diatomic Li_2 molecules can be detected in the gaseous state. They are much less stable than H_2 molecules because of the larger size of the lithium atoms and the greater nuclear repulsion between them. Compare the bond dissociation energies of H_2 and Li_2 in Table 25.2.

Beryllium, $1s^2 2s^2$. Of the eight electrons available for a Be_2 molecule, the first four are the nonbonding inner electrons. The next four fill the σ_{2s} and σ^*_{2s} orbitals. This gives a bond order of $\frac{1}{2}(2-2)=0$. As predicted, Be_2 molecules are not stable.

Boron, $1s^2 2s^2 2p^1$. The B_2 molecule has six outer electrons to be distributed in molecular orbitals. Four electrons fill the σ_{2s} and σ^*_{2s} orbitals, leaving two electrons for the π orbitals. Following the Hund principle, the two electrons enter the π_{py} and π_{pz} orbitals singly. This gives B_2 a bond order of one.

The conclusions that may be drawn from the molecular orbital diagram for gaseous B_2 (see Figure 25.7) are that it will exist and that it will exhibit a paramagnetism showing the presence of two unpaired electrons. Both of these predictions are correct. Note that the two-electron paramagnetism of boron is evidence for the lower energy of the π_{2p} orbitals. If the σ_{2p} orbital had lower energy, the last two electrons would be paired and B_2 would be diamagnetic.

Table 25.2
Diatomic Molecules of First- and Second-Period Elements
In general, bond length decreases and bond energy increases with bond order. Note these trends in the series B_2, C_2, N_2, and O_2^+, O_2, O_2^-.

Species	Total number of electrons	Bond order	Bond length (nm)	Bond dissociation energy (kJ/mol)
H_2	2	1	0.074	432
"He_2"	4	0	—	—
Li_2	6	1	0.267	108
"Be_2"	8	0	—	—
B_2	10	1	0.159	292
C_2	12	2	0.124	590.
N_2	14	3	0.109	942
O_2^+	15	$2\frac{1}{2}$	0.112	636
O_2	16	2	0.121	494
O_2^-	17	$1\frac{1}{2}$	0.130	394
F_2	18	1	0.141	154
"Ne_2"	20	0	—	—
CO	14	3	0.113	1070
NO	15	$2\frac{1}{2}$	0.115	628

Carbon, $1s^2 2s^2 2p^2$. Eight electrons must be placed into the molecular orbitals for the C_2 molecule. In addition to the pairs of electrons in the σ_{2s} and σ^*_{2s} orbitals, pairs of electrons are also placed in the π_{2p_y} and π_{2p_z} orbitals. The C_2 molecule has a bond order of $\frac{1}{2}(6 - 2) = 2$, and the higher bond order is reflected in a high bond dissociation energy (see Table 25.2). All electrons are paired, and gaseous C_2 is diamagnetic. The C_2 molecule exists at high temperatures or in the presence of an electrical discharge.

Nitrogen, $1s^2 2s^2 2p^3$. A nitrogen molecule has ten outer electrons to be placed in molecular orbitals. The first eight assume the same configuration as in C_2, and the remaining two enter the σ_{2p_x} orbital (see Table 25.1). With eight bonding electrons and two antibonding electrons, the bond order is three, in agreement with the Lewis formula that we have been writing for N_2, which is :N≡N:. All electrons are paired, in agreement with the observed diamagnetism of the N_2 molecule. The very large bond dissociation energy results from the high bond order.

Oxygen, $1s^2 2s^2 2p^4$. One of the early successes of molecular orbital theory was in explaining the paramagnetism of oxygen. The theory also nicely accounts for the relative bond dissociation energies of O_2 and O_2^+. The first ten outer electrons from the oxygen atoms fill the molecular orbitals up to the π_{2p} orbitals as in N_2, but note that the π_{2p} and σ_{2p} orbitals are reversed in order (see Figure 25.7). The next highest energy level has two equal π^*_{2p} antibonding orbitals available. The final two electrons of O_2 should enter these π^*_{2p} orbitals singly. The paramagnetism of O_2 shows that this does indeed happen—two unpaired electrons are present in an oxygen molecule. Only the molecular orbital theory was able to explain both the paramagnetism and the bond order of O_2, which is two. A demonstration of the paramagnetism of O_2 and the diamagnetism of N_2 is illustrated in Figure 25.8.

Removal of one electron from O_2 to give O_2^+ increases the O—O bond energy (see Table 25.2). Taking away an antibonding electron has strengthened the bond, which is exactly what molecular orbital theory predicts.

Fluorine, $1s^2 2s^2 2p^5$. In the F_2 molecule, the 14 available electrons are placed in the molecular orbitals of increasing energy up to the π^*_{2p} energy level (see Figure 25.7). All electrons are paired and the bond order, in agreement with the Lewis formula, :F̤̈—F̤̈:, is one.

Neon, $1s^2 2s^2 2p^6$. The Ne_2 molecule is not stable. All electrons are paired, and with all the molecular orbitals in Figure 25.6 filled, the bond order is zero.

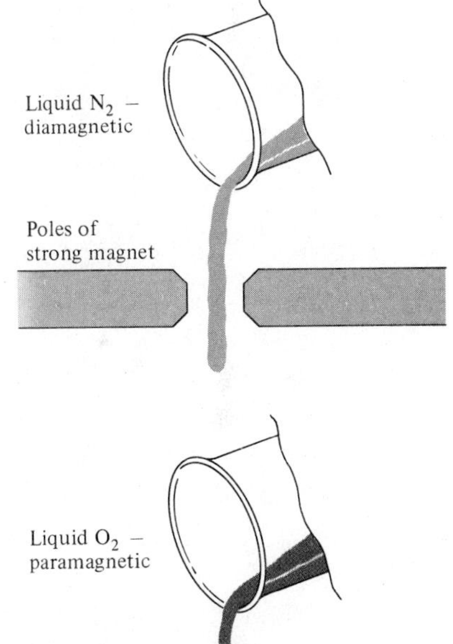

Figure 25.8
Paramagnetism of Oxygen
Liquid oxygen (b.p. −183°C) is held between the poles of a strong magnet (9 kilogauss) until it evaporates. Liquid nitrogen (b.p. −196°C) is not held by such a magnet. Note that these liquefied gases must be handled with great care.

Source: For details of this demonstration, see Bassam Z. Shakhashiri et al., *Chemical Demonstrations* (University of Wisconsin Press, 1983).

EXAMPLE 25.2 Molecular Orbital Electron Configuration; Bond Order

(a) Using the molecular orbital energy diagram given in Figure 25.6, write the electron configuration for O_2^+ and O_2^-. (b) Comment on the relative bond orders, magnetic properties, bond lengths, and bond dissociation energies of O_2^-, O_2, and O_2^+ (discussed above).

(a) In the O_2^+ molecule there is a total of 15 electrons. The four 1*s* electrons will not be involved in the bonding because they lie very close to the oxygen nuclei and we simply write the atomic configurations for these four electrons: $1s^2 1s^2$. The remaining 11 electrons are placed in the orbitals shown in the figure using the usual rules for filling orbitals: $\sigma_{2s}{}^2 \sigma^{*2}_{2s} \sigma^2_{2p_x} \pi^2_{2p_y} \pi^2_{2p_z} \pi^{*1}_{2p_y}$. The molecular orbital electron configuration for O_2^+ is

$$\mathbf{1s^2 1s^2 \sigma_{2s}{}^2 \sigma^{*2}_{2s} \sigma^2_{2p_x} \pi^2_{2p_y} \pi^2_{2p_z} \pi^{*1}_{2p_y}}$$

The electron configuration for O_2^-, which has two more electrons, is

$$\mathbf{1s^2 1s^2 \sigma_{2s}{}^2 \sigma^{*2}_{2s} \sigma^2_{2p_x} \pi^2_{2p_y} \pi^2_{2p_z} \pi^{*2}_{2p_y} \pi^{*1}_{2p_z}}$$

(b) To form O_2^- from O_2, an additional electron must enter one of the π^*_{2p} orbitals. This reduces the paramagnetism from that of two unpaired electrons in O_2 to that of one unpaired electron, the same as for O_2^+. **The bond order of O_2^- is $\frac{1}{2}(8 - 5) = 1\frac{1}{2}$. This**

decrease in bond order from that of O_2 should lead to an increase in bond length and a decrease in bond strength. For O_2^+, with a bond order of $2\frac{1}{2}$, the bond length should be less than that in O_2 and the bond strength greater. The data in Table 25.2 confirm these predictions.

EXERCISE Comment on the (a) relative bond orders, (b) magnetic properties, (c) bond lengths, and (d) bond dissociation energies of N_2, N_2^+, and N_2^-.

Answers (a) 3 for N_2, $2\frac{1}{2}$ for N_2^+ and N_2^-; (b) N_2 is diamagnetic, N_2^+ and N_2^- are paramagnetic; (c) N_2 is less, N_2^+ and N_2^- are about the same; (d) N_2 is largest, N_2^+ and N_2^- are about the same.

25.8 HETERONUCLEAR DIATOMIC MOLECULES

Within limits, the energy-level diagram of Figure 25.6 can be used to determine the molecular orbital configuration of gaseous diatomic molecules formed between atoms of different elements—heteronuclear diatomic molecules—in the second period. **Heteronuclear** means "of different nuclei"; that is, it refers to atoms of different elements.

The elements C and O, and N and O are not very different in atomic number, and the energies with which these atoms hold electrons are similar. Molecular orbitals can therefore form between these atoms by overlap of comparable atomic orbitals (as illustrated by Figure 25.6).

Consider carbon monoxide, CO, with 14 electrons (six from C, eight from O). Placing these electrons in the energy levels in order according to the usual rules gives a molecule with the same configuration as the nitrogen molecule (see Figure 25.7 and Table 25.1), and therefore a bond order of three, :C≡O:.

The nitric oxide molecule, NO, has a total of 15 electrons. Its configuration should be like that of nitrogen (see Figure 25.7), but with one more electron added in a π^*_{2p} orbital. Nitric oxide does indeed have the paramagnetism of a molecule with one unpaired electron. Like the O—O$^+$ case described above, removal of an antibonding electron (the π^*_{2p} electron) from NO produces a positive ion, NO^+, more stable than its neutral parent.

When the atoms in a heteronuclear diatomic molecule have large differences in atomic number and in ionization energies, overlap occurs between orbitals of similar energy level but not necessarily of similar description (i.e., not necessarily 1*s* with 1*s*, 2*p* with 2*p*). Such a situation is illustrated for HF in Figure 25.9. The 1*s* orbital of hydrogen has approximately the same energy as a 2*p* orbital of fluorine, allowing the formation of a σ_{sp} molecular orbital.

EXAMPLE 25.3 Molecular Orbital Electron Configuration

Assume that the molecular orbital diagram given in Figure 25.6 is valid for BO. Write the electron configuration for this molecule. What is the bond order?

There are 13 electrons in this molecule. The complete configuration is

$$1s^2 1s^2 \sigma_{2s}^2 \sigma_{2s}^{*2} \pi_{2py}^2 \pi_{2pz}^2 \sigma_{2px}^1$$

where we have written $1s^2 1s^2$ for the four 1*s* electrons that are not involved in the bonding. The bond order is $\frac{1}{2}(7 - 2) = \mathbf{2\frac{1}{2}}$.

EXERCISE Is the BO molecule paramagnetic or diamagnetic?

Answer Paramagnetic.

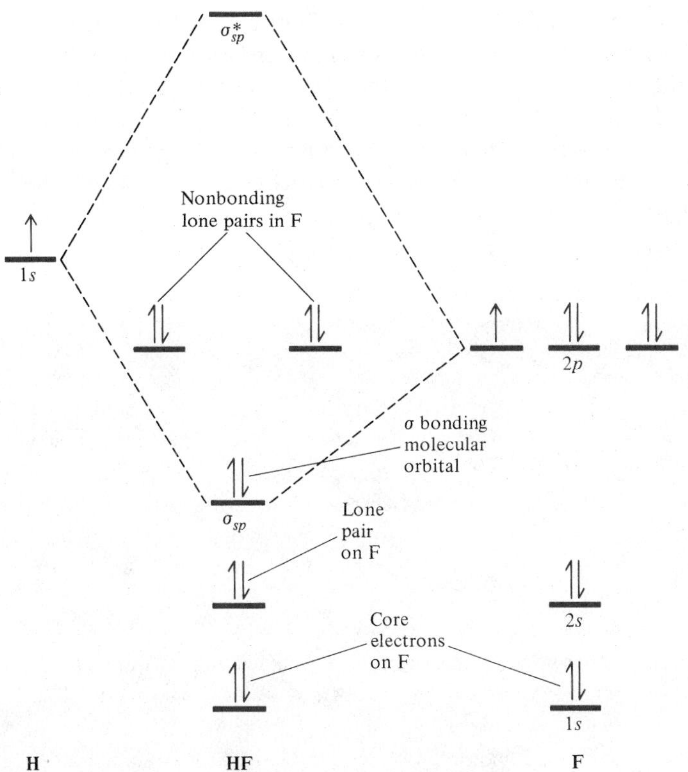

Figure 25.9
Molecular Orbital Energy Diagram for HF
Drawn for unhybridized orbitals on fluorine. (Only the $2p$ orbitals of F are sufficiently close in energy to the hydrogen $1s$ orbital to allow overlap and molecular orbital formation. Of these, only the $2p_z$ orbital has the proper symmetry.)

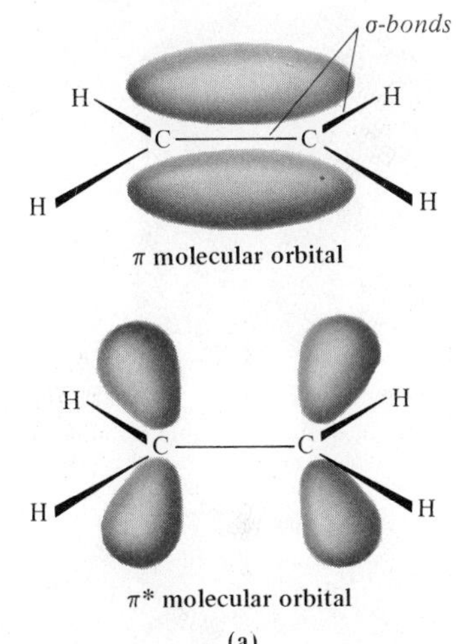

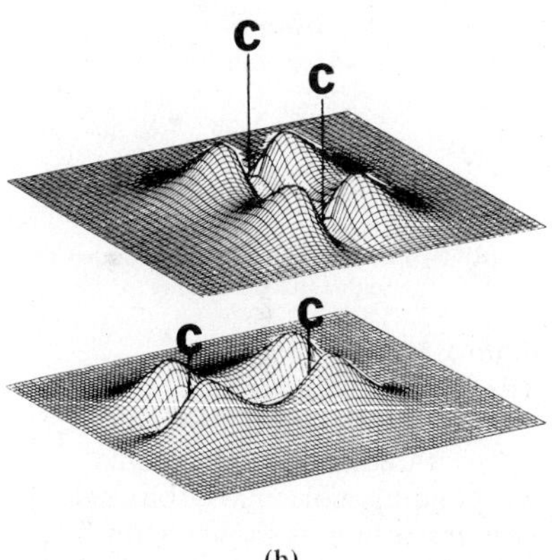

Figure 25.10
Molecular Orbital Views of Ethylene
(a) Conventional drawing of ethylene π and π^* molecular orbitals; (b) π-bonding orbital electron density distribution in ethylene. The plot, viewed from two different angles, shows electron density in a plane perpendicular to the plane of the C—H skeleton. The C atoms lie in the "valleys," as indicated; the H atoms (not shown) would be above and below the plane of the plot.

Source: J. R. Van Wazer and I. Absar, *Electron Densities in Molecules and Molecular Orbits* (New York: Academic Press, 1975), p. 33.

25.9 MULTIPLE BONDS AND DELOCALIZED ORBITALS

The molecular orbital picture of the carbon–carbon bond in ethylene (Figure 25.10) is not very different from the valence bond picture (see Figure 11.7). σ Bonds determine the geometry of the carbon–hydrogen skeleton, and π bonds lie above and below the carbon–carbon bond axis. Three valence electrons from each carbon atom enter the σ orbitals, leaving one valence electron from each carbon atom for the π orbitals. Two p orbitals, one from each carbon atom, form two π molecular orbitals, one bonding and occupied by the two electrons, and one antibonding and unoccupied in the ground state of ethylene.

When a molecule or ion has more than one multiple bond, it is often possible to write several Lewis structures that are in agreement with the known properties of the molecule or ion. We have dealt with such species by the concept of resonance (Section 9.17) or the concept of delocalization of π electrons. In molecular orbital theory, the π molecular orbitals are delocalized—spread over the entire molecule.

Previously (Section 11.3) we described the bonding in benzene in terms of the classical resonance structures

$$\text{(benzene)} \longleftrightarrow \text{(benzene)}$$

that account for the equivalence of the carbon–carbon bonds. To build up a molecular orbital picture of a benzene molecule, a total of 42 electrons from the six carbon atoms and six hydrogen atoms must be placed in orbitals. The first 12 electrons remain in $1s$ atomic orbitals on the carbon atoms. The next 24 electrons occupy the orbitals that form the carbon–hydrogen skeleton of benzene: 12 in carbon–hydrogen bonds and 12 in carbon–carbon bonds. This leaves six electrons for the next highest energy level molecular orbitals, which must be formed from the single p orbital on each carbon atom not involved in σ bonds. These six p orbitals combine to form three π orbitals and three π^* orbitals. The three π orbitals add up to a region of uniform electron density

(a) σ-bonded C–H skeleton

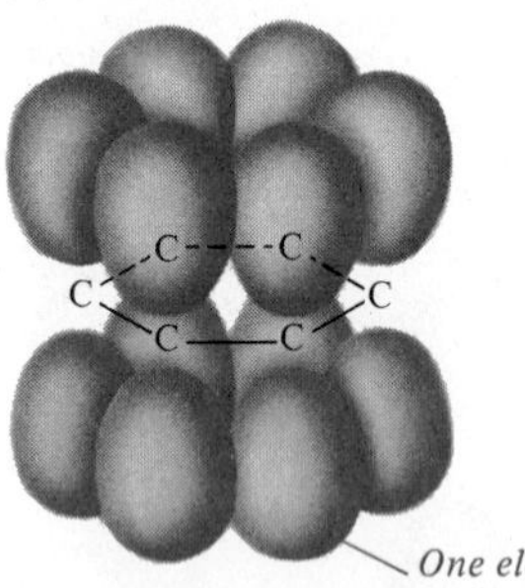

(b) p atomic orbitals available for molecular orbitals

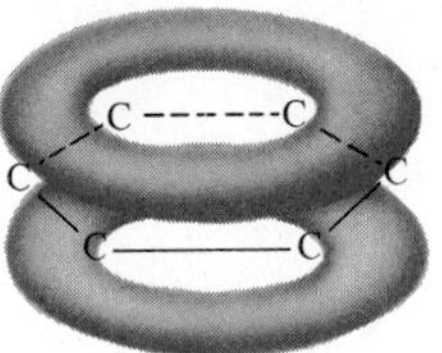

(c) Lowest energy molecular orbital

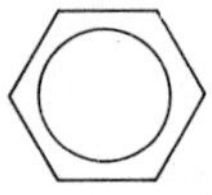

(d) Symbol for benzene that emphasizes delocalized bonding

Figure 25.11
The Benzene Molecule
The six atomic orbitals (b) combine to give three bonding and three antibonding molecular orbitals. One bonding orbital is shown in (c). The sum of the molecular orbitals, or delocalized bonding, is represented by (d).

above and below the ring of carbon atoms (Figure 25.11) and are occupied by the six electrons in pairs. The three π^* orbitals remain unoccupied.

This bonding picture accounts for the equivalence of all the carbon–carbon bonds in benzene and for the thermal stability of the molecule. As we have pointed out, electron energy is lowered when an electron is given a larger space to move about in. The six electrons in the *delocalized* π orbitals can be anywhere within the electron cloud shown in Figure 25.11.

SUMMARY OF SECTIONS 25.1–25.9

Molecular orbital theory explains bond formation as the occupation by electrons of orbitals characteristic of bonded atoms rather than individual atoms. The combination of two atomic orbitals produces two molecular orbitals: a bonding orbital, in which most of the electron density is located between the nuclei of the bonded atoms, and an antibonding orbital, in which most of the electron density is located away from the space between the two nuclei. Electrons in bonding orbitals form bonds; electrons in antibonding orbitals can weaken bonds formed between the same atoms by other electrons. The types of molecular orbitals formed from s and p atomic orbitals are shown in Figure 25.3, and the energy-level diagram for the molecular orbitals of second-period elements is shown in Figure 25.6.

By assigning electrons to bonding and antibonding orbitals in accordance with a set of rules (Section 25.4), it is possible to determine the order of a bond without resort to Lewis structures and resonance (Equation 25.1). In general, the strength of a bond increases and its length decreases with increasing bond order. The value of molecular orbital theory for predicting bond formation, bond energies, bond lengths, and the magnetic properties of molecules is demonstrated in Section 25.7 for homonuclear diatomic molecules. One of the successes of molecular orbital theory is the explanation of the two unpaired electrons in the O_2 molecule. The concept of delocalized π orbitals has its basis in the formation of molecular orbitals that encompass a number of atoms.

SIGNIFICANT TERMS (with section references)

antibonding molecular orbital (25.2)
bond order (25.6)
bonding molecular orbital (25.2)
heteronuclear (25.8)
homonuclear (25.7)
molecular orbital (25.1)
molecular orbital theory (25.1)

QUESTIONS AND PROBLEMS

Molecular Orbital Theory

25.1 What is a molecular orbital? What two types of information can be obtained from molecular orbital calculations? How do we use such information to describe the bonding within a molecule?

25.2 What is the relationship between the number of molecular orbitals and the number of atomic orbitals from which they are formed?

25.3 What is the relationship between the energy of a bonding molecular orbital and the energies of the original atomic orbitals? What is the relationship between the energy of an antibonding molecular orbital and the energies of the original atomic orbitals?

25.4 Draw an energy-level diagram for the formation of molecular orbitals from two atomic orbitals of equal energy. Identify the bonding and antibonding molecular orbitals.

25.5 Prepare sketches of the bonding and antibonding molecular orbitals formed by combining (a) two s atomic orbitals, (b) two p atomic orbitals in end-on approach, (c) two p atomic orbitals in parallel approach, (d) an s atomic orbital and a p atomic orbital interacting so that they approach along the axis of the p orbital (e.g., p_z along the z axis).

25.6 Complete the following energy-level diagram for the formation of molecular orbitals:

$2_{p_x}2_{p_y}2_{p_z}$		$2_{p_x}2_{p_y}2_{p_z}$
atomic orbitals	*molecular orbitals*	*atomic orbitals*

Assume the bonding axis to be the x axis. Identify the bonding and antibonding molecular orbitals that are formed.

25.7 State the three rules for placing electrons in molecular orbitals.

25.8 What is meant by the term "bond order"? How is the value of the bond order calculated?

Homonuclear Diatomic Molecules

25.9 Using the molecular orbital energy diagram given in Figure 25.4, write the electron configuration for each of the following molecules and predict which molecules would exist: (a) H_2^+, (b) H_2, (c) H_2^-, and (d) H_2^{2-}. Calculate the bond order of each of the species.

25.10 Repeat Question 25.9 for He_2^+ and He_2 using Figure 25.5.

25.11 Using the molecular orbital energy diagram given in Figure 25.6, write the electron configurations for (a) Be_2, (b) C_2, (c) Ne_2. Which of these molecules are predicted to exist? Calculate the bond order of the species.

25.12 Repeat Question 25.11 for (a) Li_2, (b) B_2, (c) F_2.

25.13 The standard state enthalpy of formation at 0 K is 1164.70 kJ/mol for $O_2^+(g)$, 1560.716 kJ/mol for $O^+(g)$, and 246.785 kJ/mol for $O(g)$. Calculate the bond dissociation energy for O_2^+

$$O_2^+(g) \longrightarrow O^+(g) + O(g)$$

25.14 Using the data given in Problem 25.13, calculate the bond dissociation energy for O_2

$$O_2(g) \longrightarrow 2O(g)$$

Which is more stable, O_2^+ or O_2? Why?

25.15 The standard state enthalpy of formation at 0 K is 1503.378 kJ/mol for $N_2^+(g)$, 470.842 kJ/mol for $N(g)$, and 1873.156 kJ/mol for $N^+(g)$. Calculate the bond dissociation energy for N_2^+.

25.16 The standard state heats of formation at 0 K are 155 kJ/mol for $N_2^-(g)$, 478 kJ/mol for $N^-(g)$, and 470.842 kJ/mol for $N(g)$. Calculate the bond energy of N_2^-.

25.17 List the following species in order of decreasing bond strength: O_2, O_2^+, O_2^-, O_2^{2+}, and O_2^{2-}. Which has the shortest bond length? Calculate the bond order of each of the species.

25.18 Repeat Question 25.17 for N_2, N_2^+, and N_2^-.

Heteronuclear Diatomic Molecules

25.19 Assuming that the molecular orbital diagram given in Figure 25.6 is valid for NO, write the molecular orbital description for this molecule. Would NO be paramagnetic? Would you predict that NO^+ would be more stable? Why?

25.20 Assuming that the molecular orbital diagram in Figure 25.6 is valid for CN, CN^+, and CN^-, write the molecular orbital descriptions for these species. Which would be most stable? Why?

25.21 The heat of formation is 67 kJ/mol for $CN^-(g)$, 1793 kJ/mol for $CN^+(g)$, and 434 kJ/mol for $CN(g)$ at 0 K. Calculate the carbon–nitrogen bond energies in CN^-, CN, and CN^+. Assume that the following equations describe the dissociation processes:

$$CN^-(g) \longrightarrow C(g) + N^-(g)$$
$$CN(g) \longrightarrow C(g) + N(g)$$
$$CN^+(g) \longrightarrow C^+(g) + N(g)$$

The heats of formation are 711 kJ/mol for $C(g)$, 471 kJ/mol for $N(g)$, 1798 kJ/mol for $C^+(g)$, and 478 kJ/mol for $N^-(g)$.

25.22 Prepare a molecular orbital energy diagram similar to Figure 25.9 for HF using a fluorine atom that has undergone sp^3 hybridization. Compare this diagram with Figure 25.9. Is there any difference in the number of bonding or antibonding orbitals? Is there any difference in the number of nonbonding orbitals? Is there a difference in the pattern of the nonbonding orbitals?

CHAPTER

26

Metals and Metallurgy; The *s*- and *p*- Block Metals

METALS

26.1 PROPERTIES OF METALS

We think of "metals" as dense, high melting, malleable and ductile, and good conductors of heat and electricity. Some metals, but not all, possess all these properties. Metals appear in each family of the periodic table except the halogen and noble gas families, and therefore exhibit a wide range of properties. For example, the density of lithium is 0.534 g/cm^3, whereas that of platinum is 21.45 g/cm^3. Mercury is a liquid at ordinary temperatures, but osmium melts at 3045 °C. Sodium and potassium are soft enough to be cut easily with a knife, but chromium is very hard.

Yet metals do have certain properties in common. All metals tend to form positive ions, although with varying ease. All are good conductors of heat and electricity, though, again, to quite different degrees. All, when freshly cut, have a bright luster. Many metals, however, quickly lose their luster when exposed to air because of oxides, carbonates, or sulfides that form on their surfaces.

The common properties of metals depend upon the ready availability of electrons. Cations form easily by electron loss. In metallic bonding (Section 10.3 and Section 29.16), outer electrons are freely mobile, rather than bound to individual atoms. These mobile electrons are responsible for the electrical conductivity, thermal conductivity, and luster of metals.

Keep in mind that in the periodic table no sharp line separates metals from nonmetals. The semiconducting elements, which fall between the metals and nonmetals (see Figure 9.3), are neither distinctly metallic nor distinctly nonmetallic. To the right of the semiconducting elements lie the nonmetals. To their immediate left lie the *p*-block metals, some of which have both nonmetallic and semiconducting properties. The *d*-transition metals include such elements as copper, iron, gold, and silver that best exhibit the properties commonly thought of as metallic. At the far left are the *s*-block metals—their atoms lose electrons more easily than atoms of any other elements, although in some ways (e.g., softness and high reactivity) the *s*-block metals are not typical metals.

In this chapter we first discuss some of the ways that metals are prepared from the materials in which they are found in the Earth's crust. Then we look at the chemistry of the metals of the *s* and *p* blocks in the periodic table. In a later chapter we consider the chemistry of the transition metals.

26.2 THE WHOLE EARTH

The Earth is, at least at the present time, the only source we have for all of the raw materials needed to sustain life and support our lifestyle. We draw gases from the atmosphere, water and some elements and chemical compounds from the hydrosphere, and fossil fuels, metals, and structural materials from the solid surface.

The portion of the solid Earth available to us is a very small part of the whole (less than 1% by mass). The deepest mine extends only 3.8 km beneath the surface. Geologists define the Earth's crust as the region from the surface to a depth of 30–50 km.

Three major types of rocks are found in the crust: igneous rocks, formed by solidification of molten substances; sedimentary rocks, formed by deposition of material by the oceans and rivers; and metamorphic rocks, formed by the action of heat and pressure on existing rocks. The most abundant substances in rocks (Table 26.1) are silicates, which are composed of metal cations and SiO_4 tetrahedra (Section 29.13).

A **mineral** is a naturally occurring inorganic substance with a characteristic crystal structure and composition. The compositions of many minerals, particularly the silicates, vary within the limit set by their crystal structures, because a variety of metal cations can occupy similar sites in the anionic framework.

Each year the Bureau of Mines of the U.S. Department of the Interior publishes a large volume of "Mineral Facts and Problems." Turning its data-filled pages brings to mind a picture of the Whole Earth that is our resource. The minerals needed for production of metals are buried all over the globe, some in countries that rarely come to our attention.

Most of the nickel is in New Caledonia and Zimbabwe. Most of the chromium is in Botswana and Turkey. Our industrialized society dictates the interdependence of countries on all continents. Almost all of the titanium essential for our jet engines, space vehicles, and missiles is imported into the United States from Australia and India. And 97% of the manganese that we use, a metal necessary for steelmaking, is imported from Gabon, Brazil, the Republic of South Africa, France, and a few other countries. It is obvious that there are practical, political, and economic, as well as scientific, implications in such data.

Table 26.1
Average Composition of the Crust of the Earth (Igneous Rocks)[a]

Element	Mass %
Oxygen	46.6
Silicon	27.72
Aluminum	8.13
Iron	5.0
Calcium	3.63
Sodium	2.83
Potassium	2.59
Magnesium	2.09
Titanium	0.44
Phosphorus	0.118
Manganese	0.10
Other nonmetals (excluding noble gases)	$\sim 10^{-2}$–10^{-4} each
Other metals (nonradioactive)	$\sim 10^{-2}$–10^{-7} each

[a] Source: Therald Moeller, *Inorganic Chemistry: A Modern Introduction* (New York: Wiley, 1982), pp. 24, 25.

26.3 THE OCCURRENCE OF METALS

The form in which metals occur in the Earth's crust depends chiefly upon their reactivity, the solubilities of their salts, and the ease with which their salts react with water or are oxidized. Very unreactive metals, such as gold, silver, and platinum, are usually found in nature in the elemental state. The somewhat more reactive metals are present in the form of sulfides (e.g., CuS, PbS, ZnS) which, because of their extremely low solubility, have resisted both oxidation and reactions with water. The still more reactive metals have been converted to oxides (e.g., MnO_2, Al_2O_3, TiO_2).

The most reactive metals have formed salts. Calcium and magnesium occur widely as carbonates, sulfates, and silicates (often containing both Ca and Mg). Magnesium sulfate is soluble in water and is found in many mineral springs. Calcium sulfate ($CaSO_4 \cdot 2H_2O$, gypsum) is only slightly soluble, but enough so to contribute to the hardness of natural waters. In the presence of dissolved carbon dioxide, calcium carbonate is dissolved as the hydrogen carbonate, also contributing to the hardness of water. Aluminum occurs largely in the form of aluminosilicates such as muscovite ($KAl_2(OH)_2Si_3AlO_{10}$), one form of mica, and kaolin ($H_4Al_2Si_2O_9$), a clay. Sometimes it occurs as the insoluble oxide, $Al_2O_3 \cdot xH_2O$, and the complex fluoride, $Na_3[AlF_6]$.

Sodium and potassium are found in nature either as salts dissolved in the ocean and mineral springs or in insoluble, unreactive aluminosilicates such as albite ($NaAlSi_3O_8$) and orthoclase ($KAlSi_3O_8$). These silicates are widely distributed in all parts of the world, but since they are very stable, they are not used as sources of the metals which they contain. The slow weathering of orthoclase is important, however, for it liberates potassium ions, which are essential for all plant growth.

SUMMARY OF SECTIONS 26.1–26.3

Metals vary widely in properties. However, all metals have a tendency to lose electrons. The freely mobile electrons of metallic bonding are responsible for the luster when freshly cut, and the thermal and electrical conductivity of all metals.

In the Earth's crust, the unreactive metals (e.g., Au, Ag, Pt) are usually found in the elemental state. The somewhat more reactive metals (e.g., Cu, Pb, Zn) are found as sulfides of very low solubility. The still more reactive metals (e.g., Mn, Cr, Ti) are usually present as oxides. The elements of the beryllium family (e.g., Mg, Ca) and also aluminum are found as carbonates, or silicates, or sulfates. Sodium and potassium occur either as soluble salts in the oceans and mineral springs or in insoluble, unreactive aluminosilicates.

THE PREPARATION OF METALS

26.4 METALLURGY

Metallurgy is the science and technology of metals—their production from ores, their purification, and the study of their properties and uses. Here we are interested in how pure or relatively pure metals are produced from their ores.

Most metals occur in nature as crystalline minerals scattered through the surrounding rock or in veins in the rock. If the combination of rocky matrix and mineral can profitably be mined and treated for extraction of the metal it is called an **ore.** The percentages of metals in ores vary over a wide range—a typical iron ore may contan 50% iron, whereas some valuable copper ores contain less than 1% of the metal. The determining factor is a balance between the economic value of the metal in the ore and the ease with which the metal can be won from the ore. As the supplies of ores that contain higher percentages of metals are exhausted, industry must turn to lower and lower grade ores.

The procedures for production of pure metals from ores fall into three general categories: (1) concentration of the ore, (2) extraction of the metal from the concentrated ore (which includes reduction of the metal), and (3) refining of the crude metal.

a. Concentration It is desirable to separate the valuable mineral from as much as possible of the unwanted rock, called the *gangue* (pronounced "gang"). Usually the first step is crushing or grinding the ore into smaller pieces. Then the separation can be carried out by *flotation*—concentration of the metal-bearing mineral in a froth of bubbles that can be skimmed off. The finely crushed ore is placed in a large tank containing water, a wetting agent such as pine oil that will wet the metal-bearing mineral but not the unwanted silicate rock particles, a surface-active agent, and possibly also a frothing agent. The surface-active agent functions like a soap or detergent molecule (Section 15.17); it has a polar end that is adsorbed on the mineral surface and a hydrophobic, hydrocarbon end that is drawn into the bubble, carrying the mineral into the froth. The mixture is vigorously stirred and a strong stream of air is blown through the tank, carrying the coated mineral particles to the surface with the air bubbles. Most of the unwanted rock sinks to the bottom of the tank. A concentrated ore is obtained in this way.

b. Extraction Extraction of the metal from the concentrated ore requires reduction of the metal from a positive oxidation state to the free metal. Before reduction, other operations may be necessary. A finely divided ore may undergo **sintering**—heating without melting to cause the formation of larger particles. Or a carbonate or an oxide ore may undergo **calcining**—heating to drive off a gas, for example

$$4FeCO_3(s) + O_2(g) \xrightarrow{\Delta} 2Fe_2O_3(s) + 4CO_2(g) \qquad \textbf{(26.1)}$$

In **roasting,** a term usually applied to sulfide ores, the ore is heated below its melting point in air or oxygen to convert sulfides to oxides, for example

$$2PbS(s) + 3O_2(g) \longrightarrow 2PbO(s) + 2SO_2(g) \qquad \textbf{(26.2)}$$

The processes for extracting and reducing metals, and also those used in metal refining, fall into three general areas of metallurgy: pyrometallurgy, electrometallurgy, and hydrometallurgy.

Pyrometallurgy employs chemical reactions carried out at high temperatures. For example, in **smelting,** reduction of a mineral yields the molten metal, which in the process is separated from unwanted rock. The reducing agent in smelting is usually carbon or another metal. The oxides produced by roasting sulfide ores are generally reduced by smelting with carbon.

$$ZnO(s) + C(s) \xrightarrow{\Delta} Zn(l) + CO(g) \qquad \textbf{(26.3)}$$

Ordinarily, ore concentration does not make a complete separation of the gangue from the mineral. The remaining gangue is removed during smelting by adding something that will react with it, called a **flux,** to give the **slag,** a material that is liquid at the temperature of the smelting furnace. Most slags are silicates, for example

$$\underset{\substack{\text{rock} \\ \text{gangue}}}{SiO_2(s)} + \underset{\substack{\text{limestone} \\ \text{flux}}}{CaCO_3(s)} \xrightarrow{\Delta} \underset{\substack{\text{calcium silicate} \\ \text{slag}}}{CaSiO_3(l)} + CO_2(g) \qquad \textbf{(26.4)}$$

The molten metal and the slag form separate layers in the furnace and are drawn off at different points. The slag solidifies to a glassy mass that may be discarded or may be used to make Portland cement (Section 29.14c). Pyrometallurgical methods are used in the production of copper and zinc (Section 26.6) and iron and steel (Section 26.7).

Electrometallurgy (Section 26.5) is the general term for processes that utilize electrical energy in the reduction of minerals or in the refining of metals. Sodium and aluminum (Section 26.5) are produced by electrometallurgy.

Hydrometallurgy is the general term for processes that utilize aqueous solutions in the extraction and reduction of metals. In **leaching,** the metal or a compound of the metal is dissolved out of the ore or directly out of the ore deposit by an aqueous solution. After purification of the solution, a pure compound of the metal may be isolated or reduction to the metal may be carried out in the solution. Often the leach solution is regenerated and can be used again.

Copper can be leached by sulfuric acid in the presence of oxygen

$$\underset{\text{copper ore}}{2CuFeS_2(s)} + \underset{\text{leach solution}}{H_2SO_4(aq)} + 4O_2(g) \longrightarrow 2CuSO_4(aq) + Fe_2O_3(s) + 3S(s) + H_2O(l)$$

and gold by a cyanide solution in the presence of oxygen.

$$\underset{\text{gold ore}}{4Au(s)} + \underset{\text{leach solution}}{8CN^-} + O_2(g) + 2H_2O(l) \longrightarrow 4[Au(CN)_2]^- + 4OH^-$$

The latter was one of the first hydrometallurgical processes; it was introduced in the late nineteenth century.

After conversion to an ion in solution, the metal is reduced through displacement by a more reactive metal or by reaction with some other type of reducing agent. For copper and gold the reactions are

$$CuSO_4(aq) + Fe(s) \longrightarrow FeSO_4(aq) + Cu(s)$$
$$Zn(s) + 2[Au(CN)_2]^- \longrightarrow 2Au(s) + [Zn(CN)_4]^{2-}$$

Hydrometallurgy offers several advantages: (1) The ore does not have to be concentrated. It is only broken into small pieces and the leach solution is allowed to run through it. (2) The use of large amounts of coal and coke to roast the ore and then to reduce it is avoided. (3) Pollution of the atmosphere by such by-products of pyrometallurgy as sulfur dioxide, arsenic(III) oxide, and furnace dust is avoided. (4) It is more effective for low-grade ores than older methods. For these reasons, especially (3) and (4), pyrometallurgical methods are being replaced by hydrometallurgical methods.

c. Refining The **refining,** or purification, of crude metals is important for two reasons: First, the impurities may make the metal unfit for use. For example, even a small fraction of a percent of arsenic (a common impurity) in copper lowers the electrical conductivity by 10% to 20%. Second, the impurities in the metal may be valuable. Most silver is a by-product of the metallurgy of lead and copper.

Methods for refining crude metal include electrolytic purification (as for copper; Section 26.6a), oxidation of impurities to be removed (as for iron; Section 26.7b), or distillation of low-boiling metals such as mercury or zinc.

26.5 ELECTROMETALLURGY

a. Sodium To obtain the free alkali metals from their compounds, the following reduction must be carried out in the absence of water:

$$M^+ + e^- \longrightarrow M$$

One convenient way to supply the energy required by this reaction is with electricity. Most sodium is produced by electrolysis of molten sodium chloride by the Downs process (Figure 26.1).

$$\textit{cathode} \quad 2Na^+ + 2e^- \longrightarrow 2Na(l)$$
$$\textit{anode} \quad 2Cl^- \longrightarrow Cl_2(g) + 2e^-$$

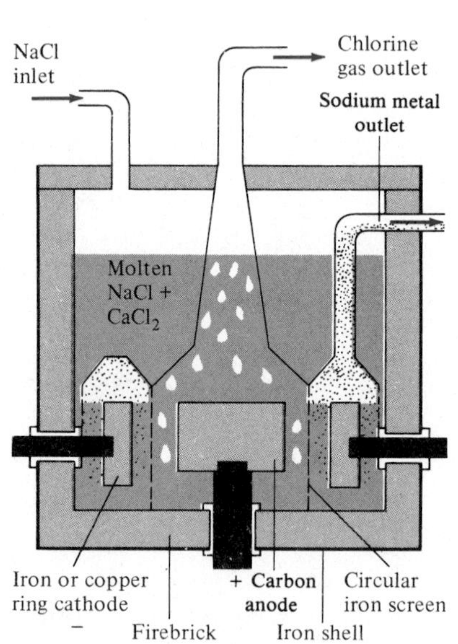

Figure 26.1
The Downs Cell for the Production of Sodium

Because sodium chloride melts at 808 °C, which is close to the boiling point of sodium, calcium chloride is added to lower the melting point to about 600 °C. This also permits the electrolysis to be carried out more economically. Resistance to the flow of electric current provides the heat required to keep the salt mixture molten. Note in Figure 26.1 that the cell is constructed so that the sodium and chlorine are kept apart and, therefore, cannot react.

b. Aluminum The source of practically all aluminum is bauxite, which is primarily hydrated aluminum oxide, $Al_2O_3 \cdot xH_2O$. Although bauxite is an abundant mineral, the best deposits are rapidly being exhausted or are far from the United States. The development of hydrometallurgical methods for obtaining aluminum is under way, for eventually new sources of aluminum will be needed. At present, however, there is no economical method for the extraction of aluminum from silicate minerals and clays.

Bauxite generally contains large amounts of silica (SiO_2) and iron(III) oxide (Fe_2O_3), and these substances must be removed. The purification of bauxite is accomplished by the Bayer process, which takes advantage of the differences in acid–base properties of the oxides. Aluminum oxide is amphoteric, whereas iron(III) oxide is basic and silica is a relatively inert acidic oxide. The crude bauxite is digested under pressure with hot sodium hydroxide solution, which dissolves the aluminum oxide as sodium tetrahydroxoaluminate, $Na[Al(OH)_4]$. Iron(III) oxide and other insoluble materials are removed by filtration. The filtrate is then diluted with water and cooled, precipitating the aluminum hydroxide. The precipitate is separated by filtration and converted to the pure anhydrous oxide, Al_2O_3, by heating. The sodium hydroxide solution is concentrated and used again.

Aluminum is obtained from the anhydrous oxide by an electrolytic method known in this country as the Hall process. While he was a student at Oberlin College, Charles Martin Hall, inspired by a professor's remark that a fortune awaited the man who could invent a cheap process for producing aluminum, set this as his goal. In 1886 at the age of 21, shortly after he graduated, he succeeded. Hall did become a rich man, and a memorial to him in the form of an aluminum statue stands on the Oberlin campus.

In the electrolytic cells (Figure 26.2) for the Hall process, steel boxes lined with graphite are the cathodes. The electrolyte is molten cryolite, Na_3AlF_6 (m.p. 1000 °C), in which the purified aluminum oxide is soluble. Carbon rods, which serve as anodes, are suspended in the molten solution, and electrolysis is carried out at 900–1000 °C. During electrolysis, Al^{3+} ions from the oxide migrate toward the cell lining (the cathode) where they are reduced to the liquid metal, which collects at the bottom of the cell. (The AlF_6^{3-} of the electrolyte is not reduced because of its great stability.) At the anodes, O^{2-} ions are oxidized to molecular oxygen that reacts with the carbon anodes. The anodes are gradually consumed and must be replaced periodically. The Hall process yields aluminum of a purity between 99.0 and 99.9%.

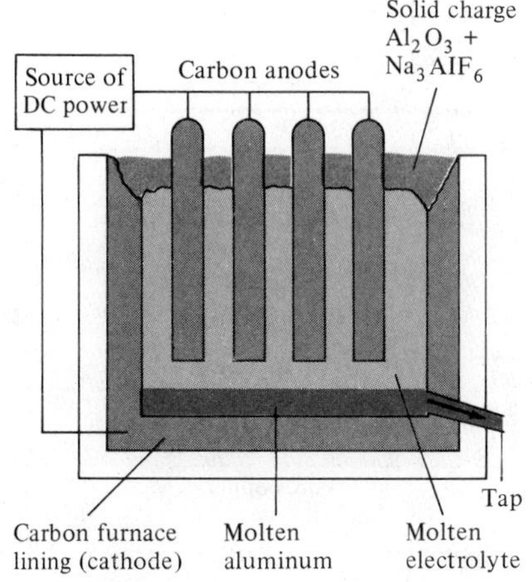

Figure 26.2
Electrolytic Cell for Production of Aluminum

26.6 METALS FROM SULFIDE ORES

a. Copper After concentration by crushing and flotation, copper ore is roasted and smelted in a multistep process that separates the iron and copper sulfides present in most copper ores (chalcocite, Cu_2S; chalcopyrite, $CuFeS_2$). The ore may first be roasted to drive off some of the sulfur as sulfur dioxide and sulfur trioxide. Next, heating in a furnace with a silica flux converts iron oxides and some iron sulfide to slag, producing a molten mixture of copper and iron sulfides with the iron silicate slag floating on top. Some of the reactions that take place in the furnace are

$$FeS_2(l) + O_2(g) \xrightarrow{\Delta} FeS(l) + SO_2(g)$$
$$3FeS(l) + 5O_2(g) \xrightarrow{\Delta} Fe_3O_4(l) + 3SO_2(g)$$
$$2CuFeS_2(l) + O_2(g) \xrightarrow{\Delta} Cu_2S(l) + 2FeS(l) + SO_2(g)$$
$$Fe_3O_4(l) + FeS(l) + 4SiO_2(s) + O_2(g) \xrightarrow{\Delta} \underset{\textit{iron silicate slag}}{4FeSiO_3(l)} + SO_2(g)$$

The molten sulfide mixture is transferred to a converter where it is smelted with silica in the presence of oxygen that is blown through the mixture. Here, the remaining iron is separated in an iron silicate slag and the final reduction to copper metal is accomplished.

$$2Cu_2S(l) + 3O_2(g) \longrightarrow 2Cu_2O(l) + 2SO_2(g)$$
$$2Cu_2O(l) + Cu_2S(l) \longrightarrow 6Cu(l) + SO_2(g)$$

Smelters in some localities in the United States have been ordered by the government to remove the sulfur dioxide from stack gases and convert it to nonpolluting products. This can be done either by catalytically oxidizing the sulfur dioxide in the stack to sulfuric acid (via sulfur trioxide) or by reducing it to elemental sulfur by passing it over red hot carbon.

$$SO_2(g) + 2C(s) \longrightarrow S(l) + 2CO(g)$$

The products of these reactions (H_2SO_4 and S) are useful materials, but commonly they cannot be sold for enough to pay for their recovery. However, the pollution problem in the vicinity of smelters is so great that removal of the sulfur dioxide from the stack

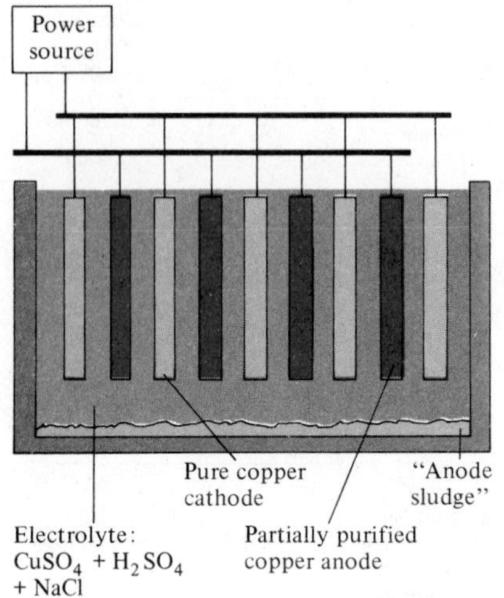

Figure 26.3
Electrolytic Cell for Refining Copper

gases is necessary. Also, the sulfur dioxide may contribute to acid rain hundreds of miles from the smelter. To avoid the high cost of pollution control, some copper producers are turning to direct leaching followed by electrolysis.

The copper from smelting a sulfide ore is quite impure, the chief impurities being silver, gold, iron, zinc, lead, arsenic, sulfur, copper(I) oxide, and bits of slag. By heating the molten metal in a stream of air, most of the arsenic and sulfur are converted to the volatile oxides and escape. The other impurities are removed by electrorefining. Bars of the crude copper serve as anodes in the electrolysis and plates of pure copper as the cathodes; a mixture of dilute sulfuric acid, sodium chloride, and copper(II) sulfate is the electrolyte (Figure 26.3). By careful control of the voltage across the cell, only copper and the more electropositive metal impurities (e.g., iron, zinc, lead) in the anode are oxidized and dissolved. Metallic impurities less electropositive than copper, such as silver and gold, are largely unaffected and drop from the anode as it disintegrates. (Any Ag^+ that forms is precipitated as AgCl.) Valuable metal by-products are recovered from these anode sludges. Voltage control is so maintained that only copper, the least electropositive of the various metals that dissolve from the anode, is plated out on the pure copper cathode. The refining process yields electrolytic copper, which has a purity of greater than 99.9%.

EXAMPLE 26.1 Electrochemistry: Copper Metallurgy

In electrorefining copper, the copper is separated from iron, lead, zinc, gold, and silver. The electrolytic cells employ anodes of impure copper and cathodes of pure copper. Based on the following standard reduction potentials, discuss what happens in this type of cell:

	E°
$Zn^{2+} + 2e^- \longrightarrow Zn$	−0.763 V
$Pb^{2+} + 2e^- \longrightarrow Pb$	−0.126 V
$Cu^{2+} + 2e^- \longrightarrow Cu$	+0.337 V
$Ag^+ + e^- \longrightarrow Ag$	+0.799 V
$Au^+ + e^- \longrightarrow Au$	+1.691 V

We know that oxidation—the loss of electrons—takes place at an anode. The series of electrode potentials is written above in the order of increasing ease of reduction and decreasing ease of oxidation. The voltage must be such that copper is dissolved from the impure copper anode

$$Cu \longrightarrow Cu^{2+} + 2e^-$$

which means that **lead and zinc, which are more easily oxidized than copper, will also be oxidized to cations and dissolve.** Gold and silver are less readily oxidized than the other metals, and at the appropriately controlled voltage they will be unaffected. **As the anode dissolves, gold and silver drop to the bottom as the anode sludge.**

Reduction—the gain of electrons—occurs at the cathode. Of the metal ions in solution—Cu^{2+}, Zn^{2+}, and Pb^{2+}—the copper(II) ion has the most positive standard reduction potential and is most readily reduced. Therefore, **the voltage can be controlled so that only copper will be deposited at the cathode.**

b. Zinc The most common zinc ores are sphalerite or zinc blende (ZnS) and smithsonite ($ZnCO_3$). Others are zincite (ZnO) and franklinite, $(Zn, Mn)O \cdot xFe_2O_3$, in which the ratios of Zn, Mn, and Fe_2O_3 are variable. The low boiling point of zinc (907 °C) allows it to be distilled away from molten rock and in some cases refined by distillation. The metallurgy of franklinite ore is interesting, for upon reduction at a high temperature, it yields zinc, manganese, and iron. The zinc distills from the mixture and the manganese–iron alloy that remains is used directly in the manufacture of alloy steels.

Most zinc ores are treated by roasting to convert the sulfide to the oxide, followed by high-temperature reduction with carbon to give zinc as a gas, which is then condensed and refined.

$$ZnO(s) + C(s) \xrightarrow{\Delta} Zn(g) + CO(g)$$

Zinc is also extracted by hydrometallurgical methods. For example, a solution of zinc sulfate can be obtained by leaching the roasted sulfide ore with sulfuric acid in the presence of oxygen. The overall reaction is

$$2ZnS(s) + O_2(g) + 2H_2SO_4(aq) \longrightarrow 2ZnSO_4(aq) + 2S(s) + 2H_2O(l)$$

Zinc dust is stirred into the zinc sulfate solution to reduce and precipitate any metals that are more easily reduced than zinc. The solution is then filtered and electrolyzed to give pure metallic zinc.

EXAMPLE 26.2 Chemical Reactions: Compounds of Metals

The following reactions take place in the roasting and hydrometallurgical extraction of zinc:

(a) $ZnS(s) + 2O_2(g) \longrightarrow ZnSO_4(s)$

(b) $2ZnS(s) + 3O_2(g) \longrightarrow 2ZnO(s) + 2SO_2(g)$

(c) $ZnO(s) + H_2SO_4(aq) \longrightarrow ZnSO_4(aq) + H_2O(l)$

(d) $Zn(s) + CdSO_4(aq) \longrightarrow ZnSO_4(aq) + Cd(s)$

How can each of these reactions be classified? (Consult Table 17.8.)

(a) One of the reactants is an element. This is therefore a **redox combination reaction**—the combination of an element and a compound.

(b) This is also a redox reaction of an element; elemental oxygen combines with both elements in a binary compound in a **redox combination reaction.**

(c) One of the reactants, aqueous sulfuric acid, contains sulfate ions. The reaction is a **nonredox partner-exchange reaction** between a basic oxide and an acid.

(d) Elemental zinc is a reactant. This is a **redox displacement reaction** in which a more active metal displaces a less active metal.

26.7 IRON AND STEEL

a. Sources and Uses of Iron Like copper and zinc, iron occurs in nature as the sulfide (Fe_2S). However, this mineral is not used as an ore because it is difficult to remove the last traces of sulfur, which make steel brittle. Iron is also found in large amounts as oxides. Hematite (Fe_2O_3) is most abundant, and magnetite (Fe_3O_4, which is $FeO \cdot Fe_2O_3$) is also a valuable ore because of its higher percentage of iron content. As the name implies, magnetite is attracted by a magnet; some samples, referred to as lodestones, can also act as magnets.

Siderite ($FeCO_3$) is present in many soils and contributes to hardness in water through its ready conversion to the soluble hydrogen carbonate.

$$FeCO_3(s) + CO_2(g) + H_2O(l) \rightleftharpoons Fe(HCO_3)_2(aq)$$

In air, iron(II) hydrogen carbonate in solution is oxidized to the insoluble iron(III) oxide.

$$4Fe(HCO_3)_2(aq) + O_2(g) \longrightarrow 2Fe_2O_3(s) + 8CO_2(g) + 4H_2O(l)$$

This accounts for the brown stain so often seen under dripping faucets and in other places where hard water is in contact with air.

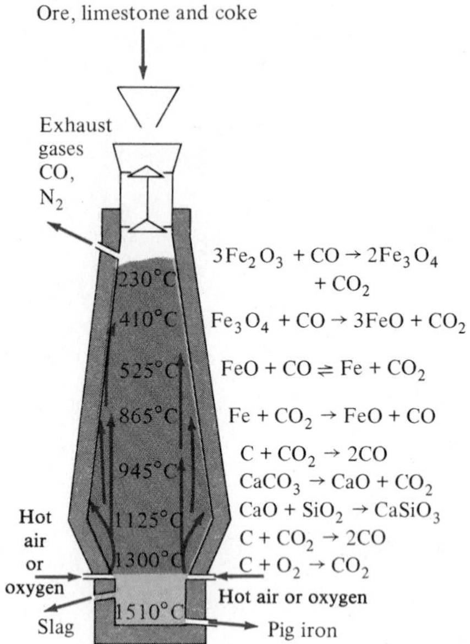

Figure 26.4
The Blast Furnace
The hot gases rise; the solids increase in temperature as they descend.

Taconite ores, which are chiefly iron oxides containing silica, are now increasingly used in the United States as sources of iron. They are extremely hard and difficult to handle, but metallurgical research has overcome most of the problems.

The term **steel** is used for alloys of iron that contain carbon and often other metals as well. Without carbon, iron is not strong enough or hard enough for many modern applications. The properties of steel depend upon the percentage of carbon present, the heat treatment of the steel, and the alloying metals present. Low-carbon, or mild, steel contains up to 0.2% carbon. It is malleable and ductile and is used in making wire, pipe, and sheet steel. Medium steel (0.2–0.6% carbon) is used in rails, boiler plate, and structural pieces. High-carbon steel (0.6–1.5% carbon) is hard, but lacks ductility and flexibility. It is used for tools, springs, and cutlery. Cast iron, which is not classified as a steel, contains more than 1.5% carbon and is more brittle than steels. Stainless steel is a medium steel that contains 4% or more of chromium.

b. Ironmaking: The Blast Furnace The raw materials for ironmaking are (1) iron ore that has been concentrated, (2) coke, and (3) limestone ($CaCO_3$), which serves as a flux. Crude iron, called pig iron or cast iron, is produced in a blast furnace—a tower about 100 ft high and 25 ft in diameter that is lined with special refractory (heat-resistant) brick.

The furnace is charged from the top with a mixture of iron ore, coke, and limestone (Figure 26.4). A strong (about 350 mph) blast of very hot air or oxygen is blown in at the bottom, where the coke is converted to carbon monoxide, which is the reducing agent. The charge is heated gradually as it descends. First the moisture is driven off. Then the ore is partially reduced by carbon monoxide. In the hotter part of the furnace the reduction of the ore to metallic iron is completed, and the limestone loses CO_2 and reacts with the impurities in the ore (mainly silicon dioxide, but also manganese and phosphorus oxides) to produce the molten slag. The molten iron and slag are immiscible and form separate layers at the bottom of the furnace.

The reduction reactions are reversible, and complete reduction takes place only if the carbon dioxide formed is destroyed. This is effected by reduction to carbon monoxide with an excess of coke.

$$CO_2(g) + C(s) \longrightarrow 2CO(g)$$

Relatively pure oxygen can be used in place of air in blast furnaces. This allows for a smaller furnace and somewhat higher temperatures. The advantages largely overcome the obvious disadvantage of the cost of the oxygen.

EXAMPLE 26.3 Chemical Reactions: Ironmaking

Classify the following reactions that take place in the iron blast furnace:

(a) $C(s) + O_2(g) \longrightarrow CO_2(g)$
(b) $C(s) + CO_2(g) \longrightarrow 2CO(g)$
(c) $CaO(s) + SiO_2(s) \longrightarrow CaSiO_3(s)$
(d) $CaCO_3(s) \longrightarrow CaO(s) + CO_2(g)$
(e) $FeO(s) + CO(g) \longrightarrow Fe(l) + CO_2(g)$

(a) Combination of elements. **Redox combination reaction.**
(b) Combination of element and compound. **Redox combination reaction.**
(c) Combination of compounds. **Nonredox combination reaction.**
(d) Thermal decomposition of a single compound. **Nonredox decomposition reaction.**
(e) **Redox reaction** between compounds. (Carbon monoxide is the reducing agent.)

c. Steelmaking The pig iron withdrawn from the blast furnace contains small amounts of carbon, sulfur, phosphorus, silicon, manganese, and other impurities. At this stage, the iron is so brittle that it is useless for most purposes. Ironmaking is a reduction process, but steelmaking is an oxidation process: the impurities are oxidized.

The two objectives in steelmaking are to burn out the unwanted impurities from pig iron and to add the exact amounts of metals or other materials needed to impart desirable properties.

The manganese, phosphorus, and silicon in molten pig iron are converted by air or oxygen to oxides, which react with appropriate fluxes to give slags. Sulfur enters the slag as sulfide, and carbon is burned to carbon monoxide or carbon dioxide. If the chief impurity is manganese, an acidic flux—the oxide of a nonmetal—must be used. Silicon dioxide is the usual acidic flux.

$$MnO(s) + SiO_2(s) \longrightarrow MnSiO_3(l)$$

If the chief impurity is silicon or phosphorus (the more common case), a basic flux must be used. This is usually magnesium oxide or calcium oxide, which give reactions such as

$$SiO_2(s) + MgO(s) \xrightarrow{\Delta} MgSiO_3(l)$$

$$P_4O_{10}(s) + 6CaO(s) \xrightarrow{\Delta} 2Ca_3(PO_4)_2(l)$$

The steelmaking furnace is lined with brick made of the fluxing material, and this lining absorbs part of the oxide which is to be removed.

Most steel in the United States is made by the basic oxygen process, which can provide 300 tons of steel in one hour from each furnace (Figure 26.5). The molten pig iron, along with scrap iron and the materials needed to form a slag, is contained in a barrel-shaped furnace that may hold up to 300 tons of material. A blast of high-purity oxygen under a pressure of 150–180 lb/sq. inch is directed against the surface of the liquid, and the barrel can be tipped and rotated to bring fresh material to the surface. The oxidation of the impurities is very rapid, and the escape of gaseous products so agitates the mass that even the iron at the bottom of the vessel is brought into reaction. The temperature of the material rises almost to the boiling point of the iron without the application of any external heat. At such a temperature, the reactions are extremely rapid, and the entire process is completed in an hour or less, leaving a product that is uniform and of high quality. The molten purified iron is converted to steel by addition of the correct amount of carbon and metals such as vanadium, chromium, titanium, manganese, and nickel.

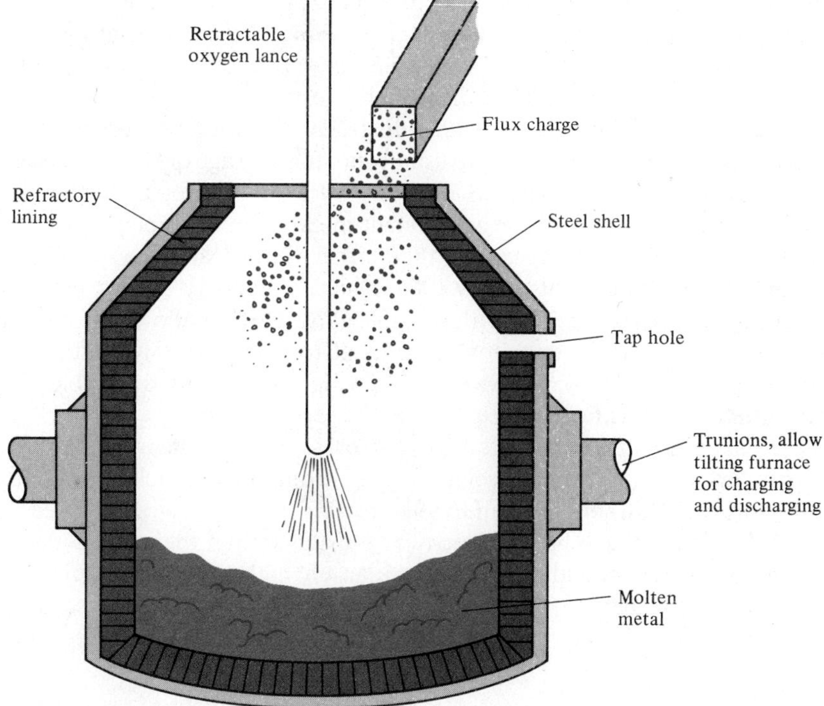

Figure 26.5
Basic Oxygen Process Steelmaking Furnace

The electric arc method, once used only for making small amounts of specialty steels, is now carried out on a larger scale and is increasing in use. An electric arc between carbon electrodes in the roof over a large saucer containing the molten metal provides the heat. Oxygen is added in controlled amounts and also released from impurities and oxides of alloying metals that are added at the beginning of the process. In this process, the amount of oxide impurities in the steel can be controlled and there is less loss of desirable added materials, some of which are quite expensive.

d. Heat Treatment of Steel At high temperatures, iron and carbon combine to form iron carbide (Fe_3C) called cementite. The reaction is reversible, but unlike most combination reactions it is endothermic.

$$3Fe + C + heat \rightleftharpoons Fe_3C$$

The stability of cementite thus increases as the temperature rises, at least in the temperature range involved in the heat treatment of steel. When steel containing cementite is cooled slowly, the equilibrium shifts toward the formation of iron and carbon, and the carbon separates as minute flakes of graphite. These give the metal a gray color. If, however, the steel is cooled very rapidly, equilibrium is not attained and the carbon remains largely in the form of cementite, which is light in color. At ordinary temperatures the decomposition of cementite is so slow that, for all practical purposes, it does not take place at all. Steel containing cementite is harder and much more brittle than that containing graphite.

By tempering, that is, by heating the steel to a suitable temperature for a short time and then cooling it rapidly, the ratio of carbon present as graphite and as cementite can be adjusted within rather wide limits. It is also possible to vary the total amount of carbon in different parts of a single piece of steel, and thus modify its properties. Ball bearings, for example, are made of a medium steel to give them toughness and strength, but the surface is *case hardened* by heating the bearings in a bed of carbon to give a thin surface coating containing cementite.

26.8 ALLOYS

When two or more metals, or metals plus nonmetals, are intimately mixed to yield a substance with metallic properties, they form an **alloy.** Alloys are usually prepared by melting the constituents together, but sometimes they are obtained by the simultaneous deposition of metals on electrodes in an electrolytic cell.

Alloys are extremely important in an industrial civilization because of the greater diversity of their properties—which are usually *not* the average of the properties of the combined metals. The melting points of many alloys are lower or higher than the average of the melting points of the component metals. For example, common solder (tin–lead) melts at a lower temperature than either tin or lead. The so-called fusible alloys, many with melting points below 100 °C, have a diversity of uses. For example, the cavities in metal parts are filled with a fusible alloy which is allowed to harden. The parts can be machined while the alloy prevents distortion of the shape; the parts are then immersed in hot water to remove the alloy, which can be reused. Fusible alloys are also present in automatic fire extinguishers and electrical fuses.

An alloy may show greater resistance to corrosion or other chemical attack than its constituents. Stainless steel, which is chiefly iron, chromium and nickel, is not attacked by hydrochloric acid, although the individual metals in it are easily dissolved.

Some alloys have much greater electrical resistance than their components. This property makes nichrome and chromel (both largely nickel and chromium) useful in electrical resistance heaters (for example, in toasters, space heaters). These alloys also resist atmospheric corrosion—they do not "rust away" even when heated to redness many times.

Other alloys have remarkable magnetic properties. Highly magnetic devices are made of Alnico (chiefly iron, aluminum, nickel, and cobalt).

Steel is greatly hardened by alloying it with manganese, vanadium, and other metals. These alloys are used in the jaws of rock crushers and bank vaults.

The color of an alloy may also be quite different from the colors of its component metals. Alloys of gold with copper are red or yellow, those with palladium or nickel are white, those with silver or cadmium are green, those with iron are blue, and those with aluminum are purple.

Alloys are of many different kinds. Some are heterogeneous mixtures—polycrystalline materials with small areas of different metals in contact. Others are solid solutions—homogeneous mixtures in which the atoms of one metal are randomly distributed among the atoms of another metal. Still others are actually intermetallic compounds.

Alloys that are heterogeneous mixtures result when crystals of more than one composition form as a mixture of molten metals cools. In such a case, the solubility of one metal in the other varies with the temperature. Consider, for example, melted metal A dissolved in melted metal B: A may crystallize out as the temperature is lowered, leaving B and some A still liquid. On further cooling, this liquid, too, will solidify. A heterogeneous alloy formed in this way will contain microscopic crystals of A embedded in a matrix of B or of A and B.

The requirements for solid solution formation are (1) similar chemical properties, (2) similar crystal structures, and (3) appropriate atomic radius ratios. A **solid solution** is an alloy in which atoms of one metal replace atoms in the crystal lattice of the other. Metals with radii that differ from each other by no more than 15% and are soluble in each other can form solid solutions in all proportions. For example, the United States "nickel" coin is 25% nickel and 75% copper. This alloy is harder and less easily corroded than either nickel or copper. Although it is mostly copper, it is white. As metal atoms become more different in size, solubility decreases and ultimately disappears.

In **interstitial solid solutions** small atoms occupy holes in the crystal structure of a metal. Very high-melting, hard, brittle, electrically conducting, and chemically inert alloys result when certain *d*-transition metals (e.g., Ti, Zr, Hf, V, Nb, Ta, Cr, Mo, W, or Fe) are united with nitrogen, carbon, or boron at elevated temperatures. The nitrides and carbides often have or closely approach 1:1 atomic ratios with the metal. Structural analyses show that they are formed by distortions of the metallic crystal structure by the entry of the small atoms (N, radius = 0.074 nm; C, 0.077 nm) into holes (Sections 13.9, 13.10). The borides often have parallel layers of metal and boron atoms. Hydrogen combines with a number of the transition metals to give interstitial hydrides (Section 16.6e).

Some metals, when melted together, react to form **intermetallic compounds.** These compounds have reproducible stoichiometry based not on the usual concepts of ionic or covalent bonding, but upon maintaining a constant (or nearly constant) ratio of the total number of valence electrons to the total number of atoms. For example, copper and zinc form three distinct compounds—CuZn, Cu_5Zn_8,and $CuZn_3$. These rather surprising formulas are rationalized by the Hume-Rothery rule, which relates the formulas of the compounds to the ratio of the total number of valence electrons to the total number of atoms in the compound. Counting one valence electron for copper and two for zinc, these valence electron/atom ratios are as follows in terms of the formulas of the three compounds or in terms of the number of atoms for every 21 electrons.

CuZn (*β-brass*)	3 electrons/2 atoms	or	21 electrons/14 atoms
Cu_5Zn_8 (*γ-brass*)	21 electrons/13 atoms	or	21 electrons/13 atoms
$CuZn_3$ (*ϵ-brass*)	7 electrons/4 atoms	or	21 electrons/12 atoms

As the ratio of electrons to atoms increases, the metal atoms are bound more tightly together, so hardness increases and malleability and ductility decrease. ϵ-Brass is so brittle that it shatters like glass when it is struck with a hammer.

SUMMARY OF SECTIONS 26.4–26.8

The steps necessary to obtain metals from their ores include (a) concentration, in which unwanted materials are separated from the ore; (b) extraction of the metal from the concentrated ore, often combined with reduction of the metal; and (c) refining the crude metal.

In pyrometallurgy, metals are extracted and reduced (e.g., from ores containing sulfides or oxides) by reactions at high temperatures. Copper, zinc, and iron are obtained by pyrometallurgy. In electrometallurgy, metals are obtained from ores or refined in electrochemical cells. Sodium and aluminum are obtained from their ores and copper is refined by electrometallurgy. Hydrometallurgy utilizes reactions in aqueous solution for the extraction and reduction of metals. Hydrometallurgy is energy-efficient, can be pollution-free, and is effective for low-grade ores.

Molten crude iron is produced in a blast furnace at high temperatures from a mixture of iron ore, coke, and limestone; carbon monoxide is the reducing agent. In steelmaking, molten iron is treated with oxygen at high temperatures to oxidize the impurities in crude iron, and alloying materials are introduced. Steel is heat-treated to modify its properties by controlling the amount of iron carbide (Fe_3C) present.

Alloys are mixtures that have metallic properties. Alloys may be heterogeneous mixtures, solutions, or intermetallic compounds. The properties of alloys are often quite different from those of their components.

THE *s*-BLOCK METALS

In reviewing the descriptive chemistry of the *s*-block elements (and all other elements), one looks at the similarities and differences between these elements and those in other parts of the periodic table, at the physical properties of the atoms (e.g., radii and ionization energy) and the elements (e.g., density and hardness), at the chemical reactivity of the elements and their simple compounds, and at the properties and uses of certain compounds.

The elements of Representative Groups I and II, the *s* block, are all metals. These are the only two groups of elements that have the same oxidation state (+1 and +2, respectively) in virtually all compounds. They are also two of the four groups (the others are the halogens and the noble gases, discussed in the next chapter) in which properties vary most predictably with increasing atomic number.

26.9 THE ALKALI METALS

The metals of Representative Group I, commonly known as the alkali metals, are the most electropositive and most reactive metals. Each has the largest atomic radius and the lowest ionization energy of the elements in its period. In Table 26.2, it can be seen that each alkali metal is lower melting and lower boiling, less dense, and of larger atomic radius and lower ionization energy than the alkaline earth metal in the same period. The expected variations in properties such as radii and ionization energy are also evident.

Table 26.2
Properties of the s-Block Metals

Alkali metals (ns^1): Alkaline earth metals (ns^2):	Li Be	Na Mg	K Ca	Rb Sr	Cs Ba
Melting point (°C)	179	97.5	63.5	39.0	28.4
	1283	650	851	757	704
Boiling point (°C)	1372	892	774	679	690
	~1500	1107	1487	1384	1640
Density (g/cm^3)	0.53	0.70	0.86	1.53	1.87
	1.85	1.74	1.54	2.58	3.65
Atomic radius (nm)	0.152	0.186	0.227	0.248	0.265
	0.111	0.160	0.197	0.215	0.217
Ionic radius (nm) M^+	0.068	0.097	0.133	0.147	0.167
M^{2+}	0.035	0.066	0.099	0.112	0.134
First ionization energy	520	496	419	403	376
(kJ/mol, 0 K)	900	738	590	550	503
Electronegativity	1.0	0.9	0.8	0.8	0.7
	1.5	1.2	1.0	1.0	0.9
Standard reduction potential (V)[a]	−3.05	−2.71	−2.93	−2.93	−2.92
$M^n + ne^- \longrightarrow M(s)$	−1.85	−2.36	−2.87	−2.89	−2.91
Flame color	bright red	yellow	violet	purple	blue
	—	—	brick red	crimson	green

[a] $n = 1$ for alkali metals; $n = 2$ for beryllium family elements.

The alkali metals are all strong reducing agents. In general, ease of electron loss and strength as reducing agents are expected to increase with atomic radius. However, as shown by the reduction potentials in Table 26.2, lithium, although it has the smallest atomic radius, is the strongest reducing agent among the alkali metals. To explain this, we must remember that reduction potentials measure reactivity in aqueous solution. The contributions to the energy of electron loss in aqueous solution are represented by the following steps. (M = a metal atom.)

$$M(s) \longrightarrow M(g) \qquad \textit{sublimation} \qquad \textbf{(26.5)}$$

$$M(g) \longrightarrow M^{n+}(g) + ne^- \qquad \textit{ionization} \qquad \textbf{(26.6)}$$

$$M^{n+}(g) + ne^- \xrightarrow{H_2O} M^{n+}(aq) + ne^-(aq) \qquad \textit{hydration} \qquad \textbf{(26.7)}$$

$$M(s) \longrightarrow M^{n+}(aq) + ne^-(aq) \qquad \textbf{(26.8)}$$

Sublimation and ionization are always endothermic, and hydration—the process of an ion attracting and being surrounded by water molecules—is always exothermic. The more strongly an ion attracts the water molecules, the larger the hydration energy. The Li^+ ion, being smaller than the other alkali metal ions and having a higher charge-to-size ratio, releases more energy upon hydration than the others. As a result, the reduction reaction (Equation 26.8) is more favorable for lithium.

The single, easily removed *ns* valence electrons in alkali metal atoms are responsible for many of the distinctive properties of these elements. Radiation of the correct wavelength can liberate these electrons (the photoelectric effect; Section 8.3). Cesium, the most electropositive of all elements, is frequently used in photoelectric devices.

The valence electrons of these large atoms are also easily excited to higher energy levels. In falling back to their ground state levels, valence electrons give up this energy by emitting radiation of characteristic wavelengths in the visible region of the spectrum. The color of the radiation emitted by sodium is seen in a sodium vapor light. The characteristic colors listed in Table 26.2 appear when salts of these elements are placed in a Bunsen burner flame. By analyzing the wavelength and intensity of such emitted radiation in a spectrometer, concentrations of these elements can be determined with great accuracy.

In compounds, the alkali metals have the +1 oxidation state. The cations have low polarizing ability, and the bonding in simple compounds of these elements is ionic.

Table 26.3
Differences between Lithium and Other Members of the Lithium Family (Na, K, Rb, Cs)
Lithium is quite similar to magnesium.

Property	Comments
Strongest reducing agent in family in aqueous solution	Li^+ has large hydration energy
Some salts (e.g., Li_2CO_3, Li_3PO_4) less soluble than those of other family members	Li^+ is of high charge density, contributing to large lattice energy
Reacts with nitrogen to give Li_3N_2	Only alkali metal that reacts directly with nitrogen
Carbonate and hydroxide undergo thermal decomposition	Only alkali metal carbonate and hydroxide to do so

Most such compounds are water soluble. However, again because of the smaller size of Li^+, lithium compounds display certain differences in properties, leading to its diagonal similarity to magnesium rather than the elements in its group (Table 26.3). For example, some lithium salts (e.g., LiF, Li_2CO_3, Li_3PO_4) are only slightly soluble, presumably because the high charge density of the Li^+ ion contributes to a large lattice energy. In such solubilities and in other ways lithium resembles magnesium, its diagonal neighbor of similar atomic size.

The alkali metals are the least dense of the metals and are so soft that all except lithium can be cut with a knife. The freshly cut surfaces have a metallic luster, but soon become dull due to reaction with atmospheric oxygen and water vapor. To prevent these reactions, the alkali metals are usually stored under oxygen-free organic liquids such as mineral oil.

Because of their great reactivity, the alkali metals are found in the Earth's crust only in compounds or as ions dissolved in seawater, brine wells, or salt lakes. Sodium (Section 26.5) and lithium are produced by electrolysis of molten salts. Potassium is made by reaction of molten potassium chloride (the mineral sylvite) with sodium vapor in the absence of air

$$KCl(l) + Na(g) \xrightarrow{\Delta} K(g) + NaCl(l)$$

This reaction is driven toward products by the continuous removal of the more volatile potassium vapor. Reduction of chlorides to produce free metals is one of the major uses of sodium metal. Other uses of sodium and the alkali metals are listed in Table 26.4.

Sodium and potassium ions are present in all body fluids and play essential roles in the function of living cells. Disturbances in their concentrations in the body have devastating effects on physical and mental well-being. The success of lithium carbonate (Li_2CO_3) in controlling the symptoms of manic depression may result from the influence of Li^+ on the Na^+/K^+ balance in the body. Potassium is also an essential nutrient for plants.

Table 26.4
Major Uses of the Alkali Metals
Each metal is used in the manufacture of numerous industrial chemicals.

Lithium	Lithium–magnesium alloys Lithium battery Compounds for use in lubricating greases, enamels, and glasses Aluminum production (in electrolyte)	**Sodium**	Coolant (liquid) Na vapor lights Reducing agent
		Potassium	Production of KO_2 for life-support systems
Rubidium and Cesium	Photoelectric cells Scavengers[a]		

[a] React with and remove traces of oxygen, etc.

26.10 THE ALKALINE EARTH METALS

The Representative Group II, or alkaline earth metals—like the alkali metals—display the expected periodic trends in properties (see Table 26.2), including increasing strength as reducing agents with increasing atomic number. Because of this strength as reducing agents, the elements are generally freed from their ores by electrometallurgy. In compounds, these elements have oxidation state +2 only. Beryllium and magnesium, because of their smaller radii, resemble the alkali metals less than do calcium, strontium, and barium.

Beryllium atoms are the smallest metal atoms, and the chemistry of beryllium is significantly different from that of the other elements in the group, as shown in Table 26.5. The beryllium ion (Be^{2+}) would have a very high polarizing ability and, as expected, beryllium often forms highly covalent bonds. Also, unlike the other alkaline earth metal ions, the beryllium ion attracts electron pairs from other molecules or ions strongly enough to form stable complex ions. Beryllium is more similar to its diagonal neighbor, aluminum, than to the other alkaline earth metals.

The polarizing ability of magnesium ions is much less than that of beryllium ions, but sufficiently high that in some compounds magnesium forms covalent or partially covalent bonds. (In organic compounds known as Grignard reagents, e.g., C_2H_5MgBr, magnesium is covalently bonded to one carbon atom and one halogen atom.)

Calcium, strontium, and barium—the alkaline earth metals of larger radii—are second only to the alkali metals in reactivity, their reactivity increasing with atomic number. All are strong reducing agents (see Table 26.2) and are flammable solids—calcium reacts with oxygen when heated and barium is spontaneously flammable in air. In compounds, calcium, barium, and strontium give up both *ns* electrons to form +2 ions. These three elements, like the alkali metals, have characteristic flame colors (see Table 26.2).

Each alkaline earth metal is higher melting and boiling, and denser than the corresponding alkali metal. None is found free in nature. Beryllium is present in the gemstones beryl ($Be_3Al_2Si_6O_{18}$) and emerald (beryl with 2% chromium), and in various ores. The free metal can be produced by electrolysis; most elemental beryllium is used in alloys (Table 26.6). Calcium, an abundant element, is present as calcium carbonate in limestone, marble, chalk, eggshells, pearls, and seashells, and as calcium sulfate in gypsum and alabaster.

Beryllium compounds are quite toxic. By contrast, calcium is essential to bone formation in animals. Magnesium is present in chlorophyll and is essential to the life of all green plants. Magnesium is also essential to humans as a component of bone and as Mg^{2+} in many enzymes. Most likely, beryllium ion is toxic because it can displace magnesium ion in enzymes and cause them to malfunction.

Table 26.5
Differences between Beryllium and the Other Members of the Beryllium Family (Mg, Ca, Sr, Ba)
Beryllium is similar in some ways to aluminum.

Property	Comments
Covalent in most binary compounds	A result of smaller size and stronger attraction for electrons.
A weaker reducing agent than other family members	In this, as in other properties, Be resembles its diagonal neighbor, Al.
Oxide and hydroxide are amphoteric	Be is the least "metallic," most electronegative element in the family.
Hydrated cation is hydrolyzed to give acidic aqueous solution	Another consequence of small size and stronger attraction for electrons.
Able to form stable complexes	Although all metals form complexes, other Be family metals form less stable complexes than Be^{2+} because cations are larger.

Table 26.6
Major Uses of the Alkaline Earth Metals
Strontium is seldom used commercially because calcium and barium serve the same purposes and are less expensive.

Beryllium	Alloys (small amounts as a hardening agent; in nonsparking tools) Shielding nuclear reactors (captures neutrons) Heat sink for aerospace vehicles Ceramics Windows in X-ray tubes	**Calcium**	Deoxidizer Reducing agent
		Barium	Alloys Scavenger (vacuum and TV tubes)
Magnesium	Structural metal: aircraft, portable tools, racing-class bicycles, industrial machinery Deoxidizer and desulfurizer Batteries Flashbulbs, flares, rocket propellants Cathodic protection of other metals		

Magnesium salts are recovered from the oceans and brine lakes, and magnesium metal is then produced by electrolysis of the molten salts. Magnesium is the lightest metal that can be used for structural applications, and is a component of many alloys that are easily fabricated and welded. Burning magnesium emits a brilliant white light, a property utilized in flares, fireworks, and flashbulbs.

26.11 CHEMICAL REACTIONS OF s-BLOCK METALS AND THEIR COMPOUNDS

a. Combination with Other Elements The *s*-block metals combine with oxygen (Section 16.10) and all halogens in reactions that, except for those of beryllium, are vigorous at room temperature. At higher temperatures, combination with many other nonmetals takes place. With an ample supply of oxygen, sodium and barium form peroxides (containing O_2^{2-}) and potassium, rubidium, and cesium form superoxides (containing O_2^{-}). Other combination reactions generally yield the expected ionic compounds.

b. Reactions with Water, Acids, and Bases The alkali metals and calcium, strontium, and barium are sufficiently active to displace hydrogen from water at room temperature [e.g., $2Na(s) + 2H_2O(l) \longrightarrow 2NaOH(aq) + H_2(g)$]. With most alkali metals, the reaction is so exothermic that it may occur with explosive violence as a result of the ignition of the hydrogen that is liberated. Beryllium does not react with water, and magnesium (free of its protective oxide coating) reacts only with boiling water. All of the *s*-block metals displace hydrogen from acids. It should be emphasized that active metals react even more violently with solutions of acids than they do with water. Beryllium, like other metals that form amphoteric oxides and hydroxides, also reacts with strongly alkaline solutions to form the hydroxo anion and hydrogen (Table 26.7, reaction a).

c. Reactions of Oxides and Hydroxides The alkali metals are so named because their oxides all react with water to give strongly alkaline solutions:

$$M_2O(s) + H_2O(l) \longrightarrow 2MOH(aq) \qquad M = \text{Li, Na, K, Rb, Cs}$$

Solutions of alkali metal hydroxides absorb carbon dioxide from the air to form carbonates or hydrogen carbonates.

$$2MOH(aq) + CO_2(g) \longrightarrow M_2CO_3(aq) + H_2O(l)$$
$$MOH(aq) + CO_2(g, xs) \longrightarrow MHCO_3(aq)$$

Magnesium oxide reacts slowly and not very energetically with water. The oxides of calcium, strontium, and barium are progressively more reactive toward water.

$$MO(s) + H_2O(l) \longrightarrow M(OH)_2 \qquad M = \text{Mg, Ca, Sr, Ba}$$

Table 26.7
Typical Reactions of an Amphoteric Metal and Its Compounds
Similar reactions to those illustrated here for beryllium are undergone (under appropriate conditions) by Al, Ga, Sn, and Pb from the *p* block and also Zn.

Metal with acid and base	
(c)	$Be(s) + 2H^+ \longrightarrow Be^{2+} + H_2(g)$
(d)	$Be(s) + 2OH^- + 2H_2O(l) \longrightarrow [Be(OH)_4]^{2-} + H_2(g)$
Oxide with acid and base	
(a)	$BeO(s) + 2H^+ \longrightarrow Be^{2+} + H_2O(l)$
(b)	$BeO(s) + 2OH^- + H_2O(l) \longrightarrow [Be(OH)_4]^{2-}$
Ion with water (hydrolysis)	
(e)	$[Be(H_2O)_4]^{2+} + H_2O(l) \rightleftharpoons [Be(OH)(H_2O)_3]^+ + H_3O^+$
or	$Be^{2+} + H_2O(l) \rightleftharpoons BeOH^+ + H^+$
Ion with base (precipitate, then dissolution in excess base)	
(f)	$Be^{2+} + 2OH^- \longrightarrow Be(OH)_2(s) \xrightarrow{xs\ OH^-} [Be(OH)_4]^{2-}$

Recall that ionic oxides are basic, covalently bonded oxides are acidic, and those with intermediate bonding are amphoteric. Among the oxides of the *s*-block metals, only beryllium oxide has sufficiently covalent bonding to be amphoteric, as shown by its reactions with both acids and bases (Table 26.7, reactions c and d).

d. Thermal Decomposition of Hydroxides and Carbonates There is a significant difference between the thermal decomposition of the carbonates and hydroxides of the alkali metals and those of most other metals. Except for lithium hydroxide and to a certain extent lithium carbonate, alkali metal hydroxides and carbonates do *not* decompose when heated. The alkaline earth hydroxides do undergo thermal decomposition. For the carbonates, the necessary temperature for decomposition increases from magnesium to barium.

$$M(OH)_2(s) \xrightarrow{\Delta} MO(s) + H_2O(g) \qquad M = Be, Mg, Ca, Sr, Ba$$
$$MCO_3(s) \xrightarrow{\Delta} MO(s) + CO_2(g) \qquad M = Mg\ (400\ ^\circ C), Ca\ (900\ ^\circ C), Sr\ (1175\ ^\circ C), Ba\ (1500\ ^\circ C)$$

e. Ions in Aqueous Solution Ions of the *s*-block metals, except for Be^{2+}, do not hydrolyze or form complex ions to any significant extent. The Be^{2+} ion forms an acidic solution (Table 26.7, reaction e). Addition of OH^- ion to Be^{2+} in aqueous solution first causes precipitation of the hydroxide and then, with excess OH^- ion, hydroxide dissolution by formation of a hydroxo complex ion (Table 26.7, reaction f), another reaction typical of ions of amphoteric metals.

The alkali metal ions form very few precipitates with simple anions. Therefore, in qualitative analysis these ions must be precipitated as higher molecular mass compounds such as sodium zinc uranyl acetate, $Na[Zn(UO_2)_3(CH_3COO)_9]$.

26.12 COMPOUNDS OF THE s-BLOCK METALS

Some of the commercially important compounds of *s*-block metals are listed in Tables 26.8 and 26.9. The uses of many of these compounds depend upon their alkalinity. Sodium hydroxide, sodium carbonate, calcium hydroxide, and calcium oxide are among the top industrial chemicals in terms of the quantities produced each year (see Table 26.10), and the chemistry of these industrial alkalis is discussed in the next section.

The alkali metal hydroxides are white solids soluble in water with the evolution of heat. The alkaline earth hydroxides are much less soluble, the solubility increasing from $Be(OH)_2$ to $Ba(OH)_2$ in line with decreasing attraction between the larger cations and the hydroxide ions.

Table 26.8
Some Commercially Important Alkali Metal Compounds and Their Uses

Compound	Uses
$LiAlH_4$	Reducing agent in organic chemistry
Li_2CO_3	Drug for treating manic depression
$NaOH$[a] (lye, caustic soda)	Industrial chemical Pulp, paper, and rayon Extraction of Al from ore S removal from petroleum Soaps and detergents Food processing
Na_2CO_3[a] (soda ash)	Glass Industrial chemical Detergents and cleansers Water softening Pulp and paper industry
$NaHCO_3$ (baking soda)	Food industry Household use Industrial chemical Fire extinguishers
Na_2O_2	Bleaching agent Preserving bacon, ham, etc.
$NaNO_2$	Metal treatment
$Na_2SO_4 \cdot 10H_2O$ (Glauber's salt)	Paper manufacture Glass
KOH (caustic potash)	K_2CO_3 manufacture Soaps, and detergents Fertilizer manufacture
K_2CO_3 (potash)	Chemical manufacture Glass and ceramics Dyes and pigments
KNO_3	Gunpowder and matches
KCl (muriate of potash)	Fertilizer

[a] A large-volume industrial chemical

Table 26.9
Some Commercially Important Alkaline Earth Metal Compounds and Their Uses

Compound	Uses
MgO (magnesia)	Refractory Insulation Paper manufacture Animal food Flocculant Antacid
$MgSO_4 \cdot 7H_2O$ (epsom salt)	Leather tanning Fabric treatment Medicine
CaO[a] (lime, quicklime) and $Ca(OH)_2$ (slaked lime)	Metallurgy Mortar, plaster, and cement Industrial alkali Pulp and paper industry Bleaching powder Pollution control Water treatment Glass manufacture
$CaHPO_4$ and $Ca(H_2PO_4)_2$	Fertilizers
$CaSO_4 \cdot 2H_2O$ (gypsum)	Sheetrock (drywall)
$CaSO_4 \cdot \frac{1}{2}H_2O$ (plaster of paris)	Plaster
Ca silicates	Glass, portland cement, ceramics
$Sr(NO_3)_2$	Roadside flares
$BaTiO_3$	Sonar and electronic equipment

[a] A large-volume industrial chemical.

Sodium peroxide (Na_2O_2) and potassium superoxide (KO_2) are powerful oxidizing and bleaching agents which form hydrogen peroxide (H_2O_2) on contact with cold solutions of acids or with an excess of cold water.

$$Na_2O_2(s) + 2H_2O(l, xs) \longrightarrow H_2O_2(aq) + 2NaOH(aq)$$
$$2KO_2(s) + 2H_2O(l, xs) \longrightarrow H_2O_2(aq) + O_2(g) + 2KOH(aq)$$

Potassium superoxide is employed in "breathing" equipment used in rescue work in mines and other areas where the air is so deficient in oxygen that an artificial atmosphere must be generated. The moisture of the breath reacts with the oxide to liberate oxygen, and at the same time the potassium hydroxide formed removes carbon dioxide as it is exhaled.

$$4KO_2(s) + 2H_2O(l) \longrightarrow 4KOH(s) + 3O_2(g)$$
$$2KOH(s) + CO_2(g) \longrightarrow K_2CO_3(s) + H_2O(l)$$

The alkali metal carbonates, except for lithium carbonate, are quite soluble in water, giving alkaline solutions due to hydrolysis of the carbonate ion. The alkaline earth metal carbonates, like most other carbonates, are insoluble in water, but dissolve in water containing dissolved carbon dioxide:

$$MCO_3(s) + H_2O(l) + CO_2(g) \longrightarrow M^{2+} + 2HCO_3^-$$

Groundwater always contains carbon dioxide, and therefore can dissolve limestone and dolomite, for example

$$\underset{\textit{dolomite}}{CaCO_3 \cdot MgCO_3(s)} + 2H_2O(l) + 2CO_2(aq) \rightleftharpoons Ca^{2+} + Mg^{2+} + 4HCO_3^-$$

The calcium and magnesium ions thus brought into solution make the water "hard" (Section 14.13)

The dissolution of carbonate minerals in groundwater has an interesting effect—if the mineral is below the surface of the Earth and is covered by rocks that do not dissolve, a cave may be formed. Groundwater may then seep into the cave through cracks in the roof. The water contains calcium and magnesium hydrogen carbonates that it has dissolved elsewhere. When it enters the cave, where the partial pressure of the carbon dioxide is lower, the carbon dioxide escapes from solution by a reversal of the reaction shown above. Insoluble calcium and magnesium carbonates are thus produced and deposit around the opening through which the water is trickling into the cave, forming an icicle-like structure called a *stalactite.* Some of the hydrogen carbonate remains in solution until drops of water fall from the tip of the stalactite. As the drops fall, the hydrogen carbonate decomposes and forms carbonate deposits on the floor of the cave, building up another "icicle," called a *stalagmite.* After many centuries, the tips of the stalactite and stalagmite meet and the two "icicles" grow into a column that one might think was left there to support the roof of the cave.

Sodium carbonate forms a number of hydrates, the most common of which is washing soda (sodium carbonate decahydrate, $Na_2CO_3 \cdot 10H_2O$). Solutions of this salt are effective cleaning agents because of the alkalinity resulting from hydrolysis of the carbonate ion. Sodium hydrogen carbonate ($NaHCO_3$), or bicarbonate of soda, is made from pure sodium carbonate.

$$Na_2CO_3(aq) + H_2O(l) + CO_2(g) \longrightarrow 2NaHCO_3(s)$$

It is used in baking powders as a *leavening agent,* a substance that produces gas bubbles in dough. This gives rise to its common name, baking soda. An acid substance must be present in the dough to liberate carbon dioxide:

$$HCO_3^- + H^+ \longrightarrow CO_2(g) + H_2O(l)$$

The evolving gas causes the dough to rise and gives the product the appropriate lightness and texture. Sour milk can be the source of the acid. Some recipes call for use of a baking powder. These powders are mixtures of sodium hydrogen carbonate and an acidic substance, for example, sodium alum (sodium aluminum sulfate), $NaAl(SO_4)_2 \cdot 12H_2O$ (acidic due to reaction of hydrated Al^{3+} with water); calcium dihydrogen phosphate, $Ca(H_2PO_4)_2$; or potassium hydrogen tartrate, $K(HC_4H_4O_6)$. "Double acting" baking powder is made possible by coated crystals of $Ca(H_2PO_4)_2$ that release about half of their hydrogen ion during mixing and half during baking.

EXAMPLE 26.4 Chemical Reactions: Alkali Metal Compounds

Complete and balance the following equations (a reaction does occur in each case):

(a) $K_2CO_3(s) + C(s) \xrightarrow{>1000\,°C}$

(b) $NaHCO_3(s) \xrightarrow{\Delta}$

(c) $Li^+ + HPO_4^{2-} + OH^- \longrightarrow$

(a) One of the reactants is an element, so this will be a redox reaction. Combination is not possible. At elevated temperature elemental carbon is a reducing agent and can be expected to "displace" potassium to yield elemental potassium. The carbon will be oxidized, most likely to carbon monoxide at this high temperature.

$$\mathbf{K_2CO_3(s) + 2C(s) \longrightarrow 2K(g) + 3CO(g)}$$

(b) With one reactant, this could be a redox or a nonredox decomposition reaction, or a disproportionation. Hydrogen carbonates undergo nonredox decomposition upon heating to form carbonates, carbon dioxide, and water.

$$\mathbf{2NaHCO_3(s) \longrightarrow Na_2CO_3(s) + CO_2(g) + H_2O(g)}$$

Sodium carbonate, like most of the alkali metal carbonates, will not undergo further thermal decomposition.

(c) This is a reaction of ions in aqueous solution. Although with three ions present, the reaction pattern will not be exactly that of a partner-exchange reaction, we should still first look for the possible formation of solids, gases, or nonionized compounds. The OH^- ion can react with the acidic HPO_4^{2-} ion.

$$HPO_4^{2-} + OH^- \longrightarrow PO_4^{3-} + H_2O(l)$$

A further reaction is possible because lithium forms several only slightly soluble salts, including the phosphate. The overall reaction should be

$$\mathbf{3Li^+ + HPO_4^{2-} + OH^- \longrightarrow Li_3PO_4(s) + H_2O(l)}$$

26.13 INDUSTRIAL ALKALIS

The hydroxides, oxides, and carbonates of sodium, potassium, magnesium, and calcium are interrelated industrial chemicals of importance. In the chemical industry, "alkali" is a general term for any compound that produces hydroxide ion in aqueous solution.

a. The Chloralkali Industry Sodium hydroxide and chlorine are co-produced by the electrolysis of saturated aqueous sodium chloride solutions and are two of the primary products of what is referred to as the *chloralkali industry*. The third major chloralkali chemical is sodium carbonate.

The overall chemical reaction in the electrolysis of saturated sodium chloride solutions (known as "brines")

$$2NaCl(aq) + 2H_2O(l) \xrightarrow{\text{electrolysis}} 2NaOH(aq) + H_2(g) + Cl_2(g)$$

simultaneously produces two large-volume industrial chemicals (Table 26.10)—chlorine (Cl_2) and sodium hydroxide (NaOH)—and also hydrogen.

Three major types of electrochemical cells are used in the chloralkali industry—diaphragm cells, membrane cells, and mercury cells. The objective in each is to keep chlorine, which is produced at the anodes, from mixing with either hydrogen or sodium hydroxide. The reactions that would result,

$$2NaOH(aq) + Cl_2(g) \longrightarrow NaOCl(aq) + NaCl(aq) + H_2O(l)$$
$$H_2(g) + Cl_2(g) \longrightarrow 2HCl(g)$$

Table 26.10
The Top 15 Industrial Chemicals in a Recent Year[a]

		Billions of kilograms produced
1	Sulfuric acid (H_2SO_4)	33.43
2	Nitrogen (N_2)	22.07
3	Oxygen (O_2)	15.00
4	Ethylene ($CH_2{=}CH_2$)	14.90
5	Lime [CaO, $Ca(OH)_2$]	13.77
6	Ammonia (NH_3)	12.72
7	Sodium hydroxide ($NaOH$)	9.99
8	Chlorine (Cl_2)	9.52
9	Phosphoric acid (H_3PO_4)	8.36
10	Propylene ($CH_3CH_2{=}CH_2$)	7.87
11	Sodium carbonate (Na_2CO_3)	7.81
12	Ethylene dichloride ($ClCH_2CH_2Cl$)	6.60
13	Nitric acid (HNO_3)	5.96
14	Urea [$CO(NH_2)_2$]	5.48
15	Ammonium nitrate (NH_4NO_3)	5.04

[a] Of these top 15 industrial chemicals in 1986, the alkalis are lime [CaO and $Ca(OH)_2$], sodium hydroxide, and sodium carbonate; the acids are sulfuric acid, phosphoric acid, and nitric acid. The organic chemicals on the list—ethylene, propylene, ethylene dichloride, and urea are used extensively in making polymers (see Chapter 33).

are undesirable because product is lost, and also because hydrogen–chlorine mixtures can explode.

In diaphragm and membrane cells the electrode reactions are

$$\textit{cathode} \quad 2H_2O + 2e^- \longrightarrow H_2 + 2OH^- \qquad E^\circ = -0.83\ \text{V}$$
$$\textit{anode} \quad 2Cl^- \longrightarrow Cl_2 + 2e^- \qquad E^\circ = 1.36\ \text{V}$$

The cathode in a diaphragm cell is a metal mesh that supports an asbestos diaphragm (Figure 26.6a). In membrane cells (Figure 26.6b), the asbestos diaphragm is replaced by an ion-exchange membrane. The sodium ions from the anode compartment are allowed by the membrane to enter the cathode compartment, where OH^- ions are being generated, but the Cl^- ions are *not* allowed to pass by the membrane. In this way a concentrated sodium hydroxide solution can be produced.

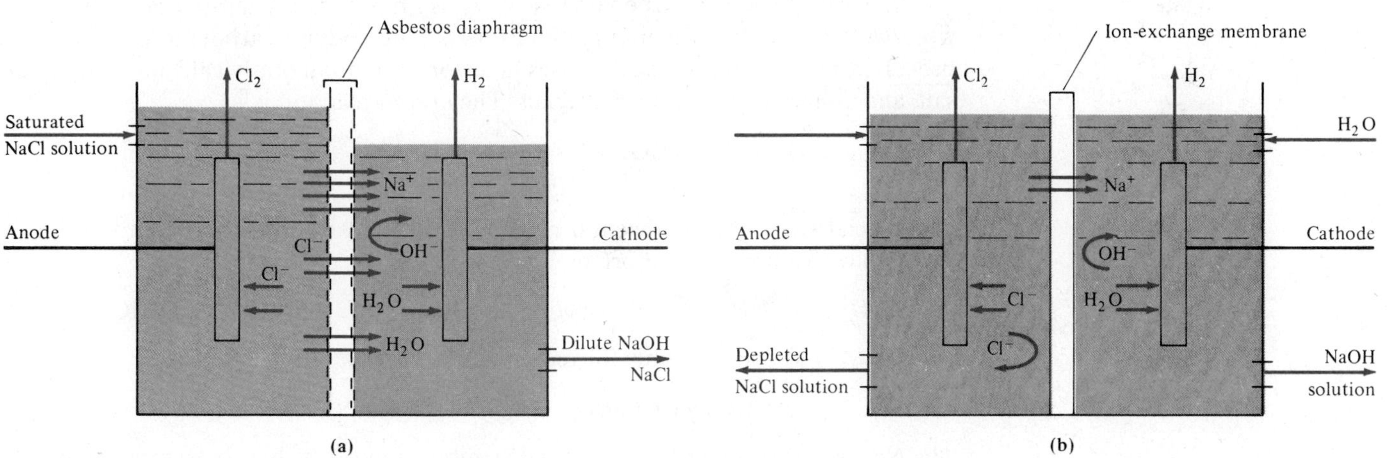

Figure 26.6
Schematic Drawings of the Diaphragm (a) and Membrane (b) Cells for the Electrolysis of Sodium Chloride Solutions
(a) Sodium and chloride ions can pass through the wet-asbestos diaphragm; hydrogen and chlorine gases cannot pass through. A positive pressure of the solution on the anode side prevents the return flow of hydroxide ions from the cathode compartment. The sodium hydroxide solution produced is contaminated with unreacted salt, which precipitates as the solution is concentrated and can be removed by filtration. (b) The ion exchange membrane allows the passage of sodium ions but prevents the passage of both chloride and hydroxide ions (and also hydrogen and chlorine gases). The product solution is not contaminated by chloride ions and is more concentrated than that from the diaphragm cells.

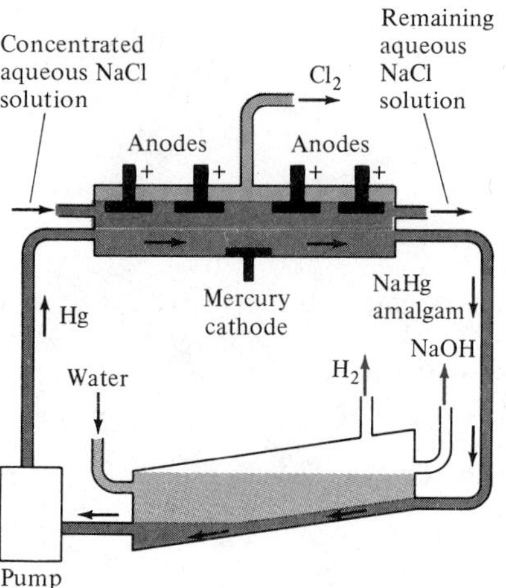

Figure 26.7
The Mercury Cell for the Electrolysis of Sodium Chloride
This cell is more expensive to build and operate than the diaphragm or membrane cells. The toxic nature of the mercury released to the environment is another drawback. For these reasons, the use of mercury cells is being discontinued.

Flowing mercury is used as the cathode in the mercury cells (Figure 26.7). Sodium ion is reduced at the mercury cathode and forms sodium amalgam

$$Na^+ + e^- + x\mathrm{Hg}(l) \longrightarrow \mathrm{NaHg}_x(l)$$

which reacts with water in a separate vessel to give a concentrated sodium hydroxide solution that is free from NaCl. The recovered mercury is recycled.

$$2\mathrm{NaHg}_x(l) + 2\mathrm{H_2O}(l) \longrightarrow \mathrm{H_2}(g) + 2\mathrm{NaOH}(aq) + 2x\mathrm{Hg}(l)$$

Caustic soda, as sodium hydroxide is known in industry, is sold as a concentrated solution, and in solid forms such as pellets and flakes. The addition of water to concentrated or solid sodium hydroxide releases heat and must be done with caution. Half of the caustic soda manufactured each year is used in the production of other chemicals. Five percent is consumed in the production of alumina from bauxite ores (Section 26.5). A small but growing application of caustic soda is in commercial food processing—caustic soda solutions are used to peel vegetables. Other uses of caustic soda are listed in Table 26.7.

Anhydrous sodium carbonate (Na_2CO_3) is known industrially as **soda ash.** On a worldwide basis, the principal method for making sodium carbonate is the Solvay process, a multi-step process that uses limestone (predominantly calcium carbonate), salt, and ammonia as the raw materials. The *overall* reaction is

$$\underset{\textit{limestone}}{\mathrm{CaCO_3}} + \underset{\textit{brine}}{2\mathrm{NaCl}} \longrightarrow \underset{\textit{soda ash}}{\mathrm{Na_2CO_3}} + \mathrm{CaCl_2}$$

The crucial steps are the reaction of CO_2 (generated by thermal decomposition of $CaCO_3$) with ammonia and brine

$$2\mathrm{CO_2}(g) + 2\mathrm{NH_3}(aq) + 2\mathrm{NaCl}(aq) + 2\mathrm{H_2O}(l) \longrightarrow 2\mathrm{NH_4Cl}(aq) + 2\mathrm{NaHCO_3}(s)$$

and recovery of the ammonia for recycling in the process

$$2\mathrm{NH_4Cl}(aq) + \mathrm{CaO}(s) \longrightarrow \mathrm{CaCl_2}(aq) + 2\mathrm{NH_3}(g) + \mathrm{H_2O}(l)$$

The Na_2CO_3 is produced by thermal decomposition of the hydrogen carbonate.

In the United States the production of natural soda ash has replaced the Solvay process. Large deposits of the mineral trona ($Na_2CO_3 \cdot NaHCO_3 \cdot 2H_2O$) are present at Searles Lake in California and in the residue from a huge lake that once covered southwestern Wyoming. Trona is heated to convert the sodium hydrogen carbonate to sodium carbonate. More than half the soda ash produced in the United States is used in making glass. Other important uses are in the pulp and paper industry and in soaps and detergents. Most dishwasher detergents contain soda ash.

b. Lime Chemicals derived from limestone, most importantly calcium oxide (CaO) and calcium hydroxide ($Ca(OH)_2$) are central to another segment of the alkali industry. Many chemicals are known commercially by common names. For example, in industry **lime** refers to either calcium oxide, specifically known as **quicklime** (or **unslaked lime**), or calcium hydroxide, specifically known as **hydrated lime** (or **slaked lime**).

Limestone is the raw material for the production of quicklime and hydrated lime. As the least expensive sources of alkalinity, these are industrial chemicals with a great diversity of uses. Various grades of commercial "lime" contain different proportions of hydroxides, oxides, carbonates, and water. Many are derived from dolomitic limestone and contain magnesium as well as calcium compounds. The following equations represent the major classifications of lime and their preparation. The addition of water to quicklime is called "slaking"; it is a very exothermic reaction.

$$\underset{\text{high-calcium limestone}}{CaCO_3(s)} \xrightarrow{\Delta} \underset{\text{high-calcium quicklime}}{CaO(s)} + CO_2(g)$$

$$\underset{\text{dolomitic limestone}}{CaCO_3 \cdot MgCO_3(s)} \xrightarrow{\Delta} \underset{\text{dolomitic quicklime}}{CaO \cdot MgO(s)} + 2CO_2(g)$$

$$CaO(s) + H_2O(l) \longrightarrow \underset{\text{high-calcium hydrated lime}}{Ca(OH)_2(s)}$$

$$CaO \cdot MgO(s) + 2H_2O(l) \xrightarrow{\text{pressure}} \underset{\text{dolomitic hydrated lime}}{Ca(OH)_2 \cdot Mg(OH)_2(s)}$$

The largest use of lime is as a flux in steelmaking, where it reacts with and helps to remove such impurities as phosphorus and sulfur. Other major uses are in the construction industry and in water treatment (see Table 26.9).

SUMMARY OF SECTIONS 26.9–26.13

Of the *s*-block elements, lithium (Table 26.3), magnesium, and more so, beryllium (Table 26.5) differ somewhat from the others (due mainly to smaller atomic size). The remaining *s*-block elements have many properties in common. The properties of these elements are summarized in Table 26.11. Knowing the properties in this table will provide a good foundation for understanding the behavior of the *s*-block elements. Similar tables are presented for each group of elements and compounds discussed in Chapters 26–30.

Table 26.11
Outstanding Properties of the *s*-Block Metals and Their Compounds

Li–Cs, ns^1 Be–Ba, ns^2	Valence electrons in *s* orbitals only
Li^+–Cs^+ Be^{2+}–Ba^{2+}	One oxidation state only; ions in all compounds (except many Be and some Mg compounds)
Li, Na, K, Rb, Cs, Ca, Sr, Ba	Most reactive metals; combine with all nonmetals (except, in Group I, only Li + $N_2 \longrightarrow$ nitride)
Li–Cs	Not stable in air; soft, low-melting, least dense metals
$M \longrightarrow M^+ + e^-$ $M' \longrightarrow M^{2+} + 2e^-$	Strong reducing agents; Li *very* strong
Li	Differs in properties from other family members (Table 26.3)
Be	An amphoteric element (Table 26.7); covalent in many binary compounds; differs in properties from other family members (Table 26.5); poisonous
M_2O, $M'O$ MOH, $M'(OH)_2$(M = Li–Cs; M' = Mg–Ba)	Strongly basic oxides and hydroxides; MOH, water soluble and thermally stable; $M'(OH)_2$, slightly soluble and less thermally stable
M_2CO_3(M = Na–Cs)	Very stable thermally
$M_n^+A^{n-}$ (M = Na–Cs; A = common anions)	Water-soluble compounds; hydrated cations not hydrolyzed
NaOH (caustic soda) Na_2CO_3 (soda ash) $Ca(OH)_2$ (hydrated lime) CaO (quicklime)	Top industrial chemicals and alkalis

Table 26.12
Common Oxidation States of the p-Block Metals
The more stable oxidation states are shown in color.

Al		+3
Ga	+1	+3
In	+1	+3
Tl	+1	+3
Sn	+2	+4
Pb	+2	+4
Bi	+3	+5

THE *p*-BLOCK METALS

The seven *p*-block metals fall into Representative Groups III (aluminum, gallium, indium, and thallium), IV (tin and lead), and V (bismuth). Consequently, the properties of *p*-block metals vary more than those of the *s*-block metals. They are generally less reactive than the *s*-block metals and more likely to form covalent bonds. Except for aluminum, each *p*-block metal commonly exhibits two oxidation states, one equal to the group oxidation state and one equal to the group oxidation state minus 2 (Table 26.12). The metals of highest atomic number in each group tend to be more stable in their lower oxidation states (Tl, +1; Pb, +2; Bi, +3). Aluminum is the *p*-block metal of greatest interest because of its abundance (see Table 26.1) and its versatility in applications. (The preparation of aluminum by electrometallurgy was discussed in Section 26.5.)

26.14 ALUMINUM AND OTHER GROUP III METALS

Aluminum has the +3 oxidation state in stable compounds, indicating involvement of the ns^2 and np^1 valence electrons in bonding. Removal of all three electrons to form Al^{3+} requires a large amount of energy and yields an ion of moderate polarizing ability. Therefore, in most simple aluminum compounds the bonding is predominately covalent. However, because the heat of hydration of Al^{3+} is large, the ion forms readily in aqueous solution and is also present in crystalline hydrates of aluminum compounds.

Aluminum is a silvery-white metal of low density. It is rather soft and weak, but its strength can be increased considerably by alloying with other metals, such as copper or magnesium. The electrical conductivity of aluminum is comparable, weight for weight, with that of copper, and it has been used in electrical wiring. Without proper installation, however, aluminum wiring can be hazardous. It must not be connected to standard brass terminals, because couple action results in corrosion of the aluminum and formation of an oxide on its surface. Oxide also forms if the insulation wears away, exposing the aluminum wire. The outcome of oxide formation is an increase in electrical resistance, the generation of heat, and the possibility of igniting surrounding materials.

Aluminum has a host of uses (Table 26.13) that take advantage of its many desirable properties, especially its combination of lightness and strength. Aluminum can be cast in massive, complex shapes, drawn into fine wires, or rolled into thin sheets. Anodized aluminum has a thick, durable coating of aluminum oxide, a coating that can be permanently colored by introduction of a dye.

Table 26.13
Uses of Aluminum
In 1934 aluminum was proclaimed the "theme metal of the twentieth century."

- Structural material
  - Buildings
  - Airplanes
  - Boats
  - Cars and trucks
  - Electrical wire
  - Kitchen utensils
  - Foil wrap
  - Beverage cans
  - Containers and packaging
- Alloys
  - Alnico (magnetic)
  - Duraluminum (light, but strong)
- Welding (thermite reaction)
- Pigments and paints
- Fireworks, flares, and rocket fuel
- Alumina (Al_2O_3) (α-alumina, γ-alumina)
  - Watch bearings (synthetic ruby or sapphire)
  - Dehydrating agent ("activated")
  - Refractory bricks
  - Catalyst support (catalyst deposited on surface of finely divided alumina)
- Aluminum sulfate
  - Water purification
  - Pulp and paper sizing

Gallium and indium, like aluminum, form compounds in the +3 oxidation state with largely covalent bonding and +3 ions in solution. Unlike aluminum, gallium and indium also form compounds in the +1 oxidation, in which only the np^1 electron is involved in bonding. However, these +1 ions are not stable in aqueous solution.

All the *p*-block metals except for aluminum are influenced by the higher effective nuclear charges caused by intervention of the transition elements. Note in Table 26.14 that gallium has the smallest atomic radius of the Group III metals and thallium has a higher ionization energy than indium.

Gallium and indium are important to the solid-state electronics industry. They are used primarily in compound semiconductors (Sections 29.18 and 29.19). The intense red light in some digital displays is provided by gallium–arsenic–phosphorus light-emitting diodes. More than 80% of all gallium and indium produced is used in the electronics industry.

Thallium, by contrast to the other Group III metals, is more stable in both aqueous solutions and compounds as the Tl^+ ion. As a result, in its +3 oxidation state thallium is a strong oxidizing agent.

Thallium compounds are excellent rat and ant poisons, but care must be exercised in their use, because thallium and its compounds are also *highly toxic* to human beings.

Table 26.14
Properties of the Representative Group III Metals
The group electron configuration is ns^2np^1.

	Al	Ga	In	Tl
Melting point (°C)	658	29.75	155	303.5
Boiling point (°C)	1800	1700	>1450	1650
Density (g/cm³)	2.70	5.9 (solid)	7.30	11.5
Atomic radius (nm)	0.143	0.122	0.163	0.170
Ionic radius, M^{3+} (nm)	0.051	0.062	0.081	0.095 (Tl^+ 0.147)
First ionization energy (kJ/mol, 0 K)	578	579	558	590
Electronegativity	1.5	1.8	1.5	1.4
Standard reduction potential (V) $M^{3+} + 3e^- \longrightarrow M(s)$	−1.66	−0.53	−0.343	0.719[a]

[a] $Tl^{3+} + 2e^- \rightleftharpoons Tl^+ \quad E° = 1.25$ V

26.15 TIN, LEAD, AND BISMUTH

The +2 and +4 oxidation states of tin and lead represent involvement of the np^2 or the ns^2np^2 electrons in bonding, respectively. For bismuth, the +3 and +5 oxidation states similarly represent involvement of the np^3 or ns^2np^3 electrons. Bonding in compounds of these elements in their lower oxidation states is predominately ionic; in the higher oxidation states it is more nearly covalent. Species containing lead and bismuth in their less-stable higher oxidation states, such as PbO_2 or BiO_3^-, are strong oxidizing agents.

The ancient Egyptians and Babylonians were familiar with tin and lead. A record of bismuth goes back to the 1400s, but it may have been isolated long before that time and confused with lead. The aqueducts that brought water to ancient Rome were lined with lead. Even though they have been known and used for so long, these elements are not abundant in nature.

Tin is obtained by heating cassiterite with charcoal or coke.

$$\underset{\text{cassiterite}}{SnO_2(s)} + 2C(s) \longrightarrow Sn(l) + 2CO(g)$$

The crude tin is partially purified by placing it on a hot sloping table and permitting the molten metal to flow away from the higher melting impurities. Further refinement is carried out by an electrolytic method.

Tin is a soft, low-melting metal that exists in three allotropic forms of different densities (Table 26.15). The form stable at ordinary temperatures is white tin (Figure 26.8), which is distinctly metallic in character. Below 13.2 °C, it changes slowly into an amorphous gray powder, gray tin, which is a semiconductor and is less dense than the metallic form. The gradual crumbling away of tin as it changes to gray tin when

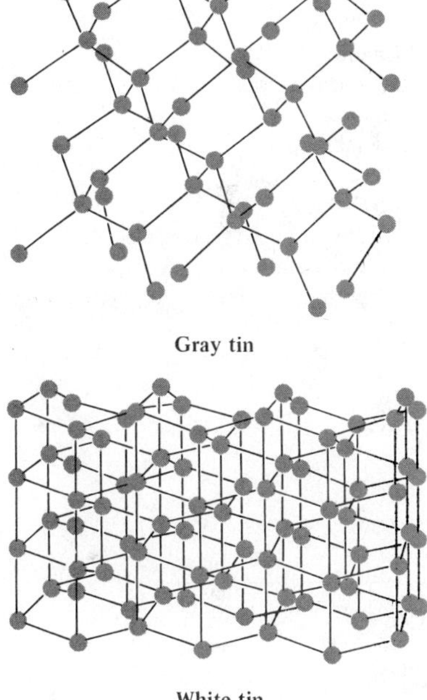

Figure 26.8
Crystal Structures of Gray and White Tin

Table 26.15
Properties of Tin, Lead, and Bismuth

	Sn	Pb	Bi
Configuration	$[Kr]4d^{10}5s^25p^2$	$[Xe]4f^{14}5s^25p^65d^{10}6s^26p^2$	$[Xe]4f^{14}5s^25p^65d^{10}6s^26p^3$
Melting point (°C)	232	327	271
Boiling point (°C)	2270	1620	1560
Density (g/cm³)	5.75 (gray) 7.28 (white) 6.52 (brittle)	11.29	9.80
Atomic radius (nm)	0.141	0.175	0.155
Ionic radius (nm)	0.093 (Sn^{2+})	0.120 (Pb^{2+})	0.096 (Bi^{3+})
First ionization energy (kJ/mol, 0 K)	709	716	703
Electronegativity	1.7	1.6	1.9
Standard reduction potential (V)			
$M^{2+} + 2e^- \longrightarrow M(s)$	−0.136	−0.126	—
$BiO^+ + 2H^+ + 3e^- \longrightarrow Bi(s) + H_2O$	—	—	0.32

Table 26.16
Major Uses of Tin, Lead, and Bismuth

Tin
Tinplate for beverage cans
Alloys, e.g.,
soft solders (Sn, Pb)
Babbit metal (Sn, Cu, Pb) used in bearings
bronzes (Cu, Sn)
pewter (>90% Sn, Sb, Cu)
Lead
Alloys, e.g., in lead storage batteries, solders, plumbing
Litharge, PbO, used in rubber, ceramics, etc.
Red lead, Pb_3O_4, rust-inhibiting pigment
X-ray shield
Bismuth
Acoustical barriers
Radiation shields
Low-melting alloys, e.g.,
Wood's metal, m.p. 70–72 °C (50% Bi, 25% Pb, 12.5% Sn, 12.5% Cd) used in, e.g., electrical fuses, automatic sprinklers, safety plugs in gas cylinders

exposed to low temperatures is called "tin disease." When the metal is heated to 161 °C it changes to brittle tin, a material that shatters when it is struck with a hammer.

The major uses of tin are in the manufacture of tinplate—low-carbon steel with a thin coating of tin—and alloys (Table 26.16). Tinplate is made by electrolytic deposition of tin on the steel; the coating improves such properties of the metal as its workability and ease of soldering, and protects it from corrosion.

The extractive metallurgy of lead is a multistep process similar in many ways to that of copper (Section 26.6). The sulfide ore is sintered, roasted, and smelted. The crude lead may be refined by pyrometallurgical or electrometallurgical methods. At all steps, the methods used must be designed to separate lead from the other elements (e.g., As, Sb, Sn, Bi, Ag, Au, Pt) usually also present in its ores, many of them metals worthy of recovery because of their own value. One method of separating silver and gold from crude lead takes advantage of the immiscibility of lead and zinc, and the solubulity of silver and gold in zinc. In the Parkes process, crude lead and a small amount of zinc are melted together. The silver and gold, along with some lead, dissolve in the zinc and the mixture rises to the surface and solidifies. Further refining is necessary to separate the silver and gold.

Lead, like tin, is soft and easily melted. It is quite dense. When freshly cut, lead has a high luster, but upon exposure to the air, it quickly becomes dull as the result of the formation of a thin coating of oxide or carbonate. This coating adheres tightly to the metal and protects it from further corrosion. For this reason, the metal has long been used for roofing, gutters, downspouts, and sewer lines. It should never be used to conduct water intended for human consumption, for lead is slowly corroded and dissolved by water (particularly by soft water), and it is toxic (see Aside 26.1: Metals as Poisons).

Bismuth is the main group metal of highest atomic number and is the least "metallic" in its properties—it is brittle rather than malleable and has the lowest electrical conductivity of any metal except mercury. The recovery of bismuth from its native ores is a simple process. The ore is heated at least to the melting point of bismuth (271 °C) and the metal is permitted to flow away from the impurities. Oxide and sulfide ores are roasted in the air and then reduced with carbon. A large amount of bismuth is obtained in the United States as a by-product of copper and lead smelting and refining.

Most of the bismuth that is used commercially appears in low-melting alloys for electrical fuses, automatic sprinkler systems, safety plugs in compressed gas cylinders, and other such devices. Bismuth alloys are also useful where accurate dimensions in the product are important, for example, in type metal. Bismuth and gallium are the only metals that increase in volume upon solidification.

26.16 CHEMICAL REACTIONS OF THE *p*-BLOCK METALS AND THEIR COMPOUNDS

a. Combination with Other Elements The *p*-block metals, except for thallium, are generally stable in dry air (i.e., they do not corrode or rust away). They are either unreactive or form a protective coating.

At high temperatures the *p*-block metals all combine with oxygen, halogens, and other nonmetals. The Group III metals form oxides and halides in their +3 oxidation states, (e.g., Al_2O_3 and $AlCl_3$), except that thallium tends to be oxidized first to its +1 oxidation state.

Tin is oxidized to the +4 state by both halogens and oxygen to give SnX_4 and SnO_2. Lead reacts with oxygen below 500 °C to give PbO and at higher temperatures to give Pb_3O_4, which is known as red lead and contains both Pb(II) and Pb(IV). With halogens, lead gives only PbX_2.

Bismuth reacts with both oxygen and halogens to give the +3 compounds, BiX_3 and Bi_2O_3, with the exception that at high temperatures BiF_5 can be produced.

b. Reactions with Water, Acids, and Bases Aluminum (freed from its protective coating of Al_2O_3) is the most reactive *p*-block metal and will displace hydrogen from steam. The *p*-block metals all displace hydrogen from nonoxidizing acids. Lead reacts only superficially with dilute sulfuric acid because the lead (II) sulfate produced is insoluble in the acid and forms a protective coating. All the *p*-block metals also react with oxidizing acids. However, the reaction of aluminum with nitric acid is not extensive because the metal is soon protected by an oxide layer. (The activity of aluminum can be demonstrated by first washing a piece of aluminum in dilute acid to dissolve the oxide, then dipping the aluminum into a solution of $HgCl_2$ to form an amalgam coating to which Al_2O_3 does not adhere. Whiskers of Al_2O_3 grow rapidly from the surface of aluminum treated in this way and the temperature of the system increases sharply. The treated aluminum will also liberate hydrogen from water and react with acids.)

Tin is oxidized to the +4 state by strongly oxidizing acids

$$Sn(s) + 4HNO_3(aq) \longrightarrow SnO_2(s) + 4NO_2(g) + 2H_2O(l)$$

whereas lead and bismuth are oxidized to their lower oxidation states of +2 and +3, respectively.

Four of the seven *p*-block metals—aluminum, gallium, tin, and lead—are amphoteric; like beryllium (Table 26.7, reaction e), they react with strong bases to form hydroxo anions.

c. Reactions of Oxides, Hydroxides, and Carbonates The oxides of the four amphoteric *p*-block metals (Al, Ga, Sn, Pb) react with both acids and bases (Table 26.7, reactions c and d).

Bismuth (III) oxide is a weakly basic oxide that is insoluble in water. Its basicity is shown by its reaction with acids to give salts plus water.

The hydroxides and carbonates of the *p*-block metals all undergo the expected thermal decomposition reactions to give water and carbon dioxide, respectively.

d. Ions in Aqueous Solution The Al^{3+}, Ga^{3+}, Sn^{2+}, and Pb^{2+} ions all react with water to give acidic solutions. These ions also form hydroxide precipitates, which then dissolve in excess OH^-. (See Table 26.7, reactions e and f.)

The simple Bi^{3+} ion is not found in aqueous solution, for it immediately reacts with water to give the bismuthyl ion, BiO^+.

$$Bi^{3+} + H_2O(l) \longrightarrow BiO^+ + 2H^+$$

$$BiO^+ + Cl^- \longrightarrow \underset{\text{bismuthyl chloride}}{BiOCl(s)}$$

Tin(IV), lead(IV), and bismuth(V) are not found as the Sn^{4+}, Pb^{4+}, and Bi^{5+} ions in aqueous solution; their water-soluble compounds immediately undergo hydrolysis or reduction.

e. Aluminothermic Reactions At high temperatures, aluminum burns in air in a strongly exothermic reaction to form the oxide.

$$4Al(s) + 3O_2(g) \longrightarrow 2Al_2O_3(s) \qquad \Delta H^\circ = -3350 \text{ kJ}$$

The large heat of this reaction makes finely divided aluminum an excellent reducing agent for metal oxides that can be reduced only at high temperatures (e.g., SrO). The reactions of aluminum with metal oxides are called **aluminothermic reactions.** So much heat is liberated in such reactions that the reduced metal is often obtained in the molten state. The Thermit process utilizes the reaction of iron(III) oxide with aluminum to provide both the high temperature and the molten iron needed for welding.

$$Fe_2O_3(s) + 2Al(s) \longrightarrow 2Fe(l) + Al_2O_3(s) \qquad \Delta H^\circ = -851.4 \text{ kJ}$$

EXAMPLE 26.5 Chemical Reactions: Elemental Sn

Write equations for the reactions of metallic tin with dilute hydrochloric acid, hot concentrated sulfuric acid, and concentrated sodium hydroxide.

Tin is a sufficiently active metal to displace hydrogen from dilute hydrochloric acid (Table 17.3). We would expect the formation of tin(II) rather than tin(IV) with this nonoxidizing acid.

$$Sn(s) + 2HCl(aq) \longrightarrow SnCl_2(aq) + H_2(g)$$

With a strong oxidizing acid tin would be oxidized to tin(IV). The usual products of the oxidation of a metal by sulfuric acid are the metal sulfate and sulfur dioxide.

$$Sn(s) + 4H_2SO_4(conc) \xrightarrow{\Delta} Sn(SO_4)_2(aq) + 2SO_2(g) + 4H_2O(l)$$

Tin is one of the metals that forms amphoteric oxides and reacts with hydroxide bases to form hydroxo ions and hydrogen. With no strong oxidizing agent present, tin(II) will be formed. Specific knowledge would be required to predict how many hydroxide ions would be present in the hydroxo ion that is formed.

$$Sn(s) + (n-2)OH^- + 2H_2O(l) \longrightarrow [Sn(OH)_n]^{(n-2)-} + H_2(g)$$

26.17 COMPOUNDS OF THE p-BLOCK METALS

a. Compounds of Aluminum Salts of aluminum with strong acids are obtained as hydrates from aqueous solution under ordinary conditions (e.g., $AlCl_3 \cdot 6H_2O$, $Al_2(SO_4)_3 \cdot 18H_2O$) a reflection of the strong tendency of the aluminum ion to combine with water. These salts contain the $[Al(H_2O)_6]^{3+}$ ion. Any additional water molecules are associated with the anions or held in crystal lattices.

Anhydrous aluminum chloride ($AlCl_3$) is a white, deliquescent substance. It fumes in moist air as it hydrolyzes to HCl. The compound is prepared by reaction of molten aluminum with chlorine. The aluminum chloride sublimes out of the hot reaction mixture. Many organic reactions are catalyzed by anhydrous aluminum chloride.

In benzene solution and in the vapor state, the anhydrous chloride, bromide, and iodide form Al_2X_6 molecules in which each aluminum atom is surrounded tetrahedrally by four halogen atoms (Figure 26.9) thereby completing its octet.

Aluminum oxide (Al_2O_3), known as alumina, occurs in nature in the hydrated form as *bauxite* and in the anhydrous form as *corundum*. Corundum is used as an abrasive and a *refractory* (a material that is unchanged by high temperatures); it is hard and comparatively inert toward chemical attack. Some deposits of corundum include crystals that are colored by the presence of small amounts of oxides of other metals and are of value as gemstones. In the ruby the impurity is Cr_2O_3; in the sapphire, FeO and TiO_2; in the oriental amethyst, Mn_2O_3; and in the oriental topaz, Fe_2O_3. Synthetic alumina gemstones practically identical with the natural ones are manufactured by melting aluminum oxide and the appropriate metal oxides in an oxyhydrogen flame and allowing the melts to crystallize.

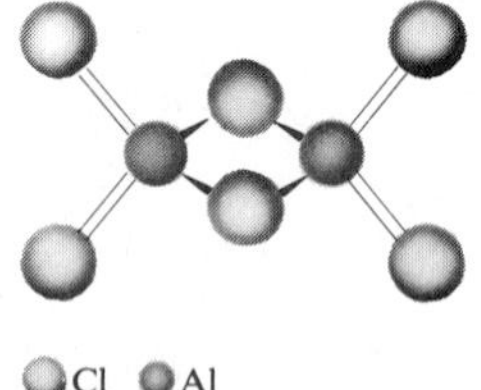

Figure 26.9
Aluminum Chloride (Al_2Cl_6) in the Vapor State

Treatment of an aqueous solution of an aluminum salt with a weak base, such as aqueous ammonia, yield a white gelatinous precipitate of variable composition usually formulated as $Al(OH)_3$ or $Al_2O_3 \cdot xH_2O$ and referred to as aluminum hydroxide. The hydroxide is converted to the anhydrous oxide at about 800 °C.

Corundum and the anhydrous aluminum oxide obtained by dehydrating the hydroxide have the same formula, Al_2O_3, but they have different crystal structures and different properties. Corundum—the form of the oxide called **α-alumina (α-Al_2O_3)**—is quite inert. The other form of the oxide, **γ-alumina,** is an amphoteric substance that will dissolve in either acids or strong hydroxide bases. Heating γ-alumina above 1000 °C converts it to α-alumina. Finely divided alumina has a very high surface area, a property that makes it useful as a support for catalysts and as a dehydrating agent. For some uses of the aluminas, see Table 26.13.

Aluminum sulfate ($Al(SO_4)_3$), which can be obtained directly from bauxite by the action of sulfuric acid, is the aluminum salt of greatest commercial use, primarily in the pulp and paper industry, where it is used in the sizing process (in which rosin renders paper resistant to water), to adjust the pH of the stock, and to treat waste effluent. The second most important use of aluminum sulfate is in the treatment of municipal water supplies and sewage. An essential step in water treatment is the precipitation of aluminum hydroxide by an alkaline substance, such as lime, naturally present in the water or added with the aluminum sulfate, for example

$$Al_2(SO_4)_3(aq) + 3Ca(OH)_2(aq) \longrightarrow 2Al(OH)_3(s) + 3CaSO_4(aq)$$

The aluminum hydroxide is gelatinous and adsorbs and entangles suspended and colloidal impurities, including bacteria. The "floc" of hydroxide plus impurities is allowed to settle and the clarified water is then filtered. Because the Al^{3+} ion is an astringent (it causes body tissues to contract, thereby cutting down on secretions such as blood or perspiration), the aluminum sulfate hydrate is used in styptic pencils and aluminum chloride hexahydrate is used in antiperspirants.

When solutions containing aluminum sulfate and potassium sulfate in equimolar concentrations are permitted to evaporate, a double salt known as alum, $KAl(SO_4)_2 \cdot 12H_2O$, crystallizes. This substance is one of a general class of compounds, the alums, of the formula $M^+M^{3+}(SO_4)_2 \cdot 12H_2O$, where M^+ can be any one of a large number of singly charged cations and M^{3+} one of several triply charged ions. Alums of Na^+, K^+, and NH_4^+ and of Al^{3+}, Cr^3, and Fe^{3+} are the most common.

b. Compounds of Tin, Lead, and Bismuth The known halides of tin, lead, and bismuth clearly illustrate the influence of electronegativity and polarizing ability on bonding and properties. All of the possible tin(II) and tin(IV) halides are known. A few examples are given in Table 26.17. The tin(IV) halides, except for SnF_4, are volatile compounds—$SnCl_4$ is a fuming liquid (b.p. 114 °C), and $SnBr_4$ and SnI_4 boil at 202 °C and 340 °C, respectively. These compounds are more molecular than ionic, as expected from the large charge-to-size ratio that an Sn^{4+} ion would have. They undergo complete reaction with water, rather than dissolution. The high electronegativity of fluorine contributes to the lesser volatility of SnF_4 (it sublimes at 705 °C).

The tin(II) halides are all significantly less volatile than the comparable tin(IV) compounds; these are considered to be ionic compounds. Tin(II) chloride, a common laboratory chemical, is a weak reducing agent. Its aqueous solutions are slowly oxidized by air.

All the lead(II) halides are known compounds. They are much less volatile than any of the tin halides, and molten $PbCl_2$ conducts electricity—a demonstration of its ionic nature. The dihalides are not soluble in water (except that $PbCl_2$ dissolves in hot water), and they can be precipitated by combining solutions containing lead(II) and halide ions. Of the possible lead(IV) halides, only PbF_4 and $PbCl_4$ are known. When heated above 100 °C, $PbCl_4$ decomposes to give $PbCl_2$ and Cl_2. This behavior and the nonexistence of lead(IV) bromide and iodide are consistent with the strength of lead(IV) as an oxidizing agent and the increasing ability down their family for the halide ions to act as reducing agents. For similar reasons, BiF_5 is the only known bismuth(V) halide. It is a potent fluorinating agent, readily supplying fluorine and being converted to the trifluoride. All four bismuth(III) halides are known, but only BiF_3 is a true salt.

Lead(II) oxide (PbO), known as *litharge*, is used in the glaze for decorative ceramic objects (but not for objects that come into contact with food). The molten oxide is an

Table 26.17
Some Compounds of Tin, Lead, and Bismuth, and Their Uses

Compound	Name and use
$SnCl_2$	Tin(II) chloride Tinplating
SnF_2	Tin(II) fluoride Toothpaste
$SnCl_4$	Tin(IV) chloride A fuming liquid
$SnCl_4 \cdot 5H_2O$	Tin(IV) chloride pentahydrate Perfume and soap
SnO_2	Tin(IV) oxide White enamel, white glass
PbO	Lead(II) oxide "Litharge," pottery glaze
PbO_2	Lead(IV) oxide A nonstoichiometric compound; lead storage battery
Pb_3O_4	Red lead Anticorrosion paint for structural steel
$PbCrO_4$	Lead chromate Chrome yellow pigment
BiF_3	Bismuth(III) fluoride
BiF_5	Bismuth(V) fluoride Fluorinating agent
Bi_2O_3	Bismuth(III) oxide Yellow enamel
BiOCl	Bismuthyl chloride Cosmetics

excellent solvent medium for growing garnet crystals useful in electronic devices and as gems. A paste of litharge and glycerine hardens on standing and yields a cement that is stable toward water and is often used to seal drain pipes to sinks.

Lead(IV) oxide (PbO_2) does not exist in nature. It may be obtained as a dark brown material by the oxidation of lead(II) in alkaline solution with very strong oxidizing agents such as hypochlorite ion. The compound isolated never has the stoichiometric formula PbO_2, the atomic ratio of oxygen to lead ordinarily being about 1.89. This is an example of a compound with a lattice defect. In this case, the crystal has some vacancies where there should be oxygen atoms. Because of this defect, lead(IV) oxide conducts electricity as the oxide ions move from hole to hole. This property allows PbO_2 to function as an electrode in the lead storage battery (Section 24.16).

For many years, white lead, a basic carbonate of the approximate composition $Pb_3(OH)_2(CO_3)_2$, was the most important white paint pigment. However, small children who tend to eat flaking paint can develop lead poisoning from it. Moreover, the compound darkens on exposure to air as a result of the formation of black lead sulfide. It has therefore been replaced by less toxic and more stable compounds such as titanium dioxide. Lead compounds still in use as pigments include the yellow chromate $PbCrO_4$, the red compound $PbO \cdot PbCrO_4$, and the red oxide Pb_3O_4, known as red lead, a component of paints used to protect structural steel from corrosion.

SUMMARY OF SECTIONS 26.14–26.17

The seven *p*-block metals vary more in properties than the *s*-block elements. The major properties of *p*-block metals are given in Table 26.18.

Table 26.18
Outstanding Properties of the *p*-Block Metals and Their Compounds

Al, Ga, In, Tl (Group III) Sn, Pb (Group IV) Bi (Group V)	All but Tl stable in air
Al, Ga, Sn, Pb	Amphoteric metals (Table 27.6)
$\overset{+3}{Al}\ \overset{+3}{Ga}\ \overset{+3}{In}$	One common oxidation state
$\overset{+1+3}{Tl}\ \overset{+2+4}{Sn}\ \overset{+2+4}{Pb}\ \overset{+3+5}{Bi}$	Two common oxidation states; lower state more stable for all but Sn; lesser ionic character in higher oxidation states
Tl^{3+}, $\overset{+4}{PbO_2}$, $\overset{+5}{BiO_3^-}$	Strong oxidizing agents
Al	Most reactive *p*-block metal; light and strong; myriad uses
Al + metal oxide $\longrightarrow$ Al_2O_3 + metal	Very exothermic reactions
p-block metals + O_2 $\not\longrightarrow$	All stable in dry air
Bi	Least reactive, least metallic *p*-block element
Pb compounds	Cumulative poison
Halides	Lesser ionic character in higher oxidation states; only known Pb(IV) halides are fluoride and chloride; only known Bi(V) halide is fluoride
Al_2O_3	Bauxite mineral (hydrated); corundum, an abrasive (anhydrous); α and γ aluminas; finely divided; many uses

Aside 26.1 METALS AS POISONS

Evaluating the danger from metals in the environment is difficult for several reasons. The concentrations involved may be small and not easily measured. Wastewater treatment and solid waste treatment usually do not remove metal ions, meaning that constant return of treated water and solids to the environment can cause a gradual increase in metal concentration in the environment. Also, there are wide variations in amount of exposure and the reaction to exposure of different individuals. The only testing possible is with animals, and much is yet to be learned about how much a human reaction will be like that of, for example, a rat. In addition, as experiences with mercury have shown, the natural cycles of metals must be better understood before the fate of metals in the environment can be predicted.

Lead and mercury are two metals under suspicion as environmental poisons. Other metals that may be of concern are beryllium, cadmium, and nickel, particularly as nickel carbonyl ($Ni(CO)_4$).

a. Lead Some historians believe that lead poisoning contributed to the fall of the Roman Empire. The Romans stored their wine in pottery vessels glazed with lead compounds, creating acidic conditions that were sure to leach lead into the wine. They also received their water from lead-lined aqueducts. Upper-class Romans, who could afford glazed pottery vessels, apparently suffered from high rates of stillbirth and brain damage, which may have contributed to their downfall. The use of lead glazes is now banned in the United States, but care must be taken with old pottery or pottery made in other countries where such glazes are still in use.

Since soluble lead salts are cumulative poisons, the indiscriminate use of the metal and its compounds represents a serious health hazard. A daily intake of more than 1 mg of the element for a prolonged period apparently can be dangerous. Early stages of lead poisoning are characterized by constipation, anemia, loss of appetite, and pain in the joints. Unfortunately, these symptoms may not immediately be associated with lead poisoning. Later stages of the disease include paralysis of the extremities and mental damage.

Concentrations of environmental lead rose dramatically beginning in the 1940s (see Figure 26.10), mainly because of the use of tetraethyllead in gasoline. Studies have shown that with the decrease in the use of "leaded" gasoline, there has been a decrease in blood lead levels in children.

b. Mercury Pure metallic mercury is not as toxic as mercury vapor and soluble mercury compounds, which are highly poisonous. However, mercury metal as well as its compounds must be handled with care.

Mercury poisoning is most often a local problem where mercury concentrations are high. The mercury reaches human beings mainly in food. It was thought for years that the discharge of metallic mercury into, say, a lake, was harmless because the mercury would sink and remain at the bottom as part of the sediment. Now it is known that bacteria can convert metallic mercury into methylmercury ion (CH_3Hg^+), a form in which it is soluble and highly poisonous. Mercury is concentrated up the food chain as, for example, bigger fish eat smaller fish that have eaten still smaller fish that contained mercury.

The symptoms of mercury poisoning (which, like lead poisoning, is hard to diagnose in its early stages), include loss of muscle control and blurred vision, leading ultimately to paralysis and kidney failure. Erratic behavior and mental deterioration are also likely to occur. (You probably remember the "Mad Hatter" from *Alice in Wonderland*. This character is based on experience, not fantasy. In the 19th century, mercury compounds were used to process felt in the manufacture of hats. As a result, it was not unusual for hatters to develop mercury poisoning, and the phrase "mad as a hatter" was in common use.) Because the body has a natural mechanism for eliminating mercury, mercury is not a great threat as a low-level cumulative poison.

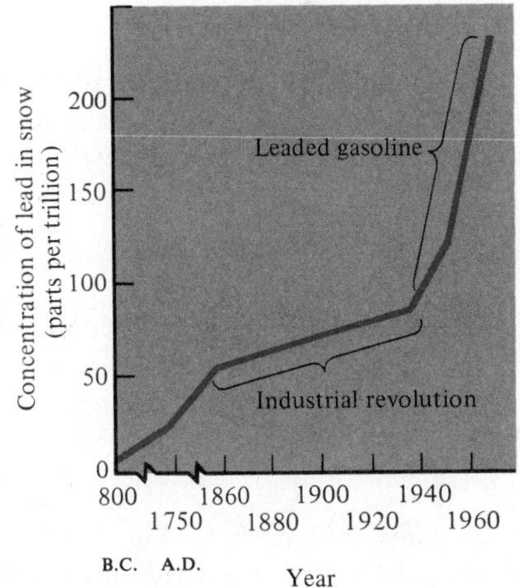

Figure 26.10
Environmental Lead
Lead levels in isolated Greenland glaciers have been growing since 800 B.C. The actual data are a series of scattered points; the curve represents the best average line through these points. In addition, there is considerable seasonal variation in lead levels. The trend is significant, however.

Source: G. Tyler Miller, Jr, *Living in the Environment* (Belmont, CA: Wadsworth, 1975), p. 97.

SIGNIFICANT TERMS (with section references)

α-alumina (26.17)
alloy (26.8)
aluminothermic reactions (26.16)
calcining (26.4)
caustic soda (26.13)
electrometallurgy (26.4)
flux (26.4)
γ-alumina (26.17)
hydrated lime, slaked lime (26.13)
hydrometallurgy (26.4)
intermetallic compounds (26.8)
interstitial solid solutions (26.8)
leaching (26.4)
lime (26.13)
metallurgy (26.4)
mineral (26.2)
ore (26.4)
pyrometallurgy (26.4)
quicklime, unslaked lime (26.13)
refining (26.4)
roasting (26.4)
sintering (26.4)
slag (26.4)
smelting (26.4)
soda ash (26.13)
solid solution (26.8)
steel (26.7)

QUESTIONS AND PROBLEMS

Metals

26.1 List the chemical and physical properties that we usually associate with metals.

26.2 Explain briefly why some metals have lustrous surfaces but others have dull surfaces.

26.3 Define "mineral." Does every mineral have a definite, fixed composition? Explain.

26.4 In what types of minerals do we generally find (a) unreactive metals, (b) the slightly reactive metals, (c) the more reactive metals, (d) the highly reactive metals, (e) the most reactive metals?

26.5 Chromite, (Mg, Fe)Cr_2O_4, is useful for making refractories (substances resistant to high temperatures) and is the only important chromium ore. What is the mass fraction of Cr in pure "ferrochromite" ($FeCr_2O_4$) and in pure "magnesiochromite" ($MgCr_2O_4$)? If a sample of chromite contains about equal numbers of Mg and Fe ions, what is the average Cr content, expressed in mass percent?

26.6 Seawater contains 0.13 mass % Mg^{2+}. What mass of seawater would have to be processed to yield 1.0 ton of the metal if the recovery process is 75% efficient?

The Preparation of Metals

26.7 Define the term "metallurgy." What does the study of metallurgy include?

26.8 How does an ore differ from a mineral? Name the three general categories of procedures needed to produce pure metals from ores.

26.9 Briefly describe one method by which gangue can be separated from the desired mineral during the concentration of an ore.

26.10 Name three general ways in which a metal can be extracted from an ore. What preliminary treatment steps might be used before a concentrated ore undergoes extraction? What does each of these steps accomplish?

26.11 How does pyrometallurgy differ from the other ways of reducing minerals? What substances are commonly used as the reducing agents in smelting? What happens to the remaining gangue during smelting?

26.12 Give details for three different refining processes.

26.13 The following equations represent reactions used in some important metallurgical processes:

(a) $Fe_3O_4(s) + CO(g) \longrightarrow Fe(l) + CO_2(g)$
(b) $MgCO_3(s) + SiO_2(s) \longrightarrow MgSiO_3(l) + CO_2(g)$
(c) $Au(s) + CN^- + H_2O(l) + O_2(g) \longrightarrow [Au(CN)_2]^- + OH^-$

Balance each unbalanced equation and classify each as (i) roasting, (ii) calcining, (iii) leaching, (iv) adding a flux, or (v) reduction.

26.14 Repeat Question 26.13 for

(a) Al_2O_3 (cryolite solution) $\xrightarrow{\text{electrolysis}} Al(l) + O_2(g)$
(b) $PbSO_4(s) + PbS(s) \longrightarrow Pb(l) + SO_2(g)$
(c) $TaCl_5(g) + Mg(l) \longrightarrow Ta(s) + MgCl_2(l)$

Electrometallurgy: Sodium and Aluminum

26.15 Write the equation that describes the electrolysis of a brine solution to form NaOH, Cl_2, and H_2. What mass of each substance will be produced in an electrolysis cell for each mole of electrons passed through the cell? Assume 100% efficiency.

26.16 In the preparation of sodium by the Downs process, calcium chloride is added to the sodium chloride to lower the melting point. A student raised the question as to why calcium metal does not form instead of sodium, since calcium is below sodium in the electromotive series. What did he overlook?

26.17 Briefly describe the Hall process for the commercial preparation of aluminum.

26.18 Write a balanced ion-electron equation for each electrode reaction in the Hall process. What amount of time is needed to produce 1.00 kg of molten aluminum using a current of 1015 A? (Assume the process to be 91% efficient.)

26.19 What is bauxite? What are the usual impurities that are found in this ore? Briefly describe how the ore is purified.

26.20 Even though clays and other minerals are more abundant sources of aluminum than bauxite, they are not of commercial importance. Why?

Metals from Sulfide Ores: Copper and Zinc

26.21 Name the undesirable gaseous product formed during the smelting of copper and other sulfide minerals. Why is it undesirable? How can the pollutant effect be overcome?

26.22 What are the common impurities in copper metal produced by smelting? How is each of these eliminated?

26.23 What would happen during the electrolytic purification of copper if the voltage across the cell were (a) too high or (b) too low?

26.24 The following reactions take place during the extraction of copper from copper ore:

(a) $2Cu_2S(l) + 3O_2(g) \longrightarrow 2Cu_2O(l) + 2SO_2(g)$
(b) $2Cu_2O(l) + Cu_2S(l) \longrightarrow 6Cu(l) + SO_2(g)$

Identify the oxidizing and reducing agents. Show that each equation is correctly balanced by demonstrating that the increase and decrease in oxidation numbers are equal.

26.25 Assuming complete recovery of metal, identify which ore would yield the larger quantity of copper on a mass basis: (a) an ore containing 3.80 mass % azurite, $Cu(OH)_2 \cdot 2CuCO_3$, or (b) an ore containing 4.85 mass % chalcopyrite, $CuFeS_2$.

26.26 What mass of copper could be electroplated from a solution of $CuSO_4$ using an electrical current of 3.00 A flowing for 5.00 h? (Assume 100% efficiency.)

26.27 Write chemical equations describing the roasting of sphalerite and the subsequent high-temperature reduction by carbon to give zinc metal.

26.28 Briefly describe the hydrometallurgical method used to produce zinc metal from zinc sulfide.

Iron and Steel

26.29 Name some of the common minerals that contain iron. Write the chemical formula for the iron compound in each. What is the oxidation number of iron in each substance?

26.30 The principal iron ore in Minnesota is hematite, Fe_2O_3. This ore is calcined to magnetite, Fe_3O_4, before being shipped to the mills in Illinois, Indiana, Michigan, Ohio, and Pennsylvania. Write the chemical equation for the calcination step. What mass of magnetite can be obtained from a metric ton of hematite?

26.31 During the operation of a blast furnace, coke reacts with the oxygen in air to produce carbon monoxide, which, in turn, serves as the reducing agent for the iron ore. Assuming the formula of the iron ore to be Fe_2O_3, calculate the mass of air needed for each ton of iron produced. Assume air to be 21% O_2 by mass.

26.32 The reaction

$$FeO(s) + CO(g) \longrightarrow Fe(s) + CO_2(g)$$

takes place in the blast furnace at a temperature of 800 K. (a) Calculate ΔH°_{800} for this reaction using $\Delta H^\circ_{f,800} = -268$ kJ/mol for FeO, -111 kJ/mol for CO, and -394 kJ/mol for CO_2. Is this a favorable enthalpy change? (b) Calculate ΔG°_{800} for this reaction using $\Delta G^\circ_{f,800} = -219$ kJ/mol for FeO, -182 kJ/mol for CO, and -396 kJ/mol for CO_2. Is this a favorable free energy change? Using your values of ΔH°_{800} and ΔG°_{800}, (c) calculate ΔS°_{800}.

26.33 What are some of the usual impurities found in pig iron? How is each of these removed from the iron? Write the chemical equations for reactions of "fluxes" with MnO, SiO_2, and P_4O_{10}.

26.34 What are the properties of steels that contain cementite? How can "tempering" give steel desirable properties? Briefly discuss the equilibrium involving Fe, C, and cementite.

Alloys

26.35 Define the term "alloy." Give three classifications of alloys. Compare the general properties of alloys to those of the individual metals.

26.36 What criterion can be used to determine the empirical formula of possible intermetallic compounds? Show that Au_3Sn can be classified as an ϵ-alloy.

26.37 Silver and cadmium form three intermetallic compounds. Analyses show these compounds to contain 24.2% Ag, 49.0% Ag, and 37.5% Ag by mass, respectively. Determine the empirical formulas of these compounds and label each as a β-, γ-, or ϵ-alloy.

26.38 A sample containing 40 g of Ca and 60 g of Al is thoroughly mixed and heated to 1200 °C in the absence of air, resulting in the formation of a liquid solution. The solution is cooled to about 1050 °C, and crystals that contain 43 mass % Ca and 57 mass % Al start to form. As the mixture is cooled further, a second solid phase appears at 72.5 °C. It contains 32 mass % Ca and 68 mass % Al. Each of these solid phases corresponds to an intermetallic compound formed between aluminum and calcium. Find the empirical formulas of these substances.

The s-Block Metals

26.39 Are the elements in Groups I and II found in the free state in nature? What are the primary sources for these elements? How are the metals obtained?

26.40 Briefly contrast the physical properties of the elements in Groups I and II.

26.41 Write the general outer electron configurations for atoms of the *s*-block metals. What oxidation state(s) would you predict for these elements? What types of bonding would you expect in most of the compounds of these elements? Why?

26.42 Where do the metals of Groups I and II fall in the activity series with respect to H_2? What does this tell us about their reactivities with water and acids?

26.43 Write chemical equations describing the reactions of O_2 with each of the alkali and alkaline earth metals. Account for differences within each family.

26.44 What type of solution is formed by dissolving the oxides of Groups I and II in water? How does BeO differ from the others in this behavior? Account for differences in behavior.

26.45 Write balanced equations and give the experimental conditions necessary for preparing solid (a) NaOH, (b) Na_2CO_3, (c) $NaHCO_3$, (d) CaO, (e) $Ca(OH)_2$, each from a readily available substance.

26.46 Write balanced chemical equations for the following:

(a) $NaOH(s) \xrightarrow{\Delta}$

(b) $BeO(s) + NaOH(s) + H_2O(l) \longrightarrow$

(c) $NaCl(aq, conc) + H_2O(l) \xrightarrow{\text{electrolysis}}$

(d) $NH_4Cl(aq) + Ca(OH)_2(s) \longrightarrow$

26.47 Contrast the behavior upon heating of the hydroxides and carbonates of Group I elements to that of the hydroxides and carbonates of Group II elements.

26.48 Write the chemical equation for slaking quicklime. The production of calcium oxide from calcium hydroxide is highly endothermic. What can be said about the heat of reaction for the slaking process?

26.49 The standard heat of formation is -426.73 kJ/mol for NaOH(*s*), -469.23 kJ/mol for NaOH(*aq*, 6 M), and -469.10 kJ/mol for NaOH(*aq*, 0.1 M). (a) Calculate the heat of solution to form a 6 M NaOH solution from solid NaOH and water. Comment on your answer. (b) Calculate the heat of dilution to prepare 0.1 M NaOH from 6 M NaOH.

26.50 The percentage of limestone in a sample of unknown composition can be determined by dissolving the sample in strong acid, precipitating the Ca^{2+} as calcium oxalate (CaC_2O_4), and titrating the $C_2O_4^{2-}$ by using MnO_4^- in an acidic solution. Write the chemical equations describing this process.

The p-Block Metals

26.51 Repeat Question 26.41 for the *p*-block metals. Explain why each *p*-block metal may have more than one positive oxidation state.

26.52 Although aluminum is a reactive metal, it is not noticeably corroded in air. Why? Why is this property important in determining various uses for the metal?

26.53 Predict whether aqueous solutions of aluminum nitrate are neutral, acidic, or alkaline and give a reason for your answer.

26.54 Elemental tin reacts with concentrated nitric acid to form SnO_2 and lead reacts to form $Pb(NO_3)_2$. Explain this difference and write the chemical equations for these reactions.

26.55 Aluminum, like beryllium, is an amphoteric metal. Write equations for (a) reactions of Al with acid and strong base and (b) reactions of Al_2O_3 with acid and base.

26.56 As a student added a solution of NaOH dropwise to a test tube containing a solution of Al^{3+}, he noticed that a gelatinous precipitate was formed that disappeared upon the addition of more of the alkaline solution. Explain his observations and write ionic chemical equations for the reactions involved.

26.57 Explain why in most simple compounds between aluminum and nonmetals the bonding is predominantly covalent and yet the ion is readily formed in aqueous solution.

26.58 Choose which of the following halides are essentially molecular and prepare three-dimensional sketches of their molecular structures: (a) $PbCl_2$, (b) $BiCl_3$, (c) SnF_4, (d) BiF_3, (e) $SnCl_2$, (f) SnI_4.

26.59 Write the chemical equation for a thermit-type reaction of PbO_2 with Al. $\Delta H_f^\circ = -277.4$ kJ/mol for PbO_2 and -1675.7 kJ/mol for Al_2O_3. Calculate ΔH° for this reaction. Is the energy change favorable?

26.60 What mass of "white lead," $Pb(OH)_2 \cdot 2PbCO_3$, can be made from 10.0 g of Pb assuming 100% conversion?

26.61 Briefly mention uses of (a) Al(*s*), (b) $AlCl_3$, (c) Al_2O_3, (d) $Al_2(SO_4)_3$. For each, indicate a property that determines the use.

26.62 Write the formula for lead(IV) oxide. How is this compound prepared? Why can it conduct electricity? What is the major use of this compound?

26.63 A voltaic cell consists of a tin electrode dipping into a 1 M $Sn(NO_3)_2$ solution and a lead electrode dipping into a 1 M $Pb(NO_3)_2$ solution. The half-cells are connected by a $NaNO_3$ salt bridge. Which electrode is the anode? The standard state reduction potentials are -0.136 V for Sn^{2+} and -0.126 V for Pb^{2+}. What voltage will the cell generate? How can the cell voltage be increased?

26.64 Lead is known to crystallize in the cubic system with a unit cell length of 0.495 nm. The density of lead is 11.29 g/cm^3. Is the unit cell primitive, body-centered, or face-centered?

ADDITIONAL QUESTIONS

26.65 Which of the following are redox reactions? Identify the oxidizing and reducing agents in each of the redox reactions.

(a) $Sn(s) + O_2(g) \longrightarrow SnO_2(s)$
(b) $MgCO_3(s) \longrightarrow MgO(s) + CO_2(g)$
(c) $2K(s) + 2H_2O(l) \longrightarrow 2KOH(aq) + H_2(g)$
(d) $NaOH(aq) + CO_2(g) \longrightarrow NaHCO_3(aq)$

26.66 Repeat Question 26.65 for

(a) $Al_2O_3(s) + 2NaOH(aq) + 3H_2O(l) \longrightarrow 2NaAl(OH)_4(aq)$
(b) $2Al(s) + 3MnO(s) \xrightarrow{\Delta} 3Mn(l) + Al_2O_3(s)$
(c) $Mg(OH)_2(s) + 2HCl(aq) \longrightarrow MgCl_2(aq) + 2H_2O(l)$
(d) $2NaCl(l) \xrightarrow{\text{electrolysis}} 2Na(l) + Cl_2(g)$

26.67 Classify each of the following reactions according to the reaction types listed in Tables 17.2 and 17.7:

(a) $2NaHCO_3(s) \longrightarrow Na_2CO_3(s) + CO_2(g) + H_2O(g)$
(b) $CaO(s) + H_2O(l) \longrightarrow Ca(OH)_2(s)$
(c) $Al_2O_3(s) + 2NaOH(aq) + 3H_2O(l) \longrightarrow 2NaAl(OH)_4(aq)$

26.68 Repeat Question 26.67 for

(a) $Cl_2(g) + H_2O(l) \longrightarrow HOCl(aq) + H^+ + Cl^-$
(b) $Na(g) + KCl(l) \longrightarrow NaCl(l) + K(g)$
(c) $NH_4HCO_3(aq) + NaCl(aq) \longrightarrow$
$NaHCO_3(s) + NH_4Cl(aq)$

26.69 Predict the major products of the following reactions:

(a) $NaOH(aq) + H_2O(l) + CO_2(g, \text{excess}) \longrightarrow$
(b) $Cl_2(g) + NaOH(aq) \longrightarrow$
(c) $MgCO_3 \cdot CaCO_3(s) \xrightarrow{\Delta}$
(d) $Na_2O(s) + H_2O(l) \longrightarrow$

26.70 Repeat Question 26.69 for

(a) $PbO(s) + H_2O(l) + NaOH(aq) \longrightarrow$
(b) $BiCl_3(aq) + H_2O(l) \longrightarrow$
(c) $Al_2O_3(s) + H_2SO_4(aq) \longrightarrow$
(d) $Pb(s) + O_2(g) \xrightarrow{\Delta}$

26.71 An abbreviated activity series for some of the representative metals studied in this chapter is $Mg/Sn/Pb/H_2$ with the most "active" element on the left and the least "active" one on the right. Use this series to predict whether each of the following reactions will occur or not under the same conditions that prevailed when the series was determined:

(a) $Sn + HCl \longrightarrow$
(b) $Pb^{2+} + Mg \longrightarrow$
(c) $Sn^{2+} + Mg^{2+} \longrightarrow$

For those that you predicted to occur, complete and balance the equations.

CHAPTER

27

Nonmetals: Halogens and Noble Gases

27.1 PERIODIC RELATIONSHIPS: HALOGENS AND NOBLE GASES

The halogens (Representative Group VII) and the noble gases form the only two complete families of nonmetals and include all but six of the elements classified as nonmetals. Like all nonmetals other than hydrogen and helium (the first member of the noble gas family), the halogens (fluorine, chlorine, bromine, iodine, and astatine) and the other noble gases (neon, argon, krypton, xenon, and radon) fall in the *p* block of the periodic table. By contrast with metals, which commonly have only positive oxidation states, the halogens (except for fluorine) have both positive and negative oxidation states, as do many other nonmetals.

Due to the differences in the electron configurations of their atoms, the noble gases and the halogens are remarkably different in their properties. Noble gas atoms have the filled ns^2np^6 configuration (except $1s^2$ for He). Halogen atoms are one electron short of the filled configuration—all have the ns^2np^5 configuration (Table 27.1).

The noble gases exist as monatomic molecules. They have essentially zero electronegativity and neither gain nor lose electrons easily. No reactions are known for helium, neon, or argon. The heavier noble gases—krypton, xenon, and radon—apparently combine directly only with fluorine.

The halogens, by contrast, are very reactive elements. The word "halogen" comes from the Greek meaning "salt former." Halogen atoms readily gain electrons to form halide ions (X^-). They also form many compounds in which they achieve electron octets by sharing electrons. The free halogens exist as diatomic molecules in which two atoms share their single unpaired electrons

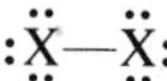

Table 27.1
Electron Configurations of Halogen and Noble Gas Atoms

Halogens	Noble gases
—	Helium $_2He$ $1s^2$
Fluorine $_9F$ $[He]2s^2 2p^5$	Neon $_{10}Ne$ $[He]2s^2 2p^6$
Chlorine $_{17}Cl$ $[Ne]3s^2 3p^5$	Argon $_{18}Ar$ $[Ne]3s^2 3p^6$
Bromine $_{35}Br$ $[Ar]3d^{10} 4s^2 4p^5$	Krypton $_{36}Kr$ $[Ar]3d^{10} 4s^2 4p^6$
Iodine $_{53}I$ $[Kr]4d^{10} 5s^2 5p^5$	Xenon $_{54}Xe$ $[Kr]4d^{10} 5s^2 5p^6$
Astatine $_{85}At$ $[Xe]4f^{14} 5d^{10} 6s^2 6p^5$	Radon $_{86}Rn$ $[Xe]4f^{14} 5d^{10} 6s^2 6p^6$

HALOGENS

27.2 PROPERTIES OF THE HALOGENS

At room temperature and pressure, fluorine and chlorine are gases, bromine is a liquid (the only liquid nonmetallic element), and iodine is a metallic-appearing, low-melting solid that sublimes readily. The intensity of their colors increases from a very pale yellow for fluorine to violet-black for iodine (Table 27.2).

Bromine and iodine have high vapor pressures, and even at room temperature dark red vapor is always present above liquid bromine and violet vapor above solid iodine. The halogens all have pungent and irritating odors, and attack the skin and flesh. Bromine is particularly nasty; it causes burns that are very slow to heal. Astatine is radioactive and its most stable isotope has a half-life of 8.3 h.

Fluorine, perhaps more so than any other element, illustrates the *differences* in properties expected of first family members compared to other elements in their families (Table 27.3). For fluorine these differences are due largely to the small size and high electronegativity of the fluorine atom, which combine to make fluorine the most reactive nonmetal.

The free halogen molecules are quite stable to dissociation. Breaking the X—X bond becomes easier down the family from Cl_2 to Br_2 to I_2, reflecting the increasing size of the atoms. Fluorine, however, has a surprisingly low bond dissociation energy (see Table 27.2)—apparently because the small fluorine atoms are so strongly

Table 27.2
Properties of the Halogens

	Fluorine	Chlorine	Bromine	Iodine
Color and state (at 1 atm, 25 °C)	Pale yellow gas	Yellow-green gas	Red-brown liquid	Violet-black solid
Melting point (°C)	−220	−101	−73	113
Boiling point (°C)	−188	−34	59	184
X—X bond dissociation energy (kJ/mol)	159	243	192	151
Atomic radius (nm)	0.071	0.099	0.114	0.133
Ionic radius, X^- (nm)	0.136	0.181	0.196	0.220
Ionization energy (0 K) (kJ/mol)	1681	1251	1140	1008
Electronegativity	4.0	3.0	2.8	2.5
Electron affinity (0 K) (kJ/mol)	−322	−349	−325	−295
Standard reduction potential, $E°$ (V) $X_2 + 2e^- \longrightarrow 2X^-$	2.87	1.36	1.07	0.536

Table 27.3
Differences between Fluorine and the Other Members of the Fluorine Family (Cl, Br, I)

Property	Comments
One of the strongest chemical oxidizing agents known	No other element has as great a tendency to add electrons
The only element that combines directly with some of the noble gases	Forms fluorides with Kr, Xe, and Rn
Low bond dissociation energy compared to Cl_2 and Br_2; value out of line with family trend	Large repulsive forces between small, highly electronegative atoms allow easy breaking of F—F bond
F^-, the only halide ion that cannot function as a reducing agent	F^- must be oxidized to F_2 electrolytically
No positive oxidation states	No oxohalo anions comparable to, e.g., ClO^- and ClO_4^- are known
HF strongly hydrogen bonded, causing high m.p., b.p., and heats of fusion and vaporization	Hydrogen bonding also partly accounts for hydrofluoric acid being a weak acid

electronegative and so close together that their nuclei and electrons can repel each other. The ease with which the fluorine–fluorine bond is broken contributes to the extraordinary reactivity of elemental fluorine. However, bonds between fluorine atoms and atoms of other elements are stronger than comparable bonds of other halogen atoms.

The trends in properties expected in periodic table groups are clearly apparent for the halogens. Atomic and ionic radii increase down the family. Removing an electron (more difficult than for any other elements except the noble gases) becomes easier, and adding an electron (more favorable than for any other elements) is less exothermic down the family, again except for fluorine.

The high electronegativities of the halogens reflect the strong ability of halogen atoms to attract electrons in compounds. Covalent bonds in halogen compounds are polar, often with partial negative charges on the halogen atoms. Fluorine is assigned an oxidation state of -1 in all compounds. The other halogens—in addition to the -1 oxidation state in binary ionic and molecular compounds—often have the positive oxidation states of $+1$, $+3$, $+5$, and $+7$ when covalently bonded to more electronegative elements, notably to oxygen. Some common compounds and ions of chlorine in various oxidation states are listed in Table 27.4.

The extremely reactive elemental halogens are never found in nature. Their salts are found in the Earth's crust and in solution in the oceans or in natural brines (Table 27.5).

Table 27.4
Oxidation States and Some Compounds and Ions of Chlorine

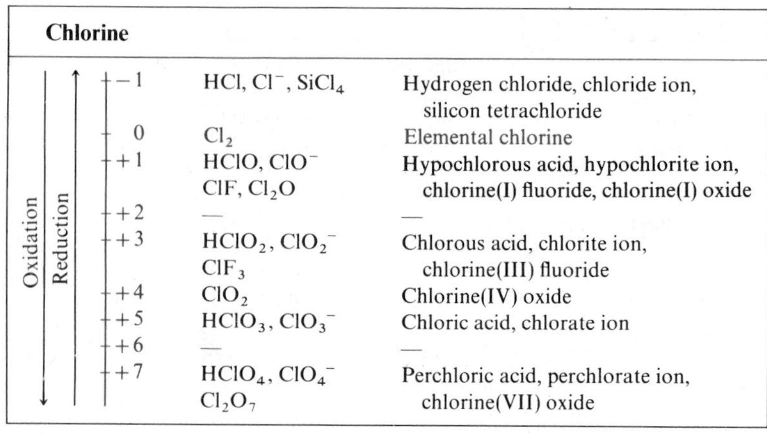

Chlorine		
-1	HCl, Cl^-, $SiCl_4$	Hydrogen chloride, chloride ion, silicon tetrachloride
0	Cl_2	Elemental chlorine
$+1$	HClO, ClO^- ClF, Cl_2O	Hypochlorous acid, hypochlorite ion, chlorine(I) fluoride, chlorine(I) oxide
$+2$	—	—
$+3$	$HClO_2$, ClO_2^- ClF_3	Chlorous acid, chlorite ion, chlorine(III) fluoride
$+4$	ClO_2	Chlorine(IV) oxide
$+5$	$HClO_3$, ClO_3^-	Chloric acid, chlorate ion
$+6$	—	—
$+7$	$HClO_4$, ClO_4^- Cl_2O_7	Perchloric acid, perchlorate ion, chlorine(VII) oxide

Table 27.5
Natural Occurrence of the Halogens

	Percent in Earth's crust (by mass)	Sources
Fluorine	$6–9 \times 10^{-2}$	Fluorspar, CaF_2 Cryolite, $Na_3(AlF_6)$ Fluorapatite, $Ca_5(PO_4)_3F$
Chlorine	3.14×10^{-2}	NaCl Salt beds, brine wells, seawater (2.8%) Carnallite, $KCl \cdot MgCl_2 \cdot 6H_2O$
Bromine	1.6×10^{-4}	NaBr, KBr, $MgBr_2$ Brine wells Seawater
Iodine	3×10^{-5}	$NaIO_3$, $NaIO_4$ $NaNO_3$ deposits in Chile Iodides Brine wells

EXAMPLE 27.1 Physical Properties: Halogens

What is responsible for the increasing melting and boiling points down the halogen family?

Increasing melting and boiling points indicate that greater intermolecular forces are present. The halogens form diatomic molecules that are symmetrical and do not have dipole moments. Therefore, **the increasing force of attraction between the molecules of the successive halogens must be due solely to the increased induced dipole forces (London forces)** that go with larger size, increased polarizability, and greater numbers of electrons.

27.3 CHEMICAL REACTIONS OF THE HALOGENS

a. Redox Reactions of X_2 and X^- As expected from their strong tendencies to react by adding electrons, all elemental halogens are oxidizing agents. Fluorine is one of the strongest known oxidizing agents, the strongest of all the elements. The oxidizing ability of the halogens decreases down the family from fluorine to iodine (see $E°$ values in Table 27.2). The relative strengths of the halogens as oxidizing agents make it possible for each to oxidize the anions of the halogens following it in the family (the basis for the activity series of the nonmetals discussed in Section 17.4). Fluorine will oxidize chloride ions

$$\begin{array}{rcll} F_2(g) + 2e^- & \longrightarrow & 2F^- & E° = 2.87 \text{ V} \\ 2Cl^- & \longrightarrow & Cl_2(g) + 2e^- & E° = -1.36 \text{ V} \\ \hline F_2(g) + 2Cl^- & \longrightarrow & Cl_2(g) + 2F^- & E° = 1.51 \text{ V} \end{array}$$

Similarly, chlorine will oxidize bromide ions, and bromine will oxidize iodide ions. (These are redox displacement reactions.)

$$\begin{array}{rcll} Cl_2(g) + 2Br^- & \longrightarrow & Br_2(aq) + 2Cl^- & E° = 0.29 \text{ V} \\ Br_2(l) + 2I^- & \longrightarrow & 2Br^- + I_2(s) & E° = 0.53 \text{ V} \end{array}$$

In the presence of excess halogen, further oxidation takes place; for example, bromide ion can be oxidized to bromate ion.

$$3Cl_2(g) + Br^- + 3H_2O(l) \longrightarrow BrO_3^- + 6HCl(aq)$$

Table 27.6
Combination of the Halogens with Other Elements
Compounds of fluorine and chlorine with oxygen and nitrogen, and of chlorine with carbon, do form by other methods.

	With metals	With semiconducting elements	With nonmetals
F_2	All	All	All except O_2, N_2, He, Ne, Ar
Cl_2	Most	All	All except O_2, N_2, C, and noble gases
Br_2	Most	Most	Only halogens, H_2, P, S
I_2	Most	Most	Only halogens, H_2, P

b. Combination with Other Elements In combination reactions with other elements, the halogens are reduced to the −1 oxidation state to give halides.

Fluorine combines directly, in most cases at ordinary temperatures, with all of the elements except oxygen, nitrogen, helium, neon, and argon (Table 27.6). In contact with fluorine, charcoal bursts into flame; hydrogen–fluorine mixtures are also self-igniting. At elevated temperatures, most metals burn in fluorine. However, some metals, including aluminum, copper, and nickel, form protective fluoride films on their surfaces, inhibiting further reaction. Fluorine is often shipped and handled in containers made of such metals.

Because of their small size, more fluorine atoms than atoms of other halogens can fit around a central atom in a molecule. Combined with the strength of fluorine as an oxidizing agent, this ability allows many elements to form compounds in their highest oxidation states in combination reactions with fluorine. For example, for arsenic from Group V and sulfur from Group VI

$$2As(s) + 5F_2(g) \xrightarrow[\text{temperature}]{\text{room}} 2AsF_5(g)$$

$$S(s) + 3F_2(g) \xrightarrow{\text{bluish flame}} SF_6(g)$$

Chlorine is also very reactive (see Table 27.6) and many metals and semiconducting elements burn on contact with chlorine (e.g., copper, iron, antimony, boron).

Bromine and iodine are less vigorous in their combination reactions, and tend to take other elements to lower positive oxidation states than do fluorine and in some cases chlorine.

c. Combination with Molecular Compounds A **halogenation** reaction is the introduction of a halogen atom or atoms into a molecular compound. The free halogens react with many compounds in this way. In each of the following examples the free halogen acts as an oxidizing agent.

$$\underset{\text{sulfur dioxide}}{\overset{+4}{S}O_2(g)} + \overset{0}{X_2} \longrightarrow \underset{\text{sulfuryl halide}}{\overset{+6}{S}O_2\overset{-1}{X_2}(g \text{ or } l)} \qquad X = F, Cl$$

$$\underset{\text{carbon monoxide}}{\overset{+2}{C}O(g)} + \overset{0}{X_2} \longrightarrow \underset{\text{carbonyl halide}}{\overset{+4}{C}O\overset{-1}{X_2}(g \text{ or } l)} \qquad X = Cl, Br$$

Fluorine is so reactive that it often behaves differently in such reactions and gives products other than those of halogenation.

Halogenation reactions are of great importance in the chemical industry in the production of halogen-substituted hydrocarbons. Halogens add across carbon–carbon double bonds

$$\underset{\text{ethylene}}{H_2C{=}CH_2(g)} + X_2(g) \longrightarrow \underset{\text{dihaloethane}}{CH_2X{-}CH_2X} \qquad X = Cl, Br \qquad \textbf{(27.1)}$$

or replace hydrogen atoms, for example

$$\underset{\text{methane}}{CH_4(g)} + Cl_2(g) \longrightarrow \underset{\text{methyl chloride}}{CH_3Cl(g)} + HCl(g) \qquad \textbf{(27.2)}$$

POLYMERS AND PLASTICS

■ **GLASS FIBER-REINFORCED POLYURETHANE** *above* A sample is being made in the laboratory by pouring molten polyurethane over glass fiber cloth. Fiber-reinforced plastics are a large and growing family of materials with many desirable properties such as strength, durability, heat resistance, and light weight. (Polyurethane is a polymer with $-O-C(=O)-NH-$ links between units.)

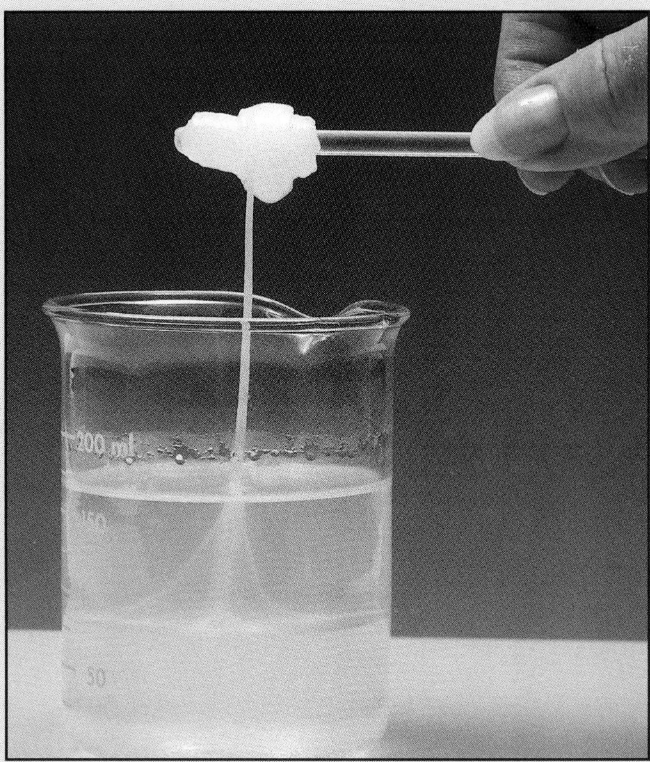

■ **NYLON** *above* Nylon, a thermoplastic polyamide, is being pulled from the interface between solutions of adipoyl chloride ($ClCOC(CH_2)_4COCl$) and hexamethylenediamine ($NH_2(CH_2)_6NH_2$), where it forms. In nylon, units are linked by $-C(=O)-NH-$ groups.

■ **VELCRO** *above* The two sides of a Velcro closing are shown pulled apart at one end. One side consists of thick nylon loops that are slit open. The other consists of thinner, closed nylon loops that slip through the slits when the two sides are pressed together.

■ **TEFLON-COATED WIRE** *above* In a process called melt-extrusion, a wire is being coated with Teflon (polytetrafluoroethylene, $(F_2CCF_2)_n$), which is an almost ideal insulator because it is chemically inert, does not deteriorate with age, and is a nonconductor over a wide temperature range. Such wires are used in, for example, telephone systems.

CHEMISTRY AT WORK

■ **SULFURIC ACID PLANT** *below* A stockpile of sulfur waits outside this large sulfuric acid plant in New Jersey. The sulfur is burned to produce sulfur dioxide (SO_2), which is then oxidized to sulfur trioxide (SO_3), from which sulfuric acid is produced.

■ **PHOSPHATE PLANT** *below* Phosphate rock is used directly in the manufacture of both phosphoric acid (H_3PO_4) and phosphate-containing fertilizers. This phosphate plant is in Idaho, one of the states with major deposits of phosphate rock (another is Florida).

■ **TNT EXPLOSION** *above* The rapid decomposition of certain nitrogen compounds in redox reactions that produce nitrogen and other gases makes these compounds valuable explosives. Explosion of TNT, which is trinitrotoluene ($(NO_2)_3C_6H_2CH_3$), completely converts the compound to nitrogen, hydrogen, carbon monoxide, and carbon.

■ **WATER PURIFICATION** *above* To remove organic matter that is decomposed by oxygen-using bacteria, sewage is exposed to a healthy bacteria population while air is vigorously blown through the water (*left*). Chlorine (Cl_2) plays essential roles in both sewage treatment, in which water is chlorinated before returning it to the environment, and in municipal treatment of water being supplied to consumers (*right*).

■ **BUTYL RUBBER** *below* The plant supervisor is standing in front of a unit that manufactures isobutene ($(CH_3)_2C{=}CH_2$); his arm is resting on bales of the butyl rubber produced from isobutene. Butyl rubber is used in tractor tires, inner tubes, hoses, and mechanical fittings.

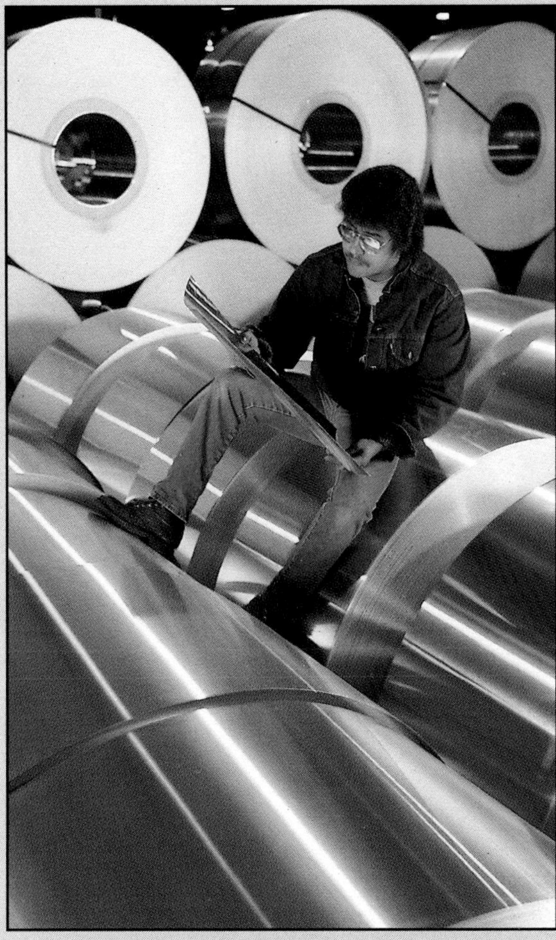

■ **ALUMINUM RECYCLING** *below and right* Aluminum is shown arriving at the recycling plant (*below, left*), entering a furnace (*below, right*), and as the finished recycled product (*right*). Recycling becomes increasingly important as natural resources are depleted, sites for disposing of waste become scarce, and natural beauty is destroyed by carelessly discarded waste materials.

CHEMISTRY IN THE ENVIRONMENT

■ **MINERAL DEPOSITS IN HOT WATER PIPE** *above* Hot water pipes and boilers can become completely clogged by boiler scale. Rock-like materials including calcium and magnesium carbonates are formed by reaction of the metal ions present in hard water with hydrogen carbonate ion (HCO_3^-).

■ **TARNISH ON SILVER** *above* Silver reacts with hydrogen sulfide in the atmosphere to form a thin coating of silver sulfide on its surface.

■ **RUST ON A SHIPWRECK** *above* On exposure to a moist atmosphere, iron corrodes in an electrochemical process that produces red-brown iron oxide. Because the oxide crumbles away from the surface, fresh iron is exposed and rust can eventually destroy an entire object.

■ **CARLSBAD CAVERNS** *above* Stalactites and stalagmites grow when ground water carrying minerals dissolved as hydrogen carbonates seeps into a cave. As water drips down and carbon dioxide escapes from the solution (due to lower CO_2 partial pressure in the cave), the minerals are deposited in the icicle-like forms.

■ **EFFECTS OF ACID RAIN** *below and right* Industrial activity that produces sulfur dioxide gas is a principal cause of acid rain. The sulfur dioxide is converted to sulfuric acid by reactions in the atmosphere, and the acid is carried by rain to the surfaces of plants, trees, statuary, and buildings, where it causes serious damage. The photos show acid-rain effects in a forest in North Carolina (*right*) and on a plaque in Washington, D.C. (*left*).

d. Reactions with Water and Bases Fluorine reacts rapidly with water at room temperature, displacing oxygen.

$$2F_2(g) + 2H_2O(l) \longrightarrow 4HF(aq) + O_2(g) \qquad \textbf{(27.3)}$$

Other oxidation products, such as oxygen difluoride (OF_2) and hydrogen peroxide (H_2O_2), are also formed.

Chlorine, bromine, and iodine disproportionate in aqueous solution to give the hydrohalic (HX) and hypohalous (HXO) acids, for example

$$\overset{0}{Cl_2}(aq) + H_2O(l) \longrightarrow \underset{\textit{hydrochloric acid}}{H\overset{-1}{Cl}(aq)} + \underset{\textit{hypochlorous acid}}{H\overset{+1}{Cl}O(aq)} \qquad \textbf{(27.4)}$$

In such reactions, the halogen is both oxidized to the +1 and reduced to the −1 oxidation states.

Aqueous solutions of chlorine and bromine are usually stored in brown bottles because the decomposition of the hypohalous acids is accelerated by sunlight:

$$2HXO(aq) \xrightarrow{h\nu} O_2(g) + 2H^+ + 2X^- \qquad \textbf{(27.5)}$$

Iodine dissolves readily in solutions that contain the iodide ion because of the formation of the brown triiodide ion, I_3^-.

Two reactions can occur when chlorine, bromine, or iodine is dissolved in an alkaline solution. The first is disproportionation to give the hypohalite (XO^-) and halide (X^-) ions:

$$\overset{0}{X_2} + 2OH^- \longrightarrow \overset{-1}{X^-} + \overset{+1}{X}O^- + H_2O(l) \qquad \textbf{(27.6)}$$

The second, which occurs readily in warm solutions, is the disproportionation of the hypohalite ion:

$$3\overset{+1}{X}O^- \longrightarrow 2\overset{-1}{X^-} + \overset{+5}{X}O_3^- \qquad \textbf{(27.7)}$$

EXAMPLE 27.2 Chemical Equilibria: Halogens

How might the concentration of HBrO in a saturated aqueous bromine solution be (a) increased, (b) decreased?

In a saturated bromine solution the equilibrium is

$$Br_2(l) + H_2O(l) \rightleftharpoons HBrO(aq) + H^+ + Br^-$$

(a) The concentration of HBrO could be increased by adding **a cation that forms a slightly soluble salt with Br^-**, for example

$$Br_2(l) + H_2O(l) + AgNO_3(aq) \longrightarrow HBrO(aq) + AgBr(s) + H^+ + NO_3^-$$

(b) The concentration of HBrO could be decreased by **adding an acid to increase the H^+ concentration or a soluble bromide to increase the Br^- concentration,** in either case displacing the equilibrium in favor of $Br_2(l)$ and $H_2O(l)$.

27.4 FLUORINE

No known chemical oxidizing agent is strong enough to oxidize fluoride ion to elemental fluorine. Since all natural sources of fluorine contain fluoride ions, elemental fluorine must be prepared by electrochemical oxidation. Hydrogen fluoride is obtained first by treating the abundant mineral fluorspar with concentrated sulfuric acid:

$$CaF_2(s) + H_2SO_4(\text{conc}) \xrightarrow{\Delta} CaSO_4(s) + 2HF(g) \qquad \textbf{(27.8)}$$

Table 27.7
Some Fluorine-Containing End Products

Refrigerants
Aerosol propellants
Polyfluorocarbon resins (e.g., Teflon)
Textile finishes (e.g., Scotchgard)
Toothpaste
Electrical insulating medium (SF_6)

Fluorine is then produced by electrolysis of hydrogen fluoride, which is dissolved in molten potassium fluoride because the reaction must be carried out in the absence of water:

$$2HF(\text{in KF}) \xrightarrow{\text{electrical energy}} H_2(g) + F_2(g)$$

The K^+ and HF_2^- ions that form in the molten salt solution are the current carriers. The anode is graphite and the cathode is Monel metal or steel.

Before the 1940s there was no commercial production of elemental fluorine, and fluorine-containing chemicals were just beginning to find uses as refrigerants. The great reactivity of fluorine had made chemists reluctant to handle it. Fluorine technology developed when, in the drive to produce the atomic bomb, it was found that the uranium isotopes ^{238}U and ^{235}U could be separated by diffusion of their gaseous hexafluorides. The principal use of elemental fluorine today is the production of UF_6 for isotope separation. Another important use is in production of sulfur hexafluoride (SF_6)—an ideal substance for insulating electrical equipment because it is an extraordinarily stable and unreactive gas that is nonflammable, nontoxic, and an excellent insulator. Some of the major fluorine-containing end products are listed in Table 27.7.

Thanks to the advertising industry, it seems impossible for anyone in the United States not to know that fluorides "fight dental cavities." When the cause of spotted and darkened teeth of residents of several towns in the western United States was investigated, fluoride ion was found to be dissolving from natural mineral deposits into municipal water supplies. Persistent U.S. Public Health dentists soon found that along with the discolored teeth went an unusually low rate of tooth decay, and this too was attributed to fluoride ion.

Teeth contain calcium in the form of the mineral apatite ($Ca_5(PO_4)_3OH$). Fluoride ions are thought to replace hydroxide ions in the apatite structure and make it resistant to attack by the acids that cause tooth decay. Fluoride toothpastes reach the surface of teeth in order to build a protective coating there. Tin(II) fluoride (stannous fluoride, SnF_2), sodium fluoride (NaF), and sodium monofluorophosphate (Na_2PO_3F, known as "MFP") are added to toothpaste to provide fluoride ions. However, the benefits from fluoride ion are believed to be greatest when it is incorporated throughout the entire tooth. Such incorporation is most likely achieved when fluoride ions in drinking water are ingested regularly by children and carried by the bloodstream to growing teeth.

27.5 CHLORINE

Chlorine is a top ten industrial chemical and a product of the chloralkali industry (Section 26.13a). About 80% of all chlorine produced is used in the manufacture of other chemicals. Some of the major products containing chlorine compounds are herbicides and pesticides, poly(vinyl chloride) plastics, dry-cleaning fluids, pharmaceuticals, and refrigerants.

Elemental chlorine and oxo compounds of chlorine in its lower oxidation states are valuable for applications that require bleaching, removing undesirable odors, and destroying disease-causing bacteria (disinfecting). Hypochlorous acid (HClO) formed in aqueous solution is the active ingredient in such applications. For example, household bleach (e.g., Chlorox) is an alkaline solution of sodium hypochlorite.

$$Cl_2(g) + H_2O(l) \rightleftharpoons H^+ + Cl^- + HClO(aq) \qquad \textbf{(27.9)}$$

$$\underset{\textit{sodium hypochlorite}}{NaClO(aq)} + H_2O(l) \rightleftharpoons Na^+ + OH^- + HClO(aq) \qquad \textbf{(27.10)}$$

Unwanted color in paper pulp, textiles, household laundry, and swimming pools is most often due to colored organic compounds. Color results when radiation in the visible region of the spectrum is absorbed by alternating single and double carbon–carbon bonds (Section 32.26). If some double bonds are broken, radiation is no longer absorbed and the objectionable color disappears. In the reaction between hypochlorous acid and a double bond, OH adds to one carbon atom and Cl to the other, producing a single bond and thereby breaking up the color-absorbing system.

Hypochlorous acid is a wonderfully useful chemical because it not only bleaches and oxidizes but acts as a disinfectant. Contrary to popular belief, it is not chlorine, but hypochlorous acid formed by reaction (27.9) that kills bacteria in city water and swimming pools (see Aside 27.1). The acid is thought to act by diffusing through the cell walls and destroying enzymes essential to the health of the bacteria. Epidemics of such water-borne diseases as dysentery, hepatitis, and typhoid are rare in areas where water is chlorinated.

Note, however, that **bleaches, scouring powders, or other cleaning agents that contain chlorine or hypochlorite must never be mixed with household ammonia.** The ammonia can be oxidized to produce nitrogen trichloride (NCl_3), an explosive and extremely toxic gas.

EXAMPLE 27.3 Electrochemistry: Sodium Chloride

If a 1 M solution of sodium chloride were electrolyzed, would the products be the same as in electrolysis of a concentrated brine? The following are the reduction potential values for the possible electrode reactions:

$$\begin{array}{llll} \text{(i)} & Na^+ + e^- \longrightarrow Na & & E^\circ = -2.71\ V \\ \text{(ii)} & 2H_2O + 2e^- \longrightarrow H_2 + 2OH^-(10^{-7}\ M) & & E = -0.41\ V \\ \text{(iii)} & O_2 + 4H^+(10^{-7}\ M) + 4e^- \longrightarrow 2H_2O & & E = 0.82\ V \\ \text{(iv)} & Cl_2 + 2e^- \longrightarrow 2Cl^- & & E^\circ = 1.36\ V \end{array}$$

The reduction reactions that might take place at the cathode are reactions (i) and (ii). The likely reaction will be reaction (ii), for it has the more positive reduction potential.

The oxidation reactions possible at the anode are

$$\begin{array}{ll} 2H_2O \longrightarrow O_2(g) + 4H^+(10^{-7}\ M) + 4e^- & E = -0.82\ V \\ 2Cl^- \longrightarrow Cl_2 + 2e^- & E^\circ = -1.36\ V \end{array}$$

Here, the first reaction is the one with the greater thermodynamic likelihood of occurring. **The overall reaction expected on the basis of the thermodynamic data is simply the decomposition of water.**

$$2H_2O(l) \xrightarrow[\text{electrical energy}]{Na^+Cl^-} 2H_2(g) + O_2(g)$$

27.6 BROMINE

In the United States bromine is prepared mainly by oxidation of the bromide ion in natural brine from Arkansas, using elemental chlorine as the oxidizing agent. Like chlorine, elemental bromine is used widely in the production of chemical compounds for many different applications.

Until recent years the major industrial use of bromine has been in the synthesis of ethylene dibromide ($C_2H_4Br_2$), which is added to leaded gasoline. The compound reacts with lead oxide and lead sulfate, combustion products that would otherwise be deposited in the engine. The reaction product, lead(II) bromide, is volatile and is carried away with the exhaust. As antipollution laws limiting the amount of lead in gasoline have taken effect, this use of bromine has declined and will eventually be eliminated.

In agriculture, organic bromine compounds (e.g., methylbromide, CH_3Br) are important pesticides used for killing rodents, insects, fungus, and weeds. Other uses of bromine compounds rely upon their fire-retarding properties (e.g., highly brominated organic compounds added to plastics), their high density (e.g., in hydraulic fluids), and the sensitivity to light of silver bromide (see Aside 30.1 on photography).

27.7 IODINE

In the United States brine wells in Oklahoma are the main source of iodine, which is produced by oxidation of iodide ion by chlorine in a process similar to that for bromine. Outside the United States, a major source of iodine is the sodium iodate ($NaIO_3$) in the solution left after removal of sodium nitrate (saltpeter) from Chilean ores. Reduction with hydrogen sulfite ion yields free iodine in a two-step process:

$$2NaIO_3(aq) + 6NaHSO_3(aq) \longrightarrow 2NaI(aq) + 3Na_2SO_4(aq) + 3H_2SO_4(aq)$$
$$5NaI(aq) + NaIO_3(aq) + 3H_2SO_4(aq) \longrightarrow 3I_2(s) + 3Na_2SO_4(aq) + 3H_2O(aq)$$

(What are the oxidizing and reducing agents in the reaction immediately above?)

Iodine is used in animal feed, pharmaceutical chemicals, catalysts, and chemicals for photography. Iodide ion is necessary for the production of thyroxine in the thyroid gland. Insufficient iodide ion in the diet leads to a condition known as goiter, which is an enlargement of the thyroid gland. To assure the presence of iodide ion in the diet, sodium or potassium iodide is added to table salt, which is sold as "iodized" salt.

SUMMARY OF SECTIONS 27.1–27.7

The halogens fluorine, chlorine, bromine, and iodine are reactive nonmetals found in many common and important compounds. The outstanding properties of the halogens are summarized in Table 27.8.

Table 27.8
Outstanding Properties of the Halogens

$:\ddot{\underset{..}{X}}\cdot$	Representative Group VII (ns^2np^5)
$:\ddot{\underset{..}{X}}:\ddot{\underset{..}{X}}:$	Diatomic molecules; strong bonds
Na^+Cl^- ← *ionic* PCl_3 ← *covalent* ClF_5 ← *covalent, expanded octet*	Highly reactive and electronegative halogens form compounds on entire continuum from ionic to covalent; halides form with almost all elements
Cl Br I ↓	Regular trends in properties; increasing radii, m.p., b.p.; decreasing electronegativity
$\overset{\delta-}{X}-\overset{\delta+}{E}$	Polar covalent bond to other elements (E) in many compounds
F_2 ← *only H, He, and Ne atoms are smaller than F atoms*	Most reactive, most electronegative, most strongly oxidizing element; differs from other family members (Table 27.2)
EF_n	Many elements have highest oxidation states in fluorides
$\overset{+1}{X}, \overset{+3}{X}, \overset{+5}{X}, \overset{+7}{X}$	Cl, Br, I have many positive oxidation states
$X'_2 + 2X^- \longrightarrow X_2 + 2X'^-$	One X_2 displaces another from halides of following halogens: $Cl_2 > Br_2 > I_2$
$X_2 + H_2O \rightleftharpoons H^+ + X^- + HOX$ (*hypohalic acid*) $X^- + 2OH^- \rightleftharpoons X^- + XO^- + H_2O$ (*hypohalite ion*)	Cl_2, Br_2, I_2 disproportionate
X_2 + hydrocarbon	Halogenation; X_2 adds across C=C bond or replaces H
F_2, Cl_2	Prepared by electrolytic oxidation of X^-
Br_2, I_2	Prepared by oxidation of X^- by Cl_2
Cl_2	Top industrial chemical; used in chemical manufacture, and as bleach and disinfectant

Aside 27.1 SWIMMING POOL CHEMISTRY

A freshly filled swimming pool is an aqueous equilibrium system open to the atmosphere and (if it is outdoors) to sunlight. When swimmers enter the pool, bacteria, dirt, and organic by-products are mixed into the system. To maintain a healthy and beautifully clear swimming pool requires careful balancing of the pool chemistry.

The majority of pools are treated with chemicals that release hypochlorous acid (Equations 27.9 and 27.10). As this acid is the active ingredient for oxidizing odor-causing organic chemicals, bleaching, and disinfecting, it is desirable to keep the concentration of the acid high. Because hypochlorous acid (and hypochlorite ion) participates in several equilibria in aqueous solution (Equations 27.4, 27.6, 27.7), its concentration is clearly pH dependent.

The pool pH is, in turn, dependent upon the natural "alkalinity" of the water—the concentration of hydrogen carbonates, carbonates, and hydroxides dissolved in the water. Hydrogen carbonates can provide a buffering action that helps to maintain a constant pH. In a pool with low alkalinity, very small additions of acid or base can cause undesirably wide swings in pH, and in such cases sodium hydrogen carbonate must be added to the pool to correct this condition. A pH range of 7.2–7.6 is ideal, for in this range hypochlorous acid has its maximum bactericidal effect.

Each pool has its own "chlorine demand" based on the amount of contaminants brought into it, the temperature, the alkalinity, and the amount of sunlight it receives, for sunlight accelerates the decomposition of hypochlorous acid (Equation 27.5). In regions where sunlight exposure is high, pools can be "chlorinated" and protected from sunlight by the same chemical. For example, trichloroisocyanuric acid yields hypochlorous acid and cyanuric acid upon hydrolysis.

The cyanuric acid absorbs ultraviolet radiation (note the double bonds alternating with single bonds), thereby screening the hypochlorous acid in a swimming pool from the sun.

$$\text{trichloroisocyanuric acid} + 3H_2O(l) \longrightarrow \text{cyanuric acid} + 3HClO(aq)$$

trichloroisocyanuric acid *cyanuric acid*

An additional factor that influences swimming pool chemistry is the hardness of the water (Section 14.13). A bad balance among alkalinity, temperature, pH, and water hardness can lead at one extreme to precipitation of calcium carbonate on the pool walls and filtering system, or at the other extreme, to corrosion of metal equipment exposed to the pool water (Section 24.18). These conditions are corrected by adding hydrochloric acid or sodium hydrogen carbonate to adjust the pH and the alkalinity as needed.

One more interesting bit of swimming pool chemistry—an unpleasant odor and eye irritation are often attributed to too much chlorine in a pool. The cause of this condition is actually *too little* "chlorine" in the pool. The irritating chemicals are chloramines (compounds with nitrogen–chlorine bonds), which are formed by the reaction of hypochlorous acid with amines and urea (from sweat and urine), for example

$$NH_3(aq) + HClO(aq) \longrightarrow \underset{\textit{monochloramine}}{NH_2Cl(aq)} + H_2O(l)$$

In the presence of additional hypochlorous acid, the chloramines are destroyed by oxidation to nitrogen, for example

$$2NH_2Cl(aq) + HClO(aq) \longrightarrow N_2(g) + H_2O(l) + 3HCl(aq)$$

The next time you dive into a sparkling swimming pool, you might remember that you are about to disturb a number of simultaneous aqueous equilibria.

COMPOUNDS OF HALOGENS

27.8 HYDROGEN HALIDES AND HYDROHALIC ACIDS

At atmospheric pressure above 20 °C, all the hydrogen halides are colorless gases. They fume in moist air as they react to form droplets of the hydrohalic acids. The hydrogen halides are strong irritants to the mucous membranes, and hydrogen fluoride is a particularly dangerous material in this respect.

Table 27.9
Properties of the Hydrogen Halides

Property	HF	HCl	HBr	HI
Melting point (°C)	−83.1	−114.8	−86.9	−50.7
Boiling point (°C)	19.5	− 84.9	−66.8	−35.4
Heat of fusion at melting point (kJ/mol)	4.58	1.99	2.41	2.87
Heat of vaporization at boiling point (kJ/mol)	30.3	16.2	17.6	19.7
Percent dissociation into elements at 1000 °C	0	0.014	0.5	33
H—X bond length (nm)	0.0917	0.127	0.141	0.161
Bond dissociation energies (kJ/mol)				
$HX(g) \longrightarrow H(g) + X(g)$	568	432	356	298

In the solid and liquid states, and even in the gaseous state if the temperature is not too high, hydrogen fluoride (HF) molecules remain hydrogen-bonded in chains and rings. The effect of this strong hydrogen bonding is apparent in the properties given in Table 27.9. Compare the melting and boiling points, and heats of fusion and vaporization of hydrogen fluoride with those of the other hydrogen halides. Note also that the hydrogen–halogen bond is strongest and shortest in hydrogen fluoride. As expected, the bond becomes weaker and longer with increasing size of the halogen atom.

The hydrogen halides other than hydrogen fluoride can all act as reducing agents, their strength increasing in the order HCl < HBr < HI. Hydrogen iodide is a strong reducing agent. In redox reactions the hydrogen halides usually are oxidized to elemental halogens.

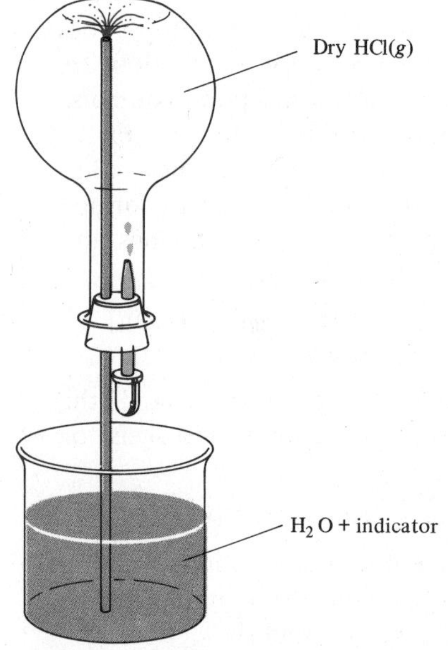

Figure 27.1
$HCl(g) + H_2O(l)$
A flask (2 L) filled with dry HCl gas is closed with a stopper fitted with a long glass tube drawn to a nozzle at the end in the flask and with a medicine dropper filled with water. The flask is up-ended, with the end of the glass tube in a large beaker of water containing an acid–base indicator. Squirting a few drops of water into the flask dissolves so much gas that the reduction in pressure causes a fountain to flow. The color change of the water in the fountain shows that the reaction to produce H^+ ions has occurred. (This experiment can also be conducted using gaseous ammonia, which yields OH^- ions.)

The combination of hydrogen with fluorine to give hydrogen fluoride is violent, even at room temperature. Commercially, hydrogen fluoride is prepared by the reaction of fluorspar with sulfuric acid (Equation 27.8). Hydrogen fluoride is used in making fluorine compounds and the cryolite needed for aluminum production (Section 26.5).

Hydrogen chloride, $HCl(g)$, can be produced in a two-step process known as the salt–sulfuric acid process. The first reaction, a partner exchange, proceeds to completion at a low temperature as hydrogen chloride escapes from the reaction mixture.

$$NaCl(s) + H_2SO_4(conc) \longrightarrow HCl(g) + NaHSO_4(s) \qquad \textbf{(27.11)}$$

Adding more sodium chloride and heating to a high temperature gives additional hydrogen chloride.

$$NaHSO_4(s) + NaCl(s) \xrightarrow{\Delta} HCl(g) + Na_2SO_4(s) \qquad \textbf{(27.12)}$$

However, most hydrogen chloride used in the United States is obtained as a by-product in the chlorination of organic compounds.

The reactions of bromine and iodine with hydrogen are slow at room temperature but can be made to go at practical rates at elevated temperatures in the presence of a catalyst such as platinum. Hydrogen bromide and hydrogen iodide can also be prepared by reduction of the elemental halogens by phosphorous or sulfur in aqueous solution, for example

$$S(s) + 3Br_2(l) + 4H_2O(l) \longrightarrow H_2SO_4(aq) + 6HBr(g)$$

Hydrogen bromide and hydrogen iodide cannot be prepared by the salt–sulfuric acid method because in the presence of concentrated sulfuric acid the hydrogen halide is oxidized, for example

$$H_2SO_4(conc) + 2HBr(g) \longrightarrow Br_2(g) + SO_2(g) + 2H_2O(l)$$

When dissolved in water, the hydrogen halides give the solutions that we know as the hydrohalic acids—hydrofluoric, hydrochloric, hydrobromic, and hydroiodic acids (Figure 27.1). Due in large part to the strength of the hydrogen–fluorine bond, hydrofluoric acid is a weak acid, comparable in strength to acetic acid. The other three hydrohalic acids are strong nonoxidizing acids.

Although used in lesser volume than the big three industrial acids (H_2SO_4, H_3PO_4, HNO_3; Section 28.6), hydrochloric acid, $HCl(aq)$, is also an essential industrial chemical. Major uses are in the manufacture of other chemicals, in removal of oxides from the surface of steel (known as *pickling*), in the extractive metallurgy of numerous metals, and for pH control in many processes. In the food industry, hydrochloric acid is utilized to digest corn starch and produce corn syrup. In our stomachs, hydrochloric acid serves the same type of function—breaking down large molecules to smaller ones by hydrolysis (Section 33.7).

27.9 METAL HALIDES

Halides are known for all of the metals, and preparation of most of them is possible by the direct combination of the elements. Other preparative methods include partner-exchange reactions or displacement of hydrogen from halogen acids by active metals.

Because metals vary considerably in electropositive character and halogens vary in electronegative character, bonding in metal halides spans the entire continuum from ionic to covalent. In general, *s*-block metals form ionic halides; *p*-block and transition metals in +3 or higher oxidation states, nonmetals, and semiconducting elements form molecular halides.

The greater the ionic character of the metal–halogen bonds, the higher the melting and boiling points of metal halides. For metals with two or more oxidation states, the halide with the metal in the lower state is the more ionic and has the higher melting and boiling points.

Because of the small size, low polarizability, and high electronegativity of the fluoride ion, fluorides often differ in properties from other comparable halides. For example, aluminum fluoride (AlF_3) is an ionic compound, whereas the other halides of aluminum have predominately covalent bonding. The fluorides of lithium, magnesium, calcium, strontium, barium, and the lanthanides are relatively insoluble in water. By contrast, the other halides of these elements, and of most metals, are water soluble (see Table 17.1).

27.10 INTERHALOGENS

Molecular compounds formed by the combination of two different halogens are known as **interhalogens.** The interhalogens are either diatomic molecules (e.g., ICl) or molecules of the type AB_n, where A is the larger halogen atom and is surrounded by three, five, or seven atoms of the smaller halogen. Note in Table 27.10 that in compounds with the small fluorine atom, chlorine, bromine, and iodine are able to form more than four covalent bonds by utilizing *d* orbitals. In interhalogens, the smaller, more electronegative halogen is assigned an oxidation state of −1, giving the central atom a positive oxidation state.

The interhalogens are all volatile, reactive compounds. Several of them are very corrosive and actively attack the skin. They are also all strong oxidizing agents. Mixtures of halides are produced in reactions between interhalogens and most other elements. With water, the more electronegative halogen forms the hydrohalic acid and the less electronegative element forms the oxoacid; for example, for bromine(I) chloride (BrCl), which is used in water treatment

$$BrCl(g) + H_2O(l) \longrightarrow HCl(aq) + HBrO(aq)$$

Note that this is a partner-exchange reaction.

A large number of anions in which several halogen atoms have combined—polyhalide ions—are also known. Some of the many polyhalide ions that form stable

Table 27.10
Interhalogen Compounds

ClF	*colorless gas*
ClF_3	*colorless gas*
ClF_5	*colorless gas*
BrF	*red gas*
BrF_3	*colorless liquid*
BrF_5	*colorless liquid*
BrCl	*red gas*
IF_3	*yellow solid*
IF_5	*colorless liquid*
IF_7	*colorless gas*
ICl	*dark red liquid; black crystals*
ICl_3	*yellow solid*
IBr	*brown-black solid*

crystalline salts are I_2Br^-, $IClBr^-$, ICl_4^-, BrF_2^-, and several polyiodide ions, I_3^-, I_5^-, and I_7^-.

EXAMPLE 27.4 Oxidation Numbers: Halogens

In iodine monochloride (ICl), what oxidation numbers should be assigned to chlorine and iodine? What reaction illustrates that these oxidation number assignments are correct?

The more electronegative element, in this case chlorine, is always assigned the negative oxidation number, so in ICl, **Cl is assigned − 1 and I is assigned + 1. In the hydrolysis of this compound**

$$\overset{+1\,-1}{ICl}(s) + H_2O(l) \longrightarrow H\overset{-1}{Cl}(aq) + H\overset{+1}{I}O(aq)$$

chlorine forms the hydrohalic acid and iodine the hypohalous acid with no change in the oxidation numbers of the halogens, demonstrating that the chlorine is indeed the more electronegative element.

27.11 OXOACIDS OF THE HALOGENS AND THEIR SALTS

The properties of the significant oxoacids of the halogens are summarized in Table 27.11. Only iodic acid (HIO_3), perchloric acid ($HClO_4$), and the two periodic acids (HIO_4 and H_5IO_6) have been isolated. The others are known in solution and by their salts. All but paraperiodic acid (H_5IO_6) are monoprotic acids. They contain one hydrogen atom in an OH group bonded to the halogen atom. The halogen–oxygen bonds all have double-bond character that is attributed to π bonding by interaction of oxygen *p* orbitals and halogen *p* or *d* orbitals.

Oxoacids and oxoanions of the halogens in their lower oxidation states disproportionate in varying degrees, particularly in alkaline solutions. The acids in the higher oxidation states are stable to disproportionation.

Table 27.11
Oxoacids of Halogens

Oxidation no. of halogen	Known acids	Can be isolated	Aqueous acid strength	Salts (example of common stable salt)
+1	**Hypohalous acids**			
	HClO	No	Weak ($K_a = 2.9 \times 10^{-8}$)	Hypochlorites ($Ca(OCl)_2$)
	HBrO	No	Weak ($K_a = 2.2 \times 10^{-9}$)	Hypobromites[a]
	HIO	No	Weak ($K_a = 2.3 \times 10^{-11}$)	Hypoiodites[a]
+3	**Halous acids**[b]			
	$HClO_2$	No	Moderately strong ($K_a = 1.1 \times 10^{-2}$)	Chlorites[a]
+5	**Halic acids**			
	$HClO_3$	No	Strong	Chlorates ($KClO_3$)
	$HBrO_3$	No	Strong	Bromates ($AgBrO_3$)
	HIO_3	Yes	Moderately strong ($K_a = 1.6 \times 10^{-1}$)	Iodates ($Ba(IO_3)_2$)
+7	**Perhalic acids**			
	$HClO_4$	Yes	Strong	Perchlorates ($KClO_4$)
	$HBrO_4$	No	Strong	Perbromates ($KBrO_4$)
	HIO_4	Yes	Weak	Periodates (KIO_4)
	H_5IO_6	Yes	Weak	Paraperiodates (Na_5IO_6)

[a] Salts are not stable. Chlorites explode readily.
[b] The bromine and iodine acids are not known. Bromites (e.g., $NaBrO_2 \cdot H_2O$) have been prepared.

Table 27.12
Standard Reduction Potentials for Oxochloro Species

Reaction			E°_{298} (V)
Acidic media			
$ClO_4^- + 2H^+ + 2e^-$	$\rightleftharpoons$	$ClO_3^- + H_2O$	1.19
$ClO_3^- + 3H^+ + 2e^-$	$\rightleftharpoons$	$HClO_2 + H_2O$	1.21
$HClO + H^+ + e^-$	$\rightleftharpoons$	$\frac{1}{2}Cl_2 + H_2O$	1.63
$HClO_2 + 2H^+ + 2e^-$	$\rightleftharpoons$	$HClO + H_2O$	1.65
Alkaline media			
$ClO_3^- + H_2O + 2e^-$	$\rightleftharpoons$	$ClO_2^- + 2OH^-$	0.33
$ClO_4^- + H_2O + 2e^-$	$\rightleftharpoons$	$ClO_3^- + 2OH^-$	0.36
$ClO_2^- + H_2O + 2e^-$	$\rightleftharpoons$	$ClO^- + 2OH^-$	0.66
$ClO^- + H_2O + 2e^-$	$\rightleftharpoons$	$Cl^- + 2OH^-$	0.89

All oxochloro acids and their anions are oxidizing agents. In Table 27.12, note that the weaker acids ($HClO$ and $HClO_2$) are present in acidic solution as the undissociated acids and in alkaline solution as the anions. Hypochlorous acid ($HClO$) and its salts, and sodium chlorate ($NaClO_3$) are valuable as bleaching agents (Section 27.5).

Comparisons of oxidizing strength at standard state conditions, that is, in terms of the E° values, reveal a trend toward decreasing oxidizing strength with increasing halogen oxidation state. Hypochlorous acid and chlorous acid are the strongest oxidizing agents. Aqueous perchlorate ion is the weakest oxidizing agent at room temperature *under standard state conditions.* However, such comparisons are misleading because the oxidizing strength of these species is strongly dependent upon pH, concentration, and the reaction conditions. At high temperatures or in concentrated solutions, perchloric acid becomes a very strong oxidizing agent and a hazardous material, frequently exploding on contact with oxidizable substances. A classic laboratory accident is a fire or explosion caused by hot concentrated perchloric acid coming into contact with a rubber stopper or some rubber tubing. **Vigorous oxidation reactions are a possibility in any system containing halogen oxoacids or oxoanions.**

Many preparations of the oxochloro acids are either disproportionation reactions or reactions of salts with nonvolatile acids, often sulfuric acid. At room temperature about 30% of the chlorine in an aqueous Cl_2 solution is present as hypochlorous acid (Equation 27.4).

Chlorous acid ($HClO_2$), an extremely unstable substance, is formed along with chloric acid ($HClO_3$) when chlorine dioxide (also unstable) is passed into water:

$$2ClO_2(g) + H_2O(l) \longrightarrow HClO_2(aq) + HClO_3(aq)$$

(What type of reaction is this?)

Chloric acid and perchloric acid ($HClO_4$) both can be prepared by the reactions of salts with concentrated sulfuric acid (chloric acid is known only in aqueous solution). Perchloric acid is distilled at low pressure from a mixture of potassium perchlorate and the acid:

$$Ba(ClO_3)_2(aq) + H_2SO_4(conc) \longrightarrow 2HClO_3(aq) + BaSO_4(s)$$

$$KClO_4(s) + H_2SO_4(conc) \xrightarrow[\text{low pressure}]{\Delta} KHSO_4(s) + HClO_4(g)$$

(What type of reactions are these? What drives them to "completion"?) Pure perchloric acid, a colorless hygroscopic liquid (m.p. −112 °C), is a highly unstable and dangerous substance. Many salts of perchloric acid (the perchlorates) and of chloric acid (the chlorates) are more stable substances, and most are water-soluble compounds. Ammonium perchlorate (NH_4ClO_4), however, is a dangerous substance that must be handled with care. When heated, this salt of ammonia with an oxidizing anion decomposes explosively (Aside 28.1).

$$2NH_4ClO_4(s) \xrightarrow{\Delta} N_2(g) + Cl_2(g) + 2O_2(g) + 4H_2O(g)$$

Commercially, perchlorates are prepared by the electrolysis of hot concentrated chloride solutions, for example

$$KCl(aq) + 4H_2O(l) \xrightarrow[\text{electrolysis}]{\Delta} KClO_4(aq) + 4H_2(g)$$

The first preparation of perbromic acid ($HBrO_4$) and perbromates provides an interesting contrast with the discovery of noble gas compounds (Section 27.14). The noble gas compounds were not expected to exist, and for a long time they were not actively sought. However, by analogy with perchlorates, it seemed reasonable that perbromates should form and be relatively stable compounds. Repeated attempts were therefore made to prepare them, and some elaborate explanations were put forth to rationalize the failure of these efforts.

In 1968 it was found that perbromic acid and perbromates *could* be prepared from bromates under strongly oxidizing conditions. Interestingly, xenon difluoride was one of the first strong oxidizing agents utilized to produce a perbromate.

$$NaBrO_3(aq) + XeF_2(aq) + H_2O(l) \longrightarrow \underset{\textit{sodium perbromate}}{NaBrO_4(aq)} + 2HF(aq) + Xe(g)$$

Once produced, the perbromates were indeed found to be reasonably stable. They are strong oxidizing agents, but the reactions are slow. It has been suggested that these compounds were elusive for kinetic reasons—their formation appears to have a large energy of activation.

27.12 FLUORINATED AND CHLORINATED HYDROCARBONS

The introduction of fluorine and/or chlorine atoms into hydrocarbon molecules makes possible many desirable and useful modifications in properties. For example, successive replacement of hydrogen atoms in methane (CH_4) produces four industrially important compounds: (1) methyl chloride (CH_3Cl), used in making silicones and CH_3-containing chemicals; (2) methylene chloride (CH_2Cl_2), used in paint removers and in extracting caffeine from coffee; and (3) chloroform ($CHCl_3$) and (4) carbon tetrachloride (CCl_4), both used in manufacturing other halogen-containing hydrocarbons. Other compounds worthy of note are trichlorethylene ($ClCH{=}CCl_2$, known as "trichlor") and 1,1,1-trichlorethane (Cl_3CCH_3), both used to degrease metal parts during their manufacture (and both of which must be handled so that human exposure is minimized). Several completely fluorinated hydrocarbons (e.g., perfluorodecalin, $C_{10}F_{22}$) are under investigation as blood substitutes for use during surgery.

Polymers made from fluorocarbons, such as Teflon, are inert and resistant to high temperatures—no doubt, due in part to the great strength of the carbon–fluorine bond. Such polymers can be used in contact with very reactive chemicals and as nonstick coatings for cooking pans.

$$n\ F_2C{=}CF_2 \xrightarrow{\text{polymerization}} \underset{\textit{Teflon}}{\left(\begin{array}{cc} F & F \\ | & | \\ -C & -C- \\ | & | \\ F & F \end{array}\right)_n}$$

many repeating units

Chlorofluorocarbon is a general term for hydrocarbons in which some or all of the hydrogen atoms have been replaced by fluorine and chlorine atoms. The low molecular mass chlorofluorocarbons, known by the trade name Freons, are nontoxic, stable at high temperatures, inert, odorless, and nonflammable. They are ideal for use as refrigerants and aerosol propellants because they are easily liquefied by pressure alone and have large enthalpies of vaporization.

In the 1970s concern arose over the possible destruction of ozone in the stratosphere by reactions of two Freons. Dichlorodifluoromethane and trichlorofluoromethane are thought to yield chlorine atoms by photodecomposition in the stratosphere.

$$CF_2Cl_2(g) \xrightarrow{h\nu} CF_2Cl(g) + Cl(g)$$
$$CFCl_3(g) \xrightarrow{h\nu} CFCl_2(g) + Cl(g)$$

The reactive chlorine atoms may then destroy ozone in the following reaction sequence, which can be repeated over and over, since the chlorine atoms are regenerated in the second reaction with oxygen atoms that are always present in the stratosphere.

$$Cl(g) + O_3(g) \longrightarrow ClO(g) + O_2(g)$$
$$ClO(g) + O(g) \longrightarrow Cl(g) + O_2(g)$$

After extensive investigation and debate, the use of these two Freons in aerosol sprays in the United States was banned. However, their use as refrigerants and as aerosol propellants outside of the United States continues.

The chemistry of the upper atmosphere is complex and is the subject of intensive ongoing investigation. The global concentration of chlorofluorocarbons (and other chlorine-substituted hydrocarbons) does appear to be increasing and the concentration of ozone decreasing. A complete disappearance of ozone over the Antarctic at certain times of the year has been discovered. As of the 1980s, efforts are being made toward an international agreement on the control of chlorofluorocarbon use, as well as the study of other effects of global pollution of the upper atmosphere, in particular, global warming (see Aside 29.1).

27.13 CHEMICAL REACTIONS OF HALOGEN COMPOUNDS

a. Thermal Decomposition of Compounds Containing Halogen Oxoanions Metal chlorates and perchlorates decompose when heated to liberate oxygen and form the halides, for example

$$2NaClO_3(s) \xrightarrow{\Delta} 2NaCl(s) + 3O_2(g) \qquad \textbf{(27.13)}$$
$$KClO_4(s) \xrightarrow{\Delta} KCl(s) + 2O_2(g) \qquad \textbf{(27.14)}$$

At moderate temperatures the perchlorate is an intermediate in the decomposition of chlorates, for example

$$4KClO_3(s) \xrightarrow{\Delta} 3KClO_4(s) + KCl(s) \qquad \textbf{(27.15)}$$

Alkali metal bromates and iodates also decompose thermally to oxygen and the halide, while compounds of less active metals give other products, often the halogen and oxygen, for example

$$2Zn(BrO_3)_2(s) \xrightarrow{\Delta} 2ZnO(s) + 2Br_2(g) + 5O_2(g)$$

b. Ions in Aqueous Solution The halide ions are all found in many soluble salts. The Cl^-, Br^-, and I^- anions simply become hydrated in aqueous solution. The F^- ion reacts to give weakly alkaline solution ($K_b = 1.5 \times 10^{-11}$).

Of the possible anions of the known halogen oxoacids (see Table 27.11), those most commonly encountered are the hypochlorite ion (ClO^-); the anions with halogens in the +5 state, namely the chlorate (ClO_3^-), bromate (BrO_3^-), and iodate (IO_3^-) ions; and the perchlorate (ClO_4^-) ion. The basicity of these anions decreases with increasing oxidation number of the halogen. Hypochlorous acid is a weak acid and the hypochlorite ion can hydrolyze to give a weakly alkaline solution:

$$ClO^- + H_2O(l) \rightleftharpoons HClO(aq) + OH^- \qquad \textbf{(27.16)}$$

(In the presence of additional OH^-, the ClO^- ion disproportionates; Equation 27.7.) Chloric acid ($HClO_3$) and perchloric acid ($HClO_4$) are strong acids, and their anions do not react with water to a significant extent. The oxoanions containing halogens all can act as oxidizing agents.

c. Hydrolysis of Halides Many halides with covalent or partially covalent bonding undergo hydrolysis reactions (partner exchange with water; Section 17.3d). The products are an oxygen-containing compound from the more electropositive element plus the binary halogen acid; for example

$$PBr_3(l) + 3H_2O(l) \longrightarrow H_3PO_3(aq) + 3HBr(aq) \quad \textbf{(27.17)}$$

$$SiCl_4(l) + 4H_2O(l) \longrightarrow Si(OH)_4(s) + 4HCl(aq) \quad \textbf{(27.18)}$$

$$SbCl_3(s) + H_2O(l) \longrightarrow SbOCl(s) + 2HCl(aq) \quad \textbf{(27.19)}$$

EXAMPLE 27.5 Reactions of Oxohalo Anions

(a) Predict the products of reaction between potassium chlorate and potassium iodide in an acidic aqueous solution. (You will find Tables 17.6–17.8 helpful.) (b) Balance the net ionic equation for the reaction by the half-reaction method.

(a) There are three reactive ions in solution—chlorate ion (ClO_3^-), iodide ion (I^-), and hydrogen ion (H^+). The chlorate ion is a relatively strong oxidizing agent in acidic medium; the iodide ion is the element in its lowest oxidation state and therefore should be oxidized readily by the chlorate ion. **The usual oxidation product of iodide ion is elemental iodine (I_2). The chlorate ion is commonly reduced to either Cl^- or Cl_2. However, as Cl_2 is formed, it will oxidize I^- and be reduced to Cl^-. As in most redox reactions, the H^+ will be converted to H_2O.**

(b) The unbalanced equation for the reaction is

$$ClO_3^- + I^- \longrightarrow Cl^- + I_2$$

The unbalanced ion-electron equations are

$$ClO_3^- \longrightarrow Cl^-$$

$$I^- \longrightarrow I_2$$

Balancing each equation for atoms and then for charge gives

$$ClO_3^- + 6H^+ \longrightarrow Cl^- + 3H_2O$$

$$ClO_3^- + 6H^+ + 6e^- \longrightarrow Cl^- + 3H_2O$$

$$2I- \longrightarrow I_2$$

$$2I^- \longrightarrow I_2 + 2e^-$$

Adjusting the half-reactions so that electrons gained equal electrons lost gives

$$ClO_3^- + 6H^+ + 6e^- \longrightarrow Cl^- + 3H_2O$$

$$6I^- \longrightarrow 3I_2 + 6e^-$$

The balanced equation is found by adding the half-reactions and canceling where appropriate:

$$\mathbf{ClO_3^- + 6I^- + 6H^+ \longrightarrow Cl^- + 3I_2 + 3H_2O}$$

EXERCISE Why isn't ClO_2^- formed when ClO_3^- reacts as an oxidizing agent?

Answer The chlorite ion is a stronger oxidizing agent than the chlorate ion.

SUMMARY OF SECTIONS 27.8–27.13

The outstanding properties of some halogen compounds are summarized in Table 27.13.

Table 27.13
Outstanding Properties of Halogen Compounds and Ions

HX	Gases
HX(*aq*)	Hydrohalic acids; HCl, HBr, HI—strong nonoxidizing acids; HI—strong reducing agent
HF(*aq*)	Weak acid
HCl(*aq*)	Important industrial chemical
$M(X)_n$	M = *s*-block metal—ionic compound; M = +3 metal, nonmetal, semiconducting element—covalent bonding
$M(X)_n + H_2O$	Molecular halides hydrolyze to give HX + oxygen-containing species
Interhalogens	Highly reactive
$HOX(O)_{1-3}$ X = Cl, Br	Monoprotic oxoacids; oxidizing agents (Table 27.11)
$HClO_4$, HIO_3, HIO_4, H_5IO_6	Only isolable halogen oxoacids
ClO^-, BrO_3^-, IO_3^-, ClO_4	Most common of the halogen oxoanions
HOCl, ClO_2, NaOCl, $Ca(OCl)_2$, $NaClO_3$	Bleaching agents
HOCl	Disinfectant
X–Hydrocarbon	Many useful compounds
Chlorates, perchlorates	Thermal decomposition to halide + O_2

Aside 27.2 THOUGHTS ON CHEMISTRY: "OXYMURIATIC ACID"

What is oxymuriatic acid? Try to decide as you read John Dalton's observations of its properties.

The highly interesting compound, now denominated oxymuriatic acid, was discovered by Scheele, in 1774. It may be procured by applying a moderate heat to a mixture of muriatic acid [HCl] *and oxide of manganese or red lead; a yellowish coloured gas ascends, which may be received over water; it is oxymuriatic acid gas. But this gas, which is largely obtained for the purposes of bleaching, is usually got from a mixture of equal weights of common salt (muriate of soda), oxide of manganese, and a dilute sulphuric acid of the strength 1.4; . . .*

Some of its properties are:

1. It has a pungent and suffocating smell, exceeding most other gases in these respects, and it is highly deleterious. . . .

2. Oxymuriatic acid gas is absorbed by water, but in a very small degree compared with muriatic acid gas. . . .

3. Water impregnated with the gas is called liquid oxymuriatic acid. It has the same odour as the gas, and an astringent, not acid, taste. When exposed to the light of the sun, the liquid acid is gradually decomposed, as was first observed by Berthollet, into its elements, muriatic acid and oxygenous gas; the former remains combined with the water, and the latter assumes the gaseous form. Neither light nor heat has been found to decompose the acid gas.

4. This acid, in the gaseous state or combined with water, has a singular effect on colouring matter. Instead of converting vegetable blue into red, as other acids do, it abstracts colours in general from bodies, leaving them white or colourless. The oxygen combines with the colouring principle, and the muriatic acid remaining dissolves the compound. Hence the use of this acid in bleaching. . . .

. . .

6. Oxymuriatic acid seems to combine readily with the fixed alkalis and the earths when dissolved in water; but it decomposes ammonia. . . .

7. . . . Upon mixing hydrogen and oxymuriatic acid in a strong phial capable of containing 600 grams of water, and exposing the mixture to the solar rays, an explosion almost instantly took place with a loud report, just as if it had received an electric spark. If the stopper was well closed, a vacuum nearly was formed, which was instantly filled with water when the stopper was drawn out under water; but it generally happened that the stopper was expelled with violence.

It remains now to point out the constitution of this acid. All experience shews, that it is a compound of muriatic acid and oxygen; but the exact proportion has not hitherto been ascertained.

Source: John Dalton, *A New System of Chemical Philosophy* (London: Bickestaff, Strand, 1808), pp. 297ff.

NOBLE GASES

27.14 PROPERTIES AND COMPOUNDS OF THE NOBLE GASES

The expected periodic trends are evident for the noble gases. Down the family from helium to radon, the atomic radii increase (Table 27.14) and the ionization energies decrease. Melting and boiling points also increase down the family.

Table 27.14
Properties of the Noble Gases
All of the noble gases are colorless, odorless, and monatomic. Values of atomic radii for elements for which compounds are not known have been found by extrapolation.

	He	Ne	Ar	Kr	Xe	Rn
Melting point (°C)	−272	−249	−189	−157	−112	−71
Boiling point (°C)	−269	−246	−186	−153	−108	−62
Atomic radii (nm)	0.050	0.065	0.095	0.110	0.130	0.145
Ionization energy at 0 K (kJ/mol)	2372	2081	1521	1351	1170.	1037

Thus far we have referred frequently to the single most outstanding property of the noble gases, their lack of chemical reactivity. For more than 60 years after the discovery of radon—the last noble gas to be identified—it was believed by most chemists and taught in most chemistry courses that these were the "inert gases." Because of their stable filled outer electron energy levels, noble gas atoms were thought to neither gain, lose, nor share electrons and to form no chemical compounds. However, since a limited number of compounds has now been prepared, we know that these elements are "noble" but not "inert." Isaac Asimov has pointed out a fine semantic distinction. Chemists were thinking and writing, "The noble gases cannot form compounds under any conditions," when they should have been saying, "As far as we know, the noble gases do not form compounds."

The 1962 preparation of xenon hexafluoroplatinate, $Xe[PtF_6]$, amazed many chemists. Neil Bartlett, who had just made $O_2[PtF_6]$, looked, "quite by chance" he said, at a chart of ionization energies plotted against atomic number (a similar chart appears in most general chemistry textbooks; see Figure 9.9). Bartlett saw that the ionization energy of xenon is almost equal to that of molecular oxygen (1170 kJ for Xe and 1175 kJ for O_2). He had the thought that xenon might undergo the same type of reaction with PtF_6 as did O_2.

The experiment worked—Xe and PtF_6 combined. Furthermore, Bartlett found that xenon tetrafluoride (XeF_4) can be made quite simply by heating xenon and fluorine together at 750 °C for one hour. In fact, if xenon and fluorine are mixed in a flask and the flask is allowed to stand in the sunlight, some of the compound eventually forms. Xenon tetrafluoride turned out to be a quite ordinary crystalline substance which can be melted, recrystallized, and stored in a bottle on the shelf. A barrage of research was set off by the initial discoveries of xenon compounds, and within a year 50 publications on noble gas compounds had appeared in scientific journals.

Thus far, most known noble gas compounds contain xenon bonded to oxygen or fluorine or both. This is reasonable, since xenon has a relatively low ionization energy, and fluorine and oxygen are highly electronegative.

Xenon combines with fluorine to give fluorides in which it has oxidation states of +2, +4, and +6 (Table 27.15). As far as we know, xenon does not combine directly with oxygen, but xenon–oxygen compounds are prepared from the xenon–fluorine compounds, for example

$$XeF_6(s) + 3H_2O(l) \longrightarrow \underset{\substack{\textit{highly}\\ \textit{explosive}}}{XeO_3(s)} + 6HF(g)$$

A few krypton compounds, notably KrF_2, have been prepared, as has a very small amount of a radon fluoride. Although radon has an even lower ionization energy than

Table 27.15
Some Xenon Compounds

Compound	M.p. (°C)	Form; properties
XeF_2	129	Colorless crystals; stable; reacts with H_2O
XeF_4	117	Colorless crystals; stable
$XeOF_2$	31	Colorless crystals; unstable
XeF_6	49.6	Colorless crystals; stable; reacts with H_2O
Cs_2XeF_8	—	Yellow solid; stable to 400 °C
XeO_3	—	Colorless crystals; explosive
XeO_4	—	Colorless gas; explosive

xenon and should form stable compounds, its radioactivity and scarcity make it very difficult to study. As yet, no comparable compounds of helium, neon, or argon have been prepared.

27.15 SOURCES AND USES OF THE NOBLE GASES

Natural gas from some wells contains up to 6% of helium. Helium is separated from the other components of natural gas by liquefaction, followed by fractional distillation. Most of the helium used commercially is obtained from this source. The other noble gases (except for radon) are obtained from liquefied air.

Several of the major uses of the noble gases take advantage of their lack of reactivity. Helium and argon are used in various ways in metallurgy to protect metals that are being melted or heated from reacting with oxygen or nitrogen in the air. For example, argon is used as a shielding gas in welding. In laboratories, both helium and argon provide protection in working with highly reactive materials.

Common incandescent light bulbs are filled with an 88% argon–12% nitrogen mixture. Argon is better than pure nitrogen for this purpose in two ways: Because they are heavier, argon molecules move more slowly and conduct heat less readily than nitrogen molecules. Thus, the filament in the bulb reaches a higher temperature and glows more brightly in an atmosphere of argon. Also, the life of the filament is prolonged because it sublimes more slowly in the heavier gas. Some nitrogen must be present in the bulb, however, or an electrical arc would be formed. Krypton is even better than argon for the purposes described because it is still heavier, but krypton is too expensive for routine use in light bulbs.

When an electric current is passed through neon under low pressure in a closed tube, the neon glows bright red. All signs made from such gas-filled tubes are called "neon" signs, though some contain mixtures of neon and argon or other gases. Helium produces a yellow-white light; argon a blue light; a helium–argon mixture, an orange light; and a neon–argon mixture, a deep lavender light. By using appropriate gas combinations, colored glass tubes, and the addition of a bit of mercury, almost any color can be produced.

Argon is of major importance in the steel industry, where an argon–oxygen mixture is used to help remove carbon impurities from the molten metal. Helium and argon provide the necessarily inert and very pure atmosphere needed for the growth of single crystals for semiconductors.

The largest use of helium is in a variety of applications that take advantage of its unusual properties at low temperatures. The possible importance of helium in future energy-related applications has caused concern over the need to conserve helium by separating it from natural gas before the gas is used as fuel. Otherwise, during combustion of the fuel helium goes up the chimney. Because helium is so light, it then escapes from the Earth's atmosphere and cannot be reclaimed.

SUMMARY OF SECTIONS 27.14–27.15

The noble gases show the expected periodic variations in physical properties. All are present in the atmosphere, radon (which is radioactive and short-lived) only near deposits of radioactive minerals. Helium is obtained from natural gas wells; argon, xenon, and krypton are separated from liquefied air. Applications take advantage of the lack of reactivity of noble gases, the colors they emit in "neon" signs, the very low temperatures they can maintain, and, for helium especially, the unusual properties at low temperatures. The discovery of xenon compounds with fluorine and oxygen has necessarily forced a change in the view that because of their filled energy levels noble gas atoms form no compounds.

SIGNIFICANT TERMS (with section references)

halogenation (27.3)
interhalogens (27.10)

QUESTIONS AND PROBLEMS

Halogens—General Properties

27.1 Write the electron configuration for each of the atomic halogens. Draw the Lewis symbol for a halogen atom, X. What is the usual oxidation state of the halogens in binary compounds with metals, semiconducting elements, and most nonmetals?

27.2 Draw the Lewis structure of a halogen molecule, X_2. Describe the bonding in the molecule. What is the trend of bond length and strength going down the family from F_2 to At_2?

27.3 What types of intermolecular forces are found in molecular halogens? What is the trend in these forces going down the family from F_2 to At_2? Describe the physical state of each molecular halogen at room temperature and pressure.

27.4 Choose the properties that increase in going down the fluorine family in the periodic table: (a) atomic radius, (b) melting and boiling points, (c) bond dissociation energies of X_2, (d) ionic radius of X^-, (e) electron affinities, (f) ionization energies, (g) electronegativities, (h) values of the standard reduction potential for X_2.

27.5 Although astatine has been detected in various minerals, most properties of this element have been observed from artificially produced samples or predicted from periodic relationships. Using Table 27.2, predict the (a) physical state, (b) melting point, (c) ionic radius, (d) bond energy of this element.

Halogens—Chemical Reactions of the Halogens

27.6 Write balanced equations for any reactions that occur in aqueous mixtures of (a) NaI and Cl_2, (b) NaCl and Br_2, (c) NaI and Br_2, (d) NaBr and Cl_2, (e) NaF and I_2.

27.7 An aqueous solution contains either NaBr or a mixture of NaBr and NaI. Using only aqueous solutions of I_2, Br_2, and Cl_2 and a small amount of CCl_4, describe how you might determine what is in the unknown solution.

27.8 Use the Nernst equation to predict the ratio of $[Cl^-]$ to $[Br^-]$ at which liquid bromine could liberate gaseous chlorine at 1 bar pressure from a solution of Cl^-. $E° = 1.0652$ V for the $Br_2(l)/Br^-$ couple and 1.3595 V for the $Cl_2(g)/Cl^-$ couple.

27.9 When powdered antimony is sprinkled into a bottle containing chlorine, tiny sparkles are generated. The white powder formed contains 47 mass % chlorine. Find the empirical formula of the chloride and write an equation for the reaction.

27.10 Calculate the standard enthalpy of formation of HI(g) from the following data: $\Delta H_f^\circ = 62.438$ kJ/mol for $I_2(g)$ and average bond energy = 435 kJ/mol for H—H, 297 kJ/mol for H—I, and 151 kJ/mol for I—I.

27.11 The rate of formation of HBr(g) from $H_2(g)$ and $Br_2(g)$ between 200 °C and 300 °C is given by the equation

$$\text{rate} = \frac{k[H_2][Br_2]^{1/2}}{1 + k'\dfrac{[HBr]}{[Br_2]}}$$

How would you adjust the concentrations in order to make the reaction proceed as fast as possible?

27.12 What is a "halogenation reaction"? Write chemical equations for X_2 reacting with SO_2, PX_3, CO, and $H_2C{=}CH_2$.

27.13 Compare the reactivities of the halogens with water. Write chemical equations for the reactions. Which halogen is quite different from the others in its reaction?

27.14 Write chemical equations describing the two different reactions that can occur when chlorine, bromine, or iodine is dissolved in an alkaline solution. Briefly compare the products of the reactions for the various halogens.

27.15 Determine the mass of $KClO_3$ theoretically obtained by the reaction of 50.0 L of $Cl_2(g)$, measured at 25 °C and 1.00 atm, with hot, concentrated KOH(aq). What mass of KClO would be theoretically obtainable by the reaction of the same quantity of $Cl_2(g)$ with cold, dilute KOH(aq)?

27.16 The equilibrium constant at 25 °C for the reaction

$$Cl_2(g) + H_2O(l) \rightleftharpoons H^+ + Cl^- + HClO(aq)$$

is 4.4×10^{-4}. Assuming that the HClO does not ionize appreciably, what will be the pH of a solution of "chlorine water" at a chlorine pressure of 0.5 bar?

Halogens—Specific Properties and Reactions

27.17 Table 27.5 shows that the halogens occur in nature as halide ions, except for iodine, which also occurs as $NaIO_3$. Explain this difference.

27.18 Write chemical equations representing the commercial preparations of F_2, Cl_2, Br_2, and I_2.

27.19 Fluorine is considered to be a highly reactive substance. Why can it be safely stored in certain metal containers?

27.20 Briefly discuss the use of "fluoride" in drinking water to prevent tooth decay.

27.21 A convenient laboratory preparation of chlorine, bromine, or iodine is to allow a metal halide or hydrohalic acid to react with $MnO_2(s)$ in an acidic solution:

$$MnO_2(s) + 4H^+ + 4X^- \longrightarrow Mn^{2+} + 2X^- + X_2(g) + 2H_2O(l)$$

An excess of concentrated hydrochloric acid is added to 0.100 mol MnO_2. Calculate the mass and volume of chlorine collected at 27 °C and 765 Torr.

27.22 In what forms is chlorine valuable as a bleach? Briefly discuss the bleaching process.

27.23 Commercial bleaching powder is a mixture of $Ca(OCl)Cl$, $CaCl_2$, and $Ca(OCl)_2$ in approximately equimolar quantities. (a) Name these substances.

The powder is prepared by passing chlorine over dry calcium oxide

$$CaO(s) + Cl_2(g) \longrightarrow Ca(OCl)Cl(s)$$
$$2CaO(s) + 2Cl_2(g) \longrightarrow Ca(OCl)_2(s) + CaCl_2(s)$$

leaving a small amount of CaO as an impurity. (b) Determine the oxidation state of chlorine in the compounds given in the above equations and identify the (c) oxidizing agent and (d) reducing agent in each reaction.

The chlorine needed for bleaching is produced in aqueous solution by the reaction between water, OCl^-, and Cl^-, represented by the equation

$$OCl^- + Cl^- + H_2O(l) \longrightarrow Cl_2(g) + 2OH^-$$

What species in the above equation is being (e) oxidized and (f) reduced? (g) What happens to the pH of the solution as the chlorine is being generated?

27.24 The chlorine content of the water in a swimming pool is to be determined. A 45.0 mL sample (density = 1.02 g/mL) of this water was treated with excess KI and the liberated iodine titrated with 26.0 mL of 0.075 M $Na_2S_2O_3$ (sodium thiosulfate). The unbalanced chemical equations for the processes are

$$Cl_2 + I^- \longrightarrow Cl^- + I_2$$
$$I_2 + S_2O_3^{2-} \longrightarrow S_4O_6^{2-} + I^-$$

Balance these equations and find the percent chlorine by mass in the water.

27.25* The vapor in equilibrium with a sample of liquid bromine at 25 °C is red-brown in color, and the intensity of the color might lead one to think that the vapor pressure of bromine is quite high. After a 20.65 g sample of argon was bubbled through liquid bromine, the combined argon-bromine mixture at 734 Torr weighed 57.46 g. Using Dalton's law of partial pressures, calculate the vapor pressure of bromine.

27.26* The average concentration of bromine in seawater is 75 ppm, calculated as Br^- ion. Calculate (a) the volume of seawater in cubic feet that is required to produce one ton of liquid bromine and (b) the volume of chlorine gas, in liters, measured at STP, required to displace all of this bromine. The density of seawater is 64.5 lb/ft³.

27.27 Bromine can be prepared in the laboratory by treating NaBr with a mixture of MnO_2 and H_2SO_4

$$NaBr(s) + H_2SO_4(aq) \longrightarrow HBr(aq) + NaHSO_4(aq)$$
$$2HBr(aq) + MnO_2(s) + H_2SO_4(aq) \longrightarrow Br_2(l) + MnSO_4(aq) + 2H_2O(l)$$

What mass of bromine can be produced from the reaction of 100.0 g of NaBr with excess H_2SO_4 and MnO_2?

27.28* Only one stable isotope of iodine is found in nature, $^{127}_{53}I$. Calculate the neutron-proton ratio for this stable isotope. The artificial radioisotope $^{131}_{53}I$, in the form of I^-, is commonly used to decrease thyroid gland activity. What mode of decay would be predicted for this isotope? Write the nuclear equation for the decay.

27.29 What is "iodized salt"? What role does iodine play in human nutrition?

27.30 In acidic solution, the dichromate ion, $Cr_2O_7^{2-}$, oxidizes iodide ion to elemental iodine and is itself reduced to chromic ion, Cr^{3+}. (a) Write the equations for the half-reactions and calculate $E°$ for the reaction using the following half-cell potentials: 0.5355 V for the I_2/I^- couple and 1.33 V for the $Cr_2O_7^{2-}/Cr^{3+}$ couple. (b) Would the reaction become more or less favorable if the concentrations of all ions were 0.1 M instead of 1M?

27.31* The element astatine was first definitely prepared by the nuclear reaction represented by $^{209}_{83}Bi(\alpha, 2n)\ ^{211}_{85}At$. (a) Write the nuclear equation for this reaction. The half-life of this isotope is 7.21 hr, with 40.9% decaying by α emission and 59.1% by electron capture. (b) Write the nuclear equations for the decay processes. (c) What amount of ^{211}At would remain at the end of a week if a researcher started with 0.05 μg?

27.32 What would be the predicted common oxidation states of astatine? Which of these oxidation states do not occur in the following species: AtI, AtBr, AtCl, HAt, AgAt, AtO^-, and AtO_3^-? Name these substances.

Compounds of the Halogens

27.33 Briefly describe the chemical and physical properties of gaseous and aqueous HCl, HBr, and HI. How does HF differ significantly in behavior from the other hydrogen halides? Why?

27.34 Because there is considerable hydrogen bonding in HF, hydrogen fluoride is polymeric. Draw a Lewis structure illustrating this behavior. Also illustrate the structure of the $[HF_2]^-$ anion.

27.35 The standard state enthalpy of formation at 25 °C is −320.08 kJ/mol for HF(*aq*), −229.994 kJ/mol for OH^-, −332.63 kJ/mol for F^-, and −285.830 kJ/mol for $H_2O(l)$. Find the enthalpy of neutralization of hydrofluoric acid.

$$HF(aq) + OH^- \longrightarrow F^- + H_2O(l)$$

Using the following thermochemical equation:

$$H^+ + OH^- \longrightarrow H_2O(l) \qquad \Delta H° = -55.835 \text{ kJ}$$

find the enthalpy change for the reaction

$$HF(aq) \longrightarrow H^+ + F^-$$

27.36 The free energy of formation at 25 °C is −95.299 kJ/mol for HCl(*g*) and −228.572 kJ/mol for $H_2O(g)$. Determine whether the reaction

$$4HCl(g) + O_2(g) \longrightarrow 2Cl_2(g) + 2H_2O(g)$$

is spontaneous or not.

27.37 Predict the type of bonding between atoms of (a) Na and F, (b) Ca and F, (c) I and Cl, and (d) Br and Br.

27.38 Aluminum fluoride is considered to be an ionic substance, but aluminum iodide is considered to be molecular. Explain this difference.

27.39 Consider an element that can exhibit multiple oxidation states such as iron or phosphorus. Compare the oxidation states that result when the element combines with the various halogens.

27.40 Hydrogen chloride can be produced by treating a metal chloride with concentrated sulfuric acid. Write the chemical equations for this process. What happens when concentrated sulfuric acid reacts with metal bromides and iodides? Write chemical equations illustrating these reactions.

27.41 What is the oxidation number of Cl in each of the following interhalogens: (a) ClF, (b) BrCl, (c) ICl, (d) ClF_3, (e) ClF_5, (f) ICl_3, (g) ClF_2^-, (h) ClF_4^-, (i) ICl_2^-, (j) ICl_4^-, (k) $BrCl_2^-$, (l) $IClBr^-$, (m) $IFCl_3^-$?

27.42 Iodine is not very soluble in water, but readily dissolves in solutions containing I^-. Why? Draw the Lewis structure and discuss the geometry of the species formed between I^- and I_2.

27.43 Draw the Lewis structures and discuss the geometries of (a) BrF, (b) BrF_3, (c) BrF_5, (c) BrF_5, (d) BrCl, (e) IBr molecules. Identify which of the molecules are polar.

27.44 Write the Lewis structures and discuss the geometries of the following oxoanions of the halogens: (a) XO^-, (b) XO_2^-, (c) XO_3^-, (d) XO_4^-.

27.45 Unlike the other halogens, fluorine forms no oxoacids nor oxoanions. Why?

27.46 Choose the strongest acid from each group: (a) HClO, HBrO, HIO; (b) HClO, $HClO_2$, $HClO_3$, $HClO_4$; (d) HIO, $HBrO_3$, $HClO_4$.

27.47 Do the oxochloro acids and anions generally act as oxidizing or do they act as reducing agents? What is the trend in reduction potential with increasing oxidation state?

27.48 Freon-12 is dichlorodifluoromethane (CCl_2F_2). Draw the Lewis structure of this molecule and describe its geometry. Will the boiling point of Freon-12 be higher than that of Freon-22 (monochlorodifluoromethane, $CHClF_2$)?

27.49 Briefly compare the products formed by the thermal decompositions of alkali metal and less reactive metal salts containing halogen oxoanions.

27.50 Arrange the oxoanions of chlorine in order of increasing thermal stability and write chemical equations describing the decomposition of the sodium salt of each ion as the salt is heated.

27.51* A 10.00 g sample of $KClO_3$ is heated. Part of it undergoes the reaction

$$2KClO_3 \longrightarrow 2KCl + 3O_2$$

and part of it undergoes the reaction

$$4KClO_3 \longrightarrow 3KClO_4 + KCl$$

If the residue weighs 7.00 g, what proportion of the $KClO_3$ reacted according to each equation?

27.52 Choose the ions which will react with water to give an alkaline solution: (a) F^-, (b) Cl^-. (c) Br^-, (d) I^-, (e) ClO^-, (f) ClO_3^-, (g) BrO_3^-, (h) IO_3^-, (i) ClO_4^-. Write chemical equations for those that react.

27.53 The solubility in water at 0 °C is 59.5 g/100 g H_2O for $CaCl_2$ and 35.7 g/100 g H_2O for NaCl. In a saturated solution, which salt is more effective in lowering the freezing point of water? $K_f = 1.86$ K/m for water.

27.54 What intermolecular forces are present in (a) Cl_2, (b) HF, (c) $HClO_4$, (d) ICl_3, (e) HBrO, (f) Cl_2O_7?

ADDITIONAL QUESTIONS—Halogens

27.55 Based on your knowledge of halogen chemistry, suggest a reaction for preparing (a) $KBrO_3(aq)$ from $Br_2(l)$; (b) $KBrO_3(aq)$ from $KBrO(aq)$; (c) $BrF_5(g)$ from $Br_2(l)$ (d) $Br_2(aq)$ from Br^-. Write a balanced ionic equation (or equations) for each of the preparations.

27.56 Write balanced equations for the following chemical reactions: (a) preparation of Cl_2 from aqueous KCl by electrolysis, (b) any reaction of Br_2 as an oxidizing agent, (c) disproportionation of KClO when it is heated, (d) displacement of one halogen by another in aqueous solution, (e) formation of an iodine chloride from the elements.

27.57 Identify which of the following are redox reactions. Identify the oxidizing and reducing agent in each of the redox reactions.

(a) $BrO_3^- + 5Br^- + 6H^+ \longrightarrow 3Br_2(l) + 3H_2O(l)$
(b) $I^- + Br_2(aq) \longrightarrow IBr(aq) + Br^-$
(c) $NaHSO_4(s) + NaCl(s) \xrightarrow{\Delta} Na_2SO_4(s) + HCl(g)$
(d) $2KClO_3(l) \xrightarrow{\Delta} 2KCl(s) + 3O_2(g)$

27.58 Repeat Question 27.57 for:

(a) $4NaClO_3(s) \xrightarrow{\Delta} 3NaClO_4(s) + NaCl(s)$
(b) $3Br_2(l) + 6OH^- \longrightarrow BrO_3^- + 5Br^- + 3H_2O(l)$
(c) $2NaF(s) + H_2SO_4(conc) \xrightarrow{\Delta} Na_2SO_4(s) + 2HF(g)$
(d) $2Cl_2(g) + HgO(s) + H_2O(l) \longrightarrow HgCl_2(aq) + 2HOCl(aq)$

27.59 Classify each of the following reactions according to the reaction types listed in Tables 17.2 and 17.7:

(a) $Cl_2(g) + 2I^- \longrightarrow I_2(s) + 2Cl^-$
(b) $K_2CO_3(s) + 2HClO_4(aq) \longrightarrow 2KClO_4(s) + H_2O(l) + CO_2(g)$
(c) $Hg(NO_3)_2(aq) + 2KI(aq) \longrightarrow HgI_2(s) + 2KNO_3(aq)$
(d) $S_8(s) + 24F_2(g) \longrightarrow 8SF_6(g)$

27.60 Repeat Question 27.59 for:

(a) $HBrO_4(aq) + KOH(aq) \longrightarrow KBrO_4(aq) + H_2O(l)$
(b) $2KClO(s) \xrightarrow{\Delta} 2KCl(s) + O_2(g)$
(c) $3KClO(s) \xrightarrow{\Delta} KClO_3(s) + 2KCl(s)$
(d) $TiI_4(g) \xrightarrow{\Delta} Ti(s) + 2I_2(g)$

27.61 Predict the major products of the following reactions:

(a) $KHF_2(l) \xrightarrow[\text{energy}]{\text{electrical}}$

(b) $S_8(s) + Br_2(l) + H_2O(l) \longrightarrow$

(c) $IO_3^- + I^- + H^+ \longrightarrow$

(d) $Fe(s) + Cl_2(g) \longrightarrow$

27.62 Repeat Question 27.61 for:

(a) $Sn(s) + Cl_2(g) \longrightarrow$

(b) $Cl_2(g) + F_2(g) \xrightarrow{\Delta}$

(c) $KI(s) + H_2SO_4(conc) \xrightarrow{\Delta}$

(d) $Cl_2(g) + KOH(aq, conc) \longrightarrow$

The Noble Gases

27.63 What are the general trends for the (a) atomic radii, (b) ionization energies, (c) gas densities, (d) boiling points of the noble gases down the noble gas family?

27.64 Which noble gases are known to form chemical compounds? With which elements are the noble gas atoms bonded?

27.65 Write the electron configuration for Xe. Why might Xe be predicted to be "inert"? Write the Lewis structure for XeF_4. How can we explain this exception to the octet rule?

27.66 Write the Lewis structures for XeF_2 and XeF_6. Name these compounds using the Stock system.

27.67 Based on your answers to Questions 27.65 and 27.66, predict the geometries of XeF_2 and XeF_4. What are the important intermolecular forces in each substance?

27.68 Argon crystallizes in a face-centered cubic unit cell at −235 °C. Determine the radius of an argon atom using 0.543 nm as the unit cell length.

27.69 Xenon(VI) fluoride can be produced by the combination of xenon(IV) fluoride with fluorine. Write a chemical equation for this reaction. What mass of XeF_6 will be produced from 3.62 g of XeF_4?

27.70 Essentially complete conversion of Xe to XeF_4 takes place when 1.00 mol of Xe reacts with 5.00 mol of F_2 at 400.°C and 6.0 atm for a few hours in a nickel container. After the reaction, what gas remains in the container and at what pressure?

27.71 Xenon hexafluoride reacts rapidly with the SiO_2 in glass or quartz containers to form $XeOF_4(l)$ and $SiF_4(g)$. What will be the pressure of SiF_4 in a 1.00 L container at 25 °C after 1.00 g of XeF_6 decomposes?

27.72 The heat of formation is −402 kJ/mol for $XeF_6(s)$ and −261.5 kJ/mol for $XeF_4(s)$. Calculate the heat of reaction at 25°C for the preparation of XeF_6 from XeF_4 and $F_2(g)$.

27.73 Assume that the heats of sublimation of XeF_4 and XeF_6 are nearly the same. What is the value of the Xe—F bond energy calculated using the F—F bond energy as 159 kJ/mol and the answer to Problem 27.72?

27.74* In most periodic tables the atomic mass for radon is given as (222), which means that the isotope with the longest half-life is ^{222}Rn. This isotope is generated by the α decay of another nuclide. (a) Identify this radioactive nuclide and write the nuclear equation for the production of ^{222}Rn. The half-life of ^{222}Rn, an α emitter, is 3.8235 day. (b) Write the equation for the decay and calculate the fraction of ^{222}Rn remaining after 1.0 week.

CHAPTER

28

Nonmetals: Nitrogen, Phosphorus, and Sulfur

28.1 PERIODIC RELATIONSHIPS: NITROGEN, PHOSPHORUS, AND SULFUR

Chapter 27 presented the chemistry of noble gases and halogens (Group VII), all nonmetals. Moving to the left in the periodic table brings us to the *p*-block groups within which properties vary more widely. In Representative Groups IV, V, and VI, the lighter elements are nonmetals, the heaviest elements are metals, and the semiconducting elements lie in between. In this chapter we explore the chemistry of the nonmetals nitrogen and phosphorus from Group V, and sulfur from Group VI (oxygen, also from Group VI, was discussed in Chapter 16). The semiconducting elements and the inorganic chemistry of carbon are considered in Chapter 29.

Nitrogen, phosphorus, and sulfur have in common their presence in numerous industrially important compounds. Each of these elements plays an essential role in animal and plant chemistry, and therefore each is also essential in fertilizers. Nitrogen, as the first member of its group, differs significantly from other members of the group, as summarized in Table 28.1.

Nitrogen and phosphorus atoms each have five valence electrons in the ns^2np^3 configuration, and the sulfur atom has six valence electrons in the ns^2np^4 configuration. Consequently, the maximum oxidation state for both nitrogen and phosphorus is $+5$, and for sulfur it is $+6$. All three elements exhibit negative oxidation states in their compounds with hydrogen (NH_3, PH_3, and H_2S) and in compounds in which their atoms complete octets by forming anions—nitride ion, N^{3-}; phosphide ion, P^{3-}; and sulfide ion, S^{2-}. The most common oxidation states of nitrogen are -3, $+3$, and $+5$, but nitrogen is unique among the elements in forming compounds in all oxidation states from -3 to $+5$. Because of the numerous possible oxidation states of nitrogen, its compounds undergo a number of disproportionation reactions. The $+3$ and $+5$ oxidation states are most common for phosphorus; for sulfur, in addition to the negative oxidation state, the $+4$ and $+6$ states are most important.

Table 28.1
Differences between Nitrogen and the Other Members of the Nitrogen Family (P, As, Sb, Bi)

Property	Comments
Occurs free in nature to a large extent	A reflection of the relative inertness of N_2 at ordinary temperatures; other elements in the group occur chiefly in combined forms
A gas at ordinary temperatures	Other elements form a variety of solid allotropes at ordinary temperatures; simplest form of phosphorus is P_4; no discrete molecules present in solid forms of the other elements
N_2 molecule has a very high bond dissociation energy	Other elements have lower bond energies; because of thermal stability of N_2, many compounds containing more than one nitrogen atom decompose to give N_2, some explosively; certain nitrogen-rich compounds find use in explosives
Can form no more than four covalent bonds	Only *s* and *p* orbitals available for bonding; other elements in family can use *d* orbitals, for example, in PF_5, PF_6^-
Only element of the group that combines directly with H_2	$N_2 + 3H_2 \rightleftharpoons 2NH_3$
Only element of the group that forms a stable cation with hydrogen	NH_4^+, the ammonium ion; PH_4^+ is known, but is unstable in aqueous solution, immediately reacting to form PH_3
All possible oxidation states from −3 to +5 are well defined	For P, As, Sb, and Bi the +3 and +5 states are most common (the −3 and +1 states are of some importance for phosphorus)

In their positive oxidation states, nitrogen, phosphorus, and sulfur atoms form covalent bonds. The oxides of these elements are all acidic oxides. Nitrogen and phosphorus utilize their unpaired *p* electrons in forming three covalent bonds, as in $:PCl_3$ or $:NH_3$, leaving a lone electron pair on the central atom. In NH_4^+ the bonding electrons are pictured as occupying four sp^3 hybrid orbitals. Nitrogen atoms, which have no *d* orbitals, are limited to formation of four covalent bonds.

Sulfur atoms form two covalent single bonds by sharing two unpaired electrons, as in H_2S or SCl_2. They form SO_4^{2-} or SO_3^{2-} by utilizing four equivalent sp^3 hybrid orbitals. Phosphorus and sulfur atoms can also enter into five and six covalent bonds, attributed to combinations of *s*, *p*, and *d* orbitals via hybridization.

Nitrogen, phosphorus, and sulfur each form a variety of compounds in which multiple bonding and resonance play important roles, notably the nitrogen oxides and the oxoacids of all three elements. Often the bonds of nitrogen, sulfur, or phosphorus with oxygen have properties intermediate between those of single and double covalent bonds. For phosphorus and sulfur, these properties are attributed to π bonding due to involvement of *p* and *d* orbitals. Although the structures of compounds containing such intermediate bonds can be written correctly with single bonds, they often are seen written with double bonds between the central elements and oxygen.

As the free elements, nitrogen is a diatomic gas (N_2), and phosphorus and sulfur each have numerous solid allotropic forms. From the summary of properties given in Table 28.2, you see that nitrogen is the most electronegative of these elements, and that nitrogen atoms have the smallest radius and the largest ionization energy. Only oxygen and fluorine are more electronegative than nitrogen.

Table 28.2
Properties of Nitrogen, Phosphorus, and Sulfur

Property	Nitrogen	Phosphorus	Sulfur
Configuration	[He]$2s^2 2p^3$	[Ne]$3s^2 3p^3$	[Ne]$3s^2 3p^4$
Common oxidation states	−3, +3, +5	+3, +5	−2, +4, +6
Formula of molecule	N_2	P_4	S_8
Melting point (°C)	−210.0	44.1 (white)	112.8 (rhombic) 119.0 (monoclinic)
Boiling point (°C)	−195.8	280 (white)	444.6
Atomic radius (nm)	0.074	0.110	0.103
Bond dissociation energy (kJ/mol)	946	209	264
Ionization energy (0 K) (kJ/mol)	1402	1012	1000
Electronegativity	3.0	2.1	2.5

Table 28.3
Uses of Elemental Nitrogen

Major use
- Manufacture of ammonia

Other uses
- Manufacture of nitrogen compounds, e.g.,
  - Calcium cyanamide (a fertilizer)
  - Nitrides (used in cutting and grinding tools)
  - Hydrazine (a rocket fuel)
- Blanketing applications, e.g.,
  - For reactive chemicals
  - For foods
  - For electrical equipment
  - For blowing bubbles into foamed polymers
- Low-temperature applications, e.g.,
  - Condensing gases
  - Preserving biological materials
  - Cryosurgery (selective destruction of tissue)
  - Freezing food
  - Shrink fitting molded parts
  - Embrittling

N, P, S: THE ELEMENTS AND THEIR INDUSTRIAL ACIDS

28.2 NITROGEN

Nitrogen is a colorless, odorless, nontoxic gas. In the nitrogen molecule (N_2), the atoms are joined by a very strong triple bond, $:N{\equiv}N:$. Consequently, molecular nitrogen is not very reactive at ordinary temperatures and is quite stable thermally. Because of this stability, some nitrogen compounds have positive heats of formation and decompose rapidly to produce molecular nitrogen (e.g., lead azide, $Pb(N_3)_2$, $\Delta H_f^\circ = 434$ kJ/mol, used as a detonator). The rapid rates of decomposition and the sudden expansion of the hot gaseous nitrogen ideally suit certain of these compounds for use as explosives (see Aside 28.1).

The Earth's atmosphere is 78% nitrogen (by volume), and nitrogen is obtained industrially from liquefied air. Among the many uses of this top-ten industrial chemical (see Table 26.9), the most important is in the manufacture of ammonia by the Haber process (see Aside 19.2).

Many other uses of nitrogen (Table 28.3) take advantage of either its lack of reactivity or its low boiling point. (Of the elements, only liquid hydrogen, helium, and neon have lower boiling points.) Chemicals that would react violently with oxygen or decompose in the presence of moist air are often "blanketed" with nitrogen to protect them from chemical change. Nitrogen is frequently swept through equipment to drive out air before oxygen- or moisture-sensitive chemicals are admitted. In the food industry, replacing air with nitrogen prevents the breakdown of food by oxidation, and also prevents mold growth and kills dormant insect eggs. The low temperatures provided by liquid nitrogen are used to fast-freeze foods and for the refrigeration of both food and biological specimens. The brittleness of materials frozen in liquid nitrogen allows, for example, spices to be ground more finely and unwanted sharp edges to be broken off molded plastic parts.

28.3 PHOSPHORUS

a. Allotropic Forms At ordinary temperatures, phosphorus usually is found as either white or red phosphorus (it has many other allotropes). White phosphorus is a soft, low-melting waxy solid that often appears yellow (sometimes called "yellow phosphorus") due to the formation of small amounts of red phosphorus on its surface. In the gaseous state below 800 °C, in the liquid state, and in the solid white form, phosphorus consists of P_4 molecules with P atoms at each corner of a regular tetrahedron (Figure 28.1). Above 800 °C, P_4 molecules are in equilibrium with P_2 molecules ($:P{\equiv}P:$). Condensation and cooling of gaseous phosphorus causes formation of white phosphorus.

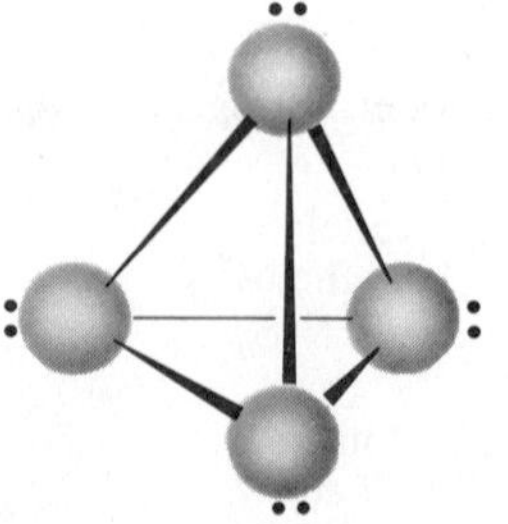

Figure 28.1
The P_4 Molecule of White Phosphorus

White phosphorus can ignite spontaneously in air and is usually stored under water, in which it is insoluble. When the supply of oxygen is extremely limited and moisture is present, oxidation of white phosphorus is slow and is accompanied by phosphorescence. White phosphorus is highly toxic. Small amounts can cause severe irritation of the lungs and gastrointestinal tract, and about 50 mg inhaled or ingested can be fatal. It should *never* be touched, for its ignition temperature is about the same as the temperature of the skin, and it can cause painful, slow-healing burns.

The story of the discovery of phosphorus is worth retelling. In the seventeenth century in Hamburg, Germany, Hennig Brand, a physician and alchemist, had what seems a very curious idea. He would try to obtain from urine a liquid with which he could convert silver into gold. Brand evaporated fresh urine and let the residue stand until it had putrefied. Then he heated this potent substance vigorously and collected in water the vapors that were given off. We have no way of knowing what he expected to see, but the white, flammable, waxy solid which glowed in the dark must surely have been a surprise. Brand was the first to discover an element not among those known since antiquity—he had distilled pure white phosphorus.

The second common allotropic form of phosphorus—red phosphorus—is obtained when the white variety is heated to about 250 °C in the absence of air or is exposed to light. Red phosphorus differs radically from the white allotrope. It is relatively nonpoisonous, is insoluble in those solvents that dissolve the white form (diethyl ether, benzene, carbon disulfide), melts at about 600 °C, and is denser. Red phosphorus is stable toward atmospheric oxidation at ordinary temperatures, but it ignites when heated to about 400 °C. In general, red phosphorus undergoes the same reactions as white phosphorus, but requires higher temperatures. Red phosphorus has a complex structure that is a polymer of P_4 molecules.

Heating white phosphorus under pressure converts it into the most thermodynamically stable form of phosphorus, black phosphorus, which is more highly polymeric and denser than red phosphorus, and is a semiconductor.

b. Sources, Preparation, and Uses of Phosphorus Phosphorus, the tenth most abundant element, occurs only in combined form, mainly as phosphates in various minerals. The most common phosphate-bearing minerals are fluorapatite, $Ca_5(PO_4)_3F$, and phosphorite, which is hydroxyapatite, $Ca_5(PO_4)_3OH$.

Elemental phosphorus is produced by treating **phosphate rock,** as the mineral deposits containing the calcium phosphates and silica are called, with coke in electric furnaces (Figure 28.2). In the overall reaction, phosphate is reduced to elemental phosphorus and coke is oxidized to carbon monoxide, with calcium being removed as calcium silicate slag.

$$2Ca_3(PO_4)_2(s) + 6SiO_2(s) + 10C(s) \xrightarrow{1200-1450\ ^\circ C} 6CaSiO_3(l) + 10CO(g) + P_4(g) \qquad \Delta H = 3050\ \text{kJ}$$

At the same time, the fluoride ion usually present in phosphate rock is converted by reaction with SiO_2 to volatile SiF_4, a toxic gas that must be removed from the vapor stream. Iron oxide, usually also present in phosphate rock to a small extent, reacts with phosphorus to give iron phosphide, known as "ferrophos."

Most phosphate rock is used directly in the manufacture of phosphate fertilizers rather than in the production of the element (see Aside 28.3). Only about 5% of all phosphorus used in a year is in elemental form (Table 28.4). Most of this is burned to give phosphorus(V) oxide, which is converted to pure phosphoric acid (Section 28.6b) and other chemicals.

Table 28.4
Uses of Elemental Phosphorus

Major use
Manufacture of P_4O_{10} and H_3PO_4
Other uses
Synthesis of phosphates and other chemicals
Matches, bombs, and fireworks
Rodent poisons
Alloying agent

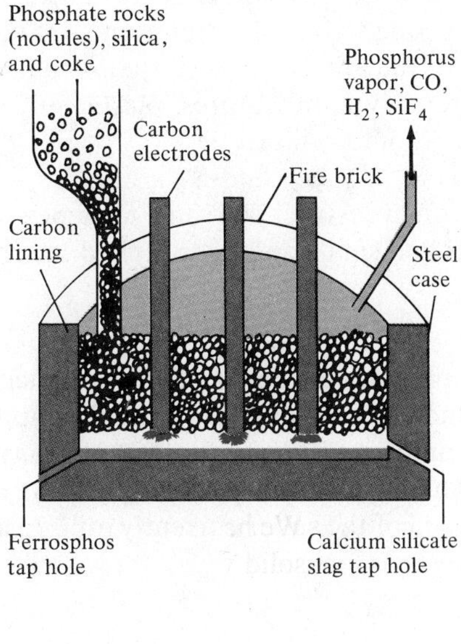

Figure 28.2
An Electric Phosphorus Furnace
A high voltage electric arc produces temperatures of 1200 °–1450 °C in the furnace. Molten ferrophos is heavy and sinks to the bottom, where it can be drawn off. Molten calcium silicate is less dense than the ferrophos and forms a second liquid layer that can be drawn off separately.
Source: A. D. F. Toy, *Phosphorus Chemistry in Everyday Living* (Washington, D.C.: American Chemical Society, 1976).

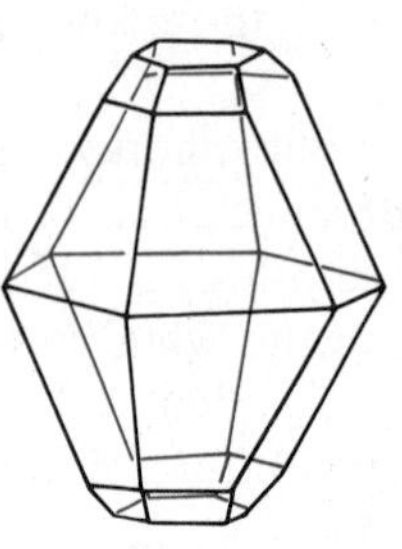

(a) Rhombic sulfur crystal

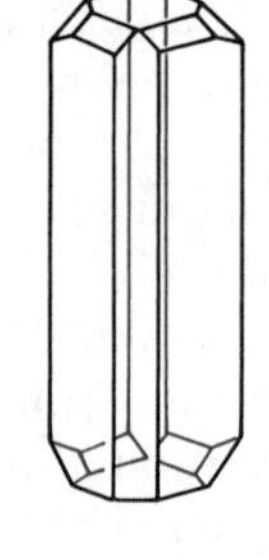

(b) Monoclinic sulfur crystal

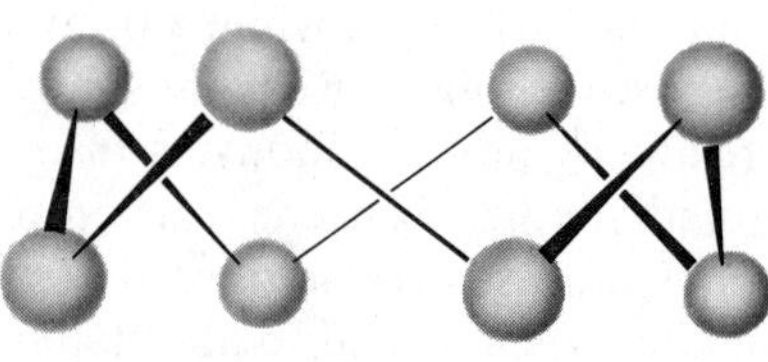

(c) S_8 ring structure

Figure 28.3
Forms of Sulfur

28.4 SULFUR

a. Allotropic Forms Like phosphorus, sulfur is a solid at room temperature and exists in various allotropic forms in which atoms are covalently bonded to each other. There are many more allotropes of sulfur (more than 30) than of phosphorus. Although elemental sulfur is reactive and combines directly with most elements, in none of its forms is it as difficult to handle or as toxic as white phosphorus. Elemental sulfur has been used medically as a laxative and in various preparations applied to the skin.

The form of sulfur stable at ordinary temperatures is rhombic sulfur (Figure 28.3a), which is crystalline, bright yellow in color, odorless, and tasteless. It is practically insoluble in water, but dissolves freely in carbon disulfide (CS_2). When heated to 95.6 °C rhombic sulfur undergoes a slow transition to monoclinic sulfur. The change is so slow that if rhombic sulfur is heated rapidly to its melting point of 112.6 °C, very little change takes place. Monoclinic sulfur (Figure 28.3b) is made up of long, needle-like crystals of m.p. 119 °C. In other properties, monoclinic sulfur is much the same as rhombic sulfur; both are composed of molecules in which eight sulfur atoms form a ring (Figure 28.3c).

All forms of solid sulfur melt to give a mobile straw-colored liquid. The melting process is accompanied to a very small extent by the breaking of S_8 rings. Between the melting point and about 160 °C there is little change in the mobility of the liquid sulfur, but between 160 °C and 187 °C, there is a *10,000-fold increase* in viscosity and the material turns brown. The increased viscosity is due to the formation of long chains of sulfur atoms that entangle each other. Above 200 °C, the viscosity decreases, probably as a result of the breaking of the long chains, and at 444.6 °C, the normal boiling point, the sulfur is again a mobile liquid. If the liquid material at about 200 °C is cooled very rapidly by pouring it into cold water, a soft rubbery substance known as plastic sulfur forms. In plastic sulfur, the long chains of sulfur atoms are in the form of coils, and the rubbery character is due to the ability of these chains to uncoil and coil again under stress. On standing at ordinary temperatures, plastic sulfur, and all other forms of sulfur, slowly revert to the rhombic allotrope.

Sulfur vapor contains S_8, S_6, S_4, and S_2 molecules, with the relative amounts varying with temperature and pressure. The same tendency of sulfur atoms to bond to each other in short chains and small rings is exhibited in species such as S_2Cl_2, S_2^{2-}, S_4^{2-}, and H_2S_6.

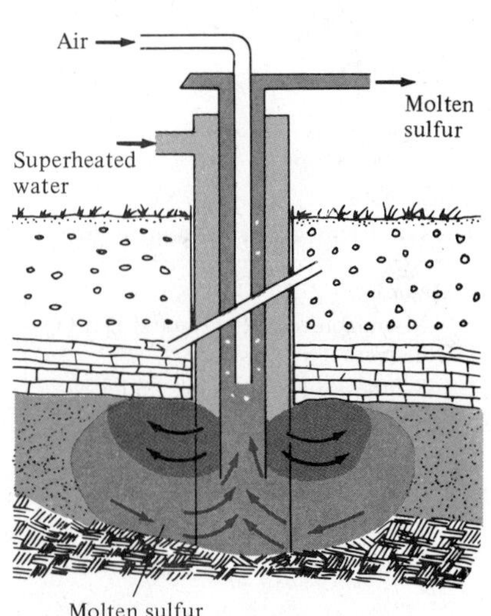

Figure 28.4
The Frasch Process for Sulfur Extraction
Three concentric pipes are sunk from the surface into the bed of sulfur. Water at a temperature of about 170 °C and a pressure of 100 lb/sq inch is forced down the outer pipe to melt the sulfur. Hot compressed air is pumped down the innermost pipe and mixes with the molten sulfur, forming a froth of water, air, and sulfur. This froth rises to the surface through the middle pipe and is collected in large bins. Removal of the water leaves sulfur with a purity of about 99.5%.

b. Sources, Preparation, and Uses of Sulfur Sulfur is much less abundant than oxygen and one-half as abundant as phosphorus, making up 0.052% (by mass) of the Earth's crust. It is found mainly as the free element; in heavy metal sulfides (e.g., PbS, ZnS, $CuFeS_2$), many of which are metal ores (Section 26.6); and in light metal sulfates (e.g., $MgSO_4$, $CaSO_4$). Much of the sulfur currently utilized in this country is obtained from vast deposits of the free element in Texas and Louisiana. The sulfur occurs in veins and pockets in beds of limestone and is covered with several hundred feet of quicksand and rock. It is brought to the surface by a process developed at the turn of this century and called the Frasch process after its inventor, the engineer Herman Frasch (the process is described in Figure 28.4; see also Aside 28.1).

Growing concern over pollution by sulfur dioxide (see Aside 28.4) has led to the removal of sulfur from natural gas and petroleum. In addition, sulfur-containing compounds are being removed from the effluent and stack gases of metal-smelting plants that process sulfide ores. Elemental sulfur from these operations is now coming into the market in increasing quantities. The major use of sulfur is in the manufacture of sulfuric acid (Table 28.5; Section 28.6a).

Table 28.5
Uses of Elemental Sulfur

Uses
Major use
Manufacture of sulfuric acid —roughly 90% of all elemental sulfur goes into sulfuric acid.
Other uses
Pulp and paper industry
Manufacture of carbon disulfide
Rubber
Fungicides, insecticides, and medicine

Aside 28.1 THOUGHTS ON CHEMISTRY: FRASCH DESCRIBES HIS FIRST SUCCESS

Herman Frasch, a German-born petroleum engineer, first demonstrated his process for recovering sulfur from natural deposits in Louisiana in about 1894. By 1900 the process had been developed for large-scale use. Following this, in just five years the production of sulfur in the United States expanded thirty-fold. Frasch himself described his first success with the process as follows:

After permitting the melting fluid to go into the ground for twenty-four hours, I decided that sufficient material must have been melted to produce some sulphur. The pumping engine was started on the sulphur line, and the increasing strain against the engine showed that work was being done. More and more slowly went the engine, more steam was supplied, until the man at the throttle sang out at the top of his voice, "She's pumping." A liquid appeared on the polished rod, and when I wiped it off I found my finger covered with sulphur. Within five minutes the receptacles under pressure were opened and a beautiful stream of the golden fluid shot into the barrels we had ready to receive the product. After pumping for about fifteen minutes, the forty barrels we had supplied were seen to be inadequate. Quickly we threw up embankments and lined them with boards to receive the sulphur that was gushing forth; and since that day no further attempt has been made to provide a vessel or a mold into which to put the sulphur.

When the sun went down we stopped the pump to hold the liquid sulphur below until we could prepare to receive more in the morning. The material on the ground had to be removed, and willing hands helped to make a clean slate for the next day. When everything had been finished, the sulphur all piled up in one heap, and the men had departed, I enjoyed all by myself this demonstration of success. I mounted the sulphur pile and seated myself on the very top. It pleased me to hear the slight noise caused by the contraction of the warm sulphur, which was like a greeting from below—proof that my object had been accomplished. Many days and many years intervened before financial success was assured, but the first step towards the ultimate goal had been achieved. We had melted the mineral in the ground and brought it to the surface as a liquid. We had demonstrated that it could be done.

Source: Herman Frasch, reprinted in part from "Address of Acceptance, Perkin Medal Award," in *The Journal of Industrial and Engineering Chemistry*, February 1912, pp. 131–140; also reprinted in *CHEM TECH*, February 1976, pp. 99–105. Published 1912 and 1976 by the American Chemical Society.

28.5 SOME CHEMICAL REACTIONS OF NITROGEN, PHOSPHORUS, AND SULFUR

a. Combination with Oxygen At elevated temperatures nitrogen combines with oxygen to form nitrogen(II) oxide (NO, nitric oxide) in an equilibrium that favors the uncombined elements

$$N_2(g) + O_2(g) \rightleftharpoons 2NO(g) \qquad \textbf{(28.1)}$$

This reaction occurs in the air during any combustion at high temperature or when lightning strikes, and plays a role in the formation of smog (Aside 28.4).

White phosphorus ignites spontaneously in air, whereas red phosphorus burns in air only when heated to 400 °C.

$$4P(s) + 5O_2(g, xs) \xrightarrow[\text{to 400 °C}]{\text{white P, spontaneous in air; red P, ignites when heated}} \underset{\textit{phosphorus(V) oxide}}{P_4O_{10}(s)} \qquad \textbf{(28.2)}$$

Sulfur, once ignited, burns with a blue flame to produce mainly sulfur dioxide, together with some sulfur trioxide:

$$S(s) + O_2(g) \longrightarrow SO_2(g) \qquad \Delta H° = -296.8 \text{ kJ} \qquad \textbf{(28.3)}$$

$$2SO_2(g) + O_2(g) \longrightarrow 2SO_3(g) \qquad \Delta H° = -197.8 \text{ kJ} \qquad \textbf{(28.4)}$$

Although we know that phosphorus and sulfur exist as $P_4(s)$ and $S_8(s)$ molecules, for simplicity it is customary to write P(*s*) and S(*s*) in equations, as above.

b. Combination with Other Elements The only element with which nitrogen combines at room temperature is lithium, the strongest elemental reducing agent.

$$6Li(s) + N_2(g) \longrightarrow 2Li_3N(s) \qquad \textbf{(28.5)}$$

At higher temperatures, presumably because it is partially dissociated into atoms, nitrogen is more reactive and combines with a number of different elements to form nitrides such as magnesium nitride (Mg_2N_3) or boron nitride (BN) (Section 28.7). We have discussed the combination with hydrogen to form ammonia (the Haber process) as an example of the roles of thermodynamics and kinetics in industrial processes (Aside 19.2).

Phosphorus and sulfur combine with most elements. In reactions with the halogens, phosphorus yields either PX_3 or PX_5, except that the pentaiodide is not known. Apparently five large iodine atoms cannot fit around the smaller phosphorus atom. The products of the combination of sulfur with the halogens (again, except for iodine) vary with the conditions and sulfur–sulfur bonds can be formed, for example

$$2S(s) + Cl_2(g) \xrightarrow{\Delta} \underset{\textit{disulfur dichloride (Cl—S—S—Cl)}}{S_2Cl_2(l)}$$

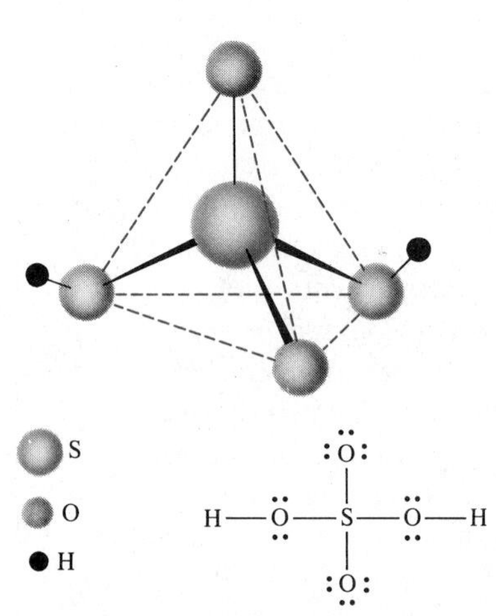

(a) Sulfuric acid, H_2SO_4

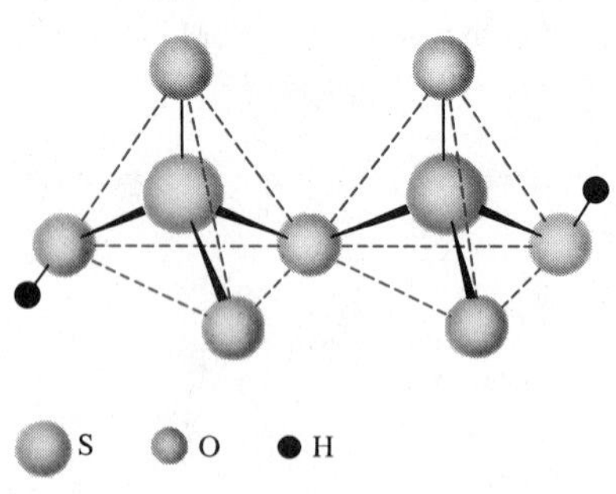

(b) Pyrosulfuric acid, $H_2S_2O_7$

Figure 28.5
Sulfur (VI) Oxoacids
Pyrosulfuric acid (disulfuric acid) is equivalent to two H_2SO_4 molecules combined by the loss of an H_2O molecule between them.

28.6 INDUSTRIAL ACIDS

Acids, like alkalis (Section 26.13), are essential industrial chemicals. They are used for adjusting the pH in industrial processes and for properties specific to the individual acids. Every year sulfuric acid is the largest-volume industrial chemical. It is the least expensive nonvolatile acid and is also valuable for its dehydrating and redox properties. It is often said that the amount of sulfuric acid a country uses is a direct measure of its industrial progress. Next after sulfuric acid, phosphoric and nitric acids are the two acids of greatest industrial importance (see Table 26.10).

a. Sulfuric Acid Pure sulfuric acid (H_2SO_4)(Figure 28.5a) is a viscous, oily liquid which begins to boil at 290 °C with decomposition into sulfur trioxide (which escapes as a gas) and water (some of which remains in the solution). Once the composition of 98.3% H_2SO_4 and a boiling point of 338 °C are reached, the composition remains unchanged, giving the usual concentrated laboratory acid, which is 18 M. When the concentrated acid is added to water, a large amount of heat is evolved, partly due to the exothermic formation of hydrates.

Sulfuric acid is produced by the catalyzed oxidation of sulfur dioxide followed by hydration of the resultant sulfur trioxide. The sulfur dioxide may be obtained by burning sulfur (Equation 28.3) or from sulfur-containing by-products of other operations. The successful oxidation of sulfur dioxide, like the reaction of nitrogen and hydrogen to give ammonia, must achieve a compromise between the kinetics and thermodynamics of the reaction. The combination of oxygen with sulfur dioxide is exothermic and reaches an equilibrium. High yields of sulfur trioxide are therefore favored by low temperatures. However, the reaction rate at low temperatures, even in the presence of a catalyst, is too slow to be economically practical. One compromise is to pass the SO_2–O_2 mixture over the catalyst at a temperature at which the rate is sufficiently high (575 °C) to give an 80% yield of the trioxide and then raise the yield by subsequent passes at lower temperatures at which the percentage of conversion is higher.

The sulfur trioxide is not converted to sulfuric acid by direct addition to water. Gaseous sulfur trioxide reacts with water to form a mist of sulfuric acid, which is absorbed only slowly in the liquid water. To avoid this problem, sulfur trioxide is passed into 98% sulfuric acid, in which it is extremely soluble, to form pyrosulfuric acid, $H_2S_2O_7$ (Figure 28.5b). By addition of water, sulfuric acid of the desired concentration is obtained.

$$\underset{\text{sulfur trioxide}}{SO_3(g)} + \underset{\text{sulfuric acid}}{H_2SO_4(l)} \rightleftharpoons \underset{\text{pyrosulfuric acid}}{H_2S_2O_7(l)} \quad (28.6)$$

$$H_2S_2O_7(l) + H_2O(l) \longrightarrow 2H_2SO_4(aq) \quad (28.7)$$

Over two-thirds of the sulfuric acid produced in the United States is used in making phosphate fertilizers (Aside 28.3). Some of the other uses, of which there are many, are listed in Table 28.6.

One distinctive property of sulfuric acid is the ease with which it forms hydrates (e.g., $H_2SO_4 \cdot H_2O$, $H_2SO_4 \cdot 2H_2O$). This ability makes concentrated sulfuric acid an extremely effective drying agent for wet gases and liquids, provided that the gas or liquid does not react with the acid. Sulfuric acid can also remove the elements of water from organic compounds that themselves contain no free water. Table sugar, or sucrose, has the formula $C_{12}H_{22}O_{11}$. Although no water is present in the compound, the hydrogen-to-oxygen ratio is the same as that in water. When concentrated sulfuric acid is added to sugar, these elements are removed in the form of water, and a residue of carbon is left.

$$C_{12}H_{22}O_{11}(s) + 11H_2SO_4(conc) \longrightarrow 12C(s) + 11(H_2SO_4 \cdot H_2O)$$

The same type of reaction accounts for the charring of wood, paper, cotton, and wool and the destruction of skin by the concentrated acid.

Hot concentrated sulfuric acid is a moderately strong oxidizing agent, although not as strong as nitric acid. The sulfur atom is usually (though not always) reduced from its oxidation state of +6 to +4, for example

$$Cu(s) + 2H_2SO_4(conc) \longrightarrow CuSO_4(aq) + SO_2(g) + 2H_2O(l)$$
$$C(s) + 2H_2SO_4(conc) \longrightarrow CO_2(g) + 2SO_2(g) + 2H_2O(l)$$

At lower temperatures concentrated sulfuric acid is a weaker oxidixing agent, and in dilute solutions its reactions are like those of any nonoxidizing acid.

b. Phosphoric Acid Pure phosphoric acid (H_3PO_4, also called orthophosphoric acid) is a white solid that melts at 42.35 °C to a viscous liquid with a strong tendency to supercool. In structure, H_3PO_4 is like H_2SO_4 (see Figure 28.5a), but with an H atom on each of three O atoms. Phosphoric acid is one of the stronger weak acids. ($K_{a1} = 7.5 \times 10^{-3}$.) It is a poor oxidizing agent and is useful when oxidizing properties are undesirable.

The two major industrial processes for phosphoric acid yield products of quite different purity. Thermal process phosphoric acid is manufactured from phosphorus produced by the electric furnace method, which has been distilled and is therefore quite pure. In the two-step process the elemental phosphorus is oxidized and then the oxide is allowed to react with water.

$$4P(s) + 5O_2(g) \longrightarrow P_4O_{10}(s) \quad (28.8)$$
$$P_4O_{10}(s) + 6H_2O(l) \longrightarrow 4H_3PO_4(aq) \quad (28.9)$$

Table 28.6
Uses of Sulfuric Acid
A few of the most important of the hundreds of ways of using sulfuric acid are listed.

Uses of Sulfuric Acid
Major uses
Manufacture of fertilizers, e.g.,
Ammonium sulfate
Superphosphates
Other uses
Synthesis of other chemicals
Pigments and paints
Petroleum refining
Iron and steel
Storage batteries
Metal production and processing
Paper and pulp processing
Rayon production

The phosphoric acid produced in this way is pure enough for most industrial uses that require relatively high purity, although small amounts of arsenic present as H_3AsO_4 must be removed if the acid is destined for use in food or pharmaceuticals.

Thermal process phosphoric acid is used in making detergent components (tripolyphosphates; Section 28.13), in metal treatment, in refractories (substances that withstand high temperature), in catalysts, and in foods and beverages. Dilute phosphoric acid is not toxic and has a pleasant sour taste similar to that of the citric acid and acetic acid present in many foods. For this reason it is widely used as a tart flavoring agent, especially in carbonated beverages. It has a number of other applications in the food industry, for example, as a buffer in jams and jellies, and as a cleaning agent for dairy equipment.

A less pure product, wet process phosphoric acid, is made by the reaction between the calcium phosphate in phosphate rock and sulfuric acid.

$$Ca_3(PO_4)_2(s) + 3H_2SO_4(aq) \longrightarrow 2H_3PO_4(aq) + 3CaSO_4(s) \qquad \textbf{(28.10)}$$

The acid from this process is generally concentrated by boiling off water to give an 85% aqueous solution, which is known as "syrupy" phosphoric acid because of its viscous nature. Wet process phosphoric acid, which represents over 90% of the phosphoric acid produced worldwide, is used directly in the production of fertilizers and animal feed supplements.

c. Nitric Acid Nitric acid (HNO_3) is a colorless, fuming liquid with a choking odor. The pure acid is rarely encountered, as it is difficult to prepare and decomposes rapidly. Both the pure acid and its solutions decompose in sunlight to give nitrogen dioxide.

$$4HNO_3(aq) \xrightarrow{hv} 4NO_2(aq) + 2H_2O(l) + O_2(g) \qquad \textbf{(28.11)}$$

Old solutions of nitric acid are usually yellow or brown because of the presence of dissolved nitrogen dioxide.

Almost all nitric acid is produced by the oxidation of ammonia. Anhydrous ammonia and air are the sole raw materials in the three-step process, known as the Ostwald process:

$$4NH_3(g) + 5O_2(g) \xrightarrow[\sim 900\ ^\circ C]{Pt-Rh\ catalyst} 4NO(g) + 6H_2O(l) \qquad \textbf{(28.12)}$$

$$2NO(g) + O_2(g) \longrightarrow 2NO_2(g) \qquad \textbf{(28.13)}$$

$$3NO_2(g) + H_2O(l) \longrightarrow 2HNO_3(aq) + NO(g) \qquad \textbf{(28.14)}$$

The first step, the catalytic oxidation of ammonia, is rapid. Once the reaction has been initiated, the ammonia burns spontaneously to give a 90% yield of nitrogen(II) oxide. The second step is a reaction that becomes less favorable at high temperatures. The hot gases from reaction (28.12) are cooled and more air is added to produce nitrogen dioxide in this step. In the third step, the nitrogen dioxide is absorbed by water to produce nitric acid in a disproportionation reaction. The nitrogen(II) oxide formed in this step is recycled. In modern plants the entire process is carried out under pressure.

Like so many of the chemicals discussed in this chapter, nitric acid has its major use in the manufacture of fertilizers. It is also important in the nitration reactions that produce many organic nitrates used in explosives, plastics, dyes, and lacquers.

In any concentration higher than about 2 M, nitric acid can function as an oxidizing agent. It is the strongest oxidizing agent of the three major industrial acids. As we saw in our study of oxidizing agents (Section 17.5a), prediction of the nitrogen-containing reduction products from nitric acid is difficult. There are many possible oxidation states and the products vary with the conditions. The most common are nitrogen dioxide (NO_2) or nitrogen(II) oxide (NO). Other possible reduction products are NH_4^+, N_2, HNO_2, and N_2O.

All metals except the least reactive ones, such as gold and platinum, are oxidized by nitric acid. In some cases the reaction must be initiated by heat, but once started it proceeds vigorously. In general, the stronger the reducing agent and the more dilute the acid, the lower will be the oxidation state of the nitrogen in the reduction product. For example, zinc, a reasonably active metal, reduces dilute nitric acid all the way to ammonium ion.

$$4Zn(s) + 10H^+ + \overset{+5}{N}O_3^- \xrightarrow[]{\text{dil HNO}_3} 4Zn^{2+} + \overset{-3}{N}H_4^+ + 3H_2O(l) \qquad \textbf{(28.15)}$$

In somewhat less dilute acid, elemental nitrogen is produced in the reaction of zinc with nitric acid. With *concentrated* nitric acid, nitrogen dioxide is the main reduction product, for example

$$Cu(s) + 4H^+ + 2\overset{+5}{N}O_3^- \xrightarrow[]{\text{conc HNO}_3} Cu^{2+} + 2\overset{+4}{N}O_2(g) + 2H_2O(l) \qquad \textbf{(28.16)}$$

With warm dilute nitric acid, the main reduction product in the reaction with copper is nitrogen(II) oxide (NO) (Equation 28.23).

A number of metals that are attacked by dilute nitric acid—for example, iron and chromium—are inert to the concentrated acid. These metals are said to be *passive* toward the acid; they are protected by oxide coatings.

A 3-to-1 by volume mixture of concentrated HCl and HNO_3 is known as **aqua regia** ("royal water") because of its ability to dissolve such noble metals as gold and platinum, which are inert to either acid alone. The redox reaction between concentrated nitric and hydrochloric acids produces water, elemental chlorine, and nitrosyl chloride.

$$3HCl(aq) + HNO_3(aq) \longrightarrow Cl_2(g) + \underset{\text{nitrosyl chloride}}{NOCl(g)} + 2H_2O(l) \qquad \textbf{(28.17)}$$

The solvent action of aqua regia probably results mainly from the action of the nitrosyl chloride and chlorine, which convert the metals initially to chlorides. The metal chlorides are then transformed to stable complex anions by reaction with chloride ion. The overall reaction for gold is

$$Au(s) + 4H^+ + 4Cl^- + NO_3^- \longrightarrow [AuCl_4]^- + NO(g) + 2H_2O(l) \qquad \textbf{(28.18)}$$

The properties of sulfuric acid, phosphoric acid, and nitric acid are summarized in Table 28.7.

Table 28.7
Properties of Sulfuric Acid, Phosphoric Acid, and Nitric Acid

Sulfuric acid

$$H_2SO_4(aq) + H_2O(l) \longrightarrow HSO_4^- + H_3O^+ \quad \textit{strong acid}$$
$$HSO_4^- + H_2O(l) \rightleftharpoons SO_4^{2-} + H_3O^+ \quad K_a = 1.0 \times 10^{-2}$$

M.p. pure acid 10.4 °C; 98% acid, 3 °C
B.p. ~290 °C with decomposition ($SO_3 + H_2O$)
Density of 98% acid: 1.84 g/mL
Concentrated acid: 18 M, 98% H_2SO_4 by mass
Hot, concentrated H_2SO_4 is a strong oxidizing acid. Dilute H_2SO_4 is not an oxidizing acid.

Phosphoric acid

$$H_3PO_4(aq) + H_2O(l) \rightleftharpoons H_2PO_4^- + H_3O^+ \quad K_a = 7.5 \times 10^{-3}$$
$$H_2PO_4^- + H_2O(l) \rightleftharpoons HPO_4^{2-} + H_3O^+ \quad K_a = 6.6 \times 10^{-8}$$
$$HPO_4^{2-} + H_2O(l) \rightleftharpoons PO_4^{3-} + H_3O^+ \quad K_a = 1 \times 10^{-12}$$

M.p. 42.35 °C; at 200 °C $\longrightarrow H_4P_2O_7(l)$
Density: 100% acid, 1.87 g/mL; 85% acid, 1.69 g/mL
Concentrated acid: 14.8 M, 85% H_3PO_4 by mass
H_3PO_4 is *not* an oxidizing acid.

Nitric acid

$$HNO_3(aq) + H_2O(l) \longrightarrow NO_3^- + H_3O^+ \quad \textit{strong acid}$$

M.p. −41.6 °C
B.p. 83 °C decomposes in sunlight (to NO_2)
Density: pure acid, 1.50 g/mL; 70.4% acid, 1.42 g/mL
Concentrated acid: 15.9 M, 70.4% HNO_3 by mass
In any concentration >2 M HNO_3 can be an oxidizing acid.

Table 28.8
Outstanding Properties of Nitrogen, Phosphorus, and Sulfur

Nitrogen	
N [He] $2s^2 2p^3$ (*5 valence electrons*)	Member of Representative Group V; cannot form more than four covalent bonds; differs from other family members (Table 28.1)
$:N\equiv N:(g)$ (*very strong bond*)	Second highest bond dissociation energy for any diatomic molecule (only C≡O is higher); unreactive molecule at ordinary temperatures; many uses based on inertness; obtained from liquid air
$N_2(l)$	Many uses based on low b.p.
$\overset{-3}{N}$ to $\overset{+5}{N}$	Forms compounds in all possible oxidation states from −3 to +5
N⋯O	Many N—O bonds are intermediate between single and double bonds
$2Li(s) + 3N_2(g) \longrightarrow 2LiN_3(s)$	Only combination at room temperature
Phosphorus	
P [Ne]$3s^2 3p^3$ (*5 valence electrons*)	Member of Representative Group V; can form more than four covalent bonds, e.g., PF_5
P_4	Formula for white phosphorus in all states; highly reactive; red and black allotropic forms are polymeric, less reactive; obtained from phosphate rock
$\overset{-3}{P}$, $\overset{+3}{P}$, $\overset{+5}{P}$	Most common oxidation states
P⋯O	Many P—O bonds are intermediate between single and double bonds
P(s) + E ⟶	Combines with many elements
Sulfur	
S [Ne]$3s^2 3p^4$ (*6 valence electrons*)	Member of Representative Group VI; can form more than four covalent bonds, e.g., SF_6
S_8 (*cyclic*)	Many allotropic forms, some with S_8 rings; strong tendency of sulfur atoms to self-link; obtained by Frasch process from free sulfur deposits
$\overset{-2}{S}$, $\overset{+4}{S}$, $\overset{+6}{S}$	Most common oxidation states
S⋯O	Many S—O bonds are intermediate between single and double bonds
S(s) + E ⟶	Combines with many elements

SUMMARY OF SECTIONS 28.1–28.6

Nitrogen and phosphorus are the nonmetals of Representative Group V; sulfur and oxygen (Chapter 16) are the nonmetals of Representative Group VI. Nitrogen differs significantly from other group members (Table 28.1). Nitrogen, phosphorus, and sulfur are important in many industrial chemicals and in the chemistry of living things. Sulfuric acid, phosphoric acid, and nitric acid (Table 28.7) are the three largest-volume industrial acids. The outstanding properties of nitrogen, phosphorus, and sulfur are summarized in Table 28.8.

COMPOUNDS OF NITROGEN

Table 28.9 gives examples of the compounds of nitrogen in all of its oxidation states from −3 to +5.

Table 28.9
Oxidation States and Some Compounds of Nitrogen

Oxidation state	Compound	Name
−3	NH_3	Ammonia
	Li_3N, Ca_3N_2	Nitrides
	NH_4^+	Ammonium ion
−2	N_2H_4	Hydrazine
−1	NH_2OH	Hydroxylamine
0	N_2	Nitrogen
+1	N_2O	Nitrogen(I)oxide (nitrous oxide)
+2	NO	Nitrogen(II)oxide (nitric oxide)
+3	N_2O_3	Nitrogen(III)oxide
	HNO_2, NO_2^-	Nitrous acid, nitrite ion
+4	NO_2	Nitrogen dioxide
	N_2O_4	Dinitrogen tetroxide
+5	N_2O_5	Nitrogen(V)oxide (dinitrogen pentoxide)
	HNO_3, NO_3^-	Nitric acid, nitrate ion

(Oxidation ↓ ; Reduction ↑)

28.7 NITRIDES

Nitrides are of several different types. The salt-like nitrides are ionic compounds that contain the nitride ion (N^{3-}). These are white, crystalline, high-melting solids formed by lithium, beryllium family metals, cadmium, and zinc. The nitride ion is a very strong base, and the salt-like nitrides react vigorously and completely with water, with the formation of ammonia and OH^-.

$$N^{3-} + 3H_2O(l) \longrightarrow NH_3(aq) + 3OH^- \qquad \textbf{(28.19)}$$

Some covalent nitrides are molecular compounds formed with nonmetals. Many of these compounds are volatile and unstable (e.g., Cl_3N, which is very unstable, and S_4N_4, which is detonated by shock or temperatures above 30 °C). Other covalent nitrides are very hard, high-melting network covalent compounds known as diamond-like nitrides. Silicon, boron, and the other boron family elements form diamond-like nitrides. Boron nitride (BN) is nearly identical in structure with graphite (see Figure 29.16), with boron and nitrogen atoms alternating throughout. Just as graphite can be converted to diamond (Section 29.3), boron nitride is converted by heat and pressure to a very hard, diamond-like material called borazon, which is used as an abrasive.

The interstitial nitrides formed by many of the *d*-transition metals (e.g., titanium, iron, tungsten) are hard, high-melting alloys that retain the metallic properties of luster and conductivity. As in the interstitial hydrides (Section 16.6) and carbides (Section 29.5e), the small nonmetal atoms occupy holes in the metal crystal structure. The metal nitrides and the diamond-like nitrides have many applications that make use of their properties of hardness and resistance to temperature and chemical attack.

28.8 AMMONIA, AMMONIUM SALTS, AND HYDRAZINE

Ammonia (NH_3) is a colorless gas with a pungent odor familiar to everyone who has used household ammonia as a cleaning agent. In such use, it should be kept in mind that ammonia is a powerful heart stimulant and that excessive inhalation may produce serious effects, even death.

In the gaseous state, ammonia exists as discrete polar molecules of triangular pyramidal shape (see Figure 11.2). In the liquid and solid states, its molecules are extensively hydrogen-bonded, as would be expected from the high electronegativity of nitrogen.

Many of the chemical properties of ammonia result from the unshared pair of electrons on the nitrogen atom. The ammonia molecule is an electron-pair donor (a Lewis base), and by sharing its electron pair with metal ions becomes a ligand in many complex ions. It is also a weak Brønsted-Lowry base, gaining a proton from water to give an alkaline solution (thus making it a useful cleaning agent):

$$NH_3(aq) + H_2O(l) \rightleftharpoons NH_4^+ + OH^-$$

Although aqueous ammonia solutions are often labeled "ammonium hydroxide," the compound NH_4OH has neither been isolated nor observed in solution.

Like nitrogen, ammonia is a top-ten industrial chemical (see Table 26.10). More than 80% of the ammonia produced each year is used either as fertilizer or in the manufacture of other fertilizers. Ammonia is also the raw material for the production of nitric acid (Section 28.6c), many explosives, and most other nitrogen-containing compounds.

Ammonia reacts with acids to give ammonium salts, for example

$$NH_3(g) + HCl(g) \longrightarrow NH_4Cl(s)$$

Most ammonium salts are very soluble in water, and the resulting solutions are acidic due to hydrolysis of the ammonium ion, which is a weak Brønsted-Lowry acid

$$NH_4^+ + H_2O(l) \rightleftharpoons NH_3(aq) + H_3O^+$$

As we saw in our study of chemical reactions, all ammonium salts undergo thermal decomposition. Salts with nonoxidizing anions give ammonia and the parent acids (Section 17.3b). Ammonium hydrogen carbonate (NH_4HCO_3) decomposes to some extent even at room temperature (to NH_3, CO_2, and H_2O). For this reason the compound is used in smelling salts, once popular for reviving ladies prone to fainting spells.

Ammonium salts containing oxidizing anions undergo redox decomposition when heated (Section 17.4b). For example, ammonium nitrate (NH_4NO_3) decomposes as follows at 170–260 °C.

$$\underset{\textit{ammonium nitrate}}{\overset{-3}{N}H_4\overset{+5}{N}O_3(s)} \xrightarrow{170-260\ ^\circ C} \overset{+1}{N_2}O(g) + 2H_2O(g)$$

In this reaction, the nitrogen in the ammonium ion is oxidized, whereas that in the nitrate ion is reduced. Ammonium nitrate is a dangerous solid and should be handled with great care. Decomposition, particularly in a limited space or at higher temperatures, may occur with explosive violence, with N_2O being reduced to elemental nitrogen (see Equation 28.22) in the overall reaction

$$2\overset{-3}{N}H_4\overset{+5}{N}O_3(s) \xrightarrow{>300\ ^\circ C} 2\overset{0}{N_2}(g) + 4H_2O(g) + O_2(g)$$

Mixtures of ammonium nitrate and oxidizable materials are also dangerous. Accidental exposure to elevated temperatures may initiate oxidation, and this in turn may increase the temperature to the point of detonation. With the intentional addition of a detonator, ammonium nitrate is used as an explosive. Proper precautions, however, allow ammonium nitrate to be handled safely. It is widely used as a fertilizer and is an important industrial chemical (see Table 26.10).

Anhydrous hydrazine (H_2NNH_2) is a fuming colorless liquid and a potent reducing agent. It is converted mainly to nitrogen by strong oxidizing agents such as hydrogen peroxide, liquid oxygen, or nitrogen tetroxide. For example

$$H_2NNH_2(l) + O_2(g) \longrightarrow N_2(g) + 2H_2O(g) \qquad \Delta H^\circ = -543.3\ \text{kJ} \qquad \textbf{(28.20)}$$

$$2H_2NNH_2(l) + N_2O_4(g) \longrightarrow 3N_2(g) + 4H_2O(g) \qquad \Delta H^\circ = -1078\ \text{kJ} \qquad \textbf{(28.21)}$$

Hydrazine ($\Delta H_f = 50$ kJ/mol) is representative of compounds that have positive heats of formation and thus decompose with the release of large amounts of heat. Hydrazine and substituted hydrazines are extensively used as rocket fuels. A reaction like (28.21) powered the Apollo lunar excursion vehicle. Reaction (28.20) allows hydrazine to be used for removing corrosion-causing oxygen from the water in high-pressure boilers.

28.9 NITROGEN OXIDES

There is an oxide of nitrogen for each oxidation state from +1 to +5 (see Table 28.9). The properties of the more common nitrogen oxides are summarized in Table 28.10. These oxides usually are referred to by their common names.

Nitrogen(I) oxide (nitrous oxide, N_2O) was the first synthetic anesthetic to be discovered and is still in use as a light anesthetic, especially in dentistry. When inhaled in low concentrations, the gas produces mild euphoria—the basis for another of its names, laughing gas. This oxide was also the first aerosol propellant. It is quite soluble in fat and was introduced before World War II as the propellant in whipped cream dispensers. When the pressure inside the can is lowered by pushing the nozzle, the N_2O passes through the cream as it escapes and whips the cream. At high temperatures N_2O decomposes to give nitrogen and oxygen, a reaction which can occur with explosive force:

$$2N_2O(g) \longrightarrow 2N_2(g) + O_2(g) \qquad \Delta H^\circ = -164.1 \text{ KJ} \qquad \textbf{(28.22)}$$

Nitrogen(II) oxide (nitric oxide, NO) can be prepared by the redox reaction between copper and warm, dilute nitric acid.

$$3Cu(s) + 8H^+ + 2NO_3^- \longrightarrow 3Cu^{2+} + 2NO(g) + 4H_2O(l) \qquad \textbf{(28.23)}$$

It has an unpaired electron and is a fairly reactive substance, combining with all of the halogens except iodine to give nitrosyl halides (NOX). However, NO does not react with water.

In the presence of oxygen at ordinary temperatures, NO adds an oxygen atom to give the nitrogen(IV) oxides, which are always in equilibrium with each other and thus difficult to study separately. They are more descriptively known by their common names, nitrogen dioxide (NO_2) and dinitrogen tetroxide (N_2O_4):

$$\underset{\text{nitrogen dioxide, red-brown gas}}{2NO_2(g)} \rightleftharpoons \underset{\text{dinitrogen tetroxide, colorless gas}}{N_2O_4(g)} \qquad \Delta H^\circ = -57.20 \text{ kJ} \qquad \textbf{(28.24)}$$

Dinitrogen tetroxide is a **dimer**—a molecule formed by the combination of two identical molecules. Two molecules of nitrogen dioxide combine to form dinitrogen tetroxide by sharing the unpaired electrons on their nitrogen atoms. The nitrogen(IV) oxides are both strong oxidizing agents. Rather than simply combining with water,

Table 28.10
The Common Oxides of Nitrogen

Formula and common name	Oxidation state of nitrogen	Structure	Properties
N_2O Nitrous oxide	I	$:N\equiv N-\ddot{\underset{\cdot\cdot}{O}}: \longleftrightarrow :\underset{\cdot\cdot}{N}=N=\ddot{O}:$	Colorless gas; quite stable; anesthetic (laughing gas); aerosol propellant; released during nitrification by bacteria in soil
NO Nitric oxide	II	$:\dot{N}=\ddot{O}:$	Colorless gas; one unpaired electron; fairly reactive; air pollutant (by-product of any combustion in air); combines with O_2 to give NO_2 at ordinary temperatures; formed by oxidation of NH_3 in HNO_3 production
NO_2 Nitrogen dioxide	IV	$:\ddot{O}=\dot{N}-\ddot{\underset{\cdot\cdot}{O}}: \longleftrightarrow :\ddot{\underset{\cdot\cdot}{O}}-\dot{N}=\ddot{O}:$	Red-brown gas; one unpaired electron; reactive; air pollutant; always in equilibrium: $NO_2(g) \rightleftharpoons N_2O_4$; intermediate in HNO_3 production
N_2O_4 Dinitrogen tetroxide	IV	$(:\ddot{\underset{\cdot\cdot}{O}}-)(:\ddot{O}=)N-N(-\ddot{\underset{\cdot\cdot}{O}}:)(=\ddot{O}:) \longleftrightarrow$ 3 other resonance forms	Colorless gas or liquid; present with NO_2 in all states; amount greater at lower temperatures

nitrogen dioxide (NO_2) disproportionates. In cold water it forms both nitrogen oxoacids

$$2NO_2(g) + H_2O(l) \longrightarrow HNO_3(aq) + HNO_2(aq) \quad \textbf{(28.25)}$$

At a higher temperature, in a reaction that is used in the commercial preparation of nitric acid (see Equation 28.14), it disproportionates to nitric acid plus nitric oxide (NO). (The role of NO and NO_2 in air pollution is discussed in Aside 28.4.)

Nitrogen(III) oxide (N_2O_3) is the least stable nitrogen oxide and is never pure, but is always in equilibrium with the nitrogen(II) and (IV) oxides

$$\underset{\textit{nitric oxide}}{NO(g)} + \underset{\textit{nitrogen dioxide}}{NO_2(g)} \overset{-20\ ^\circ C}{\rightleftharpoons} \underset{\textit{dinitrogen trioxide}}{N_2O_3(l)} \quad \textbf{(28.26)}$$

Nitrogen(V) oxide (dinitrogen pentoxide, N_2O_5), the only solid nitrogen oxide (m.p. 30 °C), is the anhydride of nitric acid

$$N_2O_5(s) + H_2O(l) \longrightarrow 2HNO_3(aq) \quad \textbf{(28.27)}$$

28.10 NITROGEN OXOACIDS AND THEIR SALTS

Nitrous acid (HNO_2) is a weak acid known only in aqueous solution

$$HNO_2(aq) \rightleftharpoons H^+ + NO_2^- \qquad K_a = 7.2 \times 10^{-4} \text{ at } 25\ ^\circ C$$

Even in cold solutions some decomposition occurs to give nitric acid and nitric oxide, another nitrogen disproportionation reaction.

$$3HNO_2(aq) \longrightarrow HNO_3(aq) + 2NO(g) + H_2O(l) \quad \textbf{(28.28)}$$

Nitrous acid can be prepared by passing equal molar amounts of nitric oxide (NO) and nitrogen dioxide (NO_2) into water, a reaction which is equivalent to adding water to nitrogen (III) oxide (see Equation 28.26).

Because of the intermediate +3 oxidation state of nitrogen, nitrous acid can act as either an oxidizing or reducing agent. It is a strong oxidizing agent in acidic solution, with the reduction product depending upon the strength of the reducing agent. It is a relatively weak reducing agent and is oxidized (to nitric acid) only by such strong oxidizing agents as chlorine and permanganate ion.

Nitrites, the salts of nitrous acid, are colorless or pale yellow solids that are stable to heat (note the difference from nitrates). The alkali metal nitrites are very soluble in water, and those of calcium, strontium, and barium are moderately soluble in water. The nitrites are made either by passing an equal molar mixture of nitric oxide and nitrogen dioxide into a solution of metal hydroxide, for example

$$NO(g) + NO_2(g) + 2KOH(aq) \longrightarrow 2KNO_2(aq) + H_2O(l)$$

or by reduction at elevated temperatures of a metal nitrate by a reducing agent such as lead or iron powder.

$$NaNO_3(s) + Pb(s) \xrightarrow{\Delta} NaNO_2(l) + PbO(l)$$

The nitrite ion is weakly basic. Acidifying a solution containing NO_2^- results in disproportionation and the production of nitric oxide (NO) and nitrogen dioxide (NO_2). In alkaline solution, the nitrite ion is a rather strong reducing agent.

Sodium nitrate ($NaNO_3$) and sodium nitrite ($NaNO_2$) together are used as preservatives and curing agents in meats such as frankfurters, bacon, and bologna. They are especially effective at preventing the growth of the microorganism that causes the serious food poisoning known as botulism.

Frequently, molecular nitrogen is released when species containing nitrogen in a higher and lower oxidation state are brought together. Warming an aqueous solution saturated with both ammonium chloride and sodium nitrite has been used as a laboratory preparation of nitrogen.

$$NH_4^+ + NO_2^- \xrightarrow{\Delta} N_2(g) + 2H_2O(g)$$

Nitric acid (HNO_3) is a large-volume industrial acid and its properties were discussed earlier (Section 28.6c).

Nitrates, the salts of nitric acid, are easily prepared by the reaction between nitric acid and metal oxides, hydroxides, or carbonates. With oxides and hydroxides, the reaction is simply neutralization and goes to completion because of the formation of water. With carbonates the reaction is complete because of the evolution of carbon dioxide gas and the formation of water, for example

$$CaCO_3(s) + 2HNO_3(aq) \longrightarrow Ca(NO_3)_2(aq) + CO_2(g) + H_2O(l)$$

Metal nitrates are very soluble in water. The few that appear to be insoluble, such as bismuth nitrate, $Bi(NO_3)_3$, and mercury (II) nitrate, $Hg(NO_3)_2$, actually react with water with the formation of insoluble basic salts that contain the O or OH group.

$$\underset{\textit{bismuth nitrate}}{Bi(NO_3)_3(s)} + H_2O(l) \rightleftharpoons \underset{\substack{\textit{basic bismuth nitrate} \\ \textit{(bismuthyl nitrate)}}}{BiO(NO_3)(s)} + 2HNO_3(aq)$$

In our study of chemical reactions (Section 17.4b), we saw that all nitrates undergo thermal decomposition, the products depending upon the reactivity of the metal. For example

$$Mg(NO_3)_2(s) \xrightarrow{\Delta} Mg(NO_2)_2(s) + O_2(g)$$
$$2Pb(NO_3)_2(s) \xrightarrow{\Delta} 2PbO(s) + 4NO_2(g) + O_2(g)$$

EXAMPLE 28.1 Chemical Reactions: N Compounds

Predict the products of the following reactions of nitrogen compounds.

(a) $Ba_3N_2(s) + H_2O(l) \longrightarrow$ (d) $TiN(s) + H_2O(l) \longrightarrow$

(b) $Ca(NO_2)_2(s) \xrightarrow{\Delta}$ (e) $AgNO_3(s) \xrightarrow{\Delta}$

(c) $NO(g) + Br_2(l) \longrightarrow$

(a) This is a reaction between two compounds. Barium is an active metal. Therefore barium nitride is a saltlike nitride and contains the strongly basic nitride ion N^{3-}, which reacts completely with water.

$$Ba_3N_2(s) + 6H_2O(l) \longrightarrow \mathbf{3Ba(OH)_2(aq) + 2NH_3(aq)}$$

(b) Thermal decomposition is the only possible reaction. However, *nitrites,* as opposed to nitrates, do not decompose when heated. **No reaction will occur.**

(c) This is a reaction between an element and a compound. Combination is possible because nitrogen can have a higher oxidation state.

$$2NO(g) + Br_2(l) \longrightarrow \mathbf{2NOBr(l)}$$

(d) Like (a) this is a reaction between two compounds. However, titanium is *not* an active metal and titanium nitride is most likely an interstitial nitride. **No reaction will take place** because the nitride ion is not present.

(e) As in (b) a thermal decomposition is possible, and nitrates of heavy metals like silver react to form oxygen, nitrogen(IV) oxide, and the metal oxide. However, because the oxides of the least active metals also decompose thermally to form the metal and oxygen, the overall reaction is as follows:

$$2AgNO_3(s) \xrightarrow{\Delta} \mathbf{2Ag(s) + 2NO_2(g) + O_2(g)}$$

Aside 28.2 EXPLOSIVES

A big bang, heat, the power to shatter solid objects, possibly flames—these are the effects that we associate with explosions. In chemical terms, just what are explosions and explosives?

All chemical explosions are very fast redox reactions which result in the release of large volumes of gases. A chemical explosion can be initiated by high temperatures, a sudden increase in pressure, or a physical shock. Once the activation energy has been provided, a sufficient amount of heat is generated for the explosive reaction to be self-sustaining.

Not all chemical compounds that explode—in fact, very few—are suitable for practical use as explosives. A substance that explodes with the slightest shock or temperature increase cannot be fabricated into an artillery shell or transported to the blasting site where a road is being built. The components of military and industrial explosives are substances that are safe to handle, and explode only under controlled conditions. Millions of kilograms of such substances are produced annually with an accident rate no higher than those in other industrial operations.

The essential characteristic of an explosion-causing redox reaction is a rapid reaction rate. The release of a large amount of energy as heat is less important. A comparison of the combustion of gasoline and the explosion of gunpowder illustrates this difference. Burning 1 g of gasoline in an internal combustion engine produces about 46 kJ of thermal energy and requires about 0.01 s. One gram of gunpowder releases only 3.3 kJ in propelling an artillery shell, but does so in 0.0005 s.

Explosives are divided into two general classes based on the way in which the chemical reaction moves through the mass of the material while it is exploding. *High explosives* undergo detonation—the redox reaction moves directly through the body of a solid at a rapid rate (2000 to 9000 m/s) as the reaction, thought to be carried by a chain mechanism, spreads. *Low explosives* "burn" by a more slow-moving reaction of material close to the surface (less than 0.25 m/s) rather than by detonation. Neither high nor low explosives react with oxygen from the air. The oxidized and reduced atoms are either part of the same compound, which undergoes an internal redox reaction, or they are present in an intimate mixture of oxidizing agent and fuel.

Most high explosives are compounds that contain nitrogen, combined with oxygen, hydrogen, and/or carbon. During the explosion, the combined nitrogen is reduced to elemental nitrogen. In addition, the other elements are all converted to gases—carbon to carbon dioxide or carbon monoxide, and hydrogen and oxygen to water or to molecular hydrogen and oxygen. The rapidly produced large volume of hot gases is responsible for the surge of pressure and the damaging shock wave that accompany an explosion.

High explosives are described as either primary or secondary, depending upon their sensitivity. Primary explosives can be set off by a spark or a blow and must be handled with great care. Most primary explosives are inorganic salts. Secondary explosives, such as nitroglycerin, are less sensitive than primary explosives, but more powerful. Once set off, they react with great heat and shattering power (known as *brisance*). Most secondary explosives are organic compounds that contain nitro (NO_2) groups, for example, TNT (trinitrotoluene). The

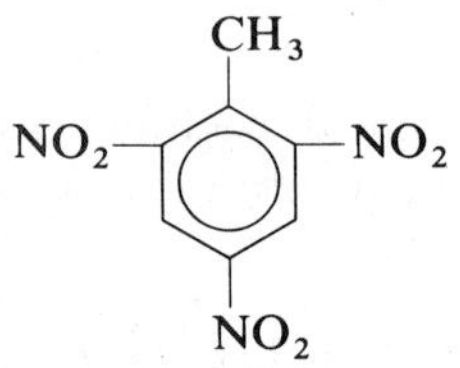

CH_3 / NO_2 / NO_2 / NO_2 (benzene ring with CH_3 and three NO_2 groups)

trinitrotoluene
(TNT)

explosion of a secondary explosive is often initiated by a primary explosive in, for example, a detonator cap. Alfred Nobel's idea that the detonation of a small amount of mercury fulminate could trigger the explosion of nitroglycerin was the first step in the development of modern high explosives. Dynamite is a mixture of nitroglycerin with a solid inert support and other reactive substances.

$$Hg(ONC)_2(s) \longrightarrow Hg(l) + 2CO(g) + N_2(g)$$

mercury fulminate
(a primary explosive)

Heat of explosion at constant volume: 1.8 kJ/g
Specific pressure (the pressure exerted by explosion of 1 kg in a 1 L volume): 5212 kg/cm²
Velocity of detonation wave: 3920 m/s
Explosion temperature: 4105 °C

$$\begin{array}{l} H_2C{-}ONO_2 \\ \;| \\ HC{-}ONO_2(l) \longrightarrow 3CO_2(g)+\tfrac{5}{2}H_2O(g)+\tfrac{3}{2}N_2(g)+\tfrac{1}{4}O_2(g) \\ \;| \\ H_2C{-}ONO_2 \end{array}$$

nitroglycerin
(a secondary explosive)

Heat of explosion at constant volume: 6.4 kJ/kg
Specific pressure: 9835 kg/cm²
Velocity of detonation wave: 8500 m/s
Explosion temperature: 3360 °C

SUMMARY OF SECTIONS 28.7–28.10

The outstanding properties of some nitrogen compounds are summarized in Table 28.11. The properties of the four common nitrogen oxides have been summarized in Table 28.10.

Table 28.11
Outstanding Properties of Nitrogen Compounds and Ions

Nitrides	May be ionic, molecular, network covalent, or interstitial
$N^{3-} + 3H_2O(l) \longrightarrow NH_3(aq) + 3OH^-$	Vigorous, complete reaction of ionic nitrides
lone pair $H{-}\ddot{N}{-}H$ (with H below N)	A Lewis base; weakly basic in aqueous solution; a top industrial chemical; important in fertilizer manufacture; ligand in complex ions
NH_4^+	Acidic aqueous solution; most NH_4^+ salts are water soluble; all salts decompose on heating
hydrazine N_2H_4	Powerful reducing agent; used in rocket fuels
Nitrogen oxides	Known in +1 to +5 oxidation states; N_2O, NO, NO_2, N_2O_4 most common (Table 28.10)
HNO_2	Known only in solution; strong oxidizing agent and weak reducing agent
NO_2^-	Moderately strong reducing agent in alkaline solution
HNO_3	An oxidizing agent; a strong acid; an industrial chemical
$M(NO_3)_n$	Nitrates are water soluble; undergo thermal decomposition
Nitrogen compound $\xrightarrow{\Delta} N_2(g)$	Many such reactions used in explosives and propellants
Disproportionation	Many nitrogen compounds disproportionate

COMPOUNDS OF PHOSPHORUS

Table 28.12 gives examples of compounds of phosphorus in its oxidation states of −3, +1, +3, and +5.

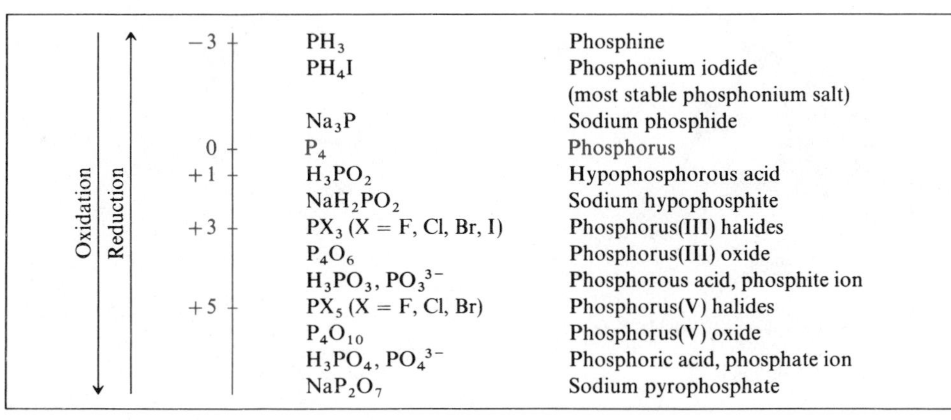

Oxidation state	Compound	Name
−3	PH_3	Phosphine
	PH_4I	Phosphonium iodide (most stable phosphonium salt)
	Na_3P	Sodium phosphide
0	P_4	Phosphorus
+1	H_3PO_2	Hypophosphorous acid
	NaH_2PO_2	Sodium hypophosphite
+3	PX_3 (X = F, Cl, Br, I)	Phosphorus(III) halides
	P_4O_6	Phosphorus(III) oxide
	H_3PO_3, PO_3^{3-}	Phosphorous acid, phosphite ion
+5	PX_5 (X = F, Cl, Br)	Phosphorus(V) halides
	P_4O_{10}	Phosphorus(V) oxide
	H_3PO_4, PO_4^{3-}	Phosphoric acid, phosphate ion
	NaP_2O_7	Sodium pyrophosphate

(Arrows: Oxidation ↓, Reduction ↑)

Table 28.12
Oxidation States and Some Compounds of Phosphorus

28.11 PHOSPHIDES AND PHOSPHINE

The compounds of phosphorus in the −3 oxidation state—the phosphides and phosphine—are similar in some ways to the comparable nitrogen compounds, the nitrides and ammonia. In other oxidation states the compounds of these two elements are quite different.

Ionic phosphides are known for most of the *s*-block metals. Like the nitride ion, the phosphide ion (P^{3-}) is such a strong base that it reacts completely with water, forming OH^- and phosphine (PH_3), a volatile molecular compound

$$P^{3-} + 3H_2O(l) \longrightarrow PH_3(g) + 3OH^- \qquad \textbf{(28.29)}$$

Other phosphides, like the nitrides, vary widely in type of bonding and stoichiometry.

Phosphine (PH_3) is a colorless gas that has an offensive, garliclike odor and is extremely toxic. The pure compound ignites when heated in air at about 150 °C. However, phosphine frequently also contains diphosphine (P_2H_4), which is very reactive and renders the mixture spontaneously flammable. Phosphine is much less soluble in water than ammonia and does not form alkaline solutions. Like ammonia, phosphine is an electron-pair donor. It can function as a ligand in complex ions and it forms salts containing the phosphonium ion (PH_4^+). Unlike the ammonium ion, the phosphonium ion hydrolyzes completely:

$$PH_4^+ + H_2O(l) \longrightarrow PH_3(g) + H_3O^+ \qquad \textbf{(28.30)}$$

28.12 PHOSPHORUS OXIDES AND HALIDES

The oxidation of white phosphorus at about 100 °C in a limited supply of air gives phosphorus(III) oxide (P_4O_6), a white, crystalline substance, as the main product. The P_4O_6 molecule is a tetrahedron in which an oxygen atom has been added between each pair of phosphorus atoms (Figure 28.6a).

With an excess of air, the phosphorus(V) oxide (P_4O_{10}), is formed. It is a white solid that exists in a number of crystalline modifications. When heated, it sublimes as discrete P_4O_{10} molecules (Figure 28.6b), which are related in structure to the P_4O_6 molecules by addition of an oxygen atom to each phosphorus atom. (Sometimes this oxide is incorrectly referred to as phosphorus pentoxide and given the formula P_2O_5.) Phosphorus(V) oxide is the anhydride of phosphoric acid, and its most important chemical property is its tremendous affinity for water. It reacts with water to form the acid (see Equation 28.9) and is one of the most useful drying agents. When phosphorus burns in air, P_4O_{10} is produced as a cloud of very white particles of colloidal

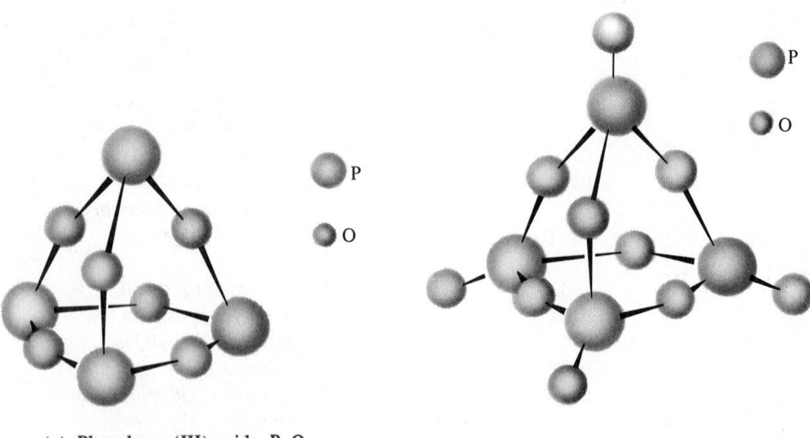

(a) Phosphorus(III) oxide, P_4O_6
m.p. 23.8°C
b.p. 175.4°C

(b) Phosphorus(V) oxide, P_4O_{10}
sublimes at 358°C

Figure 28.6
Phosphorus Oxides

dimensions. The particles are kept in suspension in the air by Brownian motion. The resulting very dense smoke has been used as a military smoke screen.

The phosphorus(III) halides (PX_3) are simple molecular substances that have the same pyramidal molecular geometry as ammonia and phosphine. At room temperature the fluoride is a colorless gas; the chloride and bromide are volatile, fuming liquids; and the iodide is a red, low-melting solid that is not stable.

In water, the trihalides are completely and irreversibly hydrolyzed (PF_3 reacts quite slowly).

$$PX_3 + 3H_2O(l) \longrightarrow \underset{\textit{phosphorous acid}}{H_3PO_3(aq)} + 3HX(aq) \qquad \textbf{(28.31)}$$

All the phosphorus(V) halides (PX_5) except the iodide are known. The PF_5 and PCl_5 molecules in the vapor have a triangular bipyramidal shape. In the solid state the pentachloride is a saltlike substance consisting of tetrahedral $[PCl_4]^+$ and octahedral $[PCl_6]^-$ ions.

Like the trihalides, the pentahalides react rapidly with water. With a limited quantity of water, two of the halogen atoms are removed to give phosphoryl halides (POX_3) and hydrohalic acids.

$$PX_5 + H_2O(l) \longrightarrow POX_3 + 2HX(aq)$$

With an excess of water, all of the halogen atoms react and the pentahalides are converted to phosphoric acid.

$$PX_5 + 4H_2O(l) \longrightarrow H_3PO_4(aq) + 5HX(aq)$$

28.13 PHOSPHORUS OXOACIDS AND THEIR SALTS

Phosphorous acid (H_3PO_3) is commonly prepared by hydrolysis of phosphorus trichloride (Equation 28.31). [Note one of the great pitfalls in the study of chemistry: The third "*o*" appears *only* in the spelling of the phosphorous(I) and (III) acids.] The H_3PO_3 molecule is unusual for an acid in that one hydrogen atom is bonded directly to the phosphorus atom (Figure 28.7). Because the hydrogen–phosphorus bond is essentially nonpolar (P and H have about the same electronegativities), this hydrogen atom is not acidic and therefore phosphorous acid is a diprotic acid. It is a moderately strong acid with respect to its first dissociation ($K_{a1} = 3 \times 10^{-2}$; $K_{a2} = 1.6 \times 10^{-7}$). Phosphorous acid is a strong reducing agent, forming phosphoric acid as its oxidation product. Phosphites are salts of phosphorous acid containing either the dihydrogen phosphite ion ($H_2PO_3^-$) or the hydrogen phosphite ion (HPO_3^{2-}). Both ions are weakly basic and can act as reducing agents.

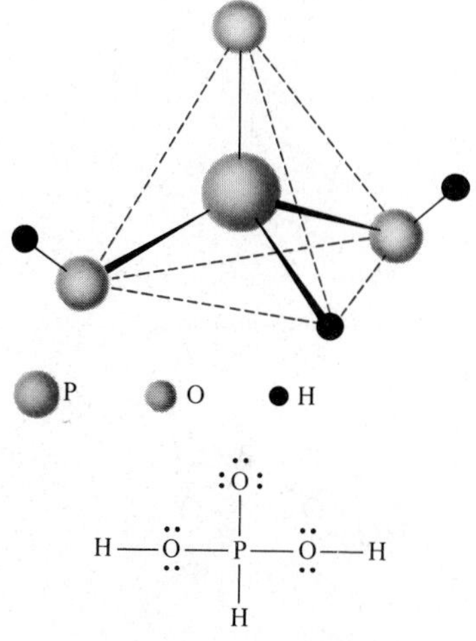

Phosphorous acid, H_3PO_3
m.p. 73.6°C
b.p. 200°C (dec)

Figure 28.7
Phosphorous Acid

The formulas and names of the phosphorus(V) acids are included in Table 28.13. In all of these acids and their salts, each phosphorus atom is surrounded by four oxygen atoms and in none is a phosphorus atom bonded directly to another phosphorus atom.

Phosphoric acid (H_3PO_4), as we have discussed (Section 28.6b), is a major industrial chemical. The other phosphorus(V) acids are products of the elimination of water between successive molecules of phosphoric acid. The elimination of one molecule of water between two molecules of phosphoric acid yields diphosphoric, or pyrophosphoric acid ($H_4P_2O_7$), which has the same structure as pyrosulfuric acid (see Figure 28.5)

elimination of H_2O

```
  O            O             O     O
  |            |             |     |
HOP(OH     H)OPOH  ---->  HOP—O—POH + H2O
  |            |             |     |
  OH           OH            OH    OH
```

phosphoric acid — *pyrophosphoric acid, $H_4P_2O_7$*

Table 28.13
Phosphorus Acids

Phosphorus(I)	
H_3PO_2	Hypophosphorous acid, monoprotic
Phosphorus(III)	
H_3PO_3	Phosphorous acid, diprotic
Phosphorus(V)	
H_3PO_4	Phosphoric acid (orthophosphoric acid), triprotic
$H_4P_2O_7$	Pyrophosphoric acid (diphosphoric acid) tetraprotic
$H_5P_3O_{10}$	Triphosphoric acid (tripolyphosphoric acid) pentaprotic
$(HPO_3)_n$	Metaphosphoric acid ($n = 3, 4, \ldots$)

Pyrophosphoric acid, a colorless glassy solid, is extremely soluble in water, in which it slowly reverts to phosphoric acid. Triphosphoric acid, or tripolyphosphoric acid, is formed by loss of two molecules of water among three molecules of phosphoric acid. Theoretically this process can continue indefinitely. Polyphosphoric acids are used in the production of fertilizers and other chemicals, as catalysts in the petroleum industry, and as dehydrating agents.

"Metaphosphoric acid" is a mixture of polymeric acids of empirical formula HPO_3, obtained when phosphoric acid is heated.

$$nH_3PO_4(l) \xrightarrow{325-350\ ^\circ C} \underbrace{(HPO_3)_n}_{n\,=\,\text{a variety of whole numbers}} + nH_2O(g)$$

The metaphosphoric acids are difficult to separate. They contain either rings or long chains of PO_4 tetrahedra. (The structures of two simple cyclic metaphosphates are shown in the margin.) Sodium metaphosphate $(NaPO_3)_n$, which forms stable complexes with the +2 metal ions in hard water, is widely used as a water softener.

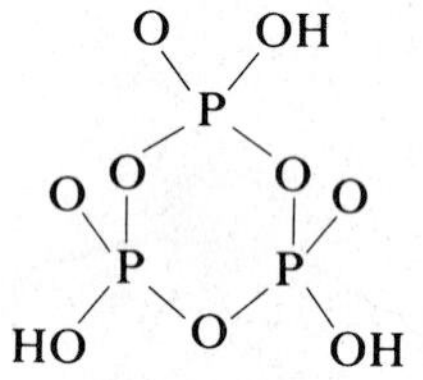

trimetaphosphoric acid, $H_3P_3O_9$

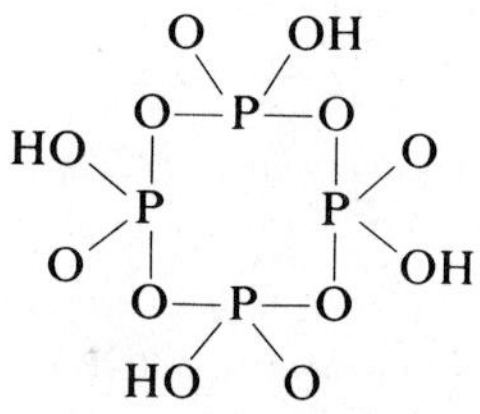

tetrametaphosphoric acid, $H_4P_4O_{12}$

The possibilities for making phosphates with desirable properties are extensive—any or all of the hydrogen atoms in any of the phosphoric acids can be replaced by metal ions. Some phosphates yield acidic aqueous solutions, others yield alkaline aqueous solutions, and others are only slightly soluble—all properties that have been put to use. Like phosphoric acid, many phosphates are nontoxic and therefore can be used in the food industry, where their ability to act as buffers is often useful.

The sodium salts derived from phosphoric acid

NaH_2PO_4	Na_2HPO_4	Na_3PO_4
sodium dihydrogen phosphate (or monosodium phosphate)	*disodium monohydrogen phosphate (or disodium phosphate)*	*trisodium phosphate (or normal sodium phosphate)*

all have numerous and varied uses. Sodium dihydrogen phosphate (NaH_2PO_4) which gives an acidic aqueous solution, is used in acidic cleaning agents. Disodium monohydrogen phosphate (Na_2HPO_4) can prevent separation of fat and water, and serves this purpose in pasteurized process cheese. In addition, it is used in cured hams, instant cereals, and evaporated milk. By contrast, trisodium phosphate (Na_3PO_4) is not appropriate for use in foods because it gives a strongly alkaline aqueous solution; it appears in heavy-duty alkaline cleaning agents and in water-softening agents, where it functions in the removal of +2 metal ions by precipitation of insoluble phosphates. The hydrate $Na_3PO_4 \cdot 12H_2O$ is a crystalline compound sold as a cleaning compound and paint remover.

Calcium phosphates, in contrast to the sodium phosphates, are not very soluble in water. The dihydrate of calcium monohydrogen phosphate is used as a polishing agent in toothpaste.

Sodium tripolyphosphate (a salt of tripolyphosphoric acid)

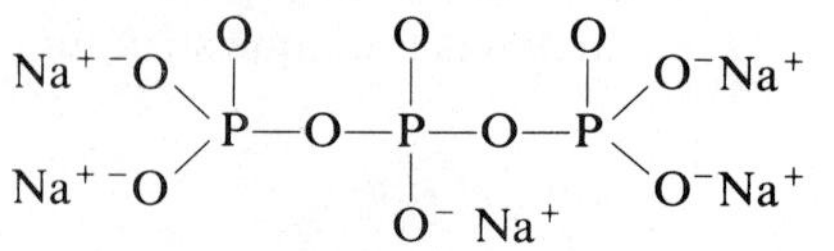

sodium tripolyphosphate

dominates the market as a builder in synthetic detergents, the largest volume application for phosphates other than fertilizer applications. A builder aids the function of a detergent by removing hard water ions, increasing the alkalinity of the solution, and

Table 28.14
Uses of Phosphates
In food applications, the calcium phosphates are most common. In cleaning applications, the sodium phosphates are most common.

Food uses	Cleaning agents	Other uses
Food supplements for humans and animals	Builders in detergents	Water softeners
Carbonated beverages (for tartness)	Strongly alkaline cleaning agents	Flame retardants
Leavening agents	Acidic cleaning agents	In ore flotation
Quick-cooking cereals	Abrasives in toothpastes	In plating plastics with metals
Emulsifiers		Dishwasher detergents
Buffers		
Consistency control		

removing dirt and preventing its redisposition. The tripolyphosphate ion sequesters hard water ions; that is, it forms complex ions with them. Other nonfertilizer uses of phosphates are summarized in Table 28.14.

SUMMARY OF SECTIONS 28.11–28.13

The phosphate compounds and properties that have been discussed in these sections are summarized in Table 28.15.

Table 28.15
Outstanding Properties of Phosphorus Compounds and Ions

Compound or ion	Properties
$P^{3-} + 3H_2O(l) \longrightarrow PH_3(g) + 3OH^-$	Vigorous, complete reaction of ionic phosphides
H—P̈—H with H below (*phosphine, weak Lewis base*)	Much less stable than NH_3; very toxic; ligand in complex ions
PH_4^+	Decomposes completely in water to give PH_3 and H^+
$P_4O_6 \xrightarrow{H_2O} H_3PO_3$ (*phosphorous acid*)	Acidic oxides
$P_4O_{10} \xrightarrow{H_2O} H_3PO_4$ (*phosphoric acid*)	
PX_3	Complete series of compounds with the halogens known; hydrolyze to give $H_3PO_3 + HX$ (X = F, Cl, Br, I)
PX_5	All but the iodide known; with an excess of H_2O, gives $H_3PO_4 + HX$
H_3PO_3	Moderately strong *diprotic* acid; powerful reducing agent
$H_2PO_3^-$, HPO_3^{2-}	Strong reducing agents
H_3PO_4	Important industrial chemical; triprotic acid; weak acid
PO_4^{3-}	Strongly basic
$H_2PO_4^-$	Weakly acidic
$H_4P_2O_7$ (*pyrophosphoric acid*)	Formed by elimination of H_2O from two H_3PO_4 molecules; contains P—O—P bonds; lowest of series of polyphosphoric acids
$Na_5P_3O_{10}$ (*sodium tripolyphosphate*)	Used in detergents as a builder
$(HPO_3)_n$ (*metaphosphoric acid*)	Cyclic and linear acids containing PO_4 tetrahedra
Phosphates	Many uses, e.g., in foods and cleaning agents

Aside 28.3 PLANT NUTRIENTS AND CHEMICAL FERTILIZERS

The major end uses of nitrogen, phosphorus, and sulfur, and of their most important industrial products, ammonia, phosphoric acid, and sulfuric acid, are in the production of chemical fertilizers. In plants and animals, nitrogen is present mainly in the reduced form (oxidation state -3) as ammonium ion or as $-NH_2$ groups in organic compounds. But plants cannot use atmospheric nitrogen, presumably because of the large amount of energy needed to break the nitrogen–nitrogen bond. Molecular nitrogen must undergo **nitrogen fixation**—the combination of molecular nitrogen with other atoms—before it can be taken up by plants from the soil, usually in the form of nitrate ion. In nature, bacteria and algae that live free in soil, ponds, and lakes, or that live in a mutually dependent relationship with specific plants, fix nitrogen by converting it to ammonia. By a process called nitrification, other microorganisms convert ammonia to nitrites and nitrates—water-soluble compounds that can be taken up by plants. During nitrification, some nitrous oxide (N_2O) is released to the atmosphere (see Aside 29.1).

Nitrogen, phosphorus, sulfur, and most other elements can be traced in various forms in their passage through the land; the rivers, oceans, and other bodies of water; and the atmosphere, if they form gaseous compounds. Nitrogen fixation and nitrification are two of several steps in the nitrogen cycle, which is shown schematically in Figure 28.8. The activities of human beings enter into the nitrogen, phosphorus, and sulfur cycles when we burn fuel and garbage, spray with pesticides, or spread fertilizer.

Nitrogen and phosphorus are two of the three major nutrients (Table 28.16) supplied to plants through the soil.

Table 28.16
Plant Nutrients
Carbon, hydrogen, and oxygen are obtained by plants from the air and the soil. All other nutrients are obtained only from the soil. Primary nutrients are needed in largest amounts (up to 200 pounds per acre), secondary nutrients are needed in smaller amounts (up to 50 pounds per acre), and others are needed only in small amounts (<1 pound per acre).

Plant Nutrients
Needed in large amounts
Carbon
Hydrogen
Oxygen
Nitrogen
Phosphorus
Potassium
Needed in moderate amounts
Calcium
Magnesium
Sulfur
Some of those needed in trace amounts
Boron
Copper
Iron
Manganese
Zinc
Molybdenum
Chlorine

Figure 28.8
The Nitrogen Cycle

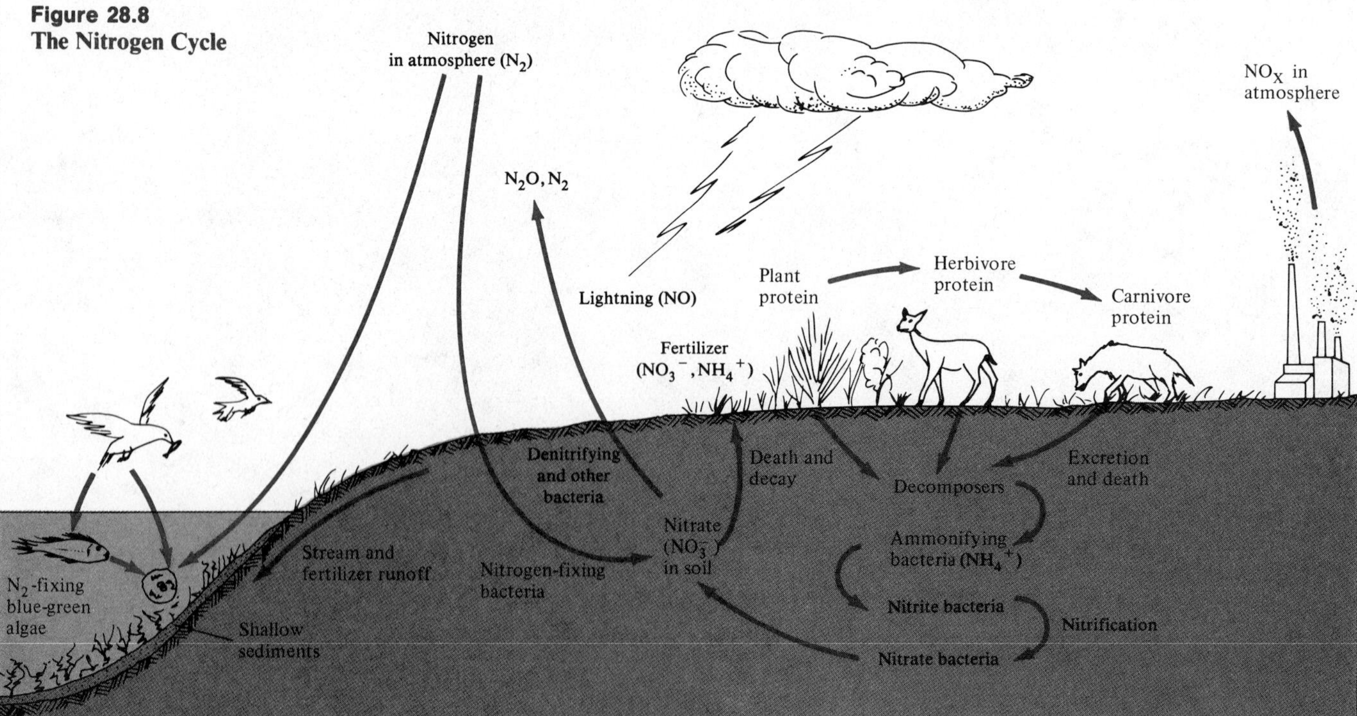

Table 28.17
Manufacture of Major Chemical Fertilizers
Natural raw materials appear in black and manufactured raw materials in color in the left-hand column.

Sources	Process	Product
Phosphate rock	Grind to about 0.1 millimeter	Ground phosphate rock
Sulfur Air Water	Catalytic oxidation and hydration	Sulfuric acid
Phosphate rock Sulfuric acid	React and cure	Single superphosphate
Phosphate rock Sulfuric acid	Dissolve and filter	Phosphoric acid Gypsum ($CaSO_4 \cdot 2H_2O$)
Phosphate rock Phosphoric acid	React and cure	Triple superphosphate
Hydrocarbons Steam Nitrogen from air	Re-form to hydrogen Synthesize	Ammonia
Ammonia Sulfuric acid	React and crystallize	Ammonium sulfate
Ammonia Phosphoric acid	React and crystallize or granulate	Ammonium phosphates
Air Water	Catalytic oxidation Absorption	Nitric acid
Phosphate rock Nitric acid Ammonia	React Ammoniate Filter	Nitrophosphates Calcium nitrate
Ammonia Nitric acid	React and crystallize or make into pellets	Ammonium nitrate
Ammonia Carbon dioxide	React and crystallize or make into pellets	Urea (H_2NCONH_2)

Source: Adapted from Christopher J. Pratt, "Chemical Fertilizers," *Scientific American,* **212**(6), June 1965.

Sulfur plays a double role in the fertilizer industry—it is necessary to plants, although in smaller amounts than nitrogen and phosphorus, and it is vitally important in the sulfuric acid used in the manufacture of chemical fertilizers (Table 28.17), particularly the superphosphates. Ammonia, also essential to fertilizers, is manufactured by the reaction of nitrogen and hydrogen in the Haber process—an industrial method of fixing nitrogen. Because natural gas and oil are the sources of hydrogen for the Haber process, countries rich in supplies of these fossil fuels are potentially big producers of fertilizers.

Nitrogen atoms are needed for amino acids—the building blocks of proteins—and for chlorophyll, as well as for many other biochemically important molecules. The rate of growth and yield from crops is directly influenced by the nitrogen available, and it has been shown that the amount of protein in, for example, corn is increased by nitrogen fertilizers.

About 50 years ago there was a thriving business in supplying bacteria that farmers plowed into the soil where clover or other legumes were growing. The business died out when chemical fertilizers became available at lower prices. But now interest is reviving in bacterial nitrogen fixation, and it has become an active area of research. With greater understanding of the natural fixation process, it is hoped that bacteria can again be put directly to work in the fields, cutting down the need for chemical fertilizers, which require large amounts of energy in their manufacture. Possibly better plant–microorganism combinations might be designed, or possibly a nonbiological system for nitrogen fixation that resembles the natural process might be developed.

The primary role of phosphorus in biological systems is in energy transfer—plants and animals store energy in the phosphorus–oxygen bonds. Good suppliers of phosphorus in the soil hasten the growth of young plants. Most of the phosphorus used by plants is in the form of phosphate ions, particularly the dihydrogen phosphate ion, $H_2PO_4^-$.

The raw material for phosphate fertilizers is phosphate rock. The rock is converted directly to single superphosphate—a mixture of monocalcium phosphate and calcium sulfate (gypsum).

$$2Ca_5(PO_4)_3F(s) + 7H_2SO_4(aq) + 17H_2O(l) \longrightarrow 3Ca(H_2PO_4)_2 \cdot H_2O(s) + 7CaSO_4 \cdot 2H_2O + 2HF(g)$$

or it is converted directly to triple superphosphate, which contains no calcium sulfate and therefore delivers a higher percentage of phosphorus

$$Ca_5(PO_4)_3F(s) + 7H_3PO_4(aq) + 5H_2O(l) \longrightarrow 5Ca(H_2PO_4)_2 \cdot H_2O(s) + HF(g)$$

Finely ground phosphate rock can also be used as a fertilizer, or phosphoric acid can be produced and soluble salts made from the acid (Table 28.17). The ammonium phosphates, $(NH_4)H_2PO_4$ (MAP) and $(NH_4)_2HPO_4$ (DAP), are manufactured economically from wet process phosphoric acid, and in the 1970s became the leading phosphate fertilizers. Ammonium phosphates are advantageous because they supply *both* nitrogen and phosphorus.

COMPOUNDS OF SULFUR

Table 28.18 gives examples of compounds of sulfur in its oxidation states of −2, +4, and +6.

Table 28.18
Oxidation States and Some Compounds of Sulfur

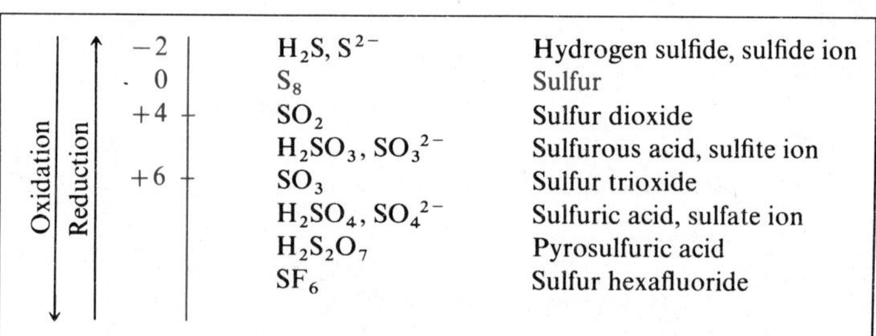

Oxidation state	Compound	Name
−2	H_2S, S^{2-}	Hydrogen sulfide, sulfide ion
0	S_8	Sulfur
+4	SO_2	Sulfur dioxide
	H_2SO_3, SO_3^{2-}	Sulfurous acid, sulfite ion
+6	SO_3	Sulfur trioxide
	H_2SO_4, SO_4^{2-}	Sulfuric acid, sulfate ion
	$H_2S_2O_7$	Pyrosulfuric acid
	SF_6	Sulfur hexafluoride

28.14 HYDROGEN SULFIDE AND OTHER SULFIDES

Hydrogen sulfide (H_2S) is an extremely poisonous colorless gas with an obnoxious odor resembling that of rotten eggs. **A few breaths of the concentrated gas can be fatal.** At lesser concentrations, headaches and dizziness are produced. Hydrogen sulfide is especially hazardous because it deadens the sense of smell so that increasing concentrations are not detected.

Many metal sulfides are only slightly soluble (Table 28.19) and the precipitation of metal sulfides plays an important role in qualitative analysis schemes (Section 22.8) At one time, many laboratories were recognizable from great distances by the strong smell of H_2S in the air as it leaked from generators that produced the gas. However, H_2S is now more safely generated directly in solution by heating an aqueous solution of thioacetamide (CH_3CSNH_2):

Table 28.19
Solubility Product Constants (K_{sp}) of Some Metal Sulfides

Metal sulfide	K_{sp}
MnS	2.3×10^{-13}
FeS	4.2×10^{-17}
ZnS	2×10^{-24}
SnS	3×10^{-27}
CdS	2×10^{-28}
PbS	1×10^{-28}
CuS	6×10^{-36}
HgS	4×10^{-53}

$$\underset{\text{thioacetamide}}{CH_3CSNH_2(aq)} + 2H_2O(l) \xrightarrow{\Delta} \underset{\text{acetate ion}}{CH_3COO^-} + \underset{\text{ammonium ion}}{NH_4^+} + \underset{\text{hydrogen sulfide}}{H_2S(aq)} \quad \textbf{(28.32)}$$

Hydrogen sulfide is a good reducing agent, both in the pure state and in aqueous solution. In an excess of oxygen (or air), hydrogen sulfide burns, when ignited, to give sulfur dioxide and water.

$$2H_2S(g) + 3O_2(g) \longrightarrow 2SO_2(g) + 2H_2O(g)$$

If the amount of air is limited, the oxidation product is elemental sulfur.

In aqueous solution, H_2S is a weak diprotic acid (hydrosulfuric acid, Section 21.7), and the S^{2-} and HS^- ions are both weakly basic. Adding acid to S^{2-} in aqueous solution generates H_2S, which will dissolve up to 0.1 M and then begin to bubble out of the solution. Slightly soluble metal sulfides can be dissolved by addition of acid to form hydrogen sulfide and a salt in solution.

28.15 SULFUR OXIDES AND HALIDES

Sulfur dioxide (SO_2) is a colorless gas which has a characteristic suffocating odor, and is very irritating to the eyes and respiratory tract. It is formed in the combustion of sulfur-containing fuels and in the smelting of sulfide ores, and it is a primary air pollutant (see Aside 28.4). The reactive chemicals on the tip of a strike-anywhere match are usually P_4S_3 and an oxidizing agent. When such a match is struck, one can often smell the sulfur dioxide that is formed:

$$P_4S_3(s) + 8O_2(g) \longrightarrow P_4O_{10}(s) + 3SO_2(g)$$

Sulfur dioxide is a strong reducing agent. In addition to its essential role in the production of sulfuric acid (Section 28.6a), it is extensively used as a bleach for textiles and food, a disinfectant, a mold inhibitor in dried fruits, and in the production of paper pulp.

Sulfur trioxide (SO_3) is usually gaseous when formed, but solidifies on cooling. It has three different solid modifications (m.p.'s, 62.3 °C, 32.5 °C, 16.8 °C). Liquid sulfur trioxide fumes in moist air as it forms sulfuric acid.

Sulfur hexafluoride (SF_6) is the most stable of the sulfur–halogen compounds and one of the most chemically unreactive compounds known (Section 27.4). It does not react with water, but most other sulfur–halogen compounds do so with sufficient vigor to make these substances hazardous to handle. For example, sulfur tetrafluoride hydrolyzes rapidly upon exposure to moist air.

$$SF_4(g) + 2H_2O(l) \longrightarrow 4HF(g) + SO_2(g)$$

28.16 SULFUR OXOACIDS AND THEIR SALTS

Sulfur dioxide is very soluble in water. Its aqueous solutions are acidic and this acidity is traditionally described as the result of the following equilibria:

$$SO_2(g) + H_2O(l) \rightleftharpoons H_2SO_3(aq)$$
$$H_2SO_3(aq) + H_2O(l) \rightleftharpoons HSO_3^- + H_3O^+ \qquad K_a = 1.43 \times 10^{-2}$$
$$HSO_3^- + H_2O(l) \rightleftharpoons SO_3^{2-} + H_3O^+ \qquad K_a = 5.0 \times 10^{-8}$$

Sulfurous acid (H_2SO_3) is thus considered to be a diprotic acid. In reality, most of the dissolved sulfur dioxide is simply in solution as a hydrate and there is no clear-cut evidence for the existence of sulfurous acid. However, stable salts containing both the hydrogen sulfite and sulfite ions are well known. Like sulfur dioxide, the sulfites and hydrogen sulfites are reducing agents.

The properties and preparation of sulfuric acid (H_2SO_4), which is a strong acid and the number one industrial chemical, have already been discussed (Section 28.6a).

Pyrosulfuric acid ($H_2S_2O_7$) is also known as "oleum" or "fuming sulfuric acid," because it evolves white fumes on contact with moist air. The fumes are caused by the immediate reaction of the escaping SO_3 with water to form ordinary sulfuric acid as a mist. Pyrosulfuric acid is a more powerful oxidizing agent than sulfuric acid. It is formed by dissolving SO_3 in H_2SO_4. As is true for phosphorus in its +5 oxoacids and salts, each sulfur atom in the +6 oxygen acids and salts is surrounded tetrahedrally by four oxygen atoms (see Figure 28.7).

Peroxosulfuric acids contain peroxo groups $—\ddot{\underset{..}{O}}—\ddot{\underset{..}{O}}—$ (as in hydrogen peroxide), for example

$$\begin{array}{c} :\ddot{O}: \quad\quad :\ddot{O}: \\ | \quad\quad\quad\quad | \\ H—\ddot{\underset{..}{O}}—S—\ddot{\underset{..}{O}}—\ddot{\underset{..}{O}}—S—\ddot{\underset{..}{O}}—H \\ | \quad\quad\quad\quad | \\ :\underset{..}{O}: \quad\quad :\underset{..}{O}: \end{array}$$

Peroxodisulfuric acid ($H_2S_2O_8$)

As you might expect, these acids are strong oxidizing agents. They are also potentially explosive and their salts are unstable.

$$2K_2S_2O_8(s) \xrightarrow{\Delta} 2K_2SO_4(s) + 2SO_3(g) + O_2(g)$$

When an aqueous solution containing sulfite ion is boiled with elemental sulfur, the thiosulfate ion ($S_2O_3^{2-}$) is produced.

$$SO_3^{2-} + S(s) \longrightarrow \underset{\text{thiosulfate ion}}{S_2O_3^{2-}}$$

The two sulfur atoms in the thiosulfate ion are not chemically equivalent. One is the central atom in an AB_4 configuration, and the three oxygen atoms and the other sulfur atom are bonded to it. Thiosulfates can be thought of as salts of thiosulfuric acid, $H_2S_2O_3$, which (like sulfurous acid) is of doubtful existence. The most common thiosulfate is the sodium salt, which is obtained from solution as the pentahydrate, $Na_2S_2O_3 \cdot 5H_2O$. This substance is known as "hypo" and is used in photography (see Aside 30.1).

The thiosulfate ion is unstable in acidic solution, immediately decomposing to give free sulfur and sulfur dioxide.

$$S_2O_3^{2-} + 2H^+ \longrightarrow S(s) + SO_2(g) + H_2O(l)$$

The ion is a fairly strong reducing agent. Strong oxidizing agents, such as chlorine, oxidize thiosulfate to sulfate ion.

$$S_2O_3^{2-} + 4Cl_2(g) + 5H_2O(l) \longrightarrow 2SO_4^{2-} + 10H^+ + 8Cl^-$$

Moderately strong oxidizing agents, such as iodine, convert $S_2O_3^{2-}$ to $S_4O_6^{2-}$.

$$2S_2O_3^{2-} + I_2(s) \longrightarrow S_4O_6^{2-} + 2I^-$$

This reaction occurs quantitatively and rapidly, and is useful for the analytical determination of iodine and in other analytical procedures.

EXAMPLE 28.2 Chemical Properties: SF_6

Sulfur hexafluoride, SF_6, is a remarkable compound. In spite of its rather large molar mass (146.05 g), this covalent substance is a gas at room temperature. Despite the fact that sulfur is in its maximum oxidation state of +6, the compound is chemically inert at ordinary temperatures and is not an oxidizing agent even at red heat. These properties make it useful as an insulator in high-voltage generators. What reasons can be given for the properties described?

Here is an excellent case where properties can be related to structure. The SF_6 molecule is octahedral, and the **sizes of the sulfur and fluorine atoms are such that the latter essentially envelop the sulfur atom completely.** As a result, **attractions between individual molecules are weak** (the solid sublimes at −63 °C) and **reagents cannot penetrate the fluoride barrier to react with the sulfur atom.**

[It is worth noting that the hexafluoride of selenium, the element lying below sulfur in Representative Group VI, is more reactive than SF_6. The larger selenium atom is less well shielded by the six fluorine atoms.]

EXAMPLE 28.3 Chemical Reactions: P and S Compounds

Categorize the following reactions of phosphorus and sulfur compounds.

(a) $P_4O_6(s) + 6H_2O(l) \longrightarrow 4H_3PO_3(aq)$

(b) $H_2S(g) + 2NaOH(aq) \longrightarrow Na_2S(aq) + 2H_2O(l)$

(c) $2ZnS(s) + 3O_2(s) \longrightarrow 2ZnO(s) + 2SO_2(g)$
(reaction used in smelting metals; produces pollution)

(d) $Ca_3(PO_4)_2(s) + 3H_2SO_4(aq) \longrightarrow 2H_3PO_4(aq) + 3CaSO_4(s)$
(preparation of wet process phosphoric acid)

(a) This is a **nonredox combination of compounds** and the reaction of an acidic oxide with water.

(b) An acid (H_2S) and a base (NaOH) combine to give water and a salt (Na_2S). This is a **neutralization reaction**—a **partner-exchange** reaction driven by the formation of water.

(c) This reaction of an element and a compound is the **redox displacement** of a nonmetal (S) by a more active nonmetal (O). Elemental oxygen is reduced and sulfide ion from ZnS is oxidized to the +4 state in SO_2.

(d) While not a reaction between ions in aqueous solution, this reaction follows the pattern of a **partner-exchange reaction.** In this case a strong acid reacts with a salt to produce a weaker acid.

EXAMPLE 28.4 Chemical Reactions: P and S Compounds

Predict the products of the following reactions:

(a) $ZnS(s) + HCl(aq) \longrightarrow$

(b) $Ba(H_2PO_2)_2(aq) + H_2SO_4(aq) \longrightarrow$

(c) $Na_3PO_4(s) + H_2O(l) \longrightarrow$

(a) Metal sulfides dissolve in strong acid solutions with the formation of hydrogen sulfide

$$ZnS(s) + 2HCl(aq) \longrightarrow \mathbf{ZnCl_2(aq) + H_2S(aq)}$$

This is a partner-exchange reaction.

(b) This is a reaction between ionic compounds in aqueous solution. A partner-exchange reaction can be driven by the formation of slightly soluble barium sulfate. The other product is hypophosphorous acid. [This reaction is a preparative method for hypophosphorous acid.]

$$Ba(H_2PO_2)_2(aq) + H_2SO_4(aq) \longrightarrow \mathbf{BaSO_4(s) + 2H_3PO_2(aq)}$$

(c) Like all sodium salts, sodium phosphate is water soluble. The PO_4^{3-} anion, the anion of a weak acid (HPO_4^{2-}), will react with water and the solution will be alkaline.

$$Na_3PO_4(s) \xrightarrow{H_2O} \mathbf{3Na^+ + PO_4^{3-}}$$
$$PO_4^{3-} + H_2O(l) \rightleftharpoons \mathbf{HPO_4^{2-} + OH^-}$$

SUMMARY OF SECTIONS 28.14–28.16

The outstanding properties of some sulfur compounds are summarized in Table 28.20.

Table 28.20
Outstanding Properties of Sulfur Compounds and Ions

H_2S	A weak diprotic acid
S^{2-}	Weakly basic
M_xS_y	Many metal sulfides are insoluble and used in qualitative analysis; dissolve in acid with generation of H_2S
SO_2	Formed in combustion of sulfur and sulfur-containing fuels and smelting of sulfide ores; air pollutant; strong reducing agent; acidic oxide
SO_3	Fumes in moist air as it forms H_2SO_4 mist
SX_n	Most sulfur halides hydrolyze, e.g., in moist air (except very stable SF_6)
HSO_3^-	Weakly acidic; reducing agent
SO_3^{2-}	Reducing agent; salts give SO_2 with acid
H_2SO_4	Largest volume industrial chemical; strong acid in first dissociation; hot, concentrated acid is strong oxidizing agent and dehydrating agent
HSO_4^-	Acidic in aqueous solution
$H_2S_2O_7$ *pyrosulfuric acid*	Intermediate in production of H_2SO_4
$S_2O_3^{2-}$ *thiosulfate ion*	Reducing agent
$Na_2S_2O_3 \cdot 5H_2O$ *sodium thiosulfate*	"Hypo"; used in photography

Aside 28.4 AIR POLLUTION, SMOG, AND ACID RAIN

Some pollutants, such as the insecticide DDT, are unknown in nature. Others, such as NO, occur naturally (NO is formed in lightning flashes and bacterial decay), but are considered to be pollutants when they are added to the environment in excessive amounts by human activities. Still other substances, known as secondary pollutants, are harmful materials formed by chemical reactions, for example, the sulfuric acid formed in the atmosphere from the oxides of sulfur.

The five major air pollutants are carbon monoxide, hydrocarbons, the nitrogen oxides, the sulfur oxides, and particulates—airborne solid particles and liquid droplets. The sources and effects of the nitrogen and sulfur oxides are considered here. The other air pollutants are discussed in Aside 29.1.

Nitric oxide (NO) and nitrogen dioxide (NO_2)—referred to as NO_x, "nox"—are produced by oxidation of atmospheric nitrogen (and nitrogen-containing compounds, as in coal).

$$N_2 + O_2 \xrightarrow{\Delta} \underset{\text{nitric oxide}}{2NO}$$

$$2NO + O_2 \xrightarrow{\Delta} \underset{\text{nitrogen dioxide}}{2NO_2}$$

The first of these reactions takes place at high temperatures (about 1200–1750 °C); therefore, NO is a by-product of any high-temperature combustion in the presence of air. Most fossil fuel combustion in both motor vehicles and stationary sources achieves the necessary temperatures, and almost 90% of the pollutant NO is introduced in this way. Pollution by NO_x reaches its highest levels over highly industrialized cities with high automobile populations.

Sulfur dioxide, SO_2, is produced when sulfur-containing coal and fuel oil are burned. Very little sulfur dioxide pollution can be blamed on the automobile, because in the production of gasoline most of the sulfur is removed. One-half of the annual sulfur dioxide emission comes from power plants, and the second most substantial contribution is made by industrial processes, especially metal smelting, in which sulfur is removed from sulfides ores by oxidation (Section 26.4). Sulfur trioxide is formed in the air by oxidation of the lower oxide.

On their own, neither of the nitrogen oxides is a health hazard at present pollution levels. Nitrogen dioxide is potentially the more dangerous of the two, and in high enough concentration can damage the lungs. Sulfur dioxide and sulfur trioxide—referred to as SO_x, "sox"—are both strongly irritating to the respiratory tract. At SO_2 concentrations of 5 ppm almost everyone suffers throat and eye irritation, and SO_3 at a concentration of 1 ppm can cause severe discomfort. Elderly persons and those with respiratory diseases are most seriously affected.

The hazards from NO_x and SO_x of greatest immediate concern are twofold—the smog produced when high concentrations accumulate over industrial areas, and the acid rain resulting from the generally rising concentrations of these pollutants throughout the atmosphere.

Nitrogen oxides are essential to the unpleasant mixture of gases and particulates that make up photochemical smog. The ingredients necessary to form **photochemical smog** are sunlight, nitrogen oxides, and hydrocarbons.

The intervention of hydrocarbons in the natural cycle of NO_2 reactions in the atmosphere starts the trouble (Figure 28.9). Hydrocarbons react with oxygen atoms to form highly reactive free radical intermediates which contain unpaired electrons. The free radicals (symbolized here by a dot on the formula, e.g., RCO·) initiate a variety

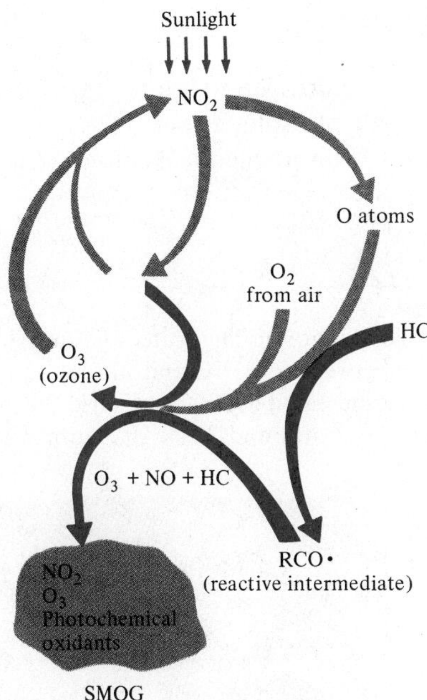

Figure 28.9
Interaction of Hydrocarbons (HC) with Natural NO_2 in Photochemical Reaction Cycle
The natural cycle is shown in gray.

of reactions, among which may be the following:

$$RCO\cdot + O_2 \longrightarrow RCO_3\cdot$$
$$RCO_3\cdot + \text{hydrocarbons} \longrightarrow R_2C{=}O, RCHO$$
$$RCO_3\cdot + NO \longrightarrow RCO_2\cdot + NO_2$$
$$RCO_3\cdot + O_2 \longrightarrow RCO_2\cdot + O_3$$
$$RCO_3\cdot + NO_2 \longrightarrow \underset{\textit{peroxyacylnitrates (PAN)}}{RCO_3NO_2}$$

As a result, concentrations of ozone, the peroxyacylnitrates (PAN), and aldehydes (RCHO) build up. These substances are all irritants to the respiratory system and can also damage many materials. Ozone is particularly destructive to rubber and fabrics, and is harmful to crops and ornamental plants. Nitrogen dioxide also accumulates and is responsible for the characteristic brown color of photochemical smog. Airplane pilots are familiar with the sight of a brown pall hanging over cities.

The word "smog" is a combination of "smoke" and "fog." It is something of a misnomer for photochemical smog, which includes nitrogen oxides, ozone, PAN, aldehydes, hydrocarbons, and carbon monoxide, but not smoke or fog. The term "smog" is more accurate for **London smog,** or gray smog, which is initiated by a mixture of sulfur oxides, particulates, and high humidity. Many of the chemicals in particulates can catalyze the formation of SO_3 from SO_2, and high humidity assures the formation of a fog containing many sulfuric acid droplets formed by reaction of SO_3 with water. The sulfuric acid also coats the surface of particulates and is drawn into the lungs with them. The 1952 London smog is blamed for 4000 deaths.

Unlike smog, acid rain is a global rather than an urban problem, affecting the wildlife in remote wilderness areas as well as stone buildings and monuments. "Pure" rainwater is defined as having a pH of 5.6—slightly acid due to equilibrium with the natural concentration of carbon dioxide in the atmosphere. With rising air pollution has come rising acid concentration that is washed out of the atmosphere by rain. Today, rainfall in the entire eastern half of the United States has a pH of 5 or below. In the Northeast, rain of pH 4.2 or less is commonplace. (The pH of tomato juice is 4.2.)

Nitric acid and sulfuric acid in varying proportions are present in acid rain. Current studies suggest the formation of nitric acid is linked to photochemical reactions of hydroxyl radicals, ·OH, generated by reaction of photochemically excited oxygen atoms with water.

$$O_3(g) \xrightarrow{h\nu} O^*(g) + O_2(g)$$
$$O^*(g) + H_2O(g) \longrightarrow 2\cdot OH(g)$$
$$\cdot OH(g) + NO_2(g) \longrightarrow HNO_3(g)$$

Sulfuric acid is formed by the oxidation of SO_2 by ozone or hydrogen peroxide. The reactions are believed to occur while the gases are also dissolved in water droplets.

Nitric and sulfuric acids are deposited not only by rainfall, but also by fog and snow, as well as on the surface of particulates that fall to the ground. Some of the worrisome effects of the resultant rising acidity of natural waters are declining fish populations (70% of Norwegian lakes with pH levels of less than 4.5 have *no* fish); the dissolution by acid runoff and the transportation into natural waters of metals such as aluminum (toxic to fish) and mercury (toxic to humans by concentration up the food chain); possible destructive effects on plants in wilderness areas as well as on farms; and problems in the lumber industry due to the deaths of trees. Also, stone buildings and monuments weather rapidly under acidic conditions (Figure 28.10).

(a) Acid rain containing H_2SO_4 falls

(b) Crust forms as calcite ($CaCO_3$) is transformed to gypsum ($CaSO_4\cdot 2H_2O$)

(c) Crust washes off as the more soluble gypsum dissolves

(d) Layer of stone is removed

Figure 28.10
Effect of Acid Rain on Limestone

SIGNIFICANT TERMS (with section references)

aqua regia (28.6)
dimer (28.9)
London smog (28.4)
nitrogen fixation (Aside 28.3)
phosphate rock (28.3)
photochemical smog (Aside 28.4)

QUESTIONS AND PROBLEMS

Periodic Relationships: N, P, and S

28.1 Write the electron configuration for a nitrogen atom. Assuming that the atom will gain enough electrons to attain a noble gas configuration, what would be the most stable negative oxidation number for N? If the *p* electrons can be removed or shared more easily than the *s* electrons, what positive oxidation numbers would be predicted for N? Repeat this question for P and S atoms.

28.2 Draw the Lewis structures of the N_2, P_4, and S_8 molecules and describe the geometry of each molecule. What intermolecular forces are important for these molecules?

28.3 The N≡N bond energy is 946 kJ/mol and the N—N bond energy is 159 kJ/mol. Predict whether four gaseous nitrogen atoms would form two nitrogen molecules or a tetrahedral molecule similar to P_4, basing your prediction on the amount of energy released as the molecules are formed. Repeat the calculations for phosphorus using 485 kJ/mol for P≡P and 243 kJ/mol for P—P.

28.4 The average atomic mass of N is 14.0067 u. There are two isotopes which contribute to this average: ${}^{14}_{7}N$ (14.00307 u) and ${}^{15}_{7}N$ (15.00011 u). Calculate the percentage of ${}^{15}_{7}N$ atoms in a sample of naturally occurring nitrogen.

28.5* Using molecular orbital notation, (a) write the electron configuration for a molecule of nitrogen. (b) What types of bonds and how many of each would you predict the molecule to have? (c) Prepare a three-dimensional sketch of a molecule of N_2. (d) Will the molecule be paramagnetic or diamagnetic?

28.6 What is meant by the term "allotropy"? Name some of the allotropic forms of phosphorus and sulfur. List some of the physical properties of N_2, P_4, and S_8.

28.7 At 502 Torr and 750 °C, 1.00 L of sulfur vapor is found to weigh 0.5350 g. What is the major component of the vapor?

28.8 The enthalpy change for dissolving sulfur in six moles of CS_2 at 25 °C

$$S(s) + 6CS_2(l) \longrightarrow S(\text{in } 6CS_2)$$

is 1695 J/mol for rhombic sulfur and 1360. J/mol for monoclinic sulfur. Calculate $\Delta H°$ for the phase transformation

$$S(\text{rhombic}) \longrightarrow S(\text{monoclinic})$$

Which form of sulfur would you predict to be more stable at this temperature?

28.9* The phase diagram for sulfur is shown in the accompanying figure. Describe what happens to a sample for sulfur (a) that is heated very slowly at 1 atm from 25 °C to 500 °C; (b) that is heated at 1420 atm from 25 °C to 153 °C; (c) that has been melted and is then poured slowly into boiling water at 1 atm; (d) that is in the gaseous phase between 95.31 °C and 115.18 °C and that has the pressure upon it increased to 2000 atm; (e) in the monoclinic form between 5.1×10^{-6} atm and 3.2×10^{-5} atm when it is heated.

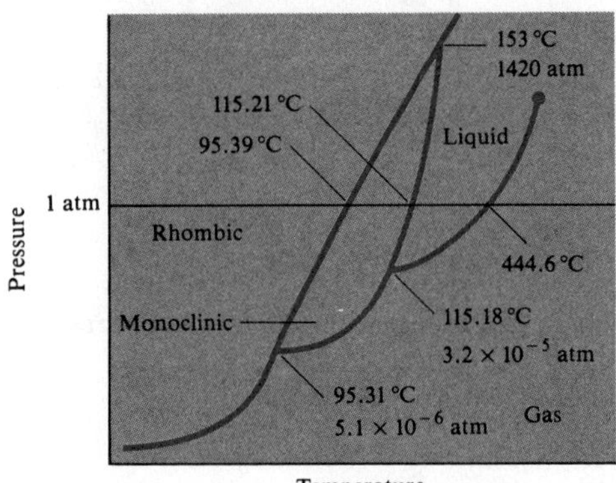

28.10 Describe the commercial preparations of elemental nitrogen, phosphorus, and sulfur. Be sure to give any pertinent chemical equations.

28.11 List some of the major uses of elemental nitrogen, phosphorus, and sulfur.

28.12 Briefly describe the types of bonds that N, P, and S atoms will form with metals, semiconducting elements, and nonmetals.

28.13 Compare the reactivity of N, P, and S in combination reactions with other elements. Does the existence of allotropic forms of phosphorus make a difference in the reactivity of this element? Explain.

28.14 Although nitrogen and phosphorus are in the same periodic table family, phosphorus forms a pentachloride, PCl_5, but the analogous nitrogen compound, NCl_5, does not exist. Why is this not surprising?

28.15 Compare the chemical reactivities of the nitride, phosphide, and sulfide ions toward water. Write chemical equations describing these reactions.

28.16 Nitrogen and oxygen occur together in the atmosphere, but do not react to form NO or NO_2 under ordinary conditions. Why not? Under what conditions do they react in the atmosphere?

Industrial Acids

28.17 Briefly describe the industrial preparations of sulfuric, nitric, and phosphoric acids.

28.18 Name three main classifications of reactions undergone by sulfuric acid. When zinc metal is added to dilute H_2SO_4, H_2 is generated; when zinc metal is added to concentrated H_2SO_4, H_2S is formed. Write chemical equations for these reactions and explain why there is a difference in the reactions.

28.19 What is the pH of a 0.010 M solution of H_2SO_4? What is the concentration of each sulfur-containing species? $K_{a2} = 1.0 \times 10^{-2}$ for H_2SO_4.

28.20 What mass of H_2SO_4 can be produced in the process given below if 1.00 kg of FeS_2 is used? The unbalanced equations for the process are

$$FeS_2(s) + O_2(g) \longrightarrow Fe_2O_3(s) + SO_2(g)$$
$$SO_2(g) + O_2(g) \longrightarrow SO_3(g)$$
$$SO_3(g) + H_2SO_4(l) \longrightarrow H_2S_2O_7(l)$$
$$H_2S_2O_7(l) + H_2O(l) \longrightarrow H_2SO_4(aq)$$

Assume complete reactions.

28.21 Briefly discuss the oxidizing properties of nitric acid. What determines the oxidation state of the nitrogen in the reduction product?

28.22 Commercial concentrated HNO_3 contains 69.5 mass % HNO_3 and has a density of 1.42 g/mL. What is the molarity of this solution? What volume of the concentrated acid should a stockroom attendant use to prepare 10.0 L of dilute HNO_3 at a concentration of 6.0 M?

Compounds of Nitrogen

28.23 Name three classes of nitrides and briefly discuss the bonding in each.

28.24 Describe the molecular geometry of the ammonia molecule. How does this geometry affect the physical properties of the substance? Will ammonia act as a Lewis acid or as a Lewis base? Illustrate this acid–base behavior by completing the following equations: (a) $NH_3(aq) + H^+ \longrightarrow$, and (b) $AgCl(s) + NH_3(aq) \longrightarrow$.

28.25 Write the Lewis structures for (a) N_2O, (b) NO, (c) HNO_2, (d) HNO_3, (e) NO_2^-, (f) NO_3^-. Name each species and identify the hybridization of the outermost orbitals of the nitrogen atom in each species.

28.26 Why can nitrous acid act as both an oxidizing agent and a reducing agent? Write the chemical equation for the disproportionation of nitrous acid.

28.27 Why does NO_2, a brown gas, lose its color when cooled to 0 °C?

28.28 Calculate the pH of each of the following solutions: (a) 0.05 M HNO_3; (b) 1 M NH_3, $K_b = 1.6 \times 10^{-5}$; (c) 0.1 M HNO_2, $K_a = 7.2 \times 10^{-4}$.

28.29 What would be the pH of a buffer solution prepared using equal volumes of 0.20 M HNO_2 and 0.20 M KNO_2? $K_a = 7.2 \times 10^{-4}$ for HNO_2.

28.30 Describe the chemical behavior of ammonium salts upon heating. Contrast the reaction products of ammonium salts containing anions that can act as oxidizing agents with the reaction products of those salts containing anions that do not act as oxidizing agents.

28.31 The detonator-induced thermal explosion of ammonium nitrate yields $N_2(g)$, $H_2O(g)$, and $O_2(g)$. Calculate the total volume of gas, measured at 1.00 atm and 827 °C, theoretically released in the explosive decomposition of 1.00 kg of $NH_4NO_3(s)$.

28.32 Some nitrogen-containing compounds used in explosives are ammonium nitrate (NH_4NO_3), hydrazine (N_2H_4), sodium azide (NaN_3), and nitroglycerin ($C_3H_5N_3O_9$). What common characteristic do these substances share? How does this relate to their use?

28.33* Although the compound HN_3, hydrazoic acid, is very explosive in the pure state, it can be studied in aqueous solution. The acid is prepared by the reaction of hydrazine with nitrous acid.

$$N_2H_4(l) + HNO_2(aq) \longrightarrow 2H_2O(l) + HN_3(aq)$$

(a) What are the oxidation states of nitrogen in the compounds in this reaction? (b) What is the oxidizing agent in the reaction? Hydrazoic acid is a rather weak acid with $K_a = 2.4 \times 10^{-5}$ at 25 °C. (c) Calculate the pH of a 0.01 M solution of HN_3.

28.34* Two acceptable resonance forms can be written for the HN_3 molecule

$$\mathrm{H{-}\ddot{N}{=}N{=}\ddot{N}} \rightleftharpoons \mathrm{H{-}\ddot{N}{-}N{\equiv}N{:}}$$

(a) Use the following thermodynamic data to calculate the heat of formation of $HN_3(g)$ for each resonance form: $\Delta H_f^\circ = 218.0$ kJ/mol for H(g) and 472.7 kJ/mol for N(g), and bond energy = 389 kJ/mol for N—H, 159 kJ/mol for N—N, 473 kJ/mol for N=N, and 946 kJ/mol for N≡N. The lengths of nitrogen–nitrogen bonds are 0.145 nm for N—N, 0.123 nm for N=N, and 0.109 nm for N≡N. (b) Estimate the bond lengths of each of the nitrogen–nitrogen bonds in HN_3. (c) Prepare a three-dimensional sketch of the molecule showing the delocalized bonding.

Compounds of Phosphorus

28.35 What structural feature do molecules of P_4, P_4O_6, and P_4O_{10} have in common? What is the common structural feature for all of the acids containing phosphorus(V)?

28.36 Draw Lewis structures for (a) PCl_3, (b) $[PCl_4]^+$, (c) PCl_5, (d) $[PCl_6]^-$. What is the hybridization of the phosphorus atom in each species? Prepare three-dimensional sketches of these species and identify the molecular geometry of each.

28.37 Phosphoric acid, H_3PO_4, forms sodium salts corresponding to replacement of all three hydrogen atoms—NaH_2PO_4, Na_2HPO_4, and Na_3PO_4—but only two analogous salts for phosphorous acid, H_3PO_3, can be obtained—NaH_2PO_3 and Na_2HPO_3. Give a reasonable explanation for this difference in the behavior of the two acids.

28.38 A sample of phosphorus was burned in air and produced a product containing 56 mass % P and 44 mass % O. Find the simplest formula of the oxide formed. The density of this gaseous oxide at 200 °C and 1 atm pressure is 5.7 g/L. What is the molecular formula of this oxide?

28.39 A laundry additive contains 9.5 mass % P in the form of $Na_5P_3O_{10}$. What mass of $Na_5P_3O_{10}$ is in a one-pound box of the additive?

28.40* The following series of balanced equations shows the reactions used to convert pure hydroxyapatite, $Ca_5(PO_4)_3(OH)$, to sodium tripolyphosphate hexahydrate, $Na_5P_3O_{10}\cdot 6H_2O(s)$:

$$Ca_5(PO_4)_3(OH) + 5H_2SO_4 \longrightarrow 5CaSO_4 + H_2O + 3H_3PO_4$$
$$3H_3PO_4 \longrightarrow H_5P_3O_{10} + 2H_2O$$
$$H_5P_3O_{10} + 5NaOH \longrightarrow Na_5P_3O_{10} + 5H_2O$$
$$Na_5P_3O_{10} + 6H_2O \longrightarrow Na_5P_3O_{10}\cdot 6H_2O$$

Classify each reaction using the scheme presented in Chapter 17. Assuming no loss of phosphorus in any of the reactions, calculate the mass of hydroxyapatite required to produce 10.0 kg of this builder for detergents.

28.41* A careless stockroom attendant prepared a liter each of the following solutions: 0.1 M Na_3PO_4, 0.1 M Na_2HPO_4, and 0.1 M NaH_2PO_4, but forgot to label the containers. Quickly the attendant realized that each solution had a unique value of pH, and all that was needed was to calculate the theoretical values of pH from the values of K_a ($K_{a1} = 7.5 \times 10^{-3}$, $K_{a2} = 6.6 \times 10^{-8}$, $K_{a3} = 1 \times 10^{-12}$), measure the pH values of the solutions using a pH meter, and properly label each container. What were the values of pH that the attendant calculated?

Compounds of Sulfur

28.42 Write Lewis structures for the following species: (a) sulfur dioxide, (b) sulfuric acid, (c) sulfur trioxide, (d) pyrosulfuric acid, (e) peroxodisulfuric acid, (f) sulfite ion, (g) sulfate ion, (h) thiosulfate ion, (i) sulfide ion, (j) molecular sulfur. What is the oxidation number of sulfur in each of these species?

28.43 Which metal sulfides are considered to be water soluble? Will solutions containing the soluble sulfides be alkaline, neutral, or acidic?

28.44* A sample of powdered sulfur was divided into two parts—one about twice the size of the other. The smaller sample was heated until the sulfur melted and began to burn. The gaseous product (a) of this reaction was collected and mixed with water to form (b). The larger sulfur sample was mixed with iron filings and heated in a crucible, producing a dark solid (c). This solid was placed in a container, and HCl was added, producing a gas (d), which was allowed to mix with (b), giving a yellow to white finely divided precipitate (e). Identify the five substances mentioned above and write chemical equations describing the reactions.

28.45 A gaseous mixture at 300 °C in a 1.00 L vessel originally contained 1.00 mol SO_2 and 5.00 mol O_2. Once equilibrium conditions were attained, 81% of the SO_2 had been converted to SO_3. What is the value of the equilibrium constant (K_c) for this reaction at this temperature?

28.46* Assume that a molecule of SO_3 contains one double bond and two single bonds between the sulfur and oxygen atoms. Calculate the heat of formation of $SO_3(g)$ using the following data: $\Delta H_f^\circ = 279$ kJ/mol for S(*g*) and 249 kJ/mol for O(*g*), and bond energy = 423 kJ/mol for S—O and 523 kJ/mol for S=O. Because of the delocalized π bonding in the molecule, SO_3 is more stable than predicted by this calculation. Calculate this difference in energy (known as the "resonance energy") given that the measured heat of formation of $SO_3(g)$ is −395.72 kJ/mol.

28.47 Common copper ores in western United States contain the mineral chalcopyrite, $CuFeS_2$. Assuming that an average commercially attractive ore contains 0.2 mass % Cu (as chalcopyrite) and that all the sulfur ultimately appears in the smelter stack gases as SO_2, calculate the mass of sulfur dioxide generated by the conversion of 1.0 kg of the ore.

ADDITIONAL QUESTIONS AND PROBLEMS

28.48 Compare the acidic or alkaline natures of solutions containing NO_3^-, PO_4^{3-}, HPO_4^{2-}, $H_2PO_4^-$, SO_4^{2-}, or HSO_4^-. Write chemical equations for any pertinent reactions.

28.49 Which of the following are redox reactions? Identify the oxidizing and reducing agent in each of the redox reactions.

(a) $P_4(s) + 3OH^- + 3H_2O(l) \longrightarrow 3H_2PO_2^- + PH_3(g)$
(b) $3NO_2(g) + H_2O(l) \longrightarrow 2HNO_3(aq) + NO(g)$
(c) $2S_2O_3^{2-} + I_3^- \longrightarrow S_4O_6^{2-} + 3I^-$
(d) $3S_8(s) + 48OH^- \longrightarrow 16S^{2-} + 8SO_3^{2-} + 24H_2O(l)$

28.50 Repeat Question 28.49 for

(a) $2Ca_3(PO_4)_2(s) + 6SiO_2(s) + 10C(s) \xrightarrow{\Delta} 6CaSiO_3(l) + 10CO(g) + P_4(g)$
(b) $P_4O_{10}(s) + 6H_2O(l) \longrightarrow 4H_3PO_4(aq)$
(c) $Au(s) + 4H^+ + 4Cl^- + NO_3^- \longrightarrow [AuCl_4]^- + NO(g) + 2H_2O(l)$
(d) $3H_3PO_4 \longrightarrow H_5P_3O_{10} + 2H_2O(l)$

28.51 Classify each of the following reactions according to the reaction types listed in Tables 17.2 and 17.7:

(a) $4NH_3(g) + 5O_2(g) \xrightarrow{\Delta} 4NO(g) + 6H_2O(g)$
(b) $Cu^{2+} + H_2S(g) \longrightarrow CuS(s) + 2H^+$
(c) $Na_2SO_3(s) + S(s) + 5H_2O(l) \longrightarrow Na_2S_2O_3\cdot 5H_2O(s)$
(d) $(NH_4)_2Cr_2O_7(s) \xrightarrow{\Delta} Cr_2O_3(s) + 4H_2O(g) + N_2(g)$

28.52 Repeat Question 28.51 for

(a) $HNO_3(l) + H_2O(l) \longrightarrow H_3O^+ + NO_3^-$
(b) $Mg_3N_2(s) + 6H_2O(l) \longrightarrow 3Mg(OH)_2(s) + 2NH_3(g)$
(c) $2NaNO_3(s) \xrightarrow{\Delta} 2NaNO_2(s) + O_2(g)$
(d) $3S_8(s) + 48OH^- \longrightarrow 8SO_3^{2-} + 16S^{2-} + 24H_2O(l)$

28.53 Predict the major products of the following reactions:

(a) $Li_3N(s) + H_2O(l) \longrightarrow$
(b) $NH_4NO_3(s) \xrightarrow{\Delta}$
(c) $HNO_2(aq) + I^- + H^+ \longrightarrow$
(d) $H_2SO_4(18\ M) + H_2S(g) \longrightarrow$

28.54 Repeat Question 28.53 for

(a) $HNO_3(l) + P_4O_{10}(s) \longrightarrow$
(b) $Cu(s) + HNO_3(6\ M) \longrightarrow$
(c) $PBr_3(l) + H_2O(l) \longrightarrow$
(d) $HSO_3^- + H_2O(l) \longrightarrow$

28.55 Write balanced equations for the following chemical reactions: (a) formation of sulfur hexafluoride; (b) combustion of hydrogen sulfide, using excess oxygen; (c) laboratory preparation of nitric oxide; (d) thermal decomposition of solid disodium monohydrogen phosphate.

28.56 Which solution has a larger mole fraction of solute: concentrated H_2SO_4 (which is 96.0 mass % H_2SO_4) or concentrated "ammonium hydroxide" (which is 58.6 mass % NH_3)?

28.57 The combustion products of P_4S_3 are P_4O_{10} and SO_2. Using the enthalpy of formation of P_4S_3 as -154 kJ/mol, of P_4O_{10} as -2984 kJ/mol, and of SO_2 as -297 kJ/mol, calculate the enthalpy of combustion of P_4S_3—the reaction that takes place when a match tip ignites.

CHAPTER

29

Carbon and the Semiconducting Elements

29.1 PERIODIC RELATIONSHIPS: CARBON AND THE SEMICONDUCTING ELEMENTS

So far in our study of the elements we have examined the *s*- and *p*-block metals; the nonmetals of Representative Groups V, VI, and VII; and the noble gases. Seven elements—boron (Group III), silicon and germanium (Group IV), arsenic and antimony (Group V), and selenium and tellurium (Group VI)—lie between the representative metals and the nonmetals in the *p* block of the periodic table. These elements are classified as semiconducting elements, and are also known as *metalloids* or *semimetals*. The semiconducting elements resemble metals in appearance but are more like the nonmetals in their chemical properties. They conduct electric current, but much less effectively than metals. It is the nature of the electrical conductivity of semiconducting elements that puts them in a class by themselves; their conductivity increases under conditions that decreases the conductivity of metals. This distinctive property is explained in terms of bonding (Section 29.17).

The inorganic chemistry of carbon, the only nonmetal in Representative Group IV, is included in this chapter along with the chemistry of the semiconducting elements. There are a number of reasons for doing this. In its elemental state and in certain aspects of its inorganic chemistry, carbon is intermediate in properties between metals and nonmetals and is similar to silicon and germanium, the semiconducting members of Group IV. Like all the semiconducting elements in their thermodynamically most stable forms, elemental carbon is crystalline and has network covalent bonding. By contrast, all the other elemental nonmetals are composed of discrete

Table 29.1
Differences between Carbon and the Other Members of the Carbon Family (Si, Ge, Sn, Pb)

Property	Comments
Only nonmetal in the family	Silicon and germanium are semiconducting elements; tin and lead are metals
Can form no more than four covalent bonds	The $n = 2$ energy level can hold only 8 electrons and only s and p orbitals are available for bonding; other elements have d orbitals available and can form species such as SiF_6^{2-}, $SnCl_6^{2-}$
Only family member to form anions containing only the element; C_2^{2-} and C^{4-} are known	Such anions form by combination of carbon with very reactive metals; e.g., in Be_2C, CaC_2, Al_4C_3
Has the greatest tendency of all elements to self-link	Gives rise to the realm of organic chemistry
C—C bond is strong (331 kJ/mol)	Strongest nonmetal–nonmetal single bond except for H_2 (other family members, Si—Si, 209 kJ/mol; Ge—Ge, 160 kJ/mol; Sn—Sn, 140 kJ/mol)
Forms multiple bonds with itself, oxygen, nitrogen, sulfur, and phosphorus	Very few multiple bonds formed by other family members (Si=Si bonds are known)

molecules, such as Cl_2, He, O_2, P_4, and S_8. Also, carbon as graphite has a metallic luster and is a fairly good electrical conductor, yet its chemical properties are those of a nonmetal.

As is true for most first members of representative element families, there are significant differences between carbon and the other members of its family, most importantly the ability of carbon atoms to undergo extensive self-linkage (*catenation*). The field of organic chemistry (Chapter 32) is based on the very great tendency of carbon atoms to bond strongly to each other (matched by no other element) and to form strong covalent bonds with hydrogen atoms. A summary of the differences between carbon and the other members of Group IV is given in Table 29.1.

CARBON AND ITS INORGANIC COMPOUNDS

29.2 PROPERTIES OF CARBON

In most compounds, carbon atoms (electron configuration $1s^2 2s^2 2p^2$) form four equivalent single bonds. A carbon atom can also form a double covalent bond (along with two single bonds) or a triple covalent bond (along with one single bond). Only in compounds with the most electropositive metals do carbon atoms form anions (C_2^{2-} or in a few cases C^{4-}). In its inorganic compounds, carbon has oxidation states of -4 (in C^{4-}), -1 (in C_2^{2-}), $+2$ (in CO), or $+4$ (e.g., in CCl_4, CO_2). [In organic compounds, carbon may have any oxidation state from -4 to $+4$.]

Elemental carbon is not very reactive and its reactions require high temperatures. Carbon is less electronegative than nitrogen, oxygen, fluorine, chlorine, and bromine, other nonmetals with which it forms many compounds. With respect to other members of its own family, however, carbon has the highest ionization energy and is the most electronegative element (Table 29.2).

Table 29.2
Properties of Carbon

Melting point (°C)	
Diamond	>3550
Graphite	3652–3697 (sublimes)
Ionization energy (kJ/mol, 0 K)	1086
Electron affinity (kJ/mol, 0 K)	123
Electronegativity	2.5
Density (g/cm^3)	
Diamond	3.51
Graphite	2.25
C—C single bond length (nm)	0.154
Atomic radius (nm)	0.077
Bond energy (kJ/mol)	C—C 331 C=C 590 C≡C 812

29.3 DIAMOND, GRAPHITE, AND OTHER FORMS OF CARBON

Carbon is found naturally in two allotropic forms that are both crystalline—diamond and graphite. There are also numerous natural and man-made amorphous forms of carbon, including coke and many varieties of finely divided industrial carbon known as, for example, carbon black, lampblack, animal charcoal, and activated carbon.

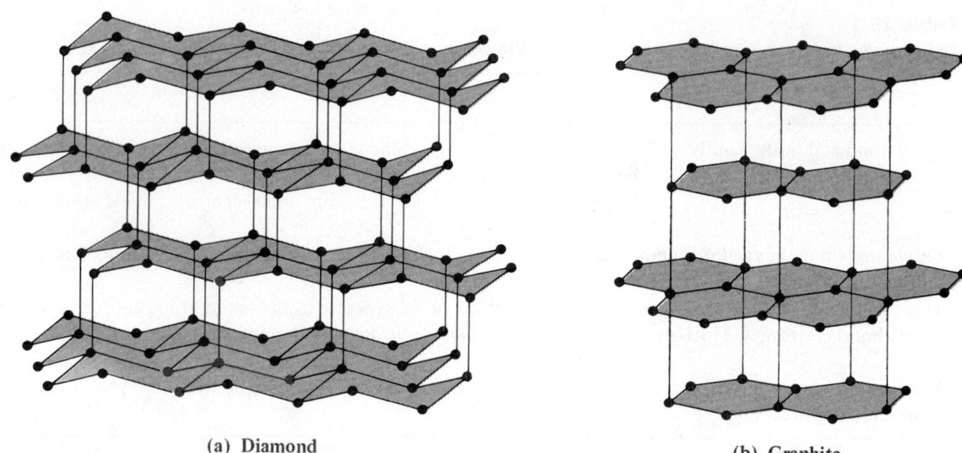

Figure 29.1
Comparison of the Crystal Structures of Diamond and Graphite, Showing the "Layers" in Each

Diamond and graphite differ greatly in properties as a result of the differences in their crystal structures. Diamond is clear and colorless and is the hardest substance known. It has a very high melting point, is extremely brittle, and when struck, breaks into many pieces. It does not conduct electricity. These properties reflect the strength of network covalent bonding (Section 10.5)—each carbon atom in the crystal shares its four valence electrons with four other carbon atoms, which surround it tetrahedrally (Figure 29.1a).

The ability of diamond to refract light rays is high, and when properly cut and polished, a diamond reflects light in an array of many colors. On this account, and because of its durability, diamond is highly prized as a gemstone. Impure samples are often black; these have no value as gems, but when crushed are used in abrasives. In rare cases a diamond has an attractive color because of traces of impurities; such a diamond, because it is unique, is especially valuable as a gem. The famous Hope diamond, for example, is blue.

Diamonds are found embedded in large "pipes" of volcanic rock. To separate the diamonds, the rock is crushed and passed in a stream of water over greased belts to which the diamonds adhere. Most gem quality diamonds come from South Africa. Zaire and the Soviet Union provide the bulk of the natural industrial-quality diamonds.

Graphite is black and soft, and is much less dense than diamond (see Table 29.2). Like diamond, graphite melts only at an extremely high temperature. It feels smooth and slippery to the touch and is an excellent lubricant. The carbon atoms are arranged in planar layers in the graphite crystal (Figure 29.1b). Within each layer, each carbon atom is bonded to three other carbon atoms by covalent single bonds. The fourth valence electron of each carbon atom participates in delocalized π bonding. The atoms in each layer are tightly bonded together (bond order, ~ 1.3; bond energy, 477 kJ/mol), but the binding force between layers is weak (17 kJ/mol), allowing the layers to slip over each other. The delocalized electrons give graphite metallic properties. It has a dull luster and conducts electricity moderately well. The lubricating properties of graphite depend not only on slippage between the planes, but also on a film of moisture or of gas molecules adsorbed on the surface of the graphite layers. The adsorbed substance decreases friction as the layers slide past each other. (Dry graphite in a vacuum is not slippery.)

In addition to its use as a lubricant, graphite is employed on a large scale in electrodes, in molds and crucibles to be used with hot metal, and in fabricated parts for the aerospace industry (e.g., nose cones). The control rods in nuclear reactors are often made of graphite. Graphite is the black ingredient in pencils, a pencil "lead" being a baked rod of graphite mixed with clay.

Comparison shows that the crystal structures of graphite and diamond are closely related (see Figure 29.1). If alternate carbon atoms in each layer of graphite could be raised above the plane of the layer and the others depressed below it, and the layers could then be pushed a little closer together, the diamond structure would result. According to the phase diagram for carbon (Figure 29.2), this should be possible at a high enough temperature and pressure. The difficulty of compressing graphite, however, was recognized by the famous physicist Percy Bridgeman, who said, "Graphite is nature's strongest spring," and by an unidentified scientist who put it this way: "It is easy to squeeze carbon atoms together, but very difficult to keep them squz."

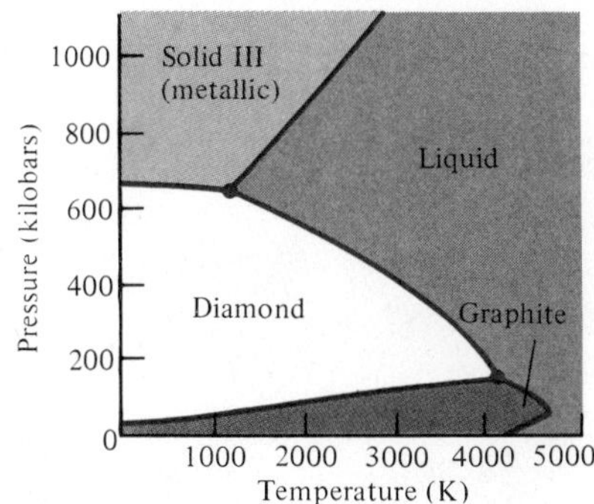

Figure 29.2
Phase Diagram for Carbon
One kilobar is approximately equal to 1000 atm.

Synthetic diamonds were first made in 1955 at the General Electric Company laboratory. The process requires chambers able to maintain, at the minimum, a 50 kbar pressure and 1500 K for long periods of time. A molten metal catalyst enhances the reaction rate. Synthetic diamonds have a variety of industrial applications that mainly utilize the properties of hardness and abrasiveness in various cutting and polishing tools. For example, concrete can be cut like wood with a blade coated with tiny diamond crystals.

Gem-quality diamonds have not yet been made synthetically. The diamond-like synthetic gems sold at less-than-diamond prices are "cubic zirconia" (ZrO_2), titanium dioxide (TiO_2), strontium titanate ($SrTiO_3$), or yttrium aluminum garnet ($Y_3Al_5O_{12}$), called "YAG." Titanium dioxide and strontium titanate give stones of fiery brilliance, but they are softer than diamond. "YAG" is less brilliant than the others, but is harder (Figure 29.3). Cubic zirconia is the closest to diamond in its properties.

The most familiar amorphous forms of carbon are coke, made by heating coal at a high temperature to drive off volatile inorganic and organic substances, and charcoal. The charcoal burned in stoves and fireplaces is made from wood by heating it in the absence of air. Finely powdered forms of carbon, sometimes also referred to as charcoal of various types, are made similarly from such materials as animal bones, coconut shells, or sugar. Finely divided carbon "activated" by heating with steam to clean its surface is an excellent adsorbent. Such activated carbon has a very high surface area—ranging from 600 to 2000 m^2/g. Activated carbon is used to remove bad-smelling or dangerous vapors from the air and to remove colored or bad-tasting impurities from water or other liquids. Many municipal water-treatment plants pass water through beds of activated carbon, and it is used to purify soft drinks, fruit juices, honey, and vodka.

Carbon black, usually made by the thermal decomposition of hydrocarbons in an open flame, is a very finely divided, very pure form of carbon. The human eye can distinguish 260 shades of black, and 10 nm particles of carbon black are the blackest of substances. Carbon black is used in printing ink and as a paint pigment, and large quantities are used to reinforce and color rubber—an automobile tire is one-fourth carbon black.

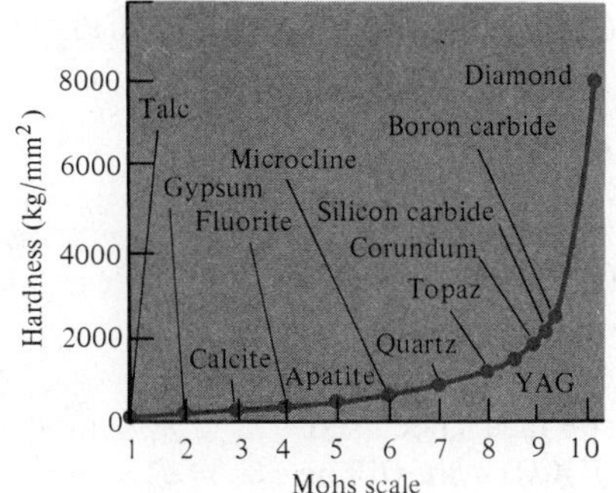

Figure 29.3
Hardness of Diamond and Other Materials
The Mohs scale for hardness is based on the properties of a series of minerals. A value of 10 is assigned to diamond (hardest) and 1 to talc (softest). (Corundum is aluminum oxide.) Note that the Mohs scale is nearly linear for values from 1 to 8 but deviates sharply for higher values.

29.4 SOME CHEMICAL REACTIONS OF CARBON AND INORGANIC CARBON COMPOUNDS

Many of the chemical reactions of carbon and carbon compounds that are common in inorganic chemistry have been discussed in earlier chapters. These and the additional reactions discussed here are summarized in Table 29.3.

Only fluorine among the halogens combines with elemental carbon and it does so at ordinary temperatures (reaction a, Table 29.3). At elevated temperatures, the reactivity of carbon depends upon its form. Diamond combines with oxygen above 800 °C, while natural graphite oxidizes slowly above 450 °C. A number of carbides (Section 29.5e) are made by the combination of finely divided carbon with metals at high temperatures (e.g., reaction c, Table 29.3). (Hydrogen is often added so that it will react with any oxygen present and prevent the oxidation of carbon.)

The combustion of carbon (reactions d, e), hydrocarbons (e.g., reaction h), and fossil fuels is exothermic and thermodynamically spontaneous (negative ΔG values). Why is it then that coal, petroleum, and dead trees do not burst into flames? These substances are *kinetically stable*—their reactions with oxygen have large energies of activation. Carbon monoxide also burns at elevated temperatures.

$$2CO(g) + O_2(g) \longrightarrow 2CO_2(g) \qquad \textbf{(29.1)}$$

The relative amounts of CO and CO_2 formed in the combustion of carbon compounds depend on the temperature and the supply of oxygen. With a limited oxygen supply, carbon monoxide is favored; with an excess of oxygen, carbon dioxide is favored. At high temperatures the reduction of carbon dioxide by carbon, an endothermic reaction, can also take place.

$$CO_2(g) + C(s) \longrightarrow 2CO(g) \qquad \textbf{(29.2)}$$

Carbon and carbon monoxide (CO) are both good reducing agents at elevated temperatures. In reactions that are important in the synthesis of fuels (Section 32.5), carbon or carbon monoxide can reduce water to hydrogen (reaction i, j). The reducing

Table 29.3
Some Reactions of Carbon and Its Compounds and Ions

Elemental carbon

(a) $C(s) + 2F_2(g) \xrightarrow{\text{room temperature}} CF_4(g)$

(b) $C(s) + 2S(s) \xrightarrow[\text{slightly} > 1\ \text{atm}]{700\text{–}900\ °C} CS_2(g)$

(c) $C(s) + W(s) \xrightarrow[H_2]{1400\text{–}1600\ °C} WC(s)$

(d) $2C(s) + O_2(g) \xrightarrow{\Delta} 2CO(g)$ $\quad \Delta H° = -221$ kJ $\quad \Delta G° = -274$ kJ

(e) $C(s) + O_2(g) \xrightarrow{\Delta} CO_2(g)$ $\quad \Delta H° = -394$ kJ $\quad \Delta G° = -394$ kJ

(f) $C(s) + 2H_2SO_4(conc) \xrightarrow{\Delta} CO_2(g) + 2SO_2(g) + 2H_2O(g)$

(g) $C(s) + 4HNO_3(conc) \xrightarrow{\Delta} CO_2(g) + 4NO_2(g) + 2H_2O(g)$

Hydrocarbon combustion

(h) $CH_4(g) + 2O_2(g) \xrightarrow{\Delta} CO_2(g) + 2H_2O(l)$ $\quad \Delta H° = -890$ kJ $\quad \Delta G° = -817$ kJ

Carbon and carbon monoxide as reducing agents

(i) $C(s) + H_2O(g) \xrightarrow{1000\ °C} H_2(g) + CO(g)$

(j) $CO(g) + H_2O(g) \xrightarrow[\textit{catalyst}]{250\ °C} H_2(g) + CO_2(g)$

(k) $C(s) + ZnO(s) \xrightarrow{\Delta} Zn(g) + CO(g)$

(l) $3CO(g) + Fe_2O_3(s) \xrightarrow{\Delta} 3CO_2(g) + 2Fe(s)$

Ions in aqueous solution

(m) $CO_3^{2-} + H_2O(l) \rightleftharpoons HCO_3^- + OH^-$ $\quad K_b = 2.1 \times 10^{-4}$

(n) $HCO_3^- + H_2O(l) \rightleftharpoons H_2CO_3(aq) + OH^-$ $\quad K_b = 2.2 \times 10^{-8}$

(o) $CN^- + H_2O(l) \rightleftharpoons HCN(aq) + OH^-$ $\quad K_b = 1.6 \times 10^{-5}$

properties of carbon (often as coke) and carbon monoxide are valuable in freeing metals from their oxide ores (e.g., reactions k and l). In sunlight, or in the presence of an appropriate catalyst, carbon monoxide combines with chlorine to give the highly toxic gas carbonyl chloride ($COCl_2$), better known as phosgene. By sharing the electron pair on the carbon atom, carbon monoxide forms coordinate covalent bonds with many metals in such compounds as $Fe(CO)_5$ and $Cr(CO)_6$. Known as metal carbonyls, these are coordination compounds in which the carbon monoxide molecules are ligands, as in complex ions.

The commonly encountered carbon-containing anions are the carbonate and hydrogen carbonate ions (CO_3^{2-} and HCO_3^-) and the cyanide ion (CN^-), which are anions of weak acids and give alkaline aqueous solutions.

The addition of acid to solutions of CO_3^{2-} or HCO_3^- generates carbon dioxide and addition of acid to CN^- in solution generates (very poisonous) hydrogen cyanide.

Cyanide ion forms very stable complex ions with many metals, in some cases by first forming an insoluble cyanide that then dissolves due to complex formation, for example

$$\begin{aligned} Ag^+ + CN^- &\longrightarrow AgCN(s) \\ AgCN(s) + CN^- &\longrightarrow [Ag(CN)_2]^- \end{aligned} \qquad (29.3)$$

29.5 SOME INORGANIC CARBON COMPOUNDS

a. Carbon Monoxide Carbon monoxide (CO) is a colorless, odorless gas which is insoluble in water and most other liquids. It is toxic because, when inhaled, it combines with the hemoglobin of the blood, displacing oxygen needed by the cells. In many of its physical properties carbon monoxide closely resembles nitrogen, with which its molecules are isoelectronic.

strongest bond in a diatomic molecule → :C≡O: *bond energy, 1075 kJ/mol*   :N≡N: *bond energy, 946 kJ/mol*

Like nitrogen, carbon monoxide is unreactive at ordinary temperatures. The current importance of carbon monoxide as an industrial raw material lies primarily in its reactions with hydrogen in the preparation of numerous organic compounds (Section 32.5).

b. Carbon Dioxide, Carbonic Acid, and Carbonates Carbon dioxide (CO_2) is a colorless, nontoxic gas with what has been described as a "faintly pungent" odor. The carbon dioxide molecule is linear, with double bonds between the carbon and oxygen atoms, $\ddot{\underset{..}{O}}=C=\ddot{\underset{..}{O}}$. The gas is readily condensed to the liquid state by cooling and compression, and upon further cooling the liquid freezes to a white solid ("dry ice"). This substance does not melt upon warming, but sublimes at −78.5 °C. Solid CO_2 leaves no trace when it evaporates (except for the water that condenses on its surface). It is very convenient for use in cooling ice cream, reactive chemicals, and many other things. At high temperatures (above 1700 °C) carbon dioxide decomposes to give carbon monoxide and oxygen (an endothermic reaction which is the reverse of reaction 29.1).

The atmosphere contains only 0.0325% carbon dioxide by volume, but atmospheric carbon dioxide plays an important role in photosynthesis and the cycling of carbon and oxygen through the environment. On the other hand, life is impossible in an atmosphere that contains too much carbon dioxide. A CO_2 concentration of 1% by volume in air can cause headaches, 10% can cause severe distress, and over 30% causes unconsciousness and death.

Because carbon dioxide is inert and does not support combustion, and because it is heavier than air, allowing it to blanket a fire, it is an excellent fire-extinguishing agent.

Table 29.4
Some Uses of Carbon Dioxide

Some Uses of Carbon Dioxide
Raw material in production of organic chemicals
Recovery of oil from depleted wells[a]
Refrigeration and food preservation
Beverage carbonation
Fire extinguishing

[a] Carbon dioxide is forced into wells, dislodging some of the remaining petroleum.

Liquid carbon dioxide (held in tanks at elevated pressure) is used in fire-extinguishing systems in airplanes and ships, and in chemical plants and other industrial installations. Hand-held extinguishers also may contain carbon dioxide under pressure. When the pressure is released by opening a valve, the liquid carbon dioxide escapes and immediately evaporates. Expansion of the gas causes a dramatic drop in temperature. As a result, the CO_2 freezes and forms a blanket of CO_2 "snow." Other uses of carbon dioxide are listed in Table 29.4.

In pure water a saturated CO_2 solution at 1 atm pressure and 25 °C contains about 0.034 mol/L of CO_2 gas, most of which is simply dissolved in the water. Only approximately one out of every 400 CO_2 molecules in solution reacts with a water molecule to give carbonic acid (H_2CO_3). This acid cannot be isolated, but does exist in small concentrations in aqueous CO_2 solutions.

The salts derived from carbonic acid are the carbonates, such as Li_2CO_3 and $CaCO_3$, and the hydrogen carbonates, such as $NaHCO_3$. Carbonates other than those of the alkali metals and ammonium ions are generally insoluble in water.

Minerals containing carbonate ions are plentiful in the Earth's crust, and more than half of the rock mined each year contains carbonates. The principal carbonate minerals are calcite ($CaCO_3$), which we know as limestone and marble, magnesite ($MgCO_3$), siderite ($FeCO_3$), and dolomite, which is a calcite with about half of the calcium ions replaced by magnesium ions ($(Ca, Mg)CO_3$). Calcium carbonate is also the major component of animal by-products such as pearls, eggshells, and coral.

c. Carbon Disulfide Carbon disulfide (CS_2) is a volatile, flammable liquid. It is made by the combination of hardwood charcoal and sulfur or the reaction of methane (CH_4) with sulfur at high temperatures. Carbon disulfide is important to the chemical industry because it is an excellent solvent for waxes, greases, hydrocarbons, and other nonpolar substances. Also, the reaction of carbon disulfide with natural cellulose is an essential step in the conversion of that material to rayon. The major disadvantages of using carbon disulfide are its toxicity and extremely high flammability.

d. Carbon Tetrachloride Carbon disulfide reacts with chlorine to give carbon tetrachloride (CCl_4) and disulfur dichloride (sometimes erroneously called sulfur monochloride).

$$\underset{}{CS_2(l)} + 3Cl_2(g) \longrightarrow \underset{\textit{carbon tetrachloride}}{CCl_4(l)} + \underset{\textit{disulfur dichloride}}{S_2Cl_2(l)} \qquad \textbf{(29.4)}$$

The reaction products are separated by distillation. The major use of carbon tetrachloride is as a reactant and solvent within the chemical industry. Because of its toxicity, the use of the compound in consumer products has been banned by the U.S. Food and Drug Administration.

e. Carbides Carbides are binary compounds of carbon with metals and semiconducting elements. There are three types of carbides. The ionic, or salt-like, carbides are formed with cations of the alkali and alkaline earth metals, and in most cases contain the C_2^{2-} ion, $[:C{\equiv}C:]^{2-}$. Carbides containing the C_2^{2-} ion react readily with water to produce acetylene (Section 32.2),

$$\underset{\textit{calcium carbide}}{CaC_2(s)} + 2H_2O(l) \longrightarrow Ca(OH)_2(aq) + \underset{\textit{acetylene}}{HC{\equiv}CH(g)} \qquad \textbf{(29.5)}$$

and are sometimes referred to as *acetylides*.

Calcium carbide (CaC_2), which is a typical salt-like carbide, is a white crystalline material prepared industrially by the reduction of calcium oxide (quicklime) with coke at a very high temperature.

$$\underset{\textit{calcium oxide (quicklime)}}{CaO(s)} + \underset{\textit{coke}}{3C(s)} \longrightarrow \underset{\textit{calcium carbide}}{CaC_2(s)} + CO(g) \qquad \textbf{(29.6)}$$

Calcium carbide produced by this reaction is the starting material in a commercial method for the production of acetylene via reaction (29.5).

Interstitial carbides, or metallic carbides, result when carbon atoms fill open spaces in the cubic or hexagonal close-packed crystal structures of transition metals. Such interstitial carbides as those of titanium, tungsten, tantalum, and niobium are very hard, heat-resistant materials and are used in cutting tools. Interstitial carbides are unreactive, and the ability of the metal to conduct electricity is retained.

Boron and silicon form extremely hard and inert carbides in which the bonding is fully covalent (the covalent carbides). Silicon carbide (SiC) is a network covalent substance similar to diamond, in which alternate atoms of carbon are replaced by silicon atoms. Boron carbide (B_4C) is composed of groups of three covalently bonded carbon atoms that fit into the crystal lattice of boron. Boron carbide and silicon carbide (also known as carborundum) come close to diamond in hardness (see Figure 29.3), and like industrial diamonds, are used in abrasives and cutting tools.

f. Cyanides Hydrogen cyanide (HCN) is an extremely poisonous, highly volatile liquid (b.p. 26 °C), which has an odor of bitter almonds. Hydrogen cyanide is made by passing a mixture of methane, ammonia, and air over a catalyst at 800 °C.

$$\underset{\textit{methane}}{2CH_4(g)} + 2NH_3(g) + 3O_2(g) \xrightarrow[\text{catalyst}]{800^\circ} \underset{\substack{\textit{hydrogen}\\\textit{cyanide}}}{2HCN(g)} + 6H_2O(g) \qquad \textbf{(29.7)}$$

Aqueous solutions of HCN are known as hydrocyanic acid. Neutralization with a base (e.g., NaOH) gives cyanide salts (e.g., NaCN). The cyanide ion, like the C_2^{2-} ion, contains triply bonded carbon, $[:C{\equiv}N:]^-$. Hydrogen cyanide and all cyanides are highly toxic. They inactivate enzymes essential to the production of energy by cellular oxidation.

EXAMPLE 29.1 Chemical Reactions: C Compounds

Following are the descriptions of three qualitative chemical tests. Write chemical equations for the reactions that occur in each test. Identify the reactions as either redox or nonredox reactions. (a) Test for elemental carbon: A solid sample is melted together with a small piece of elemental sodium. Cooling gives a white solid. Addition of water causes evolution of a gas which ignites spontaneously as a result of the heat developed. (b) Test for CO: A piece of filter paper moistened with a dilute, acidic solution of palladium(II) chloride, when held in the stream of gas to be tested, turns black. (c) Test for CO_3^{2-}: Dilute sulfuric acid is added to a crystalline solid. A gas evolves. Passage of the gas into a solution containing barium hydroxide yields a white precipitate.

(a) Carbon and the active metal sodium combine to form sodium carbide (a **redox combination reaction**). Addition of water yields acetylene (**nonredox reaction** of C_2^{2-} with water), which burns in air (**redox combination** of oxygen with a compound).

$$2Na(s) + 2C(s) \longrightarrow Na_2C_2(s)$$
$$Na_2C_2(s) + 2H_2O(l) \longrightarrow 2NaOH(aq) + C_2H_2(g)$$
$$2C_2H_2(g) + 5O_2(g) \longrightarrow 4CO_2(g) + 2H_2O(g)$$

(b) Carbon monoxide is a reducing agent; the black stain is elemental palladium (**redox reaction**).

$$PdCl_2(aq) + CO(g) + H_2O(l) \longrightarrow Pd(s) + CO_2(g) + 2HCl(aq)$$

(c) Addition of acid to a carbonate produces carbon dioxide (**nonredox partner exchange**). The carbon dioxide, an acidic oxide, reacts with the alkaline solution to form an insoluble salt and water (**nonredox**).

$$CO_3^{2-} + H_2SO_4(aq) \longrightarrow SO_4^{2-} + H_2O(l) + CO_2(g)$$
$$CO_2(g) + Ba(OH)_2(aq) \longrightarrow BaCO_3(s) + H_2O(l)$$

SUMMARY OF SECTIONS 29.2–29.5

The properties of carbon and its inorganic compounds discussed in these sections are summarized in Table 29.5.

Table 29.5
Outstanding Properties of Carbon and Some Common Inorganic Carbon Compounds

$-\overset{\vert}{\underset{\vert}{C}}-$ $\;>C=C<\;$ $-C\equiv C-$	Forms 4 single covalent bonds; also double and triple bonds
$-C-C-C-C-$ 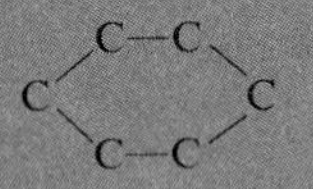 ; six-membered ring of C atoms	Catenates (self-bonds) more readily than any other element; basis of organic chemistry; C—C bond is strong
Diamond, graphite	Very different allotropes; diamond is very hard, brittle, nonconducting; graphite is soft, slippery, conducting
C(*s*), ordinary temperatures	Not very reactive at room temperature; attacked only by F_2 and oxidizing acids
$C(s) + O_2(g) \longrightarrow CO, CO_2$ $\text{Hydrocarbons} + O_2 \longrightarrow CO, CO_2, H_2O$	Thermal energy sources
C(*s*) and CO(*g*) at high temperatures	Good reducing agents; used in metallurgy
CO_2	Acidic oxide; H_2CO_3 formed in H_2O, but only by small percentage of molecules
CO_3^{2-}, HCO_3^-, CN^-	Weakly basic anions
HCN, all cyanides	*Very* poisonous; instantly fatal
$M(CO)_m$ (*metal carbonyl*), $[M(CN)_n]^{x+,x-}$	CO and CN^- form many complexes and complex ions with transition metals
$Ca^{2+}C_2^{2-}$ (*ionic*), WC (*metallic*), SiC and B_4C (*covalent*)	Three types of carbides; B_4C and SiC, only covalent carbides, are very hard

Aside 29.1 CARBON COMPOUNDS IN THE ATMOSPHERE; GLOBAL WARMING

Carbon monoxide and hydrocarbons are primary pollutants added to the atmosphere by the activities of civilization. The major source of both is the internal combustion engine. The concentration of carbon monoxide in automobile exhaust rises when the mixing of fuel and oxygen is poor and when the temperature of combustion is lower than it should be. Levels of carbon monoxide as high as 100 ppm have been measured in downtown urban areas at peak traffic hours. Since concentrations of 50–100 ppm are sufficient to slow human responses, and higher concentrations lead to headaches and nausea, rising carbon monoxide concentrations are clearly undesirable.

Hydrocarbons enter the atmosphere as partially unburned fuel emitted with the exhaust, and gasoline also escapes from fuel tanks and engines. Little hydrocarbon pollution comes from stationary fuel-burning sources, since the combustion temperatures are usually higher and the combustion more complete. Harmful physiological effects from hydrocarbon pollutants are not likely at present concentrations. The potential for damage due to hydrocarbons comes from the crucial role that they play in the formation of photochemical smog (see Aside 28.4).

The Clean Air Act of 1970 called for a 90% reduction in automobile emission of three pollutants—hydrocarbons, carbon monoxide, and nitrogen(II) oxide—between 1970 and 1975. The amounts of emission of each pollutant were to be decreased in stages, with the toughest standard enforceable in 1976 model automobiles. A variety of problems has caused continuing postponement of enforcement of the final set of standards. However, reductions have been achieved. By changing engine design to improve combustion efficiency and to better control the air-to-fuel ratio, losses of unburned gasoline have been cut back. To decrease the concentration of hydrocarbons and carbon monoxide in the exhaust gases major reliance thus far has been placed upon catalytic converters. Exhaust gases pass over a catalyst containing a metal such as platinum or rhodium, and carbon monoxide and hydrocarbons are oxidized to carbon dioxide and water by additional air taken in at the converter. Thus far, control of nitrogen(II) oxide emission has been less successful.

Particulates, the fifth of the five major air pollutants (the others are CO, hydrocarbons, NO_x, and SO_x) are airborne solid particles and liquid droplets, which often contain carbon or carbon compounds. They range greatly in size, origin, and composition. There are tiny particles of soot (amorphous carbon) and droplets of sulfuric acid mist. Particles produced in burning coal, called *fly ash*, can be up to 500,000 nm in diameter; they consist chiefly of oxides, such as SiO_2, Al_2O_3, MgO, Fe_2O_3, and TiO_2. Man-made particulates are contributed to the atmosphere in roughly equal amounts by stationary fuel combustion, industrial processes, and fires, such as forest fires and agricultural burning of wastes.

An untold amount of damage is done each year by particulates settling out on buildings and within homes. Costs of cleaning go up and corrosion is often accelerated. Particulates are also suspected of influencing weather patterns by serving as nuclei for cloud formation, and by altering the amount of radiation reaching the Earth's surface.

To decrease pollution by particulates, the particles must be captured before they enter the atmosphere. Various devices are in use to wash out, settle out, or precipitate out pollutants before waste gases leave power plants or industrial plants. As is usually the case with pollution control measures, they add to the cost of the process or the product being manufactured, a factor that retards the introduction of such measures.

Carbon dioxide is not an air pollutant in the sense that it causes immediate damage to living or material things. However, carbon dioxide, together with methane, nitrous oxide (N_2O), and the chlorofluorocarbons (Section 27.12) increasingly are recognized as sources of potential serious worldwide problems due to global warming—an increase in the average global temperature.

The temperature at the surface of the Earth is regulated through the trapping of heat between the surface and the upper atmosphere by carbon dioxide, water, and other gases. The surface of the Earth is warmed by absorption of solar radiation of many wavelengths. A constant temperature is maintained by reradiation of this energy back into the atmosphere. Because the Earth is cooler than the Sun, reradiation occurs at longer wavelengths than the incoming radiation, chiefly in the infrared region of the spectrum. The so-called "greenhouse gases" absorb in the infrared region and trap this energy in the atmosphere, preventing it from being lost to outer space. The result of this effect, known as the "greenhouse effect," is maintenance of the energy balance and surface temperature of the Earth to which we are accustomed.

Carbon dioxide, water vapor, and ozone are natural greenhouse gases. However, due mainly to fossil fuel combustion, carbon dioxide concentration in the atmosphere is rising steadily at the rate of about 1.5 ppm per year. Chlorofluorocarbon (Section 27.12), methane, and nitrous oxide (N_2O) concentrations are also rising, and these too are greenhouse gases. Methane is produced by bacterial decay and also by the activities of mankind. Nitrous oxide is mainly a by-product from bacterial action that converts ammonia in the soil to nitrates (see Figure 28.8). The concentration of nitrous oxide in the atmosphere may be on the rise because of the increased use of ammonia-containing fertilizers. The study of how the small molecules of these gases interact in the upper atmosphere is difficult because each may react with the others in many different ways. Data are being gathered and the problem is being approached through computer modeling of the global climate. The possible scientific, political, and economic results of global warming promise to be of continuing interest.

THE SEVEN SEMICONDUCTING ELEMENTS AND THEIR COMPOUNDS

29.6 GENERAL PROPERTIES OF THE SEMICONDUCTING ELEMENTS AND THEIR COMPOUNDS

Some properties of the semiconducting elements are listed in Table 29.6. The atomic radii and ionization energies of these elements follow the general trends expected from their positions in the periodic table and are intermediate between those of the metals and nonmetals (see summary in Table 10.12).

Each semiconducting element has the group oxidation state as its maximum oxidation state. Boron and silicon have only this one positive oxidation state (+3 and +4, respectively). The Group V (As, Sb) and Group VI (Se, Te) elements in their +5 and +6 oxidation states are readily reduced to +3 and +4 and thus are present in

Table 29.6
Properties of the Semiconducting Elements

Group and element	Configuration	Common oxidation states[a]	Atomic radius (nm)	Melting point (°C)	Boiling point (°C)	Ionization energy (kJ/mol)	Electronegativity
Group III							
B	$[He]2s^2 2p^1$	−3, +3	0.080	2300	2500	801	2.0
Group IV							
Si	$[Ne]3s^2 3p^2$	−4, +4	0.118	1420	2355	786	1.8
Ge	$[Ar]3d^{10}4s^2 4p^2$	−4, +2, +4	0.123	958	2700(?)	762	1.8
Group V							
As	$[Ar]3d^{10}4s^2 4p^3$	−3, +3, +5	0.125	814 (32 atm)	610 (sublimes)	947	2.0
Sb	$[Kr]4d^{10}5s^2 5p^3$	−3, +3, +5	0.145	630.5	1325	834	1.9
Group VI							
Se	$[Ar]3d^{10}4s^2 4p^4$	−2, +4, +6	0.116	217 (gray form)	685	941	2.4
Te	$[Kr]4d^{10}5s^2 5p^4$	−2, +4, +6	0.143	450	1390	869	2.1

[a] The most stable states are shown in color. For many of the unusual boron compounds, the concept of oxidation state has little meaning.

Table 29.7
Some Halides of the Semiconducting Elements

Compound	M.p. (°C)	B.p. (°C)
BCl_3	−107	13
SiF_4	−90.2	−86
$SiCl_4$	−70	57.6
Si_2Cl_6	−1	145
$GeCl_4$	−49.5	84
AsF_5	−79.8	−52.8
$AsCl_3$	−8.5	63
$SbCl_5$	4	140[a]
SeF_4	−10	101
SeF_6	−35 (2 atm)	−47[b]
TeI_4	280	100[a]

[a] Decomposes
[b] Sublimes

oxidizing agents (like lead and bismuth in their higher oxidation states). All semiconducting elements are assigned negative oxidation states in their compounds with hydrogen or with more electropositive metals.

The bonding of semiconducting elements in compounds is generally covalent. Table 29.7 gives the properties of some halides of these elements. The covalent nature of the bonding is shown by the low melting and boiling points.

Germanium, arsenic, antimony, and selenium all exist in a number of different forms, both amorphous and crystalline. In their pure crystalline forms, these elements have a dark metallic luster. The amorphous forms tend to show greater chemical reactivity than the crystalline forms, although none of the semiconducting elements is highly reactive at ordinary temperatures and pressures.

29.7 SOME CHEMICAL REACTIONS OF THE SEMICONDUCTING ELEMENTS AND THEIR COMPOUNDS

a. Combination with Other Elements All of the semiconducting elements react with the halogens and oxygen. As is to be expected, the reactions with fluorine are the most vigorous and those with iodine are the least vigorous. The elements best achieve their higher oxidation states in reactions with fluorine (unless the amount of fluorine is deficient). In general, the reactions with oxygen require elevated temperatures. Some reactions which illustrate these principles are the following:

$$2B(s) + 3F_2(g) \longrightarrow 2BF_3(g) \qquad Se(s) + O_2(g) \xrightarrow{\Delta} SeO_2(s)$$

$$2B(s) + 3I_2(s) \xrightarrow{\Delta} 2BI_3(s) \qquad Te(s) + 3F_2(g) \longrightarrow TeF_6(g)$$

Combination reactions of the semiconducting elements with metals produce binary compounds called borides, arsenides, silicides, and so on. In some cases, the products are true compounds in which the semiconducting elements are assigned the negative oxidation numbers. Such compounds are formed by arsenic, selenium, and tellurium with the most electropositive metals; for example, in potassium selenide (K_2Se) the oxidation state of Se is −2. In other cases, solid solutions or various types of alloys are formed. In many borides the small boron atoms occupy interstices in the metal lattice.

MODERN SCIENCE AND TECHNOLOGY

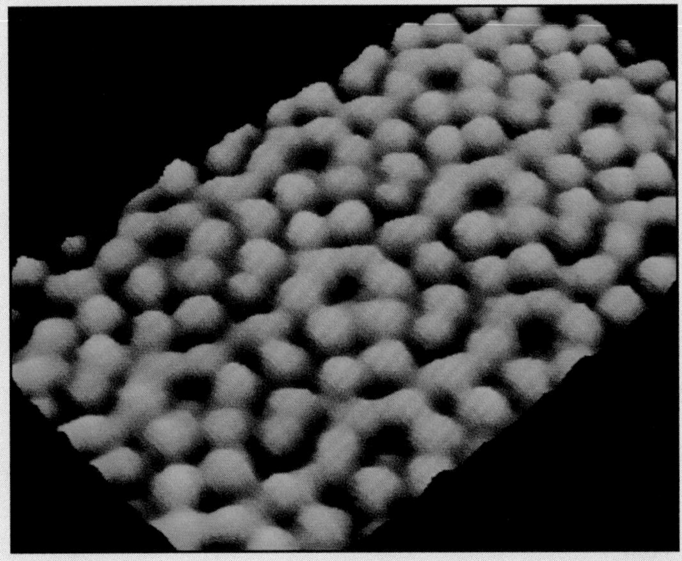

■ **SILICON ATOMS** *left* The scanning tunneling microscope (STM), developed in the 1980s at the IBM Zurich Research Laboratory, makes it possible to "see" atoms and molecules at surfaces. Magnifications up to 1×10^8 are possible. The surface contours are mapped by a tiny probe that moves back and forth at a constant distance of 1 nm above the electron clouds of the surface atoms. The contour lines, when processed by a computer, are transformed into an image like that of silicon atoms at a surface shown here. In 1986, Heinrich Roher and Gerd Binnig were awarded the Nobel prize in physics for development of the STM, which opens the door to new understanding of many physical, chemical, and biological phenomena.

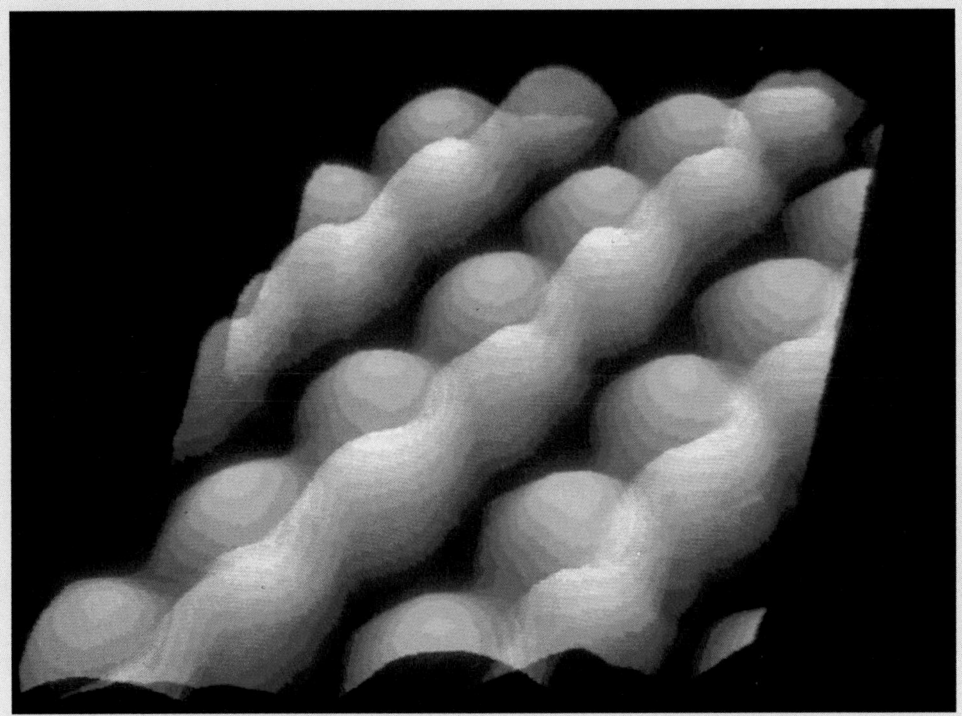

■ **GALLIUM ARSENIDE** *left* An STM photo of layers of gallium (blue) and arsenic (red) atoms at the surface of gallium arsenide (color added). The semiconducting and light-emitting properties of gallium arsenide (GaAs) are utilized in light-emitting diodes, lasers, microwave generation, and many other electronic devices.

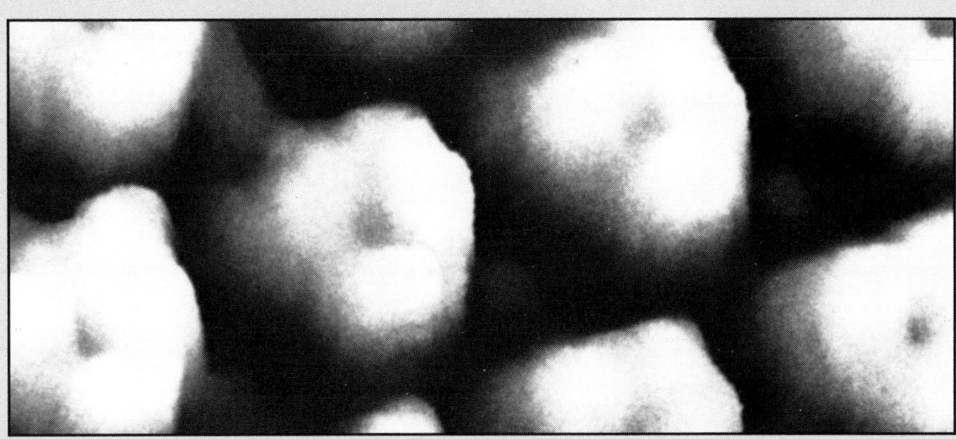

■ **BENZENE MOLECULES** *right* In 1988 this first-ever picture of individual benzene molecules was taken with the STM. The image clearly reveals a ring structure for each benzene molecule. During creation of the image, the benzene molecules were immobilized by adsorption on a clean rhodium surface. To prevent the benzene molecules from moving about and blurring the image, carbon monoxide molecules were also adsorbed to the surface, but their images are difficult to detect.

COMPUTER RECORDING HEAD *below* Information is stored permanently on a computer disk by magnetizing areas of iron oxide on the surface. To record data, a magnetic field is created by passing current through a conducting coil surrounding a magnetic material in a "read-write" head. Data are read by the same head in the reverse process. The recording head shown, which is the size of the period at the end of this sentence, has a "coil" that is a thin film of copper deposited on a silicon ceramic material; the magnetic core is a nickel-iron alloy.

INTEGRATED CIRCUIT CHIP *above* In the center of this microchip package is a gallium arsenide chip called a decision circuit. It is used for retiming and regenerating digital data signals in high-speed photonic information transmission systems. The chip measures less than $1 mm^2$, rests on a glass insulator, and is surrounded by silicon-based transistors. The leads will connect the chip to light-emitting diodes that transmit the data as light.

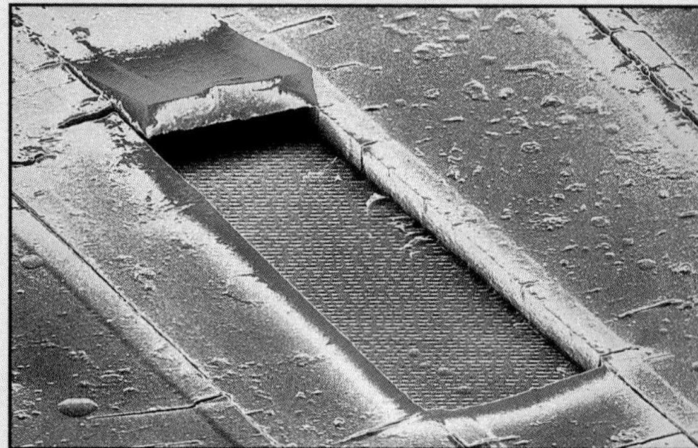

COMPACT DISC SURFACE *above* In the center of this photo, the protective transparent plastic layer of a compact disc has been cracked off to reveal the pattern of bumps that record part of Mozart's 40th Symphony. The bumps, which vary in length from about 0.8 μm to 3.5μm, are coated with a thin layer of reflective aluminum and are "read" by a laser beam. Because the size of the bumps is close to the wavelength of visible light, they refract light to produce a rainbow of colors. Magnification 333 ×.

FIBER OPTICS *right* In the photonic transmission of information, pulses of light travel down optical fibers like those pictured. Such silica (SiO_2) fibers contain only a few parts per billion of contaminants. Addition of boron oxide (B_2O_3) or germanium dioxide (GeO_2) to the glass allows control of the refractive index across the fiber diameter so that when light rays approach the surface they are reflected back into the fiber. A transatlantic optical telephone cable can carry 40,000 conversations simultaneously.

A SUN-POWERED CAR *left* The General Motors Sunraycer is powered by a 90-sq-ft array of solar cells of the same type used in satellites. In 1987, the Sunraycer won the first World Solar Challenge race in Australia by traveling almost 2,000 miles in 5½ days at an average speed of 41 mph.

SOLAR CELLS *left* Solar cells take advantage of the photovoltaic effect, in which radiation promotes valence band electrons into the conduction band at the junction between a metal and silicon or some other semiconductor. The outcome is conversion of the Sun's radiant energy to electrical energy. The photo shows the essential component of a solar cell, a thin sheet of doped silicon.

HIGH-TEMPERATURE SUPERCONDUCTOR *above and right* In 1988, great excitement was generated by discovery of materials that become superconductors — have zero resistance to the flow of electricity — at higher temperatures (above -196 °C) than any previously known material. Superconductivity at more easily achieved temperatures opens the door to a host of applications in electricity generation and storage, smaller computers, and superpowerful magnets for medicine and research. At the left is a photo of (*top, left to right*) the elements yttrium (Y), barium (Ba), copper (Cu) and (*bottom*) a thin film of one of the first high-temperature superconductors, which has the composition $YBa_2Cu_3O_{7-x}$, where $x = 0.1 - 0.2$. At the right is a computer model of the structure of this material (Ba, green; Y, silver; Cu,blue; O, red).

■ ADVANCED MATERIALS — ALUMINUM HONEYCOMB *left* "Advanced materials" are new materials that withstand extreme conditions, often in mechanical, electrical, and environmental applications for which they are specifically created. The honeycomb shown, created by adhesively bonding aluminum foil sheets to form hexagonal cells, is used in sandwich panels that have the highest known strength-to-weight ratio of any structural material.

■ ADVANCED MATERIALS — FIBERS *left and above* Aramids are aromatic polyamides (*para*-substituted) that, because of their large, linear molecules, form fibers of higher tensile strength than any other known material. Their strength, toughness, low density, and wear and temperature resistance are leading to a wide range of uses that include protective wear like fire-resistant gloves (*left*) and commercial fishnets (*above*), and composite materials.

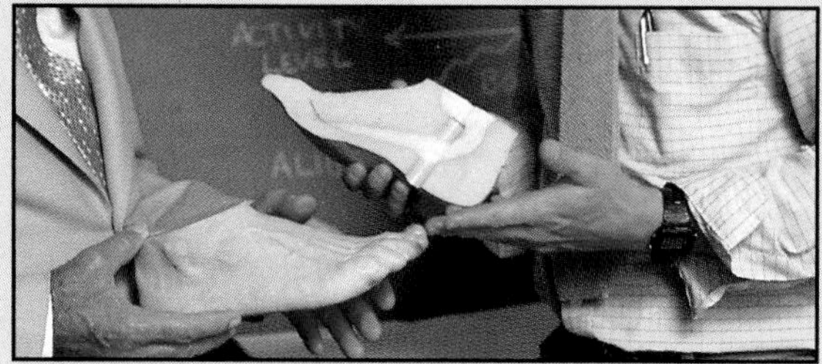

■ ADVANCED MATERIALS — PLASTICS *above* Advanced materials are providing exactly the properties needed to solve many specific problems. For example, the artificial foot shown has greatly improved the quality of possible movement because of the extreme resilience and flexibility of the acetal polymer (Delrin) from which the body of the foot is made.

■ ADVANCED MATERIALS — COMPOSITES *above* Composites are materials that physicially combine two solids that retain their identity — for example, fiber glass reinforced plastics. Strong, low-density, low-cost composites are being produced by combining aramid or carbon fibers with thermoplastics. Composites like those shown are used in aircraft, with a weight savings of up to 1,000 lb per plane for commercial airliners.

b. Reactions with Water, Acids, and Bases Unlike the more metallic elements, none of the semiconducting elements react with water or with nonoxidizing acids. With oxidizing acids, they give oxoacids or oxides, usually in their higher oxidation states:

$$B(s) + 3HNO_3(conc) \longrightarrow \overset{+3}{H_3BO_3}(s) + 3NO_2(g)$$

$$2As(s) + 5H_2SO_4(conc) \xrightarrow{\Delta} \overset{+5}{2H_3AsO_4}(aq) + 5SO_2(g) + 2H_2O(l)$$

$$2Sb(s) + 10HNO_3(conc) \longrightarrow \overset{+5}{Sb_2O_5}(s) + 10NO_2(g) + 5H_2O(l)$$

Silicon does not react with concentrated nitric acid, but it does dissolve in a mixture of nitric and hydrofluoric acids to form fluorosilicic acid.

$$Si(s) + 4HNO_3(conc) + 6HF(aq) \longrightarrow H_2SiF_6(aq) + 4NO_2(g) + 4H_2O(l)$$

Boron reacts with molten alkali metal hydroxides with the liberation of hydrogen and the formation of borates:

$$2B(s) + 6NaOH(l) \longrightarrow 2Na_3BO_3(s) + 3H_2(g)$$

Silicon and germanium (like aluminum, tin, and lead) react similarly to give hydrogen plus silicates or germanates. Arsenic, selenium, tellurium, and antimony react differently—they disproportionate. For example, for arsenic the products are an arsenide and an arsenite.

$$2\overset{0}{As}(s) + 6NaOH(l) \longrightarrow Na_3\overset{-3}{As} + Na_3\overset{+3}{As}O_3(s) + 3H_2O(g)$$

c. Reactions of Oxides The oxides of the semiconducting elements are either amphoteric (As_2O_3, Sb_2O_3, TeO_2) or acidic. None are basic oxides, again demonstrating the chemical similarity of these elements to the nonmetals. The oxides of boron and selenium react with water to give acids (these are nonredox combination reactions).

$$B_2O_3(s) + 3H_2O(l) \longrightarrow \underset{\textit{boric acid}}{2H_3BO_3(aq)}$$

$$SeO_3(s) + H_2O(l) \longrightarrow \underset{\textit{selenic acid}}{H_2SeO_4(aq)}$$

The amphoteric oxides react with both acid and base (see Table 26.7), as illustrated by the reactions of antimony(III) oxide (also nonredox reactions):

$$Sb_2O_3(s) + 3H_2SO_4(aq) \longrightarrow Sb_2(SO_4)_3(aq) + 3H_2O(l)$$
$$Sb_2O_3(s) + 6NaOH(aq) \longrightarrow 2Na_3SbO_3(aq) + 3H_2O(l)$$

All of the oxides of the semiconducting elements react with strong bases to give salts plus water.

d. Hydrolysis of Halides All of the halides of the semiconducting elements react rapidly with water. The final products are oxoacids or oxides, depending upon the solubility of the oxide and the amount of water present. Some of the reactions are quite violent.

$$SiX_4 + 2H_2O(l) \longrightarrow SiO_2(\text{hydrated})(s) + 4HX(aq)$$
$$AsX_3 + 3H_2O(l) \longrightarrow H_3AsO_3(aq) + 3HX(aq)$$
$$AsF_5 + 4H_2O(l) \longrightarrow H_3AsO_4(aq) + 5HF(aq)$$

Antimony trichloride gives an insoluble oxochloride, SbOCl, which then slowly reacts with additional water to give the oxide SbO_3. The hydrolyses of halides are all nonredox partner-exchange reactions. (The partner-exchange pattern is somewhat obscured by the way that we write the formulas of acids and by the formation of oxides rather than oxoacids in some cases.)

e. Ions in Aqueous Solution Unlike the metals and more like the nonmetals, the semiconducting elements rarely are found as hydrated monatomic cations in aqueous solution. Aside from the oxocation of antimony (SbO^+) and a few complex ions (e.g., $[BF_4]^-$, $[SiF_6]^{2-}$, $[SbCl_4]^-$), the semiconducting elements occur in aqueous solutions as oxoanions or monatomic anions (Se^{2-}, Te^{2-}). Most of these anions give alkaline solutions because they are the anions of weak acids. For example, soluble salts containing the tetraborate ion ($B_4O_7{}^{2-}$) in aqueous solution give boric acid:

$$B_4O_7{}^{2-} + 7H_2O(l) \rightleftharpoons 4H_3BO_3(aq) + 2OH^- \qquad \textbf{(29.8)}$$

Sodium silicate, which is present in glass, also reacts slowly with water to give a mixture of various hydrated silicic acids plus sodium hydroxide.

$$Na_4SiO_4(s) + 3H_2O(l) \rightleftharpoons H_2SiO_3(\text{hydrated}) + 4NaOH(aq) \qquad \textbf{(29.9)}$$

29.8 BORON

Boron occurs in the crust of the Earth mainly as boric acid (H_3BO_3), known as the mineral *sassolite,* and borates. The borates most important as boron ores are borax ($Na_2B_4O_7 \cdot 10H_2O$), kernite ($Na_2B_4O_7 \cdot 4H_2O$), and colemanite ($Ca_2B_6O_{11} \cdot 5H_2O$).

The free element is obtained as an impure microcrystalline brown powder, "amorphous" boron, by the reduction of boric oxide (B_2O_3) with magnesium at high temperatures.

$$B_2O_3(s) + 3Mg(s) \longrightarrow 2B(s) + 3MgO(s)$$

Pure boron, in the form of black lustrous crystals, can be prepared by reduction of boron trichloride with hydrogen above 1000 °C.

$$2BCl_3(g) + 3H_2(g) \xrightarrow{\Delta} 2B(s) + 6HCl(g)$$

Elemental crystalline boron is an extremely hard, high melting (2300 °C) substance that exists in at least three forms. All have giant-molecule structures in which the fundamental structural unit is an icosahedral (twenty-sided) arrangement of twelve boron atoms (Figure 29.4) united by strong covalent single bonds. Boron, like other first elements in representative groups, differs significantly from the other group members (Table 29.8) and tends to resemble its diagonal neighbor, silicon.

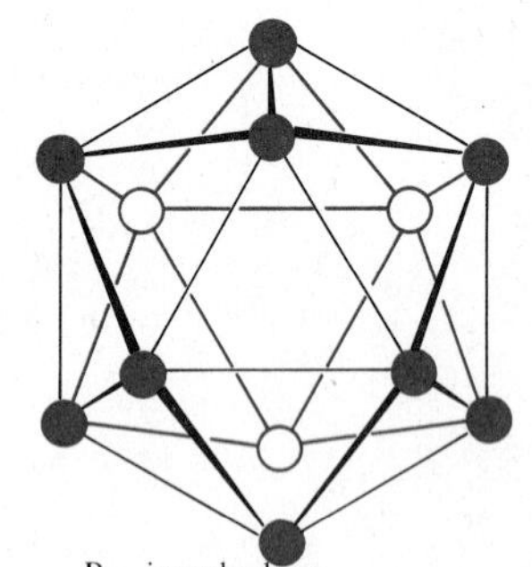

B_{12} icosahedron

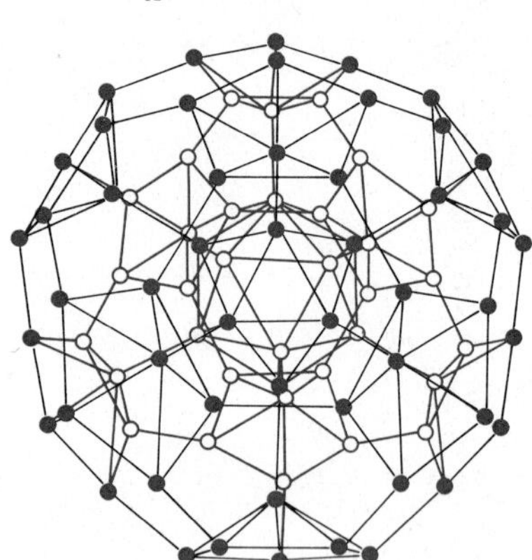

84-atom unit of β-rhombohedral boron

Figure 29.4
Elemental Boron
The B_{12} icosahedron is the fundamental structural unit of boron.

Elemental boron is added in small amounts to aluminum and to steel, where it aids in hardening. It is also used as a neutron absorber in the control rods of light-water nuclear reactors (Section 12.17). A newer application of boron is the production of fibers for use in fiber-reinforced materials. Pure boron fibers or boron-coated tungsten fibers embedded in aluminum or magnesium produce a light, very stiff, and strong material used in aircraft parts. Boron is less used in electronic devices than are the other semiconducting elements. It is limited to use as an additive in low concentrations (a dopant; Section 29.17).

Table 29.8
Differences between Boron and the Other Members of the Boron Family (Al, Ga, In, Tl)

Property	Comments
The only nonmetal in the boron family	Boron is a semiconducting element; its chemical behavior is essentially that of a nonmetal; other family members are all metals
Resembles Si more than Al in its chemistry	A consequence of similar electronegativity and effective nuclear charge
Oxide (B_2O_3) is acidic	Al_2O_3 and Ga_2O_3 are amphoteric, In_2O_3 and Tl_2O_3, basic; B_2O_3 is the anhydride of a number of boric acids, the best characterized being H_3BO_3 and HBO_2
Forms a series of unique hydrides (e.g., B_2H_6, B_4H_{10})	The hydrides are covalent compounds with unusual, three-center bonding; aluminum forms a polymeric hydride; there is little evidence for hydrides of other family members

Boron is an unusual element, and the structure and bonding in many boron compounds are not predictable on the basis of what is known of any other elements and their compounds. From the $2s^2 2p^1$ configuration of a boron atom, a hydride of the composition BH_3 would be expected. However, the simplest of the large family of boranes—boron–hydrogen compounds—is diborane (B_2H_6; Table 29.9). More than 20 other boranes of the general formula B_nH_m are known (e.g., B_4H_{10}, B_5H_9, B_5H_{11}, B_6H_{12}). The lower molecular mass boranes are gases or low-boiling liquids, with boiling points that increase with mass as they do for hydrocarbons. Boranes are highly reactive and have found many uses in chemical synthesis. They decompose when heated in the absence of air to give boron and hydrogen; many are spontaneously flammable; and they are decomposed by water to yield hydrogen and boric acid.

The nature of the bonding in the boron hydrides has been a matter of great interest. The boranes are **electron-deficient compounds**—they possess too few valence electrons for the atoms to be held together by ordinary electron-pair covalent bonds. In the diborane molecule, for example, there are 12 valence electrons, 3 from each of the two boron atoms and 1 from each of the six hydrogen atoms. For each boron atom to be joined covalently to three hydrogen atoms and to the other boron atom by covalent single bonds would require 14 electrons. Therefore, the configuration

```
   H  H
   |  |
H—B—B—H
   |  |
   H  H
```

not a possible configuration

which is found in the organic compound ethane (C_2H_6) is impossible in diborane.

Table 29.9
Some Compounds of Boron

B_2H_6	Diborane (m.p. −166 °C, b.p. −92.5 °C)
H_3BO_3	Boric acid (at 169 °C $\longrightarrow$ HBO_2)
HBO_2	Metaboric acid (m.p. 236 °C)
B_2O_3	Boric oxide (m.p. 460 °C, b.p. ~1860 °C)
B_4C	Boron carbide (m.p. 2350 °C, b.p. >3500 °C; very hard)
BN	Boron nitride (~3000 °C, sublimes; very hard)
$Na_2B_4O_7 \cdot 10H_2O$	Sodium tetraborate decahydrate (borax) (m.p. 320 °C)
CaB_6	Calcium boride (m.p. 2235 °C, very hard)

The experimentally determined molecular structure of diborane is shown in Figure 29.5. The boron atoms and the four terminal hydrogen atoms lie in one plane; the two remaining hydrogen atoms lie above and below the plane and connect the two boron atoms. Each "bridging" hydrogen atom is bonded to the two boron atoms by a **three-center bond**—a bond in which a single pair of electrons bonds three atoms covalently. In the higher boranes there are both B—H—B and B—B—B three-center bonds, leading to molecular shapes in which boron and hydrogen atoms are joined in three dimensions in interconnecting polyhedra.

Boric acid (H_3BO_3 or $B(OH)_3$) is a weak *mono*protic acid that yields acidic solutions, not by proton donation but by the following reaction:

$$B(OH)_3(aq) + 2H_2O(l) \rightleftharpoons [B(OH)_4]^- + H_3O^+ \qquad \textbf{(29.10)}$$

A dilute boric acid solution is a mild antiseptic, so mild that it often is used as an eyewash. Boric acid also is used as a preservative, to protect wood against insect damage, and as a fire retardant in paper and cotton.

Boric acid is produced by acidification of a cold aqueous solution of the mineral borax. The reaction amounts to driving the hydrolysis of the tetraborate ion to completion as the OH^- ion reacts with the acid (see Equation 29.8).

Borax has long been used as a cleansing agent and water softener because of the alkalinity of its solutions, which is due to hydrolysis of the tetraborate anion. Borax is also necessary in the manufacture of Pyrex glass (Section 29.14a) and fiberglass. In recent years the increased interest in energy conservation has led to a great demand for fiberglass insulation, and thus for borax.

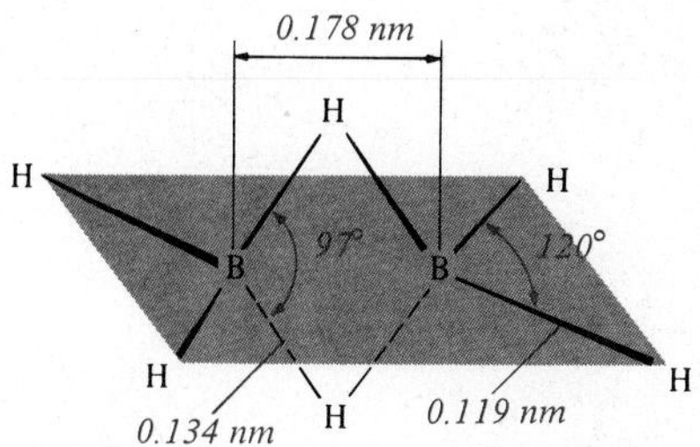

Figure 29.5
Structure of Diborane

29.9 SILICON AND GERMANIUM

Silicon is the second most abundant element in the Earth's crust; germanium, however, is among the less abundant elements. Silicon is not found naturally as the free element, but it is widely distributed in the form of silica (SiO_2), or **silicates,** compounds containing SiO_4^{4-} groups and metals. Roughly 85% of the Earth's crust is composed of silica and silicate minerals. Elemental silicon of about 98% purity is obtained by reduction of sand, which is largely SiO_2, with coke in an electric furnace.

$$SiO_2(s) + 2C(s) \xrightarrow{3000\ ^\circ C} Si(l) + 2CO(g)$$

Silicon made in this way is used in alloys, for example, in spring steel (22% silicon), in corrosion-resistant iron alloy (about 15%), and in aluminum alloys for fine casting (about 17%). It is also the starting material in the manufacture of **silicones**—polymers composed of silicon, carbon, hydrogen, and oxygen (Section 29.15). The preparation of the ultrapure silicon and germanium needed in semiconductor devices is discussed in Section 29.18.

Germanium is found mainly as a sulfide in association with other metal sulfides, for example, those of lead and zinc (see Example 29.2). Germanium is also recovered from coal ashes and flue dust. Pure germanium, which has several crystalline forms, is metallic in appearance, but brittle, like glass. Both silicon and germanium have the diamond crystal structure (see Figure 29.2a).

The first semiconductor devices were made of germanium; today the major use of germanium is in the electronics industry. Germanium is ideally suited for fabricating semiconducting devices. It can be prepared in very pure form with clearly defined crystal structure and crystal defects. It has been said that because of the interest in its electronic properties, germanium is the best understood of all of the elements. Certainly this is true of its solid-state properties. Applications of germanium outside the electronics industry include infrared-transmitting glass, low-melting gold–germanium alloys (sometimes used in dental work), and substances that exhibit red fluorescence when struck by light of appropriate wavelengths.

Silicon and germanium form both monoxides and dioxides, but only the dioxides are stable. The natural and man-made silicon–oxygen compounds are so numerous and important that they are discussed separately (Sections 29.12–29.15). Germanium dioxide (GeO_2), the major commercially available germanium compound, is a solid that exists in two crystalline forms, one more inert and less soluble than the other; both melt above 1000 °C.

Silicon and germanium atoms, like the carbon atom, can form four covalent single bonds using the four ns^2np^2 valence electrons in each atom. In compounds, atoms of these elements do not self-link to the great extent that carbon atoms do.

Silicon–hydrogen compounds are called silanes. The simplest is monosilane (SiH_4); other known members of the series of the general formula Si_nH_{2n+2} include Si_2H_6, Si_3H_8, and Si_4H_{10}—all compounds that contain silicon–silicon bonds. The higher silanes are unstable and decompose. In addition, most silanes are quite flammable.

The germanium hydrides, Ge_nH_{2n+2}, called germanes, are generally less flammable and less easily hydrolyzed than the silanes.

The silicon and germanium tetrahalides are volatile substances; their boiling points increase with the atomic mass of the halogen in each series. Silicon also forms a number of halides of the composition Si_nX_{2n+2} in which, as in the silanes, there are silicon–silicon bonds.

EXAMPLE 29.2 Chemical Reactions: Germanium Compounds

A process for the production of elemental germanium includes the following steps: (a) Ore containing germanium(IV) sulfide is treated with a mixture of concentrated sulfuric acid and nitric acid to convert the sulfide to germanium(IV) oxide. (b) The oxide-containing product is dispersed in hydrochloric acid, and upon heating, germanium(IV) chloride distills out of the mixture. (c) Treating the chloride with water yields the pure oxide. (d) Elemental germanium is produced by the reaction between the oxide and hydrogen. How does each step in this process take advantage of the distinctive properties of the substances involved? Where possible, write chemical equations.

(a) **A mixture of concentrated sulfuric acid and nitric acid is a strongly acidic and strongly oxidizing medium. Many sulfides are soluble in acids. The sulfide ion is oxidized,** but it would be difficult to write a single equation for the reaction in this mixture of acids and ore.

(b) **Halides of the semiconducting elements are covalent and many of them are volatile** [$GeCl_4$, b.p. 83.1 °C]. The reaction

$$GeO_2(s) + 4HCl(aq) \xrightarrow{\Delta} GeCl_4(g) + 2H_2O(l)$$

takes advantage of the removal of a volatile product to allow a nonredox partner-exchange reaction to proceed in the direction desired.

(c) This reaction takes advantage of a property common to halides of the semiconducting elements—their **ready reaction with water, in this case to form an oxide:**

$$GeCl_4(l) + 2H_2O(l) \longrightarrow GeO_2(s) + 4HCl(aq)$$

(d) **Hydrogen reduces oxides of many elements other than those in the lithium and beryllium families to the free elements** (Table 16.7):

$$GeO_2(s) + 2H_2(g) \xrightarrow{\Delta} Ge(s) + 2H_2O(g)$$

29.10 ARSENIC AND ANTIMONY

Arsenic and antimony are not abundant elements. Their principal minerals are sulfides, including orpiment (As_2S_3), which was known to the alchemists. Arsenic and antimony are also found in the free state in nature—like bismuth, the heaviest element in Group V, but unlike the lighter phosphorus, which is more reactive.

Elemental arsenic and antimony each have several allotropes. The common semiconducting forms, often referred to as the metallic forms, are gray, lustrous, and crystalline, while the amorphous allotropes are yellow. Yellow arsenic is very unstable and reverts quickly to the semiconducting form.

Yellow arsenic and arsenic compounds are highly poisonous, leading quickly in large doses to convulsions and death. Chronic arsenic poisoning causes fatigue and a variety of unpleasant effects, including hair loss, visual disturbances and blindness, a garlic odor on the breath, paralysis, and anemia. By taking small doses of arsenic over an extended period, tolerance to arsenic can be built up. The "arsenic" of detective stories is the oxide, As_2O_3, which is soluble, odorless, tasteless, and fatal in doses of 0.1 g or more.

Arsenic and antimony are both of use in hardening lead alloys destined for lead shot, bullets, bearings, battery grids, and cable sheathings. Antimony has been a component of the metal used to make type for printing since the fifteenth century. Because it expands on solidifying, antimony helps to produce type that has sharp edges and prints clear images.

Arsenicals is a general term for arsenic compounds used in human and veterinary medicine, in agriculture, and as pesticides. The major use of arsenic compounds is in agriculture, especially in the defoliation of cotton before harvesting. A number of organic arsenicals are valuable in the treatment of chronic skin diseases.

Antimony compounds are much less toxic than arsenic compounds, but more so than bismuth compounds. Antimonials, like the arsenicals, are used in medicine and as pesticides. They provided the first cures for two quite dreadful parasitic diseases prevalent in tropical climates, schistosomiasis and leishmaniasis, bringing them under control as they had never been before.

Arsine (AsH_3) and stibine (SbH_3) (which takes its name from the Latin name for antimony), are unstable gases that are easily decomposed into their elements by heat. They are much less thermally stable than ammonia (NH_3) and phosphine (PH_3), the hydrides of the nonmetallic members of their periodic table family. Neither arsine nor stibine appears to be significantly basic, and cations analogous to the ammonium and phosphonium ions are not known. Arsine and stibine are both very poisonous.

Arsenic and antimony combine with oxygen to give amphoteric +3 oxides. Both exist in the gaseous state as molecular compounds, As_4O_6 and Sb_4O_6. In the solid state these units combine in polymers. Arsenic (III) oxide (As_4O_6) is the source of arsenic compounds used as pesticides, wood preservatives, and glasses. As the free metal, antimony is used primarily as a component of alloys for bearings, solder, ammunition, type metal, and in batteries.

Arsenic(V) oxide (As_2O_5) and antimony(V) oxide (Sb_2O_5) are acidic oxides obtained by oxidation of the lower oxides with concentrated nitric acid. The arsenic compound dissolves in water to give a solution from which arsenic acid (H_3AsO_4) can be isolated. This acid, like phosphoric acid, is decomposed by heat to give pyro ($H_4As_2O_7$) and meta ($HAsO_3$) acids. Antimony(V) oxide is relatively insoluble in water, and no free antimony(V) acid has been prepared.

In acidic solution, arsenic(V) and the antimony(V) are moderately strong oxidizing agents, but not as strong as bismuth(V).

$$H_3AsO_4(aq) + 2H^+ + 2e^- \rightleftharpoons HAsO_2 + 2H_2O(l) \qquad E^\circ_{298} = 0.56\text{ V}$$
$$Sb_2O_5(s) + 6H^+ + 4e^- \rightleftharpoons 2SbO^+ + 3H_2O(l) \qquad E^\circ_{298} = 0.58\text{ V}$$

Arsenic and antimony form complete series of trihalides which are liquids or low-melting solids, and in the vapor state, at least, exist as simple molecules with trigonal pyramidal structures. They are rapidly hydrolyzed (Section 29.7d).

29.11 SELENIUM AND TELLURIUM

Selenium and tellurium are among the ten or fifteen rarest elements. No ores are mined specifically for their selenium or tellurium content. These elements generally occur as selenides (e.g., PbSe, $(AgCu)_2Se$, Ag_2Se) and tellurides (e.g., PbTe, Cu_4Te_3, Ag_3AuTe_2) associated with sulfide ores of such heavy metals as copper and lead. The principal commercial source of both selenium and tellurium is the anode mud formed in electrolytic copper refining (Section 26.6).

Selenium, like sulfur, which precedes it in Representative Group VI, is known in several allotropic forms, including two crystalline forms. One is red selenium (m.p. 170–180 °C), which is soluble in carbon disulfide and does not conduct electricity. It is generally obtained when selenium compounds are reduced to the free element in solution. The selenium molecule is octaatomic and has the same puckered ring structure as the S_8 molecule. When red selenium is heated below its melting point, it changes to the more stable gray semiconducting form (m.p. 217.4 °C), called metallic selenium. This substance consists of spiral chains of selenium atoms held by covalent bonds; the chains are parallel to each other in the crystal (Figure 29.6). The electrical conductivity of gray selenium increases with the intensity of the light striking it. This property makes selenium valuable in photosensitive semiconducting devices.

The photosensitivity of selenium is also the basis for the process of dry-paper copying (*xerography*). A thin layer of amorphous selenium is coated on the surface of a drum. As the drum turns, the following events take place:

1. The photoreceptor layer (the selenium) is given a uniform electrostatic charge.
2. During exposure to an image, the selenium conducts away the charge where

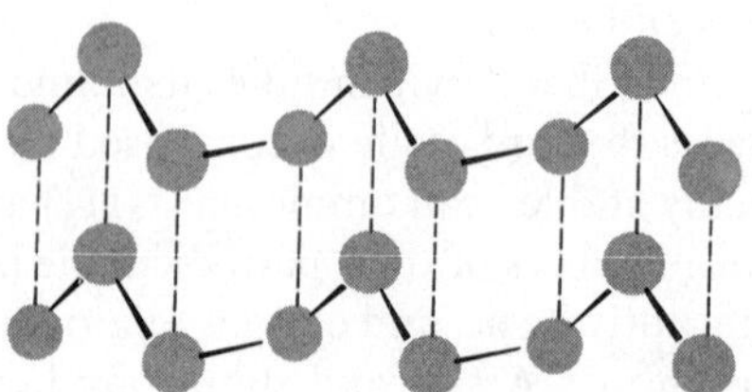

Figure 29.6
Gray Selenium, Elemental Tellurium
Both have this structure.

light strikes it, creating a "latent" image (like that on photographic film before it is developed).
3. Black particles (toner) adhere to the surface where the charge remains (where the "original" was dark).
4. The black particles are attracted from the photoreceptor surface to a piece of paper. [Further steps fix the image on the paper and clear the photoreceptor for the next exposure.]

Selenium also has a variety of uses in alloys. It is a decolorizer for glass at low concentrations, a coloring agent for ruby glass at high concentrations, and a vulcanizing agent for rubber.

Crystalline tellurium, unlike sulfur and selenium, has no allotropic forms. It is a silvery white solid that has the same crystal structure as gray selenium (Figure 29.6). The electrical conductivity of tellurium is little affected by light. Tellurium is a component of many different alloys, including various steels, malleable cast iron, and copper, lead, and tin alloys used in automotive bearings.

Selenium and tellurium form hydrogen selenide (H_2Se) and hydrogen telluride (H_2Te). In their chemistry, these compounds closely resemble hydrogen sulfide (Section 28.14). They are gaseous, emit vile odors, are poisonous, and are stronger reducing agents than hydrogen sulfide.

Selenium and tellurium form dioxides and trioxides. In contrast to SO_2, which is a colorless gas at room temperature, SeO_2 and TeO_2 are white solids. Selenium dioxide (sublimes at 315 °C) has a molecular structure of polymeric chains:

```
      O       O
 \   / \     / \   /
  Se     Se     Se
  |      |      |
  O      O      O
```

Tellurium dioxide (m.p. 733 °C) is known in several allotropic modifications, none of which is based upon discrete molecules.

The oxoacids of selenium and tellurium in their +6 oxidation states provide an interesting comparison with sulfuric acid. Selenic acid (H_2SeO_4), like sulfuric acid, has a tetrahedral structure in which the selenium atom is surrounded by four oxygen atom (see Figure 28.5a). Also like sulfuric acid, it is a strong diprotic acid. In telluric acid, which can be written as either H_6TeO_6 or $Te(OH)_6$, the tellurium atom is surrounded by the six oxygen atoms. Apparently this structure is possible because of the larger size of the tellurium atom. Telluric acid is also a diprotic acid, but much weaker than either the sulfur or selenium acids. Both selenic and telluric acids, however, are much stronger oxidizing agents than sulfuric acid. (Recall that in general higher oxidation states become less stable down representative element groups.) Salts of both these acids, such as $NaHSeO_4$, Na_2SeO_4, NaH_5TeO_6, and NaH_4TeO_6, are known. Many salts, as well as other compounds of selenium and tellurium, are quite toxic.

EXAMPLE 29.3 Chemical Reactions: Halide Preparations

Given below are the incomplete equations for several reactions (other than combination) by which some of the halides of the semiconducting elements can be prepared in the laboratory. Classify these reactions as best you can, identify any oxidizing and reducing agents, and complete the equations. Why are none of these reactions carried out in aqueous solutions?

(a) $BaGeF_6(s) \xrightarrow[N_2]{>700\ ^\circ C} GeF_4(g) + \cdots$

(b) $As_2O_3(s) + S(s) + Br_2(l) \xrightarrow[7\ h]{\Delta} AsBr_3(s) + \cdots$

(c) $AlBr_3(s) + BF_3(g) \xrightarrow{\Delta} BBr_3(g) + \cdots$

(a) Germanium is in the +4 oxidation state in both $BaGeF_6$ and GeF_4. Therefore this is a **nonredox thermal decomposition reaction** in which a gaseous product is formed. Driving off the GeF_4 by heating will leave BaF_2, a high-melting salt.

$$BaGeF_6(s) \longrightarrow \mathbf{GeF_4(g) + BaF_2(s)}$$

(b) Arsenic is in the +3 state in both As_4O_6 and the product $AsBr_3$, but bromine has been reduced to bromide ion. Therefore this is a **redox reaction, clearly not a simple one. Bromine is the oxidizing agent. Sulfur must be the reducing agent,** which indicates that an oxidation product of sulfur must also be formed, most likely SO_2.

$$2As_2O_3(s) + 3S(s) + 6Br_2(l) \longrightarrow \mathbf{4AsBr_3(s) + 3SO_2(g)}$$

(c) No oxidation numbers change here. This **nonredox reaction** between pure compounds must follow the pattern of a **partner-exchange reaction,** indicating that the other product will be AlF_3.

$$AlBr_3(s) + BF_3(g) \longrightarrow \mathbf{BBr_3(g) + AlF_3(s)}$$

The halides of the semiconducting elements cannot be prepared in aqueous solution, for they would immediately undergo hydrolysis in the presence of water.

SUMMARY OF SECTIONS 29.6–29.11

The properties of the semiconducting elements discussed in these sections are summarized in Table 29.10. Bonding and the semiconducting properties of these elements are discussed next in Sections 29.12–29.15. The properties of natural and synthetic silicon–oxygen compounds are then discussed in Sections 29.16–29.19.

Table 29.10
Outstanding Properties of the Semiconducting Elements and Their Compounds

"Metalloids"	Intermediate in properties between metals and nonmetals; form mostly covalent bonds; network covalent in most stable form; most are used in semiconductors
B(III), Si(IV), Ge(IV)	One common positive oxidation state
As, Sb (III, V); Se, Te (IV, VI)	Two common positive oxidation states; higher states, oxidizing agents (e.g., H_3AsO_4, Sb_2O_5)
SiH_4, AsH_3,..., Na_2Se, Mg_2Si,...	Negative oxidation states in compounds with electropositive elements
Free elements + H_2O or + HX $\not\longrightarrow$ NR	Inert to water and nonoxidizing acids
Free elements + HNO_3, H_2SO_4(*conc*)	React with oxidizing acids to give oxoanions, oxides
Free element + molten alkali metal	B, Si, Ge give H_2 + salt of oxoanion; As, Se, Te, Sb disproportionate
$EX_n (n = 3 - 6)$	Halides covalent; formed by combination of elements; many are volatile
$EX_n \xrightarrow{H_2O}$ oxides, oxoacids	Halides all readily hydrolyzed
As_2O_3, Sb_2O_3, TeO_2	Amphoteric oxides; others acidic
$B_4O_7^{2-}(aq)$, $HAsO_3^{2-}(aq)$, $AsO_4^{3-}(aq)$,...	In aqueous solution, various oxoanions; no cations except SbO^+; a few complex ions
B_nH_m — *boranes*	Volatile; highly reactive; have unusual three-center bonds
H_3BO_3 — *boric acid*	Weak *mono*protic acid
Si	Second most abundant element; SiO_4 tetrahedra form framework for many rocks and minerals
Ge	Essential to semiconductors
Si_nH_{2n+2} — *silanes*	Volatile and flammable
As, Se, Te, and their compounds	All toxic; As and its compounds, highly toxic
Se + hν	Photosensitivity of Se is basis for xerography

SILICON–OXYGEN COMPOUNDS

29.12 NATURAL SILICA

The common crystalline form of silica, SiO_2, is quartz, which is present in most igneous, sedimentary, and metamorphic rocks. Sand, flint, and agate are other familiar natural forms of silica. All of the cyrstalline modifications of silica are three-dimensional polymeric substances consisting of linked SiO_4 tetrahedra, with each silicon atom joined to four oxygen atoms and each oxygen atom attached to two silicon atoms. Indeed, all silicon-containing substances in the Earth's crust have the SiO_4 tetrahedron as the fundamental structural unit.

At high temperatures (about 1600 °C for quartz), silica melts to a viscous liquid that has a strong tendency to supercool and form a glass (Section 29.14). Quartz glass undergoes very small changes in volume with changes in temperature and is highly transparent to both ultraviolet and visible light. Because of these properties, it is useful for chemical apparatus where large temperature changes occur and for optical instruments where transparency to ultraviolet radiation is necessary. (Ordinary glass absorbs ultraviolet light.)

Silica is a substance of considerable chemical stability [ΔH_f° (quartz) = −910.94 kJ/mol; ΔG_f° = −856.64 kJ/mol]. It is inert toward all of the halogens except fluorine and all acids but hydrofluoric. The following is one of the reactions by which silica is etched by hydrofluoric acid.

$$SiO_2(s) + 4HF(aq \text{ or } g) \longrightarrow SiF_4(g) + 2H_2O(l)$$

At high temperatures, silica is reduced by active metals and also by carbon, with which it gives either elemental silicon or silicon carbide (SiC).

$$\underset{\textit{silica}}{SiO_2(s)} + 3C(s) \longrightarrow \underset{\substack{\textit{silica carbide}\\\textit{(carborundum)}}}{SiC(s)} + 2CO(g)$$

Hot concentrated sodium hydroxide slowly converts silica to water-soluble silicates

$$xSiO_2(s) + 2NaOH(aq) \longrightarrow Na_2O \cdot xSiO_2(aq) + H_2O(l)$$

In the mixture of silicates formed, the ratio of sodium to silicon ranges from 0.5 to 4. Concentrated, syrupy solutions of the silicate mixtures are sold under the name of water glass. The solutions are useful as adhesive, cleansing, waterproofing, and fireproofing agents. Treatment of a solution of a soluble silicate with dilute acid yields a gelatinous material, $SiO_2 \cdot xH_2O$, called silicic acid. Although silica is an acidic oxide, no protonic acids derived from it are known.

A variety of anions have been detected in the soluble silicates, including SiO_4^{4-}, $Si_2O_7^{6-}$, $Si_3O_{10}^{8-}$, and others. The soluble silicates are relatively simple substances compared with the insoluble silicates.

29.13 NATURAL SILICATES

The simplest naturally occurring silicates contain SiO_4^{4-} ions and are known as orthosilicates. For example, the gemstone zircon, $ZrSiO_4$, is an orthosilicate. In SiO_4^{4-} ions, the silicon atom is at the center of a tetrahedron of oxygen atoms (Figure 29.7).

The ions of many different metals occur in natural silicates. The most common ones are Al^{3+}, Fe^{3+}, Ti^{4+}, Mg^{2+}, Ca^{2+}, Fe^{2+}, Mn^{2+}, Li^+, Na^+, and K^+. Few silicate minerals are homogeneous substances, one reason being that metal ions of appropriate sizes can substitute for each other in the crystal lattices. For example, olivine, $(Mg,Fe)_2SiO_4$, also a simple silicate, can be found with magnesium and iron present in all proportions.

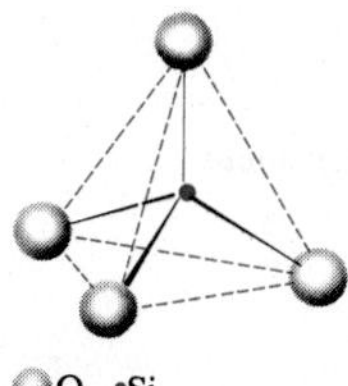

Figure 29.7
A Silicate Anion, SiO_4^{2-}

In structure, most silicates are polymeric. They are easily classified according to the way SiO_4 tetrahedra are linked to each other. Beginning with the disilicate ion, $Si_2O_7^{6-}$, in which two tetrahedra have a common oxygen corner

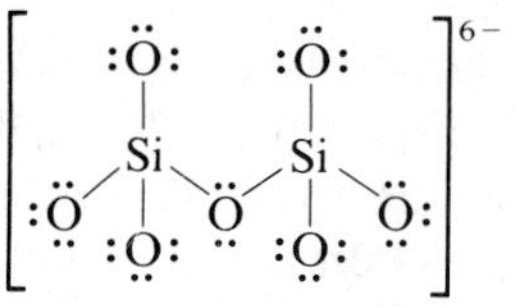

$$\left[\begin{array}{c} :\ddot{O}: \quad :\ddot{O}: \\ | \qquad\quad | \\ :\ddot{O}-Si-\ddot{O}-Si-\ddot{O}: \\ | \qquad\quad | \\ :\ddot{O}: \quad :\ddot{O}: \end{array}\right]^{6-}$$

and which occurs naturally in the rare scandium mineral thortveitite, $Sc_2Si_2O_7$, they range through rings, chains, and sheets to three-dimensional networks like that of silica. These variations are simply represented by drawing the SiO_4 tetrahedra as in Table 29.11, which summarizes some of the varieties of silicate minerals.

Table 29.11
Silicate Structures

Structurual type	Number of vertices shared	Anion composition	General class	Anion structure	Examples
Discrete anionic groups	0	SiO_4^{4-}	Orthosilicates		Be_2SiO_4 (phenacite) $(Mg, Fe)_2SiO_4$ (olivine)
	1	$Si_2O_7^{6-}$	Pyrosilicates	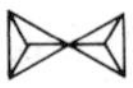	$Sc_2Si_2O_7$ (thortveitite)
	2	$Si_3O_9^{6-}$ $Si_6O_{18}^{12-}$	(Rings)	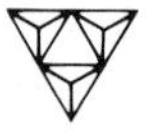 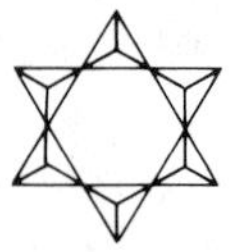	$BaTiSi_3O_9$ (benitoite) $Be_3Al_2Si_6O_{18}$ (beryl)
One-dimensional chains	2	$[SiO_3]_n^{2n-}$	Pyroxenes (linear chains)	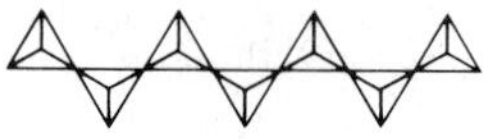	$CaMgSi_2O_6$ (diopside)
	2	$[(Si_4O_{11})_n]^{6n-}$	Amphiboles (double chains)	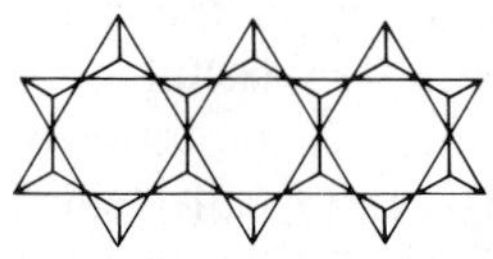	$Ca_2(OH)_2Mg_5(Si_4O_{11})_2$ (tremolite, an asbestos)
Two-dimensional sheets	3	$[Si_2O_5]^{2-}$	Mica and talc; clays	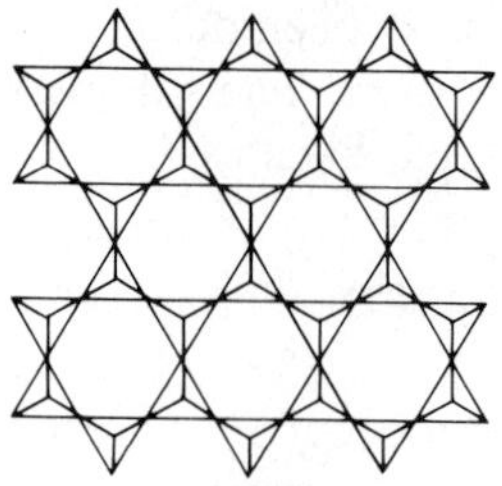	$Mg_3[Si_4O_{10}](OH)_2$ (talc) $Al_2[Si_4O_{10}](OH)_2$ (pyrophyllite)
Three-dimensional networks	4	SiO_2	Silica	—	Quartz tridymite, cristobalite
	4	$[AlSi_3O_8]^-$ $[Al_2Si_2O_8]^{2-}$	Feldspars and zeolites	—	$NaAlSi_3O_8$ (albite) $CaAl_2Si_2O_8$ (anorthite)

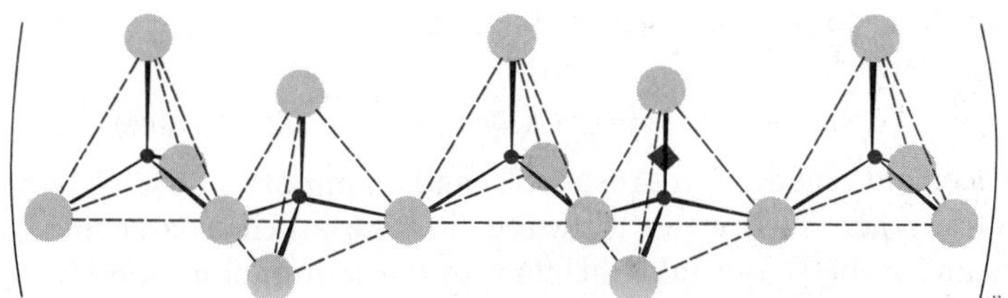

Figure 29.8
Silicon–Oxygen Chain

Since the oxygen atoms that surround each silicon atom are arranged tetrahedrally, the SiO_4 chains are not straight, but zigzag (Figure 29.8). Although each silicon atom is attached to four oxygen atoms, the composition of the chain is expressed by the formula $(SiO_3)_n^{2n-}$, for two of the four oxygen atoms are shared with other silicon atoms.

Double silicate chains are also known (see Table 29.11). In these, half of the silicon atoms share three oxygen atoms with other silicon atoms, whereas the other half share only two. The repeating anionic silicate unit is $(Si_4O_{11})^{6-}$. Minerals containing this group, the *amphiboles*, are complex in structure, for most of them contain two or more cations, and all of them have hydroxide groups linked to the cations. The asbestos minerals, such as tremolite, $Ca_2(OH)_2Mg_5(Si_4O_{11})_2$, are typical members of the amphibole family. The internal structure of these minerals is reflected in their highly fibrous nature. The strong Si—O bonds in the chains remain intact, but the weaker bonds to the ions between the chains can be broken. Asbestos is not flammable and the fibers have been incorporated in many materials, ranging from wallboard to fireproof fabrics. However, airborne asbestos fiber is a cancer-causing agent and care must be exercised that the asbestos material is not worn down so that minute fibers can be released to the atmosphere. Because of this concern, old asbestos materials are being removed from public buildings.

There are many silicate minerals in which the anions form sheets by extension of the double silicate chains. Each silicon atom shares three of its four oxygen atoms with adjacent silicon atoms. Talc, some clay minerals, and some micas contain sheets of this sort, but the structures of these minerals are complicated by the presence of hydroxide groups, which are bonded to the silicate sheets through magnesium or aluminum ions. The bonds between the hydroxide groups and the metal ions are rather weak and easily broken. This accounts for the ease with which these minerals can be split into thin layers.

The mica phlogopite is an aluminosilicate, a silicate in which aluminum atoms replace some of the silicon atoms. Many clays and feldspars are also aluminosilicates. The replacement of silicon atoms by aluminum atoms leaves a charge imbalance, for each aluminum atom has only three valence electrons rather than four. For each aluminum atom a +1 cation must also enter the aluminosilicate structure, or for every two aluminum atoms, one +2 cation must be added. The ratio of aluminum atoms to silicon atoms may vary over a wide range, but the ratio of positive ions to aluminum atoms is fixed by the requirement for neutrality. Typical natural aluminosilicates are $KAlSi_3O_8$ (orthoclase) and $CaAl_2Si_2O_8$ (anorthite).

The **zeolites** are a family of aluminosilicates with much more open structures than the clays and feldspars. The crystals contain channels which are from 0.5 to 0.13 nm in diameter. Sodium ions or other cations reside in the channels and are not attached to the rigid aluminosilicate structure. The diameters of the openings in zeolites vary with the sizes of the cations. If structures with openings of the correct size are chosen, zeolites can be used to absorb small molecules selectively, a property for which zeolites are referred to as *molecular sieves*. Such small molecules as ammonia, carbon dioxide, and ethyl and methyl alcohols are reversibly absorbed by zeolites. Molecular sieves are used industrially in drying gases and liquids, in separating nitrogen from air, and in separating straight and branched-chain hydrocarbons. Synthetic zeolites (Figure 29.9) can be produced with the properties needed for specific applications.

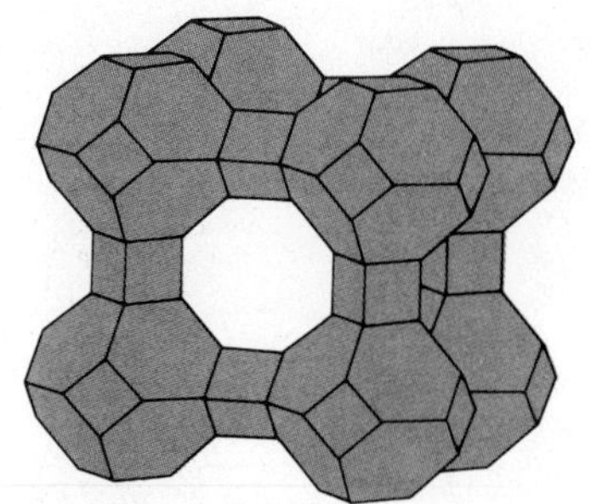

Figure 29.9
A Synthetic Zeolite
The polyhedra are units containing a total of 24 aluminum and silicon atoms; the lines connect the aluminum and silicon atoms. An oxygen atom (not shown) lies between each aluminum and silicon atom (see Figure 29.8). The molecular formula of this "molecular sieve" is $Na_{12}(Al_{12}Si_{12}O_{48})\cdot 27H_2O$.

The ions in a zeolite can be replaced readily by other ions of equivalent charge, for example

$$2NaZ + Ca^{2+} \rightleftharpoons CaZ_2 + 2Na^+ \qquad (Z = \text{zeolite})$$

These displacement reactions are reversible, and an important use of zeolites depends upon this fact. When hard water is passed through a bed of a zeolite, the calcium, magnesium, and iron(II) ions take the place of the sodium ions, thereby softening the water. Many domestic water-softening systems contain zeolites. (Zeolites are now being added to synthetic detergents for the same purpose.) A fixed bed of spent zeolite can be regenerated by passing a concentrated solution of sodium chloride through it.

Zeolites are also extremely important in the chemical industry because the sodium ions can be replaced by ions which are catalytically active, such as those of various transition metals. Zeolite catalysts are used in petroleum refining, in making gasoline from a hydrogen–methane mixture, and in other ways.

29.14 SYNTHETIC SILICATE MATERIALS

a. Glass Many silicate minerals, once melted, do not crystallize when cooled, but form hard, noncrystalline, transparent substances called glasses. (Other materials also form glasses, but the term is most often used for the silicate glass of everyday use.) Since glasses are extremely viscous, rearrangement of the structure into the more stable crystalline forms proceeds only very slowly, and may require centuries.

Common window glass is a mixture of sodium and calcium silicates. It is made by melting sodium carbonate, calcium carbonate (or oxide), and silicon dioxide together.

$$xNa_2CO_3(l) + xSiO_2(s) \longrightarrow Na_{2x}(SiO_3)_x(l) + xCO_2(g)$$
$$xCaCO_3(s) + xSiO_2(s) \longrightarrow Ca_x(SiO_3)_x(l) + xCO_2(g)$$

Countless variations in glass composition are possible. For example, the replacement of part of the sodium by potassium makes the glass harder and raises the temperature at which it softens. The addition of transition metal compounds to the glass mix forms silicates that are colored—chromium(III) oxide gives a deep green glass, cobalt(II) oxide gives a blue one, and so on. The inclusion of colloidal materials may give a glass a color (gold gives ruby glass) or make it translucent or opaque (SnO_2 gives opaque glass). The addition of lead(II) oxide increases the refractive index of the glass so that it gives a play of colors when exposed to the rays of white light. Lead glass (flint glass) is used for cut-glass dishes, lenses and artificial gems.

If part of the silicon dioxide is replaced by boric oxide in the form of borax, the resultant glass has a very low coefficient of expansion and can undergo rapid changes in temperature without breaking. Pyrex, the trademarked glass common in laboratories, is a borosilicate glass.

Ordinary glass is only slightly soluble in water, but it is soluble enough that water that has stood in a vessel of such glass for some time gives an alkaline reaction and, upon evaporation, yields a weighable residue (e.g., Equation 29.9). Alkali and alkaline earth metal hydroxides attack glass markedly, for the polymeric silicate anions are degraded by alkalies. Acids, except hydrofluoric, are almost without action on most glasses. Hydrofluoric acid attacks glass rather rapidly and is used for etching designs on objects made of glass.

$$CaSiO_3(s) + 6HF(aq) \longrightarrow CaF_2(s) + SiF_4(g) + 3H_2O(l)$$

b. Ceramics Glass is noncrystalline and has sand as its major silicate ingredient. In contrast, most of the products that we think of as ceramics—brick, tile, earthenware, pottery, and porcelain—have clay as their major ingredient and are mixtures of crystalline and glass phases. The term "ceramics" is hard to define precisely. One encyclopedia includes as ceramics all solid materials that are not organic or metallic. A common definition is materials made from clay.

Clays are formed by the weathering of aluminosilicate minerals and generally contain the original mineral and sand as impurities. Most clays also have iron(III) oxide as an impurity, and the reddish color of many clays and ceramic products is due to this oxide. Clays of greatest importance to the ceramics industry contain kaolin group minerals ($Al_2Si_2O_5(OH)_4$, for example) as the main components.

Ceramic objects are made by shaping, drying, and then heating to high temperature plastic mixtures of clay, additives such as other silicate minerals or quartz, and water. The changes that occur on firing are complex and not fully understood.

A vitreous coating or glaze is often put on ceramic pieces after a preliminary period of firing. The glaze material—it may be a metal oxide or a salt, or a metal silicate, or a mixture of these—is applied to the surface, and the object is reheated to a temperature at which the additive either melts or reacts with clay to form a glassy coating. The appearance and properties of ceramic ware may vary widely depending upon the type of clay employed, the nature and quantity of the additives, the nature of the glaze material, the time and temperature of firing, and the presence of an oxidizing or reducing atmosphere during firing.

c. Cement The main raw materials of cement, a mixture often called portland cement (because of its similarity to a stone native to the Isle of Portland in England) or hydraulic cement (because it sets even under water), are limestone ($CaCO_3$) and clay. A powdered mixture of these materials is heated in a rotary kiln to 1400–1600 °C. Carbon dioxide is evolved from the limestone, and the resulting mixture sinters together into small lumps called "clinker." This material, a mixture of calcium aluminates and silicates with lime (CaO), is mixed with a little gypsum ($CaSO_4 \cdot 2H_2O$) and powdered.

When cement is mixed with water, hydrates are formed. This gives a plastic mass. When the material first hardens, within about 24 hours, it is still reasonably soft and can even be cut with a knife for a day or two. The gypsum extends the hardening time, permitting a longer period for working with the cement. The process of hardening appears to involve hydrolysis of the calcium aluminate to give calcium hydroxide and hydrated aluminum oxide, which react with the calcium silicates to form calcium aluminosilicates. Hardening continues for many years. The compounds first formed when cement sets apparently undergo slow hydration, followed by crystallization of the hydrates.

In quick-hardening cements, aluminum oxide in the form of bauxite is substituted for some of the clay. These cements set more slowly but harden much more rapidly than portland cement. In use, cement is usually mixed with sand and gravel or crushed rock. This prevents excessive shrinkage and gives a hard, strong material known as concrete.

29.15 SILICONES

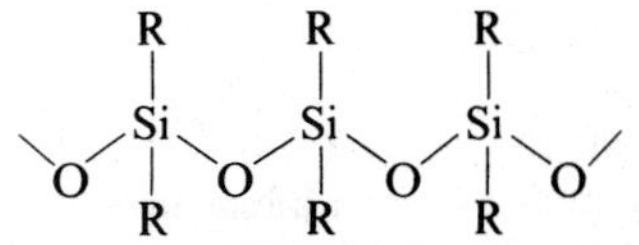

The usefulness of the mineral silicates and similar synthetic substances is largely attributable to their high stability to heat and their relative inertness toward the action of other chemicals—properties related to the great strength of the silicon–oxygen bond (368 kJ/mol). Polymeric substances that combine the —Si—O—Si— framework with hydrocarbon groups have been synthesized and have found many valuable applications. Silicones are polymers of the general formula $(R_2SiO)_n$, where R is a hydrocarbon group such as —CH_3 (methyl), —C_2H_5 (ethyl), or —C_6H_5 (phenyl). The simplest silicones are linear molecules. However, they may be cyclic or there may be cross-linking of the linear polymers, as illustrated by the structures in the margin. By varying the nature of the hydrocarbon groups, the degree of cross-linking, and the length of the chains, a great variety of silicone polymers is obtained.

linear silicone

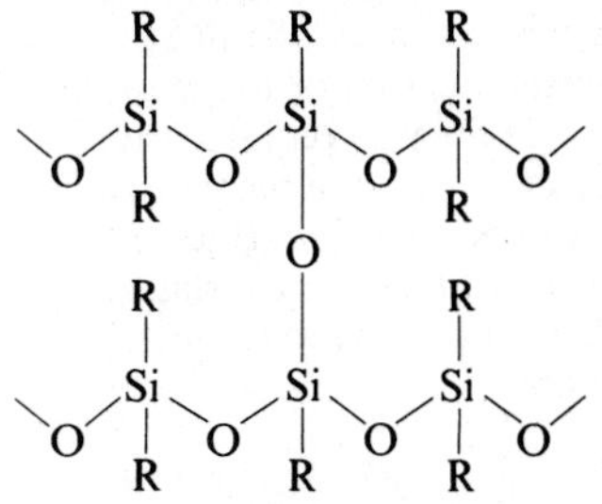

cross-linked silicone

These polymers have properties that are to a considerable extent a combination of those of hydrocarbons and of the —Si—O—Si— framework. Being nontoxic, they are harmless if taken internally and nonirritating to skin and eyes. Silicones used as heat-transfer fluids can withstand temperatures up to 750 °C for ten years. They are not susceptible to atmospheric oxidation and are not wet by water. Their inertness has

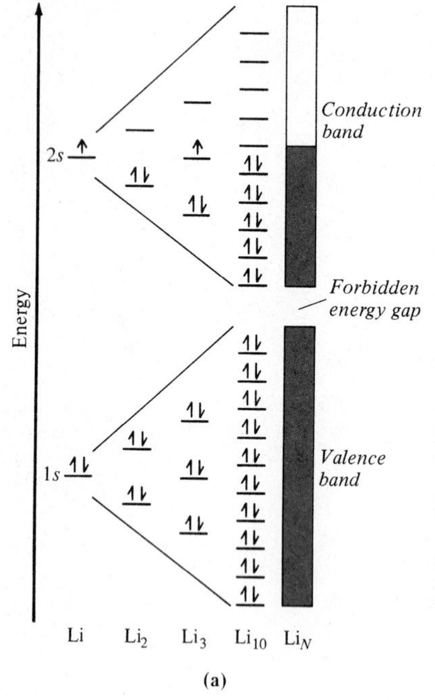

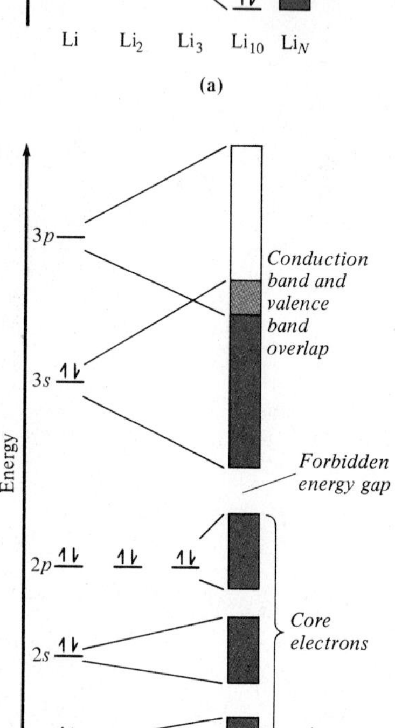

Figure 29.10
Formation of Electron Bands by Lithium and Magnesium
The band structures of all alkali metals resemble that of lithium; the band structures of all alkaline earth metals resemble that of magnesium. Metallic conduction requires the presence of either (a) a partially filled conduction bond (as in lithium) or (b) a valence band that overlaps an empty conduction band (as in magnesium).

made silicones useful in surgical applications. Those silicones that are liquid are used as brake fluids and as water repellants, those that are greaselike are good lubricants, the plastic ones have special applications in place of rubber and also in chewing gum, and the cross-linked, rigid ones are employed as insulators.

SUMMARY OF SECTIONS 29.12–29.15

Silicon is present in most rocks and minerals, in which SiO_4 tetrahedra, by sharing oxygen atoms, are connected in various ways (Table 29.11). Zeolites are aluminosilicates that, because of their open channels and replaceable cations, can be used as molecular sieves, catalysts, and in water softening. Glass, ceramics of all types, and cement are man-made silicate materials. Silicones are polymers of the general formula $(R_2SiO)_n$, which have widely varying properties and a vast number of applications. Note the following distinctions:

silicon	Si (the element)
silica	SiO_2
silicates	Materials composed of metal cations and SiO_4 tetrahedra
silicones	$(R_2SiO)_n$, synthetic polymers

BONDING IN METALS AND SEMICONDUCTORS

29.16 METALLIC BONDING

Many of the characteristic properties of metals depend upon the presence of freely mobile electrons with a continuous distribution of energies (Section 10.3). The molecular orbital description of bonding (Chapter 25) provides a picture of how free electrons become available.

When many atoms are brought together in a metal, many molecular orbitals are possible. Instead of electrons being confined to orbitals that encompass two, or three, or a dozen atoms in a molecule, some electrons reside in orbitals that encompass a great many atoms.

Consider the combination of lithium atoms. Two lithium atoms, each of configuration $1s^22s^1$, combine to give four molecular orbitals occupied by six electrons as shown at the top in Figure 29.10a. Three lithium atoms produce six molecular orbitals, three at the lower energy level and three at the upper energy level. The process of combination is pictured as continuing up to N lithium atoms that combine to give N molecular orbitals in each energy level.

As the energy levels are divided into more and more molecular orbitals, the energy differences between any two orbitals become smaller and smaller. The point is reached where the molecular orbitals in an energy level are so close together that they are essentially continuous. A continuous energy level produced by a large number of molecular orbitals is called a band, or **energy band.**

The bands formed by the lithium 1*s* and 2*s* atomic orbitals are delocalized over a great many atoms, much as six atomic orbitals are delocalized over a whole benzene molecule. The lower energy band in lithium metal is full, just as the 1*s* atomic orbital in a gaseous lithium atom is full. The upper band, to which each lithium atom contributes only one electron, is one-half full. Such delocalized orbitals in metals hold the electrons of the electron sea on which our earlier explanation of metallic bonding was based.

The highest energy electrons within the half-full energy band of elemental lithium are close to empty levels in the other half of the band that are not much higher in energy. Therefore, they can easily move from their own energy level to another that is only slightly higher. Once in an energy level that is not full, these electrons are available as current carriers. An energy band in which electrons are free to flow and therefore to conduct electricity is a **conduction band.** The electrons in a conduction band that have enough energy so that they are not held back by attraction to the positive ions are the conduction electrons. The electrons at lower levels in a conduction band need a larger amount of energy to reach empty orbitals and in general do not participate in conduction. When an electric field is applied, the conduction electrons are accelerated in the direction of the field and the net result is a flow of electrons.

Electrons contribute to conduction only if they are in a partially filled band. In a full band with no adjacent empty orbitals, all that the electrons can do is to change places with each other. The net result in the presence of an electric field is no conduction, since equal numbers of electrons are going in both directions.

At first glance, an element such as magnesium (Figure 29.10b) looks as if it should be a nonconductor, since the highest energy band is filled by two 3*s* electrons contributed by each atom. However, magnesium (and beryllium, calcium, zinc, and other metals that form +2 ions) is metallic and does conduct electricity. The empty 3*p* orbitals form an energy band that, because of the closeness of the atoms, overlaps the 3*s* band (Figure 29.10b). Therefore, electrons can move up out of the full 3*s* band and travel in the empty 3*p* band. The highest completely filled band in a metal is called the **valence band.** In divalent metals, the valence and conduction bands overlap. Electrons below the valence band, such as the 1*s*, 2*s*, and 2*p* electrons of magnesium, are the *core electrons.* They are held tightly by the nuclei.

A **forbidden energy gap,** or **energy gap**—an area of forbidden electron energies—can lie between energy bands. The energy gap is a consequence of the quantum mechanical nature of electrons. As discussed in the next section, the energy gap between the valence and conduction band plays an important role in the differences among conductors, semiconductors, and insulators.

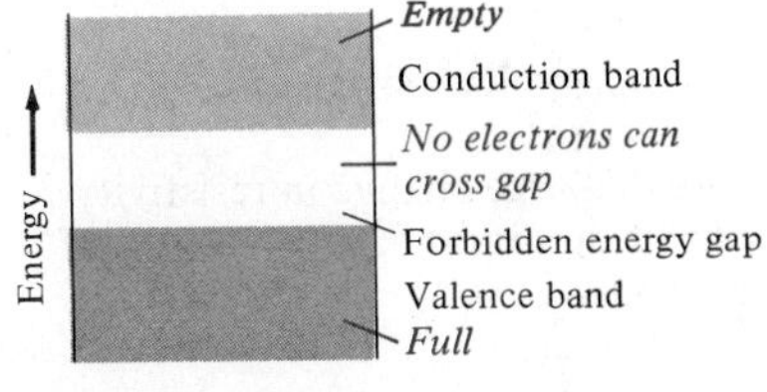

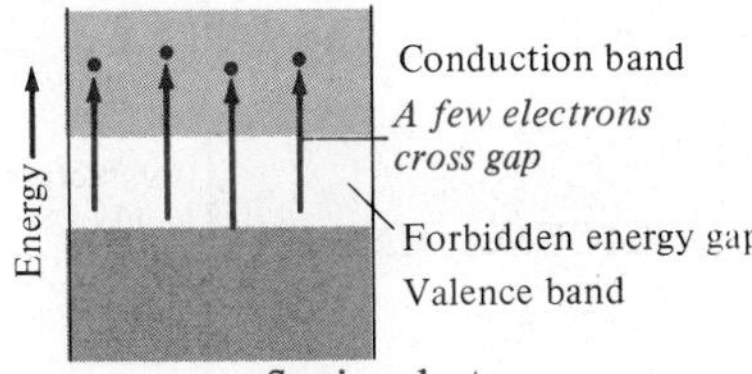

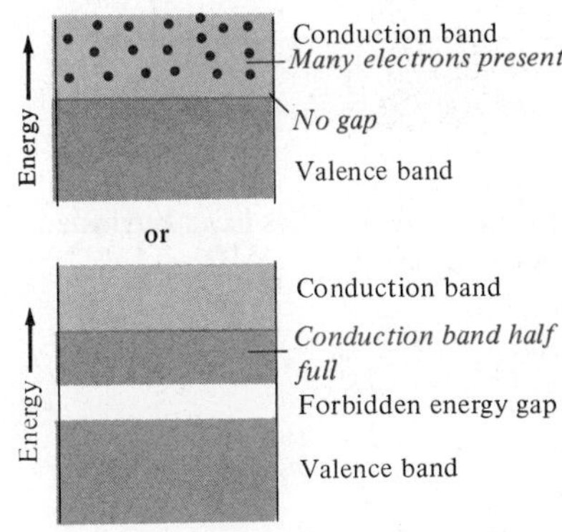

Figure 29.11
Energy Bands in Insulators, Semiconductors, and Metals

29.17 BONDING AND SEMICONDUCTIVITY

The energy band picture of bonding can be applied to the solid state of any material, for in the solid state the orbitals of the many individual atoms are brought close enough to overlap. The size of the energy gap between the valence band and the conduction band varies in different materials. In an **insulator**—a substance that does not conduct electricity—the energy gap is so large that electrons from the valence band cannot cross it (Figure 29.11). And since the valence band is full, no conduction occurs in an insulator because no net flow of electrons is possible.

In a semiconducting element, an energy gap between the valence band and the conduction band is also present, but it is smaller than in an insulator. Even at room temperature, a few electrons have enough energy to jump the gap and enter the conduction band, where they are free to move. Some energy gaps are given in Table 29.12.

A **semiconductor** is a solid, crystalline material with an electrical conductivity intermediate between that of a metal and that of an insulator. Semiconductors are not as conductive as metals, because fewer electrons are available in conduction bands. Put another way, we say that semiconductors are more resistive to the passage of electrical current. Electrical resistance is measured in ohms. The *resistivity* of a substance is the electrical resistance per centimeter of a conductor of 1 cm^2 cross-sectional area and has units of ohm centimeters (ohm cm). *Conductivity* is the reciprocal of resistivity and has units of ohm^{-1} cm^{-1}.

Aluminum, a typical metal, has an electrical resistivity of 2.7×10^{-6} ohm cm at 20 °C. Pure silicon has a resistivity of 10^5 ohm cm, while pure diamond, an insulator, is highly resistive—10^{14} ohm cm at 15 °C. The resistivity range for semiconductors is roughly 10^{-3} to 10^8 ohm cm.

Table 29.12
Energy Gaps
The energy gap for diamond is 502 kJ/mol.

	Energy gap (kJ/mol)
Semiconducting elements	
B	320
Si	100
Ge	67
As (gray)	120
β-Sb	10
Te	37
Some semiconducting compounds	
InP	130
GaAs	140
InSb	20
CdTe	140

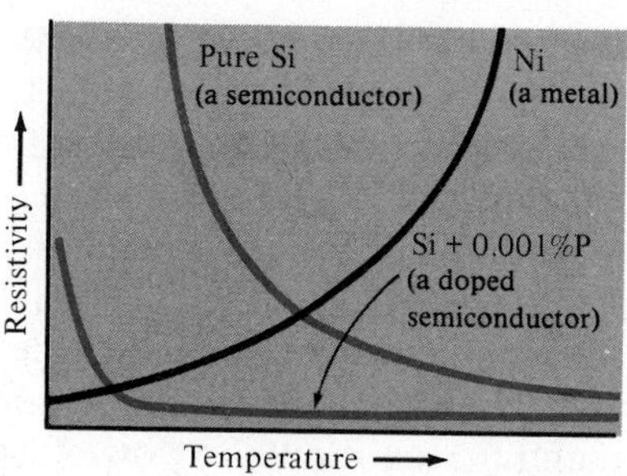

Figure 29.12
Variation of Resistivity with Temperature
Note the large change in resistivity with the addition of only 0.001% of an impurity.

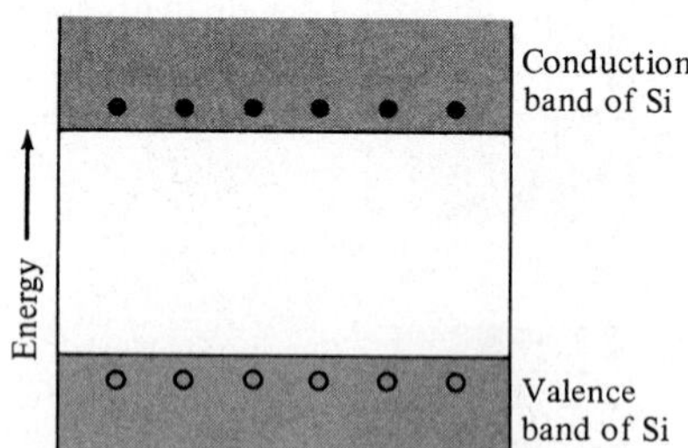

Figure 29.13
Electron–Hole Pairs in an Intrinsic Semiconductor

An increase in temperature causes the ions in a metallic crystal lattice to vibrate more within their lattice positions. This increases the chances for an electron moving through the metal under the influence of an electric field to collide with the ions. For a metal, the net result of an increase in temperature is an *increase* in resistivity (Figure 29.12).

Now we come to the essential difference between metals and semiconductors. With an increase in temperature, more electrons in a semiconductor gain the energy needed to jump out of the valence band and into the conduction band. Therefore, with rising temperature, the resistivity of a semiconductor *decreases* (Figure 29.12). The amount of decrease is different for each semiconductor. At low enough temperatures, the conductivity of semiconductors is the same as that of insulators, and at high enough temperatures it is like that of metals.

The conductance of a semiconductor is aided both by the free flow of electrons in the conduction band and by what is thought of as the migration of holes in the valence band in a direction opposite to that of the electron flow. Understanding this may take a moment's thought. Look at it this way. An electron jumping out of the valence band leaves benind a hole, in the same way as a person getting up from his seat in a theater leaves a vacant seat. Suppose that the empty seat is on the end of a row. The person in the second seat can move over to the end seat. Then the person in the third seat from the end can decide to move into the second seat. If everyone in the row moves over by one seat the vacant seat—the hole—will have moved across the entire row. Under the influence of an electric field, holes move in this way through the valence band. The net effect of the motion of the holes is that of positive charges moving in the opposite direction from the conduction electrons. The positive charge is the result of the electron deficiency.

In a pure semiconductor at room temperature, the number of electrons in the conduction band and the number of holes in the valence band are equal. An **intrinsic semiconductor** contains equal numbers of current-carrying holes and electrons; the conduction is an intrinsic property of the material (Figure 29.13). At high enough temperatures, even insulators such as diamond can become intrinsic semiconductors.

In an **extrinsic semiconductor** the number of current-carrying holes and electrons is not equal, and conduction depends on extrinsic materials—on the addition of appropriate impurities. The impurities are of two types—donors and acceptors. A donor provides electrons in the following way. In pure silicon and germanium, each atom is joined to four neighboring atoms by covalent bonds. An atom of a Representative Group V element—for example, phosphorus, arsenic, or antimony—enters the germanium or silicon crystal lattice without distorting the lattice very much and bonds to its four germanium atom neighbors. Since the atoms of the Group V element have the ns^2np^3 configuration, one electron is left over for every atom. These extra electrons enter an occupied valence band, called a **donor level,** that usually lies slightly below the conduction band of the host semiconductor (Figure 29.14). Electrons in the donor level are easily promoted to the host semiconductor conduction band. This greatly increases the conductivity.

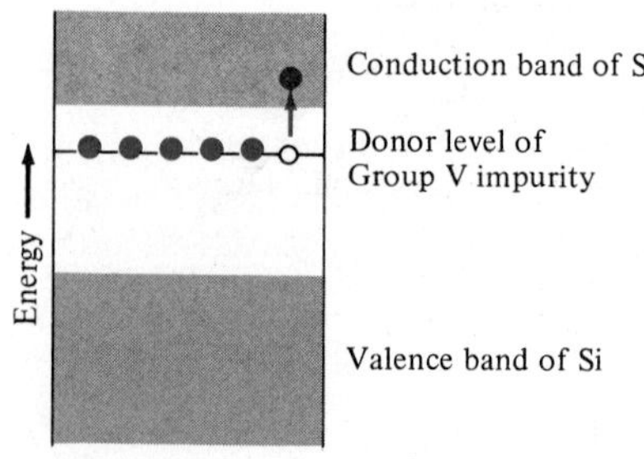

n-type

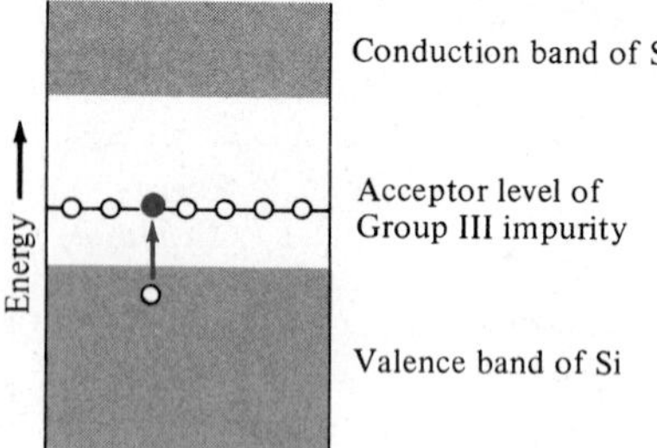

p-type

Figure 29.14
***n*- and *p*-Type Semiconductors**
In an *n*-type semiconductor, the energy difference between the impurity donor level and the conduction band is small. In a *p*-type semiconductor, the energy difference between the impurity acceptor level and the valence band is small.

In a semiconductor, a **donor impurity** contributes electrons to the conduction band and it does so without leaving holes in the valence band. The addition of controlled amounts of impurities to semiconducting elements is called **doping.** The impurity, called a **dopant,** is usually added at concentrations of 100 to 1000 parts per million. A crystal doped with a donor impurity is called an ***n*-type semiconductor**—negative electrons are the majority of the current carriers. In an *n*-type semiconductor, for each electron that enters the conduction band, a positive ion is left in the crystal structure.

An atom of a Representative Group III element—for example, boron, aluminum, or indium—has the ns^2np^1 configuration, and therefore when added to silicon or germanium can bond to only three of its neighboring atoms. One neighboring atom is left with only three bonds and a single electron that has no partner. This situation contributes an electron deficiency, in other words, a hole. An **acceptor impurity** contributes holes to a vacant **acceptor level** of energy, which is slightly above the valence band (Figure 29.14). The acceptor level is created when electrons from the valence band move to fill the vacancies in the acceptor atoms, leaving behind a band of holes. Electrons are easily promoted from the valence band to the acceptor level. In such a semiconductor, called a ***p*-type semiconductor,** positive holes are the majority of the current carriers. In a *p*-type semiconductor, for each acceptor atom, a negative ion is left in the crystal structure.

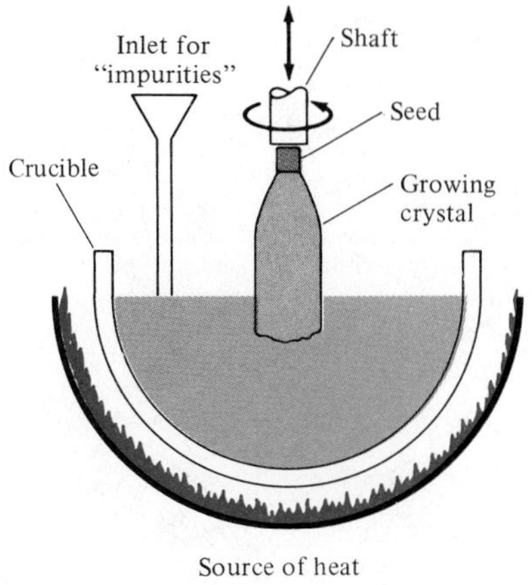

Figure 29.15
Apparatus for Pulling a Single Crystal

29.18 PREPARATION OF SEMICONDUCTOR MATERIALS

Silicon and germanium are the elements most extensively used as host semiconductors. As starting materials for semiconductor device manufacture, the elements must contain less than one part per *billion* of impurities. This is comparable to one pinch of salt in ten tons of potato chips. Conversion of the elements as they are first isolated to materials of such purity requires exacting procedures.

The 98% pure metallurgical silicon (Section 29.9) is purified by a series of chemical and physical processes such as the following. The crude silicon is converted to chloride, $SiCl_4$, by reaction with elemental chlorine. The chloride is purified by repeated fractional distillation and is then reduced to elemental silicon by reaction with hydrogen or very pure magnesium. For further purification the volatile iodide (SiI_4) is prepared from the silicon and then decomposed by heat. The silicon is then formed into ingots for the remaining steps in the purification process. Elemental germanium ingots ready for further purification are obtained directly from the preparation of the free metal as described in Example 29.2.

Zone refining, or zone melting, of the ingots is the next step. **Zone refining** is a method for purifying solids in which a melted zone carries impurities out of the solid. A rod of the solid is heated at a point near one end until it melts. The melted zone is moved slowly along the rod (either the heater or the rod is moved). Because dissolved impurities lower the melting point of the solid, the melt is always less pure than the solid. Impurities therefore collect in the molten zone and eventually are concentrated at the end of the rod.

Not only must materials for semiconducting devices be ultrapure, but they must also be free of crystalline imperfection. Boundaries between microcrystals would act as barriers to the desired free flow of electrons and holes. A semiconducting device is usually made from a piece of a single crystal. Among the many techniques of growing single crystals, one of the most successful is the "pulling" of a large crystal from a melt of the same composition as the crystal. A small seed crystal is touched to the melt and gradually withdrawn at such a rate that the melt solidifies slowly onto the seed crystal (Figure 29.15). Single crystals of germanium 5 cm in diameter and 25 cm in length and even larger silicon crystals are made by this technique.

The common semiconductor materials (Table 29.13) are silicon, germanium, and a number of binary intermetallic compounds between elements from Groups III and V (the III–V compounds) or between elements from the zinc family and Group VI (called II–VI compounds because zinc, cadmium, and mercury atoms, like those of the Group II elements, have two *s* valence electrons). In a III–V compound, for example, the

Table 29.13
Some Common Semiconductor Materials
The designation of zinc family metals as "Group II" metals is based on their similar electron configurations [Group II, or Be family, $(n-1)p^6ns^2$; Zn family, called Group IIB in some versions of the periodic table, $(n-1)d^{10}ns^2$].

Elements
Si
Ge
III–V Compounds
AlP
AlAs
AlSb
GaP
GaAs
GaSb
InP
InAs
InSb
II–VI Compounds
ZnS
ZnSe
ZnTe
CdS
CdSe
CdTe

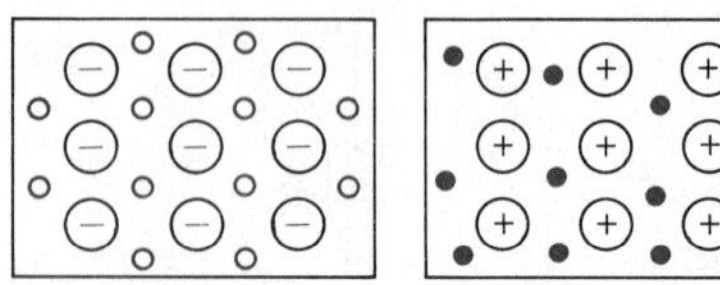

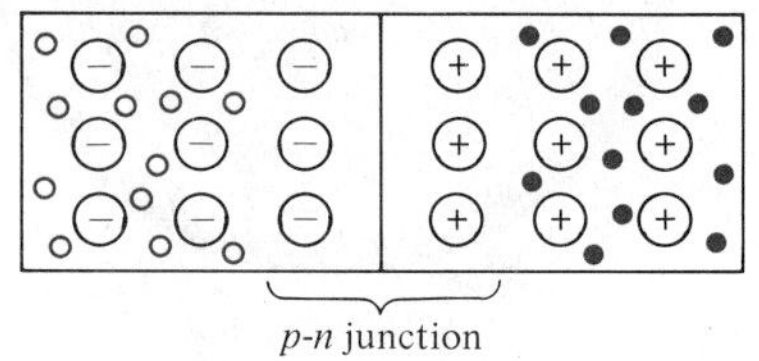

Figure 29.16
Formation of a $p-n$ Junction
The resistance of the material at the junction is increased because electrons must approach a region of negative charge and holes must approach a region of positive charge.

Group III element contributes three electrons per atom and the Group V element contributes five electrons per atom, giving an average of four electrons per atom. These compounds form crystals with a diamondlike structure similar to that of silicon and germanium, and can be doped with the same results as the elemental semiconducting materials.

Among the properties of interest in semiconducting materials are the size of the energy gap; the concentration of charge carriers, either electrons or holes; the mobility, or speed with which the charge carriers can move; and the lifetime of a charge carrier before it is annihilated by the combination of an electron with a hole. By varying the concentrations of dopants, the wide range of properties needed to create the various types of semiconductor devices is made available.

29.19 SEMICONDUCTOR DEVICES

In their operation, most semiconducting devices depend upon the properties imparted by adjacent *p*-type and *n*-type semiconductors. A ***p–n* junction** is the boundary between *n*-type and *p*-type semiconductors. The $p-n$ junction is created by different doping of adjacent areas in the same crystal.

Electrons from the *n*-type semiconductor and holes from the *p*-type semiconductor at first migrate toward the junction, where they combine (Figure 29.16). Their combination leaves excess positive ions on the *n*-type side of the junction and excess negative ions on the *p*-type side of the junction. Before long, no further migration of electrons and holes occurs, because each would have to move toward a region with a charge the same as its own. The result is a potential barrier, which makes the resistance at the $p-n$ junction greater than in the bulk of the material. By varying the characteristics of the $p-n$ junction, the current-carrying capability of the junction can be varied.

A diode is a semiconducting device that incorporates a $p-n$ junction and can perform many different functions. For example, a diode can act as a *rectifier*, converting alternating to direct current. Electrodes are attached to the ends of a rectifying diode and alternating current is passed through it. During part of the current cycle electrons from the *n*-type (electron-rich) part of the semiconductor are attracted to one electrode and the positive holes of the *p*-type part are attracted to the other electrode, leaving the junction region bare of current carriers and thus effectively stopping current flow. During the other part of the current cycle, when the polarity of the electrodes is reversed, electrons are repelled from the pole to which they were previously attracted and are now pulled to the other electrode, while the positive "holes" are attracted to the electrode from which the electrons are repelled. Now the two processes reinforce each other with respect to flow of current. The rectifier thus prevents flow of current in one direction and offers a low resistance to flow in the other (Figure 29.17).

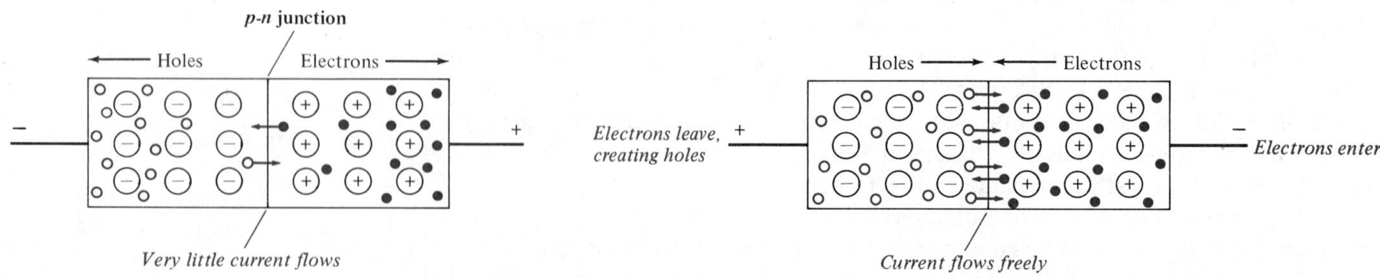

Figure 29.17
Operation of a $p-n$ Junction as a Rectifier
Alternating current is converted to direct current because current can flow freely in only one direction.

A photodiode responds to the absorption of light of energy equal to the band gap of the semiconducting material from which the diode is fabricated. Absorption of a photon raises an electron into the conduction band, creating a free electron and a hole which can migrate and allow current to flow. In a light-emitting diode, an external current causes current to flow so that electrons and holes meet at the *p-n* junction and combine. As the electrons "fall" into the holes they emit visible radiation of energy equal to that of the band gap.

Transistors consist of two diodes, back to back. The main functions performed by transistors are switching and amplifying current. Integrated circuits, in which diodes, transistors, and other necessary circuit components are combined in a single semiconductor chip, have set instrument designers free from the use of discrete components. As a result, for example, computers that filled several large cabinets have shrunk to the size of typewriters.

Aside 29.2 THOUGHTS ON CHEMISTRY: NONSENSE

The following excerpt is from a book by Gary Zukov about modern physics—the quantum revolution and its results (which are discussed in Chapter 8). The thoughts expressed apply to many discoveries in physics, chemistry, and all sciences. Suppose, for example, the young chemist trying to make aluminum and calcium by reducing their oxides with carbon had dismissed as "nonsense" the unexpected result—production of a flammable hydrocarbon. (He had made acetylene by way of calcium carbide.)

The importance of nonsense hardly can be overstated. The more clearly we experience something as "nonsense," the more clearly we are experiencing the boundaries of our own self-imposed cognitive structures. "Nonsense" is that which does not fit into the prearranged patterns which we have superimposed on reality. There is no such thing as "nonsense" apart from a judgmental intellect which calls it that....

Nonsense is only that which, viewed from our present point of view, is unintelligible. Nonsense is nonsense only when we have not yet found that point of view from which it makes sense.

In general, physicists [and chemists] do not deal in nonsense. Most of them spend their professional lives thinking along well-established lines of thought. Those scientists who establish the established lines of thought, however, are those who do not fear to venture boldly into nonsense, into that which any fool could have told them is clearly not so. This is the mark of the creative mind; in fact, this is the creative process. It is characterized by a steadfast confidence that there exists a point of view from which the "nonsense" is not nonsense at all—in fact, from which it is obvious.

In physics, as elsewhere, those who most have felt the exhilaration of the creative process are those who best have slipped the bonds of the known to venture far into the unexplored territory which lies beyond the barrier of the obvious. This type of person has two characteristics. The first is a childlike ability to see the world as it is, and not as it appears according to what we know about it.

The second characteristic of true artists and true scientists is the firm confidence which both of them have in themselves. This confidence is an expression of an inner strength which allows them to speak out, secure in the knowledge that, appearances to the contrary, it is the world that is confused and not they. The first man to see an illusion by which men have flourished for centuries surely stands in a lonely place. In that moment of insight he, and he alone, sees the obvious which to the uninitiated (the rest of the world) yet appears as nonsense or, worse, as madness or heresy. This confidence is not the obstinacy of the fool, but the surety of him who knows what he knows, and knows also that he can convey it to others in a meaningful way.

Source: Gary Zukov, *The Dancing Wu Li Masters* (New York: William Morrow, 1979), pp. 117–119. Quoted from Bantam edition, 1980. Copyright © 1979 Gary Zukov. By permission of William Morrow & Company.

SUMMARY OF SECTIONS 29.16–29.19

The property unique to the semiconducting elements is an increase in electrical conductivity with increasing temperature. This is explained as due to excitation of electrons out of the valence band, across an energy gap, and into the conduction band. An insulator has an energy gap so large that no electrons can cross it; a metal has no energy gap (Figures 29.10 and 29.11). *n*-Type semiconductors are doped so that they contain an excess of electrons as charge carriers. *p*-Type semiconductors contain an excess of "positive" holes. Semiconducting devices are made from extremely pure single crystals. Diodes contain *p–n* junctions produced by differently doping areas of the same crystal; they can function as rectifiers, light emitters, and photoactive conductors. Silicon, germanium, and III–V and II–VI compounds (where the "II" elements are from the zinc family) are used in most diodes, transistors, and integrated circuits (Table 29.13).

SIGNIFICANT TERMS (with section references)

acceptor impurity (29.17)
acceptor level (29.17)
conduction band (29.16)
donor impurity (29.17)
donor level (29.17)
dopant (29.17)
doping (29.17)
electron-deficient compounds (29.8)
energy band (29.16)
extrinsic semiconductor (29.17)
forbidden energy gap, energy gap (29.16)
insulator (29.17)
intrinsic semiconductor (29.17)
n-type semiconductor (29.17)
p-type semiconductor (29.17)
p–n junction (29.19)
semiconductor (29.17)
silicates (29.9)
silicones (29.9)
three-center bond (29.8)
valence band (29.16)
zeolites (29.13)
zone refining (29.18)

QUESTIONS AND PROBLEMS

Carbon and Its Inorganic Compounds

29.1 Write the electron configurations of the C and Si atoms. What would you predict for the formulas of the compounds formed between these elements and fluorine? Although silicon can form the $[SiF_6]^{2-}$ ion, the analogous ion for carbon, $[CF_6]^{2-}$, is unknown. Why?

29.2 Discuss the possible ways in which the carbon atom can form covalent bonds. Draw Lewis structures for each of the possibilities and predict the angles between the bonds around the carbon atom.

29.3 The largest diamond ever discovered was the Cullinan diamond, which weighed 3106 carats. Calculate the volume of this stone in cm^3. The density of diamond is 3.51 g/cm^3. One carat is equal to 200 mg.

29.4 Name the two allotropic forms of carbon. Briefly compare their properties.

29.5 The heat of combustion at 25 °C is -395.4 kJ/mol for diamond and -393.5 kJ/mol for graphite. The absolute entropies are 2.38 J/K mol and 5.74 J/K mol, respectively. Determine which allotropic form of carbon is more stable under room conditions.

29.6 Briefly describe the reactivity of carbon with (a) the halogens, (b) sulfur, (c) metals, (d) oxygen.

29.7 Describe the types of reactions in which C and CO serve as reducing agents at elevated temperatures.

29.8 Name the oxoanions of carbonic acid and write their formulas. Will aqueous solutions of salts of these anions be acidic, alkaline, or neutral? Why?

29.9 Draw Lewis structures for CO, CO_2, and CO_3^{2-}. Predict which substance has (a) the strongest carbon–oxygen bond and (b) the shortest carbon–oxygen bond. Give reasons for your answers.

29.10 What is the shape of a CO_2 molecule? What is the hybridization of the carbon atom orbitals? What intermolecular forces will be found in CO_2?

29.11 How will the concentrations of CO_2 in water depend on the temperature? How does it depend on the partial pressure of the CO_2 above the solution?

29.12 Write Lewis structures for N_2, CN^-, CO, and C_2^{2-}. Assuming that the available molecular orbitals of CN^- and CO are similar to those of N_2 and C_2^{2-}, write the electron configurations for all four species. What type of hybridization is present?

29.13 Pieces of dry ice were mixed with 100. mL of 5.0 M NaOH. Write the chemical equation for the reaction to form sodium carbonate. What mass of sodium carbonate would be formed from 2.2 g of CO_2 reacting with the NaOH solution?

29.14 What is the pH of a 0.010 M solution of $NaHCO_3$? $K_{a1} = 4.5 \times 10^{-7}$ for H_2CO_3.

The Seven Semiconducting Elements and Their Compounds

29.15 Name the seven semiconducting elements and write the chemical symbol for each. Briefly discuss the appearance, electrical conductivity, and chemical behavior of each of these elements.

29.16 List the common oxidation states of the semiconducting elements. How are these values related to their respective electron configurations?

29.17 Describe the reactivities of the semiconducting elements with water and nonoxidizing acids. What are the products formed upon reaction with oxidizing acids?

29.18 Write the formulas of the oxides of the seven semiconducting elements discussed in this chapter. Write an equation for the reaction each undergoes with water. Name each substance that is formed.

29.19 Write chemical equations illustrating how the halides of the semiconducting elements resemble the halides of nonmetals in their behavior toward water much more than they do the halides of metals.

29.20 The standard enthalpy of formation at 25 °C is 830.5 kJ/mol for $B_2(g)$ and 562.7 kJ/mol for $B(g)$. Calculate the bond dissociation energy for B_2.

29.21* Write the electron configuration for (a) an atom of boron and (b) a molecule of B_2. Based on the electron configuration for the atom, (c) predict whether boron in the gaseous atomic state is paramagnetic or diamagnetic. Based on the electron configuration for the diatomic molecule, (d) predict whether the molecules are paramagnetic or diamagnetic. (e) Describe the type of bonding present in the molecule.

29.22 Write the chemical equation for the reduction of BCl_3 by H_2. What volume of H_2 measured at STP is required to produce 1.0 g of elemental boron?

29.23 Devise a plan to prepare crystalline boron from sassolite (H_3BO_3) by using heat, hydrogen gas, chlorine gas, and magnesium metal.

29.24 The sublimation energy of Si is 456 kJ/mol at 25 °C. On the average, two bonds must be broken for each atom that undergoes sublimation. (a) Calculate the average bond energy for a Si—Si bond. Like carbon, silicon forms a diatomic molecule in the gaseous phase. The bonding consists of a σ bond and a π bond. (b) Calculate the bond dissociation energy for a Si=Si bond given that the standard heat of formation at 25 °C of $Si_2(g)$ is 594 kJ/mol and of $Si(g)$ is 456 kJ/mol. (c) Which bond energy (single or double) is larger?

29.25* What is the total mass of silicon in the crust of the Earth? Assume that the radius of the Earth is 6400 km, the crust is 50 km thick, the density of the crust is 3.5 g/cm^3, and 25.7 mass % of the crust is silicon.

29.26 The equation for the carbon reduction of SiO_2 is

$$SiO_2(l) + 2C(s) \xrightarrow{3300\text{ K}} Si(l) + 2CO(g)$$

The standard state heats of formation at 3300 K are zero for the elements, −128.8 kJ/mol for $CO(g)$ and −910.1 kJ/mol for $SiO_2(l)$. (a) Calculate ΔH°_{3300} for this reaction. Based on energy alone, would you expect this reaction to be favorable? The standard state entropies at 3300 K are 277.1 J/K mol for $CO(g)$, 109.8 J/K mol for $Si(l)$, 53.2 J/K mol for $C(s)$, and 215.7 J/K mol for $SiO_2(l)$. (b) Calculate ΔS°_{3300} for this reaction. Based on entropy alone, would you expect this reaction to be favorable? Using your values of ΔH°_{3300} and ΔS°_{3300}, (c) calculate ΔG°_{3300} for this reaction and discuss your result.

29.27 A zinc sulfide ore contains 1 part of recoverable Ge per 2,500,000 parts of ore. In the recovery process, the conversion sequence is $GeS_2(s) \rightarrow GeO_2(s) \rightarrow GeCl_4(g) \rightarrow GeO_2(s)$. Assuming complete separations and 90% recovery of germanium, calculate the quantity of GeO_2 that should be obtainable from 1000 kg of ore.

29.28 Many metals and semiconducting elements can be purified by sublimation. The vapor pressure of arsenic is 10.0 Torr at 437 °C and 40.0 Torr at 483 °C. Calculate the heat of sublimation of arsenic. If the temperature of the arsenic is increased to 518 °C, what will be the vapor pressure?

29.29 What elements make up the class of compounds known as the "boranes"? What is the formula of the simplest known borane? Draw a diagram of the molecular structure for this compound.

29.30 What does the term "three-center bond" mean? How is it different from our ordinary concept of covalent bonding?

29.31 A gaseous sample of a boron hydride had a density of 0.57 g/L at 25 °C and 0.500 atm. What is the molar mass of the compound?

29.32 A graduate student prepared some diborane using the reaction

$$4BF_3(g) + 3LiAlH_4(s) \xrightarrow{\text{ether}} 2B_2H_6(g) + 3LiF(s) + 3AlF_3(s)$$

Assuming 100% yield, what mass of B_2H_6 could be produced from the reaction of 5.0 g of BF_3 with 10.0 g of $LiAlH_4$? The student used ether that had not been carefully dried and lost some diborane to the following reaction:

$$B_2H_6(g) + 6H_2O(l) \longrightarrow 2H_3BO_3(\text{ether}) + 6H_2(g)$$

How much of the diborane would react with 0.01 g of water?

29.33 Write the Lewis structures for BCl_3 and $[BCl_4]^-$. What are the geometries of these species?

29.34 What are compounds that contain only silicon and hydrogen called? What is the general formula for this series of compounds?

29.35 Monosilane reacts vigorously with water. Write the chemical equation for this reaction. Explain why monosilane reacts vigorously with water and methane does not.

29.36 What kind of intermolecular forces would you predict to be present in the tetrahalides of silicon? Briefly describe the resulting physical properties of these compounds.

29.37 At high temperatures, silicon forms a molecular dioxide in the gas phase. Two double bonds are present in each molecule. Calculate the average bond energy for a Si=O bond given that the standard heat of formation at 25 °C is −322 kJ/mol for $SiO_2(g)$, 455.6 kJ/mol for Si(g), and 249.170 kJ/mol for O(g).

29.38 Compare the chemical properties of arsine and stibine to those of the other hydrides of Representative Group V elements.

29.39 A sample of arsenic was burned in a limited amount of oxygen to form an oxide containing 76 mass % As. (a) What is the empirical formula of this oxide? The true molar mass of this compound is 200 g/mol. (b) What is the molecular formula of the oxide? This oxide was dissolved in water to form an acid that contained 59.5 mass % As, 38.1 mass % O, and 2.4 mass % H. (c) What is the empirical formula of this acid? (d) Write chemical equations for both reactions.

29.40 Calculate the equilibrium constant for the reaction

$$H_3AsO_4(aq) + 2H^+ + 2I^- \longrightarrow HAsO_2(aq) + I_2(s) + 2H_2O(l)$$

given the following half-reactions and associated standard reduction potentials:

$$H_3AsO_4 + 2H^+ + 2e^- \longrightarrow HAsO_2 + 2H_2O \qquad E° = 0.56\text{ V}$$
$$I_2 + 2e^- \longrightarrow 2I^- \qquad E° = 0.5356\text{ V}$$

29.41 The standard state free energy of formation at 25 °C is 22.2 kJ/mol for $H_2Se(aq)$ and 44.0 kJ/mol for HSe^-. Calculate $\Delta G°$ and K for the reaction.

$$H_2Se(aq) \rightleftharpoons H^+ + HSe^-$$

29.42 Sulfuric and selenic acids have the formulas H_2SO_4 and H_2SeO_4, respectively. The formula of telluric acid, however, is H_6TeO_6. How do you account for this difference?

Silicon–Oxygen Compounds

29.43 What is the fundamental building unit in all silicon-containing substances in the crust of the Earth? How is this building block modified to form the various types of natural silicates?

29.44 How is carborundum produced? Write the equation for the reaction.

29.45 Discuss the chemical reactivity of silicon dioxide with strong aqueous acids, with strong bases, and with hydrofluoric acid. Write equations illustrating any reactions.

29.46 What is a zeolite? Name some of the properties and uses of this class of compounds.

29.47 Describe the processes involved in making ordinary glass from Na_2CO_3, $CaCO_3$, and SiO_2. Write chemical equations for any reactions that take place.

29.48 The lead content of a decorative glass was determined by allowing it to react with excess HF. The resulting suspension was evaporated to remove all of the silicon as $SiF_4(g)$. The fluoride residue was then dissolved in $HNO_3(aq)$, and the lead content determined by precipitation titration with 0.0503 M K_2CrO_4 solution. Assuming that the molecular composition of the glass is $K_2O\cdot SiO_2\cdot PbO$, (a) write equations for all the chemical reactions involved and on the basis of a 1.0000 g sample of the glass calculate (b) the volume of $SiF_4(g)$ liberated at 25 °C and 740 Torr and (c) the volume of K_2CrO_4 solution used in the titration.

29.49 How are clays formed? What is a common impurity in most clays? Briefly discuss the production of a ceramic object and subsequent glazing.

29.50 Briefly discuss the chemistry of cement.

29.51 Many times concrete is poured during the winter even though the air temperature is expected to drop a little below freezing (32 °F) that night. The workers simply cover the fresh concrete with straw and there is no problem with it freezing. Why?

29.52 What are silicones? How do they differ from silicates?

Bonding in Metals and Semiconductors

29.53 What term is used for a continuous energy level produced by a large number of closely spaced molecular orbitals? If electrons partially fill this energy level, what properties can be expected for the element?

29.54 How does the size of the energy gap between the valence band and the conduction band determine whether a substance is an insulator, a metal, or a semiconductor?

29.55 Compare the variation of electrical resistance with temperature of a metal to that of a semiconducting element.

29.56 Which elements are commonly used as donor impurities? Which elements are used as acceptor impurities?

29.57 Briefly describe how an *n*-type semiconductor device operates and how this differs from the operation of a *p*-type device.

29.58 Describe how the silicon from ordinary metallurgical reduction is treated to produce silicon suitable for semiconductor devices.

ADDITIONAL QUESTIONS

29.59 Which of the following are redox reactions? Identify the oxidizing and reducing agent in each of the redox reactions:

(a) $2CH_4(g) + 2NH_3(g) + 3O_2(g) \xrightarrow{800\,°C} 2HCN(g) + 6H_2O(g)$
(b) $AgCN(s) + CN^- \longrightarrow [Ag(CN)_2]^-$
(c) $4Au(s) + 8CN^- + O_2(g) + 2H_2O(l) \longrightarrow 4[Au(CN)_2]^- + 4OH^-$
(d) $Pd^{2+} + CO(g) + H_2O(l) \longrightarrow Pd(s) + CO_2(g) + 2H^+$
(e) $Si(s) + 4HNO_3(aq) + 6HF(aq) \longrightarrow H_2SiF_6(aq) + 4NO_2(g) + 4H_2O(l)$
(f) $3Na(s) + As(s) \xrightarrow{\Delta} Na_3As(s)$
(g) $B_2O_3(s) + 3H_2O(l) \longrightarrow 2H_3BO_3(s)$
(h) $SbCl_3(s) + H_2O(l) \longrightarrow SbOCl(s) + 2HCl(aq)$

29.60 Repeat Question 29.59 for

(a) $2KHCO_3(s) \xrightarrow{\Delta} K_2CO_3(s) + CO_2(g) + H_2O(g)$
(b) $CS_2(l) + 3Cl_2(g) \longrightarrow CCl_4(l) + S_2Cl_2(l)$
(c) $Fe_2O_3(s) + 3CO(g) \xrightarrow{\Delta} 2Fe(s) + 3CO_2(g)$
(d) $C(s) + 4HNO_3(aq, conc) \xrightarrow{\Delta} CO_2(g) + 4NO_2(g) + 2H_2O(l)$

(e) $BCl_3(l) + 3H_2O(l) \longrightarrow H_3BO_3(s) + 3HCl(aq)$

(f) $SiO_2(s) + 3C(s) \xrightarrow{\Delta} SiC(s) + 2CO(g)$

(g) $2B(s) + 6NaOH(l) \xrightarrow{\Delta} 3H_2(g) + 2Na_3BO_3(s)$

(h) $SiH_4(g) + 2O_2(g) \longrightarrow SiO_2(s) + 2H_2O(l)$

29.61 Classify each of the following reactions according to the reaction types listed in Tables 17.2 and 17.7:

(a) $CO(g) + Cl_2(g) \longrightarrow COCl_2(g)$

(b) $2CH_3OH(l) + 3O_2(g) \xrightarrow{\Delta} 2CO_2(g) + 4H_2O(g)$

(c) $Na_2CO_3(s) + SiO_2(s) \xrightarrow{\Delta} Na_2SiO_3(l) + CO_2(g)$

(d) $CdO(s) + C(s) \xrightarrow{\Delta} Cd(g) + CO(g)$

(e) $2BCl_3(g) + 3H_2(g) \xrightarrow{\Delta} 2B(s) + 6HCl(g)$

(f) $GeO_2(s) + 4HCl(aq) \xrightarrow{\Delta} GeCl_4(g) + 2H_2O(l)$

(g) $BaGeF_6(s) \xrightarrow[N_2]{\Delta} GeF_4(g) + BaF_2(s)$

(h) $Te(OH)_6(aq, conc) + 6HI(aq, conc) \longrightarrow TeI_4(s) + I_2(s) + 6H_2O(l)$

29.62 Repeat Question 29.61 for

(a) $CaCO_3(s) \xrightarrow{\Delta} CaO(s) + CO_2(g)$

(b) $MgCO_3(s) + H_2SO_4(aq) \longrightarrow MgSO_4(aq) + CO_2(g) + H_2O(l)$

(c) $CH_4(g) + 2O_2(g) \longrightarrow CO_2(g) + 2H_2O(g)$

(d) $HCO_3^- + OH^- \longrightarrow CO_3^{2-} + H_2O(l)$

(e) $B(s) + 3HNO_3(aq, conc) \longrightarrow H_3BO_3(s) + 3NO_2(g)$

(f) $SbCl_3(s) + HCl(aq) \longrightarrow H^+ + [SbCl_4]^-$

(g) $SeCl_4(s) + 3H_2O(l) \longrightarrow 4HCl(aq) + H_2SeO_3(s)$

(h) $Te(s) + 3F_2(g) \longrightarrow TeF_6(g)$

29.63 Predict the major products of the following reactions:

(a) $CaC_2(s) + H_2O(l) \longrightarrow$

(b) $Ag(s) + CN^- + O_2(g) + H_2O(l) \longrightarrow$

(c) $CH_4(g) + O_2(g) \xrightarrow{\Delta}$

(d) $C(s) + H_2SO_4(conc) \xrightarrow{\Delta}$

(e) $BBr_3(g) + H_2O(l) \longrightarrow$

(f) $BF_3(g) + NaF(aq) \longrightarrow$

(g) $Ge(s) + Cl_2(g) \longrightarrow$

(h) $SeO_2(s) + H_2O(l) \longrightarrow$

29.64 Repeat Question 29.63 for

(a) $Fe_3O_4(s) + CO(g) \xrightarrow{\Delta}$

(b) $CaCO_3(s) + H_2O(l) + CO_2(g) \longrightarrow$

(c) $CN^- + H_2O(l) \rightleftharpoons$

(d) $AlBr_3(s) + BF_3(g) \xrightarrow{\Delta}$

(e) $GeO_2(s) + H_2(g) \xrightarrow{\Delta}$

(f) $SiO_2(g) + NaOH(aq) \longrightarrow$

(g) $Na_2B_4O_7(s) + H_2O(l) \longrightarrow$

29.65 Write balanced equations for the following chemical reactions: (a) formation of $CaSiO_3(l)$ from $CaCO_3(s)$, (b) reaction of CO_3^{2-} with water, (c) formation of a metal carbonyl, (d) reaction of a solid carbonate with an aqueous acid, (e) formation of carbon monoxide from carbon dioxide.

29.66 Write balanced equations for the following chemical reactions:

(a) $H_3BO_3(s) \xrightarrow{\Delta}$

(b) $SiO_2(s) + HF(aq) \longrightarrow$

(c) $Na_2CO_3(s) + SiO_2(s) \xrightarrow{\Delta}$

(d) $B(s) + O_2(g) \xrightarrow{\Delta}$

(e) $AsCl_3(l) + H_2O(l) \longrightarrow$

29.67* In a student-designed qualitative analysis scheme, As^{3+} and Sb^{3+} are separated from other ions in the form of highly insoluble sulfide precipitates by adding $H_2S(aq)$ to the acidified solution. After filtering, these two elements are separated from each other by adding excess $HCl(aq)$ to the precipitate. The basis of this separation is that the Sb_2S_3 reacts with HCl to form the soluble $[SbCl_4]^-$ anion, whereas the As_2S_3 does not react with the HCl and remains as a precipitate. After filtering, the presence of As^{3+} in the precipitate is confirmed by forming AsO_4^{3-} by adding $H_2O_2(aq)$ in the presence of OH^- to the precipitate, and then forming the red precipitate Ag_3AsO_4 by adding Ag^+. The presence of Sb^{3+} is confirmed by adding NH_3 to remove enough H^+ to allow the orange sulfide Sb_2S_3 to precipitate again. Write chemical equations describing the above reactions.

29.68* A manufacturer wanted to prepare some semiconductor devices by using radioactive isotopes of Ge as dopants to form the Ga and As impurities upon decay. To one sample of ultrapure Ge, a sufficient amount of ^{69}Ge (a β^+ emitter) was added so that its concentration was 4 ppm. (a) Write the equation for the nuclear decay process and (b) identify the doping impurity formed by this radioisotope. (c) Will this sample of Ge be used for a *p*- or *n*-type semiconductor? To the second sample of ultrapure Ge, a sufficient amount of ^{77}Ge (a β^- emitter) was added so that its concentration was 2 ppm. (d) Write the equation for the nuclear decay process and (e) identify the doping impurity resulting from this radioisotope.

CHAPTER

30

d- and *f*-Block Elements

d-BLOCK ELEMENTS

In the preceding chapter we completed a survey of the chemistry of the representative elements. In this chapter we present an overview of the chemistry of the more than 50 remaining elements of the *d* and *f* blocks in the periodic table. The three series of *d*-block elements fall in the center of the periodic table in the fourth, fifth, and sixth periods, between the alkaline earth metals (Representative Group II) and the boron family elements (Representative Group III). Collectively they are often referred to as the *d*-transition elements or simply the transition elements.

The two series of *f*-block elements, usually placed at the bottom of the periodic table, follow lanthanum and actinium in atomic number. Collectively known as the *f*-transition elements or the inner transition elements, they include all the manmade, radioactive elements.

We first review the properties of the *d*-block elements and then look more closely at the chemistry of chromium, manganese, iron, cobalt, and nickel (from the first *d*-transition series) and of copper, silver and gold; and zinc, cadmium, and mercury (the elements of the last two *d*-block families). Finally, some notable features of the chemistry of the *f*-transition elements are presented.

30.1 PROPERTIES OF *d*-BLOCK ELEMENTS

The outermost electron configurations of the *d*-block elements are given in Table 30.1. Atoms of each (except palladium atoms) contain one or two *s* electrons in the outermost main energy level.

Because of the loosely held *s* electrons, all *d*-block elements are metals. Most are hard and strong and, with the exception of the scandium family elements, they are among the densest of elements (Figure 30.1). In the absence of surface coatings, all

Table 30.1
Outermost Electron Configurations of *d*-Block Elements
Deviations from the idealized configuration (marked with an *) occur when a different configuration is more stable.

Idealized configuration:	$(n-1)d^1 ns^2$	$(n-1)d^2 ns^2$	$(n-1)d^3 ns^2$	$(n-1)d^4 ns^2$	$(n-1)d^5 ns^2$	$(n-1)d^6 ns^2$	$(n-1)d^7 ns^2$	$(n-1)d^8 ns^2$	$(n-1)d^9 ns^2$	$(n-1)d^{10} ns^2$
First series:	$_{21}$Sc $3d^1 4s^2$	$_{22}$Ti $3d^2 4s^2$	$_{23}$V $3d^3 4s^2$	$_{24}$Cr* $3d^5 4s^1$	$_{25}$Mn $3d^5 4s^2$	$_{26}$Fe $3d^6 4s^2$	$_{27}$Co $3d^7 4s^2$	$_{28}$Ni $3d^8 4s^2$	$_{29}$Cu* $3d^{10} 4s^1$	$_{30}$Zn $3d^{10} s^2$
Second series:	$_{39}$Y $4d^1 5s^2$	$_{40}$Zr $4d^2 5s^2$	$_{41}$Nb* $4d^4 5s^1$	$_{42}$Mo* $4d^5 5s^1$	$_{43}$Tc $4d^5 5s^2$	$_{44}$Ru* $4d^7 5s^1$	$_{45}$Rh* $4d^8 5s^1$	$_{46}$Pd* $4d^{10}$	$_{47}$Ag* $4d^{10} 5s^1$	$_{48}$Cd $4d^{10} s^2$
Third series:	$_{57}$La $5d^1 6s^2$	$_{72}$Hf $4f^{14} 5d^2 6s^2$	$_{73}$Ta $4f^{14} 5d^3 6s^2$	$_{74}$W $4f^{14} 5d^4 6s^2$	$_{75}$Re $4f^{14} 5d^5 6s^2$	$_{76}$Os $4f^{14} 5d^6 6s^2$	$_{77}$Ir $4f^{14} 5d^7 6s^2$	$_{78}$Pt* $4f^{14} 5d^9 6s^1$	$_{79}$Au* $4f^{14} 5d^{10} 6s^1$	$_{80}$Hg $4f^{14} 5d^{10} 6s^2$

4f-transition elements intervene

d-block elements have a metallic luster. Most have high melting and boiling points and are good conductors of heat and electricity.

Except for those elements near the ends of the series, atoms of each *d*-block element also have an incompletely filled $(n-1)d$ energy level that is close to the *ns* energy level. Electrons in both the $(n-1)d$ and *ns* levels are, therefore, close to the outer regions of the electron clouds and are available to interact with their surroundings. The following characteristic properties are associated with "transition elements" and their compounds. These properties all result from the ready availability of the *d* electrons.

Paramagnetism The $(n-1)d$ subshell can contain up to five unpaired electrons, and many *d*-transition element compounds retain unpaired electrons.

Color Color in transition metal compounds usually (though not always) can be attributed to the presence of unpaired electrons. Unpaired electrons are so readily promoted from the *d* level to higher energy levels that absorption of visible light can provide enough energy to cause this change. When white light falls on the compound, either in the solid state or in solution, certain frequencies are absorbed, and we see a color corresponding to the frequency of the unabsorbed light. The absorbed energy is dispersed as heat, not reemitted as light. (See Section 31.11 for further discussion of color in transition metal compounds.)

Catalytic activity Ordinarily, catalytic activity is due to the ease with which electrons are lost and gained or moved from one energy level to another. Some typical applications of this property are the use of nickel as a hydrogenation

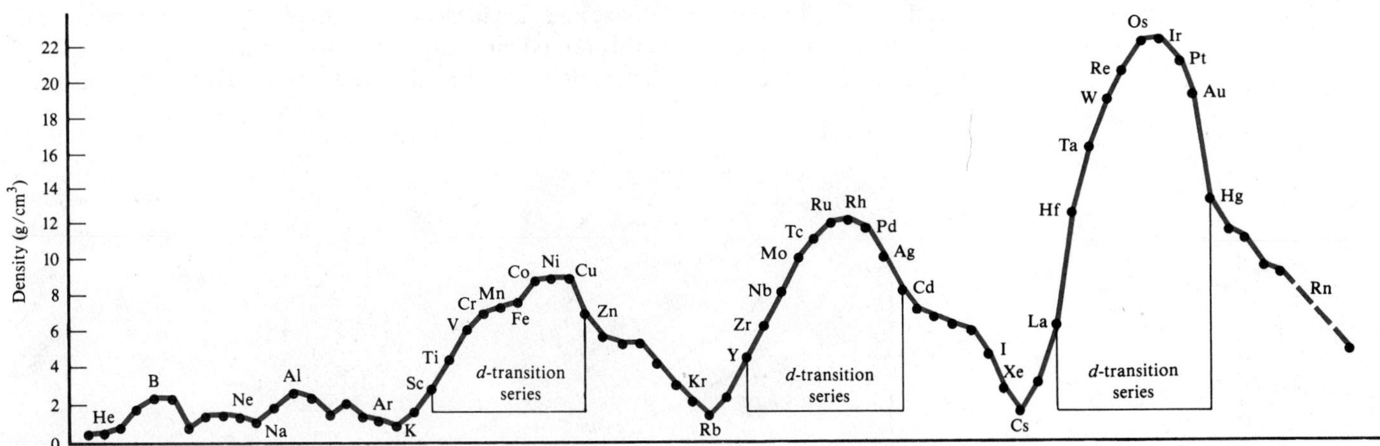

Figure 30.1
Densities of the Elements
The *f*-transition elements are not included.

Table 30.2
Some Hydrated 3*d*-Transition Metal Cations

$[Ti(H_2O)_6]^{3+}$[a]	violet
$[V(H_2O)_6]^{3+}$[a]	green
$[Cr(H_2O)_6]^{3+}$	violet
$[Mn(H_2O)_6]^{2+}$	pale pink
$[Fe(H_2O)_6]^{2+}$	pale green
$[Co(H_2O)_6]^{2+}$	pink
$[Ni(H_2O)_6]^{2+}$	green
$[Cu(H_2O)_4]^{2+}$	blue

[a] Easily oxidized.

catalyst (Section 16.6f) and the use of vanadium(V) oxide (V_2O_5) as a catalyst in the contact process for the manufacture of sulfuric acid (Section 28.6a). Many industrial processes for the production of organic compounds from hydrocarbons, particularly from unsaturated hydrocarbons, employ catalysts containing cobalt, platinum, palladium, rhodium, or titanium. The catalytic activities of the *d*-transition metals and their compounds give them great importance in the chemical industry.

Complex ion formation   The formation of complex ions (in which metal cations share electron pairs donated by ligands) plays a very important role in the chemistry of the *d*-transition metals (especially for Cr, Fe, Co, Ni; Cu family; Pt metals). For example, in aqueous solutions ions of the first-series elements (other than scandium and zinc) form hydrates that impart characteristic colors to their solutions (Table 30.2). Numerous examples of complex ion formation are mentioned in this chapter. The next chapter is devoted to coordination chemistry—the chemistry of species that contain ligands coordinated to metal atoms or ions.

Multiple oxidation states   Because varying numbers of *d* electrons can participate in bonding, many *d*-block elements are found in several different oxidation states.

The elements at either end of each series differ significantly from the other *d*-block elements. The scandium family elements (scandium, yttrium, and lanthanum) strongly resemble the alkaline earth metals that precede them—they are reactive metals that readily form ionic compounds. The zinc family elements (zinc, cadmium, and mercury) are more like representative elements in that they have fully occupied $(n-1)d$ and ns subshells, and therefore lack many of the characteristic properties due to the partially filled *d* subshells. For this reason, the elements of the zinc family are sometimes not classified as transition elements.

As we saw in our study of periodic relationships (Chapter 9), the transition metals vary less in radius, ionization energy, and electronegativity than do the representative elements. Comparison of the atomic radii of the three *d*-transition series shows a gradual decrease in size toward slight minima about halfway across each period (Figure 30.2).

In atoms of the heavier elements at the right, the increasing number of *d* electrons slightly outweighs the increasing effective nuclear charge, causing a small increase in radius. As is strikingly shown in Figure 30.2, the second and third series elements have virtually identical atomic radii, the result of the intervention of the lanthanide elements (Section 9.6). This similarity in size and also in electron configuration causes the second and third members of each family (and their compounds) to be quite similar, especially in the families at the left. The elements of the third series (La—Hg), however, have

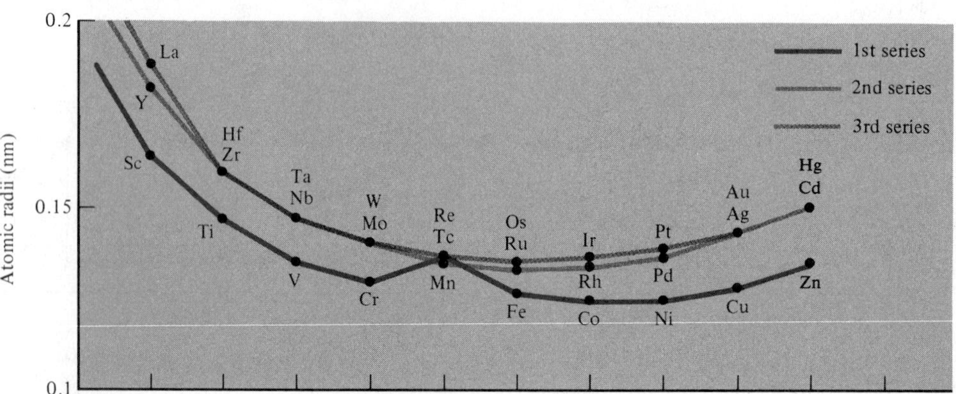

Figure 30.2
Radii of *d*-Block Metal Atoms

generally higher ionization energies than the corresponding second series elements (Y—Cd) due to their similar size but greater effective nuclear charge.

The loss of electrons in aqueous solution is also generally more difficult for elements at the right than at the left in the series. As shown by their standard reduction potentials (Table 30.3), metals at the left in the first series are better reducing agents than those at the right. Zinc, however, which we have noted is not a true transition metal, is a better reducing agent than copper.

Table 30.3
Standard Reduction Potentials of First *d*-Transition Series Metals
The more negative the E° value, the better the metal as a reducing agent.

	E° (volts)
$Sc^{3+} + 3e^- \longrightarrow Sc(s)$	−2.077
$Ti^{2+} + 2e^- \longrightarrow Ti(s)$	−1.628
$V^{2+} + 2e^- \longrightarrow V(s)$	−1.175
$Cr^{2+} + 2e^- \longrightarrow Cr(s)$	−0.913
$Mn^{2+} + 2e^- \longrightarrow Mn(s)$	−1.185
$Fe^{2+} + 2e^- \longrightarrow Fe(s)$	−0.4402
$Co^{2+} + 2e^- \longrightarrow Co(s)$	−0.277
$Ni^{2+} + 2e^- \longrightarrow Ni(s)$	−0.250
$Cu^{2+} + 2e^- \longrightarrow Cu(s)$	+0.337
$Zn^{2+} + 2e^- \longrightarrow Zn(s)$	−0.7628

30.2 OXIDATION STATES AND BONDING OF *d*-BLOCK ELEMENTS

Scandium, yttrium, and lanthanum form +3 ions of noble gas configuration in most compounds and therefore resemble the *s*-block elements. Zinc, cadmium, and mercury have the +2 oxidation state in most compounds, except that mercury also has the +1 state in Hg_2^{2+} ions.

The number of known oxidation states for the remaining *d*-block elements increases with the number of unpaired *d* electrons, reaching maxima in the manganese and iron families. Manganese has the largest number of known oxidation states (+2 to +7) of all the *d*-block elements.

For the scandium family through the manganese family, the *maximum* oxidation state equals the sum of the number of 3*d* and 4*s* electrons, for example

Sc	Ti	V	Cr	Mn
$3d^1 4s^2$	$3d^2 4s^2$	$3d^3 4s^2$	$3d^5 4s^1$	$3d^5 4s^2$
+3	*+4*	*+5*	*+6*	*+7*

Beyond the manganese family there is no correlation between maximum oxidation state and electron configuration.

The most common oxidation states of the *d*-block elements are listed in Table 30.4. For members of the first three families, the maximum oxidation states of +3, +4, and +5 are the most stable. Beyond these families, states lower than the maximum are more stable, reflecting the increasing nuclear charge, which results in the *d* electrons being less readily available for bonding. Starting with the manganese family, the +2 state is common. For chromium, manganese, and iron the highest known oxidation states are the *least* stable. Therefore, the dichromate ion ($Cr_2O_7^{2-}$ in acidic solution), the permanganate ion (MnO_4^-), and ferrate ion (FeO_4^{2-}) are all strong oxidizing agents.

Table 30.4
Common Oxidation States of the *d*-Block Elements
For the first five elements in each series, the maximum oxidation state equals the sum of the $(n-1)d$ and ns electrons. The most frequently encountered states for each element are shown in color.

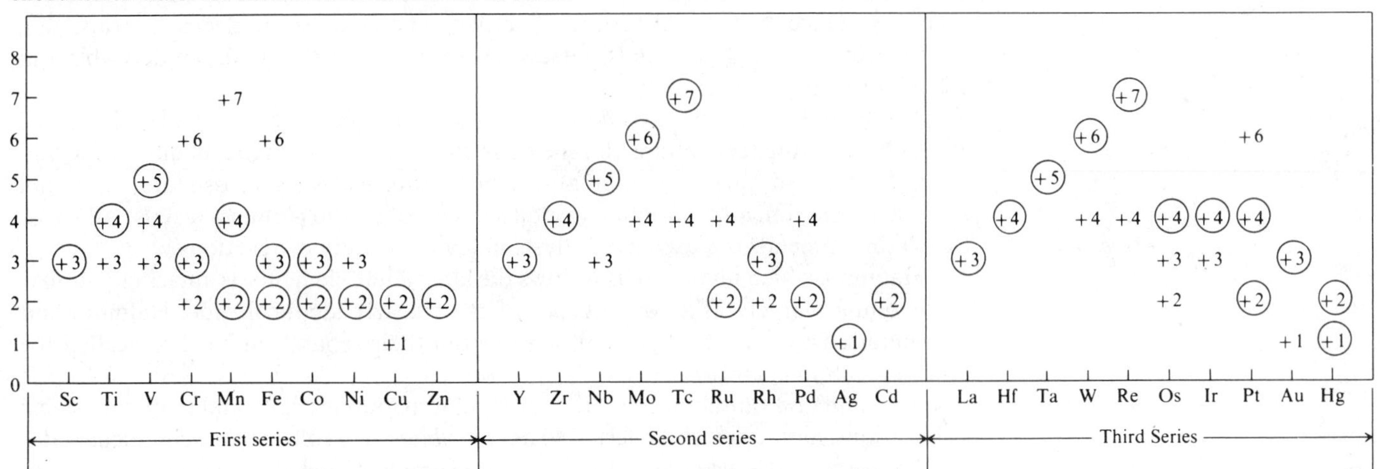

Many binary ionic compounds containing the +2 ions of the first series elements from manganese to zinc are known. In general, binary compounds containing *d*-block elements in the +3 and higher states (except for scandium family elements) are molecular compounds—the chlorides undergo hydrolysis and the oxides are acidic rather than basic. Even in +2 oxidation states, many second and third series elements form binary compounds that are not simple ionic compounds.

30.3 METALS OF THE 3*d*-TRANSITION SERIES

The 3*d*-series metals titanium, vanadium, chromium, manganese, iron, cobalt, and nickel have a great deal in common economically and industrially, as well as chemically. All of them are used in alloys in which strength, hardness, and resistance to corrosion are important. Iron, cobalt, nickel, and also gadolinium (a 4*f*-transition metal) are the only elements that are **ferromagnetic,** that is, they can exhibit magnetism in the absence of an external magnetic field. In a ferromagnetic material, large numbers of paramagnetic ions are grouped into "domains" in which their magnetic moments are aligned in the same direction. The domains, once the material has been "magnetized" by exposure to a strong magnetic field, also have their magnetic moments permanently aligned in the same direction.

Titanium is ninth in abundance of the elements in the Earth's crust and was found in a high percentage (12%) in rocks brought back from the moon by Apollo 11. The pure metal possesses excellent structural characteristics, an unusual resistance to corrosion under ordinary conditions, and a relatively low density. Most titanium is used in military and aerospace applications—there is no other metal that is a satisfactory substitute for titanium in these applications. Titanium is also widely used as titanium carbide in cutting tools. A form of titanium dioxide called rutile is the most important titanium compound; it is a bright white substance used as a pigment in paint, paper, and many other products.

The outstanding use of vanadium metal is in high-strength steels, in which small percentages of vanadium impart desirable mechanical properties. Many vanadium compounds are valuable as catalysts.

The chemistry and uses of chromium, manganese, iron, cobalt, nickel, copper, and zinc are discussed in later sections of this chapter.

30.4 METALS OF THE 4*d*- AND 5*d*-TRANSITION SERIES

Typically metallic in appearance, the second and third series metals range from the bluish gray tantalum, through platinum (a beautiful, silvery white metal that takes a high polish), to the familiar silver and gold. Whether it be hardness, corrosion resistance, resistance to high temperatures, activity as a catalyst, or electrical properties, almost every one of these metals possesses some property that makes it desirable and useful in practical applications.

Horizontal similarities exist in these series, as they do in the first series. The three elements following technetium and the three following rhenium are all quite alike and occur together in various combinations in nature. Ruthenium, osmium, rhodium, iridium, palladium, and platinum are collectively called the **platinum metals.** All six of the platinum metals are used as catalysts in a wide variety of reactions.

Hafnium, which immediately follows the lanthanide elements, is amazingly similar to zirconium, largely as a consequence of their nearly identical radii. Hafnium and zirconium always occur together in nature, and their separation has been called the most difficult in chemistry.

Niobium and tantalum also are very similar. (Commercially and in metallurgical publications, niobium is still referred to as *columbium.* This older name commemorates the discovery of the element in England, in a sample of ore that had been sent from

America.) In moving to the right across the periodic table, the second and third series elements in each family gradually become more different from each other.

Because of their great resistance to high temperatures, zirconium and zirconium oxide, along with alloys of niobium and molybdenum, are used in space vehicles that must reenter the atmosphere. Niobium and molybdenum are also important in steelmaking. Molybdenum(IV) sulfide is used as a lubricant, for example, in wheel bearings and gear oils. Tantalum is very resistant to body fluids and is used as a bone replacement; a tantalum plate can be put into the skull to replace a piece of bone that has been crushed or surgically removed.

Technetium ($Z = 43$) and promethium ($Z = 61$), a lanthanide, are unique in being the only elements with atomic numbers less than 83 that are known *only* as radioactive isotopes. Technetium, which can be recovered in reasonably large amounts from the spent uranium fuel of nuclear reactors, is used widely in diagnostic medicine. Photographic images of various body parts are rendered by a camera that detects γ radiation from injected technetium-99m.

30.5 SOME CHEMICAL REACTIONS OF *d*-BLOCK ELEMENTS

a. Combination with Elements The reactivity of the elemental transition metals in combination reactions is affected by the formation of protective coatings on the metal surfaces. In general, the combination reactions that do occur require high temperatures. For example, at elevated temperatures, manganese combines with fluorine, chlorine, nitrogen, and oxygen. As for all metals, reactions are frequently accelerated if the metal is finely divided. (A very fine form of iron, known as pyrophoric iron, spontaneously bursts into flame when exposed to air.) Many transition metals combine directly with nitrogen, carbon, and boron to form compounds that are in some cases true stoichiometric compounds and in other cases are solid solutions or interstitial compounds (Section 26.8).

b. Reactions with Acids and Bases The reactivity of many transition metals with acids is also inhibited by the formation of protective coatings. Concentrated nitric acid renders iron, cobalt, and nickel passive to further chemical attack by oxidizing the surface of the metals. Some transition metals react with nonoxidizing acids (e.g., Cr and Mn), others with oxidizing acids only, and some with concentrated aqueous alkaline solutions (e.g., Cr). With nonoxidizing acids, those that react give $+2$ ions. With oxidizing acids, some give $+2$ ions (e.g., Cu, Zn) and some give compounds with the metal in the $+3$ oxidation state. The platinum metals (Ru, Os, Rh, Ir, Pd, and Pt) are all quite resistant to reactions with acids and aqueous bases. Platinum, gold, and ruthenium are attacked only by aqua regia (a mixture of hydrochloric and nitric acids). Some of the least reactive *d*-transition metals can be oxidized in the presence of molten alkali metal hydroxides (e.g., Ru and Os). Zinc, showing its greater resemblance to the *p*-block metals than the transition metals, is amphoteric and reacts both with acids and aqueous solutions of strong bases (see Table 26.7, reactions a and b). Chromium reacts with dilute nonoxidizing acids and hot concentrated alkaline solutions.

c. Ions in Aqueous Solution The most significant aqueous chemistry of the transition metal ions involves the formation of complex ions (to which the next chapter is devoted). The redox reactions of the oxoanions of manganese and chromium are of importance and are discussed in the sections on these metals.

To varying degrees the hydrated cations of transition metals react with water to form H^+ and, in the absence of other ions that influence the pH, give acidic solutions.

A characteristic sequence of reactions for many transition metal ions is the precipitation of a salt followed by its dissolution to give a complex ion in aqueous solution. For example, the addition of aqueous ammonia to a solution containing nickel(II) ion gives an apple green precipitate of nickel(II) hydroxide, which dissolves

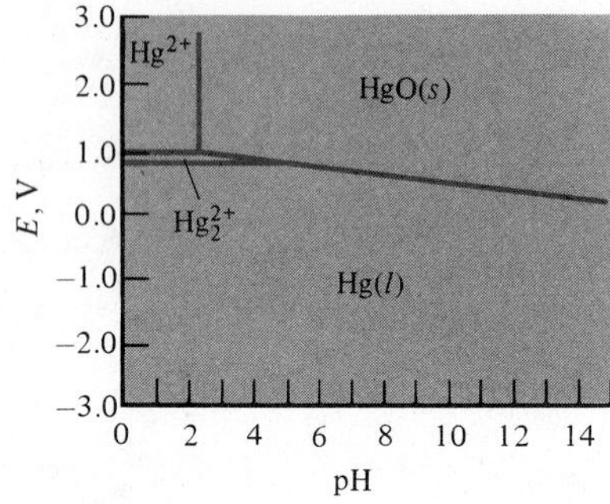

Figure 30.3
pH–Potential Diagram for Mercury

upon the addition of excess aqueous ammonia to yield the deep blue hexaamminenickel(II) ion, $[Ni(NH_3)_6]^{2+}$.

$$Ni^{2+} + 2OH^- \rightleftharpoons Ni(OH)_2(s)$$
$$Ni(OH)_2(s) + 6NH_3(aq) \rightleftharpoons [Ni(NH_3)_6]^{2+} + 2OH^-$$

In aqueous solution, mercury(I) ions establish a disproportionation equilibrium

$$Hg_2^{2+} \rightleftharpoons Hg^{2+} + Hg(l)$$

which is easily shifted in either direction according to Le Chatelier's principle. Mercury(I) disproportionates in the presence of any reagent that gives a slightly soluble or slightly dissociated mercury(II) compound, for example

$$Hg_2^{2+} + 2OH^- \longrightarrow Hg(l) + HgO(s) + H_2O(l)$$
$$Hg_2^{2+} + 4CN^- \longrightarrow Hg(l) + [Hg(CN)_4]^{2-}$$
$$Hg_2^{2+} + H_2S(aq) \longrightarrow Hg(l) + HgS(s) + 2H^+$$

The small region of stability of Hg_2^{2+} is shown in the pH–potential diagram for mercury (Figure 30.3).

SUMMARY OF SECTIONS 30.1–30.5

The outstanding properties of the *d*-transition elements and their compounds are summarized in Table 30.5.

Table 30.5
Oustanding Properties of *d*-Block Elements and Their Compounds

$ns^{1,2}$	One or two *s* electrons (except Pd)
$(n-1)d$	Incomplete *d* subshells and availability of *d* electrons responsible for characteristic properties: paramagnetism, colored and catalytically active compounds, complex ion formation, multiple oxidation states
Ionization energy, radii, electronegativity	Vary less across periods than for representative elements
Sc, Y, La	More like alkaline earth metals; +3 ions in compounds
Zn, Cd, Hg	More like representative metals
First series	Ionization energy decreases across series; metals at left better reducing agents; all industrially important
Second and third series	Very similar in size and other properties (lanthanide contraction); similarities decrease across series
Sc—Mn families	Maximum oxidation state = sum of *ns* and $(n-1)d$ electrons
Cr, Mo, Mn, Tc, Re	Stable $(n-1)d^5ns^{1,2}$ configurations; maximum number of unpaired electrons; largest number of oxidation states
+6 +7 +6 Cr, Mn, Fe	Strong oxidizing agents (e.g., in $Cr_2O_7^{2-}$, MnO_4^-, FeO_4^{2-})
Elemental metals	Typical metals (hard, lustrous, electrical conductors); some protected by surface coatings; combination reactions at higher temperatures
Fe, Co, Ni	Ferromagnetic
M + acid	Nonoxidizing $\longrightarrow$ +2 ions; oxidizing $\longrightarrow$ +2 or +3 ions; some passive; Pt metals nonreactive; Zn, amphoteric
$M(A)_n$	Molecular compounds in higher oxidation states
$[M^{2+,3+}(H_2O)_n]$	Ions hydrated in solution; many are colored; many crystalline hydrates
$[M(ligand)_n]^{x+,0,x-}$	Complex ions; especially common for Cr, Fe, Co, Ni; Cu family; Pt metals

SELECTED *d*-BLOCK ELEMENTS

30.6 CHROMIUM

a. Elemental Chromium Chromium is a bluish white, hard, and brittle metal that does not tarnish, is very resistant to corrosion, and takes a high polish. Its principal uses

(Table 30.6) are in alloys, particularly steel, and in *chrome plating*—the electrolytic deposition of a layer of chromium on another metal. Chrome plating produces a shiny, attractive surface with high resistance to wear and corrosion, commonly seen on automobile bumpers and trim. Because the chromium layer itself always has pinholes that would allow corrosion to occur, an object is first plated with copper, which provides an adherent surface, then with nickel, which imparts corrosion resistance, and then with the chromium. The chromium surface is protected by a thin layer of oxide.

The chief ore of chromium is chromite, $FeCr_2O_4$. Each iron atom is surrounded tetrahedrally by four oxygen atoms, and each chromium atom is surrounded octahedrally by six oxygen atoms. Several naturally occurring minerals have this same structure, which is indicative of its great stability. Such minerals are known as spinels—they are all represented by the general formula $M^{2+}(M^{3+})_2O_4$.

For the manufacture of stainless steel, which is mainly iron, chromium, and nickel, separation of the metals in the chromite ore is not necessary. An alloy of iron and chromium called ferrochrome is produced from chromite by reduction with carbon or silicon in an electric furnace and is used directly in steelmaking.

Chromium metal is obtained by reduction of the oxide with carbon, silicon, or aluminum as follows:

$$2Al(s) + Cr_2O_3(s) \longrightarrow 2Cr(l) + Al_2O_3(l) \qquad \Delta H^\circ = -527 \text{ kJ}$$

The chromium melts and is separated from the molten aluminum oxide by the difference in their densities.

Table 30.6
Uses of Chromium and Chromium Compounds

Chromium
Steel
Other alloys
Chrome plating
Nickel chromium heating elements
Chromium(VI) oxide(CrO_3)
Chrome plating baths
Other metal treatment
Oxidizing agents, pigments
Sodium chromate(Na_2CrO_4) and sodium dichromate($Na_2Cr_2O_7$) and other chromates
Pigments
Leather tanning
Corrosion inhibitors
Aluminum anodizing
Pigment manufacture
Chromite($FeCr_2O_4$)
Refractories

b. Chromium Oxidation States and Equilibria Chromium is known in all positive oxidation states from +2 to +6. Its common oxidation states are +2, +3, and +6. As shown in the pH–potential diagram (Figure 30.4), these are the thermodynamically stable states. The +3 state is the most common and most stable. Chromium(II) compounds are reducing agents (shown by their appearance in the negative potential region on the pH–potential diagram) and chromium(VI) compounds are oxidizing agents (shown by their appearance in the positive region on the diagram). Dichromates in acidic solution are common oxidizing agents, being reduced to Cr^{3+}, for example

$$Cr_2O_7^{2-} + 6Br^- + 14H^+ \longrightarrow 2Cr^{3+} + 3Br_2(aq) + 7H_2O(l)$$

$$\underset{\text{acetaldehyde}}{Cr_2O_7^{2-} + 3CH_3CHO(aq)} + 8H^+ \longrightarrow 2Cr^{3+} + \underset{\text{acetic acid}}{3CH_3COOH(aq)} + 4H_2O(l)$$

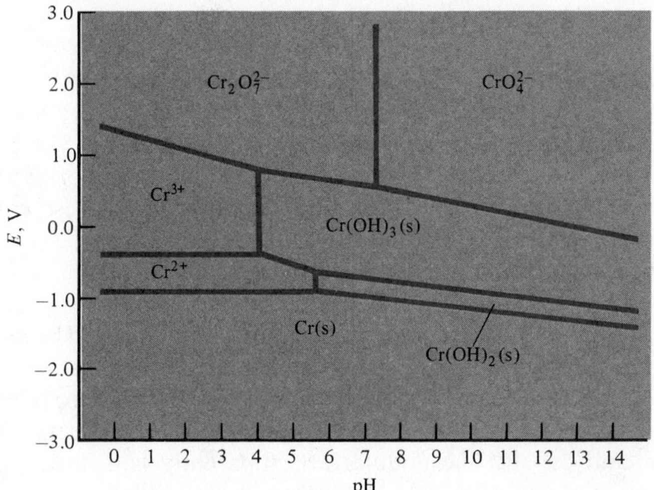

Figure 30.4
pH–Potential Diagram for Chromium

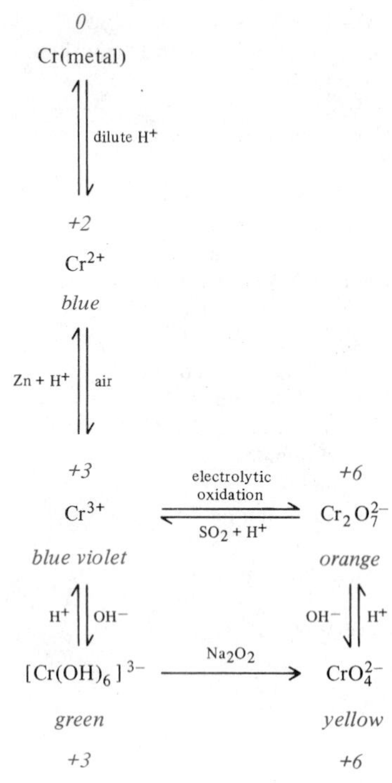

Figure 30.5
Oxidation and Reduction of Chromium

The conditions necessary for some of the reactions of chromium and its ions are shown in Figure 30.5.

Chromate ion ($CrO_4{}^{2-}$) and dichromate ion ($Cr_2O_7{}^{2-}$) exist in solution in an equilibrium that, as clearly shown in the pH–potential diagram, is pH-dependent.

$$\underset{\substack{\text{chromate}\\\text{ion}\\\text{(yellow)}}}{H^+ + CrO_4{}^{2-}} \rightleftharpoons \underset{\substack{\text{hydrogen}\\\text{chromate ion}}}{HCrO_4{}^-} \qquad 2HCrO_4{}^- \qquad \underset{\substack{\text{dichromate}\\\text{ion}\\\text{(deep orange)}}}{Cr_2O_7{}^{2-}} + H_2O(l) \quad \textbf{(30.1)}$$

c. Compounds of Chromium Many chromium compounds are colored and are used as pigments. Chromium(III) oxide (Cr_2O_3) is a green, stable, amphoteric oxide. It is used as a pigment in glass and porcelain, in fabrics, and in printing.

The crystalline chromium(III) chloride of commerce is commonly labeled $CrCl_3 \cdot 6H_2O$. It is either $[Cr(H_2O)_5Cl]Cl_2 \cdot H_2O$ (blue-green) or $[Cr(H_2O)_4Cl_2]Cl \cdot 2H_2O$ (green) or a mixture of the two, depending upon the method of manufacture. In each case, six groups (those shown inside the square brackets) have formed coordinate covalent bonds with the chromium ion. The two compounds can be distinguished from one another by their different shades of green and more certainly by the fact that when solutions of the two are treated with silver ion, one of them immediately precipitates one-third of its chlorine, and the other, two-thirds of its chlorine. The coordinated water molecules are bound tightly to the chromium ion, but the others are not bound at all; they evidently fit into holes in the crystal lattice in a stoichiometric way without being attached. The violet chloride $[Cr(H_2O)_6]Cl_3$ is also well known, but it is more difficult to prepare in crystalline form than either of the green chlorides. It is precipitated as fine crystals when an ice-cold, saturated solution of either of the green forms is treated with hydrogen chloride gas.

Chromium(IV) oxide (CrO_2) is both an electrical conductor and a ferromagnetic material. It is used as the magnetic medium in some high-quality recording tapes.

Chromium(VI) oxide (CrO_3), or chromium trioxide, is a red crystalline compound that is a strong oxidizing agent and is used in chrome plating. It is made commercially from sodium dichromate.

$$\underset{\substack{\text{sodium}\\\text{dichromate}}}{Na_2Cr_2O_7(aq)} + H_2SO_4(conc) \longrightarrow \underset{\substack{\text{chromium(VI)}\\\text{oxide}}}{2CrO_3(s)} + H_2O(l) + Na_2SO_4(aq) \quad \textbf{(30.2)}$$

Often called "chromic acid," chromium(VI) oxide gives an acidic aqueous solution.

Chromium(VI) compounds are highly toxic and are irritants to the skin and respiratory tract. Nevertheless, chromium is an essential trace element in mammals, where it plays a role in the breakdown of glucose in the blood.

30.7 MANGANESE

a. Elemental Manganese Manganese is a silvery, brittle metal with a slightly pink appearance. It is more reactive than its periodic table neighbors and has several allotropic forms that vary in brittleness and ductility. Manganese compounds are much less toxic than those of chromium. Manganese is one of the essential trace elements for both plants and animals, and manganese(II) sulfate is added to some fertilizers.

Manganese compounds are widely distributed on Earth. The primary ore is pyrolusite ($MnO_2 \cdot xH_2O$). An intriguing possible source of manganese is nodules that have been found on the ocean floor. These nodules, 1–15 cm across, contain a high percentage of manganese, together with compounds of other metals, such as iron, nickel, copper, and cobalt. The nodules are apparently laid down very slowly in concentric layers, perhaps at the rate of 1–100 mm per million years.

Manganese imparts hardness and strength to steel and is essential to the production of almost all types of steel. It is also used in other alloys, such as manganese bronze (copper and manganese) and a nonconducting alloy (with nickel and copper) called *manganin* (Table 30.7). For steelmaking, production of pure manganese is not necessary. Instead, mixed iron and manganese oxides are reduced by carbon to give ferromanganese, an iron–manganese carbide (25–30% Mn), which is used directly.

Table 30.7
Uses of Manganese and Its Compounds

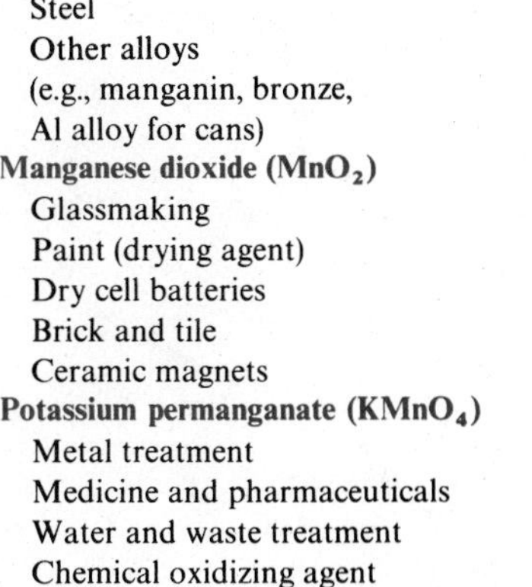

Manganese
- Steel
- Other alloys (e.g., manganin, bronze, Al alloy for cans)

Manganese dioxide (MnO_2)
- Glassmaking
- Paint (drying agent)
- Dry cell batteries
- Brick and tile
- Ceramic magnets

Potassium permanganate ($KMnO_4$)
- Metal treatment
- Medicine and pharmaceuticals
- Water and waste treatment
- Chemical oxidizing agent

b. Manganese Oxidation States and Equilibria Compounds are known containing manganese in all oxidation states from +2 to +7. The +2, +4, and +7 states are the most common ones, the +4 state mainly in MnO_2 (Table 30.8). Manganese(III) is stable only in complex ions or in compounds of small solubility.

Element and oxidation state	Compound	Comments
Cr(II)	$Cr(CH_3COO)_2 \cdot H_2O$ chromium(II) acetate monohydrate	Dark red; a slightly soluble acetate
Cr(III)	Cr_2O_3 chromium(III) oxide	Green; a pigment
Cr(IV)	CrO_2 chromium(IV) oxide	Brown-black; ferromagnetic
Cr(VI)	CrO_3 chromium(VI) oxide	"Chromic acid"; red; oxidizing agent
	$PbCrO_4$ lead chromate	Bright yellow; a pigment
	K_2CrO_4 potassium chromate	Yellow
	$K_2Cr_2O_7$ potassium dichromate	Orange; oxidizing agent
Mn(II)	$MnCl_2$ manganese(II) chloride	Pink; soluble; oxidation catalyst, like most Mn(II) salts
Mn(IV)	MnO_2 manganese(IV) oxide	Black; nonstoichiometric; one of few stable Mn(IV) compounds
Mn(VI)	K_2MnO_4 potassium manganate(VI)	Deep green
Mn(VII)	$KMnO_4$ potassium manganate(VII)	Purple; oxidizing agent (potassium permanganate)

Table 30.8
Some Compounds of Chromium and Manganese

Manganese is most stable in its +2 oxidation state (Figure 30.6). The pale pink manganese(II) salts are formed when manganese in any higher oxidation state is reduced in acidic solution. Manganese in its higher oxidation states is best known as the dark green manganate(VI) ion (MnO_4^{2-}) and the deep purple permanganate, or manganate(VII) ion (MnO_4^-), which are both strong oxidizing agents.

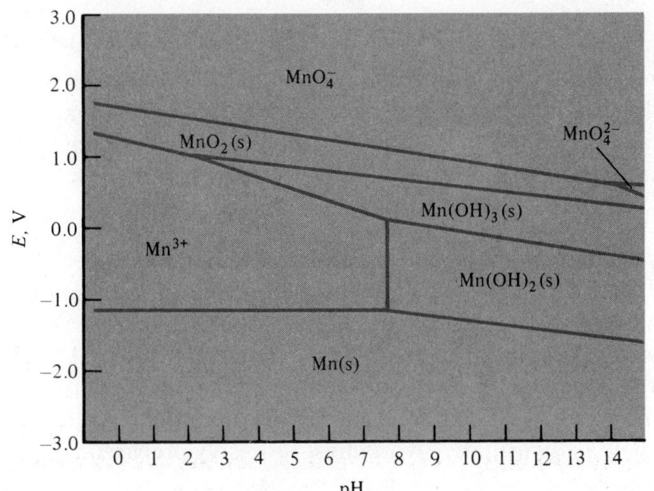

Figure 30.6
pH–Potential Diagram for Manganese

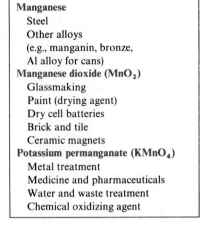

When permanganate ion acts as an oxidizing agent in very strongly alkaline solution, the manganate(VI) ion (MnO_4^{2-}) is formed, for example

$$\underset{\substack{\text{permanganate}\\ \text{ion}\\ \text{(purple)}}}{2MnO_4^-} + OH^- + \underset{\substack{\text{formate}\\ \text{ion}}}{HCOO^-} \longrightarrow \underset{\substack{\text{manganate}\\ \text{ion}\\ \text{(dark green)}}}{2MnO_4^{2-}} + CO_2(g) + H_2O(l) \qquad \textbf{(30.3)}$$

As can be seen in the pH–potential diagram for manganese (Figure 30.6), the manganate(VI) ion is thermodynamically stable *only* in strongly alkaline solution. When a solution containing this ion is acidified, disproportionation immediately takes place to give manganese(IV) oxide and permanganate ion:

$$\underset{\text{dark green}}{3MnO_4^{2-}} + 4H^+ \longrightarrow \underset{\text{purple}}{2MnO_4^-} + \underset{\text{black}}{MnO_2(s)} + 2H_2O(l)$$

In less strongly alkaline solution or in neutral solution, the reduction product of permanganate ion is manganese(IV) oxide

$$2MnO_4^- + 3H_2S(g) \longrightarrow 2MnO_2(s) + 3S(s) + 2OH^- + 2H_2O(l) \qquad \textbf{(30.4)}$$

In acidic solutions, permanganate ion is reduced to Mn^{2+}, for example

$$2MnO_4^- + 5C_2O_4^{2-} + 16H^+ \longrightarrow 2Mn^{2+} + 10CO_2(g) + 8H_2O(l) \qquad \textbf{(30.5)}$$

(unless, in weakly acidic solution, Mn^{3+} is stabilized as a complex ion such as $[Mn(PO_4)_2]^{3+}$). The endpoint in a redox titration with permanganate ion can be recognized when the deep purple permanganate ion persists instead of instantly converting to the pale pink Mn^{2+} ion.

c. Compounds of Manganese Manganese(IV) dioxide (MnO_2) is a dark brown or black, very slightly soluble, powdery substance. Stoichiometric manganese dioxide is extremely rare, if it exists at all. Analysis of the material found in nature as well as of that prepared in the laboratory shows a Mn to O ratio of 1:1.85 (approximately), the exact ratio depending on the mode of preparation. (Compare with lead dioxide; Section 26.15.) X-ray analysis shows that this material has the crystal structure calculated for pure manganese dioxide, but that several percent of the oxygen atoms are missing, leaving holes in the crystal structure. Oxide ions from adjacent sites are able to move into these holes, leaving new holes; this process makes the oxide conductive. It is both the oxidizing power and conductivity of the manganese(IV) oxide that make it valuable in dry cells. The compound has many other uses, including in brick, tile, ceramic magnets (used in, e.g., television sets), and welding fluxes.

Potassium manganate(VI) (K_2MnO_4) is an intermediate in the preparation of potassium permanganate ($KMnO_4$). Potassium manganate(VI) is prepared by fusing a compound in which manganese is in a lower oxidation state with a basic material (e.g., KOH or K_2CO_3) and an oxidizing agent (e.g., KNO_3, $KClO_3$, or air):

$$MnO_2(s) + 2KOH(l) + KNO_3(l) \xrightarrow{\Delta} K_2MnO_4(s) + KNO_2(s) + H_2O(g)$$

Potassium permanganate ($KMnO_4$) crystallizes in dark purple, almost black, needles that dissolve in water to give a purple solution. Permanganates are prepared from manganate(VI) compounds by acidification in the presence of a strong oxidizing agent, for example

$$2MnO_4^{2-} + O_3(g) + 2H^+ \longrightarrow 2MnO_4^- + O_2(g) + H_2O(l) \qquad \textbf{(30.6)}$$

Even in acidic solution permanganate ion slowly decomposes, especially when exposed to light, producing a residue of MnO_2 in the bottom of its container. The uses of potassium permanganate in treating athlete's foot and rattlesnake bite and as an antidote for poisons depend upon its strong oxidizing ability. It is also widely employed as an oxidant in organic chemistry, in quantitative analysis, and in the treatment of polluted air and water by oxidation of impurities. Because of its instability, solutions that are to be used in quantitative work must be freshly prepared and standardized by titration with a pure, stable reducing agent, such as sodium oxalate ($Na_2C_2O_4$) (Equation 30.5).

Some important compounds of chromium and manganese are listed in Table 30.8.

EXAMPLE 30.1 Chemical Reactions: Ions of Cr and Mn

Predict the major products of the following reactions (which do take place). Do not be concerned with balancing the equations.

(a) $Cr^{3+} + H_2O_2(aq) \xrightarrow{OH^-}$

(b) $Cr_2O_7^{2-} + Fe^{2+} \xrightarrow{H^+}$

(c) $MnO_4^- + Cr^{3+} \xrightarrow{OH^-\text{ (mildly alkaline)}}$

(a) Hydrogen peroxide is an oxidizing agent and water is its reduction product (Table 17.6). Chromium has only two common oxidation states, +3 and +6. We can expect it to be oxidized to a +6 oxoanion. The question arises as to whether to write CrO_4^{2-} or $Cr_2O_7^{2-}$ as the product. Because the reaction takes place in alkaline solution, the product must be CrO_4^{2-} [see Equation 30.1].

$$Cr^{3+} + H_2O_2(aq) \xrightarrow[\text{not balanced}]{} CrO_4^{2-} + H_2O(l)$$

(b) Dichromate ion is a strong oxidizing agent and is reduced to Cr^{3+}. Iron has only two common oxidation states, +2 and +3, so it will probably be oxidized to iron(III)

$$Cr_2O_7^{2-} + Fe^{2+} \xrightarrow[\text{not balanced}]{H^+} Cr^{3+} + Fe^{3+}$$

(c) Like (a), this is a reaction of Cr^{3+} with an oxidizing agent. In this mildly alkaline solution permanganate ion will be reduced to manganese(IV) oxide and Cr^{3+} will form predominantly CrO_4^{2-}:

$$MnO_4^- + Cr^{3+} \xrightarrow[\text{not balanced}]{OH^-\text{ (mildly alkaline)}} MnO_2(s) + CrO_4^{2-}$$

30.8 IRON

In many respects, our industrial civilization is built upon iron. It is fourth in abundance of all the elements. The plentiful ores are readily reduced to the metal. Pure iron, which we rarely see, is silvery white and relatively soft. By metallurgical working, iron can be made extremely strong. It can also be made soft or hard, flexible or stiff, malleable or brittle, depending on the treatment to which it is subjected and the alloying materials. The sources and metallurgy of iron and the process of steelmaking are discussed in Section 26.7. The corrosion of iron is discussed in Section 24.18.

Iron is most stable and most common in the +2 (ferrous) and +3 (ferric) states (see Figure 24.8). The most common of the soluble iron(II) salts is iron(II) sulfate ($FeSO_4$), which crystallizes from water in the form of large, light green crystals, $FeSO_4 \cdot 7H_2O$. It is a by-product of the steel industry, for sheet steel that is to be plated or enameled is often cleaned of corrosion products by treatment with dilute sulfuric acid. Cleaning metal in an acid bath is called *pickling* (the leftover acid is called pickle liquor). In this process some of the metal is dissolved, along with the rust. The iron(II) sulfate that forms is recovered as the green hydrate, the iron compound with the largest number of commercial uses. It is the starting material for the preparation of most other iron compounds and is widely used as a reducing agent, as a disinfectant, in the dyeing industry, and in such products as weed killers and wood preservatives. On exposure to air, its solutions soon become turbid and deposit precipitates of hydrated iron(III) compounds.

There are three iron oxides—FeO, Fe_2O_3, and Fe_3O_4—all of which are found in nature and all of which tend to be nonstoichiometric. In these oxides the oxygen atoms form a lattice and iron atoms occupy the holes. The crystal structures and uses of the iron oxides are summarized in Table 30.9. Rust on iron is a hydrated iron(III) oxide of variable composition, $Fe_2O_3 \cdot (H_2O)_n$.

Iron pyrite (FeS_2) a mineral, is often called "fool's gold" because of its yellow color and bright metallic luster. Upon heating, pyrite (which contains S_2^{2-}) decomposes to iron(II) sulfide (FeS) and sulfur, and it is sometimes used as a source of elemental sulfur.

Iron(II) and iron(III) salts with the common anions are generally soluble. Unlike iron(II) chloride, which is a salt, iron(III) chloride is covalent, volatile, and somewhat soluble in nonpolar solvents, demonstrating the generality that transition metals in +3 and higher oxidation states are covalently bonded. Many iron salts form crystalline hydrates. In both the +2 and +3 states, iron forms many complex ions. The addition of cyanide ion to a solution of an iron(II) salt gives a gray, slimy precipitate of iron(II) cyanide, which dissolves when excess cyanide is added to form a clear yellow solution:

$$Fe^{2+} + 2CN^- \longrightarrow Fe(CN)_2(s)$$

$$Fe(CN)_2(s) + 4CN^- \longrightarrow \underset{\textit{hexacyanoferrate(II) ion}}{[Fe(CN)_6]^{4-}}$$

Potassium hexacyanoferrate(II), $K_4[Fe(CN)_6]\cdot 3H_2O$, is a common laboratory chemical (sometimes called potassium ferrocyanide). Oxidizing agents readily change potassium hexacyanoferrate(II) to potassium hexacyanoferrate(III), $K_3[Fe(CN)_6]$ (also called potassium ferricyanide).

When $K_4[Fe(CN)_6]$ reacts with an iron(III) salt in aqueous solution *or* when $K_3[Fe(CN)_6]$ reacts with an iron(II) salt, a precipitate of Prussian blue, $KFe[Fe(CN)_6]$, is formed. The absorption of light by Prussian blue, resulting in its intense blue color, is facilitated by the ease with which electrons can migrate from one iron atom to another. This compound follows the general rule that a solid that contains atoms of the same metal in two different oxidation states is deeply colored—a phenomenon known as *interaction absorption.*

The formation of Prussian blue is utilized in making blueprints. Paper is moistened in total darkness with a solution containing iron(III) ion, hexacyanoferrate(III) ion, and ammonium citrate, a mild reducing agent. The paper is then dried and is ready for use. As long as it is kept in darkness, no reaction occurs, but upon exposure to light

Table 30.9
Iron Oxides
Fe_2O_3 occurs in two different crystal structures. All three oxides have nonstoichiometric compositions. Rust is a hydrated iron(III) oxide, FeO(OH) or $Fe_2O_3 \cdot (H_2O)_n$.

Oxide	Natural mineral name	Crystal structure	Uses
FeO Iron(II) oxide (black)	Wüstite (an ore)	Sodium chloride structure, cubic close-packed O^{2-} anions with Fe^{2+} in all octahedral holes	Green glass, ceramics, catalysts
Fe_2O_3 Iron(III) oxide (brown)	Hematite, also limonite, ($2Fe_2O_3 \cdot 3H_2O$, the most common ores)	Two structures: (a) Cubic close-packed O^{2-} anions with Fe^{3+} distributed in octahedral and tetrahedral holes (b) Hexagonal close-packed with Fe^{3+} in $\frac{2}{3}$ of octahedral holes	Red paint pigment; Magnetic recording tape
Fe_3O_4 Iron(II, III) oxide	Magnetite (an ore)	Cubic close-packed O^{2-} anions with Fe^{2+} ions in octahedral holes and Fe^{3+} ions half in octahedral and half in tetrahedral holes; a spinel	Pigment in ceramics and glass; Magnetic recording tape; Corrosion-resistant coating on steel

the citrate reduces part of the iron(III) to iron(II), and when the paper is moistened, the insoluble Prussian blue is formed. Any parts of the paper protected from the light remain unaffected. The soluble compounds that have not reacted are then washed away.

30.9 COBALT AND NICKEL

a. Elemental Cobalt and Nickel Cobalt and nickel are less abundant than chromium, manganese, or iron. The principal nickel ores are sulfides or oxides. Very few ores are valuable for their cobalt content alone. Cobalt and its salts are obtained chiefly as by-products in the metallurgy of nickel.

Cobalt is a component of many valuable alloys. Among these are Alnico (Fe, Al, Ni, Co), which is highly magnetic, Stellite (Cr, Co, W), used in surgical instruments because it is hard and corrosion resistant, and Hastelloy B (chiefly Co and Ni), which is very strong and hard, even at high temperatures.

Cobalt is much less reactive than iron and dissolves in acids only slowly. It corrodes very slightly in ordinary air. Cobalt compounds are widely used as catalysts. For example, cobalt naphthenate catalyzes the drying of paint. It is deep blue, but is used in such small concentration that, even in a white paint, the color is not perceptible. A cobalt–molybdenum–alumina catalyst is used in removing sulfur from crude oil. Also, cobalt catalysts are used in the synthesis of hydrocarbons (see Table 32.9).

The importance of nickel in modern civilization can hardly be overestimated. It is widely used as a protective plate on iron, as a catalyst for hydrogenation and petroleum refining, and as a constituent of many valuable alloys, including steel (Table 30.10). The nickel steels are hard, tough, and corrosion resistant.

Table 30.10
Uses of Nickel, Cobalt, and Their Compounds

Uses
Nickel
Stainless steel
Other alloys, e.g.,
Nichrome (Ni, Cr)
Monel metal (Ni, Cu)
Modern coinage
Hydrogenation catalyst
Nickel plating
Undercoat in chrome plating
Magnets
Ni—Cd batteries
Nickel compounds
Catalysts
Electroplating baths
Magnets
Ceramics
Dyes
Cobalt
Alloys
Alnico, and other magnetic alloys
Stellite
Hastelloy B
For cutting tools (Co–Mo–W)
Superalloys for high temperatures
Catalyst in refining
Cobalt compounds
Catalysts
Ceramics
Paints and enamels
Trace nutrient in animal feed and fertilizer

b. Compounds of Cobalt and Nickel In most of their compounds, cobalt has the oxidation state of $+2$ or $+3$, and nickel has the oxidation state of $+2$. Both form many complex ions, and $+3$ cobalt is stable in aqueous solution *only* as part of a complex ion such as $[Co(NH_3)_6]^{3+}$. Most simple cobalt salts contain the pink, hydrated ion $[Co(H_2O)_6]^{2+}$. The water is expelled from most of these salts by heating somewhat above 100 °C. In general, cobalt(II) with four ligands is blue and cobalt(II) with six ligands is pink. The color change of a hydrated cobalt complex ion is used in devices meant to show when it is going to rain. At high humidity, the blue complex ion takes up water and turns pink.

$$\underset{\textit{blue}}{[Co(H_2O)_4]Cl_2(s)} \xrightleftharpoons{\text{H}_2\text{O in atmosphere}} \underset{\textit{pink}}{[Co(H_2O)_6]Cl_2(s)}$$

The best known of the cobalt(II) salts is the pink chloride, $[Co(H_2O)_6]Cl_2$, which dissolves in water to give a pink solution and in alcohol to give a blue solution of $[Co(C_2H_5OH)_4]Cl_2$. It is the source of most other cobalt salts. Cobalt(II) oxide (CoO) gives the deep blue color to the glass known as cobalt glass.

In many soluble nickel(II) compounds, nickel is present as the light green hydrated ion $[Ni(H_2O)_6]^{2+}$. The halide salts, upon heating above 100 °C, are converted to the anhydrous salts $NiCl_2$ (yellow), $NiBr_2$ (yellow-brown), and NiI_2 (black). These dissolve readily in water, giving green solutions. Nickel(II) sulfate, nickel(II) nitrate, and other nickel(II) salts containing oxoanions retain their green color upon dehydration; this color must, therefore, be due to the coordination of oxygen atoms with the nickel(II) ion.

In nickel carbonyl $[Ni(CO)_4]$ the metal shows an oxidation state of zero. This interesting compound, which long baffled theoretical chemists, is a volatile (b.p. 43 °C), *extremely toxic* liquid. It is obtained by the action of carbon monoxide gas on nickel powder at slightly elevated temperatures. Such transition metal carbonyls are all volatile, highly poisonous, and soluble in nonpolar solvents.

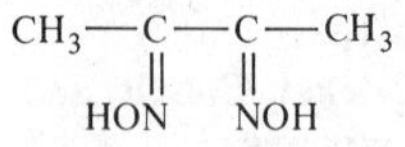

$CH_3—C—C—CH_3$ (with $\|$ to HON and NOH)

dimethylglyoxime

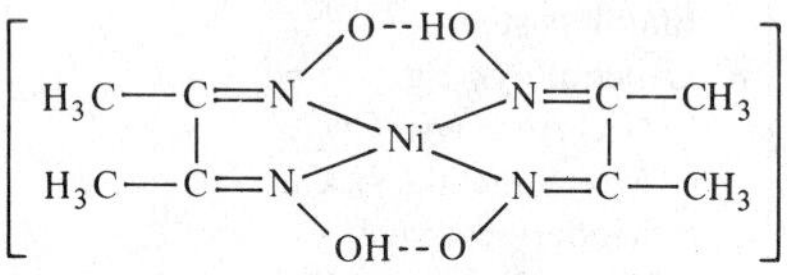

$[H_3C—C{=}N(O\text{--}HO)N{=}C—CH_3 / H_3C—C{=}N(OH\text{--}O)N{=}C—CH_3]$ with Ni

complex between nickel ion and two molecules of dimethylglyoxime

Figure 30.7
Nickel Complex with Dimethylglyoxime

One of the most striking of the nickel complexes is that formed with the organic compound dimethylglyoxime (DMG), which has the structural formula shown in Figure 30.7. When nickel(II) ion is added to a neutral solution of this material, it coordinates with the two nitrogen atoms in each of the two DMG molecules to give the nickel–dimethylglyoxime complex. At the same time one hydrogen ion escapes from each molecule of DMG, and the other forms a hydrogen bond with an oxygen atom of the other DMG molecule. This complex is only very slightly soluble in water and is deep red. Since no other common metal ion gives a precipitate with dimethylglyoxime, the formation of a red precipitate when DMG is added is a very strong indication that nickel(II) ion is present in a solution.

30.10 COPPER, SILVER, AND GOLD

a. Elemental Copper, Silver, and Gold Copper, silver, and gold together are referred to as the **coinage metals** because of their use in coins since ancient times. In properties such as malleability, ductility, and conductivity, these are the most "metallic" of all metals. If a heavy weight is hung on a fairly thin copper wire, the wire will continue to stretch slowly for several days until it becomes too small in diameter to support the weight and breaks. Gold is the most malleable of all metals; in fact, it can be rolled into sheets that are so thin as to be transparent (they transmit green light). Gold leaf is used in signs on windows and office doors, and in covering the domes of capitol buildings.

Copper, silver, and gold are excellent conductors of electricity. Copper is used in cooking vessels because of its high thermal conductivity and in electrical wiring because of its high electrical conductivity. Both silver and gold are better conductors than copper, but their higher cost rules them out for most purposes.

All three of the metals in this family are relatively soft and are often alloyed with other metals to achieve hardness, rigidity, and other desirable properties (Table 30.11). Because of their beautiful luster and color, copper, silver, and gold have been used for many centuries in ornaments, jewelry, and tableware.

The major uses of copper, silver, and gold are summarized in Table 30.12. (The metallurgy of copper was discussed in Section 26.6.)

Copper, silver, and gold all lie below hydrogen in the electromotive series. They do not react with nonoxidizing acids such as hydrochloric acid and they are not readily attacked by oxygen at ordinary temperatures. On heating with air, copper, which is the most reactive of the three, forms copper(II) oxide (CuO); silver slowly forms silver(I) oxide (Ag_2O), which, however, decomposes into its elements on strong heating. Gold does not react with oxygen.

Table 30.11
Some Nonferrous Alloys Containing *d*-Transition Metals

	Zn	Cd	Cu	Bi	Pb	Sn	Ni	In	Ag	Au
Brass	18–40	—	60–82	—	—	—	—	—	—	—
Lead–tin–yellow brass	24.0	—	72.0	—	1.0	3.0	—	—	—	—
Bronze	1–25	—	70–95	—	—	1–18	—	—	—	—
Lead–nickel brass	20.0	—	57.0	—	9.0	2.0	12.0	—	—	—
Nickel silver (German silver)	24.0	—	64.0	—	—	—	12.0	—	—	—
Wood's metal	—	12.5	—	50.0	25.0	12.5	—	—	—	—
Lipowitz metal	—	10	—	50.0	26.7	13.3	—	—	—	—
Cerrolow	—	5.3	—	44.7	22.6	8.3	—	19.1	—	—
U.S. Nickel coin	—	—	75	—	—	—	25	—	—	—
U.S. Penny[a]	97.6	—	2.4	—	—	—	—	—	—	—
Sterling silver	—	—	7.5	—	—	—	—	—	92.5	—
Gold (18 karat)[b]	—	—	5–14	—	—	—	—	—	10–20	75
Gold (14 karat)	—	—	12–28	—	—	—	—	—	4–30	58

[a] Post-1982. Pre-1982 U.S. pennies were 95% Cu and 5% Zn. The composition was changed because of the rising price of copper.
[b] White gold contains gold with varying amounts of Pd, Ni, and Zn.

Copper and silver readily react with sulfur and sulfur-containing compounds, forming either Cu_2S or CuS and Ag_2S, respectively. This reaction is particularly evident in the case of silver, which darkens rapidly when left in contact with sulfur-containing substances such as eggs, rubber, or mustard. Both metals tarnish slowly when exposed to the atmosphere, for there are nearly always traces of hydrogen sulfide in the air.

Both copper and silver readily react with oxidizing acids, such as nitric acid, copper going to the +2 state, and silver to the +1 state. Gold does not tarnish noticeably in the air or when exposed to sulfur or its compounds. It is not attacked by nitric acid, but it is dissolved by aqua regia (HCl/HNO_3).

The formation of complex ions is of great importance in the chemistry of these metals; advantage is taken of it in the metallurgy and the electroplating of all three, in analysis, and, in the case of silver, in the developing of photographic films and plates (see Aside 30.1).

Table 30.12
Uses of Copper, Silver, and Gold

Copper
Electrical applications
Pipes, plumbing, gutters
Industrial machinery
Coinage
Silver
Sterling silver tableware
Photography
Mirror backing
Heat-exchange equipment
Pharmaceuticals
Electronic devices
Coinage
Jewelry
Gold
Electronic devices
Photography
Jewelry
Reflective coating, e.g., on space-ships, windows

b. Compounds of Copper Copper(II) oxide (CuO) is a black, insoluble substance formed by gentle heating of the metal in air or by the addition of an alkaline solution to a hot solution of a copper(II) salt. An interesting use of CuO is as a black surface coating in devices that collect solar energy. A thin layer of CuO transmits infrared radiation, but not shorter wavelength radiation.

The blue-green copper(II) hydroxide is obtained either as a gelatinous material with variable water content or as a crystalline substance of the composition $Cu(OH)_2$ when a base is added to a copper(II) solution maintained at room temperature. Both the oxide and the hydroxide are predominantly basic in character and react with most acids to give solutions of copper(II) salts. Such solutions are generally blue or blue-green, the color being due to the hydrated copper(II) ion.

Copper(II) sulfide (CuS), usually obtained by the action of hydrogen sulfide on a solution of a copper(II) salt, is highly insoluble in water ($K_{sp} = 6 \times 10^{-36}$ at 25 °C) but dissolves readily in a solution of sodium sulfide because of the formation of the complex ion $[CuS_2]^{2-}$, which is even less dissociated than solid CuS.

$$Na_2S(aq) + CuS(s) \longrightarrow Na_2[CuS_2](aq)$$

This reaction is often used in analytical chemistry to separate copper from the ions of other heavy metals that do not form such complexes.

Copper(II) sulfate pentahydrate ($CuSO_4 \cdot 5H_2O$), commonly called blue vitriol, is the copper compound used commercially in the greatest quantity. Like most other hydrated copper(II) salts, this one is blue, which is characteristic of the $[Cu(H_2O)_4]^{2+}$ ion. The fifth molecule of water is held to the sulfate ion by hydrogen bonding. When the salt is dehydrated by gentle heating or placing it in a desiccator, the blue crystals crumble to a fine white powder of anhydrous copper sulfate, $CuSO_4$. On dissolution in water, this powder again gives a blue solution. Copper(II) sulfate is used to kill algae and fungi, in fertilizers, and as the starting material for the production of many other copper compounds.

Stable copper(I) compounds include the halides (e.g., CuCl, white), the sulfide (black), and the cyanide (white), all of which are anhydrous crystalline compounds. Copper(I) compounds containing oxoanions are known, but they are unstable. For example, Cu_2SO_4 can be prepared and stored in dry air, but upon exposure to moisture it disproportionates vigorously:

$$Cu_2SO_4(s) \longrightarrow Cu(s) + CuSO_4(s)$$

In general, the simple Cu^+ ion does not exist in aqueous solution, for it immediately disproportionates.

c. Compounds of Silver and Gold Silver nitrate ($AgNO_3$) (once referred to as "lunar caustic": "lunar" because in alchemical lore, silver was related to the moon and "caustic" because in the solid form, it burns flesh) is made by the reaction of metallic silver with nitric acid. It can be crystallized from the resulting solution as large, white crystals that are readily soluble in water.

Most of the other simple compounds of silver are insoluble in water and can be prepared by precipitation from a solution of silver nitrate. The formation of the white, curdy precipitate of silver chloride (AgCl) is a sensitive test for the presence of either Ag^+ or Cl^- in solution. The pale yellow bromide and yellow iodide are even less soluble than the chloride. Although silver chloride is not very soluble in water ($K_{sp} = 1.8 \times 10^{-10}$), it is dissolved by solutions of ammonia, with which it forms a stable soluble complex ($K_d = 6.2 \times 10^{-8}$), diamminesilver chloride $[Ag(NH_3)_2]Cl$.

As with copper, the simple gold(I) compounds are the halides (AuCl, yellow), the sulfide (brown-black), and the cyanide (yellow). All of these are insoluble and show a tendency to disproportionate into metallic gold and the gold(III) compound. All are readily reduced to metallic gold. The most important soluble compound of gold(I) is the cyano complex, $Na[Au(CN)_2]$, which is used in electroplating gold onto metallic objects. Among the gold(III) compounds, the commonest simple substance is red gold(III) chloride, usually written $AuCl_3$, but which is actually a dimer (Au_2Cl_6) in both the solid and vapor phases.

Certain gold compounds have been used to treat rheumatoid arthritis, although in some cases severe side effects prevent the treatment.

Aside 30.1 THE PHOTOGRAPHIC PROCESS

When exposed to light of short wavelength, silver halides are "activated," after which they are more easily reduced to the metallic state than unactivated silver halides. This is the basis of the photographic process.

$$AgX(s) + \text{light energy} \longrightarrow AgX^* \text{ (activated silver halide)}$$
$$AgX^* + \text{reducing agent} \longrightarrow Ag(s)$$

The silver halide (usually a mixture of halides) in finely divided crystalline form is suspended in gelatin, which is spread in a uniform layer on the film. When light falls on the film, electrons are released from halide ions and migrate to the crystal surfaces, where they combine with silver ions to form silver atoms. The number of silver ions reduced is proportional to the amount of light, but is too small to produce a visible image. The image is developed, or intensified, by placing the film in a solution of a weak organic reducing agent. By a catalytic mechanism not fully understood, all the silver ions in any crystal grain that contains just a few silver atoms are also reduced to silver atoms. The black crystals become visible in regions where light had struck the film.

Once the film has been developed, the unreduced silver halide must be removed before the film is taken into the light; otherwise, it will slowly turn black. This silver halide is removed by washing the film in a solution of sodium thiosulfate ($Na_2S_2O_3$), which forms a stable, soluble complex by reaction with the silver halide. This is the fixing process, and sodium thiosulfate is called *hypo*.

$$AgX(s) + 2S_2O_3^{2-}(aq) \longrightarrow [Ag(S_2O_3)_2]^{3-}(aq) + X^-(aq)$$

After rinsing in water, the negative is dried. It is black where light fell on it and clear where no light struck it. To make a print, the negative is projected onto paper coated with silver halide–gelatin. Where the negative is black, no light passes through to the paper; where the negative is transparent, the paper is exposed. Once the paper is developed, fixed, and dried, the print is ready.

Portraits can be toned by converting the silver in the picture to some color other than black. This is done by immersing the picture in a solution of a material that will react with silver and leave a suitable deposit on the paper.

$$Ag(s) + [AuCl_2]^- \longrightarrow AgCl(s) + Cl^- + Au(s) \quad \text{(sepia)}$$
$$2Ag(s) + PtCl_4^{2-} \longrightarrow 2AgCl(s) + 2Cl^- + Pt(s) \quad \text{(gray)}$$

The silver chloride formed in toning with gold or platinum is removed by treatment with hypo.

The chemistry of color photography is extremely complex, but the initial steps rely on the photosensitivity of the silver halides, just as black and white photography does. The silver halide is most sensitive to light in the blue and ultraviolet regions. In color film, the halide surface may be coated with an organic compound that absorbs, say, red light. The dye transfers energy to the silver halide, allowing it to be exposed by light of colors that it does not ordinarily "see." There are three layers of AgX–gelatin suspension, one for blue light, one for green light, and one for red light. Each layer also contains an organic dye which reacts with the silver to give a characteristic color. The silver salt formed by reaction with dye is later washed out, and the final picture contains dyes, but no silver.

30.11 ZINC, CADMIUM, AND MERCURY

a. Elemental Zinc, Cadmium, and Mercury Only the two outer *s* electrons in zinc, cadmium, and mercury are employed in bonding. Therefore, they have fewer oxidation states (+2 only for zinc and cadmium, +2 and +1 in Hg_2^{2+} for mercury) and form fewer complex ions than the other *d*-block elements. Zinc is a reasonably strong reducing agent, stronger than copper, but weaker than manganese and the elements at the left of manganese in the first transition series.

Zinc is one of the four workhorse metals of our modern civilization (the others being iron, copper, and lead). The metallurgy of zinc was discussed in Section 26.6. Zinc is used in coating iron to prevent rusting (galvanized iron) and in many important alloys, of which the most widely used are brasses, bronzes, and bearing metals (see Table 30.11). Other uses of zinc, together with uses of cadmium and mercury, are listed in Table 30.13.

The largest use of metallic cadmium is in alloys (see Table 30.11), especially those that melt at low temperatures. Wood's metal and Lipowitz metal, both of which melt at 70 °C, are used in automatic fire extinguishers and fire alarms. Cerrolow alloy, which also contains indium, melts at 47 °C. Cadmium rods are used in nuclear reactors to absorb neutrons, and thus, by moderating neutron flux, to control the chain reaction.

Mercury is less abundant in the Earth's crust than is cadmium, it is not as widely distributed, and it is not obtained as a by-product. Yet there are important uses for mercury for which there are no adequate substitutes. Hence, mercury is an expensive metal. Most of the mercury used in the United States is imported from Spain, though some is mined in California. The chief one is the red cinnabar (HgS). Roasting in the presence of oxygen yields elemental mercury.

There are, it has been estimated, more than three thousand uses of mercury (a few are given in Table 30.13). Many of them depend upon its liquid nature. The relatively great change in volume of the metal with changes in temperature makes it useful in thermometers. Its liquidity and high density account for its use in barometers, and its electrical conductivity for its use in electric switches. Mercury also conducts electricity in the vapor state, emitting the bright blue light of mercury arc lights. With most of the metals, mercury forms alloys called **amalgams.** One is used in dentistry. When the intermetallic compound Ag_3Sn is ground with mercury, it dissolves to form a semisolid amalgam which, on standing, sets to form a hard, solid mixture of the intermetallic compounds Ag_5Hg_8 (a γ intermetallic compound; Section 26.8) and Sn_7Hg. During the formation of these compounds, the amalgam expands slightly and fits the walls of a cavity so tightly that bacteria cannot easily penetrate.

Table 30.13
Uses of Zinc, Cadmium, and Mercury

Zinc
Electroplating
Alloys
Galvanized iron
Pigments, e.g., ZnO, ZnS (both white)
Rubber
Zinc oxide ointment (antiseptic)
Batteries (dry cells) and Ni–Zn batteries
Cadmium
Electroplating
Alloys, for brazing and low-melting alloys
Ni–Cd batteries
Pigments, e.g, CdS (yellow)
Fungicides
Mercury
Fungicides
Pulp and paper industry
Batteries
Leather tanning
Thermometers and other scientific instruments
Hg vapor lights
Electrical switches
Amalgams for dentistry

b. Compounds of Zinc, Cadmium, and Mercury Some compounds of zinc, cadmium, and mercury are listed in Table 30.14. All soluble compounds of zinc, cadmium, and mercury are toxic, with toxicity increasing in the order Zn < Cd < Hg (see Aside 26.1).

Table 30.14
Some Compounds of Zinc, Cadmium, and Mercury

ZnO	Zinc oxide	Chinese white, a pigment; photoconductor in copying machine; ointment base
ZnS	Zinc sulfide	In lithopone, a white pigment; TV tube phosphor
$ZnCl_2$	Zinc chloride	Deodorant; wood preservative
CdO	Cadmium oxide	Ni–Cd battery electrodes
CdS	Cadmium sulfide	Yellow to red pigment
$CdSe$	Cadmium selenide	Semiconductor
$CdTe$	Cadmium telluride	Semiconductor
HgO	Mercury(II) oxide	Red and yellow forms
Hg_2Cl_2	Mercury(I) chloride	Calomel, a drug
$HgCl_2$	Mercury(II) chloride	Corrosive sublimate, antiseptic in dilute solution
HgS	Mercury(II) sulfide	Red, the mineral cinnabar; black, precipitate from aqueous solution

Mercury(II) oxide (HgO) is either red or yellow, depending upon its method of preparation. The red and yellow forms have identical crystal structures and differ only in particle size, the yellow from being more finely divided than the red. Mercury(I) oxide is not known.

Zinc oxide (ZnO) is white at ordinary temperatures, but turns yellow when heated. On cooling, it turns white again. Cadmium oxide (CdO) varies in color from green-yellow to black, depending upon the temperature at which it is obtained. Both cadmium and zinc oxides are only slightly soluble in water.

Mercury(I) chloride (Hg_2Cl_2), known as *calomel*, has been used in medicine for a long time. It is a diuretic, a cathartic, and an antiseptic. It also kills intestinal worms and cabbage and onion maggots. The extremely poisonous nature of mercury compounds is masked in calomel by its very small solubility.

Mercury(II) chloride ($HgCl_2$), or *corrosive sublimate*, by contrast is quite soluble in water and is a *violent* poison, being corrosive to mucous membranes. In very dilute solution, it can be used as a disinfectant.

Mercury(II) sulfide (HgS) is the least soluble sulfide ($K_{sp} = 1.9 \times 10^{53}$ at 25 °C) and one of the least soluble of all simple binary compounds.

SUMMARY OF SECTIONS 30.6–30.11

Some outstanding properties of the *d*-block metals discussed in these sections are summarized in Table 30.15. The uses of the metals, and the formulas and uses of their compounds are listed in the many tables throughout these sections.

Table 30.15
Outstanding Properties of Selected *d*-Block Elements and Their Compounds

Cr	Very corrosion resistant; made by Al reduction of oxide; used in chrome plating and stainless steel
$FeCr_2O_4$	Chromite ore; reduced to ferrochrome, used in stainless steel production
Cr(II), Cr(III), Cr(VI)	Common oxidation states; Cr(III) most stable; Cr(II) compounds, reducing agents; Cr(VI) compounds, oxidizing agents (e.g., $Cr_2O_7^{2-}$ in acidic solution)
Mn	Brittle; more reactive than its neighbors; present in all steels
Mn(II), Mn(IV), Mn(VII)	Common oxidation states; Mn(II) most stable; others (e.g., MnO_4^-, MnO_4^{2-}) oxidizing agents
MnO_2, FeO, Fe_2O_3, Fe_3O_4	Usually nonstoichiometric; MnO_2 used in dry-cell batteries
Fe	Pure metal soft and silvery; properties greatly varied by alloying (steel and other iron alloys)
Ni, Co	Important in many alloys and in catalysts
Co(II), Co(III), Ni(II)	Many complex ions; often hydrated in compounds
Mn(III), Co(III)	Stable in solution only in certain complex ions
Cu, Ag, Au	Coinage metals; malleable, ductile, lustrous, good electrical conductors: used in jewelry, decorative items
Hg	Many uses based on liquidity, high density, electrical conductivity; amalgams (alloys) with most metals
Cu^+, Hg_2^{2+}	Disproportionate; Cu^+ in H_2O, Hg_2^{2+} with, e.g., OH

f-BLOCK ELEMENTS

30.12 PROPERTIES OF f-BLOCK ELEMENTS

The two series of *f*-block elements are placed at the bottom of most periodic tables. In the first series, the lanthanides—on which we concentrate our attention—successive electrons enter the 4*f* energy level. (For the configurations of the *f*-block elements, see the periodic table inside the front cover.) Lanthanum (electron configuration, $[Xe]4f^0 5d^1 6s^2$), which all lanthanides resemble in their properties, can be viewed as either the first element in the third *d*-transition series or the first element in the lanthanide series, or both. Actinium (electron configuration, $[Rn]5f^0 6d^1 7s^2$) has a similar relationship to the second *f*-transition series elements, which are known as the actinides. All actinides are radioactive elements. The first six (through plutonium) are present naturally in the Earth's crust, the best known being uranium, the nuclear reactor fuel. The remaining actinides are known only as synthetic radioisotopes.

The lanthanides are all shiny, silvery, reactive metals of moderately high melting points and densities (comparable to those of the elements of the first *d* series). All but gadolinium and lutetium tarnish readily in air from the formation of oxides. The lanthanides vary even less in properties across the period than do the *d*-transition elements.

The electrons in the incomplete *f* subshell do not participate in bonding as readily as the electrons from the incomplete *d* subshells in the *d*-transition elements. The lanthanides, therefore, are limited to oxidation states of +2, +3, and in just a few cases (e.g., Ce, Pr, Tb) +4. The +3 oxidation state is by far the most common. The +2 ions are easily oxidized. Cerium in its +4 oxidation state is a strong oxidizing agent. In most oxidation states, the atoms and ions of the lanthanides (and the actinides) contain unpaired electrons, and are therefore paramagnetic and colored.

The regular decrease in ionic radii of the lanthanides, shown in Figure 30.8, leads to a decrease in ionic character and solubility in a series of compounds with the same anion. Because of their similar atomic and ionic radii, many lanthanide compounds with a given anion are isomorphous (of the same crystal structure).

Their similar radii allow all lanthanide +3 ions to occur together in natural minerals and make separation from each other very difficult. They were first known as the "rare earths"—"earths" because of their original isolation as oxides, which were once referred to as "earths," and "rare" because of their isolation from relatively rare minerals. The "discovery" of these elements was the result of over 100 years of confusion and hard work in separations of closely similar species.

The story began in 1794 and 1803 with the separation from two different Swedish ores of two different substances, each believed to be the oxide of a newly discovered element. The oxides were named yttria (for Ytterby, Sweden) and ceria (for Ceres, a newly discovered planetoid). In 1841, after two years of work in his basement laboratory, Carl Mosander, a Swedish chemist and mineralogist, announced that ceria contained the oxides of two additional substances. He named these oxides lanthana (the hidden one) and didymia (the twin brother of lanthanum). Two years later Mosander produced three new oxides from yttria.

The secrets of the two original Swedish minerals were just beginning to unfold. There followed a period during which several investigators claimed that they had isolated new substances from Mosander's oxides, only to have others show that these also were mixtures. The poor communication of the times and the use of names such as "old terbia" and "new erbia" added to the confusion. Eventually ceria proved to be

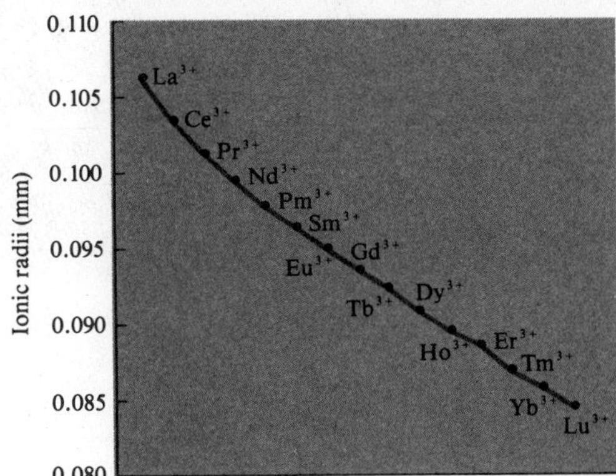

Figure 30.8
Radii of Ions of the Lanthanides

a mixture of the oxides of seven of the lighter rare earths (from lanthanum to gadolinium, with the exception of promethium). Yttria was found to be a mixture of the oxides of the eight heavier rare earths (gadolinium through lutetium) and of those of scandium and yttrium.

30.13 SOURCES AND USES OF THE LANTHANIDES

All fourteen lanthanides are present together in bastnasite, monazite, and xenotine, the major ores from which the elements are recovered. It is impractical both chemically and economically to extract only specific lanthanides from the ores as they are needed. Therefore, the relative amounts of these metals and their compounds produced each year reflect their distribution in the ores.

The difficult separation of the lanthanides was first accomplished by precipitation and repeated recrystallization (up to 3000 times) of individual salts of differing solubilities. The development of modern, highly effective separation methods was first inspired by the discovery in the 1950s that cerium(IV) oxide is a superior polishing compound for plate glass and optical glass. In the 1960s further impetus was given to lanthanide production by the demand for phosphors for color television tubes. (An yttrium–europium oxosulfide or oxide of very high purity is a bright red phosphor essential to color television.) Ion exchange or liquid–liquid extraction between water and an organic solvent containing a chelating agent (Section 31.3) are now used effectively to separate the ions from each other. The light lanthanide ions can be reduced electrolytically; the heavy ones must be reduced by high-temperature reactions with active metals, often calcium.

One of the first uses of the lanthanides was in the production of *mischmetal* for the flints in cigarette lighters. Mischmetal is an alloy of a mixture of the light rare earths (La, Ce, Pr, Nd) with iron. Some very interesting new applications for lanthanides are on the horizon. A lanthanum–nickel intermetallic compound that can reversibly store more hydrogen per unit volume than liquid hydrogen may find use in a hydrogen economy (see Aside 16.1).

$$LaNi_5(s) + 3H_2(g) \rightleftharpoons LaNi_5H_6(s)$$

Samarium–cobalt permanent magnets are five to ten times more powerful than common magnets and are making new devices possible.

SUMMARY OF SECTIONS 30.12–30.13

The lanthanides and the actinides comprise the two series of *f*-block transition elements. The lanthanides, or rare earths, are very similar in properties and occur together in nature, making their separation difficult. However, they have many specialized applications, for example, in alloys and electronic devices. The lanthanides, which are easily oxidized, form mainly ionic compounds containing +3 ions. Unpaired electrons in *f* subshells cause lanthanide compounds (and actinide compounds) to be paramagnetic and colored. However, the *f* electrons do not take part in bonding readily, and the lanthanides form fewer and less stable complexes than the *d*-transition metals.

All actinides are radioactive, and all those beyond plutonium are synthetic.

SIGNIFICANT TERMS (with section references)

amalgams (30.11)
coinage metals (30.10)
ferromagnetic (30.3)
platinum metals (30.4)

QUESTIONS AND PROBLEMS

d-Block Elements

30.1 Identify the energy levels that are being filled with electrons for each of the three *d*-transition series. Why do many chemists consider zinc, cadmium, and mercury to be representative metals?

30.2 The outer shell electron configurations of the Fe, Co, Ni, and Cu atoms are $3d^6 4s^2$, $3d^7 4s^2$, $3d^8 4s^2$, and $3d^{10} 4s^1$, respectively. Why is there a discontinuity between Ni and Cu?

30.3 Briefly describe the physical properties of the *d*-transition elements. Name some of the characteristic chemical properties of *d*-transition metals and their ions.

30.4 Zirconium and hafnium are very much alike in their chemical and physical properties even though they are in different periods in the periodic table. Why are they so similar?

30.5 Pure copper and pure zinc are both relatively soft metals, but the alloy $CuZn_3$ is very hard. Suggest a reason.

30.6 What is the relationship between the maximum oxidation states and the electron configurations for the metals of the scandium, titanium, vanadium, chromium, and manganese families?

30.7 Compare the stability of the maximum oxidation state of chromium to those of molybdenum and tungsten.

30.8 Compare the nature of the bonding in compounds formed by transition elements in their higher and lower common oxidation states.

30.9 What does the term "ferromagnetic" mean? How does ferromagnetism differ from paramagnetism?

30.10 During 1.00 h, 3.4 mg of $^{95}_{43}Tc$ underwent radioactive decay by electron capture. The original sample size was 0.1000 g. Write the equation for the nuclear reaction for the decay process and calculate the half-life of this isotope.

Selected *d*-Block Elements—Chromium

30.11 Name the chief ore of chromium. How is this ore treated to produce chromium metal? Give an equation to show the reaction involved.

30.12 What are the three common oxidation states of chromium? Which of these is most common? What properties does chromium have in the other two oxidation states?

30.13 The total surface area of a car bumper is 1750 sq inches. The bumper is to be chrome plated to a thickness of 0.0003 inch in a Cr(VI) solution in one hour. What must the current be to perform this plating assuming 20% efficiency? The density of Cr is 7.20 g/cm^3.

30.14 Write the Lewis structures for the chromate and the dichromate ions. What is the oxidation state of chromium in each of these ions?

30.15 What is wrong with the statement "Chromate ion acts as a strong oxidizing agent in acidic solution"?

30.16* The standard state free energy of formation at 25 °C is −727.75 kJ/mol for CrO_4^{2-}, −1301.1 kJ/mol for $Cr_2O_7^{2-}$, 0 for H^+, and −237.129 kJ/mol for $H_2O(l)$. What is the equilibrium constant for the reaction

$$2CrO_4^{2-} + 2H^+ \rightleftharpoons Cr_2O_7^{2-} + H_2O(l)$$

To what pH must a 0.100 M CrO_4^{2-} solution be adjusted so that the concentrations of CrO_4^{2-} annd $Cr_2O_7^{2-}$ are equal?

30.17 Write chemical equations illustrating the amphoteric behavior of chromium(III) hydroxide.

30.18 At 25 °C, $K_{sp} = 2.8 \times 10^{-13}$ for $PbCrO_4$. Calculate the solubility of this salt in pure water and in a 0.100 M Pb^{2+} solution.

Selected *d*-Block Elements—Manganese

30.19 From its general reactivity and its position in the electromotive series, would you expect manganese to occur in nature as the sulfide or the oxide? Name the primary ore of manganese.

30.20 What are the most common oxidation states of manganese? What uses are made of MnO_2 and MnO_4^-?

30.21 Why is solid MnO_2 an electrical conductor?

30.22* As the following reduction potentials indicate, the Mn^{3+} ion is a much stronger oxidizing agent in aqueous solution than the Fe^{3+} ion.

$$Mn^{3+} + e^- \longrightarrow Mn^{2+} \qquad E° = 1.51\ V$$
$$Fe^{3+} + e^- \longrightarrow Fe^{2+} \qquad E° = 0.771\ V$$

The difference in ease of gaining electrons cannot be attributed to differences in ionic radii (Mn^{3+}, 0.066 nm; Fe^{3+}, 0.064 nm). In fact, one would expect that the Fe^{3+} ion, with its larger nuclear charge and slightly smaller size, would gain electrons more easily than the Mn^{3+} ion. What electronic factors are probably significant in accounting for the difference in oxidizing power of the two species?

30.23 Outline the steps and write the chemical equations for the production of potassium permanganate from manganese dioxide.

30.24 Write the half-reaction equations and the overall equation for the disproportionation of MnO_4^{2-} in an acidic solution.

30.25 How do you explain what has happened in the following chemical changes?

(a) $MnO_4^{2-} + H^+ \longrightarrow$ purple solution + dark brown solid
(b) $MnO_4^{2-} + O_3(g) + H^+ \longrightarrow$ purple solution + gas

30.26 What is the pH of a 0.015 M solution of permanganic acid? $K_a = 2.0 \times 10^3$ for $HMnO_4$.

Selected *d*-Block Elements—Iron

30.27 What are the common oxidation states of iron? What are the common names for the simple cations having these oxidation states?

30.28 Write the formulas for the common oxides of iron. What is the oxidation state of iron in each of them?

30.29 A 1.00 g sample of iron reacted completely with 215 mL of $O_2(g)$ measured at 762 Torr and 20.0 °C. Determine the formula of the oxide formed.

30.30 Write the formula for iron pyrite. What is the oxidation number of iron in this compound? Explain.

30.31 Identify the oxidation state(s) of iron in the compound $KFe[Fe(CN)_6]$. Is there more than one answer to this question? Explain.

30.32* Solid iron is rather interesting in that it undergoes three solid–solid phase transitions before it melts at 1535 °C. Up to 760 °C, the α form is stable (body-centered cubic, unit cell length = 0.286106 nm); from 760 °C to 907 °C, the β form is stable (body-centered cubic, 0.290 nm); from 907 °C to 1400 °C, the γ form is stable (face-centered cubic, 0.363 nm); and from 1400 °C to the melting point, the δ form is stable (body-centered cubic, 0.293 nm). Show that the radius of an iron atom is the same in each of these structures.

30.33 What are the products of the following reactions?

(a) $Fe(NO_3)_2 \cdot 6H_2O(s) \xrightarrow{\text{gentle heat (140 °C)}}$

(b) [solid product of (a)] $\xrightarrow{\text{strong heat, air}}$

(c) $FeCO_3(s) \xrightarrow{\text{200 °C, air}}$

30.34* Air oxidation of solutions of Fe^{2+} presents a problem in keeping these solutions in the laboratory. Given that the $E° =$ −0.440 V for the Fe^{2+}/Fe couple and 0.771 V for Fe^{3+}/Fe^{2+} couple, show by calculation that simply putting a piece of Fe in a solution of Fe^{2+} will maintain the iron in the +2 state.

30.35 Iron(II) can be oxidized to iron(III) by MnO_4^- in acidic solution, the latter being reduced to Mn^{2+}. Write equations for the half-reactions for the oxidation and reduction processes and combine them to obtain the overall equation. A 0.302 g sample of iron ore is dissolved, the iron in it is reduced to Fe^{2+}, and the Fe^{2+} is oxidized using 42.79 mL of 0.0205 M $KMnO_4$. What is the percentage of iron in the sample?

30.36 Which solution would have a lower pH: 0.1 M $FeCl_3$ or 0.1 M $FeCl_2$? Why?

30.37 Calculate the pH of a 0.100 M $Fe(NO_3)_2$ solution. $K_a = 1.2 \times 10^{-6}$ for Fe^{2+} at 25 °C.

30.38 What is the equilibrium concentration of Fe^{2+} in a solution that is 0.050 M in $[Fe(C_2O_4)_2]^{2-}$ and 0.050 M in $C_2O_4^{2-}$? $K_d = 2 \times 10^{-8}$ for $[Fe(C_2O_4)_2]^{2-}$.

Selected *d*-Block Elements—Cobalt and Nickel

30.39 What are some of the commercial uses of cobalt and nickel?

30.40 Anhydrous nickel chloride is yellow. When this substance is dissolved in water, a green solution is formed. Upon addition of $NH_3(aq)$, an apple green precipitate forms which, upon addition of more ammonia, dissolves, giving a deep blue solution. Explain these changes and give chemical equations.

30.41* Identify and write the chemical formula for each of the species involved in the following processes: (a) anhydrous cobalt(II) chloride, which is blue, was dissolved in water to give a pink solution; (b) addition of aqueous ammonia to the pink solution caused a pink precipitate to form; (c) addition of more ammonia to the mixture dissolved the precipitate and gave a tan solution.

30.42 The major source of nickel is the mineral pentlandite, (Ni, Fe)S, in which the nickel-to-iron ratio is about one to one. The sulfides are separated by selective flotation methods, and the NiS is roasted to give the oxide. The Mond process involves passing a mixture of H_2 and CO at 50 °C over the oxide to reduce the nickel oxide and any oxides of impurities to the free metals while the CO reacts selectively with the metallic nickel to form gaseous $[Ni(CO)_4]$. The $[Ni(CO)_4]$ is heated to 200 °C to decompose it into metallic nickel and CO, which can be recycled. (a) Write chemical equations for all the processes described. (b) What mass of Ni will be obtained from 1.0 kg of pure mineral using the Mond process assuming an overall efficiency of 82%?

Selected *d*-Block Elements—Copper, Silver, and Gold

30.43 Write the outer shell electron configurations for atoms of the coinage metals. Predict the oxidation numbers expected for these elements. What are the common oxidation states observed?

30.44 To separate gold and silver from impure lead, the mixture is extracted with molten zinc. How can these metals be recovered from the zinc extract?

30.45 Compare the relative reactivities of copper, silver, and gold by discussing briefly their reactions with air, nonoxidizing acids, and oxidizing acids.

30.46 Write chemical equations describing the reaction between Cu^{2+} and $H_2S(aq)$ and the subsequent addition of $Na_2S(aq)$.

30.47 How is silver nitrate prepared? What are some of the uses for this compound?

30.48 In most activity series of metals, silver is listed above platinum. Show that the reaction between Ag and Pt^{2+} is favorable, given that standard reduction potentials for the Ag^+/Ag and Pt^{2+}/Pt couples are 0.799 V and 1.2 V, respectively. However, a student observed that Ag does not reduce Pt(II) in alkaline solution. Confirm this result using $E° = 0.16$ V for the $Pt(OH)_2/Pt$ couple and 0.345 V for the Ag_2O/Ag couple.

30.49 Write balanced chemical equations describing the reactions that occur as (a) aqueous solutions of silver nitrate and sodium hydroxide are mixed, (b) gold is treated with a mixture of concentrated nitric and hydrochloric acids to form $HAuCl_4$, (c) aqueous solutions of copper(II) ion and cyanide ion are mixed, (d) excess aqueous ammonia is added to a solution of copper(II) ion, (e) silver chloride is treated with aqueous ammonia, (f) copper(I) sulfate is mixed with water, (g) anhydrous copper(II) sulfate is treated with water.

30.50 A photograph is toned with a solution of $[AuCl_2]^-$. Write the equation for the toning process and state what color the tone will be. If 2.1 mg of silver is oxidized to Ag^+, what mass of gold is reduced to metal?

Selected *d*-Block Elements—Zinc, Cadmium, and Mercury

30.51 Write the electron configurations of Hg, Hg^+, and Hg^{2+}. Which of these should be diamagnetic? One piece of evidence that mercury(I) exists as Hg_2^{2+} is that solutions containing mercury(I) are diamagnetic. Sketch the Lewis structure for Hg_2^{2+} and explain why this ion is diamagnetic.

30.52 The vapor pressure of mercury is 0.0012 Torr at 20.0 °C and 0.2729 Torr at 100.0 °C. Calculate the heat of vaporization of liquid mercury.

30.53 There is less difference in physical and chemical properties between Zn and Cd than between Cd and Hg. Examine the periodic table and explain this fact.

30.54 When zinc sulfide ore is roasted, the major products are ZnO and $ZnSO_4$. However, when mercury(II) sulfide, cinnabar, is treated in the same way, metallic mercury is formed. Explain this difference.

30.55 Zinc can be recovered from sphalerite (ZnS) by roasting, dissolution of the roasting product in dilute sulfuric acid, and electrolysis. (a) Write a balanced equation for each chemical reaction. (b) Calculate the amount of electricity required to recover the zinc from 100. kg of ore containing 90. % by mass of ZnS. Assume that the process is 100% efficient.

30.56 Complete the following equations describing reactions of the elements in the zinc family:

(a) $M + O_2(g) \longrightarrow$ M = Zn, Cd, Hg
(b) $M + X_2 \longrightarrow$ M = Zn, Cd; X = halogen
(c) $Hg + X_2 \xrightarrow{\text{excess } X_2}$ X = halogen
(d) $M + S \longrightarrow$ M = Zn, Cd, Hg

30.57 Mercury reacts with a mixture of concentrated nitric and hydrochloric acids to produce mercury(II) chloride, nitric oxide, and water. Write equations for the oxidation and reduction half-reactions describing this reaction and write the overall equation. What is the oxidizing agent?

30.58 The standard state free energy of formation at 25 °C is 0 for Hg(*l*), 164.40 kJ/mol for $Hg^{2+}(aq)$, and 153.52 kJ/mol for $Hg_2^{2+}(aq)$. Calculate $\Delta G°$ and the equilibrium constant for the reaction

$$Hg(l) + Hg^{2+} \rightleftharpoons Hg_2^{2+}$$

Explain why a drop of Hg is added to a solution of Hg_2^{2+} to "stabilize" it toward air oxidation.

30.59 Compare the bonding in compounds of the zinc family to that in compounds of the alkaline earth family. Why is there a difference?

30.60 Explain why solid HgF_2 is more ionic than solid HgI_2.

f-Block Elements

30.61 Write the general outer shell electron configuration for atoms of the *f*-transition elements. Which electrons are primarily involved in chemical bonding?

30.62 What is the most common oxidation state for the lanthanides? Write the general formulas of the oxides, nitrides, and halides.

30.63 Compare the number of complexes formed by the *f*-transition elements to those formed by the *d*-transition elements. Why is there a significant difference? Would you expect lanthanum ion or lutetium ion to be more prone to form complexes?

30.64 Identify the transuranium element formed in each of the following alpha-particle bombardment reactions: (a) $^{238}U(\alpha, 2n)$?, (b) $^{239}Pu(\alpha, n)$?, (c) $^{242}Cm(\alpha, n)$?.

ADDITIONAL QUESTIONS AND PROBLEMS

30.65 A student placed clean samples of Mn, Fe, Ni, and Cu into test tubes containing hydrochloric acid and observed gas bubbles being formed in the test tubes containing Mn, Fe, and Ni. (a) Write chemical equations for these reactions. A student with somewhat poorer laboratory technique performed the same experiment without cleaning the surfaces of the metals and reported gas bubbles in the test tubes containing Mn, Fe, and Ni and a change of color on the Cu surface. (b) What reaction did the second student see in the test tube containing the Cu? Write an equation describing this reaction.

A surprising number of students each year insist that Cu reacts with HCl. This is incorrect. However, Cu will react with an oxidizing acid such as HNO_3. (c) Write the chemical equation for this reaction and explain the difference in reactivity.

30.66 After many years of controversy, element 41 was officially named niobium in 1950. (Many metallurgists and U.S. commercial producers still refer to it as columbium.) The major mineral is columbite (or niobite), $(Fe, Mn)(Nb, Ta)_2O_6$. The metal was first prepared by formation of the chloride by reaction of the ore with HCl(*g*) and subsequent heating of the chloride with hydrogen gas. Write chemical equations for these processes.

30.67 Which of the following are redox reactions? Identify the oxidizing and reducing agent in each of the redox reactions.

(a) $3\,MnO_4^{2-} + 4H^+ \longrightarrow 2MnO_4^- + MnO_2(s) + 2H_2O(l)$
(b) $2[Co(CN)_6]^{4-} + 2H_2O(l) \longrightarrow 2[Co(CN)_6]^{3-} + H_2(g) + 2OH^-$
(c) $(NH_4)_2Cr_2O_7(s) \xrightarrow{\Delta} Cr_2O_3(s) + N_2(g) + 4H_2O(g)$
(d) $[Cu(H_2O)_4]^{2+} + 4NH_3(aq) \longrightarrow [Cu(NH_3)_4]^{2+} + 4H_2O(l)$

30.68 Repeat Question 30.67 for

(a) $Cr(OH)_3(s) + 3OH^- \longrightarrow [Cr(OH)_6]^{3-}$
(b) $6Fe^{2+} + Cr_2O_7^{2-} + 14H^+ \longrightarrow 6Fe^{3+} + 2Cr^{3+} + 7H_2O(l)$
(c) $[Ag(NH_3)_2]^+ + Cl^- + 2H^+ \longrightarrow AgCl(s) + 2NH_4^+$
(d) $[Au(CN)_2]^- + e^- \longrightarrow Au(s) + 2CN^-$

30.69 Classify each of the following reactions according to the reaction types listed in Tables 17.2 and 17.7:

(a) $2Eu^{2+} + 2H^+ \longrightarrow 2Eu^{3+} + H_2(g)$
(b) $[Ag(NH_3)_2]^+ + 2CN^- \longrightarrow [Ag(CN)_2]^- + 2NH_3(aq)$
(c) $2La(s) + N_2(g) \xrightarrow{\Delta} 2LaN(s)$
(d) $Th(s) + 4H^+ \longrightarrow Th^{4+} + 2H_2(g)$

30.70 Repeat Question 30.69 for

(a) $Cr_2O_3(s) + 6OH^- + 3H_2O(l) \longrightarrow 2[Cr(OH)_6]^{3-}$

(b) $8Al(s) + 3Fe_3O_4(s) \xrightarrow{\Delta} 4Al_2O_3(s) + 9Fe(l)$

(c) $Fe_2O_3(s) + 3CCl_4(g) \xrightarrow{\Delta} Fe_2Cl_6(g) + 3COCl_2(g)$

(d) $Fe^{3+} + Cr^{2+} \longrightarrow Fe^{2+} + Cr^{3+}$

30.71 Predict the major products of the following reactions:

(a) $Cr_2O_7{}^{2-} + OH^- \longrightarrow$

(b) $Mn^{2+} + OCl^- + OH^- \longrightarrow$

(c) $FeS_2(s) + O_2(g) \xrightarrow{\Delta}$

(d) $MnO_4{}^- + Fe^{2+} + H^+ \longrightarrow$

30.72 Repeat Question 30.71 for

(a) $Pr(s) + HCl(aq) \xrightarrow{\Delta}$

(b) $MnO_2(s) + HBr(aq) \longrightarrow$

(c) $Fe_3O_4(s) + CO(g) \xrightarrow{\Delta}$

(d) $[Au(CN)_2]^- + Zn(s) \longrightarrow$

30.73* For each of the following pairs of transition metal ions, identify the reagent(s) that will effect a one-step separation: (a) Mn^{2+} and Fe^{2+}, (b) Cu^{2+} and Ag^+, (c) Cr^{3+} and Pr^{3+}, (d) Ni^{2+} and Cu^{2+}, (e) Fe^{2+} and Co^{2+}, (f) Y^{3+} and Ni^{2+}.

30.74* An aqueous solution contains only the cations Fe^{3+}, Cu^{2+}, and Mn^{2+}. Using only the reagents $NH_3(aq)$, $NH_4{}^+(aq)$, $HNO_3(aq)$, $KClO_3(s)$, and $NCS^-(aq)$, indicate clearly how each cation can be identified.

30.75* The presence of Fe, Co, and Ni in an alloy can be confirmed by dissolving the alloy in a nonoxidizing acid to form Fe^{2+}, Co^{2+}, and Ni^{2+} and then performing the following series of qualitative analysis separations and tests on the resulting solution: (a) The metal ions are precipitated in alkaline solution as FeS, CoS, and NiS by H_2S; (b) dilute HCl is added to dissolve the more soluble FeS, and the Fe^{2+} is thus separated from the other cations; (c) the CoS and NiS are dissolved in concentrated HNO_3, forming Co^{2+} and Ni^{2+}; (d) the Fe^{2+} is oxidized to Fe^{3+} by HNO_3; (e) a red complex, $[Fe(CNS)]^{2+}$, is formed by adding a few crystals of NH_4CNS to Fe^{3+}; (f) ammonia is added to the solution containing Co^{2+} and Ni^{2+} to form the hexaammine complexes; (g) to half of the solution of the hexaammines, dimethylgloxime is added to form the red precipitate $[Ni(DMG)_2]$; (h) to the other half of the solution of the hexaammines, HCl is added to destroy the complex and then a few crystals of NH_4CNS are added to form the blue complex $[Co(CNS)_4]^{2-}$. Write chemical equations for the reactions involved in these eight steps.

30.76* A sample of copper metal was treated as follows: (a) Concentrated nitric acid was poured on it, giving at first a green and then a blue solution and a brown gas; (b) a dilute solution of ammonia was added, forming a very dark blue solution; (c) zinc was added, producing a colorless solution and a slime of finely divided metal; (d) the metal slime was separated, dried, and then heated in the air to form a black powder; (e) the black powder was added to a dilute solution of HCl, forming a blue solution; (f) a solution of NaOH was added, forming a blue gelatinous precipitate; (g) the blue precipitate changed to a black precipitate upon boiling; (h) the black precipitate reacted with a solution of H_2SO_4 to form a blue solution; (i) the blue solution was slowly evaporated to form blue crystals; (j) the blue crystals were heated to form very light blue, nearly white crystals. Identify the form of the copper after each process and write a chemical equation for each process.

30.77 The shorthand designation for the "alkaline accumulator" (or Edison or ferro-nickel cell) is

$$steel|Fe(s)|Fe(OH)_2(s)|KOH(aq)|Ni(OH)_2(s)|NiOOH(s)|steel$$

Write equations for the half-reactions for the oxidation and reduction processes and write the overall equation for the cell. The standard reduction potential for the $Fe(OH)_2/Fe$ couple is -0.877 V and the overall potential of the cell is 1.40 V. What is the standard reduction potential for the $NiOOH/Ni(OH)_2$ couple?

C H A P T E R

31

Coordination Chemistry

STRUCTURE AND NOMENCLATURE OF COMPLEXES

31.1 SOME DEFINITIONS

Platinum (II) chloride is a stable greenish powder that is not very soluble in water. Ammonia is a simple molecular compound with covalent nitrogen–hydrogen bonds. The fact that such compounds react with each other to give completely different stable compounds was once puzzling. None of the concepts of bonding of the time could explain why compounds with all of their "valences" satisfied would combine with each other.

In 1893 Alfred Werner proposed that metal atoms had "principal" valences and "auxiliary" valences. In compounds then written, for example, $CoCl_3 \cdot 4NH_3$, Werner showed that substituents were held to cobalt by both types of valence. In modern terms, a "principal" valence is an ionic bond and an "auxiliary" valence is a coordinate covalent bond. Today the formula for this compound is written $[Co(NH_3)_4Cl_2]Cl$. One chlorine atom is held by a principal valence (ionic bond) and the remaining chlorine atoms and the ammonia groups are held by auxiliary valences (coordinate covalent bonds).

Werner's ideas initiated the modern study of complex ions and coordination compounds, or "complexes" (Sections 14.10 and 22.1), which is the basis for some of the most exciting and active areas of chemical research today. Inorganic ions in the human body are present in coordination compounds and complex ions (bioinorganic chemistry); metals form many, many coordination compounds with organic compounds (organometallic chemistry); and catalysts often function via coordination compounds or complex ions. Some coordination compounds are extremely stable, whereas others are unstable; many have colors that are quite different from the colors of their constituents; some are readily soluble in water and others are soluble only in nonpolar solvents; some are volatile and others are not. By proper tailoring of a complex ion or coordination compound, almost any property can be built into it.

To review, complexes such as

$$\underset{NH_3\ \text{molecule ligand}}{\overset{Ag^+\ \text{ion}}{[Ag(NH_3)_2]^+}} \qquad \underset{F^-\ \text{ion ligand}}{\overset{Co^{3+}\ \text{ion}}{[CoF_6]^{3-}}}$$

are formed by the combination of a central metal atom or cation and two or more molecules or ions referred to as ligands. The ligands contain nonmetal atoms that donate electron pairs (Lewis bases) and the central metal ion accepts electron pairs (Lewis acids) into unoccupied orbitals. The formation or dissociation of complex ions in aqueous solution occurs by stepwise gain or loss of the ligands. For example, for the dissociation of the complex ion $[Ag(NH_3)_2]^+$

$$[Ag(NH_3)_2]^+ \rightleftharpoons [Ag(NH_3)]^+ + NH_3(aq)$$
$$[Ag(NH_3)]^+ \rightleftharpoons Ag^+ + NH_3(aq)$$

overall dissociation

$$[Ag(NH_3)_2]^+ \rightleftharpoons Ag^+ + 2NH_3(aq)$$

Complexes may be charged or neutral, depending upon the species that combine, for example

$$2{:}NH_3 + Ni^{2+} \longrightarrow [Ag({:}NH_3)_2]^+ \quad \textbf{(31.1)}$$
$$4{:}CO + Ni \longrightarrow [Ni({:}CO)_4] \quad \textbf{(31.2)}$$
$$4{:}\ddot{C}l{:}^- + Pt^{2+} \longrightarrow [Pt({:}\ddot{C}l{:})_4]^{2-} \quad \textbf{(31.3)}$$
$$2{:}\ddot{C}l{:}^- + 2{:}NH_3 + Pt^{2+} \longrightarrow [({:}\ddot{C}l{:})_2Pt({:}NH_3)_2] \quad \textbf{(31.4)}$$
$$5{:}\ddot{C}l{:}^- + {:}NH_3 + Pt^{4+} \longrightarrow [({:}\ddot{C}l{:})_5Pt({:}NH_3)]^- \quad \textbf{(31.5)}$$
$$4{:}NO_2^- + Pt^{2+} \longrightarrow [Pt({:}NO_2)_4]^{2-} \quad \textbf{(31.6)}$$

The donor atom may be part of a molecule (Equations 31.1 and 31.2), it may be an ion (Equations 31.3–31.5), or it may be part of an ion (Equation 31.6). The charge on a complex is the sum of the charges of the constituent parts. For example, in $[PtCl_4]^{2-}$ the charge is found by adding $+2$ for the platinum and $4 \times (-1)$ for the four Cl^- ions to obtain the charge of -2 for the complex ion.

The ligand in a complex is said to be "coordinated to" the atom or ion that is in the center of the new structure. Any neutral compound that contains a metal atom and its associated ligands is called a coordination compound. Such a compound may be formed between a complex ion and other ions, for example, $[Ag(NH_3)_2]^+Cl^-$ or $(K^+)_2[Pt(NO_2)_4]^{2-}$. Or the complex itself may be neutral, for example, $[Pt(NH_3)_2(NO_2)_2]$, in which platinum has oxidation state $+2$, or $[Ni(CO)_4]$, in which nickel has oxidation state zero. The formula of the complex is usually enclosed in square brackets.

The **coordination number** of the central metal atom or ion in a complex is the number of nonmetal atoms bonded to that atom or ion (Table 31.1).

Table 31.1
Common Coordination Numbers of Cations

Ag^+	2	Hg^{2+}	4	Al^{3+}	4, 6	Pd^{4+}	6
Au^+	2, 4	Ca^{2+}	6	Au^{3+}	4	Pt^{4+}	6
Cu^+	2, 4	Co^{2+}	4, 6	Co^{3+}	6	Zr^{4+}	8
Li^+	4	Cu^{2+}	4, 6	Cr^{3+}	6	Hf^{4+}	8
Tl^+	2	Fe^{2+}	6	Fe^{3+}	6	Th^{4+}	8
		Ni^{2+}	4, 5, 6	Ir^{3+}	6		
		Pb^{2+}	4	Os^{3+}	6		
		Pd^{2+}	4	Sc^{3+}	6		
		Pt^{2+}	4				
		V^{2+}	6				
		Zn^{2+}	4				

To give a few examples—nickel(II) can have coordination numbers of four, five, or six, as in $[Ni(CN)_4]^{2-}$, $[Ni(CN)_5]^{3-}$, and $[Ni(NH_3)_6]^{2+}$. With cobalt(III), the coordination number is almost invariably six; and with zirconium (IV) and hafnium(IV), it is often eight, as in $[ZrF_8]^{4-}$ and $[HfF_8]^{4-}$. The most common coordination numbers are two, four, and six, but others ranging from three to ten are known.

EXAMPLE 31.1 Structure of Complexes

For each species listed, identify the ligands and give the coordination number and oxidation number of the central atom or ion: (a) $[Co(NH_3)_5(SO_4)]Br$, (b) $[Ag(CN)_2]^-$, (c) $[Ir(NH_3)_3Cl_3]$.

(a) In this coordination compound the complex ion is the cation $[Co(NH_3)_5(SO_4)]^+$. **The ligands are five ammonia molecules and a sulfate ion,** so **the coordination number of the central cobalt ion is 6.** To give the complex ion its +1 charge, the cobalt must be in its **+3 oxidation state.**

$$\overset{+3}{[Co}(NH_3)_5(\overset{-2}{SO_4})]^+$$

[Note that SO_4^{2-}, although it has a charge of −2, occupies only one coordination position in this complex; see Table 31.2.]

(b) In this complex ion there are **two CN^- ions as ligands,** so the coordination number of the silver ion is **2.** Silver has only one common oxidation state, **+1,** and this gives the complex ion its −1 charge.

$$[\overset{+1}{Ag}\,\overset{2\times(-1)}{(CN)_2}]^-$$

(c) In this coordination compound the absence of any other species (or of a charge) outside the bracket shows that this is a neutral molecule, not an ion. The ligands are **ammonia molecules and chloride ions.** The coordination number of the iridium is **6,** and it must have a **+3 oxidation number.**

$$[\overset{+3}{Ir}(NH_3)_3\,\overset{3\times(-1)}{Cl_3}\,]$$

EXERCISE In each of the following, (a) $[Pt(NH_3)_5Cl]Cl_3$, (b) $[Mn(H_2O)_6]^{3+}$, and (c) $[CoF_6]^{3-}$, identify the ligands and give the coordination number and the oxidation number for the central atom or ion.

Answers (a) coordination compound containing complex cation, NH_3 and Cl^-, 6, +4; (b) complex cation, H_2O, 6, +3; (c) complex anion, F^-, 6, +3.

31.2 NOMENCLATURE OF COMPLEXES

A few simple rules cover what you must know about naming complexes:

1. The ligands are named first. The prefixes di, tri, tetra, and so on are used to indicate the number of each kind of ligand present. Negatively charged ligands are given names that end in o. Sometimes the prefixes bis (two ligands), tris (three ligands), and tetrakis (four ligands) are also used, especially when the ligand name is complicated or already includes di, etc. For example, "bis" is used in the following name because it includes two "di's," dichlorobis (ethylenediamine)cobalt(III) ion, $[CoCl_2(NH_2CH_2CH_2NH_2)_2]Cl$.
2. The ligand names are given in alphabetical order. [In an older system, negative ligands were named first, neutral ones second, and positive ones last.] Some ligands have familiar names also used in naming other types of compounds (e.g., chloro, cyano); others have names special to complexes (e.g., carbonato, CO_3^{2-}; aqua, H_2O). Table 31.2 lists the names of the most common ligands. Note that the ammine group—two m's—is NH_3.

Table 31.2
Common Ligands
The donor atom in each ligand is shown in color. The NO_2^- ion can bond as a ligand through either the N atom (nitro) or the O atom (nitrito). The SCN^- group can bond as a ligand through either S or N, depending chiefly on the metal in the complex. In fact, the SCN^- or CN^- groups can attach themselves to two metal ions simultaneously.

Ligand	Name	Ligand	Name
F^-	fluoro	CO_3^{2-} [a]	carbonato
Cl^-	chloro	$C_2O_4^{2-}$ [a]	oxalato
Br^-	bromo	OSO_3^{2-}	sulfato (SO_4^{2-})
I^-	iodo	NH_3	ammine
CN^-	cyano	H_2O	aqua
NCS^-	isothiocyanato	CO	carbonyl
SCN^-	thiocyanato	NO	nitrosyl
ONO^-	nitrito	PR_3	trialkyl- or triarylphosphine
NO_2^-	nitro		
OH^-	hydroxo		

[a] Carbonate and oxalate ions usually bond to the central metal atom through two oxygen atoms (see Section 31.3).

3. The name of the central metal atom or ion followed by its oxidation state in parentheses may be given after the ligand names. The metal name is *not* separated from the ligand names by a space.
4. When a complex ion has a negative charge, the name of the central metal atom is given the ending *ate*. For some of the elements, the ion name is based on the Latin name from which the symbol is derived, for example, ferrate for iron, Fe. When naming only the ion, the word "ion" is used in the name.

Using these rules, the complexes formed in reactions (31.1)–(31.6) receive the following names:

$[Ag(NH_3)_2]^+$	diamminesilver(I) ion
$[Ni(CO)_4]$	tetracarbonylnickel(0) (commonly called "nickel carbonyl")
$[PtCl_4]^{2-}$	tetrachloroplatinate(II) ion
$[Pt(NH_3)_2Cl_2]$	diamminedichloroplatinum(II)
$[PtCl_5(NH_3)]^{3-}$	amminepentachloroplatinate(II) ion
$[Pt(NO_2)_4]^{2-}$	tetranitroplatinate(II) ion

In naming a coordination compound, the name of the cation is given first as usual, followed by name of the anion.

$K^+[Pt(NH_3)Cl_5]^-$	potassium amminepentachloroplatinate(IV)
$[Co(NH_3)_4SO_4]^+NO_3^-$	tetraamminesulfatocobalt(III) nitrate

EXAMPLE 31.2 Nomenclature of Complexes

Write the names for the following two complexes:
(a) $[Cd(CN)_4]^{2-}$ (b) $[Co(NH_3)_4(CO_3)]Cl$

(a) As a ligand, CN^- is named cyano (Table 31.2). With a total ligand charge of $4 \times (-1) = -4$ and a charge on the complex of -2, the cadmium ion must have a $+2$ charge. Naming the ligand first and changing "cadmium" to "cadmate" because the complex ion has a negative charge gives the compound name as **tetracyano-cadmate(II) ion.**
(b) In this compound, the complex ion is the cation ($+1$ charge) and is named first. The two types of ligands are CO_3^{2-}, carbonato, and NH_3, ammine. The ligand charge is $1 \times (-2) = -2$ (for the single CO_3^{2-} ion), so the cobalt ion must have a $+3$ charge. The complex name is **tetraamminecarbonatocobalt(III) chloride.**

EXERCISE Name the following: (a) $K_4[Fe(CN)_6]$, and (b) $[Co(NH_3)_2(NO_2)_4]^-$.

Answers (a) potassium hexacyanoferrate(II), (b) diamminetetranitrocobaltate(III) ion.

EXAMPLE 31.3 Nomenclature of Complexes

Write the formulas for (a) sodium tetrachlorodicyanochromate(III) and (b) pentaammineaquacobalt(III) chloride.

(a) In this coordination compound the "ate" ending shows that the complex ion is the anion. With six ligands, each with a charge of -1 (four Cl^- and two CN^-), the total ligand charge is -6. With chromate(III), Cr^{3+}, as the central ion, the complex ion must have a charge of -3. The formula is written as $\mathbf{Na_3[CrCl_4(CN)_2]}$.

(b) Here, the complex ion is the cation. The ligands are H_2O and NH_3. The cobalt is present as the $+3$ ion and, as both ligands are neutral, this is the charge of the complex ion. The formula is, therefore, $\mathbf{[Co(H_2O)(NH_3)_5]Cl_3}$.

EXERCISE Write the formulas for the following: (a) hexaamminechromium(III) chloride, (b) pentaamminechlorochromium(III) chloride, (c) tetraaquanickel(II) ion, and (d) diiodocuprate(I) ion.

Answers (a) $[Cr(NH_3)_6]Cl_3$, (b) $[Cr(NH_3)_5Cl]Cl_2$, (c) $[Ni(H_2O)_4]^{2+}$, (d) $[CuI_2]^-$.

31.3 CHELATION

In many cases, two or more atoms with unshared pairs of electrons are present in the same ion or molecule. If they are not too close together or too far apart, they may coordinate to the same metal atom or ion to form a ring. For example, oxalate ($C_2O_4{}^{2-}$) ion usually behaves in this way:

$$M \begin{matrix} \swarrow O-C=O \\ \quad | \\ \nwarrow O-C=O \end{matrix}$$

oxalato complex

Five- and six-membered rings, which are much less strained than four-membered rings, are very common. A few examples are listed in Table 31.3.

The phenomenon of ring formation by a ligand in a complex is called **chelation** and the ring formed is called a **chelate ring** (pronounced "key-late," from the Greek *kela* meaning "crab's claw"). A ligand which contains two donor atoms by which it can form

Table 31.3
Some Common Chelating Agents
The electron pairs available for donation are shown in color.

Formula	Abbreviation	Formula	Abbreviation
Form five-membered rings		**Form six-membered rings**	
$\ddot{N}H_2CH_2CH_2\ddot{N}H_2$ *ethylenediamine (bidentate)*	en	$\ddot{N}H_2CH_2CH_2CH_2\ddot{N}H_2$ *trimethylenediamine (bidentate)*	tm
$(\ddot{N}H_2CH_2C(=O)\ddot{O})^-$ *glycinate ion (bidentate)*	gly	$[CH_3-C(=\ddot{O})-CH=C(-\ddot{O})-CH_3]^-$ *acetylacetonate ion (bidentate)*	acac
$\begin{bmatrix} :O-C=O \\ \quad\ \ | \\ :O-C=O \end{bmatrix}^{2-}$ *oxalate ion (bidentate)*	ox		
$\ddot{N}H_2CH_2CH_2\ddot{N}HCH_2CH_2\ddot{N}H_2$ *diethylenetriamine (tridentate)*	dien		

(a) Ethylenediamminetetraacetic acid (edta) anion

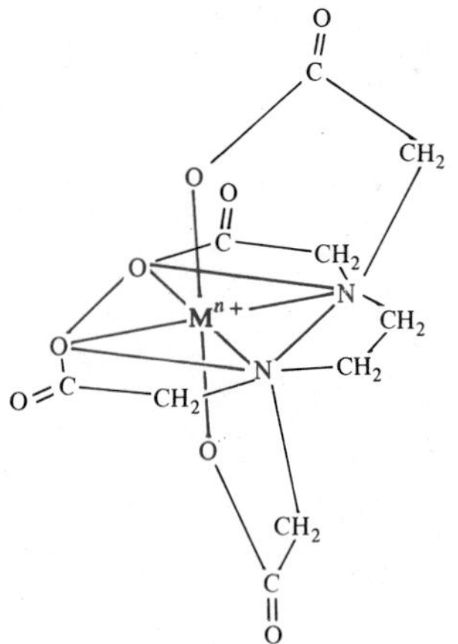

(b) edta-metal complex with five chelate rings

Figure 31.1
Ethylenediaminetetraacetic Acid and Complex Formation
(a) The donor atoms are shown in color. (b) Five rings form in the chelate, with the metal atom bonded to the six donor atoms and located at the center of an octahedron.

a chelate ring is referred to as **bidentate.** The oxalato ligand pictured above is bidentate. (There also are tridentate, quadridentate, ... ligands.)

Chelation greatly increases the stability of a complex. This is of tremendous value in altering the properties of metal ions by complexation. Chelating agents, also known as sequestering agents, have a great many practical applications, some of which are discussed in Section 31.8. One of the most powerful chelating ligands known is the anion of ethylenediaminetetraacetic acid (edta) (Figure 31.1a). This ion is able to attach itself to a single metal ion through six donor atoms—both nitrogen atoms and one oxygen atom of each —CH_2COO^- (acetate) group. Five chelate rings are formed, with one central metal ion common to all of them (Figure 31.1b). The six atoms coordinated to the metal ion are located at the corners of an octahedron (see Figure 31.2).

31.4 MOLECULAR GEOMETRY AND ISOMERISM

The central metal atom or ion and the ligands in a complex assume a definite geometry, as do the atoms in a molecule. The resulting shapes are, in many cases, like those of comparable molecules as predicted by VSEPR (see Table 11.1). However, some variations in geometry occur, partly because in complexes the bonding and nonbonding electrons are in *d* subshells rather than *p* subshells.

Complexes with two ligands (e.g., $[ClAgCl]^-$) are linear. Many complexes with four ligands are tetrahedral (e.g., $[Be(H_2O)_4]^{2+}$), but some are square-planar (e.g., $[Pt(NH_3)_4]^{2+}$), as shown in Figure 31.2. Two geometries occur with about equal frequency for five-coordinate complexes: square pyramidal (e.g., $[CuCl_5]^{3-}$) and trigonal bipyramidal (e.g., $[Ni(CN)_5]^{3+}$. Most six-coordinate complexes are octahedral, but a few are triangular prismatic.

Compounds that differ in molecular structure but have the same molecular formula are called **isomers.** The fixed positions of the ligands in a complex allow for several different types of isomerism. **Structural isomerism**—the occurrence of isomers that differ in how the atoms or groups of atoms are connected—is common because the arrangement of the ligands about the metal atom can vary. For example, ligands and ions can change places, in what is called *ionization isomerism.* In the following two isomers the Cl^- and SO_4^{2-} ions have changed places

$$[Co(NH_3)_5SO_4]Cl \qquad [Co(NH_3)_5Cl]SO_4$$

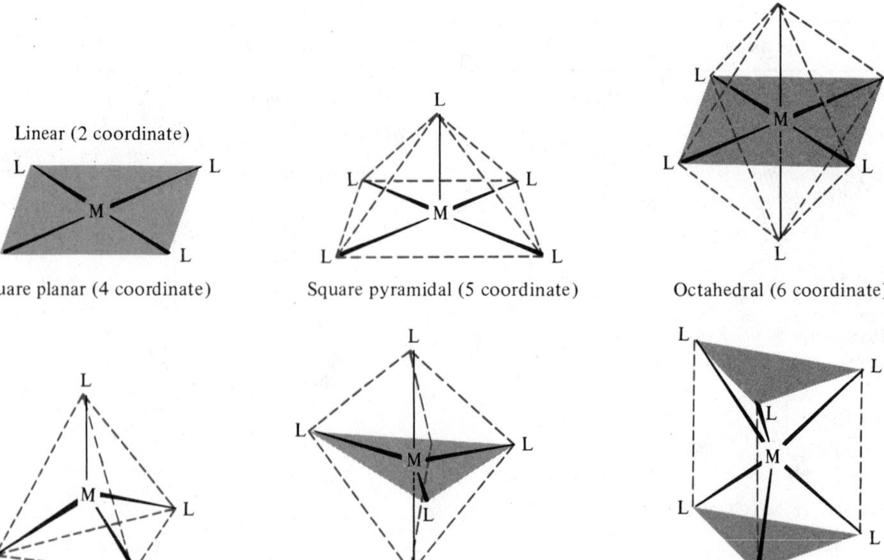

Figure 31.2
Geometry of Complexes
The square planar, octahedral, and tetrahedral geometries are the most common.

Groups that can coordinate through different atoms create *linkage isomerism,* as illustrated by $:NCS^-$ and $:SCN^-$ or $:ONO^-$ and $:NO_2^-$. Differences in how water molecules are bonded in a complex create *hydrate isomerism,* illustrated by the three forms of $CrCl_3 \cdot 6H_2O$ mentioned in Section 30.6c (in which the six coordinated groups are all H_2O molecules, five H_2O molecules plus one Cl^- ion, or four H_2O molecules plus two Cl^- ions).

***Cis–trans* isomerism** occurs when atoms or groups of atoms can be arranged in different ways on two sides of a rigid structure. For example, two identical ligands in a square-planar complex can be on either adjacent or opposite corners of the square plane:

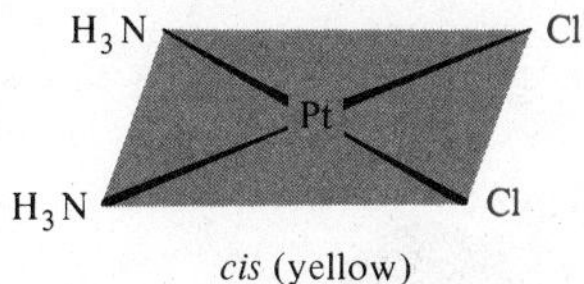

cis (yellow)

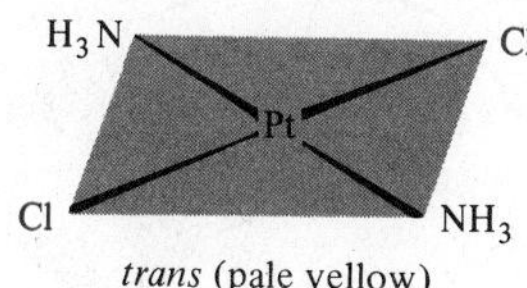

trans (pale yellow)

Diamminedichloroplatinum (II), a square-planar complex

In ***cis* isomers** the groups under consideration are on the same side of a rigid structure; in ***trans* isomers** they are on opposite sides. Note in the preceding structures that although the arrangement in space differs, all the atoms are connected in the same way in both cases—an example of **geometric isomerism.** *Cis–trans* isomerism is one form of geometric isomerism.

Many complexes, including octahedral complexes of the general formula $[Ma_4b_2]$, where a and b are different unidentate ligands, can have *cis* and *trans* isomers.

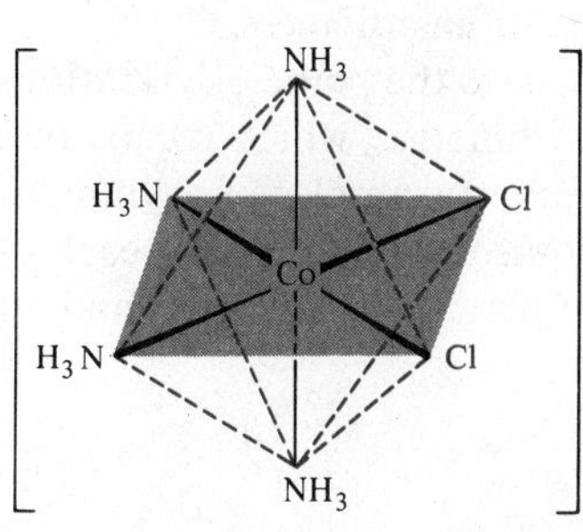

cis (violet)

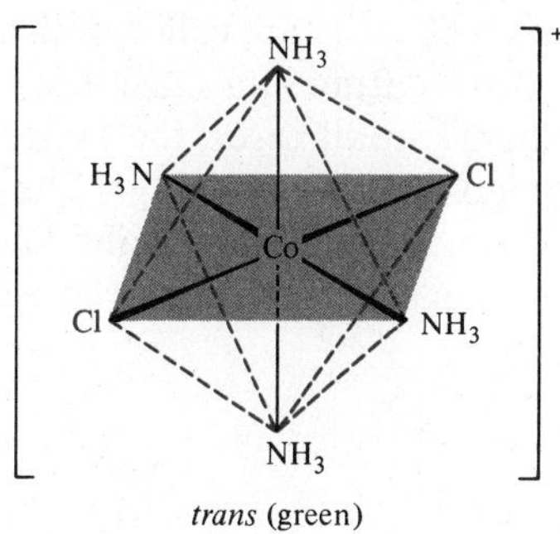

trans (green)

Tetraamminedichlorocobalt (III) ion, an octahedral complex of the type Ma_4b_2

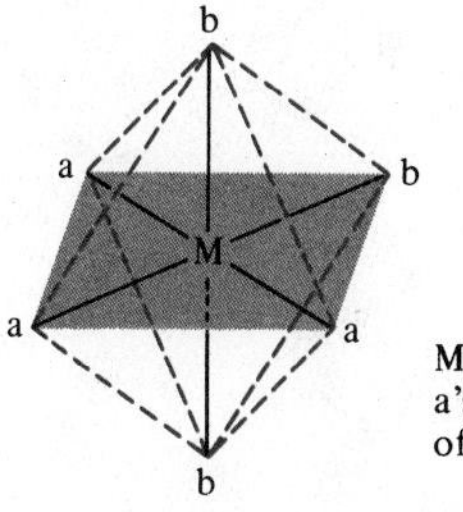

Meridional isomer – a's at three corners of the square plane

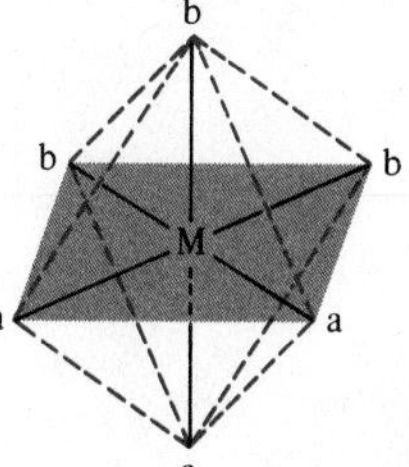

Facial isomer – a's at corners of triangular face

Figure 31.3
Geometrical Isomerism in an Octahedral Complex of the General Formula Ma_3b_3
A complex of this type is $[Co(NH_3)_3(NO_2)_3]$.

With three of each type of ligand, $[Ma_3b_3]$, a different type of geometric isomerism is also possible here, one in which the arrangements are designated as meridional (*mer*) and facial (*fac*), as shown in Figure 31.3.

The structural or geometric isomers of a given compound have different properties. In a third kind of isomerism, the isomers are almost identical in their properties and as a result are often difficult to separate. Many organic compounds and coordination compounds can exhibit this type of isomerism, which is detected by the use of polarized light.

Ordinary light rays vibrate in all planes perpendicular to the direction of travel of the rays. When such light is passed through a polarizing prism or a piece of polarizing plastic like that used in Polaroid sunglasses, the part of the light that is transmitted vibrates in one plane only and is called **plane-polarized light** (Figure 31.4). When a beam of plane-polarized light passes through certain molecules or ions (or their solutions),

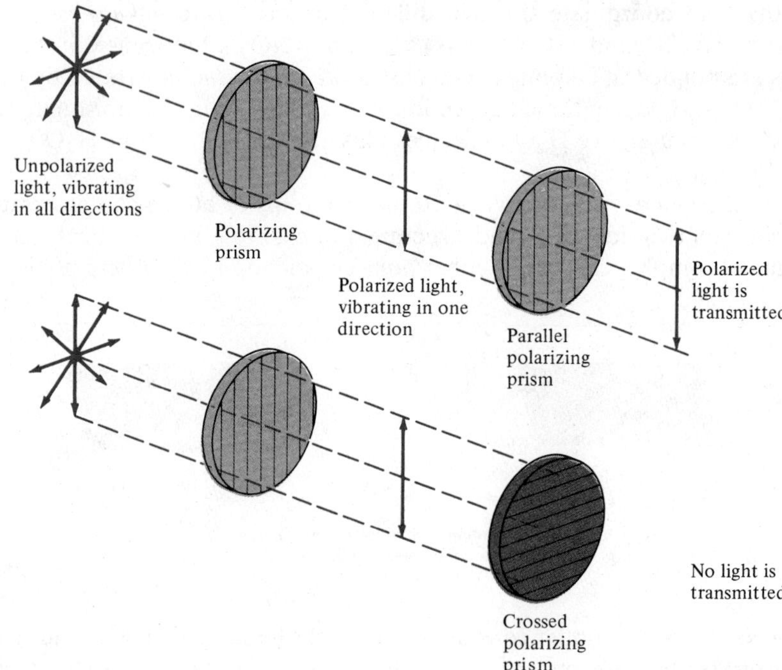

Figure 31.4
Polarized Light
The effect of parallel and crossed polarizers can be observed with a pair of lenses from polarizing sunglasses.

the plane of vibration of the light is rotated. The amount of rotation varies from compound to compound and is measured by an instrument called a polarimeter (Figure 31.5). **Optical isomerism** is the occurrence of pairs of molecules of the same molecular formula that rotate plane-polarized light equally in opposite directions. The members of such pairs are called **optical isomers,** or **enantiomers.**

What is the nature of a substance that can rotate the plane of vibration of plane-polarized light? In all cases, the molecule is asymmetric, which means that it has a nonsuperimposable mirror image. The most familiar examples of asymmetric objects are your left and right hands. Hold your two hands up, palms facing each other. You can imagine that each is the mirror image of the other. Now place one hand on the table

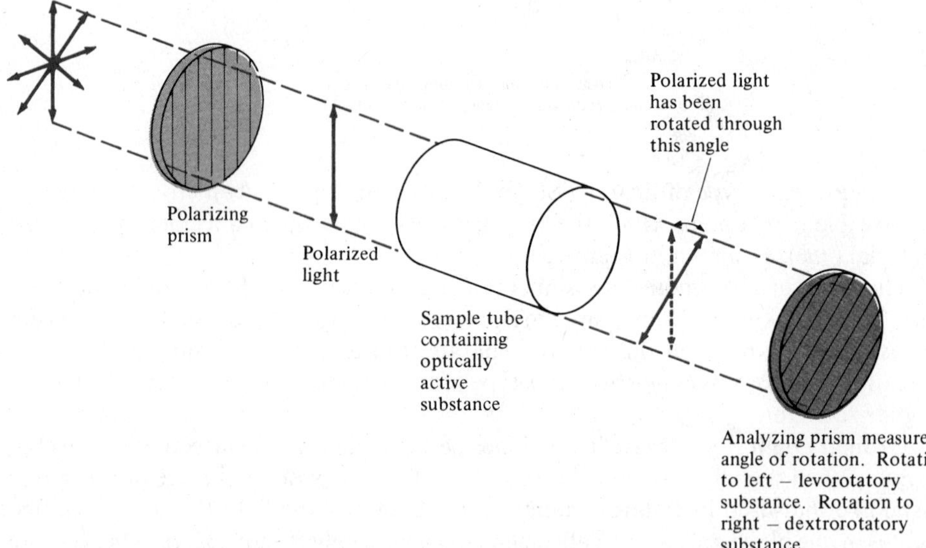

Figure 31.5
Principle of a Polarimeter
The extent to which the plane of polarization has been rotated by the sample is determined by rotating the analyzing prism until transmission of the beam is maximized.

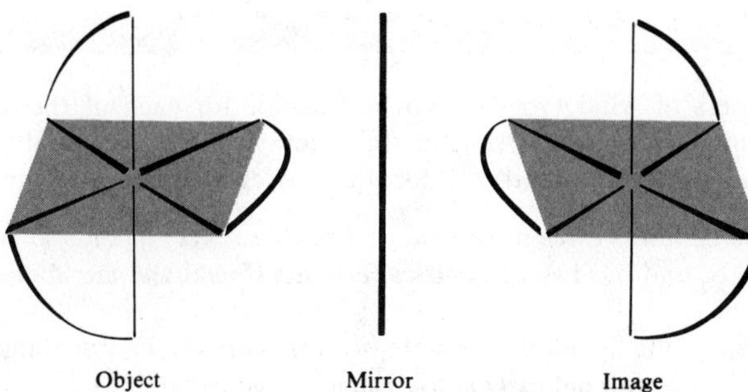

Figure 31.6
Optically Active Isomers of a Trichelate Octahedral Complex

and try to place the other on top of it so that *all* parts coincide—you will find that it can't be done.

Molecules that are asymmetric and rotate plane-polarized light are said to be **optically active.** The property of having two forms that are nonsuperimposable mirror images and are therefore optically active is called **chirality** (pronounced kī-ral-ĭ-tē). Molecules of this type are referred to as *chiral.*

Optical activity is common in complexes that contain chelate rings. Figure 31.6 represents schematically the two optically active isomers of a complex in which three chelate rings are joined to a six-coordinate central atom. The two isomers are said to have different *spatial configurations.* All six-coordinate complexes with three chelate rings are chiral.

Six-coordinate complexes that contain two chelate rings exhibit both *cis–trans* and optical isomerism, as illustrated in Figure 31.7 for a cobalt complex. The isomer in which the two nonchelated groups are in the *cis* positions is optically active. The *trans* isomer, by contrast, is not optically active—it is superimposable upon its mirror image.

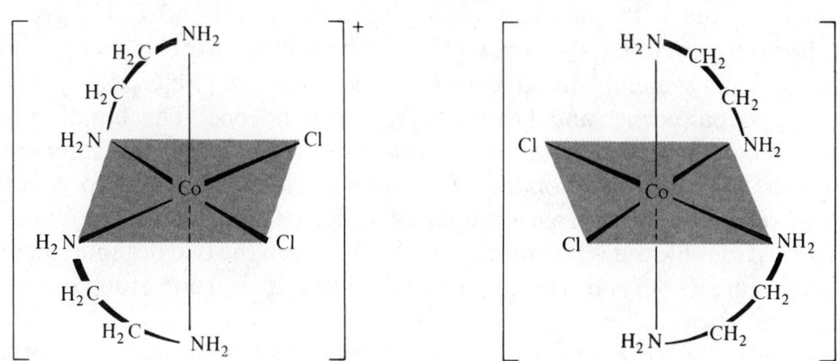

Optical isomers of the *cis* complex; both violet

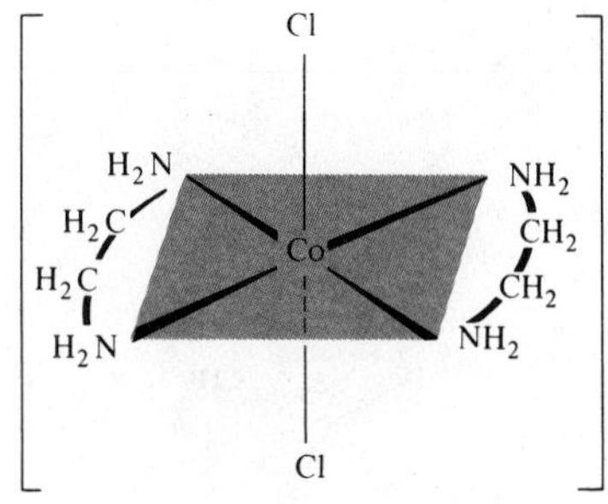

Nonchiral *trans* isomer; green

Figure 31.7
Isomers of Dichlorobis(ethylenediammine)cobalt(III) Ion, a Dichelate Octahedral Complex

EXAMPLE 31.4 Isomerism

How many isomers of what types would be possible for each of the following complexes? (a) triamminetrinitrocobalt(III) ion, (b) tris(ethylenediamine)cobalt(III) ion, (c) tetraamminediaquacobalt(III) ion, (d) dichlorobis(ethylenediamine)platinum(IV)chloride?

(a) This complex contains three nitro ligands and three NH_3 ligands. It is octahedral of the type Ma_3b_3 and has **two geometrical isomers (facial and meridional)** [see Figure 31.3].
(b) The ethylenediamine ligand is bidentate, so this is an octahedral complex with three chelate rings. It has **two optically active isomers** [see Figure 31.6].
(c) Here there are two H_2O ligands and four NH_3 ligands, to give an octahedral complex of the type Ma_2b_4. There are **two possible *cis-trans*, geometrical isomers** [see $Pt(NH_3)_2Cl_2$ structures above].
(d) The two bidentate ethylenediamine ligands make this a dichelate, octahedral complex. There are **two optical isomers—the *cis* isomers—and one *trans* isomer** [see Figure 31.7].

EXERCISE Describe the isomerism in (a) $K[Co(NO_2)_4(H_2O)_2]$, (b) $[NiBrCl(CN)_2]^{2-}$, and (c) $[Cr(C_2O_4)_3]^{3-}$.

Answers (a) *cis* and *trans* isomers of the octahedral complex, (b) *cis* and *trans* isomers of the square-planar complex, (c) two optical isomers of the octahedral complex.

31.5 METALLOCENES AND METAL CLUSTERS

Several groups of compounds are considered to be coordination compounds even though they do not exactly fit the definition given earlier because they do not have central metal atoms or ions and ligands. For example, metals and certain unsaturated hydrocarbons combine to give metallocenes, the best known of which contain the cyclopentadienyl anion, $C_5H_5^-$. Cyclopentadiene (C_5H_6; Figure 31.8a) readily loses a proton when treated with sodium to give the ionic compound $C_5H_5^-Na^+$, which contains the cyclopentadienyl group, $C_5H_5^-$ (abbreviated cp). This compound reacts with salts of many metals to give derivatives such as $[Fe(C_5H_5)_2]$, ferrocene; $[Co(C_5H_5)_2]$, cobaltocene; and $[Ru(C_5H_5)_2]$, ruthenocene. The bonding in these compounds is covalent. The metallocenes are insoluble in water but soluble in nonpolar solvents, and they are volatile. The organic rings can undergo a variety of substitution reactions, so a large number of different types of compounds can be formed. In most metallocenes the metal atom lies between the two organic rings to form a sandwich (Figure 31.8c) and is not bonded to any specific carbon atom, but is bonded equally to all of them.

Compounds that contain metal–metal bonds are also similar to coordination compounds. Mercury(I) compounds such as mercury(I) chloride (Hg_2Cl_2) have been known for many years. The two mercury atoms are joined by a single covalent bond to which each atom contributes one electron.

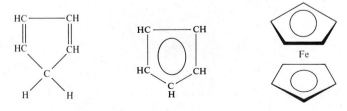

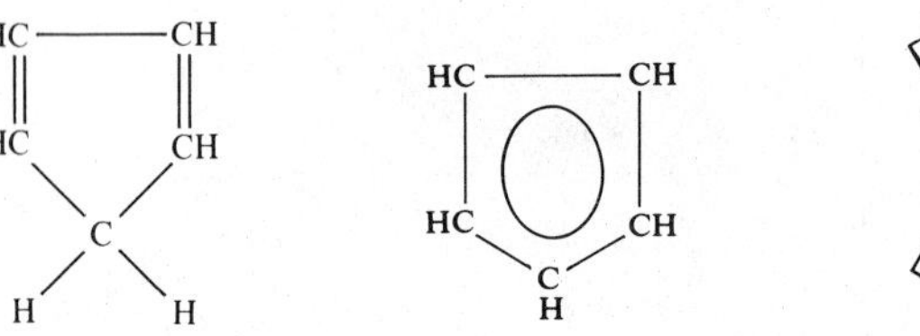

Figure 31.8
Metallocenes Derived from Cyclopentadiene

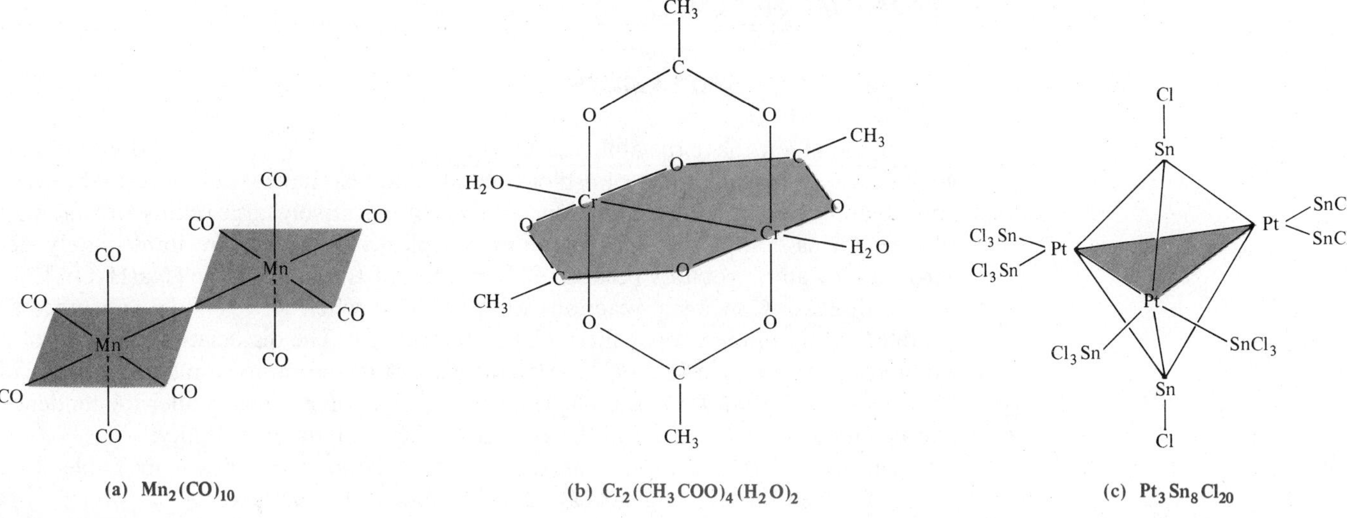

Figure 31.9
Some Simple Metal Cluster Compounds

More recently, it has been discovered that metal atoms can be bonded to each other in many ways in compounds now referred to as metal cluster compounds. In some cases, as in Hg_2Cl_2 and $Mn_2(CO)_{10}$, the metal atoms are connected solely by covalent bonds (Figure 31.9a). In others, there are also neutral or negatively charged nonmetals binding the metal atoms together. One of the simplest of these is the dimer of chromium(II) acetate hydrate, $Cr_2(CH_3COO)_4 \cdot 2H_2O$ (Figure 31.9b). The two oxygen atoms of each acetate group bind to different chromium atoms, so that the four acetate groups form a paddle wheel configuration.

In some metal clusters different kinds of metals are joined together. An example is $Pt_3Sn_8Cl_{20}$, in which each platinum atom is bonded to the other two, as well as to four tin atoms (Figure 31.9c). Some clusters that are very large have been prepared; their structures are quite beautiful (Figure 31.10). Cluster compounds have received a great deal of attention in recent years because of their usefulness as catalysts.

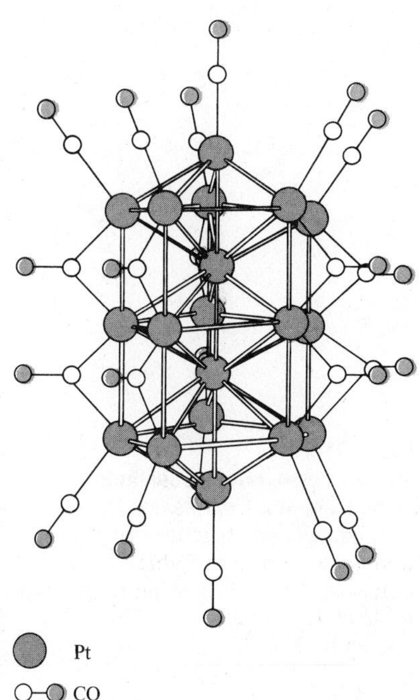

Figure 31.10
A Large Metal Cluster Anion, $[Pt_{19}(CO)_{22}]^{4-}$
The dimensions of this cluster are 80 by 110 nm.
Source: Reprinted with permission from *Chemical and Engineering News*, February 8, 1982, p. 13. Copyright 1982 American Chemical Society.

SUMMARY OF SECTIONS 31.1–31.5

A complex is an ion (a complex ion) or a neutral compound (a coordination compound) that contains a metal atom or ion surrounded by coordinately bonded ligands. The rules for naming complex ions and their compounds are given in Section 31.2. In chelation two or more atoms with unshared electron pairs in the same molecule coordinate to a single metal atom or ion.

In addition to structural isomerism, the fixed positions of the ligands in complexes create possibilities for both geometric and optical isomerism. In *cis–trans* isomerism (a form of geometric isomerism), substituents are on the same or opposite sides of a rigid structure. Optical isomerism (detected by the use of polarized light) occurs when a molecule or ion is asymmetrical and therefore has a nonsuperimposable mirror image.

Metallocenes, in most of which a metal atom or ion lies between two organic rings, and metal cluster compounds, which contain metal–metal bonds, are classified as coordination compounds.

PROPERTIES OF COMPLEXES

31.6 STABILITY AND LABILITY

All metal ions have the ability to form complex ions. Those with small radius and high charge, especially those of *d*-block metals that also have vacant *d* orbitals, do so most readily; yet even the sodium ion, with its comparatively large radius and its small charge, forms complexes. Most sodium complexes, however, are immediately destroyed by water, probably because of formation of the more stable $[Na(H_2O)_x]^+$.

In discussing chemical reactions we noted that when we refer to a substance as "stable" we are talking about thermodynamic stability. The dissociation constant of a complex (introduced in Section 22.1) is a measure of its thermodynamic stability (recall that $\Delta G^\circ = -2.303RT \log K$). The smaller the value of K_d, the smaller the tendency for a complex to dissociate and the greater its thermodynamic stability.

The overall dissociation constants of some complexes, shown in Table 31.4, illustrate several interesting facts about the stability of complexes:

1. Metal ions vary widely in the stability of the complex ions which they form, even those with the same ligand and those with the same coordination number (in Tables 31.4–2, 3, 5; 7, 8).
2. The higher the oxidation state of the metal, the more stable are the complexes it forms with a given ligand. This is illustrated by 16 and 17, 25 and 26, 27 and 28. The ligand–metal bond strength is greater for the higher metal oxidation states.
3. The stability of halide complexes often decreases in the order $I^- > Br^- > Cl^-$ (compare 7, 10, and 12).
4. The cyanide ion is like the halide ions in many respects, but far exceeds them in the stability of its complexes (compare 7, 10, and 12 with 14).
5. Chelate ring formation greatly increases the stability of complexes. Monoamines are rather poor coordinating agents (33), but diamines, if they can form five-membered rings, give extremely stable complexes (24). Much of the extra stabilization of complexes by chelation is attributable to the entropy factor (Section 23.5). Compare the K_d value for $[Cd(NH_3)_4]^{2+}$ (2), which has four unidentate ligands, with that of $[Cd(en)_2]^{2+}$ (24), which has two bidentate ligands. In the equilibrium

$$[Cd(NH_3)_4]^{2+} + 2en(aq) \rightleftharpoons [Cd(en)_2]^{2+} + 4NH_3(aq)$$

Table 31.4
Overall Dissociation Constants of Some Common Complexes
The names and structures of en, ox, and gly are given in Table 31.3. For edta, see Figure 31.1. A more extensive table of K_d values is given in Appendix V4.

Complex	K_d	Complex	K_d
1. $[Ag(NH_3)_2]^+$	6.2×10^{-8}	**18.** $[Cu(OH)_4]^{2-}$	7.6×10^{-17}
2. $[Cd(NH_3)_4]^{2+}$	1×10^{-7}	**19.** $[Zn(OH)_4]^{2-}$	5×10^{-21}
3. $[Cu(NH_3)_4]^{2+}$	1×10^{-13}	**20.** $[CaP_2O_7]^{2-}$	1×10^{-5}
4. $[Ni(NH_3)_6]^{2+}$	1×10^{-9}	**21.** $[MgP_2O_7]^{2-}$	2×10^{-6}
5. $[Zn(NH_3)_4]^{2+}$	3.46×10^{-10}	**22.** $[CuP_2O_7]^{2-}$	2.0×10^{-7}
6. $[AgCl_2]^-$	9×10^{-6}	**23.** $[Ag(en)]^+$	1×10^{-5}
7. $[CdCl_4]^{2-}$	9.3×10^{-3}	**24.** $[Cd(en)_2]^{2+}$	2.60×10^{-11}
8. $[PdCl_4]^{2-}$	6×10^{-14}	**25.** $[Co(en)_3]^{2+}$	1.52×10^{-14}
9. $[AgBr_2]^-$	7.8×10^{-8}	**26.** $[Co(en)_3]^{3+}$	2.04×10^{-49}
10. $[CdBr_4]^{2-}$	2×10^{-4}	**27.** $[Fe(ox)_3]^{4-}$	6×10^{-6}
11. $[PdBr_4]^{2-}$	8.0×10^{-14}	**28.** $[Fe(ox)_3]^{3-}$	3×10^{-21}
12. $[CdI_4]^{2-}$	8×10^{-7}	**29.** $[Cu(gly)_2]$	5.6×10^{-16}
13. $[Ag(CN)_2]^-$	1×10^{-22}	**30.** $[Zn(gly)_2]$	1.1×10^{-10}
14. $[Cd(CN)_4]^{2-}$	8.2×10^{-18}	**31.** $[Cu(edta)]^{2-a}$	1.38×10^{-19}
15. $[Au(CN)_2]^-$	5×10^{-39}	**32.** $[Zn(edta)]^{2-a}$	2.63×10^{-17}
16. $[Fe(CN)_6]^{4-}$	1.3×10^{-37}	**33.** $[Cd(CH_3NH_2)_4]^{2+}$	2.82×10^{-7}
17. $[Fe(CN)_6]^{3-}$	1.3×10^{-44}		

[a] Edta is the anion of ethylenediaminetetraacetic acid (Figure 32.1).

the formation of the chelate complex containing ethylenediamine is favored by entropy because the number of particles in the system increases. (There are three species in solution on the left in the equation above, and five on the right.)

6. The formation of fused chelate rings (rings with one or more bonds in common) gives complexes that are more stable that those containing rings that are not fused (compare 29 and 31; 30 and 32). The complexing agents in these cases are amino acids that form five-membered chelate rings, but with glycine, no fused rings are formed; edta forms five fused rings if all of its possible donor atoms coordinate.

The distinction between thermodynamic stability and kinetic stability is nicely illustrated by the behavior of certain complexes. For example, the rates of exchange of ligands in some cyano complexes have been studied by using cyano ligands labeled with radioactive carbon. The presence of the radioactive CN^- group in the complex or in solution can be detected. The relative kinetic stabilities of substances are described by the terms "labile" and "inert." A **labile complex** undergoes rapid exchange of its ligands with, say, a reaction half-life of a minute or less.

$$[Ni(CN)_4]^{2-} + 4C^*N^- \rightleftharpoons [Ni(C^*N)_4]^{2-} + 4CN^- \qquad t_{1/2} = \sim 30\text{ s}$$

As is true of this tetracyanonickelate(II) complex, a labile complex can be thermodynamically stable and dissociated to only a very small extent (the K_d value for $[Ni(CN)_4]^{2-}$ is about 1×10^{-30}), and yet the ligands of such a complex may be continuously undergoing exchange at a very rapid rate.

By contrast, an **inert complex** has a slow rate of ligand exchange, for example

$$[Cr(CN)_6]^{3-} + 6C^*N^- \rightleftharpoons [Cr(C^*N)_6]^{3-} + 6CN^- \qquad t_{1/2} = \sim 24\text{ days}$$

An inert complex which is thermodynamically unstable in acidic solution is $[Co(NH_3)_6]^{3+}$. Its thermodynamic instability is demonstrated by its spontaneous transformation to the more stable hexahydrated species cobalt(II) complex:

$$4[Co(NH_3)_6]^{3+} + 20H^+ + 26H_2O(l) \longrightarrow 4[Co(H_2O)_6]^{2+} + 24NH_4^+ + O_2(g)$$

Its inertness is demonstrated by the fact that at room temperature this reaction takes several days.

31.7 EFFECT OF COMPLEX FORMATION ON PROPERTIES

When a metal ion becomes part of a complex, most of its properties are changed. For example, when an excess of aqueous ammonia is added to a solution of a copper(II) salt, the $[Cu(NH_3)_4]^{2+}$ ion is formed, and the solution becomes dark blue. The salts of this new ion have quite different solubilities from the corresponding salts of the "simple" copper(II) ion, and they crystallize in different forms. Changes of this sort are seen whenever a complex is formed.

Consider cobalt(III) hydroxide, $Co(OH)_3$, and tris(ethylenediamine)cobalt(III) hydroxide, $[Co(en)_3](OH)_3$. The former is polymeric, insoluble, and very weakly basic (Section 16.11); the latter is easily soluble and is as strong a base as sodium hydroxide.

The reaction of tetraamminecopper(II) with cyanide ion is instructive, for it shows one ligand (ammonia) being displaced by a more strongly complexing ligand (cyanide ion). At the same time, the metal ion is reduced by the excess cyanide ion from Cu(II) to Cu(I).

$$2[Cu(NH_3)_4]^{2+} + 6CN^- \longrightarrow 2[Cu(CN)_2]^- + (CN)_2(aq) + 8NH_3(aq)$$

Such changes in oxidation state upon complex formation are not unusual. The reduction potential of a metal ion is always influenced by the nature of the complex in which it is held, though not always to the great extent that is shown by Cu^{2+},

$$Cu^{2+} + e^- \longrightarrow Cu^+ \qquad E^\circ = 0.158\ \text{V}$$
$$Cu^{2+} + 2CN^- + e^- \longrightarrow [Cu(CN)_2]^- \qquad E^\circ = 1.12\ \text{V}$$

It must be kept in mind that a complex ion in aqueous solution is in equilibrium with the simple hydrated metal cation. In many cases a solution of the complex ion contains a concentration of the metal ion that is small but sufficient for the reactions typical of that ion to take place. For example, a solution of $[Cu(NH_3)_4]^{2+}$ ($K_d = 1 \times 10^{-13}$) shows many of the chemical properties of the Cu^{2+} ion, including the precipitation of copper(II) sulfide when hydrogen sulfide is added

$$[Cu(NH_3)_4]^{2+} + H_2S(g) \longrightarrow CuS(s) + 2NH_4^+ + 2NH_3(aq)$$

31.8 COMPLEXES IN PRACTICAL APPLICATIONS AND IN NATURE

Many uses of complexing agents depend upon their ability to dissolve selectively or tie up metal ions, or to remove them from solution. The calcium, magnesium, and iron ions can be removed from hard water by complexing agents such as the tripolyphosphate ion (Section 28.13). Edta is added to foods such as mayonnaise to tie up metal ions that would otherwise accelerate spoilage. Transition metal catalysts often function by the temporary formation of coordinate covalent bonds to ligands, which then undergo reaction. Subsequent dissociation of the complex restores the catalyst for further reaction.

In analytical chemistry, an ion that would interfere with an analysis is held in solution as a complex while other ions are detected or removed as precipitates. There are "complexometric" titrations in which a complexing agent, frequently edta, is titrated into a solution of a metal ion and the endpoint detected by the color change of another complexing agent. We have already mentioned the use of complexes in hydrometallurgy (Section 26.4).

Some metals and semimetals are essential to living things, while others are poisonous, and still others (e.g., Cu, Se, As) are essential in small amounts but toxic in larger amounts. Both essential metals and poisons can be dissolved and transported as complexes.

Humic acids are organic compounds formed in the soil by the decay of organic matter. They bind metal ions needed by plants, transport them through the soil, and make them available to the plant roots. Chlorosis is a condition caused by the lack of iron and results in yellowing of the leaves of orange trees and other acid-loving plants. Sprinkling an iron salt on the soil around an orange tree will not provide the needed iron because in the moist soil the iron is coverted to $Fe(OH)_3$ or Fe_2O_3, both of which are so slightly soluble that the roots cannot absorb them. Commercial fertilizers for acid-loving plants contain "chelated iron" that stays in solution and can be absorbed.

Several of the transition metals are essential to human life. In the body they are complexed with proteins in various metalloenzymes that catalyze crucial body functions. The heme portion of hemoglobin incorporates Fe^{2+}, and vitamin B_{12} binds Co^{3+}. Both heme, which functions in the transport of oxygen in the blood, and chlorophyll *a*, which is vital to the photochemical transfer of energy in plants, are members of the same class of compounds—the porphyrins (Figure 31.11). Health food stores all have large sections devoted to "chelated minerals"—iron, magnesium, copper, zinc, and other metal ions combined with complexing agents to render them soluble.

Many chelating, or sequestering, agents, including edta, remove unwanted metal ions from the body by forming stable, soluble complexes which can be eliminated with normal waste products. The chelating agent British Anti-Lewisite (BAL), for example,

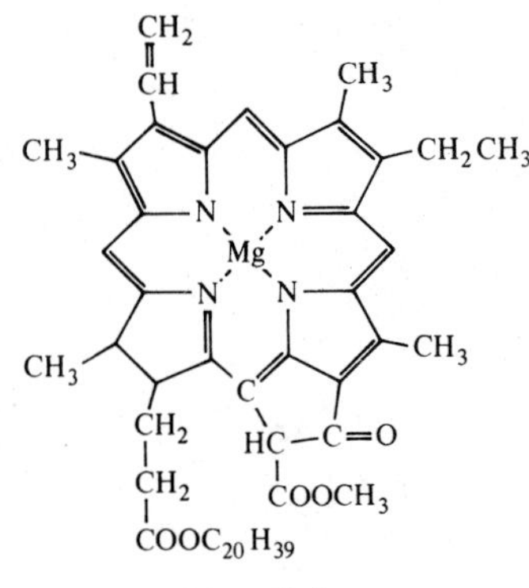

chlorophyll a

heme

Figure 31.11
Structures of Chlorophyll *a* and Heme
These compounds are both members of a class called porphyrins.

was developed during wartime as an antidote to the arsenic-containing poisonous gas called Lewisite ($ClCH{=}CHAsCl_2$). BAL, which has the formula

$$\begin{array}{ccc} H_2C- & CH- & CH_2 \\ | & | & | \\ SH & SH & OH \end{array}$$

coordinates through the sulfur atoms. BAL is now used to treat poisoning by many elements, including arsenic, mercury, gold, bismuth, antimony, thallium, tellurium, and chromium.

Cadmium has been identified as a possible cause of hypertension (high blood pressure). We are born with virtually no cadmium in our bodies. It comes into our diet as a contaminant of the zinc used in various ways in our society. For example, we inhale it in the dust from the wear and tear on tires (which contain zinc oxide) and ingest it with foods cooked in zinc-coated steel vessels. Sequestering agents for cadmium have been utilized as antihypertension drugs.

SUMMARY OF SECTIONS 31.6–31.8

Complex ions differ widely in their stability. The higher the oxidation state of the metal, the more stable are the complexes it forms with a given ligand. The CN^- ion tends to form highly stable complexes. Chelation increases the stability of a complex, and complexes with fused rings are more stable than complexes with separate rings. A labile complex undergoes rapid exchange of its ligands, while an inert complex has a slow rate of ligand exchange. (These terms refer to kinetic rather than thermodynamic stability.)

By formation of complexes, metal ions can be dissolved selectively or removed from solution, or tied up so that they are unreactive. For example, complexing agents are used to remove toxic metals from the human body. Many catalysts are complexes or function by forming complexes. Also, complexation is essential in chlorophyll and to the transportation of nutrients in plants, and in many biologically important molecules, such as hemoglobin and vitamin B.

BONDING IN COMPLEXES

Bonding in complexes, as in other compounds, is rarely strictly ionic or strictly covalent. The valence bond approach to bonding in complexes emphasizes covalent bonding, the crystal field theory emphasizes ionic bonding, and the molecular orbital theory brings about a compromise between the two. Keep in mind that in the following sections, as in our earlier discussions of bonding theory, we are describing in words concepts based primarily on mathematical models.

31.9 VALENCE BOND THEORY

In valence bond theory, metal–ligand bonds are looked upon as coordinate covalent bonds. Bonding results from the interaction of an empty orbital on a metal atom or ion with a ligand orbital that contains two electrons. The available orbitals in the metal atom are considered to be hybridized and the number of hybridized orbitals is equal to the number of bonds to ligand atoms in a given complex. Several examples of how hybridization is used to account for the properties of complexes are given in the following paragraphs.

The beryllium atom and the Be^{2+} ion have the ground-state configurations shown below. In the $[Be(H_2O)_4]^{2+}$ complex ion, the $2s$ and $2p$ orbitals are hybridized and of equal energy. Four pairs of electrons from the oxygen atoms of the four water molecules have entered the sp^3 hybrid orbitals, an interpretation of bonding consistent with the observed tetrahedral geometry of this (and other) Be^{2+} complexes and also the absence of paramagnetism.

2p — — —
$2s^2$ ⇅
$1s^2$ ⇅
Be

2p — — —
2s —
$1s^2$ ⇅
Be^{2+}

sp^3 hybrid orbitals ⇅ ⇅ ⇅ ⇅
$1s^2$ ⇅
ligand electrons
$[Be(H_2O)_4]^{2+}$

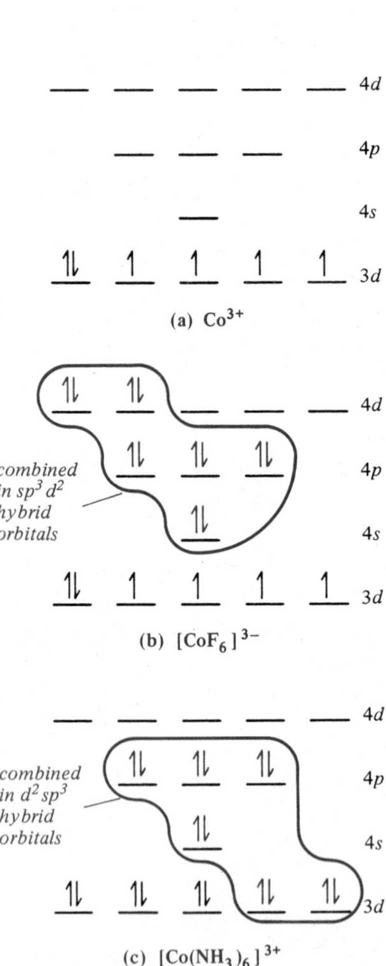

Figure 31.12
Valence Bond Picture of Configuration in Cobalt(III) Complexes
Ligand electrons are represented in color.

In most transition metal complexes, orbitals of the *d* subshells are included in the hybridization. Consider the cobalt(III) ion which, like many transition metal ions, forms complexes with six bonds to ligands and octahedral geometry. The cobalt(III) ion has six 3*d* electrons (Figure 31.12a), two paired and four unpaired, making the ion highly paramagnetic. In the $[CoF_6]^{3-}$ complex ion, the paramagnetism is unchanged, indicating that the 3*d* orbitals take no part in bonding to the fluoride ion ligands. Instead, 4*d*, or outer, orbitals are involved. The 4*s*, the three 4*p*, and two of the 4*d* orbitals interact (Figure 31.12b), giving six equivalent sp^3d^2 orbitals that are occupied by electron pairs from six ligands in what is known as an "outer orbital" complex.

The $[Co(NH_3)_6]^{3+}$ complex ion, however, is diamagnetic—it contains no unpaired electrons. Also, this complex ion is much more stable than $[CoF_6]^{3-}$, indicating that ammonia forms stronger bonds with Co^{3+} than does F^-. The additional energy released in forming such a complex is considered sufficient to cause the unpaired 3*d* electrons to pair up, leaving two 3*d* orbitals empty and available for hybridization. The 12 ligand electrons are pictured as occupying six equivalent hybrid orbitals formed by two 3*d* orbitals, the 4*s* orbital, and the three 4*p* orbitals. The resulting d^2sp^3 hybrid, known as an "inner orbital" complex, has no unpaired electrons (Figure 31.12c), in agreement with observed properties.

The possibilities for forming inner or outer orbital complex ions and for paramagnetism or diamagnetism vary with the number of *d* electrons present in a metal ion. For example, Cr^{3+} has three unpaired electrons in 3*d* orbitals and two vacant 3*d* orbitals. In this case, d^2sp^3 hybridization accounts for the observed paramagnetism due to three unpaired electrons and the octahedral geometry of complex ions such as $[Cr(NH_3)_6]^{3+}$. Some additional examples are included in Table 31.5. Most complex ions containing six ligands are octahedral whether the bonding is sp^3d^2 or d^2sp^3. In complex ions with four ligands, dsp^2 hybridization is characteristic of the square-planar configuration and sp^3 bonding is characteristic of the tetrahedral configuration (although there are some exceptions).

Table 31.5
Valence Bond Description of Electrons and Bonding in Some Transition Metal Complex Ions

Ion	Complex	No. of 3*d* electrons in metal ion	Type of hybrid	Geometry	No. of unpaired electrons in complex ion
Cr^{3+}	$[Cr(NH_3)_6]^{3+}$	3	d^2sp^3	octahedral	3
Fe^{2+}	$[Fe(CN)_6]^{4-}$	6	d^2sp^3	octahedral	0
Fe^{3+}	$[Fe(CN)_6]^{3-}$	5	d^2sp^3	octahedral	1
Fe^{3+}	$[FeF_6]^{3-}$	5	sp^3d^2	octahedral	5
Co^{3+}	$[Co(NH_3)_6]^{3+}$	6	d^2sp^3	octahedral	0
Co^{3+}	$[CoF_6]^{3-}$	6	sp^3d^2	octahedral	4
Ni^{2+}	$[Ni(CN)_4]^{2-}$	8	dsp^2	square planar	0
Cu^{+}	$[Cu(NH_3)_2]^{+}$	10	sp	linear	0
Cu^{+}	$[Cu(CN)_4]^{3-}$	10	sp^3	tetrahedral	0
Cu^{2+}	$[Cu(NH_3)_4]^{2+}$	9	dsp^2	square planar	1

EXAMPLE 31.5 Bonding in Complexes

The cyanide ion is a ligand that can cause pairing of unpaired electrons in the *d* subshell of a metal ion. Assuming d^2sp^3 hybridization for the Mn^{3+} ion in $[Mn(CN)_6]^{3-}$, determine whether $[Mn(CN)_6]^{3-}$ will be paramagnetic or diamagnetic.

The electron configurations are $[Ar]4s^2 3d^5$ for Mn and $[Ar]3d^4$ for Mn^{3+}. In order for two of the *d* orbitals from Mn^{3+} to be included in hybridization (as required for d^2sp^3 hybridization), one unpaired electron must be paired, leaving two unpaired electrons to give the following arrangement, in which color indicates ligand electrons in the d^2sp^3 hybrid orbitals

```
__ __ __ __ __   4d
   ⇅  ⇅  ⇅       4p
      ⇅          4s
⇅  ↑  ↑  ⇅  ⇅    3d
```

The hexacyanomanganate(III) ion should be **paramagnetic.** [It is.]

EXERCISE Briefly describe the bonding in $[Mn(H_2O)_6]^{3+}$ using the valence bond theory. The water molecule is not a strong ligand and no pairing of unpaired electrons in the *d* subshell of the metal ion will occur.

Answer The electron configurations for Mn and Mn^{3+} are given above. Since no pairing of electrons occurs, the ligands will use the 4*d* subshell for the sp^3d^2 hybridization. The ion will be paramagnetic.

31.10 CRYSTAL FIELD THEORY

Crystal field theory views the interaction between metal ions and ligands in complexes as electrostatic attraction and repulsion. The properties of the complex are explained by a rearrangement of the electrons of the metal ion. The bonding is considered to be completely ionic and interaction between the ligand electrons and the metal ion is *not* considered. "Crystal field theory" is so named because it derives from a theory developed for the energy relationships of ions in crystals. (A modification of crystal field theory that is corrected for small contributions from covalent bonding between the metal atom and the ligands is called ligand field theory.)

We describe here how the crystal field theory explains the formation of octahedral complexes. Application of the theory to complexes of other coordination numbers and geometry would be similar.

We begin with an isolated Ti^{3+} ion, which has one 3*d* electron. The *d* electron can reside in any of the five *d* orbitals (Figure 31.13), for they are of equal energy. As six

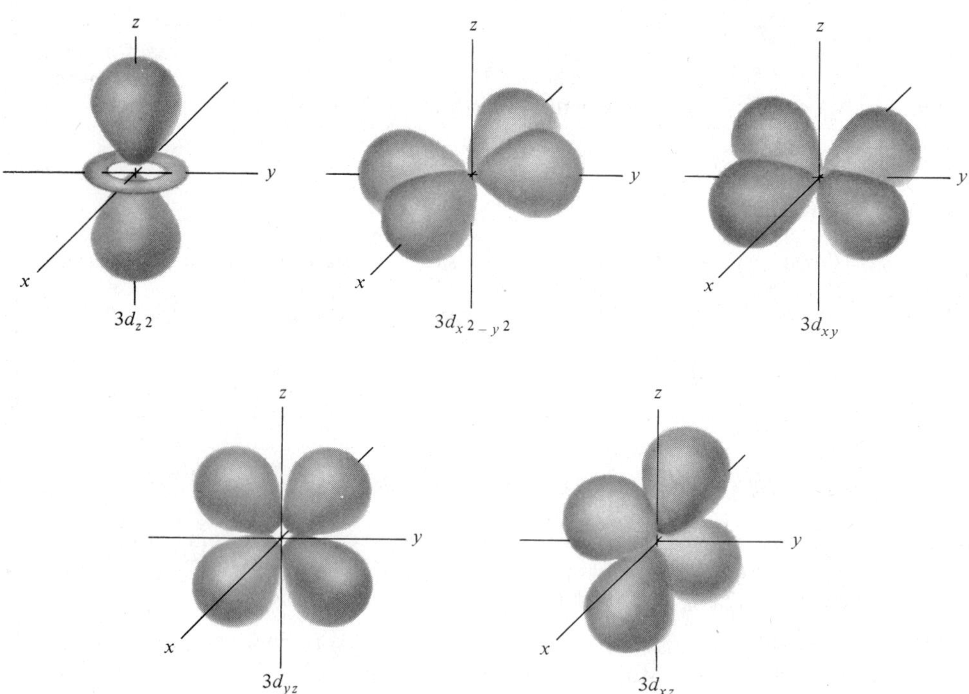

Figure 31.13
The Five *d* Orbitals

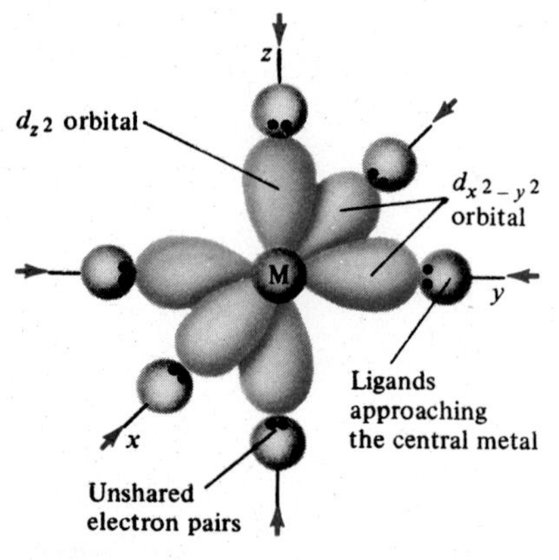

Figure 31.14
Six Ligands about to Form an Octahedral Complex

ligands bearing electron pairs approach the metal ion, it becomes more difficult for an electron to occupy the *d* orbitals, because there is a repulsion between the electrons of the metal and the approaching ligand electrons. In other words, the energy level of the *d* orbitals is raised.

If the approaching ligand electrons had a perfectly symmetrical effect on the metal ion *d* orbitals, all of the *d* orbitals would be raised in energy equally. However, the ligands approach along the x, y, and z coordinates as shown in Figure 31.14, producing an octahedral field. Comparison with the arrangement of the *d* orbitals in Figure 31.14 shows that the two lobes of the d_{z^2} orbital and the four lobes of the $d_{x^2-y^2}$ orbital point directly toward the corners of the octahedron, where the negative charge of the approaching ligands is concentrated. These orbitals are therefore raised in energy relative to their positions in a symmetrical field. (Remember that orbitals represent regions of negative charge; because like charges repel each other, an increase in potential energy occurs when they are brought together.) The other three orbitals are further from the negative charge of the donor atom and are at a lower energy than they would be in a spherical field. The three orbitals of lower energy are called t_{2g} orbitals (pronounced "t-two-g") and the two of higher energy are called e_g orbitals (pronounced "e-g"). (These names are derived from spectroscopic terms.) The changes in energy levels can be pictured as follows:

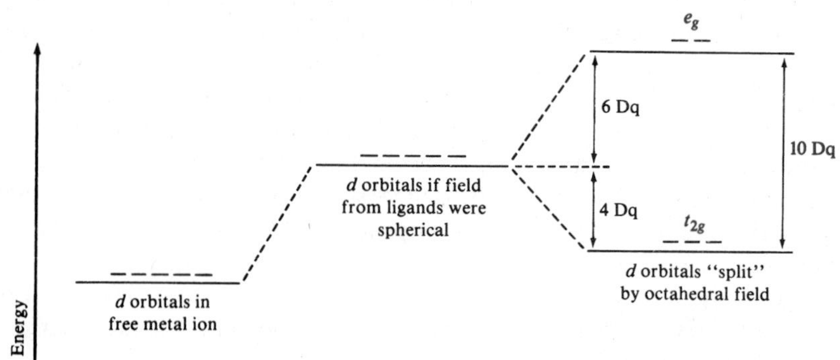

The single d electron in the Ti^{3+} ion will avoid the regions of higher energy and higher electronic repulsion at the corners of the octahedron and enter one of the t_{2g} orbitals. This increases the stability of the Ti^{3+} complex ion by lowering its energy.

The energy difference between the t_{2g} and e_g orbitals is called the *octahedral crystal field splitting* and is represented by the symbol 10 Dq or by Δ. Crystal field splitting is a measure of the "crystal field strength" of the ligand. In the formation of these two energy levels, quantum theory requires that no energy be gained or lost. The decrease in energy of the set of orbitals that lies at the lower energy level must be balanced by a corresponding increase in the other set. Therefore, in terms of the total energy, 10 Dq, the energy of each of the two e_g orbitals is 6 Dq higher, and that of each of the three t_{2g} orbitals is 4 Dq lower than if the separation had not taken place [i.e., $2(+6\text{ Dq}) = +12\text{ Dq}$ and $3(-4\text{ Dq}) = -12\text{ Dq}$].

The value of 10 Dq is measured spectroscopically. (It is inversely proportional to the wavelength of light absorbed when an electron is excited from the t_{2g} level to the e_g level: $10\text{Dq} = hc/\lambda$.) The amount of stabilization provided by splitting of the d orbitals into two levels for a given ion—the **crystal field stabilization energy** (abbreviated CFSE)—can be calculated in terms of 10 Dq. It is the algebraic sum of −4 Dq per electron in a t_{2g} level and +6 Dq for each electron in an e_g level on the central ion of a complex.

The arrangement of the electrons and the CFSE for metal ions with from one to four d electrons, d^1 to d^4 ions, in octahedral complexes are summarized in Figure 31.15. In metal ions with four d electrons there are two possible arrangements of the electrons. Three of the electrons, according to the usual rules, occupy the lower energy t_{2g} orbitals and are unpaired. The fourth electron can either go into an e_g orbital or pair up with an electron in a t_{2g} orbital. If the fourth electron goes in at the higher energy level, four d electrons occupy separate orbitals, their spins are not paired, and a **high-spin complex** has been formed.

In $[Mn(CN)_6]^{3-}$ and many other octahedral d^4 complexes, the fourth electron pairs up with one of the electrons in a t_{2g} orbital. A complex in which d electrons are

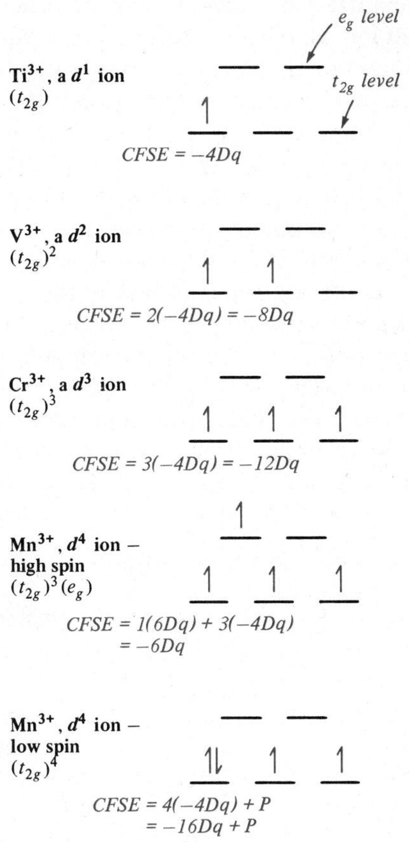

e_g level

t_{2g} level

Ti^{3+}, a d^1 ion (t_{2g})

CFSE = −4Dq

V^{3+}, a d^2 ion $(t_{2g})^2$

CFSE = 2(−4Dq) = −8Dq

Cr^{3+}, a d^3 ion $(t_{2g})^3$

CFSE = 3(−4Dq) = −12Dq

Mn^{3+}, d^4 ion – high spin $(t_{2g})^3(e_g)$

CFSE = 1(6Dq) + 3(−4Dq) = −6Dq

Mn^{3+}, d^4 ion – low spin $(t_{2g})^4$

CFSE = 4(−4Dq) + P = −16Dq + P

Figure 31.15
Crystal Field Picture of Configurations in d^1 through d^4 Ions in Octahedral Complexes
P is a factor added because of energy used in pairing electrons.

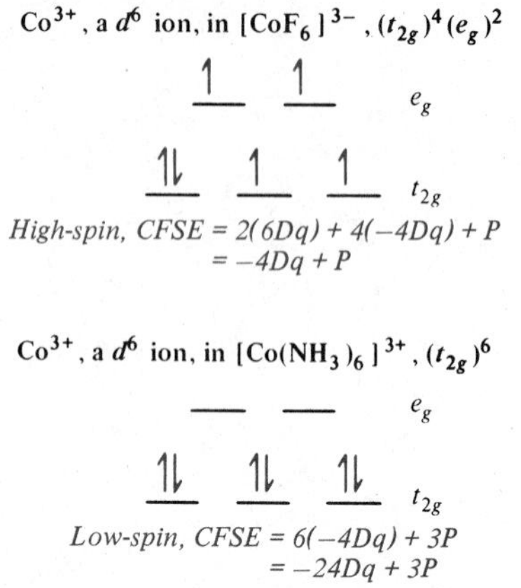

Co^{3+}, a d^6 ion, in $[CoF_6]^{3-}$, $(t_{2g})^4(e_g)^2$

1 1 e_g

1↓ 1 1 t_{2g}

High-spin, CFSE = 2(6Dq) + 4(−4Dq) + P = −4Dq + P

Co^{3+}, a d^6 ion, in $[Co(NH_3)_6]^{3+}$, $(t_{2g})^6$

— — e_g

1↓ 1↓ 1↓ t_{2g}

Low-spin, CFSE = 6(−4Dq) + 3P = −24Dq + 3P

Figure 31.16
Crystal Field Picture of Configurations in d^6 High-Spin and Low-Spin Co^{3+} Octahedral Complexes

paired wherever possible is called a **low-spin complex**. It might be expected that having *all* of the electrons in the lower energy orbitals would give added stability. However, a good deal of energy (represented by *P*; see Figure 31.15) is consumed in pairing the spins of two electrons, so some stability is lost.

The possibility of forming *either* high-spin or low-spin complexes exists for d^4 through d^7 ions. Whether a complex will be high spin or low spin depends upon the nature of the ligand. If the ligand is a strongly coordinating agent with a large Dq like CN^-—a **strong-field ligand**—its binding will provide enough energy to force a pairing of electrons and form the spin-paired system. A less strongly coordinating ligand with a small Dq such as Cl^-—a **weak-field ligand**—will form the high-spin system. It is easy to distinguish between the two cases, for the high-spin complexes are more highly paramagnetic and there is a marked difference in the position of the spectroscopic absorption bands.

To examine the crystal field approach for one more case, consider the d^6 cobalt complexes with ammonia molecules and with fluoride ion that were described from the valence bond viewpoint in the preceding section. In each case, at least two of the six *d* electrons must be paired, for there are only five orbitals. If only two electrons are paired, the complex has a high spin, $(t_{2g})^4(e_g)^2$ (where the superscripts indicate the number of electrons in each level). The $[CoF_6]^{3-}$ ion is a high-spin complex (Figure 31.16). It shows the presence of the two unpaired electrons by its paramagnetism and its CFSE is low. Most of the well-known Co^{3+} complexes are of the low-spin variety, with all of the electrons paired and occupying t_{2g} orbitals, $(t_{2g})^6$. The two e_g orbitals are empty. These low-spin Co^{3+} complexes are diamagnetic and their crystal field stabilization energy is high.

31.11 COLOR, STABILITY, AND THE SPECTROCHEMICAL SERIES

One of the triumphs of the crystal field theory is its ability to correlate the stability of complexes with their colors. To illustrate how this is possible, we can consider a series of complexes of cobalt(III) in which there are six electrons in the 3*d* subshell. In the simple, uncoordinated ion, all of these electrons have equal energy, but in the six-coordinate complexes, they are split between two levels. The difference in energy, 10 Dq, between the two levels depends upon the strength with which the ligands are bonded to the metal. In a complex containing strong-field ligands—that is, those that form strong bonds with the metal—the value of 10 Dq will be large, but if the ligands are "weak," 10 Dq will be small. Under the influence of external energy, such as light energy, an electron may be promoted from a lower level (t_{2g}) to a higher one (e_g). The smaller 10 Dq, the less the energy required. The absorbed energy, of course, is very quickly dissipated as heat as the electron falls back to the t_{2g} level. The energy of light depends upon the wavelength—the shorter the wavelength, the greater the energy. Visible light is a mixture of radiation of many wavelengths, ranging from blue (about 400 nm) to red (about 700 nm). The color that we see in a solution or a solid is due to the light which is *not* absorbed, but which is transmitted or reflected. A material which absorbs blue light (short wavelength), for example, appears to us to be orange (long wavelength) (Table 31.6).

Table 31.6
Absorbed and Transmitted Colors

Absorbed wavelength (nm)	Absorbed color	Transmitted color
410–490	violet/blue-green	yellow/red
490–530	blue-green/green	red/violet
530–580	green/yellow	violet/blue
580–680	yellow/red	blue/blue-green

Table 31.7
Color and Bond Strength in Cobalt Complexes of the Type $[Co(NH_3)_5X]^{n+}$

Compound	Color		Wavelength of light absorbed (nm)
$[Co(NH_3)_5(NO_2)]Cl_2$	Yellow	↑	455.9
$[Co(NH_3)_5(NH_3)]Cl_3$	Yellow		471.7
$[Co(NH_3)_5(H_2O)]Cl_3$	Rose-red		487.0
$[Co(NH_3)_5(NCS)]Cl_2$	Red	*Increasing*	498.3
$[Co(NH_3)_5(OH)]Cl_2$	Carmine	*strength of*	501.7
$[Co(NH_3)_5Cl]Cl_2$	Reddish purple	*Co—X bond*	521.8
$[Co(NH_3)_5Br]Br_2$	Reddish purple		540.5

The coordination compounds of the transition metals are generally colored, which we believe indicates that electrons from the t_{2g} level are raised to the e_g level when visible light falls upon them. [Other changes in the quantum states of electrons are also induced, and light of different wavelengths is absorbed, but we are interested here only in the wavelengths that affect the 3*d* electrons.] Table 31.7 lists several complexes of the type $[Co(NH_3)_5X]^{n+}$, arranged in order of decreasing strength of the Co—X bond. This order is the same as the order of decreasing energy (10 Dq) or increasing wavelength of the light absorbed.

$$Br^- < Cl^- < OH^- < NCS^- < H_2O < NH_3 < NO_2^-$$

Such a series of ligands arranged according to bond strength is called a "spectrochemical series." A tabulation including other ligands is given in Table 31.8. The $[Co(CN)_6]^{3-}$ ion, which contains six very strong ligands, is a pale yellow color. In the visible range only a small amount of radiation in the short-wavelength region is energetic enough to promote an electron in this ion from the t_{2g} level to the e_g level.

Table 31.8
Spectrochemical Series
Strongly binding ligands such as CN^- and NO_2^- are referred to as strong-field ligands. Weakly binding ligands such as Br^- and I^- are referred to as weak-field ligands.

Ligand	
CN^-	↑
NO_2^-	
ethylenediamine	
pyridine	
NH_3	
H_2O	
NCS^-	*Increasing strength of binding as a ligand*
$C_2O_4^{2-}$	
CH_3COO^-	
OH^-	
urea	
F^-	
Cl^-	
Br^-	
I^-	

Among the types of reactions that complexes undergo, one of the most important is replacement of one ligand by another, and the stability series can be related to such reactions. Consider, for example, the reactions

$$[Cu(H_2O)_4]^{2+} + 4NH_3(aq) \longrightarrow [Cu(NH_3)_4]^{2+} + 4H_2O(l)$$

$$[Ni(NH_3)_4]^{2+} + 2NH_2CH_2CH_2NH_2(aq) \longrightarrow Ni\left(\begin{matrix}-NH_2CH_2\\ |\\ -NH_2CH_2\end{matrix}\right)_2^{2+}(aq) + 4NH_3(aq)$$

$$[Zn(NH_3)_4]^{2+} + 4CN^- \longrightarrow [Zn(CN)_4]^{2-} + 4NH_3(aq)$$

In each of these reactions a weaker ligand is replaced by a stronger one (see Table 31.8).

If two complexes are of comparable stability, an equilibrium will be established which is governed by the relative concentrations of the reagents and the relative stabilities of the complexes. For example

$$[Cd(NH_3)_4]^{2+}(aq) + 4CH_3NH_2(aq) \rightleftharpoons [Cd(CH_3NH_2)_4]^{2+}(aq) + 4NH_3(aq)$$

31.12 MOLECULAR ORBITAL THEORY

The molecular orbital theory uses a somewhat different explanation of high-spin and low-spin complexes. In the formation of an octahedral complex, molecular orbitals are formed by combination of the atomic orbitals of the metal with the orbitals that contain the available electron pairs of the ligands. When six ligands approach a *d*-transition metal atom or ion, the 4*s* and 4*p* orbitals of the metal are available for overlap with the ligand orbitals. Two of the five *d* orbitals are oriented in space so that they point toward the approaching ligands (see Figure 31.14), and these two are also available for bonding. The other three *d* orbitals cannot overlap ligand orbitals.

What happens is pictured this way: the 4*s*, the three 4*p*, and the two available *d* orbitals (a total of six orbitals) combine with the orbitals that hold the bonding electron pairs in each of the six ligands to give six bonding and six antibonding molecular orbitals. The remaining three *d* orbitals from the metal are nonbonding molecular orbitals—they are not properly oriented in space, and electrons residing in these orbitals make no contribution to bonding, nor is there any change in their energy levels.

According to this picture, in the $[Co(NH_3)_6]^{3+}$ ion, all of the outer shell electrons are paired and are in bonding orbitals, giving a stable diamagnetic complex. In the $[CoF_6]^{3-}$ complex ion, however, two of the nonbonding electrons have been displaced into the two lowest level antibonding orbitals. Thus, this ion is paramagnetic and is less stable than the ammine complex.

SUMMARY OF SECTIONS 31.9–31.12

In valence bond theory, metal–ligand bonds are viewed as coordinate covalent bonds. Bonding occurs through the overlap of atomic orbitals, and the orbitals of the metal that are filled by the ligand electrons are viewed as hybridized and of equal energy, their number equal to the number of bonds to ligands.

In crystal field theory, bonding is considered to be ionic. The approach of ligands with their electron pairs causes a splitting in the energy level of the metal atom or ion orbitals; the energy of some orbitals is raised and that of others is lowered. Electrons enter the orbitals with lower energy, making the complex more stable by an amount known as the crystal field stabilization energy, or CFSE. A strong-field ligand may create a low-spin complex, in which electrons are paired in the lower-energy orbitals. A weak-field ligand will create a high-spin complex, in which some electrons are unpaired in higher-energy orbitals.

The crystal field theory makes it possible to relate the stability of complexes to their colors. The more strongly the ligands are bonded to the metal, the shorter the wavelength of light absorbed by the complex. (The color that we see is due to the wavelengths *not* absorbed.)

According to molecular orbital theory, molecular orbitals are formed by combination of the atomic orbitals of the metal with the orbitals that contain the available electron pairs of the ligands.

SIGNIFICANT TERMS (with section references)

bidentate (31.3)
chelate ring (31.3)
chelation (31.3)
chirality (31.4)
cis isomers (31.4)
cis–trans isomerism (31.4)
coordination number (31.1)
crystal field stabilization energy (31.10)
crystal field theory (31.10)
geometric isomerism (31.4)
high-spin complex (31.10)
inert complex (31.6)
isomers (31.4)
labile complex (31.6)
low-spin complex (31.10)
optical isomerism (31.4)
optical isomers, enantiomers (31.10)
optically active (31.4)
plane-polarized light (31.4)
strong-field ligand (31.10)
structural isomerism (31.4)
trans isomers (31.4)
weak-field ligand (31.10)

QUESTIONS AND PROBLEMS

Structure of Complexes

31.1 What are the two constituents of a complex? What type of chemical bonding occurs between these constituents?

31.2 Define the term "coordination number" for the central atom or ion in a complex. What values of the coordination numbers for metal ions are most common?

31.3 Identify the ligands and give the coordination number and the oxidation number for the central atom or ion in each of the following: (a) $[Co(NH_3)_2(NO_2)_4]^-$, (b) $[Cr(NH_3)_5Cl]Cl_2$, (c) $K_4[Fe(CN)_6]$, (d) $[Pd(NH_3)_4]^{2+}$.

31.4 Repeat Question 31.3 for (a) $Na[Au(CN)_2]$, (b) $[Ag(NH_3)_2]^+$, (c) $[Pt(NH_3)_2Cl_4]$, (d) $[Co(en)_3]^{3+}$.

31.5 What is the term given to the phenomenon of ring formation by a ligand in a complex? What numbers of atoms are found in stable rings?

31.6 Write a structural formula showing the ring(s) formed by a bidentate ligand such as the glycinate ion with a metal ion such as Fe^{3+}. How many atoms are in the ring?

31.7 A compound synthesized in an inorganic preparation experiment was analyzed and found to contain 17.0 mass % Fe, 21.9 mass % C, 25.5 mass % N, and 35.6 mass % K. Find the empirical formula of the compound. The osmotic pressure exerted by a 0.0010 M solution of this compound at 25 °C was 0.095 atm. How many ions are present in this compound? Write the formula of the compound as a coordination compound. Name this substance.

31.8 The yellow complex compound $K_3[Rh(C_2O_4)_3]$ can be prepared from the wine-red complex compound $K_3[RhCl_6]$ by boiling a concentrated aqueous solution of $K_3[RhCl_6]$ and $K_2C_2O_4$ for two hours and then evaporating the solution until the product crystallizes.

$$K_3[RhCl_6](aq) + 3K_2C_2O_4(aq) \xrightarrow{\Delta} K_3[Rh(C_2O_4)_3](s) + 6KCl(aq)$$

What is the theoretical yield of the oxalato complex if 1.00 g of the chloro complex is heated with 5.75 g of $K_2C_2O_4$? In an experiment, the actual yield was 0.83 g. What is the percent yield?

31.9 Consider the formation of the triiodoargentate(I) ion.

$$Ag^+ + 3I^- \longrightarrow [AgI_3]^{2-}$$

Would you expect an increase or decrease in the entropy of the system as the complex is formed? The standard state absolute entropy at 25 °C is 72.68 J/K mol for Ag^+, 111.3 J/K mol for I^-, and 253.1 J/K mol for $[AgI_3]^{2-}$. Calculate $\Delta S°$ for the reaction and confirm your prediction.

31.10 Molecular iodine reacts with I^- to form a complex ion.

$$I_2(aq) + I^- \rightleftharpoons [I_3]^-$$

Calculate the equilibrium constant for this reaction given the following data at 25 °C:

$$I_2(aq) + 2e^- \longrightarrow 2I^- \qquad E° = 0.535\ \text{V}$$
$$[I_3]^- + 2e^- \longrightarrow 3I^- \qquad E° = 0.5338\ \text{V}$$

Nomenclature of Complexes

31.11 Name the following substances: (a) $K_3[Mn(CN)_6]$, (b) $[Pd(NH_3)_4](OH)_2$, (c) $[Ag(CN)_2]^-$, (d) $[Ag(NH_3)_2]^+$, (e) $K_2[Fe(C_2O_4)_2]\cdot 2H_2O$, (f) $K_3[Co(C_2O_4)_3]\cdot 6H_2O$, (g) $[Cu(NH_3)_4]^{2+}$, (h) $Na[AgI_2]$, (i) $Na_2[PtI_4]$.

31.12 Name the following substances: (a) $Na[Au(CN)_2]$, (b) $[Ni(NH_3)_4(H_2O)_2](NO_3)_2$, (c) $[CoCl_6]^{3-}$, (d) $[Co(H_2O)_6]^{3+}$, (e) $Na_2[Pt(CN)_4]\cdot 3H_2O$, (f) $K[Cr(NH_3)_2(H_2O)_2Cl_2]$, (g) $[Pt(NH_3)_4]Cl_2$, (h) $Na[Al(H_2O)_2(OH)_4]$, (i) $[Co(NH_3)_4Cl_2][Cr(C_2O_4)_2]$.

31.13 Write formulas for the following:
(a) diamminedichlorozinc(II),
(b) tin(IV) hexacyanoferrate(II),
(c) tetracyanoplatinate(II) ion,
(d) potassium hexacyanochromate(III),
(e) tetraammineplatinum(II) ion,
(f) hexaamminenickel(II) bromide,
(g) tetraamminecopper(II) pentacyanohydroxoferrate(III).

31.14 Write formulas for the following:
(a) diamminetetrachloroplatinum(IV),
(b) hexaamminenickel(II) iodide,
(c) magnesium tetracyanoplatinate(II) heptahydrate,
(d) lead(II) tetrafluoroborate(III),
(e) iron(III) hexacyanoferrate(II),
(f) lithium hexachloroferrate(II),
(g) diamminesilver(I) nitrate,
(h) diamminetetrachlorocobaltate(III) ion.

Molecular Geometry and Isomerism

31.15 Choose which of the following formulas represent complexes that can form geometrical isomers: (a) tetrahedral Mab_3, (b) tetrahedral Ma_2b_2, (c) square planar Mab_3, (d) square planar Ma_2b_2. Name any geometrical isomers that can exist. Is it possible for any of these isomers to show optical activity? Explain.

31.16 How many geometrical isomers can be formed by complexes that are (a) octahedral Ma_2b_4 and (b) octahedral Ma_3b_3? Name any geometrical isomers that can exist. Is it possible for any of these isomers to show optical activity? Explain.

31.17 Write the structural formulas for (a) two isomers of $[Pt(NH_3)_2Cl_2]$, (b) four isomers (including linkage isomers) of $[Co(NH_3)_3(NO_2)_3]$, (c) two isomers (including ionization isomers) of $[Pt(NH_3)_3Br]Cl$.

31.18 Determine the number and types of isomers that would be possible for each of the following complexes:
(a) tetraamminediaquacobalt(III) ion,
(b) triamminetriaquacobalt(III) ion,
(c) tris(ethylenediamine) cobalt(III) ion,
(d) dichlorobis(ethylenediamine)platinum(IV) chloride,
(e) diamminedibromodichlorochromate(III) ion.

31.19* Indicate whether the complexes in each pair are identical or are isomers:

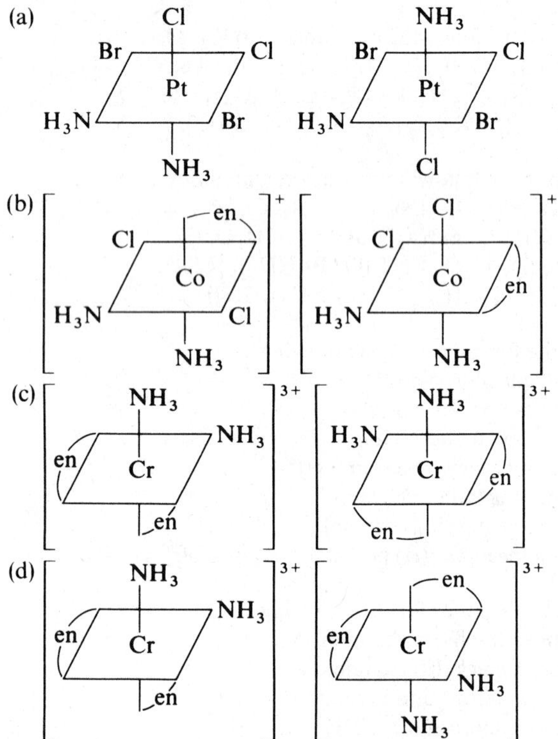

31.20* Werner studied the electrical conductance of aqueous solutions containing a series of platinum(IV) complexes having the general formula $Pt(NH_3)_xCl_4$, where x is an integer that varied from 2 to 6. His results can be summarized as

Formula of complex	Number of ions produced upon complete dissociation
$Pt(NH_3)_6Cl_4$	5
$Pt(NH_3)_5Cl_4$	4
$Pt(NH_3)_4Cl_4$	3
$Pt(NH_3)_3Cl_4$	2
$Pt(NH_3)_2Cl_4$	0

Assuming that Pt(IV) forms octahedral complexes, (a) write the formulas for the five compounds based on the dissociation results, (b) draw three-dimensional sketches of the complexes, (c) draw sketches of any isomers that are possible, and (d) name each compound.

Properties of Complexes

31.21 What properties of the central metal ion increase the thermodynamic stability of a complex? What is the trend of stability of complexes with chelation?

31.22 Does the term "labile" imply that a complex is thermodynamically unstable? Does the term "inert" imply that a complex is thermodynamically stable? Briefly explain your answers.

31.23 Why is the dissociation constant smaller for (a) $[Co(en)_3]^{3+}$ than $[Co(en)_3]^{2+}$, (b) $[Cu(gly)_2]$ than for $[Zn(gly)_2]$, (c) $[Cd(en)_2]^{2+}$ than for $[Cd(NH_3)_4]^{2+}$?

31.24 A copper penny will not react with hydrochloric acid, but will react with hydroiodic acid to give $H_2(g)$. Is this behavior explained by the fact that HI(*aq*) is a stronger acid than HCl(*aq*)? Explain your answer.

31.25 The overall dissociation constant for $[Fe(C_2O_4)_3]^{3-}$ is 3×10^{-21}. What would be the maximum concentration of Fe^{3+} in a solution containing 0.010 M $K_3[Fe(C_2O_4)]$ which also contains 1.5 M $C_2O_4^{2-}$?

31.26 Write the rate equation for the reaction

$$[Co(NH_3)_5F]^{2+} + H_2O(l) \longrightarrow [Co(NH_3)_5(H_2O)]^{3+} + F^-$$

in terms of concentrations of reactants. Experimentally it is found that doubling the concentration of $[Co(NH_3)_5F]^{2+}$ doubles the reaction rate. What is the order of reaction with respect to the complex?

31.27* Use the following standard reduction potential data to answer the questions below:

$$Co^{3+} + e^- \rightleftharpoons Co^{2+} \qquad E^\circ = 1.808\text{ V}$$

$$Co(OH)_3(s) + e^- \rightleftharpoons Co(OH)_2(s) + OH^-(aq) \qquad E^\circ = 0.17\text{ V}$$

$$[Co(NH_3)_6]^{3+} + e^- \rightleftharpoons [Co(NH_3)_6]^{2+} \qquad E^\circ = 0.108\text{ V}$$

$$[Co(CN)_6]^{3-} + e^- \rightleftharpoons [Co(CN)_5]^{3-} + CN^- \qquad E^\circ = -0.83\text{ V}$$

$$O_2(g) + 4H^+(10^{-7}\text{ M}) + 4e^- \rightleftharpoons 2H_2O(l) \qquad E^\circ = 0.815\text{ V}$$

$$2H_2O(l) + 2e^- \rightleftharpoons H_2(g) + 2OH^- \qquad E^\circ = -0.828\text{ V}$$

Which cobalt(III) species among those listed would oxidize water? Which cobalt(II) species among those listed would be oxidized by water? Explain your answers.

31.28* At 25 °C, $\Delta H^\circ = -46.4$ kJ, $\Delta G^\circ = -42.84$ kJ, and $\Delta S^\circ = -11.7$ J/K for the reaction

$$[Cu(H_2O)_4]^{2+} + 2NH_3(aq) \longrightarrow [Cu(NH_3)_2(H_2O)_2]^{2+} + 2H_2O(l)$$

and $\Delta H^\circ = -25.1$ kJ, $\Delta G^\circ = -49.0$ kJ, and $\Delta S^\circ = 79$ J/K for the reaction

$$[Cu(H_2O)_4]^{2+} + gly^- \longrightarrow [Cu(gly)(H_2O)_2]^+ + 2H_2O(l)$$

In each case, two Cu–water bonds are broken, but in the first reaction two Cu–NH_3 bonds are formed and in the second reaction a Cu–NH_2 and Cu–O bond are formed. Based on the ΔH° values, (a) which set of bonds is stronger? In each case, two water molecules are replaced—in the first reaction by two ligands and in the second reaction by one ligand. (b) Why does the entropy increase in the second reaction? Based on values of ΔG°, (c) which complex ion is more stable?

31.29 Briefly describe the bonding in metallocene complexes.

31.30 What is a metal cluster? What types of bonding are present in this type of complex?

Bonding in Complexes

31.31 Name the three theories that we use to explain the bonding in complexes. Briefly describe each of the theories.

31.32 What determines the color of a complex? How can we use color to compile a list of ligands in order of metal–ligand bond strength?

31.33 Using the valence bond theory, describe the bonding in a Co^{3+} octahedral complex with ligands that can (a) occupy only the outer orbitals on the Co^{3+} and (b) occupy inner orbitals on the Co^{3+}. What types of hybridization of the atomic orbitals on Co^{3+} are proposed for each?

31.34 Using the valence bond theory, describe the bonding in a square planar Cu^{2+} complex in which the ligands can occupy inner orbitals on the Cu^{2+}. What type of hybridization of the Cu^{2+} atomic orbitals is envisioned?

31.35 How many unpaired electrons would you predict there to be in each of the following: (a) $[Fe(CN)_6]^{3-}$, (b) $[Fe(H_2O)_6]^{3+}$, (c) $[Mn(H_2O)_6]^{2+}$, (d) $[Co(NH_3]_6]^{3+}$?

31.36 Consider the compound having the formula $[Co(NH_3)_5(H_2O)]^{3+}[Co(NO_2)_6]^{3-}$. In terms of valence bond theory, describe the bonding in each ion. Would you expect this substance to be paramagnetic or diamagnetic?

31.37 Determine the electronic distribution in (a) $[Co(CN)_6]^{3-}$, a low-spin complex ion, and (b) $[CoF_6]^{3-}$, a high-spin complex ion. Express the crystal field stabilization energy for each complex.

31.38 Determine the crystal field stabilization energy for (a) $[Mn(H_2O)_6]^{3+}$, a high-spin complex, and (b) $[Mn(CN)_6]^{3-}$, a low-spin complex.

31.39 What reactions will take place if samples of the labile complex $[Ni(H_2O)_6]^{2+}$ are mixed with solutions of (a) CN^-, (b) NH_3, (c) pyridine(C_5H_5N)?

31.40 What reactions might take place if an aqueous solution of $[Co(NH_3)_5Cl]Cl_2$ and $NaNO_2$ is allowed to stand for several hours? Would you expect a reaction to occur between $Na_3[Co(CN)_6]$ and $NaNO_2$ in aqueous solution? Explain.

31.41* Draw a sketch of the crystal field splitting for octahedral complexes and label each of the energy levels. How many t_{2g} electrons and how many e_g electrons are there in (a) $[Mn(H_2O)_6]^{2+}$, (b) $[CoF_6]^{3-}$, (c) $[Ti(H_2O)_6]^{3+}$?

31.42* Which atomic orbitals are used to form the molecular orbitals for an octahedral complex? How many molecular orbitals are formed?

CHAPTER

32

Organic Chemistry

HYDROCARBONS

During the early part of the nineteenth century the compounds found in living organisms were thought to be formed only through a subtle "vital force" present in such organisms. "Organic" chemistry was named on this basis and distinguished from "inorganic" chemistry, which covered all compounds not found in living matter. It was noted that all of the organic compounds then known contained carbon.

In 1828, German chemist Friedrich Wohler heated lead cyanate ($Pb(OCN)_2$) with an aqueous solution of ammonia, expecting to produce lead(II) oxide (PbO) and ammonium cyanate (NH_4OCN), an ionic compound. After filtering off the solid lead(II) oxide and evaporating the remaining aqueous solution, he was left with a crystalline solid that did not have any of the properties expected for ammonium cyanate. Wohler found that the product was urea, which previously had been isolated only from human urine. In the hot aqueous solution the ammonium cyanate had undergone a rearrangement reaction.

$$\underset{\textit{ammonium cyanate}}{NH_4OCN(aq)} \xrightarrow{\Delta} \underset{\textit{urea}}{H_2N{-}\overset{\overset{\displaystyle O}{\|}}{C}{-}NH_2(aq)}$$

This crucial experiment freed organic chemistry from its link to living organisms. The vital force concept gradually faded away, and organic chemistry became the chemistry of the compounds of carbon and the carbon–carbon bond. The number of organic compounds now known is in the millions. Many organic compounds do, of course, occur in nature, but even more have been made synthetically.

All organic compounds contain carbon atoms and, with a few exceptions, hydrogen atoms. Oxygen, nitrogen, the halogens, sulfur, phosphorus, silicon, and many other elements may be present also. Their frequency of occurrence decreases roughly in the order given. All organic compounds can be thought of as based upon the structures of the hydrocarbons, which are compounds containing carbon and hydrogen atoms only.

32.1 SATURATED HYDROCARBONS

Saturated hydrocarbons contain only single covalent bonds. Each carbon atom is joined to four other atoms, and each hydrogen atom is joined to a carbon atom. The simplest saturated hydrocarbon is methane (CH_4). Ethane (C_2H_6) and propane (C_3H_8) are next in the series (Table 32.1).

Saturated hydrocarbons with more than three carbon atoms all exhibit structural isomerism (Section 31.4). There are two C_4H_{10} structural isomers—both have the same molecular formula but the atoms are arranged differently, as shown by their structural formulas

```
                          H   H   H
                          |   |   |
                      H — C — C — C — H
     H   H   H   H        |   |   |
     |   |   |   |        H   |   H
 H — C — C — C — C — H    H — C — H
     |   |   |   |            |
     H   H   H   H            H
    (b.p. −0.5 °C)       (b.p. −12 °C)
```

Table 32.1
Simple Saturated Hydrocarbons

Structure	Boiling point (°C)	Systematic name	Structure and name of alkyl group	
(1) CH_4	−162	Methane	CH_3—	methyl
(2) CH_3CH_3	−89	Ethane	CH_3CH_2—	ethyl
(3) $CH_3CH_2CH_3$	−42	Propane	$CH_3CH_2CH_2$—	*n*-propyl
			$CH_3\underset{\vert}{C}HCH_3$	isopropyl
(4) $CH_3(CH_2)_2CH_3$	−0.5	Butane	$CH_3CH_2CH_2CH_2$—	*n*-butyl
			$CH_3CH_2\underset{\vert}{C}HCH_3$	*sec*-butyl
(5) $CH_3CH(CH_3)CH_3$	−12	2-Methylpropane	$CH_3CH(CH_3)CH_2$—	isobutyl
			$CH_3\overset{\vert}{C}(CH_3)CH_3$	*tert*-butyl
(6) $CH_3(CH_2)_3CH_3$	36	Pentane	$CH_3(CH_2)_4$— (two other C_5H_{11}—groups[a])	*n*-pentyl
(7) H_2C—CH_2 ring with CH_2 (cyclic C_3H_6)	−33	Cyclopropane	H_2C—CH— ring with CH_2	cyclopropyl
(8) H_2C—CH_2 / H_2C—CH_2 (cyclic C_4H_8)	13	Cyclobutane	H_2C—CH— / H_2C—CH_2	cyclobutyl
(9) cyclic $(CH_2)_6$: H_2C, CH_2, H_2C, CH_2, CH_2, CH_2	81	Cyclohexane	cyclic $(CH_2)_5$CH—	cyclohexyl

[a] The other C_5H_{11}— groups are not usually named as pentane-derived groups.

or by the more common, condensed formulas, in which it is assumed that the carbon atoms are bonded to each other.

$$CH_3CH_2CH_2CH_3 \quad \underset{\displaystyle CH_3}{CH_3\underset{|}{C}HCH_3}$$

Structural isomers have different properties and can be separated from each other by distillation or other physical methods. The number of possible structural isomers increases greatly as the number of carbon atoms in a hydrocarbon increases. For $C_{10}H_{22}$, 75 different structural isomers are possible; for $C_{40}H_{82}$, over 62 trillion isomers are possible.

The carbon atoms in a hydrocarbon may be connected in a straight chain (at the left above), in a branched chain (at the right above), or in a ring (cyclic hydrocarbons, as in compounds 7–9 in Table 32.1). All straight-chain and branched-chain saturated hydrocarbons are classified as **alkanes** and have the general molecular formula C_nH_{2n+2}. As n increases, a sequence of straight-chain alkanes is generated in which each member differs from its immediate neighbor by one CH_2 group. The first five members of the alkane series are methane, ethane, propane, butane, and pentane (1–4 and 6 in Table 32.1).

The carbon atoms in alkanes are described as sp^3-hybridized. The four bonds to each are arranged tetrahedrally and the bond angles are all close to 109.5°. As a result, a continuous chain of carbon atoms has a staggered, or zigzag, arrangement in space. Each carbon atom can rotate freely (e.g., Figure 32.1b and c) and the shape of an alkane is constantly changing.

With so many possible compounds, it is apparent that a systematic method is needed for naming hydrocarbons (and the organic compounds derived from them). Such a systematic nomenclature has been formulated by the International Union of Pure and Applied Chemistry, and is known as IUPAC nomenclature. Many organic

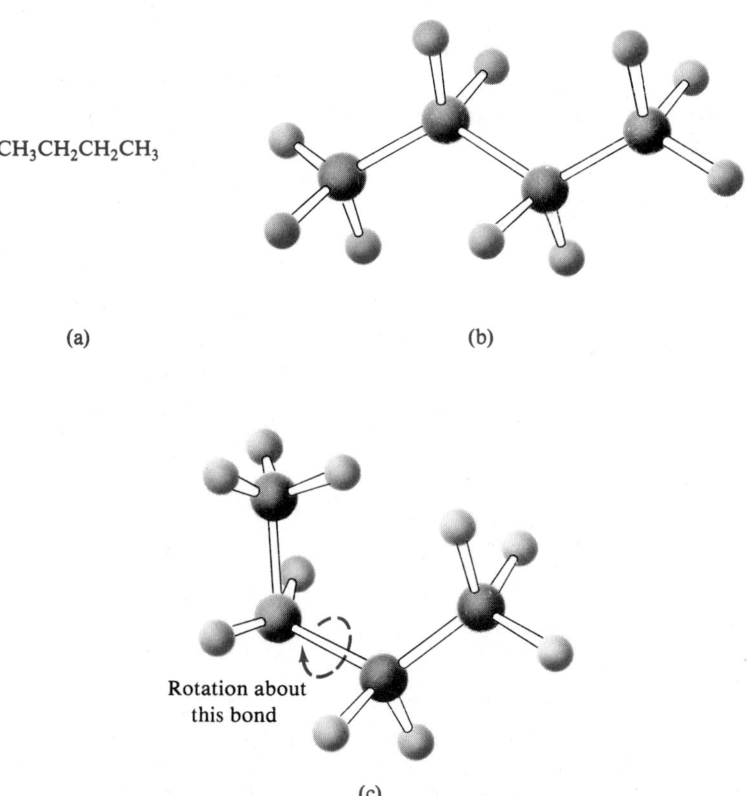

Figure 32.1
Butane, C_4H_{10}
(a) Condensed formula. (b) Ball-and-stick model illustrating zigzag arrangement of C atoms in alkanes. (c) One of many shapes possible through rotation about C—C bonds.

compounds were known long before this nomenclature was devised, and (perhaps unfortunately) certain common names persist and are encountered often. For example, the branched-chain C_4H_{10} isomer shown earlier has the common name *isobutane*. The *iso-* prefix often is used to designate a branched-chain hydrocarbon possessing the pattern

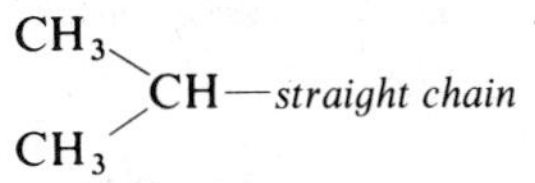

$$\begin{matrix} CH_3 \diagdown \\ \quad CH\text{—}\textit{straight chain} \\ CH_3 \diagup \end{matrix}$$

For example

$$\begin{matrix} CH_3 \diagdown \\ \quad CH\text{—}CHCH_2CH_2CH_3 \\ CH_3 \diagup \end{matrix}$$

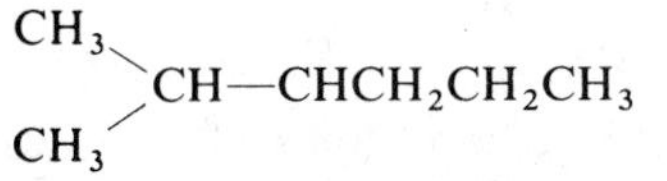

isoheptane

Common names are used often for simple, low molecular mass organic compounds.
Three rules will introduce you to the systematic nomenclature for alkanes:

1. A prefix indicates the longest chain of carbon atoms:

meth-	C_1	hex-	C_6
eth-	C_2	hept-	C_7
prop-	C_3	oct-	C_8
but-	C_4	non-	C_9
pent-	C_5	dec-	C_{10}

For a straight-chain (or *normal*) alkane, the ending *-ane* is added to this prefix, for example

$$\overset{1}{C}H_3\overset{2}{C}H_2\overset{3}{C}H_2\overset{4}{C}H_2\overset{5}{C}H_2\overset{6}{C}H_3 \quad \text{or} \quad CH_3(CH_2)_4CH_3$$

hexane

(The common name of a straight-chain alkane includes the letter *n* for normal, such as *n*-hexane.)

2. The positions and names of **substituents**—atoms or groups of atoms that have replaced hydrogen atoms in an organic compound—are indicated by adding prefixes and numbers to the name of longest carbon atom chain. The chain is numbered beginning with 1 at the end nearest the first attached group. In this way each substituent has the lowest possible number. For example, the following compound

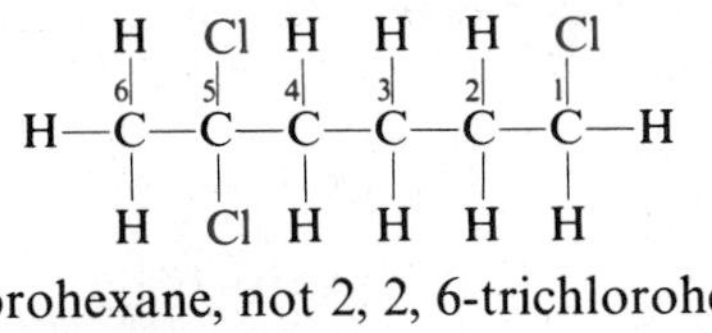

$$\begin{matrix} & H & Cl & H & H & H & Cl & \\ & | & | & | & | & | & | & \\ H— & \overset{6}{C}— & \overset{5}{C}— & \overset{4}{C}— & \overset{3}{C}— & \overset{2}{C}— & \overset{1}{C} & —H \\ & | & | & | & | & | & | & \\ & H & Cl & H & H & H & H & \end{matrix}$$

is named 1, 5, 5-trichlorohexane, not 2, 2, 6-trichlorohexane.

3. Saturated hydrocarbon substituent groups are named by replacing the *-ane* ending with *-yl*. Such groups, called **alkyl groups,** contain one less hydrogen atom than the alkane. For example, CH_4 is methane and CH_3— is a methyl group. The systematic name for isobutane is thus 2-methylpropane.

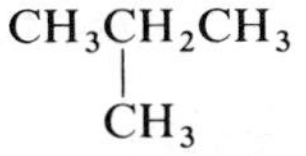

$$\begin{matrix} CH_3CHCH_3 \\ | \\ CH_3 \end{matrix}$$

2-methylpropane

Additional examples of how to assign systematic names to alkanes are given in Table 32.2 and the names of alkyl groups derived from the simple hydrocarbons are included in Table 32.1. Note the designation of two of the butyl group isomers as *tert-* and *sec-*. A carbon atom joined to only one other carbon atom is called a *primary carbon atom*, often designated by 1°; one joined to two other carbon atoms, a *secondary*

Table 32.2
Examples of Systematic Alkane Nomenclature

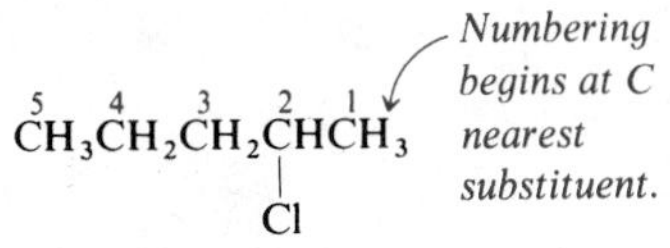

$$\overset{5}{C}H_3\overset{4}{C}H_2\overset{3}{C}H_2\overset{2}{C}H\overset{1}{C}H_3 \text{ (with Cl on C-2)}$$

Numbering begins at C nearest substituent.

2-chloropentane

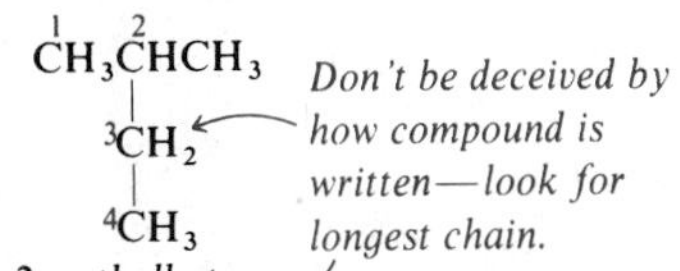

$$\begin{matrix} \overset{1}{C}H_3\overset{2}{C}HCH_3 \\ | \\ {}^3CH_2 \\ | \\ {}^4CH_3 \end{matrix}$$

Don't be deceived by how compound is written—look for longest chain.

2-methylbutane

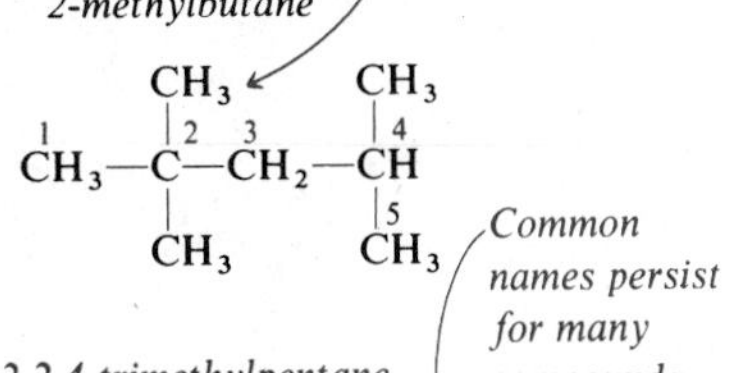

$$\begin{matrix} & CH_3 & & CH_3 \\ & |\,2 & 3 & |\,4 \\ \overset{1}{C}H_3— & C— & CH_2— & CH \\ & | & & |\,5 \\ & CH_3 & & CH_3 \end{matrix}$$

Common names persist for many compounds.

2,2,4-trimethylpentane
Common name—"isooctane"

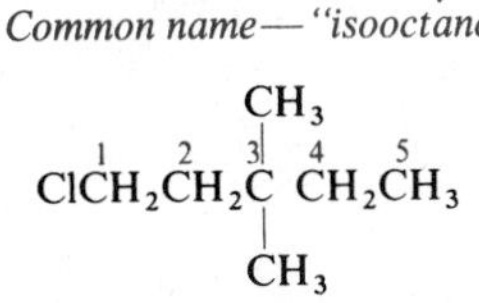

$$\begin{matrix} & & CH_3 & \\ & & |\,3 & \\ Cl\overset{1}{C}H_2\overset{2}{C}H_2 & & C & \overset{4}{C}H_2\overset{5}{C}H_3 \\ & & | & \\ & & CH_3 & \end{matrix}$$

1-chloro-3,3-dimethylpentane

(2°) *carbon atom* (as in the *sec*-butyl group); and one joined to three carbon atoms, a *tertiary* (3°) *carbon atom* (as in the *tert*-butyl group). A carbon atom joined to four other carbon atoms is a *quaternary* (4°) *carbon atom.* In the structures below, X represents an atom other than a C atom.

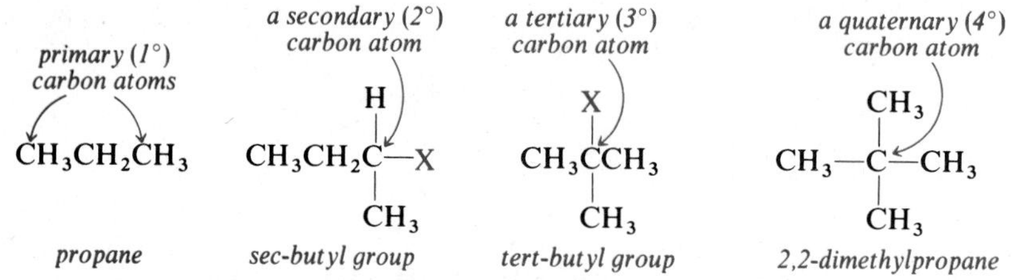

primary (1°) carbon atoms — $CH_3CH_2CH_3$ — *propane*

a secondary (2°) carbon atom — $CH_3CH_2C(H)(CH_3)$—X — *sec-butyl group*

a tertiary (3°) carbon atom — $CH_3C(X)(CH_3)CH_3$ — *tert-butyl group*

a quaternary (4°) carbon atom — CH_3—$C(CH_3)_2$—CH_3 — *2,2-dimethylpropane*

Compounds in which the carbon atoms of a saturated hydrocarbon are joined together in a ring are called cycloalkanes. **Cycloalkanes** are cyclic saturated hydrocarbons that have the general molecular formula C_nH_{2n}. The nomenclature of cycloalkanes follows the same pattern used for the noncyclic alkanes. If there are two or more substituents, one is given the number 1 and others are given the lowest possible numbers.

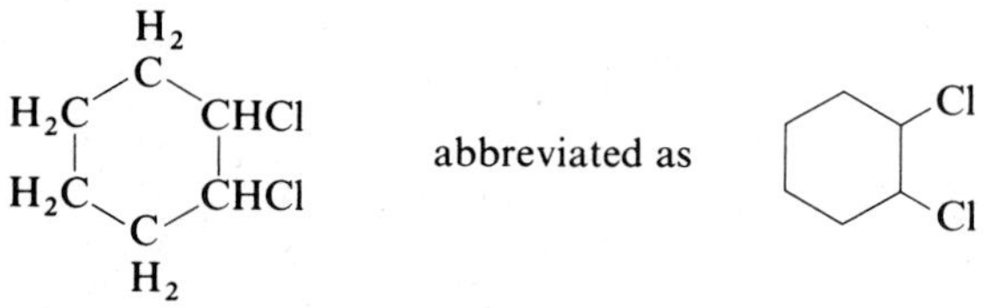

Ring: H_2C, CH_2 (H_2C), CHCl, CHCl, H_2C, CH_2 — abbreviated as [hexagon with Cl on two adjacent corners]

1,2-dichlorocyclohexane

In the abbreviated expressions, it is understood without writing them that there is a carbon atom at each corner and that sufficient hydrogen atoms are present to complete four bonds to each carbon atom.

EXAMPLE 32.1 Organic Nomenclature: Alkanes

Write the condensed formula for 2,3-dimethylbutane. Indicate the primary, secondary, tertiary, or quaternary carbon atoms.

The "butane" part of this name tells us that the longest straight carbon chain in this compound consists of four carbon atoms:

C—C—C—C

The prefix "2,3-dimethyl" indicates that there are methyl groups, CH_3—, attached to the second and third carbon atoms:

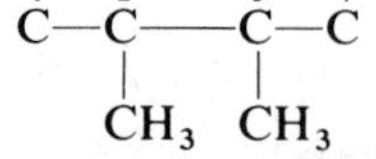

$\overset{1}{C}$—$\overset{2}{C}$—$\overset{3}{C}$—$\overset{4}{C}$, with CH_3 on C-2 and CH_3 on C-3

Filling in hydrogen atoms so that each carbon atom has four covalent bonds and labeling primary, secondary, and tertiary carbon atoms give

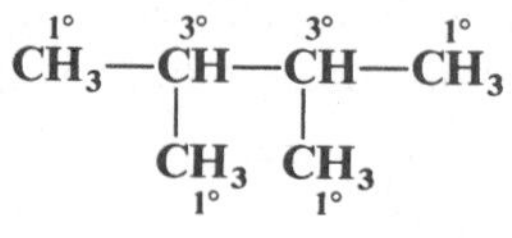

$\overset{1°}{CH_3}$—$\overset{3°}{CH}$—$\overset{3°}{CH}$—$\overset{1°}{CH_3}$, with $\underset{1°}{CH_3}$ on each CH

EXERCISE Give the IUPAC name for

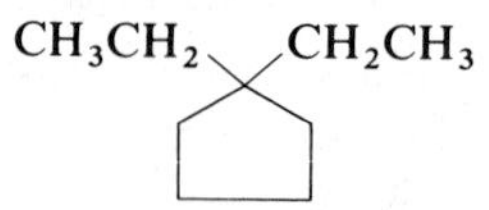

[cyclopentane ring with CH_3CH_2 and CH_2CH_3 on the same carbon]

Answer 1, 1-diethylcyclopentane.

32.2 UNSATURATED HYDROCARBONS

Hydrocarbons with covalent double or triple bonds (Section 11.9) between carbon atoms are said to be **unsaturated hydrocarbons,** "unsaturated" because the carbon atoms in such bonds can form other bonds.

In the common system of nomenclature, hydrocarbons with double covalent bonds are referred to as **olefins** and those with triple covalent bonds as **acetylenes.** The common names of olefins are formed by replacing the *-ane* ending of the saturated hydrocarbon name by *-ylene.* For example, ethane (CH_3CH_3) becomes ethylene ($CH_2{=}CH_2$). Hydrocarbons with triple covalent bonds are named as derivatives of acetylene ($CH{\equiv}CH$). The simple unsaturated hydrocarbons listed in Table 32.3 quite often are referred to by their common names.

In IUPAC nomenclature, hydrocarbons with covalent double and triple bonds are referred to as **alkenes** and **alkynes,** respectively. To derive the systematic names of alkenes, the *-ane* ending of the corresponding saturated hydrocarbon name is dropped and the *-ene* ending is added. If two double bonds are present, the *-adiene* ending is added; for three double bonds, the *-atriene* ending is added, and so on. The position of the multiple bond or bonds is indicated by numbers, starting from the end of the chain that will assign the lowest number to the first carbon atom in the multiple bond. For example

$$CH_3CH_2CH_2CH_2CH_3 \qquad \overset{1}{C}H_3\overset{2}{C}H{=}\overset{3}{C}H\overset{4}{C}H_2\overset{5}{C}H_3 \qquad \overset{1}{C}H_2{=}\overset{2}{C}H\overset{3}{C}H{=}\overset{4}{C}H\overset{5}{C}H_3$$

pentane *2-pentene* *1,3-pentadiene*

Similarly, alkynes are named by dropping the *-ane* ending and adding *-yne, -adiyne, -atriyne,* and so on.

$$\overset{1}{C}H_3\overset{2}{C}{\equiv}\overset{3}{C}\overset{4}{C}H_2\overset{5}{C}H_3 \qquad \overset{1}{C}H{\equiv}\overset{2}{C}\overset{3}{C}H_2\overset{4}{C}H_2\overset{5}{C}{\equiv}\overset{6}{C}\overset{7}{C}H_3$$

2-pentyne *1,5-heptadiyne*

As for alkanes, substituents are indicated by names and numbers that precede the name of the parent hydrocarbon (that with the longest carbon chain), for example

$$\overset{1}{C}H_3\overset{2}{C}H{=}\overset{3}{C}H\overset{4}{C}H\overset{5}{C}H_3 \quad (CH_3 \text{ on } C\text{-}4)$$

4-methyl-2-pentene

The two carbon atoms in a double bond and the four other atoms to which they are bonded all lie in the same plane. This configuration is attributed to the planar sp^2 hybrid orbitals which form the three σ bonds at each carbon atom. The region of the double bond in any molecule is thus flat, with the π bond above the plane (see

Table 32.3
Simple Unsaturated Hydrocarbons

Structure	Boiling point (°C)	Common name (systematic name)	Structure and name as substituent	
$CH_2{=}CH_2$	−102	Ethylene (ethene)	$CH_2{=}CH{-}$	vinyl
$CH_3CH{=}CH_2$	−48	Propylene (propene)	$CH_2{=}CHCH_2{-}$	allyl
$CH_3CH_2CH{=}CH_2$	−7	α-Butylene[a] (1-butene)	—	
$(CH_3)_2C{=}CH_2$	−7	Isobutylene (2-methylpropene)	—	
$CH_3CH{=}CHCH_3$	−4 (*cis*)	β-Butylene[a] (2-butene)	$CH_3CH{=}CHCH_2{-}$	crotyl
$HC{\equiv}CH$	−83	Acetylene (ethyne)	$HC{\equiv}C{-}$	ethynyl
$CH_3C{\equiv}CH$	−23	Methylacetylene (propyne)	$HC{\equiv}CCH_2{-}$	propargyl
$CH_2{=}CH{-}C{\equiv}CH$	3	Vinylacetylene (1-buten-3-yne)	—	

[a] α- and β- as used here are equivalent to 1- and 2- in the IUPAC system.

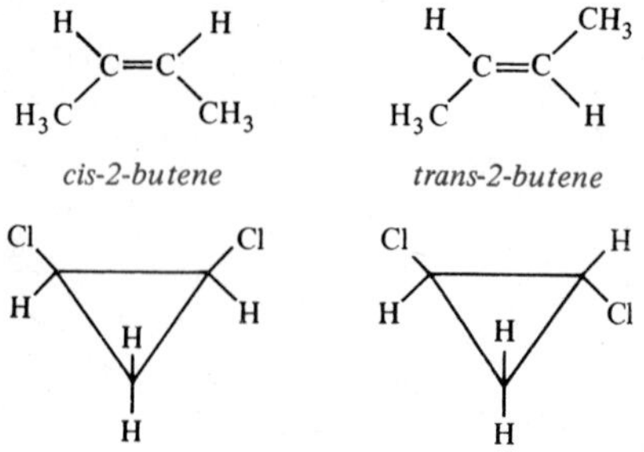

H, H, H_3C, CH_3 — C=C
cis-2-butene

H, CH_3, H_3C, H — C=C
trans-2-butene

Cl, Cl, H, H, H, H
cis-1,2-dichloro-cyclopropane

Cl, H, H, Cl, H, H
trans-1,2-dichloro-cyclopropane

Figure 32.2
***Cis–Trans* Isomerism**

Figure 11.7b). Rotation of the two carbon atoms joined by a double bond is inhibited by the π bond, which must be broken if rotation of the carbon atoms is to take place. This restricted rotation means that the placement of atoms or groups on the same or different sides of a double bond produces *cis–trans* isomers (Section 31.4), for example

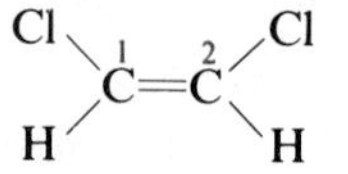

Cl, Cl, H, H — $\overset{1}{C}=\overset{2}{C}$
cis-*dichloroethylene*
m.p. −80.5 °C, b.p. 60.3 °C

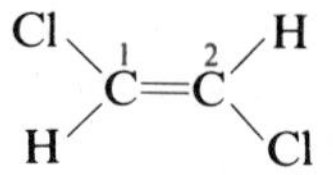

Cl, H, H, Cl — $\overset{1}{C}=\overset{2}{C}$
trans-*dichloroethylene*
m.p. −50 °C, b.p. 47.5 °C

If rotation about the double bond were easy, these two compounds would be identical. Note that in *cis–trans* isomerism involving double bonds, *each* doubly bonded carbon atom must have two *different* groups attached to it. Thus, 2-butene (Figure 32.2) also has *cis–trans* isomers. There is a third isomer of dichloroethylene, which is a structural isomer of the *cis* and *trans* compounds, but has no possibility for *cis–trans* isomerism.

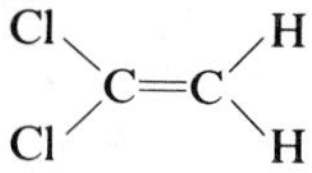

Cl, H, Cl, H — C=C
1,1-dichloroethylene

In cycloalkanes the carbon atoms also cannot rotate, so *cis–trans* isomerism exists in such compounds too, as in *cis*-1,2-dichlorocyclopropane and *trans*-1,2-dichlorocyclopropane (see Figure 32.2).

Covalent double bonds that alternate with single covalent bonds are said to be *conjugated.* For example, the long straight chain in the natural pigment β-carotene, which gives carrots their orange color, contains a series of conjugated double bonds (Figure 32.3). The color in such compounds arises because of the extensive delocalized π bonding present in the system of conjugated double bonds. These delocalized electrons are closer in energy to unoccupied energy levels than electrons in unconjugated double bonds. Therefore, less energy is required to excite them to higher levels and they absorb energy of wavelengths in the visible region, rather than in the ultraviolet region (as for unconjugated double bonds; see Aside 32.2).

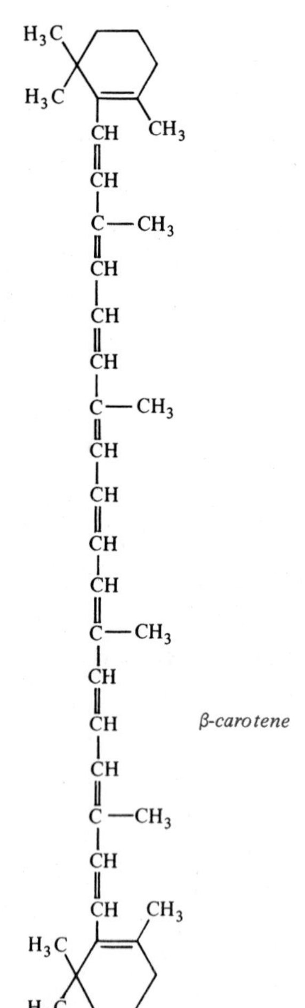

Figure 32.3
β-Carotene, the Pigment in Carrots
Also used to color processed foods (e.g., margarine).

EXAMPLE 32.2 Alkenes and Alkynes

Write the condensed formula for 3,4-dimethyl-3-hexene. Is *cis–trans* isomerism possible for this compound? If so, write the formulas showing locations of groups about the double bond in the two isomers.

The "3-hexane" part of the name tells us that the longest straight carbon chain in this compound consists of six carbon atoms with one double bond between the third and fourth carbon atoms. The "3,4-dimethyl" prefix indicates that there are two methyl groups, one attached to the third carbon atom and one attached to the fourth carbon atom.

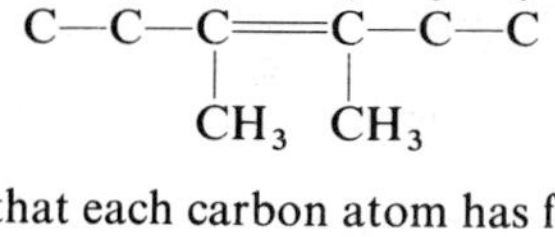

$$\overset{1}{C}-\overset{2}{C}-\underset{CH_3}{\overset{3}{C}}=\underset{CH_3}{\overset{4}{C}}-\overset{5}{C}-\overset{6}{C}$$

Filling in hydrogen atoms so that each carbon atom has four covalent bonds gives

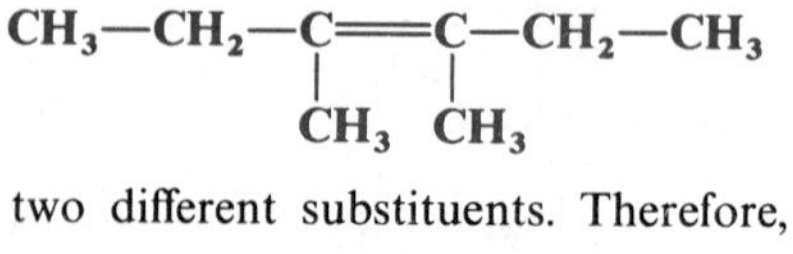

$$CH_3-CH_2-\underset{CH_3}{C}=\underset{CH_3}{C}-CH_2-CH_3$$

Each carbon atom has two different substituents. Therefore, ***cis–trans* isomerism is possible**

CH_3CH_2, CH_2CH_3, H_3C, CH_3 — C=C
cis

CH_3CH_2, CH_3, H_3C, CH_2CH_3 — C=C
trans

EXERCISE Write the IUPAC names for $CH_3C{\equiv}CCH_2CH_3$ and $ClCH_2C{\equiv}CCH_2CH_3$. Is *cis–trans* isomerism possible for alkynes?

Answers 2-pentyne, 1-chloro-2-pentyne; no (two different substituents on C atoms in triple bond not possible).

32.3 AROMATIC HYDROCARBONS

Aromatic hydrocarbons are unsaturated hydrocarbons containing planar ring systems stabilized by delocalized electrons, as in benzene. The bonding in benzene is represented by two resonance structures (Section 11.9) or the delocalization of electrons (Section 25.9).

benzene, resonance structures

benzene, structure emphasizing delocalized electrons

The pleasant odor of compounds containing such ring systems originally suggested the name "aromatic," but many not-so-sweet-smelling "aromatic" compounds are now known. Instead, their similar structures and chemical reactions are the basis for treating aromatic compounds as a group. Aromatic hydrocarbons are so distinct and different from other hydrocarbons that there is a general term for the others too. **Aliphatic hydrocarbons** are hydrocarbons that contain no aromatic rings. Saturated, unsaturated, and cyclic hydrocarbons with no aromatic rings are all aliphatic compounds.

Benzene is a colorless liquid, b.p. 80 °C, that burns with a very sooty flame, a property characteristic of aromatic hydrocarbons. Table 32.4 illustrates some aromatic hydrocarbons derived from benzene. The name of a single substituent is added to "benzene" as a prefix, as in ethylbenzene. Three structurally isomeric forms are possible for a disubstituted benzene, whether or not the substituents are the same. The three possibilities, illustrated in Table 32.4 for the xylenes, are designated *ortho* (abbreviated *o*-) for 1,2-substitution, *meta* (abbreviated *m*-) for 1,3-substitution, and *para* (abbreviated *p*-) for 1,4-substitution. Unless there is no question of what the structure is, as in hexachlorobenzene, numbers are always used to locate the substituents when three or more are present in the ring.

CH_2CH_3 / Cl

m-chloroethylbenzene or *3-chloroethylbenzene*

Cl / Cl / Cl

1,2,4-trichlorobenzene

CH_2CH_3 / Cl / Cl

2,4-dichloroethylbenzene

When the benzene ring is a substituent, it is named as a *phenyl* group. Diphenylmethane, for example, is $(C_6H_5)_2CH_2$.

Many compounds, known as *polycyclic aromatic hydrocarbons,* contain two or more aromatic rings fused together. ("Fused" rings have one side in common.) Some examples are given in Table 32.5. Several resonance forms can be written for each

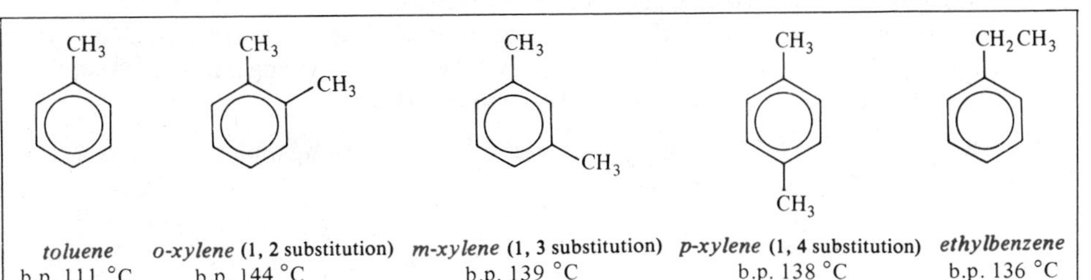

Table 32.4
Some Simple Substituted Benzene Hydrocarbons
The three xylenes illustrate *ortho, meta,* and *para* substitution.

Table 32.5
Some Simple Polycyclic Aromatic Hydrocarbons
The positions on the rings are numbered by convention as shown. Numbers are not assigned to the positions where rings are joined because substituents do not bond at these positions (there are no replaceable hydrogen atoms).

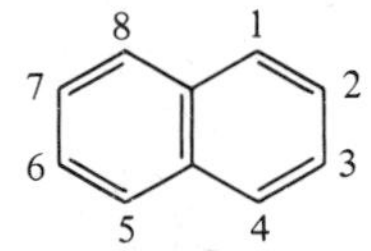

8 1 7 2 6 3 5 4

naphthalene
m.p. 80 °C

CH_3

CH_3

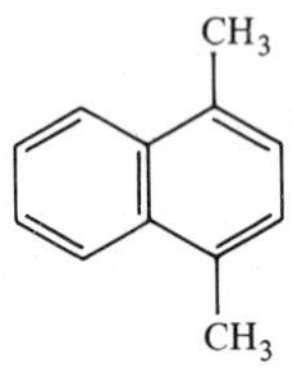

1,4-dimethylnaphthalene
(a substituted napthalene) m.p. 7.6 °C

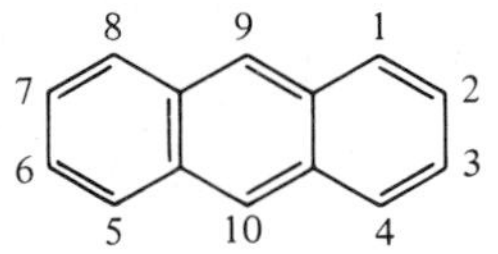

8 9 1 7 2 6 3 5 10 4

anthracene
m.p. 218 °C

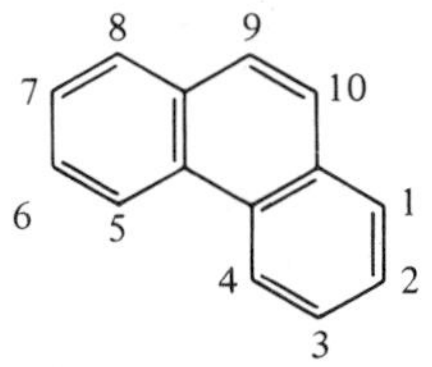

8 9 7 10 6 5 1 4 2 3

phenanthrene
m.p. 100 °C

compound of this type. Numbers are assigned (by IUPAC) to carbon atoms in such aromatic hydrocarbons containing fused rings, and the locations of substituents are identified by the assigned numbers, as illustrated in Table 32.5.

EXAMPLE 32.3 Aromatic Hydrocarbons

Give a name for

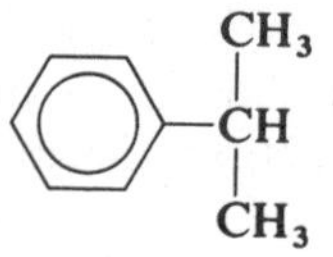

CH_3 — CH — CH_3 (attached to benzene ring at CH)

A benzene ring with a single alkyl substituent is named by adding the alkyl group name as a prefix. The name for this group is "isopropyl" (see Table 32.1) and the compound is named **isopropylbenzene.**

EXERCISE Write the structural formula for 1-methylnaphthalene (see Table 32.5).

Answer

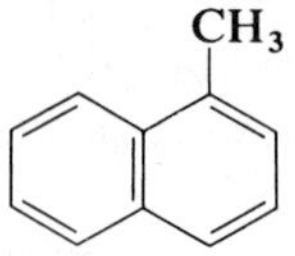

CH_3

EXAMPLE 32.4 Classification of Organic Compounds

Are the following compounds aromatic or aliphatic; saturated or unsaturated; alkanes, alkenes, or alkynes; branched or straight chains? In what other ways, if any, might these compounds each be described (other than by their names)?

(a) Cl, Br (para on benzene ring)

(b) $HC{\equiv}C{-}CH_2CH_2CH_3$

(c) $CH_3CH_2CH_2CHCH_3$ with CH_2 — CH_2 — CH_3 attached at CH

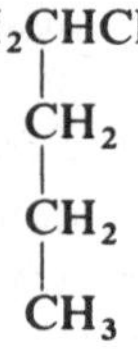

(d)

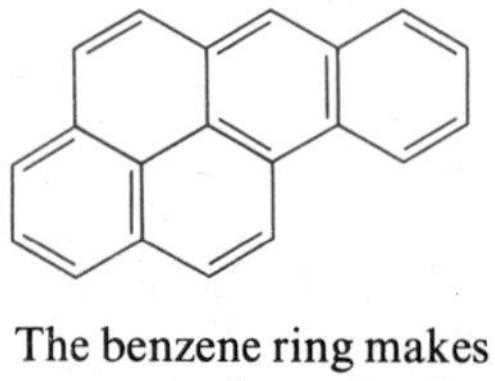

(e) $CH_3CH{=}CH_2$

(a) The benzene ring makes this an **aromatic compound.** The location of the two substituents opposite each other shows this to be a **para-substituted benzene.**
(b) With only carbon and hydrogen atoms present in a straight chain, this is an **aliphatic hydrocarbon.** The triple bond makes it an **unsaturated compound** which is an **alkyne,** or an **acetylene.** The compound can also be described as a **straight-chain compound.**
(c) This hydrocarbon is **saturated** (no double or triple bonds) and **aliphatic** (no aromatic rings), and is a **branched-chain alkane.**
(d) There are a number of fused aromatic rings with no substituents present in this compound. It is an **aromatic, polycyclic hydrocarbon.**
(e) This small molecule can be described as a **straight-chain hydrocarbon** that is **unsaturated** (it contains a carbon–carbon double bond). It is **aliphatic** (no aromatic rings) and can also be described as an **olefin** or an **alkene.**

EXERCISE Use the terms given above to describe the following molecules:

(a) Cl—(cyclohexane ring)—Br (b) $CH_3CH_2CH_2CH_3$ (c) $CH_3CH{=}CHCH{=}CHCH_3$

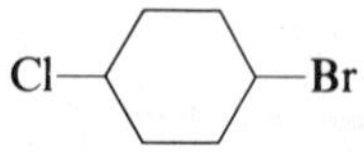

Answers (a) aliphatic, saturated, substituted cycloalkane; (b) aliphatic, saturated, alkane, straight chain; (c) aliphatic, unsaturated, alkene, straight chain (also a conjugated diene).

32.4 PROPERTIES OF HYDROCARBONS

The properties of hydrocarbons are to a large extent those imparted by the covalent bond in combination with those imparted by intermolecular forces. In solubility, the behavior of hydrocarbons reflects the fact that they are nonpolar. All but the lightest hydrocarbons are virtually insoluble in water but are soluble in each other. Most hydrocarbons are less dense than water and float on the surface of the water. Within any series of similar alkanes, alkenes, or aromatic hydrocarbons, melting points and boiling points generally increase with increasing molecular mass (as shown by the boiling points given in Tables 32.1 and 32.3). Alkanes containing eighteen or more carbon atoms are waxy solids at room temperature (the paraffins).

All hydrocarbons burn readily, and in the presence of sufficient oxygen undergo complete combustion to produce only carbon dioxide and water. Otherwise, the chemical reactivity of hydrocarbons varies with the presence or absence of multiple covalent bonds and with whether or not the compound is aromatic.

Alkanes, which contain only nonpolar C—C and C—H single covalent bonds, are generally not very reactive. They are not attacked at room temperature by acids, bases, or oxidizing agents such as potassium permanganate.

Alkanes do, however, undergo *halogenation* reactions, in which hydrogen atoms are replaced by halogen atoms. The halogen atoms are generated from the halogen molecules by absorption of energy, often from light. Such reactions—for example, the chlorination of methane—

$$Cl_2 \xrightleftharpoons{\text{light}} 2Cl\cdot$$
$$Cl\cdot + CH_4 \longrightarrow HCl + CH_3\cdot$$
$$CH_3\cdot + Cl_2 \longrightarrow \underset{\textit{methyl chloride}}{CH_3Cl} + Cl\cdot$$

are chain reactions. In the preceding example, the chlorine atom produced in the third step reacts with another methane molecule as in the second step and carries the chain reaction forward. Further chlorination produces CH_2Cl_2, $CHCl_3$, and CCl_4 (Section 27.12).

Double and triple bonds between carbon atoms in *nonaromatic* compounds undergo **addition reactions**—the addition of atoms or groups to the carbon atoms. Some of the common types of addition reactions are illustrated in Table 32.6.

Table 32.6
Addition Reactions of Unsaturated Hydrocarbons

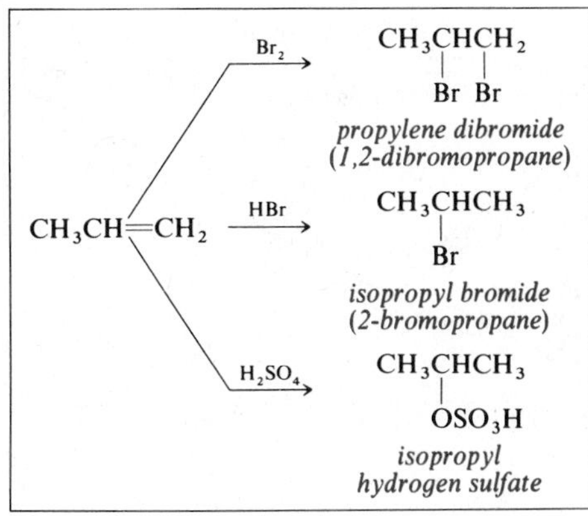

Table 32.7
Substitution Reactions of Aromatic Compounds
The reactions shown here are typical of all aromatic compounds.

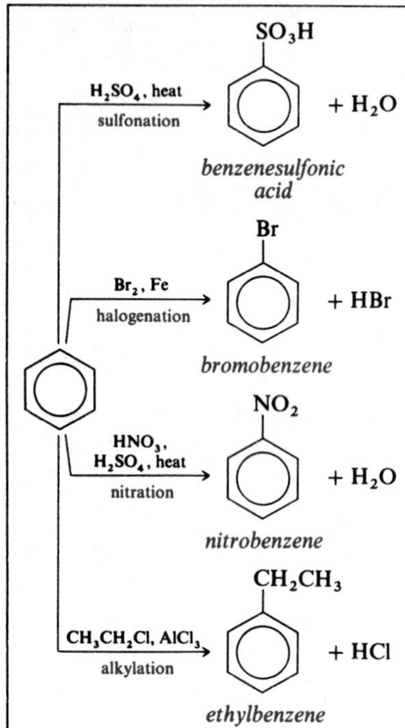

Alkenes and alkynes also react with oxidizing agents. They undergo cleavage of the double or triple covalent bond and oxidized products are formed, for example

$$CH_3CH{=}CHCH_3 \xrightarrow[H_2O]{\text{excess } KMnO_4} 2CH_3\overset{\overset{O}{\|}}{C}OH \quad \text{(32.1)}$$

acetic acid

Both the reactions with bromine (Table 32.6) and the reaction with potassium permanganate (Equation 32.1) are useful tests for the presence of double and triple bonds. The reddish color of the bromine and the purple color of aqueous permanganate are observed to disappear in the presence of unsaturated compounds.

The reactivity of aromatic compounds differs from that of nonaromatic unsaturated compounds. In contrast to the addition reactions of alkenes and acetylenes, aromatic compounds undergo **substitution reactions**—replacement of one or more hydrogen atoms of the aromatic ring. Four types of substitution reactions typical of all aromatic compounds are illustrated in Table 32.7.

The aromatic ring is so stable that it remains unchanged in most reactions, including those with oxidizing agents. For example, potassium permanganate causes oxidation of an alkyl substituent rather than of the benzene ring.

$$C_6H_5CH_3 \xrightarrow[\text{heat}]{KMnO_4} C_6H_5\overset{\overset{O}{\|}}{C}{-}OH \quad \text{(32.2)}$$

toluene → *benzoic acid*

The stability to oxidation and the reaction by substitution instead of addition are distinctive properties of aromatic compounds.

32.5 HYDROCARBONS AND ENERGY

The major sources of hydrocarbons and carbon are the fossil fuels—natural gas, petroleum, and coal. The major use of hydrocarbons and carbon is as fuel. Approximately 75% of the energy consumed in the United States is derived from natural gas and petroleum. The energy available to a society determines what that society can do and may determine what that society will do. Both the availability and the economics of hydrocarbons are changing. How long the current dominant role of naturally occurring hydrocarbons, particularly petroleum, as an energy source can, or will, continue is a question that is open to debate. That they will all eventually be gone is certain. Some predict that this will occur within your lifetime.

Many long-term energy scenarios are now used to explore possible new and changing sources of energy as the United States attempts to become less dependent on imported oil and as the present supplies of fossil fuels are exhausted. Figure 32.4

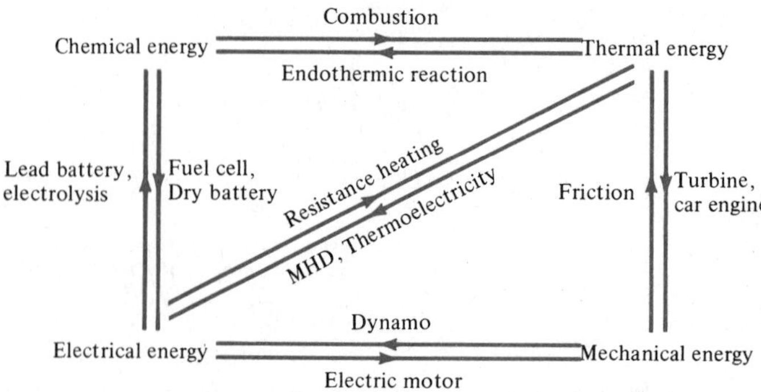

Figure 32.4
Some Energy Conversion Pathways
MHD refers to magnetohydrodynamics—the generation of electricity by a conductive fluid flowing in a magnetic field.

Source: J. H. Harker and J. R. Backhurst, *Fuel and Energy* (New York: Academic Press, 1981), p. 4.

shows some of the ways in which energy is converted from one form to another. The pathway for hydrocarbon fuels is from chemical to thermal to mechanical to electrical energy. Hydrocarbon fuels from new sources would enter the cycle of energy conversion as stored chemical energy and be utilized in the same way as the traditional fossil fuels.

Table 32.8
Composition of Natural Gas (by volume)

Methane, CH_4	70–90%
Ethane, C_2H_6	0–7%
Propane, C_3H_8	0–20%
Butane, C_4H_{10}	0–1%
Hydrogen, H_2	nil
Carbon dioxide, CO_2	0–8%
Oxygen, O_2	0–0.2%
Nitrogen, N_2	0–5%
Hydrogen sulfide, H_2S	0–5%
Noble gases, Ar, He, Ne, Xe	0–2%

a. Fossil Fuels Natural gas is a mixture of gaseous substances trapped along with petroleum and coal in the Earth's crust. The composition of natural gas varies widely (Table 32.8), but methane is always the major constituent. Natural gas is recovered from oil wells and isolated gas wells to which the gas has migrated through the rock. The "calorific" value of an average natural gas is about 38 MJ/m^3 (megajoules per cubic meter; the calorific value is the heat released when one unit—in this case one cubic meter—of the fuel undergoes complete combustion at 1 atm). Pipelines carry natural gas from large producing fields to major industrial and population centers, where it is used directly as fuel or as a chemical raw material.

Petroleum, as it is pumped from underground reservoirs, is a dark, thick, smelly liquid. Up to 95% of crude petroleum (Table 32.9) is a mixture of hydrocarbons, including alkanes, cycloalkanes, and aromatics, but only rarely, alkenes. A use has been found for virtually every component of petroleum.

Table 32.9
Approximate Composition of Crude Petroleum (by mass)

Hydrocarbons	50–95%
Alkanes, cycloalkanes, and aromatics	
Oxygen, nitrogen, and sulfur-containing compounds	0.5–8%
Resins and asphalts	5–25%
Polymers and polycyclic aromatics, among others	

The first step toward manufacture of petroleum products is separation of the crude petroleum by distillation and extraction into fractions possessing various boiling points and molecular masses. These fractions can be used as indicated in Table 32.10 or treated further to convert them into more desirable mixtures or pure compounds. For example, at least two-thirds of the molecules in the gasoline that you obtain at a service station were not present as such when the petroleum was pumped from the oil well. Isomerization, cracking, alkylation, and reforming are processes used to increase the yield of products such as gasoline from crude oil. *Petroleum isomerization,* accomplished by heat and catalysts, converts straight-chain alkanes into branched alkanes. The latter perform better as fuels. *Petroleum cracking,* also via heat and catalysts, breaks large molecules above the gasoline range into smaller molecules (alkanes and also alkenes) that are in the gasoline range. *Petroleum alkylation* combines lower molecular mass alkanes and alkenes to form molecules in the gasoline range. *Petroleum reforming* employs catalysts in the presence of hydrogen to convert noncyclic hydrocarbons to aromatic compounds. The molecules are "reformed" as, for example, in the following reaction.

$$\underset{n\text{-heptane}}{CH_3(CH_2)_5CH_3(l)} \xrightarrow[\text{Pt-Re catalyst}]{\text{Pt or}} \underset{\text{toluene}}{C_6H_5CH_3(l)} + 4H_2(g)$$

Further treatment of petroleum products depends upon their ultimate use. For example, liquid products that will be burned are usually treated to remove hydrogen sulfide and sulfur-containing organic compounds.

Table 32.10
Important Products from Crude Petroleum

	Boiling range (°C)	Composition	Source	Principal uses
Natural gas	—	Mainly methane, some nitrogen depending on source	Natural sources	Fuel gas, also reformed to synthesis gas
Liquefied petroleum gas (LPG)	—	Propane, butane	Stripped from "wet" natural gas or from cracking operations	Domestic and industrial fuel—production of coal gas, synthetic chemicals
Primary flash distillate (PFD)	varies	Propane and butane dissolved in gasoline–kerosene range of liquids	Preliminary distillation of crude petroleum	Manufacture of synthesis gas
Gasoline	25–175	Complex mixture of materials. Contains additives to improve performance but no sulfur or polymerizable components	Primary distillation, cracking and reforming processes	Spark ignition internal combustion engines
Kerosene	135–300	Paraffinic hydrocarbons with substantial proportion of aromatics, low sulfur content	Distillation, cracking	Agricultural tractors, lighting, heating and aviation gas turbines
Gas oil	175–345	Saturated hydrocarbons	Distillation, hydrodesulfurization	Diesel fuel, heating and furnaces. Feed to cracking units
Diesel fuel	175–375	Saturated hydrocarbons, often with high sulfur	Distillation, cracking	Diesel engines, furnace heating
Fuel oils	225–425	—	Residue of primary distillation, blended with distillates	Large-scale industrial heating
Lubricating oils	wide range	Three types: mainly aromatic, mainly aliphatic or mixed	Vacuum distillation of primary distillation residue; solvent extraction	Lubrication
Wax	—	Paraffins	Chilling residue from vacuum distillation	Toilet preparations, food, candles, petroleum jelly
Bitumen	—	Wide variation	Residue from vacuum distillation or oxidation of residue from primary distillation ("blown" bitumen)	Road surfacing, waterproofing

Source: J. H. Harker and J. R. Backhurst, *Fuel and Energy* (New York: Academic Press, 1981), p. 86.

Petroleum is the source of compounds that serve as starting materials for the syntheses of most industrial organic chemicals, particularly those used in synthetic fibers, plastics, coatings, and synthetic rubber. However, only 10% of the petroleum processed each year goes to the chemical industry in the form of raw materials. The remaining 90% is eventually burned, in what might be viewed as wasting a natural resource.

Coal, like natural gas, can be burned directly as a fuel. Coal contains carbon, hydrogen, nitrogen, oxygen, and sulfur in varying amounts, as well as traces of heavy metals. The carbon content of coal ranges from 45% to 95%. The second most important use of coal is in production of coke for the iron and steel industry. Coke is made by heating coal in the absence of air to drive off volatile materials. Ammonia and coal tar, from which organic chemicals, particularly aromatics, can be obtained are important by-products of this operation.

$$\text{Coal} \xrightarrow{\text{heat, no air}} NH_3 + \text{other gases } (H_2, CH_4, CO, CO_2) + \text{coke} + \text{coal tar (source of aromatic compounds)}$$

b. Nontraditional Routes to Fuels and Organic Chemicals The interest in producing fuel and organic chemicals from sources other than petroleum rises and falls with the price of oil. However, it is a topic that ultimately must grow in importance.

The term "synfuel" has come to mean any fuel derived in a nontraditional way either from natural resources or by chemical synthesis. Synfuels include, for example, liquid or gaseous fuel produced from coal (the gas is referred to as SNG, synthetic natural gas), alcohol made from sugarcane or corn (an example of the general group of "fuels from biomass"), and natural hydrocarbon fuels extracted from oil-bearing shale.

One widely applicable, nontraditional route to fuels and also organic chemicals is by reactions of carbon monoxide–hydrogen mixtures, known as synthesis gas, or syngas. A potentially very important source of carbon monoxide–hydrogen mixtures is coal, for coal supplies far exceed our petroleum supplies.

The reaction of coal or coke with steam was once used to provide household cooking gas, and is known as the water gas reaction

$$\underset{\textit{coal or coke}}{C(s)} + H_2O(g) \xrightarrow{1000\ ^\circ C} CO(g) + H_2(g) \qquad \textbf{(32.3)}$$

The $CO-H_2$ mixture produced in this manner, known as water gas, has a calorific value of about 11 MJ/m^3. The availability of natural gas at much lower prices made the use of water gas for cooking obsolete.

The reaction of carbon monoxide and steam to give hydrogen and carbon dioxide

$$CO(g) + H_2O(g) \xrightarrow{1000\ ^\circ C} CO_2(g) + H_2(g) \qquad \textbf{(32.4)}$$

known as the water gas shift reaction, is used to increase the hydrogen content of water gas. The ratio of carbon monoxide to hydrogen in the mixture can be controlled by controlling the degree of shifting. For example, a 1-to-2 carbon monoxide–hydrogen mixture is the synthesis gas for methanol production and a 1-to-3 mixture is needed for methane production. These and some of the other uses of carbon monoxide–hydrogen mixtures are illustrated in Table 32.11.

A growing and increasingly successful area of research is the production of larger and more complex organic molecules from the simple ones obtained directly from syngas. At least one functioning commercial operation is already producing gasoline from methanol. Another plant begins by making syngas from coal (known as coal gasification) and proceeds to the production of methanol, methyl acetate, and acetic

Table 32.11
Uses of Carbon Monoxide–Hydrogen Mixtures

As a fuel (water gas)

$$2H_2(g) + O_2(g) \xrightarrow{\Delta} 2H_2O(g)$$

$$2CO(g) + O_2(g) \xrightarrow{\Delta} 2CO_2(g)$$

Methanol production

$$CO(g) + 2H_2(g) \xrightarrow[\substack{300\ \text{atm} \\ \text{Ag or Cu catalyst}}]{230\text{–}400\ ^\circ C} CH_3OH(l)$$

Methane production (methanation; SNG production)

$$CO(g) + 3H_2(g) \xrightarrow[\substack{\text{Ni or} \\ \text{Co catalyst}}]{\substack{230\text{–}450\ ^\circ C \\ 1\text{–}100\ \text{atm}}} CH_4(g) + H_2O(g)$$

Hydrocarbon synthesis (Fischer–Tropsch synthesis)

$$mCO(g) + (2m + 1)H_2(g) \xrightarrow[\text{catalyst}]{\substack{150\text{–}350\ ^\circ C \\ 1\ \text{to several hundred atm}}} C_mH_{2m+2} + mH_2O(g)$$

Aldehyde production (oxo process)

$$RCH{=}CH_2 + H_2(g) + CO(g) \xrightarrow{200\text{–}250\ ^\circ C} RCH_2CH_2CHO$$

anhydride—all precursors of numerous important industrial chemicals. Addition of ammonia to syngas allows for the production of nitrogen-containing chemicals.

SUMMARY OF SECTIONS 32.1–32.5

All organic compounds can be considered derivatives of hydrocarbons, which contain only carbon and hydrogen. Saturated hydrocarbons contain only single covalent bonds; unsaturated hydrocarbons contain double and/or triple covalent bonds between carbon atoms. Aromatic hydrocarbons contain planar rings stabilized by delocalized electrons, as in benzene, and have properties different from those of nonaromatic, or aliphatic, hydrocarbons. (Aromatic properties include resistance to oxidation of the ring and reaction by substitution on the ring, Table 32.7, rather than addition across double bonds, Table 32.6.) Alkanes are relatively unreactive, but do burn readily and undergo halogenation in chain reactions with halogen atoms. Organic compounds are named systematically according to a set of rules (IUPAC nomenclature). Many also have older, common names.

Most of our energy is generated by the combustion of fossil fuels (natural gas, petroleum, and coal); petroleum (Table 32.10) is also the source of the raw materials for most manufactured organic chemicals. As petroleum resources are depleted, other sources must be found for energy and chemical raw materials. One possible alternative is based on the production of syngas ($CO–H_2$ mixtures) from coal (Table 32.11).

PROPERTIES OF ORGANIC COMPOUNDS

32.6 THE NATURE OF FUNCTIONAL GROUPS

If one of the hydrogen atoms in each of the compounds in the series of straight-chain hydrocarbons is replaced by an OH group, the following series of compounds is produced:

$$CH_3OH$$
$$CH_3CH_2OH$$
$$CH_3CH_2CH_2OH$$
$$CH_3CH_2CH_2CH_2OH$$
etc.

These compounds can all be represented by the general expression ROH, where R is an alkyl group and OH is the "hydroxyl" group.

The chemistry of these compounds is dominated by the OH group, and they exhibit similar chemical behavior, no matter how long the continuous chain of carbon atoms may be. Saturated hydrocarbons are relatively unreactive, so the hydrocarbon part of the molecule does not overshadow the characteristic reactivity of the OH group. The compounds of the family represented by the formula ROH are collectively called alcohols, and the OH group is referred to as a functional group. A **functional group** is a chemically reactive atom or group of atoms that imparts characteristic properties to the family of organic compounds containing that group.

Table 32.12 lists the most common functional groups. You should know the general formula for each type of compound listed. In speaking or writing about an organic compound that includes an OH group, for example, we might refer to the compound as having a hydroxyl group or as being an alcohol or a phenol. Large and even not so large molecules can contain many functional groups of the same or

Table 32.12
Some Common Functional Groups

General formula[a] (functional group in color)	Type of compound	Functional group name
R—X	Alkyl or aryl halide	Halo
R—OH	Alcohol	Hydroxyl
Ar—OH	Phenol	Hydroxyl
ROR	Ether	Alkoxy
RNH_2, R_2NH, R_3N	Amine[b]	Amino
R—C(H)=O	Aldehyde	Aldehyde or formyl
$R_2C=O$	Ketone	Carbonyl, or keto, or oxo
R—C(OH)=O	Carboxylic acid	Carboxyl
R—C(NH_2)=O	Amide	Amido
R—C(X)=O	Acyl halide	Carbonyl halide
R—C(=O)—O—C(=O)—R	Acid anhydride	Anhydride
R—C(OR)=O	Ester	Ester
R—NO_2	Nitroalkane or nitroaromatic	Nitro
R—SO_3H	Sulfonic acid	Sulfonic acid
R—CN	Nitrile	Cyano

[a] In these general formulas, R represents the remainder of the molecule. Ar is the abbreviation used for an aromatic group, such as the phenyl group (C_6H_5). X represents any halogen atom.
[b] RNH_2, a primary amine; R_2NH, a secondary amine; R_3N, a tertiary amine (Section 32.12).

different types. The properties of compounds containing the most common functional groups are discussed in Sections 32.9–32.17.

In organic chemistry, the power of a functional group should never be underestimated. For example, the placement of the —NO_2 or —NH_2 groups significantly changes the taste of the following three compounds.

$OCH_2CH_2CH_3$ / NO_2 / NH_2 — *tasteless*

$OCH_2CH_2CH_3$ / NH_2 / NO_2 — *4100 times sweeter than sugar*

$OCH_2CH_2CH_3$ / NO_2 / NO_2 — *bitter*

32.7 ISOMERISM IN ORGANIC MOLECULES

All three types of isomerism introduced in our discussion of coordination complexes (Section 31.4)—structural, *cis–trans*, and optical isomerism—are common in organic compounds. We have already mentioned the structural isomerism of hydrocarbons and *ortho-*, *meta-*, and *para-*substituted benzenes, and the *cis–trans* isomerism of alkenes. Introduction of functional groups into hydrocarbons produces the possibility of further structural isomer formation. Consider ethane (C_2H_6). Two chlorine atoms can replace hydrogen atoms in ethane in two different ways to give two structural isomers

$$\begin{array}{c} \text{H}\quad\text{H} \\ {}^{1}|\quad{}^{2}| \\ \text{Cl—C—C—H} \\ |\quad\;| \\ \text{Cl}\quad\text{H} \end{array} \qquad\qquad \begin{array}{c} \text{H}\quad\text{H} \\ {}^{1}|\quad{}^{2}| \\ \text{H—C—C—H} \\ |\quad\;| \\ \text{Cl}\quad\text{Cl} \end{array}$$

1,1-dichloroethane m.p. −97.0 °C, b.p. 57.3 °C — *1,2-dichloroethane* m.p. −35.4 °C, b.p. 83.5 °C

Clearly, the number of structural isomers that might be formed by introduction of functional groups increases greatly with the number of carbon atoms in a molecule.

A simple example of an optically active organic compound is one in which a single carbon atom is bonded to four different kinds of atoms or groups. For example, in 3-methylhexane

$$\begin{array}{c} \text{H} \\ | \\ CH_3CH_2CCH_2CH_2CH_3 \\ | \\ CH_3 \end{array}$$

an asymmetric carbon atom (arrow to C-3)

3-methylhexane

the number 3 carbon atom in the hexane chain is bonded to an ethyl group, a methyl group, a hydrogen atom, and a propyl group. Such an atom bonded to four different atoms or groups is said to be an **asymmetric atom.** Any molecule containing one asymmetric carbon atom is chiral— it is optically active and mirror-image isomers of that molecule can exist (Section 31.4). Figure 32.5 represents the mirror-image configurations of 3-methylhexane. If you can, examine three-dimensional models as an aid to seeing the difference between optical isomers containing asymmetric carbon atoms.

One optically active isomer rotates plane-polarized light counterclockwise, or to the left, and is called *levorotatory.* The other isomer rotates plane-polarized light clockwise, or to the right, and is called *dextrorotatory.* In solutions of equal concentrations, optical isomers rotate plane-polarized light equally in opposite directions. The individual molecules are referred to as the L-isomer and the D-isomer. In most

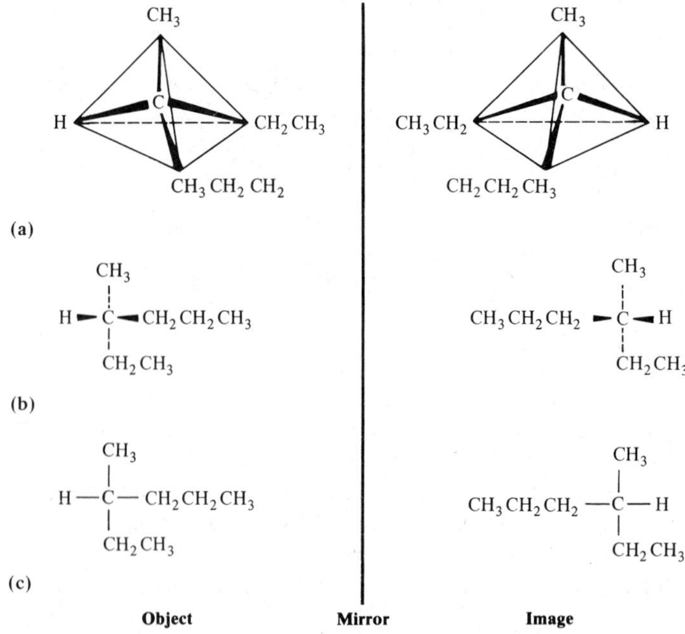

Figure 32.5
Enantiomeric Forms of 3-Methylhexane
(a) Three-dimensional representation; (b) simulated three-dimensional models; (c) Fischer projections of the models in (b). In a Fischer projection, horizontal lines represent bonds that project outward toward the reader and vertical lines represent bonds that project away from the reader.

properties, such as melting point, boiling point, and solubility, two optical isomers behave identically and appear to have the same molecular structure. However, the dextro and levo forms can differ greatly in their reactivity toward other optically active substances.

A mixture of equal parts of the levorotatory isomer and the dextrorotatory isomer of the same substance—a **racemic mixture**—shows no net rotation of polarized light as a result of the cancellation of the equal and opposite rotation of the isomers. Such a mixture is optically inactive.

Many naturally occurring substances exist only as one or the other member of an optical isomer pair. This implies that the plants or animals producing these substances must synthesize them by methods very different from those that are used in the chemistry laboratory. Furthermore, living systems are often able to distinguish between optical isomers. For example, *only* the dextrorotatory form of an optically active drug might be effective whereas the levorotatory isomer might have little or no effect or even be toxic. It is this biological specificity that requires considerable ingenuity in the synthesis of compounds that duplicate complex natural products. Often several asymmetric atoms are present in a single molecule and the number of possible optically active isomers is quite large. [Molecular asymmetry also occurs in molecules with no asymmetrically substituted atoms.]

EXAMPLE 32.5 Isomerism

Which of the following compounds can exhibit optical isomerism? Draw structural formulas (like the "Fischer projections" in Figure 32.5) which show the asymmetric carbon atoms.

(a) $CH_3CHBrCH_2CH_3$ (b) $CH_2BrCH_2CH_2CH_3$ (c) $CH_2ClCHBrCH_3$

To determine whether or not these compounds can have optical isomers, look for carbon atoms that are asymmetric—that have four different substituents. **In both compounds (a) and (c) the second carbon atom from the left is asymmetric.**

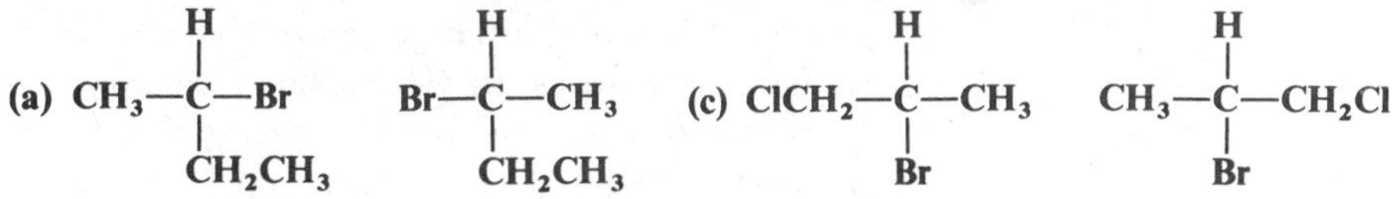

(a)
$$\begin{array}{c} \mathrm{H} \\ | \\ \mathrm{CH_3{-}C{-}Br} \\ | \\ \mathrm{CH_2CH_3} \end{array} \qquad \begin{array}{c} \mathrm{H} \\ | \\ \mathrm{Br{-}C{-}CH_3} \\ | \\ \mathrm{CH_2CH_3} \end{array}$$

(c)
$$\begin{array}{c} \mathrm{H} \\ | \\ \mathrm{ClCH_2{-}C{-}CH_3} \\ | \\ \mathrm{Br} \end{array} \qquad \begin{array}{c} \mathrm{H} \\ | \\ \mathrm{CH_3{-}C{-}CH_2Cl} \\ | \\ \mathrm{Br} \end{array}$$

Compound (b) has no asymmetric carbon atom.

EXERCISE Will the following compound exhibit optical isomerism?

$$\begin{array}{c} \mathrm{CH_2{=}CHCHCH_3} \\ \quad | \\ \quad \mathrm{Cl} \end{array}$$

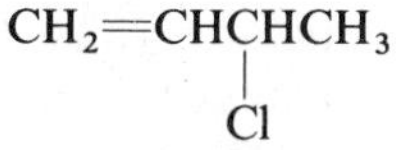

If it will, draw the Fischer projections for the enantiomeric pair.

Answer Yes.

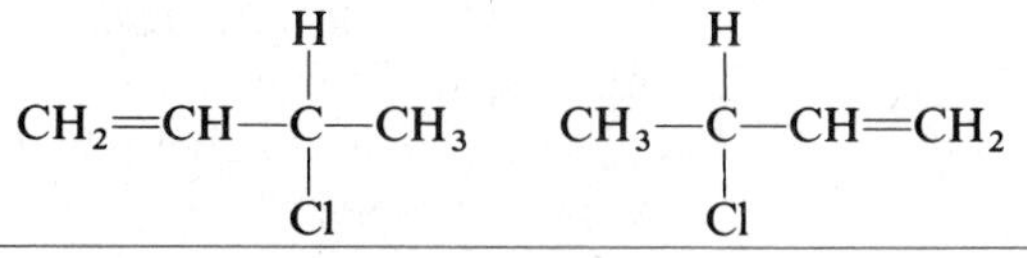

$$\begin{array}{c} \mathrm{H} \\ | \\ \mathrm{CH_2{=}CH{-}C{-}CH_3} \\ | \\ \mathrm{Cl} \end{array} \qquad \begin{array}{c} \mathrm{H} \\ | \\ \mathrm{CH_3{-}C{-}CH{=}CH_2} \\ | \\ \mathrm{Cl} \end{array}$$

EXAMPLE 32.6 Isomerism

What types of isomers and how many of each type are possible for dibromochloroethylene?

The bromine and chlorine atoms can be arranged about the double bond in two different ways to give **two structural isomers:**

$$\underset{Cl}{\overset{H}{}}\!\!>C=C<\!\!\underset{Br}{\overset{Br}{}} \qquad \underset{H}{\overset{Br}{}}\!\!>C=C<\!\!\underset{Cl}{\overset{Br}{}}$$

The second of these two structures has two different groups on each carbon atom in the double bond with the bromine atoms *cis* to each other. **A trans isomer** is also possible:

$$\underset{Br}{\overset{H}{}}\!\!>C=C<\!\!\underset{Cl}{\overset{Br}{}}$$

EXERCISE How many isomers (counting *cis* and *trans* isomers separately) are possible for monochlorobutene? *Answer* 9.

32.8 CHEMICAL REACTIONS OF ORGANIC COMPOUNDS

It is beyond the intended scope of this chapter to discuss to any great extent the many possible types of chemical reactions of organic compounds containing functional groups. However, we want to provide an example of how organic reactions are pictured and explained. We can do this by discussing one general type of organic reaction: the substitution of a halogen atom in an alkyl halide (a compound with the general formula RX, where R is an alkyl group and X is a halogen; Section 32.9) by another atom or group, OH^- in our example. The overall reaction we have chosen is

$$NaOH(aq) + \underset{\textit{methyl bromide}}{CH_3Br(l)} \longrightarrow \underset{\textit{methanol}}{CH_3OH(aq)} + NaBr(aq) \qquad \mathbf{(32.5)}$$

Most organic reactions involve making and breaking covalent bonds. The site of reaction in an organic compound is most likely to be a multiple covalent bond in a functional group or a polar single covalent bond. Atoms or groups that are electron-rich or electron-poor are potential reaction sites. A reactant that has a partial or complete positive charge and will bond with an atom that has an available electron pair is referred to as an **electrophile** ("electron-loving"). Similarly, a reactant that has a partial or complete negative charge that enables it to bond with an electron-deficient atom is referred to as a **nucleophile** ("nucleus-loving"). A given organic functional group will undergo many of the same reactions regardless of its location in different kinds of molecules.

Referring back to Equation (32.5), the carbon–bromine bond in methyl bromide is polar because the bromine atom is more electronegative than the carbon atom. This makes the carbon atom somewhat electron deficient, and therefore an electrophilic reaction site.

$$\equiv\overset{\delta+}{C}-\overset{\delta-}{Br}$$

The hydroxide ion from sodium hydroxide, with its negative charge, $:OH^-$, is a nucleophilic species.

In the reaction between hydroxide ion and methyl bromide, the nucleophilic OH^- approaches the electrophilic carbon atom in the C—Br bond. As the oxygen–carbon bond forms, the carbon–bromine bond breaks. The reaction can be represented by arrows that indicate the directions in which electron pairs move.

$$HO:^- + CH_3-Br \longrightarrow HOCH_3 + :Br^-$$

In this reaction the breaking of the C—Br bond and the formation of the C—O bond occur simultaneously. This and other similar reactions have been shown to exhibit second-order kinetics, the rate being dependent upon the concentrations of both hydroxide ion and methyl bromide. Many, but not all, alkyl chlorides, bromides, and iodides react with a variety of nucleophiles in this manner. This type of reaction is described as a "bimolecular nucleophilic substitution reaction."

In the equations for organic reactions, the formulas of simple reactants, solvents, and catalysts are often written over the arrows. The results are equations that are not stoichiometrically balanced, but show only the changes in the organic molecule or molecules of concern. The equation for the nucleophilic substitution reaction of hydroxide ion with methyl bromide, written in this way, would be

$$CH_3Br \xrightarrow{NaOH,\ H_2O} CH_3OH$$

In organic chemistry, it is common to refer to a reaction in which oxygen is added to an organic molecule or hydrogen is removed as "oxidation." Also, a reaction in which hydrogen is added to a molecule or oxygen is removed often is referred to as "reduction." For example, acetaldehyde is oxidized by the addition of an oxygen atom to give acetic acid and the same compound is reduced by the addition of two hydrogen atoms to give ethanol:

$$\underset{\textit{acetaldehyde}}{CH_3\overset{\overset{O}{\|}}{C}{-}H} \underset{\text{reduction}}{\overset{\text{oxidation}}{\rightleftharpoons}} \underset{\textit{acetic acid}}{CH_3\overset{\overset{O}{\|}}{C}OH}$$

$$\underset{\textit{acetaldehyde}}{CH_3\overset{\overset{O}{\|}}{C}{-}H} \underset{\text{oxidation}}{\overset{\text{reduction}}{\rightleftharpoons}} \underset{\textit{ethanol}}{CH_3CH_2OH}$$

There are, of course, organic redox reactions that do not involve hydrogen or oxygen, but it is useful to recognize what is meant when the terms are used as described above.

Table 32.13
Alkyl and Aryl Halides, RX[a]

CH_3Br *methyl bromide* (*bromomethane*)
b.p. 5 °C
CH_3I *methyl iodide* (*iodomethane*)
b.p. 42 °C
CH_3CH_2Br *ethyl bromide* (*bromoethane*)
b.p. 38 °C
$CH_3CH_2CH_2Br$ *propyl bromide* (*1-bromopropane*)
b.p. 71 °C
$CH_3\underset{Br}{\underset{\mid}{C}}H\underset{Br}{\underset{\mid}{C}}H_2$ *propylene dibromide* (*1,2-dibromopropane*)
b.p. 141 °C
$CH_2{=}CHCl$ *vinyl chloride* (*chloroethene*)
b.p. 14 °C
$C_6H_5{-}Cl$ *chlorobenzene*
b.p. 132 °C

[a] Common name (systematic name)

SUMMARY OF SECTIONS 32.6–32.8

The properties of organic molecules are determined to a great extent by the types of functional groups present in the molecules. The various functional groups (Table 32.12) are most often the sites of chemical reactions. Isomerism is a possibility for many organic compounds. Structural isomers increase in number with the number of atoms present. Alkenes with two different groups on each of the atoms in a carbon–carbon double bond exist as *cis–trans* isomers, which usually differ in properties. Optical isomers, which are identical in many properties, are formed by compounds that contain carbon atoms with four different substitutents (asymmetric carbon atoms).

Organic chemical reactions are often the result of interaction between electrophiles (electron-seekers) and nucleophiles (electron-rich species). The pathways of electrons in organic reactions are depicted by writing arrows in the chemical equations. "Oxidation" of an organic compound often means addition of oxygen (or removal of hydrogen); "reduction" often means addition of hydrogen (or removal of oxygen).

ORGANIC COMPOUNDS CONTAINING FUNCTIONAL GROUPS

32.9 ALKYL AND ARYL HALIDES

Fluorine, chlorine, bromine, and iodine can all be found in organic molecules. The halides of hydrocarbons may be **alkyl halides** (RX), where R is an alkyl group, or **aryl halides** (ArX), where Ar is an aromatic group. These compounds may also contain unsaturated R groups. Some common halogen derivatives of hydrocarbons are listed in Table 32.13 to illustrate their nomenclature (compare with the saturated hydrocarbons, Table 32.1) and the effect of molecular structure on boiling point. Like many simple compounds containing functional groups, these compounds are most often referred to by their common names.

Halogen atoms are more electronegative than the carbon atom, making the C—X bond polar and the carbon atom subject to attack by a nucleophilic reactant, as discussed in the preceding section. The reactions of alkyl halides with hydroxide bases (Equation 32.5), the cyanide ion, and ammonia are all of this type.

$$CH_3Br + NaCN \xrightarrow{H_2O} CH_3CN + NaBr$$

$$CH_3Br + 2NH_3 \xrightarrow{H_2O} CH_3NH_2 + NH_4Br$$

Halogen atoms bonded to aromatic rings are inert to such reactions.

Alkyl halides can also be induced to lose HX (where X is Cl, Br, or I) and form alkenes, for example

$$CH_3\underset{\displaystyle CH_3}{\underset{|}{C}}HBr + NaOH \xrightarrow[\Delta]{\text{alcohol}} CH_3CH{=}CH_2 + NaBr + H_2O$$

Aside 32.1 CHLORINATED ORGANIC COMPOUNDS IN THE ENVIRONMENT

Chlorinated hydrocarbons and pesticides in the environment present a problem. There is no doubt of this. Large amounts of such compounds have been released into the environment as a consequence of their widespread use in numerous valuable applications. Beginning in the 1940s with DDT (see below), chlorinated pesticides have been spread and sprayed around the world. In addition, many industrially important compounds, such as vinyl chloride ($CH_2{=}CHCl$, a polymer raw material; Sections 33.1, 33.2) and trichloroethylene ($ClCH{=}CCl_2$, used in degreasing metal parts), have escaped into the environment by accident or along with waste materials. Also, the suspicion has arisen that chlorinated compounds can be produced by the reaction between chlorine used in water purification and organic pollutants present in the water.

As the number of chlorine atoms in a hydrocarbon is increased, the compounds become less flammable, more dense and viscous, less soluble in water and more soluble in organic solvents, and generally more stable to chemical attack. Compounds in which chlorine in bonded to a vinyl group or an aromatic ring are especially stable due to participation of lone-pair electrons from the chlorine in electron delocalization over the entire molecule.

Because highly chlorinated compounds are not readily oxidized by air, hydrolyzed by water, degraded by sunlight, or broken down by bacterial action, those that get into the environment persist, accumulate, and can be distributed widely. DDT has been found in the polar ice and most of us have detectable amounts of DDT and PCBs (see below) in our bodies.

In recent years, one substance after another has been brought to public attention as a toxic or harmful pollutant. The following chlorine-containing insecticides, once widely used, have come under scrutiny and have been banned or restricted:

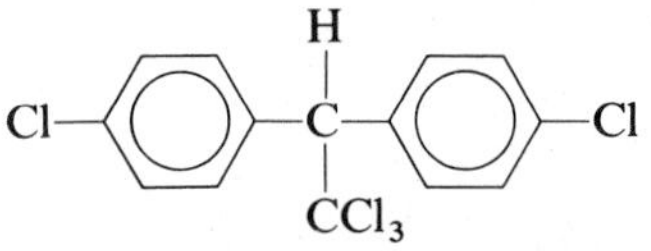

Cl—C₆H₄—C(H)(CCl₃)—C₆H₄—Cl

DDT
(now banned in U.S.)

Cl, Cl, Cl, CH₂, CCl₂, Cl, Cl

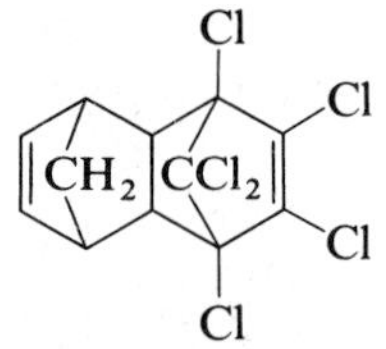

aldrin
(carcinogenic; now banned in U.S.)

Cl, Cl, O, CH₂, CCl₂, Cl, Cl

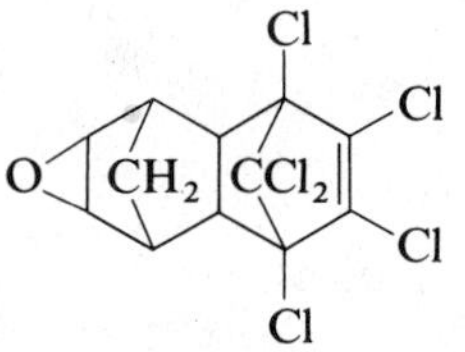

dieldrin
(carcinogenic; now banned in U.S.)

Cl, Cl, Cl, Cl, CCl₂, Cl, Cl

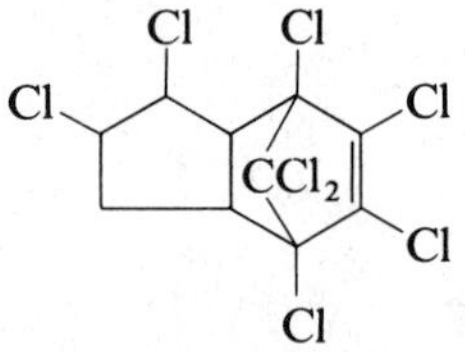

chlordane
(use in U.S. restricted)

The PCBs, in particular, have been in the news in the 1980s. The chlorination of biphenyl

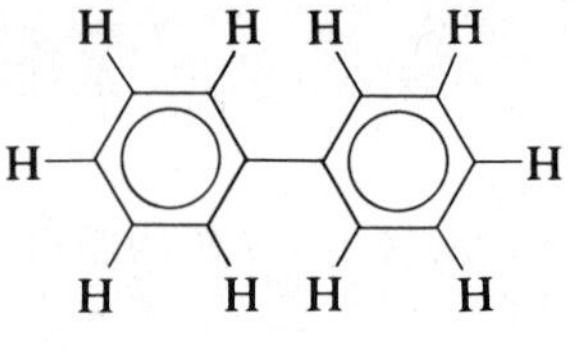

H H H H, H—C₆H₄—C₆H₄—H, H H H H

produces a mixture of *polychlorinated biphenyls* (the PCBs) containing primarily the compounds in which three to five hydrogen atoms are replaced by chlorine atoms. The PCBs are *extremely* stable—they survive municipal incineration and can be destroyed only by burning at 1000–1600 °C or by oxidation by ozone and ultraviolet radiation. The half-life of PCBs in the environment is 15 years.

The uses of the PCBs have been based on their great thermal stability and their ability to conduct heat but not electricity. They are present in most older electrical transformers and are widely used as lubricants, high-pressure hydraulic fluids, and in adhesives, inks, and paper coatings. Direct exposure to PCBs is known to cause severe skin disease. PCBs were first detected in the environment by a scientist searching for DDT and other pesticides in wildlife.

Many questions have also been raised about the toxicity in humans of the compound known in the popular press as "dioxin." (All compounds derived from the six-membered ring $C_4H_4O_2$ are dioxins.) "Dioxin" is a contaminant produced in the manufacture of a number of chlorine-containing compounds, including the defoliant 2,4,5-T, which has been widely used.

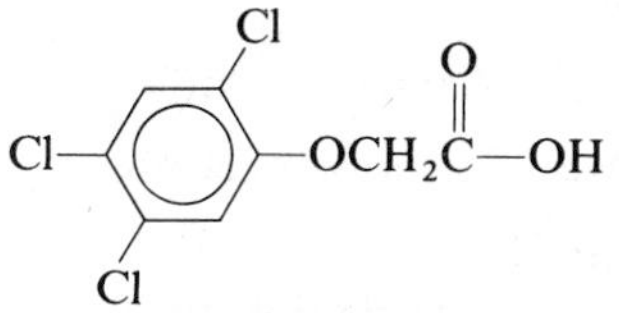

2,4,5-T

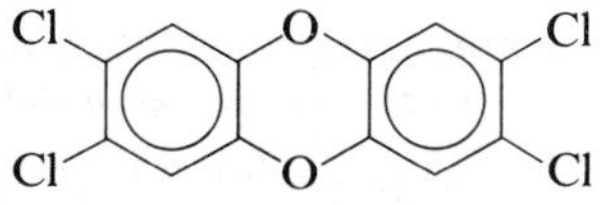

"dioxin"
*(2,3,7,8-tetrachlorodibenzo-*p*-dioxin)*

"Dioxin" has been the object of public concern because it has been found in animal tests to be extremely toxic in very low doses.

While exposure to large, obviously harmful concentrations of toxic compounds must always be avoided, the long-term effects of continued exposure to low levels of pollutants are of equally serious concern. Attention was first focused on such problems when DDT was identified as the cause of the decrease in the population of eagles, hawks, pelicans, and other large birds. DDT and other persistent compounds are concentrated by passage along a food chain (Figure 32.6).

The use of the chlorinated compounds that can persist in the environment has now been either banned or severely restricted in the United States. Chlorinated pesticides have been replaced by other types of compounds in many applications, for example, by organophosphorus compounds, which, although they are not without problems, are degraded more rapidly in the environment. The regulation of pesticides is, however, somewhat controversial. The health benefits of synthetic pesticides are great. The introduction of a DDT spraying program aimed at mosquitoes reduced the annual incidence of malaria in just one country from three million cases to about seven thousand cases. Also, the benefits to agriculture are not only desirable, but will become more and more necessary as world population grows. In agriculture, one answer lies in an approach called "integrated pest management": Conditions are controlled to favor natural predators and, rather than routine application of pesticides, pest populations are monitored so that spraying need only be done when the potential for crop damage is greatest.

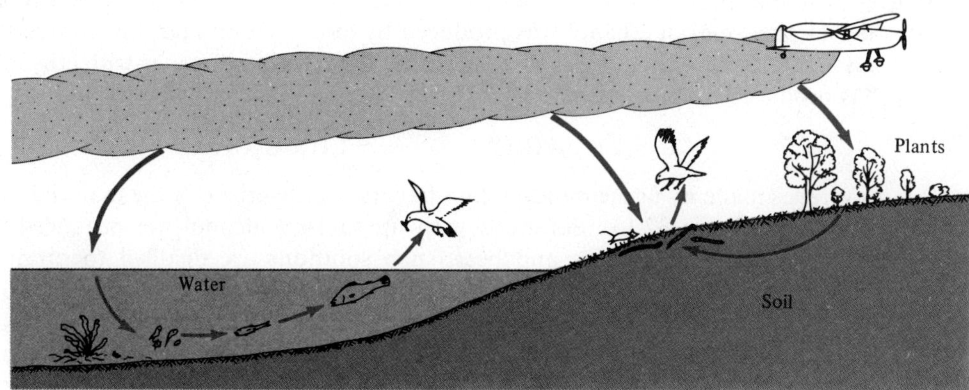

Figure 32.6
Food Chain Concentration of DDT
An example of how DDT concentration is increased by each species along a food chain. For example, in one study the following DDT concentrations were found: lake sediments, 0.0085 ppm; invertebrates, 0.40 ppm; fish feeding on invertebrates, 5 ppm; gulls feeding on fish, 3200 ppm.

32.10 ALCOHOLS AND PHENOLS

In organic compounds, the OH group usually is covalently bonded and is called the **hydroxyl group. Alcohols** have the general formula ROH. If R is aromatic, the compounds are known as **phenols** (ArOH).

Table 32.14
Alcohols, ROH[a]

CH_3OH *methyl alcohol* *(methanol)* b.p. 65 °C
CH_3CH_2OH *ethyl alcohol* *(ethanol)* b.p. 78 °C
$CH_3CH_2CH_2OH$ *propyl alcohol* *(1-propanol)* b.p. 97 °C
$CH_3CH(OH)CH_3$ *isopropyl alcohol* *(2-propanol)* b.p. 82 °C
$HOCH_2CH_2OH$ *ethylene glycol* *(1,2-ethanediol)* b.p. 197 °C
$CH_2(OH)CH(OH)CH_2(OH)$ *glycerol or glycerin* *(1,2,3-propanetriol)* b.p. 290 °C (dec)
C_6H_5–CH_2CH_2OH *(2-phenylethanol)* b.p. 219 °C

[a] Common name (systematic name)

In the common system of nomenclature for alcohols, the R group is named and the word *alcohol* follows, as in methyl alcohol (CH_3OH) and cyclohexyl alcohol ($C_6H_{11}OH$). In the IUPAC system of nomenclature, the name is derived from the longest hydrocarbon chain that includes the OH group by dropping the final *e* and adding *ol*, as in methanol, ethanol, and cyclohexanol. When necessary, a number is used to show the position of the OH group. Numbering starts at the end of the chain nearest to the OH group. Table 32.14 provides a few examples.

Like water, the alcohols exhibit hydrogen bonding, which causes their boiling points to be relatively high. The lower molecular mass alcohols are miscible with water because of hydrogen bonding between the alcohol and water molecules.

Also like water, alcohols can react as either acids or bases. As acids, they are even weaker than water, but in the presence of relatively strong bases, alcohols can give up protons. The reaction of ethyl alcohol with sodium can be used to dispose of waste sodium by converting it to sodium ethoxide, and then allowing the alkoxide to react with water.

$$\underset{\textit{ethanol}}{2CH_3CH_2OH} + 2Na \longrightarrow \underset{\textit{sodium ethoxide}}{2CH_3CH_2ONa} + H_2(g)$$

$$CH_3CH_2ONa + H_2O \longrightarrow CH_3CH_2OH + NaOH$$

In the presence of concentrated acids, alcohols act as bases in overall reactions such as

$$CH_3OH + HBr \longrightarrow CH_3Br + H_2O$$

Methanol, once known as wood alcohol because of its formation during wood distillation, is now made synthetically form carbon monoxide and hydrogen (Section 32.5b). Methanol is toxic and causes blindness or death when taken internally.

Ethylene glycol (see Table 32.14) is the most widely used permanent antifreeze and coolant for automobile engines, where its high boiling point prevents its loss through evaporation. Isopropyl alcohol is used in rubbing alcohol and is the raw material for acetone production (Section 32.13). Glycerol has a strong attraction for water and is added to, for example, lotions, creams, and foods where moisture retention is desirable. It is also a raw material for the manufacture of paints, varnishes, explosives, and many other products.

Ethanol is an important industrial chemical, being used mainly as a solvent and an intermediate in the synthesis of other chemicals. Fermentation of any natural material that contains starch or sugar, such as grains or fruits, produces ethanol. Enzymes first convert starch to sugar; other enzymes then convert the sugars to ethanol and carbon dioxide. At one time, all ethanol was produced by fermentation and this production method is still in use. Ethanol also is obtained by the reaction of water with ethylene, which is produced from petroleum.

$$CH_2{=}CH_2 + H_2O \xrightarrow[\text{pressure}]{\text{heat, catalysts,}} CH_3CH_2OH$$

Table 32.15
Alcoholic Beverages

Beverage	Starting material
Whiskey	Grains (rye, corn, wheat, oats, barley)
Rum	Molasses
Brandy	Grapes and other fruits
Gin	Grains (distilled and flavored with juniper berry)
Vodka	Potatoes and corn
Wines	Grapes
Sake	Rice
Mead	Honey
Beer	Barley, hops, and other grains
Tequila	Cactus

Beverages made by fermentation retain flavors characteristic of the material that was fermented (Table 32.15). Solutions with up to 14% alcohol are produced by fermentation. Except for wines and beer, such solutions are distilled to produce beverages with increased alcohol content. The usual 80-proof beverage is a 40% solution (by volume) of ethyl alcohol. One hundred percent, or *absolute ethyl alcohol,* is 200 proof. Ethyl alcohol for beverages is subject to a federal tax. In order to render industrial ethyl alcohol unfit for drinking, various toxic or objectionable materials (e.g., methanol) that are difficult to remove from ethyl alcohol are added. This *denatured alcohol* is tax free, hence much cheaper for industrial users.

Table 32.16 gives some examples of phenols, the simplest of which is phenol (C_6H_5OH). Like alcohols, the lower molecular mass phenols are water soluble. However, the aromatic ring in phenol permits resonance stabilization of the anion formed by loss of hydrogen from a phenol, for example, $C_6H_5O^-$. As a result, the ionization of phenols is favored and phenols are stronger acids than alcohols.

Phenol itself is poisonous and also causes blisters on the skin. The biggest use for phenol is in the manufacture of phenol–formaldehyde polymers. Another important use of phenols is in bactericidal products. "Carbolic acid" is a name given to phenol and its aqueous solutions. Joseph Lister, applying Pasteur's theories about bacteria and infection in the late 1800s, did surgery under a spray of carbolic acid.

Table 32.16
Phenols, ArOH[a]
Most phenols, unlike the lower molecular mass alcohols, are solids at room temperature.

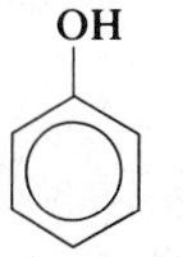

OH on benzene ring *phenol* m.p. 42 °C	OH, OH (ortho) on benzene ring *catechol* m.p. 104 °C
OH, CH_3 (para) on benzene ring *p-cresol* m.p. 36 °C	OH, OH (para) on benzene ring *hydroquinone* m.p. 169 °C
OH on naphthalene ring *α-naphthol* m.p. 94 °C	

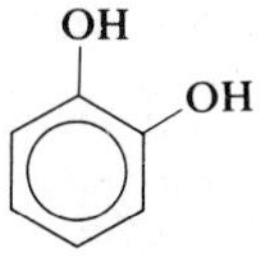

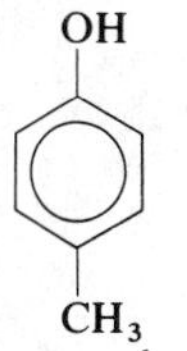

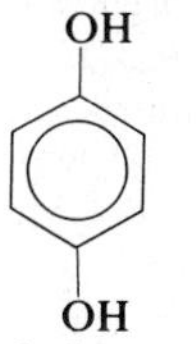

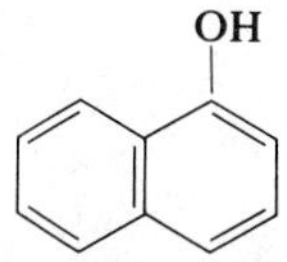

[a] These phenols are known best by their common names as given above. Phenol is also known as "carbolic acid."

32.11 ETHERS

Aliphatic **ethers** have the general formula ROR. If the two hydrocarbon groups are alike, the ether is a *simple* or *symmetrical* ether; if they are different, the ether is a *mixed* or *unsymmetrical* ether.

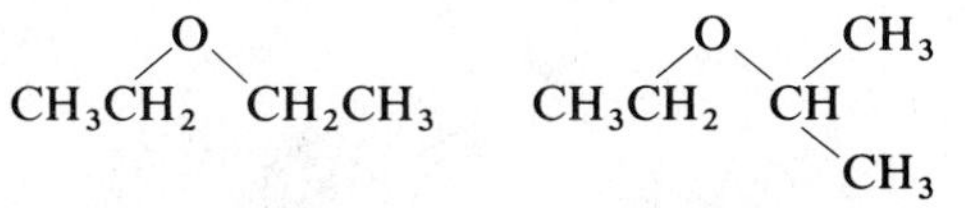

CH_3CH_2–O–CH_2CH_3
diethyl ether
a simple ether

CH_3CH_2–O–$CH(CH_3)_2$
ethyl isopropyl ether
a mixed ether

The common system of nomenclature names the groups attached to the oxygen atom followed by the word *ether,* as just shown. The RO— is an *alkoxy group.* In the IUPAC system of nomenclature, an ether is named as an alkoxy derivative of the longest chain hydrocarbon to which the alkoxy group is attached. The position of attachment is given by a number, starting at the end of the chain nearest the alkoxy group. Thus, ethyl butyl ether ($CH_3CH_2OCH_2CH_2CH_2CH_3$) is named 1-ethyoxy-butane. Table 32.17 gives several examples of ethers. Note that there are also cyclic ethers and aromatic ethers.

Ether molecules cannot form hydrogen bonds with one another. As a consequence, the boiling points of ethers are near those of hydrocarbons having the same molecular mass. The effect of hydrogen bonding is clearly illustrated by a comparison of the boiling points of ethylene glycol ($HOCH_2CH_2OH$, b.p. 197 °C) and 1,2-dimethoxy-ethane ($CH_3OCH_2CH_2OCH_3$, b.p. 85 °C).

Diethyl ether, a volatile flammable compound, is an anesthetic. The ethers are rather unreactive and are often used as solvents in carrying out reactions of organic compounds. However, ethers do have an unfortunate tendency to form peroxides (compounds containing—O—O—bonds) when exposed to the oxygen in air. These peroxides are susceptible to explosive decomposition. For this reason special precautions must always be used when working with ethers, especially ethers from containers that have been opened and allowed to stand in the air. Because aluminum reacts with peroxides, ethers often are stored in aluminum cans to minimize the hazard of peroxide formation.

Table 32.17
Ethers, ROR[a]

CH_3OCH_3 *dimethyl ether* *(methoxymethane)* b.p. −24 °C	H_2C–CH_2–O–CH_2–CH_2–O (ring) *dioxane* *(1,4-dioxin)* b.p. 101 °C
$CH_3CH_2OCH_2CH_3$ *diethyl ether* *(ethoxyethane)* b.p. 35 °C	
$CH_3OCH_2CH_2OCH_3$ *ethylene glycol dimethyl ether ("glyme")* *(1,2-dimethoxyethane)* b.p. 85 °C	benzene ring–OCH_3 *anisole* *(methoxybenzene)* b.p. 154 °C
H_2C—CH_2 bridged by O *ethylene oxide* *(oxirane)* b.p. 11 °C	

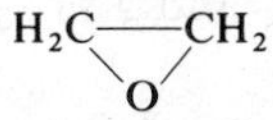

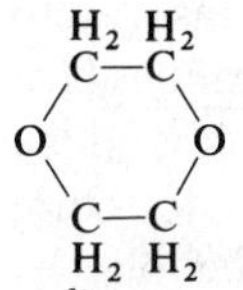

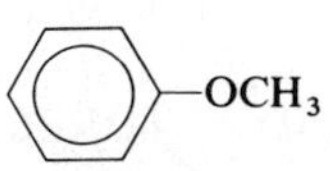

[a] Common name (systematic name)

Table 32.18
Classification of Amines

Primary amine, RNH_2
CH_3NH_2
methylamine
b.p. −6 °C
Secondary amine, R_2NH
$(CH_3)_2NH$
dimethylamine
b.p. 7 °C
Tertiary amine, R_3N
$(CH_3)_3N$
trimethylamine
b.p. 3 °C
Quaternary ammonium salt, $R_4N^+X^-$
$(CH_3)_4N^+I^-$
tetramethylammonium iodide
m.p. 230 °C (dec)

32.12 AMINES

Amines can be thought of as formed by the replacement of the hydrogen atoms of an ammonia molecule by alkyl or aryl groups. **Amines** are classified as primary (RNH_2), secondary (R_2NH), or tertiary (R_3N), according to the number of hydrogen atoms of the ammonia molecule that have been replaced. Addition of a fourth hydrocarbon group yields a quaternary ammonium ion. This classification is illustrated in Table 32.18.

Aromatic amines are those with the nitrogen atom of the amine group bonded to an aromatic ring carbon atom.

NH_2 NH_2

aniline b.p. *183 °C* *(2-naphthylamine)* m.p. *112 °C*

In cyclic amines, the nitrogen atom is part of a ring that may or may not be aromatic.

O N N H

pyridine b.p. *116 °C* *morpholine* b.p. *130 °C*

The amines are the most common organic bases. Simple amines such as the ethylamines

$CH_3CH_2NH_2$	$(CH_3CH_2)_2NH$	$(CH_3CH_2)_3N$
ethylamine	*diethylamine*	*triethylamine*
pK_b *3.33*	pK_b *3.02*	pK_b *3.28*

are somewhat stronger bases than ammonia (NH_3, pK_b 4.7). The aromatic amines are weaker bases because delocalization withdraws electron density from the nitrogen atom. For example, aniline ($C_6H_5NH_2$) has pK_b 9.38.

The salts of amines are odorless, water-soluble solids from which the free amines can be obtained by reaction with bases, for example

$$CH_3NH_2 \xrightarrow{HCl} CH_3NH_3^+Cl^- \xrightarrow{NaOH} CH_3NH_2$$

methylamine *methylammonium chloride*

By contrast, most free amines have fairly strong, unpleasant odors. Like the alcohols, amines of lower molecular mass (up to those with five carbon atoms) are water soluble.

Physiologically active amines (Table 32.19) found in plants are known as *alkaloids.* Nicotine is present in tobacco and is used as an insecticide. Morphine, a sedative and analgesic (pain killer), is a major component of opium. Codeine, another analgesic, is the methyl ether (at the phenolic hydroxyl group) of morphine. Heroin, too, is a derivative of morphine. Epinephrine, also known as adrenalin, is the stimulant released by the human adrenal glands under conditions of fear or stress.

Table 32.19
Some Physiologically Active Amines

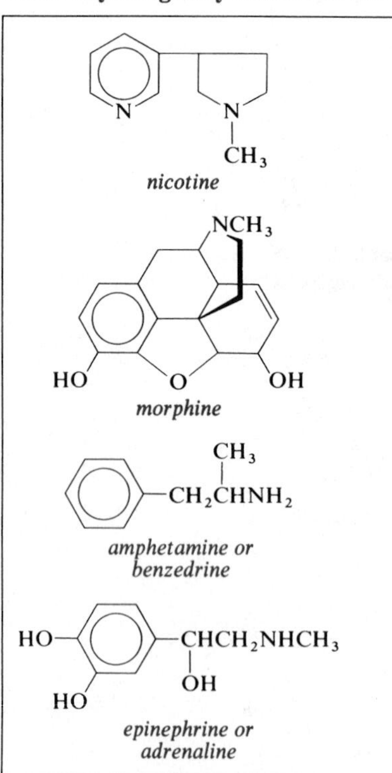

32.13 ALDEHYDES AND KETONES

The **carbonyl group** consists of a carbon atom and an oxygen atom joined by a covalent double bond.

$$\rangle C{=}O$$

a carbonyl group

The carbonyl group is present in a number of different functional groups: aldehydes and ketones, which are discussed in this section; carboxylic acids (Section 32.14); esters (Section 32.15); and, as shown in Table 32.12, also in amides ($RCONH_2$), acyl halides (RCOX, where X is a halogen), and acid anhydrides ($(RCO)_2O$). Esters, amides, acyl halides, and acid anhydrides are considered derivatives of carboxylic acids. Each of these types of compounds undergoes hydrolysis to give carboxylic acids (plus, for esters, an alcohol; for amides, NH_3, or for N-substituted amides, amines; for acyl halides, HX).

An **aldehyde** (RCHO) is a compound in which a hydrogen atom and an alkyl group are bonded to a carbonyl group (Table 32.20). The aldehyde functional group can be found only at the *end* of a molecule. The general formula for an aldehyde may be written in several ways:

$$R{-}C(=O){-}H \quad \text{or} \quad R\overset{O}{\overset{\|}{C}}{-}H \quad \text{or} \quad RCHO$$

(RCOH would look like an alcohol.)

In the simplest possible aldehyde, R is a hydrogen atom also (formaldehyde; see Table 32.20). The common names of the aldehydes are derived from the names of the carboxylic acids (next section) containing the same carbon skeleton by replacing the final *ic* by *aldehyde*. Thus, formaldehyde (HCHO) is the aldehyde corresponding to formic acid (HCOOH). The IUPAC system of nomenclature for aldehydes requires only the replacement of the final *e* of the name of the longest chain hydrocarbon including the aldehyde group by *al*. Substitutions or branches along this main chain are designated by numbers (the carbonyl carbon atom is numbered 1) and named by prefixes, for example, 3-chloropropanal ($ClCH_2CH_2CHO$).

Formaldehyde is used mainly in the manufacture of phenol–formaldehyde and urea–formaldehyde polymers. It also finds some use as a disinfectant and a preservative, often as *formalin*, a 37% aqueous solution of formaldehyde.

A **ketone** (R_2CO or RCOR′) contains two groups attached to the carbon atom of the carbonyl group (Table 32.21). The general formula for a ketone, in which R and R′ may be the same or different, takes the following forms:

$$\begin{matrix} R \\ \quad \diagdown \\ \quad C{=}O \\ \quad \diagup \\ R' \end{matrix} \quad \text{or} \quad R\overset{O}{\overset{\|}{C}}R' \quad \text{or} \quad RCOR'$$

In the common system of nomenclature, the two hydrocarbon groups are named, followed by the word *ketone*, as in methyl ethyl ketone ($CH_3COCH_2CH_3$). Dimethyl

Table 32.20
Aldehydes, RCHO[a]

HCHO *formaldehyde* *(methanal)* b.p. −21 °C	$CH_2{=}CHCHO$ *acrolein* *(propenal)* b.p. 52 °C
CH_3CHO *acetaldehyde* *(ethanal)* b.p. 21 °C	C_6H_5—CHO *benzaldehyde* b.p. 179 °C
$CH_3(CH_2)_4CHO$ *caproaldehyde* *(hexanal)*[b] b.p. 128 °C	

[a] Common name (systematic name)
[b] Note that the aldehyde group must be at the end of a chain, so it is unnecessary to write 1-hexanal.

Table 32.21
Ketones, RCOR′[a]

CH_3COCH_3 *dimethyl ketone or acetone* *(2-propanone)* b.p. 56 °C	$CH_3COCH_2CH_2CH_3$ *methyl n-propyl ketone* *(2-pentanone)* b.p. 102 °C
$CH_3COCOCH_3$ *biacetyl*[b] *(2,3-butanedione)* b.p. 88 °C	C_6H_5—$COCH_3$ *methyl phenyl ketone or acetophenone* b.p. 200 °C
$CH_3CH_2COCH_2CH_3$ *diethyl ketone* *(3-pentanone)* b.p. 102 °C	

[a] Common name (systematic name)
[b] This is one of the few colored (yellowish green) organic liquids. Practically all *pure* organic liquids are colorless; many pure organic solids are colored, however.

ketone is almost always referred to by its trivial name, *acetone*. In the IUPAC system of nomenclature, a ketone is designated by the suffix *one*, which replaces the final *e* of the name of the longest chain hydrocarbon containing the carbonyl group. A number is normally used to show the position of the carbonyl group, starting at the end of the chain nearest the carbonyl group. Thus, methyl ethyl ketone is 2-butanone in the IUPAC system. Substitutions and branches on the main chain are treated as for hydrocarbons.

Ketones are used extensively as solvents, especially in lacquers and other coatings. They frequently serve as starting materials for organic synthesis.

Alcohols are often the starting materials for the preparation of aldehydes and, especially, ketones. Sodium dichromate oxidizes a primary alcohol to an aldehyde and a secondary alcohol to a ketone.

$$\underset{\substack{\textit{ethyl alcohol}\\ \textit{a primary alcohol}}}{CH_3CH_2OH} \xrightarrow[H_2SO_4]{Na_2Cr_2O_7} \underset{\textit{acetaldehyde}}{CH_3\overset{\overset{O}{\|}}{C}{-}H} + H_2O$$

$$\underset{\substack{\textit{isopropyl alcohol}\\ \textit{a secondary alcohol}}}{CH_3\overset{\overset{OH}{|}}{C}HCH_3} \xrightarrow[H_2SO_4]{Na_2Cr_2O_7} \underset{\textit{acetone}}{CH_3\overset{\overset{O}{\|}}{C}CH_3} + H_2O$$

A ketone is resistant to further oxidation under these conditions, but an aldehyde is readily oxidized to the corresponding carboxylic acid, for example

$$\underset{\textit{acetaldehyde}}{CH_3\overset{\overset{O}{\|}}{C}{-}H} \xrightarrow{\substack{\text{oxidizing agent,}\\ \text{e.g., } KMnO_4}} \underset{\textit{acetic acid}}{CH_3\overset{\overset{O}{\|}}{C}{-}OH}$$

Even atmospheric oxygen will convert many aldehydes to the corresponding acids. A very mild oxidizing agent, $[Ag(NH_3)_2]OH$, called Tollens' reagent, provides a diagnostic test for aldehydes.

$$RCHO + 2[Ag(NH_3)_2]OH \longrightarrow 2Ag(s) + RCOONH_4 + H_2O + 3NH_3$$

When this reaction is carried out in a clean test tube, the elemental silver that forms deposits as a silver mirror on the walls of the test tube. A process for the silvering of mirrors utilizes this reaction. [Solutions from this test must never be allowed to stand because of the possibility of the formation of silver azide (AgN_3), a *highly explosive* substance.]

32.14 CARBOXYLIC ACIDS

A **carboxylic acid** (RCOOH) has a functional group that is a combination of a carbonyl group and a hydroxyl group.

$$R{-}C\begin{matrix}{\nearrow}O\\ {\searrow}OH\end{matrix} \quad \text{or} \quad R\overset{\overset{O}{\|}}{C}OH \quad \text{or} \quad RCOOH$$

The more familiar carboxylic acids have common names associated with the sources from which they were first isolated. For example, formic acid (Latin *formica* for ant; see Table 32.22) was first isolated from ant bodies and is partly responsible for the painful sting of the red ant. Acetic acid (Latin *acetum* for vinegar) is present in vinegar. And butyric acid (Latin *butyrum* for butter) is responsible for the odor of rancid butter.

In the IUPAC system of nomenclature, the ending *oic* replaces the final *e* of the hydrocarbon name that is formed from the longest continuous chain of carbon atoms including the carboxyl group. The word *acid* follows this derived name. When this chain of carbon atoms is numbered, the carboxylic carbon atom is numbered 1. Table 32.22 shows the structures and names of some of the more common carboxylic acids.

Table 32.22
Carboxylic Acids, RCOOH[a]

HCOOH *formic acid* *(methanoic acid)* b.p. 100 °C	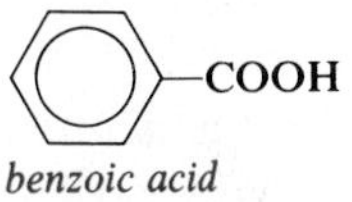 C_6H_5—COOH *benzoic acid* m.p. 121 °C
CH_3COOH *acetic acid* *(ethanoic acid)* b.p. 118 °C	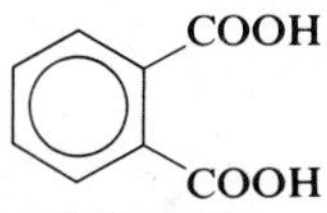 $C_6H_4(COOH)_2$ (ortho) *phthalic acid* m.p. 230 °C
CH_3CH_2COOH *propionic acid* *(propanoic acid)* b.p. 140 °C	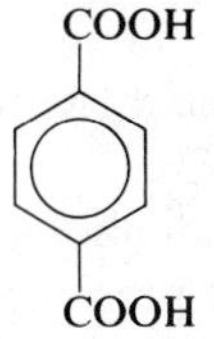 $C_6H_4(COOH)_2$ (para) *terephthalic acid* m.p. > 300 °C (sublimes)
$CH_2{=}CHCOOH$ *acrylic acid* *(propenoic acid)* b.p. 140 °C	
n-$CH_3CH_2CH_2COOH$ *butyric acid* *(butanoic acid)* b.p. 163 °C	

[a] Common name (systematic name)

Simple, unsubstituted monocarboxylic acids have pK_a values in the range 4–5; they are weaker as acids than sulfonic acids (RSO_3H) but stronger than phenols.

The ionization of a carboxylic acid, as shown, produces the *carboxylate ion* ($RCOO^-$).

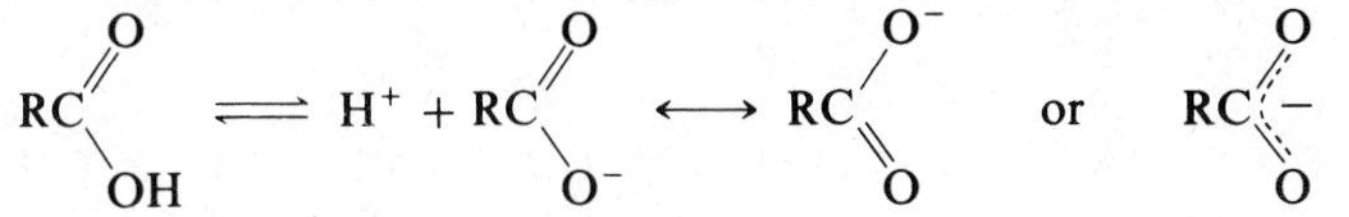

$$RC(=O)OH \rightleftharpoons H^+ + RC(=O)O^- \longleftrightarrow RC(O^-)=O \quad \text{or} \quad RC(\cdots O)(\cdots O)^-$$

a carboxylate ion, resonance structures

This ion is stabilized by the delocalization of the negative charge over the two oxygen atoms.

All of the carboxylic acids form salts when they are treated with bases (e.g., hydroxides, carbonates, or hydrogen carbonates).

$$CH_3COOH + NaOH \longrightarrow CH_3COO^-Na^+ + H_2O$$

acetic acid (ethanoic acid) — *sodium acetate (sodium ethanoate)*

The name of the salt is obtained from the name of the cation, followed by the name of the carboxylic acid with the final *ic* replaced by *ate*.

Acetic acid is produced in large quantities (up to three billion pounds per year) for use in making plastics and solvents. Phthalic acid is also widely used in the production of varnishes, paints, plastics, and synthetic fibers.

Many carboxylic acids occur in nature, either free or combined in some form. The combination of hydroxyl groups and carboxylic acid groups in the same molecule to give hydroxy acids is very common in nature. Lactic acid is found in sour milk (racemic form) and in the muscles of man and animals (dextrorotatory isomer). Tartaric acid (dextrorotatory isomer) occurs in many fruits. Citric acid occurs in both plants and animals; lemon juice contains 5–8% citric acid.

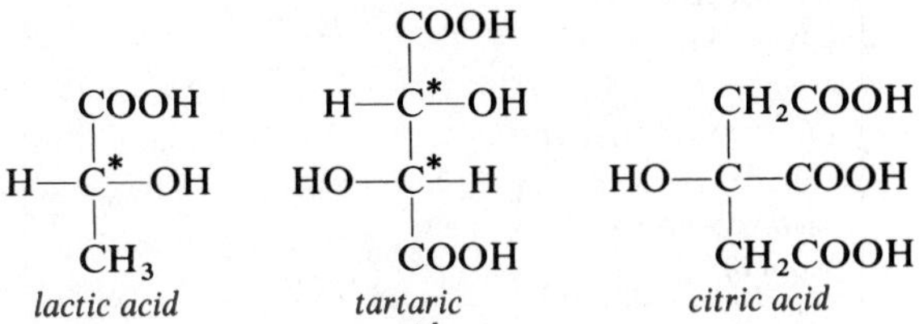

lactic acid	tartaric acid	citric acid
	COOH	
COOH	H—C*—OH	CH_2COOH
H—C*—OH	HO—C*—H	HO—C—COOH
CH_3	COOH	CH_2COOH

In the formulas above, * indicates an asymmetric carbon atom.

32.15 ESTERS

A compound in which the acidic hydrogen atom of a carboxylic acid group is replaced by an alkyl (or aryl) group is an **ester** (RCOOR′). The general formula for an ester, in which R and R′ may be the same or different, can be written

$$R{-}C(=O){-}O{-}R' \quad \text{or} \quad R\overset{\overset{\displaystyle O}{\|}}{C}OR' \quad \text{or} \quad RCOOR'$$

The common and IUPAC names for esters are derived in the same manner. In the two-word name, the first word is the name of the R′ group in RCOOR′. The second word is the name of the carboxylic acid, with the final *ic* replaced by *ate*, as in naming carboxylate anions (Table 32.23).

A carboxylic acid and an alcohol form an ester in an *esterification reaction* when they are heated together in the presence of an acid catalyst.

$$\underset{\textit{acetic acid}}{CH_3\overset{\overset{\displaystyle O}{\|}}{C}{-}OH} + \underset{\textit{ethyl alcohol}}{HOCH_2CH_3} \xrightleftharpoons{H_2SO_4} \underset{\textit{ethyl acetate}}{CH_3\overset{\overset{\displaystyle O}{\|}}{C}{-}OCH_2CH_3} + H_2O$$

The formation of an ester in such an equilibrium reaction is enhanced by the application of Le Chatelier's principle. An excess of either the carboxylic acid or the alcohol, or removal of water as it is formed, or both techniques together, are used to increase the yield of the ester.

The hydrolysis of an ester is the reversal of the reaction leading to its formation.

$$\underset{\textit{ethyl acetate}}{CH_3\overset{\overset{\displaystyle O}{\|}}{C}{-}OCH_2CH_3} + H_2O \xrightleftharpoons{H^+} \underset{\textit{acetic acid}}{CH_3COOH} + \underset{\textit{ethyl alcohol}}{CH_3CH_2OH}$$

Hot aqueous sodium hydroxide will also hydrolyze esters; the reaction is referred to as *saponification*, a term that is associated with the process for making soap (Section 15.17).

$$\underset{\textit{ethyl acetate}}{CH_3COOCH_2CH_3} + NaOH \xrightarrow{H_2O} \underset{\textit{sodium acetate}}{CH_3COONa} + \underset{\textit{ethyl alcohol}}{CH_3CH_2OH}$$

Table 32.23
Esters of Carboxylic Acids, RCO_2R'[a]

$HCOOCH_3$ *methyl formate* *(methyl methanoate)* b.p. 32 °C	$C_6H_4(COOC_2H_5)_2$ (benzene ring with two adjacent $COOC_2H_5$ groups) *diethyl phthalate* b.p. 298 °C
CH_3COOCH_3 *methyl acetate* *(methyl ethanoate)* b.p. 57 °C	*an ester of acetic acid* → $\overset{\overset{\displaystyle O}{\|}}{O}CCH_3$ on benzene ring with adjacent COOH *aspirin* *(acetylsalicylic acid)* m.p. 135 °C
$CH_3COOCH_2CH_3$ *ethyl acetate* *(ethyl ethanoate)* b.p. 77 °C	
$CH_2{=}CHCOOCH_3$ *methyl acrylate* *(methyl propenoate)* b.p. 85 °C	
$C_6H_5{-}COOCH_3$ (benzene ring–$COOCH_3$) *methyl benzoate* b.p. 198 °C	

[a] Common name (systematic name)

Low molecular mass esters, for example, ethyl acetate and butyl acetate, are used extensively as solvents. Most esters of carboxylic acids have quite pleasant odors, even when derived from rather foul-smelling acids. Many fruits and flowers owe their flavor and fragrance to the esters present. For example, methyl salicylate is responsible for the flavor of wintergreen, and the ethyl ester of butyric acid (see Table 32.22) has the flavor of pineapple.

Aspirin, which is both a fever reducer and pain killer, is an ester of acetic acid formed by reaction with the phenolic —OH group of salicylic acid (o-HOC_6H_4COOH; Table 32.23). When the aspirin reaches the alkaline environment of the intestines, the salicylic acid is liberated and absorbed by the body tissues.

EXAMPLE 32.7 Organic Compounds

Classify the following compounds according to the types of compounds listed in Table 32.24. (page 858). Identify the functional groups and the R portions in each molecule.

(a) 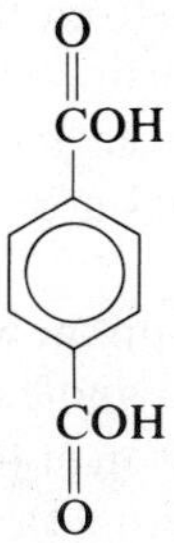 benzene ring with COH (C=O) at top and COH (C=O) at bottom

(b) $CH_3C(CH_3)(CH_3)OH$ — drawn as $CH_3\overset{CH_3}{\underset{CH_3}{C}}OH$

(c) $CH_2{=}CH\overset{O}{\overset{\|}{C}}OCH_2CH_2CH_2CH_3$

(d) 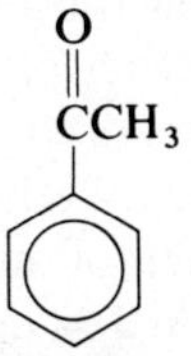 benzene ring with $\overset{O}{\overset{\|}{C}}CH_3$

The R portions of each molecule are shown in color.

(a)  benzene ring with COH (C=O) at top and COH (C=O) at bottom
a dicarboxylic acid

(b) $CH_3{-}\overset{CH_3}{\underset{CH_3}{C}}OH$
an alcohol

(c) $CH_2{=}CH\overset{O}{\overset{\|}{C}}OCH_2CH_2CH_2CH_3$
an ester

(d) 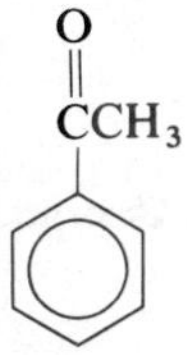 benzene ring with $\overset{O}{\overset{\|}{C}}CH_3$
a ketone

EXERCISE Classify the following compounds and identify the functional groups and the R portions in each molecule.

(a) CH_2Cl_2 (b) 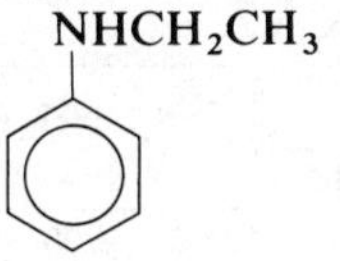 benzene ring with $NHCH_2CH_3$

Answers (a) An alkyl chloride, R is CH_2; (b) an amine (secondary), R is C_6H_5 and CH_2CH_3.

Aside 32.2 TOOLS OF CHEMISTRY: INFRARED AND ULTRAVIOLET SPECTROSCOPY

Absorption spectra arise when energy from incoming radiation is absorbed, causing atoms, molecules, ions, or electrons to move into higher energy states. Among the valuable tools of the organic chemist are infrared (IR) and ultraviolet (UV) absorption spectroscopy.

The infrared region of the electromagnetic spectrum extends from 2.5 μm to 15 μm (1 $\mu m = 1 \times 10^{-4}$ cm = 1000 nm). Infrared spectra reflect increases in the energy of motion of atoms in a molecule relative to each other. The atoms in a molecule are never motionless. The bonds may, for example, bend or stretch (Figure 32.7) or undergo other motions known as wagging, rocking, deformation, and so on. When the motion involves a change in dipole moment, and when the frequency of incoming radiation matches the frequency of the motion, absorption occurs and a "peak" appears in the infrared spectrum. Figure 32.8 shows the spectra of ethylene glycol and ethylenediamine with the origins of many of the peaks identified (str = stretching; def = deformation; wag = wagging). Customarily in infrared spectra the percent of the radiation *transmitted* (% transmittance) is plotted versus wavelength (in μm, for example) or wave number (in cm^{-1}), so the absorption peaks actually appears as valleys.

The more complicated a molecule is, the more possibilities there are for relative motions of the atoms. Therefore, infrared spectra can be quite complex, and the larger the molecule the larger the number of peaks that appear in the spectrum. In spectra with many peaks, identification of the origin of all the peaks is difficult.

Infrared spectra are most useful in two ways. One depends upon the fact that the same functional groups cause absorption at the same wavelength (or very nearly so) in the spectra of all the molecules in which they occur. From the study of the spectra of thousands of compounds, "correlation charts" have been assembled. These charts show the locations of peaks that result from the presence of specific groups of atoms. A small portion of a correlation chart is shown in Figure 32.9. This segment includes the location of the peaks of functional groups that contain a carbon–nitrogen single bond.

In the determination of the structure of an unknown compound, the infrared spectrum is used to indicate which functional groups are present by comparison of the spectrum with such charts. This is often not as simple and

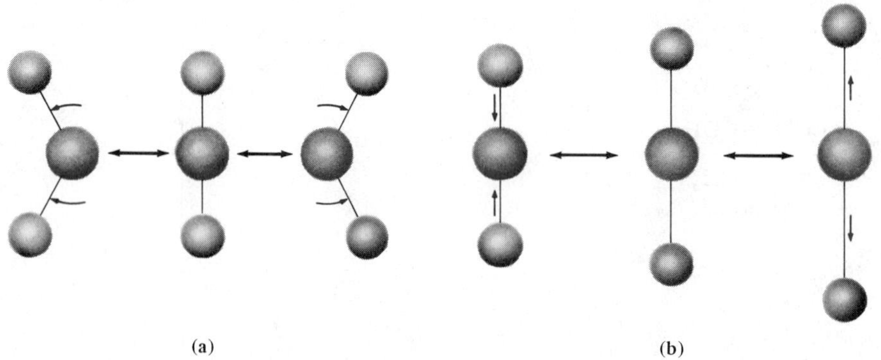

Figure 32.7
Examples of Molecular Motion Detectable by Infrared Absorption
(a) Bending. (b) Stretching.

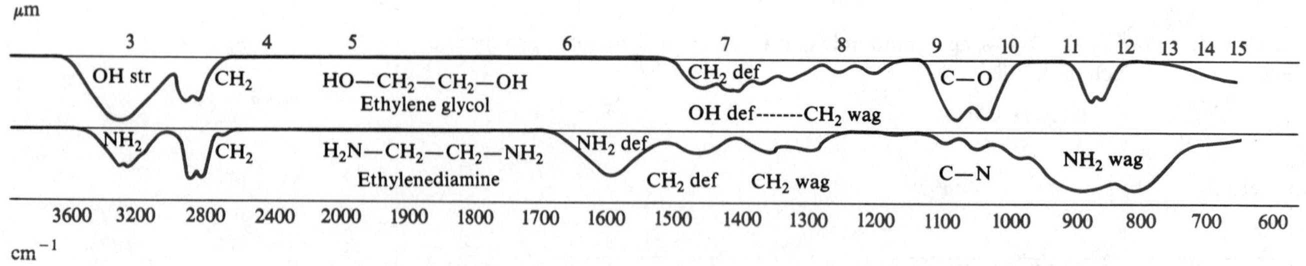

Figure 32.8
Infrared Spectra of Ethylene Glycol (1,2-Ethanediol) and Ethylenediamine (1,2-Diaminoethane)

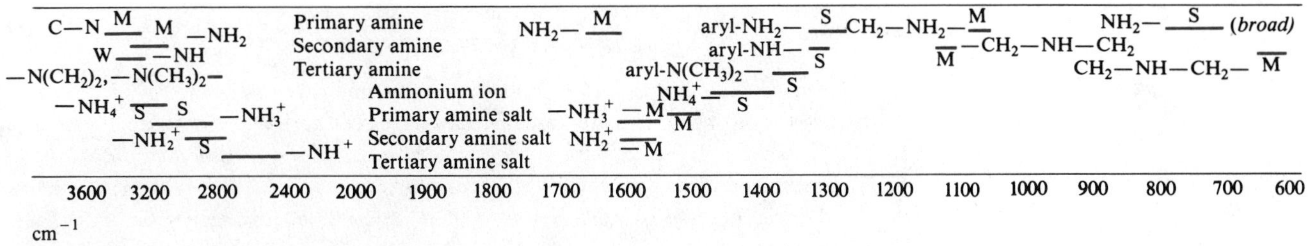

C—N M, M —NH_2, W —NH; Primary amine: NH_2— M; aryl-NH_2— S, CH_2—NH_2— M; NH_2— S (broad)
Secondary amine: aryl-NH— S; M —CH_2—NH—CH_2; CH_2—NH—CH_2— M
—$N(CH_2)_2$, —$N(CH_3)_2$—; Tertiary amine: aryl-$N(CH_3)_2$— S
Ammonium ion: —NH_4^+ S, S; NH_4^+— S
Primary amine salt: —NH_3^+; —NH_3^+ — M, M
Secondary amine salt: —NH_2^+ S; NH_2^+ — M
Tertiary amine salt: —NH^+

3600 3200 2800 2400 2000 1900 1800 1700 1600 1500 1400 1300 1200 1100 1000 900 800 700 600
cm^{-1}

Figure 32.9
Portion of an Infrared Correlation Chart
This segment shows the location of peaks for groups that contain carbon–nitrogen single bonds. The letters S, M, and W indicate strong, medium, and weak bands, respectively.

straightforward as it sounds, but with experience, much useful information about structure can be gleaned from infrared spectra. For example, an experienced observer would quickly know from the two spectra in Figure 32.8 that one compound contained OH groups and the other, NH_2 groups.

Infrared spectra, like mass spectra, are also used as fingerprints. It has been said that the infrared spectrum of a compound may be its most characteristic physical property. By comparison of the spectrum of a compound, say a reaction product that has not been identified, with the spectra of known compounds that might be products, identification can often be made. For example, if a compound was known to be either ethylenediamine or ethylene glycol, comparing its spectrum with the two known spectra in Figure 32.8 would quickly show which compound had been produced.

The near-ultraviolet region of the electromagnetic spectrum extends from 200 to 400 nm and is useful in studying organic compounds. Ultraviolet radiation is of higher energy than infrared radiation. At such energy, absorption occurs by transitions of outer π or nonbonding electrons into higher energy levels, usually into unoccupied antibonding, σ^* or π^* orbitals.

A simple saturated hydrocarbon does not absorb in the near-ultraviolet region. The presence of carbon–carbon multiple bonds or of functional groups containing multiple bonds—that is, the presence of π electrons and π orbitals—is required for UV absorption. Ultraviolet spectra are much simpler than infrared spectra. They show a single peak or a few peaks for each isolated functional group or carbon–carbon multiple bond present in the molecule. Ultraviolet spectra are described in terms of λ_{max}, the wavelength of maximum absorption. Usually measured in solution, ultraviolet spectra are presented as plots of absorption versus wavelength or, as in the pyridine spectrum shown in Figure 32.10, as the logarithm of the molar absorptivity, log ε, versus wavelength. Molar absorptivity, ε, is a function of the amount of radiation transmitted and the length of the path of the light through the sample. Ultraviolet absorption peaks are shifted to longer wavelengths when a molecule contains conjugated systems, for example, —CH=CH—CH=CH— or —CH=CH—CH=O.

Ultraviolet spectra can be used for structure analysis and compound identification in the same way that infrared spectra are used. However, their application in these ways is limited because there are fewer peaks in ultraviolet spectra and the peaks are broader. Spectra in the near-ultraviolet region are more often used in quantitative analysis, as are spectra in the visible region. The spectra are measured in solution and the intensity of the absorption varies with the concentration of the absorbing species. The relationship between the ultraviolet absorption and concentration is first determined experimentally. Then an unknown concentration can be found by comparison with the data for known concentrations. Titrations can be followed by spectral changes, and frequently, the concentration of a substance in a complex mixture can be measured by its characteristic absorption.

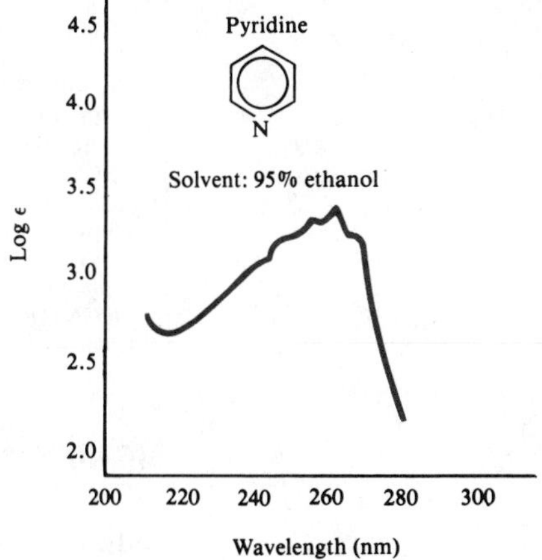

Figure 32.10
Ultraviolet Spectrum of Pyridine
Source: R. M. Silverstein et al., *Spectrometric Identification of Organic Compounds* (New York: Wiley, 1974).

SUMMARY OF SECTIONS 32.9–32.15

The structures, characteristic properties, and some uses of the classes of compounds containing the most common functional groups have been reviewed. These functional groups may be divided into those containing only single bonds (halides, alcohols and phenols, ethers, and amines) and those containing the carbonyl group, —C=O (aldehydes, ketones, carboxylic acids and the esters derived from them, acyl halides, acid anhydrides, and amides). The properties of compounds containing these functional groups are summarized in Table 32.24.

Table 32.24
Characteristic Properties of Some Common Types of Organic Compounds
As a rule of thumb, monofunctional alcohols, ethers, amines, aldehydes, ketones, acids, and esters are at least slightly water soluble up to those that contain five carbon atoms. If more than one functional group is present, more carbon atoms can be carried into solution.

Compound type	Properties
Alkyl halides RX	Water-insoluble, polar C—X bond subject to nucleophilic attack
Alcohols ROH	Hydrogen-bonded; very weak acid or base behavior
Phenols ArOH	Weak acids; lower molecular mass phenols are water soluble; aromatic ring reactive to substitution and oxidation
Ethers ROR	No hydrogen bonding; low boiling points; good solvents for organic compounds; relatively unreactive
Amines RNH_2, R_2NH, R_3N	Weak bases; foul odors; soluble in aqueous acids (salt formation)
Aldehydes and ketones RC(=O)H, RC(=O)R′	Polar C=O bond subject to nucleophilic attack; ketones good solvents for organic compounds
Carboxylic acids RC(=O)OH	Weak acids; often biting odors; form salts with bases
Esters RC(=O)OR′	Formed by reaction of carboxylic acid + alcohol; pleasant odors; good solvents for organic compounds
Acyl halides RC(=O)X	Reactive; not found in nature; lower molecular mass acyl halides hydrolyze in moist air to give RCOOH
Acid anhydrides RC(=O)OC(=O)R′	Reactive; not found in nature
Amides RC(=O)NHR′, $RC(=O)NH_2$,	Hydrogen bonded; generally solids; hydrolyze under more vigorous conditions than acyl halides

SIGNIFICANT TERMS (with section references)

addition reactions (32.4)
alcohols (32.10)
aldehyde (32.13)
aliphatic hydrocarbons (32.3)
alkanes (32.1)
alkenes, olefins (32.2)
alkyl groups (32.1)
alkyl halides (32.9)
alkynes, acetylenes (32.2)
amines (32.12)
aromatic hydrocarbons (32.3)
aryl halides (32.9)

asymmetric atom (32.7)
carbonyl group (32.13)
carboxylic acid (32.14)
cycloalkanes (32.1)
electrophile (32.8)
ester (32.15)
ethers (32.11)
functional group (32.6)
hydroxyl group (32.10)
ketone (32.13)
nucleophile (32.8)
phenols (32.10)
racemic mixture (32.7)
saturated hydrocarbons (32.1)
substituents (32.1)
substitution reactions (32.4)
unsaturated hydrocarbons (32.2)

QUESTIONS AND PROBLEMS

Hydrocarbons

32.1 What are alkanes? Write the general molecular formula for an alkane. Could a substance having the molecular formula C_6H_{14} be an alkane? Explain your answer.

32.2 What are cycloalkanes? Write the general molecular formula for a cycloalkane. Could a substance having the molecular formula C_5H_{12} be a cycloalkane? Explain your answer.

32.3 What is an alkene? Write the general molecular formula for an alkene. Is C_6H_{12} a possible molecular formula for an alkene? Explain your answer.

32.4 What is an alkyne? Write the general molecular formula for an alkyne. Is C_6H_{10} a possible molecular formula for an alkyne? Explain your answer.

32.5 What is an alkyl group? What are the general molecular formulas for the alkyl groups derived from an alkane and a cycloalkane? How is the name of an alkyl group derived from the name of an alkane?

32.6 What is a diene? What are conjugated double covalent bonds?

32.7 Write the structural formulas for the three isomeric saturated hydrocarbons having the molecular formula C_5H_{12}. Name each by the IUPAC system.

32.8 Repeat Question 32.7 for the five isomers of C_6H_{14}.

32.9 Write the structural formula for 2,2-dimethylpropane. Designate each carbon atom in the compound as primary, secondary, tertiary, or quaternary.

32.10 Give the IUPAC name for

(cyclopropane ring)—CH_3

Designate each carbon atom in the compound as primary, secondary, tertiary, or quaternary.

32.11 Draw the condensed formulas of the following compounds: (a) 1-butyne, (b) 2-methylpropene, (c) 2-ethyl-3-methyl-1-butene, (d) 3-methyl-1-butyne.

32.12 Draw the condensed formulas of the following compounds: (a) 3-hexyne, (b) 1, 3-pentadiene, (c) cyclobutene, (d) 3, 4-diethylhexane.

32.13 Write the IUPAC names for the following compounds:

(a) $CH_3C(CH_3)_2CH_2CH_3$ (b) $CH_2{=}CBr_2$

(c) $CH_3CH(CH_3)CH(CH_3)CH_3$ (d)

32.14 Repeat Question 32.13 for:

(a) (cyclopentene ring) (b) $CH_3CH(CH_3)CH_2CH_3$

(c) $CH_3C{\equiv}CCH_3$ (d) $CH_3C(CH_3){=}CHCH_3$

(e) $CH_3CH_2CH(CH_2CH_3)CH_3$ (f) $CH_3CH(CH_2CH_3)CH_2CH_3$

32.15 What is an aromatic hydrocarbon?

32.16 What is a phenyl group? How many isomeric monophenylnaphthalenes are possible?

32.17 There are three possible isomeric trimethylbenzenes. Write their structural formulas and name each of them.

32.18 How many isomeric dibromobenzenes are possible? What names are used to designate these isomers?

32.19 Write the structural formulas for the following compounds: (a) *p*-dinitrobenzene, (b) *n*-propylbenzene, (c) 1, 3, 5-tribromobenzene, (d) 1, 3-diphenylbutane.

32.20 Write the IUPAC names for the following compounds:

(a) benzene ring with Cl, Cl (b) benzene ring with CH_3, Cl

(c) naphthalene ring with Cl, Cl (d) benzene ring with Cl, Cl, Cl, Cl, Cl

32.21 The heat of formation at 25 °C is −75 kJ/mol for $CH_4(g)$, 717 kJ/mol for $C(g)$, and 218 kJ/mol for $H(g)$. (a) Calculate the average C—H bond energy. The heat of formation at 25 °C is −85 kJ/mol for $C_2H_6(g)$, 52 kJ/mol for $C_2H_4(g)$, and 227 kJ/mol for $C_2H_2(g)$. Calculate the (b) C—C, (c) C=C, (d) C≡C bond energies.

32.22 The heat of combustion at 20 °C is −3 487 kJ/mol for n-C_5H_{12}, −4 141 kJ/mol for n-C_6H_{14}, −4 811 kJ/mol for n-C_7H_{16}, and −5 450 kJ/mol for n-C_8H_{18}. Using −286 kJ/mol for the heat of formation of $H_2O(l)$ and −395 kJ/mol for $CO_2(g)$, calculate the heat of formation of each of the alkanes. Prepare a plot of $\Delta H^\circ_{f,293}$ against the number of carbon atoms in these molecules and predict the heat of formation of n-C_9H_{20}.

$$C_nH_{2n+2}(l) + \frac{(3n+1)}{2} O_2(g) \longrightarrow nCO_2(g) + (n+1)H_2O(l)$$

32.23 Write the structural formula for the organic compound or the carbon-containing compound formed in each of the following reactions:

(a) $CH_4 + O_2(\text{excess}) \xrightarrow{\Delta}$

(b) $CH_4 + Cl_2 \xrightarrow{\text{light}}$

(c) $CH_3CH{=}CH_2 \xrightarrow{H_2SO_4}$

(d) 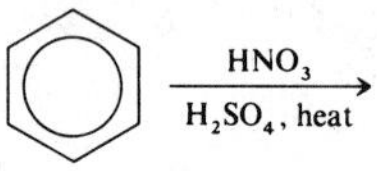 benzene $\xrightarrow[H_2SO_4,\ \text{heat}]{HNO_3}$

(e) 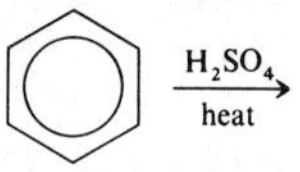 benzene $\xrightarrow[\text{heat}]{H_2SO_4}$

32.24 Repeat Question 32.23 for:

(a) $HC{\equiv}CH + O_2(\text{excess}) \xrightarrow{\Delta}$

(b) $CH_3Cl + Cl_2 \xrightarrow{\text{light}}$

(c) $CH_3CH{=}CH_2 \xrightarrow{HCl}$

(d) 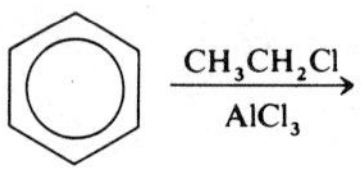 benzene $\xrightarrow[AlCl_3]{CH_3CH_2Cl}$

(e) 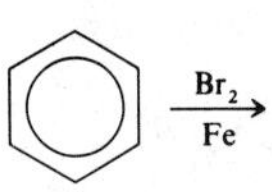 benzene $\xrightarrow[Fe]{Br_2}$

Hydrocarbons and Energy

32.25 Discuss the following processes in the petroleum industry: (a) isomerization, (b) cracking, (c) alkylation, (d) reforming.

32.26 Name some of the major components of natural gas. What does the term "synfuel" mean? Give some examples of synfuels.

32.27 Write the chemical equation for the water gas reaction. At 1000 K, $K_p = P_{CO}P_{H_2}/P_{H_2O} = 3.2$. What are the partial pressures of CO and H_2 at this temperature if $P_{H_2O} = 15.6$ bar?

32.28 Write the chemical equation for the water gas shift reaction. At 1000 °C the free energy of formation is −175.8 kJ/mol for $H_2O(g)$, −226.5 kJ/mol for $CO(g)$, and −396.2 kJ/mol for $CO_2(g)$. Calculate ΔG° for this reaction and determine whether or not it is a favorable reaction.

Properties of Organic Compounds

32.29 Name three types of isomerism possible in organic molecules.

32.30 To what do the terms levorotatory and dextrorotatory refer? How are Fischer projections used to show enantiomers?

32.31 Which of the following compounds can exist as *cis* and *trans* isomers: (a) 1-butene, (b) 2-bromo-1-butene, (c) 2-bromo-2-butene, (d) 2, 3-dimethyl-2-butene, (e) 2, 3-dichloro-2-butene?

32.32 Which of the following compounds would exhibit optical isomerism?

(a) $CH_3CH(Br)CH_3$

(b) $CH_3CH(OH)CH_2CH_3$

(c) HO—C_6H_4—CH(Cl)—C_6H_5

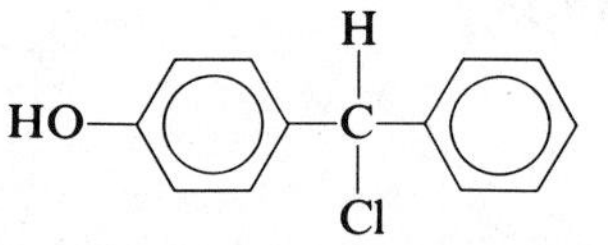

(d) $CH_3CH{=}CHCH_3$

Draw the Fischer projections for the enantiomeric pairs.

32.33 What does the term "nucleophile" mean? Which of the following could be nucleophiles: (a) CN^-, (b) I^-, (c) H^+, (d) NH_3, (e) Na^+?

32.34 What does the term "electrophile" mean? Which of the following could be electrophiles: (a) Cl^-, (b) H^+, (c) OH^-, (d) $CH_3{}^+$?

32.35 In the following reaction

$$(CH_3)_3C^+ + :\ddot{O}H_2 \longrightarrow (CH_3)_3C{-}\ddot{O}{-}H + H^+$$

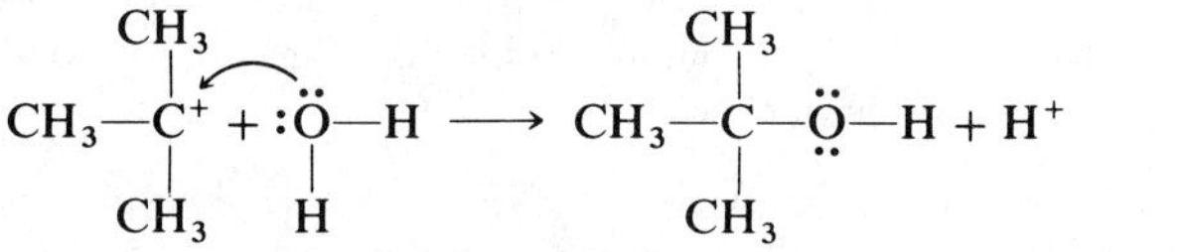

which species is considered to be a nucleophile and which is considered to be an electrophile?

32.36 Repeat Question 32.35 for

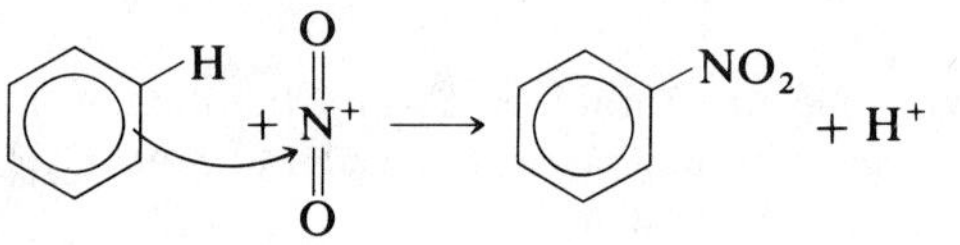

$$C_6H_6 + NO_2^+ \longrightarrow C_6H_5NO_2 + H^+$$

32.37 How are the terms "oxidation" and "reduction" often used in organic chemistry? Classify the following changes as either oxidation or reduction: (a) CH_4 to CH_3OH, (b) $CH_2{=}CH_2$ to $CH_3{-}CH_3$, (c) $CH_3CH_2CH_2OH$ to $CH_3CH_2CH_3$,

(d) 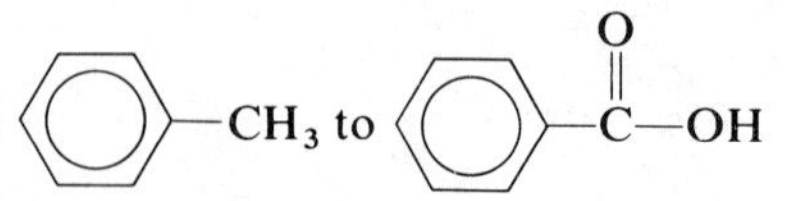 $C_6H_5{-}CH_3$ to $C_6H_5{-}C(=O){-}OH$

32.38 Classify the following changes as either oxidation or reduction: (a) CH_3OH to CO_2 and H_2O, (b) CH_3CH_2OH to CH_3CHO, (c) CH_3COOH to CH_3CHO, (d) $CH_3CH{=}CH_2$ to $CH_3CH_2CH_3$.

32.39 A laboratory procedure called for oxidizing 2-propanol to acetone using an acidic solution of $K_2Cr_2O_7$. However, an insufficient amount of $K_2Cr_2O_7$ was on hand, so the laboratory instructor decided to use an acidic solution of $KMnO_4$ instead. What mass of $KMnO_4$ was required to carry out the same amount of oxidation as 1.00 g of $K_2Cr_2O_7$?

32.40 A piece of glass 12 inches by 18 inches is to be silvered using the following reaction:

$$CH_3\overset{\overset{\displaystyle O}{\|}}{C}H(aq) + 2[Ag(NH_3)_2]OH(aq) \longrightarrow$$

$$2Ag(s) + CH_3\overset{\overset{\displaystyle O}{\|}}{C}O^-NH_4^+(aq) + 3NH_3(aq) + H_2O(l)$$

If the thickness of the silver is to be 0.0003 inch, what mass of acetaldehyde is needed for the reaction? The density of Ag is 10.5 g/cm^3.

32.41 You have three test tubes: the first contains either hexane or 2-hexene, the second contains either benzene or styrene ($C_6H_5CH{=}CH_2$), and the third contains either cyclohexene or 2-bromopropane. Describe a simple chemical test that would enable you to determine visually which compound was present in each test tube.

32.42 Write the structural formula of the organic compound formed in each of the following reactions:

(a) 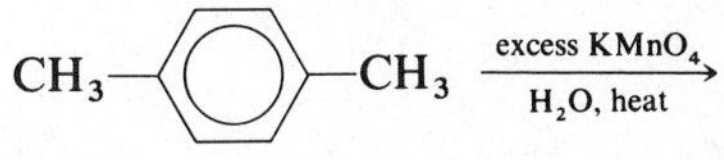 $CH_3{-}C_6H_4{-}CH_3$ (para-xylene) $\xrightarrow[H_2O,\ heat]{excess\ KMnO_4}$

(b) $CH_3CH{=}CHCH_3 \xrightarrow{HBr}$

(c) $C_6H_5CH{=}CH_2 \xrightarrow{Br_2}$

(d) 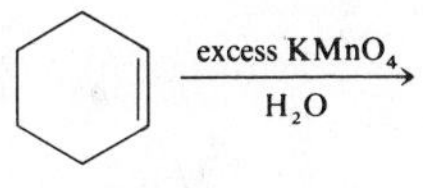 (cyclohexene) $\xrightarrow[H_2O]{excess\ KMnO_4}$

Functional Groups

32.43 The structures of many of the classes of compounds discussed in this chapter can be "derived" from that of the water molecule, $H_a{-}O{-}H_b$, by replacing one or both of the hydrogen atoms by various organic groups. Name the class of compound formed when (a) H_a is replaced by an alkyl group, (b) H_b is replaced by an aromatic group, (c) H_a and H_b are replaced by alkyl groups, (d) H_a is replaced by the $R{-}\overset{\overset{\displaystyle O}{\|}}{C}{-}$group, (e) H_a is replaced by the $R{-}\overset{\overset{\displaystyle O}{\|}}{C}{-}O{-}$ group and H_b is replaced by an alkyl group, (f) H_a and H_b are replaced by $R{-}\overset{\overset{\displaystyle O}{\|}}{C}{-}$ groups.

32.44 Consider the following oxidation processes:

(a) 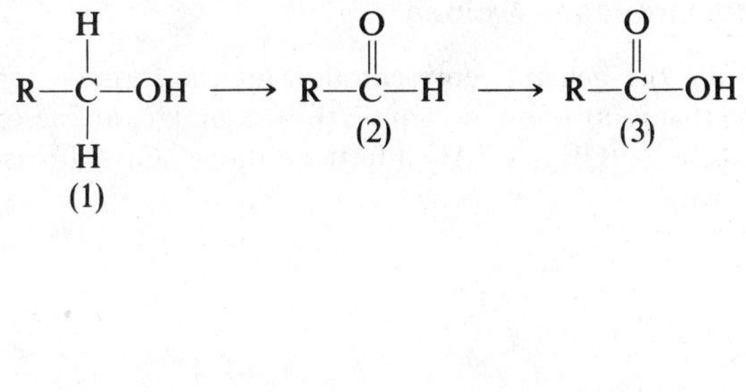 $R{-}CH_2{-}OH \longrightarrow R{-}\overset{\overset{\displaystyle O}{\|}}{C}{-}H \longrightarrow R{-}\overset{\overset{\displaystyle O}{\|}}{C}{-}OH$
(1) (2) (3)

(b) 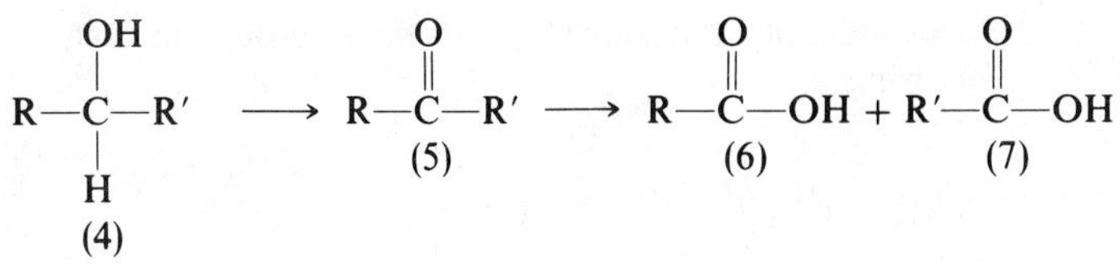 $R{-}CH(OH){-}R' \longrightarrow R{-}\overset{\overset{\displaystyle O}{\|}}{C}{-}R' \longrightarrow R{-}\overset{\overset{\displaystyle O}{\|}}{C}{-}OH + R'{-}\overset{\overset{\displaystyle O}{\|}}{C}{-}OH$
(4) (5) (6) (7)

And the following processes:

(c) 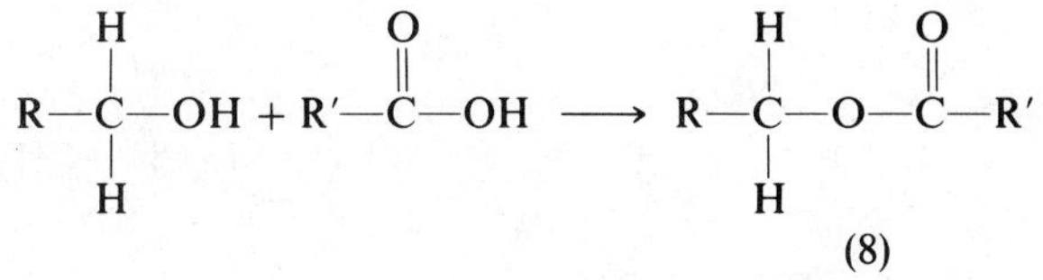 $R{-}CH_2{-}OH + R'{-}\overset{\overset{\displaystyle O}{\|}}{C}{-}OH \longrightarrow R{-}CH_2{-}O{-}\overset{\overset{\displaystyle O}{\|}}{C}{-}R'$
(8)

(d) 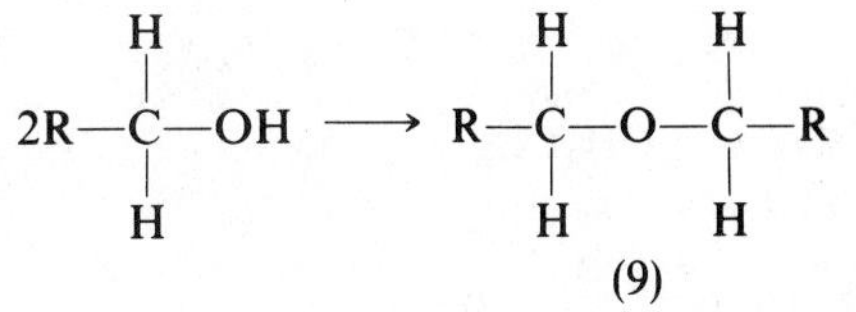 $2R{-}CH_2{-}OH \longrightarrow R{-}CH_2{-}O{-}CH_2{-}R$
(9)

(e) 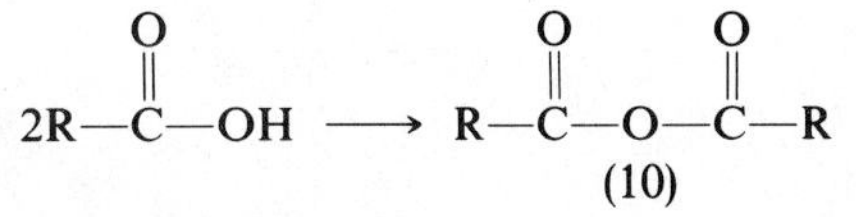 $2R{-}\overset{\overset{\displaystyle O}{\|}}{C}{-}OH \longrightarrow R{-}\overset{\overset{\displaystyle O}{\|}}{C}{-}O{-}\overset{\overset{\displaystyle O}{\|}}{C}{-}R$
(10)

Name the functional group in each of the ten numbered compounds.

32.45 Identify the class of organic compound (ester, ether, ketone, etc.) to which each of the following belongs:

(a) 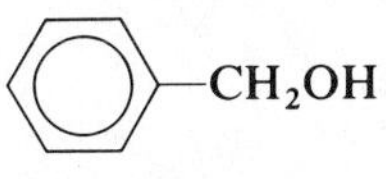 $C_6H_5{-}CH_2OH$

(b) (cyclic: $H_2C{-}C({=}O)$, $H_2C{-}C({=}O)$, joined through O)

(c) $C_6H_5{-}O{-}C_6H_5$

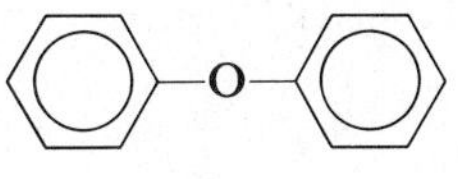

(d) $CH_3\overset{\overset{\displaystyle O}{\|}}{C}OC(CH_3)_3$

(e) 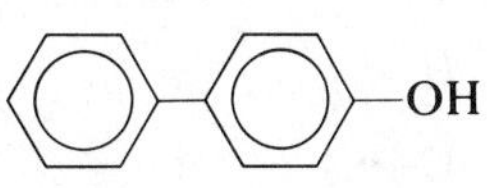 $C_6H_5{-}C_6H_4{-}OH$

(f) 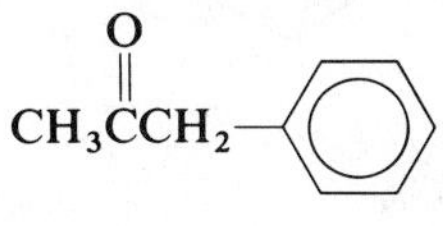 $CH_3\overset{\overset{\displaystyle O}{\|}}{C}CH_2{-}C_6H_5$

(g) 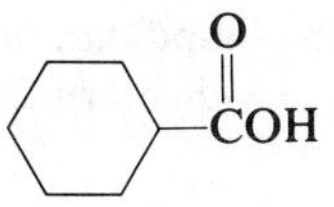 $C_6H_{11}{-}\overset{\overset{\displaystyle O}{\|}}{C}OH$

(h) $C_6H_5{-}CH_2{-}\overset{\overset{\displaystyle O}{\|}}{C}H$

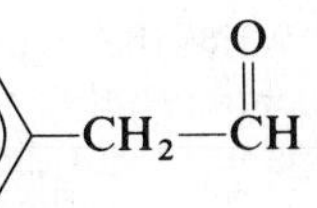

(i) 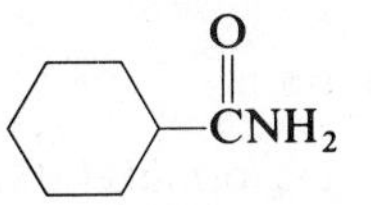 $C_6H_{11}{-}\overset{\overset{\displaystyle O}{\|}}{C}NH_2$

(j) 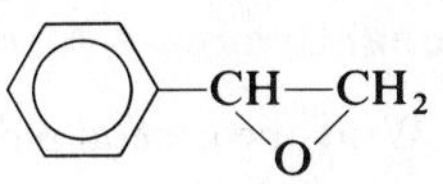 $C_6H_5{-}CH{-}CH_2$ (bridged by O)

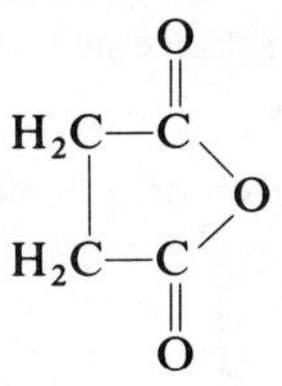

(k) 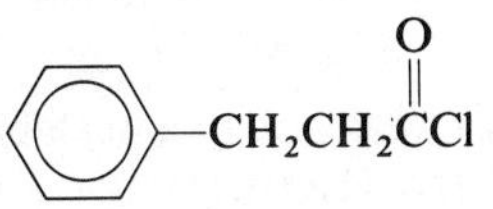 $C_6H_5{-}CH_2CH_2\overset{\overset{\displaystyle O}{\|}}{C}Cl$

32.46 Identify and name the functional groups in each of the following:

(a) 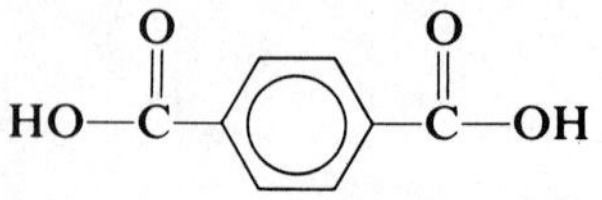 $HO-\overset{O}{\overset{\|}{C}}-C_6H_4-\overset{O}{\overset{\|}{C}}-OH$

(b) $CH_2{=}CH-\overset{O}{\overset{\|}{C}}-O-CH_2CH_2CH_2CH_3$

(c) 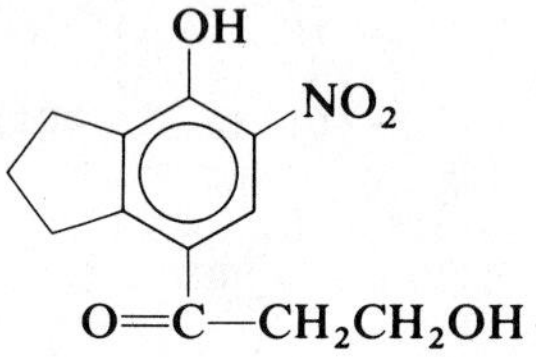 OH, NO_2, $O{=}C-CH_2CH_2OH$

(d) 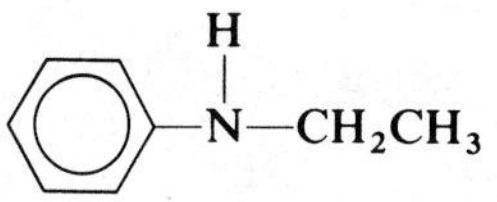 $C_6H_5-\overset{H}{\overset{|}{N}}-CH_2CH_3$

(e) $CH_3CH_2-\overset{O}{\overset{\|}{C}}-O-\overset{O}{\overset{\|}{C}}-CH_2CH_3$

(f) 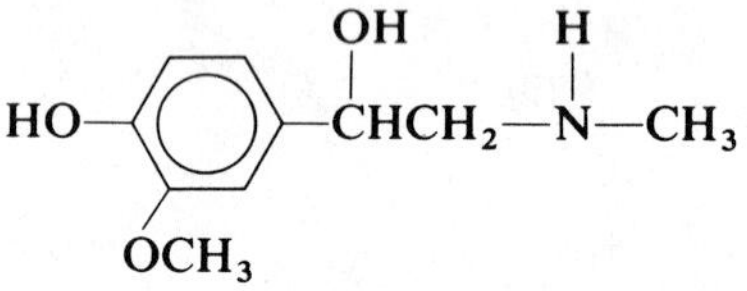 $HO-C_6H_3(OCH_3)-\overset{OH}{\overset{|}{C}}HCH_2-\overset{H}{\overset{|}{N}}-CH_3$

Functional Groups—Alkyl and Aryl Halides

32.47 Write the general representation for the formula of an alkyl halide. How does this differ from the representation for the formula of an aryl halide?

32.48* Alkyl halides readily undergo nucleophilic attack with the subsequent replacement of the halogen atom, but aryl halides do not. Explain.

32.49 Name each of the following halides:

(a) 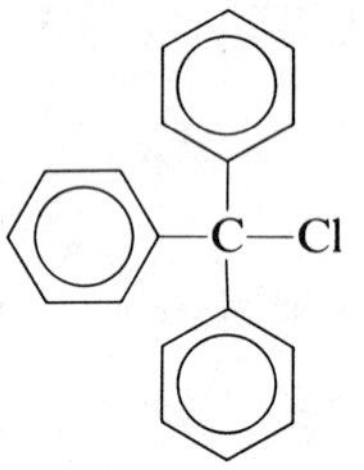 $(C_6H_5)_3C-Cl$ (b) $CH_3-\overset{CH_3}{\overset{|}{C}}H-CH_2Cl$

(c) $CHCl_3$ (d) $Cl-\overset{Cl}{\overset{|}{C}}{=}\overset{H}{\overset{|}{C}}-Cl$

32.50 Write the structural formula for each compound: (a) 2-chloropentane, (b) 4-bromo-1-butene, (c) 1,2-dichloro-2-fluoropropane, (d) 1,4-dichlorobenezne.

Functional Groups—Alcohols and Phenols

32.51 Write the general representation for the formula for (a) a primary alcohol, (b) a secondary alcohol, (c) a tertiary alcohol, (d) a phenol.

32.52 Ethanol, like water, can act as a weak acid or a weak base. Write chemical equations showing CH_3CH_2OH acting (a) as a base with HCl and (b) as an acid with Na(*s*). (c) Name the compounds formed.

32.53 Write the structural formula for each of the following compounds: (a) 1-butanol, (b) cyclohexanol, (c) 1,4-pentanediol, (d) 3-hexyn-1-ol.

32.54 Name the following compounds:

(a) $CH_3\underset{CH_3}{\underset{|}{C}}H-CH_2OH$ (b) $HC{\equiv}C-CH_2OH$

(c) $CH_3-\underset{OH}{\underset{|}{C}}H-\underset{OH}{\underset{|}{C}}H_2$ (d) $CH_3-\overset{CH_3}{\overset{|}{\underset{CH_3}{\underset{|}{C}}}}-OH$

32.55 Which of the following compounds are phenols?

(a) 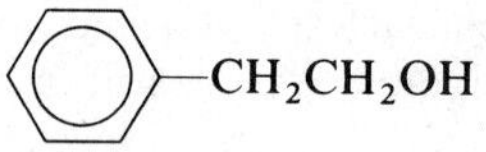 $C_6H_5-CH_2CH_2OH$ (b) $C_6H_{11}-OH$

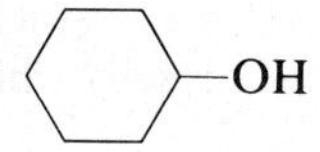

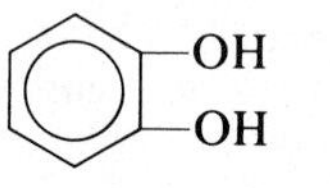

(c) $C_6H_4(OH)_2$ (d) 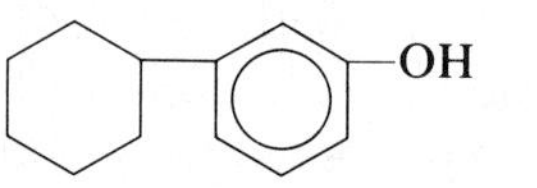 $C_6H_{11}-C_6H_4-OH$

(e) 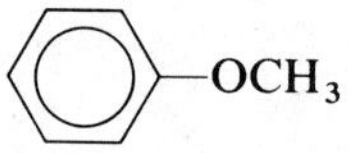 $C_6H_5-OCH_3$

Name each of these compounds.

32.56 Write the structural formula for each of the following: (a) *p*-bromophenol, (b) 4-nitro-1-naphthol, (c) *m*-nitrophenol.

Functional Groups—Ethers

32.57 What determines whether an ether is "symmetrical" or "unsymmetrical"?

32.58 Briefly describe the bonding around the oxygen atom in dimethyl ether. What intermolecular forces are found in this ether?

32.59 Write the structural formula for each of the following: (a) methoxymethane, (b) 1-ethoxypropane, (c) 1,3-dimethoxybutane, (d) ethoxybenzene, (e) methoxycyclobutane.

32.60 Name each of the following ethers:

(a) $CH_3-O-CH_2CH_3$ (b) $CH_3-O-\underset{CH_3}{\underset{|}{C}}H-CH_3$

(c) 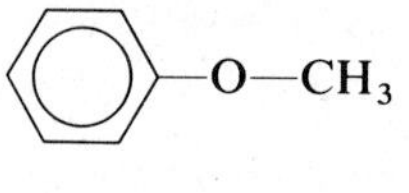 $C_6H_5-O-CH_3$ (d) 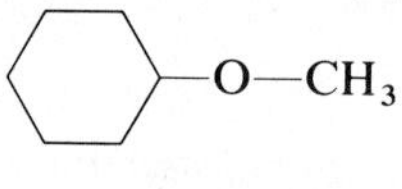 $C_6H_{11}-O-CH_3$

Functional Groups—Amines

32.61 Write the general representation for the formula for a compound that is (a) a primary amine, (b) a secondary amine, (c) a tertiary amine. Is $(CH_3)_3CNH_2$ a tertiary amine? Give a reason for your answer.

32.62 Name the following amines:

(a) 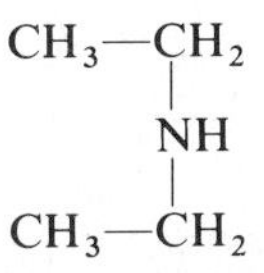 $CH_3—CH_2—NH—CH_2—CH_3$ (b) 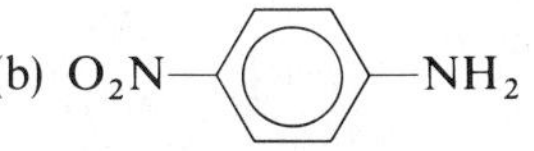 $O_2N—C_6H_4—NH_2$ (c) $C_5H_9—NH_2$ (cyclopentyl)

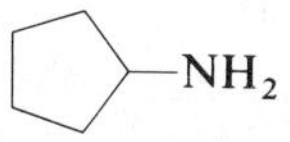

(d) 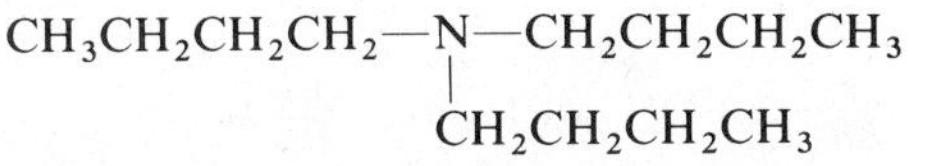 $CH_3CH_2CH_2CH_2—N(CH_2CH_2CH_2CH_3)—CH_2CH_2CH_2CH_3$

32.63 What are the equilibrium concentrations of the various species present in a 0.100 M solution of aniline? $K_b = 4.2 \times 10^{-10}$ for

$$C_6H_5NH_2(aq) + H_2O(l) \rightleftharpoons C_6H_5NH_3^+ + OH^-$$

32.64 Which solution would be the more acidic: a 0.10 M solution of aniline hydrochloride, $C_6H_5NH_3Cl$ ($K_b = 4.2 \times 10^{-10}$ for aniline, $C_6H_5NH_2$), or a 0.10 M solution of methylamine hydrochloride, CH_3NH_3Cl ($K_b = 3.9 \times 10^{-4}$ for methylamine, CH_3NH_2)?

Functional Groups—Aldehydes and Ketones

32.65 What is the fundamental arrangement of atoms that is common to both aldehydes and ketones? What is the difference between an aldehyde and a ketone?

32.66 What functional group is produced by the oxidation of a primary alcohol? Will this functional group undergo subsequent oxidation? If so, what new functional group will be produced? What functional group is produced by the oxidation of a secondary alcohol? Will this functional group readily undergo further oxidation?

32.67 Name the following compounds:

(a) 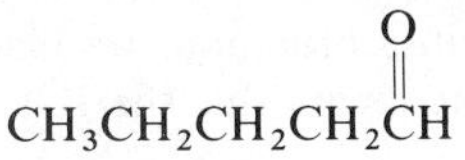 $CH_3CH_2CH_2CH_2CH{=}O$ (b) 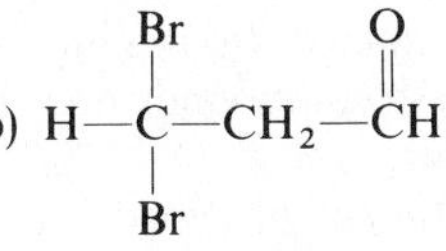 $H—CBr_2—CH_2—CH{=}O$

(c) 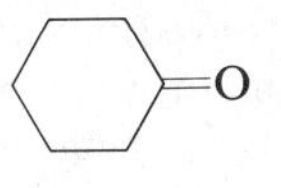 cyclohexanone ($C_6H_{10}{=}O$) (d) 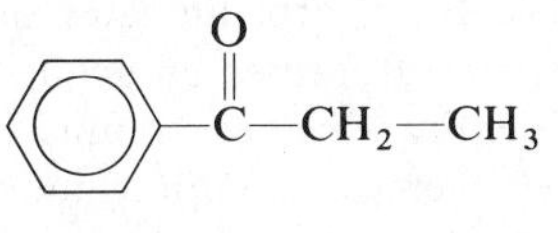 $C_6H_5—C({=}O)—CH_2—CH_3$

32.68 Write the chemical formula for each of the following: (a) 2-methylbutanal, (b) propynal, (c) *o*-methoxybenzaldehyde, (d) 2-butanone, (e) 1-bromo-2-propanone, (f) 3-hexanone.

Functional Groups—Carboxylic Acids

32.69 Write the chemical equation for the reaction of benzoic acid with sodium hydroxide. Name the salt that is produced.

32.70 Write the structural formula for each of the following: (a) 2-methylpropanoic acid, (b) 3-bromobutanoic acid, (c) *p*-nitrobenzoic acid, (d) potassium benzoate, (e) 2-aminopropanoic acid.

32.71 In aqueous solution, acetic acid exists mainly in the molecular form ($K_a = 1.745 \times 10^{-5}$). (a) Calculate the freezing point depression for a 0.10 molal aqueous solution of acetic acid neglecting any ionization of the acid. $K_f = 1.86$ °C kg/mol for water.

In nonpolar solvents such as benzene, acetic acid exists mainly as dimers

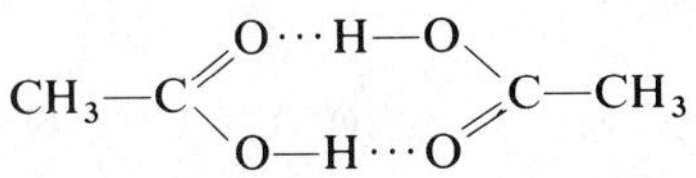

$$CH_3—C(=O\cdots H—O)(—O—H\cdots O=)C—CH_3$$

as a result of hydrogen bonding. (b) Calculate the freezing point depression for a 0.10 molal solution of acetic acid in benzene, $K_f = 4.90$ °C kg/mol for benzene.

32.72 What is the pH of a 0.10 M solution of sodium benzoate? $K_a = 6.6 \times 10^{-5}$ for benzoic acid, C_6H_5COOH. Would this solution be more or less acidic than a 0.10 M solution of sodium acetate? $K_a = 1.754 \times 10^{-5}$ for acetic acid, CH_3COOH.

Functional Groups—Esters

32.73 Name the functional groups that participate in an esterification reaction. What is the term given to the reverse of an esterification reaction?

32.74 Name the following esters:

(a) 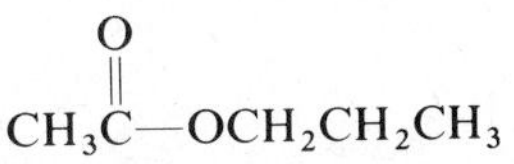 $CH_3C({=}O)—OCH_2CH_2CH_3$

(b) 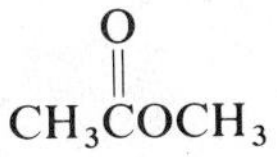 $CH_3C({=}O)OCH_3$

(c) 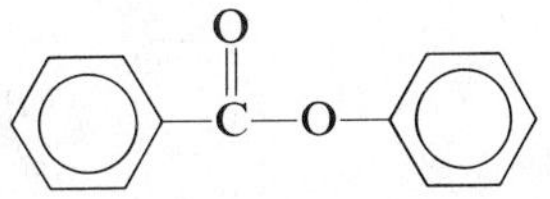 $C_6H_5—C({=}O)—O—C_6H_5$

(d)  $CH_3CH_2CH_2C({=}O)OCH_2CH_2CH_2CH_3$

(e) 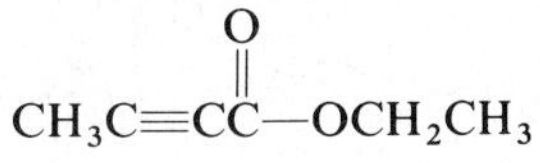 $CH_3C{\equiv}CC({=}O)—OCH_2CH_3$

ADDITIONAL QUESTIONS—Functional Groups

32.75 What is the general representation for the formula for an acyl halide? What functional group is formed by the reaction of an acyl halide and water?

32.76 What is the general representation for the formula for an acid anhydride? Write a chemical equation for the hydrolysis of acetic butyric anhydride ($CH_3CO—O—OCCH_2CH_2CH_3$) to form the original acids.

32.77 Name each of the following compounds:

(a) $CH_3CH_2CH_2CH_2OH$ (b) 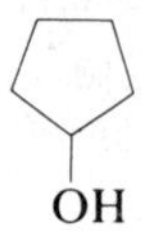 cyclopentane ring with OH

(c) $CH_3—CH(NH_2)—CH_3$ (d) $CH_3—C(Cl)=CH_2$

(e) 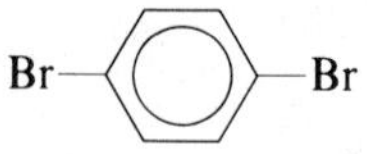 Br—C₆H₄—Br (para) (f) $(CH_3CH_2)_3N$

(g) 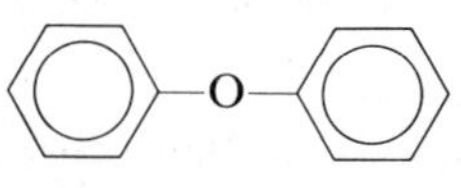 C₆H₅—O—C₆H₅ (h) 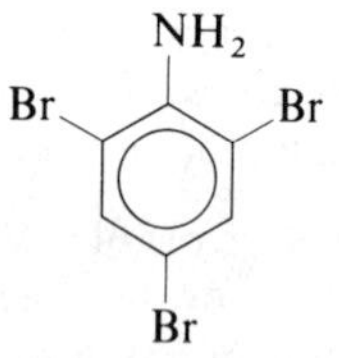 benzene ring with NH_2, Br, Br, Br

32.78 Draw the structural formula for each of the following compounds: (a) *p*-bromotoluene, (b) cyclohexanol, (c) 2-methoxy-3-methylbutane, (d) diethylamine, (e) *o*-chlorophenol, (f) 1,4-butanediol.

32.79 Name each of the following compounds:

(a) $CH_3CH(CH_3)CH_2OH$ (b) $CH_3CH_2CH_2CH_2NH_2$

(c) $CH_3CH_2CH_2CH_2CH{=}O$ (d) 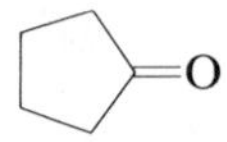 =O

(e) $CH_3CH_2CH(OCH_3)CH_3$ (f) $CH_3C(H_3C)(CH_3)—C(=O)OH$

32.80 Choose the compound that is the stronger acid in each set:

(a) $CH_3CH_2CH_2OH$ or CH_3—C₆H₄—OH

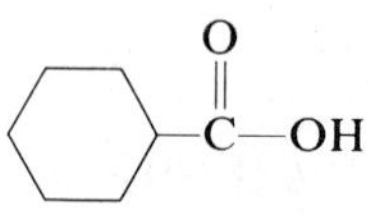

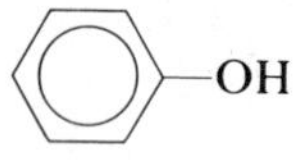

(b) CH_3CH_2OH or cyclohexyl—C(=O)—OH

(c) C₆H₅—OH or C₆H₅—C(=O)—OH

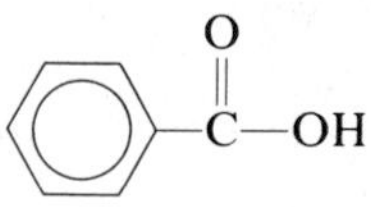

(d) 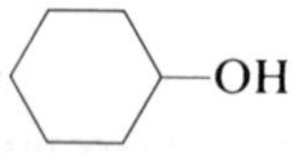 cyclohexyl—OH or C₆H₅—OH

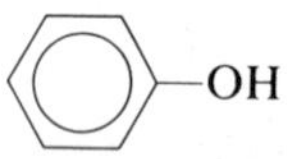

32.81 Identify the major products of each reaction:

(a) 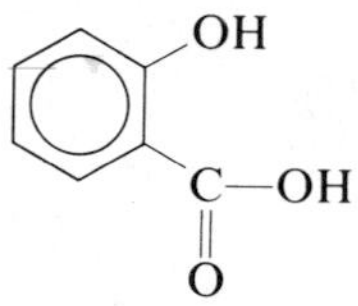 benzene ring with OH and C(=O)—OH (ortho) + 2NaOH ⟶

(b) 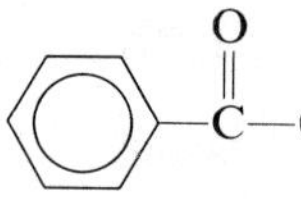 C₆H₅—C(=O)—OH + CH_3OH $\xrightarrow[\Delta]{H_2SO_4}$

(c) $CH_3CH_2CH{=}O$ $\xrightarrow[H^+]{MnO_4^-}$

32.82 Identify the major products of each reaction:

(a) 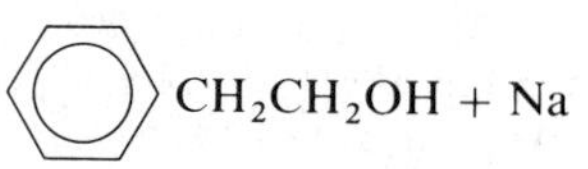 C₆H₅—CH_2CH_2OH + Na ⟶

(b) $CH_3CH_2C(=O)OCH_3$ $\xrightarrow[\Delta]{H_2O,\ conc\ HCl}$

(c) 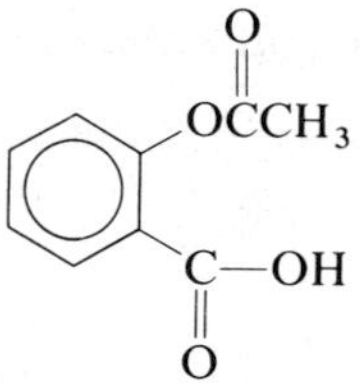 benzene ring with $OC(=O)CH_3$ and C(=O)—OH (ortho) $\xrightarrow[\Delta]{NaOH(aq)}$

32.83* Alcohols can be used as starting reagents for producing many other types of compounds: (a) aldehydes and ketones, (b) acids, (c) esters, (d) alkyl halides. Write a chemical equation illustrating the preparation of each of these types of compounds from an alcohol.

32.84* A student was given three bottles labeled A, B, and C. One of these contained acetic acid, one contained acetaldehyde, and one contained ethyl alcohol. (a) Write a structural formula for each of these compounds and (b) name them using the IUPAC nomenclature.

The student observed that substance A reacted with substance B to form an ester under certain conditions and that substance B formed an acidic solution when dissolved in water. (c) Identify which compound is in each bottle and (d) write balanced chemical equations for the reactions involving the formation of the ester and the ionization of the acid.

To confirm the identification, the student treated the compounds with a strong oxidizing agent and found that compound A required roughly twice as much oxidizing agent as compound C. (e) Write chemical equations for the reactions of compounds A and C with an acidic solution of MnO_4^-.

C H A P T E R

33

Organic Polymers and Biochemistry

ORGANIC POLYMERS

33.1 WHAT IS A POLYMER?

Plastics, synthetic fibers, rubber, cellulose, proteins—these materials are all polymers. Polymers result when large numbers of smaller molecules, called **monomers,** chemically bond to each other. The individual large polymer molecules are known as **macromolecules.** In their simplest form, polymer molecules are long chains composed of one type of repeating structural unit. If a single paper clip were a monomer, then a chain of paper clips would constitute such a polymer, or macromolecule. Linear polyethylene is a polymer of this type. The monomer is ethylene, $H_2C{=}CH_2$, and the structure of a segment of the polymer chain is as shown:

the repeating unit from the monomer

$$-CH_2CH_2CH_2CH_2CH_2\boxed{CH_2CH_2}CH_2CH_2CH_2CH_2CH_2-$$

or $$\left(CH_2CH_2\right)_n \qquad n = \text{a large number}$$

polyethylene

We encounter polyethylene almost every day in such items as flexible bottles, trash can liners, and the wrappings on food and meat in the supermarket.

The size of the macromolecules in a polymeric substance is quite variable. Polymers are characterized by the *average molecular mass* of the chains. For the majority of synthetic polymers in common use as plastics or fibers, the average molecular masses are between 1×10^4 and 1×10^6 u. The number of repeating units in such polymers, known as the *degree of polymerization,* varies roughly from 100 to 5000.

The chemistry of polymer formation and the properties of macromolecules and the materials made from them are based upon the same principles as those of smaller molecules. The differences between polymeric and nonpolymeric materials are due mainly to the much larger sizes of polymer molecules. Large, flexible molecules become tangled and twisted together, but small molecules do not. It is much more difficult for macromolecules to fall into an ordered, crystalline arrangement or to flow past each other. Also, intermolecular forces acting over the length of a macromolecule add up to a much stronger attraction than the same forces acting between small molecules.

The flexibility, impact resistance, elasticity, and other mechanical properties of polymeric materials reflect the resistance of the polymer chains to movement. Therefore, strong intermolecular forces or molecular structures that inhibit molecular motion tend to strengthen and stiffen a polymer, and also to raise its melting point. When only weak dipole–dipole forces act between macromolecules, as in polyethylene, very long chains are needed to develop useful mechanical properties in a polymeric material. If stronger intermolecular forces such as hydrogen bonding are possible between chains, a useful material can be made from smaller macromolecules. Interestingly, some polymers deform if bent slowly enough for the molecules to move to new positions, but break if bent so sharply that high local stresses occur. This behavior is unique to polymeric materials.

The properties of polymeric materials vary widely depending upon the chemical composition and structure of the macromolecules. As a result, plastics, synthetic fibers, and synthetic rubbers of many different types have found a multitude of practical applications. Polymers are also used in coatings such as paints, lacquers, and protective films, and in adhesives. Synthetic polymers are now produced in huge amounts—close to 50 billion pounds per year in the United States alone. Because of the special techniques needed to form polymers and to study their properties, polymer chemistry has developed into a distinct and separate branch of chemistry. More than half the chemists and chemical engineers in the United States are employed in fields related in one way or another to polymer chemistry.

33.2 POLYMER STRUCTURE

A long chain of repeating units bonded together as in polyethylene forms a *linear polymer* (Figure 33.1). In the natural course of polymerization or as the result of intentional control of the reaction, other, often shorter, polymer chains of the same or different composition can be attached to the main polymer chain, producing a *branched polymer*.

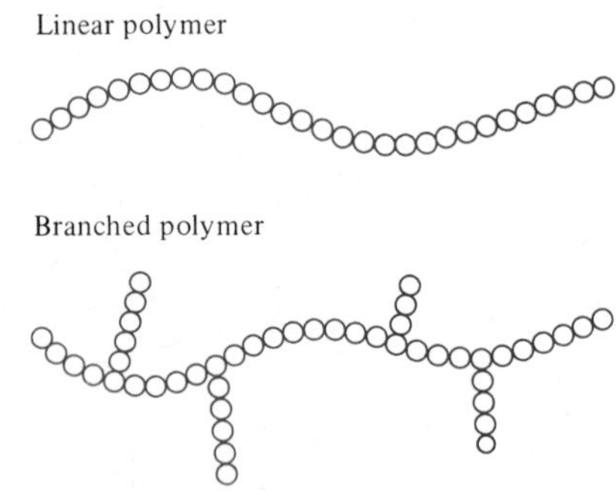

Figure 33.1
Polymer Structures
Each circle represents a single repeating unit.

The degree of branching significantly affects the properties of polymers. In linear polyethylene, the straight chains can lie close together; in branched-chain polyethylene, the branches get in the way and prevent such close packing. Therefore, the linear polymer yields a much higher density and less flexible plastic than the branched polymer. High-density polyethylene is used mainly in rigid bottles (e.g., for milk, bleach, motor oil, liquid detergents), sturdy films (e.g., in grocery bags), and pipes. Low-density polyethylene is used mainly in packaging films, coatings (e.g., on paper milk cartons), and wire and cable coverings.

By "plastics" we refer to materials composed of polymers and used in manufactured objects. All plastics are classified as either thermoplastics or thermosets. A **thermoplastic** can be softened repeatedly when heated and hardened when cooled, with little change in properties. The softening occurs as the polymer chains move more and more freely. Thermoplastics can be heated, then molded (shaped, as in making a toy or a bottle) or extruded (forced through an opening to form a continuous shape, as in making a garden hose), and then cooled to produce a desired product.

As you might expect, the difference between thermoplastics and thermosets is based upon structure. When the "branches" on one polymer chain are bonded to other polymer chains, the result is a *cross-linked polymer*—a polymer in which long chains are interconnected by covalently bonded shorter chains. The cross links hold the molecules in place so that heating does not allow them to move freely. A **thermoset plastic** is cross-linked and is permanently rigid. A thermoset plastic, if heated sufficiently, will decompose rather than melt. To form articles with the desired shapes

Table 33.1
Some Vinyl Polymers

	Monomer	Repeating unit	Uses
Poly(vinyl chloride) (PVC)	$CH_2{=}CHCl$	$\leftarrow CH_2CH(Cl) \rightarrow_n$	Pipes and pipefittings, floor tile, vinyl siding, gutters
Polystyrene	$CH_2{=}CHC_6H_5$	$\leftarrow CH_2CH(C_6H_5) \rightarrow_n$	Molded toys, dishes, kitchen equipment; insulating foam; rigid foam packaging (e.g., meat trays)
Polyacrylonitrile	$CH_2{=}CHCN$	$\leftarrow CH_2CH(CN) \rightarrow_n$	Acrylic fibers for carpets, sweaters, clothing fabric
Polypropylene	$CH_2{=}CH_2CH_3$	$\leftarrow CH_2CH(CH_3) \rightarrow_n$	Filament for rope, webbing, carpeting; molded auto and appliance parts
Poly(vinyl acetate)	$CH_2{=}CHC(=O)OCH_3$	$\leftarrow CH_2CH(OC(=O)CH_3) \rightarrow_n$	Water-based paints, adhesives, paper and textile coatings
Poly(vinyl alcohol)	—[a]	$\leftarrow CH_2CH(OH) \rightarrow_n$	Water-soluble thickening agent for foods and cosmetics; temporary protective films for metals, etc.

[a] Poly(vinyl alcohol) must be made by hydrolysis of the acetate ester group in poly(vinyl acetate), because vinyl alcohol ($CH_2{=}CHOH$) is unstable.

from thermoset plastic, the cross-linking reaction must be allowed to take place during fabrication of the article.

A **homopolymer** is formed by the polymerization of a single type of monomer. Polyethylene, whether linear or branched, is a homopolymer. A large and widely used family of homopolymers is produced from monomers structurally related to polyethylene by replacement of one hydrogen atom with a different substituent. These monomers all contain the vinyl group $H_2C{=}CH{-}$. The repeating unit in the vinyl polymers is H_2CCHX, where X varies according to the monomer (Table 33.1).

A **copolymer** is formed by the polymerization together (the "copolymerization") of two or more different monomers. Two monomers can combine in either regular fashion (although this is rare) or random fashion. Figure 33.2 illustrates some of the ways in which two monomers can combine. Careful control of the polymerization reaction makes it possible to create polymers with exactly the desired properties by combination of different monomers in various ratios and geometric arrangements.

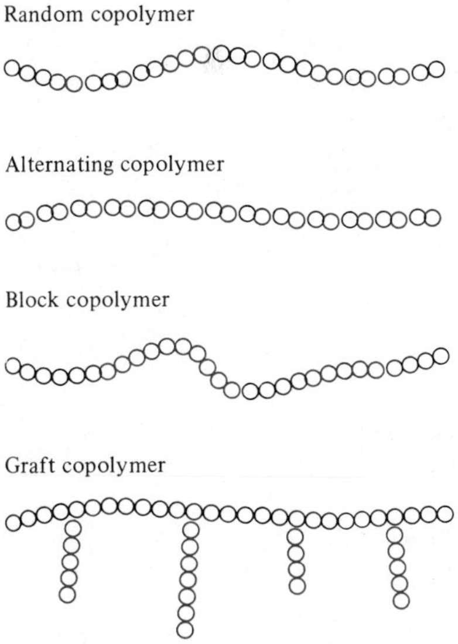

Random copolymer

Alternating copolymer

Block copolymer

Graft copolymer

Figure 33.2
Types of Copolymers
Each circle represents a single repeating unit; the black and color circles represent units from two different monomers.

33.3 CRYSTALLINITY AND THERMAL PROPERTIES OF POLYMERS

As we know, crystallinity results from the packing together of molecules or ions in a regular arrangement. On the other hand, random arrangements of molecules produce amorphous substances, which have quite different properties from crystalline

Orientation of chains in a crystalline polymer

Orientation of chains in an amorphous polymer

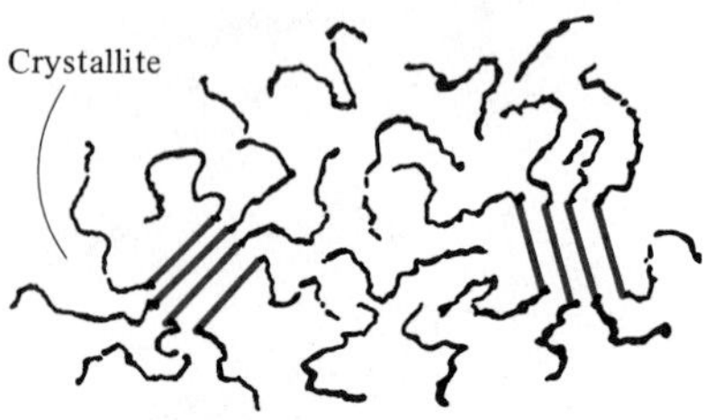

Orientation of chains in a semicrystalline polymer

Figure 33.3
Amorphous and Crystalline Polymers

ones. Most polymers are either amorphous or semicrystalline. Because of the length of polymer chains, a fully ordered arrangement is not common.

In an amorphous polymer, the macromolecules are coiled and twisted together, not unlike a group of very long earthworms twisted together in a can. Crystallinity in polymers is possible when the chains can line up close together in parallel fashion (Figure 33.3). For this to happen, the individual polymer molecules must be of reasonably regular structure and without bulky substituents or branches that would get in the way. When the polymer chains pack together, they form *crystallites*—regions within the polymer that have crystalline properties such as sharp melting points. Only a portion of a given chain is likely to become part of a crystallite. A given chain may wander from one crystallite to another, with intermediate regions being amorphous.

The thermal properties of polymers are strongly influenced by their semicrystalline composition. At low temperatures an amorphous polymer is a glass—only small groups, such as CH_3 side chains, can move at all. Glasses are rigid and brittle—they shatter if struck (like a mass of frozen worms).

As the temperature increases, it is possible for increasingly larger portions of the molecules to move. At what is called the **glass transition temperature** (T_g) there is a dramatic increase in the flexibility of a polymer as enough free volume becomes available to allow the cooperative motion of large segments of the polymer chains. Mechanical, electrical, and other properties also change at T_g.

Above their glass transition temperatures, polymers are flexible and rubbery. As the temperature increases beyond T_g, independent motion of the polymer chains becomes increasingly easier. At appropriate temperatures above T_g thermoplastic polymers can be molded and otherwise transformed into desired shapes. Complete disentanglement of the molecules occurs at the melting point (T_m) of the crystalline regions.

Rigid plastics have glass transition temperatures above the temperatures at which they are used. For example, polystyrene, a hard clear plastic used in housewares, plastic lenses, and Styrofoam, has a T_g of 100 °C, well above room temperature. On the other hand, polymers used in applications that require flexibility must have a T_g well below the temperature at which they will be used. Polyisobutylene, one of the butyl rubbers, has a T_g of -70 °C and is a tacky substance at room temperature. It is used in chewing gum and caulking compounds, as an adhesive, and as an additive for automotive lubricating oils. At ordinary temperatures, lubricating oil is not a good solvent for polyisobutylene and the polymer molecules are coiled up. At higher temperatures, the oil becomes less viscous, an undesirable property. However, it also becomes a better solvent for the polymer molecules. The molecules uncoil and counteract the decrease in viscosity of the oil, because the long chains tend to become entangled and thus resist movement past each other.

33.4 POLYMERIZATION REACTIONS

In order to participate in polymer formation, a molecule must be able to react at both ends so that the polymer chain can grow. Alkenes polymerize by addition to both ends of the double bonds and conversion of the double bonds to single bonds. Difunctional molecules polymerize by reaction at each of their two functional groups.

The two principal types of polymerization reactions are chain reaction polymerization (also referred to as addition polymerization) and step reaction polymerization (also referred to as condensation polymerization). The addition and condensation reactions involved in polymerization are no different from those of any organic molecules. However, they must be able to repeat thousands of times when the reactants are sufficiently pure.

a. Chain Reaction Polymerization **Chain reaction polymerization,** or **addition polymerization,** is a rapid polymerization characterized by three reaction steps—initiation, propagation, and termination. The polymerization begins when a molecule of an *initiator* (I) is converted to an active species (I*) that reacts with a monomer molecule (M) to give an active intermediate:

initiation

$$\mathrm{I} \longrightarrow \mathrm{I^*}$$
$$\mathrm{I^* + M} \longrightarrow \mathrm{IM^*} \qquad (33.1)$$

Propagation is the very rapid (10^{-1} to 10^{-6} s) growth of the polymer chain by the successive addition of monomer molecules.

propagation

$$\mathrm{IM^* + M} \longrightarrow \mathrm{IMM^* + M} \longrightarrow \mathrm{IMMM^* + M\cdots} \longrightarrow \mathrm{I(M)_nM^*} \qquad (33.2)$$

Propagation continues until the active end of the chain encounters a species (T) which reacts to give a molecule that is no longer active.

termination

$$\mathrm{I(M)_nM^* + T} \longrightarrow \mathrm{I(M)_nMT} \qquad (33.3)$$

Many vinyl polymers are formed by chain reaction polymerization in which the active species that initiates the reaction is a free radical (has an unpaired electron) and the chain grows by successive addition of monomer to a free radical at the end of the chain (Figure 33.4). The reactive end of the chain can also be a cation, an anion, or a coordination complex.

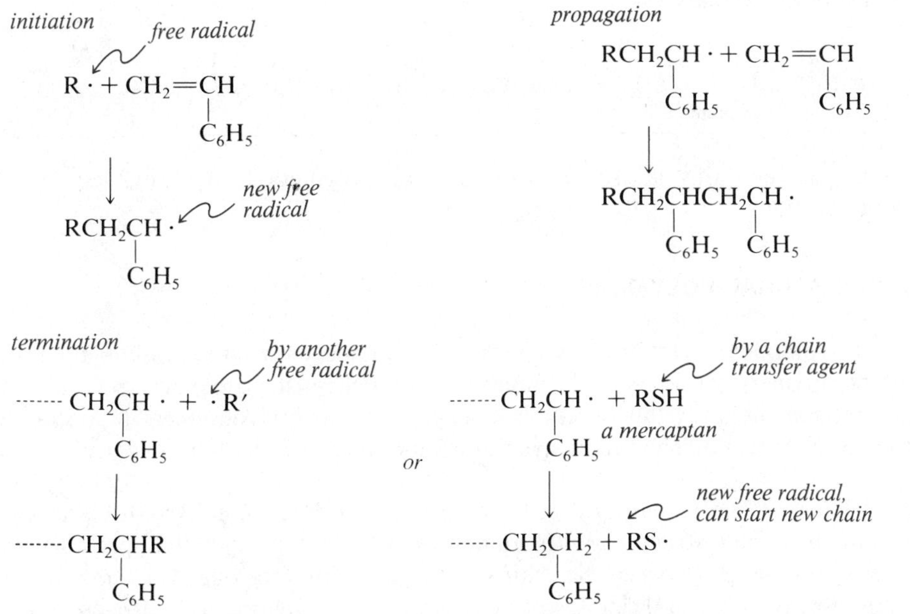

initiation

free radical

$\mathrm{R\cdot + CH_2{=}CH(C_6H_5)}$ ⟶ $\mathrm{RCH_2CH(C_6H_5)\cdot}$ *new free radical*

propagation

$\mathrm{RCH_2CH(C_6H_5)\cdot + CH_2{=}CH(C_6H_5)}$ ⟶ $\mathrm{RCH_2CH(C_6H_5)CH_2CH(C_6H_5)\cdot}$

termination

by another free radical

$\mathrm{\cdots CH_2CH(C_6H_5)\cdot + \cdot R'}$ ⟶ $\mathrm{\cdots CH_2CHR(C_6H_5)}$

or

by a chain transfer agent

$\mathrm{\cdots CH_2CH(C_6H_5)\cdot + RSH}$ (*a mercaptan*) ⟶ $\mathrm{\cdots CH_2CH_2(C_6H_5) + RS\cdot}$ *new free radical, can start new chain*

Figure 33.4
Chain Reaction Polymerization of Styrene
(1) Initiation, (2) propagation, (3) termination.

b. Step Reaction Polymerization **Step reaction polymerization,** or **condensation polymerization,** proceeds by the reaction of difunctional monomers with each other. No initiator is necessary and each "step" is the same type of chemical reaction. Usually a small molecule, such as water or methanol, is eliminated in the combination of the two functional groups and the reaction takes place at both ends of the growing chain. For example, polyesters are formed by step reaction polymerization between difunctional acids and difunctional alcohols as follows (where rectangles represent the molecules to which the functional groups are attached):

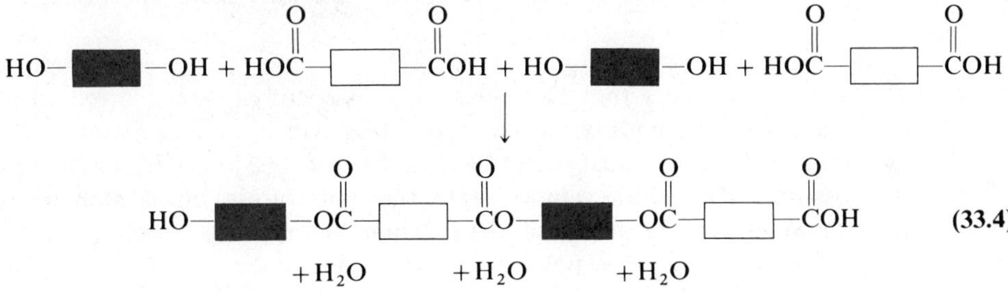

$$\mathrm{HO{-}\blacksquare{-}OH + HOC({=}O){-}\square{-}C({=}O)OH + HO{-}\blacksquare{-}OH + HOC({=}O){-}\square{-}C({=}O)OH}$$
$$\downarrow$$
$$\mathrm{HO{-}\blacksquare{-}OC({=}O){-}\square{-}C({=}O)O{-}\blacksquare{-}OC({=}O){-}\square{-}C({=}O)OH} \qquad (33.4)$$
$$+\mathrm{H_2O} \quad +\mathrm{H_2O} \quad +\mathrm{H_2O}$$

Table 33.2
Some Copolymers Formed by Step Reaction Polymerization

Nylon 66, a polyamide—used in fibers

$$\mathrm{HO\overset{\overset{O}{\|}}{C}(CH_2)_4\overset{\overset{O}{\|}}{C}OH + H_2N(CH_2)_6NH_2 \xrightarrow{\Delta} \left(\overset{\overset{O}{\|}}{C}(CH_2)_4\overset{\overset{O}{\|}}{C}NH(CH_2)_6NH\right)_n + 2nH_2O}$$

adipic acid *hexamethylenediamine*

Poly(ethylene terephthalate), a polyester—used in films (Mylar), fibers (Dacron), and clear soft-drink bottles

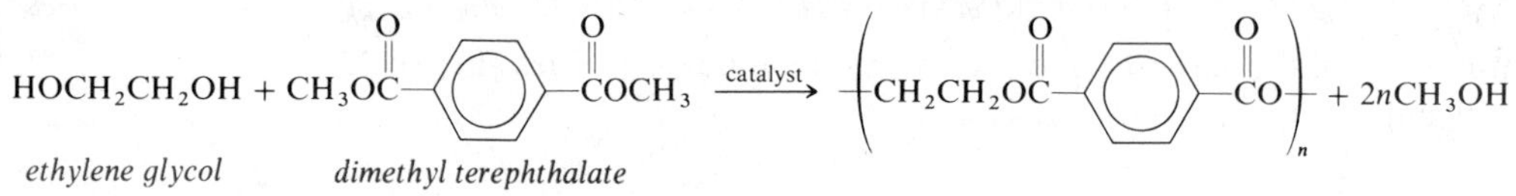

ethylene glycol *dimethyl terephthalate*

A polycarbonate, a very tough and transparent plastic (Lexan)—used in, e.g., shatter-resistant windows

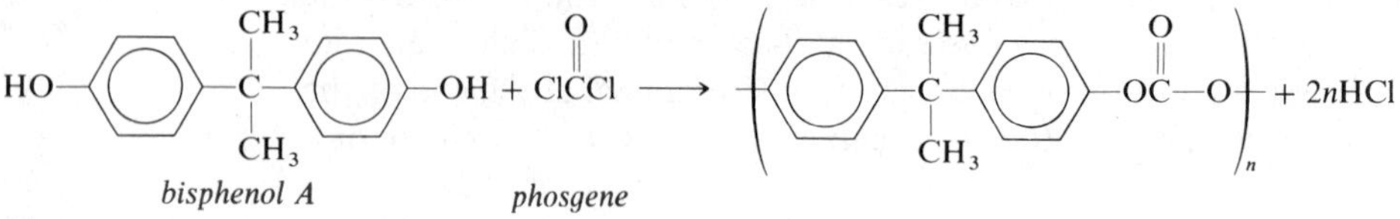

bisphenol A *phosgene*

A polyurethane—used in elastic fibers, foams, floor and surface coatings

$$\mathrm{HO(CH_2)_4OH + O{=}C{=}N(CH_2)_6N{=}C{=}O \longrightarrow \left(CH_2(CH_2)_3O\overset{\overset{O}{\|}}{C}NH(CH_2)_6NH\overset{\overset{O}{\|}}{C}{-}O\right)_n + 2nH_2O}$$

1,4-butanediol *1,6-hexanediisocyanate*

Table 33.2 contains additional examples of polymers formed by step reaction (condensation).

33.5 NATURAL POLYMERS

The unique properties of living things are in many ways the result of the properties of polymers, for many of the molecules of biochemistry are macromolecules. Two natural polymers—cellulose and rubber—were the first polymers to be put into commercial use. Each had to be structurally modified in order to modify its properties.

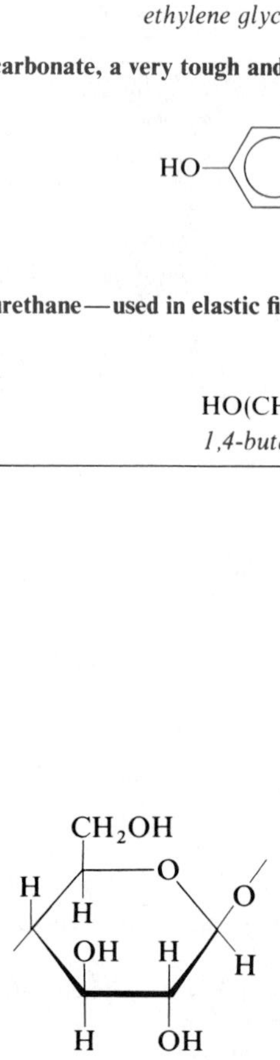

Natural cellulose repeating unit

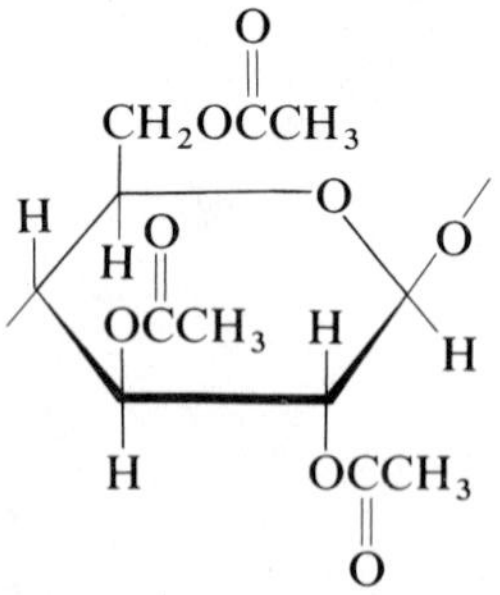

Cellulose acetate repeating unit

Figure 33.5
Repeating Units of Natural Cellulose and Cellulose Acetate, a Thermoplastic Derivative

a. Cellulose The repeating unit in natural cellulose (Figure 33.5), for example, in cotton, is a ring structure with numerous hydroxyl groups that provide strong hydrogen bonding between the chains. Microcrystalline regions also help to hold the chains together. Natural cellulose is not thermoplastic. Before the hydrogen bonds are broken and the microcrystals melted—a requirement for plastic flow—the molecule undergoes thermal decomposition. By converting some of the hydroxyl groups to functional groups that do not form hydrogen bonds, it is possible to alter the properties of the natural polymer. The nitrate ester of cellulose was prepared and used commercially in lacquers and films in the 1800s, before the nature of polymers was understood. (Due to its high flammability, the use of cellulose nitrate in such applications is now obsolete.) Cellulose acetate and other cellulose esters are still used in fibers. (Garments made of such fibers are labeled "acetate.") Rayon is also made from cellulose, not by esterification but by other treatments that alter the properties of the polymer.

b. Natural Rubber Polymers that have elasticity, like rubber, are referred to as **elastomers.** Such polymers are utilized above their glass transition temperatures. The polymer chains in an elastomer must be flexible and they must be joined by a moderate number of cross-links. In the unextended state, the molecules of an elastomer are tangled together. When a force is applied, the polymer chains move into an extended and more ordered arrangement (Figure 33.6). The cross-links are necessary to prevent the chains from slipping past each other when force is applied. When the force is released the chains return to their less ordered state, largely because of the resultant increase in entropy.

Natural rubber is an unsaturated polymer in which the chain enters and leaves each covalent double bond in a *cis* orientation. (The *trans* isomer is not rubbery.)

$$-CH_2-\overset{\overset{CH_3}{|}}{C}=CH-CH_2CH_2-\overset{\overset{CH_3}{|}}{C}=CH-CH_2CH_2-\overset{\overset{CH_3}{|}}{C}=CH-CH_2CH_2-\overset{\overset{CH_3}{|}}{C}=CH-CH_2-$$

natural rubber

Note that there are no cross-links between the polymer chains. Natural rubber becomes soft and sticky when heated and tends to become permanently deformed when stretched. However, when natural rubber is heated with sulfur and a catalyst, sulfur cross-links are formed between the chains. This process is called *vulcanization*; it was discovered accidentally by Charles Goodyear in 1839. Introduction of a moderate number of sulfur cross-links converts natural rubber into a useful elastomer. A larger number of cross-links converts it into hard rubber, or ebonite, which is not an elastomer.

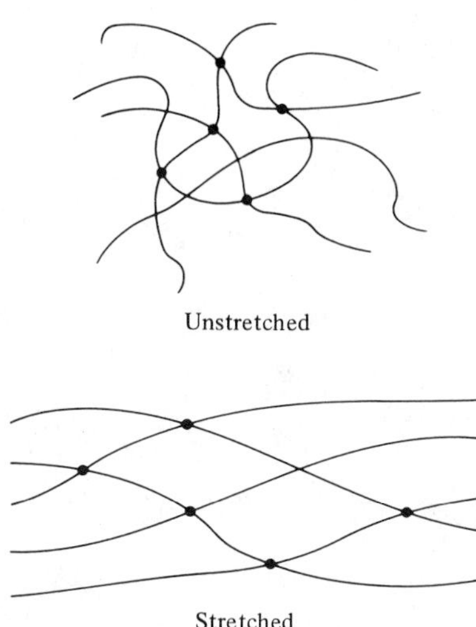

Figure 33.6
Vulcanized Rubber
The black circles represent the sulfur cross-links.

SUMMARY OF SECTIONS 33.1–33.5

Polymers, or macromolecules, are produced when up to thousands of smaller molecules (monomers) are bonded together. The physical properties of a polymer are determined by such factors as the flexibility of the macromolecules, the sizes and types of groups attached to the polymer chains, and the magnitude of intermolecular forces. Polymers may be linear, branched, or cross-linked. Copolymers are produced from two monomers combined randomly or in specific patterns.

Thermoplastics can be repeatedly softened when heated and can be hardened when cooled; thermoset plastics, which are cross-linked, are permanently rigid and decompose only if heated strongly. Most polymers are either amorphous or semicrystalline. At low temperatures an amorphous polymer is rigid and brittle. At the glass transition temperature (T_g), the flexibility of the polymer increases markedly. Rigid plastics are used at temperatures below T_g; flexible plastics are used at temperatures above T_g.

To participate in polymerization, a molecule must be able to react at both ends. The principal types of polymerization reactions are chain reaction polymerization (initiation, propagation, termination), which is undergone by, for example, vinyl monomers, and step reaction polymerization, which involves reaction between functional groups on different monomer molecules.

Cellulose with added ester groups and natural rubber with added sulfur cross-links are modified natural polymers in commercial use.

BIOCHEMISTRY

It is difficult to provide even the briefest introduction to a subject as broad and complex as biochemistry in the small space available to us here. Yet to omit this subject entirely would be a great pity. For one thing, everyone should know at least a little about the marvelous molecular machinery that not only keeps us alive, but also enables us to solve quadratic equations, play the tuba, and read books such as this one. Moreover, biochemistry has become one of the most dynamic and exciting fields of modern science—one that attracts more and more students of chemistry, and that has had an enormous impact on the lives of all of us. So, for those of you who will not have the opportunity to study this subject at another time, as well as for those of you who will pursue the subject further, we give here a short introduction to the molecules of living things and how they function.

33.6 BIOLOGICAL MOLECULES—AN INTRODUCTION

Most biologically important substances are organic compounds, built up from skeletons of carbon atoms. Many of them are very large molecules, and most of these are polymers. In some cases these polymers consist of thousands of identical repeating units. Frequently, however, they are composed of several different monomers, linked in a sequence that is highly specific and crucially important to the properties and function of such molecules.

In biological molecules we find a remarkably intimate relationship between structure and function. A seemingly small and insignificant change, involving only a few atoms in a giant molecule, can literally spell the difference between life and death for an organism. A single hemoglobin molecule, for example, contains some 8000 atoms; its molecular mass is nearly 65,000 u. The only difference between normal hemoglobin and the form of hemoglobin that causes sickle-cell disease—a painful and often fatal condition—is the following substitution at two places in the molecule:

$$-CH(CH_3)_2 \quad \text{replaced by} \quad -CH_2CH_2COOH$$

A crucial concept for any understanding of how biomolecules interact and function is the distinction between hydrophilic and hydrophobic molecules. **Hydrophilic** means, literally, "water-loving"; it refers to substances that are soluble in water and similar polar solvents. To be hydrophilic, a large organic molecule must generally have a great many charged groups or groups containing polar bonds. **Hydrophobic** means, literally, "water-fearing," and refers to substances that are insoluble in water but dissolve readily in nonpolar solvents such as benzene. Hydrophobic substances are generally composed chiefly of hydrocarbon chains, aromatic rings, and other nonpolar groups. Hydrophilic substances interact with other hydrophilic substances, and hydrophobic substances have an affinity for other hydrophobic substances, but molecules of the two classes do not readily interact with one another. In living systems, chemical reactions take place largely in aqueous solution, that is, in a hydrophilic environment. By contrast, most of the "architecture" of the living cell, including the membrane that surrounds it, is made of hydrophobic molecules. And as we shall see, some of the most important biomolecules are able to perform their special functions because they are "amphibious"—part hydrophilic and part hydrophobic.

The four major classes of organic compounds in living cells are proteins, carbohydrates, lipids, and nucleic acids. Table 33.3 shows the amounts of these substances that, together with water, inorganic ions, and smaller molecules, are present in the cells of a microorganism and an animal. The ratios vary from species to species and among different types of cells. But the carbohydrates, lipids, proteins, and nucleic acids perform similar functions in all plants and animals.

Table 33.3
Composition of Cells
Amino acids are the monomers of which proteins are composed; nucleotides are the monomers of which nucleic acids are composed.

Component	Bacterium *E. coli* (%)	Rat liver (%)
Water	70	69
Protein	15	21
Amino acids	0.4	—
Nucleic acids	7	1.2
Nucleotides	0.4	—
Carbohydrates	3	3.8
Lipids	2	6
Other small molecules	0.2	—
Inorganic ions	1	0.4

33.7 CARBOHYDRATES

Substances commonly referred to as "sugars" are carbohydrates (as are starch and cellulose). A *monosaccharide* is a simple sugar that is a polyhydroxy aldehyde (an *aldose*) or a polyhydroxy ketone (a *ketose*). Some examples are given in Table 33.4. Two simple sugar molecules combine by condensation to give a *disaccharide*. Condensation polymerization of sugars yields *polysaccharides*. **Carbohydrates** are all monosaccharides, disaccharides, or polysaccharides.

Table 33.4
Some Monosaccharides
The ketoses found in living things are almost always 2-ketoses—the second carbon atom in the chain is the carbonyl group carbon atom.

Aldoses (aldehydes)				Ketoses (ketones)	
CHO HCOH CH_2OH D-*glyceraldehyde*	CHO HCOH HCOH HCOH CH_2OH D-*ribose*	CHO HCOH HOCH HCOH HCOH CH_2OH D-*glucose*	CHO HCOH HOCH HOCH HCOH CH_2OH D-*galactose*	CH_2OH C=O HCOH HCOH CH_2OH D-*ribulose*	CH_2OH C=O HOCH HCOH HCOH CH_2OH D-*fructose*

Since most monosaccharides contain several asymmetrically substituted carbon atoms, monosaccharides and their derivatives form optically active isomers. In glyceraldehyde, for example (see Table 33.4), the middle carbon is attached to four different groups and the molecule therefore has two mirror-image enantiomers. A six-carbon aldose ($C_6H_{12}O_6$) has four such asymmetric carbon atoms and consequently forms 16 enantiomers. One of these, D-glucose, is the most biologically important simple sugar. Nearly all natural monosaccharides have the D configuration.

Sugars with five carbon atoms (*pentoses*) or six carbon atoms (*hexoses*) are more stable as cyclic structures than as the open-chain structures depicted in Table 33.4. The ring is created by a reaction between the carbonyl oxygen of the sugar molecule and one of its hydroxyl groups. As Figure 33.7 shows, for D-glucose this reaction can give rise to two different ring structures, known as α-D-glucose and β-D-glucose, depending on the configuration at carbon atom 1. (When the ring forms, carbon atom 1 also becomes an asymmetric carbon, and the α and β forms are optical isomers of one another at that carbon atom.)

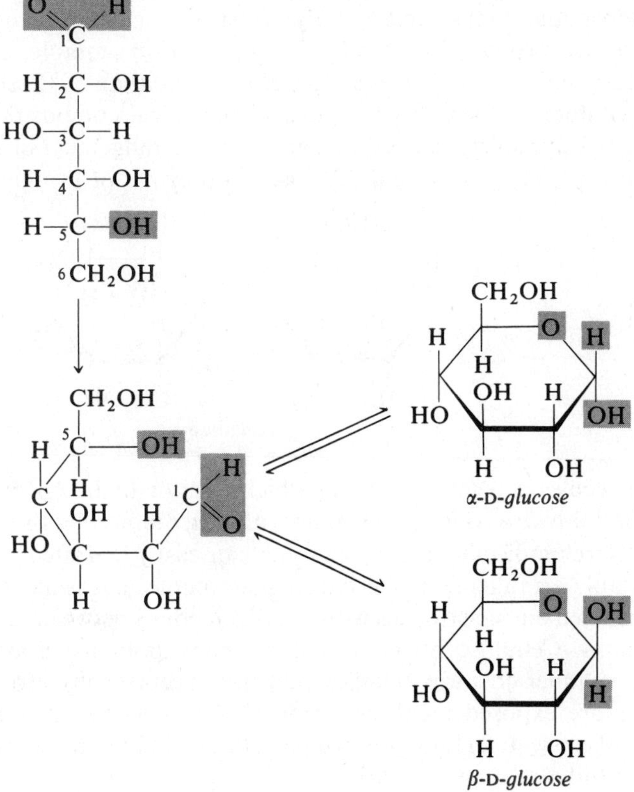

Figure 33.7
Formation of the Glucose Ring Structure
In solution the aldehyde form of glucose is unstable. The aldehyde group reacts with one of the hydroxyl groups to give a ring structure that can exist in two different configurations, the β configuration being slightly more stable.

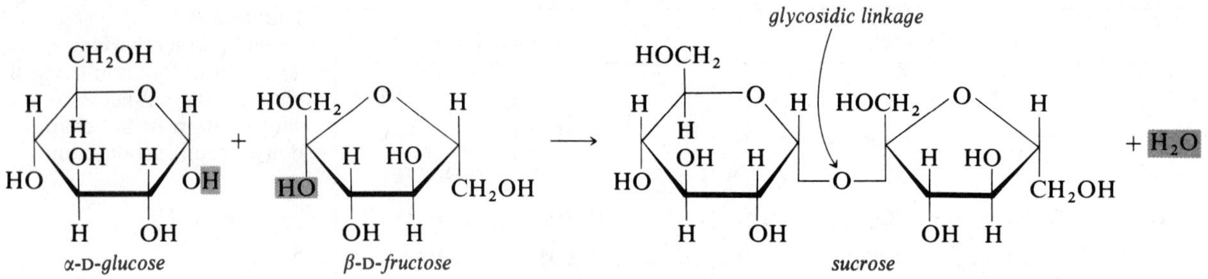

α-D-glucose + β-D-fructose ⟶ sucrose + H_2O

glycosidic linkage

Figure 33.8
Sucrose (table sugar), a Disaccharide

Disaccharides result from the loss of water between two OH groups. The result is formation of a *glycosidic linkage.* For example, sucrose, which is common table sugar, is a disaccharide of glucose and fructose (Figure 33.8). Many biopolymers are formed by condensation and the loss of water molecules. The reverse of a condensation reaction is a hydrolysis reaction, in which a bond connecting two monomers is broken by addition of the elements of water. The digestion of food containing polymeric biomolecules such as starch or protein (Section 33.11) always involves a series of hydrolysis reactions that break down the polymer into its monomers.

Polysaccharides have two principal functions: energy storage and structural support. As we will see in Section 33.11, organisms use simple sugars, particularly glucose, as their main energy source, but they do not store such sugars in large quantities. For one thing, glucose is too reactive. For another, keeping huge numbers of glucose molecules inside a cell would create tremendous osmotic pressure (Section 15.13); water would flow into the cell, causing it to swell and burst. It is much more convenient to stockpile several thousand glucose units in the form of a single macromolecule. Plants store glucose as starch, a polysaccharide. An excess of glucose in the bloodstream of animals is converted into glycogen, a highly branched polymer of α-D-glucose that is stored in the liver and muscles.

By far the most abundant structural polysaccharide is cellulose. Some 100 billion tons of cellulose are produced each year by plants. For example, cotton is 99% cellulose, and the woody parts of trees are generally more than 50% cellulose.

Although cellulose and starch have the same chemical composition, they differ greatly in properties due solely to the differences in the arrangement of their atoms—starch is a polymer of α-D-glucose and cellulose is a polymer of β-D-glucose:

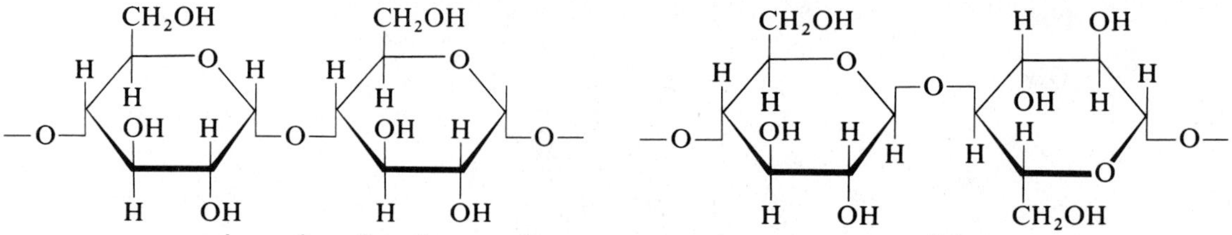

amylose, a form of starch
(polymer of α-D-glucose)

cellulose
(polymer of β-D-glucose)

Starch macromolecules coil into a helix in which interior hydroxyl groups stabilize the helix and exterior hydroxyl groups interact with water molecules through hydrogen bonding. Thus, starch molecules are water-soluble or easily hydrated.

The most stable arrangement of cellulose macromolecules is achieved when the molecules lie stretched out side by side, with hydrogen bonds between hydroxyl groups on adjacent chains. Cellulose fibers consist of many polysaccharide chains held together in this manner to form bundles that are exceptionally strong. Since few hydroxyl groups are exposed on the exterior of the fibers to interact with water, cellulose is insoluble in water. Humans can digest starch but have no enzyme capable of breaking down cellulose.

33.8 LIPIDS

"Lipid" is something of a catchall classification for biomolecules. There are many different kinds of lipids and they differ widely in structure; none are polymers, though some have quite large molecules. They have in common that they contain hydrocarbon segments and are nonpolar. **Lipids** are biochemical substances that are hydrophobic—they dissolve in nonpolar solvents such as ether, rather than in water.

One major class of lipids consists of the glycerides, which include the compounds that we know as fats and oils. Glycerides are esters (Section 32.15) formed by the condensation of the three-carbon alcohol glycerol with one or more fatty acids (Figure 33.9). Fatty acids are long, straight-chain carboxylic acids that we encountered previously in soaps (Section 15.17). Since the glycerol molecule has three OH groups, it can form ester linkages with one, two, or three fatty acids, forming a monoglyceride, a diglyceride, or a triglyceride, respectively. In glyceride molecule, the fatty acid chains may be identical or they may be different (a *mixed glyceride*). Most naturally occurring fats and oils are mixtures of several mixed glycerides.

The distinction between fats and oils is based on melting point: at room temperature, fats are solids and oils are liquids. This difference is, as you might expect, based on differences in the glyceride molecules. In fats, the long hydrocarbon chains from the fatty acids are largely saturated. The chains are similar in shape and the molecules can fit closely together, allowing for the maximum attraction between them. In oils, a large proportion of the hydrocarbon chains are unsaturated. The carbon chain enters and leaves all of the double bonds in a *cis* orientation, creating a sharp bend at each bond. Therefore, these unsaturated glyceride molecules cannot pack together efficiently, the attractions between molecules are weaker, and the melting points are lower.

Glycerides from animal sources, such as butter and lard, are usually high in saturated fatty acids, whereas those from vegetable sources, such as corn oil or safflower oil, are richer in unsaturated fatty acids. Solid shortening for cooking and baking used to be made by hydrogenating vegetable oils—adding hydrogen across carbon–carbon double bonds to produce glycerides with completely saturated fatty acid chains. Today, however, there is evidence that a diet high in saturated fats contributes to the buildup of fatty deposits in the walls of blood vessels, which may lead to high blood pressure, heart attacks, and strokes. Thus there has been a trend toward the use of *polyunsaturated* oils (those with molecules containing more than one double bond) in food products. Recent evidence indicates that monounsaturated oils (those with a single double bond in most hydrocarbon chains) may help guard against fatty deposits in blood vessels. Olive oil is a natural monounsaturated oil.

The chief function of fats in animals is energy storage. Unlike plants, animals are mobile and so must carry their energy stockpile around with them. It is therefore

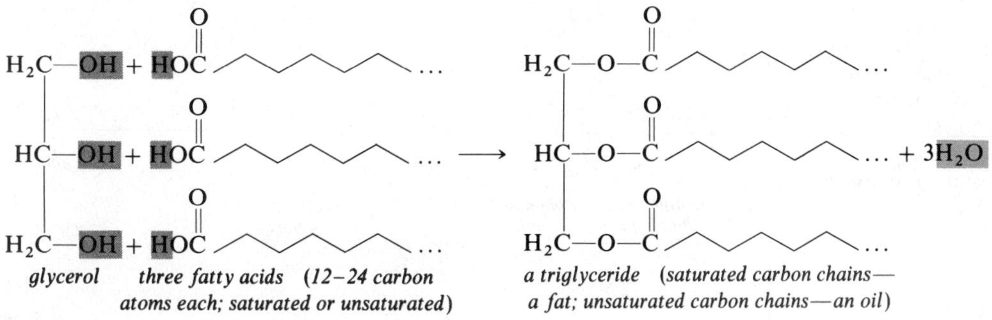

$H_2C—OH + HOC(=O)$ ... $\longrightarrow$ $H_2C—O—C(=O)$ ...
$HC—OH + HOC(=O)$ ... $\longrightarrow$ $HC—O—C(=O)$ ... $+ 3H_2O$
$H_2C—OH + HOC(=O)$ ... $\longrightarrow$ $H_2C—O—C(=O)$ ...

glycerol *three fatty acids (12–24 carbon atoms each; saturated or unsaturated)* *a triglyceride (saturated carbon chains—a fat; unsaturated carbon chains—an oil)*

Figure 33.9
Formation of a Triglyceride
Glycerol is esterified with three molecules of fatty acid. The fatty acid chains found in naturally occurring triglycerides all have an even number of carbon atoms because they are synthesized in two-carbon segments. The most common chains are 16 or 18 carbon atoms in length. When double bonds are present, the configuration of the chain with respect to the double bond is always *cis*.

critically important to achieve the highest possible energy-to-weight ratio. Oxidation of a fat yields more energy than oxidation of the same weight of a starch, which already contains many C—O and H—O bonds.

Phospholipids—molecules of enormous biological importance—represent a variation on the triglyceride structure. In a phospholipid, two of the hydroxyl groups of glycerol are esterified with fatty acids, just as in a fat or oil. The third OH group, however, forms an ester link with phosphoric acid (H_3PO_4) or a derivative of phosphoric acid—usually one containing an amine group—for example

$$(CH_3)_3\overset{+}{N}-CH_2-CH_2-O-\underset{O^-}{\overset{O}{\overset{\|}{P}}}-O-CH_2-CH(-O-C(=O)\sim\sim\sim\cdots)-CH_2-O-C(=O)\sim\sim\sim\cdots \quad (33.5)$$

hydrophilic (charged) head

hydrophobic (nonpolar) tail

a phospholipid

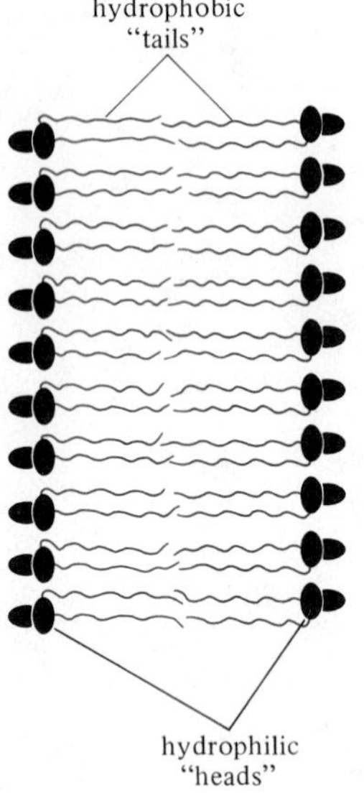

Figure 33.10
Phospholipids and the Structure of a Biological Membrane
The basis of membrane structure is a phospholipid *bilayer*. The hydrophobic tails of the phospholipid molecules form the interior of the membrane; the hydrophilic heads face outward toward the aqueous environment on either side of the membrane. The hydrophobic character of the membrane prevents the passage of most hydrophilic species such as ions or large polar molecules.

The two fatty acid chains make up a long, nonpolar "tail." The phosphate, however, carries a negative charge, while the amine group is usually in its cationic form ($—NH_3^+$ or $—NR_3^+$). Thus the molecule also has a compact, strongly dipolar "head." The overall structure of such a phospholipid is in fact very similar to that of a soap or detergent molecule (Section 15.17). Phospholipids, with their chemical "split personalities," are key constituents of biological membranes such as those that surround living cells. These membranes are thought to consist principally of a double layer of phospholipid molecules (Figure 33.10). The hydrophilic heads of the phospholipids can interact with substances in the aqueous interior of the cell and in the aqueous environment outside. These two aqueous regions, however, are separated by the hydrophobic tails of the phospholipids, which make up the interior of the membrane structure and do not easily allow the passage of ions or most polar substances.

Waxes are also esters, but their structures are less complex than those of the glycerides and phospholipids. Wax molecules are formed by condensation reactions between long-chain monohydroxy alcohols and long-chain fatty acids to give molecules 30–60 carbon atoms in length. The highly hydrophobic character of the hydrocarbon chains makes waxes useful as waterproofing materials for skin, fur, or feathers. A wax coating on the leaves and fruits of many plants helps slow the loss of water by evaporation in hot, dry conditions.

Structurally, steroids are very different from the lipids we have described so far, all of which consist in large part of hydrocarbon chains. Steroids are nevertheless classified as lipids because of their hydrophobic character. All steroids have structures based on four fused rings (Table 33.5). Very small variations in the bonding of atoms

Table 33.5
Some Steroids
Testosterone and cortisone are both hormones. Despite their similarity in structure, they have totally different biological effects. Cholesterol, like the saturated triglycerides, has been implicated in heart and circulatory disease.

in the rings and in the groups attached to them give rise to compounds that are remarkably diverse in their biological functions. Many steroids are hormones—chemical messengers that regulate various physiological processes. But notice in Table 33.5 how similar in structure are the male hormone testosterone and the adrenal hormone cortisone—compounds with drastically different biological activities. Another steroid, now rather notorious, is cholesterol, which, like saturated fats, has been implicated in heart and circulatory disease.

33.9 PROTEINS

Proteins are in a very real sense the master molecules of living things. Virtually nothing happens in living organisms without their participation. The human body probably contains at least 10,000 different kinds of protein, and the number could well be larger. Some of the diverse functions of proteins are listed in Table 33.6.

Table 33.6
Some Functions of Proteins

Enzymes	Catalyze biochemical reactions
Structural proteins	Virus coat proteins, cell wall proteins, insect silks, vertebrate protective tissues (skin, hair, feathers, scales, nails, hooves, horns, beaks, etc.), vertebrate connective tissues (bone, cartilage, tendons, ligaments, etc.)
Contractile proteins	Present in muscle fibers, cilia, flagella
Membrane proteins	Present in all cellular and intracellular membranes
Transport proteins	Bind and transport other molecules in bloodstream (e.g., hemoglobin)
Storage proteins	Release amino acids when needed, e.g., casein (milk protein), ovalbumin (egg white)
Protective proteins	Antibodies, blood-clotting agents
Hormones	Regulate growth and metabolism

a. Amino Acids All **proteins** are polymers of α-amino acids. As the name implies, an **amino acid** contains both an amine group and a carboxylic acid group. (The designation α simply means that the amine group is attached to the carbon next to the carboxylic acid.) You can most easily think of an amino acid as consisting of a central carbon atom—the *α carbon*—to which are attached four different groups: the amine group (NH_2), the carboxylic acid group (COOH), a hydrogen atom, and a fourth variable group, commonly represented by R:

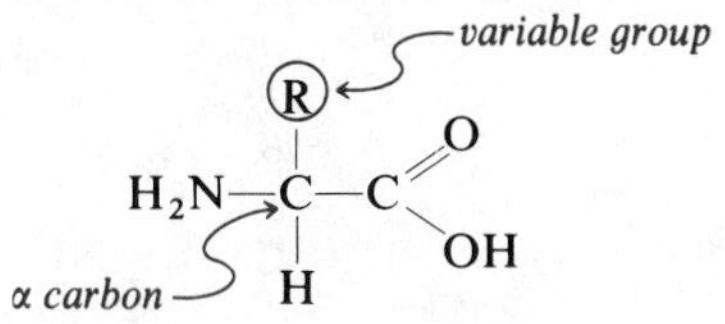

$$H_2N\text{—}\overset{\overset{\text{Ⓡ}}{|}}{\underset{\underset{H}{|}}{C}}\text{—}C\begin{smallmatrix}\diagup\!\!\!\!=O\\ \diagdown OH\end{smallmatrix}$$

variable group (Ⓡ); *α carbon* (central C)

Since the COOH group easily loses a proton and the NH_2 group easily gains one, an amino acid in solution can exist in several different ionic forms, the equilibrium concentrations varying with the pH:

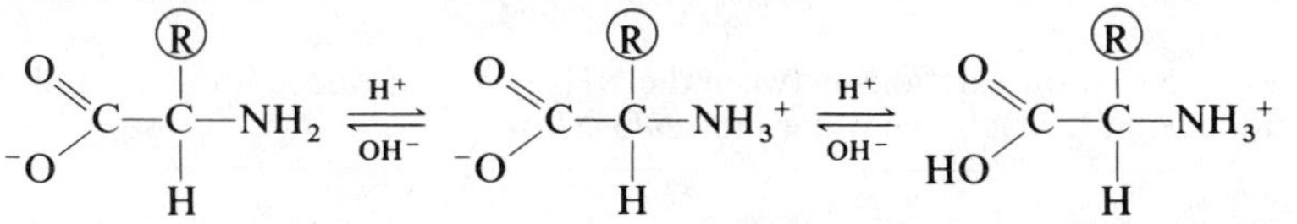

$$^{-}O(O{=})C\text{—}\overset{\text{Ⓡ}}{\underset{H}{C}}\text{—}NH_2 \underset{OH^-}{\overset{H^+}{\rightleftharpoons}} {^{-}O}(O{=})C\text{—}\overset{\text{Ⓡ}}{\underset{H}{C}}\text{—}NH_3^+ \underset{OH^-}{\overset{H^+}{\rightleftharpoons}} HO(O{=})C\text{—}\overset{\text{Ⓡ}}{\underset{H}{C}}\text{—}NH_3^+$$

In near-neutral solutions (as well as in the crystalline state) the double-ion form (in the center above) is the one generally found. The α carbon atom is asymmetric, and all amino acids (except for glycine; Table 33.7) have optically active isomers that, like those of sugars, are designated either D or L forms. Twenty different amino acids, each with a different R group, are commonly found in the proteins of living things. Since the rest of the molecule is the same in all amino acids, it is the R group that gives each amino acid its special properties. As illustrated in Table 33.7, some R groups are hydrophobic and others are hydrophilic, either because they contain polar bonds, or because they are acidic or basic and under physiological conditions exist in ionized form.

Table 33.7
Some Representative Amino Acids
Amino acids are the building blocks of all proteins. The R groups impart varying properties to the molecules. Proteins from different sources differ in their amino acid content. Some amino acids can be synthesized by the human body; others, the *essential amino acids,* cannot be made in the human body and must be obtained from the diet.

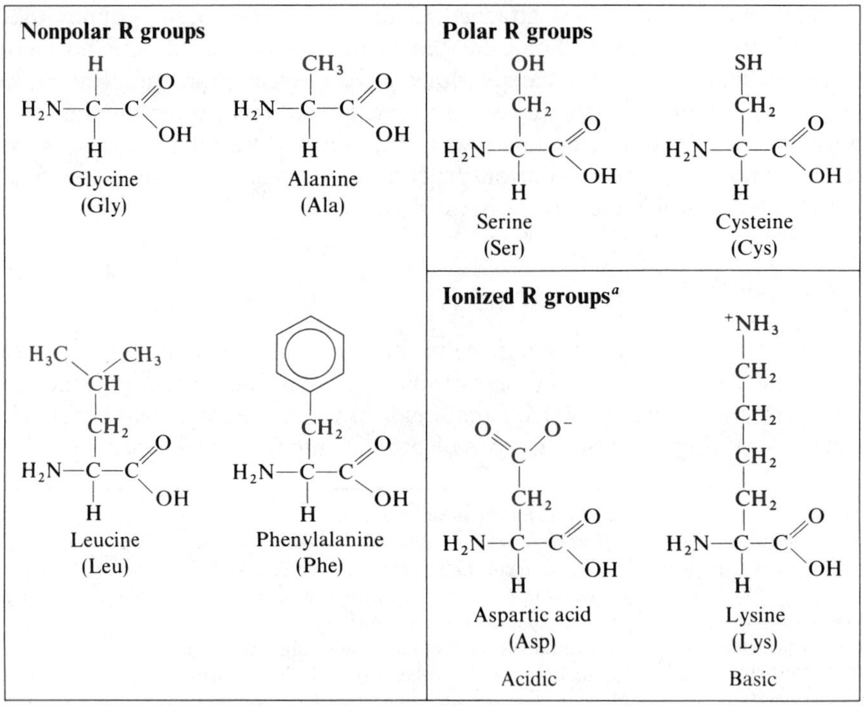

Nonpolar R groups		Polar R groups	
$H_2N-CH(H)-C(=O)OH$ Glycine (Gly)	$H_2N-CH(CH_3)-C(=O)OH$ Alanine (Ala)	$H_2N-CH(CH_2OH)-C(=O)OH$ Serine (Ser)	$H_2N-CH(CH_2SH)-C(=O)OH$ Cysteine (Cys)
		Ionized R groups[a]	
$H_2N-CH(CH_2CH(CH_3)_2)-C(=O)OH$ Leucine (Leu)	$H_2N-CH(CH_2C_6H_5)-C(=O)OH$ Phenylalanine (Phe)	$H_2N-CH(CH_2COO^-)-C(=O)OH$ Aspartic acid (Asp) Acidic	$H_2N-CH(CH_2CH_2CH_2CH_2{}^+NH_3)-C(=O)OH$ Lysine (Lys) Basic

[a] These amino acids are present in the ionized forms shown when they are in solution in water at the pH of body fluids (approx. pH 7.3).

Proteins consist of long, unbranched polymer chains made up of amino acid monomers. The link between adjacent monomers is formed by a condensation reaction between the amine group of one amino acid and the carboxylic acid group of another. As Figure 33.11 shows, a water molecule is removed, leaving the two amino acids joined by a **peptide bond.** A single chain of amino acids linked by peptide bonds is called a **polypeptide.** Some polypeptides contain several thousand amino acid molecules. Every protein molecule consists of one or more polypeptide chains.

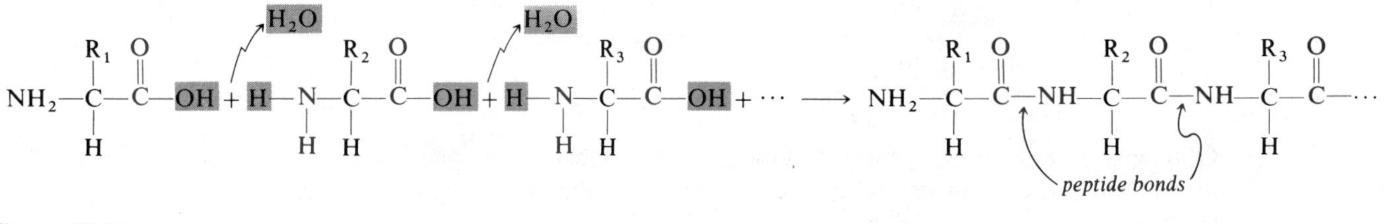

$$NH_2-CH(R_1)-C(=O)-OH + H-NH-CH(R_2)-C(=O)-OH + H-NH-CH(R_3)-C(=O)-OH + \cdots \xrightarrow{-H_2O,\ -H_2O} NH_2-CH(R_1)-C(=O)-NH-CH(R_2)-C(=O)-NH-CH(R_3)-C(=O)-\cdots$$

peptide bonds

Figure 33.11
Formation of a Polypeptide
Amino acids are linked by peptide bonds formed by condensation reactions between the NH_2 group of one amino acid and the COOH group of the next amino acid. For each peptide bond formed, one water molecule is lost.

b. Protein Structure To describe the structure of a functioning protein in an organism, it is necessary to specify the three-dimensional shape that the polypeptide chain assumes (Figure 33.12). Functioning proteins can be divided on the basis of their structure into two broad classes: fibrous proteins and globular proteins.

In fibrous proteins, the polypeptide chains are relatively extended and arranged so that they intertwine with or lie parallel to neighboring chains, like the strands of a rope or threads in a piece of fabric. Forces of various kinds hold the chains together, creating great strength. (Some fibrous proteins are as strong, for their size, as steel cable.) Most fibrous proteins have many hydrophobic R groups exposed on their surfaces, making them insoluble in water.

In globular proteins, by contrast, the polypeptide chain is wound and folded into a compact but irregular shape, like a tangled ball of yarn. Most globular proteins function in the aqueous interiors of cells or in the bloodstream, and so must be water soluble. Generally the molecules fold up so as to expose many hydrophilic R groups on the surface of the molecule.

The sequence of the amino acids combined in a peptide chain is referred to as the primary (1°) structure. This sequence also determines the three-dimensional shapes of proteins, for the chains are held in place by interactions between groups along their lengths. With one exception (the disulfide bridge), these interactions are limited to electrostatic attraction between charged groups, hydrogen bonding, and the other intermolecular forces.

You can see in Table 33.7 that the amino acid cysteine contains an R group that ends in an —SH group. When two cysteine molecules are near each other, they can react to form the double amino acid known as cystine, creating a covalent cross-link, a disulfide bridge with the structure —C—S—S—C.

The secondary (2°) structure of a protein is a regular coiling or zigzagging of polypeptide chains caused by hydrogen bonding between the NH and C=O groups of amino acids near each other in the chains. Fibrous proteins are characterized by their secondary structure. One very common secondary structure is the α helix in which the helical coiling of a chain is maintained by intrachain hydrogen bonds. Among the α-helical proteins are the keratins, which are major structural components of hair, animal skin, fur, wool, nails, hooves, and horns. When the disulfide bridges are few in number, as in the keratins of hair, the helical fibers are elastic and highly flexible (like a rubber band). A larger number of cross-links makes for a tougher, less flexible form of keratin such as is found in the horns of animals (more like a thermoset polymer).

To give hair a "permanent wave," the disulfide cross-links are broken chemically by treatment with a reducing agent and the hair is coiled on rollers to give it the desired degree of curl. New cross-links are then formed by treating the hair with an oxidizing solution; the cross-links lock in the fibers into their new shape.

If the secondary structure of a protein resembles the coils in a curly telephone cord, the tertiary (3°) structure resembles the tangling of that cord into a ball. The proteins that take part in the dynamic chemistry of living things—the enzymes, carriers, hormones, and so on—are nearly all globular proteins, and every globular protein is unique. The three-dimensional twisting and folding of the polypeptide chain in a globular protein, the tertiary (3°) structure, is created by interactions among the R groups of the amino acids. These interacting groups may be quite far from each other and the interactions may be any of the types described above. Disulfide bridges, because they are stronger than the other intramolecular interactions, are very important in determining tertiary structure.

Many globular protein molecules, especially the larger ones, consist of several polypeptide chains. The ways in which different chains are arranged and joined to one another constitute the quaternary (4°) structure of the protein. Like tertiary structure, quaternary structure usually is created by the weaker interactions, but here too disulfide bridges may be involved.

The 2°, 3°, and 4° structures of a globular protein can be disrupted easily by heat, extremes of pH, or strongly oxidizing or reducing conditions (Figure 33.13). Under such conditions the protein undergoes **denaturation**—it exhibits substantial changes in its physical properties, and its biological activity, which depends to a great extent on the maintenance of a precise spatial configuration. The most familiar example of denaturation is the change that takes place in albumin—the principal component of egg white—when you cook an egg. In this particular case the change is irreversible, but interestingly, this is not always the case. Often when the disruptive conditions are removed, a denatured polypeptide will spontaneously fold up and resume its original 3° and 4° structures. This suggests two things: that the original configuration is the most stable (i.e., lowest energy) state for that particular polypeptide, and that the polypeptide chain can "find" this configuration solely on the

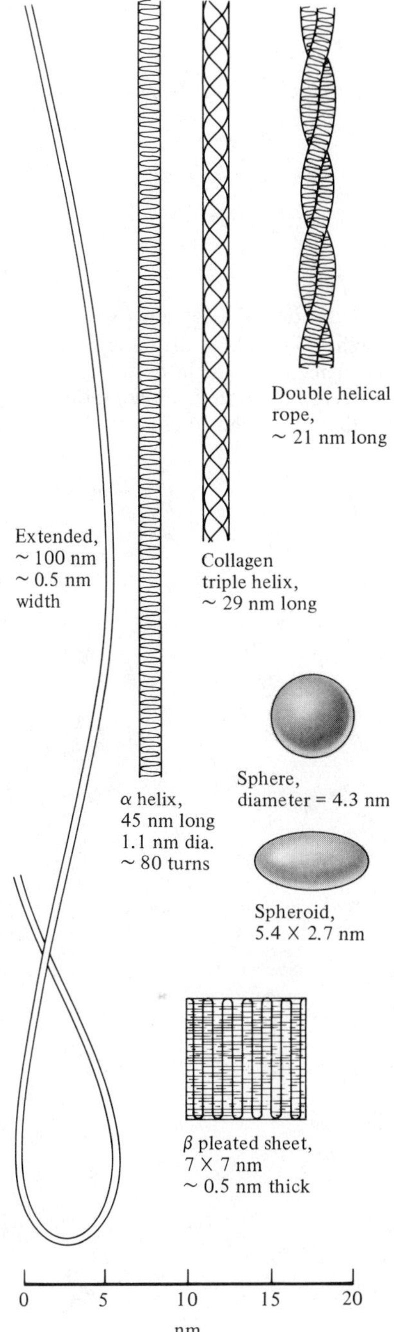

Figure 33.12
Some Shapes a Protein Molecule of 300 Amino Acids Might Assume
Source: D. E. Metzler, *Biochemistry: The Chemical Reactions of Living Cells* (Academic Press, New York, 1977), p. 76.

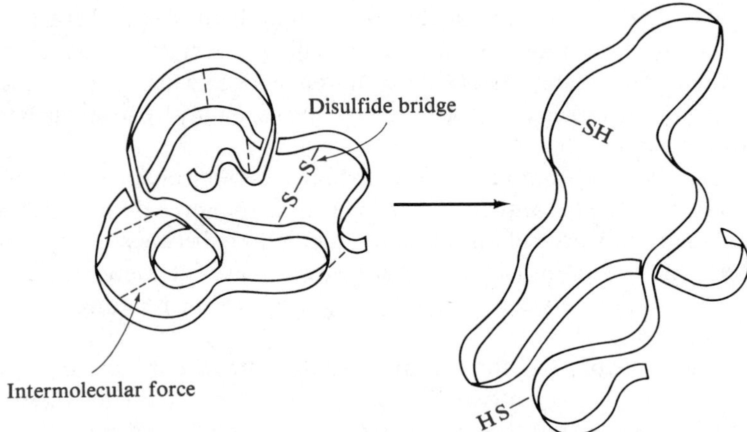

Figure 33.13
Denaturation of Globular Protein
The S—S bonds of disulfide bridges are broken and the intermolecular forces are disrupted, allowing the protein molecules to change shape.

basis of its 1° structure. Indeed, it is now thought that the instructions for the manufacture of proteins in living cells specify *only* the 1° structure of each protein—the sequence of amino acids in the polypeptide chain (Section 33.10).

c. Enzymes By far the most important of the globular proteins are the **enzymes,** or biological catalysts. Without the assistance of enzymes, there would *be* no chemistry of living things. Even the simplest cell must perform hundreds of different chemical reactions *merely to stay alive.* To appreciate what this means, suppose for a moment that you are a chemical engineer, and that you are presented with the following assignment. You must devise a way of carrying out several hundred different reactions, simultaneously. Many of them are thermodynamically unfavorable and ordinarily would not take place to any significant extent. Most of the others, though spontaneous, are too slow to be useful. Nevertheless, you must get *all* of them to occur at high rates and with high yields. You cannot use powerful oxidizing or reducing agents, or strong acids or bases, nor can you vary the temperature or pressure very much; all the reactions must be carried out at atmospheric pressure, near room temperature, and within a relatively narrow range of pH. Moreover, the number of different vessels that you can use is quite limited, so many of the reactions must take place not only at the same time but in the same solution. Despite this, they must proceed without any of the side reactions that are usually so difficult to avoid in even the simplest chemical system.

No one would blame you if, faced with such a task, you quickly switched to another field. Yet cells can do all of the things we have just described—and continuously *adjust* the rates of most of the individual reactions into the bargain! This remarkable performance is made possible by enzymes.

Typically, enzymes are large molecules, with molecular masses ranging into the millions; often they consist of several polypeptide chains. Somewhere in its elaborately folded structure, each enzyme molecule possesses a region known as the **active site.** The active site may take the form of a groove along the surface of the molecule, or a deeper crevice or channel into the interior; its shape and location are unique in each enzyme. Physically, the **substrate**—the molecule or molecules with which an enzyme interacts—fits into the active site like a hand in a glove (Figure 33.14). Even more important than the physical fit, however, is the chemical fit. The substrate molecule is held in position by interactions of various types (e.g., hydrogen bonding, electrostatic attraction) between its atoms and groups that line the active site in the enzyme.

The amino acids that make up the active site rarely lie near each other in the polypeptide chain. Indeed, they may be in different polypeptide chains. It is the folding of the protein molecule that brings these amino acids together to form the active site. This is why precise 3° and 4° structure is so essential to the biological activity of an enzyme molecule. This precision of fit, chemical as well as physical, between the substrate molecule and the active site helps to account for the remarkable specificity of

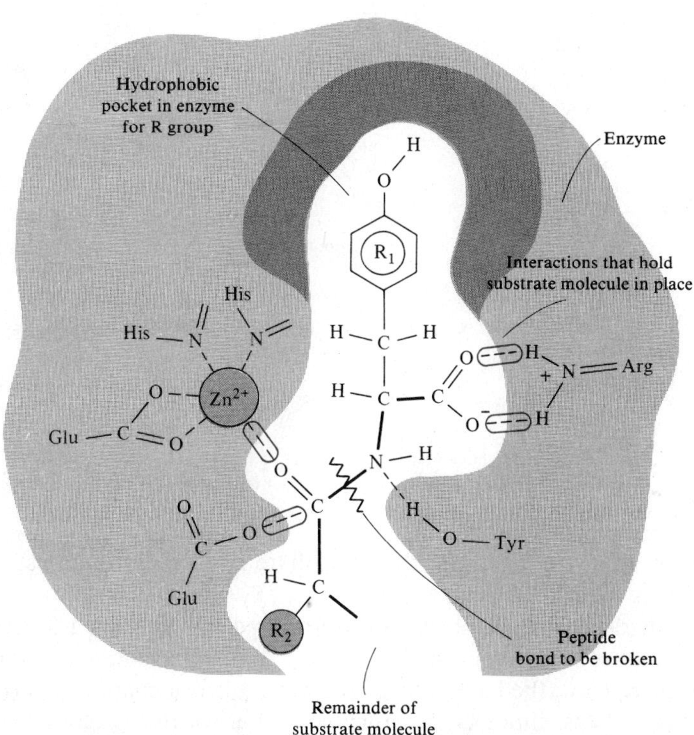

Figure 33.14
The Active Site of Carboxypeptidase A
This enzyme splits off amino acids, one at a time, from the ends of polypeptide chains. It works best when the amino acid at the end of the chain has a large hydrophobic **R** group that fits into the pocket at the top of the figure. This enzyme incorporates a Zn^{2+} ion, which helps to bind the substrate and weakens the peptide bond that is to be broken by attracting electrons from the adjacent C=O bond. Other amino acids from different parts of the enzyme molecule help to position the substrate and attack atoms on either side of the peptide bond to be broken.

many enzymes. Several factors probably are involved in catalysis by enzymes. Clearly, by binding two substrate molecules (as many enzymes do), the enzyme serves as molecular "matchmaker." That is, it brings the reactants together and ensures that the groups involved in the reaction are so oriented that the encounter will be effective (Section 18.2). Since the concentrations of reactants in living cells is often quite low, this is an important factor. Even more important, however, is the role of enzymes in weakening bonds in the substrate or substrates so as to make the reaction easier. An enzyme often undergoes a slight change in configuration upon binding the substrate. Such a change may physically strain the substrate molecule. Furthermore, binding of the substrate to the enzyme may withdraw electron density from the bond or bonds that are to be broken. An enzyme may also take a direct (though temporary) part in the reaction mechanism itself. For example, the enzyme may serve as an interim donor of an ion or group to one of the reactants (e.g., an H^+ ion in many common reactions), subsequently recovering its "loan" from another reactant. The efficiency of most enzymes is quite remarkable. Many enzyme molecules can handle thousands of substrate molecules per second, increasing reaction rates by factors of a million or more.

33.10 NUCLEIC ACIDS

Nucleic acids are the molecules of which genes are made. Their function is the preservation of the genetic information—the complete set of "blueprints" for all actual and potential characteristics received by each organism from its parent or parents and passed on from generation to generation. Nucleic acids also make up most of the molecular machinery by which these genetic blueprints are "read" and the information they contain is translated into actual traits such as red hair, perfect pitch, or color-blindness.

Nucleic acids are polymers composed of complex repeating units called nucleotides. Each **nucleotide** has three parts: (1) a five-carbon sugar—either ribose or deoxyribose, (2) one of five different nitrogen-containing organic bases (Table 33.8), (3) a phsophate group, —O—PO_3^{2-}. The base replaces the OH group normally attached

Table 33.8
The Bases of DNA and RNA
The double-ring structures are known as purines; the single-ring structures are known as pyrimidines. The nitrogen atom bonded to a sugar in DNA or RNA is shown in color (the usual abbreviations for each base given in parentheses).

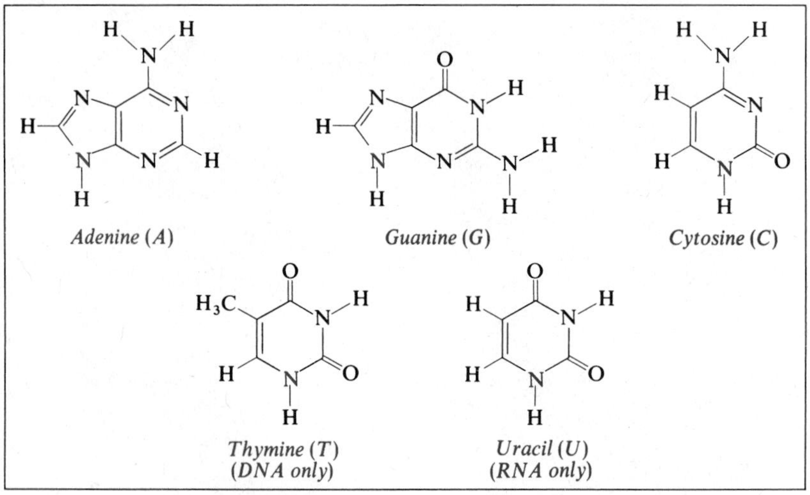

Adenine (A) *Guanine (G)* *Cytosine (C)*

Thymine (T) *(DNA only)* *Uracil (U)* *(RNA only)*

to the 1-carbon of the sugar ring; the phosphate is bonded by an ester linkage to the 5-carbon (Figure 33.15).

A nucleic acid consists of several thousand nucleotides joined together by ester linkages between the 3-carbon atom of the sugar of one nucleotide and the phosphate group of another. Thus, the backbone of the molecule is a chain of alternating sugar and phosphate groups. Since each phosphate group forms ester links to the preceding and the following sugar molecules, successive sugars are said to be joined by *phosphodiester bonds.*

There are two major types of nucleic acid: **deoxyribonucleic acid (DNA)** which carries the genetic information, and **ribonucleic acid (RNA)** which is involved in putting this information to work in the cell. They differ in three ways:

1. The sugar in RNA is ribose, while the sugar in DNA is 2-deoxyribose, in which the OH group on the 2-carbon atom in ribose has been replaced by a hydrogen atom.
2. Four different bases (see Table 33.8) are found in DNA: *cytosine* (abbreviated C) *thymine* (T), *adenine* (A), and *guanine* (G). In RNA, thymine does not occur; its place is taken by *uracil* (U).
3. DNA is nearly always double stranded (as described below), while RNA is usually single stranded.

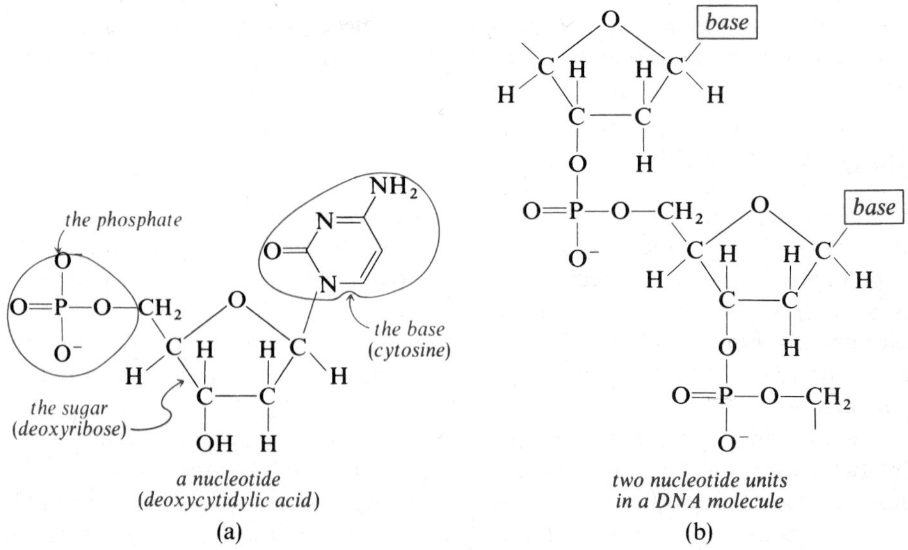

the phosphate
the sugar (deoxyribose)
the base (cytosine)
a nucleotide (deoxycytidylic acid)
(a)

two nucleotide units in a DNA molecule
(b)

Figure 33.15
Components of a Nucleic Acid
(a) A nucleotide, consisting of a sugar (ribose or deoxyribose), a phosphate group, and a nitrogen-containing base. (b) A portion of a DNA molecule, which consists of nucleotides linked by ester bonds between the sugar of one and the phosphate of the next.

You may be wondering exactly how a molecule such as DNA—indeed, any one molecule, no matter how complex—can determine specific traits of an organism. The answer is that traits are actually determined by what proteins each cell makes, and when and in what quantities it makes them. Proteins make up much of the structure of each living cell; more importantly, all enzymes are proteins, and enzymes control virtually every aspect of cell chemistry, including what *other* substances the cell can make. Nucleic acids, in turn, contain the coded information specifying the makeup of every one of the cell's thousands of different proteins. This information—*the genetic information—is incorporated in the sequence of bases of the cell's DNA.*

We already have mentioned that the 1° structure of a polypeptide seems sufficient to determine the 2°, 3°, and/or 4° structure that the protein molecule will ultimately assume. Thus the sequence of DNA bases must contain information that can be translated into a sequence of amino acids in each polypeptide chain. Since there are 20 different amino acids and only four DNA bases, it is obvious that a one-to-one correspondence between bases and amino acids is not possible. Instead, each amino acid is represented by a sequence of three consecutive bases, called a *codon*. There are 64 possible codons ($4 \times 4 \times 4$)—clearly more than enough to specify the 20 different amino acids. In fact, most amino acids can be "spelled" by any of several different codons; in addition, some codons serve as "punctuation," indicating the start and the termination of polypeptide chains.

The key to the ability of DNA to preserve genetic information and to pass it on from generation to generation is its double-stranded structure, first deduced by James Watson and Francis Crick in 1953. This was the discovery that inaugurated the field of molecular biology, which has led to today's headlines about cloning and genetic engineering. The DNA molecule has often been likened to a "spiral staircase"—an apt comparison, though strictly speaking the shape should be described as helical rather than spiral. Each strand is coiled into a helix and the two strands wind about a common axis, forming the now-famous "double helix" structure (Figure 33.16). The sides of the staircase are the sugar-phosphate backbones, which lie on the outside of the molecule. (The presence of the phosphate groups on the outside is in fact what makes the nucleic acids acidic.) The bases lie between the two backbones, in the center of the structure—they form the "steps" of the staircase. The bases, which are flat, lie almost perpendicular to the axis of the double helix. Each base is joined to a base opposite it on the other strand by hydrogen bonds. Although individually these bonds are weak, their number gives the DNA molecule great stability. The sizes of the bases, and the location of the groups that can participate in hydrogen bonding, are such that adenine can pair only with thymine and guanine can pair only with a cytosine in the opposite strands (Figure 33.16). What this means is that the base sequences of the two strands are not independent. Rather, the two strands are *complementary*—once the sequence of one strand is given, that on the other strand is completely determined. The two strands carry the same information in the sense that a photographic print and the negative from which it was made carry the same information—either one could be used to reconstruct the other.

As Watson and Crick saw, this double-stranded structure provides a mechanism whereby the genetic information can be duplicated—something that must happen every time a cell divides. With the help of a number of highly specialized enzymes, the two strands are uncoiled from one another, and a new, complementary "partner" strand is synthesized for each. The bases for the new strands are assembled in the proper sequence in accordance with the base-pairing rules described above. Wherever thymine appears in an old strand, adenine is inserted opposite it in the new complementary strand, and so on. The eventual result is two new DNA molecules, each consisting of one new and one old strand. This process is called *replication.*

The synthesis of a polypeptide in accordance with the information contained in DNA is an intricate series of events, which we can sketch only in the most general way. First, in the process known as *transcription,* a single-stranded RNA molecule—a

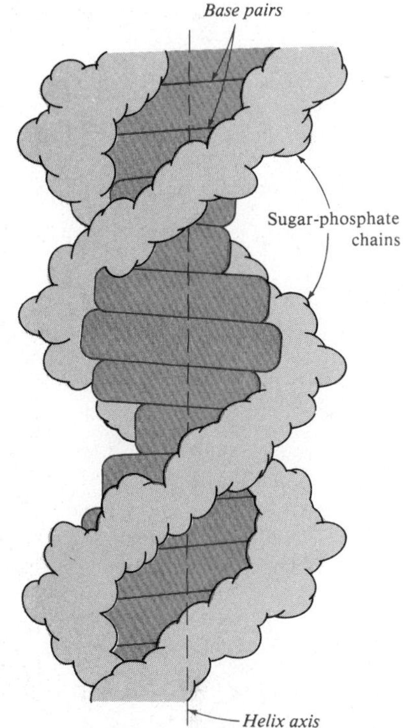

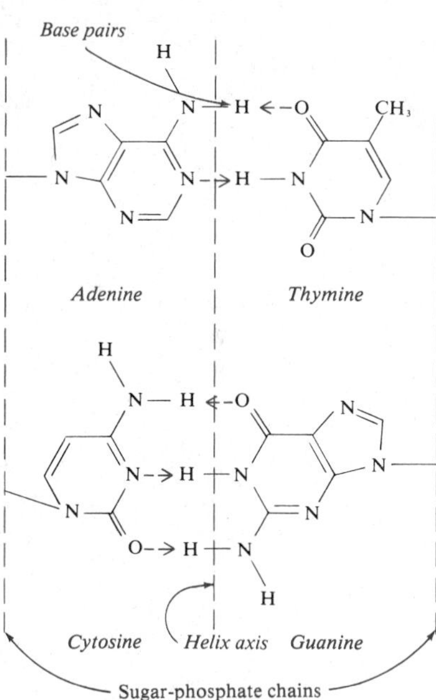

Figure 33.16
The DNA Double Helix and Base Pairing
The sugar-phosphate chains wind around the outside of the molecule: the bases are stacked nearly perpendicular to the helix axis in the center. Because of the size of the bases and the location of their hydrogen-bonding groups, adenine can pair only with thymine and cytosine can pair only with guanine.

"transcript"—is made from a DNA gene—again, in accordance with the base-pairing rules. (In the RNA molecule, U takes the place of T, pairing with A.) This RNA molecule, known as messenger RNA (mRNA), becomes attached to structures called ribosomes, which are the cell's "protein factories." The key role in protein synthesis is played by another kind of RNA molecule, transfer RNA (tRNA) which serves as a link between the codons of the mRNA and the amino acids for which they code. Each different type of tRNA molecule binds a single type of codon at one end and the amino acid molecule specified by that codon at the other end. As the tRNA molecules "read" successive mRNA codons, the appropriate amino acids are bound by enzymes into the polypeptide chain growing at the ribose surface. The "empty" tRNA molecules are then released to bind new amino acids. When a "stop" codon is reached, the polypeptide chain (which in all likelihood already has begun to fold up into its final configuration) is detached from the ribosome. The process of protein synthesis in this manner is called *translation.*

33.11 ATP, ENERGY, AND METABOLISM

The chemical processes of a living organism, collectively called **metabolism,** can be divided into two parts. Organisms are constantly breaking down large molecules (such as those in foodstuffs) into smaller, simpler ones—a process known as **catabolism.** Catabolic reactions are usually accompanied by the release of free energy. At the same time, organisms are continuously synthesizing a great variety of large molecules from simpler molecules—a process known as **anabolism.** Anabolic reactions generally cannot take place without an *input* of free energy. For simplicity, we shall refer to energy-requiring reactions as "uphill" and energy-yielding reactions as "downhill."

The overall strategy that is forced upon any organism by the laws of thermodynamics is obvious: it must somehow use the energy released by its downhill reactions to drive its uphill reactions. The question is, how? A complete answer (which would include much of what is known of biochemistry) would have to take into account a host of variations in different organisms, different types of cells, and different metabolic pathways. The basic pattern, however, is remarkably constant in all known organisms:

Whenever energy is available from the breakdown of a large molecule, it is immediately invested in making an energy-rich phosphate compound. Most commonly, it is used to add a third phosphate group to adenosine diphosphate (ADP) to make adenosine triphosphate (ATP).

Whenever a cell needs to do work—for example, to drive an uphill reaction—it obtains the necessary energy by splitting off a phosphate group from an energy-rich phosphate compound—most commonly, by removing one phosphate group from ATP to form ADP.

Adenosine diphosphate (ADP) is simply a nucleotide (Section 33.22) consisting of the sugar ribose, the base adenine, and two phosphate groups; **adenosine triphosphate (ATP)** has the same structure, but with a third phosphate group (Figure 33.17). For a number of reasons, chiefly having to do with its greater concentration of negative charge and fewer available resonance forms. ATP has a considerably higher energy content than ADP plus a free phosphate group. Stated another way, we can say that ATP has a high *free energy of hydrolysis:*

$$\text{ATP} + H_2O \longrightarrow \text{ADP} + P_i + H^+ \qquad \Delta G^\circ = -34.5 \text{ kJ (at pH 7)} \qquad \textbf{(33.6)}$$

(The symbol P_i is used to stand for "inorganic phosphate," HPO_4^{2-}.) Splitting off a phosphate group from ATP is therefore a way of releasing free energy, while adding a third phosphate group to ADP requires energy. The phosphate groups in ATP often are termed "high-energy phosphate groups," and the bonds joining these groups to

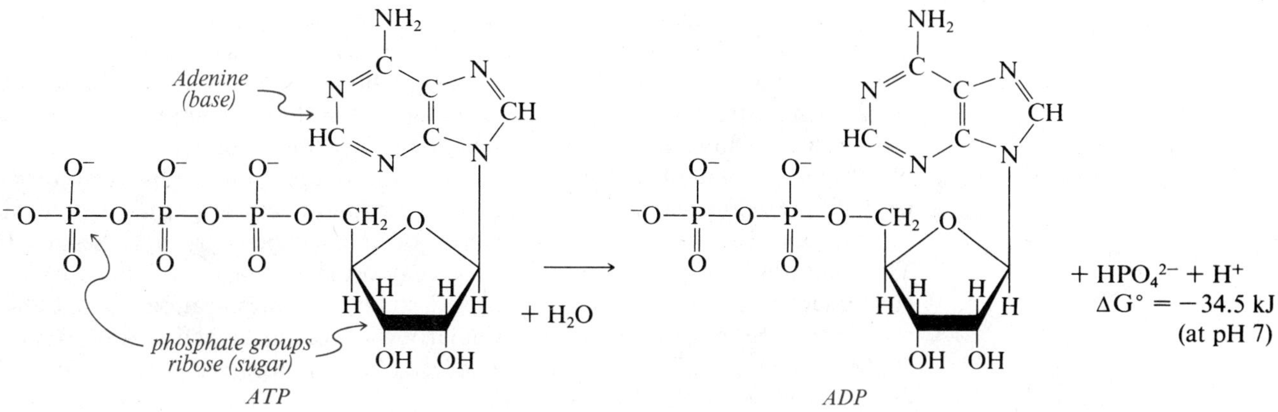

Figure 33.17
Hydrolysis of ATP
When ATP loses its third phosphate group (forming ADP), a large amount of energy is released. This reaction is used to power much of the work done by living things, including the synthesis of large biomolecules.

the rest of the molecule are referred to as "high-energy phosphate bonds." However, there is nothing unique about either the groups or the bonds; this is merely the biochemist's way of indicating the large free energy changes involved in the removal of these phosphate groups by hydrolysis.

ATP is often called the "energy currency of the living cell." When a cell must carry out an uphill reaction, it can draw on its reserves of ATP—a stockpile of energy derived from quite unrelated reactions that may have taken place at various times in different parts of the cell. For example, most cells derive energy from the process known as respiration—the oxidation of energy-rich molecules (typically glucose) to CO_2 and H_2O:

$$C_6H_{12}O_6 + 6O_2 \longrightarrow 6CO_2 + 6H_2O \qquad \Delta G° = -2872 \text{ kJ} \qquad \textbf{(33.7)}$$

In cells, this overall reaction is carried out in a complex series of carefully controlled steps. The energy released in several of the most steeply downhill steps is used to make ATP from ADP and phosphate. The net yield is some 38 moles of ATP per mole of glucose oxidized. These ATP molecules can then be "spent" at any time and in any way that the needs of the cell dictate.

How does a cell use ATP to "pay" for uphill reactions? Here again, enzymes are the key. An enzyme can *couple* two different reactions, catalyzing both simultaneously so that they become in effect steps of a single overall reaction. In living cells, enzymes often combine an uphill reaction with the hydrolysis of ATP. The net energy change of the combined reaction is the algebraic sum ofthe energy changes of the two steps. Thus if the hydrolysis of ATP provides more energy than the uphill reaction requires, the overall reaction will be thermodynamically favorable and will run downhill.

In a typical mechanism, an enzyme catalyzes the transfer of a phosphate group from ATP to a reactant molecule. This "activates" the reactant by increasing its energy. Then, in the subsequent course of the reaction, the phosphate group is split off, and the energy released in this step derives the desired reaction. Schematically, if X is the reactant and Y is the product

$$\begin{array}{lll} \text{ATP} + \text{X} & \longrightarrow & \text{ADP} + \text{X—Ⓟ} \quad \textit{activated reactant} \\ \text{X—Ⓟ} & \longrightarrow & \text{Y} + \text{Ⓟ} \\ \hline \text{ATP} + \text{X} & \longrightarrow & \text{ADP} + \text{Ⓟ} + \text{Y} \end{array}$$

The net result is clearly equivalent to the reaction X ⟶ Y accompanied by the hydrolysis of ATP to ADP.

Alternatively, enzymes can link reactions in sequences, called *metabolic pathways*, in which the products of one step are the reactants in the next. We saw in Section 23.7 that the overall free energy change of any reaction is independent of the number of steps. As in the previous case, it is the algebraic sum of the free energy changes of all the steps. Thus a downhill reaction (such as the hydrolysis of an energy-rich phosphorylated compound) *anywhere* in such a pathway can "push" subsequent reactions or "pull" preceding reactions, as long as it releases more free energy than they consume.

We can easily understand this in terms of Le Chatelier's principle (Section 17.2). The equilibrium of the reaction $B \rightleftharpoons C$ may ordinarily favor B. Nevertheless, if a subsequent reaction continously removes C, so that its concentration remains low, the reaction will be "pulled" to the right. Similarly, if a preceding reaction keeps producing B, the equilibrium will be constantly displaced by the appearance of more B, and the reaction will be "pushed" to the right.

At this point you may be wondering about the ultimate source of the energy-rich molecules that organisms use to drive their anabolic reactions and do the other work of staying alive. Cells can synthesize such molecules, of course—most cells can actually *make* glucose from simpler precursors—but only at the expense of breaking down other energy-rich substances. Such a cycle is the chemical equivalent of a perpetual motion machine. The second law of thermodynamics assures us that it is impossible to break even in this way. In fact, animals get their energy-rich molecules from eating plants (or eating other animals that have eaten plants). Plants *are* able to make energy-rich substances at a net profit, not because they can circumvent the laws of thermodynamics (for nothing can), but because they have a free, external energy source—the sun. In the process of photosynthesis, solar energy is captured in chemical form as the plant uses sunlight to derive an uphill reaction: the synthesis of sugar from CO_2 and H_2O. So virtually all the life energy on Earth ultimately comes from the sun (Figure 33.18).

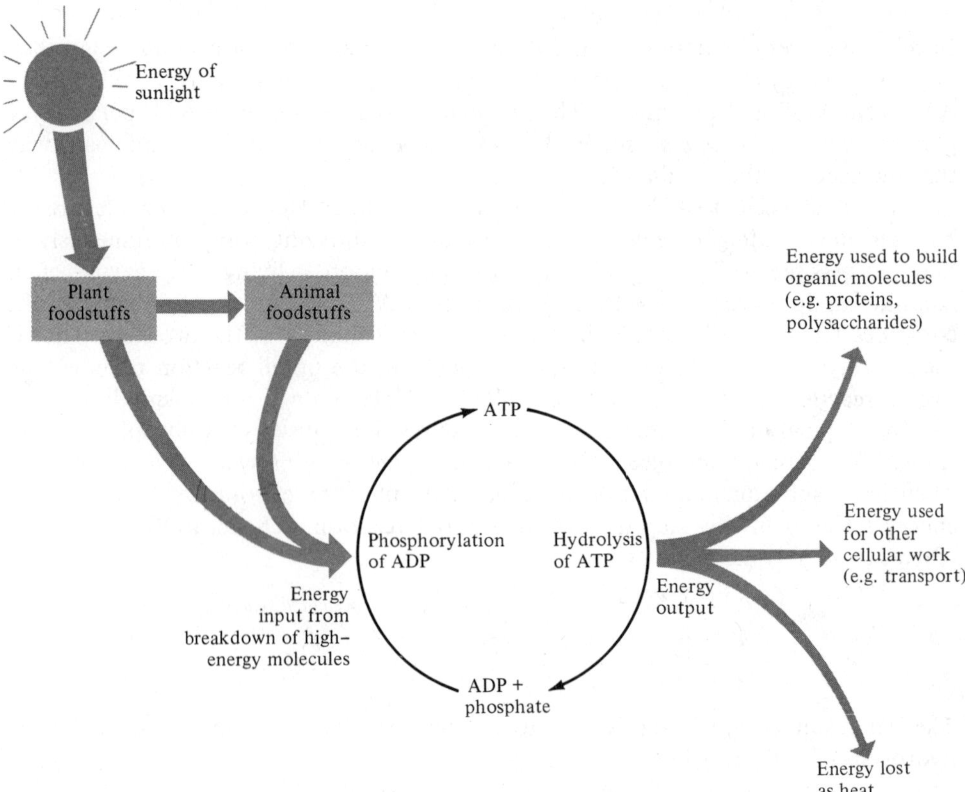

Figure 33.18
The Flow of Energy through the Living World
Through photosynthesis, plants capture the energy of sunlight to make energy-rich molecules (carbohydrates) from CO_2 and H_2O. Organisms break down these molecules and use the energy released to make ATP. Hydrolysis of ATP then supplies the energy needed to drive "uphill" reactions and for other vital functions.

SUMMARY OF SECTIONS 33.6–33.11

Most biomolecules are organic polymers. The structure and function of biological molecules are intimately related. Cellular membranes and other structural elements are composed chiefly of hydrophobic substances. Most of the substances that take part in the chemical reactions of organisms are hydrophilic.

Carbohydrates are simple sugars (monosaccharides), combinations of two sugars (disaccharides), or polymers of many sugars (polysaccharides). Glucose, like many monosaccharides, exists chiefly in a ring configuration in solution. Polysaccharides are used by organisms as storage forms for sugars (starch, glycogen) or to provide structural support (cellulose).

All *lipids* are nonpolymeric, hydrophobic substances. Among the lipids are the glycerides (fats and oils), which are esters of fatty acids, and the alcohol glycerol. Phospholipids play important roles in cellular membranes because they have polar, hydrophilic heads and nonpolar, hydrophobic tails.

Proteins consist of one or more long, unbranched polymer chains of amino acids, linked by peptide bonds (polypeptides). Fibrous proteins are generally hydrophobic, structural materials in which the polypeptide chains are relatively extended and cross-linked. Globular proteins are more compact, elaborately folded molecules, usually with hydrophilic exteriors that allow them to function in aqueous solution. Enzymes, the catalysts of living systems, are large protein molecules. Enzyme molecules possess a groove called the active site, where substrate molecules can be bound. An enzyme brings substrate molecules together in proper orientation for reaction, weakens bonds in the substrate(s), and may take a temporary part in the reaction mechanism.

Nucleic acids are polymers of nucleotides, each of which consists of a five-carbon sugar (ribose or deoxyribose), a nitrogen-containing base, and a phosphate group. DNA carries the genetic "blueprints" that specify what proteins an organism can make, while RNA is involved in "reading" the blueprints and guiding the synthesis of the proteins.

The breakdown of large molecules, which generally yields energy, is called catabolism; the synthesis of large molecules from simpler components, which generally requires energy, is called anabolism. Together these two processes make up the metabolism of the organism. Organisms use the energy released by catabolic reactions to make ATP from ADP and phosphate. Enzymes couple the reverse of this reaction—the hydrolysis of a ATP—to anabolic reactions so that the energy provided by the former can drive the latter.

SIGNIFICANT TERMS (with section references)

active site (33.9)
adenosine diphosphate (ADP) (33.11)
adenosine triphosphate (ATP) (33.11)
amino acid (33.9)
anabolism (33.11)
carbohydrates (33.7)
catabolism (33.11)
chain reaction polymerization, addition polymerization (33.4)
copolymer (33.2)
denaturation (33.9)
deoxyribonucleic acid (DNA) (33.10)
elastomers (33.5)
enzymes (33.9)
glass transition temperature (33.3)
homopolymer (33.2)
hydrophilic (33.6)
hydrophobic (33.6)
lipids (33.8)
macromolecules (33.1)
metabolism (33.11)
monomers (33.1)
nucleic acids (33.10)
nucleotide (33.10)
peptide bond (33.9)
polypeptide (33.9)
proteins (33.9)
ribonucleic acid (RNA) (33.10)
step reaction polymerization, condensation polymerization (33.4)
substrate (biochemical) (33.9)
thermoplastic (33.2)
thermoset plastic (33.2)

QUESTIONS AND PROBLEMS

Polymers

33.1 What is a macromolecule? What is the term for the smaller molecule (or molecules) that provides the repeating unit that makes up a macromolecule? What are typical molecular masses of macromolecules?

33.2 What is necessary if a molecule is to be capable of polymerization? Name three types of molecules that can polymerize.

33.3 Name the two natural polymers first used commercially. Which natural polymer is an elastomer? How can each of these polymers be modified to obtain somewhat different properties?

33.4 Describe the differences between thermosetting and thermoplastic polymers.

33.5 How would you interpret, in terms of intermolecular forces between polymer chains and intramolecular bonds, the observation that a given polymer begins to decompose before it begins to melt when it is heated?

33.6 Poly(vinyl alcohol) has a relatively high T_m of 258 °C. How would you explain this behavior? A segment of the polymer is

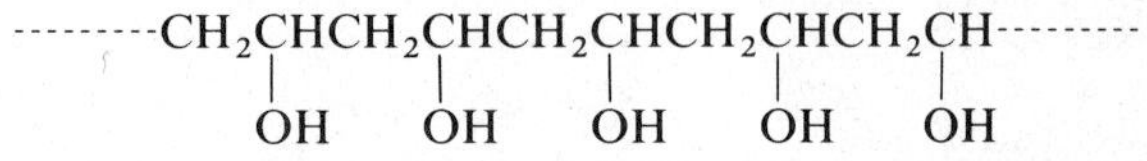

$$\cdots CH_2\underset{OH}{CH}CH_2\underset{OH}{CH}CH_2\underset{OH}{CH}CH_2\underset{OH}{CH}CH_2\underset{OH}{CH}\cdots$$

33.7 Prepare rough sketches of linear, branched, and cross-linked polymers. Which of these usually demonstrate "thermoplastic" properties?

33.8 What changes could be made in the structures of polymer molecules that would increase the rigidity of the polymer and raise its melting point?

33.9 Using A and B for two different types of monomers, write structures which show how they might combine to form copolymers.

33.10 Using letters or symbols, schematically illustrate a chain reaction polymerization and the formation of a polyester by a step reaction polymerization.

33.11 Methyl vinyl ketone ($CH_3\overset{O}{\overset{\|}{C}}CH{=}CH_2$) can be polymerized via an addition, or chain reaction, polymerization. Write the molecular structure of a five-unit segment of this polymer.

33.12 The copolymerization of styrene with a small amount of 1,4-divinylbenzene produces a cross-linked copolymer.

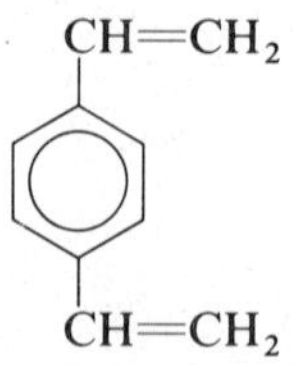

$$CH{=}CH_2 - C_6H_4 - CH{=}CH_2$$

1,4-divinylbenzene

Write the molecular structure of a portion of this copolymer that will show how the 1,4-divinylbenzene becomes involved in cross-linking the copolymer.

33.13 Suppose the following diol is used instead of ethylene glycol in preparing a polyester analogous to that shown in Table 33.2:

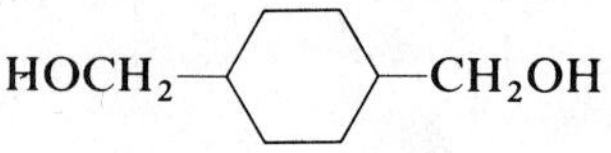

$$HOCH_2-C_6H_{10}-CH_2OH$$

What would be the molecular structure of the repeating unit of this polyester?

33.14 What is the molecular structure of the monomer that is polymerized in the formation of the following polymer?

$$\cdots CH_2\overset{CH_3}{\underset{C_6H_5}{C}}CH_2\overset{CH_3}{\underset{C_6H_5}{C}}CH_2\overset{CH_3}{\underset{C_6H_5}{C}}CH_2\overset{CH_3}{\underset{C_6H_5}{C}}CH_2\overset{CH_3}{\underset{C_6H_5}{C}}\cdots$$

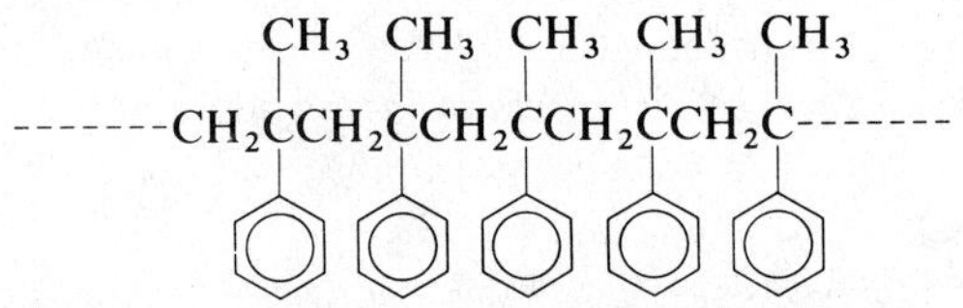

33.15 Which of the following polymers might be chosen for use in (a) polishing wax, (b) molded automobile door handles? (i) Polystyrene, $T_g = 373$ K; (ii) poly(dimethyl siloxane), $T_g = 150$ K.

33.16 Which of the following polymers might be chosen for use in (a) rope, (b) skylights? (i) Polypropylene, $T_g = 253$ K; (ii) poly(methyl methacrylate), $T_g = 378$ K.

33.17 A cellulose polymer has a molar mass of 750,000 g/mol. Estimate the number of β-glucose units in the polymer. The polymerization reaction can be represented by

$$xC_6H_{12}O_6 \longrightarrow \text{polymer} + (x-1)H_2O$$

33.18 A 2.3 g sample of poly(vinyl alcohol) was dissolved in water to give 101 mL of solution. The osmotic pressure of the solution was 55 Torr at 25 °C. What is the molar mass of the poly(vinyl alcohol)?

Biochemistry

33.19 Name the four major classes of organic compounds found in living cells, and list some of the principal functions associated with each.

33.20 Give some examples of hydrophilic and hydrophobic biomolecules. Why is this distinction so important in the chemistry of living things?

33.21 What do we call the basic structural units of carbohydrates, and what is their chemical makeup? What terms are used to describe the carbohydrates formed by combinations of these units?

33.22 Which of the following terms—carbohydrate, aldose, ketose, monosaccharide, and/or disaccharide—apply to (a) D-glucose, (b) D-fructose, (c) sucrose?

33.23 How many chiral carbon atoms are present in a molecule of each of the following hexoses:

$$CH_2OH-\overset{O}{\overset{\|}{C}}-CHOH-CHOH-CHOH-CH_2OH \qquad \text{D-}\mathit{fructose}$$

$$H\overset{O}{\overset{\|}{C}}-CHOH-CHOH-CHOH-CHOH-CH_2OH \qquad \text{D-}\mathit{glucose}$$

33.24 Write the molecular structures of the two cyclic forms that would result from ring closure (analogous to D-glucose; see Figure 33.7) between the OH on the fifth carbon atom and the carbonyl group of D-fructose.

33.25 Sorbitol, a sweet-tasting compound sometimes used instead of sucrose, is made by the catalytic reduction of D-glucose:

$$\begin{array}{c} CHO \\ | \\ HCOH \\ | \\ HOCH \\ | \\ HCOH \\ | \\ HCOH \\ | \\ CH_2OH \\ \textit{D-glucose} \end{array} \xrightarrow[\text{Ni, pressure}]{H_2} \begin{array}{c} CH_2OH \\ | \\ HCOH \\ | \\ HOCH \\ | \\ HCOH \\ | \\ HCOH \\ | \\ CH_2OH \end{array} \quad \text{or} \quad \begin{array}{c} CH_2OH \\ | \\ HOCH \\ | \\ HOCH \\ | \\ HCOH \\ | \\ HOCH \\ | \\ CH_2OH \end{array}$$

sorbitol

(a) Are the two molecular structures shown for sorbitol identical? (b) Write the molecular structure for another aldose that would form sorbitol when similarly reduced. (c) Is sorbitol considered to be a carbohydrate?

33.26 The D- and L- forms of fructose are as follows:

$$\begin{array}{c} CH_2OH \\ | \\ C{=}O \\ | \\ HOCH \\ | \\ HCOH \\ | \\ HCOH \\ | \\ CH_2OH \\ \textit{D-fructose} \end{array} \qquad \begin{array}{c} CH_2OH \\ | \\ C{=}O \\ | \\ HCOH \\ | \\ HOCH \\ | \\ HOCH \\ | \\ CH_2OH \\ \textit{L-fructose} \end{array}$$

Write the formula for the molecular structure of L-glucose, given

$$\begin{array}{c} CHO \\ | \\ HCOH \\ | \\ HOCH \\ | \\ HCOH \\ | \\ HCOH \\ | \\ CH_2OH \\ \textit{D-glucose} \end{array}$$

33.27 The equilibrium constant for the reaction

$$\beta\text{-D-glucose} \rightleftharpoons \alpha\text{-D-glucose}$$

is 0.56 at 25 °C. Find the concentration of each of the sugars at equilibrium in a solution originally containing 0.10 M β-D-glucose.

33.28 What is the osmotic pressure of a 1.00 mass % sucrose solution ($C_{12}H_{22}O_{11}$)? To what height will a column of water be supported by this pressure at 25 °C? The density of the sugar solution is 1.002 g/mL.

33.29 List the principal types of lipids. What do they have in common? How do they differ from the other large biological molecules?

33.30 Give two reasons why fats are more efficient substances for energy storage than carbohydrates.

33.31 Classify each of the following fatty acids as "saturated" or "unsaturated": (a) stearic acid (commonly found in animal and vegetable fats), $C_{17}H_{35}COOH$; (b) oleic acid (commonly found in corn oil), $C_{17}H_{33}COOH$; (c) lauric acid (commonly found in coconut oil), $C_{11}H_{23}COOH$: (d) arachidic acid (commonly found in peanut oil), $C_{19}H_{39}COOH$; (e) linolenic acid (commonly found in linseed oil), $C_{17}H_{29}COOH$.

33.32 A sample of a mixture contains 84% of the fatty acid chains from oleic acid ($C_{18}H_{34}O_2$) and 7% from palmitic acid ($C_{16}H_{32}O_2$). Is this mixture a fat or an oil? Why? Write the formula for one of the possible triglycerides in this substance.

33.33 Write the molecular structure that represents an example of each of the following types of lipids: (a) a monoglyceride, (b) a diglyceride, (c) a triglyceride, (d) a simple triglyceride, (e) a mixed triglyceride. Use saturated fatty acids containing 14, 16, or 18 carbon atoms and unbranched chains.

33.34 Write the molecular structures of three isomeric mixed diglycerides, using unbranched fatty acids for the acid components of these esters.

33.35 Write the structural formula for a triglyceride molecule (an ester formed from a glycerol molecule and three fatty acid molecules) containing the following saturated fatty acids: two molecules of $C_{12}H_{25}COOH$ and one molecule of $C_{14}H_{29}COOH$. How many mixed triglycerides are possible with these same acids?

33.36 A 10.0 g sample of a triglyceride was "saponified" (hydrolyzed) with sodium hydroxide. The products were glycerol and three moles of sodium stearate ($C_{17}H_{35}COO^-Na^+$). What mass of NaOH was required for the hydrolysis?

33.37 Describe the chemical structure of a phospholipid molecule. How is this structure related to the role of phospholipids in biological membranes?

33.38 Why are steroids classified as lipids even though there is a considerable structural difference between steroids and other types of lipids? Briefly discuss this structural difference.

33.39 Describe the structure of a natural amino acid molecule. What kind of isomerism do most amino acids exhibit? Why?

33.40 Write the structural formulas for the ionic forms of an α-amino acid in acidic, neutral, and alkaline media. In a neutral solution will the amino group be neutral or positively charged?

33.41 How many different dipeptides can be formed from the amino acids A, B, and C? Assume that each dipeptide contains two different amino acids.

33.42 Repeat Question 33.41 for the possible tripeptides. Assume each tripeptide to contain three different amino acids.

33.43 Consider two amino acids

$$\begin{array}{c} H \\ | \\ R_1{-}C{-}COO^- \\ | \\ NH_3^+ \end{array} \qquad \begin{array}{c} H \\ | \\ R_2{-}C{-}COO^- \\ | \\ NH_3^+ \end{array}$$

Write the structural formulas for the two dipeptides that could be formed containing one molecule of each amino acid.

33.44 Aspartame (trade name, NutraSweet) is a methyl ester of a dipeptide

$$\text{C}_6\text{H}_5\text{—CH}_2\text{—}\underset{\text{COOCH}_3}{\text{CH}}\text{—NH—}\overset{\text{O}}{\overset{\|}{\text{C}}}\text{—}\underset{\text{CH}_2\text{COOH}}{\text{CH}}\text{—NH}_2$$

Write the structural formulas of the two amino acids that are combined to make aspartame (neglecting optical isomerism).

33.45 How are the amino acids in a polypeptide joined together? What are the links called?

33.46 What is denaturation of a protein? What are some factors that can cause denaturation? Is this process reversible?

33.47 The sequence of amino acids in a polypeptide is frequently represented by an expression such as GlySerCysLeuPhe, where the amino acids combined are glycine, serine, cysteine, leucine, and phenylalanine (in that order). Write the molecular structure of this polypeptide.

33.48 Write the molecular structure of cystine, an amino acid formed by a S—S bond between two molecules of cysteine as a result of treating cysteine with a mild oxidizing agent.

33.49 What is the active site of an enzyme? How does a substrate molecule bind to the active site?

33.50 Explain the mechanisms involved in the catalytic function of enzymes.

33.51 What are the basic structural units of nucleic acids? Describe the composition of these units. How are they linked together in a nucleic acid?

33.52 Briefly describe the structure of the DNA molecule. Explain the role of hydrogen bonding in the structure of DNA.

33.53 At one point, Watson and Crick were considering a possible structure for DNA in which the bases were on the outside of the molecule and the phosphate groups interacted with each other in the center. What chemical and functional objections to this model can you see?

33.54 How is genetic information encoded by nucleic acid molecules? Describe briefly how this information is duplicated.

33.55 The sequence of nucleotides in a DNA molecule usually is indicated by giving the letters representing the heterocyclic bases (Table 33.8) in the many nucleotides comprising the DNA chain. Thus, a portion of a DNA chain could be given as GTGCACCCCAT. Figure 33.15b gives the pattern to be followed in writing a complete molecular structure for a DNA segment including the bases. Write the molecular structure of the portion of a DNA chain described as AGC.

33.56 Each codon in the genetic code is represented by the three letters that indicate the sequence of the three consecutive bases in the appropriate nucleic acid. For example, the sequence UCA codes for the amino acid serine, as does AGU and others also. Some codons for the amino acids in Table 33.7 are

First two letters of codon	Last letter of codon A	C	G	U
AA	Lys	—	Lys	—
GA	—	Asp	—	Asp
CU	Leu	Leu	Leu	Leu
GC	Ala	Ala	Ala	Ala
GG	Gly	Gly	Gly	Gly
UC	Ser	Ser	Ser	Ser
UG	Term[a]	Cys	—	Cys
UU	Leu	Phe	Leu	Phe

[a] Term designates the codon for the termination of a growing chain of amino acids.

(a) Give the sequence of amino acids in a polypeptide specified by the following series of codons: (i) GACGGACUGUUC and (ii) UUUUGUUUCGGGGCG. (b) Give a codon sequence for the following polypeptides: (i) GlySerCysLeuPhe and (ii) LysAsp-PheLeuAlaSerCysGlyGly.

33.57 What is ATP? Describe briefly the role of ATP in the utilization of energy by living cells.

33.58 At pH = 7.0 and 25 °C, $\Delta G° = -29$ kJ for the reaction

$$\text{ATP}(aq) + \text{H}_2\text{O}(l) \rightleftharpoons \text{ADP}(aq) + \text{HPO}_4^{2-}$$

Calculate the value of the equilibrium constant for this reaction.

APPENDIX

I

Logarithms and Graphs

I.A LOGARITHMS

Logarithms and antilogarithms are used in chemistry when working with pH and with many equations in kinetics, nuclear chemistry, thermodynamics, and electrochemistry. It is important that you know how to work with logarithms and antilogarithms even though your electronic calculator may make finding the values of these mathematical functions quite easy.

There are several commonly used "bases" such as 2, 10, 12, and e ($e = 2.718281828459$) for expressing logarithms. Logarithms to the base 10, called "common logarithms" and abbreviated "log," are used throughout this book. Sometimes the mathematics by which an equation is derived gives the equation in the form of a logarithm to the base e. Such a logarithm is called a "natural logarithm" and abbreviated "ln." In equations we can replace "$\ln x$" by "$(2.303)\log x$" because

$$\ln x = (2.302585092994)\log x$$

The *common logarithm* of a number is the power to which 10 must be raised to get that number

$$a = 10^b$$

(b: the logarithm of a; a: number)

For example, we can easily see that the logarithm of 10 is 1 because $10 = 10^1$, the logarithm of 100 is 2 because $100 = 10^2$, and the logarithm of 0.01 is -2 because $0.01 = 10^{-2}$. Similarly, the logarithm of 50.00 is 1.6690 because $50.00 = 10^{1.6990}$.

The usual way to express the logarithm of a number is to write

$$\log a = b$$

read as "the logarithm of a equals b"

Thus

$$\log 10 = 1 \qquad \log 100 = 2$$
$$\log 0.01 = -2 \qquad \log 50.00 = 1.6990$$

The logarithm of any number contains two parts called the "characteristic" and the "mantissa." The *characteristic*, the number to the left of the decimal point in the logarithm, is based upon the location of the decimal point in the original number. The *mantissa*, the number to the right of the decimal point in the logarithm, is based upon the exact value of the digits in the original number. For example

$$\log 50.00 = 1.6990 \qquad \log 500. = 2.699$$

(in each, the digit before the decimal point is the characteristic; the digits after it are the mantissa)

The mantissa should be given to the same number of significant digits as the original number, as shown above. In using a calculator to find a logarithm, keep this rule in mind:

$$\log 959 = 2.982 \qquad (\text{not } 2.981818607)$$

CALCULATIONS INVOLVING LOGARITHMS

Logarithms can be used to multiply and divide numbers, to raise a number to some power, and to find the nth root of a number. Because logarithms are exponents, the rules for exponents are followed in performing these operations. In each case the desired calculation is restated in logarithmic form as follows:

desired calculation	*logarithmic form*
$(a)(b)$	$\log(a)(b) = \log a + \log b$
a/b	$\log(a/b) = \log a - \log b$
a^n	$\log(a^n) = n \log a$
$\sqrt[n]{a}$	$\log(\sqrt[n]{a}) = (\log a)/n$

The logarithms of the numbers are found, the calculation is performed using the logarithms, and the value of the answer is found by taking the antilogarithm of the result of the calculation.

In Section 2.3 we discussed the rules of significant figures for calculations involving addition (or subtraction) and multiplication (or division). For example, we know that the answer to each of the following problems should contain three significant digits:

$$z = (16.3)(7.942) \qquad y = (6.32 \times 10^{-36})^3$$

Solving these problems using logarithms should not change this conclusion. By applying our usual rules to *only the mantissas,* we obtain the correct answers:

$$\log z = (\log \underset{123}{16.3}) + (\log \underset{1234}{7.942})$$

$$= (\underset{123}{1.212}) + (\underset{1234}{0.8999})$$

$$= \underset{123}{2.112} \quad \textit{addition rule applied only to mantissa}$$

$$z = \underset{123}{129}$$

$$\log y = (3)(\log 6.32 \times 10^{-36})$$

$$= (3)(-35.\underset{123}{199})$$

$$= -105.\underset{123}{597} \quad \textit{multiplication rule applied to mantissa}$$

$$y = \underset{123}{2.53} \times 10^{-106}$$

When the logarithm of a number appears in an equation and is to be treated like any other factor in the calculation, the usual rules of significant figures apply, once the logarithm has been supplied with the correct number of digits. For example

$$\log z = \frac{(96.5)(0.8401)}{(8.314)} = 9.75$$

$$z = 5.6 \times 10^9$$

and also

$$x = (0.8401) - \frac{(8.314)(2.303)}{(96.5)} \log(7.6 \times 10^3)$$

$$= (0.8401) - \frac{(8.314)(2.303)(3.88)}{(96.5)}$$

$$= (0.8401) - (0.770) = 0.070$$

FINDING LOGARITHMS

Most electronic calculators have a key labeled "log" (or something similar). Use of this key varies from calculator to calculator. Refer to the instruction manual for your calculator for information on how to use this key if you are unfamiliar with its use. Often the logarithm displayed by the calculator contains too many digits in the mantissa, so it must be rounded as illustrated previously. In the absence of a calculator, logarithms are found by using a log table. Such tables are contained in chemistry handbooks, other scientific handbooks, and in the appendixes of certain textbooks.

Note that logarithms of negative numbers are undefined because there is no power to which 10 can be raised that will generate a negative number.

1. The Logarithm of a Number between 1 and 10 The logarithm of 1 is zero ($10^0 = 1$) and the logarithm of 10 is one ($10^1 = 10$). The logarithm of a number between 1 and 10 is between zero and one. In other words, the characteristic is 0 and the logarithm is the same as the mantissa. For example

$$\log(4.683) = 0.6705$$

$$\log(7.2) = 0.86$$

2. The Logarithm of a Number Greater than 10 As we saw above, $\log 10 = 1$. The logarithm of a number greater than 10 is greater than 1 and has a characteristic of 1 or more. To see this, we can rewrite the number in standard scientific notation and convert this logarithm of a product into the sum of the logarithms according to the rule for multiplying with logarithms.

$$\log(a \times 10^n) = \log a + \log 10^n = (\log a) + n$$

The desired logarithm consists of a mantissa determined by $\log a$ and a characteristic equal to n. The following examples illustrate finding the characteristic and mantissa for numbers greater than 10:

$$\begin{aligned} \log(107) &= \log(1.07 \times 10^2) \\ &= \log(1.07) + \log(10^2) \\ &= 0.029 + 2 = 2.029 \end{aligned}$$

$$\begin{aligned} \log(6.022 \times 10^{23}) &= \log(6.022) + \log(10^{23}) \\ &= 0.7797 + 23 = 23.7797 \end{aligned}$$

$$\begin{aligned} \log(5 \times 10^{327}) &= \log(5) + \log(10^{327}) \\ &= 0.7 + 327 = 327.7 \end{aligned}$$

[In the above addition steps, n is an exact number.]

Most electronic calculators directly provide answers to the first two examples but display some type of error message (flashing display, "error," etc.) for the third example, because the upper limit of the calculator has been exceeded. For such calculations, it is important to remember that the characteristic is the same as n and the mantissa is determined by $\log a$. Simply write the number as a power of 10 (as illustrated), find $\log a$ on your calculator, and add the characteristic.

3. The Logarithm of a Number between 0 and 1 The logarithm is negative for a number less than 1 but greater than 0. To see this, we can rewrite the number in standard scientific notation and use logarithms as was done above.

$$\log(a \times 10^{-n}) = \log a + \log 10^{-n} = (\log a) - n$$

Because n is always greater than $\log a$, the desired logarithm is always a negative number. The following examples illustrate finding the characteristic and mantissa for numbers between 0 and 1:

$$\begin{aligned}\log(0.362) &= \log(3.62 \times 10^{-1}) \\ &= \log(3.62) + \log(10^{-1}) \\ &= 0.559 + (-1) \\ &= 0.559 - 1 = -0.441\end{aligned}$$

$$\begin{aligned}\log(4.7 \times 10^{-16}) &= \log(4.7) + \log(10^{-16}) \\ &= 0.67 + (-16) \\ &= 0.67 - 16 = -15.33\end{aligned}$$

$$\begin{aligned}\log(3.142 \times 10^{-106}) &= \log(3.142) + \log(10^{-106}) \\ &= 0.4972 + (-106) \\ &= 0.4972 - 106 = -105.5028\end{aligned}$$

Although the first two calculations are performed easily on most electronic calculators, the third probably will generate an error message because the lower limit of the calculator has been exceeded. For such calculations, it is important to remember that the logarithm is given by subtracting n from the value of $\log a$ to give a negative number. Write the number as a power of 10 (as illustrated), find $\log a$ on your calculator, and subtract the characteristic.

Some of you may have learned other ways to represent logarithms for numbers less than 1, such as $\bar{1}.559$ and $\overline{16}.67$ or $9.559 - 10$ and $4.67 - 10$. These notations are very difficult to use in the types of calculations performed in chemistry.

FINDING ANTILOGARITHMS

The *antilogarithm* (or inverse logarithm) is the value of the number to which the logarithm corresponds. For example, if $\log a = 2$, the antilogarithm of $\log a$ is 100 (i.e., $\log 100 = 2$). The value of the antilogarithm is found essentially by reversing the above procedures for finding logarithms.

Electronic calculators vary widely in the methods used to evaluate an antilogarithm. Some have a key labeled "$\log^{-1}$" or "10^x" for ease of use; however, some require the use of a sequence of key strokes to obtain the antilogarithm. Refer to the instruction manual for your calculator for information on how to find antilogarithms if you are unfamiliar with the procedure. Often the antilogarithm displayed by the calculator contains too many significant figures, so it must be rounded appropriately.

1. The Antilogarithm for log *a* between 0 and 1 The antilogarithm is a number between 1 and 10 if the logarithm is greater than 0 but less than 1. For example

$$\begin{aligned}\log x &= 0.6952 \\ x &= \text{antilog}(0.6952) = 4.957\end{aligned}$$

$$\begin{aligned}\log y &= 0.438 \\ y &= \text{antilog}(0.438) = 2.74\end{aligned}$$

2. The Antilogarithm for log *a* Greater than 1 The antilogarithm of 1 is 10 and the antilogarithm of a logarithm greater than 1 is greater than 10. For example

$$\begin{aligned}\log x &= 6.9372 \\ x &= \text{antilog}(6.9372) \\ &= \text{antilog}(0.9372 + 6) \\ &= [\text{antilog}(0.9372)][\text{antilog}(6)] \\ &= 8.654 \times 10^{6}\end{aligned}$$

$$\begin{aligned}\log y &= 173.26 \\ y &= \text{antilog}(173.26) \\ &= \text{antilog}(0.26 + 173) \\ &= [\text{antilog}(0.26)][\text{antilog}(173)] \\ &= 1.8 \times 10^{173}\end{aligned}$$

To overcome the limitations of a calculator in the second calculation, remember that a is the antilogarithm of the mantissa and that the power of 10 will be equal to the characteristic.

3. The Antilogarithm If log *a* Is Negative The antilogarithm is a number less than 1 but greater than 0 if the logarithm is negative. To find the antilogarithm of a negative logarithm, it is first necessary to restate the logarithm with a positive mantissa so that the antilogarithm of the mantissa may be determined.

The procedure is to first subtract the mantissa of the logarithm in question from 1 and then combine it with the negative number equal to the characteristic plus 1

$$-X.Y = (1 - 0.Y) - (X + 1)$$

For example

$$\begin{aligned}\log x &= -1.36 \\ x &= \text{antilog}(-1.36) \\ &= \text{antilog}(0.64 - 2) \\ &= \text{antilog}[0.64 + (-2)] \\ &= [\text{antilog}(0.64)][\text{antilog}(-2)] \\ &= 4.4 \times 10^{-2}\end{aligned}$$

$$\begin{aligned}\log y &= -17.963 \\ y &= \text{antilog}(-17.963) \\ &= \text{antilog}(0.037 - 18) \\ &= \text{antilog}[0.037 + (-18)] \\ &= [\text{antilog}(0.037)][\text{antilog}(-18)] \\ &= 1.09 \times 10^{-18}\end{aligned}$$

$$\begin{aligned}\log z &= -212.3 \\ z &= \text{antilog}(-212.3) \\ &= \text{antilog}(0.7 - 213) \\ &= \text{antilog}[(0.7) + (-213)] \\ &= [\text{antilog}(0.7)][\text{antilog}(-213)] \\ &= 5 \times 10^{-213}\end{aligned}$$

To overcome the limitations of a calculator in the third calculation, remember that a is the antilogarithm of the mantissa subtracted from 1 and that the power of 10 will be equal to the negative of 1 plus the absolute value of the characteristic.

PROBLEMS

1. Find the logarithm of (a) 7.42, (b) 5.288, (c) 0.276, (d) 4.99×10^{-132}, (e) 5×10^{26}, (f) 6.022×10^{317}. *Answer* (a) 0.870, (b) 0.7233, (c) −0.559, (d) −131.302, (e) 26.7, (f) 317.7797.

2. Find the antilogarithm of (a) 0.2635, (b) 0.9988, (c) −5.2955, (d) −104.7, (e) 5.3153, (f) 421.2. *Answer* (a) 1.834, (b) 9.972, (c) 5.064×10^{-6}, (d) 2×10^{-105}, (e) 2.067×10^{5}, (f) 2×10^{421}.

3. Perform the following calculations using logarithms: (a) (7.290)(18.26), (b) (9.435)/(0.8888), (c) $(9.35)^{3.5}$, (d) $\sqrt[5]{6.293}$.
Answer (a) 133.3, (b) 10.62, (c) 2.50×10^3, (d) 1.445.

I.B GRAPHS

Many chemical experiments are designed to determine the relationship between two measurable quantities known as variables. For example, we could gather data to describe the pressure of an ideal gas as a function of the absolute temperature, the potential of an electrochemical cell as a function of pH, or the concentration of a reactant as a function of time. If the variable y can be expressed as a function of the variable x [written mathematically as $y = f(x)$], then y is known as the dependent variable and x is known as the independent variable.

One of the best ways to communicate quickly the general behavior of the relationship between two variables is with a graph. The dependent variable is plotted along the vertical axis, the *ordinate,* and the independent variable is plotted along the horizontal axis, the *abscissa,* (see Figure I.1) [We say that "y is plotted against x" or that the figure is "a plot of y *versus* x."]

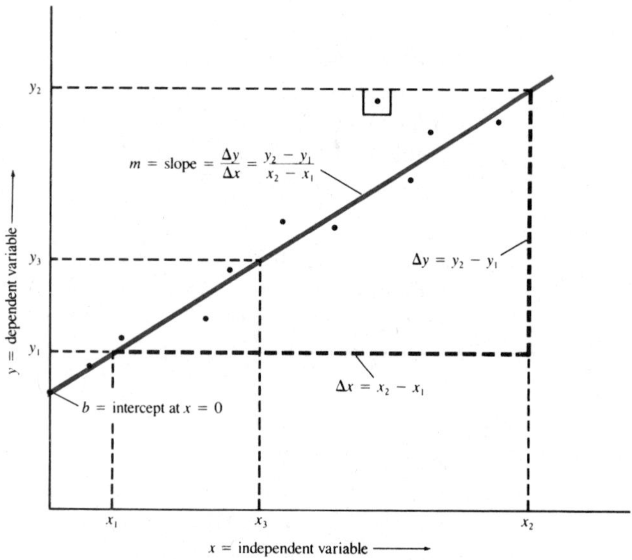

Figure I.1
Plot of y vs. x; $y = mx + b$

Here are five suggestions for making good graphs:

1. Choose a good-quality graph paper. For most graphs, paper that has 10 divisions per centimeter works nicely. Although cross-ruled notebook paper is suitable for rough plots, it should not be used for final laboratory reports or for solutions to problems. [Occasionally "semi-log" or "log-log" paper is used to construct graphs involving logarithms of variables. The logarithm scales on these styles of graph paper eliminate having to determine the logarithms of numbers before making the plot. However, when analyzing the data (for example, to determine the slope of a straight line), the actual numerical values of the logarithms must be used.]
2. The axes should be clearly marked and labeled using units. The numeric divisions along the axes should be linear (evenly spaced) and clearly indicated. Choose ranges for the numeric divisions for the variables so that the data set will occupy most of the area of the graph rather than being concentrated into one small area. It is not necessary to have an origin on the graph. Leave suitable blank margins along the edges of the graph paper.
3. All data points should be plotted accurately and marked clearly on the graph. Draw a circle or some other geometric figure around each data point so that it is highly visible. [The size of the geometric figures is sometimes used to represent the precision of the measurement.] If more than one data set appears on the graph, indicate each set by a characteristic geometric symbol or color.
4. Use mechanical drawing devices ("irregular" or drawing curves and straight edges) to draw the "best curves" through the data. Usually a "smooth curve" should be drawn through the data instead of connecting individual data points (see Figure I.1). Depending on the data, the "best curve" may not actually pass through the data points, but rather it will lie so that the points have equal positive and negative deviations from the curve.
5. A short but complete description of the graph should appear.

Once the "best curve" is drawn through the data, the graph can be used to determine the value of one of the variables for a given value of the other variable. For example, in Figure I.1 we can see that the value of y that corresponds to x_3 will be y_3.

A graph can reveal readily the presence of inaccurate data points. For example, the data point indicated by the square in Figure I.1 lies too far from the curve compared to the other data and should not be included in the analysis of the set of data. [There are statistical tests available to indicate whether or not a data point should be included, but we will not concern ourselves with them.]

Often the variables that are plotted in a graph are related by the linear function $y = mx + b$ where the value of m is the slope of the straight line and b is the intercept of the straight line with the y axis at $x = 0$. [The slope and intercept are often related to physical quantities. For example, in the study of kinetics, the slope could be related to an activation energy or a rate constant.] The graphical method for determining the slope and intercept of the line is shown in Figure I.1. To improve the accuracy in calculating the slope, the points (x_1, y_1) and (x_2, y_2) should be chosen from the "best" straight line through the data rather than chosen from among the actual data points. To improve the precision in calculating the slope, these points should be chosen near the extremes of the values of the variables. The slope should be reported with any units that Δy and Δx have.

Many electronic calculators have a linear least-squares regression program that determines the equation for the straight line through a set of data points. This program eliminates any guesswork about where the "best" straight line in a graph should be drawn and automatically gives the values of the slope and intercept. Obviously, any inaccurate data should not be included in the regression analysis.

PROBLEMS

1. The density of ethanol at various temperatures is

d, g/mL	0.805	0.797	0.788	0.782	0.777
T, °C	2.0	11.0	22.0	29.0	35.0

Prepare a plot of d against T and determine the values of a and b in the equation

$$d = a + bT$$

Using this equation, calculate the density of ethanol at 25.0 °C. *Answer* $a = 0.807$ g/mL, $b = -0.00084$ g/mL °C, 0.786 g/mL.

2. For an ideal gas, Boyle's law gives the relationship between the pressure and volume of the gas as

$$PV = k$$

where k is a constant. This equation can be rewritten as

$$V = k(1/P)$$

indicating that a plot of V against $(1/P)$ will be linear. The slope of the straight line is equal to k and the intercept is equal to 0. Using the following $P-V$ data

V, L	4.92	7.57	24.6	98.4	222
P, atm	5.00	3.25	1.00	0.250	0.111

prepare a plot of V against $(1/P)$ and determine the value of k. *Answer* $k = 24.6$ L atm.

APPENDIX

Units and Constants

Table II.1
Fundamental Constants
The numbers in parentheses represent the standard deviation in the last significant figures cited.

Symbol	Quantity	Value	Common rounded-off value
a_0	Bohr radius	$5.2917706(44) \times 10^{-11}$ m	5.29×10^{-11}
c	Velocity of light	$2.99792458(12) \times 10^{8}$ m/s	3.00×10^{8} m/s
e	Electronic charge	$1.6021892(46) \times 10^{-19}$ C	1.602×10^{-19} C
F	Faraday's constant	$9.648456(27) \times 10^{4}$ C/mol	9.65×10^{4} C/mol
g	Gravitational acceleration	9.80665 m/s^2	9.81 m/s^2
h	Planck's constant	$6.626176(36) \times 10^{-34}$ J s	6.626×10^{-34} J s
k	Boltzmann's constant	$1.380662(44) \times 10^{-23}$ J/K	1.381×10^{-23} J/K
L	Avogadro's number	$6.022045(31) \times 10^{23}$/mol	6.022×10^{23}/mol
m_e	Electron rest mass	$9.109534(47) \times 10^{-31}$ kg	9.11×10^{-31} kg
m_n	Neutron rest mass	$1.6749543(86) \times 10^{-27}$ kg	1.675×10^{-27} kg
m_p	Proton rest mass	$1.6726485(86) \times 10^{-27}$ kg	1.673×10^{-27} kg
R	Gas constant	8.31441(26)J/K mol	8.314 J/K mol
		$8.20568(26) \times 10^{-2}$ L atm/K mol	0.0821 L atm/K mol
		$8.20568(26) \times 10^{-5}$ m^3 atm/K mol	8.21×10^{-5} m^3 atm/K mol
R	Rydberg's constant	$1.097373177(83) \times 10^{7}$/m	1.097×10^{7}/m
		$2.179907(12) \times 10^{-18}$ J	2.18×10^{-18} J
$\bar{V}$	Ideal gas molar volume	22.41383(70)L/mol	22.4 L/mol
		$2.241383(70) \times 10^{-2}$ m^3/mol	0.0224 m^3/mol

Source: Values based on *J. Phys. Chem. Ref. Data*, **2,** 663–734 (1973).

Table II.2
SI Prefixes

Decimal location	Prefix	Prefix symbol
1,000,000,000,000,000,000 = 10^{18}	exa	E
1,000,000,000,000,000 = 10^{15}	peta	P
1,000,000,000,000 = 10^{12}	tera	T
1,000,000,000 = 10^{9}	giga	G
1,000,000 = 10^{6}	mega	M
1,000 = 10^{3}	kilo	k
100 = 10^{2}	hecto	h
10 = 10^{1}	deka	da
0.1 = 10^{-1}	deci	d
0.01 = 10^{-2}	centi	c
0.001 = 10^{-3}	milli	m
0.000 001 = 10^{-6}	micro	μ
0.000 000 001 = 10^{-9}	nano	n
0.000 000 000 001 = 10^{-12}	pico	p
0.000 000 000 000 001 = 10^{-15}	femto	f
0.000 000 000 000 000 001 = 10^{-18}	atto	a

Table II.3
Base Units of the International System of Units
The SI units are shown in boldface. Selected conversion factors for non-SI units are given. The numbers in parentheses represent the standard deviation in the last significant figures cited. Factors marked with an asterisk are exact.

Physical quantity Unit	Symbol	Selected conversion factors (common rounded-off conversion factor)
Length		
Meter[a]	**m**	
Ångstrom	Å	1 Å = 1×10^{-10} m*
Inch	inch	1 inch = 2.54×10^{-2} m*
Mass		
Kilogram[b]	**kg**	
Atomic mass unit	u	1 u = $1.6605655(86) \times 10^{-27}$ kg (1.6606×10^{-27} kg)
Metric ton		1 metric ton = 1×10^{3} kg*
Pound (avoir.)	lb	1 lb = 4.5359237×10^{-1} kg* (0.454 kg)
Ounce (avoir.)	oz	1 oz = $2.8349523125 \times 10^{-2}$ kg* (0.0283 kg)
Ounce (troy or apoth.)	oz	1 oz = $3.11034768 \times 10^{-2}$ kg* (0.0311 kg)
Time		
Second[c]	**s**	
Electrical current		
Ampere[d]	**A**	
Temperature		
Kelvin[e]	**K**	
Degree Celsius	°C	$T(°\text{C}) = T(\text{K}) - 273.15$*
Degree Fahrenheit	°F	$T(°\text{F}) = \left(\frac{1.8\ \text{F}°}{1\ \text{C}°}\right) T(°\text{C}) + 32$*
Amount of substance mole		
Mole[f]	**mol**	
Luminous intensity		
Candela[g]	**cd**	

Source: Values based on *J. Phys. Chem. Ref. Data,* **2,** 663–734 (1973).

[a] The *meter* is the length of the path traveled by light in vacuum during a time interval of (1/299792458) of a second.

[b] The *kilogram* is the unit of mass equal to the mass of the international platinum–iridium prototype of the kilogram kept at the International Bureau of Weights and Measures.

[c] The *second* is the duration of 9,192,631,770 periods of radiation corresponding to the transition between the two hyperfine levels of the ground state of the cesium-133 atom.

[d] The *ampere* is that constant current which, if maintained in two straight parallel conductors of infinite length, of negligible circular cross section, and placed 1 meter apart in vacuum, would produce between these conductors a force equal to 2×10^{-7} newton per meter of length.

[e] The *kelvin* is the fraction (1/273.16) of the thermodynamic temperature of the triple point of water.

[f] The *mole* is the amount of substance of a system that contains as many elementary entities as there are atoms in 0.012 kilogram of carbon-12. When the mole is used, the elementary entities must be specified and may be atoms, molecules, ions, electrons, other particles, or specified groups of such particles.

[g] The *candela* is the luminous intensity, in a given direction, of a source that emits monochromatic radiation of frequency 540×10^{12} hertz and that has a radiant intensity in that direction of (1/683) watt per steradian.

Table II.4
Derived Units of the International System of Units
The SI units are shown in boldface. Selected conversion factors for non-SI units are given. The numbers in parentheses represent the standard deviation in the last significant figures cited. Factors marked with an asterisk are exact.

Physical quantity Unit	Symbol	Selected conversion factors (common rounded-off conversion factor)
Area		
Square meter	$\mathbf{m^2}$	
Density		
Kilogram per cubic meter	$\mathbf{kg/m^3}$	
Grams per milliliter	$g/mL = g/cm^3$	$1\ g/cm^3 = 1 \times 10^3\ kg/m^3$*
Dipole moment		
Coulomb meter	**C m**	
Debye	D	$1\ D = 3.335641(14) \times 10^{-30}$ C m (3.34×10^{-30} C m)
Electrical resistance		
Ohm	$\mathbf{\Omega} = V/A = kg\ m^2/A^2\ s^3$	
Electricity, quantity		
Coulomb	**C** = A s	
Electrostatic unit	$esu = cm^{3/2}\ g^{1/2}/s$	$1\ C = 3.335641(14) \times 10^{-10}$ esu (3.34×10^{-10} esu)
Electromotive force		
Volt	**V** = $kg\ m^2/A\ s^3$	
Energy		
Joule	**J** = $kg\ m^2/s^2 = A\ V\ s = Pa\ m^3 = N\ m$	
Erg	$erg = g\ cm^2/s^2$	$1\ erg = 1 \times 10^{-7}$ J*
Calorie	cal	1 cal = 4.184 J*
Electron volt	eV	$1\ eV = 1.6021892(46) \times 10^{-19}$ J (1.602×10^{-19} J)
Liter atmosphere	L atm	$1\ L\ atm = 1.01325 \times 10^2$ J*
Wave number	cm^{-1}	$1\ cm^{-1} = 1.986477(10) \times 10^{-23}$ J (1.986×10^{-23} J)
Atomic mass unit	u	$1\ u = 1.492442(6) \times 10^{-10}$ J (1.492×10^{-10} J)
Entropy		
Joule per Kelvin	**J/K**	
Entropy unit, Gibb	cal/K	1 cal/K = 4.184 J/K*
Force		
Newton	**N** = $kg\ m/s^2$	
Dyne	$dyne = g\ cm/s^2$	$1\ dyne = 1 \times 10^{-5}$ N*
Pound force	lbf	1 lbf = 4.4482216152605 N* (4.45 N)
Frequency		
Hertz	**Hz** = $s^{-1} = /s$	
Cycles per second	cps	1 cps = 1 Hz*

Source: Values based on *J. Phys. Chem. Ref. Data*, **2**, 663–734 (1973).

Physical quantity Unit	Symbol	Selected conversion factors (common rounded-off conversion factor)
Heat—see Energy		
Power		
Watt	**W** = J/s = A V = kg m^2/s^3	
Pressure		
Pascal	**Pa** = N/m^2 = kg/m s^2	
Atmosphere	atm	1 atm = 1.01325×10^5 Pa*
Bar	bar	1 bar = 1×10^5 Pa*
Pounds per square inch	psi = lb/in^2	1 psi = $6.894757293167 \times 10^3$ Pa (6.89×10^3 Pa)
Torr, millimeter of mercury	Torr = mm Hg	1 Torr = $1.33322368421 \times 10^2$ Pa (133.3 Pa) 1 atm = 760 Torr*
Radiation, activity		
Becquerel	**Bq** = disintegration/s	
Rutherford	Rd	1 Rd = 1×10^6 Bq*
Curie	Ci	1 Ci = 3.7×10^{10} Bq*
Radiation, dosimetry		
Coulomb per kilogram	**C/kg**	
Roentgen	R	1 R = 2.57976×10^{-4} C/kg* (2.58×10^{-4} C/kg)
Radiation, energy absorbed		
Gray	**Gy** = J/kg	
Rad	rad	1 rad = 1×10^{-2} Gy*
Volume		
Cubic meter	$\mathbf{m^3}$	
Liter	L	1 L = 1×10^{-3} m^3*
Quart (US liq.)	qt	1 qt = 9.4635295×10^{-4} m^3 (9.46×10^{-4} m^3)
Milliliter	mL	1 mL = 1 cm^3*
Cubic centimeter	cm^3 = cc	1 cm^3 = 1×10^{-6} m^3*
Work—see Energy		

APPENDIX

Properties of Atoms

Table III.1
Ionization Energy and Electron Affinity at 0 K

Element Z symbol	Ionization energy (kJ/mol)	Electron affinity (kJ/mol)
1 H	1312.0	−72.7
2 He	2372.3	21
3 Li	520.2	−59.8
4 Be	899.4	240
5 B	800.6	−83
6 C	1086.4	−122.5
7 N	1402.3	0.0
8 O	1313.9	−141.4
9 F	1681.0	−322.2
10 Ne	2080.6	29
11 Na	495.6	−52.9
12 Mg	737.7	230
13 Al	577.6	−50
14 Si	786.4	−120.
15 P	1011.7	−74
16 S	999.6	−200.4
17 Cl	1251.1	−348.7
18 Ar	1520.5	35
19 K	418.8	−48.3
20 Ca	589.8	156
21 Sc	631	

Element Z symbol	Ionization energy (kJ/mol)	Electron affinity (kJ/mol)
22 Ti	658	−37.7
23 V	650.	−90.4
24 Cr	652.8	−64
25 Mn	717.4	
26 Fe	759.3	−56.2
27 Co	758	−90.3
28 Ni	736.7	−123.1
29 Cu	745.4	−123.1
30 Zn	906.4	
31 Ga	578.8	−36
32 Ge	762.1	−116
33 As	947	−77
34 Se	940.9	−195
35 Br	1139.9	−324.5
36 Kr	1350.7	39
37 Rb	403.0	−46.88
38 Sr	549.5	168
39 Y	616	
40 Zr	660.	
41 Nb	664	
42 Mo	684.9	−96
43 Tc	702	
44 Ru	711	

Element Z symbol	Ionization energy (kJ/mol)	Electron affinity (kJ/mol)
45 Rh	720.	
46 Pd	805	
47 Ag	731.0	−125.7
48 Cd	867.7	
49 In	558.3	−34
50 Sn	708.6	−121
51 Sb	833.7	−101
52 Te	869.2	−183
53 I	1008.4	−295.3
54 Xe	1170.4	41
55 Cs	375.7	−45.5
56 Ba	502.9	52
57 La	538.1	
58 Ce	528	
59 Pr	523	
60 Nd	530.	
61 Pm	535	
62 Sm	543	
63 Eu	547	
64 Gd	564	
65 Tb	564	
66 Dy	572	
67 Ho	581	

(*continues*)

Table III.1 *(continued)*

Element Z symbol	Ionization energy (kJ/mol)	Electron affinity (kJ/mol)	Element Z symbol	Ionization energy (kJ/mol)	Electron affinity (kJ/mol)	Element Z symbol	Ionization energy (kJ/mol)	Electron affinity (kJ/mol)
68 Er	589		78 Pt	870	−205.3	87 Fr		−44.0
69 Tm	596		79 Au	890.1	−222.75	88 Ra	509.3	
70 Yb	603.4		80 Hg	1007.0		89 Ac	670	
71 Lu	523.5		81 Tl	589.3	−50	90 Th		
72 Hf	680		82 Pb	715.5	−101	91 Pa		
73 Ta	761	−80	83 Bi	703.3	−101	92 U		
74 W	770.	−50	84 Po	818	−170	93 Np		
75 Re	760.	−14	85 At		−270	94 Pu	560	
76 Os	840		86 Rn	1037.0	41	95 Am	580	
77 Ir	880							

Source: Data mainly from T. Moeller, *Inorganic Chemistry* (New York: Wiley, 1982) pp. 76–79 and 81–84.

Table III.2
Atomic and Ionic Radii of the Elements, in nm

		1 H	2 He	3 Li	4 Be	5 B	6 C	7 N	8 O	9 F	10 Ne	11 Na
Anion	−4	—	—	—	—	—	0.260	—	—	—	—	—
	−3	—	—	—	—	—	—	0.171	—	—	—	—
	−2	—	—	—	—	—	—	—	0.140	—	—	—
	−1	0.208	—	—	—	—	—	—	—	0.136	—	—
Atom	0	0.037	0.05	0.152	0.111	0.080	0.077	0.074	0.074	0.071	0.065	0.186
Cation	+1	—	—	0.068	—	—	—	—	—	—	—	0.097
	+2	—	—	—	0.035	—	—	—	—	—	—	—
	+3	—	—	—	—	—	—	—	—	—	—	—
		12 Mg	13 Al	14 Si	15 P	16 S	17 Cl	18 Ar	19 K	20 Ca	21 Sc	22 Ti
Anion	−4	—	—	—	—	—	—	—	—	—	—	—
	−3	—	—	—	0.212	—	—	—	—	—	—	—
	−2	—	—	—	—	0.184	—	—	—	—	—	—
	−1	—	—	—	—	—	0.181	—	—	—	—	—
Atom	0	0.160	0.143	0.118	0.110	0.103	0.099	0.095	0.227	0.197	0.161	0.145
Cation	+1	—	—	—	—	—	—	—	0.133	—	—	—
	+2	0.066	—	—	—	—	—	—	—	0.099	—	0.080
	+3	—	0.051	—	—	—	—	—	—	—	0.081	0.076
		23 V	24 Cr	25 Mn	26 Fe	27 Co	28 Ni	29 Cu	30 Zn	31 Ga	32 Ge	33 As
Anion	−4	—	—	—	—	—	—	—	—	—	—	—
	−3	—	—	—	—	—	—	—	—	—	—	0.222
	−2	—	—	—	—	—	—	—	—	—	—	—
	−1	—	—	—	—	—	—	—	—	—	—	—
Atom	0	0.131	0.125	0.137	0.124	0.125	0.125	0.128	0.133	0.122	0.123	0.125
Cation	+1	—	—	—	—	—	—	0.096	—	—	—	—
	+2	0.088	0.083	0.080	0.074	0.072	0.069	0.072	0.074	—	—	—
	+3	0.074	0.063	0.066	0.064	0.063	—	—	—	0.062	—	—
		34 Se	35 Br	36 Kr	37 Rb	38 Sr	39 Y	40 Zr	41 Nb	42 Mo	43 Tc	44 Ru
Anion	−4	—	—	—	—	—	—	—	—	—	—	—
	−3	—	—	—	—	—	—	—	—	—	—	—
	−2	0.191	—	—	—	—	—	—	—	—	—	—
	−1	—	0.196	—	—	—	—	—	—	—	—	—
Atom	0	0.116	0.114	0.110	0.248	0.215	0.178	0.159	0.143	0.136	0.135	0.133
Cation	+1	—	—	—	0.147	—	—	—	—	—	—	—
	+2	—	—	—	—	0.112	—	—	—	—	—	—
	+3	—	—	—	—	—	0.092	—	0.070	0.067	—	0.068

(continues)

Table III.2 *(continued)*

		$_{45}$Rh	$_{46}$Pd	$_{47}$Ag	$_{48}$Cd	$_{49}$In	$_{50}$Sn	$_{51}$Sb	$_{52}$Te	$_{53}$I	$_{54}$Xe	$_{55}$Cs
Anion	−4	—	—	—	—	—	—	—	—	—	—	—
	−3	—	—	—	—	—	—	0.245	—	—	—	—
	−2	—	—	—	—	—	—	—	0.218	—	—	—
	−1	—	—	—	—	—	—	—	—	0.220	—	—
Atom	0	0.135	0.138	0.145	0.149	0.163	0.141	0.145	0.143	0.133	0.130	0.265
Cation	+1	—	—	0.126	—	—	—	—	—	—	—	0.167
	+2	—	0.080	0.089	0.097	—	0.093	—	—	—	—	—
	+3	0.068	—	—	—	0.081	—	0.076	—	—	—	—
		$_{56}$Ba	$_{57}$La	$_{58}$Ce	$_{59}$Pr	$_{60}$Nd	$_{61}$Pm	$_{62}$Sm	$_{63}$Eu	$_{64}$Gd	$_{65}$Tb	$_{66}$Dy
Anion	−4	—	—	—	—	—	—	—	—	—	—	—
	−3	—	—	—	—	—	—	—	—	—	—	—
	−2	—	—	—	—	—	—	—	—	—	—	—
	−1	—	—	—	—	—	—	—	—	—	—	—
Atom	0	0.217	0.187	0.183	0.182	0.181	≈0.181	0.180	0.199	0.179	0.176	0.175
Cation	+1	—	—	—	—	—	—	—	—	—	—	—
	+2	0.134	—	—	—	—	—	0.110	0.109	—	—	—
	+3	—	0.114	0.107	0.106	0.104	0.106	0.100	0.098	0.097	0.093	0.092
		$_{67}$Ho	$_{68}$Er	$_{69}$Tm	$_{70}$Yb	$_{71}$Lu	$_{72}$Hf	$_{73}$Ta	$_{74}$W	$_{75}$Re	$_{76}$Os	$_{77}$Ir
Anion	−4	—	—	—	—	—	—	—	—	—	—	—
	−3	—	—	—	—	—	—	—	—	—	—	—
	−2	—	—	—	—	—	—	—	—	—	—	—
	−1	—	—	—	—	—	—	—	—	—	—	—
Atom	0	0.174	0.173	≈0.173	0.194	0.172	0.156	0.143	0.137	0.137	0.137	0.136
Cation	+1	—	—	—	—	—	—	—	—	—	—	—
	+2	—	—	0.094	0.093	—	—	—	—	—	—	—
	+3	0.091	0.089	0.087	0.086	0.085	—	0.067	—	—	—	0.073
		$_{78}$Pt	$_{79}$Au	$_{80}$Hg	$_{81}$Tl	$_{82}$Pb	$_{83}$Bi	$_{84}$Po	$_{85}$At	$_{86}$Rn	$_{87}$Fr	$_{88}$Ra
Anion	−4	—	—	—	—	—	—	—	—	—	—	—
	−3	—	—	—	—	—	—	—	—	—	—	—
	−2	—	—	—	—	—	—	—	—	—	—	—
	−1	—	—	—	—	—	—	—	—	—	—	—
Atom	0	0.139	0.144	0.150	0.170	0.175	0.155	0.118	—	0.145	—	—
Cation	+1	—	0.137	—	0.147	—	—	—	—	—	0.180	—
	+2	0.080	—	0.110	—	0.120	—	0.230	—	—	—	0.143
	+3	—	0.085	—	0.095	—	0.096	—	—	—	—	—
		$_{89}$Ac	$_{90}$Th	$_{91}$Pa	$_{92}$U	$_{93}$Np	$_{94}$Pu	$_{95}$Am	$_{96}$Cm	$_{97}$Bk	$_{98}$Cf	
Anion	−4	—	—	—	—	—	—	—	—	—	—	
	−3	—	—	—	—	—	—	—	—	—	—	
	−2	—	—	—	—	—	—	—	—	—	—	
	−1	—	—	—	—	—	—	—	—	—	—	
Atom	0	0.188	0.180	0.161	0.139	—	0.151	≈0.131	—	—	—	
Cation	+1	—	—	—	—	—	—	—	—	—	—	
	+2	—	—	—	—	—	—	—	—	—	—	
	+3	0.118	0.108	0.113	0.101	0.110	0.108	0.107	0.095	0.094	0.095	

Source: Data mainly from T. Moeller, *Inorganic Chemistry*, (New York: Wiley, 1982), pp. 70–72 and 141–47.

APPENDIX IV

Thermodynamic Data

Table IV.1
ΔH_f°, ΔG_f°, S°, and C_p° for Selected Substances at 25 °C and 1 Bar

Substance	ΔH_f° (kJ/mol)	ΔG_f° (kJ/mol)	S° (J/K mol)	C_p° (J/K mol)
Aluminum				
Al(*s*)	0.0	0.0	28.33	24.35
Al_2O_3(α-solid)	−1675.7	−1582.3	50.92	79.04
AlF_3(*s*)	−1504.1	−1425.0	66.44	75.10
$AlCl_3$(*s*)	−704.2	−628.8	110.67	91.84
Al_2Cl_6(*g*)	−1290.8	−1220.4	490.	
$Al_2(SO_4)_3$(*s*)	−3440.84	−3099.94	239.3	259.41
Antimony				
Sb(solid III)	0.0	0.0	45.69	25.23
Sb(*g*)	262.3	222.1	180.27	20.79
Sb_4O_6(solid II)	−1440.6	−1268.1	220.9	
Sb_4O_6(solid I)	−1417.1	−1253.0	246.0	202.76
Sb_2O_5(*s*)	−971.9	−829.2	125.1	
$SbCl_5$(*l*)	−440.2	−350.1	301	
$SbCl_3$(*s*)	−382.17	−323.67	184.1	107.9
SbOCl(*s*)	−374.0			
Argon				
Ar(*g*)	0.0	0.0	154.843	20.786
Arsenic				
As(α-solid)	0.0	0.0	35.1	24.64
As(*g*)	302.5	261.0	174.21	20.786
As_4(*g*)	143.9	92.4	314	
AsH_3(*g*)	66.44	68.93	222.78	38.07
As_2O_5(*s*)	−924.87	−782.3	105.4	116.52
As_4O_6(monoclinic)	−1309.6	−1153.93	234	
H_3AsO_3(*aq*)	−742.2	−639.80	195.0	
H_3AsO_4(*aq*)	−902.5	−766.0	184	
$AsCl_3$(*l*)	−305.0	−259.4	216.3	

(*continues*)

Table IV.1 *(continued)*

Substance	ΔH_f° (kJ/mol)	ΔG_f° (kJ/mol)	S° (J/K mol)	C_p° (J/K mol)
Barium				
$Ba(s)$	0.0	0.0	62.8	28.07
$Ba(g)$	180	146	170.243	20.786
$BaO(s)$	−553.5	−525.1	70.42	47.78
$Ba(OH)_2(s)$	−944.7			
$BaCl_2(s)$	−858.6	−810.4	123.68	75.14
$BaCO_3(s)$	−1216.3	−1137.6	112.1	85.35
$Ba(NO_3)_2(s)$	−992.07	−796.59	213.8	151.38
$BaSO_4(s)$	−1473.2	−1362.2	132.2	101.75
$BaCrO_4(s)$	−1446.0	−1345.22	158.6	
Beryllium				
$Be(s)$	0.0	0.0	9.50	16.44
Bismuth				
$Bi(s)$	0.0	0.0	56.74	25.52
$Bi_2O_3(s)$	−573.88	−493.7	151.5	113.51
$BiCl_3(s)$	−379.1	−315.0	177.0	105
$BiOCl(s)$	−366.9	−322.1	120.5	
Boron				
$B(\beta\text{-solid})$	0.0	0.0	5.86	11.09
$B(g)$	562.7	518.8	153.45	20.799
$B_2H_6(g)$	35.6	86.7	232.11	56.90
$B_2O_3(s)$	−1272.77	−1193.65	53.97	62.93
$H_3BO_3(s)$	−1094.33	−968.92	88.83	81.38
$[B(OH)_4]^-(aq)$	−1344.03	−1153.17	102.5	
$BN(s)$	−254.4	−228.4	14.81	19.71
$BF_3(g)$	−1137.00	−1120.33	254.12	50.46
$BCl_3(l)$	−427.2	−387.4	206.3	106.7
Bromine				
$Br(g)$	111.884	82.396	175.022	20.786
$Br_2(l)$	0.0	0.0	152.231	75.689
$HBr(g)$	−36.40	−53.45	198.695	29.142
$HBr(aq)$	−121.55	−103.96	82.4	
$BrO_3^-(aq)$	−67.07	18.60	161.71	
$HBrO(aq)$	−113.0	−82.4	142	
$BrF(g)$	−93.85	−109.18	228.97	32.97
$BrCl(g)$	14.64	−0.98	240.10	34.98
Cadmium				
$Cd(\gamma\text{-solid})$	0.0	0.0	51.76	25.98
$Cd(g)$	112.01	77.41	167.746	20.786
$CdO(s)$	−258.2	−228.4	54.8	43.43
$CdS(s)$	−161.9	−156.5	64.9	
Calcium				
$Ca(s)$	0.0	0.0	41.42	25.31
$Ca(g)$	178.2	144.3	154.884	20.786
$CaO(s)$	−635.09	−604.03	39.75	42.80
$Ca(OH)_2(s)$	−986.09	−898.49	83.39	87.49
$CaC_2(s)$	−59.8	−64.9	69.96	62.72
$CaCO_3$(calcite)	−1206.92	−1128.79	92.9	81.88
$CaCO_3$(aragonite)	−1207.13	−1127.75	88.7	81.25
$Ca(NO_3)_2(s)$	−938.39	−743.07	193.3	149.37
$CaSO_4(s)$	−1434.11	−1321.79	106.7	99.66
$CaCl_2(s)$	−795.8	−748.1	104.6	72.59
$CaCl_2 \cdot 6H_2O(s)$	−2607.9			
$Ca_3(PO_4)_2(s)$	−4120.8	−3884.7	236.0	227.82
Carbon				
C(graphite)	0.0	0.0	5.740	8.527
C(diamond)	1.895	2.900	2.377	6.113
$CH_4(g)$	−74.81	−50.72	186.264	35.309
$C_2H_2(g)$	226.73	209.20	200.94	43.93
$C_2H_4(g)$	52.26	68.15	219.56	43.56
$C_2H_6(g)$	−84.68	−32.82	229.60	52.63

(continues)

Table IV.1 *(continued)*

Substance	ΔH_f° (kJ/mol)	ΔG_f° (kJ/mol)	S° (J/K mol)	C_p° (J/K mol)
$CO(g)$	−110.525	−137.168	197.674	29.142
$CO_2(g)$	−393.509	−394.359	213.74	37.11
$HCN(g)$	135.1	124.7	201.78	35.86
$CS_2(l)$	89.70	65.27	151.34	75.7
$CF_4(g)$	−925	−879	261.61	61.09
$CCl_4(l)$	−135.44	−65.21	216.40	131.75
$CH_3Cl(g)$	−80.83	−57.37	234.58	40.75
$CH_2Cl_2(l)$	−121.46	−67.26	177.8	100.0
$CHCl_3(l)$	−134.47	−73.66	201.7	113.8
$CH_3COOH(l)$	−484.5	−389.9	159.8	124.3
$CH_3COOH(aq)$	−485.76	−396.46	178.7	
$CH_3OH(l)$	−238.66	−166.27	126.8	81.6
$CH_3CH_2OH(l)$	−277.69	−174.78	160.7	111.46
cis-$CHClCHCl(l)$	−27.6	22.11	198.41	113
trans-$CHClCHCl(l)$	−23.14	27.34	195.85	113
$CH_3NH_2(l)$	−47.3	35.7	150.21	
Chlorine				
$Cl(g)$	121.679	105.680	165.198	21.840
$Cl_2(g)$	0.0	0.0	223.066	33.907
$HCl(g)$	−92.307	−95.299	186.908	29.12
$HCl(aq)$	−167.159	−131.228	56.5	
$HClO_4(aq)$	−129.33	−8.52	182.0	
$HClO(aq)$	−120.9	−79.9	142	
$ClF(g)$	−54.48	−55.94	217.89	32.05
$ClF_3(g)$	−163.2	−123.0	281.61	63.85
Chromium				
$Cr(s)$	0.0	0.0	23.77	23.35
$CrO_3(s)$	−589.5			
$Cr(OH)_3(s)$	−1064.0			
$[Cr(H_2O)_6]^{3+}$	−1999.1			
Cobalt				
Co(hexagonal)	0.0	0.0	30.04	24.81
$CoO(s)$	−237.94	−214.20	52.97	55.23
$Co_3O_4(s)$	−891	−774	102.5	123.4
$CoCl_2(s)$	−312.5	−269.8	109.16	78.49
Copper				
$Cu(s)$	0.0	0.0	33.150	24.435
$CuO(s)$	−157.3	−129.7	42.63	42.30
$Cu_2O(s)$	−168.6	−146.0	93.14	63.64
$CuS(s)$	−53.1	−53.6	66.5	47.82
$Cu_2S(s)$	−79.5	−86.2	120.9	76.32
$CuCl_2(s)$	−220.1	−175.7	108.07	71.88
$CuCl(s)$	−137.2	−119.86	86.2	48.5
$CuSO_4(s)$	−771.36	−661.8	109	100.0
$CuSO_4 \cdot 5H_2O(s)$	−2279.65	−1879.745	300.4	280.
$Cu(NO_3)_2(s)$	−302.9			
Fluorine				
$F(g)$	78.99	61.91	158.754	22.744
$F_2(g)$	0.0	0.0	202.78	31.30
$HF(g)$	−271.1	−273.2	173.779	29.133
$HF(aq)$	−320.08	−296.82	88.7	
Helium				
$He(g)$	0.0	0.0	126.150	20.786
Hydrogen				
$H(g)$	217.965	203.247	114.713	20.784
$H_2(g)$	0.0	0.0	130.684	28.824
Iodine				
$I(g)$	106.838	70.250	180.791	20.786
$I_2(s)$	0.0	0.0	116.135	54.438
$[I_3]^-(aq)$	−51.5	−51.4	239.3	
$HI(g)$	26.48	1.70	206.594	29.158
$HIO(aq)$	−138.1	−99.1	95.4	

(continues)

Table IV.1 *(continued)*

Substance	ΔH_f° (kJ/mol)	ΔG_f° (kJ/mol)	S° (J/K mol)	C_p° (J/K mol)
$IO_3^-(aq)$	−221.3	−128.0	118.4	
$IF_5(l)$	−864.8			
$ICl_3(s)$	−89.5	−22.29	167.4	
$IBr(s)$	−10.5			
Iron				
Fe(α-solid)	0.0	0.0	27.28	25.10
$FeO(s)$	−272.0			48.12
$Fe_2O_3(s)$	−824.2	−742.2	87.40	103.85
$Fe_3O_4(s)$	−1118.4	−1015.4	146.4	143.43
$Fe(OH)_2(s)$	−569.0	−486.5	88	
$Fe(OH)_3(s)$	−823.0	−696.5	106.7	
$FeCl_2(s)$	−341.79	−302.30	117.95	76.65
$FeCl_3(s)$	−399.49	−334.00	142.3	96.65
$FeSO_4(s)$	−928.4	−820.8	107.5	100.58
$FeSO_4 \cdot 7H_2O(s)$	−3014.57	−2509.87	409.2	394.47
$Fe_2(SO_4)_3(s)$	−2581.5			
Lead				
$Pb(s)$	0.0	0.0	64.81	26.44
PbO(yellow)	−217.32	−187.89	68.70	45.77
PbO(red)	−218.99	−188.93	66.5	45.81
$PbO_2(s)$	−277.4	−217.33	68.6	64.64
$Pb_3O_4(s)$	−718.4	−601.2	211.3	146.9
$PbS(s)$	−100.4	−98.7	91.2	49.50
$PbCl_2(s)$	−359.41	−314.10	136.0	
$Pb(NO_3)_2(s)$	−451.9			
$PbSO_4(s)$	−919.94	−813.14	148.57	103.207
Lithium				
$Li(s)$	0.0	0.0	29.12	24.77
$Li(g)$	159.37	126.66	138.77	20.786
$LiH(s)$	−90.54	−68.35	20.008	27.87
$LiOH(s)$	−484.93	−438.95	42.80	49.66
$LiF(s)$	−615.97	−587.71	35.65	41.59
$LiCl(s)$	−408.61	−384.37	59.33	47.99
$LiAlH_4(s)$	−116.3	−44.7	78.74	83.18
Magnesium				
$Mg(s)$	0.0	0.0	32.68	24.89
$MgO(s)$	−601.70	−569.43	26.94	37.15
$Mg(OH)_2(s)$	−924.54	−833.51	63.18	77.03
$Mg_3N_2(s)$	−460.7			
$MgCl_2(s)$	−641.32	−591.79	89.62	71.38
$MgBr_2(s)$	−524.3	−503.8	117.2	
$MgI_2(s)$	−364.0	−358.2	129.7	
$MgCO_3(s)$	−1095.8	−1012.1	65.7	75.52
$Mg(NO_3)_2(s)$	−790.65	−589.4	164.0	141.92
$Mg_3(PO_4)_2(s)$	−3780.7	−3538.7	189.20	213.47
$MgSO_4(s)$	−1284.9	−1170.6	91.6	96.48
Manganese				
Mn(α-solid)	0.0	0.0	32.01	26.32
$MnO(s)$	−385.22	−362.90	59.71	45.44
$MnO_2(s)$	−520.03	−465.14	53.05	54.14
$Mn_2O_3(s)$	−959.0	−881.1	110.5	107.65
$Mn_3O_4(s)$	−1387.8	−1283.2	155.6	139.66
$MnO_4^-(aq)$	−541.4	−447.2	191.2	
$MnO_4^{2-}(aq)$	−653	−500 7	59	
Mercury				
$Hg(l)$	0.0	0.0	76.02	27.983
$Hg^{2+}(aq)$	171.1	164.40	−32.2	
$Hg_2^{2+}(aq)$	172.4	153.52	84.5	
HgO(red)	−90.83	−58.539	70.29	44.06
HgO(yellow)	−90.46	−58.409	71.1	
$HgCl_2(s)$	−224.3	−178.6	146.0	
$Hg_2Cl_2(s)$	−265.22	−210.745	192.5	

(continues)

Table IV.1 *(continued)*

Substance	ΔH_f° (kJ/mol)	ΔG_f° (kJ/mol)	S° (J/K mol)	C_p° (J/K mol)
HgS(red)	−58.2	−50.6	82.4	48.41
HgS(black)	−53.6	−47.7	88.3	
Neon				
Ne(g)	0.0	0.0	146.328	20.786
Nickel				
Ni(*s*)	0.0	0.0	29.87	26.07
NiO(*s*)	−239.7	−211.7	37.99	44.31
NiS(*s*)	−82.0	−79.5	52.97	47.11
$NiCl_2(s)$	−305.332	−259.032	97.65	71.67
$[Ni(NH_3)_6]^{2+}(aq)$	−630.1	−255.7	394.6	
Nitrogen				
$N_2(g)$	0.0	0.0	191.61	29.125
$NH_3(g)$	−46.11	−16.45	192.45	35.06
$NH_3(aq)$	−80.29	−26.50	111.3	
$NH_4^+(aq)$	−132.51	−79.31	113.4	79.9
$N_2H_4(l)$	50.63	149.34	121.21	98.87
NO(*g*)	90.25	86.55	210.761	29.844
$NO_2(g)$	33.18	51.31	240.06	37.20
$N_2O(g)$	82.05	104.20	219.85	38.45
$N_2O_3(g)$	83.72	139.46	312.28	65.61
$N_2O_4(g)$	9.16	97.89	304.29	77.28
$N_2O_5(s)$	−43.1	113.9	178.2	143.1
$N_2O_5(g)$	11.3	115.1	355.7	84.5
$HNO_3(l)$	−174.10	−80.71	155.60	109.87
$HNO_3(aq)$	−207.36	−111.25	146.4	
$NH_4NO_3(s)$	−365.56	−183.87	151.08	139.3
$NH_4NO_2(s)$	−256.5			
$NH_4Cl(s)$	−314.43	−202.87	94.6	84.1
$NH_4Cl(aq)$	−299.66	−210.52	169.9	
$(NH_4)_2SO_4(s)$	−1180.85	−901.67	220.1	187.49
NOCl(*g*)	51.71	66.08	261.69	44.69
NOBr(*g*)	82.17	82.42	273.66	45.48
Oxygen				
O(*g*)	249.170	231.731	161.055	21.912
$O_2(g)$	0.0	0.0	205.138	29.355
$O_3(g)$	142.7	163.2	238.93	39.20
$OH^-(aq)$	−229.994	−157.244	−10.75	
$H_2O(l)$	−285.830	−237.129	69.91	75.291
$H_2O(g)$	−241.818	−228.572	188.825	33.577
$H_2O_2(l)$	−187.78	−120.35	109.6	89.1
Phosphorus				
P(g)	314.64	278.25	163.193	20.786
P(white)	0.0	0.0	41.09	23.840
P(red)	−17.6	−12.1	22.80	21.21
P(black)	−39.3			
$P_4(g)$	58.91	24.44	279.98	67.15
$PH_3(g)$	5.4	13.4	210.23	37.11
$P_4O_6(s)$	−1640.1			
$P_4O_{10}(s)$	−2984.0	−2697.7	228.86	211.71
$HPO_3(s)$	−948.5			
$H_3PO_2(s)$	−604.6			
$H_3PO_3(s)$	−964.4			
$H_3PO_4(s)$	−1279.0	−1119.1	110.50	106.06
$H_3PO_4(l)$	−1266.9			
$H_3PO_4(aq)$	−1288.34	−1142.54	158.2	
$H_4P_2O_7(s)$	−2241.0			
$PF_5(g)$	−1595.8			
$PF_3(g)$	−918.8	−897.5	273.24	58.70
$PCl_5(g)$	−374.9	−305.0	364.58	112.80
$PCl_5(s)$	−443.5			
$PCl_3(l)$	−319.7	−272.3	217.1	
$PCl_3(g)$	−287.0	−267.8	311.78	71.84

(continues)

Table IV.1 *(continued)*

Substance	ΔH_f° (kJ/mol)	ΔG_f° (kJ/mol)	S° (J/K mol)	C_p° (J/K mol)
Potassium				
$K(s)$	0.0	0.0	64.18	29.58
$K(g)$	89.24	60.59	160.336	20.786
$KOH(s)$	−424.764	379.08	78.9	64.9
$KOH(aq)$	−482.37	−440.50	91.6	
$KCl(s)$	−436.747	−409.14	82.59	51.30
$KBr(s)$	−393.798	−380.66	95.90	52.30
$KI(s)$	−327.900	−324.892	106.32	52.93
$K_2CO_3(s)$	−1151.02	−1063.5	155.52	114.53
$KNO_3(s)$	−494.63	−394.86	133.05	96.40
$K_2SO_4(s)$	−1437.79	−1321.37	175.56	131.46
$K_2SO_4(aq)$	−1414.02	−1311.07	225.1	
$KMnO_4(s)$	−837.2	−737.6	171.71	117.57
$K_2Cr_2O_7(s)$	−2061.5	−1881.8	291.2	219.24
$K_2CrO_4(s)$	−1403.7	−1295.7	200.12	145.98
$K_3[Fe(CN)_6](s)$	−249.8	−129.6	426.06	
$K_4[Fe(CN)_6](s)$	−594.1	−453.0	418.8	332.21
Selenium				
Se(black)	0.0	0.0	42.442	25.363
Se(red)	6.7			
Se(g)	227.07	187.03	176.72	20.820
$H_2Se(g)$	29.7	15.9	219.02	34.73
$SeO_2(s)$	−225.35			
$SeO_3(s)$	−166.9			
$H_2SeO_4(s)$	−530.1			
$SeF_6(g)$	−1117	−1017	313.87	110.5
$SeCl_4(s)$	−183.3			
Silicon				
$Si(s)$	0.0	0.0	18.83	20.00
$Si(g)$	455.6	411.3	167.97	22.251
$SiH_4(g)$	34.3	56.9	204.62	42.84
SiO_2(α-quartz)	−910.94	−856.64	41.84	44.43
$H_4SiO_4(s)$	−1481.1	−1332.9	192	
$H_2SiO_3(s)$	−1188.7	−1092.4	134	
$SiF_4(g)$	−1614.94	−1572.65	282.49	73.64
$SiCl_4(l)$	−687.0	−619.84	239.7	145.31
$SiBr_4(l)$	−457.3	−443.9	277.8	
SiC(cubic)	−65.3	−62.8	16.61	26.86
Silver				
Ag	0.0	0.0	42.55	25.351
$Ag_2O(s)$	−31.05	−11.20	121.3	65.86
Ag_2S(orthorhombic)	−32.59	−40.67	144.01	76.53
$AgCl(s)$	−127.068	−109.789	96.2	50.79
$[AgCl_2]^-(aq)$	−245.2	−215.4	231.4	
$AgI(s)$	−61.84	−66.19	115.5	56.82
$AgNO_3(s)$	−124.39	−33.41	140.92	93.05
$[Ag(NH_3)_2]^+(aq)$	−111.29	−17.12	245.2	
Sodium				
$Na(s)$	0.0	0.0	51.21	28.24
$Na(g)$	107.32	76.761	153.712	20.786
$NaOH(s)$	−425.609	−379.494	64.455	59.54
$NaOH(aq)$	−470.114	−419.150	48.1	
$Na_2S(s)$	−364.8	−349.8	83.7	
$NaCl(s)$	−411.153	−384.138	72.13	50.50
$NaCl(aq)$	−407.27	−393.133	115.5	
$Na_2CO_3(s)$	−1130.68	−1044.44	134.98	112.30
$NaHCO_3(s)$	−950.81	−851.0	101.7	87.61
$Na_2SO_4(s)$	−1387.08	−1270.16	149.58	128.20
$Na_2S_2O_3(s)$	−1123.0	−1028.0	155	
$NaClO_4(s)$	−383.30	−254.85	142.3	
$NaCH_3COO(s)$	−708.81	−607.18	123.0	79.9
$NaBH_4(s)$	−188.61	−123.86	101.29	86.78

(continues)

Table IV.1 *(continued)*

Substance	ΔH_f° (kJ/mol)	ΔG_f° (kJ/mol)	S° (J/K mol)	C_p° (J/K mol)
Strontium				
Sr(*s*)	0.0	0.0	52.3	26.4
SrO(*s*)	−592.0	−561.9	54.4	45.02
$SrCO_3(s)$	−1220.1	−1140.1	97.1	81.42
$SrCl_2(s)$	−828.9	−781.1	114.85	75.60
Sulfur				
S(rhombic)	0.0	0.0	31.80	22.64
S(monoclinic)	0.33			
S(*g*)	278.805	238.250	167.821	23.673
$S_8(g)$	102.30	49.63	430.98	156.44
$H_2S(g)$	−20.63	−33.56	205.79	34.23
$SO_2(g)$	−296.830	−300.194	248.22	39.87
$SO_3(s)$	−454.51	−374.21	70.7	
$SO_3(g)$	−395.72	−371.06	256.76	50.67
$H_2SO_4(l)$	−813.989	−690.003	156.904	138.91
$H_2SO_4(aq)$	−909.27	−744.53	20.1	
$SCl_2(l)$	−50			
$S_2Cl_2(l)$	−59.4			
Tellurium				
Te(*s*)	0.0	0.0	49.71	25.73
$H_2Te(g)$	99.6			
$TeO_2(s)$	−322.6	−270.3	79.5	
Tin				
Sn(white)	0.0	0.0	51.55	26.99
Sn(gray)	−2.09	0.13	44.14	25.77
SnO(*s*)	−285.8	−256.9	56.5	44.31
$SnO_2(s)$	−580.7	−519.6	52.3	52.59
$SnCl_4(l)$	−511.3	−440.1	258.6	165.3
$SnCl_2(s)$	−325.1			
$SnCl_2 \cdot 2H_2O(s)$	−921.3			
Uranium				
U	0.0	0.0	50.21	27.665
$UO_2(s)$	−1084.9	−1031.7	77.03	63.60
$UO_3(s)$	−1223.8	−1145.9	96.11	81.67
$UF_6(g)$	−2147.4	−2063.7	377.9	129.62
$UF_6(s)$	−2197.0	−2068.5	227.6	166.77
Xenon				
Xe(*g*)	0.0	0.0	169.683	20.786
$XeF_4(s)$	−261.5			
Zinc				
Zn(*s*)	0.0	0.0	41.63	25.40
ZnO(*s*)	−348.28	−318.30	43.64	40.25
$Zn(OH)_2(s)$	−643.25	−555.07	81.6	72.4
ZnS(wurtzite)	−192.63			
ZnS(sphalerite)	−205.98	−201.29	57.7	46.0
$ZnCl_2(s)$	−415.05	−369.398	111.46	71.34
$ZnSO_4(s)$	−982.8	−871.5	110.5	99.2

Source: Data mainly from "The NBS Tables of Chemical Thermodynamic Properties," *Journal of Physical and Chemical Reference Data,* **11** (2), 1982.

Table IV.2
Vapor Pressure of Water below 100 °C

Temperature (°C)	Vapor pressure (Torr)	Temperature (°C)	Vapor pressure (Torr)	Temperature (°C)	Vapor pressure (Torr)
0.0	4.579	35.0	42.175	70.0	233.7
1.0	4.926	36.0	44.563	71.0	243.9
2.0	5.294	37.0	47.067	72.0	254.6
3.0	5.685	38.0	49.692	73.0	265.7
4.0	6.101	39.0	52.442	74.0	272.2
5.0	6.543	40.0	55.324	75.0	289.1
6.0	7.013	41.0	58.34	76.0	301.4
7.0	7.513	42.0	61.50	77.0	314.1
8.0	8.045	43.0	64.80	78.0	327.3
9.0	8.609	44.0	68.26	79.0	341.0
10.0	9.202	45.0	71.88	80.0	355.1
11.0	9.844	46.0	75.65	81.0	369.7
12.0	10.518	47.0	79.60	82.0	384.9
13.0	11.231	48.0	83.71	83.0	400.6
14.0	11.987	49.0	88.02	84.0	416.8
15.0	12.788	50.0	92.51	85.0	433.6
16.0	13.634	51.0	97.20	86.0	450.9
17.0	14.530	52.0	102.09	87.0	468.7
18.0	15.477	53.0	107.20	88.0	487.1
19.0	16.477	54.0	112.51	89.0	506.1
20.0	17.535	55.0	118.04	90.0	525.76
21.0	18.650	56.0	123.80	91.0	546.05
22.0	19.827	57.0	129.82	92.0	566.99
23.0	21.068	58.0	136.08	93.0	588.60
24.0	22.377	59.0	142.60	94.0	610.90
25.0	23.756	60.0	149.38	95.0	633.90
26.0	25.209	61.0	156.43	96.0	657.62
27.0	26.739	62.0	163.77	97.0	682.07
28.0	28.349	63.0	171.38	98.0	707.27
29.0	30.043	64.0	179.31	99.0	733.24
30.0	31.824	65.0	187.54	100.0	760.00
31.0	33.695	66.0	196.09		
32.0	35.663	67.0	204.96		
33.0	37.729	68.0	214.17		
34.0	39.898	69.0	223.73		

APPENDIX

V

Equilibrium Constants

Table V.1

K_a, Ionization Constants of Acids at 25 °C

The acids are arranged in alphabetical order of the nonmetal atom.

Name	Formula	K_a	Name	Formula	K_a
Inorganic and organic acids			**Chlorine**		
			Hydrochloric	HCl	large
Arsenic			Perchloric	$HClO_4$	large
Arsenic	H_3AsO_4	6.5×10^{-3}	Chloric	$HClO_3$	large
	$H_2AsO_4^-$	1.1×10^{-7}	Chlorous	$HClO_2$	1.1×10^{-2}
	$HAsO_4^{2-}$	3×10^{-12}	Hypochlorous	$HClO$	2.90×10^{-8}
Boron			**Chromium**		
Boric	H_3BO_3	6.0×10^{-10}	Chromic	H_2CrO_4	1.8×10^{-1}
Bromine				$HCrO_4^-$	3.2×10^{-7}
Hydrobromic	HBr	large	**Fluorine**		
Hypobromous	$HBrO$	2.2×10^{-9}	Hydrofluoric	HF	6.5×10^{-4}
Carbon			**Iodine**		
Acetic	CH_3COOH	1.754×10^{-5}	Hydroiodic	HI	large
Benzoic	C_6H_5COOH	6.6×10^{-5}	Periodic	HIO_4	5.6×10^{-9}
Carbonic	H_2CO_3	4.5×10^{-7}	Iodic	HIO_3	1.6×10^{-1}
	HCO_3^-	4.8×10^{-11}	Hypoiodous	HIO	2.3×10^{-11}
Chloroacetic	$CH_2ClCOOH$	1.40×10^{-3}	**Manganese**		
Cyanic	$HNCO$	3.3×10^{-4}	Permanganic	$HMnO_4$	large
Dichloroacetic	$CHCl_2COOH$	3.32×10^{-2}	**Nitrogen**		
Formic	$HCOOH$	1.772×10^{-4}	Nitric	HNO_3	large
Hydrocyanic	HCN	6.2×10^{-10}	Nitrous	HNO_2	7.2×10^{-4}
Oxalic	$H_2C_2O_4$	5.60×10^{-2}	**Oxygen**		
	$HC_2O_4^-$	6.2×10^{-5}	Hydrogen peroxide	H_2O_2	2.2×10^{-12}
Propionic	CH_3CH_2COOH	1.3×10^{-5}	**Phosphorus**		
Thiocyanic	$HNCS$	large	Phosphoric	H_3PO_4	7.5×10^{-3}
Trichloroacetic	CCl_3COOH	2×10^{-1}			

(continues)

Table V.1 (*continued*)

Name	Formula	K_a	Name	Formula	K_a
Phosphoric—*continued*					
	$H_2PO_4^-$	6.6×10^{-8}	**Amphoteric hydroxides**		
	HPO_4^{2-}	1×10^{-12}	Aluminum hydroxide	$Al(OH)_3$	4×10^{-13}
Phosphorous	H_3PO_3	3×10^{-2}	Antimony(III) hydroxide	$SbO(OH)$	1×10^{-11}
	$H_2PO_3^-$	1.6×10^{-7}	Chromium(III) hydroxide	$Cr(OH)_3$	9×10^{-17}
Hypophosphorous	H_3PO_2	1.23×10^{-2}	Copper(II) hydroxide	$Cu(OH)_2$	1×10^{-19}
Pyrophosphoric	$H_4P_2O_7$	1.2×10^{-1}		$HCuO_2$	7.0×10^{-14}
	$H_3P_2O_7^-$	7.9×10^{-2}	Lead(II) hydroxide	$Pb(OH)_2$	4.6×10^{-16}
	$H_2P_2O_7^{2-}$	2.0×10^{-7}	Tin(IV) hydroxide	$Sn(OH)_4$	10^{-32}
	$HP_2O_7^{3-}$	4.8×10^{-10}	Tin(II) hydroxide	$Sn(OH)_2$	3.8×10^{-15}
Silicon			Zinc hydroxide	$Zn(OH)_2$	1.0×10^{-29}
Metasilicic	H_2SiO_3	3.2×10^{-10}	**Metal cations**		
	$HSiO_3^-$	1.5×10^{-12}	Aluminum ion	Al^{3+}	1.4×10^{-5}
Sulfur			Ammonium ion	NH_4^+	6.3×10^{-10}
Hydrosulfuric	H_2S	1.0×10^{-7}	Bismuth(III) ion	Bi^{3+}	1×10^{-2}
	HS^-	3×10^{-13}	Chromium(III) ion	Cr^{3+}	1×10^{-4}
Sulfuric	H_2SO_4	large	Copper(II) ion	Cu^{2+}	1×10^{-8}
	HSO_4^-	1.0×10^{-2}	Iron(III) ion	Fe^{3+}	4.0×10^{-3}
Sulfurous	H_2SO_3	1.43×10^{-2}	Iron(II) ion	Fe^{2+}	1.2×10^{-6}
	HSO_3^-	5.0×10^{-8}	Magnesium ion	Mg^{2+}	2×10^{-12}
Thiosulfuric	$H_2S_2O_3$	2.0×10^{-2}	Mercury(II) ion	Hg^{2+}	2×10^{-3}
	$HS_2O_3^-$	3.2×10^{-3}	Zinc ion	Zn^{2+}	2.5×10^{-10}

Source: Data mainly from *Stability Constants of Metal-Ion Complexes*, Special Publications **17** (1964) and **25** (1971), The Chemical Society, London.

Table V.2

K_b, Ionization Constants of Bases at 25 °C

For sparingly soluble bases, see Table V.3. The ions are arranged in alphabetical order.

Name	Formula	K_b	Name	Formula	K_b
Inorganic and organic bases			Anions—*continued*		
			Fluoride ion	F^-	1.5×10^{-11}
Ammonia	NH_3	1.6×10^{-5}	Formate ion	$HCOO^-$	5.643×10^{-11}
Aniline	$C_6H_5NH_2$	4.2×10^{-10}	Nitrite ion	NO_2^-	1.4×10^{-11}
Diethylamine	$(C_2H_5)_2NH$	9.5×10^{-4}	Oxalate ion	$C_2O_4^{2-}$	1.6×10^{-10}
Dimethylamine	$(CH_3)_2NH$	5.9×10^{-4}		$HC_2O_4^-$	1.79×10^{-13}
Ethylamine	$C_2H_5NH_2$	4.7×10^{-4}	Phosphate ion	PO_4^{3-}	1×10^{-2}
Methylamine	CH_3NH_2	3.9×10^{-4}		HPO_4^{2-}	1.5×10^{-7}
Triethylamine	$(C_2H_5)_3N$	5.2×10^{-4}		$H_2PO_4^-$	1.3×10^{-12}
Trimethylamine	$(CH_3)_3N$	6.3×10^{-5}	Phosphite ion	HPO_3^{2-}	6.3×10^{-8}
Anions				$H_2PO_3^-$	3×10^{-13}
Acetate ion	CH_3COO^-	5.701×10^{-10}	Metasilicate ion	SiO_3^{2-}	6.7×10^{-3}
Arsenate ion	AsO_4^{3-}	3.3×10^{-3}		$HSiO_3^-$	3.1×10^{-5}
	$HAsO_4^{2-}$	9.1×10^{-8}	Sulfate ion	SO_4^{2-}	1.0×10^{-12}
	$H_2AsO_4^-$	1.5×10^{-12}	Sulfite ion	SO_3^{2-}	2.0×10^{-7}
Borate ion	$H_2BO_3^-$	1.6×10^{-5}		HSO_3^-	6.99×10^{-13}
	$B_4O_7^{2-}$	10^{-3}	Sulfide ion	S^{2-}	3×10^{-2}
Carbonate ion	CO_3^{2-}	2.1×10^{-4}		HS^-	1.0×10^{-7}
	HCO_3^-	2.2×10^{-8}	Thiocyanate ion	NCS^-	1.4×10^{-11}
Chromate ion	CrO_4^{2-}	3.1×10^{-8}	Thiosulfate ion	$S_2O_3^{2-}$	3.1×10^{-12}
Cyanide ion	CN^-	1.6×10^{-5}			

Source: Data mainly from *Stability Constants of Metal-Ion Complexes*, Special Publications **17** (1964) and **25** (1971), The Chemical Society, London.

Table V.3

K_{sp}, Solubility Products of Sparingly Soluble Salts and Bases at 25 °C.
The substances within each group are arranged in order of decreasing K_{sp}.

Salt	K_{sp}	Salt	K_{sp}	Salt	K_{sp}
Acetates		**Hydroxides**		**Phosphates**	
$Ag(CH_3COO)$	4.4×10^{-3}	$Ba(OH)_2$	1.3×10^{-2}	Li_3PO_4	3×10^{-13}
$Hg_2(CH_3COO)_2$	4×10^{-10}	$Sr(OH)_2$	6.4×10^{-3}	$Mg(NH_4)PO_4$	3×10^{-13}
Arsenates		$Ca(OH)_2$	4.0×10^{-5}	Ag_3PO_4	1.4×10^{-16}
Ag_3AsO_4	1×10^{-22}	Ag_2O	2×10^{-8}	$AlPO_4$	5.8×10^{-19}
Bromides		$Mg(OH)_2$	7.1×10^{-12}	$Mn_3(PO_4)_2$	1×10^{-22}
$PbBr_2$	3.9×10^{-5}	$BiO(OH)$	1×10^{-12}	$Ba_3(PO_4)_2$	3×10^{-23}
$CuBr$	5.2×10^{-9}	$Be(OH)_2$	4×10^{-13}	$BiPO_4$	1.3×10^{-23}
$AgBr$	4.9×10^{-13}	$Mn(OH)_2$	2×10^{-13}	$Ca_3(PO_4)_2$	10^{-26}
Hg_2Br_2	5.8×10^{-23}	$Cd(OH)_2$	8.1×10^{-15}	$Sr_3(PO_4)_2$	4×10^{-28}
Carbonates		$Pb(OH)_2$	1.2×10^{-15}	$Mg_3(PO_4)_2$	10^{-32}
$MgCO_3$	1×10^{-5}	$Fe(OH)_2$	8×10^{-16}	$Pb_3(PO_4)_2$	7.9×10^{-43}
$NiCO_3$	1.3×10^{-7}	$Ni(OH)_2$	3×10^{-16}	**Sulfates**	
$CaCO_3$	3.84×10^{-9}	$Co(OH)_2$	2×10^{-16}	$CaSO_4$	2.5×10^{-5}
$BaCO_3$	2.0×10^{-9}	$Zn(OH)_2$	1.2×10^{-17}	Ag_2SO_4	1.5×10^{-5}
$SrCO_3$	5.2×10^{-10}	$SbO(OH)$	1×10^{-17}	Hg_2SO_4	6.8×10^{-7}
$MnCO_3$	5.0×10^{-10}	$Cu(OH)_2$	1.3×10^{-20}	$SrSO_4$	3.5×10^{-7}
$CuCO_3$	2.3×10^{-10}	$Hg(OH)_2$	4×10^{-26}	$PbSO_4$	2.2×10^{-8}
$CoCO_3$	1.0×10^{-10}	$Sn(OH)_2$	6×10^{-27}	$BaSO_4$	1.7×10^{-10}
$FeCO_3$	2.1×10^{-11}	$Cr(OH)_3$	6×10^{-31}	**Sulfides**	
$ZnCO_3$	1.7×10^{-11}	$Al(OH)_3$	3.5×10^{-34}	MnS	2.3×10^{-13}
Ag_2CO_3	8.1×10^{-12}	$Fe(OH)_3$	3×10^{-39}	FeS	4.2×10^{-17}
$CdCO_3$	1.0×10^{-12}	$Sn(OH)_4$	10^{-57}	NiS	3×10^{-19}
$PbCO_3$	7.4×10^{-14}	**Iodides**		CoS	4×10^{-21}
Chlorides		PbI_2	7.1×10^{-9}	ZnS	2×10^{-24}
$PbCl_2$	2×10^{-5}	CuI	1.1×10^{-12}	SnS	3×10^{-27}
$CuCl$	1.2×10^{-6}	AgI	8.3×10^{-17}	CdS	2×10^{-28}
$AgCl$	1.8×10^{-10}	HgI_2	3×10^{-26}	PbS	1×10^{-28}
Hg_2Cl_2	1.3×10^{-18}	Hg_2I_2	4.5×10^{-29}	CuS	6×10^{-36}
Chromates		**Nitrates**		Cu_2S	3×10^{-48}
$CaCrO_4$	6×10^{-4}	$BiO(NO_3)$	2.8×10^{-3}	Ag_2S	7.1×10^{-50}
$SrCrO_4$	2.2×10^{-5}	**Nitrites**		HgS	4×10^{-53}
Hg_2CrO_4	2.0×10^{-9}	$Ag(NO_2)$	6.0×10^{-4}	Fe_2S_3	1×10^{-88}
$BaCrO_4$	1.2×10^{-10}	**Oxalates**			
Ag_2CrO_4	2.5×10^{-12}	MgC_2O_4	8×10^{-5}		
$PbCrO_4$	2.8×10^{-13}	$CoCrO_4$	4×10^{-6}		
Cyanides		FeC_2O_4	2×10^{-7}		
$AgCN$	2.3×10^{-16}	NiC_2O_4	1×10^{-7}		
Ferrocyanides		SrC_2O_4	5×10^{-8}		
$KFe[Fe(CN)_6]$	3×10^{-41}	CuC_2O_4	3×10^{-8}		
$Ag_4[Fe(CN)_6]$	2×10^{-41}	BaC_2O_4	2×10^{-8}		
$K_2Zn_3[Fe(CN)_6]_2$	1×10^{-95}	CdC_2O_4	2×10^{-8}		
Fluorides		ZnC_2O_4	2×10^{-9}		
BaF_2	1.0×10^{-6}	CaC_2O_4	1×10^{-9}		
MgF_2	6.8×10^{-9}	$Ag_2C_2O_4$	3.5×10^{-11}		
SrF_2	2.5×10^{-9}	PbC_2O_4	4.8×10^{-12}		
CaF_2	2.7×10^{-11}	$Hg_2C_2O_4$	2×10^{-13}		
ThF_4	4×10^{-28}	MnC_2O_4	1×10^{-15}		
		$La_2(C_2O_4)_3$	2×10^{-28}		

Table V.4

K_d, Dissociation Constants for Complexes at 25 °C

The complexes for a given metal ion are arranged in order of decreasing K_d. The following abbreviations represent ligands: (en) is the ethylenediamine molecule, $H_2NCH_2CH_2NH_2$; (nta) is the nitrilotriacetate ion, $N(CH_2COO)_2^{3-}$; (gly) is the glycine ion, $H_2NCH_2COO^-$; and (edta) is the ethylenediaminetetraacetate ion, $(OOCCH_2)_2NCH_2CH_2N(CH_2COO)_2^{4-}$.

Complex	K_d	Complex	K_d	Complex	K_d
Aluminum		**Copper**		**Mercury**	
$[AlF_6]^{3-}$	3×10^{-20}	$[Cu(SCN)_2]$	1.8×10^{-4}	$[HgCl_4]^{2-}$	2×10^{-16}
Calcium		$[CuCl_2]^-$	1.15×10^{-5}	$[Hg(SCN)_4]^{2-}$	2.0×10^{-22}
$[Ca(P_2O_7)]^{2-}$	1×10^{-5}	$[Cu(P_2O_7)]^{2-}$	2.0×10^{-7}	**Nickel**	
$[Ca(nta)_2]^{4-}$	2.44×10^{-12}	$[Cu(C_2O_4)_2]^{2-}$	6×10^{-11}	$[Ni(NH_3)_6]^{2+}$	1×10^{-9}
Cadmium		$[Cu(NH_3)_4]^{2+}$	1×10^{-13}	**Palladium**	
$[CdCl_4]^{2-}$	9.3×10^{-3}	$[Cu(gly)_2]$	5.6×10^{-16}	$[PdBr_4]^{2-}$	8.0×10^{-14}
$[Cd(SCN)_4]^{2-}$	1×10^{-3}	$[Cu(OH)_4]^{2-}$	7.6×10^{-17}	$[PdCl_4]^{2-}$	6×10^{-14}
$[CdBr_4]^{2-}$	2×10^{-4}	$[Cu(eta)]^{2-}$	1.38×10^{-19}	**Silver**	
$[Cd(NH_3)_6]^{2+}$	1×10^{-5}	**Gold**		$[Ag(OH)_3]^{2-}$	1.7×10^{-5}
$[CdI_4]^{2-}$	8×10^{-7}	$[Au(CN)_2]^-$	5×10^{-39}	$[Ag(en)]^+$	1×10^{-5}
$[Cd(CH_3NH_2)_4]^{2+}$	2.82×10^{-7}	**Iron**		$[AgCl_2]^-$	9×10^{-6}
$[Cd(NH_3)_4]^{2+}$	1×10^{-7}	$[Fe(C_2O_4)_3]^{4-}$	6×10^{-6}	$[AgCl_4]^{3-}$	5×10^{-6}
$[Cd(en)_4]^{2+}$	2.60×10^{-11}	$[Fe(SCN)_3]$	5×10^{-7}	$[AgBr_2]^-$	7.8×10^{-8}
$[Cd(CN)_4]^{2-}$	8.2×10^{-18}	$[Fe(C_2O_4)_2]^{2-}$	2×10^{-8}	$[Ag(NH_3)_2]^+$	6.2×10^{-8}
Cobalt		$[Fe(C_2O_4)_3]^{3-}$	3×10^{-21}	$[Ag(SCN)_4]^{3-}$	2.1×10^{-10}
$[Co(NH_3)_6]^{2+}$	9×10^{-6}	$[Fe(CN)_6]^{4-}$	1.3×10^{-37}	$[Au(CN)_2]^-$	1×10^{-22}
$[Co(C_2O_4)_3]^{4-}$	2.2×10^{-7}	$[Fe(CN)_6]^{3-}$	1.3×10^{-44}	**Zinc**	
$[Co(en)_3]^{2+}$	1.52×10^{-14}	**Lead**		$[Zn(NH_3)_4]^{2+}$	3.46×10^{-10}
$[Co(en)_3]^{3+}$	2.04×10^{-49}	$[Pb(SCN)_2]$	3×10^{-3}	$[Zn(gly)_2]$	1.1×10^{-10}
		Magnesium		$[Zn(edta)]^{2-}$	2.63×10^{-17}
		$[Mg(P_2O_7)]^{2-}$	2×10^{-6}	$[Zn(CN)_4]^{2-}$	2.4×10^{-20}
		$[Mg(nta)_2]^{4-}$	6.3×10^{-11}	$[Zn(OH)_4]^{2-}$	5×10^{-21}

APPENDIX VI

Standard Reduction Potentials at 25 °C

Table VI.1
Standard Reduction Potentials in Acidic Media

Reduction half-reaction	E°, V	Reduction half-reaction	E°, V
$Li^+ + e^- \longrightarrow Li(s)$	−3.045	$U^{4+} + e^- \longrightarrow U^{3+}$	−0.607
$Rb^+ + e \longrightarrow Rb(s)$	−2.925	$As(s) + 3H^+ + 3e^- \longrightarrow AsH_3(g)$	−0.607
$K^+ + e^- \longrightarrow K(s)$	−2.925	$Ga^{3+} + 3e^- \longrightarrow Ga(s)$	−0.529
$Cs^+ + e^- \longrightarrow Cs(s)$	−2.923	$Fe^{2+} + 2e^- \longrightarrow Fe(s)$	−0.4402
$Ra^{2+} + 2e^- \longrightarrow Ra(s)$	−2.916	$Cr^{3+} + e^- \longrightarrow Cr^{2+}$	−0.408
$Ba^2 + 2e^- \longrightarrow Ba(s)$	−2.906	$Cd^{2+} + 2e^- \longrightarrow Cd(s)$	−0.4029
$Sr^{2+} + 2e^- \longrightarrow Sr(s)$	−2.888	$Se(s) + 2H^+ + 2e^- \longrightarrow H_2Se(aq)$	−0.399
$Ca^{2+} + 2e^- \longrightarrow Ca(s)$	−2.866	$Ti^{3+} + e^- \longrightarrow Ti^{2+}$	−0.369
$Na^+ + e^- \longrightarrow Na(s)$	−2.714	$PbI_2(s) + 2e^- \longrightarrow Pb(s) + 2I^-$	−0.365
$La^{3+} + 3e^- \longrightarrow La(s)$	−2.522	$PbSO_4(s) + 2e^- \longrightarrow Pb(s) + SO_4^{2-}$	−0.3588
$Ce^{3+} + 3e^- \longrightarrow Ce(s)$	−2.483	$In^{3+} + 3e^- \longrightarrow In(s)$	−0.343
$Mg^{2+} + 2e^- \longrightarrow Mg(s)$	−2.363	$Tl^+ + e^- \longrightarrow Tl(s)$	−0.3363
$H_2(g) + 2e^- \longrightarrow 2H^-$	−2.25	$PbBr_2(s) + 2e^- \longrightarrow Pb(s) + 2Br^-$	−0.284
$Sc^{3+} + 3e^- \longrightarrow Sc(s)$	−2.077	$Co^{2+} + 2e^- \longrightarrow Co(s)$	−0.277
$[AlF_6]^{3-} + 3e^- \longrightarrow Al(s) + 6F^-$	−2.069	$PbCl_2(s) + 2e^- \longrightarrow Pb(s) + 2Cl^-$	−0.268
$Be^{2+} + 2e^- \longrightarrow Be(s)$	−1.847	$V^{3+} + e^- \longrightarrow V^{2+}$	−0.256
$V^{3+} + 3e^- \longrightarrow V(s)$	−1.798	$Ni^{2+} + 2e^- \longrightarrow Ni(s)$	−0.250
$Hf^{4+} + 4e^- \longrightarrow Hf(s)$	−1.700	$AgI(s) + e^- \longrightarrow Ag(s) + I^-$	−0.1518
$Al^{3+} + 3e^- \longrightarrow Al(s)$	−1.662	$Sn^{2+} + 2e^- \longrightarrow Sn(s)$	−0.136
$Ti^{2+} + 2e^- \longrightarrow Ti(s)$	−1.628	$Pb^{2+} + 2e^- \longrightarrow Pb(s)$	−0.126
$Zr^{4+} + 4e^- \longrightarrow Zr(s)$	−1.529	$P(s) + 3H^+ + 3e^- \longrightarrow PH_3(g)$	−0.063
$V^{4+} + 4e^- \longrightarrow V(s)$	−1.50	$Fe^{3+} + 3e^- \longrightarrow Fe(s)$	−0.036
$[SiF_6]^{2-} + 4e^- \longrightarrow Si(s) + 6F^-$	−1.24	$2H^+ + 2e^- \longrightarrow H_2(g)$	0.000
$[TiF_6]^{2-} + 4e^- \longrightarrow Ti(s) + 6F^-$	−1.191	$AgBr(s) + e^- \longrightarrow Ag(s) + Br^-$	+0.0713
$Mn^{2+} + 2e^- \longrightarrow Mn(s)$	−1.185	$Si(s) + 4H^+ + 4e^- \longrightarrow SiH_4(g)$	+0.102
$V^{2+} + 2e^- \longrightarrow V(s)$	−1.175	$Hg_2Br_2(s) + 2e^- \longrightarrow 2Hg(l) + 2Br^-$	+0.1397
$Cr^{2+} + 2e^- \longrightarrow Cr(s)$	−0.913	$S(s) + 2H^+ + 2e^- \longrightarrow H_2S(aq)$	+0.142
$H_3BO_3(s) + 3H^+ + 3e^- \longrightarrow B(s) + 3H_2O(l)$	−0.869	$Sn^{4+} + 2e^- \longrightarrow Sn^{2+}$	+0.15
$SiO_2(s) + 4H^+ + 4e^- \longrightarrow Si(s) + 2H_2O(l)$	−0.857	$Sb_2O_3(s) + 6H^+ + 6e^- \longrightarrow 2Sb(s) + 3H_2O(l)$	+0.152
$Zn^{2+} + 2e^- \longrightarrow Zn(s)$	−0.7628	$Cu^{2+} + e^- \longrightarrow Cu^+$	+0.153
$Cr^{3+} + 3e^- \longrightarrow Cr(s)$	−0.744	$SO_4^{2-} + 4H^+ + 2e^- \longrightarrow H_2SO_3(aq) + H_2O(l)$	+0.172
$Te(s) + 2H^+ + 2e^- \longrightarrow H_2Te(aq)$	−0.739	$AgCl(s) + e^- \longrightarrow Ag(s) + Cl^-$	+0.2222

(continues)

Table VI.1 *(continued)*

Reduction half-reaction	E°, V	Reduction half-reaction	E°, V
$[Hg_2Br_4]^{2-} + 2e^- \longrightarrow Hg(l) + 4Br^-$	+0.223	$Br_2(l) + 2e^- \longrightarrow 2Br^-$	+1.0652
$Hg_2Cl_2(s) + 2e^- \longrightarrow 2Hg(l) + 2Cl^-$	+0.2676	$Br_2(aq) + 2e^- \longrightarrow 2Br^-$	+1.087
$Cu^{2+} + 2e^- \longrightarrow Cu(s)$	+0.337	$SeO_4^{2-} + 4H^+ + 2e^- \longrightarrow H_2SeO_3(aq) + H_2O(l)$	+1.15
$SO_4^{2-} + 8H^+ + 6e^- \longrightarrow S(s) + 4H_2O(l)$	+0.3572	$ClO_4^- + 2H^+ + 2e^- \longrightarrow ClO_3^- + H_2O(l)$	+1.19
$VO^{2+} + 2H^+ + e^- \longrightarrow V^{3+} + H_2O(l)$	+0.359	$2IO_3^- + 12H^+ + 10e^- \longrightarrow I_2(s) + 6H_2O(l)$	+1.195
$[Fe(CN)_6]^{3-} + e^- \longrightarrow [Fe(CN)_6]^{4-}$	+0.36	$Pt^{2+} + 2e^- \longrightarrow Pt(s)$	~1.2
$H_2SO_3(aq) + 4H^+ + 4e^- \longrightarrow S(s) + 3H_2O(l)$	+0.450	$ClO_3^- + 3H^+ + 2e^- \longrightarrow HClO_2(aq) + H_2O(l)$	+1.21
$Cu^+ + e^- \longrightarrow Cu(s)$	+0.521	$O_2(g) + 4H^+ + 4e^- \longrightarrow 2H_2O(l)$	+1.229
$I_2(s) + 2e^- \longrightarrow 2I^-$	+0.5355	$MnO_2(s) + 4H^+ + 2e^- \longrightarrow Mn^{2+} + 2H_2O(l)$	+1.23
$MnO_4^- + e^- \longrightarrow MnO_4^{2-}$	+0.564	$2HNO_2(aq) + 4H^+ + 4e^- \longrightarrow N_2O(g) + 3H_2O(l)$	+1.29
$Hg_2SO_4(s) + 2e^- \longrightarrow 2Hg(l) + SO_4^{2-}$	+0.6151	$Cr_2O_7^{2-} + 14H^+ + 6e^- \longrightarrow 2Cr^{3+} + 7H_2O(l)$	+1.33
$Cu^{2+} + Br^- + e^- \longrightarrow CuBr(s)$	+0.640	$Cl_2(g) + 2e^- \longrightarrow 2Cl^-$	+1.3597
$Po^{2+} + 2e^- \longrightarrow Po(s)$	+0.65	$PbO_2(s) + 4H^+ + 2e^- \longrightarrow Pb^{2+} + 4H_2O(l)$	+1.455
$[PtCl_6]^{2-} + 2e^- \longrightarrow [PtCl_4]^{2-} + 2Cl^-$	+0.68	$Au^{3+} + 3e^- \longrightarrow Au(s)$	+1.498
$O_2(g) + 2H^+ + 2e^- \longrightarrow H_2O_2(aq)$	+0.6826	$MnO_4^- + 8H^+ + 5e^- \longrightarrow Mn^{2+} + 4H_2O(l)$	+1.51
$[PtCl_4]^{2-} + 2e^- \longrightarrow Pt(s) + 4Cl^-$	+0.73	$2BrO_3^- + 12H^+ + 10e^- \longrightarrow Br_2(l) + 6H_2O(l)$	+1.52
$Fe^{3+} + e^- \longrightarrow Fe^{2+}$	+0.771	$Ce^{4+} + e^- \longrightarrow Ce^{3+}$	+1.61
$Hg_2^{2+} + 2e^- \longrightarrow 2Hg(l)$	+0.788	$2HClO(aq) + 2H^+ + 2e^- \longrightarrow Cl_2(g) + 2H_2O(l)$	+1.63
$Ag^+ + e^- \longrightarrow Ag(s)$	+0.7991	$HClO_2(aq) + 2H^+ + 2e^- \longrightarrow HClO(aq) + H_2O(l)$	+1.645
$Rh^{3+} + 3e^- \longrightarrow Rh(s)$	+0.80	$Au^+ + e^- \longrightarrow Au(s)$	+1.691
$2NO_3^- + 4H^+ + 2e^- \longrightarrow N_2O_4(g) + 2H_2O(l)$	+0.803	$H_2O_2(aq) + 2H^+ + 2e^- \longrightarrow 2H_2O(l)$	+1.776
$Cu^{2+} + I^- + e^- \longrightarrow CuI(s)$	+0.86	$Co^{3+} + e^- \longrightarrow Co^{2+}$	+1.808
$2Hg^{2+} + 2e^- \longrightarrow Hg_2^{2+}$	+0.920	$Ag^{2+} + e^- \longrightarrow Ag^+$	+1.980
$NO_3^- + 3H^+ + 2e^- \longrightarrow HNO_2(aq) + H_2O(l)$	+0.94	$S_2O_8^{2-} + 2e^- \longrightarrow 2SO_4^{2-}$	+2.01
$NO_3^- + 4H^+ + 3e^- \longrightarrow NO(g) + 2H_2O(l)$	+0.96	$O_3(g) + 2H^+ + 2e^- \longrightarrow O_2(g) + H_2O(l)$	+2.07
$Pd^{2+} + 2e^- \longrightarrow Pd(s)$	+0.987	$F_2(g) + 2e^- \longrightarrow 2F^-$	+2.87
$[AuCl_4]^- + 3e^- \longrightarrow Au(s) + 4Cl^-$	+1.00	$F_2(g) + 2H^+ + 2e^- \longrightarrow 2HF(aq)$	+3.06

Source: Data adapted from T. Moeller, *Inorganic Chemistry* (New York: Wiley, 1982), pp. 789–803.

Table VI.2
Standard Reduction Potentials in Alkaline Media

Reduction half-reaction	E°, V	Reduction half-reaction	E°, V
$Ca(OH)_2(s) + 2e^- \longrightarrow Ca(s) + 2OH^-$	−3.02	$SbO_2^- + 2H_2O(l) + 3e^- \longrightarrow Sb(s) + 4OH^-$	−0.66
$Sr(OH)_2(s) + 2e^- \longrightarrow Sr(s) + 2OH^-$	−2.88	$PbO(s) + H_2O(l) + 2e^- \longrightarrow Pb(s) + 2OH^-$	−0.580
$Ce(OH)_3(s) + 3e^- \longrightarrow Ce(s) + 3OH^-$	−2.87	$TeO_3^{2-} + 3H_2O(l) + 4e^- \longrightarrow Te(s) + 6OH^-$	−0.57
$Mg(OH)_2(s) + 2e^- \longrightarrow Mg(s) + 2OH^-$	−2.690	$Fe(OH)_3(s) + e^- \longrightarrow Fe(OH)_2(s) + OH^-$	−0.56
$BeO(s) + H_2O(l) + 2e^- \longrightarrow Be(s) + 2OH^-$	−2.613	$S(s) + 2e^- \longrightarrow S^{2-}$	−0.447
$Al(OH)_3(s) + 3e^- \longrightarrow Al(s) + 3OH^-$	−2.30	$Cu_2O(s) + H_2O(l) + 2e^- \longrightarrow 2Cu(s) + 2OH^-$	−0.358
$U(OH)_4(s) + e^- \longrightarrow U(OH)_3(s) + OH^-$	−2.20	$TlOH(s) + e^- \longrightarrow Tl(s) + OH^-$	−0.343
$U(OH)_3(s) + 3e^- \longrightarrow U(s) + 3OH^-$	−2.17	$CrO_4^{2-} + 4H_2O(l) + 3e^- \longrightarrow Cr(OH)_3(s) + 5OH^-$	−0.13
$H_2PO_2^- + e^- \longrightarrow P(s) + 2OH^-$	−2.05	$2Cu(OH)_2(s) + 2e^- \longrightarrow Cu_2O(s) + H_2O(l) + 2OH^-$	−0.080
$SiO_3^{2-} + 3H_2O(l) + 4e^- \longrightarrow Si(s) + 6OH^-$	−1.697	$Tl(OH)_3(s) + 2e^- \longrightarrow TlOH(s) + 2OH^-$	−0.05
$Mn(OH)_2(s) + 2e^- \longrightarrow Mn(s) + 2OH^-$	−1.55	$MnO_2(s) + 2H_2O(l) + 2e^- \longrightarrow Mn(OH)_2(s) + 2OH^-$	−0.05
$Cr(OH)_3(s) + 3e^- \longrightarrow Cr(s) + 3OH^-$	−1.34	$NO_3^- + H_2O(l) + 2e^- \longrightarrow NO_2^- + 2OH^-$	+0.01
$Zn(OH)_2(s) + 2e^- \longrightarrow Zn(s) + 2OH^-$	−1.245	$SeO_4^{2-} + H_2O(l) + 2e^- \longrightarrow SeO_3^{2-} + 2OH^-$	+0.05
$Te(s) + 2e^- \longrightarrow Te^{2-}$	−1.143	$HgO(s) + H_2O(l) + 2e^- \longrightarrow Hg(l) + 2OH^-$	+0.098
$PO_4^{3-} + 2H_2O(l) + 2e^- \longrightarrow HPO_3^{2-} + 3OH^-$	−1.12	$PbO_2(s) + H_2O(l) + 2e^- \longrightarrow PbO(s) + 2OH^-$	+0.247
$WO_4^{2-} + 4H_2O(l) + 6e^- \longrightarrow W(s) + 8OH^-$	−1.05	$IO_3^- + 3H_2O(l) + 6e^- \longrightarrow I^- + 6OH^-$	+0.26
$MoO_4^{2-} + 4H_2O(l) + 6e^- \longrightarrow Mo(s) + 8OH^-$	−1.05	$ClO_3^- + H_2O(l) + 2e^- \longrightarrow ClO_2^- + 2OH^-$	+0.33
$In(OH)_3(s) + 3e^- \longrightarrow In(s) + 3OH^-$	−1.00	$Ag_2O(s) + H_2O(l) + 2e^- \longrightarrow 2Ag(s) + 2OH^-$	+0.345
$PbS(s) + 2e^- \longrightarrow Pb(s) + S^{2-}$	−0.93	$ClO_4^- + H_2O(l) + 2e^- \longrightarrow ClO_3^- + 2OH^-$	+0.36
$SO_4^{2-} + H_2O(l) + 2e^- \longrightarrow SO_3^{2-} + 2OH^-$	−0.93	$O_2(g) + 2H_2O(l) + 4e^- \longrightarrow 4OH^-$	+0.401
$Se(s) + 2e^- \longrightarrow Se^{2-}$	−0.92	$IO^- + H_2O(l) + 2e^- \longrightarrow I^- + 2OH^-$	+0.485
$P(s) + 3H_2O(l) + 3e^- \longrightarrow PH_3(g) + 3OH^-$	−0.89	$NiO_2(s) + 2H_2O(l) + 2e^- \longrightarrow Ni(OH)_2(s) + 2OH^-$	+0.490
$Fe(OH)_2(s) + 2e^- \longrightarrow Fe(s) + 2OH^-$	−0.877	$MnO_4^- + 2H_2O(l) + 3e^- \longrightarrow MnO_2(s) + 4OH^-$	+0.588
$2H_2O(l) + 2e^- \longrightarrow H_2(g) + 2OH^-$	−0.8281	$BrO_3^- + 3H_2O(l) + 6e^- \longrightarrow Br^- + 6OH^-$	+0.61
$Cd(OH)_2(s) + 2e^- \longrightarrow Cd(s) + 2OH^-$	−0.809	$BrO^- + H_2O(l) + 2e^- \longrightarrow Br^- + 2OH^-$	+0.761
$Co(OH)_2(s) + 2e^- \longrightarrow Co(s) + 2OH^-$	−0.73	$ClO^- + H_2O(l) + 2e^- \longrightarrow Cl^- + 2OH^-$	+0.89
$Ni(OH)_2(s) + 2e^- \longrightarrow Ni(s) + 2OH^-$	−0.72	$O_3(g) + H_2O(l) + 2e^- \longrightarrow O_2(g) + 2OH^-$	+1.24

Source: Data adapted from T. Moeller, *Inorganic Chemistry* (New York: Wiley, 1982), pp. 789–803.

Table VI.3
Elemental Listing of Standard Reduction Potentials at 25 °C

Reduction half-reaction	E°, V
Aluminum	
$Al(OH)_3(s) + 3e^- \longrightarrow Al(s) + 3OH^-$	−2.30
$[AlF_6]^{3-} + 3e^- \longrightarrow Al(s) + 6F^-$	−2.069
$Al^{3+} + 3e^- \longrightarrow Al(s)$	−1.662
Antimony	
$SbO_2^- + 2H_2O(l) + 3e^- \longrightarrow Sb(s) + 4OH^-$	−0.66
$Sb_2O_3(s) + 6H^+ + 6e^- \longrightarrow 2Sb(s) + 3H_2O(l)$	+0.152
Arsenic	
$As(s) + 3H^+ + 3e^- \longrightarrow AsH_3(g)$	−0.607
Barium	
$Ba^{2+} + 2e^- \longrightarrow Ba(s)$	−2.906
Beryllium	
$BeO(s) + H_2O(l) + 2e^- \longrightarrow Be(s) + 2OH^-$	−2.613
$Be^{2+} + 2e^- \longrightarrow Be(s)$	−1.847
Boron	
$H_3BO_3(s) + 3H^+ + 3e^- \longrightarrow B(s) + 3H_2O(l)$	−0.869
Bromine	
$BrO_3^- + 3H_2O(l) + 6e^- \longrightarrow Br^- + 6OH^-$	+0.61
$BrO^- + H_2O(l) + 2e^- \longrightarrow Br^- + 2OH^-$	+0.761
$Br_2(l) + 2e^- \longrightarrow 2Br^-$	+1.0652
$Br_2(aq) + 2e^- \longrightarrow 2Br^-$	+1.087
$2BrO_3^- + 12H^+ + 10e^- \longrightarrow Br_2(l) + 6H_2O(l)$	+1.52
Cadmium	
$Cd(OH)_2(s) + 2e^- \longrightarrow Cd(s) + 2OH^-$	−0.809
$Cd^{2+} + 2e^- \longrightarrow Cd(s)$	−0.4029
Calcium	
$Ca(OH)_2(s) + 2e^- \longrightarrow Ca(s) + 2OH^-$	−3.02
$Ca^{2+} + 2e^- \longrightarrow Ca(s)$	−2.866
Cerium	
$Ce(OH)_3(s) + 3e^- \longrightarrow Ce(s) + 3OH^-$	−2.87
$Ce^{3+} + 3e^- \longrightarrow Ce(s)$	−2.483
$Ce^{4+} + e^- \longrightarrow Ce^{3+}$	+1.61
Cesium	
$Cs^+ + e^- \longrightarrow Cs(s)$	−2.923
Chlorine	
$ClO_3^- + H_2O(l) + 2e^- \longrightarrow ClO_2^- + 2OH^-$	+0.33
$ClO_4^- + H_2O(l) + 2e^- \longrightarrow ClO_3^- + 2OH^-$	+0.36
$ClO^- + H_2O(l) + 2e^- \longrightarrow Cl^- + 2OH^-$	+0.89
$ClO_4^- + 2H^+ + 2e^- \longrightarrow ClO_3^- + H_2O(l)$	+1.19
$ClO_3^- + 3H^+ + 2e^- \longrightarrow HClO_2(aq) + H_2O(l)$	+1.21
$Cl_2(g) + 2e^- \longrightarrow 2Cl^-$	+1.3597
$2HClO(aq) + 2H^+ + 2e^- \longrightarrow Cl_2(g) + 2H_2O(l)$	+1.63
$HClO_2(aq) + 2H^+ + 2e^- \longrightarrow HClO(aq) + H_2O(l)$	+1.645
Chromium	
$Cr(OH)_3(s) + 3e^- \longrightarrow Cr(s) + 3OH^-$	−1.34
$Cr^{2+} + 2e^- \longrightarrow Cr(s)$	−0.913
$Cr^{3+} + 3e^- \longrightarrow Cr(s)$	−0.744
$Cr^{3+} + e^- \longrightarrow Cr^{2+}$	−0.408
$CrO_4^{2-} + 4H_2O(l) + 3e^- \longrightarrow Cr(OH)_3(s) + 5OH^-$	−0.13
$Cr_2O_7^{2-} + 14H^+ + 6e^- \longrightarrow 2Cr^{3+} + 7H_2O(l)$	+1.33
Cobalt	
$Co(OH)_2(s) + 2e^- \longrightarrow Co(s) + 2OH^-$	−0.73
$Co^{2+} + 2e^- \longrightarrow Co(s)$	−0.277
$Co^{3+} + e^- \longrightarrow Co^{2+}$	+1.808
Copper	
$Cu_2O(s) + H_2O(l) + 2e^- \longrightarrow 2Cu(s) + 2OH^-$	−0.358
$2Cu(OH)_2(s) + 2e^- \longrightarrow Cu_2O(s) + H_2O(l) + 2OH^-$	−0.080
$Cu^{2+} + e^- \longrightarrow Cu^+$	+0.153
$Cu^{2+} + 2e^- \longrightarrow Cu(s)$	+0.337
$Cu^+ + e^- \longrightarrow Cu(s)$	+0.521
$Cu^{2+} + Br^- + e^- \longrightarrow CuBr(s)$	+0.640
$Cu^{2+} + I^- + e^- \longrightarrow CuI(s)$	+0.86
Fluorine	
$F_2(g) + 2e^- \longrightarrow 2F^-$	+2.87
$F_2(g) + 2H^+ + 2e^- \longrightarrow 2HF(aq)$	+3.06
Gallium	
$Ga^{3+} + 3e^- \longrightarrow Ga(s)$	−0.529
Gold	
$[AuCl_4]^- + 3e^- \longrightarrow Au(s) + 4Cl^-$	+1.00
$Au^{3+} + 3e^- \longrightarrow Au(s)$	+1.498
$Au^+ + e^- \longrightarrow Au(s)$	+1.691
Hafnium	
$Hf^{4+} + 4e^- \longrightarrow Hf(s)$	−1.700
Hydrogen	
$H_2(g) + 2e^- \longrightarrow 2H^-$	−2.25
$2H^+ + 2e^- \longrightarrow H_2(g)$	0.000
Indium	
$In(OH)_3(s) + 3e^- \longrightarrow In(s) + 3OH^-$	−1.00
$In^{3+} + 3e^- \longrightarrow In(s)$	−0.343
Iodine	
$IO_3^- + 3H_2O(l) + 6e^- \longrightarrow I^- + 6OH^-$	+0.26
$IO^- + H_2O(l) + 2e^- \qquad I^- + 2OH^-$	+0.485
$I_2(s) + 2e^- \longrightarrow 2I^-$	+0.5355
$2IO_3^- + 12H^+ + 10e^- \longrightarrow I_2(s) + 6H_2O(l)$	+1.195
Iron	
$Fe(OH)_2(s) + 2e^- \longrightarrow Fe(s) + 2OH^-$	−0.877
$Fe(OH)_3(s) + e^- \longrightarrow Fe(OH)_2(s) + OH^-$	−0.56
$Fe^{2+} + 2e^- \longrightarrow Fe(s)$	−0.4402
$Fe^{3+} + 3e^- \longrightarrow Fe(s)$	−0.036
$[Fe(CN)_6]^{3-} + e^- \longrightarrow [Fe(CN)_6]^{4-}$	+0.36
$Fe^{3+} + e^- \longrightarrow Fe^{2+}$	+0.771
Lanthanum	
$La^{3+} + 3e^- \longrightarrow La(s)$	−2.522
Lead	
$PbS(s) + 2e^- \longrightarrow Pb(s) + S^{2-}$	−0.93
$PbO(s) + H_2O(l) + 2e^- \longrightarrow Pb(s) + 2OH^-$	−0.580
$PbI_2(s) + 2e^- \longrightarrow Pb(s) + 2I^-$	−0.365
$PbSO_4(s) + 2e^- \longrightarrow Pb(s) + SO_4^{2-}$	−0.3588
$PbBr_2(s) + 2e^- \longrightarrow Pb(s) + 2Br^-$	−0.284
$PbCl_2(s) + 2e^- \longrightarrow Pb(s) + 2Cl^-$	−0.268
$Pb^{2+} + 2e^- \longrightarrow Pb(s)$	−0.126
$PbO_2(s) + H_2O(l) + 2e^- \longrightarrow PbO(s) + 2OH^-$	+0.247
$PbO_2(s) + 4H^+ + 2e^- \longrightarrow Pb^{2+} + 2H_2O(l)$	+1.455
Lithium	
$Li^+ + e^- \longrightarrow Li(s)$	−3.045
Magnesium	
$Mg(OH)_2(s) + 2e^- \longrightarrow Mg(s) + 2OH^-$	−2.690
$Mg^{2+} + 2e^- \longrightarrow Mg(s)$	−2.363
Manganese	
$Mn(OH)_2(s) + 2e^- \longrightarrow Mn(s) + 2OH^-$	−1.55
$Mn^{2+} + 2e^- \longrightarrow Mn(s)$	−1.185
$MnO_2(s) + 2H_2O(l) + 2e^- \longrightarrow Mn(OH)_2(s) + 2OH^-$	−0.05
$MnO_4^- + e^- \longrightarrow MnO_4^{2-}$	+0.564
$MnO_4^- + 2H_2O(l) + 3e^- \longrightarrow MnO_2(s) + 4OH^-$	+0.588
$MnO_2(s) + 4H^+ + 2e^- \longrightarrow Mn^{2+} + 2H_2O(l)$	+1.23
$MnO_4^- + 8H^+ + 5e^- \longrightarrow Mn^{2+} + 4H_2O(l)$	+1.51
Mercury	
$HgO(s) + H_2O(l) + 2e^- \longrightarrow Hg(l) + 2OH^-$	+0.098
$Hg_2Br_2(s) + 2e^- \longrightarrow 2Hg(l) + 2Br^-$	+0.1397
$[Hg_2Br_4]^{2-} + 2e^- \longrightarrow Hg(l) + 4Br^-$	+0.223
$Hg_2Cl_2(s) + 2e^- \longrightarrow 2Hg(l) + 2Cl^-$	+0.2676
$Hg_2SO_4(s) + 2e^- \longrightarrow 2Hg(l) + SO_4^{2-}$	+0.6151
$Hg_2^{2+} + 2e^- \longrightarrow 2Hg(l)$	+0.788
$2Hg^{2+} + 2e^- \longrightarrow Hg_2^{2+}$	+0.920

Table VI.3 (*continued*)

Reduction half-reaction	E°, V
Molybdenum	
$MoO_4^{2-} + 4H_2O(l) + 6e^- \longrightarrow Mo(s) + 8OH^-$	−1.05
Nickel	
$Ni(OH)_2(s) + 2e^- \longrightarrow Ni(s) + 2OH^-$	−0.72
$Ni^{2+} + 2e^- \longrightarrow Ni(s)$	−0.250
$NiO_2(s) + 2H_2O(l) + 2e^- \longrightarrow Ni(OH)_2(s) + 2OH^-$	+0.490
Nitrogen	
$NO_3^- + H_2O(l) + 2e^- \longrightarrow NO_2^- + 2OH^-$	+0.01
$2NO_3^- + 4H^+ + 2e^- \longrightarrow N_2O_4(g) + 2H_2O(l)$	+0.803
$NO_3^- + 3H^+ + 2e^- \longrightarrow HNO_2(aq) + H_2O(l)$	+0.94
$NO_3^- + 4H^+ + 3e^- \longrightarrow NO(g) + 2H_2O(l)$	+0.96
$2HNO_2(aq) + 4H^+ + 4e^- \longrightarrow N_2O(g) + 3H_2O(l)$	+1.29
Oxygen	
$2H_2O(l) + 2e^- \longrightarrow H_2(g) + 2OH^-$	−0.8281
$O_2(g) + 2H_2O(l) + 4e^- \longrightarrow 4OH^-$	+0.401
$O_2(g) + 2H^+ + 2e^- \longrightarrow H_2O_2(aq)$	+0.6826
$O_2(g) + 4H^+ + 4e^- \longrightarrow 2H_2O(l)$	+1.229
$O_3(g) + H_2O(l) + 2e^- \longrightarrow O_2(g) + 2OH^-$	+1.24
$H_2O_2(aq) + 2H^+ + 2e^- \longrightarrow 2H_2O(l)$	+1.776
$O_3(g) + 2H^+ + 2e^- \longrightarrow O_2(g) + H_2O(l)$	+2.07
Palladium	
$Pd^{2+} + 2e^- \longrightarrow Pd(s)$	+0.987
Phosphorus	
$H_2PO_2^- + e^- \longrightarrow P(s) + 2OH^-$	−2.05
$PO_4^{3-} + 2H_2O(l) + 2e^- \longrightarrow HPO_3^{2-} + 3OH^-$	−1.12
$P(s) + 3H_2O(l) + 3e^- \longrightarrow PH_3(g) + 3OH^-$	−0.89
$P(s) + 3H^+ + 3e^- \longrightarrow PH_3(g)$	−0.063
Platinum	
$[PtCl_6]^{2-} + 2e^- \longrightarrow [PtCl_4]^{2-} + 2Cl^-$	+0.68
$[PtCl_4]^{2-} + 2e^- \longrightarrow Pt(s) + 4Cl^-$	+0.73
$Pt^{2+} + 2e^- \longrightarrow Pt(s)$	~1.2
Polonium	
$Po^{2} + 2e^- \longrightarrow Po(s)$	+0.65
Potassium	
$K^+ + e^- \longrightarrow K(s)$	−2.925
Radium	
$Ra^{2+} + 2e^- \longrightarrow Ra(s)$	−2.916
Rhodium	
$Rh^{3+} + 3e^- \longrightarrow Rh(s)$	+0.80
Rubidium	
$Rb^+ + e^- \longrightarrow Rb(s)$	−2.925
Scandium	
$Sc^{3+} + 3e^- \longrightarrow Sc(s)$	−2.077
Selenium	
$Se(s) + 2e^- \longrightarrow Se^{2-}$	−0.92
$Se(s) + 2H^+ + 2e^- \longrightarrow H_2Se(aq)$	−0.399
$SeO_4^{2-} + H_2O(l) + 2e^- \longrightarrow SeO_3^{2-} + 2OH^-$	+0.05
$SeO_4^{2-} + 4H^+ + 2e^- \longrightarrow H_2SeO_3(aq) + H_2O(l)$	+1.15
Silicon	
$SiO_3^{2-} + 3H_2O(l) + 4e^- \longrightarrow Si(s) + 6OH^-$	−1.697
$[SiF_6]^{2-} + 4e^- \longrightarrow Si(s) + 6F^-$	−1.24
$SiO_2(s) + 4H^+ + 4e^- \longrightarrow Si(s) + 2H_2O(l)$	−0.857
$Si(s) + 4H^+ + 4e^- \longrightarrow SiH_4(g)$	+0.102

Reduction half-reaction	E°, V
Silver	
$AgI(s) + e^- \longrightarrow Ag(s) + I^-$	−0.1518
$AgBr(s) + e^- \longrightarrow Ag(s) + Br^-$	+0.0713
$AgCl(s) + e^- \longrightarrow Ag(s) + Cl^-$	+0.2222
$Ag_2O(s) + H_2O(l) + 2e^- \longrightarrow 2Ag(s) + 2OH^-$	+0.345
$Ag^+ + e^- \longrightarrow Ag(s)$	+0.7991
$Ag^{2+} + e^- \longrightarrow Ag^+$	+1.980
Sodium	
$Na^+ + e^- \longrightarrow Na(s)$	−2.714
Strontium	
$Sr^{2+} + 2e^- \longrightarrow Sr(s)$	−2.888
$Sr(OH)_2(s) + 2e^- \longrightarrow Sr(s) + 2OH^-$	−2.88
Sulfur	
$SO_4^{2-} + H_2O(l) + 2e^- \longrightarrow SO_3^{2-} + 2OH^-$	−0.93
$S(s) + 2e^- \longrightarrow S^{2-}$	−0.447
$S(s) + 2H^+ + 2e^- \longrightarrow H_2S(aq)$	+0.142
$SO_4^{2-} + 4H^+ + 2e^- \longrightarrow H_2SO_3(aq) + H_2O(l)$	+0.172
$SO_4^{2-} + 8H^+ + 6e^- \longrightarrow S(s) + 4H_2O(l)$	+0.3572
$H_2SO_3(aq) + 4H^+ + 4e^- \longrightarrow S(s) + 3H_2O(l)$	+0.450
$S_2O_8^{2-} + 2e^- \longrightarrow 2SO_4^{2-}$	+2.01
Tellurium	
$Te(s) + 2e^- \longrightarrow Te^{2-}$	−1.143
$Te(s) + 2H^+ + 2e^- \longrightarrow H_2Te(aq)$	−0.739
$TeO_3^{2-} + 3H_2O(l) + 4e^- \longrightarrow Te(s) + 6OH^-$	−0.57
Thallium	
$TlOH(s) + e^- \longrightarrow Tl(s) + OH^-$	−0.343
$Tl^+ + e^- \longrightarrow Tl(s)$	−0.3363
$Tl(OH)_3(s) + 2e^- \longrightarrow TlOH(s) + 2OH^-$	−0.05
Tin	
$Sn^{2+} + 2e^- \longrightarrow Sn(s)$	−0.136
$Sn^{4+} + 2e^- \longrightarrow Sn^{2+}$	+0.15
Titanium	
$Ti^{2+} + 2e^- \longrightarrow Ti(s)$	−1.628
$[TiF_6]^{2-} + 4e^- \longrightarrow Ti(s) + 6F^-$	−1.191
$Ti^{3+} + e^- \longrightarrow Ti^{2+}$	−0.369
Tungsten	
$WO_4^{2-} + 4H_2O(l) + 6e^- \longrightarrow W(s) + 8OH^-$	−1.05
Uranium	
$U(OH)_4(s) + e^- \longrightarrow U(OH)_3(s) + OH^-$	−2.20
$U(OH)_3(s) + 3e^- \longrightarrow U(s) + 3OH^-$	−2.17
$U^{4+} + e^- \longrightarrow U^{3+}$	−0.607
Vanadium	
$V^{3+} + 3e^- \longrightarrow V(s)$	−1.798
$V^{4+} + 4e^- \longrightarrow V(s)$	−1.50
$V^{2+} + 2e^- \longrightarrow V(s)$	−1.175
$V^{3+} + e^- \longrightarrow V^{2+}$	−0.256
$VO^{2+} + 2H^+ + e^- \longrightarrow V^{3+} + H_2O(l)$	+0.359
Zinc	
$Zn(OH)_2(s) + 2e^- \longrightarrow Zn(s) + 2OH^-$	−1.245
$Zn^{2+} + 2e^- \longrightarrow Zn(s)$	−0.7628
Zirconium	
$Zr^{4+} + 4e^- \longrightarrow Zr(s)$	−1.529

Source: Data adopted from T. Moeller, *Inorganic Chemistry* (New York: Wiley, 1982), pp. 789–803.

APPENDIX

VII

Answers to Odd-Numbered Questions and Problems*

*(Answers to Questions and Problems that require a drawing are given in the Student Solutions Manual.)

CHAPTER 2

2.1 (a) ± 1, (b) ± 0.0001, (c) probably ± 10, (d) ± 0.001, (e) probably ± 1000, (f) ± 0.0001, (g) ± 0.01, (h) probably ± 10

2.3 (a) 3, (b) 2, (c) 4, (d) 3, (e) 5, (f) 4, (g) 3 (or more)

2.5 (a) 500 (or 520 or 523), (b) 68.46, (c) 10.0, (d) 11,000, (e) 2.3, (f) 19.2, (g) 900 (or 860), (h) 850, (i) 170, (j) 42.2, (k) 18.0, (l) 127

2.7 (a) 2.432 cm, (b) 2.432 $\pm$ 0.07 cm includes all of the data, (c) -0.3% (The negative sign indicates that the experimental value is lower than the accepted value.)

2.9 (a) 6.5×10^3, (b) 4.1×10^{-3}, (c) 3.050×10^{-3}, (d) 8.10×10^2, (e) 3×10^{-7}, (f) 9.352×10^6, (g) 4.2×10^4

2.11 (a) 5260, (b) 0.00000410, (c) 500,000, (d) 3000, (e) 0.0162, (f) 9346

2.13 (a) 5.13×10^3, (b) 8.2×10^1, (c) 3.5×10^{-7}, (d) 2.2×10^{11}

2.15 0.99704 g/cm^3

2.17 length, mass, time, electric current, temperature, luminous intensity, and amount of substance; others derived from the seven base quantities

2.19 (a) ix, xvi, xvii; (b) i, vi, xiii; (c) ii, v, vii, xi, xiv; (d) iii, viii, xii, xv; (e) iv, x

2.21 (a) km, (b) MV, (c) mm, (d) TJ, (e) km^3

2.23 (a) 37.0 °C, (b) -23 °C, (c) 26 °C, (d) 120 °C, (e) 0.00 °C

2.25 (a) 310.2 K, (b) 250. K, (c) 299 K, (d) 390 K, (e) 273.15 K

2.27 (a) 212.00 °F, (b) -241.2 °F, (c) 832.5 °F, (d) 4980 °F, (e) 640. °F

2.29 (1 atm/760 Torr), (760 Torr/1 atm)

2.31 (a) 3.66 m, (b) 1×10^8 m

2.33 (a) 1.03 nm, (b) 2660 J, (c) 14.6 dm^3, (d) 1.46×10^4 g, (e) 1.48×10^6 Pa, (f) 1.9×10^{-19} J, (g) 0.9674 atm, (h) 21.65 cm^3

2.35 (a) 27 yd, (b) 13,700 kg/m^3, (c) 430 Pa, (d) 4.7 bar, (e) 6.63 kg

2.37 (a) 8.314×10^7 erg/K mol, (b) 1.987 cal/K mol, (c) 0.08205 L atm/K mol

2.39 6×10^{22} carbon atoms

2.41 2 ton sulfur dioxide

2.43 3.9×10^{-22}

2.45 (a) 3, (b) 6, (c) 4, (d) 5, (e) 3, (f) 3

2.47 1.0056×10^3 g

2.49 1.64 g/cm^3

2.51 0.8034 g/cm^3

2.53 nearly 29 diskettes

CHAPTER 3

3.1 (a) beryllium, (b) boron, (c) vanadium, (d) arsenic, (e) barium

3.3 (a) Xe, (b) Ni, (c) Mg, (d) Co, (e) Si, (f) Pb, (g) K, (h) Ag, (i) F, (j) Rn

3.5 (a) 1 phase: matter, pure substance, element; (b) 1 phase: matter, pure substance, compound; (c) 1 phase: matter, pure substance, compound, (d) 2 phases: matter, heterogeneous mixture; (e) 1 phase: matter, pure substance, compound; (f) 1 phase: matter, homogeneous mixture; (g) 2 phases: matter, heterogeneous mixture

3.7 i, j, k

3.9 see Figure 3.5; no; cathode rays are fundamental to all matter

3.11 simultaneous magnetic and electric fields were imposed upon on a beam of cathode rays in an arrangement such that the beam continued in the same direction as the initial direction; the *e*/*m* value was calculated from the field strengths; the mass value proved that the electron was a subatomic particle

3.13 little or no deflection; each atom has a small, dense central core in which virtually all the mass and positive charge of the atom are concentrated

3.15 2.3×10^{-14}

3.17 (a) smaller, (b) larger, (c) larger, (d) no change

3.19 an isotope has a certain number of protons and neutrons; different isotopes of an element have the same number of protons, but different numbers of neutrons; the hydrogen isotopes are named protium ("regular" hydrogen with no neutrons), deuterium (with 1 neutron), and tritium (with 2 neutrons)

3.21 (a) 45 protons, 55 neutrons; (b) 60 protons, 86 neutrons; (c) 35 protons, 44 neutrons; (d) 3 protons, 4 neutrons; (e) 65 protons, 94 neutrons

3.23 $^{29}_{14}Si$, 14, 15, 29, 14, 0; $^{34}_{16}S^{2-}$, 16, 18, 34, 18, −2 $^{56}_{26}Fe^{2+}$, 26, 30, 56, 24, +2; $^{188}_{79}Au^{3+}$, 79, 109, 188, 76, +3

3.25 (a) i, iii, v, vi, vii, ix; (b) ii, iv, vi, viii, x, xi; (c) A = 13 ii, iii; A = 14 iv, v; A = 16 vii, viii; A = 17 ix, x

3.27 (a) 1.29392×10^{-22} g, (b) 1.32706×10^{-22} g, (c) 1.36023×10^{-22} g

3.29 (a) 150.9196 u, (b) 152.9209 u

3.31 (a) 9.289044×10^{-23} g, (b) 9.28836×10^{-23} g, (c) 9.289207×10^{-23} g; each isotope contains a different number of protons and neutrons (the mass of a proton is slightly different than the mass of a neutron) and each isotope also has a different binding energy (the amounts of mass converted to energy)

3.33 24.31 u

3.35 69.17% ^{63}Cu, 30.83% ^{65}Cu

3.37 (a) $^{1}H^{35}Cl$, $^{1}H^{37}Cl$, $^{2}H^{35}Cl$, $^{2}H^{37}Cl$; (b) 36 u, 38 u, 37 u, 39 u, respectively; (c) $^{1}H^{35}Cl > {}^{1}H^{37}Cl > {}^{2}H^{35}Cl > {}^{2}H^{37}Cl$

CHAPTER 4

4.1 (a) iii, (b) ii, (c) iii, (d) ii, (e) iii, (f) i

4.3 (a) ii, (b) iii, (c) i, (d) ii, (e) iii, (f) iii

4.5 (a) i, (b) ii, (c) i, (d) ii, (e) i, (f) i

4.7 (a) $Ca(ClO_3)_2$, (b) $Al_2(SO_4)_3$, (c) $AuBr_3$, (d) NH_4CN, (e) K_3PO_4, (f) $Na_2Cr_2O_7$, (g) $Ca(MnO_4)_2$

4.9 (a) 70.90 u, (b) 55.85 u, (c) 55.85 u, (d) 342.3 u, (e) 122.55 u, (f) 306.78 u, (g) 616.0 u, (h) 221.1 u, (i) 98.00 u, (j) 193.1 u

4.11 (a) lithium ion, (b) cadmium ion, (c) iron(II) ion, (d) manganese(II) ion, (e) aluminum ion

4.13 (a) Na^+, (b) Zn^{2+}, (c) Ag^+, (d) Hg^{2+}, (e) Fe^{3+}

4.15 (a) nitride ion, (b) oxide ion, (c) selenide ion, (d) fluoride ion, (e) bromide ion

4.17 (a) lithium sulfide, (b) tin(IV) oxide, (c) rubidium iodide, (d) lithium oxide, (e) uranium(IV) oxide, (f) barium nitride, (g) sodium fluoride

4.19 (a) NaF, (b) ZnO, (c) BaO_2, (d) $MgBr_2$, (e) HI, (f) CuCl, (g) KI

4.21 (a) ammonium sulfate, (b) potassium dichromate, (c) iron(II) perchlorate, (d) calcium carbonate, (e) sodium nitrite, (f) potassium chromate, (g) sodium sulfite

4.23 (a) K_2SO_3, (b) $Ca(MnO_4)_2$, (c) $Ba_3(PO_4)_2$, (d) Cu_2SO_4, (e) $NH_4(CH_3COO)$, (f) $AgNO_3$, (g) $U(SO_4)_2$

4.25 (a) hydrochloric acid, (b) phosphoric acid, (c) perchloric acid, (d) nitric acid, (e) sulfurous acid, (f) phosphorous acid

4.27 carbonic acid; HCO_3^-, hydrogen carbonate ion; CO_3^{2-}, carbonate ion

4.29 (a) carbon monoxide, (b) carbon dioxide, (c) sulfur hexafluoride, (d) silicon tetrachloride, (e) iodine monofluoride

4.31 (a) B_2O_3, (b) SiO_2, (c) PCl_3, (d) SCl_4, (e) BrF_3, (f) H_2Te, (g) P_2O_3

4.33 (a) N_2O, (b) N_2 and O_2, (c) all are gases, (d) heat is required

4.35 (a) gaseous nitrogen reacts with gaseous hydrogen at 400 °C and 250 atm pressure in the presence of FeO as a catalyst to produce gaseous ammonia; (b) when gaseous carbon monoxide and gaseous oxygen are heated, gaseous carbon dioxide is formed; (c) heating solid silicon dioxide and carbon at 3000 °C gives liquid silicon and gaseous carbon monoxide

4.37 (a) $Cl_2O_7(g) + H_2O(l) \longrightarrow 2HClO_4(aq)$,
(b) $Br_2(l) + H_2O(l) \longrightarrow HBr(aq) + HBrO(aq)$,
(c) $Ca_3(PO_4)_2(s) + 3H_2SO_4(aq) \longrightarrow 3CaSO_4(s) + 2H_3PO_4(aq)$,
(d) $V_2O_4(s) + 4HClO_4(aq) \longrightarrow 2VO(ClO_4)_2(aq) + 2H_2O(l)$,
(e) $Fe_2O_3(s) + 3H_2(g) \longrightarrow 2Fe(s) + 3H_2O(l)$,
(f) $2K(s) + 2H_2O(l) \longrightarrow 2KOH(aq) + H_2(g)$,
(g) $MgCO_3(s) \xrightarrow{\Delta} MgO(s) + CO_2(g)$,
(h) $Al_2S_3(s) + 6H_2O(l) \longrightarrow 2Al(OH)_3(s) + 3H_2S(g)$,
(i) $BaO_2(s) + H_2SO_4(aq) \longrightarrow H_2O_2(aq) + BaSO_4(s)$

4.39 (a) 1.6×10^{-2} mol Cs, (b) 7.8×10^3 mol CO_2, (c) 0.271 mol $BaCl_2$, (d) 0.020 mol Cu

4.41 (a) 8.07×10^{24} Ga atoms, (b) 8.07×10^{24} $C_6H_5CH_3$ molecules, (c) 8.07×10^{24} $AgNO_3$ formula units

4.43 (a) 46.01 g/mol, (b) 171.35 g/mol, (c) 245.3 g/mol, (d) 54.94 g/mol, (e) 96.99 g/mol, (f) 28.02 g/mol, (g) 743.67 g/mol, (h) 183.14 g/mol, (i) 413.35 g/mol

4.45 (a) 64 g C, (b) 13.75 g N_2O_5, (c) 6.3×10^{-4} g $AmBr_3$, (d) 4×10^{-9} g HCl

4.47 (a) 8.0×10^{-5} mol Tc, (b) 0.986 mol NH_3, (c) 33.2 mol NH_4Br, (d) 0.027 mol PCl_5

4.49 6.022×10^{23} O_3 molecules, 1.807×10^{24} O atoms

4.51 6.0220×10^{23} formula units, 6.0220×10^{23} Au^+ cations, 6.0220×10^{23} CN^- anions, 1.8066×10^{24} atoms

4.53 (a) 0.0704 M, (b) 0.50 M, (c) 0.200 M

4.55 1.00 g NaOH

4.57 0.067 mol, 0.067 mol, 0.22 M

4.59 34.3% Fe, 65.7% Cl

4.61 (a) 43.18% K, 39.15% Cl, 17.67% O; (b) 36.70% K, 33.27% Cl, 30.03% O; (c) 31.91% K, 28.93% Cl, 39.17% O; (d) 28.22% K, 25.59% Cl, 46.19% O

4.63 2.2 g O

4.65 cuprite

4.67 $C_8H_{11}O_3N$

4.69 (a) C_2H_6O, (b) $C_4H_{12}O_2$, (c) 92.16 g/mol

4.71 C_4H_8O

4.73 (a) C_4H_9, (b) $C_4H_5N_2O$, (c) CH_2O, (d) $C_3H_4O_3$

4.75 4 Fe atoms/molecule hemoglobin

4.77 (a) 0.0178 mol CaO, 2.14×10^{22} atoms;
(b) 0.0135 mol $Ca(OH)_2$, 4.06×10^{22} atoms;
(c) 0.0411 mol Mg, 2.48×10^{22} Mg atoms;
(d) 0.0161 mol $HOCH_2CH_2OH$, 9.70×10^{22} atoms;
(e) 0.0555 mol H_2O, 1.00×10^{23} atoms; sample of water contains the largest number of moles and atoms

4.79 51.8 u, Cr

CHAPTER 5

5.1 (a) i, (b) iv, (c) iv, (d) iii, (e) ii, (f) iii (g) ii, (h) iii

5.3 (a) no reaction, (b) reaction, (c) no reaction, (d) reaction

5.5 (a) $PbSO_4(s)$, (b) $Na^+ + CH_3COO^-$, (c) $2NH_4^+ + CO_3^{2-}$, (d) $MnS(s)$, (e) $Ba^{2+} + 2Cl^-$

5.7 (a) $2Fe^{3+} + 2I^- \longrightarrow 2Fe^{2+} + I_2(s)$, (b) $2IO_3^- + 6HSO_3^- \longrightarrow 2I^- + 6SO_4^{2-} + 6H^+$, (c) $BrCl(g) + H_2O(l) \longrightarrow H^+ + Cl^- + HBrO(aq)$, (d) $PCl_5(g) + 4H_2O(l) \longrightarrow H_3PO_4(aq) + 5H^+ + 5Cl^-$, (e) $2S_2O_3^{2-} + I_2(s) \longrightarrow S_4O_6^{2-} + 2I^-$

5.9 (a) $Hg^{2+} + S^{2-} \longrightarrow HgS(s)$, (b) $Al^{3+} + 3OH^- \longrightarrow Al(OH)_3(s)$, (c) no reaction, (d) no reaction

5.11 (a) one mol C_7H_{16} reacts with 11 mol O_2 to give 7 mol CO_2 and 8 mol H_2O; (b) one molecule of C_7H_{16} reacts with 11 molecules of O_2 to give 7 molecules of CO_2 and 8 molecules of H_2O; (c) a 100.23 g sample of C_7H_{16} reacts with 352.00 g of O_2 to give 308.07 g of CO_2 and 144.16 g of H_2O

5.13 (a) (3 mol O_2)/(2 mol ZnS),
(b) (2 mol ZnO)/(2 mol ZnS) = (1 mol ZnO)/(1 mol ZnS),
(c) (2 mol SO_2)/(2 mol ZnS) = (1 mol SO_2)/(1 mol ZnS)

5.15 (a) 1.50 mol O_2, (b) 0.500 mol O_2, (c) 0.500 mol O_2, (d) 0.500 mol O_2, (e) 2.00 mol O_2

5.17 (a) 1.88 mol O_2, (b) 1.50 mol NO, (c) 2.25 mol H_2O

5.19 (a) 0.750 mol O_2, (b) 0.500 mol N_2, (c) 1.50 mol H_2O

5.21 46.0 g Na

5.23 3.87 g AgCl

5.25 13.4 g $CoCl_2$, 4.13 g HF

5.27 1.75 g F_2

5.29 0.651 g Br_2

5.31 67.6%

5.33 69.1 g H_2TeO_3

5.35 357 g $KClO_3$

5.37 (a) 0.56 ton CaO, (b) 0.74 ton $Ca(OH)_2$

5.39 3.0 L

5.41 0.1096 M

5.43 64.43%

5.45 11.7 g NaCl, Cl_2, Na

5.47 51.2 g $PbCl_2$, 8.1 PCl_3 remains

5.49 8.54 g HNF_2, 6.96 g HCl, 1.6 g $ClNF_2$, 0.50 g NH_4Cl, 0.37 g HF

5.51 50. g PCl_5

5.53 $2AgNO_3(s) \xrightarrow{\Delta} 2Ag(s) + 2NO_2(g) + O_2(g)$, 99.2%

5.55 0.0100 M SO_4^{2-}, 0.090 M HSO_4^-, 0.110 M H^+

CHAPTER 6

6.1 (1) molecules are relatively far apart (not valid for a liquid or a solid); (2) molecules are in constant motion; (3) average speed and average kinetic energy are proportional to temperature; (4) molecules in all ideal gases have the same average kinetic energy at the same temperature; (5) molecular collisions are elastic

6.3 0.856

6.5 (a) ii, (b) i, (c) i, (d) i

6.7 0.33 atm

6.9 11 L

6.11 1.004 L

6.13 94 °C

6.15 79 K

6.17 1.49 atm

6.19 13.2 L

6.21 1340 psi

6.23 9.2 K

6.25 (a) 2 L, (b) 2 L, (c) 2 L, (d) 4 L

6.27 250. mL water vapor

6.29 719 mL

6.31 44.6 mol

6.33 2.69×10^{22} molecules

6.35 137 g/mol

6.37 0.0409 mol

6.39 0.00118 mol

6.41 40.0 atm

6.43 3.1×10^4 K

6.45 1.30 L

6.47 1.0×10^{10} molecules

6.49 1.3 L

6.51 250. g/mol

6.53 7.4 g/L

6.55 6.52 atm

6.57 0.27

6.59 32.0 L

6.61 5.00 L N_2

6.63 11.2 L, 12.5 L

6.65 13.9 L HF, 3.48 L O_2, 6.95 L Xe

6.67 6 g $KClO_3$, the reaction must be heated, MnO_2 is a catalyst

6.69 1.4070

6.71 1.004

6.73 the ideal volume is 11.3 L; no, volume is less than ideal

6.75 16.6 atm, 12.3 atm

6.77 (a) 14.4 psi, (b) 745 mmHg, (c) 29.3 in Hg, (d) 99,300 Pa, (e) 0.980 atm, (f) 33.5 ft H_2O

6.79 CN, $(CN)_2$

6.81 (a) 2 mL, (b) Dalton's law, (c) volume decreases, (d) volume increases, (e) $XeF_4 < Xe < F_2$, (f) XeF_4

CHAPTER 7

7.1 (a) $\Delta E < 0$; (b) $\Delta E > 0$; (c) ΔE is > 0, $= 0$, or < 0 depending on the numerical values of q and w

7.3 -40, 300 J

7.5 $\Delta V = -0.24$ L

7.7 (a) negligible, (b) on the system, (c) on the system

7.9 $\Delta H = q_p > 0$, $\Delta E > 0$, $w > 0$

7.11 -20 J

7.13 340 J for Cl_2, 280 J for Na, 510 J for NaCl; NaCl requires the most

7.15 28.0 J/K mol, 0.140 J/K g

7.17 $\Delta T = -35$ K

7.19 24.3 J/K mol for Al, 16.4 J/K mol for Be, 23.9 J/K mol for Cr, 25.2 J/K mol for Fe, 26.9 J/K mol for Pb, 26.8 J/K mol for Sn; Be is an exception because it has a very small atomic mass; average (excluding Be) = 25.4 J/K mol; the vibration of the atoms in the crystal structure accounts for the heat capacity

7.21 225 J/K

7.23 7940 J/K

7.25 -466 kJ

7.27 (a) 0.13 J/K g, W; (b) yes, 0.26 J/K, Mo

7.29 (a) solid, (b) liquid, (c) liquid, (d) gas

7.31 (a) $O_2(g)$, (b) C(*s*, graphite), (c) Hg(*l*), (d) $I_2(s)$

7.33 (a) 106.12 kJ, (b) -212.24 kJ, (c) 212.24 kJ

7.35 -1.30×10^7 J

7.37 (a) 95.1 kJ, (b) 143 kJ, (c) 2.97 kJ, (d) 2.97 kJ

7.39 -201.34 kJ

7.41 -220.1 kJ

7.43 (a) -443.5 kJ, (b) -235.8 kJ, (c) 25.594 kJ

7.45 -55.836 kJ

7.47 $C(s) + 2H_2(g) \longrightarrow CH_4(g)$, $\Delta H_f^\circ = -85.1$ kJ

7.49 (a) 10.519 kJ/mol, (b) 310.114 kJ/mol, (c) 320.633 kJ/mol

7.51 liquid water

7.53 d < e < f < c < a < b

7.55 0.0036 K

7.57 4490 J

7.59 -284.33 kJ

CHAPTER 8

8.1 photoelectric effect and radiation by hot bodies

8.3 (a) 4.469×10^{14} s^{-1}, (b) 1.491×10^4 cm^{-1}, (c) 2.961×10^{-19} J

8.5 $\nu = 8.8 \times 10^{14}$ s^{-1}, $E = 5.8 \times 10^{-19}$ J $= 3.5 \times 10^5$ J/mol

8.7 $\nu = 5.5 \times 10^{14}$ s^{-1}, $E = 3.6 \times 10^{-19}$ J, 30 photons

8.9 only heating occurs, nothing else but more heating; electrons are dislodged, larger number of dislodged electrons

8.11 wavelength is in the order of magnitude of atomic size, wavelengths are too small to be observed

8.13 0.18 m

8.15 (1) radiation source, excites sample; (2) analyzer, divides beam; (3) sample holder, contains sample, (4) detector, measures quantities; (5) display device, makes results visible

8.17 incident radiation excites sample to higher energy levels, emission of energy returns system to lower energy state

8.19 helium was discovered by Janssen during a solar eclipse in 1868 by the detection of a new line in the solar system

8.21 $\bar{\nu} = 97{,}491.6$ cm^{-1}, $\lambda = 102.573$ nm, ultraviolet

8.23 d

8.25 $\bar{\nu} = 109{,}678$ cm^{-1}, $\lambda = 91.1760$ nm

8.27 (1) electron moves about nucleus in fixed circular orbits, (2) angular momentum is quantized as a whole-number multiple of $h/2\pi$, (3) electron does not radiate energy while in same orbit; theory fails to describe multielectron atoms, violates Heisenberg uncertainty principle, and could not explain spectral changes in magnetic fields

8.29 0.529 nm, 0.212 nm, 0.476 nm, 0.846 nm, 1.32 nm; size increases as n^2

8.31 $n = 1, 2, 3\ldots$; $1 \le n \le 7$ for known elements; $n = \infty$ is removal of electrons to form an ion; n defines energy and average distance from the nucleus for an electron

8.33 symbol is m_l; $m_l = 0, \pm 1, \pm 2, \ldots \pm l$; describes orientation of orbitals in space

8.35 (a) 2, (b) 3, (c) 0, (d) 1, (e) 4

8.37 (a) 0, ± 1, ± 2; (b) 0; (c) 0, ± 1

8.39 d

8.41 $1s < 2p < 2s < 3d < 3p < 3s$

8.43 0.0529 nm is most probable distance, electron can be found closer and farther as well

8.45 no two electrons can have same numerical values for quantum numbers; c

8.47 orbitals of equal energies occupied by single electron before pairing occurs; b

8.49 (a) K: $1s^2 2s^2 2p^6 3s^2 3p^6 4s^1$; same as in Table 8.7; paramagnetic
(b) Sc: $1s^2 2s^2 2p^6 3s^2 3p^6 3d^1 4s^2$; same as in Table 8.7; paramagnetic
(c) Si: $1s^2 2s^2 2p^6 3s^2 3p^2$; same as in Table 8.7; paramagnetic
(d) F: $1s^2 2s^2 2p^5$; same as in Table 8.7; paramagnetic
(e) U: $1s^2 2s^2 2p^6 3s^2 3p^6 3d^{10} 4s^2 4p^6 4d^{10} 4f^{14} 5s^2 5p^6 5d^{10} 5f^4 6s^2 6p^6 7s^2$; it is different from that in Table 8.7 for two subshells, $5f^3 6d^1$ rather than $5f^4$; paramagnetic
(f) Ag: $1s^2 2s^2 2p^6 3s^2 3p^6 3d^{10} 4s^2 4p^6 4d^9 5s^2$; it is different from that in Table 8.7 for two subshells, $4d^{10} 5s^1$ rather than $4d^9 5s^2$; paramagnetic
(g) Mg: $1s^2 2s^2 2p^6 3s^2$; same as in Table 8.7; diamagnetic
(h) Fe: $1s^2 2s^2 2p^6 3s^2 3p^6 3d^6 4s^2$; same as in Table 8.7; paramagnetic
(i) Pr: $1s^2 2s^2 2p^6 3s^2 3p^6 3d^{10} 4s^2 4p^6 4d^{10} 4f^3 5s^2 5p^6 6s^2$; same as in Table 8.7; paramagnetic

8.51 (a) As, (b) Unq, (c) Po, (d) Tc, (e) V

8.53 $1s^2 2s^2 2p^6 3s^2 3p^6 3d^9 4s^2$, $1s^2 2s^2 2p^6 3s^2 3p^6 3d^{10} 4s^1$; no differences in magnetic properties

8.55 (a) 0, (b) 4, (c) 2, (d) 2, (e) 5

8.57 A: $1s^2 2s^2 2p^6 3s^2 3p^6 3d^{10} 4s^1$, B: [Kr]$4d^2 5s^2$, C: [Kr]$4d^{10} 5s^2 5p^6$, D: [Xe]$4f^{14} 5d^{10} 6s^2 6p^3$, E: $1s^2 2s^2$

8.59 $1s^2 2s^2 2p^6 3s^2 3p^6 3d^{10} 4s^2 4p^6 4d^{10} 4f^{14} 5s^2 5p^6 5d^{10} 5f^{14} 6s^2 6p^6 6d^{10} 7s^2 7p^6 8s^1$, alkali metal family

CHAPTER 9

9.1 (a) nitrogen family, Representative Group V; (b) alkali metal family, Representative Group I; (c) *f*-transition elements

9.3 (a) ns^2, (b) $(n-1)d^{1-10}ns^{1,2}$, (c) ns^2np^5

9.5 (a) Ca, *s*-block representative element (alkaline earth element); (b) Rh, *d*-transition element; (c) Os, *d*-transition element; (d) Er, *f*-transition element (lanthanide); (e) Xe, noble gas element; (f) Pb, *p*-block representative element

9.7 (a) B; (b) H; (c) A; (d) C, F, G, I; (e) A, D; (f) I; (g) B, J; (h) D; (i) E, H, K; (j) G

9.9 (a) 1, (b) 5, (c) 6, (d) 8, (e) 7

9.11 (a) $\cdot\dot{\text{Tl}}:$, (b) $\cdot\dot{\text{As}}:$, (c) Mg:, (d) $\cdot\ddot{\underset{\cdot\cdot}{\text{F}}}:$, (e) $:\dot{\underset{\cdot\cdot}{\text{O}}}:$

9.13 effective nuclear charge, principal quantum number; decrease; increase

9.15 intervention of the *d*-transition elements increases the nuclear charge of Ga enough to balance the effect of the larger principal quantum number; the elements are very similar in chemical behavior

9.17 (a) iii, (b) i, (c) ii, (d) iv

9.19 $Cl^+ < Cl < Cl^-$

9.21 $Be^{2+} < Mg^{2+} < Na^+ < F^- < Cl^- < S^{2-}$

9.23 Cl, 0.099 nm; P, 0.105 nm; P–F, 0.176 nm

9.25 0.12459 nm

9.27 (a) $[Ne]3s^2$, +2, [Ne]; (b) $[Ne]3s^23p^1$, +3, [Ne]; (c) $[Ar]3d^14s^2$, +2, $[Ar]3d^1$ and +3, [Ar]

9.29 (a) $[He]2s^22p^3$, −3, [Ne]; (b) $[Ne]3s^23p^4$, −2, [Ar]; (c) $[Ne]3s^23p^3$, −3, [Ar]

9.31 $[Mg]^{2+}$, $[Al]^{3+}$, $[Sc]^{3+}$, $[:\ddot{\underset{\cdot\cdot}{\text{N}}}:]^{3-}$, $[:\ddot{\underset{\cdot\cdot}{\text{S}}}:]^{2-}$

9.33 increase across period, decrease down family

9.35 smaller, metals form cations and nonmetals do not

9.37 less, p^3 configuration is stable

9.39 −334.38 kJ/mol

9.41 more difficult to remove a negatively charged electron from a positively charged ion than from a neutral atom

9.43 (a) 30 u, (b) 15, (c) 6, (d) nonmetal

CHAPTER 10

10.1 force that acts strongly enough between two atoms or groups of atoms to hold them together in a different species that has measurable properties; electrostatic attraction between oppositely charged species

10.3 lose outermost electron to form a +1 ion

10.5 ionic between Ba^{2+} and OH^-; polar covalent between O and H

10.7 each atom gives up valence electrons to become cations in an "electron sea"; valence electrons of nonmetals are too difficult to remove

10.9 specific bonds need not be broken as layers of cations move

10.11 electrons are transferred from metal to nonmetal, resulting in electrostatic attraction between ions

10.13

(a) Zn:	+	$2\cdot\ddot{\underset{\cdot\cdot}{\text{F}}}:$	$\longrightarrow$	$[Zn]^{2+}$	$2\,[:\ddot{\underset{\cdot\cdot}{\text{F}}}:]^-$
$[Ar]3d^{10}4s^2$		$[He]2s^22p^5$		$[Ar]3d^{10}$	[Ne]
(b) Ca:	+	$\cdot\ddot{\underset{\cdot}{\text{O}}}:$	$\longrightarrow$	$[Ca]^{2+}$	$[:\ddot{\underset{\cdot\cdot}{\text{O}}}:]^{2-}$
$[Ar]4s^2$		$[He]2s^22p^4$		[Ar]	[Ne]

10.15 (a) $LaCl_3$, (b) CuF and CuF_2, (c) CsBr

10.17 (a) Ca_3N_2, (b) none, (c) K_2Se, (d) FeS and Fe_2S_3

10.19 ions surrounded by ions of opposite charge; charges and relative sizes of ions

10.21 (a) 2, (b) 4, (c) 6

10.23 coordinate covalent bond; N, B

10.25 a, d, e

10.27 upon undergoing physical changes, the molecules remain intact

10.29 distance between nuclei; atoms vibrate in the bond; decreases

10.31 463.459 kJ/mol

10.33 367.69 kJ/mol

10.35 799 kJ/mol

10.37 (a) −560 kJ, (b) −62 kJ, (c) −970 kJ

10.39 yes, both reactions involve the same number of C—C and C—H bonds

10.41 ordinarily the actual bonding is not fully any one of the classifications

10.43 unequal sharing of electrons; a, d, e

10.45 Al^{3+}

10.47 (a) 10., (b) 28; cation with higher charge has higher charge-to-radius ratio; Cu^{2+}

10.49 ability of an atom to attract electrons; nonpolar covalent bonds formed if small difference, polar covalent bonds formed if moderate differences, and ionic bonds formed if large difference (≥ 2.0)

10.51 d < c < b < a

10.53 (a) iii; (b) i, ii; (c) iv

10.55 (a) metallic; (b) nonpolar covalent; (c) ionic; (d) nonpolar covalent; (e) polar covalent; (f) :Ö=Ö:; (g) $[Cu]^{2+}$[:Ö:]$^{2-}$; (h) H—H; (i) H—Ö—H; (j) H_2, O_2; (k) Cu

10.57 numbers equal to the charge of ions or the charge that ion would have if the compound were ionic; no

10.59 *s* block, scandium family, Zn, Cd, Al, F; heavier metals in B, N, and O families

10.61 (a) K is +1, H is −1; (b) Mn is +2, Cl is −1; (c) N is −3, H is +1; (d) P is 0; (e) Cl is −1; (f) S is +4, O is −2; (g) Na is +1, O is −1; (h) Mn is +3, F is −1; (i) I is +3, Cl is −1; (j) H is +1, Se is −2

10.63 (a) N is −3 in NH_3 and 0 in N_2, Cu is +2 in CuO and 0 in Cu, O is −2 in CuO and H_2O, H is +1 in NH_3 and H_2O; N and Cu are undergoing changes in oxidation numbers; (b) H is +1 in H_2SO_4, NaOH, and H_2O, S is +6 in H_2SO_4 and Na_2SO_4, O is −2 in H_2SO_4, NaOH, Na_2SO_4, and H_2O, Na is +1 in NaOH and Na_2SO_4; there are no changes in oxidation numbers

10.65 (a) nitrogen(III) oxide, (b) iodine(III) chloride, (c) carbon(IV) oxide, (d) sulfur(IV) oxide, (e) boron(III) fluoride, (f) nitrogen(V) oxide, (g) silicon(IV) chloride, (h) carbon(IV) chloride

10.67 (a) BN, (b) CSe_2, (c) BrCl, (d) N_2O_3, (e) OF_2, (f) SF_4, (g) NO, (h) PCl_5

CHAPTER 11

11.1 arrangement of atoms and bonds; bond lengths, bond angles, bond strengths, molecular geometry

11.3 (a) [:Ö—Cl(—Ö:)(—Ö:)—Ö:]$^-$ (b) Ö=N̈—F̈: (c) :F̈—Xe(—F̈:)(—F̈:)—F̈: (d) $[CrCl_6]^{3-}$ with six :C̈l: about Cr (e) :O: double bonded to C with :C̈l—C—C̈l: (f) :F̈—C̈l(—F̈:)—F̈:

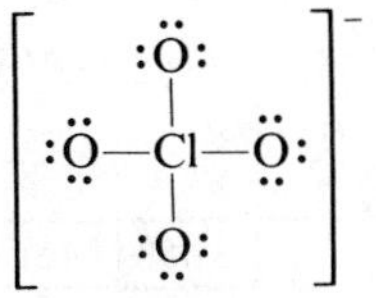

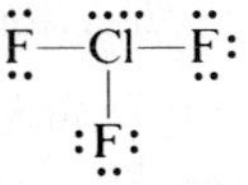

11.5 (a) H—Ö—N̈(—H)—H (b) eight-membered ring of :S̈: atoms (S—S—S—S—S—S—S—S) (c) H—Si(—H)(—H)—H (d) :F̈—Ö—Ö—F̈: (e) :C≡O: (f) :C̈l atoms (six) about Se: $SeCl_6$

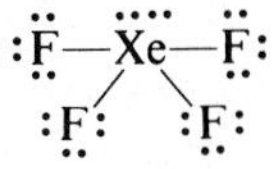

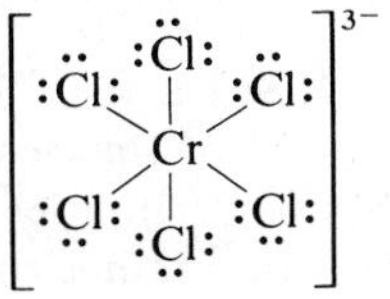

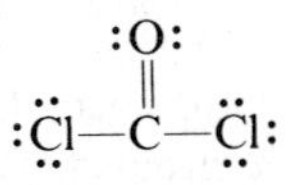

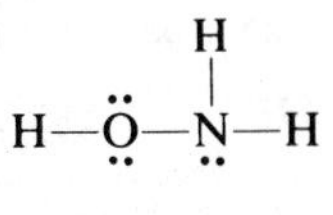

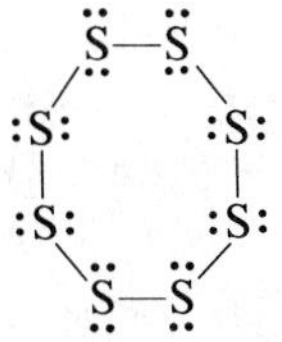

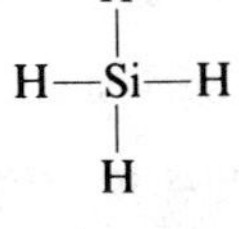

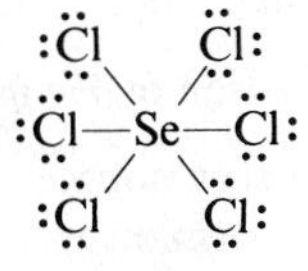

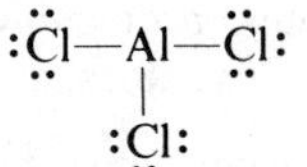

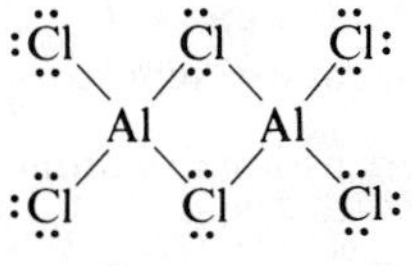

11.7 :C̈l—Al(—C̈l:)—C̈l: ; :C̈l and C̈l bridging two Al atoms, each Al with two terminal :C̈l:

11.9 [H—C(=Ö:)—Ö:]$^-$ ⇌ [H—C(—Ö:)=O:]$^-$

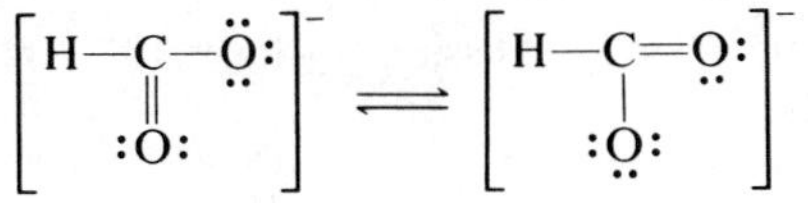

11.11 shorter

11.13 (a) linear, (b) linear, (c) triangular planar, (d) triangular bipyramidal

11.15 LP—LP > LP—BP > BP—BP

11.19 (a) linear, type A_2; (b) bent, type AB_2E_2; (c) bent, type AB_2E_2; (d) triangular pyramidal, type AB_3E; (e) square planar, type AB_4E_2; (f) linear, type AB; (g) linear, type AB_2E_3

11.21 (a) linear, type AB_2; (b) tetrahedral, type AB_4; (c) tetrahedral, type AB_4; (d) octahedral about Cr, type AB_6; (e) triangular pyramidal, type AB_3E; (f) octahedral, type AB_6; (g) bent, type AB_2E

11.23 109.47° ideal; N—O—H, H—N—H, and H—N—O should be less

11.27 σ bond formed by overlap of partially filled 2*p* orbitals

11.29 new set of orbitals formed by mixing atomic orbitals; to explain geometry

11.31 (a) sp^2, (b) sp^3, (c) sp^3, (d) sp^3d, (e) sp^3

11.33 (a) *sp*, (b) sp^2, (c) sp^3, (d) sp^3, (e) sp^3, (f) sp^3, (g) sp^3d, (h) sp^2, (i) sp^3

11.35 (a) *sp*, (b) *sp*, (c) sp^2; (a) and (b); (c)

11.37 :Ö=Ö:; :Ö=Ö—Ö: ⇌ :Ö—Ö=Ö:, sp^2

11.39 intermolecular forces; dipole–dipole interactions and London forces

11.41 electrostatic attraction resulting from momentary shifts in symmetry of electron cloud; size and geometry of molecules, ease of polarization of electron clouds

11.43 c, d, e

11.45 b, c

11.47 each H_2O molecule is involved in four hydrogen bonds; each HF molecule is involved in two hydrogen bonds

11.49 (a) London forces; (b) London forces; (c) London forces, dipole–dipole interactions; (d) London forces; (e) London forces, dipole–dipole interactions; (f) London forces; (g) London forces, dipole–dipole interactions

11.51 (a) S_8, (b) SO_2, (c) F_2, (d) *n*-octane, (e) CCl_4

11.53 NH_3: dipole–dipole interactions, hydrogen bonding, London forces; CH_4: London forces; CH_4; NH_3

11.55 first structure, more overlap between molecules

11.57 342.4 kJ/mol, 327 kJ/mol

11.59 (a) i, iii; (b) i, ii, iii, (c) i, ii; (d) i; (e) ii

11.61 (a) :F̤̈—Xe—F̤̈: sp^3d; XeF_4 (square planar) sp^3d^2; XeF_6 sp^3d^3

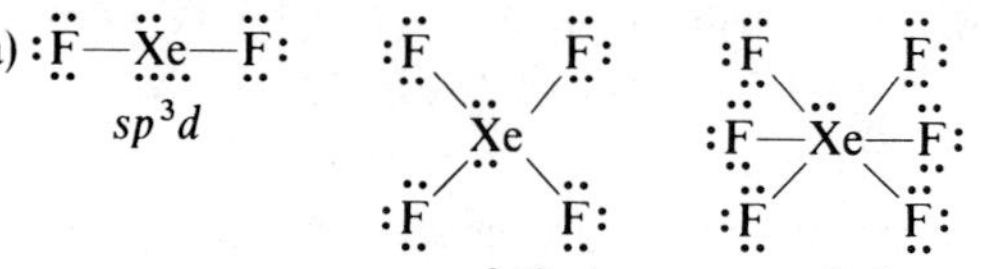

CHAPTER 12

12.1 protons and neutrons; same; $A = Z + N$

12.3 energy that would be released in the combination of nucleons to form a nucleus; apply the mass–energy relationship to the mass defect

12.5 (a) -1.6769×10^{-11} J, -1.1978×10^{-12} J/nucleon; (b) -7.887×10^{-11} J, -1.408×10^{-12} J/nucleon; (c) -1.7551×10^{-10} J, -1.3501×10^{-12} J/nucleon; $^{56}_{26}Fe$

12.7 small values for very light elements; maximum values for $40 < A < 100$; smaller values for heavier elements

12.9 (a) α particles drawn toward negative electrode, β particles toward positive electrode, γ unaffected; (b) α particles and β particles drawn to opposite portions of the field, γ unaffected; (c) α radiation reduced by paper, α radiation and β radiation reduced by concrete; α particles are $^4_2He^{2+}$, β particles are e^-, γ radiation is very high-energy electromagnetic radiation

12.11 same electronic structure

12.13 probability of e^- to be captured is a function of chemical environment

12.15 88.5%

12.17 12.2 yr

12.19 (a) $5.41 \times 10^{-10}\ yr^{-1}$, (b) $8.08 \times 10^{21}\ K^+$ ions, (c) $1620\ s^{-1}$

12.21 5100 yr

12.25 H and He are most abundant; elements with even values of Z are more abundant; lower values of Z are more abundant

12.27 (a) $^{63}_{28}Ni \longrightarrow {}^{0}_{-1}e + {}^{63}_{29}Cu$; (b) $2\ ^2_1H \longrightarrow {}^3_2He + {}^1_0n$; (c) $^{10}_{5}B + {}^1_0n \longrightarrow {}^7_3Li + {}^4_2He$, the unknown nuclide is $^{10}_{5}B$; (d) $^{14}_{7}N + {}^1_0n \longrightarrow 3\ ^4_2He + {}^3_1H$

12.29 $^{227}_{90}Th$

12.31 A is ^{13}N, B is ^{13}C, C is ^{14}N, D is ^{15}O, E is ^{15}N, F is ^4He

12.33 (a) (i) unchanged, (ii) unchanged, (iii) unchanged; (b) (i) decreases by 2, (ii) decreases by 4, (iii) increases; (c) (i) increases by 1, (ii) unchanged, (iii) decreases; (d) $^{91m}_{39}Y \longrightarrow {}^{91}_{39}Y + \gamma$; (e) $^{205}_{84}Po \longrightarrow {}^4_2He + {}^{201}_{82}Pb$; (f) $^{215}_{83}Bi \longrightarrow {}^{0}_{-1}e + {}^{215}_{84}Po$

12.35 β^- decay; n/p ratio decreases

12.37 (a) 1.6, β^- decay; (b) 1.13, β^+ decay or electron capture; (c) 1.585, β^- decay

12.39 β^- emission $^{159}_{64}Gd$

0.59 MeV β^-
0.89 MeV β^-
0.95 MeV β^-
0.362 MeV γ
0.058 MeV γ
$^{159}_{65}Tb$

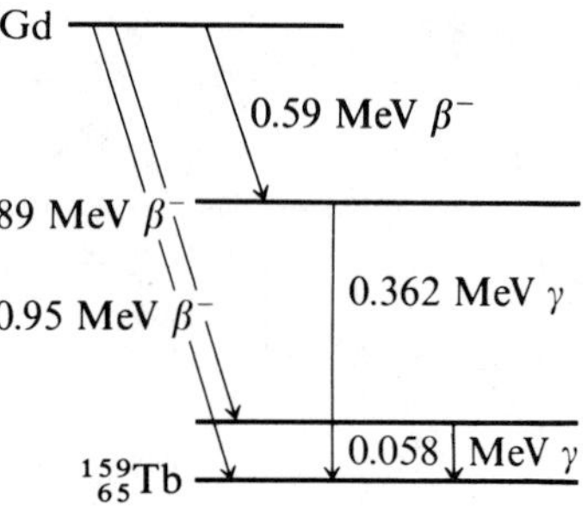

12.41 (a) -1.1964×10^{-12} J/nucleon, (b) -1.2779×10^{-12} J/nucleon, (c) -1.2418×10^{-12} J/nucleon, (d) -1.2445×10^{-12} J/nucleon, (e) -1.213×10^{-12} J/nucleon; $^{16}_{8}O$

12.43 moving particle strikes target to form unstable nucleus which decays; target(bombarding particle, product particle(s))-product

12.45 (a) $^{14}_{7}N + {}^4_2He \longrightarrow {}^{17}_{8}O + {}^1_1H$, (b) $^{106}_{46}Pd + {}^1_0n \longrightarrow {}^{106}_{45}Rh + {}^1_1H$, (c) $^{23}_{11}Na + {}^1_0n \longrightarrow {}^{0}_{-1}e + {}^{24}_{12}Mg$, $X = {}^{24}_{12}Mg$

12.47 (a) $^6_3Li(n,\alpha)^3_1H$, (b) $^{31}_{15}P(d,p)^{32}_{15}P$, (c) $^{238}_{92}U(n,\beta^-)^{239}_{93}Np$

12.49 (a) $^{14}_{7}N + {}^4_2He \longrightarrow {}^1_1H + {}^{17}_{8}O$, (b) 1.92×10^{-13} J

12.51 $\alpha < \beta < \gamma$; energy is transferred in very small distance

12.53 heavy isotope splits into two nuclides of intermediate mass and several neutrons, ^{235}U and ^{239}Pu

12.55 stars, formation of He from H

12.57 (1) fuel, energy source; (2) moderator, slow down neutrons; (3) central system, keep reaction under critical conditions; (4) cooling system, heat transfer; (5) shielding, protection from heat and radiation; spent fuel reprocessing, waste storage

12.59 produces more fissionable atoms than consumes, $^{238}_{92}U + {}^1_0n \longrightarrow {}^{239}_{92}U$, $^{239}_{92}U \longrightarrow {}^{239}_{93}Np + {}^{0}_{-1}e$, $^{239}_{93}Np \longrightarrow {}^{239}_{94}Pu + {}^{0}_{-1}e$

12.61 -1.395×10^{-13} J/u for fission, -1.5×10^{-13} J/u for fusion; fusion process

CHAPTER 13

13.1 gases and liquids flow because of movement of molecules; rigid structure of solid does not allow it to flow

13.3 flow; temperature and size, shape, and chemical nature of molecules; decreases

13.5 bottom, top

13.7 anisotropic, cleaved into pieces with planar faces, sharp melting points; isotropic, cleaved into pieces with nonplanar faces, softens

13.9 dynamic; more energetic molecules leave the restricted state at the same rate as the less energetic molecules in the freer state return to the restricted state

13.11 226 kJ/mol

13.13 38 g

13.15 (a) C_6H_6, (b) H_2CO, (c) Ga, (d) He

13.17 359.4 Torr

13.19 234 Torr

13.21 4.4 Torr, 45.9 Torr

13.23 72.2%

13.25 vapor pressure is not a linear function of temperature

13.27 99.5 °C

13.31 N_2, Ar, O_2

13.33 temperature above which no amount of pressure will cause liquefaction; no, sufficient pressure must be applied

13.35 3, some ice melts

13.37 remains solid, melts

13.39 each single layer is hexagonal; hexagonal closest packing is ABAB... stacking of layers, cubic closest packing is ABCABC... stacking

13.41 most convenient small part of a space lattice that, if repeated in three dimensions, will generate the entire lattice; see Figure 13.21

13.43 (a) 3.47×10^{-22} g, (b) 3.79×10^{-23} cm^3, (c) 9.16 g/cm^3

13.45 0.409 nm

13.47 2.3297 g/cm^3

13.51 2.1635 g/cm^3

13.53 energy liberated as gaseous ions combine to give a crystalline ionic substance; $M^{2+}(g) + X^{2-}(g) \longrightarrow MX(s)$; increases, decreases

13.55 −787.38 kJ/mol, −717.88 kJ/mol; lattice energy increases with decreasing interionic distance

13.57 theoretical density will be high if crystal imperfections are present; theoretical density could be high or low if impurities are present

13.59 ability of waves to bend around corners or to spread out; comparable in size; X-rays, neutrons and electrons

CHAPTER 14

14.1 $H-\ddot{\underset{\cdot\cdot}{O}}-H$, polar covalent

14.3 109.47°, smaller, lone pair-lone pair and lone pair-bonded pair repulsions

14.5 as temperature increases, volume decreases as more hydrogen bonds break and volume increases as normal thermal expansion occurs giving the maximum density at 3.98 °C; ice contains extended organized lattice that is not present in the liquid

14.7 12 g

14.9 0.907 g/cm^3

14.11 forms favorable dipole–dipole and dipole–ion interactions; no

14.13 ionic size and charge; Zn^{2+}

14.15 (a) (solvent–solvent) + (solute–solute) > (solvent–solute), (b) (solvent–solvent) + (solute–solute) < (solvent–solute)

14.17 25 kJ, cool to the touch

14.19 (a) chromium(III) acetate monohydrate, (b) cadmium nitrate tetrahydrate, (c) ammonium oxalate monohydrate, (d) lithium bromide dihydrate

14.21 (a) 51.17%, (b) 8.861%

14.23 230. g $CaCl_2$

14.25 $CuCl_2 \cdot 2H_2O$

14.27 H_3O^+, $\left[\begin{matrix} H-\ddot{O}-H \\ | \\ H \end{matrix}\right]^+$

14.29 H^+ and OH^- are always in equilibrium

14.31 (a) $H_2O(g) \xrightarrow{\Delta} H(g) + OH(g)$,
(b) $H_2O(l) \rightleftharpoons H^+ + OH^-$,
(c) $CaCl_2 \cdot H_2O(s) + 5H_2O(l) \longrightarrow CaCl_2 \cdot 6H_2O(s)$

14.33 (a) iv, (b) iv, (c) i, (d) ii, (e) iii

14.35 yes, relatively few ions formed by self-ionization

14.37 (a) K^+, NO_3^-; (b) H^+, ClO_4^-; (c) K^+, OH^-; (d) Ag^+, Cl^-; (e) HCOOH, H^+, $HCOO^-$; (f) CH_3OH

14.39 (a) $H_2SO_4(l) \xrightarrow{H_2O} H^+ + HSO_4^-$ or $H_2SO_4(l) + H_2O(l) \longrightarrow H_3O^+ + HSO_4^-$ and

$HSO_4^- \xrightarrow{H_2O} H^+ + SO_4^{2-}$ or
$HSO_4^- + H_2O(l) \rightleftharpoons H_3O^+ + SO_4^{2-}$
(b) $NH_4I(s) \xrightarrow{H_2O} NH_4^+ + I^-$
(c) $Sr(OH)_2(s) \xrightleftharpoons{H_2O} Sr^{2+} + 2OH^-$
(d) $HCN(g) \xrightleftharpoons{H_2O} H^+ + CN^-$ or
$HCN(g) + H_2O(l) \rightleftharpoons H_3O^+ + CN^-$
(e) $AgCl(s) \xrightleftharpoons{H_2O} Ag^+ + Cl^-$

14.41 an acid contains hydrogen and yields H^+ in solution, a base contains OH^- and yields OH^- in solution

14.43 an acid containing more than one ionizable hydrogen atom per molecule; H_2SO_4; $H_2SO_4(aq) \xrightarrow{H_2O} H^+ + HSO_4^-$ or $H_2SO_4(aq) + H_2O(l) \longrightarrow H_3O^+ + HSO_4^-$, $HSO_4^- \xrightleftharpoons{H_2O} H^+ + SO_4^{2-}$ or $HSO_4^- + H_2O(l) \rightleftharpoons H_3O^+ + SO_4^{2-}$

14.45 (a) $HOOCCOOH(s) \xrightleftharpoons{H_2O} H^+ + HOOCCOO^-$ or
$HOOCCOOH(s) + H_2O(l) \rightleftharpoons H_3O^+ + HOOCCOO^-$
$HOOCCOO^- \xrightleftharpoons{H_2O} H^+ + OOCCOO^{2-}$ or
$HOOCCOO^- + H_2O(l) \rightleftharpoons H_3O^+ + OOCCOO^{2-}$
(b) $HCN(g) \xrightleftharpoons{H_2O} H^+ + CN^-$ or
$HCN(g) + H_2O(l) \rightleftharpoons H_3O^+ + CN^-$
(c) $HNO_3(l) \xrightarrow{H_2O} H^+ + NO_3^-$ or
$HNO_3(l) + H_2O(l) \longrightarrow H_3O^+ + NO_3^-$

14.47 HBr requires 0.494 g NaOH, $HClO_4$ requires 0.398 g NaOH

14.49 4.06 L

14.51 central metal atom or cation with ligands (anions or molecules) attached

14.53 (a) −1, (b) +1, (c) −1

14.55 (a) flushed by air, (b) settle in pools and lakes or filtered by soil, (c) precipitated, (d) bacteria and other microorganisms decompose byproducts

14.57 0.54 mol/L

14.59 metal ions present that form precipitates with soap or upon boiling; temporary hardness (HCO_3^-) can be removed by boiling, permanent hardness cannot be softened by boiling

14.61 0.033 g Na_2CO_3, 0.086 g $Ca(OH)_2$

14.63 pathogens (viruses and bacteria); oxygen-demanding organic waste (from dead or live animals and plants); specific chemicals (e.g., metals, detergents)

CHAPTER 15

15.1 (a) salt water, NaCl (solute) in H_2O (solvent); (b) air, O_2 (solute) in N_2 (solvent); (c) hydrochloric acid, HCl (solute) in H_2O (solvent); (d) vinegar, CH_3COOH (solute) in H_2O (solvent); (e) brass, Cu (solute) in Zn (solvent)

15.3 dynamic equilibrium; "tag" the solute in one phase

15.5 oil-water forces are not strong enough to overcome oil-oil and water-water forces

15.7 all intermolecular forces should be identical; c

15.9 5.0×10^{-6}

15.11 1.40×10^{-3} g N_2/100 g H_2O, 7.88×10^{-4} g O_2/100 H_2O; greater

15.13 solubility increases with increasing temperature; exceptions usually include the SO_4^{2-}, SeO_4^{2-}, SO_3^{2-}, AsO_4^{3-} and PO_4^{3-} anions

15.15 molarity

15.17 mass percent $= [s/(s + 100)](100)$ where s is the solubility in units of (g solute/100 g H_2O)

15.19 1.54 M

15.21 0.0221

15.23 19 g NaCl, 141 g H_2O

15.25 0.88 m

15.27 14.8 g NaCl in 996 g H_2O

15.29 50.0 mL of 4.38 M NaOH, dilute to 250 mL with H_2O

15.31 −21 kJ, warm to the touch

15.33 1.23 M, 1.29 m, 0.0228

15.35 4.38 m, 0.0730, 70.0 mass%

15.37 (a) same, (b) vapor phase is richer in A, (c) vapor phase is richer in B

15.39 liquids exert independent vapor pressures; boiling occurs when total vapor pressure equals atmospheric pressure

15.41 207 Torr acetone, 118 Torr chloroform, $P_t = 325$ Torr

15.43 23.07 Torr

15.45 100.0640 °C, −0.233 °C

15.47 (a) methyl alcohol, (b) same

15.49 110 g/mol

15.51 0.03 atm

15.53 12.2 K; no, solution would be frozen

15.55 89 mass%

15.57 strong electrolytes $\gg$ weak electrolytes $>$ nonelectrolytes

15.59 7.4 atm

15.61 0.151 mol/kg, forms 4 ions ($3K^+$, $Fe(CN)_6^{3-}$)

15.63 751.0 Torr

15.65 59 g/mol, 112 g/mol; CH_3COOH dimerizes in benzene

15.67 0.55446 m Cl^-, 0.47576 m Na^+, 0.02859 m SO_4^{2-},

0.05419 m Mg^{2+}, 0.0103 m Ca^{2+}, 0.0101 m K^+, 0.00241 m HCO_3^-, 0.00084 m Br^-; −2.11 °C, 100.582 °C; 27.8 atm

15.69 suspended particles in a colloid are larger than molecules or ions

15.71 gases exist only as molecules and form only true solutions; use colloid of opposite charge or Cottrell precipitator

15.73 (a) aerosol, (b) foam

15.75 384.2 atm/(g CO/100 g H_2O), 0.106 atm/(g SO_2/100 g H_2O); 9.293×10^{-4} m CO, 1.47 m SO_2; −0.00173 °C, −2.73 °C; SO_2 reacts with water to form H_2SO_3

CHAPTER 16

16.1 a, b, f

16.3 (a) O in O_2^{2-} is both oxidizing and reducing agent; (b) Sn is reducing agent, H^+ is oxidizing agent

16.5 (a) Sn^{2+} is reducing agent ($+2 \longrightarrow +4$), Fe^{3+} is oxidizing agent ($+3 \longrightarrow +2$); (b) MnO_2 is oxidizing agent (Mn, $+4 \longrightarrow +2$), HCl is reducing agent (Cl, $-1 \longrightarrow 0$); (c) XeF_2 is oxidizing agent (Xe, $+2 \longrightarrow 0$), H_2O is reducing agent (O, $-2 \longrightarrow 0$); (d) I^- is reducing agent ($-1 \longrightarrow +1$), ClO^- is oxidizing agent (Cl^-, $+1 \longrightarrow -1$); (e) HNO_3 is both oxidizing agent and reducing agent (N, $+5 \longrightarrow +4$); (O, $-2 \longrightarrow 0$)

16.7 (a) $8SO_2(g) + 16H_2S(g) \longrightarrow 3S_8(s) + 16H_2O(l)$, (b) $2Ca_3(PO_4)_2(s) + 10C(s) + 6SiO_2(s) \longrightarrow 6CaSiO_3(l) + P_4(g) + 10CO(g)$, (c) $3SO_2(g) + Cr_2O_7^{2-} + 8H^+ \longrightarrow 2Cr^{3+} + 3H_2SO_4(aq) + H_2O(l)$, (d) $3Cl_2(g) + 6OH^- \longrightarrow ClO_3^- + 5Cl^- + 3H_2O(l)$, (e) $3S(s) + 6OH^- \longrightarrow 2S^{2-} + SO_3^{2-} + 3H_2O(l)$, (f) $2MnO_4^- + 3IO_3^- + H_2O(l) \longrightarrow 2MnO_2(s) + 3IO_4^- + 2OH^-$, (g) $Sn(s) + 2H_2O(l) + 2OH^- \longrightarrow 3IO_3^- + [Sn(OH)_4]^{2-} + H_2(g)$

16.9 $1s^1$; 0, ±1

16.11 0.015%

16.13 $H_2(g) + 2Li(s) \longrightarrow 2LiH(s)$, $H_2(g) + CuO(s) \xrightarrow{\Delta} Cu(s) + 2H_2O(g)$

16.15 electronegativity of H > metals, ionic bonding; electronegativity of H ≈ nonmetals, polar covalent bonding

16.17 (a) $N_2(g) + 3H_2(g) \longrightarrow 2NH_3(g)$, reducing agent; (b) $I_2(g) + H_2(g) \longrightarrow 2HI(g)$, reducing agent; (c) $Ca(s) + H_2(g) \longrightarrow CaH_2(s)$, oxidizing agent; (d) $SnO(s) + H_2(g) \longrightarrow Sn(s) + H_2O(g)$, reducing agent

16.19 steam reforming of hydrocarbons, $CH_4(g) + H_2O(g) \xrightarrow{\Delta} CO(g) + 3H_2(g)$; water gas reaction, $C(s) + H_2O(g) \xrightarrow{\Delta} H_2(g) + CO(g)$; electrolysis of water, $2H_2O(l) \longrightarrow 2H_2(g) + O_2(g)$

16.21 −678 kJ, 2.80 times as great

16.23 $1s^2 2s^2 2p^4$; 0, −2, +4, +6

16.25 F; F is more electronegative than O

16.27 element will form oxides in lower oxidation state; regular oxides will form

16.29 (a) $P_4(s) + 5O_2(g) \longrightarrow P_4O_{10}(s)$, (b) $C(s) + O_2(g) \longrightarrow CO_2(g)$, (c) $2SO_2(g) + O_2(g) \longrightarrow 2SO_3(g)$, (d) $2SnO(s) + O_2(g) \longrightarrow 2SnO_2(s)$

16.31 (a) $MgO(s) + H_2O(l) \longrightarrow Mg(OH)_2(aq)$, (b) $P_4O_6(s) + 6H_2O(l) \longrightarrow 4H_3PO_3(aq)$, (c) $CO_2(g) + H_2O(l) \longrightarrow H_2CO_3(aq)$, (d) $SO_2(g) + H_2O(l) \longrightarrow H_2SO_3(aq)$, (e) $Na_2O(s) + H_2O(l) \longrightarrow 2NaOH(aq)$

16.33 MO, most alkaline because M is in lowest oxidation state

16.35 7.6 L

16.37 $:\ddot{O}{=}\ddot{O}{-}\ddot{O}: \rightleftharpoons :\ddot{O}{-}\ddot{O}{=}\ddot{O}:$, bent

16.39 formation: $O_2(g) \xrightarrow{hv} 2O(g)$, $O(g) + O_2(g) + M(g) \longrightarrow O_3(g) + M(g)$; decomposition: $O_3(g) \xrightarrow{hv} O_2(g) + O(g)$; ultraviolet radiation will not be absorbed

16.41 19.21 mg O_3

16.43 $H{-}\ddot{O}{-}\ddot{O}{-}H$; two O—O—H planes roughly perpendicular to each other; London forces, dipole–dipole interactions, hydrogen bonding

16.45 oxidation–reduction; O is oxidized from −1 to 0 and reduced from −1 to −2; $2H_2O_2(l) \longrightarrow 2H_2O(l) + O_2(g)$

16.47 44.01 kJ/mol, 51.47 kJ/mol; intermolecular forces in H_2O_2 are stronger

16.49 O_2, H_2

16.51 12.1 atm H_2, 2.45 atm O_2, 14.6 atm total; 7.25 atm H_2 and H_2O

CHAPTER 17

17.1 not necessarily; chemical equilibrium is dynamic

17.3 (a) gas formation, (c) precipitate formation, (d) H_2O_2 is strong reducing agent

17.5 $[Ag(NH_3)_2]^+$ is more stable than AgCl in the presence of excess ammonia

17.7 (a) $2NH_3(g) + H_2SO_4(aq) \longrightarrow (NH_4)_2SO_4(aq)$, (b) $CaO(s) + SiO_2(l) \longrightarrow CaSiO_3(l)$

17.9 (a) $Na_2CO_3(l) + SiO_2(s) \longrightarrow Na_2SiO_3(l) + CO_2(g)$, (b) $[Cu(H_2O)_4]^{2+} + 4NH_3(aq) \longrightarrow [Cu(NH_3)_4]^{2+} + 4H_2O(l)$, (c) $[AgCl_2]^- + 2NH_3(aq) \longrightarrow [Ag(NH_3)_2]^+ + 2Cl^-$

17.11 (a) $NaCl(aq)$, $H_2S(g)$; (b) $Na_2SO_4(aq)$, $CdS(s)$; (c) $NH_4CH_3COO(aq)$, $PbCrO_4(s)$; (d) $As_2S_3(s)$, $HCl(aq)$

17.13 (a) nonredox decomposition, $Ca(OH)_2(s) \xrightarrow{\Delta} CaO(s) + H_2O(g)$; (b) nonredox displacement, $[Cu(H_2O)_4]Cl_2(aq) + 4NH_3(aq) \longrightarrow [Cu(NH_3)_4]Cl_2(aq) + 4H_2O(l)$; (c) nonredox partner exchange, $CuCl_2(aq) + 2NaOH(aq) \longrightarrow Cu(OH)_2(s) + 2NaCl(aq)$; (d) nonredox partner exchange, $PCl_3(l) + 3H_2O(l) \longrightarrow 3HCl(aq) + H_3PO_3(aq)$; (e) nonredox combination, $SO_2(g) + CaO(s) \longrightarrow CaSO_3(s)$; (f) nonredox decomposition, $Be(OH)_2(s) \xrightarrow{\Delta} BeO(s) + H_2O(g)$; (g) nonredox combination, $CO_2(g) + LiOH(aq) \longrightarrow LiHCO_3(aq)$

17.15 (c) $TiCl_4$ is oxidizing agent, Mg is reducing agent

17.17 (a) $CO_2(g)$, (b) $PCl_5(s)$, (c) $TiCl_4(l)$, (d) $Mg_3N_2(s)$, (e) $Fe_2O_3(s)$, (f) $NO_2(g)$, (g) $P_4O_{10}(s)$

17.19 (a) no, no reduction; (b) no, no oxidation; (c) no, Zn is stronger reducing agent than H_2; (d) yes

17.21 (a) $SO_3^{2-} + Br_2(aq) + H_2O(l) \longrightarrow SO_4^{2-} + 2Br^- + 2H^+$, (b) $2SO_3^{2-} + O_2(g) \longrightarrow 2SO_4^{2-}$, (c) $SO_3^{2-} + H_2O_2(aq) \longrightarrow SO_4^{2-} + H_2O(l)$, (d) $5SO_3^{2-} + 2MnO_4^- + 6H^+ \longrightarrow 5SO_4^{2-} + 3H_2O(l) + 2Mn^{2+}$

17.23 (a) $PCl_3(l)$ or $PCl_5(s)$; (b) $PCl_5(s)$; (c) $CuCl_2(s)$; (d) Cl^-,$I_2(aq)$

17.25 (a) more complex, $MnO_2(s) + 2Cl^- + 4H^+ \longrightarrow Mn^{2+} + Cl_2(g) + 2H_2O(l)$; (b) electron transfer between monatomic ions, $Sn^{2+} + 2Fe^{3+} \longrightarrow Sn^{4+} + 2Fe^{2+}$; (c) combination of elements, $16K(s) + S_8(s) \longrightarrow 8K_2S(s)$; (d) more complex, $N_2H_4(l) + O_2(g) \longrightarrow N_2(g) + 2H_2O(l)$; (e) decomposition (disproportionation) $Cu_2SO_4(s) \xrightarrow{H_2O} Cu(s) + CuSO_4(aq)$; (f) decomposition (disproportionation) $Hg_2Cl_2(s) \longrightarrow Hg(l) + HgCl_2(s)$

17.27 163 g $K_2Cr_2O_7$

17.29 (a) nonredox, partner exchange between aqueous ions; (b) redox, displacement of one element in a compound by another; (c) nonredox, decomposition to give compounds; (d) nonredox, combination of compounds; (e) redox, more complex; (f) nonredox, displacement; (g) redox, complex; (h) redox, combination of element with compound to give compound; (i) redox, decomposition; (j) redox, electron transfer

17.31 (a) redox displacement, $LiOH(aq)$, $H_2(g)$; (b) redox decomposition, $Ag(s)$, $O_2(g)$; (c) nonredox combination; $LiOH(aq)$; (d) redox decomposition, $H_2(g)$, $O_2(g)$; (e) no reaction, (f) no reaction, (g) redox decomposition, $NaNH_2(s)$, $O_2(g)$

17.33 (a) $NH_4NO_3(s) \xrightarrow{\Delta} N_2O(g) + 2H_2O(g)$, (b) $H_2S(g) + Pb(CH_3COO)_2(aq) \longrightarrow PbS(s) + 2CH_3COOH(aq)$, (c) $Ba(OH)_2(aq) + H_2SO_4(aq) \longrightarrow BaSO_4(s) + 2H_2O(l)$ (d) $2Fe^{3+} + Sn^{2+} \longrightarrow 2Fe^{2+} + Sn^{4+}$, (e) $2FeCl_2(aq) + Cl_2(g) \longrightarrow 2FeCl_3(aq)$, (f) $2Hg(l) + O_2(g) \longrightarrow 2HgO(s)$, (g) $CaCl_2(aq) + H_2SO_4(l) \longrightarrow CaSO_4(s) + 2HCl(aq)$, (h) $SO_2(g) + 2NaOH(aq) \longrightarrow Na_2SO_3(aq) + H_2O(l)$

17.35 (a) no, only strong electrolytes formed; (b) no, Cu is below Fe in activity series; (c) no, only strong electrolytes formed; (d) yes, $BaCl_2(aq)$, $H_2O(l)$; (e) no, Zn is below Al in activity series

17.37 $2Mg(s) + O_2(g) \xrightarrow{\Delta} 2MgO(s)$, redox, combination of elements to give a compound; $3Mg(s) + N_2(g) \xrightarrow{\Delta} Mg_3N_2(s)$, redox, combination of elements to give a compound; $Mg_3N_2(s) + 6H_2O(l) \longrightarrow 3Mg(OH)_2(aq) + 2NH_3(g)$, nonredox, partner exchange; $Mg(OH)_2(s) \xrightarrow{\Delta} MgO(s) + H_2O(g)$, nonredox, decomposition to give compounds

CHAPTER 18

18.1 reaction that occurs in a single step exactly as written; mechanism consists of all elementary steps in a single pathway showing changes that occur at the molecular level

18.3 orientation, penetration of electron shells, stable resulting structure; (a) penetration, (b) orientation, (c) stable structure

18.5 (a) i, (b) v, (c) iv, (d) ii, (e) iii

18.9 bimolecular, bimolecular

18.11 (a) Rate $= \Delta[O_2]/\Delta t$, (b) Rate $= -\Delta[HCl]/\Delta t$ or Rate $= \Delta[CaCl_2]/\Delta t = \Delta[CO_2]/\Delta t$, (c) Rate $= -\Delta[S^{2-}]/\Delta t = \Delta[HS^-]/\Delta t = \Delta[OH^-]/\Delta t$, (d) Rate $= \Delta[CO_2]/\Delta t$, (e) Rate $= -\Delta[H_2O]/\Delta t = \Delta[H_2]/\Delta t$ or Rate $= \Delta[O_2]/\Delta t$

18.13 Rate $= k[A]$; $(\text{time})^{-1}$; $\log[A] = -(k/2.303)t + \log[A]_0$; plot $\log[A]$ against t

18.15 Rate $= k[NO]^2[H_2]$, third order overall

18.17 Rate $= k[H_2][C_2H_4]^{-1}$

18.19 (a) 6.3×10^{-5} mol/L min, (b) 6.3×10^{-6} mol/L min

18.21 1.24×10^{-2} mol/L, 29 day

18.23 15.4 min, 290 min

18.25 Rate $= k\,P_{NO}^2\,P_{H_2}$

18.27 Rate $= k[A]^2[B]$

18.29 plot of $\log[CH_2N_2]$ against t is linear, $k = 0.055\ \text{min}^{-1}$

18.31 (b) 0.97 mol/L, 0.68 mol/L, 0.48 mol/L; (c) 6.5×10^{-4} mol/L s; (d) 4.3×10^{-4} mol/L s, 3.1×10^{-4} mol/L s, yes; (e) first order with $k = 6.5 \times 10^{-4}\ s^{-1}$

18.33 first step

18.35 (a) A; (b) C, D; (c) Rate $= k[B][C]$; (d) Rate $= k''[B][A]$; (e) second order overall

18.37 150 kJ/mol

18.39 2.0×10^5 J/mol

18.41 10 times faster

18.45 (a) double, (b) double, (c) increase, (d) slow down, (e) quadruple, (f) increase

18.47 (b) first order; (c) 2.30×10^{-2} s^{-1}; (d) 2, 4

18.49 (a) $x = -1$, $y = 2$; (b) first order, 1.9×10^{-4} s^{-1}, 3.1×10^{-4} s^{-1}; (c) yes; (d) 1.8×10^{-6} s^{-1} $Torr^{-1}$

18.51 1.0×10^5 J/mol

CHAPTER 19

19.1 dynamic equilibrium; rates of forward and reverse reactions become identical

19.3 (a) $K_C = [SO_2][H_2O]/[SO_3][H_2]$,
(b) $K_C = [NO]^4[H_2O]^6/[NH_3]^4[O_2]^5$,
(c) $K_C = [CO_2]^3[H_2O]^4/[C_3H_8][O_2]^5$

19.5 $K_C = [H^+][H_2PO_4^-]/[H_3PO_4]$; the additional H^+ produces more H_3PO_4 and less $H_2PO_4^-$ once equilibrium is reestablished

19.7 b, d

19.9 CrO_4^{2-}

19.11 1.1×10^{-5}

19.13 0.1

19.15 0.030 mol/L, 130

19.17 (a) $K_P = P_{SO_2}P_{H_2O}/P_{SO_3}P_{H_2}$,
(b) $K_P = P_{NO}^4P_{H_2O}^6/P_{NH_3}^4P_{O_2}^5$, (c) $K_P = P_{CO_2}^3P_{H_2O}^4/P_{C_3H_8}P_{O_2}^5$

19.19 0.01813

19.21 193

19.23 (a) $K_C = [Ag^+]^2[S^{2-}]$; (b) $K_P = P_{CO_2}$, $K_C = [CO_2]$;
(c) $K_P = P_{CO}^2$, $K_C = [CO]^2$; (d) $K_C = [Br^-]^2/[I^-]^2[Br_2]$

19.25 6.13×10^{51}

19.27 1.6×10^{10}

19.29 $K = K_a^2K_bK_c/K_d^2$

19.31 same form; Q is used for nonequilibrium conditions

19.33 $Q_C = 700$, ClF will decompose

19.35 (a) 2.0×10^9, (b) no

19.37 if a system at equilibrium is subjected to a stress, the system will react in a way that tends to relieve the stress; concentration changes, temperature changes, total pressure changes; no change, rates of forward and reverse reactions are changed equally

19.39 additional BrF_5 will decompose

19.41 moist, pink color is formed by gaseous H_2O reacting with $[Co(H_2O)_4]Cl_2$

19.43 (a) i, (b) i, (c) i, (d) i

19.45 the approximation $C \pm x \approx C$ is made if $C > K$; check approximation using 5% error rule

19.47 4.7×10^{-7} bar

19.49 3.90×10^{-31} mol/L

19.51 0.042 mol/L

19.53 2.7×10^{-5} mol/L

19.55 0.048 mol/L

19.57 0.002 mol Br_2/L, 1.13 mol F_2/L, 0.246 mol BrF/L

19.59 2.4×10^{-23} mol/L

CHAPTER 20

20.1 proton donor; all water-ion acids are proton donors;
(a) $HNO_3(aq) \longrightarrow H^+ + NO_3^-$, NO_3^-;
(b) $HSO_4^- \longrightarrow H^+ + SO_4^{2-}$, SO_4^{2-};
(c) $H_2SO_4(aq) \longrightarrow H^+ + HSO_4^-$, HSO_4^-;
(d) $HCl(aq) \longrightarrow H^+ + Cl^-$, Cl^-;
(e) $H_2O(l) \longrightarrow H^+ + OH^-$, OH^-

20.3 (a) $HCO_3^- + H_2O(l) \rightleftharpoons H_3O^+ + CO_3^{2-}$, $HCO_3^- + H_3O^+ \rightleftharpoons H_2CO_3(aq) + H_2O(l)$; (b) CO_3^{2-} is the conjugate base of HCO_3^-; H_2CO_3 is the conjugate acid of HCO_3^-

20.5 (a) iv, v, vi, vii; (b) i, iii, v, vi

20.7 proton transfer; (a) $H_2O(l) + NH_3(aq) \longrightarrow OH^- + NH_4^+$, H_2O and OH^-, NH_3 and NH_4^+; (b) $H_3O^+ + OH^- \longrightarrow 2H_2O(l)$, H_3O^+ and H_2O, OH^- and H_2O; (c) $NH_4^+ + CO_3^{2-} \longrightarrow NH_3(aq) + HCO_3^-$, NH_4^+ and NH_3, CO_3^{2-} and HCO_3^-; (d) $HNO_3(aq) + H_2O(l) \longrightarrow NO_3^- + H_3O^+$, HNO_3 and NO_3^-, H_2O and H_3O^+

20.9 (a) no reaction, (b) $H_2S(aq) + CN^- \longrightarrow HS^- + HCN(aq)$, (c) no reaction

20.11 (a) hydrofluoric acid, (b) hydrobromic acid, (c) hydrosulfuric acid, (d) hydroselenic acid

20.13 (a) phosphoric acid, phosphorous acid, hypophosphorous acid, (b) arsenic acid, arsenous acid; (c) antimonic acid, antimonous acid

20.15 loss of one or more water molecules from ortho acid yields a meta acid; $H_3PO_4 \longrightarrow H_2O + HPO_3$

20.17 (a) borate ion, (b) arsenite ion, (c) nitrite ion, (d) bismuthate ion, (e) selenite ion, (f) hypoiodite ion, (g) perchlorate ion

20.19 (a) gallium(III) selenate, (b) cobalt(II) phosphate, (c) zinc bromate, (d) thallium(I) nitrate, (e) thorium(IV) sulfate

20.21 (a) $PbSeO_4$, (b) $CoSO_3$, (c) $Cd(IO_3)_2$, (d) $CsClO_4$, (e) Cu_2SO_4

20.23 radius, H_2Te

20.25 (a) NH_3, (b) HI, (c) HBr, (d) H_2S

20.27 acid strength increases with increasing oxidation state, H_2SO_4

20.29 (a) HIO_3, (b) $Sb(OH)_3$, (c) HNO_3

20.31 molar mass divided by number of equivalents of H^+ or OH^- ions supplied; for H_2SO_4, equivalent mass = 98 g/equiv if only one proton reacts, 49 g/equiv if both protons react

20.33 (a) 1 equiv/1 mol, (b) 3 equiv/1 mol, (c) 2 equiv/1 mol

20.35 (a) 0.065 mol/L, (b) 0.13 mol/L, (c) 0.13 mol/L

20.37 $2H_2O(l) \rightleftharpoons H_3O^+ + OH^-$; $K_W = [H^+][OH^-]$; $[H^+] = [OH^-]$; acidic $[H^+] > [OH^-]$, alkaline $[OH^-] > [H^+]$

20.39 (a) 7.7×10^{-11} mol/L, (b) 1×10^{-6} mol/L

20.41 (a) 5.02, (b) 3.8321, (c) 2.693, (d) 10.92, (e) −1.00; a, b, c, e

20.43 (a) 10.54, (b) 2.04, (c) 7.65, (d) 15.0; a, c, d

20.45 (a) 13.3755, (b) 1.438, (c) 4.38, (d) 12.917

20.47 $HA(aq) + H_2O(l) \longrightarrow H_3O^+ + A^-$, $K_a = [H^+][A^-]/[HA]$

20.49 2.8×10^{-10} mol/L

20.51 $[H^+] = [F^-] = 3.7 \times 10^{-3}$ mol/L, $[HF] = 0.021$ mol/L, pH = 2.43

20.53 1.8×10^{-5}

20.55 (a) 1.32%, (b) 4.18%

20.57 (a) $M(OH)(aq) \rightleftharpoons M^+ + OH^-$, $K_b = [M^+][OH^-]/[MOH]$; (b) $B(aq) + H_2O(l) \rightleftharpoons BH^+ + OH^-$, $K_b = [BH^+][OH^-]/[B]$; K_b

20.59 4.0×10^{-4}

20.61 $[NH_4^+] = [OH^-] = 1.3 \times 10^{-3}$ mol/L, $[NH_3] = 0.10$ mol/L

20.63 1.6×10^{-5}

20.65 11.15

20.67 1.1×10^{-4}

20.69 $K_W = K_a K_b$

20.71 4.5×10^{-6}

20.73 CN^-

20.75 electron-pair acceptor; (a)

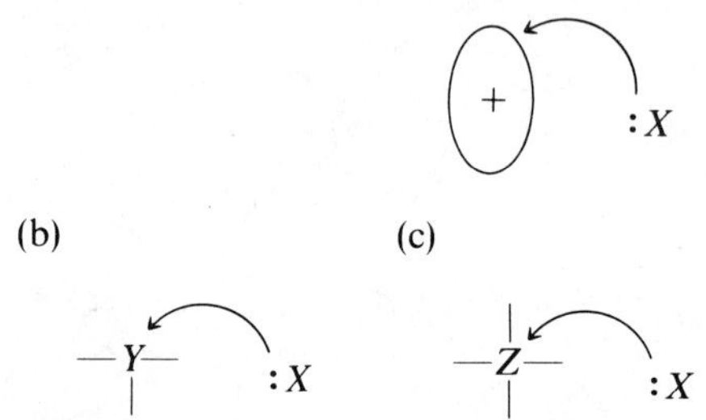

(b) (c)

—Y— :X —Z— :X

20.77 formation of coordinate covalent bond; all; (a) Ag^+, (b) I_2, (c) H_3O^+, (d) H_2O

20.79 3.101×10^{-7} mol/L, 6.5085; no, $[H^+] = [OH^-]$

20.81 −12.55 kJ for HF, −78.8 kJ for HCl, HCl

20.83 (a) 1.35×10^{-15}; (b) $[D_3O^+] = [OD^-] = 3.67 \times 10^{-8}$ mol/L; (c) 7.435; (d) pD + pOD = 14.869; (e) pD < 7.435; (f) CH_3COOH, HD_2O^+; (g) none; (h) CH_3COOH, HD_2O^+; (i) D_2O, CH_3COO^-; (j) 28 mol/L

CHAPTER 21

21.1 weaker acid than water, none

21.3 (b) $NH_4^+ + H_2O(l) \rightleftharpoons NH_3(aq) + H_3O^+$, (c) $[Al(H_2O)_6]^{3+} + H_2O(l) \rightleftharpoons [Al(OH)(H_2O)_5]^{2+} + H_3O^+$, (d) $[Fe(H_2O)_6]^{3+} + H_2O(l) \rightleftharpoons [Fe(OH)(H_2O)_6]^{2+} + H_3O^+$

21.5 6.3×10^{-10}

21.7 weaker base than water, none

21.9 (a) $S^{2-} + H_2O(l) \rightleftharpoons HS^- + OH^-$, (b) $HS^- + H_2O(l) \rightleftharpoons H_2S(aq) + OH^-$, (c) $F^- + H_2O(l) \rightleftharpoons HF(aq) + OH^-$, (e) $NO_2^- + H_2O(l) \rightleftharpoons HNO_2(aq) + OH^-$

21.11 1.5×10^{-11}

21.13 any hydrolysis of ions will change pH

21.15 (a) i, ii, v; (b) iv; (c) iii

21.17 11.80

21.19 decreases

21.21 1.3×10^{-6} mol/L

21.23 2.9×10^{-8} mol/L

21.25 1.00, no

21.27 5.46

21.29 3.14, 3.14

21.31 0.066 mol/L

21.33 0.13 mol/L

21.35 (a) 4.757, (b) 4.754, (c) 4.757, (d) 4.757

21.37 2.282, 2.5; 3.866, 3.889

21.39 $H_2SO_4(aq) + H_2O(l) \longrightarrow HSO_4^- + H_3O^+$, $HSO_4^- + H_2O(l) \rightleftharpoons SO_4^{2-} + H_3O^+$; HSO_4^-, SO_4^{2-}

21.41 (a) $[SO_3^{2-}] = 5.0 \times 10^{-8}$ mol/L, $[H^+] = [HSO_3] =$ 0.0313 mol/L, $[H_2SO_3] = 0.069$ mol/L

21.43 1.04

21.45 $H^+ + OH^- \rightleftharpoons H_2O(l)$, same net reaction

21.47 62.0 mL

21.49 1.731 mol/L

21.51 30.8%

21.53 4.12%, yes

21.55 5.36

21.61 (a) 0.602, (b) 0.824, (c) 1.125, (d) 1.777, (e) 7.000, (f) 12.097

21.63 (a) 2.177, (b) 3.032, (c) 3.575, (d) 4.353, (e) 8.303, (f) 12.097

21.65 (a) red, (b) yellow, (c) green

21.67 8.3, *m*-cresol purple or thymol blue

21.69 9.3, thymol blue or phenolphthalein

CHAPTER 22

22.1 $[Fe(C_2O_4)_3]^{3-} \rightleftharpoons [Fe(C_2O_4)_2]^+ + C_2O_4^{2-}$, $K_{d_1} = [Fe(C_2O_4)_2^-][C_2O_4^{2-}]/[Fe(C_2O_4)_3^{3-}]$; $[Fe(C_2O_4)_2]^- \rightleftharpoons [Fe(C_2O_4)]^+ + C_2O_4^{2-}$, $K_{d_2} = [Fe(C_2O_4)^+][C_2O_4^{2-}]/[Fe(C_2O_4)_2^-]$; $[Fe(C_2O_4)]^+ \rightleftharpoons Fe^{3+} + C_2O_4^{2-}$, $K_{d_3} = [Fe^{3+}][C_2O_4^{2-}]/[Fe(C_2O_4)^+]$; $K_d = K_{d_1}K_{d_2}K_{d_3}$

22.3 2×10^{-22} mol/L

22.5 5.1×10^{-9} mol/L

22.7 (a) $K_{sp} = [Ag^+]^2[SO_4^{2-}]$, (b) $K_{sp} = [Fe^{3+}]^2[S^{2-}]^3$, (c) $K_{sp} = [Sr^{2+}][CrO_4^{2-}]$

22.9 (a) 1.7×10^{-10}, (b) 3.7×10^{-5}, (c) 2.0×10^{-16}

22.11 2.0×10^{-13}

22.13 (a) 26 g/L, (b) 0.55 g/L

22.15 4.9×10^{-11} mol/L

22.17 (a) 8.5×10^{-5} mol/L, (b) 2.5×10^{-8} mol/L, (c) 7.9×10^{-6} mol/L

22.19 0.4 g Ca^{2+}/L

22.21 $Q_i = 0.005$, yes

22.23 $Q_i = 1.00 \times 10^{-7}$, no

22.25 $[S^{2-}]$ is decreased because of the presence of H^+ to form $H_2S(aq)$ and HS^-; no change

22.27 use concentrated acid, $Fe(OH)_3(s) + 3H_3O^+ \longrightarrow Fe^{3+} + 6H_2O(l)$

22.29 7.5 for $Fe(OH)_2$, 5.04 for $Cu(OH)_2$

22.31 AgCl, (b) 5×10^{-7} mol/L

22.33 1.2×10^{-3} mol/L

22.35 5.3 to 12.3

22.37 2.1×10^{-7} mol/L; $Q_i = 4 \times 10^{-15}$, no

22.39 −0.6 to 5.0

CHAPTER 23

23.1 entropy; b < c < a

23.3 heating gives more kinetic energy (or movement)

23.5 decrease; cooling gives less kinetic (or vibrational) energy

23.7 7.6 J/K mol

23.9 22.0 J/K mol, 109 J/K mol; increase in randomness is greater for vaporization

23.11 gas has more disorder than liquid

23.13 (a) O_3, (b) $FeCl_3$, (c) $H_2O(g)$

23.15 (1) S_0°, (2) γ-solid from 0 K to 24 K, (3) γ-solid to β-solid at 24 K, (4) β-solid from 24 K to 44 K, (5) β-solid to α-solid at 44 K, (6) α-solid from 44 K to 54 K, (7) fusion at 54 K, (8) liquid from 54 K to 90.2 K, (9) vaporization at 90.2 K, (10) gas from 90.2 K to 298 K

23.17 191.2 J/K mol

23.19 (a) i, (b) iii, (c) i, (d) i, (e) ii, (f) ii, (g) i

23.21 (a) 144.56 J/K, (b) 100.89 J/K; net increase of one mol of gas

23.23 (a) −274.87 J/K, (b) −345.7 J/K; neither reaction

23.25 245.45 J/K, yes

23.27 133.2 J/K, yes

23.29 $G = H - TS$; (a) negative, (b) positive, (c) zero

23.31 (a) iii; (b) ii; (c) i, iv

23.33 −326.4 kJ, yes, both

23.35 −53.47 kJ/mol

23.37 −44.43 J/K mol

23.39 liquid

23.41 -4.78 kJ, yes, ΔH°

23.43 -61.86 kJ

23.45 85.8 kJ

23.47 -1387.8 kJ

23.49 -27.42 kJ

23.51 90 kJ, -210 kJ; CuO

23.53 $\Delta H = T\,\Delta S$

23.55 6.74; yes, $\Delta G^\circ < 0$

23.57 -100.2 kJ, more favorable

23.59 (a) 91.12 kJ, (b) 1.1×10^{-16}, (c) 1.1×10^{-8} bar

23.61 more favorable, more favorable; CO reaction has larger temperature dependence

23.63 211.5 kJ

23.65 6.6×10^{-9}

23.67 -102 kJ/mol

23.69 (a) -137.55 J/K, decrease; (b) 285.4 kJ, unfavorable; (c) 326.4 kJ, 163.2 kJ/mol, no

23.71 (a) 55.836 kJ; (b) 79.885 kJ; (c) 1.0×10^{-14}; (d) 2.0×10^{-14}; (e) 6.85, no, $[H^+] = [OH^-]$

CHAPTER 24

24.1 oxidation is loss of electrons, reduction is gain of electrons; equal

24.3 (a) $H_2O_2(aq)/H_2O(l)$, $I_2(s)/I^-$; $H_2O_2(aq) + 2H^+ + 2e^- \longrightarrow 2H_2O(l)$, $2I^- \longrightarrow I_2(s) + 2e^-$, $H_2O_2(aq) + 2I^- + 2H^+ \longrightarrow 2H_2O(l) + I_2(s)$; (b) $I_2(s)/I^-$, $NO_2^-/NO(g)$; $NO_2^- + 2H^+ + e^- \longrightarrow NO(g) + H_2O(l)$, $2I^- \longrightarrow I_2(s) + 2e^-$, $2NO_2^- + 4H^+ + 2I^- \longrightarrow 2NO(g) + I_2(s) + 2H_2O(l)$; (c) $Al^{3+}/Al(s)$, $H_2SO_4(conc)/SO_2(g)$; $Al(s) \longrightarrow Al^{3+} + 3e^-$, $H_2SO_4(conc) + 2H^+ + 2e^- \longrightarrow SO_2(g) + 2H_2O(l)$, $2Al(s) + 3H_2SO_4(conc) + 6H^+ \longrightarrow 2Al^{3+} + 3SO_2(g) + 6H_2O(l)$; (d) $[Zn(OH)_4]^{2-}/Zn(s)$, $NO_3^-/NH_3(aq)$, $Zn(s) + 4OH^- \longrightarrow [Zn(OH)_4]^{2-} + 2e^-$, $NO_3^- + 6H_2O(l) + 8e^- \longrightarrow NH_3(aq) + 9OH^-$, $4Zn(s) + NO_3^- + 6H_2O(l) + 7OH^- \longrightarrow 4[Zn(OH)_4]^{2-} + NH_3(aq)$

24.5 (a) $I_2(s) + 2e^- \longrightarrow 2I^-$, $H_2S(aq) \longrightarrow S(s) + 2H^+ + 2e^-$, $H_2S(aq) + I_2(s) \longrightarrow 2I^- + S(s) + 2H^+$; (b) $PbO_2(s) + 2Cl^- + 4H^+ + 2e^- \longrightarrow PbCl_2(s) + 2H_2O(l)$, $2Cl^- \longrightarrow Cl_2(g) + 2e^-$, $PbO_2(s) + 4Cl^- + 4H^+ \longrightarrow PbCl_2(s) + Cl_2(g) + 2H_2O(l)$; (c) $2Cu(s) + 2OH^- \longrightarrow Cu_2O(s) + H_2O(l) + 2e^-$, $Br_2(aq) + 2e^- \longrightarrow 2Br^-$, $2Cu(s) + Br_2(aq) + 2OH^- \longrightarrow Cu_2O(s) + 2Br^- + H_2O(l)$; (d) $S^{2-} + 8OH^- \longrightarrow SO_4^{2-} + 4H_2O(l) + 8e^-$, $Cl_2(g) + 2e^- \longrightarrow 2Cl^-$, $S^{2-} + 4Cl_2(g) + 8OH^- \longrightarrow SO_4^{2-} + 8Cl^- + 4H_2O(l)$; (e) $MnO_4^- + 2H_2O(l) + 3e^- \longrightarrow MnO_2(s) + 4OH^-$, $IO_3^- + 2OH^- \longrightarrow IO_4^- + H_2O(l) + 2e^-$, $2MnO_4^- + 3IO_3^- + H_2O(l) \longrightarrow 2MnO_2(s) + 3IO_4^- + 2OH^-$

24.7 (a) $[Fe(CN)_6]^{4-} \longrightarrow [Fe(CN)_6]^{3-} + e^-$, $2CrO_4^{2-} + 5H_2O(l) + 6e^- \longrightarrow Cr_2O_3(s) + 10OH^-$, $6[Fe(CN)_6]^{4-} + 2CrO_4^{2-} + 5H_2O(l) \longrightarrow 6[Fe(CN)_6]^{3-} + Cr_2O_3(s) + 10OH^-$; (b) $I_2(s) + 12OH^- \longrightarrow 2IO_3^- + 6H_2O(l) + 10e^-$, $ClO^- + H_2O(l) + 2e^- \longrightarrow Cl^- + 2OH^-$, $I_2(s) + 2OH^- + 5ClO^- \longrightarrow 5Cl^- + 2IO_3^- + H_2O(l)$; (c) $CN^- + 2OH^- \longrightarrow CNO^- + H_2O(l) + 2e^-$, $[Fe(CN)_6]^{3-} + e^- \longrightarrow [Fe(CN)_6]^{4-}$, $2[Fe(CN)_6]^{3-} + CN^- + 2OH^- \longrightarrow 2[Fe(CN)_6]^{4-} + CNO^- + H_2O(l)$; (d) $Cu(OH)_2(s) + 2e^- \longrightarrow Cu(s) + 2OH^-$, $N_2H_4(aq) + 4OH^- \longrightarrow N_2(g) + 4H_2O(l) + 4e^-$, $2Cu(OH)_2(s) + N_2H_4(aq) \longrightarrow N_2(g) + 2Cu(s) + 4H_2O(l)$

24.9 voltaic (galvanic) and electrolytic cells; voltaic cell generates electrical energy from spontaneous redox reaction, electrolytic cell uses electrical energy to cause reaction

24.13 $H_2(g) \longrightarrow 2H^+ + 2e^-$, $Fe^{3+} + e^- \longrightarrow Fe^{2+}$, $2Fe^{3+} + H_2(g) \longrightarrow 2H^+ + 2Fe^{2+}$

24.15 32.69 g

24.17 39 g

24.19 3.1 L

24.21 0.055 A

24.23 $+3$

24.25 (a) $K^+ < Na^+ < Fe^{2+} < Cu^{2+} < Cu^+ < Ag^+ < F_2$; (b) Ag^+, F_2

24.27 Eu > Ra > Pu > Rh; Eu; Eu, Ra, Pu; Rh

24.29 several products can form and these products can undergo various further reaction

24.31 (a) Mg; (b) Mg, Al; (c) Mg, Al, Zn

24.33 $Zn(s) + Cl_2(g) \longrightarrow Zn^{2+} + 2Cl^-$, $E^\circ = 2.12$ V, yes

24.35 (a) $E^\circ = 0.907$ V, spontaneous; (b) $E^\circ = 0.036$ V, spontaneous; (c) $E^\circ = -0.03$ V, nonspontaneous; (d) $E^\circ = 0.79$ V, spontaneous

24.37 at cathode: Na^+, H^+, H_2O or SO_4^{2-} reduction; H^+ reduction most likely; at anode: H_2O or SO_4^{2-} oxidation, H_2O oxidation more likely

24.39 H^+ reduced to H_2; H_2 is flammable

24.41 -171.4 kJ

24.43 1.229 V

24.45 10. kJ, 3.5 kJ

24.47 0.40 V

24.49 3×10^{-8} mol/L

24.51 $H_2(g) \longrightarrow 2H^+ + 2e^-$, $O_2(g) + 4H^+ + 4e^- \longrightarrow 2H_2O(l)$, $2H_2(g) + O_2(g) \longrightarrow 2H_2O(l)$, $E^\circ = 1.229$ V; $E = 1.264$ V

24.53 -0.470 V

24.55 (a) $2Hg(l) \longrightarrow Hg_2^{2+} + 2e^-$, $Fe^{3+} + e^- \longrightarrow Fe^{2+}$, $2Hg(l) + 2Fe^{3+} \longrightarrow 2Fe^{2+} + Hg_2^{2+}$; (b) (inert)|$Hg(l)$|$Hg_2^{2+}$||$Fe^{2+}$, Fe^{3+}|(inert); (c) $E° = -0.017$ V, no; (d) $E = 0.131$ V, more favorable

24.57 6×10^{70}

24.59 2

24.61 5×10^{-13}

24.63 -1.207 V

24.65 0.233 V

24.67 $Fe(OH)_3$ will form, $Fe(s) + 3OH^- \longrightarrow Fe(OH)_3(s) + 3e^-$

24.69 favorable redox reaction during discharge, electrolysis during recharge

24.71 no, amount of electrical work that will be produced

24.73 (a) $Ag(s) + Cl^- \longrightarrow AgCl(s) + e^-$, $Cr_2O_7^{2-} + 14H^+ + 6e^- \longrightarrow 2Cr^{3+} + 7H_2O(l)$, $6Ag(s) + 6Cl^- + Cr_2O_7^{2-} + 14H^+ \longrightarrow 2Cr^{3+} + 7H_2O(l) + 6AgCl(s)$; (b) $E° = 1.11$ V; (c) 1.10 V; (d) 10^{113}

24.75 (a) 0.358 V, (b) 0.429 V, (c) 0.518 V, (d) 0 V, (e) 1×10^{12}, (f) 6×10^3 mol/L

24.77 (a) +4, (b) +4, (c) 0, (d) Mg, (e) UF_4, (f) 20.5 A, (g) 0.411 L, (h) yes

CHAPTER 25

25.1 space in which electron with specific energy is likely to be found between bonded nuclei; electron density maps, energy level diagrams; maps used for structures, diagrams used for bond description and spectral studies

25.3 lower, higher

25.7 (1) use lowest energy orbitals, (2) maximum of two electrons in an orbital, (3) parallel spins in equal energy orbitals before pairing

25.9 (a) σ_{1s}^1, exists, 1/2; (b) σ_{1s}^2, exists, 1; (c) $\sigma_{1s}^2\sigma_{1s}^{*1}$, exists, 1/2; (d) $\sigma_{1s}^2\sigma_{1s}^{*2}$, does not exists, 0

25.11 (a) $1s^2 1s^2 \sigma_{2s}^2 \sigma_{2s}^{*2}$, does not exist, 0; (b) $1s^2 1s^2 \sigma_{2s}^2 \sigma_{2s}^{*2} \pi_{2p_y}^2 \pi_{2p_z}^2$, exists, 2; (c) $1s^2 1s^2 \sigma_{2s}^2 \sigma_{2s}^{*2} \sigma_{2p_x}^2 \pi_{2p_y}^2 \pi_{2p_z}^2 \pi_{2p_y}^{*2} \pi_{2p_z}^{*2} \sigma_{2p_x}^{*2}$, does not exist, 0

25.13 642.80 kJ/mol

25.15 840.620 kJ/mol

25.17 $O_2^{2+} > O_2^+ > O_2 > O_2^- > O_2^{2-}$; O_2^{2+}; 2, $2\frac{1}{2}$, $1\frac{1}{2}$, 3, 1

25.19 $1s^2 1s^2 \sigma_{2s}^2 \sigma_{2s}^{*2} \pi_{2p_y}^2 \pi_{2p_z}^2 \sigma_{2p_x}^2 \pi_{2p_y}^{*1}$; yes; yes; contains one less antibonding electron

25.21 1122 kJ/mol for CN^-, 748 kJ/mol for CN, 476 kJ/mol for CN^+

CHAPTER 26

26.1 dense, high melting, malleable, ductile, good thermal and electrical conductors; form cations and salts

26.3 naturally occurring inorganic substance with characteristic structure and composition; no, various metal cations can be present in an anionic framework

26.5 46.4%, 54.2%, 50.3%

26.7 science and technology of metals—production, purification, and properties

26.9 crushing ore; adding water, wetting agent, and frothing agent; collecting from froth

26.11 use of high temperatures; carbon or metals; form slag

26.13 (a) $Fe_3O_4(s) + 4CO(g) \longrightarrow 3Fe(l) + 4CO_2(g)$, (v) reduction; (b) $MgCO_3(s) + SiO_2(s) \longrightarrow MgSiO_3(l) + CO_2(g)$, (iv) adding a flux; (c) $4Au(s) + 8CN^- + 2H_2O(l) + O_2(g) \longrightarrow 4[Au(CN)_2]^- + 4OH^-$, (iii) leaching

26.15 $2NaCl(aq) + 2H_2O(l) \longrightarrow H_2(g) + Cl_2(g) + 2NaOH(aq)$; 40.00 g, 1.01 g H_2, 35.45 g Cl_2

26.17 purified Al_2O_3 is dissolved in molten electrolyte, electrolysis produces Al at cathode, O_2 consumes carbon anodes

26.19 ore containing high concentration of hydrated Al_2O_3; SiO_2, Fe_2O_3; digest ore under pressure with hot NaOH, filter, dilute with water, cool, filter

26.21 SO_2; odor, kills plants, poisonous, forms H_2SO_4; catalytic oxidation to H_2SO_4 or reduction to S by hot C

26.23 (a) Ag and Au might be oxidized or water decomposed, (b) Cu would not be oxidized

26.25 a

26.27 $2ZnS(s) + 3O_2(g) \longrightarrow 2ZnO(s) + 2SO_2(g)$, $ZnO(s) + C(s) \xrightarrow{\Delta} Zn(g) + CO(g)$

26.29 pyrite (FeS_2), +2; hematite (Fe_2O_3), +3; magnetite (Fe_3O_4), +2 and +3; siderite ($FeCO_3$), +2

26.31 2.0 T

26.33 C, burned to CO or CO_2; S, slag as sulfide; P, Si, and

Mn, air converted to oxides in slag; $MnO(s) + SiO_2(s) \longrightarrow MnSiO_3(l)$, $SiO_2(s) + MgO(s) \xrightarrow{\Delta} MgSiO_3(l)$, $P_4O_{10}(s) + 6CaO(s) \xrightarrow{\Delta} 2Ca_3(PO_4)_2(l)$

26.35 mixture of two or more metals, or metals plus nonmetals, giving a substance with metallic properties; heterogeneous mixtures, solid solutions, intermetallic compounds; unique to alloy

26.37 $AgCd_3$, ε-alloy; AgCd, β-alloy; Ag_5Cd_8, γ-alloy

26.39 no; seawater, brines; electrolysis of molten chlorides

26.41 ns^x; 0, $+x$; ionic; low electronegativity

26.43 $4Li(s) + O_2(g) \longrightarrow 2Li_2O(s)$, $2Na(s) + O_2(g) \longrightarrow Na_2O_2(s)$, $M(s) + O_2(g) \longrightarrow MO_2(s)$ where M = K, Rb, Cs; $2M(s) + O_2(g) \longrightarrow 2MO(s)$ where M = Be, Mg, Ca, Sr, $Ba(s) + O_2(g) \longrightarrow BaO_2(s)$; electronegativity, size, charge

26.45 (a) $2NaCl(aq) + 2H_2O(l) \xrightarrow{\text{electricity}} Cl_2(aq) + H_2(g) + 2NaOH(aq)$, evaporation; (b) $CaCO_3(s) + 2NaCl(aq) \longrightarrow Na_2CO_3(s) + CaCl_2(aq)$ using Solvay process; (c) $Na_2CO_3(aq) + H_2O(l) + CO_2(g) \longrightarrow 2NaHCO_3(s)$; (d) $CaCO_3(s) \xrightarrow{\Delta} CaO(s) + CO_2(g)$; (e) $CaO(s) + H_2O(l) \longrightarrow Ca(OH)_2(s)$

26.47 LiOH, Li_2CO_3, Group II hydroxides, and Group II carbonates decompose upon heating

26.49 (a) −42.50 kJ/mol, $\Delta T = 50$ K; (b) 130 J/mol

26.51 $ns^2\,np^x$; 0, $+x$, $+(x+2)$, and $-(6-x)$; ionic with metals, covalent with nonmetals; moderate electronegativity; loss of the *p* electrons and then additional loss of *s* electrons

26.53 acidic, Al^{3+} hydrolyzes

26.55 (a) $2Al(s) + 6H^+ \longrightarrow 2Al^{3+} + 3H_2(g)$, $2Al(s) + 2OH^- + 6H_2O(l) \longrightarrow 2[Al(OH)_4]^- + 3H_2(g)$; (b) $Al_2O_3(s) + 6H^+ \longrightarrow 2Al^{3+} + 3H_2O(l)$, $Al_2O_3(s) + 2OH^- + 3H_2O(l) \longrightarrow 2[Al(OH)_4]^-$

26.57 high hydration energy

26.59 $3PbO_2(s) + 4Al(s) \longrightarrow 3Pb(s) + 2Al_2O_3(s)$, −2519.2 kJ, yes

26.61 (a) structural materials, wire, utensils, foil; (b) catalysts, antiperspirants; (c) abrasives, jewels, catalyst support; (d) water purification, pulp and paper sizing

26.63 tin electrode, 0.010 V, make $[Sn^{2+}] < [Pb^{2+}]$

26.65 (a) oxidizing agent; O_2, reducing agent: Sn; (c) oxidizing agent: H_2O, reducing agent: K

26.67 (a) nonredox—decomposition to give compounds, (b) nonredox—combination of compounds, (c) nonredox—combination of compounds

26.69 (a) $NaHCO_3(aq)$; (b) $NaOCl(aq)$, $NaCl(aq)$, $H_2O(l)$; (c) $MgO(s)$, $CaO(s)$, $CO_2(g)$; (d) $NaOH(aq)$

26.71 (a) $Sn(s) + 2HCl(aq) \longrightarrow SnCl_2(aq) + H_2(g)$, (b) $Pb^{2+} + Mg(s) \longrightarrow Pb(s) + Mg^{2+}$

CHAPTER 27

27.1 F is $[He]2s^2 2p^5$, Cl is $[Ne]3s^2 3p^5$, Br is $[Ar]3d^{10}4s^2 4p^5$, I is $[Kr]4d^{10}5s^2 5p^5$, At is $[Xe]4f^{14}5d^{10}6s^2 6p^5$; $:\ddot{X}:$; −1 with metals, semiconducting metals, and most nonmetals, positive values (except F) such as +1, +3, +5, +7 in some compounds such as interhalogens and oxoacids

27.3 London forces; strength increases going down the family; F_2 is a pale yellow gas, Cl_2 is a yellow-green gas, Br_2 is a red-brown liquid, I_2 is a violet-black solid

27.5 (a) solid, (b) 300 °C, (c) 0.25 nm, (d) 100 kJ/mol

27.7 add Br_2 and CCl_4; if any NaI were present, I_2 would form, as indicated a purple color in the CCl_4 layer

27.9 $SbCl_3$, $2Sb(s) + 3Cl_2(g) \longrightarrow 2SbCl_3(s)$

27.11 keep $[H_2]$ and $[Br_2]$ as high as possible and keep [HBr] as low as possible

27.13 F_2 is different, $2F_2(g) + 2H_2O(l) \longrightarrow 4HF(aq) + O_2(g)$; other halogens are similar, $X_2 + H_2O(l) \longrightarrow H^+ + X^- + HXO(aq)$

27.15 83.3 g $KClO_3$, 185 g KClO

27.17 iodine occurs as the oxoanion because it can share electrons more readily than the other halogens

27.19 fluorine forms protective fluoride coatings with certain metals

27.21 7.09 g, 2.45 L

27.23 (a) Ca(OCl)Cl is calcium hypochlorite chloride, $CaCl_2$ is calcium chloride, $Ca(OCl)_2$ is calcium hypochlorite; (b) 0 in Cl_2, +1 in OCl^- of Ca(OCl)Cl and $Ca(OCl)_2$, −1 in Cl^- of Ca(OCl)Cl and $CaCl_2$; (c) Cl_2; (d) Cl_2; (e) Cl; (f) Cl; (g) increases

27.25 226 Torr

27.27 77.66 g Br_2

27.29 table salt to which I^- has been added; thyroid gland production of thyroxine

27.31 (a) ${}^{209}_{83}Bi + {}^{4}_{2}He \longrightarrow 2{}^{1}_{0}n + {}^{211}_{85}At$; (b) ${}^{211}_{85}At \longrightarrow {}^{4}_{2}He + {}^{207}_{83}Bi$, ${}^{211}_{85}At + {}^{0}_{-1}e \longrightarrow {}^{211}_{84}Po$; (c) 5 fg

27.33 HCl, HBr, and HI are gases at room temperature and pressure, and fume in moist air to form droplets of strong nonoxidizing acids; from HCl to HBr to HI, melting points, boiling points, heats of fusion and vaporization, strengths as reducing agents, and bond lengths increase whereas bond energies decrease; HF has extensive hydrogen, giving significantly higher values of melting point, boiling point, and heats of fusion and vaporization than HCl, and acts as a weak acid in aqueous solution

27.35 −68.39 kJ, −12.56 kJ

27.37 (a) ionic, (b) ionic, (c) polar covalent, (d) nonpolar covalent

27.39 element will form highest oxidation state with F, lower oxidation states with Br or I

27.41 (a) +1, (b) −1, (c) −1, (d) +3, (e) +5, (f) −1, (g) +1, (h) +3, (i) −1, (j) −1, (k) −1, (l) −1, (m) −1

27.43 (a) :Br—F: linear, polar

(b) :F—Br—F: (with :F: bonded below Br) T-shape, polar

(c) four :F: around Br with :F: below (:F—Br—F:, two :F: above, :F: below) square pyramidal, polar

(d) :Br—Cl: linear, polar

(e) :I—Br: linear, polar

27.45 fluorine forms only a −1 oxidation state

27.47 oxidizing agents; reduction potential decreases with increasing oxidation state

27.49 alkali metal oxoanions form O_2 and alkali metal halide salt; less reactive metals form O_2, X_2, and metal oxide

27.51 76.6% undergoes reaction to form O_2

27.53 $CaCl_2$

27.55 (a) $3Br_2(l) + 6OH^- \longrightarrow BrO_3^- + 5Br^- + 3H_2O(l)$, (b) $3BrO^- \xrightarrow{\Delta} BrO_3^- + 2Br^-$, (c) $Br_2(l) + 5F_2(g) \longrightarrow 2BrF_5(g)$, (d) $2Br^- + Cl_2(aq) \longrightarrow Br_2(aq) + 2Cl^-$

27.57 redox reactions: (a), (b), (d); oxidizing agent: (a) BrO_3^-, (b) Br_2, (d) $KClO_3$; reducing agent: (a) Br^-, (b) I^-, (d) $KClO_3$

27.59 (a) redox—displacement of one element from a compound by another element; (b) nonredox—partner-exchange; (c) nonredox—partner-exchange between ions in aqueous solution; (d) redox—combination of two elements to give a compound

27.61 (a) redox decomposition to give $H_2(g) + F_2(g) + KF(s)$, (b) redox combination to give $H_2SO_4(aq) + HBr(g)$, (c) redox reaction of ions in aqueous solution to give $I_2(s) + H_2O(l)$, (d) redox combination $FeCl_3(s)$

27.63 (a) increase, (b) decrease, (c) increase, (d) increase

27.65 $[Kr]4d^{10}5s^2 5p^6$, predicted to be inert because of complete octet,

:F—Xe—F: (with :F: above and below Xe), empty $5d$ orbitals are used for the extra electrons

27.67 XeF_2 is linear, XeF_4 is square planar; London forces

27.69 $XeF_4(s) + F_2(g) \longrightarrow XeF_6(s)$, 4.29 g XeF_6

27.71 0.0499 atm

27.73 150. kJ/mol

CHAPTER 28

28.1 $[He]2s^2 2p^3$, ±3 and +5; $[Ne]3s^2 3p^3$, ±3 and +5; $[Ne]3s^2 3p^4$, −2, +4, and +6

28.3 (a) two molecules would be preferred (−1890 kJ to −954 kJ), (b) one tetrahedral molecule would be preferred (−1460 kJ to −970 kJ)

28.5 (a) $1s^2 1s^2 \sigma_{2s}^2 \sigma_{2s}^{*2} \pi_{2p_y}^2 \pi_{2p_z}^2 \sigma_{2p_x}^2$, (b) one σ and 2 π bonds, diamagnetic

28.7 S_2

28.9 (a) rhombic form increases in temperature, changes to monoclinic form at 95.39 °C, increases in temperature, melts at 115.21 °C, increases in temperature, vaporizes at 444.6 °C, increases in temperature; (b) rhombic form increases in temperature, melts, some liquid crystallizes in monoclinic form, resulting in an equilibrium mixture of monoclinic solid, rhombic solid, and liquid; (c) liquid cools, solidifies at 115.21 °C to monoclinic solid, cools to 100 °C; (d) gas deposits as monoclinic solid, monoclinic solid changes to rhombic solid; (e) monoclinic form increases in temperature, sublimes

28.11 see Tables 28.3, 28.4, and 28.5

28.13 S is more reactive than P or N; S and P combine with most elements; S combines with certain metals at room temperature to form sulfides, burns in air to form SO_2, and combines with halogens; reaction conditions determine whether +3 or +5 for P forms; N combines only with Li at room temperature; N combines with reactive metals and certain p-block elements at elevated temperatures; white P is much more reactive than the other allotropic forms

28.15 N^{3-} and P^{3-} are much stronger bases than S^{2-}; $N^{3-} + 3H_2O(l) \longrightarrow NH_3(aq) + 3OH^-$, $P^{3-} + 3H_2O(l) \longrightarrow PH_3(g) + 3OH^-$, $S^{2-} + H_2O(l) \rightleftharpoons HS^- + OH^-$

28.17 sulfuric acid: $2SO_2(g) + O_2(g) \longrightarrow 2SO_3(g)$ using V_2O_5 catalyst, $SO_3(g) + H_2SO_4(l) \rightleftharpoons H_2S_2O_7(l)$, $H_2S_2O_7(l) + H_2O(l) \longrightarrow 2H_2SO_4(aq)$; nitric acid: $4NH_3(g) + 5O_2(g) \longrightarrow 4NO(g) + 6H_2O(g)$ at 700 °C using Pt–Rh catalyst, $2NO(g) + O_2(g) \longrightarrow 2NO_2(g)$; $3NO_2(g) + H_2O(l) \longrightarrow 2HNO_3(aq) + NO(g)$; phosphoric acid: $P_4(s) + 5O_2(g) \longrightarrow P_4O_{10}(s)$, $P_4O_{10}(s) + 6H_2O(l) \longrightarrow 4H_3PO_4(aq)$ or $Ca_3(PO_4)_2(s) + 3H_2SO_4(aq) \longrightarrow 2H_3PO_4(aq) + 3CaSO_4(s)$

28.19 1.85; $[SO_4^{2-}] = 4 \times 10^{-3}$ M, $[HSO_4^-] = 6 \times 10^{-3}$ M

28.21 HNO_3 is an oxidizing agent in solutions of greater than 2 M concentration; product formed depends on the concentration of the acid, the temperature, and nature of reducing agent

28.23 salt-like, ionic; covalent, covalent either in molecules or network; interstitial, metallic

28.25 (a) dinitrogen monoxide, nitrogen(I) oxide, or nitrous oxide

$$:\ddot{N}{=}N{=}\ddot{O}: \longleftrightarrow :N{\equiv}N{-}\ddot{O}:$$

sp^2 sp sp sp

(b) nitrogen monoxide, nitrogen(II) oxide, or nitric oxide

$$:\dot{N}{=}\ddot{O}:$$

sp^2

(c) nitrous acid

$$H{-}\ddot{O}{-}\ddot{N}{=}\ddot{O}:$$

sp^2

(d) nitric acid

$$H{-}\ddot{O}{-}\overset{:O:}{\overset{\|}{N}}{-}\ddot{O}: \longleftrightarrow H{-}\ddot{O}{-}\overset{:\ddot{O}:}{\overset{|}{N}}{=}\ddot{O}:$$

sp^2 sp^2

(e) nitrite ion

$$\left[:\ddot{O}{-}\ddot{N}{=}\ddot{O}:\right]^- \longleftrightarrow \left[:\ddot{O}{=}\ddot{N}{-}\ddot{O}:\right]^-$$

sp^2 sp^2

(f) nitrate ion

$$\left[\overset{:\ddot{O}:}{\overset{|}{:\ddot{O}{-}N{=}\ddot{O}:}}\right]^- \longleftrightarrow \left[\overset{:\ddot{O}:}{\overset{|}{:\ddot{O}{=}N{-}\ddot{O}:}}\right]^- \longleftrightarrow \left[\overset{:O:}{\overset{\|}{:\ddot{O}{-}N{-}\ddot{O}:}}\right]^-$$

sp^2 sp^2 sp^2

28.27 NO_2/N_2O_4 equilibrium lies toward colorless N_2O_4 at low temperatures

28.29 3.14

28.31 3950 L

28.33 (a) -2 in N_2H_4, $+3$ in HNO_2, $-1/3$ (average) in HN_3, (b) HNO_2; (c) 3.4

28.35 tetrahedron of P atoms; tetrahedron of O atoms around P atom

28.37 all three protons on H_3PO_4 are ionizable because they are bonded to O atoms; only two protons on H_3PO_3 are ionizable because one H is bonded directly to the P atom and is not acidic

28.39 170 g $Na_5P_3O_{10}$

28.41 12.5 for PO_4^{3-}, 10.0 for HPO_4^{2-}, 4.1 for $H_2PO_4^-$

28.43 sulfides of alkali metals, ammonium ion, calcium, strontium, and barium; alkaline solutions resulting from hydrolysis of S^{2-} ion

28.45 4.0

28.47 4.0 g SO_2

28.49 all four reactions are redox; oxidizing agents: (a) P_4, (b) NO_2, (c) I_3^-, (d) S_8; reducing agents: (a) P_4, (b) NO_2, (c) $S_2O_3^{2-}$, (d) S_8

28.51 (a) redox combination of element with compound to give other compounds, (b) nonredox partner-exchange reaction between ions in aqueous solution, (c) redox combination of element with compound to give another compound and nonredox combination of compounds, (d) redox decomposition

28.53 (a) nonredox partner-exchange reaction, $Li_3N(s) + 3H_2O(l) \longrightarrow 3LiOH(s) + NH_3(g)$; (b) redox decomposition reaction, $NH_4NO_3(s) \xrightarrow{\Delta} N_2O(g) + 2H_2O(g)$ or $2NH_4NO_3(s) \xrightarrow{\Delta} 2N_2(g) + 4H_2O(g) + O_2(g)$; (c) redox reaction of ions in aqueous solution, $2HNO_2(aq) + 2I^- + 2H^+ \longrightarrow I_2(s) + 2NO(g) + 2H_2O(l)$; (d) redox reaction, $H_2SO_4(18\ M) + H_2S(g) \longrightarrow S(s) + SO_2(g) + 2H_2O(l)$

28.55 (a) $S(s) + 3F_2(g) \longrightarrow SF_6(g)$ or $S_8(s) + 24F_2(g) \longrightarrow 8SF_6(g)$; (b) $2H_2S(g) + 3O_2(g, \text{excess}) \xrightarrow{\Delta} 2SO_2(g) + 2H_2O(g)$; (c) $3Cu(s) + 8H^+ + 2NO_3^- \longrightarrow 3Cu^{2+} + 2NO(g) + 4H_2O(l)$; (d) $2Na_2HPO_4(s) \xrightarrow{\Delta} Na_4P_2O_7(s) + H_2O(g)$

28.57 -3721 kJ

CHAPTER 29

29.1 $1s^22s^22p^4$, $1s^22s^22p^63s^23p^4$; CF_4, SiF_4; two 3*d* orbitals on Si are used, no *d* orbitals are available on C

29.3 177 cm^3

29.5 $\Delta G° = -2.9$ kJ for diamond $\longrightarrow$ graphite, graphite is more stable

29.7 reduce water to H_2 in fuel synthesis, reduce metal oxides in metallurgy

29.9 $:C{\equiv}O:$ $\quad :\ddot{O}{=}C{=}\ddot{O}:$ $\quad \left[\overset{O}{\overset{\|}{O{\cdots}C{\cdots}O}}\right]^{2-}$

(a) CO, (b) CO

29.11 concentration decreases with increasing temperature, concentration increases as partial pressure increases

29.13 5.3 g Na_2CO_3

29.15 boron (B), silicon (Si), germanium (Ge), arsenic (As), antimony (Sb), selenium (Se), tellurium (Te); resemble metals in appearance, conduct electricity much less effectively than metals, more like nonmetals than metals in chemical properties

29.17 no reaction; form oxoacids or oxides, usually in higher oxidation states

29.19 $BX_3 + 3H_2O(l) \longrightarrow H_3BO_3(aq) + 3HX(aq)$, $SiX_4 + 2H_2O(l) \longrightarrow SiO_2(s) + 4HX(aq)$, $AsX_3 + 3H_2O(l) \longrightarrow H_3AsO_3(aq) + 3HX(aq)$

29.21 (a) $1s^2 2s^2 2p^1$, (b) $1s^2 1s^2 \sigma_{2s}^2 \sigma_{2s}^{*2} \pi_{2p_y}^1 \pi_{2p_z}^1$, (c) paramagnetic, (d) paramagnetic, (e) π bonding

29.23 $2H_3BO_3(s) \xrightarrow{\Delta} 3H_2O(g) + B_2O_3(s)$, $B_2O_3(s) +$

$3Mg(s) \longrightarrow 2B(s) + 3MgO(s)$, $2B(s) + 3Cl_2(g) \longrightarrow 2BCl_3(g)$, $2BCl_3(g) + 3H_2(g) \xrightarrow{\Delta} 2B(s) + 6HCl(g)$

29.25 3×10^{25} g Si

29.27 0.5 g GeO_2

29.29 boron and hydrogen, B_2H_6, H H H / B B / H H H (bridged structure)

29.31 28 g/mol

29.33 :Cl̤—B—C̤l: with :Cl̤: below B — triangular planar; $[BCl_4]^-$ — tetrahedral

29.35 $SiH_4(g) + (2 + x)H_2O(l) \longrightarrow SiO_2 \cdot xH_2O(aq) + 4H_2(g)$; hydrated silica oxide is very stable, methane does not form hydrated oxide

29.37 638.0 kJ/mol

29.39 (a) As_2O_3, (b) As_2O_3, (c) H_3AsO_3, (d) $4As(s) + 3O_2(g) \longrightarrow 2As_2O_3(s)$, $As_2O_3(s) + 3H_2O(l) \longrightarrow 2H_3AsO_3(aq)$

29.41 21.8 kJ, 1.5×10^{-4}

29.43 SiO_4 tetrahedron; tetrahedra are linked to each other to form various polymers (rings, chains, sheets, and networks)

29.45 inert to acids, except HF(*aq*); hot concentrated NaOH converts silica to water-soluble silicates; $SiO_2(s) + 4HF(aq \text{ or } g) \longrightarrow SiF_4(g) + 2H_2O(l)$, $xSiO_2(s) + 2NaOH(aq) \longrightarrow Na_2O \cdot xSiO_2(aq) + H_2O(l)$

29.47 $xNa_2CO_3(l) + xSiO_2(s) \longrightarrow Na_{2x}(SiO_3)_x(l) + xCO_2(g)$, $xCaCO_3(s) + xSiO_2(s) \longrightarrow Ca_x(SiO_3)_x(l) + xCO_2(g)$; replacement of Na by K makes harder glass and raises the softening temperature, replacement of SiO_2 with B_2O_3 gives a glass with a very low coefficient of expansion

29.49 formed by the weathering of aluminosilicate minerals; iron(III) oxide is a common impurity; ceramic objects are shaped, dried, heated, glazed, reheated

29.51 hardening process is exothermic, straw acts as insulator

29.53 energy band, element will conduct electricity

29.55 increase in temperature causes an increase in electrical resistance of a metal but a decrease in resistance in semiconductors

29.57 *n*-type semiconductor operates by electrons from the donor impurities serving as the current carriers, *p*-type semiconductor operators by movement of positive holes contributed by the acceptor impurities

29.59 redox reactions: (a), (c), (d), (e), and (f); oxidizing agents: (a) O_2, (c) O_2, (d) Pd^{2+}, (e) HNO_3, and (f) As; reducing agents: (a) NH_3, (c) Au, (d) CO, (e) Si, and (f) Na

29.61 (a) redox—combination of an element with a compound to give another compound, (b) redox—combination of an element with a compound to give other compounds, (c) nonredox—displacement, (d) redox—displacement of one element from a compound by another element, (e) redox—displacement of one element from a compound by another element, (f) nonredox—partner-exchange, (g) nonredox—decomposition to give compounds, (h) redox—not one of the categories listed

29.63 (a) $CaC_2(s) + 2H_2O(l) \longrightarrow Ca(OH)_2(s) + C_2H_2(g)$, (b) $4Ag(s) + 8CN^- + O_2(g) + H_2O(l) \longrightarrow 4[Ag(CN)_2]^- + 4OH^-$, (c) $CH_4(g) + 2O_2(g) \xrightarrow{\Delta} CO_2(g) + 2H_2O(g)$ or $2CH_4(g) + 3O_2(g) \xrightarrow{\Delta} 2CO(g) + 4H_2O(g)$, (d) $C(s) + 2H_2SO_4(\text{conc}) \xrightarrow{\Delta} CO_2(g) + 2SO_2(g) + 2H_2O(l)$, (e) $BBr_3(g) + 3H_2O(l) \longrightarrow H_3BO_3(s) + 3HBr(aq)$, (f) $BF_3(g) + NaF(aq) \longrightarrow Na[BF_4](aq)$, (g) $Ge(s) + 2Cl_2(g) \longrightarrow GeCl_4(l)$, (h) $SeO_2(s) + H_2O(l) \longrightarrow H_2SeO_3(aq)$

29.65 (a) $CaCO_3(s) + SiO_2(s) \xrightarrow{\Delta} CaSiO_3(l) + CO_2(g)$, (b) $CO_3^{2-} + H_2O(l) \rightleftharpoons HCO_3^- + OH^-$, (c) $Ni(s) + 4CO(g) \longrightarrow [Ni(CO)_4](g)$, (d) $CaCO_3(s) + 2H^+ \longrightarrow Ca^{2+} + H_2O(l) + CO_2(g)$, (e) $CO_2(g) + C(s) \xrightarrow{\Delta} 2CO(g)$ or $2CO_2(g) \xrightarrow{\Delta} 2CO(g) + O_2(g)$

29.67 $2As^{3+} + 3H_2S(aq) \longrightarrow As_2S_3(s) + 6H^+$, $2Sb^{3+} + 3H_2S(aq) \longrightarrow Sb_2S_3(s) + 6H^+$, $Sb_2S_3(s) + 8HCl(aq) \longrightarrow 2[SbCl_4]^- + 3H_2S(aq) + 2H^+$, $As_2S_3(s) + 2H_2O_2(aq) + 12OH^- \longrightarrow 2AsO_4^{3-} + 8H_2O(l) + 3S^{2-}$, $3Ag^+ + AsO_4^{3-} \longrightarrow Ag_3AsO_4(s)$, $2[SbCl_4]^- + 8NH_3(aq) + 2H^+ + 3H_2S(aq) \longrightarrow Sb_2S_3(s) + 8NH_4Cl(aq)$

CHAPTER 30

30.1 3*d* for Sc—Cu, 4*d* for Y—Ag, 5*d* for La—Au; $d^{10}s^2$ configuration give them many properties of representative elements

30.3 hard, strong, and dense, metallic luster, high melting and boiling points, good conductors of heat and electricity; large catalytic activity, formation of complex ions

30.5 crystal structure and electron structure of alloy are different from those of the individual metals

30.7 less stable

30.9 ferromagnetic elements can exhibit magnetism in absence of external magnetic field; paramagnetism requires the presence of external magnetic field

30.11 chromite, $FeCr_2O_4$; reduction of the oxide with carbon, silicon, or aluminum; $2Al(s) + Cr_2O_3(s) \longrightarrow 2Cr(l) + Al_2O_3(l)$

30.13 900 A

30.15 the dichromate ion is the major species present in acidic solution

30.17 $Cr(OH)_3(s) + 3H^+ \longrightarrow Cr^{3+} + 3H_2O(l)$,
$Cr(OH)_3(s) + 3OH^- \longrightarrow [Cr(OH)_6]^{3-}$

30.19 oxide, pyrolusite

30.21 oxide ions can move between holes in crystal structure

30.23 fusing manganese dioxide with basic material and an oxidizing agent, followed by acidification in the presence of a strong oxidizing agent; $MnO_2(s) + 2KOH(l) + KNO_3(l) \xrightarrow{\Delta} K_2MnO_4(s) + KNO_2(s) + H_2O(g)$, $2MnO_4^{2-} + O_3(g) + 2H^+ \longrightarrow 2MnO_4^- + O_2(g) + H_2O(l)$

30.25 MnO_4^{2-} disproportionates to MnO_4^- and MnO_2, (b) MnO_4^{2-} oxidized to MnO_4^-

30.27 +2, iron(II) ion, ferrous ion; +3 iron(III) ion, ferric ion

30.29 FeO

30.31 +2, +3; yes, electrons can migrate from one iron atom to another

30.33 (a) $Fe(NO_3)_2$; (b) NO_2, O_2, Fe_2O_3; (c) CO_2, FeO

30.35 81.1%

30.37 3.46

30.39 alloys, catalysts; nickel is protective plate on iron

30.41 (a) $[Co(H_2O)_6]^{2+}$, (b) $Co(OH)_2$, (c) $[Co(NH_3)_6]^{2+}$

30.43 $(n-1)^{10}ns^1$; +1, +2, +3; +2 for Cu, +1 for Ag, and +1 and +3 for Au

30.45 copper forms CuO, silver forms Ag_2O, gold does not react; no reaction with nonoxidizing acids; copper forms Cu^{2+} and silver forms Ag^+ with oxidizing acids, and gold will react only with aqua regia to form $[AuCl_4]^-$

30.47 metallic silver and nitric acid; source of other silver compounds

30.49 (a) $2NaOH(aq) + 2AgNO_3(aq) \longrightarrow Ag_2O(s) + H_2O(l) + 2NaNO_3(aq)$,
(b) $Au(s) + 3HNO_3(aq) + 4HCl(aq) \longrightarrow HAuCl_4(aq) + 3NO_2(g) + 3H_2O(l)$,
(c) $2Cu^{2+} + 4CN^- \longrightarrow 2CuCN(s) + (CN)_2(g)$,
$CuCN(s) + CN^- \longrightarrow [Cu(CN)_2]^-$,
(d) $Cu^{2+} + 4NH_3(aq) \longrightarrow [Cu(NH_3)_4]^{2+}$,
(e) $AgCl(s) + 2NH_3(aq) \longrightarrow [Ag(NH_3)_2]^+ + Cl^-$,
(f) $Cu_2SO_4(s) \xrightarrow{H_2O} Cu(s) + CuSO_4(aq)$,
(g) $CuSO_4(s) + 5H_2O(l) \longrightarrow CuSO_4 \cdot 5H_2O(l)$

30.51 $[Xe]4f^{14}5d^{10}6s^2$, diamagnetic; $[Xe]4f^{14}5d^{10}6s^1$, paramagnetic; $[Xe]4f^{14}5d^{10}$, diamagnetic; $[Hg—Hg]^{2+}$; σ bond makes this ion diamagnetic

30.53 Zn and Cd differ by 18 in atomic number, Cd and Hg differ by 32 in atomic number

30.55 (a) $2ZnS(s) + 3O_2 \xrightarrow{\Delta} 2ZnO(s) + 2SO_2(g)$,
$ZnO(s) + H_2SO_4(dil) \longrightarrow ZnSO_4(aq) + H_2O(l)$,
$Zn^{2+} + 2e^- \longrightarrow Zn(s)$;
(b) 1.7×10^8 C

30.57 $Hg + 2Cl^- \longrightarrow HgCl_2 + 2e^-$, $3e^- + NO_3^- + 4H^+ \longrightarrow NO + 2H_2O$, $3Hg(l) + 2NO_3^- + 8H^+ + 6Cl^- \longrightarrow 3HgCl_2(s) + 2NO(g) + 4H_2O(l)$; NO_3^- is the oxidizing agent

30.59 zinc family compounds are more covalent because their ions are smaller in size

30.61 $(n-2)f^x\ (n-1)d^y\ ns^2$; 5*d* and 6*s* for lanthanides, 6*d* and 7*s* for actinides

30.63 *f*-transition elements form fewer complexes because bonding is mainly ionic and does not involve *f* electrons; lutetium ion is slightly smaller and is more prone to form complexes

30.65 (a) $M(s) + 2HCl(aq) \longrightarrow MCl_2(aq) + H_2(g)$ where M = Mn, Fe, Ni; (b) reaction of oxide coating with the acid, $CuO(s) + 2HCl(aq) \longrightarrow CuCl_2(aq) + H_2O(l)$; (c) $3Cu(s) + 8HNO_3(aq) \longrightarrow 3Cu(NO_3)_2(aq) + 2NO(g) + 4H_2O(l)$, copper does not react with a nonoxidizing acid because copper lies below hydrogen in the electromotive series

30.67 (a), (b), (c); oxidizing agents: (a) MnO_4^{2-}, (b) H_2O, (c) $Cr_2O_7^{2-}$; reducing agents: (a) MnO_4^{2-}, (b) $[Co(CN)_6]^{4-}$, (c) NH_4^+

30.69 (a) redox—electron transfer between cations in aqueous solution, (b) nonredox—displacement, (c) redox—combination of two elements to give a compound, (d) redox—displacement of one element from a compound by another element

30.71 (a) $Cr_2O_7^{2-} + 2OH^- \longrightarrow 2CrO_4^{2-} + H_2O(l)$,
(b) $Mn^{2+} + OCl^- + 2OH^- \longrightarrow MnO_2(s) + Cl^- + H_2O(l)$,
(c) $4FeS_2(s) + 11O_2(g) \longrightarrow 2Fe_2O_3(s) + 8SO_2(g)$
(d) $MnO_4^- + 5Fe^{2+} + 8H^+ \longrightarrow Mn^{2+} + 5Fe^{3+} + 4H_2O(l)$

30.73 (a) HNO_3, $KClO_3$, Δ; (b) HCl; (c) NaOH; (d) H_2S, H^+; (e) NH_3, NH_4^+; (f) NH_3, NH_4^+

30.75 (a) $M^{2+} + H_2S(aq) \longrightarrow MS(s) + 2H^+$ where M = Fe, Co, Ni;
(b) $FeS(s) + 2HCl(aq) \longrightarrow FeCl_2(aq) + H_2S(aq)$;
(c) $MS(s) + 2HNO_3(aq) \longrightarrow M^{2+} + 2NO_3^- + H_2S(aq)$ where M = Co, Ni;
(d) $Fe^{2+} + 2HNO_3(aq) \longrightarrow Fe^{3+} + NO(g) + H_2O(l) + NO_3^-$;
(e) $Fe^{3+} + NH_4CNS(s) \longrightarrow [Fe(CNS)]^{2+} + NH_4^+$;
(f) $M^{2+} + 6NH_3(aq) \longrightarrow [M(NH_3)_6]^{2+}$ where M = Co, Ni;
(g) $[Ni(NH_3)_6]^{2+} + 2DMG^- \longrightarrow [Ni(DMG)_2] + 6NH_3(aq)$;
(h) $[Co(NH_3)_6]^{2+} + 6HCl(aq) + 4NH_4CNS(s) \longrightarrow [Co(CNS)_4]^{2-} + 10NH_4^+ + 6Cl^-$

30.77 $Fe(s) + 2OH^- \longrightarrow Fe(OH)_2(s) + 2e^-$,
$NiOOH(s) + H_2O(l) + e^- \longrightarrow Ni(OH)_2(s) + OH^-$,
$Fe(s) + 2NiOOH(s) + 2H_2O(l) \longrightarrow Fe(OH)_2(s) + 2Ni(OH)_2(s)$; 0.52 V

CHAPTER 31

31.1 central metal cation or atom, ligands; coordinate covalent bonding

31.3 (a) NH_3, NO_2^-, 6, +3; (b) NH_3, Cl^-, 6, +3; (c) CN^-, 6, +2; (d) NH_3, 4, +2

31.5 chelation, 5- and 6-membered rings

31.7 $FeC_6N_6K_3$, 4 ions, $K_3[Fe(CN)_6]$, potassium hexacyanoferrate(III)

31.9 decrease, −153.5 J/K

31.11 (a) potassium hexacyanomanganate(III),
(b) tetraamminepalladium(II) hydroxide,
(c) dicyanoargentate(I) ion, (d) diamminesilver(I) ion,
(e) potassium dioxalatoferrate(II) dihydrate,
(f) potassium trioxolatocobaltate(III) hexahydrate,
(g) tetraamminecopper(II) ion, (h) sodium diiodoargentate(I),
(i) sodium tetraiodoplatinate(II)

31.13 (a) $[Zn(NH_3)_2Cl_2]$, (b) $Sn[Fe(CN)_6]$, (c) $[Pt(CN)_4]^{2-}$,
(d) $K_3[Cr(CN)_6]$, (e) $[Pt(NH_3)_4]^{2+}$, (f) $[Ni(NH_3)_6]Br_2$,
(g) $[Cu(NH_3)_4]_3[Fe(CN)_5(OH)]_2$

31.15 (d) *cis* and *trans* isomers only

31.19 (a) identical, (b) geometric isomers, (c) optical isomers, (d) identical

31.21 small radius and high charge of central metal ion, stability increases

31.23 (a) higher oxidation state, (b) smaller ionic size, (c) ring formation

31.25 9×10^{-24} mol/L

31.27 Co^{3+}; $[Co(CN)_5]^{3-}$

31.29 sharing of the π electrons by metal atom and carbon atoms

31.31 valence bond approach, covalent bonding using hybridized atomic orbitals; crystal field theory, ionic bonding; molecular orbital theory, electrons in molecular orbitals

31.33 (a) six pairs of electrons from ligand go into 4*s*, 4*p*, and 4*d* orbitals, sp^3d^2; (b) six pairs of electrons go into 3*d*, 4*s*, and 4*p* orbitals, d^2sp^3

31.35 (a) 1, (b) 5, (c) 5, (d) 0

31.37 (a) $(t_{2g})^6(e_g)^0$, −24 Dq + 3P; (b) $(t_{2g})^4(e_g)^2$, −4 Dq + P

31.39 (a) $[Ni(H_2O)_6]^{2+} + 4CN^- \longrightarrow [Ni(CN)_4]^{2-} + 6H_2O(l)$,
(b) $[Ni(H_2O)_6]^{2+} + 4NH_3(aq) \longrightarrow [Ni(NH_3)_4]^{2+} + 6H_2O(l)$,
(c) $[Ni(H_2O)_6]^{2+} + 4\text{pyridine}(aq) \longrightarrow [Ni(\text{pyridine})_4]^{2+} + 6H_2O(l)$

31.41

e_{2g} — —

6 Dq

−4 Dq

t_{2g} — — —

(a) $(t_{2g})^3(e_g)^2$, (b) $(t_{2g})^4(e_g)^2$, (c) $(t_{2g})^1$

CHAPTER 32

32.1 straight and branched saturated hydrocarbons; C_nH_{2n+2}; yes, if $n = 6$

32.3 hydrocarbon with double bond; C_nH_{2n} with $n \geq 2$; yes, if $n = 6$ (or could be cycloalkane)

32.5 group containing one less hydrogen atom than an alkane; C_nH_{2n+1}, C_nH_{2n-1}; replace *ane* with *yl*

32.7 $CH_3—CH_2—CH_2—CH_2—CH_3$, *n*-pentane;
$CH_3—CH_2—CH(CH_3)—CH_3$, 2-methylbutane;
$CH_3—C(CH_3)_2—CH_3$, 2,2-dimethylpropane

32.9 $CH_3—C(CH_3)_2—CH_3$ (each CH_3 1°, central C 4°)

32.11 (a) $CH{\equiv}C—CH_2—CH_3$ (b) $CH_2{=}C(CH_3)—CH_3$
(c) $CH_2{=}C(CH_2CH_3)—CH(CH_3)—CH_3$ (d) $CH{\equiv}C—CH(CH_3)—CH_3$

32.13 (a) 2,2-dimethylbutane (b) 1,1-dibromoethene
(c) 2,3-dimethylbutane (d) 1,3-cyclohexadiene

32.15 unsaturated hydrocarbons containing ring systems stabilized by delocalized electrons

32.17

1,2,3-trimethylbenzene

1,2,4-trimethylbenzene

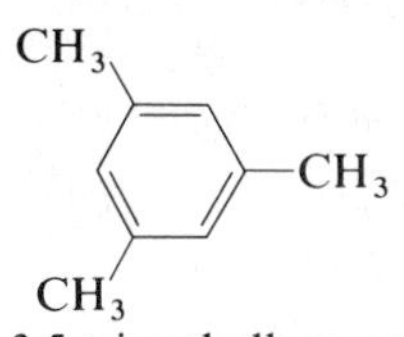

1,3,5-trimethylbenzene

32.19 (a) $O_2N-C_6H_4-NO_2$

(b) $C_6H_5-CH_2-CH_2-CH_3$ (c) Br, Br, Br on a benzene ring (1,3,5)

(d) 1,3-diphenylbutane is

$C_6H_5-CH_2-CH_2-CH(C_6H_5)-CH_3$

32.21 (a) 416 kJ/mol, (b) 330 kJ/mol, (c) 590 kJ/mol, (d) 810 kJ/mol

32.23 (a) CO_2; (b) CH_3Cl or CH_2Cl_2, etc.; (c) $CH_3-CH(OSO_3H)-CH_3$

(d) $C_6H_5-NO_2$ (e) $C_6H_5-SO_3H$

32.25 (a) converts straight-chain alkanes into branched alkanes, (b) breaks large molecules into smaller ones, (c) combines smaller molecules into larger ones, (d) converts noncyclic hydrocarbons to aromatic compounds

32.27 $H_2O(g) + C(s) \longrightarrow CO(g) + H_2(g)$, 7.1 bar

32.29 structural, *cis–trans*, and optical isomerism

32.31 c, e

32.33 reagent having partially or completely negative charge that bonds to an electron-deficient atom; a, b, d

32.35 H_2O, $(CH_3)_3C^+$

32.37 loss or gain of hydrogen or oxygen atoms; (a) oxidation, (b) reduction, (c) reduction, (d) oxidation

32.39 0.645 g

32.41 2-hexene, styrene, and cyclohexene will undergo additions with Br_2, in which the red-brown color of bromine will disappear

32.43 (a) alcohol, (b) phenol, (c) ether, (d) carboxylic acid, (e) ester, (f) carboxylic acid anhydride

32.45 (a) alcohol, (b) carboxylic acid anhydride, (c) ether, (d) ester, (e) phenol, (f) ketone, (g) carboxylic acid, (h) aldehyde, (i) amide, (j) cyclic ether, (k) acyl chloride

32.47 RX where R is a saturated hydrocarbon group, ArX where Ar is an aromatic group

32.49 (a) triphenylchloromethane, (b) 1-chloro-2-methylpropane (or isobutyl chloride), (c) trichloromethane (or chloroform), (d) trichloroethene (or trichloroethylene)

32.51 (a) $R-CH_2-OH$ (b) $R-CH(R')-OH$

(c) $R-C(R'')(R')-OH$ (d) $Ar-OH$

32.53 (a) $CH_3-CH_2-CH_2-CH_2-OH$, (b) $C_6H_{11}-OH$ (cyclohexane ring)

(c) $CH_3-CH(OH)-CH_2-CH_2-CH_2-OH$

(d) $CH_3-CH_2-C{\equiv}C-CH_2-CH_2-OH$

32.55 (c) catechol, (d) *m*-cyclohexylphenol

32.57 symmetrical ethers have like hydrocarbon groups attached to the oxygen atom, unsymmetrical ethers contain different hydrocarbon groups

32.59 CH_3-O-CH_3
(b) $CH_3-CH_2-O-CH_2-CH_2-CH_3$
(c) $CH_3-O-CH_2-CH_2-CH(O-CH_3)-CH_3$

(d) $CH_3-CH_2-O-C_6H_5$ (e) CH_3-O-(cyclobutane ring)

32.61 (a) $R-NH_2$; (b) $R-NH(R')$; (c) $R-N(R')-R''$; no, only one hydrogen atom has been replaced by one hydrocarbon group

32.63 $[C_6H_5NH_3^+] = [OH^-] = 6.5 \times 10^{-6}$ mol/L, $[C_6H_5NH_2] = 0.100$ mol/L

32.65 $-C(=O)-$; aldehydes contain a hydrogen atom and a hydrocarbon group, ketones contain two hydrocarbon groups

32.67 (a) pentanal, (b) 3,3-dibromopropanal, (c) cyclohexanone, (d) phenylethyl ketone

32.69

$C_6H_5-C(=O)-OH + NaOH \longrightarrow$

$C_6H_5-C(=O)-O^-Na^+ + H_2O$, sodium benzoate

32.71 (a) 0.19 °C, (b) 0.25 °C

32.73 carboxylic acid, alcohol; hydrolysis

32.75 $R-C(=O)-X$, carboxylic acid

32.77 (a) 1-butanol, (b) cyclopentanol, (c) 2-aminopropane (or isopropylamine or 2-propylamine), (d) 2-chloropropene, (e) 1,4-dibromobenzene (or *p*-dibromobenzene),

(f) triethylamine, (g) diphenyl ether (or phenoxybenzene)
(h) 2,4,6-tribromoaminobenzene (or 2,4,6-tribromoaniline)

32.79 (a) 2-methyl-1-propanol (or isobutyl alcohol), (b) *n*-butylamine, (c) pentanal, (d) cyclopentanone, (e) 2-methoxybutane, (f) 2,2-dimethylpropanoic acid

32.81 (a) $C_6H_4(O^-Na^+)(COO^-Na^+)$ (ortho: benzene ring bearing O^-Na^+ and $—C—O^-Na^+$, with C=O)

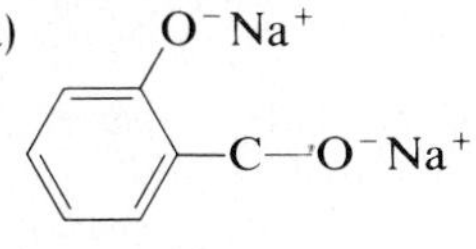

(b) $C_6H_5—\overset{O}{\overset{\|}{C}}—O—CH_3$, H_2O (c) $CH_3—CH_2—\overset{O}{\overset{\|}{C}}—OH$

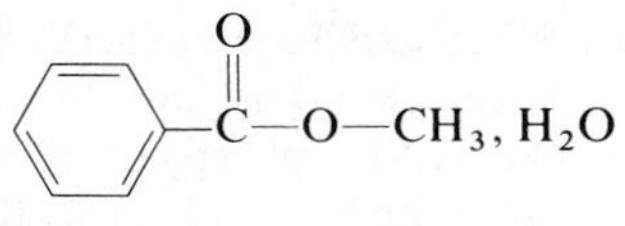

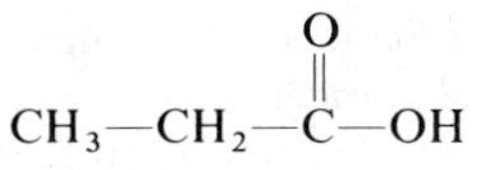

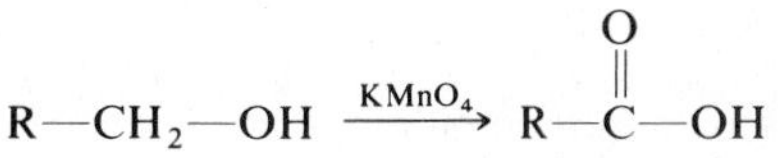

32.83 (a)

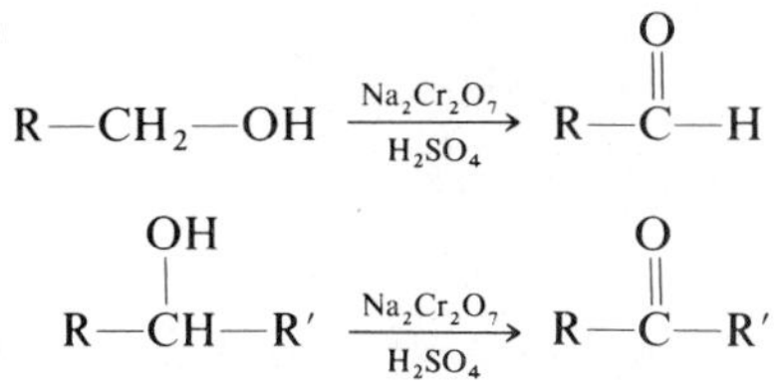

$$R—CH_2—OH \xrightarrow[H_2SO_4]{Na_2Cr_2O_7} R—\overset{O}{\overset{\|}{C}}—H$$

$$R—\overset{OH}{\overset{|}{C}H}—R' \xrightarrow[H_2SO_4]{Na_2Cr_2O_7} R—\overset{O}{\overset{\|}{C}}—R'$$

(b)

$$R—CH_2—OH \xrightarrow{KMnO_4} R—\overset{O}{\overset{\|}{C}}—OH$$

(c)

$$R—OH + R'—\overset{O}{\overset{\|}{C}}—OH \longrightarrow R'—\overset{O}{\overset{\|}{C}}—O—R + H_2O$$

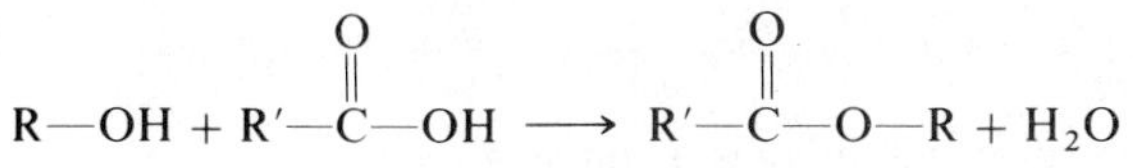

(d) $R—OH + HX \longrightarrow R—X + H_2O$

CHAPTER 33

33.1 polymer, monomer, 10^4 to 10^6 u

33.3 cellulose, natural rubber; natural rubber; convert some hydroxyl groups to other functional groups or cross-link some molecules

33.5 melting requires thermal energy to overcome intermolecular attractions; decomposition occurs if intramolecular bonds break before intermolecular attractions are overcome

33.7 linear polymer

———————

branched polymer

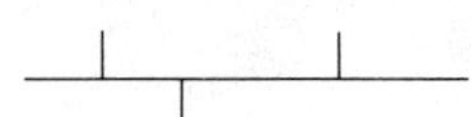

cross-linked polymer

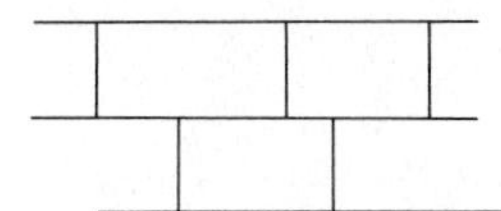

linear and branched polymers

33.9 random AAABBAABAAABBAA
alternating ABABABABAB
block AAAABBBBBAAAAABBBB

graft AAAAAAAAAAAAAA (with side chains of B: BBBB, BBBBBB, BBB)

33.11 $—CH_2CHCH_2CHCH_2CHCH_2CHCH_2CH—$, each CH bearing $C{=}O$ with CH_3 (five $—C(=O)CH_3$ groups)

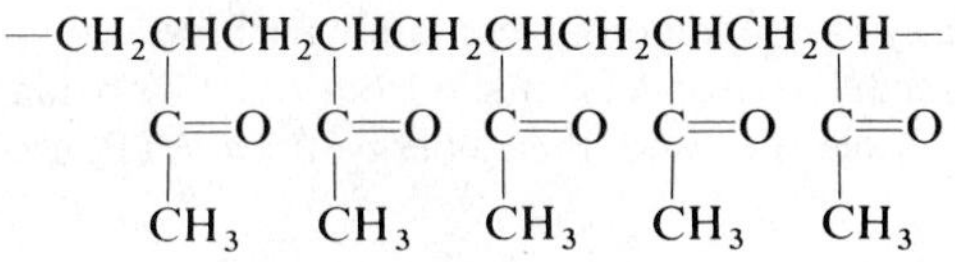

33.13 $\left(—CH_2—C_6H_{10}—CH_2O\overset{O}{\overset{\|}{C}}—C_6H_4—\overset{O}{\overset{\|}{C}}O—\right)_n$

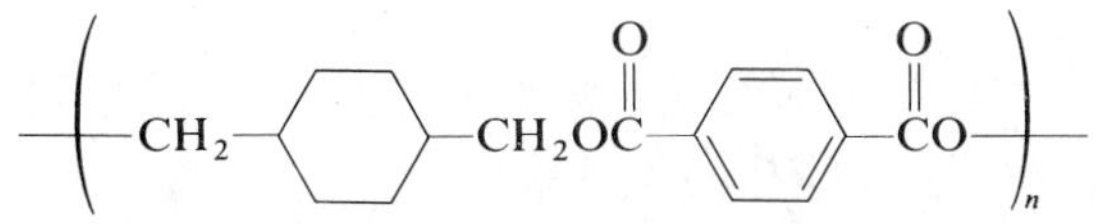

33.15 (a) ii, (b) i

33.17 4600

33.19 carbohydrates, energy storage and structural support; fats, energy storage; proteins, see Table 33.6; nucleic acids, genetic information

33.21 simple sugars (monosaccharides or disaccharides) or polymers (polysaccharides); monosaccharides are polyhydroxyl aldehydes or polyhydroxyl ketones

33.23 3, 4

33.25 (a) yes; (b) CH_2OH–HCOH–HOCH–HCOH–HCOH–CHO or CHO–HOCH–HOCH–HCOH–HOCH–CH_2OH; (c) no, it is a polyol

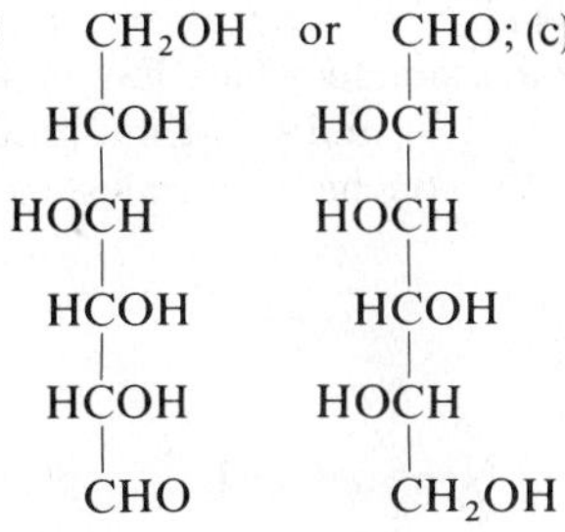

33.27 [α-D-glucose] = 0.036 mol/L, [β-D-glucose] = 0.064 mol/L

33.29 glycerides, phospholipids, waxes, steroids; hydrophobic, nonpolar compounds containing large hydrocarbon segments; nonpolymeric

33.31 (a) saturated, (b) unsaturated, (c) saturated, (d) saturated, (e) unsaturated

33.33 (a) CH_2OH–CHOH–$CH_2O\overset{O}{\overset{\|}{C}}CH_2CH_2CH_2CH_2CH_2CH_2CH_2CH_2CH_2CH_2CH_2CH_2CH_2CH_3$

(b)

$$\begin{array}{l} CH_2OCCH_2CH_2CH_2CH_2CH_2CH_2CH_2CH_2CH_2CH_2CH_2CH_2CH_3 \text{ (C=O)} \\ CHOH \\ CH_2OCCH_2CH_2CH_2CH_2CH_2CH_2CH_2CH_2CH_2CH_2CH_2CH_2CH_3 \text{ (C=O)} \end{array}$$

(c)

$$\begin{array}{l} CH_2OCCH_2CH_2CH_2CH_2CH_2CH_2CH_2CH_2CH_2CH_2CH_2CH_2CH_3 \text{ (C=O)} \\ HCOCCH_2CH_2CH_2CH_2CH_2CH_2CH_2CH_2CH_2CH_2CH_2CH_2CH_3 \text{ (C=O)} \\ CH_2OCCH_2CH_2CH_2CH_2CH_2CH_2CH_2CH_2CH_2CH_2CH_2CH_2CH_3 \text{ (C=O)} \end{array}$$

(d) the triglyceride in (c) is a simple glyceride (all of the fatty acids chains are identical)

(e)

$$\begin{array}{l} CH_2OCCH_2CH_2CH_2CH_2CH_2CH_2CH_2CH_2CH_2CH_2CH_2CH_2CH_3 \text{ (C=O)} \\ HCOCCH_2CH_2CH_2CH_2CH_2CH_2CH_2CH_2CH_2CH_2CH_2CH_2CH_2CH_2CH_3 \text{ (C=O)} \\ CH_2OCCH_2CH_2CH_2CH_2CH_2CH_2CH_2CH_2CH_2CH_2CH_2CH_2CH_2CH_2CH_2CH_2CH_3 \text{ (C=O)} \end{array}$$

33.35

$$\begin{array}{l} CH_2—O—\overset{O}{\overset{\|}{C}}—C_{12}H_{25}\text{, two} \\ CH—O—\overset{O}{\overset{\|}{C}}—C_{12}H_{25} \\ CH_2—O—\overset{O}{\overset{\|}{C}}—C_{14}H_{29} \end{array}$$

33.37 two hydroxyl groups of glycerol are esterified with fatty acids and the third hydroxyl group forms an ester link with phosphoric acid or a derivative (usually containing an amine group); hydrophilic heads (the phosphate/amine groups) interact with substances in aqueous interior of cell and aqueous outside layer, hydrophobic tails (fatty acid chains) do not permit passage of ions or polar substances

33.39 an α carbon atom to which four different groups are attached: NH_2, COOH, H, and an organic group (except glycine, to which a second H is attached); optical isomerism

33.41 6

33.43

$$R_1—\underset{NH_3^+}{\overset{H}{C}}—\overset{O}{\overset{\|}{C}}—\underset{H}{N}—\underset{H}{\overset{R_2}{C}}—COO^-$$

$$R_2—\underset{NH_3^+}{\overset{H}{C}}—\overset{O}{\overset{\|}{C}}—\underset{H}{N}—\underset{H}{\overset{R_1}{C}}—COO^-$$

33.45 condensation reaction of amine group and carboxylic acid group; peptide bonds

33.47

$$H_2NCH_2\overset{O}{\overset{\|}{C}}NH\underset{CH_2OH}{CH}\overset{O}{\overset{\|}{C}}NH\underset{CH_2SH}{CH}\overset{O}{\overset{\|}{C}}NH\underset{CH_2CH(CH_3)_2}{CH}\overset{O}{\overset{\|}{C}}NH\underset{CH_2C_6H_6}{CH}COOH$$

33.49 a groove along surface of molecule or deeper crevice or channel inside; hydrogen bonds, hydrophobic interactions, electrostatic attraction between charged groups

33.51 a pentose sugar with a nitrogeneous base replacing the OH attached to carbon 1 and a phosphate group joined by an ester link to carbon 5; ester links between the phosphate of one nucleotide and carbon 3 of the sugar in the following nucleotide

33.53 no need for base pairing by hydrogen bonding, and two strands of DNA would not have to be complementary; DNA would not be acidic

33.55

A: adenine nucleotide — CH_2–deoxyribose (O; H, H, H, H, H) with adenine (NH_2, N, H) attached; 3′ O linked to $O—P(=O)—O^-$

G: guanine nucleotide — CH_2–deoxyribose (O; H, H, H, H, H) with guanine (O, N–H, NH_2) attached; 3′ O linked to $O—P(=O)—O^-$

C: cytosine nucleotide — CH_2–deoxyribose (O; H, H, H, H, H) with cytosine (NH_2, H, H, N, O) attached; 3′ O linked to $O—P(=O)—O^-$

33.57 energy-rich compound containing ribose, adenine and three phosphate groups; ATP made from ADP by oxidation of energy-rich molecules and then energy from ATP used later

Glossary (with section/aside references)

Glossary terms referenced to Chapters 34 and 35 appear only in *Chemistry with Inorganic Qualitative Analysis* by Moeller, Bailar, Kleinberg, Guss, Castellion, and Metz.

α decay the emission of an α particle by a radionuclide (12.9)

α-alumina the inert form of aluminum oxide, α-Al_2O_3; also called corundum (26.17)

α particle a helium ion with a charge of $+2$, He^{2+}; given off by some radioactive elements (3.8)

β decay electron emission, positron emission, or electron capture by a radionuclide (12.9)

γ decay the emission of a γ ray by a radionuclide (12.9)

γ-alumina the reactive form of aluminum oxide, γ-Al_2O_3 (26.17)

π bonds bonds that concentrate electron density above and below the bond axis and have a plane of zero electron density passing through the bond axis (11.9)

σ bonds bonds in which the region of highest electron density surrounds the bond axis (11.7)

absolute entropy (S) the entropy of a pure substance (23.4)

absolute temperature scale a scale that takes absolute zero as its zero point (6.5)

absolute zero -273.15 °C, the lowest possible temperature (6.5)

absorption the incorporation of one substance into another at the molecular level (16.6)

absorption spectrum the spectrum produced by the radiation that is not absorbed by a sample (Aside 8.1)

acceptor atom the atom that accepts the electron pair for sharing in a coordinate covalent bond (10.5)

acceptor impurity a dopant that contributes holes to a vacant acceptor level in a semiconductor (29.17)

acceptor level a vacant energy level above the valence band in a semiconductor (29.17)

acetylenes hydrocarbons with covalent triple bonds; also called alkynes (32.2)

acid (water-ion) a substance that contains hydrogen and yields hydrogen ions in aqueous solution (14.8)

acid ionization constant (K_a) the equilibrium constant for the ionization of an acid, $K_a = [H^+][A^-]/[HA]$ (20.9)

acid salt a salt with the hydrogen-containing anion from a polyprotic acid (17.3)

acidic anhydride (acidic oxide) a nonmetal oxide that combines with water to give an acid (16.11)

acidic aqueous solution a solution that contains a greater concentration of H^+ ions than of OH^- ions; $[H^+] > 1 \times 10^{-7}$ mol/L (14.8, 20.7)

actinides the elements of the second f-transition series in which the $5f$ subshell is being filled (9.2)

activated complex the short-lived combination of reacting atoms, molecules, or ions that is intermediate between reactants and products; also called transition state (18.2)

activation enery (E_a) the minimum energy that reactants must have for reaction to occur (18.3)

active site the region in an enzyme that binds the substrate in a biochemical reaction (33.9)

activity (radioactive) the number of disintegrations in a given unit of time (12.5)

actual yield the weighed mass or measured volume of product formed in a reaction (5.8)

addition polymerization a rapid polymerization characterized by three reaction steps—initiation, propagation, and termination; also called chain reaction polymerization (33.4)

addition reactions the addition of atoms or groups to the carbon atoms in a double or a triple covalent bond (32.4)

adenosine diphosphate (ADP) a nucleotide consisting of the sugar ribose, the base adenine, and two phosphate groups; involved in metabolic functions (33.11)

adenosine triphosphate (ATP) a nucleotide that has the same structure as adenosine diphosphate but with a third phosphate group; involved in metabolic functions (33.11)

adsorption the attraction of a substance to the surface of a solid (15.15)

aerosol a colloid in which a gas is the dispersing medium (15.16)

alcohols ROH; organic compounds containing a hydroxyl group (32.10)

aldehyde RCHO; an organic compound containing an aldehyde group (—CHO) (32.13)

aliphatic hydrocarbons hydrocarbons that contain no aromatic rings (32.3)

alkali metals the first representative element family; the lithium family; Group I (9.1)

alkaline aqueous solution a solution that contains a greater concentration of OH^- ions than of H^+ ions; $[OH^-] > 1 \times 10^{-7}$ mol/L (14.8, 20.7)

alkaline earth metals the second representative element family; the beryllium family; Group II (9.1)

alkanes the straight- and branched-chain saturated hydrocarbons of general formula C_nH_{2n+2} (32.1)

alkenes hydrocarbons with covalent double bonds; also called olefins (32.2)

alkyl groups groups containing one less hydrogen atom than alkanes (32.1)

alkyl halide RX; an organic compound in which a halogen atom is bonded to an alkyl group (32.9)

alkynes hydrocarbons with covalent triple bonds; also called acetylenes (32.2)

allotropes different forms of the same element in the same state (16.13)

alloy an intimate mixture of metals, or of metals plus nonmetals, that yields a substance with metallic properties (26.8)

alpha decay α decay; the emission of an α particle by a radionuclide (12.9)

alpha-alumina α-alumina; the inert form of aluminum oxide, α-Al_2O_3; also called corundum (26.17)

alpha particle α particle; a helium ion with a charge of +2, He^{2+}; given off by some radioactive elements (3.8)

aluminothermic reactions the reactions of aluminum with metal oxides (26.16)

amalgams alloys of mercury (30.11)

amines RNH_2, R_2NH, R_3N; organic compounds containing the amino (—NH_2) or substituted amino (—NHR, —NR_2) groups (32.12)

amino acid a biomolecule that contains both an amine group and a carboxylic acid group (33.9)

amorphous solid a solid in which the atoms, molecules, or ions assume a random and nonrepetitive three-dimensional arrangement (13.3)

amphoteric capable of acting as either an acid or a base (16.11)

anabolism the synthesis of large molecules from small molecules by organisms (33.11)

angular momentum the product of the mass times the velocity times the radius for a body in motion; a measure of the tendency of a body to keep moving on a curved path (8.7)

anions negatively charged ions (4.1)

anode the electrode at which oxidation occurs (24.3)

antibonding molecular orbital an orbital in which most of the electron density is located away from the space between the nuclei (25.2)

aqua regia a 3-to-1 by volume mixture of concentrated HCl and HNO_3 (28.6)

aqueous solution a solution of any substance in water (3.4)

aromatic hydrocarbons unsaturated hydrocarbons containing planar ring systems stabilized by delocalized electrons, as in benzene (32.3)

aryl halide ArX; an organic compound in which an aryl group is bonded to a halogen atom (32.9)

asymmetric atom an atom bonded to four different atoms or groups (32.7)

atom the smallest particle of an element (3.3)

atomic mass (atomic weight) the average mass in atomic mass units of the atoms of the naturally occurring mixture of isotopes (3.12)

atomic mass unit (u) 1/12 of the mass of one carbon-12 atom (3.12)

atomic number (Z) the number of protons in the nucleus of each atom of an element (3.10)

atomic orbital the region in which an electron with a specific energy will most probably be located (8.8)

atomic radii an internally consistent set of radii based on the atoms in comparable chemical bonds (9.4)

Avogadro's law equal volumes of gases, measured under the same conditions of temperature and pressure, contain equal numbers of molecules (6.7)

Avogadro's number the number of atoms in exactly 12 g of carbon-12; 6.022×10^{23} (4.7)

azeotropes constant-boiling mixtures that distill without change in composition (Aside 15.2)

band of stability range of neutron–proton ratios that are stable for each nuclear charge (12.4)

band spectra spectra in which groups of closely spaced lines form bands; produced by the absorption or emission of energy by molecules (Aside 8.1)

barometer an instrument used to measure atmospheric pressure (Aside 6.1)

base (water-ion) a substance that contains hydroxide ions and dissociates in water to give hydroxide ions (14.8)

base ionization constant (K_b) the equilibrium constant for the ionization of a base, $K_b = [BH^+][OH^-]/[B]$ (20.11)

basic anhydride (basic oxide) a metal oxide that yields a hydroxide base with water, also called a basic oxide (16.11)

beta decay β decay; electron emission, positron emission, or electron capture by a radionuclide (12.9)

bidentate a ligand that contains two donor atoms and can form a chelate ring (31.3)

bimolecular reaction an elementary reaction that has two reactant particles (18.4)

binary compound a compound that contains atoms or ions of only two elements (4.4)

binding energy per nucleon the nuclear binding energy of a nucleus divided by the number of nucleons in that nucleus (12.2)

biological sciences the study of living matter (Chapter 1)

boiling point the temperature at which the vapor pressure of a liquid in an open container equals the pressure of the gases above the liquid and bubbles of vapor form throughout the liquid (13.6)

bombardment reactions reactions in which electromagnetic radiation or fast-moving particles are captured by a nucleus to form an unstable nucleus that subsequently decays (12.8)

bond angle the angle between the bonds that join one atom to two other atoms (11.1)

bond axis the line between the nuclei of two atoms bonded to each other (11.6)

bond dissociation energy the enthalpy change per mole for breaking exactly one bond of the same type per molecule (10.6)

bond energy the average enthalpy change per mole for breaking one bond of the same type per molecule (10.6)

bond length the distance between the nuclei of two bonded atoms (9.4)

bond order the number of effective bonding electron pairs shared between the two atoms of a covalent bond (25.6)

bonding molecular orbital an orbital in which most of the electron density is located between the nuclei of the bonded atoms (25.2)

Boyle's law at constant temperature, the volume of a given mass of an ideal gas is inversely proportional to the pressure upon the gas (6.4)

breeder reactor a nuclear reactor that produces at least as many fissionable atoms as it consumes (12.15)

Brønsted-Lowry acid any molecule or ion that can act as a proton donor (20.1)

Brønsted-Lowry acid–base reaction a reaction in which a proton is transferred from a proton donor to a proton acceptor (20.1)

Brønsted-Lowry base any molecule or ion that can act as a proton acceptor (20.1)

buffer solution a solution that resists changes in pH when small amounts of acid or base are added to it (21.5)

calcining the heating of an ore to drive off a gas; for example, to convert a carbonate to an oxide or to convert a hydrate to an oxide (26.4)

calorimeter a device for measuring the heat absorbed or released during a chemical or physical change (Aside 7.2)

canal rays positively charged ions flowing from the anode to the cathode in a gas-discharge tube (3.7)

carbohydrates polyhydroxy aldehydes or ketones and their dimers and polymers, including sugar, starch, and cellulose (33.7)

carbonate hardness hardness imparted to water by HCO_3^- anions together with the Ca^{2+}, Mg^{2+}, and Fe^{2+} ions; also called temporary hardness (14.13)

carbonyl group C=O; a carbon atom and an oxygen atom joined by a covalent double bond (32.13)

carboxylic acid RCOOH; an organic compound that contains a carboxyl group (–COOH) (32.14)

catabolism the breakdown of large molecules into smaller, simpler molecules by organisms (33.11)

catalyst a substance that increases the rate of a reaction but can be recovered chemically unchanged after the reaction is complete (4.5)

cathode the electrode at which reduction occurs (24.3)

cathode rays streams of electrons flowing from the cathode toward the anode in a gas-discharge tube (3.6)

cations positively charged ions (4.1)

caustic soda sodium hydroxide (26.13)

cell potential (cell voltage) (*E*) a measure of the potential difference between two half-cells; also called electromotive force (emf) (24.6)

cell reaction the overall chemical reaction that occurs in an electrochemical cell (24.3)

centrifugate the solution that lies above the precipitate after centrifugation (34.1)

chain reaction a series of reactions in which the products of the reactions initiate other similar reactions, so that the series can be self-sustaining (12.13)

chain reaction polymerization a rapid polymerization characterized by three reaction steps—initiation, propagation, and termination; also called addition polymerization (33.4)

changes of state interconversions between the solid, liquid, and gaseous states (3.5)

Charles' law at constant pressure, the volume of a given mass of an ideal gas is directly proportional to the absolute temperature (6.5)

chelate ring a ring formed by chelation (31.3)

chelation ring formation by a ligand in a complex; bonding occurs between two or more ligand atoms and the central metal atom or ion (31.3)

chemical adsorption adsorption in which the forces between the adsorbate and the surface are of the magnitude of chemical bond forces; also called chemisorption (18.5)

chemical bond a force that acts strongly enough between two atoms or groups of atoms to hold them together in a different species that has measurable properties (10.1)

chemical change a change in the composition of matter (Chapter 1)

chemical compound a substance of definite, fixed composition in which atoms of two or more elements are chemically combined (3.4)

chemical equation representation of the total chemical change that occurs in a chemical reaction by symbols and formulas (4.5)

chemical equilibrium equilibrium in which the rates of the forward and the reverse chemical reactions are the same and the amounts of the species present do not change with time (14.5)

chemical formula representation of a chemical compound or polyatomic ion by the symbols for the elements combined, with subscripts indicating how many atoms of each element are included (4.2)

chemical kinetics study of the rates and mechanisms of chemical reactions (18.1)

chemical nomenclature collective term for the rules and regulations that govern naming chemical compounds (4.4)

chemical properties properties of a substance that can be observed only in chemical reactions (3.4)

chemical reaction any process in which the identity and composition of at least one substance is changed (3.4)

chemical reactivity the tendency to undergo chemical reactions (5.1)

chemical stability resistance to chemical change (8.13)

chemistry the study of matter, the changes that matter can undergo, and the laws that describe these changes (Chapter 1)

chirality the property of having two forms that are nonsuperimposable mirror images (31.4)

chromatography the distribution of a solute between a stationary phase and a mobile phase (Aside 19.1)

***cis* isomers** isomers in which the groups under consideration are on the same side of a double bond or other rigid structure (31.4)

***cis–trans* isomerism** the occurrence of isomers in which atoms or groups of atoms can be arranged in different ways on either side of a double bond or some other rigid structure, such as that in a cyclic compound (31.4)

coinage metals copper, silver, and gold; used in coins since ancient times (30.10)

colligative property any property of a solution that depends only on the relative numbers of solute and solvent particles and not on their identity (15.10)

colloid a substance made up of suspended particles larger than most molecules but too small to be seen in an optical microscope (15.15)

colloidal dispersion a mixture in which small particles remain dispersed almost indefinitely (15.15)

combination reaction a chemical reaction in which two reactants combine to give a single product; A + B → C (5.1)

combined gas law $P_1V_1/T_1 = P_2V_2/T_2$ for a fixed mass of an ideal gas (6.6)

combustion any chemical change in which heat and light are produced; usually refers to burning in the presence of oxygen (Aside 7.2)

common ion an ion added to a solution that already contains some of that ion (21.4)

common ion effect a displacement of an ionic equilibrium by an excess of one or more of the ions involved (21.4)

complex ion an ion consisting of a central metal atom or cation to which are bonded one or more molecules or anions (14.10)

concentration a quantitative statement of the amount of solute in a given amount of solvent or solution (4.8)

condensation the movement of molecules from the gas phase to the liquid phase; also called liquefaction (13.4)

condensation polymerization a polymerization that proceeds by reaction between difunctional monomers, often with elimination of a small molecule between each two reacting groups; also called step reaction polymerization (33.4)

conduction band an energy band in which electrons are free to flow (29.16)

constant a quantity with a numerical value that does not change (6.3)

continuous spectra spectra produced by radiation that is emitted at all wavelengths in a region of the electromagnetic spectrum (Aside 8.1)

coordinate covalent bond a single covalent bond in which both electrons in the shared pair come from the same atom (10.5)

coordination compounds compounds that contain metal ions or atoms and their surrounding ligands (14.10)

coordination number (complex) the number of nonmetal atoms bonded to the central atom or ion in a complex (31.1)

coordination number (crystal) the number of nearest neighbors of an atom, ion, or molecule in a crystal (13.8)

copolymer a polymer formed by the polymerization of two or more different monomers (33.2)

Coulomb force the force of electrical attraction or repulsion between charged particles (Aside 3.3)

covalent bonding the attraction between two atoms that share electrons (10.5)

critical mass the smallest mass of fissionable material that will support a self-sustaining chain reaction under a given set of conditions (12.13)

critical point the point at which the densities of a liquid and its vapor become equal, and the boundary between the two phases disappears (13.5)

critical pressure the pressure that will liquefy a substance at its critical temperature (13.5)

critical temperature the temperature above which a substance cannot exist as a liquid no matter how great the pressure (13.5)

crystal a solid that has a shape bounded by plane surfaces intersecting at fixed angles (13.3)

crystal field stabilization energy (CFSE) the amount of stabilization provided by splitting of the *d* orbitals into two levels for a given ion (31.10)

crystal field theory bonding theory that views the interaction between metal ions and ligands in complexes as electrostatic attraction and repulsion (31.10)

crystal structure the complete geometric arrangement of the particles that occupy the space lattice (13.9)

crystalline solid a solid in which the atoms, molecules, or ions assume a characteristic, regular, and repetitive three-dimensional arrangement (13.3)

cubic closest packing the closest-packed arrangement of layers of atoms in an ABCABCABC... sequence (13.8)

cycloalkanes cyclic saturated hydrocarbons that have the general molecular formula C_nH_{2n} (32.1)

d^{10} configuration a noble gas core with an outer d^{10} configuration; also called pseudo-noble gas configuration (10.4)

Dalton's law of partial pressures in a mixture of ideal gases, the total pressure exerted is the sum of the pressures that each gas would exert if it were present alone under the same conditions (6.11)

decanting carefully pouring off a liquid to separate it from a solid or another liquid; often used to separate a precipitate from the centrifugate after centrifugation (34.1)

decomposition reaction a chemical reaction in which a single compound breaks down to give two or more other substances; $C \longrightarrow A + B$ (5.1)

deliquescent takes up enough water from the air to dissolve in the water (14.4)

delocalized electrons electrons that occupy a space spread over three or more atoms (11.9)

denaturation substantial changes in the physical properties and biological activity of a protein caused by disruption of 2°, 3°, or 4° structure due to heat, pH change, and so forth (33.9)

density (*d*) the mass per unit volume of a substance (2.8)

deoxyribonucleic acid (DNA) nucleic acid that carries the genetic information (33.10)

descriptive chemistry the description of the elements and their compounds, their physical states, and how they behave (Chapter 1)

deuterium hydrogen-2 or "heavy hydrogen" (3.11)

diamagnetism the property of repulsion by a magnetic field shown by substances that do not contain unpaired electrons (8.12)

diatomic molecules molecules made of two atoms (4.1)

diffraction the spreading of waves as they pass obstacles or openings comparable in size to their wavelength (8.2)

diffusion the mixing of molecules by random motion and collisions until the mixture becomes homogeneous (6.13)

dimer a molecule formed by combination of two identical molecules (28.19)

dipole a pair of opposite charges of equal magnitude at a specific distance from each other (10.7)

dipole moment (μ) the magnitude of the plus and minus charges of a dipole times the distance between them (11.10)

dipole–dipole interaction electrostatic force of attraction between molecules with dipole moments (11.11)

displacement reaction a chemical reaction in which the atoms or ions of one substance take the place of other atoms or ions in a compound; $A + BC \longrightarrow AC + B$ (5.1)

disproportionation reaction a reaction in which an element in one oxidization state is both oxidized and reduced (17.4)

dissociation constants (complex ions) (K_d) equilibrium constants for the dissociation of complex ions (22.1)

dissociation of an ionic compound the separation of the ions of a neutral ionic compound, usually by dissolution in water (4.5)

dissolution the process of one substance dissolving in another (3.4)

distillate a product of distillation (Aside 15.2)

distillation the heating of a liquid or a solution to boiling, followed by collecting and condensing the vapors (Aside 15.2)

distribution coefficient the equilibrium constant for the distribution of a solute between two immiscible solvents (Aside 19.1)

donor atom the atom that provides the electron pair for sharing in a coordinate covalent bond (10.5)

donor impurity a dopant that contributes electrons to the conduction band in a semiconductor without leaving holes in the valence band (29.17)

donor level an occupied valence band, usually just below the conduction band in a semiconductor (29.17)

dopant an impurity added to a crystalline semiconductor (29.17)

doping the addition of controlled amounts of impurities to semiconducting elements (29.17)

double covalent bond a bond in which two electron pairs are shared between the same two atoms (10.5)

dynamic equilibrium a state of balance between exactly opposite changes occurring at the same rate, as in a system at chemical equilibrium (6.11)

effective nuclear charge the portion of the nuclear charge that acts on a given electron (9.5)

efflorescence the loss of water by a hydrate upon exposure to air (14.4)

effusion the escape of molecules in the gaseous state one by one, without collisions, through a hole of molecular dimensions (6.13)

elastomers polymers that have elasticity like rubber (33.5)

electrochemical cell any device in which an electrochemical reaction occurs (24.3)

electrochemistry the chemistry of oxidation–reduction reactions that either produce or utilize electrical energy (24.3)

electrode potential the potential for an electrochemical half-reaction (24.7)

electrode reaction the half-reaction that occurs in a half-cell (24.3)

electrodes conductors through which electrical current enters or leaves a conducting medium (3.6)

electrolysis the process of driving a nonspontaneous redox reaction to occur by means of electrical energy (24.3)

electrolytes pure substances or substances in solution that conduct electricity by the movement of ions (14.7)

electrolytic cell a cell that uses electrical energy from outside the cell to cause a redox reaction to occur (24.3)

electrometallurgy the use of electrical energy for the extraction, reduction, or refining of metals (26.4)

electromotive force a measure of the potential difference between two half-cells; also called cell potential (24.6)

electron a fundamental, subatomic, negatively charged particle (3.6)

electron affinity the enthalpy change for the addition of one electron to an atom or ion in the gaseous state (9.9)

electron capture the capture by a radionuclide nucleus of one of its own inner orbital electrons (12.9)

electron configuration the distribution among the subshells of all electrons in an atom (8.11)

electron-deficient compounds compounds that possess too few valence electrons for the atoms to be held together by ordinary covalent bonds (29.8)

electronegative atom an atom that acquires a partial negative charge in a covalent bond or that forms a negative ion (10.9)

electronegativity the ability of an atom in a covalent bond to attract electrons to itself (10.9)

electronic structure the arrangement of electrons in atoms, ions, or molecules (8.1)

electrophile a reactant that has a partial or complete positive charge and will bond with an atom that has an available electron pair (32.8)

electropositive atom an atom that acquires a partial positive charge in a covalent bond or that forms a positive ion (10.9)

elementary reaction a reaction that occurs in a single step exactly as written (18.1)

elements substances composed solely of atoms of the same atomic number (3.4)

emf a measure of the potential difference between two half-cells; also called cell potential (24.6)

emission spectrum the spectrum of radiation emitted by a substance that has absorbed energy (Aside 8.1)

empirical formula an experimentally determined simplest formula (4.10)

empirical relationship a relationship based solely on experimental facts or derived without the use of any theory or explanation of the facts (4.10)

emulsion a colloid in which particles of a liquid are dispersed in another liquid (15.16)

enantiomers a pair of optically active isomers; one rotates plane-polarized light counterclockwise, the other rotates plane-polarized light clockwise; also called optical isomers (31.4)

end point the point at which an indicator changes color in a titration (21.8)

endothermic absorbs heat from the surroundings (7.2)

energy band a continuous energy level produced by a large number of adjacent molecular orbitals (29.16)

energy gap an area of forbidden electron energies between energy bands; also called forbidden energy gap (29.16)

enthalpy (H) a thermodynamic property of a system defined so that a change in enthalpy (ΔH) is equal to the amount of heat gained or lost in processes that occur at constant external pressure (7.4)

entropy (S) a measure of disorder (23.1)

enzymes biological catalysts (33.9)

equilibrium constant (K) a constant for a chemical reaction equal to the product of the concentrations of the reaction products, each raised to the power equal to its stoichiometric coefficient, divided by the product of the concentrations of the reactants, each raised to the power equal to its stoichiometric coefficient (19.2)

equilibrium vapor pressure the pressure exerted by the vapor over a liquid (or solid) when evaporation and condensation are at equilibrium (6.11)

equivalence point the point at which the chemically equivalent, or stoichiometric, amounts of reactants have reacted (21.8)

equivalent mass of a base the mass of a base that reacts with one mole of H^+ ions (20.6)

equivalent mass of an acid the mass of an acid that donates one mole of H^+ ions (20.6)

ester RCOOR; an organic compound in which the acidic hydrogen atom of a carboxylic group is replaced by a hydrocarbon group (32.15)

ether ROR; an organic compound in which the carbon atoms of two organic groups are bonded to an oxygen atom (32.11)

evaporation the escape of molecules from the surface of a liquid in an open container to the gas phase (13.4)

exact numbers numbers with no uncertainty; they arise by directly counting whole items or by definition (2.1)

excited states energy levels for electrons, atoms, molecules (or other systems) higher than the ground state (8.7)

exothermic releases heat to the surroundings (7.2)

exponential notation notation in which the significant figures of a number are retained, usually as a factor from 1 to 9.99..., and the location of the decimal point is indicated by a power of 10, e.g. 1.0×10^9; also called scientific notation (2.4)

extrinsic semiconductor a semiconductor in which the number of current-carrying holes and electrons is not equal and conduction depends upon the addition of appropriate impurities (dopants) (29.17)

family (periodic table) the elements in a single vertical column in the periodic table; also called group (8.13)

faraday (*F*) the amount of electrical charge represented by 1 mol of electrons, equivalent to 96,484.56 C (24.5)

ferromagnetic exhibits magnetism in the absence of an external magnetic field (30.3)

filtrate the liquid that passes through a filter (34.1)

first law of thermodynamics the energy of the universe is constant (7.4)

fluorescent emits radiation during exposure to light or to some other form of energy (3.6)

flux a substance that forms a molten material by combining with unwanted rocks during smelting (26.4)

foam a colloid consisting of tiny bubbles of gas dispersed in a liquid (15.16)

forbidden energy gap an area of forbidden electron energies between energy bands; also called energy gap (29.16)

force any interaction that can cause a change in the motion or state of rest of a body (2.8)

formation constants (complex ions) equilibrium constants for the formation of complex ions (22.1)

formula unit the simplest unit indicated by the formula of a nonmolecular compound (4.2)

fossil fuels natural gas, coal, and petroleum (16.7)

fractional distillation a process that separates liquid mixtures into fractions that differ in boiling points (Aside 15.2)

free energy change (ΔG) the energy that is available to do useful work as the result of a chemical or physical change; $\Delta G = \Delta H - T\Delta S$ (23.6)

free radical a highly reactive species that contains an unpaired electron (18.2)

frequency (ν) the number of complete waves passing a given point in a unit of time (8.2)

fuel cells cells that produce electrical energy directly from the air oxidation of a fuel continuously supplied to the cell (24.17)

functional group a chemically reactive atom or group of atoms that imparts characteristic properties to organic compounds containing that group (32.6)

fundamental particle a particle that is present in all matter (3.6)

fusion melting of a solid (13.4)

galvanic cell a cell that generates electrical energy from a spontaneous redox reaction; also called voltaic cell (24.3)

gamma decay γ decay; the emission of a γ ray by a radionuclide (12.9)

gamma-alumina γ-alumina; the reactive form of aluminum oxide, γ-Al_2O_3 (26.17)

gas-discharge tubes glass tubes that can be evacuated and into which electrodes are sealed (3.6)

Gay-Lussac's law of combining volumes when gases react and/or gaseous products are formed, the ratios of the volumes of the gases involved, measured at the same temperature and pressure, are small whole numbers (6.7)

gel colloid in which solid particles (usually very large molecules) unite in a random and intertwined structure that gives rigidity to the mixture (15.16)

geometric isomerism isomerism in which all atoms are connected in the same way but arrangement in space differs (31.4)

glass transition temperature (T_g) the temperature at which the flexibility of a polymer increases dramatically because enough free volume becomes available to allow the cooperative motion of large segments of the polymer chains (33.3)

Graham's law the relative rates of diffusion and effusion of two gases at the same pressure and temperature are inversely proportional to the square roots of their densities (or their molar masses) (6.13)

ground state the lowest energy state of an electron, atom, molecule (or other system) (8.7)

group (periodic table) the elements in a single vertical column in the periodic table; also called a family (8.13)

group reagent (qualitative analysis) a reagent that reacts simultaneously with all cations in a particular group in cation analysis (34.1)

half-cell an electrode and its surrounding electrolyte in an electrochemical cell (24.3)

half-life ($t_{1/2}$) the time it takes for one half of any quantity of a reactant to undergo a reaction; often refers to the nuclei in a sample of a radioactive substance (12.5)

half-reactions reactions that represent either oxidation only or reduction only (24.1)

halogenation a reaction in which one or more halogen atoms are introduced into a molecular compound (27.3)

halogens the seventh representative element family; the fluorine family; Group VII (9.1)

hard water water containing metal ions (principally Ca^{2+}, Mg^{2+}, and Fe^{2+}) that form precipitates with soap or upon boiling (14.13)

heat (q) the energy transferred between objects or systems at different temperatures (7.2)

heat capacity the amount of heat required to raise the temperature of a given amount of a substance by one kelvin (7.5)

heat of combustion the standard enthalpy of combustion (7.11)

heat of formation the standard enthalpy of formation (7.9)

heat of reaction the total amount of heat released or absorbed between the beginning of a reaction and the return of the substances present to the original temperature; for reactants and products in standard states, equal to the standard enthalpy of reaction (7.7)

heat of solution the enthalpy change for the dissolution of one mole of substance in a given amount of solvent (14.3)

Heisenberg uncertainty principle it is impossible to know simultaneously both the exact momentum and the exact position of a particle such as an electron (8.5)

Henry's law the partial pressure of a gas over a solution is directly proportional to the concentration of the gas in that solution at a given temperature; $P_A = kC$ (15.4)

Hess's law the enthalpy change of a chemical reaction is the same whether the reaction takes place in one step or several steps (7.10)

heterogeneous catalyst a catalyst present in a phase different from that of the reactants and products (18.15)

heterogeneous mixture a mixture in which the individual components of the mixture remain physically separate and can be seen as separate components (3.4)

heterogeneous reaction a reaction between substances in different phases (18.14)

heteronuclear refers to atoms of different elements (25.8)

hexagonal closest packing the closest-packed arrangement of layers of atoms in an ABABAB... sequence (13.8)

high-spin complex a complex ion in which d electrons remain unpaired (31.10)

homogeneous catalyst a catalyst present in the same phase as the reactants (18.15)

homogeneous mixture a mixture that has a uniform composition and appearance throughout (3.4)

homogeneous reaction a reaction between substances in the same phase (18.14)

homonuclear refers to atoms of the same element (25.7)

homopolymer a polymer formed by the polymerization of a single type of monomer (33.2)

hybridization the mixing of the atomic orbitals on a single atom to give a new set of orbitals (hybrid orbitals) on that atom (11.8)

hydrated surrounded by water molecules (14.2)

hydrated lime calcium hydroxide; also called slaked lime (26.13)

hydrates chemical compounds that incorporate water molecules (14.2)

hydration the association of water molecules with ions or molecules of another substance (14.2)

hydrocarbons compounds of carbon and hydrogen; saturated hydrocarbons contain only single covalent bonds; unsaturated hydrocarbons contain multiple bonds (16.7, 32.1, 32.2)

hydrogen bond the attraction of a hydrogen atom covalently bonded to an electronegative atom for a second electronegative atom (11.13)

hydrogenation a chemical reaction in which hydrogen atoms are added to a molecular compound (16.6)

hydrolysis a general term for reactions in which the water molecule is split (14.6)

hydrolysis of an ion the reaction of an ion with water to give either H_3O^+ or OH^-, plus whatever reaction product is formed by the ion (21.3)

hydrometallurgy the use of aqueous solutions in the extraction and reduction of metals (26.4)

hydrophilic soluble in water and similar polar solvents (literally, "water-loving") (33.6)

hydrophobic insoluble in water but readily soluble in nonpolar solvents such as benzene (literally, "water-fearing") (33.6)

hydroxyl group a covalently bonded OH group (32.10)

hygroscopic capable of taking up water from the air (14.4)

ideal gas a gas assumed to be composed of molecules that occupy no volume and exert no forces of attraction on each other (6.2)

idal gas law $PV = nRT$ (6.9)

ideal solution of a molecular solute a solution in which the forces between all particles of both the solvent and the solute are identical (15.3)

ideal solution of an ionic solute a solution in which solute ions are independent of each other and attracted only to solvent molecules (15.3)

immiscible describes liquids that are mutually insoluble or very nearly so (15.3)

indicator (acid–base) a compound that changes color in a specific pH range (21.8)

inert complex a complex that has a slow rate of ligand exchange (31.6)

infinitely miscible describes substances that are completely soluble in each other (15.2)

inhibitors substances that slow down a catalyzed reaction (18.15)

initial reaction rate the instantaneous reaction rate at $t = 0$; the instant when a chemical reaction begins (18.6)

inorganic chemistry the chemistry of all the elements and their compounds, except compounds of carbon with hydrogen and their derivatives (Chapter 1)

inorganic qualitative analysis a systematic method that allows the identification of the cations and anions present in an aqueous solution of unknown composition; takes advantage of similarities and differences in acid–base properties, redox properties, solubilities of salts, and stabilities of complex ions (22.8)

instantaneous reaction rate the rate at any specific instant of time during a chemical reaction (18.6)

insulator a substance that does not conduct electricity (29.17)

interhalogens molecular compounds formed by the combination of two different halogens (27.10)

intermediate a reactive species that is produced during the course of a reaction but always reacts further and is not among the final products (18.1)

intermetallic compounds metal–metal alloys of reproducible stoichiometry (26.8)

intermolecular forces the forces of attraction and repulsion between individual molecules (7.3)

internal energy (E) all energy contained within a chemical system (7.3)

internal energy change (ΔE) the amount of energy exchanged with the surroundings during a chemical or physical change of a system (7.3)

internal redox reactions reactions in which the oxidized and reduced elements originate in the same compound (17.4)

interstitial solid solutions alloys in which small atoms occupy holes in the crystal of a metal (26.8)

intramolecular forces the forces of attraction and repulsion between atoms in the same molecule (7.3)

intrinsic semiconductor a semiconductor that contains equal numbers of current-carrying holes and electrons (29.17)

ion–electron equations balanced equations that include only the electrons and other species directly involved in the oxidation or reduction of a given atom, molecule, or ion (24.1)

ion exchange the process of replacement of one ion by another (14.13)

ion product (Q_i) the reaction quotient for the dissolution of an ionic solid (22.5)

ion product constant for water (K_w) the equilibrium constant for the ionization of pure water, $K_w = [H^+][OH^-]$ (20.7)

ionic bonding the attraction between positive and negative ions (10.4)

ionic radii the radii of anions and cations in crystalline ionic compounds (9.4)

ionization formation of ions from a molecular compound or from atoms (4.5)

ionization energy the enthalpy change for the removal of the least tightly bound electron from an atom or an ion in the gaseous state (9.7)

ions positively or negatively charged atoms or other species formed by the loss or gain of electrons (3.6)

isoelectronic having identical electron configurations (9.6)

isomeric transition delayed γ-ray emission by a radionuclide (12.9)

isomers compounds that differ in molecular structure but have the same molecular formula (31.4)

isomorphous having the same crystal structure (13.10)

isotopes forms of the same element with different mass numbers (3.11)

Kelvin temperature scale the absolute temperature scale based on the Celsius scale; 0 K is equal to −273.15 °C (6.5)

ketone R_2CO; an organic compound in which two hydrocarbon groups are bonded to the carbon atom of a carbonyl group (32.13)

kinetic energy the energy of a moving body (6.2)

labile complex a complex that has a rapid rate of ligand exchange (31.6)

lanthanide contraction the gradual decrease in size across the first *f*-transition element series from lanthanum to lutetium (9.6)

lanthanides elements of the first *f*-transition series in which the 4*f* subshell is being filled (9.2)

lattice energy the energy liberated when gaseous ions combine to give one mole of a crystalline ionic compound (13.11)

leaching any process that dissolves a metal or a compound of a metal out of an ore or directly out of an ore deposit by the use of an aqueous solution (26.4)

Le Chatelier's principle if a system at equilibrium is subjected to a stress, the system will react in a way that tends to relieve the stress (17.2)

Lewis acid a molecule or ion that can accept one or more electron pairs (20.13)

Lewis acid–base reaction a reaction in which an electron pair from one atom is donated to another atom, thus forming a covalent bond (20.13)

Lewis base a molecule or ion that can donate an electron pair (20.13)

Lewis structures structures in which Lewis symbols are combined so that the bonding and nonbonding outer electrons are indicated (10.5)

Lewis symbol a symbol in which the outer electrons of an atom are shown by dots arranged around the atomic symbol (9.1)

ligands the molecules or ions bonded to the central atom or cation in a complex ion or coordination compound (14.10)

lime calcium oxide or calcium hydroxide (26.13)

limiting reactant the reactant that determines the amount of product that can be formed (5.7)

line spectra spectra consisting of lines at specific wavelengths; produced by the absorption or emission of energy by electrons in atoms (Aside 8.1)

lipids nonpolar biochemical molecules that are hydrophobic, including fats, oils, phospholipids, waxes, and steroids (33.8)

liquefaction the movement of molecules from the gas phase to the liquid phase; also called condensation (13.4)

London forces forces of attraction between fluctuating dipoles in atoms and molecules that are very close to each other (11.12)

London smog smog produced in the presence of a mixture of sulfur oxides, particulates, and high humidity; also called gray smog (Aside 28.4)

lone electron pairs pairs of valence electrons not involved in bonding; also called nonbonding electron pairs or unshared pairs (10.5)

low-spin complex a complex ion in which *d* electrons are paired (31.10)

macromolecules individual large polymer molecules (33.1)

magic numbers certain numbers of nucleons (2, 8, 20, 28, 50, 82, 126) that impart particularly great nuclear stability (12.7)

main group elements the elements of the *s* and *p* blocks in which the inner subshells are occupied fully, and the outer *s* and *p* subshells are filling; also called representative elements (9.1)

manometer a device used to measure gas pressure in closed systems (Aside 6.1)

mass (*m*) the physical property that represents the quantity of matter in a body (2.8)

mass defect the difference between the mass of an atom and the masses of the nucleons and electrons present in that atom (12.2)

mass number (*A*) the sum of the number of neutrons and the number of protons in an atom (3.11)

mass percent the concentration of a solution (or other mixture) expressed as the mass of solute divided by the mass of solution and multiplied by 100; also called weight percent (15.6)

matter everything that has mass and occupies space (Chapter 1)

melting point the temperature at which the solid and liquid phases of a substance are at equilibrium (13.4)

metabolism the sum of all chemical processes of a living organism (33.11)

metals elements that are good conductors of electricity (9.3)

metallic bonding the attraction between positive metal ions and surrounding, freely mobile electrons (10.3)

metallurgy the science and technology of metals—their production from ores, their purification, and the study of their properties (26.4)

mineral a naturally occurring inorganic substance with a characteristic crystal structure and composition (26.2)

miscibility the mutual solubility of two substances in the same phase (both liquids, both solids, or both gases) (15.2)

mixture matter composed of two or more substances that retain their separate identities (3.4)

molality (*m*) the concentration of a solution expressed as the number of moles of solute per kilogram of solvent (15.7)

molar heat capacity the amount of heat required to raise the temperature of one mole of substance by one kelvin (7.5)

molar mass the mass in grams of one mole of a substance (4.7)

molar solubility the solubility expressed in moles per liter of solution (22.3)

molar volume molar volume is the volume occupied by one mole of a substance, normally given for gases at STP (6.8)

molarity (*M*) the concentration of a solution expressed as number of moles of solute per liter of solution (4.8)

mole (*n*) a number of anything equal to Avogadro's number (4.7)

mole fraction (X_i) the number of moles of a component in a mixture divided by the total number of moles of all components in the mixture (6.11)

molecular formula a formula representing the actual number of atoms of each element that are combined in each molecule of a compound (4.11)

molecular geometry the spatial arrangement of the atoms in a molecule (11.1)

molecular mass (molecular weight) the sum of the atomic masses, in atomic mass units, of all atoms in the formula of a compound (4.3)

molecular orbital the space in which an electron with a specific energy is most likely to be found in the vicinity of two or more nuclei that are bonded together (25.1)

molecular orbital theory a theory that explains bond formation as the occupation by electrons of orbitals characteristic of bonded atoms rather than of individual atoms (25.1)

molecularity the number of reactant particles involved in an elementary reaction (18.4)

molecule the smallest particle of a pure substance that has the composition of that substance and is capable of independent existence (4.1)

momentum the product of the mass times the velocity for a body in motion (8.5)

monatomic ions cations or anions formed by single atoms of elements (4.1)

monomers molecules that can undergo reaction with each other to form polymers (33.1)

multiple covalent bond a bond in which more than one pair of electrons are shared between the same two atoms (10.5)

multiple unit cell a unit cell that contains other lattice points in addition to those at the corners (13.9)

net ionic equation a chemical equation that shows only the species involved in the chemical change and excludes spectator ions (5.2)

network covalent substance a substance composed of three-dimensional arrays of covalently bonded atoms (10.5)

neutral solution a solution in which $[H^+] = [OH^-]$ (20.7)

neutralization the reaction of an acid with a base (14.9)

neutron a fundamental subatomic particle that has a mass almost the same as the mass of a proton and has no charge (3.9)

neutron number (*N*) the number of neutrons in a nucleus (3.11)

nitrogen fixation the combination of molecular nitrogen with other atoms (Aside 28.3)

noble gases elements consisting of atoms in which all the energy sublevels that are occupied are completely filled and the outer configuration is ns^2np^6 (except for He, which is $1s^2$) (8.13)

noble gas configuration outer-shell electron configuration of ns^2np^6 (or $1s^2$) (9.8)

nonbonding electron pairs pairs of valence electrons not involved in bonding; also called lone electron pairs (10.5)

noncarbonate hardness hardness imparted to water by Ca^{2+}, Mg^{2+}, and Fe^{2+} ions essentially in the absence of HCO_3^- ions; also called permanent hardness (14.13)

nonelectrolytes substances that dissolve in water to give solutions that do not conduct electricity (14.7)

nonmetals elements that do not conduct electricity (9.3)

nonpolar covalent bond a covalent bond in which the electrons are shared equally (10.7)

nonstoichiometric compounds compounds in which atoms of different elements combine in other than whole-number ratios (13.12)

normal boiling point the temperature at which a liquid boils at 760 Torr pressure (13.6)

normal freezing point the temperature at which a liquid freezes at 760 Torr pressure (13.7)

normality (*N*) the concentration of a solution expressed as the number of equivalents of solute per liter of solution (20.6)

***n*-type semiconductor** a semiconductor crystal doped with a donor impurity; a semiconductor in which electrons are the majority of current carriers (29.17)

nuclear binding energy the energy that would be released in the combination of nucleons to form a nucleus (12.2)

nuclear chemistry the study of nuclei and of reactions that cause changes in nuclei (12.2)

nuclear fission the splitting of a heavy nucleus into two lighter nuclei of intermediate mass numbers (and sometimes other particles as well) (12.2)

nuclear force the force of attraction between nucleons (12.1)

nuclear fusion the combination of two light nuclei to give a heavier nucleus (12.2)

nuclear reactions reactions that lead to changes in the atomic number, mass number, or energy states of nuclei (12.8)

nuclear reactor equipment in which fission is carried out at a controlled rate (12.13)

nucleic acid a biopolymer in which the repeating units are nucleotides; involved in transmitting genetic information (33.10)

nucleons a collective term for protons and neutrons (12.1)

nucleophile a reactant that has a partial or complete negative charge and will bond with an electron-deficient atom (32.8)

nucleotide a biomolecule composed of a sugar, a nitrogen-containing base, and a phosphate group; present in nucleic acids (33.10)

nucleus a central region, very small by comparison with the total size of an atom, in which virtually all the mass and the positive charge of the atom are concentrated (3.8)

nuclide any isotope of any element (12.1)

octet rule atoms tend to combine by gain, loss, or sharing of electrons so that the outer energy level of each atom holds or shares four pairs of electrons (10.4)

olefins hydrocarbons with covalent double bonds; also called alkenes (32.2)

optical isomerism the occurrence of pairs of molecules of the same molecular formula that rotate plane-polarized light equally in opposite directions (31.4)

optical isomers a pair of optically active isomers, which rotate plane-polarized light equally in opposite directions; also called enantiomers (31.4)

optically active the property of rotating the plane of vibration of plane-polarized light (31.4)

ore a combination of rocky matrix and mineral that can be mined profitably and treated for the extraction of a metal (26.4)

organic chemistry the chemistry of compounds of carbon with hydrogen, and of their derivatives (Chapter 1)

osmosis the passage of solvent molecules through a semipermeable membrane from a more dilute solution into a more concentrated solution (15.13)

osmotic pressure the external pressure exactly sufficient to oppose and stop osmosis from occurring (15.13)

outer electron configurations the electron configurations in the occupied energy level of highest principal quantum number (8.13)

overall reaction order the sum of the exponents of the concentration terms in the rate equation (18.1)

overvoltage a collective term for several effects that add to the voltage required by an electrochemical reaction (24.9)

oxidation any process in which an oxidation number increases (16.2)

oxidation numbers (oxidation states) numerical values assigned to atoms in compounds, equal either to the charges of ions or to the charges the atoms would have if the compound were ionic (10.10)

oxidation–reduction reactions chemical reactions in which oxidation and reduction take place; also called redox reactions (16.2)

oxidizing acids acids that are oxidizing agents (17.5)

oxidizing agent an atom, molecule, or ion that causes an increase in the oxidation state of another substance and is itself reduced (16.3)

oxidizing anion an anion that is capable of being reduced (17.4)

oxoacid an acid that contains hydrogen and oxygen atoms and a central atom of a third element (20.3)

paramagnetism the property of attraction to a magnetic field shown by substances that contain unpaired electrons (8.12)

partial pressure (P_i) the pressure of a single gas in a mixture (6.11)

partner-exchange reaction a chemical reaction in which two compounds interact as follows: $AC + BD \rightarrow AD + BC$, where A, B, C, and D may be atoms, monatomic ions, or polyatomic ions (5.1)

peptide bond a bond formed by the loss of a water molecule between two amino acids (33.9)

percent yield the extent to which product in a chemical reaction has been formed: [(actual yield)/(theoretical yield)] (100) (5.8)

percentage composition the percentage by mass of each element in a compound (4.9)

period (periodic table) a horizontal row in the periodic table (8.13)

periodic law the properties of the elements vary periodically with their atomic numbers (9.3)

periodic table a table that groups the elements in order of increasing atomic number so that elements with similar properties fall near each other (8.1)

permanent hardness hardness imparted to water by Ca^{2+}, Mg^{2+}, and Fe^{2+} ions essentially in the absence of HCO_3^- ions; also called noncarbonate hardness (14.13)

pH the negative logarithm of the hydrogen ion (or hydronium ion) concentration, $pH = -\log [H^+]$ (20.8)

phase a homogeneous part of a system in contact with, but separate from, other parts of the system (3.5)

phenols ArOH; organic compounds in which a hydroxyl group is bonded to an aryl group (32.10)

phosphate rock mineral deposits containing various calcium phosphates and silica (28.3)

photochemical smog smog produced in the presence of sunlight, nitrogen oxides, and hydrocarbons (Aside 28.4)

photoelectric effect the release of electrons by certain metals (particularly Cs and other alkali metals) when light shines on them (8.3)

photon a quantum of radiant energy (8.3)

physical adsorption adsorption in which the forces between the adsorbate and the surface are van der Waals forces; also called physisorption (18.15)

physical change a change in which the composition of matter is unaltered (Chapter 1)

physical properties properties than can be measured or observed without changing the composition and identity of a substance (3.4)

physical sciences the study of natural laws and processes other than those peculiar to living matter (Chapter 1)

pi bond π bond; a bond that concentrates electron density above and below the bond axis and has a plane of zero electron density passing through the bond axis (11.9)

plane-polarized light light that vibrates in one plane only (31.4)

platinum metals ruthenium, osmium, rhodium, iridium, palladium, and platinum (30.4)

p–n junction the boundary between an n-type and a p-type semiconductor (29.19)

pOH the negative logarithm of the hydroxide ion concentration, $pOH = -\log [OH^-]$ (20.8)

polar covalent bond a covalent bond in which the electrons are shared unequally (10.7)

polarization of an ion the distortion of the electron cloud of an ion by an ion of opposite charge (10.8)

pollutant an undesirable substance added to the environment (14.11)

polyatomic ions ions containing more than one atom (4.2)

polyatomic molecule a molecule containing more than two atoms (4.1)

polymers very large molecules formed by linking together many smaller molecules (4.1)

polymorphous able to crystallize in more than one crystal structure (13.10)

polypeptide a single chain of amino acids linked by peptide bonds (33.9)

polyprotic acids acids that contain more than one ionizable hydrogen atom (14.8)

positron a particle identical to an electron in all its properties except for the charge, which is +1 rather than −1 (12.9)

potential energy energy of an object by virtue of its position, that is, because of a force acting on it that could cause it to move (Aside 3.3)

precipitate a solid that forms during a reaction in solution (5.1)

precipitation the formation of a solid when a reaction takes place in solution (5.1)

pressure (P) force exerted per unit area (2.8)

primitive unit cell a unit cell in which only the corners are occupied (13.9)

principles of chemistry explanations of chemical facts; for example, by theories and mathematics (Chapter 1)

products the substances that are produced in a chemical reaction (4.5)

promoters substances that make a catalyst more effective (18.15)

proteins biopolymers composed of α-amino acids; function as enzymes, membranes, structural molecules, and so forth (33.9)

proton a fundamental subatomic particle with a positive charge equal in magnitude to the negative charge of the electron (3.7)

pseudo-noble gas configuration a noble gas core with an outer d^{10} configuration; also called d^{10} configuration (10.4)

***p*-type semiconductor** a semiconductor doped with an acceptor impurity; a semiconductor in which positive holes are the majority of the current carriers (29.17)

pure substance a form of matter that has the same composition and the same properties, no matter what its source (3.4)

pyrometallurgy the use of high temperatures for the extraction and reduction of metals (26.4)

qualitative analysis the identification of substances, often in a mixture (Chapter 1)

quantitative analysis the determination of the amounts of substances in a compound or mixture (Chapter 1)

quantized restricted to amounts that are whole-number multiples of a basic unit, or quantum, for the particular system (8.3)

quantum the definite amount of energy of a small particle (8.3)

quantum mechanics the study of the motion of entities that are small enough and move fast enough to have both observable wavelike and particle-like properties (8.5)

quantum number a whole-number multiplier that specifies an amount of energy (or anything else that is quantized) (8.3)

quantum theory a general term for the idea that energy is quantized and for the consequences of that idea (8.3)

quicklime calcium oxide; also called unslaked lime (26.13)

racemic mixture a mixture of equal parts of the levorotatory isomer and the dextrorotatory isomer of the same substance (32.7)

radioactivity the spontaneous emission by unstable nuclei of particles or electromagnetic radiation, or both (12.3)

radionuclides (radioisotopes, radioactive nuclides) nuclides that spontaneously break down (12.3)

Raoult's law the vapor pressure of a liquid above a solution is equal to the mole fraction of that liquid in the solution times the vapor pressure of the pure liquid (15.9)

rare earth elements the elements in the scandium family and the sixth period elements from lanthanum to lutetium (9.2)

rate constant (*k*) the proportionality constant that relates the reaction rate to some function of the reactant concentrations raised to various powers (18.7)

rate-determining step the slowest step in a reaction mechanism (18.11)

rate equation the mathematical relationship between the reaction rate and the concentration of one or more of the reactants (18.7)

reactants the substances that are changed in a chemical reaction (4.5)

reactant reaction order the exponent on the term for a given reactant in the rate equation (18.7)

reaction mechanism all elementary steps in a single reaction pathway (18.1)

reaction quotient (*Q*) a value found from an expression that has the same form as the equilibrium constant but is used for a reaction not at equilibrium (19.7)

reaction rate the change in concentration of a reactant or a product in a unit of time (18.5)

reagent any chemical compound or mixture used to bring about a desired chemical reaction (34.1)

redox couple the oxidized and reduced species that appear in an ion–electron equation (24.1)

redox reactions chemical reactions in which oxidation and reduction take place; also called oxidation–reduction reactions (16.2)

reducing agent an atom, molecule, or ion that causes a decrease in the oxidation state of another substance and is itself oxidized (16.3)

reduction any process in which an oxidation number decreases (16.2)

refining the purification of crude metals (26.4)

refluxing the process of vapor moving up a distillation column, condensing, and trickling down the column (Aside 15.2)

representative elements the elements of the *s* and *p* blocks, in which the inner subshells are occupied fully and the outer *s* and *p* subshells are filling; also called main group elements (9.1)

residue what is left after part of a solid has dissolved or after a solution has been evaporated to dryness (34.1)

resonance the arrangement of valence electrons in molecules or ions for which more than one Lewis structure can be written (11.3)

resonance hybrid the structure of a molecule or ion for which resonance structures can be written (11.3)

ribonucleic acid (RNA) nucleic acid that puts genetic information to work in the cell by controlling amino acid sequence in protein synthesis (33.10)

roasting the heating of an ore below its melting point in air or oxygen, usually to convert sulfides to oxides (26.4)

salts ionic compounds formed between cations and the anions of acids (4.4)

saturated hydrocarbons hydrocarbons that contain only covalent single bonds (32.1)

saturated solution a solution in which the amount of dissolved solute is equal to that which would be in equilibrium with undissolved solute at a given temperature (15.1)

scientific notation notation in which the significant figures of a number are retained, usually as a factor from 1 to 9.99..., and the location of the decimal point is indicated by a power of 10, e.g., 1.0×10^9 (2.4)

screening effect the decrease in the nuclear charge acting on an electron due to the effects of other electrons (9.5)

second law of thermodynamics the entropy of the universe is increasing constantly (23.2)

selective precipitation the separation of ions by controlling concentrations so that one compound precipitates while others do not (22.7)

semiconducting elements elements that fall between the metals and the nonmetals in the periodic table (B, Si, Ge, As, Se, Sb, Te) and have properties intermediate between those of metals and nonmetals (9.3)

semiconductor a solid crystalline material with an electrical conductivity intermediate between that of a metal and an insulator (29.17)

semipermeable membranes membranes that allow the passage of some molecules but not others (15.13)

sigma bond σ bond; a bond in which the region of highest electron density surrounds the bond axis (11.7)

significant figures (significant digits) all digits that are known with certainty, plus the first digit to the right that has an uncertain value (2.1)

silicates compounds containing silicon–oxygen groups and metals (29.9)

silicones polymers composed of silicon, carbon, hydrogen, and oxygen (29.9)

simplest formula the simplest whole-number ratio of atoms in a compound (4.10)

single covalent bond a bond in which two atoms are held together by sharing two electrons (10.5)

sintering the heating of an ore without melting, causing the formation of larger particles (26.4)

slag a material that is liquid at the temperature of the furnace in smelting or other pyrometallurgical processes (26.4)

slaked lime calcium hydroxide; also called hydrated lime (26.13)

smelting the reduction of a mineral to yield a molten metal, which in the process is separated from unwanted rock (26.4)

soda ash anhydrous sodium carbonate (26.13)

sol a colloid consisting of solid particles suspended in a liquid (15.16)

solid solution an alloy in which atoms of one metal replace atoms in the crystal lattice of another (26.8)

solubility the amount of a substance that will dissolve in a given solvent to produce a saturated solution (15.1)

solubility product (K_{sp}) the equilibrium constant equation for a solid electrolyte in equilibrium with its ions in solution (22.2)

solute the component of a solution usually present in the smaller amount (3.4)

solution any homogeneous mixture of two or more substances (3.4)

solvation the association of solvent molecules with the ions or molecules of a solute (14.2)

solvent the component of a solution usually present in the larger amount (3.4)

space lattice an orderly array of points representing sites with identical environments in the same orientation in a crystal (13.9)

specific heat the amount of heat required to raise the temperature of one gram of a substance by one kelvin (7.5)

spectator ions ions that are present during a reaction in aqueous solution, but are unchanged in the reaction (5.2)

spectrometer an instrument that records the intensity and frequency of absorbed or emitted radiation (Aside 8.1)

spectroscopy the study of spectra (Aside 8.1)

spectrum **(spectra, *pl.*)** an array of radiation or particles spread out according to the increasing or decreasing magnitude of some physical property (3.10)

spontaneous chemical change a chemical change that takes place without any continuing outside influence (23.6)

standard electrode potential the electrode potential established relative to the potential of 0.000 V for the H^+/H_2 couple under standard state conditions; also called standard reduction potential (24.7)

standard enthalpy changes ($\Delta H°$) enthalpy changes for chemical reactions or other transformations of substances in their standard states (7.8)

standard enthalpy of combustion ($\Delta H_c°$) the heat for the reaction of one mole of a substance in its standard state with oxygen (7.11)

standard enthalpy of formation ($\Delta H_f°$) the heat of formation of one mole of a compound in its standard state by the combination of elements in their standard states at a specified temperature (7.9)

standard enthalpy of reaction the enthalpy change for the transformation of reactants in their standard states to products in their standard states (7.8)

standard molar volume the volume occupied by one mole of a substance at standard temperature and pressure (6.8)

standard reduction potential the electrode potential established relative to the potential of 0.000 V for the H^+/H_2 couple under standard state conditions; also called standard electrode potential (24.7)

standard state the physical state in which a substance is most stable at 1 bar pressure and a specified temperature (7.8)

standard temperature and pressure (STP) 0 °C (273 K) and 760 Torr (1 atm) (6.6)

states of matter the gaseous state, the liquid state, and the solid state (3.5)

steel alloys of iron that contain carbon and often other metals as well (26.7)

step reaction polymerization a polymerization that proceeds by reaction between difunctional monomers, often with elimination of a small molecule between each two reacting groups; also called condensation polymerization (33.4)

stoichiometric amount the exact amount of a reactant required or produced as determined by a balanced chemical equation (5.7)

stoichiometry the calculation of the quantitative relationships in chemical changes (5.3)

storage cells (storage batteries) cells that can be recharged by using electrical energy from an external source to reverse the initial oxidation–reduction reactions (24.16)

strong electrolytes compounds that are 100% ionized or dissociated into ions in aqueous solution (14.7)

strong-field ligand a strongly coordinating ligand; can form a spin-paired complex (31.10)

structural isomerism the occurrence of isomers that differ in how atoms or groups of atoms are connected (31.4)

subatomic particle a particle smaller than the smallest atom (3.6)

sublimation the vaporization of a solid (13.4)

substituents atoms or groups of atoms that have replaced hydrogen atoms in organic compounds; also, atoms or groups that have replaced atoms in any original molecule (26.8)

substitution reactions reactions in which one or more hydrogen atoms of an aromatic ring are replaced (32.4)

substrate (biochemical) the molecule or molecules with which an enzyme interacts (33.9)

supercooled cooled below the freezing point without the occurrence of freezing (13.7)

superheated heated to a temperature above the boiling point without the occurrence of boiling (13.6)

supernatant solution the solution that lies above the precipitate after centrifugation (34.1)

supersaturated solution a solution that holds more dissolved solute than would be in equilibrium with undissolved solute (15.1)

surface tension the property of a surface that imparts membrane-like behavior to the surface (13.2)

surroundings everything in the universe that is not part of a system under study (7.2)

system a portion of the universe under study (7.2)

temporary hardness hardness imparted to water by HCO_3^- anions together with the Ca^{2+}, Mg^{2+}, and Fe^{2+} ions; also called carbonate hardness (14.13)

termolecular reaction an elementary reaction that requires the interaction of three reactant particles (18.4)

theoretical density the mass of a unit cell divided by its volume (13.9)

theoretical yield the maximum amount of a product that can, according to the balanced chemical equation, be obtained from known amounts of reactants (5.8)

theory a unifying principle or group of principles that seeks to explain a body of facts or phenomena (Chapter 1)

thermochemical equation a chemical equation that includes ΔH for the balanced equation as written (7.7)

thermochemistry the study of the thermal energy changes that accompany chemical and physical changes (7.2)

thermodynamics the study of energy transformations (7.2)

thermonuclear reactions nuclear fusion reactions, which occur at very high temperatures (12.14)

thermoplastic a plastic that softens when heated (33.2)

thermoset plastic a plastic composed of a cross-linked polymer that is permanently rigid (does not soften with heat) (33.2)

three-center bond a bond in which a single pair of electrons bonds three atoms covalently (29.8)

titration measurement of the volume of a solution of one reactant that is required to react completely with a measured amount of another reactant (21.8)

titration curve (acid–base) a plot of pH versus volume of acid or base added in an acid–base titration (21.9)

***trans* isomers** isomers in which the groups under consideration are on opposite sides of a double bond or other rigid structure (31.4)

transition elements the elements in which the *d* or *f* subshells are filling (9.2)

transition state the short-lived combination of reacting atoms, molecules, or ions that is intermediate between reactants and products; also called activated complex (18.2)

transuranium elements the elements in the seventh period in the periodic table of atomic numbers greater than uranium (9.2)

triple covalent bond a bond in which three electron pairs are shared between the same two atoms (10.5)

triple point a point at which three phases are in equilibrium (13.7)

tritium hydrogen-3 (3.11)

unit cell the most convenient small part of a space lattice that, if repeated in three dimensions, will generate the entire lattice (13.9)

unknown (analytical) a solution or solid of unknown composition to be analyzed (34.1)

unsaturated hydrocarbons hydrocarbons that possess covalent double or triple bonds between carbon atoms; alkenes and alkynes (32.2)

unsaturated solution a solution that can still dissolve more solute (15.1)

unshared pairs pairs of valence electrons not involved in bonding; also called lone electron pairs (10.5)

unslaked lime calcium oxide; also called quicklime (26.13)

valence band the highest energy band completely filled by electrons (29.16)

valence bond theory theory that describes bond formation as the interaction, or overlap, of atomic orbitals (11.6)

valence electrons the electrons that are available to take part in chemical bonding (9.1)

valence-shell electron-pair repulsion (VSEPR) theory theory for prediction of molecular geometry based on repulsion between electron pairs (11.4)

van der Waals forces the collective name for intermolecular forces other than hydrogen bonding; include dipole–dipole forces and London forces (11.10, intro)

vapor pressure the pressure exerted by the vapor over a liquid (or solid) when evaporation and condensation are at equilibrium; also called equilibrium vapor pressure (6.11)

vapor pressure lowering the difference between the vapor pressure of a pure solvent and the total vapor pressure over a solution of a nonvolatile solute (15.10)

vaporization the escape of molecules from the liquid or solid phase to the gas phase (13.4)

variables measurable properties that can change (6.3)

viscosity the resistance of a substance to flow (13.2)

volatile forms a vapor very readily (6.1)

voltaic cell a cell that generates electrical energy from a spontaneous redox reaction; also called a galvanic cell (24.3)

water softening the removal of the ions that cause hardness in water (14.13)

wave number ($\bar{\nu}$) the number of wavelengths per unit of length (8.2)

wavelength (λ) the distance between any two similar points on adjacent waves (8.2)

weak electrolytes substances that are only partially ionized in aqeuous solution (14.7)

weak-field ligand a weakly coordinating ligand; can form a high-spin complex (31.10)

weight the force exerted on a body by the pull of gravity on the mass of that body (2.8)

weight percent the concentration of a solution (or other mixture) expressed as the mass of solute divided by the mass of solution and multiplied by 100; also called mass percent (15.6)

work (w) a force moving an object over a distance (7.4)

yield, actual the weighed mass or measured volume of product formed in a reaction (5.8)

yield, percent the extent to which product in a chemical reaction has been formed: [(actual yield)/(theoretical yield)] (100) (5.8)

yield, theoretical the maximum amount of a product that can, according to the balanced chemical equation, be obtained from known amounts of reactants (5.8)

zeolites a family of aluminosilicates with much more open structures than the clays and feldspars; used for selective absorption (29.13)

zone refining a method for purifying solids in which a melted zone carries impurities out of the solid (29.18)

ILLUSTRATION CREDITS

PHOTO ESSAY ONE: ENERGY 1 (*clockwise from top left*): NASA; Los Alamos National Laboratory; University of California, Lawrence Livermore Laboratory; University of California, Lawrence Livermore Laboratory. 2 (*clockwise from top left*): Courtesy Wabash Instrument Corp.; © Fundamental Photographs; © Robert W. Metz. 3 (*clockwise from top*): © Robert W. Metz; © Sam C. Pierson, Photo Researchers; © Robert W. Metz. 4 (*clockwise from top left*): Courtesy Northeast Utilities; Courtesy Northeast Utilities; U. S. Department of Energy, Science Photo Library, Photo Researchers; Courtesy Northeast Utilities, G. Betancourt Photo.

PHOTO ESSAY TWO: NATURAL RESOURCES 1 (*clockwise from top left*): NASA; Courtesy AMAX, Inc.; Courtesy AMAX, Inc.; E. R. Degginger. 2 (*clockwise from top left*): Courtesy AMAX, Inc.; Courtesy AMAX, Inc.; Courtesy AMAX, Inc.; Paul Silverman, © Fundamental Photographs. 3 (*top and center*): Courtesy American Petroleum Institute; (*bottom*): © Richard Megna, Fundamental Photographs; © Richard Megna, Fundamental Photographs; © Richard Megna, Fundamental Photographs. 4 (*top*): Paul Silverman, © Fundamental Photographs.

PHOTO ESSAY THREE: ELEMENTS AND COMPOUNDS 1 (*right*): © Robert W. Metz; (*left, top to bottom*): Andrew McClenaghan, Science Photo Library, Photo Researchers; E. R. Degginger; E. R. Degginger; Paul Silverman, © Fundamental Photographs. 2: Richard Megna, © Fundamental Photographs. 3: Paul Silverman, © Fundamental Photographs. 4: © Robert W. Metz.

PHOTO ESSAY FOUR: COLOR IN TRANSITION METAL COMPOUNDS 1: © Robert W. Metz.

PHOTO ESSAY FIVE: VISUALIZING MOLECULES 1 (*right*): Evans & Sutherland images generated by PS 390 computer graphics system; (*left*): © 1979 Joel Gordon. 2 (*right*): Courtesy of Molecular Design, Inc. Lenox Hill Station, NY, NY; (*left*): Evans & Sutherland images generated by PS 390 computer graphics system.

PHOTO ESSAY SIX: INORGANIC QUALITATIVE ANALYSIS 1: © Robert W. Metz.

PHOTO ESSAY SEVEN: CHEMICAL REACTIONS 1 (*clockwise from top left*): © Robert W. Metz; © Robert W. Metz; © Richard Megna, Fundamental Photographs; E. R. Degginger; © Robert W. Metz. 2 (*all*): © Robert W. Metz—except (*top center and right*): E. R. Degginger. 3 (*top left*): © Robert W. Metz; (*top center*): Paul Silverman, © Fundamental Photographs; (*top right*): © Robert W. Metz; (*bottom*): © Richard Megna, Fundamental Photographs. 4 (*clockwise from top left*): © Richard Megna, Fundamental Photographs; E. R. Degginger; © Kristen Brochman, Fundamental Photographs; © Kristen Brochman, Fundamental Photographs.

PHOTO ESSAY EIGHT: CHEMICAL PRINICPLES 1 (*clockwise from top left*): © Robert W. Metz; © Richard Megna, Fundamental Photographs; © Kip Peticolas, Fundamental Photographs. 2 (*top right*): © Robert W. Metz; (*bottom right*): © Lawrence Migdale, Science Source, Photo Researchers; (*left*): Dr. Jeremy Burgess, Science Photo Library, Photo Researchers. 3: © Robert W. Metz. 4 (*all*): © Robert W. Metz—except (*top left*): © Richard Megna, Fundamental Photographs.

PHOTO ESSAY NINE: POLYMERS AND PLASTICS 1 (*clockwise from top left*): E. R. Degginger, FPSA; E. R. Degginger, FPSA; Courtesy Du Pont; Manfred Kage, © Peter Arnold.

PHOTO ESSAY TEN: CHEMISTRY AT WORK 1 (*clockwise from top left*): E. R. Degginger, FPSA; E. R. Degginger, FPSA; © George J. Gaspar, Taurus Photos; © Peter Angelo, Phototake; © Tom Hollyman, Photo Researchers. 2 (*clockwise from top left*): © Burk Uzzle, Archive Pictures; Brian Payne, courtesy Adolf Coors Co.; © L. L. T. Rhodes, Taurus Photos; © L. L. T. Rhodes, Taurus Photos.

PHOTO ESSAY ELEVEN: CHEMISTRY IN THE ENVIRONMENT 1 (*clockwise from top left*): © Richard Megna, Fundamental Photographs; E. R. Degginger; © Robert W. Metz; © Phil Degginger; E. R. Degginger; © Fundamental Photographs.

PHOTO ESSAY TWELVE: MODERN SCIENCE AND TECHNOLOGY 1: Courtesy IBM. 2 (*clockwise from top left*): Courtesy IBM; Courtesy AT&T; Dr. Jeremy Burgess, Science Photo Library, Photo Researchers; © Charles Falco, Photo Researchers. 3 (*clockwise from top left*): Courtesy General Motors; Arthur Singer, © Phototake; Courtesy IBM. Computer model generated by WINSOM software, IBM UK Scientific Centre. 4 (*clockwise from top left*): Courtesy American Cyanamid Company; Courtesy Du Pont; Courtesy American Cyanamid Company; Courtesy Du Pont; Courtesy Du Pont.

INDEX

Page numbers 891–964 refer only to *Chemistry with Inorganic Qualitative Analysis*, Third Edition, Chapters 34 and 35.

A 8
B 9
C 0
D 1
E 2
F 3
G 4
H 5
I 6
J 7

Table of Atomic Masses and Electron Configurations Listed by Atomic Number
Scaled to the relative atomic mass $^{12}C = 12$ exactly. A number in parentheses is the atomic mass number of the isotope of longest known half-life.

Atomic number	Element	Symbol	Atomic mass	Electron configuration
1	Hydrogen	H	1.0079	$1s^1$
2	Helium	He	4.00260	$1s^2$
3	Lithium	Li	6.941	$1s^2 2s^1$
4	Beryllium	Be	9.01218	$1s^2 2s^2$
5	Boron	B	10.81	$1s^2 2s^2 2p^1$
6	Carbon	C	12.011	$1s^2 2s^2 2p^2$
7	Nitrogen	N	14.0067	$1s^2 2s^2 2p^3$
8	Oxygen	O	15.9994	$1s^2 2s^2 2p^4$
9	Fluorine	F	18.998403	$1s^2 2s^2 2p^5$
10	Neon	Ne	20.179	$1s^2 2s^2 2p^6$
11	Sodium	Na	22.98977	$1s^2 2s^2 2p^6\ 3s^1$
12	Magnesium	Mg	24.305	$1s^2 2s^2 2p^6\ 3s^2$
13	Aluminum	Al	26.98154	$1s^2 2s^2 2p^6\ 3s^2\ 3p^1$
14	Silicon	Si	28.0855	$1s^2 2s^2 2p^6\ 3s^2\ 3p^2$
15	Phosphorus	P	30.97376	$1s^2 2s^2 2p^6\ 3s^2\ 3p^3$
16	Sulfur	S	32.06	$1s^2 2s^2 2p^6\ 3s^2\ 3p^4$
17	Chlorine	Cl	35.453	$1s^2 2s^2 2p^6\ 3s^2\ 3p^5$
18	Argon	Ar	39.948	$1s^2 2s^2 2p^6\ 3s^2\ 3p^6$
19	Potassium	K	39.0983	$1s^2 2s^2 2p^6\ 3s^2\ 3p^6\ 4s^1$
20	Calcium	Ca	40.08	$1s^2 2s^2 2p^6\ 3s^2\ 3p^6\ 4s^2$
21	Scandium	Sc	44.9559	$1s^2 2s^2 2p^6\ 3s^2\ 3p^6 3d^1\ 4s^2$
22	Titanium	Ti	47.88	$1s^2 2s^2 2p^6\ 3s^2\ 3p^6 3d^2\ 4s^2$
23	Vanadium	V	50.9415	$1s^2 2s^2 2p^6\ 3s^2\ 3p^6 3d^3\ 4s^2$
24	Chromium	Cr	51.996	$1s^2 2s^2 2p^6\ 3s^2\ 3p^6 3d^5\ 4s^1$
25	Manganese	Mn	54.9380	$1s^2 2s^2 2p^6\ 3s^2\ 3p^6 3d^5\ 4s^2$
26	Iron	Fe	55.847	$1s^2 2s^2 2p^6\ 3s^2\ 3p^6 3d^6\ 4s^2$
27	Cobalt	Co	58.9332	$1s^2 2s^2 2p^6\ 3s^2\ 3p^6 3d^7\ 4s^2$
28	Nickel	Ni	58.69	$1s^2 2s^2 2p^6\ 3s^2\ 3p^6 3d^8\ 4s^2$
29	Copper	Cu	63.546	$1s^2 2s^2 2p^6\ 3s^2\ 3p^6 3d^{10} 4s^1$
30	Zinc	Zn	65.38	$1s^2 2s^2 2p^6\ 3s^2\ 3p^6 3d^{10} 4s^2$
31	Gallium	Ga	69.72	$1s^2 2s^2 2p^6\ 3s^2\ 3p^6 3d^{10} 4s^2\ 4p^1$
32	Germanium	Ge	72.59	$1s^2 2s^2 2p^6\ 3s^2\ 3p^6 3d^{10} 4s^2\ 4p^2$
33	Arsenic	As	74.9216	$1s^2 2s^2 2p^6\ 3s^2\ 3p^6 3d^{10} 4s^2\ 4p^3$
34	Selenium	Se	78.96	$1s^2 2s^2 2p^6\ 3s^2\ 3p^6 3d^{10} 4s^2\ 4p^4$
35	Bromine	Br	79.904	$1s^2 2s^2 2p^6\ 3s^2\ 3p^6 3d^{10} 4s^2\ 4p^5$
36	Krypton	Kr	83.80	$1s^2 2s^2 2p^6\ 3s^2\ 3p^6 3d^{10} 4s^2\ 4p^6$
37	Rubidium	Rb	85.4678	Krypton core $5s^1$
38	Strontium	Sr	87.62	$5s^2$
39	Yttrium	Y	88.9059	$4d^1\ 5s^2$
40	Zirconium	Zr	91.22	$4d^2\ 5s^2$
41	Niobium	Nb	92.9064	$4d^4\ 5s^1$
42	Molybdenum	Mo	95.94	$4d^5\ 5s^1$
43	Technetium	Tc	(98)	$4d^5\ 5s^2$
44	Ruthenium	Ru	101.07	$4d^7\ 5s^1$
45	Rhodium	Rh	102.9055	$4d^8\ 5s^1$
46	Palladium	Pd	106.42	$4d^{10}$
47	Silver	Ag	107.868	$4d^{10}\ 5s^1$
48	Cadmium	Cd	112.41	$4d^{10}\ 5s^2$
49	Indium	In	114.82	$4d^{10}\ 5s^2 5p^1$
50	Tin	Sn	118.69	$4d^{10}\ 5s^2 5p^2$
51	Antimony	Sb	121.75	$4d^{10}\ 5s^2 5p^3$
52	Tellurium	Te	127.60	$4d^{10}\ 5s^2 5p^4$
53	Iodine	I	126.9045	$4d^{10}\ 5s^2 5p^5$
54	Xenon	Xe	131.29	$4d^{10}\ 5s^2 5p^6$
55	Cesium	Cs	132.9054	$4d^{10}\ 5s^2 5p^6\ 6s^1$